Erickson

# ANIMAL SCIENCE

## (Animal Agriculture Series)

by

M. E. Ensminger, B.S., M.A., Ph.D.

Formerly: Assistant Professor in Animal Science
University of Massachusetts

Chairman, Department of Animal Science
Washington State University

Consultant, General Electric Company
Nucleonics Department (Atomic Energy Commission)

Currently: President, Consultants-Agriservices
Clovis, California

President, Agriservices Foundation
Clovis, California

Collaborator
U.S. Department of Agriculture

Adjunct Professor
California State University, Fresno

Adjunct Professor
The University of Arizona, Tucson

Distinguished Professor
University of Wisconsin, River Falls

Ninth Edition

INTERSTATE PUBLISHERS, INC.
Danville, Illinois

ANIMAL SCIENCE, Ninth Edition. Copyright © 1991 by Interstate Publishers, Inc. All rights reserved.

Printed in the United States of America.

Editions:

First ............1950
Second ..........1951
Third ...........1955
Fourth ..........1960
Fifth ............1962
Sixth ............1969
Seventh .........1977
Eighth ...........1983
Ninth ...........1991

Order from

Interstate Publishers, Inc.
510 North Vermilion Street, P.O. Box 50
Danville, IL 61834-0050
Phone: (800) 843-4774
FAX: (217) 446-9706

*Translations:* Translated into Spanish under the direction of Dr. Mauricio B. Helman, Professor, Veterinary Sciences, Catholic University of Argentina; and published by El Ateneo, Florida 34–344, Buenos Aires, Argentina. Translated into Chinese under the direction of the Chinese Academy of Agricultural Sciences; and published in Beijing, China.

*Library of Congress Catalog Card No. 90–82364*

ISBN 0-8134-2887-4

Dr. M. E. Ensminger (right), the author of *Animal Science*, is shown seated in his study conferring with Dr. G. A. Bogdanov, distinguished Russian Scientist and Director of the Ukraine Agricultural Academy, Kiev, U.S.S.R., who traveled to Clovis, California to seek Dr. Ensminger's permission to translate *Animal Science* and his other books into Russian. Dr. Ensminger happily accorded the permission, and waived all royalties in the interest of scientific and cultural exchange between the U.S.S.R. and the United States; a policy which he has followed all over the world, and which has resulted in his books being translated into several languages and the whole world being his classroom.

**Cover Picture:** The cover picture is from an original painting by the noted artist, Tom Phillips. It portrays the artist's conception of what *Animal Science* is all about. It pictures how farmers and ranchers produce plants, which have their tops in the sun and their roots in the soil, then feed these plants to animals which convert them into bountiful and nutritious foods for all the people, many of whom live in distant cities.

To the Memory
of my father, the late Jacob Ensminger, who
was my first instructor in animal science
and the best stockman I have ever known,
this book is dedicated.

# Preface to the Ninth Edition of *Animal Science*

Animal science refers to the total store of knowledge relative to the breeding, feeding, care, and management of animals and marketing and processing of animals and their products, as gained through practical experience and research methods. That's what this book, titled *Animal Science*, now in several languages and used all over the world, is all about.

Better to serve the animal science team—students, teachers, animal producers, and researchers—this ninth edition of *Animal Science* has been made 21st-century new. This involved the following three-step approach, which I considered again and again as I revised chapter by chapter, and section by section, and of which all members of the animal science team should be knowledgeable in the decades to come: (1) Where are we now, (2) where are we headed, and (3) how do we get there?

## WHERE ARE WE NOW?

The global perspective of our present status follows:

1. **The world's fastest population increase is going on now.** During the 1990s, more people will be added to the world total than in any previous—or possibly future—decade. Currently, between 90 and 100 million people are being added to the world's population every year. Between 1990 and 2000, the world will add population equivalent to another China.

2. **People are living longer and healthier.** Life expectancy throughout the world is going up and up. This calls for more food; and, in many cases, it calls for government support to augment inadequate personal incomes.

3. **Biotechnology and genetic wizardry have arrived.** Through genetic wizardry and gene transfer, we now have the tools through which to redesign plants and animals—to tailor-make all living creatures.

4. **People are aware of good nutrition.** Millions of people throughout the world have discovered the importance of good nutrition. But heightened interest has not necessarily led to better nutrition. Instead, in altogether too many cases it has led to a rash of quick solutions by fads, foibles, and trade secrets.

5. **There is more trade between countries.** The ceasing of isolation of countries is making for increased trade between them, which will expand unabated.

6. **There is more scientific, cultural, and educational exchange among nations.** Today, people all over the world are learning from one another. In science, culture, and education, we are building bridges across barriers between countries as we race to the 21st century—and beyond.

7. **There is no need for hunger.** Currently, the world is producing enough food for all the people, yet many people are going to bed hungry each night (a) because of area famines resulting from droughts and pests, and/or (b) because of lack of money or of transportation and storage facilities.

## WHERE ARE WE HEADED?

Based on facts, figures, and historical records, along with my native intuition, I picture futuristic animal science as follows:

1. **Improved environment and sustainable agriculture will be in vogue.** In the future, we shall no longer operate like there is no tomorrow. Instead, we shall be good stewards of nature, which we shall preserve for our children and posterity. The accolades of the future will be accorded for such things as pollution control, soil and water conservation, clean air, and preservation of rain forests.

2. **Food production emphasis will shift from quantity to quality.** In the future, food producers, processors, and marketers will extol the safety and nutritive values of foods, rather than quantity.

3. **People will have more leisure time.** With a shorter work week and more money to spend, people will have more lesiure time, much of which will be used in recreation and travel.

4. **There will be more convenience foods and more eating away from home.** With more women working, their kitchen time is limited, so, the current food trend will continue, unabated, (a) with more ready-to-cook and ready-to-serve foods, and (b) with more eating away from home.

5. **The keepers of herds and flocks will have more knowledge of animal behavior, and will improve animal welfare.** In the future, improved knowledge of animal behavior will make for improved animal welfare and animal profits, for all three—behavior, welfare, and profits—are on the same side of the ledger.

6. **There will be more and better visual aids.** The younger generation has been brought up on color TV, with the result that they desire that their educational material be presented graphically—and preferably in color. This calls for more illustrations in textbooks of the future, preferably augmented by color.

## HOW DO WE GET THERE?

My greatest challenge in preparing the ninth edition of this work was to assist the members of the animal science team—students, teachers, animal producers, researchers—in getting from hither to yon, to get from where we are now to where we are headed in the 21st century—and beyond.

So, as I revised and updated each part of *Animal Science*, and as a I added each new section and illustration to it, I asked myself this simple question: Will it be helpful? But the simple question generated complex answers! To assist the readers of *Animal Science* in following the blueprint for the future, I made the following major changes in the ninth edition of *Animal Science*:

1. **Revised and updated it.** The extent of these changes can best be described by the word *completely*, which tells it all.

2. **Made it 21st-century new.** This involved expanding many parts such as those on sustainable agriculture and animal behavior, adding new sections on such pertinent matters as the environment, and adding new and complete chapters on rabbits and fish.

3. **Added more visual aids.** More visual aids were added throughout the book. But, in selecting and adding each of them, I posed the question: Will it be helpful? Also, a beautiful and informative eight-page colored picture section was added.

4. **Made it more scientific, yet kept it practical.** In recognition that we are entering the age of biotechnology, which will make for more changes than in any period of history, *Animal Science* was made more scientific, yet it was kept practical.

If this ninth edition of *Animal Science* helps the members of the animal science team—students, teachers, animal producers, and researchers—to make a great leap into the 21st century—and beyond, I shall be amply rewarded.

In my writings, I rely on many people for help. I am especially indebted to the following persons who greatly pursued this revision of *Animal Science*: Audrey H. Ensminger, who shepherded it from my Missouri hieroglyphics stage in the manuscript to camera-ready; Dr. Lawrence A. Duewer, Agricultural Economist, USDA, ERS, who provided many of the statistical facts and figures; Joan and Lynn Wright, who did, or supervised, the typing of the entire manuscript; Margo Williams, who did the art and pasteup work; Randall and Susan Rapp, who set the type; and Jean Nelson and Deanna Ross, who did the proofreading. Additionally, a host of individuals, associations, and companies provided pictures, served as reviewers, or made other notable contributions, which are gratefully acknowledged at appropriate places throughout the book. Indeed, *Animal Science* was a dedicated effort for the animal science team—students, teachers, animal producers, and researchers.

*M. E. Ensminger*

Clovis, California
1991

# References

The following books are by the same author and publisher as *Animal Science:*

> *Animal Science Digest*
> *Beef Cattle Science*
> *Dairy Cattle Science*
> *Sheep and Goat Science*
> *Swine Science*
> *Poultry Science*
> *Horses and Horsemanship*
> *The Stockman's Handbook*
> *Stockman's Handbook Digest*

*Animal Science Digest* is a condensation of *Animal Science*. Each of the books devoted to a specific class of farm animals contains much wider and more complete coverage for the species indicated by name than is possible in *Animal Science*.

*The Stockman's Handbook* is a modern know-how, show-how book which contains, under one cover, the pertinent things that a livestock producer needs to know in the daily operation of a farm or ranch. It covers the broad field of animal agriculture, concisely and completely, and whenever possible, in tabular and outline form. *Stockman's Handbook Digest* is a condensation of *The Stockman's Handbook*.

# Other Selected References

| Title of Publication | Author(s) | Publisher |
|---|---|---|
| *Animal Agriculture* | H. H. Cole<br>M. Ronning | W. H. Freeman and Company, San Francisco, CA, 1974 |
| *Animal Science and Industry,* Fourth Edition | Duane Acker | Prentice-Hall, Englewood Cliffs, NJ, 1991 |
| *Introduction to Livestock Production* | H. H. Cole *et al.* | W. H. Freeman and Company, San Francisco, CA, 1966 |
| *Livestock and Poultry Production* | R. V. Diggins<br>V. W. Christensen | Prentice-Hall, Inc., Engelwood Cliffs, NJ, 1975 |
| *Modern Livestock & Poultry Production,* Second Edition | J. R. Gillespie | Delmar Publishers, Inc., Albany, NY, 1983 |
| *Science of Animal Husbandry, The* | J. Blakely<br>D. H. Bade | Reston Publishing Company, Inc., Reston, VA 1976 |
| *Science of Animals That Serve Mankind, The,* Third Edition | J. R. Campbell<br>J. F. Lasley | McGraw-Hill Book Company, New York, NY, 1985 |
| *Scientific Farm Animal Production,* Third Edition | R. E. Taylor<br>R. Bogart | Macmillan Publishing Company, New York, NY, 1988 |

# Animal Science Goes "High Tech"

*Fusion and Copy*, the first two cloned heifer calves reported, are the result of taking a single cell from a 16-cell embryo and transferring it into a one-cell egg from which the genetic material has been removed. The research project was sponsored by W. R. Grace & Co., and carried out jointly by the University of Wisconsin-Madison and American Breeders Service, a GRACE Animal Services company. (Courtesy, Dr. Robert Walton, American Breeders Service, DeForest, WI)

# Contents

Seeds of hope! (Courtesy, *Ceres*, FAO, United Nations, Rome, Italy; and Kongskilde of Denmark)

# FOOD AND ANIMALS— A GLOBAL PERSPECTIVE[1]

## Chapter 1

[1]Grateful appreciation is expressed to all those who assisted with this chapter by responding so generously to my call for authoritative information, facts, figures, and pictures. Very special thanks are due to the following: United Nations—Food and Agriculture Organization (FAO), Rome, Italy; and the U.S. Department of Agriculture, Washington, DC

It is important that both producers and consumers consider food and animals in the global perspective. The keepers of herds and flocks need to know which countries are potential competitors and which are likely markets ahead—to 2000 and beyond. Consumers—all—need to know from whence their meat, milk, eggs, and fish will come and how much to expect.

## WORLD FOOD DISTRIBUTION

Diogenes, the Greek philosopher, when asked about the proper time to eat, replied, "If a rich man, when you will; if a poor man, when you can." Today, well-fed nations are scrambling to eat better, and hungry nations to eat at all. One-third of all the people on earth go to bed hungry each night.

Famines are nothing new, of course. They are as old as the twelfth chapter of Genesis, where it is recorded, "And there was a famine in the land; and Abram went down into Egypt to sojourn there; for the famine was grievous in the land." Later, during the seven lean years, Joseph was in Egypt when "famine was over all the face of the earth" (Genesis 41:56). Also, famines were frequent in subsequent years. Every school child has read of the potato famines of Ireland. All these were *food famines,* caused by sudden failures in the production of food due to weather, water, disease, or pests. Today, the world faces the prospect of a new kind of famine—a famine caused by unequal distribution of food and lack of purchasing power, interspersed with droughts.

For 200 years, we proved Malthus[2] wrong because, as the population increased, new land was brought under cultivation; and machinery, chemicals, new crops and varieties, and irrigation were added to step up the yields. Now, science has given us the miracle of better health and longer life; and world population is increasing at the rate of about 240,000 people a day, or 87.5 million per year; and it is predicted that world population will be 6.2 billion by the year 2000, and be doubled by the year 2045 (see Fig. 1–1).

Despite an increase of 1.8 billion people in the world during the 25-year period 1961–63 to 1983–85, people as a whole were better fed than previously. On the average, the food available per capita rose from 2,320 calories to 2,660 calories. *But the exceptions were many!* In the low income countries as a group, apart from China and India, per capita food supplies in 1983–85 were no higher than 15 years earlier. More disturbing yet, some 500 million people in the world remained seriously undernourished, further widening the gap between the "haves" and the "have nots," between the rich and the poor.[3]

To match population growth and feed people adequately, world food production needs to increase at an *average* rate of about 2.5% per year. This is feasible. Yet people in parts

---

[2]Thomas Robert Malthus was an English clergyman. In 1798, he prophesied that world population grows faster than people's ability to increase food production.

[3]Those concerned with world food and population matters commonly classify countries on the basis of economies as developed, developing, or centrally planned. Among the developed economies are the U.S.A., Canada, West Germany, Japan, Australia, and New Zealand. The 94 countries with developing economies are located principally in Southeast Asia, Africa, and South America. The centrally planned economies include the Soviet Union and North Korea.

Classifying a large number of countries as "developing" tends to obscure the level of their income and the nature of their food situation. Some of the least developed countries are very poor. Others, such as oil-rich Kuwait and Saudi Arabia, are very wealthy; they, and other members of the OPEC (Organization of Petroleum Exporting Countries), are quite able to pay for cereal grain imports. Some developing countries are self-sufficient in food; and a few—Argentina and Thailand, for example—are net exporters of grain.

## THE POPULATION EXPLOSION

| World Population | 1980 | 1990 | 2000 |
|---|---|---|---|
| | *(millions)* | | |
| P.R. of China ..... | 983 | 1,115 | 1,242 |
| India ........... | 689 | 850 | 1,013 |
| Africa .......... | 493 | 660 | 887 |
| Latin America ..... | 365 | 445 | 551 |
| Europe ......... | 484 | 499 | 510 |
| U.S.S.R. ........ | 266 | 291 | 312 |
| North America .... | 252 | 277 | 296 |
| (United States) .... | 227 | 250 | 268 |
| Oceania ......... | 23 | 26 | 30 |
| **World Total** .... | **4,476** | **5,320** | **6,241** |

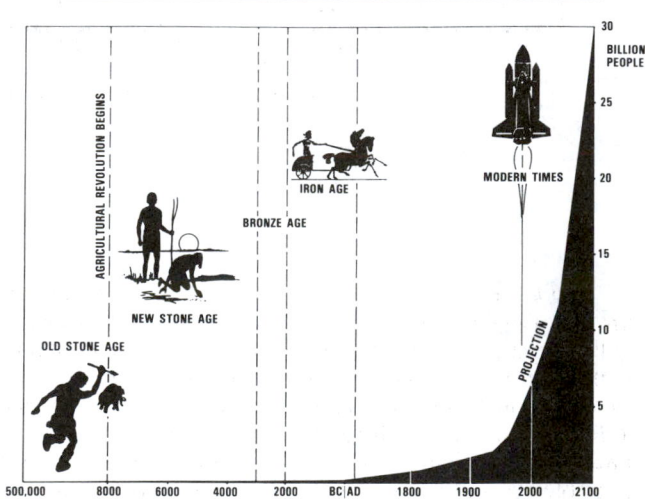

Fig. 1–1. World human population—past, present, and future.

of the world will continue to starve to death, because of lack of food or lack of money. For them, a world average increase in food of 2.5% per year does nothing to satisfy their hunger pangs when they are far below the average.

## WORLD NUTRITION DEFICIENCIES

In addition to food shortages (undernutrition), there is the equally vital issue of the nutritional adequacy of available supplies within countries (see Fig. 1–2) and the extent of malnutrition (caused by deficiency of a nutrient or nutrients). Over a billion people—⅓ of the world's population—are seriously short of calories in their diets; ⅔ of the people lack good quality protein (see Fig. 1–35).

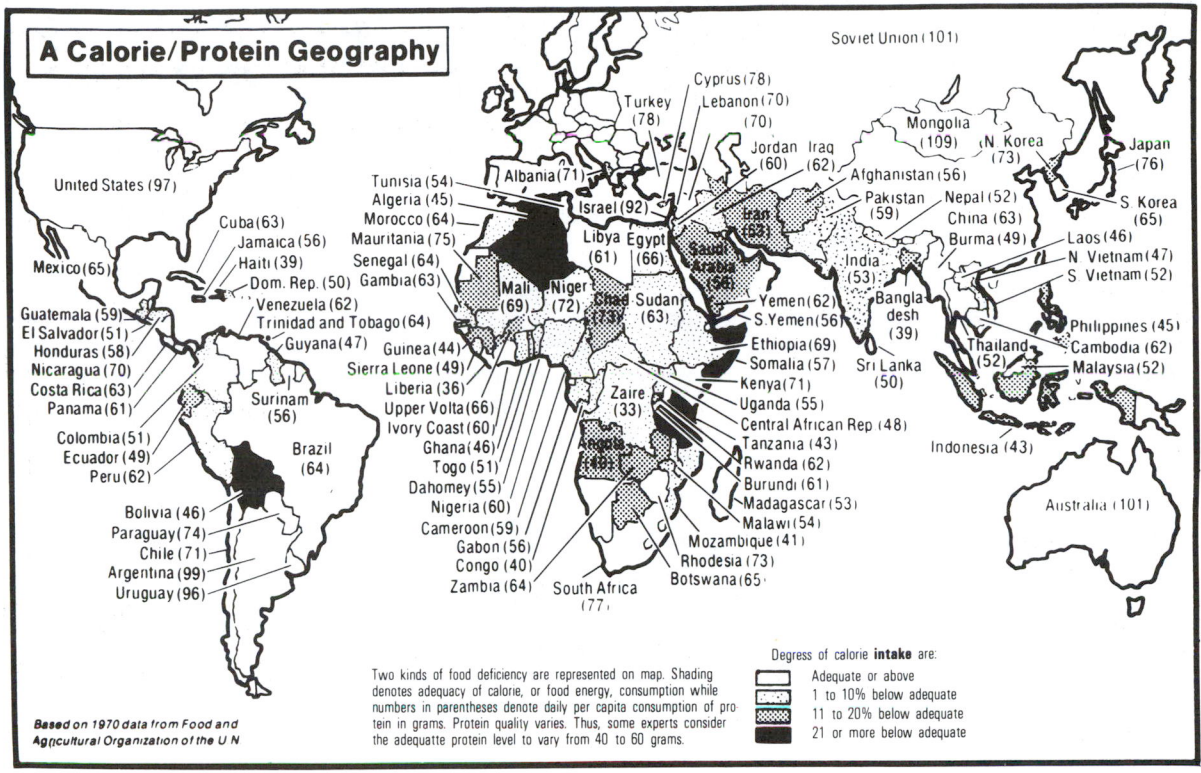

Fig. 1-2. World geography of calories and proteins. (Courtesy, *The New York Times*)

There are, of course, many adverse physical conditions brought about by prolonged periods of malnutrition, especially the absence of sufficient quantities of critical compounds in the diet. Rickets, scurvy, beriberi, pellagra, vitamin A deficiency (a major cause of blindness in many countries), anemia, iodine deficiency (which often leads to goiter and cretinism), kwashiorkor, and marasmus are but a few of the many classical examples of nutritional deficiency diseases. When malnutrition begins at an early age and persists over a substantial period of time, a variety of defects may develop involving bone and body structures and mental condition, all of which may be irreversible. Unfortunately, in the developing countries, the basic problem is one of gross dietary inadequacies which may be compounded by specific vitamin or protein insufficiency.

In Latin America, where protein-poor cereal grains are the imperfect staff of life, 82 out of every 1,000 children die before their first birthday; another 12 die before they reach the age of 4. Even the survivors may envy the dead; often brain-damaged, they become the adults who are the most in need of help and least able to help themselves.

It is not easy to present a precise picture of the extent of undernutrition and malnutrition in the world. Intake of energy and protein over a period of time appears to be the best indicator of dietary sufficiency. This information is presented in Fig. 1-2 and Table 1-1 which shows the per capita calories and proteins per day of the geographic areas of the world.

An insufficient intake of energy and/or protein needed to maintain the body functions—for activity, for growth, and for the reproductive cycle—can be manifested in the following ways: from the slight impairment of growth or thinness seen in mildly undernourished children to the gross alterations shown by persons suffering from kwashiorkor or marasmus; and between these two extremes the spectrum includes apathy, emaciation, incapacity to perform the required physical work, weight loss, inadequate weight gain during pregnancy, and low birth weight.

**TABLE 1-1**

**PER CAPITA CALORIES AND PROTEIN PER DAY IN GEOGRAPHICAL AREAS OF THE WORLD[1]**

| Area | Per Capita Calories Per Day | Per Capita Grams of Protein Per Day |
|---|---|---|
| | (no.) | (g) |
| Europe | 3,397 | 101.5 |
| U.S.S.R. | 3,394 | 105.6 |
| North and Central America | 3,370 | 94.8 |
| Oceania | 3,126 | 91.4 |
| South America | 2,622 | 66.2 |
| Asia | 2,485 | 60.7 |
| Africa | 2,299 | 57.4 |
| World Total | 2,694 | 70.3 |

[1]Source: *FAO Production Yearbook 1988*, United Nations, Rome, Italy, Vol. 42, pp. 291–294. Data for 1984–86.

## PROTEIN SHORTAGES

Protein malnutrition is the most serious and common cause of infant mortality and general debility in developing countries, and among the poor in developed countries. Inadequacy of proteins in the diet accounts for the premature deaths of as high as 40% of all children born in certain of the developing areas of the world. The diseases kwashiorkor (primarily protein deficiency) and marasmus (severe undernourishment) directly or indirectly account for 3 to 10 times greater infant mortality and as much as 20 to 50 times higher death rate among 1- to 4-year-old children in certain African countries than in the industrialized regions of the world. Children below 5 years of age may account for 40% of the total mortality. The low protein reserves of the body and generally poor nutritional status cannot sustain those children who develop fevers from respiratory and gastrointestinal infections, which markedly increase the body losses of protein.

The telltale symptoms of protein deficiency, which can readily be seen in the people of many of the developing countries of the world, are: stunted growth, bellies bloated by edema, diarrhea, brittle hair with an abnormal reddish color, and retarded mentality.

Food and Agriculture Organization (FAO) figures show that there are sufficient protein supplies in the world to meet minimum needs if they were properly distributed according to the individual needs of each person. But food supplies of different countries vary widely.

Although grain will remain the basic ingredient of the world's food supply, rising incomes and a strong desire for a more affluent life-style will increase the demand for more animal protein in the future.

## FOOD AND ANIMALS—NOW, AND IN THE YEARS AHEAD

The Food and Agriculture Organization of the United Nations, Rome, Italy, recently completed a big and scientific study of the agriculture of the 94 developing countries and the 34 developed countries, now and in the year 2000. The author briefed this important 263-page report plus appendix,[4] which is herewith presented.

- **Food and Animals—in the 1980s**— The status in the 1980s was as follows:

  1. **For Both Developing and Developed Countries.**
     a. **Human population growth rates were modest.** For the developing countries, the average population growth rate was 3.5% per year; for the developed countries, the annual growth rate was 2.5%.

     In the year 1989, world human population rose 87.5 million, and total world population reached 5.2 billion.

     b. **People as a whole were better fed.** Despite an increase of 1.8 billion people during the 25-year period 1961–63 to 1983–85, people as a whole were better fed. On the average, the food available per capita rose from 2,320 calories to 2,660 calories. *But the exceptions were many!* In the low income countries as a group, apart from China and India, per capita food supplies in 1983–85 were no higher than 15 years earlier. Some 335 to 500 million

people in the world remained seriously undernourished.
     c. **Fish catch slowed.** After almost doubling in the 1950s and 1960s, the world fish catch slowed. In 1985, world fish production totaled 85 million tons.
  2. **For Developing Countries.**
     a. **Hunger persisted.** Most of the 335 to 500 million undernourished people in the world in 1979–81 were in the developing countries, where hunger persisted.

     b. **Food supplies per capita did not improve.** During the 15-year period 1970 to 1985, the food supplies per capita in the developing countries did not improve significantly.

     c. **Productivity of food and labor increased.** During the 15-year period 1970 to 1985, biotechnology transformed the agriculture of the developing countries; it upped the rice yields by 41%, wheat yields by 77%, and labor productivity by 50%.

     d. **Rural poverty worsened.** In 1980, it was estimated that some 780 million people in the developing countries were living in absolute poverty, 90% of whom were rural people dependent wholly or partly on agriculture. Moreover, the study revealed that poverty is associated with (1) illiteracy, (2) undernourishment, and (3) high infant mortality.

- **Food and Animals—toward 2000 and beyond**— The projections/goals/needs toward 2000 and beyond follow:

  1. **For Both Developing and Developed Countries.**
     a. **More food for more people.** We need to provide food for one billion more people by the year 2000 than in the mid-1980s.

     b. **Production and consumption of food will be affected by new forces.** In the years ahead, the following forces will significantly influence trends in the production and consumption of food: (1) rise in income (purchasing power), (2) international trade, (3) attitude toward agriculture, and (4) improved technology.

     c. **Increased production will be achieved by science and technology.** Only limited new areas can be brought under cultivation. So, most of the food increase in the future will be achieved by the application of science and technology.

     d. **Biotechnology will make for new opportunities, but it will take time.** Recent advances in biotechnology offer hope and help for all people. It is now possible to transfer genes from one species to another without taking unwanted genes; and embryo transfer and cloning will become practical in the years ahead. But the widespread use of biotechnology will take time; it takes about 15 years from discovery to application—and even more time for widespread use.

     e. **Genetic potential of food crops will be maximized.** Despite the great strides that have been made in the past 50 years in increasing the yields of the major food crops, the potential for further improvement has not been exhausted. The development and use of high-yielding varieties of crops, worldwide, will continue to be an important part of technological progress.

     f. **Forests will be vital.** In addition to the economic value of timber, in the years ahead forests will be vital for the conservation of soil and water, and the security of agricultural production.

     g. **Fish demand and production will continue strong.** The demand for fishery products to the end of the century

---

[4]*Agriculture: Toward 2000,* Food and Agriculture Organization of The United Nations, Rome, Italy, 1987.

will remain strong. In the year 2000, the production of fish may exceed 100 million tons.

h. **The addition of soil nutrients will remain essential.** Technologies of the future must continue to provide for the application, with discretion, of chemical fertilizers and/or manure to the soil.

i. **The biggest challenge will be concern about the environment.** Meeting, by the year 2000, the 60% increase in agricultural production in the developing countries, and the 20% increase in the developed countries, will inevitably press more heavily on the natural resources, which are already under heavy strain. Thus, the challenge ahead is to increase agricultural production so as to feed more people, and to feed them well, without seriously damaging the world's natural resources. This calls for increased pest control, but with pesticides used in combination with biological pest control and other methods; the use of chemical fertilizers and manure in moderate amounts; the development of sustainable agriculture—and more.

Concern about the environment and saving the planet reached a new peak in 1989 prompted by (1) the earth's human population increasing 87.5 million that year, making for a total world population of 5.2 billion; (2) the spill of nearly 262,000 barrels of crude oil into the pristine waters of Alaska's Prince Williams Sound; (3) the deforestation, and burning of fossil fuels spewing an estimated 19 billion tons of carbon dioxide into the atmosphere, aggravating the global warming process that could cause the average worldwide temperature to rise as much as 8°F within the next 60 years; (4) the destruction of 28 million acres of tropical rain forests; (5) acid rain; and (6) the ozone hole over Antarctica remaining alarmingly large. Whether or not all the dire predictions come to pass, they do underscore the message: *The planet is in grave trouble.* Thus, nations need to take drastic action now; otherwise, the earth could one day be unfit for human habitation.

2. **For Developing Countries.**

a. **Elimination of hunger.** The primary responsibility for producing enough food to achieve this goal rests with the developing countries.

b. **Increased purchasing power.** Increased food supplies will not alleviate hunger unless accompanied by increased purchasing power of the poor.

c. **Nutrition will improve, but not everywhere.** For the 94 developing countries, average per capita food availability for direct human consumption will rise from 2,420 calories in 1983–85 to 2,620 calories in the year 2000. But, some low income countries with low per capita food consumption will remain low; in particular, this will be true of sub-Saharan Africa.

d. **Agricultural production in the developing countries will rise.** It is projected that agricultural production in the developing countries will rise about 3% from 1983–85 to the year 2000.

e. **Livestock production in the developing countries will increase.** The production of livestock products will grow the fastest. Most of the increase will come from pigs, poultry, and to a lesser extent, from dairy cattle. The projected methods of achieving increased meat production to the year 2000 are: 46% of the meat production will result from higher yields per animal, 20% from greater numbers, and 34% from improved pasture carrying capacity, health, and feed.

f. **Yield increases will be the major source of increased crop production.** Yield increases will be the major source of increased crop production in the developing countries, with 63% of the increase coming therefrom. Of the remaining 37% increased crop production, 22% will come from increased arable land, and 15% from multiple cropping during a given year.

g. **Continued increase in cereal grains for livestock feeds.** The use of cereal grains for animals in the developing countries, including China, is expected to more than double between the mid-1980s and the year 2000; with most of it fed to pigs, poultry, and dairy cattle.

h. **Increased irrigation.** The share of irrigated crops in relation to the total value of crop production is projected to rise by a fifth, to 43%, by the year 2000. Almost one-fifth of the arable land in the 93 developing nations (not including China) will then be irrigated; and two-thirds of all wheat and rice produced will be grown under irrigation.

i. **Fertilizer use will almost double.** Between 1982–84 and the year 2000, the use of chemical fertilizer is projected to more than double, with wheat and rice accounting for 55 to 60% of all fertilizer use.

j. **Other off-farm imports will continue to rise.** Mechanization (including tractors), improved seed, and the use of pesticides and herbicides will continue to increase in the developing countries.

k. **Employment in agriculture will continue to be large.** In the year 2000, the agricultural sector will still employ half of the labor force in the developing countries.

3. **For Developed Countries.**

a. **Decreasing demands, accompanied by a slowing down of their overproduction.** The developed countries will encounter decreased demands, both domestic and in exports, in the years ahead. So, they need to slow down their overproduction of foods and feeds, much of which is subsidized.

b. **Livestock products and cereals will continue to account for three-fourths of the developed market economies (DMEs).** But, in order to avoid surpluses, the meat exports of the developed countries need to be reduced from the 1.2% of their DMEs during the period of 1980–85 to approximately 0.9% in the year 2000; and the cereal grain exports need to be reduced from the 2.3% of their DMEs in 1980–85 to approximately 0.9% in the year 2000.

c. **Exports of milk and dairy products will continue to grow.** It is projected that there will be continued growth in the exports of milk and dairy products from the developed countries, although some low income countries will import these products only in the form of concessionary sales or food aid.

Both the World Health Organization and the U.S. Department of Agriculture voice grave concerns similar to the preceding FAO report. In summary form, their respective reports follow.

• **World Health Organization (WHO)**—According to WHO, 10 million children under the age of 5 are now chronically and severely malnourished, and 90 million more are moderately affected. While undernourished children may remain alive, they are extremely vulnerable to minor infectious diseases. WHO figures also show that of all the deaths in the poor countries, more than half occur among children under 5, and that the vast majority of these deaths, perhaps as many as 75%, are due to malnutrition complicated by infection.

According to WHO, in the first comprehensive study of the global toll of disease, issued in 1989, about 1.3 billion people, or more than 20% of the world's population, are sick or malnourished. The most affected areas are south and east Asia, where an estimated 500 million people, or about 40% of the population, suffer from malnutrition or diseases such as malaria, measles, diarrhea, and respiratory illness. Health problems are also severe in sub-Saharan Africa, where malnutrition and disease, including AIDS, are believed to affect 30% of the population, or 160 million people.

• **The U.S. Department of Agriculture (USDA)**— The USDA is authority for the following noteworthy statistics pertaining to world nutrition:[5]

1. In the developing countries, 62% of the calories come from direct consumption of cereals, compared to one-third of the calories from direct grain consumption in the developing countries.

2. The average person in the developed countries consumes more than four times as much meat and about six times as much milk and eggs as the average person in developing countries.

3. To close the nutritional gap between the developed and developing countries in terms of calories would require the addition of 500 calories per day per person to the diet of the developing countries, or only about 2% of the world's grain production annually, or about 25 million tons of cereals. But the mere production of 25 million tons more grain would not solve the problem of the world's malnourished, for the most difficult task of all would be to collect and deliver the grain to the malnourished people.

4. The value of food production per capita in the developed countries is more than five times as large as in the developing countries. This difference reflects (a) the higher level of income in developed countries which permits consumption of high-value food products such as meat, milk, and eggs; and (b) the much higher level of agricultural productivity per person.

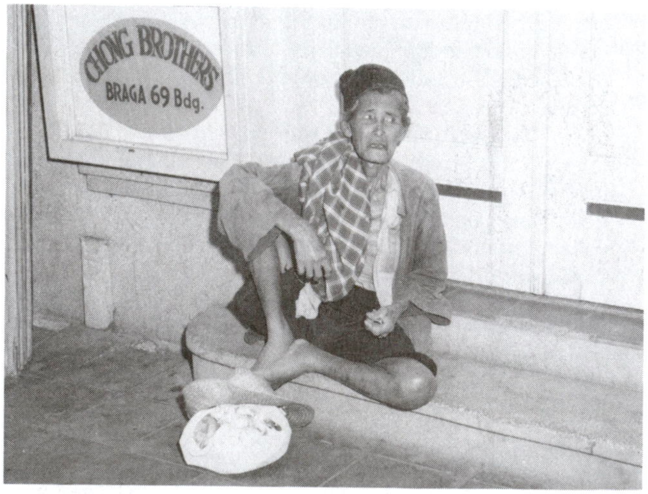

Nutritional deficiency diseases, which afflict both animals and humans, offer a very dramatic and vivid way in which to tell the story of undernutrition and malnutrition. A select few of these are presented in Figs. 1-3 through 1-10. It is noteworthy that nutritional deficiency areas throughout the world generally affect all species within the area, and that the manifestations of most deficiency diseases are similar regardless of species. For example, soils of an iodine-deficient area produce iodine-deficient crops and result in many big-necked (goiterous) animals and people. It is noteworthy, too, that much of our knowledge of human nutrition came the animal route.

Fig. 1-4. Cretinism, a condition originating during pregnancy or early infancy characterized by stunted physical and mental development, caused by severe thyroid deficiency. (Courtesy, FAO, Rome, Italy)

Fig. 1-3. Beriberi, a deficiency disease due to lack of vitamin B-1 (thiamin) in the diet. In the Far East, it is largely due to the almost exclusive use of polished rice. Note his cracked skin and swollen legs (edema). (Courtesy, FAO, Rome, Italy)

[5]*The World Food Situation and Prospects to 1985,* For. Ag. Econ. Rpt. No. 98, Economic Research Service, USDA, pp. 48–51.

Fig. 1-5. Goiter (big neck), caused by iodine deficiency. The enlarged thyroid gland (goiter) is nature's way of attempting to make sufficient thyroxin under conditions where a deficiency exists. (Courtesy, FAO, Rome, Italy)

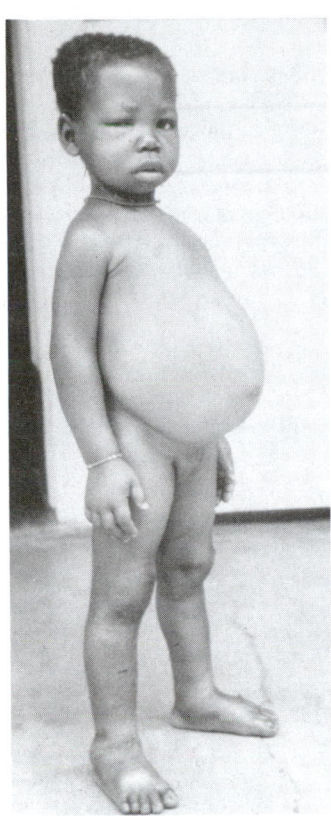

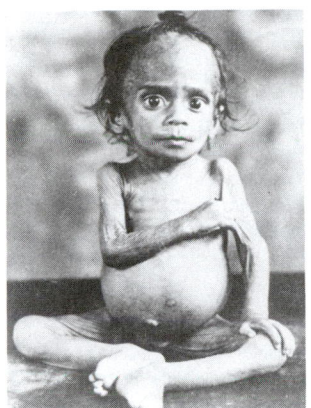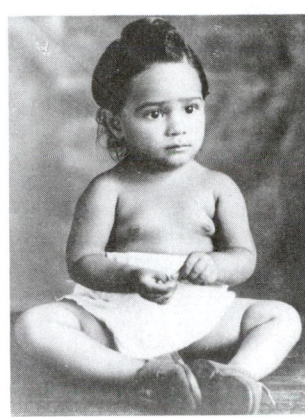

Fig. 1–7. Marasmus, a condition that occurs mainly in infants, resulting primarily from a lack of calories (energy foods) for a considerable period of time. *Left:* A 2-year-old girl, weighing 11.5 lb. *Right:* Same little girl after 10 months of treatment. (Courtesy, FAO, Rome, Italy)

Fig. 1–6. Kwashiorkor, a protein deficiency disease. Note characteristic bloated belly. Other usual symptoms are: stunted growth, diarrhea, brittle hair with an abnormal reddish color, and retarded mentality. (Courtesy, FAO, Rome, Italy)

Fig. 1–8. Pellagra, caused by a deficiency of niacin. Note the dermatitis (inflammation of the skin marked by reddening, swelling, oozing, crusting, and/or scaling). The disease is also characterized by a fiery red tongue, loss of appetite, nausea, and other symptoms. (Courtesy, FAO, Rome, Italy)

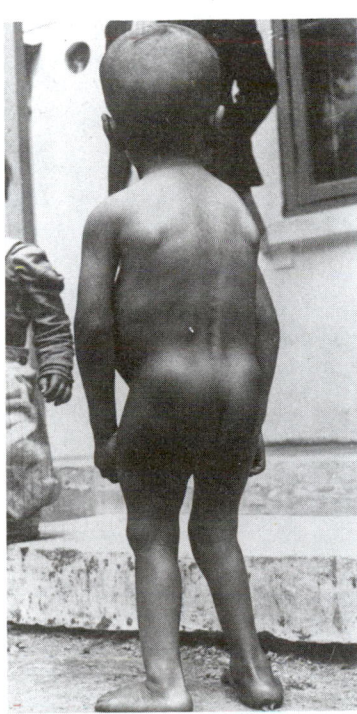

Fig. 1–9. Rickets, caused by lack of either calcium, phosphorus, or vitamin D; or an incorrect ratio of the two minerals. This back view shows curvature of the spine, enlarged knees, and crooked legs. (Courtesy, FAO, Rome, Italy)

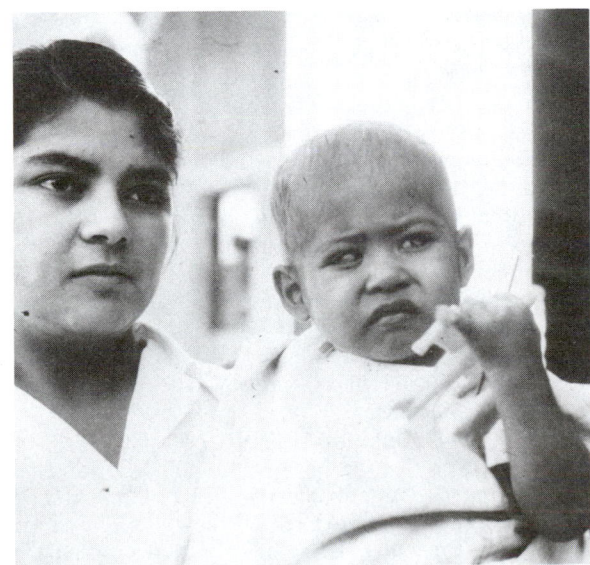

Fig. 1–10. Vitamin A deficiency, resulting in keratomalacia, characterized by a softening and ulceration of the cornea of the eye. Unless carotene or vitamin A is provided, blindness may follow. (Courtesy, FAO, Rome, Italy)

## FACTORS AFFECTING THE SUPPLY OF FOOD

The world's ability to supply food depends on (1) the availability and use of land; (2) water; (3) irrigation; (4) fertilizer; (5) yield, including the Green Revolution; (6) weather and climate; and (7) incentives to producers. The efficiency of food marketing and distribution systems and the size, organization, and management of agricultural enterprises also influence food supply.

## LAND

Only limited new land areas can be brought under cultivation.

Table 1-2 shows, by areas: (1) the acres/hectares of cultivated land per person, ranked in descending order; (2) the total land area; (3) the total cultivated land; (4) the percent of total land area cultivated; and (5) the human population.

It is noteworthy that Asia has the least cultivated land per person (only 0.38 acre) and Oceania (Australia and New

**TABLE 1-2**
**CULTIVATED LAND PER PERSON BY GEOGRAPHICAL AREAS[1]**

| Geographical Area | Cultivated Land Per Person | | Total Land Area | | Total Cultivated Land | | Percent of Total Land Area Cultivated | Human Population |
|---|---|---|---|---|---|---|---|---|
| | | | (1,000) | | (1,000) | | | |
| | (acres) | (hectares) | (acres) | (hectares) | (acres) | (hectares) | (%) | (1,000) |
| Oceania | 4.68 | 1.89 | 2,081,783 | 842,827 | 120,684 | 48,860 | 6 | 25,757 |
| U.S.S.R. | 2.01 | 0.82 | 5,501,184 | 2,227,200 | 574,448 | 232,570 | 11 | 285,993 |
| North & Central America | 1.63 | 0.66 | 5,280,356 | 2,137,796 | 676,417 | 273,853 | 13 | 417,276 |
| South America | 1.23 | 0.50 | 4,331,078 | 1,753,473 | 350,671 | 141,972 | 8 | 285,024 |
| Africa | 0.75 | 0.30 | 7,320,158 | 2,963,627 | 457,997 | 185,424 | 6 | 609,922 |
| Europe | 0.70 | 0.29 | 1,168,211 | 472,960 | 346,047 | 140,100 | 30 | 496,812 |
| Asia | 0.38 | 0.15 | 6,616,273 | 2,678,653 | 1,113,772 | 450,920 | 16 | 2,994,005 |
| World | 0.70 | 0.28 | 33,070,966 | 13,389,055 | 3,640,037 | 1,473,699 | 11 | 5,114,788 |

[1]Source: *FAO Production Yearbook 1988*, United Nations, Rome, Italy, Vol. 42, pp. 47-79. Data for 1987-88.

Zealand) the most (4.68 acres). It is noteworthy, too, that on the basis of ''percent of total land area cultivated,'' Europe is the highest (30%), and Africa and Oceania are the lowest (6%).

Of course, some of the land not cultivated presently could be brought under the plow with simple technology, whereas other parts of it could be brought under cultivation only with modern and costly technology. So, although the world now uses slightly less than half of the land area potentially suitable for crop production, most of the additional suitable land lies outside the densely populated countries. For example, India, Bangladesh, and Egypt have precious little new land suitable for bringing under cultivation. Also, the potentially cultivatable new lands are not surplus acres available merely for the taking for crop production. Many areas are inaccessible or cannot be developed economically at this time. Still other areas are in permanent pastures or forested; hence, putting them under the plow would adversely affect the supplies of animal and forest products. Thus, a good part of future food production gains will have to come from yield-increasing techniques—more irrigation, more fertilizer, improved seed varieties, and better cultural practices.

In developing countries, size of land area is especially critical in food production because human labor and farm-produced capital—draft animals, manure, homemade equipment, ditches, and wells—are often the only resources available to farmers to augment their land's basic production capabilities.

Size of land area becomes less important as people learn about and can afford means of increasing output through modern techniques.

## WATER

It has long been assumed that water is abundant and available at little cost. But water suitable for human needs and for irrigating crops is becoming scarce in many parts of the world.

As the population increases, the demands for water will increase for direct use by people, for industry and for agriculture. Competition for the available supply will become acute, areas of critical shortage will increase in number and size, and limitations on the availability of water for agricultural uses may affect food production.

It is estimated that the theoretical minimum water requirement to sustain life is about 300 gal per person per day. This is based on the approximate amount of water necessary to produce 2½ lb of bread. But people cannot live by bread alone. To provide a person with 2 lb of bread plus 1 lb of beef per day increases the daily water requirement to about 2,500 gal.

Farmers have increased the availability of water, while at the same time conserving the soil by terracing, contour farming, construction of dams, and other techniques designed to slow and reduce water runoff. But much more water and soil conservation work needs to be done.

In the future, water shortage may well become as critical as land shortage in limiting food production and lowering the living standard in many areas. Thus, the time has come when we need to recognize that water is precious and should be conserved. Also, we need to intensify research in an attempt to find an economically feasible method of desalting sea water.

## IRRIGATION

Development and utilization of irrigation water has not reached its full potential around the world. In many areas, irrigation is both a crucial input and a prerequisite for improved production.

To gauge both the impact of future droughts on production and the possibilities of future agricultural progress, it is useful to examine which countries have the most irrigation (Table 1–3) and which countries are most intensively irrigated (Table 1–4). It is noteworthy that the top 5 countries in terms of irrigated areas—China, India, U.S.S.R., United States, and Pakistan—have over 62% of the world's irrigated area, with China alone accounting for almost 20%.

Fig. 1–11. A Persian waterwheel with buckets, powered by a camel, raising water for irrigation in Pakistan. (Courtesy, International Bank for Reconstruction and Development, International Development Association, Washington, DC)

### TABLE 1–3
### MAJOR IRRIGATING COUNTRIES, RANKED BY IRRIGATED AREA[1]

| Country | Irrigated Area | | Cultivated Area[2] | | Percentage Irrigated |
|---|---|---|---|---|---|
| | (1,000 acres) | (1,000) hectares | (1,000 acres) | (1,000 hectares) | (%) |
| China | 110,738 | 44,833 | 239,531 | 96,976 | 47 |
| India | 103,987 | 42,100 | 417,406 | 168,990 | 25 |
| U.S.S.R. | 50,598 | 20,485 | 574,448 | 232,570 | 9 |
| United States | 44,712 | 18,102 | 469,090 | 189,915 | 10 |
| Pakistan | 39,718 | 16,080 | 51,278 | 20,760 | 77 |
| Indonesia | 18,278 | 7,400 | 52,414 | 21,220 | 35 |
| Iran | 14,178 | 5,740 | 36,631 | 14,830 | 39 |
| Mexico | 12,558 | 5,084 | 61,022 | 24,705 | 21 |
| Thailand | 9,871 | 3,996 | 49,524 | 20,050 | 20 |
| Romania | 8,314 | 3,366 | 26,395 | 10,686 | 32 |
| Spain | 8,077 | 3,270 | 50,450 | 20,425 | 16 |
| Italy | 7,534 | 3,050 | 30,053 | 12,167 | |
| World | 560,957 | 227,108 | 3,640,037 | 1,473,699 | 15 |

[1]Source: *FAO Production Yearbook 1988*, United Nations, Rome, Italy, Vol. 42, pp. 47–59.

[2]Cultivated area is arable land plus land under permanent crops.

### TABLE 1–4
### MAJOR IRRIGATING COUNTRIES, RANKED BY PERCENTAGE OF AREA IRRIGATED[1]

| Country | Percentage Irrigated | Cultivated Area[2] | | Irrigated Area | |
|---|---|---|---|---|---|
| | (%) | (1,000 acres) | (1,000) hectares | (1,000 acres) | (1,000 hectares) |
| Egypt | 100 | 6,324 | 2,560 | 6,324 | 2,560 |
| Pakistan | 77 | 51,278 | 20,760 | 39,718 | 16,080 |
| Israel | 63 | 1,082 | 438 | 687 | 278 |
| Japan | 61 | 11,629 | 4,708 | 7,139 | 2,890 |
| Korea, Rep. of | 59 | 5,294 | 2,143 | 3,113 | 1,260 |
| Albania | 57 | 1,764 | 714 | 1,011 | 409 |
| China | 47 | 239,531 | 96,976 | 110,738 | 44,833 |
| Iran | 39 | 36,631 | 14,830 | 14,178 | 5,740 |
| Indonesia | 35 | 52,414 | 21,220 | 18,278 | 7,400 |
| Peru | 33 | 9,201 | 3,725 | 3,039 | 1,230 |
| Iraq | 32 | 13,462 | 5,450 | 4,323 | 1,750 |
| Romania | 32 | 26,395 | 10,686 | 8,314 | 3,366 |
| World | 15 | 3,640,037 | 1,473,699 | 560,957 | 227,108 |

[1]Source: *FAO Production Yearbook 1988*, United Nations, Rome, Italy, Vol. 42, pp. 47–59.

[2]Cultivated area is arable land plus land under permanent crops.

In judging a country's agricultural productivity and ability to withstand drought, the percentage of land irrigated may be more important than the total area irrigated. Egypt, Pakistan, Israel, Japan, and Korea are the top five countries in percentage of land irrigated (Table 1–4).

The People's Republic of China ranks first in area irrigated and seventh in the percentage of cultivated area irrigated. This explains how, to a large extent, China has been able to produce enough food for over one billion people on 10% of their total available land. In contrast to China, India—which is second in irrigated area (Table 1–3)—is not among the top 12 countries in proportion of the crop area irrigated (Table 1–4). Thus, it is not surprising to find that Indian yields for rice, a principal irrigated crop in both countries, is less than half (44%) of China's yields.

## FERTILIZER

Although more land can be brought into production, studies of world food production conclude that, except for Africa and Latin America, yield-increasing techniques will be the primary source of future growth. Fertilizer, which is in part a substitute for more land, is a key factor in yield increases, although it must be combined with improved varieties of seeds and improved cultural practices if it is to have much impact on yields.

World fertilizer production and consumption has increased rapidly, but unevenly, since 1950. In 1987/88, world fertilizer consumption totaled 140.5 million metric tons (see Table 1–5).

**TABLE 1–5**
**WORLD PRODUCTION AND CONSUMPTION OF COMMERCIAL FERTILIZERS[1]**

| | Nitrogen | | Phosphate | | Potash | | Total NPK | |
|---|---|---|---|---|---|---|---|---|
| | Production | Consumption | Production | Consumption | Production | Consumption | Production | Consumption |
| | (thousand metric tons) | | | | | | | |
| Africa | 1,856 | 1,885 | 1,888 | 1,146 | — | 446 | 3,744 | 3,477 |
| North & Central America | 16,384 | 12,712 | 10,402 | 4,997 | 8,839 | 2,350 | 35,625 | 23,058 |
| South America | 1,443 | 1,973 | 1,603 | 2,235 | 37 | 1,907 | 3,083 | 6,115 |
| Asia | 28,570 | 31,600 | 8,806 | 11,107 | 2,014 | 4,159 | 39,390 | 46,866 |
| Europe | 18,167 | 15,655 | 7,137 | 7,799 | 8,745 | 8,396 | 34,050 | 31,850 |
| Oceania | 288 | 428 | 932 | 1,100 | — | 229 | 1,221 | 1,757 |
| U.S.S.R. | 15,538 | 11,787 | 8,840 | 8,564 | 10,888 | 7,052 | 35,266 | 27,403 |
| World total | 82,246 | 76,040 | 39,609 | 36,948 | 30,523 | 27,539 | 152,379 | 140,526 |

[1]Source: *1988 FAO Fertilizer Yearbook*, Vol. 38, United Nations, Rome, Italy. Data for 1987–88.

The major fertilizer-exporting countries, in order, are Canada, U.S.A., and the U.S.S.R.; while the developing regions of Africa, Latin America, and Asia have been large importers. The largest exporters of each of the plant nutrients are: for nitrogen, U.S.A. and U.S.S.R.; for phosphate, the United States; and for potash, Canada.

## YIELD

The most effective means of increasing the total supply of any farm product is the adoption of new technology, thereby increasing the production per acre or per animal.

The continuing impact of improved production technology is evident in the increasing grain yields in the developed countries and in the achievements of the high-yielding varieties, especially wheat and rice, in the developing countries.

Doubts have arisen as to whether future crop yields will increase at the rates that have been achieved in the past. Attention has been focused on an apparent slowdown in the rate of yield increases for some crops in certain developed countries and on the apparent loss of momentum of the Green Revolution in developing countries. But we must not forget that it took 36 years for hybrid corn to achieve 95% acceptance in the United States.

In 1988, the developed countries produced 2,493 lb of cereals per acre, whereas the developing countries produced only 2,027 lb of cereals per acre. During the 30-year period, 1950–1980, U.S. wheat yield doubled, corn yield increased by

2.4 times, and sorghum yields tripled, due primarily to greater use of fertilizer, improved seed varieties, and better cultivation practices. There is every reason to believe that similar production increases are possible in the developing world.

## THE GREEN REVOLUTION

To most people, the Green Revolution refers to the adoption of high-yielding varieties of wheat and rice. It all started with a short-strawed wheat developed by Norman Borlaug, a Rockefeller Foundation scientist stationed in Mexico, who evolved with new spring wheats that helped Mexico and Southeast Asia close the food gap, a development that brought him fame and a Nobel Prize. The first of these improved wheat varieties was released in 1948.

Actually, the Green Revolution is a combination of techniques, including short-strawed varieties of wheat and rice, along with proper irrigation, fertilizer, and management.

Today, more than 90% of the wheat of Mexico consists of "shorty" varieties; and half, or more, of the wheat and rice of the Asian developing nations are of the new high-yielding varieties. It is expected that fuller adoption of the new varieties of wheat and rice will be achieved in Asia, but it will take time, just as it took from one to two decades for high-yielding varieties of wheat to be accepted throughout Mexico. Also, it is recognized that new agricultural technologies are accepted much more quickly and fully in progressive regions than in poorer regions.

But high-yielding varieties are only one component of the Green Revolution; the other major ingredients are (1) improved water control, (2) increased use of farm chemicals for fertilization and plant protection, and (3) improved management practices such as seedbed preparation, seeding rates, weed control, and timing of fertilizer applications. Also, if the Green Revolution is to be maintained or expanded, newer varieties must constantly be developed and distributed, with much of this work done in the area where they are to be used.

The Green Revolution bought the world extra time in which to get food and people into balance.

## WEATHER AND CLIMATE

Most of the world's food supply still depends on the weather. Persistent, widespread droughts; heavy flooding; changes in the severity of winter weather; and shifts in monsoons give rise to claims that global shifts in climate are in progress. But there is no reputable scientific evidence either to support or refute such claims.

Climate—the average weather variables over a considerable period of years—is much more stable than the short-period fluctuations in the weather. Yet, evidence of long-term climatic change has been obtained from descriptive historical records and natural phenomena, such as the width of tree rings, vegetative layers in peat bogs, pollen samples in lake sediment, glacial deposits, fossil remains of plants and animals, carbon dating, and core samples from ocean floors and the polar ice caps.

Projections for future food production levels generally rest on the assumption that *normal* weather can be expected to prevail. But the policies and programs for expanding food production should include provisions for the possibility that weather conditions could be either less favorable or more favorable than normal. This underscores the need for flexible world food policies to adapt to changes in conditions and to provide a margin of security against sudden or unexpected changes.

## INCENTIVES TO PRODUCERS

Lack of profit incentive has been a major impediment to increasing food production in many countries of the world. Policies designed to maintain low and stable food prices to consumers have dampened the farmer's incentive to produce, and been partially responsible for the large grain imports of certain countries.

## FACTORS AFFECTING THE DEMAND FOR FOOD

The demand for food depends primarily upon (1) human population growth, (2) income growth (disposable income), (3) the proportion of income spent for food, (4) income distribution, (5) food preferences and trends, and (6) the nutritive qualities.

## HUMAN POPULATION GROWTH

Population growth is the major determinant of demand for food.

The population of the world first topped one billion about the year 1830. It took from the dawn of humankind until nearly 1,600 years after the birth of Christ for the number of people in the world to build up to that point. Disease had a major role in holding population down. It took another 100 years for the world to add its second billion people, which we reached about 1930. It took a mere 30 years more to reach 3 billion. In 1989, world population reached an estimated 5.2 billion.

Currently, world population is compounding at the rate of about 1.7% a year—a deceptively small number, but remember that this increase is like compound interest. In actual numbers, the population increase amounts to 5.55 additional people per second, some 240,000 more per day, over 7.3 million per month, and about 87.5 million per year. The demographers predict that the world will have 6.2 billion people in the year 2000; 7.2 billion in the year 2010; 8.2 billion in the year 2020; 9.1 billion in the year 2030; and 10.8 billion people in the year 2050.[6] But the developing countries will account for most of the increase. For every birth in the developed countries, there will be 3 in the developing countries.

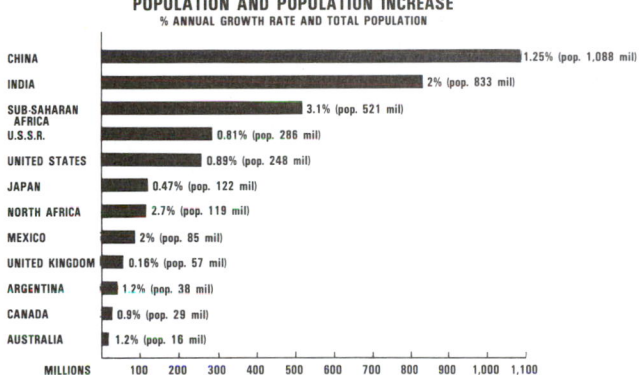

**POPULATION AND POPULATION INCREASE**
% ANNUAL GROWTH RATE AND TOTAL POPULATION

| | |
|---|---|
| CHINA | 1.25% (pop. 1,088 mil) |
| INDIA | 2% (pop. 833 mil) |
| SUB-SAHARAN AFRICA | 3.1% (pop. 521 mil) |
| U.S.S.R. | 0.81% (pop. 286 mil) |
| UNITED STATES | 0.89% (pop. 248 mil) |
| JAPAN | 0.47% (pop. 122 mil) |
| NORTH AFRICA | 2.7% (pop. 119 mil) |
| MEXICO | 2% (pop. 85 mil) |
| UNITED KINGDOM | 0.16% (pop. 57 mil) |
| ARGENTINA | 1.2% (pop. 38 mil) |
| CANADA | 0.9% (pop. 29 mil) |
| AUSTRALIA | 1.2% (pop. 16 mil) |

MILLIONS   100  200  300  400  500  600  700  800  900  1,000  1,100

Fig. 1-12. 1989 population and annual growth rate of selected countries. (Source: *World Population Growth by Country and Region, 1950–86, and Projections to 2050*, USDA, ERS, 1988)

Do these population figures foretell the fulfillment of the doomsday prophecy of Thomas Malthus, made nearly two centuries ago, that the world population grows faster than humankind's ability to increase food production? The author's answer to this question is "not immediately, but perhaps eventually." The Reverend Malthus did not foresee scientific-technological and sociological developments which would both speed food production increases and slow population growth. For example, the Green Revolution, which bought extra time, and the biotechnological era now in progress, were outside Malthus' reckoning. Nor did Malthus foresee the use of "the pill," which has reduced the population growth in Europe, Japan, Oceania, and North America to a fraction of the biological potential which he cited. Nevertheless, as we race to the year 2000, there will be famine in different areas of the world from time to time, and it will be increasingly difficult to feed a growing population beyond 2000. We must remember, too, that, no matter how rigid and effective, there is a time lag of 15 to 20 years following any policy changes designed to curb population.

[6]Urban, F. and P. Rose, *World Population by Country and Region, 1950–86, and Projections to 2050*, U.S. Department of Agriculture, ERS, Agriculture and Trade Analysis, April 1988, p. 3.

## INCOME GROWTH (DISPOSABLE INCOME)

Rising disposable incomes around the world are making for higher standards of living and increased demands for the good things of life, including food. This, along with more people to feed, is causing supply to lag behind. Rising incomes in the developing countries, which are unequally distributed, result in demand for more grain.

Rising incomes in the developed countries result in consumption of less grain directly and the conversion of more grain to meat, milk, and eggs; hence, the consumption of animal protein is closely related to income.

Since the early 1970s, competition between food and feed uses of grain has been apparent, as has competition between the world's rich and poor for all foods.

The above trend is expected to prevail in the years ahead, mildly relieved or accentuated in keeping with the relative abundance or the world's harvest.

Although incomes in the developing nations have improved in recent years, they are still very low. The developed countries pay their workers 10 to 20 times as much as the workers are paid in the developing countries.

## PROPORTION OF INCOME SPENT FOR FOOD

As income rises, consumers spend a smaller proportion of their disposable income for food (Fig. 1–13). In most low-income countries, consumers spend about one-half of their income for food; whereas the proportion drops less than one-seventh in the highest income countries. The data contained in Fig. 1–13 show that consumers spent the following proportion of their income for food:

- **Percent of income spent for food by consumers in the six countries spending least for food:** U.S., 10.4%; Canada, 11.5%; U.K., 13.7%; Netherlands, 14.4%; Luxembourg, 15.5%; and Australia, 15.5%.

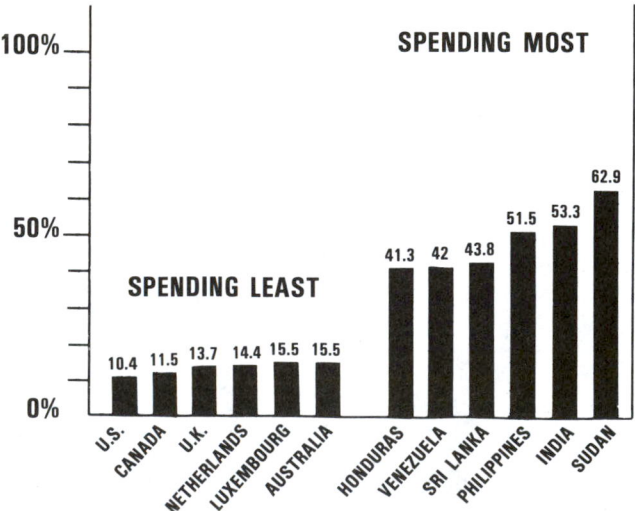

## PERCENT OF INCOME SPENT ON FOOD

Fig. 1–13. Proportion of income spent for food. As income rises, consumers spend a smaller proportion of their wages for food. Most countries are experiencing a decline in the amount of income spent for food due to greater efficiency in production, processing, and distribution. (Source: *National Food Review*, Oct.-Dec. 1989, USDA, p. 29) Data for 1986.

- **Percent of income spent for food by consumers in the six countries spending most for food:** Honduras, 41.3%; Venezuela, 42%; Sri Lanka, 43.8%; Philippines, 51.5%; India, 53.3%; and Sudan, 62.9%.

Also, food preferences, or eating habits, change with income. The hierarchy of preferences, from lowest income to highest income is: lowest for roots and tubers, a little higher for coarse grains (corn, sorghum, etc.) for human consumption, and progressively higher for other cereals (wheat and rice), pulses (the edible seeds of leguminous crops, like peas and beans), fruits and vegetables, and animal products.

Low-income consumers spend a high proportion of their budget for direct cereal grain consumption. High-income consumers spend less of their food budget for cereals and more of it for livestock products. In the United States, consumers spend about 2.5% of their disposable income for red meat and poultry.

## INCOME DISTRIBUTION

World income distribution is very unequal, as evidenced by the gross national product (the gross domestic product at purchaser values [market prices] plus net factor income from abroad) per capita in U.S. dollars in 1987: low-income countries (low-income Africa, China, and India, and other low-income Asia), $290; developing countries as a whole, $720; industrialized countries (Japan, United States, and European Community), $14,677.[7]

Two-thirds of the world's population—about 3.5 billion people—live in countries in which the gross national product averages about $720 per capita. While their incomes are expected to increase, they will change slowly and little to the year 2000—and perhaps beyond. Thus, the character of their demand for food cannot be expected to change very much—they will continue to be heavily dependent on grains, tubers, and root crops. Since a large proportion of these 3.5 billion people have incomes substantially below the $720 average level, and since the proportion of people in this group is likely to rise, the demand for food will be strongly influenced by demands of the lower income groups; hence, this demand will be largely for grain for direct consumption. For this reason, the food situation for a large part of the world's population will continue to be a problem of grain—its availability and price.

## FOOD PREFERENCES AND TRENDS

The selection of foods is based on taste preferences and relative availability/price, with different foods competing with one another for a place in the consumer's diet. In prosperous times and with increased buying power, the preferred foods increase in demand and price. Because cereal and bakery products are not luxury foods, their per capita consumption is little affected by buying power.

Based on per capita consumption by countries, people show the following meat preferences: Argentina and Uruguay favor beef; Hungary and Germany favor pork; New Zealand and Australia favor lamb; the United States and Israel favor poultry; and Japan favors fish.

[7]*World Tables, 1988–89 Edition,* from the data files of the World Bank, published by the World Bank, The Johns Hopkins University Press, Baltimore and London, 1989, pp. 16–17.

Worldwide, two-income families indicate that consumers have more money to spend for food and less time for preparing food. This has led to an increase in both convenience foods and in eating away from home—at restaurants, cafeterias, and other eating places. Also, health conscious people are eating more poultry and less red meat, more fruit and vegetables, and more low-fat dairy products instead of whole milk.

## NUTRITIVE QUALITIES

Today, enlightened people throughout the world are diet conscious. To them, animal products are not only good, but they're good for them. To them, meat, milk, eggs, and fish are far more than just very tempting and delicious foods. To them, these products contain the nutritive essentials of an adequate diet—high-quality protein, plus calories, minerals, and vitamins.

Among those who know, the nutritive qualities of animal products are a major factor in determining demand. Knowledgeable and financially able people are willing to pay a premium for meat, milk, eggs, and fish.

Additional pertinent facts about the nutritive qualities of animal products follow:

1. **Protein**. The word *protein* comes from the Greek word *proteios,* meaning *in first place.* It merits this definition because it is recognized as the most important body-builder. Next to water, it's the most plentiful substance in the human body. The major portion of body protein is found in the muscles. However, there is some protein in the blood, bones, teeth, hair, eyes, skin, and nails. In fact, protein is found in every living cell.

Protein is needed to build new tissue and to maintain and repair the old. It's not a fixed, unchanging substance; rather, it's in a constant state of exchange. While some is breaking down, more is being built as replacement. The need for new tissue is greatest in the young, but there is a continuing need for the adult, also. The hair and nails continue to grow, and the outer layer of the skin scales off and must be replaced.

The hormones, the chemical regulators of body processes, and the enzymes, the spark plugs of chemical reactions, are protein, too. Also, protein is one of the factors which control the fluid movement in and out of cells and to and from the bloodstream.

Animal products are much higher in protein content than cereal grains. On a dry basis (which is the only true comparable basis), the comparison is as follows: fish (carp), 81.0%; eggs, 47.0%; beef (Choice grade, total edible, trimmed to retail level, raw), 30.7%; milk (whole), 26.4%; wheat (whole flour, hard wheat), 15.1%; and rice (milled or polished), 7.6%.

In addition to supplying more protein than the cereals, animal protein is of higher quality than cereal protein. It is a complete protein—it contains all the essential amino acids, or building stones, to maintain, repair, and build body tissue; and the proportion of amino acids in animal protein almost exactly parallels that in human protein. Certain amino acids, of which there are 23, can be made by the body to satisfy its needs. Others cannot be formed fast enough to supply the body's needs, and therefore, are known as essential amino acids. These must be supplied in the food.

An incomplete protein will maintain life, but it will not support growth. it does not contain all the essential amino acids. Generally speaking, vegetable proteins are incomplete, although there are exceptions, such as the protein from soybeans.

2. **Calories, minerals, vitamins, and digestibility.** Nutritionally, people eat animal products primarily for their protein content. However, meat, milk, eggs, and fish are rich sources of other nutrients.

Generally speaking, animal products are moderate from the standpoint of calorie content. For example, a medium-sized egg contains about 77 calories. The energy value of meat, milk, and fish varies according to the amount of fat. A pound of moderately fat beef provides about 1,350 calories, or nearly half the daily energy requirement of the average adult.

Animal products are good sources of minerals, but there are differences in their content. Meat and eggs are an especially good source of phosphorus and iron. Milk is an excellent source of calcium and phosphorus; but it is a poor source of iron, copper, and iodine. Fish is high in phosphorus and fair in calcium and iron.

Animal products contain all the vitamins required by humans. They are a rich source of the important B vitamins, including vitamin B-12.

Finally, in considering the nutritive qualities of animal products, it should be noted that they are highly digestible. For example, about 97% of meat protein is digested, whereas wheat protein is only 79% digestible.

## WORLD PRODUCTION AND CONSUMPTION OF ANIMAL PRODUCTS

In general, meat production and consumption are highest in countries that have extensive grasslands, temperate climates, well-developed livestock industries, and sparse population. In many of the older and more densely populated regions of the world, insufficient grain is produced to support the human population even when consumed directly. This lessens the possibility of keeping animals, except for consuming forages and other humanly inedible feeds. Certainly, when it is a choice between the luxury of meat and animal by-products or starvation, people will elect to accept a lower standard of living and go on a grain diet. In addition to the available meat supply, food habit and religious restrictions affect the kind and amount of meat produced and consumed.

Table 1-6 (see next page) shows the meat and egg production and fish catch in selected countries. Note the favored position of beef in the United States, which produces 10.8 million metric tons annually, or about 22% of the world production of beef. Note, too, that worldwide, on a total tonnage produced basis, meats rank as follows in descending order: pork, beef, poultry, and mutton/lamb/goat meat. In egg production, the three leading countries, by rank, are China, U.S.S.R., and the United States. In fish catch, the three leading countries by rank are: Japan, U.S.S.R., and China.

Fig. 1-14 shows the total per capita consumption of all red meat in certain specified countries, whereas, Figs. 1-15 through 1-20 show the world's leading consumers of beef and veal; pork; mutton, lamb, and goat meat; poultry meat; milk; and eggs, respectively.

From 1965 to 1968, New Zealand was the world's largest per capita consumer of red meat, followed closely by Australia. In 1969, Australia took the lead. In 1989, the leading countries in red meat consumption by rank were: East Germany, Uruguay, Czechoslovakia, New Zealand, Australia, and the United States (see Fig. 1-14).

In 1989, Argentina led in per capita beef consumption, followed in order, by Uruguay, United States, Canada,

### TABLE 1-6
### MEAT, EGG, AND MILK PRODUCTION IN SPECIFIED COUNTRIES; FISH CATCH

| Country (leading countries, by rank, of all meats, 1988) | Total Red Meat Production[1] | Beef and Veal Production[1] | Pork Production[1] | Mutton, Lamb, and Goat Meat Production[1] | Poultry Meat Production[1] | Egg Production[1] | Fish Catch[2] | Cow Milk |
|---|---|---|---|---|---|---|---|---|
| | (1,000 metric tons) | | | | | | | |
| United States .... | 27,935 | 10,854 | 7,109 | 152 | 9,517 | 4,046 | 5,736 | 66,010 |
| China .......... | 24,996 | 631 | 20,134 | 781 | 2,656 | 6,685 | 9,346 | 3,845 |
| U.S.S.R. ....... | 19,213 | 8,600 | 6,324 | 875 | 3,127 | 4,656 | 11,160 | 105,950 |
| France ....... | 5,477 | 1,832 | 1,740 | 160 | 1,429 | 912 | 844 | 27,510 |
| W. Germany ..... | 5,428 | 1,608 | 3,342 | 29 | 407 | 726 | 202 | 23,978 |
| Brazil ......... | 4,735 | 2,447 | 1,050 | 62 | 1,836 | 1,280 | 793 | 13,200 |
| Italy ........... | 3,822 | 1,144 | 1,287 | 68 | 1,047 | 706 | 554 | 10,869 |
| Japan ......... | 3,654 | 570 | 1,578 | — | 1,480 | 2,409 | 11,841 | 7,608 |
| Argentina ...... | 3,494 | 2,650 | 224 | 89 | 434 | 293 | 559 | 6,450 |
| U.K. .......... | 3,396 | 964 | 1,106 | 321 | 1,079 | 790 | 665 | 14,981 |
| Spain ......... | 3,287 | 447 | 1,680 | 236 | 826 | 758 | 1,393 | 6,620 |
| World total ...... | 163,540 | 48,834 | 64,381 | 8,991 | 36,862 | 34,880 | 92,693 | 468,362 |

Source: *FAO Production Yearbook 1988*, United Nations, Rome Italy, Vol. 42. Data for 1988.

[2]Source: *FAO Fishery Statistics Yearbook 1987*, United Nations, Rome Italy, Vol. 64, pp. 97, 101. Data for 1988.

Australia, and New Zealand (see Fig. 1–15). Hungary led in per capita pork consumption, followed by East Germany, Denmark, Czechoslovakia, and West Germany (see Fig. 1–16). New Zealand and Australia, which rank number one and two, respectively, held a commanding lead in per capita mutton and lamb consumption (see Fig. 1–17).

In 1988, the world's leading poultry eaters by rank were: United States, Israel, Hong Kong, Canada, Australia, and Hungary (see Fig. 1–18). That same year, the leading milk consumers by rank were: New Zealand, Ireland, Netherlands, Switzerland, Finland, and Austria (see Fig. 1–19). In 1988, the leading egg eaters by rank were: West Germany, France, United States, Japan, United Kingdom, and Poland (see Fig. 1–20).

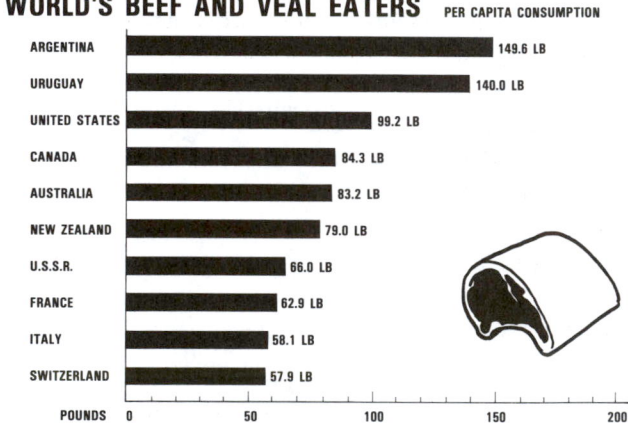

**WORLD'S BEEF AND VEAL EATERS** PER CAPITA CONSUMPTION

| | |
|---|---|
| ARGENTINA | 149.6 LB |
| URUGUAY | 140.0 LB |
| UNITED STATES | 99.2 LB |
| CANADA | 84.3 LB |
| AUSTRALIA | 83.2 LB |
| NEW ZEALAND | 79.0 LB |
| U.S.S.R. | 66.0 LB |
| FRANCE | 62.9 LB |
| ITALY | 58.1 LB |
| SWITZERLAND | 57.9 LB |

POUNDS 0   50   100   150   200

Fig. 1–15. Leading beef and veal consumers, in pounds per capita, 1989. (Source: *World Livestock Situation*, USDA, FAS, Nov. 1989)

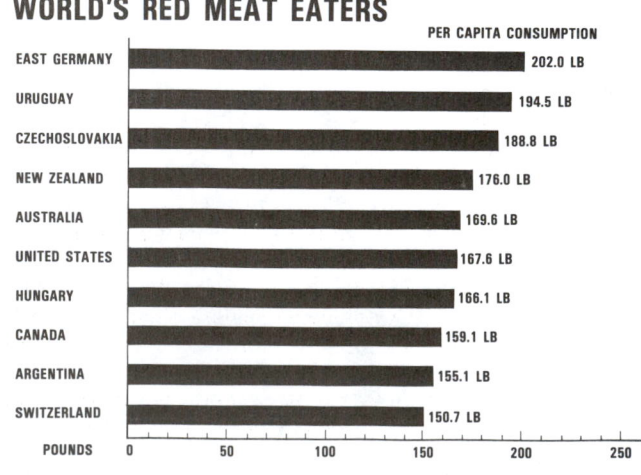

**WORLD'S RED MEAT EATERS** PER CAPITA CONSUMPTION

| | |
|---|---|
| EAST GERMANY | 202.0 LB |
| URUGUAY | 194.5 LB |
| CZECHOSLOVAKIA | 188.8 LB |
| NEW ZEALAND | 176.0 LB |
| AUSTRALIA | 169.6 LB |
| UNITED STATES | 167.6 LB |
| HUNGARY | 166.1 LB |
| CANADA | 159.1 LB |
| ARGENTINA | 155.1 LB |
| SWITZERLAND | 150.7 LB |

POUNDS 0   50   100   150   200   250

Fig. 1–14. Total per capita consumption of all red meat—beef and veal, pork, and mutton, lamb, and goat meat—in certain specified countries, 1989. (Source: *World Livestock Situation*, USDA, FAS, Nov. 1989)

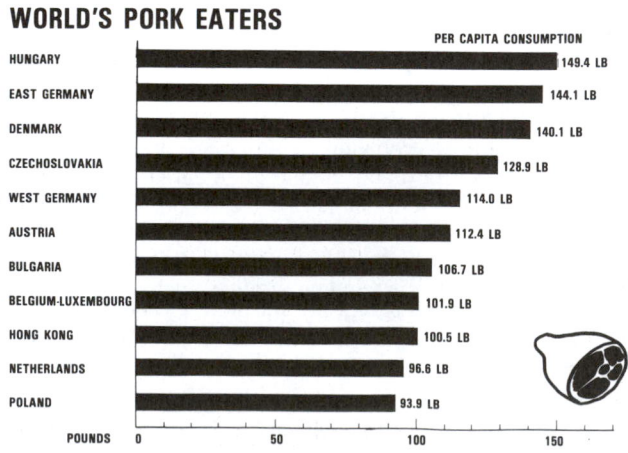

**WORLD'S PORK EATERS** PER CAPITA CONSUMPTION

| | |
|---|---|
| HUNGARY | 149.4 LB |
| EAST GERMANY | 144.1 LB |
| DENMARK | 140.1 LB |
| CZECHOSLOVAKIA | 128.9 LB |
| WEST GERMANY | 114.0 LB |
| AUSTRIA | 112.4 LB |
| BULGARIA | 106.7 LB |
| BELGIUM-LUXEMBOURG | 101.9 LB |
| HONG KONG | 100.5 LB |
| NETHERLANDS | 96.6 LB |
| POLAND | 93.9 LB |

POUNDS 0   50   100   150

Fig. 1–16. Leading pork consumers, in pounds per capita, 1989. (Source: *World Livestock Situation*, USDA, FAS, Nov. 1989. Data for 1989)

## WORLD'S MUTTON, LAMB, AND GOAT EATERS

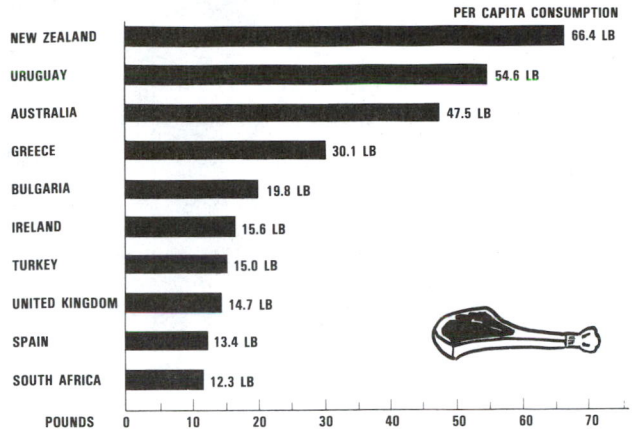

PER CAPITA CONSUMPTION

| Country | Pounds |
| --- | --- |
| NEW ZEALAND | 66.4 LB |
| URUGUAY | 54.6 LB |
| AUSTRALIA | 47.5 LB |
| GREECE | 30.1 LB |
| BULGARIA | 19.8 LB |
| IRELAND | 15.6 LB |
| TURKEY | 15.0 LB |
| UNITED KINGDOM | 14.7 LB |
| SPAIN | 13.4 LB |
| SOUTH AFRICA | 12.3 LB |

POUNDS 0 10 20 30 40 50 60 70

Fig. 1–17. Leading mutton, lamb, and goat meat consumers, in pounds per capita, 1989. (Source: *World Livestock Situation,* USDA, FAS, Nov. 1989. Data for 1989)

## WORLD'S POULTRY EATERS

PER CAPITA CONSUMPTION

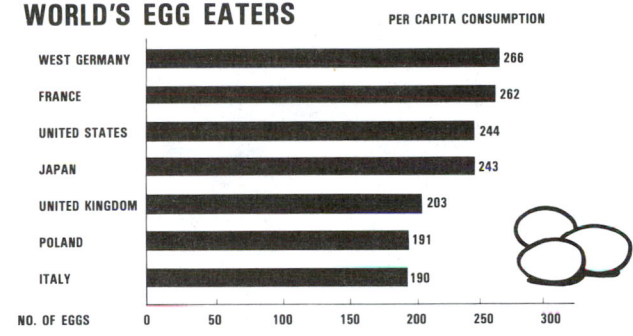

| Country | Pounds |
| --- | --- |
| UNITED STATES | 81.3 LB |
| ISRAEL | 79.0 LB |
| HONG KONG | 62.8 LB |
| CANADA | 59.4 LB |
| AUSTRALIA | 54.1 LB |
| HUNGARY | 50.2 LB |
| SPAIN | 49.4 LB |
| FRANCE | 42.8 LB |
| ITALY | 38.9 LB |
| JAPAN | 31.3 LB |

POUNDS 0 10 20 30 40 50 60 70 80

Fig. 1–18. Leading poultry meat (includes chicken meat, turkey meat, ducks, geese, and guinea fowl, ready-to-cook basis) consumers, in pounds per capita, 1988. (Source: USDA, Foreign Agricultural Service)

## WORLD'S MILK CONSUMERS

PER CAPITA CONSUMPTION

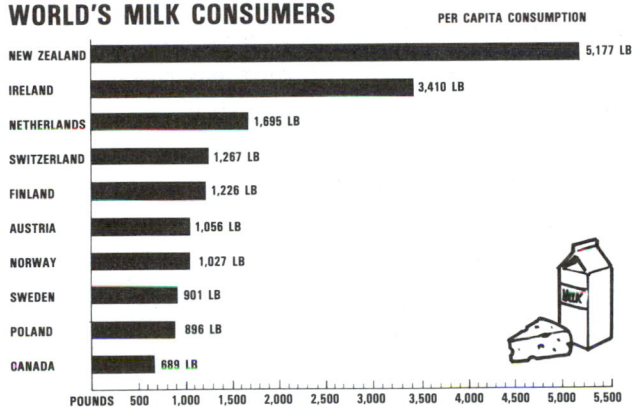

| Country | Pounds |
| --- | --- |
| NEW ZEALAND | 5,177 LB |
| IRELAND | 3,410 LB |
| NETHERLANDS | 1,695 LB |
| SWITZERLAND | 1,267 LB |
| FINLAND | 1,226 LB |
| AUSTRIA | 1,056 LB |
| NORWAY | 1,027 LB |
| SWEDEN | 901 LB |
| POLAND | 896 LB |
| CANADA | 689 LB |

POUNDS 500 1,000 1,500 2,000 2,500 3,000 3,500 4,000 4,500 5,000 5,500

Fig. 1–19. Milk consumption in selected countries, in pounds per capita, 1988. (Sources: *FAO Production Yearbook 1988,* Vol. 42, p. 272, Table 99, and *World Population by Country and Region, 1950-86, and Projections to 2050,* USDA, p. 3)

## WORLD'S EGG EATERS

PER CAPITA CONSUMPTION

| Country | No. of Eggs |
| --- | --- |
| WEST GERMANY | 266 |
| FRANCE | 262 |
| UNITED STATES | 244 |
| JAPAN | 243 |
| UNITED KINGDOM | 203 |
| POLAND | 191 |
| ITALY | 190 |

NO. OF EGGS 0 50 100 150 200 250 300

Fig. 1–20. Egg consumption in selected countries, in number of eggs per capita, 1988. (Source: USDA, Foreign Agricultural Service)

## FACTORS THAT LIMIT PRODUCTION AND CONSUMPTION OF ANIMAL PRODUCTS

Many factors interact to determine how much and what kinds of food humans eat in any particular country. Differences between countries in animal protein consumption are much greater than for calories or total protein, because countries differ much more in their ability to produce animal products, or to purchase such products from other countries, or both, than in their ability to produce food in general.

Without attempting to list them in any order of priority, among the major factors that limit the production and consumption of animal products are the following:

• **Level of income**—Animal protein consumption shows a rather close relationship to the level of income, because animal products are more expensive than other foods. People with low incomes consume relatively low-priced carbohydrates. With higher incomes, high-quality foods, notably animal products, are eaten. Thus, per capita animal product consump-

tion is a good barometer of the gross domestic product (GDP) and standard of living of a country. Indeed, animal products are a status symbol.

It is emphasized, however, that national demand for animal products is not a simple matter of multiplying the human population times the per capita gross domestic product. For example, meat prices are clearly a factor in determining variations in consumption levels.

Major meat exporters such as Uruguay, Argentina, New Zealand, and Australia have lower meat price levels; and their people consume more meat in relation to income levels than the rest of the world. Also, in the South American countries, exports have been restricted to ensure adequate supplies, hold domestic prices down, and keep traditional high consumption levels intact.

Countries such as the United States, Canada, and the United Kingdom, which have internal grain prices at world levels and generally free access to their meat markets, can be considered to have meat consumption levels in undistorted relation to their income levels. Here, beef and pork prices are influenced by world grain prices, and meat imports from other sources can compete freely.

Countries with sufficient protection in the grain and/or meat sector to put consumer meat prices above world levels—such as the European Economic Community (EC) countries and Switzerland—have consumption levels below what disposable income would indicate.

Some countries maintain very tight import controls, often through quotas and/or high tariffs; and the resultant very high meat prices offset higher income levels, causing per capita meat consumption to lag. Japan is such a country.

• **Religious beliefs, social customs, and traditional eating habits**—These forces dictate the kinds of food eaten and may even prevent the consumption of meat and meat products. Although India is the leading cattle country of the world in numbers, to the Hindu (approximately 80% of India's population) the cow is regarded as a mother and an object of reverence; hence, the eating of beef is taboo.

African tribes, such as the Masai, are much more concerned with numbers than with productivity. They do not evaluate cattle in price per pound; rather, they think in terms of the number of head required to purchase a bride.

In general, people in the western world, including Argentina and Uruguay, prefer beef. The Australians and New Zealanders consume large quantities of mutton and lamb, in addition to beef. The Europeans include pork as a significant portion of their diets. The Japanese favor fish, with pork consumption ranking second, and beef third.

• **Lack of integration of animal and plant agriculture**—In many countries, animal and plant agriculture are not well integrated. The western range and the Corn Belt of the United States are a classical example of desirable integration, giving rise to a two-phase system of cattle and sheep production. Feeder cattle and lambs are produced on the vast rangelands of the West, then finished on corn or grain sorghum in the Corn Belt and in the irrigated areas.

There are some rather large areas of the earth in which nomadism is still a way of life. This includes the northern portion of Africa, much of the Near East, and some parts of central Asia, such as Tibet and Mongolia. If the animals that these nomads maintain are to produce anywhere near their maximum potential, ways must be found to combine nomadism and farming, so that feed supplies may be available during the winter, in periods of droughts, and for finishing for market.

Also, if there is to be greater production of meat, milk, and eggs in many of the developing countries, there must be a more adequate feed supply. This calls for growing, harvesting, and storing forages and coarse grains for livestock and poultry, with these feeds used to supplement the residues of crops, by-product feeds, and nonprotein nitrogen (such as urea).

• **Lack of art and science in animal production**—Neither the art nor the science of animal production is well developed in most of the emerging countries where animal protein is generally in short supply. Moreover, there is lack of trained and competent personnel to provide specialized teaching, research, and extension in animal science. As a result of this lack of expertise and leadership in animal science, farmers continue to depend upon and devote their time to their crops.

Fig. 1–21. Tibet, in some parts of which nomadism is still a way of life. This shows a typical Tibetan tent (including ropes) made of yak hair. Traditionally, nomads overgraze an area, then move tent and herd to a new pasture. No feed is stored for either droughts or winter.

• **Research work with animals is both slow and costly**—Research work with livestock and poultry is inherently slower and more costly than research with plants. As a result, there is a paucity of research work on which to base animal improvement programs in the poorer countries.

• **Lack of capital**—Generally speaking, the capital requirements for a livestock program are greater than for a crop program. This mitigates against establishing, expanding, and/or improving livestock programs in all countries, especially in the poor countries.

• **Low productivity of animals**—Even though many of the animal product-deficient countries are reasonably well stocked in terms of animal numbers, the contribution of those animals to the supply of meat, milk, and eggs is quite low. This is due to lack of selection and breeding for genetic superiority, as well as to deficiencies in husbandry.

• **Diseases and parasites**—Diseases and parasites limit animal production in developing countries. Considerable attention has been given to major killer diseases, such as rinderpest. However, many less devastating diseases, and also parasites, do a great deal of damage. Foot-and-mouth disease is accepted as an endemic fact of life in many countries, where it causes substantial losses and makes for a major barrier to trade in livestock and meat for many countries. There are few cattle in large sections of Africa because of tsetse flies and the trypanasomes they transmit. Recently, African horse sickness and African swine fever have shown up in some of the developed countries.

• **Lack of facilities for marketing, processing, and distributing animals and their products**—Even if animals are produced, there yet remains the problem of marketing, processing, and distributing their products under the conditions which prevail in most developing countries. Limitations in transportation facilities, lack of organized markets, and lack of refrigeration are but a few of the roadblocks to producing more animal products in the emerging countries.

• **Orientation of in-country agricultural policies**—There is a tendency of many countries to orient their agricultural development policies—in subsidies, in tariffs and quotas, and in other ways—in the direction of plant production, and to overlook some of the very great potentials for improvement in animal production.

## WORLD ANIMAL PRODUCT DEMAND AHEAD

The three major factors that will influence the demand for animal products in the years ahead are (1) the population increase; (2) the increase in per capita income, which determines buying power, and (3) food preferences and trends.

Of course, feeding more people calls for more food. Calories—merely satisfying hunger—will have first call on income. Above this point, a part of the income will be spent on increased quantities of animal products. Hence, in projecting animal product demand ahead, it is important to project income—to determine how much people will have to spend.

In recent years, the gross domestic product (GDP), which is highly related to domestic buying power, has risen in almost every nation on the globe. It is expected that this trend will continue in the years ahead, especially in developed nations. But the rate of growth of the national income in the developing nations is expected to be less than in the developed nations. It follows that the demand for animal products, in the developed countries will continue to grow more rapidly than the supply. It is expected that agricultural policies in the richer countries will almost certainly encourage further expansions in animal production, but with the expansion limited to the available feeds.

In North America and Western Europe, the demand for beef will remain strong, but the hunger for beef regardless of price has been blunted. Many people are substituting chicken, turkey, fish and shellfish for steaks, roasts, and hamburgers. Pork will compete for the consumer's meat dollar on the basis of price and convenience. Poultry consumption will continue to rise, with the poultry industry capitalizing on diet and health issues and offering consumers a variety of tasty items at a very reasonable cost. The demand for fish and seafoods will continue to rise. Dairy product consumption will increase rather slowly.

The major meat importers in the years ahead will be the United States, the EC-12 (European Economic Community), Japan, and the Soviet Union.

## MEETING ANIMAL PRODUCT DEMAND

As the world demand for animal products increases, the major questions are (1) how will this demand be met, and (2) will this demand be met by the developing nations or by the developed nations?

It will be met by all classes of animals—by livestock, poultry, fish, and rabbits, with the choice of species determined primarily by available feeds, efficiency of feed conversion, animal product demand, and environment.

Throughout the world, cattle, sheep, and goats will increasingly become "roughage burners," utilizing the vast quantities of available grass and other humanly inedible forages, along with by-product feeds. However, considerable feed grains will continue to be fed to finishing cattle and sheep, with the quantity dependent upon the supply and price. Although the added finish from grain feeding improves the eating qualities of meat, it should be noted that it actually lowers the percentage protein slightly. Thus, when trimmed to retail level, Choice grade carcass beef runs 17.4% protein vs 19.4% for Standard grade.

Cattle will continue to furnish most of the world's milk. In 1987, West Germany exported 35.6% of the total world dairy products; and Belgium/Luxembourg ranked second in the ex-

Fig. 1–22. Cattle will continue to furnish most of the milk of the world. But milk shortages in the developing countries will limit the nutritional programs, even for children and pregnant-lactating mothers. This shows a woman milking a cow in North Senegal, Africa. In order to induce milk let-down, cows are first sucked briefly by a calf; then the woman goes ahead with milking. (Courtesy, FAO of the U.N., Rome, Italy)

port of dairy products. Some additional quantities of milk will come from buffalo, goats, and sheep (like the Awassi breed of sheep in Israel).

Swine will continue to be an important source of animal protein. Among the four-footed animals, the pig is the most efficient converter of feed into products for human consumption. Also, it has the added virtues of early maturity, short period of production, large litters (see Fig. 1–23), small space requirements, and relatively small capital requirements. Increasingly, pigs will be fitted into a production system that utilizes a maximum of refuse and by-product feeds and a minimum of humanly edible products. This is the system that has been developed so well by the Chinese.

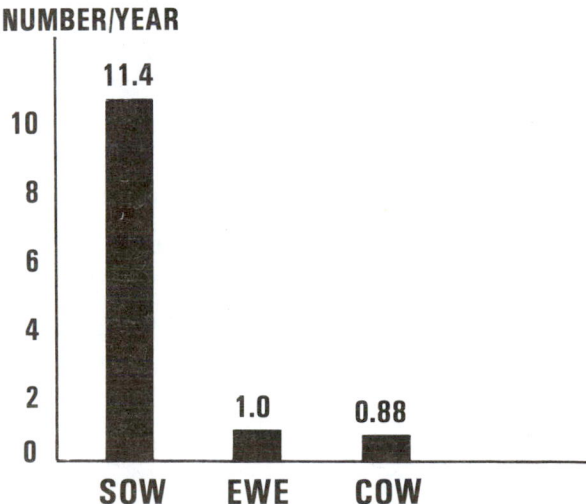

Fig. 1–23. Number of young raised to weaning age per year per breeding female. As shown, swine are the most prolific of all four-footed animals, Also, they are the most efficient and the best scavengers of all red meat producers. Hence, they will continue to be an important source of animal protein in many parts of the world.

Chickens will continue to expand in numbers and produce more eggs and meat. Also, more turkeys, ducks, geese, guinea fowl, and Japanese quail will augment the output of animal protein. Like swine, poultry require a relatively short production period and a minimum amount of capital per unit, virtues which make them quite adaptable to low-income areas.

Surprising as it may seem, on a worldwide basis the average number of pounds of fish per person per year is higher than the amount of beef; fish run about 39.3 lb vs 20.7 lb for beef. It is projected that more fish will be available in the future. Some of the catch presently used for animal feeding will likely be diverted to human use. Also, it is expected that freshwater fish, including fish production in small ponds and paddy fields, will receive increasing attention.

Rabbits, which require little space, reproduce rapidly, and feed primarily on products that do not compete seriously for human food, will be kept in more backyards and furnish additional animal protein in the future.

Both the developing and developed nations will mobilize to meet the animal product demand of the future. But, because their approaches and results will differ, they are treated in the separate sections that follow.

• **Developing nations**—Future nutritional improvement in the developing countries will depend to a very large extent on the modernization of their livestock and fish industries. All projections point to an increase in the demand for animal proteins—meat, milk, eggs, and fish. Even now, these products are generally in short supply, with the result that prices are out of reach of the lower income groups; and milk shortages limit the nutritional programs of young children and pregnant-lactating mothers.

It is predicted that the developing countries will greatly increase their poultry production (meat and eggs), because of the speed with which a flock can be increased, and because of the proven rapidity with which modern technology can be applied to the poultry industry. Also, there will be increases in pork. Yet it is on cattle that the developing nations will rely for about a third of their meat supplies and virtually all their milk requirements in the years ahead.

Right off, it would appear that there is great potential for increasing meat production in many of the developing nations—in Africa, Latin America, and the Far East. Most of them have a surplus of forages. Also, they desperately need more high-quality protein in the diet; and they need to improve their income. Projections show that the demand for livestock products in the developing countries will rise to the year 2000 and beyond. However, there are several bottlenecks which restrict rapid increases in these areas; among them (1) lack of infrastructure—roads, railroads, marketing facilities, slaughterhouses, and refrigeration; (2) periodic droughts; (3) diseases and parasites which take huge tolls of the animal population and/or restrict expansion; (4) lack of soil testing facilities, irrigation, and fertilizers; (5) inadequate agricultural extension service and coordination between teaching, research, and extension; (6) insufficient transportation, storage (including refrigeration), and markets; and (7) insufficient investment capital.

It is difficult to make rapid increases in beef production in the developing countries. Due to the extremely low fertility and late maturity of the cattle in many of these countries, they are forced to slaughter some of the heifers in order to maintain an adequate offtake rate to meet demand rather than being able to keep these heifers to build up their cattle population.

The low levels of efficiency which characterize animals in the developing countries, with a few exceptions, offer enormous opportunities for improvement. The main efforts to increase beef and mutton production in these countries should be directed toward accomplishing the following: (1) increasing the fertility of the breeding animals, (2) decreasing death losses, (3) decreasing the age at slaughter, and (4) increasing the average carcass weight. This calls for the following improvements:

1. **Genetic improvement.** This can be achieved through breeding and selecting adapted animals.

2. **Feed and nutrition improvement.** The following three approaches for improving the feed base are recommended: (a) improving native grasslands, (b) producing and utilizing more crop roughages, and (c) providing supplemental feeds.

3. **Animal health improvement.** Improved animal health is a major potential for increasing livestock production in the developing areas. This may be achieved through better veterinary service, better extension service, and better animal health products, with such health programs extending across national boundaries where major epidemics are involved.

Some areas of the world live with foot-and-mouth disease, primarily because wild animals serve as constant reservoirs and spreaders. In the high-rainfall belt of tropical Africa, the dreaded tsetse fly has kept large areas out of agricultural production. It is estimated that if trypanosomiasis, a disease borne by the tsetse fly, were brought under control, the Savannah pastures of the tsetse fly infested area would carry a cattle population of 140 million head, in excess of the total cattle population of the United States. Successful campaigns to control diseases and parasites can be carried out as evidenced by the eradication of rinderpest.

The above discussion points up the enormous possibilities for increasing the livestock production of the developing countries—in numbers, genetics, feed, management, and health. In instances where unproductive animals are pressing upon the too meager feed supplies, the first step would be to decrease numbers. In other areas, much of the world's unused land could be mobilized quickly for raising cattle and sheep. Hand in hand, much of the world's livestock could become far more productive through breeding, nutrition, and health programs. Such improvements require capital and knowledgeable personnel, both of which are generally in short supply in the developing nations.

• **Developed nations**—Rapid mobilization for increased meat production can come about in the developed countries more readily than in the developing countries because they have the necessary infrastructure, the capital for investment, adequate-sized breeding herds, and the management support and knowledge necessary for bettering production. An example of this mobilization was the dramatic growth in cattle feeding in the United States following World War II, where, in a period of 26 years (1947 to 1973), the percentage of all slaughter cattle grain fed increased from 35 to 77%.

In summary, it may be said that increased meat will come from both the developing and the developed nations. Among the developing nations having high potential for increasing meat production are Argentina, Brazil, Uruguay, Paraguay, Columbia, and Mexico. However, it must be remembered that due to the presence of foot-and-mouth disease in many countries, the major importing countries do not wish to take their beef unless it is deboned, cooked, or pasteurized. Consequently, the countries which have an immediate advantage in increasing exports are Australia, New Zealand, the Central American countries, Mexico, Ireland, the United States, and Canada.

## FOOD VS FEED

Vegetable products (largely cereal grains) are the most important single component of the world's food supply, accounting for 84% of the calories, worldwide. They are the major, and sometimes almost exclusive, source of food for many of the world's poorest people, supplying more than 90% of the total calories many of them consume (see Fig. 1–24). However, in many developed countries, more grain is fed to animals than is consumed directly by humans. Under these circumstances, during a period of food scarcity it is inevitable that some will suggest that grain be diverted from livestock and poultry feeding

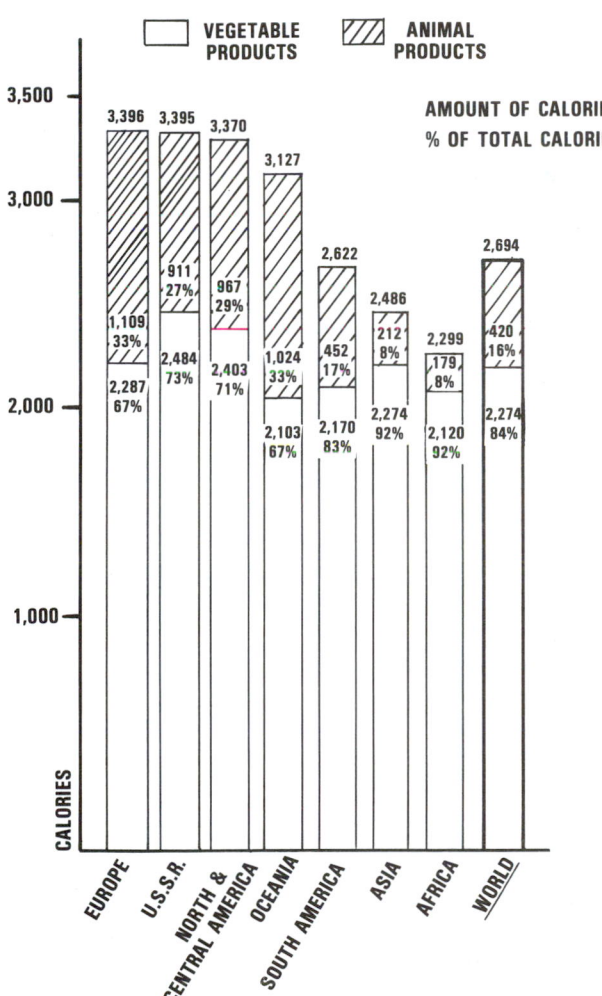

## WHERE THE CALORIES COME FROM

Fig. 1-24. Calories per person per day from vegetable vs animal products, by geographical areas. Worldwide, 84% of the calories come from vegetable products (largely cereal grains) and 16% from animal products. But there is great variation between areas; the developing areas (Asia and Africa) receive a much higher proportion of their calories from vegetable products than do the developed areas (Europe, Oceania, North and Central America, and the U.S.S.R.). (Source: *FAO Production Yearbook 1988*, U.N., Rome, Italy, Vol. 42, p. 291, Table 106. Data for 1984–86)

in the developed countries—that some will challenge the efficiency of animals in converting feed to food and the place of animals in the economical production of human food. This happened in 1972 when a world food crisis, originating from a combination of long-term problems, suddenly emerged in pronounced form due to a fall in the world output of cereals by 33 million tons.

Some government officials, politicians, economists, commentators, and others claim that livestock production is wasteful and inefficient. The implication is that, as a matter of public policy, the U.S. livestock industry should be curtailed or eliminated and crops should be consumed directly by people. This might be desirable if animals subsisted entirely on cereal grains and other humanly edible foods, or if animals and feeds were grown on land needed to produce foods for human consumption. However, much of the land of the world is unsuited to the production of food crops. In temperate climates, this nonarable land is used to grow cattle, sheep, and goats. These ruminants can eat pasture and coarse roughages that have no value except as animal feed. Thus, they contribute high-quality protein (meat), which is greatly prized for its nutritive qualities, plus its flavor and appetite appeal. Cattle produce most of the meat consumed throughout the world.

The following oversimplified example illustrates the plight that people throughout the world face in times of food shortages.

If two people were suddenly to become shipwrecked and find themselves on a barren island on which there was a pregnant cow and a bag of corn as the only food in sight, no doubt they would kill the cow and eat both the meat and the corn. If however, they found on the island a considerable amount of humanly inedible material, such as grass and other roughages, which would be suitable for the cow, and if only a little of the corn would be required as a supplement for the animal, sound judgement and ingenuity would probably result in a compromise program—their sharing the corn with the cow and the propagation of more cattle.

But instead of involving two people, one cow, one bag of corn, and some humanly inedible feed, the food vs feed question is of grave concern to 5.2 billion people; and the decision involves all four-footed animals and poultry, and a good many cereal grains and other products that they eat but which are equally well suited for human consumption.

What, then, will hungry people do relative to animals? Will they retain an animal agriculture, or will they go on a cereal grain diet?

Of course, when people are hungry enough—when there is famine—they'll consume most grains directly. They'll emulate the shipwrecked people on a barren island; they'll eat both the cow and the grain. However, where considerable humanly inedible feeds are available—such as vast acreages of grasses, and other forages, and numerous by-product feeds—they'll work out a compromise program; they'll share certain feeds with animals and propagate the latter. But the hungrier people get, the more carefully they'll scrutinize the animals with which they are sharing their precious grain; and if the situation becomes desperate enough, they'll eye each other—as cannibals do.

Sporadic food shortages and famine in different parts of the world give rise to the following recurring questions: Who should eat grain—people or animals? Shall we have food or feed?

These are straightforward questions that deserve unvarnished answers. However, in searching for answers, it should be recognized that such terms as *malnutrition, starvation,* and *famine* make for sensationalism and news, and that intellectual seduction and headlines can be obtained from a story about how the starving children of the world could eat if only Americans would give up one hamburger per week. Indeed, because of intellectual laziness or the desire to find simple solutions to a complex problem, those who favor going on a grain diet often fall into the pit of overgeneralization and easy substitution of moral indignation for knowledge.

## FAVORING DIRECT GRAIN CONSUMPTION

Historically, the people of new and sparsely populated countries have been meat eaters, whereas the people of the older more densely populated areas have been vegetarians. The latter group has been forced to eliminate most animals and to consume plants and grains directly in an effort to avoid famine.

Forgetting for a moment the high nutritive value of meats, there can be no question that more hunger can be alleviated with a given quantity of grain by completely eliminating animals. About 2,000 lb of concentrates must be supplied to livestock in order to produce enough meat and other livestock products to support a person for a year, whereas 400 lb of grain (corn, wheat, rice, soybeans, etc.) eaten directly will support a person for a year. Thus, a given quantity of grain eaten directly will feed 5 times as many people as it will if it is first fed to livestock and then is eaten indirectly by humans in the form of livestock products. This is caused by the unavoidable nutrient losses in all animal feeding and the fact that no return is received from that portion of the animal's feed which goes for maintenance (which amounts to approximately ½). This is precisely the reason the people of the Orient have become vegetarians.

Among the arguments sometimes advanced by the prophets of doom in favor of direct grain consumption for feeding the starving masses are the following:

1. **On a calorie or protein conversion basis, it's not efficient to feed grain to animals and then to consume the livestock products**—This fact is pointed up in Table 1-7.[8]

The inefficiency of animals is further pointed up in the following figures: Fig. 1-25, feed efficiency; Fig. 1-26, energy (calorie) efficiency conversion; and Fig. 1-27 (see page 22), protein efficiency conversion.

Of course, in any valid evaluation of calorie or protein conversion of feed to food, such as presented in Table 1-7, cognizance must be taken of the following facts:

a. Where ruminants are supplied with urea as a nitrogen supplement, they may yield far more product protein than is supplied to them in the feed because they are capable of manufacturing protein. Under such circumstances, the feed-to-food protein conversion may exceed 100%.

The efficiency of conversion of both energy and protein may exceed 100% if consideration is given only to the quantities of plant products consumed by ruminants that are digestible by humans.

Animal products are far more than empty calories and protein; they also provide all the essential amino acids, minerals, and vitamins, along with digestibility and palatability.

2. **More people can be fed**—Grain consumed directly by people will feed about 5 times as many people as could be fed were it converted to livestock products and then consumed. Thus, in the developing countries, where the population explosion is most dramatic, about 400 lb of grain are available per person per year. Virtually all of this is eaten as grain; precious little of it is converted to animal products. In the United States and Canada, however, the average person uses

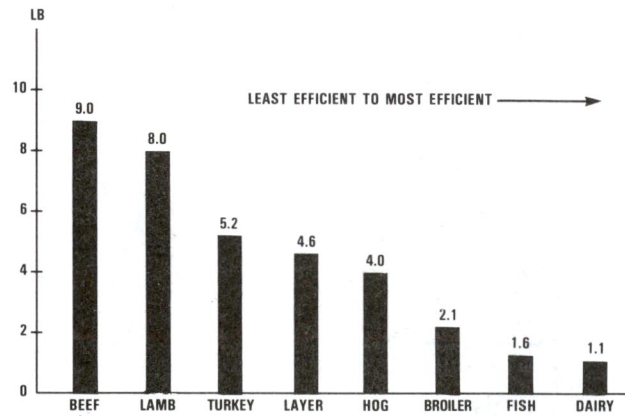

**FEED EFFICIENCY**
**POUNDS OF FEED REQUIRED TO PRODUCE ONE POUND OF PRODUCT**

Fig. 1-25. Pounds of feed required to produce 1 lb of product. This shows that it takes 9 lb of feed to produce 1 lb of on-foot beef, whereas it takes only 1.11 lb of feed to produce 1 lb of milk. (Source: Table 1-7 this chapter)

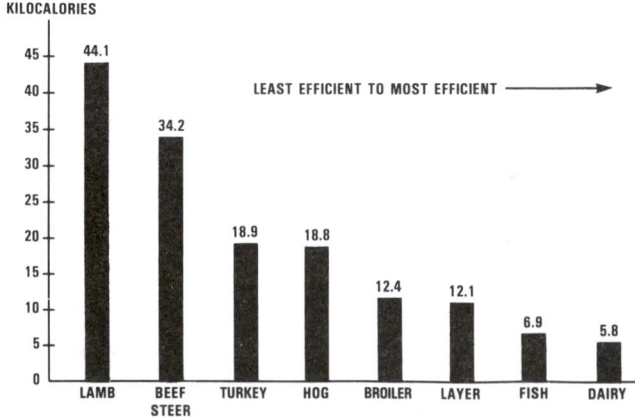

**ENERGY (CALORIE) EFFICIENCY CONVERSION**
**KILOCALORIES OF FEED REQUIRED TO PRODUCE ONE KILOCALORIE OF PRODUCT**

Fig. 1-26. Kilocalories in feed required to produce 1 kcal of product. This shows that it takes 44.1 kcal in feed to produce 1 kcal in lamb, whereas only 5.8 kcal in feed will produce 1 kcal in milk. (Source: Table 1-7 this chapter)

---

[8]It could be argued that Table 1-7 makes no provisions for the feed used by the sires and dams of these animals—the animals that gave birth to these producers. Others may be critical of using a yearling steer without making provisions to get him to the feedlot stage. Finally, it may be contended that any such comparison should be between animals of like age; for example, between broilers and veal calves. Having raised these questions, the author submits Table 1-7, which in his judgement is as fair a rating on feed to food efficiency as can be made.

## TABLE 1-7
### FEED TO FOOD EFFICIENCY RATING BY SPECIES OF ANIMALS, RANKED BY PROTEIN CONVERSION EFFICIENCY
(Based on Energy as TDN or DE and Crude Protein in Feed Eaten by Various Kinds of Animals Converted into Calories and Protein Content of Ready-to-Eat Human Food)

| Species | Unit of Production (on foot) | Feed Required to Produce One Production Unit | | | | Dressing Yield | | Ready-to-Eat; Yield of Edible Product (meat & fish deboned & after cooking) | | | | Feed Efficiency[4] (lb feed to produce one lb product) | | Efficiency Rating | | | |
| | | Pounds | TDN[1] | DE[2] | Protein | Percent | Net Left | As % of Raw Product (carcass) | Amount Remaining from One Unit of Production | Calorie[3] | Protein[3] | | | Calorie Efficiency[5] | | Protein Efficiency[6] | |
| | | (lb) | (lb) | (kcal) | (lb) | (%) | (lb) | (%) | (lb) | (kcal) | (lb) | (%) | (ratio) | (%) | (ratio) | (%) | (ratio) |
|---|---|---|---|---|---|---|---|---|---|---|---|---|---|---|---|---|---|
| Broiler | 1 lb chicken | 2.1[7] | 1.7[8] | 3,400 | 0.21[8] | 72[13] | 0.72 | 54[14] | 0.39 | 274 | 0.11 | 47.6 | 2.1:1 | 8.1 | 12.4:1 | 52.4 | 1.9:1 |
| Fish | 1 lb fish | 1.6[9] | 0.98 | 1,960 | 0.57 | 65[10] | 0.65 | 57[11] | 0.37 | 285 | 0.27 | 62.5 | 1.6:1 | 14.5 | 6.9:1 | 47.6 | 2.1:1 |
| Dairy cow | 1 lb milk | 1.11[7] | 0.98 | 1,800 | 0.1[8] | 100 | 1.0 | 100 | 1.0 | 309 | 0.037 | 90.0 | 1.11:1 | 17.2 | 5.8:1 | 37.0 | 2.7:1 |
| Turkey | 1 lb turkey | 5.2[7] | 4.21[8] | 8,420 | 0.46[8] | 79.7[13] | 0.797 | 57[15] | 0.45 | 446 | 0.146 | 19.2 | 5.2:1 | 5.3 | 18.9:1 | 31.7 | 3.2:1 |
| Layer | 1 lb eggs (8 eggs) | 4.6[7] | 3.73[8] | 7,460 | 0.41[8] | 100 | 1.0 | 100[12] | 1.0[12] | 616 | 0.106 | 21.8 | 4.6:1 | 8.3 | 12.1:1 | 25.9 | 3.9:1 |
| Hog (birth to market weight) | 1 lb pork | 4.0[18] | 3.2 | 6,400 | 0.36 | 70[18] | 0.70 | 44[17] | 0.31 | 341 | 0.088 | 0.25 | 4.0:1 | 5.3 | 18.8:1 | 24.4 | 4.1:1 |
| Rabbit | 1 lb fryer | 3.0[18] | 2.20 | 4,400 | 0.48 | 55[19] | 0.55 | 79[19] | 0.43 | 301 | 0.08 | 35.7 | 2.8:1 | 6.8 | 14.6:1 | 16.7 | 6.0:1 |
| Beef steer (yearling finishing period in feedlot) | 1 lb beef | 9.0[18] | 5.85 | 11,700 | 0.90 | 58[18] | 0.58 | 49[17] | 0.28 | 342 | 0.085 | 11.1 | 9.0:1 | 2.9 | 34.2:1 | 9.4 | 10.6:1 |
| Lamb (finishing period in feedlot) | 1 lb lamb | 8.0[18] | 4.96 | 9,920 | 0.86 | 47[18] | 0.47 | 40[17] | 0.19 | 225 | 0.052 | 12.5 | 8.0:1 | 2.3 | 44.1:1 | 6.0 | 16.5:1 |

[1] TDN pounds computed by multiplying pounds feed (column to left) times percent TDN in normal rations. Normal ration percent TDN taken from M. E. Ensminger's books and rations, except for the following: dairy cow, layer, broiler, and turkey from *Agricultural Statistics 1974*, p. 358, Table 518. Fish based on averages recommended by Michigan and Minnesota Stations and U.S. Fish and Wildlife Service.

[2] Digestible Energy (DE) in this column given in kcal, which is 1 Calorie (written with a capital C), or 1,000 calories (written with a small c). Kilocalories computed from TDN values in column to immediate left as follows: 1 lb TDN = 2,000 kcal.

[3] From *Lessons on Meat*, National Live Stock and Meat Board, 1965.

[4] Feed efficiency as used herein is based on pounds of feed required to produce 1 lb of product. Given in both percent and ratio.

[5] Kilocalories in ready-to-eat food = kilocalories in feed consumed, converted to percentage. Loss = kcal in feed ÷ kcal in product.

[6] Protein in ready-to-eat food = protein in feed consumed, converted to percentage. Loss = pounds protein in feed ÷ pounds protein in product.

[7] *Agricultural Statistics 1974*, p. 358, Table 518. Pounds feed per unit of production is expressed in equivalent feeding value of corn.

[8] Pounds feed (column No. 2) per unit of production (column No. 1) is expressed in equivalent feeding value of corn. Therefore, the values for corn were used in arriving at these computations. No. 2 corn values are TDN, 81%; protein, 8.9%. Hence, for the dairy cow 81% × 1.11 = 0.9 lb TDN; and 8.9% × 1.11 = 0.1 lb protein.

[9] Data from report by Dr. Phillip J. Schaible, Michigan State University, *Feedstuffs*, April 15, 1967.

[10] *Industrial Fishery Technology*, edited by Maurice E. Stansby, Reinhold Pub. Corp., 1963, Ch. 26, Table 26-1.

[11] *Ibid.* Reports that "Dressed fish averages about 73% flesh, 21% bone, and 6% skin." In limited experiments conducted by A. Ensminger, it was found that there was a 22% cooking loss on filet of sole. Hence, these values—73% flesh from dressed fish, minus 22% cooking losses—give 57% yield of edible fish after cooking, as a percent of the raw, dressed product.

[12] Calories and protein computed basis per egg; hence, the values herein are 100% and 1.0 lb, respectively.

[13] *Marketing Poultry Products*, 5th Ed., by E. W. Benjamin et al., John Wiley & Sons, 1960, p. 147.

[14] *Factors Affecting Poultry Meat Yields*, University of Minnesota Sta. Bull. 476, 1964, p. 29, Table 11 (fricassee).

[15] *Ibid.* Page 28, Table 10.

[16] Ensminger, M. E., *The Stockman's Handbook*, 6th Ed., Sec. XII.

[17] Allowance made for both cutting and cooking losses following dressing. Thus, values are on a cooked, ready-to-eat basis of lean and marbled meat, exclusive of bone, gristle, and fat. Values provided by National Live Stock and Meat Board (personal communication of June 5, 1967, from Dr. Wm. C. Sherman, Director, Nutrition Research, to the author), and based on data from *The Nutritive Value of Cooked Meat*, by Ruth M. Leverton and George V. Odell, Misc. Pub. MP-49, Appendix C, March 1958.

[18] Estimates by the author.

[19] Based on information in *Commercial Rabbit Raising*, Ag. Hdbk. No. 309, USDA, 1966, and *A Handbook on Rabbit Raising*, by H. M. Butterfield, Washington State University Ext. Bull. No. 411.

**PROTEIN EFFICIENCY CONVERSION**
**POUNDS OF FEED PROTEIN REQUIRED TO PRODUCE**
**ONE POUND OF PRODUCT PROTEIN**

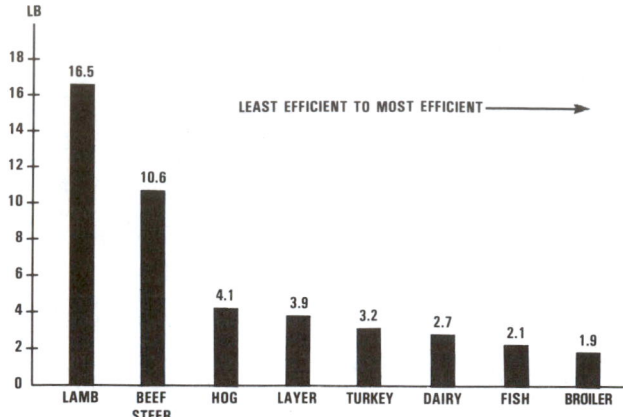

LB

LEAST EFFICIENT TO MOST EFFICIENT ⟶

| LAMB | BEEF STEER | HOG | LAYER | TURKEY | DAIRY | FISH | BROILER |
|------|------|------|------|------|------|------|------|
| 16.5 | 10.6 | 4.1 | 3.9 | 3.2 | 2.7 | 2.1 | 1.9 |

Fig. 1–27. Pounds of feed protein required to produce 1.0 lb of product protein. This shows that it takes 16.5 lb of feed protein to produce 1.0 lb of lamb protein, whereas only 1.9 lb of feed protein will produce 1.0 lb of broiler protein. (Source: Table 1–7. See column headed "Protein Efficiency")

Fig. 1–28. Vast areas throughout the world, such as this rough terrain, are not suited to cultivation. Hence, their only use is for grazing or forest.

nearly 2,000 lb of grain per year. All but about 150 lb of it is consumed as meat, milk, and eggs. This shows 2 things: (1) About 5 times as many people can be fed on a cereal diet as can be fed on a diet consisting chiefly of animal products; and (2) as people become more affluent, they use more grain, with most of it converted into animal products. It's noteworthy, too, that no nation appears to have reached an affluence level where its per capita grain requirement has stopped rising.

## FAVORING SHARING GRAIN WITH ANIMALS, THEN CONSUMING THE ANIMAL PRODUCTS

Among some social reformists, the charge persists that much of the world goes hungry because of the substitution of meat, milk, and eggs for direct grain consumption. A response to this accusation requires far more than a simple denial.

The following facts are presented in favor of sharing grain with animals, then consuming the animal products.

1. **Much of the world's land is not cultivated.** More land throughout the world can, and will be, brought under cultivation and used for crop production. But, like the western range of the United States, vast acreages throughout the world—including arid and semiarid grazing lands; and brush, forest, cutover, and swamplands—are unsuited to the production of bread grains or any other type of farming; their highest and best use is, and will remain, for grazing.

In the United States, only 21% of the land area of the 50 states is cultivated; 29.1% of the total land area, including Alaska and Hawaii, is devoted to pasture and rangeland. The enormous productivity of this vast area becomes apparent from the following figures: Every 22 lb of usable forage (grass, shrubs, and other plants) eaten by a ewe-lamb combination will produce about 1 lb of lamb; every 26 lb of usable forage eaten by a cow-calf combination will yield about 1 lb of calf; and every 10 lb of forage eaten by a calf will produce about 1 lb of calf.

In China, only 10.0% of the land is cultivated; yet this 6.8% of the world's cultivated land sustains a fourth of the world's population. North of China's Great Wall, life centers on pastoral areas; large flock and herds of cattle, sheep, and horses roam these vast grasslands.

2. **Forages provide most of the feed for livestock.** Pastures and other roughages—feeds not suitable for human consumption—provide most of the feed for livestock, especially for ruminants (four-stomached animals such as cattle, sheep, goats, buffalo and certain wild species including deer, antelope, and elk), throughout the world. Fortunately, the uniqueness of the ruminant's stomach permits it to consume forages, and, through bacterial synthesis, to convert such inedible (to humans) roughages into high-quality proteins—meat and milk. Hence, cattle and sheep manufacture human food from nonedible forage crops. Not only that, they serve as the primary means of storing such forage from one season to the next.

In Australia, New Zealand, and Argentina, pastures are relatively abundant and little grain is fed to cattle and sheep. Despite grains being relatively plentiful in the United States, forages provide the bulk of animal feeds; pastures and other roughages account for 93.8% of the total feed of sheep and goats, 84.5% of the feed of beef cattle, and 61.7% of the feed of all livestock.

Fig. 1–29. New Zealand, where grass is abundant and little grain is fed to animals. This shows cattle and sheep on pasture in the Waikato District. (Courtesy, National Publicity Studios, Wellington, New Zealand)

Even feedlot cattle consume relatively little grain in total. Generally speaking, feeder cattle, raised on milk and grass and that are to be grain fed, are put into the feedlot at weights of around 600–700 lb, to be fed to weights of about 1,050 lb. This means that they attain 60 to 65% of their weight gain before entering the feedlot. In the feedlot, it takes 9 lb of feed to make 1 lb of gain, with 6 lb of this consisting of grains and by-product feeds and 3 lb of roughage. Assuming a feeding period of 140 days and a gain of 450 lb in the lot, the total market weight (1,050 lb) would represent 2.57 lb of feed grain expended for each pound of gain (450 × 6 = 2,700. Then, 2,700 ÷ 1,050 = 2.57). So, on a birth-to-market basis, it takes only 2 to 3 lb of grain per pound of weight gained. Less grain is consumed during those times when grains are scarce and high in price, at which times cattle are grazed longer and kept in the feedlot a shorter time. For example, had the steer in the above example been kept on pasture longer, had he been short fed for 90 to 100 days (instead of 140 days), and had he been fed to the same weight, but marketed at Good grade instead of Choice, each pound of on-foot weight would have required only about 1.8 lb of grain, which is comparable to the feed efficiency of broilers.

A Choice steer weighing 1,050 lb on foot will produce 454 lb of salable beef (processed, cut, and trimmed). As noted above, this steer can be finished on 2,700 lb of grain. This means that it requires slightly less than 6 lb of grain to produce a pound of beef for the retail meat counter (2,700 ÷ 454 = 5.95).

Of course, not all beef is grain fed; about one-third of it is strictly grass fed. Besides, when grain is scarce and high in price, feeder steers are generally carried on grass to heavier weights (maybe 800 lb, instead of 650 lb) before being fed for about 100 days (short fed), rather than 140 days. This lessens the grain consumed per pound of salable beef.

3. **Animals convert chemical energy in plants into a form available to people.** Life on earth is supported by a finite amount of chemical energy produced by photosynthesis in plants using solar energy. But only 1% of the total solar energy reaching the planet is captured as chemical energy during photosynthesis; and only 5% of this captured energy is fixed in a form suitable as food for people. To supply more food, we must be able to convert a greater percentage of total energy fixed as chemical energy in plants—the other 95%—into a form available to humans. Ruminants are the answer. They can convert energy from human inedible materials into foods for humans.

4. **Some grain feeding of ruminants is in the best interest of consumers.** Cattle and sheep can be fed entirely on grass and other forages. But the public should be aware of the consequences of such a program.

Without any feedlot feeding of cattle, total beef supplies would become more limited because of reduced cattle numbers, a longer time to market, and less meat yield per animal, resulting in higher prices for a less palatable product. Also, supplies would be more seasonable and less uniform, with much greater fluctuation in availability and price to consumers. Feeding no grain would have the same effect on lamb supplies. Likewise, without some grain, dairy cows would produce less milk which would cost more.

For these reasons, the U.S. animal industry will continue to feed grain in the foreseeable future, with the proportion of grain to roughage in rations varied in keeping with (a) feed supplies and prices, and (b) prices of animal products.

Fig. 1-30. Ruminants—cattle, sheep, and goats—convert the photosynthetic energy derived from solar energy and stored in the grass into food for humans. This shows Hereford steers on bromegrass pastures in Nebraska. (Courtesy, C.B. & Q. Railroad, Chicago, IL)

5. **Food and feed grains are not synonymous.** Animals do not compete to any appreciable extent with the hungry people of the world for food grains such as rice and wheat. Instead, they eat feed grains—like field corn, grain sorghum, barley, and oats—for which there is little or no demand for human use in most countries, plus forages and grasses—fibrous stuff that people can not eat. For example, in the United States only 3% of the corn—the major animal feed grain—is used for human food. Also, it is noteworthy that the feed grains which the United States ships overseas are used almost entirely for livestock and poultry production abroad.

6. **Ruminants utilize low-quality roughages.** Cattle, sheep, and goats efficiently utilize large quantities of coarse, high-cellulose roughages, including crop residues, straw, and coarse low-grade hays. Such products are indigestible by humans, but from 30 to 80% of the cellulose material is digested by ruminants.

Fig. 1-31. Cattle can utilize efficiently large quantities of coarse, humanly inedible roughages, like cornstalks. This shows cows feeding on corn residue which had been harvested by mechanical means. Grazing cornstalks limits utilization to about 15% of the total available dry matter, whereas harvesting increases utilization to about 60%. Combining the two methods increases utilization to about 70%. (Courtesy, Iowa State University)

Of all U.S. crop residues, the residue of corn (cornstalks and husklage) is produced in greatest abundance and offers the greatest potential for expansion in cow numbers. Corn usually produces an amount of residue equal to the quantity of the grain produced. So, the more than 200 million tons of corn grain produced each year result in more than 200 million tons of corn residue. That's more than 200 million tons of potential cow feed, enough to winter more than 150 million pregnant cows. Mature cows are physiologically well adapted to utilizing such roughage. Moreover, when corn residue is used to the maximum as cow feed, acreage which would otherwise be used to pasture the herd is liberated to produce more corn and other crops. Also, there are many other crop residues which, if properly utilized, could increase the 151.5 million head figure given above.

7. **Animals provide an outlet for by-product feeds.** Animals provide a practical outlet for a host of by-product feeds not suited for human consumption, including corncobs, cottonseed hulls, gin trash, oilseed meals, beet pulp, citrus pulp, molasses (cane, beet, citrus, and wood), wood by-products (sawdust), rice bran and hulls, wheat milling by-products, and fruit, nut, and vegetable refuse. It is estimated that each year ruminants convert more than 9 million tons of industrial by-products into human food.

Fig. 1–32. Chinese hogs in Kwangtung Province, China. Their ration consisted of two by-products—rice mill feed and bagasse (the pith of sugarcane), along with water hyacinth—all of which the pigs ate with relish. In the People's Republic of China, swine utilize millions of tons of otherwise wasted crop residues and by-products. (Photo by A. H. Ensminger)

8. **Animals provide elasticity to grain production.** Livestock feeding provides a large and flexible outlet for the year-to-year changes in grain supplies. When there is a large production of grain, more can be fed to livestock, with the animals carried to heavier weights and higher finish. On the other hand, when grain supplies are low, herds and flocks can be maintained by reducing the grain that is fed and by increasing the grasses and roughages. Thus, livestock give stability to grain farming.

In the United States, this flexibility of animals from the standpoint of grain supplies came clearly into focus in cattle finishing following World War II. In 1947, the majority of slaughter cattle were grass finished—only 6.9 million head of cattle were grain fed, representing 30.1% of the slaughter cattle. Grain bins bulged with surpluses, and consumers were able and willing to pay for Choice, grain-fed beef, which was ably promoted and merchandised by self-service chain stores. The

time was ripe for the chain reaction that spawned the era of grain-fed cattle. Commercial cattle feeders, acutely attuned to consumer demands, set out to fill the need. Cattle feeding went big—and modern. High-energy rations were the order of the day. By 1973, 77% of the U.S. beef production came from grain finished cattle, and only 23% from grass cattle. Then, suddenly there were grain shortages—and there was a rude awakening. The U.S. cattle feeding industry was experiencing a prelude to world food shortages. The message was loud and clear: When grain is scarce and high in price, cattle and other animals cannot compete with humans for cereals. By 1980, 68% of the U.S. beef production came from grain finished cattle, and 32% from grass cattle. In a period of 33 years—from 1947 to 1980—the percentage of all slaughter cattle grain fed increased from 30 to 68%.

When grains are in short supply, fewer slaughter cattle are grain fed—more are grass finished. In the years ahead, depending on future grain supplies and prices, it is predicted that two-thirds, or less, of the U.S. domestic beef supply will come from feedlot cattle. Also, during periods of high-priced grains, heavier feeder cattle will go into feedlots, and they will be fed for a shorter period on less grain and more roughage than when grains are more abundant and cheaper.

In the future, animals will increasingly be "roughage burners," with the proportion of grain to roughage determined by grain supplies and prices.

Beef, dairy, and sheep producers will more and more rely upon the ability of the ruminant to convert coarse forage, grass, and by-product feeds, along with a minimum of grain, into palatable and nutritious food for human consumption, thereby competing less for humanly edible grains. The longtime trend in animal feeding will be back to roughages; increasingly, flesh will be grass.

9. **Animals stimulate grain production and avoid surpluses.** Lessening or eliminating animals would reduce the demand for grain and make for lower grain prices. In turn, lower grain prices would provide less incentive to produce grain, with the result that less grain would be produced. Fewer animals might even contribute to a return to grain surpluses and their accompanying familiar problems.

10. **Animals provide products that meet consumer preferences.** Most people who can afford to do so eat a portion of their food in the form of livestock products simply because of preference—because they like them. Thus, when they have the money, they consume meat, milk, and eggs. In turn, when the price ratios between grain and livestock products are favorable, more grain is fed to more animals.

11. **Animals convert feed grains to higher protein-content products.** Grains, such as corn, are much lower in protein content in cereal form than after conversion into meat, milk, or eggs. On a dry basis, the protein contents of selected products are: corn, 10.45%; beef (Choice grade, total edible, trimmed to retail level, raw), 30.7%; milk, 26.4%; and eggs, 47.0%.[9]

12. **Ruminants convert nonprotein nitrogen to protein.** Ruminant animals (cattle, sheep, and goats) can use nonprotein nitrogen, like urea, to produce protein for humans in the form of meat and milk.

13. **Animals produce proteins of higher value than plants.** Proteins from animals sources (meat, milk, and eggs) have a higher value than proteins from plant sources because they have every amino acid needed for growth, including lysine and methionine, which are deficient in vegetable sources. Also, they

[9]*Composition of Foods*, Ag. Hdbk. No. 8, Agricultural Research Service, USDA.

are an excellent source of zinc, iron, and many other trace minerals. Thus, animals improve protein quality by converting low quality plant protein to high quality, balanced animal protein.

The Food and Agriculture Organization of the United Nations is on record as advocating that the world's diet have animal protein in amounts up to ⅓ of the total protein requirement. The high value of animal protein becomes apparent when it is realized that one 3½-oz serving of cooked, lean beef is equal to 14 oz of cooked dried beans or 7½ tbs of peanut butter.

Sometimes folks are prone to compare cereal grains and animal proteins, pound for pound, without considering the digestibility of each. The fallacy of such reasoning is pointed up in the example that follows.

A ton of corn (dent, No. 2) contains 8.9% cereal protein, or 178 lb (2,000 × .089 = 178) of available protein. That same ton of corn will finish a 700-lb milk-fed and grass-raised steer to Standard grade and a weight of 1,100 lb. Following slaughter, this steer will yield 660 lb of carcass containing 19.4% beef protein, or 128 lb of available protein. Offhand, this appears to favor people eating grain directly—178 vs 128 lb, or 50 lb more protein. But it's net protein utilization (which is a measurement of both digestibility and biological value of a protein) by the human body that counts. Hence, one further step is necessary.

In the net protein utilization (NPU) scale, eggs rank at the top, with a score of 94%, followed by cows' milk (whole) at 82%. Beef protein has a score of 73%, whereas corn scores only 53%.[10] This means that a higher proportion of the beef protein than of the corn protein is assimilated by the human body.

Thus, 178 lb (the available protein in a ton of corn) × 53 (net protein utilization of corn) = 94.34 lb of protein actually available to humans who eat the cereal (corn). By using the same arithmetic, the beef story is: 128 lb (the available protein in the beef carcass) × 73 (net protein utilization of beef) = 93.44 lb of protein actually available to people who eat the beef. That makes it a standoff—less than 1 lb difference in net protein between consuming the corn directly or putting it through a steer. But, given the choice, most people would rather eat beef. No cereal platter, no matter how well disguised, can ever inspire the toast or impart the status symbol of a roast of beef or a sizzling steak.

**14. Animals step up the nutritive quality and energy of food.** Unquestionably, more people can be fed by consuming grain directly, rather than by putting it through animals (see Table 1–8).

### TABLE 1–8
### CALORIES IN FOOD INTAKE PER PERSON, PER DAY,
### UNITED STATES VS INDIA[1]

|  | Total Calories | Cereals | Roots & Plantains | Live-stock | Other |
|---|---|---|---|---|---|
| U.S.A. ...... | 3,652 | 781 | 116 | 1,159 | 1,596 |
| India ........ | 2,161 | 1,382 | 43 | 118 | 618 |
| Difference .... | 1,491 | 601 | 73 | 1,041 | 978 |

[1]Source: *Agriculture: Toward 2000*, United Nations, Rome, Italy, p. A24, Table 4. Data for 1983–85.

[10]*Nutrition and Physical Fitness*, by Bogert, Briggs, and Callaway, 1973, p. 89, Table 5–2, from FAO Nutrition Studies.

Thus, as shown in Table 1–8, on the basis of calories per person per day, the comparison between the United States and India is not, as generally assumed, 3,652 vs 2,161, or a difference of 1,491 calories. Rather, the following differences in the sources of the calories in the U.S.A. vs India are noteworthy: (1) India gets 601 more calories from cereals, and (2) the U.S.A. gets 1,041 more calories from livestock products. Of course, in any valid comparison, consideration must be given to (a) the total nutrients furnished in the different diets (not energy alone)—to the superior quality of the proteins, and to the vitamins and minerals obtained from animal products in comparison with a cereal grain diet; and (b) the stepped-up energy levels made by animals through utilizing grasses and legumes which collect energy from the sun by photosynthesis, and through eating humanly inedible by-product feeds such as cornstalks, straw, and bagasse (sugarcane residue).

**15. Animals provide medicinal and other products.** Animals are not processed for meat alone. They are the source of hundreds of important by-products, including some 100 medicines.[11]

Besides medicines, many familiar products are derived from animals, including leather, shoe polish, photographic film, soap, lubricants, candles, glue, buttons, and bone china, to name a few.

## HOW A HUNGRY WORLD WILL MEET THE FOOD VS FEED DILEMMA

Practicality dictates that a hungry world should and will, proceed in about the following order in meeting the food vs feed dilemma:

1. Consume a higher proportion of humanly edible grains and seeds, and their by-products, directly—without putting them through animals, simply because approximately five times more people can be fed by doing it this way. This will make for a gradual lessening of animal agriculture, just as it has in the Orient for years.

2. Utilize a higher proportion of roughages to concentrates in animal rations as increasing quantities of cereal grains are needed for human consumption.

3. Retain more of those species that can utilize a maximum of humanly inedible feeds and a minimum of products suitable for human consumption. This would favor cattle and sheep, provided they are fed a maximum of pasture and other roughages. Both poultry and swine may compete with people for grains. Nevertheless, it is expected that further increases in poultry will come, primarily because of their efficiency as converters of protein from feed to food; and it is expected that there will be further increases in swine, especially in China, where pigs are scavengers and manure producers par excellence.

4. Propagate the most efficient feed to food species converters (see Table 1–7). This means dairy cows, fish, and poultry. Because beef cattle and sheep are at the bottom of the totem pole when it comes to feed efficiency, the pressure will be to eliminate them, except as roughage consumers.

[11]The count on the number of medicines derived from animals varies, perhaps due to (a) whether or not certain derivatives are counted, and (b) whether or not experimental products, as well as commercial, are included. Swift and Company lists 90 such products, the American Meat Institute states that over 100 different pharmeceuticals come from cattle alone, while the National Live Stock and Meat Board pegs the number of different medicines coming from animals at 134.

5. Increase the within-species efficiency of all animals and eliminate the inefficient ones. This calls for more careful selection and more rigid culling than ever before.

6. Improve pastures and ranges. Good pasture will produce 200 to 400 lb of beef or lamb per acre annually (in weight of young weaned, or in added weight of older animals); superior pastures will do much better.

## POWER, NUTRIENTS, AND MANURE

Animals are the main sources of agricultural power, animal proteins, and manure in many parts of the world, especially in populous China and India, and in Nepal and Thailand. They will continue to fill these roles for many years to come.

Thanks to the Green Revolution, the people of these developing countries will soon have enough rice and wheat—carbohydrates. But, for the most part, they must rely on animals (1) to provide needed power, (2) to manufacture the needed animal proteins, and (3) to produce the needed manure for fertilizing the fields and fueling their homes.

### ANIMALS AS A SOURCE OF POWER

A century ago, muscles provided 94% of the world's energy needs; coal, oil, and waterpower provided the other 6%. Today, the situation is reversed in the developed nations. They now obtain 94% of their energy needs from coal, oil, natural gas, and waterpower, and only 6% from the muscle power of people and animals. However, in developing nations, cattle, water buffalo, and horses still provide much of the agricultural power. In this capacity, they contribute to the food supply of people from plant sources. Such draft animals are a part of the agricultural scene of Asia, Africa, the Near East, Latin America, and parts of Europe—areas characterized by small farms, low incomes, abundance of human labor, and lack of capital. But animals have certain advantages. They can be fueled on roughages to produce power, a most important consideration in time of energy shortage; and both cattle and water buffalo are "triple threat animals"—they're used for work, milk, and meat. Also, when it comes to tilling wet, muddy rice paddies, water buffalo are without a peer; and, under adverse conditions, they will outproduce cattle in power, milk yield, and butterfat.

Although the general trend in the world is toward more and more mechanization, animals will likely continue to provide much of the agricultural power in many of the developing countries.

Fig. 1–33.  Oxen pulling a stick (one-handled) plow. Draft animals are a part of the agricultural scene in most of the developing countries of the world. (By Mr. Burton Holmes, from Ewing Galloway)

### ANIMALS AS A SOURCE OF NUTRIENTS

Perhaps most people consume animal products—meat, milk, eggs, and fish—simply because they like them. They derive a rich enjoyment and satisfaction therefrom. For flavor, variety, and appetite appeal, they are unsurpassed.

But animal products are far more than just very tempting and delicious foods. From a nutritional standpoint, they contain certain essentials of an adequate diet. This is important, for how we live and how long we live are determined in large part by our diet.

It is estimated that the average American gets the percentages of his food nutrients shown in Table 1–9 and Fig. 1–34 from animal products. Foods of ruminant origin (meat, milk, and their various by-products) are especially important in the American diet; they provide ⅔ of the total protein, about ⅓ of the total energy, ⅘ of the calcium, ⅔ of the phosphorus, and significant amounts of the other minerals and vitamins needed in the human diet.

In addition to the nutrients listed in Table 1–9, meat, dairy products, and eggs are a rich source of vitamin B-12, which does not occur in plant foods—only in animal sources and fermentation products. Also, it is noteworthy that the availability of iron in beef is twice as high as in plants.

**TABLE 1–9**
**FOOD NUTRIENTS: PERCENTAGE OF TOTAL CONTRIBUTED**
**BY LIVESTOCK AND POULTRY PRODUCTS[1]**

|  | Food Energy | Protein | Fat | Carbo-hydrates | Calcium | Phos-phorus | Iron | Vitamin A Value | Thiamin | Ribo-flavin | Niacin | Vitamin B-6 | Vitamin B-12 |
|---|---|---|---|---|---|---|---|---|---|---|---|---|---|
|  | | | | | | | (%) | | | | | | |
| **M**eat, fish & poultry ........ | 19 | 43 | 31 | — | 4 | 29 | 24 | 26 | 26 | 24 | 46 | 41 | 75 |
| **E**ggs ........... | 2 | 4 | 2 | — | 2 | 4 | 4 | 3 | 1 | 4 | — | 2 | 5 |
| **D**airy products, excluding butter . | 10 | 21 | 12 | 6 | 77 | 36 | 2 | 16 | 8 | 35 | 2 | 11 | 18 |
| **T**otal ........... | 31 | 68 | 45 | 6 | 83 | 69 | 30 | 45 | 35 | 63 | 48 | 54 | 98 |

[1]Source: *Agricultural Statistics 1988*, USDA, p. 493, Table 679. Data for 1985.

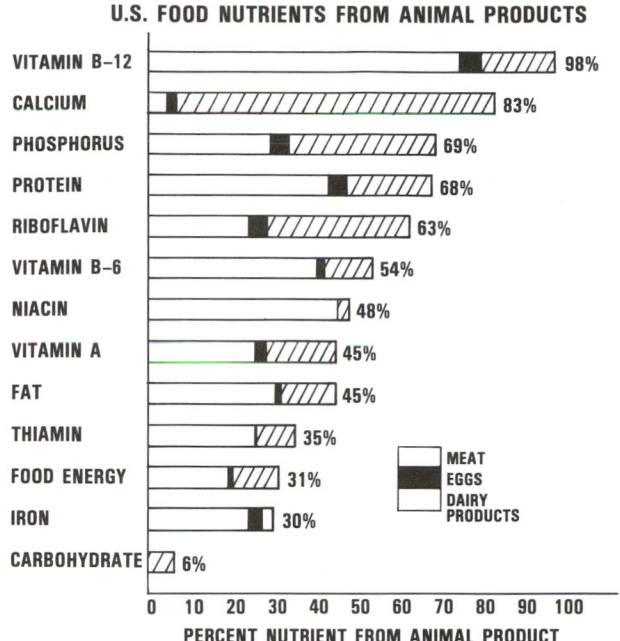

**U.S. FOOD NUTRIENTS FROM ANIMAL PRODUCTS**

| Nutrient | Percent |
|---|---|
| VITAMIN B-12 | 98% |
| CALCIUM | 83% |
| PHOSPHORUS | 69% |
| PROTEIN | 68% |
| RIBOFLAVIN | 63% |
| VITAMIN B-6 | 54% |
| NIACIN | 48% |
| VITAMIN A | 45% |
| FAT | 45% |
| THIAMIN | 35% |
| FOOD ENERGY | 31% |
| IRON | 30% |
| CARBOHYDRATE | 6% |

MEAT
EGGS
DAIRY PRODUCTS

0  10  20  30  40  50  60  70  80  90  100
**PERCENT NUTRIENT FROM ANIMAL PRODUCT**

Fig. 1–34.  Percentage of food nutrients contributed by animal products of the total nutrient supply in the United Staes in 1985. (Source: Table 1–9)

## ANIMALS AS A SOURCE OF PROTEIN

About two-thirds of the world's protein supply is provided from plant sources, one-third from animal sources. Since the Food and Agriculture Organization of the United Nations reports that the world's diet needs animal protein in amounts equivalent to one-third of the total protein, *provided* it were equally distributed. But is isn't (see Fig. 1–35).

The people of Europe have 4.7 times as much high quality animal protein per person as the people of Africa.

The most important role of animal protein is to correct the amino acid deficiencies of the cereal proteins, which supply about two-thirds of the total protein intake, and which are notably deficient in the amino acid, lysine. The latter deficiency can also be filled by soybean meal, fish, protein concentrates and isolates, synthetic lysine, or high-lysine corn. But such

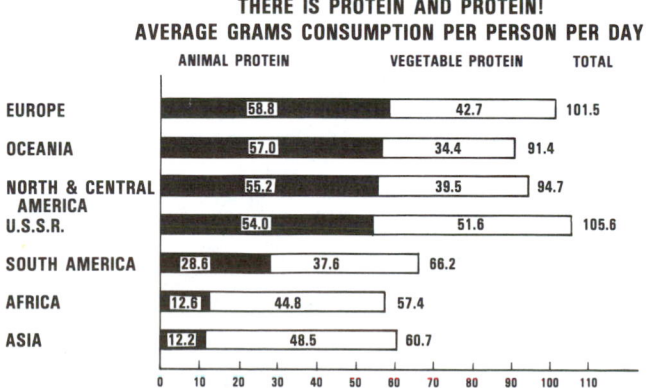

**THERE IS PROTEIN AND PROTEIN!**
**AVERAGE GRAMS CONSUMPTION PER PERSON PER DAY**

| | ANIMAL PROTEIN | VEGETABLE PROTEIN | TOTAL |
|---|---|---|---|
| EUROPE | 58.8 | 42.7 | 101.5 |
| OCEANIA | 57.0 | 34.4 | 91.4 |
| NORTH & CENTRAL AMERICA | 55.2 | 39.5 | 94.7 |
| U.S.S.R. | 54.0 | 51.6 | 105.6 |
| SOUTH AMERICA | 28.6 | 37.6 | 66.2 |
| AFRICA | 12.6 | 44.8 | 57.4 |
| ASIA | 12.2 | 48.5 | 60.7 |

0  10  20  30  40  50  60  70  80  90  100  110

Fig. 1–35.  Average grams protein consumption per person per day, with a breakdown into animal and vegetable protein, by geographic areas. (Source: *FAO Production Yearbook 1988*, p. 293, Table 107. Data for 1984–86)

Fig. 1–36.  A woman milking a water buffalo in India. Because of the large proportion of vegetarians in India (35 to 40%), milk is by far the nation's most important animal protein food. More than 50% of the milk produced in India is buffalo milk. In comparison with cow's milk, buffalo milk is higher in fat content (7.5% vs 4%) and sells at a higher price. (Courtesy, FAO of the U.N., Rome, Italy)

Fig. 1–37.  Tibetan nomad milking sheep. Note the lineup. Sheep provide about 1.5% of the world's milk supply. (Courtesy, Mr. Ray G. Johnson)

products have neither the natural balance in amino acids nor the appetite appeal of animal protein.

As soon as people get enough calories—as they achieve higher incomes, as they approach affluence—they start turning away from a starch oriented diet to one based on animal protein. This has happened in the United States, Canada, New Zealand, and Sweden. The affluent do not necessarily eat more animal protein products for nutritional reasons. Rather they consume more meat, milk, eggs, and fish because they like them—because they derive a rich enjoyment and satisfaction therefrom.

It is noteworthy that 8.6%—slightly less than one-twelfth—of the world's milk supply comes from buffalo, goats, and sheep.[12] However, in 16 Asian countries for which data are available, only 51.5% of the milk is from cows, while water buffalo, goats, and sheep produce 11.1, 15.2, and 22.2%, respectively.[13]

---

[12]Phillips, R.W., *Agriculture and the World's Food Supply*, Institute of Agriculture, University of Minnesota, 1974, p. 16.
[13]*Ibid.*

## ANIMALS AS A SOURCE OF MANURE

Animals provide manure for fields, a fact which was often forgotten during the era when chemical fertilizers were relatively abundant and cheap. One ton of average manure contains 10 lb of nitrogen (N), 5 lb of phosphoric acid ($P_2O_5$), and 10 lb of potassium ($K_2O$). Thus, it is worth $5.40 per ton when nutrients are computed at the following prices per pound: N = 29¢, $P_2O_5$ = 30¢, and $K_2O$ = 12¢ (see Table 5-9).

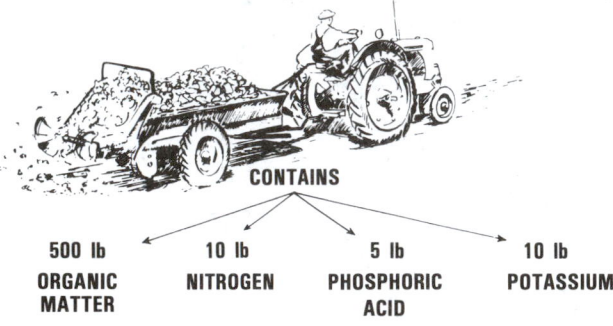

## ONE TON OF AVERAGE MANURE

CONTAINS

| 500 lb ORGANIC MATTER | 10 lb NITROGEN | 5 lb PHOSPHORIC ACID | 10 lb POTASSIUM |

Fig. 1-38. The content of one ton of average manure.

On the average general farm, with various classes and ages of animals, probably 80% of the fertilizing value of the feed is excreted in the feces and urine. With proper conservation, therefore, this fertility value may be returned to the soil.

When the author was a boy on a Missouri farm, we fed livestock to produce manure, to grow more crops, to feed more livestock, to produce more manure. But time was! The use of chemical fertilizer expanded manyfold; labor costs rose to the point where it was costly to conserve and spread manure on the land; more animals were raised in confinement; and a predominantly urban population didn't appreciate what they referred to as "foul smelling, fly-breeding, stuff." As a result, what to do with manure became a major problem on many livestock establishments.

China has kept its soils productive for thousands of years, primarily through the use of night soil (human waste) and every other kind of manure, applied to the land in primitive, but effective, fashion. All over China, a familiar saying is: "The more pigs, the more manure; and the more manure the more grain." Indeed, manure is very precious in China; it is carefully conserved and added to the land. Manure is used as a way in which to increase yields of farmland already under cultivation.

The energy crisis has prompted concern that farmers will not have sufficient fertilizers at reasonable prices in the years ahead. Since nitrogenous fertilizers are oil- or petroleum-based, there is cause for concern. As a result, a growing number of U.S. farmers are returning to organic farming—they are using more manure—the unwanted barnyard centerpiece of the past 40 years. They are discovering that they are just as good reapers of the land and far better stewards of the soil.

### AMOUNT AND VALUE OF MANURE

Currently, the United States is producing manure (exclusive of bedding) at the rate of about 1.35 billion tons annually. That is sufficient manure to add ¾ ton each year to every acre of the total land area (1.9 billion acres) of mainland United States.

Based on equivalent fertilizer prices (see section on "Animals as a Source of Manure"), the yearly manure crop is worth $10.3 billion (see Table 5-10). That's a potential income of more than $4,905 for each of the nation's 2.1 million farms.

The value of manure varies according to (1) the class of animals, (2) the kind of feed consumed and the kind of bedding used, (3) the method of handling, (4) the rate and method of application, and (5) the kind of soil and crops on which it is used.

Of course, the value of manure cannot be measured alone in terms of increased crop yields and equivalent cost of a like amount of commercial fertilizer. It has additional value for the organic matter which it contains, which almost all soils need, and which farmers and ranchers cannot buy in a sack.

Also, it is noteworthy that, due to the slower availability of its nitrogen and to its contribution to the soil humus, manure produces rather lasting benefits which may continue for many years. Approximately ½ of the plant nutrients in manure are available to and effective upon the crops in the immediate cycle of the rotation to which the application is made. Of the unused remainder, about ½, in turn, is taken up by the crops in the second cycle of the rotation; ½ of the remainder in the third cycle, etc., etc. Likewise, the continuous use of manure through several rounds of a rotation builds up a backlog which brings additional benefits and a measurable climb in yield levels.

In the future, as fertilizer and feed become increasingly scarce and expensive, the economic value of animal manure will increase, and it will be looked upon as a resource and not as a waste that presents a disposal problem. More and more feedlot manure will be recycled and either (1) incorporated in a grower ration, or (2) fed to range cattle during periods when range supplementation is beneficial, with the residues distributed over grazing areas where they would have fertilizing value.

Livestock producers sometimes fail to recognize the value of this barnyard crop because (1) it is produced whether or not it is wanted, and (2) it is available without cost. Most of all, no one is selling it. Whoever heard of a traveling manure salesman?

### HOW MUCH MANURE CAN BE APPLIED TO THE LAND?

With today's heavy animal concentration in one location, the question is being asked: How much manure can be applied to the land without depressing crop yields, making for salt problems in the soil, making for nitrate problems in feed, or contributing excess nitrate to groundwater or surface streams?

Based on earlier studies in the midwestern United States, before the rise of commercial fertilizers, it would appear that one can apply from 5 to 20 tons of manure per acre, year after year, with benefit.

Heavier applications can be made, but probably should not be repeated every year. With rates higher than 20 tons per annum, there may be excess salt and nitrate buildup. Excess nitrate from manure can pollute streams or groundwater and result in toxic levels of nitrate in crops. Without doubt the maximum rate at which manure can be applied to the land will vary widely according to soil type, rainfall, and temperature.

## THE WORLD FOOD SITUATION—WHAT'S AHEAD, AND WHAT TO DO ABOUT IT

The world supply of food will keep up with the demand, and perhaps exceed it, from now to the year 2000—and likely well beyond. But there will be substantial and troublesome disparities between the developed and developing nations, between the rich and the poor, between the "haves" and the "have nots," and between the well-fed and the hungry.

Hunger and malnutrition will continue to be widespread despite global abundance of food supplies. It is estimated that 500 million people will continue to suffer from severe under-nutrition, and that additional millions will be unable to acquire enough food to enjoy an active and productive life. At the same time, some countries will be striving to reduce food production in order to bring stocks to more manageable levels. Against this background, the following forces will affect the world food situation in the years ahead: (1) the persistence of large imbalances in trade and international payments; (2) protectionist trade policies; (3) low export prices for some agricultural commodities; (4) the burden of debt, particularly in Africa, Latin America, and the Caribbean; (5) the age-old spectre of droughts, locusts and grasshoppers; (6) the need for developed countries to adjust their agricultural support policies to bring supply more in line with demand; (7) closing the vast gulf between the level of science and technology employed by the modern commercial operator in the industrial world and the subsistence farmer in many developed countries, yet recognizing that much of the new science and technology will be beyond the reach of most of the latter; (8) greater concern for the environment, including the greenhouse effect of increased emissions of carbon dioxide and other gases on global temperature, and the long-term damage being wrought by some industrial gases on the world's protective ozone layer; (9) the development of sustainable agriculture; and (10) mobilizing an increasingly complex, independent, and competitive world to give priority to assisting the poor and deprived.

From a purely production standpoint, worldwide food shortages are unlikely. Nevertheless there will be famine in different areas over the land, primarily because of crop failures and/or lack of money with which to buy food.

Throughout much of the preceding portion of this chapter, reference has been made to general ways and means of alleviating world hunger. In the sections that follow, specific methods of improving the world food situation are discussed.

## CURB POPULATION GROWTH

Members of the animal kingdom other than humans have their numbers held in check by the many factors encompassed in the term *balance of nature*. People are different! Their strong propensity is to overpopulate the earth and to create conditions which threaten their very existence—their food supply, the water they drink, the environment in which they live, and the very air they breathe.

Without doubt, people will continue to live longer. Hence, curbing population growth will be required to maintain the balance between production and demand for food. It will be necessary to bring the number of people and their supply of food into proper balance.

It is hoped that reason will prevail—that people will not recklessly reproduce themselves into oblivion, with too many mouths nibbling away at natural resources faster than the earth

can combine the energy of the sun, the rains of the heaven, and the minerals of the soil, to produce food. The unhappy alternative: Accept the fact that in some nations the chosen methods of population control are starvation, disease, and/or war.

## INCREASE FARM PRICES AND PROFITS

People do those things which are most profitable to them; and farmers are people. Farmers in the developing countries of the world, can produce more, but higher prices than have existed in the past will be necessary to assure this.

Worldwide, farmers have always demonstrated their willingness to respond to incentives—prices and profits.

## BRING MORE ARABLE LAND UNDER CULTIVATION

Ever since people stopped living a nomadic life, they have been hunting for arable (cultivable) land. Today, only limited new areas can be brought under cultivation. More than half of the potential, but presently unused, arable land is in the tropics, and about one sixth of it is in the humid tropics—the largest areas being Africa and South America.

New land will be brought under cultivation by plowing grasslands, building dams, irrigating arid lands, terracing, and controlling wind and water erosion. The People's Republic of China has done a marvelous job of converting many once uncultivable areas into highly productive farmlands.

## DEVELOP MORE IRRIGATION

The value of irrigation for increasing crop yields is generally known. Yet, a study of the world's 20 major irrigating countries, in areas irrigated, showed that only 15% of their total cultivated areas was irrigated (see Table 1–3, page 9). Thus, the potential to increase crop yields through irrigation is very great.

The new high-yielding varieties of rice and wheat have a proven potential. But utilizing their high potential requires both proper timing and amount of water application. Thus, there is need to develop more and better irrigation throughout the world.

## INCREASE CROP YIELDS

While more land can be brought under cultivation than is currently being used, all recent studies of world food production conclude that, outside of Africa and Latin America, yield-increasing techniques—irrigation, fertilizer, new seeds, and improved technology—will be the primary source of future food increases.

In addition to irrigation, fertilizer is a key factor in yield increases, although it must be combined with improved varieties of seeds and improved cultural practices if it is to have much impact on yields. As evidence of the soundness of the fertilizer approach, it is noteworthy that almost half of the 50% gain in crop output per acre in the United States since 1940 is attributed to the increased use of fertilizers. From this, it may be concluded that increased use of fertilizer could increase world food output by 50% in the years ahead.

Hand in hand with increased irrigation and fertilization, new seeds and techniques will continue to make for increased crop yields in the years ahead. The Green Revolution, spawned by new varieties of wheat and rice, should be expanded wherever these crops are grown; and new varieties of crops adapted to other parts of the world, along with new techniques, should be developed.

## IMPROVE PASTURES AND RANGES

Some sparsely populated areas of the world, such as Australia, New Zealand, Argentina's pampas area, and the western range areas of the United States and Canada, are now important sources of livestock products. But there are still vast areas of sparsely settled grasslands where the production of livestock products is small; among them, large portions of Africa, the highlands of central Asia, some portions of the Andean area of South America, and the nomadic grazing areas of the Near East. In these areas, subsistence is the goal and animal numbers are generally regarded as being more important than the yield of salable products. Nevertheless, the potential for increased production in these areas is considerable. Also, and most important, the only practical way of harvesting human food from many of these areas is through livestock. So, grass—the world's largest crop—should no longer be taken for granted. The contribution of properly managed grazing lands in terms of food and fiber production needs to be pursued. No other program offers so much potential to increase the world's food production capacity quickly and at so little cost; this is especially true of the grasslands in the tropics and subtropics.

More and more grains will be used for direct human consumption. As a result, there will be increased reliance on grass for meat, milk, and wool production. Petroleum is not needed to make wool, and animals do not require fuel to graze the land and recover the energy that is stored in the grass. Also, animals are completely recyclable; they produce a new crop each year and perpetuate themselves through their offspring. But it takes thousands of years to create coal, oil, and natural gas; and when they're gone, they're gone forever.

Most grazing areas can be improved by seeding new and better varieties of grasses and legumes, by fertilizing, and by management, including scientifically controlled grazing, avoiding overgrazing by both domestic livestock and wild animals, and supplemental feeding.

Fig. 1–39. *Bos indicus* cattle of India in an open field. Due to constant overgrazing, most such areas have been reduced to mere "gymnasiums" for livestock. (Photo by A. H. Ensminger)

## FEED MORE ROUGHAGE AND LESS GRAIN

The 25-year trend in the United States toward feeding more concentrate and less roughage to cattle is now reversing itself. Beginning in 1950, concentrate feeding rose markedly as the price of corn and other high-energy feeds went down. From a 14% share of the feed unit intake of beef cattle in the 1950s, concentrates grew to comprise close to a quarter of the 1970 ration. The grain-fed cattle binge ended with the world grain shortages and high-priced grains of the early 1970s.

In the future, cattle and sheep will increasingly be "roughage burners." Producers will rely upon the ability of the ruminant to convert coarse forage, grass, and by-product feeds, along with a minimum concentrate, into palatable and nutritious food for human consumption, thereby competing less for humanly edible grains. Increasingly, the steer of tomorrow will be produced on a maximum of milk and grass and minimum of grain.

Ruminants can make the transition to more roughage with ease. For them, it is merely a return to nature, for they evolved as consumers of forage.

## PRODUCE LEANER MEATS

Leaner meat is higher in protein content than fat meat. For example, on a carcass basis, trimmed to retail level, Standard grade beef runs 19.4% protein vs 17.4% for Choice grade—that's 2% higher. Besides, leaner beef can be produced with much less grain.

Consumer preferences and costs of production underlie the relative prices of fat and lean beef, lamb, pork, and poultry, but changes in U.S. grading standards help consumers adjust their consumption patterns. For this reason, when grain prices are high, producers exert pressure to have the federal-grading system changed so as to reduce the amount of grain fed.

## LESSEN ANIMAL NUMBERS

As the world's food situation worsens, animal numbers will come under scrutiny—particularly where they compete for potential human food. Today, there are over 3 billion 4-footed animals in the world and about 3.2 billion domesticated fowl—more than twice as many animals and birds as people in the world.

## DEVELOP MORE EFFICIENT ANIMALS

Improved genetics, along with improved feeding and management, have made for more meat, milk, and eggs. Yet, much further improvements are possible and needed, especially in the developing countries. Table 1–10 shows the pounds of production per animal per year for different classes of animals (1) in the developed countries, and (2) in the developing countries. Note that productivity in the developing countries is very low compared with that of the developed countries. Fig. 1–41 shows the wide difference in the productivity of cattle for beef production in selected countries.

Fig. 1-40. Unimproved sheep. The primary challenge throughout the world is to improve the great masses of animals. In the years ahead, there must be (1) improved reproduction—a higher percent young crop, fewer death losses from birth to weaning, earlier weaning ages, and heavier weights; (2) increased productivity per animal; (3) greater efficiency of feed utilization; and (4) improved market products.

### TABLE 1-10
### POUNDS OF PRODUCTION PER ANIMAL PER YEAR[1]

|  | Production in Developed Countries | | Production in Developing Countries | |
|---|---|---|---|---|
|  | (lb) | (kg) | (lb) | (kg) |
| Milk | 7,810 | 3,550 | 1,722.6 | 783 |
| Beef & Veal, carcass | 514.8 | 234 | 354.2 | 161 |
| Pork, carcass | 178.2 | 81 | 145.2 | 66 |
| Mutton & Lamb, carcass | 35.2 | 16 | 30.8 | 14 |
| Goat, carcass | 26.4 | 12 | 26.4 | 12 |
| Eggs | 9.5 | 4.3 | 6.0 | 2.7 |

[1]Source: *FAO Production Yearbook 1988*, United Nations, Rome, Italy, Vol. 42, pp. 252, 255, 259, 262, 265, 274, & 282.

Although two thirds of the animals of the world are raised in the developing countries, these nations produce only 37% of the world's meat, milk, and eggs.[14] This low productivity is largely due to the failure to utilize the scientific principles of husbandry and disease control.

The object of livestock production is to convert the production of the land—grass and/or crops—into the maximum of animal products. This calls for animal improvements in the following outputs: in the yield of meat, milk, or eggs of acceptable quality; and, in four-footed animals, a higher percent of young and more pounds of offspring produced per dam. More efficient animals can be selected through modern production-testing techniques designed to obtain maximum production on minimum feed. Also, new and potentially useful breeds should be evaluated in each of the countries in which it is planned to propagate them.

Without doubt, there is more opportunity to improve the food and nutrition of the human population in the developing countries by increasing the productivity of their ruminant livestock than in any other way. Also, more efficient animals

[14]Source: *Commodity Review and Outlook*, Food and Agriculture Organization of the United Nations, Rome, Italy, 1988, p. 61, Table 2.14.

## BEEF PRODUCTION PER INVENTORY OF CATTLE PER YEAR

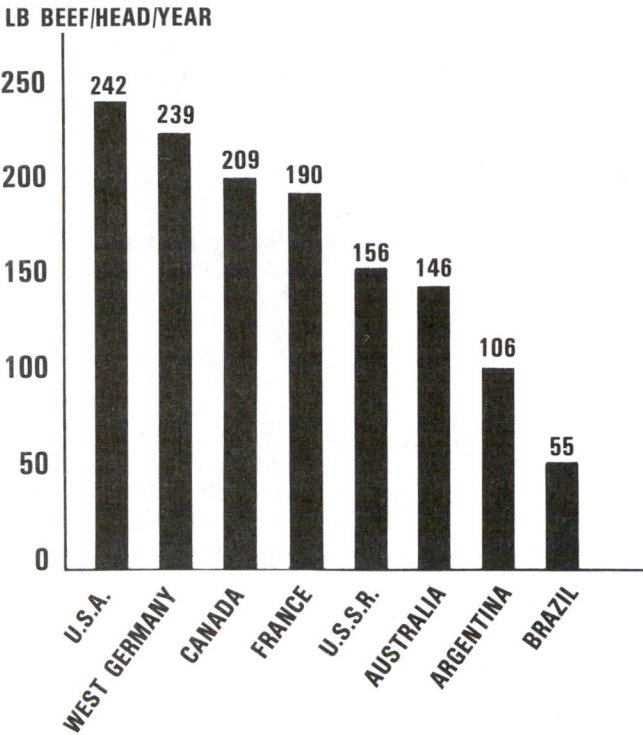

Fig. 1-41. Beef production per head of cattle population per year in selected countries. This is not a precise measure of productivity of cattle, because no account is taken of (1) herd increase or decrease, or (2) international movement of cattle numbers. Nevertheless, this figure does provide a reasonable indication of the efficiency of the cattle of a country. (Source: *Agricultural Statistics 1988*, USDA, p. 262, Table 389, and *FAO Production Yearbook 1988*, U.N. Rome, Italy, Vol. 42, p. 253, Table 92. Data for 1988.)

afford a logical approach for providing more food in these developing countries for the following reasons: (1) Animals provide a way to store a considerable supply of food in good condition *on the hoof*; (2) the farmers have a history of livestock use—hence, it should be easier to improve the productivity of their animals than to institute an entirely different agricultural system; and (3) ruminants do their own harvesting.

## CONTROL DISEASES AND PARASITES

Diseases in farm animals reduce the supply of meat and other animal products by an unknown, but large, quantity, and add substantially to the cost of food and fiber.

Deaths of animals take a tremendous toll. Even greater economic losses—hidden losses—result from failure to reproduce live young, and from losses due to retarded growth and poor feed efficiency, carcass condemnations and decreases in meat quality, and labor and drug costs. Also, considerable cost is involved in keeping out diseases that do not exist in a country, such as keeping foot-and-mouth disease out of the United States. Quarantine of a diseased area may cause depreciation of land values or even restrict whole agricultural programs. Additionally, and most importantly, it is recognized that some 200 different types of infectious and parasitic diseases can be transmitted from animals to human beings; among them, such dreaded diseases as brucellosis (undulant fever), leptospirosis, anthrax, Q fever, rabies, trichinosis, tuberculosis, and tularemia. Thus, rigid meat and milk inspection is necessary for the protection of human health. This is added expense which the producer, processor, and consumer must share.

One of the principal causes of low livestock productivity in the developing countries is the incidence of animal disease. The most important of these is trypanosomiasis, caused by the dreaded tsetse fly, prevalent throughout tropical and subtropical Africa in an area about the same size as the United States. Were it not for the tsetse fly and trypanosomiasis, some 70% of the area would be suitable for livestock production and mixed agriculture, and 140 million more cattle—or the equivalent number of smaller ruminants—could be kept.

Thus, the potential throughout the world of providing more food through animal and disease control is very great. The level of animal health attained in the advanced countries shows that tremendous scope exists for improvement in most developing areas.

The prevention and control of internal and external parasites through the use of the most effective anthelmintic or insecticide is a quick, cheap, and dependable method of increasing meat, milk, and egg production with no extra animals, no additional feed, and little more labor.

## IMPROVE AND INCREASE PROTEIN SOURCES

It is generally recognized that diet customs are somewhat emotional in character—that many people will put synthetic clothes on their backs long before they will put synthetic food in their stomachs. Yet, when people are hungry or suffering from malnutrition, they're not finicky about the "pedigree" of their food.

New protein sources are being used, and others will be discovered. However, people like animal products, so they will pay more for them than for plant foods. There is an old saying in the southwestern part of the United States that "Thin beef is better than fat beans." Nevertheless, there is no known nutritive essential in meat which cannot be provided from fish, vegetable sources, or by synthesis, although animal sources are the surest way of meeting nutritive needs, without the hazard of imbalances.

Researchers are attempting to bolster traditional protein sources and to develop entirely new proteins, with their efforts centered around the following approaches and protein sources:

• **Traditional protein sources**—In many countries where proteins are in short supply, the growing of grain legumes and pulses and the greater use of nuts and protein concentrates made from protein-rich residues of oil manufacturing are receiving increasing attention.

Attempts are being made to bolster plant protein sources by (1) increasing through genetic means the content and nutritional quality of the protein in cereal grains; (2) improving the quality of cereal protein by fortification with chemically produced amino acids (principally lysine; sometimes with tryptophan and methionine added); and (3) adding protein concentrates from fish, oilseeds, and other sources to cereal foods.

• **Soybean protein**—The protein of soybean, after suitable processing, has the highest nutritive value of any plant protein source. Foods made from soybeans include soy flour, soy milk, spun fiber, soy sauce, tofu and tempeh (fermented cheese and curd), and soy butter. Also, the soybeans may be cooked as green or dried beans, canned in sauce, or roasted (like peanuts).

Soy flour may be used in many baked products. Some six to eight percent of soy flour added to wheat flour used for bread and pastries will significantly improve the protein value of the product without making much change in texture and appearance.

Soy milk has been consumed for centuries in China and in some other countries of the Far East. A 200-milliliter bottle of soy milk will supply at least 6 grams of high-quality protein, 50% of a child's daily requirement. Soy milk is made by soaking soybeans 4 to 5 hours, grinding them in a hot water slurry (1 part of soaked beans to 3 parts of water), then straining out the insoluble residue.

Soybean isolates can be spun into fibers, flavored, colored, and fabricated into meatlike products, including beef steaks, chicken, pork chops, lamb chops, ham, bacon, and sausage—all difficult to distinguish from the real product.

• **High protein/high amino acid corn**—It has been known for many years that corn, the world's third most important human food after rice and wheat, is nutritionally inadequate. In 1914, researchers at the Connecticut Agricultural Experiment Station induced starvation in laboratory rats by feeding them generous helpings of corn. Further, it was found that rats could be restored to health by supplementing the high corn diet with two protein fractions—the amino acids lysine and tryptophan.

Although normal corn contains about 10% protein, half of it is locked up in the fraction zein, which is useful in the manufacture of textiles, plastics, and other things, but totally indigestible by single-stomached creatures—including humans. Moreover, normal corn is especially poor in lysine and tryptophan, essential amino acids that the human body cannot manufacture and must get from food.

This deficiency of corn shows up in people wherever corn is a major source—if not the only source—of protein in the diet. Known by the exotic name, kwashiorkor, this nutritional deficiency disease is the leading cause of mortality among infants and children in many parts of the world.

For years, plant scientists assayed the world's corn varieties one by one, looking for a strain with more nutritionally balanced protein. Finally, in 1963, a Purdue University team headed by biochemist Edwin T. Mertz analyzed an odd group of corns characterized by soft, floury endosperm inside an opaque, chalk-white kernel. The Purdue scientists found that the opaque characteristics of corn, which had been noted for years without exciting much scientific interest, is associated with a recessive gene that replaces some of the kernel's humanly useless zein with needed lysine and tryptophan. The mutant—routinely labelled Opaque-2, or $O_2$ for short—had a lysine content of 3.4%, compared to 2.0% for normal corn. Additionally, Opaque-2 showed higher levels of tryptophan and other amino acids.

But the millenium that seemed so near with the discovery of Opaque-2 has remained frustratingly out of reach. Although the nutritional value of the high-lysine corn is recognized, two major hurdles between research discovery and application must yet be overcome: (1) The mutant gene is linked to Opaque-2's soft, floury kernel, which is both light in weight and vulnerable to pest attacks, producing lower yields for farmers; and (2) Opaque-2 has not been accepted by the majority of consumers, who are accustomed to the harder *flint* or *dent* kernels with a deeper, translucent color. But the need is great—human lives are at stake. So, the goal of plant breeders is to develop corn varieties that better meet the demands of farmers and consumers.

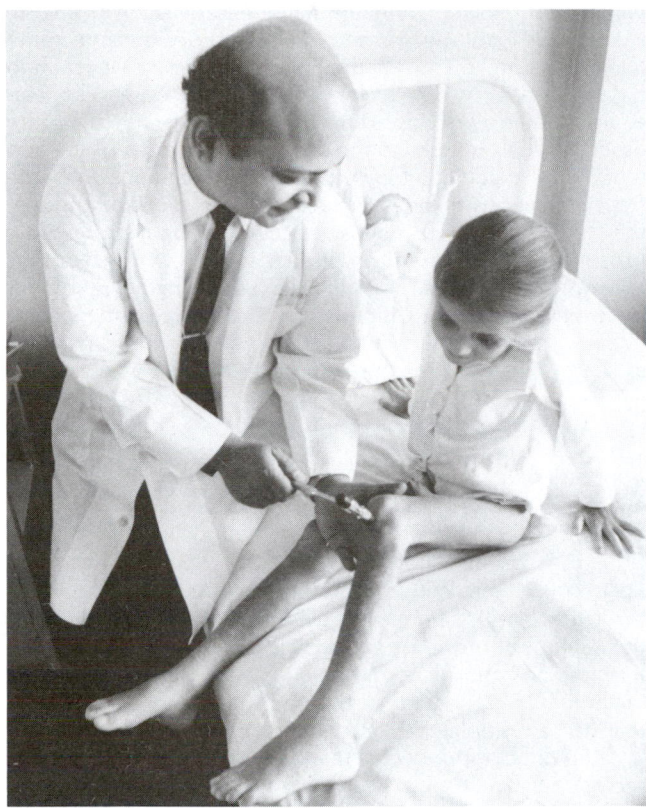

Fig. 1–42.  Lysine made the difference! There are two ways to starve: (1) lack of food, or (2) lack of one or more nutrients. This little South American girl was taken to Valle Hospital, in Colombia, to die (left). She showed all the symptoms of advanced protein malnutrition—even her hair turned from black to coarse red. She recovered on a high-lysine corn diet (right). (Courtesy, The Rockefeller Foundation, New York, NY)

In 1989, University of Minnesota researchers announced the discovery of a gene in corn that controls the level of protein produced, which can be used to produce corn with 3% more protein and 20% more methionine and lysine than normal corn, without lowering the yield or producing soft kernels. A patent application on the process is pending at this writing.

• **Fish protein**—The present world fish catch runs about 97 million metric tons per year, with an average protein content of about 15%. Fish fillets contain over 20% protein; hence, they are one of the richest sources of protein. Each year, some 25 million metric tons of fish are processed into fish meal for animal feed. With improved handling and processing techniques, some of this fish could be converted into fish protein concentrate (FPC) suitable for human consumption. This product is produced from types of fish that are not popular in the usual channels of fresh fish trade. The fish are extracted to remove oil, dried, and ground to make a bland meal containing about 80% protein, 0.2% fat, and 13% mineral. Fish protein concentrate furnishes high-quality protein and is inexpensive. However, it lacks appetite appeal—it is bland, tasteless, and odorless.

• **Single-celled protein (SCP)**—Some single-celled protein types such as yeasts, algae, and bacteria can be useful sources of protein and vitamins for human and animal feeding. The safety of these foods depends on the organisms selected, the quality of substrate used, and the conditions of growth. Of course, yeasts and bacteria have been used for centuries in the baking, brewing, and distilling industries, in making cheese and other fermented foods, and in the storage and preservation of foods.

Dried brewer's yeast, a residue from the brewing industry, and Torula yeast, resulting from the fermentation of wood residue and other cellulose sources, have been marketed as animal feeds for years. With proper processing, they are also suitable for human foods.

Bacteria grow faster than yeasts under favorable conditions, doubling their mass in a matter of minutes, rather than hours. Dried bacterial cells contain at least 55% protein.

Various bacteria and yeasts can be selected and cultured to grow on organic wastes. These include animal wastes, sewage, many different chemical residues from industrial plants, petroleum by-products, sawdust, and other fibrous residues. Petroleum companies in several countries have built factories to produce bacterial protein for the animal feed market. Also, considerable research is in progress to convert manure from poultry and other animals through bacterial fermentation into animal protein feed. This recycling process could produce much protein and help solve a pollution problem.

Algae are single-celled plants which may contain 20 to 60% protein on a dry basis. They synthesize proteins by the use of solar energy. In northern Nigeria, these plants are dried and eaten for human food. A noted University of California scientist states that with the sunshine available in southern California, one acre of algae would feed 500 people for one year. Although algae grow widely on the earth's water surfaces, problems of harvesting and processing them into acceptable food products remain unsolved.

## TAP THE SEA FOR MORE FOOD

The ocean, which covers 70.73% of the earth's surface and, therefore, receives a proportionate amount of all the solar energy reaching this planet, is one of the most promising potentials for providing added food for the world's spiraling human population. It's an immense reservoir of food which man has only lightly tapped. Hence, there is growing interest in the seas as a source of food supply, including both fish and vegetables.

## INCREASE FISH CULTURE

Fish production for human food should be greatly expanded. Today, fish contribute less than 1% of world food supplies in terms of dietary energy, 5% of total protein, and 14% of animal protein, although it is of much greater importance in some individual countries.

Fish are very efficient food converters. One and six-tenths pounds of fish food will produce one pound of fish (see Table 1–7). For this reason, plus the fact that fish do not compete with crop production, it is expected that both freshwater and brackish-water fisheries, including fish production in small ponds and paddy fields, will receive increasing attention in the future.

Fig. 1–43. A hog-fish combination in Kaw Yao County, in China. Note that the hog barn is located so that it can be flushed directly into the fish ponds. (Photo by A. H. Ensminger)

The Chinese have ingeniously fed manure to fish since the days of the Ming Dynasty; hence, the use of manure in fish culture is of long standing in that country. In Kwangtung Province, in the heart of the land of fish and rice, the author was impressed with a hog-fish combination on a commune where they produced 49,700 pigs and had a fish catch of 863,000 lb. The fish were fed solely on pig manure flushed from the nearby hog barns directly into the fish ponds, plus grass clippings from the pond banks. This unique operation provided a method of handling manure plus pollution control and, at the same time, recycled and used the feed twice—first through the hogs, and second through the fish. It is noteworthy, too, that the Chinese were feeding manure to fish without the hazard of nitrate-nitrite toxicity and/or of oxygen starvation of the fish.

The Chinese Academy of Agriculture and Forestry Sciences recommends the following practices when feeding manure to fish: (1) mixing the manure with plant material, then composting it before feeding (although manure may be, and is, fed to fish

without composting, pollution may be lessened by first composting); (2) fertilizing and fermenting the composted manure; and (3) feeding manure to the little fingerlings, rather than to big fish. But, according to the fish experts of China, the feed of fish should vary according to age and species (carp are especially well adapted to the use of animal waste). They also report that, in addition to manure, freshwater fish of China are fed a great variety of foods, including silk worm cocoons, silage, the leaves of such crops as sweet potatoes and turnips, soybeans, bean cake and curd, distiller's grains, and wheat. It's noteworthy, too, that, in addition to supplying needed food, the freshwater fish of China lessen the mosquito menace.

## CONSERVE ENERGY

Fossil fuels—the stored photosynthates of previous millennia—are like a bank account. There is nothing wrong with drawing upon either of them, but neither is inexhaustible. It is highly imprudent not to be aware of big withdrawals and not to cover them. Within a short span of a few years, the world made the transition from a positive energy balance based upon the capture of the sun's energy via green plants, crops, and forests to an imbalance, or even a negative balance, by resorting primarily to the bank of trapped sun energy of fossil fuels that had accumulated over millions of years.

The direct input of fuel into food production is of rather recent origin. It all began in a very small way about 1840, when fuel-powered ships transported fertilizer (guano, and later bone meal) from South America to Europe. Then, after 1910, transportation vehicles relied almost exclusively on fossil fuels. But the direct use of fossil fuels in agriculture started with the manufacturing of a chemical fertilizer beginning about 1922. Following closely in period of time, tractors replaced horses, mules, and oxen—eventually almost completely eliminating them.

In addition to food production on the farm, there are two other important steps in the food line as it moves from the producer to the consumer; namely, processing and marketing, both of which require higher energy inputs than to produce the food on the farm (see Table 1–11).

### TABLE 1–11
### MODERN FOOD PRODUCTION IS INEFFICIENT IN ENERGY UTILIZATION—THE STORY FROM PRODUCER TO CONSUMER[1]

| Year | On the Farm | Food Processing | Marketing and Home Cooking | Total/ Person/ Year |
|---|---|---|---|---|
| 1940[2] | | | | |
| Million kcal | 0.9 | 2.2 | 2.1 | 5.2 |
| Percent | 18.0 | 42.0 | 40.0 | 100.0 |
| 1990[3] | | | | |
| Million kcal | 2.8 | 5.7 | 4.6 | 13.1[4] |
| Percent | 21.4 | 43.5 | 35.1 | 100.0 |
| Increase, times, 1940–1990 | 3.1 | 2.6 | 2.2 | 2.5 |

[1]Energy in million kcal used per capita to produce one million kcal of food in the U.S.

[2]Values from Borgstrom, G. "The Price of a Tractor," *Ceres*, FAO of the U.N., Rome, Italy, Nov.-Dec., 1974, p. 18, Table 3.

[3]Author's estimate based on several reports detailing trends in energy usage.

[4]This means that in 1990, it required 13.1 million kcal to produce 1 million kcal of food for each person, a daily consumption of 2,740 kcal (1,000,000 ÷ 365 = 2,740).

Table 1–11 points up the increasing drain that modern food production is putting on the energy supply. In 1990, U.S. farms put in 2.8 calories of fuel per calorie of food grown, 3.1 times more than the on-farm energy input in 1940.

Table 1–11 also shows that, in the United States in 1990, a total of 13.1 calories were used in the production, food processing, and marketing-cooking for every calorie of food consumed, with a percentage distribution of the total cost of energy at each step from producer to consumer as follows: on the farm, 21.4% food processing, 43.5%; and marketing and home cooking, 35.1%. In 1940, it took only 5.2 calories—about 40% of the 1990 figure—to get 1 calorie of food on the table. It is noteworthy, too, that more energy is required for food processing and marketing-home cooking than for growing the product; and that, from 1940 to 1990, the on-the-farm energy requirement increased by 3.1 times, in comparison with an increase of 2.6 and 2.2 times for each of the other steps—processing and marketing-home cooking.

Fig. 1–44. An Oriental wet rice peasant, using animal power (water buffalo), expends only 1 calorie of energy to produce each 50 calories of food. By comparison, the average U.S. farmer, using mechanical power (tractors), expends 2.8 calories of fuel energy to produce 1 calorie of food. (Courtesy, International Bank for Reconstruction and Development, Washington, DC)

Prior to the advent of machines and fuel in crop production, 1 calorie of energy input on the farm produced about 16 calories of food energy. Today, on the average, U.S. farms put in about 2.8 calories of fuel per calorie of food grown; hence, to produce a daily intake of 3,000 calories of edible food from cultivated crops may require 8,400 calories of energy from fossil fuels—an exhaustible source. It's more surprising yet—and thought-provoking—to know that, even today in the poorer or developing countries, it takes only 1 calorie to produce each 10 calories of food consumed. The Oriental wet rice peasant uses only 1 unit of energy to produce 50 units of food energy. This gives the Orientals a favorable position among the major powers as the energy crisis worsens.

The following additional points are pertinent to any energy conservation program:

1. **Photosynthesis fixes energy.** Photosynthesis is by far the most important energy-producing process. But currently only about 1% of the solar energy falling on an area is fixed by photosynthesis; and only 5% of this captured energy is fixed in a form suitable as food for man. Thus, (a) manipulating plants for increased efficiency of solar energy conversion, and (b) converting more total energy fixed as chemical energy in plants

(the other 95%) into a form available to people would appear to hold great promise in solving the future food problems of the world.

2. **Animals step up energy.** The increase in the energy level through animal products—through animals consuming the photosynthetic energy in crops—almost equals the energy subsidies at each of the two steps after the product leaves the farm (in food processing and marketing-home cooking—see Table 1–11).

Fig. 1–45. Animals conserve energy. For the most part, they consume photosynthetic energy obtained from the sun and stored in grasses and other forages; and, unlike synthetic fibers, it does not require fossil energy to produce wool.

3. **Crop residues contain energy.** Crop residues left in the field, above or below the soil surface, may well constitute four to five times more energy than is harvested. Increasingly, this potential source of added feed, organic fertilizer, and energy will be utilized in the future.

4. **Increased yields; the law of diminishing returns in energy.** Modern intensive farming has markedly increased crop yields per acre and per man hour—by as much as 50- to 100-fold. But this has been done at the cost of large inputs of fuel (including electricity).

For a surprising number of modern cropping systems, a 10- to 50-fold increase in the energy output merely doubles or triples the food energy. Substantial expenditures fail to produce corresponding increases in yields. Thus, the law of diminishing returns prevails.

High petroleum costs have spurred a search for other energy sources and for means of conserving energy. Higher productivity of the agriculture of tomorrow must be achieved through ingenious approaches in order to reverse the present lopsided energy balance. In obtaining increased food yields, we must consider the use of energy to produce energy. We must remember that photosynthesis is by far the most important energy-producing process; indeed, that it is the only basic food-manufacturing process in the world. We must remember, too, that grazing animals do not require fuel outside of their own body use to harvest the energy and other nutrients of grass (solar energy converted into chemical energy by grass), a renewable resource.

## CONTROL POLLUTION

Pollution is the issue of the decade. It matters little whether pollution is due to agriculture or factories. Everything that defiles, desecrates, or makes impure or unclean streams or atmosphere must be controlled.

In agriculture, we need to give particular attention to the pollution caused by manure, fertilizer, insecticides, herbicides, and growth promotants.

One cow produces as much waste as 16 humans. Hence, with 20,000 steers in a feedlot, the disposal problem is equal to a city of 320,000 people. In addition to being used as fertilizer, manure is now being recycled as a livestock feed and serving as a source of energy (methane gas).

Of course, there is no one best manure management system for all situations. But, one way or another, science and technology must evolve with ways of disposing of 1.5 billion tons of manure annually; and this must be accomplished without polluting streams or the atmosphere or being offensive to the neighbors.

We must ever be mindful that life, beauty, wealth, and progress depend upon how wisely man uses nature's gifts—the soil, the water, the air, the minerals, and the plant and animal life.

## ESTABLISH GRAIN RESERVES

People have attempted to prevent widespread starvation through grain reserves since biblical times, when Joseph interpreted Pharaoh's dream about 7 years of famine befalling ancient Egypt. Pharaoh directed Joseph to garner 20% of the crops harvested during the 7 fat years as a reserve for the 7 lean years.

World security involves both short-run and long-run arrangements. In the short run, grain reserves are the logical way to deal with floods, drought, pests, and similar emergency problems. Over the long run, food security can best be assured by improving the ability of the world to feed itself, through increased purchasing power and the application of science and technology. For sheer survival, each nation will have to assume the major responsibility for its own food supply—whether by production, barter, or purchase—and for controlling its own population.

In view of the uncertain nature of weather and world food supplies, it would appear wise to ensure against major, short-run shortages through maintaining an ever normal granary—by storing up food in times of plenty and dispensing in times of adversity. Hand in hand with a food reserve, the world needs to develop an early warning system to detect food shortages.

Food reserves to provide short-term emergency famine relief would involve relatively small amounts of grain, perhaps around 10 million tons. The question of how to support a famine relief program would have to be decided. The consensus is that rich and poor alike would provide commodities or financial assistance for this food aid reserve, probably in proportion to their gross national product (GNP). If 10 million tons of grain were prorated among the developed countries on the basis of GNP, the United States would need to provide about 3 million tons as its share. The cost of such a program would depend largely on how frequently the reserve would be tapped, but it would be small.

In considering grain reserves, it is noteworthy that, today, the United States exports 61% of the cereal grain exports of the world. Argentina ranks second, accounting for 13% of cereal grain exports.

Regardless of the organization for and method of providing food reserves, the United States does care and will do its share—and likely more. On this count, the record speaks for itself. From 1954 to 1980, $11.4 billion was contributed to needy countries by the United States.

## WASTE NOT; WANT NOT

Waste of food supplies will increasingly nag the consciences and pocketbooks of all people—producers and consumers alike.

Pests cause an estimated 30% annual loss in the worldwide potential production of crops, livestock, and forests.[15] Every part of our food, feed, and fiber supply—including marine life, wild and domestic animals, field crops, horticultural crops, and wild plants—is vulnerable to pest attack. Obviously, if these losses could be prevented, or reduced, world food supplies would be increased. Think of it!—The potential to increase food supplies by nearly one-third through eliminating wanton waste. The problems are complex, but the stakes are high.

Remember that a worldwide annual loss of 30% potential food productivity occurs despite the use of advanced farming technology and mechanized agriculture. Remember, too, that in many of the developing countries losses greatly exceed this figure.

Pests of many kinds attack plants during all stages of their growth, and they attack food and food products after harvest—in storage, during transportation to market, in warehouses, in elevators, in ships, in supermarkets, and in homes after purchase. Here are a few notable pest losses:

- **Plant diseases and insects**—Disease organisms kill plants, cause rotting and blemishing of food products, and reduce crop yields and quality.

  Insects devour growing crops, lower yields and quality, and attack grains and other food products in storage and during transport. Also, insects harbor and transmit diseases to plants, animals, and humans.

  More than 160 bacteria, 250 viruses, and 8,000 fungi are known to cause plant diseases. In the United States alone, approximately 10,000 species of insects are destructive enough to be called *enemies*; and about four-fifths of them are injurious to crops.

- **Weeds**—Weeds reduce yields and quality by competing with crops for water, nutrients, light, and space. Also, they poison livestock, interfere with harvesting, and slow the flow of water for irrigation and drainage. In the United States, some 2,000 species of weeds and brush cost farmers and ranchers an estimated $8 billion annually, with reductions in quantity and quality of crops heading the list.[16]

- **Rats**—Each year, rats consume feed, damage additional feed, destroy property, and spread disease, for a total of $18 per rat. Thus, the yearly keep on U.S. rats totals nearly $2.8 billion.[17]

[15]Ennis, Jr., W. B., W. M. Dowler, and W. Klassen, "Crop Production to Increase Food Supplies," *Science*, Vol. 188, No. 4188, May 9, 1975, pp. 593–598.

[16]Ensminger, M. E., *The Stockman's Handbook*, 6th ed., The Interstate Printers & Publishers, Inc., Danville, IL, 1983, p. 757.

[17]*Ibid.*, p. 758.

Although they are not as damaging as rats, mice, gophers, and other rodents should also be controlled—and for the same reasons.

• **Birds**—Birds are gluttonous and filthy. They consume and contaminate much feed and spread many diseases. Hence, they should be controlled. In a study of a 12,000-head cattle feedlot in California, University of California researchers found that the 10,000 to 20,000 birds that came to "dinner" ate an average of about 350 lb of feed each day, for a total of 57,750 lb during the 5½-month winter season. Iowa cattle feeders figure that starlings add $3 to $4 to the cost of each steer marketed. Some western feedlot operators estimate that starling nuisance and feed costs add 2¢ to the cost of each pound of gain. The quelea, a grain-eating bird of Africa and Asia, causes serious losses of rice, millet, sorghum, and wheat crops; in the Sudan area alone, daily consumption by these birds has been estimated at 3,000 tons of grain.

A few notable statistics pointing up the enormous economic damage caused by pests follow:

• **In the Near East**—Preharvest losses of food and industrial crops due to insects and diseases in 11 developing countries in the Near East run an estimated 23%.[18]

• **In Chile**—The average losses in yield of potatoes due to blight exceed 20% per year.

• **In India**—Insects and rodents that attack grain in the field and in storage cause annual losses of 13 million tons; such an amount of wheat would have supplied 77 million families with one loaf of bread per day for a year.

In addition to the direct losses caused by pests, there are hidden, or indirect, losses: losses in efficiency, and losses in the input of energy involved in crop production—wasted energy. And losses in suffering!

Indeed, science and technology have been the great multipliers. Together, they have upped the ounce to the pound, the pint to the bushel, and the dozen to the gross. Despite this accomplishment, the very existence of people is threatened; not because they cannot produce enough food, but because they have not protected that which they have produced from the ravages of pests—from losses in food and fiber, in lowered efficiency, and in deaths. Like a thief in the night, each year pests steal away billions of dollars in losses to farmers and increased cost to consumers; and for the most part, they go unrectified. Only when millions of people die from starvation are steps taken to control them.

By applying the science and technology that we already have, food and fiber losses can be reduced substantially—perhaps by as much as 30 to 50%. The net result would be an increase of 10 to 15% in the world food supply, with no new land required. In no other way can the hunger gap be filled so quickly and at so little cost.

## INCREASE SCIENTIFIC AND CULTURAL EXCHANGE BETWEEN COUNTRIES

There will always be international boundaries. But scientists and agriculturalists the world over can, and will, work together through the *tie that binds*—their desire to help humanity. Since no nation has a corner on all the brains, scientific and cultural exchange between countries offers our best hope for survival.

## INCREASE RESEARCH, EDUCATION, AND EXTENSION

The best long-run solution to the problem of world food and nutrition shortages is to develop research, education, and extension programs that provide farmers in the developing countries with the necessary tools and techniques to increase their production, augmented by fortification of foods and education of consumers.

Research gave the world Mendel's laws of inheritance; hybrid corn and high-yielding varieties of wheat and rice; the vitamins, antibiotics, and hormones; a host of insecticides, vermifuges, and herbicides; many disease controls; new breeds of livestock and new varieties of plants; frozen semen, estrus control, embryo transfer, gene transfer, and cloning; fixation of nitrogen from the air; radioisotopes and radiation; and automation. Farmers have combined these products of research to produce more food and fiber.

A backlog of basic and applied research, supported by programs through which the results of research can be quickly moved into the field to increase production, is an important insurance against long-run food shortages. It is especially important that agricultural research and application be directed toward the problems of increasing productivity in the developing countries, where modern scientific agriculture is in its infancy and where fancy technology will do little to ease the needs of hungry people.

Many of today's products and techniques were developed from ideas and research done some years earlier. There is a time lag between research breakthrough and its application; hence, it is urgent that more research be initiated now. The story of hybrid corn points up both the time lag and the benefits of research. The basic research that gave the world hybrid seed was done prior to 1920. Yet, in 1933—13 years later—only 2.0% of the corn acreage of the Corn Belt was planted to hybrid seed. But today, 98% of the corn grown in the corn-producing sections of the United States comes from hybrid seed; and it's yielding 8 to 10 bushels more per acre than the open pollinated varieties, or over one-half billion bushels more annually. The same time lag in application is being experienced in the Green Revolution. The first of the improved wheat varieties was released in Mexico in 1948. Yet, 35 years later, only 35% of the wheat of the Asian developing nations, exclusive of the centrally planned economies, is of the new high-yielding varieties. Thus, research started now will be appearing 15 to 25 years hence. Certainly, new techniques will be needed then, as now.

Hand in hand with research, there is need for improved teaching and research application in the developing nations. All too often, the elite of these countries are trained, only to go home and put on a white laboratory coat, far removed from the buffalo that pulls the plow. Too often the results of research are never gotten out to and applied by practicing farmers.

Teaching, research, and extension—especially adult education through short courses—should be expanded and implemented by imaginative and bold approaches because, over the long pull, they afford the most logical way in which to provide adequate food and clothing for the world's expanding population.

[18]*Report of the President's Science Advisory Committee Panel on the World Food Supply*, Vol. 3, U.S. Government Printing Office, Washington, DC, 1967, p. 130–175.

## SUMMARY

The world will continue to grow richer and more populous. Sociological developments which slow population growth, and technological developments which speed production increases will make for more meat, milk, and eggs in the future. Until 2000, and perhaps beyond, more food will produced per person and food production will be generally adequate to meet demand for the world as a whole. However, substantial malnutrition, punctuated by famine in areas and periods of crop failure, will likely persist among low-income groups and in the developing countries, with the degree and extent of undernourishment and starvation determined by the presence and effectiveness of special national and international food programs.

In the period beyond 2000, it will be increasingly difficult to mobilize the necessary human, institutional, and economic resources rapidly enough to keep pace with food demand in countries where the need will continue to be most critical. Meeting the food production needs of the 21st century will depend upon accelerated adoption of known methods, as well as development of new technology.

Feeding the hungry people in the developing countries in the years ahead is of worldwide concern to all people of all races, all religions, all colors, and all political and economic philosophies.

Indeed, concern about food and animals—ages old and 21st century new—will be the major global issue in the decades to come. But the author is bullish about the future! Agricultural scientists and farmers (peasants) throughout the world will go on researching, discovering, creating, and advancing. Then, by sharing and applying their science and know-how, each of us, and the whole world, will have a brighter tomorrow. Our dreams will come true—faster and more abundantly, with more food and animals in our future.

## QUESTIONS FOR STUDY AND DISCUSSION

1. Who was Thomas Robert Malthus? What, and when, did he prophesy relative to world population and food?

2. In the 1970s, many experts predicted that Malthus would be proven right before the end of the 20th century. What forces and developments prevented this from happening?

3. Name, and describe the symptoms of, the most common nutritional deficiency diseases affecting humans.

4. In 1990, many countries were burdened with food surpluses. Yet, at the same time, it was estimated that 500 million people in the world remained seriously malnourished. How could this be?

5. Table 1–1 shows the following per capita calories: Europe, 3,397; Africa, 2,299. Why is there this wide disparity of 1,098 calories per person per day? How can this difference be lessened?

6. Describe the symptoms of protein deficiency disease known as Kwashiorkor.

7. According to the FAO report entitled *Agriculture: Toward 2000*, which the author briefed in this chapter, the projections/goals/needs for food and animals toward the year 2000 and beyond differ for (a) the developing countries, and (b) the developed countries. What are the major differences between the two groups?

8. Is the world as a whole running out of land which it would be practical to develop and cultivate? Justify your answer.

9. Table 1–2 shows that Asia has 0.38 acres of cultivated land per person, whereas Africa has 0.75 acres per person—twice as much. Yet, there is much more hunger and malnutrition in Africa than in Asia. What is the explanation of this difference?

10. How do you account for the fact that, among the nations of the world, the People's Republic of China ranks first in total acres irrigated and seventh in percentage of area irrigated?

11. Why is it said that fertilizer is a substitute for land?

12. What is the Green Revolution? What did Dr. Norman Borlaug have to do with it; and what honor is bestowed upon him as a result of it?

13. Explain how each of the following factors affects the demand for food: (a) human population growth, (b) income growth (disposable income), (c) proportion of income spent for food, (d) income distribution, (e) food preferences and trends, and (f) nutritive qualities.

14. Why are animal proteins much sought?

15. What factors account for the world leadership of the United States in (a) total red meat production and (b) beef and veal production?

16. List, by rank, the three leading countries of the world in fish catch. How do you account for these countries dominating fish catch?

17. What factors favor the following countries ranking first among the nations of the world in per capita consumption of the animal products listed: *total red meat*, East Germany; *beef*, Argentina; *pork*, Hungary; *mutton and lamb*, New Zealand; *poultry*, United States; *milk*, New Zealand; and *eggs*, West Germany?

18. Explain how each of the following factors limits the production and consumption of animal products: (a) level of income, (b) religious beliefs, (c) lack of integration of animal and plant agriculture, (d) lack of art and science in animal production, (e) the high cost of animal research work, (f) lack of capital, (g) low productivity of animals, (h) diseases and parasites, (i) lack of facilities for marketing, processing, and distributing animals and their products, and (j) orientation of in-country agricultural policies.

19. What nations will be the major meat importers in the years ahead? How do you account for the fact that each of these countries will be importing so much meat rather than producing it, in the years ahead?

20. What do you foresee in product demand during the final decade of the 20th century for beef, pork, poultry, fish, and dairy products?

21. As the world demand for animal products increases, (a) how can this demand be met? and (b) will this demand be met by the developing nations or by the developed nations?

22. Worldwide, 84% of the calories consumed by people come from vegetable products and 16% from animal products.

How do the developed and the developing countries differ as to their sources of calories?

23. List and discuss the factors favoring direct grain consumption by people.

24. Table 1–7 shows that on a calorie or protein conversion basis, it's not efficient to feed grain to animals and then consume the livestock products. Evaluate this table.

25. Discuss the relative (comparative) (a) feed efficiency, (b) energy efficiency, and (c) protein efficiency of different species of farm animals.

26. List and discuss the factors favoring sharing grain with animals.

27. If there should be worldwide famine, how would a hungry world meet the food vs feed dilemma?

28. In this day and age is there any justification for using animals as a source of power anywhere in the world?

29. What nutrients can best be secured from animal proteins, rather than from plant proteins?

30. Should manure be looked upon as (a) a valuable fertilizer, or (b) an unwanted barnyard centerpiece?

31. List and discuss the forces which will affect the world food situation in the years ahead.

32. List and discuss specific methods for improving the world food situation.

33. Study Table 1–10. Why is there so much difference between the developed and developing nations in the production per animal per year of the following products: milk, beef/veal, pork, mutton and lamb, and eggs?

34. Fig. 1–41 shows a wide difference in the production per inventory of cattle per year. What is the reason for this?

35. Of what importance is high-lysine corn?

36. On the average, a U.S. farmer uses 2.8 calories of fuel per calorie of food produced, whereas an Oriental wet rice peasant uses only 1 unit of energy to produce 50 units of food energy. Does this give the Orientals a favorable position among the major powers of the world as the energy crisis worsens? Does this indicate that the whole world should do away with tractors and go to animal power?

37. Why has pollution become such an issue?

38. For protection against crop failure and famine, grain reserves are needed. This gives rise to the following pertinent questions: (a) How large should these reserves be? (b) who should hold them—should they be stored with the countries that produced them, or should they be held by a group of nations (perhaps the U.N.), or the private sector (farmers, grain traders, processors)? (c) who should pay for them? and (d) by whom, and how, should they be managed?

39. Pests cause an estimated 30% annual loss in worldwide potential production of crops, livestock, and forests. How can this happen in an era of advanced farming technology and mechanized agriculture?

40. How can world food shortages be lessened through (a) inreased scientific and cultural exchange between countries, and (b) increased research, education, and extension?

41. Will world food production be adequate to meet world demand to the year 2010 A.D.?

42. What solution do you propose for the world food problems?

## SELECTED REFERENCES

| Title of Publication | Author(s) | Publisher |
|---|---|---|
| *Agricultural Commodity Projections, 1970–1980,* Vols. 1 and 2 | | Food and Agriculture Organization of the United Nations, Rome, Italy, 1971 |
| *Agricultural Statistics 1988* | USDA staff | U.S. Department of Agriculture, Washington, DC, 1988 |
| *Agriculture: Toward 2000* | FAO staff | Food and Agriculture Organization of the United Nations, Rome, Italy, 1987 |
| *Animal Agriculture and the World's Food Supply* | R. W. Phillips | Institute of Agriculture, University of Minnesota, St. Paul, MN, 1974 |
| *Animal Production and World Food Needs,* Spec. Pub. 12 | H. Degraff, et al. | College of Agriculture, University of Illinois, Urbana-Champaign, IL, 1968 |
| *By Bread Alone* | L. R. Brown<br>E. P. Eckholm | Praeger Publishers, New York, NY, 1975 |
| *Commission of the European Communities, Report for 1973,* Parts 3 and 4 | | The European Community Information Service, Washington, DC, 1973 |
| *Commodity Review and Outlook* | FAO staff | Food and Agriculture Organization of the United Nations, Rome, Italy, 1988 |
| *Demographic Yearbook 1971,* 23rd Edition | | Department of Ecomonic and Social Affairs, United Nations, New York, NY, 1972 |
| *Energy and Protein Requirements,* Report of a Joint FAO/WHO Ad Hoc Expert Committee | | Food and Agriculture Organization, and World Health Organization, United Nations, Rome, Italy, 1971 |

*(Continued)*

## SELECTED REFERENCES (Continued)

| Title of Publication | Author(s) | Publisher |
|---|---|---|
| *FAO Yearbook, Fertlizer, Vol. 38* | FAO Staff | Food and Agriculture Organization of the United Nations, Rome, Italy, 1988 |
| *FAO Yearbook, Fishery Statistics, Vol. 65* | FAO staff | Food and Agriculture Organization of the United Nations, Rome, Italy, 1988 |
| *FAO Yearbook, Production, Vol. 42* | FAO staff | Food and Agriculture Organization of the United Nations, Rome, Italy, 1988 |
| *FAO Yearbook, Trade Commerce, Vol. 41* | FAO staff | Food and Agriculture Organization of the United Nations, Rome, Italy, 1987 |
| *Food and Agriculture in the Common Market: Policitical and Economic Aspects* | | The European Research Bureaux, Oxford, England, 1973 |
| *Food and Nutrition, Vol. 1, No. 1* | | Food and Agriculture Organization of the United Nations, Rome, Italy, 1975 |
| *Food Consumption, Prices, and Expenditures, 1966–87, Statistical Bul. 773* | J. J. Putnam | U.S. Department of Agriculture, ERS, Washington, DC, 1989 |
| *Foods & Nutrition Encyclopedia* | A. H. Ensminger, et al. | Pegus Press, Clovis, CA, 1983 |
| *Handbook on Human Nutritional Requirements* | R. Passmore B. M. Nicol M. N. Rao | Food and Agriculture Organization of the United Nations, Rome, Italy, 1974 |
| *Manual of Tropical Veterinary Parasitology* | M. Shah-Fischer | CAB International, Wallingford, U.K., 1989 |
| *State of Food and Agriculture, The: 1987–88* | FAO staff | Food and Agriculture Organization of the United Nations, Rome, Italy, 1987–88 |
| *Statistical Abstracts of the United States*, 109th edition | Staff | U.S. Department of Commerce, Bureau of the Census, Washington, DC, 1989 |
| *Strategy for the Conquest of Hunger: Proceedings of a Symposium* | | The Rockefeller Foundation, New York, NY, 1968 |
| *Strategy for Plenty, A* | | Food and Agriculture Organization of the United Nations, Rome, Italy, 1970 |
| *United Nations World Food Conference: Assessment of the World Food Situation, Present and Future* | | United Nations, Rome, Italy, 1974 |
| *United Nations World Food Conference: The World Food Problem; Proposals for National and International Action* | | United Nations, Rome, Italy, 1974 |
| *Working Papers: Conference of International Development Strategies for the Sahel* | | The Rockefeller Foundation, New York, NY, 1975 |
| *Working Papers: Food Production and the Energy Dilemma* | R. W. Cummings, Jr. | The Rockefeller Foundation, New York, NY, 1974 |
| *Working Papers: Perspectives on Aquaculture* | | The Rockefeller Foundation, New York, NY, 1974 |
| *World Animal Review, No. 4* | | Food and Agriculture Organization of the United Nations, Rome, Italy, 1972 |
| *World Almanac, 1989* | Edited by M. S. Hoffman | Scripps Howard Company, New York, NY, 1989 |
| *World Food Production, Demand, and Trade* | L. L. Blakeslee E. O. Heady C. F. Framingham | Iowa State University Press, Ames, IA, 1973 |
| *World Food Situation and Prospects to 1985, The* | | Economic Research Service, USDA, Washington, DC, 1974 |
| *World Population by Country and Region, 1950–86, and Projections to 2050* | F. Urban P. Rose | U.S. Department of Agriculture, ERS, April 1988 |
| *World Tables, 1988–89 edition* | World Bank staff | The World Bank, The Johns Hopkins University Press, Baltimore and London, 1989 |
| *Yearbook of International Trade Statistics 1972–73* | | United Nations, New York, NY, 1974 |

Food for the table! Animal products are good—and good for you. (Courtesy, National Live Stock and Meat Board, Chicago, IL)

# ANIMAL SCIENCE, U.S.A.

# Chapter 2

*Animal science refers to the total store of knowledge relative to the breeding, feeding, care, and management of animals and the marketing and processing of animals and their products, as gained through practical experience and research methods.*

Animal agriculture is essential to a well-nourished and happy people. If we would look in the family refrigerator, we would find a lot of animal products—foods derived from beef and dairy cattle, pigs, sheep, and poultry. Behind the livestock, we would see vast expanses of pasture and range land, feed grains, and such by-product feeds as cull potatoes, beet by-products, and surplus citrus fruit—all being utilized as animal feeds. Back of the feeds are the soil resources, spring rains, and the energy of the sun. With calloused hands, the farmer and rancher combine these to produce a tasty platter of meat for the table, cream for the peaches, butter for the biscuit, and cheese for the macaroni—all derived from the land via animals. In addition, leather, fats, wool, grease for soap, glands for adrenalin and other essential medical products, glues, gelatins, important organic chemicals, and countless other materials come from animals raised on American farms and ranches.

Back of present-day successful animal production has gone years of experience and scientific research—progress in animal breeding; feeding; physiology; disease and parasite control; management; marketing; and processing, storing, and distribution of meats, milk, and eggs. This progress is the result of studies and experiences that have extended from the farms and ranches to the nation's kitchens.

## THE BIRTH OF ANIMAL SCIENCE

The field of animal science had its humble beginnings with the domestication of animals, for from this remote day forward it was necessary to give attention to their breeding, feeding, and care and management.

The domestication of animals also marked the first step toward civilization of the most primitive people—the transformation from the savage to the civilized way of life. Savages hunted animals as sources of food and raiment. They lived on what roots, berries, and seeds they could find and on such insects, animals, and fish as they could catch. In addition, they were only too likely to include their kin and kind in their hunting and eating.

Fig. 2-1. Ancient drawing of a bison on a rock, dating to the Old Stone Age. Even prior to their domestication, people revered animals, according them a conspicuous place in the art of the day. (Courtesy, The Bettmann Archive, Inc.)

From very early times, primitive people had domesticated the dog, using it to assist in their hunting and to provide protection by night. Perhaps even more important, the presence of the dog furnished animal companionship, thus filling a deep-rooted want which has always existed in human beings.

Fig. 2-2. Neolithic (New Stone Age) people continued to hunt animals as a source of food and raiment, even though at this remote period many animals had been domesticated. (Photo from a painting by Charles Knight, obtained through the courtesy of the American Museum of Natural History, New York, NY)

Gradually, people adopted a more settled mode of life, and with this came the desire to safeguard the food supply for times when hunting was poor and to have the food close at hand. At this stage, seeds were stored for winter use; and nearly all of our modern animals were tamed and confined or, as we say, domesticated.

In addition to using domestic animals as a more certain supply of food and clothing, ingenious caretakers soon began to employ them for purposes of pack and draft. Through selection and controlled matings, they also molded animal types better to serve specific needs.

Fig. 2-3. Photograph of the painting, *The Sacrifice of Noah*, by Bernardo Cavallino. This shows an animal being served up as an offering to the deities. (Courtesy, National Gallery of Art, Kress Collection)

But the contribution of animals extended far beyond their utility value. People revered them; they accorded them a conspicuous place in the art of the day and made them the chief objects of worship and myth and the sacrificial offerings of many a religious ceremony. Also, from the day of domestication forward, the herding of animals became indicative of the superiority of one tribe over another, and the great livestock countries of the world have always supported the most advanced civilizations and been the most powerful. Down through the ages, therefore, animals have been the most useful helpmates of people—contributing richly to their food, clothing, power, recreation, and inspiration.

## DISTRIBUTION OF ANIMALS AND EARLY AMERICAN IMPORTATIONS

People followed the migration of animals prior to their domestication in order to keep near their food supply. But with their taming and confinement, they took the initiative in migrating and moving animals with them. It was in this very manner that domestic animals were brought to the Western Hemisphere; for only the llama, alpaca, guinea pig, and turkey were native to the Americas at the time of Columbus' first landing in 1492. Yet today, the United States possesses the largest and most varied domestic animal population of all countries.

It is generally believed that cattle, pigs, sheep, goats, and horses were first brought to the West Indies by Christopher Columbus on his second voyage in 1493. Cattle, horses, and sheep were brought to Mexico by Cortez in 1519, whereas pigs and horses were first introduced directly into what is now the United States by Hernando de Soto in 1539.

## TRANSFORMATION OF THE U.S LIVESTOCK INDUSTRY

The domestic animals that were first brought to America by the Spanish explorers were of low efficiency in comparison with those of present standards. Nevertheless, they furnished the sturdy foundation stock that was subsequently improved and developed into a great industry.

As cities grew and animals increased in numbers and pushed inland from the Eastern Seaboard, there was the development of trailing and marketing, the enlargement of the local slaughterhouse, and eventually the birth of the modern packing industry. Following closely in period of time came the development of railroads, the luxury of rail transportation of animals, and the perfection of artificial refrigeration and the refrigerator car. Such developments marked a new era in animal production. In the meantime, there was the opening of the fertile Ohio Valley and the western range.

With rapid development and expansion of the livestock industry came the demand for improvement in breeding. Progressive breeders, ever awake to their opportunity, proceeded to make large importations of animals from the mother countries—especially from Great Britain, France, and Spain. Thus, through the years, the Texas Longhorns were replaced by prime bullocks; the Arkansas Razorbacks were replaced by improved meat-type swine, and the black, brown, and spotted sheep were replaced by modern mutton and wool-type animals. Hand in hand with improved breeding came improved feeding and management.

## U.S. ANIMAL AND HUMAN POPULATION TRENDS

Despite all the improvements that have been wrought in the husbandry of animals and in the marketing, processing, and distribution of animals and their products, U.S. animal numbers have failed to keep pace with human population increases. Table 2–1 gives the comparative U.S. animal and human population in 1840 and 1988, whereas Fig. 2–4 shows their respective population trends during this same period of time.

On January 1, 1988, for every hundred people in this country there were 40 cattle and calves, compared with a peak of 97 in 1888; 4 sheep and lambs, compared with 124 in the peak year of 1867; and 22 hogs, compared with an all-time high of 94 in 1872.

**TABLE 2–1**
**U.S. ANIMAL AND HUMAN POPULATION IN 1840 AND 1988[1]**

| Human and Animal Population | 1840, or as Designated | 1988 |
|---|---|---|
| Human population | 17,069,000 | 245,602,000 |
| All cattle and calves | 14,971,000 | 98,994,000 |
| Milk cows (milk cows that have calved) (1850) | 6,385,000 | 10,307,000 |
| All sheep and lambs | 19,311,000 | 10,774,000 |
| Hogs and pigs | 26,301,000 | 53,795,000[2] |
| Horses and mules | 4,336,000 | 10,840,000 |
| Poultry (1880) | 102,272,000[3] | 5,620,799,000[2] |
| Broilers | | 5,002,934,000[2] |
| Chickens | | 377,516,000[2] |
| Turkeys | | 240,349,000[2] |

[1]Sources: human population from *Statistical Abstract of the United States 1989*, p. 7, No. 2; all cattle and calves, milk cows, sheep and lambs, hogs and pigs, and poultry from *Agricultural Statistics 1988*, USDA, Tables 386, 402, 400, 500, 510, and 520. Horses and mules from *Horses and Horsemanship*, p. 14, Table 2–1, by M. E. Ensminger, published by Interstate Publishers, 1990; data for 1986.

[2]Figures for 1987.

[3]1880 figure for chickens and turkeys only.

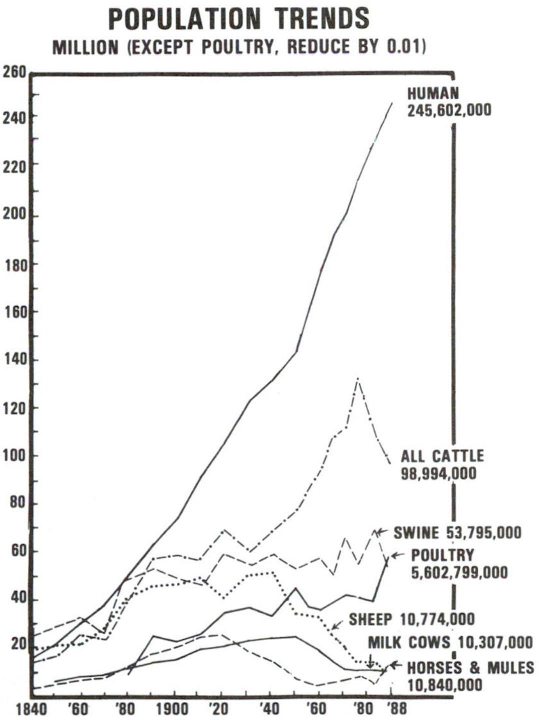

**POPULATION TRENDS**
MILLION (EXCEPT POULTRY, REDUCE BY 0.01)

HUMAN 245,602,000
ALL CATTLE 98,994,000
SWINE 53,795,000
POULTRY 5,602,799,000
SHEEP 10,774,000
MILK COWS 10,307,000
HORSES & MULES 10,840,000

Fig. 2–4. U.S. animal and human population trends. Note that animal numbers have failed to keep pace with human population increases. (Source: U.S. Bureau of the Census and USDA)

But animal numbers alone do not tell the whole story. Productivity per animal unit has been greatly accelerated. Meat animals now mature more rapidly than formerly—the quicker turnover actually meaning more products from fewer animals. Dairy cattle and poultry also produce more milk and eggs, respectively.

Further human population increases are inevitable. Hand in hand with it, science and technology will continue to make for greater efficiency of production. Fewer farms and farm workers will produce more food and fiber (see Figs. 2–5 and 2–6).

## FARM AND NONFARM POPULATION

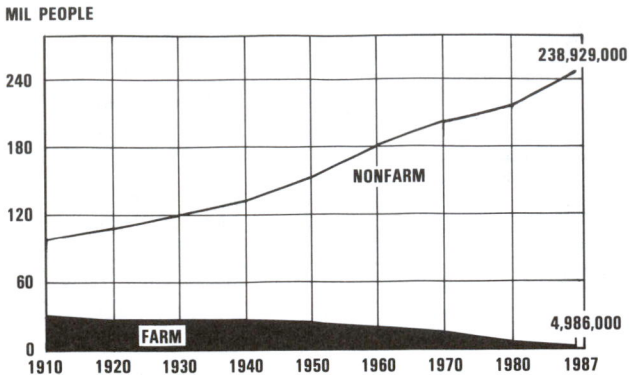

Fig. 2–5.  In 1987, the U.S. population totaled 243,915,000, with a breakdown as follows: 238,929,000 were nonfarm, and 4,986,000 were farm. **NOTE WELL:** Farm population consists of persons living in rural areas on places of 10 or more acres with at least $50 or more of agricultural product sales sold in the reporting year, and under 10 acres with at least $250 in agricultural sales. (Source: *Agricultural Statistics 1988*, USDA, Washington, DC)

## PERSONS SUPPLIED PER FARM WORKER

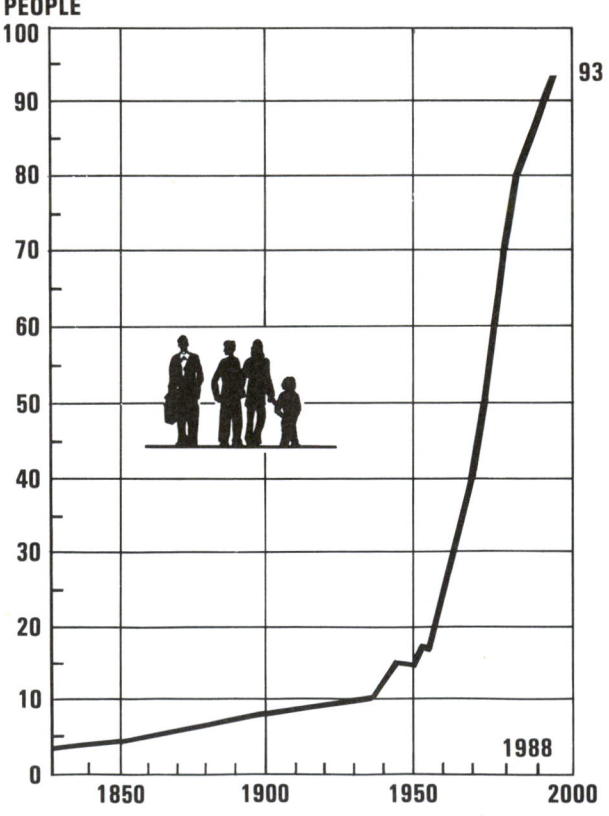

Fig. 2–6.  Productivity of farm worker; the number of persons supplied farm products by one farm worker. In 1820, each farm worker supplied farm products for 4.12 persons, including self. In 1952, the figure was 17.32; in 1966, it was 39.6; in 1975, it was 55; and in 1988, it was 93 people. Never have so many people been dependent on so few. (Source: U.S. Department of Agriculture)

In 1988, one farm worker supplied enough food and fiber for 93 people. That year (1988), each U.S. resident consumed an average of 142 lb of beef, veal, pork, lamb, and mutton; 72 lb of chicken and turkey; 88 lb of fresh fruits; 79 lb of fresh vegetables; 290 lb of dairy products; and 50 lb of potatoes; so, one U.S. farm worker produced a whopping 93 times this amount of food in 1988. Not only that; farmers also produce large quantities of farm products for export.

In order that the increased U.S. human population of the future will be well fed, inefficient animals will be culled. This elimination is inevitable if this nation is to have a long-range meat program that will supply the nutritional requirements of an expanding population on a basis that will be profitable to producers. Also, in the years ahead, increased emphasis will be placed on such things as higher percentage calf, lamb, and pig crops, and on lessening losses due to diseases and parasites.

## MAGNITUDE OF THE U.S. LIVESTOCK INDUSTRY

The far-flung livestock business comprises one of the largest industries in the United States. Providing America's livestock products requires 2.2 million farms and ranches, 6,753 federally inspected meat packers and processors, 181,000 meat retailers, and 727,000 food service outlets.[1] Additionally, it requires immense investments in land, feedlots, herds, processing plants, and vast networks of transportation lines and distribution facilities. In the sections that follow, the magnitude of the livestock industry will be shown through several important criteria.

## LAND AREA AND NUMBER OF FARMS AND RANCHES DEVOTED TO ANIMAL PRODUCTION

Animals are the largest users of the nation's land. Table 2–2 and Fig. 2–7 show land use, land ownership, and the number of farms and ranches in the United States.

Although land use is not constant, it is noteworthy that 46% of the total land area of the United States is devoted to the production of feeds and foods, including 29.2% of the U.S. lands used for grazing purposes and an additional 16.9% devoted to the production of hay and other forage crops and grain. Both publicly and privately owned lands are used in the production of feeds and foods.

Most U.S. lands are privately owned, but an astounding 885 million acres, or 39.1%—over 1 acre in 2½—of all U.S. lands, are under public ownership. The federal government, with 730 million acres, or 32.2% of all U.S. lands, controls most of the publicly owned lands. State and local governments own an additional 155 million acres. The publicly owned lands are, for the most part, effectively used by private farmers and ranchers for livestock production under a grazing permit system. It is unfortunate, however, that 60.3% of the federal holdings of the 50 states is located in the Western Region (see

---

[1]U.S. Department of Agriculture, ERS, sources; 1987 data, except for number of farms and ranches which is 1988.

### TABLE 2-2
### LAND USE, LAND OWNERSHIP, AND NUMBER OF FARMS AND RANCHES IN THE UNITED STATES[1]

| | | Area (million acres) | Area (million hectares) | Percentage Total (%) |
|---|---|---|---|---|
| Total land area of the United States, including Alaska and Hawaii | | 2,265 | 917 | 100.0 |
| I Land Use | A. Area devoted to production of feeds and foods | 1,045 | 423 | 46.0 |
| | Pasture and grazing land | 662 | 268 | 29.2 |
| | Cropland used for forages and grains | 383 | 155 | 16.9 |
| | B. Area not devoted to production of feeds and foods including forests not pastured, urban areas, roads, farmsteads, crop failure, and fallow and idle areas | 1,220 | 493.9 | 53.9 |
| II Land Ownership | A. Private ownership | 1,329 | 538.1 | 58.7 |
| | B. Indian land | 51 | 20.6 | 2.3 |
| | C. Public ownership | 885 | 358.3 | 39.1 |
| | Federal | 730 | 293.1 | 32.2 |
| | State and local governments | 155 | 62.8 | 6.8 |
| III In Farms and Ranches | A. Land in nation's 2,159,000 farms and ranches (1988) | 999 | 404.5 | 44.1 |
| | B. Land not in farms and ranches | 1,266 | 512.6 | 55.9 |

[1]Sources: *Statistical Abstract of the United States 1989*, USDA, pp. 193, 628. **Note:** Number of U.S. farms is for 1988. Rest of data for 1982.

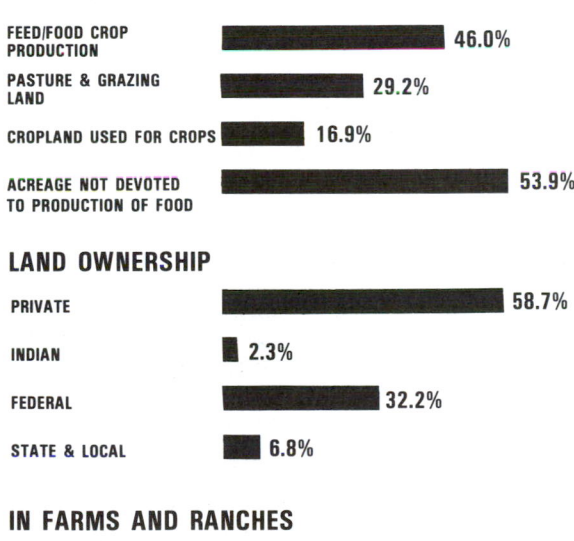

## LAND USE, OWNERSHIP, AND NUMBER OF FARMS AND RANCHES IN THE U.S.

### TOTAL LAND AREA OF THE UNITED STATES 2,265 MILLION ACRES

**LAND USE**

- FEED/FOOD CROP PRODUCTION — 46.0%
- PASTURE & GRAZING LAND — 29.2%
- CROPLAND USED FOR CROPS — 16.9%
- ACREAGE NOT DEVOTED TO PRODUCTION OF FOOD — 53.9%

**LAND OWNERSHIP**

- PRIVATE — 58.7%
- INDIAN — 2.3%
- FEDERAL — 32.2%
- STATE & LOCAL — 6.8%

**IN FARMS AND RANCHES**

- IN NATION'S FARMS & RANCHES — 44.1%
- NOT IN FARMS AND RANCHES — 55.9%

Fig. 2-7. Land use and ownership of the total land area of the United States, including Alaska and Hawaii. It is noteworthy (1) that 46% of the total U.S. land area is devoted to the production of feeds and foods, and (2) that 41.9% of the total U.S. land area is under non-private ownership (Indian, federal, or state owned). (Data based on Table 2-2 of this book.)

Fig. 2-9). By contrast, in the Northeastern Region only 3.4% of the land is federally controlled. Naturally, the character of the land area, rather than state boundaries as such, has been the chief criterion in arriving at what lands shall remain under federal ownership.

Fig. 2-8. On the range! A total of 29.2% of the U.S. land in the 50 states is used for pasture and grazing, much of it unsuited for cultivation. (Source: *The Sheepman's Production Handbook*, Denver, CO)

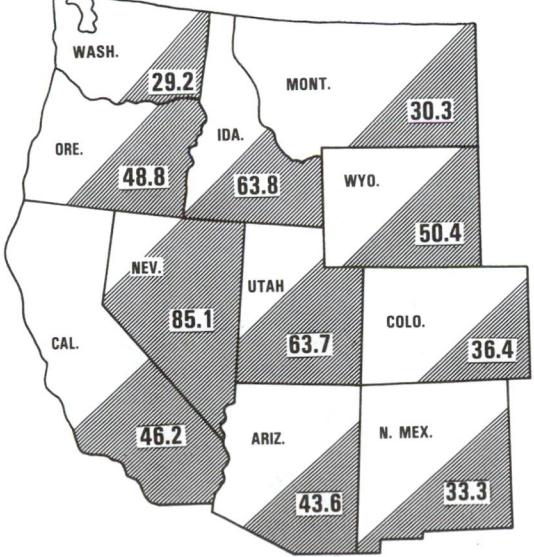

Fig. 2-9. Proportion of all land owned by the U.S. government in each of the 11 western states. (Source: U.S. General Services Administration, *Inventory Report on Real Property Owned by the United States Throughout the World*, Annual. Data for 1986.)

Table 2–2 also shows that 44.1% of the total land area of the United States is operated in the nation's 2,159,000 farms and ranches. The remaining 55.9% of nonfarm and ranch areas consists chiefly of publicly owned lands, cities and towns, and areas not suitable or available for private operations, but much of this area is grazed by animals.

It is also important that livestock producers be informed relative to the major uses of land in different regions of the United States. This information is presented in Fig. 2–10.

Fig. 2–10 classifies land use into four major categories: (1) cropland; (2) grassland, pasture and range (including cropland used only for pasture); (3) forest land (exclusive of forest reserved for parks and other uses); and (4) other land. Note (1) that in total (for the United States as a whole), "grassland, pasture, and range" constitutes the major land use; and (2) that the Mountain and Southern Plains regions rank first and second, respectively, as the largest grazing areas in the nation.

## MAJOR USES OF LAND BY REGION

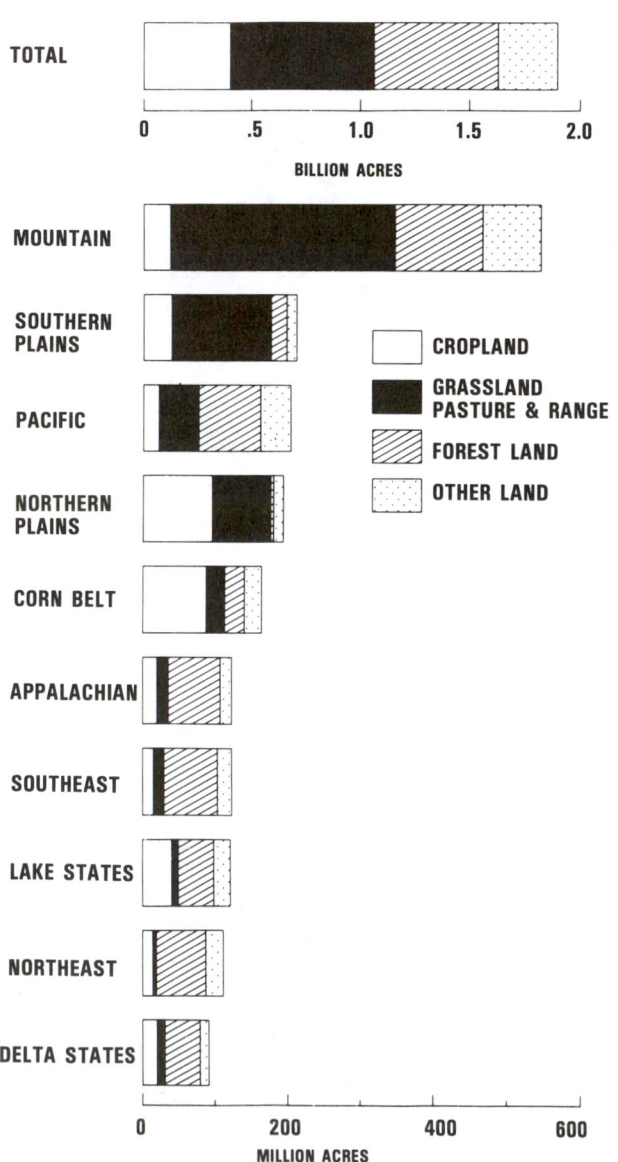

Fig. 2-10. Major use of land: (1) in total, excluding Alaska and Hawaii; and (2) by regions. (Source: *1988 Agricultural Chartbook*, U.S. Department of Agriculture, Agricultural Handbook No. 673, p. 41, Chart 46, 1988)

## COMPARATIVE CASH INCOME DERIVED FROM DIFFERENT AGRICULTURAL PURSUITS

Fig. 2–11 shows the cash receipts derived from U.S. livestock and crops in 1987. The pie diagram on the left shows the proportion of livestock income accounted for by each class of livestock. The dominant position of beef cattle (cattle and calves) is quite obvious; they accounted for 44.4% of all the income derived from livestock and livestock products in 1987.

The chart on the right shows the relative importance of U.S. livestock and crops in terms of cash receipts. It is noteworthy that livestock and their products accounted for 55.2% of the cash income received by U.S. farmers and ranchers, with beef cattle accounting for nearly one-fifth of the total cash income.

In the livestock sector, dairy products ranked second—accounting for 12.9% of the income; poultry ranked third—accounting for 8.3% of the income; hogs came next—accounting for 7.5% of the income; and sheep were last—accounting for only 0.41% of the income. Thus, of each $100 worth of farm products sold, animals and their products accounted for $55.20 whereas, grains, cotton, vegetables, fruits and nuts, tobacco, etc., brought in $44.80.

The figures below the two pie diagrams show the 1987 cash receipts in millions of dollars.

As would be expected, the proportions are somewhat changeable from year to year, depending upon the relative value of the various farm products and the amount produced. In addition to the cash income aspects, it is important to remember that the available home supply of animal products greatly improved the diets of farm people.

## NUMBER AND VALUE OF LIVESTOCK ON FARMS AND RANCHES

Table 2–3 gives the numbers and total value of livestock on U.S. farms and ranches, by classes. As noted, there were more than 792 million farm animals, including chickens and turkeys, but excluding commercial broilers, with an aggregate value of $59.3 billion, exclusive of horses, for which no evaluation was available. Thus, animals represent a huge investment on the part of farmers and ranchers.

**TABLE 2–3**
**NUMBER AND VALUE (VALUE/HEAD AND TOTAL VALUE)**
**OF LIVESTOCK AND POULTRY ON U.S. FARMS, BY CLASSES[1]**

| Class of Livestock and Poultry | Number on Farms and Ranches | Farm Value | |
|---|---|---|---|
| | | Per Head | Total |
| | (thousands) | ($) | ($1,000) |
| **C**attle ................. | 98,994 | 523.00 | 51,807,580 |
| **H**ogs ................... | 53,795 | 76.20 | 4,096,647 |
| **S**heep and lambs ........ | 10,774 | 89.90 | 968,918 |
| **H**orses and mules[2] | 10,840 | — | — |
| **C**hickens, excluding commercial broilers ..... | 377,516 | 1.87 | 706,079 |
| **T**urkeys ................ | 240,349 | 6.96 | 1,701,137 |
| Total ................. | 792,268 | | 59,280,361 |

[1]Source: *Agricultural Statistics 1988*, USDA, pp. 258, 271, 282, 348, and 363. Data for 1987–88.
[2]Source: *Horses and Horsemanship*, p. 14, Table 2–1, by M. E. Ensminger, published by Interstate Publishers, 1990. Data for 1986.

# CASH RECEIPTS FROM LIVESTOCK AND CROPS

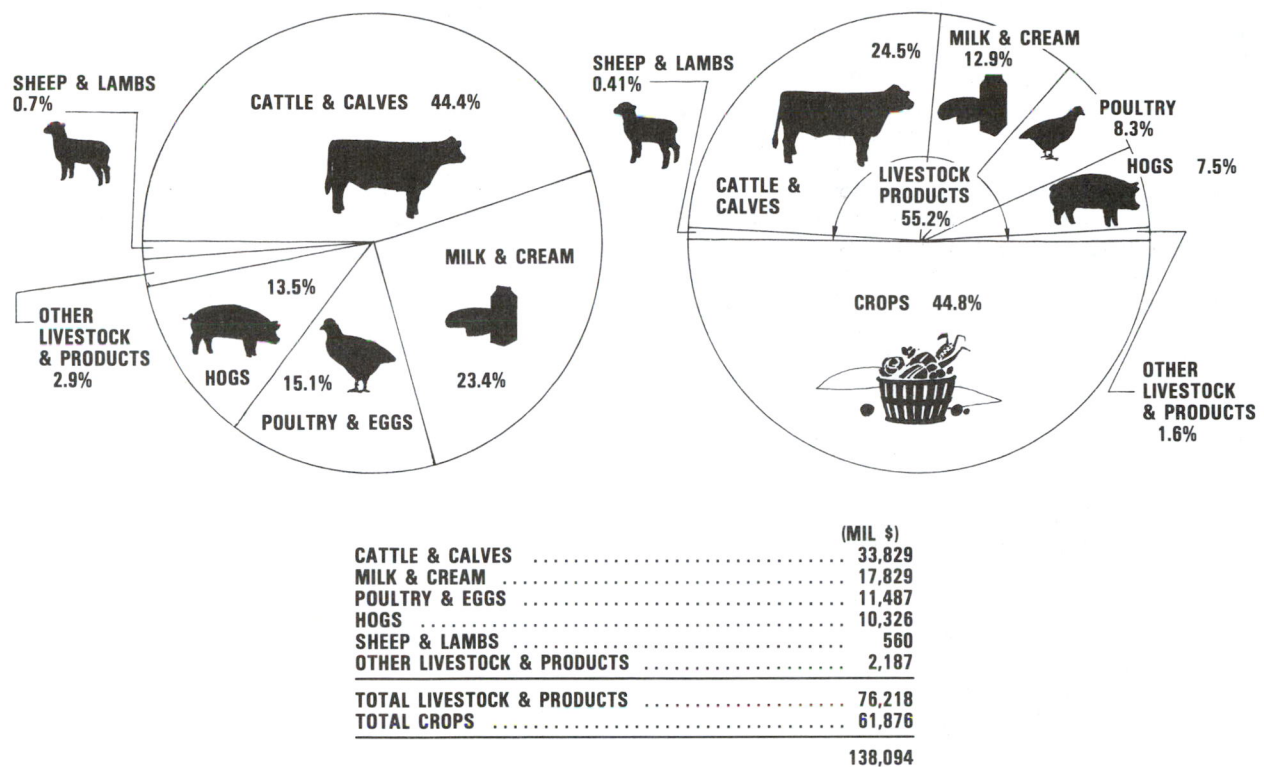

| | (MIL $) |
|---|---|
| CATTLE & CALVES | 33,829 |
| MILK & CREAM | 17,829 |
| POULTRY & EGGS | 11,487 |
| HOGS | 10,326 |
| SHEEP & LAMBS | 560 |
| OTHER LIVESTOCK & PRODUCTS | 2,187 |
| TOTAL LIVESTOCK & PRODUCTS | 76,218 |
| TOTAL CROPS | 61,876 |
| | 138,094 |

Fig. 2–11. Comparative cash income derived from different agricultural pursuits in 1987. (Source: *Agricultural Statistics 1988*, USDA, p. 409, Table 581)

## SIZE OF THE MEAT-PACKING INDUSTRY

In 1986, the U.S. meat packing industry slaughtered 37.6 million cattle, 3.5 million calves, 5.8 million sheep and lambs, and 80 million hogs; and produced 39.3 billion lb of red meat. Additionally, poultry processors produced (ready to cook) 14.9 billion lb of chicken and 3.3 billion lb of turkey. Also, the meat-packing industry employed 143,000 people in 1987. Meat-packing, therefore, is one of the nation's leading industries.

## OTHER ORGANIZATIONS ENGAGED IN THE LIVESTOCK INDUSTRY

In addition to the millions of people and the many facilities engaged in raising livestock and in slaughtering, processing, and distributing the products therefrom, additional personnel and facilities are equally essential in the operation of the far-flung livestock industry. This includes such essential operations as (1) transportation and communication; (2) marketing; (3) commercial feed companies; (4) livestock magazines; (5) purebred registry associations; and (6) research, teaching, extension, promotional, and regulatory organizations and officials.

## FUNCTIONS OF ANIMALS IN THE UNITED STATES

The average person is aware, at least in part, of the basic utility functions of animals in contributing food, clothing, power, and recreation. But few recognize that, because of their added functions, animals are an integral part of a sound, mature, and permanent agriculture.

In the sections that follow, the functions of animals in the United States are pointed up. The functions of animals in the global perspective are presented in Chapter 1; hence, the reader is referred thereto for a discussion of the worldwide aspects of the subject.

## ANIMALS CONTRIBUTE FOOD, CLOTHING, POWER, AND RECREATION

The primary utility function of animals in the United States is the production of food, clothing, power, and recreation—products upon which humans became more and more dependent with each succeeding step in their advancing civilization.

• **Food**—The development of railroads, the perfection of artificial refrigeration, the growth of livestock markets and packinghouses, the improvement and extension of highways, and the invention of the motor truck all richly enhanced the food supply, especially the quality and quantity of available animal products. Today, nearly half of the total food supply in the United States is contributed by mammalian, avian, and aquatic life. The list of food products of animal origin includes meat from domestic and wild animals, fowl and eggs from domestic and wild birds, fish of many kinds, and milk from cows and goats. Research has given positive proof of the leadership of these animal foods as rich sources of those nutrients that are so essential for good health and proper nutrition.

Fig. 2–12. Animals contribute food. (Courtesy, American Dairy Association, Chicago, IL)

In general, the consumption of food products of animal origin is limited by their cost, which in turn is governed by supply and demand. Naturally, the comparative prices of other food products and competition is also a major factor. On the average, U.S. consumers spend about 2.5 to 4.0% of their total income, or about 15 to 25% of their food budget, for meat. Table 2–4 shows the U.S. per capita consumption of selected food products.

### TABLE 2–4
### ANNUAL PER CAPITA CONSUMPTION OF SELECTED FOOD PRODUCTS[1]

| Product | Per Capita/Year | |
|---|---|---|
| | (lb) | (kg) |
| **Meats** (total red meat) | 135.4 | *61.5* |
| Beef | 73.4 | *33.4* |
| Veal | 1.5 | *0.7* |
| Lamb and mutton | 1.3 | *0.6* |
| Pork (excluding lard) | 59.2 | *26.9* |
| **Fish** (edible weight) | 15.4 | *7.0* |
| **Poultry products** | | |
| Eggs | 31.6 | *14.4* |
| Chicken (ready-to-cook) | 62.7 | *28.5* |
| Turkey (ready-to-cook) | 15.1 | *6.9* |
| **Dairy products** | | |
| Cheese | 24.0 | *10.9* |
| Condensed and evaporated whole milk | 3.8 | *1.7* |
| Fluid milk and cream | 235.7 | *107.1* |
| Ice cream | 18.3 | *8.3* |
| **Fats and oils** | 62.7 | *28.5* |
| Butter | 4.6 | *2.1* |
| Margarine | 10.5 | *4.8* |
| Lard | 1.8 | *0.8* |
| Shortening | 21.3 | *9.7* |
| Other edible fats and oils | 26.2 | *11.9* |
| **Fruits** | | |
| Fresh | 101.7 | *46.2* |
| Processed (canned, frozen, and dried) | 27.6 | *12.5* |
| **Vegetables** (fresh, canned, and frozen) | 188.5 | *85.7* |
| **Potatoes and sweet potatoes** | 27.6 | *12.5* |
| **Sugar** (refined) | 68.8 | *31.3* |

[1]Source: *Agricultural Statistics 1988*, USDA, p. 494, Table 680; p. 184, Table 252; p. 231, Table 337. Data for 1987.

• **Clothing**—Today, the chief contributions of animal life to clothing for people are in the forms of wool, leather, hair, and furs. In addition to serving as sources of clothing, each of these

Fig. 2–13. Clothing—fashion in wool. A broad-shouldered, full-skirted, navy, knit dress over a maize merino wool polo sweater. (Courtesy, The Wool Bureau, Inc., New York, NY)

articles has many other uses. For example, leather not only provides boots and shoes, jackets and helmets, gloves and mittens, and belts; but it is also used in the manufacture of harness, belting, scabbards, holsters, gaskets, and countless other articles.

• **Power**—In the United States, we commonly think of horses and mules as the only source of animal power, but in different parts of the world and at different times oxen, buffalo, reindeer, elephants, camels, goats, and dogs have all been brought under submission and used as sources of power for different purposes.

With the invention and extension of mechanical power during the 19th and 20th centuries, there was a steady, but marked, decline in the use of animals as a source of power. Thus, in the United States the use of horses for draft purposes reached a peak in numbers in 1915, and mules reached a peak in 1925, following which their use for power declined to the point where they are almost a novelty in this country today.

Fig. 2-14. How it used to be done! This shows a big team hitch drawing a combine in the famous Palouse wheat area of eastern Washington, prior to the advent of tractor power. (Courtesy, Washington State University, Pullman)

• **Recreation**—Throughout the ages, animal life has contributed to recreation and sport. It is reasonable to surmise that a certain amount of enjoyment must have accompanied primitive hunting and fishing, though in times of scarcity it was a grim business. Today, wild animals and aquatic life contribute richly to the enjoyment of people. Domestic animals, notably the horse and the dog, are used extensively in recreation and sport. Also, the animals in the zoo and the circus are a source

Fig. 2-15. Animals contribute to recreation and sport. Picture of a trail ride. (Courtesy, American Quarter Horse Assn., Amarillo, TX)

of recreation for old and young alike. In recent years, many people of wealth have established outstanding purebred herds and flocks as a means of recreation and enjoyment. In brief, no mechanical device has had such wide appeal as animals, for they serve as sources of recreation for people in all walks of life.

## ANIMALS CONVERT INEDIBLE FEEDS INTO VALUABLE PRODUCTS

About two-thirds of the feed used by U.S. livestock is not suited for human consumption. In this category are hay, pasture, coarse forages (such as straws, fodders, etc.); certain grains; such by-products as are obtained from mills, packinghouses, and food-processing plants; and damaged grains and foods, and garbage. These are converted into meat, eggs, milk, and wool.

Such well-known grains as corn, oats, and barley would have only limited value if restricted solely to direct human consumption; but because they can eventually ride to market as animal products, their value is immensely greater. A distinction needs to be made, therefore, between food grains for people and feed grains for livestock. It is also well to remember that farm animals—especially sheep and beef cattle—exist largely on such roughages as pasture, hay, and silage (see Table 2-5).

#### TABLE 2-5
#### PERCENTAGE OF FEED OF DIFFERENT CLASSES OF U.S. LIVESTOCK DERIVED FROM (1) CONCENTRATES AND (2) ROUGHAGES, INCLUDING PASTURE[1]

| Class of Animal | Concentrates | Roughages |
|---|---|---|
| | (%) | (%) |
| Beef cattle | 15.5 | 84.5 |
| Dairy cattle | 41.3 | 58.7 |
| Sheep and goats | 6.2 | 93.8 |
| Swine | 95.7 | 4.3 |
| Horses and mules | 27.0 | 73.0 |
| Poultry | 100.0 | 0.0 |
| All livestock | 38.3 | 61.7 |

USDA, Economic Research Service. Data for the feed year 1983-84.

Fig. 2-16. Dairy cows obtain 58.7% of their feed from forage (see Table 2-5). This shows Jersey cows on pasture. (Courtesy, The American Jersey Cattle Club, Columbus, OH)

Much of the forage utilized in U.S. livestock production is produced on land not suited for the raising of bread grains or gardens. Then, too, in addition to the character of the land itself, labor, tools, storage, processing plants, and transportation facilities would be limiting factors in any rapid shift away from utilizing the present inedible feeds by animals.

## ANIMALS MAKE FOR DESIRABLE DIVERSIFICATION

Diversification refers to having more than one type of enterprise or source of income. There can be diversification within a specialty. For example, beef cattle producers might maintain a breeding herd and produce calves, buy and grow out stocker cattle, and finish out cattle—either in their own feedlot or on a custom basis.

In contrast to *one-crop* farming, as applied to either plants or animals, a well-diversified livestock program has the following advantages:

1. It distributes the risks.
2. It establishes a more desirable basis for credit.
3. It makes for better distribution of labor—providing employment throughout the year.
4. It provides for more effective utilization of homegrown feeds.
5. It makes for a more flexible farm program.
6. It allows for the development of superior markets.
7. It affords an opportunity for the application of wider experience and superior managerial ability.

Without attempting to elaborate upon each point, it should be readily apparent that a well-managed diversified livestock program makes for a sound and stable agriculture. Also, over a long period of time, it makes for a more profitable enterprise; for, after all, though the need of the general public for animal products creates a demand for them, farmers would not engage in livestock production without a profit incentive.

## OTHER FUNCTIONS OF ANIMALS

Animals have the following functions in addition to those detailed in the preceding sections: (1) They serve as the sheet anchor in erosion control, (2) they maintain soil fertility, (3) they serve as an important companion of grain production, and (4) they provide medicinal and other products.

(Also see Chapter 1, section entitled, "Favoring Sharing Grain with Animals, Then Consuming the Animal Products.")

## SOME FACTORS TO CONSIDER IN ESTABLISHING THE LIVESTOCK ENTERPRISE

It is not intended that this section shall cover all of the multiple factors that should be considered in establishing a livestock enterprise. Rather it is hoped that thought may be focused on some of the primary factors that should be considered if the venture is to be successful.

## SYSTEMS OF FARMING

Three broad systems of farming are practiced in the United States: (1) crop farming, (2) livestock farming, and (3) crop and livestock combined. Because considerable variation exists within each classification, some farm management specialists prefer further and more refined divisions. For example, instead of placing crop farming in one category, there are those who list one-crop farms and diversified-crop farms as two separate systems of farming. For purposes of this discussion, however, the three broad classifications will be used, with each system of farming defined as follows:

1. **Crop farming** is that system in which crops are grown that are useful for food, feed, or clothing—with the income derived from the sale of cash crops.
2. **Livestock farming** is that system in which the crops are used chiefly or entirely as feed for animals—with the income derived from the sale of animals, meat, milk, eggs, and/or wool.
3. **Crop and livestock farming combined** is that system that combines significant amounts of crop and livestock farming—with the income derived both from the sale of cash crops and animals and their by-products.

## LEADING LIVESTOCK STATES OF THE UNITED STATES

In a country as large and variable as the United States, it is but natural to find that some areas and states are better adapted to livestock production than others. Though the size of land area within each of the 50 states varies considerably and, consequently, is a major factor in determining the total livestock population of each, the cash receipts of livestock and livestock products by states is one of the best available criteria of the general livestock adaptation of the respective states. This information is presented in Table 2–6.

**TABLE 2–6**
**LEADING LIVESTOCK STATES BY RANK BASED ON CASH RECEIPTS OF LIVESTOCK AND LIVESTOCK PRODUCTS**

| State | Cash Receipts of Livestock and Livestock Products |
|---|---|
| | (1,000 dollars) |
| Texas | 6,059,064 |
| Iowa | 5,270,416 |
| Nebraska | 4,847,955 |
| California | 4,740,981 |
| Wisconsin | 4,221,897 |
| Kansas | 3,914,297 |
| Minnesota | 3,644,652 |
| Colorado | 2,321,104 |
| Pennsylvania | 2,319,173 |
| Illinois | 2,261,845 |
| United States Total | 76,218,300 |

¹Source: *Agricultural Statistics 1988*, USDA, p. 413, Table 584. Data for 1987.

Over a long period of time, farmers and ranchers do those things that are most profitable to them. Thus, there is perhaps considerable logic in assuming that there exists in these leading livestock states a set of conditions making for successful livestock production. This fact is of importance to those who wish to get established in the livestock business in an adapted area.

## KIND OF LIVESTOCK ADAPTED TO THE REGION

Some regions or areas of the United States possess a combination of feed and economic conditions that make them best adapted to a particular kind or kinds of livestock. Available feeds are determined chiefly by climate, soil, and topography; whereas economic conditions are determined by markets, available labor, transportation facilities, and perishability of product.

Because of a combination of feed and economic conditions, the following kinds of livestock farming predominate in different regions or areas of the United States.

• **The Dairy Region**—This region includes the densely populated northeastern part of the United States, extending from southeastern Minnesota north and east. The rainfall and climate of the area are well adapted to the production of pasture, hay, silage crops, and oats—important crops for the dairy ration. In addition, there are excellent and nearby markets for milk, a highly perishable product.

- **The Corn Belt**—This region refers to the seven Corn Belt states of Iowa, Illinois, Nebraska, Missouri, Kansas, Indiana, and Ohio. The climate, soil, and topography of the area are well suited to the production of corn, the principal crop; and corn is preeminently adapted to the finishing of hogs and cattle. Also, the Corn Belt feedlots are in close proximity to large livestock markets.

- **The Cotton Belt**—This region, which derives its name from its predominant crop, is confined to the southern United States. Corn is the second most important crop of the area. Additionally, there is a considerable acreage of peanuts, sweet potatoes, and other crops, part or all of which are fed to animals. more important, from the standpoint of efficient livestock production, year-round pastures are possible in this area. The combination of a mild climate, a variety of crops, and a long grazing season makes the Cotton Belt well adapted to most types of livestock production. From an economic standpoint, the region has an abundance of comparatively cheap labor, but satisfactory outlets are in need of further developments.

- **The Western Range**—This region comprises the great ranching area of the United States which lies west of the 100th meridian. The extensive grazing areas of this territory are especially suited to the production of feeder cattle, and lambs and wool—highly concentrated and nonperishable products that may be economically and successfully transported great distances to market.

Although regional adaptations are important and should always be given careful consideration when selecting the livestock enterprise, it is well to keep in mind that some regions are almost equally well suited to other kinds of livestock than those for which they are most noted. For example, the Corn Belt states are well adapted to support a considerable number of animals other than hogs and beef cattle—including dairy cattle, sheep, poultry, and horses. Likewise, poultry successfully competes with dairy cattle in the northeastern United States, and some of the more isolated grazing areas of the Dairy Belt are best adapted to beef production. Also, some of the irrigated valleys of the Far West are noted for their large cattle and sheep feeding yards—primarily because of their abundant supply of by-product feeds and forages.

## KIND OF LIVESTOCK ADAPTED TO THE FARM OR RANCH

Under some conditions a hog farmer may have one neighbor who is a cattle feeder, a second who operates a dairy, a third who keeps a sizable flock of sheep, a fourth who produces light horses for recreation and sport, and a fifth whose chief source of income is from poultry. All may be successful and satisfied with their respective livestock enterprises. This indicates, therefore, that several types of livestock farming may be nearly equally well adapted to an area or region. This means that the selection of the dominant type of livestock enterprise should be analyzed from the standpoint of the individual farm or ranch.

Usually a combination of several factors suggests the livestock enterprise or enterprises best adapted to a particular farm or ranch. One of these factors is the labor requirement. Table 2-7 may be of assistance in arriving at a decision as to the kind or kinds of livestock best suited to the individual farm or ranch.

**TABLE 2-7**
**ANIMAL LABOR REQUIREMENTS**

| Animal or Product | Work Hours/Hundredweight Production[1] | | No. Head Cared for in 1989 by One Worker in the Most Efficient Operations[2] |
|---|---|---|---|
| | 1935–39 | 1982–86 | |
| | (hours) | (hours) | |
| Beef cattle | 4.2 | 0.9 | Cow-calf (brood cows) .........300–500<br>Feedlot cattle .........2,500 |
| Milk cows | 3.4 | 0.2 | Milk cows .........150 |
| Sheep | 6.1 | 1.0[2] | Farm flock ewes .........1,000<br>Range ewes .........1,000–2,000<br>Feedlot lambs .........7,500–10,000 |
| Hogs | 3.2 | 0.3 | Sows .........200 |
| Eggs | 1.7/100 eggs | 0.2/100 eggs | Laying hens (cage) .........40,000[3]<br>Laying hens (floor) .........15,000[3] |
| Broilers | 8.5 | 0.1 | .........75,000 |
| Turkeys | 23.7 | 0.2 | .........40,000 |

[1]Source: Data on "work hours/hundredweight production," except sheep, from *Agricultural Statistics 1988*, USDA, p. 395, Table 569.
[2]Estimates by M. E. Ensminger.
[3]Does not include time devoted to egg processing.

## REQUISITES OF A SUCCESSFUL LIVESTOCK PRODUCER

The first and most important requisite of successful livestock producers is that they must possess a great love for animals.

This appears to be an inborn trait, for some people never acquire a natural ability to work with animals—no matter how long or how hard they try. When such love for stock exists, animals are more docile and easier to handle, for the caretakers' feelings are relayed to their charges. Also, a great love for animals appears to be essential if caretakers are to feed them regularly, cheerfully, and with enjoyment, without regard to long hours and Sundays or holidays; if they are to provide clean, dry bedding, despite the fact that a driving storm may make it necessary to repeat the same operation the next day; if they are to serve as nursemaids to newborn or sick animals, though it may mean the loss of sleep and working with cold, numb fingers; and if they are to remain calm and collected, though striking an animal or otherwise giving vent to their feelings might at first appear warranted.

Next to having a great love for animals, it is important that successful livestock producers have adequate knowledge of the broad field of animal science. In addition, good producers must be well versed relative to soil and crops, for in most successful livestock enterprises the feeds are homegrown. In general, this means that livestock farmers and ranchers should be equally as competent as grain farmers from the standpoint of raising feeds, and in addition, they must be thoroughly competent in the production and marketing of animals. Successful livestock farming requires great skill.

Finally, experience, industry, and good judgment are very necessary requisites of successful producers. These words carry the same connotation in all industries and are self-explanatory.

## NEW ANIMAL FRONTIERS THROUGH RESEARCH

In the past, livestock producers could be successful if they merely bred, fed, and managed their animals. Today, this is not enough. For profit and survival, their operations must be predicated on more and better research followed by prudent application.

Research gave us Mendel's laws of inheritance, an understanding of nutrient requirements, the antibiotics, insecticides, and a host of other discoveries. In brief, it is not exaggeration to say that we are eating better and living healthier and longer as a result of science and technology.

Because we have made so much progress in the past, some people may be of the opinion that all of the worthwhile things in the field of animal science have been done. Not so! We need new technology to produce livestock and their products more efficiently, and we need to adjust and adapt the industry to changes in other parts of our economy. We need to increase efficiency of reproduction—to bring reproductive rates and seasons under more precise control, and to reduce losses, both before and after birth; we need to improve the quality of meat, milk, and fiber produced; we need new management practices, facilities, and equipment to reduce labor and increase productive efficiency; and we need to conserve energy and control pollution. All these challenges, and more, lie ahead.

## QUESTIONS FOR STUDY AND DISCUSSION

1. Define the term *animal science*.

2. Why is *animal agriculture* essential to a well-nourished and happy people?

3. What prompted the birth of animal science?

4. How did the domestication of animals contribute to the civilization of primitive people?

5. Why have the great livestock countries of the world supported the most advanced civilizations and been the most powerful?

6. What was the chief motive for bringing the first animals into the Western Hemisphere? What animals are native to the Americas?

7. Columbus, Cortez, and de Soto introduced animals to the Western Hemisphere. For each of them, give the year when they came, where they landed, and what animal species they brought.

8. What developments prompted the early improvement of animals in the United States?

9. As shown in Table 2-1 and Fig. 2-4, U.S. animal numbers have failed to keep pace with human population increases. Yet, the per capita consumption of animal products has increased. How has the latter been accomplished?

10. Table 2-1 and Fig. 2-4 also reveal that "all cattle and calves" increased much more than "hogs and pigs" from 1840 to 1988. How do you account for this? The same table and figure also show that "all sheep and lambs" decreased during this same period. What caused the decrease in sheep and lamb numbers?

11. Fig. 2-5 shows that the U.S. nonfarm human population has increased while the farm population has declined. What caused the farm population to decline?

12. Fig. 2-6 shows that productivity per farm worker—the number of persons supplied farm products by one farm worker—increased from 4.12 persons in 1820 to 93 persons in 1988. What were the main forces back of this increase? Did this increase contribute to the energy crisis?

13. In the future, what practical production methods will be employed for the purpose of maintaining high per capita red meat, milk, and egg consumption for our expanding population?

14. Show the magnitude of the U.S. livestock industry, using all possible criteria.

15. Fig. 2-7 and Table 2-2 show that 29.2% of the U.S. land in the 50 states is in pasture and grazing land in comparison with 16.9% in cropland. Would you expect any shifts in these percentages in the future? If so, what shifts would you expect and why would you expect them?

16. Fig. 2-7 and Table 2-2 show that 39.1% of the U.S. land in the 50 states is under public ownership. Should this land be sold to private individuals?

17. Why is the highest percentage of federal lands (60.3% of it) located in the Western Region?

18. Fig. 2-10 classifies land into four major categories. Identify these categories. What two regions have the largest grazing areas in the nation? Why should livestock producers be well informed relative to the major land uses in different regions of the nation?

## SELECTED REFERENCES *(Continued)*

| Title of Publication | Author(s) | Publisher |
|---|---|---|
| *New Technologies in Animal Breeding* | B. G. Brackett<br>G. E. Seidel, Jr.<br>S. M. Seidel | Academic Press, Inc., New York, NY, 1981 |
| *Physiology of Reproduction and Artificial Insemination of Cattle*, Second Edition | G. W. Salisbury<br>N. L. Van Dewark | W. H. Freeman & Co., Publishers, San Francisco, CA, 1978 |
| *Principles of Genetics*, Second Edition | I. H. Herskowitz | The Macmillan Company, New York, NY, 1977 |
| *Principles of Genetics*, Seventh Edition | E. J. Gardner<br>D. P. Snustad | John Wiley & Sons, Inc., New York, NY, 1984 |
| *Reproduction in Domestic Animals*, Third Edition | H. H. Cole<br>P. T. Cupps | Academic Press, Inc., New York, NY, 1981 |
| *Reproduction in Farm Animals*, Fourth Edition | Edited by<br>E. S. E. Hafez | Lea & Febiger, Philadelphia, PA, 1980 |
| *Reproductive Physiology of Mammals and Birds*, Third Edition | A. V. Nalbandov | W. H. Freeman & Co., Publishers, San Francisco, CA, 1976 |
| *Science of Genetics, The*, Fourth Edition | G. W. Burns | The Macmillan Company, New York, NY, 1980 |

Jersey bull, *Brownys Masterman Jester*, Grand Champion Bull at the National Jersey Show five consecutive years. Owned by Haven Hill Farms, Lake Placid Club, New York, NY. (Courtesy, American Jersey Cattle Club, Reynoldsburg, OH)

Chicken breeding farm, showing individually caged birds, designed to control egg production, egg quality, egg weight, and feed intake of each individual hen. (Courtesy, Euribrid B. V., Boxmeer, Holland)

A high percentage lamb crop is desirable. Multiple births are affected by heredity, environment, and age. (Courtesy, Sandy Petersen, American Polypay Sheep Assn., Sidney, MT)

Vitamin B–12 made the difference! Both chicks are shown at 3½ weeks of age. The little chick on the left was fed a ration deficient in vitamin B–12. The big, perky chick on the right received plenty of B–12. The little chick weighed 157 grams. The big chick weighed 280 grams—nearly twice as much as his smaller companion. (Courtesy, Merck Chemical Division, Rahway, NJ)

# Chapter 4

## FEEDING LIVESTOCK

Animals inherit certain genetic possibilities, but how well these potentialities develop depends upon the environment to which they are subjected; and the most important influence in the environment is the feed. In turn, all feeds come directly or indirectly from plants which have their roots in the soil. Thus, we have the cycle as a whole—from the soil, through the plant, thence to the animal and back to the soil again.

Lush pastures and well-cured, dry forages produced on fertile soils, together with a multitude of grains and by-product feeds—many of which are inedible by people—constitute the basis of successful livestock production. In this category are hay, pasture, certain grains, millfeeds, and other by-products which are converted into meat, eggs, milk, and wool. Moreover, 29.1% of the land area of the United States is devoted to pasture and range land—more than any other land use. Much of this forage is produced on lands unsuited for the growth of bread grains or gardens. Fig. 4–1 shows the relative importance of the principal U.S. livestock feeds.

Also, feeding is important from an economic standpoint; it is the major item of expense in the production of livestock. For example, feed accounts for approximately the following proportions of the cost of livestock production: finishing cattle, 70%; feedlot lambs, 50%; pork, 65 to 75%; milk production, 55%; and poultry, 55 to 75%, with the production of eggs toward the lower side of this range and the production of broilers and turkeys toward the upper side.

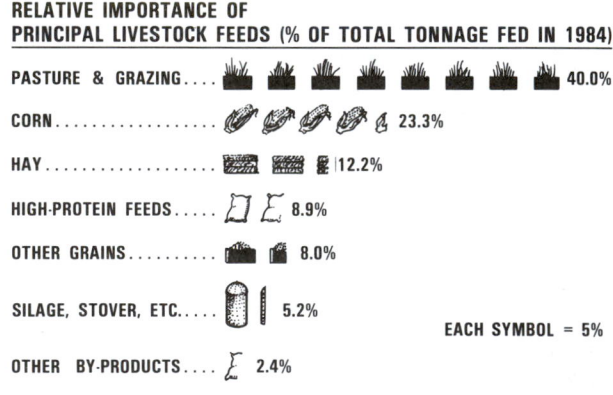

**RELATIVE IMPORTANCE OF PRINCIPAL LIVESTOCK FEEDS (% OF TOTAL TONNAGE FED IN 1984)**

| | |
|---|---|
| PASTURE & GRAZING | 40.0% |
| CORN | 23.3% |
| HAY | 12.2% |
| HIGH-PROTEIN FEEDS | 8.9% |
| OTHER GRAINS | 8.0% |
| SILAGE, STOVER, ETC. | 5.2% |
| OTHER BY-PRODUCTS | 2.4% |

EACH SYMBOL = 5%

Fig. 4–1. These principal livestock feeds are converted into meat, eggs, milk, and wool. (Source of data: USDA, Economic Research Service)

It is important, therefore, that feeding practices on the farm be as efficient and economical as possible. To this end, every livestock producer should endeavor to provide a ration that is both satisfactory and inexpensive—a ration which makes for the maximum production of quality product per unit of feed consumed.

## DIGESTION

Animals cannot utilize most nutrients in the forms found in feeds. Complex nutrients must be broken down into simpler compounds before they can be absorbed. This is accomplished by a combination of mechanical and enzymatic processes known as digestion. Digestion also includes absorption of food from the digestive tract into the bloodstream. Most of the unused portion of the feed is eliminated in the feces, although a considerable portion is also given off as gas through the mouth and nose.

Based on the kind of feed eaten, animals are classed as follows:

1. **Herbivores.** These are the vegetarians. They depend entirely upon plants for their feed supply.

2. **Carnivores.** These are the flesh eaters. They feed almost entirely upon the flesh of other animals.

3. **Omnivores.** These consume both flesh and plants.

## THE DIGESTIVE SYSTEM

The digestive system, or alimentary canal, consists of a tube which courses internally from the lips to the anus. At prescribed intervals, it becomes specialized in regions called the mouth, esophagus, stomach, small intestine, large intestine, rectum, and anus. Protruding along the way are the liver and the pancreas, which provide essential secretory products for digestion.

There are major differences in the anatomy and physiology of the organs of the digestive tract of different animal species. These differences are of great nutritional significance, as they affect both the nature of the digestion process and the kind of feed that can be utilized. Based on the type of digestive system, animals may be grouped as (1) monogastric (simple-stomached), (2) avian (poultry), (3) polygastric (ruminants), or (4) pseudo-ruminants (those with functional ceca).

An understanding of species differences in types of digestive systems and kinds of feed eaten is essential to intelligent feeding. This information is presented in Fig. 4–2. Table 4–1 (next page) gives average capacity figures of digestive tracts.

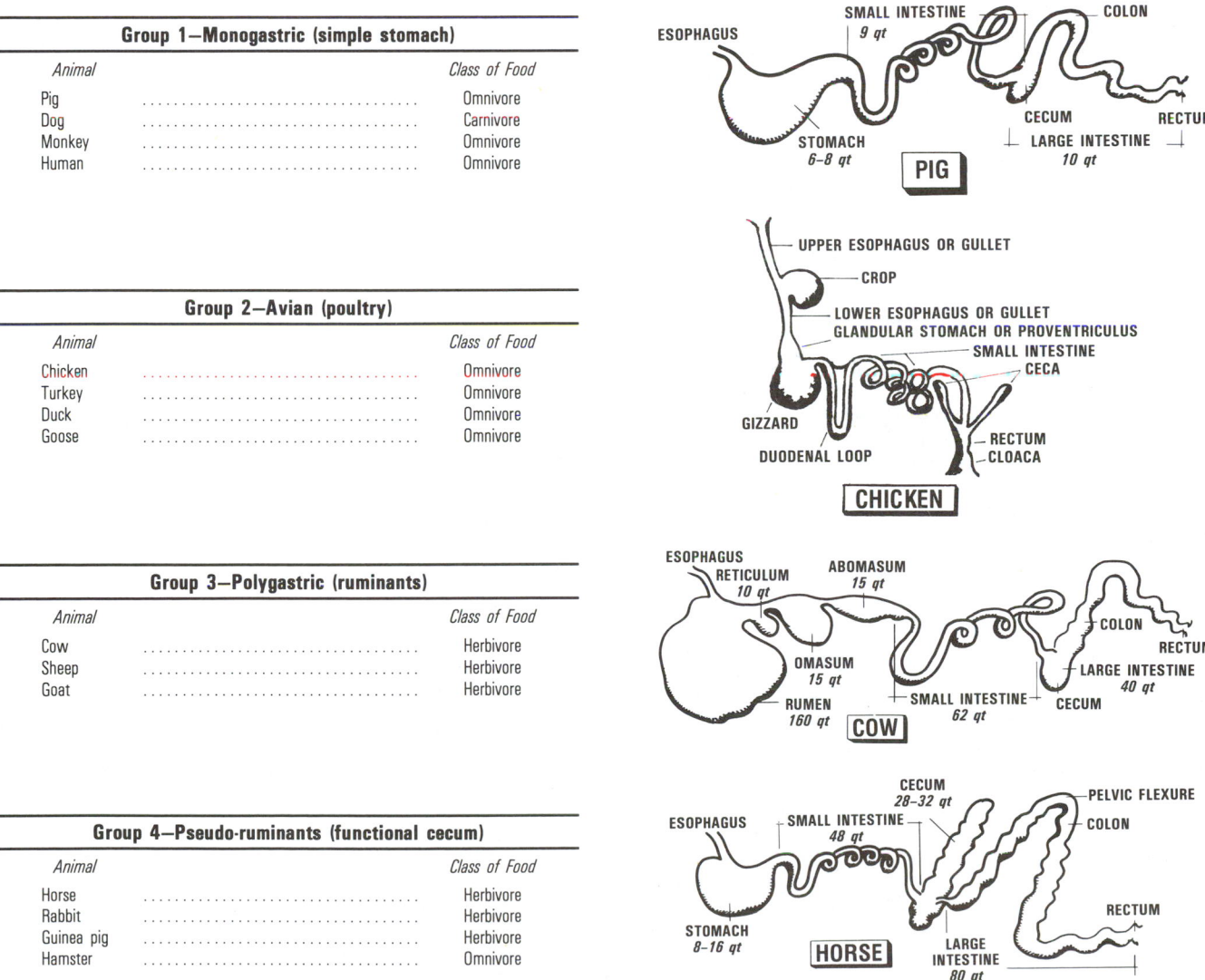

| Group 1—Monogastric (simple stomach) | |
| --- | --- |
| *Animal* | *Class of Food* |
| Pig | Omnivore |
| Dog | Carnivore |
| Monkey | Omnivore |
| Human | Omnivore |

| Group 2—Avian (poultry) | |
| --- | --- |
| *Animal* | *Class of Food* |
| Chicken | Omnivore |
| Turkey | Omnivore |
| Duck | Omnivore |
| Goose | Omnivore |

| Group 3—Polygastric (ruminants) | |
| --- | --- |
| *Animal* | *Class of Food* |
| Cow | Herbivore |
| Sheep | Herbivore |
| Goat | Herbivore |

| Group 4—Pseudo-ruminants (functional cecum) | |
| --- | --- |
| *Animal* | *Class of Food* |
| Horse | Herbivore |
| Rabbit | Herbivore |
| Guinea pig | Herbivore |
| Hamster | Omnivore |

Fig. 4–2. Classification of animals according to type of digestive system and class of feed, with schematic diagram of digestive tracts of pig, cow, horse, and chicken.

## TABLE 4-1
### PARTS AND AVERAGE CAPACITIES OF DIGESTIVE TRACTS OF SELECTED ANIMALS[1]

| | Animal Species | | | | | | | | | | | |
| | Cattle | | Sheep or Goat | | Horse | | Pig | | Dog | | Human | |
| | (gal) | *(liter)* | (gal) | *(liter)* | (gal) | *(liter)* | (gal) | *(liter)* | (gal) | *(liter)* | (gal) | *(liter)* |
| **Gastric Compartment:** | | | | | | | | | | | | |
| Rumen (paunch) | 53.4 | *202* | 6.2 | *23.4* | | | | | | | | |
| Reticulum (honeycomb) | 2.0 | *8* | 0.5 | *2.0* | | | | | | | | |
| Omasum (manyplies) | 5.0 | *19* | 0.2 | *0.9* | | | | | | | | |
| Abomasum (true stomach) | 6.1 | *23* | 0.9 | *3.3* | 4.8 | *18.0* | 2.1 | *8.0* | 1.1 | *4.3* | 0.3 | *1.0* |
| **Subtotal** | 66.5 | *252* | 7.8 | *29.6* | 4.8 | *18.0* | 2.1 | *8.0* | 1.1 | *4.3* | 0.3 | *1.0* |
| Small intestine | 17.4 | *66* | 2.4 | *9.0* | 16.9 | *63.8* | 2.4 | *9.2* | 0.4 | *1.6* | 1.0 | *4.0* |
| Cecum | 2.6 | *10* | 0.3 | *1.0* | 8.9 | *33.5* | 0.4 | *1.6* | 0.03 | *0.1* | | |
| Large intestine | 7.4 | *28* | 1.2 | *4.6* | 25.4 | *96.0* | 2.3 | *8.7* | 0.2 | *0.9* | 0.3 | *1.0* |
| **Total** | 93.9 | *356* | 11.7 | *44.2* | 56.0 | *211.3* | 7.2 | *27.5* | 1.73 | *6.9* | 1.6 | *6.0* |

[1]Source: Adapted from Melvin J. Swenson (ed.), *Dukes' Physiology of Domestic Animals*, eighth ed., by Cornell University,by permission of Cornell University Press.

## MONOGASTRIC SPECIES (SIMPLE STOMACH)

Monogastric animals have the simplest of all digestive systems. It consists of the mouth and associated glands, esophagus, stomach, small intestine, large intestine, pancreas, and liver. This is the type of gastrointestinal tract found in the pig, dog, monkey, and human. It is characterized by limited capacity, chemical secretory factors playing a very important role in the digestion of foods, and limited microbial action and fiber digestion. It follows that monogastric animals are better adapted to the use of concentrated feeds, such as grains, than the use of large quantities of roughages.

## AVIAN (POULTRY) DIGESTIVE SYSTEM

The digestive system of poultry differs considerably from that of other monogastric animals. Birds have no teeth, hence, there is no chewing. The esophagus empties directly into the crop, where the food is stored and soaked. From the crop, the food passes to the proventriculus (or grandular stomach), the thick walled organ immediately in front of the gizzard, where it is stored temporarily and digestive juices are copiously secreted and mixed with it. Thence, it passes to the gizzard, a very muscular organ, which normally contains stones or grit, and which functions like teeth. Here it is crushed and ground. Then it moves through the small intestine, the cecum, and the large intestine.

Digestion in the fowl is rapid. It requires only about 2½ hours in the laying hen, and 8 to 12 hours in the nonlaying hen, for the feed to pass from the mouth to the cloaca.

## POLYGASTRIC SPECIES (RUMINANTS)

Cattle, sheep, and goats are polygastric animals, or ruminants. They differ from simple-stomached animals in the following important ways:

1. **Mouth.** Ruminants have no upper incisor or canine teeth. Thus, they depend on the upper dental pad and lower incisors, along with the lips and tongue, for the intake of feed into the mount (called prehension).

2. **Four stomach compartments.** Ruminants possess four stomach compartments—rumen, reticulum, omasum, and abomasum (true stomach)—whereas monogastrics only have one. Such a digestive system makes for two primary nutritional differences between ruminants and simple-stomached animals:

a. **More space.** They have the necessary space for processing large quantities of bulky forages to provide their nutrients. The cow, for example, when compared to a human on a proportion-to-weight basis, has about nine times the digestive tract capacity.

b. **More microorganisms.** The rumen provides a highly desirable environment for the enormous population of microorganisms. Typical counts of rumen bacteria range from 25 to 50 billion/ml, and typical counts of protozoa range from 200,000 to 500,000/ml. The number of rumen bacteria varies according to the nature of the diet, feeding regimen, time of sampling after feeding, species differences, individual animal differences, season, availability of green feed, and the presence or absence of ciliate protozoa.

Rumen microorganisms serve two important functions:

(1) They make it possible for ruminants to utilize roughage—to digest the fiber therein. They break down the cellulose and pentosans of feeds into usable organic acids, chiefly acetic, propionic, and butyric acid—commonly called the volatile fatty acids (VFA). The VFAs are largely absorbed through the rumen wall and provide the ruminant 60 to 80% of its energy needs. Microbial digestion is of great practical importance in the nutrition of ruminants; it is the fundamental reason why they can be maintained chiefly on roughages.

(2) In exchange for their rumen-housing privileges, the microbes synthesize nutrients for their host, in a true type of symbiotic relationship. They synthesize, or manufacture, all the B-complex vitamins and all the essential amino acids. The latter can even be made from nonprotein nitrogen compounds (NPN), such as urea or ammoniated products, or from proteins that are deficient in one or more of the amino acids. Finally, the microorganisms give their life to their host in payment for food and shelter, being digested further along in the gastrointestinal tract.

3. **Rumination.** A placid cow lying under a tree slowly chewing her cud conveys a special sense of contentment, symbolic of the tranquility of the countryside. But this activity, or

phenomenon, which is peculiar to ruminants, is of great practical significance.

During rumination, the animal regurgitates and rechews a soft mass of coarse feed particles, called a bolus. Each bolus is chewed for about a minute, then swallowed again. Ruminants may spend eight hours or more per day in rumination, the amount of time varying according to the nature of the diet. Coarse, fibrous diets result in more time ruminating. Rechewing does not improve digestibility. Rather, rumination has an important bearing on the amount of feed the animal can eat and utilize. Feed particle size must be reduced to allow passage of the material from the rumen. It follows that high-quality forages require much less rechewing and pass out of the rumen at a faster rate; hence, they allow the cow to eat more.

4. **Eructation (belching of gas).** Substantially more gas is produced in digestion by ruminants than by simple-stomached animals. The microbial fermentation in the rumen results in the production of large amounts of gases (primarily $CO_2$ and methane) which must be eliminated, otherwise bloat results. Normally, these gases are expelled quite freely by eructation (belching) and, to a lesser extent, by absorption into the blood, from which they are eliminated through exhaled air from the lungs.

5. **Stomach of the newborn.** When the calf is born, the rumen is small and the fourth stomach is by far the largest of the compartments. Thus, digestion in the young calf is more like that of a single-stomached animal than that of a ruminant. The milk which the calf normally consumes bypasses the first two compartments and goes almost directly to the fourth stomach in which the rennin and other compounds for the digestion of milk are produced. If the calf gulps too rapidly, or gorges itself, the milk may go into the rumen where it is not digested properly and may upset the calf's digestive system. As the calf nibbles at hay, small amounts of material get into the rumen. When certain bacteria become established, the rumen develops and the calf gradually becomes a full-fledged ruminant.

## PSEUDO-RUMINANTS

The digestive tract of the pseudo-ruminant—the horse, rabbit, guinea pig, and hamster—is anatomically and physiologically quite different from both the ruminant and the monogastric. The horse, a herbivore, possesses a much larger digestive system than the pig and is able to utilize large amounts of roughages because of the greatly enlarged cecum and large intestine.

Some basic differences between the stomach of the horse and the stomach of the ruminant are:

1. The ruminant has four compartments, whereas the horse has one.

2. The stomach capacity of the horse is much smaller—less than 5 gal for the mature horse as opposed to about 66 gal for the mature. Because of its small stomach, if the horse is fed too much roughage, labored breathing and quick tiring may result. Actually, the horse's stomach is designed for almost constant intake of small quantities of feed, rather than large amounts at any one time.

3. The cecum (the horse's fermentation vat) is located on the distal end of the gut, a fact that is of far greater significance than its size. It follows the small intestine, with the result that the ingesta pass from the cecum directly to the large intestine. By contrast, the anatomical arrangement of the cow is such that the ingesta pass from the rumen (the cow's fermentation vat) to the small intestine, thence to the large intestine.

4. At the time of eating, feed passes through the horse's stomach vary rapidly—so much so that feed eaten at the beginning of the meal passes to the intestine before the last part of the meal is completed. Without feed, the horse's stomach will empty completely in 24 hours, whereas it takes about 72 hours (3 times as long) for the ruminant's stomach to empty.

The anatomical and physiological differences in the horse are of great importance nutritionally, for the following reasons:

1. There is less microbial activity in the horse than in the ruminant. As a result—

a. The horse does not break down more than about 30% of the cellulose of feed, whereas the ruminant breaks down 60 to 70%. Hence, horses cannot handle as much roughage as ruminants. Also, higher quality (lower cellulose content) forages must be fed to horses.

b. The horse synthesizes only limited proteins, B vitamins, and vitamin K, whereas the ruminant synthesizes sizable quantities of these. Thus, the addition of B vitamins to the ration of the horse (along with vitamins A and D, which are dietary essentials and not synthesized in the digestive tract of any class of farm animals) is good insurance, especially where horses are under stress and high-quality feeds are not being fed.

2. The efficacy of absorption of nutrients synthesized by the microorganisms in the cecum is questioned. This is so because the small intestine of the horse never gets a chance at the ingesta from the cecum, for the latter empties directly into the large intestines.

It may be concluded, therefore, that in comparison with the cow, the horse should be fed less roughage, higher quality proteins (and not such non-protein nitrogenous products as urea), and added B vitamins and vitamin K. These facts, along with the stress and strain to which most modern light horses are subjected, lead the author to the conclusion that the nutritive requirements of the horse more nearly parallel those of the pig than the cow.

## ENZYMES OF THE DIGESTIVE TRACT

*Enzymes are organic catalysts produced by certain cells within the body which speed biochemical reactions at ordinary body temperatures without being used up in the process.* Enzymatic activity is responsible for most of the chemical changes occurring in feeds as they move through the digestive tract. A summary of enzymes involved in the digestive process of farm animals is presented in Table 4–2 (see next page).

## KIND OF FEED IN RELATION TO TYPE OF DIGESTIVE SYSTEM

From the preceding discussion, it becomes clear that monogastric animals, including poultry, must eat a large percentage of grains and other concentrates and depend almost entirely on digestive enzymes to break down these compounds. On the other hand, ruminants, with their four stomach compartments and the help of microorganisms, can subsist largely, or entirely, on bulky, high-fiber forages which, because of their low energy per unit weight of dry matter, must be consumed in large quantities to supply their nutrient needs. The

**TABLE 4–2**
**DIGESTIVE PROCESSES IN FARM ANIMALS AND POULTRY**

| Region | Secretion (Secreted by) | Enzyme | Enzyme Acts On, or Function | End Product of Digestion | Comments |
|---|---|---|---|---|---|
| **Mouth** | Saliva (salivary glands). | Amylase (ptyalin). Maltase. | Starch, dextrins. Maltose. | Maltose and dextrins. Glucose. | Saliva lubricates food. In ruminants, the buffer salts of saliva help control acidity in the stomach. |
| **Crop** (birds) | Mucus. | | Lubricates and softens food. | | |
| **Rumen** | | Enzymes from microorganisms. | Cellulose, polysaccharides, starches, sugars, fats, proteins (urea). | Volatile fatty acids. Microbial protein. B vitamins. Vitamin K. | |
| **Stomach** (abomasum) in animals; proventriculus in birds | Gastric juice and acids (chiefly HCl)(walls of stomach). Mucus. | Pepsin. Lipase (in carnivores). Amylase. | Protein. Fat. | Proteoses, polypeptides, peptides. Higher fatty acids and glycerol. Coating of stomach lining and lubrication of food. | |
| Nursing animals. | Gastric juice (walls of stomach). | Renin. | Milk protein (casein). | Coagulates milk protein. | |
| **Gizzard** (birds) | | | Grinding. | Ground foods. Reduced particle size. | |
| **Duodenum** (small intestine) | Pancreatic juice (pancreas). | Trypsin. Chymotrypsin. Amylopsin (amylase). Steapsin (lipase). Carboxypeptidase. Collagenase. Cholesterol esterase. | Proteins, proteoses, peptones, and peptides. Starch, dextrins. Fats. Peptides. Collagen. Cholesterol. | Peptones, peptides. Amino acids. Maltose, dextrins. Higher fatty acids and glycerol. Amino acids and peptides. Peptides. Cholesterol esterified with fatty acids. | Low in ruminants. |
| | Bile (liver). | | Fats. | Emulsion of fats (soap, glycerol). | |
| **Small intestine** | Intestinal juice (secreted by intestinal wall). | Peptidase (erepsin). Sucrase (invertase). Maltase. Lactase. Polynucleotidase. | Peptides. Sucrose. Maltose. Lactose. Nucleic acid. | Amino acids and dipeptides. Glucose and fructose. Glucose. Glucose and galactose. Mononucleotides. | Very low in ruminants. Low in ruminants. High in young mammals. |
| **Large intestine** (cecum and colon) | | Cellulase from microorganisms. | Cellulose, polysaccharides, starches, sugars. | Volatile fatty acids. Microbial protein. B vitamins. Vitamin K. | |

horse, because of its greatly enlarged cecum and large intestine, can utilize quantities of roughage intermediate between simple-stomached and ruminant animals.

## THE FUNCTIONS OF FEEDS

The feed consumed by animals is used for a number of different purposes, the exact usage varying somewhat with the class, age, and productivity of the animal. With all animals, a certain part of the feed is used for the maintenance of bodily functions aside from any useful production. This is known as the maintenance requirement. In addition, the various classes of animals use feed to take care of the functions for which they are kept. Thus, young, growing animals need nutrients suitable for building muscle tissue and bone; finishing animals need a surplus of energy feeds for the formation of fat; breeding females require feed for the development of their fetuses, and following parturition, for the production of milk; whereas work (or running) animals use feed to supply energy for productive work. Still other classes of animals require feed for the production of eggs and wool. Each of these needs will be discussed separately.

## MAINTENANCE

An animal differs from an engine in that the latter has no fuel requirement when idle; whereas an animal requires fuel every second of the day, whether it is idle or not.

*The maintenance requirement may be defined as a ration which is adequate to prevent any loss or gain of tissue in the body when there is no production.* Although these requirements are relatively simple, they are essential for life itself. A mature animal must have heat to maintain body temperature, sufficient energy to cover the internal work of the body and the minimum movement of the animal, and a small amount of proteins, vitamins, and minerals for the repair of body tissue.

No matter how quietly an animal may be lying in the stall, it still requires a certain amount of fuel, and the least amount on which it can exist is called its basal maintenance requirement. The animal (cattle and sheep, but not horses) requires about 9% more fuel when standing than when lying and still more for any movement it may make. This explains why it is desirable for economic reasons that finishing animals should eat and then lie down as much as possible. Even under the best conditions, about one-half of all the feed consumed by animals is used in meeting the maintenance requirements.

## GROWTH

*Growth may be defined as the increase in size of the muscles, bones, internal organs, and other parts of the body.* Naturally, the growth requirements become increasingly acute when animals are forced for early usage, such as the training and racing of a two- or three-year-old horse.

Growth has been referred to as the foundation of animal production. Young cattle, sheep, and swine will not make the most economical finishing gains unless they have been raised to be thrifty and vigorous. Likewise, breeding females may have their productive ability seriously impaired if they have been raised improperly. Nor can one expect the most satisfactory yields of milk from dairy cows or eggs from layers, unless they were well developed when young. Workhorses and mules cannot perform the maximum amount of work, and running horses do not possess the desired speed and endurance, if their growth has been stunted or if their skeletons have been injured by inadequate rations during the growth period.

## FINISHING (OR SHOW-RING FITTING)

This is the laying on of fat, especially in the tissues of the abdominal cavity and in the connective tissues just under the skin and between the muscles.

The composition of a ration for fitting may be the same as for maintenance, but it must be supplied in larger quantities.

In practical fitting rations, higher condition in mature animals is usually obtained through increasing the allowance of feeds high in carbohydrates and fats—a more liberal allowance of grains. Any surplus of protein may also serve for the production of fat, but usually such feeds are more expensive and are not used for economy reasons. In fitting mature animals, very little more proteins, minerals, and vitamins are required than for maintenance. In fitting young, growing animals, however, it is essential that—in addition to supplying more carbohydrates and fats—ample protein, minerals, and vitamins be provided.

In general, the more feed a growing or finishing animal consumes, the more economical will be its gains. For example, it may be assumed that a calf requires 6 lb of feed daily to maintain itself, when making no gain. With an additional 3 lb of feed, or a daily ration of 9 lb, it gains 1 lb daily in weight. If the ration is increased by another 3 lb, bringing the daily allowance to 12 lb, it gains 2 lb daily in weight. Thus, each 3 lb of feed over and above the maintenance requirement (6 lb per day for maintenance) produces a gain of 1 lb in liveweight. On the basis of total feed consumed, however, the first pound of gain requires a total of 9 lb of feed (the first 6 lb for maintenance and an additional 3 lb for gain), whereas the next pound of gain only requires an additional 3 lb of feed. These facts, as illustrated in the oversimplified way, are the basis for the statement that for economical finishing the feeder should get every possible ounce of feed "under their hides." The chief difference between good and poor feeders is in the amount of feed above the maintenance requirement that they are able to consume. This also shows why it is necessary to have animals with ample feed capacity.

## REPRODUCTION (EGGS) AND LACTATION

Regular and normal reproduction is the basis for profit on any farm or ranch. Despite this undeniable fact, it has been estimated that from 20 to 50% of all matings are infertile, that 25% of all cows culled from dairy herds are removed because of reproductive inefficiency, that the overall average U.S. calf crop of all cattle (beef and dairy combined) is only 88%, that 5% of all ewes are sterile, that only 50% of all mares bred actually produce foals, and that 15% of all sows bred fail to produce litters. With birds, the development of the egg is the chief part of reproduction. Certainly, there are many causes of reproductive failure, but most scientists are agreed that inadequate nutrition is a major one.

With all species, most of the growth of the fetus occurs during the last third of pregnancy, thus making the reproductive requirements most critical during this period. The ration of the pregnant female should supply sufficient amounts of protein, minerals, and vitamins.

With females of all species, the nutritive requirements for moderate to heavy milk production are much more rigorous than the pregnancy requirements. There is special need for a rather liberal protein, mineral, and vitamin allowance.

In the case of the young, growing, pregnant females, additional protein, minerals, and vitamins, above the ordinary requirements, must be provided; otherwise, the fetus will not develop properly or milk will be produced at the expense of the tissues of the dam.

It is also known that the ration exerts a powerful effect on sperm production and semen quality. Too fat a condition can even lead to temporary or permanent sterility. Moreover, there is abundant evidence that greater fertility of herd sires exists under conditions where a well-balanced ration and plenty of exercise are provided.

## WORK (RUNNING)

In many respects, work requirements are similar to the needs for finishing, both functions requiring high-energy feeds. The function of work (or running) is, for the most part, limited to horses, though in certain parts of the world oxen furnish the chief source of power.

For mature workhorses, not in reproduction, work is performed primarily at the expense of the carbohydrates and fats in the ration—energy that can be supplied in the form of additional grain. Theoretically, the protein is not drawn upon so long as the other nutrients are present in adequate amounts. From a practical standpoint, however, it is usually desirable to feed more proteins than the maintenance requirement, merely to ensure that the animal can make efficient use of the remainder of the nutrients in the ration. When a ration too low in protein is fed, more feed is required because the animal is unable to utilize the ration efficiently. For work animals, the mineral and vitamin requirements are practically the same as for comparable idle animals—except for the greater need for salt because of increased perspiration.

## WOOL/MOHAIR

Wool and mohair are high-protein products. Thus, a shortage of protein in the ration will lessen wool and mohair production, even though the total amount of the ration is adequate. It is also known that both the quality and quantity of fiber may be materially lowered if the animals are subjected to unfavorable thrift or health. If such periods are of relatively short duration, tender spots (weak spots) appear in the growth of the fiber.

## NUTRITIVE NEEDS

*Nutrients are the chemical substances found in feed materials that can be used, and are necessary for the maintenance, production, and health of animals.* The chief classes of nutrient substances are carbohydrates, fats, proteins, minerals, vitamins, and water. Nutrients are needed by the animal in definite amounts, with the quantities varying according to the class and age of animal, and the purpose for which it is being fed. A deficiency in a nutrient can be, and often is, a limiting factor in animal production.

## ENERGY NEEDS

Energy is required for practically all life processes—for the action of the heart, maintenance of blood pressure and muscle tone, transmission of nerve impulses, ion transport across membranes, reabsorption in the kidneys, protein and fat synthesis, the secretion of milk, and the production of eggs and wool.

A deficiency of energy is manifested by slow or stunted growth, body tissue losses, and/or lowered production of meat, milk, eggs, or fiber, rather than by specific signs such as those which characterize many mineral and vitamin deficiencies. For this reason, energy deficiencies often go undetected and unrectified for extended periods of time.

It is common knowledge that a ration must contain carbohydrates, fats, and proteins. Although each of these has specific functions in maintaining a normal body, they can all be used to provide energy for maintenance, for work, or for finishing. From the standpoint of supplying the normal energy needs of animals, however, the carbohydrates are by far the most important, more of them being consumed than any other compound, whereas the fats are next in importance for energy purposes. Carbohydrates are usually more abundant and cheaper, and they are very easily digested, absorbed and transformed into body fat. Also, carbohydrate feeds may be more easily stored than fats in warm weather and for longer periods of time. Feeds high in fat content are likely to become rancid, and rancid feed is unpalatable, if not actually injurious in some instances.

## CARBOHYDRATES

*The carbohydrates are organic compounds composed of carbon, hydrogen, and oxygen.* This group includes the sugars, starch, cellulose, gums, and related substances. They are formed in the plant by photosynthesis as follows: $6CO_2 + 6H_2O$ + energy from sun = $C_6H_{12}O_6$ (glucose) + $6O_2$. On the average, the carbohydrates comprise about three-fourths of all the dry matter in plants, the chief source of animal feed. They form the woody framework of plants as well as the chief reserve food stored in seeds, roots, and tubers. When consumed by animals, carbohydrates are used as a source of heat and energy, and any excess of them is stored in the body as fat.

No appreciable amount of carbohydrate is found in the animal body at any one time, the blood supply being held rather constant at about 0.05 to 0.1% for most animals. However, this small quantity of glucose in the blood, which is constantly replenished by changing the glycogen of the liver back to glucose, serves as the chief source of fuel with which to maintain the body temperature and to furnish the energy needed for all body processes. The storage of glycogen (so-called animal starch) in the liver amounts to 3 to 7% of the weight of that organ.

• **Carbohydrate feeds consist of nitrogen-free extract and fiber**—From a feeding standpoint, the carbohydrates consist of nitrogen-free extract (N.F.E.) and fiber. The nitrogen-free extract includes the more soluble, and, therefore, the more digestible, carbohydrates—such as starches, sugars, hemicelluloses, and the more soluble part of the celluloses and pentosans. Also, N.F.E. contains some lignin. The fiber is that woody portion of the plants (or feeds) which is not dissolved out by weak acids and alkalies. Fiber, therefore, is less easily digested. It includes cellulose, hemicellulose, and lignin.

• **Use of fibrous roughages possible because of bacterial action**—The ability of animals to utilize roughages—to digest the fiber therein—depends chiefly on microbial action. This is confined largely to the first three compartments of the stomach of ruminants, to the cecum and colon of the horse, and to a lesser extent in the large intestine of all animals. This bacterial digestion breaks down cellulose and pentosans of feeds into usable organic acids (chiefly acetic, propionic, and butyric acid). These volatile fatty acids are absorbed directly through the ruminal wall, and furnish much of the maintenance energy requirements of the animal.

• **Fiber of young grasses and early cut hay more digestible**—The fiber of growing pasture grass, fresh or dried, is more digestible than the fiber of most hay. Likewise, the fiber of early cut hay is more digestible than that of hay cut in the late-bloom or seed stages. The difference is due to both chemical and physical structure, especially in the presence of certain encrusting substances (notably lignin) which are deposited in the cell wall with age.

• **Optimum fiber content of ration variable**—Young stock of all classes, finishing steers and lambs, high-producing dairy cows, swine, poultry, and working horses (running horses) must have rations in which a large part of the carbohydrate content of the ration is low in fiber and is in the form of nitrogen-free extract. On the other hand, a considerable amount of fiber or bulk in the ration is believed desirable for mature breeding animals of all classes of livestock, especially when too high a condition is not desired. Likewise, with young animals being developed for breeding purposes, the increased fiber will tend to develop more growth and not so much fat.

To promote normal physiological activity of the gastrointestinal tract, one must feed a certain amount of coarse roughage to all classes of farm animals, except possibly to swine.

## FATS

Lipids (fat and fatlike substances), like carbohydrates, contain the three elements: carbon, hydrogen, and oxygen. Fats are soluble in such organic solvents as ether, chloroform, and benzene. For this reason, grease spots on clothing are usually removed with one of these solvents. As livestock feeds, fats function much like carbohydrates in that they serve as a source of heat and energy and for the formation of fat. Because of the larger proportion of carbon and hydrogen, however, fats liberate more heat than carbohydrates when digested, furnishing approximately 2.25 times as much heat or energy per pound on oxidation as do the carbohydrates. A smaller quantity of fat is required, therefore, to serve the same function.

• **Fats vary in melting point and degree of unsaturation**—The physical and chemical properties of fats are quite variable.

From a chemical standpoint, a molecule of fat consists of a combination of three molecules of certain fatty acids with one molecule of glycerol. Fats differ in their melting points and other properties, depending on the particular fatty acids that they contain. Thus, because of the high content of unsaturated acids (such as oleic and linoleic) and acids of low molecular weight, corn fat is a liquid at ordinary temperatures; whereas, because of the high content of stearic and palmitic acids, beef fat is solid at ordinary temperatures.

Some fatty acids are unsaturated, which means that they have the ability to take up oxygen or certain other chemical elements. Chemically, these unsaturated acids contain one or more pairs of double-bond carbon atoms. These characteristics are important. The value of linseed oil and varnish is due to their high content of unsaturated fatty acids, by virtue of which oxygen is absorbed when they are exposed to air, resulting in a tough, resistant coating. Likewise, because of their high content of unsaturated fats, soybeans or peanuts, when fed liberally, will produce soft pork.

An unsaturated fat readily unites with iodine, two atoms of this element being added for each double bond. Thus, in experimental work, the iodine number (the number of grams of iodine absorbed by a hundred grams of fat) is an excellent criterion of the degree of unsaturation. In the past, the iodine test was commonly applied to studying the soft pork problem. At the present time, the chief measure used in such determinations is the refractive index, as determined by the refractometer.

- **Rancidity of fats**—Because of their unsaturation, fats often become rancid through oxidation or hydrolysis, resulting in disagreeable flavors and odors which lessen their desirability as feeds. The development of rancidity may be retarded through proper storage or by adding antioxidants. The hydrogenation of fats (adding hydrogen to the double bonds) also lessens rancidity. The latter process has long been effectively used in improving the keeping qualities of vegetable shortenings and lard.

- **Feed fats affect body fats**—The nature of the body fat in animals is markedly influenced by the character of the fat in feeds. This phenomenon is of considerable practical importance, as the degree of hardness of the fat is a major factor in determining carcass quality and value. This is particularly true of hogs in which, because of the nature of the feeds consumed in certain areas, the soft-pork problem exists. Likewise, the kind of fat consumed by cows has a similar influence on the nature of the milk fat.

- **Some fats in ration desirable**—Common belief to the contrary, animals can tolerate a rather high fat content in the ration. As evidence of this, it is to be noted that sucklings normally handle a relatively large amount of fat, for milk contains 25 to 40% of this nutrient on a dry matter basis. Also, except for the soft-pork problem, no apparent difficulty is encountered in feeding hogs on a rather high fat diet, such as results when large quantities of peanuts or soybeans are fed.

A small amount of fats in the ration is desirable, as these fats are the carriers of the fat-soluble vitamins. There is evidence that some species (humans, swine, rats, and dogs) require certain of the fatty acids. Fortunately, normal farm rations contain ample quantities of these nutrients.

- **Animal fats vs vegetable oils**—The comparative food value of animal fats and vegetable oils has long been a stormy issue, particularly in regard to the relative merits of butter vs oleomargarine or of lard vs vegetable shortenings. In general, except for the vitamins which they carry, there is no conclusive experimental work to indicate that, as a source of fatty acids, animal fats are superior to vegetable oils. There is reason to believe that margarine, when fortified with vitamins A and D, is—from a nutritional standpoint—just as effective as butter in promoting growth, good health, reproduction, and lactation. There is, however, evidence that some species (humans, swine, rats, and dogs) require certain of the fatty acids, but it is quite likely that these differences are unimportant when a normal mixed diet is used. Of course, many people still prefer butter, and many desire to patronize it as a more genuinely and naturally home produced product than margarine.

## PROTEIN NEEDS

For more than a century, proteins and their structural units, the amino acids, have been studied and recognized as important dietary constituents.

*Proteins are complex organic compounds made up chiefly of amino acids, which are present in characteristic proportions for each specific protein.* This nutrient always contains carbon, hydrogen, oxygen, and nitrogen, and in addition it usually contains sulfur and frequently phosphorus. Proteins are essentially in all plant and animal life as components of the active protoplasm of each living cell.

Crude protein refers to all the nitrogenous compounds in a feed. It is determined by finding the nitrogen content and multiplying the result by 6.25. The nitrogen content of protein averages about 16% ($100 \div .16 = 6.25$).

In plants, the protein is largely concentrated in the actively growing portions, especially the leaves and seeds. Plants also have the ability to synthesize their own proteins from such relatively simple soil and air compounds as carbon dioxide, water, nitrates and sulfates. Thus, plants, together with some bacteria which are able to synthesize these products, are the original source of all proteins.

Proteins are much more widely distributed in animals than in plants. Thus, the proteins of the animal body are primary constituents of many structural and protective tissues—such as bones, ligaments, hair, hoofs, skin, and the soft tissues which include the organs and muscles. The total protein content of the bodies of animals ranges from about 10% in very fat mature animals to 20% in thin, young animals. By way of further contrast, it is also interesting to note that, except for the bacterial action in the rumen, animals lack the ability of the plant to synthesize proteins from simple materials. They must depend upon plants or other animals as a source of dietary protein. In brief, except for the high-quality proteins built by the bacterial action in the paunch of ruminants, the animal must have amino acids or more complete protein compounds in the ration.

Animals of all ages and kinds require adequate amounts of protein of suitable quality; for maintenance, growth, finishing, reproduction, work, and wool production. Of course, the protein requirements for growth, reproduction, and lactation are the greatest and most critical.

### QUALITY OF PROTEINS

In addition to supplying an adequate quantity of proteins, it is essential that the character of proteins be thoroughly understood. Proteins are very complex compounds with each molecule made up of hundreds of thousands of amino acids combined with each other. The amino acids, of which some 23 are known, are sometimes referred to as the building stones of proteins. Certain of these amino acids can be made by the animal's body to satisfy its needs. Others cannot be formed fast enough to supply the body's needs and, therefore, are known as essential (or indispensable) amino acids. These must be supplied in the feed. Thus, rations that furnish an insufficient amount of any of the essential amino acids are said to have proteins of poor quality, whereas those which provide the proportions of the various necessary amino acids are said to supply proteins of good quality.

In general, animal proteins are superior to plant proteins for monogastric animals (including man) because they are better balanced in the essential amino acids. For example, zein (a corn protein) is an incomplete plant protein. It is deficient in the essential amino acids lysine and tryptophan. On the other hand, animal proteins are excellent sources of lysine, and many of them (especially milk and eggs) are abundant in tryptophan.

Fig. 4–3. Lysine made the difference! Big pig (B) received high-lysine corn. Little pig (C), a littermate, was fed a lysine-deficient ration. (Courtesy, Cornell University, Ithaca, NY)

The necessity of each amino acid in the diet of the experimental rat has been thoroughly tested, but less is known about the requirements of large animals or even the human. According to our present knowledge, based largely on work with the rat, the following division of amino acids as essential and nonessential seems proper:

| Essential (indispensable) | Nonessential (dispensable) |
|---|---|
| Arginine | Alanine |
| Histidine | Asparagine |
| Isoleucine | Aspartic acid |
| Leucine | Cysteine |
| Lysine | Cystine |
| Methionine (may be replaced in part by cystine) | Glutamic acid |
|  | Glutamine |
| Phenylalanine | Glycine |
| Threonine | Hydroxyproline |
| Tryptophan | Proline |
| Valine | Serine |
|  | Tyrosine |

Arginine is regarded as essential for animals, whereas it is not for humans; most young mammals cannot synthesize it in sufficient amounts to meet their needs for growth.

In practical animal nutrition, the amino acids most likely to be deficient are lysine, methionine, and tryptophan. This stems from the fact that cereal grains, which are primary energy feeds, are quite low in these amino acids. So, it follows that rations based on a high percentage of these grains usually require supplementation with proteins which contain higher levels of these amino acids.

Fortunately, the amino-acid content of proteins from various sources varies. Thus, the deficiencies of one protein may be improved by combining it with another, and the mixture of the two proteins often will have a higher feeding value than either one alone. It is for this reason that a considerable variety of feeds in the ration is usually recommended.

The feed proteins are broken down into amino acids by digestion. They are absorbed and distributed by the bloodstream to the body cells, which rebuild these amino acids into body protein.

## BIOLOGICAL VALUE OF PROTEIN

Most chemical and microbiological tests for nutrient substances give information about the total amount of a nutrient present in a particular feedstuff or ration. However, they tell nothing about the digestibility and utilization of the feedstuff or ration in the digestive tract of the animal. Hence, biological tests directly involving animals are required to establish the true usefulness of feed supplying the nutrient needs of animals. These biological tests are particularly important in evaluating protein. (They are also important in evaluating energy-yielding nutrients like carbohydrates and fats.)

*The biological value of a protein is the percentage of the digestible protein of a feed or feed mixture which is usable as a protein by the animal.* It can be determined by a balance experiment in which a measured intake of protein is compared to the measured undigested protein in the feces of the animal. Thus, the biological value of a protein is a reflection of the kinds and amounts of amino acids available to the animal after digestion. If the amino acids available to the animal closely match those needed for body protein formation, the biological value of the protein is high. If, on the other hand, there are excesses of certain amino acids and deficiencies of other amino acids as a result of digestion, the biological value of the protein is low because of the increased number of amino acids which must be excreted via the kidney.

The biological values of animal proteins are much higher than those of plant proteins. For example, the biological values of whole eggs and milk are 94 and 85, respectively, compared to whole corn and navy beans (cooked) at 60 and 38, respectively. Because of this situation, the nutritional value of plant proteins (corn, white bread, navy beans, etc.) is substantially increased by consuming them with animal products rich in the amino acids in which vegetables are deficient. Thus, because of its high biological value, milk protein is of special value in the treatment of kwashiorkor, a type of protein malnutrition found among young children, in areas where people subsist largely on plant foods.

## PROTEINS AS AN ENERGY SOURCE

In general, high protein-content feeds are more expensive than those high in carbohydrates or fats. For this reason, there is the temptation to feed too little of them, rather than too much. On the other hand, when protein feeds are the cheapest, excess quantities of them may be fed as energy feed without harm, provided that the ration is balanced out in all other respects. Any amino acids that are left over, after the protein requirements of the body have been met, are deaminated, or broken down, in the liver. In this process, a part of each amino acid is turned into energy and the remainder (the $NH_2$ group split off from the amino acid molecule) is excreted via the kidneys as urea in the urine. Because only a part of the original energy value of the protein is retained, as sources of energy, fats and carbohydrates are more efficient. It is estimated that carbohydrates yield approximately 95% of their potential energy to the body, whereas only about 70% of the potential energy of protein is useful in meeting the energy needs of the body.

## PROTEIN NEEDS OF RUMINANTS

Ruminants are thought to have just as rigid physiological requirements for the essential amino acids as nonruminants. Yet, high-quality protein—protein containing all the essential amino acids—need not be fed to ruminants. The microbes (bacteria and protozoa) in the rumen synthesize the essential amino acids from nonessential amino acids and from certain nonprotein nitrogen-containing materials, like urea. The microbes produced in the rumen are then digested further along in the digestive tract (in the abomasum and intestines) to provide the ruminant with its amino acid needs. Because of this phenomenon, nonprotein nitrogen, like urea, may be used as a substitute for protein in ruminant rations.

## MINERAL NEEDS

Animal bodies contain small amounts—only 2 to 5%—of inorganic elements, called minerals. But these constituents play an important role in animal nutrition. In addition to furnishing structural material for bones and teeth, as constituents of the soft tissues, the blood, the fluids of the body, and certain of the secretions, they regulate many of the vital processes.

Although acute mineral-deficiency diseases and actual death losses are relatively rare, inadequate supplies of any one of the 18 essential mineral elements may result in lack of thrift, poor gains, inefficient feed utilization, lowered reproduction, and decreased production of meat, milk, eggs, wool, and work. Thus, like a thief in the night, subacute mineral deficiencies in farm animals each year steal away millions of dollars in profits from the farmers and ranchers of America, and, for the most part, go unnoticed. Only when the mineral deficiency reaches such proportions that it results in excess emaciation, reproductive failure, or death, is it likely to be detected.

To avoid deficiencies, minerals are usually provided, either free choice or added to the ration. Such supplements should supply only the specific minerals that are deficient—and in the quantities necessary. Excesses and mineral imbalances should be avoided.

Special attention needs to be given to trace minerals in areas where there is a deficiency of one or more of them, when poor quality roughage is fed, or when high- or all-concentrate rations are fed.

Eighteen mineral elements are known to be required by at least some animal species. They can be divided into two groups based on the quantity required in the ration.

• **Major or macrominerals**—These elements are required in amounts ranging from a few tenths of a gram to one or more grams per day.

• **Trace or microminerals**—These elements are required in minute quantities, ranging from a millionth of a gram (microgram) to a thousandth of a gram (milligram) per day.

The terms—major/macromineral and trace/micromineral—do not imply any lesser role for the latter group; rather, they represent quantity designation based on the amounts needed by animals.

The two groups follow:

| Major/Macrominerals | Trace/Microminerals | |
|---|---|---|
| Salt (sodium & chlorine, NaCl) | Chromium (Cr) | Molybdenum (Mo) |
| Calcium (Ca) | Cobalt (Co) | Selenium (Se) |
| Phosphorus (P) | Copper (Cu) | Silicon (Si) |
| Magnesium (Mg) | Fluorine (F) | Zinc (Zn) |
| Potassium (K) | Iodine (I) | |
| Sulfur (S) | Iron (Fe) | |
| | Manganese (Mn) | |

The percentages of the principal mineral constituents of the body are indicated by the following data, showing the average analyses of 18 steers of varying ages exclusive of the contents of the digestive tract.[1]

| Element | Percent |
|---|---|
| Calcium | 1.33 |
| Phosphorus | 0.74 |
| Potassium | 0.19 |
| Sodium | 0.16 |
| Sulfur | 0.15 |
| Chlorine | 0.11 |
| Magnesium | 0.04 |
| Iron | 0.01 |
| Total | 2.73 |

The dominant position of calcium and phosphorus in the above data becomes apparent when it is realized that calcium accounts for 49% of the total mineral; phosphorus, 27%; and all other minerals, 24%.

The general functions of minerals are as follows:

1. Give rigidity and strength to the skeletal structure.
2. Serve as constituents of the organic compounds, such as protein and lipid, which make up the muscles, organs, blood cells, and other soft tissues of the body.
3. Activate enzyme systems.
4. Control fluid balance—osmotic pressure and excretion.
5. Regulate acid-base balance.
6. Exert characteristic effects on the irritability of muscles and nerves.
7. Engage in mineral-vitamin relationships.

A summary of each mineral is presented in Table 4–3, Animal Mineral Chart on pages 98 to 101, with the minerals listed alphabetically within each of two classifications: (1) Major/Macrominerals, or (2) Trace/Microminerals. Additionally, because of their dominant position and certain relationships, a narrative discussion of salt, calcium, and phosphorus follows.

---

[1] Hogan, A. G., and J. L. Nierman, *Studies of Animal Nutrition—VI, The Distribution of the Mineral Elements in the Animal Body as Influenced by Age and Condition*, Missouri Ag. Exp. Res. Bull., No. 107.

TABLE 4–3
ANIMAL MINERAL CHART

| Mineral | Major Functions | Some Deficiency Symptoms | Major Interrelationships; Toxicities | Good Sources for Animals | Comments |
|---|---|---|---|---|---|
| *Major or Macrominerals* | | | | | |
| Salt (NaCl) | Salt serves as both a condiment and a nutrient. Sodium and chlorine help maintain osmotic pressure in body cells, upon which depends the transfer of nutrients to the cells and the removal of waste materials. Sodium is associated with muscle contraction and is important in making bile, which aids in the digestion of fats and carbohydrates. Chlorine is required for the formation of hydrochloric acid in the gastric juice so vital to protein digestion. | Reduced growth and efficiency of feed utilization in growing animals; reduced milk production and weight loss in adults. Lowered reproduction (infertility in males, and delayed sexual maturity in females). Craving for sodium, evidenced by such things as drinking urine. In laying hens, a deficiency of sodium results in lowered production, loss of weight, and cannibalism. Chicks on chlorine-deficient diet exhibit nervous symptoms induced by sudden noise. | Salt toxicity, which is accentuated with restriction of water intake, readily occurs in nonruminants. It is characterized by a staggering gait, blindness, and other nervous disorders. Excess Na results in hypertension. Excess Cl is not likely. | Salt: free choice, or added to the ration at a level of 0.25 to 0.50%. | In practice, Na and Cl are supplied together as common salt. The body's requirement for Cl is approximately half that of Na. The body contains approximately 0.2% sodium. |
| Calcium (Ca) | Bone and teeth formation and maintenance; nerve function; muscle contraction; blood coagulation; cell permeability. Essential for milk production and for formation of eggshell in poultry. | Rickets in young. Osteomalacia in adults. Tetany (hypocalcemia). Milk fever in dairy cows is the classical example of Ca tetany. Hens: Thin-shelled eggs, drop in egg production, and lowered hatchability. | Calcium-phosphorus ratio is important. For non-ruminants, it should be 1 to 2 parts Ca to 1 part P. For ruminants, it may be anywhere from 1:1 to 7:1. Vitamin D is involved. If adequate vitamin D is present, the ratio of calcium to phosphorus is less important. Excess Ca reduces the absorption and utilization of Zn. In swine, this causes parakeratosis. Excess Mg decreases Ca absorption, replaces Ca in the bone and increases Ca excretion. | Oystershells. Limestone. Dicalcium phosphate. Defluorinated phosphate. Protein supplements of animal origin, legume forages, and rape. Milk. Bone meal. | Only 20 to 30% of the calcium in the average ration is absorbed from the intestinal track and taken into the bloodstream. Over 70% of the ash of the body consists of Ca and P. Approximately 99% of the Ca in the body is present in the bones and teeth. |
| Phosphorus (P) | Bone and teeth formation and maintenance; a component of phospholipids which are important in lipid transport and metabolism and cell-membrane structure. Milk secretion. In energy metabolism. A component of RNA and DNA, the vital cellular constituents required for protein synthesis. A constituent of several enzyme systems. | Rickets in young. Osteomalacia in adults. Depraved appetite (pica), but this is not specific for phosphorus deficiency. Breeding problems. Hens: Reduced egg production in poultry. | Ratio of Ca-P is important; somewhere between 1 to 2 parts of Ca to 1 part of P. Sufficient vitamin D is necessary for P assimilation and utilization. Excess Ca and Mg cause decrease in P absorption. In ruminants, excess P in relation to Ca is likely to cause calculi. | Monosodium phosphate. Diammonium phosphate. Dicalcium phosphate. Defluorinated phosphate. Bone meal. Most cereal grains and their by-products (notably wheat bran) are high in P. | Phosphorus is more efficiently absorbed than calcium; about 70% of the ingested phosphorus is absorbed. Approximately 80% of the P of the body is present in the bones and teeth. Excess P may result in lameness and spontaneous fracture of long bones. High P has a laxative effect. |
| Magnesium (Mg) | Essential for normal skeletal development; as a constituent of bones and teeth; enzyme activator, primarily in glycolytic system; involved in activating certain peptidases in protein digestion; relaxes nerve impulses; serves as a ruminant alkalizer and buffer. | Vasodilation, with resulting reduction in blood presure (manifested outwardly by a flushing of the skin). Hyperirritability. Tetany (grass tetany, or grass staggers) characterized by loss of appetite (anorexia), hyperemia, convulsions, and death. | Excess of Mg upsets Ca and P metabolism. Mg toxicity from feeding has not been demonstrated. | Magnesium sulfate or oxide, mixed with salt or small amount of feed | Deficiencies of Mg may be encountered with suckling calves and pigs. |
| Potassium (K) | Major cation intracellular fluid where it is involved in osmotic pressure and acid-base balance. Relaxes the heart muscle. Involved in secretion of insulin, in enzyme reactions of the phosphorylation of creatine, in carbohydrate metabolism, and in protein synthesis. | Growth retardation, unsteady gait, general muscle weakness, pica, diarrhea, distended abdomen, emaciation, enlargement of the heart and kidneys, followed by death. | Magnesium deficiency results in failure to retain potassium; hence, it may lead to K deficiency. Excessive levels of potassium interfere with magnesium absorption. Excessive use of salt depletes the body's potassium. | Potassium chloride, kelp. Molasses (beet and cane), beet tops. Roughages usually contain ample potassium. | Potassium deficiency may occur in drylot finishing cattle or sheep on a high-concentration ration. |

*(Continued)*

**TABLE 4-3** *(Continued)*

| Mineral | Major Functions | Some Deficiency Symptoms | Major Interrelationships; Toxicities | Good Sources for Animals | Comments |
|---|---|---|---|---|---|
| **Sulfur** (S) | Required as a component of sulfur-containing amino acids, cystine and methionine.<br>As a component of biotin, sulfur is important in lipid metabolism.<br>As a component of thiamin and insulin, it is important in carbohydrate metabolism.<br>As a component of coenzyme A, it is important in energy metabolism.<br>As a component of hair, wool, and feathers. | Retarded growth, primarily due to not meeting the sulfur amino acid requirement for protein synthesis.<br>Sheep fed nonprotein N to replace protein without S supplementation show reduced wool growth (wool contains approximately 4% sulfur). | Sulfur is related to the amino acids, cystine and methionine, and to biotin, thiamin, and coenzyme A (see column to left, "Major Functions").<br>Sulfur toxicity is not a practical problem. | Nonruminants should be provided sulfur-containing proteins.<br>Ruminants and horses may be provided sulfur in protein, as elemental sulfur or as sulfate sulfur. | The body contains approximately 0.15% sulfur.<br>Sulfur requirements are primarily those involving amino acid nutrition.<br>Ruminants fed urea as a source of protein nitrogen may benefit from supplemental sulfur. |
| *Trace or microminerals* | | | | | |
| **Chromium** (Cr) | In glucose metabolism.<br>Activator of certain enzymes.<br>Stabilizer of nucleic acids.<br>Stimulation of the synthesis of fatty acids and cholesterol in the liver. | Impaired glucose tolerance.<br>Disturbance of lipid and protein metabolism. | Diets high in carbohydrates may cause the supply of GTF-chromium to be depleted. | There is no evidence that practical animal rations need to be supplemented with Cr. | The importance of Cr in glucose metabolism of other animals (other than the rat) and humans has not been established to date. |
| **Cobalt** (Co) | As a component of vitamin B-12.<br>Rumen microorganisms use Co for the synthesis of vitamin B-12 and the growth of rumen bacteria. | Deficiency of Co in cattle and sheep produces symptoms similar to a deficiency of vitamin B-12.<br>Ruminants grazing in Co-deficient areas show loss of appetite, reduced growth, and loss in body weight, followed by emaciation, anemia, and eventually death. Frequently a depraved appetite is noted.<br>The disease called *salt sick* in Florida is due to Co deficiency associated with Cu deficiency.<br>In different parts of the world, Co deficiency is known as Denmark disease, coast disease, enzootic marasmus, bush sickness, wasting disease, Nakuritis, and pining disease. | Related to vitamin B-12.<br>Cobalt toxicity is not likely. | Cobaltized mineral mixture made by adding Co at rate of 0.2 oz/100 lb of salt as cobalt chloride, cobalt sulfate, cobalt oxide, or cobalt carbonate. Also, several good Co-containing commerical minerals are on the market.<br>Grazing animals may be given pellets composed of cobalt oxide and iron administered orally with a balling gun. The pellets lodge in the rumen and are gradually dissolved over a period of months.<br>Poultry by-product meal, soybean meal, meat meal, rice bran, and blackstrap molasses. | The Co content of the leaves of the catalpa tree is regarded as a good indicator of the adequacy of cobalt in an area.<br>Co-deficient areas have been reported in Australia, western Canada, and in the U.S. in the states of Florida, Michigan, Wisconsin, Massachusetts, New Hampshire, Pennsylvania, and New York. |
| **Copper** (Cu) | Along with iron and vitamin B-12, copper is necessary for hemoglobin formation, although it forms no part of the hemoglobin molecule (or red blood cells).<br>Essential in enzyme systems, hair development and pigmentation, bone development, reproduction, and lactation. | Fading hair coat; light wool growth and straight, hairlike fibers, known as steely wool.<br>A condition called *swayback* (enzootic ataxia) in newborn lambs.<br>Lameness, swelling of joints, and fragility of bones.<br>Nutritional anemia, commonly called *salt sick*. | Copper is involved in iron metabolism.<br>An excess of molybdenum in the presence of sulfate causes a condition which can be cured by administering copper.<br>Excess copper is toxic; it accumulates in the liver, and death may result.<br>In high molybdenum areas, the Cu level for horses and cattle should be about 5 times higher than normal. | Trace mineralized salt containing copper sulfate or copper carbonate.<br>Cane or blackstrap molasses, liver meal, brewers' grains, and gluten feed and meal. | A variable store of copper is located in the liver and spleen.<br>Milk is low in Cu; hence, young animals raised almost exclusively on milk may develop anemia.<br>Copper deficiencies are common in Australia and New Zealand,, and in southern U.S. |
| **Fluorine** (F) | Necessary for sound bones and teeth. | Excesses of fluorine are of more concern than deficiencies in livestock production. | Large amounts of calcium, aluminum, or fat will lower the absorption rate of fluorine.<br>High dietary Ca depresses F uptake of bone. F is a cumulative poison; hence, the toxic effects may not be noticed for some time.<br>High levels result in enlarged bones; softening, mottling, and irregular wear of the teeth; roughened hair coat; delayed maturity; and less efficient utilization of feed. | No need to supplement livestock with fluorine has been demonstrated. Should such supplementation be necessary, 1 ppm in the drinking water should suffice. | Fluorine in excess of 20 to 40 ppm of the dry matter of the diet (depending on the species of animal, age, and rate of production) may show a progressive severe toxicity. |

*(Continued)*

**TABLE 4–3** *(Continued)*

| Mineral | Major Functions | Some Deficiency Symptoms | Major Interrelationships; Toxicities | Good Sources for Animals | Comments |
|---|---|---|---|---|---|
| Iodine (I) | Needed by the thyroid gland for making thyroxin, an iodine-containing hormone which controls the rate of body metabolism or heat production. | Goiter (big-neck) in humans, calves, lambs, and kids; stillbirths and weak young; hairless pigs; woolless lambs at birth.<br>There is no satisfactory treatment for animals that have developed pronounced I-deficiency symptoms.<br>Iodine deficiency in young animals is called cretinism. In adults it is known as myxedema. | Feeds of the cabbage family contain goitrogens, which interefere with the use of thyroxin and may produce goiter.<br>Long-term chronic intake of large amounts of I reduces thyroid uptake of I.<br>Marked species differences exist in tolerance to high intakes of I. | Stabilized iodized salt containing 0.01% potassium iodide (0.0076%I).<br>Calcium iodate.<br>Ethylenediamine dihydriodide (EDDI).<br>Whey, marine by-products, poultry by-products, blackstrap molasses, and meat meal. | Enlargement of the thyroid gland (goiter) is nature's way of trying to make enough thyroxin (an I-containing hormone) when there is insufficient I in the feed.<br>Mature animal body contains less than 0.00004% I.<br>I deficiencies are worldwide. In the U.S., the Northwest, the Pacific Coast, and the Great Lakes regions are goiter areas. |
| Iron (Fe) | Iron is a constituent of hemoglobin, the iron-containing compound that transports oxygen.<br>Also, iron plays a role in cellular oxidations, being a component of certain enzymes concerned with oxygen transfer. | Fe-deficiency anemia, characterized by smaller than normal number of red cells and less than normal amount of hemoglobin. | Iron is related to hemoglobin.<br>Cu is required for proper Fe metabolism.<br>Pyridoxine deficiency decreases the absorption of Fe.<br>Too much iron may be deleterious—interfering with phosphorus absorption by forming an insoluble phosphate. | Ferrous sulfate administered orally, or iron dextran infection.<br>Leafy portions of plants, meat by-products, legume seeds, cereal grains, and cane molasses.<br>Trace mineralized salt. | The body contains only about 0.004% iron. Thus, a mature human contains only about 1/10 ounce of this mineral.<br>Iron is stored in the liver, spleen, and kidneys.<br>Young animals are born with a store of iron. But milk is low in iron. So, when young animals are continued on milk for a long time, particularly under confined conditions and with little or no supplemental feed, nutritional anemia will likely develop. |
| Manganese (Mn) | Essential for normal bone formation (as a component of the organic matrix), and growth of other connective tissues.<br>Blood clotting.<br>Insulin action.<br>Activator of enzyme systems in the metabolism of carbohydrates, fats, proteins, and nucleic acids. | Poor growth.<br>Lameness, shortening and bowing of the legs, and enlarged joints. *Knuckling over* in calves.<br>In pigs, crooked legs and enlarged hocks.<br>Impaired reproduction (testicular degeneration of males; defective ovulation of females).<br>Slipped tendons (perosis) in poultry. | Excess Ca and P decreases absorption.<br>Mn is not toxic in moderate excesses. | Trace mineralized salt containing 0.25% manganese (or more).<br>Rice, wheat, hays, blackstrap molasses, cottonseed hulls. | The manganese content of plants is dependent on soil content.<br>Plants grown on alkali soils may be abnormally low in manganese. |
| Molybdenum (Mo) | As a component of three different enzyme systems involved in the metabolism of carbohydrates, fats, proteins, sulfur-containing amino acids, nucleic acids, and iron.<br>As a component of the enzyme xanthine oxidase especially important in poultry for uric acid formation.<br>Stimulates action of rumen organisms. | Toxic levels of Mo are of greater practical concern than deficiencies. | Molybdenum utilization is reduced by excess copper, sulfate, and tungsten.<br>Mo is related to uric acid formation in poultry and microbial action in ruminants.<br>Mo as a toxic mineral affects cattle and sheep grazing pastures grown on soils high in Mo content.<br>Toxic levels of Mo interfere with copper metabolism; hence, increase copper requirements. | No Mo supplementation of normal rations is necessary. | Mo toxicity results in severe scours and loss of condition. |
| Selenium (Se) | Not completely known. But involved in vitamin E absorption and/or retention. Also, a required nutrient in its own right. Se prevents degeneration and fibrosis of the pancreas in chicks.<br>Component of the enzyme glutathione peroxidase, which protects against oxidation of polyunsaturated fatty acids.<br>Protects tissue against certain poisonous substances, such as arsenic, cadmium, and mercury.<br>Interrelation with vitamin E. | Nutritional muscular dystrophy, called *white muscle disease* in calves and *stiff lamb disease* in lambs.<br>Exudative diathesis in poultry.<br>Liver necrosis in pigs. | Selenium is closely related to vitamin E and the sulfur-containing amino acids.<br>Animals consuming forage or grain produced on seleniferous soils develop blind staggers or alkali disease, characterized by emaciation, loss of hair, soreness and sloughing of hooves, lameness, anemia, excess salivation, grinding of the teeth, blindness, paralysis, and death.<br>In poultry, egg production and hatchability are reduced and deformities are common, including lack of eyes and deformed wings and feet. | Sodium selenate, sodium selenite.<br>Marine by-products, cereal grains, wheat by-products, and plants grown on selenium-rich soils. | In 1987, FDA approved a maximum of 0.3 ppm selenium in complete feed for all classes of animals. |

*(Continued)*

**TABLE 4–3** *(Continued)*

| Mineral | Major Functions | Some Deficiency Symptoms | Major Interrelationships; Toxicities | Good Sources for Animals | Comments |
|---|---|---|---|---|---|
| **Silicon** (Si) | Necessary for normal growth and skeletal development of the chick and rat. | Deficiency in chicks and rats results in retarded growth and skeleton deformities, especially in the skull. | From a practical standpoint, adverse effects of high Si intake, rather than Si deficiency, appear to be of concern. | One of most abundant elements on earth. Present in large amounts in soils and plants. | On purified diets, the addition of Si has increased the growth rate of chicks and rats. |
| **Zinc** (Zn) | Zinc is needed in normal skin, bones, hair, and feathers.<br>Zinc is a component of several enzyme systems, including peptidases and carbonic anhydrase.<br>Also, Zn is required for normal protein synthesis and metabolism and is a component of insulin.<br>Zinc imparts bloom to the hair coat. | Loss of appetite and stunted growth.<br>Poor hair or feather development; slipping of wool.<br>Rough and thickened skin in swine, known as parakeratosis. | Excess Ca reduces the absorption and utilization of Zn, precipitating parakeratosis in swine.<br>Excess Zn interferes with Cu metabolism and may cause anemia. | Zinc carbonate.<br>Zinc sulfate.<br>Fish meal.<br>Corn gluten feed and meal.<br>Poultry by-products.<br>Distillers' solubles. | Zinc availability is affected adversely by phytates. |

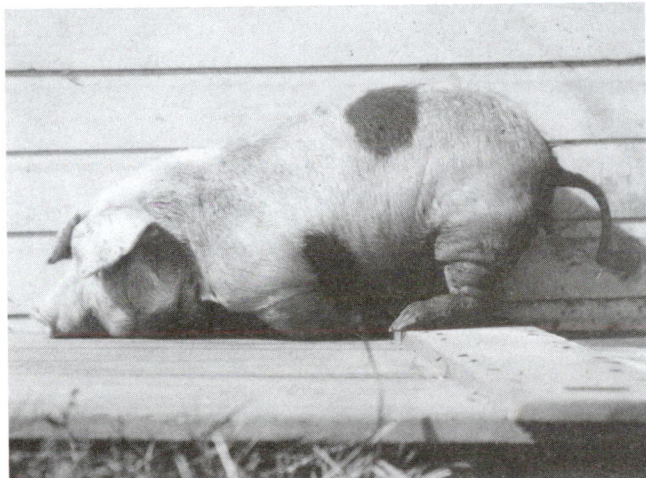

Fig. 4–4. Growing swine need plenty of calcium. *Top:* Pig with severe rickets due to a calcium deficiency. *Bottom:* Healthy pig that received adequate calcium. (Courtesy, Ohio Agricultural Research and Development Center, Wooster, OH)

Fig. 4–5. Calcium, phosphorus, and/or vitamin D deficiency, resulting in rickets. Note the bowed front legs, enlarged knee joints, and difficulty in standing. (Courtesy, USDA)

Fig. 4–6.  Cobalt deficiency. Note thin, emaciated condition and loss of wool. (Courtesy, Cornell University, Ithaca, NY)

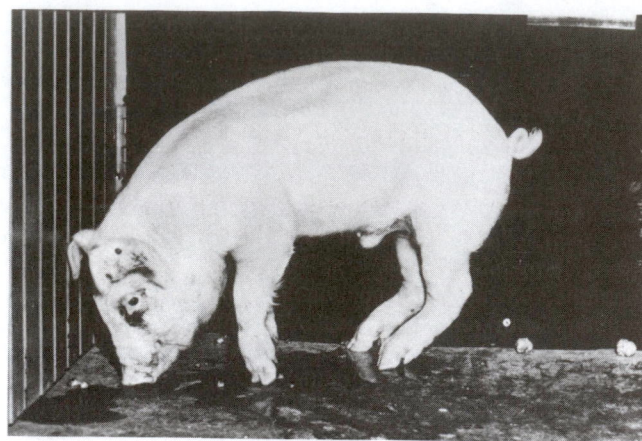

Fig. 4–7.  Copper deficiency. Note the drawing under of the rear legs and the crooked front legs. (Courtesy, Hormel Institute, Austin, MN)

Fig. 4–8.  Iron deficiency, resulting in nutritional anemia, characterized by a swollen condition about the head and neck and paleness of the mucous membranes. (Courtesy, University of Illinois, Urbana)

Fig. 4–9.  Manganese deficiency, resulting in perosis, or slipped tendon. A deficiency of either choline or biotin will also cause perosis. (Courtesy, Cornell University, Ithaca, NY)

## SALT (SODIUM CHLORIDE)

Salt, which serves as both a condiment and a nutrient, is needed by all classes of animals, but more especially by herbivora (grass-eating animals). It may be provided in the form of granulated, rock, or block salt. In general, the form selected is determined by price and availability. It is to be pointed out, however, that very hard block and rock salt are difficult for stock to eat, often resulting in sore tongues and inadequate consumption. Also, if there is much competition for the salt block, the more timid animals may not get their requirements.

Both sodium and chlorine are essential for animal life. The blood contains 0.25% chlorine, 0.22% sodium and 0.02 to 0.22% potassium; thus, the chlorine content is higher than that of any other mineral in the blood. The salt requirements are greatly increased under conditions which cause heavy sweating, thereby resulting in large losses of this mineral from the body. Unless replaced, fatigue will result. For this reason, when engaged in hard work and perspiring profusely, both horses and humans should receive liberal allowances of salt.

• **Salt starved animals**—Salt can be fed free choice to cattle, sheep, swine, and horses provided that they have not previously been salt starved. If animals have not been fed salt for a considerable length of time, they may overeat, resulting in digestive disturbances and even death. Salt starved animals should first be hand fed salt, and the daily allowance should be increased gradually until they start leaving a little in the mineral box. When this point is reached, self-feeding may be followed. The Indians and the pioneers of this country handed down many legendary stories about the huge number of buffalo and deer that killed themselves by gorging at a newly found *salt lick* after having been salt starved for long periods of time.

## CALCIUM AND PHOSPHORUS

Farm animals are more likely to suffer from a lack of calcium and phosphorus than from any of the other minerals except common salt. These 2 minerals comprise about 70% of the mineral content of the animal body and from 1/3 to 1/2 of the minerals of milk. About 99% of the calcium and over 80% of the phosphorus are found in the bones and teeth.

Liberal allowances of calcium and phosphorus are especially important for young, growing animals, for those that are pregnant, and for those that are producing milk.

The following general characteristics of feeds in regard to calcium and phosphorus are noteworthy:

1. The cereal grains and their by-products and straws, dried mature grasses, and protein supplements of plant origin are low in calcium.

2. The protein supplements of animal origin, legume forage, and rape are rich in calcium.

3. The cereal grains and their by-products are fairly high or even rich in phosphorus, but a large portion of the phosphorus is not readily available.

4. Most all protein-rich supplements are high in phosphorus. But, here again, plant sources of phosphorus contain much of this element in a bound form.

5. Beet by-products and dried, mature, non-leguminous forages (such as grass hays and fodders) are likely to be low in phosphorus.

6. The calcium and phosphorus content of plants can be increased through fertilizing the soil upon which they are grown.

## CALCIUM-PHOSPHORUS RATIO AND VITAMIN D

In considering the calcium and phosphorus requirements of animals, it is important to realize that the proper utilization of these minerals by the body is dependent upon three factors: (1) an adequate supply of calcium and phosphorus in an available form, (2) a suitable ratio between them, and (3) sufficient vitamin D to make possible the assimilation and utilization of the calcium and phosphorus.

Generally speaking, nutritionists have recommended a calcium-phosphorus ratio somewhere between 1 to 2 parts of calcium to 1 part of phosphorus for non-ruminants (hogs and horses), and anywhere from 1:1 to 7:1 for ruminants.

If plenty of vitamin D is present (as provided either by sunlight or through the ration), the ratio of calcium to phosphorus becomes less important. Also, less vitamin D is needed when there is a desirable calcium-phosphorus ratio.

## RECOMMENDED CALCIUM AND PHOSPHORUS SUPPLEMENTS

Table 4–4 gives several sources of calcium and phosphorus and the typical analysis of each.

### TABLE 4–4
### TYPICAL ANALYSIS OF CALCIUM AND PHOSPHORUS SUPPLEMENTS (MOISTURE-FREE BASIS)[1]

| Compound | Calcium Content | Phosphorus Content | Sodium Content | Protein Equivalent N × 6.25 | Fluorine Content |
|---|---|---|---|---|---|
| | (%) | (%) | (%) | (%) | (%) |
| **Calcium compounds:** | | | | | |
| Calcium carbonate | 38.13 | 0.04 | 0.07 | — | 0 |
| Limestone, ground | 37.22 | 0.22 | 0.06 | — | — |
| Oystershells, ground | 36.27 | 0.10 | 0.21 | 0.7 | — |
| **Defluorinated phosphates manufactured from defluorinated phosphoric acid:** | | | | | |
| Defluorinated phosphate | 32.10 | 17.13 | 3.27 | — | 0.18 |
| Diammonium phosphate | 0.52 | 20.54 | 0.04 | 115.5 | 0.16 |
| Dicalcium phosphate | 22.67 | 19.00 | 1.61 | — | 0.10 |
| Monoammonium phosphate | 0.39 | 24.99 | 0.08 | 71.0 | 0.19 |
| Monocalcium phosphate | 18.80 | 21.27 | 0.06 | — | 0.14 |
| **Defluorinated phosphates manufactured from furnace phosphoric acid:** | | | | | |
| Dicalcium phosphate | 23.71 | 19.07 | 0.08 | — | 0.19 |
| Disodium phosphate | — | 22.32 | 32.00 | — | — |
| Feed-grade phosphoric acid | 0.18 | 27.84 | 0.23 | — | 0.25 |
| Monocalcium phosphate | 22.92 | 23.96 | — | — | 0.03 |
| Monosodium phosphate | 0.04 | 25.60 | 19.63 | — | — |
| Sodium tripolyphosphate | — | 25.38 | 31.23 | — | 0.03 |
| Tricalcium phosphate | 31.44 | 17.34 | 0.17 | — | 0.05 |
| **High-fluoride phosphates:** | | | | | |
| Ground low-fluorine rock phosphate | 36.00 | 14.00 | — | — | 0.45 |
| Ground rock phosphate, raw | 35.00 | 13.00 | 0.03 | — | 3.70 |
| Soft rock phosphate | 16.09 | 9.05 | 0.10 | — | 1.21 |
| **Packinghouse by-products:** | | | | | |
| Bone charcoal (bone black) | 34.08 | 15.85 | — | — | — |
| Bone meal, steamed | 27.31 | 12.40 | 0.42 | 19.5 | 0.07 |

[1]Adapted by the author from data provided especially for this book by the former International Feedstuffs Institute, now the Feed Composition Data Bank, National Agricultural Library, USDA, Beltsville, MD.

Where calcium alone is needed, ground limestone or oystershell flour are commonly used, either free choice or added to the ration in keeping with nutrient requirements.

Where phosphorus alone is needed, monosodium phosphate or diammonium phosphate may be used.

## PRECAUTIONS RELATIVE TO CALCIUM AND PHOSPHORUS SUPPLEMENTS

Earlier experiments cast considerable doubt on the availability of phosphorus when it was largely in the form of phytin. Although wheat bran is very high in phosphorus, containing 1.32%, there was some question as to its availability due to the high phytin content of this product. More recent studies, however, indicate that cattle, and perhaps mature swine, can partially utilize phytin phosphorus. Cattle can utilize about 60% of the total phosphorus from most plant sources, whereas swine can utilize only about 50%. It must be emphasized, however, that phosphorus availability depends to a large extent on phosphorus sources, dietary supplies of calcium, and adequate vitamin D. Recent work indicates that high calcium levels enhance the formation of the insoluble phytic acid, whereas high vitamin D levels aid materially in the utilization of phosphorus in the form of phytin. On the other hand, in the case of humans and poultry, the evidence seems clear that phytin phosphorus is a less satisfactory source of phosphorus.

Likewise, for humans, the availability of the calcium of certain leafy materials is impaired by the presence of oxalic acid—the acid precipitating the calcium and preventing its absorption. Thus, it was a great consolation to many people (including the author) to discover that, due to the high oxalic acid content of spinach, this food possesses very questionable value in human nutrition from the standpoint of calcium. On the other hand, the deleterious effects of oxalic acid are reduced in the ruminant because of its apparent ability to metabolize oxalic acid in the body.

During World War II, the shortage of phosphorus feed supplements led to the development of defluorinated phosphates for feeding purposes. Raw, unprocessed rock phosphate usually contains from 3.25 to 4.0% fluorine, whereas steamed bone meal normally contains only 0.05 to 0.10%. Fortunately, through heating at high temperatures under conditions suitable for elimination of fluorine, the excess fluorine of raw rock phosphate can be removed. Such a product is known as defluorinated rock phosphate.

The Association of American Feed Control Officials has established maximum fluorine content for (1) mineral substances, and (2) total ration (see Table 4–5).

### TABLE 4–5
### MAXIMUM FLUORINE CONTENT FOR (1) MINERAL SUBSTANCES AND (2) TOTAL RATION[1]

| Class of Animal | Maximum Fluorine Content of Any Mineral or Mineral Mixture Which Is to Be Used Directly for the Feeding of Animals Shall Not Exceed— | Fluorine Content of Rock Phosphate (or other ingredients) Shall Be Such That the Maximum Fluorine Content of the Total Ration (exclusive of roughage) Shall Not Exceed— |
|---|---|---|
| | (%) | (%) |
| Cattle | | |
| slaughter | 0.30 | 0.009 |
| breeding and dairy | 0.20 | 0.004 |
| Sheep | 0.30 | 0.006 |
| Lambs | 0.35 | 0.01 |
| Swine | 0.45 | 0.015 |
| Poultry | 0.60 | 0.03 |

[1]*Feed Control*, official publication, Association of American Feed Control Officials, 1988, p. 73.

## VITAMIN NEEDS

Until early in the 20th Century, if a ration contained proteins, fats, carbohydrates, and minerals—together with a certain amount of fiber—it was considered to be a complete diet. True enough, the disease known as beriberi made its appearance in the rice-eating districts of the Orient when milling machinery was introduced from the West, having been known to the Chinese as early as 2600 B.C.; and scurvy was long known to occur among sailors fed on salt meat and biscuits. However, for centuries these diseases were thought to be due to toxic substances in the digestive tract caused by pathogenic organisms, rather than food deficiencies; and more time elapsed before the discovery of vitamins. Of course, there was no medical profession until 1835, the earlier treatments having been based on superstition rather than science.

Largely through the trial-and-error method, it was discovered that specific foods were helpful in the treatment of certain of these maladies. In 1747, Lind, a British naval surgeon, showed that the juice of citrus fruits was a cure for scurvy. Lunin, as early as 1881, had come to the conclusion that certain foods, such as milk, contain, besides the principal ingredients, small quantities of unknown substances essential to life. Eijkman, working in Java in 1897, had satisfied himself that the disease beriberi was due to the continued consumption of a diet of polished rice. At a very early date, the Chinese used a concoction rich in vitamin A as a remedy for night blindness. Also, codliver oil was used in treating or preventing rickets long before anything was known about the cause of the disease.

The significance of these observations relative to diet, however, was not fully appreciated until scientists found it desirable in many types of investigations to use the biological approach, with purified diets to supplement chemical analyses in measuring the value of feeds. These rations were made up of relatively pure nutrients—proteins, carbohydrates, fats, and minerals—from which the unidentified substances were largely excluded. With these purified rations, all investigators shared a common experience; the animals limited to such diets not only failed to thrive, but they even failed to survive if the investigations were continued for any length of time. At first, many investigators explained such failures on the basis of unpalatability and monotony of the rations. Finally, it was realized that these purified rations were lacking in certain substances, minute in amount and the identity of which was unknown to science. These substances were essential for the maintenance of health and life itself and the efficient utilization of the main ingredients of the food. With these findings, a new era of science was ushered in. The modern approach to nutrition was born.

Funk, a Polish scientist working in London, first referred to these nutrients as *vitamines* in 1912. Presumably, the name vitamines alluded to the fact that they were essential to life, and they were assumed to be chemically of the nature of amines—nitrogen-containing (the chemical assumption was later proved incorrect, with the result that the "e" was dropped; hence the word *vitamin*).

The actual existence of vitamins, therefore, has been known only since 1912, and only within the last few years has it been possible to see or touch any of them in a pure form. Previously, they were merely mysterious invisible "little things" known only by their effects. In fact, most of the present fundamental knowledge relative to the vitamin content of both human foods and animal feeds was obtained through measuring their potency in promoting growth or in curing certain disease conditions in animals—a most difficult and tedious method. For the most part, small laboratory animals were used, especially the rat, guinea pig, pigeon, and chick.

The lack of vitamins in a ration may, under certain conditions and in a more limited way, be more serious than a short supply of feed and may result in serious economic losses. On the other hand, such vitamin deficiencies are less widespread throughout the world than hunger itself, for actual starvation has always stalked across much of the world, being referred to as famine only when the numbers dying approach the millions.

Under practical conditions, the rations of farm animals usually contain adequate quantities of each of the several vitamins. However, deficiencies may occur during periods (1) of extended drought or in other conditions of restriction in diet, (2) when production is being forced, (3) when large quantities of highly refined feeds are being fed, or (4) when low-quality forages are utilized. Also, deficiencies may occur as a result of lack of availability of vitamins or because of the presence of antimetabolites. Both are important concepts. For example, analyses show corn to be adequate in niacin. Yet, due either to an antimetabolite or unavailability, there may be niacin deficiencies when corn is fed—deficiencies that can be remedied by niacin supplementation.

The absence of one or more vitamins in the ration may lead to a failure in growth or reproduction, or to characteristic disorders known as deficiency diseases. In severe cases, death itself may follow. Although the occasional deficiency symptoms are the most striking result of vitamin deficiencies, it must be emphasized that, in practice, mild deficiencies probably cause higher total economic losses than do severe deficiencies. It is relatively uncommon for a ration, or diet, to contain so little of a vitamin that obvious symptoms of a deficiency occur. When one such case does appear, it is reasonable to suppose that there must be several cases that are too mild to produce characteristic symptoms but which are sufficiently severe to lower the state of health and the efficiency of production. It is also to be emphasized that different species of animals vary in their needs for the vitamins. Also, not all animals suffer from the same deficiency diseases. Thus, humans, monkeys, and guinea pigs react severely to the absence of vitamin C in the ration, whereas fowl and ruminants are unaffected.

Initially, vitamin nutrition was dependent upon incorporating into the ration feeds that were known to contain a high natural content of the needed vitamins. But this was not too satisfactory, for the vitamin content of feeds varies considerably according to soil, climate conditions, and curing and storing. Through the years, however, methods for the laboratory synthesis of various vitamins were developed. So, today, most of the vitamins are available in pure crystalline form at prices that make them economical for use in the vitamin supplementation of livestock.

## THE VITAMINS

*Vitamins are substances that are required in minute amounts by one or more animal species for normal growth, production, reproduction, and/or health.*

The omission of a single vitamin from the diet of a species that requires it will produce specific deficiency symptoms. Many of the vitamins function as coenzymes (metabolic catalysts); others have no such role, but perform certain essential functions.

Many phenomena of vitamin nutrition are related to solubility—vitamins are soluble in either fat or water. Consequently, it is important that both nutritionists and producers be well informed about solubility differences in vitamins and make use of such differences in programs and practices. Based on solubility, vitamins may be grouped as follows:

| The Fat-Soluble Vitamins | The Water-Soluble Vitamins |
|---|---|
| Vitamin A | Biotin |
| Vitamin D | Choline |
| Vitamin E | Folacin (folic acid) |
| Vitamin K | Inositol |
| | Niacin (nicotinic acid, nicotinamide) |
| | Pantothenic acid (vitamin B-3) |
| | Para-aminobenzoic acid (PABA) |
| | Riboflavin (vitamin B-2) |
| | Thiamin (vitamin B-1) |
| | Vitamin B-6 (pyridoxine, pyridoxal, pyridoxamine) |
| | Vitamin B-12 (cobalamins) |
| | Vitamin C (ascorbic acid, dehydro-ascorbic acid) |

It is noteworthy that vitamin C is the only member of the water-soluble group that is not a member of the B family.

The two groups of vitamins exhibit the following differences that distinguish them both chemically and biologically.

The fat-soluble vitamins contain only carbon, hydrogen, and oxygen, whereas the water-soluble B vitamins contain these three elements plus nitrogen and occasionally sulfur.

Vitamins originate primarily in plant tissues; with the exceptions of vitamins C and D, they are present in the animal tissues only if an animal consumes feed containing them or harbors microorganisms that synthesize them. Fat-soluble vitamins can occur in plant tissue in the form of a provitamin (or precursor of a vitamin), which can be converted into a vitamin in the animal body. Also, the B vitamins are universally distributed in all living tissues, whereas the fat-soluble vitamins are completely absent from some.

The fat-soluble vitamins are stored in appreciable quantities in the body, whereas the water-soluble vitamins are not. Any of the fat-soluble vitamins can be stored wherever fat is deposited; and the greater the intake, the greater the storage. By contrast, the water-soluble B vitamins are not stored in any appreciable amount. Moreover, the large amounts of water which pass through the body daily tend to carry out the water-soluble vitamins, thereby depleting the supply. Hence, they should be supplied in the diet on a daily basis. However, because all living cells contain all the B vitamins, and because the body conserves nutrients that are in short supply by using them only in vital reactions, deficiency symptoms do not appear immediately following their removal from the diet.

Table 4-6, pages 106 to 108, contains a list of the 16 vitamins, the existence of which is undisputed. It is believed that most of the vitamins essential to animal nutrition have now been identified.

Each of the vitamins is as much a distinct chemical compound as is cane sugar, for example. All of them contain carbon, hydrogen, and oxygen. In addition, all of the B vitamins except inositol contain nitrogen. Certain of the B vitamins also contain one or more of the mineral elements of their molecules. Even when added to the diet in very small amounts, vitamins are extraordinarily potent.

In Table 4–6, only a brief presentation is made relative to (1) animals most affected, (2) the function of the vitamins, (3) some deficiency symptoms, and (4) good animal sources of each of the vitamins. In reviewing this, it must be remembered that single, uncomplicated vitamin deficiencies are the exception rather than the rule. Multiple deficiencies are altogether too common, making diagnosis difficult even to the trained observer. In addition to the summary in Table 4–6, pertinent information pertaining to certain vitamins is contained in the sections that follow, and a summary of each of the vitamin deficiency diseases is included in Table 4–12 along with nutritional diseases and ailments from causes other than vitamin deficiencies.

**TABLE 4–6**
**ANIMAL VITAMIN CHART**

| Name of Vitamin | Animals Most Affected | Functions | Some Deficiency Symptoms | Good Sources for Animals | Comments |
|---|---|---|---|---|---|
| *Fat-soluble Vitamins* | | | | | |
| **Vitamin A** | Affects all farm animals, including poultry. | Bone growth. Night vision (formation of visual purple in the eye). Prevents xerophthalmia. Essential for body growth, bone growth, and normal tooth development. Epithelial tissue maintenance—respiratory, urogenital and digestive tracts, and the skin. | Stunted growth or loss in weight and loss of appetite, xerophthalmia (an eye disease), night blindness, nervous incoordination as shown by a staggering gait, unsound teeth, rough, dry skin, and sterility in males and females or young which are born weak or dead. Chicks: Wobbly gait. Hens: Reduced egg production and hatchability. | Vitamin A can be provided as the synthetic vitamin or as its precursor, carotene. Rich sources of carotene follow: Green, leafy hays, not over 1 year old. Grass silages. Lush, green pastures. Yellow corn. Green and yellow peas. Fish oils. Carrots. Whole milk. Dehydrated alfalfa meal. | Vitamin A is found only in animals; plants contain the precursor, carotene. Animals are able to store considerable vitamin A, but because of their greater requirements and less storage, young animals suffer from a deficiency much sooner than those that are mature. Both carotene and vitamin A are readily destroyed by oxidation, thus resulting in considerable losses in processing and storing (as in making or storing of hay). |
| **Vitamin D** | Affects all farm animals, including poultry. | Aids in the assimilation and utilization of calcium and phosphorus and necessary in normal bone development of animals, including the bone of the fetus. Promotes sound teeth. | Rickets in young. Osteomalacia in adults. Tetany, characterized by muscle twitching, convulsions, and low serum calcium. Chicks: Reduced growth, soft bones (rickets), leg deformities. Hens: Poor eggshells and lowered hatchability. | Vitamin $D_2$ (irradiated ergosterol), the plant form. Vitamin $D_3$, the animal form. Sunlight. Sun-cured hays. Cod and certain other fish-liver oils. Irradiated yeast. | Most mammals can use either $D_2$ or $D_3$, but birds require vitamin $D_3$. When animals are exposed sufficiently to direct sunlight, the ultraviolet light in the sunlight penetrates the skin and produces vitamin D from traces of certain cholesterols in the tissues. Tissue storage is very limited. The vitamin D requirement is less when a proper balance of calcium and phosphorus exists. |
| **Vitamin E** | Calves, sheep, horses, poultry, rats, and perhaps certain other animals. | Antitoxidant. A sparer of selenium. As an essential factor for the integrity of red blood cells. Essential in cellular respiration, primarily in heart and skeletal muscle tissue. Regulator in the synthesis of DNA, vitamin C, and coenzyme Q. | Muscular dystrophy (stiff-lamb disease in lambs and white muscle disease in calves). Reproductive failure. Steatitis. Chicks: Encephalomalacia (crazy chick disease). Hens: Poor hatchability. | Alpha-tocopherol. Rice polishings. Wheat germ meal. Alfalfa leaf meal. Green grass. Early cut hay. | Vitamin E is widely distributed in all natural feeds. Utilization of vitamin E is dependent on adequate selenium. |
| **Vitamin K** | All species, but ruminants have the advantage of microbial synthesis. | Essential for blood clotting. | Prolonged blood clotting time, generalized hemorrhages, and death in severe cases. | Menadione (vitamin $K_3$). Green pastures. Well-cured hays. Fish meal. In general, this factor is widely distributed in normal farm rations. Also, all classes of farm animals synthesize it. | Vitamin K has definite value in human therapy where clotting of the blood is impaired due to a deficiency of the vitamin. Menadione is widely used commercially as a source of vitamin K. Well-known antagonists of vitamin K are dicoumarol and warfarin. |

*(Continued)*

**TABLE 4–6** *Continued*

| Name of Vitamin | Animals Most Affected | Functions | Some Deficiency Symptoms | Good Sources for Animals | Comments |
|---|---|---|---|---|---|
| *Water-soluble Vitamins* | | | | | |
| **Biotin** | Required by all species. | Biotin is required in many reactions in the metabolism of carbohydrates, fats, and proteins.<br><br>Biotin serves as a coenzyme for transferring $CO_2$ from one compound to another.<br><br>Biotin also serves as a coenzyme for deamination (removal of $NH_2$) of amino acids for the production of energy. | Pigs exhibit spasticity of the hind legs, cracks in the feet, and a dermatitis. There is also lowered efficiency of feed utilization.<br><br>Chicks and turkey poults show dermatitis and perosis.<br><br>In hens, hatchability is severely reduced.<br><br>In mink, biotin deficiency makes for abnormal fur. | Synthetic biotin.<br>Alfalfa leaf meal (dehy.)<br>Rice polishings.<br>Yeast.<br>Distillers' solubles.<br>Safflower meal.<br>Cottonseed meal.<br>Blackstrap molasses. | Ordinary farm rations probably contain ample biotin, or farm animals synthesize all they need.<br><br>Biotin is rendered unavailable by raw egg white. |
| **Choline** | Swine, rats, and poultry. | Choline is involved in the prevention of fatty livers, in transmitting nerve impulses, and in the metabolism of fat. | Poor growth and fatty livers in most species.<br><br>In chickens and turkeys, slipped tendon (perosis).<br><br>In swine, abnormal gait in growing pigs and reproductive failure in adult females. | Choline chloride or choline dihydrogen.<br>Rice polishings.<br>Soybean lecithin.<br>Wheat germ.<br>Yeast.<br>Rapeseed (canola) meal.<br>Poultry by-products.<br>Fish meal. | With a high-protein diet, enough choline is synthesized from certain precursors and amino acids. Deficiency symptoms are more readily obtained as the protein content is lowered. |
| **Folacin** (folic acid) | All animals and birds may be affected. | Folacin enzymes are responsible for the following important functions: (1) the formation of purnes and pyrimidines; (2) the formation of heme; (3) the interconversion of the amino acid serene to the amino acid glycine; (4) the formation of the amino acids tyrosine from phenylalanine and glutamic acid from histidine; (5) the formation of the amino acid methionine from hemocystene; (6) the synthesis of choline from ethanolamine; and (7) the conversion of nicotinamide to N-methylnicotinamide. | Macrocytic anemia (of young) and macrocytic anemia (of pregnancy).<br><br>In chicks, retarded growth and depigmentation of colored feathers.<br><br>In humans and dogs, a sore, red, smooth tongue, disturbance of the digestive tract, and poor growth. | Synthetic folacin.<br>Wheat germ.<br>Yeast.<br>Soybean meal.<br>Alfalfa hay.<br>Cottonseed meal.<br>Linseed meal. | Folic acid is widely distributed in both plants and animals. It was given this name because of the abundance of the factor in plant leaves. |
| **Inositol** | Chicks, fish, swine, guinea pigs, hamsters, rats, and mice. | Not known. But it appears to aid in the metabolism of fats and helps reduce blood cholesterol.<br><br>In combination with choline, it prevents hardening of arteries and protects the heart.<br><br>As a precursor of phosphoinosities, which is found in various tissues. | Not demonstrated in animals. | Synthetic inositol.<br>Wheat germ.<br>Yeast.<br>Liver meal.<br>Citrus meal.<br>Blackstrap molasses. | Widely distributed in animal feeds.<br>Synthesized in intestines. |
| **Niacin** (Nicotinic Acid, Nicotinamide) | It is a dietary essential of pigs, chickens, monkeys, and humans. Apparently synthesized in the digestive tract of ruminants (sheep and cattle) and the horse. | Constituent of two coenzymes, which are necessary in cell respiration; in the release of energy from carbohydrates, fats, and protein; and in biological oxidation-reduction systems. | Reduced growth and appetite.<br><br>Swine exhibit diarrhea, vomiting, dermatitis, unthriftiness, and ulcerated intestine.<br><br>Chicks show poor feathering, scaly dermatitis, and sometimes, a "spectacled eye."<br><br>Dogs show a darkening of the tongue (black tongue) and mouth lesions.<br><br>Humans develop pellagra characterized by a bright red tongue, mouth lesions, anorexia, and nausea. | Synthetic niacin.<br>Rice polishings.<br>Yeast.<br>Rice bran.<br>Marine by-products.<br>Liver meal.<br>Animal by-products.<br>Green alfalfa is a fair source. | Niacin is the most stable of the B-complex vitamins.<br><br>Niacin present in most cereal grains is not available to the pig and other simple-stomached animals.<br><br>Niacin can be synthesized in the body from surplus tryptophan.<br><br>Mature ruminants do not need dietary niacin under most conditions because of synthesis of rumen microflora. |
| **Pantothenic Acid** (vitamin B–3) | Rats, dogs, pigs, chickens, and turkeys. Synthesized in rumen of cow and sheep; perhaps the horse also synthesizes it. | Component (1) of coenzyme A, required for energy metabolism; and (2) of acyl carrier protein (ACP).<br><br>ACP, along with CoA, is required by the cells in the biosynthesis of fatty acids. | All species exhibit reduced growth, loss of hair, and enteritis.<br><br>Pigs develop "goose-stepping" gait.<br><br>Chicks show dermatitis and embryonic death.<br><br>Dogs vomit and show fatty infiltration of liver.<br><br>Mature ruminants synthesize pantothenic acid in rumen. Signs of deficiency in calves are rough coat, dermatitis, anorexia, and loss of hair around eyes. | Calcium pantothenate.<br>Rice polishings.<br>Yeast.<br>Safflower meal.<br>Whey.<br>Fish solubles.<br>Blackstrap molasses.<br>Alfalfa meal. | Grain is very deficient in pantothenic acid.<br><br>Of all B vitamins, it is most likely to be deficient under drylot conditions.<br><br>Pantothenic acid is commonly added to commercial swine and poultry rations. |

*(Continued)*

**TABLE 4–6** *Continued*

| Name of Vitamin | Animals Most Affected | Functions | Some Deficiency Symptoms | Good Sources for Animals | Comments |
|---|---|---|---|---|---|
| **Para-aminobenzoic Acid** (PABA) | Essential growth factor for certain microorganisms. | For higher animals, PABA functions as an essential part of the folacin molecule. | Not demonstrated in animals. | Synthetic para-aminobenzoic acid. Lecithin. Wheat germ. Yeast. Fish meal. Soybean meal. Peanut meal. Blackstrap molasses. | Abundantly synthesized in intestines. |
| **Riboflavin** (vitamin B-2) | Thought to be required by all animals, but deficiency symptoms not observed in ruminants, perhaps due to rumen synthesis. Deficiency symptoms noted in poultry, swine, and horses. | Promotes growth and functions in the body as a constituent of several enzyme systems and as such is important in carbohydrate, fatty acid, and amino acid metabolism. | Retarded growth in most species, with a wide variety of other symptoms somewhat variable with the species. Periodic ophthalmia (moon blindness) in horses; reproductive failure in the sow, and slow growth, anemia, diarrhea, unthrifty appearance, eye opacities, and an abnormal gait in the young pig; and curled toe paralysis in birds. | Synthetic riboflavin. Yeast. Skim milk. Whey. Liver meal. Alfalfa hay. Grass (immature/green). Poultry by-product meal. | Grains are poor source of riboflavin. Many common rations are borderline or deficient in riboflavin, especially swine and poultry rations. Riboflavin is destroyed by light or heat. |
| **Thiamin** (vitamin B-1) | All animals must have a dietary source, unless there is rumen synthesis, as in cattle and sheep. | As a coenzyme in energy metabolism. In the functioning of the peripheral nerves. Maintains (1) normal appetite, (2) tone of the muscles, and (3) healthy mental attitude. | Reduction in appetite (anorexia) and loss in weight. Cardiovascular disturbances. Beriberi (in humans). Lowered body temperature. Chicks: Polyneuritis (retraction of the head). Hens: Lowered egg production. | Thiamin hydrochloride. Thiamin mononitrate. Rice polishings. Wheat germ meal. Yeast. Rice bran. Wheat and wheat by-products. Cottonseed meal. | Fats exhibit a thiamin-sparing effect. |
| **Vitamin B-6** (pyridoxine, pyridoxal, pyridoxamine) | B-6 is a dietary essential for the rat, pig, chick, and dog. It is synthesized in the rumen of cattle and sheep and perhaps in the cecum of the horse; thus, no deficiency symptoms in these species have been reported. | As coenzyme in protein and nitrogen metabolism. Involved in red blood cell formation and in absorption of amino acids. Involved in carbohydrate and fat metabolism. Involved in clinical problems, including (1) anemia that is iron-resistant, (2) kidney stones, and (3) physiological demands of pregnancy. | All species exhibit convulsions. Pigs show anorexia, poor growth, and convulsions. Chicks show retarded growth and abnormal feathering. Hens show lowered egg laying and hatchability. Rats develop a specific dermatitis. | Safflower meal. Fish solubles. Pasture (green). Meat meal and tankage. Wheat and wheat by-products. Alfalfa hay. | Normally, animal rations are not lacking in vitamin B-6. |
| **Vitamin B-12** (cobalamins) | Swine, rats, poultry, and humans. Ruminants synthesize B-12 unless cobalt is deficient. | Vitamin B-12 functions in two enzyme forms: coenzyme B-12 and Methyl B-12. Also, the role of B-12 is closely related to other vitamins. | All animals show retarded growth. Pigs show uncoordinated hind leg movements; and there is reproductive failure in sows. Eggs from B-12-deficient hens fail to hatch. | Synthetic B-12. Protein supplements of animal origin. Fermentation products. | B-12 is apt to be lacking in swine and breeder poultry rations. |
| **Vitamin C** (ascorbic acid, dehydroascorbic acid) | Dietary need is limited to humans, guinea pigs, monkeys, fur-eating bats, and bulbul birds. Probably required by other species but synthesized in the body. | Collagen formation. Metabolism of the amino acids, tyrosine and tryptophan. Absorption and movement of iron. Metabolism of fats and lipids, and cholesterol control. Sound teeth and bones. Strong capillary walls and healthy blood vessels. Metabolism of folic acid. As a general antioxidant. Requirement increases in periods of stress. | Scurvy; swollen, bleeding and ulcerated gums, loosening of teeth, and weak bones. | Vitamin C (ascorbic acid). Acerola cherry. Rose hips. Citrus pulp. Well-cured hay. Green pasture. | Ordinary farm rations and body synthesis provide adequate vitamin C. |

Fig. 4–10. Vitamin A made the difference! *Left:* Normal chick that received adequate vitamin A. *Right:* Vitamin A deficient chick, showing poor growth, watery eyes, unsteady gait, and ruffled feathers. (Courtesy, University of Maryland, College Park)

Fig. 4–11. Vitamin D deficiency, resulting in rickets in this young lamb. (Courtesy, College of Veterinary Medicine, University of Illinois, Champaign-Urbana)

Fig. 4–12. Vitamin E deficiency, resulting in encephalomalacia (crazy chick disease). Note the head retraction and loss of control of the hind legs. (Courtesy, Cornell University, Ithaca, NY)

Fig. 4–13. Choline deficiency. Note the spraddled hind legs of these pigs, characteristic of a choline deficiency. (Courtesy, Washington State University, Pullman)

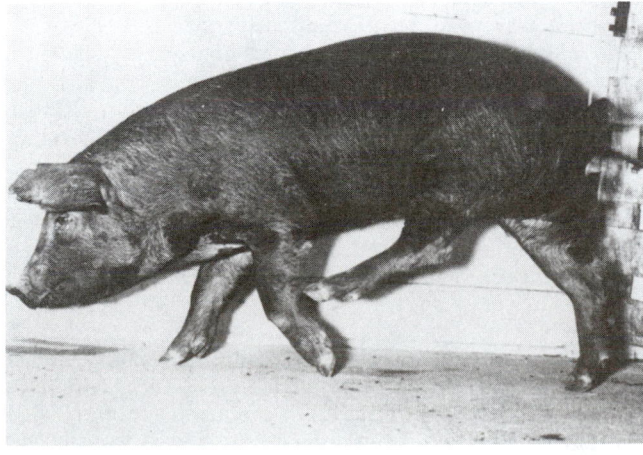

Fig. 4–14. Pantothenic acid deficiency. Note the high, goose-stepping gait. (Courtesy, University of California, Davis)

Fig. 4–15. Riboflavin deficiency. All pigs from this sow that received a riboflavin-deficient ration during gestation were either born dead or died within 48 hours. (Courtesy, Washington State University, Pullman)

## VITAMIN A AND CAROTENE

Vitamin A is required by all farm animals. It is strictly a product of animal metabolism, no vitamin A being found in plants. The counterpart in plants is known as carotene, which is the precursor of vitamin A. Because the animal body can transform carotene into vitamin A, this compound is often spoken of as *provitamin A.*

Carotene is the yellow-colored, fat-soluble substance that gives the characteristic color to carrots and butterfat (vitamin A in nearly a colorless substance). Carotene derives its name from the carrot, from which it was first isolated over 100 years ago. Although its empirical formula was established in 1906, it was not until 1919 that Steenbock discovered its vitamin A activity. Though the yellow color is masked by the green chlorophyll, the green parts of plants are rich in carotene; hence, they have a high vitamin A value. Also, the degree of greenness in a roughage is a good index of its carotene content provided it has not been stored too long. Early cut, leafy green hays are very high in carotene.

Aside from yellow corn, practically all of the cereal grains used in livestock feeding have little carotene or vitamin A value. Even yellow corn has only about $\frac{1}{10}$ as much carotene as well-cured hay. Dried peas of the green and yellow varieties, carrots, yellow sweet potatoes, yellow pumpkins, and squash are also valuable sources of carotene.

On most farms and ranches, adequate vitamin A can be supplied for all classes of animals by allowing access to green pastures in the grazing season and through providing green, leafy hay (ground legumes for swine) not over one year old, or good quality grass or legume silage for winter feeding. The circumstances most conducive to vitamin A deficiencies are (1) extended periods of drought, resulting in the pastures becoming dry and bleached; (2) a long winter feeding period on bleached hays or straws, especially overripe cereal hays and straws; (3) using feeds which have lost their vitamin A potency through extended storage (for example, it has been found that alfalfa leaf meal may lose $\frac{9}{10}$ of its vitamin A value in a year's storage); or, (4) the drylot feeding of swine predominantly on cereals, especially if yellow corn is not included in the ration. There is reason to believe that mild deficiencies of vitamin A, especially in the winter and early spring, are fairly common.

The vitamin A potency (whether due to the vitamin itself, to carotene, or to both) of feeds is usually reported in terms of IU or USP units. These two units of measurement are the same. They are based on the growth response of rats, in which several different levels of test product are fed to different groups of rats, as a supplement to vitamin A-free diet which has caused growth to cease. A USP or IU is the vitamin A value for rats of 0.30 microgram of pure vitamin A alcohol, or of 0.60 microgram of pure beta-carotene. The carotene or vitamin A content of feeds is commonly determined by colorimetric or spectroscopic methods.

## VITAMIN D

Like vitamin A, vitamin D is required by all farm animals and humans. Most of the commonly used feeds contain little or no vitamin D, yet there is no wide-spread need for special supplements containing this factor. Fortunately, the skin of animals and many feeds contain the provitamins in certain forms of cholesterol and ergosterol, respectively, which, through the action of ultraviolet light (light of such short wavelengths that it is invisible) from the sun, is converted into vitamin D. These certain forms of cholesterol and ergosterol themselves have no antirachitic effect.

Of all the known vitamins, vitamin D has the most limited distribution in common feeds. Very little of this factor is contained in the cereal grains and their by-products, in roots and tuber, in feeds of animal origin, or in growing pasture grasses. The only important natural sources of vitamin D are sun-cured hay and other roughages. The chief vitamin D-rich concentrates include sun-cured hay, cod-liver and other fish oils, irradiated cholesterol and ergosterol, and irradiated yeast. Vitamin $D_3$, the animal form, is more active for poultry and should be used instead of vitamin $D_2$, the plant form of the vitamin.

As might be suspected from the preceding discussion, artificially dehydrated hay contains little vitamin D.

The effectiveness of sunlight is determined by the length and intensity of the ultraviolet rays which reach the body. It is more potent in the tropics than elsewhere, more potent at noon than earlier or later in the day, more potent in the summer than in the winter, and more potent at high altitudes. The ultraviolet rays are largely screened out by clothing, window glass, clouds, smoke, or dust.

During cloudy, winter weather, or when animals are confined, vitamin D should always be added to the ration.

## THE B VITAMINS

Originally, the anitberiberi factor was designated as vitamin B, to distinguish it from the antinightblindness factor (called vitamin A) which had been found in carrots and butterfat of milk. As research continued, it was found that vitamin B actually consisted of several factors, the total of which has risen to 11 to date. This had made the terminology rather confusing. As a whole, this group has come to be known as the *B complex.* Some members of this group are referred to by numbers as vitamin B-1, B-2, etc.; others are known by their chemical names; still others have both a number and a chemical designation.

The last new vitamin discovery was made in 1948, when Merck and Company isolated the antianemia factor in liver, which became vitamin B-12.

Under normal circumstances, most, if not all, of the B vitamins (and vitamin K) are synthesized by ruminants (cattle and sheep) in sufficient quantities to meet their requirements. They are formed in the rumen and lower gastrointestinal tract as metabolic by-products of microbial fermentations. In pseudo-ruminants, such as the horse and rabbit, similar microbial synthesis occurs in the cecum and the colon.

Unlike ruminants, however, pigs and poultry have one stomach (and no large cecum like the horse). As a result, they do not synthesize enough of certain of the B vitamins. Consequently, these factors must be provided regularly in the ration in adequate amounts if deficiencies are to be averted.

This means that the nutritionist and the caretaker should provide a dietary source of water-soluble B vitamins on a daily basis for simple-stomached animals—pigs, poultry, and man, but they need not be too concerned about providing the B vitamins to cattle, sheep, goats, horses, and rabbits.

It is also to be emphasized that subacute deficiencies can exist although the actual deficiency does not appear. In fact, borderline deficiencies are both the most costly and the most difficult with which to cope, going unnoticed and unrectified. Such borderline deficiencies result in poor and expensive gains.

Also, under farm conditions one will usually not find a vitamin deficiency which involves only a single vitamin. In other words, deficiencies usually represent a combination of factors, and usually the deficiency symptoms will not be clear-cut.

## UNIDENTIFIED FACTORS

In addition to the vitamins listed in Table 4–6, certain unidentified or unknown factors are important in animal nutrition. They are referred to as *unidentified* or *unknown* because they have not yet been isolated or synthesized in the laboratory. Nevertheless, rich sources of these factors and their effects have been well established. A diet that supplies the specific levels of all the known nutrients but which does not supply the unidentified factors is inadequate for best performance. There is evidence that the growth factors exist in dried whey, marine and packinghouse by-products, distillers' solubles, antibiotic fermentation residues, alfalfa meal, and certain green forages. There is also evidence that at least one unknown hatchability factor is in fish solubles and green forage. Most of the unidentified factor sources are added to the diet at a level of one to three percent.

## WATER NEEDS

Water is one of the most vital of all nutrients. In fact, animals can survive for a longer period without feed than they can without water. Fortunately, under most conditions, it can be readily provided in abundance and at little cost. In addition to what animals drink, water is found in all feeds, ranging from about 10% in air-dry feeds to over 80% in fresh green forage.

Water is one of the largest single constituents of the animal body, varying in amount from 40% in fat hogs to 80% in newborn pigs, 50% in a 1,000 lb steer to 70% in a newborn calf, and 50% in a fat lamb to 80% in a newborn lamb. In general, the percentage of water in the bodies of animals varies with their species, condition, and age. The younger the animal, the more water it contains. Also, the fatter the animal, the lower the water content. Thus, as an animal matures, it requires proportionately less water on a weight basis, because it consumes less feed per unit of weight and the water content of the body is being replaced by fat. This accounts for the fact that gains in older animals are more costly than those in younger animals.

Water performs the following important functions in animals:

1. It is necessary to the life and shape of every cell and is a constituent of every body fluid.

2. It acts as a carrier for various substances, serving as a medium in which nourishment is carried to the cells and waste products are removed therefrom.

3. It assists with temperature regulation in the body, cooling the animal by evaporation from the skin as perspiration.

4. It is necessary for many important chemical reactions of digestion and metabolism.

5. As a constituent of the synovial fluid, it lubricates joints; in the cerebrospinal fluid, it acts as a water cushion for the nervous system; in the perilymph in the ear, it transports sound; and in the eye it is concerned with sight and provides a lubricant for the eye.

Surplus water is excreted from the body, principally in the urine, and to a slight extent in the perspiration, feces, and water vapor from the lungs.

The specific water requirements of each class of animals will receive further consideration in the sections devoted to the respective species. In general, however, under practical conditions, the needs for water can best be taken care of by allowing the animals free access to plenty of clean, fresh water at all times.

## FEEDS

*Feed (or feedstuff) is any ingredient, or material, fed to animals for purposes of sustaining them.* Most feedstuffs provide one or more nutrients, but nonnutritive products may be fed for such purposes as providing flavor, color, or other factors related to palatability, adding bulk, or preserving feeds.

A wide variety of feedstuffs can be, and are, used for animal feeding throughout the world. More than 2,000 different products have been classified as animal feeds, not counting varietal, grade, and stage of maturity differences. However, as shown in Table 4–7, relatively few of these products make up the great bulk of the U.S. feed supply.

### TABLE 4–7
### ANIMAL FEEDS CONSUMED IN THE UNITED STATES[1]

|  | Acreage Harvested | Used for Feed | Yield Per Acre |
|---|---|---|---|
|  | (1,000 acres) | (1,000 tons) | (tons) |
| **Hay** (all) | 60,748 | 149,302 | 2.46 |
| Alfalfa | 25,535 | 84,794 | 3.32 |
| All other hay | 35,215 | 64,508 | 1.83 |
| **Silage:** |  |  |  |
| Corn | 5,829 | 84,468 | 14.5 |
| Sorghum | 424 | 5,157 | 12.2 |
|  |  | (1,000 bu) | (bu) |
| **Grains:** |  |  |  |
| Corn | 59,208 | 7,072,073 | 119.4 |
| Sorghum | 10,604 | 739,249 | 69.7 |
| Barley | 10,057 | 529,530 | 52.7 |
| Oats | 6,925 | 374,000 | 54.0 |
|  |  | (1,000 metric tons) |  |
| **High-Protein:** |  |  |  |
| *Oilseed meals:* |  |  |  |
| Soybean | | 19,410 | |
| Cottonseed | | 1,429 | |
| Sunflower | | 390 | |
| Canola | | 205 | |
| Linseed | | 120 | |
| Peanut | | 100 | |
| *Animal proteins:* | | | |
| Tankage and meat meal | | 2,471 | |
| Fishmeal and solubles | | 481 | |
| Milk products | | 364 | |
| *Grain protein feeds:* | | | |
| Gluten feed and meal | | 1,321 | |
| Distillers' dried grains | | 1,046 | |
| Brewers' dried grains | | 120 | |
| **Miscellaneous feeds:** | | | |
| Wheat millfeeds | | 5,659 | |
| Molasses | | 1,605 | |
| Fats and oils | | 888 | |
| Dried and molasses beet pulp | | 662 | |
| Alfalfa meal | | 554 | |
| Rice millfeeds | | 548 | |
| Urea[2] | | 265 | |
| Miscellaneous by-product feeds | | 1,267 | |

[1]USDA sources. 1987 data.

[2]Estimate for the feed year 1986–87 from *Feed Management*, June, 1987, p. 18.

# CLASSES OF FEEDS

The number of feedstuffs is so great that it is impossible to cover each of them in this book. Rather, we shall classify them, then discuss their nutritional properties by groups. For convenience, the commonly used feeds are herein classified as (1) roughages, (2) concentrates, (3) by-product feeds, (4) protein supplements, (5) minerals, (6) vitamins, (7) special feeds, and (8) additives. For application of this knowledge, readers may refer to the feeding chapters in this book devoted to each class of livestock.

## ROUGHAGES

*Roughages are bulky feeds that are low in weight per unit of volume, contain more than 18% crude fiber, and are low in energy.* They are the natural feeds of all herbivorous animals, including ruminants and horses. Although swine can survive solely on roughages, productivity is too low to be economical.

Roughages include pasture, dry forages, and silages. Fig. 4-1 shows that these three feeds account for 57.4% of all U.S. livestock feeds. Of course, the proportion of roughage-to-concentrate consumption varies widely according to class of animal. As shown in Chapter 2, Table 2-5, sheep and goats head the list of "roughage burners," with 93.8% of their total feed coming from roughage, including pasture. Beef cattle obtain 84.5% of their feed supply from roughages. Poultry consume the least roughage, only an insignificant amount of their feed coming therefrom.

From a feeding standpoint, the following general characteristics of roughages are pertinent, although some well-known roughage can be cited as an exception to each characteristic (for example, on a dry basis well-eared corn silage runs 18% crude fiber, but the TDN is high—about 70%):

1. **Bulk.** They are bulky feeds with low weight per unit of volume.

2. **Fiber and energy.** They contain more than 18% crude fiber, and they are lower in energy than the concentrates.

3. **Digestibility.** They are generally lower in digestibility than concentrates, due to lignin content.

4. **Minerals.** They are generally higher in calcium, potassium, and trace minerals than most concentrates, but phosphorus content is apt to be moderate to low.

5. **Vitamins.** They are higher in fat-soluble vitamins than most concentrates. Legumes are good sources of B vitamins.

6. **Protein.** They are variable in protein content. Legumes may run 20% or more crude protein, whereas other roughages, such as straws, may have only 3 to 4% crude protein.

From an overall nutrition standpoint, roughages may range from very good nutrient sources (such as lush young grass, legumes, and high-quality silage) to very poor feeds (such as straws, hulls, and some browse). Nevertheless, all of them can be used advantageously, provided (1) they are properly prepared and supplemented, and (2) the feeder uses judgement in selecting the species and classes of animals in which the particular roughage is fed.

In the sections that follow, the following groups of roughages will be discussed: pasture, hay, crop residue, silage, haylage (low-moisture silage), green chop, and other roughages.

## PASTURE

*A pasture is an area of land on which there is a growth of forage that animals may graze.* Pasture and rangeland account for 29.2% of the total land area of the United States, including Alaska and Hawaii (see Chapter 2, Fig. 2-7).

Fig. 4-16. Stocker cattle grazing irrigated pasture in the Texas Panhandle. (Courtesy, *The Progressive Farmer*, Birmingham, AL)

Broadly speaking, all U.S. pastures may be classified as either (1) seeded pastures, or (2) native pastures. Although no sharp line of demarcation exists between the two groups, seeded pastures include those which either receive more than approximately 20 in. of rainfall annually or are irrigated. They are the seeded (cultivated) pastures of the Corn Belt, the South, the East, and the irrigated areas, and smaller and scattered moderate-to-high rainfall areas throughout the West. The native pastures include those range pastures which receive less than 20 in. of rainfall annually.

Pasture may be further classified as—

1. **Permanent pastures.** Those which, with proper care, last for many years. They are most commonly found on land that cannot be used profitably for cultivated crops, mainly because of topography, moisture, or fertility. The vast majority of the farms of the United States have one or more permanent pastures, and most range areas come under this classification.

2. **Semipermanent or rotation pastures.** Those that are used as a part of the established crop rotation. These are seeded pastures that are generally used for 2 to 7 years before plowing.

3. **Temporary and supplemental pastures.** Those that are used for a short period, usually annuals, such as Sudan grass, sorghum, millet, rye, barley, wheat, oats, rape, or soybeans. They are generally seeded for the purpose of providing supplemental grazing during the season when the regular permanent or rotation pastures are relatively unproductive.

Pastures vary greatly in quality, depending on type, soil, growing conditions, and stage of maturity. Mature grasses, especially those that are leached and bleached, are low in palatability, digestible energy, protein, and carotene, and in some of the minerals. Grasses are usually adequate in Ca, Mg, and K, but they are apt to be borderline or deficient in P, and they may be low in some of the trace minerals.

Beef cattle, dairy cattle, and sheep compete for many of the grazing areas of the United States. Horses and hogs also relish some of the same kind of pastures, but competition from them is relatively minor.

It is estimated that 84.5% of the total feed supply of all U.S. beef cattle is derived from forage; in season, this means pasture (see Chapter 2, Table 2–5). Good pasture alone will produce 200 to 500 lb of beef per acre annually (in weight of calves weaned, or in added weight to older cattle); superior pastures will do even better.

It is estimated that 58.7% of the nation's milk is produced from forages—pastures, hay, silage, and miscellaneous forages (see Chapter 2, Table 2–5). Good pasture alone will provide cows with sufficient nutrients for body maintenance and for the production of about 20 lb of milk daily; superior pasture will provide for maintenance and more than 40 lb of milk daily.

It is estimated that 93.8% of the total feed supply of all U.S. sheep is derived from forages; for the most part this means pasture (see Chapter 2, Table 2–5). No class of farm animals is so well adapted to the utilization of maximum quantities of pastures as sheep. They are unique in that the vast majority of their young are marketed as milk-fat animals directly off grass.

Prior to 1950, pastures were considered essential for successful swine production. But the importance of pastures for hogs declined with increased knowledge of nutrition and more confinement production. Today, only 4.3% of the U.S. swine feed is derived from forage, including pasture. Most modern swine operations do not use pastures for lactating sows and growing-finishing pigs. However, many operators still use pastures for pregnant sows.

## HAY

*Hay is forage harvested during the growing period and preserved for drying and subsequent use.* It is the most important harvested roughage for U.S. livestock, and ranks third among all livestock feeds, being exceeded only by pasture and corn (see Fig. 4–1). The importance of the nation's crop is further attested to by the fact that more than 60 million acres of hay, producing more than 140 million tons, worth more than $9 billion, are harvested annually.

Fig. 4–17. Alfalfa hay (long), corral fed. Hay is the most important source of energy for U.S. dairy cows. (Courtesy, Ford-New Holland, New Holland, PA)

Hay varies more in nutritive value than any other feed, primarily because of (1) differences in the crop from which it is made, (2) stage of cutting, (3) handling, and (4) possible weather damage during curing. Average quality hay will run 25 to 35% crude fiber and 45 to 55% TDN.

Hays are made from legumes, grasses, or cereal crops. In terms of total tonnage produced annually, alfalfa accounts for approximately 40% of the nation's hay production. Many different kinds of hay make up the other 60% of the country's hay supply; among them, cereal hays made from oats, barley, wheat, and rye, and grass hays made from Bermuda grass, prairie grass, redtop, Johnson grass, orchard grass, and timothy.

The object of haymaking is to (1) harvest the crop at the optimum stage of maturity which will provide the maximum yield of nutrients per acre without damage to the next crop, and (2) cure properly, which involves lowering the water content of the green herbage from 65 to 85% moisture to 20% or less.

Hay is primarily a cattle, sheep, and horse feed, although dehydrated alfalfa may be included in swine and poultry rations.

## CROP RESIDUES

*Crop residues are the portions of crops that are normally left in the field following harvest.* Among such crop residues are: corn stalks and husklage, sorghum stalks, soybean refuse, small grain straws and chaff, and legume and grass seed straws. Crop residues must be fed to the right class of animals, and they must be properly supplemented.

Of all the crop residues, the residue of corn is produced in greatest abundance in the United States and offers the greatest potential for expansion in animal numbers. Corn usually produces an amount of residue equal to the quantity of the grain produced. So, the more than 200 million tons of corn grain produced each year results in more than 200 million tons of corn residue, an amount approximately equal to all other kinds of crop residue combined. Each year, that's more than 200 million tons of potential cow feed, enough to winter more than 150 million dry pregnant cows.

Mature cows are physiologically well adapted to utilize crop residues. Such feeds will meet the daily energy (TDN) needs of dry pregnant cows, but they are slightly deficient in protein, and low in phosphorus and carotene.

## SILAGE

*Silage is fermented forage plants.*

Silage making is one of the 3 common methods of utilizing forage crops, the other 2 methods being pasturing and haymaking. Pasturing is the least expensive of the 3 methods, but it is seasonal in nature. In the spring and early summer, forage plants generally grow faster than they can be utilized by normal grazing, and become dormant in cold weather.

The importance of silage in this country is evidenced by the fact that over 170 million tons of it are made annually. Further, there is ample evidence that silage making is on the increase. It is estimated that 2,000 to 3,000 silos are constructed in the U.S. each year, with tower silos increasing more rapidly than other types.

Most silage in the United States is made from either corn or sorghum, with corn silage far in the lead—over 16 times as much corn as sorghum silage is made. At the present time, it is estimated that 65% of the nation's silage is made from corn and sorghum and 35% from grasses, legumes, and other feeds. Among the other feeds made into silage are small grains, waste from food processing (sweet corn, green beans, and green peas), root crops, and various vegetable residues.

Silage is primarily a beef and dairy feed, where it is used as part or the only roughage in the ration. It is also a good sheep feed. Sometimes it is fed to brood sows. Very little silage is fed to horses.

From 2½ to 3 lb of silage are required to replace 1 lb of hay, due to the lower dry matter content (usually only 25 to 35%) of the silage.

### HAYLAGE (LOW-MOISTURE SILAGE)

*Haylage is made from grass and/or legume that is wilted to 40 to 60% moisture content before ensiling.* Properly made haylage has a pleasant aroma and is a palatable, high-quality feed. Animals usually receive more dry matter and net feed value in haylage than in silage made from the same cut.

Haylage is easy to prepare and preserve in a gas-type silo where air is excluded. But it can be made in a conventional silo provided certain precautions designed to keep out the air are taken.

Haylage is growing in popularity, especially as a dairy feed. Its nutritive value depends on the stage of the growth of the crop when cut and the percentage of dry matter in it.

### GREEN CHOP (SOILAGE)

*Green chop, or soilage is fresh herbage that is cut and chopped in the field, then transported and fed to animals in confinement.* Legumes, Sudan grass, and corn are sometimes used in this manner. With tall growing crops, 50% more feed value may be realized from a given area than can be obtained by any other method of harvesting. However, green chop requires special equipment and harvesting every day. Also, there are harvesting problems in wet weather, and there is an inevitable change in feed quality as the season progresses.

Most green chop is fed to lactating dairy cows, usually in combination with hay or silage because the total intake tends to be greater.

### OTHER ROUGHAGES

Among the other roughages (other than pasture, hays, crop residue, silage, haylage, and green chop) used for livestock are cottonseed hulls, corncobs, sawdust and other wood products, oat hulls, beet tops, root crops, peanut hay, newspapers, and a host of others. When properly (1) used for the right species and class of animal, (2) combined with a high-quality legume roughage, and/or (3) supplemented with the necessary protein, minerals, and vitamins, all of them are excellent feeds. Availability, costs, and results should be the determining factors in their use, just as the economics of the situation should determine the use of any other feed ingredient.

### CONCENTRATES

*Concentrates are feeds that are high in energy and low in fiber (under 18%).*

Many different kinds of concentrate feeds can be, and are, used as animal feeds. Availability and price are the two most important factors determining the choice of concentrates. Consideration of the latter factor—price—necessitates that feeders be keen students of values. They must change the formulations of their rations in keeping with comparative feed prices. Fig. 4–18 shows the tonnage of feed concentrates fed to U.S. livestock and poultry from 1977 to 1987.

Corn is the most common grain fed to livestock; about 9 times as much corn as sorghum is fed in the United States. It is palatable and rich in the energy-producing carbohydrates

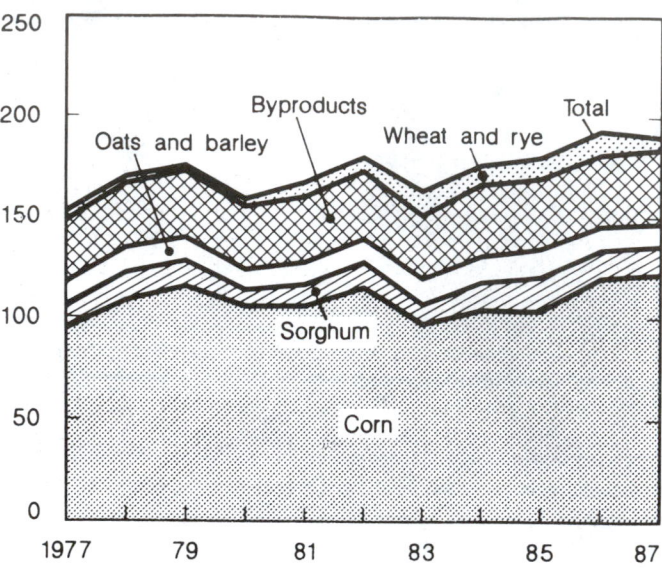

## Feed Concentrates Fed to Livestock and Poultry

Fig. 4–18. This shows the concentrates, including by-product feeds (oilseed meals, animal protein feeds, and milk by-products only), fed to U.S. livestock and poultry, 1977–87. This figure shows (1) that corn is by far the most important livestock feed, (2) that by-products rank second, and (3) that wheat is of minor importance as a livestock feed. (From: 1988 *Handbook of Agricultural Charts,* USDA, p. 97)

and fats (80% TDN), and low in fiber. Also, corn is easily stored, with only moisture and carotene being lost over a period of time. Corn-and-cob meal (consisting of 20 to 25% cob) is excellent for finishing cattle. However, corn has certain very definite limitations—it is low in protein (and in the essential amino acids lysine, methionine, and tryptophan) and calcium.

The grain sorghums are assuming an increasingly important role in livestock feeding, particularly in the fringe areas of the Corn Belt, and in the South and Southwest where moisture conditions are less favorable. New and high-yielding varieties have been developed and have become popular. As a result, more and more grain sorghums are being fed to livestock. The chemical composition of sorghum (milo) is similar to corn except that the protein content is generally higher and more variable. Its feeding value is greatly enhanced by proper processing.

Although corn and sorghum are by far the most common feed grains, such grains as barley, rye, oats, wheat, and triticale, are used in many sections of the United States and Canada. The small grains are excellent when properly prepared and used.

### BY-PRODUCT FEEDS

*By-product feeds are concentrates and roughages other than the primary products from animal and plant processing and from industrial manufacturing.*

Innumerable by-products—both roughages and concentrates—from plant and animal processing, and from manufacturing—are standard and valuable livestock feeds; among them, the following: milling by-products from the cereal grains and oilseeds, root crops (cull potatoes and by-products of potato processing, turnips, mangels, swedes, fodder beets, carrots, and parsnips), dried beet pulp, and beet tops (from sugar beet processing), distillery and brewing by-products, unused bakery products, and by-products from numerous fruits and nuts. (Also, see section on "Protein Supplements," which follows.)

Fig. 4–19. Drylot steers eating field cured dry beet tops. (Courtesy, The Great Western Sugar Co., Denver, CO)

As is true of any ration ingredient, the requisites to effective and profitable use of each by-product feed are (1) that it be bought at a favorable price, nutritive composition considered; (2) that its proximate composition be known, and that it be incorporated in a balanced ration; (3) that it be palatable and consumed in adequate quantity; and (4) that it not adversely affect carcass quality, particularly from the standpoint of harmful chemical residues from pesticides applied to crops. Generally speaking, the use of by-product feeds calls for ingenuity and experience in handling them, special knowledge relative to their nutritive qualities and use in balanced rations, and relatively high labor costs. As a result, many feeders are not interested in using them, whereas others find it a lucrative business.

## PROTEIN SUPPLEMENTS

*Protein supplements are feedstuffs that contain more than 20% protein or protein equivalent.* At least 23 amino acids have been identified and may occur in combinations to form an almost limitless number of proteins.

High-protein feeds are usually named and classified according to their origin and method of processing. On the basis of origin, they are usually grouped into two general categories as follows:

1. **Animal proteins.** Animal protein supplements are derived from inedible tissues from meat-packing or rendering plants, from surplus milk or milk products, and from marine sources. They include proteins from meat, fish, poultry, eggs, milk, and their products. With hogs and chickens, one of these protein sources was formerly a must. With the discovery and general availability of vitamin B–12, high-protein feeds of animal origin became less essential for swine and poultry.

Not all animal proteins are of high quality. For example, feather meal, a by-product from poultry processing, runs about 85% protein, but protein is very poorly digested by monogastrics and must be hydrolyzed for good utilization. Even then, not more than 3 to 5% should be used in swine rations.

2. **Plant proteins.** This group includes the common oilseed by-products—soybean meal, cottonseed meal, linseed meal, peanut meal, safflower meal, sunflower meal, rapeseed meal (canola meal), and coconut (or copra) meal. They vary in protein content and feeding value, depending on the seed from which they are produced, the amount of hull and/or seed coat included, and the method of oil extraction used.

In practical feeding operations, hogs and chickens are usually provided with some protein feeds of animal origin in order to supplement the proteins found in grains and forages. Protein quality is less important with ruminants and pseudo-ruminants, because of microbial synthesis. In feeding mature cattle, sheep, and horses, a safe plan to follow is to provide a liberal supply of high-quality legume hay or lush young pasture along with the concentrates. Also, the quality of the proteins in a ration is likely to be higher if a variety of feeds is combined.

In addition to the oilseed meals, numerous good commercially manufactured protein supplements are available. Usually, they are prepared for a particular class of livestock. They are generally blends of animal and vegetable protein ingredients, with urea added for ruminants. They may also include minerals, vitamins, and/or antibiotics.

### High-Protein Feed Use

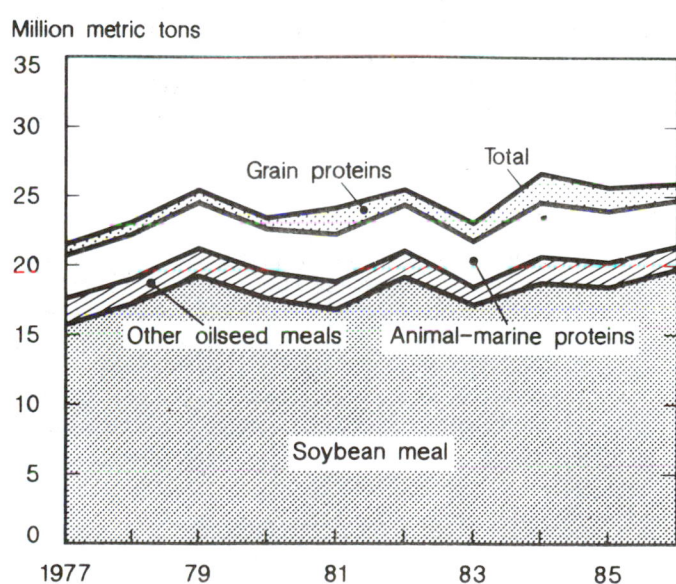

Fig. 4–20. This shows the high-protein feeds fed to U.S. livestock and poultry, 1977–86, in 44% protein soybean meal equivalent. The following explanation is pertinent: Grain proteins include gluten feed and meal, and brewers' and distillers' dried grains; animal-marine proteins include tankage, meat meal, marine by-products, and milk products; other oilseed meals include cottonseed, linseed, peanut, sunflower, and copra. This figure shows that soybean meal is by far the most important high-protein feed. (From 1988 *Handbook of Agricultural Charts*, USDA, p. 97)

## NONPROTEIN NITROGEN (NPN) SOURCES

Certain nonprotein nitrogen sources may be substituted for all or much of the supplemental protein required in most ruminant rations, provided such rations are adequate in minerals and readily available carbohydrates. Among such products are urea, ammoniated molasses, ammoniated beet pulp, ammoniated cottonseed meal, ammoniated citrus pulp, and ammoniated rice hulls.

The rumen microorganisms—which are a low form of plant life and are able to use inorganic compounds much like plants utilize chemical fertilizers—build proteins of high quality in their cells from sources of inorganic nitrogen that nonruminants cannot use. Since the life-span of these microorganisms is short, further on in the digestive tract the ruminant digests the bacteria and obtains good protein therefrom. In ruminant nutrition, therefore, even such nonprotein sources of nitrogen as urea and ammonia have a protein replacement value. An exception is the very young ruminant in which the rumen and its ability to synthesize are not yet well developed.

## Urea

Urea is a white, crystalline, odorless, nonprotein nitrogen compound of the formula $N_2H_4CO$. It is manufactured in chemical plants that produce anhydrous ammonia by fixing some of the nitrogen of the air, some of the ammonia gas is combined with gaseous carbon dioxide to produce the white crystalline solid urea which is quite stable. In addition to feeds, urea is used as a fertilizer, either dry or in solutions, and in making plastics. Also, it is noteworthy that urea occurs as the principal end product of nitrogen metabolism in nearly all mammals; it is found in the urine of all farm animals and humans.

Initially, the protein equivalent value of feed grade urea was 42% (nitrogen) times 6.25 (common protein factor), or 262% protein. Today, more concentrated 45% nitrogen (45 × 6.25 = 281% protein) urea has replaced most of the 42% grade, at a lower unit cost.

Approximately 265,000 metric tons of urea are fed annually in the United States, as a source of protein for cattle, sheep, and goats. In recent years there has been increasing interest in feeding urea to ruminants, due primarily to the following circumstances:

1. **Shortage of oil meal proteins.** The scarcity and high price of oil meal protein feeds, due mainly to increasing use of these products for (1) human consumption, and (b) monogastric animals.

2. **Progress in fundamental ruminant nutrition.** Through basic studies, scientists have established many of the nutrient requirements of rumen microorganisms.

These factors, plus meeting the needs of a rapidly expanding human population, are likely to continue to accentuate the interest of feed manufacturers and livestock producers in utilizing urea and other nonprotein nitrogen sources.

Urea may constitute up to one-third of the total protein of the ration of ruminants, provided additional energy is added in the form of molasses or grain to compensate for the lack of energy in the urea, in order to feed properly the rumen bacteria. By total protein is meant the protein intake of the entire ration—including forage, grain, and protein supplements.

Common guidelines relative to the use of urea for beef cattle are given in Table 4–8.

### TABLE 4–8
### COMMON GUIDELINES TO THE USE OF UREA FOR CATTLE

| | For Finishing Cattle | For Grower (stocker) Cattle | For Wintering Pregnant and Lactating Cows |
|---|---|---|---|
| Percent of total protein in ration from urea ....... (%) | 33⅓ | 25.0 | 25.0 |
| Maximum urea/animal/day .................... (lb) | 0.22 (100 g) | 0.15 (68 g) | — |
| Percent of urea, by weight, of total air-dry feed consumed ............... (%) | 1.0 | 1.0 | 1.0 |
| Percent of urea, by weight, of concentrate mix (grain plus protein supplement)[1] .................. (%) | 2.0–3.0 | 3.0 | 3.0 |
| Percent of urea, by weight, of the protein supplement ............... (%) | 20–30[2] | 10.0[3] | 10.0 |
| Percent of supplemental nitrogen in high-protein supplement from urea[4] .... (%) | 60–90[5] | 30.0 | 30.0 |
| Pounds of urea added/ ton of corn silage at ensiling time[6] ........... (lb) | 10.0 (4.5 kg) | 10.0 (4.5 kg) | 10.0 (4.5 kg) |

[1]Feed intake may be depressed if over 1% is used. Yet, many beef producers are successfully using 2%.

[2]High-urea supplements are best fed in complete mixed rations, which are *thoroughly* mixed. Supplements containing 20–30% urea require extreme caution when being hand-fed.

[3]A protein supplement containing 10% urea provides 28.1% of the protein equivalent (281% × .10) from nonprotein nitrogen.

[4]This means that as much as 60–90% of the protein value of the supplement may come from nonprotein sources. However, because such a supplement will constitute only 2–5% of the total ration fed, the first rule of thumb given in Table 4–8 still applies; namely, only ¼–⅓ of the total protein in the ration will be supplied from a nonprotein source.

[5]In a feedlot ration, this may be equivalent to 25–40% of the total nitrogen from all sources.

[6]On a dry matter basis, corn silage ensiled at the well-dented stage contains about 8% protein. The addition of 10 lb of urea per ton (or *5 kg/1,000 kg*) of silage increases the protein content from 8–13%. However, there is loss of flexibility in feeding such a ration, and the rate of gain will be less than can be secured from higher energy, more dense rations. Also, it is extremely important that the urea be well mixed in the silage; otherwise, there is hazard of toxicity.

### Slow-Release Urea Products

Several products in which urea is bound in a slow-release complex have been developed in recent years; among them, urea combined with starch from grain, and urea combined with the sugars in molasses through heat and chemical treatment. These products are designed to decrease the solubility of urea in the rumen and thereby slow the release of ammonia. Slow ammonia release, or a more uniform ammonia level in the rumen throughout the day, is desirable, especially for urea used in low-energy rations. Additionally, there should be less danger of urea toxicity from overconsumption with slow-release products.

### Single-Celled Protein (SCP)

*Single-cell protein (SCP) is protein obtained from single-cell organisms, such as yeast, bacteria, fungi, and algae, that have been grown on specially prepared growth media.*

Some single-celled protein types can be useful sources of protein and vitamins for animal feeding. The safety of these feeds depends on the organisms selected, the quality of substrate used, and the conditions of growth. Of course, yeast and bacteria have been used for centuries in the baking, brew-

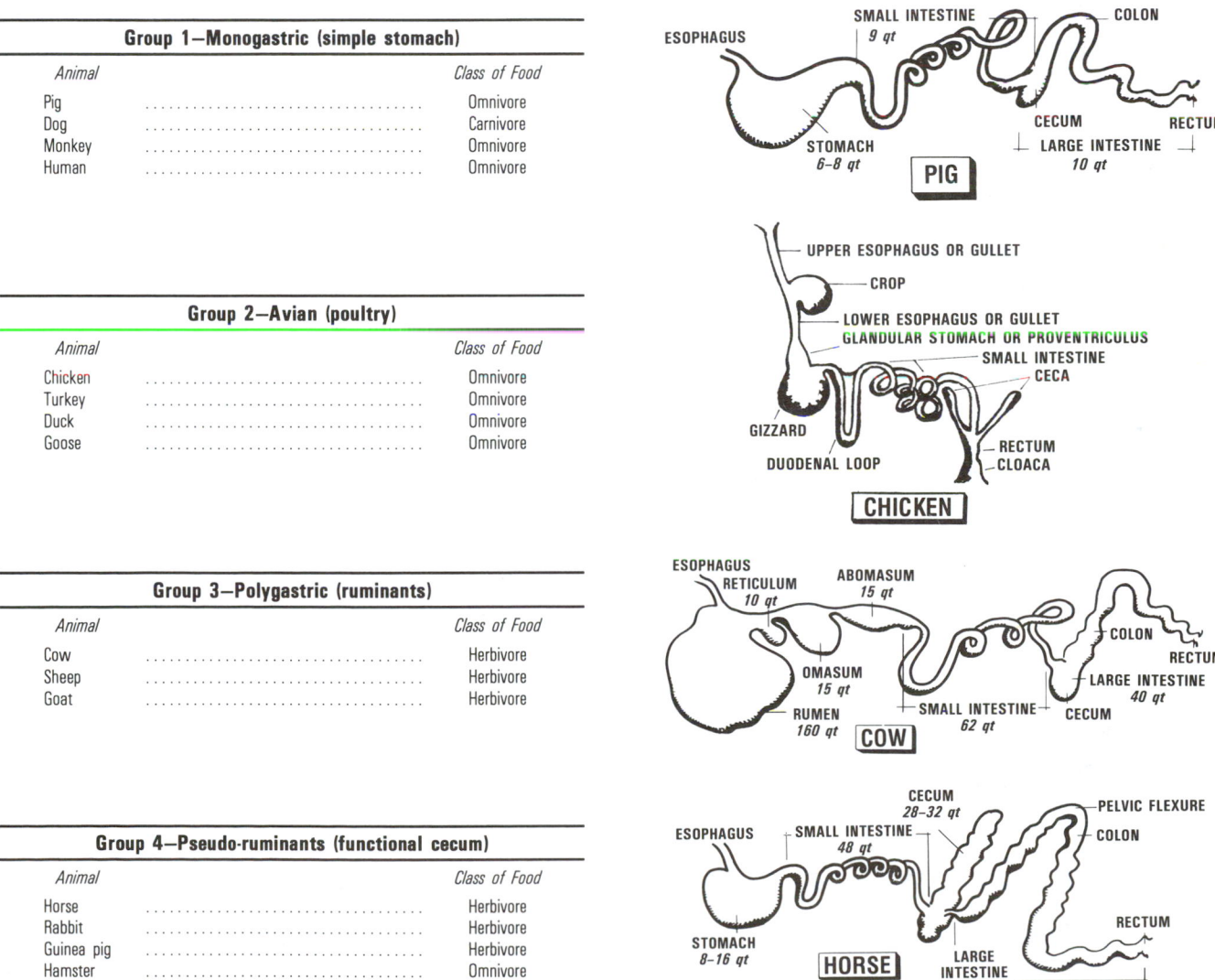

ing, and distilling industries, in making cheeses and other fermented foods, and in storage and preservation of foods.

Dried brewers' yeast, a residue from the brewing industry, and Torula yeast, resulting from the fermentation of wood residues and other cellulose sources, have long been used as animal feeds.

Bacteria grow faster than yeasts under favorable conditions, doubling their mass in a matter of minutes, rather than hours. Dried bacterial cells contain at least 55% protein.

Various bacteria and yeasts can be selected and cultured to grow on organic wastes. These include animal wastes, sewage, many different chemical residues from industrial plants, petroleum by-products, sawdust, and other fibrous residues. Petroleum companies in several countries have built factories to produce bacterial protein for the animal feed market. In this process, certain yeasts that can use the carbon from petroleum by-products as a source of energy are selected. A culture of yeast is propagated under controlled conditions, by blending a nitrogen source (such as ammonia), a suitable mixture of inorganic minerals, and petroleum hydrocarbons. The yeast culture multiplies at a prodigious rate and produces large yields. With improvement in technology, large-scale production of yeast protein will likely become an important animal protein supplement.

Algae, single-celled plants which contain about 50% protein on a dry basis, offer an attractive possibility as a protein source. They synthesize proteins by the use of solar energy. Preliminary results with cultivated freshwater algae indicate that they will produce about 10 times as much protein per unit of land area as soybeans.

Considerable research is in progress to convert manure from poultry and other animals through bacterial fermentation into animal protein feed. This recycling process could produce much protein and help solve a pollution problem.

## MINERAL SUPPLEMENTS

*Mineral supplements are rich sources of one or more of the inorganic elements needed to perform certain essential body functions.*

Only the specific minerals that are deficient, and in the quantities necessary, should be provided. Excesses and mineral imbalances should be avoided. It follows that needed supplementary minerals will vary according to the animal species, age, production, ration, and the mineral content of the soils and crops in the area where grown. Most feeds provide minerals in addition to basic organic nutrients, although fat and urea are marked exceptions. Nevertheless, most rations require more concentrated sources of one or more mineral elements.

Generally, the macrominerals of concern are NaCl (common salt), Ca, P, Mg, and sometimes S; and the trace elements that may be deficient are Cu, Fe, I, Mn, and Zn, and in some places Co and Se.

Needed mineral mixes may be either home mixed or provided by a commercial mineral. Commercial mineral mixes are minerals mixed by manufacturers who specialize in the commercial mineral business, either handling minerals alone or in combination with a feed business. Because mineral mixes have become more complicated with the recognition of the importance of trace elements and interrelationships, and because most farmers and ranchers do not have the equipment with which to mix minerals properly, commercial minerals are finding a place of increasing importance in all livestock feeding.

Minerals may be either incorporated in the ration or self-fed.

When livestock are fed a mixed feed, totally or in part, the needed minerals are generally incorporated into the ration in keeping with known requirements. This is usually accomplished by adding 0.25 to 0.50% trace mineralized salt to the total ration, plus calcium and phosphorus (and any other minerals that are in short supply) as needed to balance the ration. In addition, where the lower level of salt (0.25%) is incorporated in the ration, trace mineralized salt is usually provided free choice.

Where animals are fed an unmixed ration or are on pasture, minerals may be provided as follows:

1. **Where animals are on liberal grain feeding.** Provide free access to a 2-compartment mineral box, with (a) trace mineralized salt in one side, and (b) in the other side, a mixture of ⅓ trace mineralized salt (salt included for purposes of palatability), ⅓ defluorinated phosphate or steamed bone meal, and ⅓ ground limestone or oystershell flour.

2. **Where animals are primarily on roughage (pasture, hay, and/or silage).** Provide free access to a 2-compartment mineral box, with (a) trace mineralized salt in one side, and (b) in the other side, a mixture of ⅓ trace mineralized salt (salt included for purposes of palatability), and ⅔ defluorinated phosphate or steamed bone meal.

As noted, no limestone or oystershell flour (source of calcium only) is needed in the latter mix, because forages are generally more deficient in phosphorus than in calcium.

## VITAMIN SUPPLEMENTS

*Vitamin supplements are rich synthetic or natural feed sources of one or more of the complex organic compounds, called vitamins, that are required in minute amounts by animals for normal growth, production, reproduction, and/or health.*

Formerly, a wide variety of feed ingredients were added to livestock rations for their vitamin content. But it was found that the vitamin concentration of feedstuffs varied tremendously, being affected by plant species and part (leaf, stalk, or seed), harvesting, storing, and processing. Generally speaking, vitamins are easily destroyed by heat, sunlight, oxidation, and mold growth. So, today, nutritionists rely on vitamin supplements, which in many cases are chemically pure sources that need to be used only in very minute amounts. In modern feed formulation, premixes often represent the common sense approach to providing vitamins.

For adult ruminants, vitamins A, D, and E are of concern, with A being the one most likely to be deficient. Under ordinary circumstances, ruminants synthesize adequate B vitamins, and vitamins C and K. Unless they are kept indoors, they usually receive sufficient exposure from direct sunlight to meet their needs for vitamin D.

Because of the greater prevalence of confinement feeding, along with limited gastrointestinal synthesis, swine are more apt to suffer from vitamin deficiencies than ruminants. Under practical conditions, special consideration should be given to the need for supplementing swine rations with the following vitamins: A, D, E, riboflavin, niacin, pantothenic acid, B–12, and choline.

Vitamins A, D ($D_3$ for poultry), B–12, and riboflavin are commonly low in poultry rations. Also, it is in the nature of good insurance to add the following vitamins to poultry rations, as they may be deficient: E and K, and the rest of the B vitamins (in addition to vitamin B–12 and riboflavin, already mentioned). Because only limited amounts of the water-soluble vitamins can be stored, they should be fed regularly in the ration in adequate amounts.

## SPECIAL FEEDS

Among the special feeds used by livestock producers are molasses, and fats and oils.

### MOLASSES

Molasses (including cane or blackstrap, beet, citrus, and wood molasses) is extensively used as a livestock feed. 1,605,000 metric tons are used for animal feeds in the United States, annually (see Table 4–7).

Cane and beet molasses are by-products of the manufacture of sugar from sugarcane and sugar beets, respectively. Citrus molasses is produced from the juice of citrus wastes. Wood molasses is a by-product of the manufacture of paper, fiberboard, and pure cellulose from wood; it's an extract from the more soluble carbohydrates and minerals of the wood material. Cane or blackstrap molasses is by far the most extensively used type.

When used at levels of 10 to 15% of the ration, molasses has about three-fourths the energy value of corn. However, molasses has added value as an appetizer, to reduce dustiness of a ration, as a binder for pelleting, to stimulate rumen microbial activity, and as a source of unidentifiable factors. Also, cane molasses is a good source of certain minerals.

*Brix is a term used to express molasses quality, as reflected by the relative level of sugar present.* It is arrived at by first determining specific gravity. Then by use of conversion tables, the degrees Brix, or level of sucrose present is obtained.

The different types of molasses may also be available in dehydrated form.

### FATS AND OILS

Feeding of fats was promoted in an effort to find a profitable outlet for surplus packinghouse and rendering plant fats. For the most part fats were formerly used for soapmaking, but they are not used extensively in detergents. Thus, with the rise in the use of detergents in recent years, they became a "drug on the market."

In 1987, a total of 888,000 metric tons of fats and oils were used in animal feeds in the United States.

Animal and vegetable fats seem to be equally effective additions to rations; thus, selection should be determined solely by comparative price. Ordinarily, animal fats are much cheaper than such vegetable oils as soybean oil or cottonseed oil. Vegetable oils are generally priced out of the animal feed market, for use in margarine, paint, and other industrial uses.

Several different fat products are used as animal feed; among them, acidulated soap stock (foots), tallows, greases (white and yellow), blended feeding fat, house grease, brown grease, sewer grease, and modified yellow grease. Each of them should be bought by specifications and guarantees.

Fat serves the following three practical functions when added to livestock rations:

1. **It increases the caloric density of the ration.** Since fat contains approximately 2¼ times as much energy as soluble carbohydrate, it is possible to increase the energy content with little increase of the bulk of the ration. Thus, with the same feed intake, energy intake is higher.

2. **It controls dust.** Thus, the addition of one to two percent fat materially lessens the dust involved in feed processing. Also, it is well known that dusty rations are not consumed readily by animals; hence, fat enhances consumption from this standpoint.

3. **It lessens the wear and tear on feed-mixing equipment.** This is important, because both break-downs and new equipment are costly.

If the price is favorable, fat may be added to rations at the following levels: for swine and poultry, 5 to 10%; and for cattle, 2 to 6%. Higher levels of fat usually result in drastically lowered feed consumption. When fed at the levels recommended above, the energy value of fat is approximately 2¼ times that of the grains. When corn is the major source of grain, fat additions can be expected to be less useful than with the small grains. This is understandable when it is realized that corn contains approximately 4% fat as compared to 1 to 1½ for the other feed grains.

Higher levels of fat than indicated above may be used for young ruminants in milk replacers; depending on the purpose, replacers may contain 15 to 30% added fat.

One of the more exciting areas, which is still in the experimental stage, is the so-called rumen protected fat system, a research development of Australia. In this system, the fat is emulsified with a protein, then the protein is treated with formaldehyde. The product cannot be digested in the rumen due to the formaldehyde linkage on the protein. However, the pH of the abomasum is such that the formaldehyde linkage of the protein is broken, with the result that the product will then be digested similarly to what takes place in monogastric animals. Studies reveal that the utilization of fat by the ruminant can be improved by the use of this system. Also, and most significant, by the use of the protein protected fat system it is possible to alter the ratio of the saturated to the unsaturated fatty acids in beef depot fat. The same is true of butterfat.

Through this technique, the digestion of fat in the ruminant is delayed. It bypasses the rumen, or first stomach, where, normally, bacteria break it down and turn it into saturated fat. Instead, it isn't digested until it reaches the cow's fourth stomach, where the polyunsaturated fats are released, digested, and absorbed into saturated fats. Imagine choice polyunsaturated beef! The idea is to reduce a person's level of cholesterol, suspected by some, but not proven, to be the cause of certain types of heart disease.

## ADDITIVES, IMPLANTS, AND INJECTIONS

More than 1,000 drug products are approved by the Food and Drug Administration (FDA) for use by livestock and poultry producers. This includes additives, implants, and injectables, along with other drugs that are used to fight disease and protect animals from infections. Two other statistics which point up the important role of drugs in animal production are: (1) 8 out of every 10 animals raised for food in the United States receive some drugs during their lifetime; and (2) chemicals that regulate growth, modify the rumen's activity, and/or improve feed efficiency increase U.S. meat, milk, and egg production approximately 15% each year. Used properly, these drugs enable livestock producers to provide safe and wholesome meat, eggs, and milk to consumers at lower costs than would otherwise be possible. Used improperly, however, these drugs can be hazardous to consumers.

Consumers are little bothered about what goes on their backs, but they are much concerned about what goes in their stomachs. While they enjoy the price and supply benefits of modern food production technology, they want to be assured of the safety of the food they eat.

Thus, livestock producers have the task of choosing the right drug(s) to maximize rate and efficiency of production, while, at the same time, observing FDA regulations and protecting the consumer. Under such circumstances, they should carefully analyze all the information presented by each company in support of its product. Also, they should study the results of unbiased experimental work, as reported in both scientific literature and popular articles; and they should sound out reliable users of the product. Finally, in the United States, they must comply with FDA regulations.

Feed additives and implants constitute a diverse group—so diverse that it is difficult to classify all of them as to mode of action or function. For this reason, some are grouped whereas others are merely listed alphabetically in the sections that follow.

• **Abortifacients**—*An abortifacient is a drug or other agent that induces abortion.* In the livestock industry, the primary use of abortifacients is to abort feedlot heifers. Prostaglandins and prostaglandin analogues are the abortifacients of choice during the first 150 days of pregnancy; beyond 150 days pregnancy, additional products, such as dexamethasone or estradiol, may be used. By pregnancy testing and the use of an abortifacient, the termination of early pregnancy in feedlot heifers can be brought about.

• **Activated carbon (charcoal)**—In recent years, sporadic cases of pesticide contamination in livestock have been reported. In a number of these cases, activated carbon, or charcoal, has been fed to the contaminated animals to facilitate excretion of the pesticides. It has been suggested that the activated carbon absorbs the pesticide as it is excreted into the digestive tract and recycled in the bile.

• **Additives that enhance the market value**—Traditional methods of breeding and selecting animals to meet changing consumer demands take time—lots of time. So, the race is on to speed up this transition, with genetic engineering and additives leading the way. Moreover, it no longer suffices to have additives that stimulate growth and improve feed efficiency. The additives of tomorrow must also improve the composition and quality of the product produced—the meat, milk, and eggs. Among the types of product-enhancing additives currently being used or tested experimentally are the following:

1. **Xanthophylls and carotenoids in poultry.** Many consumers believe that a deep yellow color of broiler skin/shanks and egg yolks is indicative of top quality. Xanthophylls or carotenoid additives are commonly used for this purpose.

2. **More lean/less fat.** Several products for the purpose of producing more muscle/protein and less fat in beef cattle, lambs, hogs, and broilers are being tested.

3. **Lower cholesterol.** Consumers are cholesterol conscious; so, scientists are developing and testing products that will lower the cholesterol in meat, milk, and eggs.

• **Additives that physically aid digestion.** In some types of livestock—notably poultry and ruminants—the physical characteristics of the feed can markedly alter its digestibility. Two products which have received considerable attention as physical aids to digestion are grit for poultry and roughage substitutes for ruminants.

1. **Grit.** Since poultry do not have teeth to facilitate grinding of feed, most grinding takes place in the thick-muscled gizzard. The more thoroughly feed is ground, the more surface area is provided for digestion and subsequent absorption. Hence, when hard, coarse, or fibrous feeds are fed to poultry, grit is sometimes added to supply additional surface for grinding within the gizzard. Additionally, grit serves to break down the ingested feathers and litter which can sometimes lead to gizzard impaction.

2. **Roughage substitutes.** *Roughages are bulky, coarse feeds which are high in fiber (cellulose and related compounds) and thus low in digestibility*; such as hay, straw, silage, wheat bran, corncobs, and cottonseed hulls. In ruminants, roughage promotes rumination and rumen function and the production of milk of normal fat content.

Polyethylene and oystershell have been tested relative to their respective efficacies as roughage substitutes. While several studies have indicated some improvement from the use of these products, some nutritionists feel that they are of little or no value. The most common practice in feeding high-energy feeds is to incorporate in the ration some source of natural roughage, such as hay, cottonseed hulls, almond hulls, or ground corncobs.

• **Anthelmintics (wormers; vermifuges)**—*Anthelmintics are drugs used to control worms.*

Knowing what kind of worm(s) is present in an animal is the first requisite to the choice of the proper anthelmintic. Since no one drug is appropriate or economical for all conditions, the next requisite is to select the right one; the one which, when used according to directions, will be most effective and produce a minimum of side effects for the animal treated.

Each livestock establishment should, in cooperation with the local veterinarian, evolve with a parasite control program and schedule. Also, a schedule of treatments should be prepared based on knowledge of the life cycle of the various parasites.

From time to time, new vermifuges, or wormers, are approved and old ones banned or dropped. When parasitism is encountered, therefore, it is suggested that the producer obtain from local authorities the current recommendation relative to the choice and concentration of the vermifuge to use. This information can be obtained from a county extension agent, entomologist, veterinarian, or agricultural consultant.

• **Antibiotics**—*Antibiotics are substances which are produced by living organisms (molds, bacteria, fungi, or green plants) and which have bacteriostatic or bactericidal properties.* They are the most widely used of the microbial drugs.

In addition to their use as growth stimulators, antibiotics are used as nutritional stimulants to promote better feed efficiency in ruminants and swine, and to increase egg production, hatchability, and shell quality in poultry. They are also added to feed in substantially higher quantities to remedy pathological problems.

The following six mechanisms by which antibiotics increase rate of gain and feed efficiency have been suggested:

1. **Disease control.** There is evidence that antibiotics exert a "disease defense effect" by suppressing microorganisms which might otherwise produce subclinical diseases in the animals.

2. **Nutrient-sparing effect.** When antibiotics are fed, the populations of microorganisms in the animal's digestive tract change. Studies indicate that antibiotics have a sparing effect on some vitamins and amino acids.

3. **Metabolic effect.** According to this theory, there is a *metabolic effect*, by which the antibiotic affects the body functions.

4. **Feed and water intake.** Antibiotics usually increase feed and/or water intake.

5. **Toxic waste products or toxins.** Antibiotics may inhibit the growth of organisms which produce toxic waste products or toxins.

6. **Digestion and absorption.** Antibiotics may improve the digestion and subsequent absorption of certain nutrients.

More than likely, all six modes of action apply to antibiotics in general, but not to any specific antibiotic—for antibiotics differ. Also, there are dose-related responses. Since antibiotics are most effective in stressful conditions where animals are more likely to be exposed to disease and more susceptible to it, the disease-control effect is probably the most important of the six modes of action.

Antibiotics are used in the following two ways in livestock and poultry production:

1. **Low levels in feeds.** Low, subtherapeutic doses of antibiotics are included in livestock and poultry feeds to increase growth rate, and/or feed efficiency, and help prevent bacterial diseases.

2. **High (therapeutic) levels in feeds.** High levels of antibiotics are used to treat disease, just as they are in human medicine. Used for short periods of time, high levels have been quite effective in treating anaplasmosis and controlling shipping fever in cattle, bacterial enteritis in swine, and respiratory diseases, diarrhea, fowl cholera, typhoid, and breast blisters in poultry. Also, high levels of antibiotic have been useful for preventing and treating stresses associated with transporting animals and their adjustment to a new environment.

Most antibiotics are approved for cattle, swine, chickens, and turkeys. Few are approved for sheep, horses, rabbits, ducks, pheasants, and quail. No approvals have been given for goats, geese, dogs, and cats. The reason for this situation is the cost of obtaining approval in relation to potential sales.

The list of antibiotics and antibiotic combinations approved as feed additives is long—and growing longer; hence, space limitations in this book will not permit a listing of each of them, along with giving their FDA status, dosage, and indications of use. For up-to-date and detailed information on this subject, feed manufacturers and livestock producers (1) are referred to the *Feed Additive Compendium,* published annually and updated monthly, by The Miller Publishing Company, Minnetonka, MN; or (2) should seek the advice of their veterinarian, nutritionist, and/or county extension agent.

• **Antioxidants**—*Antioxidants are compounds that prevent oxidative rancidity of polyunsaturated fats.*

Many of the most useful products in the feed industry are readily subject to autoxidation; among them, fat sources, fish by-products, animal-rendered products, essential vitamins, and pigmentation sources. Feeds which are high in unsaturated fatty acids are especially prone to autoxidation and subsequent rancidity.

Oxidation reactions are accelerated by high temperature, light (ultraviolet and blue), ionizing radiation, peroxides (including oxidized fats), lipoxidase enzymes, organic ion catalysts (hemoglobin), and trace metals (copper).

Oxidation reactions are inhibited by refrigeration, exclusion of light, exclusion of oxygen, destruction of enzymes, metal deactivators, and antioxidants.

• **Arsenicals**—Arsenilic acid and sodium arsanilate are the most widely used growth-promoting arsenicals. The FDA-approved for use alone or in certain drug combinations for chickens, turkeys, and swine. When used alone according to directions, they increase rate of gain and improve feed efficiency of chickens, turkeys, and swine; improve pigmentation in growing chickens and turkeys; increase egg production in layers; and prevent coccidiosis in chickens and dysentery in swine.

• **Bloat control products**—Three products are approved by FDA for bloat control; namely, poloxalene (trade name, Bloat Guard), oxytetracycline (trade names, Terramycin or Neo-Terramycin), or laureth-23 (Enproal Bloat Blox).

Also, two ionophores (lasalocid [trade name, Bovatec] and monensin [trade name, Rumensin]) inhibit gas formation and decrease methane production in the rumen, thereby reducing feedlot bloat. (See section on "Ionophores" in this chapter.)

• **Buffers**—*Buffers are the substances which lessen the change in hydrogen ion concentration produced by adding acids or alkalis to ruminant rations.* The counterpart substances for humans are called antacids.

When used in beef, dairy, and sheep rations, a buffer chemically maintains a balanced pH in the animal's digestive system.

High-concentrate rations, smaller feed particle size, fermented feeds, and rapid shifts from high-roughage to high-concentrate rations all lower the pH of the rumen. Since many of the microorganisms in the rumen cannot tolerate low pH concentrations, the normally heterogeneous, balanced population of the microbes becomes skewed, favoring the acidophilic (acid-loving) bacteria. This condition often leads to upsets. However, the addition of feed buffers can prevent dramatic changes of pH in the rumen, thereby stabilizing the microbial population.

Many buffer products are available. However, the following six products make up the most common ingredients: sodium bicarbonate, magnesium oxide, sodium bentonite, sodium sesquicarbonate, limestone, and whey.

• **Chemotherapeutics**—*Chemotherapeutics are organic compounds with bacteriostatic or bactericidal properties similar to those of antibiotics.* But, unlike antibiotics, these compounds are produced chemically rather than microbiologically. The chemotherapeutics are used primarily for disease control. However, arsenicals, carbadox, furaxolidone, and roxarsone are also used as growth promotants and to improve feed conversion efficiency.

The arsenicals and the sulfas, two of the most widely used chemotherapeutics, have been around for a very long time. Pertinent information about each of them follows:

1. **Arsenicals.** Various compounds containing arsenic, when used at carefully-controlled levels, have been found to increase rate of gain and improve feed efficiency in chickens, turkeys, and swine; and to prevent coccidiosis in chickens and turkeys and prevent dysentery (bloody scours) in swine.

2. **Sulfonamides.** The therapeutic use of the sulfonamides precedes that of antibiotics. The coming of the antibiotics, many of which were more specific in action and easier to administer than the sulfas, markedly reduced the use of the sulfas. In time, however, bacteria emerged that were resistant to many of the antibiotics, which led to a resurgence in the use of the sulfonamides, in feed or water.

Although many of the antibiotics and chemotherapeutics have similar properties, the expertise of the nutritionist should be sought when increased growth rate and improved feed efficiency is the primary objective, and the veterinarian should be consulted where disease control and/or treatment is involved. Correctly (1) determining the objective and/or problem and (2) matching the additive to the objective and/or problem are the first and most important decisions when choosing a feed additive.

It is the responsibility of the producer to comply with regulations and use chemotherapeutics properly. To this end, the producer should—

1. Read and follow directions on the tag or label.
2. Follow the withdrawal period.

• **Copper (CU)**—Copper is widely used as a swine feed growth additive in Europe, where numerous experiments have shown growth response and increased feed efficiency comparable to those obtained with antibiotics. A supplemental level of 175 to 200 ppm copper is recommended.

Copper carbonate, copper chloride, copper oxide, copper sulfate, copper glycinate, and copper methionine are effective sources of copper, with copper sulfate the preferred form. They cost less than antibiotics.

• **Electrolytes**—*An electrolyte is a substance which when dissolved in water enables the resulting solution to conduct an electric current.* The most common electrolytes in the animal body are salts of such minerals as sodium, potassium, magnesium, calcium, phosphorus, sulfur, and chlorine.

The most commonly used electrolyte solution is 0.9% sodium chloride, which is also called *physiological saline* because it has the same solute strength (tonicity) as the body fluids. Some solutions also contain potassium and magnesium salts because these minerals are also highly essential in the maintenance of a variety of vital functions.

Solutions of electrolytes are administered when it is necessary to replace the mineral salts and water that have been lost under circumstances such as dehydration, diarrhea, hemorrhage, excess urination, and vomiting.

The volume of the electrolyte administered must be adequate, with the veterinarian determining the amount. Up to 7 to 10% of the body weight may be administered over a 24-hour period.

The oral route of administering the electrolyte should be chosen whenever the condition of the animal so permits. The IV route of administration is indicated in life-threatening situations (severe vomiting, diarrhea, impending circulatory failure). Subcutaneous or intraperitoneal administration may be used effectively if circulation is adequate to ensure absorption.

• **Enzymes**—*Enzymes are complex protein compounds produced in living cells which cause changes in other substances without being changed themselves. They are organic catalysts.*

Normally, the enzymatic output of the digestive system of animals is adequate for maximum digestion of the starches, fats, and proteins. However, repeated experiments in many laboratories with western U.S. barley have shown that the metabolizable energy value of barley for poultry produced in the West under semiarid conditions is improved either by soaking in water before being fed or by the addition of enzyme preparations derived from fungal fermentations.

Results from feeding enzymes to species other than poultry have been negative to slightly favorable.

• **Flavoring agents**—*Flavoring agents are feed additives that are designed to increase palatability and feed intake.*

Hand in hand with improved animal nutrition and care have come feed palatability problems. Antibiotics, wormers, growth stimulants, and waste products often taste bad. Besides, today's animals are in forced production and stressed. There is a place, therefore, for feed flavors and aromas that will overcome palatability problems, attract animals to feed and keep them on feed, and increase feed consumption and performance. To meet this need, a wide array of feed flavor products is on the market; and they are available in both dry and liquid forms, usually under alluring brand names.

• **Hormone and hormonelike compounds**—*Hormones are chemicals released by a specific area of the body that are transported to another region within the animal where they bring about a physiological response.*

Scientists have identified many of the hormones of the body and have successfully synthesized several hormone or hormonelike compounds which produce the same physiological responses as those from naturally produced hormones. In some cases, this has made it possible to administer these products to obtain a specific response; for example, increased growth, milk production, or meat production.

Each hormone or hormonelike compound elicits a different response. In separate sections that follow, the nutrition-related responses of the following products are presented: Somatotropin (growth hormones), including the bovine somatotropin (BST) for dairy cattle and porcine somatotropin (PST) for swine; Melengestrol acetate (MGA).

1. **Somatotropin (growth hormone).** Somatotropin is secreted naturally by the anterior pituitary gland, located in the skull at the base of the brain of all vertebrates. It is a peptide, and it is species-specific. Thus, bovine somatotropin (BST) and porcine somatotropin (PST) are distinctly different.

Many problems must be resolved before growth hormones become commercially available and widely used, including a practical method of administering somatotropin to animals. Since it is a protein, as is insulin, it cannot be taken orally in the feed because it would be digested in the gut.

Currently, scientists are working feverishly away perfecting and testing growth hormones. For dairy cattle, somatotropin is being tested as a stimulator of milk production, and as a growth promotant for replacement heifers. For beef cattle, sheep, swine, and poultry, somatotropin is being tested as a promotant to improve gains and feed efficiency, and, hopefully, produce leaner meat.

The full impact of these products on the livestock industry won't be felt until they have been used for a decade. It appears, however, that growth hormones will be one of the early significant commercial successes of DNA technology; and that they will usher in a new era in livestock production.

a. **Bovine somatotropin (BST) for dairy cattle.** In 1985, Bauman and Eppard of Cornell University, and DeGeeter and Lanza of the Monsanto Company in St. Louis, reported responses of high-producing dairy cows to long-term treatment with pituitary somatotropin and recombinant somatotropin.[2] Cows injected with either type of somatotropin produced from 16 to 41% more 3.5% fat-corrected milk (FCM) than the control group. Cows injected with somatotropin consumed more feed on a body weight

---

[2]Bauman, Dale E., et al., "Responses of High-Producing Dairy Cows to Long-Term Treatment with Pituitary Somatotropin and Recombinant Somatotropin," *Journal of Dairy Science*, Vol. 68, No. 6, 1985, p. 1352.

basis than the control group, but their relative feed energy efficiency was higher. The increased feed efficiency could be accounted for by the smaller proportion of total intake required to meet the cows' maintenance energy requirements.

Before the commercial use of bovine somatotropin can become a reality, a practical method of administering the product must be developed and FDA approval must be secured.

Some additional pertinent information relative to bovine somatotropin follows:

(1) **Human safety**—BST is a natural protein produced by all cows, which is always present in milk. It is not a steroid or sex hormone, and it is not relative to the steroid hormones which are used as growth promotants. There is no evidence that the BST normally found in milk is harmful when consumed by humans. Moreover, milk levels do not increase significantly following treatment. As a peptide, when the BST molecule is consumed by humans, it is broken down and inactivated in the digestive tract in the same manner as any other protein. Both the U.S. Food and Drug Administration and England's Milk Marketing Board have approved the sale of milk from BST-treated cows. Thus, there does not appear to be any negative effect on human health.

(2) **How to administer**—Because BST is a peptide and would be broken down in the digestive tract, it cannot be fed. But daily injections don't seem practical. (In the experimental stage, it was injected daily, in the hip area.) So, a long lasting time-released injection or implant appears to be the answer.

(3) **Increased feed efficiency and milk production**—In a commercial dairy and over a full lactation, bovine growth hormone may be expected to increase feed efficiency from 10 to 20%, and milk production by 10 to 25%. But scientists still don't know exactly how it works.

(4) **Growth rate of heifers**—Bovine growth hormone will increase the growth rate of heifers by 8 to 10% and stimulate the development of secretory tissue in the mammary gland.

(5) **No adverse effect on cows**—It appears that the growth hormone is a safe product to use on dairy animals; treated cows have about the same somatic cell count, disease incidence, post-calving behavior, and rebreeding performance as untreated cows. However, because of their higher production, treated cows will require greater feed intake and more intensive management and observation.

b. **Porcine somatotropin (PST) for swine.** Porcine somatotropin (PST) is the scientific name for the growth hormone in swine, which is the counterpart of bovine somatotropin (BST) in dairy cattle; and it appears to be equally as effective. Both products are normally produced by the anterior pituitary at the base of the brain. But each is species specific; that is, PST works only on swine, and BST works only on cattle. PST is present in all pigs. It stimulates protein synthesis and growth in most tissues of the body, but it causes breakdown of fat deposits in adipose tissue.

When porcine somatotropin is injected, lactating sows produce more milk and wean heavier pigs; and growing-finishing hogs grow faster and more efficiently, and pro-

duce leaner pork. Under field conditions, it is estimated that the use of PST will produce 15 to 20% (1) more rapid gain, (2) greater feed efficiency, and (3) increase in lean muscle.

A major deterrent to the commercial use of PST is lack of an appropriate method of administering the product to swine. Daily injections, as used in early experimental studies, are not practical in commercial hog production. A suitable method of administering PST will be developed; the only question is when and by whom. Also, FDA approval will be necessary.

2. **Melengestrol acetate (MGA).** Melengestrol acetate (MGA), a synthetic progestogen hormone, is approved by the FDA for use as a feed additive for nonpregnant heifers.

Melengestrol acetate is similar in structure and activity to progesterone, the naturally-occurring hormone of pregnancy. It suppresses estrus (heat) and ovulation and promotes growth.

MGA is fed at a very low level to feedlot heifers—0.25 to 0.40 mg/head/day. On the average, it will increase daily rate of gain by 10% and feed efficiency by 6%. A 48-hour withdrawal period prior to slaughter is required.

• **Implants**—*An implant is a small pellet that is deposited underneath the skin behind the ear of an animal for the purpose of promoting growth.*

A list of the presently available and FDA-approved implants, all of which have been shown to improve rate of gain and feed efficiency, follows:

1. **Compudose.** This long-lasting implant is approved by the FDA for use in steers of any age and weight.

2. **Finaplex.** This implant, chemically known as trenbolene acetate (TBA), was cleared by the FDA in 1987, to be used in growing and finishing feedlot steers. It is a synthetic androgen with much greater anabolic potency than testosterone.

3. **Ralgro.** This implant is derived from a mold produced on corn by the organism, *Gibberella zeae*. The active ingredient is zeranol, an anabolic agent—not a hormone. Ralgro improves the rate of gain and feed efficiency of cattle of either sex in the suckling, growing, or finishing stages, in either the feedlot or on pasture. Ralgro is also approved as an implant for sheep.

4. **Steer-oid and Heifer-oid.** These products were developed in Canada. Essentially, Steer-oid, which contains estradiol benzoate and progesterone, is similar to Synovex-S; and Heifer-oid, which contains estradiol benzoate and testosterone, is similar to Synovex-H.

5. **Synovex-S, Synovex-H, and Synovex-C.** Synovex-S and Synovex-H are sex-specific implants; ''S'' is for steers, and ''H'' is for heifers. Synovex-C is for calves of either sex 45 days of age or older.

When using implants, follow label instructions for implanting technique, level of application, and withdrawal period prior to slaughter.

• **Ionophores**—*Ionophores are feed additives that change the metabolism within the rumen by altering the rumen microflora to favor propionic acid production.*

Currently, two ionophores—Bovatec (lasalocid) and Rumensin (monensin)—are approved for cattle. Both products are antibiotics that have been around for a long time. However, they were initially approved as anticoccidial drugs for poultry; monensin was marketed under the trade name Coban, and lasalocid as Avatec.

Lasalocid and monensin, are very effective in altering the rumen microflora, resulting in (1) increasing propionic acid production and decreasing acetic acid, methane, and carbon dioxide production, and (2) decreasing the breakdown of natural protein by rumen bacteria (making for rumen bypass). During the 1980s, the ionophores were fed to the majority of feedlot cattle.

Bovatec (lasalocid) was approved by the Food and Drug Administration for use in feedlot cattle, both to improve feed efficiency and daily gain. Lasalocid is a polyether antibiotic produced by *Streptomyces lasaliensis*. Research has shown that lasalocid is an effective coccidiostat in both cattle and sheep. Lasalocid is toxic to horses, but higher doses are required than of monensin to cause death.

Rumensin, an antibiotic fermentation product produced by a strain of *Streptomyces cinnamonensis*, alters the metabolism within the rumen. Extensive testing has shown Rumensin to be a highly effective improver of feed efficiency in feedlot cattle and in stocker and feeder cattle and beef and dairy replacement heifers on pasture. Rumensin-fed cattle eat less feed per day, but gain about the same amount of weight per day as cattle not fed Rumensin.

• **Isoacids (IsoPlus)**—*Isoacids provide three branched-chain fatty acids (isobutyric acid, isovaleric acid, and 2-methyl butyric acid) and valeric acid—the same fatty acids that are made by ruminant bacteria and are present naturally in the rumen of cattle.* Isoacids are essential for the growth of some rumen organisms that digest fiber. Adding isoacids to the cow's ration results in higher milk production.

The U.S. Food and Drug Administration has approved a blend of isoacids as a feed additive for lactating cows, sold as the calcium salts, under the trade name IsoPlus, manufactured by Eastman Chemical Co., a division of Eastman Kodak.

• **Kelp**—Kelp, or seaweed, grows in the sea. Botanically, it is a member of the algae family. For centuries, kelp has been promoted as a natural feed for animals and food for humans, prized for its minerals and vitamins. Kelp is always a rich source of iodine; dried kelp contains more than eight times as much iodine as iodized salt.

Kelp has long been promoted for its therapeutic properties for both animals and people, but with few of these claims substantiated by properly conducted and controlled experiments.

• **Medicated feeds**—*Medication is the administration of remedies for the prevention or healing of disease.* It follows that medicated feeds are feeds that contain remedies to prevent or heal disease.

Medicated feeds should be used in keeping with the instructions on the label, especially as they pertain to (1) the drug withdrawal period from animals in advance of marketing products; and (2) proper mixing, handling, and storage of medicated feeds.

• **Mold (fungi) inhibitors**—*Molds are fungi distinguished by the formation of mycelium (a network of filaments or threads), or by spore masses.*

*Mold inhibitors are substances that prevent the growth of molds.*

Certain fungi, most notably the organism *Aspergillus flavus*, produce toxins; the toxin produced by *Aspergillus flavus* is known as aflatoxin. Aflatoxin, which has clearly been shown to be a carcinogen (cancer-producing), causes much trouble in livestock. But it is not the only mycotoxin to be feared.

Mycotoxins affect all species, especially the young. Generally, ruminants appear to tolerate higher levels of mycotoxins over longer periods of intake than simple-stomached animals. Growing chickens are markedly less susceptible to aflatoxins than ducklings, gosling, pheasants, or turkey poults. Fish are probably one of the most susceptible species of animals to aflatoxin poisoning.

Of all the mold inhibitors currently available at an economical price, propionic acid is the most efficacious. It is Generally Recognized As Safe (GRAS), by FDA, available in liquid or dry form, of low toxicity to animals, and economical for addition to feeds at an effective level. Sorbic acid is also very effective in preventing mold growth. Mold inhibitors should be applied in keeping with the directions of the manufacturer.

The toxicity of aflatoxin-contaminated feed can be reduced when irradiated by ultraviolet light or exposed to anhydrous ammonia under pressure.

• **Pellet binders**—*Pellet binders are products that enhance the firmness of pellets.* Several feed additives are known to produce a marked increase in the firmness of pellets; among them (1) sodium bentonite (clay), (2) cellulose products from the wood pulp industry, (3) lignin derivatives, and (4) grain industry byproducts.

Molasses or fat are sometimes added to feed as an aid in pelleting, as well as being a concentrated source of energy.

• **Probiotics (microbial enhancers)**—*Probiotics are substances that contain desirable gastrointestinal microbial cultures and/or ingredients that enhance the growth of desirable gastrointestinal microbes.* They establish a desirable balance of gastrointestinal organisms and/or the substances which contribute toward the balance.

The concept of probiotics is not unlike the use of bacterial cultures as a silage preservative or the inoculation of legume seeds.

As knowledge of the types and functions of microorganisms in the digestive tract unfolded, there evolved with it the concept of inoculating animals with beneficial microorganisms and/or giving them substances to encourage the growth of beneficial microorganisms.

Without doubt, young animals and animals under stress—adverse weather, changes in ration, weaning, transporting, comingling, or other stressful situations—will be more likely to respond to probiotics. It follows that the greater the need, the greater the response from probiotics. They should be used in accordance with the label of the manufacturer.

• **Steroids**—Steroids are a group of fat-related organic compounds. They include cholesterol, 7-dehydrocholesterol and ergosterol, bile acids, and steroid hormones.

Various types of steroids have been used to assist human and equine athletes increase muscle mass and eliminate pain from creaky joints. Additionally, steroids are sometimes used on horses to assist in the healing of broken bones, to fight against parasitic anemias, to correct low blood counts, to combat respiratory infections, to increase appetites, and to improve performance of race horses.

But the use of anabolic steroids may reduce fertility in both stallions and mares. The use of corticosteroids carries the risk of the horse using the injured part more than it should and making the injury worse. Also, on most U.S. race tracks the use of steroids is forbidden within a certain period prior to a race (usually 48 hours of a race). Because horses may be used for human consumption in a number of European countries, notably in France and Belgium, the use of steroid drugs is banned in the European Economic Community (EEC).

While anabolic steroids and corticosteroids are controversial because of some of their side effects, no such argument surrounds the proper use of progesterone and estrogen.

• **Tranquilizers**—Several drugs such as reserpine, aspirin, ethylene glycol, and other tranquilizers, have been fed to animals for the purpose of quieting and curbing activity.

Tranquilizers are sometimes added to poultry feeds to quiet birds being moved from place to place, to reduce the incidence of cannibalism, or to calm flocks affected with hysteria. But the use of these drugs as feed additives for other animals species is essentially nonexistent.

## FEED PROCESSING

Feed is the major cost in animal production. Hence, it is economically important that it be processed in such a manner as to make for maximum efficiency (1) in handling, from a mechanical standpoint; and (2) in feed efficiency, from the standpoint of the animal.

Feed preparation can influence the nutritive value of a feed. For example, fine grinding and pelleting of forages tend to increase rate of passage through the gut, which lowers fiber digestibility. However, overall animal response to pelleted forages is usually increased over the same forage fed in long or chopped form, because the slightly lower digestibility is more than offset by increased feed consumption.

Generally speaking, the higher the level of feeding and the greater the production desired, the more important proper feed preparation becomes. This is so because (1) the higher the level of feeding, the more selective animals become in their eating habits; and (2) in ruminants, digestibility decreases as level of feeding increases, primarily because the feed does not remain in the digestive tract long enough for maximum effect of the various digestive processes.

Most of the recent technology in feed preparation has been with feedlot cattle. It came in with the development of large commercial feedlots. But much of it is applicable to all ruminants. Feed preparation for swine and poultry has remained relatively simple as compared with the variety of methods available and in use for ruminant feeds. The major change in horse feed preparation has been the increased use of all-pelleted rations (hay and grain combined).

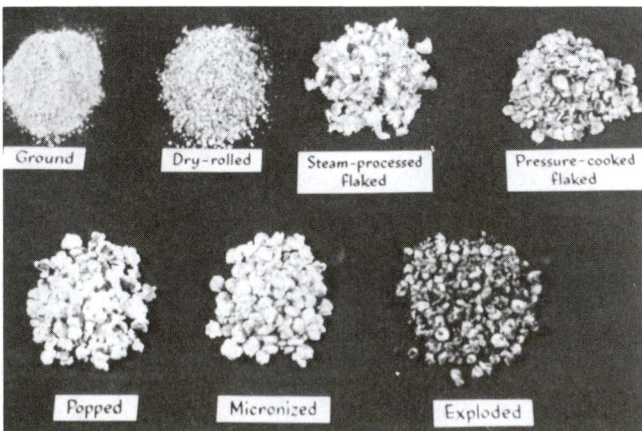

Fig. 4–21. Milo (grain sorghum) processed by different methods. (Courtesy, Dr. W. H. Hale, Department of Animal Science, The University of Arizona, Tucson)

## CONCENTRATE PROCESSING METHODS

Several concentrate processing methods have evolved. Some are physical, others are chemical; some are dry processing, others are wet processing.

It is recognized that any grouping of processing methods cannot be precise, for two or more processing treatments may be involved in a feed; for example, in making pellets, grinding is followed by adding heat and moisture, then pressure. Despite some overlapping, the author evolved with the following classification of grain processing methods:

**Mechanical alterations**
    Dehulling
    Extruding (gelatinization)
    Grinding
    Rolling
    Dry rolling (cracking, crushing)
    Steam rolling (crimping, steam crimping)

**Heat treatments**
    Dry heat processing
        Micronizing
        Popping—Jet-sploding
        Roasting
    Moist heat processing
        Cooking
        Exploding
    Flaking
        Steam Flaking
        Pressure Flaking
    Pelleting
        Crumbling

**Moisture alterations**
    Bran mash
    Drying (dehydration)
    High-moisture grain (early harvested)
    Reconstituted grain
    Watered feeds

**Blocks**

**Liquid Supplements**

**Fermenting**

**Hydroponics (sprouted grain)**

**Unprocessed (whole) corn**

## MECHANICAL ALTERATIONS

The mechanical methods by which the grain kernel is broken vary, but generally speaking, they involve either shearing, cutting, or mashing. In the milling of grains, there is also the abrasive action to scrub off the outer coats in processes referred to as burring, pearing, polishing, dehulling, and other similar terms.

When any of the dry processing methods are used, it is important that the kernel be broken, but that there be coarseness and relative freedom from fines.

• **Dehulling**—*Dehulling is the process of removing the outer coat of grain, nuts, and some fruits.* The hulls are high in fiber and low in digestibility by swine, poultry, and other monogastric animals.

- **Extruding (gelatinization)**—*Extruding is a process by which feed is pressed, pushed, or protruded through constrictions under pressure.*

Extruding usually involves grinding the grain, followed by heating with steam in order to soften it, then forcing the material through a steel tube by an auger. The softened material is then extruded through cone-shaped holes which are smaller where the feed enters and gradually enlarge where the feed is expelled. The expansion causes disruption, or granulation, of the starch granules. Various factors, including moisture of the grain, influence the character of the final product.

- **Grinding**—*Grinding is that process by which a feedstuff is reduced in particle size by impact, shearing, or attrition.* It may change the digestibility of cellulose and protein. Grinding is the most common, cheapest, and simplest method of feed preparation. It is usually accomplished by means of a hammer mill, which, by impact, reduces the particle size of the grain until it passes through a screen of a certain size.

Feed manufacturers commonly refer to feed mixtures in which all the ingredients are ground as a mash or meal.

Fig. 4–22.  Hens eating mash. (Courtesy, Ralston Purina Company, St. Louis, MO)

- **Rolling**—*Rolling refers to the process by which grain is compressed into flat particles by passing it between rollers.* The rolling may be accomplished without the addition of water (dry rolling) or after subjecting the grain to steam (steam rolling). A brief description of each method follows.

1. **Dry rolling (cracking, crushing).** *Dry rolling, which is also called cracking or crushing, refers to passing grain, without steam, between a closely fitted set of steel rollers which are usually grooved on the surface.* It breaks the hull and/or seed coat and results in an end product resembling coarsely ground grain.

2. **Steam rolling (crimping, steam crimping).** *Steam rolling, which is also called crimping or steam crimping, refers to exposing grain to steam for a short period of time, usually 1 to 8 minutes, followed by rolling.* The steam softens the kernel, producing a more intact, crimped appearing product than that produced by dry rolling.

## HEAT TREATMENTS

Excess heating damages some nutrients, such as the amino acids, and vitamins, whereas proper heating of protein sources (such as soybeans) and of carbohydrate sources (such as cereal grains, potatoes, and beans) result in better availability of nutrients. Heating soybeans destroys the trypsin inhibitor or a possible active protein fraction in raw soybeans, increasing the amino acid availability, results in better availability of the fat, and increases metabolizable energy.

Proper heating of cereal grains, such as corn, barley, and milo, will make for partial gelatinization and improve rate and efficiency of gains of cattle.

In general, heat treatments do not improve the nutritional value of most feedstuffs for monogastrics. However, they are the most successful of the newer feed processing techniques for ruminants.

- **Dry heat processing**—*Dry heat processing consists of surrounding the feed with dry air.*

The common methods of processing by dry heat are micronizing, popping, and roasting. Details follow.

1. **Micronizing.** *Micronizing is a coined word used to describe a dry heat treatment of grain by microwaves emitted from infrared burners.* In micronizing, grain is heated to 300°F by gas-fired infrared generators as it passes along an oscillating steel plate or skillet, thence is dropped into knorling rolls. Micronized grain is not popped.

Micronized grain sorghum compares favorably to steam flaked grain sorghum, from the standpoint of rate and efficiency of gain. However, cost of processing favors the micronizing technique over steam flaking because of a lower initial cost in equipment.

2. **Popping (jet-sploding).** *Popping is the exploding, or puffing, of grain resulting from the rapid application of dry heat.* Popping grain for livestock involves the same principle as processing popcorn for people, and the end results are similar.

All grains can be processed by this method, but it appears that it is especially effective in processing sorghum grain.

3. **Roasting.** *Roasting is a simple process of heating feed to the desired temperature in some form of oven for a period of time.* It is another method of heat treatment.

The effects of roasting are not fully understood. But it appears to increase the availability of nutrients, possibly as a result of changes in the starch (perhaps gelatinization) due to the heat, along with some effect on the proteins.

Corn and soybeans are the principle feeds that are processed by roasting.

- **Moist heat processing**—Moist heat processing consists in surrounding the feed by water or steam and (1) cooking either in a conventional vessel or under pressure, or (2) compressing.

The common moist heat processing methods are cooking, exploding, flaking (steam flaking, pressure flaking), pelleting, and crumbling.

1. **Cooking.** *Cooking is processing by applying heat.*

Farmers have long known that potatoes, beans, and soybeans should be cooked for pigs.

In general, cooking feedstuffs does not improve their nutritional value for monogastrics—swine and poultry. However, soybeans are an exception. Their nutritional value is greatly enhanced by heating.

Garbage is cooked to control *Trichinella,* which causes trichinosis in humans.

2. **Exploding.** *Exploding is the swelling of grain, produced by steaming under pressure followed by releasing the air.* The product resembles puffed breakfast cereals. Excellent quality control and uniformity of product are possible with this process.

California workers compared exploded milo with steam flaking. The puffed material produced feed intake, gain, and feed efficiency comparable to the best performing flaked grain treatment.

3. **Flaking.** *Flaking is a modification of steam rolling in which the grain is subjected to steam either for a longer period of time or under pressure.* The end product has a distinct and pleasant aroma, resembling cooked cereal. Proper flaking of grains renders the starch fraction more readily available to rumen microorganisms and enzyme degradation than conventional methods of steam or dry rolling.

Flaking is rolling into flat pieces following either (1) steaming at atmospheric pressure, or (2) steaming under pressure.

　　a. **Steam flaking.** This was the first modern technique which markedly increased feed efficiency and rate of gain in the case of milo. This process differs from steam rolling or crimping in that the grain is subjected to steam under atmospheric conditions for a longer period of time, usually 15 to 30 minutes, prior to rolling.

　　b. **Pressure flaking.** In pressure flaking, the grain is subjected to steam under pressure for a short time, such as 50 psi for 1 to 2 minutes. In comparison with steam flaking, flakes produced by pressure are less brittle and less subject to fragmenting during the mixing and feeding operation.

4. **Pelleting.** *Pelleting is the agglomerating of feed by compacting and forcing it through die openings by a mechanical process.* Pellets can be made into small chunks or cylinders of different diameters, lengths, and degrees of hardness. Large pellets—especially those large enough to be fed on pasture or range—are commonly called range cubes.

Fig. 4-23　Broilers eating bite-sized pellets. (Courtesy, Ralston Purina Company, St. Louis, MO)

Grains and other concentrates are pelleted for the purposes of (1) facilitating mechanization in handling; (2) eliminating fines and dust, and increasing palatability; (3) alleviating separation of ingredients and sorting; (4) increasing feed density—thereby lessening transportation and labor costs; (5) reducing storage space; (6) making it possible to feed on the ground or in windy areas with little loss; and (7) improving the nutritional value of certain feedstuffs through the instantaneous heat and pressure.

Fig. 4-24.　Range cubes fed to replacement heifers wintered on low-quality pasture. (Courtesy, Ralston Purina Company, St. Louis, MO)

On rations containing a low level of crude fiber, there is no advantage in pelleting feed for beef cattle or swine. However, with more fibrous feeds—especially barley—there is a decided advantage in pelleting feed for swine. Pelleted feeds are popular with horse owners and caretakers; and this includes pelleted concentrates, pelleted hay, and pelleted complete feeds (concentrates and hay combined).

Pelleting broiler and turkey feeds improves performance. But there is no advantage from pelleting layer feeds, with the result that layers are usually fed mash feeds. Pelleting results in marked improvement in the nutritional value for chicks and poults of certain feedstuffs, such as wheat bran, wheat germ meal, dehydrated alfalfa meal, rye, rapeseed oil meal, and field peas.

The following concentrates may be pelleted: (1) the entire concentrate, (2) the fines only, (3) the protein supplement, and (4) range supplements. Also, concentrates may be combined with a roughage(s) to make a complete feed, and the entire mix may be pelleted.

　　a. **Crumbling.** *Crumbles are crushed pellets.* They are made by crushing pellets into coarse, granular form. In comparison with pellets, crumbles are preferred by many poultry producers and are better adapted to mechanical feeders. Crumbles retain the heating and density advantages of pellets, but alleviate the sometimes disadvantages of pellets being difficult to chew, swallow, and digest. In comparison with ground feeds, crumbles have the advantage of being dust-free, irregular, and granular.

## MOISTURE ALTERATIONS

Water is important in feed preparation and processing. Sometimes the water content of a feed must be altered for proper feed storage, and sometimes it must be changed for feeding purposes.

Some feeds must be stored dry; others must be stored wet. Feeds carrying more than 14% moisture cannot be stored in bulk, for they will likely mold. For safe storage, therefore, grains with higher moisture content must be dried, ensiled, or acid treated.

The moisture content of forages that are to be preserved for ensiling is also of importance, since it affects the ease with which ensiling can be effected. Grass-legume forages must frequently be wilted to reduce moisture content to about 60 to 67%. On the other hand, mature forages often require the addition of water during the ensiling process.

Very dry feeds are often very dusty following grinding or dry rolling. Animals universally dislike dusty feeds; consequently, powdery rations are not eaten well. Dry, dusty rations may be improved by adding small quantities of water, by steaming, or by feeding the product in wet form.

- **Bran mash**—*A bran mash is steamed wheat bran.* It is the traditional feed for use in regulating the bowels of horses on idle days and at such other times as required. The wet mash is prepared by filling a 2- to 2½-gal bucket with wheat bran, pouring enough boiling water over it to make it the consistency of breakfast oatmeal, covering the bucket with a blanket and allowing it to steam until cooked, then feeding it to the horse.

- **Drying (dehydrating)**—*Drying is the removal of moisture by artificial or natural means.* To avoid spoilage in storage, grains must be dry enough to prevent the growth of bacteria and molds.

Generally speaking, shelled or threshed grains stored in unventilated bins should not have more than about 14% moisture; preferably, it should not exceed 10 to 12%. Grain may be dried (1) by the use of fuel—artificially; (2) by natural air drying; or (3) by a combination of the two methods. Artificial air drying is usually accomplished by running the grain through a heated chamber at a rate that will ensure its being adequately dried when it passes from the drier. The amount of heat and the drying time will vary with the amount of moisture to be removed. The process is expensive.

- **High-moisture grain**—*High-moisture grain refers to grain that is harvested at a moisture level of 22 to 40% and stored without drying.* Optimum conditions for ensiling high-moisture grain appear to be 25 to 32% moisture content. Correctly speaking, high-moisture grain does not involve moisture alteration.

High-moisture grain may be successfully stored in three ways:

1. It may be ensiled (fermented) in an oxygen-limiting silo.
2. It may be ensiled in unsealed storage (in conventional upright silos, or in horizontal silos).
3. It may be preserved by the addition of an organic acid, most commonly propionic acid (or a mixture of propionic acid with either acetic acid or formic acid), or ammonia.

- **Reconstituted grain**—*Reconstituted grain is mature grain that is harvested at the normal moisture level (10 to 14% moisture), following which water is added to bring the moisture level to 25 to 30% and the wet product is stored in a suitable structure for 15 to 21 days prior to feeding.* Thus, reconstituted grain involves processing that resembles soaking, and which results in an end product similar to high-moisture grain.

When stored in upright silos, the grain is stored whole, then rolled or ground at the time of removal. Reconstituted grain cannot be satisfactorily stored in horizontal silos as compaction cannot be obtained.

Properly reconstituted milo and steam processed flaked milo give similar results with fattening cattle. Corn is also greatly improved by reconstituting, but there appears to be less advantage from reconstituting barley or wheat.

- **Watered feeds**—Water is frequently added to feed, with the amount varying from just enough for dust control to making a slop.

Ground and dried rolled grains, and finely ground alfalfa, tend to be dusty. The palatability of such feeds may be improved by adding a small amount of water at the time of feeding.

1. **Soaking.** Sometimes hard grains that are not mechanically processed are soaked for 12 to 24 hours. The soaking softens and swells the grain. Also, dried beet pulp and soybean flakes may be fed in wet form.

2. **Liquid and paste feeding.** Liquid feeding usually involves mixing predetermined amounts of feed and water prior to, or at the time of, feeding. When properly used, this method can practically eliminate feed dust in the feeding area and minimize wastage. Ratios of feed and water can be varied to produce a free-flowing liquid or a thick paste.

Some swine producers feed a slop (slurry, gruel, or swill), especially to early weaned pigs and pigs being fitted for show or sale, feeling that they get greater feed consumption and gains thereby.

## BLOCKS

*Blocks are compressed packages, generally weighing from 30 to 50 lb each, although high-energy blocks (high in fat content) weighing up to 500 lb are now available.* Mineral blocks have been used for a very long time. These were followed by the development of protein blocks, primarily for supplementing cattle on the range and horses on pastures or in corrals. More recently, high-energy blocks evolved.

Fig. 4–25. Block in use on pasture—a means of lessening the labor attendant to the daily feeding of supplement on pasture or range. (Courtesy, Moorman Manufacturing Co., Quincy, IL)

Blocks may be placed in grazing areas where cattle have frequent access to them, with one block provided to 15 mature cattle. Intake will vary with the feed supply and the type of block. Generally, it is planned to limit feed consumption to about 2 lb per head per day by hardness of block and salt and/or fat content.

Range cattle producers use blocks as a means of (1) lessening the labor attendant to the daily feeding of a range supplement, (2) alleviating the loss that accompanies feeding a meal, and (3) distributing cattle on the range.

## LIQUID SUPPLEMENTS

*Liquid supplements are supplements in liquid form.* Many of them contain water, molasses, and urea, usually with added trace minerals and vitamins. This is a convenient way of feeding supplements to cattle on pasture or in a corral. Also, liquid supplements are sometimes added to complete ration mixes, either as part of the mix or as a top dressing.

The amount of molasses in most liquid supplements varies from 50 to 70% of the total weight. Most liquid supplements contain ½ to 2% phosphorus, often phosphoric acid. Other compounds that may be present in liquid feed supplements are fat, either animal or vegetable, to increase the amount of energy; alcohols—both ethyl alcohol and propylene glycol are used; and/or a product(s) to govern consumption.

Liquid supplements in a "lick" tank can be offered free choice. This is a convenient and satisfactory way in which to supply protein, energy, and other nutrients, so long as the cattle do not consume more than they need.

## FERMENTING

Two fermentation processes are of practical importance in livestock feeding: (1) ensiling; and (2) improvement of the nutritional value of feeds, either by fermenting the feedstuff itself, or by fermenting other materials that may be used as feed additives to supplement the original feed.

- **Ensiling**—The earliest use of fermentation in animal feeding, and still the most extensive one, involves the ensiling process which takes place when certain feeds with sufficient moisture are stored in a silo in the absence of air. The entire ensiling process requires 2 to 3 weeks, during which time, a small amount of oxygen is deleted with aerobic respiration, and anaerobic fermentation occurs.

- **Improvement of feed nutrient content by fermentation**— The age-old practice of slop-feeding pigs was a continuous-batch fermentation process. Today, more sophisticated and better controlled fermentation techniques are being used; and fermentation is being employed to produce amino acids and proteins, certain vitamins, antibiotics, and certain enzymes.

Beyond these accomplishments there exists an area of significant impact of fermentative processes in converting by-products and waste materials into livestock feeds, and, at the same time, lessening the pollutants in the environment of humans. For example, considerable research is in progress to convert manure from poultry and other animals through bacterial fermentation into animal protein feed. This recycling process could produce much more protein and help solve a pollution problem.

## HYDROPONICS

*Hydroponics is the growing of plants with their roots immersed in an aqueous solution containing the essential mineral nutrient salts, instead of soil.* This means that the plants are produced with water and chemicals, but without soil.

The Wisconsin Alumni Research Foundation chemically analyzed and compared the composition of oat grain and 5-day oat grass on a dry matter basis (see Table 4–9).

### TABLE 4-9
### COMPOSITION OF OAT GRAIN AND 5-DAY OAT GRASS, MOISTURE-FREE BASIS[1]

| Constituent | | Oat Grain | Oat Grass |
|---|---|---|---|
| Dry Matter | (%) | 100.00 | 100.00 |
| Protein | (%) | 15.00 | 21.00 |
| Ether extract (fat) | (%) | 4.21 | 5.20 |
| Nitrogen-free extract | (%) | 65.86 | 42.79 |
| Fiber | (%) | 11.71 | 26.11 |
| Ash | (%) | 3.22 | 3.90 |
| Calcium | (%) | 0.063 | 0.238 |
| Phosphorus | (%) | 0.360 | 0.509 |
| | | (mg/kg) | (mg/kg) |
| Carotene[2] | | 0 | 39.067 |
| Vitamin E | | 17.95 | 48.87 |
| Niacin | | 7.18 | 103.96 |
| Riboflavin | | 1.96 | 22.29 |
| Thiamin | | 3.14 | 12.86 |
| Vitamin C | | 0 | 218.3 |

[1]Analyses by Wisconsin Alumni Research Foundation.

[2]Each mg of beta carotene was considered to be equivalent to 1,556 IU of vitamin A.

As shown in Table 4–9, the 5-day oat grass is a better source than oat grain of calcium, phosphorus, carotene, vitamin E, the B vitamins (riboflavin, thiamin, and niacin), and vitamin C.

Based on studies conducted by the different universities, sprouting results in an average loss of 83% of the dry matter of oat grain. One study showed a reduction in TDN from 75.7% in the oat grain to 70.2% for the sprouted oats. Also, the digestibility of dry matter, energy, protein, ether extract, and nitrogen-free extract was lower for the sprouted oats than for the oat grain. The composition of hydroponically grown forage will vary according to the growth stage of the plants, the temperature, the nutrients in the aqueous solution, and several other variables.

In arriving at a decision whether to produce feeds hydroponically, consideration should be given to (1) the needs of different classes of livestock for each nutrient, and (2) the cost of supplying these nutrients hydroponically.

Although hydroponically produced forages are high-quality feeds from the standpoint of certain minerals and vitamins, the need for supplemental quantities of such nutrients in common rations for livestock is questionable.

Without doubt, sprouted grains will give an assist when added to poor rations—and the poorer the ration, the bigger the boost. However, with our present knowledge of nutrition, it should be recognized that balanced rations can be formulated without the added equipment and labor costs of producing forages hydroponically.

## UNPROCESSED (WHOLE) CORN

*Unprocessed (whole) corn refers to shelled corn, the kernels of which have not been broken.*

It is generally recognized that young cattle (both beef and dairy animals under 6 months of age) masticate their feed well. Thus, although the digestibility of corn may be increased when it is processed for young bovines, the increased feeding value may not be sufficient to offset the added cost of processing.

With the exception of young cattle, it has been assumed that corn should be ground, or otherwise processed, for cattle. Recent experiments at a number of experiment stations have indicated that there are exceptions—that the proportion of concentrate to forage is a factor in determining whether corn should be processed for cattle. Cattle fed dry, whole shelled corn gain an average of 5% faster and require 7% less feed per pound of gain than cattle fed ground, rolled, or crimped corn *when high concentrate rations are fed.* However, processing appears to have some value for dry shelled corn in rations with 20% or more roughage content or when corn is very dry—less than 12% moisture.

Corn kernels appearing in the feces of cattle fed whole-shelled corn have caused some feeders to think that whole corn is much less digestible than ground corn. Some of the ground grain passes through the digestive tract, too, but it is not as noticeable in the feces as whole grain. Tests at Ohio State University showed very little difference in the digestibility of whole- and ground-shelled corn fed to steers on high-concentrate rations.

Eliminating processing costs is the main advantage of feeding whole corn.

## FORAGE PROCESSING METHODS

The common methods of forage preparation are chopping, grinding, shredding, cubing (wafering), drying, ensiling, and pelleting.

### CHOPPING, GRINDING, OR SHREDDING

Chopping, grinding, or shredding result in forages divided into smaller particles; but they differ in how they section it, and in the size of the particles. In comparison with a similar forage fed in long form, a forage subjected to any one of these three processes (1) is easier to handle and mechanize, (2) can be stored in a smaller area at less cost, (3) is fed with less feed refusal and waste, and (4) may make for slightly greater production. Low-quality, coarse forages are usually improved more from processing than high-quality, fine forages.

• **Chopping**—Refers to cutting forage not less than 2 in. in length. (The 2 in. refers to the set of the choppers. Some of the material will be cut longer than this, and some shorter).

Fig. 4–26. Power-bale feeder moving bales into portable on-the-farm mill for chopping and mixing. (Courtesy, Gehl Company, West Bend, WI)

Chopping has the disadvantage of being dusty. Also, there may be considerable leaf loss, or shattering, in field chopping because the hay must be drier than when it is baled or put up as long hay.

• **Grinding**—Refers to processing forage less than 1 in. in length. Usually grinding is accomplished by means of a hammer mill, in which the forage is beaten by revolving metal hammers until it is small enough in size to pass through the screen placed in the grinder.

Fine grinding is more costly than coarse chopping; hence, it is less appealing from a practical standpoint. Yet, fine grinding is sometimes desirable when the material (either sun-cured or dehydrated) is to be incorporated in the rations of swine or poultry. Ground forages are less digestible for ruminants because they pass through the paunch more rapidly, with only limited bacterial action. When finely ground hay is fed to lactating cows, the fermentation in the rumen produces less acetic acid and more propionic acid than when coarse forage is fed, and, in turn, this results in the fat content of the milk being substantially reduced.

When it is advantageous to use ground hay in a ration, the addition of molasses, fat, or water will lessen the dustiness and reduce the air pollution by nutrients. Some commercial mills spray a small amount of liquid fat on bales of hay just before they enter the grinder. Fat is easier to work with in a grinder or mixer than molasses, for the latter has a tendency to be sticky and "gum up" the equipment.

• **Shredding**—Is similar to chopping, except shredding tends to separate the stems longitudinally rather than cut them crosswise. Coarse forages, such as fodder and stover, are better suited to shredding than to chopping and grinding.

### CUBING[3] (WAFERING)

When applied to forages, the term cubing (wafering) refers to the practice of compressing long or coarsely cut hay into cubes about 1¼ in. square and 2 in. long, with a bulk density of 30 to 32 lb per cubic foot. They do not necessitate fine grinding, and they facilitate automation in both haymaking and feeding.

This method of haymaking is increasing, because it offers most of the advantages of pelleted forages, with few of the disadvantages. Because cubed forage is relatively coarse, it lowers milk fat percentage only slightly, if at all.

It is noteworthy, however, that horses occasionally choke when fed cubes.

### DRYING

For safe storage, the moisture of hay must be lowered to the following levels: loose hay, 25%; baled hay, 20 to 22% (the lower figure for larger bales); field chopped hay, 18 to 20%; and cubes (wafers), 16 to 17%. These figures must be modified according to temperature; higher temperatures necessitate lower moisture.

*Artificial dehydrating refers to that process in which forage is taken from the field as soon as it is cut (or in some instances after wilting), put through a hay chopper or silage cutter, and*

---

[3]There is overlapping in the use of the word *cube.* The compressed long or coarsely cut hay packages about 1¼ in. square and 2 in. long are known as cubes. Also, pellets that are large enough to be fed on pasture or range are commonly called cubes.

dried in large rotating drum driers of different types. For the most part, this method of drying is limited to processing forage for swine and poultry feeds. The most popular type of artificial dehydrator in use in this country is one that uses a high initial heat (1,200 to 1,400°F), and which is usually heated by natural gas or fuel oil. Due to high equipment and fuel cost, along with the added cost of moving heavy high-moisture forage from field to dryer and in operating the dehydrator, artificially dried forages must command a premium price over field-cured forage.

## ENSILING

*Ensiling refers to the changes which take place when forage or feed with sufficient moisture to allow fermentation is stored in a silo in the absence of air.* The entire ensiling process requires 2 to 3 weeks, during which time a small amount of oxygen is deleted with aerobic respiration, and anaerobic fermentation occurs.

Ensiling is notable for its versatility. It can be conducted in facilities ranging from simple to sophisticated, and it can be applied to a wide variety of feedstuffs. Its greatest use is in preserving forages, with acetic, lactic, and other of the lower acids formed. The addition of grains (at the rate of about 150 lb per ton) as preservatives in ensiling forage crops also involves the principle of fermentation. Likewise, when high-moisture grain is stored in an oxygen-limiting silo, it undergoes a fermentation process.

## PELLETING[4]

*When applied to forages, the term pelleting refers to the process of forcing ground forage (usually with some added moisture) through a thick steel die and compressing it into a circular or rectangular mass which is cut at predetermined lengths.* They can be formed into shapes of varying thickness, length, and hardness. The larger shapes, commonly fed to cattle and sheep on the range, are referred to as cubes.

Binding agents are sometimes added to feedstuffs to regulate the hardness of pellets, especially forage pellets which bind less than concentrates.

The two biggest deterrents to pelleting forages are (1) fine grinding, and (2) cost.

Since the pelleting of forages usually involves quite fine grinding prior to the pelleting process, and since finely ground feeds cause lowered butterfat, it follows that pelleted forage is not suitable for lactating cows.

Pelleted feeds for horses (hay, concentrate, and complete feed) have become fairly popular in recent years.

On the average, cattle fed high-roughage (above 80% roughage) or all-roughage rations will eat about ⅓ more pellets than long or chopped hay, make about ½ to ¾ lb faster daily gains, and require 2 to 2½ lb less feed per lb of gain. Also, it is recognized that low-quality roughages are improved most by pelleting.

Both cubing (wafering) and pelleting forages will (1) simplify haymaking, (2) lessen transportation costs and storage space, (3) reduce labor, (4) make automatic hay feeding feasible, (5) decrease nutrient losses, and (6) eliminate dust.

---

[4]Pellets may refer to (a) the entire concentrate in pellet form, (b) the fines of the concentrate in pellet form, which are usually added back to the grain for feeding, (c) the forage in pellet form, (d) the protein supplement, or (e) the range supplements in pellet form.

With cubing or pelleting, the spread between high- and low-quality roughage is narrowed; that is, the poorer the quality of the roughage, the greater the improvement from cubing or pelleting. This is so because such preparation assures complete consumption of the roughage. Also, cubing or pelleting, especially the latter, tends to speed the passage of roughage through the digestive system.

## MISCELLANEOUS PROCESSING METHODS

There is hardly any limit to the number of processing methods—some old, others new. Some preserve quality, others increase consumption and lessen labor, and still others change the chemical composition and feeding value. In addition to the processing methods already covered, several miscellaneous, but important, methods are discussed in the sections that follow.

### AMMONIATION

Ammonium salts and anhydrous ammonia (gas or liquid) have been used for ammoniating feeds that contain high levels of carbohydrates and low levels of nitrogen. Among such ammoniated feeds are: citrus pulp, beet pulp, molasses, sugarcane bagasse, and rice hulls. Also, low quality roughages may be ammoniated.

### ANIMAL WASTE (MANURE) PROCESSING

Animal waste (manure) has nutritive value for ruminants because these animals are capable of utilizing nonprotein nitrogen and fiber. So, proper processing is important.

Broiler and layer litter has been successfully used as an ingredient of cattle feed for many years. However, wastes from all species may be, and are, used. Among the methods employed to process animal wastes prior to feeding are: deep-stacking, ensiling (fermentation), dehydration, and pelleting. The two most common and practical methods of processing are:

1. **Deep-stacking.** In this method, the litter is deep-stacked for several weeks, during which it generates temperatures of 160°F or higher, which render it free of any potentially pathogenic microorganisms that might be present (Pathogenic bacteria do not grow at temperatures over 80°F, and they are killed at 145°F in a matter of minutes.) It follows that there have been no documented animal health problems associated with feeding broiler or layer litter processed in this manner.

2. **Ensiling (fermentation).** Ensiling is a controlled fermentation process during which carbohydrates in the mixture are converted to lactic, acetic, and other acids. Once sufficient acids are produced, bacterial action ceases and the ensilage is stable. During the fermentation, heat is generated, thereby diminishing the hazard from certain pathogenic organisms that might be present.

Dehydration and pelleting of animal wastes are excellent processing methods as such. However, current energy costs make them uneconomical.

In December 1980, the U.S. Food and Drug Administration published a document leaving regulation of feeding animal waste to the individual states.

## FAT ADDED

Typical livestock rations contain relatively small quantities of fat. Although most animals require minimal amounts of certain fatty acids in their diets, these minimals are low and easily met by normal rations. Nevertheless, fat serves the following practical functions when added to livestock rations:

1. It increases the caloric density of the ration.
2. It improves palatability.
3. It facilitates absorption of vitamins A and D and provides fatty acids.
4. It delays the sensation of hunger.
5. It controls dust.
6. It lubricates feed processing equipment.
7. It improves handling qualities.

However, the following problems are inherent in the incorporation of fats in feeds:

1. Animal fats tend to solidify in cold weather.
2. Fats can coat and clog mixing and distribution equipment.
3. High levels of fat in pelleted feeds can cause soft pellets.
4. Fats can become rancid.

If the price is favorable, fat may be added to rations at the following levels: for swine and poultry, 5 to 10%; and for cattle, 2 to 6%. Higher levels of fat usually result in drastically lowered feed consumption. When fed at the levels recommended above, the energy value of fat is approximately $2\frac{1}{4}$ times that of the grains. When corn is the major grain, fat additions can be expected to be less useful than with the small grains. This is understandable when it is realized that corn contains approximately 4% fat as compared to 1 to $1\frac{1}{2}$% for the other feed grains.

Animal and vegetable fats seem to be equally effective additions to rations, provided they are stable and nonreactive (noncorrosive); hence, selection should be determined solely by comparative price.

## FREEZING

A considerable amount of mink food, consisting of meat and fish, is frozen. Freezing inhibits bacterial growth and slows the enzymatic processes which can destroy the product.

## IRRADIATION

For many years, it was known that both ultraviolet light (from the sun) and cod-liver oil had identical effects in healing of rickets. In 1924, Steenbock of Wisconsin and Hess of Columbia University, independently announced that certain food materials could be made antirachitic by exposing them to ultraviolet light.

Upon irradiation, ergosterol, a plant sterol, yields ergocalciferol, commonly known as vitamin $D_2$.

The ultraviolet radiation in sunlight serves as a source of radiant energy necessary to convert 7-dehydrocholesterol (an animal sterol stored beneath the skin surface) into biologically active vitamin $D_3$.

Vitamin $D_2$, the plant form of the vitamin, and vitamin $D_3$, the animal form, have the same antirachitic value for the rat, dog, pig, ruminant, and human, but vitamin $D_3$ is more active for poultry.

Sun-cured hay is a good natural source of vitamin D for four-footed animals, but not for poultry. Fortified fish oils and irradiated sterols are good sources of vitamin D for both four-footed animals and poultry.

## MOLASSES ADDED

Molasses (including cane or blackstrap, beet, citrus, wood and starch molasses) is extensively used as a livestock feed. When used at levels of 5 to 15% of the ration, it has about $\frac{3}{4}$ the energy value of corn. However, molasses has added value as an appetizer, to reduce dustiness of a ration, as a binder for pelleting, to stimulate rumen microbial activity, and as a source of unidentified factors. Also, cane molasses is a good source of certain trace minerals.

In hot, humid areas, molasses should be limited to 5% of the ration; otherwise, mold may develop. Where mustiness is a problem, it may be controlled by adding calcium propionate to the feed according to the manufacturer's directions.

## ORGANIC ACIDS

The proper use of organic acids provides another way in which to preserve high-moisture grains. The organic acid treatment involves the application of 1 to $1\frac{1}{2}$% acid (i.e. propionic, acetic, formic, ammonium isobutyric, etc.) at time of harvest, followed by storage in a pile.

Acid treatment inhibits the growth of molds and bacteria. Research has shown that propionic acid alone or a mixture of 75% acetic and 25% propionic acid are quite effective. Acetic acid should not be used alone. Limited research indicates that sodium propionate, formalin, ammonium isobutyrate and citric acids have been successful, as well as combinations of propionic acid and formic acid or formalin.

Experimental studies indicate that acid-treated grain has approximately the same feeding value as high-moisture grain stored in an oxygen-limiting silo. Also, it alleviates the cost of drying. Thus, organic acid treatment of grain may be a practical way in which to preserve high-moisture grains.

## PRESERVATIVES

*A preservative is a material added at the time of mixing or storing to enhance the keeping qualities of a feed. A brief description of hay and silage preservatives follows:*

• **Hay preservatives**—Preservatives are available commercially which can be applied to hay. Usually the directions (1) recommend the addition of 1 to 3 lb of these products for each ton of damp hay, and (2) claim that there will be no heating or molding.

More experimental work is needed relative to chemical hay preservatives. But, available data indicate that propionic acid is the hay preservative of choice. Missouri workers report that most hay (28% moisture at baling) treated with an organic acid was equivalent in digestion to dry hay, whereas hay baled moist and not treated had significantly lower digestibility than treated or control hays.

Anhydrous ammonia is one of the most recent materials being studied as a hay preservative. In Indiana trials, applying this material at the rate of 1.0% to hay baled at 30% moisture successfully prevented molding, heating, and quality deterioration.

- **Silage preservatives**—Two types of additives have generally been used in silage making: (1) feed additives, and (2) chemical additives.

Feed additives supply a readily available source of carbohydrates for bacterial fermentation of the silage. Some feed additives, such as corn-and-cob meal, when mixed with high-moisture forages, also absorb water and help to reduce run-off. When used as preservatives, approximately 75 to 85% of the feed nutrients added may be recovered as feed.

A large number of chemical additives have been used in silage making, with variable results.

## SELF-FEEDING GOVERNORS

The commonly used self-feeding governors are (1) bulky fibrous feeds; (2) salt-feed or fat-feed mixtures; (3) fat content of block; and (4) liquid supplements.

- **Bulky, fibrous feeds**—Bulk can be used as a self-feeding governor. This consists in adding to the bulkiness of the ration, such as can be achieved by increasing the amount of chopped hay and lessening the concentrate. Actually, this is a way in which to lower the energy content of the ration. Since an animal can hold only so much, it is an effective control of feed intake.

- **Salt-feed mixtures**—The practice of using salt as a governor to limit feed consumption of pasture or range has been used for a very long time. It was ushered in as a labor-saving device for cattle and sheep in inaccessible and rough areas. Today, salt-feed mixtures are used in either meal or block form.

- **Fat content of block**—Since animals tend to eat until a certain caloric intake is reached, they consume less total weight when fed high-fat rations. Thus, pounds of feed consumed can be governed by the amount of fat in a block. It is noteworthy, too, that fat serves as a needed feed nutrient, whereas consuming more salt than required (as happens when a salt-feed mixture is used) makes for a waste of salt.

- **Liquid supplements**—When self-fed, the consumption of liquid supplements is generally controlled by (1) the use of a lick tank, and/or (2) incorporating in the formulation phosphoric acid, beet solubles, and/or citrus peel liquor.

## SLOW-RELEASE AND RUMEN BYPASS TREATMENTS

Two feed processing techniques—slow-release nonprotein nitrogen, and rumen bypass protein—are designed to delay digestion.

- **Slow-release nonprotein nitrogen**—Among the slow-release nonprotein nitrogen products that liberate nitrogen slowly are a combination of urea and gelatinized starch, and urea combined with gelatinized corn.

- **Rumen bypass**—This refers to bypass protein (also known as protected or escaped protein) in feed that escapes digestion in the rumen and passes into the lower digestive tract where it is digested and absorbed. Feed processors have developed treatments through which the bypass proteins in certain feeds can be increased; among them, heat and pressure treatment, treatment with tannins, treatment with formaldehyde or other aldehydes, lipid (fat) treatment, complexing with bentonite clay, use of amino acid analogs, increasing microbial metabolism in the rumen, and adding ionophores.

## TREATMENT OF HIGH-CELLULOSE FEEDS

High feed prices and more stringent burning regulations have spurred research to find a practical method of improving the feeding value of several high-cellulose products, such as rice, wheat, barley and oat straws; bagasse; tree bark; corncobs; gin trash; newspaper; and seed hulls.

In their natural state, these products make poor feedstuffs because little lignin or silica, or a combination of the two, (1) encrust the energy-rich carbohydrates, cellulose, and hemicellulose; and (2) keep the microbes in the ruminant's stomach from breaking them down to release energy.

The answer to this problem lies in some treatment that opens up the fibers enough to permit increased digestion in the rumen. Several methods of chemical and/or physical treatment are being investigated; among them, alkali treatment, ammoniation, hydrogen peroxide treatment, and high pressure steam.

- **Alkali treatment**—Sodium hydroxide is the common alkali treatment of high-cellulose products (crop residues), although calcium hydroxide and potassium hydroxide are sometimes used. The effectiveness of alkali treatment depends on the residue or waste being treated and the technique employed. Treatment level ranges from about 2 to 10% of the chemical based on the total dry matter content of the material being treated; and treatment time is about 24 hours.

On the basis of efficacy of the treatment, the cereal straws rank as follows: wheat straw, barley straw, and oat straw.

- **Ammoniation treatment**—This method involves placing the high-cellulose product (crop residues) in an air-tight enclosure (such as black plastic) and adding either anhydrous gas or liquid ammonia. It is important to add the correct amount of ammonia; too much makes for unnecessary expense, and too little makes for poor quality feed. Optimum treatment level appears to be 3.0 to 3.5% anhydrous ammonia based on the total dry matter content of the material being treated; and optimum treatment time is about 20 days vs 24 hours for the sodium hydroxide treatment. The two major advantages of using ammonia in comparison with the alkali treatment are: (1) it adds nonprotein nitrogen to the product, and (2) no mineral residue remains that might be detrimental to the animal or to the soil to which the manure is added. Ammoniation produces the following benefits:

1. It increases the crude protein equivalent by 3 to 10%.
2. It increases the digestible energy (TDN) by 3 to 23%.
3. It increases animal intake by 20 to 27%.
4. It prevents molding of crop residues and high-moisture forages.

CAUTIONS relative to ammoniation: Anhydrous ammonia in liquid form is very toxic to the skin and eyes; so, when a person's skin/eyes come in contact with anhydrous ammonia, it should be flushed away with water immediately; otherwise, serious injury may result. Ammonia can be flammable and even explosive; so, never smoke or light a flame near it. Also, there is hazard of ammonia toxicity among some cattle and sheep receiving ammoniated feeds, and among some young that are suckling their mothers that are fed ammoniated feeds, characterized by hyperexcitability, circling, convulsions, and some deaths.

- **High pressure steaming**—High pressure steaming, with or without added chemicals, has been used to a limited extent in treating crop residues and wood. Aspen, which is the most digestible of the woods, has been shown to reach a digestibility of 56.6% after steaming for 2 hours at 165°C, with the treated product readily accepted by sheep at up to 60% of the ration and producing normal weight gains and carcass yields. The treating of wood products is limited because: (1) the cost of treatment is high, (2) conventional feedstuffs need to be relatively high priced before treated wood residues can compete in the marketplace, and (3) the lack of a steady market for the treated products.

Decisions as to type and amount of chemicals and treatment systems must be based on evaluation of processing costs in relation to the value of the finished feedstuff.

## COMPLETE (ALL-IN-ONE) RATIONS

Most experiments and experiences have not shown any difference between mixed rations and the feeding of roughage and concentrates separately insofar as efficiency and production are concerned. However, a mixed ration has the following advantages:

1. It makes for greater efficiency in feeding and lessens the sorting at the feed bunk.
2. When the roughage is relatively unpalatable, a mixed ration forces consumption.
3. When it is desired to limit concentrate consumption, mixing with the roughage is desirable.
4. A mixed ration makes it easier to get animals on full feed.

Thus, each feeder must decide on the matter of mixed feed vs feeding roughage and concentrate separately, with relative costs and other factors considered. Most large cattle and sheep feedlots use completely mixed rations. Also, the trend is toward complete feeds for both dairy cows and swine, primarily because such complete feeds (1) lend themselves better to automation, and (2) provide better control of nutrient intake.

• **All-pelleted rations (grain and forage combined)**— Increasingly, complete pelleted rations are being used for horses, swine, and fish. Among the virtues ascribed to all-pelleted rations are (1) they prevent selective eating—if properly formulated, each mouthful is a balanced diet; (2) they alleviate waste; (3) they eliminate dust (thereby lessening heaves in horses); (4) they lessen labor and equipment; (5) they lessen storage; and (6) they facilitate automation.

## CHOICE OF PROCESSING METHOD

The choice of a processing method is highly dependent on the feedstuff to be fed. It is clear that a given processing technique may be very desirable for one grain, but quite detrimental to another. Corn may be fed without any processing, but not milo. Pressure treating appears to be desirable for milk, but harmful to wheat.

Comparison of grain processing techniques are difficult because there are a number of interactions between processing technique and roughage level or type of ration fed. For example, data from Ohio State University have shown that whole shelled corn was superior to crimped corn in very low-roughage rations, whereas crimped corn was clearly superior in high-roughage rations.

## FEED PROCESSING TABLE

Table 4–10 is a summary of pertinent information relative to the preparation of feeds for each class of livestock.

### TABLE 4–10
### PREPARATION OF FEEDS

| Class of Animal | Concentrates | Forages | Comments |
|---|---|---|---|
| Beef cattle | Extruding, flaking, micronizing, popping, roasting, or high-moisture grain—with choice determined by cost—are preferable, especially for full-fed animals on a high-grain ration. But such equipment is costly to purchase and operate; hence, a large-volume operation is required to cover fixed costs.<br>Dry or steam roll or grind coarsely for most beef cattle, especially those not full fed high-grain rations and those in smaller operations.<br>On high-concentrate rations (those with 80% or more concentrate), whole corn need not be processed.<br>Grain (except for very hard seeds) need not be processed for calves under 6 months of age, for young calves masticate feed thoroughly.<br>Cubes (large pellets) preferred for feeding on pasture or range.<br>Professional caretakers often cook feed (especially barley) for show cattle to increase palatability. | Long hay is satisfactory for most cattle other than commercial feedlot operations.<br>Chopped (2″ length), cubed (wafered), or pelleted forage should be used (1) in commercial feedlots or when quality of hay is poor; and (2) in all cattle operations from standpoints of ease of handling and lessening wastage.<br>Shredding fodders and stovers (corn or milo) makes them easier to handle and lessens waste. | Fine grinding grain increases incidence of hyperkeratosis (ruminal parakeratosis) in feedlot cattle.<br>Dry or steam rolling or coarse grinding of grains are of about equal value for most beef cattle. Either method is just as satisfactory as more expensive methods (like flaking) when grain intake is relatively low.<br>Chopping or pelleting low-quality hay is more advantageous than chopping high-quality hay.<br>Finely ground hay not recommended, as it decreases digestibility. |
| Dairy cattle | Grinding is the simplest and the most widely used grain processing method for dairy cattle. Cracking, steam rolling, and pelleting are the other popular procedures.<br>Exploding, extruding, flaking, micronizing, popping, roasting, or high-moisture grain are preferable for high-producing lactating cows, but they are not widely used.<br>Dry or steam roll or grind coarsely for dry cows, young stock, and low-producing cows.<br>Grain (except for very hard seeds) may be fed whole to young calves under 6 months of age. | Long hay or cubes. Cubes lend themselves to automation; and lower milk fat percentage only slightly, if at all. | Butterfat is depressed unless the ration contains some threshold level of coarse material.<br>Finely ground or pelleted roughage will result in reduced rumen acetate production and lower milk fat percentage. |

*(Continued)*

**TABLE 4–10** (Continued)

| Class of Animal | Concentrates | Forages | Comments |
|---|---|---|---|
| **Sheep and goats** | Processing grains not necessary unless seeds are hard (like sorghum or millet) or the teeth are poor.<br>Hard seeds (like sorghum) may be prepared by exploding, extruding, flaking, micronizing, popping, roasting, or high-moisture grain, with cost determining the choice.<br>Pellets are increasingly being used by lamb feeders.<br>Cubes or pellets preferred for feeding on pasture or range.<br>Professional shepherds prefer flaked grain for show sheep, as the ration is lighter and there are fewer digestive disturbances. | Chop (2″ in length), pellet, or cube.<br>Many lamb feeders are using all-pelleted rations (hay and grain combined). | Sheep and goats masticate grain more thoroughly than cattle, with the result that feed preparation for them is of less value than for cattle.<br>A high incidence of parakeratosis—a degeneration of the rumen papilla—appears to result from feeding pellets, especially when low forage-high concentrate pellets are used. Hence, breeding sheep should not be fed pellets for extended periods without any long or chopped forage. |
| **Swine** | Corn, barley, grain sorghum, and oats should be finely ground for swine. Medium to coarse grinding is best for wheat, because fine grinding makes it pasty and less palatable.<br>Pelleting corn-soybean rations generally improves feed utilization and increases rate of gain by at least 4–5%.<br>Cook Irish potatoes, beans, soybeans, and garbage.<br>Cooking (except for the feeds listed above), soaking, or fermenting are not of value when swine are on full feed.<br>Liquid and paste feeding give inconsistent results in feed consumption and rate of gain; hence, they should be evaluated on the basis of a mechanical means of dispensing feed. However, slop (slurry or gruel) is desirable for early weaned pigs, and perhaps for pigs being fitted for show or sale.<br>High-moisture corn does not result in any improvement of efficiency for swine; hence, the value of high-moisture corn as compared to regular corn should be computed on a dry matter basis. | Alfalfa (or other legume) that is to be incorporated in mixed feeds should be ground.<br>Rations containing considerable amounts of fiber are improved by pelleting because of increased consumption, improved carbohydrate digestibility, and reduced sorting and wastage compared to meal rations. | Fine grinding will cause some bridging in self-feeders. Also, finely ground feed is associated with increased incidence of stomach ulcers in swine. |
| **Poultry** | Grains for poultry are prepared in 3 forms—mash, pellets, crumbles.<br>1. *Mash:* grind medium fine.<br>2. *Pellets:* composed of mash feeds that are pelleted. Birds usually consume more of a pelleted ration than the same ration in mash form. Usually there is more cannibalism with pellets than with mash or crumbles.<br>3. *Crumbles:* produced by rolling pellets. | Grind hay that is to be included in poultry feeds. | Proper heating of protein sources will result in better availability of nutrients if temperature and time of heating are not excessive.<br>Heating soybean meal destroys trypsin inhibitor and possibly other factors which limit protein digestion especially in growing chicks.<br>Methionine addition to heated meals increases chick growth. |
| **Horses** | Flaking is the preferred method of grain preparation for horses; it makes for a light ration and few digestive disturbances.<br>For horses with good teeth, the value of oats is increased only 5% by processing. | Either feed long hay or an all-pelleted ration (grain and hay combined). | Cubes (wafers) sometimes cause horses to choke.<br>Horses are very sensitive to moldy feed.<br>Horses should not be fed dusty feed, because of the hazard of heaves. |
| **Rabbits** | Nutritionally complete pellets containing 50–60% concentrate and 40–50% roughage, ⅛–³⁄₁₆″ diameter and ⅛–¼″ long, of two types:<br>1. Production rations, high in protein and energy.<br>2. Maintenance rations, medium in protein and energy.<br>Grains may be fed whole, but to maximize digestibility oats and barley should be rolled and corn should be cracked.<br>Protein supplements should be in cake, pellet, or crumble form. | Roughages should be cut to lengths of 3″ and fed in a rack. | Rabbits are very sensitive to moldy and dusty feeds, so they should always be fed clean, high-quality feeds.<br>Protein supplements in mash form are not desirable, because they may settle out from the rest of the feed and be wasted.<br>Uncut roughage will result in wastage. |
| **Mink** | Traditionally, mink rations are high in moisture (60–70%) and are fed as a semisolid mass (with a consistency not unlike that of a hamburger) on the wire mesh constituting the top of the pen.<br>**Fresh or frozen animal products and by-products from:**<br>1. Horses (now in limited supply).<br>2. Meat-packing houses.<br>3. Poultry processing.<br>And many other animal products.<br>**Fresh or frozen fish and fish by-products.**<br>**Processed grains and cereal by-products:**<br>1. From breakfast food industry.<br>*Limit to*—10–15% in breeding diets.<br>        10–25% in production diets.<br>2. Supplemental cereals (same as No. 1 above) plus skim milk, wheat germ meal, alfalfa meal, yeast, and meat and fish meals.<br>*Limit to*—15–20% in breeding diets.<br>        15–35% in production diets.<br>3. Dried bakery products.<br>*Limit to*—15–20% of diet.<br>**Vegetables:**<br>*Limit to*—8% of diet.<br>**Completely dehydrated mixed mink feeds.**<br>Such feeds are now available in bags. They do not require refrigeration. The ultimate goal of the large mink producer, which is now being realized, is to supply the complete diet as dry pellets that can be dispensed from a self-feeder. | | Fresh meat and fish should be chilled immediately, put in moistureproof bags not to exceed 4–5″ thick, quick frozen at −10°F (−23°C), stored at 0°F (−18°C), and not held more than 6 months.<br>Dehydration is an alternative to freezing for longtime storage, but protein quality is reduced.<br>Fish containing thiaminase should be cooked.<br>Cooking grain improves digestibility of carbohydrates but reduces protein digestibility. However, there may be diarrhea when fed raw. |

(Continued)

TABLE 4–10 (Continued)

| Class of Animal | Concentrates | Comments |
|---|---|---|
| Fish | Fish diets may be either wet or dry.<br>*Wet diets:*<br>1. Natural diets from the immediate environment. No feed preparation involved.<br>2. Artificial diets consisting of various organs, meats, and by-products from animals, poultry, and fish. These products must be refrigerated to prevent spoilage.<br>*Dry diets:*<br>Fed in pellet form, with the size of the pellet altered according to the size of the fish.<br>Pellets are of 2 kinds:<br>1. Pellets that *float* (they're spongelike) for surface feeders, like rainbow trout.<br>2. Pellets that *sink* for bottom feeders, like brown trout.<br>Eels are fed mixed diets in paste form. | Some researchers recommend that at least 50% of the ration of carp consist of natural ingredients. Maximum production of natural feeds usually calls for fertilization with organic or inorganic fertilizers.<br>Catfish ponds should be fertilized to establish an optimum level of plankton growth.<br>Fish products should be heated to destroy thiaminase.<br>Normally, catfish are bottom feeders, but they can be taught to feed at the surface. |

## EVALUATING FEEDSTUFFS

Some feeds are more valuable than others; hence, measures of their relative usefulness are important. Among such methods of evaluating feeds are the following:

1. Physical evaluation of feedstuffs.
2. Cost per unit of nutrients.
3. Proximate analysis.
4. Digestion (or metabolism) trials.
5. Measuring energy value of feedstuffs.
   a. Total digestible nutrients (TDN).
   b. Calorie system.
6. Other feed requirements.
7. Feeding trials.

## PHYSICAL EVALUATION OF FEEDSTUFFS

In order to produce or buy superior feeds, producers need to know what constitutes feed quality, and how to recognize it. They need to be familiar with those recognizable characteristics of feeds which indicate high palatability and nutrient content. If in doubt, the animals will tell them, for they like and thrive on high-quality feed.

The physical evaluation of feedstuffs, especially forages, is based largely on eye and smell appeal. Does it look good and smell good? The easily recognizable characteristics of hay of high feeding value are:

1. It is made from plants cut at an early stage of maturity, thus assuring the maximum content of protein, minerals, and vitamins, and the highest digestibility.
2. It is leafy, thus giving assurance of high protein content.
3. It is bright green in color, thus indicating proper curing, a high carotene or provitamin A content, and palatability.
4. It is free from foreign material, such as weeds and stubble.
5. It is free from mold and dust.
6. It is fine stemmed and pliable—not coarse, stiff, and woody.
7. It has a pleasing fragrant aroma; it "smells good enough to eat."

## COST PER UNIT OF NUTRIENTS

One method of arriving at the best buy in feeds is to compute and compare the cost per unit of nutrients, based on feed composition. Where chemical analysis of a specific feed is not available, feed composition tables may be used. Thus, feed composition tables may serve as a basis of feed purchasing and merchandising, as well as for ration formulation.

The use of the cost per unit of nutrients method can best be illustrated by the examples that follow:

If 44% protein (crude) soybean meal is selling at $9.88 per 100 lb whereas 35% protein (crude) linseed meal sells for $6.25 per 100 lb, which is the better buy? Divide $9.88 by 44 to get 22.5¢ per pound of crude protein for the soybean meal. Then divide $6.25 by 35 and get 17.9¢ per pound of crude protein for the linseed meal. Thus, at these prices linseed meal is the better buy—by 4.6¢ (22.5 − 17.9 = 4.6) per pound of crude protein.

When buying energy feed, one can compare the cost per pound of total digestible nutrients (TDN). For example, if corn is priced at $3.63 per 100 lb and has a TDN of 91% divide $3.63 by 91 and the result is 3.99¢ per pound of TDN. If milo with 86% TDN sells for $3.25 per 100 lb, divide $3.25 by 86, and the price is 3.78¢ per pound of TDN. Thus, milo would be the better buy by 0.21¢ (3.99 − 3.78 = 0.21) per pound of TDN.

Of course, it is recognized that many other factors affect the actual feeding value of each feed, such as (1) palatability, (2) grade of feed, (3) preparation of feed, (4) ingredients with which each feed is combined, and (5) quantities of each feed fed. It follows that, from the standpoint of the producer, the most important measurement of a feed's usefulness is in terms of *net returns* rather than cost per bag or cost per ton. To a swine producer, for example, cost per pound or per ton of feed, and pounds of feed required to produce a pound of pork, are important only as they reflect or affect the cost per unit of pork produced. Thus, if the cost of a growing-finishing ration is 6¢ per pound and 4 lb of the ration are required to produce 1 lb body weight, then the feed cost per pound of body weight can be arrived at by multiplying the above figures (6 × 4), which gives a feed cost of 24¢ per pound of pork. Obviously when rations are compared, the ration that produces a pound of pork at the lowest total feed cost is the most desirable from an economic point of view.

## PROXIMATE ANALYSIS

For more than 100 years, feeds have been analyzed by a method developed by two scientists, Henneberg and Stohmann, at the Weende Experiment Station in Germany. This method is called the proximate analysis, or the Weende System of feed analysis. Feeds are evaluated in terms of 6 components: (1) moisture, (2) ash, (3) crude protein, (4) ether extract, (5) crude fiber, and (6) nitrogen-free extract (see Table 4–11, next page).

### TABLE 4-11
### THE FRACTIONS OF PROXIMATE ANALYSIS

| Fraction | Procedure [1] | Major Components |
|---|---|---|
| 1. Moisture (dry matter). | Heat sample to constant weight at temperature just above boiling point of water. Loss in weight equals water. | Water and any volatile compounds (100% − $H_2O$ = DM%). |
| 2. Ash (mineral matter). | Burn at 500° to 600°C for 2 hours. | Mineral elements. |
| 3. Crude protein (protein averages 16% N; hence, N × 6.25 = crude protein). | Determine nitrogen by Kjeldahl process. | Proteins, amino acids, non-protein nitrogen. |
| 4. Ether extract (fat). | Extraction with diethyl ether. | Fats, oils, waxes, resins, pigments. |
| 5. Crude fiber (CF). [2] | Residue after boiling in weak acid and weak alkali. | Cellulose, hemicellulose, lignin. |
| 6. Nitrogen-free extract (NFE). [2] | Remainder; i.e., 100 minus sum of the other fractions. | Starch, sugars, some cellulose, hemicellulose, and lignin. |

[1]Each procedure can be applied to a separate sample, of standard weight, of the feedstuff to be analyzed; or a single sample can be used to determine dry matter, crude fat, and crude fiber. In the latter case, separate samples would be run for ash and crude protein.

[2]Carbohydrates (CHO = CF + NFE).

Today, feeds are being analyzed routinely through highly sophisticated chemical procedures. Many agricultural experiment stations, as well as most large feed companies, have facilities to analyze feeds for both the prevention and diagnosis of nutritional problems.

A chemical analysis gives a solid foundation on which to start in the evaluation of feeds. Thus, feed composition tables serve as a basis for ration formulation and for feed purchasing and merchandising. Commercially prepared feeds are required by state law to be labeled with a list of ingredients and a guaranteed analysis. Although state laws vary slightly, most of them require that the feed label (tag) show in percent the minimum crude protein and fat; and maximum crude fiber and ash. Some feed labels also include maximum salt, minimum TDN, and/or minimum calcium and phosphorus. These figures are the buyer's assurance that the feed contains the minimal amounts of the higher cost items—protein and fat; and not more than the stipulated amounts of the lower cost, and less valuable, items—the crude fiber and ash.

In addition to proximate analysis, methods are available for assaying individual vitamins—biological assays for some, chemical determinations for others.

Despite the recognized value of a chemical analysis, it is not the total answer for the following reasons:

1. **Feeds vary widely.** Because of wide variations in the composition of feeds, feed composition tables ("book values") should be considered as excellent, but not precise, guides. For example, the protein and moisture content of hay is quite variable, and the phosphorus and iodine content of soils affect plant composition. So, whenever possible, especially in large operations, it is best to take a representative sample of each major feed ingredient and have a chemical analysis made of it.

2. **It does not go far enough.** A chemical or proximate analysis does not provide any information relative to digestibility, palatability, texture, toxicity, digestive disturbances, the associated effect of feedstuffs, or nutritional adequacy. Neither does it tell anything about the soil on which the feed was grown, despite the fact that soils high in molybdenum and selenium

affect the composition of feeds produced. Other similar soil-plant-animal relationships exist and are important. Thus, further steps need to be taken to evaluate a feed.

## DIGESTION (OR METABOLISM) TRIAL

Animals are not able to extract all the nutrients present in feeds. The actual value of ingested nutrients is dependent upon the use which the body is able to make of them. The first consideration here is digestibility, since undigested nutrients do not get into the body proper.

A digestion trial is made by determining the percentage of each nutrient in the feed through chemical analysis; giving the feed to the test animal for a preliminary period (usually 7 to 10 days for ruminants), so that all residues of former feeds will pass out of the digestive tract; giving weighed amounts of the feed during the test period (7 to 10 days for ruminants); collecting, weighing, and analyzing the feces; determining the difference between the amount of the nutrient fed and the amount found in the feces; and computing the percentage of each nutrient digested. The latter figure is known as the *digestion coefficient* for that nutrient in the feed.

Various techniques and equipment may be used to make the fecal collections; among them; a specially designed digestion stall (see Fig. 4-27); collection harness and bag; markers (such as carmine, ferric oxide, chromic oxide, or soot), fed with the ration at the beginning and the end of the collection period; and indicators of inert reference subject.

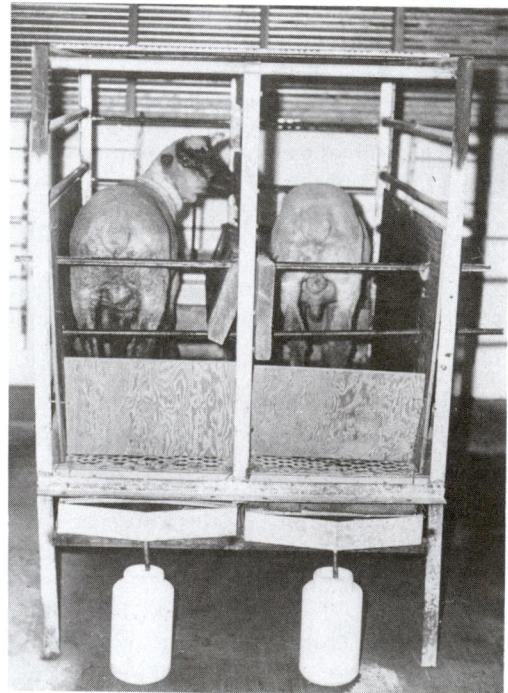

Fig. 4-27. Metabolism stalls for digestibility studies with sheep. Note the pans at the bottom for collecting feces and the containers on the floor for collecting urine. (Courtesy, Irrigated Agriculture Research and Extension Center, Washington State University, Pullman)

The digestibility of a feedstuff is affected by four factors: the species of animal (type of digestive tract), the chemical composition of the feed, the way in which the feed was processed, and the individuality of the animal (some animals have more efficient digestive systems than others).

## MEASURING AND EXPRESSING ENERGY VALUE OF FEEDSTUFFS

One nutrient cannot be considered as more important than another, because all nutrients must be present in adequate amounts if efficient production is to be maintained. Yet, historically, feedstuffs have been compared or evaluated primarily on their ability to supply energy to animals. This is understandable because (1) energy is required in larger amounts than any other nutrient, (2) energy is most often the limiting factor in livestock production, and (3) energy is the major cost associated with feeding animals.

Our understanding of energy metabolism has increased through the years. With this added knowledge, changes have come in both the methods and terms used to express the energy value of feeds.

The methods of measuring the energy value of feedstuffs currently employed in the United States are:

1. Total digestible nutrient (TDN)
2. Calorie System, including—
   a. Gross energy (GE)
   b. Digestible energy (DE)
   c. Metabolizable energy (ME)
   d. Net energy (NE)

Each system has its advantages and advocates. Also, both the difficulty in determining energy values of feeds according to the different systems of measurement and the accuracy of the results increase in the order that they are listed above. Nevertheless, more and more feedstuffs are being evaluated in calories, with net energy being the method of choice.

Although net energy is the most precise measure known of the real value of a feed, it is very difficult to determine. It requires either (1) the measurement of all forms of energy loss, or (2) the actual amount of energy retained by the animal or produced as a useful product. The first determination may be accomplished by balance methods for determining energy, and the second by the comparative slaughter technique.

## TOTAL DIGESTIBLE NUTRIENT (TDN) SYSTEM

*Total Digestible Nutrients (TDN) is the sum of the digestible protein, fiber, nitrogen-free extract, and fat × 2.25.* It has been the most extensively used measure for energy in the United States.

Back of TDN values are the following steps:

1. **Digestibility**—The digestibility of a particular feed for a specific species is determined by a digestion trial.

2. **Computation of digestible nutrients**—Digestible nutrients are computed by multiplying the percentage of each nutrient in the feed (protein, fiber, nitrogen-free extract [NFE], and fat) by its digestion coefficient. The result is expressed as digestible protein, digestible fiber, digestible NFE, and digestible fat. For example if dent corn contains 8.9% protein of which 77% is digestible, the percent of digestible protein is 6.9.

3. **Computation of total digestible nutrients (TDN)**—The TDN is computed by using the following formula:

% TDN = % DCP + % DCF + % DNFE + (% DEE × 2.25)

where DCP = digestible crude protein; DCF = digestible crude fiber; DNFE = digestible nitrogen-free extract; and DEE = digestible ether extract.

TDN is ordinarily expressed as a percent of the ration or in units of weight (lb or kg), not as a caloric figure.

The main **advantage** of the TDN system is that it has been used for a very long time and many people are acquainted with it.

The main **disadvantages** of the TDN system are:

1. It is really a misnomer, because TDN is not an actual total of the digestible nutrients in a feed. It does not include the digestible mineral matter (such as salt, limestone, and defluorinated phosphate—all of which are digestible); and the digestible fat is multiplied by the factor 2.25 before being included in the TDN figure, because its energy value is higher than carbohydrates and protein. As a result of multiplying fat by the factor 2.25, feeds high in fat will sometimes exceed 100 in percentage TDN (a pure fat with a coefficient of digestibility of 100% would have a theoretical TDN value of 225%—100% × 2.25).

2. It is an empirical formula based upon chemical determinations that are not related to actual metabolism of the animal.

3. It is expressed as a percent or in weight (lb or kg), whereas energy is expressed in calories.

4. It takes into consideration only digestive losses; it does not take into account other important losses, such as losses in the urine, gases, and increased heat production (heat increment).

5. It overevaluates roughages in relation to concentrates when fed for high rates of production, due to the higher heat loss per pound of TDN in high-fiber feeds.

Because of these several limitations, in the United States the TDN system is gradually being replaced by other energy evaluation systems, particularly net energy. However, due to the voluminous TDN data on many feeds and long-standing tradition, it will continue to be used by many people for a long time to come.

## CALORIE SYSTEM

Calories are used to express the energy value of feedstuffs. *One calorie (always written with a small "c") is the amount of heat required to raise the temperature of 1 g of water 1°C precisely from 14.5°C to 15.5°C).*

To measure this heat energy, an instrument known as the bomb calorimeter is used, in which the feed (or other substance) to be tested is placed and burned in the presence of oxygen.

Through various digestive and metabolic processes, much of the energy in feed is dissipated as it passes through the animal's digestive system. About 60% of the total combustible energy in grain and about 80% of the total combustible energy in roughage is lost as feces, urine, gases, and heat. These losses are illustrated in Figs. 4–28 and 4–29 (see next page).

As shown in Figs. 4–28 and 4–29, energy losses occur in the digestion and metabolism of feed. Measures that are used to express animal requirements and the energy content of feeds differ primarily in the digestive and metabolic losses that are included in their determination. Thus, the following terms are used to express the energy value of feeds:

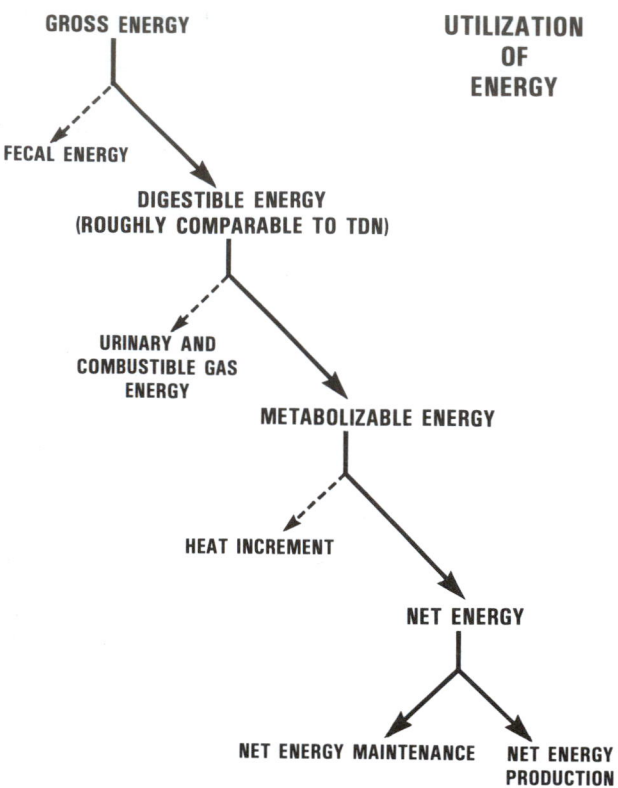

UTILIZATION
OF
ENERGY

Fig. 4-28.  Utilization of energy

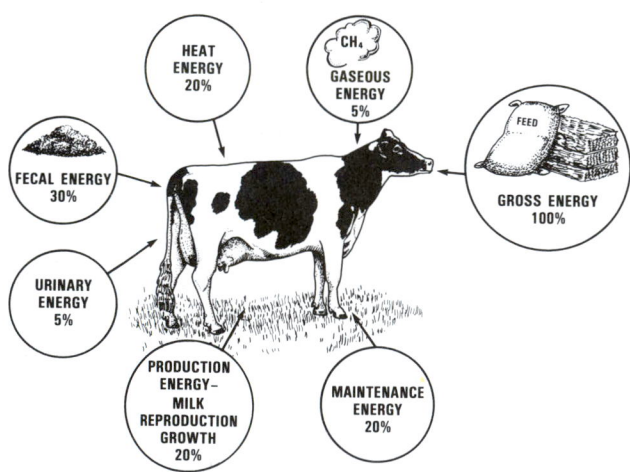

Fig. 4-29.  Energy utilization by a lactating cow showing average partition of feed energy by the animal.

Fig. 4-30.  Bomb calorimeter for the determination of gross energy (caloric content). (Courtesy, Parr Instrument Co., Moline, IL)

- **Digestible energy (DE)**—*Digestible energy is that portion of the GE in a feed that is not excreted in the feces.*

- **Metabolizable energy (ME)**—*Metabolizable energy represents that portion of the GE that is not lost in the feces, urine, and gas.* Although ME more accurately describes the useful energy in the feed than does GE or DE, it does not take into account the energy lost as heat.

- **Net energy (NE)**—*Net energy represents the energy fraction in a feed that is left after the fecal, urinary, gas, and heat losses are deducted from the GE.* The net energy, because of its greater accuracy, is being used increasingly in ration formulations, especially in computerized formulations for large operations.

Although net energy is a more precise measure of the real value of the feed than other energy values, it is much more difficult to determine.

Two systems of net energy evaluation are presently being used. Lofgreen and Garrett[5] developed a system whereby the net energy requirements are listed as dictated by physiological functions—for example, net energy for maintenance (NE$_m$) and net energy for gain (NE$_g$). Also, Moe and Flatt[6] developed a net energy system that compares the physiological function to that of lactation through the use of regression analysis. This value, NE$_{lactation}$, is applicable for all physiological functions.

## CALORIMETRIC SYSTEMS OF MEASURING ANIMAL HEAT

Measurement of animal heat began 200 years ago, when Lavoisier and LaPlace, the great French scientists, enclosed a guinea pig in a chamber surrounded by ice. The amount of ice melted by the heat of the animal multiplied by the latent heat of the ice indicated the heat given off by the guinea pig, thus giving rise to direct calorimetry. They also noted that the melting of the ice was directly related to the amount of carbon dioxide given off by the animal. This was the basis of in-

- **Gross energy (GE)**—*Gross energy represents the total combustible energy in a feedstuff.* It does not differ greatly among feeds, except for those high in fat. For example, 1 lb of corncobs contains about the same amount of GE as 1 lb of shelled corn. Therefore, GE does little to describe the useful energy in feeds for finishing animals.

[5]Lofgreen, G. P., and W. N. Garrett, "A System for Expressing Net Energy Requirements and Feed Values for Growing and Finishing Beef Cattle," *Journal of Animal Science*, Vol. 27, 1968, p. 793.

[6]Moe, P. W., and W. P. Flatt, "Net Energy of Feedstuffs for Lactation," *Journal of Dairy Science*, Vol. 52, 1969, p. 928.

direct calorimetry. Since that time, more sophisticated types of equipment for measuring body heat have been developed, but the basic principles remain the same.

## DIRECT CALORIMETRY

Armsby, while Director of the Agriculture Experiment Station at Pennsylvania State College in the late 1800s, felt that there was a great need for more sophisticated techniques in the evaluation of feeds. At the time, feeds were evaluated solely on their digestibility. So, together with two other Pennsylvania State professors, Fries and Osmond, Armsby constructed a respiration calorimeter capable of directly measuring heat production in large animals. It took six people at any one time to operate the machine, plus an additional person to handle the animal being studied. Through the use of this apparatus, Armsby became one of the foremost authorities in the research of energy utilization in animals. Today, newer and cheaper methods of measuring energy utilization in animals have rendered the machine obsolete; but Armsby's calorimeter still stands, now housed in a museum commemorating the monumental research which he conducted.

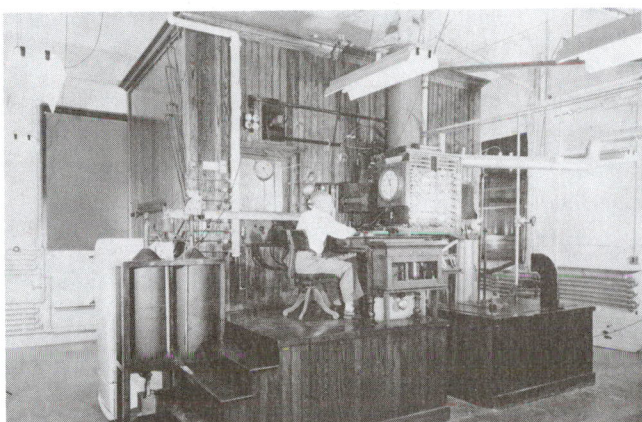

Fig. 4–31. Dr. R. W. Swift, former head of the Department of Animal Nutrition at Pennsylvania State University, is shown operating the respiration calorimeter built by H. P. Armsby. (Courtesy, Pennsylvania State University, University Park)

In direct calorimetry, the animal is confined to a well-insulated chamber and the heat losses (by radiation, convection, and conduction from the body surface; by evaporation of water from the skin and lungs; and by excretion of urine and feces) are measured either by (1) the increase in temperature of a known volume of water, or (2) electrical current generated as heat passes across thermocouples (gradient layer calorimetry). Brody[7] estimated that about one-fourth of the heat lost by the body resulted from moisture vaporization. The remaining three-fourths of the body heat is lost by radiation, conduction, and convection.

Direct calorimetry has been useful in the determination of the heat increment fraction of energy utilization. This parameter includes the heat of fermentation and the heat of nutrient metabolism. The heat produced by feeding a particular ration is first measured. Then, the intake of that feed is increased and the heat produced by the animal is measured. The increase in heat due to the increase in feed intake is subsequently termed *heat increment.*

Direct calorimetry is the most accurate method of measuring the heat production of animals, but it is costly and arduous. The machine itself is very expensive to build and operate; and considerable labor is involved in controlling the animal, running the machine, and analyzing the results. For these reasons, indirect calorimetry is usually the method of choice for the evaluation of energy in feeds.

## INDIRECT CALORIMETRY

In indirect calorimetry, heat of production is calculated from measurement of the respiratory exchange—the $O_2$ consumption, and usually the $CO_2$ production—of the animal. This method is based on the fact that $O_2$ consumption and $CO_2$ production are closely related to heat production. The ratio of carbon dioxide produced to oxygen consumed, which is known as the respiratory quotient (RQ), is distinctive for each compound; hence, it serves to indicate the type of nutrient being metabolized. Respiratory quotient can be determined through the following equation:

$$RQ = \frac{CO_2 \text{ produced}}{O_2 \text{ consumed}}$$

Therefore, if we look at the equation for the catabolism of carbohydrates, we can calculate a respiratory quotient.

1. $C_6H_{12}O_6 + 6\,O_2 \longrightarrow 6\,CO_2 + 6\,H_2O + \text{heat}$

2. $RQ = \dfrac{6\,CO_2}{6\,O_2} = 1$

The RQ for fat would be less than 1, as seen in the following example:

1. Palmitic acid $(C_{16}H_{32}O_2) + 23\,O_2 \longrightarrow$
   $16\,CO_2 + 16\,H_2O + \text{heat}$

2. $RQ = \dfrac{16\,CO_2}{23\,O_2} = .70$

Most mixed fats have RQ values of about .7, while short-chain fatty acids have higher values (about .8). Proteins have values intermediate between carbohydrates and fats—about .81.

It should be noted that in stress conditions, the RQ value can be greater than one. This can occur when there is hyperventilation where large quantities of carbon dioxide are exhaled while an oxygen debt exists in the body. Metabolic acidosis creates an excess exhalation of carbon dioxide as a compensatory mechanism.

The total heat production of the animal may be computed by either (1) measuring RQ along with oxygen, then making readings from tables; or (2) using a single equation relating heat production to the respiratory exchange.

Comparative measurements of direct and indirect calorimetry reveal that the two methods give results that are in close agreement.

The actual measurement of the respiratory exchange of animals may be accomplished by several different types of apparatus; among them (1) the open circuit gravimetric system, (2) the open circuit chamber system involving gas analysis, (3) the open circuit mask system, and (4) the closed circuit mask system (spirometer).

1. **Open circuit gravimetric system.** This system is by far the simplest and easiest to set up, maintain and operate. Also, it is readily acceptable to measure heat production from small

---

[7]Brody, S. B., *Bioenergetics and Growth*, Reinhold Publishing Company, New York, NY, 1945.

animals (poultry, mice, rats, and rabbits). The needed materials are an enclosed holding chamber, scales, water and carbon dioxide absorbants, and an air pump. A schematic diagram of the setup is presented in Fig. 4–32.

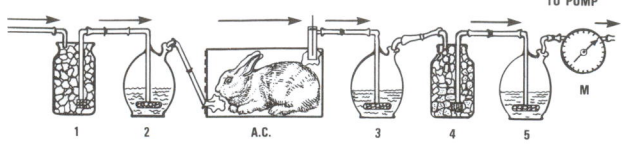

Fig. 4–32. Haldane respiration apparatus for the open circuit gravimetric system of indirect calorimetry: A.C., animal chamber. M, meter for measuring rate of ventilation. Bottles 1 and 4 contain soda lime (or caustic alkali) for absorption of carbon dioxide. Bottles 2, 3 and 5 contain sulfuric acid for absorption of water. Air entering the animal chamber is freed of carbon dioxide and moisture by passing through bottles 1 and 2. The animal gives off carbon dioxide and moisture, and these are collected in the bottles of the outgoing chain. Bottle 5 is necessary because soda lime gives off moisture. The gain in weight of bottles 3, 4 and 5 minus the loss in weight of the animal and chamber represents the oxygen consumption.

2. **Open circuit chamber system.** This system of indirect calorimetry is very similar to the open circuit gravimetric system with a few exceptions. The open circuit chamber system enables researchers to use large animals. Since the chamber is too large to allow for accurate weighing procedures, and since large quantities of $CO_2$ are being exhaled, modifications of the open circuit gravimetric system are employed.

The composition of the air that is pumped into the chamber is carefully analyzed. The air circulation within the chamber is measured with a gas flow meter; and at given periods, air is sampled and analyzed to determine the decrease of oxygen and the increase of $CO_2$.

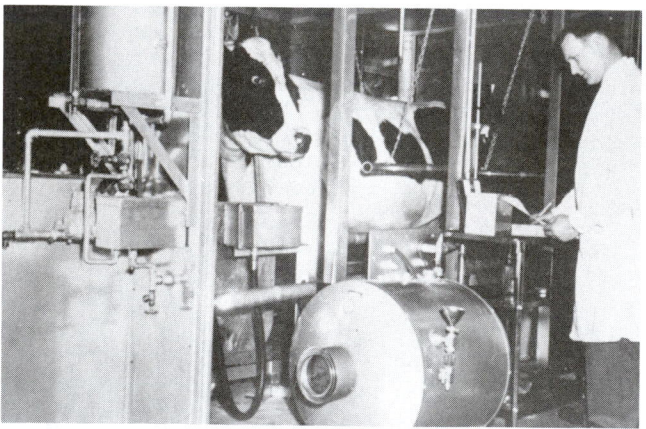

Fig. 4–33. A cow in an open circuit respiration chamber. The gas meter to the left of the chamber is used to measure the respiratory exchange of the cow. These data, plus the gas composition, provide the information needed to calculate the heat production (HP) of the cow. The HP of an animal consuming feed in a thermoneutral environment is composed of the heat increment (heat of fermentation plus heat of nutrient metabolism) plus heat used for maintenance (basal metabolism plus voluntary activity). (Courtesy, USDA)

3. **Open circuit mask system.** This type of indirect calorimetry is commonly used in human research. The subject is told to exhale in a mask for a short period, and all of the exhaled air is collected in a bag or spirometer for analysis. Because large animals exhale large amounts of air, a special collection and sampling apparatus is utilized.

4. **Closed circuit mask system (spirometer).** This system is commonly used in both human and animal research. Doctors frequently use this method of indirect calorimetry to measure the basal metabolic rate of humans. The subject, either human or animal, breathes into a mask which is connected to a spirometer—an apparatus which is filled with oxygen. A one-way valve enables the subject to inhale the oxygen from within the bell of the spirometer, and the rate of oxygen consumption is then recorded on a kymograph. Once the rate of oxygen consumption has been determined, the researcher can calculate the utilization of energy through the use of a number of correction factors for sex, weight, and age of the subject, as well as atmospheric temperature and barometric pressure.

## COMPARATIVE SLAUGHTER METHOD OF DETERMINING NET ENERGY

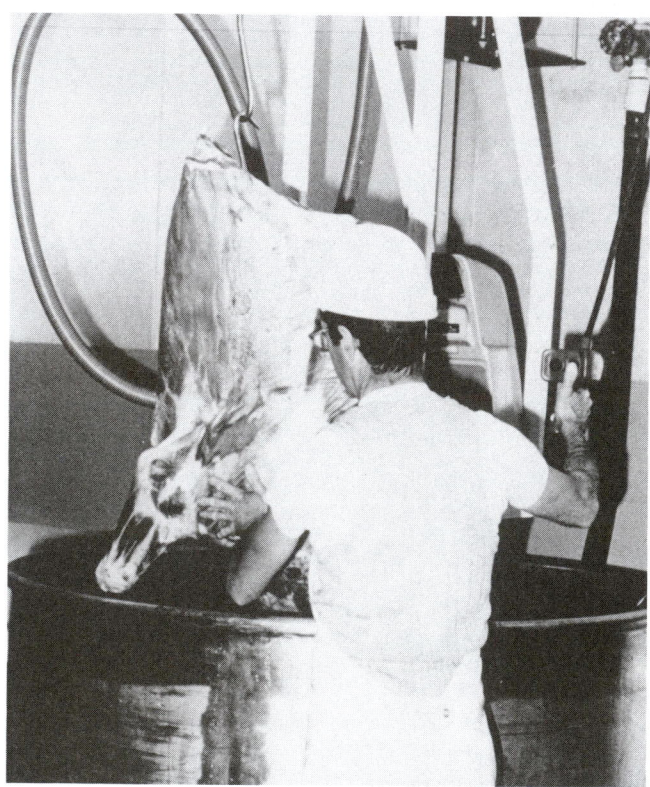

Fig. 4–34. This shows the dipping procedure being used to obtain carcass density from which specific gravity is computed, which, in turn is used to estimate carcass fat content. By use of an initial and a final slaughter group of animals in each feeding trial, the energy gain can be measured, giving a more accurate measure of the true feed value than does just liveweight gain. Further, the method provides a measure of energy content without grinding and chemical analysis of the carcass. (Courtesy, University of California, Davis)

The comparative slaughter method of determining net energy (energy storage and heat production) is an old technique with a new look. It was first employed by Mitchell and coworkers, of the Illinois Agricultural Experiment Station, in 1926. But, for the most part, it was discontinued, because chemically analyzing the body (carcass) is slow, tedious, and expensive. Then, in 1959, Lofgreen and Garrett, of the California Agricultural Experiment Station, reported an ingenious modification of the comparative slaughter technique. By making use of established relationships between carcass density and the composition of the animal, they were able to estimate the energy content of the carcass without analyzing the body chemically. The density of the carcass is determined by weighing in water—using a dipping procedure (see Fig. 4–34), which can be done quickly and without affecting the sale value of the carcass.

The comparative slaughter method is especially well suited to studies involving growing and fattening animals—cattle, lambs, hogs, and broilers, in which the amount of energy stored in the carcass can be measured. However, it is not adapted to use with dairy animals.

The comparative slaughter method requires relatively large numbers of animals. A random sample, or check group, is selected and slaughtered at the beginning of an experiment to determine the initial body composition. Then, at the close of the experiment, the remaining animals are slaughtered and analyzed. The difference in the calorie content of the two groups represents the energy storage or gain, which is a far more accurate measure of the true value of feed than liveweight gain.

The comparative slaughter method of feed evaluation is unique in the following respects:

1. It provides a relatively inexpensive way in which to determine net energy values.

2. The animals can be kept under more natural conditions, similar to those found in a commercial enterprise.

3. Feeds can be assigned $NE_m$ (net energy for maintenance and $NE_g$ (net energy for gain) values in keeping with the efficiency of metabolizable energy utilization for these different physiological processes.

The modified comparative slaughter technique has had a major impact on feed evaluation and ration formulation throughout the United States.

## ENERGY DEFINITIONS AND CONVERSIONS

Some pertinent definitions and conversions of energy terms follow:

- **Calorie (cal)**—The amount of energy as heat required to raise the temperature of 1 gram of water 1°C (precisely from 14.5 to 15.5°C). It is equivalent to 4.184 joules. Although *not preferred*, it is also called a *small calorie* and so designated by being spelled with a lower case "c." **Note well:** In popular writings, especially those concerned with human caloric requirements, the term calorie is frequently used erroneously for the kilocalorie.

- **Kilocalorie (kcal)**—The amount of energy as heat required to raise the temperature of 1 kilogram of water 1°C (from 14.5 to 15.5°C). Equivalent to 1,000 calories. In human nutrition, it is referred to as a kilogram calorie or as a *large Calorie* and is so designated by being spelled with a capital "C" to distinguish it from the *small calorie*.

- **Megacalorie (Mcal)**—Equivalent to 1,000 kilocalories or 1,000,000 calories. Also, referred to as a *therm*, but the term megacalorie is preferred.

- **British Thermal Unit (BTU)**—The amount of energy as heat required to raise 1 pound of water 1°F; equivalent to 252 calories. This term is seldom used in animal nutrition.

- **Joule**—A proposed international unit (4.184J = 1 calorie) for expressing mechanical, chemical, or electrical energy, as well as the concept of heat. In the future, energy requirements and feed values will likely be expressed by this unit.

- **Converting TDN to Mcal**—One pound of TDN = 2.0 Mcal or 2,000 kcal. It is recognized, however, that the roughage component in a ration affects its energy value. Thus, when converting all-roughage rations from TDN to Calories, some scientists figure that 1 pound of TDN = 1,500 kcal, instead of 2,000.

- **Hay equivalent (HE)**—This is the energy equivalent of 1 ton of hay which, on the average, contains 800 Mcal of net energy. With an Animal Unit Month (AUM) being equivalent to 320 Mcal of net energy, 2.5 AUM are required to furnish the same amount of energy as 1 ton of hay.

## OTHER FEED REQUIREMENTS

In addition to being nutritionally complete, the following feed requirements are important:

1. **Palatability.** If they don't eat it, they won't produce; and if they don't eat enough, feed efficiency will be poor. The relationship of feed consumption to feed efficiency becomes clear when it is realized that the maintenance requirement of an animal producing at a low rate represents a much greater percent of the total feed required than for an animal producing at a more rapid rate.

Palatability is the result of the following factors sensed by the animal in locating and consuming feed: appearance, odor, taste, texture, temperature, and, in some cases, auditory properties of the feed (like the sound of pigs eating corn). These factors are affected by the physical and chemical nature of the feed.

2. **Variety.** Some variety in the ration is desirable, particularly from the standpoints of assuring (a) increased palatability, and (b) balance of nutrients—for example, all the essential amino acids.

3. **Digestive disturbances.** Bloat, colic, scours, and constipation are the bane of all feeders. The choice of feeds can give a big assist in minimizing such disturbances. For example, bloat in cattle and colic in horses can be lessened by avoiding lush or frosted pastures; scours can be lessened by proper feeding; and constipation can be corrected by feeding alfalfa, wheat bran, linseed meal, and molasses.

4. **Bulk.** The amount of bulk in the ration will vary. Ruminants can consume bulkier feeds than monogastric animals; the younger the animal, the less bulk; and the higher the production desired, the less bulky the ration. Also, the relative cost of feeds—concentrates vs roughages—will influence the relative amount of bulk in the ration.

5. **Cost.** Cost is important. But even more important is net returns; hence, it may well be said that it is net returns rather than cost per ton, or per bag, that counts.

6. **Poisonous plants and feeds.** Poisonous plants and feeds should be avoided. The livestock producer should know the poisonous plants common to the area, and avoid them. Also, the following poisons should be avoided: prussic acid, hydrocyanic acid, ergot, scabbed grain, smut on grain, spoiled or moldy feed, botulism, aflatoxins, selenium poisoning, nitrate poisoning, lead poisoning, and mercury poisoning.

## FEEDING TRIALS

Each method of evaluating feedstuffs, discussed earlier in this chapter, has a place and is valuable. But none of them takes into consideration all the factors which determine the true value of any feed for a particular class of livestock. The "court of last judgment" for determining the true value of a feedstuff is the animal. How well do the animals eat the "stuff"? How does it affect their health and well-being? How are they producing? Answers to these questions call for feeding the ingredient or ration under controlled conditions to the particular class of livestock.

The U.S. Department of Agriculture and the state experiment stations have conducted numerous experiments to determine the feeding value of specific feeds and of different rations, for each class of livestock. The results of these studies have paid handsome dividends.

Generally speaking, it is in the best interest of livestock producers that experiment stations continue to assume responsibility for the majority of research, including the evaluation of feedstuffs. In comparison with private industry, they have more trained research personnel; generally their studies involve greater accuracy and more controls; and their results are unbiased and unquestioned. But even experiment stations need to bear in mind that adequate animal numbers are important; that there are species and individual animal differences; that feeds differ widely; and that the results must be repeatable.

## CONDUCTING APPLIED TESTS ON THE FARM

When carefully conducted and properly interpreted and used, feeding trials can be a valuable adjunct in many of today's large livestock operations. Among their virtues, the operator can study area and feed differences. Among their limitations, usually there is less accuracy and there are fewer controls than in most university conducted experiments. For the latter reason, most of them should be looked upon as applied tests or demonstrations *per se*, rather than carefully controlled, basic experiments; terminology which doesn't detract from their value, but which does place them in proper perspective.

Applied feed tests have been conducted extensively in many large commercial cattle feedlots.

## FEEDING STANDARDS

*Feeding standards are tables showing the amounts of one or more nutrients needed by different species of animals for different purposes, such as growth, finishing, and lactation.* They serve as guides in balancing rations and feeding practices. Most feeding standards are expressed in (1) quantities of nutrients required per day, and/or (2) percent of the ration; the first type being used for animals given exact quantities of a ration, and the second type used when animals are fed free choice (*ad libitum*).

The first feeding standard was developed by Thaer, a German scientist, in 1810. He took meadow hay as his standard, compared the extractable nutrients of other feeds to it, then assigned them *hay values*. In 1859, Grouven made use of analyses of protein, fat, and carbohydrate to formulate the first feeding standard for farm animals. In 1864, Wolff devised a standard based on digestible nutrients obtained from feeding trials. In 1897, Lehmann modified Wolff's standards. Other systems followed. Among them, the TDN system by Henry, published in the first edition of his book, *Feeds and Feeding*,

in 1898[8]; the starch values of Kellner, a German scientist, in 1907; the Scandinavian feed-unit system (Woll, 1912); the dairy cow standards by Haecker, of Minnesota, in 1914; the net energy values of Armsby, of Pennsylvania, in 1915; the productive units developed by Mollgaard of Denmark, in 1939; and the productive energy values computed by Fraps of the Texas Station (1937, 1941).

Today, the most widely used feeding standards in the United States are those published for each animal species by the National Research Council (NRC) of the National Academy of Sciences. These requirements were adapted for, and presented in, *Feeds & Nutrition* and *Feeds & Nutrition Digest*, books by M. E. Ensminger, *et al.*, published by Ensminger Publishing Company; hence, the reader is referred thereto. In England, similar standards are issued by the Agricultural Research Council (ARC). Other countries have similar bodies which make recommendations on the nutritive requirements of animals.

In the United States, the TDN system is giving way to other energy evaluation systems, particularly net energy. England uses metabolizable energy (ME), adjusted according to the efficiency with which a feedstuff or diet is used for a particular purpose.

Other European standards are based on starch equivalents, Scandinavian Feed Units, and other methods of evaluation.

Although feeding standards are excellent and needed guides, there are still many situations where nutrient needs cannot be specified with great accuracy for animals. Also, in practical feeding operations, economy must be considered; for example, dairy producers are interested in obtaining that level of milk production which will make for the largest net returns in light of current feed costs and the market price of milk. Moreover, feeding standards tell nothing about the palatability, physical nature, or possible digestive disturbances of a ration. Neither do they give consideration to animal differences, management differences, effects of such stresses as weather, disease, parasitism, surgery (dehorning, castrating, etc.). Thus, there are many variables that alter the nutrient needs and utilization of animals—variables that are difficult to include quantitatively in feeding standards, even when feed quality is well known.

## BALANCED RATIONS

To supply all the needs—for maintenance, growth, finishing, reproduction, lactation, work (or running), and/or wool—the different classes of animals must receive sufficient feed to furnish the necessary quantity of energy (carbohydrates and fats), proteins, minerals, vitamins, and water. Perhaps under certain conditions feed additives may be desirable, although it is not likely that they are essential. A ration that meets all these needs is said to be balanced. More specifically, by definition, *a balanced ration is one which provides an animal the proper proportions and amounts of all the required nutrients for a period of 24 hours.*[9]

---

[8]Henry, W. A., (1850–1932) was Professor of Agriculture, and later Dean, at the University of Wisconsin. His textbook, *Feeds and Feeding*, bore the subtitle, "A Handbook for the Student and Stockman." Frank B. Morrison, Dean Henry's son-in-law, eventually became junior author, then in 1936 he became sole author of *Feeds and Feeding*, long the leading textbook in the field.

[9]Although Webster defines the noun *ration* as "the amount of food supplied to an animal for a definite period, usually for a day," to most livestock producers the word implies the feeds fed to an animal or animals, without limitation to the time in which they are consumed. In this and other chapters of *Animal Science*, the author accedes to the common usage of the word, rather than to dictionary correctness.

When in confinement, animals have access only to the feeds provided by the caretaker. This points up the importance of balanced rations.

Several suggested rations for different classes of livestock are given in the feeding chapters of this book which are devoted to the respective classes of livestock. Generally these rations will suffice, but it is recognized that rations should vary with conditions, and that many times they should be formulated to meet the conditions of a specific farm or ranch, or to meet the practices common to an area.

Also, good livestock producers should know how to balance rations. Then, if the occasion demands, they can do it. Perhaps of even greater importance, they will then be able more intelligently to select and buy rations with informed appraisal; to check on how well their manufacturer, dealer, or consultant is meeting their needs; and to evaluate the results.

## HOW TO BALANCE A RATION

Ration formulation consists of combining feeds to make a ration that will be eaten in the amount needed to supply the daily nutrient requirements of the animal. This may be accomplished by the methods presented later in this chapter, but first the following pointers are necessary:

1. In computing rations, more than simple arithmetic should be considered, for no set figures can substitute for experience. Compounding rations is both an art and a science—the art comes from animal know-how and experience, and keen observation; the science is largely founded on chemistry, physiology, and bacteriology. Both are essential for success.

2. Before attempting to balance a ration, the following major points should be considered:

a. **Availability and cost of the different feed ingredients**—Preferably, cost of ingredients should be based on delivery after processing—because delivery and processing costs are quite variable.

b. **Moisture content**—When considering costs and balancing rations, feeds should be placed on a comparable moisture basis; usually, an air-dry basis, or 10% moisture content, is used. This is especially important in the case of high-moisture grain or silage.

c. **Composition of the feeds under consideration**—Feed composition tables (*book values*), or average analysis, should be considered as good guides, but not precise, because of wide variations in the composition of feeds. For example, the protein and moisture contents of sorghum, hay, and silages are quite variable. Whenever possible, especially with large operations, it is best to take a representative sample of each major feed ingredient and have a chemical analysis made of it for the more common constituents—protein, fat, fiber, nitrogen-free extract, and moisture; and often calcium, phosphorus, and carotene. Such ingredients as oil meals and prepared supplements, which must meet specific standards, need not be analyzed so often, except as quality-control measures.

Despite the recognized value of a chemical analysis, it is not the total answer. It does not provide information on the availability of nutrients to the animal; it does not tell anything about the associated effect of feedstuffs—for example, the apparent way in which beet pulp enhances the value of ground milo; and it does not tell anything about taste, palatability, texture, or undesirable physiological ef-

fects such as bloat and laxativeness. Nevertheless, a chemical analysis does give a basis on which to start the evaluation of feeds. Also, with chemical analysis at hand, and bearing in mind that it's the composition of the total feed (the finished ration) that counts, the person formulating the ration can more intelligently determine the quantity of protein to buy, and the kind and amounts of minerals and vitamins to add.

d. **Soil analysis**—If the origin of a given feed ingredient is known, a soil analysis or knowledge of the soils of the area can be very helpful; for example, (1) the phosphorus content of soils affects plant composition, (2) soils high in molybdenum and selenium affect the composition of the feeds produced, (3) iodine- and cobalt-deficient areas are important in animal nutrition, and (4) other similar soil-plant-animal relationships exist.

e. **The nutrient allowances**—This should be known for the particular class of animals for which a ration is to be formulated. Also, it must be recognized that nutrient requirements and allowances must be changed from time to time, as a result of new experimental findings.

3. In addition to providing a proper quantity of feed and to meeting the protein and energy requirements, a well-balanced and satisfactory ration should be:

a. Palatable and digestible.

b. Economical. Generally speaking, this calls for the maximum use of feeds available in the area, especially forages.

c. Adequate in protein content, but not higher than is actually needed. Generally speaking, medium and high-protein feeds are in scarcer supply and higher in price than high-energy feeds.

d. Well fortified with the needed minerals, or free access to suitable minerals should be provided; but mineral imbalances should be avoided.

e. Well fortified with the needed vitamins.

f. So formulated, where ruminants are involved, as to nourish the billions of bacteria in the paunch in order that there will be satisfactory (1) digestion of roughages, (2) utilization of lower quality and cheaper proteins and other nitrogenous products (thus, it is possible to use urea to constitute up to one-third of the total protein of the ration of ruminants, provided care is taken to supply enough carbohydrates and other nutrients to assure adequate nutrition for rumen bacteria), and (3) synthesis of B vitamins.

This means that rumen microorganisms must be supplied adequate (1) energy, including small amounts of readily available energy such as sugars or starches; (2) ammonia-bearing ingredients such as proteins, urea, and ammonium salts; (3) major minerals, especially sodium, potassium, and phosphorus; (4) cobalt and possibly other trace minerals; and (5) unidentified factors found in certain natural feeds rich in protein or nonprotein nitrogenous constituents.

g. One that will enhance, rather than impair, the quality of the product (meat, milk, eggs, or wool/mohair) produced.

4. In addition to considering changes in availability of feeds and feed prices, ration formulation should be altered at stages to correspond to changes in weight and productivity of animals.

The above points are pertinent to the balancing of rations, regardless of the mechanics of computation used.

## METHODS OF FORMULATING RATIONS

In the sections that follow, four different methods of ration formulation are presented: (1) the square method, (2) the trial-and-error method, (3) the net energy method, and (4) the computer method. Despite the sometimes confusing mechanics of each system, if done properly, the end result of all four methods is the same—a ration that provides the desired allowance of nutrients in correct proportions economically (or at least cost), but more important, so as to achieve the greatest net returns—for it is net profit, rather than cost, that counts. Since feed usually represents the greatest cost item in livestock production, the importance of balanced rations is evident.

An exercise in ration formulation follows for purposes of illustrating the application of each of these four methods:

1. **Square method,** applied to a swine ration.
2. **Trial-and-error method,** applied to a lactating cow ration.
3. **Net energy method,** applied to a cattle finishing ration.
4. **Computer method,** applied to a layer ration.

### SQUARE (OR PEARSON SQUARE) METHOD

The square method is a simple, direct, and easy way in which to figure proportions between two ingredients. It permits quick substitution of feed ingredients in keeping with market fluctuations, without disturbing the protein content.

In balancing rations by the square method, it is recognized that one specific nutrient alone receives major consideration. Correctly speaking, therefore, it is a method of balancing one nutrient requirement, with no consideration given to the other nutritive requirements.

To compute rations by the square method, or by any other method, it is first necessary to have available both feeding standards and feed composition tables. These may be obtained from *Feeds & Nutrition,* by Ensminger, *et al.,* or from the nutrient requirement publications of the National Academy of Sciences, of which there are separate publications for each species of animal.

The following example shows how to use the square method in formulating a swine ration:

**Example.** *A swine producer has 40–lb pigs to which it is desired to feed a 16% protein ration until they reach 120 lb weight. Corn containing 8.9% protein is on hand. A 36% protein supplement, which is reinforced with minerals and vitamins, can be bought. What percent of the ration should consist of corn and of the 36% protein supplement?*

Step by step, the procedure in balancing this ration is as follows:

1. Draw a square, and place the number 16 (desired protein level) in the center.
2. At the upper left-hand corner of the square, write *protein supplement* and its protein content (36); at the lower left-hand corner, write *corn* and its protein content (8.9).
3. Subtract diagonally across the square (the smaller number from the larger number), and record the difference at the corners on the right-hand side (36 − 16 = 20; 16 − 8.9 = 7.1). The number at the upper right-hand corner gives the parts of concentrate by weight, and the number at the lower right-hand corner gives the parts of corn by weight to make a ration with 16% protein.

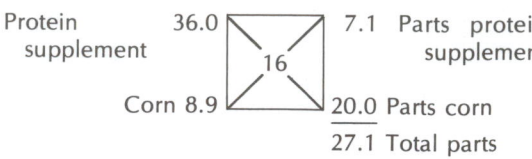

4. To determine what percent of the ration would be corn, divide the parts of corn by the total parts and multiply by 100:20.0 ÷ 27.1 × 100 = 73.8% corn. The remainder, 26.2%, would be supplement.

### TRIAL-AND-ERROR METHOD

In the example that follows, the trial-and-error method is used, with consideration given to energy and protein. Also, crude protein rather than digestible protein is used because (1) this is what feed manufacturers want to know as they plan feed formulas, and (2) this is what livestock producers see on the tag when they purchase feed. In most mixed feeds, approximately 80% of the total protein is digestible.

**Example.** *Let's assume that a dairy producer has a 1,433–lb cow producing 65 lb of milk testing 4.0% fat. The producer is feeding 14 lb of alfalfa hay and 40 lb of corn silage per day. Corn, oats, and soybean meal are available. What concentrate mix should the producer use to meet the needs of this lactating cow, from the standpoint of energy and protein?*

The available feeds have approximately the following composition (as-fed basis):

|  | TDN | Crude Protein |
|---|---|---|
|  | (%) | (%) |
| Alfalfa hay, all analyses ......... | 51.0 | 16.0 |
| Corn silage, all analyses ........ | 18.0 | 2.2 |
| Corn, all analyses .............. | 80.0 | 9.9 |
| Oats, all analyses .............. | 69.0 | 11.9 |
| Soybean meal, solv extd, 44% ... | 76.0 | 44.4 |

Here are the steps in balancing this ration:

**Step 1.** The daily TDN and crude protein requirements of this cow (1,433 lb body weight, 65 lb of milk testing 4% fat) are:

Requirements of cow for—

|  | TDN | Crude Protein | |
|---|---|---|---|
|  | (lb) | (lb) | (g) |
| Maintenance ............ | 9.94 | 0.94 | 428 |
| Milk production .......... | 20.9 | 5.87 | 2,665 |
| Total ................ | 30.84 | 6.81 | 3,093 |

**Step 2.** The forage (14 lb alfalfa hay, 40 lb corn silage) is supplying:

|  | TDN | Crude Protein |
|---|---|---|
|  | (lb) | (lb) |
| Alfalfa hay, 14 lb ............. | 7.14 | 2.24 |
| Corn silage, 40 lb ............. | 7.20 | 0.88 |
| Total from forage ............ | 14.34 | 3.12 |

**Step 3.** Remainder, to be supplied by concentrate:

| | TDN | Crude Protein |
|---|---|---|
| | (lb) | (lb) |
| | 16.5 | 3.69 |

**Step 4.** Let's try out (that's why it is called the *trial-and-error method*) a grain mix of 700 lb corn, 280 lb oats, 10 lb monosodium phosphate, and 10 lb salt, and determine the amounts of TDN and crude protein in 1,000 lb of the grain mix:

| | TDN | Crude Protein |
|---|---|---|
| | (lb) | (lb) |
| Corn, 700 lb .................... | 560.0 | 69.30 |
| Oats, 280 lb .................... | 193.2 | 33.3 |
| Monosodium phosphate, 10 lb ..... | — | — |
| Salt, 10 lb .................... | — | — |
| Total .................... | 753.2 | 102.60 |
| or in percent ................ | 75.3% | 10.3% |

**Step 5.** Divide the TDN needed from concentrate (16.5 lb) by the percent TDN in the mixture (75.3%). Thus, feeding 21.9 lb of the concentrate will meet the energy needs.

**Step 6.** Will this level of grain mix (21.9 lb) also meet the crude protein needs? By multiplying the pounds of concentrate mixture by the percent crude protein (21.9 × 10.3%), we find that the proposed concentrate would supply 2.26 lb of crude protein, whereas 3.69 lb are needed. Therefore, a high-protein supplement must be substituted for some of the homegrown grain.

**Step 7.** Let's substitute 175 lb of soybean meal for 175 lb of corn. Hence, the concentrate mix as now proposed will consist of:

| | TDN | Crude Protein |
|---|---|---|
| | (lb) | (lb) |
| Corn, 525 lb .................... | 420.0 | 52.0 |
| Oats, 280 lb .................... | 193.2 | 33.3 |
| Soybean meal, 175 lb ............ | 133.0 | 77.7 |
| Monosodium phosphate, 10 lb ..... | — | — |
| Salt, 10 lb .................... | — | — |
| Total .................... | 746.2 | 163.0 |
| or in percent ................ | 74.6% | 16.3% |

**Step 8.** By referring back to Step 3, we can divide the pounds of TDN and crude protein needed from the concentrate, by the percentage of TDN and crude protein found in the grain mix in Step 7. We find that 16.5 ÷ .746 = 22.1 lb needed to supply 16.5 lb TDN; and 3.69 ÷ .163 = 22.63 lb needed to supply 3.69 lb crude protein. Thus, we find that the following ration will supply the needed TDN (with a slight overage) and crude protein for a 1,433-lb lactating cow producing 65 lb of milk testing 4% fat:

| | TDN | Crude Protein |
|---|---|---|
| | (lb) | (lb) |
| Alfalfa hay, 14 lb ................ | 7.1 | 2.2 |
| Corn silage, 40 lb ................ | 7.2 | 0.9 |
| Concentrate mix (Steps 7 & 8), 22.63 lb . | 16.9 | 3.7 |
| Total ..................... | 31.2 | 6.8 |

In many sections of the country, especially in grain-deficient areas and on highly specialized dairies where little or no grain is grown, the dairy producer may find it most economical to purchase a commercial dairy feed to augment the roughage that is being fed.

### NET ENERGY METHOD

In order to apply the net energy method to the feeding of livestock, the following net energy values must be available:

1. A table showing the net energy requirements of the particular class of animal.

2. A table showing the nutrient composition of feeds, with the net energy of each feed partitioned into energy used for body maintenance and for gain; thus, the net energy values in megacalories (Mcal) per unit (lb or kg) are needed for each feed for maintenance ($NE_m$) and for gain ($NE_g$).

These net energy values may be obtained from *Feeds & Nutrition*, by Ensminger, *et al.*, and from the nutrient requirement publications of the National Academy of Sciences, of which there are separate publications for each species of animals.

The two examples that follow will show how to apply the net energy method. In the first example, net energy values of feeds are used to calculate the number of pounds of a given ration that a steer would need to consume to make a specified daily gain. In the second example, net energy is used to predict average daily gain based on consuming a certain number of pounds of a specified ration. Bear in mind that the ration in both cases (in these examples, the ration in Table 4–12) must be balanced for protein, minerals, and vitamins, in order for these net energy values to have validity for calculating daily consumption and predicting average daily gain.

**TABLE 4–12**
**RATION FOR FINISHING CATTLE**

| Ration Ingredient | | | Composition of Ingredients (as-fed basis) NE_m[1] | | Ration Supplies NE_m[1] | Composition of Ingredients (as-fed basis) NE_g[2] | | Ration Supplies NE_g[2] |
|---|---|---|---|---|---|---|---|---|
| | (lb) | (kg) | (Mcal/lb)[3] | (Mcal/kg)[3] | (Mcal)[3] | (Mcal/lb)[3] | (Mcal/kg)[3] | (Mcal)[3] |
| **Shelled corn,** all analyses ......................... | 68.60 | 31.14 | 0.86 | 1.90 | 59.00[4] | 0.59 | 1.89 | 40.47[5] |
| **Soybean meal** (solvent), 44% ..................... | 4.00 | 1.82 | 0.79 | 1.74 | 3.16 | 0.53 | 1.17 | 2.12 |
| **Alfalfa hay** (mid-bloom) ......................... | 27.00 | 12.26 | 0.52 | 1.15 | 14.04 | 0.28 | 0.62 | 7.56 |
| **Salt** ......................... | 0.40 | 0.18 | — | — | — | — | — | — |
| **Total** ......................... | 100.00 | 45.4 | — | — | 76.20 | — | — | 50.15 |

[1]$NE_m$ = net energy for maintenance.    [2]$NE_g$ = net energy for gain.    [3]Mcal stands for megacalorie.    [4]68.60 lb × 0.86 = 59.00    [5]68.60 lb × 0.59 Mcal = 40.47

**Example 1.** *Using net energy values to calculate the number of pounds of the ration that must be consumed to produce a specific gain*—How many calories would a 770–lb medium-frame steer calf need to consume to gain 2.6 lb daily?

**Step 1.** Calculate the net energy for maintenance ($NE_m$) and gain ($NE_g$) values for a pound of the ration shown in Table 4–12.

By referring to a table of Composition of Feeds, it is determined that l lb of the Table 4–12 ration supplies 0.7620 megacalories of net energy for maintenance (Mcal $NE_m$) and 0.5015 megacalories of net energy for gain (Mcal $NE_g$).

**Step 2.** By referring to a table of Net Energy Requirements, we find that the requirement for a 770–lb medium-frame steer calf to gain 2.6 lb daily is as follows:

|  | Mcal/day |
| --- | --- |
| $NE_m$ ........ | 6.24 |
| $NE_g$ ........ | 5.50 |

**Step 3.** Pounds of feed to meet the daily maintenance requirement:

$$6.24 \text{ Mcal} \div .7620 \text{ Mcal} = 8.19 \text{ lb}$$

**Step 4.** Pounds of feed to meet the requirement for 2.6 lb daily gain:

$$5.50 \text{ Mcal} \div .5015 \text{ Mcal} = 10.97 \text{ lb}$$

**Step 5.** Total pounds of feed that the steer calf must eat daily to gain 2.6 lb:

$$8.19 \text{ lb} + 10.97 \text{ lb} = 19.16 \text{ lb}$$

**Example 2.** *Using net energy to predict the average daily gain of a 770–lb medium-frame steer calf that is consuming a certain number of pounds of a specified ration*—Let's assume that we have a 770–lb steer that is consuming 18 lb of the ration shown in Table 4–12. What daily gain should be expected?

**Step 1.** Pounds of feed to meet the daily maintenance requirement = 8.19 lb (see prior example).

**Step 2.** Pounds of feed left for gain:

$$18 \text{ lb} - 8.19 \text{ lb} = 9.81 \text{ lb}$$

**Step 3.** Mcal of $NE_g$ supplied by remaining feed:

$$9.81 \text{ lb} \times .5015 \text{ Mcal} = 4.92 \text{ Mcal}$$

**Step 4.** Daily gain expected from 4.92 Mcal of $NE_g$. (Value obtained from a table of Net Energy Requirements. In this case, *Feeds & Nutrition*, p. 698, Table 19–3.)

**4.51 Mcal produces 2.2 lb gain**

Therefore, 4.92 Mcal will produce 2.40 lb daily gain

$$\left[ \frac{4.92 \ (2.2)}{4.51} \right]$$

## COMPUTER METHODS[10]

Most large livestock establishments and feed companies now use computers in ration formulation. Also, many of the state universities, through their Federal-State Extension Services, are offering ration balancing computer services to farmers within their respective states on a charge basis. Consulting nutritionists are available throughout the United States and provide computer services, as well as other services. With the recent advent of the low cost personal computer, this powerful technology is available to almost everyone.

Despite their sophistication, there is nothing magical or mysterious about the use of computers in ration balancing. Their primary advantages are accuracy and speed of computation. In addition, computer programs (software) used in ration balancing provide a means of organizing needed information in a logical and systematic manner. The computer should be viewed as an extension of the knowledge and skills of the formulator.

At this time, there is no "push-button" system of feed formulation available. The degree of success realized is very dependent on the management of data put into the computer, and on the evaluation of the resulting formulations that the computer generates. In the hands of experienced users, the computer enables the producer and nutritionist to be more precise in carrying out ration formulation.

Two basic approaches to ration formulation are practiced with computers:

1. Trial-and-error formulation.
2. Linear programming (LP).

### Trial-And-Error Formulation With the Computer

For a discussion of the trial-and-error method of ration balancing, see the earlier section in this chapter headed "Trial-and-Error Method." Many ration balancing software programs written for the computer allow for trial-and-error ration balancing. Feed mill nutritionists frequently use this technique to enter into the computer rations that are given to them by other nutritionists or by a producer. The objective in this case is to confirm the nutrient values for the ration based on the specific ingredients used by the feed manufacturer. In many cases, these rations are not to be altered without permission. In other cases, the number of ingredients for a specific ration may be limited so that the trial-and-error technique is just as fast as using linear programming to arrive at the desired nutrient levels in the ration.

**NOTE:** It does not take specialized computer software to use the trial-and-error method. Spreadsheet (or Financial Spreadsheet) programs, for instance, organize data into rows and columns. Information, such as nutrient values for a feedstuff, may be entered into data cells (see Fig. 4–35). Simple and complex arithmetic operations can be controlled by the user to the extent that rather large trial-and-error method rations can be programmed and run.

Spreadsheets have been developed with specific microcomputers in mind; and there are a great number of them on the market.

[10]The section on "Computer Methods" was prepared by L. M. Larsen, Ph.D., Consultant, Nutri-Systems, 426 E. Shields, Fresno, CA 93704.

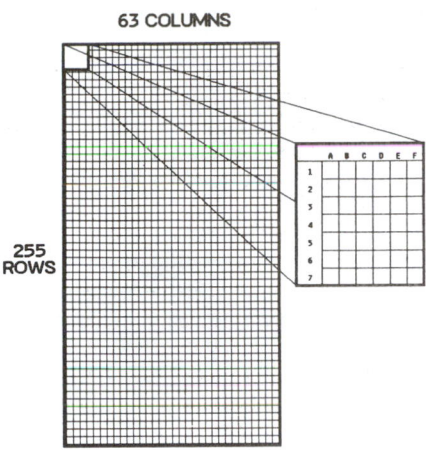

Fig. 4–35. Graphic representation of a spreadsheet (From Lane, R. J. and T. L. Cross, *Spreadsheet Applications for Animal Nutrition and Feeding,* Reston Publishing Co., Inc., Reston, VA, 1985).

## Linear Programming (LP)

The most common technique for computer formulation of rations is the linear programming (LP) technique. At times, this is referred to as *least cost* ration formulation. This designation results from the fact that most LP techniques for ration formulation have as their objective *minimization of cost.* A few LP programs are in use that solve for *maximization of income over feed costs.* Regardless, the livestock producer and nutritionist should always keep in mind that maximizing net profit is the only true objective of most ration formulations. A skilled user of the LP system will control ration quality by writing specifications that lead to rations that will maximize profit.

Briefly described, the LP program is a mathematical technique in which a large number of simultaneous equations are solved in such a way as to meet the minimum and maximum levels of nutrients and levels of feedstuffs specified by the user at the lowest possible cost. It is not necessary to understand the inner workings of the computer program to use LP, though it does take experience to use it to good advantage and to avoid certain pitfalls. The most common pitfalls are incorrectly entered or missing data and the specification of minimums and maximums that cannot be met with the feedstuffs available. The latter is called an infeasible solution. When an infeasible solution is encountered, the user must determine (1) if this is due to incorrect or missing data, or (2) if the specifications must be relaxed.

## Procedure for Use of Linear Programming (LP)

Before using the LP approach to ration formulation, the user should become familiar with the specific software package to be used. (See later section on "Selection of Computer Software and Hardware for Ration Formulation.") It is also desirable to study the LP technique as applied to feed formulation. After users are familiar with LP and their computer software, they are ready to begin using the computer for ration formulation by LP. It must first be understood that all data entered into the computer is directed to files. In most cases, these files are located on disks, or perhaps on tapes. Currently, most computers use keyboards and CRT (cathode ray tube) displays for entry of data. The necessary data files are generally created in steps as follows:

1. **Enter names of available feed ingredients, and the cost of each.** It is necessary that all of the available feeds be listed along with the unit cost. It matters little if the formulator uses cost per ton, cost per hundred weight (cwt), or cost per pound, but the same method of cost input must be used for all feeds. The computer software may call for a specific form for entering costs.

2. **Enter nutrient values for feeds.** Tables of feed composition using average or typical values may be used, but, because of the wide variation in the composition of feeds, a chemical analysis of a representative sample of each lot of feed is more precise and should be used if available. This is especially true of forages, in which composition may be affected in a major way by cultural conditions and stage of maturity.

3. **Enter ration specifications.** Ration specifications are generally broken into two parts: (1) Nutrient limits, and (2) Ingredient limits. In each case, the formulator specifies either a lower limit and/or an upper limit for each item. If no specification for the particular item is desired, it may be specified as zero (0) or left blank, depending on the circumstances. It is also appropriate to list feedstuffs available, but not currently on hand (with an upper limit of zero). Most LP solutions will then tell the user the highest cost at which such feeds would enter the solution if allowed to do so. Ratios between nutrients (such as a calcium/phosphorus ratio) or feedstuffs (corn/barley ratio) may also be specified in most LP software packages. The experienced formulator usually deals with palatability or feedstuff quality considerations by setting an upper limit on the amounts of problem feeds or a lower limit on feeds that contribute a positive quality to the ration. Nonnutritive attributes, such as bulk density, may also be programmed into the LP system. The LP technique is a very flexible and powerful ration balancing tool.

**NOTE:** Important additional items to consider when creating ration specifications are upper limits on the use of nonprotein nitrogen (or urea) and limits on the usage of feed additives, such as drugs, feed flavors, and the like.

Fig. 4–36 illustrates, by means of a worksheet, a logical method of organizing the restrictions for a ration.

### LEAST-COST FORMULATION WORKSHEET

| Specifications | Ingredient A | Ingredient B | Ingredient C | Restrictions |
|---|---|---|---|---|
| Cost | | | | Minimize |
| Total weight | | | | 1,000 lb |
| Crude protein | | | | 133 lb |
| Digestible protein | | | | 100 lb |
| Ether extract | | | | 25 to 80 lb |
| Net energy lactation | | | | 900 Mcal |
| Calcium | | | | 5 to 10 lb |
| Phosphorus | | | | 7 lb |
| Vitamin A equivalent | | | | 35,000 |
| Vitamin D | | | | 60,000 |
| Limits on ingredients | | | | |
| Minimum | | | | |
| Maximum | | | | |

Fig. 4–36. Sample worksheet for a least-cost formulation. The first column lists the specifications. The various feedstuffs to be considered are listed in the succeeding columns with their respective costs and nutritive values. The last column lists the restrictions desired on the final formulation.

4. **Submit all of the above information to the matrix building and solving portion of the LP software package.** Matrix building and solving are generally accomplished automatically by the computer software once the specifications have been entered into the computer. Mathematically, the procedure involves the solution of a complex algebraic problem, with an answer being derived in seconds or minutes. Using the LP program, the computer produces a mix that will meet the desired specifications at the lowest possible cost.

5. **Examine the solution provided by the computer software.** The end result should be feasible, both from a mathematical standpoint and from a nutritional standpoint. The feedstuff mixture should be acceptable to the animals for which it is intended. In most cases the first solution provided to the user is not acceptable. Repeat runs may be necessary to obtain the best solution.

Fig. 4–37, leghorn layer ration, is a computer printout of an LP solution. The various columns of the report have been numbered for indentification. Similar columns have been given the same number. The three sections of the report are each identified with a Roman numeral.

An explanation of the information contained in each column of Fig. 4–37 follows:

**Column (1)**—Ingredient and nutrient numbers.

**Column (2)**—Ingredient and nutrient names.

**Column (3)**—Solution *amounts* given in percentage for feed ingredients (Section I) and nutrients (Section II).

**Column (4)**—The percentage solution for ingredients has been converted to a "ton" *batch* using prespecified rounding factors for each ingredient. (The batch totals 1,992.73 lb, rather than an exact 2,000 lb because of the rounding requirements.)

**Column (5)**—Ingredient *costs* in dollars per hundred weight (cwt)

**Column (6)**—*Lower(L)* and *upper(U) limits* specified for each ingredient (Section I) and each nutrient (Section II).

**Column (7)**—A pair of values, one under the other, gives the *stable costs*. The first value is the feed cost below which the present optimal solution would no longer be valid. Similarly, the second value is the feed cost above which the present solution would no longer be valid. The *stable cost* figures let the LP user know when it is desirable to reprocess the ration.

```
16-JAN-89    ANYCO GRAIN AND MILLING                          ( 7) ID. NO. 29704
             P. O. BOX 1234                       COSTS.. ...$/CWT ...$/TON
             ANYTOWN, USA 90909    (123) 456-7890 BASE LP   8.069    161.38
  * FEASIBLE* RATION: L507 LAYER MASH, 17 PCT.    BATCH     8.030    161.60
```

**Section I**

| (1) # | (2) INGREDIENT | (3) AMOUNT | (4) BATCH | (5) COST | (6) LIMITS | (7) STABLE COSTS | (8) RANGE | (9) COST PER UNIT |
|---|---|---|---|---|---|---|---|---|
| 4 | GROUND CORN | 33.333 | 665.00 | 6.40 | | 6.15 | 64.50 | 0.0024I |
| | | | | | 10.000L | 6.47 | 25.05 | 0.0007D |
| 6 | GROUND MILO | 33.848 | 675.00 | 6.00 | | 5.94 | 43.26 | 0.0006I |
| | | | | | | 6.22 | 0.00 | 0.0022D |
| 10 | WHEAT MILLRUN | 2.085 | 40.00 | 6.00 | 15.000U | 4.84 | 28.64 | 0.0116I |
| | | | | | | 6.23 | 0.00 | 0.0029D |
| 19 | SOYBEAN MEAL,47.5 | 16.117 | 320.00 | 14.95 | | 11.04 | 18.80 | 0.0391I |
| | | | | | | 26.37 | 14.80 | 0.1142D |
| 20 | MEAT SCRAP,50 | 6.000 | 120.00 | 15.03 | 6.000U | 20.13 | 7.54 | -0.0510I |
| | | | | | | | 3.94 | 0.0510D |
| 21 | DICAL PHOS(22CA/18P) | 0.396 | 10.00 | 17.00 | | 0.41 | 2.03 | 0.1659I |
| | | | | | | 139.27 | 0.27 | 1.2227D |
| 23 | LIMESTONE | 7.584 | 150.00 | 1.45 | | 0.00 | 7.34 | 0.0309I |
| | | | | | | 12.89 | 7.53 | 0.1144D |
| 24 | SALT, PLAIN | 0.250 | 5.00 | 2.35 | 0.250U | 0.00 | 0.93 | 0.0393I |
| | | | | | 0.250L | | 0.00 | -0.0393D |
| 27 | POULTRY PREMIX | 0.250 | 5.00 | 56.37 | 0.250U | 0.00 | 0.93 | 0.5795I |
| | | | | | 0.250L | | 0.00 | -0.5795D |
| 33 | DL-METHIONINE,99 | 0.086 | 1.73 | 157.00 | | 76.28 | 0.09 | 0.8072I |
| | | | | | | 540.27 | 0.07 | 3.8327D |
| 37 | SELENIUM, 90.8 | 0.050 | 1.00 | 17.50 | 0.050U | 0.00 | 0.73 | 0.1903I |
| | | | | | 0.050L | | 0.00 | -0.1903D |
| | TOTALS | 100.000 | 1992.73 | | REQUESTED BATCH WEIGHT IS 2000.00 | | | |

**Section II**

| (1) # | (2) NUTRIENT | (3) AMOUNT | (6) LOWER | (6) UPPER | (9) COST PER UNIT DECREASE | (9) INCREASE | (8) EFFECTIVE RANGE DECREASE | (8) INCREASE |
|---|---|---|---|---|---|---|---|---|
| 1 | WEIGHT | 100.000 | 100.000 | 100.000 | 0.016 | -0.016 | 99.3204 | 104.2085 |
| 8 | CRUDE PROTEIN | 17.000 | 17.000 | | -0.243 | 0.243 | 15.8640 | 18.1362 |
| 12 | CRUDE FAT | 2.930 | 0.000 | | 0.054 | 0.053 | 2.8216 | 5.7481 |
| 13 | CRUDE FIBER | 2.308 | | 4.000 | 0.137 | 0.097 | 2.2498 | 2.3682 |
| 14 | ASH | 11.963 | 0.000 | | 0.657 | 0.033 | 11.9068 | 12.2025 |
| 15 | CALCIUM | 3.600 | 3.600 | 3.700 | -0.080 | 0.080 | 2.0008 | 3.8582 |
| 16 | PHOSPHORUS | 0.671 | 0.650 | | 0.557 | 0.938 | 0.6609 | 0.9609 |
| 17 | AVAIL. PHOSPHORUS | 0.470 | 0.470 | | -0.910 | 0.910 | 0.4480 | 0.7686 |
| 31 | M. E. (POULTRY)/LB | 1270.000 | 1270.000 | | -0.003 | 0.003 | 1204. | 1281. |
| 47 | LYSINE | 0.798 | 0.700 | | 20.036 | 1.694 | 0.6856 | 0.8010 |
| 48 | METHIONINE | 0.350 | 0.350 | | -1.359 | 1.359 | 0.3309 | 1.7876 |
| 49 | METHIONINE + CYSTINE | 0.619 | 0.600 | | 3.532 | 1.380 | 0.6172 | 2.0342 |
| 61 | LINOLEIC ACID | 1.134 | 1.000 | | 0.061 | 0.107 | 1.0381 | 2.5310 |
| 62 | XANTHOPHYLL /LB | 3.000 | 3.000 | | -0.026 | 0.026 | 1.6701 | 5.8047 |

**Section III**

| (1) # | (2) ** NOT USED ** INGREDIENT | (5) COST | (10) RELATIVE WORTH | (6) UPPER LIMIT | (9) COST/UNIT INCREASE | (8) INCREASE RANGE |
|---|---|---|---|---|---|---|
| 8 | GROUND BARLEY | 6.500 | 5.782 | | 0.0072 | 6.531 |
| 12 | ALFALFA, DEHY.,17 | 8.000 | 7.374 | 2.500 | 0.0063 | 0.932 |
| 31 | VEGETABLE FAT | 17.200 | 11.914 | | 0.0529 | 2.835 |
| 32 | CANE MOLASSES | 4.450 | 2.384 | 0.000 | 0.0207 | 1.738 |

Fig. 4–37. Example Leghorn layer ration processed by computer linear programming. See text for explanation of marked ( ) columns. (Courtesy, Nutri-Systems, Fresno, CA)

**Column (8)**—The *range* values are related to the *stable costs* and the *cost per unit* columns. The values delineate the limits over which the *stable cost* and *unit decrease/ increase* columns are applicable. (An example: If the cost of ground corn decreases to $6.163/cwt [Column 7], then the usage of corn would increase to 64.50% [Column 8]. Of course, there would be changes in the usage of other ingredients as corn increases in amount.)

**Column (9)**—Another pair of values gives the *cost per unit decrease(D)* and *increase(I)*. These values indicate how much the cost of the ration would be changed if either an ingredient or nutrient is increased or decreased by one unit in the percent solution. A positive value means that cost would be increased and a negative value means that cost would be decreased.

**Column (10)**—Section III contains information about the *ingredients not used* in the solution. The *relative worth* column indicates the cost at which each of these ingredients would enter the solution.

6. **Reformulate with LP at periodic intervals.** Changes in ingredient costs, in ingredient availability, and in the needs of animals dictate the need for reprocessing the ration. The good formulator monitors all these items on a regular basis. It is also critical to evaluate the feeding results to confirm that production goals and cost objectives are being met with the ration. Computers don't feed animals—people do!

### Selection of Computer Software and Hardware for Ration Formulation

Numerous companies market computer software for ration formulation. The software varies from the very simple and straightforward to very complex packages intended for large feed manufacturers. The latter packages include applications for formula costing, inventory control, control of usage of ingredients in limited availability, production of feed tags, etc. Most software is intended for use on a single computer model or at least a certain family of computers (IBM-PC™, for example). It is therefore most desirable to select the software desired before purchasing the computer hardware. Computer type and size of memory and disk drive storage capacity must meet the criteria of the software developer or the software may not be usable.

Directories which list software by application are a good place to start looking. Other sources of information are feed and livestock trade publication advertisements, university personnel, and nutritionists who use feed blending software. Nutritionists are a good source of information as to how well a certain software package performs.

Ration formulation software may be generalized so that it can be made applicable to all species of animals or it may be designed with the unique requirements of specific species such as poultry, dairy cattle, etc. When the software has been designed for a certain species, it may incorporate tables of nutrient requirements and tables of typical feedstuffs and their nutrient values. This can save the user time, but it does not mean that the software will run itself without the judgment of the user. No one has yet developed software that will anticipate all the conditions under which livestock will be fed. Computers are not able to assess all aspects of ingredient quality, environment, and animal management. The judgment of the producer and formulator must be imposed on the computer software. Look for the freedom to make changes as needed. When in doubt, seek advice from those with experience.

## FEED SUBSTITUTIONS

Successful livestock producers are keen students of values. They recognize that feeds of similar nutritive properties can and should be interchanged in the ration as price relationships warrant, thus making it possible at all times to obtain a balanced ration at the lowest cost.

In arriving at feed substitutions, two primary factors besides cost, chemical composition, and feeding value should be considered—namely, palatability and product quality. Also, when substituting feed ingredients, the following facts should be kept in mind:

1. Feeds differ widely in feeding value. Barley, for example, varies widely in feeding value according to the hull content and the test weight per bushel; and the same can be said relative to oats. There is a wide range in the protein content of sorghum. The forages vary widely in feeding value according to the stage of maturity at which they are cut and how well they are cured and stored.

2. Nonlegume forages may have a higher then normal value relative to legumes when the chief need of the animal is for additional energy rather than for supplemented protein. Thus, nonlegume forages of low value can be used to better advantage for wintering mature, dry beef cows and dry dairy cows than for lactating cows and young calves. On the other hand, legumes may have a higher than normal value relative to nonlegumes where the chief need is for additional protein rather than for added energy. Thus, no protein supplement is necessary for breeding beef cows, provided a good-quality legume forage is fed.

3. Based primarily on available supply and price, certain feeds—especially those of medium protein content, such as brewers' dried grains, corn gluten feed (gluten feed), distillers' dried grains, distillers' dried solubles, peanuts, and peas (dried)—may be used interchangeably as (a) grains and by-product feeds and/or (b) protein supplements.

4. The feeding value of certain feeds is materially affected by preparation. Thus, wheat must be coarsely ground or rolled for cattle.

For the reasons noted above, no comparative value of feeds can be absolute. Rather, feed substitutions should be based on the class and age of animal and quality of feed, together with experience and experiments.

## COMMERCIAL FEEDS

Commercial feeds are just what the term implies—instead of being farm mixed, these feeds are mixed by commercial feed manufacturers who specialize in the business. In 1988, a total of 103.1 million tons of primary feeds (complete feeds) were manufactured in the United States, and an additional 10 million tons of secondary feeds (supplements) were produced; making for a total of 113.1 million tons of commercial feeds. Primary feed is that which is mixed from individual ingredients, sometimes with the addition of a premix at a rate of less than 100 lb per ton of finished feed. Secondary feed is that which is mixed with one or more ingredients and a formula feed supplement (which is a primary feed); normally, the supplement is used at a rate of 300 lb or more per ton of finished feed, depending upon the protein content of the supplement and the percentage of protein content desired in the finished feed. The breakdown, percentagewise, by classes of livestock for which primary (complete) commercial feeds were used in 1988 follows: poultry, 44.5%; beef and sheep, 17.6%; dairy, 16.9%; hogs, 14.2%, and all other, 6.8%.[11]

The commercial feed manufacturer has the distinct advantages of (1) purchasing feed in quantity lots, making possible price advantages; (2) economical and controlled mixing; (3) the hiring of scientifically trained personnel for use in determining the rations; and (4) quality control. Most livestock producers have neither the know-how nor the quantity of business to provide these services on their own. Because of these several advantages, commercial feeds are finding a place of increasing importance in livestock feeding. Also, it is to the everlasting credit of reputable feed dealers that they have been good teachers, often getting livestock producers started in feeding balanced rations.

Numerous types of commercial feeds, ranging from additives to complete rations, are on the market, with most of them designed for the specific species, age, or need. Among them, are complete rations (including hay for ruminants and pseudo-ruminants), concentrates, pelleted or cubed forages, protein supplements (with or without reinforcements of minerals and/or vitamins), mineral and/or vitamin supplements, additives (antibiotics, hormones, etc.), milk replacers, starters, young stock rations, fitting rations, rations for different levels of production—for the idle (like dry cow rations) to the forced producers, and medicated feeds.

Fig. 4–38. This is the 170-ft commercial feed mill of Pennfield Feed-Grain, Lancaster, PA. This mill features safety explosion panels, fully-automated computerized batching, and a dust control system. More than 1,000 tons of grain and other ingredients are handled daily in this mill. (Courtesy, National Broiler Council, Washington, DC)

## HOW TO SELECT A COMMERCIAL FEED

There is a difference in commercial feeds! That is, there is a difference from the standpoint of what producers can purchase with their feed dollars. Enlightened operators will know how to determine what constitutes the best in commercial feeds for their specific needs. They will not rely solely on how the feed looks and smells or on the feed salesperson. The most

_____
[11]Source: American Feed Manufacturers Assn.

important factors to consider or look for in buying a commercial feed are:

1. **The reputation of the manufacturer.** This should be determined by (a) checking on who is back of it, (b) conferring with other producers who have used the particular products, and (c) checking on whether or not the commercial feed under consideration has consistently met its guarantees. The latter can be determined by reading the bulletins or reports published by the respective state departments in charge of enforcing feed laws.

2. **The specific needs.** Feed needs vary according to (a) class, age, and productivity of the animals, and (b) whether animals are fed primarily for maintenance, growth, finishing (or show-ring fitting), reproduction, lactation, or work (running). The wise operator will buy different formula feeds for different needs.

3. **The feed tag.** Most states require that mixed feeds carry a tag that guarantees the ingredients and the chemical makeup of the feed. Feeds with more protein and fat are better, and feeds with less fiber are better.

In general, if the fiber content is less than 8%, the feed may be considered as top quality, if the fiber is more than 8 but less than 12%, the feed may be considered as medium quality; while feeds containing more than 12% fiber should be considered carefully. Occasionally, a high-fiber feed is good. For example, many feeds are high in fiber simply because they contain generous quantities of alfalfa; yet they may be perfectly good feeds for the purpose intended. On the other hand, if oat hulls and similar types of high-fiber ingredients are responsible for the high-fiber content of the feed, the quality should be questioned. The latter type of fiber is poorly digested and does not provide the nutrients required to stimulate the digestion of the fiber in roughages.

4. **Flexible formulas.** Feeds with flexible formulas are usually the best buy. This is because the price of feed ingredients in different source feeds varies considerably from time to time. Thus, good feed manufacturers will shift their formulas as prices change in order to give their customers the most for their money. This is as it should be, for (a) there is no one best ration, and (b) if substitutions are made wisely, the price of the feed can be kept down and feeders will continue to get equally good results.

## STATE COMMERCIAL FEED LAWS

Nearly all states have laws regulating the sale of commercial feeds. These benefit both producers and reputable feed manufacturers. In most states the laws require that every brand of commercial feed sold in the state be licensed, and that the chemical composition be guaranteed.

Samples of each commercial feed are taken each year, and analyzed chemically in the state's laboratory to determine if manufacturers lived up to their guarantees. Additionally, skilled microscopists examine the sample to ascertain that the ingredients present are the same as those guaranteed. Flagrant violations on the latter point may be prosecuted.

Results of these examinations are generally published, annually, by the state department in charge of such regulatory work. Usually, the publication of the guarantee alongside any "short-changing" is sufficient to cause the manufacturer promptly to rectify the situation, for such public information soon becomes known to both users and competitors.

## HOME-MIXED VS COMMERCIAL FEEDS

The value of farm-grown grains—plus the cost of ingredients which need to be purchased in order to balance the ration, and the cost of grinding and mixing—as compared to the cost of commercial ready-mixed feeds laid down on the farm, should determine whether it is best to mix feeds at home or depend on ready-mixed feeds.

Fig. 4–39. A portable grinding, mixing, conveying unit for preparing home-mixed feds. (Courtesy, Gehl Company, West Bend, WI)

Feeders have the following options for purchasing and preparing feeds:

1. Purchase of a commercially prepared, complete feed.
2. Purchase of a commercially prepared protein supplement (likely reinforced with minerals and vitamins), which may be blended with local or homegrown grain.
3. Purchase of a commercially prepared mineral-vitamin premix which may be mixed with an oil meal, and then blended with local or homegrown grain.
4. Purchase of individual ingredients (including minerals and vitamins) and mixing the feed from the ground up.

In summary, it may be said that there now exist two good alternative sources of most feeds and rations—home mixed or commercial—and the able manager will choose wisely between them.

## FEEDING AS AN ART

The feed requirements of animals do not necessarily remain the same from day to day or from period to period. The age and size of the animal; the kind and degree of activity; climatic conditions; the kind, quality, and amount of feed; the system of management; and the health, condition, and temperament of the animal are all continually exerting a powerful influence in determining the nutritive needs. How well the feeder understands, anticipates, interprets, and meets these requirements usually determines the success or failure of the ration and the results obtained. Although certain principles are usually followed by good feeders, no book knowledge or set of instructions can substitute for experience and born livestock intuition. Skill and good judgment are essential when feeding. Indeed, there is much truth in the old adage that "the eye of the master fattens the animals."

## STARTING ANIMALS ON FEED

In cattle and lamb feeding operations, it is important that the animals be accustomed to feed gradually. In general, upon arriving at the feedlot, the animals may be given as much nonlegume roughages as they will consume. On the other hand, it is necessary that they be gradually accustomed to high-quality legumes, which may be too laxative. The latter can be accomplished by slowly replacing the nonlegume roughage with greater quantities of legumes. Of course, as the grain ration is increased, the consumption of roughages will be decreased.

Starting cattle and lambs on grain requires care and good judgment. With both classes of stock, it is usually advisable first to accustom them to a bulky type of ration, a starting ration with considerable oats or beet pulp being excellent for this purpose. A common *rule of thumb* in starting cattle on feed is to give them 2 lb per head the first day; increase the ration from ½ to 1 lb daily until they reach approximately the halfway mark of what is anticipated will be a full feed; and then increase the ration by 1 lb every third day until full feeding is obtained. Lambs are usually started on grain by feeding about ¼ lb per head daily, gradually increasing the allowance so that they are getting a feed of about 2 lb per head daily when on full feed about 4 weeks later.

In general, little difficulty is ever encountered in starting hogs on feed. Although the same precautions that apply to cattle and sheep should be observed, swine are less sensitive to feed changes than other classes of farm animals.

The keenness of the appetites and the consistency of the droppings of the animals are an excellent index of their capacity to take more feed. In all instances, scouring, the bane of the feeder, should be avoided.

## AMOUNT TO FEED

Many farm animals throughout the world are underfed all or some part of the year. Temporary underfeeding is most likely to occur in the winter months or during periods of extended droughts. Fortunately, during such times of restricted feed intake, animals have nutritive reserves upon which they can draw. Although they may survive for a considerable period of time under these conditions, there is an inevitable loss in body weight and condition.

Overfeeding is also undesirable, being wasteful of feeds and creating a health hazard. Animals that suffer from mild digestive disturbances are commonly referred to as off feed. When overfeeding exists, there is usually considerable leftover feed and wastage, and there is a high incidence of bloat (colic in horses), founder, scours, and even death.

When on full feed, finishing cattle will consume 1¾ to 2½ lb of concentrates and from ¾ to 2 lb of hay per 100 lb liveweight daily. Lambs on full feed will consume about 2 lb of grain per head daily and about the same amount of roughage. The proportion of grain to roughage for cattle and sheep should be varied according to their comparative price and the age and quality of the animals. Finishing hogs will consume 4 to 5 lb of feed, mostly concentrates, per 100 lb of liveweight. The pig, therefore, is capable of consuming more feed in proportion to body weight than either the cow or sheep. This is a major factor in the greater efficiency with which pigs convert feeds into meats. Regardless of the ration or class of stock, however, finishing animals should receive a maximum ration over and above the maintenance requirement.

## FREQUENCY, REGULARITY, AND ORDER OF FEEDING

In general, finishing animals are fed twice daily. With animals that are being fitted for show, where maximum consumption is important, it is not uncommon to find three or even four feedings daily. When self-fed, animals eat at more frequent intervals, though they consume smaller amounts each time.

Animals learn to anticipate their feed. Accordingly, they should be fed *with great regularity*, as determined by a timepiece. During warm weather, they will eat better if the feeding hours are early and late, in the cool of the day.

Usually, the grain ration is fed first, with the roughage following. In this manner, the animals eat the bulky roughages more leisurely.

## FEEDING SYSTEMS

The vast majority of large cattle and sheep feedlots, large swine operations, and large poultry enterprises feed complete rations, in which all of the needed nutrients are provided in the quantities necessary. Little or no feed is provided in addition, because such feeding would destroy the balance of nutrient provided by the complete ration. Even with this system of feeding, animals are frequently provided with a supplementary source of minerals, to which they are given free access.

Sometimes swine producers and poultry producers (especially farm flock owners) feed whole grain and fortified protein supplement in separate feeders. The supplement provides the extra amounts of protein, minerals, and vitamins lacking in the grain. This system enables the farmer to make maximum use of farm-grown or local grains. By estimating the approximate consumption of each the grain and the supplement, fortifying the supplement accordingly, an overall nutrient intake that is reasonably well balanced is achieved.

## HAND FEEDING VS SELF-FEEDING

Self-feeding is the most common method of feeding employed in finishing hogs and feeding poultry. Though less prevalent, the practice is increasing in cattle- and lamb-finishing operations, but it is almost never used with horses.

Self-feeding may be accomplished satisfactorily in either of two ways: (1) providing a suitable self-feeder or hopper-type of container, or (2) keeping a large-type feed bunk (trough for swine) well filled at all times. In order to self-feed cattle or sheep, they must be hand fed until they are on full feed, after which they may have free access to the self-feeder.

The principal **advantages** derived from the use of self-feeders are:

1. Less labor and time are required in the feeding operations. By using self-feeders with large bins or hoppers, a large quantity of feed—enough for a week or two—may be mixed and placed before the animal at one time. Hand feeding is a twice-a-day chore, whereas it is merely necessary to check the filled self-feeders at intervals to make certain that the feed has not clogged.

2. Slightly higher feed consumption and larger daily gains are secured. This is especially true when hand feeding is done by an inexperienced feeder or when more or less irregularity occurs in time of feeding and quantity and character of ra-

tions—conditions which sometimes prevail under ordinary farm conditions.

3. Self-fed animals are usually ready for market earlier than hand-fed animals because of their slightly more rapid gains.

4. The animals are less likely to go off feed. This is because they learn that feed is available at all times, with the result that they are inclined to eat leisurely. On the other hand, greedy steers or lambs are likely to gorge when self-fed, particularly at those times when weather conditions may cause them to come to the feed trough with very keen appetites.

5. Often self-feeders are especially well adapted to grain feeding on pastures due to the distance of the pastures from the feed storage and mixing facilities.

6. With most feeds, pigs can be relied upon to balance their own rations when fed cafeteria style; thus, considerable time can be saved in the mixing of feeds.

The principal **disadvantages** of using self-feeders are the following:

1. Unless chopped roughage is thoroughly mixed with the grain ration, self-feeding is not adaptable where it is desired to utilize the maximum quantity of roughage. This is simply due to the fact that, under self-fed conditions, the animals elect to eat a large proportion of grain. Thus, where roughages are abundant and low in price, hand feeding may be more practical.

2. The gains of self-fed cattle and sheep may cost slightly more, particularly where grains are relatively higher than roughages. However, the slightly higher feed cost is usually offset by the slightly higher selling price of the finished animals.

3. If hopper-type feeders are used, rather than bunks, there is usually a slightly higher equipment cost in self-feeding.

4. Somewhat more time and expense are usually required in preparing feeds (shelling, grinding, and mixing) for the self-feeder.

5. Practical cattle feeders usually consider that self-feeding is not adapted to an extended feeding period, whereas they feel that it is quite a satisfactory method to use for a short-feed of 90 to 120 days.

6. Unlike pigs, cattle and sheep cannot be trusted to balance their own rations when different feeds are made available on a free-choice basis.

In general, because of the greater quantities of concentrates consumed in self-feeding, it is advisable that the ration used be slightly more bulky than when hand feeding is used. The extra bulk may be obtained either by (1) adding a small amount of chopped or ground hay to the grain ration, or (2) using in the ration a considerable proportion of such bulky, fibrous feeds as oats, beet pulp, or bran. With favorable feed prices, it may be well to have as much as one-third of the ration of cattle and sheep made up of oats, beet pulp, or bran. With brood sows, a generous supply of ground alfalfa should be mixed in the ration, thus providing a bulky, less fattening ration and furnishing a desirable source for minerals, vitamins, and proteins.

## CREEP FEEDING YOUNG ANIMALS

*Creep feeding is the supplemental feeding of young nursing animals in an enclosure which is accessible to them but not to their dams.* Young gains are cheap gains. This is due to (1) the higher water and lower fat content of the young animal in comparison with older animals, and (2) the higher feed consumption per unit of weight of young animals. This has encouraged more and more creep feeding of calves, pigs, and lambs.

## FEEDS SHOULD NOT BE CHANGED ABRUPTLY

Sudden changes in diet are to be avoided, especially when changing from a less concentrated ration to a more concentrated one. When this rule of feeding is ignored, digestive disturbances result, and the animals go off feed. In either adding or omitting one or more ingredients, the change should be made gradually. Likewise, caution should be exercised in turning animals out to pasture or in transferring them to more lush grazing. If it is not convenient to accustom them to new pasture gradually, they should at least be well filled with hay (or with the former pasture) before being turned out.

## ATTENTION TO DETAILS PAYS

The successful feeder pays great attention to details. In addition to maintaining the health and comfort of the animals and filling their feed troughs, consideration is also given to their individual likes and temperaments.

It is important to avoid excessive exercise, which results in loss of energy by animals through unnecessary muscular activity. Rough treatment, excitement, and noise usually result in nervousness and inefficient use of feed. Finishing animals and milking cows should not be required to exercise any more than is deemed necessary for the maintenance of good health. Dehorning is usually necessary in order to reduce fighting and possible bruises or injury. Likewise, all males should be castrated, for they will be much quieter.

## GOOD LIVESTOCK REQUIRE GOOD SOILS

Good livestock producers can be identified by the appearance of their animals. Good animals are proof of good breeding and superior nutrition; and proper nutrition is obtained by feeding hays, grains, and pastures which are grown on fertile soils.

In the wild state, animals did not possess the many bone unsoundnesses and nutritional deficiencies common to domesticated livestock. They roved over the prairies or through the forests, gleaning the feeds provided by nature—vegetation produced on highly mineralized, unleached soils—whereas on a modern farm the range is restricted, leached, and sometimes entirely devoid of vegetation. Domestic animals have little or no choice in the selection of their diet, being able to consume what the caretaker provides or what can be grazed within the confines of fenced holdings. The condition is further aggravated by forcing early development and high production, such as is obtained in racing 2- or 3-year-olds, in producing high yields of milk and butterfat at 2 years of age, in finishing yearling cattle, in farrowing gilts at one year of age, in marketing pigs at 4 to 5 months of age, and in the production of a fat market lamb at weaning time.

Present research, together with practical observation, points to the fact that the mere evaluation of yields in terms of tons of forage or bushels of grain produced per acre is not enough.

Neither does a standard feed analysis (a proximate analysis) tell the whole story. Rather, there appears to be a direct and most important relationship of the fertility of the soil to the composition of the plant in terms of calcium, phosphorus, proteins, vitamins, and other nutrients. Moreover, the animal cannot be expected to be well nourished when forced to subsist on plants which themselves are suffering from such nutritive deficiencies. Since the need for these nutritive constituents for the development of strong bone and healthy bodies is fully understood, it may truly be said that good animals require good soils.

Unsuccessful attempts are constantly being made to overcome these nutritive deficiencies of plants by dosing the animal with patent mineral and vitamin mixtures. Unfortunately, at times, such "cure-alls" may even accentuate the difficulty, because of harmful imbalances in calcium or phosphorus and possibly other nutrients. Therefore, the wise policy consists of first improving the fertility of the soil, which in turn means well-nourished and highly nutritious forages and grains for the animal. Soil fertility can best be improved through adding crop residues, manure, and certain commercial fertilizers as required. Restoring depleted soils not only increases the mineral, protein, and perhaps vitamin content of plants; but the tonnage yield of the crop will be higher.

The rapidity of soil leaching is always greatest in areas of high rainfalls and high temperatures, which means the eastern and southern parts of the United States. It can also be stated that the failure of a soil to grow an adapted legume is an indication of lack of calcium. In such areas, the nutritive deficiencies of animals are more prevalent.

Under certain conditions and in certain areas where mineral deficiencies become very severe, deficiency areas become known. Thus, the southwestern United States is known as a phosphorus-deficient area, the northwestern United States as an iodine-deficient area, and the southeastern United States as a cobalt-deficient area.

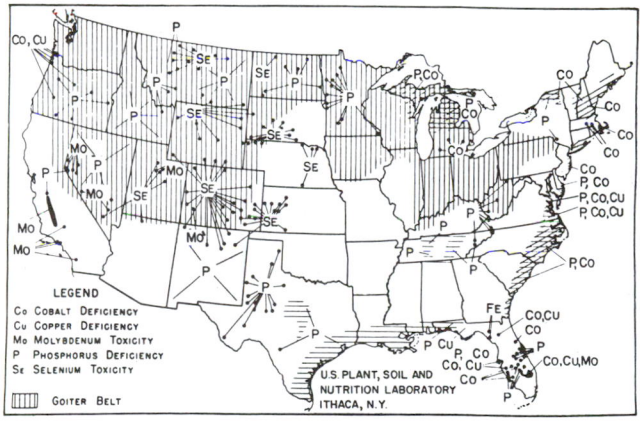

Fig. 4-40. Mineral deficiency areas of the United States and the excess selenium area of the northern and central Great Plains. (Courtesy, USDA)

## NUTRITIONAL DISEASES AND AILMENTS

More animals (and people) throughout the world suffer from hunger—from just plain lack of sufficient feed-than from the lack of a specific nutrient (or nutrients); therefore, it is recognized that nutritional deficiencies may be brought about either by (1) too little feed, or (2) rations that are too low in one or more nutrients.

Also, forced production (such as very high milk yields and finishing animals at early ages) and the feeding of forages and grains which are often produced on leached and depleted soils have created many problems in nutrition. This condition has been further aggravated through the increased confinement of stock, many animals being confined to lots or buildings all or a large part of the year. Under these unnatural conditions, nutritional diseases and ailments have become increasingly common.

Although the cause, prevention, and treatment of most of these nutritional diseases and ailments are known, they continue to reduce profits in the livestock industry simply because the available knowledge is not put into practice. Moreover, those widespread nutritional deficiencies which are not of sufficient proportions to produce clear-cut deficiency symptoms cause even greater economic losses because they go unnoticed and unrectified. Table 4–13 contains a summary of the important nutritional diseases and ailments affecting animals. Pictures showing typical symptoms of several mineral and vitamin deficiency diseases are presented in the feeding chapters pertaining to each species—Chapters 15, 24, 32, 40, 44, and 54. Pictures showing two nutritional toxicity diseases are presented on page 155.

**TABLE**
**NUTRITIONAL DISEASES**

| Disease | Species Affected | Cause | Symptoms and Signs (or age group most affected) | Distribution and Losses Caused By |
|---|---|---|---|---|
| **Acetonemia** (see Ketosis) | | | | |
| **Acidosis** (or lactic acid acidosis)—a metabolic disease of cattle and sheep | Cattle, especially feedlot cattle. Sheep, especially feedlot lambs. | Acidosis is caused by an increase in lactic acid-producing bacteria (both the d- and l-forms) and the rapid production of lactic acid. It commonly occurs where there is a sudden shift from high-roughage to high-concentrate ration. However, cattle maintained on high-energy rations are constantly in a marginal state of acidosis due to the formation of lactic acid in the rumen flora. Thus, ingredient changes, poor mixing of grain in the ration, or faulty feeding can promote acute acidosis. | Marginal acidosis is characterized by poor performance and inconsistent feed ingestion. If ingredient changes or erratic feeding persist, acute acidosis may result, creating laminitis—and eventually "ski shoe" in cattle. In severe cases, the rumen becomes immobilized, followed by increased pulse and respiration rate, variable rectal temperature, sunken eyes, loss of dermal elasticity, staggering, coma, and death. | Acidosis occurs wherever cattle and lambs are fed, especially on high-concentrate rations. The annual loss from acidosis has been estimated at about 1% of the production. |
| **Alkali disease** (see Selenium poisoning) | | | | |
| **Anemia**, nutritional | All warm-blooded animals, including humans. | Commonly an iron deficiency, but it may be caused by a deficiency of copper, cobalt, and/or certain vitamins. | Loss of appetite, progressive emaciation, and death. Most prevalent in suckling young. Pigs show listlessness, rough hair coat, wrinkled skins, drooping ears and tails, pale membranes around the mouth and eyes, labored breathing, and a swollen condition about the head and shoulders. | Worldwide. Losses consist of slow and inefficient gains, and deaths. |
| **Aphosphorosis** | Cattle; sheep to a lesser extent. | Low available phosphorus in feed. | Decreased growth rate; inefficient feed utilization; depraved appetite—chewing bones, wood, hair, rags, and other objects; stiff joints and fragile bones. Breeding problems and a high incidence of milk fever in dairy cattle. | Worldwide. Southwestern U.S. |

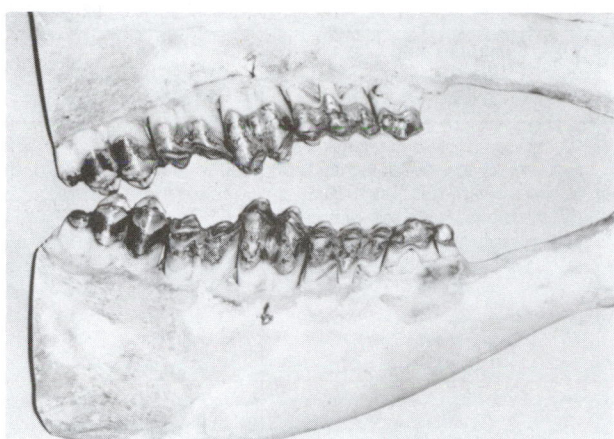

Fig. 4–41. Fluorine toxicity. Note the mottled and stained teeth, resulting from the animal ingesting too much fluorine in the feed and/or water. (Courtesy, University of Minnesota, Minneapolis)

Fig. 4–42. Selenium toxicity. Note the deformity of this lamb, traceable to the ewe ingesting too much selenium during pregnancy, resulting in selenium injury during fetal development. (Courtesy, University of Wyoming, Laramie)

**4–13**
## AND AILMENTS OF ANIMALS

| Treatment | Control and Eradication | Prevention | Remarks |
|---|---|---|---|
| **D**ifferent treatments have been used with varying degrees of success; among them: (1) removal of rumen contents and replacement by contents of an animal on a normal ration; (2) feeding a high level of penicillin (12–20 million units) to suppress lactic acid-producing bacteria; (3) drenching (or intravenous injection) with a solution of sodium bicarbonate to restore the acid base balance; (4) daily intramuscular administration of antihistamines and cortical steroids for each of several days to help prevent intoxication and laminitis; or (5) backing the animals down on both amount and kind of feed (lessening the ration, and returning to the mix that was being used before trouble was encountered). | **A**cidosis is best controlled by (1) avoiding accidental access of animals to large amounts of concentrates, and (2) changing gradually and stepwise from a low to a high proportion of concentrate in the ration. | **P**revention consists of starting animals on a high-roughage ration and gradually reducing the roughage and increasing the grain; avoiding erratic feeding; and avoiding abrupt ration changes. | |
| **P**rovide dietary sources of the nutrient or nutrients, the deficiency of which is known to cause the condition. | **W**hen nutritional anemia is encountered, it can usually be brought under control by supplying dietary sources of the nutrient or nutrients, the deficiency of which is known to cause the condition. | **S**upply dietary sources of iron, copper, cobalt, and certain vitamins (especially folacin, riboflavin, and vitamin B–6). <br> **K**eep confinement of suckling animals to a minimum and provide dry feeds at an early age. <br> **A**nemia in pigs can be prevented by providing supplemental iron in one of the following forms: | **A**nemia is a condition in which the blood is either deficient in quality or quantity. (A deficient quality refers to a deficiency in hemoglobin and/or red cells.) <br> **L**evels of iron in feed believed to be ample, since feeds contain 40 to 400 mg/lb. |

    1. Inject intramuscularly 100 to 200 mg of iron from iron dextran into baby pigs at 2 to 3 days of age. If pigs remain in confinement and do not have access to creep feed at an early age, a second injection at 2 to 3 weeks of age is desirable. Injection is the method of choice, for it assures that every pig receives its requirement.

    2. Orally administer iron dextran in a liquid or a solid preparation. To ensure daily intake by all pigs, it is important to have a preparation that is palatable and readily consumed. Also, placement of the oral preparation at the right location in the creep area is most important.

    3. Give the pigs iron tablets or paste at 2 to 3 days of age. Repeat the treatment every 7 to 10 days until the pigs are eating the creep ration adequately. If pills are given, it is important to see that the pigs swallow them and not spit them out.

    4. Place clean soil in the farrowing pen daily. Soil should not be contaminated with parasite eggs and other disease organisms. Iron sulfate can be sprinkled over the soil.

    5. Swab sow's udder daily with a solution of 1 lb ferrous sulfate dissolved in 1 gal of warm water.

    6. Provide pigs with access to a creep feed by the time they are 10 days old.

| | | | |
|---|---|---|---|
| **I**ntravenous drench with suitable phosphorus solution. <br> **A**dd phosphorus to the ration. | **C**ontrolled by feeding phosphorus, either free-choice or added to the ration. | **F**eed phosphorus in feed and/or as mineral supplement (free-choice). <br> **K**eep the calcium-phosphorus ration within the range 4:1 to 1:1. | **G**enerally caused by lack of phosphorus in the pasture. Phosphorus fertilizing may help. |

*(Continued)*

TABLE 4-13

| Disease | Species Affected | Cause | Symptoms and Signs (or age group most affected) | Distribution and Losses Caused By |
|---------|------------------|-------|------------------------------------------------|-----------------------------------|
| **Azoturia** (hemoglobinuria, Monday morning disease, blackwater) | Horses. | Sudden exercise, following a day or two of rest during which time the horse has been on full feed, resulting in partial spasm or tie-up. Azoturia is caused by an abnormal amount of glycogen stored in the muscle. As the glycogen breaks down, lactic acid is formed, which builds up in the muscle, causing severe muscle destruction and the release of myoglobin which manifests itself as partial spasm or tie-up, and wine-colored urine. | Profuse sweating, elevated temperature and pulse, wine-colored urine, stiff gait, reluctance to move due to pain, and knuckling over of the hind pasterns. Finally, animal assumes a sitting position, and eventually falls prostrate on the side. | Worldwide, but the disease is seldom seen in horses at pasture and rarely in horses at constant work. |
| **Baby pig shakes** (see Hypoglycemia) | | | | |
| **Bloat—Feedlot** | All ruminants. | High-concentrate rations, especially when finely ground, increase numbers of slime-producing bacteria in rumen. Slime traps fermentation gas and produces bloat. | Symptoms same as pasture bloat (see "Bloat—Pasture" which follows). Occurs when cattle or sheep have been fed high-concentrate, low-roughage rations for approximately 60 days or longer. | A survey of Kansas feedlots showed the following losses from bloat; 0.1% died of bloat; 0.2% bloated severely; and 0.6% bloated mildly to moderately, with animal performance affected adversely.[1] |
| **Bloat—Pasture** | All ruminants. | Most common on lush legume pastures. Incidence on wheat pasture has been increasing in recent years. Pasture bloat is a frothy bloat caused by interaction of several factors—plant, animal, and microbial. Soluble plant proteins and the presence of saponins play a prominent role in permitting stable froth formation. There may be a genetic tendency in chronic bloaters. | First observed as distension of paunch on left side in front of hipbone. This is followed by distension of right side, protrusion of anus, respiratory distress, cyanosis of tongue (bluish coloration), struggling, and death if not treated. | Widespread, although some areas appear to have more bloat than others. Often results in death. Causes average annual losses in beef and dairy cattle (including milk) of more than $100 million. |
| **Colic** | Horses. | Internal parasites are the number one cause of colic. Improper feeding, working, or watering. | Excruciating pain; and, depending on the type of colic, other symptoms are: the horse looking at its belly, distended abdomen, increased intestinal rumbling, violent rolling and pawing, profuse sweating, constipation, and refusal of feed and water. | Worldwide. Colic is the most common ailment among horses and is the leading cause of death. |
| **Crooked calves** | Cattle. | Manganese deficiency. Lupine consumption. | Calves born with crooked necks and legs. | Manganese deficiency occurs in northwestern U.S. |
| **Downer cow syndrome** | Cattle. | Downer cow syndrome commonly develops following a malady that causes recumbency (lying down); for example, anesthesia, arthritis, calving paralysis, nutritional deficiency, grass tetany, ketosis, mastitis, metritis, or milk fever. | Cow goes down and is reluctant or unable to rise voluntarily. Generally, the hindlimbs are affected—not the forelimbs. | It may be found wherever there are cattle. In afflicted beef herds, up to 10% of the cows may go down. |
| **Encephalomalacia** (crazy chick disease) | Chicken. | Vitamin E deficiency. | Chicks: Head retraction and loss of control of legs. Adults: Poor reproductive performance; prolonged vitamin E deficiency results in permanent sterility in the male and reproductive failure in the female. | Encephalomalacia occurs wherever there is vitamin E deficiency in chickens. The incidence is higher where vitamin E deficiency is accompanied by polyunsaturated fats that oxidize and become rancid. |

(Continued)

| Treatment | Control and Eradication | Prevention | Remarks |
|---|---|---|---|
| **A**bsolute rest and quiet. While awaiting the veterinarian, apply heated cloths or blankets, or hot-water bottles to the swollen and hardened muscles. *But do not move the horse.* <br> **T**he veterinarian should determine treatment. In mild cases, treatment may consist of the use of a tranquilizer or sedative. In severe cases, the veterinarian may use (1) a muscle relaxant, or (2) sodium bicarbonate in solution to readjust the acid balance in the muscles. | **W**hen trouble is encountered, decrease the concentrate ration and increase the exercise on idle days. | **R**estrict the ration and provide daily exercise when the animal is idle. Give a wet bran mash the evening before an idle day or turn the idle horses to pasture. <br> **S**ome believe that a diuretic (a drug which will increase the flow of urine) will prevent the tie-up syndrome. This is a common treatment of racehorses. Others feel that increased B vitamins will prevent the lactic acid buildup. | **T**he chances of recovery are good for horses that remain standing, are not forced to move after the signs are noticed, and whose pulse returns to normal within 24 hours. <br> **A**zoturia and colic have some similar symptoms; hence, there is danger of misdiagnosis and the wrong treatment. |
| **A**dminister a defoaming agent immediately, such as (1) 1 pint of corn oil, peanut oil, or soybean oil, or (2) poloxalene administered according to the manufacturer's directions. | **I**f feasible, increase proportion of roughage in ration. However, good quality legume hay may increase incidence of feedlot bloat. In the latter case, poloxalene or oxytetracycline will be effective preventives when used according to manufacturer's directions. | **U**se (1) poloxalene or oxytetracycline according to manufacturer's directions; and (2) proper management. | **F**eedlot bloat may occur duirng any month of year; however, it is more common during hot, humid weather. |
| **T**ime permitting, severe cases of bloat should be treated by a veterinarian. Puncturing of the paunch should be a last resort. <br> **M**ild cases may be home-treated by (1) keeping the animal on its feet and moving, and (2) drenching the cattle either with (a) 1 pint of corn oil or soybean oil, or (b) 1 to 2 oz of poloxalene. | **W**hen there is high incidence of bloat, it may be desirable to change the feed. <br> **W**here legume bloat is encountered, use poloxalene, oxytetracycline, or polyoxyethylene (23), lauryl ether (Laureth-23/Enproal Bloat Blox), according to the respective manufacturer's directions. | **T**he incidence is lessened by (1) avoiding straight legume pastures and immature legumes, (2) feeding a coarse grass hay prior to turning onto lush pasture, (3) feeding dry forage along with pasture, (4) avoiding a rapid fill from an empty start, (5) keeping animals continuously on pasture after they are once turned out, (6) keeping salt and water conveniently accessible at all times, (7) avoiding frosted pastures, and/or (8) using poloxalene, oxytetracycline, or Laureth-23 (Enproal Bloat Blox) according to manufacturer's directions, including placing blocks containing these antifoaming agents in various parts of the pasture. | **L**egume pastures, alfalfa hay, and barley appear to be associated with a higher incidence of bloat than many other feeds. <br> **L**egume pastures are particularly hazardous when moist, after a light rain or dew. |
| **C**all a veterinarian. To avoid danger of inflicting self-injury, (1) place the animal in a large, well-bedded stable, or (2) take it for a slow walk. <br> **D**epending on diagnosis, veterinarian may use one or more of following: sedatives; laxatives (such as mineral oil); drugs; or surgery. | **F**ollow a good management program, including parasite control. <br> **F**eed, work, and water horses properly. | **P**roper feeding, working, watering, and parasite control. | **C**olic is also a symptom of abdominal pain that can be caused by a number of different conditions. For example, bloodworms cause a colic due to damage in the wall of blood vessels. This results in poor circulation to the intestine. Also, mares with uterine torsions exhibit colic pain, as do mares with urinary stones in the bladder. |
| **T**here is no treatment. | **P**rovide manganese where needed. <br> **A**void pasturing pregnant cows on pasture or range containing lupines. | **F**eed manganese; 30 ppm of total feed. Either manganese carbonate or manganese sulfate may be used. | **C**rooked calf disease due to manganese deficiency can be eliminated almost completely by providing manganese where needed. |
| **G**ood nursing is the best help. <br> **M**ove animal to an area with adequate bedding and footing; turn animal from side to side frequently; feed and water. | **I**n beef cattle, separate the first calf heifers from the thinner cows and give them special feeds. <br> **I**n dairy cows, feed cows properly, especially those that are likely candidates for milk fever. | **R**ecognize the earliest signs of downer cow syndrome and treat the animals before they become recumbent. | **D**owner cow is not a good name because it lumps together all cows that become recumbent for whatever reason. |
| **N**o treatment of affected birds is effective. | **A**dd vitamin E or another antioxidant to the diet. | **S**upplement the ration with vitamin E or a suitable antioxidant. | **U**nder farm conditions in the U.S., encephalomalacia is the major vitamin E deficiency disease found in growing chicks. |

*(Continued)*

| Disease | Species Affected | Cause | Symptoms and Signs (or age group most affected) | Distribution and Losses Caused By |
|---|---|---|---|---|
| **Enterotoxemia** (overeating disease, pulpy-kidney disease) | Sheep; less frequently goats; and rarely cattle. | *Clostridium perfringens*, type D, triggered by overconsumption of high-energy feeds, milk, or lush pasture. | Sudden death. Enterotoxemia can be confirmed by laboratory tests if necropsy is performed soon after death. | Worldwide. In unvaccinated feedlot lambs, death rate averages 3 to 4%. In explosive outbreaks, losses range from 10 to 40%. |
| **Fescue foot** (fescue toxicosis) | Cattle, sheep, and horses. | A fungus, *Acremonium coenophialam*, which lives in tall fescue. | Poor conception rates, low pasture gains, and depressed milk production. | Wherever tall fescue is grown, worldwide. |
| **Fluorine poisoning** (fluorosis) | All farm animals, poultry, fish, and humans. | Ingesting excessive quantities of fluorine through either the feed, air, water, or some combination of these. | Abnormal teeth (especially mottled enamel) and bones (bones become thickened and softened), stiffness of joints, loss of appetite, emaciation, reduction in milk flow, diarrhea, delayed maturity, and salt hunger. | The water in parts of Arkansas, California, South Carolina, and Texas has been reported to contain excess fluoride. Occasionally, throughout the U.S., high-fluorine phosphates are used in mineral mixes. Areas near certain industries which heat earthy materials or burn high-fluoride coal may be a problem. |
| **Founder** (laminitis) | Horses. Cattle. Sheep. Goats. | Overeating (grain; or lush legume or grass—known as *grass founder*), overdrinking, or from inflammation of the uterus following parturition. Also intestinal inflammation. Too rapid change in the ration. | Extreme pain, fever (103 to 106°F), and reluctance to move. If neglected, chronic laminitis will develop, resulting in a dropping of the hoof soles and a turning up of the toe walls. | Worldwide. Actual death losses from founder are not very great, but usefulness may be affected. |
| **Goiter** (see Iodine deficiency) | | | | |
| **Gossipol toxicity** | Swine. Poultry. Ruminants are seldom affected. | Gossipol, a toxic yellow pigment, contained in the glands of most cottonseeds. | *Swine:* Difficult breathing, loss of appetite, retarded growth, lack of thrift, weakness, and death. *Poultry:* High-gossipol will cause discoloration in stored eggs. | Gossipol toxicity is seldom a problem in the U.S., due to: (1) inactiviting the gossipol in cottonseed by processing, (2) proper cautions in feeding cottonseed and cottonseed meal, and (3) use of new glandless cottonseed and cottonseed meal. |
| **Grass tetany** (grass staggers) | Cattle. Sometimes sheep and goats. | Magnesium deficiency. | Generally occurs during first 2 weeks of pasture season, on pasture containing less than 0.2% magnesium and more than 3% potassium and 4% nitrogen, and among lactating animals. Initial signs of grass tetany are nervousness, attentive ears, and decreased milk flow. In more severe cases, affected animals may avoid the rest of the herd, walk with a stiff gait, lose their appetite, and urinate frequently; they are nervous, have staring eyes, and keep their head and ears in an erect position; they stagger, have a twitching skin—especially about the face, ears, and flanks; lie down and get up frequently; behave aggressively (they may even charge a person); after 2 to 3 days, extreme excitement may cause convulsions, during which animal lies flat on side, saliva flows freely, breathing is labored, and the heart pounds rapidly; and if treatment is not given, usually death occurs during or after a convulsion. | Worldwide. It is generally considered to be the leading cause of cattle deaths in the U.S., killing an estimated 1 to 3% of the cattle of the temperate regions. |

*(Continued)*

| Treatment | Control and Eradication | Prevention | Remarks |
|---|---|---|---|
| None. | When an outbreak occurs in feeder lambs, (1) increase the roughage in the ration, and (2) add 200 g of chlortetracycline per ton of feed.<br><br>When an outbreak occurs in nursing lambs, inject susceptible lambs with enterotoxemia antiserum, followed by vaccination 14 to 21 days later. | Proper feeding, along with vaccination.<br>Ewes should be vaccinated with types C and D toxoid. Lambs should be vaccinated with type D only. | The disease is caused by bacteria, but it is triggered by high-energy rations, lots of milk, or lush pasture. |
| There is no effective treatment. Cattle usually recover if removed from fescue pasture or hay. | Good management, good nutrition of animals, early detection of symptoms, and/or destroying toxic pastures and reseeding with fungus (endophyte)- free seed. | Seeding of fungus-free seed.<br>Where fescue foot is a problem, remove mares from fescue pasture last 2 to 3 months of pregnancy. | Most toxic fescue pastures are several years old. |
| Any damage may be permanent, but animals which have not developed severe symptoms may be helped to some extent, if the sources of excess fluorine are eliminated. | Discontinue the use of feeds, water, or mineral supplements containing excessive fluorine. | Avoid the use of feeds, water, or mineral supplements containing excessive fluorine. The maximum safe level in the total dry ration is: beef and dairy cattle, 50 to 100 ppm; sheep and swine, 100 to 200 ppm; and chickens, 300 to 400 ppm. | Fluorine is a cumulative poison.<br>Underfluorinated rock phosphate often contains 3.5 to 4.0% (35,000–40,000 ppm) of fluorine.<br>Phosphate clays (soft phosphates) are usually too high in fluorine to be used safely unless defluorinated. |
| Pending arrival of the veterinarian, the attendant should stand the animal's feet in a cold-water bath.<br>The following treatments are used, with varying degress of success: (1) mineral oil, (2) analgesics (pain killers), (3) injectable antihistamine, (4) antibiotics, (5) sodium bicarbonate, (6) temporarily deadening the nerve supply to the feet, (7) water soaks, and/or (8) wraps applied to the affected feet. | Alleviate the causes. | Avoid (1) overeating, (2) overdrinking (especially when hot), and/or (3) inflammation of the uterus following parturition.<br>Veterinary attention should be given if mares retain the afterbirth longer than 12 hours.<br>Careful grain feeding will prevent many causes of founder in cattle, sheep, and goats. | Unless foundered animals are quite valuable, it is usually desirable to dispose of them following a case of severe founder.<br>Swine do not founder because they can unload their stomachs by vomiting. |
| Remove the cause—do not feed toxic cottonseed or cottonseed meal. | Limit the cottonseed meal content of swine rations to ½ the protein supplement of the ration.<br>Limit the free gossipol poultry ration to 0.03%, and add iron salts. | Gossipol toxicity can be alleviated in three ways: (1) limiting the amount of gossipol, (2) adding iron salts, and (3) using glandless cottonseed and cottonseed meal. | In addition to inactiviting the gossipol, ways have been devised to extract the gossipol. |
| Under range conditions, 200 cc of a sterile, saturated solution of magnesium sulfate (Epsom salts) injected under the animal's skin (inject only 50 cc at any one place). | (See Prevention.) | Grass tetany can be prevented by not turning animals to pasture, but this is not practical. Feeding hay at night during the first 2 weeks of the pasture season is helpful.<br>Prevention consists in providing magnesium daily throughout the high-risk period. The daily magnesium requirement of cattle is as follows: dry cows, 10 g daily; cows suckling calves, 20–25 g daily; dairy cows, 30 g daily; and calves, 4–8 g daily.<br>Any of the following mineral supplements are satisfactory:<br>1. A mix of ⅓ each: magnesium oxide, iodized salt, and either soybean meal or cottonseed meal.<br>2. A mineral mix of 30% magnesium oxide, 30% iodized salt, 30% bone meal, and 10% dried molasses.<br>3. A mineral mix of ⅔ magnesium oxide and ⅓ salt as the only source of salt.<br>4. A commercial mineral (block or mix) usually formulated to be fed at rate of ½ to 1½ lb/head/day, and to provide 10 to 15 g of magnesium daily. | Treated cattle may be aggressive on arising; so watch out! |

TABLE 4–13

| Disease | Species Affected | Cause | Symptoms and Signs (or age group most affected) | Distribution and Losses Caused By |
|---|---|---|---|---|
| **Heaves** | Horses. Mules. | Exact cause unknown, but it is known that the condition is often associated with the feeding of damaged, dusty, or moldy hay, and the use of dusty bedding or paddocks. It often follows severe respiratory infection such as strangles. Probably an allergy. | Difficulty in forcing air out of the lungs, resulting in a jerking of flanks (double flank action) and coughing. The nostrils are often slightly dilated and there is a nasal discharge. Heaves in horses is similar to emphysema in people. | Worldwide. Losses are negligible. |
| **Hypoglycemia** (or Baby pig shakes) | Swine. | Low blood-sugar level accompanies the trouble, but the cause of the low blood sugar is unknown. Predisposition occurs from any disease which decreases or inhibits the milk production of sows. | Shivering, weakness, failure to nurse, with no evidence of scouring. If disturbed, the pigs emit a weak, crying squeal. Hair becomes erect and rough, and the heart action slow and feeble. Without treatment, death usually comes in 24 to 36 hours after the first symptoms appear. Confined to baby pigs only. | Worldwide. Hypoglycemia accounts for 15 to 25% of total piglet mortality. |
| **Iodine deficiency** (goiter) | All farm animals and humans. | A failure of the body to obtain sufficient iodine from which the thyroid gland can form thyroxin (an iodine-containing compound) | Goiter (big neck) is the most characteristic symptom in humans, calves, lambs, and kids. Also, there may be reproductive failures and weak offspring that fail to survive. Pigs may be born hairless, and show edema of shoulders and neck. Foals may be weak. | Worldwide; wherever feeds are grown on iodine-poor soils. Highest incidence has been observed in the Alps, the Pyrenees, the Himalayas, the Thames Valley of England, parts of New Zealand, and in a number of Central and South American countries. In the U.S., from the Great Lakes to Washington. Also, reported in California and Texas. |
| **Ketosis** (also known as acetonemia in cattle and pregnancy disease in sheep) | Cattle. Sheep. Goats. | A metabolic disorder of nutritional origin, characterized by hypoglycemia (low blood sugar). If the increased nutrient requirements are not met by more feed during the high-demand period: (in cows, 1–6 weeks after calving; in ewes, 2 weeks after lambing), the animals must draw on body fat reserves. If this is done too rapidly, and without adequate carbohydrates in the ration, ketosis follows. | In cows, ketosis or acetonemia is usually observed within first 1–6 weeks after calving. Affected animals show loss in appetitie and condition, a marked decline in milk production, and the production of a peculiar sweetish chloroformlike odor of acetone that may be present in the milk and pervade the barn. In ewes and goats, ketosis or pregnancy disease generally strikes during last 2 weeks of pregnancy. Usually, affected ewes are carrying twins or triplets. Symptoms include grinding of teeth, dullness, weakness, frequent urination and trembling when exercised—with the final stage being complete collapse, followed by death in 90% of the cases. In dairy goats, lactation ketosis, which is similar to ketosis of dairy cows, may be observed in high milk producers following kidding. | Worldwide. Ketosis or acetonemia affects dairy cattle throughout the U.S. Metabolic diseases of cattle (which include ketosis, milk fever, grass tetany, lameness, and reduced resistance to infectious agents) cause estimated losses of $15 million annually. Ketosis or pregnancy disease of sheep affects farm flocks more than range bands, the losses in the former sometimes being as high as 25% |
| **Liver Abscesses** | Cattle, especially feedlot cattle. | *Fusobacterium necrophorum*, the same bacteria that causes foot rot. High concentrate (low roughage) finishing rations predispose cattle to a high incidence of liver abscesses. | Liver abscesses generally go undetected until cattle are slaughtered. However, reduced feed intake and gain near the end of the feeding period may be indicative of liver abscesses. | Liver abscesses occur in all countries where intensive beef production is practiced. The incidence of liver abscesses is highest in feedlot cattle, where an estimated 18 to 20% of the livers are affected. In a particular lot of feedlot cattle, liver abscesses may range from 1 to 90%. |
| **Manganese deficiency** (see Crooked calf disease and Grass tetany) | All farm animals and poultry. | Deficiency of manganese. | Young born with stiff, curved, or crooked necks and backs, and permanently bent forward legs caused by contracted tendons. | Reported in Washington, Montana, and Utah. |

*(Continued)*

| Treatment | Control and Eradication | Prevention | Remarks |
|---|---|---|---|
| **A**ntihistamine granules can be administered in feed to control coughing due to lung congestion. | **A**ffected animals are less bothered if turned to pasture. If used only at light work, if fed an all-pelleted ration, or if the hay is sprinkled lightly with water. | **A**void the use of damaged feeds.<br>**F**eed an all-pelleted ration, thereby alleviating dust. | **U**nlike humans, a horse cannot breathe through the mouth.<br>**B**asically, heaves is a rupture of some of the alveoli in the lungs, of which the specific cause is unknown. |
| **P**rovide heat lamps for pigs.<br>**A**t earliest symptoms either (1) force feed at frequent intervals a mixture of 1 part corn syrup diluted with 2 parts water or (2) give intraperitoneal injections of 5% glucose solution every 4 to 6 hours.<br>**C**onsult veterinarian. | **N**ot contagious. | **A**dequate rations and good care and management of the gestating sows may lessen the incidence of the disease.<br>**B**e sure there is adequate milk for baby pigs during first days of life. | **O**ne of the hazards of hypoglycemia is that the milk flow of the sow will not be stimulated or may even cease due to the inactivity of the affected pigs. In the latter case, pigs may have to be either transferred to a foster mother or hand fed. |
| **O**ccasionally borderline cases may survive; in these the moderate thyroid enlargement disappears in a few weeks.<br>**O**nce the iodine deficiency symptoms appear, no treatment is very effective. | **A**t the first signs of iodine deficiency, iodized salt should be fed to all farm animals. | **I**n iodine-deficient areas, feed iodized salt to all farm animals throughout the year.<br>**S**tabilized iodized salt containing 0.01% potassium iodide is recommended.<br>**O**rganic iodide is also a suitable source of iodine, but is usually more costly. | **T**he enlarged thyroid gland (goiter) is nature's way of attempting to make sufficient thyroxin under conditions where an iodine deficiency exists.<br>**M**ares fed excess iodine (48 mg or more) during late gestation will produce foals with hyperplastic goiter. Some mares will also develop goiter. |
| *Cattle:* ½ to 1 lb of either propylene glycol or sodium propionate daily, with the dose divided into 2 treatments for 5 to 10 days. Put treatment in grain if cow is eating; otherwise, give as drench.<br>**I**ntravenous injection of glucose (500 ml of 50% glucose solution) is rapid way of getting sugar in blood.<br>*Sheep and goats:* 3 to 4 oz of propylene glycol, given orally twice daily.<br>*Dairy goats after kidding:* 6 to 8 oz of propylene glycol, given orally twice daily. | *Cows:* Maintain relatively high energy intake before calving; increase energy intake substantially after calving.<br>*Ewes:* Avoid obesity in early pregnancy. Feed rather liberally last 6 weeks of pregnancy. | *Cows:* The incidence of ketosis can be lessened by (1) not allowing cows to be excessively fat at calving; (2) increasing the level of concentrates rapidly after calving; (3) feeding good quality roughage after calving, and avoiding abrupt changes in roughage; (4) feeding adequate proteins, minerals, and vitamins, and (5) providing comfort, exercise, and ventilation.<br>**I**n problem herds, feeding ¼ lb daily of proylene glycol or sodium propionate may be helpful.<br>*Sheep and goats:* Feed more hay and ½ to 1 lb of grain beginning a month before parturition. Good management is important too, including exercise, freedom from parasites, and minimizing stress. | **T**he clinical findings are similar in the case of affected cattle and sheep, but it usually strikes ewes just before lambing, whereas cows are usually affected within the first 1 to 6 weeks after calving. |
| **T**reatment of feedlot cattle should be left to the veterinarian, who may administer (1) sulfapyridine, or (2) an antibiotic. | **T**he low level of feeding certain antibiotics during the finishing period of cattle will markedly reduce the number of abscesses. | **I**ncidence can be reduced, but not entirely eliminated, by (1) changing from high-roughage to high-concentrate rations gradually, and (2) feeding an antibiotic (commonly chlortetracycline or tylosin) at the daily rate of 70 mg/animal. | **A**t slaughter, livers affected with abscesses are condemned for human food. |
| | (See Prevention.) | **F**eed a mineral containing manganese; 30 ppm of total feed, or 27.24 g/ton feed. | **T**he Utah station has also produced crooked calves by feeding lupine.<br>**A**lkali can tie up the manganese in water, soils, or plants. |

*(Continued)*

TABLE 4–13

| Disease | Species Affected | Cause | Symptoms and Signs (or age group most affected) | Distribution and Losses Caused By |
|---|---|---|---|---|
| **Milk fever** (parturient paresis; hypocalcemia) | Cattle. Sheep. Goats. | Low blood calcium concentration. Initiation of lactation places a severe strain on the calcium balance of the cow due to the amount of calcium secreted in the milk. | Commonly occurs within 3 days after calving and in high-producing cows. Rarely occurs at first calving. First symptoms are loss of appetite, constipation, and general depression. This is followed by nervousness and finally collapse and complete loss of consciousness. The head is usually turned back. | A common, widespread disease of dairy cows. Losses are not great, although untreated animals are likely to die. Milk fever causes estimated average animal losses in dairy cattle (including milk) of $100 million. |
| **Molybdenum toxicity** | Ruminants, especially calves and cows in milk. | As little as 6 ppm can cause toxicity, depending on the amount of copper available. | Toxic levels of molybdenum interfere with copper metabolism, thus increasing the copper requirement and producing typical copper deficiency symptoms. The physical symptoms in cattle are anemia and extreme diarrhea, with consequent loss in weight and milk yield. Black hair may turn brown. In sheep, there is depigmentation of the wool, and loss of crimp. | Canada (Manitoba), England, and the U.S. (California, Nevada, and other states). |
| **Night blindness** (nyctalopia) | All farm animals and humans. | Deficiency of vitamin A. Ability to see in dim light depends upon the rate of synthesis of rhodopsin (visual purple) in the retina of the eye. When vitamin A is deficient, rhodopsin synthesis is impaired, resulting in lessened ability to see in dim light, commonly known as night blindness. | Deficiency first manifests itself as a slow dark adaptation, then progresses to total night blindness. | Worldwide. Especially prevalent when (1) extreme drought, or (2) winter feeding on bleached grass or hay. |
| **Nitrate poisoning** (oat hay poisoning, corn stalk poisoning) | Primarily cattle. Sheep. Horses. Ruminants are most susceptible because of the conversion of nitrates to nitrites by the microorganisms in the rumen. | Forages (vegetative part) of most grain crops (oats, wheat, barley, rye, corn, sorghum), Sudangrass, and numerous weeds, especially (1) when under stress such as drought, insufficient sunlight, or after spraying with weed killer (herbicide); or (2) following heavy nitrate fertilization of soils (commercial, green manure crop, barnyard manure). Some nitrate may be formed after forage is stacked. Inorganic nitrate or nitrite salts, or fertilizers left where animals have access to them, or where they may be mistaken for salt. Pond or shallow well into which surface runoff from barnyard or well-fertilized soil might drain. | Accelerated respiration and pulse rate; diarrhea; frequent urination; loss of appetite; general weakness; trembling and staggering gait; frothing from mouth; lowered milk production; abortion; blue color of the mucous membrane, muzzle, and udder due to lack of oxygen in blood; death within 4½ to 9 hrs. after consuming nitrates. A rapid and accurate diagnosis of nitrate poisoning may be made by examining blood. Normal blood is red and becomes brighter when exposed to air, whereas blood from cows toxic with nitrates is a brown color due to formation of methemoglobin. Nitrates oxidize ferrous hemoglobin (oxyhemoglobin) to ferric hemoglobin (methemoglobin) which cannot transport oxygen. The animal essentially suffocates for lack of oxygen in tissues. When ¾ of the oxyhemoglobin is converted to methemoglobin, the animal will die. | Excessive nitrate content of feeds is an increasingly important cause of poisoning in farm animals, due primarily to more and more high nitrogen fertilization. But nitrate toxicity is not new, having been reported as early as 1850, and having occurred in semiarid regions of this and other countries for years. |
| **Oat hay poisoning** (see Nitrate poisoning) | | | | |

*(Continued)*

| Treatment | Control and Eradication | Prevention | Remarks |
|---|---|---|---|
| The standard treatment is intravenous infusion of calcium borogluconate as soon as first signs appear. *Caution:* Overdose of calcium salts can result in heart damage. Intravenous injection of calcium salts should be given slowly. | (See Prevention.) | Each of the following measures will lessen the incidence of milk fever: <br> 1. *Low calcium during the dry period.* Feeding low calcium (less than 100 g/day)—high phosphorus (more than 40 g/day) rations during the dry period is important. High calcium levels in dry cow rations aggravate the problem. So, feeding a low calcium ration (less than | The name *milk fever* is a misnomer, because the disease is not accompanied by fever, the temperature really being below normal. <br> The incidence of milk fever is highest in the Guernsey and Jersey breeds. |

0.1 lb/day) before calving has shown promise of preventing milk fever.

2. *Calcium shock treatment.* Feed low calcium, high phosphorus rations containing only 15 to 20 g of calcium per day for 2 weeks prior to the expected calving date, rather than a restricted, but somewhat higher, calcium intake during the entire dry period. This creates a mild calcium deficiency which stimulates production of the biologically active form of vitamin D in the animal's body. In turn, this form of vitamin D stimulates the bone and gut to supply more calcium and phosphorus. As a result, when the greater demand for calcium and phosphorus occurs at calving, the bone and gut are already activated and are able to meet the increased demands for calcium. Thus, milk fever is avoided.

3. *Calcium-phosphorus ratio and amounts.* Balancing cow rations to contain 0.5% calcium and 0.25% phosphorus on a dry matter basis will limit the incidence of milk fever.

4. *High vitamin D.* Feeding massive doses of 20 million I.U. of vitamin D/cow/day starting about 5 days before calving and continuing through the first day postpartum, with a maximum dosage period of 7 days, has been effective in controlling milk fever. However, difficulty in predicting calving dates accurately has reduced the effectiveness of this treatment under practical conditions.

5. *Avoid excessive fatness.* Excessive fatness, or any other conditions that reduce feed intake at calving, tends to cause more milk fever.

| Treatment | Control and Eradication | Prevention | Remarks |
|---|---|---|---|
| One gram of copper sulfate per head daily in the feed will usually cure symptoms of molybdenum toxicity. For calves up to one year of age, ½ g of copper sulfate will usually be adequate. | (See Prevention.) | When the molybdenum content of the forage is below 5 ppm, add 1% copper sulfate to the salt. <br> When the molybdenum content of the forage is above 5 ppm, add 2 to 5% copper sulfate to the salt, depending on the level of molybdenum. | When feeds are high in sulfate, toxic symptoms will be produced on lower levels of molybdenum and, conversely, higher levels of molybdenum can be tolerated with low levels of sulfate. |
| Add vitamin A or carotene to the ration. | (See Treatment and Prevention.) | Provide good sources of carotene (provitamin A) through green hay, pasture, or silage, or from yellow corn. | The vitamin A requirement varies (1) according to species, and (2) within species according to age, weight, and reproductive status. |
| A 4% solution of methylene blue (in a 5% glucose or a 1.8% sodium sulfate solution) administered by a veterinarian intravenously at the rate of 100 cc/1,000 lb liveweight. | (See Prevention.) | More than 0.5% nitrate nitrogen (dry basis) may be considered as potentially toxic. Feed should be analyzed when in question, by using a simple test to detect presence of nitrates (qualitative); if present, follow with a quantitative test to determine how much is present. <br> Nitrate poisoning may be reduced by (1) feeding high levels of grains and other high-energy feeds (molasses) and vitamin A, (2) limiting the amount of high-nitrate feeds, (3) ensiling forages which are high in nitrates, (fermentation reduces some nitrates to gas, but care must be taken to avoid nitric oxide and nitrogen dioxide released in early stages of fermentation) and avoid feeding until 3 to 4 weeks in storage. | Nitrate form of nitrogen does not appear to cause the actual toxicity. During digestion, the nitrate is reduced to nitrite, a far more toxic form (10 to 15 times more toxic than nitrates). In cows and sheep, this conversion takes place in the rumen (paunch); in horses in the cecum. <br> Lethal dose varies with (1) nutritional state, size and type of animal; and (2) the consumption of feed other than nitrate-containing material. (Nitrate over 5% of total ration is a potential source of trouble; 0.75% content nitrate forages must be fed with caution, and milk production will be lowered; and 1.5% death will likely occur). <br> Where nitrate troubles are suspected, consult the local veterinarian or county agent. |

*(Continued)*

| Disease | Species Affected | Cause | Symptoms and Signs (or age group most affected) | Distribution and Losses Caused By |
|---|---|---|---|---|
| **Osteomalacia** | All species. | Inadequate phosphorus (sometimes inadequate calcium). Lack of vitamin D in confined animals. Incorrect ration of calcium to phosphorus. | Phosphorus deficiency symptoms are: depraved appetitie (gnawing on bones, wood, or other objects, or eating dirt); lack of appetitie, stiffness of joints, failure to breed regularly, decreased milk production, and an emaciated appearance. Calcium deficiency symptoms are: fragile bones, reproductive failures, and lowered lactations. Mature animals most affected. Most of the acute cases occur during pregnancy and lactation. | Southwestern U.S. is classed as a phosphorus-deficient area, whereas calcium-deficient areas have been reported in parts of Florida, Louisiana, Nebraska, Virginia, and West Virginia. |
| **Parakeratosis** (greasy skin disease) | Swine. | Low zinc level. High calcium levels in the ration—above 0.8% | Pigs have mangy appearance, reduced appetite and growth rate, diarrhea, and vomiting. It affects pigs 6- to 16-weeks of age. | Mortality is not high; economic loss is mainly in reduced gains and lowered feed efficiency. |
| **Periodic ophthalmia** (moon blindness) | Horses. Mules. Asses. | It may be caused by (1) lack of riboflavin, (2) an autoimmune reaction, (3) an allergic reaction, (4) genetics, (5)deptoapirosis, brucellosis, and stronglyes, (6) parasitic infection, or (7) viral, fungal, chlamydial and mycoplasmal infections. | Periods of cloudy vision, in one or both eyes, which may last for a few days to a week or two and then clear up; but it recurs at intervals, eventually culminating in blindness in one or both eyes. | In many parts of the world. In the U.S., it occurs most frequently in the states east of the Missouri River, where an estimated 10 to 20% of the horses may be afflicted. |
| **Perosis** (slipped tendon) | Poultry. | Manganese deficiency. Perosis may also be caused by a deficiency of biotin or choline. | Characerized by enlargement and deformity of the hock joint and slipped tendon, twisting the shank to one side. One or both. | Perosis is seldom seen in large commercial poultry operations, where rations are scientifically formulated. However, it is sometimes seen in farm flocks. |
| **Photosensitization** | Cattle. Horses. Sheep. Swine. | Sensitization of light-colored skin to sunlight. Some feeds, forages, and certain medicines contain substances which may sensitize the skin. Products of metabolism, which are normally removed from the body, may accumulate because of faulty liver function (hepatogenous photosensitization). | The signs of the disease resemble severe sunburn. The lesions are confined to the white, or lightly pigmented, exposed areas of skin. The muzzle, eyes, face, and light areas over the back are usually affected first. Areas of the belly and udder, which are exposed to the sun when the animal is down, may also be affected. The earliest signs are redness and swelling of the skin. Later, tissue fluids ooze from the affected areas and crusting of the skin occurs, with resultant matting of the hair. In severe cases, the eyelids and nostrils may be swollen closed. In extreme cases, sloughing of the skin and gangrene result. | Worldwide. Death loss is higher in sheep than in cattle. The monetary loss is considerable, including loss in weight, damaged udders and teats, screwworms, secondary infections, eye damage, and stunted offspring. |
| **Polioencephalomalacia** (cerebrocortical necrosis, polio) | Cattle. Sheep. Goats. | Believed to be due to a thiamin (B-1) deficiency. It is noninfectious. | Sudden deaths in feedlot cattle. Sick animals are excitable, uncoordinated, and have impaired vision. On driving, these animals go down into convlusion. In sheep,the incidence is highest in feedlot lambs 5 to 8 months of age. In goats, it may strike suckling young on pasture. | Worldwide. Most common in feedlot cattle and sheep. In sheep, the morbidity rate may range from a few cases up to 10% of the flock, and 50% of the affected animals may die. |
| **Polyneuritis** | Chickens. | Deficiency of thiamin (vitamin B-1), which is required by poultry for metabolism. | Loss of appetite, followed by loss of weight, ruffled feathers, leg weakness, and unsteady gait. Head retraction. Nervous disorders, culminating in paralysis of the peripheral nerves. | Polyneuritis is not observed in poultry under practical conditions, because thiamin is found in abundance in whole grains which make up the major part of most poultry rations. |
| **Pregnancy disease in sheep** (see Ketosis) | | | | |

(Continued)

| Treatment | Control and Eradication | Prevention | Remarks |
|---|---|---|---|
| **S**elect natural feeds that contain sufficient quantities of calcium and phosphorus.<br>**F**eed a special mineral supplement or supplements.<br>**I**f this disease is far advanced, treatment will not be successful. | (See Treatment.) | **F**eed balanced rations, and allow animals free access to a suitable phosphorus and calcium supplement.<br>**I**ncrease the calcium and phosphorus content of feed through fertilizing the soils. | **C**alcium deficiencies are much more rare than phosphorus deficiencies in cattle, sheep, and horses.<br>**C**alcium deficiencies are fairly common in swine because grains, which are their chief feed, are low in this mineral. |
| **T**he amount of zinc to add to the ration will vary with the type of ration fed. The common recommendation is: Add 0.4 lb of zinc carbonate or 0.9 lb of zinc sulfate heptahydrate/ton of feed. | It is not contagious. | **M**eet the zinc requirement for swine. | **E**xcess calcium reduces the absorption and utilization of zinc. In swine, this causes parakeratosis.<br>**Z**inc carbonate, sulfate, and oxide are effective sources of zinc in prevention and treatment. |
| **A**ntibiotics, corticosteroids, and/or atropine administered promptly are helpful in some cases.<br>**I**mmediately (1) change to greener hay or grass, and (2) add riboflavin at the rate of 40 mg/day/animal. | **T**here is no sure control, because of the elusive causes of the disease. | **F**eed green grass, or well-cured green, leafy hay; or add riboflavin to the ration at the rate of 40 mg/animal/day. | **T**his disease has been known to exist for at least 2,000 years.<br>**B**ecause of the recurrent nature of the disease, it should be considered an unsoundness in the horse. |
| **A**fter the deformity has occurred, there is no treatment. | **P**rovide sufficient manganese in the ration, along with sufficient biotin and choline. | **G**rains tend to be deficient in manganese; so, prevention consists in adding manganese to poultry rations. | **P**erosis is made more severe by excessive amounts of calcium and phosphorus in the ration, and by rearing birds on wire or slotted floors. |
| **D**iscontinue the forage that is causing the trouble and keep the animal out of sunlight. In some cases, treatment of local lesions is warranted. In severe cases, the veterinarian may resort to intravenous fluids, antibiotics, and other special medicines. | **A**void animal access to offending plants, and keep animals in the shade. | **G**ood range and pasture management generally prevent the problem of photosensitization. | **P**hotosensitization should be differentiated from sunburn. |
| **I**ntramuscular or intravenous thiamin injections at a dosage of 1 to 2 mg/lb should be administered to sick animals. Twice-daily treatments may be necessary for 2 days.<br>**R**apidity of recovery relates directly to the speed of disease recognition and institution of thiamin treatment. Good nursing will help. | **C**oncentrate should be decreased, and roughage increased for 5 days, after which there should be gradual return to high energy ration. | **U**ntil the cause is discovered, little can be done to prevent the disease, except to provide a good ration. | **R**uminants normally derive adequate thiamin from symbiotic ruminal activity. The inadequacy is thought to be the result of some unknown intraruminal thiamin destructor. |
| **C**hickens suffering from a thiamin deficiency respond soon after the administration of vitamin B-1.<br>**I**njection of chickens suffering from a thiamin deficiency is necessary, because they are usually unable to eat. | **C**ontrol of the disease is not a problem, because most poultry feeds contain an abundance of thiamin. | **S**pecial sources of thiamin are not normally added to poultry rations. | **T**hiamin is the least stored of the vitamins; so, it should be supplied regularly in the ration.<br>**T**oday, polyneuritis is primarily of historic interest, because poultry rations contain adequate thiamin. |

TABLE 4–13

| Disease | Species Affected | Cause | Symptoms and Signs (or age group most affected) | Distribution and Losses Caused By |
|---|---|---|---|---|
| **Protein poisoning** | Horses in particular, but claims of protein poisoning are sometimes made relative to other classes of animals. | High levels of protein incriminated. Some horses do appear to be allergic to certain proteins or to excesses of specific amino acids. | "Protein bumps" over the body—an allergic reaction. | Evidence of the occurrence of protein poisoning is lacking. |
| **Pulmonary emphysema** (bovine pulmonary emphysema, cow asthma) | Cattle. | Caused by cattle consuming lush pasture containing high levels of tryptophan. When changed from dry feed to lush pastures, the clostridial organisms in the rumen help convert tryptophan to 3-methylindole (3-Mi). When large quantities of 3-Mi are absorbed in the blood stream, pulmonary emphysema may result.<br>Also, the disease may be caused by pneumonia, allergic reactions to lungworm larvae or inhaled fungal organisms, or inhalation of irritating gases. | Difficult breathing. In severe cases, panting, difficulty in exhaling, coughing, excess salivation, reluctance to move, extreme weakness, and rapid loss in condition. Death may follow within a few hours after onset of the disease. | Worldwide.<br>In affected herds, up to 20% of the cattle may develop emphysema and as many as 10% may die. |
| **Rickets** | All farm animals and humans. | Lack of either calcium, phosphorus, or vitamin D; or an incorrect ratio of the 2 minerals.<br>In housed animals, vitamin D deficiency is not uncommon. Grazing animals are more likely to be phosphorus deficient. | Enlargement of the knee and hock joints, and the animal may exhibit great pain when moving about. Irregular bulges (beaded ribs) at juncture of ribs with breastbone, and bowed legs.<br>Rickets is a disease of young animals—calves, foals, pigs, lambs, kids, pups, and chicks.<br>*Poultry:* The bones of growing birds become soft and rubbery. | Worldwide.<br>It is seldom fatal, but it can be severely debilitating and economically disastrous. |
| **Salt deficiency** (sodium chloride) | All farm animals and humans. | Lack of salt (sodium chloride). | Loss of appetite, retarded growth, loss of weight, a rough coat, lowered production of milk, and a ravenous appetite for salt.<br>*Horses:* Salt-deficient horses tire easily, stop sweating, and exhibit muscle spasms.<br>*Poultry:* Salt-deficient poultry exhibit cannibalism and poor growth. | Worldwide, especially among grass-eating animals. |
| **Salt poisoning** (sodium chloride) | All farm animals, but swine and sheep most frequently affected. | Brine from cured meats; wet salt.<br>Where large amounts of brine or salt have been mixed in hog slop.<br>When excess salt is fed following salt starvation.<br>When salt is improperly used to govern self-feeding of concentrate. | Sudden onset—1 to 2 hours after ingesting salt; extreme nervousness; muscle twitching and fine tremors; much weaving, wobbling, staggering, and circling; blindness; weakness; normal temperature, rapid but weak pulse, and very rapid and shallow breathing; diarrhea; death from a few hours up to 48 hours.<br>Convulsions seldom occur, except in pigs. | Salt poisoning is relatively rare, except in pigs. |
| **Salt sick** (cobalt deficiency) | Cattle.<br>Sheep.<br>Goats. | Cobalt deficiency.<br>In Florida, cobalt deficiency is associated with copper deficiency. | Loss of appetite, depraved appetite, scaliness of skin, listlessness, and lack of thrift. | Cobalt deficiency is widespread in different parts of the world.<br>In the U.S., cobalt deficiencies occur in Connecticut, Florida, Massachusetts, Michigan, New Hampshire, New York, North Carolina, Pennsylvania, Rhode Island, South Carolina, Vermont, and Wisconsin. |

(Continued)

| Treatment | Control and Eradication | Prevention | Remarks |
|---|---|---|---|
| Lower protein content of the ration. | Economics generally control the level of protein feeding. High protein feeds are more expensive than high energy feeds, with the result that there is temptation to feed too little of them. | Do not feed excessive levels of protein. | There is no proof that heavy feeding of high-protein feeds is harmful, provided (1) the ration is balanced out in other respects, (2) the animal's kidneys are normal and healthy (a large excess of protein in terms of body needs increases the work of the kidneys for the excretion of the urea), (3) any ration change to a high-protein feed is made gradually, as is recommended in any change in feed, and (4) there is adequate exercise and normal metabolism. |
| The recommended treatment: 1. Remove the animals from lush pastures and place them on hay. 2. Inject antihistamines, steroids, or other compounds to lessen the respiratory distress. 3. Use antibiotics or sulfonamides to prevent secondary bacterial infections. | The removal of cattle from lush pasture and placing them in a drylot and feeding hay will control the disease. | Prevention consists in: 1. Making a gradual transition from summer range to lush pasture. 2. Continuing to feed hay or straw while cattle are on pasture. | Pulmonary emphysema is primarily a nutritional disease, although there are other causes. |
| If the disease has not advanced too far, treatment may be successful by supplying adequate amounts of vitamin D, calcium, and phosphorus, and/or adjusting the ratio of calcium to phosphorus. | Control of rickets is usually achieved by providng a balanced ration, with special consideration given to calcium, phosphorus, and vitamin D. | Provide (1) sufficient calcium, phosphorus, and vitamin D, and (2) correct ratio of the 2 minerals. Vitamin $D_3$, rather than $D_2$, is required by the chicken. | Rickets is characterized by a failure of growing bone to ossify, or harden, properly. Hens fed rations deficient in vitamin D lay eggs with progressively thinner shells until production ceases. |
| Salt starved animals should be gradually accustomed to salt, slowly increasing the hand-fed allowance until the animals may be safely allowed free access to it. | (See Treatment and Prevention.) | Provide plenty of salt at all times, preferably by free-choice feeding. | Common salt is one of the most essential minerals for grass-eating animals, and one of the easiest and cheapest to provide. Excessive salt intake can result in toxicity if animals are deprived of water (see Salt poisoning). |
| Provide large quantities of fresh water to affected animals. Those that can and do drink seldom need additional treatment. Those unable to drink should be given water via stomach tube, by the veterinarian. | (See Prevention.) | If animals have not had salt for a long time, they should first be hand-fed salt, gradually increasing daily allowance until they leave a little in the mineral box; then self-feed. | Indians and pioneers handed down many legendary stories about huge numbers of wild animals that killed themselves simply by gorging at a newly found salt lick after having been salt-starved for long periods of time. Salt poisoning in pigs and poultry does not occur unless there is water deprivation. |
| Provide 0.2 to 0.5 oz cobalt salt/100 lb of salt—or feed a suitable trace mineral supplement. | Provide adequate cobalt in the ration—about 0.1 ppm. Deficiency symptoms appear when the level drops to the range of 0.04 to 0.07 ppm or lower. | Mix 0.2 to 0.5 oz of cobalt chloride, cobalt sulfate, or cobalt carbonate/100 lb of either (1) salt, or (2) whatever mineral mix is being used. | Cobalt is needed especially for microbial synthesis of vitamin B-12. Nonruminants must be fed preformed vitamin B-12. |

TABLE 4–13

| Disease | Species Affected | Cause | Symptoms and Signs (or age group most affected) | Distribution and Losses Caused By |
|---|---|---|---|---|
| **Selenium poisoning** (alkali disease, blind staggers) | All farm animals and humans. | Chronic selenium poisoning results when animals consume forages and grains containing 5 to 40 ppm selenium. Acute selenium poisoning results when animals consume high selenium content plants known as *selenium accumulators*. | In chronic selenium poisoning, there is loss of hair from the tail in cattle, a general loss of hair in swine, and a loss of hair from the mane and tail in horses. In severe cases, the hoofs slough off, lameness occurs, feed consumption decreases, and death may occur by starvation.<br>In acute selenium poisoning (blind staggers) vision is impaired. | In certain regions of the western U.S.—especially in Colorado, Nebraska, South Dakota, and Wyoming.<br>Also, high selenium has been reported in Alberta, Canada; and in Mexico, Ireland, Israel, and China; in Queensland, Australia; and in parts of South America.<br>Losses result primarily from unthrifty animals. But some deaths occur from the chronic and acute forms of the disease. |
| **Stiff-lamb disease** (see White muscle disease) | | | | |
| **Sweet clover disease** | Cattle; rarely affects sheep or horses. | Usually produced only by moldy or spoiled sweet clover, hay, or silage.<br>Caused by presence of dicoumarol which interferes with vitamin K in blood clotting. | Loss of clotting power of the blood. As a result, blood forms soft swellings beneath skin on different parts of body. Serious or fatal bleeding may occur at time of dehorning, castration, parturition, or following injury.<br>All ages affected. A newborn animal may also have the condition at birth. | Wherever sweet clover is grown and cured for hay. |
| **Urinary calculi** (gravel, stones, water belly, urolithiasis) | Cattle.<br>Sheep.<br>Goats.<br>Horses.<br>Humans. | Unknown, but it does seem to be nutritional. Experiments and experiences have shown a higher incidence of urinary calculi when there is (1) a high potassium intake, (2) a high phosphorus-low calcium ratio (from the standpoint of preventing urinary calculi, the Ca:P ratio should be about 2:1), (3) a high silica content in the ration, or a high proportion of high-silica grains and forages, such as native grasses, wheat straw, sugar beet leaves or pulp, sorghums, and cottonseed meal. A deficiency of vitamin A may be contributing factors. | Frequent attempts to urinate, dribbling or stoppage of the urine, pain and renal colic.<br>Usually only males affected, the females being able to pass the concretions.<br>Bladder may rupture, with death following. Otherwise, uremic poisoning may set in.<br>Urinary calculi is one of the most important diseases in feedlot cattle and sheep, particularly in steers and wethers on full feed. | Worldwide.<br>Affected animals seldom recover completely.<br>The economic loss may be considerable, since calculi formation frequently comes near the end of the feeding period. |
| **Vitamin A deficiency** (night blindness and xerophthalmia) | All animals including poultry. | Vitamin A intake too low.<br>High levels of nitrate intake from hay, silage, and/or water.<br><br>Slowed bone growth, abnormal bone shape, and paralysis.<br>Unsound teeth, characterized by abnormal enamel, pits, and decay.<br>Rough, dry, scaly skin; increased abscesses in ears, mouth, and/or salivary glands; increased diarrhea, and kidney and bladder stones.<br>Reproductive disorders, including poor conception, abnormal embryo growth, placental injury, and death of the fetus.<br>Xerophthalmia in cattle develops in the advanced stages of vitamin A deficiency. The eyes become severely affected, and blindness may follow.<br>Severe diarrhea in young calves and intermittent diarrhea in advanced stages in adults.<br>In finishing cattle, generalized edema or anasarca with lameness in hock and knee joints and swelling in the brisket area. | Night blindness, the first symptom of vitamin A deficiency, is characterized by faulty vision, especially noticeable when the affected animal is forced to move about in twilight in strange surroundings.<br>Stunted growth. | Worldwide. Especially prevalent where one of the following conditions frequently prevails: (1) extended drought, and (2) winter feeding on bleached grass cured on the stalk or on bleached hay. |
| **Weak calf syndrome** | Cattle. | Deficiencies of protein and other essential nutrients and/or weather stress. | Calves show severe depression, general weakness, arched back, red crusted muzzle, diarrhea, and inability to stand and nurse.<br>Most prevalent in calves from first calf heifers. | Most common in southwestern U.S.<br>Weak calf syndrome has resulted in calf mortality as high as 48% in some herds. In addition to calf losses, another economic loss is the reproductive failure of cows. |

*(Continued)*

| Treatment | Control and Eradication | Prevention | Remarks |
|---|---|---|---|
| **C**hronic selenium toxicity may be treated by feeding a high protein ration, by the use of trace amounts of arsenic compounds, by the oral administration of such compounds as naphthalene or bromobenzene; with such treatments under the direction of the veterinarian or consultant. **T**here is no known treatment for acute selenium poisoning. | (Control measures based on Prevention.) | **A**bandon areas wehre soils contain excess selenium, because crops produced on such soils constitute a menace to both animals and humans. **S**oils that contain more than 0.5 ppm selenium are potentially dangerous. Chronic toxicity results from rations containing as little as 8.5 ppm of selenium. Some high selenium areas may be used in pasture rotation and to produce supplemental feeds. | **T**he maximum toxic levels of selenium for farm animals are:<br><br>**Class of Animal** / **Max. Recommended by FDA** *(mg/head/day)* / **Toxic Level** *(ppm in feed)* / *(mg/head/day)*<br><br>Beef cattle .. 1.0 / 10–30 / 100–300<br>Dairy cattle .. 2.0 / 3–5 / 30–60<br>Sheep ....... 0.23 / 3–20 / 7–50<br>Swine ...... — / 5–10 / 8–16<br>Chickens .... — / 2 / —<br>All spec. 2.0 (or 2 ppm) / — / — |
| **R**emove the offending materials and administer menadione (vitamin K₃) **T**he veterinarian usually gives the affected animal an injection of plasma or whole blood from a normal animal that was not fed on the same feed. | **W**hen a case of sweet clover disease is observed in the herd, either (1) discontinue feeding the damaged product, or (2) alternate it with a better-quality hay, especially alfalfa. | **P**roperly cure any sweet clover hay or ensilage. **C**ultivars of sweet clover that are low in coumarin content, and hence safe to feed, have been developed. | **T**he disease has also been produced from feeding moldy lespedeza hay and from sweet clover pasture. |
| **W**hen calculi develop, it may be advisable to dispose of the animal, since treatments have limited success. **T**reatment: (1) add ammonium chloride at the rate of 1 oz (lambs) or 1¼ to 1½ oz (cattle) per head daily—or 50 to 60% more ammonium sulfate; (2) increase the phosphorus content of the ration (or pasture) so that it equals the calcium content (by adding monosodiumphosphate); (3) increase salt content of ration to 3 to 4% so as to increase water consumption (too much may lower feed intake); (4) incorporate 20% alfalfa in the ration; (5) administer muscle relaxants to help the passage of calculi from the bladder; or (6) surgically remove the calculi; however, males will become nonbreeders after such an operation. | **I**f severe outbreaks of urinary calculi occur in finishing steers or lambs it is usually well to dispose of them if they are carrying acceptable finish. **I**ncrease water consumption by including 3 to 4% salt in the ration. **T**he addition of a broad-spectrum antibiotic to the ration has been useful in controlling urinary calculi in some cases. | **G**ood feed and management appear to lessen the incidence. **D**elayed castration (castration of bull calves at 4–5 mo. of age) and high-salt diets of feedlot cattle (1–3% salt in the grain ration, using the upper limits in the winter months) in order to induce more water consumption are effective preventive measures. **A**void (1) high potassium or phosphorus intake, (2) an incorrect Ca:P ratio, or (3) an excessive amount of beet pulp or grain sorghum in the ration. | **C**alculi are stonelike concretions in the urinary tract which almost always originate in the kidneys. These stones block the passage of urine, resulting in the condition commonly referred to as water belly. **T**he mineral deposits may be of variable sizes, shapes, and composition. In cattle, the phosphatic type predominates under feedlot conditions and the siliceous type occurs most frequently in range cattle. **A**mmonium chloride (see Control and Eradication) appears to be the product of choice. However, ammonium sulfate may be used, at the rate of 1.7 to 2.0 oz/head/day. Add it to the ration when an outbreak occurs. |
| **T**reatment consists in correcting the dietary deficiencies and (1) adding vitamin A to the ration, or (2) injecting cattle intramuscularly or intraruminal 500,000 to 1,000,000 IU of vitamin A. | (See Prevention and Treatment.) | **P**rovide good sources of carotene (provitamin A) through green, leafy hays, silage, lush green pasture, yellow corn. **A**dd stabilized vitamin A to ration or inject slow-release vitamin A intramuscularly. | **H**igh levels of nitrates interfere with the conversion of carotene to vitamin A. **S**heep will not develop xerophthalmia on a vitamin A deficiency. |
| **I**mprove the nutrition, and minimize weather stress. | **P**rovide an adequate ration, and provide shelter where necessary. | **F**eed and manage pregnant females so that they are gaining weight prior to calving. **S**eparate first-calf heifers and older cows from the herd and feed and manage them separately. **F**eed adequate energy, protein, minerals, and vitamins. **C**ontrol internal and external parasites. | **W**eak calf syndrome results from the nutritional deficiency of pregnant cows, along with weather stress. |

*(Continued)*

TABLE 4–13

| Disease | Species Affected | Cause | Symptoms and Signs (or age group most affected) | Distribution and Losses Caused By |
|---------|------------------|-------|--------------------------------------------------|-----------------------------------|
| **White muscle disease** (Muscular dystrophy; in sheep, stiff-lamb disease) | Calves. Lambs. Foals. In lambs, it is commonly referred to as stiff-lamb disease. | Selenium deficiency, due to continuous consumption of a ration containing less than 0.02 ppm selenium. | *In calves:* White muscle disease is characterized by lameness or inability to stand, and heart failure. It most commonly affects calves 2 to 4 months of age. More calves than lambs or foals develop heart damage. *In lambs:* The symptoms and signs are a stiff, stilted way of moving, chiefly in the hind legs, although the front legs and shoulders may be involved. Young, rapidly growing lambs are especially susceptible. | White muscle disease occurs widely throughout the world. In the U.S., the disease is widely distributed, but the severity is greatest on the two coasts. Economic losses result from the deaths of severely affected calves and lambs, the unthriftiness of survivors, and the cost of the preventive programs. Death losses range up to 50%, with an average of 15%; and the mortality in untreated animals may reach 80%. |

[1]Meyer, R. M., *27th Kansas Formula Feed Conference Proceedings*, P. H1, Kansas State University, Manhattan, KS, 1972.

## NUTRITION RESEARCH OF THE FUTURE

In the present era, livestock producers have applied the latest in science and technology to the feeding of animals in order to maximize their genetic potential and increase profits. As a result, the people of the world, as a whole, are better fed and living longer than in any other period in history. But, in the production of feed and food, we have given little thought to endangered animals, endangered people, and endangered planet—presaged by ominous warning signals.

As we enter the 21st century, we are on the brink of a major scientific revolution called *biotechnology*, which will spawn many developments exceeding our fondest dreams. It will involve every facet of animal production from feeding to finished products. In the future, however, the world will no longer be able to afford excesses in the use of pesticides, chemical fertilizers, irrigation water, and subsidized overproduction. So, hand in hand with the development of the biotechnology era, there will be need for a great array of research, much of which will not be directed at maximizing production and net returns. Rather, it will embrace pollution control, animal welfare, soil conservation, sustainable agriculture, the greenhouse effect, acid rain, the ozone layer, rain forests—and more.

## QUESTIONS FOR STUDY AND DISCUSSION

1. Discuss the relative importance of the principal U.S. livestock feeds.

2. Why is knowledge of livestock feeding so important from an economic standpoint?

3. Classify animals on the basis of kind of feed eaten.

4. Discuss the nutritional significance of differences in the anatomy and physiology of the organs of the digestive tract of monogastric, avian, ruminant, and pseudo-ruminant animals.

5. What are enzymes: Name three enzymes involved in the digestive process of farm animals, and for each of them (a) identify the food nutrients(s) that it acts upon, and (b) give the end product of digestion.

6. Discuss each of the following functions of feed:
   a. Maintenance
   b. Growth
   c. Finishing (or show-ring fitting)
   d. Reproduction (eggs) and lactation
   e. Work (running)
   f. Wool/mohair

7. What are nutrients? List the chief classes of nutrients.

8. What are the chief animal feed sources of (a) energy, and (b) protein? How is a deficiency of each manifested?

9. From a feed standpoint, discuss (a) animal fats vs vegetable oils, and (b) quality of proteins.

10. What is meant by "essential amino acids"? Name the essential amino acids.

11. Why is quality of protein of particular interest in monogastric animals? Why are biological tests so important in evaluating protein?

12. List three major, or macro, minerals and give for each the (a) major functions, (b) deficiency symptoms, and (c) mineral relationships.

13. List five trace, or micro, minerals and give for each of them the (a) major functions, (b) deficiency symptoms, and (c) mineral relationships.

14. Discuss the importance of the proper ratio of calcium-phosphorus, along with adequate vitamin D, for animals.

15. Group vitamins as either (a) fat-soluble, or (b) water soluble. For each of them, list the animal species most affected, the functions, some deficiency symptoms, and give practical sources.

16. What are "unidentified factors"? List three common sources of them.

*(Continued)*

| Treatment | Control and Eradication | Prevention | Remarks |
|---|---|---|---|
| Intramuscular injection of sodium selenite/vitamin E in aqueous solution at the rate of 0.25 mg Se/lb of body weight. This may be repeated in 2 weeks, but should not exceed 4 doses. Federal law restricts injectable Se to the order of a licensed veterinarian. Do not use within 30 days of slaughter. | Control consists of meeting the selenium requirement by (1) supplementing the ration of cows and ewes during the last ⅓ of pregnancy and the first part of lactation with selenium in the form of sodium selenite at the rate of 0.3 ppm dry matter, or (2) injecting intramuscularly each cow or ewe 1 month before parturition with approved levels of selenium/vitamin E preparation. | Add selenium to the ration. In 1987, FDA approved the addition of 0.3 ppm selenium to the complete feed of cattle, sheep, swine, and poultry. Also, mineral mixes containing selenium are available for free-choice feeding in areas of known selenium deficiency. | White muscle disease is most common in calves and lambs on lush pasture or selenium deficient soils. |

17. How do you account for the fact that corn is by far the leading grain feed in the United States and that soybean meal is by a wide margin the leading U.S. protein supplement?

18. What important functions does water perform?

19. Define, then discuss the importance, characteristics, and uses of each of the following roughages: pasture, hay, crop residue, silage, haylage, and green chop.

20. Define concentrates. Name six concentrates, and tell where in the United States and/or Canada they are commonly grown and available.

21. Define *by-product feed*. Name six common by-product feeds, give the source of each, and tell how they are commonly used as feeds.

22. What is urea? What is a single-celled protein? Discuss the importance of each.

23. Give the formula for a home-mixed mineral for self-feeding (a) where animals are on liberal grain feeding, and (b) where animals are primarily on roughage?

24. Discuss each of the following feeds: (a) molasses, and (b) fats and oils.

25. What are additives, implants, and injections? Discuss the role of each of the following: abortifacients, anthelmintics, antibiotics, bloat control products, buffers, flavoring agents, somatotropin (growth hormone), implants, ionophores, kelp, and mold (fungi) inhibitors.

26. How do you explain the fact that feed processing changed so much in the era of commercial cattle feedlot expansion?

27. Describe each of the following grain processing methods: extruding (gelatinization), dry rolling, steam rolling, flaking, pelleting, bran mash, high-moisture grain, and hydroponics.

28. Describe each of the following forage processing methods: chopping, cubing, ensiling, ammoniation, animal waste (manure) processing, self-feeding governors, and treatment of high-cellulose feeds.

29. What is unique about the preparation of feeds for each of the following classes of animals: beef cattle, sheep, swine, poultry, and fish?

30. Discuss the technique, along with the advantages and limitations, of each of the following methods of evaluating feeds: (a) proximate analysis, (b) total digestible nutrients (TDN), (c) direct calorimetry, (d) indirect calorimetry, (e) comparative slaughter method, and (f) feeding trials.

31. In addition to being nutritionally complete, of what importance are the following feed considerations: palatability, variety, digestive disturbances, bulk, cost, and poisonous plants?

32. Define the term *balanced ration*. Discuss the advantages, and limitations of each of the following methods for balancing rations: (a) square method, (b) trial-and-error method, (c) net energy method, and (d) computer method.

33. When and how should feed substitutions be made?

34. What factors should be considered in choosing between (a) home-mixed feeds, and (b) commercial feeds?

35. Discuss the art of feeding from the following standpoints: (a) starting animals on feed; (b) frequency, regularity, and order of feeding; (c) creep feeding young animals; and (d) changing feeds.

36. When buying feeds, how important is it to have information about the soil (soil nutrients) on which they are grown?

37. List ten of the most prevalent animal nutritional diseases and ailments in the United States. Then, for each of them, give the species most affected, cause, symptoms, distribution and losses caused by, treatment, control, and prevention.

38. For what type of nutrition research should federal and state research dollars be spent in the future?

## SELECTED REFERENCES

| Title of Publication | Author(s) | Publisher |
|---|---|---|
| *Alternative Sources of Protein for Animal Production* | National Research Council | National Academy of Sciences, Washington, DC, 1973 |
| *Animal Feeds* | M. Gutcho | Noyes Data Corporation, Park Ridge, NJ, 1970 |
| *Animal Growth and Nutrition* | E. S. Hafez<br>I. A. Dyer | Lea & Febiger, Philadelphia, PA, 1969 |
| *Animal Nutrition*, Seventh Edition | L. A. Maynard<br>J. K. Loosli<br>H. F. Hintz<br>R. G. Warner | McGraw-Hill Book Company, New York, NY, 1979 |
| *Animal Nutrition*, Third Edition | P. McDonald | Longman, London and New York, NY, 1981 |
| *Applied Animal Feeding and Nutrition*, Third Edition | M. H. Jurgens | Kendall/Hunt Publishing Company, Dubuque, IA, 1974 |
| *Applied Animal Nutrition*, Second Edition | E. W. Crampton<br>L. E. Harris | W. H. Freeman and Co., Publishers, San Francisco, CA, 1969 |
| *Atlas of Nutritional Data on United States and Canadian Feeds* | National Research Council, U.S.A.; Committee on Feed Composition, Research Branch, Canada Department of Agriculture. | National Academy of Sciences, Washington, DC, 1971 |
| *Basic Animal Nutrition and Feeding*, Second Edition | D. C. Church<br>W. G. Pond | D. C. Church, O & B Books, Corvallis, OR, 1974 |
| *Bioenergetics and Growth* | S. Brody | Reinhold Publishing Corp. New York, NY, 1945 |
| *Body Composition in Animals and Man*, Pub. 1598 | National Research Council | National Academy of Sciences, Washington, DC, 1968 |
| *Composition of Cereal Grains and Forages*, Pub. 585 | National Research Council | National Academy of Sciences, Washington, DC, 1958 |
| *Crop Production*, Fifth Edition | R. J. Delorit<br>L. J. Greub<br>H. L. Ahlgren | Prentice-Hall, Inc., Eaglewood Cliffs, NJ, 1984 |
| *Digestive Physiology and Nutrition of the Ruminant* | Edited by<br>D. Lewis | Butterworth & Co., Ltd., London, England, 1961 |
| *Digestive Physiology and Nutrition of Ruminants*, Vols. 1, 2, and 3, Second Edition | D. C. Church | D. C. Church, O & B Books, Corvallis, OR, Vols. 1 and 2, 1979; Vol. 3, 1980 |
| *Effect of Processing on the Nutritional Value of Feeds* | National Research Council | National Academy of Sciences, Washington, DC, 1973 |
| *Energy Metabolism of Ruminants* | K. L. Blaxter | Hutchinson & Co., Ltd., London, England, 1962 |
| *Fats in Animal Nutrition* | J. Wiseman | Butterworth, London, 1984 |
| *Feed Formulations*, Third Edition | T. W. Perry | The Interstate Printers & Publishers, Inc., Danville, IL, 1982 |
| *Feeds and Feeding*, Third Edition | A. Cullison | Reston Publishing Company, Inc., Reston, VA, 1982 |
| *Feeds and Feeding*, 22nd Edition | F. B. Morrison | The Morrison Publishing Company, Ithaca, NY, 1956 |
| *Feeds and Feeding*, Abridged | F. B. Morrison | The Morrison Publishing Company, Ithaca, NY, 1956 |
| *Feeds & Nutrition*, Second Edition | M. E. Ensminger<br>J. E. Oldfield<br>W. W. Heinemann | Ensminger Publishing Co., Clovis, CA, 1990 |

## SELECTED REFERENCES *(Continued)*

| Title of Publication | Author(s) | Publisher |
|---|---|---|
| *Feeds & Nutrition Digest* | M. E. Ensminger<br>J. E. Oldfield<br>W. W. Heinemann | Ensminger Publishing Co., Clovis, CA, 1990 |
| *Feeds for Livestock, Poultry and Pets* | M. H. Gutcho | Noyes Data Corporation, Park Ridge, NJ, 1973 |
| *Fire of Life, The* | M. Kleiber | John Wiley & Sons, Inc. New York, NY, 1961 |
| *Foods & Nutrition Encyclopedia*, Vols. 1, and 2 | A. H. Ensminger, *et al.* | Pegus Press, Clovis, CA, 1983 |
| *Forages*, Fourth Edition | H. D. Hughes<br>M. E. Heath<br>D. S. Metcalf | The Iowa State College Press, Ames, IA, 1985 |
| *Fundamentals of Nutrition*, Second Edition | E. W. Crampton<br>L. E. Lloyd<br>B. F. McDonald | W. H. Freeman and Co. Publishers, San Francisco, CA, 1978 |
| *Handbook of Feedstuffs, The* | R. Seiden<br>W. H. Pfander | Springer Publishing Co., Inc., New York, NY, 1957 |
| *International Feed Nomenclature and Methods for Summarizing and Using Feed Data to Calculate Diets, An*, Bull. 479 | L. E. Harris<br>J. M. Asplund<br>E. W. Crampton | Agriculture Experiment Station, Utah State University, Logan, UT, 1968 |
| *Livestock Feeds Feeding*, Second Edition | D. C. Church | O & B Books, Inc., Corvallis, OR, 1984 |
| *Manual of Clinical Nutrition* | R. S. Goodhart<br>M. G. Wohl | Lea & Febiger, Philadelphia, PA, 1964 |
| *Mineral Nutrition of Animals* | V. I. Georvgievski, *et al.* | Butterworth, London. 1981 |
| *Mineral Nutrition of Livestock, The*, Second Edition | E. J. Underwood | Food and Agriculture Organization of the United Nations, Rome, Italy, 1981 |
| *Mineral Nutrition of Plants and Animals* | F. A. Gilbert | University of Oklahoma Press, Norman, OK, 1948 |
| *Nitrogen and Energy Nutrition of Ruminants* | R. L. Shirley | Academic Press, Inc., Orlando, FL, 1986 |
| *Nonprotein Nitrogen in the Nutrition of Ruminants* | J. K. Loosli<br>I. W. McDonald | Food and Agriculture Organization of the United Nations, Rome, Italy, 1968 |
| *Nutrition of Animals of Agricultural Importance*, Parts 1 and 2 | Edited by<br>D. Cuthbertson | Pergamon Press, London, England, 1986 |
| *Nutritional Deficiencies in Livestock* | R. T. Allmann<br>T. S. Hamilton | Food and Agriculture Organization of the United Nations, Rome, Italy, 1952 |
| *Physiology of Digestion and Metabolism in the Ruminant* | A. T. Phillipson | Oriel Press, Ltd., Newcastle upon Tyre, England, 1970 |
| *Physiology of Digestion in the Ruminant* | Dougherty, *et al.* | Butterworth, Inc., Washington, DC, 1964 |
| *Principles of Biochemistry*, Sixth Edition | A. White, *et al.* | McGraw-Hill, Inc., New York, NY, 1978 |
| *Proceedings of the Fifth International Congress on Nutrition* | National Research Council | Waverly Press, Inc., Baltimore, MD, 1961 |
| *Processed Plant Protein Foodstuffs* | Edited by<br>A. M. Altschul | Academic Press, Inc., New York, NY, 1958 |
| *Processing and Utilization of Animal By-products* | I. Mann | Food and Agriculture Organization of the United Nations, Rome, Italy, 1962 |
| *Protein Contribution of Feedstuffs for Ruminants* | E. L. Miller<br>I. H. Pike<br>A. J. H. Zanes | Butterworth, London, 1982 |
| *Proteins: Their Chemistry and Politics* | A. M. Altschul | Basic Books, Inc., Publishers, New York, NY, 1965 |
| *Recent Advances in Animal Nutrition* | W. Haresign<br>D. J. A. Cole | Butterworth, London, 1985 |

## SELECTED REFERENCES *(Continued)*

| Title of Publication | Author(s) | Publisher |
|---|---|---|
| *Rumen and Its Microbes, The* | R. E. Hungate | Academic Press, Inc., New York, NY, 1966 |
| *Science of Animals That Serve Humanity, The* | J. R. Campbell<br>J. F. Lasley | McGraw-Hill Book Company, New York, NY, 1985 |
| *Science of Nutrition of Farm Livestock* | Edited by<br>D. Culbertson | Pergamon Press, New York, NY, 1969 |
| *Science of Providing Milk for Man, The* | J. R. Campbell<br>R. T. Marshall | McGraw-Hill Book Company, New York, NY, 1975 |
| *Selenium in Nutrition* | National Research Council | National Academy of Sciences, Washington, DC, 1971 |
| *Single-Cell Protein* | Edited by<br>R. I. Mateles<br>S. R. Tannenbaum | The M.I.T. Press, Cambridge, MA, 1968 |
| *Stockman's Handbook, The*, Sixth Edition | M. E. Ensminger | The Interstate Printers & Publishers, Inc., Danville, IL, 1983 |
| *United States-Canadian Tables of Feed Composition* | J. H. Conrad, Chairman of Subcommittee | National Academy Press, Washington, DC, 1982 |
| *Urea and Non-protein Nitrogen in Ruminant Nutrition*, Second Edition | Edited by<br>H. J. Stangel | Nitrogen Division, Allied Chemical Corporation, Morristown, NJ, 1963 |
| *Urea as a Protein Supplement* | Edited by<br>M. H. Briggs | Pergamon Press, Inc., New York, NY, 1967 |
| *Use of Drugs in Animal Feeds, The*, Pub. 1679 | National Academy of Sciences | National Academy of Sciences, Washington, DC, 1969 |
| *Vitamins, The: Chemistry, Physiology, Pathology, Methods*, Vols. I–III, Second Edition | Edited by<br>W. H. Sebrell, Jr.<br>R. S. Harris | Academic Press, Inc., New York, NY, 1967, 1968, 1971 |
| *Vitamins in Feeds for Livestock* | F. C. Aitken<br>R. G. Hankin | Commonwealth Agricultural Bureaux, Farnham Royal, Bucks, England, 1970 |
| *Vitamins and Hormones*, Vol. XV | Edited by<br>R. S. Harris<br>G. F. Marrian<br>K. V. Thimann | Academic Press, Inc., New York, NY, 1957 |

A dairy farm in Tennessee. (Courtesy, American Jersey Cattle Club, Columbus, OH)

# LIVESTOCK BUILDINGS AND EQUIPMENT

# Chapter 5

Properly constructed and arranged buildings and equipment are an asset to the farm or ranch. They increase production, make for labor and feed efficiency, conserve crops and manure, provide comfort for people and animals and add to the beauty of the farm landscape. In serving these purposes, it is not necessary that buildings and equipment be either elaborate or expensive.

Fig. 5–1. Buildings and fences may add beauty to the farm landscape, as on this horse farm. (Courtesy, Kentucky Department of Travel Development, Frankfort)

Increasingly, the U.S. livestock industry has moved to more and more confinement—to production in limited quarters, with or without shelter, but with no pasture. The reasons: (1) It minimizes management time, (2) it minimizes labor, (3) it maximizes control of animals and birds, (4) it maximizes genetic potential of animals and birds, and (5) it maximizes feed conversion. These motivating forces will be accentuated in the future, with the result that confinement production will continue to increase. The trend and acceptance is toward more closed, insulated, mechanically ventilated, environmentally controlled structures—especially for poultry, swine, and dairy enterprises. But the shift to confinement structures and high-density production operations has introduced new problems and accentuated old ones, especially in the areas of animal behavior, manure management, optimum environment, and flexibility. These problem areas are accorded special and rather complete treatment in this chapter.

Because of variations in climatic conditions, sizes and types of enterprises, and systems of management, no attempt will be made herein to present detailed building and equipment plans and specifications. Rather, it is desired merely to convey suggestions regarding some of the desirable features of buildings and equipment in use in various parts of the country. For detailed plans and specifications for a particular locality, the farmer or rancher should (1) study successful buildings and equipment on neighboring farms and ranches, (2) consult the local county agricultural agent, vocational agriculture instructor, or lumber dealer, and/or (3) write to the state college of agriculture.

## LOCATION OF FARM OR RANCH HEADQUARTERS

When planning an entirely new farm or ranch headquarters, the choice of location for the buildings is the first consideration. Likewise, when appraising the desirability of an existing headquarters, the same factors should be considered. These factors are:

1. Water supply.
2. Topography.
3. Drainage.
4. Size and shape.
5. Soil type.
6. Accessibility to fields.
7. Roads.
8. Views.
9. Sun exposure and wind protection.
10. Electricity.
11. Schools and churches.
12. Telephone.
13. Mail route.
14. Service facilities.
15. Erosion control.
16. Vegetation.

## FARMSTEAD ARRANGEMENT

Fig. 5–2. Farmstead arrangement at an Iowa Angus cattle breeding farm. (Courtesy, Corn Belt Farm Dailies, Chicago, IL)

In planning a new farm or ranch headquarters or in altering an old one, buildings, fences, lots, trees, etc., should be added according to an established master plan; for, once constructed, buildings are difficult and expensive to move. In general, for conservation of space and time, the barn and other service buildings should be located around a central court and should be so arranged that most of them can be seen from the house. In arriving at the best arrangement—which means the location and arrangement of individual buildings within the site—the farmstead cannot and should not be modeled after one popular pattern. Consideration should be given to the following pertinent factors:

1. **The house location comes first.** As the farm or ranch house is the headquarters or office of the farm business as well as a home, its location is of greatest importance in farmstead arrangement. The ranch house should be located (a) on a high area which is well drained away from the house, and which will command a view of other buildings as well as one or more scenic views; (b) where it is easily accessible; (c) in the direction of the prevailing winds, but sheltered from strong winds; (d) to obtain the maximum of sunlight in the North and a minimum of sunlight in the South; (e) with access to either the front or back door but with best access to the front door; (f) where there is adequate yard which can be well landscaped; and (g) approximately 150 ft from the road.

2. **Orientation.** Fortunately, the farm or ranch headquarters need not be oriented with the compass. Although in general the farmstead plan will be developed to present the front to the road, most buildings can be turned, quarterturned, or reversed, as may be necessary to take advantage of the prevailing winds, sunlight, view, etc. In general, livestock barns are placed with the long axis north and south, whereas, when possible, livestock sheds in northern areas are faced to secure direct sunlight and yet to face away from the direction of the prevailing winds. In the South, sheds are usually oriented for maximum shade and storm protection.

3. **Direction of the wind.** The house should be located on the windward side of the headquarters, with special consideration given to summer winds. Swine barns should be located at the greatest distance from the other buildings, especially from the house and dairy barn, and so that the prevailing winds will not blow from the swine barn to the house. Unless hills form a natural windbreak, it is desirable to arrange suitable tree plantings for this purpose. Usually, a tree windbreak is located 75 to 150 ft from the buildings to be protected, with 3 to 7 rows of trees 20 to 75 ft wide.

4. **Efficiency.** The buildings should be located so as to require a minimum of walking when doing the chores. This means that those buildings in which the most time is spent—such as the dairy barn, poultry house, and machine shop—should be closest to the house and that the buildings should be near enough together to permit efficiency of labor without making a fire hazard. Likewise, animal barns should be convenient to feedlots and pastures.

5. **Corrals and lots.** The buildings and their adjacent corrals and lots should be arranged so that the buildings are accessible without walking through feedlots and corrals.

6. **Fire protection.** Farm buildings should be far enough apart so that fire will not spread easily from one building to another. In general, this means at least 100 ft apart in the case of large buildings. In acquiring added fire protection through spacing buildings farther apart, one should avoid extreme distances that will mean inefficiency in operation; fire insurance is probably cheaper than labor.

7. **Appearance.** Careful attention to the headquarters arrangement can add to the attractiveness of the entire unit. Manure piles and unsightly objects should not be visible from the main highway or house; shrubbery and trees should be planted to screen unsightly objects; fences and buildings should be repaired and painted regularly; and yard, driveways, and corrals should be kept free from rubbish, scattered farm machinery, etc.

8. **Gates and lanes.** The adoption of larger machinery has necessitated wider gates and lanes than have been commonplace in the past. Often the wider lanes can serve as pasture as well as roadway.

## REQUISITES OF LIVESTOCK BUILDINGS

Fig. 5–3. Confinement dairy barn, with automated feeding and watering in the central area and free-stalls on the right. (Courtesy, Hoard's Dairyman, Ft. Atkinson, WI)

Each farm or ranch is different, and the type and size of buildings will vary accordingly. Among the factors determining the type and size of buildings are (1) kind and fertility of the soil, (2) available markets, (3) size of farm, (4) tenant or owner operation, (5) kind and amount of livestock and crops to be grown, (6) personal preference, (7) climatic conditions of the region, and (8) storage requirements. Thus, the specific requisites of animal buildings will vary according to the needs of the region, state, community, and individual farm or ranch. There are, however, certain general requirements of animal buildings that should always be considered. Once buildings are constructed, there is a definite limit to the changes that can be made in remodeling. Consequently, it is most important that very careful consideration be given to their initial design. The general requisites which livestock buildings should meet are:

1. Reasonable construction and maintenance cost.
2. Flexible design.
3. Reduce labor.
4. Have utility value.
5. Provide protection from the elements.
6. Protect newborn animals.
7. Attractiveness.
8. Durability.
9. Dryness.
10. Well ventilated.
11. Well lighted.
12. Provide direct sunlight.
13. Sanitary.
14. Easily cleaned.
15. Provide for manure disposal.
16. Convenient.
17. Provide adequate space for animals.
18. Provide adequate space for feed and bedding storage.
19. Serve multiple use.
20. Make for minimum fire risk.
21. Safety.
22. Rodent control.
23. Surrounded by suitable corrals or lots.
24. Near water.

25. Modify extreme temperatures.
26. Keep proper humidity.
27. Properly insulated.
28. Adapted to present and future needs.
29. Protect animal health.

## SPACE REQUIREMENTS OF BUILDINGS AND EQUIPMENT

One of the first and frequently one of the most difficult problems confronting the farmer or rancher who wishes to construct a building or item of equipment is that of arriving at the proper size or dimensions. In general, too little space may jeopardize the health and well-being of animals, whereas more space than needed will make buildings and equipment more expensive than necessary.

It is not within the scope of this book to detail all the space requirements for all classes, ages, and uses of animals. Instead, the reader is referred to *The Stockman's Handbook,* a book by the same author and publisher as this book.

### RECOMMENDED MINIMUM WIDTH FOR SERVICE PASSAGES

In general, the requirements for service passages are similar, regardless of the kind of animals. Accordingly, suggestions contained in Table 5–1 are equally applicable to beef and dairy cattle, sheep, swine, poultry, and horse facilities.

### TABLE 5–1
### RECOMMENDED MINIMUM WIDTHS FOR SERVICE PASSAGES

| Kind of Passage | Use | Minimum Width |
|---|---|---|
| Feed alley ............ | For feed cart. | 4 ft |
| Driveway ............. | For wagon, spreader, or truck. | 9 ft |
| Doors and gate ........ | Drive through. | 9 ft |
| Doors and gate ........ | To small pens. | 4 ft |

### STORAGE SPACE REQUIREMENTS FOR FEED AND BEDDING

The space requirements for feed storage for the livestock enterprise may vary so widely that it is difficult to provide a suggested method of calculating space requirements applicable to such diverse conditions. The amount of feed to be stored depends primarily upon (1) length of pasture season, (2) method of feeding and management, (3) kind of feed, (4) climate, and (5) the proportion of feeds produced on the farm or ranch in comparison with those purchased. Normally, the storage capacity should be sufficient to handle all feed and grain and silage grown on the farm and to hold purchased supplies. Forage and bedding may or may not be stored under cover. In those areas where weather conditions permit, hay and straw are frequently stacked in the fields or near the barns in loose, baled, or chopped form. Sometimes, poled framed sheds or a cheap cover of waterproof paper or wild grass is used for protection. Other forms of low-cost storage include temporary upright silos, trench silos, and temporary grain bins.

Table A–5 in the Appendix of this book gives the storage space requirements for feed and bedding. This information may be helpful to the individual operator who desires to compute the barn space required for a specific livestock enterprise. This

table also provides a convenient means of estimating the amount of feed or bedding in storage.

## NATURALLY VENTILATED BUILDINGS

There is continuing interest among livestock producers in naturally ventilated buildings as opposed to environmentally controlled buildings, primarily because of their significantly lower construction and operating costs. Because no attempt is made to regulate temperature, the cost of heavy insulation, tight fitting doors and windows, and a mechanical ventilation system are averted. Of course, buildings which for management reasons must be maintained at temperatures above winter levels are not suited to natural ventilation (e.g., farrowing barns or stanchion barns).

Naturally ventilated buildings can be successfully used for most livestock housing; among them, (1) free-stall housing for dairy cattle, (2) loafing or bedded pack barns for dairy, beef, or sheep, (3) swine-finishing buildings, (4) calf barns, and (5) poultry houses.

Naturally ventilated buildings are mainly a shell to protect animals from rain and snow, and to protect the building contents (grain, hay, etc.). Winter inside temperatures will often

Fig. 5–4. Naturally ventilated, open shed for cattle. (Courtesy, Texas A&M, College Station, TX)

Fig. 5–5. Free-stalls for lactating cows in a loafing barn. (Courtesy, J. C. Allen & Son, Lafayette, IN)

be within 3 to 10°F of outside temperatures. Thus, such buildings are often referred to as cold confinement livestock buildings.

A naturally ventilated building has a continuous opening at the high point (normally the ridge) of the building for air exhaust and continuous openings or inlets along the sidewalls of the building for fresh air. The size of these openings is based on rules of thumb or experience. Air entering along the sidewalls (normally under the eaves) of the building is warmed by the heat from the animals in the building and picks up moisture as it rises toward the ridge. The continuous open ridge allows this warm, moist air to escape, thus completing the air exchange process.

During warm weather, the building should serve mainly to keep rain out and act as a sunshade. Large continuous openings in the sidewalls allow summer breezes to blow through the building.

Typical naturally ventilated buildings can be divided into two types as follows:

1. **Open front.** These buildings have one long side completely open at least one-half the height of the sidewall. The open side faces away from the direction of prevailing winter winds, normally to the south or southeast.

2. **Enclosed.** These buildings have all sides closed but provide continuous eave openings and large doors or vent panels for summer conditions. Enclosed naturally ventilated buildings offer more protection from wind and precipitation.

## ENVIRONMENTALLY CONTROLLED BUILDINGS

*Environment may be defined as all the conditions, circumstances, and influences surrounding and affecting the growth, development, and production of a living thing.* In animals, this includes air temperature, relative humidity, air velocity, wet bedding, dust, light, ammonia buildup, odors, and space requirements. Control or modification of these factors offer possibilities for improving animal performance. There is still much to be learned about environmental control, but the gap between awareness and application is becoming smaller.

The keepers of herds and flocks were little concerned with the effect of environment on animals as long as they roamed pastures and ranges. Space requirements, wet bedding, am-

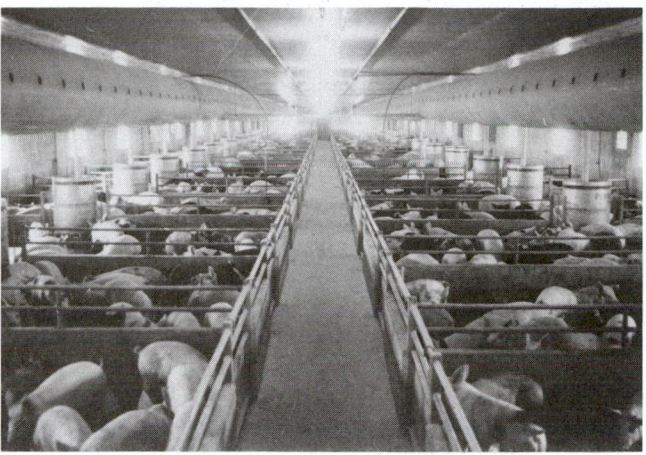

Fig. 5-6. Environmentally controlled hog finishing building. This unit, which holds 1,200 hogs, also has a partially slotted floor and scraping gutter manure handling system. (Courtesy, *National Hog Farmer*, St. Paul, MN)

monia buildup, odors, and manure disposal were no problem. But the concentration of animals into smaller places changed all this. With the shift to confinement structures and high-density production operations, building design became more critical.

Environmentally controlled buildings are costly to construct, but they make for the ultimate in animal comfort, health, and efficiency of feed utilization. Also, they lend themselves to automation, which results in a saving in labor; and, because of minimizing space requirements, they effect a saving in land cost. Today, environmental control is rather common in poultry and swine housing, and it is on the increase with other classes of livestock.

Before an environmental system can be designed for animals, it is important to know their (1) heat production, (2) vapor production, and (3) space requirements. This information is as pertinent to designing livestock buildings as nutrient requirements are to balancing rations.

### HEAT PRODUCTION OF ANIMALS

The heat production of animals is given in Table 5-2.

Table 5-2 may be used as a guide, but in doing so, consideration should be given to the fact that heat production varies

### TABLE 5-2
### HEAT PRODUCTION OF ANIMALS[1]

| Heat Source | Unit | | Heat Production, BTU/Hr | | | Heat Production, Kcal/Hr | | |
|---|---|---|---|---|---|---|---|---|
| | | | Temperature | Total | Sensible | Temperature | Total | Sensible |
| | *(lb)* | *(kg)* | *(°F)* | | | *(°C)* | | |
| Cow | 1,000 | *453.6* | 40 | 3,600 | 2,640 | *4* | *907.2* | *665.3* |
| | | | 70 | 3,000 | 1,550 | *21* | *756.0* | *390.6* |
| Calves (6–10 months) | — | — | 60 | 780 | 660 | *16* | *196.6* | *166.3* |
| | | | 80 | 720 | 420 | *27* | *181.4* | *105.8* |
| Sheep | 100 | *45.4* | 0.039-in. **fleece length** | | | 0.1-cm **fleece length** | | |
| | | | 45 | 560 | 500 | *7* | *141.1* | *126.0* |
| | | | 70 | 320 | 245 | *21* | *80.6* | *61.7* |
| | | | 3.937-in. **fleece length** | | | 10.0-cm **fleece length** | | |
| | | | 45 | 245 | 185 | *7* | *61.7* | *46.6* |
| | | | 70 | 260 | 125 | *21* | *65.5* | *31.5* |

*(Continued)*

**TABLE 5–2** (Continued)

| Heat Source | Unit | | Heat Production, BTU/Hr | | | Heat Production, Kcal/Hr | | |
|---|---|---|---|---|---|---|---|---|
| | | | Temperature | Total | Sensible | Temperature | Total | Sensible |
| | (lb) | (kg) | (°F) | | | (°C) | | |
| **H**og: | | | | | | | | |
| Sow & litter (3 weeks after farrowing) .............. | 400 | 181.4 | — | 2,000 | 1,000 | — | 504.0 | 252.0 |
| Fattening ..................... | 200 | 90.7 | 35 | 860 | 740 | 2 | 216.7 | 186.5 |
| | | | 70 | 610 | 435 | 21 | 153.7 | 109.6 |
| **L**ayer hen ..................... | 4.5 | 2.04 | 50 | 40 | 28 | 10 | 10.1 | 7.1 |
| **H**orse ........................ | | | 70 | 1,900–2,500[2] | — | 21 | 453.6¢630.0 | — |

[1]Adapted by the author from *Agricultural Engineers Yearbook*, St. Joseph, MI, except for horse. Data for horse from *Farm Buildings*, by John C. Wooley, McGraw-Hill Book Company, Inc.

[2]Armsby and Kriss, in a paper entitled, "Some Fundamentals of Stable Ventilation," published in the *Journal of Agricultural Research*, Vol. 21, p. 343, list the total heat output as follows: a 1,000-lb horse, 1,500 BTU per hour; a 1,500-lb horse, 2,450 BTU per hour.

with age, body weight, ration, breed, activity, house temperature, and humidity at high temperatures. As noted, Table 5–2 gives both total heat production and sensible heat production. Total heat production includes both sensible heat and latent heat combined. *Latent heat refers to the energy involved in a change of state and cannot be measured with a thermometer; evaporation of water or respired moisture from the lungs are examples. Sensible heat is that portion of the total heat, measurable with a thermometer, that can be used for warming air, compensating for building losses, etc.* Heat is measured in British thermal units (BTUs). *One BTU is the amount of heat required to raise the temperature of one pound of water one degree Fahrenheit.*

## VAPOR PRODUCTION OF ANIMALS

Animals give off moisture during normal respiration; and the higher the temperature the greater the moisture. This moisture should be removed from buildings through the ventilation system. Most building designers govern the amount of winter ventilation by the need for moisture removal. Also, cognizance is taken of the fact that moisture removal in the winter is lower than in the summer; hence, less air is needed. However, lack of heat makes moisture removal more difficult in the wintertime. Table 5–3 gives the information necessary for determining the approximate amount of moisture to be removed.

Since ventilation also involves a transfer of heat, it is important to conserve heat in the building to maintain desired temperatures and reduce the need for supplemental heat. In a well-insulated building mature animals may produce sufficient heat to provide a desirable balance between heat and moisture; but young animals will usually require supplemental heat. The major requirement of summer ventilation is temperature control, which requires moving more air than in the winter.

**TABLE 5–3**
**VAPOR PRODUCTION OF ANIMALS[1]**

| Vapor Source | Unit | | Temperature | | Vapor Production | | Vapor Production | |
|---|---|---|---|---|---|---|---|---|
| | (lb) | (kg) | (°F) | (°C) | (lb/hr) | (BTU/hr) | (kg/hr) | (kcal/hr) |
| **C**ow ......................... | 1,000 | 453.6 | 40 | 4 | 0.92 | 960 | 0.42 | 241.9 |
| | | | 70 | 21 | 1.38 | 1,450 | 0.63 | 365.4 |
| **C**alves (6–10 months) ............ | — | — | 60 | 16 | 0.11 | 120 | 0.05 | 30.2 |
| | | | 80 | 27 | 0.29 | 300 | 0.13 | 75.6 |
| **S**heep ........................ | 100 | 45.4 | | | 0.039-in. **fleece length** | | 0.1-cm **fleece length** | |
| | | | 45 | 7 | 0.06 | 60 | 0.03 | 15.1 |
| | | | 70 | 21 | 0.07 | 75 | 0.03 | 18.9 |
| | | | | | 3.937-in. **fleece length** | | 10.0-cm **fleece length** | |
| | | | 45 | 7 | 0.06 | 60 | 0.03 | 15.1 |
| | | | 70 | 21 | 0.13 | 135 | 0.06 | 34.0 |
| **H**og: | | | | | | | | |
| Sow & litter (3 weeks after farrowing) .............. | 400 | 181.4 | | | 0.97 | 1,020 | 0.44 | 257.0 |
| Fattening ..................... | 200 | 90.7 | 35 | 2 | 0.11 | 120 | 0.05 | 30.2 |
| | | | 70 | 21 | 0.16 | 175 | 0.07 | 44.1 |
| **L**ayer hen ..................... | 4.5 | 2.04 | 50 | 10 | 0.012 | 12 | 0.005 | 3.0 |
| **H**orse ........................ | | | 70 | 21 | 0.729 | — | 0.33 | — |

[1]Adapted by the author from *Agricultural Engineers Yearbook*, St. Joseph, MI, except for horse. Data for horse from *Farm Buildings*, by John C. Wooley, McGraw-Hill Book Company, Inc.

## RECOMMENDED ENVIRONMENTAL CONDITIONS FOR ANIMALS

The comfort of animals is a function of temperature, humidity, and air movement. Likewise, the heat loss from animals is a function of these three items.

The prime function of the winter ventilation system is to control moisture, whereas the summer ventilation system is primarily for temperature control. If air in livestock barns is sup-plied at a rate sufficient to control moisture—that is, to keep the inside relative humidity in winter below 75%—then this will usually provide the needed fresh air, help suppress odors, and prevent an ammonia buildup.

Some typical temperature, humidity, and ventilation recommendations for different classes of livestock are given in Table 5–4. This table will be helpful in obtaining a satisfactory environment in confinement livestock buildings, which require careful planning and design.

### TABLE 5–4
### RECOMMENDED ENVIRONMENTAL CONDITIONS FOR ANIMALS

| Class of Animal | Temperature | | | | Acceptible Humidity | Commonly Used Ventilation Rates[1] | | | | | Drinking Water | | | |
|---|---|---|---|---|---|---|---|---|---|---|---|---|---|---|
| | Comfort Zone | | Optimum | | | Basis | Winter[2] | | Summer | | Winter | | Summer | |
| | (°F) | (°C) | (°F) | (°C) | (%) | | (cfm) | (m³/min.) | (cfm) | (m³/min.) | (°F) | (°C) | (°F) | (°C) |
| **B**eef cow | 40–70 | 5–21 | 50–60 | 10–15 | 50–75 | 1,000 lb (or *454 kg*) | 100 | *2.8* | 200 | *5.7* | 50 | *10* | 60–75 | *15–24* |
| **S**teer, enclosed bldg. on slotted floor | 40–70 | 5–21 | 50–60 | 10–15 | 50–75 | 1,000 lb (or *454 kg*) | 100 | *2.1–2.3* | 200 | *14.2* | 50 | *10* | 60–75 | *15–24* |
| **D**airy cow | 40–70 | 5–21 | 50–60 | 10–15 | 50–75 | 1,000 lb(or *454 kg*) | 100 | *2.8* | 200 | *5.7* | 50 | *10* | 60–75 | *15–24* |
| **D**airy calves | 50–75 | 10–24 | 65 | 17 | | per 100 lb (*45 kg*) | 10 | | 25 | | | | | |
| **S**heep: | | | | | | | | | | | | | | |
| Ewe | 45–75 | 7–24 | 55 | 13 | 50–75 | | 20–25 | *.6–.7* | 40–50 | *1.1–1.4* | 40–45 | *5–8* | 60–75 | *15–24* |
| Feeder lamb | 40–70 | 5–21 | 50–60 | 10–15 | 50–75 | | 15 | *.3* | 30 | *.65* | 40–45 | *5–8* | 60–75 | *15–24* |
| Newborn lamb | 75–80 | 24–27 | | | | | | | | | | | | |
| **S**wine: | | | | | | | | | | | | | | |
| Sow, farrowing house | 60–70 | 15–20 | 65 | 17 | 60–85 | Sow and litter | 80 | *1.4* | 210 | *2.8* | 50 | *10* | 60–75 | *15–24* |
| Newborn pigs (brooder area) | 80–90 | 27–32 | 85 | 29 | 60–85 | | | | | | | | | |
| Growing-finishing hogs | 60–65 | 15–17 | 60 | 15 | 60–85 | 125 lb (or *57 kg*) | 15 | *.7* | 75 | *2.1* | 50 | *10* | 60–75 | *15–24* |
| **H**orse | 45–75 | 7–24 | 55 | 13 | 50–75 | 1,000 lb (or *454 kg*) | 60 | *1.7* | 150 | *4.5* | 40–45 | *5–8* | 60–75 | *15–24* |
| **N**ewborn foal | 75–80 | 24–27 | | | | | | | | | | | | |
| **P**oultry: | | | | | | | | | | | | | | |
| Layers | 50–75 | 10–24 | 55–70 | 13–20 | 50–75 | per bird | 2 | | 5 | | 50 | *10* | 60–75 | *15–24* |
| Broilers | 85–95 | 21–27 | 70 | 24 | 50–75 | per lb body weight | ½ | | 1 | | 50 | *10* | 60–75 | *15–24* |
| Turkeys | 95–100 (beginning poults) | 35–38 | | | | per lb body weight | ½ | | 1 | | 50 | *10* | 60–75 | *15#24* |

[1]Generally 2 different ventilating systems are provided: one for winter, and an additional one for summer. Hence, as shown in Table 5–4, the winter ventilating system in a beef cow barn should be designed to provide 100 cfm (cubic feet/minute) for each 1,000-lb cow. Then, the summer system should be designed to provide an added 100 cfm, thereby providing a total of 200 cfm for summer ventilation.

[2]In practice, in many buildings, added summer ventilation is provided by opening (1) barn doors and (2) high-up hinged walls.

[3]Provide approximately ¼ the winter rate continuously for moisture removal.

## REQUISITES OF LIVESTOCK EQUIPMENT

Generally speaking, livestock equipment refers to structures other than barns or shelters used in the care and management of animals. Much of this equipment is portable.

The size and design of livestock equipment may differ; that is, not all hayracks or self-feeders, for example, are the same. Yet there are certain fundamentals of livestock equipment that are similar regardless of the kind of equipment, the design, or the size. The requisites are:

1. Useful, practical, and efficient.
2. Simple construction.
3. Durable.
4. Dependable.
5. Low initial cost and upkeep.
6. Movable.
7. Accessible.
8. Save feed.
9. Reduce labor.
10. Conserve manure.

## AUTOMATION

*Automation is a coined word meaning the mechanical handling of materials.* Livestock producers automate to lessen labor and cut costs.

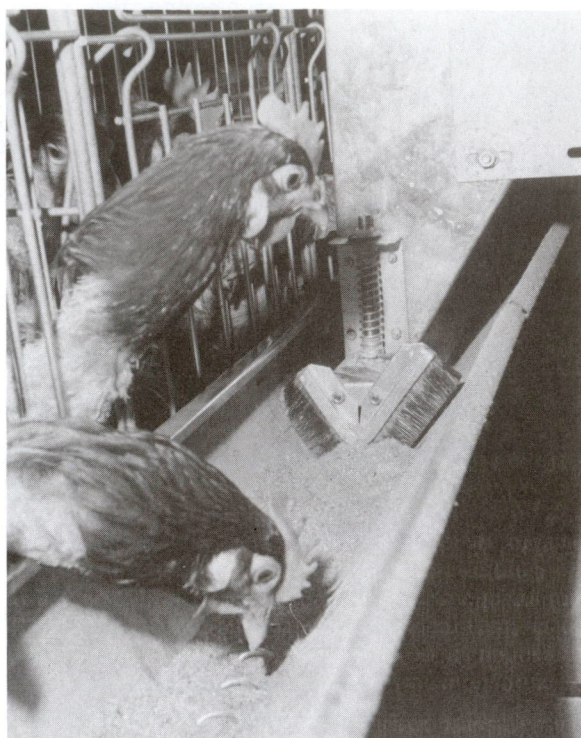

Fig. 5-7. Automated poultry feeding. Note (1) the feed dispenser, and (2) the brushes sweeping the sides of the trough; minimizing wastage. (Courtesy, SALOPIAN Industries, Ltd., Shrewsbury, England)

Modern equipment has practically eliminated the pitchfork, bucket, basket, and milk stool. Such chores as feeding, watering, milking, bedding, and barn cleaning have been, or are being mechanized. Livestock producers are using more self-unloading trucks and trailers, self-feeders, feed bunk augers and belts, labor-saving grain- and forage-processing equipment (pellets, cubes, or wafers, etc.), automatic waterers, manure disposal units, and milking machines. Automation of the livestock industry will increase.

Fig. 5-8. Automated milking. Modern polygon milking parlor in which 24 cows are milked at the same time. (Courtesy, James Tappan, Arizona Dairy Company, Higley, AZ)

## TYPES OF BARN ROOFS

Fig. 5-9 illustrates various roof shapes. The shape, style, slope, and type of roof construction selected should be based upon the function to be served, the economy of construction, strength, and appearance.

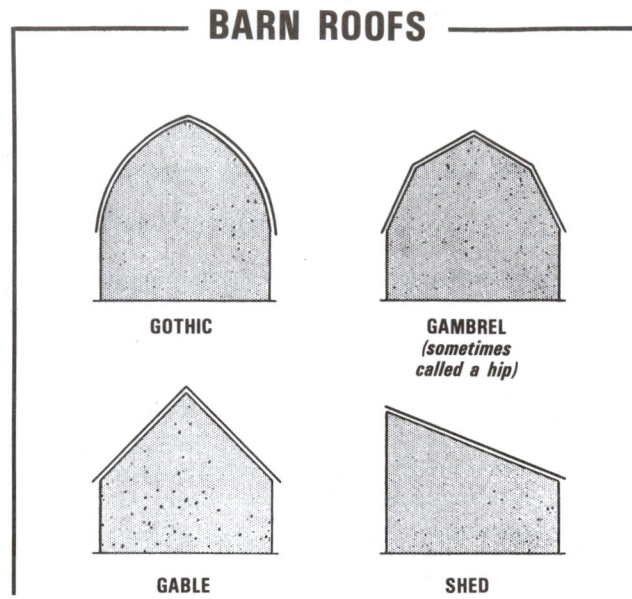

**BARN ROOFS**

GOTHIC

GAMBREL
*(sometimes called a hip)*

GABLE

SHED

Fig. 5-9. Types of barn roofs. (Drawing by Prof. R. F. Johnson)

On permanent buildings, the roofing should last 15 to 20 years without replacement. Among the more durable roofings are: cedar shingles, cement-asbestos shingles; 250 lb asphalt shingles; 28-gauge, 2 oz coated steel sheets; and aluminum.

## TYPES OF FLOORS FOR STALLS OR STABLES

Solid floors for animals may be, and are, constructed of numerous materials—including clay, clay with a concrete border, plank, concrete, concrete with board surfacing, cork brick, creosoted wooden blocks, cinders, or various combinations of these materials. Regardless of the type of flooring material, for a good dry bed there should be a combination of surface and subsurface drainage, together with a cover provided for a suitable absorbent litter.

Most livestock producers feel that a perfect flooring material has not yet been developed, as each of the existing types has certain disadvantages. Rough wooden floors furnish good traction for animals and are warm to lie upon; but they are absorbant and unsanitary. They also lack durability and often harbor rats and other rodents. Concrete floors are durable, impervious, easily drained and sanitary; but they are rigid and without resilient qualities, are slippery when wet, and are cold to lie upon. Clay floors are noiseless, springy; and they afford a firm natural footing unless wet, but they are difficult to keep clean and level. After considering both the advantages and disadvantages of the many types of flooring materials, most practical producers are agreed that, under average conditions and where solid floor is desired, concrete is the most satisfactory flooring for central hog houses and poultry houses and that clay is the most satisfactory flooring for cattle, sheep, and horse barns or shelters.

## SLOTTED FLOORS

*Slotted floors are floors with slots through which the feces and urine pass to a storage area below or nearby.* Such floors are not new; they have been used in Europe for over 200 years. More and more slotted floors are being used for swine and poultry in this country, and there is increased interest in using them for cattle and sheep.

Fig. 5-11. Layers on slotted floor. (Photo by J. C. Allen & Son, Inc., West Lafayette, IN)

Fig. 5-10. Lambs on slotted floor.

Slats may be made of wood, concrete, or metal (steel or aluminum). Table 5-5 summarizes pertinent facts pertaining to each type of material. Fig. 5-12 and Table 5-6 (see next page) present important design information relative to concrete slats. The A, B, C dimensions given in Table 5-6 correspond to the location shown in Fig. 5-12.

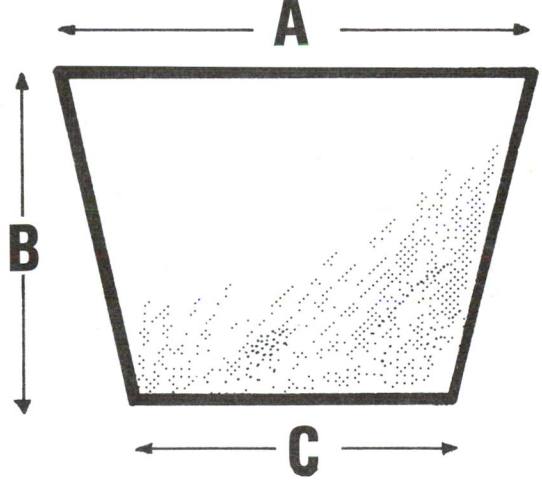

Fig. 5-12. Concrete slat.

The main **advantages** of slotted floors are (1) less space per animal is needed, (2) bedding is eliminated, (3) manure handling is reduced, (4) increased sanitation, and (5) saving in labor.

The chief **disadvantages** of slotted floors are (1) higher initial cost than conventional solid floors, (2) less flexibility in the use of the building, (3) spilled feed is lost through the slots, (4) animals raised on slotted floors resist being driven over a solid floor, and (5) environmental conditions become more critical.

### TABLE 5-5
### MATERIALS FOR FLOOR SLATS

| Material | Expected Life | Advantages | Disadvantages | Comments |
|---|---|---|---|---|
| **Wood** | 2–4 years | Low initial cost. | Difficult to maintain spacing. Not too durable. | Make from hardwood, like oak. |
| **Concrete** | 10–15 years | Long life. May be homemade. | Quality control difficult when poured at site. | |
| **Metal** (steel or aluminum) | 4–8 years | Easily cleaned. | High cost. | Special erosion-resistant metal now available. |
| **Flattened expanded steel mesh** | | Low cost. | Will not take too much weight. | Satisfactory for young lambs and for young pigs to 50 lb weight. |

## TABLE 5-6
### CONCRETE SLAT DIMENSIONS[1]

| Class of Animal | Length | A | B | C | Space Slats | Floor Space |
|---|---|---|---|---|---|---|
| | (ft) | (in.) | (in.) | (in.) | | |
| **Cattle** (beef and dairy): Designed for approx. 250 lb per linear foot live load | 6 | 6 | 6 | 3 | *Calves:* 1¼ in. apart. *Steers,* weaning to market: 1½ to 1¾ in. | For 1,000 lb: 20–25 sq ft. |
| | 8 | 6 | 6 | 3 | | |
| | 10 | 6 | 7½ | 3 | | |
| **Hogs and Sheep:** Designed for approx. 100 lb per linear foot live load | 4 | 4 | 3½ | 3 | Space slats either a uniform ⅜ in. apart or ¾ to 1 in. apart. Spaces between ⅜ and ¾ in. are not recommended because pigs' legs may get caught. Space slats 1 in. apart behind the sow to improve cleaning. | *Dry ewe:* 6 sq ft. *Ewe & lamb:* 8 sq ft. *Feeder lamb:* 4 sq ft. *Sows:* 8 sq ft. *Sow and litter:* 20–30 sq ft. *Growing-finishing:* 1. To 75 lb: 5 sq ft. 2. 75–125 lb: 6 sq ft. 3. 125 lb–market: 8 sq ft. |
| | 6 | 4 | 4 | 3 | | |
| | 8 | 5 | 4½ | 4 | | |
| | 10 | 5 | 5 | 4 | | |
| | 12 | 5 | 5½ | 4 | | |

Concrete slat

[1]For concrete slats, use a 7½ bag mix with ¾ in. maximum aggregate.

## PAVED LOTS AND FEEDING FLOORS

In those areas in which barnyard mud is a usual winter problem, a concrete lot floor for beef and dairy cattle, hogs, and sheep may constitute one of the most profitable improvements. A properly constructed paved feedlot results in (1) a saving in time and labor; (2) less waste in feed and bedding; (3) greater conservation of manure; (4) greater sanitation and fewer diseases; and (5) more animal comfort, which means greater gains and production.

The paved feedlot should be located where it will make for convenience in feeding and watering operations and will be sheltered from the prevailing winds and exposed to the sun. In most areas, these conditions are met by locating the paved lot adjacent to the building housing the animals and on the south or east side thereof.

## SILOS

The first silo built by a Caucasian in the United States is said to have been erected by F. Morris in Maryland in 1876. However, Columbus found that the Indians used pits or trenches in which to store their grain, and centuries earlier in the Old World, silos were used as a means of preserving both grain and green forage.

The value and use of silage in feeding the respective classes of livestock is fully covered in the chapters on feeding. The ensuing sections will be devoted to a discussion of types and requisites of silos—especially from structural and utility standpoints.

## TYPES OF SILOS

Broadly speaking, most silos can be classified as either upright or horizontal. The choice between the two, as well as the selection of the construction material, should be determined primarily by the suitability to the particular needs of the farm or ranch, the cost, and the silage storage losses.

Silage storage losses vary widely between kinds of silos and are generally larger than most farmers realize. (see Table 5–7).

## TABLE 5-7
### ESTIMATED AVERAGE SILAGE STORAGE LOSSES

| Type of Silo | Percent of Loss | |
|---|---|---|
| | Average | Range |
| **G**astight upright | 5 | 1–10 |
| **C**onventional upright | 6 | 2–12 |
| **H**orizontal (trench) | 15 | 8–25 |
| **O**pen stack | 20 | 12–30 |

Some pertinent information relative to each kind of silo is given in the discussion which follows, but it is not within the scope of this book to give detailed silo plans and specifications. The latter may be obtained from local authorities, from silo manufacturers, or by writing to the state agricultural colleges.

But first a word of caution: *Before entering an upright or pit silo, it is recommended that a lighted cigarette lighter, candle, or lantern be lowered into the silo. If the flame goes out, assume that the pit is dangerous to enter and replenish it with fresh air before entering.*

### UPRIGHT SILOS (TOWER SILOS)

This type of silo, which is sometimes referred to as the ''watchtower of prosperity,'' is a cylinder built above ground. It is a permanent farm structure and, as such, it should be constructed to withstand long usage. Among the materials used are concrete, brick, tile, stone, metal, and wood.

An upright silo has the following advantages: (1) Its round shape withstands pressure well and is adapted to good packing; (2) it is durable; (3) it has lower storage losses than any other type; (4) adverse weather, particularly a heavy snow, does not affect the unloading of an upright silo; and (5) it is well adapted to a conveyor feeding system. However, an upright silo is the most expensive of all types from the standpoint of initial cost per ton storage capacity.

Upright silos and conveyor feeding systems find their greatest use on dairy farms and in cattle feedlots of up to 1,000-head capacity.

Fig. 5-13. Upright, or tower, silos. (Courtesy, A. O. Smith Harvestore Products, Inc., Arlington Heights, IL)

Conventional upright silos with tight walls and well-sealed doors are fairly airtight except for the top (surface). The top surface of upright silos should be covered with a sheet of plastic to minimize spoilage.

Where two upright silos are in use, consideration should be given to storing two kinds of feedstuffs—perhaps silage in one and high-moisture corn in the other. Also, farmers may find it advantageous to use an upright silo the year around; for example, it might be filled with corn silage in the fall for winter feeding, then refilled with haylage in the spring for summer feeding.

Gastight upright silos allow for more flexibility than conventional concrete stave silos. Feeding can be stopped for a time with less concern about spoilage.

## HORIZONTAL SILOS (TRENCH SILOS)

This type of silo is horizontal; generally it is mostly underground—it is a trenchlike structure. The walls of a trench silo may or may not be lined, but for making good silage they should always be smooth. Also, a trench silo should always be wider at the top than at the bottom, and the bottom should slope away from one end in order that the excess juices may drain off.

In comparison with upright silos, horizontal silos have the following **advantages**: (1) They cost less—that is, they provide storage at a lower cost per ton; (2) they are easier to construct; (3) they cost less to fill, because a blower is not necessary; (4) they are less prone to freezing (because of being mostly underground); (5) they are better suited to self-unloading truck feeding; (6) they lend themselves to easier expansion; and (7) they are better adapted to big operations—to establishments with 1,000 or more beef or dairy cattle.

In comparison with upright silos, horizontal silos have two main **disadvantages**: (1) Packing is more of a problem, because they do not have the extra weight of a vertical silo to assist; and (2) there is a larger area to seal in order to prevent spoilage. To prevent excess spoilage losses, it is recommended that a horizontal silo be covered. When filling is completed, it may be sealed by (1) plastic weighted with old tires to keep the cover from flapping in the breeze, (2) chopped green forage or weeds, or (3) such other materials as waterproof paper, aluminum, soil, or sawdust.

## OTHER TYPES OF SILOS

In addition to the upright and horizontal types of silos, each of the following types of silos has a place under certain circumstances:

1. **Pit silos.** The pit silo is shaped like the tower silo, but inverted into the ground. It resembles a well or cistern. The walls of a pit silo may or may not be lined. Where the water table is low enough that the silo will not fill with water, such as in semiarid areas, the pit silo is very satisfactory.

2. **Self-feeder or bunker silos.** As a laborsaving measure, some operators are now constructing silos above ground (or slightly recessed)—usually with concrete floors, and sidewalls of wood, concrete, or other materials—and self-feeding silage to cattle by making use of a portable hurdle or fence.

In this method, the fence (gate, hurdle, or stanchion) through which the animals put their heads, is placed at the end (or side) of the silo and moved back as the silage is eaten. About 6 in. of space per cow is allowed.

3. **Above-ground temporary silos.** Several kinds of above-ground temporary silos are used. Generally, this kind of storage is used to meet emergencies, to supplement permanent silos, or to ensile such by-product feeds as cannery refuse, pea vines, and beet tops or pulp. Above-ground temporary silos are low in cost, can be erected on short notice, require no special foundation, and can be set up on almost any sight convenient for filling and feeding.

Perhaps most above-ground temporary silos can be classed as belonging to one of the following four kinds:

a. **Enclosed stack silos.** These are built entirely above ground, without trenches or holes. They are upright, are generally circular, and are enclosed by snow or picket fences, poles, wooden staves, heavy woven wire, or other materials. Most of them are lined with tar paper or tough fiber-reinforced paper made especially for the purpose. Because of the relatively weak walls of these silos, their height should not be greater than twice their diameter unless poles are set at four to six points around their circumference and tied together at the top.

b. **Open stack silos.** These are similar to enclosed stack silos, except that no supports or walls are used. As would be expected, greater spoilage is encountered in the open stack than in the enclosed stack, because of the greater evaporation and spoilage which accompanies the exposed sides.

c. **Modified trench stack silos.** This silo, which is intermediate between a trench and a stack silo, is adapted to areas where the groundwater level is high. It is constructed by excavating a shallow trench 12 to 18 in. deep, by piling the excavated earth on either side of the trench to support the silage and keep out surface water, by packing silage thoroughly in and over the trench to a height of 10 to 15 ft, and by covering the stack with any one of the materials recommended for covering a horizontal silo (see "Horizontal Silos").

4. **Plastic silos.** Plastic (polyethylene) is now available for use as temporary silos, and for use as covers for trench, bunker, and tower silos, and as silo liners. If not punctured, it is nearly airtight. Plastic thicknesses range from 4 to 9 mils. The thicker grades have better tear and puncture resistance, and low permeability by both air and moisture; however, they cost more and are difficult to tie tightly. Thinner grade plastics are less costly, more pliable and easier to seal.

The two common types of plastic silos are: (1) enclosed plastic bag or tube silos, and (2) round bale plastic covered silage.

a. **Enclosed plastic bag or tube silos.** These temporary silos are made of heavy plastic in the form of a tube into which forage is forced by a special machine (much like stuffing sausages). The machine needed to pack the tube is generally rented or owned cooperatively. The filled structure is 8 ft in diameter and about 100 ft long. Preservation of silage is excellent provided the ends are kept sealed and the plastic is not torn or damaged by rodents or other animals. To remove or self-feed the silage, the plastic is cut and folded back at one end to expose as much silage as needed each day. The plastic cannot be re-used.

b. **Round bale plastic covered silage.** The most common methods for using plastic material to produce round bale silage are:

(1) **Individual bags.** Bags come in various lengths, diameters, and thicknesses. A tractor-mounted spear device is needed to lift the bale while applying the bag. Then the bale is placed in storage position before it is tied off. If possible, the bales should be stacked in cord-wood fashion to reduce exposed surface area. Then, a plastic cover over the entire stack may reduce storage damage.

(2) **Plastic tubes.** These consist of several round bales stuffed by a machine into a long plastic tube which is then sealed at both ends. The filled plastic tube resembles an "Enclosed Plastic Bag or Tube Silo" described earlier, except that it consists of a row of round bales covered with plastic rather than long, continuous sausage-type silage material. Plastic tubes can be effective and time-saving, but the multiple bales stored in one package tend to increase the loss if the bag is torn, punctured or opened for feeding. However, the tube can be easily tied off into one-bale (or more) segments for feeding.

(3) **Sheet plastic.** Several round bales can be stacked under two sheets of plastic, with the plastic ends on the ground covered with soil, sand, or other effective sealing procedure. The hazard with this type of storage is that there are more possibilities for air leaks to develop, which may result in a large number of bales being spoiled.

## MANURE

*The term manure refers to a mixture of animal excrement (consisting of undigested feeds plus certain body wastes) and bedding.*

Modern livestock buildings and equipment should be designed to handle the manure produced by the animals they serve; and this should be done efficiently, with a minimum of labor and pollution, so as to retrieve the maximum value of the manure, and make for maximum animal sanitation and comfort.

## DAILY MANURE PRODUCTION AND STORAGE

Table 5–8 shows the approximate daily manure production of each class of animal.

**TABLE 5–8**
**APPROXIMATE DAILY MANURE PRODUCTION; WITHOUT BEDDING[1]**

| Animal | Cu Ft/Day Solids and Liquids[2] | Gallons/ Day[3] |
|---|---|---|
| 1,000-lb cow ........ | 1½ | 11 |
| 1,000-lb steer ........ | 1 | 7½ |
| 10 head of sheep .... | ½ | 4 |
| 10 head of hogs: | | |
| 50 lb ............ | ⅔ | 5 |
| 100 lb ............ | 1⅓ | 10 |
| 150 lb ............ | 2¼ | 17 |
| 200 lb ............ | 2¾ | 20½ |
| 250 lb ............ | 3½ | 26 |
| 1,000 5-lb layers ..... | 3 | 22½ |
| 1,000-lb horse ........ | ¾ | 5½ |

[1]Adapted by the author from Michigan State University Circular Bull. 231.
[2]There are about 34 cu ft in a ton of manure.
[3]One cu ft = 7½ gallons.

Manure may be stored in a separate tank or it may be left to accumulate in a pit under slotted floors.

Storage capacity can be computed as follows:

Storage capacity = No. of animals × daily manure production × desired storage time in days + extra water.

**Example:** 80 cows (1,000 lb each) × 1½ × 120 days = 14,400 cu ft; 7½ gal × 14,400 cu ft = 108,000-gal capacity.

Extra water must often be added to liquify the wastes. Thus, if the manure is to be pumped, ⅕ to ⅗ of the storage volume may be needed for the extra water. For irrigation, there should be about 95% water and about 5% manure. Water should be kept to a minimum if the manure is to be field spread with a tank wagon.

Generally, three to six months' storage capacity is desirable.

## YEARLY MANURE PRODUCTION, COMPOSITION, AND VALUE

The quantity, composition, and value of manure produced vary according to species, weight, kind and amount of feed, and kind and amount of bedding. The computations herein are on a fresh manure basis (exclusive of bedding). Table 5–9 gives the quantity, composition, and value of manure excreted per year per 1,000 lb liveweight by different species of farm animals. Table 5–10 gives the tonnage and value of manure excreted by U.S. livestock by species per year, along with the total tonnage and total value per year.

The data in Tables 5–9 and 5–10 are based on animals confined the year-around. Actually, the manure recovered and available to spread where desired is considerably less than indicated because (1) animals are kept on pasture and along roads and lanes much of the year, where the manure is dropped, and (2) losses in weight often run as high as 60% when manure is exposed to the weather for a considerable time.

Currently, we are producing manure (exclusive of bedding) at the rate of 1.35 billion tons annually (see Table 5–10). That is sufficient manure to add nearly 3/4 ton each year to every acre of the total land area (1.9 billion acres) of continental United States.

About 75% of the nitrogen, 80% of the phosphorus, and 85% of the potassium contained in animal feeds are returned

**TABLE 5-9**
**QUANTITY, COMPOSITION, AND VALUE OF FRESH MANURE**
**(FREE OF BEDDING) EXCRETED PER YEAR PER 1,000 LB LIVEWEIGHT BY VARIOUS KINDS OF FARM ANIMALS**

| (1) Animal | (2) Tons Excreted/ Year/1,000 Lb Liveweight[1] | (3) Excrement | Composition and Value of Manure on a Tonnage Basis[2] | | | | | | |
|---|---|---|---|---|---|---|---|---|
| | | | (4) Lb/Ton[3] | (5) Water | (6) N | (7) $P_2O_5$[4] | (8) $K_2O$[4] | (9) Value/ Ton[5] |
| | | | (lb) | (%) | (lb) | (lb) | (lb) | ($) |
| **Cow** (beef or dairy) .................. | 12 | Liquid .......... Solid .......... | 600 1,400 | 79 | 11.2 | 4.6 | 12.0 | 6.07 |
| | | Total .......... | 2,000 | | | | | |
| **Steer** (finishing cattle) ............... | 8.5 | Liquid .......... Solid .......... | 600 1,400 | 80 | 14.0 | 9.2 | 10.8 | 8.12 |
| | | Total .......... | 2,000 | | | | | |
| **Sheep** ....................... | 6 | Liquid .......... Solid .......... | 660 1,340 | 65 | 28.0 | 9.6 | 24.0 | 13.88 |
| | | Total .......... | 2,000 | | | | | |
| **Swine** ....................... | 16 | Liquid .......... Solid .......... | 800 1,200 | 75 | 10.0 | 6.4 | 9.1 | 5.81 |
| | | Total .......... | 2,000 | | | | | |
| **Horse** ....................... | 8 | Liquid .......... Solid .......... | 400 1,600 | 60 | 13.8 | 4.6 | 14.4 | 7.11 |
| | | Total .......... | 2,000 | | | | | |
| **Poultry** ....................... | 4.5 | Total .......... | 2,000 | 54 | 31.2 | 18.4 | 8.4 | 15.58 |

[1]*Manure Is Worth Money—It Deserves Good Care,* University of Illinois Circ. 595, 1953, p.4.

[2]Columns 5, 6, 7, and 8 from *Farm Manures,* University of Kentucky Circ. 593, 1964, p.5, Table 2.

[3]From *Reference Material for 1951 Saddle and Sirloin Essay Contest,* compiled by M. E. Ensminger, p. 43; data from *Fertilizers and Crop Production* by Van Slyke, published by Orange Judd Publishing Co.

[4]$P_2O_5$ can be converted to phosphorus (P) by dividing the figure given above by 2.29, and $K_2O$ can be converted to potassium (K) by dividing by 1.2.

[5]Calculated on the assumption that nitrogen (N) retails at 29¢, $P_2O_5$ at 30¢, and $K_2O$ at 12¢ per pound in commercial fertilizers.

**TABLE 5-10**
**TONNAGE AND VALUE OF MANURE (EXCLUSIVE OF BEDDING) EXCRETED BY U.S. LIVESTOCK[1]**

| Class of Livestock | Number of Animals on Farms[2] | Average Liveweight | Tons Manure Excreted/Year/ 1,000 Lb Liveweight[3] | Total Manure Production | Total Value of Manure[4] |
|---|---|---|---|---|---|
| | | (lb) | (tons) | (tons) | ($) |
| **Cattle** (beef and dairy; including steers) ...... | 102,468,000 | 900 | 11 | 1,014,433,200 | 7,202,475,720 |
| **Sheep** ......................... | 10,328,000 | 100 | 6 | 6,196,800 | 86,011,584 |
| **Swine** ......................... | 50,960,000 | 200 | 16 | 163,072,000 | 947,448,320 |
| **Chicken,** layers ............. | 280,140,000 | 4.5 | 4.5 | 5,672,835 | 88,382,769 |
| **Broilers** ..................... | 4,646,312,000 | 3.5 | 4.5 | 73,179,414 | 1,140,135,270 |
| **Turkeys** ..................... | 207,216,000 | 22 | 4.5 | 20,514,384 | 319,614,102 |
| **Horses** ...................... | 8,519,000 | 1,000 | 8 | 68,152,000 | 484,560,720 |
| | | | | 1,351,220,633 | 10,268,628,485 |

[1]In these computations, no provision was made for animals that died or were slaughtered during the year. Rather, it was assumed that their places were taken by younger animals, and that the population of each species was stable throughout the year.

[2]From USDA, *Agricultural Statistics 1987,* except horse numbers from American Horse Council, Inc., Wash., D.C. The cattle and sheep figures are for Jan. 1, 1987; the swine and layer figures are for Dec. 1, 1986; the broiler and turkey figures are number raised during the year; the horse numbers are for 1986. All figures are assumed averages throughout the year.

[3]*Manure Is Worth Money—It Deserves Good Care,* University of Illinois Circ. 595, 1953, p. 4.

[4]Computed on the basis of the value per ton given in the right-hand column of Table 12–9.

as manure. In addition, about 40% of the organic matter in feeds is excreted as manure. As a rule of thumb, it is commonly estimated that 80% of the total nutrients in feeds are excreted by animals as manure.

Naturally, it follows that the manure from well-fed animals is higher in nutrients and worth more than that from poorly fed ones. For example, the manure produced from steers liberally fed on nutritious concentrates is more valuable than that produced from cattle wintered on hay.

## MODERN WAYS OF HANDLING MANURE

Modern handling of manure involves maximum automation and a minimum loss of nutrients. Among the methods being used, with varying degrees of success, are: slotted floors emptying or pumping into irrigation systems; storage vats; spreaders (including those designed to handle liquids alone or liquids and solids together); dehydration; power loaders; conveyors; industrial-type vacuums; lagoons; and oxidation ditches.

Actually, there is no one best manure management system for all situations; rather, it is a matter of designing and using that system which will be most practical for a particular set of conditions.

## MANURE USES

Historically, manure has been used as a fertilizer. But high feed prices and shortage of fossil fuels have made for new uses. Although it is expected that manure will continue to be used primarily as a fertilizer for many years to come, increasingly it will be recycled and used as a feed and converted into energy. Other manure-based products will continue to evolve, but it is expected that they will be of minor importance.

### MANURE AS A FERTILIZER

The actual monetary value of manure can and should be based on (1) equivalent cost of a like amount of commercial fertilizer, and (2) increased crop yields. Table 5-9 gives the equivalent cost of a like amount of commercial fertilizer. Numerous experiments and practical observations have shown the measurable monetary value of manure in increased crop yields.

There was a time when farmers fed livestock to produce manure, to grow more crops, to feed more livestock. The use of chemical fertilizer expanded manyfold; labor costs rose to the point where it was costly to conserve and spread manure on the land; more animals were raised in confinement; and a predominantly urban population didn't appreciate what they referred to as "foul-smelling, fly-breeding stuff." As a result, what to do with manure became a major problem on many livestock establishments.

Based on equivalent fertilizer prices (see Table 5-9, right-hand column), and livestock numbers (see Table 5-10), the yearly manure crop is worth $10.3 billion. That's a potential income of $4,905 for each of the nation's 2.1 million farms.

The value of manure as a fertilizer varies according to (1) the class of animals, (2) the kind of feed consumed and the kind of bedding used, (3) the method of handling, (4) the rate and method of application, and (5) the kind of soil and crops on which it is used.

Livestock producers sometimes fail to recognize the value of this barnyard crop because (1) it is produced whether or not it is wanted, and (2) it is available for only the cost of handling.

- **How much manure can be applied to the land?**—State regulations differ in limiting the rate of manure application. Missouri draws the line at 30 tons per acre on pasture, and 40 tons per acre on cropland. Indiana limits manure application according to the amount of nitrogen applied, with the maximum limit set at 225 lb per acre per year. Nebraska requires only one-half acre of land for liquid manure disposal per acre of feedlot, which appears to be the least acreage for manure disposal required by any state.

One of the big problems in applying animal waste to the land as a fertilizer is knowing the plant nutrient content of the material. If this is known, the amount of manure necessary to supply the needed nutrients can be added. So, representative samples of the manure should be analyzed for nitrogen, phosphorus, potash, and moisture content. Then the application rates should be based on soil tests, crop requirements, and com-

Fig. 5-14. Big capacity manure spreader in use at a cattle feedlot. (Courtesy, Sperry-New Holland, New Holland, PA)

position of the manure sample. This relatively inexpensive procedure will avoid errors in application rate.

The amount of manure to be applied can usually be geared to the amount of nitrogen that the crop needs. Thus, if 150 lb of available nitrogen would be adequate for maximum crop production, the manure containing 300 lb of total nitrogen should be applied for the first year of use (twice the amount of nitrogen needed the first year is applied because only half the nitrogen is available the first year).

When farmers have sufficient land they should use rates of manure which supply only the nutrients needed by the crop rather than the maximum possible amounts suggested for pollution control.

Based on earlier studies in the Midwest, before the rise of commercial fertilizers, it would appear that one can apply from 5 to 20 tons of manure per acre, year after year, with benefit.

### MANURE AS A NONFEED ENERGY SOURCE

Manure can also serve as a source of nonfeed energy, which, of course, is not new. The pioneers burned dried bison dung, which they dubbed *buffalo chips*, to heat their sod shanties. In this century, methane from manure has been used for power in European farm hamlets when natural gas was hard to get. While the costs of constructing plants to produce energy from manure on a large-scale basis may be high, some energy specialists feel that a prolonged fuel shortage will make such plants economical. India now has about 10,000 anaerobic digestion plants in operation.

Methane, of course, is usable like natural gas. There is nothing new or mysterious about this process. Sanitary engineers have long known that a family of bacteria produces methane when they ferment organic material under strictly anaerobic conditions. (Our grandparents called it *swamp gas;* their city cousins called it *sewer gas.*) However, it should be added that, due to capital and technical resources needed, for some time to come, the production of methane by anaerobic digestion will likely be limited. If all animal manure were converted to energy, it has been estimated that it could produce energy equal to 10% of the petroleum requirements or 12½% of our natural gas requirements.

## MANURE AS A FEED

Recycling manure as a livestock feed is the most promising of the nonfertilizer uses. Various processing methods are being employed; and some manure is being fed without processing. More and more feedlot manure will be either (1) incorporated in a grower ration, or (2) fed to breeding herds during periods when pasture supplementation is beneficial, with the residues distributed over grazing areas where they would have fertilizing value.

Animal wastes contain several nutrients that are capable of being utilized when the material is recycled by feeding. Nitrogen, which is present in both protein and nonprotein forms, is a major constituent. Available energy is rather low. Fiber and ash are generally high. The high ash indicates that animal wastes are high in minerals; they are especially rich in phosphorus. Additionally, they contain certain vitamins synthesized in the digestive tract. The wastes possessing the highest nutritive value are broiler litter and layer waste.

One characteristic of all animals wastes is variability in composition due to diet regime, kind and amount of bedding, length of time before collecting, and processing method. The main difference in composition between raw and processed wastes is in the moisture content; many of the processed wastes are low in moisture.

The high fiber and considerable nonprotein nitrogen of animal wastes indicate that they are best suited for feeding to ruminants, since they possess a digestive tract capable of efficiently utilizing high fiber and nonprotein nitrogen. Also, because of their low energy content, they are best adapted for use in maintenance and gestating rations, rather than in lactating and growing rations.

Animal wastes processed by ensiling, dehydration, and other methods can be fed successfully to a wide range of animals. But, for best results the rations in which they become a part should be well balanced following their incorporation. Several workers have shown that the inclusion of too high levels of waste in a ration results in an excessive level of fiber and/or minerals, followed by lowered animal performance. Because of this limitation, not more than 10 to 20% waste should be

included in high-energy rations, such as in cattle finishing rations. However, much higher levels (up to 80%) can be incorporated in rations of gestating beef cows.

• **Poultry waste**—Nearly 100 million tons of poultry wastes (from layers, broilers, and turkeys) are produced annually (see Table 5–10). Because poultry production is highly intensive, with many birds in a small area, waste disposal is a major problem. Most cage-layer operations produce manure free of litter as the primary form of waste. Broiler operations, generally, produce litter.

On a moisture-free basis, cage-layer manure generally contains 25 to 35% crude protein and mineral fiber, while broiler litter contains somewhat less protein—about 18 to 30% and substantially more fiber due to the presence of absorbent materials.

Poultry litter is the most collectable and the most nutritious of all animal wastes. It follows that many experiments have been conducted with it, involving feeding trials with different species. The results of numerous experiments are summarized in Table 5–11. The mean values for waste-fed animals reported therein were obtained by averaging all levels of feeding poultry wastes in the respective categories, though some of the levels were excessive. As shown in Table 5–11 the performance of animals fed wastes was generally slightly lower than that of the controls that were fed traditional feed ingredients. But, on a dry-matter basis, animal wastes generally make for least cost rations and higher net returns.

Also, dried poultry litter has been fed successfully to dry and lactating dairy cows, to growing and breeding sheep, to growing swine, and to broilers.

## SCALES

Scales are a valuable piece of equipment for the modern stock farm or ranch; for they make it possible to determine weights of animals on production-testing studies, to secure the accurate rate of gains of animals being finished, to sell animals on the farm or ranch on a weight basis, and to buy and sell

**TABLE 5–11**
**PERFORMANCE OF ANIMALS FED RATIONS CONTAINING POULTRY WASTES[1]**

| Species of Experimental Animal Used | Kind of Poultry Waste Studied | Criteria | Control Group | Waste-Fed Group |
|---|---|---|---|---|
| Cattle | Dehydrated layer waste | Daily gain ............ lb<br>Daily feed dry-matter intake .. lb<br>Feed/gain ratio ............ | 2.35 (1.07 kg)<br>15.82 (7.19 kg)<br>7.81 | 2.31 (1.05 kg)<br>15.44 (7.02 kg)<br>7.72 |
| Lactating cows | Dehydrated layer waste | Milk yield ............ lb/day<br>Milk fat ............ %<br>Milk total solids ............ % | 41.8 (19.0 kg)<br>3.51<br>12.04 | 38.94 (17.7 kg)<br>3.63<br>12.01 |
| Sheep | Dehydrated layer waste | Daily gain ............ lb<br>Feed/gain ratio ............ | 0.42 (0.19 kg)<br>5.52 | 0.40 (0.18 kg)<br>6.66 |
| Swine | Dehydrated layer waste | Daily gain ............ lb<br>Feed/gain ratio ............ | 1.32 (0.60 kg)<br>4.12 | 1.14 (0.52 kg)<br>4.82 |
| Growing chicks | Dehydrated layer waste | Daily gain ............ grams<br>Feed/gain ratio ............ | 16.1<br>2.36 | 15.7<br>2.60 |
| Laying hens | Dehydrated layer waste | Egg production ......... % lay<br>Feed/dozen eggs ............ lb | 71.9<br>4.18 (1.90 kg) | 72.8<br>4.18 (1.90 kg) |
| Cattle | Poultry litter[2] | Daily gain ............ lb<br>Feed/gain ratio ............ | 2.2 (1.0 kg)<br>10.18 | 1.91 (0.87 kg)<br>11.58 |

[1]Adapted by the authors from *Unidentified Resources as Animal Feedstuffs*, NRC, National Academy Press, Wash., D.C., 1983, pp. 132–144, Tables 35–41.

[2]Also, dried poultry litter has been fed successfully to dry and lactating dairy cows, to growing and breeding sheep, to growing swine, and to broilers.

feed on a weight basis. For greatest usefulness, scales should be so arranged that a pen may be set up quickly when weighing mature animals or may be removed when weighing feed.

A convenient place for the farm scales is in the farm court, next to the corrals or feedlot. In this location, the scale is convenient for weighing livestock or loads of feed and supplies.

Fig. 5-15. Modern dial scales at C & B Livestock Co., Hermiston, OR. (Courtesy, Ron Baker, owner of C & B Livestock Co.)

## FENCES FOR LIVESTOCK

Good fences (1) maintain farm boundaries, (2) make livestock operations possible, (3) reduce losses of both animals and crops, (4) increase land values, (5) promote better relationships between neighbors, (6) lessen accidents from animals getting on roads, and (7) add to the attractiveness and distinctiveness of the premises.

Wire is the leading fence material, although such materials as steel, poles, boards, stone, and hedge have a place and are used under certain circumstances.

Fig. 5-16. Woven wire fence with a barbed wire top. (Courtesy, Keystone Steel & Wire, Peoria, IL)

## PAINT FOR BUILDINGS AND EQUIPMENT

The chief purposes of painting farm buildings and equipment are (1) to preserve them from the effects of the weather, and (2) to add to their attractiveness. In addition, interior painting, such as is done in most homes, makes walls and ceiling more sanitary and dark rooms lighter.

Fig. 5-17. Paint makes the difference! (Courtesy, *Corn Belt Farm Dailies*, Chicago, IL)

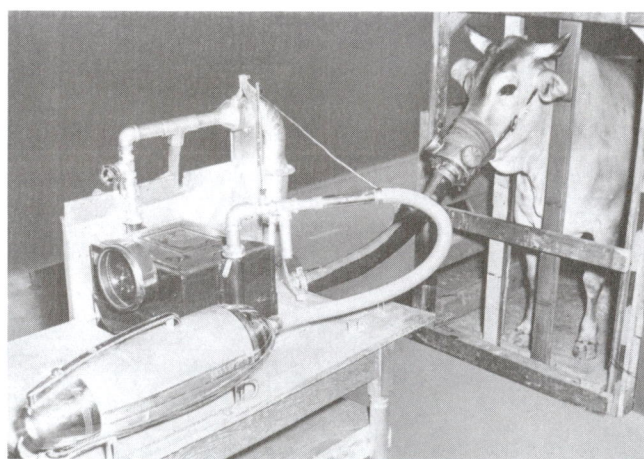

Fig. 5-18. Evaluating heat tolerance by respiratory volume. In the future, environmental control will be an important consideration in livestock buildings. (Courtesy, USDA)

# QUESTIONS FOR STUDY AND DISCUSSION

1. What purposes do farm and ranch farm buildings and equipment serve?

2. What motivating forces have made for more confinement production?

3. What new animal problems have arisen and what old problems have been accentuated by increased confinement production?

4. What factors are important in determining the location of the farm or ranch headquarters?

5. List and discuss each of the most important factors to consider in farmstead arrangement.

6. List and discuss the requisites of livestock buildings. Why is flexible design of great importance?

7. One of the first, and frequently one of the most difficult, problems confronting the farmer or rancher who wishes to construct a building or item of equipment is that of arriving at the proper size or dimensions. In planning to construct new buildings and equipment for animals, what factors and measurements for buildings and equipment should be considered?

8. List and discuss the pros and cons of (a) naturally ventilated buildings vs (b) environmentally controlled buildings.

9. Define environment. In animals, what environmental factors are involved?

10. Before an environmental system can be designed for a species of animals, it is important to know (a) their heat production, (b) vapor production, and (c) space requirements. Why is this information pertinent thereto?

11. Define the following terms: total heat, sensible heat, latent heat, BTU, and vapor.

12. The major requirement of winter ventilation is moisture removal, whereas the major requirement of summer ventilation is temperature control. Why the difference?

13. List the requisites of livestock equipment.

14. Define automation. Give some examples of automation on the farm or ranch.

15. Sketch four types of barn roof designs. What type of roof design do you prefer? Justify your preference.

16. List the common types of floor for stalls or stables, and discuss the *advantages* and *disadvantages* of each of them.

17. List and discuss the *advantages* and the *disadvantages* of slotted floors.

18. What kind of silo do you prefer? Justify your choice.

19. What are plastic silos? Describe the common types of plastic silos.

20. What is manure?

21. Calculate the manure storage capacity needed for a hundred 1,000-lb steers for six months.

22. Discuss each of the following points pertaining to manure:
    a. Yearly manure production, composition, and value.
    b. Modern ways of handling manure.

23. Why is there renewed interest in using manure as a fertilizer? How much manure can be applied to the land?

24. What is the potential of manure as a nonfeed energy source?

25. Discuss manure as a feed.

26. Why are scales so valuable on a modern farm or ranch?

27. What purposes do fences serve on a livestock operation?

28. Why paint buildings and equipment?

# SELECTED REFERENCES

| Title of Publication | Author(s) | Publisher |
|---|---|---|
| *Agricultural Engineers Yearbook* | Edited by R. H. Hahn, Jr. | American Society of Agricultural Engineers, St. Joseph, MI, annually |
| *Bibliography of Livestock Waste Management* | J. R. Miner D. Bundy G. Christenbury | Office of Research and Monitoring, U.S. Environmental Protection Agency, Washington, DC, 1972 |
| *Dairy Housing II* | R. L. Fehr, Planning Committee Chairman | American Society of Agricultural Engineers, St. Joseph, MI, 1983 |
| *Farm Builder's Handbook*, Second Edition | R. J. Lytle | Structures Publishing Company, Farmington, MI, 1973 |
| *Farm Building Design* | L. W. Neubauer H. B. Walker | Prentice-Hall, Inc., Englewood Cliffs, NJ, 1961 |
| *Farm Buildings* | R. E. Phillips | Doane-Western, Inc., St. Louis, MO, 1981 |
| *Farm Buildings*, Second Edition | J. C. Wooley | McGraw-Hill Book Company, New York, NY, 1946 |
| *Farm Buildings*, Third Edition | D. G. Carter W. A. Foster | John Wiley & Sons, Inc., New York, NY, 1947 |
| *Farm Service Buildings* | H. E. Gray | McGraw-Hill Book Company, New York, NY, 1955 |
| *Farm Structures* | H. J. Barre L. L. Sammet | John Wiley & Sons, Inc., New York, NY, 1950 |
| *Handbook of Livestock Equipment*, Second Edition | E. M. Juergenson | The Interstate Printers & Publishers, Inc., Danville, IL, 1979 |
| *Housing of Animals* | A. Matson J. Daelemans J. Lambrecht | Elsevier, Amsterdam, The Netherlands, 1985 |
| *Livestock Waste Management and Pollution Abatement* | | American Society of Agricultural Engineers, St. Joseph, MI, 1971 |
| *Livestock Waste Management System Design Conference for Consulting and SCS Engineers* | | University of Nebraska Cooperative Extension Service, Lincoln, NE, 1973 |
| *Machines for Power Farming*, Second Edition | A. A. Stone H. E. Gulvin | John Wiley & Sons, Inc., New York, NY, 1967 |
| *Midwest Plan Service Handbooks by species* | Staff/Committee | Agricultural Engineering Department, Iowa State University, Ames, IA (pub. at intervals) |
| *Practical Farm Buildings: A Text and Handbook*, Second Edition | J. S. Boyd | The Interstate Printers & Publishers, Inc., Danville, IL, 1979 |
| *Principles of Animal Environment* | M. L. Esmay | Avi Publishing Co., Westport, CT, 1969 |
| *Stockman's Handbook, The*, Sixth Edition | M. E. Ensminger | The Interstate Printers & Publishers, Inc., Danville, IL, 1983 |
| *Structures and Environment Handbook* | | Midwest Plan Service, Iowa State University, Ames, IA, 1972 |
| *Ventilation of Agricultural Structures* | M. A. Hellickson J. N. Walker | American Society of Agricultural Engineers, St. Joseph, MI, 1983 |

Plans and specifications for beef cattle buildings and equipment can also be obtained from your local country agricultural agent, your state college of agriculture, and materials and equipment manufacturers and dealers.

# ANIMAL HEALTH, DISEASE PREVENTION, AND PARASITE CONTROL

A serious case! (Courtesy, The Bettmann Archive)

by

DR. ROBERT F. BEHLOW, D.V.M., Professor Emeritus, North Carolina State University, Raleigh, North Carolina

and

Dr. M. E. ENSMINGER, Ph.D., Distinguished Professor, Wisconsin State University; Adjunct Professor, California State University at Fresno; and Collaborator, U.S. Department of Agriculture

# Chapter 6

In the natural state, animals roved over the broad prairies and through virgin forests, gleaned the feeds provided by nature, and bedded down on new sites each night. Finally, as civilization advanced and herds and flocks increased in size, animals were placed in restricted areas. Simultaneously, they were bred and fed for greater production of meat, milk, eggs, fiber, or for power and speed. Under unnatural conditions of close confinement, greater numbers, forced production, and eating and sleeping in close contact with their own body discharges, the control of diseases and parasites became of paramount importance.

Each year, American farmers and ranchers, from one end of the country to another, are robbed of their profits in livestock production—robbed by diseases and parasites. Inevitably, these losses are passed along to consumers, where they add substantially to the cost of food and fiber. It is conservatively estimated that annual U.S. losses from diseases, parasites, and pests of livestock and poultry are equivalent to about 15% of the cash receipts from marketings of livestock and livestock products.[1] Thus, with cash receipts of $76.218 billion in 1987 (see Chapter 2, Fig. 2–11 in this book), losses for that year on an equivalent basis of 15% would total $11.4 billion ($76.218 × 15% = $11.4 billion).

Animal diseases are one of the principal causes of low productivity in the developing countries, in addition to high mortality losses. In the early 1980s, moderate to high incidence of foot-and-mouth disease was reported in 30 developing countries, rinderpest in 5, trypanosomiasis in 36, theilariosis in 18, babesiosis in 34, swine fever in 5, fowl pox in 43, Newcastle disease in 63, and chronic respiratory disease in 51. Although these figures are constantly changing, they do point up the seriousness and extent of the disease/parasite problem, worldwide. Studies, including extensive surveys made by the author, reveal that American farmers and ranchers suffer the following appalling losses:

1. Twelve percent of the cows that are bred never calve.
2. Calf losses from birth to weaning run 6%.
3. Ten percent of all calves (beef and dairy combined) are afflicted by calf scours, and 18% of all dairy calves so afflicted die.
4. Diseases cost cow-calf operators $26.95 per cow in 1990.
5. Cattle feedlot losses on calves run about 2.0%; and on yearling, about 1.4%.
6. About 1.5 million head of cattle die in feedlots each year, at an estimated loss of more than $750 million.
7. Sterility and delayed breeding in dairy cattle make for an estimated yearly loss of $60 per cow, or a national total of $650 million.

8. Dairy herds average 10% breeding difficulty at any one time and 1.85 services per conception.
9. Retained placenta occurs in about 10% of the parturitions of dairy cattle.
10. Nearly 40% of all dairy cows have some form of mastitis, which causes a yearly loss of $90 to $250 per infected cow per year, depending on the severity of the infection, according to the National Mastitis Council.
11. Five percent of the ewes bred never lamb.
12. Twenty percent of the lambs born die between birthing and weaning.
13. About 2.5% of lambs on finishing rations die.
14. Fifteen percent of the sows bred never farrow.
15. Twenty-five percent of the pigs born die between birth and weaning.
16. Horse owners and caretakers are spending millions on needless concoctions.
17. One-half of all pregnant mares either abort or produce weak foals.
18. We produce only a 50% foal crop, which means that 2 mares are kept a whole year to produce one foal.
19. Six percent of the foals born die between birth and weaning.
20. Layer death losses run about 1% per month, or 10 to 12% per year.
21. Broiler losses average about 4%.
22. Turkey losses are high; 8 to 9% of poults do not survive, and 5 to 6% of breeder hens die each year.

Although deaths of animals take a tremendous toll, even greater economic losses—hidden losses—result from failure to reproduce living young, and from losses due to retarded growth and poor feed efficiency, carcass condemnations and decreases in meat quality, and in added labor and drug costs.

Also, considerable cost is involved in keeping out diseases, such as foot-and-mouth disease, that do not exist in this country. Quarantine of a diseased area may cause depreciation in land values or even restrict whole agricultural programs. Additionally, and most important, some 200 different types of infectious and parasitic diseases can be transmitted from animals to human beings; among them, such dreaded diseases as brucellosis (undulant fever), leptospirosis, anthrax, Q fever, rabies, trichinosis, tuberculosis, and tularemia. Thus, rigid meat and milk inspection is necessary for the protection of human health. This is added expense which the producer, processor, and consumer must share.

Despite all these disturbing factors, it is satisfying to know that the United States is regarded as the safest country in the world for a flourishing livestock industry. In order to ensure further progress, however, thousands of workers—including scientists with the U.S. Department of Agriculture, colleges, pharmaceutical houses, practicing veterinarians, and others—are constantly striving to make this country even healthier for both people and animals.

It is not intended that this book shall serve as a source of home remedies. Rather, enlightened livestock producers will institute programs designed to assure herd health, disease prevention, and parasite control. When animal disease troubles are encountered, they will not attempt to diagnose or treat, but will call upon their local veterinarian in exactly the same manner as they call upon the family doctor when human ill health is encountered. But well-enlightened producers will (1) be in a better position to institute programs designed to assure herd health, (2) more readily recognize any serious outbreak of disease and promptly call their veterinarian, (3) prevent un-

[1]The author arrived at the 15% by two methods:
1. **Method #1:** A report prepared for the Council of Deans, Association of American Veterinary Medical Colleges, based on information available on disease losses as of Feb. 1, 1981, and including cattle (beef and dairy), sheep, swine, poultry, horses, and fish, showed animal disease losses of $10 billion per year. In 1980, cash receipts from marketing of total livestock and products amounted to $67.991 billion (Source: *Agricultural Statistics 1988*, USDA, p. 409, Table 581). So, a $10 billion disease/parasite loss was equivalent to 14.7% of the cash receipts from marketings of livestock and livestock products that year.
2. **Method #2:** *Losses in Agriculture*, Ag. Handbook No. 291, ARS, USDA, 1965, pp. 73–82, Tables 26–32, reported estimated annual U.S. losses from the more important diseases, parasites, and pests of livestock and poultry at $2.8 billion for the period 1951–60. During this same period (1951–60), cash receipts from marketings of total livestock and products averaged $17.76 billion (Source: *Agricultural Statistics 1962*, USDA, p. 567, Table 688). So, a $2.8 billion disease/parasite loss was equivalent to 15.8% of the average annual cash receipts for the 10-year period 1951–60.

necessary suffering of sick animals, (4) be better qualified to assist the veterinarian in administering treatment, and (5) be more competent in carrying out a program designed to bring the disease under control with a minimum spread of the infection.

## SIGNS OF GOOD HEALTH

In order that caretakers may know when animal disease strikes, they must first know the signs of good health, any departure from which constitutes a warning of trouble. Some of the signs of good health are:

1. **Contentment.** Healthy animals appear contented; the cow will stretch on rising, the sheep will stand or lie quietly, the pig will curl its tail, and the horse will look completely unworried when resting.

2. **Alertness.** Healthy animals are alert and bright-eyed and will prick their ears up on the slightest provocation.

3. **Eating with relish, and cudding by ruminants.** In healthy animals, the appetite is good and the feed is attacked with relish (as indicated by eagerness to get to the trough, wagging the tail, etc.) In cattle and sheep, cudding is a sure sign of good health, and is one of the first things to disappear in sickness.

4. **Sleek coat and pliable, elastic skin.** A sleek, oily coat and a pliable and elastic skin characterize healthy animals. When the hair coat loses its luster and the skin becomes dry, scurfy, and hidebound, there is usually trouble.

5. **Bright eyes and pink eye membranes.** In healthy animals, the eyes are bright and the membranes—which can be seen when the lower lid is pulled down—are whitish pink in color and they are moist.

6. **Normal feces and urine.** The consistency of the feces varies with the diet; for example, when animals are first turned on lush grass they will be loose. Also, the consistency and dryness of the feces vary between species, but they should be firm and not dry. And there should not be large quantities of undigested feed. The urine should be clear. Both the feces and urine should be passed without effort, and should be free from blood, mucus, or pus.

7. **Normal temperature, pulse rate, and breathing rate.** Table 6–1 gives the normal temperature, pulse rate, and breathing rate of farm animals. In general, any marked and persistent deviations from these normals may be looked upon as a sign of animal ill health.

### TABLE 6–1
### NORMAL TEMPERATURE, PULSE RATE, AND BREATHING RATE OF FARM ANIMALS

| Animal | Normal Rectal Temperature | | Normal Pulse Rate | Normal Respiration Rate |
| --- | --- | --- | --- | --- |
| | Average | Range | | |
| | (degrees F) | (degrees F) | (rate/min.) | (rate/min.) |
| Cattle ....... | 101.5 | 100.4–102.8 | 60–70 | 10–30 |
| Sheep ...... | 102.3 | 100.9–103.8 | 70–80 | 12–20 |
| Goats ....... | 103.8 | 101.7–105.3 | 70–80 | 12–20 |
| Swine ...... | 102.6 | 102.0–103.6 | 60–80 | 8–13 |
| Horses ...... | 100.5 | 99.0–100.8 | 32–44 | 8–16 |
| Poultry ...... | 106.0 | 105.0–107.0 | 200–400 | 15–36 |

Every caretaker should have an animal thermometer, which is heavier and more rugged than the ordinary human thermometer. The temperature is measured by inserting the ther-

mometer full length in the rectum, where it should be left 2 to 3 minutes. Prior to inserting the thermometer, a long string should be tied to the end.

In general, infectious diseases are ushered in with a rise in body temperature, but it must be remembered that body temperature is affected by stable or outside temperature, exercise, excitement, age, feed, etc. It is lower in cold weather, in older animals, and at night.

The pulse rate indicates the rapidity of the heart action. The pulse of different farm animals is taken at the following body areas: cattle, either on the outside of the jaw just above its lower border, on the soft place immediately above the inner dewclaw, or just above the hock joint; sheep and swine, on the inside of the thigh where the femoral artery comes in close proximity to the skin; and horse, either at the margin of the jaw where an artery winds around from the inner side, at the inside of the elbow, or under the tail. It should be remembered that the younger, the smaller, and the more nervous the animal, the higher the pulse rate. Also, the pulse rate increases with exercise, excitement, digestion, and high outside temperature.

The breathing rate can be determined by placing the hand on the flank, by observing the rise and fall of the flanks, or, in the winter, by watching the breath condensate in coming from the nostrils. Rapid breathing due to recent exercise, excitement, hot weather, or stuffy buildings should not be confused with disease. Respiration is accelerated in pain and in febrile conditions.

## ANIMAL DISEASE

*Disease is defined as any departure from the state of health.* Beyond a doubt, the most serious menace threatening the livestock industry is animal ill health. There are many degrees of ill health, but by far the largest loss is a result of the diseases that are due to a common factor transmitted from animal to animal. These disorders are classed as infectious, contagious, and parasitic diseases and are considered theoretically controllable. Today, with the modern rapid transportation facilities and the dense livestock population centers, the opportunity for animals to become infected are greatly increased compared with a generation ago.

### CAUSES OF DISEASE

Any agent that may bring about an abnormal condition of any or all tissues of the body is a disease-producing entity. Among the chief causes may be listed infectious agents, such as bacteria, viruses, and parasites; and noninfectious agents, including chemicals, poisons of various types, faulty nutrition (see Chapter 4), and injuries. In addition to the actual causative agents, any of the following conditions may predispose disease; overwork, exposure to cold, and long shipments—especially in cold weather.

Diseases are classified according to the following bases:

1. **Infectiousness.** As either (a) infectious—one caused by the presence in or on an animal body of a foreign living organism that creates disturbances and leads to symptoms; or (b) noninfectious.

2. **Communicability or contagiousness.** As (a) communicable; or (b) noncommunicable.

Fig. 6–1. Germ-free animals aid research workers in fighting diseases. This lamb, being fed by an Ohio Station (Wooster) veterinary researcher, is being reared under sterile conditions for studies aimed at pinning down specific disease-causing agents and finding means of disease prevention or cure. The lamb was taken from its mother by modified hysterectomy and immediately moved to the sterile isolation unit, thus avoiding any contamination by normally present organisms. The germ-free technique has helped the Ohio scientists find causes and control techniques for several important livestock diseases. The technique has been used successfully with lambs, pigs, turkeys, and calves. (Courtesy, Agricultural Research and Development Center, Ohio State University)

3. **Manner of occurrence.** As (a) sporadic—one which occurs in isolated cases or outbreaks, like glanders or swamp fever in horses; (b) epizootic—one which appears suddenly and affects many animals over a large area at the same time, like influenza; or (c) enzootic—one which affects certain animals of a given area year after year, like goiter.

4. **Anatomic.** As (a) respiratory; (b) nervous; (c) urogenital, etc.

5. **Course and duration.** As (a) acute—one that runs a rapid course of a few days; (b) subacute—one which runs a slower course and lasts 2 or 3 weeks; or (c) chronic—one which lasts from 4 weeks to an indefinite period.

6. **Prognosis.** As (a) curable; (b) incurable; (c) malignant; or (d) benign.

7. **Origin.** As (a) inherited; (b) acquired; (c) prenatal; or (d) postnatal.

8. **Kinds or types of organisms that produce them.** As bacteria, molds, protozoa, viruses, etc. (see the section that follows).

## INFECTIOUS TYPES OF ORGANISMS

The principal types of infectious organisms that cause disease may be grouped as follows:

1. **Bacteria.** Are one of the smallest and simplest known forms of plant life. They are microscopic, possess just one cell, vary in shape, multiply by transverse fission, and possess no chlorophyll.

2. **Chalamydia** (*C. psittaci* and *C. trachomatis*). Are bacteria which lack some important mechanisms for production of metabolic energy and thus lead an intracellular existence.

3. **Flukes** (*trematodes*). Are soft, flat, leaf-shaped parasitic worms.

4. **Insect larvae.** Are the immature, wingless form of insects.

5. **Moldlike bacteria.** Are somewhat higher in the evolutionary scale than ordinary bacteria. Some have referred to them as higher bacteria.

6. **Molds** (*fungi*). Are fungi distinguished by the formation of mycelium (a network of filaments or threads), or by spore masses.

7. **Mycoplasmas** (*PPLO*). Are microscopic organisms intermediate between viruses and bacteria.

8. **Protozoa.** Are the simplest and most primitive form of animal life; they consist of one single cell.

9. **Rickettsiae.** Appear to be intermediate between the bacteria and the viruses.

10. **Roundworms** (*nematodes*). Are the unsegmented worms, usually cylindrical and elongated in shape and with tapered ends.

11. **Tapeworms** (*cestodes*). Have bodies made up of flattened segments joined together to make a chain. Each segment contains a set of male and female reproductive organs.

12. **Viruses.** May be defined as disease-producing agents that (a) are so small that they cannot be seen through an ordinary microscope (they can be seen by using an electron microscope), (b) are capable of passing through the pores of special filters which retain ordinary bacteria, and (c) propagate only in living tissue. They are generally classified according to the tissues they invade, although this is a very arbitrary method, as some viruses invade many tissues.

13. **Yeastlike fungi.** Are characterized by budding, yeastlike cells.

## PARASITES

*Broadly speaking, parasites are organisms living in, on, or at the expense of another living organism.*

Animals are attacked by a wide variety of internal and external parasites. They include fungi, protozoa (or unicellular animals), arthropods (or insects, ticks, and related forms), and helminths (or worms).

Many diseases cannot be spread unless carried by insects. They are among the most ancient afflictions of people and they have played their part in shaping human history. Malaria has influenced the rise and fall of civilizations, epidemics of plague and yellow fever have decimated populations in both the old and new worlds, and outbreaks of louse-borne typhus have often determined the outcome of military campaigns.

Fig. 6–2. Calf with *bottle jaw*. Generally this condition is indicative of a heavy infection of stomach worms. (Courtesy, School of Veterinary Medicine, Auburn University, Auburn, AL)

*Any animal that serves as a residence for a parasite is referred to as a host.* In order to complete their life-span (cycle), some parasites require only one host while others need more.

While in residence, parasites usually seriously affect the host, but there are notable exceptions. Among the ways in which parasites may do harm are (1) absorbing food, (2) sucking blood or lymph, (3) feeding on the tissue of the host, (4) obstructing passages, (5) causing nodules or growths, (6) causing irritation, and (7) transmitting diseases. These may result in death of the affected animal; or they may cause large financial loss through stunted growth, lowered production, general unthriftiness, and emaciation.

The prevention and control of parasites is one of the quickest, cheapest, and most dependable methods for increasing production with no extra animals. no additional feed, and little more labor. This is important, for, after all, the farmer or rancher bears the brunt of this reduced production, wasted feed, and damaged hides. It is hoped that the discussion that follows may be helpful in (1) preventing the propagation of parasites, and (2) causing the destruction of parasites through the use of the most effective anthelmintic or insecticide.

From time to time, new insecticides and vermifuges are approved and old ones are banned or dropped. Where parasitism is encountered, therefore, it is suggested that the livestock producer obtain from local authorities the current recommendation relative to the choice and concentration of the insecticide and vermifuge to use. This information can be obtained from the county agent, extension entomologist, agricultural consultant, or local veterinarian.

### POISONOUS PLANTS

Poisonous plants have been known since time immemorial. Biblical literature alludes to the poisonous properties of certain plants, and history records that hemlock (a poison made from the plant from which it takes its name) was administered by the Greeks to Socrates and other state prisoners.

No section of the United States is entirely free of poisonous plants, for there are hundreds of them. But the heaviest livestock losses from them occur on the western ranges because (1) there has been less cultivation and destruction of poisonous plants in range areas, and (2) the frequent overgrazing on some of the western ranges has resulted in the elimination of some of the more nutritious and desirable plants, and these have been replaced by increased numbers of the less desirable and, in some cases, poisonous species. It is estimated that poisonous plants account for 8 to 10% of all range animal losses each year; and even more in some areas.

Fig. 6–3.  Cow with white snakeroot poisoning. Marked weakness results in the *trembles* characteristic of this condition. (Courtesy, College of Veterinary Medicine, University of Illinois)

### SOME COMMON POISONOUS PLANTS

The list of poisonous plants is so extensive that no attempt is made herein to discuss them. Nevertheless, both the livestock producer and the veterinarian should have a working knowledge of the principal poisonous species in the area in which they operate. The common poisonous plants of the intermountain ranges to which cattle and/or sheep are susceptible at certain times of the grazing season are listed in Table 6–2.

**TABLE 6–2**
**TYPE OF RANGE ANIMAL SUSCEPTIBLE TO POISONOUS PLANTS AT DEFINITE SEASONS**

| Poisonous to Cattle | Time of Year | Poisonous to Sheep | Time of Year | Poisonous to Cattle & Sheep | Time of Year |
|---|---|---|---|---|---|
| Low larkspur | Spring | Death camas | Spring | Broomweed | Spring and summer |
| Oak | Spring | Greasewood | Fall | Chokecherry | Spring |
| Tall larkspur | Early summer & early fall | Horsebrush | Spring | Copperweed | Summer |
| Timber milk vetch | Spring | Rubberweed | Summer | Desert parsley | Spring |
| Water hemlock | Spring | Sneezeweed | Summer | Halogeton | All year |
| | | | | Loco | Spring |
| | | | | Lupine | Summer and fall |
| | | | | Milkweeds | Summer |
| | | | | Veratrum | Summer |

### PREVENTING LOSSES FROM POISONOUS PLANTS

With poisonous plants, the emphasis should be on prevention of losses rather than on treatment, no matter how successful the latter. The following are effective preventive measures:

1. Follow good pasture or range management.
2. Know the poisonous plants common to the area.

3. Know the symptoms that generally indicate plant poisoning.

4. Avoid turning to pasture in very early spring.

5. Provide supplemental feed during droughts, after plants become mature, and after early frost.

6. Avoid turning very hungry animals where there are poisonous plants.

7. Avoid driving animals too fast when trailing.

8. Remove promptly all animals from infested areas when plant poisoning strikes.

9. Treat promptly.

## MODES OF SPREADING DISEASES

Infectious diseases may be spread from one animal to another in a variety of ways; among them, the following:

1. **Direct contact** with diseased animals, in which the infected host actually touches the susceptible animal and transmits the disease. Venereal diseases are spread in this manner.

2. **Indirect contact,** such as (a) by susceptible animals touching infected animals' excretions or secretions, like the placenta, or aborted fetuses; or (b) by susceptible animals breathing infected droplets exhaled from the nose and mouth of infected animals.

3. **Contaminated facilities and equipment,** including vehicles used to transport animals, contaminated feed, water, cattle chutes, syringes, poultry crates, clothes used to wash cows' udders, etc. Gastrointestinal infections, such as salmonellosis, and bovine mastitis are transmitted in this manner.

4. **Vectors,** which include insects, mites, ticks, and snails. In some cases, transmission of the agent by vector is purely mechanical; for example, a biting fly serves as a "flying needle." In other cases, a stage of development, or part of the life cycle, of the infectious agent in the vector is actually necessary before it may be passed on to a new host.

5. **Carrion feeders** (flesh eaters, such as dogs, foxes, or birds) may carry bits of infected carcasses to clean farms.

The rapidity of spread of different infections, their geographical and seasonal distributions, and the relative ease of their prevention and control may all depend, at least in part, upon their means of transmission.

## MODES OF PATHOGEN BODY ENTRY

The infection of a tissue and the production of a disease by a living agent is not always easily accomplished. The agent must first gain entrance to the animal by one of the body openings or through the skin. It then usually multiplies and attacks the tissues. To accomplish this, it must be sufficiently powerful (virulent) to overcome the defenses of the animal's body. The defenses of the animal's body vary and may be weak or entirely lacking, especially under conditions of a low nutritional plane and poor management practices.

Pathogens commonly gain entrance into the body through one or more of the following channels:

1. Respiratory tract.
2. Digestive tract.
3. Genital tract, especially during mating or parturition.
4. Wounds.
5. Mucous membranes of the eye, e.g., pinkeye and leptospirosis (the latter may be acquired when the urine of an infected animal gets into the eye).
6. Teat canal, especially in lactating females.
7. Navel cord in the newborn.
8. Contaminated syringes or surgical instruments.
9. Insect bites.

## ANIMAL DISEASES TRANSMISSIBLE TO HUMANS

The progress of people, from cave to condominium, has been greatly influenced by animals. From the remote day of their domestication forward, the most advanced civilizations of the day have been the keepers of herds and flocks. Although progress walks Indian file behind animals, certain diseases follow. Many of these diseases are transmitted through meat, milk, or eggs; others are transmitted through close contact with the animal, contact with its excreta, or contact with its products—such as hides, wool, or hair; still others are carried from animals to people by insect vectors.

This group of shared diseases is known as *zoonoses*. It includes African sleeping sickness, anthrax, brucellosis, leptospirosis, Q fever, rabies, trichinosis, tuberculosis, and tularemia. It is important that we realize that many zoonotic diseases cannot be prevented or controlled in people except through their control in animals. To the latter end, Table 6–3, "Selected Animal Diseases Transmissible to Humans" is presented.

## HUMAN DISEASES TRANSMISSIBLE TO ANIMALS

Disease transmission is a two-way street! Some human diseases are transmitted from people to animals, thence back to people. Pertinent information relative to a few of these diseases follows:

| | | | | TABLE |
| | | | | SELECTED ANIMAL DISEASES |
| Disease | Species Affected | Cause | Symptoms and Signs (or age group most affected) | Distribution and Losses Caused by |
| --- | --- | --- | --- | --- |
| **Anthrax** (splenic fever, charbon)—an acute, infectious disease. | **A**ll warm-blooded animals and humans. | *Bacillus anthracis*, a large, rod-shaped organism. | **H**istory of sudden death. Sick animals are feverish, excitable, and later depressed. They carry head low and lag behind herd. Respiration is rapid. There are swellings over the body, especially around the neck region.<br>**M**ilk secretion may turn bloody or cease entirely, and there may be a bloody discharge from all body openings.<br>**C**attle are most susceptible. Most frequent in mature animals on summer pasture. | **G**eneral throughout the world in so-called anthrax districts. |

1. **Viral infections.** The viruses that cause influenza (Type A) in humans appears to be closely related to those isolated from swine, horses, avians (chickens, ducks, and turkeys), and possible cattle and sheep. This prompts much speculation relative to the role of animals in the epidemiology of human influenza, particularly as a source of new major antigenic variants. It is now generally accepted that the swine strain is a prototype of the influenza strain responsible for the widespread and severe human epidemic of 1917–18, and that swine got this strain from people. This gives rise to the following unanswered question: Does a human influenza strain have the potential to establish itself in animals, then at some later date be reintroduced into humans, when the latter's antibody status permits?

Chimpanzees are susceptible to the virus that causes the common cold and are capable of transmitting the infection back to people. Also, certain of the great apes may acquire chicken pox from children and serve as a source of infection to other children.

People may transmit the virus that causes mumps to dogs and hamsters, and possibly to cats. Present evidence also indicates a close relationship between the virus of measles and canine distemper; thus, measle virus may be used to immunize dogs against canine distemper.

2. **Bacterial infections.** *Streptococcus pyogens,* the streptococcus that causes scarlet fever in humans may be transmitted from people to cows, then be shed in milk and produce milk-borne epidemics of the disease. However, the staphylococci of people are usually less virulent in cows than in humans.

3. **Protozoa infections.** Several animal parasites, including amoebic dysentery, may be passed from people to animals (cats, dogs, monkeys, and rats), then back to people again.

## DISEASES FOR WHICH ANIMALS SERVE AS PASSIVE CARRIERS

Animals may host the spores of several pathogenic organisms in their intestinal tracts, without exhibiting any symptoms of disease. Among such diseases are the following:

**Tetanus.** *Clostridium tetani,* the tetanus organism, is commonly found in the intestine of herbivorous animals (cattle and horses) and to a lesser extent in humans. Manure, and soil that has been fertilized with manure, are prime sources of these spores. It follows that people working around horses and stables are more likely to be carriers than those engaged in other occupations.

In people and farm animals (especially horses, sheep, and goats), tetanus often follows a deep puncture wound, such as may be inflicted by a nail, a firecracker, or a gunshot. The deep puncture provides both an entry (broken skin) and anaerobic conditions for growth of the organism. After such an injury, tetanus antitoxin will give a very effective passive immunity. However, active immunizations, using tetanus toxoid, is the preferred preventive treatment.

2. **Gas gangrene.** Like tetanus, gas gangrene infection in people results from the introduction of spores through the broken skin. The clostridia organisms, especially *C. perfringens,* are frequently found in the intestinal tract of most farm animals and people. Gas gangrene is more prevalent in combat zones where people are more apt to be wounded and more likely to be in contact with soil contaminated with animal or human excreta.

3. **Botulism.** Botulism in people and animals is caused by the ingestion of food in which the organism *C. botulinum* has produced toxins. Among farm animals, horses are the most susceptible to the disease. They develop botulism from eating moldy or spoiled forages or grains in which botulism toxin has been produced. Poultry are also quite susceptible to the disease.

## DIAGNOSING DISEASE

In addition to a physical examination (the signs of good health vs the signs of disease), the veterinarian has several means of diagnosing disease; among them, the following:

1. Skin testing with antigen, e.g., tuberculosis in cattle. This test is considered sufficiently accurate to cause the slaughter of an apparently normal animal that reacts positively.

2. Agglutination tests, in which there is a characteristic clumping together of cells when serum from a positive animal is brought in contact with the specific antigen. This type of test is used in bloodtyping and in diagnosing such diseases as brucellosis, typhoid fever, and tularemia.

3. Microscopic examination of blood, used to identify the type of infectious organism and to make red and white blood cell counts.

4. Chemical tests on blood, urine, feces, milk, saliva, etc.

5. Skin scrapings, which are especially useful for identifying fungi or parasitic mite infections.

6. Microscopic and/or visual fecal examination, primarily for parasites.

7. Strip cup or other simple physical tests on milk.

8. Biopsy and/or autopsy.

**6–3**
**TRANSMISSIBLE TO HUMANS**

| Treatment | Control and Eradication | Prevention | Remarks |
|---|---|---|---|
| Early treatment with massive doses of penicillin may be effective. | Quarantine infected herds, and withhold all milk and other products from the market until the danger of disease transmission is past. All carcasses and contaminated material should be burned completely or buried deeply and covered with quicklime, preferably on the spot. Vaccinate all exposed but healthy animals, rotate pastures, and initiate a rigid sanitation program. Spray affected and normal animals to avoid fly transmission of infection. | In infected areas, vaccination should be repeated each year, usually in the spring; and there should be adequate fly control by spraying animals during the insect season. Prevention of anthrax in people depends on (1) eradication of the disease in animals, (2) elimination of industrial infections (tanneries, woolen mills, and factories utilizing animal hair), and (3) early diagnosis and prompt treatment of infected cases. | The farmer or rancher should never open the carcass of a dead animal suspected of having died from anthrax; instead the veterinarian should be summoned at the first sign of an outbreak. Control measures should be carried out under the supervision of a veterinarian. The bacillus that causes anthrax can survive for years in a spore stage, resisting all destructive agents. |

*(Continued)*

TABLE 6–3

| Disease | Species Affected | Cause | Symptoms and Signs (or age group most affected) | Distribution and Losses Caused by |
|---|---|---|---|---|
| **Brucellosis** (Bang's disease, undulant fever, Malta fever) —a hidden disease; one of the most serious and wide-spread affecting the livestock industry. | Cattle, sheep (but rare), swine, goats, and humans. The *suis, abortus*, and *melitensis* strains are seen in horses, and the *suis* and *melitensis* strains are seen in cattle, but the incidence is less frequent than *abortus*. People are suscep-tible to all 3 types. | *Brucella abortus.* *Brucella suis.* *Brucella melitensis.* | The act of abortion is the most characteristic symp-tom (especially in cattle), although not all animals that abort are affected and not all affected animals abort. In cattle, the typical symptoms are (1) abortion in the last third of pregnancy, (2) retained afterbirth, (3) several services per conception, and (4) uterine infections. In swine, abortion and sterility are not so com-mon as in cattle; infection may cause swollen joints and lameness, and swelling or atrophy of the testes, epididymus, and prostate in the male. There are no marked symptoms in goats. In people, the disease is characterized by chills, headache, fever, severe night sweats and extreme weakness. | Worldwide. It is the most important U.S. animal-human disease, and there is great economic loss in fewer animal offspring, in breeding and par-turition trouble, in lowered milk production, etc. For the U.S. as a whole, fewer than 1% of all cattle tested (including both beef and dairy animals) react. It is rather common in goats, but rare in sheep. |

Fig. 6–4. Brucellosis. Cow with aborted fetus—the most characteristic symptom. (Courtesy, USDA)

| Disease | Species Affected | Cause | Symptoms and Signs (or age group most affected) | Distribution and Losses Caused by |
|---|---|---|---|---|
| **Leptospirosis** | Cattle, sheep, swine, goats, horses, dogs, foxes, rats and other rodents, and humans. The disease is trans-ferable between species. | Several species of corkscrew-shaped organisms of the spirochete group. *Leptospira pomona* primarily affects cattle and swine. | In most herds, leptospirosis is a mild disease. However, the symptoms may vary from herd to herd. In general, the symptoms noted in *cattle* are (1) high fever, (2) poor appetite, (3) abortion anytime, (4) bloody urine, (5) anemia, and (6) ropy milk. All ages of cattle, and both sexes are affected (in-cluding steers). Swine leptospirosis is usually characterized by abor-tion, pigs born dead or weak, and unthrifty market hogs. Equine leptospirosis is usually characterized by fever, inappetence, mild depression, and occasionally jaun-dice. Human leptospirosis is characterized by abrupt onset of fever with chills, headache, vomiting, and pains in the extremities, joints, and muscles. | Leptospirosis was first observed in humans in 1915–16, in dogs in 1931, and in cattle in 1934. It was first reported in cattle and swine in the U.S. in 1944 and 1952, respectively, although it has been found in dogs in the U.S. since 1939. Bovine leptospirosis has been reported in Europe, Australia, and the U.S. Surveys indicate this disease is widespread in the cattle population of the U.S. as well as in many other parts of the world. Mortality is low in most outbreaks; however, in young calves, it may be high. The main losses are from poor growth in beef cattle and loss of milk production in dairy cows. If it were not for abortions, this disease would go undetected in many herds. |
| **Q Fever** (Nine Mile fever) | Cattle, goats, sheep, and humans. | The causative organism of Q fever is *Coxiella burnetti*. Ticks are the most important vector. People may acquire Q fever through the inhalation of con-taminated dust (including tick feces). However, most persons become in-fected through exposure to livestock, or through the ingestion of their pro-ducts (raw milk or meat of infected animals). | This disease is classified as a rickettsial disease, although the mode of infection to humans differs from that of other infections in this group. In people, the disease manifests itself by acute onset, chills, prostration, and fever. Headache is pro-nounced in most cases. The fever is continuous and lasts from a few days to 2 or 3 weeks. It resembles influenza. | Q fever was first recognized in Australia in 1935. The disease has been identified in some 35 states within the U.S. and therefore is recognized as endemic. A recent Ohio study indicates that the disease is more prevalent among large than among small dairy herds. |

(Continued)

| Treatment | Control and Eradication | Prevention | Remarks |
|---|---|---|---|
| Since there is no successful animal treatment, farmers and ranchers should not waste valuable time and money on so-called cures that are advocated by fraudulent operators.<br><br>In people, treatment should be by a medical doctor. | The nationwide cooperative federal-state brucellosis eradication program has been very successful in reducing the incidence of bovine brucellosis in the U.S. The program consists of blood testing and certifying brucellosis-free herds and areas. The certification progresses from an individual herd, thence to an area or county, thence to a state.<br><br>*Cattle:* Two principles are involved: (1) finding infected animals and eliminating them from the herd, and (2) vaccination. In heavily infected herds where valuable animals are involved, the test-and-slaughter plan is not practical. In such herds, vaccination with Strain 19 at 2 to 10 months of age is recommended.<br><br>In lightly infected herds, blood testing and removal of reactors is recommended. If there is danger of exposure, calfhood vaccination should be used as a protective measure.<br><br>A federal-state cooperative plan for the control and eradication of brucellosis is in progress in the U.S. under this program. *Certified* herds are those that are free of the disease; *Modified Certified Areas* are areas that, as a result of complete testing, are considered nearly free of the disease; *Certified Brucellosis-free* are former Modified-Certified Areas in which continued testing indicates that the disease has been completely eradicated.<br><br>*Swine:* Several plans for eradication of brucellosis in swine are followed. With an infected commercial herd, it is recommended that the entire herd be sold for slaughter. With a valuable purebred herd, blood testing and slaughter is recommended, with the separation of the pigs from the sows if there are several reactors. Vaccination of swine has not been successful and is not recommended.<br><br>*Goats:* Blood testing and the elimination of reactors is recommended. It is claimed that strains of *Br. melitensis* used as a bacterin (killed vaccine) and as a vaccine induce a high degree of immunity in sheep and goats. | Buy replacement animals that are free of the disease and that are from herds known to be free of the disease. Divert or fence off drainage from infected areas. Animals that are purchased or that are shown should be isolated for 30 days and tested before adding to the herd. Avoid visiting infected farms or premises, as the bacteria may be brought home on shoes or clothing. For the same reason, feeds should not be bought from such farms, and one should beware of used feed bags.<br><br>Do not use calfhood vaccination unless (1) there is a disease problem in the herd or in an adjoining herd, or (2) it is so required in order to ship cattle into certain states. However, in problem areas, vaccinate with Strain 19; dairy heifer calves should be vaccinated at 2 to 6 months of age and beef heifers at 2 to 10 months of age. | Brucellosis derives its name from a British Army Surgeon, Sir David Bruce, who, in 1887, discovered the bacteria later named *Brucella melitensis*. In cattle, it is called Bang's disease after a Danish veterinarian, who isolated *Br. abortus* in 1896.<br><br>Pasteurizing milk and cooking meat make these foods safe for human consumption.<br><br>There is ample evidence that boars transmit the disease. Bulls are less apt to do so.<br><br>*Tests for Brucellosis:* The following tests are used for diagnosis of the disease in cattle:<br>1. *Agglutination test,* of which there are two common methods:<br>    a. *The tube, or slow, method*—in which a blood sample is taken from the jugular vein; the blood is allowed to clot and the serum to separate; and the serum is mixed in small test tubes with a suspension of specially selected strain of *Br. abortus*. Complete agglutination in dilutions of 1:100 and higher are positive.<br>    b. *The plate or rapid test*—This is a rapid agglutination test which is done on a glass slide or plate. The antigen consists of specially selected strains of *Br. abortus* stained with gentian violet and brilliant green. |
| | | 2. *Milk ring test*—This is a modification of the agglutination test which is done with milk. The test involves mixing the antigen with fresh milk. The test depends on the fact that clumps of agglutinated organisms are carried to the surface by rising fat globules. A positive test is indicated by a purple cream layer with white milk below. The milk ring test is a highly efficient and accurate screening test for locating infected dairy herds.<br>3. *Card test*—This test involves the use of a disposable card on which blood serum or plasma is mixed with buffered whole-cell suspension of *Br. abortus* (antigen), which reacts (agglutinates) with antibodies in the blood serum of animals infected with brucellosis. | |
| Treatment of animals, which should be prescribed by a veterinarian, may include blood transfusions, administration of selected antibiotics, and good care.<br><br>Antibiotics give fairly good results if cases are promptly treated. It appears that selected antibiotics must be used to eliminate shedders.<br><br>In human leptospirosis, the M.D. should be consulted relative to treatment. | The disease is spread by infected urine; therefore, spread animals out over a large area; avoid congestion in a corral or barn.<br>Fence off waterholes or ponds of slow-running streams.<br>Isolate sick animals or new additions to the herd.<br>Discard milk from diseased cows.<br>Clean and disinfect the barns; exterminate rodents.<br>Administer leptospirosis vaccine to all female cows and sows on problem farms each year.<br>Keep different classes of livestock separated, because leptospirosis can be spread from one species to another. | Blood test animals prior to purchase, isolate for 30 days, and then retest prior to adding them to the herd.<br>Periodic vaccination of animals.<br>Vaccination of people with a suspension of killed leptospires has been employed and reported successful in several countries. | Carrier animals—animals that have had leptospirosis and survived —may spread the infection by shedding spirochetes in the urine. The infected urine may then either (1) be breathed as a mist in cow barns, or (2) contaminate feed and/or water and thus spread the infection. Breeding bulls can transmit this disease to cows.<br><br>It is known that recovered cattle can remain carriers for up to 3 months and swine can remain carriers up to 1 year.<br><br>Leptospirosis is mainly a warm-weather disease.<br><br>The spirochetes seldom survive for more than 30 days outside the animal. Stagnant water favors their survival. |
| Treatment with an antibiotic may reduce the duration of fever and illness, but the response to antibiotic therapy usually is not dramatic. | A vaccine against Q fever is not currently available.<br>Separation of pregnant animals and burning or burying reproductive products can greatly reduce the spread of the disease. | Avoid ticks, take care in aiding animals through parturition, and pasteurize milk properly.<br>Since milk-borne transmission of Q fever has occurred, pasteurization temperatures have been elevated slightly to ensure killing of the causative organism. | Occupational associations that bring people in contact with infected animals at the time of parturition or with their products (such as wool or hides) greatly increase the risk of infection. |

TABLE 6-3

| Disease | Species Affected | Cause | Symptoms and Signs (or age group most affected) | Distribution and Losses Caused by |
|---|---|---|---|---|
| **Rabies** (or hydrophobia, madness)—an acute infectious disease. Fig. 6-5. Rabies. Note the violent butting with the head, a characteristic of the furious form which is most often seen in cattle. At this stage, the animal is insane and very dangerous; it will attack other animals, and even people. (Courtesy, Pitman-Moore, Indianapolis, IN) | All warm-blooded animals and humans. | A filtrable virus which is usually carried into a bite wound by a rabid animal. | Disease manifests itself in two forms: (1) the furious form, and (2) the dumb form. In early stages of furious form, there is loss of appetite, cessation in milk secretion, anxiety, restlessness, and a change in disposition. Next there is madness, excitability, loud bellowing, pawing of the ground, inability to swallow, and violent butting of the head. At this stage, the animal is very dangerous, attacking and biting itself or other animals, or even people. Posterior paralysis strikes on the 4th or 5th day, followed by a coma and death on about the 6th day. | Fewer than 10% of the cases appear in cattle, horses, swine, or sheep. |
| **Trichinosis** *Trichinella spiralis*—a parasitic disease of humans contracted largely by consuming infested pork, eaten raw or imperfectly cooked. Fig. 6-6 shows the life history and habits of trichina. | Swine. Humans. | The parasite, *Trichinella spiralis*. Fig. 6-6 | No specific symptoms in hogs, even when the parasite is present in the muscle tissue, its usual abode. | Old studies (conducted prior to current garbage-cooking laws) showed (1) less than 1% of pork from grain-fed hogs infected with trichinosis, and (2) 5–6% infection of pork in hogs fed uncooked garbage. |
| **Trypanosomiasis** (African Sleeping Sickness—a chronic infectious disease caused by protozoa of the genus *Trypanosoma* and spread by the bite of infected tsetse fly. | All animals and humans. | Flagellates (protozoa) of the genus *Trypanosoma*. The trypanosomes live in the blood of their host, where they multiply and release poisonous by-products of metabolism. | In people and animals, the protozoa invade the nervous system, causing lethargy and finally death. The trypanosomes are spread from host to host by bloodsucking tsetse flies. When a tsetse fly withdraws blood from an infected animal, trypanosomes are sucked into its intestine, where they multiply and undergo developmental changes. They migrate to the fly's salivary glands, in which they further develop and multiply. The fly can then transfer the trypanosomes via saliva into a vertebrate host. | The diseae is confined to Africa, wherever tsetse flies (the vectors) occur. It makes large areas of Africa uninhabitable for people. If trypanosomiasis were brought under control, the Savannah pastures of the tsetse fly-infested area, which is about the size of the U.S., would carry 146 million more head of cattle—or the equivalent of smaller ruminants. |
| **Tuberculosis**—a chronic infectious disease. (Fig. 6-7 shows a positive reaction to the TB test. Note swelling.) | All animals and humans. Tuberculosis in sheep and goats is rare and of chronic character. | *Mycobacterium tuberculosis*, of which there are 3 kinds: (1) the human, (2) the bovine, and (3) the avian (bird) types. | Animals usually get tuberculosis of the lungs and lymph nodes; although in poultry the liver, spleen, and intestines are chiefly affected. In cows, the udder sometimes becomes infected and swollen in chronic cases. Many times infected animals show no outward physical signs of the disease. There may be loss in weight, swelling of joints, and a chronic cough and labored breathing. Other seats of infection are genitals, central nervous system, and the digestive system. *Sheep:* Manifest few symptoms: observed on postmortem. Coughing is prominent in goats, but not in sheep. | Worldwide. The incidence of tuberculosis in the U.S. is steadily declining. |
| **Tularemia** | Cattle, sheep, horses, swine, and humans. Cats, chipmunks, dogs, hamsters, muskrats, beavers, coyotes, foxes, mice, rats, rabbits, deer, opossums, squirrels, guinea pigs, and many other wild mammals. | It is caused by *Francisella tularensis* (closely related to the plague bacillus). The chief vectors of tularemia are ticks (especially the dog tick and the wood tick) and the bloodsucking flies (especially the deerfly and horsefly), but it may also be spread by fleas, certain mosquitoes, and lice. Infected animals that die may contaminate water and thereby spread the infection to sheep and perhaps other species. | In people, the disease is characterized by headache, chills, fever, and vomiting, accompanied by irregular fever, which lasts for several weeks. | This disease of mammals was first described in 1911 in Tulare County, CA (from whence came the name), as a "plaguelike disease of rodents." In 1919 it was described as "deerfly fever." The disease has been reported throughout North America and Eurasia. |

*(Continued)*

| Treatment | Control and Eradication | Prevention | Remarks |
|---|---|---|---|
| After the disease is fully developed, there is no known treatment. Where animals are bitten or exposed to rabies, see the veterinarian. | Persons bitten by a rabid animal should immediately report to the family doctor who will prescribe treatment. Complete eradication would be difficult to achieve because of the reservoir of infection in wild animals. | Immunize all dogs, and regulate the licensing, quarantine, and transportation of dogs. Vaccines, both live and inactivated, derived from a variety of animal cells or tissue cultures are available. Some require annual revaccination, others protect adequately for 3 years. The National Association of Public Health Veterinarians, Inc. issues an annually revised compendium of animal rabies vaccines available in the U.S., including recommendations for their use. Those responsible for the control of rabies should obtain their compendium. Control wild carnivores and bats. | Rabies is generally transmitted to farm animals by dogs, or by wild animals (skunks, foxes, etc.). Where people are bitten or exposed to rabies, they should see their local doctor, who may use a vaccine made of (1) killed virus, nervous tissue origin, or (2) killed virus, duck embryo origin. Also, new and promising vaccines have been developed and are being tested experimentally. |
| There is no practical treatment for infected hogs. Infected humans should be under the care of an M.D. | Trichinosis in swine may be lessened by: (1) destruction of all rats on the farm, (2) proper carcass disposal of hogs and other animals that die on the farm, and (3) cooking all garbage and offal from slaughter houses—it should be cooked for 30 minutes at 212°F (100°C) | Prevention of trichinosis in humans may be obtained by: (1) thoroughly cooking all pork at a temperature of 137°F (58°C) for 6 minutes before it is consumed, or (2) freezing at –13°F (–25°C) for 10 to 20 days before eating. | In people, the disease is usually accompanied by a fever, digestive disturbances, swelling of infected muscles, and severe muscular pain (in the breathing muscles as well as others). Microscopic examination of pork is the only way in which to detect the presence of trichina, but such a method is regarded as impractical in meat inspection procedure. All states have laws requiring that garbage be cooked prior to feeding. |
| Treatments have been disappointing. | It is difficult to control, because many wild animals serve as a reservoir of trypanosomes. | Prevention is dependent upon control of the tsetse fly vector by means of insect repellents, insecticides, and brush clearing. | This organism has a life cycle similar to that of the malarial parasite. Like malaria, it can be spread from human to human, but unlike malaria, it can also be carried from animals to humans. |
| In humans, tuberculosis can be arrested by hospitalization and complete rest, but in animals this method of treatment is neither effective nor practical. Also, no known medical treatment is effective with animals. | *Cattle:* Periodic testing and removal of reactors is the only effective method of control. The following is an effective control program for animals: (1) disposing of tubercular swine, cattle, and poultry; (2) applying strict sanitation, and (3) rotating feedlots and pastures. *Sheep:* Testing with avian tuberculin may be of assistance. | Removal and supervised slaughter of reactor animals, and pasteurization of milk and creamery by-products. Fig. 6–7 shows a positive reaction to the intradermic (into the skin) tuberculin test in a cow. This reaction indicates the presence of TB. Avoid pasturing or housing cattle and swine with chickens. | All states will accept for entry the following: (1) accredited herds which have been tested for TB within the past 12 months, or (2) cattle which have had individual negative tests within the past 30 days. |
| Streptomycin and tetracycline at recommended dose levels are very effective in the treatment of this disease. | Avoid picking up sick, easily caught, or dead rabbits. Wear rubber gloves when handling wild game. | Cooking readily renders the infected tissues safe for human consumption. Vaccines are effective. | Primary modes of transmissioin to humans include bites from infected ticks or other arthropods, the handling of infected animals, and the inhalation of dust or vapor containing *F. tularensis*. |

## DRUGS

*Drugs, or medicinal agents, are substances of mineral, vegetable, or animal origin used in the relief of pain or for the cure of disease.* Much superstition cloaks the reasons for the recommended use of many drugs that have been employed for centuries. An example of this is liverwort, which was heralded as a sure cure for liver disorders only because it was shaped like a liver. Unfortunately, there is no known cure-all for a large number of diseases or for the relief of a great number of different parasitisms.

Lacking the knowledge of limitations of drugs and the nature of disease, many farmers and ranchers have been sold worthless products. There is a flourishing business in various cure-alls that are sold under such names as *tonic, reconditioner, worm expeller, liver medicine, mineral mixture, mineral and vitamin mix, regulator,* and numerous others. It is poor practice to disregard the advice of reputable veterinarians and experimental workers and to rely on claims made by unscrupulous manufacturers of preparations of questionable or fraudulent nature. Most of these patent drugs are sold for fantastic prices, considering their actual cost, and most of their ingredients are never indicated. To avoid being swindled, purchases should be limited to preparations from reliable firms and confined to those recommended by the local veterinarian. Fortunately, the Food and Drug Administration has been very vigilant and has been instrumental in the disappearance of many misbranded drugs and remedies from interstate channels.

## ANIMAL HEALTH

Modern science has conceived of many artificial protective mechanisms against disease. These are important; hence, they are discussed in the sections that follow. Valuable as they are, however, there is no substitute for livestock sanitation and disease prevention. The artificial protections should merely be used as an adjunct to a high state of natural health that is built around a program of improved breeding, feeding, and management.

## IMMUNITY

*When an animal is immune to a certain disease, it simply means that it is not susceptible to that disease.*

The animal body is remarkably equipped to fight disease. Chief among this equipment are large white blood cells, called phagocytes, which are able to overcome many invading organisms.

The body also has the ability, when properly stimulated by a given organism or toxin, to produce antibodies and/or antitoxins. When an animal has enough antibodies for overcoming particular (disease-producing) organisms, it is said to be immune to that disease.

When immunity to a disease is inherited, it is referred to as a natural immunity. For example, when sheep are exposed to hog cholera they never contract the disease because they have a type of natural immunity referred to as species immunity. Likewise, people are naturally immune to Texas fever. Algerian sheep are said to be highly resistant to anthrax; this constitutes a type of natural immunity called racial immunity.

Acquired immunity or resistance is either active or passive. When the animal is stimulated in such manner as to cause it to produce antibodies, it is said to have acquired active im-

munity. On the other hand, if an animal is injected with the antibodies (or immune bodies) produced by an actively immunized animal, it is referred to as an acquired passive immunity. Such immunity is usually conferred by the injection of blood serum from immunized animals, the serum carrying with it the substances by which the protection is conferred. Passive immunization confers immunity upon its injection, but the immunity disappears within three to six weeks.

In active immunity, resistance is not developed until after one or two weeks; but it is far more lasting, for the animal apparently keeps on manufacturing antibodies. It can be said, therefore, that active immunity has a great advantage. There are exceptions, however—for example, when a disease must be checked immediately as in a virulent outbreak of swine erysipelas, when immune serum from actively immunized horses is injected.

It is noteworthy that young suckling mammals secure a passive immunity from the colostrum that they obtain from the mother for the first few days following birth.

## VACCINATION

*Vaccination may be defined as the injection of some agent (such as a bacterin or vaccine) into an animal for the purpose of preventing disease.*

In regions where a disease appears season after season, it is advised that healthy susceptible animals be vaccinated before exposure and before there is a disease outbreak. This practice is recommended not only because it takes time to produce an active immunity but also because some animals may

Fig. 6–8. Cattle in a chute being vaccinated. (Courtesy, Fort Dodge Laboratories, Fort Dodge, IA)

be about to be infected with the disease. The delay of vaccination until there is a disease outbreak may increase the seriousness of the infection. In addition, a new outbreak will "reseed" the premises with the infective agent.

In vaccination, the object, as has been previously pointed out, is to produce in the animal a reaction that in some cases is a mild form of the disease.

It is a mistake, however, to depend on vaccination alone for disease prevention. One should always ensure its success by the removal of all the interfering adverse conditions. It must also be said that varying degrees of immunity or resistance result when animals are actively immunized. Individual animals vary widely in their response to similar vaccinations. Heredity also plays a part in the determination of the level of resistance. In addition, nutritional and management practices play an important part in degrees of resistance displayed by animals.

## BIOLOGICS

*Biologics may be defined as medicinal preparations made from microorganisms (bacteria, protozoa, or viruses) and their products.* They include various vaccines, bacterins, serums, and similar preparations. These agents are one of the most valuable contributions to animal health, and they are constantly being improved. They are used essentially for rendering animals immune to various infections.

It is noteworthy, however, that not all attempts to confer immunity by biologics are successful. In some cases, it seems impossible to create an immunity against infection. The common cold is a case in point. In other cases, the animal may die from the disease or its complications, in spite of an inoculation—because of a biologic of poor quality, infection before the treatment is begun, or improper administration of the biologic.

### VACCINES

Fig. 6–9. Inoculation of eggs for the production of vaccine. Some vaccines are made by growing virus on chick embryos. For this process, the two main constituents are fertile chicken eggs and a stock culture of virus. (Courtesy, Dr. Salsbury's Laboratories, Charles City, IA)

*Vaccines are suspensions of live organisms (bacteria or virus) or microorganisms that have had their pathogenic properties removed but their antigenic properties retained.* As pointed out previously, vaccines are purposely administered to produce a mild attack of disease, thus stimulating the resistance of that animal to that specific disease, often resulting in permanent immunity. Vaccines are employed mainly in the prevention rather than in the curing of disease. Examples are Strain 19 vaccine of *Brucella abortus* and anthrax vaccine. Great care must be used in their preparation, storage, and administration. Since the improper use of vaccines may result in disease outbreaks, it is strongly recommended that a veterinarian be consulted about their use.

### BACTERINS

*Bacterins are standardized suspensions of bacteria (and their products) that have been killed by heat or chemical means and are unable to produce disease.* When introduced into the body, they stimulate the production of protective antibodies which act against subsequent attacks of organisms of the kind contained in the bacterin. They produce an active immunity.

Theoretically, bacterins should be useful in the prevention of every infectious disease in which the causative agent is known. Unfortunately, they do not always give the desired results, especially in diseases of a chronic nature.

Often a product may be a mixed bacterin; that is, it may contain more than one organism. This usually includes secondary invaders when the true causative agent is unknown.

Among the common bacterins in use are those for blackleg in cattle, enterotoxemia in sheep, and malignant edema in all animals.

### SERUMS

Serums, also known as immune blood serum or immune serum, are obtained from the blood of animals (often horses) that have developed a solid immunity from having received one or more doses of infectious organisms. They do not contain any organisms, either dead or alive. Serums are used for the protective nature of the antibodies that they contain, which stop the action of an infectious agent or neutralize a product of that agent. They give a passive immunity. Among the serums that have proved successful are those for tetanus and anthrax.

### TOXOIDS (OR ANTITOXINS)

A toxoid is a "tamed" toxin. Some bacteria, such as the germs that cause diphtheria and tetanus, produce powerful poisons or toxins. These are the substances that actually cause the damage; the bacteria themselves may produce only very mild symptoms. The same toxin is formed when the bacteria are grown in the laboratory, but it is then treated chemically. It loses the poisonous or toxic properties but still retains the power to stimulate the body cells; they form the appropriate antibody (antitoxin—a word derived from the Greek *anti*, meaning against; and *toxin*, meaning poison). Among toxoids are diphtheria toxoid and tetanus toxoid.

## OTHER ARTIFICIAL PROTECTIVE MECHANISMS AGAINST DISEASE

In addition to the vaccines, bacterins, serums, and toxoids, the following products are employed for the protection of animals against disease: modified live vaccines, sensitized bacterins, germ-free extracts, natural and artificial aggressions, and bacterial filtrates. Although these products are prepared differently, they all serve to provide protective substances against corresponding infections.

## GENERAL ANIMAL SANITATION AND DISEASE PREVENTION

In order to reduce the possibility of disease, one must adopt certain management practices relative to the environment of the animal. It has been said that domestication and increased animal numbers imply sort of a contract. Caretakers in fulfilling their obligation for services rendered, must protect their animals from the elements, parasites, and diseases, and furnish them sanitary quarters and suitable rations. Abuse leads to the reduction of profits—a case in which money and decency are on the same side of the ledger.

Animals require sanitary quarters. In the wild state, they had access to plenty of fresh air, clean feed, and plenty of room. They are naturally of clean habits and if given the choice will not voluntarily consume contaminated feed nor lie in filth.

## VENTILATION

The need for ventilation is not as great for the animal as it is for human beings, for most of the animal's life is spent out of doors where plenty of fresh air is available. Ventilation is significant only when animals are housed in crowded quarters.

*Ventilation is the act of causing the movement of air through buildings with the objective of supplanting foul air with fresh air containing needed oxygen.* Contrary to common opinion, when a feeling of discomfort is noticed, it is the result of oxygen starvation rather than carbon dioxide poisoning.

The amount of moisture in the air is important. When improper ventilation prevents proper evaporation, the moisture content of the air increases. If humidity rises too high, interfering with heat elimination, heat stroke may ensue. Moist air generally is a more favorable medium for the existence of microorganisms, thus lending itself well to the transmission of contagious diseases. When one animal is infected with a contagious disease and is closely housed with others, an epidemic will usually follow. The air may also pick up various noxious gases, such as ammonia from decomposing urine, which may cause irritation to the sensitive membranes of the mouth, eyes, nose, and respiratory tract.

Ventilation is measured in cubic feet per minute (cfm). The required ventilation differs according to species of animal, size of animal, and outside temperature. (Also, see Chapter 5.)

## HOUSING

Although housing and close confinement predispose animals to more disease, it is often very necessary. Housing must frequently be provided to facilitate handling, to combat the elements, or to furnish protection during illness or when young are arriving. Proper drainage and dryness, adequate space, and good lighting are some of the requirements for good housing. In addition, animal quarters must be of such construc-

tion as to facilitate proper cleaning, disinfection, and maintenance of sanitary conditions. This includes suitable floors, adequate waste disposal, and proper absorbent bedding. Further discussion of the requisites of livestock buildings is found in Chapter 5.

Fig. 6–10.   Visitors can bring diseases, so precautions must be taken to ensure that a healthy flock or herd stays that way. Plastic boots help ensure against disease and parasite introduction. (Courtesy, DeKalb Poultry Research, Inc., DeKalb, IL)

## ADEQUATE MANURE DISPOSAL

Situations that compel animals to live in close contact with their own body excreta are most injurious to physical well-being. Urine, feces, exhalations, and nose and mouth discharges may often contain disease-producing agents, and furnish an ideal medium for the growth of microorganisms. Livestock producers are fully aware of the miraculous recovery many animals undergo when taken from small, unsanitary enclosures to good, clean pastures.

The importance of removing excrement frequently from the immediate surroundings (enclosures, barns, and loafing sheds) cannot be stressed too much. The method of disposal of solid and liquid manure is also very important. As this manure may contain a variety of parasites and eggs, proper disposal offers an excellent opportunity for breaking the life cycles of these parasites. On the other hand, if left in an accessible place for animals, manure can be a rich, never-ending source of disease and parasitism.

In order to ensure the killing of many harmful parasites, one may store manure (for two weeks to a month) so that the heat generated will cause their death. It should be stored in a covered concrete pit and located far enough away from the buildings to prevent contamination. These enclosures should be inaccessible to all animals. Spraying manure pits with a suitable insecticide will inhibit fly development. If the manure is believed to be free of specific infectious microorganisms (for example, tuberculosis, brucellosis, and blackleg), it may be spread daily on arable land containing no animals. Here the purifying elements—such as rain, sunshine, soil, and vegetable processes—will tend to render the manure sanitary. Food and water should always be protected from contamination by manure.

## PASTURE ROTATION

Pasture rotation provides a very practical method of control of many diseases and parasites. Permanent pastures used by one species of animal may be regarded as highly dangerous for profitable endeavors. A method by which land areas for pasturage are systematically changed periodically to crop production is recommended.

As many parasites (including bacteria) are often specific for a certain host (for example, bots of horses affect no other animal), frequently pastures may be rotated between different species.

## CARCASS DISPOSAL

In the disposal of carcasses, it is a safe rule to assume that all of them are a source of some infection, and subsequently, it is important to adopt proper sanitary precautions.

The most sanitary method of destroying a carcass is to burn it, preferably at the site of death in order to prevent the contamination of surrounding ground. A trench of sufficient size should be prepared, a fire built, and the animal placed on top so that it will be consumed in its entirety. An incinerator may be used for poultry.

The most common method of large animal carcass disposal is by burial. So that this method will be effective, the carcass should be buried deep and covered with quicklime. The top of the carcass should be at least 4 feet below the surface of the ground and in soil from which there is no danger of contamination by drainage. Burial should not be near a flowing stream, for this will only serve to spread the disease downstream.

Near large centers of population, rendering plants will take carcasses, and they afford the easiest method of disposal.

When an animal dies, it is recommended that a veterinarian be called immediately to perform a postmortem examination. This is done in an attempt to study the abnormal conditions present and to determine the cause of death. It is never safe for one who is uninformed about specific disease lesions to open an animal carcass. Such practice may not only serve to spread a very highly contagious disease, but may also expose the operator to a dangerous infection.

It is also unsafe to feed the carcass to other animals. Such procedure may cause the animal consuming it to become sick, or it may serve only to spread the disease.

## DISINFECTANTS AND THEIR USE

A disinfectant is a bactericidal or microbicidal agent that frees from infection (usually a chemical agent which destroys disease germs or other microorganisms, or inactivates viruses).

The high concentration of animals and the continuous use of modern livestock buildings often result in a condition referred to as disease buildup. As disease-producing organisms—viruses, bacteria, fungi, and parasite eggs—accumulate in the environment, disease problems can become more severe and be transmitted to each succeeding group of animals raised on the same premises. Under these circumstances, cleaning and disinfection become extremely important in breaking the life cycle. Also, in the case of a disease outbreak, the premises must be disinfected.

Under ordinary conditions, proper cleaning of barns removes most of the microorganisms, along with the filth, thus eliminating the necessity of disinfection.

Effective disinfection depends on five things:

1. Thorough cleaning before application.
2. The phenol coefficient of the disinfectant, which indicates the killing strength of a disinfectant as compared to phenol (carbolic acid). It is determined by a standard laboratory test in which the typhoid fever germ often is used as the test organism.
3. The dilution at which the disinfectant is used.
4. The temperature; most disinfectants are much more effective if applied hot.
5. Thoroughness of application, and time of exposure.

Disinfection must in all cases be preceded by a very thorough cleaning, for organic matter serves to protect disease germs and otherwise interferes with the activity of the disinfecting agent. This includes the burning of all inflammable refuse and the spreading out of the remainder on arable land (cultivated land not occupied by animals). Included also is the moistening of all litter and its subsequent removal. All walls, ceilings, and woodwork must be brushed down and also washed down with water and scrubbed. Old sacks should be disinfected or destroyed. Contaminated feed should be destroyed. Having accomplished this cleaning, one is ready to choose the disinfectant.

A good disinfectant should (1) have the power to kill disease-producing organisms, (2) remain stable in the presence of organic matter (manure, hair, soil), (3) dissolve readily in water and remain in solution, (4) be nontoxic to animals and humans, (5) penetrate organic matter rapidly, (6) remove dirt and grease, and (7) be economical to use.

When using a disinfectant, *always read and follow the manufacturer's directions.*

The number of available disinfectants is large because the ideal universally applicable disinfectant does not exist. (See *The Stockman's Handbook*, a book by the same author and publisher as *Animal Science*, for a tabular summary of the limitation, usefulness, and strength of some common disinfectants.)

Unfortunately, there is no one best germ killer, nor is there anything like a general disinfectant that is effective against all types of microorganisms under all conditions. This stems partially from the fact that not all disease-producing bacteria are susceptible to the same chemical agents. A few bacteria—such as those which cause blackleg, tetanus, and anthrax—possess the ability of forming seedlike spores that can remain dormant for years and resist destruction. Others, like those organisms causing tuberculosis, are resistant to oxidizing disinfectants such as the chlorine compounds. The organisms that cause brucellosis, strangles, and some other diseases are, fortunately, very readily killed by almost any disinfectant that reaches them.

The application of heat by steam, by hot water, by burning, or by boiling is an effective method of disinfection. In many cases, however, it may not be practical to use heat.

Sunlight possesses disinfecting properties, but it is variable and superficial in its action. Heat and some of the chemical disinfectants are more effective.

## GENETIC RESISTANCE TO DISEASE

Genetics as a tool for eliminating or controlling certain diseases holds promise. In this area, plant breeders have led the way. In 1905, it was discovered that certain varieties of wheat were more resistant to mycotic stem rust than others, thereby laying the foundation for important advances in the knowledge of genetic resistance to disease. Subsequently, scientists have evolved with many varieties of plants showing genetic resistance to disease. Evidence that similar genetic resistance to disease holds for animals has been demonstrated by experiences and experiments. For example, Brahma cattle are more resistant to certain parasites, notably Texas fever, than the British breeds; Bronze turkeys have a genetic predisposition to pendulous crop, a disease which may occur in this particular breed during very hot, dry summer months, with excess water consumption which causes an irreversible stretching of the crop; and it has been demonstrated that selected lines of chickens vary in their resistance and susceptibility to fowl leukosis.

In the future, scientists may be able to genetically engineer animals resistant to some of the most costly diseases. The goal is to improve the overall health of livestock without compromising desirable production traits like reproduction, growth, or meat quality. But for this to be a reality, the genes responsible for disease resistance, or susceptibility, must be identified and understood.

Fig. 6–11. These White Leghorn roosters look alike, but their genes are different. The rooster on the left was injected with genes of avian leukosis virus when it was a 1-day-old embryo. The center and right roosters are of two succeeding generations, which directly inherited these virus genes. Viruses may someday be used as carriers for genes for leaner, tastier, bigger, and more profitable chickens. (Courtesy, USDA)

The application of genetics to disease control in animals presents greater problems than in plants; it's more expensive and time-consuming. Also, to be of greatest practical value, it would be necessary to develop strains or breeds of animals that are genetically resistant to several diseases. Nevertheless, the stakes are high and this approach is worthy of greater attention than it has received in the past.

## REGULATIONS RELATIVE TO DISEASE CONTROL

Certain animal diseases are so devastating that no individual farmer or rancher could long protect his herds and flocks against their invasion. Moreover, where human health is involved, the problem is much too important to be entrusted to individual action. In the United States, therefore, certain regulatory activities in animal disease control are under the supervision of various federal and state organizations. Federally, this responsibility is entrusted to the following agency:

Veterinary Services
Animal and Plant Health Inspection Service
U.S. Department of Agriculture
Federal Center Building
Hyattsville, Maryland 20782

## HISTORY OF FEDERAL ANIMAL DISEASE ERADICATION

The year 1884 marked the beginning of an organized cooperative effort, under legal authority, for the control and eradication of animal disease. In that year, Congress passed an act creating the Federal Bureau of Animal Industry (now the Animal Disease Eradication Division, Agricultural Research Service). It was first established to prevent exportation of diseased cattle and to eradicate contagious pleuropneumonia and other contagious diseases. In three years, it had succeeded in completely eliminating contagious pleuropneumonia, a dreaded disease.

In the famous Bulletin No. 1, the Bureau of Animal Industry revealed that Texas fever ticks were the biological bearer of the protozoa causing Texas fever. The description of this type of disease relationship laid the foundation for the subsequent work on such diseases as yellow fever and malaria of humans.

Since then great progress has been made. Bovine tuberculosis has all but been eradicated, and cattle tick fever has been reduced to negligible proportions by a very extensive eradication program. Advances are being made in the control of brucellosis. In addition, there are many ways through which federal and state agencies make less spectacular, though equally important, contributions to human and animal welfare in the United States. For example, where certain animal diseases are involved, the producer can obtain financial assistance in eradication programs through federal and state sources.

## FOOD AND DRUG ADMINISTRATION (FDA)

In 1906, the U.S. Congress enacted the Pure Food and Drug Law. Concurrently, the Federal Meat Inspection Act was passed. Both laws became effective in 1907. The U.S. Food and Drug Administration (FDA) was established as a separate unit of the U.S. Department of Agriculture in 1927. Then, in 1940, the FDA was transferred from the USDA to the Federal Security Agency, presently, the U.S. Department of Health and Human Services.

The FDA is charged with the responsibility of safeguarding American consumers against injury, unsanitary food, and fraud. It also protects industry against unscrupulous competition. It inspects and analyzes samples and conducts independent research on such things as toxicity (using laboratory animals), disappearance curves for pesticides, and long-range effects of drugs.

## U.S. DEPARTMENT OF AGRICULTURE (USDA)

The following four divisions of the U.S. Department of Agriculture have primary responsibilities in the area of animal and human health.

1. **The Animal and Plant Health Inspection Service.** This division is charged with maintaining the wholesomeness and safety of meats processed in packing plants that ship meat and meat products, including poultry and poultry products, interstate. Veterinarians and other trained personnel make the inspections. Its purpose is to protect consumers against infected meats and fraudulent and unsanitary preparation of meat products. The inspection first consists of an examination of the live animal so that any unfit beast may be removed and disposed of properly. Secondly, the carcasses and internal organs are inspected for any abnormalities of animals carrying infectious diseases. Centers of infection sources may be located, thus assisting the livestock owners in the vicinity. The records of meat inspection also serve a useful purpose to the research scientist. (Also, see Chapter 8, section entitled, "Federal Meat Inspection.")

Fig. 6–12. Animal health inspection of meat animals consists of two examinations: (a) live animals, and (b) carcasses and internal organs. This shows live hogs being inspected prior to slaughter. (Courtesy, USDA)

2. **The Labeling and Registration Section.** This section in the USDA has responsibility for the proper labeling and safe use of pesticides. Manufacturers of pesticides must present new products with their proposed labels for approval before they are authorized to sell them. The label must indicate, as a minimum, the following: the name of the product; the active and inactive ingredients, together with the percentage of each, in the formulation; the pest(s) controlled; directions for use—including the method and rate of application; any restrictions to be observed in application and handling; and an antidote—if known.

It is the responsibility of FDA, however, to set legal tolerances for pesticides on or in raw agricultural products. Also, it sets the safe interval between last application of the insecticide and the time of harvest of the crop or the slaughter of the animal.

Thus, through the cooperative supervision of the USDA and the FDA, both the pesticide user and the consumer of the product are safeguarded.

3. **The Veterinary Services Division (VSD).** This division of the USDA is responsible for programs to control and eradicate (if possible), certain diseases of livestock, e.g., brucellosis, tuberculosis, scabies, and hog cholera. It does the following things: conducts nationwide federal-state cooperative programs for the control and eradication of animal diseases; suppresses

spread of disease through control of interstate and international movements of livestock; keeps informed of the overall disease situation nationally and internationally; administers laws to ensure humane treatment of livestock and certain laboratory animals; collects and disseminates information on morbidity and mortality; and provides training for USDA employees and others in related government agencies.

4. **Stockyard Inspection.** With the advent of large public markets, public stockyards inspection was initiated. This is an addition to the regular inspection performed on animals by meat inspectors prior to slaughter. Among the principal diseases for which inspections are made are: anthrax, scabies of cattle and sheep, tick or splenetic fever, hog cholera, and erysipelas of swine.

Not only are the incoming shipments of livestock inspected, but a reinspection is made of outgoing shipments. Tests for tuberculosis and brucellosis are accomplished, and dipping for scabies is performed before shipments are allowed to return to farms and ranches. (Also see Chapter 8, section entitled, "Health-Control Officials.")

## U.S. PUBLIC HEALTH SERVICE (USPHS)

This section of the Department of Health and Human Services is concerned with the prevention and treatment of disease. It works in the areas of vector control, pollution control, and control of communicable diseases of people. A part of this important complex is the National Institute of Health (NIH), which was formed in 1930, and which is composed of the following nine sister institutes: the National Cancer Institute, the National Heart Institute, the National Institute of Allergy and Infectious Diseases, the National Institute of Arthritis and Metabolic Diseases, the National Institute of Dental Research, the National Institute of Mental Health, the National Institute of Neurological Diseases and Blindness (including multiple sclerosis, epilepsy, cerebral palsy, and blindness), the National Institute of Child Health and Human Development, and the National Institute of General Medical Science. In addition to its own research program, the USPHS provides grants for health-related research at many universities and research institutes in the United States.

## STATE VETERINARIANS, SANITARY COMMISSIONS, AND BOARDS

Most states have state veterinarians, or comparable officials, who direct the livestock sanitary and regulatory programs within their respective states. Livestock producers may secure the regulations applicable to the state in which they reside by writing their State Department of Agriculture.

## FOREIGN DISEASE PROTECTION

Distance no longer provides a buffer against the invasion of foreign diseases. More than 90% of animals imported into the United States arrive by air. A jet plane can outpace the development of clinical signs of disease in an animal that has been exposed to infection just prior to shipment. This prompts great concern for epizootic diseases capable of crippling or destroying entire livestock populations. Such diseases still exist in Asia, Africa, and Latin America; among them, dreaded diseases such as rinderpest, contagious bovine pleuropneumonia, foot-and-mouth disease, African horse sickness, African swine fever, Newcastle disease, fowl plague, trypansomiasis, East Coast fever, and piroplasmosis.

Until 1875, the importation of livestock into the United States was free and easy. But, that year the United States prohibited the importation of cattle and hides from Spain, where foot-and-mouth disease was rampant. By 1880, European countries were refusing to buy cattle or beef from the United States, for fear of getting contagious bovine pleuropneumonia. Then, in 1884, Congress established the Bureau of Animal Industry in the U.S. Department of Agriculture and gave the Secretary of Agriculture authority to enforce quarantine laws.

Today, there are stations at several entry points, where inspectors of the U.S. Department of Agriculture's Veterinary Service Division inspect all animals and poultry to be imported to the United States. If no communicable diseases are found, the animals may be quarantined for a period of time, during which time they are treated for external parasites and subjected to various tests—e.g., horses are tested for glanders and cattle are tested for brucellosis. At the end of the quarantine period, if no communicable diseases are found, they are released to the purchaser.

The Veterinary Service Division is also charged with the responsibility of safeguarding against diseases introduced by the importation of zoo animals into this country. Wild animals brought into this country must undergo an extensive quarantine period abroad, followed by a further quarantine period at the animal quarantine station at Clifton, New Jersey. Moreover, they are allowed to go only to certain approved zoos, where the zoo animals are isolated from domestic livestock and where proper measures are taken to dispose of waste to prevent the spread of diseases.

## QUARANTINE

Fig. 6–13. Cow with foot-and-mouth disease, a dreaded foreign disease, which has been kept out of the U.S. by extreme precautions, such as quarantine at ports of entry. Note that the animal is reluctant to stand because of sore feet. Note, too, the profuse flow of saliva caused by blisters in the mouth. (Courtesy, USDA)

Many highly infectious diseases are prevented by quarantine from (1) gaining a foothold in this country, or (2) spreading. *By quarantine is meant (1) segregation and confinement of one or more animals in the smallest possible area to prevent any direct or indirect contact with animals not so restrained; or (2) regulating movement of animals at points of entry.*

Many highly infectious diseases are prevented by strict enforcement of local quarantine at points of entry. By such means rinderpest, surra, and other diseases have never gained a foothold in the United States. When an infectious disease outbreak occurs, drastic quarantine must be imposed to restrict movement out of an area or within areas. The type of quarantine varies from one involving a mere physical examination and movement under proper certification to the complete prohibition against the movement of animals, produce, vehicles, and even human beings.

### HARRY S. TRUMAN ANIMAL IMPORT CENTER

A Federal Quarantine Center was authorized in Public Law 91–239, signed by the President on May 6, 1970. A 16.1-acre site for the Center was selected at Fleming Key, near Key West, Florida. In 1978, the Center was named the Harry S. Truman Import Center in honor of former President Truman; and in 1979, the Center opened under the administration of the USDA's Animal and Plant Inspection Service.

The Center holds some 400 head of cattle or other animals at one time, for a 5-month quarantine period. This maximum security station enables American livestock producers to import breeding animals from all parts of the world, while at the same time safeguarding our domestic herds and flocks from such diseases as foot-and-mouth disease, rinderpest, piroplasmosis, and others.

## INDEMNITY PAYMENTS IN DISEASE ERADICATION

Where certain animal diseases are involved, producers may obtain financial assistance in eradication programs through federal and state sources.

Information relative to indemnities paid to owners by the federal government for animals disposed of as a result of outbreaks of certain diseases is contained in Title 9, Code of Federal Regulations, Sub-chapter B, Parts 51 through 56, copies of which can be obtained from the U.S. Department of Agriculture. These reports are entitled as follows:

1. Part 51—"Cattle Destroyed Because of Brucellosis (Bang's Disease), Tuberculosis, or Paratuberculosis."

2. Part 52—"Dourine in Horses and Asses."

3. Part 53—"Foot-and-Mouth Disease, Pleuropneumonia, Rinderpest, and Certain Other Communicable Diseases of Livestock or Poultry."

4. Part 54—"Animals Destroyed Because of Scrapie."

5. Part 55—"Cattle Destroyed Because of Anaplasmosis (Hawaii Only).

6. Part 56—"Swine Destroyed Because of Hog Cholera."

Regulations pertaining to each state can be secured by writing to the respective State Department of Agriculture.

### ANIMAL DISEASE RESEARCH

There is a close relationship between the health of animals and that of humans; between the health of farm animals, and the profits realized by those who produce them; and finally, between animal health and the availability and cost of animal products—meat, milk, eggs, and wool/mohair. In the future, much research will be conducted in genetic manipulation to produce animals (1) which are resistant to disease and parasites; and (2) which also possess superior reproductive efficiency and efficiency of feed conversion, and produce high-quality meat, milk, eggs, and wool/mohair.

Like many experimental studies, animal disease research is best approached as a team effort, involving bacteriologists, biochemists, biological scientists, nutritionists, pathologists, physiologists, veterinarians, and virologists.

Figs. 6–14 to 6–19 picture some of the new and exciting animal disease research studies currently in progress.

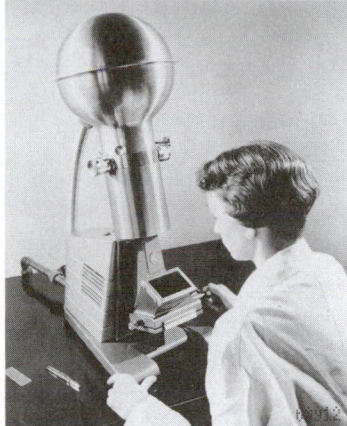

Fig. 6–14. An electron microscope. This instrument is used to study structural characteristics of microorganisms and tissue cells. This searchlight of science helps scientists to see disease-causing viruses in complex with natural antibodies. The electron microscope enables scientists to observe objects too small to be seen through conventional microscopes. (Courtesy, Radio Corporation of America, Camden, NJ)

Fig. 6–15. Inoculating chicken embryos through the eggshell against Marek's disease, a serious poultry virus. Researchers have found that chicks vaccinated in the eggshell have eight times more resistance to the disease than those vaccinated after hatching. (Courtesy, USDA)

Fig. 6–16. Two 8-week-old chickens vaccinated and exposed to Marek's disease. The difference: The bird on the left was vaccinated a few hours after hatching. The one on the right was vaccinated through the eggshell before hatching. (Courtesy, USDA)

Fig. 6–17. Injecting hybridoma-produced coccidia antibodies into poultry. This work is aimed at laying the groundwork for the commercial development of a variety of vaccines against animal diseases. (Courtesy, USDA)

Fig. 6–18. A chick being treated with D-mannos, a natural sugar that tends to block the ability of *Salmonella typhimurium* to take hold in the bird's intestine. (Courtesy, USDA)

Fig. 6–19. The use of monoclonal antibodies in serological tests for trichinosis-infected swine. Scientists are using monoclonal antibodies to probe the mysteries of immunity. (Courtesy, USDA)

# QUESTIONS FOR STUDY AND DISCUSSION

1. Why have diseases and parasites become of greater concern with animal domestication?

2. The author estimated that disease, parasites, and pests caused animal losses totalling $11.4 billion in 1987. In footnote No. 1, the author detailed the bases for the estimate. Challenge the author's two methods of arriving at this estimate. If you disagree, present a better basis for estimating annual livestock losses due to diseases, parasites, and pests.

3. Present facts and figures portraying the appalling losses suffered by American farmers and ranchers due to diseases, parasites, and other inefficiencies.

4. List and describe the signs of good animal health.

5. Define the term *animal disease*. How does the chance for an animal to become infected with a disease compare today with a generation ago?

6. List a minimum of five bases for classifying diseases.

7. List the principal types of infectious organisms that cause disease and give for each (a) the properties, and (b) some infections with which associated.

8. Define the terms *parasite* and *host*. List five ways in which parasites may do harm.

9. List and discuss five effective measures for preventing losses from poisonous plants.

10. List and discuss the common modes for spreading diseases.

11. How do pathogens enter the body?

12. List the most dreaded diseases that are transmitted from animals to humans, then tell how people may prevent them.

13. What is the tsetse fly? It is estimated that if the tsetse fly were brought under control, the Savannah pastures could carry 140 million more cattle. So, why hasn't this parasite been exterminated?

14. List and discuss human diseases that are transmissible to animals.

15. List and discuss the diseases for which animals serve as passive carriers.

16. What techniques are commonly used in diagnosing disease?

17. List and discuss the pertinent things that a livestock producer should know about (a) immunity; (b) vaccination; and (c) biologics, including vaccines, bacterins, and serums.

18. Discuss the pertinent things that a livestock producer should know about each of the following, from a disease prevention standpoint: (a) ventilation; (b) housing; (c) manure disposal; (d) pasture rotation; (e) carcass disposal; and (f) disinfectants.

19. Why don't we develop animals that are genetically resistant to diseases, much like plant breeders have developed varieties of wheat that are resistant to rust?

20. List the federal agencies that have chief responsibility for disease control and give the main area(s) of responsibility of each.

21. Sometimes folks accuse livestock producers of keeping out imports of live animals and meat as a means of lessening competition. Do you agree or disagree with this accusation? Justify your answer.

22. Give a brief history of federal animal disease eradication.

23. What are the primary responsibilities of the Food and Drug Administration (FDA)?

24. List the four divisions of the U.S. Department of Agriculture which have primary responsibilities in the area of animal and human health; and tell what each of them is responsible for.

25. What are the primary responsibilities of the U.S. Public Health Service (USPHS)?

26. Obtain the following publication from the U.S. Department of Agriculture, Animal and Health Inspection Service, Veterinary Services, Federal Center Building, Hyattsville, Maryland: Title 9, Code of Federal Regulations, Sub-chapter B, Parts 51 through 56. Also write to your State Department of Agriculture for information about indemnity payments. Then determine the indemnity payments that you could expect were you to encounter an outbreak in cattle of (a) brucellosis, or (b) foot-and-mouth disease.

27. Assume that a specific contagious disease (you name it) has broken out in your herd or flock. What steps would you take to meet the situation? (List in 1, 2, 3 order; be specific.)

28. Assume that a specific parasite (you name it) has broken out in your herd or flock. What steps would you take to meet the situation? (List in 1, 2, 3 order; be specific.)

29. Justify meat inspection as a federal and/or state expense, as opposed to letting such inspections be handled by the individual processor.

30. Select (a) a specific farm or ranch (either your own or one with which you are familiar), and (b) a specific class of farm animals (beef cattle, dairy cattle, sheep, swine, poultry, or horses); then, outline (in 1, 2, 3 order) a program of animal health, disease prevention, and parasite control.

31. Do you feel that animals that are more resistant to disease and parasites will be developed by genetic manipulation in the decades ahead?

## SELECTED REFERENCES

| Title of Publication | Author(s) | Publisher |
|---|---|---|
| Animal Agents and Vectors of Human Disease | E. C. Faust | Lea & Febiger, Philadelphia, PA, 1956 |
| Animal Disease and Human Health | J. H. Steele | Food and Agriculture Organization of the United Nations, Rome, Italy, 1962 |
| Animal Diseases: Yearbook of Agriculture, 1956 | Edited by A. Stefferud | U.S. Department of Agriculture, Washington, DC, 1956 |
| Animal Health: A Layman's Guide to Disease Control | J. K. Baker W. J. Greer | The Interstate Printers & Publishers, Inc., Danville, IL, 1980 |
| Animal Health Livestock and Pets, 1984 Yearbook of Agriculture | Staff | U.S. Department of Agriculture, Washington, DC, 1984 |
| Animal Parasitism | C. P. Read | Prentice-Hall, Inc., Englewood Cliffs, NJ, 1972 |
| Contagious Bovine Pleuropneumonia | J. R. Hudson | Food and Agriculture Organization of the United Nations, Rome, Italy, 1971 |
| Control of Ticks on Livestock, The | S. F. Barnett | Food and Agriculture Organization of the United Nations, Rome, Italy, 1968 |
| Current Veterinary Therapy | J. L. Howard, Editor | W. B. Saunders Company, Philadelphia, PA, 1981 |
| Diseases of Livestock, Sixth Edition | T. G. Hungerford | Angus & Robertson, Ltd., Sydney, Australia, 1967 |
| Diseases of Poultry, Eighth Edition | M. S. Hofstad, Editor | Iowa State University Press, Ames, IA, 1984 |
| Diseases of Swine, Fifth Edition | A. D. Leman, et al., Editors | Iowa State University Press, Ames, IA, 1981 |
| Diseases Transmitted from Animals to Man, Sixth Edition | Edited by W. T. Hubbert, et al. | Charles C. Thomas, Publisher, Springfield, IL, 1975 |
| Disinfection, Sterilization, and Preservation | C. A. Lawrence S. S. Block | Lea & Febiger, Philadelphia, PA, 1968 |
| Emerging Diseases of Animals | Veterinary Research Laboratory, Onderstepoort, South Africa | Food and Agriculture Organization of the United Nations, Rome, Italy, 1968 |
| Hagan's Infectious Diseases of Domestic Animals, Sixth Edition | D. W. Bruner J. H. Gillespie | Cornell University Press, Ithaca, NY, 1973 |
| Handbook of Pest Management in Agriculture, Vol. 1 | D. Pimentel | CRC Press, Inc., Boca Raton, FL, 1981 |
| Livestock Health Encyclopedia, Third Edition | R. Seiden | Springer Publishing Co., Inc., New York, NY, 1968 |
| Manual of Tropical Veterinary Parasitology | M. Shah-Fischer R. R. Say | CAB International, Wallingford, U.K., 1989 |
| Merck Veterinary Manual, The, Sixth Edition | Edited by O. H. Siegmund | Merck & Co., Inc., Rahway, NJ, 1973 |
| Nationwide System for Animal Health Surveillance, A | National Research Council | National Academy of Sciences, Washington, DC, 1974 |
| Practical Parasitology: General Laboratory Techniques and Parasitic Protozoa | C. J. Price J. E. Reed | United Nations Development Programme, and Food and Agriculture Organization of the United Nations, Rome, Italy, 1970 |
| Preventive Medicine and Public Health, Ninth Edition | Edited by P. E. Sartwell | Appleton-Century-Crofts, New York, NY, 1965 |
| Principles of Veterinary Science, Fourth Edition | F. B. Hadley | W. B. Saunders Company, Philadelphia, PA, 1949 |
| Some Diseases of Animals Communicable to Man in Britain | Edited by O. Graham-Jones | Pergamon Press, Ltd., London, England, 1968 |

## SELECTED REFERENCES *(Continued)*

| Title of Publication | Author(s) | Publisher |
|---|---|---|
| *Some Important Animal Diseases in Europe: Papers Presented at the Animal Disease Meeting, Warsaw, 1948* | K. V. Kesteven | Food and Agriculture Organization of the United Nations, Rome, Italy, 1952 |
| *State of Food and Agriculture, The* | Staff | Food and Agriculture Organization of the United Nations, Rome, Italy, 1987–88 |
| *Stockman's Handbook, The,* Sixth Edition | M. E. Ensminger | The Interstate Printers & Publishers, Inc., Danville, IL, 1983 |
| *Veterinary Medicine* | D. C. Blood<br>J. A. Henderson | The William & Wilkins Co., Baltimore, MD, 1960 |
| *Veterinary Parasitology,* Second Edition | G. Lapage | Charles C. Thomas, Publisher, Springfield, IL, 1963 |

In addition to the above selected references, valuable publications on different subjects pertaining to animal diseases, parasites, disinfectants, and poisonous plants can be obtained from the following sources:

1. Division of Publications, Office of Information, U.S. Department of Agriculture, Washington, DC, 20250
2. Your state agricultural college
3. Several biological, pharmaceutical, and chemical companies

Isolation stalls—a distinct asset in the control of infectious diseases. (Courtesy, College of Veterinary Medicine, University of Illinois, Urbana)

Bright eyed and bushy tailed! (Courtesy, Ruth White, White Horse Ranch, Naper, NE)

# ANIMAL BEHAVIOR/ ENVIRONMENT

## Chapter 7

Animal behavior and environment are interrelated. A good understanding of each of them will enhance the productivity of animals.

Animal behavior is the reaction of the animal to certain stimuli or its environment.

Animal environment is all of the conditions, circumstances, and influences surrounding and affecting the growth, development, and production of animals.

# PART I. ANIMAL BEHAVIOR

Those of us who remember the last round of the barn at night learned the marvels and mysteries of barnyard talk. The animals kept up a running conversation with us as we gave a handful of hay to one still-hungry cow, treated a favorite horse to an apple, took a last look at a litter of newborn pigs, closed the gate on the sheep corral, and shut the poultry house door. It was a sign language, but it spoke louder than words; it told us how the animals felt and what they wanted. Every movement and every sound conveyed a message of well-being, distress, or disease. Lack of interest, dull eyes, sluggishness, rough coat, poor appetite, and/or abnormal droppings spelled trouble. Alertness, stretching and rising, yawning, vocalizing, eating with relish, and frisking were good omens and told us that all was well in the barnyard. Whether we come from farms and hold affectionate remembrances of our animal companions or are city-bred and know them secondhand, animal behavior is one of the marvels and mysteries of life.

Human beings have always had to know something about the behavior of the animals around them. Without such knowledge, they would have become extinct long ago; their game would have eluded them, pests would have devoured their crops, and scavengers would have taken away their food stores.

The dog appears to have been the very first animal domesticated by humans. It took place in the Old Stone Age, at least 12,000 years ago,[1] long before recorded history.

Domestication probably took place through the wolves, from which dogs stem, being attracted to human camps by surplus meat wasted in times of plenty; through cubs being taken as pets by children of the tribe; and through semitame wolves (pets that retreated to the wild as they grew up) joining in the hunt with people. But why did prehistoric people choose dogs and certain other animal species for domestication from among the hundreds of thousands of species that were available? Why were not more animals domesticated in prehistoric times? Why are most young animals that are captured today and retained as pets for a short time never truly domesticated? The answers to these questions are a perplexing scientific puzzle, clues to the solution of which can be obtained by studying the behavior of our farm animals and household pets.

Written observations of animal behavior date to the writings of the ancient Greeks, Aristotle in particular, about 350 B.C. Of course, people had tamed, or domesticated, some animals much earlier, notably the dog, the cow, and the sheep. Prior to animal domestication, the very survival of the human race depended upon knowledge of animal habits and habitats as people hunted for their food. But the behavioral information needed—first to hunt the game, and later to domesticate animals—did not assume primary scientific significance for many years after the writings of Aristotle and of subsequent hunters, explorers, naturalists, and agriculturalists. Finally, in two classical books—*The Origin of Species by Means of Natural Selection,* published in 1859; and *The Descent of Man and Selection in Relation to Sex,* published in 1871, Charles Darwin laid the foundation for modern animal behavior. For many years thereafter, however, conditions were ripe for animal behavior practitioners to turn a "quick buck" as they made all sorts of claims for the reasoning powers of their charges. The most notable show-on-the-road of this type involved *Clever Hans,* a horse in Germany, about 1900, billed as the wonder horse who could add, multiply, divide, and even spell out words and sentences. *Hans* would stand in front of its trainer, and by pawing the ground with a hoof the appropriate number of times, answer questions. If asked, "How much is 2 + 2? *Hans* would paw the ground four times. If asked to spell out words or sentences, *Hans* would paw the proper number of times for each letter of the alphabet. Finally, a committee of scientists was appointed to study the celebrated horse. They found that Hans did, indeed, paw out the correct answers. But, close observation revealed that the trainer cued the horse through a slight movement of the head. Hence, by watching the trainer, the horse would always stop when it observed the head signal. Both horse and trainer were amply rewarded; the horse by treats and affection, and the trainer by another stellar performance before a large and satisfied audience. Both maintained their behavior.

Despite some charlatans along the way, people applied animal behavior from the remote day of animal domestication forward. It required knowledge of basic behavior patterns to capture, confine, and herd animals, as did breeding, feeding, watering, and sheltering them. Without this understanding, domestication would have failed and animals would not have survived. By 1900, the groundwork had been laid for the scientific work that followed; and the study of animal behavior became a distinct discipline. In recent years, it has advanced rapidly.

*Animal behavior is the reaction of the whole organism to certain stimuli, or the manner in which it reacts to its environment.* Through the years, behavior has received less attention than the quantity and quality of meat, milk, eggs, power, and fiber produced by animals. But modern breeding, feeding, and management have brought renewed interest in behavior, especially as a factor in obtaining maximum production and efficiency. With the restriction, or confinement of herds and flocks, many abnormal behaviors evolved to plague those who raise them, including cannibalism, loss of appetite, stereotyped movements, poor paternal care, overaggressiveness, dullness, degenerate sexual behavior, tail biting, cribbing, and a host of other behavioral disorders. Confinement has not only limited space, but it has interfered with the habitat and social organization to which, through thousands of years of evolution, the species became adapted and best suited.

We now know that controlled environment must embrace far more than an air-conditioned chamber, along with ample feed and water. The producer needs to concern himself more

---

[1]Reed, C. A., "Animal Domestication in the Prehistoric Near East," *Science,* Vol. 130, No. 3389, December 11, 1959, pp. 1629–1639. (Subsequent to the publication of this book, Dr. Reed and Priscilla Turnbull have discovered an authentic jawbone of a dog in a cave in Iraq, estimated to be 12,000 years old, which is the oldest known remains of a dog.)

with the natural habitat of animals. Nature ordained that they do more than eat, sleep, and reproduce. For example, studies on the behavior of swine show that they spend much of their day in active investigative behavior, primarily rooting and manipulating their movement. When free ranging, pigs may spend 40% of their day resting, 35% investigating novel surroundings, 15% eating, and 10% in other activities. What happens when pigs are confined in a building on slotted floors? How is the neutral energy dissipated that would normally be used to satisfy the drives for investigating and rooting?

Evidently, environmental deficiencies are manifested by tail biting, gastric ulcers, poor maternal care and loss of young, or other physiological functions resulting in a sudden death syndrome or tissue degeneration.

This chapter is for the purpose of presenting some of the principles and applications of animal behavior. Those who have grown up around farm animals and dealt with them in practical ways have already accumulated substantial workaday knowledge about animal behavior. Those with urban backgrounds may need to familiarize themselves with the behavior of animals, better to feed and care for them and to recognize the early signs of illness. To all, the principles and applications of animal behavior depend on understanding, which is the intent of this chapter.

## CAUSES OF ANIMAL BEHAVIOR

Animal behavior is caused by, or is the result of, three forces: (1) genetic, (2) simple learning (training and experience), and (3) complex learning (intelligence).

## GENETIC

The relation of breeding and selection to behavior becomes obvious in a group of weaning foals of mixed breeding. Upon racing across a field, some amble off in a rhythmic running walk, nodding their heads as they go; others travel high enough to clear the tops of daisies; still others break away in an easy gallop. Each of these three actions is executed with equal ease and naturalness. The first weanlings described are Tennessee Walking Horses, the second are Hackneys, and the third are Thoroughbreds. In each of these breeds, the distinctive way of going has been accomplished through years of breeding and selection—through heredity.

## SIMPLE LEARNING (TRAINING AND EXPERIENCE)

No horse—whether it be used for saddle, race, or other purposes—reaches a high degree of proficiency without an education. Thus, if the offspring of Man O'War and six of the fastest mares ever to grace the tracks had merely worked on laundry trucks until six years old, then if they were suddenly—without warning or other preparation—placed upon a racetrack, the immediate results would have been disappointing. Their natural aptitude and conformation in breeding would not have been enough. Schooling and training would still have been necessary in order to bring out their inherent abilities.

In general, the behavior of animals depends upon the particular reaction patterns with which they were born. These are called *instincts* and *reflexes*. They are unlearned forms of behavior. Thus, all horses instinctively like to run. But how well and how fast they run depends upon the training to which they

are subjected. They learn by experience. However, the training is only as effective as the inherited neural pathways will permit. Several types of learning processes are known; among them are those that follow.

## HABITUATION

*Habituation is getting used to, or ignoring, certain stimuli.* A classical example is the response of young turkeys (poults) to the danger call of the mother when there is impending danger—like the approach of a hawk. At first, they scamper for cover or to the corners of their pen, pile in clumps, and remain motionless (freeze) for several minutes. However, they gradually get used to a repetition of this call, with each response becoming less intense and producing a shorter freeze. By the time they're two months old, they practically ignore such an alarm—they merely come to attention.

Bunk breaking calves is another example. If calves are weaned without prior bunk breaking, then suddenly transferred to a cattle feedlot where there is no milk or grass, and where their feed must be obtained from a bunk, it is a traumatic experience for them. This is so because (1) there is a mother-young separation reaction, (2) they get homesick (and animals do get homesick), and (3) there is a change in feed and water. On the other hand, if they have been preconditioned, as well as bunk broken, prior to weaning, they take to the new feed bunk in the feedlot because they are used to it.

## CONDITIONING (OPERANT CONDITIONING)

*Conditioning is the type of learning in which the animal responds to a certain stimulus.* For example, upon hearing the rolling of a barn door a cow may lick her tongue and moo, even though she can see no feed; and upon hearing the rattle of a milk pail, she may let down her milk.

Artificial insemination techniques have been developed around the understanding of normal reproductive behaviors and the modifications of these behaviors. Semen collection routines are faced by behavioral responses that can change from impotence to optimum performance and high-quality semen. Proper stimulation of some bulls, for example, can increase sperm cell output by nearly 40%, compared to ejaculates after minimum stimulation.

Another example of conditioning is the use of an electric fence. When an electric fence is installed, the immediate instinct of animals is to investigate—to touch it with their noses. Upon receiving a shock, they back off and let it alone. Thereafter, the electricity can be shut off for a considerable period of time before some animal again tests it.

*Operant conditioning, or operant learning, refers to animal operation of some aspect of the environment to obtain access to feed or other animals.* It's the learning of an act that has some consequence; i.e., one that operates the environment—like pressing a bar that supplies some feed or turns off a light. An example of natural operant learning is the lifting of the cover of a self-feeder by pigs.

Broadly speaking, training is operant conditioning—it's an attempt to modify an animal's behavior. There are two types of training: (1) reinforced training, usually with positive rewards, and (2) forced training in which the animal is compelled to do certain things. Thus, the training of horses is best accomplished through the judicious employment of rewards and punishments. This doesn't mean that the animal is rewarded each time it obeys, or that it is beaten when it refuses to do something.

But horses are big and strong; hence, it's best that they want to do something, rather than have to be forced. Also, too frequent or improper use of such artificial aids as whips, spurs, reins, and bits makes them less effective; worse yet, it will likely make for a mean horse. However, horses appreciate a pat on the shoulder or a word of praise. Even better results may be obtained by working on an equine's greediness—its fondness for such things as carrots or a sugar cube. Also, treats may be used effectively as rewards to teach some specific thing such as posing, or to cure a vice like moving while the rider is mounting; but this should not be overdone.

### INSIGHT LEARNING (REASONING)

*Insight learning is the sudden adaptive reorganization of experience or sudden production of a new adaptive response not arrived at by overt trial-and-error behavior.* It replaces trial-and-error. Of course, it is difficult to be certain in such cases that the animal did not have a similar type of problem before. Even so, the immediate application of past experience to a new situation is a noteworthy capacity.

This type of learning is most prevalent in the higher animals. Some examples of insight learning are (1) a detour, or barrier, in which the animal must go away from the feed in order to reach it; (2) a chimpanzee obtaining a banana that is out of reach by stacking boxes beneath it and climbing up; and (3) a chimpanzee fitting two sticks together to pull in a piece of food that is out of reach of either stick alone.

The most important single factor to remember in training animals is that none of them (dogs included) can reason things out. An animal's mind functions by intuition, not logic. Moreover, it has no conscious sense of right and wrong. Thus, it is one of the trainer's tasks to teach an animal the difference between right and wrong—between good and bad. Although the animal cannot utilize pure reason, it can remember, and it has the ability to use the memory of one situation as it applies to another.

### IMPRINTING (SOCIALIZATION)

This is a form of early social learning which has been observed in some species. The pioneering work in this field was done by the Austrian zoologist Lorenz, with goslings. He found that if a baby gosling was exposed immediately after hatching to some moving object, especially if the object emitted sound, it would adopt that object as its parent-companion. Further studies revealed that goslings would adopt any other moving object in the same manner—dogs, cats, humans, and so forth. Also, it was found that the same principle applies to other fowl and to mammals. Thus, if a person is present during the critical period of socialization, a puppy may form a long-term association with that individual. Dog kennels use this principle to produce the most desirable pets, work dogs, and guide dogs for the blind.

Apparently, inheritance controls the time and the length of the critical period when an individual can be imprinted, the type of object to which it can be imprinted, the tendency to respond to the first object to which it is exposed, and the permanence of the attachment to the object following imprinting. For example, goslings can usually be imprinted only within the first 36 hours following hatching.

Although human-socialized animals make adorable pets, they are often a nuisance on the farm. People who have hand reared an orphan lamb, foal, or duck have found that such animals seek human companionship and never fit in well with their own kind. Animals that are socialized to people can also be dangerous, especially if they are large and attempt to respond to people in the same manner as they do to other members of their own species. For example, upon reaching maturity a deer that was bottle fed as a fawn may react toward humans in the rutting season as if they were other deer, with the result that a stag may attack a human and inflict serious injury.

### COMPLEX LEARNING (INTELLIGENCE)

*Complex learning (intelligence) is the capacity to acquire and apply knowledge—the ability to learn from experience and to solve problems.* It is the ability to solve complex problems by something more than simple trial-and-error, habit, or stimulus-response modifications. In humans, we recognize this capacity as the ability to develop concepts, to behave according to general principles, and to put together elements from past experience into a new organization.

Animals learn to do some things, whereas they inherit the ability to do others. The latter is often called instinct. Thus, ducks do not have to learn to swim—instinctively, they take to water.

Some folks judge the intelligence of animals by the size of their brain in relation to body size. Others rank them according to their ability to solve a maze (a pathway complicated by at least one blind alley, used in learning experiments and intelligence tests) in order to get food.

Generally speaking, behavioral scientists are agreed that each species has its own special abilities and capacities, and that it should only be tested on these. For example, the dog, pig, and rat, are more adept at solving a maze test than the horse. Hence, solving a maze in order to find food favors the scavengers (and the dog, the pig, and the rat are all scavengers)—they have connived for their food since the beginning of time.

However, horses, whose natural feed was the grass that lay around them, never had to develop this kind of intelligence. They were plains-living animals, highly specialized for speed as a means of escape from their enemies and with almost no powers of manipulation. Thus, horses should be good at any problem that can be solved by running, including racing, polo, pole bending, calf roping, etc. Indeed, had equines not been smart and adapted to their particular environment, they would never have made it through 58 million years.

Thus, each species is uniquely adapted to only one ecological niche. Moreover, a niche is filled by the particular species that can solve food finding therein, and that is best adapted under the conditions that prevail. It follows that intelligence comparisons between species are not meaningful, and that it is absurd to say that one species is smarter than another.

Of course, human intelligence is generally recognized. In fact, were it nor for their superior mental faculties, along with their limited muscular force, they might find themselves under the saddle or between the shafts, instead of the horse.

## METHODS OF ANIMAL COMMUNICATION

Communication between animals involves giving off by one individual of some chemical or physical signal, which, on being received by another, influences its behavior. Communication between animals need not be visual or auditory, nor must it be confined to the present.

Without doubt, this trait accounts, in part at least, for the foundation horse stock of the American Indians and the hardy bands of Mustangs—the feral horses of the Great Plains. In some mysterious manner, the abandoned and stray horses of the expeditions of de Soto and Coronado communicated and found each other; otherwise, they could not have reproduced.

## SOUND

Sound communications is of special interest because it forms the fundamental basis of human language. The gift of language alone sets humans apart from the rest of the animals and gives them enormous advantages in adaptation to their environment and in their social organization.

Sound is also an important means of communication among animals. They use sounds in many ways; among them, (1) feeding, in sounds of hunger by young, or food finding, and of hunting cries; (2) distress calls, which announce the approach or presence of an enemy and the all-clear signal following the departure of a predator; (3) sexual behavior, courting songs, and related fighting; (4) mother-young interrelations to establish contact and evoke care behavior; and (5) maintaining the group in its movements and assembly.

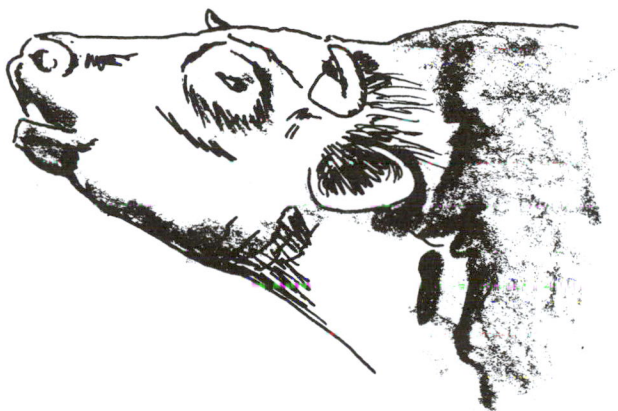

Fig. 7–1.  Communicating by sound—a bellowing bull.

Animals have a very acute sense of hearing, perceiving higher and fainter noises than the human ear.

## CHEMICAL

In mammals, females in estrus secrete a substance that attracts males. Hence, males locate females that are in heat by the sense of smell.

On the range, it has been observed that each stallion usually stakes out a territory for himself and his harem of mares, with the outside boundary thereof marked by his feces.

The sense of smell is perhaps the dog's most important weapon. It is much more highly developed than that of the human being. Police departments report that the dog's sense of smell is much greater than that of humans. Scent can be used for tracking by the direct scent of the human body, by ground scent made from human sweat deposited on the ground from feet or by footwear; and by scent, which is more persistent, created by pressure on the ground of the human foot or of any heavy article. Consequently, a police dog brought to

the scene of the crime can pick up and follow a ground scent, or can be trained to search a wide area quickly and discover persons hiding in the vicinity by picking them up by direct scent, even if they are concealed in dense vegetation or up a tree. Equally good results can be obtained in the search of large premises. Moreover, darkness is no handicap for this type of dog work. It is noteworthy, too, that the dog's keen sense of smell makes it invaluable in tracking down marijuana.

## VISUAL DISPLAYS

Birds are noted for their sexual behavior in the act of courtship. In chickens, the male typically takes the initiative in sexual behavior. There are a variety of approaches by which a cock may evoke a sexual response in the hen. The most spectacular of these is the wing flutter, waltz, or dance. These initial sexual reactions of cocks have been called *courting*. A hen may be indifferent to courting, or she may respond either negatively or positively. As a negative reaction, she may step aside, walk or run away, or struggle if captured. These types of avoidance of the male by the female may be accompanied by vocalizations varying in intensity from faint screams to loud squawks. A positive reaction to courting takes the form of a crouch, often with head low and wings spread. This behavior has been called a *sex invitation*, or *crouching*. The sexual crouch is a strong stimulus for the cock to mount and tread, particularly when the rooster approaches from the rear. The male stands on the outstretched wings, grasps the comb or hackle of the hen, and moves the feet up and down in a treading manner. Subsequently, the male rears up, the hen moves the tail to one side, and each averts the cloaca as vents meet. The male usually steps off in a forward direction and the hen shakes herself vigorously as she gets to her feet. She may run in an arc and the cock may execute a waltz.

Visual displays during courtship are less evident among four-footed animals, but they do occur to some extent.

Most animals will strike a hostile stance when they are excited or prior to fighting. Also, boars bristle—the hair on the top of their necks rises up; this serves to make them look larger and more formidable.

## HOMING AND ORIENTATION (PATH-FINDING, NAVIGATION, MIGRATION)

Through sound, scent, or some sense of which we do not know, cattle, horses, dogs, and cats often find their way back home when moved to distant places.

But the ability of pathfinding is possessed by many species, great and small. Ants that go in groups usually travel along a narrow trail that is chemically saturated by their passing; when one of them comes upon such a trail it is easily identified as the right roadway.

When baby green turtles hatch from their eggs, they must do two things: (1) dig upward through the sand in which the mother deposited the eggs; and (2) head for the water. To guide them for the first action they have a built-in urge to go uphill. But, as a rule, they must go downhill to find water. This immediately prompts the question: What cancels instruction Number 1 and supplies instruction Number 2? Scientists have established the answer—light. Water reflects light from the sky, and it flashes the baby turtle a signal that overrides its uphill orders.

But it is the wide-roaming salmon whose navigational feats are the most fantastic. It lays its eggs inland, usually far up some freshwater stream. During their second year, the young salmon move downstream to the sea, but when they reach sexual maturity, they head back toward their birthplace to spawn. To reach it, they may have to make choices at fork after fork of the waterways. But they get there! It has been well established that salmon use chemical cues to return to the rivers in which they were spawned.

Then there is the migrating and homing of birds for which no single explanation is really satisfying. Each year, an estimated 10 billion American birds, ranging from tiny hummers to giant eagles, sweep in flood tide up the face of North America, sending their ripples into every field and woodlot. It is suspected that birds use the sun and stars as compasses to guide them. But it is one of nature's best-kept secrets, for which there is no complete answer.

Both mammals and birds form attachments to particular places and attempt to return to them if removed. This is important in practical management. For example, an effective reward to a horse in training consists in letting it return to its stall immediately after it has done something right. One noted cutting horse trainer claims to have developed a champion by using this technique.

## HOW ANIMALS BEHAVE

Animals behave differently, according to species. Also, some behavioral systems or patterns are better developed in certain species than in others. Moreover, ingestive and sexual behavioral systems have been most extensively studied because of their importance commercially. Nevertheless, most animals exhibit the following nine general functions or behavioral systems, each of which will be discussed:

1. Ingestive behavior.
2. Eliminative behavior.
3. Sexual behavior.
4. Care-giving and care-seeking (mother-young behavior).
5. Agonistic behavior (combat).
6. Allelomimetic behavior.
7. Gregarious behavior.
8. Shelter-seeking behavior.
9. Investigative behavior.

## INGESTIVE BEHAVIOR

This type of behavior includes eating and drinking; hence, it is characteristic of animals of all species and all ages. It is very important because animals cannot live without feed and water.

The first ingestive behavior trait, common to all young mammals, is suckling.

Each species has its own particular method of ingesting food. The methods pertaining to each of the different animal species may be summarized as follows:

• **Cattle, sheep, and goats**—The grazing habits of these animals are determined by the lack of upper incisors. Thus, cattle wrap their tongues around grass, then jerk their heads forward so that the vegetation is cut by the lower teeth. Sheep graze very much like cattle, but their cleft upper lip allows them to graze vegetation closer to the ground. Goats graze like cattle and sheep, but they are very fond of browse—the young

Fig. 7-2. Ingestive behavior begins with suckling, in all mammals. (Courtesy, Lynn Smith, Troy, NH, reprinted with permission of *Small Farmer's Journal,* Eugene, OR)

shoots of shrubs and trees. Cattle, sheep, and goats are ruminants. Thus, they regurgitate their food and chew it again, in a process known as rumination.

*Rumination is the act of chewing the cud, characteristic of herbivorous animals with split hoofs.* It involves regurgitation of ingesta from the reticulo-rumen, swallowing of regurgitated liquids, remastication of the solids accompanied by reinsalivation, and reswallowing of the bolus. Rumination occupies about 8 hours of the cow's time each day. (In addition, the harvesting or grazing time may take another 8 hours. This means that cows may work a 16-hour day.)

• **Horses and pigs**—Both of these species possess teeth in the upper and lower jaws; hence, they bite off grass or take a mouthful of grain, then chew and swallow it. Space will not permit describing the digestive process in horses and pigs. Suffice it to say, however, that pigs have a single stomach, whereas ruminants have a four-compartment stomach. The horse is somewhat intermediate between ruminants and monogastrics, primarily due to the large *blind gut,* which is the seat of considerable bacterial action.

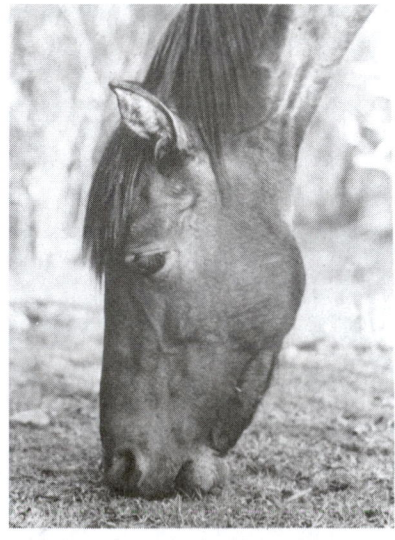

Fig. 7-3. Because of having incisor teeth in both the upper and lower jaws, horses graze vegetation closer to the ground than cattle. Because of their cleft upper lip, sheep also graze close to the ground. (Courtesy, Wild Horse Research Farm, Porterville, CA)

By nature, pigs love to root. If given the opportunity, they will stick their noses into the ground and lift forward and upward, moving earth out of the way and exposing earthworms, grubs, and roots.

- **Poultry**—Chickens and turkeys ingest their food by pecking; ducks scoop their feed with their broad, soft bills. Except for geese (which graze on grass), poultry do not eat much forage.

## ELIMINATIVE BEHAVIOR

In recent years, elimination has become a most important phenomenon, and pollution has become a dirty word. Nevertheless, nature ordained that if animals eat, they must eliminate.

A full understanding of the eliminative behavior will make for improved animal building design and give a big assist in handling manure. Right off, it should be recognized that the eliminative behavior in farm animals tends to follow the general pattern of their wild ancestors; but it can be influenced by the method of management. The eliminative behavior of different species follows:

- **Cattle**—Cattle deposit their feces in a random fashion. Although cows can defecate while walking, with the result that their feces are scattered, generally they deposit their "chips" in neat piles. Most cows hump up to urinate, whereas bulls are inclined to stand squarely on all "fours."

- **Sheep**—The eliminative behavior in sheep is very similar to that of cattle. However, ewes usually assume a squat position when they urinate.

- **Swine**—If given an opportunity, pigs are of very clean habits. They like to keep their bedding area clean and dry. Hence, they usually deposit their feces in a corner of the pen, away from the sleeping quarters. Modern methods of raising pigs in restricted quarters, which are often overcrowded, has disturbed their natural eliminative patterns.

- **Horses**—Horses tend to deposit their feces in certain locations, such as along well-traveled paths, like those leading to waterholes. Hence, if given the opportunity, they often return to these locations.

- **Chickens and turkeys**—Except when on the roosts at night, chickens and turkeys deposit their excreta at random.

- **Dogs**—Dogs tend to excrete their urine at specific locations, known as scent posts. The selection of an upright vertical target, or post, followed by sniffing and leg lifting by adult male dogs at the time of urination has been shown to be a secondary sex characteristic under hormonal control.

- **Cats**—Cats bury their feces and urine.

## SEXUAL BEHAVIOR

Reproduction is the first and most important requisite of livestock breeding. Without young being born and born alive, the other economic traits are of academic interest only. Thus, it is important that all those who breed animals should have a working knowledge of sexual behavior.

Sexual behavior involves courtship and mating. It is largely controlled by hormones, although males that are castrated after reaching sexual maturity (which among farm animals, are known as stags) usually retain considerable sex drive and exhibit sexual behavior. This suggests that psychological, or learned, as well as hormonal factors may be involved in sexual behavior.

Each animal species has a special pattern of sexual behavior. As a result, interspecies matings do not often occur. There are two notable exceptions, however: The best known cross between animal species is the mule, a hybrid, which is a cross between the horse and the ass. Also, when sheep and goats are confined, they readily mate with each other, although such matings are never fertile.

Males in most species of farm animals detect females in heat by sight or smell. Also, it is noteworthy that courtship is more intense on pasture or range than under confinement, and that captivity has the effect of producing many distortions of sexual behavior compared to wild animals. Perhaps this explains the high percentage foal crop of wild bands of mares, where conception and foaling rate of 90% or better were commonplace, in comparison with the average 50% foaling rate under domestication.

Today, livestock producers are attempting to control the sex life of animals, by bringing about ovulation at the time of choice of the owners, rather than of the female.

Additional pertinent information pertaining to the sexual behavior of each animal species follows:

- **Cattle**—Experienced producers can usually detect in-heat cows through one or more of the following characteristic symptoms: (1) restlessness; (2) mounting other cows, and standing to be mounted by another cow (standing heat appears to be the best single indicator of the proper time to breed); (3) a noticeable swelling of the labia of the vulva; (4) an inflamed appearance about the lips of the vulva; (5) frequent urination; (6) switching and raising the tail; and (7) a mucous discharge. A day or two following estrus, a bloody discharge is sometimes seen. Dry cows and heifers usually show a noticeable swelling or enlargement of the udder during estrus, whereas in lactating cows a rather sharp decrease in milk production is often encountered. When kept alone, some cows become restless, walk the fence, and bawl when they are in heat. Some may even jump the fence, or go through it, as they attempt to find a bull.

A bull can often detect a cow that is coming in heat 24 to 48 hours before she will mate, at which time he will remain in her company. Courtship of the bull consists of following the in-heat cow; licking and smelling the external genitalia, with the head extended horizontally and the lip uncurled; and chin-resting, with the chin and throat resting on the cow's rump.

Fig. 7-4. Courtship of the bull, showing chin-resting, in which the chin and throat are rested on the cow's rump.

- **Sheep**—Unlike other farm animals, the ewe shows few visible external indications of heat. The acceptance of the ram (or of a teaser with an apron) is the best method of detection.

Ovulation seems to occur late in the heat period, usually from about 24 to 30 hours after the onset of estrus.

In sheep, the display of sexual behavior of the male is more elaborate than that of the female. Typically, the ram responds to the urination of a ewe in estrus by sniffing the urine, then extending his head with his lips upcurled. He sticks his tongue in and out of his mouth as he follows the ewe, noses her external genitalia, and rubs along her side, biting her wool. A characteristic part of the sexual display, or teasing, by the ram is the raising and lowering of one front leg in a stiff-legged striking motion.

Fig. 7–5. Courtship in sheep, showing ram biting ewe's fleece.

• **Swine**—The external signs of heat in the sow are restless activity, swelling or enlargement of, and discharge from, the vulva (although these signs are not always present), mounting of other sows, frequent urination, and occasional loud grunting.

The boar often nudges the sow or gilt around the head or in the flanks with his head and nose and emits a courting song. He will then attempt to mount her.

Fig. 7–6. Courtship in swine, showing male nudging flank. (The boar also vocalizes with a courting song.)

• **Horses**—The signs of estrus in the mare are (1) the relaxation of the external genitalia; (2) frequent urination in small quantities; (3) the teasing of other mares; (4) the apparent desire for company; (5) a slight mucous discharge from the vulva; (6) allowing the stallion to smell and bite her; (7) spreading the hind legs; and (8) lifting the tail sideways. But many mares are shy breeders. Thus, when there is any question about a mare

being in season, she should be tried with the stallion. When possible, it is usually good business regularly to present mares to the teaser every day or every other day as the breeding season approaches. A systematic plan of this sort will save much time and trouble.

The courtship of the stallion is characterized by neighing; smelling the external genitalia of the mare, followed by extended head and upcurled upper lip; and pinching the mare in the croup area with his teeth.

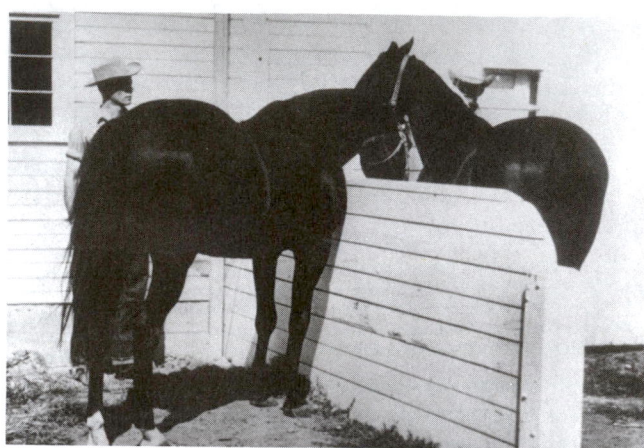

Fig. 7–7. Teasing a mare, using a solid fence for separation. (Courtesy, Washington State University)

• **Poultry**—The sexual behavior of chickens is illustrated in Fig. 7–8.

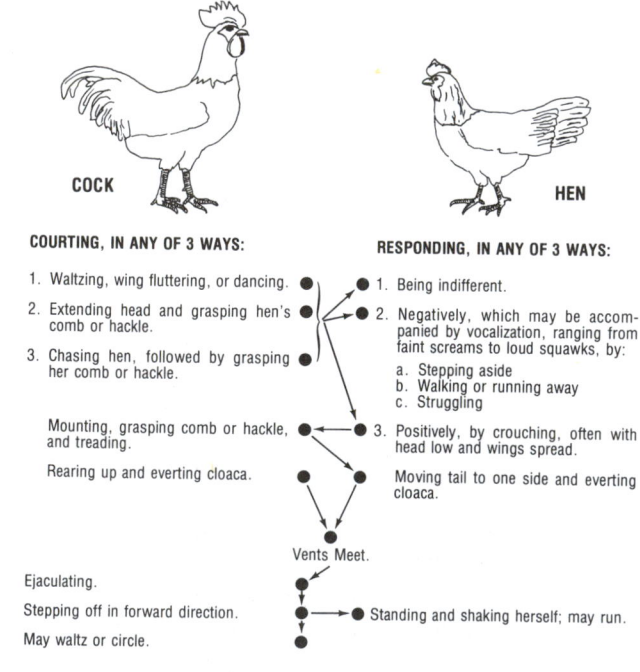

Fig. 7–8. Mating behavior in chickens, showing chain reaction between rooster (left) and hen (right).

The mating behavior in turkeys tends to follow a chain reaction similar to chickens. That is, the behavior of one sex partner elicits a specific response from the other, and, in turn, that response elicits a further response from the first partner.

Geese of the larger breeds mate best in pairs or trios, although ganders of some lighter breeds will mate satisfactorily with as many as five females. Canadian wild geese are largely monogamous and will usually mate that way for life.

The breeding habits of some other species of poultry follow:

1. **Guinea fowls** have a tendency to mate in pairs; however, one male may be mated with 3 or 4 hens.

2. **Pigeons** mate in pairs and usually remain with their mates throughout life, although the mating may be changed if desired by placing the male and female in a coup together, leaving them there for 6 to 14 days or until they become settled.

3. **Peafowl** hens are usually grouped with 4 or 5 hens to one cock bird.

4. **Swans** live in pairs and remain faithful to each other until death.

## CARE-GIVING AND CARE-SEEKING (MOTHER-YOUNG) BEHAVIOR

The care-giving behavior is largely confined to females among domestic animals, where it is usually described as *maternal*; the care-seeking behavior is normal for young animals. It begins shortly before birth and extends until the young are weaned. Care-giving and care-seeking vary widely among the different species of farm animals. These differences are pointed up in the summary that follows:

• **Cows**—Nature ordained that cows seek isolation at calving time. So, where possible, they will hide out.

Following birth, the excitement of the new mother becomes evident almost immediately. She gets up and begins to dry her newborn calf by licking it. Simultaneously, some cows "talk" to their newborn. They may become quite concerned and nervous as their "baby" first attempts to stand, takes a few footsteps—and falters. Aided by its mother's licking, and encouraged by her "talking," eventually the calf makes it to

Fig. 7–9. Care-giving and care-seeking. The mother-young bond is a remarkably strong one.

its unsteady feet and commences to search for a teat.

A newborn calf cannot see too well, but it can smell, touch, and taste. It associates everything that is good and that cares for it with its mother. This is the beginning of the herd instinct.

If on pasture, the new mother usually hides her calf. During the first day or two, the calf sleeps a great deal, while the mother grazes nearby. But a mother takes great pains not to disclose the hiding place of her calf. At intervals, she returns to feed it. If it is necessary for her to leave her calf in order to get water or supplemental feed, she does not tarry much along the way. Frequently, where there are a number of newborn calves, the cows "baby-sit" for each other. Part of the cows will leave for feed or water, but one or two will remain behind and guard all the calves. Then, when the first cows to leave have returned, the "baby-sitters" will take their turn and depart. In this manner, there are older cows with calves at all times.

When a calf in hiding is approached by a human, it will usually lie as close to the ground as possible without movement except for its eyes. If picked up, and if scared, it may bawl (cry) for its mother. If the mother hears the call, she will come running—often ready to fight. Frequently, other cows in the vicinity, especially if they have calves of their own, may join in the response. If the disturbed calf runs away, it will return to the area after the danger has passed.

By the time the calf is two days old, the mother wanders more extensively, with the calf at her side. Soon, they rejoin the herd.

Recognition between mother and calf is by smell (olfactory), sight (visual), and sound (auditory). Cows usually sniff their calves after being away for a time; and the calf recognizes its mother's call. The attachment of the mother to her calf is very strong. However, the calf accepts separation with less stress.

If a calf is stillborn, or dies soon after birth, some cows will leave the place where the fetus lies, never to return. Others may return to their dead calf at frequent intervals over a period of several days, smelling it and mooing gently.

Beef calves are normally weaned at about 7 months of age. The bond between cows and calves is very considerable, with the result that the separation is a traumatic experience. Thus, both mothers and calves bawl, often in unison, for 2 to 3 days. In all cases, however, the weaning separation should be complete and final, preferably with no opportunity for the calf to see or hear its dam again. In no case, however, should the cows and calves be turned together once the separation has been made, for it will only prolong the weaning process and may cause digestive disorders in the calf. Dairy calves are normally removed from their mothers when they are from 1 to 4 days of age, with the result that the tie between mother and offspring is soon severed.

• **Ewes**—After parturition, the ewe licks the newborn lamb, removing moisture and placental membranes. The lamb soon staggers to its feet and makes awkward efforts to find a teat to nurse. Quite often a very weak lamb will have to be held to the teat. Normally, lambs will suckle in a standing or kneeling position. While suckling, they wiggle their tails from side to side. The mother-young bond in sheep is very strong; the ewe becomes attached to her offspring, and the lamb develops an attachment to its mother. Although ewes are normally timid and easily frightened, they will defend their young even if the attacker is formidable. It is noteworthy, too, that sheep will accept and suckle orphan goats (kids), and vice versa (interspecies rearing).

• **Sows**—The sow is very protective of her pigs, especially if they squeal. She goes toward the intruder with mouth open and emits a series of sharp, barking grunts in rapid succession. She continues to mother her pigs until they are weaned, but after 2 to 3 days' separation she loses interest in them. If pigs are left with the sow for 3 to 4 months, she will usually wean them herself. Sows will readily accept pigs from another litter, provided the transfer is made the first day or two following farrowing. Exchanging of pigs among sows, in order to even out the size of the litters is a common practice in herds where many sows are farrowing about the same time.

Some nervous sows eat their pigs during or immediately after farrowing. If this trait is observed, all pigs, both live and dead, along with the placental membranes should be removed as soon as possible, before the sow has an opportunity to eat them. Once the sow has acquired a taste for flesh, she may develop a permanent pig-eating habit. Usually, such nervous sows calm down following farrowing, after which their pigs can be returned to them and they will express normal protective behavior.

• **Mares**—Mares show much the same maternal behavior toward their young as is exhibited by females of other species of farm animals. Thus, a mare calls her foal with a neigh or a whinny and exhibits nervousness and distress when her young is disturbed. When mares are separated from their foals, such as sometimes happens when they are worked or taken away for rebreeding, there is usually a noisy exchange of whinnying between mother and foal when they are put back together again and the foal is allowed to nurse.

It is noteworthy that a mare will devote as much attention and affection to a mule colt—a hybrid (ass × horse)—as she will to a horse foal.

Fig. 7-10. Mother love. (Photo by Bob Hopper, Los Angeles, CA)

• **Poultry**—Modern incubators have replaced the setting hen and precluded the need for broody hens. As a result, breeders have increased egg production by selecting against this trait. Also, with incubators, few hens and chicks are allowed to run together. Nevertheless, when this happens, the maternal behavior is intense. When foraging for food, a mother hen will cluck to her chicks and call them each time a choice morsel is found. All the chicks will come running to participate in the

find. Hens hover over their chicks by covering them with their spreading wings and nesting them close to their bodies during the night or at other periods when they rest or need protection. A hen with chicks exhibits a definite antagonistic behavior and will attack an enemy that bothers her young. To warn her chicks of danger, the hen emits a loud, shrill cry. The chicks react quickly and seek protection.

## AGONISTIC BEHAVIOR (COMBAT)

This type of behavior includes fighting, flight, and other related reactions associated with conflict. Among all species of farm mammals, males are more likely to fight than females. Nevertheless, females may exhibit fighting behavior under certain conditions. Castrated males are usually quite passive, which indicates that hormones (especially testosterone), are involved in this type of behavior. Thus, farmers have for centuries used castration as a means of producing docile males, particularly cattle, swine, and horses.

Bulls, rams, boars, and stallions that are run together from a very young age seldom fight. Perhaps they have already settled their social rank. On the other hand, bringing together sexually mature strange males of these species almost always results in a fight. The intensity of fighting depends upon the tenacity of the two combatants. Although fighting rarely results in death, it usually continues until one gives up. There is the hazard, however, that bulls will be stifled as a result of the fighting. Also, a fight among young boars on a summer day may end in death of one or both of them due to heat exhaustion.

The agonistic behavior of each species of farm animals follows:

• **Cattle**—In combat, bulls paw the ground and bellow, followed by putting their heads together and butting

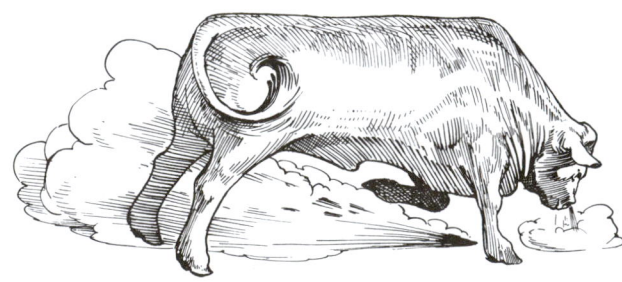

Fig. 7-11. Pawing the ground and bellowing, generally the first stage of combat in cattle.

Under range conditions, it is common for large numbers of bulls to be run together with a herd of cows. Even though many different bulls of different ages are included in the herd, fighting among them seldom occurs. Outside of the breeding season, as in the fall of the year, it is not uncommon to see bulls congregate together on the range, away from the cow herd.

Although young bulls raised together will seldom fight, a group of bulls may single out one individual and ride him to death, unless he is removed from the group.

Bringing together sexually mature strange bulls almost always results in a fight.

It is noteworthy that breeds of cattle differ in their agonistic behavior.

- **Sheep**—Rams fight by backing off and charging at each other headlong. The fight generally continues until one ram gives up, usually after both combatants have bloody noses.

Fig. 7-12. Agonistic behavior of rams.

- **Swine**—When strange boars are brought together, some fighting ensues. Sows and barrows will also fight, but they do not exhibit the jaw-clicking and saliva-producing (champing) characteristic of fighting boars. A sow will try to bite, whereas a boar will slash his opponent with his tusks.

When strange boars are first penned together, they smell one another and begin to circle as they size up each other. They frequently strut shoulder to shoulder with the hair on their crest bristled, ears cocked, and head raised in an alert, threatening position. In a serious encounter, the combatants utter deep-throated barking grunts and champ throughout the fight. As the fighting becomes intense, each boar repeatedly thrusts his head and neck sideways and upward, with his jaws open and his teeth bared. If the boars have tusks, slashes are usually inflicted on the shoulders of each other.

Fighting boars await the opportunity to discontinue shoulder contact and to nip at the ears or the neck and front legs. Sometimes, they even charge the side of the opponent with their mouths wide open.

Fighting may continue for as long as an hour, or it may end very quickly. In any event, it will continue until the dominant boar is satisfied and the loser retreats, with the winner biting and slashing him as he scampers away.

- **Horses**—Bringing together sexually mature strange stallions for the first time almost always results in a vicious fight. Stallions fight by biting, kicking, and striking. Generally they fight head to head and most of the biting is on the neck, shoulders, and front legs. Although fighting rarely results in death, it usually continues until one gives up—battle scarred by teeth and hoof marks.

Fighting among mares is less vicious than between stallions. Body biting and kicking are used as a means of establishing social order. Geldings may fight much like mares.

Jacks are unusually vicious fighters. They rely on their teeth, rather than kicking. Sometimes wild jacks killed a rival by cutting his windpipe or jugular vein. Also, it is reported that the dominant jack occasionally castrated the weaker jacks with his teeth.

Agonistic behavior is of practical importance when strange horses are first put together. One way or another, a social order must be established. Hence, there is always the potential of

injury until rank is settled. Also, agonistic behavior may create a potentially dangerous situation to both horses and riders in group riding. To reduce the hazard of such accidents, all horses should be spaced well apart when standing or moving.

Wild bands of horses and bands of domestic horses on range behave very much alike. The stallions have keen sight, hearing, and smell; and each stallion leader is very good at protecting his harem, which usually includes 10 to 20 mares. When frightened or facing danger, the stallion warns his band with snorting and restless movements and takes his place at the head of the herd, ready for battle if necessary.

- **Poultry**—In chickens, agonistic behavior includes attack, escape, avoiding, and submissive behavior. Where roosters are run with a flock of hens, fighting may continue for several days with both combatants covered with blood. Eventually, one gives up and escapes by running away. Combats between hens are much less serious than those between roosters. Submission in hens is characterized by crouching.

Breed differences in the agonistic behavior of chickens are very great. Usually the meat breeds are quite placid, whereas in game chickens (cockfighting), special strains of breeds of chickens are selected because of their agonistic behavior. In passing, it is noteworthy that the sport of cockfighting exercised a tremendous influence in the domestication of wild chickens, and also in the subsequent distribution of chickens throughout the world. Much importance was attached to the pastime of cockfighting by many human races. The literature of various nations contains many references to this sport, and it would appear that cockfighting had as much to do with the domestication of the fowl as the demand for food and that the sport was chiefly instrumental in the wide-spread distribution of chickens that followed.

When a number of strange hens are placed together in a pen, fights occur by twos until each bird has engaged all the others. The winner of each initial contest thereafter has the right to peck the loser, and the latter usually avoids the former. Some individuals give way without a fight, whereas others may challenge the winner again before dominance is settled. At subsequent meetings, one member of each pair pecks or threatens the other, definite dominance-subordination patterns become habitual, and thus the pecking order is established.

## ALLELOMIMETIC BEHAVIOR

*Allelomimetic behavior is mutual mimicking behavior.* Thus, when one member of a group does something, another tends to do the same thing; and because others are doing it, the original individual continues. Cows moving across a pasture toward a milking barn often display allelomimetic behavior. One cow starts toward the barn, and the others follow. Because the rest of the herd are following, the first cow proceeds on. Sheep walk, run, graze, and bed down together. A timid horse will follow behind a pack, in order not to be left behind.

In the wild state, this trait was advantageous in detecting the enemy, and in providing protection therefrom. In wolves and coyotes, this behavior is important in attacks on prey, since a pack working together is much more likely to be successful than animals working alone.

Under domestication, animals are usually protected from predators. Nevertheless, the allelomimetic behavior still has important consequences. By stimulating each other, animals produce the phenomenon known as *social facilitation*. Thus, the competition between animals makes for higher feed consump-

Fig. 7–13. Social facilitation. If not overcrowded, a group of steers will usually consume more feed, on the average, than one steer alone. (Courtesy, *Feedlot Management*, Minneapolis, MN)

tion. For example, one steer kept alone may eat X pounds of feed per day. However, when he is placed with other steers, his intake may be X + Y pounds. But, of course, the feed consumption advantage can be nullified when animals are placed together too closely, with the result that the agonistic behavior comes into play.

## GREGARIOUS BEHAVIOR

*Gregarious behavior refers to the flocking or herding instinct of certain species.* It is closely related to allelomimetic behavior. If animals imitate each other, they must stay together. If they stay together as a mobile group, they must use allelomimetic behavior to do so. All such behavior arises out of the process of social attachment.

Gregarious behavior differs among species; hence, each class of animals is discussed separately in the sections that follow.

• **Cattle**—Cattle tend to roam in groups of various sizes when a large herd is placed on a pasture or a range. However, there is usually considerable space between the members of the herd. Moreover, on close observation it is evident that there are several small groups within a herd, each ranging from three to five head.

• **Sheep**—The gregarious, or flocking, instinct is particularly strong in sheep. Moreover, it is more evident in some breeds than others. The Merino, and animals carrying Merino breeding, are noted for their flocking instinct. This makes it possible to herd them on the range.

During the formative period of the Merino in Spain, the powerful nobility and clergy engaged in the lucrative sheep industry. In an attempt to produce the finest staple possible, the early Spanish shepherds drove their sheep from southern to northern pastures in the spring and returned them in the fall. In this manner, it was possible to secure the most favorable grazing and climatic conditions for the flock. The early laws of the kingdom stipulated that the owners of large flocks should be allowed a path 90 paces wide through all enclosed lands. In the migration process, any animal that failed to keep with the band was left by the wayside. Presumably, this accounts for the flocking or gregarious instinct of the Merino sheep as well as their hardiness.

It is noteworthy that the gregarious instinct of sheep diminishes to some extent when they are placed within fenced holdings, instead of herded. As a result, those who handle western range bands do not try to switch back and forth from fenced range to herding, for the reason that the band becomes unmanageable from the standpoint of herding once they have been in a fenced holding for an extended period of time.

Fig. 7–14. Gregarious, or flocking, instinct exhibited by Merino sheep.

Packers use the gregarious instinct of sheep by having an old goat, appropriately called a *Judas,* lead sheep to slaughter. A well-trained Judas will lead group after group of sheep to slaughter all day long.

Fig. 7–15. Judas goat leading sheep to slaughter. (Courtesy, Monfort of Colorado, Inc., Greeley)

• **Swine**—In the wild state, swine roved through the forest in herds. Usually these wild groups consisted of 5 to 10 sows, under the leadership of a boar. The wild boar, with his large, long head and strong tusks, was a formidable match for most any enemy.

Under domestication, swine retain their gregarious nature. However, caretakers have altered it a great deal. Today, hogs are usually confined to a very limited area. Also, under domestication, they have lost most of their ferocity and are usually gentle and easily handled.

• **Horses**—In the wild state, horses ran in bands; thus, they were gregarious by nature. These bands seldom consisted of more than 40 animals; and always there was a stallion in each group.

Under domestication, horses show definite preference for their herdmates; they will even avoid certain horses in the herd. In the draft horse era, animals that were worked together usually stayed together when they were turned to pasture.

• **Poultry**—Chickens, turkeys, ducks, and geese tend to flock together. Of course, under domestication people have interfered with the normal flocking instinct of both chickens and turkeys. Even under domestication, however, ducks and geese exhibit their gregarious or flocking nature as they walk Indian file, one behind the other.

## SHELTER-SEEKING BEHAVIOR

All species of animals seek shelter—protection from the sun, wind, rain and snow, insects, and predators.

• **Cattle**—Cattle are not as sensitive to extremes in temperature—heat and cold, as are swine. Nevertheless, they do seek shelter under natural conditions—this may consist of hills, valleys, timber, and other natural windbreaks; or they may even group closely together.

Cattle seem to be able to sense the coming of a storm, at which time they may race about and act up. During a severe rain or snow storm, they turn their rear ends to the storm and tend to drift away from the direction of the wind. By contrast, bison (buffalo) face a storm head on.

Fig. 7–16. Shelter seeking behavior, showing cattle in a ravine, seeking protection from the storm. Note that they are grouped together and facing away from the storm. (Courtesy, American Hereford Assn., Kansas City, MO)

During the summer months, cattle seek either shade or a waterhole during the heat of the day. Then they graze in the cool of the evening or early morning. There are well-known breed differences in tolerance to heat. Brahma cattle can withstand more heat than the European breeds, whereas the heat tolerance of the Santa Gertrudis is intermediate.

• **Sheep**—Sheep seek shelter by moving into barns or under trees, by huddling together to keep off flies, by crowding together in extremely cold weather, and by pawing the ground and lying down. Like cattle, during a severe storm they turn their rear ends towards the wind.

When there is no shelter, there is danger of sheep massing together and smothering during a very severe storm.

• **Swine**—Hogs are very sensitive to extremes of heat and cold; hence, shelter seeking is a very important trait with them. It is particularly important that swine be provided with shade during hot weather, hogs will wallow in water if given the opportunity.

Hogs pant rapidly when they are hot; sleep stretched out full length when they are hot, so as to expose the maximum body surface to the air; and sleep curled up and huddled together during cold weather, thereby exposing minimal body surface to the air.

• **Horses**—Horses are not very sensitive to either heat or cold. When confined in cold areas, they develop a shaggy coat of hair in the wintertime and seek shelter from storms under trees and in the valleys. They paw to get their feed supply when the ground is covered with snow. Like cattle, horses face away from the direction of a severe storm.

• **Poultry**—In the wild, one of the ancestors of chickens (the red Burmese jungle fowl, ancestor of the Leghorn breed) and the ancestors of turkeys lived in forests where they had natural shelter and roosted in trees. Even today, turkeys show little tendency to seek shelter in a severe blizzard. It is not unusual, therefore, for turkey losses to be high in a severe storm.

## INVESTIGATIVE BEHAVIOR

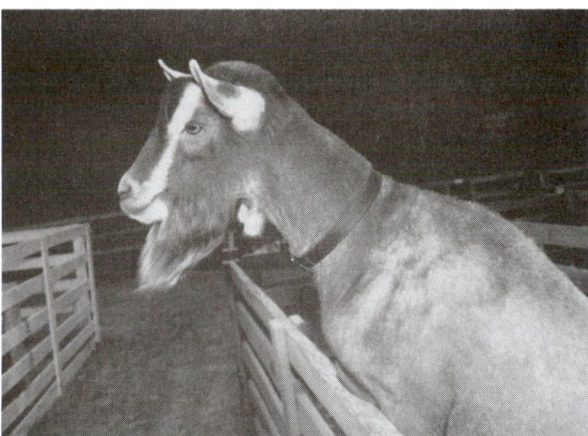

Fig. 7–17. Investigative behavior exhibited by a Toggenburg goat male. (Courtesy, University of Delaware, Newark)

All animals are curious and have a tendency to explore their environment. Investigation takes place through seeing, hearing, smelling, tasting, and touching. Whenever an animal is introduced into a new area, its first reaction is to explore it. Experienced caretakers recognize that it is important to allow animals time for investigation before attempting to work them, either when they are placed in new quarters or when new animals are introduced into the herd.

• **Cattle**—If they are not afraid, cattle investigate a strange object at close range. They proceed toward it with their ears pointed forward and their eyes focused directly upon it. As they approach the object, they sniff and their nostrils quiver. When they reach the object, sniffing is replaced by licking; and if the object is small and pliable, they may chew it or even swallow it.

Cattle exhibit investigative behavior when placed in a new pasture or in a new barn. As a result, if there is an open gate in a pasture or a hole in the fence, they usually find it, then proceed to explore the new area.

Calves are generally more curious than older cattle. Perhaps this is due to the fact that older animals have seen more objects, with the result that fewer things are new or strange to them.

• **Sheep**—Sheep investigate strange objects and quarters much like cattle. They also approach objects in the same heads-up, ears forward, and eyes-fixed manner. However, sheep are much more timid than cattle, with the result that they usually turn and run if the object moves or if something frightens them.

• **Swine**—Pigs are curious, too. When a strange person approaches a herd of hogs, an alarm, or "woof," is sounded and the animals scatter—scampering as fast as they can for a short distance. In the meantime, if the intruder remains stationary, either standing or sitting, the pigs invariably return to investigate by smelling, rooting, and nibbling. Of course, when pigs are placed in confinement, they have little area to investigate.

• **Horses**—Foals are more curious than older horses. Young equines spend much of their time looking at and sniffing objects in their pastures or stalls. As the foal grows older, it may exhibit fear of certain objects. At this stage, it may even move away from its caretaker. When this happens, the caretaker should never run after the foal. Rather, stand still; very soon, the foal's curiosity will get the best of it, and it will return. A mare frequently becomes very nervous as she watches her offspring investigate, fearful that it may get hurt in the process.

• **Poultry**—The investigative behavior exhibited by chickens and turkeys is more casual and subtle than with the four-footed animals. They will, however, walk slowly toward a new object, looking it over as they go. Usually, they stop a short distance from the object, then turn and go on their way.

## BEHAVIORAL NORMS

The producer needs to be familiar with behavioral norms of animals in order to detect and treat abnormal situations—especially illness. Many sicknesses are first suspected because of some change in behavior—loss of appetite (anorexia); listlessness; labored breathing; posture; reluctance or unusual movement; persistent rubbing or licking; and altered social behavior, such as one animal leaving the herd or flock and going off by itself—these are among the useful diagnostic tools.

Some signs of good health are:

1. Contentment.
2. Alertness.
3. Eating with relish, and cudding by ruminants.
4. Sleek coat and pliable and elastic skin.
5. Bright eyes and pink membranes.
6. Normal feces and urine.
7. Normal temperature, pulse rate, and breathing.

The above signs of good health, including normal temperature, pulse rate, and breathing rate of different species of farm animals are detailed in Chapter 6 of this book under the heading, "Signs of Good Health"; hence, the reader is referred thereto.

## HOW ANIMALS SEE

The eyes of animals are on the side of the head (the cat is the exception). This gives them an orbital, or panoramic view—to the front, to the side, and to the back—virtually at the same time. Also, this is a rounded, or globular type of vision. This leads to a different interpretation than that of the binocular type of vision of people.

Fig. 7-18. The eyes of a horse allow it to see what's going on all around it—to the front, the side, and the back—all at the same time. With the head turned very slightly to the left, the horse shown above was able to keep an eye on the photographer who was standing behind it. (Courtesy, Wild Horse Research Farm, Porterville, CA)

## HOW ANIMALS SLEEP

Normal behavior in sleep should be recognized, especially since it differs widely between species.

• **Cattle**—Cattle typically lie on their stomach or tilt to one side, with the four limbs folded under the body; one hind limb extends forward while the other protrudes toward the outside. Although cattle rest in this manner, they do not sleep in the sense that the term usually connotes. While lying down, they do shut their eyes for short periods of time. Beef bulls sometimes assume a sitting position.

Calves commonly spend up to one-half hour at a time with their heads turned back in the flank position.

• **Sheep**—The normal sleeping posture of sheep is on the stomach but tilted to the side with one front leg folded under the body and the other extended forward. Usually the head is turned to one side and the eyes are closed. Although sheep are usually inactive about half of the day, as with cattle there is considerable debate as to whether they actually sleep. Certainly, sheep do not enter the state of deep sleep that exists in horses, dogs, and cats.

• **Swine**—The resting position of swine varies according to temperature—in the summer, they sleep stretched full length; in cold weather, they sleep curled up. In any event, pigs sleep soundly—they even snore.

• **Horses**—The horse rests and sleeps standing up. This is made possible by a system of ligaments, which do not get tired like muscles, and which take the weight off muscles during rest. Sometimes horses lie down in the sun, apparently to expose the body to warmth.

In contrast to cattle and sheep which sleep very little, the horse may sleep soundly for as much as 7 hours out of each 24 hours, mostly during the warmest part of the day. But not all of the 7 hours of sleep are taken at one time; rather, it is short and irregular, depending on the degree of hunger and the climatic conditions.

• **Poultry**—Chickens and turkeys wind their claws tightly around a pole, or roost, and snuggle closely together. They like to close their eyes and hide their heads in the feathers of their wings. Ducks and geese sleep on both land and water. On land they often drowse while standing on one leg; on water they paddle every now and then, in order not to drift ashore.

## ABNORMAL BEHAVIOR

Abnormal behavior in animals develops where there is a combination of confinement, excess stimulation, and forced production with a lack of opportunity to adapt to the situation. Also, it is recognized that confinement of animals makes for lack of space; this often leads to unfavorable changes in habitat and social interactions for which the species have become adapted and best suited over thousands of years of evolution.

Homosexual behavior is common where adult mammals of one sex are confined together. Other abnormal behaviors in different species that frequently develop among domestic animals in confinement follow:

• **Cattle**—The *mean bull* complex is an example of abnormal behavior in cattle. Of course, there are inherited differences in the temperaments of cattle. Nevertheless, constant stress can change the temperament of an animal, just as it can in people. Thus, when a bull is kept for hand mating in a corral by which the cowherd passes each day, cows in heat, or coming in heat, stimulate his sexual behavior. Since he cannot respond naturally through coitus, he becomes a mean bull.

• **Sheep**—As a consultant for the Atomic Energy Commission for 20 years, the author had an opportunity to observe sheep that were kept in confinement, generation after generation, for 20 years. These animals developed a *wool-eating* habit. They didn't inflict special harm, for they only took small nibbles of wool from each other. Since these sheep were getting the most complete diet that science knew how to formulate, the only conclusion was that the wool-eating habit came about as a result of the unnatural confinement.

• **Swine**—Tail biting accompanies close confinement. It results when pigs are prevented from rooting, nibbling, and chewing—the pig's normal behavior pattern.

Docking is the best way in which to stop pigs from tail biting. Swine producers have tried all sorts of things to prevent tail biting. Some have substituted other materials for pigs to bite, such as rubber tires or chains hung near the pigpen. Others have tried spraying the tails with distasteful chemicals. But none of these methods work very well.

• **Horses**—Few animals have undergone such drastic changes through evolution as equines. Little *Eohippus* (the dawn horse of 58 million years ago) was a denizen of the swamp. Later, through evolution, horses became creatures of the prairie. Even

though their natural habitat shifted during this long predomestication period, until caretakers confined them, they gleaned the feeds provided by nature. Inevitably, this occupied their time and provided exercise. But domestication and confinement to stalls wrought many changes—changes which spawned abnormal behaviors, including balking, bolting feed, cribbing, halter pulling, kicking, tail rubbing, weaving, wood chewing (pica), backing, rearing, shying, striking with the front feet, a tendency to run away, and objection to harnessing, saddling, and grooming. Many of these vices originate with incompetent handling; nevertheless, they may be difficult to cope with or to correct. This is especially true in older animals.

Fig. 7–19.   A cribber in action. This is the vice of biting or setting the teeth against some object, such as a post or manger, while sucking in air. Such animals are also referred to as wind suckers or stump suckers. (Courtesy, Dr. George H. Waring, Department of Zoology, Southern Illinois University, Carbondale)

• **Poultry**—The most common abnormal behavior observed in chickens in confinement is cannibalism. This trait may be encountered among birds of all ages. Among baby chicks, the trouble is usually confined to toe and tail picking. With mature birds, the vent, tail, and comb are the regions most frequently picked.

The cause of cannibalism is not fully understood. It is known that it is more frequent under confined conditions. Without doubt, it may be accentuated by deficiencies in management and nutrition. Also, it may be brought on by just plain boredom or too much light.

The best way in which to control cannibalism is by debeaking. Many broiler producers, and some egg producers, have their chicks debeaked at the hatchery. Additionally, layers can be debeaked at the time of housing in the laying house.

Chickens confined to small cages in laying batteries will develop stereotyped head movements.

In turkeys, desnooding, or cutting off the snood of the young poults, is practiced by some growers to prevent injury among mature males. It is believed that this practice helps prevent the spread of erysipelas in the flock. Most hatcheries will desnood poults for a small fee upon request.

## POPULATION DENSITY

Members of the animal kingdom other than humans have their numbers held in check by the many factors encompassed in the term *balance of nature*. The principal factors operating to prevent the full realization of the potential of livestock in producing human food throughout the world can be classified into those which relate to the provision of feed, those which prevent livestock from making best use of feed resources—notably diseases and vicissitudes of climate, and those which relate to the socioeconomic context in which livestock production takes place. In addition to these factors, it seems that certain forms of social behavior which limit the rate of reproduction and prevent population size from exceeding the feed supply enter into the regulation of animal numbers.

It is generally recognized that the behavior of animals is related to population density, with the amount of fighting and other difficulties due to crowding varying according to species. Some livestock producers believe that a high density of animals affects the fertility rate. There is also evidence that population density produces more stress, which, in turn, affects maternal behavior and results in fewer young surviving. Under higher density, males show certain sexual deviations.

## SOCIAL RELATIONSHIPS

*Social behavior may be defined as any behavior caused by or affecting another animal, usually of the same species, but also, in some cases, of another species.*

*Social organization may be defined as an aggregation of individuals into a fairly well integrated and self-consistent group in which the unity is based upon the interdependence of the separate organisms and upon their responses to one another.*

The social structure and infrastructure in herds and flocks are of great practical importance. Livestock producers should be knowledgeable relative to the social relationships of each species with which they work. Then, if this social relationship is disturbed and/or modified under intensive, confined conditions, they will be better able to feed, care, and manage the animals with maximum consideration accorded to both economy of production and animal welfare.

• **Social organization among cattle**—Breed affects social stratification in cattle. For example, on pasture or range, Angus tend to dominate Shorthorns, while Herefords tend to submit to both breeds. Older cows generally dominate the younger ones, and the heavier animals (usually the older animals) tend to dominate the lighter; for this reason, 2-year-old heifers should be segregated from older cows. However, among cows of similar age and breed, the smaller and more aggressive ones are most dominant. Also, cows with more seniority in the group and cows with horns tend to be of higher social rank. Aggression in cows appears to be ritualized, with most encounters taking place in the following sequence: approach, threat, and physical contact (or fighting).

Limited studies indicate a high relationship between social status of cattle and spacing, or social distance. The higher the social rank, the more likely cows are to be found near other members of the herd. Also, dominant cows tend to allow close approach by other cows more often than subordinates.

On the range, the following rank orders are evident in large heterosexual herds of cattle: (1) adult males, (2) adult females, and (3) juveniles. Adult males dominate adult females, which in turn, dominate juveniles. However, at about 1½ years of age, young males begin to fight with adult females, and by 2½ years of age, they dominate all the females, and join the adult male rank order.

When grouped together on the range, bulls are loners, and do not organize socially.

When moving from the paddock to the milking parlor, dairy cows travel in a consistent leadership order. Mid-dominant cows tend to be in front of the group. However, the same individuals are seldom consistent leaders; instead, there is a pool of animals which tends to be in or near the lead. More consistency is found in the cows bringing up the rear of the moving herd. So, *rearship* is a more distinctive feature than *leadership.* The animals at the rear are usually the younger subordinate heifers. Also, in most herds there is a definite order in which cows enter the milking parlor; the mid-age, mid-dominant cows tend to be milked first, followed by the older cows. Social dominance orders (called *bunt order* in polled cows and *hook order* in horned cows) become more complex as herd sizes increase. Generally, not more than 100 dairy cows should be group-fed.

Fig. 7–20. When milk cows are moving to and from the milking parlor, bringing up the rear is more distinctive and consistent than leadership. Mid-dominant cows tend to be in front of the string, but their order changes. (Courtesy, USDA)

Within a corral of feedlot (finishing) cattle, a linear-tending (a linear-tending hierarchy is a type of social hierarchy in which dominance-subordinance ranking includes a triangle or some more complex hierarchy loop) *peck order*, or dominance order, can be determined by observing agonistic encounters between pairs of animals within the group. The degree of linearity is greater with increases in heterogeneity of such factors as age, sex, weight, breed, and background. Linearity and stability of the dominance order tend to increase the longer the group is together, and linearity is greater among small groups.

• **Social organization among sheep**—When grazing, large flocks (bands) of sheep generally split into subgroups and occupy separate areas. Different breeds vary in their tendency to move or flock together, with the gregarious trait being the strongest in the Merino and Rambouillet breeds.

The social structure of sheep is dependent upon visual contact and a flocking tendency. Also, group size is important. It has been shown that Merino sheep kept in pairs (just 2 sheep) gained less liveweight and produced less wool than their counterparts in groups ranging from 4 to 30 animals. The pairs spent less time grazing and more time walking along the fence line trying to keep contact with the flock in the next pasture; thus, pairing produced a level of stress which affected production. The need for visual contact during grazing may well account for this stress. It should be noted, however, that preferred group size and overall flock dispersion are dependent on breed, age, stability of flock membership, and vegetation.

Studies have not shown any correlation between the rank in which sheep reach feed troughs and their competitive ability at the troughs. However, certain individuals are constantly among the first few sheep to reach the troughs. Thus, it appears that going to feed is initiated by a few sheep, then others follow; and that dominance is not involved. Aggression is rare during normal grazing and social hierarchies; *i.e.*, dominance of one sheep over another is far from absolute.

• **Social organization among goats**—A herd of goats forms smaller groups than sheep, usually built up from the extended family group. Dominance is a relatively mild phenomenon in goat societies; when competing for feed, they may rear up and head clash with downward stroke of the head.

The principles and practices of good goat herding are very similar to those with sheep, with one exception: Rarely do sheep herders work ahead of a range band. However, it is common practice for goat herders to work in front, turning the lead goats back to avoid unnecessary travel.

• **Social organization among swine**—Wild and feral pigs prefer to live in herds in scrub brush and light forest areas, rather than on the open range. The basis of the social structure is the matriarchal herd, consisting of one or several females with their offspring. Juvenile males are tolerated in the family group. Mature boars join the matriarchal herds when sows become sexually receptive. But, apart from the breeding season, boars move about together in bachelor groups, although older ones often range as solitary animals. Under domestication, the pig has been transformed from a pugnacious, free-ranging animal to a docile animal which is readily handled in large groups under confinement. Shortly after birth, pigs show preference for the front teats of the udder. The teat order is effective until weaning, after which it is superseded by adult dominance.

• **Social organization among horses**—Studies of feral horses, along with the historical records of wild horses, give clear evidence of the social organization of horses in their natural habitat: It centers on a dominant stallion and his harem of 1 to 3 mares and their immature offspring. Each group has an alpha or leading female, and the other members respond as followers. However, the stallion maintains his patriarchal position until displaced by another adult male; sometimes he is at the front of his group, at other times he assumes a defensive position between intruders and his band, and at still other times he herds or drives his harem during the breeding season. Immature males are very submissive to the dominant stallion. The stallion and his harem are a closed society; animals not belonging to a group are rejected by either the dominant stallion or the mares.

Excess stallions live apart from the family/harem groups, either singly or in bachelor groups of up to eight males. The groups are organized with the dominant individual herding the other stallions in the same manner as the harem stallion herds mares.

Home range behavior exists, with each home range including at least one watering hole and a large grazing area. But there is considerable overlapping of the home ranges, with more than one group using the same area. The spacing between groups is controlled primarily by dominating stallions, who approach each other with threats, which occasionally result in pushing and kicking matches, during which time the animals within the groups move closer together. Following such encounters, the stallions return to their respective harems and move them apart. When approaching a watering hole, the group whinnies; if the watering site is already occupied by a band, they await their turn until the watering group moves away.

Wild and feral mares commonly foal in the spring and summer and usually mate a few days after foaling. In an amazingly short time after birth, foals run almost as fast as their mothers—and on legs almost as long. Foals stay near the sides of their mothers; and the mares are very protective of their young. Harem stallions also look after the foals and herd them back to the group when they become separated. Barren mares are sometimes protective of other mare's foals.

• **Social organization among poultry, including chickens, turkeys, ducks, and geese—**

1. **Chickens.** Among both wild and domestic chickens, a dominant male will organize a harem, prevent fighting, and mate with several females. Subordinate cocks will remain at a distance.

Chicks develop aggressive tendencies a couple of days after hatching, but the brood's dominance order is not established until sometime between six and ten weeks after hatching.

Chickens recognize each other and stratify themselves socially mainly by the configuration of the head and neck. Consequently, hens with combs and wattles trimmed have significantly more fights than do undubbed hens, because individual recognition is more difficult. Also, dubbed hens tend to be subordinate to those with combs and wattles.

Debeaking reduces injuries, but debeaked hens are able to maintain normal social relations with intact flockmates.

Dominance orders of laying hens are established more quickly when the hens are in cages than when kept on the floor. Competition for space, feed, and water is reduced in cages, primarily because the social group is smaller.

2. **Turkeys.** During the first three months after hatching, turkey poults spar; during the fourth and fifth months, this is replaced by fighting.

Like chickens, turkeys recognize each other mainly by the sight of the head and neck structure. Dominance orders are maintained by specific displays of higher ranking turkeys when they meet lower ranking flockmates. Also, dominant turkey hens peck the backs of necks of subordinates; and the larger the denuded area, the lower the social rank.

As with chickens, a gobbler will mate with several hens. Adult male and female turkeys spend much of the nonbreeding season in separate flocks, males in bachelor groups and hens in groups with their young.

3. **Ducks.** All breeds of domestic ducks, except the Muscovy, descend from the wild Mallard. Domestic ducks neither fly nor migrate. They form pair-bonds during the mating season, but the drakes are noted for being unfaithful; most pair-bonds break while the female is incubating.

4. **Geese.** All breeds of geese, except the Chinese goose, descend from the Graylag goose. Unlike ducks, geese are known for their fidelity; mates usually form pair-bonds lasting for life. The gander remains with the goose during incubation; parents and young live together until the next breeding season, at which time the gander drives the young away. No social hierarchy is apparent within a brood, but family units may interact; thus, the members of a dominant family may chase a subordinate family.

## DOMINANCE

Within most groups of farm animals of the same species, there is a well-organized social rank which is just as real as any social register or blue book of people. Animals observe this order in their relationships just as carefully as protocol demands that it be observed at a state dinner.

In chickens, in which it was first observed, the social rank order is called the *peck order*. It is established as follows: When a number of strange birds are placed together in a pen, fights ensue by twos until each bird has engaged all the others. The winner of each initial contest thereafter has the right to peck the loser, the latter usually avoiding the former. Some individuals give way without a fight and others may challenge the winner again before dominance relations are settled. At subsequent meetings, one member of each pair pecks or threatens the other, definite dominance-subordination patterns become habitual, and thus the peck order is established.

A social rank order similar to the peck order in chickens exists in most species of farm animals, although it is not always as clear-cut as in poultry. Thus the alpha animal in the herd or flock will be dominant over all other individuals and the omega animal will be subordinate to all. In between, some animals will be subordinate in some relationships and dominant in others. Moreover, once these relationships are established, they seldom change. The social rank order is usually important only in females, because mature male animals are seldom run together in groups. It is unimportant in domestic ewes.

Once the social rank order is established, it results in a peaceful coexistence of the herd or flock. Thereafter, when the dominant one merely threatens, the subordinate animal submits and avoids conflict. Of course, there are some pairs that fight every time they chance to meet. Also, if strange animals are introduced into such a group, social disorganization results in the outbreak of new fighting, as a new social rank order is established.

Fig. 7–21. Dominance. This shows a dominant cow attacking the neck of a subordinate. The latter submits and avoids a fight.

Among wild animals, social rank order is nature's way of giving mating priority to the top ranking males. Hence, they leave behind more of their progeny than do the less dominant males. Also, dominance establishes priority in feeding.

Social rank among farm animals is of little consequence as long as they are on pasture or range, and if there is plenty of feed and water; but it becomes of very great importance when animals are placed in confinement. When cows are moved into winter quarters, social dominance decrees that replacement heifers be sorted out and fed separately, that

young bulls be cared for in separate quarters, and that old cows with poor mouths be fed separately; otherwise, these animals will not get enough food.

Of course, social rank becomes of importance when a group of animals is fed in confinement; and it becomes doubly important if limited feeding is practiced. Under such circumstances, the dominant individuals crowd the subordinate ones away from the feed bunk, with the result that they may go hungry. This happens both in feedlot cattle and in breeding cattle being wintered. Several factors influence social rank; among them, (1) age—both young animals and those that are senile rank toward the bottom; (2) early experience—once a subordinate in a particular herd, usually always a subordinate; (3) weight and size; and (4) aggressiveness or timidity. Also, it is noteworthy that social rank is influenced by hormones; for example, a capon (castrated male chicken) automatically goes to the bottom of the totem pole, whereas the injection of roosters and hens with the male sex hormone, testosterone, increases their social rank.

In feedlots and other confinement operations, social facilitation is of great practical importance. Dominants should be sorted out, and, if possible, grouped together. Of course, they will fight it out until a new social order is established. In the meantime, both feed efficiency and gains will suffer. But, as a result of removing the dominants, the feed intake of the rest of the animals will be improved, followed by greater feed efficiency and profit. Among the more settled animals, social facilitation will become more evident. After the dominants have been removed, the rest of the animals will settle down into a new hierarchy, but within the limits of their dominance. Their interaction or social facilitation will be far more likely to have a calming effect on this group, to both the economic and practical advantages of the operator.

Dominance and subordination are not inherited as such, for these relations are developed by experience. Rather, the capacity to fight (agonistic behavior) is inherited, and, in turn, this determines the dominance and subordination. Hence, when combat has been bred into the herd, such herds never have the same settled appearance and docility that is desired of high-production and intensive animals

## LEADER-FOLLOWER

Leader-follower relationships are important in cattle, sheep, hogs, and horses. In each case, the young follow their mothers; hence, they continue to follow their elders. Leader-follower relationships are particularly strong in sheep, where lambs follow their mothers from birth.

It is important to distinguish leader-follower relationships from dominance; in the latter, the herd is driven, rather than led. After the dominants have been removed from the herd, the leader-follower phenomenon usually becomes more evident. It is well known that the dominant animal is not necessarily the leader; in fact, it is very rarely the leader. It pays too much attention to other matters of dominance in its relationship within the herd, with the result that it does not have the quality of leadership.

Taking a page from history and people, it is noteworthy that one of the world's greatest leaders, Napoleon, was small in stature. So it is with animals. The leader may be small, but is always intelligent. In horses, the leader is usually a mare with a well-developed investigatory sense.

Fig. 7-22. When on the move, horses always follow the leader in Indian file—never abreast. (Courtesy, Wild Horse Research Farms, Porterville, CA)

## INTERSPECIES RELATIONSHIPS

Social relationships are normally formed between members of the same species. However, they can be developed between two different species. In domestication this tendency is important (1) because it permits several species to be kept together in the same pasture or corral, and (2) because of the close relationship between people and animals. Such interspecies relationships can be produced artificially, generally by taking advantage of the maternal instinct of females and using them as foster mothers. It's not unusual, for example, to set duck eggs under a hen. All goes well until the young ducklings take to water for their first swim. Thereupon, the mother hen becomes quite excited, fearful that her babies will drown (little realizing that swimming comes naturally for young ducks).

All sorts of bizarre interspecies relationships have been arranged—including cows raising pigs; bitches (dogs) raising pigs, rabbits, and cats; and cats raising mice.

## PEOPLE-ANIMAL RELATIONSHIPS

Social relationships can also be transferred to human beings. Thus, animal caretakers usually form care-dependency relationships with the animals under their care. This is particularly true with pets—horses, dogs, and cats.

People need pets and pets need people! Both groups desire to love and be loved. This relationship is especially valuable for children, shut-ins, handicapped, and elderly people. The Delta Society is contributing richly to the happiness and well-being of people through furthering the human-animal bond and animal-facilitated therapy.

Without doubt, one of the best-known stories of a human-animal relationship pertained to groom Will Harbut and the great Thoroughbred, *Man O'War* ("Big Red"). When training, *Man O'War's* morning came early. Will Harbut gave him his first meal at 3:30 a.m.; at 7:30 a.m., he was groomed. "Big Red" was very fond of his caretaker; he liked to snatch his hat and carry it around as he showed off for visitors. Will Harbut, who had quite a way with words as well as with horses, never tired of telling the thousands of visitors who came to see *Man O'War* that "He was the mostest hoss dat ever was."

Without doubt, the most fantastic human-animal relationship of all time is the legendary story pertaining to Romulus and Remus, who were suckled by a wolf. Anulius, who was on the throne of Alban, ordered a mother to be buried alive (because she had broken her vestal vows) and her two children to be thrown into the Tiber River. The whimsical story goes on to say that the river received the babies kindly and bore them to a little bank, where they were cast ashore at the foot of a fig tree. Here they were found by a she-wolf, who cared for them until they were discovered by the shepherd Faustulus, who took them into his home and reared them. Romulus later became the legendary first king of Rome and the founder of the city.

In passing, it is also noteworthy that women have been known to suckle pups. During the 18th century, this sometimes occurred in Italy, where women suckled the pups of highly prized, delicate lapdogs. Eskimo women have also been known

Fig. 7-23. Interspecies relationship, showing a mare suckling a calf. (Courtesy, Carnation Genetics, Hughson, CA)

Fig. 7-24. She-wolf suckling Romulus and Remus; bronze sculpture by Carl Milles, in Milles Garden, Stockholm, Sweden. (Photo by A. H. Ensminger)

to rear orphaned Huskies with their own children, thereby pointing up the recognized importance of sled dogs in the Arctic from the standpoint of overall survival.

## BEHAVIOR IN THE BARNYARD

At the outset of this chapter, it was stated that the application of animal behavior depends upon understanding. The presentation to this point has been for the purpose of understanding. Let us next turn to some practical applications of animal behavior—its application in the barnyard. The following sections pertain thereto.

### BREEDING FOR ADAPTATION

The wide variety of livestock in different parts of the world reflect a continuous process of natural and artificial selection which has resulted in the survival of animals well adapted to climate and other environmental factors. Among the examples are: haired sheep (devoid of wool) in desert areas, fat-tailed sheep in arid zones, *Bos indicus* (Zebu) types of cattle in tropical areas, and *Bos taurus* cattle in temperate zones. Such adaptations relate to survival of the animals, but they do not necessarily entail maximum productivity of food for people. European cattle usually have much higher yields of milk and propensities for rapid growth than have the breeds native to Africa or India. It is understandable, therefore, why there have been many attempts to introduce improved European livestock into countries in which the productivity of native stock is low. But there are many problems in breed replacement, with the result that a large number of experimental introductions of new breeds have not been successful. Tropical Africa provides an example. Because of disease problems, poor resistance to high temperatures, and limited feed supplies, many of the attempts made by former colonial powers to improve the output of native stock by replacing them with the European breeds failed. Breed replacement or a simple crossbreeding system might seem to be a panacea for low productivity. However, unless associated with special provisions for subsequent importation of breeding stock and simultaneous improvement of the nutritional, parasitological, disease, and husbandry environments of the crossbreds, it is not likely to succeed.

But animals can be changed through heredity and selection. For example, in Israel, which has the highest average milk yields per cow of any nation in the world, the flight distance between animals approaches contiguousness in some herds; this is due to Israel having selected intensively for docility for 30 years. In other words, the animals are literally touching each other, with no antagonistic or dominant-type response. This allows them to concentrate their animals even more than they had previously, thereby giving them a higher productivity per unit area. The only problem reported by Israel is that estrus, or heat, in animals in close proximity is difficult to detect.

### QUICK ADAPTATION—EARLY TRAINING

We need to breed and select animals that adapt quickly to people-made environment—animals that not only survive, but thrive, under the conditions that people impose upon them. Actually, they may be less intelligent than their wild ancestors.

Also, early training and experience are extremely important. In general, young animals learn more quickly and easily

than adults; hence, advance preparation for adult life will pay handsome dividends. The optimum time for such training varies according to species. Thus, the ideal time to introduce a puppy to its life work is between 8 and 12 weeks of age, whereas the establishment of feed preference in chicks can be traced back to their first meal. Also, stress can be reduced or avoided entirely if animals proceed through a graduated sequence of events leading to an otherwise noxious experience. Preconditioning of cattle is an application of the latter principle to production practices. If calves are properly preconditioned (started on feed, vaccinated, treated for parasites, etc.) prior to weaning, the stress of subsequent weaning and movement to a feedlot is minimized.

### LOADING CHUTE AND CORRAL DESIGN

With knowledge of basic behavior patterns and of social habits within the herd, we have at our disposal the necessary tools for designing facilities and housing which will enable us to make animals do what we want them to do, when we want them to do it—and with a saving in both labor and tempers. As an example, let us consider the matter of putting cattle through a chute on their own accord without interference or any extra driving from the handlers.

At the outset, it is recognized that cattle will follow the leader, and that they will automatically try to escape through a gap, or opening. Hence, they will follow the leader through a curved chute much more easily than through a straight one. As the lead animal approaches the chute and realizes that there is an opening to either left or right, it goes forward with the idea of going through this gap—and escaping. Thus, if one can get the leader-type cow into the chute first, it's a simple matter to get the rest of the herd to follow through the curved chute. So, the practical application of cattle behavior and social habits to handling facilities calls for a curved chute, with a curved entrance on the outer portion of the corral and the normal straight side on the funnel portion of the corral—that is, on the inside.

The corral should always match the work requirements, labor, and herd size. The chute should always be curved; and the corral should be designed so that there is always a gap to the left or right. Then, if the animals have been selected for

Fig. 7-25. Cattle entering a curved loading chute. (Courtesy, Wm. R. Farr, Greeley, CO)

their lack of dominance, they will automatically enter the funnel portion, thence the chute, thence proceed at their own pace throughout the whole facility.

• **The corral**— By designing corral facilities in an entirely different fashion from the traditional— by abandoning the straight-sided corral—a facility can be designed that will lend itself to very rapid and adequate handling of stock, with a minimum of interference from the handlers. This has the advantages of (1) cutting down stress, (2) speeding up the work, and (3) distributing the cattle after an examination for a particular series of operations or cutouts.

Fig. 7–26.  Corral system on the Gordon Whiting Ranch, Klondike, AZ. Designed by Temple Grandin. (Courtesy, Grandin Livestock Handling Systems, Tempe, AZ)

## MANURE ELIMINATION

Body waste is a major concern; although unavoidable, it is expensive and time-consuming to handle, and it may create a major pollution problem. But manure handling can be facilitated by an understanding and application of eliminative behavior.

Cattle are indiscriminate eliminators. Even so, this trait can be used effectively. For example, if cattle are fed at the same time each day, feed is released from the rumen into the true stomach regularly, causing a gastro-colic reflex. When this happens and cattle are put under slight stress, they defecate. Knowing this, cattle can be moved to the defecating area at the right time.

Pigs can be trained to defecate in a particular area, separate and apart from their sleeping and resting areas, by: (1) keeping defecating areas cool and resting areas warm, as pigs like to defecate where it is cool and rest where it is warm; (2) locating the water near the defecating area, as a wet floor is conducive to elimination; (3) having the defecating area 1–2 in. below the feeding and resting areas seems to direct the pigs to where the defecating should occur; (4) wetting down the defecating area; and (5) feeding on the floor of the eating and sleeping areas for a few days will encourage pigs to keep these areas clean and to defecate elsewhere.

## CONTROL OF PESTS

Behavioral studies are involved in pest control. This is so because successful pest control is based on two approaches:

(1) finding the weakness of the pest, and (2) developing animals more resistant to, or able to cope with, pests. Pesticide development must consider feeding habits, motility and perception of the organisms, and pollution. Parasite control must often consider the behaviors of the host organism and of the various life forms of the parasite itself. Often more knowledge is needed relative to the lives of the afflicted animals.

Screwworm control is the classic example of applied behavior in parasite control. The eradication program, first initiated by the U.S. Department of Agriculture in 1958, takes advantage of the fact that screwworms mate only once. Here is how it works: Masses of screwworm larvae are reared on artificial media. Two days before the fly emerges, the pupae are exposed to gamma irradiation at a dosage which causes sexual sterility but no other deleterious effects. Sterile flies are then distributed over the entire screwworm infested region in sufficient quantity to outnumber the native flies, at an average rate of about 400 males per square mile per week. The female mates only once and, therefore, when mated with a sterile male does not reproduce. There is a decline in the native population each generation until the native males are so outnumbered by sterile males that no fertile matings occur, and the native flies are eliminated.

## TRAINING HORSES

There are as many successful ways to train horses as there are to train children. The author has observed top professional trainers. Each used a different technique, yet all ended up with the same result—a champion. Most of them apply the basic principles of behavior given herein.

The good trainer who has followed a program of training and educating the foal from the time it was a few days old has already eliminated the word *breaking*. The saddling and/or harnessing of the young horse is merely another step in the training program, which is done with apparent ease and satisfaction.

Each animal species has characteristic ways of performing certain functions, and rarely departs therefrom. The horse is no exception. A good understanding of horse behavior enhances horse training.

Fig. 7–27.  Training a yearling on "how to be a racehorse." Learning to gallop beside a lead pony is one of the first lessons. (Courtesy, *From Dawn to Destiny*, by Frank Jennings and Allen F. Brewer, Jr., Published by The Thoroughbred Press, Lexington, KY)

• **Memory of horses**—To a very considerable degree, the horse's aptitude for training is due to its memory. An example which substantiates the excellent memory of horses follows:

In the days of Mohammed, intelligence and obedience were the main requisites of the Arab's horse. For war purposes, only the most obedient horses were used, and they were trained to follow the bugle.

Legend has it that the Prophet Mohammed had need for some very obedient horses, so he inspected a certain herd to make personal selections.

The horses from which he wished to make selection were pastured in a large area bordering on a river. The Prophet gave orders that the animals should be fenced off from the river until their thirst became very great.

He then ordered the fence removed, and the horses rushed for the water. When they were just about to dash into the river to quench their thirst, a bugle was sounded.

All but ten of the horses ignored the call of the bugle. The obedient ten remembered—they turned and answered the call of duty, despite their great thirst. The whimsical story goes on to say that these ten head constituted the foundation of the Prophet Strain.

• **Controlling the horse**—The horse has whims and ideas of its own; but the caretaker should be boss, with the animal promptly responding as desired. With experienced equestrians, this relationship is clear-cut, for they are able to relay their feelings to the horse instantly and unmistakably. For complete control and a finished performance, the horse should have a proud and exalted opinion of itself; but, at the same time, it should subjugate those undesirable traits that make a beast of its size and strength so difficult to handle by a comparatively small and frail person. Complete control, therefore, is based on mental faculties rather than muscular force. The faculties of the horse which must be understood and played upon to obtain skillful training and control at all times are: memory, confidence and fear, association of ideas, and willingness.

## COMPANIONSHIP

Companionship in animals is of great practical importance. Except for the cat, all domestic animals are highly social and have constant need for companionship.

If not too crowded, placing animals together sometimes accomplishes two things: (1) greater feed consumption, due to the competition between them (mutual facilitation); and (2) a quieting effect. For example, nervous high-strung boars are frequently provided with a barrow as a companion.

Among horses, the two strongest sources of motivation are (1) the desire to be with other horses (companionship), and (2) the desire to go home. Thus, a pack train will always go better if the leader rides a fast horse. Then the slower ones will try to keep up, rather than be left behind. When away from companions, a horse will try to go home.

The best-known animal companionship of all pertains to high-strung racehorses and stallions, in which all sorts of companions are used—a goat, a sheep, a chicken, a duck, or a pony. Such companions are commonly referred to as *mascots.*

The expression "to get his goat" was born of the common custom of having goats for mascots. Back in the days when skulduggery was as important as form in winning races, the caretakers of one stable sometimes plotted to kidnap the goat mascot of a rival's horse. By "getting the goat" of a favorite, they cleaned up by betting against a horse that was odds-on to win, but too upset to run at its best.

The great *Stymie*, Thoroughbred winner of $918,485, became attached to a hen of nondescript breeding who came to dinner one day and never left.

Probably the most publicized mascot of all times was the pony, *Peanuts*, constant companion to the Thoroughbred racehorse *Exterminator*. *Peanuts* died three years ahead of the great old gelding. When his pony-pal failed to appear in the stall the next morning, *Exterminator* stopped eating. He would likely have died of a broken heart had his handlers not acted wisely. They left the remains of *Peanuts* in the Thoroughbred's stall one night, to demonstrate to him that his mate was dead. All night long, *Exterminator* lay with his head over the pony's body. By morning he was resigned to the situation. A new pony was brought to him, and the old warrior carried on.

## SUMMARY OF BEHAVIOR

People need animals, and animals need people. People domesticated the dog to assist in their hunting and to provide protection by night. Gradually, people adopted a more settled mode of life, and with this came the desire to safeguard their food supply for times when hunting was poor and to have their food close at hand; at this stage, nearly all our modern animals were tamed or confined, or, as we say, domesticated.

In domesticating animals, people recognized the importance of behavior; they selected those species which could both be tamed and used to satisfy their needs. However, in the breeding, care, and management that followed, behavior received less attention than the quantity and quality of the meat, milk, eggs, fiber, and power produced. The race was on for greater rate and efficiency of production. Animals in forced production were confined and automated. Then, suddenly, in sign language that spoke louder than words, animals told us that all was not well in the barnyard. They told us that something was missing—something as vital to them as an essential amino acid, mineral, or vitamin—something as important as disease prevention and environmental control. They told us that consideration of their habitat and social organization had not kept pace with advances in genetics, nutrition, environmental control, and other areas of animal care. They told us what was wrong and what they wanted through a whole host of abnormal behaviors, including cannibalism, loss of appetite, poor parental care, overaggressiveness, dullness, degenerate sexual behavior, tail biting, and cribbing. These warning signals are being heeded. Today, there is renewed interest in the study and application of animal behavior; we are trying to make it right with animals by correcting the causes of the disorders. For the time being, this calls for emulating the natural conditions of the species, including their space requirements, social organization, and training and experience. Over the long pull, it calls for breeding and selecting animals better adapted to people-made environments. It is hoped that the principles and applications of animal behavior presented in this chapter will speed the process.

# PART II.
# ANIMAL ENVIRONMENT

An animal is the result of two forces—heredity and environment. Heredity has already made its contribution at the time of fertilization, but environment works ceaselessly away until death. Since most animal traits are only 30 to 50% heritable, the expression of the rest (more than 50%) depends on the quality of all components of the environment. Thus, it is very important that the keeper of herds and flocks have enlightened knowledge of, and apply expert management to, animal environment.

*Environment may be defined as all the conditions, circumstances, and influences surrounding and affecting the growth, development, and production of animals.* The most important influences in the environment are the feed and quarters (space and shelter).

*The branch of science concerned with the relation of living things to their environment and to each other is known as ecology.*

Through the years, the domesticated animals best suited to a particular environment survived, and those that were poorly adapted either moved to a more favorable environment or perished. During the past two centuries, livestock producers have made great strides in the selection and propagation of animals suited to a particular environment, and during the past 50 years they have made progress in modifying the environment for the benefit of their animals and themselves.

It is becoming increasingly difficult to define environment, because scientists continue to discover important new environmental factors. Primitive people recognized that the sun and fire provided both heat and light, that body heat could be conserved by draping the body with animal skins, and that trees and caves provided protection from the weather. Today, it is recognized that these, along with a host of other environmental factors affect animals and people.

The keepers of herds and flocks were little concerned with the effect of environment on animals so long as they grazed on pastures or ranges. But rising feed, land, and labor costs, along with the concentration of animals into smaller spaces, changed all this. Today, most layers and broilers are on litter floors. Turkeys are shifting rapidly from range to confinement. Water is important for ducks, but even with ducks the trend is toward higher population densities and more confinement. Many swine are raised partially or totally in confinement; and confinement production is increasing with beef cattle, dairy cattle, and sheep.

Among animals, environmental control involves space requirements, light, air temperature, relative humidity, air velocity, wet bedding, ammonia buildup, dust, odors, and manure disposal, along with proper feed and water. Control or modification of these factors offers possibilities for improving animal performance. Although there is still much to be learned about environmental control, the gap between awareness and application is becoming smaller. Research on animal environment has lagged, primarily because it requires a melding of several disciplines—nutrition, physiology, genetics, engineering, and climatology. Those engaged in such studies are known as ecologists.

## EFFECT OF ENVIRONMENTAL FACTORS ON ANIMALS

Preventing disorders by merely cutting off the tails of pigs to alleviate tail biting, debeaking poultry to prevent cannibalism, and using choke collars on horses to inhibit cribbing, is not unlike trying to control malaria fever in humans by the use of drugs without getting rid of mosquitoes. Rather, we need to recognize these disorders for what they are—warning signals that conditions are not right. Correcting the cause of the disorder is the best solution. Unfortunately, this is not usually the easiest. Correcting the cause may involve trying to emulate the natural conditions of the species, such as altering space per animal and group size, providing training and experience at opportune times, promoting exercise, and gradually changing rations. Over the long pull, selection provides a major answer to correcting confinement and other behavioral problems; we need to breed animals adapted to people-made environments.

The following factors are of special importance in any discussion of animal environment:

1. Feed
2. Water
3. Weather
4. Facilities
5. Health
6. Stress

## FEED/ENVIRONMENTAL INTERACTIONS

Animals may be affected by either (1) too little or too much feed, (2) rations that are too low in one or more nutrients, (3) an imbalance between certain nutrients, or (4) objection to the physical form of the ration—for example, it may be ground too finely.

Forced production (such as growth, milk products, and racing 2-year-old horses) and the feeding of forages and grains which are often produced on leached and depleted soils have created many problems in nutrition. These conditions have been further aggravated through the increased confinement of animals, many animals being confined to stalls or lots all or a large part of the year. Under these unnatural conditions, nutritional diseases and ailments have become increasingly common.

Also, nutritional reproductive failures plague livestock operations. Generally speaking, energy supply tends to be more limiting than protein in reproduction. The level and kind of feed before and after parturition will determine how many females will show heat and conceive. After giving birth, feed requirements increase tremendously because of milk production; hence, a female suckling young needs approximately 50% greater feed allowance than during the pregnancy period. Otherwise, she will suffer a serious loss in weight, and she may fail to come in heat and conceive. This basic fact, along with other pertinent findings, was confirmed by researchers at the Montana Agricultural Experiment Station. Based on 12 years research at the Havre and Miles City Stations, they concluded that beef cattle size and milk production should be tailored to fit the environment. Big size and more milk are not better unless the range forage supply is better. The best size cow is one that fits the range conditions. Small cows do best on poor range because they can usually get 100% of their daily feed requirement for maintenance and milk production, whereas big cows on a poor range are borderline hungry all the time. Also, cows that give a lot of milk must have a good range; otherwise, they are stressed by lack of feed; and their fertility rate and calf crops drop. So, cow size and milk production should match their environment.

The next question is whether a breeding program can make maximum progress under conditions of suboptimal nutrition (such as is often found under some farm and range conditions). One school of thought is that selection for such factors as body form and growth rate in animals can be most effective only under nutritive conditions promoting the near maximum development of those characters of which the animal is capable. The other school of thought is that genetic differences affecting usefulness under suboptimal conditions will be expressed under such suboptimal conditions, and that differences observed under forced conditions may not be correlated with real utility under less favorable conditions. Those favoring the latter thinking argue, therefore, that the production and selection of breeding animals for the range should be under typical range conditions and that the animals should not be highly fitted in a box stall.

The results of a 10-year experiment conducted by the senior author and his colleagues at Washington State University, designed to study the effect of plane of nutrition on meat animal improvement, support the contention that selection of breeding animals should be carried on under the same environmental conditions as those under which commercial animals are produced.[2]

## WATER/ENVIRONMENTAL INTERACTIONS

Animals can survive for a longer period without feed than without water. Water is one of the largest constituents in the animal body, ranging from 40% in very fat, mature animals to 80% in newborn animals. Deficits or excesses of more than a few percent of the total body water are incompatible with health, and large deficits of about 20% of the body weight lead to death.

The total water requirement of animals varies primarily with the weather (temperature and humidity); feed (kind and amount); the species, age, and weight of animal; and the physiological state. The need for water increases with increased intakes of protein and salt, and with increased milk production of lactating animals. Water quality is also important, especially with respect to the content of salts and toxic compounds.

It is generally recognized that animals consume more water in summer than in winter. Based on 5 summer and 4 winter trials, the Iowa State Agricultural Experiment Station reported that yearling cattle consumed an average of 8.5 gal per day in summer vs 5 gal per day in winter.

[2]Fowler, S. H., and M. E. Ensminger, *Relationship of Plane of Nutrition to the Improvement of Swine for Meat Production Through Selection*, Tech. Bull. 34, Washington State University, Pullman, 1961.

The water content of feeds ranges from about 10% in air-dry feeds to more than 80% in fresh, green forage. Feeds containing more than 20% water are known as *wet feeds*. The water content of feeds is especially important for animals which do not have ready access to drinking water. Also, the water on the surface of plants, such as dew, may serve as an important source for cattle, sheep, and goat on arid ranges, but this supply is rarely sufficient to meet their needs.

Home range behavior exists among most wild and feral animals, with each home range including at least one watering hole—generally a stream, spring, lake or pond. The frequency of watering wild and feral animals, as well as of domestic animals on extensive pastures or ranges, is determined primarily by temperature and humidity—the higher the temperature and humidity, the more frequent the watering. However, under average conditions, the frequency of watering of different species is as follows: Cattle, 1 to 4 times per day; sheep, 1 time per day, although when grazing desert ranges in early spring, they may go for weeks without drinking water; goats, 1 time per day, although goats approach camels in their water requirements. When hand-fed, pigs generally eat all their feed, then drink; when self-fed, they alternate between eating and drinking. Horses, usually drink once each day, but during extreme heat they may return a second time. Poultry drink frequently, alternating between eating and drinking. The frequency of watering of cattle and sheep decreases as distance to water increases.

Under practical conditions, the frequency of watering is best determined by the animals, by allowing them access to clean, fresh water at all times.

## WEATHER/ENVIRONMENTAL INTERACTIONS

*Webster defines weather as a state of the atmosphere with respect to heat or cold, wetness or dryness, calm or storm, clearness and cloudiness.*

Extreme weather can cause wide fluctuations in animal performance. The difference in weather impact from one year to the next, and between areas of the country, causes difficulty in making a realistic analysis of buildings and management techniques used to reduce weather stress.

The research data clearly show that winter shelters and summer shades improve production and feed efficiency. The issue is clouded only because the additional costs incurred by shelters have frequently exceeded the benefits gained by the improved performance, particularly in those areas with less severe weather and climate.

The maintenance requirement of animals increases as temperature, humidity, and air movement depart from the comfort zone. Likewise, the heat loss from animals is affected by these three factors. Animals adapt to weather as follows:

• **In cold weather,** the heating mechanisms are employed, including (1) increased insulation from growth of hair and more subcutaneous fat; (2) increase in thyroid activity; (3) seeking protective shelter and warming solar radiations (the animals sun themselves); (4) huddling together; (5) consumption of more feed, which increases the heat increment and warms animals; and (6) increasing activity. The most important animal body heating mechanisms are amount of feed consumed and body activity, which are also evidenced in people. For example, after skiing in bitter cold weather, a skier feels comfortable after eating a beefsteak; and during a marathon race, a runner may feel quite warm when the temperature is near freezing (30°F).

- **In hot weather,** the cooling mechanisms are employed, including (1) moisture vaporization (from the skin and lungs), (2) avoidance of the heating solar radiation (the animals seek shade), (3) depression of thyroid activity, and (4) loafing (including lessening the production of meat, milk, and eggs, since they increase heat production).

## THERMONEUTRAL ZONE (COMFORT ZONE)

Fig. 7–28 and the definitions that follow are pertinent to an understanding of thermal zones.

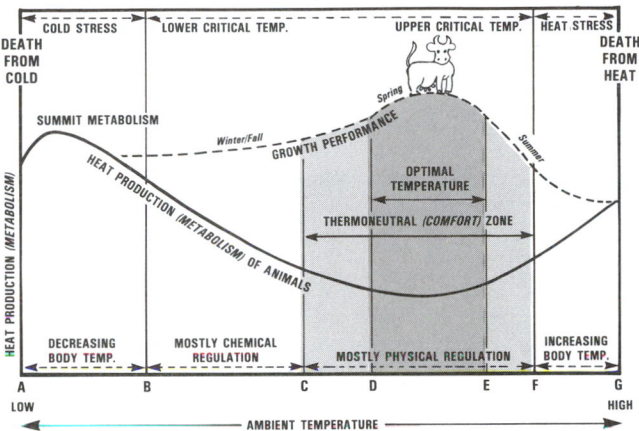

Fig. 7–28. Diagram showing (1) the influence of thermal zones and temperature on homeotherms (warm-blooded animals), and (2) the peak of milk yields in the spring, followed by the summer slump due to high (hot) summer temperature and lignification of forage.

In Fig. 7–28, *heat production (metabolism)* is plotted against *ambient temperature* to depict the relationship between chemical and physical heat regulation. Note, too, the broad range of accommodation to low (cool) temperatures in contrast to the restricted range of accommodation to high (warm) temperatures. Definitions of terms pertaining to Fig. 7–28 follow.

- **Thermoneutral (comfort) zone** (C to F) is the range in temperature within which the animal may perform with little discomfort, and in which physical temperature regulation is employed.

- **Optimum temperature** (D to E) is the temperature at which the animal responds most favorably, as determined by maximum production (gains, milk, wool, work, eggs) and feed efficiency.

- **Lower critical temperature** (B) is the low point of the cold temperature beyond which the animal cannot maintain normal body temperature. The chemical temperature regulation is employed in the zone below C. When the environmental temperature reaches below point B, the chemical-regulation mechanism is no longer able to cope with the cold, and the body temperature drops, followed by death. The French physiologist, Giaja, used the term *summit metabolism* (maximum sustained heat production) to indicate the point beyond which a decrease in ambient temperature causes the homeothermic mechanisms to break down, resulting in a decline in both heat production and body temperature and eventually death of the animal.

- **Upper critical temperature** (F) is the high point on the range of the comfort zone, beyond which animals are heat stressed and physical regulation comes into play to cool them.

The cow produces the maximum yield of milk during the spring when the temperature is optimum (D to E), and the minimum yield in the summer when it is hot (F to G).

The comfort zone, optimum temperature, and both upper and lower critical temperatures vary with different species, breeds, ages, body sizes, physiological and production status, acclimatization, feed consumed (kind and amount), the activity of the animal, and the opportunity for evaporative cooling.

The temperature varies according to age, too. For example, the comfort zone of newborn lambs is 75 to 80°F, whereas the comfort zone of mature sheep is 45 to 75°F.

Animals that consume large quantities of roughage or high-protein feeds produce more heat during digestion; hence, they have a different critical temperature than the same animals fed a high-concentrate, moderate-protein ration. Because of this, experienced cattle feeders decrease the roughage and increase the concentrate of finishing cattle during the hot summer months.

Stresses of both high and low temperatures are increased with high humidity. The cooling effect of evaporating sweat is minimized and the respired air has less of a cooling effect. As humidity of the air increases, discomfort at any temperature, and nutrient utilization, decrease proportionately

Air movement (wind) results in body heat being removed at a more rapid rate than when there is no wind. In warm weather, air movement may make the animal more comfortable, but in cold weather it adds to the stress temperature. At low temperatures, the nutrients required to maintain the body temperature are increased as the wind velocity increases. In addition to the wind, a drafty condition where the wind passes through small openings directly onto some portion or all of the animal body will usually be more detrimental to comfort and nutrient utilization than the wind itself.

## ADAPTATION, ACCLIMATION, ACCLIMATIZATION, AND HABITUATION OF SPECIES/ BREEDS TO THE ENVIRONMENT

Every discipline has developed its own vocabulary. The study of adaptation/environment is no exception. So, the following definitions are pertinent to a discussion of this subject:

*Adaptation* refers to the adjustment of animals to changes in their environment.

*Acclimation* refers to the short-term (over days or weeks) response of animals to their immediate environment.

*Acclimatization* refers to evolutionary changes of a species to a changed environment which may be passed on to succeeding generations.

*Habituation* is the act or process of making animals familiar with, or accustomed to, a new environment through use or experience.

Species differences in response to environmental factors result primarily from the kind of thermoregulatory mechanism provided by nature, such as type of coat (hair, wool, feathers), and sweat glands. Thus, hogs, which have a light coat of hair, are very sensitive to extremes of heat and cold. On the other hand, nature gave cattle an assist through growing more hair for winter and shedding hair for summer, with the result that they can withstand higher and lower temperatures than hogs. The long-haired, shaggy yak of Tibet and the wooly Scotch

Fig. 7–29. Long and wooly haired Scotch Highland cattle equipped by nature to withstand very cold weather. (Courtesy, American Scotch Highland Breeders, Assn., Edgemont, SD)

Fig. 7–30. Environmentally controlled dairy barn, with automatic feed auger, automatic waterer, and free stalls. (Courtesy, Babson Bros. Co., Chicago, IL)

Highland cattle of Scotland are as cold tolerant as the arctic-dwelling caribou and reindeer.

From time to time, American buffalo (*Bison bison*) and domestic beef cattle (*Bos taurus*) have been crossed to obtain a more hardy beast than cattle. The most publicized early work of this type was the development of the Cattalo (bison × domestic cattle), the initial cross for which was made at the Dominion Experiment station, at Scott, Saskatchewan, in Canada, in 1915.

Male fertility and female reproductive rate have remained a problem in Cattalo. Although unquestionably hardy, bison × cattle crosses can be outperformed in nearly all environments by the currently available cattle breeds or crosses; and management procedures.

Also, there are breed differences, which make it possible to select animals well adapted to specific environments. Thus, the breeds of cattle that originated in the British Isles and Northern Europe are cold tolerant, whereas the Indian-evolved Zebu, or Brahman, cattle are heat tolerant. The long-fibered Black-faced Highland sheep are cold tolerant; the haired sheep are suited to hot, desert areas; and the fat-tailed sheep are adapted to arid conditions. The Shetland Pony, native to the Shetland Isles, not more than 400 miles from the Arctic Circle, evolved in the rigors of the northland climate and on sparse vegetation, which imparted that hardiness for which the stocky breed is famed. The long-legged donkey is adapted to hot, desert areas.

In recent years, attempts have been made to combine the heat tolerance characteristics of tropical breeds with the high productive capacity of European stock. The best known of these planned beef breeds is the Santa Gertrudis, developed on the famed King Ranch of Texas, in the early 1900s, which carry approximately five-eighths Shorthorn and three-eighths Brahman breeding.

## FACILITIES/ENVIRONMENTAL INTERACTIONS

Optimum facility environments can only provide the means for animals to express their full genetic potential of production, but they do not compensate for poor management, health problems, or improper rations.

Research has shown that animals are more productive and feed-efficient when raised in an ideal environment. The primary reason for having facilities, therefore, is to modify the environment. Proper barns and other shelters, shades, sprinklers, insulation, ventilation, heating, air conditioning, and lighting can be used to approach the desired environment. Also, increasing attention needs to be given to other stress sources such as space requirements, and the grouping of animals as affected by class, age, size, and sex.

The principal scientific and practical criteria for decision making relative to the facilities for animals in modern, intensive operations is the productivity and cost of production of animals, which can be achieved only by healthy animals under minimal stress. So, the investment in environmental control facilities is usually balanced against the expected increased returns.

Temperature, humidity, and ventilation recommendations for different classes of livestock are given in Chapter 5 of this book, Table 5–4, "Recommended Environmental Control for Animals." This table will be helpful in obtaining a satisfactory environment in confinement livestock buildings, which require careful planning and design.

In recent years, there has been a trend to modify the environmental control facilities as much as possible; among such modifications designed for maximum animal comfort and efficiency of production are fans, floors, lights, shades, sprinklers/sprayers/foggers, ventilation, wallows, and windbreaks.

## HEALTH/ENVIRONMENTAL INTERACTIONS

*Health is the state of complete well-being, and not merely the absence of disease.*

*Environment embraces the forces and conditions, both physical and biological, that (1) surround animals, and (2) interact with heredity to determine behavior, growth, and development.*

*Disease is defined as any departure from the state of health.*

*Parasites are organisms living in, on, or at the expense of another living organism.*

Feed, air quality, lighting, noise, other animals, and weather are among the many factors that constitute an animal's environment. Extremes or alterations in the environment may subject an animal to stress; and stress may affect health and lead to more diseases and parasites.

The importance of good animal health is underscored by the following statistics: It is estimated that the animal diseases and parasites in the United States (1) decrease animal productivity by 15 to 20%; and (2) make for annual losses equivalent to 15% of the annual cash farm income from marketing livestock and products, on which basis the estimated livestock losses for 1987 totalled $11.4 billion. Further, there is evidence that nutrition has some involvement in 85% of the veterinary cases. In the developing countries, diseases and parasites take an even greater toll—they decrease animal productivity by 30 to 40%.

Some important health/environmental interactions not covered elsewhere in this book are discussed in the sections that follow.

## TRANSMITTING DISEASES AND PARASITES TO HUMANS

The progress of humans, from cave to condominium, has been greatly influenced by animals. From the remote day of their domestication forward, the most advanced civilizations of the day have been the keepers of herds and flocks. Although progress walks Indian file behind animals, certain diseases follow. Many of these diseases are transmitted through meat, milk, or eggs; others are transmitted through close contact with animals, contact with their excreta, or contact with their products—such as hides, wool, or hair; still others are carried from animals to humans by insect vectors.

This group of shared animal/human diseases is known as *zoonoses*. It is important to realize that zoonotic diseases cannot be prevented or controlled in people except through their control in animals. The common zoonotic diseases transmitted from animals to humans are: anthrax, brucellosis, leptospirosis, Q fever, rabies, trichinosis, trypanosomiasis (African sleeping sickness), tuberculosis, and tularemia. A summary of these diseases is presented in Chapter 6 of this book, Table 6–3.

## STRESS/ENVIRONMENTAL INTERACTIONS

*Stress is the nonspecific response of the body to any demand.*

As used herein, stress indicates an environmental condition that is adverse to an animal's well-being, either external (nutritional, weather, social) or internal (disease, parasites).

Stresses of many kinds affect animals; among them, cold stress, heat stress, drafts, poor ventilation, excitement, presence of strangers, fatigue, mixing animals, number of animals together, space, changing corral and corral mates, weaning, previous nutrition, hunger, thirst, poor sanitation, disease, parasites, surgical operations, injury, and management.

Race and show horses are always under stress; and the greater the speed and the more tired they become, the greater the stress. Also, the greater the stress, the more exacting the nutritive requirements. Thus, the ration of race and show horses should be scientifically formulated.

Animals can be prepared, or adapted, to the environment, in such a manner as to reduce stress. For example, if calves are properly *preconditioned* (started on feed, vaccinated, treated for parasites, etc.) prior to weaning, the stress of subsequent weaning and movement to a feedlot will be minimized.

In the life of an animal, some stresses are normal, and they may even be beneficial—they can stimulate favorable action on the part of an individual. Thus, we need to differentiate be-

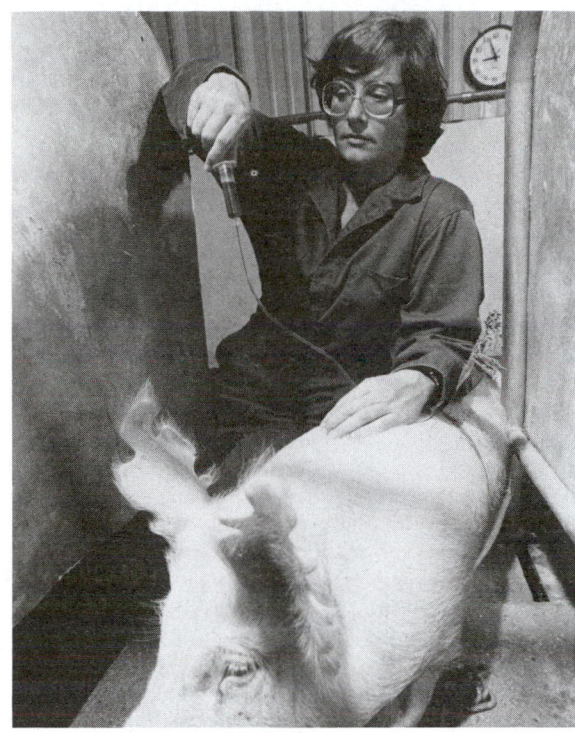

Fig. 7–31. Pigs are stressed, too. This shows a U.S. Department of Agriculture Environmental Physiologist taking a blood sample from a confined hog in an animal stress study at the U.S. Animal Research Center, Clay Center, NE. (Courtesy, USDA)

tween stress and distress. Distress—not being able to adapt—is responsible for harmful effects. The trick is to manage stress so that it doesn't become distress and cause damage and to recognize the warning signals of distress. For example, Texas Agricultural Experiment Station workers recently reported that added vitamin C, in either the feed or water, may reduce many of the health hazards associated with various kinds of stress to chickens such as hot weather, interaction with other birds in crowded conditions, and exposure to diseases.

The principal criteria used to evaluate, or measure, the well-being or stress of people are: increased blood pressure, increased muscle tension, body temperature, rapid heart rate, rapid breathing, and altered endocrine gland function. In the whole scheme, the nervous system and the endocrine system are intimately involved in the response to stress and the effects of stress.

The principal criteria used to evaluate, or measure, the well-being or stress of animals are: growth rate or production, efficiency of feed use, efficiency of reproduction, body temperature, pulse rate, breathing rate, mortality, and morbidity. Other signs of animal well-being, any departure from which constitutes a warning signal, are: contentment, alertness, eating with relish (and cudding by ruminants), sleek coat and pliable and elastic skin, bright eyes and pink eye membranes, and normal feces and urine.

Stress is unavoidable. Wild animals were often subjected to great stress; there were no caretakers to modify their weather, often their range was overgrazed, and sometimes malnutrition, predators, diseases, and parasites took a tremendous toll.

Domestic animals are subjected to different stresses than their wild ancestors, especially to more restricted areas and greater animal density. However, in order to be profitable, their stresses must be minimal.

# POLLUTION OF THE ENVIRONMENT

Pollution is the issue of the decade. Anything that defiles, desecrates, or makes impure or unclean the surroundings pollutes the environment and can have a detrimental effect on animal health and performance. Thus, gases, odorous vapors, and dust particles from animal wastes (feces and urine) in buildings directly affect the quality of the environment. Muddy lots and stray electrical voltage may also pollute the environment. For healthy and productive animals, each of these pollutants must be maintained at an acceptable level. Among the most troublesome animal pollutants are: dust, manure, muddy lots, pests and pesticides, and stray voltage.

## DUST

*Dust may be defined as a mixture of small particles of different sizes of dry matter.*

Dust is a contributing factor to both animal and human health, especially with respect to respiratory diseases. Thus, it should be considered a significant contaminant that adversely affects environmental quality of animal houses and feedlots.

Dust may be present in significant amounts both inside and outside buildings.

Cattle and sheep feedlot dust is both organic, from excreta, and inorganic, from soil. Sprinkling and increased animal density (resulting in more moisture from urine) are the most effective methods of preventing feedlot dust in dry climates or during dry seasons. In swine houses and horse barns, most of the dust comes from the feed. In poultry houses, the dust contains a considerable amount of feather and skin debris, along with particles from the feed and litter. Dust in animal buildings may also carry microbes, gases, and vapors.

## MANURE

In animal agriculture, we need to give particular attention to the pollution caused by manure. One cow produces as much waste as 16 humans. Hence, with 20,000 steers in a feedlot, the disposal problem is equal to a city of 320,000 people. In addition to being used as fertilizer, manure is now being recycled as livestock feed and serving as a source of energy (methane gas).

Of course, there is no one best manure management system for all situations. But, one way or another, science and technology must evolve with ways of disposing of 1.5 billion tons of manure annually; and this must be accomplished without polluting streams or the atmosphere or being offensive to the neighbors.

We must ever be mindful that life, beauty, wealth, and progress depend upon how wisely people use nature's gifts—the soil, the water, the air, the minerals, and the plant and animal life.

If not managed properly, animals may produce the following pollutants in troublesome quantities: manure, gases/odors, dust, and flies/other insects. Also, they may pollute water supplies.

Also, see Chapter 5, section on "Manure."

## MUDDY LOTS

Muddy lots often plague livestock producers, especially during the winter months. Mud increases scours and other

Fig. 7–32. Muddy lots like this one are a source of pollution. (Courtesy, *Wallaces' Farmers & Iowa Homestead*)

diseases in newborn animals and reduces production and feed efficiency in older animals.

California Agricultural Experiment Station studies show that mud can reduce finishing cattle gains by as much as 10 to 35%, and increase the feed required per pound of gain by a like amount. Thus, it is important that the problem be minimized, especially in high rainfall areas. Good drainage is the first essential. This should be assured at the time the lot is located and constructed.

Mounds 6 to 12 ft high, preferably perpendicular to the feed bunk, will provide finishing cattle a dry place on which to lie. Concrete aprons 10 × 12 ft wide and sloping 1 in. per foot along the bunk will provide them with solid footing on which to stand and feed. Also, lessening of cattle density during the winter months—fewer animals per lot—is an effective method of controlling the mud problem. Thus, many feedlots plan to feed fewer cattle during the muddy season.

To cope with the mud and alleviate calf scours, cow-calf operators should move the cows to a clean pasture during the calving season. If no pasture is available, dirt mounds and a coarse straw bedding are recommended.

## PESTS AND PESTICIDES

Although science and technology have been the great multipliers in increasing our food supply, potential food supplies are still destroyed by the ravages of pests. For example, in the high-rainfall belt of tropical Africa, the dreaded tsetse fly has kept large areas out of agricultural production. It is estimated that if *trypanosomiasis,* a disease borne by the tsetse fly, were brought under control the Savannah pastures of the tsetse fly infested area would carry a cattle population of 140 million head, which is more than the cattle population of the United States.

• **Pesticides**—*A pesticide is any substance that is used to control pests.* Pesticides are an integral part of modern agricultural production and contribute greatly to the quality of food, clothing, and forest products we enjoy. Also, they protect our health from disease and vermin. Pesticides have been condemned, however, for polluting the environment, and in some cases for posing human health hazards. Unfortunately, opinions relative to pesticides tend to become polarized. A report by the National Research Council summarized the situation as follows:

> Users of pesticides fear that they will be regulated to the point where pests cannot be effectively controlled, with concomitant losses of food while opponents of the use of pesticides fear that people are being poisoned and that irreversible damage is being done to the environment.[3]

No pest control system is perfect; and new pests keep evolving. So, research and development on a wide variety of fronts should be continued. We need to develop safer and more effective pesticides, both chemical and nonchemical. In the meantime, there is need for prudence and patience.

## STRAY VOLTAGE

Stray electrical voltage has caused serious problems on many dairy farms—affecting animal behavior and lowering milk production, although it may affect other animal species also. Contrary to popular belief, stray voltage is not new; it is as old as electricity itself. However, it has become a problem on many farms recently for two reasons: (1) There is more electrical load on today's farms; and (2) in the last 20 years we have used more equipment grounding for safety purposes.

*Stray voltage is excessive voltage between two animal contact points.* The conditions that cause stray voltage are, electrically, quite simple: If sufficient voltage is present, it may force a current through any available conductor, including a cow's body. Cows are good conductors because of their body design (the length from mouth to front and rear legs); cows bridge the gaps between electrically grounded objects and "true earth." The cow doesn't feel the voltage as such; she feels the tingling current running through her body.

People seldom feel the current for several reasons. Usually, caretakers wear rubber-soled shoes when in the barn, whereas the bare-footed cow stands on concrete that is often wet. Also, humans have only two legs instead of four like the cow, and human's legs touch the floor near the same vicinity.

• **Sources of stray voltage**—Any electrical condition which creates large enough voltage between two animal contact points may create a stray voltage problem. The source of stray voltage may be either "on-farm" or "off-farm."

On-farm voltage problems stem from defective equipment, faulty wiring, bad connections, or having several 120-volt motors on the same line. On-farm stray voltage can be minimized by maintaining good electrical wiring systems that meet the requirements of the National Electric Code. Also, properly balanced 120-volt circuits and conversion of larger 120-volt meters to 240 volts will reduce the effect of secondary neutral voltage drops at the farm service entrance.

Off-farm voltage comes onto the farm through the electrical supplier's lines. Voltage will vary with the load and the natural grounding ability of the area. As usage increases, so may stray voltage. Heavier loads are seen at milking time and in the fall when grain dryers may be running on many farms.

• **Signs of stray voltage**—One or more of the following signs may indicate that stray voltage exists in a dairy:

1. **Cows reluctant to enter the parlor.** When cows are subjected to stray voltage in the parlor stalls, they soon become reluctant to enter the parlor.
2. **Cows nervous in parlor.** Cows often dance or step around almost constantly while in the milking parlor.
3. **Uneven milk let-down and milk-out.** When milk let-down and milk-out are uneven, more machine stripping is required and longer milking time becomes apparent.
4. **Increased mastitis.** When milk-out is incomplete, more mastitis is likely to occur; all that is required is the presence of an infectious bacteria. In turn, this will result in increased somatic count.
5. **Reduced feed intake in the parlor.** If cows encounter stray voltage while eating from the grain feeders, a reluctance to eat and reduced feed intake usually follow.
6. **Reluctance to drink water.** If stray voltage reaches the cows in stall barns through the water supply or metal drinking cups, the animals soon become reluctant to drink.
7. **Lowered milk production.** Each of the symptoms listed above is associated with stress and reduced feed intake, followed by a drop in daily milk production.

But detection of stray voltage is not easy! Other factors such as mistreatment of animals, milking machine problems, disease, sanitation, and nutritional disorders can create problems which manifest themselves in the seven symptoms mentioned above.

• **Use voltmeter to monitor voltage**—The only sure method to determine if significant stray voltage is present is to have a qualified person perform a stray voltage survey, using approved equipment and monitoring the voltage through one, and preferably two, milkings. Point to point measurements between cow and contact points will determine if the voltage is actually getting to the cow. Generally, stray voltage is not constant throughout the day; so, readings should be taken over a long period.

Most milking machine company representatives, many power supplier employees, some milking equipment dealers, and some veterinarians and county extension agents have equipped themselves with suitable voltmeters and are prepared to lend assistance. *Someone familiar with electrical systems, wiring, and equipment should be present when measurements are made.*

## POLLUTION LAWS AND REGULATIONS

Invoking an old law (the Refuse Act of 1899, which gave the Corps of Engineers control over runoff or seepage into any stream which flows into navigable waters), the U.S. Environmental Protection Agency (EPA) launched a program to control water pollution by requiring that all cattle feedlots which had 1,000 head or more the previous year must apply for a permit by July 1, 1971. The states followed suit; although differing their regulations, all of them increased legal pressures for clean water and air. Then followed the Federal Water Pollution Control Act Amendments, enacted by Congress in 1971, charging the EPA with developing a broad national program to eliminate water pollution.

Owners/operators of animal feeding facilities with more than 1,000 animal units must apply. Animal units are computed as follows: multiply number of slaughter and feeder cattle by 1.0; multiply number of mature dairy cattle by 1.4; multiply number of swine weighing over 55 lb by 0.4; multiply the number of sheep by 0.1; and multiply the number of horses

---

[3]*Pesticide Decision Making*, Vol. VII of the Analytical Studies for the U.S. EPA, NRC, National Academy Press, 1978, pp. 14–15.

by 2.0. (See Table 7–1, footnote 1, for what constitutes 1,000 animal units.)

#### TABLE 7–1
#### SUMMARY OF REGULATIONS

| Feedlots with 1,000 or More Animal Units[1] | Feedlots with Less than 1,000 but with 300 or More Animal Units[2] | Feedlots with Less than 300 Animal Units |
|---|---|---|
| **P**ermit required for all feedlots with discharges[3] of pollutants. | **P**ermit required if feedlot— 1. Discharges[3] pollutants through an unnatural conveyance, or 2. Discharges[3] pollutants into waters passing through or coming into direct contact with animals in the confined area. **F**eedlots subject to case-by-case designation requiring an individual permit only after on-site inspection and notice to the owner or operator. | **N**o permit required unless— 1. Feedlot discharges pollutants through an unnatural conveyance, or 2. Feedlot discharges pollutants into waters passing through or coming into direct contact with the animals in the confined area, and 3. After on-site inspection, written notice is transmitted to the owner or operator. |

[1]More than 1,000 feeder or slaughter cattle, 700 mature dairy cows (milked or dry), 2,500 swine weighing over 55 lb *(24.9 kg)*, 500 horses, 10,000 sheep or lambs, 55,000 turkeys, 100,000 laying hens or broilers with continuous overflow watering, 30,000 laying hens or broilers with liquid manure handling, 5,000 ducks; or any combination of these animals adding up to 1,000 animal units.

[2]More than 300 slaughter or feeder cattle, 200 mature dairy cows (milked or dry), 750 swine weighing over 55 lb *(24.9 kg)*, 150 horses, 3,000 sheep, 16,500 turkeys, 30,000 laying hens or broilers with continuous overflow watering, 9,000 laying hens or broilers with liquid manure handling, 1,500 ducks; or any combination of these animals adding up to 300 animal units.

[3]Feedlot not subject to requirement to obtain permit if discharge occurs only in the event of a 25-year, 24-hour storm event.

## SUSTAINABLE AGRICULTURE

Fig. 7–33. Sustainable agriculture involves such ages old, and 21st century new, practices as terracing. (Courtesy, USDA, Soil Conservation Service)

Endangered species—and more! Today, it is endangered planet, endangered people and animals, and endangered agriculture. Among the deluge of warnings of environmental catastrophes are:

• Pollution-caused warming of the atmosphere, known as the *greenhouse effect*, threatening weather changes that could render large areas of the planet unproductive and uninhabitable.

• Toxic and radioactive wastes and dumped garbage that could poison drinking water and despoil the land.

• Chemical pollution that is depleting the atmosphere's protective ozone layer.

• Slashing and burning of tropical rain forests, driving thousands of species to extinction, increasing the amount of carbon dioxide in the atmosphere, and contributing to the greenhouse effect that warms the earth.

Is ¼ lb of hamburger worth ½ ton of Brazil's rain forest? Is 67 sq ft of rain forest (an area about the size of one small kitchen) too much to pay for 1 hamburger? Should we form cattle pastures to produce hamburgers in the Amazon, or should we retain the rain forest and the natural environment? These and other similar questions are being asked too little and too late to preserve much of the great tropical rain forest of the Amazon and its environment. It took nature thousands of years to form the rain forest, but it took a mere 25 years for people to destroy much of it. And when a rain forest is gone, it is gone-forever![4]

Although less dramatic, the Amazon rain forest story has been, or is being, repeated all over the world in the form of the greenhouse effect, toxicities, polluted streams, and/or other harbingers of threats to our environment. Too long we have managed our nonrenewable resources like there is no tomorrow! Now, the situation is being righted. World-wide, environmental quality and economic efficiency are in vogue. In the United States, this movement is called *Sustainable Agriculture.*

Sustainable agriculture is often described as farming that is ecologically sound and economically viable. It may be high or low input, large scale or small scale, a single crop or diversified farm, and use either organic or conventional inputs and practices. Obviously, the actual practices will differ from farm to farm. A definition follows.

*Sustainable agriculture is farming with reduced off-farm purchased inputs of pesticides, herbicides, and fertilizers, along with reduced negative impact on natural resources and improved environmental quality and economic efficiency, while producing and distributing abundant, nutritious, affordable, high-quality foods and' fibers for American and world markets.*

The development of improved crops, cropping systems, irrigation, farm management, and marketing will be needed to make farms more profitable and sustainable. Typically, such farms will rely more on biological resources and management than on nonrenewable inputs of energy and chemicals. The foundation of a sustainable farm system is a comprehensive understanding of the land, the farm resources and operations, and potential short- and long-term markets.

Many of the practices advocated under sustainable agriculture are not new; they involve such timeless agricultural practices as soil erosion control, the protection of groundwater, the use of legumes as a source of nitrogen, biological insect and weed control, and the use of pastures as a primary feed source.

## ANIMAL WELFARE/ANIMAL RIGHTS

In recent years, the behavior and environment of animals in confinement have come under increased scrutiny of animal welfare/animal rights groups all over the world. For example, in 1987 Sweden passed legislation designed (1) to phase out

---

[4]Uhl, C. and G. Parker, "Is a One-Quarter Pound Hamburger Worth a Half-Ton of Rain Forest?," *Interciencir,* 1986. Sept.-Oct., Vol. II, No. 5, p. 213.

layer cages as soon as a viable alternative can be found; (2) to discontinue the use of sow stalls and farrowing crates; (3) to provide more space and straw bedding for slaughter hogs; and (4) to forbid the use of genetic engineering, growth hormones, and other drugs of farm animals except for veterinary therapy. Also, the law provides for fining and imprisoning violators.

Animal welfarists see many modern practices as unnatural, and not conducive to the welfare of animals. In general, they construe animal welfare as the well-being, health, and happiness of animals; and they believe that certain intensive production systems are cruel and should be outlawed. The animal rightists go further; they maintain that humans are animals, too, and that all animals should be accorded the same moral protection. They contend that animals have essential physical and behavioral requirements, which, if denied, lead to privation, stress, and suffering; and they conclude that all animals have the right to live.

Livestock producers know that the abuse of animals in intensive/confinement systems leads to lowered production and income—a case in which decency and profits are on the same side of the ledger. They recognize that husbandry that reduces labor and housing costs often results in physical and social conditions that increase animal problems. Nevertheless, means of reducing behavioral and environmental stress are needed so that decreased labor and housing costs are not offset by losses in productivity. The welfarist/rightists counter with the claim that the evaluation of animal welfare must be based on more than productivity; they believe that there should be behavioral, physiological, and environmental evidence of well being, too. And so the arguments go!

But wild animals were often more severely stressed than domesticated animals. They didn't have caretakers to store feed for winter or to irrigate during droughts; to provide protection against storms, extreme temperatures, and predators; and to control diseases and parasites. Often survival was grim business. In America, the entire horse population died out during the Pleistocene Epoch. Fossil remains prove that members of the horse family roamed the plains of America (especially the area that is now known as the Great Plains of the United States) during most of tertiary period, beginning about 58 million years ago. Yet no horses were present on this continent when Columbus discovered America in 1492. Why they perished, only a few thousand years before, is still one of the unexplained mysteries. As the disappearance was so complete and so sudden, many scientists believe that it must have been caused by some contagious disease or some fatal parasite. Others feel that perhaps it was due to multiple causes, including (1) climatic changes, (2) competition, and/or failure to adapt. Regardless of why horses disappeared, it is known that conditions were favorable to them at the time of their reestablishment by the Spanish conquistadores about 500 years ago.

To all animal caretakers, the principles and application of animal behavior and environment depend on understanding; and on recognizing that they should provide as comfortable an environment as feasible for their animals, for both humanitarian and economic reasons. This requires that attention be paid to environmental factors that influence the behavioral welfare of their animals as well as their physical comfort, with emphasis on the two most important influences of all in animal behavior and environment—feed and confinement.

Animal welfare issues tend to increase with urbanization. Moreover, fewer and fewer urbanites have farm backgrounds. As a result, the animal welfare gap between town and country widens. Also, both the news media and the legislators are increasingly from urban centers. It follows that the urban views that are propounded will have greater and greater impact in the years ahead.

## FOOD SAFETY AND DIET/HEALTH CONCERNS

Many food safety and diet/health concerns are unwarranted. American consumers are prone to over-react to rumors relative to their food. They care little about what they put on their backs, but they are greatly concerned about what goes into their stomachs.

America's food supply is the safest in the world! Nevertheless, there is need for constant vigilance and improvement, especially in animal products which are subject to all the hazards of other foods (spoilage, pesticides, toxicities), plus being capable of transmitting, or serving as passive carriers, of certain diseases to humans.

In colonial times, the livestock producer slaughtered animals and processed meats, milked the cows, and gathered the eggs; then, delivered the products door-to-door to urban customers. If the products were not acceptable (spoiled meat, sour milk, cracked eggs), the matter was resolved quickly and on the spot, or the producer lost a customer. Today, the public expects the livestock team—farmers, processors, and retailers—to provide wholesome and safe products free from disease agents, toxic substances, and pesticide and drug residues.

Uptake of pesticides by animals, leading to residues in animal products, can result either from direct application of pesticides to animals or from animals ingesting feeds carrying pesticide residues. Drug residues are caused by (1) producers failing to withdraw drugs from livestock far enough in advance of marketing products; (2) contaminated feed storage, mixing, and handling equipment; and/or (3) the wastes (feces and urine) of treated animals coming in contact with untreated animals. *Reading and following the directions on the label is the key to safe pesticide and drug use.*

Dr. C. Everett Koop, former U.S. Surgeon General, is authority for the following statement:

People who are worried about pesticides fail to recognize that cancer rates in the United States have dropped remarkably over the past 40 years. During this period of time, stomach cancer has dropped more than 75%, and rectal cancer has dropped more than 65%. The only cancer going up today is cigarette-induced lung cancer.

Dr. Koop continued:

The same hysteria that sometimes accompanies food safety issues also has affected how Americans look at diet and health. However, one major issue—cholesterol—is waning, as consumers realize how much it has been oversold. The cholesterol bubble has been pricked and is slowly deflating. While cholesterol is a risk factor in coronary heart disease, scientists consider other risk factors, such as smoking, hypertension, and heredity, to be much more significant than cholesterol. Because cholesterol is manufactured in the body naturally, diet does not have the direct relationship to blood levels of cholesterol that many misled laymen assume.[5]

[5]Koop, Dr. C. Everett, *Gulf Coast Cattleman*, Vol. 55, No. 9, Dec., 1989, p.9.

Because the welfare of the nation is dependent upon the health of its people, animal (and other) products are carefully monitored by various government agencies to assure consumers that they are wholesome and safe; and because of recognizing the importance of consumers in the safety of their food, the private sector may do additional testing. The agencies most responsible for this important work are:

1. The U.S. Department of Health and Human Services, including the following agencies: The Center for Disease Control, the Food and Drug Administration (FDA), and the National Institute of Health.

2. The U.S. Department of Agriculture, including the following agencies: the Agricultural Research Service, the Animal and Plant Health Inspection Service, the Cooperative State Research Service, the Federal Extension Service, the Labeling and Registration Section, and the Veterinary Service Division.

3. State and local government agencies.

4. International organizations engaged in health and/or nutrition activities, including the World Health Organization (WHO), and Food and Agriculture Organization (FAO).

5. Private industry groups such as the National Livestock and Meat Board and the National Dairy Council.

6. Professional organizations, including dentists, dieticians, doctors, health educators, nurses, and public health workers.

7. Food processors and retailers. For example, in 1988 Lucky Stores, Inc., the largest food handler in California, initiated a testing program in cooperation with the California Department of Food and Agriculture (CDFA) to have the CDFA check their produce for pesticide residues on a regular basis, as a way of assuring their customers that the produce that they buy in Lucky Stores is completely safe. Some food handlers are using private laboratories to conduct similar tests.

(Also see Chapter 6, section on "Regulations Relative to Disease Control.")

# QUESTIONS FOR STUDY AND DISCUSSION

1. Discuss the sign language through which animals tell us how they feel and what they want.

2. What behavioral traits do dogs possess that probably account for their being the first animal domesticated by man?

3. In what way did knowledge of the behavior of animals enter into their capture and domestication?

4. Define animal behavior.

5. Why have (a) increased confinement of animals, and (b) a higher proportion of city-bred caretakers made for great interest in the subject of animal behavior?

6. Animal behavior is caused by, or is the result of, three forces: (a) heredity, (b) training and experience, and (c) intelligence. Cite an example of each. What is meant by imprinting?

7. Why are intelligence comparisons between species meaningful; and why is it absurd to say that one species is smarter than another?

8. Give examples of animal communication by each of the following: (a) sound, (b) chemicals, and (c) visual displays. Give an example of homing instinct.

9. List and discuss briefly each of the nine general functions or behavioral systems that most animals exhibit.

10. Why have ingestive and sexual behavioral systems been more extensively studied than other behavioral systems?

11. Which animal species have the cleanest habits from the standpoint of elimination?

12. Discuss the differences in courting behavior between different animal species.

13. How do you explain the fact that cattle, sheep, and horses face away from a severe storm, whereas buffalo face toward it head on?

14. Those who care for animals need to be familiar with behavioral norms of animals in order to detect and treat abnormal situations—especially illness. Describe a normal cow, sheep, pig, horse, chicken, and turkey.

15. Give one example of abnormal behavior of each species of animal which seems to have resulted from close confinement and changed habitat, and tell what you would do to rectify each condition.

16. Do you foresee the balance of nature curbing further increases in world animal population?

17. Describe the social organization in each of the following species: cattle, sheep, goats, swine, horses, chickens, turkeys, ducks, and geese. How is dominance determined by a group of animals?

18. Is social relationship between species important where cattle and sheep are grazed in the same pasture?

19. Discuss the importance of the people-animal relationship. Why is this so important in guide dogs for the blind?

20. List and discuss the significance of one example of the practical application of animal behavior in the barnyard in each of the following areas:
   a. Breeding for adaptation.
   b. Quick adaptation—early training.
   c. Loading chute and corral design.
   d. Manure elimination.
   e. Control of pests.
   f. Training horses.
   g. Companionship.

21. Should all animal science majors in colleges and universities take one animal behavior course? Justify your opinion.

22. Define the following terms: *environment* and *ecology*.

23. Why were the keepers of herds and flocks little concerned with the effect of environment on animals so long as they grazed on pastures and ranges?

24. Why is rectifying abnormal behavior by cutting off the tails of pigs to alleviate tail biting likened to trying to control malaria fever in humans by the use of drugs without getting rid of the mosquitoes?

25. Discuss how feed/environmental interactions affect animals.

26. Discuss how water/environmental interactions affect animals. Under practical conditions, how is the frequency of watering animals best solved?

27. Define each of the following terms:
    a. Thermoneutral zone (comfort zone).
    b. Optimum temperature.
    c. Lower critical temperature.
    d. Upper critical temperature.

28. Discuss how each of the following weather/environmental interactions affects animals:
    a. Cold weather animal heating mechanisms.
    b. Hot weather animal cooling mechanisms.
    c. Thermoneutral zone (comfort zone).
    d. Adaptation, acclimation, acclimatization, and habituation of species/breeds to the environment, including pertinent facts about the buffalo × domestic cattle cross made in Canada.

29. Discuss how facilities/environmental interactions affect animals.

30. Discuss how health/environmental interactions affect animals.

31. Define each of the following terms: *health, disease, parasite.*

32. Define the term *zoonoses.* List the common zoonoses.

33. Discuss how stress/environmental interactions affect animals.

34. What criteria are commonly used to evaluate, or measure, the well being or stress of (a) people and (b) animals?

35. Define the term *pollution.* Why did pollution become such an issue in the latter part of the 20th century?

36. Describe, and discuss the concerns relative to, each of the following pollutants:
    a. Dust.
    b. Manure.
    c. Muddy lots.
    d. Pests and pesticides.
    e. Stray voltage.

37. What types of feedlots are required to secure a permit under existing pollution laws and regulations? What are the requirements?

38. Define the term *sustainable agriculture.* What warning signals have prompted great interest in sustainable agriculture?

39. Define the terms *animal welfare* and *animal rights.*

40. Discuss food safety and diet/health concerns. Why are American consumers prone to over-react relative to their food?

## SELECTED REFERENCES

| Title of Publication | Author(s) | Publisher |
|---|---|---|
| *Agricultural Waste Utilization and Management Proceedings* | F. J. Humenik, Chairman Executive Committee | American Society of Agricultural Engineers, St. Joseph, MI, 1985 |
| *Agriculture and Groundwater Quality* | P. F. Pratt, Chairman | Council for Agricultural Sciences and Technology, Ames, IA, 1985 |
| *Animal Behavior*, 2nd Edition | J. P. Scott | The University of Chicago Press, Chicago, IL, 1972 |
| *Animal Behavior*, 3rd Edition | V. G. Dethier E. Stellar | Prentice-Hall, Inc., Englewood Cliffs, NJ, 1970 |
| *Animal Behavior* | N. Tinbergen | Time Incorporated, New York, NY, 1965 |
| *Animal Behavior and Its Application* | D. V. Ellis | Lewis Publishers, Inc., Chelsea, MI, 1986 |
| *Animal Behavior in Laboratory and Field* | A. W. Stokes | W. H. Freeman and Company, San Francisco, CA, 1968 |
| *Animal Behavior, The Marvels of* | T. B. Allen, Editor | National Geographic Society, Washington, DC, 1972 |
| *Animal Behaviour: A Syntheses of Ethology and Comparative Psychology* | R. A. Hinde | McGraw-Hill Publishing Co., New York, NY, 1970 |
| *Applied Animal Ethology* | W. R. Stricklin, Guest Editor | Elsevier Science Publishers B. V., Amsterdam, The Netherlands, Vol. II, No. 4, Feb., 1984 |

*(Continued)*

## SELECTED REFERENCES *(Continued)*

| Title of Publication | Author(s) | Publisher |
|---|---|---|
| *Behavior of Domestic Animals, The,* 3rd Edition | E. S. E. Hafez, Editor | The Williams and Wilkens Company, Baltimore, MD, 1975 |
| *Bibliography of Livestock Waste Management* | J. R. Miner<br>D. Bundy<br>G. Christenbury | Office of Research and Monitoring, U.S. Environmental Protection Agency, Washington, DC, 1972 |
| *Biology of Stress In Farm Animals: an integrative approach* | P. R. Wiepkema<br>P. W. M. van Adrichem | Kluwer Academic Publishers, Hingham, MA, 1987 |
| *Brazil's Imperiled Rain Forest, Rondonia's Settlers Invade* | W. S. Ellis | *National Geographic,* Vol. 174, No. 6, pp. 772–799, Dec., 1988 |
| *Development and Evolution of Behavior* | Edited by<br>L. R. Aronson, *et al.* | W. H. Freeman and Company, San Francisco, CA, 1970 |
| *Domestic Animal Behavior* | J. V. Craig | Prentice-Hall, Inc., Englewood Cliffs, NJ, 1981 |
| *Effect of Environment on Nutrient Requirements of Animals* | D. R. Ames, Chairman | NRC, National Academy Press, Washington, DC, 1981 |
| *Effects of Air Temperature, Air Humidity, and Air Movement On Heat Loss From the Pig* | T. Kamada<br>I. Notsuki | Proc. 3rd AAAP Anim. Sci. Cong. 2:1174, Seoul, S. Korea, 1985 |
| *Environmental and Functional Engineering of Agricultural Buildings* | H. J. Barre<br>L. L. Sammet<br>G. L. Nelson | Van Nostrand Reinhold Co., New York, NY, 1988 |
| *Environmental Biology* | P. L. Altman<br>D. S. Dittmer | Federation of American Societies for Experimental Biology, Bethesda, MD, 1966 |
| *Environmental Control for Agricultural Buildings* | M. L. Esmay<br>J. E. Dixon | The AVI Publishing Company, Inc., Westport, CT, 1986 |
| *Environmental Management in Animal Agriculture* | S. E. Curtis | Animal Environment Services, Mahomet, IL, 1981 |
| *Ethology of Free-Ranging Domestic Animals* | G. W. Arnold<br>M. L. Dudzinski | Elsevier Scientific Publishing Company, Amsterdam, The Netherlands, 1978 |
| *Ethology, The Biology of Behavior,* 2nd Edition | I. Eibl-Eibesfeldt | Holt, Rinehart and Winston, New York, NY, 1975 |
| *Farm Animal Manures: an overview of their role in the agricultural environment* | J. Azevedo<br>P. R. Stout | Agricultural Publications, University of California, Berkeley, CA, 1974 |
| *Guide to Environmental Research on Animals, A* | R. G. Yeck, Chairman | NRC, National Academy of Science, Washington, DC, 1971 |
| *Health Issues Related to Chemicals in the Environment: A Scientific Perspective* | A. L. Craigmill, Chairman | Council for Agricultural Sciences and Technology, Ames, IA, 1987 |
| *Impact of Stress, The,* Proceedings | Edited by<br>R. E. Moreng<br>J. R. Herbertson | Colorado State University, Ft. Collins, CO, 1986 |
| *Introduction to Animal Behavior, An,: ethology's first century,* 2nd Edition | P. H. Klopfer | Prentice-Hall, Inc., Englewood Cliffs, NJ, 1974 |
| *Kinships of Animals and Man* | A. H. Morgan | McGraw-Hill Book Company, Inc., New York, NY, 1955 |
| *Livestock Behaviour, a practical guide* | R. Kilgour<br>C. Dalton | Westview Press, Boulder, CO, 1984 |
| *Livestock Environment, Proceedings, Second International Livestock Environment Symposium* | D. S. Bundy, Planning Chairman | American Society of Agricultural Engineers, St. Joseph, MI, 1982 |

*(Continued)*

## SELECTED REFERENCES *(Continued)*

| Title of Publication | Author(s) | Publisher |
|---|---|---|
| *Mechanisms of Animal Behavior* | P. Marler<br>W. J. Hamilton, IL | John Wiley & Sons, New York, NY, 1966 |
| *Organic Farming: current technology and its role in a sustainable agriculture* | D. M. Kral, Editor | American Society of Agronomy, Madison, WI, 1984 |
| *Our Friendly Animals and Whence They Came* | K. P. Schmidt | M. A. Donohue & Co., Chicago, IL, 1938 |
| *Portraits in the Wild* | C. Moss | Houghton Mifflin Company, Boston, MA, 1975 |
| *Poultry Welfare, Proceedings* | R. M. Wegner, Editor | German Branch, The World Poultry Science Association, Federated Agricultural Research Centre, Braunschweig-Volkenrode, Germany, 1985 |
| *Principles of Animal Behavior* | W. N. Tavolga | Harper & Row, New York, NY, 1969 |
| *Principles of Animal Environment* | M. L. Esmay | The AVI Publishing Company, Inc., Westport, CT, 1978 |
| *Readings in Animal Behavior* | T. E. McGill, Editor | Holt, Rinehart and Winston, New York, NY, 1973 |
| *Safe and Effective Use of Pesticides, The* | P. J. Marer | University of California, Publications, Oakland, CA, 1988 |
| *Scientific Aspects of the Welfare of Food Animals* | F. H. Baker, Chairman | Council for Agricultural Science and Technology, Ames, IA, 1981 |
| *Social Hierarchy and Dominance* | Edited by<br>M. W. Schein | Dowden, Hutchinson & Ross, Inc., Stroudsburg, PA, 1975 |
| *Social Space for Domestic Animals* | R. Zayan, Editor | Kluwer Academic Publishers, Hingham, MA, 1985 |
| *Social Structure in Farm Animals* | G. J. Syme<br>L. A. Syme | Elsevier Scientific Publishing Co., Amsterdam, The Netherlands, 1979 |
| *Stray Voltage: Proceedings of the National Stray Voltage Symposium* | O. M. Majerus<br>R. O. Martin<br>R. A. Peterson | American Society of Agricultural Engineers, St. Joseph, MI, 1984 |
| *Stray Voltages in Agriculture, Proceedings* | H. J. Hansen<br>L. J. Endahl, Co-Chairman | American Dairy Science Association, Champaign, IL, 1983 |
| *Stress Physiology in Livestock*<br>Vol. 1, *Basic Principles*<br>Vol. 2, *Ungulates*<br>Vol. 3, *Poultry* | M. K. Yousef | CRC Press, Inc., Boca Raton, FL, 1985 |
| *Structures and Environment Handbook* | | Midwest Plan Service, Iowa State University, Ames, IA, 1972 |
| *Utilization, Treatment, and Disposal of Waste on Land, Proceedings* | E. C. A. Runge, President of Society | Soil Science Society of America, Inc., Madison, WI, 1986 |
| *Wild Animals in Captivity* | H. Hediger | Dover Publications, Inc., New York, NY, 1964 |

They also serve those who watch and wait! This shows a band of ewes during a blizzard on the range exhibiting two behavioral systems: (1) shelter-seeking, by making use of a tree shelter; and (2) the gregarious (flocking) behavior, for which sheep are noted. (Courtesy, Charles Belden, Pitchfork, WY)

How it used to be done! This shows an early-day cattle car which was equipped with feed and water facilities that could be filled from the outside. (Woodcut, 1880. Courtesy, The Bettmann Archive, New York, NY)

# Chapter 8

# MARKETING LIVESTOCK

Fig. 8–1. Two Egyptian herders taking an ox to market. (An old photograph from Bas Relief found in Sakara in the tomb of King Ephto Stoptep.) At first, meat animals were bartered for crafted articles. (Courtesy, The Bettmann Archive, New York, NY)

Livestock marketing is the end of the line. From the producer's standpoint, it is that part which gives point and purpose, and profit or loss, to all that has gone before. Market receipts constitute the only source of reimbursement to producers for their work; market day is producers' payday—hence, it is the most important single day of operation to them.

Livestock marketing embraces those operations beginning with loading animals out on the farm, ranch, or feedlot and extending until they are sold to go into processing channels. In this chapter, the term is conceived to include processing as it reflects on livestock marketing. Treatment is limited to the marketing of four-footed animals—cattle, sheep, and hogs. Because of their distinct and different market channels and procedures, separate chapters are accorded to marketing milk and dairy products (Chapter 26) and marketing poultry and eggs (Chapter 49).

In the past, livestock producers could be successful if they knew how to breed, feed, and manage their stock. Today, this is not enough; preconsidered, if not prearranged, markets are essential.

## IMPORTANCE OF LIVESTOCK MARKETING

The importance of livestock marketing is attested to by the following facts and figures:

1. In 1987, a total of 152,050,000 head of meat animals (cattle, calves, hogs, sheep, and lambs) were marketed in the United States (see Fig. 8–2); cash receipts from livestock marketed totaled $44.7 billion (see Fig. 8–3); 121,103,000 head of meat animals were slaughtered under federal inspection; and 38.3 billion lb of red meat were produced.[1]

2. Some animals are marketed, or change ownership, more than once (for example, feeder calves may pass through 2 auctions within a week). As a result of this turnover, along with a more complete coverage of small livestock markets, the actual volume and value of livestock marketed is higher than indicated in point 1, and in Figs. 8–2 and 8–3.

3. In 1987, U.S. farmers and ranchers received $76.2 billion, or 55.2% of their cash income, from livestock and livestock products (see Chapter 2, Fig. 2–11).

[1]USDA sources.

## VOLUME OF LIVESTOCK MARKETED BY SPECIES

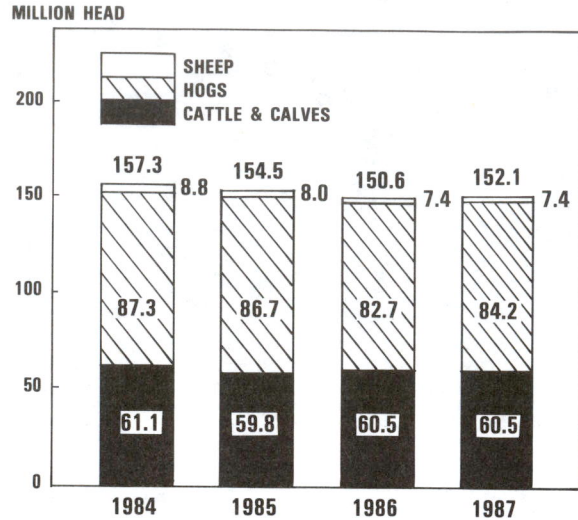

Fig. 8–2. Volume of livestock marketed by species, 1984–1987. (Source: *Agricultural Statistics 1988*; USDA; pp. 265, 276, 287; 1988)

## CASH RECEIPTS FROM LIVESTOCK MARKETED BY SPECIES

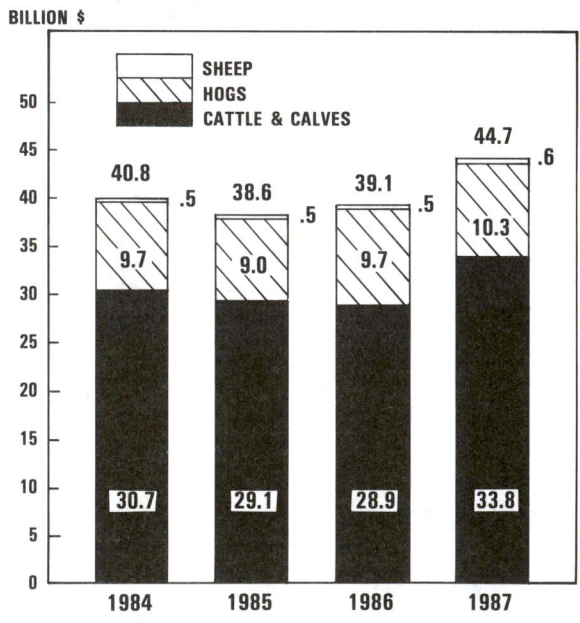

Fig. 8–3. Cash receipts from livestock marketed by species, 1984–1987. (Source: *Agricultural Statistics 1988*; USDA; pp. 265, 276, 287; 1988)

4. Livestock markets establish values for all animals, including those back on the farm or ranch. Thus, livestock markets were primarily responsible for livestock (cattle, hogs, sheep and lambs, chickens—exclusive of commercial broilers, and turkeys) having an aggregate value of $59.2 billion in 1987 (see Chapter 2, Table 2–3).

5. In 1987, the U.S. livestock market industry was served by 207 terminal markets (markets with more than one market agency selling on commission) and 469 auctions (markets with

only one agency), selling livestock;[2] and by 6,753 federally inspected meat packers and processors, buying and processing livestock.[3] Additionally, several thousand small locker plants and processors performed certain meat-marketing services.

6. The Corn Belt and the Northern Plains are major surplus red meat production areas; and the Southern Plains has a surplus of beef. The Northeast and the Atlantic Coast are red meat deficit areas; and the Far West is a pork-deficit area. This points up the enormity of the task of marketing the nation's livestock, keeping in mind that marketing is the process through which livestock and their products are moved from where they are produced to where they are consumed.

7. The marketing bill on farm foods comes high, as shown in Fig. 8-4. In 1987 consumers spent $377.1 billion for all farm foods, of which $283.2 billion (75%) went for marketing, and $93.9 billion (25%) went to farmers. That same year (1987), marketing (the cost of transporting, processing, and distributing) accounted for 53% of the cost of red meat (beef, pork, and lamb) to the consumer, whereas marketing accounted for a whopping 92% of the consumer cost of bakery and cereal products. The lower marketing cost on red meat than on bakery and cereal products is primarily due to meats being a more valuable product, volume for volume, and pound for pound.

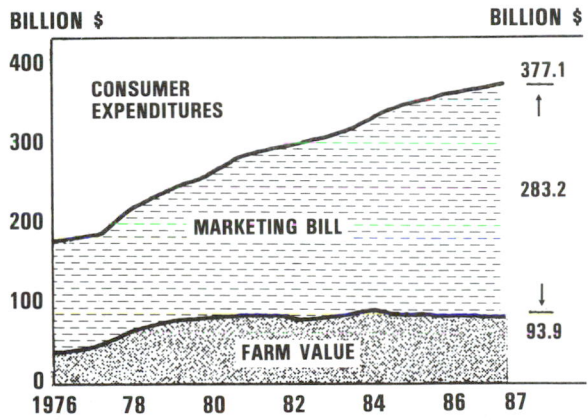

## MARKETING BILL, FARM VALUE, AND CONSUMER EXPENDITURE FOR FARM FEEDS

Fig. 8-4. Farm value, marketing bill, and consumer expenditures for farm foods, 1976 to 1987. In 1987, consumers spent $377.1 billion for all farm foods, of which $283.2 billion (75%) went for marketing, and $93.9 billion (25%) went to farmers. Since 1976, the farm value of foods has changed very little, while the marketing bill has increased greatly. (Source: 1988 *Agricultural Chartbook*, USDA, Agricultural Handbook No. 673, p. 63, Chart 121, 1988)

8. But marketing meat animals is far more than dollars. It involves the millions of people engaged in transporting, processing, and distributing meats. It includes everything that takes place from the time market animals leave the farm, ranch, or feedlot until the end product—meat—is sold over the counter. Although figures covering the marketing of meat animals alone are not available, the U.S. Department of Commerce does present a breakdown of numbers of workers, and types of work involved, in marketing food and fiber as a whole. In 1987, a total of 18.1 million workers were engaged in the food and fiber

sector after the products left the nation's farms and ranches. Of these 18.1 million workers, 1.3 million were employed in food processing, 2.7 million in manufacturing, 6.6 million in transportation, trade, and retailing, 3.7 million in eating establishments, and 3.9 million in other related work.[4] But these figures do not tell the whole story! There are 2.0 million farm workers, and millions more are employed in the production and distribution of farm production supplies—things the farmer buys to operate. In total, more than 20.1 million people, about 16.7% of those gainfully employed in the United States, make their living from agriculture (they are in agribusiness); either they are producing, marketing, or processing farm commodities, or they are supplying production goods used on the farm.

## THE DEVELOPMENT OF MARKETING AND MEAT PACKING

The development of the livestock business has truly gone hand in hand with the transformation and extension of the nation itself, and progress in each phase of livestock marketing and processing has unwittingly stimulated the whole. Thus, the growth of cities brought about the demand for meats. Progressively, this was followed by farm slaughtering for barter, the establishment of the town butcher, the operation of the local slaughterhouse, the trailing of animals and the drover, the establishment of private markets, the advent of rail transportation, the development of refrigeration, further opening up of the western range, the shifting of livestock production and leading markets to the West, improved highways, and shipment by motor truck. These changes were not alone, for the local slaughterhouse was gradually increasing is size and eventually we had the birth of the modern meat packing industry.

Fig. 8-5. Hough House, headquarters for the early-day Chicago stock yards. (Courtesy, *Chicago Daily Drovers Journal*)

---

[2]*Packers and Stockyards' Statistical Report,* 1987 reporting year, USDA, Packers and Stockyards Administration, P&SA Statistical Report No. 88-1, 1989.

[3]USDA, ERS sources (see Chapter 2, section on "Magnitude of the U.S. Livestock Industry").

[4]*Statistical Abstracts of the United States 1989*, U.S. Department of Commerce, Bureau of the Census, p. 626, No. 1074.

# THE HISTORY OF LIVESTOCK MARKETING

Fig. 8–6.   A herd of Texas Longhorns being driven to Dodge City, Kansas, an early-day cattle marketing rendezvous. (Courtesy, The Bettmann Archive, New York, NY)

Fig. 8–7.   Early Abilene, Kansas shipping point. This end of the Texas cattle trail in eastern Kansas, on the Kansas Pacific Railroad, was established in 1867, for the purpose of providing safe transportation of the cattle to the East, unmolested by the Ozark outlaws. (Courtesy, Abilene Chamber of Commerce)

Fig. 8–8.   Truck transportation of animals to market had its beginning in 1911. This shows trucks at the unloading dock of the East St. Louis, Illinois, stockyards. The most important factor in the decentralization of meat packing was the development and use of motor trucks, along with hard-surfaced roads. (Courtesy, Producers Live Stock Marketing Association, East St. Louis, IL)

Fig. 8–9.   Early day Swift and Company refrigerator car. The first refrigerator car was invented in 1868, but successful rail shipment of fresh meat by refrigeration was not achieved until 1871. Artificial refrigeration revolutionized the livestock marketing and meat packing industry. It made it possible to slaughter the live animals near the area of production and to ship the fresh dressed carcasses to the more distant consuming centers. (Courtesy, Swift & Company)

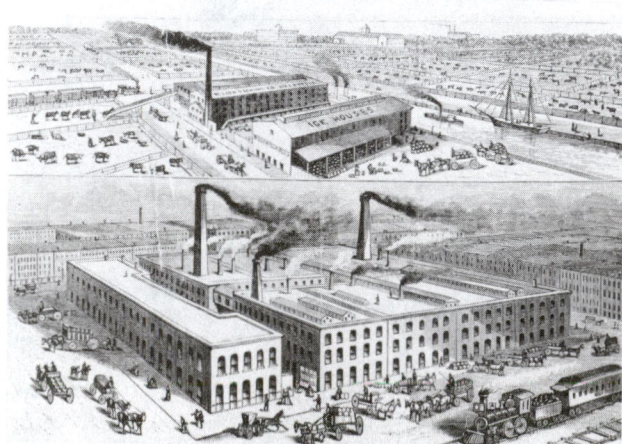

Fig. 8–10.   Early-day view of the Union Stock Yards, Chicago. Note the ice houses and Wilson Packing Company in the distance. (Courtesy, Chicago Historical Society, Chicago, IL)

Fig. 8–11.   Stone gate entrance to the Union Stock Yards, Chicago. In 1865, three railroads purchased 320 acres at 39th and Halsted Streets, and founded what was destined to become the world's greatest livestock market. (Courtesy, Chicago Historical Society, Chicago, IL)

## THE DEVELOPMENT OF LIVESTOCK TRANSPORTATION

In no phase of livestock marketing have the developments been more phenomenal than in transportation. With the humble beginning of marketing on foot or by a combination of trailing and boat shipment, transportation progressed to the luxury of the railroad, to the development of the refrigerator car designed to carry and preserve fresh meats, and finally to the construction of modern highways and the use of motor trucks.

Fig. 8–12.   Texas cattle fording a stream in trailing north. A few hundred miles' travel meant nothing to the drovers—those who drive slaughter animals overland to eastern markets. Their lives were filled with adventure and hardship. (After a woodcut; courtesy, The Bettmann Archive, New York, NY)

Today, trucks transport virtually all market livestock. In moving animals from farms, ranches, and feedlots to markets, they have several advantages over rail shipments. They minimize handling, provide protection against adverse weather, and reduce transit time. The speed and flexibility of trucks have made them dominant in both short- and long-haul shipment of livestock.

Fig. 8–13.   A truck unloading/loading dock, roof-covered pens, good lighting, self-feeders, and concrete water troughs are features of this modern cattle market facility at the Chicago-Joliet, Illinois, market. This facility will accommodate up to 8,000 cattle daily. (Courtesy, The Drovers Journal, Kansas City, KS)

## MODERN MEAT PACKING

Today the name packing is a misnomer, for the barreled, or pickled, pork from whence the name originated is now only a memory. But modern meat slaughtering and processing developed slowly, and the original name was so well established that there has never been any attempt to rename the industry.

It may be said that there have been four distinct eras in the evolution of the modern meat packing industry: (1) the era of cured pickled pork and dry salted beef which correctly resulted in the early day meat processors being designated as packers; (2) the era of rail transportation and the development of artificial refrigeration and the refrigerator car; (3) the era of complete utilization of by-products; and (4) the era of truck transportation of livestock and meat products, and of decentralization.

Fig. 8–14.   Old-time farm slaughter scene. When nearly all the people lived on the land—prior to the growth of cities and the rise of the town butcher—each family did its own slaughtering. (Courtesy, Swift and Company)

The first era extended from the time when operators in the eastern cities first devoted the major share of their time to meat curing to approximately the time of the Civil War, a span of nearly 200 years. The second era occupied a relatively short period of time, spanning only about two decades during and immediately following the Civil War. The era of the utilization of by-products was somewhat less rapid and devoid of much of the glamour of the first two periods. Truck transportation had its beginning in 1911.

At first, the packing plants were local in nature and small in size. But as transportation facilities improved and the services were broadened, the plants increased in both numbers and volume of business. Today, there are 6,753 federally inspected meat packers and processors—large and small, national, sectional, and local—in all parts of the country. It is also interesting to note that the meat-packing concerns regularly provide about one-fourth of the nation's manufactured food products. The American meat-processing industry, humbly originating with the home slaughtering and curing by the colonists, has thus risen to the status of big business. This position has not been achieved by mere chance, but because of the services that these plants render to both the producer and the consumer.

Fig. 8–15. The first Swift and Company packing plant. Small, local slaughterhouses of this type were the forerunners of modern packing-houses. (Courtesy, Swift and Company)

## EFFICIENCY AND LOW COST OF OPERATION

In addition to rendering these many valuable services, the modern meat-packing industry has become great because of the efficiency and relatively low cost at which it has operated over a period of years. The net profit on each dollar of sales of meat and other products is generally less than 1.0 cent; in 1985, it was 0.83 cent.

As further evidence of efficiency and low cost of operation in the meat-packing industry, it is noteworthy that the net profit of packing companies per 100 lb liveweight of animals is generally about 50 cents. Thus, based on averages (and recognizing that there may be species differences, and that the net profit of species shifts from time to time), for each 1,000-lb steer purchased, the packer nets about $5 in profits; for each 200-lb hog, $1; and for each 100-lb lamb, 50 cents. Of course, the volume of sales (the number of hams sold in a given year), the efficiency in operations, and the utilization of by-products make it possible for the industry to operate on these comparatively small margins.

Obviously, the net profits in the meat-packing industry have never been large enough to have any appreciable effect upon either meat or livestock prices. The gross margin is too small. Consequently, there is no justification for the housekeeper blaming the meat packer for high meat prices. Neither is there any basis for the producer blaming the packer for low market livestock prices. This situation is largely true because of the keen competition within the industry and the thousands of meat-slaughtering and processing plants located throughout the country.

It should also be realized that, however much the consumer may complain about the high price of meat, the price would be higher if it were not for the careful attention that has been given to the manufacture and sale of by-products. Sometimes the packer actually pays out for a live animal more than is received from the sale of the dressed carcass. Under

these conditions, the 30 to 55% nonmeat portion of the liveweight of a slaughter animal must be handled most efficiently in order to come out with any profit at all from the operations.

## GEOGRAPHY OF MEAT PRODUCTION AND CONSUMPTION

Cattle, sheep, and hogs are produced in every state. However, the density of population of each species varies from area to area. For example, the greatest density of hogs is in the Corn Belt. Likewise, the density of human population varies; most consumers are located east of the Mississippi River. Fig. 8–15 shows the geography of meat production and consumption. It shows that—

1. About 51% of the cattle are raised west of the Mississippi River, whereas about 73% of the beef is consumed east of the Mississippi.
2. About 69% of the lambs are raised west of the Mississippi River, whereas about 78% of the lamb is consumed east of the Mississippi.
3. About 84% of the hogs are raised east of the Mississippi River, whereas about 73% of the pork is consumed east of the Mississippi.

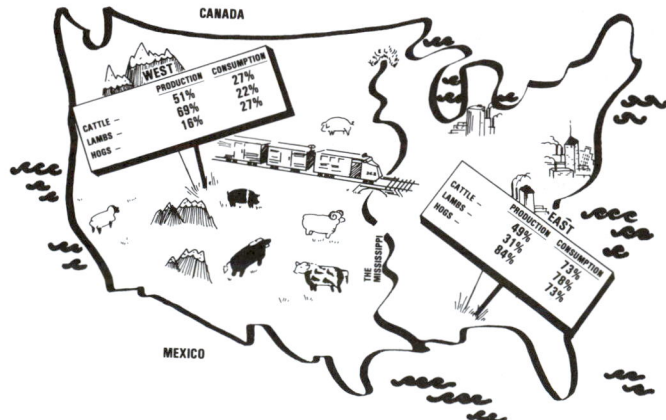

Fig. 8–16. This shows the geography of meat production and consumption. Using the Mississippi River as the dividing line, the percentage of production-consumption is given for each side of the river. Note that the majority of the consumption is in the East—thus making for a transportation problem.

4. The West produces a surplus of beef and lamb, but it has a deficit of pork.
5. The East is a deficit beef and lamb area, but it has a surplus of pork.
6. The Corn Belt and Northern Plains are the major surplus red meat areas. The Corn Belt produces the most surplus pork, and the Northern Plains produces the most surplus beef. The Lakes States and Mountain regions are the only other areas with a surplus of both beef and pork. The Southern Plains has a surplus of beef.
7. Beef and lamb movement is from west to east; beef movement is especially heavy from the Northern Plains to the eastern seaboard; and pork shipments out of the Corn Belt are especially heavy, with movement to both the East and the West.

Under these conditions, either livestock or meats, or both, must be moved from the surplus-producing regions to the deficit-producing areas. From a practical standpoint, it has been found to be more economical to slaughter animals in plants located in close proximity to the surplus-producing sections, to save the by-products, and then to transport the meat in refrigerator trucks and cars to the consuming sections than to ship the live animals. Also, it is not economical to ship the live animals farther than necessary on account of deaths, injuries, bruises, shrinkage, and freight costs. With this procedure, the average pound of meat is moved about 1,000 miles from where it is produced to where it is consumed.

## THE LARGE NATIONAL MEAT COMPANIES

Certain names loom large in the history of the packing industry. In 1863, Philip D. Armour, in partnership with a Mr. Plankinton, established a meat-packing enterprise at Milwaukee; and 4 years later Armour and Company set up business in Chicago. Gustavus F. Swift first opened a retail meat market at Eastham, Massachusetts, in 1859; gradually, he moved westward and opened up the Swift and Company plant in Chicago in 1877. The names of Armour and Swift, therefore, have long been landmarks in the meat-packing industry. Later the Cudahy Packing Company and Wilson and Company rounded out what became known as the "Big Four."

The 10 largest meat companies (ranked by 1988 sales), and the headquarters location of each, are:[5]

1. Iowa Beef Processors (IBP), Inc.
   Dakota City, NE
2. Con Agra, Inc.
   Omaha, NE
3. Excel Corp.
   Wichita, KS
4. Sara Lee Corp.
   Memphis, TN
5. Oscar Mayer Foods Corp.
   Madison, WI
6. Geo. A. Hormel & Co.
   Austin, MN
7. John Morrell & Co.
   Cincinnati, OH
8. Swift Independent Packing Co., Inc.
   Greeley, CO
9. BeefAmerica, Inc.
   Omaha, NE
10. Swift-Eckrich, Inc.
    Oakbrook, IL

## DECENTRALIZATION OF MEAT PACKING

Following World War I, a trend toward geographical decentralization of the meat-packing industry became apparent. This represented attempts to seek the optimum plant location between supplies of livestock and the market for meat. These new plants are commonly referred to as interior plants.

This gradual shift in location of slaughter decreased the relative importance of terminal markets. The major packers, located on the old established central markets, countered this move through the decentralization of their own operations,

Fig. 8–17. With the decentralization of meat packing, one-story, streamlined, interior plants like this one in western Kansas, replaced the old-fashioned multi-story plants of previous generations.

chiefly through direct buying and purchase or construction of interior plants.

Without doubt, the most important single factor contributing to the decentralization of meat packing was the development and use of motor trucks, together with the extension of hard-surfaced roads. Other factors that favored the development of interior plants were: improved market information, increased feed and livestock production in certain areas, saving in transportation rates by marketing closer to home, and more favorable labor costs in certain areas.

## METHODS OF MARKETING LIVESTOCK

The producer of livestock is confronted with the perplexing problem of where and how to market his animals. Usually there is a choice of market outlets, and the one selected often varies with different species of livestock and among sections of the country. The methods of marketing also differ between slaughter and feeder animals, and all of these differ from the marketing of purebreds.

Prior to the advent of terminal public markets in 1865, country selling accounted for virtually all sales of livestock. But sales of livestock in the country declined with the growth of terminal markets, until the latter method reached its peak at the time of World War I. Country selling was reactivated by the large nationwide packers beginning about 1920, in order to meet the increased buying competition of the small interior packers. The decline in the proportion of all livestock moving through terminal public markets was largely accounted for by the growth in country selling until the late 1930s and by the growth of both country selling and auctions since.

In 1987, meat packers (722 of them) purchased their animals through the following channels:[6]

|  | Cattle | Calves | Sheep | Hogs |
|---|---|---|---|---|
|  | (%) | (%) | (%) | (%) |
| Nonpublic markets ... | 80.2 | 61.0 | 81.4 | 88.8 |
| Auction markets ...... | 15.6 | 35.8 | 13.0 | 4.9 |
| Terminal markets ..... | 4.2 | 3.2 | 5.5 | 6.3 |

[5]*Meat Facts*, American Meat Institute, 1988, p. 19.

[6]*Packers and Stockyard's Statistical Report*, USDA, Packers and Stockyards Administration, P&SA Statistical Report No. 88–1, Jan. 1989, p. 9.

Most U.S. livestock are marketed through four channels—nonpublic markets, auctions, terminals, or carcass grade and weight basis.

## NONPUBLIC MARKETS

*Nonpublic markets include all purchases from sources except from public markets—from all sources except from terminals and auctions.* The term *nonpublic markets* evolved to replace the term *direct markets* because some questions had been raised as to whether *directs* included *only* purchases direct from sellers to plants or to packer buyer stations.

Nonpublic markets do not involve a recognized market. The selling usually takes place at the farm, ranch, feedlot, or some other nonmarket buying station or collection yard.

Prior to the advent of public stockyards in 1865, direct selling, as it was known at the time, accounted for virtually all sales of livestock. Sales of livestock in the country declined with the growth of public stockyards until the latter method reached its peak of selling at the time of World War I. Country selling was accelerated by the large nationwide packers following World War I in order to meet the increased buying competition of the small interior packers.

Nonpublic market selling is similar to terminal market selling with respect to price determination; both are by private treaty and negotiation. But it permits producers to observe and exercise some control over selling while it takes place, whereas consignment to distant terminal markets usually represents an irreversible commitment to sell. Larger and more specialized livestock farmers feel competent to sell their livestock by direct negotiation with buyers.

Improved highways and trucking facilitated the growth of direct selling. Farmers were no longer tied to outlets located at important railroad terminals or river crossings. Livestock could move in any direction. Improved communications, such as the radio and telephone, and an expanded market information service, also aided in the development of country selling of livestock, especially in sales direct to packers.

The out-of-pocket cost to producers for nonpublic market selling is zero. Their only selling expense is their time, which, of course, is not a direct out-of-pocket cost.

## AUCTION MARKET METHOD

*Auction markets (also referred to as sales barns, livestock auction agencies, community sales, and community auctions) are trading centers where animals are sold by public bidding to the buyer who offers the highest price per hundredweight or per head.* Auctions may be owned by individuals, partnerships, corporations, or cooperative associations.

This method of selling livestock in this country is very old, apparently being copied from Great Britain where auction sales date back many centuries.

The auction method of selling was used in many of the colonies as a means of disposing of property, imported goods, secondhand household furnishings, farm utensils, and animals.

According to available records, the first public livestock auction sale was held in Ohio in 1836 by the Ohio Company, whose business was importing English cattle. This event also marked the first sale of purebred cattle ever held in America.

The auction market method of selling is similar to the terminal market in that both markets (1) are an assembly or collection point for livestock being offered for sale, (2) furnish or provide all necessary services associated with selling activity, (3) are supervised by the federal government in accordance with the provisions of the National Packers and Stockyards Act, and (4) are characterized by buyers purchasing their animals on the basis of visual inspection.

But there are several important differences between terminals and auctions; among them, auction markets (1) are not always terminal with respect to livestock destination; (2) are generally smaller; (3) sell by bid, rather than by offer and counter-offer; and (4) are completely open to the public with respect to bidding, and all buyers present have an equal opportunity to bid on all livestock offered for sale.

Fig. 8–18. A modern auction facility, with scales serving as the auction ring. As cattle enter the ring (scales), they are weighed and an electric scoreboard indicates the gross and average weights. (Courtesy, USDA, Washington, DC)

The cost to the producer of using the auction market is the combined cost of selling, yardage, feed, and services. Auction market charges are generally somewhat higher than terminal charges.

## TERMINAL OR CENTRAL MARKET METHOD

*Terminal or central markets are livestock trading centers which generally have several commission firms and an independent stockyards company.* Formerly, terminal markets were synonymous with private treaty selling. Today, however, many terminal markets operate their own sale ring and all, or almost all, of their livestock are sold by auction.

The terminal or central market method entail the following distinct steps:

1. The producer must decide when to sell animals.

2. The producer must deliver animals to the terminal.

3. The producer must consign animals to a commission firm.

4. The commission firm must accept the animals upon their arrival at the terminal.

5. The commission firm must pen, water, feed, sort, sell, and attend to all the necessary tasks from the time the animals arrive until they are sold.

6. The commission firm collects from the buyer and pays the seller.

7. The buyer takes title to the animals when they are weighed.

Fig. 8–19.   Trading on a central market on private treaty, or person-to-person basis, showing seller and prospective buyer discussing the offering and arriving at an agreed upon price. (Courtesy, *The Drovers Journal*, Kansas City, KS)

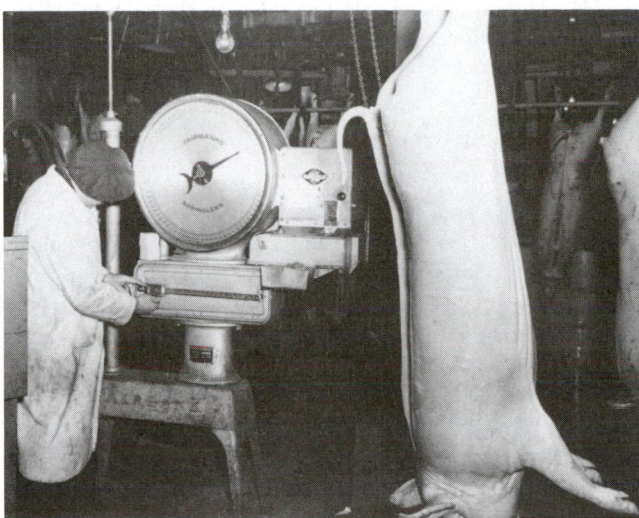

Fig. 8–20.   Weighing Canadian hog carcasses. Selling on the basis of carcass grade and yield is the most common method of marketing hogs in Denmark, Sweden, and Canada. (Courtesy, Department of Agriculture, Ottawa, Canada)

The stockyards company performs a number of useful services concerning the physical operation of the market. But, the stockyards company is a service company. Generally, it takes no active part in either buying or selling animals.

The direct out-of-pocket cost to the producer is the combined cost of yardage, sales, commission, feed, and service. These costs vary somewhat from market to market. Commission fees are the largest terminal market cost. They usually account for one-half of the total charges, with yardage plus feed and bedding accounting for the other half.

## CARCASS GRADE AND WEIGHT BASIS

It is generally agreed that there is need for a system of marketing which favors payment for a high cutout value of primal cuts and a quality product. Selling on the basis of carcass grade and weight fulfills these needs.

This is the most common method of marketing hogs in Denmark, Sweden, and Canada. The bargaining is in terms of price to be paid per 100 lb dressed weight for carcasses that meet certain grade specifications. It is the most accurate and unassailable evaluation of the value of a carcass. From the standpoint of the packer, this procedure is more time-consuming than the conventional basis of buying, and there is less flexibility in the operations.

In general, farmers who produce superior animals benefit from selling on the basis of carcass grade and weight, whereas the producers of lower quality animals usually feel that this method unjustly discriminates against them. In countries where rail grading has been used extensively, there has been an unmistakable improvement in the breeding and feeding of swine. Denmark, Sweden, and Canada have effectively followed this type of program in producing high-quality bacon, chiefly for export to the London market.

The **factors favorable** to selling on the basis of carcass grade and weight may be summarized as follows:

1. It encourages the breeding and feeding of quality animals.

2. It provides the most unassailable evaluation of the product.

3. It eliminates wasteful filling on the market.

4. It makes it possible to trace losses from condemnations, bruises, and soft pork to the producer responsible for them.

5. It is the most effective approach to animal improvement.

The **factors unfavorable** to selling on the basis of carcass grade and weight are:

1. The procedure is more time-consuming than the conventional basis of buying.

2. There is less flexibility in the operations.

3. The physical difficulty of handling the vast volume of U.S. animals on this basis is great.

The first data for carcass purchasing of livestock was published by the U.S. Department of Agriculture for the year 1961. For the United States (48 contiguous states) as a whole, the following percentages of total slaughter were purchased on a carcass grade and weight basis in 1987:

| All Cattle | Calves | Sheep and Lambs | Hogs |
|---|---|---|---|
| (%) | (%) | (%) | (%) |
| 30.4 | 36.7 | 35.9 | 13.5 |

The U.S. Department of Agriculture guidelines, wherein meat packers buy livestock on the basis of carcass grade, carcass weight, or a combination of the two, are:

1. Make known to the seller, before the sale, the significant details of the purchase contract.

2. Maintain the identity of each seller's livestock and carcass.

3. Maintain sufficient records to substantiate settlement for each purchase transaction.

4. Make payment on the basis of actual carcass weight before carcasses are shrouded.

5. Use hooks, rollers, gambrels and similar equipment of uniform weight in weighing carcasses from the same species of livestock in each packing plant; and include only the weight

of this equipment in the actual weight of the container.

6. Make payments on the basis of final USDA carcass grades or furnish the seller with detailed written specifications for any other grades used in determining final payment.

7. Grade carcasses by the close of the second business day following the day of slaughter.

## SELLING PUREBRED ANIMALS

Selling purebred animals is a highly specialized and scientific business. Purebred animals are usually sold at private treaty directly to other purebred breeders or commercial producers or through auctions which may be sponsored by either one or a few breeders (joint sales or consignment sales).

In general, the vast majority of males of all species saved for breeding purposes go into commercial herds and flocks. Only the elite sires are retained with the hope of effecting further breed improvement in purebred herds. On the other hand, the sale of purebred females is fairly well restricted to meeting the requirements for replacement purposes in existing purebred herds or for establishing new purebred herds.

Most consignment sales are sponsored by a breed association, either local, statewide, or national in character. Such auctions, therefore, are usually limited to one breed. Purebred auction sales are conducted by highly specialized auctioneers. In addition to being good salespersons, such auctioneers must have a keen knowledge of values and be familiar with blood lines of the breeding stock.

Fig. 8–21. A purebred auction sale in progress, with a Maine-Anjou bull in the ring. (Courtesy, International Maine-Anjou Assn., Kansas City, MO)

## PREPARING AND SHIPPING LIVESTOCK

Improper handling of livestock immediately prior to and during shipment may result in excess shrinkage; high death, bruises, and crippling losses; disappointing sales; and dissatisfied buyers. Unfortunately, many owners who do a superb job of producing animals dissipate all the good things that have gone before by doing a poor job of preparing and shipping to market. Generally speaking, such omissions are due to lack of know-how, rather than any deliberate attempt to take advantage of anyone. Even if the sale is consummated prior to delivery, negligence at shipping time will make for a dissatisfied

customer. Buyers soon learn what to expect from various producers and place their bids accordingly.

In addition to the important specific considerations covered in the sections which follow, these general considerations should be accorded in preparing livestock for shipment and in transporting them to market:

1. Select the best-suited method of transportation.

2. Feed and water properly prior to loading. Withhold grain feeding of all classes of livestock 12 hours before loading (omit one feed). Cattle and sheep may be allowed free choice to dry, well-cured grass hay up to loading time, but they should not be allowed access to water within 2 to 3 hours of shipment.

3. Keep animals quiet. Remember that "easy does it."

4. Comply with the requirements for health certificates, permits, and brand inspections where interstate shipments are involved.

5. Comply with the federal 28-hour law[7] in rail shipments. This prohibits transporting livestock by rail for a longer period than 28 consecutive hours without unloading, feeding, watering, and resting for 5 consecutive hours before resuming transportation. On request of the owner, the period can be extended to 36 hours.

6. Feed or graze cattle or sheep in transit if advantageous. This refers to a provision of railroads whereby livestock producers may be granted permission to graze or finish animals for a period of up to 12 months, at some intermediate stop between their point of origin and the market to which they will be consigned at the end of the finishing period.

7. Use partitions in the truck or car when necessary to separate species, sexes, or age groups.

8. Avoid shipping during extremes in weather, either when it is very cold or very hot.

## BRUISES, CRIPPLING, AND DEATH LOSSES

Losses from bruising, crippling, and death that occur during the marketing process represent a part of the cost of marketing livestock; and, indirectly, the producer foots most of the bill.

Most bruises are caused by horned cattle; by projections in feed lots, motor trucks, cars, and stockyards; by failure to partition off different kinds and classes of livestock in cars or trucks properly; by overloading or underloading; and by rough and careless handling. A careful analysis of the cause of bruises indicates that most of this loss could be prevented. Unfortunately, unless cuts or bumps are present, most bruises cannot be detected until after the animals have been slaughtered. This being true, and because market animals are purchased on the basis of liveweight in the United States, slaughter buyers take the probable loss into consideration at the time of purchase. This procedure is entirely justified because the bruises render the particular spot unfit for human food.

In a nationwide survey, involving 775,000 hogs and 163,000 cattle, Livestock Conservation, Inc., found that 8.5% of all market hogs and 6.4% of all market cattle showed unmistakable and costly carcass bruises.[8]

The same factors that are responsible for bruising are also responsible for much of the loss from death and crippling. The following observations have been made relative to crippling and death losses:

---

[7]No such law applies to truck transportation of animals.

[8]Estimate made by Livestock Conservation, Inc., in a letter to the author.

1. Crippling in hogs is more common than in other species; the incidence of crippling in calves, cattle, and sheep ranks after hogs in the order named.

2. Death losses in calves are higher than those encountered in any other kind or class of livestock. From the standpoint of species, sheep have the highest death loss in marketing, followed by hogs and cattle.

3. In-transit losses from crippling and death are higher among thin, emaciated, or overfinished animals than among strong, vigorous animals.

4. Extreme heat or cold, or sudden temperature changes, lead to increased crippling and death losses.

5. The nutrition of the animals during the growing-finishing period is a major factor.

Although the connotation of the word *crippled* varies somewhat from market to market, usually an animal that arrives in such a condition that it cannot walk into pens unassisted is recorded as a cripple. Badly crippled animals are generally bought subject to inspection.

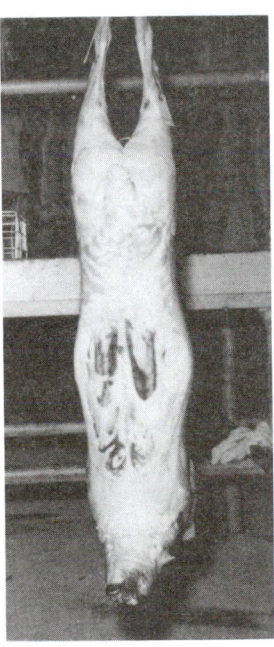

Fig. 8–22. A bruised lamb carcass showing trimming losses. Total losses from bruises trimmed out in U.S. slaughter plants cost millions of dollars, and, indirectly, the producer foots most of the bill.

## NUMBER OF ANIMALS IN A RAILROAD CAR AND IN A TRUCK

Overcrowding of market animals causes heavy losses. Sometimes a railroad car or a truck is overloaded in an attempt to effect a saving in hauling charges. More frequently, however, it is simply the result of not knowing space requirements.

Normally, railroad cars are either 36 or 40 feet in length, but truck beds are variable in size. The size of the car or truck, and the class and size of animals, determine the number of head that should be loaded therein. For comfort in shipping, the car or truck should be loaded heavily enough so that the animals stand close together, but both overloading and underloading should be avoided. Table 8–1 and 8–2 give some indication as to the number of market animals that may be loaded in a railroad car or truck.

**TABLE 8–1**
**NUMBER OF ANIMALS FOR SAFE LOADING IN A TRUCK[1]**

| Length of Truck Floor | Kind and Weight of Animals | | |
|---|---|---|---|
| | Cattle | Lambs | Hogs & Calves |
| (ft) | (1,000 lb) (454 kg) | (100 lb) (45.4 kg) | (222 lb) (100.8 kg) |
| 8 (2.4 m) ...... | 4 | 20 | 16 |
| 12 (3.7 m) ...... | 7 | 31 | 24 |
| 15 (4.6 m) ...... | 9 | 40 | 30 |
| 20 (6.1 m) ...... | 12 | 54 | 40 |
| 24 (7.3 m) ...... | 15 | 65 | 48 |

[1]Recommendations of Livestock Conservation, Inc., Chicago, IL.

**TABLE 8–2**
**ANIMALS PER RAILROAD CAR[1]**

| Car Size | Kind and Weight of Animals | | |
|---|---|---|---|
| | Cattle | Lambs | Hogs |
| (ft) | (1,000 lb) (454 kg) | (100 lb) (45.4 kg) | (225 lb) (102 kg) |
| 36 (11 m) ...... | 26 | 110 | 73 |
| 40 (12 m) ...... | 28 | 125 | 81 |

[1]Recommendations of Western Weighing and Inspection Bureau, Chicago, IL. In loading double-deck cars, the upper deck should contain 10% fewer than the lower deck.

## SHRINKAGE IN MARKETING ANIMALS

*The shrinkage (or drift) refers to the weight loss encountered from the time animals leave the farm or feedlot until they are weighed over the scales at the market.* Thus, if a steer weighed 1,000 lb at the feedlot and had a market weight of 970 lb, the shrinkage would be 30 lb or 3.0%. Shrink is usually expressed in terms of percentage. Most of this weight loss is *excretory shrink,* or in the form of feces and urine, and the moisture of the expired air. On the other hand, there is some *tissue shrinkage,* which results from metabolic breaking-down changes.

Shrinkage is of importance because the carcass meat is the most valuable portion of the animal. For this reason, dressed yield is one of the most important factors taken into consideration by packers in buying livestock for slaughter.

The most important factors affecting shrinkage are:

1. **Fill.** The fill refers to the amount of feed and water consumed by animals upon their arrival at the market and prior to selling. Normally, the fill of hogs consists of a feed or some grain (small grains are ground) common to the area, and water; whereas the fill of cattle and sheep consists of hay and water, although grain-fed animals may be given some concentrate. Naturally, the larger fill animals take, the smaller the shrinkage.

Because animals transported via motor truck may not have remained off feed too long, there is an increasing tendency not to feed and water them prior to selling on the market. This saving in feed and expense seems to be economically sound.

2. **Time and distance in transit.** The longer the animals are in transit and the greater the distance, the higher the total shrinkage. Also, the shrink takes place at a rapid rate during the first part of the haul, and then decreases as time in transit progresses.

3. **Truck vs rail transportation.** Based on practical experience and observation, most livestock shippers are of the opinion (1) that truck shipments result in less shrinkage than rail shipments for short distances, and (2) that rail shipments result in less shrinkage than truck shipments for long distances.

4. **Season.** Extremes in temperature, either very hot or very cold weather, result in higher shrinkage.

5. **Age and weight.** Young animals of all species shrink proportionally more than older animals, because of their lower carcass yield caused by less body fat and greater amount of fill in proportion to liveweight. Likewise, feeder cattle shrink about 25% more than finished cattle.

6. **Overloading.** Overloading always results in abnormally high shrinkage.

7. **Rough ride, abnormal feeding, and mixed loads.** Each of these factors will increase shrinkage.

On the average, the following shrinkage is obtained on market animals:

Cattle .............................. from 3 to 6%
Sheep .............................. from 6 to 10%
Hogs .............................. from 1 to 2%

## DOCKAGE

*Dockage refers to deductions made in the liveweight of market animals because of excessive dressing losses, or because part of the product is of low quality.* Some common dockages on livestock markets are:

1. **Piggy sows.** Usually docked 40 lb, but it may range from 0 to 50 lb, depending on the market.

2. **Stags (hogs).** Usually docked 70 lb, but it may range from 40 to 80 lb, depending on the market.

3. **Cattle with lumpy jaw.** Usually bought subject to the amount of wastage.

## LIVESTOCK MARKET NEWS SERVICES

Accurate market news is essential to the efficient marketing of livestock, both from the standpoint of the buyer and the seller. In the days of trailing, the meager market reports available were largely conveyed by word of mouth. Moreover, the time required to move livestock from the farm or ranch to market was so great that detailed market information would have been of little benefit even if it had been available. With the speed of transportation afforded by railroads and trucks, late information on market conditions became important.

### FEDERAL MARKET NEWS SERVICE

The Federal Market News Service was initiated by the U.S. Department of Agriculture beginning in 1916. This service was established for the purpose of providing unbiased and uniformly interpretable market information.

Today, livestock markets are covered by livestock, meat, and wool field offices maintained throughout the United States. This network provides coverage of terminal livestock markets, auctions, and major direct sales areas. Because of limitations in funds and personnel, market reporting in an era of increased nonpublic markets and auction sales has been difficult, if not inadequate. The problem is accentuated because the Federal Market News Service depends on voluntary cooperation in gathering information. There is no legal compulsion for buyers and sellers to divulge purchase and sale information. Reports on direct sales are obtained largely by telephone and teletypewriter, augmented by interviews made at feedlots, packing plants, and ranches.

The Federal Market News Service utilizes leased wire facilities. At scheduled times, market information is fed into a leased teletypewriter system for dissemination to the press, radio, television, and trade sources. Several offices have automatic, self-answering, taped telephone reports. In addition, mimeographed and printed reports are sent out by mail.

The Federal Market News Service relies upon local and privately owned newspapers, radio stations, and TV stations, merely supplying them with the market reports. Because at least a part of the readers or listeners are interested in this type of information, the local papers and radio stations are usually glad to serve as media for disseminating these reports.

The chief contributions of the Federal Market News Service may be summarized as follows: (1) A common terminology has come to be established from market to market, thus a slaughter steer which will grade good at Chicago would be designated the same grade at every other market in the country; and (2) facilities for the dissemination of market information have been provided.

## MARKET INFORMATION PROVIDED BY FARM AND TRADE MAGAZINES AND MARKET AGENCIES

Farm magazines are one of the most important sources of livestock market information. Trade magazines, oriented to a particular sector of the industry, such as meat or wool, also carry market reports.

Many market agencies—i.e., commission firms, auction markets, and related organizations—prepare and distribute market information. By means of weekly market newsletters or cards, they commonly emphasize the price and market conditions of the particular market they serve.

### CATTLE-FAX

The National Cattlemen's Association has been particularly effective in utilizing privately developed cattle market information. Under a separate corporation, known as Cattle Marketing Information Service, Inc. (CMIS), it operates its cattle marketing service, appropriately dubbed *Cattle-Fax.* CMIS has developed a system for determining numbers of cattle on feed, potential supplies at future dates, current prices, market conditions, and other market factors. It assembles, analyzes, and interprets this information centrally, then transmits it by phone and mail back to its members. Members utilize the information individually (or in local pooled arrangements) in negotiating direct sales with packers. *Cattle-Fax* subscribers pay a monthly fee, at a fixed minimum plus a charge per head up to a stipulated maximum total fee The *Cattle Fax* service is rather expensive for the small farmer-breeder, but pooled arrangements can be set up wherein several small operators share the cost of installation, which may be located in a bank or other business establishment convenient to the group.

### OTHER PRIVATELY DEVELOPED MARKET INFORMATION SERVICES

In addition to *Cattle-Fax,* sponsored by the National Cattlemen's Association, a number of other producer organizations are active in marketing work and have varying degrees of market information services, usually on a smaller scale than *Cattle-Fax*. The following are cited, but space limitation will not permit elaborating upon them:

American Farm Bureau Federation, which sponsors the American Marketing Association
American Sheep Producers Council, Inc.
National Farmers Organization (NFO)
National Live Stock Producers Association
National Pork Producers Council
National Wool Growers Association

## FEDERAL AND STATE LIVESTOCK MARKETING AGENCIES

With the growth of the far-flung livestock-marketing system and meat slaughtering, processing, and distribution, it soon became apparent that federal and state supervision was necessary to prevent unfair and discriminatory trade practices, to protect the public health, to ensure humane methods of handling animals, and to establish fair and equitable rates and charges for all agencies operating on a given public market. The various services rendered, legislative acts, and state and federal organizations carrying out these functions will be discussed briefly.

### FEDERAL MEAT INSPECTION

The federal government requires supervision of establishments which slaughter, pack, render, and prepare meats and meat products for interstate shipment and foreign export; it is the responsibility of the respective states to have and enforce legislation governing the slaughtering, packaging, and handling of meats shipped intrastate, but state standards cannot be lower than federal levels. The meat inspection laws do not apply to farm slaughtering for home consumption, although all states require inspection if the meat is sold.

The meat inspection service of the U.S. Department of Agriculture was inaugurated, and is maintained under, the Meat Inspection Act of June 30, 1906. This act was updated and strengthened by the Wholesome Meat Act of December 15, 1967. The latter statute (1) requires that state standards be at least to the levels applied to meat sent across state lines; and (2) assures consumers that all meat sold in the United States is inspected either by the federal government or by an equal state program. The Animal and Plant Health Inspection Service of the U.S. Department of Agriculture is charged with responsibility of meat inspection.

The purposes of meat inspection are (1) to safeguard the public by eliminating diseased or otherwise unwholesome meat from the food supply, (2) to enforce the sanitary preparation of meat and meat products, (3) to guard against the use of harmful ingredients, (4) to prevent the use of false or misleading names or statements on labels. Personnel for carrying out the provisions of the act are of two types: Professionals or veterinary inspectors who are graduates of accredited veterinary colleges, and non-professional food inspectors who are required to pass a Civil Service examination. In brief, the inspections consist of the following two types:

1. **Antemortem (before death)** inspection is made in the pens or as the animals move from the scales after weighing. The inspection is performed to detect evidence of disease or any abnormal condition that would indicate a disease. Suspects are provided with a metal ear tag bearing the notation "U.S. Suspect No . . . .," and are given special postmortem scrutiny. If in the antemortem examination there is definite and conclusive evidence that the animal is not fit for human consumption, it is *condemned*, and no further postmortem examination is necessary.

2. **Postmortem (after death)** inspection is made at the time of slaughter and includes a careful examination of the carcass and the viscera (internal organs). All good carcasses are stamped "U.S Inspected and Passed," whereas the inedible carcasses are stamped "U.S. Inspected and Condemned." The latter are sent to the rendering tanks, the products of which are not used for human food.

In addition to the antemortem and postmortem inspections referred to, the government meat inspectors have the power to refuse the application of the mark of inspection to meat products produced in a plant that is not sanitary. All parts of the plant and its equipment must be maintained in a sanitary condition at all times. In addition, plant employees must wear clean, washable garments, and suitable lavatory facilities must be provided for hand washing.

Fig. 8–23. Antemortem (before death) inspection of hogs being made by a federal veterinarian. Animals that are clearly diseased, emaciated, or otherwise unfit for human food are destroyed. Their carcasses may be used only in making inedible grease, fertilizer, or other nonfood products. Animals that appear slightly abnormal on foot are tagged "U.S. Suspect," and are given special postmortem scrutiny. (Courtesy, Technical Staff, Animal and Plant Health Inspection Service, USDA)

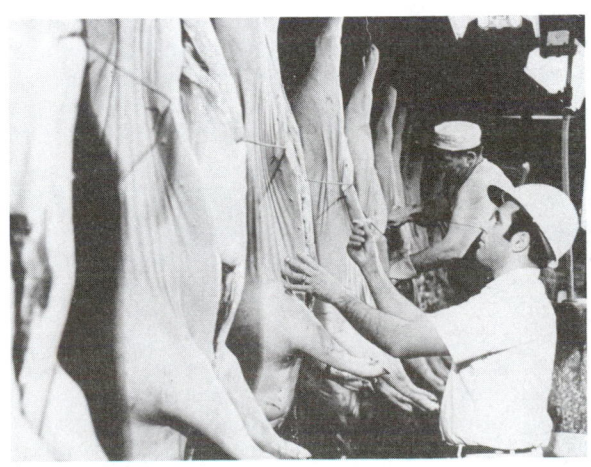

Fig. 8–24. Postmortem (after death) inspection. (Courtesy, Technical Staff, Animal and Plant Health Inspection Service, USDA)

Meat inspection regulations require the condemnation of all or affected portions of carcasses of animals with various disease conditions, including pneumonia, peritonitis, abscesses and pyemia, uremia, tetanus, rabies, anthrax, tuberculosis, various neoplasms (cancer), arthritis, actinobacillosis, and many others.

Most of the larger meat packers are under federal inspection; hence, they are allowed to ship interstate. Fig. 8–25 shows the proportion of the total U.S. meat slaughter that was produced in (1) federally inspected plants, (2) nonfederally inspected plants, and (3) farm slaughter in 1987.

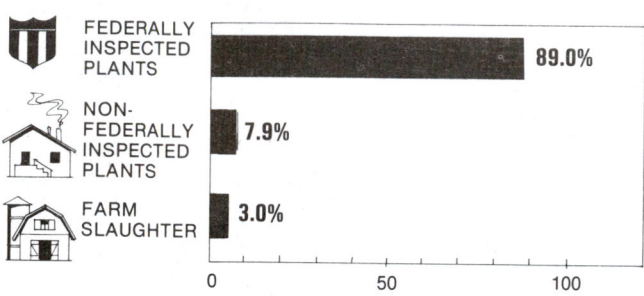

**PROPORTION OF TOTAL U.S. MEAT SLAUGHTER PRODUCED IN:**

FEDERALLY INSPECTED PLANTS — 89.0%

NON-FEDERALLY INSPECTED PLANTS — 7.9%

FARM SLAUGHTER — 3.0%

Fig. 8–25. Proportion of total U.S. meat slaughter produced in (1) federally inspected plants, (2) nonfederally inspected plants, and (3) farm slaughter. (Source: *Agricultural Statistics 1988*, USDA, p. 299, Table 442)

(Also see Chapter 6, section entitled, "U.S. Department of Agriculture.")

## STATE MEAT INSPECTION

States have varying legislation governing the slaughtering and further processing of meats produced for intrastate commerce. However, the Wholesome Meat Act of 1967 requires that the state standards be equal to the federal standards. Inspection that was often formerly conducted under local ordinances is now conducted, with one exception, by state employees. The one exception is the city of Chicago, where city employees conduct inspection under the overall supervision of the state.

The Wholesome Meat Act gave the states the option of either conducting their own inspection service, or turning the responsibility over to the federal government. In most of these states, the service is administered by the State Department of Agriculture. Quite frequently, they simply apply the federal regulations.

## PACKERS AND STOCKYARDS ACT OF 1921

On August 15, 1921, Congress passed the Packers and Stockyards Act which gives the Secretary of Agriculture supervision of packers and all others engaged in business at public stockyards. This law was enacted in order to prevent packers from engaging in unfair, discriminatory, or deceptive practices tending to restrain trade or create a monopoly, or from indulging in practices having the effect of manipulating or controlling prices. The law also decrees that all rates charged by stockyards companies and market agencies shall be just, reasonable, and nondiscriminatory. Fines are levied as penalties for violations, but the accused has the recourse of the courts, which

must uphold any ruling of the Secretary of Agriculture before the action is considered final.

## HEALTH CONTROL OFFICIALS

At every large central market, two groups of health control officials are represented: the U.S. Department of Agriculture and regulatory officials of the state department of agriculture. These health control officials are vested with the responsibility of preventing the spread of animal diseases. The federal government has the jurisdiction over interstate shipments, and the laws governing such movement apply uniformly to all parts of the country. The several states have separate and different laws governing the inshipment of livestock from other states and the control of livestock diseases within the state.

Although the act has been considerably modified and extended, the first federal legislation designed to control the spread of animal diseases was passed in 1884. State regulations have concerned themselves primarily with the eradication of tuberculosis, hog cholera, and brucellosis. Because of lack of uniformity in state laws, there has been considerable controversy in the regulations governing the control of the latter disease, particularly when calfhood vaccinations are involved.

## BRAND INSPECTION

Brand inspection of western cattle is undertaken primarily to prevent the stealing (rustling) of animals on the range and their subsequent shipment and sale to innocent parties at the big markets. This service is also very useful in recovering stray animals that have become mixed in with other cattle and cattle that, through oversight, have been shipped to market by parties other than their real owners. Brand inspection is supported by, and is under the supervision of, the different stock growers' associations of the range states. In Montana, however, the state government manages the work. Brand inspection is taken care of by inspectors expert in the reading and deciphering of brands. They examine each individual in every shipment of branded cattle.

## DEVELOPMENT OF MARKET CLASSES AND GRADES OF LIVESTOCK

With the turn of the century, the importance of having some uniform system of marketing livestock according to their relative merits or values was clearly recognized. Sensing this need, investigators of the Illinois Agricultural Experiment Station made a careful study of markets and published their findings and recommendations relative to market classes and grades in three different bulletins which appeared successively between 1902 and 1908.[9]

The need for uniform nationwide standards also soon became evident. Thus, in 1916, the U.S. Department of Agriculture was authorized to report market conditions and transactions at the various livestock markets in the United States. Using the earlier publication of the University of Illinois as a basis—and with the help of market specialists, representatives of producer organizations, and land-grant colleges—

[9]Mumford, H. W., *Market Classes and Grades of Cattle*, IL Ag. Exp. Sta. Bull. No. 78, 1902; Dietrich, W., *Market Classes and Grades of Swine*, IL Ag. Exp. Sta. Bull. No. 97, 1904; and Coffey, W. C., *Market Classes and Grades of Sheep*, IL Ag. Exp. Sta. Bull. No. 129, 1908.

the U.S. Department of Agriculture undertook to define tentative standards for uniform market classes and grades of cattle, sheep, and hogs. Through the years, these standards have been further refined, and more and more people have come to interpret them in a like manner regardless of the market or section of the country.

In addition to being rather specific and uniform—for example, so that a Choice lightweight, yearling, slaughter steer would be so designated on the several markets throughout the United States—it was recognized that the terminology must be easily understood and reasonably permanent in order to be workable. Moreover, the terminology must serve the market needs of producers, selling agencies, and buyers.

## USE MADE OF MARKET CLASSES AND GRADES

Live animal grades are not officially applied in the marketing of livestock, but the U.S. Department of Agriculture Market News Service publishes price quotations by grade.

Market classes and grades of animals have demonstrated their practicality in a number of ways. They have been especially valuable in rendering the following services:

1. In providing a means of selling animals according to their values.

2. As the standard vocabulary in market reporting, thus making possible more accurate and useful market reports.

3. As a guide to producers in shaping their breeding and feeding operations to meet market demands.

4. As a standard to ensure fair dealing when animals are purchased on order.

5. As a means of keeping various markets fairly well in line with each other.

6. As a means of ensuring that direct buying prices are on an equitable basis to those prevailing at terminal and auction markets.

7. In providing a producer with a method of selecting the market in which he may secure the best returns. The producer may review market reports based on standard classes and grades, and, with the flexibility of modern truck transportation, take advantage of the best market.

8. In providing a basis for compiling important statistical data used in analytical studies of supplies, demands, prices, and movements of livestock.

## DEFINITIONS OF MARKET CLASSES AND GRADES

Broadly speaking, *the market class is the use to which animals are put,* whereas *the market grade is a measure of how well the animal fulfills the requirements for the class.* More accurately, however, the market class is determined by all of those factors affecting the use and value of the animal, except the final grade. Thus, in cattle, the market class is determined by whether the animals are cattle or calves; by the general use to which the animals are put (slaughter cattle, stocker and feeder cattle, milkers and springers, vealers, slaughter calves, and stocker and feeder calves); by the sex (steers, heifers, cows, bulls, or stags); by the age (yearling or two-year-old or over steers) and by weight.

*Grading livestock is the act of sorting, dividing, or designating animals of similar classes and grades.* The grade is the final subdivision in the classification process. It indicates the relative degree of excellence of an animal or group of animals. When grading is properly and expertly done, each individual of a specific class and grade group is quite similar to other individuals in that group, regardless of whether the animals are in the same pen or in separate markets hundreds of miles removed from each other.

## FACTORS DETERMINING MARKET GRADES

Market grades are determined by attributes associated with market preferences and valuation. But the relevant attributes differ widely among species—in numbers, in range of variability, and in ease of objective measurement. Among species grade differences are the following:

1. Eight grades are used to cover the range in quality of steer and heifer carcasses, in comparison with only three for ready-to-cook poultry.

2. In addition to quality grades, five separate yield or, cutability grades (i.e., yield of boneless, closely trimmed retail cuts) are used in conjunction with quality grades of both beef and lamb.

Until 1989, meat packers could choose to grade or not to grade beef. But, if they graded, they were *required* to grade for both yield and quality. In 1989, the law was changed, separating quality and yield grades of beef; and allowing packers to choose whether beef carcasses are graded for quality, for yield, or for both quality and yield. But packers could continue to choose to grade or not to grade.

Quality grades, such as Prime, Choice, and Select, gauge differences in the taste of beef, primarily by examining marbling and maturity of the meat. Yield grades, numbered 1 to 5, identify differences in the percentage of lean meat obtained from a carcass.

3. Pork grades incorporate yield and quality considerations without separate quality and yield designations. Pork carcass grades rely heavily on objective measurements to determine expected combined yield of lean cuts, used in conjunction with a subjective determination of *acceptable* or *unacceptable* quality of the lean.

4. Poultry grades are based entirely on quality.

Thus, drawing up standards or specifications for a system of grades is a complex undertaking. Some of the attributes upon which grades are based can be evaluated directly; others must be evaluated indirectly, through indicators. For example, the yield of lean cuts of pork is related to (or indicated by) backfat thickness, carcass weight, and carcass length.

## SUPPLY AND DEMAND DETERMINE PRICES OF MARKET LIVESTOCK

The price that prevails for slaughter animals on foot is primarily determined by supply and demand.

Because animals normally provide about half of the entire cash income of U.S. farmers and ranchers, the importance of market prices is self-evident. But livestock raised for market are valuable only to the extent to which the meat and by-products obtainable from them are valuable.

The price of meat and animal by-products is largely determined by supply and demand, with the demand being affected primarily by buying power and competition from other products.

Contrary to the opinion held by some, packers do not control livestock and meat prices. Rather, like most commodities, they are dominated chiefly by supply and demand forces. That is, fluctuations in livestock prices are due to either changes in the demand for meat or to changes in the supply of slaughter animals. The amount of meat that people will buy is affected by a wide variety of circumstances, such as industrial conditions and buying power, the competition of other foods in season, holidays, and weather. On the other hand, the number of animals available for slaughter and meat supply is largely determined by whether the animals return a profit or incur a loss to the producer, and by the extent of each.

## FACTORS WHICH INFLUENCE MARKET SUPPLIES AND PRICES

Studies reveal that livestock prices change in keeping with a somewhat regular pattern according to the period of time. These types of price changes are (1) longtime trends, (2) cyclical trends, (3) seasonal changes, (4) short-time changes, and (5) Jewish holidays.

### LONGTIME (OR SECULAR) TRENDS

Longtime price trends are those which operate so slowly that they may not be detected for a number of years, perhaps for a period of 50 years or more. They are caused by such factors as changes in the standard of living or earning power of the masses of people in a country, by major population increases or decreases, or by longtime changes in production costs. The longtime price trends of cattle, sheep, and hogs have been very similar. In general, prices tended downward from the close of the Civil War to about 1896, after which price trends started upward. With the replacement of draft horses and mules by mechanical power, there was a steady decrease in the demand for work animals accompanied by a longtime decrease in horse and mule prices. Aside from the wartime years, the increase in numbers of meat animals has not kept pace with the increase of human population

### CYCLICAL TRENDS

The price cycle as it applies to livestock may be defined as that period of time during which the price for a certain kind of livestock advances from a low point to a high point and then declines to a low point again. In reality, it is a change in animal numbers that represents the producer's response to prices. Until about 1970, the price cycle of the different classes of animals was about as follows: hogs, 4 years; sheep, 9 to 10 years; and cattle, 10 years. (See Fig. 8–26.)

The species cycles were a direct reflection of the rapidity with which the numbers of each class of farm animals may be shifted under practical conditions to meet consumer meat demands. Litter-bearing and early-producing swine could be increased in numbers much more rapidly than either sheep or cattle. When market hog prices were favorable, established swine enterprises were expanded, and new herds were founded, so that about every 4 years, on the average, the market was glutted and prices fell, only to rise again because too few hogs were being produced to take care of the demand for meats.

It is noteworthy that cattle cycles were formerly 15 years or longer, but that they were shortened to about 10 years, due to the earlier maturity of modern cattle and the marketing of cattle at younger ages.

But big operations, confinement production, more year around births of young, and less seasonal production, may have outmoded price cycles.

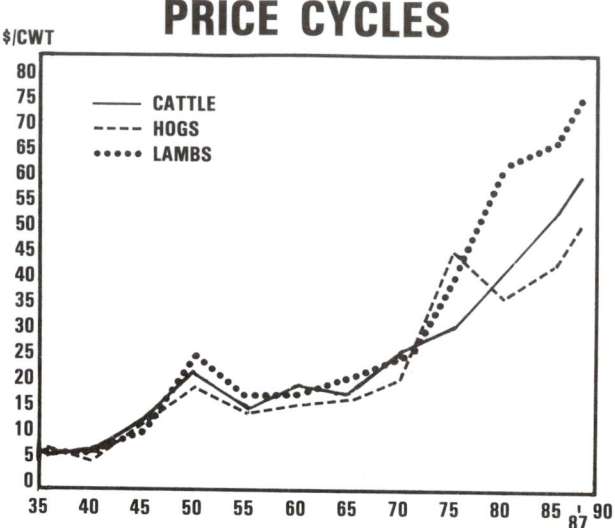

Fig. 8–26. Average price received by U.S. farmers for each class of livestock, 1935–87. Until about 1970, this showed that the price cycle of each class of animals was approximately as follows: hogs, 4 years; sheep, 9 to 10 years; and cattle, 10 years. But price cycles may be outmoded. (Source: Data for 1935–60 from *Agricultural Statistics 1962*, USDA, pp. 371, 387, 402. Data for 1965–70 from *Agricultural Statistics 1975*, USDA, pp. 306, 315. Data for 1975–1987 from *Agricultural Statistics 1988*, USDA, pp. 265, 276, 287)

### SEASONAL CHANGES

In recent years, seasonal patterns have not been as marked as they used to be. Year-round finishing of cattle and lambs in large commercial feedlots, year-round farrowing of sows in confinement, and control of estrus by light and hormones, have made for more uniform marketing throughout the year, and lessened seasonality in livestock marketing, in both receipts and prices. Thus, when arriving at livestock forecasts and marketing advice, proper reservation should be exercised in considering seasonal patterns.

Also, it must be realized that the normal seasons of high and low market prices may be changed by such factors as (1) federal farm programs and controls, (2) business conditions and general price levels, (3) feed supplies and weather conditions, (4) wars, etc.

Then too, it is not always wise to plan production to hit the highest market, for sometimes that would push up production costs more than enough to offset the gains from higher prices. Nevertheless, a careful study of normal seasonal receipts and seasonal prices will serve as useful guides.

Fig. 8–27 shows the seasonality of livestock receipts.

Fig. 8–27 shows the following seasonality in receipts of (1) hogs, (2) cattle, and (3) sheep and lambs:

1. The highest receipts of hogs is in March and October, and the lowest in May to August.

2. The highest receipts of cattle are from June to October—during the pasture season; and the lowest receipts are in November to March.

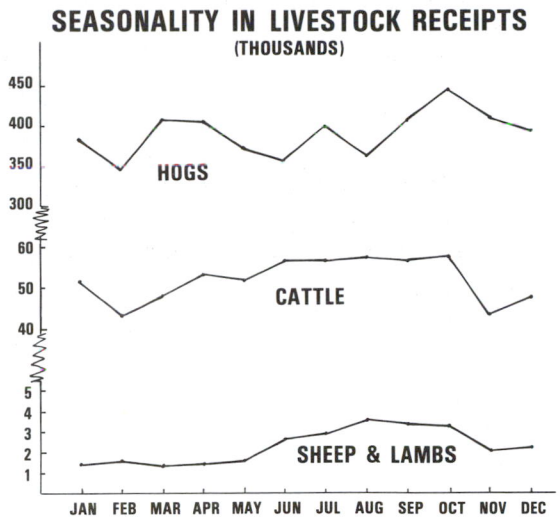

**SEASONALITY IN LIVESTOCK RECEIPTS**
(THOUSANDS)

Fig. 8-27. Seasonality in Livestock Receipts. This shows the average monthly receipts for the combined years of 1977, 1980, 1985, and 1987, of (1) hogs, (2) cattle, and (3) sheep and lambs, in the interior Iowa and southern Minnesota area. (Source: Chart based on data from *Livestock and Meat Statistics,* 1984-85, USDA, ERC, Statistical Bulletin No. 784, p. 64, Table 59, 1989)

3. The highest receipts of sheep and lambs occur in June to October, when spring lambs are marketed; and the lowest receipts occur in January to May.

Fig. 8-28 shows the seasonality in cattle prices.

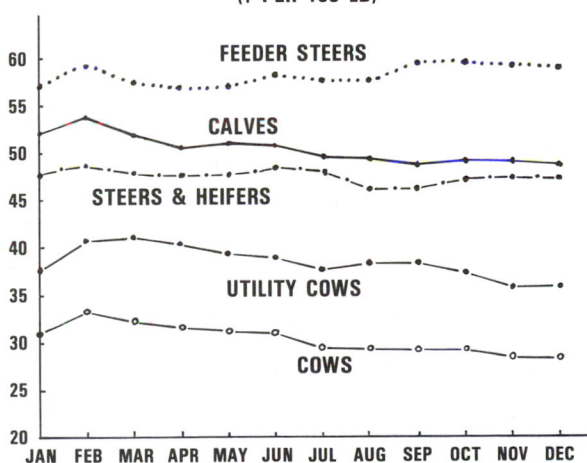

**SEASONALITY IN CATTLE PRICES**
($ PER 100 LB)

Fig. 8-28. Seasonality in cattle prices. This shows average monthly prices of (1) feeder steers, (2) calves, (3) steers and heifers, (4) utility cows, and (5) cows in the United States. The feeder steer prices are based on eight markets, for the combined years 1975, 1980, 1985, and 1987. The utility cow prices are based on the Omaha market, for the combined years 1976, 1980, 1985, and 1987. The prices of calves, steers and heifers, and cows are based on markets throughout the United States, for the combined years of 1970, 1975, 1980, and 1985. (Source: Chart based on data from *Livestock and Meat Statistics, 1984-88,* USDA, ERC, Statistical Bulletin No. 784: feeder steers, p. 182, Table 109; utility cows, p. 191, Table 115; calves, steers and heifers, and cows, p. 221, Table 130)

Fig. 8-28 shows the following seasonality in cattle prices:

1. Feeder steer prices change very little throughout the year, reflecting the year-round buying of feeder cattle by big commercial feedlots.

2. Calf prices peak in February, and are lowest from August to December—when the great bulk of the U.S. calf crop is weaned and marketed.

3. Steer and heifer prices show little seasonal change throughout the year, reflecting year-round marketing of finished cattle by big commercial feedlots.

4. Utility cows and cows are highest in February to May—at breeding/calving time, and lowest in October to January—during the period of winter feeding.

Fig. 8-29 shows the seasonality in lamb prices.

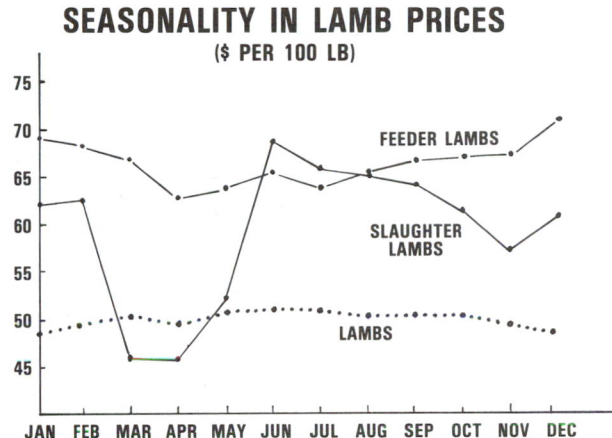

**SEASONALITY IN LAMB PRICES**
($ PER 100 LB)

Fig. 8-29. Seasonality in lamb prices. This shows average monthly prices of (1) feeder lambs, (2) slaughter lambs, and (3) lambs in the United States. The feeder lamb prices are based on the South St. Paul market, for the combined years 1977, 1980, 1985, and 1987. The slaughter lamb prices are based on the Omaha market, for the combined years 1977, 1980, 1985, and 1987. The lamb prices are based on markets throughout the United States, for the combined years 1970, 1975, 1980, and 1985. (Source: Chart based on data from *Livestock and Meat Statistics, 1984-88,* USDA, ERC, Statistical Bulletin No. 784: feeder lambs, p. 215, Table 124; slaughter lambs, p. 215, Table 125; lambs, p. 222, Table 132)

Fig. 8-29 shows the following seasonality in lamb prices:

1. Feeder lambs, which consist of young animals under one year of age that carry insufficient finish for slaughter, are highest from November to March—when few feeder lambs are available, and feeder lamb prices are lowest from April to August—when the bulk of the spring lamb crop is being marketed.

2. Slaughter lambs, which consist of young animals under one year of age that carry sufficient finish for immediate slaughter, show the greatest spread in seasonality between high and low prices of any species. The highest slaughter lamb prices are in May and June—when slaughter lambs are in short supply, and the lowest prices are in March and April—when many lambs finished in feedlots are marketed.

3. Lamb prices, which consist of all lambs under one year of age, and includes feeder lambs, slaughter lambs, and shearer lambs, shift very little throughout the year, but they are slightly higher in May to July, and slightly lower in December.

### SHORT-TIME CHANGES

Day-to-day variations in livestock prices usually are caused by uneven distribution of receipts on a given market because of such factors as weather, interference with transportation, strikes, uncertain or threatened federal policies, and stock market fluctuations. Although such changes are not large, the shrewd producer and market specialist is quick to take advantage of fluctuations of financial interest.

## JEWISH HOLIDAYS

The Jewish abstinence from eating meat during certain holidays exerts a minor influence on livestock prices. The important thing is that no kosher slaughtering is done on certain of these holidays. Likewise, producers should know that no animals are kosher slaughtered on Saturdays (the Jewish Sabbath). Normally, therefore, it is not good business to have too many kosher-type cattle, calves, or lambs on the eastern markets just prior to or during these holidays.

## PROMOTIONAL, RESEARCH, AND REGULATORY ORGANIZATIONS OF IMPORTANCE TO THE LIVESTOCK AND MEAT INDUSTRY

There are a number of nationwide promotional and research organizations whose business it is to work for the betterment of the livestock industry, especially in the areas of livestock marketing and/or meats. Among these are the following:

American Farm Bureau Federation
225 West Touhy Avenue
Park Ridge, IL 60068

American Meat Institute
P.O. Box 3556
Washington, DC 20007

American Sheep Industry Association
6911 South Yosemite Street
Englewood, CO 80112

Beef Promotion and Research Board
P.O. Box 3316
Englewood, CO 80155

Chicago Board of Trade
141 West Jackson Boulevard
Chicago, IL 60604

Chicago Mercantile Exchange
30 South Wacker Drive
Chicago, IL 60606

Livestock Marketing Association
7509 Tiffany Springs Parkway
P.O Box 901402
Kansas City, MO 64190–1402

National Cattlemen's Association
5420 South Quebec Street
P.O. Box 3469
Englewood, CO 80155

National Farmers Union
12025 East 45th Avenue
Denver, CO 80251

National Independent Meat Packers Association
734 - 15th Street, N.W.
Washington, DC 20005

National Live Stock and Meat Board
444 North Michigan Avenue
Chicago, IL 60611

National Live Stock Producers Association
4851 Independence Street #200
Wheat Ridge, CO 80033

National Pork Producers Council
P.O. Box 10383
Des Moines, IA 50306

Packers and Stockyards Agricultural Marketing Service
U.S. Department of Agriculture
Washington, DC 20250

Washington States Meat Association
88 First Street
San Francisco, CA 94105

## FUTURES AND OPTIONS TRADING

Futures and options markets offer ways to provide (1) an insurance medium in the marketing field, and (2) the facilities and machinery for underwriting price risks.

Trading in live cattle futures opened on the Chicago Mercantile Exchange in November 1964. Trading in live hogs began 15 months later. Since then, livestock futures trading has grown enormously. Between 1965 and 1988, the volume of cattle futures contracts increased from 59,219 to 5,477,205. Trading in hog futures contracts increased from 8,063 in 1966 to 2,008,750 in 1988.

## THE FUTURE OF LIVESTOCK MARKETING

The following transformation appears to be inevitable in future livestock marketing:

More animals will be marketed, for we shall need more meat to feed our expanding population; fewer, but larger, farms and feedlots, more ably managed and more deft in the marketplace; fewer, but bigger, meat packers, due to mergers; increased nonpublic market (direct) buying; more controls of futures trading; increased contract marketing and specifications; increased competition from the EC (Economic Community) markets; the continued age-old fight on freight rates, on carcasses vs live animals; increased government in business, including more supervision and antitrust action; unionization of more areas in the agribusiness chain; more futures trading; and increased and more effective meat promotion; and increased competition from broilers and turkeys.

A real challenge and a real opportunity lie ahead, provided livestock producers are not stymied by traditions and old allegiances—if they have the "tomorrow mind" instead of the "yesterday mind." The improvement of livestock marketing is in the hands of the producer, the seller, the processor, and the retailer. To meet this challenge, there must be understanding, patience, and an open mind; and all segments of the industry must work together. The future belongs to those who make wise and timely changes in a changing world.

# QUESTIONS FOR STUDY AND DISCUSSION

1. Why is market day the most important single day of operation to producers?

2. Present facts and figures that attest to the importance of livestock marketing.

3. Fig. 8–2 shows the volume of livestock marketed by species. Why are so many hogs and so few sheep/lambs marketed?

4. Fig. 8–3 shows the cash receipts from livestock marketed by species. Discuss (a) the present status as portrayed in this figure, and (b) the future changes that you foresee in cash receipts by species.

5. Enumerate the transition stages through which early-day marketing passed, finally culminating in the development of terminal markets.

6. In succession, market animals were transported by (a) drovers and water, (b) rail, and (c) truck. What factors caused each of these types of transportation, except trucking, to decline in importance.

7. Obviously, the term *packer* no longer connotes the total functions of meat processors. Should it be changed? If so, what name would you propose?

8. Are packer profits big? Justify your answer.

9. How does (a) the competition between meat slaughtering and processing plants, and (b) the utilization of by-products help to keep meat prices down?

10. Discuss the geography of meat production and consumption as portrayed in Fig. 8–15.

11. Who were the "Big Four" in early-day meat packing? Name five of the largest meat companies from among the ten largest in operation today.

12. List the primary factors which caused a decentralization of meat-packing. Are these factors likely to cause further decentralization? How have the major packers countered this move?

13. What method or methods of marketing (what market channel) do you consider most advantageous? Why do you consider it most advantageous?

14. What forces caused the phenomenal growth of non-public markets (direct marketing) during the last quarter of the 20th century?

15. What are the primary similarities, and what are the primary differences, between terminal market and auction selling?

16. Why aren't more U.S. animals sold on a carcass grade and weight basis? Why are fewer hogs sold on a carcass grade and weight basis than cattle, calves, sheep and lambs?

17. How does selling of purebreds differ from the marketing of slaughter animals?

18. Is there adequate assurance of honesty, of sanitation, and of humane treatment of animals in their shipment and marketing?

19. What steps would you take in preparing and shipping cattle, sheep, and swine (a) by truck and (b) by rail?

20. Define (a) shrinkage, (b) fill, and (c) dockage. Is the dockage of piggy sows and stags justified?

21. Do you consider today's livestock market news services adequate, especially from the standpoint of reporting non-public (direct) sales?

22. Define (a) antemortem inspection, and (b) postmortem inspection. Are both necessary?

23. What are the primary functions of federal and state meat inspections? Which is preferred? What proportion of U.S. meat slaughter is produced in (a) federally inspected plants, (b) nonfederally inspected plants, and (c) farm slaughter (see Fig. 8–25)?

24. What is the "Packers and Stockyards Act of 1921"? What are its purposes?

25. What are the responsibilities and purposes of (a) health control officials at livestock markets, and (b) brand inspectors?

26. Brief the historical development of market classes and grades of livestock.

27. Define on-foot market (a) classes, and (b) grades, and tell of their value.

28. What factors determine market grades? What grade differences exist between species?

29. Do packers control livestock and meat prices?

30. Today, longtime price cycles are less marked and less sharp than prior to 1970. What forces were back of this change?

31. What does Fig. 8–27 show relative to the seasonality in market receipts of each (a) hogs, (b) cattle, and (c) sheep and lambs?

32. What does Fig. 8–28 show relative to the seasonality in prices of each of the following market classes of cattle: (a) feeder steers, (b) calves, (c) steers and heifers, (d) utility cows, and (e) cows?

33. What does Fig. 8–29 show relative to the seasonality in prices of each of the following market classes of sheep: (a) feeder lambs, (b) slaughter lambs, and (c) lambs?

34. How do each of the following factors influence market supplies and prices: (a) short-time changes, and (b) Jewish holidays?

35. Name six promotional, research, and regulatory organizations of importance in the livestock and meat industry, and tell what each of them does.

36. List the livestock marketing areas in which you feel more research is needed, and give the economic importance of each.

37. What help, direct or indirect, does the producer obtain from the meat promotional organizations?

38. What is futures and options trading? Should a cattle feeder or a hog feeder use futures trading? If so, how?

39. Discuss the future of livestock marketing.

## SELECTED REFERENCES

| Title of Publication | Author(s) | Publisher |
|---|---|---|
| *Economics of Futures Trading* | T. A. Hieronymus | Commodity Research Bureau, Inc., New York, NY, 1971 |
| *Essentials of Marketing Livestock* | R. C. Ashby | (For the National Live Stock Exchange) Morningside College, Sioux City, IA |
| *Evolution of Futures Trading* | H. S. Irwin | Mimir Publishers, Inc., Madison, WI, 1954 |
| *Futures Trading in Livestock: Origins and Concepts* | Edited by H. H. Bakken | Mimir Publishers, Inc., Madison, WI, 1970 |
| *Futures Trading Seminar: History and Development*, Vol. I | H. H. Bakken, *et al.* | Mimir Publishers, Inc., Madison, WI, 1960 |
| *Livestock Marketing* | A. A. Dowell K. Bjorka | McGraw-Hill Book Company, New York, NY, 1941 |
| *Livestock and Meat Marketing* | J. H. McCoy | Avi Publishing Co., Westport, CT, 1972 |
| *Marketing: A Farmer's Problem* | B. F. Goldstein | The Macmillan Company, New York, NY, 1928 |
| *Marketing of Livestock and Meat, The*, Second Edition | S. H. Fowler | The Interstate Printers & Publishers, Inc., Danville, IL, 1961 |
| *Marketing: The Yearbook of Agriculture, 1954* | Edited by A. Stefferud | U.S. Department of Agriculture, Washington, DC, 1954 |
| *Organization and Competition in the Livestock and Meat Industry*, Tech. Study No. 1 | National Commission on Food Marketing | Superintendent of Documents, U.S. Government Printing Office, Washington, DC, June, 1966 |
| *Problems and Practices of American Cattlemen*, Wash. Ag. Exp. Sta. Bull. No. 562 | M. E. Ensminger M. W. Galgan W. L. Slocum | Washington State University, Pullman, WA, 1955 |
| *Stockman's Handbook, The*, Sixth Edition | M. E. Ensminger | The Interstate Printers & Publishers, Inc., Danville, IL, 1983 |

# HOW MEAT REACHES THE TABLE

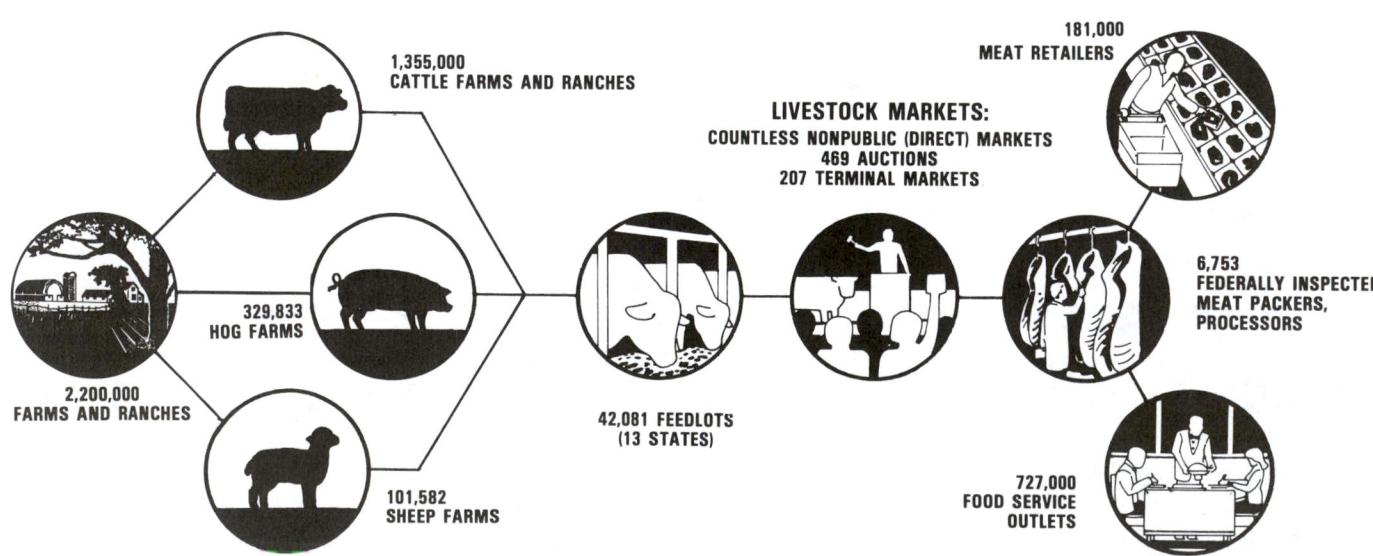

From whence meat comes. Back of meat on the table are 2,200,000 farms, of which 1,355,000 are cattle farms and ranches, 329,833 are hog farms, and 101,582 are sheep farms. All statistics are from the USDA, and for 1988, except no. of cattle farms, no. of hog farms, and no. of sheep farms are from 1982 Agriculture Census.

Broiler—meat animal of the decade! This male broiler weighed 4¾ lb at 42 days of age. (Courtesy, Hubbard Farms, Walpole, NH)

# MEAT AND ANIMAL BY-PRODUCTS

# Chapter 9

# HOW IT USED TO BE DONE!

Fig. 9-1. *Left:* Expeditionary Force sent to Persia and India in 1670. Formerly, live animals had to be carried along to be slaughtered as the army proceeded into uninhabited regions. Today, frozen, dehydrated, and canned meat simplifies the problem. It is also interesting to note that the search for spices for use in meat preservation prompted much of the early-day sailing. (Courtesy, The Bettmann Archive)

Fig. 9-2. Indians drying and smoking venison. (Photo from a watercolor original by Ernst Smith. Owned by the Rochester Museum of Arts and Sciences)

Fig. 9-3. *Below:* Smoking meats in early colonial days. This method of preserving meats is very old. Smoking is still a common practice in the homecuring of pork, and modern meat packers smoke many of their products. (Courtesy, Swift and Company)

Fig. 9-4. *Above:* Early-day (about 1840) butcher shop, from a painting by W.S. Mount (1807–1869). There were neither standard wholesale cuts nor grades. (Courtesy, The Bettmann Archive)

# HOW IT IS DONE TODAY

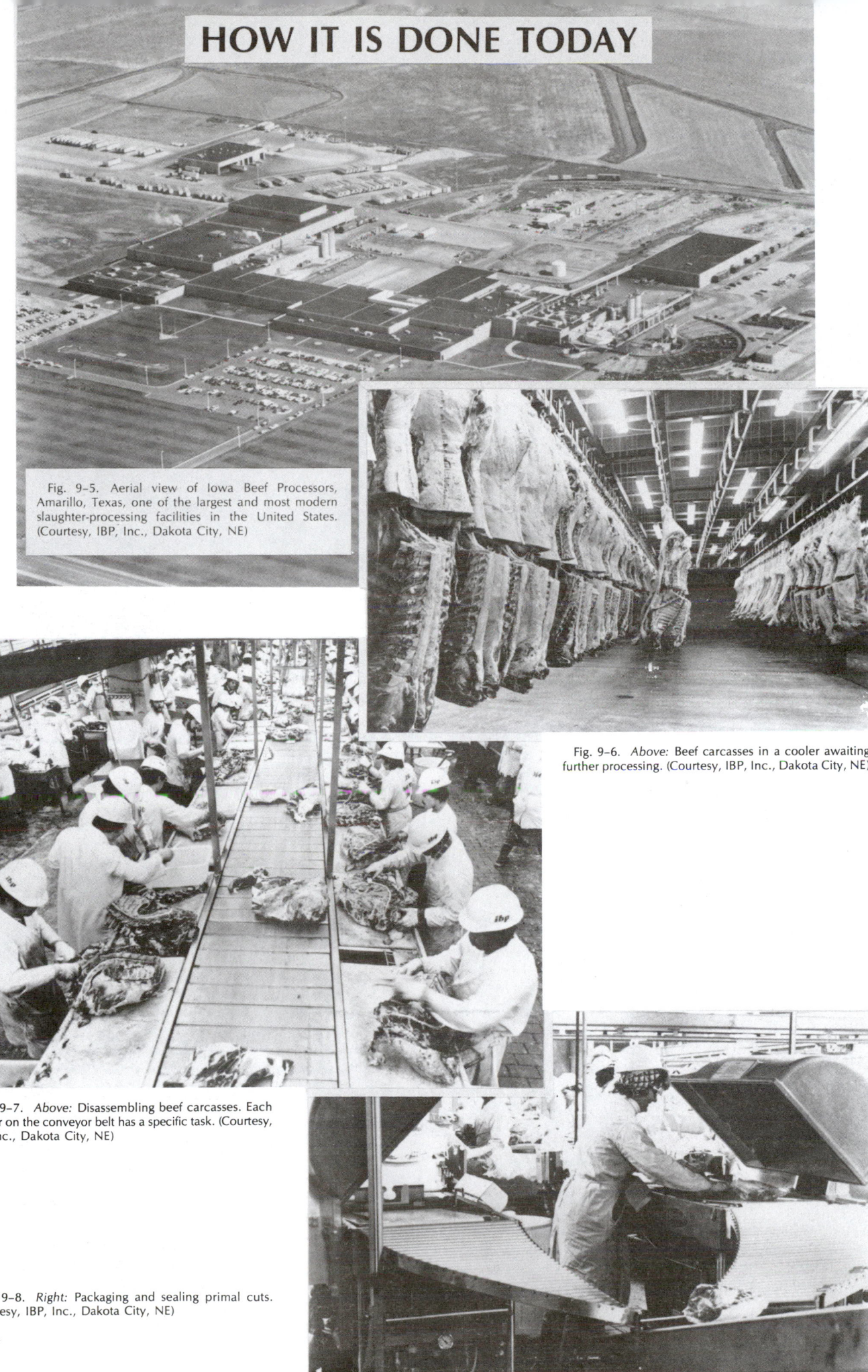

Fig. 9-5. Aerial view of Iowa Beef Processors, Amarillo, Texas, one of the largest and most modern slaughter-processing facilities in the United States. (Courtesy, IBP, Inc., Dakota City, NE)

Fig. 9-6. *Above:* Beef carcasses in a cooler awaiting further processing. (Courtesy, IBP, Inc., Dakota City, NE)

Fig. 9-7. *Above:* Disassembling beef carcasses. Each worker on the conveyor belt has a specific task. (Courtesy, IBP, Inc., Dakota City, NE)

Fig. 9-8. *Right:* Packaging and sealing primal cuts. (Courtesy, IBP, Inc., Dakota City, NE)

Fig. 9–9. Meat on the table—a roast of beef. (Courtesy, National Live Stock & Meat Board, Chicago, IL)

*Meat may be defined as the edible flesh of mammals used for food.* In a broad sense, the term meat may also include the flesh of fish, shellfish, poultry, and game. In this chapter, the discussion of meat is limited to red meat—to beef, lamb, and pork. Poultry meat is treated in the poultry section.

*Meat by-products include all products, both edible and inedible, other than the carcass meat.* The edible glands and organs are usually classed as by-products, but lard is usually grouped along with pork.

Although this chapter is devoted primarily to the final animal product—meat—it must be remembered that the top grades of this important food constituent represent the culmination of years of progressive breeding, the best in nutrition, vigilant sanitation and disease prevention, superior care and management and modern marketing, slaughtering, processing, and distribution. In brief, the efficient availability of the highest quality meat is dependent upon the well-coordinated operation of the whole field of animal science. Much effort and years of progress have gone into the production of luscious steaks or chops.

## MEAT OVER THE BLOCK
## THE ULTIMATE OBJECTIVE

Except for the wool produced by sheep, meat over the block is the ultimate objective in producing cattle, sheep, and swine. To be sure, this has not always been true. Time was, even in colonial America, when the greatest utility value of cattle was as work oxen. The production of meat and milk was purely incidental and hides were the most valuable product of slaughter. Yet, as feeds and animals increased in abundance, the colonists showed their fondness for beef, as well as for pork and lamb.

Of course, the type of animals best adapted to the production of meat over the block has changed in a changing world. Thus, in the early history of this country, the very survival of animals was often dependent upon their speed, hardiness, and ability to fight. Moreover, long legs and plenty of bone were important attributes when it came time for animals to trail hundreds of miles as drovers took them to market. The Texas Longhorn, Arkansas Razorback, and multicolored sheep of the Navajo Indians were adapted to these conditions.

Fig. 9–10. A modern meat counter. (Courtesy, Hyde Park Cooperative, Chicago, IL)

With the advent of rail transportation and improved care and feeding methods, the ability of animals to travel and fight diminished in importance. It was then possible, through selection and breeding, to produce meat animals better suited to the needs of more critical consumers. With the development of large cities, artisans and their successors in industry required fewer calories than those who were engaged in the more arduous tasks of logging, building railroads, etc. Simultaneously, the American family decreased in size. The demand shifted, therefore, to smaller and less fatty cuts of meats; and, with greater prosperity, high-quality steaks and roasts were in demand. To meet the needs of the consumer, the producer gradually shifted to the breeding and marketing of younger animals with maximum cut-out value of the primal cuts. The need was for a muscular animal with a minimum of external fat. Instead of marketing large, ponderous, fat, 3- to 5-year-old steers, yearlings came into prominence; the 600- to 1000-lb packing hogs were transformed into 200-lb meat-type barrows; and instead of marketing mature sheep as unrelished mutton, the marketing of lamb increased in importance.

Thus, throughout the years, consumer demand has always exerted a powerful influence upon the type of animals produced. To be sure, it is necessary that such production factors as prolificacy, economy of feed utilization, rapidity of gains, size, longevity, etc., receive due consideration along with consumer demands. But once these production factors have received due weight, the producers of all meat animals—whether they be purebred or commercial producers—must remember that meat over the block is the ultimate objective.

## QUALITIES IN MEATS DESIRED BY THE CONSUMER

Until World War II, big "lardy" hogs were preferred; families were large and engaged in strenuous outdoor occupations, there was a lively export for lard, and lard was in demand for use as shortening and for the manufacture of soaps and munitions. But times have changed! Vegetable oils have largely replaced lard as a shortening and we have lost much of our export market. From a position of minor importance in 1946, synthetic detergents captured the major part of the combined soap-detergent market. As a result of these changes, there has been a rather constant widening of the gap between the prices of the primal lean cuts of pork (hams, loins, picnics, and butts) and of fat for lard. Whereas fat was worth nearly as much as lean cuts in the early part of the century, in 1990 few fat cuts were wanted at any price.

But excess lard is not the only problem facing red meat producers. From 1980 to 1988, the percent of income spent for red meat dropped as follows: beef, from 2.16% to 1.29%; pork, from 1.13% to 0.82%; and all red meat from 3.28% to 2.11%. During this same period of time (1980 to 1988) the per capita consumption of poultry increased, yet, due to the increased efficiency of the American poultry producer, the percent of income spent for poultry decreased from 0.51%; to 0.42% (see Fig. 9–11).

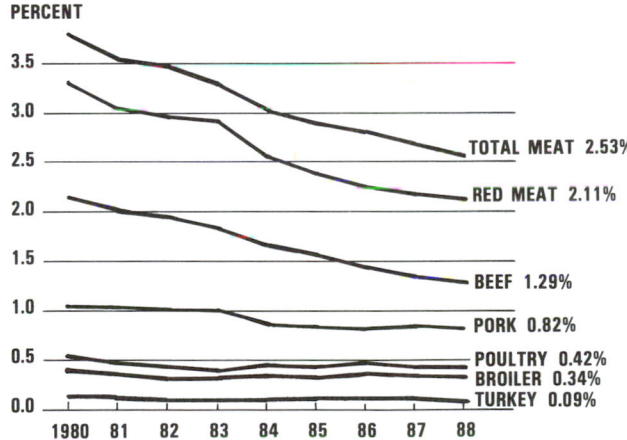

**PERCENT OF INCOME SPENT FOR RED MEAT AND POULTRY**

Fig. 9–11. Percent of U.S. disposable income spent for red meat and poultry 1980 to 1988. Red meat includes beef and pork only; poultry includes broilers and turkeys only; total meat includes beef, pork, broilers, and turkeys only. Note that the percent of income spent for red meat has declined, while percent of income spent for poultry meat has remained about the same. Since poultry meat consumption has increased markedly during this period, this indicates that poultry prices have decreased. (Source: USDA)

Because consumer preference is such an important item in the production of meats, it is well that the farmer and rancher, the packer, and the meat retailer be familiar with these desired qualities, which are summarized as follows:

1. Palatability.
2. Attractiveness, eye appeal.
3. Minimum amount of fat.
4. Tenderness.
5. Small cuts.
6. Ease of preparation.
7. Repeatability.

If these qualities are not met by meats, other products will. Recognition of this fact is important, for competition is keen for space on the shelves of a modern retail food outlet.

## FEDERAL GRADING

A scientific study of meat grading was first undertaken by the University of Illinois, beginning in 1902. The results of these studies were used as a basis by the U.S. Department of Agriculture workers in setting up tentative standards for classes and grades of beef carcasses. They were first used in market news reporting work in 1916. As would be expected in such pioneering and in a changing industry, several revisions of the early meat standards have been made through the years, thus keeping them in line with consumer demands and production practices.

Federal grading of beef was first started as a special service to U.S. Steamship Lines in 1923; and on February 10, 1925, the 68th U.S. Congress passed an act setting up a federal meat grading service. But commercial meat grading was not inaugerated until 1927, at which time Prime and Choice grades of steer and heifer beef were stamped at Boston, New York City, Philadelphia, Washington, Chicago, Omaha, and Kansas City. A year later, the Good grade was added, and finally the service broadened to include beef of all classes and grades.

Fig. 9–12. Federal grader shown rolling (grading or stamping) beef with an edible vegetable dye. (Courtesy, *Livestock Breeder Journal,* Macon, GA)

At first, federal grading of meats was limited to beef, but it now includes mutton, lamb, calf, veal, and pork carcasses.

Government grading, unlike meat inspection, is not compulsory. Official graders are subject to the call of anyone who wishes their services (packer, wholesaler, or retailer) with a charge per hour made. Government meat graders are appointed from a list of eligibles submitted by the Civil Service Commission. To qualify, a candidate must have had at least six years of suitable practical experience in wholesale meat marketing or grading. Beginners are also trained for a period of time under experienced graders.

Fig. 9–13 shows the proportion of the total U.S. meat slaughter (including federally inspected, nonfederally inspected, and farm slaughter) which was quality graded in 1988. It is noteworthy (1) that 93.6% of all lamb and mutton was quality graded, (2) that 56.4% of the beef was quality graded, and (3) that only negligible amounts of pork were quality graded. No farm-produced meat is federally graded.

## PROPORTION OF U.S. COMMERCIAL MEAT SLAUGHTER QUALITY GRADED

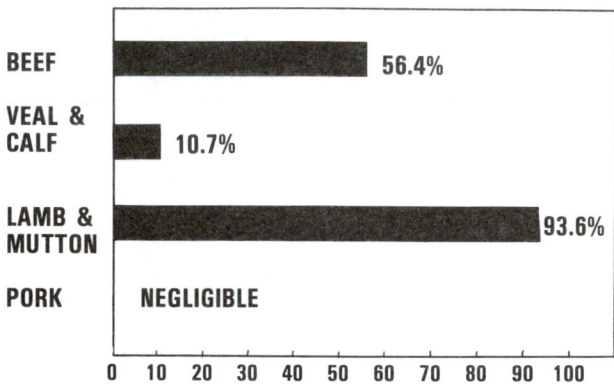

Fig. 9–13. Proportion of U.S. commercial meat production federally graded in 1988. (Source: USDA, Agricultural Marketing Service, Livestock and Seed Division, Meat Grading and Certification Branch)

## THE FEDERAL GRADES OF MEATS

_The grade of meat may be defined as a measure of its degree of excellence based on quality, or eating characteristics of the meat, and the yield, or total proportion of primal cuts._ Naturally, the attributes upon which the grades are based vary between species. Nevertheless, it is intended that the specifications for each grade shall be sufficiently definite to make for uniform grades throughout the country and from season to season, and that on-hook grades shall be correlated with on-foot grades.

Both producers and consumers should know the federal grades of meats and have a reasonably clear understanding of the specifications of each grade. From the standpoint of producers, this is important, for, after all, meat over the block is the ultimate objective. From the standpoint of consumers, especially those who buy most of the meat, this is important, because (1) in these days of self-service, prepackaged meats there is less opportunity to secure the counsel and advice of the meat cutter when making purchases, and (2) the average consumer is not the best judge of the quality of the various kinds of meats on display in the meat counter.

Federally graded meats are so stamped (with an edible vegetable dye) that the grade will appear on the retail cuts as well as on the carcass and wholesale cuts. These are summarized in Table 9–1.

In addition to the quality grades given in Table 9–1, there are the following yield grades for beef and lamb (mutton):

Grade #1
Grade #2
Grade #3
Grade #4
Grade #5

### TABLE 9–1
### QUALITY GRADES OF MEATS, BY CLASSES[1]

| Beef[2] | Veal | Mutton and Lamb | Pork |
|---|---|---|---|
| 1. Prime[3] | 1. Prime | 1. Prime[4] | 1. U.S. No. 1 |
| 2. Choice | 2. Choice | 2. Choice | 2. U.S. No. 2 |
| 3. Select | 3. Select | 3. Good | 3. U.S. No. 3 |
| 4. Standard | 4. Standard | 4. Utility | 4. U.S. No. 4 |
| 5. Commercial | 5. Utility | 5. Cull[5] | 5. U.S. Utility |
| 6. Utility | | | |
| 7. Cutter | | | |
| 8. Canner | | | |

[1]In rolling meat, the letters "U.S." precede each federal grade name. This is important, as only the government-graded meat can be so marked. For convenience, however, the letters "U.S." are not used in this table or in the discussion which follows.

[2]In addition to the quality grades given herein, there are the following yield grades for beef and lamb (and mutton) carcasses: Yield Grade 1, Yield Grade 2, Yield Grade 3, Yield Grade 4, and Yield Grade 5.

[3]Cow beef is not eligible for the Prime grade.

[4]Limited to lamb and yearling carcasses.

[5]Limited to mutton carcasses.

Quality grades are designed to gauge differences in taste (palatability), whereas yield grades identify the percentage of lean meat obtained from a carcass.

Until 1989, meat packers could choose to grade or not to grade beef. But, if they graded, they were _required_ to grade for both yield and quality. In 1989, the law was changed, separating quality and yield grades of beef; and allowing packers to choose whether beef carcasses are graded for quality, for yield, or for both quality and yield. But packers could continue to choose between to grade or not to grade.

Because of the different grade designation and the smaller number of grades in comparison with beef, the following elucidation is given relative to the grades of pork: Pork is more uniform in quality than any other class of meat and, therefore, there is need for fewer grades. Also, the grades of barrow and gilt carcasses are based on two general considerations: (1) the quality-indicating characteristics of the lean, and (2) the expected combined yields of the four lean cuts (ham, loin, picnic, shoulder, and Boston butt). Although the quality of the lean is best evaluated by a direct observation of its characteristics in a cut surface, such observations are impractical in grading carcasses. Thus, in carcasses, the quality of the lean is evaluated indirectly, on the basis of such quality-indicating characteristics as (1) firmness of fat and lean; (2) amount of feathering between the ribs; (3) color of lean; and (4) belly thickness, determined primarily by the thickness of the belly pocket. Research has shown that the actual average thickness of the backfat in relation to carcass length is a rather reliable guide to the yield of the four lean cuts. Therefore, in determining the grade of pork carcasses, the actual thickness of backfat and the carcass length are considered.

Some additional and pertinent facts relative to the federal grades of meat are:

1. **There is no differentiation between steer, heifer, and cow beef.** Federal grades make no distinction between steer, heifer, and cow beef. It is not intended that this should be construed to imply that there is no carcass difference between these sex classes. As a matter of fact, it is generally recognized that there is a pronounced difference between the sex classes of cattle, with a lesser difference between the sex classes of sheep and hogs. This step in simplification was taken so that it might be easier for the buyer or consumer to purchase meat on the

basis of quality, without the added confusion of a more complicated system.

2. **Bull and bullock beef are identified.** Bull and bullock beef are identified by class as *bull* beef and *bullock* beef, respectively. Within the bullock class, there are the following five grades: (1) Prime, (2) Choice, (3) Select, (4) Standard, and (5) Utility. However, no designated grade of bullock beef is necessarily comparable in quality to a similarly designated grade of beef obtained from steers, heifers, or cows. Neither is the yield in a designated yield grade of bull or bullock comparable to a similarly designated grade of steer, heifer, or cow beef.

3. **Lower grades seldom sold as retail cuts.** It is seldom that the lower grades—Cutter and Canner beef, Utility veal, Cull mutton and lamb, and Utility pork—are sold as retail cuts. The consumer, therefore, only needs to become familiar with the upper grades of each kind of meat.

As would be expected, in order to make the top grade in the respective classes, the carcass or cut must possess a very high degree of the attributes upon which grades are based. The lower grades of meats are deficient in one or more of these grade-determining factors. Because each grade is determined on the basis of a composite evaluation of all factors, a carcass or cut may possess some characteristics that are common to another grade. It must also be recognized that all of the wholesale cuts produced from a carcass are not necessarily of the same grade as the carcass from which they are secured. Fig. 9–14 gives the percentage distribution by grades of the total beef graded during 1988.

### PERCENTAGE DISTRIBUTION BY GRADE OF TOTAL BEEF QUALITY GRADED

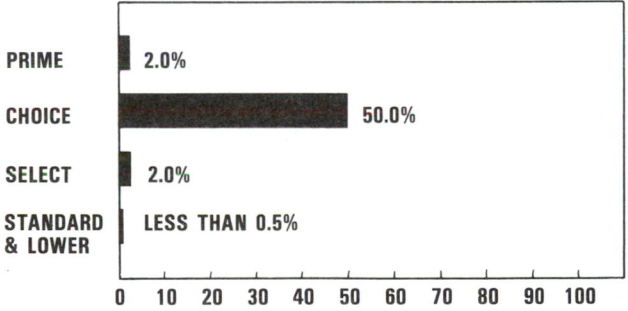

Fig. 9–14. Percentage distribution of beef by quality grades, in 1988. About 56.4% of the commercial beef slaughter was quality graded in 1988. (Source: USDA)

## ADVANTAGES AND DISADVANTAGES OF FEDERAL MEAT GRADING

Any system of evaluating products is certain to possess advantages and disadvantages, and meat grading is no exception.

Some of the **advantages** of competent federal grading are:

1. Because the federal grade is applied by an independent agent, rather than an employee of the packinghouse, the average consumer is less likely to question it. Moreover, it reduces possible misunderstanding between buyers and sellers of meats.

2. The average consumer is not a good judge of the quality of the various kinds of meat that are on display. However, if meats are correctly graded and labeled, consumers are in a position to select the grade that meets their particular requirements.

3. Because there are few federal grades, the consumer will become familiar with their significance more readily than with the more numerous packer brands.

4. Federal grading facilitates large-scale purchasing by federal, state, county, and city institutions and by hotels, restaurants, dining cars, and other large users of meats, chiefly because they can place large orders through one or several channels and obtain uniform products. Such large orders are usually on a contract basis, with grade specifications set forth.

5. The federal grades of meats correspond rather closely to the on-foot grades. They, therefore, provide a very important correlation for the benefit of producers. The payment of a premium for a quality product is a definite stimulation to improve breeding and feeding.

6. There is hardly any limit to the packer brands as established by different companies, whereas the federal grades are fairly comparable regardless of the area or ownership of the plant.

7. It is but natural that there should be less area or seasonal variation in appraisal of meat value or grades when a large overall agency is functioning than when graders are trained by many organizations, with only limited effort toward standardization between them.

8. Federal grading avoids any suspicion or temptation of upgrading, an alleged practice when demand for top grades exceeds the supply.

9. It makes meat market reports more intelligible.

10. Purchasers need not inspect carcasses or wholesale cuts to secure uniform products of merit to meet their requirements.

Some of the **disadvantages** of federal grading are:

1. It ignores sex classes that are important. It would hardly be expected that a cow carcass grading U.S. Select could be comparable to a steer carcass of the same grade.

2. It does not provide sufficiently narrow classifications for a critical trade.

3. Many feel that packer brand names are more alluring to buyers than the rather unglamorous federal grades. Naturally, this feeling has been accentuated through advertising.

4. A meat packer's reputation is an individual proposition more than a collective one. It is argued, therefore, that packers are more likely to uphold the reputation of their respective brands.

5. The national packers prefer advertising their brand names, claiming that to advertise federal grades places them in the position of helping the small packer, who does little or no advertising but who would greatly benefit therefrom.

## PACKER BRAND NAMES

Practically all packers identify their higher grades of meats with alluring private brands so that the consumer as well as the retailer can recognize the quality of a particular cut. Thus, the top brand names synonymous with the names of the big four early day meat packers and applied to their highest quality beef, pork, and lamb were: Armour's *Star*, Swift's *Premium*, Wilson's *Certified*, and Cudahy's *Puritan*.

A meat packer's reputation depends upon consistent standards of quality for all meats that carry the company brand names. The brand names are also effectively used in advertising campaigns.

## KOSHER MEATS

Meat for the Jewish trade—known as kosher meat—is slaughtered and processed according to ancient Biblical laws, called *Kashruth,* dating back to the days of Moses, more than 3,000 years ago. The Hebrew religion holds that God issued these instructions directly to Moses, who, in turn, transmitted them to the Jewish people while they were wandering in the wilderness near Mount Sinai.

The Hebrew word *kosher* means *fit or proper,* and this is the guiding principle in the handling of meats for the Jewish trade. Also, only those classes of animals considered clean— those that both chew the cud and have cloven hooves—are used. Thus, cattle, sheep, and goats—but not hogs—are koshered.[1] Poultry is also koshered.

Contrary to common belief, both forequarters and hindquarters of kosher slaughtered meat may be used by orthodox Jews. But because all veins must be removed before the meat is delivered to the customer, the Jewish trade usually confines itself to the forequarters, from which the veins can be removed with a minimum of tearing of the flesh. In order to devein a hindquarter, it is necessary to cut it up into such small pieces that it is very unattractive and unsalable for anything but ground meat or stews. Because the forequarters do not contain such choice cuts as the hinds, the kosher trade attempts to secure the best possible fores; thus, there is a secondary reason the kosher trade demands well-finished beef; namely, the ban against the use of lard in cooking, and the need for beef fat therefor.

Kosher meat must be sold by the packer or the retailer within 72 hours after slaughter, or it must be washed (a treatment known as *begiss,* meaning to wash) and reinstated by a representative of the synagogue every subsequent 72 hours. At the expiration of 216 hours after the time of slaughter (after begissing 3 times), however, it is declared *trafeh,* meaning forbidden food, and is automatically rejected for kosher trade. It is then sold in the regular meat channels. Because of these regulations, kosher meat is moved out very soon after slaughter. Also, it is easier to devein the meat while it is still warm than after it has been chilled.

Kosher sausage and prepared meats are made from meats which are deveined, soaked in water one-half hour, sprinkled with salt, allowed to stand for an hour, and washed thoroughly. This makes them kosher indefinitely.

The Jewish law also provides that before kosher meat is cooked, it must be soaked in water for one-half hour. After soaking, the meat is placed on a perforated board in order to drain off the excess moisture. It is then sprinkled liberally with salt. One hour later, it is thoroughly washed. Such meat is then considered to remain kosher as long as it is fresh and wholesome.

As would be expected, the volume of kosher meat is greatest in those Eastern Seaboard cities where the Jewish population is most concentrated. New York City alone uses about one-fourth of all the beef koshered in the United States.

While only about half of the total of six million U.S. Jewish population is orthodox, most members of the faith are heavy users of kosher meats.

## THE NUTRITIVE QUALITIES OF MEATS

Perhaps most people eat meat simply because they like it. They derive rich enjoyment and satisfaction therefrom. For flavor, variety, and appetite appeal, meat is unsurpassed.

But meat is far more than just a very tempting and delicious food. From a nutritional standpoint, it contains certain essentials of an adequate diet. This is important, for how we live and how long we live are determined in part by our diet.

It is estimated that the average American gets the following percentages of food nutrients from meat, poultry, and fish:[2]

| | |
|---|---|
| 77% of the vitamin B-12 | 30% of the phosphorus |
| 51% of the zinc | 27% of the iron |
| 46% of the niacin | 26% of the thiamin (B-1) |
| 44% of the protein | 25% of the riboflavin |
| 43% of the vitamin B-6 | |

The nutritive qualities of meats may be summarized as follows:

1. **Proteins.** The word *protein* is derived from the Greek word *proteios,* meaning *in first place.* Protein is recognized as a most important body builder. Fortunately, meat contains the proper quality and quantity of protein for the building and repair of body tissues. On a fresh basis, it contains 15 to 20% protein. Also, meat contains all of the amino acids, or building stones, which are necessary for the making of new tissue; and the proportion of amino acids in meat is similar to that in human protein.

2. **Calories.** The energy value of meat is largely dependent upon the amount of fat it contains. In turn, the fat content of meat is affected by breeding, feeding, and trimming—all of which have changed markedly in recent years. For example, since 1986, the external fat trim of beef has changed as shown in Fig. 9–15.

### THE EXTERNAL FAT TRIM ON BEEF HAS CHANGED

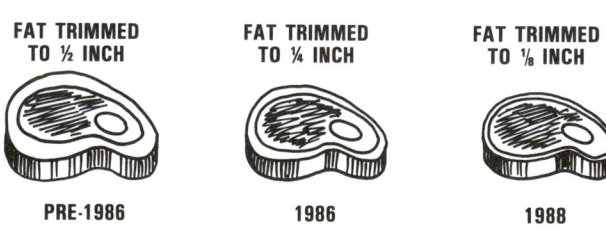

**FAT TRIMMED TO ½ INCH**     **FAT TRIMMED TO ¼ INCH**     **FAT TRIMMED TO ⅛ INCH**

PRE-1986          1986          1988

Fig. 9–15.   This shows the changes that have occurred in the external fat trim of beef since 1986.

A composite average serving of 3 oz of cooked and trimmed beef, pork, lamb, or veal will provide about 10% of the recommended calorie intake for an adult.

3. **Minerals and vitamins.** Minerals are necessary in order to build and maintain the body skeleton and tissues and to regulate body functions. Meat is a rich source of several minerals, but it is especially good as a source of phosphorus and iron.

In 1934, Doctors Minot, Murphy, and Whipple were jointly awarded the Nobel Prize in Medicine for the discovery that liver was effective in the treatment of anemia, a disease which once was regarded as fatal. The average adult would be assured an adequate supply of iron if two servings of meat were taken daily along with one serving of liver each week.

---

[1]*Deuteronomy 14:4–5 and Leviticus 11:1–8.*

[2]*Agricultural Statistics 1988,* p. 492, Table 678.

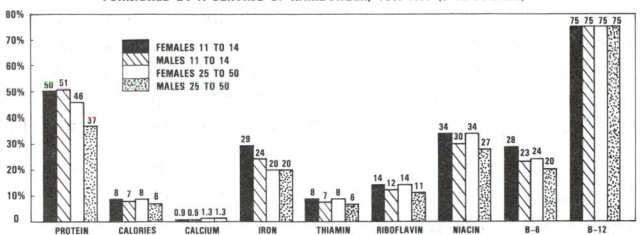

Fig. 9–16.  Percentage of daily recommended nutrient allowances furnished by a 3-oz serving of hamburger, 10% fat. Note that from left to right, each nutrient is given for the following sex-age groups: females, 11–14 years of age; males, 11–14 years of age; females, 25–50 years of age; and males, 25–50 years of age. (Source: *Recommended Dietary Allowances,* National Academy of Sciences, National Research Council, Washington, DC, 10th edition, 1989, p. 285. Nutrients in meat from *Foods and Nutrition Encyclopedia,* by Ensminger, *et al.,* published by Pegus Press, Clovis, CA)

## MEAT IS A GOOD SOURCE OF PHOSPHORUS

 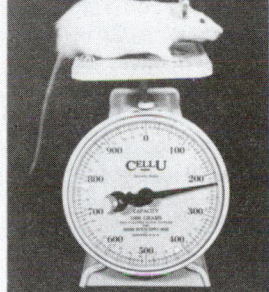

LOW PHOSPHORUS DIET (NO MEAT)
WT. 168 GRAMS

SAME DIET PLUS MEAT
WT. 225 GRAMS

Fig. 9–17.  Shows that meat is a good source of phosphorus. Minerals are necessary in order to build and maintain the body skeleton and tissues and regulate body functions. Although meat is a rich source of several minerals it is especially good as a source of phosphorus and iron. (Demonstration conducted by the National Live Stock and Meat Board in the laboratory of Rush Medical College.)

## MEAT IS A GOOD SOURCE OF THIAMIN (B–1)

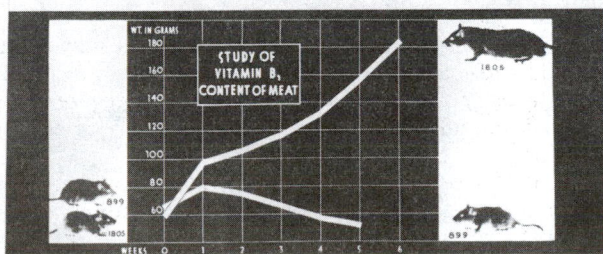

RAT NO. 899 RECEIVED ADEQUATE DIET EXCEPT FOR THIAMIN (B–1)
RAT NO. 1805 RECEIVED SAME DIET + 2% DRIED PORK HAM WHICH IS RICH IN THIAMIN (B–1)

Fig. 9–18.  Meat is a good source of thiamin (vitamin B–1). (Studies by the University of Wisconsin; supported by the National Live Stock and Meat Board)

## MEAT IS A GOOD SOURCE OF NIACIN

DOG ON A PELLAGRA-PRODUCING DIET, THAT IS A DIET LOW IN NIACIN.

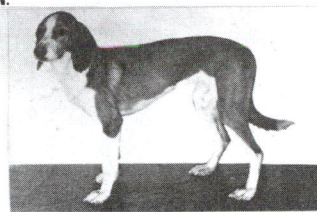

SAME DOG AFTER HAVING BEEN FED MEAT—A GOOD SOURCE OF NIACIN—FOR ONLY TWO WEEKS.

Fig. 9–19.  Meat is a good source of niacin which prevents and cures pellagra in people and "black tongue" in dogs. (Studies by the University of Wisconsin; supported by the National Livestock and Meat Board)

### TABLE 9–2
### VITAMIN CONTENT OF MEATS[1]

| Meat Product | Thiamin | Ribo-flavin | Niacin | Vitamin B-12 |
|---|---|---|---|---|
|  | (mg/100 g) | (mg/100 g) | (mg/100 g) | (mcg/100 g) |
| Beef liver ........... | .26 | 4.19 | 16.5 | 111.34 |
| Beef—ground w/10% fat ........ | .09 | .23 | 6.0 | 1.8 |
| Lamb, composite of cuts, good, 79% lean ........ | .15 | .21 | 4.9 | 2.15 |
| Pork, all cuts, thin class, separable lean, roasted ....... | .63 | .28 | 5.6 | 3.00 |
| Weiners or frank-furters w/non fat dehy. milk & cereal | .18 | .19 | 2.8 | 1.68 |

[1]*Foods & Nutrition Encyclopedia,* 1st ed., Ensminger *et al.,* Pegus Press, pp. 912 to 936.

As early as 1500 B.C. the Egyptians and Chinese hit upon the discovery that eating livers would improve one's vision in dim light. We now know that liver furnishes vitamin A, a very important factor for night vision. In fact, medical authorities recognize that night blindness, glare blindness, and poor vision in dim light are all common signs pointing to the fact that the person so affected is not getting enough vitamin A in the diet.

Meat is one of the richest sources of the important B group of vitamins, especially thiamin, riboflavin, niacin, and vitamin B–12 (see Table 9–2)—the vitamins now being used to reinforce certain foods and which are indispensable in our daily diet.

4. **Digestibility.** Finally, in considering the nutritive qualities of meats, it should be noted that this food is highly digestible. About 97% of meat proteins and 95% of meat fats are digested.

We come to realize, therefore, the important part that meat is playing in the nutrition of the nation. It should play an even greater part, for we are told that 35 to 40% of our population is now failing to receive an adequate diet to maintain health and vigor.

## MEAT MYTHS

Much has been said and written linking the consumption of meat to certain diseases. Also, the sustained annual increase in per capita and total meat consumption over a period of years may have lulled the industry into a sense of false security. A summary of three incorrect statements along with the facts follows:

• **Myth: Meat fats cause coronary heart disease**—Much has been said about the relationship of meat fats to heart disease, prompted by the extremely high mortality rate from coronary heart disease in the United States in recent years. Since 1965, deaths from cardiovascular disease in the United States have approximated one million, over half the total deaths from all causes. Some have conjectured that cholesterol and saturated fat, found in animal fat, raise blood cholesterol and cause a high incidence of coronary heart disease.

**Fact**—It is tempting to ascribe the high incidence of coronary heart disease to a single factor; for example, to blame either saturated fatty acids, cholesterol, calories, stress, heredity, hypertension, or lack of exercise. It seems far more prudent, however, to recognize that many factors are involved, and perhaps they are involved to different degrees under various sets of conditions.

Researchers at the University of Georgia have found that only one saturated fatty acid—caproic acid—raises serum cholesterol; but there is no caproic acid in meat fat. The Georgia studies also revealed that stearic acid, one of the main saturated fatty acids in meat fats, lowers both cholesterol and blood pressure.

Dr. Michael E. DeBakey, noted heart transplant surgeon and President of Baylor College of Medicine, Houston, Texas, reports that only about 30% of heart patients have any form of abnormality in their cholesterol. While Dr DeBakey has not ruled out the importance of diet, he has concluded that it is not the specific cause of heart disease. Said he, "We don't know the cause and we need to take a much saner attitude towards diet in relation to the disease since it is obvious that diet, so far as 65% to 70% of the patients are concerned, has not been related to or associated with the disease in our experience."

Dr. T. C. Huang (one of the author's former students, of whom he is very proud), in research that he conducted at Timken Mercy Hospital, Canton, Ohio, found that rats fed on diets containing saturated animal fats supplemented with cholesterol grew better and lived longer than rats given unsaturated fats.

Dr. Raymond Reiser, Distinguished Professor of Biochemistry and Biophysics at Texas A&M University, in a review of all pertinent research studies which have been the basis for the popular concept that saturated fats (largely those of animal origin) raise blood cholesterol and cause heart disease, concluded that saturated fats do not, by any criterion, elevate cholesterol to high risk levels, "if indeed they raise it at all."

In a study of the Masai tribesman of Africa, Dr. George V. Mann, Vanderbilt University School of Medicine, Nashville, Tennessee, found that these nomadic people, whose diet is made up almost entirely of meat and milk, had unusually low blood cholesterol levels and were virtually free of coronary heart disease.

Based on the above, it may be concluded that until more is known relative to the causes of heart disease, restraint should be exercised in making unsubstantiated recommendations relative to prevention and treatment lest there be created more problems than those for which attempt is being made to solve.

• **Myth: Beef causes bowel cancer**—In 1973, a "scare story" appeared on the front pages of a number of metropolitan newspapers, quoting the National Cancer Institute to the effect that, "There is now substantial evidence that beef consumption is a key factor in determining bowel cancer."

**Fact**—The study to which reference is made pertained to Hawaiian residents of Japanese descent. The truth of the matter is that string beans, peas, and macaroni all had more of a statistical relationship to colon cancer than beef. Also, the Japanese in Hawaii changed more than their eating habits when they left Japan. They changed their life-styles, too. Eating more beef wasn't the only change. They also ate less of some Japanese foods which could have been functioning as cancer suppressants.

Others have conjectured that colon cancer is predisposed by (1) lack of crude fiber or roughage in the diet, or (2) high amounts of fat. But the truth of the matter is that the real culprit is unknown.

In each of the above myths—that meat fats cause heart disease, and that beef causes bowel cancer—the facts support the thesis that the primary role of the researcher should remain that of helping people live longer and healthier, based on research facts, rather than advancing phobias.

• **Myth: U.S. meat consumption will continue unabated**—The optimistic view is that *come what may*, U.S. consumption, both in total and on a per capita basis, is elastic and unlimited.

**Fact**—This is not so. Neither beef consumption nor production is elastic *at any price*. They are subject to the old law of supply and demand. Cattle producers can, and will, produce enough beef if it is profitable. Price at the meat counter and consumption are, and will continue to be, governed by what consumers as a group are able and willing to pay.

Also, remember that the rate of U.S. human population increase has slowed; hence, there will be fewer meat eaters in the years ahead. Finally, cognizance should be taken of the fact that eating habits have changed. (See Figs. 9–24, 9–25, 9–26, and 9–27.)

## WHAT DETERMINES THE PRICE OF MEAT?

During those periods when meat is high in price, especially the choicest cuts, there is a tendency on the part of the consumer to blame either or all of the following: (1) the farmer or rancher, (2) the packer, (3) the meat retailer, or (4) the government. Such criticisms, which often have a way of becoming quite vicious, are not justified. Actually, meat prices are determined by the laws of supply and demand; that is, the price of meat is largely dependent upon what the consumers as a group are willing to pay for the available supply.

## THE AVAILABLE SUPPLY OF MEAT

Because the vast majority of meats are marketed on a fresh basis rather than cured and because meat is a perishable product, the supply of this food is very much dependent upon the number and weight of cattle, sheep, and hogs available for slaughter at a given time. In turn, the number of market animals is largely governed by the relative profitability of livestock enterprises in comparison with other agricultural pursuits. That is to say, farmers and ranchers—like any other good business people—generally do those things that are most profitable to them. Thus, a short supply of market animals at any given time usually reflects the unfavorable and unprofitable production factors that existed some months earlier and which caused curtailment of breeding and feeding operations.

Historically, when short meat supplies exist, meat prices rise, and the market price on slaughter animals usually advances, making livestock production profitable. But, unfortunately, livestock breeding and feeding operations cannot be turned on and off like a spigot. For example, a heifer cannot be bred until she is about 1½ years of age; the pregnancy period requires another 9 months; for various reasons only an average of 92 out of 100 cows bred in the United States conceive and give birth to young; and finally, the young are usually grown and fed until at least 1½ years of age before marketing. Thus, under the most favorable conditions, this production process, which is controlled by the laws of nature, requires about four years in which to produce a new generation of market cattle.

History also shows that if livestock prices remain high and feed abundant, producers will step up their breeding and feeding operations as fast as they can within the limitations imposed by nature, only to discover when market time arrives that too many other producers have done likewise. Overproduction, disappointingly low prices, and curtailment in breeding and feeding operations are the result.

Nevertheless, historically, the operations of livestock farmers and ranchers do respond to market prices, producing so-called cycles. Thus, the intervals of high production, or cycles, in cattle occur about every 10 years. In sheep, they occur about every 9 to 10 years, and in hogs—which are litter bearing, breed at an earlier age, have a shorter gestation period, and go to market at an earlier age—they occur every 4 years. Since about 1970, however, these cycles have been less pronounced.

## THE DEMAND FOR MEAT

The demand for meat is primarily determined by buying power and competition from other products. Stated in simple terms, demand is determined by the spending money available and the competitive bidding of millions of homemakers who are the chief home purchasers of meats. On a nationwide basis, a high buying power and great demand for meats exist when most people are employed and wages are high.

It is also generally recognized that in boom periods—periods of high personal income—meat purchases are affected in three ways: (1) More total meat is desired; (2) there is a greater demand for the choicest cuts; and (3) because of the increased money available and shorter working hours, there is a desire for more leisure time, which in turn increases the demand for those meat cuts of products that require a minimum time in preparation (such as steaks, chops, and hamburger). Thus, during periods of high buying power, people not only want more meats, but they compete for the choicer and easier prepared cuts of meats—porterhouse and T-bone steaks, lamb and pork chops, hams, and hamburger (chiefly because of the ease of preparation of the latter).

Because of the operation of the old law of supply and demand, when the choicer and easier prepared cuts of meat are

in increased demand, they advance proportionately more in price than the cheaper cuts. This results in a great spread in prices, with some meat cuts very much higher than others. While porterhouse steaks, or pork or lamb chops, may be selling for four or five times the cost per pound of the live animal, less desirable cuts may be priced at less than the cost per pound of the animal on foot. This is so because a market must be secured for all the cuts.

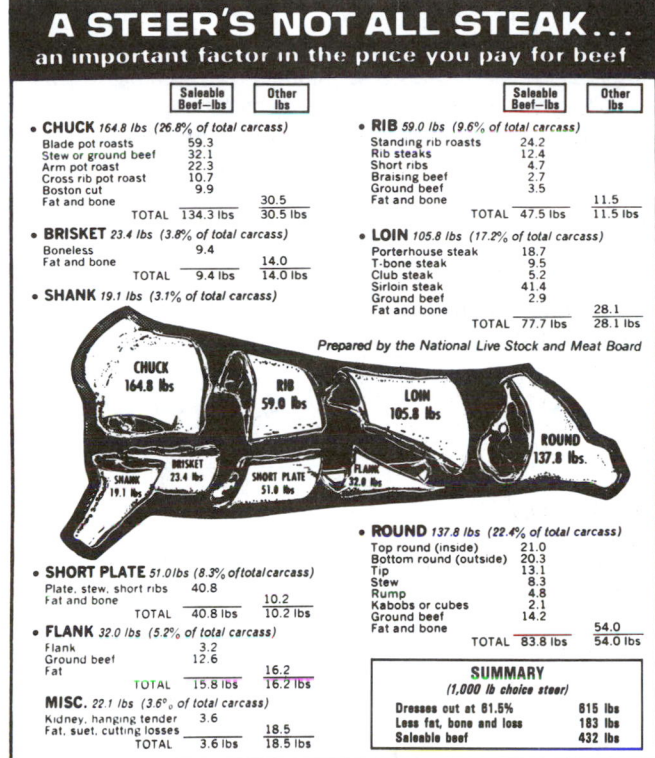

Supply and Demand are not the only factors in the price you pay for beef. For instance, today's modern-type 1,000 lb choice steer produces an approximate 615 lb carcass which the packer sells to a retailer who trims away 183 lbs of fat, bone and waste . . . ending up with only 432 lbs of beef that is cut, wrapped, and sold to customers.

Of that a surprisingly small amount is steak and a much larger quantity is roasts as shown in the chart above. Retail stores put a higher price on steak and a lower price on pot-roasts and ground beef so that they sell it all . . . not end up with only less in-demand cuts like pot-roasts and short ribs left in the cooler.

Fig. 9–20. Cattle are not all beef, and beef is not all steak! It is important therefore, that those who produce and slaughter animals and those who purchase wholesale and/or retail cuts know the approximate (1) percentage yield of chilled carcass in relation to the weight of the animals on foot, and (2) yield of different retail cuts. This figure illustrates these points. As noted, an average 1,000-lb steer grading Choice will yield about a 615-lb carcass or 432 lb of retail cuts, only 33.4 lb of which will be porterhouse, T-bone, and club steaks. (Courtesy, National Live Stock and Meat Board, Chicago, IL)

But the novice may wonder why these choice cuts are so scarce, even though people are able and willing to pay a premium for them. The answer is simple: Nature does not make many choice cuts or top grades, regardless of price. Moreover, a hog is born with 2 hams only, a lamb with 2 hind legs; and only 2 loins (a right and a left one) can be obtained from each carcass upon slaughter. In addition, not all weight on foot can be cut into meat. For example, the average steer weighing 1,000 lb on foot and grading Choice will only yield 432 lb of retail cuts (the balance consists of hide, internal organs, etc.). Secondly, this 432 lb will cut out only about 33.4 lb of porterhouse, T-bone, and club steaks. The balance of the cuts are equally wholesome and nutritious; and, although there are other steaks, many of the cuts are better adapted for use as roasts, stews, and soup bones. To make bad matters worse, not all cattle are of a quality suitable for the production of steaks. For example, the meat from most worn-out dairy animals and thin cattle of beef breeding is not sold over the block. Also, if the moneyed buyer insists on buying only the top grades of meat—namely U.S. Prime or U.S. Choice—it must be remembered that not all slaughter cattle produce carcasses of these top grades. To be sure, the lower grades are equally wholesome, but they are simply graded down because the carcass is somewhat deficient in conformation, finish, and/or quality.

Thus, when the national income is exceedingly high, there is a demand for the choicest but limited cuts of meat from the very top but limited grades. This is certain to make for high prices, for the supply of such cuts is limited, but the demand is great. Under these conditions, if prices did not move up to balance the supply with the demand, there would be a marked shortage of the desired cuts at the retail counter.

It must also be remembered that meats must compete with other food products for the consumer's dollar. On the average, consumers spend about 2.5% of their disposable income, of about 20% of their food budget, for meat.[3] In addition to preference, therefore, relative prices are an important factor in determining food selection.

Meat must also compete with certain nonfood items, for there are people who would literally go hungry in order to be able to spend their money for other purposes.

## WHERE THE CONSUMER'S FOOD DOLLAR GOES

Food is the nation's largest industry. A total of $411 billion,[4] approximately 12% of the disposable personal income, was spent for food in 1987. This vast sum included the bill for about 235,800 retail food stores; a large majority of the nation's 2,176,000 farms; thousands of wholesalers, brokers, eating establishments, and other food firms; and the transportation, equipment, and container industries. In 1987, the food and fiber sector employed 20.1 million people.[5]

In recognition of the importance of food to the nation's economy and the welfare of its people, it is important to know where the consumer's food dollar goes—the proportion of it that goes to the farmer, and the proportion that goes to the middle-person.

---

[3]It is noteworthy that food expenditures as a percent of income have decreased as follows:

| Year | Percent of Income |
|------|-------------------|
| 1937–39 | 23.1 |
| 1947–49 | 24.7 |
| 1957–59 | 20.6 |
| 1960 | 20.0 |
| 1965 | 18.0 |
| 1970 | 13.9 |
| 1975 | 13.9 |
| 1980 | 13.6 |
| 1985 | 12.5 |
| 1987 | 12.1 |

Source: 1937–1965 from "Report to the National Commission on Food Marketing," *Food from Farmer to Consumer*, June 1966, p. 9. Data: from *Food Consumption, Prices, and Expenditures, 1960–80,* Stat. Bull. No. 672, USDA, p. 99, Table 88. Data for 1985 and 1987 from *Food Consumption, Prices, and Expenditures, 1966–87,* USDA, ERC, Stat. Bull. No. 773, p. 106, Table 86, 1989.

[4]*Meat Facts,* 1989 Edition, American Meat Institute, Washington, DC, 1989, p. 34. Data for 1988.

[5]*Statistical Abstract of the United States 1989,* U.S. Department of Commerce, pp. 626 and 377.

---

Table 9–3 reveals that of each food dollar in 1987, the farmer's share was only 30 cents, the rest—70 cents—went for processing and marketing. This means that considerably more than one-half of today's food dollar goes for preparing, processing, packaging, and selling—and not for the food itself.

### TABLE 9–3
### WHERE THE CONSUMER'S FOOD DOLLAR GOES
### AS SHOWN BY THE MARKET BASKET[1]

| Year | Retail Price | Farm Value | Farm-to-Retail Spread | Farm Value Share of Retail Price |
|------|-------------|-----------|----------------------|----------------------------------|
| | ◄——— (Index, 82–84 = 100) ———► | | | (%) |
| 1980 | 88 | 97 | 83 | 37 |
| 1985 | 104 | 96 | 108 | 32 |
| 1987 | 112 | 97 | 119 | 30 |

[1]The market basket contains the average quantities of U.S. farm food products purchased annually per household. The farm value is the return to the farmer for the farm product equivalent of foods in the market basket. The spread between the retail cost and farm value represents charges for processing and marketing the product.

Source: *Food Cost Review, 1987,* USDA, ERS, Agricultural Economic Report No. 596, p. 10, Table 5, 1988.

Fig. 9–21 gives a further breakdown, by selected food items, of where the consumer's food dollar goes. Among the products shown, in 1986, the farmer's share of the consumer's dollar ranged from 7 cents per dollar for white bread to 62 cents per dollar for eggs, to 54 cents for Choice grade beef, and 46 cents per dollar for pork. The farmer's share of the retail price of eggs, beef, and pork is relatively high because processing is simple and inexpensive, and the transportation costs are low due to the concentrated nature of the products. Also, in the case of beef, retailer's handling costs are moderate because stores frequently sell beef at special, low prices—as an attraction. On the other hand, the farmer's share of white bread is low due to its bulky, costly transportation and space requirement and perishability in the retail store.

## FARM VALUE SHARE OF RETAIL FOOD PRICES

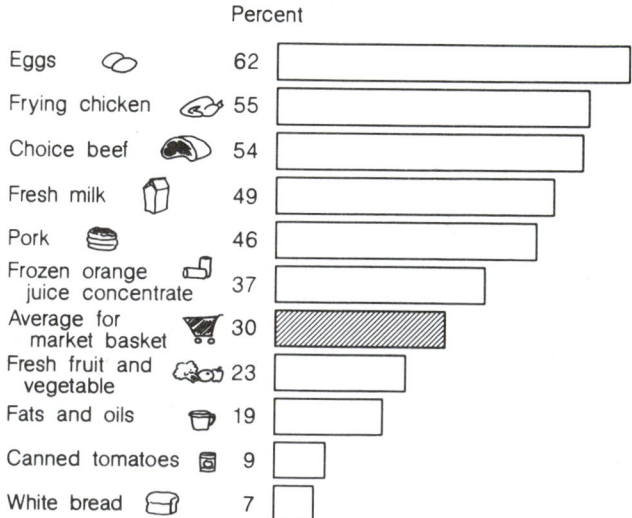

Fig. 9–21. Farmer's share of the market basket of selected foods, 1986. The farmer's share represents the price farmers receive for the raw commodity. The remainder of the dollar goes for processing, distributing, and retailing costs. (Source: *1988 Agricultural Chartbook,* USDA, Agricultural Handbook No. 673, p. 62, Chart 119.)

Among the reasons farm prices of foods fail to go up, thereby giving the farmer a greater share of the consumer's food dollar, are (1) rapid technological advances on the farm, making it possible for the farmer to stay in business despite small margins; (2) overproduction; and (3) the relative ease with which cost pressures within the marketing system can be passed backward rather than forward.

Over the years, processing and marketing costs have increased primarily because consumers have demanded, and gotten, more and more processing and packaging—more built-in services. For example, few consumers are interested in buying a live hog—or even a whole carcass. Instead, they want a pound of pork chops—all trimmed and packaged, and ready for cooking. Likewise, few homemakers are interested in buying flour and baking bread. But consumers need to be reminded that, fine as these services are, they cost money—hence, they should be expected to pay for them. Without realizing it, American consumers have millions of people working for them. They are the people who work on the food line from the time it leaves the farms and ranches until it is on the tables. They're the people who make it possible to choose between quick-frozen, dry-frozen, quick-cooking, ready-to-heat, ready-to-eat, and many other conveniences. Although figures covering the marketing of meat animals alone are not available, the U.S. Department of Commerce does present a breakdown of numbers of workers, and types of work involved, in marketing food and fiber as a whole. In 1987, a total of 18.1 million

workers were engaged in the food and fiber sector after the products left the nation's farms and ranches. Of these 18.1 million workers, 1.3 million were employed in food processing, 2.7 million in manufacturing, 6.6 million in transportation, trade, and retailing, 3.7 million in eating establishments, and 3.9 million in other related work.[6] These millions of workers engaged in transporting, processing, and distributing foods must be paid, for they have to eat, too. Of course, both men and women make up this work force. The proportion of the nation's labor force made up of women rose from 28% in 1947 to 52.5% in 1987.[7]

## U.S. MEAT PRODUCTION

The U.S. production of meat (total and by kinds) is shown in Table 9–4 and Fig. 9–22.

Table 9–4 and Fig. 9–22 reveal that (1) beef accounts for an astounding 61% of all U.S. red meat production, (2) beef production is increasing faster than second-ranking pork, and (3) veal and lamb-mutton production account for a very small proportion of the total red meat production.

Table 9–5 shows the number of animals of each of the respective classes slaughtered in the United States in the year 1987, and the proportions of these animals slaughtered in federally inspected plants, nonfederally inspected plants, and farm slaughter. It is noteworthy that hog slaughter numbers are about 2¼ times that of cattle.

### TABLE 9–4
### U.S. PRODUCTION OF MEAT[1]

| Year | Beef (million) | | U.S. Meat Prod. | Veal (million) | | U.S. Meat Prod. | Lamb and Mutton (million) | | U.S. Meat Prod. | Pork (excluding lard) (million) | | U.S. Meat Prod. | All Meats (excluding lard) (million) | |
|------|------|------|------|------|------|------|------|------|------|------|------|------|------|------|
| | (lb) | (kg) | (%) | (lb) | (kg) | (%) | (lb) | (kg) | (%) | (lb) | (kg) | (%) | (lb) | (kg) |
| 1965 | 18,699 | 8,482 | 59 | 1,020 | 463 | 3.3 | 651 | 295 | 2.1 | 11,132 | 5,049 | 35 | 31,502 | 14,288 |
| 1970 | 21,652 | 9,821 | 60 | 588 | 267 | 1.6 | 551 | 250 | 1.5 | 13,429 | 6,091 | 37 | 36,220 | 16,429 |
| 1975 | 23,974 | 10,874 | 65 | 873 | 396 | 2.4 | 410 | 186 | 1.1 | 11,503 | 5,218 | 31 | 36,760 | 16,674 |
| 1980 | 21,643 | 9,838 | 56 | 400 | 182 | 1.0 | 318 | 145 | 0.8 | 16,617 | 7,553 | 43 | 38,978 | 17,717 |
| 1985 | 23,728 | 10,785 | 60 | 515 | 234 | 1.3 | 359 | 163 | 0.9 | 14,807 | 6,730 | 38 | 39,409 | 17,913 |
| 1987 | 23,566 | 10,712 | 61 | 429 | 195 | 1.1 | 315 | 143 | 0.8 | 14,374 | 6,534 | 37 | 38,648 | 17,567 |

[1]Source: USDA.

## MEAT PRODUCTION

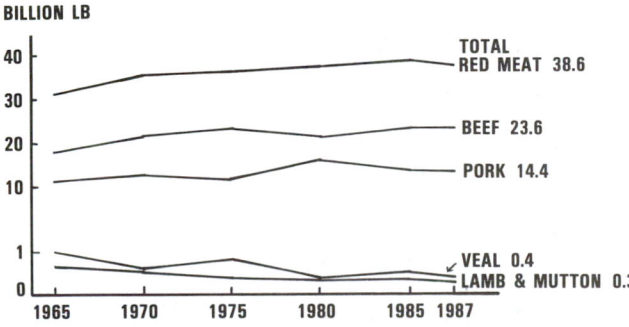

Fig. 9–22.  U.S. red meat production by kinds. (Source: USDA)

### TABLE 9–5
### UNITED STATES MEAT SLAUGHTER
### (FEDERALLY, NONFEDERALLY INSPECTED, AND FARM SLAUGHTER)
### BY CLASSES OF ANIMALS[1]

| Federal, Nonfederal, or Farm Slaughter | Cattle | Calves | Sheep and Lambs | Hogs |
|------|------|------|------|------|
| | (1,000 head) | | | |
| Federally inspected .... | 34,468 | 2,680 | 5,042 | 78,913 |
| Nonfederal ......... | 1,179 | 135 | 158 | 2,168 |
| Farm ............. | 243 | 87 | 112 | 341 |
| Total ............. | 35,890 | 2,902 | 5,312 | 81,422 |

[1]Data for 1987. Source: *Livestock and Meat Statistics, 1984–88*, USDA, ERC, Statistical Bulletin No. 784, p. 79, Table 64.

[6]*Statistical Abstract of the United States 1989*, U.S. Department of Commerce, Bureau of the Census, p. 626, No. 1074.

[7]*Statistical Abstract of the United States 1989*, U.S. Department of Commerce, pp. 626 and 377.

## PER CAPITA MEAT CONSUMPTION IN THE UNITED STATES

Although comprising only 5% of the world's population, the people of the United States consume 11.4% of the total world production of meat. The amount of meat consumed in this country varies from year to year (see Fig. 9–23). In 1989, the average per capita meat consumption, exclusive of lard, was 219.8 lb, with distribution by types of meat as shown in Table 9–6.

**MEAT CONSUMPTION PER PERSON**

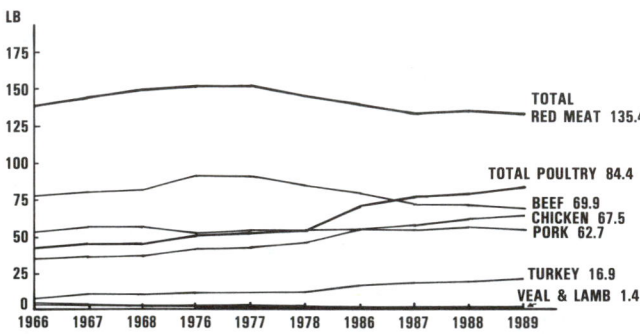

Fig. 9–23. U.S. per capita meat consumption on a retail basis, by kind of meat. Note that per capita beef consumption has gradually declined, while poultry consumption has increased dramatically—almost doubling since 1966. (Source: *National Food Review*, USDA, ERC, April-June, 1989, Vol. 12, Issue 2, p. 2, Table 1)

**TABLE 9–6**
**UNITED STATES PER CAPITA CONSUMPTION OF MEAT, BY KINDS, RETAIL WEIGHT BASIS[1]**

| Type of Meat | 1989 Annual Per Capita Consumption (retail weight basis) | |
|---|---|---|
| | (lb) | (kg) |
| **Red Meat:** | | |
| Beef | 69.9 | 31.8 |
| Veal | 1.4 | 0.6 |
| Lamb | 1.4 | 0.6 |
| Pork | 62.7 | 28.5 |
| Total | 135.4 | 61.5 |
| **Poultry:** | | |
| Chicken | 67.5 | 30.7 |
| Turkey | 16.9 | 7.7 |
| Total | 84.4 | 38.4 |

[1]Data for 1989. Source: *National Food Review*, USDA, ERC, April-June 1989, Vol. 12, Issue 2, p. 2, Table 1.

For the most part, meat consumption in this country is on a domestic basis, with only limited amounts being either imported or exported. Although cured meats furnish somewhat of a reserve supply—with more meats going into cure during times of meat surpluses—meat consumption generally is up when livestock production is high. Also, when good crops are produced and feed prices are favorable, market animals are fed to heavier weights. On the other hand, when feed-livestock ratios are unfavorable, breeding operations are curtailed, and animals are marketed at lighter weights. But during the latter periods, numbers are liquidated, thus tending to keep the meat supply fairly stable.

## TRENDS IN PER CAPITA CONSUMPTION OF FOODS

The selection of foods is based on taste preference and relative prices, with different foods competing for a place in the consumer's diet. In prosperous times and with increased buying power, the preferred foods increase in demand and price. Because cereal and bakery products and fresh fruit and vegetables are not luxury foods, their per capita consumption is little affected by buying power.

Figs. 9–24, 9–25, 9–26, and 9–27, show U.S. trends in the per capita consumption of foods since 1967[8] These charts clearly point up the following U.S. trends in eating habits: (1) The consumption of vegetables, poultry, fish, dairy products, sugars and sweeteners, and fats and oils is increasing; and (2) the consumption of red meat and eggs is decreasing.

**Per Capita Consumption of Food**

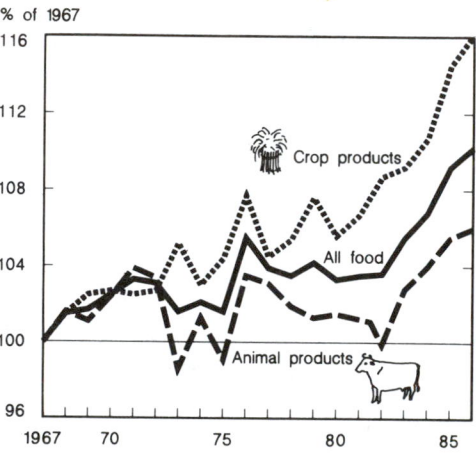

Fig. 9–24. Per capita consumption of (1) crop products (vegetables and cereal grains), (2) animal products, and (3) all food (crop and animal combined). Note that the consumption of crop products has increased more than the consumption of animal products.

**Per Capita Consumption of Meat, Poultry, and Fish**

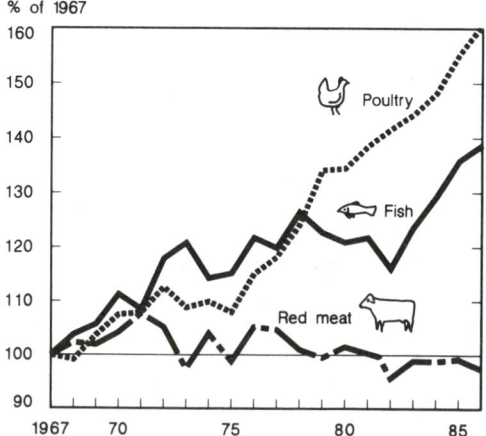

Fig. 9–25. Per capita consumption of (1) red meat, (2) fish, and (3) poultry. Note the marked change in eating habits; the consumption of both poultry and fish has increased sharply, while the consumption of red meat has decreased.

[8]These four charts are from *1988 Agricultural Chartbook*, USDA, Ag. Hdbk. No. 673, p. 64.

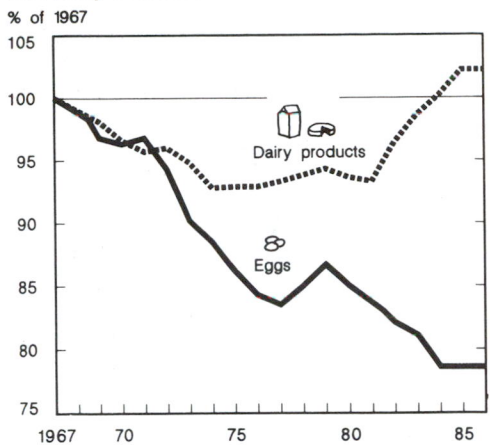

**Fig. 9–26.** Per capita consumption of (1) eggs, and (2) dairy products. Note that the consumption of dairy products has increased, while the consumption of eggs has dropped sharply.

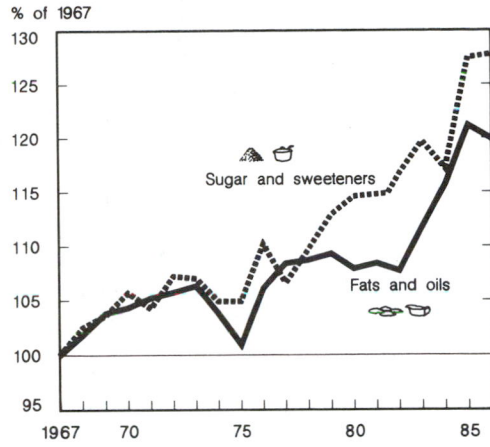

**Fig. 9–27** Per capita consumption of (1) sugars and sweeteners, and (2) fats and oils. Note that the consumption of both types of products has increased.

## MEAT IMPORTS AND EXPORTS

Livestock producers are prone to ask why the United States, which produces about one-fourth of the world's meat, buys meat from abroad. Conversely, consumers sometimes wonder why we export meat. Occasionally, there is justification for such fears. Table 9–7 gives a comparison of U.S. meat imports and production. Figs. 9–28 and 9–29 give a breakdown by kinds of meat imports and exports.

### TABLE 9–7
### U.S. IMPORTS OF MEATS COMPARED WITH U.S. MEAT PRODUCTION[1]
(Average 1983–87, Annual 1987)

| Commodity | 5-Year Average | | | | | 1987 | | | | |
|---|---|---|---|---|---|---|---|---|---|---|
| | Imports (1983–87) | | Production (1983–87) | | Imports Compared with Production | Imports | | Production | | Imports Compared with Production |
| | (million) | | (million) | | | (million) | | (million) | | |
| | (lb) | (kg) | (lb) | (kg) | (%) | (lb) | (kg) | (lb) | (kg) | (%) |
| Beef and veal ............... | 2,044 | 929 | 23,701 | 10,773 | 8.6 | 2,269 | 1,031 | 23,566 | 10,712 | 9.6 |
| Pork ........................ | 1,020 | 464 | 14,651 | 6,659 | 7.0 | 1,195 | 543 | 14,374 | 6,534 | 8.3 |
| Lamb and mutton ............. | 32 | 14 | 353 | 160 | 9.1 | 44 | 20 | 315 | 143 | 14.0 |
| Total red meat .............. | 3,096 | 1,407 | 38,705 | 17,592 | 8.0 | 3,508 | 1,594 | 38,255 | 17,389 | 9.2 |

[1]Source: USDA.

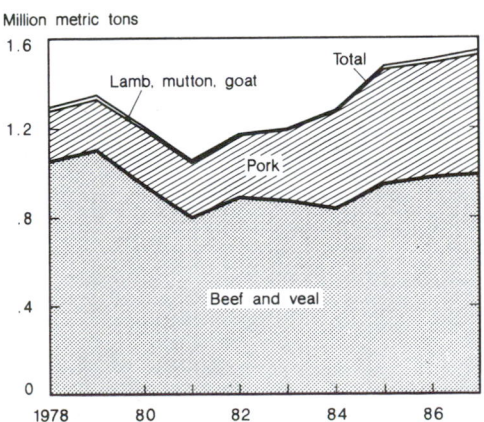

**Fig. 9–28.** U.S. imports of red meats, 1978 to 1987. Note the dominant position of beef and veal. (Source: *1988 Agricultural Chartbook,* Ag. Hdbk. No. 673, USDA, p.88, Chart 188)

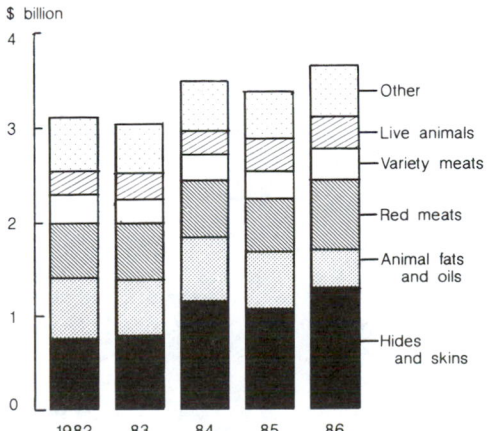

**Fig. 9–29.** U.S. exports of meat, livestock products, and live animals, 1982 to 1986. (Source: *1988 Agricultural Chartbook,* Ag. Hdbk. No. 673, USDA, p.88, Chart 187)

Table 9–8 reveals that the United States imports more meat than it exports, but, as shown in Table 9–7, total meat imports actually constitute a very small percentage of the available U.S. supply; imports constituted only 8% of the total U.S. red meat supply during the 5-year period 1983–1987.

## TABLE 9–8
## U.S. MEAT IMPORTS AND EXPORTS[1]

| Year | Imports | | | | | | | | | | Exports | | | | | | | | | |
|---|---|---|---|---|---|---|---|---|---|---|---|---|---|---|---|---|---|---|---|---|
| | Beef and Veal[2] | | Pork[3] | | Lamb, Mutton, and Goat | | Other Meals[4] | | All Meats[5] | | Beef and Veal[6] | | Pork[7] | | Lamb, Mutton, and Goat | | Other Meals[8] | | All Meats[9] | |
| | (million) | | | | | | | | | | | | | | | | | | | |
| 1983 | 472 | *215* | 139 | *63* | 19 | *8.6* | 10.5 | *4.8* | 2,021 | *919* | 68.4 | *31.1* | 31.2 | *14.2* | 1.4 | *0.64* | 166 | *75.5* | 859 | *390* |
| 1984 | 437 | *199* | 196 | *89* | 19 | *8.6* | 11.0 | *5.0* | 2,148 | *976* | 82.0 | *37.3* | 23.2 | *10.5* | 1.9 | *0.86* | 164 | *74.5* | 856 | *389* |
| 1985 | 498 | *226* | 234 | *106* | 34 | *15.5* | 12.8 | *5.8* | 2,500 | *1,136* | 80.5 | *36.6* | 18.0 | *8.2* | 1.0 | *0.45* | 187 | *85.0* | 894 | *406* |
| 1986 | 518 | *235* | 235 | *107* | 37 | *16.8* | 13.1 | *6.0* | 2,568 | *1,167* | 135.5 | *61.6* | 12.0 | *5.5* | 1.0 | *0.45* | 189 | *85.9* | 1,034 | *470* |
| 1987 | 548 | *249* | 254 | *115* | 38 | *17.3* | 16.3 | *7.4* | 2,748 | *1,249* | 155.1 | *70.5* | 15.4 | *7.0* | 1.5 | *0.68* | 177 | *80.5* | 1,075 | *489* |

[1]Adapted by the author from *Agricultural Statistics 1988*, USDA, p. 305, Table 450, and p. 303, Table 447.

[2]Includes fresh or frozen; canned, including sausage; other prepared or preserved, including pickled and cured.

[3]Includes fresh or frozen; canned hams, shoulders, and bacons; other prepared or preserved, including pickled and cured; and sausage, all types, including fresh and cured sausage.

[4]Includes mixed sausage; other meats, mostly mixed luncheon meats, and variety meats, fresh or frozen.

[5]May not add due to rounding.

[6]Includes fresh or frozen; canned, pickled or cured.

[7]Includes fresh or frozen; hams and shoulders, cured or cooked; bacon; and other pork, pickled, salted, or otherwise cured; and other pork, canned.

[8]Includes sausage, bologna, and frankfurters; variety meats, fresh or frozen, including edible animal organs; and other meats, including sausage ingredients, cured (excluding canned), meat and meat products, canned; and baby food.

[9]May not add due to rounding.

The quantity of meat and live animals imported from abroad depends to a substantial degree on (1) the level of U.S. meat production, (2) consumer buying power, (3) livestock prices, (4) quotas and tariffs, and (5) the need for manufacturing-type beef (the kind that is boned and used in making hamburgers, franks, sausages, and bologna). When animal prices are high, more meat is imported. Actually, there may be some virtue in *judiciously* increasing imports of meat and animals during times of scarcity and high prices, as an alternative to pricing meat out of the market.

Because of the restrictions designed to prevent the introduction of foot-and-mouth disease, neither fresh nor salted refrigerated beef can be imported to the United States from South America; beef importations from these countries must be canned or fully cured (i.e., corned beef).

Our meat imports come from many countries, but our chief sources, ranked on the basis of 1987 data, are:

**Beef and veal**—from Argentina, Brazil, and Uruguay.
**Lamb and mutton**—from Australia and New Zealand.
**Pork**—from Canada, Denmark, Poland, Hungary, and Yugoslavia.
**Total meat** (beef and veal, lamb and mutton, and pork)—from Australia, Canada, New Zealand, Denmark, and Poland.

Fig. 9–29 shows U.S. exports of meat, livestock products, and live animals.

Some animals of all classes are imported from a number of countries for breeding purposes,[9] but the vast majority of on-foot importations consist of feeder cattle shipments from Mexico and Canada (see Table 9–9). Relatively few animals are destined for immediate slaughter.

The quantity of meat and animals exported from this country is dependent upon (1) the volume of meat and the number

## TABLE 9–9
## U.S. LIVE ANIMAL IMPORTS[1]
## (1983–1987)

| Year | Cattle | Sheep and Lambs | Hogs |
|---|---|---|---|
| | (no.) | (no.) | (no.) |
| 1983 | 920,807 | 7,128 | 447,465 |
| 1984 | 753,483 | 16,285 | 1,322,017 |
| 1985 | 836,018 | 24,199 | 1,226,571 |
| 1986 | 1,406,524 | 22,434 | 503,728 |
| 1987 | 1,200,484 | 26,640 | 446,056 |

[1]*Livestock and Meat Statistics, 1984–1988*, Stat. Bull. No. 784, USDA, p. 244, Table 148.

of animals produced in the United States, (2) the volume of meat and the number of animals produced abroad, and (3) the relative vigor of international trade, especially as affected by buying power and trade restrictions.

Our exports of animal products consist largely of those by-products which are surplus in the United States. As shown in Fig. 9–29, hides and skins lead in importance, followed by red meats, animal fats and oils, variety meats, and live animals. The "other" category includes such products as wool and mohair. The major markets for each of these products, ranked on the basis of 1987 data, are: **Lard**—Mexico, Canada, and Belize. **Tallow**—Mexico, Egypt, the Republic of Korea, Spain, and the Netherlands. **Hides**—Republic of Korea, Japan, Mexico, and Taiwan. **Skins**—Italy and Japan.

The United States exports more cattle than any other class of animal, followed, in order, by sheep, lambs, goats, and hogs.

## MEAT IMPORT LAW (PUBLIC LAW 88–482)

Livestock producers recognize that, because of cheaper labor (and often cheaper feed supplies), farmers in the surplus meat producing countries can produce meat at a lower cost than the American producer. Transportation distances and costs

---

[9]Certified (registered) purebred animals are permitted free entry into the United States.

are not prohibitive in obtaining meat from these countries. It would appear, therefore, that only protective walls—tariffs, quotas, and embargo legislation enacted by the U.S. government—can stand in the way of increased meat competition from foreign sources.

But the question of regulating the supply of meat, like regulating any other commodity, raises extremely complex issues and makes for conflicts of interest between producers and consumers. The U.S. quotas and tariffs of live animals and meats that existed in 1987 are given in Table 9–10.

In the early 1960s cattle producers became deeply concerned over rising imports of beef. Instead of holding at the historical 1 to 5% level of U.S. beef production, imports spiraled to more than 10% of U.S. production. A number of factors caused the increased imports, the most significant of which was that the U.S. market was the most lucrative outlet in the world. Although the U.S. State Department favored voluntary agreements with major exporters, American cattle producers insisted on more stringent regulations. Finally, Congress enacted the Meat Import Law (Public Law 88–482) in August 1964, to become effective January 1, 1965. This bill provided for import quotas, based on a formula, for fresh, chilled, and frozen beef, veal, mutton (not lamb), and goat meat—including both carcasses and boneless meat. It did not include lamb or canned meats. The law was for the purpose of limiting annual imports of the specified meats to a level comparable to the designated base period 1959–1963, with an adjustment, or *growth factor*, based on changes in domestic production relative to the base period.

The base quota was established as the average annual quantity imported during the base period (1959–1963), which was 725,400,000 lb, or 4.6% of domestic production during those years. Each year, the growth factor is determined by calculating the percentage by which the estimated U.S. commercial production of the specified meats in the current calendar year and the 2 preceding years (i.e., a 3-year-average) exceeds (or falls short of) the average annual U.S. production during the base period. Thus, to determine the growth factor at the beginning of a given year, it is necessary to estimate production for that year. Then, the calculated growth factor is multiplied times the base quantity to determine the amount of increase (or decrease) in the base. The increase (or decrease) added to the base gives an adjusted base quota. The Act allows a 10% leeway above the adjusted base quota before quotas are applied to individual countries. Thus, a "quota trigger point" is determined at 110% of the adjusted base quota.

**TABLE 9–10**
**U.S. QUOTAS AND TARIFFS OF LIVE ANIMALS AND MEATS[1]**

| Import Item | Quotas (no. head/year)[2] | Tarrif (or duty) | | |
|---|---|---|---|---|
| | | 1[3] | Special[4] | 2[5] |
| **Animals for breeding**[6] | None | Free | | Free |
| **Cattle:** | | | | |
| Cattle weighing: | | | | |
| under 200 lb | 200,000 | 1¢/lb | Free (E, I) | 2.5¢/lb |
| between 200 and 700 lb | | 1¢/lb | Free (E, I) | 2.5¢/lb |
| Dairy cattle weighing: | | | | |
| over 700 lb | | Free | | 3¢/lb |
| Other cattle[7] | 400,000 | 1¢/lb | Free (E, I) | 3¢/lb |
| Beef and veal (fresh, chilled, or frozen) | (See footnote 8) | 2¢/lb | Free (E, I) | 6¢/lb |
| **Sheep:** | | | | |
| Live sheep | | Free | | $3/head |
| Mutton | | 1.5¢/lb | Free (E, I) | 5¢/lb |
| Lamb | | 0.5¢/lb | Free (E, I) | 7¢/lb |
| **Goats:** | | | | |
| Live goats | | $1.50/head | Free (E, I) | $3/head |
| Goat meat | | Free | | 5¢/lb |
| **Swine:** | | | | |
| Live hogs | | Free | | 2¢/lb |
| Pork | | Free | | 2.5¢/lb |
| **Horses:** | | | | |
| Valued under $150/head | | Free | | $30/head |
| Valued over $150/head | | Free | | 20% *ad valorem* |
| **Mules:** | | | | |
| Valued under $150/head | | $15/head | Free (E, I) | $30/head |
| Valued over $150/head | | 10% *ad valorem* | Free (E) | 20% *ad valorem* |

[1] *Tariff Schedules of the United States Annotated (1987)*, USITC Publication 1317, United States International Trade Commission, Washington, DC.

[2] Includes Canada, Mexico, and all other countries.

[3] Products of Canada and all other countries not designated LDDC or 2.

[4] Products accorded special consideration.

[5] Products of communist countries.

[6] Must be purebreds of a recognized breed and registered in a recognized registry book.

[7] For not over 400,000 head entered in the 12-months period beginning April 1, in any year, of which not over 120,000 shall be entered in any quarter beginning April 1, Oct. 1, or Jan. 1.

[8] Legislation of Aug. 1964 establishes a basic limit of 725.4 million lb plus an added factor based on U.S. production.

The Secretary of Agriculture is required to estimate at the beginning of each quarter year the quantity of prospective imports. If the estimated quantity of prospective imports exceeds the trigger point, the President is required to invoke a quota on imports of these meats. In case quotas are imposed, the total import quota is allocated among the countries from whom the United States is importing on the basis of shares supplied by those countries during a representative period.

The law does contain provisions under which the President is empowered to suspend or increase quotas when it is deemed that it is in the best interest of the nation to do so because (1) of overriding economic or national security interest, (2) the supply of meat is inadequate to meet domestic demand at reasonable prices, or (3) international agreements have been entered into which will have the same effect as the Act.

It appears that voluntary agreements will be the chief means of controlling future imports. But the fact that a law exists may have considerable psychological effect on negotiations.

## MEAT PRESERVATION

From time immemorial, one of the major food problems has been that of preserving meats over a period of time and in a condition suitable as a food. Fundamentally, meat preservation is a matter of controlling putrefactive bacterial action. Various methods of preserving meats have been practiced through the ages, the most common of which are (1) drying, (2) smoking, (3) salting, (4) freezing, (5) canning, and (6) making into sausages.

In this country, meat curing is largely confined to pork, primarily because of the keeping qualities and palatability of cured pork products. Considerable beef is dried or corned and some lamb and veal are cured, but none of these is of such great importance as cured pork.

No phase of meat preservation and merchandising has received greater interest in recent years than frozen meats. For this reason, further discussion is devoted to this subject in the section that follows.

Fig. 9–30. Canned meats displayed in the second and third tiers of the meat counter of a supermarket. The preservation of meats by canning was first initiated by Arthur Libby in 1874. (Courtesy, National Association of Retail Grocers, Chicago, IL)

## FREEZING MEATS

Freezing is not a new method of meat preservation, for in arctic regions meats have been frozen since time immemorial. But special freezing methods are important recent developments. The rapid growth of freezer lockers, and the increased popularity of food freezers, have made this method of meat preservation available in homes throughout America.

When properly prepared and stored, frozen meats resemble fresh meats in appearance, flavor, appetite appeal, and food value; and they furnish a welcome diversion to the familiar stocks of canned, salted, and cellar stored food.

Among the reasons for increased interest in frozen meats in recent years are the following:

1. Uniform supplies of meats can be available throughout the year by freezing. This is particularly advantageous for such products as lamb, which is highly seasonal.

2. The consumer is accepting frozen foods in greater quantities, especially frozen prepared foods, fruits, and vegetables. Hence, consumer confidence in frozen food has improved; and with it, frozen meat is becoming more acceptable.

3. Home freezers are in widespread use. With frozen food space available, there is a tendency to fill it up by purchasing greater quantities of food in frozen form.

4. Improved quality control is possible in frozen meat merchandising. Cutting and packaging can be done at scheduled times and under strict supervision, rather than as dictated by the need to refill the fresh meat display case.

5. Central fabricating (cutting) is facilitated by frozen meat merchandising, thereby eliminating or greatly reducing the need for in-store cutting.

6. Improved utilization of the entire carcass results from frozen meat merchandising. Cuts do not depreciate in retail display cases as is true in fresh meat merchandising; and shrinkage and spoilage is practically eliminated. As a result, the retailer is able to merchandise all of the cuts to the best advantage.

7. No case of botulism from frozen foods is on record, and 3 weeks of zero storage will kill any trichinae (the parasites that cause trichinosis) in pork.

## MODERN MEAT MERCHANDISING

With the advent of the refrigerator car, meat was shipped in exposed halves, quarters, or wholesale cuts, and divided into retail cuts in the back rooms of meat markets. But this traditional procedure leaves much to be desired from the standpoints of efficiency, sanitation, shrinkage, spoilage, and discoloration. To improve this situation, more and more packers are fabricating and packaging (boxing) meat in their plants, thereby freeing the back rooms of supermarkets.

## FABRICATED; PACKAGED MEAT

Today, modern packing plants are fabricating and packaging (boxing) meat. After chilling, the carcass is subjected to a disassembly process, in which it is fabricated or broken into counter-ready cuts; vacuum sealed; moved into storage by an automated system; loaded into refrigerated trailers; and shipped to retailers across the nation.

The following benefits accrue from central fabricating, or cutting, of meats:

Fig. 9–31. In modern meat merchandizing, meat is fabricated and boxed in the meat packing plant. This shows boxed beef being placed in the meat counter of a supermarket. (Courtesy, Iowa Beef Processors, Inc., Dakota City, NE)

1. Twenty-five percent of the weight of the carcass is removed at the packing plant. This results in a reduction in shipping costs because bones and trim do not have to be transported.

2. Twenty to thirty percent of the carcass is bone and fat, which ends up as waste in the retail store. Where fabricating is done at the packing plant, trimmings, fats, and tallow are federally inspected and can be used as edible products; for example, fat and trim can be used in ground meats.

3. Processing equipment cost at the retail store level is usually high because of the low volume and idle time. Hence, counter-ready cuts make for a saving in equipment cost at the retail store.

4. Central fabricating permits a regulated aging process.

5. Central fabricating aids in achieving uniformity of quality standards.

6. Central fabricating improves merchandising through more uniformity of cuts, new cuts, more variety of cuts, matching cuts to area preferences, creating a more tenable situation for meat retailers who lack expertise and interest in cutting, and freeing meat managers from butchering—thereby permitting them to devote their time to personal selling and customer relations.

## MEAT AND WOOL/MOHAIR PROMOTION

Effective meat promotion—which should be conceived in a broad sense and embrace research, educational, and sales approaches—necessitates full knowledge of the nutritive qualities of the product. To this end, we need to recognize that (1) meat contains 15 to 20% high-quality protein, on a fresh basis; (2) meat is a rich source of energy, the energy value be-

ing dependent largely upon the amount of fat it contains; (3) meat is a rich source of several vitamins, but it is especially good as a source of phosphorus and iron; (4) meat is one of the richest sources of the important B group of vitamins, especially thiamin, riboflavin, niacin, and vitamin B–12, and (5) meat is highly digestible, with about 97% of meat proteins and 95% of meat fats being digested. Thus, meat is one of the best foods with which to alleviate human malnutrition, a most important consideration in light of the estimation that 35 to 40% of the U.S. population is now failing to receive an adequate diet.

Also, it is noteworthy that the per capita consumption of red meat in five countries exceeds that of the United States: by rank, based on 1989 meat consumption, these are: (1) Germany, East, 202 lb; (2) Uruguay, 194.5 lb; (3) Czechoslovakia, 188.8 lb; (4) New Zealand, 176 lb; and (5) Australia, 169.6 lb.

Thus based on (1) its nutritive qualities and (2) per capita consumption in those five countries exceeding us, it would appear that there is a place and a need for increased meat promotion, thereby increasing meat consumption and price.

There will always be controversy as to how meat promotion funds should be raised and spent. There can be no doubt, however, that all segments of the meat industry would benefit by working together in a unified approach.

Several beef, pork, lamb, and wool/mohair check-off/promotion programs are in operation. Each program is under different sponsorship; some nationwide, others statewide.

• **Beef promotion**—The 1985 Farm Bill contained the Beef Promotion and Research Act—enabling legislation which made it possible for the beef industry to establish a uniform national checkoff of $1.00 per head to fund promotion and research programs. So, beginning October 1, 1986, $1.00 per head was collected on all cattle sold or imported into the United States. It is estimated that this checkoff program will generate $60 million per year. A 113-member Cattlemen's Beef Promotion and Research Board—made up of representatives of the different states—has basic responsibility for the program, with 10 persons elected therefrom serving as the Beef Promotion and Operating Committee.

The Beef Promotion and Research Board has promoted effectively the slogan:

*Beef, real food for real people.*

The headquarters/address is:

Beef Promotion and Research Board
P.O. Box 3316
Englewood, CO 80155

• **Pork promotion**—The National Pork Producers Council (NPPC), organized in 1954, is today the largest commodity organization in the nation with an identified membership. The NPPC is the pork producer's voice and advocate to solve problems efficiently for the industry. Headquartered in Des Moines, Iowa, with an additional office in Washington, DC, the Council's purpose is to improve and increase the quality, production, distribution, and sales of pork and pork products.

The NPPC is funded by a voluntary 20¢ checkoff on market hogs and a 10¢ checkoff on feeder pigs. These funds are collected by NPPC which retains 60%. The remaining 40% is divided between the National Live Stock and Meat Board (20%) and member states (approximately 20%). The National Pork Board approved a $26.2 million checkoff-funded budget for 1990.

The National Pork Board has scored well with its promotion of:

*Pork, the other white meat.*

The headquarters/address is:
National Pork Producers Council
P.O Box 10383
Des Moines, IA 50306

• **Wool and lamb promotion**—Through passage, and subsequent extension, of the National Wool Act (first passed in 1954), Congress recognized wool as an essential and strategic commodity which is not produced in the United States in sufficient quantity to meet our domestic needs.

The incentive payments are financed from the duties collected on the imports of wool. Also, the Act authorized an industry self-help program for the purpose of developing and conducting advertising and sales promotion programs for lamb and wool.

The incentive program for shorn wool and unshorn lambs follows:

1. **Payment for shorn wool.** Payments are made to wool producers after the marketing year is over, to bring the national average price received by all producers, for wool sold, up to the incentive price.

The rate of payment for the marketing year is announced after the year is over, and the average price received by all producers is known. The rate paid will be the percentage required to bring the national average price, received by all producers, up to the incentive price.

The percentage rate recognizes quality production and encourages producers to improve the quality and marketing of their wool. The higher the price the individual producer receives, the greater the payment.

2. **Payment for unshorn lambs.**—The payment on unshorn lambs is at a comparable rate to the shorn wool payment and is designed to encourage the normal practice of marketing lambs without shearing the wool. The quantity of wool on unshorn lambs sold for slaughter averages around 5 lb per 100 lb of animal, liveweight.

The headquarters/address is:
American Sheep Industry Association
6911 S. Yosemite Street
Englewood, CO 80112–1414

## MEAT PACKERS AND RETAILERS COURT CONSUMERS

Among the factors—some old, with new wrinkles added: others new—used by packers and retailers in courting consumers are the following:

• **Meat inspection stamp**—The primary purpose of Federal Meat Inspection is protection of the consumer by guaranteeing that all meat so inspected and passed is from healthy animals, slaughtered and passed under sanitary conditions; that the meat is entirely suitable for consumption when it leaves the processing establishment; and that no labels carrying misleading statements appear on it. Meat products which pass federal inspection standards are marked or stamped in abbreviated form, "U.S. Inspected and Passed."

• **Vacuum packaging**—This refers to the use of a moisture-vapor-proof film to protect the meat from the time it is fabricated until it reaches the consumer. Such packaging reduces weight loss and surface spoilage for 2 to 3 weeks.

• **Attractive display in the supermarkets**—The open refrigerated cases in a modern supermarket or meat specialty store present a panorama of appetizing meats and meat products. Generally, meats are displayed wrapped in polyvinyl chloride (PVC) film that is sealed, on which there is an affixed label stamped with the new weight, total price, price per pound, and the name of the cut.

The arrangement for displaying meats is rather standard. They are segregated as to kind (beef, veal, pork, lamb, poultry) and type of cut (steak, roast, ground beef, etc.) Smoked meats have separate display space as do fish, liver, and other specialty items. Usually a clerk behind the display case adds a personal touch and furnishes a source of information.

• **Specials**—Most meat retailers feature a good many meat specials. Experienced homemakers watch for these announcements and take advantage of them. When a home freezer is available, it may be possible to effect additional savings by buying extra large amounts of meat during special sales, then storing them.

## METHODS OF MEAT COOKERY

The method used in meat cookery depends on the nature of the cut to which it is applied. In general, the types of meat cookery may be summarized as follows:

### DRY-HEAT COOKING

Dry-heat cooking is used in preparing the more tender cuts, those that contain little connective tissue. This method of cooking consists of surrounding the meat by dry air in the oven or under the broiler. The common methods of cooking by dry heat are (a) roasting, (b) broiling, and (c) panbroiling.

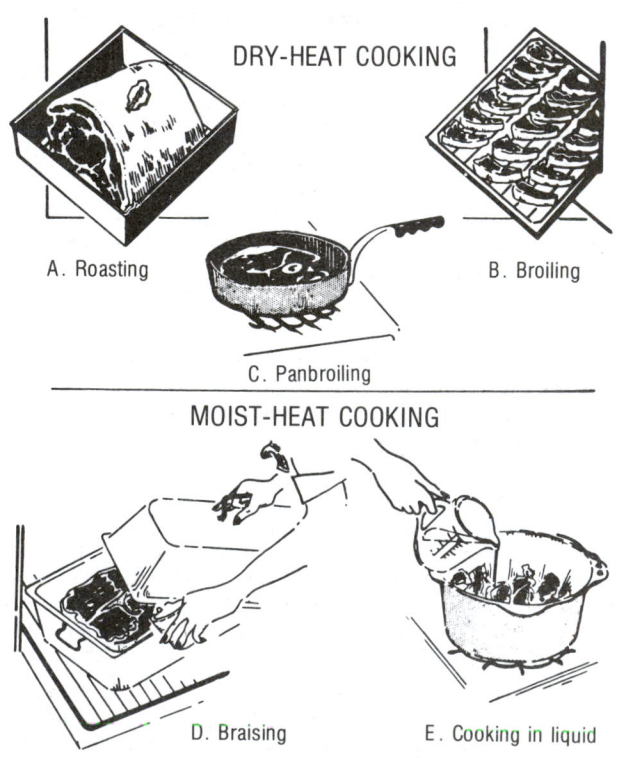

Fig. 9–32. Common methods of meat cookery: *Dry-heat cooking:* A, roasting; B, broiling; and C, panbroiling; *moist-heat cooking:* D, braising; and E, cooking in liquid. (Drawing by Prof. R. F. Johnson)

## MOIST-HEAT COOKING

Moist-heat cooking is generally used in preparing the less tender cuts, those containing more connective tissue that require moist heat to soften them and make them tender. In this type of cooking, the meat is surrounded by hot liquid or by steam. The common methods of moist-heat cooking are (a) braising, and (b) cooking in water (called stewing when small pieces of meat are so cooked.

## PACKINGHOUSE PRODUCTS FROM SLAUGHTER

The meat or flesh of animals is the primary object of slaughtering. The numerous other products are obtained incidentally. Thus, all products other than the carcass meat are designated as by-products, even though many of them are wholesome and highly nutritious articles of the human diet. Yet it must be realized that, upon slaughter, cattle, sheep, and hogs produced in the United States yield an average of 40, 50.5, and 32%, respectively, of products other than carcass meat.[10] When meat packers buy a steer, lamb, or hog, they buy far more than the cuts of meat that will eventually be obtained from the carcass; that is, only about 50% of a meat animal is meat.

In the early days of the meat-packing industry, the only salvaged by-products were hides, wool, tallow, and tongue. The remainder of the offal was usually carted away and dumped into the river or burned or buried. In some instances, packers even paid for having the offal taken away. In due time, factories for the manufacture of glue, fertilizer, soap, buttons, and numerous other by-products sprang up in the vicinity of the packing plants. Some factories were company owned, and others were independent industries. Soon much of the former waste material was being converted into materials of value.

Naturally, the relative value of carcass meat and by-products varies, both according to the class of livestock and from year to year. In the late 1980s, packers retrieved about the following percentages of the live cost of different classes of slaughter animals from the value of the by-products: 1,050-lb Choice steer, 12%; 100-lb Choice lamb, 13.50%; and 240-lb hog, 10.50%. Cattle hides and sheep pelts alone accounted for 70% and 60%, respectively, of the value of lamb and beef-by-products in 1989. Livestock prices are related mainly to meat values, but are also influenced by by-product values.

In contrast to the four early-day by-products—hide, wool, tallow, and tongue—modern cattle slaughter alone produces approximately 80 by-products which have a great variety of uses. Although many of the by-products from cattle, sheep, and hogs are utilized in a like manner, there are a few special products which are peculiar to the class of animals (e.g., wool and catgut from sheep).

The complete utilization of by-products is one of the chief reasons large packers are able to compete so successfully with local butchers. Were it not for this conversion of waste material into salable form, the price of meat would be much higher than under existing conditions. In fact, under normal conditions, the wholesale value of the carcass is about the same as the cost of the animal on foot. The returns from the sale of by-products cover all operating costs and return a reasonable profit.

It is not intended that this book should describe all of the by-products obtained from animal slaughter. Rather, Fig. 9–33

is presented in order to show some of the more important items, and a few select ones are discussed:

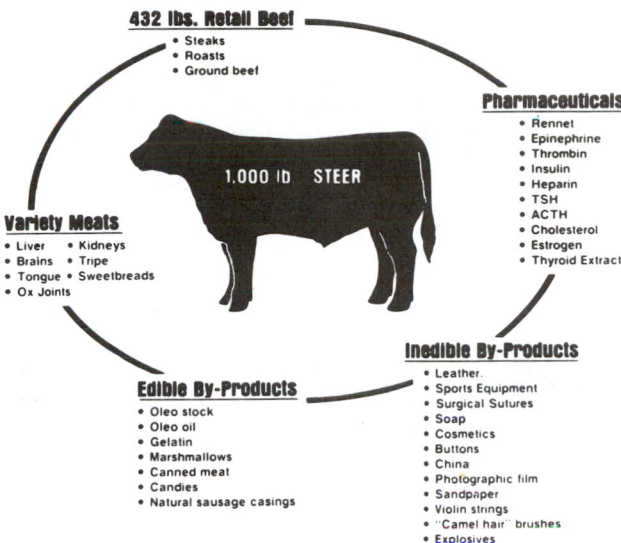

Fig. 9–33. Many good things come from cattle in addition to about 432 pounds of steaks, roasts, and hamburger normally yielded by a 1,000-lb steer. Several of these products are shown in the above figure. (Courtesy, National Live Stock and Meat Board, Chicago, IL)

1. **Hides.** Hides are particularly important as a by-product of the cattle slaughter. In the mid 1960s, cattle hides represented 2.5 to 3.0% of the total on-foot value of steers. In the 1980s, cattle hides rose to 8.5% of the value of steers.

The leather from animal hides is used for shoes, harness and saddles, belting, traveling bags, razor straps, footballs, baseball mitts, "sheepskins" for diplomas, sweatbands for hats, gloves, and numerous other leather goods.

Fig. 9–34. Brine curing cattle hides. (Courtesy, Iowa Beef Processors, Dakota City, NE)

---

[10]USDA figures.

2. **Sheep pelts.** The sheep pelt is the most valuable by-product of sheep slaughtering. Sheep skins with short wool, ¾ inch or less in length, are usually tanned with the wool on and are used for coats, robes, rugs, felts, slippers, and other articles. Pelts with longer wool are sent to the pullery. Usually they are temporarily preserved by the addition of salt until the wool is removed. The pulling process consists of applying a depilatory solution (made of sodium sulfide, slaked lime, and water) to the skin side of the pelt and then pulling the wool loose from the skin after the chemical action has loosened the hold of the fibers.

3. **Fats.** Next to hides and pelt, the fats (not including lard) are the most valuable by-product derived from slaughtering. Products rendered from them are used in the manufacture of oleomargarine, soaps, animal feeds, lubricants, leather dressings, candles, fertilizers, etc.

Oleomargarine, which is one of the better known of the products in which rendered animal fat is incorporated, is usually a mixture of vegetable oils and select animal fat.[11] Oleo oil, one of the chief animal fats of this product, is obtained from beef, mutton, or lamb.

4. **Variety meats.** The heart, liver, brains, kidneys, tongue, cheek meat, tail, feet, sweetbreads (thymus and pancreatic glands), and tripe (pickled rumen of cattle and sheep) are sold over the counter as variety meats or fancy meats.

5. **Glands.** Various glands of the body are used in the manufacture of numerous pharmaceutical preparations (See Fig. 9–35).

Proper preparation of glands requires quick chilling and skillful handling. Moreover, a very large number of glands must be collected in order to obtain any appreciable amount of most of these pharmaceutical products. For example, the glands from more than 100,000 lambs are necessary to produce one pound of adrenalin, a powerful heart stimulant; and it takes pancreatic glands from 1,500 cattle or 7,500 hogs to produce one precious ounce of insulin. But, fortunately, only minute amounts of insulin are required—the insulin from 2 hogs per day, or from 750 hogs per year, will suffice for each diabetic.

Thus, in a modern packing plant, there is no waste; literally speaking, "everything but the squeal" is saved. These by-products benefit the human race in many ways. Moreover, their utilization makes it possible to slaughter and process beef, lamb, and pork at a much lower cost. But this is not the end of accomplishment! Scientists are continually striving to find new and better uses for packinghouse by-products in an effort to increase their value.

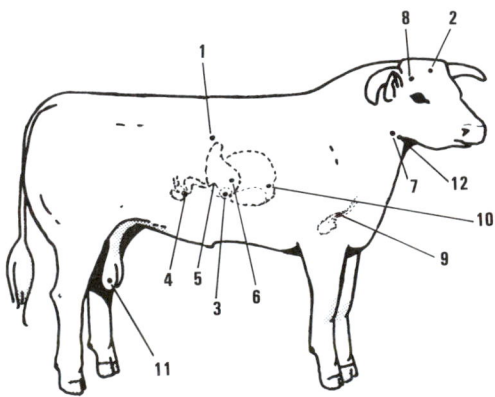

Fig. 9–35.  Meat animals are the source of more than 100 medicines and medical products (the count on the number of animal-derived medicines varies according to [1] whether or not certain derivatives are counted, and [2] whether or not experimental, as well as commercial, products are included. Swift and Company lists 90 such products; the National Live Stock and Meat Board pegs the number of different medicines coming from animals at 134; and the American Meat Institute refers to over 100 different pharmaceuticals coming from cattle alone), which doctors and veterinarians administer daily to millions of people and animals to save lives, battle disease, relieve pain, and restore health. This figure shows the approximate location of a few of the glands and other tissues used in the manufacture of some of the pharmaceutical products of human and veterinary medicine.

1. **Adrenal (suprarenal).** Source of (a) epinephrine (used for asthma, hay fever, allergies and shock), and (b) adrenal cortex extract (used for Addison's Disease, and in postsurgical and burn shock).

2. **Brain.** Source of kephalin or cephalin, used on oozing surfaces to check bleeding.

3. **Gallbladder.** Source of (a) bile salts, (b) dehydrocholic acid—used for gallbladder disturbances and abnormalities of fat digestion, and (c) cortisone (used for rheumatic fever, arthritis, various allergies, inflammatory eye diseases, etc.).

4. **Intestines.** Lamb intestines are used for surgical sutures.

5. **Liver.** Source of (a) liver extract (used for pernicious anemia) and (b) heparin (used to delay clotting of shed blood of ulcers following surgery).

6. **Pancreas.** Source of (a) insulin (the only substance known to medical science which can control diabetes), (b) trypsin (the protein-digesting enzyme), (c) amalase (the starch-splitting enzyme), and (d) lipase (the fat-splitting enzyme). Each enzyme is used for digestion of these respective nutrients; and trypsin is also used to soften scar tissue or digest necrotic tissue in wounds and ulcers.

7. **Parathyroid.** Parathyroid extract is used for tetany, which follows removal of these glands.

8. **Pituitary.** Source of (a) posterior pituitary extract (used to increase blood pressure during shock, to promote uterine contraction during and after childbirth, and to control excessive urination of diabetes insipidus), and (b) ACTH (used for rheumatic fever, arthritis, acute inflammation of eyes and skin, acute alcoholism, severe asthma, and hay fever and other allergy conditions).

9. **Red bone marrow.** Bone marrow concentrates used in treatment of various blood disorders.

10. **Stomach.** Source of rennin, used to aid milk digestion.

11. **Testes.** Source of the enzyme hyaluronidase.

12. **Thyroid.** Thyroid extract is used for malfunctions of the thyroid gland (some goiters, cretinism, and myxedema).

13. **Blood.** Source of thrombin, applied locally to wounds to stop bleeding.

15. **Lard.** With 1% benzoin added, it produces benzoinated lard, which is used as a medical ointment base.

---

[11] Oleomargarine is of 2 kinds: (1) a mixture of 50 to 80% animal fat and 20 to 50% vegetable oil, churned with pasteurized skimmed milk; or (2) 100% vegetable oil, churned with pasteurized skimmed milk. Oleomargarine was first perfected in 1869 by the Frenchman, Mege, who won a prize offered by Napoleon III for a palatable table fat which would be cheaper than butter, keep better, and be less subject to rancidity.

# QUESTIONS FOR STUDY AND DISCUSSION

1. Define the terms: (a) meat, and (b) meat by-products.

2. How did meat prompt early sailing? If so, why?

3. Explain how consumer demand has influenced the type of animals produced.

4. Why has the percent of income spent for red meat declined in recent years?

5. The per capita consumption of poultry meat has increased markedly in recent years, yet, the percent of income spent for poultry meat has remained about the same. How could this be?

6. What qualities in meats are desired by consumers; what affects will present consumer preferences exert on the livestock production and meat processing of the future?

7. Trace the development of federal meat grading, and explain the forces or motives which prompted each step.

8. What is the explanation of the wide species difference in the proportion of U.S. commercial meat slaughter federally graded; ranging from a negligible amount of pork graded to 93.6% of lamb and mutton graded (see Fig. 9–13)?

9. Define (a) grade of meat, (b) quality grade, and (c) yield grade.

10. What's the difference between quality grades and yield grades of beef and lamb? Which is the more important?

11. Why should both producers and consumers be familiar with the federal grades?

12. List five **advantages** of federal meat grading, and discuss the advantage which you feel is most important.

13. List three **disadvantages** of federal meat grading, and discuss the disadvantage which you feel is most important.

14. Should we have only one system of grading; if so, would you recommend federal grades or packer brands? Justify your answer to this two-pronged question.

15. Trace the history of koshered meats. Wherein does koshering differ from customary slaughtering and processing procedure?

16. Why should meat be included in the diet from the standpoint of its nutritive qualities? Summarize the nutrient qualities of meats.

17. Confirm or refute each of the following meat myths:
    a. Meat fats cause coronary heart disease.
    b. Beef causes bowel cancer.
    c. U.S. meat consumption will continue unabated.

18. Choose and debate either the affirmative or negative of each of the following:
    a. Meat prices are controlled by (1) the producer, (2) the packer, or (3) the retailer.
    b. Excessive profits are made by (1) the producer, (2) the packer, or (3) the retailer.

19. If a steer is not all steak, what is the rest of it?

20. Footnote No. 3 shows the U.S. food expenditures as a percent of income have decreased dramatically since 1937. How has this been accomplished?

21. Table 9–3 shows that for each consumer food dollar in 1987, the farmer's share was only 30 cents, whereas 70 cents went for processing and marketing. Why this great difference?

22. Table 9–4 and Fig. 9–22 show the dominant position of beef in U.S. meat production, and that we produce little veal and lamb/mutton. Should we encourage U.S. farmers to produce more veal, and lamb/mutton?

23. Fig. 9–23 shows that per capita beef consumption has gradually declined, while poultry production has increased dramatically—almost doubling since 1966. What forces caused beef consumption to decline and poultry consumption to increase?

24. Study with care Figs. 9–24, 9–25, 9–26, and 9–27. What U.S. trends in eating habits do these charts show?

25. Study Table 9–7 and Fig. 9–28 and analyze them from the standpoints of quantities and kinds of meats imported compared with production.

26. Fig. 9–29 shows that hides and skins lead by quite a margin in dollars derived from U.S. exports of livestock products. Why is there such a great demand for U.S. hides?

27. Compare U.S. meat imports vs exports, by kinds and quantities (see Table 9–8).

28. From what countries do our different kinds of meats come?

29. Why is the U.S. importing so many live animals (see Table 9–9)? Where do the bulk of these animals come from?

30. What is the Meat Import Law (Public Law 88–482); what is it designed to accomplish, and how does it accomplish its purpose?

31. What is meant by (a) quota, and (b) tariff?

32. What factors favor the increased use of freezing as a means of meat preservation?

33. What factors favor fabricating and packaging meats in packing plants, rather than in the back rooms of meat markets?

34. What promotional organizations have the primary responsibility for promoting each (a) beef, (b) pork, and (c) wool/lamb; how is each of these organizations funded; and what type of promotion do they promulgate?

35. In what ways do meat packers and retailers court consumers?

36. List the common methods of (a) dry-heat cooking, and (b) moist-heat cooking; and tell how each is commonly used.

37. What are packinghouse products from slaughter? Discuss the relative value of carcass meat and by-products. Discuss the value and use of each of the following by-products: (a) hides, (b) fats, (c) variety meats, and (d) glands.

## SELECTED REFERENCES

| Title of Publication | Author(s) | Publisher |
|---|---|---|
| *Adventures in Diet* | V. Stefansson | Reprinted from *Harper's Magazine* by American Meat Institute, Washington, DC |
| *By-products in the Packing Industry* | R. A. Clemen | University of Chicago Press, Chicago, IL, 1927 |
| *Food For Health—a nutrition encyclopedia* | A. H. Ensminger, *et al.* | Pegus Press, Clovis, CA, 1986 |
| *Food from Farmer to Consumer* | | National Commission on Food Marketing, Washington, DC, 1966 |
| *Foods & Nutrition Encyclopedia*, Vols. 1 & 2 | A. H. Ensminger, *et al.* | Pegus Press, Clovis, CA, 1983 |
| *Hides and Skins* | National Hide Association | Jacobsen Publishing Co., Chicago, IL, 1970 |
| *Lessons on Meat*, Second Edition | | National Live Stock and Meat Board, Chicago, IL, 1972 |
| *Livestock and Meat Marketing*, Second Edition | J. H. McCoy | Avi Publishing Co., Westport, CT, 1979 |
| *Marketing of Livestock and Meat, The*, Second Edition | S. H. Fowler | The Interstate Printers & Publishers, Inc., Danville, IL, 1961 |
| *Meat Reference Book* | | American Meat Institute, Washington, DC |
| *Meat Facts* | Department of Economic Research | American Meat Institute, Washington, DC, 1989 |
| *Meat for the Table* | S. Bull | McGraw-Hillk Book Company, New York, NY, 1951 |
| *Meat We Eat, The*, Twelfth Edition | J. R. Romans, *et al.* | The Interstate Printers & Publishers, Inc., Danville, IL, 1985 |
| *Stockman's Handbook, The*, Sixth Edition | M. E. Ensminger | The Interstate Printers & Publishers, Inc., Danville, IL, 1983 |

Also, literature on meats may be secured by writing to meat packers and processors and trade organizations; in particular the following two trade organizations:

American Meat Institute
P.O. Box 3556
Washington, DC 20007

National Live Stock and Meat Board
444 North Michigan Avenue
Chicago, IL 60611

Loin lamb chops (Courtesy, National Live Stock & Meat Board, Chicago, IL)

Business and management aspects will be increasingly important in future livestock production.

This shows the modern electronic data processing equipment which provides the necessary information for the control of delicate blending rations for feedlot cattle. This system (1) delivers data on magnetic tape for detailed computer analysis, and (2) prints the data on paper tape for immediate use by mill personnel. (Courtesy, Farr Feeders, Inc., Greeley, CO)

# BUSINESS ASPECTS OF ANIMAL PRODUCTION

# Chapter 10

Agriculture, with assets totaling $813.1 billion in 1987, (1) ranks as the nation's biggest single industry, and (2) has a value roughly equal to 40% of the total capital assets of all manufacturing corporations in the United States. The scope and importance of U.S. agriculture is further attested by the fact that more than 21 million people are employed in some phase of agriculture—from the farmers and ranchers that produce the food and fiber to the people in the farm supply and processing industries—to the workers in the local supermarket. Broadly speaking, about one job out of five in the U.S. is somehow farm/food related.

From 1935 to 1988, within a span of 53 years, the number of farms decreased from 6.8 million to 2.2 million, and the size of farms increased from 154.8 acres to 463 acres.[1] Thus, within 53 years, nearly 68% of the farms disappeared from American agriculture, and the average size of farms tripled. With this transition, herds, flocks, and feedlots became bigger.

In 1952, each farm worker supplied enough food and fiber for 17 persons, including self; by 1988, each farm worker supplied enough food and fiber for 93 persons, including self.

Farmers are big buyers, too. Each year, they generally spend over $6 billion for tractors, trucks, cars, and machinery, plus another $9.5 billion to fuel, lubricate, and maintain the fleet. Additional major annual outlays by farmers include about $19 billion for feed and seed and nearly $6 billion for fertilizer and lime.

The above trends to bigness will continue. Hopefully, agriculture will become more profitable, too, for historically, farm earnings have been low. During the year 1987, the average return on assets (the money value of property) was 5.5%.[2] This prompts the question: Would investors buy stock on the New York Stock Exchange or put money in an apartment house if they knew in advance that they would realize only 5.5% on their investment?

The business and management aspects of animal production will be increasingly important in the future. Changes in the type of business organization and in financial management will come. More capital will be required, more money will be borrowed, competent managers will be in demand, better and more complete records will be necessary, futures trading will increase, and livestock producers will become more knowledgeable relative to tax management, estate planning, and liability. The net result will be that those engaged in the business of agriculture will treat it as the big business that it is, and become more sophisticated and efficient; otherwise, they won't be in business very long.

## TYPES OF BUSINESS ORGANIZATION

The success of today's farming is very dependent on the type of business organization. No one type of organization is superior under all circumstances; rather, each situation must be considered individually. The size of the operation, the family situation, the enterprises, the objectives—all these, and more, are important in determining the best way in which to organize the farming business.

Three major types of business organizations are commonly found among farming enterprises: (1) the sole proprietorship; (2) the partnership; and (3) the corporation.

---

[1]The Census definition of a farm is as follows: "A place that sells $1,000 a year in ag products."

[2]*National Financial Summary, 1987,* USDA, ERC, ECIFS7–1, 1988, p. 6, in section headed "Rate of Return."

## PROPRIETORSHIP (INDIVIDUAL)

*The sole proprietorship is a business which is owned and operated by one individual.*

This is the most common type of business organization in U.S. farming—90% of the nation's farms are individually or family owned. Under the sole proprietorship, or individual (or family) ownership, one person controls the business. Proprietors may not provide all the capital used in the business; in fact, they usually do not. However, they have sole management and control of the operation, although this may be modified and delegated somewhat through lease agreements, contracts, etc. Basically, sole proprietors get all the profits of the business; likewise, they must absorb all the losses.

In comparison with other forms of organization, the sole proprietorship has two major limitations: (1) it may be more difficult to acquire new capital for expansion; and (2) not much can be done to provide for continuity and to keep the present business going as a unit, with the result that it usually goes out of existence with the passing of the owner.

## PARTNERSHIP (GENERAL PARTNERSHIP)

*A partnership is an association of two or more persons who, as co-owners, operate the business.* About 13% of U.S. farms are partnerships.

The basic idea of two or more persons joining together to carry out a business venture can be traced back to the syndicates that were used in major centers in western Europe in the Middle Ages. Many of the early efforts to colonize the New World were also partnerships, or *companies* which provided venture capital, ships, provisions, and trade goods to induce settlement of large land grants.

Most farm partnerships involve family members who have pooled land, machinery, working capital, and often their own labor and management to operate a larger business than would be possible if each member limited the operation to his/her own resources. It is a good way in which to bring a son, or daughter, who is usually short on capital, into the business, yet keep the parents in active participation. Although there are financial risks to each member of such a partnership, and potential conflicts in management decisions, the existence of family ties tends to minimize such problems.

In order for a partnership to be successful, the enterprise must be sufficiently large to utilize the abilities and skills of the partners and to compensate each adequately in keeping with his/her contribution to the business.

A partnership has the following **advantages:**

1. **Combining resources.** A partnership often increases returns from the operation due to combining resources. For example, one partner may contribute labor and management skills, whereas another may provide the capital. Under such an arrangement, it is very important that the partners agree on the value of each person's contribution to the business, and that this be clearly spelled out in the partnership agreement.

2. **Equitable management.** Unless otherwise agreed upon, all partners have equal rights, regardless of financial interest. Any limitations, such as voting rights proportionate to investments, should be a written part of the agreement.

3. **Tax savings.** A partnership does not pay any tax on its income, but it must file an informational return. The tax is paid

as part of the individual tax returns of the respective partners, usually at lower tax rates.

4. **Flexibility.** Usually, the partnership does not need outside approval to change its structure or operation—the vote of the partners suffices.

Partnerships may have the following **disadvantages**:

1. **Liability for debts and obligations of the partnership.** In a partnership, each partner is liable for all the debts and obligations of the partnership.

2. **Uncertainty of length of agreement.** A partnership ceases with the death or withdrawal of any partner, unless the agreement provides for continuation by the remaining partners.

3. **Difficulty of determining value of partner's investment.** Since a partner owns a share of every individual item involved in the partnership, it is often very difficult to judge value. This tends to make transfer of a partnership difficult. This disadvantage may be lessened by determining market values regularly.

The above is what is known as a partnership or general partnership. It is characterized by (1) management of the business being shared by the partners, and (2) each partner being responsible for the activities and liabilities of all of the partners, in addition to personal activities within the partnership.

## LIMITED PARTNERSHIP

*A limited partnership is an arrangement in which two or more parties supply the capital, but only one partner is involved in the management.* This is a special type of partnership with one or more general partners and one or more limited partners. The limited partnership avoids many of the problems inherent in the general partnership and has become the chief legal device for attracting outside investor capital into farm and ranch ventures. Although this device has been widely used in the oil and gas industry, and for acquiring income-producing urban real estate for a number of years, its application to agricultural ventures on a national scale is quite new. As the term implies, the financial liability of each partner is limited to the original investment, and the partnership does not require, and in fact prohibits, direct involvement of the limited partners in management. In many ways, a limited partner is in a similar position to a stockholder in a corporation.

A limited partnership must have at least one general partner who is responsible for managing the business and who is fully liable for all obligations.

The **advantages** of a limited partnership are:

1. It facilitates bringing in outside capital.
2. It need not dissolve with the loss of a partner.
3. Interests may be sold or transferred.
4. The business is taxed as a partnership.
5. Liability is limited.
6. It may be used as a tax shelter.

The **disadvantages** of a limited partnership are:

1. The general partner has unlimited liability.
2. The limited partners have no voice in management.

## CORPORATIONS

*A corporation is a device for carrying out a farming or ranching enterprise as an entity entirely distinct from the persons who are interested in and control it.* Each state authorizes the existence of corporations. As long as the corporation complies with the provisions of the law, it continues to exist—irrespective of the changes in its membership.

Until about 1960, few farms and ranches were operated as corporations. In recent years, however, there has been increased interest in the use of corporations for the conducting of farm and ranch business. Even so, only about 2% of U.S. farms use the corporate structure.

From an operational standpoint, a corporation possesses many of the privileges and responsibilities of a real person. It can own property; it can hire labor; it can sue and be sued; and it pays taxes.

Separation of ownership and management is a unique feature of corporations. The owners' interest in a corporation is represented by shares of stock. The shareholders elect the board of directors who, in turn, elect the officers. The officers are responsible for the day-to-day operation of the business. Of course, in a close family corporation, shareholders, directors, and officers can be the same persons.

The major **advantages** of a corporate structure are:

1. It provides continuity despite the death of a stockholder.
2. It facilitates transfer of ownership.
3. It limits the liability of shareholders to the value of their stock.
4. It may make for some savings in income taxes.

The major **disadvantages** of a corporation are:

1. It is restricted to doing only what is specified by its charter.
2. It must register in each state.
3. It must comply with stipulated regulations which involve considerable paperwork and expense.
4. It is subject to the hazard of higher taxes.
5. It is possible to lose control.

Still another type of corporation is family owned (privately owned). It enjoys most of the advantages of its generally larger outside investor counterpart, with few of the disadvantages. The chief **advantages** of the family owned corporation over a partnership arrangement are:

1. **It alleviates unlimited liability.** For this reason, a lawsuit cannot destroy the entire business and all the individual partners with it.
2. **It facilitates estate planning and ownership transfer.** It makes it possible to handle the estate and keep the business in the family and going if one of the partners should die. Each heir can be given shares of stock—which are easy to sell or transfer and can be used as collateral to borrow money—while leaving the management of the enterprise to those heirs interested in operating it, or even outsiders.

## CAPITAL NEEDS

In 1987, farm assets-investments in land, improvements, machinery, equipment, animals, feed and supplies—totalled $813.1 billion, while farm debt totaled $153.3 billion. Thus, in the aggregate, farmers had 81.2% equity in their business and 18.8% borrowed money (debts). The balance sheet of U.S. farming from 1970 through 1987 is shown in Fig. 10–1.

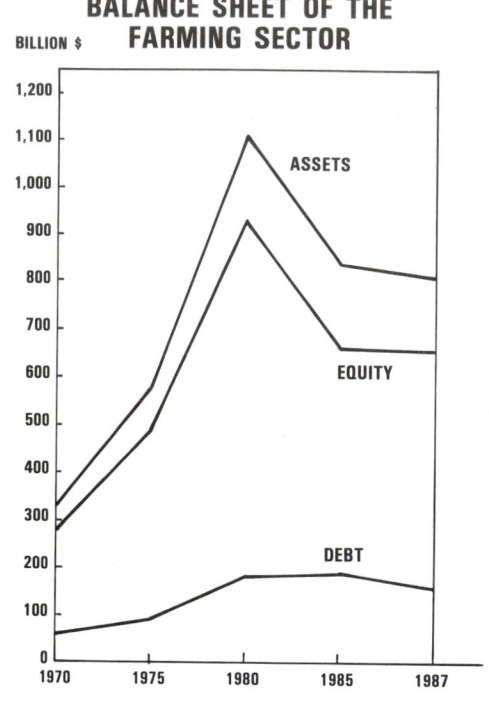

**BALANCE SHEET OF THE FARMING SECTOR**

Fig. 10–1. Balance sheet of U.S. farming, showing (1) assets, (2) debts, and (3) equities. (Source: *Statistical Abstracts of the United States 1989,* U.S. Department of Commerce, p. 632, No. 1089)

Perhaps agriculturists have been too conservative, for it is estimated that ¼ to ⅓ of American farmers could profit from the use of more credit in their operations.

Another statistic which points up the enormity of capital needs is that it takes about $17.58 in farm assets to produce $1 of net farm income.[3]

## CREDIT IN THE LIVESTOCK BUSINESS

Credit is an integral part of today's livestock business. Wise use of it can be profitable, but unwise use of it can be disastrous. Accordingly, livestock producers should know more about it. They need to know something about the lending agencies available to them, the types of credit, and how to go about obtaining a loan.

## TYPES OF CREDIT OR LOANS

Getting the needed credit through the right kind of loan is an important part of sound financial farm management. The following three general types of agricultural credit are available, based on length of life and type of collateral needed:

• **Short-term loans**—This type of loan is made for operating expenses and is usually for one year or less. It is used for the

[3]Based on 1987 figures, when farm assets were $813.1 billion and net farm income was $46,264 billion ($813.1 ÷ $46.26 = $17.58). Source: *Statistical Abstracts of the United States 1989,* U.S. Department of Commerce, P 636, No. 1099.

purchase of feeders or birds, feed, seed, fertilizer, gasoline, and family living expenses. Security, such as a chattel mortgage on the feeders, birds, or crop, may be required by the lender; and the loan is repaid when the animals or crops are sold.

• **Intermediate-term loans**—These loans are used to buy equipment and breeding stock, for making land improvements, and for remodeling existing buildings. They are paid back in one to seven years. Generally, they are secured by a chattel mortgage on livestock and machinery.

• **Long-term loans**—These loans are secured by mortgage on real estate and are used to buy land or make major improvements to farmland and buildings or to finance construction of new buildings. They may be for as long as 40 years. Usually they are paid off in regular annual or semiannual payments. The best sources of long-term loans are: an insurance company, the Federal Land Bank, the Farmers Home Administration, or an individual.

## CREDIT SOURCES

Table 10–1 shows where farmers borrow and the amount of loans from each source.

**TABLE 10–1**
**WHERE FARMERS BORROW (1987)[1]**

| Type and Source of Loan | Amount of Loan | Percent of Total |
|---|---|---|
|  | ($ million) | (%) |
| **Real Estate Mortgage Loans:** | | |
| Federal Land Banks | 32,332 | 37.0 |
| Individuals and others | 20,600 | 23.6 |
| Commercial banks | 14,455 | 16.5 |
| Farmers Home Administration | 10,083 | 11.5 |
| Insurance companies | 9,986 | 11.4 |
| Total | 87,412 | 100.0 |
| **Nonreal Estate Loan:** | | |
| Commercial banks | 29,041 | 36.0 |
| Farmers Home Administration | 16,049 | 19.9 |
| Commodity Credit Corp. | 14,695 | 18.2 |
| Individuals and others | 11,139 | 13.8 |
| Production Credit Associations | 9,486 | 11.8 |
| Federal Intermediate Credit Banks | 165 | 0.2 |
| Total | 80,575 | 100.0 |
| Total loans | 167,987 | |
| Percent real estate | 52.0 | |
| Percent nonreal estate | 48.0 | |

[1]Data from *Agricultural Statistics 1988*, USDA, p. 424, Table 596, and p. 429, Table 603.

But, agricultural financing is changing, and it will continue to change even more in the years ahead. Today, farmers are tapping the vast supply of farm equity or risk capital that is constantly seeking investment opportunities—nonfarm equity capital is being used in agriculture.

Sometime or other most farmers and ranchers find it necessary to borrow money to buy land; to construct buildings and other improvements; to purchase equipment, seed, and livestock; and/or to pay for seasonal labor. They should know something, therefore, about the lending organizations available to them in order that they may determine which one will best serve their needs. The leading sources of farm credit are summarized in Table 10–2.

TABLE 10–2
MAJOR SOURCES OF CREDIT, AND THE CHARACTERISTICS OF EACH

| Lenders | Sources of Funds | Limitations on Agricultural Lending | Loans Offered to Farmers | Comments |
|---|---|---|---|---|
| **Commercial Banks** | Demand and time deposits, bank stock, retained earnings. Funds from correspondent banks, the Federal Reserve, and participation with PCAs or FmHAs. | Commercial banks may prefer more profitable alternatives, such as installment loans. Also, legal reserve laws limit the volume of deposits available for loans. | (1) Short-term operating loans repayable within 1 year. (2) Intermediate-term loans, repayable in 1 to 7 years. (3) Some banks make long-term or real estate loans (7 to 25 years). | A financial statement is required by bank examiners. Some commercial banks have special agricultural representatives who are qualified to assist the borrower in many ways. |
| **Farm Credit System**, which embraces the following 3 federal lending units, all under the supervision of the Farm Credit Administration, an independent federal agency: | Sale of its bonds and discount notes in the private money market. | | | The Federal Land Banks and PCAs supply nearly ¼ of the credit used by farmers, and the Banks for Cooperatives provide nearly ⅔ of the borrowed capital used by farmer cooperatives. |
| 1. Federal Land Banks (FLB) | Sell bonds publicly on the national money markets. FLBs also draw on some money available from capital stock and retained earnings. | To borrow, you must buy stock in the local FLBA equal to 5% of the requested loan amount. Congress currently limits the loan to 85% or less of the appraised property value. | Real estate loans, amortized over periods ranging up to 40 years. FLBs secure loans for improvements by mortgaging the improved real estate. | The Federal Land Bank is actually a farmer's cooperative. Loans are on first mortgage only. When making loans, consideration is given to market value of the real estate, plus the income and the management ability of the borrower. |
| 2. Production Credit Associations (PCAs) | PCAs borrow from the Federal Intermediate Credit Bank, which gets funds by selling short-maturity debentures on the national money market. Also capital stock, and retained earnings. | Borrower must buy stock in the local PCA equal to 5% (sometimes 10%) of the loan. You can also get loans for farm-related services, such as rural home construction. | Only short-term and intermediate-term loans. PCAs do not offer long-term and real estate loans. | These are local cooperative lending organizations. The loan limit varies with individual cases, but it can be up to 100% of cost. However, 70–30 is most common, with the borrower providing 30% margin. |
| 3. Banks for Cooperatives | | | | Banks for Cooperatives provide the majority of financing for the nation's farm supply, marketing, and business service cooperatives. |
| **Farmers Home Administration** (FmHA) | Congressional appropriations FmHA also secures insured loans from other lenders, and from emergency or revolving funds. | Eligible only to farmers who can't reasonably borrow elsewhere. Farm ownership loan may not exceed $200,000, but FmHA will guarantee ownership loans as high as $300,000 from other credit sources. Farm operating loans made by FmHA may not exceed $200,000, but the agency can guarantee loans to farmers from other credit sources as high as $400,000 | Short-, intermediate-, and long-term, FmHA provides supervision for its loans. | Applicants who are veterans and have farm experience receive preference. Where a natural disaster has occured, under the Emergency Loan Program, a borrower may borrow up to 80% of the loss, but not to exceed $500,000. |
| **Individuals** | Personal loans, made by one individual to another. | Compared to the other sources, individuals frequently offer lower interest rates and down payments. However, the repayment period may be shorter. | There's an infinite variety of conditions and interest rates. Range all the way from real estate contracts or mortgages to personal unsecured loans. | One disadvantage of a loan from an individual is that the arrangement may be complicated by the lender's death unless adequate provision has been made for this eventuality. |
| **Insurance Companies** | Premiums received on insurance policies. They also draw funds from reserves held to pay insurance claims, and from capital and retained earnings. | Prefer long-term loans. Higher yielding, more secure investment opportunities lure money away from agricultural lending. May limit to selected geographic areas. | Typically, insurance companies limit themselves to real estate loans up to 30 to 40 years, and improvement loans which they secure through real estate mortgages. | Generally insurance companies will make loans up to 60% of the appraised value of the farm or up to 50% of the sale value. |
| **Merchants and Dealers** | Borrow from lending institutions, capital and retained earnings. Their supplier, distributor, or manufacturer may extend similar credit to them. | Limited credit, because they must keep a certain level of liquidity. Rates are higher than other lenders. Also limited by their supplier and other sources of capital. | Open account or sales contract. Short-term loans on equipment and machinery, intermediate-term loans on equipment. May charge add-on interest. Some other discounts if you pay cash. | It must be realized that merchants and dealers extend credit to farmers and ranchers primarily for the purpose of promoting the sale of products and services, and that their profits come from both sales and interest. |

# MODERN LIVESTOCK ENTERPRISES ARE BIG AND SOPHISTICATED!

It follows that they should be treated as such; otherwise, those engaged in them won't be in business very long.

Fig. 10–2. Cattle feedlot, with shades. This is the Benedict Cattle Feedlot, Casa Grande, Arizona. (Courtesy, James Benedict)

Fig. 10–3. Dairy cows in an environmentally controlled barn—completely enclosed, with controlled air circulation, temperature, humidity and light. (Courtesy, Sperry-New Holland, New Holland, PA)

Fig. 10–4. Western range ewes. Such a band usually consists of 1,000 to 3,000 ewes. (Courtesy, Ralston Purina Company, St. Louis, MO)

Fig. 10–5. Finishing hogs in confinement, on a partially slotted floor, equipped with automatically filled self-feeders. Each such finishing unit may accommodate up to 1,000 or 1,200 hogs. (Courtesy, *National Hog Farmer*, St. Paul, MN)

Fig. 10–6. Poultry breeding. Today, modern poultry breeding begins with the foundation breeder and involves much record keeping. (Courtesy, Hy-Line International, Division of Hy-Line Indian River Company, Johnston, IA)

Fig. 10–7. A horse auction in progress. Today, much horse breeding and marketing is centered in large horse establishments. (Courtesy, California Thoroughbred Breeders Association, Arcadia, CA)

## HELPFUL HINTS FOR BUILDING AND MAINTAINING A GOOD CREDIT RATING

Livestock producers who wish to build up and maintain good credit are admonished to do the following:

1. **Keep credit in one place, or in a few places.** Generally, lenders frown upon split financing. Borrowers should shop around for a creditor (a) who is able, willing, and interested in extending the kind and amount of credit needed, and (b) who will lend at a reasonable rate of interest, then stay with that borrower.

2. **Get the right kind of credit.** Don't use short-term credit to finance long-term improvements or other capital investments. Also, use the credit for the purpose intended.

3. **Be frank with the lender.** Be completely open and above board. Mutual confidence and esteem should prevail between borrower and lender.

4. **Keep complete and accurate records.** Complete and accurate records should be kept by enterprises. By knowing the cost of doing business, decision making can be on a sound basis.

5. **Keep annual inventory.** Take an annual inventory for the purpose of showing progress made during the year.

6. **Repay loans when due.** Borrowers should work out a repayment schedule on each loan, then meet payments when due. Sale proceeds should be promptly applied on loans.

7. **Plan ahead.** Analyze the next year's operation and project ahead.

## BORROW MONEY TO MAKE MONEY

Livestock producers should never borrow money unless they are reasonably certain that it will make or save them money. With this in mind, borrowers should ask, "How much should I borrow?" rather than, "How much will you lend me?"

## MANAGER

According to Webster, a manager is one who conducts business affairs with economy, and management is the act, or art, of managing, handling, controlling or directing.

Three major ingredients are essential to success in the livestock business: (1) good animals, (2) good feeding, and (3) good management. A manager can make or break any livestock enterprise. Unfortunately, this fact was long overlooked, primarily because the accent was on scientific findings, automation, and new products.

Management gives point and purpose to everything else. The skill of the manager materially affects how well animals are bought and sold, the quality of the animals, the health of the animals, the results of the ration, the stress of the stock, efficiency of production, the performance of labor, the public relations of the establishment, and even the expression of the genetic potential. Indeed, livestock managers must wear many hats—and they must wear each of them well.

The bigger and the more complicated the operation, the more competent the management required. This point merits emphasis because currently (1) bigness is a sign of the times, and (2) the most common method of attempting to bail out of an unprofitable business venture is to increase its size. Although it's easier to achieve efficiency of equipment, labor,

purchases, and marketing in big operations, bigness alone will not make for greater efficiency, as some owners have discovered to their sorrow, and others will experience. Management is still the key to success. When in financial trouble, owners should have no illusions on this point.

In manufacturing and commerce, the importance and scarcity of top managers are generally recognized and reflected in the salaries paid to persons in such positions. Unfortunately, agriculture as a whole has lagged; and altogether too many owners still subscribe to the philosophy that the way to make money out of the livestock business is to hire a manager cheap, with the result that they usually get what they pay for—a "cheap" manager.

## TRAITS OF A GOOD MANAGER

There are established bases for evaluating many articles of trade, including animals, hay, and grain. They are graded according to well-defined standards. Additionally, we chemically analyze feeds and conduct feeding trails with them. But no such standard or system of evaluation has evolved for livestock managers, despite their acknowledged importance.

The author has prepared the "Livestock Manager Checklist," given in Table 10–3, which (1) students may use for guidance as they prepare themselves for managerial positions, (2) employers may find useful when selecting or evaluating a manager, and (3) managers may apply to themselves for self-improvement purposes. No attempt has been made to assign a percentage score to each trait, because this will vary among livestock establishments. Rather, it is hoped that this checklist will serve as a useful guide (1) to the traits of a good manager and (2) to what the boss wants.

**TABLE 10–3**
**LIVESTOCK MANAGER CHECKLIST**

☐ **Character**
Has absolute sincerity, honesty, integrity, and loyalty; is ethical.

☐ **Industry**
Has enthusiasm, initiative, and aggressiveness; is willing to work, work, work.

☐ **Ability**
Has livestock know-how and experience, business acumen—including ability systematically to arrive at the financial aspects and convert this information into sound and timely management decisions—knowledge of how to automate and cut costs, common sense, organization, imagination, and growth potential.

☐ **Plans**
Sets goals; prepares organization chart and job description; plans work and works plans.

☐ **Analyzes**
Identifies the problem, determines the pros and cons, then comes to a decision.

☐ **Courage**
Has the courage to accept responsibility, to innovate, and to keep on keeping on.

☐ **Promptness and Dependability**
Is a self-starter; has "T.N.T." which means doing it "today, not tomorrow."

☐ **Leadership**
Stimulates subordinates and delegates responsibility.

☐ **Personality**
Is cheerful; not a complainer.

## ORGANIZATION CHART AND JOB DESCRIPTION

It is important that all workers know to whom they are responsible and for what they are responsible; and the bigger and the more complex the operation, the more important this becomes. This should be written down in an organization chart and job description. A sample Organization Chart is given in Fig. 10–8 and a sample Job Description is given in Table 10–4.

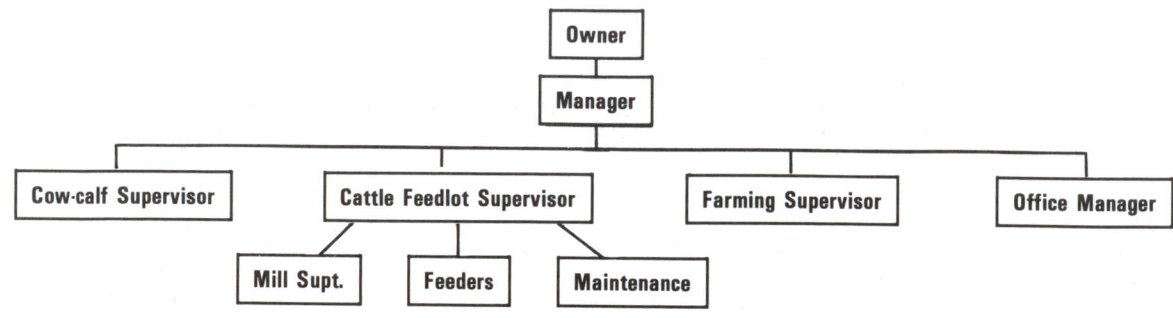

### ORGANIZATION CHART OF BAR-NONE RANCH

Fig. 10–8. A suggested farm-ranch Organization Chart.

**TABLE 10–4**
**SUGGESTED JOB DESCRIPTIONS OF**
**BAR-NONE RANCH**

| Owner | Manager | Cattle Feedlot Supervisor | Cattle Feeder No. 1 |
|---|---|---|---|
| **R**esponsible for:<br><br>1. Selecting management.<br>2. Making policy decisions.<br>3. Borrowing capital.<br>4. (List others) | **R**esponsible for:<br><br>1. Supervising all staff.<br>2. Preparing proposed long-time plan.<br>3. Budgets.<br>4. (List others) | **R**esponsible for:<br><br>1. Directing feedlot staff.<br>2. Buying and selling cattle.<br>3. Processing incoming cattle.<br>4. Animal health.<br>5. Feedlot rations.<br>6. (List others) | **R**esponsible for:<br><br>1. Morning and evening feedings.<br>2. Cleaning water troughs.<br>3. (List others) |

## AN INCENTIVE BASIS FOR THE HELP

Big farms and ranches must rely on hired labor, all or in part. Good help—the kind that everyone wants—is hard to come by; it's scarce, in strong demand, and difficult to keep. Moreover, the farm labor situation is going to become more difficult in the years ahead. There is need, therefore, for some system that will (1) give a big assist in getting and holding topflight help, and (2) cut costs and boost profits. An incentive basis that makes hired help partners in profit is the answer.

Many manufacturers have long had an incentive basis. Executives are frequently accorded stock option privileges, through which they prosper as the business prospers. Common laborers may receive bonuses based on piecework or quotas (number of units, or pounds produced). Also, most factory workers get overtime pay and have group insurance and a retirement plan. A few industries have a true profit-sharing arrangement based on net profit as such, a specified percentage of which is divided among employees. No two systems are alike. Yet, each is designed to pay more for labor, provided labor improves production and efficiency. In this way, both owners and laborers benefit from better performance.

Family owned and family operated farms have a built-in incentive basis; there is pride of ownership, and all members of the family are fully cognizant that they prosper as the business prospers.

Fig. 10–9. A good incentive basis makes hired help partners in profit.

Many different incentive plans can be, and are, used. There is no best one for all operations. The various plans in Table 10–5 are intended as guides only.

**Note well:** The various plans given in the tables and narrative that follow are for the purpose of showing how different incentives work; and not as indicators of amounts. The type and amount of incentive chosen, along with the provisions, should be tailored to fit the specific operation, with consideration given to kind and size of operation, extent of owner's supervision, present and projected productivity levels, mechanization, and other factors.

For most livestock operations, the author favors a production sharing and prevailing price type of incentive.

**TABLE 10–5**
**INCENTIVE PLANS FOR LIVESTOCK ESTABLISHMENTS**

| Types of Incentives | Pertinent Provisions of Some Known Incentive Systems in Use | Advantages | Disadvantages | Comments |
|---|---|---|---|---|
| **Bonuses** | **A** flat, arbitrary bonus; at Chirstmastime, year-end, or quarterly or other intervals.<br>**A** tenure bonus such as (1) 5 to 10% of the base wage or 2 to 4 weeks' additional salary paid at Christmastime or year-end, (2) 2 to 4 weeks vacation with pay, depending on length and quality of service or (3) $10 to $20/week set aside and to be paid if employee stays on the job a specified time. | It's simple and direct. | **N**ot very effective in increasing production and profits. | |
| **Equity-building plan** | **E**mployee is allowed to own a certain number of animals. In breeding operations, these are usually fed without charge. | It imparts pride of ownership to the employee. | **T**he hazard that the owner may feel that employees accord their animals preferential treatment; suspected if not proved. | |
| **Production sharing** | **$**2 to $6/calf weaned, $1/cwt on gain of feeder cattle; 50¢ to $2/head on fed cattle marketed; $1 to $2/pig marketed above 7 pigs/litter; 50¢ to $1/lamb weaned; $1/cwt of gain on lambs fed; 20¢ to $1/head on fed lambs marketed; so much per day for meeting certain levels of milk production/cow (for example, a 20¢ bonus/cow/month for 45.0 to 51.9 lb milk/cow/day; and 70¢ for 52.0 to 58.9 lb), with the bonus graduated upward with higher production in order to reflect the difficulty of increasing milk production in a herd where milk yield is already high; so much per 100 dozen eggs or 100 lb of broilers. | It's an effective way to achieve higher production. | **N**et returns may suffer. For example, a higher rate of gain than is economical may be achieved by feeding stockers more concentrated and expensive feeds than are practical. This can be alleviated by (1) specifying the ration and (2) setting an upper limit on the gains to which the incentive will apply.<br>**I**f a high performance level already exists, further gains or improvements may be hard to come by. | **I**ncentive payments for production above certain levels—for example, above 450 lb calf weaned/cow bred—are more effective than paying for all units produced. |
| **Profit sharing**<br>1. Percent of gross income<br>2. Percent of net income | **1**% to 2% of the gross.<br>**1**0% to 20% of the net after deducting all costs. | **N**et income sharing works better for managers, supervisors, and other administrators than for common laborers because fewer hazards are involved to opening the books to them.<br>**I**t's an effective way to get hired help to cut costs.<br>**I**t's a good plan for a hustler. | **P**ercent of gross does not impart cost of production consciousness. Both (1) percent of gross income and (2) percent of net income expose the books to workers, who may not understand accounting principles. This can lead to suspicion and distrust.<br>**C**ontroversy may arise (1) over accounting procedures—for example, from the standpoint of the owner a fast tax write-off may be desirable on new equipment, but this reduces the net shared with the worker, and (2) because some owners are prone to overbuild and over-equip, thereby decreasing net. | **T**here must be prior agreement on what constitutes gross or net receipts, as the case may be, and how it is figured. |
| **Production sharing and prevailing price** | *Cow-calf, ewe-lamb, or sow operation:* Basis (1) percent offspring weaned, and (2) weaning weight (which means pounds offspring weaned/female bred).<br>*Finishing cattle, or finishing hogs:* Basis (1) pounds feed/lb gain, and (2) daily rate of gain.<br>*Horse breeding establishment:* Basis (1) percent foal crop weaned, and (2) price of yearlings.<br>**I**n each of the above, establish break-even point(s), then split profit(s) beyond this point(s) basis (1) 80% (owner) and 20% (help), or (2) use escalator arrangement, giving help greater percentage as profits rise. With a dairy, the following production-sharing basis will work:<br>*Dairy:* Basis (1) udder health (for example, 20¢/cow/month for a somatic cell count of 500,000 to 600,000, graduated upward to 50¢/cow/month for a somatic cell count of 400,000 or less); or, (2) calving interval (for example, $20/month for a herd calving interval of 13.0 to 13.4 months, graduated upward to $50 bonus/month for a calving interval of 12.5 months or under). | It embraces the best features of both production sharing and profit sharing, without the major disadvantages of each. It (1) encourages high productivity and likely profits, (2) is tied in with prevailing prices, (3) does not necessitate opening the books, and (4) is flexible—it can be split between owner and employee on any basis desired, and the production part can be adapted to a sliding scale or escalator arrangement—for example, the incentive basis can be higher for the quarter pound feedlot gain made in excess of 2¾ lb than for a quarter-pound gain in excess of 2¼ lb | It is a bit more complicated than some other plans, and it requires more complete records. | **W**hen properly done, and all factors considered, this is the most satisfactory incentive basis for livestock enterprises. |

## INDIRECT INCENTIVES

Normally, we think of incentives as monetary in nature—as direct payments or bonuses for extra production or efficiency. However, there are other ways of encouraging employees to do a better job. The latter are known as indirect incentives. Among them are (1) good wages; (2) good labor relations; (3) adequate house plus such privileges as the use of the farm truck or car, payment of electric bill, use of swimming pool, hunting and fishing, use of a horse, and furnishing meat, milk, and eggs; (4) good buildings and equipment; (5) vacation time with pay, time off, and sick leave; (6) group health insurance; (7) security; (8) the opportunity for self-improvement that can accrue from working for a top person; (9) the right to invest in the business; (10) an all-expense-paid trip to a short course, show, or convention; and (11) year-end bonus for staying all year. These indirect incentives will be accorded to the help of more and more establishments, especially the big ones.

## FARM RECORDS AND ACCOUNTS

Modern farming necessitates adequate records and accounts, the chief functions of which are:

1. To provide information from which the farm business may be analyzed, with its strong and its weak points ascertained. From the facts thus determined, the operators may adjust current operations and develop a more effective plan of organization.

2. To provide profit and loss statements.

3. To provide a net worth statement, showing financial progress during the year.

4. To provide cash flow projections, showing when money is needed, and showing loan repayability.

5. To furnish an accurate, but simple, net income statement for use in filing tax returns.

6. To keep production records on livestock and crops.

7. To aid in making a credit statement when a loan is needed.

8. To keep a complete historical record of financial transactions for future reference.

Farmers can make their own record book by simply ruling off the pages of a bound notebook to fit their specific needs, but the saving is negligible. Instead, it is recommended that they obtain a copy of a farm record book prepared for and adapted to their area. Such a book may usually be obtained at a nominal cost from the agricultural economics department of each state college of agriculture. Also, certain commercial companies distribute very acceptable farm record and account books at no cost.

## BUDGETS IN THE LIVESTOCK BUSINESS

*A budget is a projection of records and accounts and a plan for organizing and operating ahead for a specified period of time.* A short-time budget is usually for one year, whereas a long-time budget is for a period of years. The principal value of a farm budget is that it provides a working plan through which the operation can be coordinated. Changes in prices, droughts, and other factors make adjustments necessary. But these adjustments are more simply and wisely made if there is a written budget to use as a reference.

It's unimportant whether a printed form (of which there are many good ones) is used, or one made up on an ordinary ruled 8½ × 11 in. sheet placed sidewise. The important things are (1) that a budget is kept, (2) that it be on a monthly basis, and (3) that the operator be comfortable with whatever forms or system evolved.

No budget is perfect. But it should be as good an estimate as can be made—despite the fact it will be affected by such things as droughts, diseases, markets, and many other unpredictables.

## COMPUTERS IN THE LIVESTOCK BUSINESS

Accurate and up-to-the-minute records and controls have taken an increasing importance in all agriculture, including the livestock business, as the investment required to engage in farming and ranching has risen and profit margins have narrowed. Today's successful farmers and ranchers must have, and use, as complete records as any other business. Also, records must be kept current; it no longer suffices merely to know the bank balance at the end of the year.

Big and complex agricultural enterprises have outgrown hand record-keeping. It's too time-consuming, with the result that it doesn't allow management enough time for planning and decision making. Additionally, it does not permit an all-at-once consideration of the complex interrelationships which affect the economic success of the business. This has prompted a new computer technique known as linear programming.

Fig. 10–10. Computer. Modern livestock producers augment a walk through the feedlot with a computer screen. (Courtesy, USDA, Washington, DC)

Linear programming is similar to budgeting, in that it compares several plans simultaneously and chooses from among them the one likely to yield the highest returns. It is a way in which to analyze a great mass of data and consider many alternatives. It is not a managerial genie; nor will it replace decision-making managers. However, it is a modern and effective tool in the present age, when just a few dollars per head or per acre can spell the difference between profit and loss.

There is hardly any limit to what computers can do if fed the proper information. Among the difficult questions that they can answer for a specific farm or ranch are:

1. **How is the entire operation doing so far?** It is possible to obtain a quarterly or monthly progress report, often making it possible to spot trouble before it's too late.

2. **What farm enterprises are making money; which ones are freeloading or losing?** By keeping records by enterprises—cow-calf, cattle feedlot, dairy, hogs, broilers, layers, wheat, corn, etc.—it is possible to determine strengths and weaknesses, then either to rectify the situation or shift labor and capital to a more profitable operation. Through *enterprise analysis*, some operators have discovered that one part of the farm business may earn $10 or more per hour for labor and management, whereas another may earn only $2 per hour, and still another may lose money.

3. **Is each enterprise yielding maximum returns?** By having profit, or performance, indicators in each enterprise, it is possible to compare these (a) with the historical average of the same farm or ranch or (b) with the same indicators of other farms or ranches.

4. **How does this ranch stack up with its competition?** Without revealing names, the computing center (local, state, area, or national) can determine how a given ranch compares with others—either the average, or the top (say, 5%).

5. **How to plan ahead?** By using projected prices and costs, computers can show what moves to make for the future—they can be powerful planning tools. They can be used in determining when to plant, when to schedule farm machine use, etc.

6. **How can income taxes be cut to the legal minimum?** By keeping an accurate record of expenses and figuring depreciations accurately, computers make for a saving in income taxes on most farms and ranches.

For providing answers to these questions and many more, computer accounting costs an average of about 1% of the gross farm income. By comparison, it is noteworthy that city businesses pay double this amount.

There are three requisites for linear programming a farm or ranch:

1. Access to a computer.

2. Computer know-how, so as to set the program up properly and be able to analyze and interpret the results.

3. Good records.

The pioneering computer services available to farmers and ranchers were operated by universities, trade associations, and government—most of them were on an experimental basis. Subsequently, others have entered the field, including commercial data-processing firms, banks, machinery companies, feed and fertilizer companies, and farm suppliers. They are using it as a service sell, as a replacement for the days of hard sell.

Information on "How to Balance a Ration by the Computer Method" is contained in *The Stockman's Handbook*, a book by the same author and publisher as *Animal Science*.

## FUTURES TRADING

Futures trading is not new. It is a well-accepted, century-old procedure used in many commodities for protecting profits, stabilizing prices, and smoothing out the flow of merchandise. For example, it has long been an integral part of the grain industry; grain elevators, flour millers, feed manufacturers, and others, have used it to protect themselves against losses due to price fluctuations. Also, a number of livestock products—hides, tallow, frozen pork bellies, hams, eggs, and turkey toms—were traded on the futures market before the advent of live cattle and hog futures. Many of these operators prefer to forego the possibility of making a high speculative profit in favor of earning a normal margin or service charge through efficient operation of their business. They look to futures markets to provide (1) an insurance medium in the marketing field and (2) the facilities and machinery for underwriting price risks.

A commodity exchange is a place where buyers and sellers meet on an organized market and transact business on paper, without the physical presence of the commodity. The exchange neither buys nor sells; rather, it provides the facilities, establishes rules, serves as a clearinghouse, holds the margin money deposited by both buyers and sellers, and guarantees delivery on all contracts. Buyers and sellers either trade on their own account or are represented by brokerage firms. Except for dealing in futures, and in paper contracts instead of live animals, futures trading is very similar to a terminal livestock market.

The unique characteristic of futures markets is that trading is in terms of contracts to deliver or to take delivery, rather than on the immediate transfer of the physical commodity. In practice, however, very few contracts are held until the delivery date. The vast majority of them are cancelled by offsetting transactions made before the delivery date.

Many farmers and ranchers have long contracted their cattle for future delivery without the medium of an exchange. They contract to sell and deliver to a buyer a certain number and kind of cattle at an agreed upon price and place. Hence, the risk of loss from a decrease in price after the contract is shifted to the buyer, and, by the same token, the seller foregoes

Fig. 10–11. A view of the trading floor of the Chicago Mercantile Exchange. The price quotation systems, telecommunications, and other news and information technologies are the most up-to-date systems possible. (Courtesy, Chicago Mercantile Exchange, Chicago, IL)

the possibility of a price rise. In reality, such contracting is a form of futures trading. Unlike futures trading on an exchange, however, actual delivery of the cattle is a must. Also, such privately arranged contracts are not always available, the terms may not be acceptable and the only recourse to default on the contract is a lawsuit.

The Chicago Mercantile Exchange initiated trading in live beef cattle on November 30, 1964. Shortly thereafter, they added dressed beef carcasses, as did the Chicago Board of Trade. Then, early in 1966, the Chicago Board of Trade announced plans for trading in live cattle, which they initiated October 4, 1966. The Chicago Mercantile Exchange (1) first offered hog futures on February 28, 1966, and (2) first offered feeder cattle futures on November 30, 1971.

## TAX MANAGEMENT AND REPORTING

Good tax management and reporting consists of complying with the law, but of paying no more tax than is required. It is the duty of revenue agents to see that taxpayers do not pay less than they should, but it is the business of the taxpayers to make sure that they do not pay more than is required. From both standpoints, it is important that farmers and ranchers familiarize themselves with as many of the tax regulations as possible.

The cardinal principles of good tax management are (1) maintenance of adequate records so as to assure payment of taxes in amounts no less or no more than required by law, and (2) conduction of business affairs to the end that the tax required by law is no greater than necessary.

Also, farmers and ranchers need to recognize that good tax management and good farm management do not necessarily go hand in hand. In fact, they may be in conflict. When the latter condition prevails, the advantages of the one must be balanced against the disadvantages of the other to the end that there shall be the greatest net return.

Under the cash system, farm income includes all cash or value of merchandise or other property received during the taxable year. It includes all available receipts from the sale of items produced on the farm and profits from the sale of items which have been purchased, exclusive, generally speaking, of one-half of the profits received from the sale of property used by the farmer in trade or business, such as breeding stock and farm machinery. It does not include the value of products sold or service performed for which payment was not actually available during the taxable year.

The accrual basis necessitates that complete annual inventories be kept. On the accrual basis, tax is paid on all income earned during the taxable year regardless of whether payment is actually received, and on increases of inventory values of livestock, crops, feed, produce, etc., at the end of the year as compared with the beginning of the year. All expenses incurred during the year's business are deducted from gross income regardless of whether payment is actually made, and deductions are made for any decrease in inventory values of livestock, etc., during the year.

Farmers are permitted to choose between the cash method and the accrual method.

Consultation with a tax specialist is recommended when tax problems are a bit out of the ordinary, and, like a visit to the family doctor, can be most effective when aid is sought before it is too late.

## ESTATE PLANNING

• **Special use valuation**—Owners of farms and small businesses have been granted an estate planning advantage by means of what is called *special use valuation*. Under this concept, a farm or ranch can escape valuation for estate tax purposes at the highest and best use. Thus, a farm located in an area undergoing development may be considerably more valuable to developers than it is as a farm. Nevertheless, if the family is willing to continue the farming or ranching use for ten years, the farm can be included in the estate at its value as a farm. The aggregate reduction in fair market value cannot exceed $750,000.

In order to qualify for special use valuation, the decedent must have been a U.S. citizen or resident and the farm must be located in the U.S. The farm must have been used by the decedent or a family member at the date of the decedent's death. A lease to a nonfamily member, if not dependent on production, will not satisfy this requirement. At least 50% of the value of the decedent's estate must consist of the farm and more than 25% of the estate must consist of the farm real property. It may be possible to split up a farm and take the special valuation for only part of it, but this part must involve real property worth at least 25% of the estate.

The property must be passed to a qualified heir, including ancestors of the decedent, the decedent's spouse and lineal descendants, lineal descendants of the decedent's spouse or parents, and the spouse of any lineal descendant. Aunts, uncles, and first cousins are excluded. Legally adopted children are included.

The property must have been owned by the decedent or a family member for five of the eight years preceding the decedent's death and used as a farm in that period. The decedent or a family member must have participated in the farming operation for such a period prior to the decedent's death or disability.

• **Electing special use valuation**—Though the procedures are clear as to how special use valuation is elected, the frequency with which mistakes are made indicates the importance of having a competent tax attorney or CPA firm prepare the estate tax return. A procedural failure denying the estate the considerable savings that can be gained by the election may give sufficient grounds for a malpractice suit against the return preparer.

• **Recapture tax**— If the farm ceases to be operated by the heir or a family member within ten years, an additional estate tax will be imposed and the advantage of the election will be substantially lost. Partition among qualified heirs will not bring about recapture. When heirs granted oil leases on a family farm, the portion of the land devoted to the oil rigs was subject to recapture. A recent change allows the surviving spouse of the decedent to lease a farm on a net cash basis to a family member without being subject to the recapture tax.

• **Payment extension**—Estates eligible for special use valuation may often be able to defer the payment of estate taxes. Where more than 35% of an estate of a U.S. citizen or resident consists of a farm, the estate tax liability may be paid in up to ten annual installments beginning as late as five years from when the tax might otherwise be due. If any portion of the farm is disposed of before the final payment, a corresponding portion of the amount deferred will come due.

## LIABILITY INSURANCE; WORKER'S COMPENSATION INSURANCE

Most farmers are in such financial position that they are vulnerable to damage suits. Moreover, the number of damage suits arising each year is increasing at an almost alarming rate, and astronomical damages are being claimed. Studies reveal that about 95% of the court cases involving injury result in damages being awarded.

Several types of liability insurance offer a safeguard against liability suits brought as a result of injury suffered by another person or damage to another's property.

Comprehensive personal liability insurance protects farm operators who are sued for alleged damages suffered from an accident involving their property or family. The kinds of situations from which a claim might arise is quite broad, including suits for injuries caused by animals, equipment, or personal acts.

Both worker's compensation and employer's liability insurance protect farmers against claims or court awards resulting from injury to hired help. Worker's compensation usually costs slightly more than straight employer's liability insurance, but it carries more benefits to the worker. An injured employee must prove negligence by the employer before the company will pay a claim under employer's liability insurance, whereas worker's compensation benefits are established by state law and settlements are made by the insurance company without regard to who was negligent in causing the injury. Conditions governing participation in worker's compensation insurance vary among the states.

# QUESTIONS FOR STUDY AND DISCUSSION

1. Show the trends to bigness of American agriculture by applying each of the following criteria: (1) assets, (2) number and size of farms, (3) number of persons for whom each farm worker supplies food and fiber, (4) purchases and operation of tractors, trucks, cars, and machinery.

2. Why will the business and management aspects of animal production become increasingly important in future years?

3. List the common types of business organizations, and give the advantages and disadvantages of each. For your farm or ranch, or for a farm or ranch with which you are familiar, what type of business organization would you choose, and why would you choose it?

4. U.S. farmers have more than 81% equity in their business and operate on less than 19% borrowed money. By contrast, most apartment house owners borrow about 80% of the cost of the unit, and use only about 20% of their own money. Why the difference?

5. It takes more than $17 in farm assets to generate $1 of net farm income. Wouldn't most farmers be better off to invest their money in some nonagricultural enterprise?

6. Assume that you are going to enter the livestock business, and that you have decided on the particular kind (with you making the decision between beef, cow-calf, dairy, broilers, etc.). What types of credit may be needed, and what source would you use?

7. List and discuss five good rules for building and maintaining a good credit rating.

8. Define the term *manager*. What three major ingredients are essential to success in the livestock business?

9. What are the traits of a good manager? How may a student acquire these traits?

10. For your home farm, or for a farm with which you are familiar, prepare and present (a) an organizational chart, and (b) job description.

11. List and discuss the four major types of incentives.

12. Take your own farm or ranch, or one with which you are familiar, and develop a workable incentive basis for the help.

13. List and discuss the chief functions of farm records and accounts.

14. Develop a yearly budget for your own farm or ranch, or for one with which you are familiar.

15. How may computers be used, on a practical basis, for a livestock enterprise?

16. Define the term *futures trading*. Under what circumstances would you use futures trading?

17. What constitutes good tax management and reporting?

18. Explain the main difference between (a) the cash system, and (b) the accrual basis of tax management and reporting.

19. Discuss the importance of each of the following: (a) estate planning, and (b) worker's compensation insurance and employer's liability insurance.

## SELECTED REFERENCES

| Title of Publication | Author(s) | Publisher |
|---|---|---|
| *Agricultural Statistics, 1988* | Staff | U.S. Department of Agriculture, Washington, DC, 1988 |
| *Cowboy Arithmetic, Cattle as an Investment,* Third Edition | H. L. Oppenheimer | The Interstate Printers & Publishers, Inc., Danville, IL, 1971 |
| *Cowboy Economics: Rural Land as an Investment,* Third Edition | H. L. Oppenheimer | The Interstate Printers & Publishers, Inc., Danville, IL, 1976 |
| *Cowboy Litigation: Cattle and the Income Tax,* Second Edition | H. L. Oppenheimer<br>J. D. Keast | The Interstate Printers & Publishers, Inc., Danville, IL, 1972 |
| *Fact Book of Agriculture 1989* | Staff | U.S. Department of Agriculture, Washington, DC, 1989 |
| *Farm Management Economics* | E. O. Heady<br>H. R. Jensen | Prentice-Hall, Inc., Englewood Cliffs, NJ, 1955 |
| *Introduction to Agribusiness Management, An,* Second Edition | W. J. Wills | The Interstate Printers & Publishers, Inc., Danville, IL, 1979 |
| *Land Speculation: An Evaluation and Analysis* | H. L. Oppenheimer | The Interstate Printers & Publishers, Inc., Danville, IL, 1972 |
| *Microcomputing in Agriculture* | J. Legacy<br>T. Stitt<br>F. Reneau | Reston Publishing Company, Inc., Reston, VA, 1984 |
| *Spreadsheet Applications For Animal Nutrition and Feeding* | R. J. Lane<br>T. L. Gross | Reston Publishing Company, Inc., Reston, VA, 1985 |
| *Statistical Abstracts of the United States 1989* | Staff | U.S. Department of Commerce, Washington, DC, 1989 |
| *Stockman's Handbook, The,* Sixth Edition | M. E. Ensminger | The Interstate Printers & Publishers, Inc., Danville, IL, 1983 |

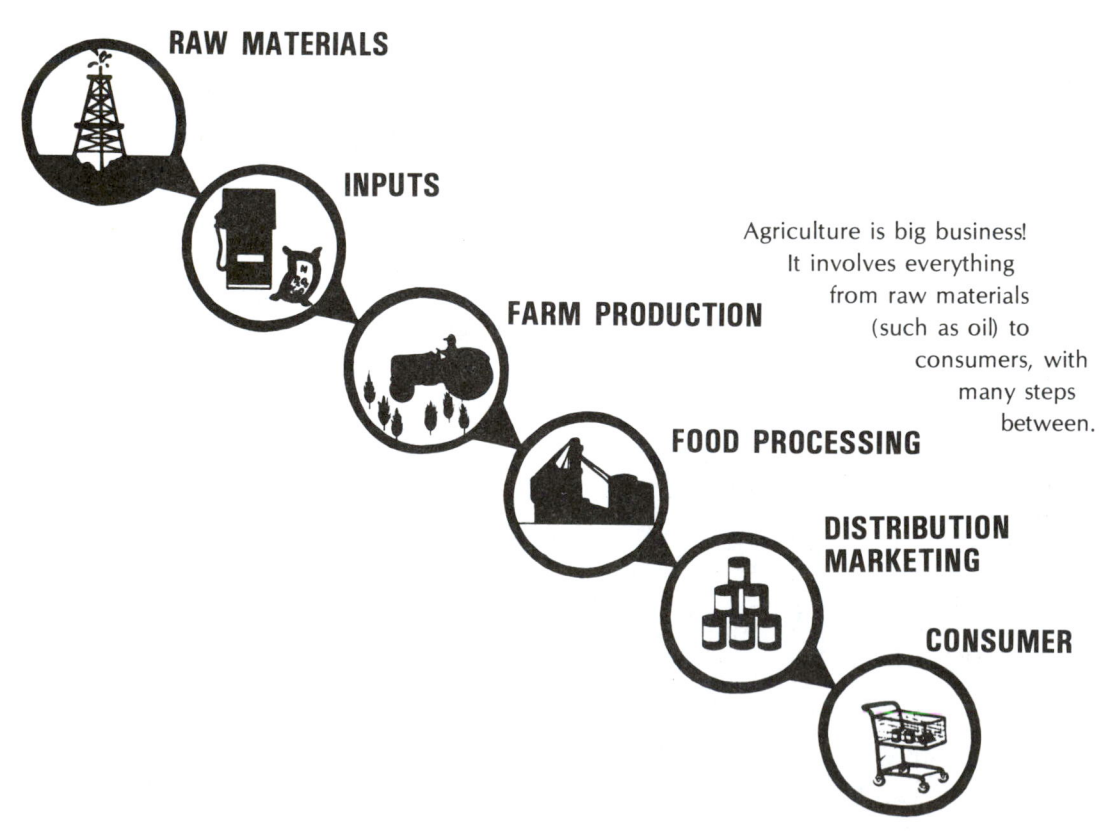

Agriculture is big business! It involves everything from raw materials (such as oil) to consumers, with many steps between.

# ANIMAL SCIENCE

Plate 1. **Business aspects,** involving computers and record keeping, are an integral part of modern livestock production. (Courtesy, University of Illinois, Urbana)

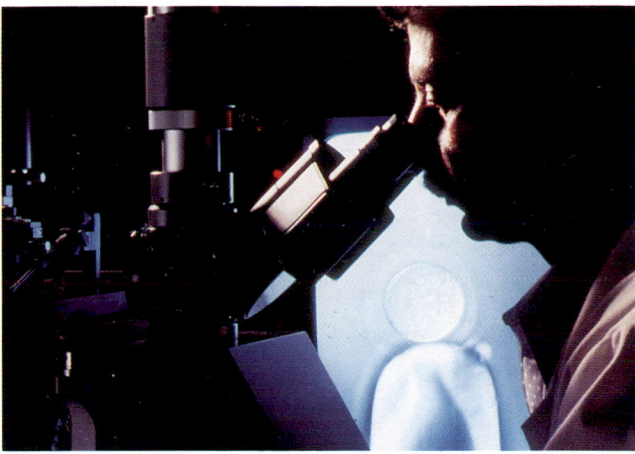

Plate 2. **Microscope** being used in genetic engineering—the redesigning of animals. (Courtesy, Granada Land & Cattle Co., Inc., Wheelock, TX)

Plate 3. **Embryo transfer**. Twenty progeny produced by one superior female. (Courtesy, Granada Land & Cattle Co., Inc., Wheelock, TX)

Plate 4. **Foal resulting from a frozen embryo**. (Courtesy, Dr. George Seidel, Colorado State University, Ft. Collins, CO)

Plate 5. **Food**. A dinner featuring roast turkey, fresh cranberry sauce, creamed onions with mushrooms and carrots, maple walnut sweet potatoes, and steamed apple pudding. (Courtesy, United Fresh Fruit & Vegetable Assn., Alexandria, VA)

Plate 6. **Children need pets**. Boys and girls, our best products, have treasured animals as pets throughout history. (Courtesy, The American Hampshire Sheep Assn., Ashland, MO)

# ANIMAL SCIENCE

Plate 7. **Polled Hereford cows watering** at a woodland pond. (Courtesy, American Polled Hereford Assn., Kansas City, MO)

Plate 8. **Livestock farm**, showing beautiful buildings, white fences, well-manicured pastures, and cattle and sheep. (Photo by J. C. Allen & Son, Inc., West Lafayette, IN)

Plate 9. **Holstein cows on pasture** in British Columbia, Canada. (Courtesy, *Holstein World*, Sandy Creek, NY)

Plate 10. **Twin Lambs**. (Photo by J. C. Allen & Son, Inc., West Lafayette, IN)

Plate 11. **Sow and litter in a farrowing crate**. (Courtesy, American Yorkshire Club, Inc., West Lafayette, IN)

Plate 12. **Chief Joseph Trail Ride**, over the trail taken by Chief Joseph when he fled from the U.S. Cavalry in the late 1800s. (Courtesy, Appaloosa Horse Club, Moscow, ID)

Plate 13. **Californian doe and litter of eight**. This is a major meat breed. (Courtesy, Alabama Agricultural and Mechanical University, Normal)

Plate 14. **Hybrid Striped Bass**. (Courtesy, Dr. D. Gatlin, Texas A&M University, College Station)

# BEEF INDUSTRY

Plate 15. **Cattle grazing in meadow** on Harlan Ranch, Indian Valley, Quincy, Californnia. (Courtesy, Floyd and Marilyn Harlan, Clovis, CA)

Plate 16. **Shorthorn cattle on the range**. (Courtesy, American Shorthorn Assn., Omaha, NE)

Plate 17. **Hereford bull in cacti (dry) country**. (Courtesy, American Hereford Assn., Kansas City, MO)

Plate 18. **Texas Longhorn cow and calf**. (Courtesy, Larry and Sandra Southard, Porterville, CA)

Plate 19. **Finishing cattle at feed bunk**. (Courtesy, American Polled Hereford Assn., Kansas City, MO)

Plate 20. **Show champions** at the Junior National Polled Hereford Show. (Courtesy, American Polled Hereford Assn., Kansas City, MO)

# DAIRY INDUSTRY

Plate 21. **Dairy farm**. (Courtesy, Milk Marketing, Inc., Strongsville, OH)

Plate 22. **Maddox Dairy**, the "Taj Mahal" of dairies, Riverdale, California. (Photo by A. H. Ensminger)

Plate 23. **Cows feeding** at Arizona Dairy, Higley, AZ, where they milk 4,500 cows, do their own embryo transfer work, and produce electricity from the manure. (Courtesy, James M. Tappan, Managing Partner)

Plate 24. **Lactating cows going to pasture**. (Courtesy, Milk Marketing, Strongsville, OH)

Plate 25. **Milk with a choice!** (Courtesy, American Dairy Assn., Rosemont, IL)

Plate 26. **Cheeses with a variety!** (Courtesy, American Dairy Assn., Rosemont, IL)

# SHEEP AND GOAT INDUSTRY

Plate 27. **Made of wool!** (Courtesy, the Wool Bureau, New York, NY)

Plate 28. **Hampshire sheep**. (Courtesy, Marvin and Elinor Heupel, Santa Maria, CA )

Plate 29. **Range band of ewes**, near Babb, Montana, with Sherburne Peak in the background. (Courtesy, Ernst Peterson, Hamilton, MT)

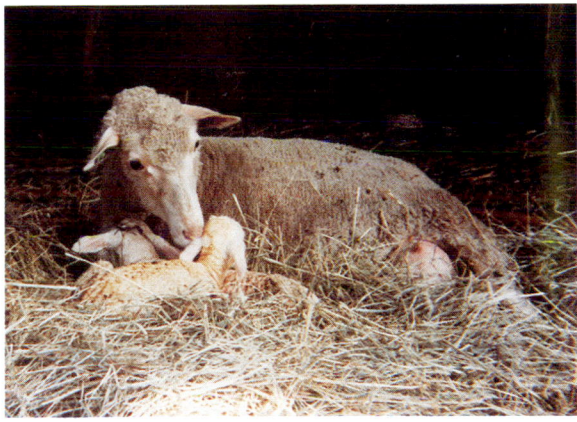

Plate 30. **Polypay ewe and newborn lamb**. (Courtesy, American Polypay Sheep Assn., Sidney, MT)

Plate 31. **Angora billy goats**. (Courtesy, Texas A&M University, College Station)

Plate 32. **Dairy doe and kids**. (Photo by Cindy Schneider, Sacred Heart, MN)

# SWINE INDUSTRY

Plate 33. **Swine confinement complex**—an aerial view. (Courtesy, University of Illinois, Urbana)

Plate 34. **Duroc boar**. (Courtesy, United Duroc Swine Registry, Peoria, IL)

Plate 35. **Sow and litter in farrowing crate**. (Courtesy, Compart's Boar Store, Nicollet, MN)

Plate 36. **Grower pen**, with 15–20 pigs per pen. (Courtesy, Compart's Boar Store, Nicollet, MN)

Plate 37. **Finishing barn**. (Courtesy, *Hog Farm*, Minnetonka, MN)

Plate 38. **Youth show** at National Show, Bowling Green, KY. (Courtesy, National Hereford Hog Assn., Flandreau, SD)

# POULTRY INDUSTRY

Plate 40. **Baby chicks**. (Courtesy, Monsanto Agricultural Company, St. Louis, MO)

Plate 39. **Laying hens in a modern cage facility**. Note the several tiers of cages, the feed troughs, and the automatic egg collecting belts. (Courtesy, L. S. Jensen, Ph.D., Dept. of Poultry Science, the University of Georgia, Athens)

Plate 42. **Large White turkeys**. (Courtesy, National Turkey Federation, Reston, VA)

Plate 41. **Broilers**, 6 weeks old. (Courtesy, California Polytechnic University, San Luis Obispo)

Plate 43. **White Pekin Ducks**. (Courtesy, California Polytechnic University, San Luis Obispo)

Plate 44. **Fried chicken**. (Courtesy, Gold Kist, Inc., Atlanta, GA)

# HORSE INDUSTRY

Plate 45. **Royal Canadian Mounted Police**. (Courtesy, RCMP, Ottawa, Canada)

Plate 46. **Mares on the range**. (Courtesy, Ernst Peterson, Hamilton, MT)

Plate 47. **Lexington Stakes, 1988**; showing *Risen Star* outside and *Forty Niner* inside. (Photo by Anne M. Eberhardt. Courtesy, *The Blood Horse*, Lexington, KY)

Plate 48. **Morgan pair**, drawing a *Stanhope Phaeton*. (Courtesy, American Morgan Horse Assn., Shelburne, VT)

Plate 49. **Horse talk**. (Courtesy, Gold Kist, Inc., Atlanta, GA)

Plate 50. **Miniature Horse**, *Little Boy Blue*, a stallion standing 30 in. tall. (Courtesy, Flying W Farms, Piketon, OH)

Robert Bakewell of Dishley (1726–1795), the first great improver of cattle in England. The American poet, Emerson, said of him: "He created cows, sheep, and horses to order—the cow is sacrificed to her bag, the ox to his sirloin."

# THE
# BEEF CATTLE
# INDUSTRY

═══════════════════════════════════════
## Chapter 11
═══════════════════════════════════════

Cattle are the most important domesticated animal, and, next to the dog, the most ancient. There are about 1.3 billion cattle in the world.

The word *cattle* seems to have the same origin as *chattel*, which means possession. This is a very natural meaning, for, when Rome was in her glory, a person's wealth was often computed in terms of cattle possessions, a practice which still persists among primitive people in Africa and Asia. That the ownership of cattle implied wealth is further shown by the fact that the earliest known coins bear an ox head; and the Roman word *pecunia* for money (preserved in our adjective *pecuniary*) was derived from the Latin word *pecus,* meaning cattle. It is also noteworthy that the oldest known treatise on agriculture, written by the Greek poet Hesiod, referred to cattle. Apparently having had some disturbing experiences with young oxen, Hesiod advised: "For draught and yoking together nine-year-old oxen are best because, being past the mischievous and frolicsome age, they are not likely to break the pole and leave the plowing in the middle."

## ORIGIN AND DOMESTICATION OF CATTLE

It seems probable that cattle were first domesticated in Europe and Asia during the New Stone Age. In the opinion of most authorities, today's cattle bear the blood of either or both of two ancient ancestors—namely *Bos taurus* and *Bos indicus*.

Other species or subspecies were frequently listed in early writings, but these are seldom referred to today. Perhaps most, if not all, of these supposedly ancestral species were also descendants of *Bos taurus,* or *Bos indicus* or crosses between the two.

## BOS TAURUS

*Bos taurus* includes those domestic cattle common to the more temperate zones, and it, in turn, appears to be derived from a mixture of the descendants of the Aurochs (*Bos primigenius*) and Celtic Shorthorns (*Bos longifrons*).

Most cattle, including the majority of the breeds found in the United States, are believed to have descended mainly from the massive Aurochs (also referred to as Uri, Ur, or Urus). This was the mighty wild ox that was hunted by our forefathers. It roamed the forests of central Europe down to historic times, finally becoming extinct about the year 1627. About the year 65 B.C., Caesar mentioned this ox in his writings, but it was domesticated long before (perhaps early in the Neolithic age), probably south of the Alps or in the Balkans or Asia Minor. Caesar referred to these animals as "approaching the elephant in size but presenting the figure of a bull." Although this is somewhat of an exaggeration as to the size of Aurochs, it was a tremendous beast, standing 6 to 7 ft high at the withers, as is proven by complete skeletons found in the bogs.

Fig. 11–1  Artist's conception of an Aurochs (*Bos primigenius*) based on historical information. This was the mighty wild ox that was hunted by our ancestors. Most cattle are believed to have descended mainly from the Aurochs.

In addition to the Aurochs, another progenitor of some of our modern breeds and earliest known domestic race of cattle was the Celtic Shorthorn or Celtic Ox. These animals, which have never been found except in a state of domestication, were the only oxen in the British Isles until 500 A.D., when the Anglo-Saxons came, bringing with them animals derived from the Aurochs of Europe. The Celtic Shorthorn was of smaller size than the Aurochs and possessed a dished face. It may have had a still different wild ancestor, or may have been an independent domestication from the Aurochs.

## BOS INDICUS

*Bos indicus* includes those humped cattle common to the tropical countries that belong to the Zebu (or Brahman) group. They are wholly domestic creatures, no wild ancestors having been found since historic times. It has been variously estimated that cattle of this type were first domesticated anywhere from 4000 to 2100 B.C. The Zebu is characterized by a hump of fleshy tissue over the withers (which sometimes weighs as much

Fig. 11-2. Zebu (*Bos indicus*). These wholly domestic animals were the ancestors of the humped cattle common to the tropical countries.

as 40 to 50 lb), a very large dewlap, large drooping ears, and a voice that is more of a grunt than a low. These peculiar appearing animals seem to have more resistance to certain diseases and parasites and to heat than the descendants of *Bos taurus*. For this reason, they have been crossed with some of the cattle of Brazil and of the southern states of this country, especially in the region bordering the gulf of Mexico.

## POSITION OF THE OXEN IN THE ZOOLOGICAL SCHEME

Domesticated cattle belong to the family *Bovidae*, which includes ruminants with hollow horns. Members of this family possess one or more enlargements for food storage along the esophagus, and they chew their cuds. In addition to what we commonly call oxen or cattle, the family *Bovidae* (and the subfamily *Bovinae*) includes the true buffalo, the bison, musk-ox, banteng, gaur, gayal, yak, and Zebu.

The following outline shows the basic position of the domesticated cow in the zoological scheme:

Kingdom *Animalia*: Animals collectively; the animal kingdom.

Phylum *Chordata*: One of the approximately 21 phyla of the animal kingdom, in which there is either a backbone (in the vertebrates) or the rudiment of a backbone, the chorda.

Class *Mammalia*: Mammals are warm-blooded, hairy animals that produce their young alive and suckle them for a variable period on a secretion from the mammary glands.

Order *Artiodactyla*: Even-toed, hoofed mammals.

Family *Bovidae*: Ruminants having polycotyledonary placenta; hollow, nondeciduous, up-branched horns; and nearly universal presence of a gall bladder.

Genus *Bos*: Ruminant quadrupeds, including wild and domestic cattle, distinguished by a stout body and hollow, curved horns standing out laterally from the skull.

Species *Bos taurus* and *Bos indicus*: *Bos taurus* includes the ancestors of the European cattle and of the majority of the cattle found in the United States; *Bos indicus* is represented by the humped cattle (Zebu) of India and Africa and the Brahman breed of America.

## USE OF CATTLE IN ANCIENT TIMES

Like other animals, cattle were first hunted and used as a source of food and other materials. As civilization advanced and humans turned to tilling the soil, it is probable that the domestication of cattle was first motivated because of their projected value for draft purposes. Large, well-muscled, powerful beasts were in demand; and any tendency to fatten excessively or to produce more milk than was needed for a calf was considered detrimental rather than desirable. Not all cattle were used for work purposes, however, in the era following their domestication. Instead of planting seeds, some races of people chose a pastoral existence—moving about with their herds as they required new pastures. These nomadic people lived mainly on the products of their herds and flocks.

As populations became more dense, feed became more abundant, and cattle more plentiful, people became more interested in larger production of meat and milk. The pastoral people adopted a more settled life and began selecting out those animals that possessed the desired qualities—including rapid growth, fat storage, and milk production. Following this transformation, Biblical and other literature referred to milk cows, the stall-fed ox, and the fatted calf.

# HISTORY AND CATTLE

Fig. 11-3. Herder tending cattle. Painting by Albert Cuyp (1620–1691), a Dutch painter.

On the crest of a rocky hill, a herd of 7 cows is silhouetted against the evening sky. The herder stands in the foreground. (Courtesy, National Gallery of Art, Washington, DC, Mellon Collection)

Fig. 11-4. Danish shell mound people eating *Bos urus* (Aurochs), their favorite food. After a painting by Ernst Griset. *Bos urus* was the mighty wild ox that was hunted by our ancestors. (Courtesy, Smithsonian Institute)

Fig. 11-5. Trail herd—point riders. Etching by Edward Borein, 1922. Note the bull in the lead. Famous cattle trails were a part of the Old West. (Courtesy, E. N. Wentworth)

Fig. 11-6. Oxen hauling logs to the saw mill. Oxen were more prized by the colonists than beef.

## BAKEWELL'S IMPROVEMENT OF ENGLISH CATTLE

Robert Bakewell of Dishley (1726–1795)—an English farmer of remarkable sagacity and hard, common sense—was the first great improver of cattle in England. His objective was to breed cattle that would yield the greatest quantity of good beef rather than to obtain great size. Bakewell had the imagination to picture the future needs of a growing population in terms of meat and set about creating a low-set, blocky, quick-maturing type of beef cattle. He paid little or no attention to fancy points. Rather, he was intensely practical, and no meat animal met with his favor unless it had the ability to put meat on the back.

Bakewell's efforts with cattle were directed toward the perfection of the English Longhorn, a class of cattle common to the Tees River area. He also contributed greatly to the improvement of the Leicester breed of sheep, and the Shire horse. Success crowned his patient skill and unwearied efforts. But success in breeding was not mere happenstance in Bakewell's program. Careful analysis of his methods reveals that three factors were paramount: (1) a definite goal as evidenced by the joints that he preserved in pickle and the skeletons of the more noted animals that adorned his halls, (2) a breeding system characterized by "breeding the best to the best" regardless of relationship rather than crossing breeds as was the common practice of the time, and (3) a system of proving sires by leasing them at fancy prices to his neighbors rather than selling them. Because of Bakewell's methods and success, he often has been referred to as the founder of animal breeding.

Bakewell's experiments were the top news of the day; and his successes were the subject of much comment, both oral and written. The American poet Emerson, for example, said of the British farmer, "He created sheep, cows, and horses to order . . . the cow is sacrificed to her bag, the ox to his sirloin."

By the beginning of the Napoleonic Wars, Bakewell's methods were widely practiced in England, and sheep and cattle were raised more for their flesh than formerly. A new era in livestock improvement was born. As an indication of this change, it is interesting to observe the increase in weights of animals at the famous Smithfield market. In 1710, beeves had averaged 370 lb, calves 50 lb, sheep 28 lb, and lambs 18 lb; whereas in 1795 they had reached 800, 148, 80, and 50 lb, respectively. Although the effect of improved agriculture is not to be minimized, the main influence in this transformation can be attributed to Robert Bakewell, whose imagination, initiative, and courage put a firm foundation under improved methods of livestock breeding.

## THE INTRODUCTION OF CATTLE TO AMERICA

Cattle were not native to the Western Hemisphere. They were first brought to the West Indies by Columbus on his second voyage in 1493. According to historians, these animals were intended as work oxen for the West Indies colonists. Cortez took cattle from Spain to Mexico in 1519. Then, beginning about 1600, other Spanish cattle were brought over for work and milk purposes in connection with the chain of Christian missions which the Spaniards established among the Indians in the New World. These missions extended from the east coast of Mexico up the Rio Grande, thence across the mountains to the Pacific Coast. Here, in a land of abundant feed and water, these Longhorns multiplied at a prodigious rate. By 1833, the Spanish priests estimated that their missions owned a total of 424,000 head of cattle,[1] many of which were running in a semiwild state. The hardy Texas Longhorns, animals of Spanish extraction, were of little commercial value except for their hides.

The colonists first brought cattle from England in 1609. Other English importations followed, with Governor Edward Winslow bringing a notable importation to the Plymouth Colony in 1623. The latter shipment included 3 heifers and a bull. Three years later, at a public court, these animals and their progeny—and perhaps some subsequent importations—were apportioned among the Plymouth settlers on the basis of 1 cow to 6 persons. It is further reported that 3 ships carried cattle to the Massachusetts Bay Colony in 1625. Other colonists came to the shores of New England bringing with them their oxen from the mother country. As would be expected, the settlers brought along the kind of cattle to which they had been accustomed in the mother country. This made for considerable differences in color, size, and shape of horns, but all of these cattle imported by the colonists possessed ruggedness and the ability to perform work under yoke.

Fig. 11–7. Texas Longhorn Steer. (Photo by J. Frank Dobie. Courtesy, N. H. Rose Collection, San Antonio, TX)

For a number of years, there were very few cattle in the United States. Moreover, those animals that the colonists did possess went without winter feed and shelter, and the young suffered the depredations of the wolves. It was difficult enough for the settlers to build homes for themselves, and they could barely raise enough corn in their fields to sustain human life.

Conditions presently changed for the better. The cattle of earlier importations multiplied, new shipments were received, and feed supplies became more abundant. Cambridge, Massachusetts, enjoyed the double distinction of being the seat of Harvard College, the first institution of higher learning in what later came to be the United States, and the most prosperous cattle center in early New England. In order to provide ample grass and browse for the increased cattle population, it was necessary that the animals range some distance from the commons (the town pasture). Thus, the tale that the streets of Boston were laid out along former cowpaths is not legend but fact. Usually in their travels, the cattle were under the supervision of a paid *cowkeeper* whose chief duty consisted of safely escorting the cattle to and from pasture.

[1]*Yearbook of Agriculture, 1921,* USDA, p. 233.

In the village economy, the bull was an animal of considerable importance. Usually the town fathers selected those animals that they considered most desirable to retain as sires, and those citizens who were so fortunate as to own animals of this caliber were paid an approved service fee on a per head basis.

## DRAFT OXEN MORE PRIZED BY COLONISTS THAN BEEF

From the very beginning, the colonists valued cattle for their work, milk, butter and hides; but little importance was attached to their value for meat. In fact, beef was considered as much a by-product as hides are today. After all, wild game was plentiful, and the colonists had learned to preserve venison, fish, and other meats by salting, smoking, and drying. So necessary were cattle for draft purposes that, in some of the early-day town meetings, ordinances were passed making it a criminal offense to slaughter a work ox before he had passed the useful work age of seven or more years. The work requirement led to the breeding of large rugged cattle, with long legs, lean though muscular bodies, and heavy heads and necks. Patient oxen of this type were well adapted for clearing away the forest and turning the sod on the rugged New England hillsides, for hauling the harvested produce over the rough roads to the seaport markets, and for subsisting largely on forages.

Fig. 11-8. Oxen pulling a prairie schooner (a type of covered wagon). Draft oxen were more prized by the colonist than beef. (Courtesy, The Bettmann Archive—from an engraving)

## THE FAR WEST EXPANSION OF THE CATTLE INDUSTRY

From the very beginning, cattle raising on a large scale was primarily a frontier activity. As the population of the eastern United States became more dense, the stock-raising industry moved farther inland. The great westward push came in the 19th century. By 1800, the center of the cow country was west of the Alleghenies, in Ohio and Kentucky; in 1860, it was in Illinois and Missouri; and by the 1880s, it was in the Great Plains. The ranches and cowboys of the Far West were the counterpart of the New England commons and cowdrivers of the 17th and 18th centuries.

The western range was recognized as one of the greatest cattle countries that the world had ever known. Plenty of water and unlimited grazing area were free to all comers, and the market appeared to be unlimited. Fantastic stories of the fabulous wealth to be made from cattle and ranching caused

a rush comparable to that of the gold diggers of 1849. All went well until the severe winter of 1886. It was the type of winter that is the bane of the cattle owners. Then, but all too late, it was realized that too many cattle had been kept and too little attention had been given to storing up winter feed supplies. The inevitable happened. With the melting of the snow in the spring of 1887, thousands of cattle skeletons lay weathering on the western range, a grim reminder of overstocking and inadequate feed supplies. Many ranchers went broke, and the cattle industry of the West suffered a crippling blow that plagued it for the next two decades. Out of this disaster, however, the rancher learned the never-to-be-forgotten lessons of (1) avoiding overexpansion and too-close grazing, and (2) the necessity of an adequate winter feed supply.

## STATUS OF THE U.S. CATTLE INDUSTRY

There was a lower trend in United States cattle numbers during the last quarter of the 20th century. As shown in Fig. 11-9, this applied to the number of beef cows, the number of dairy cows, and the total number of cattle and calves. More than any other forces, this downward trend in cattle numbers was caused by a change in eating habits, with poultry meat (1) more favorably priced than beef, and (2) capitalizing on diet and health issues.

### Cattle on Farms

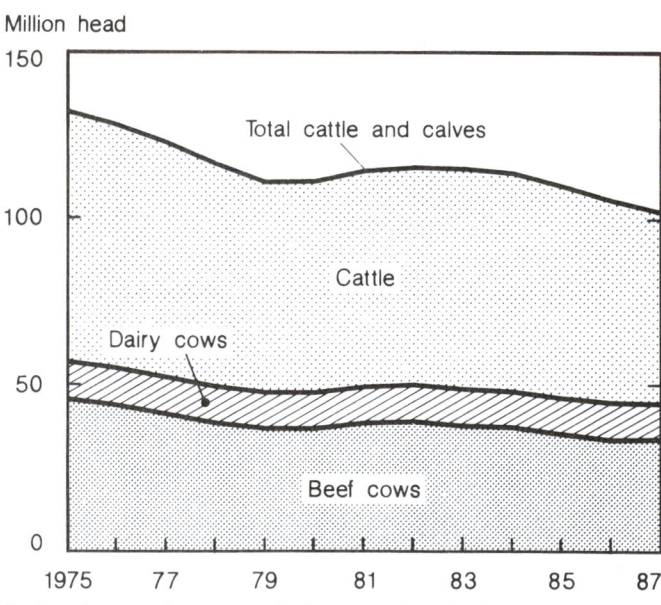

Cattle on farms as of January 1. Beef cows are those that have calved.

Fig. 11-9. Cattle on U.S. Farms, 1975 to 1987. Note that beef cows, dairy cows, and total cattle and calves have decreased since about 1975. (Source: *1988 Agriculture Chartbook*, USDA, Agriculture Handbook No. 673, p. 88, Chart 189)

Although cattle numbers trended downward beginning about 1975, the productive rate of cattle increased during the last quarter of the 20th century. Cattle received better care. There was a marked improvement in feeding; more newborn animals were saved; more attention was given to sanitation, disease prevention, and parasite control; animals were more adequately housed; waste resulting from death, crippling, and bruises in transit decreased; and feedlot cattle were grain fed for a shorter period and marketed with less finish.

## WORLD CATTLE AND BEEF

It is important that cattle producers and those who counsel them be well informed concerning worldwide beef production in order to know which countries are potential competitors. Like the price of all commodities in a free commerce, the price of beef is determined chiefly by supply and demand—that is, by the demand existing in those countries that do not produce enough to meet their domestic needs and by the supply which can be spared by those nations producing a surplus.

The production of beef cattle is worldwide. Table 11–1 gives the size and density, by rank, of the cattle population of the 12 leading cattle-producing countries of the world. As noted, world cattle numbers total 1.3 billion head; this is about one cow for every 4 people, or 22 head per square mile.

But cattle numbers alone do not tell the whole story. It is the production of beef and veal that counts. For example, large numbers of cattle are kept for work and for milk production in India. Besides, cattle are sacred to the Hindus of India, who comprise about 80% of the population.

### TABLE 11–1
### LEADING CATTLE-PRODUCING COUNTRIES OF THE WORLD

| Country | Cattle Population[1] | Human Population[2] | Size of Country[2] | | Cattle Per Capita | Cattle per | |
|---|---|---|---|---|---|---|---|
| | | | (sq mi) | (sq km) | | (sq mi) | (sq km) |
| India | 193,000,000 | 833,422,000 | 1,266,595 | 3,280,481 | 0.23 | 152.4 | 58.8 |
| Brazil | 134,133,000 | 153,992,000 | 3,286,470 | 8,511,957 | 0.90 | 40.8 | 15.8 |
| U.S.S.R. | 120,593,000 | 287,015,000 | 8,649,496 | 22,402,195 | 0.42 | 13.9 | 5.4 |
| United States | 98,994,000 | 247,498,000 | 3,618,770 | 9,372,614 | 0.40 | 27.4 | 10.6 |
| China, People's Republic of | 73,963,000 | 1,069,628,000 | 3,705,390 | 9,596,960 | 0.07 | 20 | 7.7 |
| Argentina | 50,782,000 | 32,617,000 | 1,065,189 | 2,758,840 | 1.56 | 47.7 | 18.4 |
| Mexico | 31,200,000 | 88,087,000 | 761,604 | 1,972,554 | 0.35 | 41 | 15.8 |
| Ethiopia | 31,000,000 | 47,709,000 | 471,776 | 1,221,900 | 0.65 | 65.7 | 25.4 |
| Colombia | 24,307,000 | 31,821,000 | 439,735 | 1,138,914 | 0.76 | 55.3 | 21.3 |
| Australia | 23,500,000 | 16,090,000 | 2,966,200 | 7,682,458 | 1.46 | 7.9 | 3.1 |
| Bangladesh | 22,789,000 | 112,757,000 | 55,598 | 143,999 | 0.2 | 409.9 | 158.3 |
| France | 21,100,000 | 55,813,000 | 220,668 | 571,530 | 0.38 | 95.6 | 36.9 |
| World Total | 1,263,584,000 | 5,055,000,000 | 57,800,000 | 149,702,000 | 0.25 | 21.9 | 8.4 |

[1]*FAO Production Yearbook 1988*, FAO, United Nations, Rome, Italy, Vol. 42, p. 244, Table 89. Data for 1988.
[2]*The World Almanac and Book of Facts 1989*, The Fresno Bee, 1989.

## BEEF PRODUCTION IN INDIA

India, land of sacred cows and native home of the U.S. Brahman breed, is the leading cattle country in the world in numbers and density. There are 152 cattle per square mile vs 27 in the United States; that's more than 5 times greater density in India than in the United States. But India's cattle are of very negligible importance from the standpoint of meat production, due to the large number that are either sacred or used for draft or milk purposes. The humped cattle of India pillage crops in rural areas and roam the streets of villages and cities—gentle and traffic-wise. Some are homeless, others are turned loose by owners who do not wish to pay for their keep, and still others are just AWOL. To the Hindu, the cow is regarded as a mother and an object of reverence; and the eating of beef is taboo. Although India's cattle population puts a serious drain on the nation's resources, no politician dares twist a cow's tail, or even flick a hair. To do so would not be unlike an American politician campaigning against apple pie and motherhood.

## BEEF PRODUCTION IN AUSTRALIA AND NEW ZEALAND

Beef production in Australia and New Zealand increased sharply in the late 1960s and early 1970s in response to high beef prices and low wool prices. Of all the beef-producing countries of the world, Australia and New Zealand have the best potential for increase.

Australia is a natural cattle country, and it is free from foot-and-mouth disease. Most of the cattle are grazed year-round on unfenced ranges, herded by jackaroos—the counterpart of the American cowboys. Slaughter animals consist of 2- to 4-year-old steers which are grass finished, although there is a growing trend to market younger animals and to grain finish. The vast majority of the cattle operations in Australia are very large, ranging in size from 5,000 acres in the more developed southeastern part of Australia to over 3 million acres in the Northern Territory, and with 10,000 to 50,000 cattle per unit. In 1988, Australia had 23,500,000 head of cattle.

Shorthorns, Herefords, and Angus are the leading breeds. In the tropical areas of the North, Brahman and Santa Gertrudis have been introduced and are increasing in numbers. Cross-

Fig. 11–10. Scene in India, showing cattle in a village. Usually Indian cattle are herded by either the young or the old. (Photo by the Ford Foundation)

breeding is practiced widely. Many stations (we call them ranches) are inadequately fenced and watered, and there is room for improvement in their nutrition and husbandry. As evidence of the latter statement, it is noteworthy that in northwestern Queensland, Northern Territory, and parts of Western Australia, only a 45 to 55% calf crop is raised to branding age. Also, on some properties it is standard practice to write off a 12 to 15% mortality each year.

The beef industry of Australia is subjected to recurrent droughts. Until recently, the chief obstacle to further expansion of the nation's beef production was the great distance to the consumer markets of Europe and the United States. However, improved technology in processing and transporting beef is gradually overcoming this handicap. Also, a new and relatively nearby market for beef in Japan has opened up. Thus, the improved market for beef, along with government policies favorable toward the development of the cattle industry of the country, indicates a bright future for the expanding beef industry of Australia. Also, it is noteworthy that the cost of beef production in Australia is much lower than in the United States.

New Zealand is a small, picturesque country about the size of Colorado. The climate is temperate with plentiful sunshine, adequate rainfall, and no great extremes of heat or cold. Year-round grazing is available on fenced holdings. To the advantage of each other, cattle and sheep share many areas, with cattle utilizing the coarser vegetation and sheep, the finer grasses and legumes.

In 1988, there were 8,062,000 cattle in New Zealand. Very few farmers devote themselves exclusively to beef production. In general, the raising and fattening of beef is carried on in conjunction with sheep farming. The dairy sector provides a large contribution from its cull cows and surplus calves.

Fig. 11–11.   Australian *jackaroos*, counterpart of American cowboys rounding up cattle in northern Queensland, Australia. They may drive them over trails that were pioneered 100 years before, when no one knew what lay beyond the next horizon—or the next. (Courtesy, Australian Consulate-General, San Francisco, CA)

## BEEF PRODUCTION IN EUROPE, INCLUDING THE ECONOMIC COMMUNITY (EC-12)

Beef production throughout Europe is largely from dual-purpose cattle, animals bred to produce both milk and meat. This poses the problem of how to increase beef output from such herds without pushing up milk surpluses.

In the original EC-6 (Belgium, France, West Germany, Luxembourg, the Netherlands, and Italy), about 45% of the beef

and veal production came from cull dairy cows and milk-fed calves. But cull dairy cows and calves in the 6 more recent member countries (the United Kingdom, Ireland, Denmark, Greece, Portugal, and Spain) account for only about 20% of the total beef and veal production. By comparison, in the United States, dairy cows and calves merely account for about 15% of the federally inspected beef and veal production.

In addition to obtaining beef from cull dairy cows, bull calves (not steers) are fattened out and slaughtered as yearlings throughout Europe.

In much of Europe, dairy cattle are selected for their beef qualities. For example, stud bulls that are widely used in artificial insemination are often selected on the basis of rate and efficiency of gain, very much as beef bulls are selected on performance test in the United States.

The EC-12 has had several programs to pay farmers to convert from dairy to beef cattle, but payments were not large enough to equal current income from milk sales. The only way EC farmers could substantially increase beef production at a profit would be through confined feeding, provided the price of feed grains were favorable.

All indications are that the EC-12 will continue to be a net importer of beef, even if local seasonal surpluses, especially the cheaper cuts, occur from time to time.

## BEEF PRODUCTION IN CANADA

Canada is still a frontier type of country with almost unlimited opportunities for expansion of the beef cattle industry. In general, Canadian cattle are noted for their size, scale, and ruggedness. This is due to the fact that in the great expanses of frontier agriculture, cattle production is on a cost-per-head rather than on a cost-per-pound basis; that is, it costs little more to produce a sizable beast than to produce a small one. The main obstacles to increased beef production in Canada are (1) the long severe winters in much of the cattle country centered primarily in the eastern and western provinces, where up to seven month's feeding is required; (2) the high duty and frequently closed borders for exports to the United States, the most natural potential market; and (3) the need for a permanent outlet for stocker and feeder cattle, as Canada has no finishing area comparable to the Corn Belt.

In 1988, Canada had 12,060,000 head of cattle. In 1987 (most recent year available), Canada exported 267,707 head of cattle, mainly feeders, to the United States.

Fig. 11–12.   Cattle roundup in Alberta, Canada. (Courtesy, *Cattlemen,* Winnipeg, Canada)

The cattle producers of Canada appear to be optimistic about the future of the industry. It is predicted that more and more cattle will be finished on the small grains which are produced in great abundance.

## BEEF PRODUCTION IN MEXICO

Mexico, with 31.2 million cattle in 1988, ranks seventh among the leading cattle countries of the world (Table 11–1).

Since January 1, 1955, Mexico has been free of foot-and-mouth disease, and the border has been open, subject to (1) Mexico's quotas, or not permitting exports at intervals, and (2) to U.S. quotas and duties.

Factors unfavorable to beef production in Mexico are (1) the ravages of parasites, particularly the Texas tick; (2) lack of improved breeding, which is made difficult because of the susceptibility of newly imported cattle to disease and parasites; (3) frequent droughts; and (4) political uncertainties and government policies unfavorable to the development of cattle units of adequate size that would permit practical and economic operation in the present era.

Despite all the difficulties now existing in Mexico, the fact remains that cattle are afforded a long grazing season, and labor is cheap and abundant. Cattle can be produced very cheaply. Also, in recent years, the better cattle producers of Mexico have made marked progress in improving both the quality of their cattle and the efficiency of their production.

Fig. 11–13. Part of a fine herd of Herefords on the ranch of Guillermo Finan, Hacienda Valle Columbia, Muzquiz, Coahuila, Mexico. This herd would be considered outstanding anywhere—in Mexico, in the United States, or in Canada. (Photo by A. H. Ensminger)

Each year, Mexico provides several thousand head of feeder cattle for growing on the ranges of the Southwest or finishing in U.S. feedlots. The United States received 940,000 feeders from Mexico in 1987. However, Mexico has a growing domestic market. Already Mexico is beginning to feel the drain of live cattle to the United States. As a result, Mexico will not be in a position to increase significantly beef or cattle exports in the near future. Rather, beef and cattle will continue to be controlled through quotas established by the government.

## BEEF PRODUCTION IN SOUTH AMERICA

Of the South American countries, Argentina, which ranks sixth in world cattle numbers, is recognized as the outstanding beef producer. In fact, taken as a whole, Argentine cattle pro-bably possess better breeding, and show more all-around beef excellence than do the cattle of any other country in the world. The excellence of Argentine cattle can be attributed to two factors—their superior breeding and the lush pastures of the country. Beginning in 1850 and continuing to the present time, large numbers of purebred animals have been imported from England, Scotland, and the United States. No price has been considered too high for bulls of the right type; and, again and again, British and American breeders have been outbid by Argentine *estancieros* in the auction rings of Europe. These bulls and their progeny have been crossed on the native stock of Spanish extraction (Criollo cattle). Today, Herefords, Angus, and Shorthorns are the most numerous breeds of the country.

Fig. 11–14. Well-bred cattle on lush pastures in Argentina.

The finest cattle pastures of Argentina are found along the La Plata River, in the region known as the Pampas, a vast, fertile plains area embracing about 250,000 square miles, which slopes ever so gently toward the sea. It's a dreamland of cattle and grass, and the "beef basket" of South America. Much of this fertile area is seeded to alfalfa upon which cattle are pastured year-round. Instead of finishing cattle largely on grains, as we do, the producers in Argentina finish their stock on alfalfa pastures. The corn of the Pampas region, which represents an acreage one-half as great as that devoted to alfalfa, is largely exported. Usually 2- and 3-year-old steers are finished by turning them into lush alfalfa pasture for a period of 4 to 8 months prior to marketing. The surplus beef of Argentina is marketed as frozen or chilled beef to the European countries, especially to Great Britain. None of the frozen or chilled beef from Argentina is admitted into the United States because of the hazard of foot-and-mouth disease; it must be canned or fully cured (i.e., corned beef).

Other South American countries of importance in beef production are Brazil, Colombia, Uruguay, and Paraguay.

Generally speaking, Brazil, which is slightly larger than the United States, produces hardy cattle of rather low quality, predominantly of Zebu breeding.

Colombia is handicapped by lack of improved breeding, poor transportation facilities, and limited refrigeration, although beef production is one of the nation's principal industries.

Uruguay, which is but little larger than the state of Missouri, is noted (1) as an ideal cattle country (because of its rich pastures, abundant water supply, and temperate climate), (2) for Hereford and Shorthorn cattle of good breeding, although they are not equal in quality to the cattle in Argentina, (3) as one of the most highly specialized beef cattle countries in the world, and (4) as a beef exporting country, despite its small size (80% of the nation's exports consist of animal products).

Paraguay, which is about 2½ times larger than Uruguay, produces cattle of similar breeding and quality to those in Brazil.

As in Argentina, year-round grazing constitutes the basis of the beef cattle industry of the South American countries. Virtually no grain is used in finishing animals, except for those being fitted for show. No attempt is made to finish steers until they are fully mature.

In general, the foremost obstacles, or unfavorable factors, affecting South American beef production are:

1. The ever-present foot-and-mouth disease, which, though seldom fatal, results in enormous economic losses through retarded growth and emaciation and limits the foreign sale of both beef and cattle on foot.

2. Droughts are rather frequent in many of the cattle sections, and they are likely to be of rather long duration.

3. Parasites and certain diseases other than foot-and-mouth disease are rather prevalent in the warmer sections.

4. Prices are very much dependent upon the export trade, thus making for an uncertain market.

5. Local markets are often unsatisfactory; modern packing plants are not plentiful; and refrigeration facilities are limited. Many of the cattle slaughtered in the more isolated areas of South America, especially in Brazil and Paraguay, are still made into jerked or salted beef.

6. Transportation facilities are few and far between.

7. Except for the cattle of Argentina and Uruguay, much improvement in breeding is needed; but the introduction of improved blood is difficult because of the heavy infestation of diseases and parasites to which the native and Zebu cattle are more resistant.

Because of the glowing reports about the cattle industry of Argentina, many young people from the United States have, from time to time, been interested in establishing a cattle enterprise in South America. Without exception, experienced U.S. cattle producers who have visited in South America in person, and who know whereof they speak, point out the almost impossible odds of success in such a venture. In the first place, the land is in the hands of a comparatively few families who hold a monopoly on the cattle industry; and, secondly, the political unrest in these countries is usually not conducive to such private foreign investments in land or cattle.

Fig. 11–15. Part of a herd of 800 head of Criollo × Brahman cattle in a corral at Santa Clara Ranch. (Courtesy, Sr. Jorge Cordero, Cordero Ranches, Guatemala City, Guatemala. Photo by A. H. Ensminger)

ly Criollo, although efforts have been made to improve their beef cattle in recent years by importing breeding stock from the United States, primarily Brahman and Santa Gertrudis.

Livestock are concentrated chiefly along the Pacific Coast, where they fatten on grass. Pasture is abundant, except during the dry season, when it is short for about three months. Some of the more progressive ranchers are (1) ensiling grass, corn, and/or sorghum, or (2) irrigating pastures to provide forage during the dry season.

Corn is high in price, because of the demand for human consumption. Whole cottonseed, cottonseed hulls, and cane molasses are relatively cheap. Rising land values are prompting interest in supplemental pasture finishing.

Among the factors **favorable** to beef production in Central America and Caribbean are: abundant grass, relatively cheap land and labor, freedom from foot-and-mouth disease, and tax deferment (up to 15 years).

Among the **unfavorable** factors are: scarce and high-priced cereal grains; heavy insect infestation—especially flies and ticks; each country controlling its live cattle exports more carefully each year (for example, since 1972 Guatemala has prohibited feeder cattle exports because of the desire to increase their weight in order to take full advantage of beef exports); little culling; and low percentage calf crop.

## BEEF PRODUCTION IN CENTRAL AMERICA AND CARIBBEAN

Central America and Caribbean comprise the tropical land mass connecting North America and South America, consisting of Costa Rica, Dominican Republic, El Salvador, Guatemala, Honduras, and Panama. On an individual basis, none of these countries produces or exports sufficient beef to be much of a factor. However, as a group, in 1988 they had 12,307,000 cattle; and in 1987, they exported 124,469,511 lb of beef and veal to the United States. Moreover, it is estimated that their exports will increase in the future.

Most of the cattle are of unimproved breeding, low in both yields and quality, but high in resistance to ticks and other environmental handicaps of the region. Cattle are predominant-

## BEEF PRODUCTION IN THE UNITED STATES

The dominant position of beef production in the United States is attested to by the following statistics:

1. It has 4.9% of the world's human population.
2. It has 8.0% of the world's cattle population.
3. It produces 22.2% of the world's beef and veal.
4. It consumes 16.6% of the world's beef and veal.

Thus, 8.0% of the world's cattle population produces 22.2% of the world's beef. This points up the tremendous efficiency of the U.S. cattle industry. Without doubt, much of this increased beef production on a per head basis has come about as a result of increased cattle feeding and decreased calf slaughter. In 1947, 3.6 million cattle were grain fed in the United States; in 1987, 26.3 million head were grain fed.

The present and future importance of beef cattle in the agriculture of the United States rests chiefly upon their ability to convert coarse forages, grass, and by-product feeds, along with a minimum of grain, into a palatable and nutritious food for human consumption. In 1987, 74% of U.S. commercial beef cattle slaughter consisted of grain-fed steers and heifers, and 26% came from nongrain-fed (grass) cattle.

The proportion of U.S. grain-fed slaughter cattle will likely decrease slightly in the future due to (1) consumer preference for lean beef, and (2) scarcer and higher priced grain (due to increased demand for grain for the increased world population). Also, there will be fewer long-fed and more short-fed cattle.

The production of beef cattle differs from that of most other classes of livestock in that the operation is frequently a two-phase proposition: (1) the production of stockers and feeders; and (2) the finishing of cattle. In general, each of these phases is distinctive to certain areas.

## AREAS OF BEEF PRODUCTION

Fig. 11–16 shows the geographic location of the nation's beef brood cows. Some 48% of the U.S. beef cows are in the West North Central (Iowa, Kansas, Minnesota, Missouri, Nebraska, North Dakota, South Dakota) and in the South and East areas (from Maryland, West Virginia, and Kentucky south and westward to and including Arkansas and Louisiana). The greatest expansion since 1950 has been in the West North Central and South and East areas.

There are some rather characteristic production practices common to each cattle area.

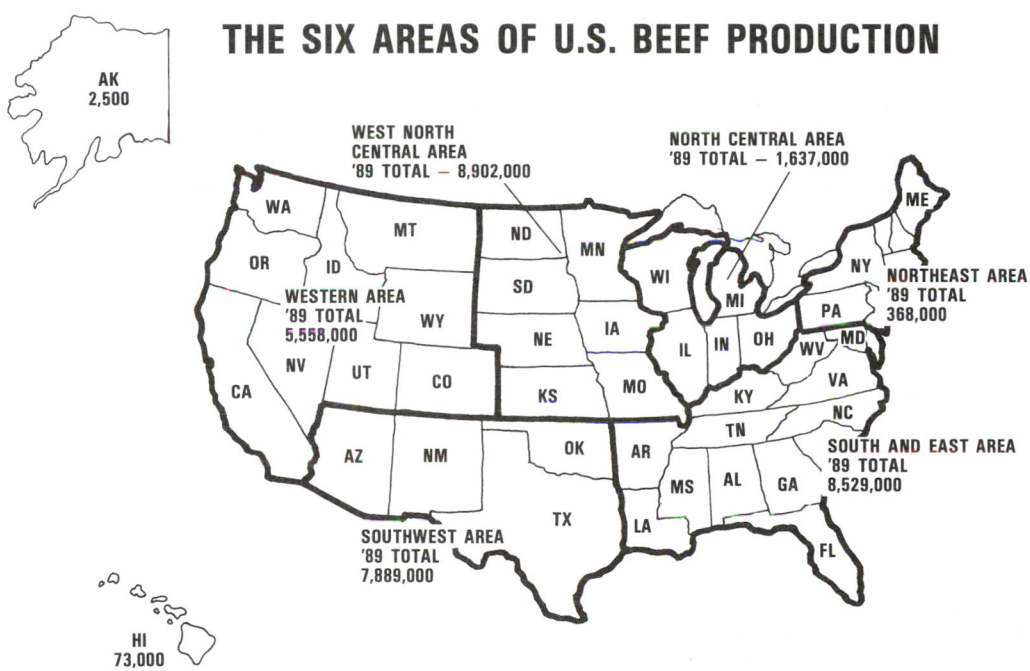

**THE SIX AREAS OF U.S. BEEF PRODUCTION**

Fig. 11–16. The six areas of U.S. beef production, and the 1989 beef cows that have calved in each area. In each area, the methods followed are determined largely by the climate, feed supply, and economic condition. (Source: USDA)

## LEADING STATES IN BEEF PRODUCTION

A ranking of the 10 leading states in beef cow numbers, together with total numbers for the United States, is given in Table 11–2. As noted, Texas is far in the lead. The large cattle numbers in the state of Texas may be attributed to the following: (1) the fact that this state represents truly great range cattle country; (2) the immense size of the state; and (3) increased cattle finishing. Oklahoma, which ranks third in cattle numbers, is also located in the Southwest Area, along with Texas. Five of the top 10 states (Kansas, Nebraska, Iowa, Missouri, and South Dakota) are in the West North Central Area.

**TABLE 11–2**
**TEN LEADING STATES IN BEEF COW NUMBERS, BY RANK, 1988[1]**

| State | No. Beef Cows |
|---|---|
| Texas | 5,260,000 |
| Missouri | 1,866,000 |
| Oklahoma | 1,842,000 |
| Nebraska | 1,680,000 |
| Kansas | 1,466,000 |
| South Dakota | 1,448,000 |
| Montana | 1,275,000 |
| Iowa | 1,201,000 |
| Florida | 1,086,000 |
| Kentucky | 1,017,000 |
| United States Total (50 states) | 32,958,000 |

[1]*Agricultural Statistics 1988*, USDA, p. 260.

## FACTORS FAVORABLE TO BEEF PRODUCTION

Some of the special advantages of beef cattle production as compared to other kinds of livestock enterprises on the farm or ranch are:

1. It utilizes land not suited for grain production.
2. It utilizes low-quality roughages.
3. It provides a profitable outlet for by-product feeds.
4. It uses homegrown feeds.
5. It provides an elastic outlet for grain.
6. It maintains fertility.
7. It requires a minimum of labor.
8. It distributes labor.
9. It requires small investment in buildings and equipment.
10. It entails little death risk.
11. It is not normally a source of pollution.
12. It produces the maximum amount of meat from milk and grass.
13. It provides flexibility, with the cow-calf operator choosing between (a) selling weaners, (b) holding calves over and selling yearlings, or (c) finishing calves or yearlings for slaughter.
14. It is suited to part-time farming.
15. It produces a nutritious product—beef.
16. It results in 54% of the consumer's beef dollar for Choice grade beef going to the producer.
17. It need not compete with people for grains.
18. It is in a favorable export situation.
19. It imparts pride of ownership.
20. It makes for a favorable balance of trade.

## FACTORS UNFAVORABLE TO BEEF PRODUCTION

Some factors which, under certain conditions, may be unfavorable to beef cattle production are:

1. It requires considerable land.
2. It requires much capital for equipment, cattle, and operation.
3. It requires fencing and water.
4. It requires considerable knowledge and management ability.
5. It isn't easy to comply with grazing regulations on public land.
6. It makes for high sire costs in boom periods.
7. It is characterized by great spread between classes and grades.
8. It cannot be expanded or liquidated quickly.
9. It propagates slowly.
10. It is subject to the hazard of such dreaded foreign diseases as foot-and-mouth disease.
11. It is inefficient in converting feed.
12. It is inefficient in converting either energy or protein.
13. It is costly to assemble feeder cattle.
14. It is not very responsive to price and cost changes.
15. It is not very responsive to technological advances.
16. It overreacts to the economy.
17. It will constantly be threatened by poultry, and by soybeans and other meat substitutes.

## FUTURE OF THE AMERICAN BEEF INDUSTRY

The author's crystal ball shows the following ahead for the U.S. beef industry:

1. **Beef will encounter increasing competition for the consumer's dollar.** The demand for beef in the United States and for export will remain strong, but the hunger for beef regardless of price has been blunted. Many people are substituting chicken, turkey, fish, and shellfish for steaks, roasts, and hamburgers. Pork will compete for the consumer's dollar on the basis of price and convenience. Poultry consumption will continue to rise, with the poultry industry capitalizing on diet and health issues and offering consumers a variety of tasty items at a very reasonable cost. The demand for fish and seafoods will continue to rise. Dairy product consumption will increase rather slowly.

2. **Biotechnology will give the beef industry a big assist, but it will take time.** It is now possible to transfer genes from one species to another without taking unwanted genes; and embryo transfer and cloning will become practical in the years ahead. But the widespread use of biotechnology will take time; it takes about 15 years from discovery to application—and even more time for widespread use.

3. **Greatest U.S. potential cow-calf expansion in the future.** The greatest future potential expansion of cow-calf operations is in the (1) West North Central, (2) North Central, and (3) South and East areas, primarily through more effective utilization of low-quality roughages, such as cornstalks, in the first two areas, and through more improved pastures and year-round grazing in the South and East.

4. **Greatest U.S. potential cattle feeding expansion in the future.** The Southern Plains and the Central Plains are potential future growth areas in cattle feeding. But future expansion in cattle feeding will be slowed because of (a) the shortened finishing period (thereby making for more annual turnover in feedlot use), and (b) consumer demand for leaner beef (with the result that more grass finished beef will be marketed). Also, the Southern Plains may encounter limitations in irrigation water and feed grains.

5. **Beef imports will pose no great threat to U.S. cattle feeders.** Limited grain feeding potential in other countries precludes any real foreign threat to U.S. cattle feeders from the standpoint of importations of high-quality beef. But we shall continue to import considerable quantities of lean, frozen, grass finished beef, known as manufacturing beef; and importation of any beef, regardless of quality, competes for the consumer's dollar.

6. **Beef shortages and high prices will goad governments to apply all conceivable, and some inconceivable, means of increasing domestic supplies.** Beef shortages and rising prices have goaded, and will continue to goad, governments to invoke all the known methods, plus some new ones, in dealing with the situation, some of which will affect the U.S. beef industry. Governments, depending on their individual position in the world of beef trade, have instituted consumer price ceilings and freezes—and higher producer price supports; consumption subsidies—and meatless days and weeks; lower tariffs—and higher export taxes; freer import quotas—and tighter export quotas; import subsidies—and export embargoes. All of these are for the purpose of increasing domestic supplies at lower consumer prices—and all are counterproductive on a worldwide basis.

7. **There will be pressure for liberalizing or removing U.S. quotas and tariffs.** American consumers will exert more and more pressure for liberalizing or removing quotas and tariffs. Also, when we are (1) eyeing exports for top grade beef, and (2) importing manufacturing beef suitable for hamburgers, hot dogs, and luncheon meats, there is less need for quotas and tariffs than in earlier years.

8. **The United States will export high quality beef.** The United States will continue to develop the high-quality portion of the Japanese beef market, because American grain-fed beef closely resembles the highly marbled Kobe beef which is so coveted in Japan.

With increased buying power in Japan and Europe, and with the price of high-quality beef in these countries higher than in the United States, it is inevitable that U.S. beef producers would like to export finished beef.

Finished beef, produced in U.S. feedlots and fabricated, packaged, and frozen in U.S. packing plants, can be transported via refrigerated jet freight and marketed in Japan and Europe at higher prices than can be secured at home.

All of the above could be realized on a free market. However, there are trade barriers in both Japan and Europe. Japan restricts U.S. beef by quotas; and the EC countries have a standard 20% duty on incoming beef in addition to a variable levy.

9. **Ever-lengthening delivery times.** A growing proportion of world beef trade will be in chilled form as modern technology—vacuum packing and temperature controlled containers—makes possible ever-lengthening delivery times.

10. **Agricultural exports will make for trade balance.** Agricultural exports will continue to be the great white hope for a favorable trade balance in the United States. Exporting a high-value product, like grain-fed beef, gives a big assist.

11. **Beef exporting countries will find it increasingly difficult to satisfy their own people.** All people are demanding more animal products. As a result, exporting countries will run into more and more difficulty satisfying their own people.

12. **We shall import more manufacturing beef.** We shall import more and more manufacturing-type beef—suitable for hamburgers, hot dogs, and luncheon meats—to meet the growing demands of the younger generation and those with moderate incomes.

13. **Mexico and Canada will limit cattle and beef exports to the United States.** Our neighbors will restrict exports of both feeder cattle and beef in order to meet their ever-increasing domestic demands.

14. **Trade barriers can make inefficient producers.** The more trade barriers (quotas, tariffs, subsidies, etc.)—the more protectionism—built around the beef industry (or any other agricultural economy), the more apt it is to become inefficient.

15. **More animal protein, especially beef, will be demanded.** Population growth and rising per capita disposable income will make for greater demand for animal protein, including beef.

16. **Beef can be priced out of the market basket.** High prices for beef will do two things: (1) decrease demand; and (2) increase competition from broilers and pork.

17. **Competition from simulated meats will increase.** The simulated meats (synthetic meats, or meat analogs) will likely become more competitive with beef in the future, as their price becomes relatively more favorable and their taste and texture are improved. To meet this type of competition, the cattle industry of the future will place increasing emphasis on the palatability and nutritive qualities of beef.

18. **Increasing health consciousness and warnings against unsaturated fats.** There will continue to be nationwide concern about nutrition, including warnings against consumption of foods high in unsaturated fats. Health faddists are telling people to eat fish instead of steak. This kind of thinking and publicity gives cause for cattle producers to keep a wary eye on longtime beef consumption trends.

19. **Inflation will increase demand for beef.** There will be continued inflation around the world, with the result that many countries will import more farm products to relieve the pressures of inflation.

20. **The energy shortage will affect beef.** Energy shortages in many parts of the world will make for a decline in gross national product, which, in turn, will lessen buying power and demand for beef.

21. **More pollution control.** People will force more and more pollution controls.

22. **Production and profits must go together.** The American farmers and feeders can and will produce more beef, providing the business is profitable. Remember that people do things which are most profitable to them. Remember, too, that cattle producers are people.

23. **Increased human population.** The population of the United States continues to expand, even though it is at a slower rate. Therefore, it is reasonable to surmise that—as has happened in the older and more densely populated areas of the world—gradually less meat per capita will become available; and more and more grains will be consumed directly as human foods. This does not mean that people of the United States are on the verge of going on an Asiatic grain diet. Rather, history often has an uncanny way of repeating itself—even though such changes come about ever so slowly. Certainly, these conditions would indicate the desirability of eliminating the less efficient animals.

24. **Beef cattle will increasingly be "roughage burners."** Beef cattle will increasingly be expected to rely upon their ability to convert coarse forage, grass, and by-product feeds, along with a minimum of grain, into palatable and nutritious food for human consumption, thereby not competing for humanly edible grains.

Fig. 11-17. Vast areas throughout the world, such as this rough terrain, are not suited to cultivation; hence, their only use is for grazing or forest. Increasingly, beef cattle will be "roughage burners." (Courtesy, American Hereford Association, Kansas City, MO)

When put into the feedlot as calves or short yearlings and long fed on a high-concentrate ration, their lifetime total feed (forage and grain combined) conversion is on the order of 10 to 1 liveweight.

In 1987, 74% of U.S. commercial beef cattle slaughter consisted of grain-fed steers and heifers, and 26% of it was grass fed. There will be less grain-fed and more grass-fed beef in the future.

25. **Increased productivity per animal unit must come.** Increased productivity per animal unit to offset higher production costs, plus the continued output of a more desirable kind of beef, will be necessary for the prosperity and survival of America's number one agricultural industry—beef production.

26. **"The tie that binds" will cross national boundaries.** There will always be political boundaries between nations. But cattle producers everywhere will continue to work together through the "tie that binds"—their love for their cattle.

# QUESTIONS FOR STUDY AND DISCUSSION

1. Discuss the origin of the word *cattle.*

2. Tell of the origin and domestication of cattle.

3. How do you account for the fact that most of the cattle in the United States are descendants of *Bos taurus* rather than *Bos indicus?*

4. Present in outline form the basic position of the domesticated cow in the zoological scheme.

5. Throughout the ages, and in many sections of the world, cattle have been used for work purposes more than horses and mules. Why has this been so?

6. Compare Robert Bakewell's breeding methods with those used in modern production testing programs. What three factors contributed most to his success as an animal breeder?

7. Recite the story of the introduction of cattle to America, and their early history in this new land.

8. Why were draft oxen more prized by colonists than beef?

9. Recite the story of the far west expansion of the cattle industry.

10. Describe the status of the U.S. beef industry in the last quarter of the twentieth century.

11. List by rank the five leading beef cattle countries of the world in cattle numbers, and give their cattle per capita and cattle per square mile.

12. Describe the beef production in each of the following countries or areas: India, Australia, New Zealand, Europe and the EC-12, Canada, Mexico, South America, Central America and Caribbean.

13. The United States has 8.0% of the world's cattle population. Yet, it produces a fantastic 22.2% of the world's beef. What's the explanation of this situation?

14. In the United States, what has caused much of the increased beef production on a per head basis?

15. The greatest expansion in U.S. cow-calf operations since 1950 has been in the West North Central and in the South and East areas. What's the explanation of the expansion in these two areas?

16. Assuming that young people had no "roots" in a particular location, in what area—(a) the West North Central area,

(b) the South and East, (c) the Southwest, (d) the Western area, (e) the North Central area, or (f) the Northeast—would you recommend that they establish a beef cattle enterprise? Justify your answer.

17. Why does Texas have such a commanding lead in beef cow numbers?

18. List 10 factors that are favorable to beef production, and discuss in detail the factor which you consider to be most important.

19. List 10 factors that are unfavorable to beef production, and discuss in detail the factor you consider to be most important.

20. Detail the type and extent of competition for the consumer's dollar that beef will encounter in the years ahead.

21. How may biotechnology give the beef industry an assist in the decades to come? Will the interval between discovery and application negate such developments?

22. What are the potential U.S. growth areas of the future for expansion of (a) cow-calf operations, and (b) cattle feeding? Why do these areas possess expansion possibilities?

23. In the years ahead, U.S. beef imports will consist primarily of manufacturing beef, whereas U.S. beef exports will consist primarily of high quality beef. Why this difference?

24. Why will agricultural exports continue to be our great white hope for a favorable trade balance?

25. Can beef be priced out of the market basket? Justify your answer.

26. Should cattle producers be concerned about increased competition from simulated meats?

27. Will more pollution control mitigate against beef production in the future?

28. Will U.S beef cattle of the future increasingly be roughage burners, and will there be less emphasis on carcass quality? Justify your answer.

29. How can U.S. cattle producers achieve increased productivity per cow unit?

30. On the whole, do you feel that the future of U.S. beef cattle production warrants optimism or pessimism? Justify your answer.

## SELECTED REFERENCES

| Title of Publication | Author(s) | Publisher |
| --- | --- | --- |
| American Cattle Trails, 1540–1900 | G. M. Brayer<br>H. O. Brayer | Western Range Cattle Industry Study and American Pioneer Trails Association, Bayside, NY, 1952 |
| Beef Cattle, Seventh Edition | A. L. Neumann | John Wiley & Sons, Inc., New York, NY, 1977 |
| Beef Cattle Production | K. A. Wagnon<br>R. Albaugh<br>G. H. Hart | The Macmillan Company, New York, NY, 1960 |
| Beef Cattle Production | J. F. Lasley | Prentice-Hall, Inc., Englewood Cliffs, NJ, 1981 |
| Beef Cattle Production | N. T. Yeates<br>P. J. Schmidt | Butterworths Pty., Ltd., Brisbane, Australia, 1974 |
| Beef Cattle Science, Sixth Edition | M. E. Ensminger | The Interstate Printers & Publishers, Inc., Danville, IL, 1987 |
| Beef Cattle Science Handbook | Edited by<br>M. E. Ensminger | Agriservices Foundation, Clovis, CA, pub. annually since 1964 |
| Beef Production and Distribution | H. DeGraff | University of Oklahoma Press, Norman, OK, 1960 |
| Beef Production and The Beef Industry: A Beef Producer's Perspective | R. E. Taylor | Burgess Publishing Company, Minneapolis, MN, 1984 |
| Beef Production in the South, Modified Edition | S. H. Fowler | The Interstate Printers & Publishers, Inc., Danville, IL, 1979 |
| Cattle and Men | C. W. Towne<br>E. N. Wentworth | University of Oklahoma Press, Norman, OK, 1955 |
| Commercial Beef Cattle Production, Second Edition | Edited by<br>C. C. O'Mary<br>I. A. Dyer | Lea & Febiger, Philadelphia, PA, 1978 |
| History of Livestock Raising in the United States, 1607–1860 | J. W. Thompson | U.S. Department of Agriculture, Washington, DC, 1942 |
| Our Friendly Animals and Whence They Came | K. P. Schmidt | M. A. Donohue & Co., Chicago, IL, 1938 |
| Principles of Classification and a Classification of Mammals, The, Vol. 85 | G. G. Simpson | American Museum of Natural History, New York, NY, 1945 |
| Stock Raising in the Northwest, 1884 | H. O. Brayer<br>G. Weis | The Branding Iron Press, Evanston, IL, 1951 |
| World Cattle, Vols. I, II, and III | J. E. Rouse | University of Oklahoma Press, Norman, OK, Vols. I and II, 1970; Vol. III, 1973 |
| Yearbook of Agriculture | | U.S. Department of Agriculture, Washington, DC |

Good cattle producers, good cattle, and good grass go hand in hand in the United States.

Cattle fashions have changed! This shows "The White Heifer That Traveled," painting by Thomas Weaver, 1811. This noted female was born about 1804. The painting shows her at 7 years of age and a weight of 2,300 lb. She was widely exhibited through England to advertise the beef-making qualities of the Shorthorn breed. (Courtesy, Harding and Harding, Geneva, IL)

# TYPES AND BREEDS
# OF BEEF/DUAL-PURPOSE
# CATTLE[1]

================================================================

# Chapter 12

================================================================

[1]Sometimes people construe the write-up of a breed of livestock in a book or in a U.S. Department of Agriculture bulletin as an official recognition of the breed. Nothing could be further from the truth, for no person or office has authority to approve a breed. The only legal basis for recognizing a breed is contained in the Tariff Act of 1930, which provides for the duty-free admission of purebred breeding stock provided they are registered in the country of origin. But the latter stipulation applies to imported animals only.

In this book, no *official* recognition of any breed is intended or implied. Rather, the author has tried earnestly, and without favoritism, to present the factual story of the breeds in narrative and picture. In particular, such information relative to the new and/or less widely distributed breeds is needed, and often difficult to obtain.

Early in the progress of cattle improvement, especially during the development of fenced holdings in England, the keepers of herds began to select certain animals for specific purposes and to plan matings with such uses in mind. Fortunately, because of the diversity of genes carried by the parent stock, it was possible, through selection, to mold certain types, each of which proved superior to the common cattle or even to other types for specific purposes. Thus, there evolved beef-type, dairy-type, dual-purpose-type, and draft-type cattle, and the several breeds of each.

## TYPES OF CATTLE

*Type may be defined as an ideal or standard of perfection combining all the characteristics that contribute to the animal's usefulness for a specific purpose.* It should be noted that this definition of type does not embrace breed fancy points. These have certain value as breed trademarks and for promotional purposes, but in no sense can it be said that they contribute to an animal's utility value. There are four distinct types of cattle: beef-type, dairy-type, dual-purpose-type, and draft-type.

Beef Cow

Dual Purpose Cow

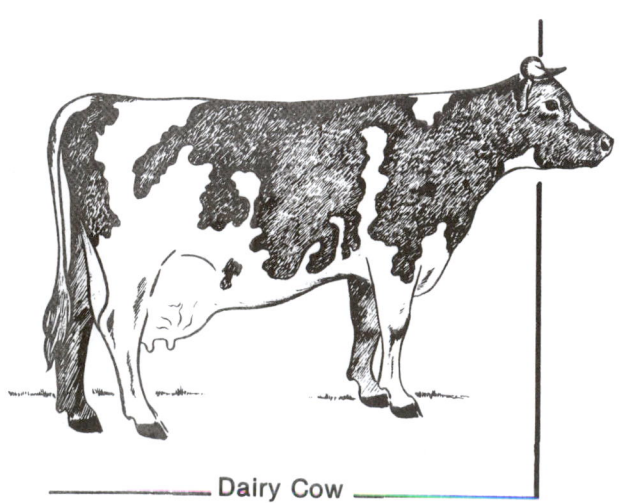

Dairy Cow

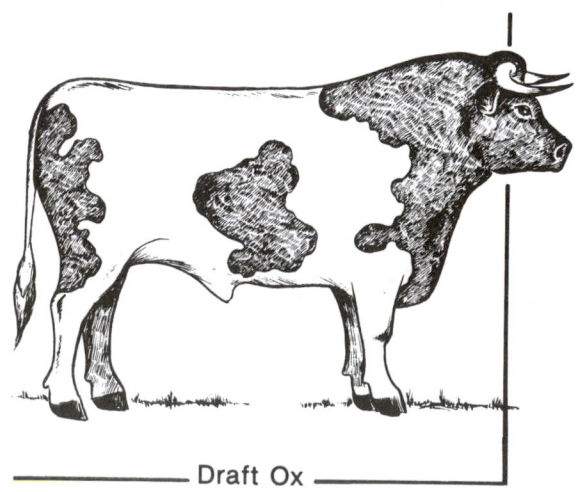

Draft Ox

Fig. 12–1. Beef-type cow (above), characterized by bred-in meat quality. Dairy-type cow (below), characterized by a lean, angular form and a well-developed mammary system.

Fig. 12–2. Dual-purpose-type cow (above), intermediate between the beef-type and dairy-type in conformation. Draft-type ox (below), characterized by great size and ruggedness with considerable length of leg.

- **Beef-type** animals are characterized by meatiness. Their primary purpose is to convert feed efficiently into the maximum of high-quality meat for human consumption.

- **Dairy-type** animals are characterized by a lean, angular form and a well-developed mammary system. Their type is especially adapted to converting feed efficiently into the maximum of high-quality milk.

- **Dual-purpose-type** animals are intermediate between the beef-type and dairy-type in conformation and also in the production of both meat and milk.

Although many breeders have the dual-purpose-type clearly in mind and although many fine specimens of the respective breeds have been produced, there is lack of uniformity in dual-purpose cattle. This is as one would expect when two important qualities, beef and milk, are being combined.

- **Draft-type** animals, when true to form, are characterized by great size and ruggedness with considerable length of leg. Although oxen are seldom seen in the United States, it must be remembered that these patient, steady, plodding beasts are still the chief source of power in many parts of the world.

On a worldwide basis, the four distinct types of cattle—beef, dairy, dual-purpose, and draft—may be found. However, with (1) the coming of the newer exotic breeds to America, and (2) the emphasis on rapid growth, milk production, and leaner beef, type differences between beef-, dual-purpose-, and draft-type cattle have become minimal. Today, there is more difference within beef breeds than between them from the standpoint of beef type.

## BREEDS OF BEEF/DUAL-PURPOSE CATTLE

*A breed may be defined as a group of animals having a common origin and possessing certain well-fixed and distinctive characteristics not common to other members of the same species; these characteristics are uniformly transmitted.* A breed may come about as a result of planned matings, or, as has been more frequently the case, it may be purely a happenstance. Once a breed has evolved, a breed association is usually organized.

The breeds of beef/dual-purpose cattle in the United States are:

| | |
|---|---|
| American Breed | Charbray |
| Amerifax | Charolais |
| Angus | Chianina |
| Ankina | Devon |
| Ankole-Watusi | Dexter |
| Barzona | Fleckvieh |
| Beefalo | Galloway |
| Beef Friesian | Gelbvieh |
| Beefmaster | Hays Converter |
| Belted Galloway | Hereford |
| Blonde d'Aquitaine | Indu Brazil (Zebu) |
| Braford | Limousin |
| Brahman | Lincoln Red |
| Bralers | Maine-Anjou |
| Brangus | Marchigiana |

| | |
|---|---|
| Murray Grey | Salers |
| Normande | Santa Gertrudis |
| Norwegian Red | Scotch Highland |
| Piedmontese | Shorthorn |
| Pinzgauer | Simbrah |
| Polled Hereford | Simmental |
| Polled Shorthorn | South Devon |
| Ranger | Sussex |
| Red Angus | Tarentaise |
| Red Brangus | Texas Longhorn |
| Red Poll | Welsh Black |
| Romangola | White Park |

## NEW BREEDS; THE EXOTICS

From the time cattle of improved breeding began to make their influence evident in North America, the majority of the genetic material used was of British origin. During the early period, the most significant changes were in the proportions of the British breeds composing the national cow herd; first came the Shorthorns, then the Herefords, and finally the Angus.

Next came the exotics; first the Brahman, followed by the Charolais, thence a whole tidal wave of new breeds. According to Webster, the word *exotic* means, "from another country; not native to the place where found. Having the appeal of the unknown—mysterious, romantic, picturesque, glamorous. Strikingly unusual in color or design." Exotic cattle were all these things—and more.

The introduction of the Zebu type of cattle in the early part of the present century significantly improved the efficiency of beef production in subtropical and desert range areas of the southern United States. Also, the Brahman was used in crossbreeding and in developing the Santa Gertrudis, Beefmaster, Charbray, and Braford breeds.

The Charolais found its way into the United States from Mexico in the late 1930s, thence spread north into Canada. In test stations, Charolais crosses demonstrated a higher lean growth rate than straightbred or crossbred British breeds. Through active promotion based on performance facts, the enchantment of a new breed, and the momentum of a new registry, the Charolais was used widely enough to be regarded as a serious threat to the established breeds.

The first Simmental was imported into Canada in 1967, and the first Limousin entered Canada in 1968.

Back of the rage for the exotics was the desire to use them in crossbreeding. Among commercial cattle producers, they engendered enthusiasm and excitement such as had not been seen in years. But among some purebred breeders of the established breeds, they produced animosities reminiscent of the range wars in the days of intruding range sheep operators and nesters. The establishment was riled because crossbreeding was being extolled, and bulls of these stark newcomers were being used. Arguments waxed hot; many old-time purebred breeders became emotional and explosive. This was surprising because the practice of crossbreeding had been common in this country since the birth of the cattle business. The commercial cattle producers had always exploited the benefits of heterosis—first by using Shorthorns or Longhorns, thence followed by Herefords and Angus. Moreover, at one time or another, all breeds of cattle in North America, including the Longhorn and British breeds, were exotics—none of them were indigenous to this country. They are no more native than North American people, except the Indians. But there were two great differences about the recent exotics: (1) They came in more quickly, and (2) they came in greater numbers. Indeed, the influx of the exotics disturbed the placid tranquility of the pastoral scene—it never will be the same.

## POPULARITY OF BREEDS

Table 12-1 shows the annual (for most recent year available) and total registrations to date of the breeds of beef/dual-purpose cattle, ranked in descending order by the most recent annual registration numbers. In these changing times, the recent annual figures are probably more meaningful than the all-time registrations, although it is recognized that one year's data fail to show trends. Of course, some breeds show limited registrations because of (1) their recent introduction to the United States, or (2) their recent origin in the United States. Also, some breed registries are inactive; still others may cease to exist in the future.

## TABLE 12-1
### REGISTRATIONS OF BEEF CATTLE IN U.S. BREED REGISTRY ASSOCIATIONS

| Breed | 1988 Registrations (or for year specified in left column) | Total Registrations to Date |
|---|---|---|
| **A**ngus | 143,520 | 11,215,047 |
| **H**ereford | 98,105 | 19,000,005 |
| **P**olled Hereford | 74,937 | 5,822,164 |
| **S**immental | 71,261 | 1,208,171 |
| **L**imousin | 53,165 | 623,672 |
| **B**eefmaster[1] | 40,723 | 447,250 |
| **C**harolais | 38,732 | 1,695,865 |
| **B**rangus | 27,742 | 442,006 |
| **S**horthorn and Polled Shorthorn (1987) | 19,557 | 3,864,132 |
| **S**alers | 18,482 | 70,871 |
| **S**anta Gertrudis | 18,003 | 653,421 |
| **G**elbvieh | 17,545 | 204,700 |
| **B**rahman | 16,483 | 968,071 |
| **R**ed Angus | 14,004 | 350,000 |
| **T**exas Longhorn | 13,092 | 159,834 |
| **S**imbrah | 8,850 | 58,554 |
| **T**arentaise | 7,000 | 70,500 |
| **C**hianina | 6,984 | 192,016 |
| **M**aine-Anjou | 4,910 | 103,450 |
| **R**ed Brangus (1989) | 4,575 | 62,000 |
| **B**raford (1989) | 2,259 | 22,000 |
| **B**arzona | 1,638 | 24,103 |
| **R**ed Poll (1989) | 1,493 | 183,844 |
| **P**inzgauer | 1,280 | 25,238 |
| **B**raler | 1,191 | 5,388 |
| **A**merifax (1989) | 755 | 9,204 |
| **W**hite Park | 408 | 1,604 |
| **A**nkina (1983) | 325 | 2,465 |
| **D**exter | 269 | 3,830 |
| **M**archigiana | 239 | 5,043 |
| **D**evon | 149 | 59,626 |

[1]Beefmaster Breeders Universal 1988 registrations 35,485, total registrations 375,250; Foundation Beefmaster Assn. 1988 registrations 5,238, total registrations 72,000.

**Note well:** Where more than one registry association exists for a given breed, some animals of that breed are generally registered in more than one registry association; hence, when breed registry numbers are combined, registration numbers will likely exceed numbers of animals actually registered.

## CHARACTERISTICS OF BREEDS

The characteristics of the different breeds of beef/dual-purpose cattle are summarized in Table 12-2. The future of all breeds will be determined primarily by (1) how well they perform (their efficiency in producing quality beef), and (2) their promotion. Most of the breeds listed in Table 12-2 will be around for a very long time, but some will fall by the wayside.

# CATTLE FASHIONS HAVE CHANGED!

*Firly,* a prize ox of Britain in 1835, shown at 4 years and 8 months of age and weighing 3,000 lb. The near animal is the growthy, meaty, productive, useful type in vogue today.

Representative animals of the 54 breeds that fit the mold and standards of today's producers and consumers are pictured in Figs. 12-3 to 12-56 which follow on pages 337 to 345.

**TABLE 12-2**
**BREEDS OF BEEF/DUAL-PURPOSE CATTLE AND THEIR CHARACTERISTICS**

| Breed | Place of Origin | Color | Distinctive Head Characteristics | Other Distinguishing Characteristics | Disqualifications; Comments |
|---|---|---|---|---|---|
| **American Breed** (½ Brahman, ¼ Charolais, ⅛ buffalo, ¹⁄₁₆ Hereford, and ¹⁄₁₆ Shorthorn. The name, according to Art Jones, the breed founder, reflects the mixture of breeds from which the American breed originated. (See Fig. 12-3) | On Art Jones' Cactus Ranch, Portales, NM. | Color characteristics of the breed defy description. Some of the cattle are white like their Charolais ancestors, whereas others are dark or marked with various color patterns, showing their buffalo, Hereford, and Shorthorn background. | The head is not distinctive. | The American Breed is moderate in size; mature bulls weigh from 1,800 to 2,400 lb *(818 to 1,091 kg)*, and mature cows from 1,100 to 1,500 lb *(500 to 682 kg)*. The American Breed was selected for (1) doing well on alkaline range grass, (2) ability to travel long distances for water, (3) high fertility, (4) easy calving, (5) high percentage calf crop of small calves with high weaning weights, (6) high natural immunity to most diseases and parasites, and (7) carcass cutability and quality. | Lack of performance is the only disqualification. Art Jones, founder of the breed, considers the infusion of ⅛ buffalo blood to be one of the most important ingredients in the mixture of breeds from which the American Breed originated. |
| **Amerifax** (⅝ Angus, ⅜ Beef Friesian) (See Fig. 12-4) | U.S.; in Kansas, Nebraska, South Dakota, and Wyoming. | Solid black or red. | Polled. | Moderate mature cow weight of about 1,000 lb *(454 kg)*. | Amerifax must pass an inspection for registration. Name stands for American Friesian × Angus cross. |
| **Angus** (See Fig. 12-5) | Scotland; in the northeastern counties of Aberdeen, Angus, Kincardine, and Forfar. | Black. | Polled. | Comparatively smooth coat of hair. Somewhat cylindrical body. | Horns, scurs, or buttons. Red color. A noticeable amount of white above the underline, or in front of the navel, or on one or more legs. Calves from females less than 18 mo. of age when calf was dropped, or from bulls less than 9 mo. of age at the time of service. |
| **Ankina** (⅝ Angus, ⅜ Chianina) (See Fig. 12-6) | Clayton, Ohio. The Ankina Breeders, Inc., was founded in 1975. | Black or dark brown. | Polled; scurs acceptable. | | Any color other than black or dark brown; horns; failing to meet minimum performance requirements; genetic defect in sire or dam. |
| **Ankole-Watusi** (See Fig. 12-7) | Ankole district of Uganda, also Rwanda, and Kenya, Africa. Imported to the U.S. in 1960 from Sweden to the Catskill Game Farm, NY. The Ankole-Watusi International Registry was formed in 1983. | Dark red, black, white, gray, brown, yellow dun, also spotted with small spots or large white splashes. | Ankole-Watusi cattle have the largest horns of any cattle in the world. They have large, uprising, outswept horns, 5 ft *(1.5 m)* long and 16 in. *(41 cm)* in diameter. Up to 6 ft *(1.8 m)* between horn tips. Head long and face straight or slightly dished. | Slight hump; moderate dewlap; good depth of body; strong topline; moderate to fine bone. | A variety of deleterious genetic recessives, such as arthrogryposis, double muscling, hydrocephalus, dwarfism, syndactylism, cryptorchidism; polled animals. |
| **Barzona** (¼ each Africander, Hereford, Angus, and Santa Gertrudis) (See Fig. 12-8) | U.S.A., on the Bard Ranches of Kirkland, Arizona; hence, the name *Barzona* (a contraction of *Bard* and *Arizona*). The foundation of the breed, which was laid in 1942, consisted of Africander, Hereford, Santa Gertrudis and Angus. The Barzona Breeders Assn. of America was founded on March 28, 1968. | Red. | Horned, long head, straight profile. | Well adapted to arid and semi-arid ranges (the Bard Ranches where the breed was developed, have an average rainfall of 12.76 in. *[80 cm]*). In the formative stage of the breed, production records were maintained and rigid selectivity was carried out on fertility, rate of gain, and mothering characteristics. Disqualifications: Any heritable defects or deformities; solid black color; white on bull other than underline, white on cow other than head or underline. | The breeding program followed by Bard Ranches to create the Barzona breed consisted of (1) forming a large genetic pool by crossing breeds, then breeding within the herd, and (2) using records to eliminate undesirable genes and retain desirable genes. |
| **Beefalo** (⅜ buffalo [bison] and ⅝ domestic cattle) (See Fig. 12-9) | U.S.A. by D. C. Basolo, Tracy, CA. It is claimed that the first fertile Beefalo bull was produced in 1966. | Variable. | The head is not distinctive. | The breed is promoted for its foraging ability, adaptability, hardiness, calving ease, low maintenance cost, and long life. | Earlier bison × domestic cattle crosses in Canada were not successful. Foundation sires are those developed by Basolo and those approved by the Association. The Association registers full-blooded American Beefalo and percentage of offspring. |
| **Beef Friesian** (See Fig. 12-10) | U.S.A.; based on dual-purpose Friesians brought from Europe, beginning with an importation from Ireland in 1971. In Europe, the Friesian has always been a dual-purpose animal, whereas the American descendent, the Holstein-Friesian, has been developed exclusively as a dairy breed. The Beef Friesian Society was organized in 1973. | Black and white. Beef Friesians × Angus cattle are generally black. | Horned or polled. Broad muzzle, open nostrils, strong jaw, broad and moderately dished forehead, straight bridged nose. | Rate and efficiency of gains comparable to the exotics; little calving difficulty; good milking ability. | Beef Friesians were developed by 3 approaches: (1) from purebred Beef Friesians, based on European stock; (2) through crossing Beef Friesian bulls on Holstein females, thence grading up; and (3) from Beef Friesian × Angus crosses. |

*(Continued)*

**TABLE 12-2** *Continued*

| Breed | Place of Origin | Color | Distinctive Head Characteristics | Other Distinguishing Characteristics | Disqualifications; Comments |
|---|---|---|---|---|---|
| **Beefmaster** (approx. ½ Brahman, and ¼ each Shorthorn and Hereford) (See Fig. 12–11) | U.S.A.; Lasater Ranch, Falfurias, TX, beginning in 1908. In 1931, Tom Lasater took over his father's herd and carried forward. In 1949, he moved the herd to its present location near Matheson, CO. | Red is the dominant color, but color is variable and disregarded in selection. | The majority are horned, although a few are naturally polled. | Good milk producers under range conditions; heavy weaning weights. | In order that each Beefmaster may be permanently identified with the breeder thereof, breeders must use a prefix name such as "Jones Beefmaster," etc., to designate their cattle. Thus, in a unique way, the responsibility for the continued improvement of the breed is placed squarely upon the individual breeder. |
| **Belted Galloway** (See Fig. 12–12) | Scotland; in the southwestern district of Galloway. First imported into the U.S. in 1950. The Belted Galloway Society, Inc. was established in 1962. | Black with a brownish tinge, or dun; with a white belt completely encircling the body between the shoulders and the hooks. | Polled. | Striking white belt; heavy coat of hair. | Red color; incomplete belt, other white marks, or scurs. |
| **Blonde d'Aquitaine** (See Fig. 12–13) | Southwest France; in 1961, when three French strains of similar background—Garonne, Quercy, and Pyrenee—combined. Also, there were infusions of Shorthorn, Charolais, and Limousin blood. The American Blonde d'Aquitaine Assn. was established in 1973. | Yellow, brown, fawn, or wheat colored. | Horned. | The breed is long bodied, long rumped, and relatively fine boned. There is little calving difficulty, due to the width and shape of the pelvis. The breed is considered to be large. | In France, Blonde d'Aquitaines are usually performance and progeny tested. Generally, the top third of the bulls in a performance test are subsequently progeny tested. |
| **Braford** (approx. ⅝ Hereford and ⅜ Brahman) (See Fig. 12–14) | U.S.A., on Adams Ranches, Fort Pierce, FL, beginning about 1948. Breed registry formed in 1973. Evolved from crossing Brahmans and Herefords. | Red or brindle, with white markings on the head and pigmentation around the eyes. | Both horned and polled strains. | Short haired; heat tolerant; only a slight hump; fertile; good milk production. Mature bulls weigh 1,500 to 2,000 lb *(682 to 909 kg)* and cows 1,000 to 1,500 lb *(454 to 682 kg)*. | Offspring cannot be registered if they are from cows that have not calved annually, that required veterinary assistance at calving, or that have bad udders. For registration, the Association requires pedigree, performance records, and that the animal pass an inspection. |
| **Brahman** (See Fig. 12–15) | Brahmans are the sacred cattle of India. In India, 30 or more breeds or varieties of *Indicus* cattle exist. In the U.S., the Brahman is an amalgamation of several Indian types, probably with a small infusion of European breeding. | Gray or red preferred; either solid color, or a gradual blending of the two. However, there are brown, black, white, and spotted Brahmans. | Drooping ears, and long face. | Prominent hump over the shoulders. An abundance of loose, pendulous skin under the throat and along the dewlap. A voice that resembles a grunt rather than a low. | Brindle grulla (a smutty or blackish red), or albino color. Cyrptorchid bull. Freemartin heifer. Inherited lameness. Dwarf or midget characteristics. Brahman are well adapted to hot, insect-infested areas, and to sparse vegetation. |
| **Bralers** (⅝ Salers, ⅜ Brahman) (See Fig. 12–16) | In the United States, on R-NOL-D Farms, Brenham, TX. | Dark mahogany. | Horned. Long head, straight profile. | Efficient feed utilization, high fertility, and ease of calving. | Breed rules and disqualifications are the same as for Brangus. |
| **Brangus** (⅜ Brahman and ⅝ Angus) (See Fig. 12–17) | U.S.A.; on Clear Creek Ranch, Welch, OK, owned by Frank Buttram, beginning in 1942. | Black. | Polled, with evidence of Brahman influence. | Slight crest over the neck. Smooth, sleek coat. | Horns; any color other than black; white ahead of navel; small for age; extremely nervous; too fine boned; too long ears and too loose hide; or thin hided and short hair coat. |
| **Charbray** (¾ Charolais × ¼ Brahman to ⅞ Charolais × ⅛ Brahman) (See Fig. 12–18) | U.S.A.; in the Rio Grande Valley of Texas, beginning in the late 1930s. From Charolais × Brahman crosses. | Light tan at birth, but usually change to a creamy white in a few weeks. | | A slight hint of the Brahman dewlap remains. The Charbray has the growth thrust of the Charolais and the heat-insect tolerance of the Brahman. | Purebred Charbrays must have a minimum of ⅝ to a maximum of ⅞ Charolais breeding; and the balance must be Brahman (Zebu) breeding. |
| **Charolais** (spelled Charollais in France) (See Fig. 12–19) | France; in the province of Charolles, in Central France. Later in the province of Nivernais. Breed society founded in France in 1887. Two earlier associations merged in 1957 under the current name American International Charolais Association. | Light tan at birth. Changes to cream white in a few weeks. | Horned. | Pink skin and mucous membranes. Noted for large size, growth thrust, and bred-in red meat. | The Association disqualifies any animal that (1) has a dark nose, (2) is spotted, or (3) has excessive dark skin pigmentation. |

*(Continued)*

**TABLE 12-2** *Continued*

| Breed | Place of Origin | Color | Distinctive Head Characteristics | Other Distinguishing Characteristics | Disqualifications; Comments |
|---|---|---|---|---|---|
| **Chianina** (pronounced *Key-a-nee-na*) (See Fig. 12-20) | Central Italy; in the Chiana Valley (from which they take their name), in the province of Tuscany. Of very ancient origin, going back to the days of the Roman Empire, when they were used for draft. The Chianina first came to the United States via Canada. The American Chianina Association began registering cattle in 1972. | Porcelain white hair, black switch, and dark skin. Calves are born tan colored, which gradually turns to white at about 60 days of age. | Horned. Narrow head with black pigmentation around eyes, black tongue, and black nose and palate. | The largest breed of cattle in the world. Mature bulls stand about 6 ft *(1.83 m)* (18 hands) high at the withers and weigh up to 4,000 lb *(1,814 kg)*. Mature cows weigh up to 2,400 lb *(1,089 kg)*. The breed is also noted for trimness of middle; fineness of head, horn, and bone; absence of excessive dewlap and brisket; and being poor milkers. | Calving difficulties are infrequent, perhaps due to the rather small heads and long, narrow bodies of the newborn. The growth rate and leanness of the breed give Chianina bulls an important role as a terminal cross in a crossbreeding program. |
| **Devon** (See Fig. 12-21) | England; in the counties of Devon and Somerset. Red cattle from the Devon area were brought to the Plymouth Colony in America in 1623. Vol. 1 of the *Devon Record* was published in 1881. | Red. A rich dark red is preferred; hence the name *Ruby Red*. | Creamy white horns with black tips. Also, there are polled strains. | Switch varies from whitish red to nearly white at tip. Skin is orange-yellow with pigment especially noticeable around eyes and muzzle. | Double muscling. Dwarfism. Excessive white color. |
| **Dexter** (See Fig. 12-22) | Ireland; in the southern and southwestern parts. These animals were named after their founder, a Mr. Dexter. | Black or red. | Horned. Head is rather long. | Small size and short legs, smallness accentuated by shortness of legs from knees and hocks down. Mature bulls should not exceed 1,000 lb *(454 kg)* and mature cows 800 lb *(364 kg)*. Some mature animals are less than 40 in. *(102 cm)* high. | Animals having white other than on the belly, switch, udder, or scrotum are disqualified for registry. Bulldog calves, a lethal condition, occurs in some animals of the breed; but it is a rarity. |
| **Fleckvieh** (German Simmental) (See Fig. 12-23) | Southern Germany; where it has been bred since 1895. Evolved from Simmental cattle, which originally came from Switzerland. | Generally red and white spotted, with a white face. The red varies from dark to a more common diluted, almost yellow shade. | Horned. | In Germany, it is considered a dual-purpose breed, with emphasis on beef. Mature bulls average about 2,550 lb *(1,157 kg)* and cows about 1,550 lb *(703 kg)*. | The progeny testing and selection program of the Fleckvieh breed in Bavaria is, without doubt, the best in the world. To be selected for licensing as a dam of a herd sire, a cow must be in the top 8% of the breed on milk production, must be classified by a committee for size and conformation, and must meet rigid standards for calving intervals, calving ease, milking ease, disposition, and pedigree. |
| **Galloway** (See Fig. 12-24) | Scotland; in the southwestern province of Galloway. In 1878, the Galloway Cattle Society of Great Britain was organized. The Galloway was first imported into Canada in 1853. The first importation of registered Galloways into the U.S. came via Canada in 1866. The American Galloway Breeders' Assn. was established in 1882. | Black, dun, belted, white, or red. | Polled. | Thick, wavy hair; hardiness and ability to rustle in cold weather. Easy calving. | Scurs or horns. |
| **Gelbvieh** (German Yellow) (See Fig. 12-25) | Germany; in Bavaria. Descended from red-brown Keltic-German Landrace, on which Simmental and Shorthorn were crossed in early 1800s. Actually came into being when 4 breeds of German cattle—Franconian, Glan-Donnersberg, Lahn, and Limpurg—amalgamated around 1920. In 1971, the American Gelbvieh Association was established. Gelbvieh semen was first imported to North America in 1972. | Golden red to rust. Solid color. | Horned. | Large, long-bodied, well-muscled, fast-gaining, and high-quality carcasses. Mature bulls weigh from 2,300 to 2,800 lb (average 2,500), *(1,043 to 1,270 kg [average 1,134])*, and cows 1,400 to 1,800 lb (average 1,500) *(635 to 816 kg [average 680])*. | In Germany, no Gelbvieh bull is put into general A.I. service until he is 6 years old and his progeny have proven him superior to his contemporaries. Performance data are required for registration. |
| **Hays Converter** (See Fig. 12-26) | In Canada; by the former Minister of Agriculture, Harry Hays, beginning in 1957. Foundation breeds were Hereford, Brown Swiss, and Holstein. Claimed that the breed converts feed into profit, hence the name *Converter*. | Predominant color is black with a white face, white feet, and a white tail. About 30% are red with white faces. Color is not a factor in selection. | | Traits upon which Sen. Hays built the breed are: (1) growth; (2) fertility; (3) minimum calving problems; (4) well-attached udders; (5) abundant milk; (6) sound feet and legs; and (7) pigmentation. The Hays Converter is a beef breed. Mature bulls weigh about 2,200 lb *(999 kg)* and cows 1,400 lb *(636 kg)*. | The system and steps used to produce the breed were: (1) selected from different breeds the important characteristics needed; (2) combined the genes into one large breeding population; (3) selected intensely for important characteristics and culled ruthlessly for several generations; and (4) measured the resulting animals after hybrid vigor was no longer important, |

Only performance tested animals are used.

to determine the transmissible genetic superiority.

**TABLE 12–2** *Continued*

| Breed | Place of Origin | Color | Distinctive Head Characteristics | Other Distinguishing Characteristics | Disqualifications; Comments |
|---|---|---|---|---|---|
| **Hereford** (See Fig. 12–27) | England; in the county of Hereford. The first importation of Hereford cattle into the U.S. was made by Henry Clay of Kentucky, in 1817. | Red with white markings; white face and white on the underline, flank, crest, switch, breast, and below the knees and hocks. White back of the crops, high on the flanks, or too high on the legs is objectionable. Likewise, dark or smutty noses and red necks are frowned upon. | Horned. The white face is the distinctive trademark of the breed. | A thick coat of hair. | Calves from females less than 21 mo. of age when calf was dropped, or from bulls less than 12 mo. of age when service producing the calf occured, cannot be registered. |
| **Indu-Brazil** (Zebu) (See Fig. 12–28) | Brazil. | Light gray to silver gray; dun to red. | Metrical horns drawing upward and to the rear. Prominent forehead and long, drooping ears. | Prominent hump over the shoulders. An abundance of loose, pendulous skin under the throat and along the dewlap. A voice that resembles a grunt rather than a low. | Brindle color combinations. White markings on the nose or switch. Absence of loose, thick, mellow skin. Weak and improperly formed hump. |
| **Limousin** (See Fig. 12–29) | Southwestern France; in the 19th century. Presently, found in largest numbers around Limoges. The breed takes its name from the Limousin Mountains. First imported to North America in 1969. The North American Limousin Foundation was started in 1968. | Wheat to rust red. | Horned. | Modern meat-type cattle—long, relatively shallow, with moderate to heavy muscling. Mature bulls average about 2,400 lb *(1,089 kg)*, and cows about 1,300 lb *(590 kg)*. Noted for ease of calving and high carcass quality. | The Limousin is one of the new European breeds raised primarily for meat production, rather than the dual-purpose of meat and milk. |
| **Lincoln Red** (See Fig. 12–30) | England; in Lincolnshire—the rugged coast in England. The name *Lincoln Red* was acquired in 1960. | Deep cherry red, with occasional white markings. | There are both horned and polled strains, with the polled predominating in their native land. | A long body; light birth weights and ease of calving; pigmentation; excellent milk production; fast growth rate; and good fertility. In England, they are a dual-purpose breed, with separate strains for beef and dairy. | It appears that the main use for the breed in America will be in crossbreeding to produce $F_1$ females, since they make excellent brood cows. |
| **Maine-Anjou** (See Fig. 12–31) | Western France; in the provinces of Maine and Anjou, from which it takes its name. The French Herd Book was established in 1919. The Maine-Anjou first came to Canada, and shortly thereafter to the U.S. In 1972, the first Maine-Anjous were registered in the U.S. | Dark red with white underline, often with small white patches on the body. Also, dark roans are found. | Most heads are either red, or the eyes are surrounded by red. | The Maine-Anjou is the largest of the French breeds. Mature bulls weigh 2,500 lb *(1,134 kg)* or above and cows 2,000 lb *(907 kg)* or more. Considered a dual-purpose breed, with emphasis on beef. Cows average about 5,000 lb *(2,268 kg)* of milk per lactation, testing 3.7%. They are long, rather up-standing, have a particularly long rump, and are noted for rapid growth. | The logical place for Maine-Anjou in American Crossbreeding systems is as maternal sires, although use as terminal sires may occur to some extent. |
| **Marchigiana** (pronounced *Mar-key-jayna*) (See Fig. 12–32) | Italy; in the Marche region, around Rome. With the fall of the Roman Empire in the 5th century, nomadic cattle were crossed with the two native Italian breeds of the time—the Chianina and the Romagnola. Out of these crosses evolved the basic foundation stock for the Marchigiana. The *Herd Book* was established in Italy in 1957. The American International Marchigiana Society was founded in 1973. | Grayish white, although bulls may be darker. Dark skin pigmentation, and dark muzzle, switch, and below or around the eyes. Calves are born tan, but turn white at about 2 months of age. | Horns that appear small in proportion to the size of the cattle. | Ability to do well under adverse conditions. Mature bulls weigh 2,650 to 3,100 lb *(1,203 to 1,407 kg)*. Mature cows weigh from 1,400 to 1,800 lb *(636 to 817 kg)*. | In Italy, Marchigianas have been very popular in crossbreeding programs with dairy cattle. |

*(Continued)*

**TABLE 12–2** *Continued*

| Breed | Place of Origin | Color | Distinctive Head Characteristics | Other Distinguishing Characteristics | Disqualifications; Comments |
|---|---|---|---|---|---|
| **Murray Grey** (See Fig. 12–33) | Australia; from a mating of a very light roan (almost white) Shorthorn cow and an Angus bull, first made by the Sutherlands on *Thologolong* in Murray Valley, near Wodonga, Victoria, Australia, in 1905. Because of the use of Angus bulls following the first cross, the Murray Grey is predominantly Angus. The Murray Grey Beef Cattle Society of Australia was formed in 1962; and the American Murray Grey Assn., Inc., was organized in 1970. | Silver-gray color, which adapts them to sunny areas, as well as colder areas. | Polled. | Ease of calving, because of small calves at birth; dark skin pigmentation, which lessens cancer eye; superior carcass; good dispositions. Bulls weigh around 2,000 lb *(909 kg)* at maturity, and females from 1,100 to 1,300 lb *(500 to 590 kg)*. | In the American Murray Grey Assn., Inc., females with ⅞ Murray Grey blood can be registered. Bulls are eligible for registry with ¹⁵/₁₆ Murray Grey blood. In addition, Recordation Certificates can be obtained on any crosses of ½ or more Murray Grey breeding. A ranch prefix (the owner's last name, the ranch name, or whatnot) is required of each breeder. |
| **Normande** (See Fig. 12–34) | France; in the area of Normandy, Brittany, and Maine. The breed was established in 1883. The American Normande Assn. was formed in 1974. | Primarily dark red and white. Colored patches around the eyes give them a "bespectacled" appearance and resistance to cancer eye and pinkeye; and dark pigmentation on the udder prevents sunburn. | Bespectacled eyes, due to dark coloring. | The Normande is known as a dual-purpose breed in France. Mature bulls in good condition average about 2,425 lb *(1,101 kg)* although weights up to 2,850 lb *(1,294 kg)* have been reported. Cows weigh 1,550 to 1,750 lb *(704 to 795 kg)* and produce an average of about 8,800 lb *(3,895 kg)* of milk per year. | For purebred registration, females cannot have less than ⅞ Normande blood and bulls not less than ¹⁵/₁₆. |
| **Norwegian Red** (See Fig. 12–35) | Norway, where they are known as Norwegian Red-and-Whites. The first importation of the breed into the U.S. was made by the Southern Cattle Corporation, Memphis, TN, in 1973. | Red; red and white. | Horned. | Abundant milk production; excellent feed conversion; and good carcasses. Mature bulls weigh from 2,200 to 2,640 lb *(999 to 1,199 kg)* and mature cows from 1,210 to 1,430 lb *(549 to 649 kg)*. | In Norway, the Norwegian Red-and-White is a dual-purpose breed, kept for both milk and beef. Today the Norwegian Red-and-White is the dominant breed of Norway; there are only 200 cattle of other breeds. It is noteworthy, too, that 58% of the cattle of Norway are registered purebreds. |
| **Piedmontese** (Piedmont) (See Fig. 12–36) | Italy, where they are the most popular breed. | White or pale gray with black points. | | About 80% of the bulls of the Piedmontese breed are double muscled to some degree. In Italy, Italian breeders report 9% higher dressing percentage and twice the steaks from double muscled cattle over normal cattle. Piedmontese cattle command a very considerable premium on the Italian market. Originally, the Piedmontese was considered to be a dual-purpose breed in Italy, but today it is selected and bred for beef qualities. | In Italy, under their system of intensive care, breeding and calving problems of double-muscled Piedmontese cattle do not appear to be serious. Double-muscled Piedmontese cattle mean to the cattle industry of Italy what broad-breasted turkeys and Cornish cross broilers mean to the U.S. poultry industry; all are meat producers *par excellence*. |
| **Pinzgauer** (pinzgau) (See Fig. 12–37) | In the Pinz Valley of Austria, and in adjacent areas of Italy and Germany; in the Alpine region. The American Pinzgauer Association was established in 1973. | Chestnut brown sides, with a white top-line and underline, and usually white feet. Deep orange pigment around eyes and on udder. | Horned. | A "beefy" breed, more so than most of the exotics, although it is classed as a dual-purpose breed in its native land. Mature bulls weigh 2,200 to 2,900 lb *(998 to 1,315 kg)*, and mature cows from 1,300 to 1,650 lb *(590 to 748 kg)*. Breed is noted for hardiness, longevity (the oldest cows and bulls reach 17 to 18 years of age), fertility and foraging ability. | In Austria, all animals are subjected to, and must pass, a rigid conformation test before they can be registered. Additionally, performance of dams and daughters in milk production and butterfat content is a criterion in the selection of breeding bulls. |
| **Polled Hereford** (See Fig. 12–38) | U.S.A.; in Iowa, by Warren Gammon, from a mutation (polled) that appeared in the horned Hereford breed. In 1901, Gammon contacted 2,500 members of the American Hereford Association to locate some naturally hornless purebred Herefords. Eventually, he located 10 females and 1 bull which became the foundation for the Polled Hereford. | Red with white markings, white face and white on underline, flank, crest, switch, breast, and below the knees and hocks. White back of the crops, high on the flanks, or too high on the legs is objectionable. Likewise, dark or smutty noses are frowned upon. | Polled, with a white face. | of age at the time of conception, and its dam at least 20 mo. of age at the time of calving. In the formative period of the breed, Polled Herefords recorded in both The American Hereford Assn. and the American Polled Hereford Assn. were known as double standard. Those that could be recorded only in the American Polled Hereford Assn. were called single standard. This distinction is no longer made, and present-day Polled Herefords trace their ancestral lines to the original English herd books of Hereford cattle. | Horned animals are disqualified. No calf is eligible for registration unless its sire was at least 10 mo. |

*(Continued)*

**TABLE 12–2** *Continued*

| Breed | Place of Origin | Color | Distinctive Head Characteristics | Other Distinguishing Characteristics | Disqualifications; Comments |
|---|---|---|---|---|---|
| **Polled Shorthorn** (See Fig. 12–39) | U.S.A.; in the north-central states, chiefly Ohio and Indiana. | **R**ed, white, or any combination of red and white. **A** smutty nose or dark nose is objectionable. | **P**olled. | **O**ther than being polled, they resemble Shorthorns, except there are more spotted animals among them. | **H**orned animals are disqualified. **U**ntil 1919 Polled Shorthorns were known as Polled Durhams, because most of these early polled individuals came from the Shorthorn Mulley cross (*Mulley* means hornless). Strains of |
| | | animals developed from this cross were called single standard animals, because they were eligible for registry in the American Polled Shorthorn Herd Book, but not in the American Shorthorn Herd Book. Today, all Polled Shorthorns are double standards; that is, they are the polled offspring of parents both of which are registered in the American Shorthorn Herd Book. **A**nimals must be under 18 mo. of age at the time of registration; otherwise, they must be approved by the Executive Committee. **B**oth Shorthorns and Polled Shorthorns are recorded in the same Herd Book. | | | | |
| **Ranger** (See Fig. 12–40) | U.S.A.; beginning in 1950, on the following three ranches: Barnes Livestock Company, Riverton, WY; W. W. Ritchie and Family, Buffalo, WY; and Watson Cattle Company, Cedarville, CA. | **T**hey run the gamut of cattle colors, including both solid and broken colors. | | **M**edium size; hardy; fertile—animals have been selected to calve at an early age, at yearly intervals or less, and without adequate and persistent milk production; heavy weaning weight; good carcass quality. | **T**he developers of the breed refer to it as, "a 'cow' breed for the producer who must have a profitable commercial operation." |
| | **T**he following breeds were used in developing the Ranger: Hereford, Milking Shorthorn, Red Angus, Shorthorn, Beefmaster, Scotch Highland, and Brahman. **T**he name *Ranger* was selected because the breed was developed on, and is adapted to, the range areas of the west. | | | | |
| **Red Angus** (See Fig. 12–41) | **S**cotland.[1] The Red Angus evolved as a result of a specific color in an established breed. | **R**ed. | **P**olled. | **S**imilar to black Angus, except for recessive red. | **D**isqualifications include any of the following: horns, scurs, or any hornlike growth; white any place other than on underline; dwarf; double muscle or other genetic defects. |
| **Red Brangus** (See Fig. 12–42) | U.S.A.; on Paleface Ranch, Spicewood, TX, from Brahman × Angus cross (about 50% of each), made in 1946. **R**egistry chartered in 1956. | **R**ed pigmentation around the eyes. | **B**road head with slightly curved forehead and straight profile; with medium sized, moderately drooping ears. | **M**ales have crest immediately forward of the shoulders. Smooth, sleek coat. **M**ature bulls weigh 1,800 to 2,200 lb *(816 to 998 kg)* and cows 1,200 to 1,400 lb *(544 to 635 kg)*. | **H**ernia, one testicle, or malformed genitals; small infantile vulva; wry nose, wry tail, double muscling, undershot jaw, overshot jaw, dwarfism, mulefoot, marble bone disease; large, pendulous sheath, long and nonretractable prepuce, lack of sheath; under- |
| | | developed teats and/or udder; very large teats, meaty udder, loosely attached udder; extremely wild or nervous; extremely small cattle for their age; animals that are extremely compact, rangy, or light boned; hard horns; or excessive Brahman or Angus type. | | | |
| **Red Poll** (A dual-purpose breed) (See Fig. 12–43) | **E**ngland; in the eastern middle coastal counties of Norfolk and Suffolk. **T**he Red Poll Cattle Club of America was organized in 1883. | **R**ed, varying from light to dark red. **A**ny white except in the switch is discriminated against. **A**lso, a smoky nose or dark spots on the nose are objectionable. | **P**olled. | | **D**isqualifications include bulls with only one testicle, scurs or any horny growth, or total blindness. |
| **Romagnola** (See Fig. 12–44) | **I**taly; in the lower Po Valley, from crossing the Podolic and native cattle. | **S**olid off-white to light gray, varying from light to dark. The bulls have a characteristically darker color of hair about the shoulders, black color around the eyes, and a black switch. | **H**ead appears to be short. **H**orns are longer and sharper than in the other white breeds, and they grow upward and outward (lyra horns). | **I**n Italy, the Romagnola is a dual-purpose breed, used for both meat and milk. **S**kin is pigmented; and the tongue, muzzle, body orifices, tail, and hooves are black. **M**ature bulls average 2,500 lb *(1,134 kg)* and mature cows 1,500 lb *(680 kg)*. At birth, bull calves average about 110 lb *(50 kg)* and heifers 95 lb *(43 kg)*. | |
| **Salers** (pronounced *Sa-lair*) (See Fig. 12–45) | **F**rance; in the south-central area—a mountainous region. **T**he name *Salers* was first applied to the cattle of this area in 1840, after a small town located in the center of the province. **T**he Salers Herd Book was started in 1906. **T**he American Salers Association was formed in 1974. | **S**olid, deep cherry red, with a white switch and sometimes white spots under the belly. | **H**orned, although there are polled strains. | **I**n its native land, the breed is noted for rapid gain, hardiness, and adaptability. **M**ature bulls have an average weight of 2,530 lb *(1,148 kg)*, mature cows average about 1,540 lb *(699 kg)*. **T**he Salers was founded as a dual-purpose breed. | **A**merican Salers Association, the breed registry, makes the claim that no genetical defect and no double muscling have ever been reported in the Salers breed. |

**TABLE 12-2** *Continued*

| Breed | Place of Origin | Color | Distinctive Head Characteristics | Other Distinguishing Characteristics | Disqualifications; Comments |
|---|---|---|---|---|---|
| **Santa Gertrudis** (⅝ Shorthorn and ⅜ Brahman) (See Fig. 12–46) | U.S.A.; on the King Ranch in Texas, based on a Shorthorn × Brahman cross. The foundation sire, *Monkey*, was born in 1920. Named from the Santa Gertrudis Land Grant, granted by the Crown of Spain, on which the breed evolved, now the headquarters division of King Ranch. In 1951, Santa Gertrudis Breeders International was incorporated. | Red or cherry red. | Generally horned, but there are polled strains. | Hair should be short, straight, and slick. Hide should be loose, with surface area increased by neck folds and sheath or navel flap. But neither should be excessive. | Disqualifications include: hernia; cryptorchid; malformed genitalia; wry nose; wry tail; double muscling; undershot and overshot jaw; nonretractable parietal prepuce; loose, 90° pendulous, close to the ground; excessive sheath development and large skinfold forward of orifice. Also, an animal is disqualified for white spots out of the underline, white underline exceeding 50%, fawn or cream color, brindling or roan, or solid black skin. |
| **Scotch Highland** (or **Highland**) (See Fig. 12–47) | Scotland. The first importation into the U.S. was made in 1883. The American Scotch Highland Breeders Assn. was established in 1948. | Red, yellow, silver, white, dun, black, or brindle. | Short head; long, widespread horns; and heavy foretop. | Long, shaggy hair, short legs; hardiness and ability to rustle in cold weather. | Polled and spotted animals are disqualified. They should be solid in color except for an occasional white tip on switch, or white on the underline or udder. |
| **Shorthorn** (See Fig. 12–48) | England; in the northeastern counties of Durham, Northumberland, York, and Lincoln. The Shorthorn was the first breed of cattle to have a herd book—the *Coates Herd Book*, founded by George Coates in 1822. The name *Shorthorn* stems from the fact that, through breeding and selection, the early improvers of the breed shortened the horns of the native cattle. The American Shorthorn Breeder's Association was established in 1882. | Red, white or any combination of red and white. A smutty nose or dark nose is objectionable. | Rather short, refined, in-curving horns. | | No calf is eligible for registration unless its sire and dam were at least 18 mo. of age at the birth date of the calf. Animals must be under 18 mo. of age at the time of registration (otherwise, they must be approved by the Executive Committee). |
| **Simbrah** (Purebred Simbrah must be ⅝ Simmental and ⅜ Brahman) (See Fig. 12–49) | In the southern part of the U.S., mostly in Texas. A few producers experimented by combining Simmental and Brahman in the late 1960s, but the first registration of Simbrah occurred in 1977. | Red, straw colored, or gray with white intermixed. | Simbrah may be polled, scurred, or horned. | The breed exhibits some characteristics of both Brahman and Simmental, but less of the Brahman extremes for sheath, hump, and looseness of skin folds. Simbrahs are fertile, hardy, adaptable, disease and parasite resistant, and fast-gaining. | In addition to purebred Simbrahs, the registry records percentage Simbrahs with a minimum of ⅜ Simmental, ¼ Brahman, and a maximum of ⅜ combination of other breeds. |
| **Simmental** (See Fig. 12–50) | Western Switzerland; in the Simme Valley, from which it derives its name. It is much older than the herd register, which was set up in Bern, Switzerland in 1806. The first Simmental was brought to North America in 1967, by a group of southern Alberta producers headed by Travers Smith of Cardston, Alberta, Canada. The American Simmental Association was established in 1968. | Generally red-and-white spotted, although some are nearly solid in color. The red varies from dark to a more common diluted, almost yellow, shade. A white face, which, like the Hereford, appears to be dominant in inheritance. | Horned, although polled strains exist. | The Simmental was first developed as a dual-purpose breed. The animals combine meat and milk to an unusually high degree, along with rapid growth rate. Mature bulls average about 2,300 to 2,400 lb *(1,043 to 1,089 kg)*, and cows about 1,600 to 1,700 lb *(726 to 771 kg)*. The breed milk production average is about 8,000 lb *(3,629 kg)* with a 4% butterfat test. | Genetic unsoundnesses. To qualify for registration, animal must be at least ½ Simmental. Any animal whose sire cannot be determined through blood typing cannot be registered. As the breed spread through Europe, it became known by several different names; *Pie Rouge* in France, *Austrovich* in Austria, and *Fleckvich* in Germany. |
| **South Devon** (See Fig. 12–51) | England; originated in southern Devonshire, through infusion of Guernsey blood into the Devon breed. The South Devon has had its own *Herd Book* in England since 1891. Henry Wallace, former U.S. Vice President, brought the first shipment of South Devons to the U.S. in 1936. The North American South Devon Assn. was formed in 1968. | Medium light red color. | Horned. | The South Devon is a dual-purpose breed. Mature bulls weigh 2,000 to 2,800 lb *(909 to 1,271 kg)* and mature cows weigh 1,200 to 1,700 lb *(545 to 772 kg)*. Cows are heavy milkers; they average about 6,550 lb *(2,974 kg)* of milk per lactation, with 4.2% fat. | The South Devon is the only breed in England that both (1) receives a Milk Marketing Board premium for rich milk, and (2) qualifies for the British Beef Subsidy. The breed is extolled as a *dam breed*, superior in maternal traits. During 1969 and 1970, Big Beef Hybrids, Stillwater, MN, imported nearly 200 purebred South Devons. |

*(Continued)*

**TABLE 12-2** *Continued*

| Breed | Place of Origin | Color | Distinctive Head Characteristics | Other Distinguishing Characteristics | Disqualifications; Comments |
|---|---|---|---|---|---|
| **Sussex** (See Fig. 12–52) | England; descended from the indigenous red cattle that inhabited Sussex and Kent Counties at the time of the Norman Conquest (1066).<br>First registered in England in 1840.<br>First imported into the U.S. in 1883. | Deep mahogany red, white switch, and ivory horns with dark tips. | A high percentage of Sussex cattle are polled. The head is moderately long, and the nose is flesh colored. | In England, the breed has earned the reputation as the "butcher's beast," because of the evenness of fleshing, predominance of lean meat, and high dressing percentage. | In the U.S., the Sussex Cattle Assn. of America registers cattle in the *English Herd Book*. American entries in the *English Herd Book* commenced in 1967.<br>Sussex cattle were immortalized by Rudyard Kipling in the poem "Alnascher and the Oxen." |
| **Tarentaise** (Tarine) (See Fig. 12–53) | France; in the Alps, where the cattle were known as Tarentaise since 1863.<br>The Herd Book was started in 1888. | Solid wheat-colored, ranging from a cherry to dark blond. Bulls tend to darken around the neck and shoulders with maturity. Frequently, bulls have a darker dorsal stripe. | Black pigmentation of the muzzle and around the eyes. | Noted for easy calving, due to adequate pelvic capacity and small calves; vigorous calves at birth; hardiness; black hair around the eyes and pigmented udders and teats, thereby making cancer eye and sunburned teats rarities; good fertility; and good milking ability, with cows averaging about 8,000 lb *(3,629 kg)* per lactation.<br>Smaller than most of the exotics. Mature bulls average about 1,800 lb *(816 kg)* and cows about 1,150 lb *(522 kg)*. | In France, the breed registry advocates eliminating (although it does not *disqualify*) widespread patches of white hairs or badger grey coloring, bright red or mahogany overall color, a stripe on the back lighter than the general coloring; very dark or black parts of the coat (cheeks, dewlap, shoulders, etc.); total absence of black pigmentation on mucous membranes and extremities; poor general conformation, particularly a crest-shaped tail.<br>Calves are small and vigorous at birth; hence, Tarentaise bulls are well suited for use on virgin heifers. |
| **Texas Longhorn** (See Fig. 12–54) | U.S.A.; from cattle of Spanish extraction. On the second voyage in 1493, Columbus brought Spanish cattle to Santo Domingo.<br>By 1900, the Texas Longhorn was driven to near extinction, replaced by the European breeds—the Shorthorn, Hereford, and Angus.<br>The Texas Longhorn Breeders Assn. of America was organized in 1964. At that time, there were only approximately 1,500 head of genuine Texas Longhorn cattle in existence. | Texas Longhorns are characterized by a great array of colors, in all degrees of richness, and in all possible combinations and patterns. | Large, spreading horns that curve upward.<br>Long head, with small ears.<br>Longer hair between the horns than on the body. | Long legs and high shoulders. Noted for fertility, ease of calving, hardiness, resistence to many common diseases, good rustling ability, good feet and legs, longevity, and adaptation to a wide variety of environmental conditions. | Registry disqualifications include any of the following: (1) any hereditary deformities that occur in cattle, such as hernia, cryptorchid, or malformed genitalia; (2) extra large sheath or navel flap; (3) any evidence of a hump over the shoulder region; (4) extremely large, droopy ears; (5) wry nose, overshot or undershot jaw; (6) horns under 24 in. *(61 cm)* tip to tip at 4 to 5 years of age; (7) wry tail; or (8) double muscling. |
| **Welsh Black** (See Fig. 12–55) | Wales; where they have long been bred as dual-purpose cattle.<br>The Canadian Welsh Black Cattle Society was formed in 1970.<br>The U.S. Welsh Black Cattle Assn. was formed in 1974. | Black. | Horned, although there is a polled strain in Wales. | High fertility; little calving difficulty; good milk production; adapted to harsh conditions of climate and forage; longevity; relative freedom from sunburned udders and cancer eye.<br>Mature bulls weigh 1,800 to 2,000 lb *(817 to 909 kg)* and cows from 1,000 to 1,300 lb *(454 to 590 kg)*.<br>Cows give 6,000 to 7,700 lb *(2,724 to 3,495 kg)* of milk per lactation. | The major impact of the breed is expected to be on brood cows—as maternal sires in a crossbreeding program.<br>In Wales, white except on udder or scrotal area is a disqualification. |
| **White Park** (See Fig. 12–56) | England. The ancestors of the breed were brought to England by the Romans, in 55 B.C. They were domesticated in the 18th century. The name *White Park* came from large game preservelike enclosures in which these cattle were kept after being domesticated. | White, with black or red pigmentation and markings around the eyes, ears, nose, feet, legs, teats, and anal area. | Predominantly polled.<br>White head with black ears, eyes, and nose. | Comparable in size to Angus, Hereford, and Shorthorn cattle.<br>Strong maternal instinct; ease of calving; excellent milking ability; docile and easy to handle. | Overmarking and undermarking are undesirable. |

During World War II, 5 females and 1 bull of the White Park breed were shipped to the U.S. to insure preservation of the seed stock.

The White Park Cattle Assn. of America was formed in 1975.

[1]In England and Scotland, both reds and blacks are registered in the same association, without distinction. In the U.S., however, red-colored animals have been barred from registry in the American Angus Association since 1917. The Red Angus Association of America was organized in 1954.

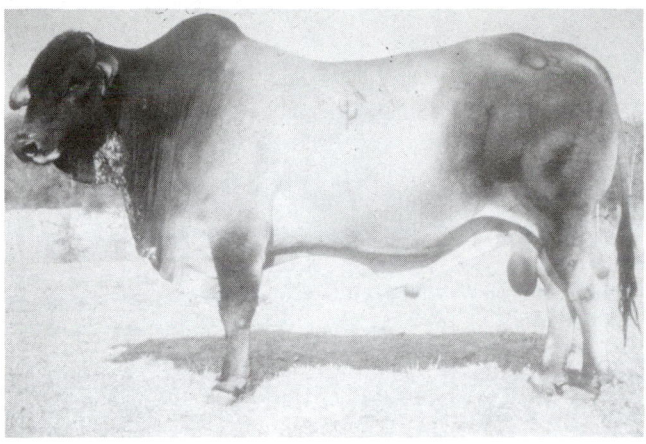

Fig. 12-3. American Breed bull, *Ron 103*, shown at 5 years of age and weighing 2,750 lb. (Courtesy, Art Jones, Cactus Ranch, Portales, NM)

Fig. 12-4. Amerifax bull. (Courtesy, Amerifax Cattle Assn., Hastings, NE)

Fig. 12-5. Angus bull, of modern type. (Courtesy, American Angus Assn., St. Joseph, MO)

Fig. 12-6. Ankina bull calf. His adjusted 205-day weight was 656 lb. (Courtesy, Ankina Breeders, Clayton, OH)

Fig. 12-7. Ankole-Watusi bull at 5 years of age. (Courtesy, Ankole-Watusi International Registry, Hebron, ND)

Fig. 12-8. Barzona bull. A good representative of the breed. (Courtesy, Barzona Breeders Assn. of America, Prescott, AZ)

Fig. 12–9. Beefalo bull. (Courtesy, American Beefalo World Registry, Louisville, KY)

Fig. 12–10. Beef Friesian bull. (Courtesy, Beef Friesian Society, Denver, CO)

Fig. 12–11. Beefmaster cow. There are both horned and polled Beefmasters. (Courtesy, Beefmaster Breeders Universal, San Antonio, TX)

Fig. 12–12. Belted Galloway bull, *Burnside Great Scot,* A Supreme Champion Belted Galloway at the Royal Highland Show, Aberdeen, Scotland. Imported by Mr. A. H. Chatfield, Jr., Rockport, ME. (Courtesy, Mr. Chatfield)

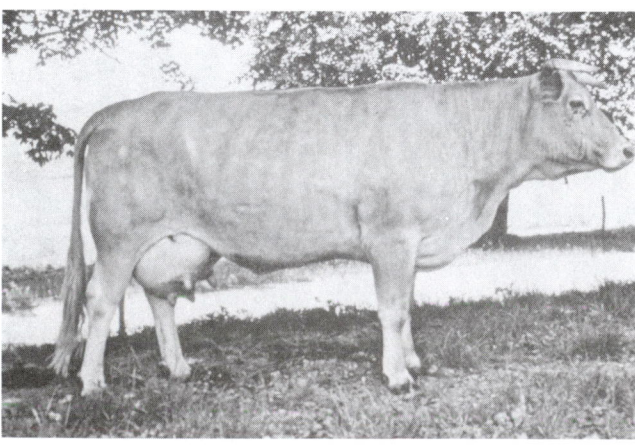

Fig. 12–13. Blonde d'Aquitaine cow. (Courtesy, American Blonde d'Aquitaine Assn., Grandview, ID)

Fig. 12–14. Braford bull on pasture in Florida. (Courtesy, Adams Ranch, Inc., Fort Pierce, FL)

Fig. 12–15. Brahman cow. (Courtesy, American Brahman Breeders Assn., Houston, TX)

Fig. 12–16. Braler heifer. (Courtesy, Brandt Ranches, Inc., Houston, TX)

Fig. 12–17. Brangus bull. The breed was developed from ⅜ Brahman and ⅝ Angus. (Courtesy, International Brangus Breeders Assn., Inc., San Antonio, TX)

Fig. 12–18. Charbray cow and calf. (Courtesy, American-International Charolais Assn., Kansas City, MO)

Fig. 12–19. Charolais bull. (Courtesy, American-International Charolais Assn., Kansas City, MO)

Fig. 12–20. Chianina bull, imported into Canada from Italy. (Courtesy, *American Chianina Journal*, Platte City, MO)

Fig. 12–21. Devon cow (polled). (Courtesy, Devon Cattle Assn., Inc., Alexandria, VA)

Fig. 12–22. Dexter bull, *J. F. Black Warrior,* owned by Peerless Herd, Decorah, IA. (Courtesy, American Dexter Cattle Assn., Concordia, MO)

Fig. 12–23. Fleckvieh bull, *Tattenhall Achilles.* This bull had an adjusted weaning weight of 775 lb and an adjusted yearling weight of 1,175 lb.

Fig. 12–24. Galloway bull. *Grange Bounty,* at Castle-Douglas, Scotland, sold for export to Round Mountain Ranch, Round Mountain, GA. (Courtesy, Round Mountain Ranch)

Fig. 12–25. Gelbvieh bull. (Courtesy, American Gelbvieh Assn., Denver, CO)

Fig. 12–26. Hays Converter bull, *Eric The Red.* (Courtesy, American Breeders Service Div., W. R. Grace & Co., DeForest, WI)

Fig. 12–27. Hereford bull, *BB Hi Tech 8205 1ET*, Reserve Grand Champion Bull at the 1989 Nugget National Hereford Show, Reno. Bill Bennett, owner, BB Cattle Co, Connell, WA. (Courtesy, American Hereford Assn., Kansas City, MO)

Fig. 12–28. Indu Brazil bull. (Courtesy, *Farm & Ranch*, Mexia, TX)

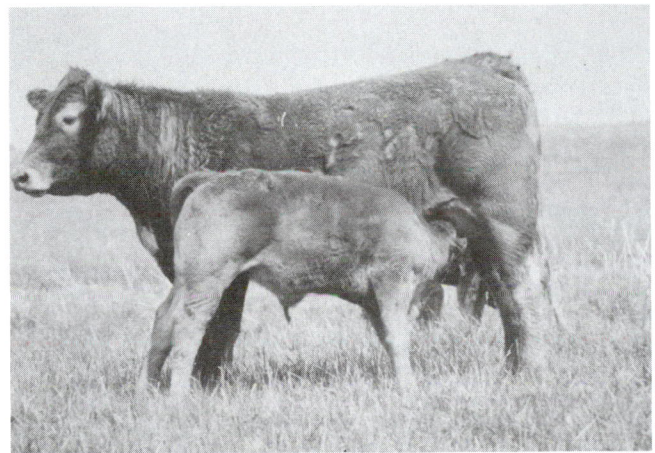

Fig. 12–29. Limousin cow and calf. (Courtesy, North American Limousin Foundation, Denver, CO)

Fig. 12–30. Lincoln Red bull, *Cockerington Lord*. Weight, 2,275 lb. (Courtesy, Carnation/Genetics, Hughson, CA)

Fig. 12–31. Maine-Anjou cow. (Courtesy, International Maine-Anjou Assn., Kansas City, MO)

Fig. 12–32. Marchigiana heifer. (Courtesy, Marky Cattle Assn., Walton, KS)

Fig. 12–33. Murray Grey bull, *Jilba Morgan,* imported from Australia. Weighed 1,055 lb at 16 months of age. (Courtesy, American Breeders Service, DeForest, WI)

Fig. 12–34. Normande bull. (Courtesy, American Normande Assn., Kearney, MO)

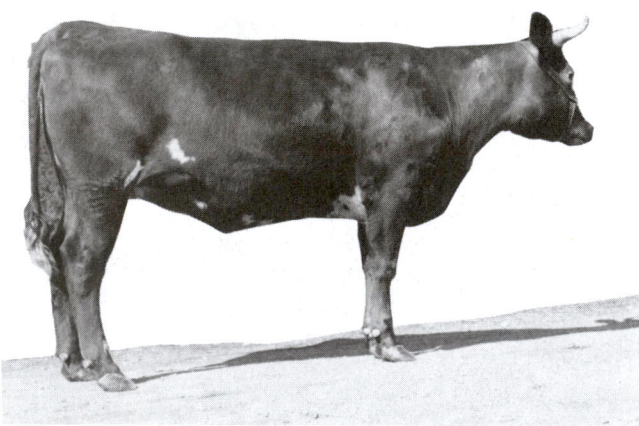

Fig. 12–35. Norwegian Red cow, *91 Mari,* imported from Norway by the Southern Cattle Corporation, Memphis, TN. (Courtesy, Southern Cattle Corp.)

Fig. 12–36. Piedmontese bull calf, showing double muscling; owned by Dr. Ugo and Gianni Galassi, Aimonetta Farm, Carentino, Italy. (Photo by the author)

Fig. 12–37. Pinzgauer cow. (Courtesy, American Pinzgauer Assn., Kelly, IA)

Fig. 12–38. Polled Hereford cow and calf. (Courtesy, American Polled Hereford Assn., Kansas City, MO)

Fig. 12-39. Polled Shorthorn cow and calf. (Courtesy, American Shorthorn Assn., Omaha, NE)

Fig. 12-40. Ranger foundation sire, *Educator 309*. (Courtesy, Mr. Frank G. Watson, Ranger Cattle Company, Denver, CO)

Fig. 12-41. Red Angus cow. (Courtesy, Red Angus Assn. of America, Denton, TX)

Fig. 12-42. Red Brangus cow and her calf on the range. (Courtesy, Mike Levi, Paleface Spicewood, TX)

Fig. 12-43. Red Poll cow. (Courtesy, Red Poll Cattle Club of America)

Fig. 12-44. Romagnola bull, *Monello.* Weighed 1,450 lb at 21 months of age. (Courtesy, American Breeders Service, DeForest, WI)

Fig. 12–45.  Salers cow and calf. (Courtesy, American Salers Assn., Englewood, CO)

Fig. 12–46.  Santa Gertrudis cow. (Courtesy, Santa Gertrudis Breeders International, Kingsville, TX)

Fig. 12–47.  Scotch Highland bull, *L. E. Loretta's King,* Golden Certified Meat Sire. (Courtesy, American Scotch Highland Breeders Assn., Remer, MN)

Fig. 12–48.  Shorthorn cow (horned), *Melbourne Rosanna 12th,* owned by John Alexander & Sons, Melbourne Farm, Big Rock, IL. (Courtesy, *The Shorthorn World,* Brandon, SD)

Fig. 12–49.  Simbrah, the result of combining the Simmental and Brahman breeds. (Courtesy, American Simmental Assn., Bozeman, MT)

Fig. 12–50.  Simmental bull. (Courtesy, American Simmental Assn., Bozeman, MT)

Fig. 12–51. South Devon bull. (Courtesy, North American South Devon Assn., Lynnville, IA)

Fig. 12–52. Sussex cow on pasture, at Refugio, TX. (Courtesy, Mr. Lawrence Wood, Refugio, TX)

Fig. 12–53. Tarentaise bull, *Boby,* owned by C-C Ranch, New Rockford, ND. (Courtesy, American Tarentaise Assn., New Rockford, ND)

Fig. 12–54. Texas Longhorn cow at 11 years of age. (Courtesy, Texas Longhorn Breeders Assn. of America, Fort Worth, TX)

Fig. 12–55. Welsh Black cow. (Courtesy, United States Welsh Black Cattle Assn., Wahton, MN)

Fig. 12–56. White Park bull. (Courtesy, White Park Cattle Assn., Madrid, IA)

# QUESTIONS FOR STUDY AND DISCUSSION

1. Define the term *type*. Describe each of the following distinct types of cattle: (a) beef-type, (b) dairy-type, (c) dual-purpose-type, and (d) draft-type.

2. Define the term *breed*. What generally causes a breed and a breed registry to evolve?

3. Must a new breed of cattle be approved by someone, or can anyone start a new breed (see footnote No. 1)?

4. Define the term *exotic*. Why were Shorthorn, Hereford, and Angus cattle brought to America? Were they exotics?

5. Why did supporters of the British breeds (Shorthorns, Herefords, and Angus) not consider the introduction of the Brahman a serious threat, whereas the introduction of the Charolais disturbed them greatly?

6. What breeds would you classify as among the recent exotics? What prompted the bringing in of these exotics? How do you account for the fact that the recent exotics came quickly and in large numbers?

7. Why, in recent years, have we brought to North America so many dual-purpose and draft-type cattle?

8. Why have beef-type and dairy-type (a) live animals, and (b) carcasses become more alike in recent years?

9. What makes a breed popular? Why are certain breeds more numerous than others?

10. List in order of 1988/89 registration numbers the six leading beef and dual-purpose breeds of cattle in the U.S.

11. List 10 breeds of beef or dual-purpose cattle, and for each of these breeds give (a) the place of origin, and (b) the color and other distinguishing characteristics.

12. What factors should be considered in selecting the breeds for a crossbreeding program?

13. Justify any preference that you have for one particular breed of beef or dual-purpose cattle.

14. Obtain breed registry association literature and a sample copy of a magazine of your favorite beef or dual-purpose breed of cattle. (See Appendix Tables A–8 and A–9 for addresses.) Evaluate the soundness and value of the material that you receive.

## SELECTED REFERENCES

| Title of Publication | Author(s) | Publisher |
| --- | --- | --- |
| *Aberdeen-Angus Breed: A History, The* | J. R. Barclay<br>A. Keith | The Aberdeen-Angus Cattle Society, Aberdeen, Scotland, 1958 |
| *Beef Cattle Science*, Sixth Edition | M. E. Ensminger | Interstate Publishers, Inc., Danville, IL, 1987 |
| *Birth of a Breed* | O. K. Sweet | The Lowell Press, Kansas City, MO, 1975 |
| *Breeds of Cattle* | H. R. Purdy | Paul Steiner, Chanticleer Press, Inc., New York, NY, 1987 |
| *Breeds of Livestock, The* | C. W. Gay | The Macmillan Company, New York, NY, 1916 |
| *Breeds of Live Stock in America* | H. W. Vaughan | R. G. Adams and Co., Columbus, OH, 1937 |
| *Hereford in America, The* | D. R. Ornduff | The author, Kansas City, MO, 1957 |
| *Hereford Heritage* | B. R. Taylor | The author, Univ. of Arizona, Tucson, AZ, 1953 |
| *History of Linebred Anxiety 4th Herefords, A* | J. M. Hazelton | Assoc. Breeders of Anxiety 4th Herefords, Graphic Arts Building, Kansas City, MO, 1939 |
| *Lasater Philosophy of Cattle Raising, The* | L. M. Lasater | Texas Western Press, U. of Texas, El Paso, 1972 |
| *Modern Breeds of Livestock*, Fourth Edition | H. M. Briggs<br>D. M. Briggs | The Macmillan Company, New York, NY, 1980 |
| *Santa Gertrudis Breed, The* | A. O. Rhoad | Inter-American Institute of Agriculture Sciences, Turrialba, Costa Rica, 1949 |
| *Santa Gertrudis Breeders International Recorded Herds* | R. J. Kleberg, Jr. | Santa Gertrudis International, Kingsville, TX, 1953 |
| *Shorthorn Cattle* | A. H. Sanders | Sanders Publishing Co., Chicago, IL, 1918 |
| *Stockman's Handbook, The*, Sixth Edition | M. E. Ensminger | Interstate Publishers, Inc., Danville, IL, 1983 |
| *Story of the Herefords, The* | A. H. Sanders | *Breeders Gazette*, Chicago, IL, 1914 |
| *Study of Breeds in America, The* | Y. Shaw | Orange Judd Co., New York, NY, 1900 |
| *Types and Breeds of African Cattle* | N. R. Joshi<br>E. A. McLaughlin<br>R. W. Phillips | Food and Agriculture Organization of the United Nations, Rome, Italy, 1957 |
| *Types and Breeds of Farm Animals* | C. S. Plumb | Ginn and Company, Boston, MA, 1920 |
| *World Cattle*, Vols. I, II, and III | J. E. Rouse | University of Oklahoma Press, Norman, OK, Vols. I and II, 1970; Vol. III, 1973 |
| *World Dictionary of Breeds, Types and Varieties of Livestock, A* | I. L. Mason | Commonwealth Agricultural Bureaux, Slough, Bucks, England, 1951 |
| *Zebu Cattle of India and Pakistan* | N. R. Joshi<br>R. W. Phillips | Food and Agriculture Organization of the United Nations, Rome, Italy, 1953 |

Also, breed literature pertaining to each breed may be secured by writing to the respective breed registry associations (see Table A–8, Appendix, for the name and address of each association).

Cattle come in different sizes, colors, and shapes, thereby making for choices—or selection. Shown here is *Satan,* a Brahman × Shorthorn cross, billed as the "world's largest steer"; owned by Jack and Penny Mericle. At 6 years of age, and still growing, *Satan* tipped the scales at 4,000 lb, stood more than 6 ft at the withers, had a heart girth of 11 ft 3 in., measured 13 ft 7 in. from nose to tailhead, had a horn spread of 39 in., and consumed 140 lb of hay per day. (Courtesy, Mr. and Mrs. Jack Mericle, Mericle Ranches, Kuna, ID)

# ESTABLISHING THE BEEF HERD; SELECTING AND JUDGING BEEF CATTLE

# *Chapter 13*

Whether establishing or maintaining a herd, cattle producers must constantly appraise or evaluate animals; they must buy, sell, retain, and cull. Where the beef cattle herd is neither being increased nor being decreased in size, each year about 46% of the heifers, on the average, are retained in order to replace about 20% of the old cows.[1] In addition, bulls must be selected and culled, and steers and other surplus animals must be marketed. Thus, in normal operations, producers are constantly called upon to cull out animals, to select replacements, and to market surpluses. Each of these decisions calls for an evaluation or appraisal.

Cattle producers are ever aware of market demands as influenced by consumer preferences. Also, the great livestock shows throughout the land have exerted a powerful influence in molding cattle types.

It must be realized, however, that only a comparatively few animals on the farms and ranches are subjected annually to the scrutiny of market specialists or experienced show-ring judges. Rather, the vast majority of purebred animals and practically all commercial herds are evaluated by practical producers—operators who select their own foundation or replacement stock and conduct their own culling operations. Such producers have no interest in the so-called breed fancy points. These practical operators may not be able to express fluently their reasons for selecting certain animals while culling others, but usually they become quite deft in their evaluations. Whether young animals are being raised for market or for breeding stock, successful livestock operators are generally good judges of livestock.

## FACTORS TO CONSIDER IN ESTABLISHING THE HERD

Except for the comparatively few persons who keep animals merely as a hobby, farmers and ranchers raise cattle because, over a period of years, they have been profitable, provided the production and marketing phases were conducted in an enlightened and intelligent manner. Therefore, after it has been ascertained that the feeds and available labor are adapted to cattle production, and that suitable potential markets exist, the next assignment is that of establishing a herd that is efficient from the standpoint of production, and that meets market demands. This involves a number of considerations.

## PUREBRED OR COMMERCIAL CATTLE

Broadly classed, cow-calf producers are either (1) purebred breeders, or (2) commercial producers. Purebred breeders are a small, but select, group. The vast majority of farmers and ranchers are commercial producers. An estimated 96% of the cattle of America are nonpurebreds.

Purebred breeders produce "seedstock" for other purebred breeders, and both bulls and purebred females for $F_1$ heifer programs of commercial producers.

Purebred breeders need to be more than good caretakers. They should be knowledgeable relative to breeding systems, pedigrees and registration, production testing, advertising, sales and other special marketing methods, and perhaps fitting and

showing. Also, they should be thoroughly knowledgeable relative to breeding commercial cattle the modern way in order that the needs of the commercial producer will be reflected in their breeding programs. Indeed, both types—purebred and commercial—are interdependent.

For the person with experience and adequate capital, the breeding of purebreds offers unlimited opportunities. It has been well said that honor, fame, and fortune are all within the realm of possible realization in the purebred business, but it should also be added that only a few achieve this high calling.

The goal of most commercial beef operations is to convert the production of the land—grass and crops—into dollars through the traditional cow-calf operation. Usually the product is marketed at the weaning stage, although some commercial producers carry them to the yearling stage, or even finish them for market. More and more commercial herds will be crossbreds, simply because the economics favor crossbreeding accompanied by complementary genes and heterosis, and because the crossbreds can be used to produce beef according to specification.

Fig. 13–1. A commercial bull buyer looking over yearling bulls. (Courtesy, American Polled Hereford Assn., Kansas City, MO)

As a group, commercial producers are intensely practical. No animal meets with their favor unless it produces meat over the block at a profit. The commercial cattle business requires less outlay of cash than the purebred business on a per animal basis, and less knowledge relative to the many facets of the purebred business—pedigrees, promotion, etc.

## SELECTION OF THE BREED OR CROSS

No one breed of cattle can be said to excel all others in all points of beef production for all conditions.

For purebred breeders, the selection of a particular breed is most often a matter of personal preference, and usually the breed that they like is the one with which they will have the greatest success. Where no definite preference exists, however, it is well to choose the breed that is most popular in the community—if any one breed predominates. If this procedure is followed, it is often possible to arrange for an exchange of animals, especially bulls. Moreover, if a given community is noted for producing good cattle of a particular breed, there are many advantages from the standpoint of advertising and sales.

---

[1]A herd of 100 cows will produce about 44 heifers each year where there is an 88% calf crop. Hence, 46% of the heifers will be needed to replace 20% of the cows (44 × .46 = 20).

Germ plasm choice for the commercial cattle producer is becoming increasingly difficult because of the large number of breeds and breed cross combinations now available. With only 3 main breeds (Shorthorns, Herefords, and Angus), there were 3 single-cross combinations and three 3-way cross combinations from which to choose. However, with 10 breeds there are 45 single-cross combinations and 360 possible 3-way combinations from which to select. Of course, there are more than 10 breeds; 54 breeds of beef/dual-purpose cattle are listed in Chapter 12 of this book, and more breeds will evolve.

## MILKING ABILITY

Weaning weight is the most important trait affecting net income in a cow-calf operation; and weaning weight of beef calves is influenced more by the dam's milk production than by any other single factor. For this reason, the pressure is on to increase the milk production of beef cows. For this reason, also, this subject—"Milking Ability"—is fully covered herein.

Fig. 13-2. "Lotta" milk—"lotta" calf. Weaning weight of calves is influenced more by the dam's milk production than by any other single factor. This shows a Beefmaster cow with her growthy heifer calf. (Courtesy, Beefmaster Breeders Universal, San Antonio, TX)

Performance-testing programs which emphasize weaning weight automatically result in selection for higher milk production. But, to increase beef production more rapidly, dual-purpose or dairy breeding are being infused into many commercial beef herds.

Research has shown a strong correlation between the level of milk production of cows and the weaning weight of their calves. Also, conversion of milk to beef is rather efficient—on the order of 10 lb of milk to one additional lb of weaned calf, although conversion may not be quite as efficient at higher levels of milk production.

Researchers are just now beginning to study carefully the milk production of beef cows. There may well be several answers to the 2-pronged question, "How much milk should a beef cow produce, and what's the best way to increase milk production in a beef herd?" On a poor range where feed is sparse, a relatively low level of milk production may be necessary to allow good reproduction, while on improved pastures a very high level of milk production may be desirable. Also, more study needs to be given to selecting the high-

producing strains of beef cattle and to the use of *dam breeds*. Reasonable goals for the more successful beef herds are (1) a cow averaging 20 lb of milk per day (about double the present level); and (2) weaning off a 600 lb calf. Cattle producers can be sure of one thing: "Little milk—little calf; lotta milk—lotta calf."

## SIZE OF THE HERD

No minimum or maximum figures can be given as to the best size for the herd. Rather, each case is one for individual consideration. It is to be pointed out, however, that labor costs differ very little whether the herd numbers 100 or 500. The cost of purchasing and maintaining a herd bull also comes rather high when too few females are kept. Other efficiencies can be achieved through size, provided the operation is under competent management. For this reason, bigness in every kind of business, including the cattle business, is a sign of the times.

The extent and carrying capacity of the pasture, the amount of hay and other roughage produced, and the facilities for wintering stock are factors that should be considered in determining the size of herd for a particular farm unit. The system of disposing of the young stock will also be an influencing factor. For example, if the calves are disposed of at weaning time or finished as baby beef, practically no cattle other than the breeding herd are maintained. On the other hand, if the calves are carried over as stockers and feeders or are finished at an older age, more feed, pasture, and shelter are required.

Then, too, whether the beef herd is to be a major or minor enterprise will have to be decided upon. Here again, each case is one for individual consideration. In most instances, replacements should be made from heifers raised on the farm.

## UNIFORMITY

Uniformity in a herd has reference to the animals looking alike—"like peas in a pod"—as a cattle producer is prone to remark, particularly from the standpoints of size, type, and color.

Fig. 13-3. A Hereford × Angus cross results in an animal with a black body and a white face—a black baldie. All such crosses will look much alike. (Courtesy, American Polled Hereford Assn., Kansas City, MO)

Uniformity in color is still important in purebred herds. For the most part, however, the desire for uniformity in color of commercial herds went out as crossbreeding came in. As long as a quality product is produced efficiently, consumers and most cattle feeders have no interest in color of hair. It should be added for the benefit of the color-conscious, however, that it is still possible to obtain uniformity of color even in crossbreds through making certain crosses; for example, uniform-colored animals with black bodies and white faces can be produced by crossing Herefords and Angus.

Size and type in any given lot of cattle are still important to cattle feeders and packer buyers. Cattle of uniform size and type feed better in a lot, and make it possible for packers to provide their retail outlets with carcasses and cuts that meet their exacting specifications and grades. But buyer appeal, for both the feedlot and packer buyer, can be imparted amazingly well by sorting and grading prior to offering cattle for sale. Properly done, this practice can make for more uniformity faster in a few minutes than can be achieved through years of selective breeding.

## HEALTH

All animals selected should be in a thrifty, vigorous condition and free from diseases and parasites. They should give every evidence of a life of usefulness ahead of them. The cows should appear capable of producing good calves, and the bull should be able to withstand a normal breeding season. Tests should be made to make certain of freedom from both tuberculosis and contagious abortion, and perhaps certain other diseases in some areas. In fact, all purchases should be made subject to the animals being free from contagious diseases. With costly purebred animals, a health certificate should be furnished by a licensed veterinarian. Newly acquired animals should be isolated for several days before being turned with the rest of the herd.

## CONDITION

Fig. 13–4. Angus feeders. It takes a unique ability to project the end result of feeding a few hundred pounds of grain or hay to an animal. (Courtesy, American Angus Assn., St. Joseph, MO)

Although an extremely thin and emaciated condition which may lower reproduction is to be avoided, it must be remembered that an overfat condition may be equally harmful from the standpoint of reproduction.

It takes a unique ability to project the end result of feeding a few hundred pounds of grain or hay to a thin animal, and fortunate indeed is the cattle producer who possesses this quality. This applies alike to both the purebred and the commercial producer. In fact, it is probably of greater importance to the commercial producer, for replacement females and stocker and feeder steers are usually in very average condition.

## AGE AND LONGEVITY

In establishing the herd, it is usually advisable to purchase a large proportion of mature cows (cows 4 to 5 years of age) that have a record of producing uniformly high-quality calves. Perhaps it can be said that not over one-half of the newly founded herd should consist of untried heifers. Aside from the fact that some of the heifers may prove to be nonbreeders, they require more assistance during calving time than do older cows. Perhaps the best buy of all, when they are available, consists in buying cows with promising calves at side and rebred to a good bull—a three-in-one proposition.

Once the herd has been established, replacement females should come from the top heifers raised on the farm or ranch. Old cows, irregular breeders, and poor milkers sell to best advantage before they become thin and "shelly."

A sound practice in buying a bull is to seek one of the serviceable age that is known to have sired desirable calves—a proven sire. However, with limited capital, it may be necessary to consider the purchase of a younger bull. Usually a wider selection is afforded with the latter procedure, and, also, such an individual has a longer life of usefulness ahead. Naturally, the time and number of services demanded of the bull will have considerable bearing on the age of the animal selected.

Since most beef females do not reproduce until they are 2 or 3 years of age, their regular and prolonged reproduction thereafter has an important bearing upon the overhead cost of developing breeding stock in relation to the number of calves produced. The longer the good, proven, producing cows can be kept without sacrifice of the calf crop or too much decrease in salvage value, the less the percentage replacement required. Moreover, the proportion of younger animals that can be marketed is correspondingly increased. Selection and improvement in longevity are possible in all breeds and should receive more attention.

In a survey made by Washington State University, it was found that old cows are culled or removed from the beef breeding herd at an average age of 9.6 years, and bulls at 6.3 years.[2] It is recognized that the severity of culling will vary somewhat from year to year, primarily on the basis of whether cattle numbers are expanding or declining; and that purebred cattle are usually retained longer than commercial cattle.

[2]Ensminger, M. E., M. W. Galgan, and W. L. Slocum, *Problems and Practices of American Cattlemen,* Wash. Ag. Exp. Sta. Bull. No. 562, Washington State University, 1955.

## ADAPTATION

As has already been indicated—except in those localities where a certain breed predominates, thus making possible the exchange of breeding stock and joint benefits in selling surplus stock—one will usually do best to select that breed for which the producer may have a decided preference. On the other hand, there are certain areas and conditions wherein the adaptation of the breed or class of animals should be given consideration. For example, in the South, Brahman cattle and certain breeds with Brahman blood are able to thrive despite the extreme heat, heavy insect infestation, and less abundant vegetation common to the area. Because of this, Brahman blood has been added to many herds of the South and Southwest, and new strains of beef cattle have evolved.

The cattle producer must always breed for a strong constitution—the power to live and thrive under the adverse conditions to which most animals are subjected sometimes during their lifetime. Under natural conditions, selection occurs for this characteristic by the elimination of the unfit. In domestic herds, however, the constitution of foundation or replacement animals should receive primary consideration.

## PRICE

With a commercial herd, it is seldom necessary to pay much in excess of market prices for the cows. However, additional money paid for a superior bull, as compared to a mediocre sire, is always a good investment. In fact, a poor bull is high at any price.

With the purebred breeder, the matter of price for foundation stock is one of considerable importance. Though higher prices can be justified in the purebred business, sound judgment should always prevail.

## SELECTION BASES

In simple terms, *selection in cattle breeding is an attempt to secure or retain the best of those animals in the current generation as parents of the next generation.* Obviously, the skill with which selections are made is all important in determining the future of the herd. It becomes perfectly clear, therefore, that the destiny of herd improvement is dependent upon the selection for breeding purposes of those animals which are genetically superior. Making the wrong selections and using genetically inferior animals for breeding purposes has ruined many a herd. Under the latter circumstances, the producer would be better off to let the cattle decide on the breeding program by random sampling.

The ultimate objective of beef production is selling beef over the block. Thus, fads or fancies in beef cattle selection that stray too far from this objective will, sooner or later, bring discredit and a penalty.

Strictly from the standpoint of the consumer, a beef animal should produce a carcass which has a high proportion of lean meat, no excess fat, and a minimum of bone—plus "eating quality," which includes tenderness, flavor, and juiciness. Additionally, for efficiency of production under practical farm or ranch conditions, the producer must have animals that produce regularly throughout a long life, and that utilize feed efficiently. From this it may be deducted that the profitability of any one animal, or of a herd, is determined by the following two factors:

1. **Individuality.** This factor is based upon the ability of the animal to produce beef for a discriminating market.

2. **Performance or efficiency of production.** This means the ability to reproduce regularly and utilize feed efficiently.

Four methods of selection are at the disposal of the cattle producer: (1) selection based on individuality or appearance, which may involve either the traditional scorecard system or the Functional Scoring System; (2) selection based on production testing; (3) selection based on pedigree; and (4) selection based on show-ring winnings. Since each method of selection has its place, a producer, especially a purebred breeder, may make judicious use of more than one of them.

## SELECTION BASED ON INDIVIDUALITY OR APPEARANCE/MEASUREMENTS

In starting a new herd, certain matters pertinent thereto must be decided—like the breed or cross, whether to start with open heifers or bred cows, etc. Of equal importance to the success of the operation is the selection of the individuals—the choice of the cows and bulls that constitute the foundation herd, and the selection of replacement heifers, usually from within the herd.

Visual appearance (evaluation or appraisal) has been, and is, the basis of both feeder cattle and slaughter cattle trade. Likewise, individuality or appearance (scored or unscored) is the usual method followed by both commercial and purebred breeders. For the most part, it has been responsible for the transformation of the Texas Longhorn to the present-day bullock.

In making selections based on individuality, it must be borne in mind that the characteristics found in the parents are likely to be reflected in the offspring, for here, as in any breeding program, a fundamental principle is that "like tends to produce like." From a practical standpoint, this points up two things:

1. Only those animals which are at least average, or preferably better than average, should be used for breeding purposes.

2. A cow's inheritance will influence only one calf each year, whereas the herd bull may influence as many as 25 to 50 animals in a given season. Hence, in any selection based on individuality, the selection of the herd bull merits maximum attention.

Different methods of scoring individual animals have evolved; among them are (1) the traditional scorecard system, and (2) the Functional Scoring System presented in this chapter. A discussion of each of these two methods of scoring will follow. But first it should be recognized that both methods are based on visual appraisal. This point bears emphasis because cattle producers and students often get the erroneous impression that, just because some visual scoring system (scoring systems based on visual appearance, in contrast to actual weights, measurements, etc.) is recommended for or used in conjunction with a production-testing program, it must be more accurate than all other scoring systems. This isn't true. All are visual methods, and the score resulting from the use of any of them is no better than the person making it. Some method of selecting all animals by score, preferably on a systematic and written down basis, is the important thing. Any one of several methods may be used. It is the author's contention that the Functional Scoring System herein presented is the most complete and accurate, yet the most simple, of all methods of scoring, and that it is admirably adapted for use in all types of selection programs, including for use in scoring animals on performance test.

## THE TRADITIONAL SCORECARD SYSTEM

A scorecard is a listing of the different parts of an animal, with a numerical value assigned to each part according to its relative importance. It is a standard of excellence. The use of the scorecard involves studying each part—head, neck, back, loin, rump, round, fore and rear flanks, etc.—and assigning a score to each, with a total of 100 for a perfect score. Most of the breed registry associations have scorecards for their respective breeds, which are available on request.

A scorecard is a valuable teaching aid for beginners. It systematizes judging and avoids any part of the animal being overlooked. However, the traditional scorecard has the following limitations: (1) It is not adapted to scoring a great number of animals, or to comparative or show-ring judging, because of the time involved in using it; (2) a near worthless animal may score quite highly—for example, an animal that is so structurally unsound that it can hardly walk may have a rather high total score; (3) it evaluates each part of an animal, rather than the systems—the skeletal system, the muscular system, etc; (4) it is based almost entirely on consumer needs, on the end product—meat; and (5) it accords precious little consideration as to whether, or how, an animal can be changed better to conform to the needs and desires of those it serves.

The Functional System, which follows, is a new and improved method of making individual or appearance selection, on a systematic and recorded basis.

## THE FUNCTIONAL SCORING SYSTEM

Many progressive cattle producers and scientists have long felt that there should be a better way of evaluating animals than either (1) unsystematic and unrecorded visual appraisal, or (2) the traditional scorecard system—that there is need for a system that relates structure to the function of producing quality beef more efficiently. Out of this thinking, several related research findings and concepts based on anatomy have evolved in different parts of the world; and out of this thinking evolved the Functional Scoring System herein presented. This system relates structure (anatomy) to desired function and usefulness—*it is concerned with the composition of gain* (with the quality and yield of the carcass), with eating value. It is limited to 6 traits—all of economic importance, and it is simple and easy to apply. It involves scoring each animal for the following 6 traits on a point basis, with a total, or combined, score of 100.

Hence, the highest and lowest scores that an animal can receive are:

| Trait | Highest Score | Lowest Score |
|---|---|---|
| Reproductive efficiency . . . . | 20 | 1 |
| Muscling . . . . . . . . . . . . . . . | 20 | 1 |
| Size . . . . . . . . . . . . . . . . . . | 15 | 1 |
| Freedom from waste . . . . . . | 15 | 1 |
| Structural soundness . . . . . . | 15 | 1 |
| Breed type . . . . . . . . . . . . . | 15 | 1 |
| **Total** | **100** | **6** |

If desired, one can establish a minimum score for each character, then cull those animals that fall below this score for the particular character. For example, it might be decided to cull all animals that score less than 5 in muscling. When used in this manner, the Functional Scoring System is somewhat similar to the system of selection known as "Establishing minimum culling standards for each character, and selecting simultaneously, but independently, for each character," described in Chapter 3 of this book under the heading "Systems of Selection," except that the latter usually implies an objective measure (like pounds of feed, or pounds of gain) whereas the Functional Scoring System is visual. Of course, the minimum standards should increase slightly with each generation if progress is being made.

Like the computer, the Functional Scoring System is no better than the people back of it—the scorers. For best results, it is recommended (1) that the scorers completely disregard any facts about the animal that would color their thinking and affect its score—such as pedigree, purchase price, and show-ring record—and concentrate on the particular characteristic that they are scoring at the time; (2) that only one animal be scored at a time; and (3) that all cattle being scored be in the same corral or arena, on firm level footing, and at approximately the same location with reference to the scorer.

It is important that the same person score the animals of a given herd.

For convenience, the author has prepared the All Breed Functional Scorecard which follows (Fig. 13–5).

| | Perfect Score | Animal | | | | |
|---|---|---|---|---|---|---|
| | | No. 1 | No. 2 | No. 3 | No. 4 | Etc. |
| **REPRODUCTIVE EFFICIENCY:** ...................................... | 20 | | | | | |
| *Highly fertile female*—Feminine—long body, lean, smooth muscled; refined, feminine head; lean cheek, jaw, neck, brisket, shoulder, and hindquarters; and a good functional udder (or promise of udder development in a heifer). | | | | | | |
| *Avoid lowly fertile female*—Steery appearance—coarse, heavy front, masculine rather than feminine; protruding brisket; bristly hair on neck and top of shoulders; rounded hindquarters; and fat deposits on the face, brisket, shoulders, hips, rump, pins, below the vulva, and in front of the udder. | | | | | | |
| *Highly fertile bull*—Masculine—*he's on the look*, with head up and ears cocked; well-developed crest; muscles well developed and clearly defined especially in the regions of the neck, loin, and thigh; and well-developed genitalia, with testicles of equal size and well defined, and a proper neck to the scrotum. | | | | | | |
| *Avoid lowly fertile bull*—Lacking masculinity—ears not alert; undeveloped crest; muscles lacking development and not clearly defined; testicles small, unbalanced, or with one carried high; scrotum that is twisted or filled with fat. | | | | | | |
| **MUSCLING:** ........................................................ | 20 | | | | | |
| *Well muscled*—Bulging in those areas least affected by fatness—the arm, forearm, gaskin, and stifle, muscles move and bulge as animal walks. Look for curved loin and round; crease in thigh; well-defined groove down topline, with loin eye bulging on each side of backbone. Look for calves with long, smooth muscling, indicating continued growth. Since muscling is a masculine trait, it is more important in bulls and steers than in heifers. | | | | | | |
| *Avoid coarse shoulders* in breeding cattle, because it is usually associated with calving problems. | | | | | | |
| **SIZE:** ............................................................... | 15 | | | | | |
| *Adequate size*—As indicated by height over the hip and body length from shoulders to tailhead. Young breeding animals and steers should be long, tall, and not excessively fat—indications that they will continue to grow. *Avoid* bulls showing signs of early sexual maturity; they are not likely to make continued rapid growth and reach large mature size. | | | | | | |
| **FREEDOM FROM WASTE:** ........................................ | 15 | | | | | |
| *Freedom from waste; trimness*—In both breeding and slaughter cattle. Excessively fat breeding cattle usually have lowered reproduction. Excessively fat slaughter cattle have reduced carcass value. | | | | | | |
| *Avoid loose hide* that is filled, or will fill, with fat. Look for loose hide on the throat, dewlap, brisket, foreflank, navel or sheath, and twist. Look for fat over back ribs, point of shoulder, and along backbone; since no muscle should be found at these places, if you feel something, it is fat. | | | | | | |
| **STRUCTURAL SOUNDNESS:** ........................................ | 15 | | | | | |
| *Structurally sound*—Legs straight, true, and squarely set; feet large, wide, and deep at the heel, with toes of equal size and shape that point straight ahead; hock and knee joints correctly set and clean. | | | | | | |
| *Avoid* sickle-hocked, post-legged, back at the knees (calf-kneed), over at the knees (buck-kneed), or puffiness or swelling of knee or hock joints. | | | | | | |
| **BREED TYPE:** ...................................................... | 15 | | | | | |
| *Characteristics true to breed*—Breed distinguished by color and markings, shape of head, presence or absence of horns (and shape of horns if present), set of ears, body shape, and size. | | | | | | |
| *Avoid* breed characteristics associated with undesirable traits. Commercial producers can disregard this trait. Likewise, it is unimportant in steers. | | | | | | |
| **TOTAL** ............................................................. | 100 | | | | | |

Fig. 13-5. All Breed Functional Scorecard.

## TESTICULAR SIZE IN BULLS

Testicular size is influenced by age, breed, and condition. But, within a given breed and at the same age and condition, testicular size and semen production are highly correlated.

The recommended method for taking scrotal circumference measurements is shown if Fig. 13–6. Note that the tape measurement is made at the largest circumference. Recommended minimum scrotal circumferences for young bulls of different ages in good condition are as follows:

| Age | Scrotal Circumference | |
|---|---|---|
| (months) | (in.) | (cm) |
| 6 .................... | 7.9 | 20.0 |
| 12 .................... | 12.6 | 32.0 |
| 18 .................... | 13.2 | 33.5 |
| 24 .................... | 13.8 | 35.0 |

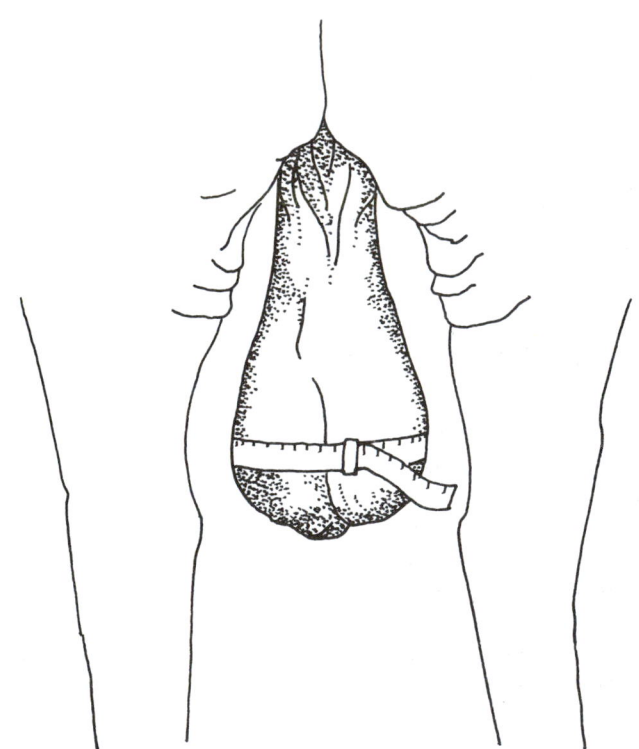

Fig. 13–6. Method of determining scrotal circumference; using a tape and making the reading at the largest circumference.

## MEASURING CATTLE

In recent years, measurements have become a descriptive supplement to many herd testing programs. Adjusted weights and weight ratios accompanied by linear measurements have added another dimension to evaluating the fat:lean ratio of an individual animal in a performance measure program.

Linear measurements are objective. They serve as supplemental information for comprehensive performance testing. How much emphasis breeders should place on linear measurement information should depend on their goals relative to shape and growth patterns, the extent to which certain shape relationships may be important to them, and any advantage these shape relationships give them in marketing beef cattle.

A linear measurement should never be interpreted as a replacement for the weight of an animal at a given age. Instead, linear measurements should be used with growth information as a supplement to selection. No one frame size for an animal will be best for all feed resources, breeding systems, and feed costs. Reproductive efficiency and market weight will determine the optimum frame size range within a given set of feed resources, breeding systems, and production costs.

The most common measurement is over the hip, halfway between the hook and the pin. Some producers also measure height at the shoulders (the highest point over the shoulders), and body length from the shoulders (highest point) to the tailhead. Fig. 13–7 shows these points, which may be measured by tape or calipers. When making height measurements, the animals should be stood squarely on a level area.

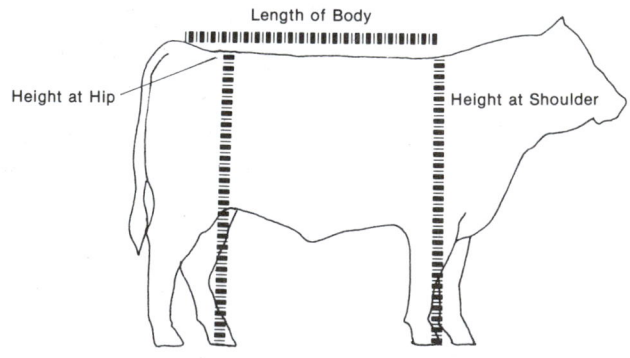

Fig. 13–7. Measurement points pertinent to cattle.

The hip measurement is adjusted to relatively logical production end points at 205 days and 365 days (within the Beef Improvement Federation [BIF] ranges currently used for adjusted weights). It is recommended that the actual hip height and adjusted hip height be printed in the sire summary, but that no height ratio be calculated. The following height adjustments are recommended for bulls and heifers:

# ADJUSTED WEANING HEIGHTS

1. **Sex adjustment factors** for heights at weaning are:

   **Bulls** = 0.033 in. *(10.084 cm)*; **Heifers** = 0.025 in. *(0.063 cm)*

   To adjust heights for sex, (a) multiply the number of days *under* 205 by 0.033 for bulls or 0.025 for heifers, and *add* to the actual height; or (b) multiply the number of days *over* 205 by 0.033 for bulls or 0.025 for heifers and *subtract* the result from the actual height.

2. **Age-of-dam adjustment factors** for heights at weaning are:

| Age of Dam | Adjustment Factors for | | | |
|---|---|---|---|---|
| | Bulls | | Heifers | |
| (years) | (in.) | (cm) | (in.) | (cm) |
| 2 or 13+ | 1.02 | 2.59 | 1.02 | 2.59 |
| 3 or 12 | 1.015 | 2.58 | 1.025 | 2.60 |
| 4 or 11 | 1.01 | 2.57 | 1.01 | 2.57 |
| 5 through 10 | . . . . . . . . . . . . . . . . . . . . . . . . . . no adjustment . . . . . . . . . . . . . . . . . . . . . . . . . . | | | |

To adjust for age of dam, multiply the adjusted height for sex by the age-of-dam factor.

- **Example—Heifer**

| Age | Weight | Height |
|---|---|---|
| Born January 1, 1984 | 70 lb *(32 kg)* | |
| *Weaned July 21, 1984* | *470 lb (214 kg)* | 38 in. *(96.5 cm)* |
| 201 days old | Adjusted 205-day weight = 496 lb *(225 kg)* | |

**Calculations:**

205 days – 201 days = 4 days
38 in. *(96.5 cm)* + (4 × 0.025) = 38.1 in. *(96.8 cm)*
*38.1 in.* × 1.015 *(25.4 mm)* = 38.67 in. *(98.2 cm)*

*38.67 in. (98.2 cm)* is the 205-day height adjusted for sex of calf and age of dam

# ADJUSTED POSTWEANING HEIGHTS

| | Bulls | Heifers* |
|---|---|---|
| *Under 365 days* | Actual height *plus* (number of days *under* 365 × 0.033) = Adjusted Height | Actual height *plus* (number of days *under* 365 × 0.025) = Adjusted Height |
| *Over 365 days* | Actual height *minus* (number of days *over* 365 × 0.033) = Adjusted Height | Actual height *minus* (number of days *over* 365 × 0.025) = Adjusted Height |

*The great variation in the nutritional levels of rations which heifers are fed between weaning and 1 year old can vary their growth rate slightly. However, the above adjustments are fairly accurate for all heifers fed to gain in a range from 0.75 to 2.0 lb *(0.19 to 0.9 kg)* per day.

**Example—Bull**

*Age:* 359 days old          *Weight:* Adjusted 365-day          *Height:* 49.5 in. *(19.5 cm)*
          weight = 1,050 lb *(47.7 kg)*

          365 – 359 days = 6 days

49.5 in. *(19.5 cm)* + (6 × 0.033) = 49.7 in. *(19.6 cm)* is the Adjusted Postweaning Height

## SELECTION BASED ON PERFORMANCE TESTING

The modern era in beef cattle breeding was ushered in with the weighing of animals and the keeping of written production records. This led to performance-testing programs, most of which were aimed at improvement of growth rate and feed efficiency. The ease of measuring these two characteristics made performance testing acceptable to producers.

Simply stated, *performance testing is a record-keeping system for the purpose of collecting data to be used in selection*. It has been an important selection tool in the hands of cattle breeders. It brought an awareness that some animals are more efficient than others, and that such characteristics as rate of gain and feed efficiency are at least partially under genetic control and can be passed on to offspring.

However, a performance-testing program based only on rate of gain and feed efficiency is not adequate, because composition of gain cannot be ignored and still meet consumer demand. We need to know if the animal is converting its feed to red meat or fat. We need to know the pounds of feed required to produce a pound of consumer accepted lean beef.

Rapid growth favors large size, but there is likely a point beyond which large size is not entirely compatible with other production traits. Also, within a breed, and at a given age, cattle with the higher weight per day of age tend to be fatter and contain a lower percentage of muscle. At a given weight, however, the greater the weight per day of age, the less the total fat and the greater the amount of total muscling.

Production-testing programs that stop too early (*i.e.*, under 1 year of age, or at less than 900 lb weight) may lead to the selection of some animals which do not have the inheritance to continue to grow lean meat to heavier weights.

Thus, the goals of performance testing need to be reappraised, and measuring techniques need to be improved and become more sophisticated. To growth rate and feed efficiency, we need to add composition of gain as determined by some such methods as (1) the Functional Scoring System on live animals, and/or (2) carcass studies on slaughter animals.

The Beef Improvement Federation (BIF), whose membership is made up of United States organizations with an interest in performance testing, was formed on February 1, 1968. One of its main objectives is to establish accurate and uniform procedures for measuring and recording data concerning the performance of beef cattle.

## SELECTION BASED ON PEDIGREE

Selection based on pedigree refers to the selection of animals to be the parents of the next generation, based upon their ancestors. This method is used in most purebred herds, usually in combination with one or more of the other bases of selection: individuality, production test, or show-ring winnings.

Pedigree fads as such should be avoided, especially if there has not been rigid culling and selection based on utility value; and in all instances, poor individuals should be culled, regardless of the excellence of the relatives. Likewise, one should not be misled by or overestimate the value of such pedigree information as the following:

1. Family names or favorite animals many generations removed.

2. Percentage of blood, including—

    a. The use of "blood" fractions instead of percentages.

    b. Setting an arbitrary fraction for purebreds.

For maximum genetic progress, the pedigree should be more than a mere listing of birth dates and names of ancestors. A pedigree should combine genealogy and performance. Such a pedigree should contain a complete listing of an animal's performance record and its ancestors' performance and progeny records.

The Beef Improvement Federation (BIF) recommends that a performance pedigree contain the following basic information, with the format of the pedigree left to each recording organization:[3]

1. **Animal's individual record:**

    a. Birth weight

    b. 205-day adjusted weaning weight

    c. Weaning weight ratio (the individual weight divided by the average weight of its contemporaries times 100)

    d. Number of contemporaries, weaning

    e. 365-, 452-, or 550-day adjusted yearling weight

    f. Yearling weight ration (computed same as "b" above)

    g. Number of contemporaries, yearling

2. **Progeny of each individual in pedigree:**

    a. Number of progeny and average weight ratios

3. **Additional considerations:**

    a. Breeding values may be added to any traits that are considered important.

## SELECTION BASED ON SHOW-RING WINNINGS

For years, many cattle producers (purebred breeders and commercial producers alike) looked favorably upon and used show-ring winnings as a basis of selection. Purebred breeders were quick to recognize this appeal and to extol their champions through advertising. In most instances, the selection of foundation and replacement cattle, and herd bulls, on the basis of show-ring winnings and standards was for the good. On some occasions, however, purebred and commercial breeders alike came to regret selection based on show-ring winnings. A case at point was the period from 1935 to 1946, when the smaller, earlier-maturing, blockier, and smoother types of cattle were

---

[3]*Guidelines for Uniform Beef Improvement Programs,* Beef Improvement Federation Recommendation.

winning. Among many, this debacle brought disrepute, from which livestock shows may never fully recover. This would indicate that some scrutiny should be exercised relative to the type of animals winning in the show, especially to ascertain whether they are the kind that are efficient from the standpoint of the producer, and whether, over a period of years, they will command a premium on a discriminating market.

Perhaps the principal value of selections based on show-ring winnings lies in the fact that shows direct the attention of new breeders to those types and strains of cattle that are meeting with the approval of the better breeders and judges.

## HERD IMPROVEMENT THROUGH SELECTION

Once the herd has been established, the primary objective should be to improve it so as to obtain the maximum production of quality offspring. In order to accomplish this, there must be constant culling and careful selection of replacements. The breeders who have been most constructive in such a breeding program have usually used great breeding bulls and then have obtained their replacements by selecting some of the outstanding, early-maturing heifers from the more prolific families.

Improvements through selection are really twofold: (1) the immediate gain in increased calf production from the better animals that are retained; and (2) the genetic gain in the next generation. The first is important in all herds, whereas the second is of special importance in purebred herds and in all herds where replacement females are raised. Most of the immediate gain is attained in selecting the cows, which are more numerous than the bulls; whereas the majority of the genetic gain comes from the careful selection of bulls. The genetic gain is small, but it is permanent and can be considered a capital investment.

Many good cattle breeders consider it a sound practice to make about a 20% replacement each year. Under such a system of management, 46% of the heifer calves are retained each year (see footnote No. 1).

## BULL GRADING

Bull grading programs have exerted a powerful influence in improving the commercial cattle on the western ranges of this country. Perhaps this movement received its greatest impetus in those areas where several owners run herds on unfenced public grazing lands. Formerly, those progressive ranchers who believed that only purebred beef bulls of high quality should be used could do nothing to prevent the presence of inferior bulls on these public ranges. The producers who bought superior bulls got no more use from them than their neighbors who turned out scrubs because they could buy them cheaply. This problem was finally solved when groups of cattle producers using common ranges decided to have their bulls classified and to use only bulls meeting certain grades. Today, grazing permits are sometimes refused or delayed because of ranchers refusing to use graded bulls.

In some cases in the West, all animals consigned to range bull sales are graded, and individual ranchers grade their young bulls before turning them with the cow herd. In some consignment sales, bulls must be of a certain specified minimum grade in order to be sold. Grading of sale bulls is especially popular with most buyers, but some sellers object to it.

Many different systems of bull grading have been used in the past. Today, most grading is on a numbering system, in which numbers from 0 to 17 are used.

## JUDGING BEEF CATTLE

Fig. 13–8. In livestock shows, a great number of animals may be ranked, or placed in each class. This shows Simmental cattle in the showing at the National Western Stock Show. (Courtesy, National Western Stock Show, Denver, CO)

Livestock judging is the art of visual appraisal, or the making of a subjective evaluation, of an animal. Cattle producers use it every day to select herd sires and replacement females for their breeding programs and to determine when animals are ready for market.

In addition to individual merit, the word judging implies the comparative appraisal or placing of several animals. In most judging contests, 4 animals are used in each class; and they are numbered 1, 2, 3, and 4, left to right as viewed from the rear. In livestock shows, a great number of animals may be ranked, or placed, in each class.

Judging beef cattle is an art, the rudiments of which must be obtained through patient study and long practice. The master breeders throughout the years have been competent livestock judges.

## PARTS OF AN ANIMAL

The first step in preparation for judging beef cattle consists of mastering the language that describes and locates the different parts of the animal. This information is set forth in Fig. 13–9.

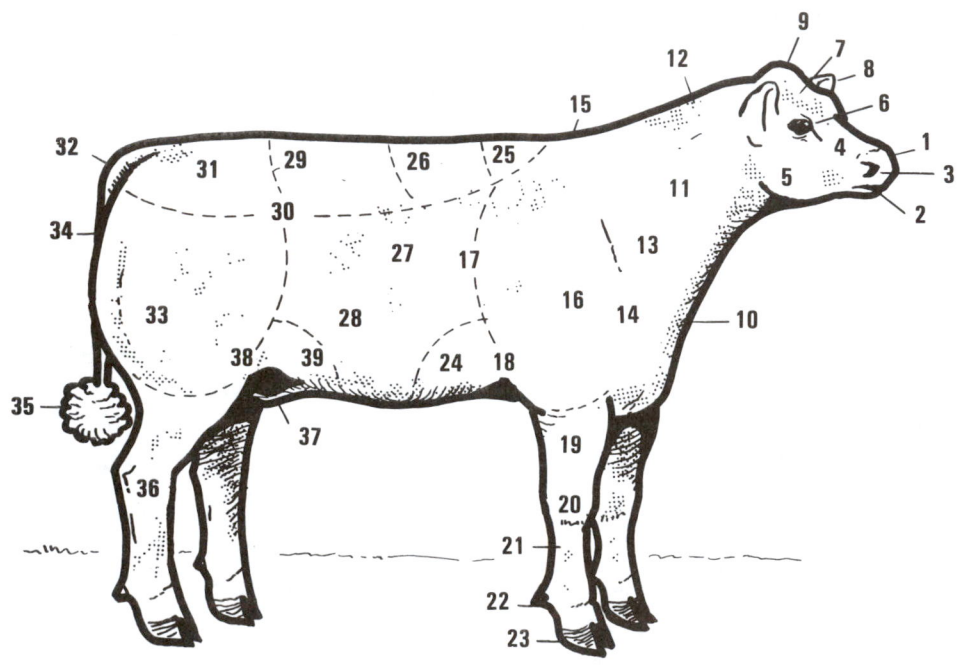

Fig. 13–9. Parts of a steer. The first step in preparation for judging beef cattle consists in mastering the language that describes and locates the different parts of the animal. (Drawing by Prof. R. F. Johnson)

| | | | |
|---|---|---|---|
| 1. Muzzle | 11. Neck | 21. Shank | 31. Rump |
| 2. Mouth | 12. Crest | 22. Dew claw | 32. Tailhead |
| 3. Nostril | 13. Shoulder vein | 23. Foot | 33. Round |
| 4. Face | 14. Point of shoulder | 24. Fore flank | 34. Tail |
| 5. Jaw | 15. Top of shoulder | 25. Crops | 35. Switch |
| 6. Eye | 16. Shoulders | 26. Back | 36. Hock |
| 7. Forehead | 17. Foreribs or heart girth | 27. Ribs | 37. Cod |
| 8. Ear | 18. Elbow | 28. Belly | 38. Stifle |
| 9. Poll | 19. Arm | 29. Loin | 39. Rear flank |
| 10. Dewlap | 20. Knee | 30. Hip or hook | |

## IDEAL BEEF TYPE AND CONFORMATION

The next requisite in judging is to have clearly in mind a standard or ideal. Presumably, this ideal should be based on a combination of (1) the efficient performance of the animal from the standpoint of the producer, and (2) the desirable carcass characteristics of market animals as determined by the consumer.

Expert livestock judges endeavor to select cattle that are most efficient at producing quality beef. Steers that are growthy, muscular, and lack excessive finish (external fat) are usually placed at the top. Judges look for the same things in breeding classes, but, in addition, they look for structural soundness and

sex character—traits that indicate that they will be efficient producers of offspring in the future.

Both breeding cattle and steers should be growthy. They should be long, tall, and not excessively fat if they are to continue to grow. Above all, avoid selecting a young animal that looks mature (that looks old for its age) and is overfat, because such characteristics are indicative of an animal that will not grow enough to be competitive in the show-ring.

Muscling, being a masculine trait, is more important in bulls and steers than in heifers. Long smooth muscling is preferred in young animals because they will usually grow for a longer period of time and get thicker as they get older. Truly muscular cattle are not smooth all over. They show some creases and indentations between the muscles; they are prominent in the forearm and stifle; and they are slightly narrower through the heart girth and loin than through the shoulder and round. Breeding cattle with coarse shoulders (very heavily muscled shoulders) should be avoided since this condition is frequently associated with calving problems.

Trimness, or freedom from predisposition to waste, is important in both breeding and slaughter cattle. Excessively fat breeding cattle will usually have poorer reproduction than trimmer ones; and very fat slaughter cattle will have reduced carcass value. Usually, prospective calves are not fat to begin with; hence, the exhibitor must estimate whether they are the kind that are likely to get overfat. Generally, calves that have large briskets and that are deep in the flanks and twist will have a tendency to be overfat when fitted to mature weight, and should be avoided.

Structural soundness is especially important in breeding cattle, but only moderately so in show steers. Nevertheless, the skeletal structure of steers should be sufficiently correct that they will not develop any serous unsoundnesses as fitting progresses. Breeding cattle must be structurally sound so that they will calve easily and be able to travel well on the pasture or range. Proper structural soundness involves a big foot, a deep heel, toes that are the same size and point straight ahead, clean joints, and correctly set legs. Avoid cattle that are sickle-hocked or post-legged. Breeding cattle should be wide through the pins.

Signs of fertility and reproductive efficiency are extremely important in breeding cattle. Bulls should have two testicles that are of normal size, and that are well defined in the scrotum, rather than surrounded by excess fat. A mature bull should be masculine in front. But bull calves should not show extreme masculinity at a young age, because such animals are apt to mature too early and stop growing.

When selecting females for show, it is well to look for femininity at all ages. Feminine females are trim in the jaw, throat, and dewlap, and smooth shouldered. Avoid masculine females—the kind that are coarse, heavy fronted, and excessively muscular.

In summary, when selecting prospective show steers, emphasize growthiness, skeletal size, muscling, and trimness. Add masculinity and soundness to the list when selecting bulls; and add femininity and structural soundness when selecting females.

The Functional Scoring System lists the requisites of an ideal animal (if it scores 100) in terms of the following six traits of economic importance: (1) reproductive efficiency, (2) muscling, (3) size, (4) freedom from waste, (5) structural soundness, and (6) breed type. (See earlier section in this chapter headed "The Functional Scoring System.")

The great breeders of the past visualized an ideal better than the standards current at the time, and they were not stampeded by passing fads. They possessed both a goal and a program.

It must be recognized, however, that the perfect specimen has never been produced. Each animal possesses one or several faults. In appraising an individual animal, therefore, its good points and its faults must be recognized, weighed, and evaluated in terms of an imaginary ideal. In comparative judging—that is, in judging a class of animals—the good points and the faults of each animal must be compared with the good points and the faults of every other animal in the class. In no other manner can they be ranked.

In addition to recognizing the strong and weak points in an animal, it is necessary that the successful judge recognize

the degree to which the given points are good or bad. A sound evaluation of this kind requires patient study and long experience. Fig. 13–10 shows the ideal beef type vs some of the common faults.

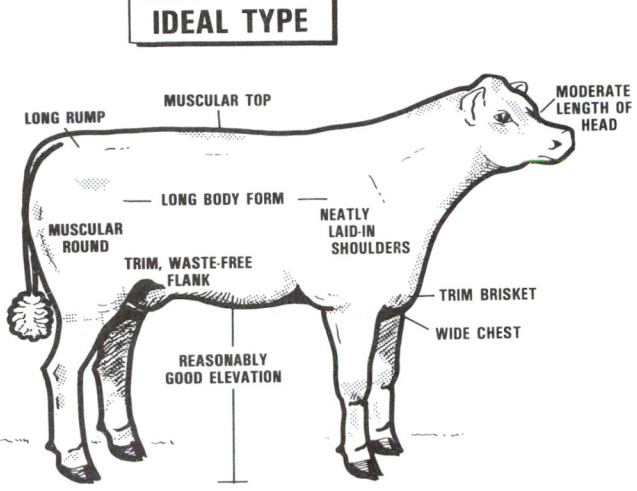

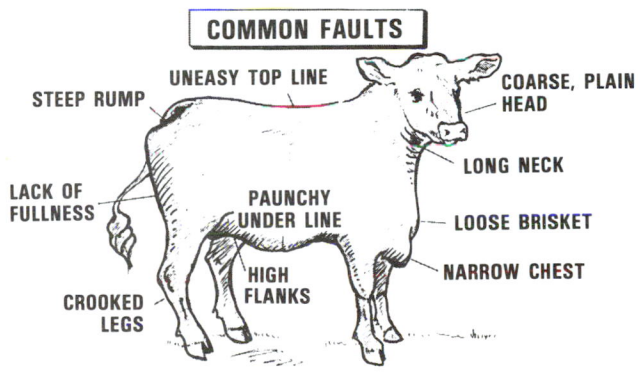

Fig. 13–10. Ideal beef type vs common faults. After mastering the language that describes and locates the different parts of an animal, the next requisite in judging is to have clearly in mind a standard or ideal. In brief, the successful beef cattle judge must know what to look for, and be able to recognize and appraise the common faults. (Drawing by Prof. R. F. Johnson)

## THE JUDGE

The judge is the person chosen by the show management to determine the relative merits of the animals entered in the show. Judging is hard work and a great responsibility. Not only do judges pick the winners, but they lead or mislead many people. For better or worse, they may be the cause of changing breeding programs which can affect the traits of the entire breed.

The essential qualifications that a good judge of beef cattle must possess, and the recommended procedure to follow in selecting or judging are as follows:

1. **Knowledge of the parts of cattle and the relative importance of each.** This consists of mastering the language that describes and locates the different parts of cattle (Fig. 13–10). In addition, it is necessary to know which of these parts are of major importance in terms of form relating to function. In a slaughter animal, the latter necessitates knowledge of cut-

ability and quality grades. In breeding animals, the Functional Scoring System may serve as a useful guide.

2. **A clearly defined ideal or standard of perfection.** Successful cattle judges must know for what they are looking; that is, they must have in mind an ideal or standard of perfection based on a combination of (a) the efficient performance of the animal from the standpoint of producers, and (b) the desirable carcass characteristics of market animals as determined by consumers.

The trained judge can evaluate an animal in a matter of seconds, using an objectively developed mental index.

3. **Keen observation and sound judgment.** The good judge possesses the ability to observe both good conformation and defects, and to weigh and evaluate the relative importance of the various good and bad features.

4. **Honesty and courage.** The good judge of any class of livestock must possess honesty and courage, whether it be in making a show-ring placing or conducting a breeding and marketing program. For example, it often requires considerable courage to place a class of cattle without regard to (a) placings in previous shows; (b) ownership; and (c) public applause. It may take even greater courage and honesty with oneself to discard from the herd a costly animal whose progeny has failed to measure up.

5. **Tact.** In discussing either (a) a show-ring class, or (b) animals on a producer's farm or ranch, it is important that the judge by tactful. Owners are likely to resent any remarks that imply that their animals are inferior.

Having acquired the knowledge referred to above, long hours must be spent in patient study and practice in comparing animals. Even this will not make expert and proficient judges in all instances, for there may be a grain of truth in the statement that "the best judges are born and not made." Nevertheless, training in judging and selecting cattle is effective when directed by a competent instructor or experienced cattle producer.

## METHOD OF EXAMINING

As in examining any class of livestock, in judging beef cattle the examination should be systematic and thorough. First, the animal (or animals) should be viewed from a distance, obtaining views from the side, rear, and front. Finally, the animal should be handled.

It makes little difference as to the order of the views in inspecting cattle, but it is important that the same procedure be followed each time. Though good judges differ, perhaps as logical a method of examining as any is illustrated in Fig. 13–11.)

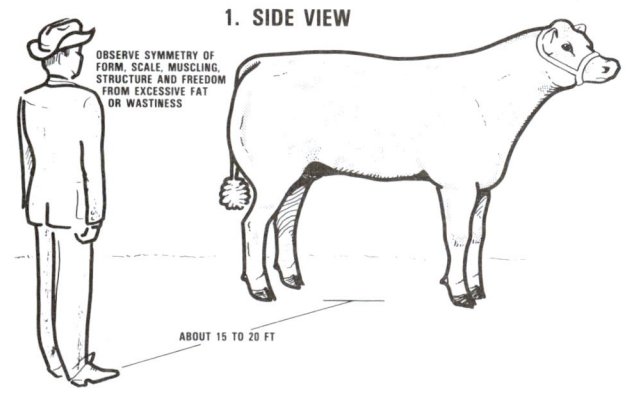

**1. SIDE VIEW**

OBSERVE SYMMETRY OF FORM, SCALE, MUSCLING, STRUCTURE AND FREEDOM FROM EXCESSIVE FAT OR WASTINESS

ABOUT 15 TO 20 FT

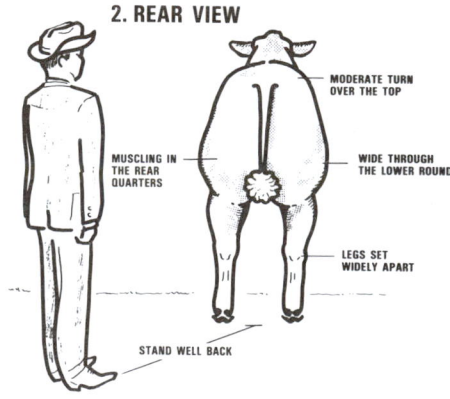

**2. REAR VIEW**

MODERATE TURN OVER THE TOP

MUSCLING IN THE REAR QUARTERS

WIDE THROUGH THE LOWER ROUND

LEGS SET WIDELY APART

STAND WELL BACK

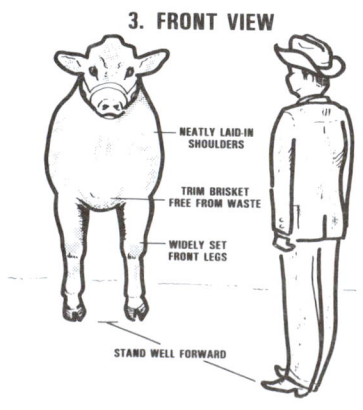

**3. FRONT VIEW**

NEATLY LAID-IN SHOULDERS

TRIM BRISKET FREE FROM WASTE

WIDELY SET FRONT LEGS

STAND WELL FORWARD

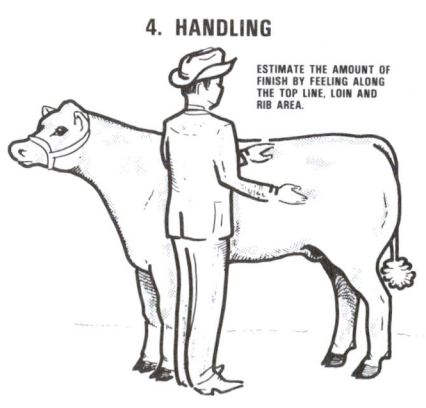

**4. HANDLING**

ESTIMATE THE AMOUNT OF FINISH BY FEELING ALONG THE TOP LINE, LOIN AND RIB AREA.

Fig. 13–11. Steps in examining (1) side view, (2) rear view, (3) front view, and (4) handling. It is important that the judge follow a logical procedure in looking over animals. (Drawing by Prof. R. F. Johnson)

# QUESTIONS FOR STUDY AND DISCUSSION

1. Cattle producers are constantly called upon (a) to cull out animals, (b) to select replacements, and (c) to market surpluses. Each of these decisions calls for a different emphasis in the evaluation or appraisal. List, then explain, why these differences.

2. What are the primary differences between purebred breeders and commercial producers from the standpoint of (a) goals, and (b) breeder/producer requirements?

3. Select a certain farm or ranch (either your home farm or ranch, or one with which you are familiar). Assume that there are no beef cattle on this establishment at the present time. What breed or cross would you select? Justify your selection.

4. How would you go about getting more milking ability (a) in a purebred herd, and (b) in a commercial herd?

5. What factors should be considered in determining the size of the beef herd for a particular farm unit?

6. Is uniformity of color important in a commercial herd where a crossbreeding program is being followed? If so, how could you obtain uniformity?

7. When establishing a new herd, what precautions should be taken relative to herd health?

8. When establishing a new herd, how important is condition?

9. Discuss the economics of longevity in a cow herd. Use as the examples two cows dropping their first calves as 2-year-olds, each producing a calf each year thereafter—but one producing through age 8, and the other producing through age 10.

10. Why should selection be from among animals kept under an environment similar to that under which the producer expects them and their offspring to perform?

11. What factors determine price when establishing a herd?

12. Define *selection*. Why and how is the profitability of any one animal, or of a herd, determined by (a) individuality, and (b) performance or efficiency of production?

13. List the four methods of selection that are at the disposal of the cattle producer.

14. Discuss the methods and importance of selecting beef cattle based on individuality or appearance.

15. What is meant by "the traditional scorecard system?" What are the virtues and what are the limitations of the scorecard system?

16. Under the Functional Scoring System, six traits are listed. Rank these traits in order of economic importance. Justify your ranking.

17. Evaluate the All Breed Functional Scorecard which is presented as Fig. 13–5. How would you improve it?

18. What factors influence testicular size in bulls? Why is testicular size important?

19. Why and how are some cattle being measured? What are the most common measurements?

20. Define *performance testing*. What bases are used in performance testing beef cattle?

21. How would you improve a performance-testing program that is based primarily upon rate and efficiency of gain.

22. How could pedigrees be made more useful from a selection standpoint?

23. Cite examples as proof of the fact that show-ring standards have not always been practical.

24. What is the twofold objective in making herd improvements through selection?

25. Under what conditions is bull grading particularly valuable? Why is bull grading generally more popular with buyers than sellers?

26. When judging beef cattle, why is it important that judges know the parts of an animal?

27. Describe the ideal beef-type animal. List and describe the common faults.

28. List and discuss the essential qualifications that a good judge of beef cattle must possess.

29. What procedure in examining beef cattle would you recommend?

# SELECTED REFERENCES

| Title of Publication | Author(s) | Publisher |
|---|---|---|
| *Beef Cattle*, Seventh Edition | A. L. Neumann | John Wiley & Sons, Inc., New York, NY, 1977 |
| *Beef Cattle Production* | J. F. Lasley | Prentice-Hall, Inc., Englewood Cliffs, NJ, 1981 |
| *Beef Cattle Science*, Sixth Edition | M. E. Ensminger | The Interstate Printers & Publishers, Inc., Danville, IL, 1987 |
| *Beef Production in the South*, Modified Edition | S. H. Fowler | The Interstate Printers & Publishers, Inc., Danville, IL, 1979 |
| *Breeding and Improvement of Farm Animals*, Sixth Edition | V. A. Rice, *et al.* | McGraw-Hill Book Company, New York, NY, 1967 |

*(Continued)*

## SELECTED REFERENCES *(Continued)*

| Title of Publication | Author(s) | Publisher |
|---|---|---|
| *Commercial Beef Cattle Production*, Second Edition | Edited by<br>C. C. O'Mary<br>I. A. Dyer | Lea & Febiger, Philadelphia, PA, 1978 |
| *Genetics of Livestock Improvement*, Third Edition | J. F. Lasley | Prentice-Hall, Inc., Englewood Cliffs, NJ, 1978 |
| *Guidelines for Uniform Beef Improvement Programs*, Program Aid 1020 | | Beef Improvement Federation, USDA, Washington, DC, 1972 |
| *Improvement of Livestock* | R. Bogart | The Macmillan Company, New York, NY, 1959 |
| *Improving Reproductive Efficiency in Beef Cattle* | J. R. Beverly, *et al.* | Glidwell Printers, Bryan, TX, 1972 |
| *Lasater Philosophy of Cattle Raising, The* | L. M. Lasater | Texas Western Press, The University of Texas, El Paso, TX, 1972 |
| *Livestock Judging, Selection and Evaluation*, Second Edition | R. E. Hunsley<br>W. M. Beeson<br>J. E. Nordby | The Interstate Printers & Publishers, Inc., Danville, IL, 1978 |
| *Proceedings of Second World Conference on Animal Production* | R. E. Hodgson, *et al.* | American Dairy Science Association, Urbana, IL, 1969 |
| *Stockman's Handbook, The*, Sixth Edition | M. E. Ensminger | The Interstate Printers & Publishers, Inc., Danville, IL, 1983 |

Cattle must eat—come rain, snow, or sleet! This shows cattle eating hay on the snow on the LaVaughn Herrick ranch, Cascade, Idaho. (Courtesy, USDA)

Being born and born alive isn't easy! This shows a newborn calf covered with snow. (Courtesy, National Cattlemen's Assn., Englewood, CO)

# BREEDING BEEF CATTLE

## Chapter 14

Reproduction is the first and most important requisite of beef cattle breeding. Without a calf being born—and born alive, the other economic traits are of academic interest only. Yet, 8% of the nation's cows never calve, and there is an appalling calf loss of 6% at birth.

Although it is true that exactly the same laws of nature govern heredity in all animals, special breeding problems are peculiar to each class of stock, and cattle are no exception. These will be discussed in this chapter.

The reproductive organs of the bull and the cow are pictured and described in Chapter 3 of this book; hence, the reader is referred thereto for information on this subject.

## NORMAL BREEDING HABITS OF COWS

In general, cattle that are bred when out on pasture or range are mated under environmental conditions approaching those which existed in nature prior to domestication. Less breeding trouble is generally encountered among such animals than among beef or dairy animals that are kept in confined conditions and under forced production.

## AGE OF PUBERTY

The normal age of puberty in cattle is 8 to 12 months. It is recognized, however, that the age at which puberty is attained varies according to (1) breeds, with the smaller breeds having an earlier onset of puberty than the larger, slower maturing ones; and (2) nutritional and environmental factors, with puberty occurring when animals have reached about one-third of their adult size.

## AGE TO BREED HEIFERS

The age at which to breed heifers will vary with their growth and development. However, when heifers are reasonably well grown and weigh 600 to 650 lb, a safe rule is to breed at the first breeding season after they are 13 to 14 months old. Some breed registry associations will not register a calf born to a heifer under a certain stipulated age; thus, purebred breeders need to be informed relative to such rules.

There is an increasing tendency among producers to breed heifers at an early age. Some operators sort the young bred heifers out from the rest of the herd in order to provide the choicest pasture or range for them. Others breed first-calf heifers to a bull that is known to sire relatively small calves at birth. Certainly, if the dam and the calf are not adversely affected, breeding at an early age would be advantageous from the standpoint of cutting production costs.

## BREEDING HEIFERS TO CALVE AS TWO-YEAR-OLDS

The following are some of the advantages and disadvantages of having heifers calve for the first time as 2-year-olds:

### Advantages—

1. On a lifetime basis, it will result in the production of about one more calf and an added calf weight of approximately 350 lb.

2. Cow cost per hundred pounds of weaned calf will be lower.

Fig. 14–1. Two-year-old Beefmaster heifers with their first calves. (Courtesy, Tom Lasater, Lasater Ranch, Matheson, CO)

### Disadvantages—

1. The conception rate of young heifers bred when just reaching puberty is lower than in older animals. This results in spreading the calving season over a longer period, with accompanying greater inconvenience and expense.

2. The percentage calf crop of heifers calving as 2-year-olds will be about 10% lower than of older cows.

3. More calving troubles and a higher death loss of both mothers and calves will be encountered.

4. Early calving heifers will likely be somewhat undersized until they reach 4 to 5 years of age.

5. The calves will wean at 25 to 50 lb lighter weights.

6. It may not be desirable for the purebred breeder from the standpoints of (a) compliance with breed registry rules relative to minimum age at calving time (check them out; different breed registries have different rules), (b) selling of open heifers, (c) having well-grown, young cows for visitors to see, and (d) distribution of birth dates so as to fill more show classifications; but, of course, these points do not affect commercial producers.

From the preceding, it may be concluded that more producers can, and should, breed yearling heifers to calve as 2-year-olds. But, in doing so, the following rules should be observed:

1. Breed only well-developed heifers, weighing 600 to 650 lb (depending on the breed) at 13 to 14 months of age. Size at breeding is more important than age. Also, some breeds come in heat and mature a little earlier than others.

2. Use young, small-headed, refined-type bulls; preferably, bulls known to sire calves that are smaller than average at birth—Angus bulls are widely used for this purpose because (a) the calves tend to be somewhat smaller at birth, and (b) when crossed on other breeds, hybrid vigor is also obtained (of course, similar hybrid vigor applies to other crosses too).

3. Feed a well-balanced ration, for a continuous gain of about 1 lb daily during the last 100 to 120 days before calving.

4. Give special care at calving time.

5. Provide superior nutrition—well balanced and rather liberal—during the lactation period. This requires a good ration—one containing adequate energy and proteins, and fortified with the necessary vitamins and minerals. In season, usually this can be accomplished by keeping these heifers on *good pastures,* with or without supplemental feeding, during both pregnancy and lactation. When good grass is not available—in the winter, early and late, or during droughts—proper feeding must be relied upon.

6. If practical, wean early; at around 4 to 6 months of age, rather than the normal 7 months.

7. Try it out on half your replacement heifers to start with; make sure that you are ready before going all out.

For the average breeder, it is best not to have the heifers calve until they are about 30 months of age. This practice can be followed if calves are being dropped in both spring and fall. When the management practice calls for either spring or fall calving and no departure therefrom, it is necessary to have the heifers calve when either 2 or 3 years of age rather than somewhere between these limits.

## HEAT PERIODS

The period of duration of heat—that is, the time during which the cow will take the bull—is very short, usually not over 16 to 20 hours, although it may vary from about 6 to 30 hours. Cows tend to have a pattern of external behavior; for example, they come in heat during the morning hours, and go out of heat in the evening or early part of the night.

The heat period recurs approximately at 21-day intervals, but it may vary from 19 to 23 days. In most cases, cows do not show signs of estrus until some 6 to 8 weeks after parturition, or in some instances even longer. Occasionally, an abnormal condition develops in cows that makes them remain in heat constantly. Such animals are known as *nymphomaniacs.* This condition is due to the development of cysts and the failure of the follicle to rupture. Treatment, which may or may not be successful, consists of rupturing the cysts via rectum or hormone injection.

### SIGNS OF ESTRUS

Experienced cattle producers can usually detect in-heat cows because they generally exhibit one or more of the following characteristic symptoms: (1) nervousness, (2) attempts to mount other members of the herd and standing to be mounted by another cow, (3) a noticeable swelling of the labia of the vulva, (4) an inflamed appearance about the lips of the vulva, (5) frequent urination, and (6) a mucous discharge. Dry cows and heifers usually show a noticeable swelling or enlargement of the udder during estrus, whereas in lactating cows a rather sharp decrease in milk production is often noted.

Standing to be mounted by another cow appears to be the best single characteristic of the heat period. Using this as a guide, it is recommended that cows be bred during the final 10 hours of standing heat, or during the first 10 hours after the end of standing heat.

## FERTILIZATION

Fertilization is the union of male and female germ cells, sperm and ovum. The sperm are deposited in the vagina at the time of service and from there ascend the female reproductive tract. Under favorable conditions, they meet the egg and one of the sperm cells fertilizes it in the upper part of the oviduct near the ovary.

In cows, fertilization is an all-or-none phenomenon, since only one ovum is ordinarily involved. Thus, the breeder's problem is to synchronize ovulation and insemination, to ensure that large numbers of vigorous, fresh sperm will be present in the fallopian tubes at the time of ovulation.

## GESTATION PERIOD

The average gestation period of cows is 283 days, or roughly about 9½ months. Though there may be considerable breed and individual variation in the length of the gestation period, it is estimated that two-thirds of all cows will calve between 278 and 288 days after breeding.

## FERTILITY IN BEEF CATTLE

Fertility refers to the ability of the male or female to produce viable germ cells capable of uniting with the germ cells of the opposite sex and of producing vigorous, living offspring. Fertility is lacking in very young animals, manifests itself first at puberty, increases for a time, then levels out, and finally recedes with the onset of senility. In cattle, as with other classes of farm animals, fertility is determined by heredity and environment.

Of course, the final test for fertility is whether young are produced, but unfortunately this test is both slow and expensive. Through evaluation of the quality of semen, it is possible to make a fairly satisfactory appraisal of the male's fertility, but no comparable measure of the female's relative fertility has yet been devised.

## METHODS OF MATING

Two methods of mating beef cattle are as follows: (1) hand mating, and (2) pasture mating. Each of these will be discussed.

## HAND MATING

In hand mating, the bull is kept separate from the cows at all times, except when an individual cow is to be bred and is turned in with him for this purpose. As a rule, in hand mating only a single service is allowed, the cow being removed immediately after service. In the breeding of purebred cattle, when breeding records are so important, this method is frequently followed. Hand mating allows for a more accurate check on whether the bull is settling the cows. It also permits a larger number of cows to be served by a bull, an especially important consideration with a proven sire.

## PASTURE MATING

In this system the bull is turned in with the herd, either throughout the entire year or during the breeding season. Even with pasture breeding, when it is desired to have the calves all come within a few weeks of each other, thereby assuring more uniformity in size and offspring, the herd bull should be separated from the cows except during the breeding season. Uniformity in size is very important from the standpoint of marketing the calves advantageously. Furthermore, by having the calves come as nearly as possible at one time, closer observation may be given the herd at the time of parturition.

Pasture breeding is most often followed with a commercial herd. As a rule, this system requires less labor, and there is less danger of missing cows when they are in heat. However, the convenience of pasture mating should not result in neglect to check whether the cows are being settled during the breeding season.

## PREGNANCY TEST

Absence of heat is not always a sign of pregnancy, but a positive diagnosis can be made. By about the second month in heifers and the third month in cows, the uterus becomes enlarged, especially in the pregnant horn, and drops into the abdominal cavity. An experienced technician can ascertain this sign of pregnancy by *feeling with the hand through the rectal wall* (see Fig. 14–2). Application of this method depends upon the recognition of changes in tone, size, and location of the uterine horns and changes in the uterine arteries. This is the most common test of pregnancy.[1] It is popular because it affords early diagnosis, and there is little hazard when performed by an experienced operator. It is recommended that cows be pregnancy tested by this method about two months after the bulls have been removed. With convenient facilities—corrals and squeeze—an experienced operator can pregnancy test 800, or more, cows per day.

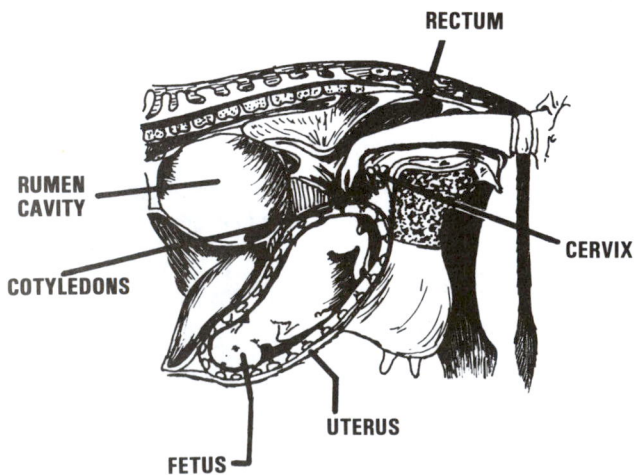

Fig. 14–2.   Rectal method for determining pregnancy in the cow.

## CARE AND MANAGEMENT OF THE BULL

Outdoor exercise throughout the year is one of the first essentials in keeping the bull virile and in a thrifty, natural condition. The finest and easiest method of providing such exercise is to arrange for a well-fenced, grassy paddock (about 2 acres is a good size for one bull). Many valuable sires have been ruined through close confinement in a small stall—or more likely yet—through being kept knee-deep in mud within a small, filthy enclosure. In addition to the valuable exercise obtained in the grassy paddock, the animal gets succulent pasture, an ideal feed for the herd bull.

A satisfactory and inexpensive shelter should be provided for the bull. The most convenient arrangement is to have this within or adjacent to the paddock, so that the bull may run in and out at will. Sufficient storage space for feed, along with materials and equipment for caring for the bull, should be provided in this building. Normally, purebred bulls are kept in separate stalls and enclosures, though some successful purebred breeders regularly run several valuable bulls in one enclosure. Bulls used in commercial herds are usually run together, both on the range and when separated out from the cows. Because of their scuffling and fighting, there is more injury hazard when bulls are handled in a group.

Under range conditions, it is rather difficult to give the bulls much attention during the breeding season. Usually the proper number of bulls is simply turned with the cow herd. During the balance of the year, however, the bulls are usually kept separate. Thus, if the producer desires calves that are dropped from February 1 to June 1, the bulls are turned with the cows about May 1 and are removed September 1.

The feeding of the herd bull is fully covered in Chapter 15. In brief, it may be said that the feeding program should be such as to keep the bull in a thrifty vigorous condition at all times.

## AGE AND SERVICE OF THE BULL

Table 14–1 gives pertinent information relative to the use of the bull, including consideration that should be given to age and method of mating. In a nationwide survey,[2] Washington State University found that, on the average, one bull was used for every 21.5 cows and heifers bred.

**TABLE 14–1**
**HANDY BULL MATING GUIDE**

| Age | No. of Matings/Yr. | | Comments |
| --- | Hand Mating | Pasture Mating | --- |
| Yearling ............ | 10–12 | 8–10 | **M**ost producers use 1 bull to 20 to 25 cows. |
| Two-year-old ........ | 25–30 | 20–25 | **A** bull should remain a vigorous and reliable breeder up to 10 years or older; up to 6 to 7 years under range conditions. |
| Three-year-old or over .. | 40–50 | 25–40 | |

---

[1]For other pregnancy tests, see *Beef Cattle Science;* by the same author and publisher as *Animal Science.*

[2]Ensminger, M. E., M. W. Galgan, and W. L. Slocum, Wash. Ag. Exp. Sta. Bull. 562, p. 50.

Should the bull prove to be an uncertain breeder, he should be given rest from service, forced to take plenty of exercise, and then placed in proper condition—neither fat nor thin. Sometimes a bull that is being let down in condition following showing will be sterile temporarily during the reducing process. Even though this lack of fertility may last for a year, usually such animals bounce back.

## CARE OF THE PREGNANT COW

The nutritive requirements of the pregnant cow are less rigorous than those of the lactating cow. In general, pregnant cows should be provided a year-round pasture as nearly as possible. During times of inclement weather or when deep snows or droughts make supplemental feeding necessary, dry roughages and silage are the common feeds. If produced on fertile soils, such forage will usually provide all the needed nutrients for reproduction. Further discussion of the nutritive needs of pregnant cows is contained in Chapter 15.

No shelter is necessary except during periods of inclement weather. Normally, the cows will prefer to run outdoors. This desire is to be encouraged—in order to provide exercise, fresh air, and sunshine. Where and when shelter is necessary, it should be neither elaborate nor expensive. An open shed facing away from the direction of prevailing winds is quite as satisfactory for the protection of dry cows as a warm bank barn with individual box stalls—and it is far less expensive. The chief requirements are that the shelter be tight overhead, that it be sufficiently deep to afford protection from inclement weather and remain dry (depths of 34 to 36 ft are preferred), that it be well drained, and that it be of sufficient size to allow the animals to move about and lie down in comfort.

## CARE OF THE COW AT CALVING TIME

As has been previously indicated, the period of gestation of a cow is about 283 days, but it may vary a few days in either direction. The careful and observant caretaker, therefore, will be ever alert and will make definite preparations in ample time. It is especially important that first-calf heifers be watched at calving time, for frequently they will need some assistance. Older cows that habitually have trouble in parturition may well be culled from the herd.

### SIGNS OF APPROACHING PARTURITION

Perhaps the first sign of approaching parturition is a distended udder, which may be observed some weeks before calving time. Near the end of the gestation period, the content of the udder changes from a watery secretion to a thick, milky colostrum. As parturition approaches, there generally will be a marked shrinkage or falling away of the muscular parts in the region of the tailhead and pinbones, together with a noticeable enlargement and swelling of the vulva.

The immediate indications that the cow is about to calve are extreme nervousness and uneasiness, separation from the rest of the herd, and muscular exertion and distress.

### PREPARATION FOR CALVING

Fig. 14–3. Calving on pasture is ideal when the weather is good. (Courtesy, American Angus Assn., St. Joseph, MO)

At the time the signs of approaching parturition seem to indicate that the calf may be expected within a short time, arrangements for the place of calving should be completed.

During the seasons of the year when the weather is warm, the most natural and ideal place for calving is a clean, open pasture away from other livestock. Hogs should not be allowed in the same place with the cow, for they are likely to injure or kill the young calf. They have even been known to injure the cow.

Under pasture conditions, there is decidedly less danger of either infection or mechanical injury to the cow and calf. In commercial range operations, it is common practice to ride the range more frequently at calving time. A better procedure consists in having a small pasture adjoining headquarters into which heavy springing cows are placed a few days before calving. With the added convenience of such an arrangement, the animals can be given more careful attention.

During inclement weather, the cow should be placed in a roomy (10- or 12-ft square), well-lighted, well-ventilated, comfortable box stall or maternity pen which should first be carefully cleaned, disinfected, and bedded for the occasion.

### NORMAL PRESENTATION

Labor pains in a mild form usually start some hours before actual parturition. After a time, the water bag appears on the outside, usually increasing in size until it ruptures from the weight of its own contents. This is followed closely by the appearance of the amniotic bladder (the second water bag), with the fetus. With the rupture of the second water bag, the straining becomes more violent, and presentation soon follows. Most commonly in presentation, the front feet come first followed by the nose which is resting on them, then the shoulders, the middle, the hips, and then the hind legs and feet (see Fig. 14–4, next page).

With posterior presentation (hind feet first) there is likely to be difficulty in calving. Moreover, there is considerably more danger of having the calf suffocate through rupture of the umbilical cord and strangulation.

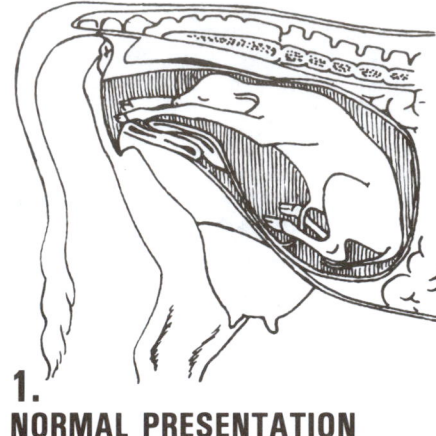

**1.**
**NORMAL PRESENTATION**

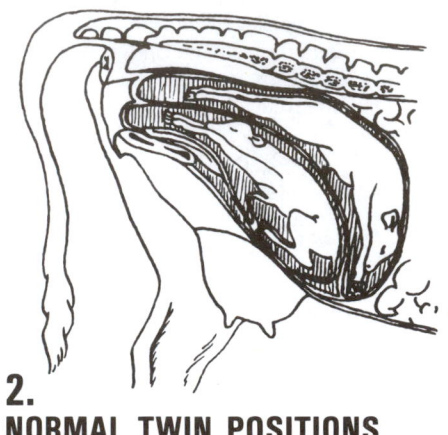

**2.**
**NORMAL TWIN POSITIONS**

Fig. 14–4.  1. Normal single presentation; the back of the fetus is directly toward that of the mother, the forelegs are extended toward the vulva, and the head rests between the forelegs. If it is necessary to render assistance, apply ropes above the ankle joints and pull alternately downward on each leg as the cow strains.
2. Normal twin positions. If delivery does not proceed normally, this is a case for a veterinarian.

## RENDERING ASSISTANCE

A good rule for the attendant is to be near but not in sight. If presentation is normal and within an hour or two after the onset of signs of calving, no assistance will be necessary. On the other hand, if the cow has labored for some time with little progress or is laboring rather infrequently, it is usually time to give assistance. Such aid will usually consist of fastening small ropes around the pasterns and pulling the young outward and downward as the cow strains. This should be done by an experienced caretaker or a competent veterinarian. It is always well to be reminded that rough, careless, or unsanitary methods at such a time may do more harm than good.

## THE NEWBORN CALF

If parturition has been normal, the cow can usually take care of the newborn calf, and it is best not to interfere. However, in unusual cases, it may be necessary to wipe the mucus from the nostrils to permit breathing; or, more rarely yet, artificial respiration methods should be applied to some calves. This may be done by blowing into the mouth, working the ribs, rubbing the body rather vigorously, and permitting the calf to fall gently. The cow should be permitted to lick the calf dry.

With calves born in sanitary quarters or out on clean pastures, there is little likelihood of navel infection. To lessen the danger of such infection, the navel cord of the newborn calf should be treated at once with a 10% solution of tincture of iodine.

A vigorous calf will attempt to rise in about 15 minutes and usually will be nursing in half an hour to an hour. The weaker the calf, the longer the time before it will be able to be up and nursing. Sometimes it may even become necessary to assist the calf by holding it up to the cow's udder.

The colostrum (the milk yielded by the mother for a short period following the birth of the young) is most important for the well-being of the newborn calf. Experiments have shown that it is almost impossible to raise a calf that has not received any colostrum. Aside from the difference in chemical composition, compared with later milk, the colostrum has the following functions:

1.  It contains antibodies which temporarily protect the calf against certain infections, especially those of the digestive tract.
2.  It serves as a natural purgative, removing fecal matter which has accumulated in the digestive tract.
3.  It contains a very high content of vitamin A, from 10 to 100 times that of ordinary milk. This provides the young calf, which is born with little body storage of this vitamin, with as much vitamin A on the first day as it would secure in some weeks from normal milk.

Usually it is best to keep the cow and calf in a small pasture for a few days. After this, they may be turned back with the main herd. Nothing is better for the cow at calving time than plenty of grass, and both the cow and calf will be helped by an abundance of fresh air and sunshine. The cow may deliberately hide the calf for the first few days, and the job may be so thoroughly done as to require considerable cleverness on the part of the caretaker to find it.

## THE AFTERBIRTH

Under normal conditions, the fetal membranes (placenta or afterbirth) are expelled from 3 to 6 hours after parturition. Should they remain as long as 24 hours after calving, competent assistance should be given by an experienced caretaker or a licensed veterinarian. The operation of removing a retained afterbirth requires skill and experience; and, if improperly done, the cow may be made a nonbreeder. Furthermore, before doing this, the fingernails should be trimmed closely; the hands and arms should be thoroughly washed with soap and warm water, disinfected, and then lubricated with petroleum jelly or linseed oil. In no case should a weight be tied to the placenta in an attempt to force removal.

As soon as the afterbirth is ejected, it should be removed and burned or buried in lime, thus preventing the development of bacteria and foul odors. This step is less necessary on the open range, where animals traverse over a wide area.

## NORMAL BREEDING SEASON AND TIME OF CALVING

The season at which the cows are bred depends primarily on the facilities at hand, taking into consideration the feed supply, pasture, equipment, labor, and weather conditions; whether the cattle are being produced for ordinary commercial or for purebred purposes; and whether they are strictly beef or dual-purpose cattle.

## ADVANTAGES OF SPRING CALVES

The production of spring calves has the following **advantages**:

1. In producing spring calves, the cows are bred during the most natural breeding season—at a time when they are on pasture, gaining in flesh, and more likely to conceive. The calving percentage is usually higher, therefore, with a system of spring calving.

2. The calves will be in shape to sell directly from the cows in the fall, at which time there is a good demand for feeder calves.

3. If the calves are to be sold as yearlings, one wintering is saved; or if they are to be sold at weaning time, no wintering is required.

4. Because of greater utilization of cheap roughage, dry cows may be wintered more cheaply.

5. Less labor and attention is required in caring for the calves the first winter.

6. Spring calves require less grain and utilize the maximum amount of pasture and roughage.

## ADVANTAGES OF FALL CALVES

The production of fall calves has the following **advantages**:

1. The cows are in better condition at calving time.
2. The cows give more milk for a longer period.
3. The calves make better use of the grass during their first summer.
4. The calves escape flies, screwworms, and heat while they are small (this is especially important in the South).
5. Upon being weaned the following spring, the calves can be placed directly on pasture instead of in a dry lot; or, if the desire is to sell, they usually find a ready market ahead of the influx of fall feeder calves from the range area.
6. When the intention is to sell market milk from dual-purpose cows, fall calves are usually best. The greater flow of milk is obtained during the period of highest prices.

## CROSSBREEDING SYSTEMS

Without a planned breeding program, crossbreeding will almost inevitably end up with (1) a motley collection of females and progeny varying in type and color, and (2) minimum benefits from hybrid vigor or heterosis.

Several different systems of crossbreeding may be used. Among them are the following:

1. **Two-breed cross**. This consists of mating purebred bulls to purebred or high-grade cows of another breed; for example, using Angus bulls on Hereford cows, to give crossbred Angus × Hereford offspring—black baldies. This system of crossing has been used with success by cattle producers for many years.

Fig. 14–5. A steer produced from a two breed cross—Brahman × Hereford. Note the great muscling and trim middle. (Courtesy, American Brahman Breeders Assn., Houston, TX)

In the 2-breed cross, only the calves are crossbred—the breeding of the sires and dams remains the same. Hence, the 2-breed cross imparts hybrid vigor only in the calf. On the average, it gives about an 8 to 10% increase in pounds of calf weaned per cow bred, plus another 2 to 3% advantage in rate of gain in the feedlot.

The 2-breed cross is relatively simple. However, it has one major deficiency; it does not make use of the crossbred cow.

2. **Two-breed backcross or crisscross.** This system involves the use of bulls of breed A on cows of breed B, then backcrossing the progeny to bulls of either breed A or B. The rotation is accomplished by using bulls of the breed least related to the particular set of cows. For example, if Charolais bulls are mated to Hereford cows, the crossbred Charolais × Hereford heifers could be retained and bred to either a Charolais or a Hereford bull. If Hereford bulls were used, the calves produced would be ¼ Charolais and ¾ Hereford. Later, if the heifers of this breeding are saved, they should be bred to a Charolais bull. The 2-breed backcross results in about 67% of the maximum heterosis being attained in the crossbred calves. But since crossbred cows are used, overall performance should be a little better in pounds of calf weaned per cow bred than in the 2-breed cross.

3. **Three-breed rotation cross.** This system calls for the selection of 3 breeds (e.g., breeds A, B, and C, which might represent Herefords, Angus, and Charolais), possessing the combination of maternal, carcass traits, and growth desired in the crossbred cows and the slaughter cattle produced. Crossbred females, selected for growth rate, are retained for breeding and bred to a purebred bull of one of the 3 breeds. Each new generation of crossbred females is retained for breeding and mated to a purebred bull until bulls of all 3 breeds have been used in rotation. Thus, such a system would operate as follows: Mate the existing B cow herd continuously to bulls of breed A; select crossbred heifers for growth rate and mate them continuously to bulls of breed C; mate the selected C (AB) females to bulls of breed B. After the rotation of bulls from the 3 breeds is completed, the rotation of purebred sires begins all over again. Thus, mate the selected BX (ABC) females to bulls of breed A.

Fig. 14–6. Three-breed cross calf. Angus × Hereford crossbred cow (a black baldie) with a calf sired by a Charolais bull. (Courtesy, American Breeders Service, DeForest, WI)

Continue the same system indefinitely, always selecting the best performing crossbred females to be mated to the breed of sire in the program to which they are least related.

In addition to the genetic advantages of this system, commercial producers select their own replacement; hence, the only outside cattle purchases are production tested bulls. The major disadvantage is that after the first four years it is necessary to maintain bulls of all three breeds simultaneously (unless A.I. is used).

A three-way rotation system results in about 87% of the maximum heterosis being attained.

4. **Three-breed fixed or static cross (terminal cross).** In this system, crossbred cows from a two-breed cross ($F_{1S}$) are used as females and are mated to a bull of a third breed. All offspring from this cross are sold. When replacement females are needed, they are purchased. Thus, crossbred cows are used and crossbred calves with a fixed percentage of inheritance from three breeds are always produced.

In addition to realizing 100% of the maximum heterosis in each calf crop, this system allows the selection of maternal breeds to go into the production of the crossbred female and the selection of growthy breeds having desirable carcasses for the terminal cross sire breed. It allows the breeds to be used for their strong points without regard to some of their weaker points. A breeder can tailor-make the crossbred market animal,

putting together in one animal desirable traits of several breeds. Such specification is not possible in the rotational system because all breeds contribute to maternal performance and calf performance.

The mechanics of this system consists in selecting three breeds for crossbreeding—two breeds (A and B) that will produce crossbred cows with outstanding maternal characteristics for fertility, milking ability, mothering ability, and adaptation; and a third breed (breed C) with rapid, efficient, postweaning muscle growth rate. Breed C would be considered a *terminal* sire breed. All crossbred progeny of bull C are marketed for slaughter.

The problem with this system is the acquisition of production tested, crossbred ($F_1$) heifers for replacements in such a program, since all the three-way crosses are marketed. The system is perpetuated by having specialized multipliers produce crossbred ($F_1$) replacement females.

Four or more breeds may be used in a rotation crossbreeding system if the commercial producer so desires. However, the maximum hybrid vigor is usually realized with the three-breed cross.

It is noteworthy that all of these crossbreeding systems rely upon the use of purebred bulls. Additionally, the two-breed cross relies on the use of purebred females, and the three-breed fixed or static cross relies on purebred females to produce the $F_1$ heifers necessary for the program.

Crossbreeding is no magic or cure-all, but it will give a powerful assist to the pocketbook if properly used. Also—and this point bears emphasis—sound management and sound selection of breeding stock based on performance, potential carcass characteristics, and overall productivity are just as important in crossbreeding as in any other breeding program.

## BUFFALO × CATTLE HYBRIDS[3]

From time to time, American buffalo (*Bison bison*) and domestic beef cattle (*Bos taurus*) have been crossed, in both Canada and the United States. Out of such crosses have evolved cattalo (cattle of less than ½ bison parentage), beefalo (⅜ buffalo, ⅜ Charolais, and ¼ Hereford), and the American Breed (⅛ buffalo, ½ Brahman, ¼ Charolais, ¹⁄₁₆ Durham, and ¹⁄₁₆ Hereford). These breeds are variously extolled, on the basis of their adaptability to cold, snowy climates; ability to thrive on weeds, shrubs, and other vegetation which domestic cattle pass up; small birth weights (straight buffalo calves weigh only about 25 lb at birth); and leaner and more flavorful meat.

Pertinent information relative to the reproductive ability of the American buffalo (*Bison bison*) × domestic cattle (*Bos taurus* hybrids follows:

1. Bison and domestic cattle interbreed.

2. Fewer maternal calving losses occur when domestic bulls are used on bison cows, although the reciprocal mating may be made.

3. Half-buffalo bull calves ($F_1$ hybrids) show normal sexual behavior, but are always sterile. The scrotum is held close to the body cavity, as in the bison.

4. The half-buffalo heifers ($F_1$ hybrids) are fertile.

5. A few backcross bull hybrids have produced semen containing some sperm.

[3]Buffalo breeders are banded together in the National Buffalo Assn., Box 995, Pierre, SD, 57501)

Fig. 14–7. Cattalo (⅛ buffalo and ⅞ domestic cattle) cow. The initial cattalo breeding experiment was started by the Dominion Experimental Station, Scott, Saskatchewan, Canada, in 1915. The foundation herd consisted of 16 female and 4 male hybrids. (Courtesy, Research Station, Canada Department of Agriculture, Lethbridge, Alberta, Canada)

6. Reproductive ability improves in both sexes of further generations as the percentage of domestic blood increases.

It is possible that animals carrying a small percentage of buffalo breeding may have a place under certain conditions. However, more scientific research on the subject is needed.

## MULTIPLE BIRTHS

A review of the literature reveals that, on the average, multiple births occur at the following frequencies in beef cattle: beef cattle—all breeds, 0.44%; Angus, 0.81% (based on 1,111 births); and Simmental, 4.61% (based on 12,625 births).

Most breeders prefer single births to twins, for the following reasons:

1. There is a high incidence of stillbirths in twins. Herefords on the range show 3.6% stillbirths among singles vs 15.7% stillbirths among twins.
2. About 85% of all heifers born twin with a bull are apt to be freemartins (sterile heifers).
3. Twin calves average 20 to 30% lighter weights at birth than singles.
4. There is a tendency in cows that have produced twins to have a lowered conception rate following twinning.

## ABNORMAL DEVELOPMENT OF SEX IN CATTLE

A peculiar form of sterility usually occurs in heifers born twin to a bull. Such sterile heifers are called *freemartins*. This condition prevails in about 85% of twin births when a calf of each sex is involved. The fetal circulations fuse, and the male hormones get into the circulation of the unborn female, where they interfere with the normal development of sex and modify the female embryo in the direction of the male. In about 15% of heifers born twin to a bull, fusion of the circulation does not occur, and the animals are normal and fertile. As, on the average, only 15 out of every 100 heifers born twin to a bull

are normal, it is usually considered best to sell such animals for slaughter, unless very valuable purebred cattle are involved, or it has been ascertained that such heifers are not freemartins. An experienced person can, at the time of birth, determine if the circulatory systems were fused. Also, there are blood tests for detecting freemartins (see Chapter 3, section on "Blood Typing").

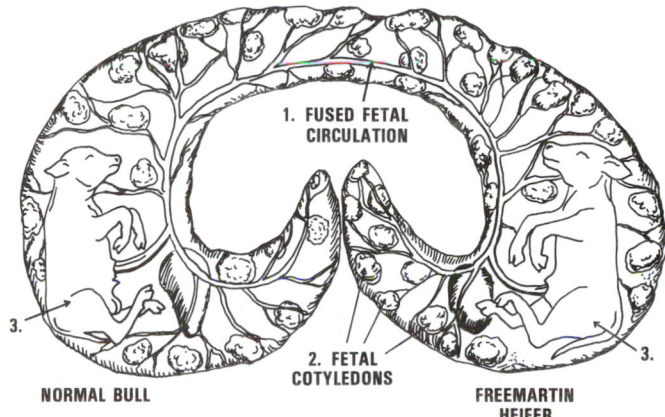

Fig. 14–8. Diagram showing fused fetal circulation of twin calves of opposite sex. Note (1) the fetal circulation of the male fused with that of the female, (2) fetal cotyledons, and (3) normal bull on the left and freemartin heifer on the right.

In addition to freemartins, other abnormal sexual developments occasionally occur in animals; namely, hermaphrodites and intersexes. Hermaphrodites are animals that possess characteristics of both sexes, whereas intersexes do not possess definite characteristics of either sex. These conditions occur sporadically when something happens to the embryo to disturb normal development.

## LETHALS AND OTHER HEREDITARY ABNORMALITIES[4]

Lethal characters in beef cattle or in any class of animals are caused by the presence of hereditary factors in the germ plasm that produce an effect so serious as to cause the death of the individual, either at birth or later in life. Breeding animals possessing hereditary lethals should be culled from the herd.

## DWARFISM IN CATTLE

Beginning about 1940, a disturbing condition known as dwarfism appeared among beef cattle, probably in all breeds. Though very small (usually weighing about half as much as normal calves), these calves are exceedingly stocky. The eyes protrude, giving a characteristic "popeyed" appearance. Some dwarfs are weak and unsteady in gait at birth. Others appear to be strong enough, but soon develop a large stomach, heavy shoulders, crooked hind legs, and sometimes labored breathing. Survival is somewhat lower than with normal calves, though most purebred breeders make no attempt to raise them.

[4]For a complete summary of "Lethals and Other Abnormalities in Cattle," see *Beef Cattle Science*, a book by the same author and publisher as *Animal Science*.

It is known (1) that the dwarf condition is of genetic origin, and (2) that it is inherited as a simple autosomal recessive (the word autosomal meaning that the factor is not carried on the sex chromosome). One or the other of the conditions (or perhaps both conditions) shown in Figs. 14–10 and 14–11 prevail in any herd of cattle in which dwarf-carrying animals are being used.

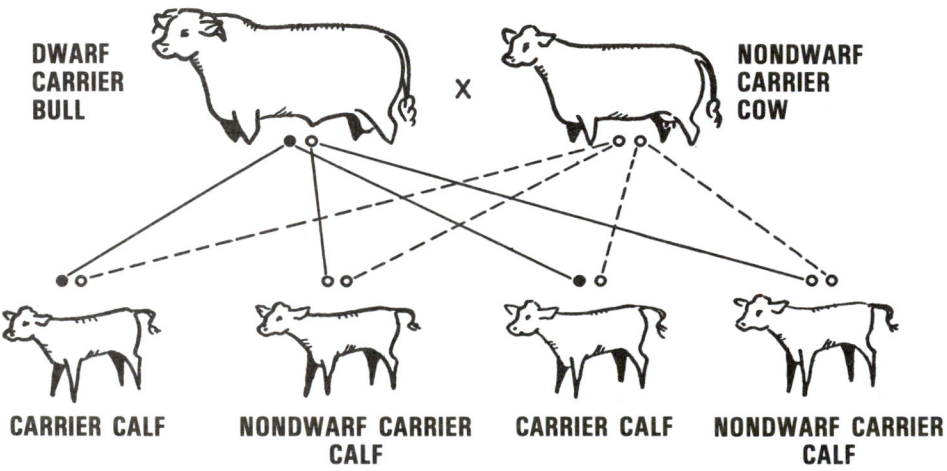

Fig. 14–10. Diagrammatic illustration of the inheritance of dwarfism, showing what to expect when a carrier (heterozygous) bull(s) is mated to a noncarrier (homozygous normal) cow(s); or the sexes may be reversed. As shown, carrier × noncarrier matings will, *on the average,* produce calves of which (1) 50% are carriers, although not dwarfs, and (2) 50% are noncarriers and nondwarfs. Unfortunately, the two groups look alike and cannot be detected by sight. (Drawing by Prof. R. F. Johnson)

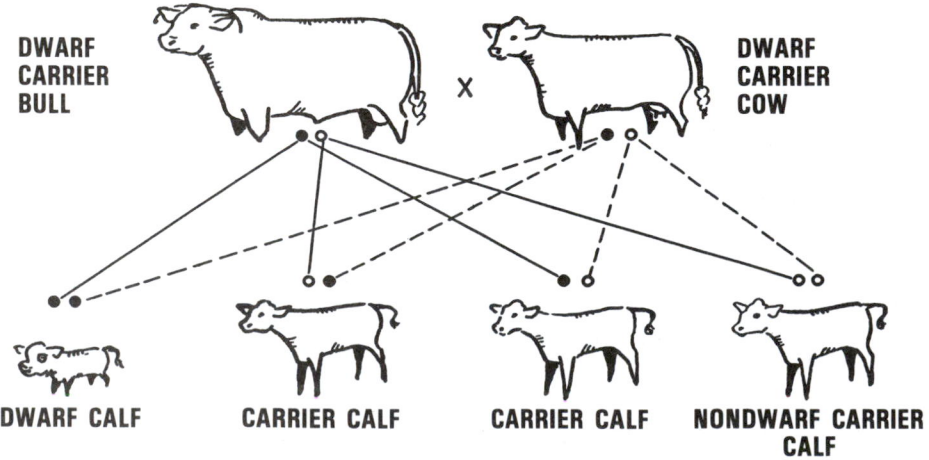

Fig. 14–11. Diagrammatic illustration of the inheritance of dwarfism, showing what to expect when a carrier (heterozygous) bull(s) is mated to a carrier (heterozygous) cow(s); or the sexes may be reversed. As shown, carrier × carrier matings will, *on the average,* produce calves of which (1) 25% are dwarfs, (2) 50% are carriers, although not dwarfs, and (3) 25% are noncarriers and nondwarfs. Unfortunately, only the dwarfs can be detected by sight; the two nondwarf groups look alike and cannot be distinguished by sight. (Drawing by Prof. R. F. Johnson)

# DOUBLE MUSCLING (MUSCULAR HYPERTROPHY)

Double muscling refers to cattle characterized by bulging muscles of the shoulder and thigh, a very rounded rear end (as viewed from the side), a wide but shallow body throughout, appearance of intermuscular grooves, and fine bones.

Double muscling is really a misnomer. Likewise, the scientific name, muscular hypertrophy, is incorrect because it implies increased size of fibers in each muscle, which is not the case. Rather, it has been shown that double-muscled cattle have more fibers, not larger fibers.

Double muscling has been around for a very long time—for at least 100 years; and it has been reported in Europe, Australia, and the United States. Also, the condition has been observed in almost every breed. Since beef cattle are produced primarily for their muscle, it's logical that selection should be centered around muscularity. Hence, when selecting the heavier-muscled animals for breeding purposes, more and more carriers may be used; and more and more double-muscled cattle may show up.

Fig. 14–12. The double-muscled Piedmont (Piedmontese) bull, *Barolo*. In Italy, where 80% of the Piedmont bulls are double muscled, these animals are highly prized and command a premium price on the market because of their high dressing percentage and high proportion of desirable steaks.

Double muscling is a genetically controlled character. It appears to be caused by a single recessive gene, which tends to be masked by the dominant gene in the heterozygous carriers. Other examples of a character controlled by one pair of genes are: polledness and hornedness, and dwarfism. The genetics, therefore, are relatively simple. Since each animal has two genes for such characters, all cattle can be classified as follows:

**DM DM**—Homozygous normal; two dominant normal genes—a normal animal.

**DM dm**—Heterozygous; one dominant normal gene (DM) which tends to cover up the one recessive gene (dm)—these are called carriers. This cover up is not complete; hence, there is a tendency toward double muscling.

**dm dm**—Homozygous recessive; two recessive double-muscled genes—a double muscled animal.

The progeny from a sire and dam of all these three genotypes are predictable, on the average, but not necessarily for any one offspring. The possible matings and progeny are shown in Table 14–2.

## TABLE 14–2
### POSSIBLE MATINGS AND PROGENY OF DOUBLE-MUSCLED CATTLE

| Sire | Dam | Progeny |
|---|---|---|
| DM DM | DM DM | DM DM All normal. |
| DM DM or DM dm | DM dm or DM DM | ½ DM DM, ½ DM dm All normal, but ½ carriers. |
| DM dm | DM dm | ¼ DM DM, ½ DM dm, ¼ dm dm Of the ¾ normal, 2 out of 3 are carriers; the remaining ¼ are double muscled. |
| DM DM or dm dm | dm dm or DM DM | DM dm ALL carriers. |
| DM dm or dm dm | dm dm or DM dm | ½ DM dm, ½ dm dm ½ carriers, ½ double muscled. |
| dm dm | dm dm | dm dm ALL double muscled. |

Obviously, the problem is to determine if an animal is DM dm (a carrier), rather than DM DM (a normal animal). There are two ways to do this: (1) appearance; and (2) breeding tests. Detection by appearance is not 100% sure, but the experienced observer doesn't make many mistakes.

If double muscling is caused by a single recessive gene, as seems to be the case in the British breeds (Angus, Hereford, and Shorthorn), (1) mating phenotypically normal (noncarrier) cows to a homozygous, double-muscled bull would produce 100% heterozygotes (carriers), none of which would be double muscled, and (2) mating known carrier cows (heterozygotes) to a homozygous, double-muscled bull would produce 50% homozygous, double-muscled calves and 50% heterozygous carriers. There are disadvantages as well as advantages to double muscling. Hence, cattle producers should be familiar with the characteristics and genetics of double muscling, and its side effects.

• **Advantages of double-muscled cattle**—Here are some of the pluses in favor of double-muscled over normal cattle:

1. The calves grow more rapidly up to one year of age.
2. They convert feed more efficiently; it requires fewer pounds of feed to produce a pound of beef.
3. They (a) have a higher dressing percentage, and (b) yield a higher proportion of the more desirable cuts—more steaks. In comparison with normal cattle, double-muscled cattle have a larger rib eye, produce less brisket, plate, and flank; and produce less kidney and pelvic fat.

Thus, double-muscled cattle are superior to normal cattle in (1) rate and efficiency of gain to one year of age, and (2) general carcass desirability.

• **Disadvantages of double-muscled cattle**—It is recognized that there may be breed differences when it comes to the advantages and disadvantages of double-muscled cattle. Nevertheless, here are some of the disadvantages to double muscling that have been reported in different countries:

1. The conception rate is lower, due to (a) the infantile reproductive tracts or slow sexual maturity of some animals, and (b) the flat vulva, which makes copulation difficult.
2. The gestation period is about 10 days longer.
3. There is more calving difficulty (Caesarean section; pulling calves; stillborn) due to heavier calves at birth (Piedmontese

double-muscled calves average 108 lb at birth vs 99 lb for normal calves), along with the enlarged rump and round regions.

4. Double-muscled calves are more difficult to raise, due to such things as (a) enlarged tongues (macroglossia), and (b) greater susceptibility to disease.

5. Double-muscled cows are poor milkers; they produce 30 to 50% less milk than normal cows.

6. Double-muscled cattle must be fed a higher proportion of concentrate to roughage, simply because they cannot utilize roughage effectively.

7. There is less marbling.

It's unlikely that all of the above disadvantages will occur in any one herd at any one time. Moreover, the degree to which they occur among breeds, and within breeds, will vary according to the "background" genes or modifying genes. Nevertheless, cattle producers should be apprised of the possibilities.

## PRODUCTION TESTING BEEF CATTLE[5]

Fig. 14–13. Performance tested bulls in a commercial herd. (Courtesy, American Polled Hereford Assn., Kansas City, MO)

In comparison with that of chickens or even swine, production testing of beef cattle is slow, and, like most investigational work with large animals, it is likely to be expensive. Even so, in the realization that such testing is absolutely necessary if maximum improvement is to be made, progressive purebred beef cattle breeders will have their herds on production test. Widely used and highly satisfactory record forms for production testing are shown in Figs. 14–14, 14–15a, and 14–15b (see pp 375, 376, and 377).

The following items appear to be of greatest importance in evaluating the profitability of beef cattle.[6]

1. **Calving interval—fertility** (approximate heritability of character: 10%[7]). Fertility is economically the most important trait in beef cattle. Without a calf being born, and born alive, cattle are self-eliminating.

2. **Calving ease** (approximate heritability of character: 10%). Calving losses at birth are the second most important reason for lower percent calf crops. Calving difficulty (dystocia) accounts for most calf deaths within the first 24 hours after calving, and most calving difficulty occurs in 2-year-old heifers.

3. **Birth weight** (approximate heritability of character: 40%). Birth weight is associated with calf survival. Also, it has a positive correlation of .39 with growth rate. However, selecting for increased birth weight is generally avoided because of the likeliness of increased calving difficulty.

4. **Weaning weight** (approximate heritability of character: 20%). Heavy weaning weight is important because (a) it is indicative of the milking ability of the cow; (b) gains made before weaning are cheaper than those made after weaning; and (c) those who sell calves at weaning usually make more profit due to the heavier weight available to sell.

5. **Cow mothering ability** (approximate heritability of character: 40%). Mothering ability is important in beef cows, because it contributes to calf survival and weaning weight.

6. **Feedlot gain** (approximate heritability of character: 45%). Daily rate of gain is important because (a) it is highly correlated with efficiency of gain; and (b) it makes for a shorter time in reaching market weight and condition, thereby effecting a savings in labor and making for a more rapid turnover in capital.

7. **Pasture gain** (approximate heritability of character: 30%). Most beef animals spend a good part of their lives on grass; hence, pasture gain is important.

8. **Efficiency of gain** (approximate heritability of character: 40%). Efficiency of feed conversion is expressed as pounds of feed intake per 100 lb of gain. It is seldom measured in performance and progeny tests, because a positive relationship exists between rate and efficiency of gain. Hence, selection for rate of gain automatically selects for efficiency of gain.

9. **Yearling weight—feedlot** (approximate heritability of character: 50%). Final feedlot weight is usually referred to as *weight per day of age*. It is generally computed at one year of age or at the end of the performance test. It is probably the most important measurement of the estimated value of a beef bull. It is composed of birth weight, weaning weight, and postweaning gain.

10. **Yearling weight**—pasture.

11. **Conformation score** (approximate heritability of character: (a) at weaning, 30%; (b) at yearling, 40%). These scores should be based on skeletal soundness and indications of carcass desirability. Structural soundness, especially of the feet and legs, is most important in breeding animals.

---

[5]For further information on this subject see the section in Chapter 3 devoted to "Selection Based on Production Testing."

[6]The heritability figures given in this section for each of the economically important traits are from the following source: Gregory, *Beef Cattle Breeding*, Ag. Info. Bull. No. 286 (Revised), USDA.

[7]The rest is due to environment. The heritability figures given herein are averages based on large numbers, thus, some variations from these may be expected in individual herds.

## GET OF SIRE RECORD

Calf Crop for Year of _____   Sire's Name _____   Reg. No. _____
Sex of Get[1] _____   Date of Birth _____
Owner and Address _____

| Herd No. of Calf | Calf Data | | | | | | | | Yearling Data | | | | | | Dam Data | | | | Remarks |
|---|---|---|---|---|---|---|---|---|---|---|---|---|---|---|---|---|---|---|---|
| | Date of Birth | Weaning Date | Weaning Age in Days | Weight in Lb | Daily Gain from Birth Weight, Lb | Adj. 205-Day Weaning Weight, Lb | Weanling Weight Ratio[2] | Confor- mation Score | Date Weighed | Weight, Lb | Wt. Adj. to days | Yr. Wt. Ratio[2] | Confor- mation Score | Herd No. | Age This Year | Mature Weight, Lb | Confor- mation Score | |
| | | | | | | | | | | | | | | | | | | | |
| | | | | | | | | | | | | | | | | | | | |
| | | | | | | | | | | | | | | | | | | | |
| | | | | | | | | | | | | | | | | | | | |
| | | | | | | | | | | | | | | | | | | | |
| | | | | | | | | | | | | | | | | | | | |
| | | | | | | | | | | | | | | | | | | | |
| | | | | | | | | | | | | | | | | | | | |
| | | | | | | | | | | | | | | | | | | | |
| | | | | | | | | | | | | | | | | | | | |
| | | | | | | | | | | | | | | | | | | | |
| | | | | | | | | | | | | | | | | | | | |
| Totals | | | | | | | | | | | | | | | | | | | |
| Averages | | | | | | | | | | | | | | | | | | | |

[1]One sheet should be used to record all the bull calves and another sheet to record all the heifer calves by the same sire.

[2]Ratio calculated as follows:

$$\frac{\text{Individual record}}{\text{Av. of all calves on same farm and same season}} \times 100$$

Fig. 14-14. Get of Sire Record Form.

## INDIVIDUAL COW RECORD

Tattoo, Horn Brand, and/or Neck Chain No. _____

Name _____    Reg. No. _____

Bred by _____    Birth Date _____

Purchased from _____    Birth Wt., Lb _____

Address _____    Weaning Wt., Lb _____ Age _____ Conf. Score _____

Sire _____

Purchase Date _____ Price, $ _____    Yearling Wt., Lb _____ Age _____ Conf. Score _____

Disposition _____ Price, $ _____    Two Year Wt., Lb _____ Age _____ Conf. Score _____

Reason for Disposal _____    Av. Daily Gain Weaning to 1 yr., Lb _____

Dam _____    Feed Efficiency _____ lb feed/100 lb gain

_____ Date _____    Temperament _____

Faults & Abnormalities _____

## PRODUCE OF DAM RECORD

| Birth Date | Sex | Tattoo | Sire | Birth Wt., Lb | Vigor at Birth[1] | Weaning Age Days | Weaning Wt., Lb Act. | Weaning Wt., Lb 205 day Adj. | Weaning Wt. Ratio[2] | Weaning Cond. | Conf. Score | Yr.,Wt., Lb Date | Yr.,Wt., Lb Adj. | Yr.,Wt., Lb Days | Yearling Wt. Ratio[2] | Conf. Score[1] | Days on Feed | Av. Daily Gain, Lb | Gain Ratio[2] | Lb Feed /100 Lb Gain | Disposition; Price Remarks |
|---|---|---|---|---|---|---|---|---|---|---|---|---|---|---|---|---|---|---|---|---|---|
| | | | | | | | | | | | | | | | | | | | | | |
| | | | | | | | | | | | | | | | | | | | | | |
| | | | | | | | | | | | | | | | | | | | | | |

(Column groups: Calf Data | Yearling Data | Production Testing)

[1]0=dead at birth; 1=definitely undersized at birth; 2=unthrifty, definite indications of disorders; 3=moderately thrifty, slight indications of disorders; 4=thrifty, no signs of disorders, dry hair coat; 5=thrifty, no signs of disorders, sleek hair coat; 6=very large, healthy, and vigorous

[2]Ratio calculated as follows:

$$\frac{\text{Individual record}}{\text{Av. of all calves on same farm and same season}} \times 100$$

Fig. 14-15a. Individual Cow Record (see Fig. 14-15b for reverse side of record form).

## IMMUNIZATION AND TEST RECORD

| | Immunizations | | | | Health Tests | | | | | | | Remarks |
|---|---|---|---|---|---|---|---|---|---|---|---|---|
| Date[1] | Blkg. | M. Edema | Bangs | Misc. | TB-Bangs | Johnes | Lepto. | Anaplas. | Vib. | Trich. | Misc. | |
| | | | | | | | | | | | | |
| | | | | | | | | | | | | |
| | | | | | | | | | | | | |
| | | | | | | | | | | | | |
| | | | | | | | | | | | | |
| | | | | | | | | | | | | |
| | | | | | | | | | | | | |
| | | | | | | | | | | | | |

[1] Indicate vaccinations by check in appropriate column opposite date given; indicate test results by P (positive), N (negative), or S (suspect) opposite date of test.

## GENERAL INFORMATION

Record all facts pertinent to the history of this cow, viz.: veterinary treatment (except immunizations), udder condition, mothering instinct, calving peculiarities, etc.

| Date | Remarks |
|---|---|
| | |
| | |
| | |
| | |
| | |
| | |
| | |
| | |
| | |
| | |
| | |
| | |
| | |

Fig. 14–15b.  Individual Cow Record. This is the reverse side of the record form shown in Fig. 14–15a.

12. **Carcass traits**. Quality of product and quantity of edible portion are the basic factors of carcass merit.

Where breeding animals are involved, and are not to be slaughtered, carcass quality may be evaluated by either (a) ultrasonic measurements, or (b) the $K^{40}$ counter. Ultrasonics can be used to measure rib eye area and outside fat cover. The $K^{40}$ counter evaluates the entire animal; it provides an effective method of measuring the total lean content of the live animal.

13. **Carcass grade** (approximate heritability of character: 30%). High carcass grade is important because it determines selling and eating quality. BIF recommends that USDA yield Grade (Nos. 1 to 5) also be used as a basis for evaluation of carcasses.

14. **Rib eye area** (approximate heritability of character: 50%). The rib eye (the large muscle which lies in the angle of the rib and vertebra) is indicative of the bred-in muscling of the entire carcass. Thus, a large area of rib eye is much sought.

15. **Tenderness** (approximate heritability of character: 50%). Warner-Bratzler shear test and taste panel test are recommended as methods of measuring tenderness.

16. **Thickness of outside carcass fat** (approximate heritability of character: 30%). Fat thickness is taken at the twelfth rib.

17. **Cancer eye susceptibility** (approximate heritability of character: 30%). There is indication that susceptibility to cancer eye is hereditary.

Differences among animals in traits of economic value are, to a considerable extent, inherited differences. Thus, systematic measurement of these differences, the recording of the measurements, and the use of records in selection will increase the rate of genetic improvement in the individual herds, and eventually in the breed and in the total cattle population.

Research has shown that when cattle are kept under nearly the same conditions and their performance records are adjusted for known environmental differences—such as age, age of dam, and sex—genetically superior animals can be identified.

The rate of improvement in a herd, breed, and population is dependent on (1) the percentage of observed differences between animals that is due to heredity (heritability), (2) the difference between selected individuals and the average of the herd or group from which they come (selection differential), (3) the genetic association among traits upon which selection is based (genetic correlations), and (4) the average age of parents when the offspring are born (generation interval).

The essentials of effective record of performance programs are:

1. All animals of a given sex and age are given equal opportunity through uniform feeding and management.

2. Systematic written records are kept of important traits of economic value of all animals.

3. Records are adjusted for known sources of variation, such as age of dam, age of calf, and sex.

4. Records are used in selecting replacements (bulls and heifers) and in culling poor producers.

5. Nutritional program and management practices are practical and uniform for the entire herd and are similar to those where progeny of the herd are expected to perform.

# QUESTIONS FOR STUDY AND DISCUSSION

1. Eight percent of the nation's cows never calve, and there is an appalling calf loss of 6% at birth. Discuss the economics and the causes of this situation.

2. Why are fewer breeding troubles encountered when cattle are bred on pasture or range rather than in confinement?

3. List the advantages and the disadvantages of having heifers calve for the first time as 2-year-olds.

4. List the important rules that should be observed when breeding yearling heifers to calve as 2-year-olds.

5. List the signs of estrus of cattle. What is the best single indicator of the heat period?

6. Define *fertilization* and tell of its importance.

7. What is the average gestation period of cows?

8. Define *fertility*. What is the final test of fertility?

9. Describe each of the two common methods of mating beef cattle. What factors determine which method a cow-calf producer should follow?

10. Why should cow-calf producers have their cows pregnancy tested?

11. Discuss (a) the care and management of the bull, and (b) the age and service of the bull.

12. Discuss (a) the care of the pregnant cow, (b) the care of the cow at calving time, (c) the signs of approaching parturition, (d) preparation for calving, (e) normal presentation, and (f) rendering assistance.

13. What is colostrum? What are its functions?

14. How should a retained afterbirth be removed?

15. Discuss the advantages of (a) spring calves, and (b) fall calves.

16. Select a given cow-calf operation (either your own, or one with which you are familiar), then outline a planned crossbreeding program for it. Justify the program that you select over alternative crossbreeding programs.

17. Do you feel that buffalo × cattle hybrids have a place? If so, under what circumstances?

18. Do you conceive of any circumstances under which multiple births in beef cattle would be desirable?

19. Define the following: (a) freemartin, (b) hermaphrodite, and (c) intersex.

20. How would you purge a herd of dwarfs (or any other undesirable recessive)?

21. Give the expected genetic picture of dwarfism of 100 offspring from mating of (a) carrier bulls × noncarrier cows, and (b) carrier bulls × carrier cows. What steps can be taken to get rid of dwarfism?

22. Is double muscling good or bad? Are carrier (DM dm) steers for double muscling dominating the American show-ring at the present time?

23. Why are record forms important in a production testing program? What are the primary differences between (a) a get of sire record, and (b) a produce of dam record?

24. In production testing beef cattle, what traits are important? What is the approximate heritability of each of these traits?

25. The Beef Improvement Federation no longer recommends that cattle be graded for type or individuality. Do you agree with them on the elimination of grading or type in production testing? Justify your position.

26. List the essentials of record of performance programs.

## SELECTED REFERENCES

| Title of Publication | Author(s) | Publisher |
|---|---|---|
| Battle of Bull Runts, The | L. P. McCann | L. P. McCann, Columbus, OH, 1974 |
| Beef Cattle, Seventh Edition | A. L. Neumann | John Wiley & Sons, Inc., New York, NY, 1977 |
| Beef Cattle Production | K. A. Wagnon<br>R. Albaugh<br>G. H. Hart | The Macmillan Company, New York, NY, 1960 |
| Beef Cattle Production | J. F. Lasley | Prentice-Hall, Inc., Englewood Cliffs, NJ, 1981 |
| Beef Cattle Production | N. T. Yeates<br>P. J. Schmidt | Butterworths Pty., Ltd., Brisbane, Australia, 1974 |
| Beef Cattle Science Handbook | Edited by<br>M. E. Ensminger | Agriservices Foundation, Clovis, CA, pub. annually since 1964 |
| Beef Production in the South, Modified Edition | S. H. Fowler | The Interstate Printers & Publishers, Inc., Danville, IL, 1979 |
| Breeding Difficulties in Dairy Cattle | S. A. Asdell | Ag. Exp. Sta. Bull. 924, Cornell University, Ithaca, NY, 1957 |
| Cattle Fertility and Sterility | S. A. Asdell | Little, Brown & Company, Boston, MA, 1968 |
| Commercial Beef Cattle Production, Second Edition | Edited by<br>C. C. O'Mary<br>I. A. Dyer | Lea & Febiger, Philadelphia, PA, 1978 |
| Crossbreeding Beef Cattle, Series 2 | M. Koger<br>T. J. Cunha<br>A. C. Warnick | University of Florida Press, Gainesville, FL, 1973 |
| Factors Affecting Reproductive Efficiency in Dairy Cattle, Ky. Ag. Exp. Sta. Bull. 605 | D. Olds<br>D. M. Seath | University of Kentucky, Lexington, KY, 1954 |
| Genetics of Livestock Improvement | J. F. Lasley | Prentice-Hall, Inc., Englewood Cliffs, NJ, 1972 |
| Lasater Philosphy of Cattle Breeding, The | L. M. Lasater | Texas Western Press, The University of Texas, El Paso, TX, 1972 |
| Prenatal and Postnatal Mortality in Cattle | H. Marsh, Subcommittee Chairman | National Academy of Sciences, Washington, DC, 1968 |
| Problems and Practices of American Cattlemen, Wash. Ag. Exp. Sta. Bull. 562 | M. E. Ensminger<br>M. W. Galgan<br>W. L. Slocum | Washington State University, Pullman, WA, 1955 |
| Reproduction and Infertility | | Michigan Agricultural Experiment Station, Michigan State University, East Lansing, MI, 1955 |
| Robert Bakewell: Pioneer Livestock Breeder | H. C. Pawson | Crosby Lockwood & Son, Ltd., London, England, 1957 |

Black and ''black baldy'' cows on the Mabee Ranch in Texas. (Courtesy, *Livestock Weekly*, San Angelo, TX)

All flesh is grass!

# FEEDING AND MANAGING BEEF CATTLE

## Chapter 15

381

Fig. 15-1.  Under the old system of unforced production and marketing at 3 to 5 years of age—or older, reasonably good pasture and hay sufficed. Not so today! Forced production, frequently in confinement, and growing and finishing simultaneously, have made the nutritive requirements more critical than formerly—especially the protein, minerals, and vitamins. (Courtesy, *West Texas Livestock Weekly*, San Angelo, TX)

The feeding of beef cattle constitutes the greatest single cost item of their production. It is important, therefore, that the feeding practices be as satisfactory as possible.

Pastures and other roughages, preferably with a maximum of the former, are the very foundation of successful beef cattle production. In fact, it may be said that the principal function of beef cattle is to harvest vast acreages of forages, and, with or without supplementation, to convert these feeds into more nutritious and palatable products for human consumption. It is estimated (1) that 85.7% of the total feed of beef cattle is derived from roughages (see Table 2–5), and (2) that 31% of the land area of the continental United States is grazed (pastured) all or part of the year, with much of this area utilized by beef cattle. If produced on well-fertilized soils, green grass and well-cured, green, leafy hay can supply all of the nutrient requirements of beef cattle except the need for common salt and whatever energy-rich feeds may be necessary for additional conditioning or drylot finishing.

## DIGESTIVE SYSTEM OF CATTLE

The most unique anatomical feature of a ruminant is the large, four-compartment stomach (Fig. 15–2). The digestive tract

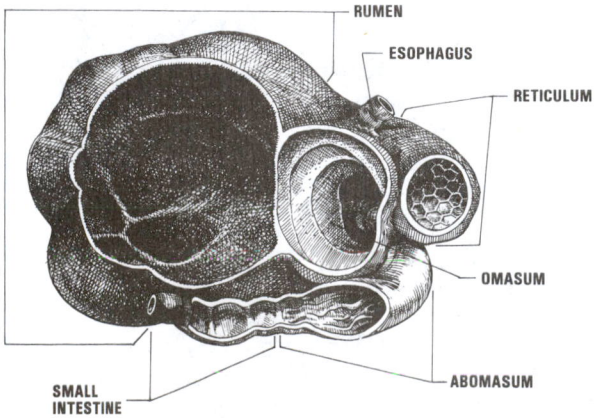

Fig. 15-2.  Stomach of mature cow.

of a mature cow comprises about ¼ of the liveweight of the animal. The rumen, or paunch, which accounts for about ¾ of the total stomach capacity of the mature cow, serves as a "fermentation vat" that enables cows, and other ruminants, to digest cellulose—a major constituent of plant tissue, utilize urea and other nonprotein nitrogen compounds, and synthesize B vitamins and vitamin K. The rumen hosts billions of bacteria and protozoa. It is estimated that one teaspoonful of rumen content may contain 200 billion microorganisms. Forage remains in the rumen and reticulum, the first two stomach compartments in which fermentation occurs, for 48 to 60 hours, whereas feed remains in the omasum and abomasum only about 8 and 3 hours, respectively.

As a result of the unique digestive system, especially the rumen, the cow is able to perform wonders for people to convert feeds that are inedible by humans, and which would otherwise go to waste, into highly nutritious meat and milk.

## NUTRITIVE NEEDS OF BEEF CATTLE

The nutritive requirements of beef cattle have become more critical with the shift in beef production practices. Steers were formerly permitted to make their growth primarily on roughages—pastures in the summertime and hay and other forages in the winter. After making moderate and unforced growth for 2 to 4 years, usually the animals were either turned into the feedlot or placed on more lush pastures for a reasonable degree of finishing. With this system, the growth and finishing requirements of cattle came largely at two separate periods in the life of the animal.

Under the old system of moderate growth rate, reasonably good pastures and good quality hay fully met the protein requirements, as well as the mineral and vitamin needs. As the feeding period was not so long with these older cattle, and the stress was not so great in comparison with the period required in the finishing of calves or yearlings in a drylot, there also was less tendency for vitamin deficiencies to develop in the feedlot; and the protein requirements were less important during the finishing period.

In recent years, the introduction of crossbreeding and the exotic breeds has produced faster-gaining calves, later-maturing cattle, and heavier-milking cows. Hand in hand with this development, scarce and high-priced grains, compared to roughages, have caused feeder cattle to be carried to heavier weights on grass and other roughages before going into feedlots, then grain fed for a shorter period than formerly. Also, more and more heifers are being bred to calve as 2-year-olds. In this revision, provisions have been made for the nutritive needs created by these changes.

As feeds represent by far the greatest cost item in beef production, it is important that there be a basic understanding of the nutritive requirements. The nutritive requirements of beef cattle have been established by the National Academy of Sciences—NRC, and published in *Nutrient Requirements of Beef Cattle*, sixth revised edition, 1984. These nutritive requirements may be obtained from the NRC publication. Also, these requirements were adapted for, and presented in, *Feeds & Nutrition* and *Feeds & Nutrition Digest*, books by M. E. Ensminger, *et al.*, published by Ensminger Publishing Company; hence the reader is referred thereto. The nutritive needs of animals as a whole are discussed in Chapter 4, "Feeding Livestock"; hence, the reader is referred thereto for general information on the subject. The specific nutritive needs of beef cattle follow.

## ENERGY

Carbohydrates, which constitute about 75% of all the dry matter of plants, are the chief source of energy in cattle feeds. Next to carbohydrates, fats are important as energy sources.

A relatively large portion of the feeds consumed by beef cattle is used in meeting the energy needs, regardless of whether the animals are merely being maintained (as in wintering) or fed for growth, finishing, or reproduction.

The first and most important function of feeds is that of meeting the maintenance needs. If there is not sufficient feed, as is frequently true during perods of drought or when winter rations are skimpy, the energy needs of the body are met by the breakdown of tissue. This results in loss of condition and body weight.

After the energy needs for body maintenance have been met, any surplus energy may be used for growth, finishing, reproduction, or lactation. When cattle are finished at early ages, growth and finishing are in most instances simultaneous, and, therefore, not easily separated.

Through bacterial action in the rumen, cattle are able to utilize a considerable portion of roughages as sources of energy. Yet it must be realized that with extremely bulky rations, the animal cannot consume sufficient quantities to produce the maximum amount of fat. For this reason, finishing rations contain a considerable proportion of concentrated feeds, mostly cereal grains. On the other hand, when the energy requirements are primarily for maintenance, roughages are usually the most economical sources of energy for beef cattle.

### SYMPTOMS OF ENERGY DEFICIENCY (UNDERFEEDING)

Many cattle throughout the world are underfed all or some part of the year. In fact, lack of sufficient total feed is probably the most common deficiency suffered by beef cattle, although it is recognized that underfeeding is frequently complicated by a concomitant shortage of protein and other nutrients. Restricted rations often occur during periods of drought, when pastures or ranges are overstocked, or when winter rations are skimpy. Also, many range producers regularly plan that cows in good flesh should lose some condition during the winter months; they feel that it is uneconomical to feed enough to retain the fleshy condition. Fortunately, during such times of restricted feed intake, animals have nutritive reserves upon which they can draw. Although they may survive for a considerable period of time under these conditions, there is an inevitable loss in body weight and condition; and, varying with the degree of underfeeding, there may be a slowing or cessation of growth (including the skeletal growth), failure to conceive, and increased mortality. Low feed intake also commonly results in increased deaths from toxic plants and from lowered resistance to parasites and diseases.

## PROTEIN

The protein allowance for beef cattle, regardless of age or system of production, should be ample to replace the daily breakdown of the tissues of the body including the growth of hair, horns, and hoofs. In general, the protein needs are greatest for the growth of the young calf and for the gestating-lactating cow.

As protein supplements ordinarily cost more per ton than grains, normally beef cattle should not be fed larger quantities of these supplements than actually needed to balance the ration.

With stocker cattle, or in the maintenance of the beef breeding herd, it usually does not pay to add a protein supplement when a legume hay is fed. With feedlot cattle on high-concentrate rations, or when the breeding herd is being wintered on a nonlegume roughage, sufficient protein supplement—usually 1 to 2 lb daily—should be added to the ration.

• **Bypass protein (undegraded intake protein)**—This refers to protein in the feed that escapes digestion in the rumen and passes into the lower digestive track where it is digested and absorbed. (Also see Chapter 4 of this book, section on "Slow-Release and Rumen Bypass Treatments."

### SYMPTOMS OF PROTEIN DEFICIENCY

Depressed appetite is the primary symptom of protein deficiency in beef cattle rations. Depressed appetite may, in turn, lead to an inadequate intake of energy; hence, protein deficiency and energy deficiency often occur together.

Other symptoms of protein deficiency are loss of weight, poor growth, irregular or delayed estrus, and reduced milk production.

### NONPROTEIN NITROGEN SOURCES

Certain nonprotein nitrogen sources may be substituted for all or much of the supplemental protein required in most beef cattle rations, provided such rations are adequate in minerals and readily available carbohydrates. Among such products are urea, ammoniated molasses, ammoniated beet pulp, ammoniated cottonseed meal, ammoniated citrus pulp, and ammoniated rice hulls. The possibility exists that other products will be forthcoming. Each such product should be evaluated by a controlled feeding trial.

(Also see Chapter 4 of this book, section on "Nonprotein Nitrogen [NPN] Sources.")

## MINERALS

Beef cattle are susceptible to the usual inefficiencies and ailments when exposed to (1) prolonged and severe mineral deficiencies, or (2) excesses of fluorine, selenium, or molybdenum. (See Table 4–13 for a summary of "Nutritional Diseases and Ailments of Animals.")

Needed minerals may be incorporated in beef cattle rations or in the water. In addition, it is recommended that all classes and ages of cattle be allowed free access to a two-compartment mineral box, with (1) salt (iodized salt in iodine-deficient areas) in one side, and (2) a suitable mineral mixture in the other side. Free-choice feeding is in the nature of cheap insurance, with the animals consuming the mineral if they are needed.

• **Salt**—Most ranchers compute the yearly salt requirement on the basis of about 25 lb for each cow. Mature animals will consume 3 to 5 lb of salt per month when pastures are lush and succulent, and 1 to 1½ lb per month during the balance of the season.

• **Calcium**—In contrast to phosphorus deficiency, calcium deficiency in beef cattle is relatively rare and mild and the symptoms much less conspicuous. In general, when the forage of cattle consists of at least one-third legume (legume hay, pasture, or silage), ample calcium will be provided. But even nonlegume forages contain more calcium than cereal grains. This indicates that a mineral source of calcium is less important when large quantities of roughage are being consumed. Also, plants grown on calcium-rich soils contain a higher content of this element.

As finishing cattle consume a higher proportion of grains to roughages—and the grains are low in calcium—they have a greater need for a calcium supplement than do beef cattle that are being fed largely on roughages. This is especially true of younger cattle and where a long feeding period is involved.

When the ration of beef cattle is suspected of being low in calcium, the animals should be given free access to a calcium supplement, with salt provided separately; or a calcium supplement may be added to the daily ration in keeping with nutrient requirements. (See Table 15–1.)

• **Phosphorus**—In some sections of the United States and other countries, the soils are so deficient in phosphorus that the feeds produced thereon do not provide enough of this mineral for cattle or other classes of stock. As a result, the cattle produced in these areas may have depraved appetites, may fail to breed regularly, and may produce markedly less milk. Growth and development are slow, and the animals become emaciated and fail to reach normal adult size. Death losses are abnormally high.

In range areas where the soils are either known or suspected to be deficient in phosphorus, cattle should always be given free access to a suitable phosphorus supplement.

To be on the safe side, the general recommendation for beef cattle on both the range and in the finishing lot is to allow free choice of a suitable phosphorus supplement in a mineral box, or to add a phosphorus supplement to the ration in keeping with nutrient requirements. (See Table 15–1.)

• **Cobalt**—Deficiencies of cobalt in cattle are costly, for the

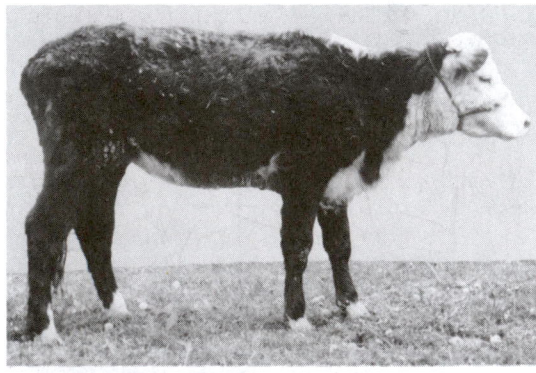

Fig. 15–3. Cobalt deficiency: The upper picture shows a heifer suffering from a cobalt deficiency. Anemia, loss of appetite, and roughness of hair coat characterize the malady. The lower picture illustrates the remarkable recovery in the same animal brought about by the administration of cobalt. (Courtesy, Michigan State University)

affected animals become weak and emaciated and eventually die. Cattle in affected areas should have access to a cobaltized mineral mixture, made by mixing 0.2 oz of cobalt chloride, cobalt sulfate, cobalt oxide, or cobalt carbonate per 100 lb of either (1) salt, or (2) mineral mix.

• **Copper**—Copper is sometimes deficient in the soils of certain areas, notably the state of Florida. In such areas, 0.25 to 0.5% of copper sulfate or copper oxide should be incorporated in the salt or mineral mixture. In addition to being an area disease, copper deficiencies have occurred in beef calves kept on nurse cows for periods extending beyond the normal weaning age.

• **Iodine**—Iodized salt should always be fed to cattle in iodine-deficient areas (such as the northwestern United States and the Great Lakes region). This can be easily and cheaply accomplished by providing stabilized iodized salt containing 0.01% potassium iodide (0.0076% iodine).

• **Magnesium**—Certain pastures in early spring are inadequate in magnesium, with the result that grass tetany or grass staggers may occur in cattle grazed on such pastures. Lactating cows are most commonly affected. In problem areas, as high as 0.7 oz of supplemental magnesium per head daily may be required to prevent this malady.

• **Manganese**—A deficiency of manganese exists in some areas of the northwestern United States, where it has been shown to be one cause of *crooked calves*—calves born with enlarged joints, stiffness, twisted legs, *overknuckling*, and weak and shortened bones.

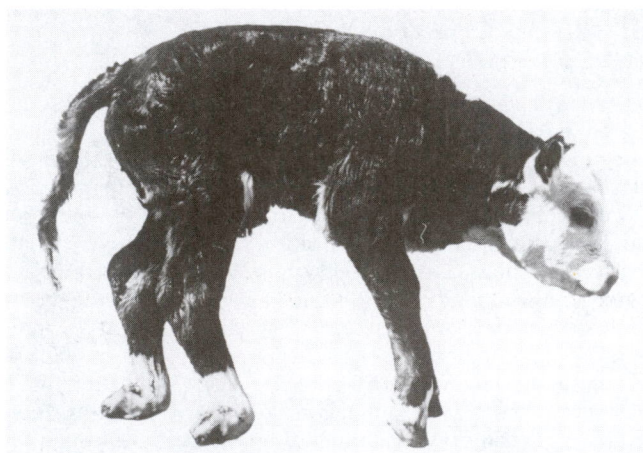

Fig. 15–4. Manganese deficiency in a newborn calf. Note weak legs and overknuckling. (Courtesy, Washington State University)

• **Molybdenum**—Molybdenum deficiencies have not been demonstrated in cattle. The greatest concern about molybdenum is its toxicity, which has been observed in areas where pastures are grown on high-molybdenum soils—known as *teart pastures* in England, Canada, and the United States. Excess molybdenum interferes with copper metabolism.

• **Potassium**—Potassium deficiencies are rare, but they occasionally occur in drylot finishing cattle fed high-concentrate rations. Forages are extremely good sources of potassium. For this reason, potassium is not generally added to feeds for cattle.

• **Selenium**—Cows grazing on low-selenium pastures may be affected as follows, in comparison with cows grazing similar pastures supplemented with selenium: (1) produce more calves with nutritional muscular dystrophy (white muscle disease), (2)

---

Fig. 15–5a. White muscle disease in a calf. This shows the generalized weakness of muscles, lameness, and difficulty in locomotion of an afflicted calf. Calf is about three months old. (Courtesy, Oregon Agricultural Experiment Station, Corvallis)

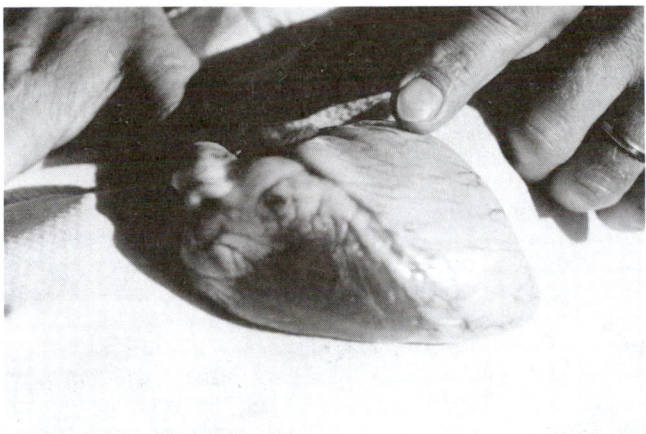

Fig. 15–5b. This shows abnormal white areas in the heart muscles of a six-week old calf afflicted with white muscle disease. (Courtesy, Oregon Agricultural Experiment Station, Corvallis)

have higher death losses, (3) have a higher incidence of retained placenta, and (4) wean off lighter calves. Also, it has been shown that the performance of feedlot cattle fed a selenium deficient ration is improved by adding selenium.

- **Sulfur**—Sulfur is a component of protein, some vitamins, and several important hormones. The common sulfur-containing amino acids are methionine, cysteine, and cystine. Also, the following amino acid derivatives contain sulfur: cystathionine, taurine, and cysteic acid. Methionine is a key amino acid, because all other sulfur compounds, except the B vitamins thiamin and biotin, can be synthesized from methionine.

All feeds contain some sulfur, but the amount usually depends on the protein content of the feed; generally speaking, the higher the protein content, the higher the sulfur content. Availability of the sulfur in the feed to microbial reduction in the rumen may be of as much concern as the actual amount that is present.

- **Zinc**—Added zinc intake has been shown to increase the rate and efficiency of gains of feedlot cattle in certain areas. This may be due to the relationship between (1) phytic acid and zinc, and (2) calcium and zinc, the improper ratios of which may create a need for supplemental zinc.

## MINERAL REQUIREMENTS

The National Research Council mineral requirements of beef cattle are summarized in Table 15–1.

When buying and home mixing minerals, or when buying commercial mineral mixes, the cattle producers should first determine their needs, based on (1) available feeds, (2) area (for example, the Northern Great Plains and the Southwest are phosphorus-deficient areas), and (3) the age and reproduction status (pregnancy and lactation make a difference) of the animals for which the mineral mix is intended.

Cattle pastured on native grass should be offered a free-choice mineral mix consisting of 40% dicalcium phosphate or bone meal and 60% trace mineralized salt. Minerals that are self-fed on pastures or in corrals should be in boxes protected from the weather.

Salt should always be available on a free-choice basis in addition to whatever mineral mix is provided.

### TABLE 15–1
### MINERAL REQUIREMENTS AND MAXIMUM TOLERABLE LEVELS FOR BEEF CATTLE[1]

| Mineral | Requirements | | Maximum Tolerable Level[3] | Mineral | Requirements | | Maximum Tolerable Level[3] |
| | Suggested Value | Range[2] | | | Suggested Value | Range[2] | |
|---|---|---|---|---|---|---|---|
| Calcium ............. (%) | — | Breeding 0.15–0.43 Growing/finishing 0.17–1.38 | 2 | Phosphorus .......... (%) | — | Breeding 0.15–0.28 Growing/finishing 0.15–0.53 | 1 |
| Cobalt ............. (ppm) | 0.10 | 0.07–0.11 | 5 | Potassium ............ (%) | 0.65 | 0.5–0.7 | 3 |
| Copper ............ (ppm) | 8 | 4–10 | 115 | Selenium .......... (ppm) | 0.20 | 0.05–0.30 | 2 |
| Iodine ............. (ppm) | 0.5 | 0.20–2.0 | 50 | Sodium ............. (%) | 0.08 | 0.06–0.10 | 10[4] |
| Iron ............... (ppm) | 50 | 50–100 | 1,000 | Chlorine ............ (%) | — | — | — |
| Magnesium ......... (%) | 0.10 | 0.05–0.25 | 0.40 | Sulfur ............. (%) | 0.10 | 0.08–0.15 | 0.40 |
| Manganese ......... (ppm) | 40 | 20–50 | 1,000 | Zinc .............. (ppm) | 30 | 20–40 | 500 |
| Molybdenum ........ (ppm) | — | 6 | | | | | |

[1]Adapted from *Nutrient Requirements of Beef Cattle*, 6th revised edition, National Research Council–National Academy of Sciences, 1984, p. 43.

[2]The listing of a range in which requirements are likely to be met recognizes that requirements for most minerals are affected by a variety of dietary and animal factors (body weight, sex, rate of gain). Thus, it may be better to evaluate rations based on a range of mineral requirements and for content of interfering substances than to meet a specific dietary value.

[3]From National Research Council (1980). Maximum tolerable levels are given on the basis of the ration dry matter.

[4]10% sodium chloride.

## VITAMINS

Vitamin deficiencies in cattle may occur as a result of lack of availability of vitamins or because of the presence of anti-metabolites. Both are important concepts. For example, analyses show corn to be adequate in niacin. Yet, due either to an antimetabolite or unavailability, there may be niacin deficiencies when corn is fed—deficiencies which can be remedied by niacin supplementation.

• **Vitamin A**—The vitamin most likely to be deficient in beef cattle rations is vitamin A. True vitamin A is a chemically formed compound, which does not occur in plants. It is furnished in most beef cattle rations in the form of its precursor, carotene. However, plants are a variable and often undependable source of carotene, due to oxidation. Also, cattle are relatively inefficient converters of carotene to vitamin A. The latter fact was taken into consideration in the development of international standards for vitamin A, which are based on the rate at which the rat converts beta-carotene to Vitamin A. The conversion rate for the rat is one mg of beta-carotene to 1,667 IU of vitamin A, whereas it is estimated that one mg of beta-carotene is equal to 400 IU of vitamin A in cattle. Moreover, the conversion rate for cattle varies under different conditions; it is influenced by type of carotenoid, breed, individual differences in animals,

Fig. 15–6. Same bull before (upper) and after (lower) vitamin A (carotene) feeding. In the upper picture, the bull shows advanced stages of vitamin A deficiency—note the dejected appearance and rough hair coat. In the bottom picture, the bull shows a general improvement in appearance and male characteristics following vitamin A feeding. (Courtesy, USDA)

and level of carotene intake. Stress conditions—such as extremely hot weather, viral infections, and altered thyroid function—have also been suggested as causes for reduced conversion.

Under practical feeding conditions, producers should consider (1) previous feeding as it influences body stores of vitamin A; (2) vitamin A destruction during processing or when mixed with oxidizing materials; and (3) carotene destruction in feeds during storage.

• **Vitamin D**—When exposed to enough direct sunlight, beef cattle normally acquire their vitamin D needs, for the ultraviolet rays in sunlight penetrate the skin and produce vitamin D from traces of sterols in the tissues. Also, cattle obtain vitamin D from sun-cured roughages. However, the addition of vitamin D to the ration is important where cattle, especially calves, are kept in the barn most of the day, where there is limited sunshine, where the calcium:phosphorus ratio leaves much to be desired, and/or where little or no sun-cured hay is fed. Vitamin D helps build strong and sturdy frames.

• **Vitamin E**—Added vitamin E may be necessary under certain conditions because of its relationship to vitamin A utilization and the prevention of white muscle disease.

• **Vitamin K**—Under normal conditions, adequate vitamin K is synthesized in the rumen of cattle. However, symptoms of inadequacy (a bleeding syndrome known as *sweet clover disease*) occur when moldy sweet clover hay, high in dicoumarol content, is fed.

• **B vitamins**—Dietary requirements for the B vitamins (thiamin, biotin, niacin, pyridoxine, pantothenic acid, riboflavin, and vitamin B–12) have been demonstrated experimentally for the young calf during the first eight weeks of life, prior to the development of the functioning rumen. At this stage in life, these requirements are usually met by the milk of the dam. Later, the B vitamins appear to be synthesized in sufficient quantities by rumen bacterial fermentation. However, inadequacy of protein or other nutrients in the ration may impair rumen fermentation, with the result that sufficient quantities of the B vitamins will not be synthesized.

## VITAMIN REQUIREMENTS

The National Research Council vitamin requirements of beef cattle are summarized in Table 15–2.

**TABLE 15–2**
**VITAMIN REQUIREMENTS OF BEEF CATTLE**
**(IN AMOUNT PER KILOGRAM OF DRY RATION)[1]**

| Nutrient | Growing and Finishing Steers and Heifers | Dry Pregnant Cows | Breeding Bulls and Lactating Cows |
|---|---|---|---|
| Vitamin A activity ....IU[2] | 2,200 | 2,800 | 3,900 |
| Vitamin D ..........IU | 275 | 275 | 275 |
| Vitamin E ..........IU | 15–60 | — | 15–60 |

[1]From: *Nutrient Requirements of Beef Cattle*, Sixth Revised Edition, National Research Council, National Academy of Sciences, Washington, DC, 1984. Requirements given in IU/kg may be converted to IU/lb by dividing by 2.2.

[2]May be vitamin A or provitamin A equivalent.

## WATER

Beef cattle should have an abundant supply of water before them at all times. Mature cattle will consume an average of about 11 gal of water per head daily, with younger animals requiring proportionately less. The water requirement is influenced by several factors, including rate and composition of gain, pregnancy, lactation, activity, type of ration, feed intake, and environmental temperature.

Saline water containing 1% soluble salts may be toxic. Excessive nitrates or alkalinity may make water unsatisfactory for cattle.

In the northern latitudes, heaters must be provided to make the water available, but they are not needed to warm the water further.

Fig. 15–7. Watering facility typical of the western range—well, windmill, and tank.

## FEED ALLOWANCE AND SOME SUGGESTED RATIONS[1]

Some general rules of feeding may be given, but it must be remembered that "the eye of the expert feeder fattens the cattle." Nevertheless, the beginner may well profit from the experience of successful feeders. It is with this hope that the suggested rations are herewith presented.

Table 15–3 (see pages 388 and 389) is a handy beef cattle feeding guide for different classes and ages of cattle. All of these are merely intended as general guides. Variations can and should be made in the rations used. The feeder should give consideration to (1) the supply of homegrown feeds, (2) the availability and price of purchased feeds, (3) the class and age of cattle, (4) the health and condition of the animals, and (5) the length of the grazing season.

In using Table 15–3 as a guide, it is to be recognized that feeds of similar nutritive properties can and should be interchanged as price relationships warrant. Thus, (1) the cereal

---

[1]Insofar as possible, these rations were computed from the requirements as reported by the National Research Council and applied by the author.

grains may consist of corn, barley, wheat, oats, and/or sorghum; (2) the protein supplement may consist of soybean, cottonseed, peanut, sunflower, and/or linseed meal; (3) the roughage may include many varieties of hays and silages, and (4) a vast array of by-product feeds may be utilized.

## FEEDS FOR BEEF CATTLE

Beef cattle feeding practices vary according to the relative availability of grasses, dry roughages, and grains. Where roughages are abundant and grain is limited, as in the western range states, cattle are primarily grown out or finished on roughages. On the other hand, where grain is relatively more abundant, as in the Corn Belt and in the High Plains area of Texas and Oklahoma, finishing with more concentrates is common.

## PASTURES

*Pastures include all crops that are harvested directly by animals.*

Good pasture is the cornerstone of successful beef cattle production. In fact, there has never been a great beef cattle country or area which did not produce good grass. It has been said that good cattle producers can be recognized by the character of pastures and that good cattle graze good pastures. Thus, the three go hand in hand—good producers, good pastures, and good cattle. The relationship and importance of cattle and pastures has been further extolled in an old Flemish proverb which says, "No grass, no cattle; no cattle, no manure; no manure, no crops."

Fig. 15–8. Cattle and grass go together. This shows a typical Texas Gulf Coast commercial herd. (Courtesy, *Livestock Weekly*, San Angelo, TX)

Approximately 31% of the total land area of the continental United States is devoted to pasture and grazing lands. Much of this area, especially in the Far West, can be utilized only by beef cattle or sheep. Although the term pasture usually suggests growing plants, it is correct to speak of pasturing stalk and stubble fields.

**TABLE**
**DAILY RATIONS**
**(As-Fed**

| Suggested Rations<br><br>With all rations and for all classes and ages of cattle, provide free access in separate containers to (1) salt (iodized salt in iodine-deficient areas), and (2) a suitable mineral mixture. | Wintering Mature Pregnant Beef Breeding Cows (avg. wt. 1,100 lb or *499 kg*) | | Wintering Mature Lactating Beef Breeding Cows (avg. wt. 1,100 lb or *499 kg*) | | Wintering Replacement Heifers (weighing 400–500 lb or *181–227 kg* start of wintering) | |
|---|---|---|---|---|---|---|
| | **Per Day** | | **Per Day** | | **Per Day** | |
| | **(lb)** | **(kg)** | **(lb)** | **(kg)** | **(lb)** | **(kg)** |
| **1.** Legume hay or grass-legume mixed hay, good quality ........... | 18–20 | *8.2–9.1* | 30 | *13.6* | 13–15[3] | *5.9–6.8*[3] |
| Grain ..................................................... | — | — | — | — | 2–3 | *0.91–1.36* |
| Protein supplement ........................................ | — | — | — | — | — | — |
| **2.** Grass hay or other nonlegume dry roughage ................. | 18–20 | *8.2–9.1* | 24–26 | *10.9–11.8* | 12–18[3] | *5.4–8.2*[3] |
| Grain ..................................................... | — | — | 2 | *0.91* | 2½–4½ | *1.13–2.04* |
| Protein Supplement ........................................ | ½–1 | *0.23–0.45* | 3 | *1.36* | 1¼–1½ | *0.57–0.68* |
| **3.** Legume hay or grass-legume mixed hay, good quality ........... | 7–11 | *3.2–5.0* | 26–28 | *11.8–12.7* | 8–12[3] | *3.6–5.4*[3] |
| Grass hay or other nonlegume dry roughage ................. | 9–11 | *4.1–5.0* | — | — | 4–6 | *1.8–2.7* |
| Grain ..................................................... | — | — | 1 | *0.45* | 2½–4 | *1.13–1.81* |
| Protein supplement ........................................ | — | — | 1 | *0.45* | ½–1 | *0.23–0.54* |
| **4.** Corn or sorghum silage ................................... | 50–55 | *22.7–25* | 55 | *25* | 25–40 | *11.3–18.2* |
| Grain ..................................................... | — | — | 2 | *0.91* | — | — |
| Protein supplement ........................................ | 0–½ | *0–0.23* | 3 | *1.36* | 1½–1¾ | *0.68–0.79* |
| **5.** Grass silage, half or more legume ....................... | 50 | *22.7* | 50 | *22.7* | 25–40 | *11.3–18.2* |
| Grain ..................................................... | — | — | 4 | *1.81* | 3–4 | *1.36–1.81* |
| Protein supplement ........................................ | — | — | — | — | ½ | *0.23* |
| **6.** Silage (corn or sorghum silage fed with legume hay or legume silage fed with grass hay) ......................... | 35 | *15.9* | 40 | *18.1* | 15–30 | *6.8–13.6* |
| Hay ...................................................... | 5–6 | *2.3–2.7* | 10 | *4.5* | 3–4 | *1.4–1.8* |
| Grain ..................................................... | — | — | — | — | 1–2 | *0.45–0.91* |
| Protein supplement ........................................ | 0–½ | *0–0.23* | — | — | ½–1 | *0.22–0.45* |

[1]If stocker calves are late or the roughage is fair to poor quality, it may be desirable to add 2–4 lb *(0.91–1.81 kg)* of grain per head daily. If farm scales are available, monthly weights may be used as the criterion for grain feeding. Keep in mind that calves should gain ¾–1 lb *(0.34–0.45 kg)* daily.

[2]In general, the experienced feeder plans that cattle on full feed shall consume (1) feeds in amounts (daily; air-dry basis) equal to about 2.5–3.0% of their liveweight, (2) 70–90% concentrates, and (3) a minimum of 2–4 lb *(0.9–1.8 kg)* roughage for each 100 lb *(45 kg)* liveweight. In areas where roughage is more abundant and comparatively cheaper than grain, the proportions of roughage to grain should be somewhat higher than indicated. In computing roughage consumption, 3 lb *(1.36 kg)* of silage are considered equivalent to 1 lb *(0.45 kg)* of hay.

The type of pasture, as well as its carrying capacity and seasonable use, varies according to topography, soil, and climate. Because of the hundreds of species of grasses and legumes that are used as beef cattle pastures, each with its own best adaptation, no attempt is made to discuss the respective virtues of each variety. Instead, it is recommended that farmers or ranchers seek the advice of their local country agricultural agent, or write to their state agricultural college.

No method of harvesting has yet been devised that is as cheap as that which can be accomplished through grazing animals. Accordingly, successful beef cattle management necessitates as nearly year-round grazing as possible. In the northern latitudes of the United States, the grazing season is usually of about six months' duration, whereas in the deep South, yearlong grazing is approached. In many range areas of the West, the breeding herds obtain practically all their forage the year-round from the range, being given supplemental roughage only if the grass or browse is buried deep in snow.

During the winter months, and in periods of drought, the pasture utilized by beef cattle may consist of dried grass cured on the stalk. On a dry basis, the crude protein content of mature, weathered grasses may be 3% or less. To supplement such feed, producers commonly feed cake or cubes. The use of cake or cubes instead of meal reduces losses from wind blowing, an especially important factor on the range.

In some instances, cattle on pasture fail to make the proper growth or gain in condition because the soil is seriously deficient in fertility or the pasture has not been well managed. In such instances, striking improvement will result from proper fertilization and management.

## HAYS AND OTHER DRY ROUGHAGES

Hay is the most important harvested roughage fed to beef cattle, although many other dry roughages can be utilized.

The dry roughages are all high in fiber and therefore lower than concentrates in total digestible nutrients. Hay averages about 28% fiber and straw approximately 38%, whereas such concentrates as corn and wheat contain only 2 to 3% fiber.

**15-3**
**FOR BEEF CATTLE**
Basis)

| Wintering Stocker Calves Roughed Through Winter and Grazed the Following Summer. Fed for winter gain of ¾–1 lb (0.34–0.45 kg) per head daily (weighing 400–500 lb or 181–227 kg start of wintering)[1] | | Finishing Calves in Drylot, Generally in Winter (weighing 400–500 lb or 181–227 kg start of feeding and 750–850 lb or 340–386 kg at marketing)[2] | | Wintering Yearlings; Roughed Through the Winter, and Generally Pasture Finished the Following Summer. Fed for winter gains of 1–1¼ lb or 0.45–0.57 kg per head daily (weighing about 600 lb or 227 kg start of wintering) | | Finishing Yearlings in Drylot, Generally in Winter (weighing about 600 lb or 272 kg start of feeding, and 900–1,050 lb or 409–477 kg at marketing)[2] | | Finishing Long-yearling Steers in Drylot Generally in Winter (weighing about 850 lb or 386 kg start of feeding and 1,000–1,100 lb or 454–499 kg at marketing)[2] | |
|---|---|---|---|---|---|---|---|---|---|
| Per Day | | Per Day | | Per Day | | Per Day | | Per Day | |
| (lb) | (kg) | (lb) | (kg) | (lb) | (kg) | (lb) | (kg) | (lb) | (kg) |
| 12–18³ | 5.4–8.2 | 4–6 | 1.8–2.7 | 16–24 | 7.2–10.9 | 4–8 | 1.8–3.6 | 6–12 | 2.7–5.4 |
| — | — | 12–15 | 5.4–6.8 | — | — | 15–19½ | 6.80–8.8 | 16–22 | 7.2–10.0 |
| — | — | 1–1½ | 0.45–0.68 | — | — | 1–1½ | 0.45–0.68 | | |
| 12–18³ | 5.4–8.2 | 4–5 | 1.8–2.3 | 16–24 | 7.2–10.9 | 4–8 | 1.8–3.6 | 6–12 | 2.7–5.4 |
| — | — | 12–15 | 5.4–6.8 | — | — | 15–20 | 6.8–9.1 | 16½–22¾ | 7.5–10.3 |
| ¼–1½ | 0.57–0.68 | 1¾–2 | 0.79–0.91 | 1½–1¾ | 0.68–0.79 | 1½–2½ | 0.68–1.1 | 1½–1¾ | 0.68–0.79 |
| 8–12³ | 5.4–8.2 | 2–3 | 0.91–1.36 | 6–8 | 2.7–3.6 | 2–4 | 0.91–1.81 | 3–6 | 1.4–2.7 |
| 4–6 | 1.8–2.7 | 2–3 | 0.91–1.36 | 10–16 | 4.5–7.2 | 2–4 | 0.91–1.81 | 3–6 | 1.4–2.7 |
| — | — | 12–15 | 5.4–6.8 | — | — | 15–19¾ | 6.8–9.0 | 16–22 | 7.2–10.0 |
| ¼–1 | 0.11–0.45 | 1½–1¾ | 0.68–0.79 | 1–1½ | 0.45–0.68 | 1¼–1¾ | 0.57–0.79 | ½–¾ | 0.23–0.34 |
| 25–40 | 11.3–18.1 | 6–16 | 2.7–7.3 | 40–55 | 18.2–24.9 | 6–25 | 2.7–11.3 | 6–35 | 2.7–5.9 |
| — | — | 8–12 | 3.6–5.4 | — | — | 11–16 | 5.0–7.3 | 15–21 | 6.8–9.5 |
| 1–1¼ | 0.45–0.57 | 2 | 0.91 | 1¼–1½ | 0.57–0.68 | 2 | 0.91 | 1¼–1½ | 0.57–0.68 |
| 25–40 | 11.3–18.1 | 6–16 | 2.7–7.3 | 40–55 | 18.1–24.9 | 6–25 | 2.7–11.3 | 6–35 | 2.7–15.9 |
| 2–3 | 0.91–1.36 | 8–12 | 3.6–5.4 | 4–5 | 1.8–2.3 | 11–16 | 5.0–7.3 | 15–21 | 6.8–9.5 |
| ½ | 0.23 | 1–2 | 0.45–0.91 | ½ | 0.23 | 1–1½ | 0.45–0.68 | 1 | 0.45 |
| 15–30 | 6.8–13.6 | 3–8 | 1.4–3.6 | 20–35 | 9.1–15.9 | 3–15 | 1.4–6.8 | 3–15 | 1.4–6.8 |
| 3–4 | 1.4–1.8 | 1–3 | 0.45–1.4 | 7 | 3.2 | 1–4 | 0.45–1.8 | 1–7 | 0.45–3.2 |
| 1–2 | 0.45–0.91 | 8–12 | 2.6–5.4 | — | — | 11–16 | 5.0–7.2 | 15–21 | 6.8–9.5 |
| ½ | 0.23 | 1–2 | 0.45–0.91 | ½–¾ | 0.23–0.34 | 1–1¾ | 0.45–0.79 | 1–1¼ | 0.45–0.57 |

³With calves (both replacement heifers and stockers) an extra 2 lb (0.91 kg) of hay daily, over and above requirements, are herewith indicated to allow for wastage. Practical operators generally feed stemmy or other hay left over by calves to the cow herd.

Fortunately, cattle are equipped to handle large quantities of roughages. In the first place, the paunch of a mature cow has a capacity of about 53 gal, thus providing ample storage for large quantities of less concentrated feeds. Secondly, the billions of microorganisms in the rumen attack the cellulose and pentosans of the fibrous roughages, such as hay, breaking them down into available and useful nutrients. In addition to providing nutrients at low cost, the roughages add needed bulk to cattle rations.

Roughages, as well as concentrates, may be classified as carbonaceous or nitrogenous, depending on their protein content. The principal dry carbonaceous roughages used by cattle include hay from grasses, the straws and hays from cereal grains; corncobs; the stalks and leaves of corn and the grain sorghums; and cottonseed hulls. Cured nitrogenous roughages include the various legume hays such as alfalfa, the clover hays, peanut hay, soybean hay, cowpea hay, and velvet bean hay.

Although leguminous roughages are preferable, weather conditions and soils often make it more practical to produce nonlegumes. Also, in many areas, such feeds as dry grass cured

Fig. 15–9.  Haying has gone modern! This shows a tightly packed, virtually waterproof round bale. Such bales shed water and may be left in the field with a minimum loss of quality. (Courtesy, International Harvester Company, Chicago, IL)

on the stalk, cereal straws, corncobs, and cottonseed hulls are abundantly available and cheap. Under such circumstances, these feeds had best be used as part of the ration for wintering beef cows, for wintering stockers that are more than one year of age, or for finishing beef cattle.

In comparison with good quality legume hays, the carbonaceous roughages are lower in protein content and quality of proteins, lower in calcium, and generally deficient in carotene (provitamin A). Thus, where nonlegume roughages are used for extended periods, these nutritive deficiencies should be corrected; this is especially true with the gestating-lactating cow or the young, growing calf. To the end that the feeding value of some of the common nonlegumes may be enhanced for beef cattle, the following facts are pertinent:

1. The feeding value of nonlegume hays can be increased by cutting them at an early stage of maturity and curing so as to retain as much of the carotene content as possible.

2. Where dry and bleached pastures are grazed for an extended period of time, or where there is an unusually long winter, it is important that at least part of the roughage be a legume, either silage or hay, or that vitamin A either be added to the ration or injected intramuscularly; and that suitable energy, protein, and mineral supplements be provided.

3. Potentially, corncobs—which were formerly considered a waste product and of little worth—have a feeding value approaching that of hay. However, their energy cannot be utilized unless they are fortified with certain nutrients which help the bacteria and other organisms of the rumen break them down into a form which can be digested. Also, corncobs are low in palatability.

4. Cereal straws and cottonseed hulls may be incorporated in the wintering ration of pregnant cows or in the rations of finishing cattle provided their fundamental characteristics and nutritional limitations are recognized and corrected.

## SILAGES AND ROOT CROPS

Silage is an important adjunct to pastures in beef cattle production, it being possible to use a combination of the two forages in furnishing green succulent feeds on a year-round basis. Extensive use of silage for beef cattle dates back only to about 1910. Prior to that time, it was generally thought of as a feed for dairy cows. Even today, only a relatively small percentage of the beef cattle of the United States is fed silage.

Fig. 15–10. Upright (tower) silos used for storing feed for beef cattle. (Courtesy, *Livestock Breeder Journal*, Macon, GA)

Corn was the first and still remains the principal crop used in the making of silage, but many other crops are ensiled in various sections of the country (see Table 4–7). The sorghums are the leading ensilage crop in the Southwest, and grasses and legumes are the leading ensilage crops in the Northeast. Also, in different sections of the country to which they are adapted, the following feeds are ensiled: cereal grains, field peas, cowpeas, soybeans, potatoes, and numerous fruit and vegetable refuse products. A rule of thumb is that crops that are palatable and nutritious to animals as pasture, as freshly harvested feed, or as dry forage, also make palatable and nutritious silage. Conversely, crops that are unpalatable and nonnutritious as pasture, as green feed, or as dry forage, also make unpalatable and nonnutritious silage.

Grass silage can be produced in those areas where the climate is too cool and the growing season too short for corn or sorghum silage. It is generally higher in protein and carotene, but lower in total digestible nutrients and vitamin D than corn or sorghum silage. Generally, grass silage contains about 80% as much total digestible nutrients (TDN) as corn silage, but it is equal in TDN when 150 lb of grain per ton have been added as a preservative. Thus, grass silage usually requires the addition to the ration of less protein supplement but more total concentrates than corn or sorghum silage. This would indicate that corn or sorghum silage would be slightly preferable to grass silage in high-roughage finishing rations for beef cattle, whereas grass silage would be preferable in high-roughage rations for young, growing beef cattle.

When silage is fed to cattle, it must be remembered that, because of its high moisture content, about 3 lb of silage are generally considered equivalent to 1 lb of dry roughage of comparable quality. A ration of 55 to 60 lb of corn silage plus ½ to ¾ lb of a protein concentrate daily will carry a dry cow through the winter. The ration may be improved, however, by replacing ⅓ to ½ of the silage with an equivalent amount of a dry roughage, adding 1 lb of dry roughage for each 3 lb of silage replaced.

Silage may be successfully used for finishing steers. Long-yearling steers will eat 25 to 35 lb a day at the beginning of the feeding period, the larger amounts being consumed when no hay is fed with it. Better results are obtained, however, if hay is included in the rations. The amount of silage is gradually decreased as the concentrates are increased. At the end of the feeding period, the cattle should be getting around 4 to 6 lb of silage and 1 or 2 lb of hay. Because of their more limited digestive capacity, the allowance of silage fed to calves should be correspondingly less.

Usually, silage provides a much cheaper succulent feed for beef cattle than roots. For this reason, the use of roots for beef cattle is very limited, being confined almost entirely to the northern areas.

## CONCENTRATES

The concentrates include those feeds which are low in fiber and high in energy. For purposes of convenience, concentrates are often further classified as (1) carbonaceous feeds, and (2) nitrogenous feeds.

In general, the use of concentrates for beef cattle is limited to (1) the finishing of cattle, (2) the development of young stock, and (3) use as limited supplements in the winter ration. Over most of the United States, the cereal grains are the chief concentrates fed to beef cattle—these grains being combined, if necessary, with protein supplements to balance the ration.

The chief carbonaceous concentrates used for beef cattle are the cereal grains and such processed feeds as hominy feed, beet pulp, and molasses. The choice of the particular feeds is usually determined primarily by price and availability.

For best results, the feeder should correct the nutritive deficiencies of the cereal grains. All of them are low in protein, low in calcium, and lacking in vitamin D. All except yellow corn are also deficient in carotene. Regardless of whether the cereal grains are fed to growing, breeding, or finishing animals, their nutritive deficiencies can be corrected in a very effective and practical way be adding either (1) a good quality legume hay to the ration, or (2) a protein concentrate plus suitable minerals.

## OTHER FEEDS AND ADDITIVES

In addition to the beef cattle feeds discussed in the preceding sections, the following feeds and additives of importance to beef cattle feeding are presented in this book in Chapter 4, "Feeding Livestock"; hence, the reader is referred thereto:

Protein supplements
Urea
By-product feeds
Feed substitutions
Mineral supplements
Vitamin supplements
Additives, implants, injections

## FEED PREPARATION

The physical preparation of cereal grains for cattle by soaking and cooking has been practiced by exhibitors for a very long time. In recent years, many sophisticated techniques for the processing of grains have been developed, especially for feedlot cattle. Basically, however, grain is either soaked, cooked, ground, or rolled (wet or dry), and hay is either cut, shredded, ground, pelleted, or cubed.

(Also see Chapter 4 of this book, section on "Feed Processing," including Table 4–10.)

## NUTRITIONAL DISEASES AND AILMENTS

Nutritional deficiencies may be brought about by either (1) too little feed, or (2) rations that are low in one or more nutrients.

Also, forced production (such as finishing animals at early ages) and the feeding of forages and grains which are often produced on leached or depleted soils have created many problems in nutrition. This condition has been further aggravated through the increased confinement of cattle, many animals being confined to lots or buildings all or a large part of the year. Under these unnatural conditions, nutritional diseases and ailments have become increasingly common.

Although the cause, prevention, and treatment of most of these nutritional diseases and ailments are known, they continue to reduce profits in the livestock industry simply because the available knowledge is not put into practice. Moreover, those widespread nutritional deficiencies which are not of sufficient proportions to produce clear-cut deficiency symptoms cause even greater economic losses because they go unnoticed and unrectified.

(Also see Chapter 4 of this book, section on "Nutritional Diseases and Ailments.")

## FEEDING AND MANAGING BROOD COWS

Feed affects total profit and cow productivity. It accounts for 65 to 70% of the total cost of keeping cows, and it exerts a powerful influence on cow fertility and calf weaning weight—the two biggest success factors in the cattle business.

Fig. 15–11. A 100% calf crop and heavy weaning weights are the two biggest success factors in the cattle business. This shows cows and calves on pasture in West Virginia.

## NUTRITIONAL REQUIREMENTS

Experiments and practical observation reveal that the period during which calf crop percentage is affected most by nutrition extends from 30 days before calving until 70 days after calving—until after rebreeding; a period of approximately 100 days. This, then, is the most critical period in the cow-calf business. It's when life begins—that period within which one calf is born and another is conceived. The needs for the cow during this most critical production period are approximately equal to her needs for the remainder of the year.

The average estimated energy requirement of a 1,000-lb beef cow during her 12-month reproductive cycle is about 14.5 lb TDN (24 megacalories); based on a 90-day calving season and a 500-lb calf at 7 months of age. However, the requirements are above the average for nearly 6½ months of the year—during the lactation period. This means that, for reasons of economy, the calving season should be timed so that much of the feed can be supplied by pasture and other economical sources of homegrown energy and protein.

A second important requisite of a sound beef cattle nutrition program is to feed animals according to their requirements. It is impossible to feed the herd properly where calving occurs the year around, or when dry pregnant cows, replacement heifers, and cows nursing calves are run together. This point becomes apparent from the following nutritional differences of (1) cows nursing calves, and milking well; (2) yearling heifers, last 3-4 months of pregnancy; and (3) dry mature cows, last 2-3 months of pregnancy.

1. The requirements of cows nursing calves are higher and more critical than those (a) of yearling heifers the last 3-4 months of pregnancy, or (b) of dry mature cows the last 2-3 months of pregnancy in total feed consumed, in energy and protein of the ration, and in calcium and phosphorus. After a cow calves, her energy needs jump about 50%, her protein needs double, and her calcium and phosphorus needs triple.

2. The requirements of yearling heifers the last 3-4 months of pregnancy are higher than those of dry mature cows the last 2-3 months of pregnancy in energy and protein, and in calcium and phosphorus.

Weight also makes a difference, as shown in Table 15–4, which gives the daily nutrient requirements at various weights of (1) dry pregnant cows and (2) cows nursing calves.

### TABLE 15–4
### DAILY NUTRIENT REQUIREMENTS OF BEEF COWS[1]

| Body Weight | | TDN | | Total Protein | | Calcium | Phosphorus |
|---|---|---|---|---|---|---|---|
| (lb) | (kg) | (lb) | (kg) | (lb) | (kg) | (g) | (g) |
| Dry pregnant mature cows (middle third of pregnancy) | | | | | | | |
| 770 | 350 | 9.02 | 4.1 | 1.34 | .61 | 20 | 15 |
| 880 | 400 | 9.68 | 4.4 | 1.45 | .66 | 22 | 16 |
| 990 | 450 | 10.56 | 4.8 | 1.55 | .70 | 23 | 18 |
| 1,100 | 500 | 11.22 | 5.1 | 1.64 | .74 | 25 | 20 |
| 1,210 | 550 | 11.88 | 5.4 | 1.74 | .79 | 26 | 21 |
| 1,320 | 600 | 12.54 | 5.7 | 1.83 | .83 | 28 | 23 |
| 1,430 | 650 | 13.20 | 6.0 | 1.92 | .87 | 30 | 25 |
| Cows nursing calves, first 3 to 4 months after calving (superior milking ability) | | | | | | | |
| 770 | 350 | 11.2 | 5.1 | 2.22 | 1.01 | 36 | 24 |
| 880 | 400 | 13.0 | 5.9 | 2.42 | 1.10 | 37 | 25 |
| 990 | 450 | 14.1 | 6.4 | 2.61 | 1.19 | 39 | 26 |
| 1,100 | 500 | 15.0 | 6.8 | 2.74 | 1.24 | 40 | 28 |
| 1,210 | 550 | 15.6 | 7.1 | 2.86 | 1.30 | 42 | 30 |
| 1,320 | 600 | 16.5 | 7.5 | 2.97 | 1.35 | 43 | 31 |
| 1,430 | 650 | 17.2 | 7.8 | 3.07 | 1.39 | 45 | 33 |

[1]Adapted by the author from *Nutrient Requirements of Beef Cattle*, sixth revised edition, National Research Council-National Academy of Sciences, Washington, DC, 1984, pp. 45 and 46; with U.S. Customary added by the author.

Heavy grain feeding is uneconomical and unnecessary for the beef breeding herd. The nutrient requirements should be adequate merely to provide for maintenance, growth (if the animals are immature), and reproduction. Fortunately, these requirements can largely be met through the feeding of roughages.

## NUTRITIONAL REPRODUCTIVE FAILURE

Since producers largely determine their own destiny when it comes to feeding, it is important that they know the causes of nutritional reproductive failure and how to rectify them.

A review of the literature clearly points to 3 important reproductive difficulties: (1) the small number of cows in heat and bred the first 21 days of the breeding season, (2) the low conception rate at first service, and (3) the excessive calf losses at birth or within the first 2 weeks of age. Also, it is noteworthy that each of the causes is more marked in young cows (first-calf heifers) than in mature cows.

Research throughout the country gives ample evidence that the real cause of most beef cow reproductive failures is a deficiency of one or more nutrients just before and immediately following calving—nutritive deficiencies during that critical 100-day period when life begins—a deficiency of energy, protein, minerals and/or vitamins.

Based on a review of literature, the author concluded as follows:

1. Energy is more important than protein in reproduction.

2. Beef cows receiving inadequate energy reproduce at a low level.

3. Phosphorus supplementation of cows on range areas deficient in phosphorus increases the calf crop.

4. Administering additional vitamin A to heifers grazing dry forage increases the calf crop.

5. The level and kind of feed before and after calving will determine how many cows will show heat—and conceive. After calving, feed requirements increase tremendously because of milk production; hence, when a cow is suckling a calf, she needs approximately 50% greater feed allowance than during the pregnancy period (see Table 15–4). Otherwise, she will suffer a serious loss in weight and fail to come in heat and conceive.

6. Cows in average condition should gain a minimum of 100 lb during the pregnancy period, followed by a gain of ½ to ¾ lb daily after calving and extending through the breeding season. If they are on the thin side at calving time, they should gain 1½ to 2 lb daily after they drop calves. This calls for 7 to 12 lb of TDN daily before calving (which can be provided by feeding 14 to 22 lb of average quality hay), and 10 to 17 lb of TDN after calving (which can be provided by feeding 14 to 28 lb of hay plus 4 lb grain), with the lactating requirement dependent on both cow weight and milking ability. Additionally, there must be adequate protein, minerals, and vitamins.

## WINTER FEEDING

In a country as large and diverse as the United States, wide variations exist in both the length of the winter season and the available feeds. But the same principles are applicable to all areas and enterprises, and the chief objective remains the same—economically to produce high-percentage calf crops with heavy birth and weaning weights.

Winter feeding is the most expensive time in cow-calf operations. From an economics standpoint, therefore, it is important that wintering practices be both knowledgeable and wise. The cheaper homegrown roughages should constitute the bulk of the winter ration for dry pregnant cows. Most of the grain and the higher class roughages may be used for other classes of livestock. A practical ration may consist of silage and/or dry roughage (legume or grass hays) combined with a small quantity of protein-rich concentrates (such as soybean meal or cottonseed meal). With the use of a leguminous roughage, the protein-rich concentrate may be omitted. Dusty or moldy feed and frozen silage should be avoided in feeding all cattle—especially in the case of the pregnant cow, for such feed may produce complications and possible abortion.

Except during the winter months, pastures constitute most of the feed of beef cattle. By fall, however, grass is usually in short supply and relatively poor as a source of protein, certain minerals (especially phosphorus), and carotene (provitamin A). To overcome these deficiencies, the producer must resort to either (1) supplemental feeding on pasture, or (2) drylot feeding.

At no other time in the operations is a possible profit so likely to be dissipated and replaced by a loss.

Fall feeding should not be delayed so long that animals begin to lose weight. The reason the cattle often eat and get poor on dry, weathered grass is that it is low in energy, protein, carotene, phosphorus, and perhaps certain other minerals. These deficiencies become more acute and increase in severity as winter advances. Cattle simply cannot consume sufficient quantities of such bulky, low-quality roughage to meet their needs; and the younger the animal the more acute the problem. Under such circumstances, the maintenance needs are met by the breakdown of body tissues, and accompanied by the observed loss in weight and condition. Young animals fail to grow; it makes for lightweight calves. Also, reproduction is affected adversely; serious underfeeding results in lowered calf crops. Supplementing fall grass with a concentrated type of supplement is the practical and ideal way in which to alleviate such nutritive deficiencies.

Dry pregnant cows in average condition should make sufficient gain in weight to account for the growth of the fetus (60 to 90 lb), plus sufficient increase in weight and condition to carry them through the suckling period. In total, they should gain 100 to 150 lb during the pregnancy period, or at the rate of approximately ½ lb daily. Of course, the size and condition of the cow is the best gauge as to the feed allowance and desired gain. As previously noted, dry cows require less supplementation than cows suckling calves. Nevertheless, they should not be permitted to lose too much flesh, unless of course, they are overfat. Also, it is recognized that it requires less feed to keep cattle from losing flesh than it does to restore them to proper condition after they have become thin. Thus, it is good economy to start feeding before the cows show any signs of malnutrition. Unless a good quality legume roughage is fed, the concentrate should provide protein, energy, and needed minerals and vitamins.

Noteworthy, too, is the makeup of a calf at birth. An average 70-lb calf at birth is about 75% water, 20% protein, and 5% ash. The calf's 70-lb weight represents about 17.5 lb dry matter. From this, it is apparent that the dry gestation period does not create a heavy nutritional drain. Thus, this is a period when producers may economize by utilizing crop residues and winter pasture.

Cows with calves at side should be fed for the production of milk, for which the requirements are more rigorous than those during pregnancy. This is important because, until weaning time, the growth of the calf is determined chiefly by the nourishment available through the milk of its dam. The principal part of the calf's ration, therefore, may be cheaply and safely provided by giving its mother the proper feed for the production of milk. To stimulate milk flow, most beef cows need a concentrate during the winter months; and the poorer or the more limited the roughage, the higher the supplement requirement. On the average, cows that calve in the fall should be fed a minimum of 4 to 6 lb concentrate daily throughout the winter.

## RATIONS FOR DRY PREGNANT COWS

When winter grazing is not possible, one of the rations in Table 15–5 may be used to meet the daily needs for energy and protein of a 1,100-lb, dry, pregnant cow. A combination of legume roughage with lower quality roughage (such as stalklage, straw, corncobs, or cottonseed hulls) will meet both the energy and protein requirements without the use of a supplement.

**TABLE 15–5**
**WINTERING RATIONS FOR A 1,100-LB (500-KG)**
**DRY PREGNANT COW**

| | Rations | | | | |
|---|---|---|---|---|---|
| | 1 | 2 | 3 | 4 | 5 |
| | ← ((lb *[kg]* per day)) → | | | | |
| Legume-grass hay .......... | 18 *(8)* | | | | 10 *(4.5)* |
| Legume-grass haylage[1] ...... | | 30 *(12)* | | | |
| Corn or grain sorghum silage | | | 35 *(15)* | | |
| Stalklage or husklage ....... | | | | 45 *(20)* | |
| Straw, cobs, or cottonseed hulls ................. | | | | | 10 *(4.5)* |
| Supplement[2] ............. | | .5 *(0.2)* | | 1 *(0.45)* | |

[1]Haylage figured at 55% dry matter, corn or grain sorghum silage at 35% dry matter, stalklage or husklage at 45% dry matter.

[2]Supplement figured at 48% crude protein. Quantity to be adjusted in keeping with the protein content of the supplement. For example, if a 24% crude protein supplement is fed, the quantity of supplement should be doubled.

## CALVING CONTROL (DAYTIME CALVING)

The following benefits would accrue if the majority of cows would calve during the daylight hours:

1. Improved calf survival due to birth during warmer daylight hours and readily available assistance.
2. Reduced nighttime labor from fewer cows calving.

Limited research supports the theory of altering calving time by late feeding (feeding 5 p.m. to 9 p.m., starting about two weeks before calving), with the bulk of calves born during daylight. Scientists at the Fort Keogh Livestock and Range Research Laboratory, Miles City, Montana, report that heifers fed between 8 and 9 p.m. had 17% more daytime births than heifers fed between 8 and 9 a.m. Other researchers report a 10 to 15% increase in daytime calving from nighttime feeding. The reason late feeding causes daytime calving has not been established.

Daytime calving has many advocates in the cow-calf industry. As a managing practice, it does not require capital outlay, and, most important, it can make for increased profits from more live calves. But further experiments and experiences on the effect of feeding time on calving time are needed. In the meantime, producers may try nighttime feeding, but they shouldn't eliminate their night calving crews, because calving at night has not been totally eliminated.

## RATIONS FOR COWS NURSING CALVES

The energy requirement of a cow nursing a calf is about 50% higher than that of a dry pregnant cow; and the protein, calcium, and phosphorus requirements are about double. Since the vast majority of the nation's cows with calves at side are on pasture most, if not all, of the lactation period, the only supplemental need is for salt and other minerals, unless the pasture is insufficient in quantity or quality of feed to support adequate milk production. The rations in Table 15–6 may be used for drylot feeding of beef cows nursing calves. Of course, the daily levels shown in this table should be approached gradually so that nutritional scours will not develop in baby calves.

### TABLE 15–6
### WINTERING RATIONS FOR A 1,100-LB *(500-KG)* COW NURSING A CALF

| | Rations | | | | |
|---|---|---|---|---|---|
| | 1 | 2 | 3 | 4 | 5 |
| | ((lb *[kg]* per day)) | | | | |
| Legume-grass hay | 30 *(13)* | | | 20 *(9)* | 10 *(4.5)* |
| Legume-grass haylage[1] | | 50 *(22)* | | | |
| Corn or grain sorghum silage | | | 60 *(27)* | | 40 *(18)* |
| Grain | | | | 5 *(2)* | |
| Supplement[2] | | | 1.5 *(0.6)* | | |

[1]Haylage figured at 55% dry matter; corn or grain sorghum silage figured at 35% dry matter.

[2]Supplement figured at 48% crude protein. Quantity to be adjusted in keeping with the content of the supplement. For example, if a 24% crude protein supplement is fed, the quantity of the supplement should be doubled.

## CROP RESIDUES AND WINTER PASTURES

Two requisites are important in wintering the cow herd; (1) bringing them through the winter in proper condition for calving, and (2) keeping feed costs to the minimum consistent with nutritional demands. Meeting these requirements has prompted increased use of crop residues and winter pastures for brood cows. As the ever-increasing human population of the world consumes a higher proportion of grains and seeds, and their by-products, directly, cattle will utilize increasing amounts of crop residues and pastures and a minimum of products suitable for human consumption. Thus, more and more farmers with crops will include a beef herd in their operations and realize a fair return from feeds which would otherwise be wasted.

### CROP RESIDUES

*Crop residues are the parts of forages that remain after harvesting a grain or seed crop.* Among such crop residues are cornstalks and husklage, sorghum stalks, soybean refuse, small grain straws and chaff, and legume and grass seed straws. Crop residues left in the field, above or below the soil surface, may well constitute 4 to 5 times more energy than is harvested. This potential source of added feed, organic fertilizer, and energy will be increasingly utilized in the future.

### Corn Residues

Of all crop residues, that from corn is produced in greatest abundance and offers the greatest potential for expansion in cow numbers. Of course, knowing the feeding value and proper supplementation of corn residues is pertinent to their profitable use.

Fig. 15–12. Cow grazing cornstalks. (Courtesy, Iowa State University, Ames, IA)

Table 15–7 lists the daily nutritive requirements of a dry pregnant cow, middle third of pregnancy, weighing 1,100 lb. Table 15–8 gives the nutritive composition of air-dry corn stover and husklage.

### TABLE 15–7
### NUTRITIVE REQUIREMENTS OF A DRY PREGNANT COW (MIDDLE THIRD OF PREGNANCY) WEIGHING 1,100 LB *(500 KG)*[1]

| | |
|---|---|
| Dry matter, daily | 20.9 lb *(9.5 kg)* |
| TDN, daily | 11.2 lb *(5.1 kg)* |
| Total protein | 7.8% |
| Calcium | 0.26% |
| Phosphorus | 0.21% |
| Vitamin A | 27,000 IU |

[1]*Nutrient Requirements of Beef Cattle*, sixth revised edition, National Research Council-National Academy of Sciences, Washington, DC, 1984.

### TABLE 15–8
### ANALYSIS OF AIR-DRY CORN STOVER AND HUSKLAGE[1]

| | Corn Stover | Husklage |
|---|---|---|
| | (%) | (%) |
| TDN | 48 | 57 |
| Crude Protein | 4.5 | 3.4 |
| Calcium | 0.4 | 0.02 |
| Phosphorus | 0.07 | 0.05 |
| Vitamin A | — | — |

[1]*Cow-Calf Information Roundup*, University of Illinois, 1971, p. 10, Table 2.

Studies show that a 1,100-lb cow will eat approximately 22–24 lb per day of palatable, air-dry stover, or about 2 lb or more of air-dry stover per 100 lb body weight per day. She will eat slightly larger amounts of husklage. This consumption, along with the information presented in Tables 15–7 and 15–8, suggests that stover and/or husklage rations will meet the daily energy (TDN) needs of dry pregnant cows, but such rations will be slightly deficient in protein, and low in phosphorus and vitamin A. Nevertheless, the highest and best use for corn residue is for dry pregnant cows for the period following conception to about 30 days before calving.

For corn refuse feeding, mature cows should be in medium to good condition at the start of the winter feeding period; and they should not be permitted to lose over 10 to 15% of their weight from fall through calving. Heifer weight losses should be under 5%. When weight loss approaches this limit, it's time to feed some grain or silage.

Corn residues provide adequate energy to maintain dry pregnant cows, but they must be supplemented with additional energy when fed to cows nursing calves or to young growing animals.

The crude protein content of corn stover is on the low side for dry pregnant cows. It averages about 4.5% (Table 15-8, whereas a dry pregnant cow requires 7.8% crude protein (Table 15-7). Thus, it is recommended that ½ lb per head per day of a 30 to 40% crude protein equivalent (CPE) supplement be provided.

It follows that the protein content of corn refuse is much too low to support either productivity or growth. For example, a 1,100-lb lactating cow requires a daily allowance of 2.7 lb of crude protein. However, a daily consumption of 29 lb of stover will provide only about 1.1 lb—less than half that needed.

For nursing cows, the protein deficiency of stover and/or husklage may be corrected by supplementation, on a per head per day basis, with 2 lb of a 40% protein supplement, or 6 lb of a good legume hay. If desired, the protein supplement may be provided in the form of protein blocks, with one block provided for each 15 cows. Where hay is fed, it should be taken to the field, rather than fed in a feedlot, as this will encourage the cows to stay in the field and graze the cornstalks.

Phosphorus should be provided to all cattle fed corn residue. Calcium may be deficient, especially for lactating cows. Also, some of the trace elements may be deficient. Hence, it is recommended that all cattle on high corn refuse have free access to a complete mineral supplement. A mineral mixture with a Ca:P ratio of 1:2 is recommended for gestating cows, and a 1:1 ratio for lactating cows.

Corn residue is deficient in vitamin A, which should be supplemented. The precalving and postcalving (heavy milking) needs of approximately 27,000 and 39,000 IU per head per day, respectively (NRC-1984), may be met by feeding vitamin A supplement, by intramuscular injection of vitamin A solution, or by feeding adequate levels of green, leafy hay.

It is important that corn residue be tailored to match the cow's nutritional needs. This is relatively simple with dry pregnant cows, where supplementation with a high-phosphorus mineral and vitamin A will usually suffice. Beginning four to six weeks before calving and continuing through the lactation period, much heavier supplementation is necessary; in addition to phosphorus and vitamin A, protein must be added, and preferably some energy and calcium for nursing cows.

## WINTER PASTURE

Where feasible, winter pasture offers cattle producers a means of reducing costs. By accumulating the feed in the field, rather than harvesting, storing, and handling the forage, the cost and labor of winter feeding can be substantially reduced. Also, cost of bedding and manure hauling can be eliminated.

Tall fescue is used as a winter pasture in the area to which it is adapted—Missouri, Illinois, Indiana, and Ohio. Usually, the new regrowth is baled in late June into round bales and left in the field. The round bales shed rain and snow and, together with the regrowth, make excellent late fall and winter grazing. Experience shows that field-stored forage has adequate quality to maintain beef cows in good condition.

Fescue is a cool season grass which keeps growing during the winter in the areas to which it is adapted; and it is more palatable during the fall and winter than any other season because of the high concentration of soluble sugars. Trampling during the fall, winter, and spring months does not injure the turf.

The Ohio Station researchers reported that tall fescue winter pasture—including both standing growth and baled hay—carried 2 cows per acre for a 4-month period. The use of electric fence to strip graze the bales and regrowth increases carrying capacity by 50 to 60% over permitting the herd across the entire field.

Fig. 15-13. Cows on winter fescue pasture, supplemented with round bales of fescue harvested the previous June and left in the field. (Courtesy, University of Illinois, Urbana)

## RANGE CATTLE SUPPLEMENTATION

Improved range should be the first goal of cattle producers, without using supplemental feeding as a substitute for good grass or as a crutch for poor range. Instead, the two—good range and proper supplemental feeding—go hand in hand.

### RANGE NUTRIENT DEFICIENCIES

Every cattle producer worthy of the name forces young stock for an early market; most soils are deficient in certain nutrients, which, in turn, affect the plants and the animals feeding thereon; during droughts and early and late in the season, feed may be in short supply (thereby limiting energy and other nutrients); early spring pastures are washy and lacking in energy; during droughts and late in the season, grasses become mature, leached and bleached—they increase in fiber and decrease in protein, phosphorus, and carotene. To meet these conditions, a supplemental source of energy, protein, phosphorus, and vitamin A is necessary.

- **Energy**—Hunger, due to lack of feed, is the most common deficiency on the western range. In particular, there may be a shortage of energy during droughts, late in the season, or early in the spring when grass is washy. Under such energy-deficient circumstances, animals lose weight and condition and young animals fail to grow. Also, reproduction is adversely affected; serious underfeeding results in the failure of some cows to show heat, more services per conception, lowered calf crops, and lightweight calves. Supplemental feeding is the practical way in which to alleviate such energy deficiencies.

- **Protein**—Mature, weathered native range grass is almost always deficient in protein—being as low as 3%, or less. Protein leaching losses due to fall and winter rains may range from 37 to 73%.

When on mature, weathered grass, cows should receive about 2 lb of concentrate supplement daily—the exact amount depending upon the nutrient content of the supplement and other factors.

The protein and energy requirements are closely interdependent; hence, it follows that energy rather than feed intake should be the dietary component relative to which the nutrient needs are adjusted.

- **Minerals**—Growth, reproduction, and lactation require adequate minerals.

Salt should be available at all times, on the basis of about 25 lb per range cow annually.

Phosphorus deficiencies are rather common among range beef cattle. A severe phosphorus deficiency will result in depraved appetite, emaciation, retarded growth and development, failure to reach normal adult size, failure to breed regularly, lowered calf crop, lowered milk production, and high death losses. The New Mexico station reported (1) phosphorus losses in grasses of 49 to 83% during the winter period, (2) increased average annual calf production per range cow of 53 lb through proper mineral supplementation, and (3) that the phosphorus supply should be continuous throughout the year, and not limited to the winter months.

Iodine, copper, and cobalt supplements should be added in those areas where deficiencies of these minerals are known to exist.

- **Vitamins**—Under normal conditions, vitamin A is the vitamin most likely to be deficient in cattle rations, because dry, bleached range grass is very low in carotene (the precursor of vitamin A).

Inadequate amounts of vitamin A (carotene) during pregnancy may cause cows to abort or give birth to dead or weak calves. Extreme deficiencies may also impair the ability of cows to conceive. Bulls receiving insufficient vitamin A show a decline in sexual activity and semen quality.

In low-sunshine areas, especially during the winter months, it is recommended that vitamin D be added to the ration.

## PASTURE AND RANGE SUPPLEMENTS

Where dried grass cured on the stalk is grazed, or where insufficient pasture is available—perhaps due to drought or overstocking—supplemental feeding is necessary. Also, supplemental feeding is a way in which to extend the grazing season, both early and late.

## TYPES AND SYSTEMS OF PASTURE AND RANGE SUPPLEMENTATION

There is no one best and most practical pasture or range supplement for any and all conditions. Many different feeds may be, and are, used; among them (1) ranch or locally produced hay, (2) alfalfa pellets or cubes, with or without fortification, and (3) supplements of various kinds.

Also, farmers and ranchers can lessen the labor attendant to the daily feeding of a pasture or range supplement by (1) hand feeding cubes at intervals, rather than daily, (2) use of protein blocks, (3) use of liquid protein supplements, or (4) self-feeding salt-feed mixtures. Where these feeding systems do not result in the neglect of the herd, there is no effect upon the health and weight of the cows, percent calf crop, or weaning weight of calves.

- **Range cubes or pellets**—Traditionally, cattle have been supplemented either once or twice daily on pasture or range, with the cubes scattered on the ground.

Fig. 15–14.  Range cubes fed on pasture or range. Many cattle producers prefer this method of supplementation, primarily for reasons of convenience and reducing losses from wind blowing. (Courtesy, Ralston Purina Company, St. Louis, MO)

Urea-containing supplements, particularly those containing high levels of urea, should not be fed at intervals on the range because (1) range forages are relatively low in energy, and (2) urea is extremely soluble and its nitrogen becomes available very quickly in the rumen. Where nonprotein nitrogen is used in a range cube or pellet, a slow-release product is safest. Two suggested formulations for pasture-range cubes or pellets are given herein: one without urea, Table 15–9; and the other with urea, Table 15–10.

## TABLE 15-9
### RANGE CUBE OR PELLET, WITHOUT UREA (AS-FED BASIS)

| Ingredient | Percent | Per Ton |
|---|---|---|
| | (%) | (lb) |
| Soybean meal, 44% (or cottonseed meal) .......... | 72.7 | 1,454 |
| Alfalfa meal, 15% ................................ | 15.0 | 300 |
| Molasses (sugarcane) ........................... | 8.5 | 170 |
| Dicalcium phosphate, or equivalent .............. | 2.0 | 40 |
| Salt ............................................ | 1.0 | 20 |
| Trace minerals[1] ................................. | .5 | 10 |
| Vitamin A[2] (30,000 IU/g potency) ................ | .3 | 6 |
| **Total** ........................................ | 100.0 | 2,000 |
| **Proximate analysis:** | (%) | |
| Crude protein ............................ | 35.9 | |
| Fat ..................................... | 1.2 | |
| Fiber ................................... | 8.3 | |
| Calcium ................................. | 1.01 | |
| Phosphorus .............................. | .9 | |
| TDN ..................................... | 68.7 | |

[1]Trace minerals should be in keeping with Table 15-1. Generally, trace minerals can best be provided by a mineral or feed company, rather than home mixed.

[2]In low-sunshine areas, also add 6 million IU of vitamin D/ton of finished feed.

## TABLE 15-10
### RANGE CUBE OR PELLET, WITH UREA
### (PREFERABLY A SLOW-RELEASE PRODUCT; AS-FED BASIS)

| Ingredient | Percent | Per Ton |
|---|---|---|
| | (%) | (lb) |
| Corn #2 (barley, wheat, oats, and/or milo) ............. | 34.7 | 694 |
| Soybean meal, 44% (cottonseed, linseed[1], and/or peanut meal) | 32.5 | 650 |
| Alfalfa meal, 15% ................................. | 15.0 | 300 |
| Molasses, sugarcane ............................. | 10.0 | 200 |
| Urea, 45% grade (use slow-release product) ............ | 4.0 | 80 |
| Dicalcium phosphate, or equivalent ................ | 2.0 | 40 |
| Salt ............................................. | 1.0 | 20 |
| Trace minerals[2] ................................. | .5 | 10 |
| Vitamin A[3] (30,000 IU/g potency) ................ | .3 | 6 |
| **Total** ........................................ | 100.0 | 2,000 |
| **Proximate analysis:** | (%) | |
| Crude protein[4] .............................. | 31.8 | |
| Fat ..................................... | 2.2 | |
| Fiber ................................... | 6.6 | |
| Calcium ................................. | .9 | |
| Phosphorus .............................. | .7 | |
| TDN ..................................... | 67.5 | |

[1]If linseed is used, limit to 6% of the ration.

[2]Trace minerals should be in keeping with Table 15-1. Generally, trace minerals can best be provided by a mineral or feed company, rather than home mixed.

[3]In low-sunshine areas, also add 6 million IU of vitamin D/ton of finished feed.

[4]This includes not more than 11.24% equivalent protein from nonprotein nitrogen; 34.9% of the total protein is furnished by urea (use a slow-release product).

- **Hand feeding at intervals, rather than daily**—Based on a four-year study done at the Texas Station, plus observations and experiences, the author recommends feeding nonurea range supplements twice weekly, allocating in each of the two feedings one-half as much supplement as would have been fed in a week on a daily feeding basis.

Protein cubes may be scattered on the ground—two or three times a week. This offers a method of checking the animals because they are attracted by the sight or sound of the vehicle when they know that there is something to eat.

Twice-weekly feeding has two distinct advantages over the use of salt-feed mixes: (1) It alleviates the cost of using excess salt, which has no nutritive value when so used; and (2) it forces inspection of the herd two times per week, which is as infrequent as is desirable.

- **Protein blocks**—Protein blocks are just what the designation implies. They are compressed protein blocks, generally weighing from 50 to 500 lb each.

Blocks may be placed in grazing areas where cattle have frequent access to them, with one block provided to 15 cows. Intake will vary with the feed supply and the type of block. Generally, it is planned to limit feed consumption to about 2 lb per head per day by hardness of the block and salt and/or fat content.

Fig. 15-15. Beef cattle eating protein blocks on the range.

- **Liquid protein supplements**—Liquid supplement in a *lick tank* can be offered free choice. This is a convenient and satisfactory way in which to supply protein, energy, and other nutrients, so long as the cattle do not consume more than they need.

Fig. 15-16. Liquid protein supplement fed free-choice in a *lick tank*, with consumption limited by a tongue-turning plastic wheel; on Sr. Guillermo Osuna's Infante Ranch, Muzquiz, Coahuila, Mexico. (Photo by A. H. Ensminger)

• **Self-feeding salt-feed mixture**—The practice of using salt as a governor to limit feed consumption on pasture or range has been around a very long time. It was ushered in as a labor-saving device for cattle and sheep in inaccessible and rough areas.

The proportion of salt to feed may vary anywhere from 5 to 40% (with 30 to 33⅓% salt content being most common), with the actual intake of feed supplements limited to 1 to 2½ lb daily. By varying the proportion of salt in the mixture, it is possible to hold the consumption of feed supplement to any level desired. In some range areas, a reduction of the salt level from 33⅓ to 24% will increase consumption by about 50%. When a liberal feeding of grain on pasture is desired, 5% salt may be sufficient.

Two suggested salt-meal supplements (salt-cottonseed or soybean meal, 41%; *do not pellet*) follow:

| Ingredient | Salt-Meal Mix No. 1 (lb) | Salt-Meal Mix No. 2 (lb) |
|---|---|---|
| Salt | 665 | 499 |
| Meal (either 41% cottonseed or soybean meal) | 1,331 | 1,497 |
| Vitamin A (30,000 IU/g) | 4 | 4 |
|  | 2,000 | 2,000 |
| **Consumption level:** | approx. 1½ lb daily | approx. 2 lb daily |
| **Guarantee:** | | |
| Crude protein | min. 27% | min. 30% |
| Salt | max. 35% | max. 27% |
| Vitamin A | 24,000 IU/lb | 24,000 IU/lb |

## CONFINEMENT (DRYLOT) BEEF COWS

*Confining (drylotting) beef cows refers to the practice of confining beef cows to small quarters—to drylots, all or part of the year.*

From a feeding standpoint, the following points are pertinent in a drylot beef cow operation:

1. All feed must be mechanically harvested and moved to the feedlot, rather than being harvested directly by the cows.

2. An assured, adequate, and economic feed supply must be available. The capital tied up in stored feeds may be quite large.

3. More knowledge of beef cow nutrition and ration formulation is needed.

(Also see Chapter 16 of this book, section on "Confinement [Drylot] Beef Cows.")

### RATIONS FOR DRYLOT COWS

Rations for drylot cows generally consist of cheap roughages—such as crop refuse, straw, cottonseed hulls, and gin trash—supplemented with protein, grain, vitamins, and minerals as required. Where available, higher quality roughages—such as silage, hays and haylages—may be used, especially (1) during the critical 100 days, beginning 30 days before calving and extending to 70 days after calving, and (2) for heifers calving as 2-year-olds. Also, during the summer and fall, green chop is frequently fed.

## SEMICONFINEMENT (OR PARTIAL CONFINEMENT) COW HERDS

A semiconfinement (or partial confinement) operation is one which takes advantage of grazing during part of the year, such as winter grazing of corn or sorghum stalks or seasonal grazing of pastures. In addition to providing low-cost feed and allowing the animals to do their own harvesting, breeding may be timed so that the calves will be dropped on clean pasture as a means of (1) preventing calf scours, and (2) stimulating milk flow.

## FEEDING AND MANAGING BULLS

Frequently, little thought is given to the feeding and management of bulls except during the breeding season. Instead, the feeding program for herd bulls should be such as to keep them in a thrifty, vigorous condition at all times. They should neither be overfitted nor in thin, run-down condition. Also, exercise is necessary for the normal well-being of the bull.

The feeding and management of bulls differ according to age and condition. For this reason, sale bulls, young bulls, and mature bulls are treated separately in the sections that follow.

Fig. 15–17. Hereford bull on the range. Note cacti in front of the bull. (Courtesy, American Hereford Assn., Kansas City, MO)

## FEEDING SALE BULLS

Bull sales are generally held in late winter and early spring, at which time mostly yearling and 2-year-old bulls are sold. In order to attract buyers, they have usually been grain fed since calfhood. Most bull buyers—especially commercial producers in rougher range areas—would rather have their new bulls in less than fitted sale condition. They find that such bulls are more fertile and more apt to range with the cows when turned to pasture during the breeding season.

Handling highly conditioned sale bulls during the critical period—after the sale is over, and just ahead of the breeding season—is all important. Experienced producers "let them down" and yet retain strong vigorous animals. They do this successfully by (1) providing plenty of exercise, (2) increasing the amount of bulky feeds, such as oats, in the ration, (3) cutting down gradually on the grain allowance, and (4) retaining the succulent feeds and increasing the pasture and hay.

## FEEDING YOUNG BULLS

Lack of fertility in a bull may often be traced back to his early care and feeding. From weaning to three years of age, bulls should be kept separate by age groups. Young bulls should be fed more liberally than mature bulls because their growth requirements must be met before any improvement in condition can take place.

Following weaning, bulls should be fed and developed sufficiently to show their inherited characteristics, but without excessive finishing. Simultaneously, they should be given plenty of exercise. Overfeeding and lack of exercise are apt to result in infertility, low-quality sperm, and unsound feet and legs.

To achieve proper development, young bulls should gain at least 2½ lb daily from weaning to 12 to 15 months of age. This will necessitate a daily feed allowance equal to about 2½% of their body weight, with a ration comprised of 50% or more concentrate. From 15 months to 3 years old, they should make a daily gain of 2 to 2¼ lb and receive a feed allowance equal to 2 to 2¼% of their body weight, with the proportion of roughage increased after the first year.

Without a doubt, the least laborious and most convenient management arrangement in handling young bulls consists in allowing a group, not exceeding 10 to 15 head of uniform size and age, the run of a pasture or enclosure of ample size, thereby providing (1) exercise, and (2) pasture in season. Of course, wherever possible, bulls should be performance tested while being developed. Ideally, this calls for individual feeds and body weight records, although group feeding plus individual weight records will suffice.

Bulls handled as recommended above will generally attain half their mature weight by the time they are 14 to 15 months of age and may be used in limited service.

During the breeding season, young bulls should be fed a grain ration consistent with pasture quality and number of cows to be bred in order to promote proper growth and development. Drought, overpasturing, and poor quality pastures are situations in which grain supplementation is particularly needed. Heavy service and poor pasture with no supplemental feeding may shorten the breeding career of a young bull.

After the breeding season, yearling bulls generally need 5 to 6 lb of grain along with good roughage.

## FEEDING MATURE BULLS

Winter is the proper time to condition bulls for the next breeding season. Bulls that have been running on pasture with the cows are likely to be thin; thus, they require sufficient concentrate to put them in proper flesh. Mature bulls will consume daily amount of feeds equal to 1½ to 3% of their liveweight, depending upon condition and individuality.

Outdoor exercise is also essential in keeping bulls virile and thrifty. The finest and easiest method of providing such exercise is to allow them the run of a well-fenced pasture. About two acres is a good size for one bull. Where several bulls are run together, the enclosure should be larger.

Mature bulls should be fed all the legume hay they will eat plus 3 to 5 lb of ground or rolled grain and 1 lb of a 32% protein supplement (or equivalent) per head per day. Also, free access to a suitable mineral mixture should be provided. About 60 days before the bulls are turned out with the cows, the concentrate allowance should be increased by 25 to 50% with the amount of the increase determined by the condition of the bulls.

Mature herd bulls need no additional feed when running with the cow herd on good summer pasture.

## FEEDING AND MANAGING CALVES

Cattle producers as a whole, have lagged in applying much of what we know about feeding and managing calves. They're inclined to let mother cows and mother nature fend for the calves. Indeed, more good proven practices, based on both successful operations and research, need to be put to use in feeding and handling calves.

## FEEDING ORPHAN AND MULTIPLE BIRTH CALVES

Occasionally a cow dies during or immediately after parturition, leaving an orphan calf to be raised. Also, there are times when cows fail to give a sufficient quantity of milk for the newborn calf. Sometimes, there are multiple births.

If there are only a few orphans, usually they can be grafted onto other cows (or adopted)—either cows that have lost their calves or that give sufficient milk to raise two calves. When such calves cannot be grafted, they must be raised by artificial methods—without a cow.

Regardless of whether orphans are grafted or raised artificially, the problem will be simplified if the calf receives colostrum, the first milk produced by a cow after giving birth to a calf, during the first 24 hours, and preferably for the first 3 days of life—from its mother, from another fresh cow, or from frozen-stored colostrum. Colostrum is higher than normal milk in dry matter, protein, vitamins, and minerals. Also, it contains antibodies (a modified type of serum globulin) that gives newborn calves a passive immunity against common calfhood diseases.

Fortunately, orphan calves can now be raised successfully on milk replacer and calf starter ration, using them as directed. The milk replacer may be fed by using a bottle or bucket equipped with a rubber nipple, or the calf may be taught to drink from a pail. It is important that all receptacles be kept absolutely clean and sanitary (clean and scald each time) and that feeding be at regular intervals. Dry feed should be started at the earliest possible time; not later than one week of age. With proper management, healthy calves may be switched entirely to a suitable dry feed at four to five weeks of age.

A suggested feeding program for calves is given in this book in Chapter 24, in the section headed "Dairy Calves"; and six starter rations are presented in the same chapter, in Tables 24–15 and 24–16.

## FEEDING EARLY WEANED CALVES

*Early weaning refers to the practice of weaning calves earlier than the usual weaning age of about 7 months, usually within the range of 35 days to 5 months of age. Although it is not common practice among U.S. beef producers, dairy producers have been weaning 3-day-old calves for years. Also, early weaning has long been an integral part of many of the beef programs of Europe.*

Currently, there is much interest in early weaning because (1) it fits into a drylot cow-calf management system, and (2) it can give a big assist in getting females, especially 2-year-old heifers, to rebreed in a short period of time.

Considering the low efficiency involved in converting supplemental energy to milk and in converting milk to meat, it is apparent that a more efficient use of feed could be achieved by giving the supplemental feed directly to the calf. A lactating cow requires about 50% more feed than a dry cow. So, rather than give her the additional feed, it is more efficient to give feed directly to the calf, thereby favoring early weaning.

Where early weaning is successful, the only responsibility of the beef cow is to produce a calf and give it a good start in life for a brief period, then go on a maintenance ration the rest of the year.

Like many good things in life, early weaning does have some disadvantages. To be successful, superior nutrition and management are essential; and the earlier the weaning age, the more exacting these requirements.

From 35 days of age on, early-weaned calves can be fed any good starter ration. Six such rations are given in Chapter 24 of this book, in Tables 24–15 and 24–16. Of course, the starter ration should be made available to the calves well ahead of weaning in order that they will be accustomed to it, thereby avoiding any setback.

## CREEP FEEDING

*Creep feeding is the supplementation of calves while they are nursing their dams.* It increases weaning weight. The basis for this response is related to the lactation curve of beef cows, the increasing nutrient requirements of the calf during the nursing period, and the decline in feed quality and quantity typical of most pastures or ranges which support the cows and calves during lactation. Studies reveal that milk production of dairy cows increases up to the fourth to sixth month following freshening, then declines gradually. By contrast, maximum milk production of beef cows occurs during the first two months after calving, then declines.

Fig. 15–18 shows why creep feeding is important. From birth to weaning, the protein and energy requirements of a

growing calf increase well beyond the ability of most beef cows to meet those needs. For example, to meet the protein and energy requirements for growth, a 100-lb calf needs 10 lb of milk, whereas a 500-lb calf needs 50 lb of milk. Since the average beef cow gives only 13 lb of milk per day throughout a seven-month suckling period, a 500-lb calf lacks 37 lb of getting enough milk from its dam at this stage of lactation to meet its needs—that's the *hungry-calf gap.*

To fill the *hungry-calf gap*—the nutrient requirements over and above that provided by 13 lb of milk—would require the consumption of 50 lb of green grass daily. Of course, that's a physical impossibility, because a 500-lb calf simply cannot hold that much bulk. So the best way to fill the *hungry-calf gap* is to creep feed with concentrate mixes.

## THE CREEP

*A creep is an enclosure or feeder for feeding purposes which is accessible to the calves but through which the cows cannot pass.* It allows for the feeding of the calves but not their dams.

Fig. 15–19. Movable calf creep (note runners), with openings that will permit the calves to enter and keep the cows out. (Courtesy, *Livestock Breeder Journal,* Macon, GA)

## CREEP RATIONS; FEEDING DIRECTIONS

Creep-fed calves need special rations. They are bovine babies; and they are both in forced production and finishing. The calf's body is expected simultaneously to lay on fat and grow in protein tissues and skeleton. Consequently, their ration requirements are for feed high in protein; rich in readily available energy; fortified with vitamins, minerals, and unidentified factors; and with all the nutrients in proper balance. Also, the ration must be very palatable. This calls for an exacting ration. To meet these needs, more and more producers are finding it practical to buy a commercially prepared complete creep feed, or a well-fortified and highly concentrated supplement to add to locally available feeds, rather than purchase individual ingredients and mix from the ground up.

Tables 15–11 and 15–12 show two creep rations, formulated by the author, that have been widely and successfully used. A simple, yet very satisfactory, creep ration may be made by grinding and pelleting 75% alfalfa and 25% cereal grain.

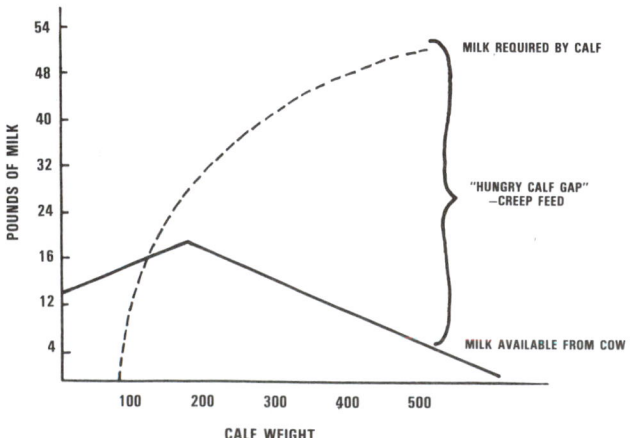

Fig. 15–18. *The hungry-calf gap—the difference between (1) the milk required by the calf, and (2) the milk available from the cow.*

## TABLE 15–11
## CALF CREEP RATION #1[1] (AS-FED BASIS)

| Ingredient | Precent | Per Ton | |
|---|---|---|---|
| | (%) | (lb) | (kg) |
| Oats | 39.60 | 800.0 | 363.2 |
| Corn #2 | 14.80 | 300.0 | 136.2 |
| Barley | 8.90 | 177.5 | 80.7 |
| Wheat bran | 9.90 | 200.0 | 90.8 |
| Dried molasses beet pulp | 9.90 | 200.0 | 90.8 |
| Soybean meal, 44% | 9.90 | 200.0 | 90.8 |
| Molasses | 4.90 | 100.0 | 45.4 |
| Salt | .50 | 10.0 | 4.5 |
| Dicalcium phosphate | .50 | 10.0 | 4.5 |
| Trace minerals[2] | .04 | 1.0 | 0.45 |
| Vitamin A (30,000 IU/g) | .06 | 1.5 | 0.68 |
| **Total** | 100.00 | 2,000.0 | 907.2 |

| Proximate analysis: | (%) |
|---|---|
| Crude protein | 14.30 |
| Fat | 3.20 |
| Fiber | 8.30 |
| Calcium | .32 |
| Phosphorus | .50 |
| TDN | 69.60 |

[1]Feed preparation: Preferably ⅛- or ³/₁₆-in. pellets. Otherwise, steam roll and flake grains, or grind grains coarsely.

[2]See Table 15-1 for recommended trace mineral levels. Follow manufacturer's directions.

## TABLE 15–12
## CALF CREEP RATION #2 (AS-FED BASIS)

| Ingredient | Percent | Per Ton | |
|---|---|---|---|
| | (%) | (lb) | (kg) |
| Corn # 2 | 24.25 | 485 | 220.2 |
| Alfalfa meal, 15% | 22.50 | 450 | 204.3 |
| Oats | 20.00 | 400 | 181.6 |
| Alfalfa hay (all analyses) | 10.00 | 200 | 90.8 |
| Soybean meal, 44% | 6.20 | 124 | 56.3 |
| Bran | 5.00 | 100 | 45.4 |
| Linseed meal, 35% | 5.00 | 100 | 45.4 |
| Molasses | 5.00 | 100 | 45.4 |
| Dicalcium phosphate | 2.00 | 40 | 18.2 |
| Trace minerals[1] | .05 | 1 | 0.45 |
| Vitamin A (325,000 IU/g)[2] | — | 63 g | |
| **Total** | 100.00 | 2,000 | 908.0 |

| Proximate analysis: | (%) |
|---|---|
| Crude protein | 15.10 |
| Fat | 3.00 |
| Fiber | 12.70 |
| Calcium | 1.04 |
| Phosphorus | .73 |
| TDN | 64.90 |

[1]See Table 15-1 for recommended trace mineral levels. Follow manufacturer's directions.

[2]When 4 lb/head/day of the calf creep ration is consumed, 40,950 IU of vitamin A will be obtained in the feed.

Calves will consume approximately 500 lb of creep feed per head from one month of age to weaning. In years of lush pasture, it will be less; in dry years, more.

**NOTE WELL:** Instead of allowing creep fed calves to consume all they will eat, limited creep feeding, described in the section that follows, is gaining in popularity.

• **Salt-limited creep feeding**—Generally, limiting creep feed to about 3 lb/head/day is recommended. This will supply enough supplemental energy and protein to the dam's milk and forage to meet the requirements for normal growth of young calves. Feeding more than 3 lb/head/day may result in excessive fat deposition instead of skeletal and muscle growth.

Limited creep feeding can be accomplished by hand-feeding. But, the disadvantages of this practice are: (1) the larger calves tend to overeat and the smaller ones not to get enough, unless adequate bunk space is provided.

A common salt-limited creep feed consists of a mixture of 5 to 10% salt and 90 to 95% cottonseed meal. Also, either of the creep rations listed in Tables 15–11 and 15–12 may be limit-fed by adding approximately 10% salt.

## CREEP GRAZING

*Creep grazing refers to the practice of grazing nursing calves on separate pastures from their dams.* They may either graze before the cows do, getting first choice of the more succulent, highly nutritious pastures; or they may have access to special pastures. The calves enter the special pastures through gates with openings large enough for calves, but too small for the cows to get through. In an alternative method, electric fences are positioned high enough (36 to 42 in.) for calves to pass under, but low enough to keep cows out. Limited studies indicate that as much as ½ lb extra weight gain per day may be obtained by creep grazing.

## WHY CREEP FEED?

Unquestionably, the best yardstick for measuring performance in a beef breeding herd is pounds of calf weaned per cow bred. This fact, along with the demand for healthy, gain-ready feeder calves and the prices being paid, is causing cow-calf operators to take a new "pencil pushing" look at the immediate and residual benefits of creep feeding.

The three major reasons for creep feeding are:

1. **It provides a way to fill the *hungry-calf gap.*** Creep feeding provides a logical and practical way to compensate for insufficient milk which usually characterizes the following conditions: (a) the normal fall off in milk production of beef cows about two months after freshening; (b) periods of unfavorable feed conditions—droughts, overgrazing, early and late in the season, and fall calving herds that are maintained on a low plane of winter nutrition—when the calf is shortchanged on both milk and pasture; (c) first-calf heifers whose milk production is generally lower than mature cows; (d) "shelly" and poorly doing cows; and (e) poor milkers. Under such conditions, creep feeding not only makes for heavier weights, but the calf crop is more uniform—the hungry calves eat more feed.

2. **It makes for heavier weaning weights.** Creep feeding results in 50 to 75 lb heavier weaning weight per calf, at no extra cost for the capital investment in land and cows.

3. **It usually pays.** The potential profitability of creep feeding depends upon (1) the price of cattle, and (2) the price of feed.

The following rule of thumb may be used to determine whether or not it will pay to creep feed: It pays to creep feed when the selling price per hundred pounds of calf is greater than the cost of one-fourth ton (500 lb) of feed.

One producer of the author's acquaintance summed up the economics of creep feeding in this way: "If your cost of gains runs 40¢ per pound, and you can sell the calves at 70¢ per pound, it's almost as good as finding a money tree." Assuming a $1.00/cwt selling advantage for noncreep-fed calves, because they're lighter and thinner, most producers feel that from creep feeding, they'll net $10 to $15 per calf more after feed and other costs are deducted.

Other reasons for, or benefits from, creep feeding are: It facilitates fall calving; the calves are more uniform; the calves achieve full genetic growth potential; young gains are cheap gains; it is more efficient to feed calves directly than to feed cows too liberally; it makes for attractive purebred calves; it makes it easy to reinforce and improve milk; it helps control parasites; it simplifies weaning; it facilitates early weaning; it makes for marketing flexibility; it narrows the price between heifers and steers; it makes for better lifetime reproductive performance of heifers; creep-fed calves are bunk broke; it makes for earlier cycling and conception of the dams; the cows are in better condition; and it gives first calf heifers a big assist.

## LIMITATIONS OF CREEP FEEDING

Like many good things, creep feeding does have its limitations; among them, the following: It isn't always profitable; it lowers subsequent feedlot gains and efficiency; it makes for less desirable stockers, because the latter are normally placed on less nutritious growing rations consisting predominantly of roughages; it mitigates against selecting cows for milk production; it is difficult to do in remote areas; and it cannot be done where there are hogs, sheep, or goats.

## FEEDING AND MANAGING REPLACEMENT HEIFERS

The feed and management program of replacement heifers will have a lifelong effect on their productivity. It will determine how young they may be bred, whether they calve early or late, whether they are good milkers or poor milkers, the weaning weight of their calves, and how long they remain in the herd. Also, feed accounts for 40 to 70% of the cost of rais-

Fig. 15–20. Replacement heifers on pasture in West Virginia. (Courtesy, American Polled Hereford Assn., Kansas City, MO)

ing replacement heifers; hence, it is important to know whether it is possible to effect savings on feed and during the growing period without affecting reproduction adversely. It is even more important to know whether their performance as adult animals can be enhanced by proper nutrition and management.

The pregnancy requirements of replacement heifers are really not too great. The body of an 80-lb newborn calf contains only about 12 lb of protein, 3.0 lb of fat, and 3.6 lb of mineral matter. But the lactation requirements are much more rigorous. If a 2-year-old heifer gives her calf an average of 1¾ gal of milk per day over a seven-month suckling period, she will produce in that milk a total of 93 lb of protein, 107 lb of fat, 133 lb of sugar, and 20 lb of minerals.

Hence, the comparison: 12 lb of protein in the fetus vs 93 lb in the milk during the suckling period. This means that nearly 8 times more protein is required for 7 months of lactation than for 9 months of pregnancy.

Also, when breeding yearlings to calve as 2-year-olds, producers should be aware that nature has ordained that the growth of the fetus, and the lactation which follows, shall take priority over the maternal requirements. Hence, when there is a nutritive deficiency, the young mother's body will be deprived, or even stunted, before the developing fetus or milk production will be materially affected.

## NUTRIENT REQUIREMENTS

Meeting the nutrient requirements of heifers from weaning to first calving is of great importance. The requirements of heifers of different body weights and growth rates are given in Table 15–13.

**TABLE 15–13**
**DAILY NUTRIENT REQUIREMENTS**
**OF MEDIUM-FRAME GROWING HEIFERS[1]**

| Body Weight | | Daily Gain | | TDN[2] | Protein[2] | Calcium | Phosphorus |
|---|---|---|---|---|---|---|---|
| (lb) | (kg) | (lb) | (kg) | (lb) | (lb) | (g) | (g) |
| 400 | 182 | 2.00 | 0.9 | 7.7 | 1.29 | 26 | 13 |
| 500 | 227 | 2.00 | 0.9 | 9.1 | 1.34 | 24 | 13 |
| 600 | 273 | 2.00 | 0.9 | 10.4 | 1.40 | 23 | 14 |
| 700 | 318 | 2.00 | 0.9 | 11.7 | 1.45 | 22 | 15 |
| 800 | 364 | 1.40 | 0.6 | 10.4 | 1.60 | 25 | 16 |
| 900 | 409 | 1.40 | 0.6 | 11.3 | 1.60 | 26 | 18 |

[1]The above requirements for 800- and 900-lb (364- and 409-kg) weights are for pregnant yearling heifers last third of pregnancy. Adapted by the authors from *Nutrient Requirements of Beef Cattle*, sixth revised edition, National Research Council-National Academy of Sciences, Washington, DC, 1984.
[2]Pounds protein and TDN can be converted to kg by dividing by 2.2.

## SUGGESTED RATIONS FOR REPLACEMENT HEIFERS

In season, good pasture plus mineral supplements fed free choice will meet the nutrient requirements for proper growth and development of heifers.

On the winter range, when dry forage is of low quality, and sometimes not too abundant, 1 to 2 lb of a protein supplement should be provided in the form of cubes, blocks, meal-salt, or liquid. When consumed at the intended level, the supplement should contain sufficient vitamin A to meet the requirements. Mineral supplements should also be provided, preferably free-choice.

Where winter grazing is not available, heifers must be drylotted and fed a complete ration. Sufficient nutrients should be provided to meet the requirements and to keep heifers in a thrifty condition, neither too fat nor too thin.

The wintering rations in Table 15–14 for 500 lb heifer calves should result in a rate of gain of 1 to 1.5 lb per day.

The wintering rations in Table 15–15 for 800- to 900-lb bred yearling heifers should allow a gain of 0.75 to 1 lb per day during the wintering period prior to calving.

## TABLE 15–14
### DAILY RATIONS FOR HEIFER CALVES (500 LB [227 KG]) (AS-FED BASIS)

| | Rations | | | | | | | | | |
|---|---|---|---|---|---|---|---|---|---|---|
| | 1 | | 2 | | 3 | | 4 | | 5 | |
| | (lb) | (kg) | (lb) | (kg) | (lb) | (kg) | (lb) | (kg) | (lb) | (kg) |
| Legume-grass haylage | 25 | 11 | | | | | | | | |
| Legume-grass hay | | | 10 | 4.5 | 10 | 4.5 | | | 5 | 2.3 |
| Corn or sorghum silage | | | | | | | 30 | 13.6 | 20 | 9.1 |
| Ground ear corn | | | 4 | 1.8 | | | | | | |
| Corn, grain sorghum, or barley | | | | | 3 | 1.4 | | | | |
| Supplement[1] | | | | | | | | | 1 | .45 |

[1]Supplement contains 48% crude protein. Quantity to be adjusted in keeping with the protein content of the supplement.

## TABLE 15–15
### RATIONS FOR BRED YEARLING HEIFERS (800–900 LB [364–409 KG]) (AS-FED BASIS)

| | Rations | | | | | | | | | |
|---|---|---|---|---|---|---|---|---|---|---|
| | 1 | | 2 | | 3 | | 4 | | 5 | |
| | (lb) | (kg) | (lb) | (kg) | (lb) | (kg) | (lb) | (kg) | (lb) | (kg) |
| Corn or sorghum silage | 45 | 20.5 | 25 | 11 | | | | | 15 | 2.3 |
| Legume-grass hay | | | 10 | 4.5 | 20 | 9.1 | | | | |
| Legume-grass haylage | | | | | | | 35 | 15.9 | | |
| Corn, grain sorghum, or barley | | | | | | | | | 3 | 1.4 |
| Supplement[1] | 1.5 | 0.7 | | | | | | | | |

[1]Supplement contains 48% crude protein. Quantity to be adjusted in keeping with the protein content of the supplement.

Replacement heifers should be fed rather liberally—more so than stocker cattle which are being grown for the feedlot, to the end that they will acquire most of their growth and development before calving. With limited feeding, they will not have enough weight for age to breed when they are 15 months old; and it is best not to have them calve until they are 30 months of age.

Occasionally, a replacement animal is injured by overfeeding or by fitting for the show, but such losses are insignificant compared with those resulting from the thousands of undersized, poorly developed animals that are grossly underfed.

## SEPARATE HEIFERS BY AGE

The nutritive requirements for heifers differ according to body weight and expected daily gain (Table 15–13). Consequently, the recommended ration for a 500-lb heifer calf (Table 15–14) differs from that of an 800- to 900-lb bred heifer (Table 15–15). It is important, therefore, that replacement heifers be separated by ages for wintering, with coming yearlings in one group and coming twos in another.

## SUMMARY RELATIVE TO CALVING TWO-YEAR-OLDS

From the above, it may be concluded that more producers can, and should, breed yearling heifers to calve as 2-year-olds. But, in doing so, the following practices should be observed in order to lessen calving difficulties.

1. Select the heaviest and highest scoring individual heifers at weaning. Weight at weaning is a means of evaluating the dam's milking ability, provided the calves have not been creep fed.

2. Keep heifers separate from older cows.

3. Start with 50% more weaner replacement heifers than needed if it is the intent to maintain the same size herd—with no provision for expansion whatsoever. This means that for every 100 cows in the herd, 20 replacement heifers are actually needed each year in order to maintain the same size herd. (There is about a 20% replacement in each herd each year.) However, 30 weaner replacement prospects (50% more than actually needed) should be held simply because, based on averages, 10 of them will either die or have to be culled before they replace older cows.

4. Give consideration to the increased nutritional requirements of heifers with increased growth rates.

5. Replacement heifers should be fed for gains of approximately 1 lb/head/day from weaning to first breeding. Following the breeding season, heifers should be managed to assure continued growth and achieve 80 to 85% of expected mature weight at the time of first calving. From breeding until calving, 1¼ lb gain/day is about right.

6. Select yearling and coming 2-year-old heifers on the basis of individuality and rate of gain. Also, cull heifers with small pelvic openings; those with large pelvic openings (above 34 sq in.) have less calving difficulty. Avoid excessively fat heifers.

7. Breed only well-developed heifers, weighing 700 to 750 lb (depending on breed) at 13 to 14 months of age. Size at breeding is more important than age. Also, some breeds come in heat and mature a little earlier than others.

8. Breed heifers 20 days earlier than the cow herd and restrict the breeding season to 45 days. This gives a short concentrated calving period, therefore, proper attention and help can be given heifers at calving time.

9. ''Flush'' feed heifers to gain approximately 2.0 lb per head daily beginning 20 days before the start of and continuing through the breeding season.

10. Breed heifers to a bull known to sire small calves at birth.

11. Feed a well-balanced ration, and feed for continuous gain of 1 to 1.25 lb during the pregnancy period; but don't get them too fat.

12. Feed heifers to weigh at least 800 lb by 120 days before calving.

13. Feed heifers to gain 100 to 120 lb the last 120 days prior to calving. Heifers should weigh at least 875 lb just before calving and approximately 775 lb shortly after calving.

14. Give heifers special care at calving time. This should include—

a. Providing adequate facilities, including (1) a pull stall, and (2) small pens, each suitable for confining a heifer and her calf for approximately 24 hours of "mothering up."

b. Moving each heifer into the calving area approximately two weeks before the expected calving date.

c. Checking heifers for calving at two-hour intervals.

d. Rendering assistance quickly and expertly when it is needed.

e. Removing heifer and calf from calving area within 24 hours after birth and putting them into a clean, dry pasture or other similar area.

15. Provide superior nutrition—well balanced, and rather liberal—during the lactation period, because a heifer's nutritional requirements double after calving. This requires a good ration—one containing adequate energy and proteins, and fortified with the necessary minerals and vitamins. In season, usually this can be accomplished by keeping these heifers on good pastures, with or without supplemental feeding, both during pregnancy and lactation. When good grass is not available—in the winter, early and late, or during droughts—proper feeding must be relied upon.

16. If practical, wean early; at 2 to 6 months of age, rather than the normal 7 months. Otherwise, creep feed the calves.

17. Run heifers that calved as 2-year-olds in a separate herd until after they have had their second calf.

18. Try it (calving 2-year-olds) out on half of your replacement heifers to start with; make sure that you know what is involved before going all out.

Of course, the below-average breeder—the person who has lightweight, poorly developed heifers, and who wouldn't think of staying up nights and having cold, numb fingers while being nursemaid to a heifer and a newborn calf—should take another year and stick to calving out 3-year-olds. But progressive, commercial cattle producers should calve out more 2-year-olds from the standpoint of cutting production costs and increasing profits.

## FITTING RATIONS FOR SHOW CATTLE

Fig. 15–21. Cy Hedrick, Columbus, Michigan, keeps his eyes on the judge while showing his heifer in the National Junior Polled Hereford Show. (Courtesy, American Polled Hereford Assn., Kansas City, MO)

All animals intended for show purposes, including both breeding animals and steers, must be placed in proper condition. To accomplish this, a suitable ration must be selected and the animal or animals must be fed with care over a sufficiently long period. The rations listed in Table 15–16 have been used by successful exhibitors. They are higher in protein content than rations normally used in commercial-finishing operations, but

### TABLE 15–16
### FITTING RATIONS FOR SHOW AND SALE CATTLE
#### (As-Fed Basis)

**Rations 1 to 5** are bulky. They are recommended for use (1) by the inexperienced feeder, and (2) in starting prospective show animals on feed.

**Rations 6 to 11** are less bulky and higher in energy. They are recommended for use (1) by the experienced feeder, and (2) during the latter part of the fitting period.

| Ration No. 1 | (lb) | (kg) | Ration No. 8 | (lb) | (kg) |
|---|---|---|---|---|---|
| Rolled barley | 50 | 22.7 | Flaked corn | 40 | 18.1 |
| Crushed oats | 20 | 9.1 | Rolled barley | 20 | 9.1 |
| Wheat bran | 20 | 9.1 | Crushed oats | 10 | 4.5 |
| Protein supplement[1] | 10 | 4.5 | Dried beet pulp | 10 | 4.5 |
| | | | Wheat bran | 10 | 4.5 |
| **Ration No. 2** | | | Protein supplement[1] | 10 | 4.5 |
| Rolled barley | 30 | 13.6 | | | |
| Flaked corn | 20 | 9.1 | | | |
| Crushed oats | 20 | 9.1 | **Ration No. 9** | | |
| Wheat bran | 20 | 9.1 | | | |
| Protein supplement[1] | 10 | 4.5 | Crushed oats | 25 | 11.3 |
| | | | Rolled barley | 20 | 9.1 |
| **Ration No. 3** | | | Rolled wheat | 20 | 9.1 |
| | | | Flaked corn | 20 | 9.1 |
| Flaked corn | 40 | 18.1 | Wheat bran | 10 | 4.5 |
| Crushed oats | 30 | 13.6 | Protein supplement[1] | 5 | 2.3 |
| Wheat bran | 20 | 9.1 | | | |
| Protein supplement[1] | 10 | 4.5 | | | |
| | | | **Ration No. 10** | | |
| **Ration No. 4** | | | | | |
| | | | Rolled barley | 35 | 15.9 |
| Crushed oats | 30 | 13.6 | Crushed oats | 20 | 9.1 |
| Rolled barley | 30 | 13.6 | Rolled wheat | 20 | 9.1 |
| Wheat bran | 20 | 9.1 | Dry beet pulp | 15 | 6.8 |
| Flaked corn | 10 | 4.5 | Protein supplement[1] | 10 | 4.5 |
| Protein supplement[1] | 10 | 4.5 | | | |
| | | | | | |
| **Ration No. 5** | | | | | |
| | | | **Ration No. 11** | | |
| Flaked corn | 55 | 25.0 | | | |
| Crushed oats | 30 | 13.6 | Rolled barley | 20 | 9.1 |
| Protein supplement[1] | 15 | 6.8 | Flaked corn | 20 | 9.1 |
| | | | Crushed oats | 20 | 9.1 |
| **Ration No. 6** | | | Whole barley (dry wt. basis but cooked before feeding) | 13 | 5.9 |
| Flaked corn or sorghum | 50 | 22.7 | Commercial supplement | 8 | 3.6 |
| Rolled barley | 40 | 18.1 | Linseed meal | 8 | 3.6 |
| Protein supplement[1] | 10 | 4.5 | Wheat bran | 6 | 2.7 |
| | | | Beet pulp, dried molasses | 4 | 1.8 |
| **Ration No. 7** | | | Salt | 1 | .5 |
| Flaked corn | 55 | 25.0 | | | |
| Crushed oats | 20 | 9.1 | | | |
| Dried beet pulp | 10 | 4.5 | | | |
| Protein supplement[1] | 15 | 6.8 | | | |

[1]The protein supplement may consist of linseed, soybean, cottonseed, or peanut meal. With most caretakers, linseed meal is the preferred protein supplement. It gives the animal a sleek hair coat and a pliable hide. Because it is a laxative feed, however, caution should be used in feeding it. Although it is true that an animal getting good clover or alfalfa hay needs less protein supplement than does one eating nonleguminous roughage, it is not possible to supply all the needed protein with hay and still get enough grain into young animals to finish them quickly.

most experienced fitters feel that by such means they get more bloom. In general, when show animals are being force-fed on any one of these concentrate mixtures, experienced caretakers prefer to feed a grass hay or a grass-legume mixed hay to a straight legume, because of the laxative effect and possible bloat hazard of the latter.

Ration 11, which the author has used extensively in fitting show steers, is prepared as follows: The whole barley is processed by (1) adding water in the proportion of 2 to 2½ gal to each gallon of dry barley, and (2) cooking until the kernels are thoroughly swelled and can be easily squashed between the thumb and forefinger. Then it is mixed with the balance of the ration in about the proportions (on a dry basis) indicated. Each steer also receives 4 lb daily of a supplement high in milk by-products. As the animal approaches show finish, the ration is changed by decreasing the rolled barley to 7 lb and increasing the rolled oats by 5 lb and the wheat bran by 2 lb.

## RULES OF FEEDING SHOW CATTLE

The selection of the fitting ration should be made largely on the basis of availability and price of feeds and the results obtained. Other important points in compounding the show ration and feeding the animal are:

1. Use care in getting the animal on feed. Avoid digestive disturbances and setbacks. A safe plan consists in feeding not more than 1 lb of grain at the first feed, or 2 lb for the day. This may be increased by approximately ¼ to ½ daily until the animal is on full feed 3 to 4 weeks later.

From the beginning, it is safe to full feed grass hay or the hay to which the animal is accustomed. Oats are the best concentrate for the beginning ration. As the grain feed is increased according to the directions given above, gradually (a) replace the oats with the mixed ration selected, and (b) decrease the hay, until the animal is eating only 3 to 6 lb of the hay daily at the end of the feeding period.

2. When on full feed, the average animal will eat from 1½ to 2½ lb of grain for each 100 lb of liveweight. Feed only as much grain as the animal will clean up in ½ to 1 hour's time.

3. Most fitters prefer flaked grains in fitting rations. But they may be coarsely ground or crushed.

4. Provide needed minerals.

5. If the droppings are too thin or there is scouring, (a) cut down on the grain allowance, and (b) clean up the quarters. If trouble still persists, cut down on the legume roughage and the protein supplement (especially linseed meal). Many experienced fitters prefer feeding grass or grass-legume mixed hay to a straight legume hay, because of some difficulty in keeping force-fed animals on feed when a legume is fed.

6. The palatability of the ration may be enhanced by adding blackstrap molasses. Make it by diluting one-half to one pint of molasses with an equal volume of water and mixing it with each grain ration just before feeding. Although blackstrap molasses is preferable, beet molasses is satisfactory.

7. Satisfactory milk replacers, which most caretakers like to include in the fitting ration, are now on the market. These products should be used in keeping with manufacturer's directions.

## QUESTIONS FOR STUDY AND DISCUSSION

1. Why is knowledge of beef cattle feeding so important from the standpoints of (a) cost, (b) quantity or roughage consumed, and (c) the nation's land area that is pastured?

2. What is unique about the digestive system of cattle?

3. Why and how have the nutritive requirements of beef cattle become more critical with (a) the marketing of cattle at earlier ages, and (b) the introduction of crossbreeding and the exotic breeds?

4. Discuss the energy needs, and the chief sources of energy, of beef cattle. What are the symptoms of energy deficiency of beef cattle?

5. Discuss the protein needs, and the chief sources of protein, of beef cattle. What are the symptoms of protein deficiency of beef cattle?

6. Discuss the impact of the expanding world human population on the use of urea and other nonprotein nitrogen sources for cattle.

7. For beef cattle, list the minerals most apt to be deficient, then give (a) some of the deficiency symptoms, and (b) practical sources of each mineral for use on a farm or ranch.

8. For beef cattle, list the vitamins most apt to be deficient; then give (a) some of the deficiency symptoms and (b) practical sources of each vitamin for use on a farm or ranch.

9. What factors influence the water requirements of beef cattle?

10. When using the suggested rations presented in Table 15–3, (a) what factors may give reason to vary the suggested rations, and (b) what feeds may be interchanged within groups as feed prices vary?

11. Discuss pastures as feeds for beef cattle from the standpoints of (a) availability/cost, (b) nutritive qualities, and (c) primary uses.

12. Discuss hays and other dry roughages as feeds for beef cattle from the standpoints of (a) availability/cost, (b) nutritive qualities, and (c) primary uses.

13. Discuss silages and root crops as feeds for beef cattle from the standpoints of (a) availability/cost, (b) nutritive qualities, and (c) primary uses.

14. Discuss concentrates as feeds for beef cattle from the standpoints of (a) availability/cost, (b) nutritive qualities, and (c) primary uses.

15. How may feed costs, which account for 65 to 75% of the total cost of keeping brood cows, be lowered?

16. How may a practical producer meet the added energy requirements of a brood cow during the critical 100 days, extending from 30 days before calving until 70 days after calving?

17. Compare and discuss the nutritive requirements of (a) dry pregnant cows, (b) cows nursing calves, and (c) yearling heifers. From a practical standpoint, does this mean that a producer should separate different classes and ages of cattle, then feed them according to needs?

18. Three important reproductive difficulties in beef cattle are: (a) the small number of cows in heat and bred the first 21 days of the breeding season, (b) the low conception rate at first service, and (c) excessive calf losses at birth. What nutritional deficiencies are the real cause of most beef cow reproductive failures?

19. Discuss the winter feeding of beef cattle from the standpoints of (a) expense, (b) delaying too long, (c) weight gain of dry pregnant cows, and (d) the nutritive requirements of lactating cows.

20. Wintering rations are presented for (a) dry, pregnant cows (Table 15–5); and (b) cows nursing calves (Table 15–6). How do these rations differ?

21. What is meant by *daytime calving*? Why is daytime calving important, and how is it accomplished?

22. Define *crop residues*. Discuss crop residues from the standpoints of (a) kinds and quantities available, and (b) the kind of beef cattle that make the highest and best use of them.

23. Define *winter pasture*. What are the virtues of fescue as a winter pasture?

24. Discuss the nutrient deficiencies of cattle that are frequently encountered on U.S. ranges. How would you rectify each one?

25. What type and system of pasture and range supplementation (range cubes or pellets, hand feeding at intervals, protein blocks, liquid protein supplements, or self-feeding salt-feed mixtures) would you recommend? Justify your answer.

26. What feeds are most commonly used for (a) confinement (drylot) cows, and (b) semiconfinement cows?

27. How would you feed and manage a heavily fitted sale bull from auction time to breeding season, which we shall assume to be a period of 60 days?

28. Wherein does the feed and management of young bulls 13 to 14 months old differ from the feed and management of mature bulls?

29. How would you raise orphan calves or multiple-birth calves? Give the feeds and the schedule of feeding for the first 120 days of a calf's life.

30. What are the advantages and disadvantages of early weaning? Outline a program for early weaning, giving the age of weaning, the feed, and the feeding schedule.

31. Define *creep feeding*. What are the advantages and disadvantages of creep feeding? Under what conditions would you recommend creep feeding? Under what conditions would you recommend against creep feeding?

32. What is meant by (a) salt-limited creep feeding, and (b) creep grazing? Why and how are these limited-feeding methods practiced?

33. How do the nutrient requirements of heifers calving as 2-year-olds differ from the nutrient requirements of heifers calving as 3-year-olds?

34. Why should heifers be separated according to age, body weight, and expected daily gain?

35. How do (a) the rations of show cattle, and (b) the rules of feeding show cattle, differ from the normal rations of commercial cattle?

## SELECTED REFERENCES

| Title of Publication | Author(s) | Publisher |
|---|---|---|
| *Animal Nutrition*, Seventh Edition | L. A. Maynard<br>J. K. Loosli<br>H. F. Hintz<br>R. G. Warner | McGraw-Hill Book Company, New York, NY, 1979 |
| *Animal Nutrition* | P. McDonald<br>R. A. Edwards<br>J. F. D. Greenhalgh | Oliver and Boyd, Ltd., Edinburgh, Scotland, 1972 |
| *Applied Animal Feeding and Nutrition*, Third Edition | M. H. Jurgens | Kendall/Hunt Publishing Company, Dubuque, IA, 1974 |
| *Applied Animal Nutrition*, Second Edition | E. W. Crampton<br>L. E. Harris | W. H. Freeman and Co., Publishers, San Francisco, CA, 1969 |
| *Basic Animal Nutrition and Feeding* | D. C. Church<br>W. G. Pond | O & B Books, Corvallis, OR, 1974 |
| *Beef Cattle*, Seventh Edition | A. L. Neumann | John Wiley & Sons, Inc., New York, NY, 1977 |
| *Beef Cattle Production* | K. A. Wagnon<br>R. Albaugh<br>G. H. Hart | The Macmillan Company, New York, NY, 1960 |
| *Beef Cattle Production* | J. F. Lasley | Prentice-Hall, Inc., Englewood Cliffs, NJ, 1981 |
| *Beef Cattle Science*, Sixth Edition | M. E. Ensminger | The Interstate Printers & Publishers, Inc., Danville, IL, 1987 |
| *Beef Cattle Science Handbook* | Edited by<br>M. E. Ensminger | Agriservices Foundation, Clovis, CA, pub. annually since 1964 |

*(Continued)*

## SELECTED REFERENCES (Continued)

| Title of Publication | Author(s) | Publisher |
|---|---|---|
| *Beef Production and The Beef Industry* | R. E. Taylor | Burgess Publishing Company, Minneapolis, MN, 1984 |
| *Beef Production in the South,* Modified Edition | S. H. Fowler | The Interstate Printers & Publishers, Inc., Danville, IL, 1979 |
| *Commercial Beef Cattle Production*, Third Edition | Edited by C. C. O'Mary | Lea & Febiger, Philadelphia, PA, 1983 |
| *Digestive Physiology and Nutrition of Ruminants*, Vols. 1 and 3 | D. C. Church | D. C. Church, Dept. of Animal Science, Oregon State University, Corvallis, OR, 1969 and 1972 |
| *Energy Metabolism of Ruminants* | K. L. Blaxter | Hutchinson & Co., Ltd., London, England, 1962 |
| *Feed Additive Compendium* | Staff | The Animal Health Institute, Washington, DC, and The Miller Publishing Company, Minneapolis, MN, annual |
| *Feeds and Feeding*, Third Edition | A. Cullison | Reston Publishing Company, Inc., Reston, VA, 1981 |
| *Feeds & Nutrition Abridged* | M. E. Ensminger C. G. Olentine, Jr. | The Ensminger Publishing Company, Clovis, CA, 1978 |
| *Feeds & Nutrition Complete* | M. E. Ensminger C. G. Olentine, Jr. | The Ensminger Publishing Company, Clovis, CA, 1978 |
| *Feeds & Nutrition*, Second Edition | M. E. Ensminger J. E. Oldfield W. W. Heinemann | The Ensminger Publishing Company, Clovis, CA, 1990 |
| *Feeds & Nutrition Digest* | M. E. Ensminger J. E. Oldfield W. W. Heinemann | The Ensminger Publishing Company, Clovis, CA, 1990 |
| *Fundamentals of Nutrition*, Second Edition | L. E. Lloyd B. E. McDonald E. W. Crampton | W. H. Freeman and Co. Publishers, San Francisco, CA, 1978 |
| *Handbook of Feedstuffs, The* | R. Seiden W. H. Pfander | Springer Publishing Co., Inc., New York, NY, 1957 |
| *Intensive Beef Production*, Second Edition | T. R. Preston M. B. Willis | Bergman Press, Oxford, England, 1970 |
| *Livestock Feeds and Feeding*, Second Edition | D. C. Church | O & B Books, Inc., Corvallis, OR, 1984 |
| *Nonprotein Nitrogen in the Nutrition of Ruminants* | J. K. Loosli I. W. McDonald | Food and Agriculture Organization of the United Nations, Rome, Italy, 1968 |
| *Nutrient Requirements of Beef Cattle*, Sixth Revised Edition | National Research Council | National Academy of Sciences, Washington, DC, 1984 |
| *Nutrition of Animals and Agricultural Importance*, Parts 1 and 2 | Edited by D. Cuthbertson | Pergamon Press, London, England, 1969 |
| *Nutritional Deficiencies in Livestock* | R. T. Allman T. S. Hamilton | Food and Agriculture Organization of the United Nations, Rome, Italy, 1952 |
| *Physiology of Digestion in the Ruminant* | Dougherty, *et al.* | Butterworth, Inc., Washington, DC, 1964 |
| *Processing and Utilization of Animal By-products* | I. Mann | Food and Agriculture Organization of the United Nations, Rome, Italy, 1962 |
| *Rumen and Its Microbes, The* | R. E. Hungate | Academic Press, Inc., New York, NY, 1966 |
| *Single-Cell Protein* | Edited by R. I. Mateles S. R. Tannenbaum | The M.I.T. Press, Cambridge, MA, 1968 |
| *Stockman's Handbook, The*, Sixth Edition | M. E. Ensminger | The Interstate Printers & Publishers, Inc., Danville, IL, 1983 |
| *Urea and Non-protein Nitrogen in Ruminant Nutrition*, Second Edition | Edited by H. J. Stangel | Nitrogen Division, Allied Chemical Corporation, Morristown, NJ, 1963 |
| *Urea as a Protein Supplement* | Edited by M. H. Briggs | Pergamon Press, Inc., New York, NY, 1967 |

Cattle round-up in Nevada. (Bennett Photo Enterprises, Reno, NV)

Reproduction—the first and most important requisite in the cow-calf system. (Courtesy, American Hereford Assn., Kansas City, MO)

# Chapter 16

# COW-CALF SYSTEM

409

The cow-calf system refers to the breeding of cows and raising of calves. In this system, the calves run with their dams, usually on pasture, until they are weaned, and the cows are not milked. It is the very foundation of beef production. Without the cow-calf system, there would be no stocker programs and no finishing operations, for there wouldn't be any raw materials—calves and feeders. The importance of the cow-calf system in the agriculture of the nation rests chiefly upon the conversion of coarse forage and grass into palatable and nutritious food for human consumption. It is especially adapted to regions where pasture is plentiful and land is cheap; hence, it is the standard system of beef production on the western range. Reproduction—the production of calves—is the first and most important requisite of the cow-calf system, for if animals fail to reproduce, the breeder will soon be out of business. Nationally, cattle producers get about a 92% calf crop.[1]

Since the 1960s, the cow-calf business of America has changed more than it has since the days of the Texas Longhorn. One noteworthy evidence of this tremendous transition appears in the listing of breeds of beef cattle in college textbooks. A popular textbook published in 1920 listed 16 breeds of cattle, whereas the author's book, *Animal Science*, Sixth Edition, published in 1969, listed 21 breeds. Hence, only 5 more breeds were added in the 49-year period. This edition (ninth) of *Animal Science* lists 54 breeds—that's 33 new breeds, or more than double the number listed 21 years earlier in the sixth edition of this same textbook.

Fig. 16–1. Today's commercial cow-calf programs may be crossbreds, with production tested ancestry, and weaned at heavy weights off milk and grass. This shows Angus cows with cross-bred calves at Fort McKavett, Texas. (Courtesy, *Livestock Weekly*, San Angelo, TX)

Hand in hand with the coming of new breeds, breeding programs changed. Many calves no longer look like "peas in a pod"—and on the small side. Instead, more and more of them are multicolored crossbreds, with production tested ancestry, weaned at heavy weights off milk and grass.

[1]This 92% figure is based on the USDA method of computation. It is higher than the actual rate, primarily because (1) more heifers are now calving at less than 2 years of age, with the result that the calves are counted but the dams are not (the USDA figures are based on the assumption that cows don't calve until they are at least 2 years of age); and (2) cows that are pregnancy tested, found barren, and marketed for slaughter are usually not counted as having been bred. The author estimates that, nationally, cattle producers actually get about an 88% calf crop.

## KINDS OF COW-CALF OPERATIONS

Usually cow-calf operators have several options open to them. They may choose (1) a farm or ranch herd, (2) running commercial or purebred cattle, (3) selling weaners or stockers, or (4) dual-purpose production.

## FARM OR RANCH HERD

In general, beef cattle production in the farm states is merely part of a diversified type of farming. Grain and pasture crops are produced; and, on the same farm, beef cattle may compete with dairy cattle, hogs, and sheep for the available feeds. This applies to practically all the farms located to the east of the 17 western range states. In general, farm beef cattle herds are much smaller than range herds of the West, and many of them lack the uniformity which prevails in range cattle.

Fig. 16–2. Range cattle herd in Canada, winding serpentinelike over the hills and through the valleys. (Courtesy, National Film Board of Canada)

More than half of all U.S. beef cattle are produced on the western range. Because a considerable portion of the range area is not suited to the production of grains, and because sheep, and in some areas big game, offer the only other major use of the grasses, it seems evident that range beef cattle production will continue to hold a place of prominence in American agriculture.

From the foregoing, it should be concluded that geographic location largely determines whether a herd of cattle shall be operated as farm herd or range herd. Thus, the majority of beef herds in the West and Southwest, except for relatively small herds in irrigated areas, are operated as range herds, whereas the vast majority of the herds in the central and eastern parts of the United States are operated as relatively small farm herds.

## PUREBRED OR COMMERCIAL

Based on number of calves raised vs number of calves registered, the author's computations show that only about 4.0% of U.S. cattle (beef and dairy) are registered purebreds. Hence, purebreds are a small, but mighty, minority.

Fig. 16-3. Purebred Angus herd. (Courtesy, *Angus Journal*, St. Joseph, MO)

There is nothing sacred about purebreds, nor does the word itself imply any magic. It is generally agreed, however, that purebreds have been the major factor in the beef improvement of the past, and that they will continue to exert a powerful influence in the future. Although limited in numbers, purebred herds are scattered throughout the United States and include both farm and ranch herds.

The purebred cattle business is a specialized type of production. Generally speaking, few cattle producers should undertake the production of purebreds with the intention of furnishing foundation or replacement stock to other purebred breeders, or purebred bulls to the commercial producer.

All nonpurebred cattle are known as commercial cattle. This includes the vast majority of the beef cattle of the United States—probably 96%. In general, however, because of the obvious merit of using well-bred bulls, most commercial calves are sired by purebreds. Beef over the block is the ultimate article of commerce of the commercial cow-calf producer; although in the process of getting from pasture to packer, the calf may be subjected to one or two intermediate stages—as stockers, and in the feedlot.

Commercial cattle producers are intensely practical. No cow meets with their favor unless she regularly produces the right kind of calf. Experience, industry, and good judgment are requisites to success in the commercial cow-calf business. Additionally, the commercial producer is a key person in the nation's economy.

Fig. 16-4. A commercial cow herd on winter range. Today, most range cattle feeding operations are mechanized—with tractors and trucks. But horses and mules are occasionally used in inaccessible areas and in deep snow. Note the sleigh (see runners) in use in the above picture. (Courtesy, Union Pacific Railroad)

## WEANERS OR STOCKERS

Commercial cattle producers seldom adhere strictly to a cow-calf operation as such. Rather, based primarily upon the price of feed and the price of cattle, they may option to (1) sell weaners, except for replacements, (2) carry all or part of their calf crop over to the yearling stage, (3) buy additional calves, and/or (4) finish out their home produced calves in their own feedlot or in a custom lot.

## DUAL-PURPOSE PRODUCTION

For the most part, dual-purpose production has been confined to small farmers who live on the land and who make their living therefrom. Cows of dual-purpose breeding are often referred to as the *farmer's cow*. In this type of production, an attempt is made to obtain, simultaneously, as much beef and milk as possible. That is to say, in its truest form, this type of management cannot be classified as either beef or dairy production.

One of the chief virtues of dual-purpose production is the flexibility which it affords. When labor is available and dairy products are high in price, the herd may be managed for market milk production. On the other hand, when labor is scarce and dairy products are low in price, calves may be left running with their dams, and emphasis may be placed on beef production.

Because of the very nature of dual-purpose production, it is not adapted to the extensive ranches of western and southwestern United States. Rather, it is practiced on a limited number of small farms scattered throughout the humid area of central and eastern United States. It is noteworthy, however, that many of the exotic breeds recently introduced into the United States and Canada, primarily for crossbreeding purposes in an effort to secure cows that will produce more milk for their calves, are known as dual-purpose breeds in the countries of their origin. Noteworthy, too, is the fact that these breeds are being used for crossbreeding in both farm and ranch herds.

## CONSIDERATIONS WHEN ESTABLISHING OR EXPANDING THE COW HERD

Whether or not cows and calves should be produced on a particular farm or ranch should be determined by a careful analysis of (1) the available resources (land, labor, capital, and managerial skills), and (2) the relative profitability of beef cattle in comparison with alternative enterprises.

Choices of alternate enterprises are limited on many ranches of the West and Southwest. Much of the land is suited only for grazing cattle or sheep. If beef cattle are selected rather than sheep, the only decision that remains is whether it should be strictly cow-calf, with all calves except replacements marketed at weaning time; a combination of cow-calf and stockers; or stockers only.

Farmers in the central and eastern United States have more options; hence, the choice of the enterprise becomes more difficult. They must first decide between grain and animal production, or some combination of the two. Additionally, if they decide to go the animal route, they must decide between beef cattle, dairy cattle, hogs, and sheep, or some combination of them.

Beef cattle do have certain advantages. Likewise, they have their disadvantages. These follow.

## FACTORS FAVORABLE TO COW-CALF PRODUCTION

The special advantages of cow-calf production as compared to other kinds of livestock are listed in Chapter 11, "The Beef Cattle Industry," under the heading, "Factors Favorable to Beef Production"; hence, the reader is referred thereto.

## FACTORS UNFAVORABLE TO COW-CALF PRODUCTION

Factors which, under certain conditions may be unfavorable to cow-calf production are listed in Chapter 11, "The Beef Cattle Industry," under the heading, "Factors Unfavorable to Beef Production"; hence the reader is referred thereto.

## AVAILABLE RESOURCES FOR COW-CALF PRODUCTION

Basically, when establishing or expanding the cow herd, farmers and ranchers need to give consideration to the availability to them of the following four resources: land, labor, capital, and managerial skills. In the years ahead, the most successful cattle producers will put these together in such a manner as to maximize profits, followed by increased cow numbers. The cow-calf resource requirements, compared to alternate enterprises, are (1) large acreages of land for which the highest and best use is pasture, along with considerable quantities of comparatively low-quality winter roughage; (2) available labor during the calving season (preferably with a liking for and a knowledge of cattle); (3) adequate capital; and (4) able management commensurate with the size and sophistication of the operation.

## PROJECTED BEEF NEEDS AHEAD

Pertinent information pertaining to past, present, and future cattle raising follows:

• **Beef consumption and price trend**—The demand for beef at any price has been blunted. Since 1966, the per capita consumption of beef has gradually declined, while poultry consumption has increased dramatically—almost doubling (see Chapter 9, Fig. 9–23). The increased consumption of poultry will continue, stiffer competition from pork is predicted, and more fish will be consumed. The price differential between beef, pork, and poultry will continue to cause consumers to exercise their choice; for every pound of beef that they buy, they can buy 1.5 lb of pork, or 3.5 lb of poultry. Increasingly, beef prices over the counter will be determined (1) by pork and poultry prices, and (2) by the nation's economy and level of employment.

• **Beef eaters and cattle numbers**—In 1990, there were 248 million people to feed in the U.S. By the year 2000, it is estimated that there will be 268 million, an increase of 20 million people. It is projected that the beef consumed by the increased human population will counterbalance the decline in per capita beef consumption, thereby resulting in cattle numbers remaining about the 1990 level—about 100 million head.

• **Increased beef production per cow**—From 1960 to 1990, beef production per cow increased about 200 lb (from 320 to 520 lb). This was a 62% increase, or an average annual increase of 2.1%—the end result of improved genetics and nutrition. As we enter the era of biotechnology, further, but gradual increases in beef production per cow may be expected. It follows that greater beef production per cow will call for fewer cows.

## TRENDS IN COW-CALF OPERATIONS

Prior to 1950, there were more milk cows than beef cows in the United States. Beef cow numbers first exceeded milk cow numbers in 1954.

Total cattle (beef and dairy) numbers peaked in 1975 with a record 132 million head, then dropped to about 100 million head in 1988.

Beef cow numbers (exclusive of dairy) grew from 16.7 million in 1950 to 45.7 million in 1975, then declined to about 33 million in 1988.

Due to the increased consumption of poultry, and perhaps of pork, too, U.S. consumption of beef will likely trend downward, and the U.S. export market for beef will be limited. It follows that any future increase in U.S. beef cattle numbers will be determined primarily by human population increase, economy, and level of employment.

## GEOGRAPHIC TRENDS

Fig. 16–5, which summarizes changes in calf numbers from 1977 to 1987, indicates geographic trends. It is noteworthy that beef calf numbers decreased in all regions from 1977 to 1987, presaging the downward trend in cattle numbers nationally. But the percentage decrease in calf numbers was least in those areas with an abundance of roughage suitable for beef cattle. Thus, the decrease in calf numbers was low in the West North Central Area because of the grazing areas of the Great Plains, along with low quality roughage, such as cornstalks, for winter feed. The decrease in calf numbers in the North Central Area was low because of the effective utilization of the abundant low quality roughages, especially cornstalks, in the area. It was low in the Southwest Area, where the highest and best use for the extensive ranges is for cattle.

### REGIONAL SUMMARY OF CHANGE IN CALF NUMBERS 1977–1987

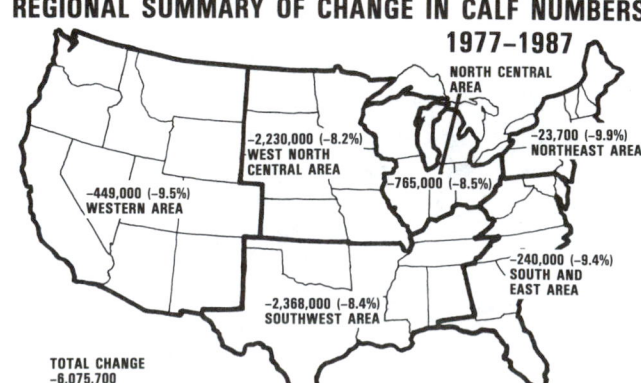

Fig. 16–5. Geographic trends in calf numbers from 1977 to 1987, in the six areas of U.S. beef production. (Source: *Agricultural Statistics 1978*, USDA, p. 305, *Ibid.* 1988, USDA, p. 263)

## TRENDS IN NUMBERS OF HERDS AND HERD SIZE

Contrary to popular belief, the average U.S. beef cattle herd is relatively small. In 1987, the census showed that 31.6 million beef cows were owned by 841,778 farmers and ranchers, and that the average herd size was 37.5 cows.[2] Further, 97.5% of the cow-calf operators had fewer than 200 cows; and 80% had fewer than 100 head. Most of these were part-time farmers or those not totally dependent on cattle for a living. But the 2.5% (25,000 operators) were big; they controlled 31% of the total cattle numbers.[3]

Increases in part-time farming activities and continued farm consolidation may both be important in shaping future increases in beef supplies. Part-time farming favors a beef cattle enterprise which can be managed on a labor-extensive basis. As farms consolidate, small holdings of pasture are brought under one management, which makes it possible to support a profitable beef enterprise.

The preponderance of small beef cow herds is likely to affect the beef industry in the future in two ways: (1) They won't respond dramatically to price and cost changes—that is, part-time farmers are less responsive to prices than commercial farmers; and (2) they may limit technological change, simply because they cannot justify the cost of the technique.

Also, this means that (1) many producers will continue to lack sufficient volume to justify economically the use of many available technologies, and (2) assembling uniform lots of feeder cattle from many small producers for movement to a few large feedlots will continue to present structural problems for the entire beef industry. Nevertheless, the rising number of producers with larger herds can be expected to lead the way in the adoption of improved management and technology.

## MORE BEEF PRODUCTION PER COW

Fig. 16–6. Milk and grass produce heavy calves. (Courtesy, International Braford Assn., Ft. Pierce, FL)

Beef cows have become more productive in recent years. In 1990, each cow produced about 200 more pounds of beef (520 versus 320 lb) than in 1970, 20 years earlier—the end result of improved genetics and nutrition. The cow of tomorrow will

be even more productive—she will be the product of biotechnology.

In the future, several potential changes in productivity per cow may be effected which will directly improve the supply of beef obtainable from a given inventory of cows, among them, those which follow.

• **Increased biotechnology**—In the future, biotechnology will give the beef industry a big assist. Embryo transfer and cloning will become more common; genes will be transferred from one species to another without taking unwanted genes; and scientists may be able to genetically engineer cattle resistant to some of the most costly diseases. But the widespread use of biotechnology will take time; it takes 15 years from discovery to application—and even more time for widespread use.

• **Increased percent calf crop**—There is room for improvement in the percent calf crop born. The U.S. Department of Agriculture reports the following percent calf crop born: In 1980, 89%[4]; in 1990, 92%. This figure is higher than the actual rate, primarily because (1) more heifers are now calving at less than 2 years of age, with the result that the calves are counted but the dams are not (the USDA figures are based on the assumption that cows don't calve until they are at least 2 years of age); and (2) cows that are pregnancy tested, found barren, and marketed for slaughter are usually not counted as having been bred. Nevertheless, the U.S. Department of Agriculture figures are the best that we have; hence, they serve as a useful guide. The author estimates that the actual percent calf crop for the United States is about 88%. No great improvement is seen in calf crop percentage ahead, primarily because gains become more difficult as calving percentages approach perfection—or 100%. The largest gains are expected in the Southeast and Southwest where rates have been the lowest.

Calf crop percentage on some cow-calf operations will be increased through selection; management practices, such as controlled breeding seasons (by use of hormones), fertility testing of bulls, and pregnancy testing of cows; improved nutrition, including improved pastures and ranges; and calving heifers at two years of age, instead of three.

Superovulation (twinning and multiple births) could dramatically improve production per cow, if some of the obstacles could be overcome.

Fig. 16–7. This cow has produced three sets of twins in five births. Producing twins can increase the efficiency of beef production by 20 to 30%. (Courtesy, USDA)

---

[2]*Beef Today*, March, 1990, p. 34.

[3]Report by Topper Thorpe, *Cattlefax*, National Cattlemen's Assn., Englewood, CO)

[4]*Cattle*, Statistical Recording Service, USDA, Jan. 29, 1982, p. 3. The percent calf crop born is based on (a) cows and heifers that have calved, and (b) calves born.

But the increased application of several practices that have an adverse effect on calving rate will partially offset gains. This includes artificial insemination; the use of the larger breeds of cattle, which will make for more calving difficulty; shifts in the calving period; and more confinement cow operations.

- **Reduced calf losses**—There is much room for improvement in the percent calf crop weaned. The U.S. Department of Agriculture reports the following appalling calf losses from birth to weaning: In 1979, 8.6%; in 1987, 7.7%.[5] Young calf losses have been reduced to some extent through the hybrid vigor of crossbreds and improved nutrition and management. However, these gains have been partially offset by increased confinement calving, which has been accompanied by more calf scours and other diseases. By 1990, it appeared that calf losses from birth to weaning had been lowered to 6%.

- **Use of dairy calves**—The average milk cow breaks down or is sold because of poor production after being in the milk string 4 years. To maintain a 100-cow dairy, therefore, 25 first-calf heifers must replace their elders on the milk line each year. But not all dairy heifer calves become tomorrow's cows! There are calf losses, and others must be culled for one reason or another. To maintain status quo in a milking herd, therefore—with no provision whatsoever for expansion—each year a dairy producer must start with a minimum of 3 heifer calves for every 10 cows in the milking string. That's 30% replacement. Of course, cull dairy cows and replacement heifers end up as beef, along with dairy steers and bulls.

In recent years, part of the demand for feeder calves has been met by increasing the proportion of the dairy calf crop fed out as dairy beef rather than vealed or kept for replacements. (Of course, fewer replacements percentagewise, are kept when dairy cow numbers are being reduced instead of expanded.) Also, the need for dairy bulls has declined with increased artificial insemination. Although milk cow numbers halved in the 1950s and '60s, and some modest declines will occur in the years ahead, it is estimated that dairy beef, from cull dairy calves that are fed out, will provide about 15% of the nation's total annual beef supply to the year 2000.

In 1990, an estimated 30% of the nation's dairy calf crop was used for veal. This will change drastically, due to a continuing gain in demand for feeder calves relative to that for veal. By 2000, the author predicts that of all dairy calves born, 30% will be retained as replacements, 60% will be finished out as dairy beef, and only 10% will be vealed. Each calf sent to the feedlot means about 800 lb more liveweight at slaughter than would have been obtained had the animal been vealed; hence, such a shift in disposition of dairy calves will increase production per cow.

- **Weights and ages of cull breeding stock**—Cull cows contribute significantly to the total supply of beef, and account for nearly all nonfed beef produced in the United States. About 20% of beef cows and 25% of milk cows are culled each year and sent to slaughter. Beef cows are culled at an average of 7 to 9 years of age. No significant changes in age of culling are expected in the years ahead. Milk cows average only 4 years on the production line; they have a shorter life of usefulness than beef cows.

The trend is toward heavier weights of cull cows. Today, cull cows in most areas average around 1,000 lb in weight, except in the Southeast and Southwest where they are 100 to

150 lb lighter. By 2000, the author predicts an average of 75-lb heavier cow culling weights, as a result of (1) improved nutrition, (2) increasing numbers of the larger exotics, and (3) conversion of more dairies to Holsteins, with lessening of the lighter breeds.

- **Production testing and crossbreeding**—Production testing and crossbreeding combined, which are out of the experimental and in the practical realm, can increase beef yield per cow maintained over that of straightbreds by 15 to 25%, depending on the choice of breeds and the breeding system. The 15 to 25% is achieved in two ways: (1) through selection, based on production testing, of the purebreds used in the crossbreeding program; and (2) through heterosis increase of the crossbreds.

Fig. 16–8. Production testing begins with birth weight, which is associated with calf survival and growth rate. But too heavy birth weight may make for calving difficulty. (Courtesy, USDA)

Through production testing, it is possible to achieve (1) heavier weaning weights (29% heritability)—and young gains off milk and grass are very efficient; (2) higher daily gains from weaning to marketing, making for shorter time in reaching market weight and condition, thereby effecting a saving in labor and making for a more rapid turnover in capital; and (3) greater efficiency of feed utilization, thereby making it possible to feed more cows and calves on a given quantity of feed.

A 2-breed cross (in which only the calves are crossbred) gives an 8 to 10% increase in pounds of calf weaned per cow bred. Through a 3-breed cross, it is possible to achieve even greater production per cow.

- **Calf weights**— Cow-calf producers will produce heavier calves in their efforts to increase efficiency and obtain a larger share of the total returns to the beef industry. Of course, heavier calves at weaning add to the total beef supply only if they are eventually carried to heavier slaughter weights. When calves are sold at around 400 lb weight, there is need for a stocker or backgrounding stage because most cattle feeders prefer cattle weighing 700 to 800 lb. Weaning heavier calves off milk and grass, especially if weights approximating 600 lb are achieved, will have a tendency to eliminate or shorten the stocker stage.

---

[5]*Agricultural Statistics 1988*, USDA, p. 265, Table 393.

Larger cattle and heavier milking strains will result in heavier weaning weights in the years ahead. Much of this transition will come from selection of British breeding stock for more size and more milk, from improved nutrition of cows and suckling calves, and from greater use of production records to select breeding stock with superior growth rate potential.

The author predicts that the combination of breeding (improved British breeds, along with infusion of exotic and dairy breeding) plus improved nutrition, will result in an average of 550-lb weaning weights at 7 to 8 months of age by 2000. This means that the better herds will be weaning off calves weighing 650 lb or more. The greatest increase will occur in the Southeast, but it should be added that this area has the most room for improvement, because its weaning weights have always been lower than other sections of the country.

• **Yearling weights**—Some producers have long grown calves beyond weaning weight, with the prevalence of the practice varied by years according to available resources—particularly feed.

Short yearlings are weaner calves held over because they are too light to sell at weaning time and/or because adequate forage is available. Usually, they are sold at just under one year of age, although they may range up to 14 months in Arizona and New Mexico. Long yearlings average about 16 months of age.

Market weights of each group will gradually move upward, due primarily to high-priced grains and shorter feedlot finishing periods, along with selecting breeding stock for more size. The author predicts that by 2000, short yearlings will weigh around 700 lb, and long yearlings will weigh around 800 lb.

## INCREASINGLY, COW-CALF OPERATIONS WILL CONTRACT, PARTNERSHIP, OR INTEGRATE

Competition from the poultry and swine industry will force more and more organizational changes in cow-calf operations. The traditional four stages—(1) cow-calf, (2) stocker, (3) feedlot, and (4) meat packer; with commissions and expenses at each stage—will give way to more one person or group control from cow-calf production to retail, either by contract, partnership, or ownership/integration. For cow-calf operators to survive this transition, the key essentials are: (1) good records, (2) flexibility, (3) discipline, (4) being well-informed, and (5) least cost of production.

## FACTORS INFLUENCING THE NUMBER OF BEEF COWS

Over a long period of time, economics determines the number of beef cows. Until 1975, economic conditions favored cattle production. As a result, beef cow numbers increased greatly in all areas of the U.S., peaking at 45.7 million in 1975. Since 1975, beef cow numbers have declined, dropping to 33 million in 1988.

## CATTLE PRICES

Cattle prices are expected to decline in the years ahead, thereby making retail beef prices more competitive with pork and poultry. In 1990, 1.5 lb of pork or 3.5 lb of poultry could be bought for the price of a pound of beef.

As shown in Fig. 16–9, there hasn't been much incentive to increase cow numbers in recent years.

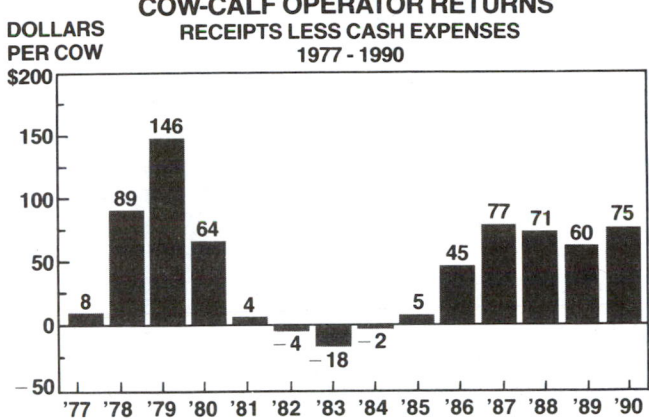

Fig. 16–9. Cow-calf returns. (Source: *Successful Farming*, Feb., 1990, p. 13, from newsletter "As I See It," by Bill Helming, The Helming Group, Overland Park, KS)

Fig. 16–9 shows cow-calf returns (cash receipts, less cash expenses) for the period 1977 to 1990. But the cash expenses do not include land charges. When land costs are included, the cow-calf returns are about 70% less than shown. Note that, even without land charges, cash receipts did not cover cash expenses in 1982, 1983, and 1984; and that cash receipts barely covered cash expenses in 1981 and 1985. Thus, there has been little profit motive to increase cow numbers in recent years.

In the decade of the 1990s, total cattle numbers (including beef cows, dairy cows, heifers, steers, bulls, and calves) are expected to remain at about 100 to 105 million head, with some decline in per capita beef consumption being counterbalanced by increased human population. However, if a relatively severe recession is encountered, some cattle liquidation may be expected.

## SHIFTS IN CROP PRODUCTION

Traditionally, land for which the highest and best use is for pasture and hay has been used for beef cows. This practice will continue in the years ahead. However, forage crops produced with known technology will become more competitive with grain production. Forage for beef cows grown on land now occupied by wheat and cotton is uncertain, and largely dependent upon the world market for food and fiber. Also, should such acreage be released, the potential for soybeans and newly introduced cultural practices for raising corn will act as a counterforce.

## SHIFTS IN LIVESTOCK ENTERPRISES

Beef cow numbers already occupy much of the land formerly used by dairy cows, particularly in the humid regions. In 1987, U.S. per capita milk consumption in all forms (all dairy products, milk equivalent, fat content basis) was 598.2 lb, up from 560.1 lb in 1972. But milk production per cow rose from 10,259 in 1972, to 13,786 in 1987. Thus, even with 33.9 million more people to feed in 1987 than in 1972, and an increase in per capita milk consumption of 38.1 lb within this same

15-year period, the total milk needs in 1987 were met with 1,366,000 fewer dairy cows (11,700,000 cows in 1972 versus 10,344,000 in 1987). Hence, from 1972 to 1987, sufficient land resources were freed by dairy cows to feed 1.4 million more beef cows. Additionally, during this 15-year period more dairy enterprises were moved into year-round confinement that utilized only harvested feeds, thereby freeing some permanent pastures.

Nationally, beef cows have been substituted for sheep through the years. Although sheep numbers appear to have stabilized, nationally, some further decreases in sheep numbers in the West and the Southwest may occur, due primarily to difficulty in obtaining labor, followed by a conversion from sheep to beef.

With higher grain prices, it is expected that the Midwest will lessen cattle feedlot operations and go more to cow-calf production and push for heavier calves as a means of utilizing more roughage (especially cornstalks). In the Southeast, more calves will be grown out to the yearling stage and finished on pasture. Cow-calf systems are expected to decrease somewhat in some of the southwestern states and to be replaced by more cow-yearling programs. In the Northern Plains, the trend will be for more cow-calf programs and fewer cow-yearling and stocker programs.

## FORAGE PRODUCTION, HARVESTING, AND UTILIZATION

Most ranges are now fully stocked; some are overgrazed. Hence, increased beef cow numbers are dependent upon increased forage production. Available forage will be increased by—

1. Application of forage technology, including fertilization of pasture and hay crops, improvement of forage plant mixes, controlled grazing, renovation and reseeding of existing pastures and ranges, and the control of undesirable plants.
2. Irrigation of forage crops in some of the western dryland areas.
3. Improvements in methods of salvaging crop residues, especially *husklage* and *stalklage* from corn.

## PUBLIC POLICES

Public polices will change from time to time, but, for the most part, it is anticipated that they will not be adverse to the cattle industry. It is unlikely that there will be any longtime depressing effect on beef cattle numbers to accrue from crop control programs, from banning drugs, from environmental control, from livestock wastes, from imports and exports of beef, or from changes in federal grades of beef.

## POPULARITY OF CATTLE RAISING

Being a cattle raiser serves as a status symbol for many people, farmers and nonfarm investors alike. Considerable romanticism has been attached to cattle raising through the years; and it has been enhanced by the recent publicity on crossbreeding and the exotic breeds. Ownership of cattle carries more prestige than ownership of any other kind of livestock.

## SHIFTS IN LAND USE

There will be increased demands for grazing lands for industrial, residential, and recreational uses, all of which will mitigate against the expansion of cow-calf operations. In particular, pressures for recreational development and from the environmentalists will increase in the western range area.

## ELDERLY AND PART-TIME FARMERS

The increasing age of farmers and the growth of part-time farming favor cow-calf operations. Farmers, like other people, are living longer. Many of them choose to pull up the reins gradually by shifting to less labor intensive enterprises. A beef cow herd is often the enterprise of choice. Rising costs and scarcity of hired labor also favor beef cattle. More part-time cattle farming seems likely.

## THE CATTLE FARM OR RANCH

The farm or ranch is the most important basic resource to the success of the cow-calf operation.

Many people are interested in buying a cattle spread. Tenants are climbing the ladder to ownership. Present owners are mechanizing their holdings and borrowing more money, with the result that they want more acres. Farms and ranches are being bought for sons and daughters. City folks want to fulfill their dreams by semiretiring on a cattle ranch. Investors, who have been disillusioned with the stock market, who are concerned about continued inflation, and who feel that manufacturing profits are on the wane, are looking for cattle farms and ranches. Some of these buyers will be happy with their purchases. Others will rue the day that they made the decision. How well the farm or ranch is selected, bought, and managed will make the difference. Indeed, no decision in the cattle business has greater consequences for the individual than selecting and purchasing the farm or ranch.

Although a cattle farm or ranch may be resold, most owners plan to operate the place that they acquire for a lifetime. This means that the fields and buildings are to be the workshop of the purchasers, and that the alternatives open thereafter are greatly reduced. They can no longer consider operation in another area, even if the climate, feed, and market are more favorable. In many cases, they are committed to a narrow range of alternatives, such as (1) cattle only, as on some holdings of the Southwest; (2) the particular combination of livestock—for example, hogs cannot be run on the western range; and (3) adapted crops and grasses. Moreover, except for an absentee owner, the farm or ranch selected will be the family home. As such, the family will develop community ties that may last for years or even for generations, since many cattle farms and ranches are passed on from one generation to the next.

Indeed, the ultimate success of a cattle operation is determined by the careful selection of a farm or ranch; the proper combination of cattle, feed, and sometimes other livestock; the weather, disease, price fluctuations, and market demands; and the well-being and happiness of the family.

# TYPES OF CATTLE FARMS OR RANCHES

When selecting a cattle farm or ranch, the first major consideration should be the purpose of the ranch. For what use will it be put? The most common types of cattle farms and ranches are listed and discussed very briefly in the sections that follow.

## FULL-TIME CATTLE FARM OR RANCH

A full-time cattle farm or ranch is one in which the operator devotes his full time to the enterprise and depends entirely upon cattle for his income. As indicated later in this chapter (see section headed "Carrying Capacity; Size of Herd or Ranch), (1) carrying capacity, not acreage, determines whether a unit is big enough to constitute a full-time operation, and (2) the author estimates that a farm or ranch with a minimum carrying capacity of 300 head of brood cows (or equivalent; for example, two yearling stockers may be substituted for one cow) is necessary to be an economic unit for a full-time cattle operation. Without doubt, exceptions can be cited where operators with smaller herds are making a good living, with no other source of income. However, most operators with fewer than 300 brood cows are either employed part-time off the farm or ranch or have another sizable farming enterprise in combination with the cattle.

## PART-TIME CATTLE FARM OR RANCH

In 1990, 97.5% of the cow-calf operators had fewer than 200 cows; and 80% had fewer than 100 head. Most of these were either employed part-time off the farm or ranch or semi-retired and had another source of income.

Part-time cattle enterprises will increase in the future. More and more folks will own a little cattle farm or ranch, as a source of some income and as a way of life. But they will derive most of their income from off-farm employment. An additional, and growing, number of part-time cattle farmers and ranchers will consist of senior citizens and semiretired folks. They are more interested in an enterprise that is relatively free from labor problems, and in the good life, than in monetary gain, for most of them have already made it.

## NOT STRICTLY COW-CALF

Traditionally, few beef cattle farms have been strictly cow-calf enterprises, limited to selling calves at weaning time. Instead, operators exercise one or more of the options open to them, based primarily upon the availability and price of feed and cattle. Some cow-calf producers sell part of the calf crop and hold the remainder over for sale as short- or long-yearlings, depending on the forage supply. Others buy additional calves and sell them as yearlings. Still others finish their home-raised calves out to slaughter weight. Additionally, there are cattle operations that own no cows; instead they buy and sell stocker cattle. Once weaned, a calf competes with beef cows for feed and forage and, consequently, acts as a constraint on production of additional calves.

## CATTLE AND CASH CROP COMBINATIONS

Cattle and cash crop combination farms and ranches are preferred by some operators. It makes for desirable diversification, and, hopefully, cattle and the cash crop(s) will not be down in price at the same time.

## CATTLE AND BY-PRODUCTS AND SPECIALTY CROP COMBINATIONS

Fig. 16–10. Cattle efficiently utilize large quantities of by-product feeds, such as cornstalks. This shows cows feeding on corn residue which has been harvested by mechanical means. (Courtesy, Iowa State University, Ames, IA)

Although they are not likely to be primary factors in determining the selection of a ranch, or how much will be paid for it, certain other cattle-crop combinations should not be overlooked. This includes such by-products as small grain stubble fields, cornstalks, cotton fields following harvest, cull potatoes, cottonseed hulls, corncobs, cull citrus, cannery refuse, beet tops, and a host of other products.

Among the specialty crops sometimes grown with a cattle combination are: cotton, tobacco, hay, and timber.

## CATTLE AND OTHER LIVESTOCK COMBINATIONS

In many areas, and on many farms and ranches, beef cattle and one or more other classes of livestock may be combined to advantage, thereby using the resources more efficiently and increasing income. For example, farmers in the central states long ago recognized the advantages of combining beef cattle and hogs; many cattle producers would do well to add sheep to their enterprises; cattle and horses are found on many farms and ranches; and combinations of beef cattle and poultry are occasionally seen, primarily as a means of disposing of poultry manure.

## IRRIGATED FARM OR RANCH

Rising land and labor costs favor more irrigation. Where the cost per cow-calf carrying capacity is cheaper under irrigation than a dryland operation, irrigation will increase. Likewise, intensive cow-calf operations under irrigation usually require less labor per cow-calf unit than more extensive dryland operations. Of course, once an area is irrigated, new crop alternatives are opened up, with the result that a determination will have to be made as to which will be the most profitable—cattle or crops.

## ATTACHED GOVERNMENT AND PRIVATE LEASES

In the West, much of the grazing land that ranchers rely upon to maintain their cattle is built into operating units by leasing or obtaining use permits from several federal and state agencies. A total of 60.3% of the federal lands of the 50 states is located in the 11 western states. This land is made available through permit or lease to nearby ranch operators, usually at a fixed annual fee per head of livestock.

The bulk of federal land is administered by the Bureau of Land Management and the U.S. Forest Service. Thus, where a cattle ranch is being acquired in the West, it is important that the rancher have knowledge of these particular agencies.

• **Bureau of Land Management**—The Bureau of Land Management of the U.S. Department of the Interior administers about 42% of federally owned land. From the standpoint of the rancher the most important function of the Bureau of Land Management is its administration of the grazing districts established under the Taylor Grazing Act of 1934 and of the unreserved public land situated outside of these districts which are subject to grazing lease under Section 15 of the act.

Fig. 16–11. Cattle of a grazing permittee on a Bureau of Land Management range near Conrow Creek, Montana. Picture shows mahogany, juniper, and pine on rough, rocky slopes. (Courtesy, Bureau of Land Management, U.S. Department of the Interior)

Grazing privileges are allocated to individual operators, associations, and corporations, and a fee is charged for grazing privileges. In 1990, the Bureau of Land Management charged $1.81 per animal unit month.

• **U.S. Forest Service**—Over 51.5 million acres of national forests are used for grazing under a system of permits issued to local farmers and ranchers by the Forest Service of the U.S. Department of Agriculture. These grazing allotments provide grazing for about 2.5 million livestock and about 3.4 million head of big-game animals.

The Forest Service issues 10-year permits to producers who hold preferences and annual permits to those who hold tem-

porary use. Among other things, the permit prescribes the boundaries of the range which they may use, the maximum number of animals allowed, and the season when grazing is permitted.

Forest Service grazing fees are based on a formula which takes into account livestock prices over the past 10 years, the quality of the forage on the allotment, and the cost of ranch operation. In 1990, the charge came to $1.81 per animal unit month.

## HOBBY FARM

In recent years, many people of wealth have established outstanding cattle herds, especially purebred herds. Most of these folks operate such cattle ranches as money-making enterprises. They are just as "money hungry" in the cattle business as they were in the industry from which they made their initial wealth. Also, conducting the cattle operation as a business is more of a challenge to them than if it were a hobby. Besides, the income tax regulations today are such as to make it impossible for many to afford not to operate a cattle farm or ranch as a business.

Internal Revenue Service agents are prone to attack the *farmer* status of absentee owners. Most challenges are raised where the taxpayer earns substantial off-farm or ranch income and is showing farm or ranch losses for a particular year or over a period of years. Though the concept is similar to the *hobby farm* challenge, the distinction exists in that the disallowance or challenge relates to capitalization of expenses. In hobby farm situations, the expenses are considered personal, and thus not deductible; nor can they be capitalized.

Of course, if a ranch is to be purchased and run as a hobby, the net return in terms of investment is likely to be of less concern to that individual.

## SELECTING AND BUYING A BEEF CATTLE FARM OR RANCH

The first and most important requisite for success in the cattle business is proper selecting and buying of the beef cattle farm or ranch. The fundamental considerations will be discussed in the sections that follow.

## EFFECT OF HERD SIZE ON COW-CALF RETURNS

Fig. 16–12. A full-time economic cow-calf unit should have 300 to 500 cows. (Courtesy, American Hereford Assn., Kansas City, MO)

*Size of herd is defined as the number of units (one cow, plus a suckling calf—if there is a calf; or one heifer two years or over) a farm or ranch will carry on a year-round basis. This includes the winter feed.*

Table 16–1 shows the effect of herd size on returns (cash receipts, less cash expenses).

#### TABLE 16–1
#### EFFECT OF HERD SIZE ON COW-CALF RETURNS[1]

| Size of Operation | Cash Receipts | Cash Expenses | Returns |
|---|---|---|---|
| | ($/cow) | ($/cow) | ($/cow) |
| All sizes ............... | 312.94 | 255.84 | 57.10 |
| 100 or fewer cows ..... | 312.95 | 274.11 | 41.84 |
| 100 to 499 cows ...... | 314.33 | 237.52 | 76.81 |
| 500 or more cows ..... | 299.90 | 200.83 | 99.07 |

[1]Source: *Economic Indicators of the Farm Sector, Cost of Production, 1987*, USDA, Economic Research Service, ECIFS 7–3, pp. 89–92. Data for 1987.

Table 16–1 shows that, in 1987, size of U.S. cow-calf operation affected per cow returns as follows:

1. All sizes of U.S. cow-calf operations averaged returns of $57.10 per cow cash receipts, less cash expenses.

2. The bigger the operation (500 or more cows vs 100 to 499) (a) the lower the cash receipts per cow, (b) the lower the cash expenses per cow, and (c) the greater the returns per cow. **Note well:** In computing the returns, consideration was not given to such items as taxes, insurance, and land charges; hence, it should not be construed as net profit per cow.

Based on the above information, plus observation, the author estimates that a farm or ranch with a carrying capacity of 300 to 500 head of brood cows (or equivalent; for example, two yearling stockers may be substituted for one cow) is necessary to be an economic unit for a cattle rancher who devotes full time to the operation and depends entirely upon cattle for income.

### EFFECT OF GEOGRAPHICAL AREA ON COW-CALF RETURNS

Geographical area may affect land use, cattle, people, and profit; hence, it may determine the kind of cattle operation in a particular area. In addition to soil, it takes moisture and reasonably warm weather to grow grass; and the longer the growing season, the more grass. Also, winter feeding is always more costly than summer grazing; hence, an area with mild winters has considerable advantage.

In the Corn Belt and Northern Plains, more than half the land is cropland, from which the crop residues, especially residues from corn, provide abundant winter feed. In the Great Plains and the Southwest, pasture and rangeland predominate. In the Northeast, Appalachian, Southeast, Delta States, and the Pacific Coast, forest and forest grazing constitute the major use of land.

Table 16–2 shows the effect of geographical area on cow-calf returns (cash receipts, less cash expenses).

#### TABLE 16–2
#### EFFECT OF U.S. GEOGRAPHICAL AREA ON COW-CALF RETURNS[1]

| Geographical Area | Cash Receipts | Cash Expenses | Returns |
|---|---|---|---|
| | ($/cow) | ($/cow) | ($/cow) |
| Great Plains ........... | 352.01 | 271.65 | 80.36 |
| North Central .......... | 346.88 | 265.65 | 81.23 |
| South ................. | 282.89 | 267.55 | 13.34 |
| West ................. | 320.93 | 223.65 | 97.28 |

[1]Source: *Economic Indicators of the Farm Sector, Cost of Production, 1987*, USDA, Economic Research Service, ECIFS 7–3, pp. 93–96. Data for 1987.

Table 16–2 shows that, in 1987, U.S. geographical area affected per cow returns as follows:

1. The Great Plains area had the highest cash receipts per cow of the four areas.

2. The West had by far the lowest cash expenses per cow of the four areas.

3. The West had, by a considerable margin, the highest returns per cow of the four areas. **Note well:** The returns should not be construed as net profit per cow, because, in computing it, consideration was not given to such items as taxes, insurance, and land charges.

### CAPITAL REQUIREMENTS

Those thinking of becoming cow-calf operators inevitably ask, "How much money will it take, and what can I make?" Operators need to take stock of their financial situation. How much money do they have available now, how much credit do they have, and how much debt are they willing to assume?

Table 16–3 shows the capital requirements in two different areas of Nebraska: (1) the Sandhills ranching area, and (2) the general farming area of eastern Nebraska. Thus, based on these figures, the capital requirement for a 500 cow outfit in the Nebraska Sandhills would be $2,000,000 (500 × $4,000); while a 500 cow crop and livestock farm in eastern Nebraska would require a capital investment of $1,527,500 (500 × $3,055).

#### TABLE 16–3
#### ESTIMATED CAPITAL REQUIREMENT ON A PER COW BASIS IN TWO AREAS OF NEBRASKA[1]

| | Sandhills Ranches in Nebraska | | Crop and Livestock Farms in Eastern Nebraska | |
|---|---|---|---|---|
| Cow ...................... | | $ 515 | | $ 515 |
| Heifer replacement ........... | (⅙) | 85 | | 85 |
| Bull (share) ................. | (1 to 25) | 50 | | 50 |
| Pastureland ................. | (15–16 acres) | 2,800 | (3–4 acres) | 2,000 |
| Hayland ................... | (1¼–1¾ acres) | 500 | (½ acre) | 325 |
| Buildings and equipment ...... | | 50 | | 80 |
| Total ...................... | | $4,000 | | $3,055 |

[1]Estimates submitted by NE Beef and Agricultural Economics Specialists: Guyer and Jose. Data for 1981–82.

It should be noted that ranches made up principally of deeded land sell for substantially more than those made up mainly of government leased land. A rule of thumb is that, on a per animal unit basis, all deeded land is worth twice as much as comparable land of which only one-half is deeded. Of course, the type of government lease, as well as the way the leased area lies in relationship to the deeded land, can make a tremendous difference in value. Yet, no matter what the history of the lease or what the old-timers say, a government lease can be cancelled quickly and without compensation.

## OTHER CONSIDERATIONS WHEN SELECTING THE LOCATION

Choosing the right location and the right farm or ranch are very important requisites for success. Nearness to friends and relatives is a major, but unmeasurable, factor in determining location, although it is less important than formerly with the development of more rapid transportation (improved highways and air travel) between areas. However, many nonsentimental factors should be considered in the selection of the particular location and the specific farm or ranch; among them, the following:

1. Area and climate.
2. Soil and topography.
3. Improvements
4. Wind direction, windbreaks, and natural shelters.
5. Service facilities, community, and markets.
6. Water, water rights.
7. Mineral rights.
8. Timber.
9. Easements, property lines.
10. Nearness to factories or cities, with the possible advantages of (a) off-farm employment, and (b) big-city attractions; and likely disadvantages of (a) air and noise pollution, and (b) limited expansion possibilities.
11. Expansion possibilities.

## WAYS OF ACQUIRING A FARM OR RANCH

Getting a suitable farm is a big problem, particularly for the beginners. They must compete for available farms with established farmers as well as with other beginners. Many established farmers need more land to enlarge their operations. Others move during the year, getting a better or more suitable farm. Some simply move to a new locality for personal reasons.

Farms or ranches may be acquired through gift or inheritance, by marriage, by renting or leasing, or by purchasing.

## FACILITIES

The investment in beef cattle buildings and equipment should be kept to a minimum. The farmstead should be neat and attractive, particularly where purebred cattle are sold for breeding purposes.

Buildings and yards should be located on a well-drained area; and plans should be made for the efficient feeding and management of the herd.

- **Shelter**—Buildings for beef cows need not be elaborate or expensive. Allow the herd to be outside during the grazing season and most other times—even in winter. Beef cattle naturally grow long, thick hair coats in the fall. Except where it is extremely cold and windy, the most shelter they need is a wooded area or a hill for a windbreak. Generally speaking, the cattle producer should give more attention to feed storage and to saving labor in feeding and manure handling than in the necessity of getting the cows inside. Also, it should be recognized that a combination of drafts, dampness, poor ventilation, and lack of sunlight creates hazards.

If a shed or barn is used, have it open to the south or east (away from the direction of the prevailing winds), with an adjoining lot to permit the cattle to stay indoors or run out at will. Mature brood cows require 59 to 60 sq ft of shelter, yearlings 35 to 40 sq ft, and weaned calves 25 to 30 sq ft. If the cattle are fed roughage under the shelter, more space will be needed—60 to 75 sq ft per cow, and a little more for young stock than the figures given.

Sheds more than 20 ft deep are preferred. The greater the depth, the warmer and drier the building.

Pole-type barns or sheds are excellent for beef cattle. They should be built on high ground, so that there will be good drainage away from them; they should have dirt floors; and they should be built high enough for convenience in removing manure—10 ft from the floor to the plate is sufficient.

Pole-type barns can be arranged so that roughage can be stored in the back of the shed, and so that, by use of movable racks, the cattle are permitted to eat their way back during the winter.

- **Fencing**—All corral fences should be built of heavy board material, pole-, or rail-type construction.

For holding cattle, 3 or 4 barbed wires are sufficient. Four wires should be used along roads and boundary lines, and 3 wires are sufficient for other areas. Woven wire is satisfactory, but more expensive. The life of the fence depends to a great extent on the life of the posts and the stability of the corner posts.

Electric fences are satisfactory for temporary fencing or rotation grazing.

- **Cattle-handling facilities**—Lack of adequate facilities for handling beef cattle prevents cattle producers from carrying out practices which would otherwise be routine and would increase their returns from the beef cattle operation. Time and money spent in planning and developing handling facilities for cattle will return dividends in terms of added profits and greater efficiency. Cattle-handling facilities should include the following:

1. A permanent corral or holding pen located near the main livestock buildings, with a working alley and several attached smaller catch pens to help in sorting cattle.
2. A head-gate, with a chute leading thereto.
3. A portable corral for use in pastures that are a considerable distance from headquarters, or a permanent corral constructed in such pastures.
4. A stationary or portable loading chute.
5. A scale, or ready access to a scale.
6. Cattle stocks where cattle are fitted for shows and sales, and for use in trimming feet.

# CONFINEMENT (DRYLOT) BEEF COWS

Confinement of cattle is not new. The pioneers planted trees and shrubs, such as osage orange, or built rail or stone fences, to hold their animals. Then in 1873, Joseph F. Glidden of DeKalb, Illinois, invented barbed wire, or the "devil's rope," as it was dubbed by those who considered the barbs inhumane or who opposed its use because it marked the beginning of the end of the open range and free grazing. But, for the most part, the wire that fenced the West confined cattle to large areas—pastures.

Today, on a very limited commercial basis, beef cows are being confined to small quarters—to drylots, all or part of the year. The practice will increase. More and more feeder calves will be produced under an intensive system.

Historically, as countries become more densely populated, land values and tax rates rise, necessitating that the highest and best use of land be made in order to pay taxes and yield a satisfactory return on investment. It's the same principle that spawns multistory buildings—as urban land prices mount, skyscrapers appear.

Although confinement production as a management technique in beef cow-calf production is relatively new in the United States, the concept is very old. To gain efficiency and reduce cost of production, American dairy, poultry, and swine producers have, of necessity, invested in intensive production units. Also, confinement cattle production has long been traditional in China, Japan, and Europe—in the more densely populated areas of the world, where land is scarce and high in price.

# ADVANTAGES AND DISADVANTAGES OF CONFINING BEEF COWS

Based on experiments and experiences, the following advantages and disadvantages are inherent in most confinement cow-calf operations in comparison with conventional pasture systems.

The **advantages:**

1. They require less investment in land per cow unit.
2. Cow numbers can be increased without obtaining more land.
3. They maximize feed production per acre of land, through utilizing harvested forages instead of grazing.
4. Hazards of drought and adverse weather conditions may be minimized through storing adequate feed reserves for future use.
5. Precise breeding programs can be designed and carried out more easily.
6. Control of estrus and use of artificial insemination programs become more practical.
7. Fewer services per conception are required in drylot than on pasture.
8. Individual performance records can be kept more easily.
9. Selection and culling are easier, because cattle can be observed more closely.
10. Plane of nutrition can be accurately known and controlled in accordance with the age and production needs of the cow; for example, nutrients can be increased during the critical 100 days, extending from 30 days before calving to 70 days after calving. After a calf is born, the cow's nutritional requirements rise approximately 50% over maintenance alone.
11. They make it easier to flush cows prior to the breeding season.
12. By-products and low-quality feeds may be utilized to advantage.
13. Partial environmental control may be achieved.
14. They permit close observation of cattle. Hence, illness and injuries are quickly detected and may be promptly treated.
15. Calving at various times of the year is more practical than in pasture handling.
16. They make for maximum flexibility in creep feeding. For example, steers and heifers can be fed separately.
17. Confinement produced calves wean more easily and go on finishing rations more readily.

The **disadvantages:**

1. They usually require a higher investment in buildings and equipment.
2. They require more labor.
3. Labor must be provided seven days per week. Hence, it is more confining to the operator, unless additional help is provided.
4. Disease problems, especially among calves, are generally more acute.
5. All feed must be harvested and moved to the feedlot, rather than being harvested directly by the cows.
6. An assured, adequate, and economic feed supply must be available. The capital tied up in stored feeds may be quite large.
7. More able management is required.
8. More knowledge of beef cow nutrition and ration formulation is needed.
9. Risks due to blizzards, hurricanes, tornadoes, floods, and lightening are increased due to the heavy concentration of animals.

# SEMICONFINEMENT (OR PARTIAL CONFINEMENT) COW HERDS

A semiconfinement (or partial confinement) operation is one which takes advantage of grazing during part of the year, such as winter grazing of corn or sorghum stalks or seasonal grazing of pastures. In addition to providing low-cost feed and allowing the animals to do their own harvesting, breeding may be timed so that the calves will be dropped on clean pasture as a means of (1) preventing calf scours, and (2) stimulating milk flow.

# MANAGEMENT PRACTICES

Beef cattle management practices vary widely between both areas and individual farmers and ranchers. In a general sort of way, however, the principles of good management of farm and range herds, and of purebred and commercial herds, are much alike. The main differences arise from the sheer size of the range cattle enterprise, which means that things must be done in a big way. Without attempting to cover all management practices, some facts relative to—and methods of accomplishing—some simple beef cattle management practices follow.

## IDENTIFICATION

All calves should be identified as soon after birth as possible, and not later than three days of age. A combination of flexible plastic ear tags and tattoo numbers is recommended, thereby securing both ease of reading and permanency.

On the western range, branding is the primary method of establishing ownership and/or age.

Fig. 16–13. Branding range calves with a hot iron. (Courtesy, Iowa Beef Processors, Inc., Dakota City, NE)

## DEHORNING

Dehorning is an economic necessity, because horned calves usually bring lower prices. In addition, dehorned and naturally polled animals do less damage to facilities and other animals than cattle that have horns.

All naturally horned animals should be dehorned, preferably before they are two months old to minimize the shock effects of the operation. At that time, the blood vessels in the horn area are very small, which means less blood loss and minimum shock.

Dehorning may be accomplished by any of the following means:

1. **Mechanical.** A tube, spoon, or Barnes-type dehorner.
2. **Electrical.** Electrically heated irons have been used with good results.
3. **Chemical.** Liquids, pastes, or caustic sticks can be used for dehorning, but all of them should be used with caution and according to directions.

## CASTRATING

Castration is recommended for all bull calves destined to be sold as feeders or finished in the feedlot. Bull calves and staggy-looking steer calves will not be accepted in many feeder calf sales. If sold as feeder animals, they usually bring a reduced price.

Castration time will vary according to the method employed and management program, and it will be different for a commercial than for a purebred operation. Some producers use elastrator bands when calves are only a few days old. Others use a knife or burdizzo (clamp) when the calves are three to four months old. Still others wait until weaning time to castrate. Bull calves will weigh more at weaning than steer calves; however, younger calves are easier to restrain for castration and suffer less shock therefrom than older animals.

## WEANING

Calves should be weaned when they are seven to eight months old. Weaning at this age fits well with most performance-testing programs. Also, calves will be about the right age and weight for fall feeder calf sales.

Weaning earlier than seven or eight months may be necessary in years when pastures are short or when calves are from first-calf heifers.

The best way to wean is to remove the calves from their dams and keep them out of sight of each other. Cows and calves should never be turned together once the separation has been made. Such a practice will only prolong the weaning process, and it may also cause digestive disorders in the calf. Provide calves with plenty of water, free-choice hay, and 3 to 4 lb of grain per head per day. If calves were creep fed, continue their rations during the weaning period.

During the weaning process, calves should be confined to a small area to cut down on walking and shrinkage. In bad weather, protection should be provided from cold wind and rain; they should have access to a shed, wooded area, gorge, or other shelter.

With higher-milking strains of beef cattle, when drying up cows, beef producers will have the same concerns as dairy producers—that of avoiding "spoiled" udders. To alleviate this problem, the following procedure is recommended:

1. Do not feed milk-stimulating feeds at weaning time. Either put the cows on poorer pastures or feed a nonlegume forage.

2. Let back pressure in the udder build up. Examine the udder at intervals, *but do not milk it out.* If the bag fills up and gets tight, rub an oil preparation (such as camphorated oil or a mixture or lard and spirits of camphor) on it, *but do not milk it out.* At the end of five to seven days, when the bag is soft and flabby, what little secretion remains (perhaps not more than half a cup) may be milked out if so desired.

## PRECONDITIONING

Fig. 16–14. Preconditioning, along with production testing, will be the trademark of the producer of reputation feeder calves of the future. (Courtesy, Ralston Purina Co., St. Louis, MO)

*Preconditioning is a way of preparing the calf to withstand the stress and rigors of leaving its mother, learning to eat new kinds of feed, and shipping from the farm or ranch to the feedlot.* To the cow-calf producer, it is a program of management, nutrition, and immunization.

The term *preconditioning* is new, but the concept has long been recommended. It consists of administering generous amounts of TLC (tender loving care), along with immunological practices and treatment for parasites.

Changed environment; excitement of sorting, loading, and shipping; long periods without feed and/or water; movement through one or more assembly points; change of feed; and exposure to disease—all add up to *fatigue, stress, shrink,* and *lowered disease resistance.*

Preconditioning is the answer. The steps used in preconditioning may, and should, vary somewhat among areas, farms, and ranches. But generally it involves the following:

1. Handling calves quietly, with a minimum of excitement.

2. Dehorning and castrating well ahead of weaning time—about two months of age is best.

3. Weaning calves 30 days ahead of shipment.

4. Starting calves on a ration that will be similar to the starting ration that they will receive when they arrive in the feedlot.

5. Vaccinating in keeping with the advice of your veterinarian.

6. Treating for parasites, internal and/or external, before weaning.

7. Reducing time, stress, and exposure to infection from the farm or ranch to the feedlot.

8. Providing a preconditioning certificate, which is a record of all husbandry, nutritional and medical histories provided by the seller of feeder cattle to the buyer.

## MARKET ALTERNATIVES

The following marketing alternatives are available to most cow-calf producers:

- **Selling feeder calves**—Spring dropped calves may be weaned and sold as feeder calves in the fall when they are 7 to 8 months old. Steer calves normally weigh 400 to 600 lb. Usually heifer calves weigh about 5% less than steer calves.

- **Selling yearling feeders**—Some producers have always grown calves beyond weaning weights, with the prevalence of the practice varied by years according to feed. This practice is most common in southwestern United States.

- **Finishing cattle**—This alternative is open to all cow-calf producers, either in their own facilities if they have them, or in custom feedlots. Calves may go directly into the feedlot at weaning to be marketed as finished cattle at 12 to 16 months of age; or they may be grown out as stockers, then finished out at slightly older ages and heavier weights.

- **Replacement heifers**—The best source of replacement heifers is from the breeder's own herd, regardless of whether the operation is purebred or commercial. This is so because more is known about them—their age, immunization history, health status, and performance—than replacements from any other source. Thus, at weaning time, the top-performing (heaviest) heifer calves should be retained—keeping 50% more than will actually be needed.

Fig. 16–15. Replacement Shorthorn heifers. (Courtesy, Maple Leaf Mills, Ltd., Ontario, Canada)

There is one exception where saving replacements would not be practical—that's a crossbreeding program where specialized $F_1$ females are mated to a terminal sire breed. In such a program, all calves are marketed at weaning after finishing. Thus, it is necessary to purchase $F_1$ heifers from an outside source.

# QUESTIONS FOR STUDY AND DISCUSSION

1. Describe the cow-calf system.

2. Why does the importance of the cow-calf system in the agriculture of the nation rest chiefly upon the conversion of coarse forage and grass into palatable and nutritious food for human consumption?

3. What are the pros, what are the cons, and what is your choice between (a) a farm or ranch herd, (b) purebred or commercial cattle, (c) selling weaners or stockers, and (d) dual-purpose production?

4. How do you account for the fact that probably fewer than 4.0% of U.S. cattle are registered purebreds?

5. What are the alternative enterprises from which a farm or ranch owner may choose (a) in the West and Southwest, (b) in the Midwest, (c) in the Northeast, and (d) in the Southeast?

6. Basically, farmers and ranchers have four resources at their disposal with which to change cow numbers—land, labor, capital, and managerial skills. In the years ahead, how will the most successful cattle producers put these together to maximize profits and increase cow numbers? How do the cow-calf resource requirements compare to the alternate enterprises?

7. What changes do you foresee in cattle raising in the years ahead from the standpoints of (a) beef consumption and beef price trends, (b) beef eaters and cattle numbers, and (c) increased production per cow and increased carcass weight?

8. Discuss trends in cow-calf operations from the standpoints of (a) cattle numbers, (b) competition from poultry and pork, and (c) export market for beef.

9. During the period of decline in calf numbers shown in Fig. 16–5, what factors resulted in the lowest decrease in the following three areas: the West North Central Area, the North Central Area, and the Southwest Area?

10. How do you account for the fact that, in 1990, 97.5% of the beef cattle operators had fewer than 200 cows? Is the preponderance of small beef cow herds good or bad?

11. In 1990, each cow produced about 200 more pounds of beef than in 1970—20 years earlier. How was this achieved?

12. In the future, several potential changes in productivity per cow may be effected which will directly improve the supply of beef obtainable from a given inventory of cows. List and discuss five of these potential changes.

13. In the decade ahead, how will beef cattle numbers be affected by (a) the decline in per capita beef consumption, and (b) increased human population.

14. In 1990, 1.5 lb of pork or 3.5 lb of poultry could be bought for the price of 1 lb of beef. How are these retail prices apt to influence the future price of live beef cattle?

15. How will each of the following factors influence the number of beef cows:
    a. Shifts in crop production.
    b. Shifts in livestock enterprises.
    c. Forage production, harvesting, and utilization.
    d. Public policies.
    e. Popularity of cattle raising.
    f. Shifts in land use.
    g. Elderly and part-time farmers.

16. Why is the ultimate success of a cattle operation so dependent upon the selection of the proper farm or ranch?

17. What are the primary differences between a full-time and a part-time cattle farm or ranch?

18. Why are the vast majority of U.S. beef cattle herds small—with fewer than 200 cows?

19. What factors determine whether a beef operation shall be (a) strictly cow-calf, with all calves except replacements marketed at weaning time, (b) a combination cow-calf and stocker operation, (c) a cow-calf operation that finished its home-raised calves out to slaughter, or (d) stocker only?

20. What factors are likely to favor an irrigated cattle farm or ranch over a dryland operation?

21. Discuss the function and method of operation of (a) the Bureau of Land Management, and (b) the U.S. Forest Service.

22. What is a *Hobby Farm?*

23. Discuss the effect of herd size on returns and expenses as shown in Table 16–1.

24. Discuss the effect of geographical area on returns and expenses as shown in Table 16–2.

25. Discuss the capital requirements of a farm or ranch as shown in Table 16–3.

26. What are the advantages and disadvantages of acquiring a farm by (a) gift or inheritance, (b) marriage, (c) rent or lease, or (d) purchase?

27. When buying a farm or ranch, what consideration should be given to facilities?

28. What factors favor increased confinement cow-calf production in the future?

29. List the advantages of confinement cow-calf production.

30. List the disadvantages of confinement cow-calf production.

31. Discuss the advantages and disadvantages of (a) a semiconfinement vs (b) a year-round (total) confinement cow-calf operation.

32. Outline a management program for a cow-calf operation.

33. How would you go about weaning calves, from the standpoint of both the cows and the calves?

34. Who benefits the most from a preconditioning program, the cow-calf producer or the cattle feeder?

35. Detail a preconditioning program.

36. What marketing alternative—selling feeder calves, selling yearling feeders, or finishing cattle—should a cow-calf producer select?

# SELECTED REFERENCES

| Title of Publication | Author(s) | Publisher |
|---|---|---|
| *Beef Cattle*, Seventh Edition | A. L. Neumann | John Wiley & Sons, Inc., New York, NY, 1977 |
| *Beef Cattle Production* | K. A. Wagnon<br>R. Albaugh<br>G. H. Hart | The Macmillan Company, New York, NY, 1960 |
| *Beef Cattle Production* | J. F. Lasley | Prentice-Hall, Inc., Englewood Cliffs, NJ, 1981 |
| *Beef Cattle Science*, Sixth Edition | M. E. Ensminger | The Interstate Printers & Publishers, Inc., Danville, IL, 1987 |
| *Beef Cattle Science Handbook* | Edited by<br>M. E. Ensminger | Agriservices Foundation, Clovis, CA, pub. annually since 1964 |
| *Beef Production and Management*, Second Edition | G. L. Minish<br>D. G. Fox | Reston Publishing Co., Reston, VA, 1982 |
| *Beef Production and the Beef Industry* | R. E. Taylor | Burgess Publishing Co., Minneapolis, MN, 1984 |
| *Beef Production in the South*, Modified Edition | S. H. Fowler | The Interstate Printers & Publishers, Inc., Danville, IL, 1979 |
| *Cattle Raising in the United States* | R. N. Van Arsdall<br>M. D. Skold | Economic Research Service, USDA, Washington, DC, 1973 |
| *Commercial Beef Cattle Production*, Second Edition | Edited by<br>C. C. O'Mary<br>I. A. Dyer | Lea & Febiger, Philadelphia, PA, 1978 |
| *Cowboy Arithmetic: Cattle as an Investment*, Third Edition | H. L. Oppenheimer | The Interstate Printers & Publishers, Inc., Danville, IL, 1971 |
| *Cowboy Economics: Rural Land as an Investment*, Third Edition | H. L. Oppenheimer | The Interstate Printers & Publishers, Inc., Danville, IL, 1976 |
| *Cowboy Litigation, Cattle and the Income Tax* | H. L. Oppenheimer<br>J. D. Keast | The Interstate Printers & Publishers, Inc., Danville, IL, 1972 |
| *Do It Right the First Time* | J. D. Keast *et al.* | Doane Agricultural Service, Inc., St. Louis, MO, 1973 |
| *Land Speculation: An Evaluation and Analysis* | H. L. Oppenheimer | The Interstate Printers & Publishers, Inc., Danville, IL, 1972 |
| *Stockman's Handbook, The*, Sixth Edition | M. E. Ensminger | The Interstate Printers & Publishers, Inc., Danville, IL, 1983 |

A 100% calf crop should be the goal of every cow-calf operation. (Courtesy, American Breeders Service, DeForest, WI)

Cowboy on a motorcycle! This shows a cow-calf operation on the George Landers Farm of Dadeville, MO. (Courtesy, USDA)

Yearling stocker cattle. (Courtesy, Pennsylvania Millers and Feed Dealers Assn., Ephrata, PA)

# STOCKER
# (FEEDER) CATTLE

## Chapter 17

Stocker cattle and stocker cattle programs have changed with the passing of time. Until the early 1900s, stockers involved growing purchased or homegrown calves or yearlings on grass and hay until they were 3 to 4 years of age. As calf weaning weights increased and finished slaughter cattle weights decreased, the amount of time and gain required to grow calves from weaning until the beginning of the finishing period was substantially shortened. The stocker cattle industry became a calf-yearling industry, usually starting with 300- to 500-lb calves and ending with a yearling sold to a feeder at 600 to 700 lb.

The development of large feedlots and year-round feeding increased the demand for feeders ready to go on high-concentrate rations. The main reason that the larger feedlots like to purchase feeders ready to go on high-energy rations is that roughage is used more efficiently in growing cattle, whereas it is usually an expensive item to use in large feedlots.

Today, the stocker stage is changing again, as a result of forces working in opposite directions, with one force favoring lengthening of the stocker stage and the other favoring shortening it:

1. Scarce and high-priced grain favor more roughage feeding and less grain feeding, resulting in carrying stockers to older ages and heavier weights, followed by a shorter feedlot period.

2. Heavier milking cows and heavier weaning weights, coupled with high-priced land, favor shortening the stocker stage, or even eliminating it, as 600-lb or heavier, weaning weights are achieved.

In the future, both types of stocker operations will prevail, with the choice determined primarily by the price of grain and the weaning weight of the calves. Heavy weaned calves will likely go directly into the feedlot for slaughter. Calves with light to average weaning weights will likely be carried as stockers to 700- to 800-lb weights, thereby shortening the feedlot period and lessening grain feeding.

Thus, the growing of calves from weaning until placing on finishing rations is not new. However, in recent years some new "wrinkles" have been added to the methods of conducting it. Today, stockers are grown according to two systems: (1) Calves or light yearlings are either roughed through the winter, followed by grazing, or grazed only, then sold as feeders in late summer and fall; or (2) calves or yearlings are fed harvested roughage and grain in drylot, and then transferred to another location for finishing. Also, some new terms have evolved. Definitions of presently used stocker (feeder) terms follow.

• **Stockers** *are calves and yearlings, both steers and heifers,*

Fig. 17–1. Stocker steers on a western range, with stacks of hay for winter feeding. (Courtesy, U.S. Forest Service)

*that are intended for eventual finishing and slaughtering and which are being fed and cared for in such a manner that growth rather than finishing will be realized. They are generally younger and thinner than feeder cattle.*

• **Feeders** *are calves and yearlings, both steers and heifers, carrying more weight and/or finish than stockers, which are ready to be placed on high-energy rations for finishing and slaughtering.*

• **Replacement heifers** *are the top end of the heifer calves selected to replace the older cows that are culled from the herd.*

• **Preconditioning** *refers to preparing the calf to withstand the stress and rigors of leaving its mother, learning to eat new kinds of feeds, and shipment from the farm or ranch to the feedlot or stocker grower.*

• **Backgrounding** *is an old practice with a new emphasis and a new name. Actually,* backgrounding *and* the stocker stage *are one and the same. Both refer to that period in the life of a calf from weaning to around an 800-lb weight, when they are ready to go on a high-energy finishing ration.* However, the term *backgrounding,* which was ushered in with the development of large feedlots, indicates a shift in emphasis. The term *stocker stage* connotes emphasis on marketing roughages through thin cattle, whereas *backgrounding* connotes emphasis on growing out feeder calves ready to go on a high-energy finishing ration. Backgrounding may be done on pasture or in the drylot, or some combination of both. At its best, the animals should be in good health, bunk broke, and ready to go on full feed.

From the above, it may be concluded that in the variable period of a calf's life between weaning and finishing, it is usually classed as either a stocker, a feeder, or a replacement heifer. Prior to weaning, calves may or may not be preconditioned.

The dividing line between stockers and feeders is not always as clear cut as the above definitions would indicate. That is, not all thin cattle are suitable for stockers. For example, very large yearlings and most heifers are usually sold as feeders, to be placed on high-energy feeds. Also, "Okie-" type cattle are usually backgrounded for 50 to 60 days, then placed on a finishing ration.

## TYPES OF STOCKER PROGRAMS

Sometimes the stocker operation is the only cattle enterprise on the farm or ranch. More frequently, however, it is conducted in conjunction with a cow-calf operation or it precedes the finishing program.

When the stocker enterprise is the only cattle enterprise on a farm or ranch, it is usually conducted in one of the following ways:

1. Calves or light yearlings are bought in the fall to be wintered on high-roughage rations in drylot and sold in the spring to buyers either (a) to go on grass for the summer, or (b) to go on a drylot finishing program.

2. Lightweight calves are bought in the fall to be wintered on roughage rations, then, under the same ownership, grazed throughout the following pasture season and sold in the fall. Under this plan, usually lighter weight calves are acquired and they are wintered at a lower rate of gain than in plan 1.

3. In Kansas, Oklahoma, and Texas, calves or light yearlings are bought in the fall and grazed on winter small grains, chiefly wheat. Good wheat pastures will produce very accept-

Fig. 17-2. Stocker calves being wintered on a high-roughage ration. (Courtesy, *Livestock Breeder Journal*, Macon, GA)

able stocker gains. The main disadvantage to the program is that, due to weather conditions, winter wheat pasture cannot always be counted upon. When it fails, the stockers must either be sold or fed a higher cost roughage.

4. In the southeastern states, which is primarily a cow-calf area, winter oats and fescue are used extensively in stocker programs. This area is turning to stocker programs in order to utilize winter pasture profitably, and to satisfy the demand of cattle feedlots for 600- to 800-lb feeder steers.

There is a trend for more and more calves (not yearlings) to be handled according to plan 1—that is, bought in the fall, wintered on roughage, and sold directly into a finishing program. This trend will be accelerated because of heavier calves being weaned in the fall, and because it is more profitable either to use presently available pasture areas (1) for brood cows to produce more calves, or (2) for crop production.

Fig. 17-3. Yearling stockers on pasture. (Courtesy, USDA)

The most common type of operation is a combination stocker-feeder program, typical of the Corn Belt and the irrigated sections of the West, where high yielding corn and sorghum crops are produced for silage. In these areas, cattle feeders usually purchase steer calves or light yearlings in the fall or late winter; fall-graze stalk fields and small grain stubble where available; move into the drylot for the winter and feed corn or sorghum silage, supplemented with a legume hay or protein supplement; then finish on a high-energy ration either in the drylot or on pasture and sell for slaughter in the summer or fall.

An increasing number of feeders are grown on contract for, and delivered to, a feedlot for finishing. This trend has been prompted by the competition between feedlots. It is their way of assuring a continuous supply of feeders of the desired weights and quality. As a further inducement, many of the feedlots finance the grower (backgrounding) operation.

## ADVANTAGES AND DISADVANTAGES OF A STOCKER PROGRAM

A stocker enterprise has both advantages and disadvantages in comparison with a cow-calf or a cattle finishing operation. These should always be weighed and balanced, especially where there is a choice.

A stocker program has the following **advantages** over other types of cattle programs:

1. **Flexibility.** A stocker operation is more flexible than a cow-calf operation from the standpoint of adjusting to feed supplies and cost, labor, and economic outlook. The number of stockers purchased each year may be altered accordingly.

2. **Efficient gains.** Stocker operators have the cattle when they make the most efficient gains.

3. **Low labor requirement.** Stocker cattle have a lower labor requirement in comparison with a cow-calf operation conducted on the same amount of land.

4. **Distribution of labor.** The peak labor requirement in wintering stockers in the drylot is completed ahead of spring and summer farm work.

5. **Quick returns.** Where a stocker program is limited either to wintering or pasturing only, returns come quickly, within 4 to 6 months. In some cases, this quick turnover permits handling 2 to 3 droves of stockers per year.

6. **Adapted to areas lacking accessibility to fat-cattle markets or sources of grain.** A stocker operation is better adapted than a cattle-finishing operation to areas lacking accessibility to slaughter cattle markets or source of grain. Of course, such areas are also well suited to cow-calf operations; hence, a choice must be made.

7. **Utilize roughage and salvage feeds.** Stockers are adapted to the use of roughage and salvage feeds.

8. **Investment in buildings and equipment may be less.** If the stocker program is limited to grazing winter or summer pastures, a minimum of buildings and equipment is required.

9. **Contract basis requires little capital.** If a grower contract is arranged with a feedlot, as is sometimes possible, little capital is required.

There are also **disadvantages** to a stocker program including the following:

1. **High risk.** It is a high-risk venture because of (a) the seasonal and yearly price fluctuations in feeder cattle, and

(b) the fact that total gains are not large in proportion to the weight purchased. Also, stocker operators who rely on winter wheat or oat pastures or summer grazing may have to purchase cattle when the price is high and sell them when the market is flooded with similar cattle. Moreover, lack of rain may make for small gains and high costs.

2. **Cost of gain may have to offset negative margins.** Cost of gain must be kept down to offset negative margins that may prevail.

3. **High buying and selling skills are required.** Buying and selling skills are extremely important because (a) there is no established market such as exists with slaughter cattle, and (b) the original weight purchased is a high percentage of the weight sold. It follows that any mistake made in buying and selling, such as mistakes in judging the quality and health of the stockers, has a greater influence on profits or losses than in a finishing program. Moreover, the entire livestock inventory is bought and sold at least once per year.

4. **Buying, selling, and shrink costs must be absorbed by a limited amount of gain.** Shrink on both ends, along with buying and selling costs, can wipe out any economical, but relatively small, gains that can be made.

5. **High land, labor, and interest costs mitigate against stocker programs.** Because of the relatively small gains made by stockers, high land, labor, and interest costs mitigate against such programs. For this reason, growing calves at rates of less than one pound per head daily becomes increasingly difficult to justify.

6. **Disease can make for heavy losses.** The weaning calf is at the most susceptible stage in life to contagious diseases. Also, the limited weight gains in growing operations leave little opportunity to recover severe losses.

7. **High transportation costs.** Often the transportation cost is high because of a long distance from the source of supply and/or feedlots.

## COMPENSATORY GROWTH

Compensatory growth is making up for a bad start in life. It is common practice for stocker cattle to be roughed through the winter as cheaply as possible, with limited daily gains. Then, in the spring, the animals are turned to lush spring pasture or put in a feedlot on a high-energy ration. Animals so managed exhibit the phenomenon of *compensatory growth;* that is, on the high-energy ration they gain faster and more efficiently than similar cattle which were fed more liberally during the wintering period. Feedlot operators were quick to sense this situation, and to take advantage of it. This is the chief reason for the popularity of Okie-type cattle. Usually, they are animals whose growth has been held back to less than their genetic potential. When fed more liberally, they exhibit a surge in growth rate and feed efficiency. Large compensatory growth usually indicates that someone (the stocker operator) has lost money while someone else (the feeder) has made money. It is noteworthy that Holsteins and the larger exotics should never be handled so as to exhibit compensatory gains. If they're held back in the winter, they're too heavy when they finish.

## MARKET CLASSES AND GRADES OF STOCKERS

Stocker cattle are of many kinds, displaying a wide range of combinations of the various characteristics such as breeding,

sex, age, weight, size, conformation, and consolidation. Fortunately, there is a market for each kind, with variation in prices reflecting both the supply of and the demand for each kind and the degree of suitability for the intended purpose. The use of uniform terms and descriptions of all classes and grades of cattle by all members of the beef team—cow-calf producers, feeders, packers, and buyers and sellers—throughout the country has contributed much to the orderly marketing of all cattle, including stockers. The market classes and grades of stocker cattle are fully covered in Chapter 19 of this book; hence, the reader is referred thereto.

## SOURCES OF STOCKERS

Sources of stockers vary from area to area. Nationwide, most stockers are secured through auctions and direct purchase from cow-calf producers. In the Corn Belt, however, most stocker cattle are secured through public stockyards, followed by local dealers, producers, and auction markets.

Stockers may be purchased to advantage in an auction, but there are problems. They have already been stressed by being loaded, hauled, unloaded, and handled; in most cases, the buyer knows little or nothing of the origin of the cattle; and they may either have come straight off native grass (which is good) or they may be making their third sale in the last 10 days (which is very bad).

The best-doing stockers are usually those coming directly from cow-calf producers—preconditioned, and without passing through a market facility.

## NATIVE (HOMEGROWN OR LOCAL) OR WESTERN STOCKERS

Fig. 17-4. Native stockers of mixed breeding. (Courtesy, Elanco Products Company, Indianapolis, IN)

Native stockers are those coming from the farms of the Corn Belt, Great Lakes, and Southeast, whereas western stockers are the branded animals coming from the western ranges. For the most part, native cattle come from comparatively small farm herds. Calves produced in these herds are often fed out on the farms where they are produced. Others are offered for sale locally. There are both advantages and disadvantages to buying native stocker cattle.

Native stockers usually have the following **advantages** in comparison with western stockers:

1. Fewer disease problems.
2. Lower freight and buying costs.
3. Lower shrink.
4. They are acclimated.
5. They are accustomed to the feed.

Native stockers usually have the following **disadvantages** in comparison with western stockers:

1. They lack uniformity.
2. They may be of lower quality.
3. They are fleshier.
4. Small lots must be combined.
5. They may not be available in large enough numbers.

## METHOD OF BUYING STOCKERS

Stocker cattle may be purchased by stocker operators themselves, or by salaried buyers, order buyers, commission firms, or cattle dealers. Although owners or managers of stocker operations should be knowledgeable enough to buy stocker cattle, if they are large operators, they are generally too busy to do so if they are doing a proper job of running their outfits. Accordingly, the trend is to shift the buying responsibility to specialists—primarily order buyers.

## TRANSPORTATION AND SHRINK OF STOCKER CATTLE

Improper handling of stocker cattle immediately prior to and during shipment may result in (1) excess shrinkage; (2) high death, bruise, and crippling losses; (3) disappointing sales; and (4) dissatisfied buyers. Unfortunately, many cow-calf operators who do a superb job of producing stockers dissipate all the good things that have gone before by doing a poor job of preparing and shipping.

## PRECONDITIONING; HANDLING NEWLY ARRIVED CATTLE

These important subjects are fully covered under similar headings in Chapters 16 and 18, respectively, of this book; hence, the reader is referred thereto.

## FEEDING STOCKERS

For a stocker operation to be profitable, the grower must be ever aware of the following reasons back of it, then feed stockers accordingly: (1) to provide a supply of the kind of cattle desired by finishing lots at the time needed; (2) to utilize the roughages and other low-cost feeds; and (3) to "cheapen down" the cattle.

Because of the very nature of the operation, the successful feeding of the stockers requires the maximum of economy consistent with normal growth and development. This necessitates cheap feed—either pasture or range grazing or such cheap harvested roughage as hay, straw, fodder, and silage. In general, the winter feeds for stockers consist of the less desirable and less marketable roughages. It is important, therefore, that the high-roughage rations of young stockers be properly supplemented from the standpoints of proteins, minerals, and vitamins.

Of course, too small gains may be unprofitable to the grower. Besides, young animals can be stunted. To make maximum growth without fattening—just to maintain condition—calves of the British breeds and crossbreds should gain 1.25 lb daily, and yearlings should gain 0.9 lb daily.

Table 15–3 contains some recommended rations for stocker cattle. Variations can and should be made in the rations used. The grower should give consideration to (1) the supply of homegrown feeds, (2) the availability and price of purchased feeds, (3) the class and age of cattle, (4) the health and condition of animals, and (5) the kind of feeder cattle in demand by feedlots.

In using Table 15–3 as a guide, it is to be recognized that feeds of similar nutritive properties can and should be interchanged as price relationships warrant. Thus, (1) the cereal grains may consist of corn, barley, wheat, oats, and/or sorghum; (2) the protein supplement may consist of soybean, cottonseed, canola, peanut, linseed, safflower, and/or sunflower meal; (3) the roughage may include many varieties of hays and silages; and (4) a vast array of by-product feeds may be utilized.

The following points are pertinent to feeding stocker cattle and should be kept in mind:

- **Recommended nutrient allowances**—Where grower rations are formulated on the basis of percentage of nutrients in the ration, the following allowances are recommended:

**Protein**
For up to 1.5 lb daily gain . . . . . . . . . . . . . . . . . 10.0%
For 1.5 lb daily gain or more . . . . . . . . . . . . . . 10.5%

**Calcium and Phosphorus**
For up to 500 lb liveweight . . . . . . . . . . . . . . . . 0.30%
For over 500 lb liveweight . . . . . . . . . . . . . . . . 0.25%

**Vitamin A**
Air-dry feed (10% moisture) . . 800 to 1,000 IU per lb
. . . . 10,000 IU daily per head

- **Level of wintering**—The level of wintering stockers affects the gains in the next stage. Thus, calves gaining the most during the winter make the least gains on pasture the following summer.

Calves wintered to gain 1.0 lb daily make satisfactory summer pasture gains. This level is recommended for calves to be grazed season-long the following summer, provided the same ownership is retained all the way through. A daily gain of 1 to 2 lb during the winter is usually desirable if calves (1) are to be sold in the spring, (2) will be on full feed 2 to 3 months after going to grass, (3) will be receiving a limited feed of grain when on grass, or (4) are replacement heifers that are to be bred at 13 to 15 months of age.

Since yearlings are not growing as rapidly as calves, they may be fed for smaller gains than calves, and yet show comparable condition. Thus, for maximum growth without fattening (for just holding their condition) calves should gain approximately 1.25 lb daily, whereas yearlings need to gain only 0.9 lb daily.

## WINTER PASTURES

Fig. 17–5. Block supplementation on winter pasture. (Courtesy, Elanco Products Company, Indianapolis, IN)

Wherever possible, stocker calf operations are planned around a winter pasture program. Weanling calves or lightweight, thin yearlings are purchased in the fall. In some cases, homegrown calves are retained and developed under this system for sale as yearlings. As would be expected, winter pasturing of stockers is largely limited to the southern part of the United States, with the kind of pasture varying from area to area.

• **Winter wheat pastures**—Winter wheat pastures are widely used for stocker cattle in Kansas, Oklahoma, and Texas. When such pastures are good, cattle make very acceptable gains on them. However, wet weather or droughts make winter wheat pasture unreliable, with the result that it is important that there be flexibility in the stocker program, both in numbers and season of use.

• **Other cool-season pastures**—In the Southern states, extensive use is made of oats, rye, ryegrass, vetch, and fescue—a perennial grass that remains green throughout the winter. This area is turning more and more to winter grazing, as a means of making profitable year-round use of their land and labor and providing 600- to 800-lb feeder cattle in greatest demand by feedlots.

## STOCKERS AND GROWER CONTRACTS

Hand in hand with the development of big feedlots and year-round feeding came the need for an assured supply of feeder cattle of the desired kind and on a continuous basis. To meet this need, more and more feedlots have turned to contractual agreements with stocker growers, with numerous kinds of contracts. Usually, the cattle are owned by the feedlot, most of which are large and in a stronger financial position than the majority of stocker growers. The two most common kinds of contracts are based on either (1) a fixed cost for the gain, or (2) an agreed feed cost plus an extra charge for labor and lot rental. Usually, there is provision for adjusting for death loss. Such contracts should always be in writing, with all provisions, including weighing conditions, spelled out.

Although the use of stocker and grower contracts has increased in recent years, the concept is not new. Many of the Kansas bluestem pasture owners have long grown out yearlings owned by Iowa and other Corn Belt feeders.

Today, many corn farmers in the fertile irrigated area near Greeley, Colorado, produce corn silage and feed stocker cattle on a contract basis for one of several large feedlots in the vicinity. Stocker cattle are also being grown under contract on the wheat pastures of Kansas, Oklahoma, and Texas; on hay and other roughages in the irrigated valleys of the West; and on sorghum silage and stalk fields throughout the Southwest.

Fig. 17–6. Stocker cattle on lush seeded pasture. (Courtesy, *Livestock Breeders Journal*, Inc., Macon, GA)

# QUESTIONS FOR STUDY AND DISCUSSION

1. How and why have stocker programs changed through the years?

2. Define the following terms: (a) stockers; (b) feeders; (c) replacement heifers; (d) preconditioning; and (e) backgrounding.

3. What are the common types of stocker programs, and what are the characteristics of each?

4. What are the advantages and disadvantages of a stocker program?

5. Do you feel that, for the most part, the stocker stage will be lengthened or shortened? Justify your answer.

6. What is compensatory growth?

7. For farmers or ranchers who produce their own stocker calves, and who finish them out on a custom basis, is compensatory growth good or bad? Justify your answer.

8. What are the usual sources of stockers?

9. What are the advantages and disadvantages of (a) native stockers, and (b) western stockers?

10. What are the alternative methods of buying stockers? What method of buying stockers do you prefer? Justify your choice.

11. Discuss rations and rate of gain of (a) stocker calves, and (b) stocker yearlings.

12. What are the primary differences between stocker and finishing rations?

13. Tell how the level of wintering stockers affects the next stage.

14. Discuss the favorable and unfavorable factors of winter wheat pastures.

15. Discuss the favorable and unfavorable factors of the common cool-season pastures of the Southeast.

16. What provisions should be incorporated in a stocker and grower contract?

# SELECTED REFERENCES

| Title of Publication | Author(s) | Publisher |
| --- | --- | --- |
| *Beef Cattle*, Seventh Edition | A. L. Neumann | John Wiley & Sons, Inc., New York, NY, 1977 |
| *Beef Cattle Production* | J. F. Lasley | Prentice-Hall, Inc., Englewood Cliffs, NJ, 1981 |
| *Beef Cattle Science*, Sixth Edition | M. E. Ensminger | The Interstate Publishers, Danville, IL, 1987 |

Yearling stockers on grass in Virginia.

Stockers at water hole in Texas. (Courtesy, *Livestock Weekly*, San Angelo, TX)

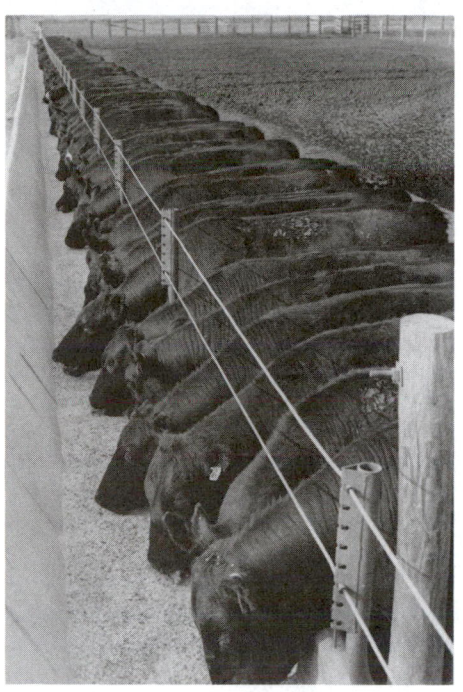

Finishing cattle the modern way. (Courtesy, American Angus Assn., St. Joseph, MO)

# FINISHING CATTLE: FEEDLOT/PASTURE

## Chapter 18

*The finishing of cattle is what the name implies, the laying on of fat.* The ultimate aim of the finishing process is to produce beef that will best answer the requirements and desires of the consumer. This is accomplished through an improvement in the flavor, tenderness, and quality of the lean beef which results from marbling.

In a general way, there are two methods of finishing cattle for market: (1) cattle feedlots, including confinement (sheltered) finishing; and (2) pasture finishing. Prior to 1900, the majority of fat cattle sent to market were 4- to 6-year-old steers that had been finished primarily on grass. Even today, the utilization of pasture continues to play an important role in all types of cattle-finishing operations; and grass finishing will increase in the future.

The center of cattle feeding has shifted from the Corn Belt to the West and Southwest. Today, about 85% of the fed cattle marketed in the U.S. is fed in 13 states. The leading cattle-feeding states by rank are: Texas, Nebraska, Kansas, Colorado, Iowa, and California.

## MODERN CATTLE FEEDING

Fig. 18–1. Truck unloading the precise ration for a pen of cattle. (Courtesy, Farr Feeders, Inc., Greeley, CO)

Cattle feeders are commonly classed as either (1) commercial feeders, or (2) farmer-feeders, based largely on numbers. From the standpoint of statistical reporting, the U.S. Department of Agriculture commonly draws the line at 1,000 head. A commercial cattle-feeding operation is defined as one having a capacity of 1,000 head or more, at any one time.

The traditional farmer-feeders evolved with Corn Belt farming, in the North Central region of the United States. Generally speaking, they market their crop, usually corn, through cattle (or hogs, or lambs), and spread the manure on the land. The purchase of feeder cattle for this type of enterprise is generally in the fall, with the actual feeding done during the winter months when labor is available due to limited field work. This traditional farmer-feeder type of operation has persisted to the present time, although it has been modernized through the years.

In addition to being larger, commercial cattle feeders generally differ from farmer-feeders in the following respects: (1) They usually feed cattle on a year-round basis, rather than during the winter months only; (2) they may grow little, or none, of their feed; (3) they are highly mechanized; and (4)

they are knowledgeable of costs and returns, skillful buyers and sellers, and aware of market trends. Today, commercial feedlots with more than 1,000-head capacity dominate cattle feeding.

### Fed Cattle Marketed by Feedlot Capacity

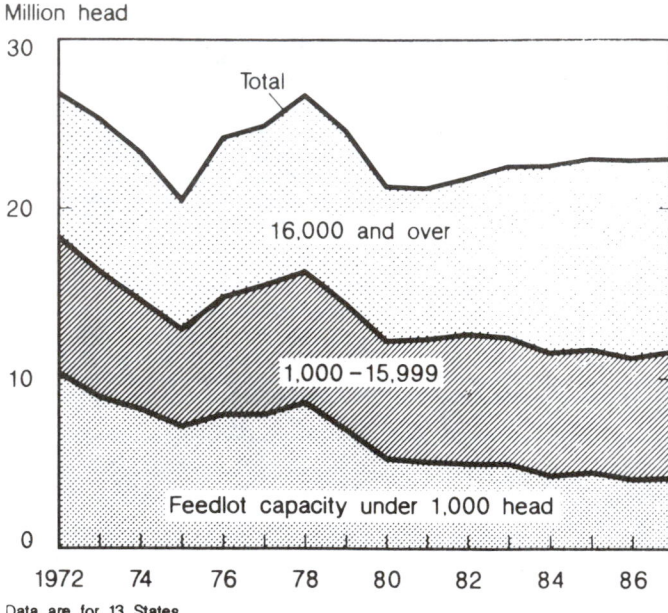

Fig. 18–2. Feedlots have gotten bigger! In 1962, feedlots with more than 1,000 head marketed 40% of the fed cattle. By 1987, 25 years later, feedlots with 1,000 head and over marketed 82% of the nation's fed cattle. (Source: *1988 Agricultural Chartbook*, Ag. Hdbk. No. 673, USDA, p. 90, Chart 194)

Feedlots exceeding 100,000 capacity are now in operation in Arizona, California, Colorado, and Texas. Even in the Corn Belt, feedlots are getting bigger.

## GROWTH OF CATTLE FEEDING; COMPOSITION OF CATTLE SLAUGHTER

The growth of the cattle-feeding industry in the United States has been spectacular since World War II. In 1947, only 6.9 million head of market cattle were grain fed, representing 30.1% of the slaughter cattle that year. The number of grain-fed cattle peaked in 1978, with 24,321,000 head marketed that year. Thereafter, the number of grain-fed cattle marketed declined to 22.5 to 23 million head in the 1980s.

During the fifties and sixties, grain bins bulged with surpluses, and consumers were able and willing to pay for Choice, grain-fed beef, which was ably promoted and merchandised by self-service chain stores. The time was ripe for the chain reaction that spawned the era of grain-fed cattle. Commercial cattle feeders, acutely attuned to consumer demands, set out to fill the need. Cattle feeding went big—and modern. Then, suddenly, in the early 1970s, there were grain shortages—and there was a rude awakening. The message was loud and clear: When grain is scarce and high in price, cattle and other animals cannot compete with humans for the cereals; fewer cattle are grain fed, more are grass finished.

Fed beef will continue to be a vital part of the American cattle industry, but it will not return to the preponderance that it occupied in the early 1970s.

The primary reasons for the phenomenal growth of the cattle feeding industry after World War II and extending through the 1970s were:

1. Abundant feeds, reasonably priced.
2. Increased human population.
3. Increased beef consumption.
4. Beef enjoyed the status of the preferred meat.
5. Increased disposable income.
6. Cattle feeding was profitable.
7. Consumer demand for beef grading U.S. Choice, with repeatability (the shoppers wanted the same quality of meat that they purchased last week).
8. Modern, commercial cattle feeders were attuned to consumer needs.
9. Finishing of cattle on a year-round basis.
10. Dual grading improved the product.
11. Increased mechanization and labor efficiency.
12. Additives and hormones improved performance.
13. New methods of grain processing improved efficiency.
14. Marketing of finished cattle became more efficient.
15. Beef packing became more efficient.
16. The advent of futures trading.

## CHARACTERISTICS OF MODERN CATTLE FEEDING

The characteristics of the modern cattle-feeding industry are:

1. It is unique to the United States.
2. It is highly concentrated in a relatively few states.
3. It gave rise to a new "beef belt" extending from the southern Great Plains through Nebraska.
4. It is a high risk business.
5. Facilities and man-hours required for feeding vary.
6. Feed costs alone account for about 70% of the variable costs.
7. Several methods of financing are being used.
8. Type of legal ownership varies with size.
9. Large amounts of investor money are involved.
10. There is considerable vertical integration of lots, but little horizontal integration. (In 1987, packers fed only 4.9% of the nation's fed cattle.)[1]
11. Larger lots do more custom feeding.
12. Most feeder cattle are bought through auctions or by direct purchase.
13. Most finished cattle are sold directly to packing plants.
14. Most big lots buy every week; sell every week.
15. Nutrition and animal health consultants are used.

## KINDS OF CATTLE TO FEED[2]

All kinds of cattle may be, and are, fed. But, for the maximum success, it is imperative that the right kind of cattle be selected for a particular feedlot. The cattle should match the operator's available feed, labor, shelter, and credit. Also, it is imperative that there be a suitable market outlet following finishing; for example, it would be unwise to feed lightweight heifers in an area where the strongest slaughter market is for heavy steers; nor should one finish out heavy Holstein steers where the primary interest of packers is for Choice grade beef. But, assuming that a satisfactory slaughter outlet exists for different kinds of cattle, the following general guides will be helpful in determining what kind of cattle to feed in a given lot.

Fig. 18–3. Yearling cattle on pasture. (Courtesy, J. C. Allen and Son, West Lafayette, IN)

## AGE AND WEIGHT OF CATTLE

A generation ago the term *feeder steer* signified to both the rancher and the Corn Belt feeder a 2½- to 3-year-old animal weighing approximately 1,000 lb. Today, cattle are referred to by ages as calves, yearlings, and 2-year-olds. This shift to younger cattle has been brought about primarily by consumer demand for smaller and lighter cuts of meat and improved feeding and management practices.

The age of cattle to feed is one of the most important questions to be decided upon by every practical cattle feeder. The following factors should be considered in reaching an intelligent decision on this point:

• **Rate of gain**—When cattle are fed liberally from the time they are calves, the daily gains will reach their maximum the first year and decline with each succeeding year thereafter. On

---

[1]*Packers and Stockyards,* Statistical Report, 1987 reporting year, USDA, Packers and Stockyards Administration, P & SA Statistical Report No. 88–1, Jan. 1989, p. 32, Table 14.

---

[2]Tables 18–1 to 18–5 are a consensus (or judgment) of several knowledgeable feedlot consultants and college animal scientists, based on a survey made by the author. A consensus was resorted to for the reasons that (1) no extensive, nationwide, scientific study of this sort has ever been made, and (2) this information is much needed by cattle feeders and those who counsel with them. No claim is made relative to the scientific accuracy of the data; rather, it is presented (1) because it is the best, if not the only, information of the kind presently available, and (2) with the hope that it will stimulate needed research along these lines.

the other hand, when in comparable condition, thin but healthy 2-year-old steers will make more rapid gains in the feedlot than yearlings; likewise, yearlings will make more rapid gains than calves. Table 18–1 illustrates this situation.

### TABLE 18–1
### EFFECT OF AGE OF CATTLE ON DAILY GAINS
### OF MEDIUM-FRAME CATTLE

| | Daily Gains | | | |
|---|---|---|---|---|
| | Average of U.S. Feedlots | | Top 5% of U.S. Feedlots | |
| Age | (lb) | (kg) | (lb) | (kg) |
| Calves | 2.4 | 1.09 | 2.7 | 1.23 |
| Yearlings | 2.8 | 1.27 | 3.2 | 1.45 |
| Two-year-olds | 2.9 | 1.32 | 3.4 | 1.54 |

• **Economy of gain**—Calves require less feed to produce 100 lb of beef than do older cattle. This may be explained as follows:

1. The increase in body weight of older cattle is largely due to the deposition of high-energy fat, whereas the increase in body weight of young animals is due mostly to the growth of muscles, bones, and organs. Thus, the body of a calf at birth usually consists of more than 70% water, whereas the body of a fat 2-year-old steer will contain only 45% moisture. In the latter case, a considerable part of the water has been replaced by fat.

For feedlot efficiency figures to be very meaningful, it is necessary to know what kind of ration was fed; otherwise, on a poundage basis, it is not unlike comparing steaks and carrots in the human diet. The more concentrated and the higher the energy value of the ration, the fewer the pounds of feed required to produce 100 lb of gain. Yet, many times it is in the nature of good business to feed rations that necessitate more pounds of feed to produce 100 lb of gain simply because lower cost gains can be produced thereby. This applies to feeding corn silage, potato waste, and many other by-product feeds.

2. Calves consume a larger proportion of feed in proportion to their body weight than do older calves.

3. Calves masticate and digest their feed more thoroughly than older cattle. Despite the fact that calves require less feed per 100 lb gain—because of the high energy value of fat—older cattle store as much energy in their bodies for each 100 lb of total digestible nutrients consumed as do younger animals.

From the above, it is apparent that age of cattle affects the pounds of feed required to produce 100 lb of gain—that the younger the cattle, the greater the feed efficiency.
Table 18–2 points up the effect of age of cattle on feedlot efficiency.
A more accurate measure of feed efficiency than pounds of feed per pound gain can be obtained through the use of energy conversion—the TDN or Mcal required to produce a pound of beef. It alleviates much of the inevitable disadvantage to which a relatively low-energy, bulky ration (such as a high-silage ration) is put when it is compared on a poundage basis to a more highly concentrated feed (such as an all-concentrate ration). Thus, energy requirements (in both TDN and Mcal per pound of gain) are given in Table 18–2.

### TABLE 18–2
### EFFECT OF AGE ON FEED EFFICIENCY
### OF MEDIUM-FRAME STEERS[1]

| | Average of U.S. Feedlots | | | Top 5% of U.S. Feedlots | | |
|---|---|---|---|---|---|---|
| Age | Feed/Lb Gain | TDN/Lb Gain | Mcal/Lb Gain[2] | Feed/Lb Gain | TDN/Lb Gain | Mcal/Lb Gain[2] |
| | (lb) | (lb) | (lb) | (lb) | (lb) | (lb) |
| Calves | 7.5 | 6.0 | 9.86 | 7.0 | 5.6 | 9.20 |
| Yearlings | 8.0 | 6.4 | 10.52 | 7.5 | 6.0 | 9.86 |
| Two-year-olds and over | 8.5 | 6.8 | 11.17 | 8.0 | 6.4 | 10.52 |

[1]Air-dry basis (approximately 98% dry matter).
[2]Mcal metabolizable energy (ME) was calculated by assuming 1.6434 Mcal ME = 1 lb TDN.

• **Flexibility in marketing**—Calves will continue to make satisfactory gains at the end of the ordinary feeding period, whereas the efficiency of feed utilization decreases very sharply when mature steers are held past the time that they are finished. Therefore, under unfavorable market conditions, calves can be successfully held for a reasonable length of time, whereas prolonging the finishing period of older cattle is usually unprofitable.

• **Length of feeding period**—Calves require a somewhat longer feeding period than older cattle to reach a comparable finish. To reach Choice condition, steer calves are usually full-fed about 7 to 8 months; yearling 4 to 5 months; and 2-year-olds only about 3 months. Table 18–3 points up this situation. The longer finishing period required for calves is due to the fact that they are growing as well as finishing.

### TABLE 18–3
### EFFECT OF AGE OF CATTLE ON LENGTH OF
### FEEDING PERIOD OF MEDIUM-FRAME CATTLE

| Age | Average of U.S. Feedlots | Top 5% of U.S. Feedlots |
|---|---|---|
| | (days) | (days) |
| Calves | 230 | 210 |
| Yearlings | 140 | 130 |
| Two-year-olds | 110 | 100 |

• **Total gain required to finish**—In general, calves must put on more total gains in the feedlot than older animals to attain the same degree of finish. In terms of initial weight, calves practically double their weight in the feedlot. On the average, yearlings increase in weight about 400 lb, and 2-year-olds increase their initial feedlot weight about 320 lb. Table 18–4 illustrates this situation.

### TABLE 18–4
### EFFECT OF AGE OF CATTLE ON TOTAL GAIN
### REQUIRED TO FINISH MEDIUM-FRAME CATTLE

| Age | Average of U.S. Feedlots | Top 5% of U.S. Feedlots |
|---|---|---|
| | (days) | (days) |
| Calves | 550 | 525 |
| Yearlings | 400 | 380 |
| Two-year-olds | 320 | 300 |

[1]To convert lb to kg, divide by 2.2.

- **Total feed consumed**—Because of their smaller size, the daily feed consumption of calves is considerably less than for older cattle. However, as calves must be fed for a longer feeding period, the total feed requirement for the entire finishing period is approximately the same for cattle of different ages.

- **Experiences of the feeder**—Young cattle are bovine *babies*. As such, they must be fed more expertly. Thus, the inexperienced feeder had best feed older cattle.

- **Kind and quality of feed**—Because calves are growing, it is necessary that they have more protein in the ration. Since protein supplements are higher in price than carbonaceous feeds, the younger the cattle the more expensive the ration. Also, because of smaller digestive capacity, calves cannot utilize as much coarse roughage, pasture, or cheap by-product feeds as older cattle.

Calves also are more likely to develop peculiar eating habits than older cattle. They may reject coarse, stemmy roughages or moldy or damaged feeds that would be eaten readily by older cattle. Calves also require more elaborate preparation of the ration and attention to other small details designed to increase their appetite.

- **Comparative costs and selling price**—Calves cost more per 100 lb as feeders than do older cattle.

- **Dressing percentages and quality beef**—Older cattle have a slightly higher dressing percentage than calves or baby beef. Moreover, many consumers have a decided preference for the greater flavor of beef obtained from older animals.

From the above discussion, it should be perfectly clear that there is no best age of cattle to feed under any and all conditions. Rather, each situation requires individual study and all factors must be weighed and balanced.

## SEX OF CATTLE

More steers than heifers are fed, simply because more of them are available. A portion of the heifers is held back for replacement purposes. In the future, more young bulls will be fed. Thus, feedlot operators must give consideration to the sex of cattle fed. First, and foremost, they should consider their market outlets.

Table 18–5 shows the effect of sex on rate of gain and feed efficiency. It is noteworthy that bulls gain more rapidly on less feed than steers, and that steers gain more rapidly on less feed than heifers.

**TABLE 18–5**
**EFFECT OF SEX ON GAIN OF MEDIUM-FRAME CATTLE[1]**

| Sex | Average of U.S. Feedlots | | Top 5% of U.S. Feedlots | |
| | Daily Gain | Feed/Lb Gain[2] | Daily Gain | Feed/Lb Gain[2] |
|---|---|---|---|---|
| | (lb) | (lb) | (lb) | (lb) |
| Heifers | 2.4 | 7.5 | 2.7 | 6.9 |
| Steers | 2.8 | 6.9 | 3.2 | 6.6 |
| Bulls | 3.0 | 6.7 | 3.4 | 6.4 |

[1]To convert lb to kg, divide by 2.2.
[2]Air-dry basis (approximately 90% dry matter).

## STEERS VS HEIFERS

On the market, cattle are divided into five sex classes; steers, heifers, cows, bullocks, and bulls. The sex of feeder cattle is important to the producer from the standpoint of cost and selling price (or margin), the contemplated length of feeding period, quality of feeds available, and ease of handling. The consumer is conscious of sex differences in cattle and is of the impression that it affects the quality, finish, and conformation of the carcass.

Steers are by far the most important of the sex classes on the market, both from the standpoint of numbers and their availability throughout the year, whereas heifers are second.

The relative merits of steers vs heifers, both from the standpoint of feedlot performance and the quality of carcass produced, has long been a controversial issue. Based on experiments and practical observations, the following conclusions and deductions seem to be warranted relative to this question:

- **Length of feeding period**—Heifers mature earlier than steers and finish sooner, thus making for a shorter feeding period. In general, heifers may be ready for market 30 to 40 days earlier than steers of the same age started on feed at the same time.

- **Market weight**—The most attractive heifer carcasses are obtained from animals weighing 650 to 900 lb on foot, showing good condition and finish but not patchy and wasty.

- **Rate and economy of gain**—Because of their slower daily gains and lower feed efficiency, the feedlot gains made by heifers are usually somewhat more costly than those made by steers of the same age.

- **Price**—Because of existing prejudices, feeder heifers can be purchased at a lower price per pound than steers, but they also bring a lower price when marketed. Thus, the net return per head may or may not be greater with heifers.

- **Carcass quality**—Carefully controlled experiments have now shown conclusively that when heifers are marketed at the proper weight and degree of finish, sex makes no appreciable difference in the dressing percentage, in the retail value of the carcass, or in the color, tenderness, and palatability of the meat.

- **Ease of handling in the feedlot**—Because of disturbances at heat periods, many feeders do not like to handle heifers in the feedlot. Of course, the incidence of estrus can be lowered by feeding the additive MGA.

- **Flexibility in marketing**—If the market is unfavorable, it is usually less advisable to carry heifers on feed for a longer period than planned because (1) of possible pregnancies, and (2) they become too patchy and wasty.

- **Effect of pregnancy**—Packer buyers have long insisted that they are justified in buying finished heifers at a lower price than steers of comparable quality and finish because (1) most heifers are pregnant and have a lower dressing percentage; and (2) pregnant heifers yield less desirable carcasses. In realization that the packer will lower the price anyway, many feeders make it a regular practice to turn a bull with heifers about 3 to 4 months before the market period. Such feeders contend that the animals are then quieter and will make better feedlot gains.

The economic loss that accrues from pregnant feedlot heifers was pointed up by two different studies made and reported by Monfort of Greeley, Colorado.[3]

---

[3]Bennett, Bill, *Animal Nutrition and Health,* "The Pregnant Feedlot Heifer," May 1985.

As a result of a 1983 survey of a number of cattle feeders and packers, Monfort reported the following relative to feedlot heifers: (1) a pregnancy rate of 16.5% on incoming feedlot heifers, (2) feeders placed a lower value of $30.32 per head on pregnant vs open heifers, (3) lower dressing percentage at slaughter of 3.3% on pregnant vs open heifers, and (4) a cost of $5.29 per head to pregnancy check and attempt an abortion on each incoming pregnant heifer. Additionally, in an actual study of 10,000 head of heifers and 1,000 heiferettes (heiferettes are large, heavy heifers possessing nearly the size and development of a mature cow) slaughtered at Monfort's Greeley, Colorado, plant in 1983–84, it was found that pregnancy lowered the dressing percentage on heifers by 5.6%, and on heiferettes 6.1%. Thus, the bottom line is that feeding pregnant heifers is costly.

In light of the preceding facts, the following management options may be considered when feeding heifers of unknown pregnancy status:

1. Feed heifers like steers, and meet the calving problems (difficult births, and caring for newborn calves) as they occur.

2. Buy only open or spayed heifers, the supply of which is limited.

3. Pregnancy examine all heifers and use an abortive agent on the pregnant ones, according to directions.

By pregnancy testing and the use of abortifacient, the termination of early pregnancy can be brought about. However, the cost and setback in performance may not justify such action.

## SPAYED HEIFERS

Spaying prevents heifers from becoming pregnant and eliminates the necessity of separating heifers from bulls or steers. Also, some buyers pay a slight premium for spayed heifers. However, most experiments have shown that spayed heifers make less rapid gains and require more feed per 100 lb gain than open (unspayed) heifers. The latter fact, plus the attendant danger of the operation, generally does not justify spaying unless the sale price is sufficiently higher. However, two new methods of sterilizing heifers without lowering performance appear promising.

## BULLS

The feeding of bulls (uncastrated males) instead of steers has been standard practice throughout Europe for many years. For example, since about 1954 West Germany has, for the most part, fed out and slaughtered bulls as yearlings, instead of steers, because they obtain 10 to 15% greater rate and feed efficiency thereby. The practice will increase in the United States now that carcasses from young bulls are federally graded as bullock beef (the use of the term *bullock beef* to identify meat from young bulls became effective July 1, 1973) rather than *bull beef,* thereby removing the connotation that the meat is inferior to or different from steer or heifer beef.

Fig. 18–4. Young bulls on feed in a farm feedlot, the carcasses of which will be graded *bullock beef.* (Courtesy, Oklahoma State University, Stillwater)

The carcasses from older bulls are still labeled *bull beef,* to differentiate them from the carcasses of younger bulls. Bullock beef from young bulls is graded according to the same quality standards as beef from steers and heifers.

Also, the economics of the situation favors the feeding of bulls instead of steers. The male hormones secreted by the testicles are excellent growth stimulants and will improve gain and efficiency by 10 to 15%. Also, bulls will produce more lean meat than steers, and research has shown that bull meat is equal in value, quality, and palatability to steer meat.

Now that the carcasses from young bulls are differentiated from older bulls, only consumer acceptance remains. Without doubt, beef shortages will speed acceptance of *bullock beef.*

The following guidelines are recommended in the feeding of bulls:

1. Start young bulls on full feed at weaning age (6 to 7 months) and feed out as rapidly as possible to a market weight of 1,100 lb.

2. Use high-energy rations for bull feeding, because they tend to grow rapidly and lay down less fat than steers.

3. Feed out bulls so that they are finished for market before 18 months of age.

4. Do not add new bulls to the pen after the weanling bulls are started on feed, because this tends to encourage fighting and riding, and results in reduced gains.

5. Keep bulls separate from other cattle when marketing them. If possible, do not permit the bulls to stand in the pen overnight before slaughter.

6. Bulls of beef breeding are less nervous and ride less than bulls of dairy breeding.

## GRADE OF CATTLE

The most profitable grade of cattle to feed will generally be that kind of cattle in which there is the greatest margin between the purchase price as feeders and the selling price as fat cattle. As can be readily understood, one cannot arrive at this decision by merely comparing the existing price between the various grades at the time of purchase. Rather, it is necessary to project the differences that will probably exist, based on past records, when the animals are finished and ready for market.

As fewer grain-fed cattle are marketed in the summer and fall, the spread in price between Good and Choice fed cattle and those of the lower grades is usually the greatest during this season. On the other hand, the spread between the grades is likely to be least in late winter and early spring, when a large number of well-finished cattle are coming to market from the feedlots. However, such seasonal effects are minimal today, due to large feedlots and year-round feeding.

The length of the feeding period and the type of feed available should also receive consideration in determining the grade of cattle to feed. Thus, for a long feed and when a liberal allowance of grain is to be fed, only the better grades of feeders should be purchased. On the other hand, when a maximum quantity of coarse roughage is to be utilized and a short feed is planned, cattle of the medium or lower grades are most suitable. Thus, successful cattle feeders match the quality of the cattle selected with the quality of the available feed; the better the feed, the higher the grade of cattle.

Cattle of the lower grades should be selected with very special care to make certain that only thrifty animals are bought. Ordinarily, death losses are much higher among low-grade feeder cattle, especially when the low-grade animals are calves. The death loss in handling average or high-grade feeders seldom exceeds 1 to 2% whereas with cull or *dogie* cattle, it frequently is 2 to 3 times this amount. Many low-grade cattle are horned, and dehorning further increases the risk—in addition to the added labor and shrinkage resulting therefrom.

No given set of rules is applicable under any and all conditions in arriving at the particular grade of cattle to feed, but the following factors should receive consideration.

1. The feeding of high-grade cattle is favored when:
   a. The feeder is more experienced.
   b. A long feed with a maximum of grain in the ration is planned.
   c. Conditions point to a wide spread in price between grades at marketing time. Such conditions normally prevail in the late summer or early fall.

2. The feeding of average or low-grade cattle is favored when:
   a. The feeder is less experienced.
   b. A short feed with a maximum of roughage or cheap by-products is planned.
   c. Conditions point to a narrow spread in price between grades at marketing time. Such conditions normally prevail in the spring.

3. In addition to the profit factors enumerated above, it should be pointed out that with well-bred cattle the following conditions prevail:
   a. Well-bred cattle possess greater capacity for consuming large quantities of feed than steers of a more common grade, especially during the latter part of the feeding period.
   b. The higher the grade of cattle, the higher the dressing percentage and the greater the proportionate development of the high-priced cuts.
   c. The higher the grade of the cattle, the greater the opportunities for both profit and loss.
   d. There is a great sense of pride and satisfaction in feeding well-bred cattle.

Certainly producers who raise their own feeder cattle should always strive to breed high-quality cattle, regardless of whether they finish them themselves or sell them as feeders.

On the other hand, the purchaser of feeder steers can well afford to appraise the situation fully prior to purchasing any particular grade.

## BREEDING AND TYPE OF CATTLE

Although the supporting data are rather limited, it is fully realized that there is considerable difference between individual animals insofar as rate and economy of gain is concerned. It is to be emphasized that these differences are greater within breeds than between breeds.

### CROSSBREDS

Good crossbreds will likely show 2 to 4% improvement over the average of the parent breeds for rate and efficiency of feedlot gains. Additionally, even larger advantages accrue to the cow-calf producer. Thus, it is inevitable that an increasing number of crossbred feedlot cattle will be seen. The primary characteristics desired in feedlot cattle are the same, whether they are crossbreds or straightbreds; namely, (1) high rate of gain, (2) efficiency of feed conversion, (3) high cutout percent, and (4) tender palatable beef.

Fig. 18–5. Crossbred steers being finished in an Arizona feedlot. (Courtesy, Benedict Feeding Co., Casa Grande, AZ)

### DAIRY BEEF

Dairy beef accounts for about 20 to 25% of the beef consumed in this country, with these animals marketed as veal calves, cull dairy cows and bulls, and finished dairy heifers and steers. Improvements in the science and technology of feeding and processing favor growing and finishing dairy beef, and minimum slaughter of veal calves.

*Dairy beef is just what the term implies—beef derived from cattle of dairy breeding, or from dairy × beef crossbreds.*

Modern cattle feeders are primarily concerned with rate and efficiency of gains, and net returns. As a result, most of them would just as soon feed steers of dairy breeding, either purebreds or crossbreds; some actually prefer them. Consumers demand beef that has a maximum amount of lean, with a minimum amount of waste fat, and which is tender and flavorful; and they couldn't care less whether it comes from a critter that was black, white faced, roan, pink, yellow, or polka dot.

As a result, more and more steers of dairy breeding are going the feedlot route, rather than as veal. As evidence of this transition, during the 45-year period 1942 to 1987, U.S. per capita veal consumption declined from 8.2 to 1.8 lb, while per capita beef consumption increased from 61.2 to 103.3 lb in this same period of time. Also, the shift in consumer demand to more lean and less fat has been reflected in the changed federal grades of beef. As a result when properly fed, Holstein steers will make Choice, Select, or Standard grades.

Fig. 18–6. Dairy beef in the making. This shows Holstein steers on feed in a California feedlot.

Some pertinent points relative to dairy beef follow.

• **High growth thrust essential**—For a dairy beef program to be most successful, scientists and producers in both Great Britain and the United States are agreed that the animals should have a high growth potential, as evidenced by heavy birth weight and heavy weight at maturity. Since Holsteins are heavy at birth and mature out at around 1,400 lb in comparison with mature weights of 1,000 to 1,200 lb of the European beef breeds, it can be readily understood that Holsteins are ideal when it comes to producing dairy beef.

• **High-energy rations; light market weights**—If dairy steers are to be slaughtered at young ages and light weights, high-energy (low-roughage) rations are imperative. Under this system, usually young calves of either dairy or dairy × beef breeding are fed in confinement—in barns; and fed milk replacers from 1 to 4 days of age to 200 to 300 lb; and are full-fed a high-concentrate ration from about 300 lb to market weight of 750 to 950 lb.

Crowding for market at any early age takes advantage of the fact that growth is generally most economical when most rapid, and that young gains are cheap gains. Also, experience shows that when Holstein calves are started on super-energy rations at around 350 to 450 lb weight and marketed under 1,100 lb, (1) there is excellent marbling with very little bark (outside fat), and (2) many of these animals will grade Choice.

• **High-roughage rations; heavy market weights**—If roughages are relatively more abundant and cheaper than concentrates, then it may be more remunerative to feed dairy beef more roughage and market at heavier weights—and with it to expect slower and less efficient gains. In any event, it is net returns that count, rather than rate of gain and pounds of feed required per pound of gain.

Under the high-roughage system, steers of dairy breeding are grown on maximum roughage to 600 to 750 lb weight, following which the ratio of concentrate to roughage is increased. Most dairy steers fed according to this system are marketed at weights of 1,150 to 1,400 lb, grading Select or Commercial. (Most of them are too old to grade Standard, and lack the necessary conformation and marbling to grade Choice.)

• **Dairy beef has good potential**—There is ample evidence that male calves of the larger dairy breeds (Holstein and Brown Swiss,) along with dairy beef crosses, have the potential for producing acceptable beef with good feed efficiency, under a system of either (1) full feeding from an early age on a high-energy ration, or (2) growing and finishing on a maximum of roughage and marketing at older ages and heavier weights. In the final analysis, therefore, the system selected should be determined by net returns. Both methods necessitate the rearing of young, one- to four-day-old calves to weights of around 300 lb, with such early rearing done by either a calf-raising specialist or by the cattle feeder who will do the ultimate finishing. Such calves must usually be obtained from over a wide area, and be of variable ages and sizes; hence, they are difficult to come by. Also, death losses are frequently high and discouraging.

Both commercial cattle feeders and dairy producers are showing increased interest in producing dairy beef, with the result that there is competition between them. More dairy beef will be produced in the future.

## BUSINESS ASPECTS OF CATTLE FEEDING

Cattle feeding is a business, and as in other businesses, cattle feeders hope to obtain a reasonably good return for the use of their capital, labor, and management. To this end, their business aspects must become more sophisticated and efficient; they must—

1. Compute break-even prices prior to buying feeder cattle, especially if they do not buy and sell each week.
2. Buy feeder cattle of the right size, quality, and price.
3. Sell the cattle to the best advantage.
4. Integrate when possible.
5. Feed cattle to weight and grade.
6. Evaluate performance.
7. Obtain economies with size.
8. Finance the feedlot and cattle properly and adequately.

Of course, the above eight points represent a great oversimplification of a complex business, but they do clearly set forth the main requisites for profitable cattle feeding. Anyone who wishes to make money feeding cattle must have expertise in these eight areas.

Business aspects outweigh all other factors—feed additives, crossbreds, pollution control, etc.—producing change in cattle feeding. It is important, therefore, that feeders and those who counsel with them be thoroughly grounded in each of these areas.

Other factors than the cost of feed, amount of gain, and price of slaughter cattle affect the price that a feeder can afford to pay for feeder cattle. Among them, are the following:

1. Condition of the cattle. Thin cattle, if in good health, will make faster gains than fleshy cattle.

2. Growthy cattle—cattle that are big framed and on the rangy order—make better gains than the little, compact kind; and they may be carried to heavier weights. If the feeder cannot obtain cattle backed by production records, ''eyeballing'' will help.

3. Younger, lighter weight cattle tend to make more efficient gains.

4. Cattle of known, superior, ancestry with gaining ability are worth more.

5. Higher grade cattle are worth more. This is so because better grades generally bring a higher selling price, and, therefore, a higher price is obtained on their gains made in the feedlot.

6. The higher the cost of feed, the greater the necessary margin between the cost of feeder cattle and the selling price of finished animals. This is so because of the high cost of gains as compared to their selling price.

7. Feeder steers are generally worth approximately $3 to $4 per cwt more than heifers. This is because they gain about 10% faster, require 5 to 10% less feed, and bring from $.50 to $1.50 per cwt more than heifers when finished. Additionally, there is no pregnancy problem.

8. Good crossbreds will make 2 to 4% more rapid and efficient gains than the average of the parent breeds.

## MARGIN

A positive margin exists when feeder cattle cost less than finished cattle. A negative margin exists when feeder cattle cost more than finished cattle. Cattle feeders will pay more for feeder cattle than they expect to receive for them as finished animals when there appears to be a favorable margin on the gain in weight; that is, when they can sell the gain in weight for considerably more than it cost to produce it.

Profits in cattle feeding come from two different kinds of margins—price margin and feeding margin.

*Price margin is the difference between the cost per cwt of the feeder animal and the selling price per cwt of the same animal when finished.*

For example, if a feeder pays $70 per cwt for a 600-lb steer and sells it for $65 per cwt, the price margin is a negative $5. This means that the cattle feeder would take a $30 loss on the original 600 lb bought.

*Feeding margin is the difference between the cost of putting on 100 lb of gain and the selling price per cwt of the same animal when finished.* Thus, if it cost $55 per cwt to put gain on yearling steers, and if finished cattle sell for $65 per cwt, the feeding margin would be $10 per cwt. Assuming a market weight of 1,000 lb, or 400 lb gain, the feeder could expect to make about $40 on the feeding margin.

The amount a cattle feeder makes as a result of a good feeding margin can more than offset the losses accruing from a negative price margin, but it doesn't always work that way. It depends on many different things—the selling price, the cost of gain, the price paid for feeder animals, and other factors.

In the example just cited, the feeding margin amounted to $40 per animal. This is not to suggest, however, that cattle feeders should always put more gain on yearling steers. How much gain a cattle feeder should put on depends upon the kind and condition of the feeder cattle when they go into the lot, the rate of gain, and several other factors. Research and experiences have clearly demonstrated that costs of gain go up pretty fast if cattle are fed much beyond Choice slaughter grade.

The principal of profits from price margin and feeding margin applies to feeder calves, also. But the relative importance of price margin vs feeding margin is not quite the same for calves as for yearlings. Let us analyze the situation further: The feeder who is feeding yearlings buys 600 lb of the 1,000 lb which will be marketed at the end of the feeding period. So, getting cattle bought right is pretty important. Thus, if the feeder pays $1 per cwt too much for feeders, that takes $6 off the potential profit from price margin and from total profits. On the other hand, the feeder who buys calves is more interested in costs of gain and feeding margin than in price margin, because about 60% of the weight which will be sold is from the gain put on in the feedlot. Thus, if the feeder pays $1 per cwt too much for a 400-lb feeder calf, it hurts, but not quite so much—$4 compared to $6 per cwt per head.

If farmer-feeders just manage to balance gains from feeding margin with their losses from price margin, this does not necessarily mean that they should not feed cattle. Actually, they are not in too bad shape. They are getting paid market price for their feed, a going wage for their labor, around 12% on their own capital, and they are getting enough to cover all fixed costs like depreciation, taxes, etc., on their facilities and equipment. With commercial feeders, it's another story. They usually buy most of their feed, and they operate on borrowed capital. Thus, to stay in business, they must turn in a profit over and above these costs.

## CUSTOM (CONTRACT) FEEDING

*Custom cattle feeding is the feeding of cattle for a fee, usually without taking ownership of the animals.*

Capital requirements, periods of severe economic conditions (like scarce money and high interest), times of depressed feeder cattle prices, and adverse pasture conditions caused custom feeding to grow following World War II. These same forces, along with the need for high occupancy (full feedlots) and increased integration, have resulted in further expansion of custom feeding.

Most custom feeders have developed large, highly mechanized, and very efficient plants. Usually, they have on their staffs highly trained nutritionists who are charged with the responsibility of formulating rations and of obtaining maximum gains and feed efficiency at the lowest possible cost. Through custom feeding, they sell the use of their facilities, services, and know-how to cattle owners, usually with profit to each party.

The proportion of custom-fed to cattle owned by the feedlot varies (1) in period of time—it increases in times of financial stress (when cattle feeding is not profitable, money is scarce, and interest is high); (2) according to area—for example, there is more custom feeding in California than Colorado; (3) according to size of feedlot—generally speaking, the larger the feedlot, the greater the percentage of custom feeding. Some feedlots do not do any custom feeding whatsoever; others are almost wholly on a custom basis; but most lots have part of each. Feedlots that do both—those in the dual role of custom feeding and owning cattle—vary in the proportion of cattle in each category, but most of them seem to prefer about ⅔ custom-fed cattle and ⅓ ownership. It's a good bread-and-butter division; in times when fed cattle lose money, such a feedlot has sufficient assured income to pay its bills.

The ownership of custom-fed cattle is diverse. It includes (1) cow-calf producers (farmers and ranchers) who wish to retain ownership of the cattle that they produce through the feedlot phase, (2) packers, and (3) investors, including limited partnerships, corporations, cattle buyers, cattle dealers, and others.

Custom feeding contracts should always be detailed and in writing, for a good understanding is the best way to avoid a misunderstanding. Also, contracts should be fair to both parties—to both the feedlot owner and the cattle owner.

## TYPES OF CUSTOM FEEDING CONTRACTS

The services rendered vary from feedlot to feedlot and according to the type of contract. In some instances, the services may be so complete that the customer never sees the cattle. The feedlot operators may buy the feeder cattle, feed them, market them, and send the customer (the client) a check for the balance, after deducting input costs, interest charges, and custom feeding charges. Less complete services are usually available to suit the customer.

Both the feedlot owner and the cattle owner should analyze the different types of contracts and determine which best fits their respective circumstances. Some feedlots offer several types of contracts, thereby according the cattle owner a choice.

Competition may dictate the type of contract and the charges made. But by knowing the variables and managing them correctly, feedlot owners can write and carry out contracts that will be fair to themselves and to their customers.

Generally speaking, contracts with fixed charges are the most satisfactory and the most common, primarily because there is less room for misunderstanding.

Although there are many types of custom cattle-feeding contracts, and many variations of each kind exist, most of them can be classified under one of the following types:

• **Feed cost plus daily yardage fee per head**—This type of contract is based on the cost of feed plus an additional 15 to 21 cents per head per day to cover handling, yardage, feed grinding, and similar expenses. Generally, an additional charge of $5.50 to $7.50 per head is made to cover medication, vaccination, branding, dehorning, and dipping. The customer finances purchase of the cattle.

• **Feed cost plus markup**—This type of contract calls for reimbursement on cost of feed plus a feed markup on (1) a flat rate (so many dollars per ton), (2) a percent of cost, or (3) a percent of moisture added in steam processing.

With a flat rate markup, a $12 to $21 charge per ton above feed cost is made to cover feed handling, grinding, and labor costs. An additional assessment is made to cover chute charges and medication, and the customer finances the purchase of the cattle. Since actual feed milling costs (for labor, power, insurance, mill maintenance, etc.) run $6 to $10 per ton, profit to the feedlot accrues from having a higher markup than the milling cost. A flat markup per ton of feed favors the feedlot when prices fall, and the cattle owner when prices rise.

Also markup on feed may be a percentage of cost basis. With this arrangement, higher feed costs favor the feedlot, whereas lower feed costs reduce the actual return to the feedlot.

Any system of feed markup will be more profitable to the feedlot with heavy, high-performing cattle than with light, slow-gaining cattle, simply because the former eat more.

With the "feed cost plus markup contract," the feedlot is essentially a feed manufacturer processing and delivering feed to its customers—the owners of the cattle.

• **Feed cost plus (1) daily yardage fee per head, and (2) markup per ton of feed**—This is a combination of the first two types. Those feedlots that charge the higher yardage rates per head daily add a smaller markup per ton of feed above the actual ingredient prices; conversely, those that charge the lower yardage rates per head per day make a higher feed charge over and above actual ingredient cost. As a rule of thumb, feed markup is generally lowered by $1 per ton for each 1¢ per head per day of yardage charged.

• **Agreement to purchase contract**—In this plan the cattle feedlot operator buys the feeder cattle, usually with the client required to make a down payment of 20 to 30% of the purchase price. The client then executes an agreement to buy the cattle when they are ready for slaughter, including the original purchase price of the feeder cattle (less any down payment made) and all feeding, handling, and interest charges. (Interest charges are tax deductible.)

• **Payment for weight gained**—This plan is based on a charge per pound of gain. In this arrangement, the feeder is reimbursed on the basis of the gain in weight put on the cattle, at an agreed price of so many dollars per hundred.

Payment for weight gained may be used in growing and backgrounding operations, where it may be desirable to specify both the minimum and the maximum rate of gain. Also, it may be used on cattle that are pastured for a time before being sent to the feedlot.

• **The incentive basis contract**—This is another system of charging on a payment-for-gain basis that some cattle owners like because it gives an incentive for the feedlot to produce rapid daily gains. It consists of paying the feedlot for all feed plus a charge arrived at by "multiplying the average daily gain times itself or times some factor." Thus, if the cattle being finished should average 3 lb gain daily over the entire feeding period, the per head per day basis of payment to the feeder would be as follows:

$$3 \text{ (gain)} \times 6 = 18¢ \text{ per head per day.}$$

This factor might be varied in keeping with the economic conditions. There is no reason it must be identical to the average daily gains. It may be higher or lower according to economic conditions.

## FUTURES AND OPTIONS TRADING IN CATTLE[4]

The three big uncertainties in the cattle feeding business, any one of which can cause a cattle feeder to suffer heavy losses, are prices of (1) feeder cattle, (2) feed, and (3) finished cattle. Through futures and options contracts, cattle feeders can now hedge all three. In advance of feeding, they can protect their prices of feeder cattle, feed, and finished cattle.

---

[4]This section was authoritatively reviewed by, and many helpful suggestions were received from, James Graham, Director, Commodity Marketing Education, Chicago Mercantile Exchange, Chicago, IL)

This discussion is devoted primarily to live (slaughter) beef cattle futures and options as they apply to cattle feedlot operators, because it is the highest risk phase of the cattle business, as well as the least flexible. Unless feeders contract ahead, they have no assurance of what their finished cattle will bring when they are ready to go. Moreover, there is little flexibility in market time, for the reason that excess finish is costly and unwanted by the consumer. As a result of this uncertainty of market price, and in realization of the high risks involved, sleepless nights are rather commonplace among cattle feeders; they find it difficult to concentrate on the business at hand—the efficient feeding and management of cattle. Live (slaughter) beef cattle futures and options provide a means through which cattle feeders can protect their selling price before the cattle are ready to be marketed.

*A futures contract is a standardized, legally binding paper transaction in which the seller promises to make delivery and the buyer promises to take delivery of a specified quantity and type of a commodity at a specified location(s) during a specified future month.* The buying and selling are done through a third party (the exchange clearing member) so that the buyer and seller remain anonymous; the validity of the contract is guaranteed by reputable and well-financed exchange clearing members; and either buyer or seller can readily liquidate their respective positions by simply offsetting sale or purchase.

The Chicago Mercantile Exchange specifications of a live (slaughter) cattle contract read as follows:

Delivery and acceptance of 40,000 lb of USDA yield grade 1, 2, or 3 Choice grade steers (approximately 37 head) within the weight range of 1,050 to 1,125.5 lb and yielding 62%, or within the weight range of 1,125.6 to 1,200 lb, and yielding 63%; stated discounts and tolerances including substitutions in estimated grade, weight, yield, fat thickness, and other details; and delivery to: Peoria, Illinois; Dodge City, Kansas; Omaha, Nebraska; Sioux City, Iowa; Amarillo, Texas; and Greeley, Colorado.

The commission fee on finished cattle contract, covering both purchase and sale (called a round turn), is negotiable between the brokerage firm and customer. In 1989, the minimum initial hedge margin was $600 and the minimum initial speculative margin was $800. The margin deposit may be increased by the broker if the value of the contract should change unfavorably.

Cattle feeders may use a futures contract for hedging purposes. *Hedging is an offsetting transaction by which purchases or sales of a commodity in the cash market are counterbalanced by sales or purchases of an equivalent quantity of futures contracts in the same commodity.*

The following example shows how a cattle feeder can hedge to lock in price:

*Example* (see Table 18–6): It is now November, and the cattle feeder has just purchased feeder cattle to place in the feedlot. Based on past experience, the feeder is quite confident that these cattle should be ready for market the following April. Through good record keeping, the feeder is also quite confident that production (including labor) and marketing costs would be about $62.50/cwt. The feeder decides to hedge the cattle with the April futures contract which at the time is selling for $67.45/cwt. The feeder has also estimated that the basis will be about –$2.10/cwt in April. So, this figure is subtracted from the April futures price and a localized futures price of $65.35/cwt is obtained, or an estimated $2.85/cwt profit; hence, the feeder sells April futures.

### TABLE 18-6
### EXAMPLE OF A CATTLE FEEDER USING A SHORT HEDGE TO LOCK IN A PRICE

| Cash Market | | Futures Market | | Basis |
|---|---|---|---|---|
| | Per Cwt | | Per Cwt | Per Cwt |
| **Nov. 15:** | | | | |
| Expects to receive in April | $65.35 | Sells April futures at | $67.45 | $–2.10 expected |
| **April 10:** | | | | |
| Sells cattle on cash market at | $62.35 | Buys April futures at | $63.85 | $–1.50 actual |
| Futures Gain | $ 3.60 | | | |
| Realized Price | $69.95 | Gain | $ 3.60 | Gain $ .60 |

The cattle feeder sold the finished cattle on the cash market for $62.35/cwt which, after subtracting the production costs of $62.50/cwt gives a loss of $0.15/cwt. However, the feeder realized a profit of $3.60/cwt on the futures transaction, so that the total profit was $3.45/cwt.

This example illustrates what could happen on a declining market. The feeder still showed a profit, even though the cattle were sold in the cash market for a price lower than the production costs, because this loss was offset by a larger profit in the futures market. This was true because, as the cash price declined, the futures price also declined.

If, however, the cash and futures prices had risen, the feeder still could make a profit, this time in the cash market. But because of a loss in the futures market, the total profit would have been less without hedging. Nevertheless, the feeder still received the price protection desired, which was the main purpose in hedging.

Although very few contracts, usually fewer than 3% are consummated by actual delivery of the commodity, a hedger should consider delivery as one of the alternatives, particularly when the cash and futures prices are out of line with each other. However, due consideration must be given to the costs of delivering or receiving delivery, since such costs may be of such magnitude as to offset the differences between the cash and futures prices.

It is not the function of the futures market to provide an alternative source of supply nor an alternate means of disposal of surplus commodities. The purpose of delivery is merely to serve as a safeguard to be used when all else fails.

Feedlot operators who buy feeder cattle and sell finished cattle on a regularly scheduled basis—weekly, biweekly, or monthly—throughout the year are doing their own hedging, provided feeder prices roughly parallel fed cattle prices.

*Live cattle options* were introduced by the Chicago Mercantile Exchange in October, 1984; followed by *feeder cattle options* in January, 1987. Options manage against price risk, but leave in price opportunity. The volume of live cattle *puts* and *calls* exceeded 1,000,000 contracts in 1988.

• **Calls or Puts**—There are two different types of options—*calls* and *puts*, which offer opposite pricing alternatives. *A call option gives the buyer the right to buy a futures contract at a fixed price level on or before an expiration date.* Conversely, *a put option gives the buyer the right to sell a futures contract at a fixed price level, on or before an expiration date.*

An easy way to remember the difference between calls and puts is: With a call option one can *call away* or purchase a futures contract; with a put option, one can *put to* or sell a contract.

Generally, when people buy a call option they are bullish or optimistic about the underlying futures price; if they're bearish about the underlying futures price, they would probably buy a put option.

Calls and puts are separate option contracts. They are not the opposite sides of the same transaction. For every purchase of a call option, there is a corresponding sale of the same call option. This is also true for put options: one buyer—one seller for each put option transaction.

### TABLE 18–7
### OPTION POSITION SUMMARY

| Call Option — Buyer | Call Option — Seller |
|---|---|
| Bullish | Bearish/Neutral |
| Right to buy futures | Obligation to sell futures |
| Pays premium | Receives premium |

| Put Option — Buyer | Put Option — Seller |
|---|---|
| Bearish | Bullish/Neutral |
| Right to sell futures | Obligation to buy futures |
| Pays premium | Receives premium |

There are three alternatives once an option is purchased: (1) let the option expire; (2) offset the option at the current premium value; or (3) exercise the option. Typically, the purchaser would offset the option prior to or at expiration and receive the current premium value. Prior to expiration, the premium value could be higher or lower than the original purchase price depending on how the underlying futures price had changed. The third alternative—to exercise the option—would be used if the option buyer desired to have the underlying futures position or actually wanted to make or take delivery on the underlying futures contract.

There are also three alternatives once you write (sell) an option. The option may be (1) offset at the current premium value; (2) exercised by the buyer, obligating the writer to accept a futures position at the price specified in the contract; or (3) allowed to expire. The writer can either choose to offset the option or wait for expiration of the option. Only the buyer can exercise the option.

## FEEDLOT FINISHING OF CATTLE

*Feedlot finishing refers to feeding cattle in a restricted area, with the feed conveyed to the animals; and it may involve either (1) an open pen or feedlot, or (2) confinement (sheltered) feeding.*

## FEEDLOT FACILITIES AND EQUIPMENT

Cattle feeding facilities and equipment are a manufacturing plant, wherein animate objects (cattle) convert feed into beef. Hence, they merit the same level of competence in planning and design as any other sophisticated manufacturing plant.

Some preliminary feedlot planning suggestions follow:

1. Decide on the type of facilities: (a) feedlot (open pen), (b) cold confinement, or (c) warm confinement.

2. Decide on the number of cattle and the feed and storage requirements with provision for expansion.

3. Determine the justifiable investment in cattle feeding facilities.

4. Select the facilities, equipment, and arrangement that best fit the management program chosen; for example, (a) fence-line bunks and a central feed-processing plant, (b) upright storage with distributors and bunks, or (c) self-feeders.

5. Design a system that is practical, laborsaving, environmentally suitable for economical gains of cattle, and attractive.

## OPEN PEN FEEDLOT

An open pen feedlot is, as indicated by the name, a lot in which the cattle are in the open—usually it is without shelter (except for such natural protection as may be afforded by trees, hills, or wind fences, or perhaps a roof over the feed bunks or shades).

An open lot without shelter is the cheapest type of feedlot construction of any. In the Southern Plains area, where the weather is mild and shelters are unnecessary, investment costs range from $100 to $125 per head of capacity.[5]

Fig. 18–7. Open pen cattle feedlot, with shades—the cheapest and most common type of feedlot. This is the Alta Verde Industries feedlot near Eagle Pass, Texas. (Courtesy, *West Texas Livestock Weekly*, San Angelo, TX)

---

[5]Estimates made by the author.

Housing increases costs, and the more elaborate the housing the greater the cost. It is estimated that, on the average, the cost per head capacity where housing is involved is about as follows for three types of feedlot facilities[6]:

| Type of Facility | Sq Ft per Animal | Cost per Animal Capacity |
|---|---|---|
| | | ($) |
| Open shed ........... | 20 | 140 |
| Cold confinement ....... | 17 | 185 |
| Warm confinement (heated) | 17 | 300 |

## CONFINEMENT FEEDING; SLOTTED (OR SLATTED) FLOORS

Currently, there is much interest in cattle confinement feeding and slotted floors. The main deterrent is cost; construction costs vary with type of structure and may range up to $300 per steer space.

*Confinement cattle feeding refers to feeding in limited quarters, generally 20 to 25 sq ft per yearling animal, which is about ⅛ the space normally allotted to a yearling in an unsurfaced lot and ⅓ that of a paved lot.* The confinement is usually under roof on slotted floors.

*Slotted floors are floors with slots through which the feces and urine pass to a storage area immediately below or nearby.*

Interest in confinement feeding and slotted floors was ushered in for the purpose of (1) automating and saving labor, (2) cutting down on bedding and facilitating manure handling, (3) lessening mud, dust, odor, and fly problems, (4) increasing gains and saving feed, (5) lessening land requirements, and (6) lessening pollution.

Fig. 18–8. Slotted floor, through which manure falls into a collection pit. (Courtesy, USDA)

Research has shown conclusively that cattle fed during the winter months in cold areas gain faster and more efficiently if they are sheltered. However, as pointed out earlier in this chapter under the section headed "Open Pen Feedlot," the per head cost is much higher for confined or sheltered cattle. Thus, the decision on whether or not cattle confinement can be justified, even in the northern part of the United States,

should be determined by economics. Will the cattle in confinement quarters gain sufficiently more rapidly and efficiently to justify the added cost? Of course, manure disposal and pollution control should also be considered.

## COLD CONFINEMENT[7]

Fig. 18–9. Finishing cattle at feed in an open shed—cold confinement. (Courtesy, *Feedlot Management*, Minneapolis, MN)

Cold confinement refers to a more or less open shed for confining cattle; hence, winter temperatures therein are within a few degrees of outdoor temperatures. Open sheds should be faced away from the direction of the prevailing winds. Additionally, doors or other openings in the closed walls should be provided for summer ventilation.

## WARM CONFINEMENT[8]

Warm confinement refers to a confinement building for cattle which is sufficiently insulated and ventilated to maintain inside winter conditions about 35°F in severe weather, and in the range of 50° to 60°F most of the time.

---

[6]Ibid.

[7]The terms *cold confinement* and *warm confinement* refer to winter conditions. Without mechanical cooling, both systems are warm during the summer months.

[8]Ibid.

## DESIGN REQUIREMENTS

The figures that follow, based on information and experiences presently available, may be used as guides.

• **Floor space**—Allot 15 to 30 sq ft per animal exclusive of the bunk and alley, with an average of 20 to 25 sq ft for a 1,000-lb animal.

• **Animals per pen**—25 to 100 head per pen, with 25 to 30 being most common.

• **Bunk**—Allow 6 to 18 in. of linear bunk space per animal, with the amount of feeding space determined by frequency of feeding and size of animal.

• **Waterers**—Locate one waterer per 25 head at the back (opposite feed bunk) of each pen, preferably in the center.

• **Slats**—Reinforced concrete, steel, or aluminum may be used. Most concrete slats are 5 to 6 in. wide across the top, 6 to 7 in. deep, tapered to 3 to 4 in. wide at the bottom, and placed so as to provide a slot width of 1½ to 1¾ in.

• **Manure production and storage**—Manure production will vary with size of animal and kind of feed, but it will be approximately as follows:

| Animal | Cu Ft/Day Solids & Liquids | % Water | Gallons/Day |
|---|---|---|---|
| 1,000-lb steer | 1–1½ | 80–90 | 7½–10¾ |

A rule of thumb is that when the pit occupies the entire area beneath the cattle, it will fill at a rate of 8 to 10 in. per month.

A newer and less costly system than slotted floors, which seems to be gaining favor among some feedlot operators, consists of a concrete floor sloped so that manure drains into gutters spaced equidistant throughout the building. Periodically, water is flushed through the gutter, carrying the waste material to pits.

## FEEDING FINISHING (FATTENING) CATTLE

The principles of nutrition are covered in Chapter 4; hence, they will not be repeated in this chapter. Instead, the application of nutrition to cattle finishing (fattening) will be covered.

Digestion of feedstuffs in ruminants is primarily a fermentation process that occurs in the rumen. This allows ruminant animals to use both roughages and grains as sources of carbohydrates for energy. Part of the carbohydrates pass through the rumen and are digested in the abomasum and small intestine. Most carbohydrates in feeds are converted to either acetic, propionic, or butyric acid by rumen bacteria and protozoa. These short chain fatty acids are then absorbed through the rumen wall into the bloodstream and are eventually used for energy in body tissues.

The major nutritional requirements of finishing cattle are: energy, protein, minerals, vitamins, and water.

About 75% of the cost of finishing cattle, exclusive of the purchase price of the feeders, is feedstuffs—grain, hay, silage, and miscellaneous wastes and by-products. The greatest need is for energy. Of course, net profit depends on how much of that energy can be converted to pounds of gain—and how efficiently.

## RATION FORMULATION

Feedlot cattle have access only to the rations provided by the caretaker. It is important, therefore, that cattle feedlot rations be balanced, and that they make for maximum net returns.

In addition to considering changes in availability of feeds and feed prices, ration formulation should be altered at stages to correspond to weight increases in the cattle.

Some suggested rations that may serve as useful guides are given in Chapter 15, Table 15–3.

Ration formulation consists of combining feeds to make a ration that will be eaten in the amount needed to supply the daily nutrient requirements of the animal. Instructions on "How to Balance a Ration" are given in Chapter 4, "Feeding Livestock"; hence, the reader is referred thereto.

## FEEDS

The growth of the cattle-feeding industry of America has gone hand in hand with the production and feeding of more grains and by-product feeds. Such feeds are high in energy and low in fiber. Hence, their availability and price influence the extent, location, and type of feeding program. Roughages, which require relatively more energy to digest and metabolize than grains, are used at low levels in most finishing rations. However, they are important in grower programs or in warm-up rations and they are even more important for maintenance of the breeding herd.

## ADDITIVES, IMPLANTS, AND INJECTABLES

Additives, implants, and injectables constitute a diverse group of products that are used to improve feed efficiency, regulate growth, protect animals from diseases and infections, and to provide safe and wholesome products to consumers at lower costs than would otherwise be possible. These products have been used extensively in cattle finishing since 1952, at which time Iowa State University researchers announced the results of cattle-feeding trials indicating a major breakthrough in lowering feed usage and increasing weight gains by feeding the compound diethylstilbestrol (DES). In 1979, the use of this product was banned by the FDA, because it was found to be cancer-producing in laboratory animals. But, subsequently, a host of new and safe additives, implants, and injectables have been developed, and have become widely used in cattle finishing. These products are listed and discussed in Chapter 4 of this book in the section headed "Additives, Implants, and Injections."

## OTHER METHODS OF IMPROVING RATE AND EFFICIENCY OF GAIN

Several other methods, in addition to additives, implants, and injectables, can be used to increase the rate of gain of feedlot cattle; among them, the following:

1. Feed young bulls (uncastrated males) instead of steers. The male hormones secreted by the testicles are excellent growth stimulants and will improve gain and feed efficiency by 10 to 15%. Alternatives to bulls that merit consideration are short-scrotum bulls (induced cryptorchidism), and Russian castrates. With these methods, testosterone is produced, yet, in comparison with bulls, the animals are easier to handle and may be carried to advanced ages without being labeled *bull beef*.

2. Take advantage of the genetic improvement of beef cattle by crossbreeding and introduction of genes from exotic breeds. This offers one of the most permanent ways of increasing the weaning weight of calves and improving performance in feedlot cattle. The selection of fast gaining and efficient cattle within the straightbreds, and in crossbreeding, may improve the efficiency of performance of feedlot cattle by 10% or more.

3. Reduce the cost of producing beef by improving the quality of cattle rations through grain processing and nutritionally balanced protein supplements.

4. Eliminate internal and external parasites, and protect cattle against the common diseases. This will save millions of dollars for both cattle feeders and consumers.

5. Keep abreast of new developments, including the discovery of new growth stimulants.

## FEED PREPARATION

The preparation of feeds for cattle is fully covered in Chapter 4, section entitled, "Feed Processing," including Table 4–10 therein; hence, the reader is referred thereto.

## FEED CONSUMPTION; RATE OF GAIN

If they don't eat it, they won't gain. But feed consumption and rate of gain are affected by many things.

The daily consumption of dry matter (feed) by cattle is primarily dependent upon the following:

1. **Size.** Large feedlot cattle consume more feed, animal for animal, than small cattle. But they may or may not be more efficient than small cattle when consideration is given to (a) the production efficiency of their dams, and (b) carrying them to the same degree of finish.

2. **Age and condition.** Older and more fleshy feedlot cattle consume less feed per unit of liveweight than do younger, leaner animals. Mature animals in good condition may be expected to consume amounts of dry matter equal to 2% or more of their liveweight, whereas thin animals eating high-quality roughage should eat amounts equal to 3% of their liveweight per day.

3. **Digestible nutrient content (energy density).** As digestible nutrient content (energy density) increases, consumption of feed dry matter is usually reduced. It follows that feed efficiency is improved in high- and all-concentrate rations, due to their high energy. In the final analysis, however, the comparative price of concentrates and roughages—the economics of the situation—will be a major determining factor.

4. **Environmental stress.** Cattle producers have long known that environmental stress caused by high and low temperatures, mud, and other adverse environmental factors can affect the voluntary consumption of feed. For example, feedlot cattle consume less during very hot weather than in cold weather.

The rate of gain of feedlot cattle is influenced by the following factors:

1. **Sex.** Under feedlot conditions, at comparable weight and finish, bulls can be expected to make about 10% greater gains than steers, and steers can be expected to make about 10% greater gains than heifers.

2. **Implants and growth stimulants.** The use of certain implants and growth stimulants in finishing steers and heifers usually increases gains by 8 to 12%.

## WATER

Water is the cheapest feed! Thus, cattle should have access to plenty of clean, fresh water at all times. They will consume 7 to 12 gal per head per day. In cold climates, waterers should be equipped with heaters. Where the water supply is not limited by cost or volume, continuous-flow waterers are excellent. In order to keep the pathogen and algae content at a minimum, water tanks should be cleaned at least once a week in the winter and twice a week in the summer. In sick pens and pens of new cattle, the water tanks should be cleaned daily.

## MANAGEMENT OF FEEDLOT CATTLE

Although it is not possible to arrive at any overall, certain formula for success in operating a cattle feedlot, those operators who have made money have paid close attention to the details of management.

There are many facets of cattle management. Only those that are unique to cattle feedlots will be discussed in the sections that follow.

### BACKGROUNDING

*Backgrounding is the preparation of cattle from weaning until placing on finishing rations.* It involves maximum roughage consumption and moderate gains.

The growing of calves from weaning until placing on finishing rations is not new. Only the term *backgrounding* is new. It evolved with the development of large commercial feedlots. However, growing of calves to the yearling stage for placement in feedlots was, and still is, known as growing stockers. (Also, see Chapter 17 of this book for further discussion of backgrounding.)

### FEED BUNK MANAGEMENT

Feed bunk management is a combination of management factors involved with obtaining a maximum performance, minimum digestive disorders, and keeping cattle on feed. Feed bunk management and quality control are directly involved with obtaining maximum and economical performance from cattle. It should be every feeder's goal to obtain maximum feed intake of a consistently high-quality ration, since both rate and efficiency of gain are directly related to nutrient intake.

### AMOUNT TO FEED; FULL VS LIMITED FEEDING

Feed intake is one of the key factors affecting feedlot performance. Perhaps no other factors have such overriding importance in determining rate and efficiency of gain, and, ultimately the profit derived from feeding cattle. Of course, the reason for emphasis on high feed intake is that once a sufficient amount of the ration is consumed to meet the maintenance needs of a finishing animal, the remainder is converted to gain with remarkable efficiency. Thus, by adding 4 lb to the daily feed intake of a 600-lb steer, rate of gain may be increased by $1\frac{1}{10}$ lb per day. Conversely, poor feed intake results in too high a percentage of the total nutrients being expended for maintenance.

Thus, finishing cattle should receive a maximum ration over and above the maintenance requirements. In general, they will consume daily an amount (on an air-dry basis) equal to 2.5 to 3.0% of their liveweight. Feed intake will vary according to the condition of the cattle, the palatability of the feeds, the energy of the ration (in general, animals eat to meet their energy needs), the weather conditions, and the management practices. For example, older and more fleshy cattle consume less feed per hundred-weight than do younger animals carrying less condition; thus, mature, over-finished steers will consume feeds in amounts equal to about 1.5% of their liveweight, whereas thin steers under 2 years of age will consume fully twice as much feed per unit liveweight.

Overfeeding is also undesirable, being wasteful of feeds and creating a health hazard. When overfeeding exists, there is usually considerable leftover feed and wastage, and there is a high incidence of bloat, founder, scours, and even death. Animals that suffer from milk digestive disturbances are commonly referred to as *off feed*.

Limited feeding means just what the name indicates—not giving the animals all they want. Limited feeding generally decreases the rate of gain, adversely affects feed conversion, and increases cost of gains. Under most conditions, cattle should be full fed throughout the finishing period.

## MUD PROBLEM

University of California studies show that mud can reduce cattle gains by as much as 25 to 35%. Thus, it is important that the problem be minimized, especially in high rainfall areas. Good drainage is the first essential. This should be assured at the time the feedlot is located and constructed. Mounds, preferably perpendicular to the feed bunk, will provide cattle a dry place on which to lie down. Concrete aprons along the bunk will provide them with solid footing on which to stand and feed. Also, lessening of cattle density during the winter months—fewer animals per lot—is an effective method of controlling the mud problem. Thus, many feedlots plan to feed fewer cattle during the muddy season.

## BIRD CONTROL

Birds are gluttonous and filthy; hence, they should be controlled. Nationwide, starlings constitute the major feedlot bird problem. Other feed-consuming bird species commonly identified in feedlots are: brewer blackbirds, redwing blackbirds, and cowbirds.

Some large commercial feedlots estimate their starling population at 100,000 per lot. Iowa feeders figure that starlings add $5 to $8 to the cost of each steer marketed. Some western feedlot operators compute the cost for overwintering each 1,000 starlings at $150; others estimate that starling nuisance and feed costs add 2¢ to the cost of each pound of gain.

In addition to feed consumption, birds contaminate feed and spread diseases—to both animals and humans. The starling has been incriminated in the spread of coccidiosis among animals, transmissible gastric enteritis (TGE) in swine, and histoplasmosis in humans.

Recordings of distressed bird calls, carbide cannons, and harassment or killing with guns achieve only partial control.

Many chemicals and baits have been tested, and a few have been found to be effective. However, some states do not allow the use of chemicals in bird control. Therefore, before using any chemical, the cattle feeder should check with the appropriate federal, state, and local departments of health.

## FLY CONTROL

The housefly is the most common type of fly found around cattle feedlots. It is a scavenger and does not feed on animals, but it does cause irritation and annoyance. Stable flies, which are blood feeders, may also be present in certain areas and certain feedlots.

Effective housefly control requires proper animal waste management and good feedlot sanitation. The basic objective in fly control is to eliminate possible sources of fly development. This can be accomplished by the following steps: (1) Provide proper drainage and avoid wet spots; (2) remove manure immediately after a pen is vacated; and (3) remove manure and spilled feed at important fly-breeding areas such as fence lines, feed bunks, hospital pens, horse pens, truck-washing stations, and receiving and shipping areas. Chemical control should be used in conjunction with the proper waste management techniques, and not as a sole means of control. Residual and space sprays aid in reduction of adult flies; and larvicides may be applied to areas of intense larval development such as manure stockpiles, hospital, and horse pens.

## FEEDLOT POLLUTION CONTROL

In recent years, there has been a worldwide awakening to the problem of pollution of the environment (air, water, and soil) and its effect on human health and on other forms of life. Much of this concern stemmed from the amount of manure produced by the sudden increase of animals in confinement. Certainly, there have been abuses of the environment (and it hasn't been limited to agriculture). There is no argument that such neglect should be rectified in a sound, orderly manner, but it should be done with a minimum disruption of the economy and lowering of the standard of living.

Pollution control is a most critical factor in site selection and operation of a cattle feedlot. Remoteness from urban development is recommended because of dust and odor. Also, before constructing a cattle feedlot the owner should become familiar with both state and federal regulations. Federal regulations apply to feedlots (1) with 1,000, or more, feeder cattle; or (2) in which animal wastes either empty directly into a stream that crosses the feedlot, or are conveyed directly into a nearby waterway by a pipe, ditch or other means. The state regulations can be secured from the state water board. They differ from state to state, but most states require a catch basin (detention pond) sufficient to contain the runoff from a storm of the magnitude of the largest rainfall during a 48-hour period of the most recent 10 years. A feedlot may minimize runoff by locating near the top of the slope and, if necessary, by using diversion embankments to divert runoff from other areas.

Cattle feedlots located near centers of populations are having an increasing number of complaints lodged against them because of manure, dust, and odor. Lawsuits, based on the nuisance law, are being filed against them.

## HEALTH OF FEEDLOT CATTLE

Loss from disease is greater in feedlot operations than in any other type of cattle enterprise. The movement of cattle, stress conditions, methods of purchase, feeding of concentrated feeds, population density, sometimes unsanitary conditions, and the bigness and complexity of the operation all contribute to disease incidence; and disease incidence is directly proportional to population density.

About 675,000 head of feedlot cattle, equivalent to 2.6% of the fed cattle marketings, valued at $600 per head, die each year at an estimated total loss of $405 million.[9] It is further estimated (1) that losses from sickness run 2 to 5 times greater than actual death losses, and (2) that the combined losses from death and sickness add $40 per head onto the cost of every feedlot finished animal. This situation can be greatly improved by (1) preconditioning, (2) moving cattle directly from the producer's farm or ranch to the feedlot (fewer than 20% of the cattle now move directly from producer to feedlot), (3) reducing the time between ranch and feedlot, (4) lessening the amount of stress and exposure to infection during the marketing and transportation periods, (5) providing the feedlot receiving the cattle with more adequate medical and nutritional history of the cattle (6) handling incoming feedlot cattle properly, and (7) diagnosing and treating sick cattle early.

## HANDLING NEWLY ARRIVED CATTLE

The most critical period for feeder cattle is the first 28 days in the feedlot. The following recommendations pertaining to incoming cattle will minimize death losses and maximize performance:

• **Provide clean, dry, comfortable quarters**—Whether it be an open lot or a building, incoming cattle should be provided with clean, dry, comfortable quarters. A dry and comfortable bed for resting is very essential because cattle are tired and have a low resistance to respiratory diseases.

• **Process upon arrival**—The relative merits of processing calves (1) at point of origin, (2) upon arrival at destination, or (3) 2 to 3 weeks after arrival are often debated.

In a well-designed experiment, involving 358 calves that originated in Texas, the University of California provided the answer to this question. Based on (1) rate of gain, (2) disease resistance, and (3) cost per pound of gain for feed, processing, and medication, the California study showed that processing at arrival is best, and that processing at point of origin is preferable to delayed processing.

**Note well:** If the cattle are very tired and weak when they get off the truck, it may be advisable to delay processing for 24 hours, but processing should not be delayed more than 72 hours.

• **Provide clean, fresh water**—Give the cattle easy access to clean, fresh water because they are usually dehydrated and thirsty upon arrival and will drink water before they eat feed. Open water tanks are preferable to automatic water bowls because most farm and ranch cattle are accustomed to drinking from tanks or ponds.

---

• **Provide a palatable ration**—Feeding a palatable ration—one that cattle will start eating soon after they are unloaded in the feedlot—will reduce the incidence of shipping fever and make the cattle recover their weight loss more rapidly.

1. **Roughage**—The best roughage for newly arrived feedlot cattle is *long grass* hay, because it is very similar in composition and taste to the grass to which most feedlot cattle have been accustomed. Thus, cattle will usually eat long grass hay more quickly than any other roughage. In areas where grass hays are not available, or are too expensive to feed, any other nonlegume roughage can be fed, such as corn silage, sorghum silage, cottonseed hull, corncobs, or grass-legume hay that contains more grass than legumes. Above all, do not feed high-quality alfalfa hay because it is too laxative and it will cause scouring which will trigger shipping fever. The same may be said relative to alfalfa haylage or alfalfa silage.

Corn silage of approximately 65% moisture content is an excellent feed for new cattle. If cattle do not eat the corn silage too well at the outset, the feeder should sprinkle a little grass hay on the top of it to encourage them to start eating.

2. **Concentrate**—Incoming cattle may be fed approximately 4 lb of concentrate per head daily, consisting of 2 lb of grain and 2 lb of protein supplement. The protein supplement should be fortified so as to provide 50,000 IU of vitamin A daily. For heavily stressed cattle, the protein supplement should also contain a high level of antibiotic, or a combination of antibiotic and a bactericidal agent such as sulfamethazine. The following level of antibiotic-sulfamethazine is recommended:

Feed 350 mg of Aureomycin plus 350 mg of sulfamethazine per head daily to newly arrived cattle for a period of 28 days. With the antibiotic-sulfamethazine treatment, shipping fever is practically alleviated.

Fig. 18–10. Incoming cattle may be started on about 4 lb of concentrate/head/day, consisting of 2 lb of grain and 2 lb of protein supplement. (Courtesy, Ralston Purina Company, St. Louis, MO)

Do not feed urea for the first 28 days after the cattle arrive. Starvation destroys the ability of the rumen to utilize urea or other nonprotein nitrogen and makes cattle more sensitive to urea toxicity. Therefore, it is not wise to put extra stress on cattle by using urea during this adjustment period.

• **Satisfy mineral hunger**—Incoming cattle are usually hungry for minerals, especially if they have been on dry range forage.

Thus, they should have access either to a mineral mixture consisting of two parts of dicalcium phosphate and one part of salt, or to a good commercial mineral.

• **Observe, isolate, and treat sick animals**—Newly arrived cattle should be observed at least twice daily. Sick animals should be removed and treated. Treating sick animals promptly, rather than waiting until tomorrow, may mean the difference between life and death. Animals that show clinical signs of shipping fever—sunken eyes, runny nose, drooling at the mouth, labored breathing, and/or weaving (unsteady gait)—should be isolated in a separate sick pen or hospital.

Rest, fresh water, good feed, proper medication, and TLC (tender loving care) are the cardinal essentials for preventing shipping fever and death losses.

## OVERFINISHING

Excessive finishing is undesirable, both from the standpoint of the producer and the consumer. Experienced cattle feeders are fully aware of the fact that to carry finishing cattle to an unnecessarily high finish is usually prohibitive from a profit standpoint. This is true because the gains in weight then consist chiefly of fat but little water. In addition, a very fat animal eats less heartily, with the result that a small proportion of the nutrients, over and above the maintenance requirement is available for making body tissue.

Fig. 18–11 shows that the heavier the cattle, the more expensive the gains. This points up (1) the importance of topping out finished cattle, rather than waiting until the entire lot is ready; and (2) the reason it is generally wise to sell cattle when they are ready to go, rather than to hold for a higher market.

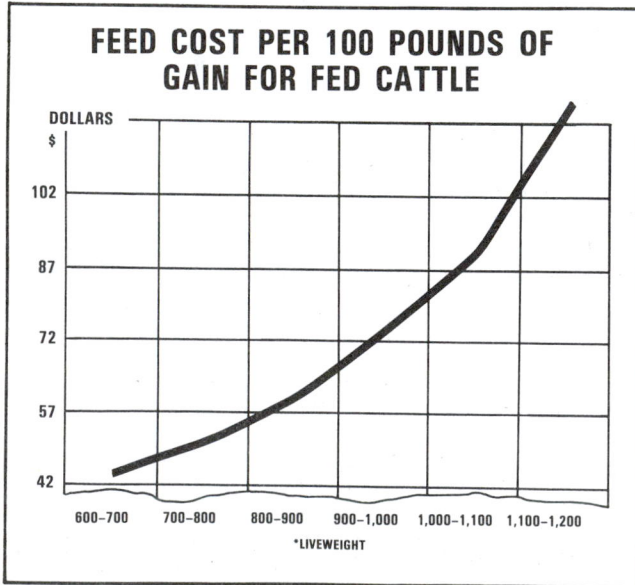

**FEED COST PER 100 POUNDS OF GAIN FOR FED CATTLE**

DOLLARS
$

102

87

72

57

42

600–700  700–800  800–900  900–1,000  1,000–1,100  1,100–1,200
*LIVEWEIGHT

Fig. 18–11. This graph illustrates changes in feed conversion efficiency for cattle from normal feeder weights to slaughter weights. Note that feed costs per 100 lb gain about double from 600–700 to 1,000–1,100 lb and that the conversion efficiency ratio changes even more sharply when cattle pass 1,100 lb.

## PASTURE FINISHING CATTLE

When grains are scarce and high in price, more cattle are grass finished. But, because young cattle grow and do not reach market finish under usual pasture conditions, it is impossible to finish them at early ages and light weights without either supplemental feeding on pasture and/or lot finishing at the end of the grazing season.

Generally speaking, no cheaper method of harvesting forage crops has been devised than is afforded by harvesting directly by grazing animals. Moreover, even most seeded pastures last several years; thus, seeding cost may be distributed over the entire period. Naturally, the cash income to be derived from pastures will vary from year to year and from place to place depending upon such factors as market price levels, class of animals, soil, season, and the use of adapted varieties.

Fig. 18–12. Hereford steers on pasture in Indiana. (Courtesy, J. C. Allen and Son, West Lafayette, IN)

## ADVANTAGES AND DISADVANTAGES OF GRAIN FEEDING ON PASTURE

The **advantages** of grain feeding cattle on pasture, compared to strictly feedlot finishing are:

1. Pasture gains are cheaper because (a) less grain is required per 100 lb of gain; (b) grass is a cheaper roughage than hay or silage; and (c) less expensive protein supplement is required. Generally speaking, self-feeding on pasture vs drylot indicates that pasture saves about 100 lb of dry feed per 100 lb of gain. Thus, if we assume that it requires 500 lb of gain to finish a steer then each steer would require 500 lb less feed on pasture than in drylot. If feed costs 7¢ per pound, that's a savings of $35 per steer on feed cost.

2. Less labor is required because the cattle gather their own roughage and the labor required for feeding roughage is eliminated. In brief, grass-finished cattle do their own harvesting. Furthermore, it may be possible to get satisfactory results with but one grain feeding each day in finishing on pasture, or the animals may be self-fed, with the caretaker merely filling the feeder at intervals.

3. Handling of manure is eliminated, the maximum fertility value of the manure is conserved, and there is no pollution problem. When pastures are utilized by livestock, approx-

imately 80% of the plant nutrients of the crop is returned to the soil.

4. Pasture finishing eliminates any requirement for buildings.

5. Finishing cattle on pasture is especially adapted to the small feeder and to areas where some of the land should be kept in permanent pasture.

The **disadvantages** of finishing cattle on pasture, compared to strictly feedlot finishing are:

1. Most feeder cattle are marketed in the fall rather than in the spring. Therefore, feeder steers purchased in the spring and intended for pasture finishing are usually scarce and high in price.

2. Though less labor is required, less labor is available. The cropping season is a rush season.

3. During the midsummer, the combination of heat and flies may cause much discomfort to the animals and reduce the gains made.

4. Pastures may become dry and parched, reducing the gains made during dry seasons.

5. The manure is usually dropped on permanent pastures year after year, which may result in the neglect of the other fields.

6. In many pastures, availability of shade and water does not present a problem. However, some areas are less fortunate in this regard.

After both the advantages and disadvantages of pasture finishing are considered, the availability of cheap, rough pastureland and the price of concentrates will usually be the determining factors in deciding upon the system to follow.

## SYSTEMS OF PASTURE FINISHING

When cattle are finished on pasture, any one of the following systems may be employed:

1. Finishing on pastures alone—no concentrates being fed.
2. Limited grain allowance during the entire pasture period.
3. Full feeding during the entire pasture period.
4. Full or limited grain feeding on pasture following the period of peak pasture growth.
5. Short feeding (60 to 120 days) in the feedlot at the end of the pasture period.

The system of pasture finishing that will be decided upon will depend upon the age of the cattle, the quality of the pasture, the price of concentrates, the rapidity of gains desired, and the market conditions.

## BASIC CONSIDERATIONS IN UTILIZING PASTURES FOR FINISHING CATTLE

The following points are basic in utilizing pastures for finishing cattle:

• **Moderate winter feeding makes for most effective pasture utilization**—The more liberally beef cattle are fed during the winter, the less will be their effective utilization of pasture the following summer—the less the compensatory gains. Generally speaking, for maximum utilization of pasture, stocker calves should be fed for winter gains not in excess of 1.25 lb per head daily, and yearlings not in excess of 0.9 lb.

• **Early pastures are "washy" but high in protein**—Cattle should not be turned on pasture too early. The first growth is extremely "washy," possessing little energy. However, the crude protein content of the forages is high during the early stages of growth and rapidly decreases as the forages mature. This would indicate the importance of pasturing rather heavily during the period of maximum growth in the spring and early summer.

• **Sudden changes are to be avoided**—Changes from drylot to pastures or from less succulent to more succulent pastures should be made with care; for grass is a laxative, and the cattle may shrink severely. Also, bloat may occur.

• **Time of starting grain feeding on pasture is determined by condition of cattle and quality of pastures**—Cattle that have been fed grain rather liberally through the winter and are in good condition should usually be fed grain from the beginning of the grazing period. On the other hand, if they have been roughed through the winter, it may be just as well to feed the grain only during the last 80 to 120 days of the grazing season, after the season of peak pasture growth. The latter recommendation is made because it is sometimes difficult to get animals to consume grain when an abundance of palatable forage is available. At peak pasture growth, the animals should be started on feed and brought to full feed as rapidly as possible.

• **Grain supplements on pastures usually make for larger daily gains and earlier marketing**—Young cattle (calves and yearlings) on summer pasture usually do not grow at their maximum potential due to energy and protein deficiencies in the feed at various times of the season. Thus, the addition of a grain supplement for cattle on pasture makes for larger daily gains and earlier marketing—either directly off grass or with a shorter drylot finishing period. The owner thus avoids late fall competition and lower prices of strictly grass cattle. Also, because cattle that are grain fed on pasture can be marketed over a wider period of time, there is greater flexibility in the operations. However, cattle that are supplemented on summer pasture often sell for less to go into feedlots because feedlot operators fear that they may not gain as rapidly as cattle that are not supplemented on pasture; they feel that more rapid gains on pasture make for less rapid and efficient gains in the feedlot—for loss of compensatory growth.

• **Whole corn preferred to rolled corn**—When self-feeding steers on pasture, whole corn is preferred to rolled corn for the following reasons: (1) Slightly less feed is required per 100 lb gain; (2) it alleviates processing cost; and (3) it results in less incidence of founder and rumen parakeratosis because whole corn supplies some roughness factor in the ration to stimulate the rumen.

• **Protein supplement not needed on good pasture**—As long as pasture is green and growing, no supplemental protein is required. During drought periods and late fall when the grass matures, extra protein is needed. At such times, it is good business to add 1 lb of protein supplement to each 8 to 12 lb of grain. Usually this will increase the rate and efficiency of gain.

• **Carrying capacity of pastures will vary**—The carrying capacity of pastures will vary with the amount of grain supplement, the quality of pasture, and the age and condition of the cattle. Because of these factors, the acreage per steer will vary all the way from 1 to 10.

• **Age is a factor**—Young cattle (yearlings) tend to grow as well as to fatten. Thus, older cattle (2 years or older) will reach a high degree of finish on pastures alone. As fine as pastures may be, it must be remembered that grass is still roughage.

• **Minerals for cattle on pasture**—Salt is especially necessary when grass is being utilized. Finishing steers consume from ¾ to 1½ oz of salt per head daily. Also, cattle on pasture should have free access to a mineral mixture composed of 2 parts of dicalcium phosphate and 1 part of iodized salt.

• **Species of grasses or legumes will vary**—The most desirable species of grasses or legumes or grass-legume mixtures to be seeded will vary according to the area, especially according to the soil and climatic conditions. Pasture yields vary greatly from area to area and season to season.

Temporary or supplemental pastures, such as Sudan grass or millet, are used for a short period and are usually more productive and palatable than permanent pastures. They are seeded for the purpose of providing supplemental grazing during the season when the regular permanent or rotation pastures are relatively unproductive.

Fig. 18–13. Yearling steers on grass pasture. (Courtesy, USDA)

• **Grass vs grass-legume mixtures should be considered**— In general, where adapted legumes can be successfully grown— either alone or with grass mixtures—the results are superior to yields obtained from pure stands of the grasses. At Pullman, Washington, in a study of pure species of smooth brome and crested wheatgrass vs grass-alfalfa mixtures, it was found that (1) the grasses produced an average of 87 lb of beef per acre, whereas the grass-alfalfa mixtures averaged 223 lb of beef per acre; (2) when based on forage yields at monthly intervals, the same mixtures produced 3 times as much oven-dry forage per acre as the pastures seeded to grasses alone; (3) the grass-legume mixtures provided a slightly longer grazing season; (4) the grass-legume mixtures provided a higher carrying capacity in terms of animals per acre; (5) the erosion-resisting characters of the soil were improved by the fibrous grass roots, both while the crop was growing and after the seeding had been plowed under; and (6) the addition of grasses to legumes tended to keep out cheatgrass and other undesirable plants.[10] The two latter points are based merely on careful observation, whereas the rest of the points are proved experimentally.

• **Grain feeding will lengthen the grazing season**—At the Washington Agricultural Experiment Station, grain feeding cattle on pasture lengthened the grazing season by an average of 57 days.

---

[10]Ensminger, M. E., et al., *Wash. Ag. Exp. Sta. Bull.* No. 444.

• **Self-feeding vs hand feeding on pasture**—Self-feeding on pasture has generally proved superior to hand feeding, as the animals consume more feed, make more rapid gains, and return more profit.

• **Economy of grain feeding on pasture**—Whether or not it will be profitable to feed grain on pasture will depend primarily upon the price of grain, the premium paid for cattle of higher finish and grade, the season in which it is desired to market, and the area and quality of pasture.

• **Pasture bloat can be controlled**—The following practices will be helpful in reducing the bloat hazard:

1. Give a full feed of hay or other dry roughage before the animals are turned to legume pastures, to prevent the animals from filling too rapidly on the green material.

2. After the animals are once turned to pasture, they should be left there continuously. If they must be removed overnight or for longer periods, they should be filled with dry roughage before they are returned to pasture.

3. Mixtures that contain approximately half grasses and half legumes should be used.

4. Water and salt should be conveniently accessible at all times.

5. The animals should not be allowed to become empty when they congregate in a drylot for shade or insect protection and then be allowed to gorge themselves suddenly on the green forage.

6. Many practical producers feel that the bloat hazard is reduced by mowing alternate strips through the pasture, thus allowing the animals to consume the dry forage along with the pasture. Others keep in the pasture a rack well filled with dry hay or straw.

7. Because of the many serrations on the leaves, Sudan hay appears especially effective in preventing bloat when fed to cattle on legume pastures.

Where legume bloat is encountered, use poloxalene (Bloat Guard), oxytetracycline (antibiotic), or polyoxyethylene (23) lauryl ether (Laureth-23/Enproal Bloat Blox) according to the respective manufacturer's directions.

## FUTURE OF CATTLE FINISHING (FATTENING)

Fig. 18–14. Future cattle finishing will be sophisticated, integrated, big, and paced by the broiler and pork industries. (Courtesy, National Cattlemen's Assn., Englewood, CO)

The author's crystal ball shows that the cattle-feeding industry in the future will change; it will be characterized by the following:

1. Cattle price will be pressured downward by cheaper broilers and pork.
2. Cycles will almost disappear.
3. Lower profits per animal will be normal.
4. Financing will be available for well managed outfits.
5. More custom feeding will be on a guaranteed cost of gain basis.
6. The trend will be to more grass and other roughage; heavier feeders; shorter feedlot periods.
7. Leaner beef will be produced.
8. Feed grain supplies and prices will be unstable and volatile.
9. More beef will be produced from grass and milk.
10. There will be more pasture finishing.
11. More by-product feeds will be utilized in feedlots.
12. More nonprotein nitrogen (NPN) and single-cell protein (SCP) will be fed.
13. More lots will be operating to capacity.
14. There will be more integrated feedlots, both horizontally and vertically.
15. There will be lessened tax shelter advantages.
16. Feedlot managers with superior business acumen will become more important.
17. There will be increased computerized management.
18. Greater emphasis on public relations will develop.
19. Pollution control will be the order of the day.
20. Recycling manure will be perfected.
21. Increased mechanization will be necessary.
22. More sheltered or confined feeding will be done.
23. The growth areas ahead will be the Southern Plains and the Central Plains.
24. Feedlots will spread further over the country so as to be located near available feeds and cattle.
25. More feeders will be produced by cows kept for beef.
26. More crossbreds will be fed.
27. More efficient cattle will be fed.
28. Most feeders will be preconditioned.
29. New additives will be developed.
30. More bulls will be fed.
31. Feedlots will be producing beef to more exacting specifications.
32. More beef being fabricated, boxed, and branded in packing plants.
33. Increased use of frozen beef.
34. More beef tenderizers being used.
35. Inroads made by simulated meats (synthetic meats, or meat analogs).
36. Improved market efficiency.
37. Beef futures trading specifications that are more exacting and more regulated.
38. Beef imports causing constant apprehension among beef producers.
39. More finished beef will be exported to Japan and Europe.
40. Less emphasis on carcass quality.
41. Broilers and pork playing major roles in the retail price of beef, the price of cattle on foot, and cattle numbers.

# QUESTIONS FOR STUDY AND DISCUSSION

1. What factors should determine the following alternative choices of a cattle feeder: (a) cattle feedlot (including confinement feeding) vs pasture finishing; and (b) farmer-feeder vs commercial cattle feeder?

2. What were the primary reasons for the shift in the center of the geography of cattle feeding from the Corn Belt to the West? Which force was the most important in this shift?

3. List the six leading cattle-feeding states of the nation. What factors are favorable to cattle feeding in each of them?

4. What factors caused cattle feedlots to get so big that feedlots with 1,000 head and over marketed 82% of the nation's cattle in 1987? (See Fig. 18-2.)

5. Detail the reasons back of the phenomenal growth of cattle feeding following World War II.

6. List and discuss the characteristics of modern cattle feeding.

7. Discuss the effect of age of cattle on each of the following: (a) daily gains, (b) feed efficiency, (c) length of feeding period, (d) total gain required to finish. (See Tables 18-1 to 18-4.)

8. How will each of the following factors affect the choice between feeding calves versus feeding yearlings: (a) experience of the feeder, and (b) kind and quality of feed?

9. Discuss how each of the following enters into the choice of the kind of cattle to feed in a given lot; (a) steers, heifers, spayed heifers, or bulls; (b) grade of cattle; and (c) breeding and type of cattle.

10. Would you recommend that a cattle feeder breed all heifers about 3 to 4 months before marketing? What pertinent facts did the Monfort study show relative to the effects of pregnancy on feedlot heifers?

11. Will more young bulls be fed in the U.S. in the future? Justify your answer. List five guidelines that should be followed when feeding bulls in a feedlot.

12. What advantages may accrue from feeding crossbreds rather than straightbreds?

13. Will more dairy beef be fed in the future? Who will feed the dairy beef of the future—the dairy producer or the specialized feedlot feeder?

14. What breed of cattle is particularly suited to the production of dairy beef?

15. When feeding dairy beef, what will determine the choice between: (a) high-energy rations and light market weight cattle; vs (b) high-roughage rations and heavy market weight cattle?

16. What business aspects of cattle feeding need to become more sophisticated and efficient?

17. List factors other than the cost of feed, amount of gain, and price of slaughter cattle that affect the price that a feeder can afford to pay for feeder cattle.

18. Define (a) price margin, (b) feeding margin, (c) positive margin, and (d) negative margin. Why isn't the relative importance of price margin vs feeding margin the same for calves as for yearlings?

19. Discuss each of the types of custom feeding contracts.

20. Discuss each of the following points pertinent to futures trading in finished cattle: (a) specifications; (b) commission fees and margin requirements; (c) how a futures contract works; (d) delivery against contract; (e) advantages and limitations of live (slaughter) cattle futures; (f) live cattle options, including calls or puts, and (g) how to go about hedging beef futures.

21. Under what circumstances would you recommend that a cattle feeder use beef futures? Under what circumstances would you recommend that a cattle feeder not use beef futures?

22. Select a certain area for feeding cattle. Then, give for that particular area the pros and cons for each of the following: (a) an open pen feedlot, (b) cold confinement, and (c) warm confinement. Finally, give your recommendation.

23. What are the (a) advantages, and (b) disadvantages of slotted floors?

24. What are additives, implants, and injectables? When and what ushered in the era of additives in cattle feeding? What methods in addition to additives, implants, and injectables can be used to increase the rate and efficiency of gain of feedlot cattle?

25. List and discuss the factors upon which the daily amount of feed is primarily dependent.

26. Discuss each of the following management aspects of feedlot cattle: (a) backgrounding, (b) feed bunk management, and (c) full vs limited feeding.

27. How may a feedlot mud problem be lessened?

28. How would you control (a) birds, and (b) flies in a cattle feedlot?

29. What can a feedlot do to lessen pollution?

30. What's a detention pond? How does it work?

31. It is estimated that the combined losses from death and sickness adds $40 per head onto the cost of every feedlot finished animal. Outline how this cost could be lowered.

32. Detail a program for handling newly arrived cattle that will minimize death losses and maximize performance.

33. Why is overfinishing undesirable?

34. Under what conditions would you recommend pasture finishing of cattle rather than feedlot finishing?

35. What are the (a) advantages, and (b) disadvantages of grain feeding on pasture?

36. In the future, is it likely that more cattle will be grass finished, perhaps by supplemental grain feeding?

37. Discuss the alternate systems of pasture finishing.

38. List and discuss the basic considerations in utilizing pastures for finishing cattle.

39. Using current feed prices, compute the value of grass on a per steer basis if it effects a saving of 100 lb of dry feed per 100 lb of gain. What additional advantages accrue from grain feeding cattle on pasture, compared to drylot finishing?

40. Discuss each of the following basic points as each applies to pasture finishing: (a) moderate winter gains; (b) early, "washy" pasture, (c) whole corn vs rolled corn; and (d) control of pasture bloat.

41. Discuss the future of cattle finishing.

## SELECTED REFERENCES

| Title of Publication | Author(s) | Publisher |
|---|---|---|
| *Beef Cattle*, Seventh Edition | A. L. Neumann | John Wiley & Sons, Inc., New York, NY, 1977 |
| *Beef Cattle Production* | K. A. Wagnon<br>R. Albaugh<br>G. H. Hart | The Macmillan Company, New York, NY, 1960 |
| *Beef Cattle Science*, Sixth Edition | M. E. Ensminger | The Interstate Printers & Publishers, Inc., Danville, IL, 1987 |
| *Beef Cattle Science Handbook* | Edited by<br>M. E. Ensminger | Agriservices Foundation, Clovis, CA, pub. annually since 1964 |
| *Beef Housing and Equipment Handbook* | Staff | Midwest Plan Service, Iowa State University, Ames, IA, 1975 |
| *Beef Production in the South*, Modified Edition | S. H. Fowler | The Interstate Printers & Publishers, Inc., Danville, IL, 1979 |
| *Diseases of Feedlot Cattle*, Third Edition | R. Jensen<br>D. R. Mackey | Lea & Febiger, Philadelphia, PA, 1971 |
| *Feeding Beef Cattle* | J. K. Matsushima | Springer-Verlag, New York, NY, 1979 |
| *Feedlot, The*, Third Edition | Edited by<br>C. B. Thompson<br>C. C. O'Mary | Lea & Febiger, Philadelphia, PA, 1983 |
| *Futures Markets for Livestock: Value as Marketing and Management Tools*, Res. Rpt. 63 | W. D. Dobson | Research Division, The University of Wisconsin, Madison, WI, 1970 |
| *Stockman's Handbook, The*, Sixth Edition | M. E. Ensminger | The Interstate Printers & Publishers, Inc., Danville, IL, 1983 |

King Charles II knighting the sirloin of beef, after a painting by John Gilbert. According to an old English legend, ecstatic with a platter of beef which was served at one of his feasts, King Charles II ceremoniously arose, touched his sword to the steaming platter, and proclaimed: "A noble joint—it shall have a title. Loin, I dub thee Knight—henceforth thou shall be *Sir Loin.*" (Courtesy, Radio Times Hulton Picture Library, London, England)

# MARKETING AND SLAUGHTERING CATTLE AND CALVES

# *Chapter 19*

Fig. 19–1. To market! To market! Cattle roundup for marketing. (Courtesy, Iowa Beef Processors, Inc., Dakota City, NE)

The general subject of livestock marketing is covered in Chapter 8 of this book, whereas this chapter is devoted specifically to the marketing and slaughtering of cattle and calves.

## MARKET CHANNELS FOR CATTLE

Most cattle are sold through the following channels: (1) public stockyards (terminals, some of which conduct auctions, also); (2) livestock auction markets; (3) nonpublic markets (direct), which means that they go directly to buyers and do not move through organized marketplaces such as terminals or auctions; and (4) on the basis of carcass grade and weight.

Channels through which cattle move to market have changed since 1960. Packers have shifted away from public stockyards (terminal markets) for their cattle purchases, to direct buying. Between 1960 and 1987, the proportion of cattle purchased by packers at terminal markets declined from 46% to 4.2%, while direct purchases increased from 39% to 80.2%. The remaining 15.6% were purchased at auction markets.

In 1987, packers bought a mere 3.2% of their calves at terminal markets. Auction markets were the source of 35.8% of the calves purchased by packers. The majority of their calves (61%) were purchased from nonpublic markets (direct).

In 1987, packers purchased their cattle and calves through the following market channels:

|  | Cattle | Calves |
|---|---|---|
|  | (%) | (%) |
| Nonpublic markets (direct) . . . . . . . . . . | 80.2 | 61.0 |
| Auction markets . . . . . . . . . . . . . . . . . . . | 15.6 | 35.8 |
| Terminal markets . . . . . . . . . . . . . . . . . . | 4.2 | 3.2 |

As noted, most cattle (80.2%) and most calves (61%), were sold direct to packers.

The rise in direct cattle purchase by packers went hand in hand with two developments:

1. The movement of plants away from terminal markets and closer to production areas, made possible by improved roads and truck transportation.

2. The growth of large cattle feedlots, which lend themselves to directs sales. The adaptation of large feedlots to direct marketing is pointed up by the following figures showing (a) steers and heifers, and (b) cows and bulls separated according to market channels, in 1987:

|  | Steers and Heifers | Cows and Bulls |
|---|---|---|
|  | (%) | (%) |
| Nonpublic markets (direct) . . . . | 91.1 | 31.3 |
| Auction markets . . . . . . . . . . . . . | 5.1 | 68.6 |
| Terminal markets . . . . . . . . . . . | 3.8 | 0.1 |

Steers and heifers are mostly fed cattle that represent the primary output of specialized cattle feeders. Cows and bulls, on the other hand, represent important by-products from beef-breeding herds and dairy enterprises.

The basis of sale of livestock as they move through market channels has changed, also. Cattle purchased by packers on carcass grade and weight increased from 5% in 1961 to 30.4% in 1987. Purchase of calves on the basis of carcass grade and weight increased from 3.9% in 1969 to 36.7% in 1987.

## PUBLIC STOCKYARDS

Fig. 19–2. Cattle being sold at auction in the Omaha, Nebraska public stockyards. (Courtesy, *The Drovers Journal,* Kansas City, KS)

*Public stockyards are livestock trading centers where animals are consigned to commission firms for private treaty selling (or they may be sold at auction in some public stockyards).*

Even though public stockyards have declined in importance in relation to other outlets, they are still a powerful and vital influence as evidenced by the following:

1. In 1987, 7.8 million cattle and calves (6.8 million cattle and 989,000 calves) passed through public stockyards.

2. They provide a price barometer in establishing values (a) for cattle sold direct, and (b) for cattle down on the farm.

3. They provide a desirable market outlet for (a) cattle producers with small numbers, or (b) producers without the necessary expertise to deal with highly professional skilled buyers.

4. Their marketing conditions and charges are known and standardized. Sales at public stockyards are made under standard conditions, on the basis of actual weights, and with the marketing charges that will be deducted from sale proceeds known.

Price quotations on cattle sold through terminal markets have the disadvantage of not representing the bulk of the sales, though they are often taken as representative. On the other hand, price quotations on direct sales have the disadvantage of being difficult to obtain and interpret.

## LEADING PUBLIC STOCKYARDS

Although public stockyards vary from year to year in total receipts, Table 19–1 shows the largest public stockyards in cattle receipts. It is noteworthy that Oklahoma City leads by a wide margin.

As would be expected, many market calves are of dairy breeding, especially the surplus bull calves that are not needed for breeding purposes. Of the remainder, a considerable number are culled out from beef herds because of undesirable type or breeding from the standpoint of future development. It can be expected, therefore, that the leading calf markets would not coincide with the leading cattle markets.

**TABLE 19–1**
**CATTLE RECEIPTS (EXCLUDING CALVES AND VEALERS) OF 10 LEADING PUBLIC STOCKYARDS, BY RANK, 1987[1]**

| Market | 1987 |
|---|---|
| Oklahoma City, OK | 687,299 |
| Sioux Falls, SD | 398,080 |
| La Junta, CO | 370,462 |
| So. St. Paul, MN | 347,430 |
| Torrington, WY | 324,977 |
| Dodge City, KS | 307,033 |
| Lexington, KY | 291,594 |
| Omaha, NE | 289,914 |
| Louisville, KY | 241,404 |
| Springfield, MO | 217,039 |
| Total of 36 public markets | 6,809,000 |

[1]*Livestock and Meat Statistics 1984–88*, Stat. Bull. No. 784, USDA, p. 61.

Table 19–2 shows the 10 leading public stockyards in calf receipts in the U.S. South St. Paul, Minnesota is primarily a market for veal calves.

**TABLE 19–2**
**CALF RECEIPTS OF 10 LEADING PUBLIC STOCKYARDS, BY RANK, 1987[1]**

| Market | 1987 |
|---|---|
| So. St. Paul, MN | 17,875 |
| So. St. Joseph, MO | 12,877 |
| Lancaster, PA | 7,032 |
| Oklahoma City, OK | 6,832 |
| Springfield, MO | 5,690 |
| Joplin, MO | 4,006 |
| Louisville, KY | 2,494 |
| Tulsa, OK | 1,882 |
| Memphis, TN | 1,551 |
| Indianapolis, IN | 755 |
| Total of 12 markets | 989,000 |

[1]*Livestock and Meat Statistics 1984–88*, Stat. Bull. No. 784, USDA, p. 61.

It is natural that the leading feeder cattle markets should be conveniently located between the cow-calf areas and the feeding areas. Table 19–3 gives the eight leading public stockyards in feeder cattle outshipments, by rank. As noted, Oklahoma City has a commanding lead, due primarily to its proximity to the High Plains feeding areas.

**TABLE 19–3**
**FEEDER CATTLE AND CALVES SHIPPED FROM PUBLIC STOCKYARDS, BY RANK, 1987[1]**

| Market | 1987 |
|---|---|
| Oklahoma City, OK | 407,253 |
| Dodge City, KS | 269,694 |
| Amarillo, TX | 154,158 |
| Sioux Falls, SD | 131,853 |
| Kansas City, MO | 98,312 |
| So. St. Joseph, MO | 85,548 |
| So. St. Paul, MN | 48,209 |
| Sioux City, IA | 31,399 |

[1]*Livestock and Meat Statistics 1984–88*, Stat. Bull. No. 784, USDA, pp. 65, 66. The figures include steers, heifers, and calves.

In recent years, there has been an increasing tendency to market feeder cattle direct, without passing them through a public stockyard. Also, an increasing number of producers are arranging to have their feeders custom fed.

## CATTLE MERCHANDISING

Because of the preponderance of finished cattle sold direct, the National Cattlemen's Association has prepared the following set of marketing guidelines to which they recommend that the industry adhere:

1. **Presentation of cattle.** Present only those cattle to each packer buyer which the feeder might reasonably expect that person to buy.

2. **Delivery time.** All cattle should be sold for delivery within seven days of the sale date; and time extensions should not be allowed.

3. **Prompt payment.** Cattle should be paid for by check on the day of delivery to the buyer's account, or wire payment should be made to the seller's bank by the close of the next business day after the transfer.

4. **Weighing conditions.** Cattle should be weighed early on the morning of delivery, without prior feeding that morning.

5. **Mud.** Excessive mud on cattle should not affect weighing conditions. Mud should be considered as a price factor and not a weight factor.

6. **Grade and yield selling.** Unless producers and feeders are paid a premium for superior meat-type animals, cattle should not be sold on a grade and yield basis, and all such transactions should have an agreement between packer and feeder on an established price. The yield should be on a hot weight basis and the packer should furnish a certified copy of the weight receipt.

7. **Rail killing.** Under no conditions should cattle be sold on carcass weight basis without a fixed price being established at the time of sale.

8. **Condemnation claims.** Any claims of packers should be supported by a previous agreement with the feeder.

9. **Point of change of ownership.** Ownership shall pass to the buyer when cattle exit scale at time of weigh-up.

10. **Maintenance of** current **status at feedlot.** Feeders are encouraged to market cattle when they reach optimum weight and grade. They should never attempt to "bull" the market.

11. **Sales reporting.** Feeders should make a prompt (within the hour) report of all sales, both to *Cattle-Fax* and the nearest USDA market news office, in order to keep the industry accurately informed on prices.

## MARKET CLASSES AND GRADES OF CATTLE

The generally accepted market classes and grades of live cattle are summarized in Table 19–4. The first five divisions and subdivisions include those factors that determine the class of the animal or the use to which it will be put. The grades indicate how well the cattle fulfill the requirements to which they are put. Figs. 19–3, 19–4, 19–5, and 19–6 portray the commonly used grades.

**TABLE 19–4**
**THE MARKET CLASSES AND QUALITY GRADES OF CATTLE**

| Cattle or Calves | Use Selection | Sex Classes | Age | Wt. (Group) | Weight Divisions (lb) | Commonly Used Quality Grades[1] |
|---|---|---|---|---|---|---|
| Cattle | Slaughter cattle[1] | Steers | Yearlings | Light / Medium / Heavy | 750 down / 750–950 / 950 up | Prime, Choice, Select, Standard, Utility, Cutter, Canner |
| | | | 2-year-olds and over | Light / Medium / Heavy | 1,100 down / 1,100–1,300 / 1,300 up | Prime, Choice, Select, Standard, Commercial, Utility, Cutter, Canner |
| | | Heifers | Yearlings | Light / Medium / Heavy | 750 down / 750–900 / 900 up | Prime, Choice, Select, Standard, Utility, Cutter, Canner |
| | | | 2-year-olds and over | Light / Medium / Heavy | 900 down / 900–1,050 / 1,050 up | Prime, Choice, Select, Standard, Commercial, Utility, Cutter, Canner |
| | | Cows | All ages | All weights | | Choice, Select, Standard, Commercial, Utility, Cutter, Canner |
| | | Bullocks | 24 mo. & under | All weights | | Prime, Choice, Select, Standard, Utility |
| | | Bulls | | All weights | | None (yield graded only) |
| | Feeder cattle | Steers | Yearlings | Light / Medium / Heavy / Mixed | | Prime, Choice, Select, Standard, Utility, Inferior |
| | | | 2-year-olds and over | Light / Medium / Heavy / Mixed | | Prime, Choice, Select, Standard, Commercial, Utility, Inferior |
| | | Heifers | Yearlings | Light / Medium / Heavy / Mixed | | Prime, Choice, Select, Standard, Utility, Inferior |
| | | | 2-year-olds and over | Light / Medium / Heavy / Mixed | | Prime, Choice, Select, Standard, Commercial, Utility, Inferior |
| | | Cows | All ages | All weights | | Choice, Select, Standard, Commercial, Utility, Inferior |
| | | Bullocks | 24 mo. & under | All weights | | Prime, Choice, Select, Standard, Utility, Inferior |
| | | Bulls | 24 mo. & under | All weights | | None |
| | Milkers & springers | Cows (milkers or springers) | All ages | All weights | | None |
| Calves | Vealers | No sex class (Sex characteristics of no importance at this age) | Under 3 mo. | Light / Medium / Heavy | 110 down / 110–180 / 180 up | Prime, Choice, Select, Standard, Utility |
| | Slaughter calves | Steers / Heifers / Bulls | 3 mo. to 1 year | Light / Medium / Heavy | 200 down / 200–300 / 300 up | Prime, Choice, Select, Standard, Utility |
| | Feeder calves | Steers / Heifers / Bulls | Usually 6 mo. to 1 year | Light / Medium / Heavy / Mixed | | Prime, Choice, Select, Standard, Utility, Inferior |

[1]In addition to the quality grades, there are the following yield grades for all slaughter cattle, except bulls: Yield Grade 1, Yield Grade 2, Yield Grade 3, Yield Grade 4, and Yield Grade 5; with Yield Grade 1 representing the highest cutability, and Yield Grade 5 the lowest. Thus, slaughter cattle may be graded for quality and/or yield.

# SLAUGHTER STEERS
## U.S. QUALITY GRADES

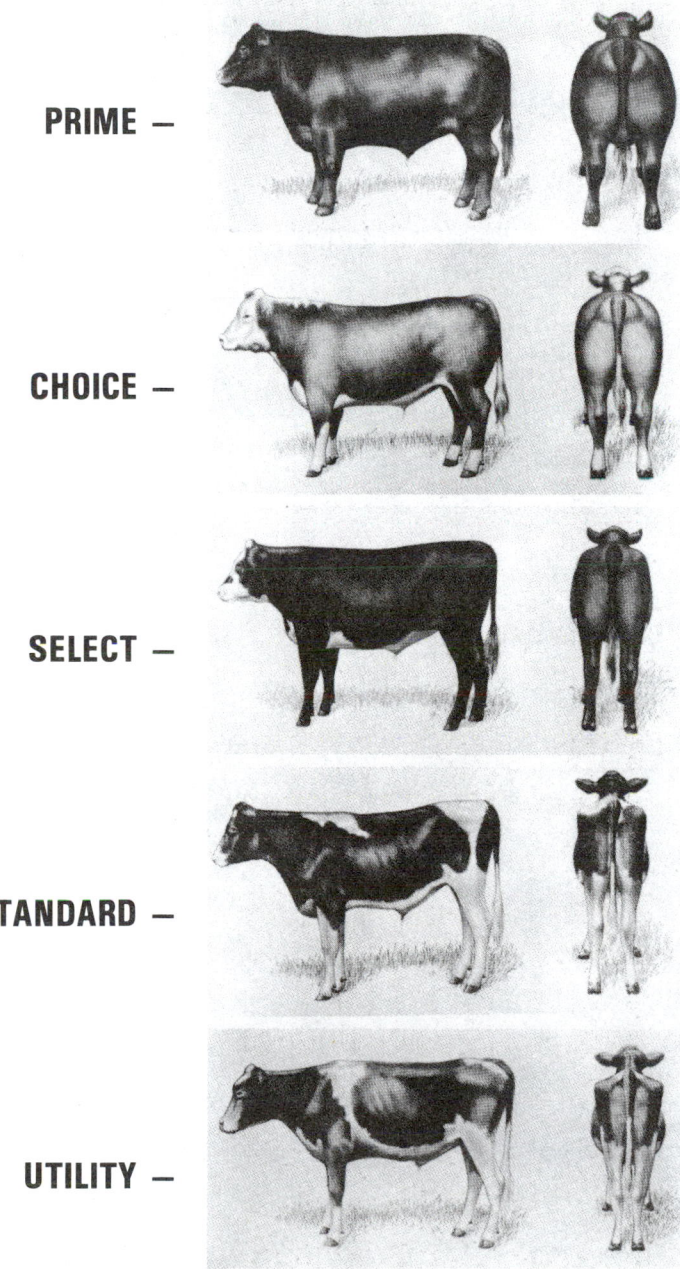

PRIME —

CHOICE —

SELECT —

STANDARD —

UTILITY —

COMMERCIAL, CUTTER, AND
CANNER GRADES ARE OMITTED

Fig. 19-3. The market *quality* grades of slaughter steers, as adapted by the author (Commercial, Cutter, and Canner not shown). (Courtesy, USDA, Agricultural Marketing Service, Livestock Division)

# SLAUGHTER STEERS
## U.S. YIELD GRADES

**YIELD GRADE 1 —**

**YIELD GRADE 2 —**

**YIELD GRADE 3 —**

**YIELD GRADE 4 —**

**YIELD GRADE 5 —**

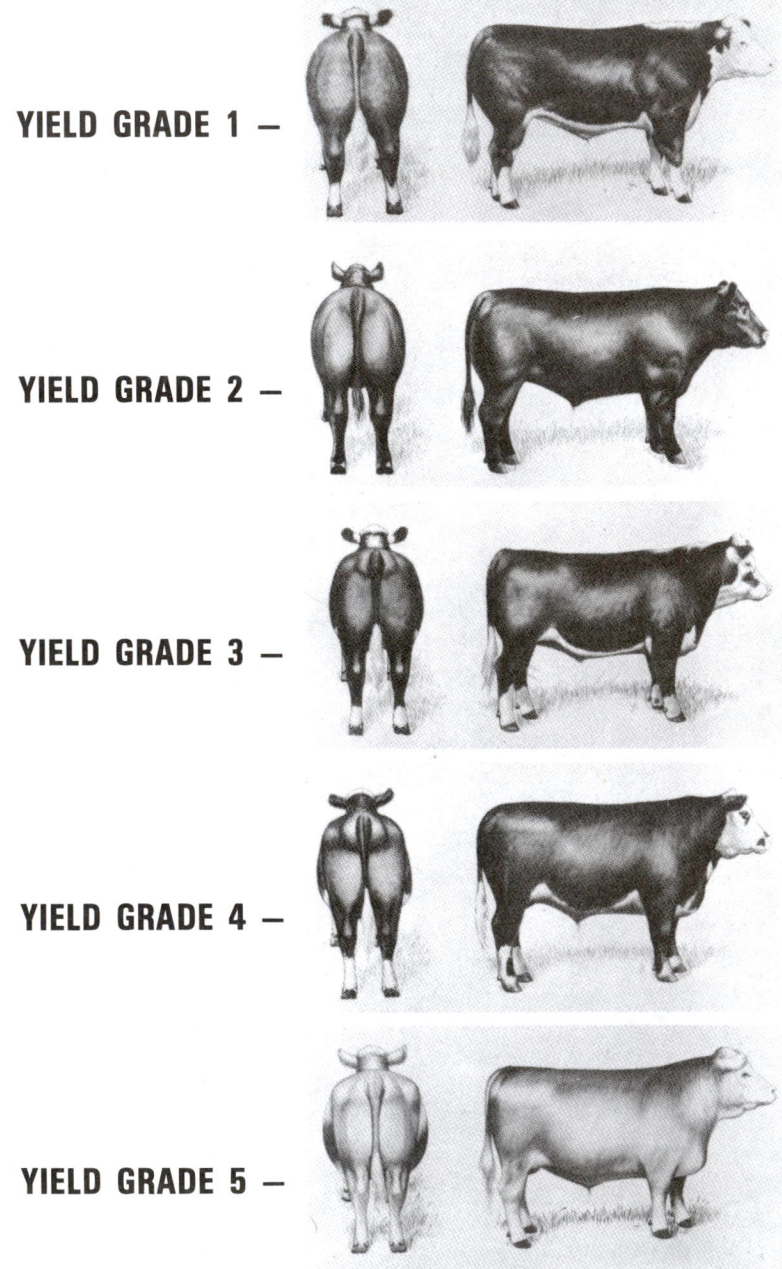

Fig. 19–4.  The market *yield* grades of slaughter steers, as adapted by the author. (Courtesy, USDA Agricultural Marketing Service, Livestock Division)

# FEEDER STEERS
## U.S. GRADES

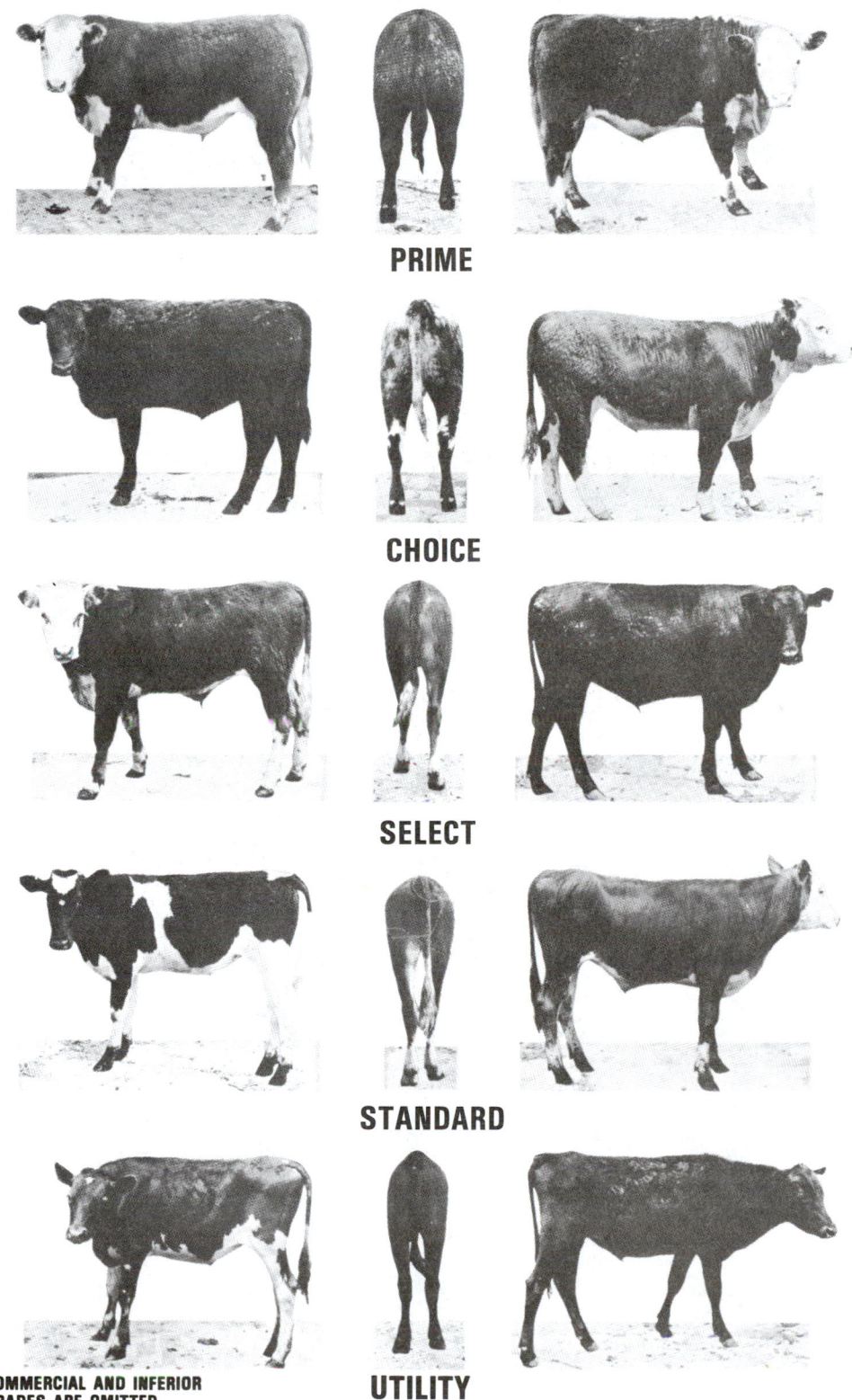

**PRIME**

**CHOICE**

**SELECT**

**STANDARD**

COMMERCIAL AND INFERIOR
GRADES ARE OMITTED

**UTILITY**

Fig. 19–5.   The market grades of feeder steers, as adapted by the author (Commercial and Inferior not shown). (Courtesy, USDA Agricultural Marketing Service, Livestock Division)

# FEEDER STEERS (CALVES)
## U.S. GRADES

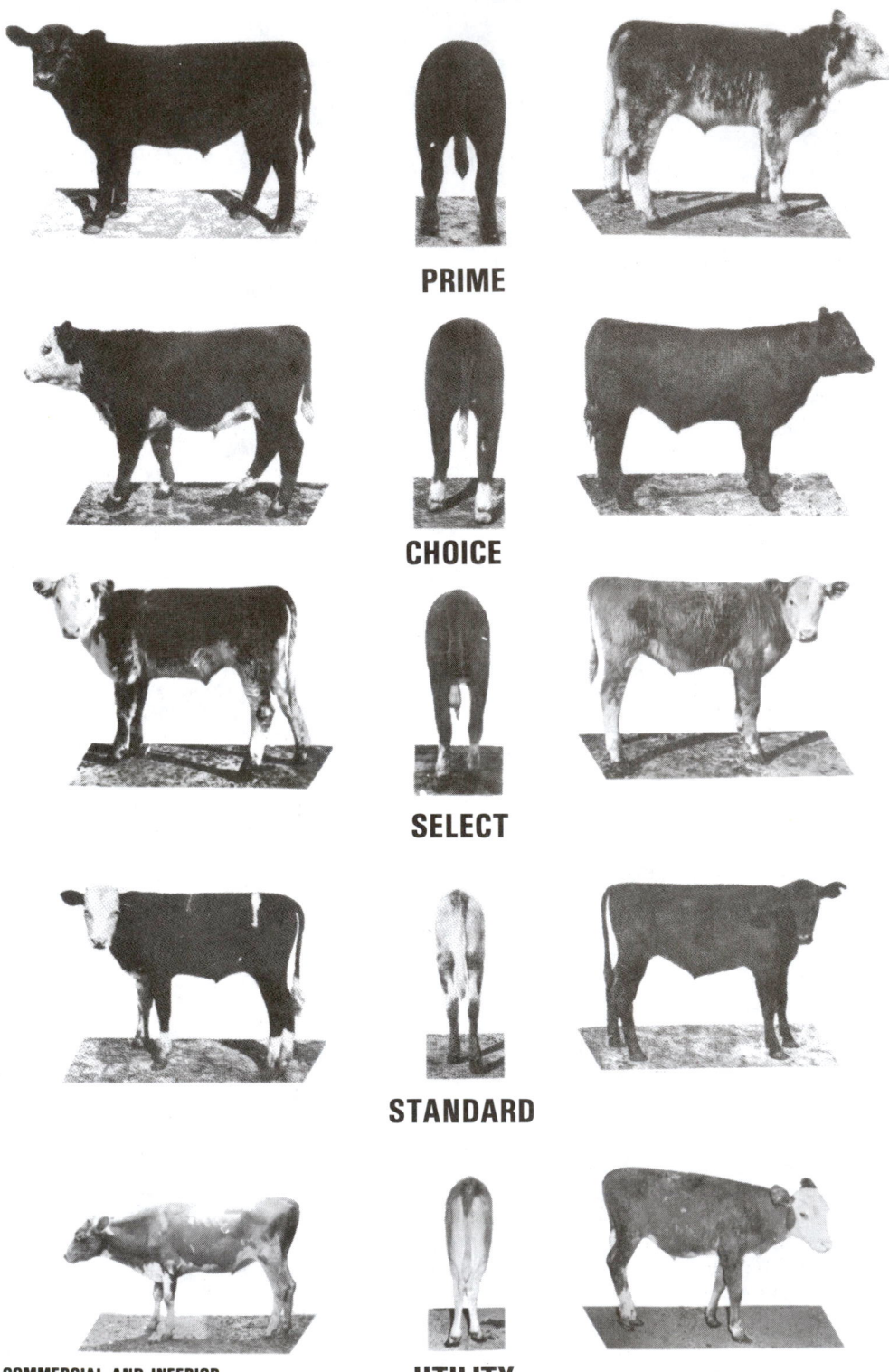

**PRIME**

**CHOICE**

**SELECT**

**STANDARD**

COMMERCIAL AND INFERIOR
GRADES ARE OMITTED

**UTILITY**

Fig. 19-6.   The market grades of feeder steer calves as adapted by the author (Commercial and Inferior not shown). (Courtesy, USDA Agricultural Marketing Service, Livestock Division)

## FACTORS DETERMINING MARKET CLASSES

The market class of cattle is determined by (1) use selection, (2) sex, (3) age, and (4) weight.

### CATTLE AND CALVES

All members of the bovine family are designated as calves until they are about one year of age, after which they are known as cattle. At public stockyards, cattle under approximately 400 lb in weight are designated as calves. Of the cattle and calves marketed in 1987, 93% were cattle and only 7% were calves.

### USE SELECTION OF CATTLE AND CALVES

The cattle group is further divided into three use divisions, each indicating something of the purpose to which the animals will be put. These divisions are (1) slaughter cattle, (2) feeder cattle, and (3) milkers and springers. Slaughter cattle include those which are considered suitable for immediate slaughter; feeders are those which are to be taken back to the country and grown for a time or finished; and milkers and springers include those cows recently freshened or soon due to calve and which are sold for milk purposes.

The calf group is also subdivided into 3 classes: (1) vealers, including milk-fat animals under 3 months of age which are sold for immediate slaughter; (2) slaughter calves that are between the ages of 3 months and one year, which have usually received grain in addition to milk, and which are fat enough for slaughter; and (3) feeder calves which are of weaning age and are sold to go back into the country for further growing or finishing.

In the selection of feeder cattle or calves, the sex, age, weight, and grade are of importance. In addition, consideration should be given to the following factors: (1) constitution and thrift, (2) natural fleshing, (3) breeding, (4) uniformity, (5) absence of horns, and (6) temperament and disposition.

As can be readily understood, the use to which animals are put is not always clear-cut and definite. Thus, when feed is abundant and factors are favorable for cattle finishing, feeders may outbid packer buyers for some of the animals that would normally go for slaughter purposes. On the other hand, slaughterers frequently outbid feeders for some of those animals that would normally go the feeder route.

### THE SEX CLASSES

Cattle are divided into five sex classes: steers, heifers, cows, bullocks, and bulls. Each of these groups has rather definite and easily distinguishable characteristics. In slaughter cattle, the sex groups are related to the commercial value of the carcass; in older animals, it affects carcass quality, finish, and conformation. In feeder cattle, sex is important in determining the suitability of animals as feeders. The definition of each sex class follows:

1. **Steer.** A male bovine castrated when young and which has not begun to develop the secondary physical characteristics of a bull.

2. **Heifer.** An immature female bovine that has not developed the physical characteristics typical of cows.

3. **Cow.** A female bovine that has developed through reproduction or with age, the relatively prominent hips, large middle, and other physical characteristics typical of mature females.

4. **Bullock.** A young (under approximately 24 months of age) male bovine (castrated or uncastrated) that has developed or begun to develop the secondary physical characteristics of a bull.

5. **Bull.** A mature (approximately 24 months of age or older) uncastrated male bovine. However, for the purpose of these standards, any mature, castrated male bovine which has developed or begun to develop the secondary physical characteristics of an uncastrated male also will be considered a bull.

Calves are merely divided into three sex classes: steers, heifers, or bulls. Because the secondary sex characteristics are not very pronounced in this group, the sex classes are of less importance for slaughter purposes than in older cattle. On the other hand, bull calves are not preferred as feeders because castration involves extra trouble and risk of loss.

### AGE GROUPS

Because the age of cattle does affect certain carcass characteristics, it is logical that age groups should exist in market classification. The terms used to indicate approximate age ranges for cattle are: vealers, calves, yearlings, 2-year-olds, and older cattle. As previously indicated, vealers are under 3 months of age,[1] whereas calves are young cattle between the vealer and yearling stage. Yearlings range from 12 to 24 months in age, and 2-year-olds from 24 to 36 months. Older cattle are usually grouped along with the 2-year-olds as "2-year-olds and over."

### WEIGHT DIVISIONS

It is common to have three weight divisions: light, medium, and heavy. When several weight divisions are included together, they are referred to as mixed weight. The usual practice is to group animals by rather narrow weight divisions because purchasers are frequently rather choosy about weights, and market values often vary quite sharply with variations in weights.

## THE MARKET GRADES

While no official grading of live animals is done by the U.S. Department of Agriculture, market grades do form a basis for uniform reporting of livestock marketing. The grade is the final step in classifying any kind of market livestock. It indicates the relative degree of excellence of an animal or group of animals. The three factors of primary importance in determining cattle grades—conformation, finish, and quality—apply to all classes of market animals. Since they have been discussed in the general chapter on marketing livestock—Chapter 8—the reader is referred thereto for additional information on the subject.

Table 19–4 lists the commonly used grades of cattle by classes; Figs. 19–3, 19–4, 19–5, and 19–6 portray the commonly used grades. As noted, the number of grades varies somewhat between classes chiefly because certain groups of animals present a wider range of variations in conformation, finish, and quality than do other groups.

---

[1]Vealers must also be over 21 days of age for slaughter. Underage veal calves are called *deacons* or *bob veal*.

Slaughter steers and heifers are divided into 8 grades: Prime, Choice, Select, Standard, Commercial, Utility, Cutter, and Canner. However, only 7 grades apply to slaughter cows, the grade Prime being deleted chiefly because of deficient conformation, finish, and quality. Bullocks have 5 grades: there are no Cutter and Canner grade bulls. There are no quality grades for bulls; they are yield graded only.

The terms *Cutter* and *Canner* are applied to the two lowest grades of slaughter cattle. Cutter cattle are so poor in form and lacking in muscle and fat covering that only such wholesale cuts as the loin and round are cut out and sold over the block. The balance of the carcass is boned out and used in sausage and canned meat products. Canners are almost entirely processed as ground and canned meats.

The grades of feeder cattle are: Prime, Choice, Select, Standard, Commercial, Utility, and Inferior. These grades are based on two value-determining characteristics—logical slaughter potential and thriftiness. The logical slaughter potential of an animal is the slaughter grade at the time the animal's carcass quality grade and carcass conformation grade become equal.

Thriftiness in feeder cattle refers to the ability of the animal to gain weight and finish rapidly and efficiently.

There are no U.S. grades for feeder bulls, although those animals are occasionally used as feeders.

As would be expected, the higher grades of slaughter cattle usually carry more weight, and the lower grades are lighter and usually underfinished.

Because the production of better grades of cattle usually involves more expenditure in the breeding operations (due to the need for superior animals) and feeding to a higher degree of finish, there must be a price spread and market grades in order to make the production of the top grades profitable. Fig. 19-7 shows the 17-year average wholesale price per hundredweight of dressed (carlots) steer beef, Choice grade 700–800 lb vs Select grade 600–700 lb, 1970–1987. This price spread in carcass grades of beef is reflected in on-foot grades.

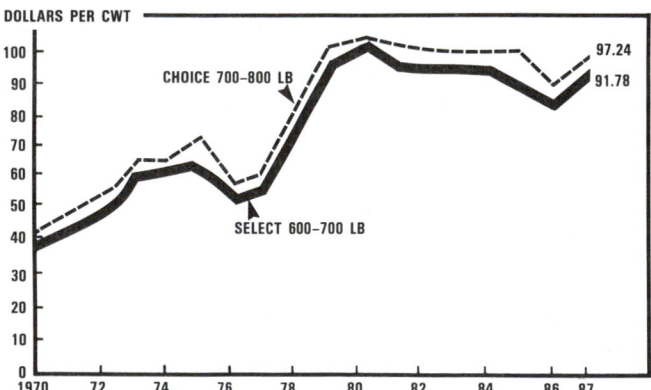

Fig. 19-7. Average wholesale price per 100 lb dressed (carlots) steer beef, Choice grade 700–800 lb vs Select grade 600–700 lb Chicago prices through 1975; Midwest prices starting in 1976. (Source: *Agricultural Statistics 1981,* USDA, p. 312, Table 461 and *Livestock and Meat Statistics 1984–88,* USDA, p. 230)

On March 6, 1975, the U.S. Department of Agriculture announced beef grading changes to become effective April 14, 1975. Although subsequent court action halted the implementation of the new grade standards for nearly a year, they became effective on February 23, 1976. Basically, the revised standards were designed to accomplish the following:

1. Reduce slightly the marbling required up to 30 months of age for Prime and Choice beef, thereby making it possible for leaner and younger beef to qualify for the two top grades.

2. Establish a more restrictive (narrower) Select (formerly Good) grade.

3. Eliminate conformation as a factor when carcasses are graded for quality.

4. Make yield grading mandatory; that is, require that all graded beef be dual graded—graded for both quality and yield. Quality grades gauge differences in taste (palatability), whereas yield grades identify the percentage of lean meat obtained from a carcass.

Until 1989, meat packers could choose to grade or not to grade beef. But, if they graded, they were *required* to grade for both yield and quality. In 1989, the law was changed, separating quality and yield grades of beef; and allowing packers to choose whether beef carcasses are graded for quality, for yield, or for both quality and yield. But packers could continue to choose between to grade and not to grade.

## PACKER SLAUGHTERING AND DRESSING OF BEEF CATTLE

Cattle and calves are either (1) commercially slaughtered, meaning they are slaughtered in federally inspected and other wholesale and retail establishments, or (2) farm slaughtered. In 1987, 99% of all cattle and 97% of all calves were slaughtered commercially.

Although farm slaughter procedures may differ somewhat from commercial slaughter, the ultimate objective is always the same. Because of the greater total numbers of cattle involved in this system of handling, only packer slaughtering and dressing—the kind that is done commercially—will be discussed.

### STEPS IN SLAUGHTERING AND DRESSING CATTLE

Upon reaching the packinghouse, cattle rapidly pass through the operation of killing and dressing. Unlike most manufacturing, meat-packing is primarily a disassembly process wherein the manufacturing operation starts with a complete unit that is progressively broken down into its component parts. In most of the larger and newer slaughtering plants, cattle are processed by the endless chain method of dressing, similar to that used for dressing calves, hogs, and sheep. Although the procedure differs somewhat between plants, in general the endless chain method of slaughtering and dressing cattle involves the following steps, carried out in rapid succession:

1. **Rendering insensible.** The cattle are rendered insensible.[2]

The following methods are accepted as humane for cattle and calves: captive bolt stunners, gunshot, or electric current. Carbon dioxide may also be used for calves.

2. **Shackling, hoisting, sticking, and bleeding.** The animal is next shackled, hoisted by the hind legs, stuck, and bled. The head is then skinned and removed.

---

[2]By federal law (known as the Humane Slaughter Act) passed in 1958 and effective June 30, 1960, unless packers use humane slaughter methods, they forfeit the right to sell meat to the government. The law lists the following methods as humane:

1. By rendering insensible to pain by a single blow or gunshot or an electrical, chemical, or other means that is rapid and effective, before being shackled, hoisted, thrown, cast, or cut.

2. By slaughtering in accordance with the ritual requirements of the Jewish faith or any other religious faith.

3. **Skinning.** The shanks are skinned and removed at the knees and hocks; beef hooks are inserted on the gam cord; the hide is opened along the median line of the belly and is removed from the belly and sides; then hide pullers are used for removing the rest of the hide in most of the newer plants. The breast and aitch (rump) bones are split by sawing.

4. **Removing viscera.** All internal organs are removed except the kidneys. If the plant is under federal inspection, the carcass and viscera are examined at this stage in the slaughtering process.

5. **Splitting carcass and removing tail.** The carcass is then split through the center of the backbone and the tail is removed.

6. **Washing and drying.** The split carcass or sides are washed with warm water under pressure.

7. **Shrouding.** The better carcasses are shrouded tightly with cloth so that they may have a smoother appearance following chilling.

8. **Sending to coolers.** Following slaughtering, the sides are sent to the coolers where they are kept at a temperature of about 34°F for a minimum period of 24 hours before ribbing.

Fig. 19–8. Beef in the coolers. (Courtesy, Iowa Beef Processors, Inc., Dakota City, NE)

## HOW SLAUGHTERING OF VEAL CALVES DIFFERS

Because of their smaller size, calves are easily dressed by the endless chain method. A wheel hoist is used in lifting the shackled calves to the rail. They are then stuck, bled, dressed, and washed. Because of the high moisture content of veal, the hide is usually left on for the purpose of reducing evaporation. This also produces a more desirable carcass color. When the hide is left on, it is thoroughly washed before dressing.

## KOSHER SLAUGHTER

Meat for the Jewish trade must come from animals slaughtered according to the rules of *Shehitah* (the ancient dietary rules). Although we usually think in terms of cattle when kosher slaughtering is mentioned, calves, sheep, lambs, goats, and poultry are slaughtered in a similar manner.

The killing is performed by a rabbi of the Jewish synagogue or a specially trained representative; a person called *shohet* or *shocket,* meaning slaughterer.

In kosher slaughter, the animal is hoisted without stunning and is cut across the throat with a special razor-sharp knife, known as a *chalaf.* With one quick, clean stroke the throat is cut through the jugular vein and other large vessels, together with the gullet and windpipe. Two reasons are given for using

this method of killing instead of the more conventional method of stunning and sticking; namely, (1) it produces more instant death with less pain, and (2) it results in more rapid and complete bleeding, which Orthodox Hebrews consider essential for sanitary reasons.

The shohet also makes an inspection of the lungs, stomach, and other organs while dressing. If the carcass is acceptable, it is marked on the brisket with a cross inside a circle. The mark also gives the date of slaughter and the name of the inspector.

Since neither packers nor meat retailers can hold kosher meat longer than 216 hours (and even then it must be washed at 72-hour intervals; see Chapter 9, section under "Kosher Meats"), rapid handling is imperative. This fact, plus the heavy concentration of Jewish folks in the eastern cities, results in large numbers of live cattle being shipped from the markets farther west to be slaughtered in or near the eastern consuming areas.

## THE DRESSING PERCENTAGE

*Dressing percentage may be defined as the percentage yield of chilled carcass in relation to the weight of the animal on foot.* For example, a steer which weighed 1,200 lb on foot and yielded a carcass weighing 729 lb may be said to have a dressing percentage of 60. The offal—so-called because formerly (with the exception of the hide, tallow, and tongue) the offal (waste) was thrown away—consists of the blood, head, shanks, tail, hide, viscera, and loose fat.

A high carcass yield is desirable because the carcass is much more valuable than the by-products. Although the packers have done a marvelous job in utilizing by-products, about 88% of the income from cattle and calves is derived from the sale of the carcass and only 12% from the by-products. Thus, the estimated dressing percentage of slaughter cattle is justifiably a major factor in determining the price or value of the live animal.

The chief factors determining the dressing percentage of cattle are (1) the amount of fill, (2) the finish or degree of fatness, (3) the general quality and refinement (refinement of head, bone, hide, etc.), and (4) the size of udder. The better grades of steers have the highest dressing percentage, with thin Canner cows showing the lowest yield. Table 19–5 gives the dressing percentages that may be expected for different grades of cattle and calves.

**TABLE 19–5**
**DRESSING PERCENTAGE OF CATTLE AND CALVES, BY GRADE[1]**

| Cattle | | |
|---|---|---|
| Grade | Range | Average |
| **P**rime | 62–67 | 64 |
| **C**hoice | 59–65 | 62 |
| **S**elect | 58–62 | 60 |
| **S**tandard | 55–60 | 57 |
| **C**ommercial | 54–62 | 57 |
| **U**tility | 49–57 | 53 |
| **C**utter | 45–54 | 49 |
| **C**anner | 40–48 | 45 |

| Calves and Vealers | | |
|---|---|---|
| Grade | Range | Average |
| **P**rime | 59–65 | 62 |
| **C**hoice | 59–60 | 58 |
| **S**elect | 52–57 | 55 |
| **S**tandard | 47–54 | 51 |
| **U**tility | 40–48 | 46 |

[1]From USDA sources.

The highest dressing percentage on record was a yield of 76.75% made by a spayed Angus heifer at the Smithfield Fat Stock Show in England.

The average liveweights of cattle and calves dressed by commercial meat-packing plants, and their percentage yields in meats, for the year 1987 are given in Table 19–6. As shown, cattle average 60% and calves 61%.

### TABLE 19–6
### AVERAGE LIVEWEIGHT, CARCASS YIELD, AND DRESSING PERCENTAGES OF ALL CATTLE AND CALVES COMMERCIALLY SLAUGHTERED IN THE UNITED STATES IN 1987[1]

| | Average Liveweight | | Dressed Weight | | Dressing Percentage |
|---|---|---|---|---|---|
| | (lb) | (kg) | (lb) | (kg) | (%) |
| Cattle ............ | 1,105 | 502 | 659 | 300 | 60 |
| Calves ............ | 249 | 113 | 151 | 69 | 61 |

[1]Source: *Livestock and Meat Statistics, 1984–88*, Stat. Bull. No. 784, USDA, pp. 107, 132.

## AGING BEEF

Except for veal, fresh beef is not at optimum tenderness immediately after slaughter. It must undergo an aging or ripening process before it really becomes tender. This process consists of the dissolution of the connective tissue (collagen) by the action of the enzymes. Beef should be aged from 2 to 6 weeks at temperatures ranging from 34° to 38°F, but only the better grades can be aged for the longer periods. Beef must have a covering of fat to protect the meat from bacterial action by sealing it from the air. With well-finished beef, some trimming is necessary anyway and the removal of the mold does not constitute any additional loss. The aging process may be hastened by the use of ultraviolet lights in high-temperature coolers with controlled humidity.

## DISPOSITION OF THE BEEF CARCASS

Fig. 19–9. Processing beef. Each worker on the conveyor belt has a specific task. (Courtesy, Iowa Beef Processors, Dakota City, NE)

Beef carcasses are disposed of in one of three ways: (1) block beef; (2) fabricated, boxed beef; or (3) processed meats.

1. **Block beef**—Block beef refers to beef that is suitable for sale over the block. Such beef is purchased by the retailer in sides, quarters, or wholesale cuts. Block beef may enter regular channels of trade either as fresh chilled or fresh frozen. Fresh chilled beef is chilled at temperatures ranging from 34° to 38°F for a minimum of 24 hours before moving out to the retail trade, or it may be held longer if aging is desired. Frozen beef is subjected to temperatures of 0°F or below and is frozen solid, in which form it can be kept for a period of several months. U.S. consumers prefer fresh chilled beef, although frozen beef is increasing in acceptance and quantity. The bulk of imported beef is frozen, rather than chilled.

2. **Fabricated, boxed beef**—Formerly, beef was shipped in exposed halves, quarters, or wholesale cuts and divided into retail cuts in the back rooms of meat retailers. This procedure left much to be desired from the standpoints of efficiency, sanitation, shrink, spoilage, and discoloration. To improve this situation, more and more packers are fabricating and boxing beef in their plants. In the Iowa Beef Processors' ultramodern plant at Dakota City, Nebraska, beef is fabricated and handled as follows: After chilling, the carcass is subjected to a disassembly process, in which it is fabricated or broken into subprimal cuts; vacuum sealed; boxed; moved into storage by a fully automated, computer controlled system; loaded by computer-automation into trailers; and shipped to retailers across the nation. The fabrication and boxing of beef at packing plants will increase in the future.

3. **Processed meats**—Beef that is not suitable for sale over the block is boned out and disposed of as boneless cuts, is canned, is made into sausage, or is cured by drying and smoking. It is estimated that about one-fifth of all slaughter cattle are disposed of as processed meats.

## BEEF AND VEAL CUTS, AND HOW TO COOK THEM

There was a time when each area of the U.S. had its traditional cuts of beef and veal. However, increased central processing and boxed beef prompted the need for greater uniformity in cutting and labeling, among both packers and retailers. Out of this need arose a new nationwide standardized identification-labeling system, coordinated by the National Live Stock and Meat Board and adopted by the industry. The names for various cuts of beef, pork, and lamb sold in U.S. food stores were reduced from more than 1,000 to about 300. As a result of this system of uniform labeling, a *rib eye steak* is a *rib eye steak*—not a Delmonico steak at one place, a filet steak someplace else, or a Spencer steak or a beauty steak in still other stores, depending on where you live in the United States—or even where you shop in the same city.

Whether a beef carcass is cut up in the home or by an expert, it should always be cut across the grain of the muscle tissue, and the thick cuts should be separated from the thin cuts and the tender cuts from the less tender cuts.

Fig. 19–10 shows the wholesale and retail cuts of beef and gives the recommended method(s) of cooking each. Fig. 19–11 presents similar information for veal.

In order to buy and/or process beef and veal wisely, and to make the best use of each part of the carcass, the consumer should be familiar with the types of cuts and how each should be processed.

Every grade and cut of meat can be made tender and palatable provided it is cooked by the proper method. Also, it is important that meat be cooked at low to moderate temperatures, usually between 300° and 325°F for roasting. At this temperature, it cooks slowly, and as a result is juicier, shrinks less, and has a better flavor than when cooked at high temperatures.

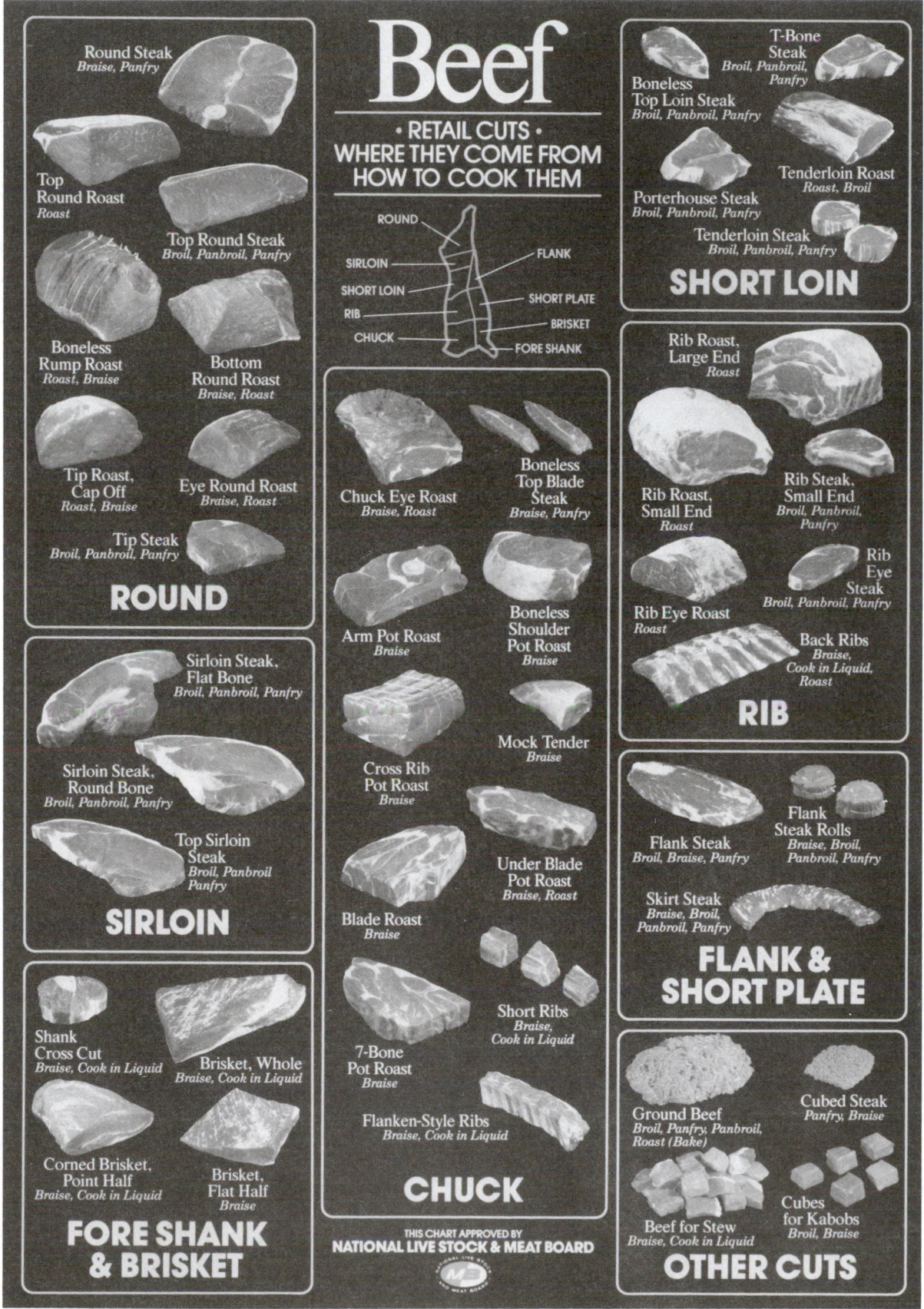

Fig. 19–10.   Beef. The retail cuts of beef; where they come from; and how to cook them. (Courtesy, National Live Stock and Meat Board, Chicago, IL)

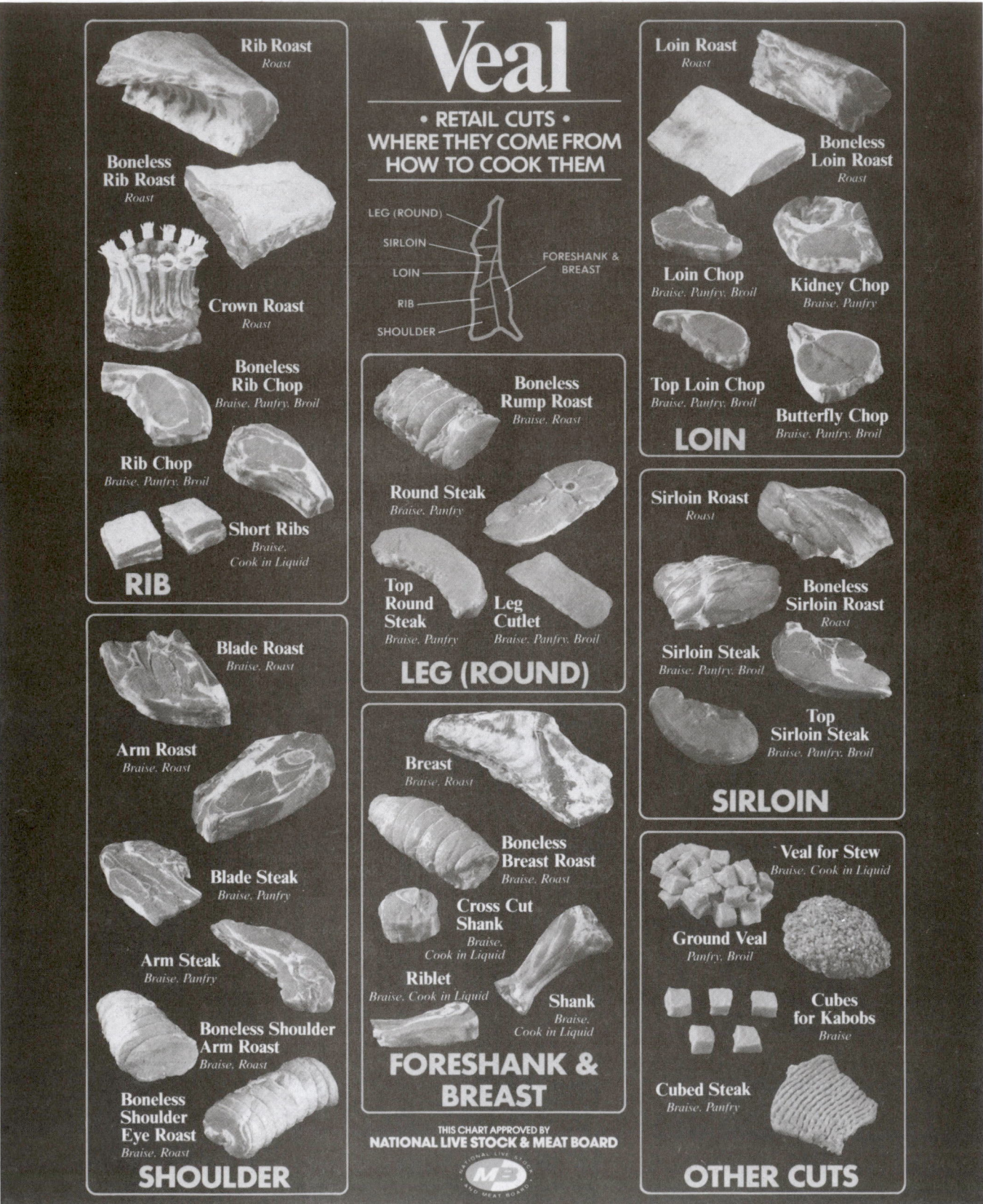

Fig. 19–11.  Veal. The retail cuts of veal; where they come from; and how to cook them. (Courtesy, National Live Stock and Meat Board, Chicago, IL)

# QUESTIONS FOR STUDY AND DISCUSSION

1. What forces cause packers to increase their purchases of cattle and calves through nonpublic markets (direct) and to decrease their purchases of cattle and calves through auction and terminal markets?

2. What method of marketing (what market channel) do you consider most advantageous for the cattle sold off your home farm or ranch, or off a farm or ranch with which you are familiar? Explain why.

3. Why are more than two-thirds of the cows and bulls sold at auctions, whereas only about 5% of steers and heifers are sold at auctions?

4. How do you account for the recent increase in cattle purchased by packers on carcass grade and weight?

5. What are public stockyards? Why are they still a powerful and vital force in livestock marketing?

6. How do you account for the fact that the leading cattle markets (Table 19–1), the leading calf markets (Table 19–2), and the leading feeder cattle markets (Table 19–3) do not coincide?

7. Why do the following markets lead in receipts, by quite a margin, in their respective class of cattle marketed: (a) Oklahoma City in cattle, (b) St. Paul in calves, and (c) Oklahoma City in feeder cattle and calves?

8. Why is it important that cattle producers know the leading markets for each class of cattle?

9. What prompted the National Cattlemen's Association to issue guidelines for the marketing of slaughter cattle direct? List and discuss these guidelines.

10. Why is it important that cattle producers know the market classes and grades of cattle and what each implies?

11. List on-foot market classes of live cattle, and describe each class.

12. List on-foot grades of live cattle, and describe each of them.

13. Since there is a rather uniform difference in the selling price of the different grades of cattle, with the top grades bringing the higher prices, why don't more growers produce the top grades?

14. What do beef quality grades gauge, and what do yield grades gauge? Do you favor using quality grades, yield grades, or both quality and yield grades?

15. List and describe briefly the usual steps in packinghouse slaughtering and dressing of cattle?

16. How does the slaughtering of veal calves differ from the slaughtering of cattle?

17. How does kosher slaughter differ from regular slaughter?

18. Define *dressing percentage.*

19. What effect does (a) grade, and (b) age (cattle vs calves) have on dressing percent?

20 Define *aging beef.* Why is beef aged?

21. What is fabricated, boxed beef? Is it likely to increase in importance in the future?

22. Discuss the value of the nationwide beef and veal identification-labeling system, coordinated by the National Live Stock and Meat Board and adopted by the industry.

## SELECTED REFERENCES

| Title of Publication | Author(s) | Publisher |
|---|---|---|
| *Lessons on Meat,* Second Edition | | National Live Stock and Meat Board, Chicago, IL, 1972 |
| *Livestock and Meat Marketing,* Second Edition | J. H. McCoy | Avi Publishing Co., Westport, CT, 1979 |
| *Marketing of Livestock and Meat, The,* Second Edition | S. H. Fowler | The Interstate Printers & Publishers, Inc., Danville, IL, 1961 |
| *Meat Handbook,* Fourth Edition | A. Levie | Avi Publishing Co., Westport, CT, 1979 |
| *Meat Reference Book* | | American Meat Institute, Washington, DC |
| *Meat for the Table* | S. Bull | McGraw-Hill Book Company, New York, NY, 1951 |
| *Meat We Eat, The,* Twelfth Edition | J. R. Romans, *et al.* | The Interstate Printers & Publishers, Inc., Danville, IL, 1985 |
| *Problems and Practices of American Cattlemen,* Wash. Ag. Exp. Sta. Bull. 562 | M. E. Ensminger M. W. Galgan W. L. Slocum | Washington State University, Pullman, WA, 1955 |
| *Stockman's Handbook, The,* Sixth Edition | M. E. Ensminger | The Interstate Printers & Publishers, Inc., Danville, IL, 1983 |
| *Uniform Retail Meat Identity Standards* | Industrywide Cooperative Meat Identification Standards Committee | National Live Stock and Meat Board, Chicago, IL, 1973 |
| *Using Information in Cattle Marketing Decisions, A Handbook,* WEMC Pub. No. 5 | | Western Extension Marketing Committee Task Force on the Economics of Marketing Livestock, Fort Collins, CO, 1973 |

Four ''helpers'' (working cow dogs) corralling market cattle. (Courtesy, *Livestock Weekly*, San Angelo, TX)

Foster mother of the human race. (Courtesy, Pennsylvania Millers and Feed Dealers Assn.)

# THE DAIRY INDUSTRY

## Chapter 20

Fig. 20–1. A modern dairy. (Courtesy, United Cooperative Farmers, Inc., Fitchburg, MA)

Under natural conditions, wild mammals produce only enough milk for their offspring. However, long before recorded history, people found that milk was good—and good for them—with the result that they domesticated milk-producing animals and began using and selecting them for higher production for their use. For the most part, this included the cow, the buffalo, and the goat—although the ewe, the mare, the sow, and other animals have been used for producing milk for human consumption in different parts of the world. The importance of the cow in milk production is attested to by her well-earned designation as, "the foster-mother of the human race."

Records exist of cows being milked as early as 9000 B.C. The Bible contains many references to milk, one of the best remembered of which is from Exodus 3:8—"milk and honey." Also, Sanskrit writings, thousands of years old, relate that milk was one of the most essential of all foods. Hippocrates recommended milk as a medicine five centuries before Christ.

When Christopher Columbus came to America, there were no cows on the American continent. But on his second voyage, in 1493, he brought cattle and other farm animals to the West Indies.

The Pilgrims did not bring any cows with them. As a result, lack of milk is said to have contributed to the high death rate of the colonists, particularly of the children.

The first cows in the United States were brought over to Jamestown in 1611, and the first cows arrived at the Plymouth Colony in 1624.

Throughout the colonial period, and until past the middle of the 19th century, dairying was limited to a few cows cared for by family labor. Management practices were poor when compared with today's standards. Also, the perishable nature of milk and the difficulty in transporting it made large-scale dairy operations impractical.

Starting soon after 1850, the following developments paved the way for the modern U.S. dairy industry of today: Cattle, which were the foundation of our present-day breeds were brought to this country; milk was pooled by neighboring farm families in a cooperative effort to make cheese; condensed milk was developed by Borden in 1856; the centrifugal cream sep-

arator was invented in 1878; the Babcock test for fat evolved in 1892, followed by the adaptation of pasteurization to milk, mechanical refrigeration, homogenization, and modern packaging and transportation.

But even more dramatic changes in the dairy industry occurred between 1929 and 1988; during this 60-year span, the only thing that remained the same was *milk* itself. These changes are pointed up in Table 20–1, which portrays dairying in 1929 vs 1988.

**TABLE 20–1**
**U.S. DAIRYING 1929 VERSUS 1988**

| Comparison | In 1929 | In 1988 |
|---|---|---|
| **N**umber of dairy farms | 4,500,000 | 220,800 |
| **N**umber of cows per herd, average | 5 | 47.2 |
| **M**ilk production per cow per year, lb | 4,000 | 13,623 |
| **T**ime required to produce 100 lb milk | 3 hrs. & 15 min. | 15 min. |
| **T**otal U.S. milk production, billion lb | 100 | 150 |
| **D**elivery/marketing | **V**ia horse-drawn milk wagons, in qt glass bottles, delivered to door-steps of consumers. | **V**ia refrigerated motor trucks, in gallon and half gallon plastic containers, delivered to, and sold by, supermarkets. |

Four major developments propelled the enormous transition of the U.S. dairy industry from 1929 to 1988, as summarized in Table 20–1; namely, (1) rural electrification, which made it possible to use milking machines to replace hand-milking, (2) bulk handling and refrigeration of milk on the farm, (3) improved transportation of milk, and (4) artificial insemination.

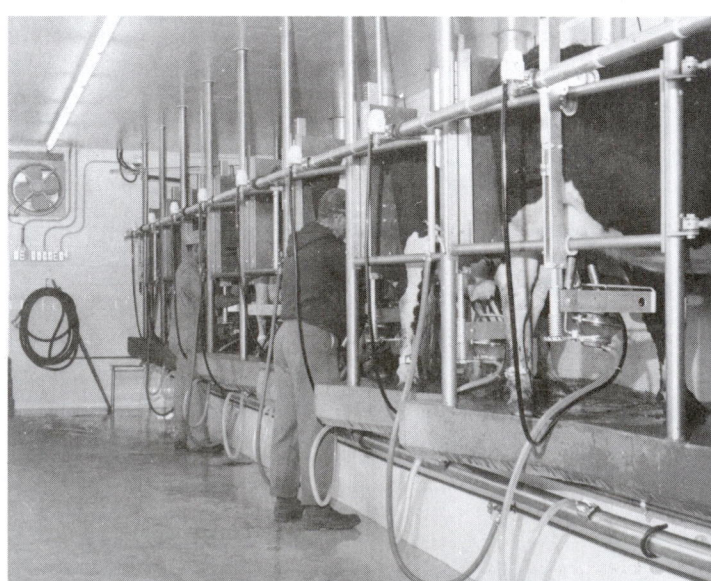

Fig. 20–2. Rural electrification and milking machines were major factors in automating the U.S. dairy industry. This shows a single row eight herringbone parlor with 2 in. low line and breaker cup milkers. (Courtesy, Babson Bros., Oak Brook, IL)

# IMPORTANCE AND USES OF MILK

Milk is an important constituent of the American diet in terms of (1) per capita consumption (Table 20–2), and (2) a rich source of needed nutrients (Table 20–3).

In 1987, the total milk equivalent per capita consumption of all dairy products was 598 lb, up from 560 lb per capita in 1972.[1]

Table 20–4 shows the quantity of milk going into different channels, and Fig. 20–3 shows the relative importance of each outlet.

### TABLE 20–2
### PER CAPITA SALES OF FLUID MILK, CREAM, AND MANUFACTURED DAIRY PRODUCTS, 1987[1]

|  | Lb |
|---|---|
| **Fluid Milk Products, Total** | 220.7 |
| Fresh whole milk | 103.8 |
| Lowfat milk | 89.3 |
| Skim milk | 14.0 |
| Flavored milk drinks | 9.7 |
| Buttermilk | 4.0 |
| **Cream and Specialty Products, Total** | 12.2 |
| Yogurt | 4.6 |
| Half and half | 3.1 |
| Sour cream and dips | 2.4 |
| Heavy cream | 1.1 |
| Light cream | 0.5 |
| Eggnog | 0.5 |
| **Manufactured Products, Total** | 69.0 |
| Evaporated and condensed whole milk | 3.6 |
| Butter | 3.7 |
| Cheese: | |
| American | 10.4 |
| Other | 11.4 |
| Cottage | 3.9 |
| Dry whole milk | 0.5 |
| Nonfat dry milk | 1.9 |
| Evaporated and condensed: | |
| Skim milk | 4.2 |
| Whole milk | 3.6 |
| Frozen desserts: | |
| Ice cream product | 18.3 |
| Ice milk product | 7.5 |

[1]*Milk Facts*, Milk Industry Foundation, Washington, DC, 1988, pp. 14–17.

### TABLE 20–4
### HOW THE U.S. MILK SUPPLY WAS USED IN 1987[1]

| Product | Milk Equivalent |
|---|---|
|  | (mil. lb) |
| Fluid milk and cream (26.4 billion qt) | 56,703 |
| Cheese | 42,851 |
| Creamery butter | 21,072 |
| Frozen dairy products[2] | 13,456 |
| Evaporated and condensed milk | 2,163 |
| Used on farms where produced | 2,203 |
| Other uses | 7,536 |

[1]*Milk Facts*, Milk Industry Foundation, Washington, DC, 1988, p. 31. Includes Alaska and Hawaii.
[2]Plus 2,125 million lb of milk equivalent in other manufactured dairy products used in production of frozen dairy products.

A total of 37.4% of the milk marketed by farmers in 1987 was consumed in fluid form (See Fig. 20–3). In the 1930s, 54% of the milk supply was used for butter compared to 14.8% in 1987. Today, more milk is used for making cheese than for making butter.

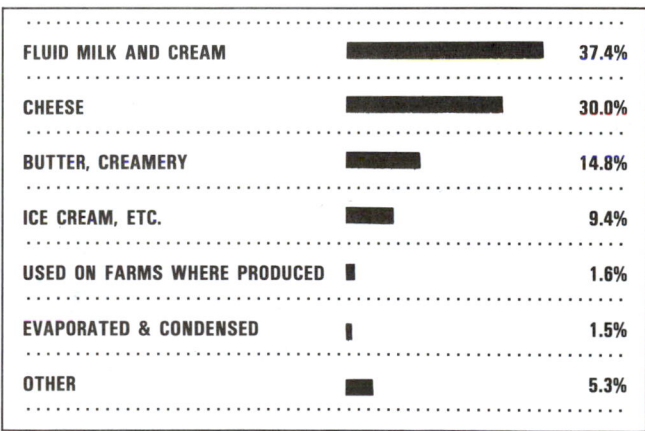

| | |
|---|---|
| FLUID MILK AND CREAM | 37.4% |
| CHEESE | 30.0% |
| BUTTER, CREAMERY | 14.8% |
| ICE CREAM, ETC. | 9.4% |
| USED ON FARMS WHERE PRODUCED | 1.6% |
| EVAPORATED & CONDENSED | 1.5% |
| OTHER | 5.3% |

Fig. 20–3. The uses of milk, in percentages. (From *Milk Facts*, Milk Industry Foundation, Washington, DC, 1988, p. 30)

### TABLE 20–3
### PERCENTAGE OF TOTAL NUTRIENTS CONTRIBUTED BY DAIRY FOODS (EXCLUDING BUTTER) TO THE U.S. DIET[1]

| Nutrient | Contributed by Dairy Foods |
|---|---|
|  | (%) |
| Energy | 10 |
| Protein | 21 |
| Calcium | 76 |
| Phosphorus | 36 |
| Magnesium | 20 |
| Riboflavin | 35 |
| Vitamin A value | 10 |
| Vitamin B–6 | 11 |
| Vitamin B–12 | 20 |
| Thimain | 9 |

[1]*Milk Facts*, Milk Industry Foundation, Washington, DC, 1988, p. 38.

# WORLD DISTRIBUTION AND PRODUCTION OF DAIRY CATTLE

Cattle furnish most of the milk of the world; only about 8.6% (less than one-twelfth) of the world's milk supply comes from buffalo, goats, and sheep. Table 20–5 shows the 10 leading milk cow producing countries of the world, ranked by cow numbers. As noted, the U.S.S.R. has a commanding lead. The United States ranks fourth in milk cow numbers.

[1]*National Food Review*, USDA, ERC, April-June 1989, Vol. 12, Issue 2, p.4, Table 4.

Table 20–5 also shows the proficiency of cows in the principal countries of the world. It is noteworthy that the United States ranks third in average milk production per cow, being outproduced by both Israel and Japan.

## TABLE 20–5
### LEADING COUNTRIES OF THE WORLD IN (1) MILK COWS ON FARMS, AND (2) MILK PRODUCTION PER COW[1]

| Country | No. Milk Cows on Farms | Country | Average Milk Production per Cow | |
|---------|------------------------|---------|--------|--------|
| | (1,000 head) | | (lb) | (kg) |
| U.S.S.R. | 42,400 | Israel[2] | 19,085 | 8,675 |
| India | 27,500 | Japan | 15,481 | 7,037 |
| Brazil | 15,300 | United States | 13,633 | 6,197 |
| United States | 10,430 | Denmark | 13,296 | 6,044 |
| Canada | 7,431 | Sweden | 13,111 | 5,960 |
| France | 6,359 | Canada | 12,317 | 5,599 |
| Mexico | 6,300 | Norway | 12,158 | 5,526 |
| Germany, West | 5,150 | Netherlands | 12,024 | 5,465 |
| Poland | 4,940 | Finland | 11,338 | 5,154 |
| United Kingdom | 3,311 | Switzerland | 10,631 | 4,832 |

[1]*Agricultural Statistics 1988*, USDA, p. 322. Data for 1987.

[2]Israel average milk production per cow is from *FAO Production Yearbook 1988*, Vol. 42, FAO, Rome,, Italy, p. 273, Table 99.

Some countries produce more dairy products than they can use, whereas others are importers. New Zealand accounts for more than one-third (37% in 1987) of the total world dairy exports. The United Kingdom is an especially heavy importer of butter, cheese, and condensed milk.

Fig. 20–4. Typical New Zealand dairy scene—Jersey cows on lush pasture. New Zealand accounts for more than one-third of the total world dairy exports. (Courtesy, New Zealand Government)

Among the factors that determine the present development of the dairy industry in different countries are: the character and preferences of the people; the adaptation of the country to dairying—dairying is not adaptable to areas that are excessively hot or cold, or where soils are poor; the relative size of urban and rural population; and the extent and effectiveness of dairy research and education.

# MAGNITUDE OF THE U.S. DAIRY INDUSTRY

Table 20–6 shows the importance of the U.S. dairy industry.

## TABLE 20–6
### U.S. DAIRY FARM DATA[1]

| | | |
|---|---|---|
| Milk cows on farms (not including heifers not fresh) | (no.) | 10,334,000 |
| Farm milk production | (lb) | 142,462,000,000 |
| Average production per cow | (lb) | 13,786 |
| Cash farm income from milk and cream: Value | ($) | 17,723,430,000 |

[1]*Agricultural Statistics 1988*, USDA, pp. 320, 324. Data for 1987.

As noted, farm income from milk and cream totaled $17.7 billion in 1987, which was 12.9% of the cash farm income that year (see Chapter 2 of this book, Fig. 2–11, for the latter statistic). Fig. 2–11 also shows that milk and cream ranked second only to cattle and calves in percent of cash income received by farmers and ranchers.

In addition to the income derived from milk, it is estimated that the sale of cows, heifers, steers, and veal calves from dairy herds accounts for an additional 3% of farm cash receipts, or a total of approximately $4.1 billion in 1987.

Another indication of the magnitude and importance of the dairy industry is that consumers spent $54 billion for dairy products in 1987, which represented about 12% of their total expenditures for food.

# LEADING STATES IN MILK PRODUCTION AND COW NUMBERS

Milk is produced in every state of the union. However, the greatest concentration of dairy cows is found in those areas with the most dense human population. This is as one would expect from the standpoint of the demand for and the marketing of fresh milk.

Table 20–7 gives the 10 leading states in dairy cattle numbers, by rank (see left half of table). Human population centers, which provide a large market for milk, have been a major factor in determining the intensity of dairying. Also, climate, land, and feed exert a considerable influence.

## TABLE 20–7
### TEN LEADING STATES IN (1) MILK COWS ON FARMS, AND (2) MILK PRODUCTION PER COW[1]

| State | No. Milk Cows on Farms | State | Average Milk Production per Cow |
|-------|------------------------|-------|--------|
| | (1,000 head) | | (lb) |
| Wisconsin | 1,795 | Washington | 18,091 |
| California | 998 | California | 17,970 |
| New York | 858 | New Mexico | 16,879 |
| Minnesota | 823 | Arizona | 15,911 |
| Pennsylvania | 721 | Oregon | 15,989 |
| Ohio | 370 | Colorado | 15,893 |
| Michigan | 361 | Utah | 15,149 |
| Texas | 329 | Idaho | 14,937 |
| Iowa | 302 | Delaware | 14,632 |
| Missouri | 221 | Nevada | 14,632 |

[1]*Agricultural Statistics 1988*, USDA, p. 320. Data for 1987.

Some states have a well-managed dairy industry, but, because of their small size, they do not have a large total dairy cattle population. Therefore, proficiency of dairy management is best indicated by average annual production per cow, as shown in Table 20–7 (see right half of table).

## PRODUCTION PER COW, MILK COW NUMBERS, AND HUMAN POPULATION

The average annual production per cow has steadily in-creased, from 3,138 lb in 1920 to 13,623 lb per cow in 1988; simultaneously, cow numbers have declined and human population has spiraled (Table 20–8). This increase in average production per cow represents progress in improved breeding, feeding, disease control, and management. Also, increased milk production has made for greater efficiency.

The cows which were in record-keeping dairy herd im-provement associations, and which obviously were the more efficient ones, produced an average of 17,008 lb of milk and 625 lb of butterfat in 1987.

### TABLE 20–8
### U.S. MILK PRODUCTION PER COW, MILK COW NUMBERS, AND HUMAN POPULATION

| Year | Milk Production per Cow[1] | Number of Milk Cows[2] | Human Population[3] | Ratio of Milk Cows to People |
|------|-----------------------------|--------------------------|----------------------|-------------------------------|
| | (lb) | | | (no. cows to no. people) |
| 1920 | 3,138 | 21,455,000 | 105,710,620 | 1: 4.9 |
| 1930 | 4,508 | 23,032,000 | 122,775,046 | 1: 5.5 |
| 1940 | 4,625 | 24,940,000 | 131,669,275 | 1: 5.6 |
| 1950 | 5,314 | 23,853,000 | 150,697,361 | 1: 6.6 |
| 1960 | 7,002 | 19,527,000 | 180,684,000 | 1: 9.3 |
| 1970 | 9,385 | 12,483,000 | 202,711,000 | 1:16.2 |
| 1980[4] | 11,889 | 10,810,000 | 221,700,000[5] | 1:20.5 |
| 1987 | 13,633[6] | 10,430,000[6] | 243,830,000 | 1:23.3 |

[1]*Agricultural Statistics*, USDA.

[2]*Livestock and Meat Statistics, 1957*, Stat. Bull. Nos. 230 and 333.

[3]*Statistical Abstracts of the United States 1949*, and *The World Almanac and Book of Facts, 1966, 1970*, and *1989*

[4]Except for human population, 1980 data from *Milk Production, Disposition and Income, 1979–81*, USDA, p. 3.

[5]*Statistical Abstracts of the United States 1980*, p. 6.

[6]*Agricultural Statistics, 1988*, p. 322. Data for 1987.

## KINDS AND SIZES OF DAIRY FARMS

Fig. 20–5. Dairy farms have become bigger, fewer, and confined. This shows open-air free-stall housing with a milking parlor in the background. (Courtesy, Babson Bros. Co., Oak Brook, IL)

Through the years, dairy farms have become larger and more specialized. Separating cream, making butter, process-ing market milk, growing and mixing concentrates, and keep-ing bulls have largely disappeared from the average dairy farm; and this trend will continue. Dairy producers tend to buy more of their feed and replacements, instead of investing in more land. A few produce part or all of their forage requirements. Some have specialized heifer-raising operations.

Most dairy farms in the major dairy areas—especially the Lake States, Corn Belt, and Northeast—continue to be operated as family enterprises. However, as dairy farms become multi-ple units, operating agreements—such as father-son agree-ments—become more important. Through a partnership, or by incorporating, it is possible (1) to assure continuity of the enter-prise by transfer of ownership from one generation to the next without excessively heavy tax penalty, and (2) to arrange for the necessary capital for a prospective producer who cannot finance an operation.

A limited number of dairy farms are owned and operated by milk-processing plants, are under corporations, or are in cow pools and cooperatives.

The trend in recent years has been toward fewer but larger herds (see Fig. 20–6 and Fig. 20–7).

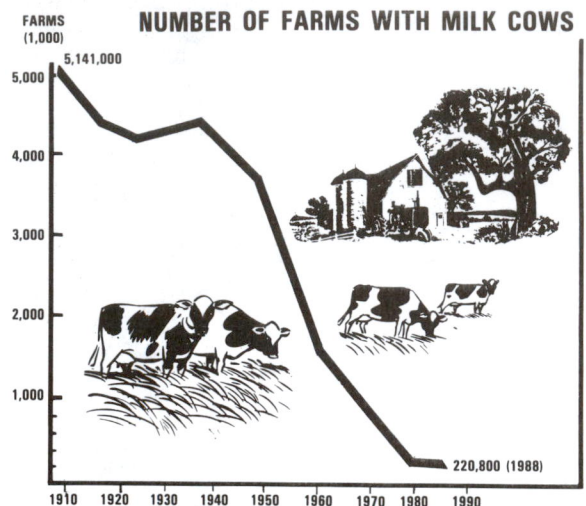

## NUMBER OF FARMS WITH MILK COWS

FARMS (1,000)

5,141,000

220,800 (1988)

Fig. 20–6. Trend in number of dairy farms. Note that in 1910 there were 5,141,000 farms with milk cows, and in 1988 there were only 220,800 farms with milk cows. (Source: USDA Agricultural Census as reported in *1988 Dairy Producer Highlights*, p. 4)

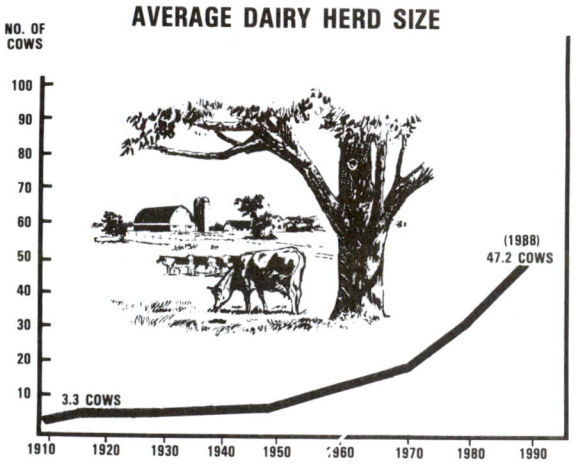

## AVERAGE DAIRY HERD SIZE

NO. OF COWS

(1988) 47.2 COWS

3.3 COWS

Fig. 20–7. Trend in dairy herd size—in number of cows per herd. Note that in 1910 there were 3.3 cows per herd, and in 1988 there were 47.2 cows per herd. (Source: USDA)

From Figs. 20–6 and 20–7, along with other authoritative sources, the following trends in the U.S. dairy industry become apparent:

1. **Decline in dairy farms.** The number of farms reporting milk cows has declined sharply. In 1910, a total of 5,141,000 farms reported milk cows; in 1950, 3,648,000 farms reported milk cows; in 1988 there were only 220,880. So, in the 78-year period, the number of farms reporting milk cows dropped 95%.

2. **Little dairy farms went out of business; big dairy farms got bigger.** The trend toward larger dairy herds in relation to the total number of dairy herds was continued. In 1910, the average dairy herd size was 3.3 cows; in 1950, it was 5.8 cows; and in 1988, it was 47.2.

3. **Cow numbers decreased.** During the 37-year period, 1950 to 1987, the number of U.S. milk cows declined from 23,853,000 to 10,430,000.

4. **Both herd size and production per cow increased.** During the same 37-year period, 1950 to 1987, the number of milk

cows declined by 56%, cows per farm increased eight fold, and milk production per cow increased by 156%.

Fig. 20–8 puts it all together. As shown, since 1974 the number of cows has declined, but the production per cow has increased dramatically with the result that total milk production has increased.

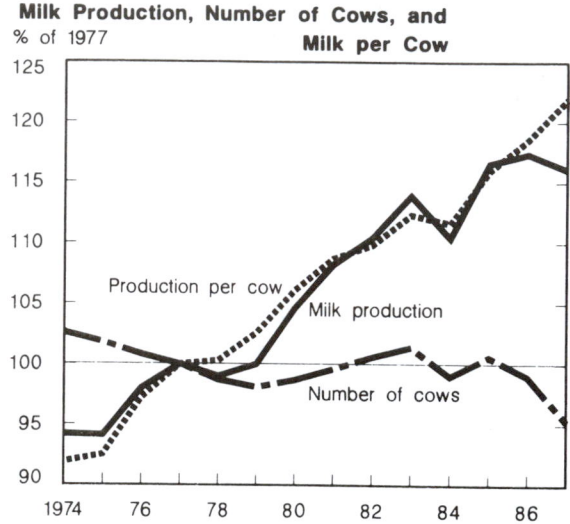

Milk Production, Number of Cows, and Milk per Cow

% of 1977

Production per cow

Milk production

Number of cows

Fig. 20–8. Although cow numbers have gone down, milk production has increased due to higher production per cow. (Source: *1988 Agricultural Chartbook*, USDA, Agricultural Handbook No. 673, p. 90, Chart 197)

## THE BUSINESS ASPECTS

On a typical dairy farm, the business may consist of one or more of the following enterprises: (1) the milking herd, (2) raising heifers for replacement purposes, (3) production of forage, and (4) production of cash crops.

As can be seen, dairy farm businesses are seldom comparable to one another, because the proportions of the enterprises vary. Hence, from the standpoint of records and analyses, it is important that each enterprise of the dairy farm be considered separately. Only by doing so, is it possible to determine which enterprises are most profitable, and which are least profitable. Such information makes it feasible for the owner to make important managerial decisions as to the future coarse—how much and what feed crop to grow, whether to enlarge or reduce the number of cows in the milking herd, whether to raise replacement heifers, and what to do to reduce costs or improve income in each enterprise in the dairy farm business.

## FACTORS FAVORABLE TO DAIRY PRODUCTION

Some of the special advantages to dairy production as compared to other livestock on the farm are:

1. **Dairying is a stable business.** Total milk production does not vary from year to year as much as the output of most other agricultural products; the change is often less than 1%, and usually not more than 2%. Nor does milk consumption vary widely. As a result, milk and other dairy products are not subject to as wide daily, seasonal, or yearly price fluctuations as are many other agricultural products.

2. **The dairy cow is unequaled as an efficient producer of human food.** A cow producing 10,000 lb of milk per year

supplies as many food nutrients as are produced by two 1,250 lb steers. Additionally, she is still available for more productive years. See Chapter 1, Table 1–7 of this book for the feed to food efficiency rating of the dairy cow compared to other species of animals.

3. **Steady income is assured.** Grain farmers, fruit producers, or vegetable growers receive income only when products are sold, usually once per year. Likewise, beef cow-calf producers secure most of their income when the calf crop is marketed, and here again, this is generally once per year. On the other hand, dairy producers receive a regular milk check at frequent intervals (biweekly or monthly) throughout the year.

4. **Steady employment is provided for labor.** Many types of agricultural work are highly seasonable, with the result that a labor force must be increased and decreased at such intervals as necessary, particularly during harvest. In the dairy enterprise, however, fairly uniform labor needs exist throughout the year. This makes it possible to keep better quality employees on a permanent basis.

5. **Dairy cows use much unsalable roughage.** Each year, considerable amounts of roughages are produced on the farms and ranches of America which would have little value if not used by dairy cattle and other ruminants. Also, much of the rolling land on which such feeds are grown is unsuited to the production of grain or other crops.

6. **Soil fertility is maintained.** By returning the manure to the land, the fertility and physical condition of the soil are preserved.

## FACTORS UNFAVORABLE TO DAIRY PRODUCTION

Among the factors which, under certain conditions, may be unfavorable to dairy production are:

1. **Considerable capital is required.** The investment for land, buildings, equipment, and cows is double, or more, what is normally required in investment per cow in a beef cow-calf enterprise.

2. **Successful dairy management necessitates superior training.** The larger the dairy enterprise, the more important this becomes. Superior managers must be knowledgeable in the basic sciences, business administration, animal physiology, and nutrition; and they must possess the necessary personality and ability to weld this knowledge into a smooth functioning, efficient production unit.

3. **The bewildering number and kinds of regulatory programs.** The federal, state, and local regulatory programs necessitate that producers be familiar with, and follow, those that are applicable to them. No matter how noble the objectives of these regulatory programs, this takes considerable time, and often it is a frustrating experience.

4. **Dairying is confining.** Unlike many agricultural endeavors, dairying must be done regularly, and without fail, particularly where it is a market milk enterprise. Thus, an owner or a manager who is interested in vacation time, short work weeks, or even short days, should not enter into a one-person dairy enterprise.

5. **Hourly returns are low.** Hourly returns to dairy farmers have, on the average, been below returns in many other types of farming, and well below the average for all U.S. manufacturing industries.

6. **The threat of imitations.** Milk—long extolled as the perfect food—is being threatened by imitations.

## FUTURE OF DAIRYING

Fig. 20–9. Peaches and cream. The future of dairying will be determined by consumers. (Courtesy, California Foods Research Institute, San Francisco, CA)

One factor above all others, assures U.S. dairy production a bright future: the efficiency of the dairy cow from the standpoint of transforming plant material into animal food products for humans (see Chapter 1, Table 1–7).

The following developments will likely characterize the dairy production of the future:

1. **Larger units, increased automation, and environmental control.** Figs. 20–6 and 20–7 show that units have become fewer but larger (more cows per herd). This trend will continue, and with it there will be more automation and more environmental control.

2. **Higher production and greater efficiency per cow.** It is evident that, as profit margins narrow and knowledge becomes greater, we shall continue to move, at an accelerated pace, to higher production and greater efficiency per cow. This will be achieved through the application of sophisticated biotechnology, breeding, feeding, milking, management, and marketing.

3. **Fewer producers, fewer cows.** To the year 2,000 and beyond, it is expected that the number of dairy farmers and the number of dairy cows will continue to decline.

4. **Milk output per caretaker will increase.** Milk output per caretaker will increase as a result of increased herd size, higher milk production per cow, and the application of labor-saving machinery and equipment.

5. **Free-stall housing and mechanized milking will increase.** More and more dairy farmers will use free-stall housing and mechanized milking systems.

6. **Family farms will continue.** It is expected that the majority of dairy farms will continue to be family owned and operated.

7. **New and useful products.** Intensive research attempts are being made to provide products (a) with the characteristics of fresh milk, but which will last longer in storage, and (b) which can be transported at lower cost. Sterile milk and concentrated milk are receiving the most attention.

8. **More dairy beef.** The dairy industry provides an estimated one-fourth of the beef consumed in this country, with these animals being marketed as veal calves, cull dairy cows and bulls, and finished dairy heifers and steers. Improvements in the science and technology of feeding and processing, along with consumer preference for less fat and more lean beef, favor the growing and finishing of more dairy beef.

9. **More training, experience, and business acumen.** The successful dairy manager of the larger and more highly specialized units of the future must have more know-how and experience, and the ability to operate the dairy establishment as a big business.

10. **Changes will be made.** Many factors will determine to what extent the above changes will occur. Feed costs will be a factor. Also, uncertainties about government policies relative to dairy product imports and grain exports will help to determine the degree and direction of change.

## QUESTIONS FOR STUDY AND DISCUSSION

1. Why are some mammals other than milk cows—such as water buffalo, goats, sheep, and mares—used for milk production in different parts of the world?

2. Summarize the dramatic changes in U.S. dairying in 1929 vs 1988, and list the four major developments that propelled these changes.

3. Discuss (a) uses of milk, and (b) the important role of milk in the diet of the average American.

4. Why does the U.S.S.R. hold such a commanding lead in milk cow numbers?

5. Israel and Japan must be doing a lot of things right in order to rank first and second, respectively, in average milk production per cow. Explain.

6. Discuss the importance of the U.S. dairy industry in terms of (a) farm income from milk and cream, (b) leading states, and (c) milk production as related to human population.

7. Table 20–8 shows that the ratio of number of milk cows to number of people has declined since 1920. What forces were back of this decline?

8. Discuss the kinds of U.S. dairy farms.

9. Discuss recent trends to fewer dairy farms, but larger herds, as pointed up in Figs. 20–6 and 20–7.

10. From the standpoint of records and analyses, why is it important that each enterprise of the dairy farm be considered separately?

11. List the factors *favorable* to dairy production.

12. List the factors *unfavorable* to dairy production.

13. List and discuss the developments that will likely characterize the dairy production of the future.

14. In the future, how is dairy production likely to fare in comparison with beef production, sheep production, swine production, and poultry production from the standpoint of increasing or decreasing?

15. What factors have prompted the trend toward more and more specialization in the dairy enterprise, in contrast to diversification?

## SELECTED REFERENCES

| Title of Publication | Author(s) | Publisher |
|---|---|---|
| *Dairy Cattle in American Agriculture* | A. R. Porter<br>J. A. Sims<br>C. F. Foreman | Iowa State University Press, Ames, IA, 1965 |
| *Dairy Cattle Feeding and Management*, Sixth Edition | W. M. Etgen<br>P. M. Reaves | John Wiley & Sons, Inc., New York, NY, 1978 |
| *Dairy Cattle Management* | J. M. Wing | Reinhold Pub. Corp., New York, NY, 1963 |
| *Dairy Cattle and Milk Production* | C. H. Eckles<br>E. L. Anthony | The Macmillan Company, New York, NY, 1956 |
| *Dairy Cattle: Principles, Practices, Problems, Profits*, Third Edition | D. L. Bath *et al.* | Lea & Febiger, Philadelphia, PA, 1985 |
| *Dairy Cattle Science*, Second Edition | M. E. Ensminger | The Interstate Printers & Publishers, Inc., Danville, IL, 1980 |
| *Dairy Cattle Selection, Feeding and Management* | W. W. Yapp<br>W. B. Nevens | John Wiley & Sons, Inc., New York, NY, 1955 |
| *Dairy Farm Management* | T. Quinn | Delmar Publishing, Inc., Albany, NY, 1980 |
| *Dairy Science* | W. E. Peterson | J. B. Lippincott Co., Philadelphia, PA, 1950 |
| *Large Dairy Herd Management* | C. J. Wilcox, *et al.* | University Presses of Florida, Gainesville, FL, 1978 |
| *Milk Production and Processing* | H. F. Judkins<br>H. A. Keener | John Wiley & Sons, Inc., New York, NY, 1963 |
| *Modern Dairy Cattle Management* | R. F. Davis | Prentice-Hall, Inc., Englewood Cliffs, NJ, 1962 |
| *Principles of Dairy Farming, The*, Sixth Edition | K. Russell<br>S. Williams | Farming Press Ltd., Fenton House, Ipswich, England, 1972 |
| *Science of Providing Milk For Man, The* | J. R. Campbell<br>R. T. Marshall | McGraw-Hill Book Company, New York, NY, 1975 |
| *Stockman's Handbook, The*, Sixth Edition | M. E. Ensminger | The Interstate Printers & Publishers, Inc., Danville, IL, 1983 |

World Champion Milk Producer, *Beecher Arlinda Ellen*. In 1975, this great cow produced 55,661 lb of milk (365–2X), with a peak of 22.6 gal in one day. *Ellen* was bred and owned by Harold L. Beecher, Rochester, Indiana.

# BREEDS OF DAIRY CATTLE

---

## Chapter 21

---

| Contents | Page |
|---|---|
| **Popularity of Breeds** | 481 |
| **Characteristics of Breeds** | 482 |
| **Milk and Butterfat Production of Breeds** | 484 |

| Contents | Page |
|---|---|
| **Programs of the Registry Association** | 484 |
| **Questions for Study and Discussion** | 486 |
| **Selected References** | 486 |

There are seven breeds of dairy cattle in the United States, all of which have proven satisfactory in different sections of the country.

## POPULARITY OF BREEDS

Table 21-1 (see next page) shows (1) when each of the breed registries was formed, and (2) the 1988/1989 and total registrations to date of the breeds of dairy cattle. Even though data for one year only fail to show trends, the recent annual figures reflect the current popularity and numbers of the respective breeds more than do the total registrations since establishing the breed registry.

**TABLE 21-1**
**REGISTRATIONS OF DAIRY CATTLE IN U.S. BREED REGISTRY ASSOCIATIONS**

| Breed | Breed Association Formed | | 1988 or 1989 Registrations | Total Registrations to Date |
|---|---|---|---|---|
| | Abroad | in U.S. | | |
| Holstein-Friesian | 1873 | 1871 | 392,883 | 15,625,900 |
| Jersey | 1833 | 1868 | 56,030 | 4,324,252 |
| Guernsey | 1814 | 1877 | 16,644 | 3,587,137 |
| Brown Swiss[1] | 1911 | 1880 | 12,376 | 973,000 |
| Ayrshire | 1877 | 1875 | 8,075 | 1,111,189 |
| Dutch Belted | — | 1886 | — | — |
| Milking Shorthorn | — | 1948 | 3,059 | 122,543 |

[1]Not including Brown Swiss beef registrations.

## CHARACTERISTICS OF BREEDS

Table 21-2 gives in summary form the place of origin and the characteristics of each of the breeds of dairy cattle.

**TABLE 21-2**
**BREEDS OF DAIRY CATTLE AND THEIR CHARACTERISTICS**

| Breed | Place of Origin | Color | Distinctive Head Characteristics | Other Distinguishing Characteristics | Disqualifications; Comments |
|---|---|---|---|---|---|
| Ayrshire | County of Ayrshire in southwestern Scotland. | Light to deep cherry red, mahogany, brown, or a combination of these colors, with white, or white alone. Black or brindle is objectionable. | Horns are widespread and tend to curve upward and outward. However, there is a polled strain. | The udders are especially symmetrical and well attached to the body. The breed is noted for its style and animation, good feet and legs, and grazing ability. | |
| Brown Swiss | The Alps of Switzerland. Brought to America in 1869. | Solid brown, varying from very light to dark. White markings are objectionable. | The nose and tongue are black, and there is a characteristic light-colored band around the muzzle. Medium-length horns. | Strong and rugged, with some tendency toward the heavy muscling characteristic of the beef breeds. Calm and unexcitable. | Spotting is undesirable. In 1971, The Brown Swiss Cattle Breeders' Association formed the Brown Swiss Beef International, Inc., for registration of beef type Brown Swiss. |
| Dutch Belted | Netherlands, prior to 17th Century. | Black and white. | Horns. Head somewhat long and dished. | White belt extending entirely around the body, from a little back of the shoulder to just in front of the hips. | No belt. |
| Guernsey | Island of Guernsey. | Fawn with white markings clearly defined; preferably a clear (buff) muzzle. | Good length of head; horns incline forward, are refined and medium in length, and taper toward the tips. | The milk is especially yellow in color; golden yellow skin pigmentation; the unhaired portions of the body are light or pinkish in color (whereas in the Jersey they are nearly black); calves are relatively small at birth. | |
| Holstein-Friesian | Netherlands and northern Germany. | Black and white or red and white. | Clean-cut, broad muzzle, open nostrils, strong jaw, broad and moderatley dished forehead, straight and bridged nose. | Large angular animals; females should weigh 1,500 lb (mature); males in breeding condition 2,200 lb. | Black and white animals are disqualified if (1) solid black, or (2) solid white. Suffix *Red* is added to name of red and white animals. Suffix *OC* is added to name of off color animals. |
| Jersey | Island of Jersey. | Jerseys vary greatly in color, but the characteristic color is some shade of fawn, with or without markings. | Forehead, broad and moderately dished with large, bright eyes. Clean-cut and proportionate to body. | Jerseys are especially known for their well-shaped udders, strong udder attachments, and ease of calving. They are also very angular and refined. | Total blindness, permanent lameness that interferes with normal function, blind quarter, freemartin heifers and animals showing signs of being operated upon or tampered with. |
| Milking Shorthorn | England: the breed traces to a milking strain of Shorthorns developed by Thomas Bates. | Red, white, or any combination of red and white. | Fine horns that are rather short. | The breed is very adaptable; it competes favorably for either milk or beef production. | No calf is eligible for registration unless its sire and dam were each at least 18 mo. of age at the birth date of the calf. |

In 1949, the Milking Shorthorn breed split off from the American Shorthorn Assn. and formed a separate breed registry—the American Milking Shorthorn Society.

In 1973, the American Shorthorn Assn. made provision to register Milking Shorthorns in its Herd Book. But there is no reciprocal arrangement for acceptance of Shorthorn (beef) blood in the American Milking Shorthorn Herd Book.

Beginning in 1969, semen of the Illawarra, a Shorthorn developed in Australia especially for milk production, was imported. Some feel that an infusion of this blood will improve udders and milk production of the Milking Shorthorn in the U.S. The American Milking Shorthorn Society must approve each importation of semen before the offspring can be registered in the Herd Book.

Fig. 21-1. Ayrshire cow, *Chestnut Ridge Starman's Gwennie*. (Courtesy, Ayrshire Breeders' Assn., Brandon, VT)

Fig. 21-2. Brown Swiss cow, *Idyl Wild Improver Jinx*, All American Aged Cow, 1985, 365D–2X 29,840 lb milk, 4.3% butterfat. (Courtesy, The Brown Swiss Cattle Breeders' Assn. of the U.S.A., Beloit, WI)

Fig. 21-3. Dutch Belted cow, *Perfection's Belle,* shown at 10 years of age. During the 4-year period, this cow averaged over 18,000 lb of milk per year. (Courtesy, Dutch Belted Cattle Assn. of America, Inc., Venus, FL)

Fig. 21-4. Guernsey cow, *Win Crest Knights Vel,* 365D–2X 30,560 lb milk, 4.3% butterfat. (Courtesy, American Guernsey Assn., Reynoldsburg, OH)

Fig. 21-5. Holstein bull, *Round Oak Rag Apple Elevation,* All-American Get/Sire, 1977, 1978, 1980, and 1981. (Courtesy, Holstein-Friesian Assn. of America, Brattleboro, VT)

Fig. 21-6. Jersey cow, *Vair-y Best Rockette*. (Courtesy, The American Jersey Cattle Club, Reynoldsburg, OH)

Fig. 21–7. Milking Shorthorn cow, *Innisfail Lobelia 72nd,* 1988 All-American 3-year-old. Owned and shown by Pinehurst Farm, Daleville, Indiana. (Courtesy, American Milking Shorthorn Society, Beloit, WI)

## MILK AND BUTTERFAT PRODUCTION OF BREEDS

There are breed differences in milk and butterfat production. Table 21–3 summarizes the averages by breeds. On the

basis of milk production, they rank in the following order: Holstein, Brown Swiss, Ayrshire, Guernsey, and Jersey. However, on the basis of the butterfat test they rank: Jersey, Guernsey, Brown Swiss, Ayrshire, and Holstein.

**TABLE 21–3**
**BREED DIFFERENCES IN MILK AND BUTTERFAT TEST AND PRODUCTION[1]**

| Breed | Milk Production | Butterfat Test | Butterfat Production |
|---|---|---|---|
| | (lb) | (%) | (lb) |
| Holstein | 15,528 | 3.65 | 566 |
| Brown Swiss | 13,063 | 3.97 | 519 |
| Ayrshire | 11,730 | 3.90 | 457 |
| Milking Shorthorn | 10,596 | 3.64 | 386 |
| Guernsey | 10,573 | 4.61 | 487 |
| Jersey | 10,150 | 4.90 | 497 |

[1]Data for Table 21–3 was compiled by the Animal Improvement Programs Laboratory, Science and Education Administration, USDA, from the 1977 records submitted by the 11 dairy records processing centers serving the National Cooperative Dairy Herd Improvement Program.

U.S. milk and butterfat record holders for each of the breeds are given in Table 21–4.

**TABLE 21–4**
**U.S. PRODUCTION RECORDS FOR EACH DAIRY BREED**

| Breed | Name of Cow | Year | Days; Times Milked | Milk Production | Butterfat Test | Butterfat Production |
|---|---|---|---|---|---|---|
| | | | | (lb) | (%) | (lb) |
| **Milk Production** | | | | | | |
| Holstein | *Beecher Arlinda Ellen* | 1975 | (365–2X) | 55,660 | 2.8 | 1,573 |
| Brown Swiss | *Century Acres Liz C.* | 1980 | (356–2X) | 37,846 | 4.4 | 1,667 |
| Ayrshire | *Leete Farms Betty's Ida* | 1976 | (305–2X) | 37,170 | 4.3 | 1,592 |
| Jersey | *Rocky Hill Favorite Deb* | 1979 | (365–2X) | 35,881 | 5.4 | 1,923 |
| Guernsey | *Willow Brook C Dena* | 1980 | (365–2X) | 35,170 | 3.9 | 1,385 |
| Milking Shorthorn | *Washita Ann's Bonnie Exp* | 1986 | (365–2X) | 30,790 | 2.9 | 902 |
| **Butterfat Production** | | | | | | |
| Holstein | *Breezewood Patsy Bar Pontiac* | 1976 | (365–2X) | 47,500 | 4.7 | 2,230 |
| Jersey | *Rocky Hill Favorite Deb* | 1979 | (365–2X) | 35,881 | 5.4 | 1,923 |
| Guernsey | *Fox Run AFC Faye* | 1971 | (365–2X) | 31,040 | 5.6 | 1,736 |
| Brown Swiss | *Letha Irene Pride* | 1959 | (365–3X) | 34,810 | 5.0 | 1,733 |
| Ayrshire | *Leete Farms Betty's Ida* | 1976 | (305–2X) | 37,170 | 4.3 | 1,592 |
| Milking Shorthorn | *Ashgrove Golden Lily 38th* | 1982 | (365–2X) | 25,749 | 4.4 | 1,144 |

## PROGRAMS OF THE REGISTRY ASSOCIATIONS

In addition to registering animals, most of the breed associations promote the following programs:

1. **Production testing in Dairy Herd Improvement Registry (DHIR).** Each breed registry association has a program for testing registered cows under the Unified Rules for Official Testing as adopted by the Purebred Dairy Cattle Association

and the American Dairy Science Association. This program is conducted cooperatively by the breed association and the Division of Dairy Herd Improvement Investigations of the U.S. Department of Agriculture (the division which, in cooperation with the states, is responsible for DHIR records). Records recognized as official by both groups are included in one program.

Under DHIR, milk testing is conducted once each month, with the tester obtaining a 24-hour milk weight and butterfat test. Also, the tester secures data on each cow that has freshened, including: feed consumption and quality, labor, price of milk, etc. All this data is sent to a central laboratory for analysis, with many states cooperating on a regional basis in the Electronic Data Processing Method (EDPM). The machine processed records are then returned to the producer, giving current information on milk yield, income over feed costs, milk produced per cow, and other pertinent information to help in making culling and managerial decisions.

In 1966-67, two breed registry production-testing programs—(a) Advanced Registry (AR), and (b) Herd Improvement Registry (HIR)—were discontinued, and the Dairy Herd Improvement Registry (DHIR) became the official milk-recording program of all the breeds.

2. **Type classification.** Since many animals are not exhibited at cattle shows, the association started a voluntary program of herd classification whereby a qualified classifier, selected by the association at the request of the owner and on a nominal charge basis, comes to the farm and classifies, for type or conformation, each milking animal in the herd. Each animal is rated according to the scorecard and placed in the corresponding category, as shown in Table 21-5.

### TABLE 21-5
### TYPE CLASSIFICATION BY BREEDS

| Nomenclature | Ayrshire | Brown Swiss | Guernsey | Holstein | Jersey | Milking Shorthorn |
|---|---|---|---|---|---|---|
| | (score in points) | | | | | |
| E-Excellent | 90-99 | 90 & over | 90 & over | 90 & over | 90 & over | 90 & over |
| VG-Very Good | 80-89 | 85-89 | 85-89 | 85-89 | 85-89 | 85-89 |
| GP-Good Plus (D-Desirable in Guernseys and Jerseys) | 70-79 | 80-84 | 80-84 | 80-84 | 80-84 | 80-84 |
| G-Good (A-Acceptable in Guernseys and Jerseys) | 60-69 | 75-79 | 75-79 | 75-79 | 75-79 | 75-79 |
| F-Fair (P-Poor in Jerseys) | 50-59 | 65-74 | 70-74 | 65-74 | 70-74 | 65-74 |
| P-Poor | — | Under 65 | Under 70 | Under 65 | — | Under 65 |
| Year when initiated | 1941 | 1944 | 1947 | 1929 | 1932 | |

The program has been increasingly utilized by breeders, and it has been highly effective in the general improvement of conformation and in merchandising cattle.

3. **Recognition awards.** Some of the breed associations have established certain programs and recognition awards for breeders, cows, and bulls; and two of them have special milk-merchandising programs. These are summarized in Table 21-6.

### TABLE 21-6
### BREED RECOGNITION AWARDS

| Breed | Breeder Award | Sire Award | Dam Award | Merchandising Program |
|---|---|---|---|---|
| Ayrshire | Constructive Breeder. | Approved Sire. | Approved Dam. | |
| Brown Swiss | None. | Superior Sire. Qualified Sire. | Certified Dam. Elite Dam. | |
| Guernsey | Gold Star Breeder. Gold Star Herd. Gold Star Breeder and Herd. | | Gold Star Dam. | Golden Guernsey Milk (started in 1923). |
| Holstein-Friesian | Progressive Breeder Registry. | Gold Medal Sire (Type & Production). | Gold Medal Dam (Production & Classification). | |
| Jersey | Constructive Breeder. Gold Star Herd. | | Tested Dam. 100,000 Pound Milk. Ton, Double, Triple Gold Awards. Hall of Fame. | All-Jersey Milk. |
| Milking Shorthorn | Progressive Breeder. | None. | Gold, Silver, and Bronze Medals | |

These awards vary somewhat between breeds, and they are revised upward from time to time so as to maintain them as worthwhile goals. For breeder recognition, the general factors taken into account are: ownership of a certain minimum number of (a) registered females, and (b) animals bred by owner; meeting established minimum production requirements under one of the breed testing programs; meeting established minimum type classification requirements; and evidence of a healthy herd, especially with respect to tuberculosis and brucellosis. Sire and dam awards are based on their apparent ability to transmit a high level of production and/or type to a specified number and percentage of offspring.

# QUESTIONS FOR STUDY AND DISCUSSION

1. What factors make for a world milk champion such as *Beecher Arlinda Ellen?*

2. Table 21–1 shows that more Holsteins are being registered annually than all the rest of the dairy breeds combined. Is this dominant position of Holsteins good or bad? Justify your answer.

3. Of what importance are the (a) distinguishing characteristics, and (b) disqualifications of the breeds of dairy cattle?

4. Select four breeds of dairy cattle; then present the following information relative to each of them: (a) place of origin, (b) color, and (c) distinguishing characteristics.

5. Is there a need for more breeds of dairy cattle than we now have in the United States?

6. How do you account for the fact that no new U.S. breed of dairy cattle has been developed in the last 100 years?

7. Are breeds of dairy cattle likely to decline in importance as happened in the poultry industry?

8. How do you account for the difference between breeds in milk and butterfat production?

9. What is the Dairy Herd Improvement Registry (DHIR)? What services do they perform?

10. Of what value are breed type classifications?

11. Of what value are breed recognition awards?

12. Justify any preference or bias that you may have for one particular breed of dairy cattle.

13. Obtain breed registry association literature and a sample copy of a magazine of your favorite breed of dairy cattle. (See Appendix for addresses.) Evaluate the soundness and value of the material that you receive.

# SELECTED REFERENCES

| Title of Publication | Author(s) | Publisher |
|---|---|---|
| *Breeds of Cattle* | H. R. Purdy | Chanticleer Press, Inc., New York, NY, 1987 |
| *Breeds of Livestock in America* | H. W. Vaughn | R. G. Adams and Co., Columbus, OH, 1937 |
| *Dairy Cattle Breeds* | R. B. Becker | University of Florida Press, Gainsville, FL, 1973 |
| *Modern Breeds of Livestock*, Fourth Edition | H. M. Briggs D. M. Briggs | The Macmillan Company, New York, NY, 1980 |
| *Types and Breeds of Farm Animals* | C. S. Plumb | Ginn and Company, Boston, MA, 1920 |
| *World Dictionary of Breeds, Types, and Varieties of Livestock, A* | I. L. Mason | Commonwealth Agricultural Bureaux, Farnham House, Farnham Royal, Slough, Bucks, England, 1951 |

Jersey cows on pasture in Canada. (Courtesy, The Canadian Jersey Cattle Club)

AYRSHIRE COW

BROWN-SWISS COW

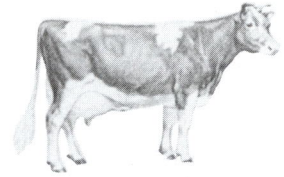

GUERNSEY COW

HOLSTEIN-FRIESIAN COW

JERSEY COW

# ESTABLISHING THE DAIRY HERD; SELECTING AND JUDGING DAIRY CATTLE

MILKING SHORTHORN COW

Breed characteristics of dairy cattle portrayed by breed registry ideals.

# Chapter 22

Whether establishing new herd or maintaining old ones, dairy producers must constantly appraise or evaluate their animals—they must buy, sell, retain, and cull.

## FACTORS TO CONSIDER IN ESTABLISHING THE HERD

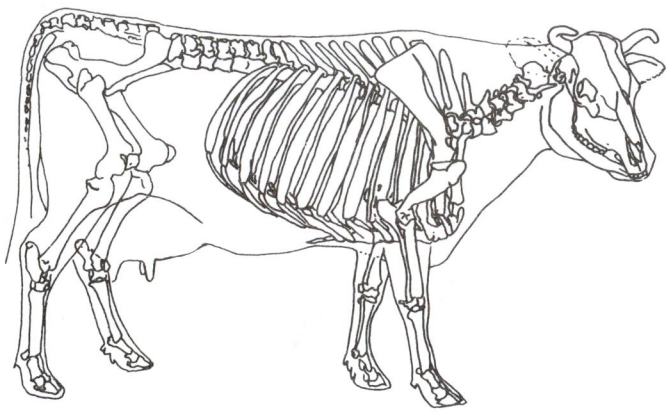

Fig. 22–1. General appearance is determined by the animal's bone structure. Proper bone structure is indicative of the cow's wearing ability and useful life.

In establishing a dairy herd, the following factors must be considered:

1. **Dairy or dual-purpose type.** Approximately 20% of the cows of the United States are kept for milk production. Of these, about 70% belong to the 5 major dairy breeds, and the remainder to the dual-purpose breeds, beef cattle, or nondescript breeding. Where dairying is highly specialized, most milk cows are strictly dairy breeding, rather than of dual-purpose type.

2. **Grades or purebreds.** Technically, a purebred animal is one that can meet ancestry requirements in one of the breed registry associations, whereas a registered animal is a purebred which has been recorded in one of the registry books. Grade animals are those that are not registered or eligible for registry; however, such animals frequently approach purebred status as a result of several generations of breeding up by sires of one breed. Thus, if a registered sire is used successively for seven generations, the final offspring will, mathematically speaking, consist of 99% registered parentage. From this, it can be reasoned that, from an inherited production potential, the gap between registered and grade animals is often very small.

In general, the person who is inexperienced in handling dairy animals, or who has a limited amount of capital, should start with grade animals, then improve them by the use of good purebred sires through artificial insemination programs. However, the person who is experienced, and who has adequate capital, may well consider the purebred business.

3. **Choice of a breed.** The choice of a breed is usually made on the basis of personal preference, prior association with the breed and the breeders, and the availability of the breed in the immediate area. Perhaps it is well to add that the choice of a breed is likely less important than the choice of good individuals within the breed, simply because there is more difference within breeds than between breeds.

4. **Buying cows, heifers, or calves.** In starting a herd, three methods are available: buying cows, buying heifers, or buying calves. Also, cows and heifers of breeding age may either be open or bred at the time of purchase. The choice between the alternatives should be determined primarily by (a) the time when it is desired to be in production, (b) available capital, and (c) experience.

Where there is no question pertaining to the honesty and integrity of the seller, and good cows with production records back of them can be acquired at what is considered to be a fair and reasonable price, this usually constitutes the best buy. In acquiring cows, it is well to keep in mind that, on the average, they remain in the milking herd for about four years only, and for the most part, they are culled from, or leave, the herd before they are seven years of age.

The purchase of heifers at breeding age, or as springers, to start a herd is a very popular method. When buying heifers, consideration should be given to the caliber of their dams and the record of their sires.

The purchase of calves requires the least initial capital of any of the methods, but it also takes more time to get into production. In many ways, however, the purchase of calves offers the best opportunity to get high-quality breeding animals.

5. **Production records.** Production records should be used in selecting individual animals, with proper consideration given to the environmental factors under which the records were made.

6. **The disease problem.** In establishing a herd, one should take every possible precaution to avoid bringing a disease, especially such diseases as tuberculosis, brucellosis, leptospirosis, trichomoniasis, and mastitis. Despite all precautions, however, it is well that newly acquired animals be isolated for a period of 30 to 60 days, and that they be retested at the end of that period before being placed in the herd.

Of course, there are other factors to consider in establishing a herd; among them, longevity. Since it normally takes the profit from the first lactation plus about one-half of the second lactation to cover the initial investment in the animal before she freshens, it becomes crystal clear that longevity is important to the owner. Other factors of importance are: uniformity, condition, reproductive ability, temperament, size, and price.

## PUREBRED BUSINESS

It is estimated that approximately 10% of the dairy cattle of the United States are purebred. However, these animals exert a powerful influence through supplying seed stock to the dairy industry—to other purebred breeders and to those who have grade dairy cattle.

## REQUISITES OF A PUREBRED HERD

The primary requisites of a purebred herd are:

1. **High production.** If purebred animals are to bring about further improvement through the dissemination of seed stock, it is imperative that they be top producers.

2. **Good type.** There is much controversy over the value of type and its relationship, or lack thereof, to production. Nevertheless, no one has proven that, on the average, cattle with good conformation produce less than cattle with poor conformation. Further, it is generally acknowledged that, among top producers, one can select animals with superior type. Also, people instinctively desire and appreciate beauty, no matter what the object; and usually they are willing to pay extra for

it. It follows, therefore, that good type is important in the purebred herd because animals must be sold.

3. **Attractive surroundings.** Although it is not necessary that the physical plant of a purebred herd be elaborate, it should be neat and attractive.

## REQUISITES OF A REGISTERED BREEDER

As is true of the commercial milk producer with grade cows, to be successful the purebred breeder must possess a love for the dairy business and the necessary knowledge and experience in the production, handling, and marketing of milk. Additionally, the registered breeder should (1) have knowledge of breeding and pedigrees, (2) pay attention to details, (3) be able to promote and sell, (4) be able to withstand disappointment, and (5) practice honesty and integrity in all dealings.

## SELECTING HERD REPLACEMENTS

Fig. 22–2. One cow in four is replaced each year in the average U.S. dairy herd.

The average dairy cow in the United States remains in the milking herd about four years. This means that about 25% of the average milking herd must be replaced each year. These replacement animals must be either raised or purchased. In either event, the animals brought in should be of herd-improving kind.

In this section, selection of the herd replacements will be limited to females. Bull selection is covered in the chapter on "Breeding Dairy Cattle."

When the major source of income is from the sale of milk, selection is simplified. The cows can be ranked from high to low on the basis of milk production, and the most profitable ones used as replacements. Where purebred breeding stock is involved, the breeder usually invokes an additional point— type or conformation.

Where heifers are being selected for eventual replacement purposes, pedigree information is very important.

## CULLING DAIRY CATTLE

Successful dairy producers are constantly checking their herds and evaluating individual cows, then culling those that do not make money. With high milk prices and low salvage prices, they often find it difficult to decide which cows to cull from the herd, and how many they should cull. Tables 22–1 and 22–2 are designed to assist in this regard. Such forms facilitate establishing minimum standards in each area of concern and importance.

Table 22–1, "Culling Guide," is simply a form on which the butterfat production of each cow can be listed (list each cow by number in the proper category), with provision for first lactation cows, second lactation cows, and mature cows. By using this form, culling decisions can be made without having to compute age-conversion factors. After listing cows in the proper columns, dairy producers can decide where to draw the line. For example, if they decide to cull all mature cows that are below the 425-lb level, they will take out all cows in columns A, B, C, and D. Similarly, they might decide to draw the line on second lactation cows at 376 lb and first lactation cows at 340 lb.

### TABLE 22–1
### CULLING GUIDE (Based on 305-Day Butterfat Production)

| | A | B | C | D | E | F | G | H | I | J | K | L |
|---|---|---|---|---|---|---|---|---|---|---|---|---|
| **Pounds of butterfat** | Under 350 | 350– 374 | 375– 399 | 400 424 | 425– 449 | 450– 474 | 475– 499 | 500– 524 | 525– 549 | 550– 574 | 575– 599 | 600– Over |
| Mature cows (3rd lactation or more) | | | | | | | | | | | | |
| **Pounds of butterfat** | Under 310 | 310– 331 | 332– 353 | 354– 375 | 376– 397 | 398– 419 | 420– 441 | 442– 464 | 465– 486 | 487– 508 | 509– 530 | 531– Over |
| Second lactation cows | | | | | | | | | | | | |
| **Pounds of butterfat** | Under 280 | 280– 299 | 300– 319 | 320– 339 | 340– 359 | 360– 379 | 380– 399 | 400– 419 | 420– 439 | 440– 459 | 460– 479 | 480– Over |
| First lactation cows | | | | | | | | | | | | |
| **Herd Total** | | | | | | | | | | | | |

After culling on the basis of productivity, producers must also cull for other reasons. The cow evaluation form, Table 22–2, is designed for this purpose. It can be applied, and those that do not measure up can be culled.

**TABLE 22–2**
**COW EVALUATION**

| Production | Retain | | Cull |
|---|---|---|---|
| | **High** | **Average** | **Low** |
| **M**ature cows | 10% above mature herd average. | Within 10% of mature herd average. | 10% or more below mature herd average. |
| **S**econd lactation cows | Above mature herd average. | Within 20% of mature herd average. | 20% or more below mature herd average. |
| **F**irst lactation cows | Above or within 10% of mature herd average. | 10 to 30% below mature herd average. | 30% or more below mature herd average. |
| **Miscellaneous Factors:** | | | |
| **E**xpected dry period | 2 months. | 3–5 months. | 6 months or more. |
| **H**ealth and injuries | Good. | Temporary. | Chronic. |
| **M**ilking qualities | Fast. | Medium. | Slow and hard. |
| **D**isposition | Quiet. | Not easily excitable. | Nervous or dull. |
| **T**ype, particularly udder conformation | Highly desirable. | Sound. | Undesirable. |

## JUDGING DAIRY CATTLE

It appears that only about 21% of the cows in lactation are tested to determine their milk-yielding capacity. Even fewer dairy animals are subjected to the scrutiny of an experienced judge in the show-ring. Thus, the only method, other than pedigree, available to evaluate the great bulk of dairy animals is by what is commonly known as judging.

Judging—as practiced in shows, in contests, or on the farm—is an attempt to place or rank animals in the order of their excellence in body type. Scoring, or type classifying, an animal accomplishes the same thing, in that the individual being scored is classified and compared to an animal that is theoretically perfect, and a rating is assigned on this basis.

Admittedly, there is considerable question as to the degree of correlation between type and production. Yet, it is generally recognized that desirable type in no way negates functional value. Moreover, it is generally recognized that attractiveness, and what we think of as desirable type, enhances the market value of purebred animals. Also, well-attached udders are less subject to injury and mastitis infection, and strong legs and feet hold up longer than weak legs and feet.

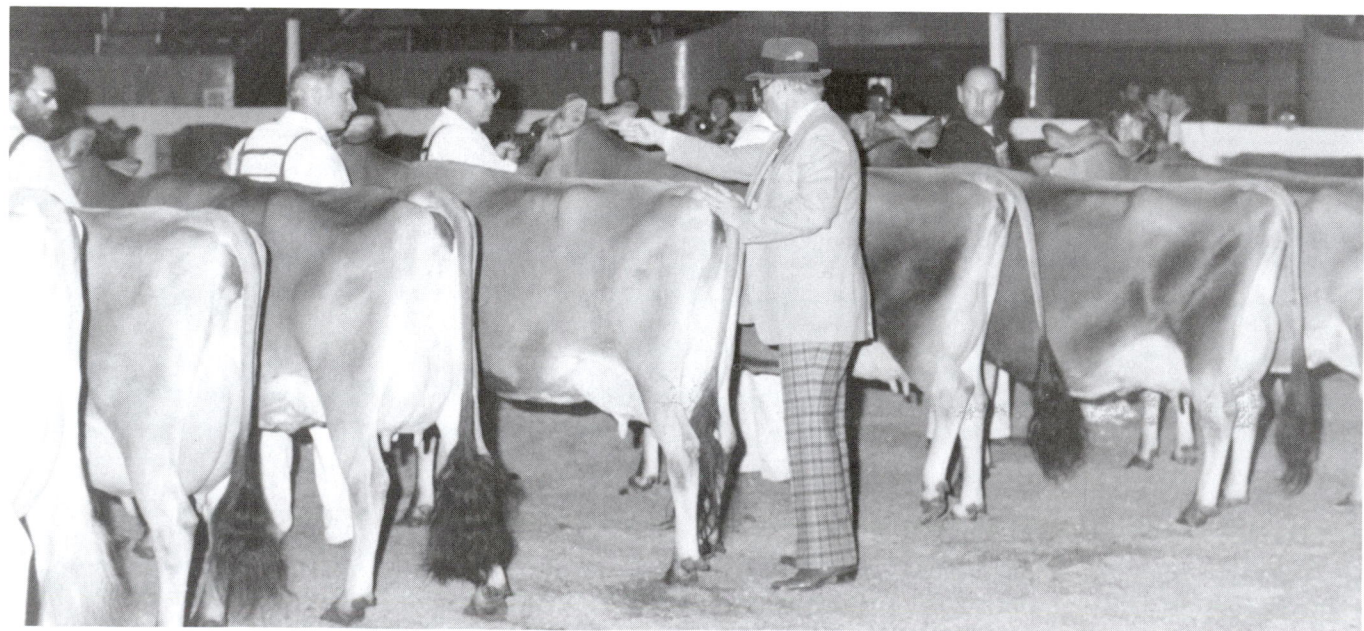

Fig. 22–3. Jersey cows being judged at the All American Jersey Show. (Courtesy, The American Jersey Cattle Club, Reynoldsburg, OH)

It is noteworthy that good and successful owners and managers are generally good judges of dairy cattle.

## PARTS OF A COW

One of the characteristics of good judges is that they possess a thorough knowledge of animals. In speaking of the characteristics of a dairy animal, we usually refer to parts rather than to the individual as a whole. It is important, therefore, to become familiar with the names of the parts. Figs. 22–5 and 22–6 show animals in outline form and identify by name the various parts of animals. These figures should be studied until each part of the animal can be easily and quickly identified by location and name. Nothing so quickly sets a professional apart from a novice as a thorough knowledge of the parts and the language commonly used in describing them.

## DAIRY TYPE; BREED TYPE

Webster defines type as, "the combination of characters appropriate to a special kind of use." Certainly, this definition is adequate for distinguishing beef-type from dairy-type cattle, or perhaps even for distinguishing Jersey type from Holstein type. However, it lacks the necessary specificity for those desiring to differentiate type within a breed. Hence, the following definition is proposed.

*Type refers to an ideal or standard of perfection combining the physical characteristics which contribute to an animal's usefulness for a specific purpose.*

Additionally, to practical milk producers the word "type" has come to express the kind of cow that is adapted to modern herd management, with emphasis placed on profitability.

Type has become more meaningful in dairy cattle since the establishment of the type classification program, first introduced in the Holstein breed in 1929. Today, each of the major dairy breed associations has an active committee on dairy types.

There is no conflict between dairy type and breed type; the latter merely adds certain distinctive breed characteristics. The breed registry associations promulgate breed type through shows, type classification programs, and models and paintings.

A monetary value is placed on type by most dairy producers, particularly by purebred breeders; hence, type may be said to be of market importance.

## UNIFIED SCORECARDS

A scorecard is a listing of the different parts of an animal, with a numerical value assigned to each according to its relative importance. Fortunately, there was a common meeting ground by the purebred dairy cattle associations in the development of the unified scorecards herewith reproduced as Figs. 22–5 and 22–6 (see pages 492 and 493). Breed characteristics may be, and are, considered in the use of the unified scorecard—among them are the picturesque style of the Ayrshire; the traditional strength and ruggedness of the Brown Swiss; the tractable disposition and milk color of the Guernsey; the size, scale, and color markings of the Holstein; and the refinement of the Jersey.

The scorecard for dairy cattle has considerable value in acquainting students and beginners with the various parts of an animal and the relative importance of each. It also has the advantage of listing the various parts of an animal under functional relationship—such as dairy character, body capacity, mammary system, etc.; and it promotes systematic observation and analysis of the points of strength and weakness in the animal being examined. The major weakness in any scorecard is that the user may not recognize that the failure of an important part of an animal may bring about the functional failure of the entire animal, and not just the part involved.

Fig. 22–4. *Mar-Ral Hi Kick,* one of the outstanding sires of the Ayrshire breed. Bull selection generally implies selecting semen and breeding AI. (Courtesy, Ayrshire Breeders Assn., Brandon, VT)

# DAIRY COW UNIFIED SCORE CARD

Copyrighted by The Purebred Dairy Cattle Association, 1943. Revised, and Copyrighted 1957 and 1971
Approved — The American Dairy Science Association, 1957

| | Perfect Score |
|---|---|
| *Breed characteristics should be considered in the application of this score card* | |

**Order of observation**

## 1. GENERAL APPEARANCE — 30

*(Attractive individuality with, feminity, vigor, stretch, scale, harmonious blending of all parts, and impressive style and carriage. All parts of a cow should be considered in evaluating a cow's general appearance)*

**BREED CHARACTERISTICS** — (see reverse side) — 10

**HEAD** — clean cut, proportionate to body; broad muzzle with large, open nostrils; strong jaws; large, bright eyes; forehead, broad and moderately dished; bridge of nose straight; ears medium size and alertly carried

**SHOULDER BLADES** — set smoothly and tightly against the body — 10

**BACK** — straight and strong; loin, broad and nearly level

**RUMP** — long, wide and nearly level from **HOOK BONES** to **PIN BONES**; clean cut and free from patchiness; **THURLS**, high and wide apart; **TAIL HEAD**, set level with backline and free from coarseness; **TAIL**, slender

**LEGS AND FEET** — bone flat and strong, pasterns short and strong, hocks cleanly moulded. **FEET**, short, compact and well rounded with deep heel and level sole. **FORE LEGS**, medium in length, straight, wide apart, and squarely placed. **HIND LEGS**, nearly perpendicular from hock to pastern, from the side view, and straight from the rear view — 10

## 2. DAIRY CHARACTER — 20

*(Evidence of milking ability, angularity, and general openness, without weakness; freedom from coarseness, giving due regard to period of lactation)*

**NECK** — long, lean, and blending smoothly into shoulders; clean cut throat, dewlap, and brisket — 20
**WITHERS**, sharp. **RIBS**, wide apart, rib bones wide, flat, and long. **FLANKS**, deep and refined. **THIGHS**, incurving to flat, and wide apart from the rear view, providing ample room for the udder and its rear attachment. **SKIN**, loose, and pliable

## 3. BODY CAPACITY — 20

*(Relatively large in proportion to size of animal, providing ample capacity, strength, and vigor)*

**BARREL** — strongly supported, long and deep; ribs highly and widely sprung; depth and width of barrel tending to increase toward rear — 10

**HEART GIRTH** — large and deep, with well sprung fore ribs blending into the shoulders; full crops; full at elbows; wide chest floor — 10

## 4. MAMMARY SYSTEM — 30

*(A strongly attached, well balanced, capacious udder of fine texture indicating heavy production and a long period of usefulness)*

**UDDER** — symmetrical, moderately long, wide and deep, strongly attached, showing moderate cleavage between halves, no quartering on sides; soft, pliable, and well collapsed after milking; quarters evenly balanced — 10

**FORE UDDER** — moderate length, uniform width from front to rear and strongly attached — 6

**REAR UDDER** — high, wide, slightly rounded, fairly uniform width from top to floor, and strongly attached — 7

**TEATS** — uniform size, of medium length and diameter, cylindrical, squarely placed under each quarter, plumb, and well spaced from side and rear views — 5

**MAMMARY VEINS** — large, long, tortuous, branching — 2

"Because of the natural undeveloped mammary system in heifer calves and yearlings, less emphasis is placed on mammary system and more on general appearance, dairy character, and body capacity. A slight to serious discrimination applies to overdeveloped, fatty udders in heifer calves and yearlings."

| *Subscores are not used in breed type classification.* | **TOTAL** | **100** |
|---|---|---|

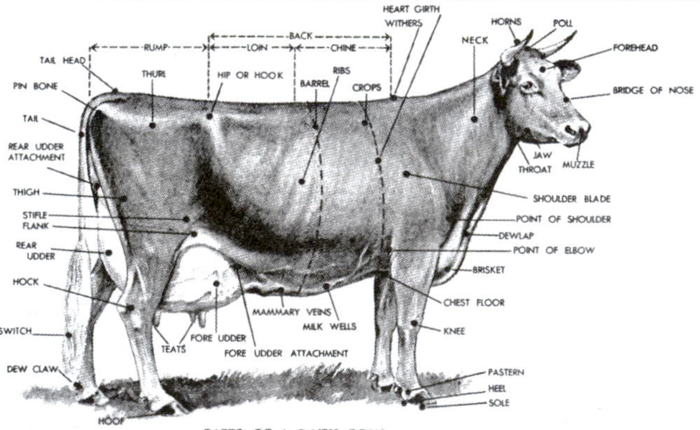

**PARTS OF A DAIRY COW**

# EVALUATION OF DEFECTS

*In a show ring, disqualification means that the animal is not eligible to win a prize. Any disqualified animal is not eligible to be shown in the group classes. In slight to serious discrimination, the degree of seriousness shall be determined by the judge.*

**EYES**
1. Total blindness: *Disqualification.*
2. Blindness in one eye: *Slight discrimination.*
3. Cross-eyes: *Slight discrimination.*

**WRY FACE**
*Slight to serious discrimination.*

**CROPPED EARS**
*Slight discrimination.*

**PARROT JAW**
*Slight to serious discrimination.*

**SHOULDERS**
Winged: *Slight to serious discrimination.*

**TAIL SETTING**
Wry tail or other abnormal tail settings: *Slight to serious discrimination.*

**LEGS AND FEET**
1. Lameness — apparently permanent and interfering with normal function: *Disqualification.*
— apparently temporary and not affecting normal function: *Slight discrimination.*

2. Bucked knees: *Slight to serious discrimination.*
3. Evidence of arthritis, crampy hind leg: *Serious discrimination.*
4. Evidence of fluid in hocks: *Slight to serious discrimination.*

**ABSENCE OF HORNS**
*No discrimination.*

**LACK OF SIZE**
*Slight to serious discrimination.*

**UDDER**
1. Blind quarter: *Disqualification.*
2. Abnormal milk (bloody, clotted, watery): *Possible disqualification.*
3. Udder definitely broken away in attachment: *Serious discrimination.*
4. A weak udder attachment: *Slight to serious discrimination.*
5. One or more light quarters, hard spots in udder, obstruction in teat (spider): *Slight to serious discrimination.*
6. Side leak: *Slight discrimination.*

**DRY COWS**
Among cows of apparently equal merit: *Give strong preference to cows in milk.*

**FREEMARTIN HEIFERS**
*Disqualification unless proved pregnant.*

**OVERCONDITIONED**
*Slight to serious discrimination.*

**TEMPORARY OR MINOR INJURIES**
Blemishes or injuries of a temporary character not affecting animal's usefulness: *Slight discrimination.*

**EVIDENCE OF SHARP PRACTICE**
1. Animals showing signs of having been operated upon or tampered with for the purpose of concealing faults in conformation, or with intent to deceive relative to the animal's soundness: *Disqualification.*
2. Uncalved heifers showing evidence of having been milked: *Serious discrimination.*

Fig. 22–5. Dairy cow unified scorecard. (Copyrighted by the Purebred Dairy Cattle Association; approved by the American Dairy Science Association)

## DAIRY BULL UNIFIED SCORE CARD

| | Perfect Score |
|---|---|
| *Breed characteristics should be considered in the application of this score card* | |

**Order of observation**

**1. GENERAL APPEARANCE** — 45

*(Attractive individuality, with masculinity, vigor, stretch, and scale, harmonious blending of all parts, and impressive style and carriage. All parts of a bull should be considered in evaluating a bull's general appearance)*

**BREED CHARACTERISTICS** — (see reverse side) — 15

**HEAD** — clean cut, proportionate to body; broad muzzle with large, open nostrils; strong jaws; large, bright eyes; forehead, broad and moderately dished; bridge of nose straight; ears medium size and alertly carried

**SHOULDER BLADES** — set smoothly and tightly against the body — 15

**BACK** — straight and strong; loin, broad and nearly level

**RUMP** — long, wide, and nearly level from **HOOK BONES** to **PIN BONES**; clean cut and free from patchiness; **THURLS**, high and wide apart; **TAIL HEAD**, set level with backline and free from coarseness; **TAIL**, slender

**LEGS AND FEET** — bone flat and strong, pasterns short and strong, hocks cleanly moulded. **FEET**, short, compact, and well rounded with deep heel and level sole. **FORE LEGS**, medium in length, straight and wide apart, squarely placed. **HIND LEGS**, nearly perpendicular from hock to pastern from the side view, and straight from the rear view — 15

**2. DAIRY CHARACTER** — 30

*(Angularity and general openness, without weakness; freedom from coarseness)*

**NECK** — long, with medium crest and blending smoothly into shoulders; clean cut throat, dewlap, and brisket. **WITHERS**, sharp. **RIBS**, wide apart, rib bones wide, flat, and long. **FLANKS**, deep and refined. **THIGHS**, incurving to flat, and wide apart from the rear view. **SKIN**, loose, and pliable

**3. BODY CAPACITY** — 25

*(Relatively large in proportion to size of animal, providing ample capacity, strength, and vigor)*

**BARREL** — strongly supported, long, and deep; ribs highly and widely sprung; depth and width of barrel tending to increase toward rear — 12

**HEART GIRTH** — large and deep, with well sprung fore ribs blending into the shoulders; full crops; full at elbows; wide chest floor — 13

*Subscores are not used in breed type classification.* **TOTAL** — **100**

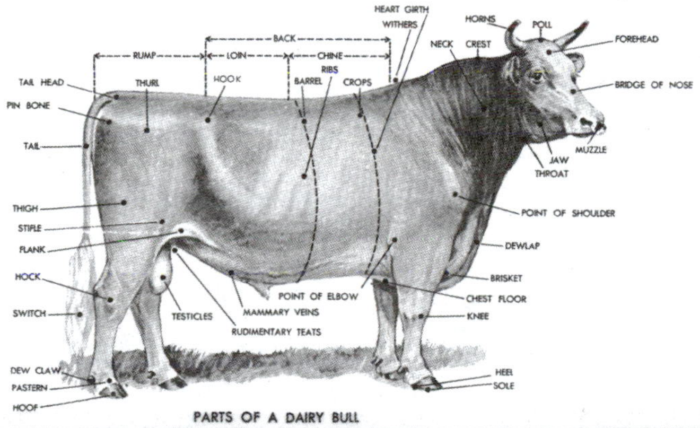

PARTS OF A DAIRY BULL

### EVALUATION OF DEFECTS

*In a show ring, disqualification means that the animal is not eligible to win a prize. Any disqualified animal is not eligible to be shown in the group classes. In slight to serious discrimination, the degree of seriousness shall be determined by the judge.*

**EYES**
1. Total blindness: *Disqualification.*
2. Blindness in one eye: *Slight discrimination.*
3. Cross-eyes: *Slight discrimination.*

**WRY FACE**
*Slight to serious discrimination.*

**CROPPED EARS**
*Slight to serious discrimination.*

**PARROT JAW**
*Slight to serious discrimination.*

**SHOULDERS**
Winged: *Slight to serious discrimination.*

**TAIL SETTING**
Wry tail or other abnormal tail settings: *Slight to serious discrimination.*

**LEGS AND FEET**
1. Lameness — apparently permanent and interfering with normal function: *Disqualification.*
— apparently temporary and not affecting normal function: *Slight discrimination.*
2. Bucked knees: *Slight to serious discrimination.*
3. Evidence of arthritis, crampy hind leg: *Serious discrimination.*
4. Boggy hocks: *Slight to serious discrimination.*

**LACK OF SIZE**
*Slight to serious discrimination.*

**TESTICLES**
Bull with one testicle or with abnormal testicles: *Disqualification.*

**OVERCONDITIONED**
*Slight to serious discrimination.*

**TEMPORARY OR MINOR INJURIES**
Blemishes or injuries of a temporary character not affecting animal's usefulness: *Slight discrimination.*

**EVIDENCE OF SHARP PRACTICE**
Animals showing signs of having been operated upon or tampered with for the purpose of concealing faults in conformation, or with intent to deceive relative to the animal's soundness: *Disqualification.*

Fig. 22–6. Dairy bull unified scorecard. (Courtesy, Purebred Dairy Cattle Association)

# QUESTIONS FOR STUDY AND DISCUSSION

1. List and discuss the factors which must be considered when establishing a dairy herd. Which factor is most important of all? How important is longevity?

2. List and discuss the primary requisites of a purebred herd.

3. List and discuss the special requisites of a registered breeder.

4. Were you to enter the dairy business, what breed would you select? Would you start with (a) grades or purebreds, or (b) cows, heifers, or calves? Justify your choice.

5. Evaluate the usefulness and completeness of Table 22–1, Culling Guide.

6. Evaluate the usefulness and completeness of Table 22–2, Cow Evaluation.

7. Define each of the following terms: *judging* and *scoring/type* classifying. Is it important that modern producers be good judges of dairy cattle?

8. How important is it for dairy producers to know the parts of a cow?

9. Of what value is (a) dairy type, and (b) breed type?

10. Should production records replace body type and evaluation?

11. Under what circumstances would you evaluate an animal by (a) use of the scorecard, (b) show-ring record, and (c) type classification sponsored by the breed registry?

# SELECTED REFERENCES

| Title of Publication | Author(s) | Publisher |
| --- | --- | --- |
| *Dairy Cattle Judging and Selection* | W. W. Yapp | John Wiley & Sons, Inc., New York, NY, 1959 |
| *Dairy Cattle Judging Techniques*, Second Edition | G. W. Trimberger | Prentice-Hall, Inc., Englewood Cliffs, NJ, 1977 |
| *Livestock Judging, Selection and Evaluation*, Second Edition | R. E. Hunsley<br>W. M. Beeson<br>J. E. Nordby | The Interstate Printers & Publishers, Inc., Danville, IL, 1978 |

Guernsey cows on pasture. (Courtesy, J. C. Allen and Son, West Lafayette, IN)

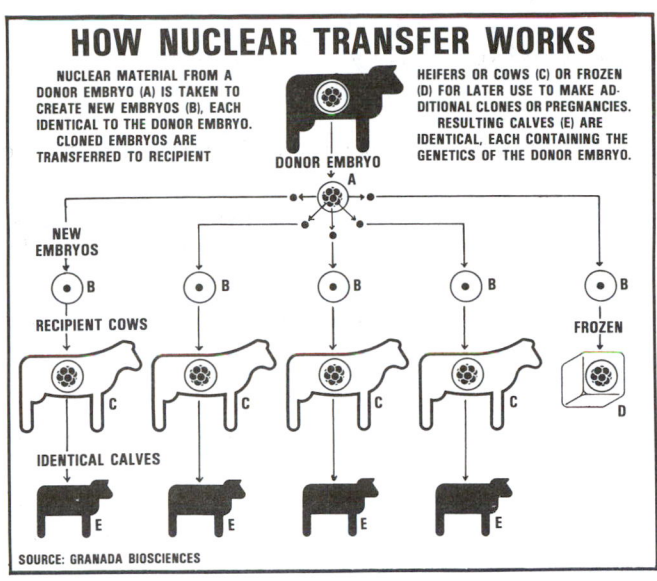

**HOW NUCLEAR TRANSFER WORKS**

NUCLEAR MATERIAL FROM A DONOR EMBRYO (A) IS TAKEN TO CREATE NEW EMBRYOS (B), EACH IDENTICAL TO THE DONOR EMBRYO. CLONED EMBRYOS ARE TRANSFERRED TO RECIPIENT

HEIFERS OR COWS (C) OR FROZEN (D) FOR LATER USE TO MAKE ADDITIONAL CLONES OR PREGNANCIES. RESULTING CALVES (E) ARE IDENTICAL, EACH CONTAINING THE GENETICS OF THE DONOR EMBRYO.

SOURCE: GRANADA BIOSCIENCES

Today's embryo/tomorrow's cow! New biotechnology (donor embryo/nuclear transfer/cloned embryos transferred to recipient heifers) will make it possible for the dairy producer of the future to be assured (1) of the sex of animals before birth, and (2) of a whole string of identical, high-producing cows. (Courtesy, Granada Biosciences, College Station, TX)

# BREEDING DAIRY CATTLE

## Chapter 23

The objective of dairy cattle breeding is to mate individuals whose offspring will possess the necessary heritability to (1) produce the maximum amount of milk of the desired composition, and (2) develop the desired body type; then to feed and manage these animals so that their maximum genetic potential will be expressed. This recognizes the fact that dairy cattle are products of heredity and environment, as are all other animals.

The economic justification for improved breeding is that good cows make more profit. This fact is clearly illustrated in Fig. 23–1. There are three main explanations for the increase in rate of production per dairy animal in recent years: (1) The productive ability of milk cows has been increased through the selection of better producing animals, (2) cows have been better fed and managed, and (3) with the decline in milk cow numbers, most of the culling has come about among the low producing cows and in the marginal herds, with the result that the remaining cows are usually among the higher producers.

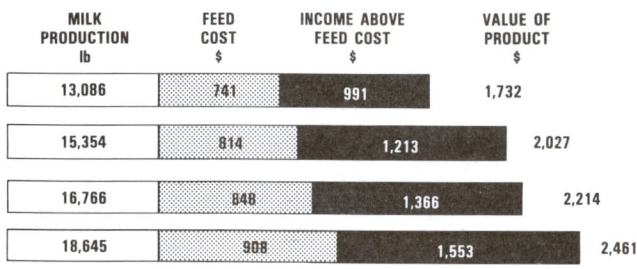

| MILK PRODUCTION lb | FEED COST $ | INCOME ABOVE FEED COST $ | VALUE OF PRODUCT $ |
|---|---|---|---|
| 13,086 | 741 | 991 | 1,732 |
| 15,354 | 814 | 1,213 | 2,027 |
| 16,766 | 848 | 1,366 | 2,214 |
| 18,645 | 908 | 1,553 | 2,461 |

Fig. 23–1.  It costs more to feed high-producing cows, but the practice pays handsome dividends. The reason: feed and overhead costs for maintenance are practically the same, regardless of the level of production. (Based on data from New York Dairy Records Laboratory; Courtesy, L. R. Brown, Extension Dairyman, The University of Connecticut, Storrs)

As noted in Fig. 23–1, it costs more to feed high producers; it costs $167 more to feed cows producing 18,645 lb of milk than cows producing only 13,088 lb. However, the income above feed costs more than compensates for the higher feed costs; the income above feed costs is $562 more for cows producing 18,645 lb of milk than for cows producing only 13,088 lb. The reason: feed and overhead costs for maintenance are practically the same, regardless of the level of production.

Today, all dairy cattle breeding may be classed as two kinds: (1) breed improvement, and (2) herd improvement, both of which come about through improvement of individual animals. As a whole, dairy producers throughout the country are greatly interested in breed improvement. However, as individuals they are primarily interested in herd improvement. If they are able to improve their individually-owned herds, they benefit as individuals. Also, any permanent herd improvement made by an individual inevitably contributes to permanent breed improvement. In fact, breed improvement through the years has largely come about through the progress made by superior breeders in their own herd improvement programs. It is important, therefore, that modern dairy breeders understand and follow a constructive breeding program.

## PHYSIOLOGY OF REPRODUCTION

Dairy producers encounter many reproductive problems, a reduction of which calls for a full understanding of physiology and the application of scientific practices therein. In fact, it may

be said that reproduction is the first and most important requisite of successful dairy cattle breeding, for if animals fail to reproduce, the breeder is soon out of business. Simply stated, milk production is a by-product of the reproductive process.

Table 23–1 gives the reasons cows leave herds. As shown, low productivity results in 32.5% of cows being culled. Reproductive problems are the second most important reason for eliminating cows from dairy herds, accounting for 26.6% of cows culled.

### TABLE 23–1
### REASONS FOR COWS LEAVING DAIRY HERDS[1]

| Reasons for Culling | Percent |
|---|---|
| | (%) |
| Cow production | 32.5 |
| Reproductive problems | 26.6 |
| Mastitis | 10.4 |
| Disease or inabilities | 7.7 |
| Teat or udder injury | 7.2 |
| Udder conformation | 5.0 |
| Accident and injury | 4.0 |
| Type | 3.1 |
| Disposition and milking ease | 2.7 |

[1]Dairy Guide, Cooperative Extension Service, The Ohio State University, Leaflet DG 300, by B. J. Conlin, University of Minnesota.

The reproductive organs of the bull and the cow are pictured and described in Chapter 3 of this book; hence, the reader is referred thereto for information on this subject.

## STERILITY AND DELAYED BREEDING

It has been estimated that sterility and delayed breeding in dairy cattle account for an average yearly loss of $60 per cow, for a national total loss of $650 million. At least half of these losses could be prevented. Thus, on the average, sterility and infertility may be expected to cause losses of the magnitude indicated in Table 23–2 (center column), at least half of which could be prevented (right column).

### TABLE 23–2
### HERD SIZE DETERMINES THE SAVINGS POSSIBLE BY MINIMIZING STERILITY

| Herd Size | Expected Yearly Loss | Possible Yearly Saving |
|---|---|---|
| | ($) | ($) |
| 40 | 2,400 | 1,200 |
| 70 | 4,200 | 2,100 |
| 100 | 6,000 | 3,000 |
| 150 | 9,000 | 4,500 |
| 200 | 12,000 | 6,000 |
| 500 | 30,000 | 15,000 |

Another noteworthy statistic is that, nationally, including both beef and dairy herds, only an 88% calf crop is produced—the other 12% of the cows abort or are sterile, temporarily or permanently.

## REPRODUCTIVE DISEASES

Brucellosis, leptospirosis, trichomoniasis, and vibriosis are the most troublesome specific genital diseases of dairy cows. In summary form, Table 23–3 gives the pertinent facts about each of them from the standpoint of sterility and delayed breeding.

**TABLE 23–3**
**FOUR MAIN REPRODUCTIVE DISEASES OF DAIRY CATTLE**

| Disease | Symptoms | Prevention |
|---------|----------|------------|
| **Brucellosis** | **A**bortion last third of pregnancy. **R**etained afterbirth. **S**everal services per conception. **U**terine infections. | **C**alfhood vaccination. **U**se artificial insemination. **P**urchase only vaccinated or tested (and clean) animals. |
| **Leptospirosis** | **H**igh fever (103° to 107°F). **P**oor appetite. **A**bortion any time. **B**loody urine. **A**nemia. **R**opy milk. | **V**accinate susceptible animals annually if disease is present in area. **K**eep different classes of livestock separated. **P**urchase clean animals, isolate for 30 days, and retest. |
| **Trichomoniasis** | **A**bortion in first third of pregnancy. **U**terine infection. **I**rregular heat periods. **S**everal services per conception. | **U**se artificial insemination with semen from bulls free of trichomoniasis. |
| **Vibriosis** | **A**bortion in middle third of pregnancy. **S**everal services per conception. **I**rregular heat periods. | **U**se artificial insemination with semen (1) from known noninfected bulls, or (2) which has been antibiotic-treated. **U**se vaccine as directed. |

## NORMAL HEAT AND GESTATION PERIODS

After heifers reach puberty, the normal heat period recurs at approximately 21-day intervals, but it may vary from 18 to 23 days. Best conception is obtained when cows are bred near the end of standing heat.

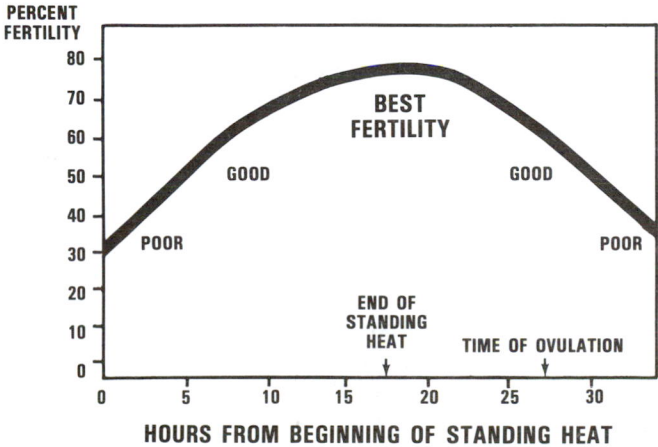

Fig. 23–2. For best conception, breed near the end of standing heat.

The normal gestation period of the dairy cow is 283 days. Based on studies made by the Ohio Experiment Station, however, it is evident that there is variation among breeds. The following figures show the gestation period by breeds, as found in this experiment: Ayrshire, 278 days; Brown Swiss, 288 days; Guernsey, 283 days; Holstein, 279 days; and Jersey, 278 days. The Ohio station also reported (1) that the gestation period of first calf heifers is about 2 days less than that of older cows of the respective breeds, and (2) that the gestation period where bull calves are born is about one day longer than that where females are born.

## GENETICS OF DAIRY CATTLE BREEDING

In all animal breeding, and dairy cattle are no exception, no new genetic material is created. Rather, it is simply a matter of sorting or rearranging the many factors already present in the male and female gametes. These factors are referred to as genes. They are contained in the chromosomes of the sperm of the male and the egg of the female.

Dairy animals have 30 pairs of chromosomes in each cell. The number of genes per chromosome is not definitely known; estimates are that there may be as many as 30,000 genes for dairy cattle. These genes are responsible for how the animal looks and produces.

When the sperm and egg unite, the new cell formed contains 30 pairs of chromosomes, or a total of 60 chromosomes, half of which come from the sperm (male) and half from the egg (female). What determines which genes and which chromosomes are to be passed on to the new cell is still a relatively dark secret.

## HERITABILITY OF CHARACTERS

The expression of a trait, such as milk production, depends upon two factors: (1) inheritance, or the ability to produce, and (2) environment, or the opportunity to express the inherited ability. Heritability is 100% when the expression of the trait varies solely because of inheritance. A trait that varies solely because of environment has a heritability of zero. Variations in most traits are neither wholly environmental nor completely hereditary. The heritability of some common dairy cattle is given in Table 23–4.

**TABLE 23–4**
**HERITABILITY ESTIMATES FOR SOME DAIRY CATTLE TRAITS**

| Trait | Heritability |
|-------|--------------|
| | (%) |
| **M**ilk yield | 30 |
| **F**at yield | 25 |
| **F**at percent | 50 |
| **P**rotein percent | 50 |
| **S**olids not fat percent | 50 |
| **M**ilking rate | 30 |
| **L**ongevity (length of life) | 5 |
| **T**ype (final rating) | 30 |
| **B**reeding efficiency | 5 |

As shown in Table 23–4, milk yield is about 30% heritable, whereas fat percentage is 50% heritable.

The following example will show how heritability can be computed: Let us assume that we have a herd that averages 440 lb of butterfat on a mature level. Further, let us select a young sire that we estimate is capable of transmitting inheritance for 500 lb of butterfat production. He is mated to select cows in the herd with production records averaging 500 lb of butterfat in a normal 305-day lactation period. Because heritability is 30%, we expect only three-tenths of the apparent superiority of the parents expressed in the offspring. The selected parents averaged 60 lb of butterfat higher than the herd. Three-tenths of the 60 equals 18 lb of butterfat. Thus, the offspring would be expected to average 458 lb of butterfat in this herd when given the same opportunity as the parents. The additional butterfat of the parents resulted from influences such as better feeding and management.

## NUMBER OF CHARACTERS SELECTED FOR SIMULTANEOUSLY

Most rapid progress can be made in a breeding program by selecting for one trait only. However, when two characteristics are inherited in close relationship, considerable progress may be achieved in both of them. This is true, for example, in total milk production and total production of fat. On the other hand, body type and high production of milk are not closely associated in inheritance, with the result that selection for both of them will result in relatively slow progress in either one. Under these circumstances, herd owners must decide which of the traits in the herd need most improvement and make their decision on the characteristics to be emphasized in selection accordingly. If good type will bring more monetary return than increased milk production, it should be emphasized. On the other hand, if higher production will increase income to a greater degree than improved body type, then it should be given greater importance in selection.

In many herds, particularly purebred herds, both type and production are important and selection for both should be made. In other words, selection for either type or production should not be made at the expense of serious loss in the other.

As a general rule, milk production and butterfat percentage are the characters usually considered in the breeding of dairy cattle, but other characters should also be considered, as many of them have economic value.

## GENETIC EVALUATION OF DAIRY CATTLE IN THE PAST

For many years, genetic improvement of the U.S. dairy cattle was focused on identifying and selecting superior cows, with little attention given to the evaluation of bulls. As improved methods were developed, dairy producers who used them gained some advantage in rate of genetic improvement over their competitors. In order, the following methods of genetic evaluation were used in the past: (1) lactation record, (2) daughter average, (3) daughter-dam comparison, (4) herdmate comparison, and (5) contemporary comparison. Each of these is briefed in the sections that follow.

### LACTATION RECORD

Because of environmental influences, knowledge of a cow's lactation record tells very little about her breeding value for milk production. A cow's production may vary as much as 50% between a well-managed herd and a poorly managed herd. On the average, about 80% of the difference between herds is due to environment and 20% is genetic.

### DAUGHTER AVERAGE

This index is based on the premise that the average production of a bull's daughter is indicative of his transmitting ability for milk production. It's main weakness is that it does not consider the production of their dams.

### DAUGHTER-DAM COMPARISON

This is one of the oldest methods of evaluating bulls; it was widely used from about 1930 to 1960. It consists in comparing the production of a bull's daughters with that of their dams. The disadvantages of a Daughter-Dam Comparison are (1) the dams and daughters may not be milking at the same time, which means that environmental differences are inevitable; (2) comparisons are usually made in one, or only a few herds, which means that only a limited range of environmental conditions prevail; and (3) many dams do not have records, with the result that there is nothing with which to compare their daughters.

In the early 1960s, the daughter-dam comparison was largely abandoned in favor of the Herdmate Comparison, which was superior in eliminating environmental effects from estimates of breeding value.

### HERDMATE COMPARISON

The Herdmate Comparison, which was developed by the U.S. Department of Agriculture and first used in 1954, was the first genetic procedure developed specifically for bulls. It was used extensively from the early 1960s, until 1974, when the U.S. Department of Agriculture introduced the Modified Contemporary Comparison.

*The term herdmates refers to other cows in the herd that were calved in the same year and season as the daughters of the sire being compared.* This method of sire evaluation has two objectives: (1) to compare performance of one animal with another which is not paternally related, and (2) to compare records made at the same time under the same environment. Under this system, all records are converted to twice-a-day milking, 305 days, and mature (5 years) age. All animals in the herd, regardless of age, which freshen within a period from 3 months before to 3 months after each daughter of the particular sire in question freshens, are included in the comparison.

### CONTEMPORARY COMPARISON

The Contemporary Comparison is based on the same principles as the Herdmate Comparison except that only first records are used. The name stems from the fact that all animals used are *contemporaries;* that is, they commenced their first records at approximately the same time. Proponents of the contemporary comparison claimed two primary advantages: (1) use of first records minimized many of the possible inaccuracies that might arise from the use of age-correction factors; and (2) the use of first records, which were made only by cows that escaped being culled; hence, were a select group. The primary disadvantage of the Contemporary Comparison is that in many herds there were too few cows that qualified as true contemporaries.

## GENETIC EVALUATION OF DAIRY CATTLE AT PRESENT

The U.S. dairy industry and consumers have benefitted greatly from the national research program on genetic improvement of dairy cattle conducted by the Animal Improvement Programs Laboratory (AIPL), Agricultural Research Service, U.S. Department of Agriculture, and from the research at the land-grant universities. Genetic evaluation of bulls (Sire Summaries) and cow (Cow Indexes) emanating from this research have been the primary source of information for identifying animals with superior genetic merit for yield.

Since 1974, national genetic evaluations of bulls and cows for milk and fat have been calculated by AIPL-USDA using the Modified Contemporary Comparison. National Sire Summaries for protein and solids-not-fat have been calculated by the Mixed Model Method since 1976. In 1983, the AIPL-USDA evolved with the first National Buck Summary for dairy goats.

Since there is no measure of a sire's individual performance, his evaluation is based on the performance of his daughters. An estimate of a sire's ability to transmit may be arrived at in two steps: (1) an overall average is calculated from the average difference from herdmates for all of his daughters; and (2) the overall average difference is weighed according to repeatability of that amount of information. Two terms are used in expressing the resulting estimate of genetic transmitting of the sire: Predicted Difference (PD), and Repeatability.

- **Predicted Difference (PD)**—This is the term applied to the genetic values which rank bulls for production traits. *Predicted Difference is the expected extra production per daughter per year, when compared to a zero PD bull.*

Fig. 23–3. *Favorite Saint,* the highest PD milk bull in the history of the Jersey breed. His PD is + 2,248 lb milk and + 3.2 for PD type. (Courtesy, The American Jersey Cattle Club, Reynoldsburg, OH)

- **Repeatability**—It is important to realize that PD values are not absolute. They change, up and down, as more daughters are added in more herds. *Repeatability is an estimate of how sure we are that the Predicted Difference reflects an individual bull's true transmitting ability.*

## MODIFIED CONTEMPORARY COMPARISON (MCC)

Since 1974, national genetic evaluations of bulls and cows for milk and fat have been calculated by AIPL-USDA using a procedure called the Modified Contemporary Comparison. To comprehend the value of the MCC over a simple average of daughter-herdmate differences, it is necessary to understand some problems that have arisen in sire evaluation.

Many owners have used high Predicted Difference sires year after year and have developed entire herds sired by these top sires. These herds have far surpassed the average dairy herd in their genetic ability for production. Any bull with daughters competing in herds of this caliber has a difficult time coming up with a high PD

By contrast, other owners have used mediocre sires, and in many cases have done little culling. Any bull whose daughters are competing in these herds looks better than he should.

The Modified Contemporary Comparison (MCC) takes into account the genetic level of herdmates competing with the daughters. This is done by the use of PDs of the sires of these herdmates.

Another frequent problem in sire evaluation is the bias due to culling that the older cows have survived. This may be a problem when a bull's summary includes older daughters as well as when the competition the heifer faces includes older cows. The MCC contains an adjustment for bias that occurs due to culling for production after the first lactation.

The MCC takes its name from the use of contemporary groups or herdmates that are of similar age. it makes use of two contemporary groups: *Contemporary Group 1* includes first lactation only; *Contemporary Group 2* includes second and later lactations. The greatest emphasis in the sire evaluation is given to the comparisons of daughters' records with contemporary information—with Group 1 records in the case of daughter's first lactation or with Group 2 records for second and later lactations of daughters. A daughter's non-contemporary herdmates are also included but are counted the equivalent of only one additional contemporary.

Another characteristic of the USDA-MCC sire summary is the use of pedigree information on the sire. The pedigree information used is the PD of the bull's sire and that of his maternal grandsire. Each bull is put into a *genetic group* based on this pedigree evaluation. The average superiority (or inferiority) of the daughter's of the bulls in this genetic group is then used in the sire's summary. This pedigree group average is weighed along with the daughter's information according to the genetic worth of each source. As daughter information (Repeatability) increases, the weight given to the pedigree group average decreases. This relationship is R to 1-R where R is the Repeatability and 1-R is the weight given to the pedigree group average.

The PD computed by USDA is adjusted to a constant genetic base such as PD 82. This means that a breed average bull in 1982 had a PD of zero. As the breed improves genetically, more and more bulls will have higher and higher PDs. The breed average of sires may be + 400, + 500 or even higher after a period of years. This genetic improvement of the breed may necessitate changing the base to a more recent year base so that an average bull is again zero and "+" sign will again denote a breed improver.

- **Pedigree evaluation**—The sire and dam contribute equally to the genetic makeup of the offspring. The estimates of these contributions are the PD of the sire and the Cow Index (CI) or *Estimated Average Transmitting Ability* of the dam. The pedigree estimate of breeding value (EBV) of a heifer or bull would therefore be:

> Pedigree EBV = Sire's PD plus Dam's CI.

As an example, suppose a heifer calf is sired by a bull with PD = + 1,200 and is from a cow with CI = + 1,000.

> Pedigree EBV = + 1200 plus + 1000 = + 2200

The best estimate of this heifer's future producing ability, above or below her competition, is + 2,200. Since she is expected to transmit only a sample half of her genetic superiority to her offspring, the best estimate of her future transmitting ability would be half her pedigree EBV, or + 1,100. The offspring is expected to be intermediate in genetic merit to her parents.

## MIXED MODEL METHOD

The term *Mixed Model* refers to the statistical properties of the effects in these models where some of the effects are random and some are fixed.

The Mixed Model Method possesses all the attributes of the Modified Contemporary Comparison—and more. The Mixed Model Method uses fixed seasons; AIPL-USDA uses two 6-month seasons—(1) January through June, and (2) July through December.

The AIPL-USDA continues to use the Modified Contemporary Comparison for national genetic evaluation, rather than the Mixed Model Method, because of the prohibitive cost of the latter method.

In addition to the Mixed Model Method being more costly to conduct than the Modified Contemporary Comparison, it is of questionable accuracy for daughters of non-A.I. bulls with most or all daughters in a single herd.

## SUMMARY OF DAIRY GENETICS

The ultimate goal for the dairy herd breeding program is to produce replacement heifers that will:

1. Produce large quantities of milk with a protein, fat and total solids content that will command a premium price.
2. Reproduce regularly and without problems, calving every 12 or 13 months.
3. Have a minimum of health problems, such as mastitis, milk fever, and ketosis.
4. Be completely mobile and require minimal care of feet.
5. Have a disposition that allows them to fit the facilities and management routine for the herd.

6. Milk out quickly, cleanly, and without special labor requirements at milking time.
7. Wear well and do the above for a long lifetime.

The most rapid avenue for attaining this goal is by selection of the best sires available through artificial insemination. Top pedigreed young sires are also a good buy and can contribute to herd improvement if a number of them are used, each sparingly. If natural services are used, they should be highly selected, with a high pedigree estimate of breeding value, and should be used sparingly and no more than one season.

Milk production is the primary trait for which to select. Fat, protein, or other total solids are also highly important and likely to increase in importance when and if nutritional pricing becomes widespread. Some attention to conformation is necessary to avoid serious problems, especially in the udder traits. The best approach is probably to select the highest production sires and then eliminate the very poorest ones based on total score of their daughters. This should avoid creating conformation problems that cannot be controlled by light culling of females in the herd.

Common sense dictates that the breeder should be aware of how production is being sacrificed each time a lower PD sire is accepted in hopes of obtaining an improvement in some other trait. If the possible improvement in that other trait is worth more than the potential loss in production, the sacrifice should be made. If too many sacrifices in production are made, the genetic improvement for production in that herd will grind to a stop.

- **Ranking of estimates of transmitting ability of bulls**—Most dairy scientists and producers rank different kinds of estimates of transmitting ability of bulls in descending order of accuracy as follows:

1. Modified Contemporary Comparison or Mixed Model Method.
2. Herdmate Comparison.
3. Daughter-Dam Comparison.
4. Daughter Average.
5. Outstanding records, show-ring winnings, or distant relatives, or other selected information.

- **Ranking of estimates of transmitting ability of cows**—Most dairy scientists and producers rank different kinds of phenotype information for estimating the transmitting ability of cows in descending order of accuracy as follows:

1. Modified Contemporary Comparison or Mixed Model Cow Index.
2. Herdmate Comparison Cow Index.
3. Daughter-Dam Comparison.
4. Lactation Record.
5. Highest Record or other selected information.

- **Further information**—All dairy farmers can improve the genetic ability of their herds to produce. Modern cow evaluations are available through the national Cooperative Dairy Herd Improvement Program in every state. Accurate and useful sire evaluations are available from the U.S. Department of Agriculture through the extension dairy specialists, artificial insemination (A.I.) associations, and breed associations. Service to top sires is available through A.I. Breed associations provide additional programs of benefit in the breeding and management of dairy herds especially through their type classification programs. The Cooperative Extension Service in every state stands ready to help with explanations and suggestions for use of all available information.

The ingredients for a successful breeding program to improve the efficiency and profitability of the herd are available. It is up to dairy producers to take the necessary time to plan, to make decisions, and to use the cow and sire evaluations for the improvement of their herds.

## RECORDS ARE NECESSARY

The foundation for a good breeding program is production records on every cow in the herd, year after year. Additionally, for the purebred breeder, type evaluation records are important. Use of these two tools—production and type evaluation records—through a careful culling and selection program tends to get rid of the undesirable genes and concentrate on those which are superior. Milk and butterfat records are also the key to scientific dairy cattle feeding.

Fig. 23-4. Performance information and breeding guidelines make it possible for dairy producers to make wise animal selections and breeding decisions. (Courtesy, Holstein-Friesian Assn. of America, Brattleboro, VT)

## KIND OF RECORDS; CHOOSING A SYSTEM

As is true in bookkeeping or cost accounting, a number of different systems and forms of dairy records are available. The important thing is to choose that system, or those systems, which will give the desired information, then stick to it.

## MILK AND BUTTERFAT RECORDS

A number of production record systems have been developed (see Chapter 21, "Breeds of Dairy Cattle," and Chapter 24, "Feeding and Managing Dairy Cattle").

## COW BREEDING, HEALTH, AND LIFETIME RECORDS

Cow breeding, health, and lifetime records are as essential as milk production records.

With proper records, here are some reasonable goals for which to strive:

1. At least 70% of the cows should conceive at first breeding.

2. At any given time, there should be no more than 10% of the cows with reproductive difficulties.

3. The herd should average no more than 1.3 services per conception.

4. Calving interval should be no longer than 12½ months.

To achieve these goals, records should be easy to keep and useful in managing the dairy herd. They should be sufficiently complete to serve the following purposes:

1. Indicate earliest date to breed cows and heifers.

2. Identify dates to turn cows dry.

3. Indicate expected calving dates.

4. Measure breeding efficiency.

5. Suggest disease problems and the need for veterinary services.

6. Evaluate (a) the fertility of the bull, and (b) the effectiveness of natural and artificial insemination.

7. Show parentage and disposition of calves.

An example of a good breeding, health, and lifetime record form is given in Figs. 23-5a and 23-5b (see pages 502 and 503).

In addition to the monthly record provided by a testing program, for genetic progress it is important to have a complete lifetime record on each individual cow. This should provide (1) complete identification of the individual animal, (2) individual lifetime lactation summaries, (3) breeding record, (4) calving record, and (5) health and veterinary record. Figs. 23-5a and 23-5b will serve this purpose. With this information available, each cow may be evaluated.

HERD NO.

| | JAN. | FEB. | MAR. | APRIL | MAY | JUNE | JULY | AUG. | SEPT. | OCT. | NOV. | DEC. |
|---|---|---|---|---|---|---|---|---|---|---|---|---|

| NAME OF COW | REG. No. | DATE OF BIRTH | BIRTH WEIGHT | CONDITION AT BIRTH |
|---|---|---|---|---|
| SIRE | REG. No. | TATTOO OR EAR TAG<br>RIGHT      LEFT | CALFHOOD VACCINATION<br>NO.     DATE | RUMEN MAGNETS |
| DAM | REG. No. | DEHORNED | EXTRA TEATS REMOVED | DATE & REASON FOR REMOVAL FROM HERD |

| MASTITIS & UDDER HEALTH | | | | | | BREEDING RECORD | | | | | | CALF RECORD | | |
|---|---|---|---|---|---|---|---|---|---|---|---|---|---|---|
| DATE | RF | LF | LR | RR | TREATMENT | HEAT DATES | DATE BRED | SERVICE BULL | PREG. CHECK | DATE DUE | DATE FRESH | SEX | NAME OR NO. | CALF DISPOSAL |
| | | | | | | | | | | | | | | |
| | | | | | | | | | | | | | | |
| | | | | | | | | | | | | | | |
| | | | | | | | | | | | | | | |
| | | | | | | | | | | | | | | |
| | | | | | | | | | | | | | | |
| | | | | | | | | | | | | | | |
| | | | | | | | | | | | | | | |
| | | | | | | | | | | | | | | |
| | | | | | | | | | | | | | | |
| | | | | | | | | | | | | | | |
| | | | | | | | | | | | | | | |

PRODUCTION RECORDS

| | AGE | DAYS IN MILK | MILK – LBS. | | TEST | FAT – LBS. | |
|---|---|---|---|---|---|---|---|
| | | | ACTUAL | M. E. | | ACTUAL | M. E. |
| 1. | | | | | | | |
| 2. | | | | | | | |
| 3. | | | | | | | |
| 4. | | | | | | | |
| 5. | | | | | | | |
| 6. | | | | | | | |
| 7. | | | | | | | |
| 8. | | | | | | | |
| 9. | | | | | | | |
| 10. | | | | | | | |

COMMENTS, NOTES & MISCELLANEOUS

SKETCH OR PICTURE

Fig. 23–5a. An example of a good breeding, health, and lifetime record form, prepared by Utah State University. This shows the front side. See Fig. 23–5b for reverse side.

## DAIRY BREEDING AND HEALTH RECORD

Prepared by Utah State University
Extension Services
Logan, Utah

### DIAGNOSIS AND TREATMENT BY VETERINARIAN

| REPRODUCTIVE DISEASES | | | OTHER DISEASES | | |
|---|---|---|---|---|---|
| DATE | CONDITION | TREATMENT | DATE | CONDITION | TREATMENT |
| | | | | | |
| | | | | | |
| | | | | | |
| | | | | | |
| | | | | | |
| | | | | | |
| | | | | | |
| | | | | | |
| | | | | | |
| | | | | | |
| | | | | | |
| | | | | | |
| | | | | | |
| | | | | | |
| | | | | | |
| | | | | | |
| | | | | | |
| | | | | | |
| | | | | | |
| | | | | | |

ADDITIONAL INFORMATION

Fig. 23–5b. An example of a good breeding, health, and lifetime record form, prepared by Utah State University. This shows the reverse side of Fig. 23–5a.

## RECORDS AFFECTED BY ENVIRONMENT

The effect of environment on dairy cattle was clearly demonstrated in an experiment in New Zealand. It involved the selection of 20 calves from low-producing herds and 20 calves from high-producing herds. All of them were sired artificially by outstanding bulls. The 40 head were assembled at the Rurakura Experiment Station, raised and milked together for the first lactation. Under these conditions, no significant difference between the production of the two groups was observed. Then, they were sent back to the respective herds from whence they came, whereupon their production was comparable to that of the cows with which they were being milked. Then, for a second time, they were returned to the Rurakura Experiment Station, where again there was no significant difference in their production. The Rurakura Station then went one step further; they confirmed these results by using identical twins, with both twins milked at the Rurakura Station, and then later divided between high- and low-producing herds for subsequent lactation.

Fig. 23-6. Production records are affected by environment, the most important factor of which is feed. This shows the corral system in use by a Jersey herd. (Courtesy, The American Jersey Cattle Club, Reynoldsburg, OH)

The New Zealand experiment points up the importance of management. No matter how good the genetics, a good environment is essential to obtain high production.

Among the environmental and other factors affecting milk production records are the following:

1. **Feeding.** The most important of all factors in determining the productivity of a cow is the quantity and quality of feed provided.

2. **Milking practices.** Good milking practices and a properly functioning milking machine are necessary for high milk production.

3. **Age of animal.** On the average, production increases each year from the time the first calf is born until the cow reaches 5 to 8 years of age, after which it declines.

4. **Size of animal.** Within the same breed and age group, the larger animals, as measured by their capacity to consume more feed, usually produce more milk than the smaller animals.

5. **Season of freshening.** Cows that freshen in the spring and summer months usually produce less than cows freshening in the winter months. Of course, this variation differs among herds and areas.

6. **Calving interval.** Cows that calve within 12 months to 14 months from last calving produce more for that lactation than cows calving at shorter intervals. Lifetime production will usually be reduced if calving intervals are longer than 12 to 14 months.

7. **Length of dry period.** Cows with a dry period of 6 to 8 weeks produce more during the following lactation than those cows with a dry period of less than 4 weeks' duration.

8. **Freedom from disease, parasites, and injury.** Any one of these may depress production, with the degree of depression determined by the severity of the ailment.

9. **Rate of maturity.** Some strains or families of cows mature at a slower rate than others.

10. **Yearly differences.** There are important yearly differences within a given area, primarily due to weather conditions and the general quality of feed available. Nevertheless, when an attempt is being made to evaluate the breeding worth of an individual animal, all of these factors may be playing an important part in the production records.

## CORRECTION FACTORS

It is frequently desirable to compare the performance of individuals or groups of animals. To do so, it is necessary to correct all records to a comparable basis. For this purpose, correction factors have been developed for each breed for (1) length of lactation, (2) the number of milkings per day, (3) age and month of calving, and (4) fat content of milk. These four adjustments are important for comparing milk and fat of cows in different environmental conditions. Each of these factors will be discussed. At the outset, however, the following point is pertinent: Although correction factors are usually necessary in order to reduce two or more records to a common basis, it is recognized that records that are comparable without factors are more reliable.

### LENGTH OF LACTATION

The most generally accepted standard length of lactation records is 305 days. When a cow is milked longer than 305 days, her yield for the first 305 days is used as the standard lactation yield. Partial lactations (those terminated in less than

305 days because of environmental influences having no relation to the cow's genetic ability to complete normal length lactations) are considered legitimate measures of the cow's performance up to the time they were terminated and are used with a correction factor to 305 days. The factors commonly used for this projection are:

| Days Milked | Factor |
|---|---|
| 95 | 2.82 |
| 125 | 2.16 |
| 155 | 1.77 |
| 185 | 1.51 |
| 215 | 1.32 |
| 245 | 1.18 |
| 275 | 1.08 |

For comparing a 365-day record, reduce this to a 305-day record equivalent by taking 85% of it.

Total lactation records, or 365-day records, are often quoted verbally and in promotional literature, with or without an adequate definition of the lactation length. Care should be taken to clarify the length of the lactation when comparing or evaluating production records.

### NUMBER OF MILKINGS PER DAY

Most cows are milked twice daily (usually referred to as 2X); hence, for most lactations no adjustment is necessary.

To convert 3-times-a-day milking to 2-times-a-day basis, multiply by 83% (0.83). For purposes of illustrating how this works, let's assume that we have a 4-year-old Holstein cow that has a 3-times-a-day, 305-day record of 14,000 lb of milk and 610 lb of fat, and that it is desired to convert it to a 2-times-a-day basis. Simply multiply the cow's record by 0.83. Hence—

14,000 lb milk × 0.83 = 11,620 lb milk on 2X basis
610 lb fat × 0.83 = 506.3 lb fat on 2X basis

### AGE AND MONTH OF CALVING

The age of a cow is always based on her age when she calved, which is when her record begins. It is estimated, on a rule-of-thumb basis, that at 2 years of age a cow produces approximately 70–80% of her mature production; at 3 years 80–90%; at 4 years, 90–95%; 5 years, 96–100%; and at 6 years, her mature record.

*Age adjustment factors* have been developed to standardize 305-day lactation records to a mature equivalent basis and to minimize environmental variation due to the month of the year in which the record began. These age and month-of-calving factors are based on a total of 4,452,332 official DHI and DHIR lactations between 1964 and 1968. These factors are more accurate than any others available for milk and fat, because they more accurately remove recent environmental effects from age and month of calving in individual breeds and regions.

Table 23–5 shows the milk and fat age adjustment factors for cows in the United States calving in the month of May, by breed, for selected ages. A complete list of adjustment factors for milk and fat by breeds, by regions (and for the United States), by month of calving and by age, is given in the following report: *USDA-DHIA Factors for Standardizing 305-Day Lactation Records for Age and Month of Calving*, ARS-NE-40, USDA.

**TABLE 23–5**
**MILK AND FAT AGE ADJUSTMENT FACTORS FOR COWS IN THE UNITED STATES**
**CALVING IN THE MONTH OF MAY, BY AGE AND BY BREED[1]**

| Age | Ayrshire | | Brown Swiss & Red Poll | | Guernsey | | Holstein & Red Dane | | Jersey | | Milking Shorthorn | |
|---|---|---|---|---|---|---|---|---|---|---|---|---|
| (months) | (milk) | (fat) | (milk) | (fat) | (milk) | (fat) | (milk) | (fat) | (milk) | (fat) | (milk) | (fat) |
| 20 | 1.33 | 1.31 | 1.54 | 1.51 | 1.29 | 1.28 | 1.40 | 1.39 | 1.37 | 1.36 | 1.49 | 1.47 |
| 24 | 1.23 | 1.21 | 1.40 | 1.37 | 1.21 | 1.21 | 1.30 | 1.29 | 1.27 | 1.26 | 1.31 | 1.29 |
| 30 | 1.16 | 1.13 | 1.29 | 1.27 | 1.13 | 1.12 | 1.21 | 1.20 | 1.17 | 1.16 | 1.19 | 1.18 |
| 36 | 1.13 | 1.12 | 1.20 | 1.18 | 1.09 | 1.08 | 1.15 | 1.15 | 1.12 | 1.11 | 1.16 | 1.16 |
| 42 | 1.09 | 1.08 | 1.14 | 1.13 | 1.06 | 1.05 | 1.10 | 1.10 | 1.07 | 1.06 | 1.14 | 1.15 |
| 48 | 1.06 | 1.06 | 1.10 | 1.09 | 1.04 | 1.04 | 1.07 | 1.07 | 1.04 | 1.04 | 1.12 | 1.12 |
| 54 | 1.04 | 1.03 | 1.06 | 1.06 | 1.02 | 1.02 | 1.04 | 1.04 | 1.02 | 1.02 | 1.09 | 1.10 |
| 60 | 1.02 | 1.02 | 1.04 | 1.04 | 1.01 | 1.02 | 1.02 | 1.03 | 1.00 | 1.01 | 1.06 | 1.07 |
| 66 | 1.01 | 1.02 | 1.03 | 1.03 | 1.01 | 1.02 | 1.01 | 1.02 | .99 | 1.00 | 1.04 | 1.05 |
| 72 | 1.01 | 1.02 | 1.02 | 1.03 | 1.01 | 1.02 | 1.01 | 1.01 | .98 | 1.00 | 1.02 | 1.03 |
| 84 | 1.00 | 1.02 | 1.01 | 1.02 | 1.01 | 1.03 | 1.01 | 1.02 | .98 | 1.00 | 1.00 | 1.01 |
| 95 | 1.00 | 1.02 | 1.00 | 1.02 | 1.01 | 1.04 | 1.01 | 1.02 | .98 | 1.01 | .98 | 1.00 |
| 110 | 1.02 | 1.04 | 1.01 | 1.03 | 1.03 | 1.06 | 1.03 | 1.05 | 1.00 | 1.03 | .98 | 1.01 |
| 120 | 1.03 | 1.05 | 1.02 | 1.05 | 1.04 | 1.08 | 1.05 | 1.07 | 1.01 | 1.04 | 1.00 | 1.02 |
| 135 | 1.04 | 1.07 | 1.04 | 1.08 | 1.06 | 1.11 | 1.08 | 1.10 | 1.03 | 1.07 | 1.01 | 1.04 |
| 140 | 1.06 | 1.09 | 1.05 | 1.09 | 1.06 | 1.11 | 1.10 | 1.12 | 1.04 | 1.08 | 1.02 | 1.05 |
| 150 | 1.08 | 1.11 | 1.07 | 1.11 | 1.08 | 1.14 | 1.12 | 1.15 | 1.06 | 1.10 | 1.03 | 1.06 |
| 160 | 1.10 | 1.14 | 1.09 | 1.13 | 1.10 | 1.16 | 1.15 | 1.18 | 1.08 | 1.12 | 1.04 | 1.07 |

[1]*USDA–DHIA Factors for Standardizing 305-Day Lactation Records for Age and Month of Calving, ARS-NE-40, Agricultural Research Service, USDA, pp. 80–91.*

The standardized yield is obtained by multiplying yield for the first 305 days of lactation by the factor corresponding to the age at calving for the appropriate breed, region of the country, season of the year, and trait (milk or fat production). For example, let's assume that we have a Guernsey cow that was 20 months old when she calved and began her lactation record in the month of May; that she was milked 2 times daily; and that her 305-day record was 11,510 lb of milk and 508 lb of fat. By referring to Table 23–5, it is observed that the age adjustment factors for a 20-month-old Guernsey cow are 1.29 for milk and 1.28 for fat. Hence—

11,510 lb milk × 1.29 = 14,847.9 lb milk on ME basis
508 lb fat × 1.28 = 650.2 lb fat on ME basis

### FAT CONTENT OF MILK (FCM)

For comparative purposes, the fat content of milk is usually based on calculating the milk and fat production to 4% fat (4% FCM), but it may be calculated to any desired fat basis. The formula for 4% FCM is:

4.0% FCM = (0.4 × milk weight) + (15 × fat weight)

### SELECTING BREEDING STOCK

For most rapid progress, the producer should have an organized program of selecting and breeding. The first step in such a program consists in (1) establishing goals—goals in milk production, milk composition, body conformation, longevity, freedom from hereditary defects, etc.; (2) determining where you are now—recording the pertinent information on each animal; and (3) determining how you will get from hither to yon—from where you are now to the goals that you have set.

## BASES OF SELECTION

The success of any breeding program depends primarily on the ability of the breeder to select properly the animals that are to be parents of the next generation. Three methods of selecting such animals are recommended:

1. **Individual merit.** This consists in selecting cows on the basis of their record of production and/or body type. It must be recognized that this basis of selection is materially affected by the environment. For this reason, it is most effective if based on more than one lactation period, or even on the basis of a lifetime average, although admittedly, the latter is too slow for most conditions. Judgment in the use of individual merit records as a basis of selection may be as essential as the records themselves.

2. **Pedigree.** The usefulness of a pedigree depends to a large extent on its completeness, and upon the understanding of the descriptive material available. There is no uniformly accepted method of reporting information on a pedigree. However, a trend is rapidly developing toward reporting information on twice-a-day milking for a a 305-day lactation, and either listing the actual age at time of freshening of each lactation or figuring all records to a mature equivalent basis of 5 to 8 years of age. When the cow's age is listed for each record, it permits an appraisal of the frequency of calving, calving interval, and to some extent, the breeding efficiency.

Of course, the ancestors close up in the pedigree are the most important ones in attempting to evaluate the breeding worth of an animal from its pedigree.

It is generally agreed that pedigree selection should be used as an accessory to individual selection. It is particularly useful when selecting young animals for traits that are sex limited or that are exhibited only after sexual maturity; for example, udder shape and attachment, milk production, etc.

3. **Progeny testing.** This method of selecting involves a study of the individual's offspring. Progeny testing is particularly valuable for selecting such quantitative characters as milk production and milk constituents. When properly used, progeny tests prevent the breeder from being deceived by the effects of environment. It is emphasized, however, that progeny testing should be used to supplement, rather than replace, the other two bases of selection.

## METHODS OF SELECTION

In addition to arriving at a basis of selection (usually a combination of individuality, pedigree, and progeny test), and the traits for which selection is to be made, the dairy cattle breeder must determine what method or methods of selection shall be used. The following three general methods are available:

1. **Cull simultaneously, but independently, for each character.** This means that culling levels are established for each trait, below which all individuals are culled, no matter how good they may be in other respects.

2. **Tandem method of selection.** In this method, one characteristic is selected for at a time until it is improved, then selection is made for a second trait, and later a third, and so on and so on. Tandem selection will result in improving one trait faster than can be achieved through any other method, but while that is being done other traits may deteriorate.

3. **Establish a selection index.** This consists in totalling the animal's score for its merits in each characteristic, then retaining those with the highest total score. The selection index is looked upon more favorably than the tandem method because it permits unusually high merit in one characteristic to make up for deficiencies in some other trait.

In practice, a combination of all three methods of selection is usually most desirable and effective.

## SELECTING COWS

When milk is the major source of income, selection is simplified. Cows in production are ranked from high to low on the basis of milk production, and the most profitable milk cows are retained and the least profitable ones sold.

The USDA annually compiles and publishes estimates of the genetic transmitting ability of cows identified (through the Dairy Herd Improvement Testing Program) as having demonstrated superior genetic merit for milk production. These cow index values are listed in the USDA-DHIA Cow Performance Index and are based on production records of the cow (modified contemporary deviation), her paternal half-siblings (sire's PD), and her dam's cow index.

A purebred breeder usually finds it desirable and profitable to select animals for type as well as production.

Where grade cows are involved, and replacement heifers are not being raised, cows can be selected or culled primarily on the basis of milk production. Where breeding animals are involved—that is, where replacement heifers or bulls are being selected for retaining in the herd or for sale purposes—cows should be selected on the bases of their milk production, pedigree, and progeny, provided all three are available.

## SELECTING REPLACEMENT HEIFERS

Fig. 23–7. Replacement heifers. (Courtesy, Holstein-Friesian Assn. of America, Brattleboro, VT)

The number of heifer replacements needed each year to maintain herd size will depend upon the number of cows eliminated from the herd because of disease, injury, low production, or poor type. Normal turnover in DHIA herds is about 25% each year. To meet this, and to allow some opportunity for culling undesirable first calf heifers, it is necessary to raise approximately one-third as many heifer calves each year as there are milking animals in the herd.

Producers who raise their replacements are in a better position to evaluate the animals genetically than operators who buy replacements, simply because their dams and close relatives are available in the same herd under similar feed and management conditions.

## SELECTING SIRES

The selection of a sire is extremely important because he becomes the parent of many more offspring than any individual cow. A superior sire may be responsible for 80% or more of the genetic improvement in a herd.

Generally speaking, the producer has three sources of herd sires: (1) artificial insemination service, (2) purchase of a herd sire, or (3) raising a herd sire. The producer must also decide between a proven sire and a young sire. The challenge is to select the herd sire that will maintain a higher level of production than the present herd average. This is essential if there is to be improvement in the herd.

Sire evaluation has become a very sophisticated procedure as a result of experience, the development of larger and faster computers, improved statistical methods, and research.

The most reliable source of superior germ plasm for the breeding program of a herd is bulls that have been accurately evaluated for a large number of traits, including yield, conformation, and calving ease. Basically, this means bulls available through A.I.

Twice annually, the USDA publishes a USDA-DHIA Sire Summary. These genetic evaluations are based on information on the bulls' daughters in herds participating in official production testing programs (DHIA and DHIR).

The USDA sire summaries are publicized and interpreted by the Cooperative Extension Service, state associations, dairy breed registries, A.I. organizations, dairy magazines, and other channels.

The *Hoard's Dairyman* Bull List of top active A.I. bulls, which was started in 1967, is excellent and widely used. Pertinent details relative to it follow.

*Hoard's Dairyman* uses three official sources: (1) USDA Sire Summaries (all production information and colored breed type information), (2) the Holstein Association (Holstein type information), and (3) the National Association of Animal Breeders (calving ease information). Bulls are ranked according to Predicted Transmitting Ability Milk-Fat-Protein Dollars (PTA MFP$). To further emphasize the importance of the ranking, the bulls are grouped according to their percentile rankings. (Those in the 90th percentile have higher MFP$ than 90% of the other active A.I. bulls of the same breed.) The bulls are listed under the following headings:[1]

| | | | Predicted Transmitting Abilities | | | | | | | | | | Type Data | | TPI | Calving Ease | |
| | | | Protein | | | | Milk and Fat | | | | | | | | | | |
| Name of Bull | Reg. Number | NAAB Code | Rel. | MFP$ | lb P | % P | CY$ | Rel. | MF$ | lb M | lb F | % F | Rel. | PTAT | PTI | Rel. | % DBH |

An explanation of the above headings, as given by *Hoard's Dairyman*, follows:

- **Name of Bull**—Bull's registered name.

- **Reg. Number**—Bull's registration number.

- **NAAB Code**—A three-part code. The number before the letter indicates the stud from which the bull's semen can be purchased. The letter indicates the breed. The number following the letter is an individual bull identification number assigned by the bull stud.

- **Rel.**—Reliability replaces Repeatability as a measure of the accuracy of the genetic evaluations. Four Reliabilities are listed . . . one for milk, fat, and proteins; one for milk and fat; one for type; and one for calving ease.

The closer Reliability is to 100, the more reliable the Predicted Transmitting Abilities (PTAs). Reliabilities on Active A.I. sires range from 45 to 99%. *They do not measure the bull's conception rate.*

- **PTA MFP$**—Weights the Predicted Transmitting Abilities (PTAs) for Milk, Fat, and Protein to reflect the gross income per lactation future daughters of bulls will earn in excess of herdmates sired by bulls having a MFP$ equal to zero when milk receives a protein premium. It is based on prices of $12.50 per hundred for milk with 3.5% fat and 3.2% protein. Differentials are 14.8 cents per point of fat and 14.3 cents per point of protein.

- **PTA LB P, % P, LB M, LB F, and % F**—PTAs for pounds protein, percent protein, pounds milk, pounds fat, and percent fat indicate how much more (or less) performance to expect from an average daughter of a bull with a PTA of zero for the same trait. The new genetic base, called PTA90, was selected so half the cow population will have positive PTAs. It was established by setting to zero the weighted average PTAs of all cows born in 1985.

- **PTA MF$**—Predicted Transmitting Ability Milk-Fat Dollars weights the PTA Milk and Fat to reflect the gross income per lactation future mature daughters of bulls will earn in excess of herdmates sired by bulls having a MF$ equal to zero. It is based on prices of $12.50 per hundred of 3.5% fat and 14.8 cents per point fat differential. This was the U.S. average milk price for 1989 minus the average hauling and assessments for promotion.

- **CY$**—Predicted Transmitting Ability Cheese Yield Dollars reflects the income per lactation future mature daughters of the bull will earn if their milk is priced according to its value in Cheddar cheese.

- **PTAT**—Predicted Transmitting Ability-Type is the expected difference in final score between daughters of the bull and breed average.

- **TPI (Holsteins)**—Type-Production Index is a value which is determined by placing an emphasis of 2 for PTA Protein, 2 for PTA Fat, 1 for PTA Type, and 1 for udder composite traits.

- **PTI (all breeds except Holsteins)**—The Production Type Index is a value which is determined by placing an emphasis of 3 for PTA$, 3 for CY$, and 1 for Predicted Transmitting Ability-Type (PTAT).

- **Calving Ease, % DBH**—This is the estimate of the Percentage of Difficult Births in Heifers when they calve the first time. Producers use this information when choosing bulls to breed heifers.

Fig. 23–8. Today, most dairy producers select semen from an A.I. bull, such as this stud bull.

[1]Reprinted with permission from the February 25, 1990 issue of *Hoard's Dairyman*, Copyright 1990 by W. D. Hoard and Sons Company, Fort Atkinson, WI)

## YOUNG SIRES

Generally speaking, producers can avail themselves of the use of proven sires through artificial insemination. However, good proven sires are not always available at a price that the individual breeder can afford to pay, particularly when it is desired to use them in natural service. Further, they are even expensive for artificial insemination associations to purchase. Additionally, it is recognized that proven sires are generally 6 to 8 years old, and that they have a remaining life expectancy of only 2 to 3 years. For these reasons, there is increasing interest in young sires.

Generally, young bulls are highly selected on pedigree and should, on the average, have high PDs when proven.

The young, sampled sires are priced lower and can be useful to the dairy farmer if used properly. It is recommended (1) that a few cows be bred to each of several young bulls, and (2) that not more than 25% of all matings be to young sires. One disadvantage with young sires is that their calving ease is an unknown factor. So, young sires should be bred to cows rather than to heifers.

## SYSTEMS OF BREEDING

In dairy cattle breeding, one of the following accepted systems of breeding is generally followed: (1) inbreeding, which embraces [a] close breeding or [b] linebreeding; (2) outcrossing; or (3) crossbreeding.

Close breeding (the mating of animals that are closely related) is usually limited to those breeders who are particularly good students of their animals, who recognize their weak and strong points, and who cull ruthlessly when the situation demands such action. When successful, it concentrates the most desirable traits and produces some outstanding individuals who transmit fairly uniformly. Close breeding is best left to the breeder who has complete records of production and type, and whose herd is at a high average level of production, and has been so for a number of years.

Linebreeding is practiced much more extensively by producers than close breeding. It is a more conservative type of inbreeding program which the vast majority of average and small dairy breeders can safely follow to their advantage.

Outcrossing (the mating of unrelated animals) is the most widely used system of breeding by the majority of producers. It offers considerable opportunity to introduce new genes into the herd, simply because a wide choice of animals can be made. It often results in producing animals which are highly desirable within themselves, but they may not transmit uniformly. However, this system does not carry the dangers that often go with inbreeding, such as possible reduction in size and scale, lack of vigor, development of possible recessive factors which may become undesirable, and greater concentration of any undesirable trait.

Crossbreeding cannot be practiced in registered herds because the offspring cannot be registered. For those producers who have no preference as to color and general appearance, crossbreeding may be followed with good results provided good proven sires are used. The aims of this system of breeding are to use the best sires available regardless of breed, and to gain hybrid vigor in the offspring (reduce calfhood mortality) and possibly more economical milk production.

## DEVELOP AND FOLLOW A BREEDING PROGRAM

Where replacement animals are raised, in either a purebred or a grade herd, a breeding program must be developed and followed if herd progress and breed progress are to be made. The following steps are pertinent to such a program:

1. Choose a suitable breed of cattle.
2. Select or purchase the best cows available, based primarily upon their production record, but with due consideration given to type and pedigrees.
3. Decide on the breeding system—inbreeding, outcrossing, or crossbreeding—to achieve the desired goals.
4. Evaluate the strong points and the weak points of the cows in the herd.
5. Obtain the services of the sire(s) which offers the greatest promise of further improvement in production and in type, but with due consideration given to the price of the sire, along with the age and health of the individual if he is to be used in natural service.
6. Enroll in the particular testing program which best meets the breeding and management programs that will be followed.
7. Follow the program of type evaluation which best meets the needs of the breeding program.
8. Arrive at the system of selection that shall be followed, choosing between individual culling levels, the tandem method, or a selection index (see Chapter 3, "Genetics and Animal Breeding," section headed, "Bases of Selection").
9. Establish and maintain reasonable standards for freedom from disease, temperament, fertility and sterility, ease and completeness of milking, and such other factors as are considered important.
10. Follow a feeding and management program which will permit the animals in the herd to express the maximum genetic potential which they possess.

## ARTIFICIAL INSEMINATION OF DAIRY CATTLE

Artificial insemination, as a means of dairy cattle improvement, is now accepted and utilized worldwide. The increased use of outstanding sires to enhance production potential, control certain genital diseases transmitted through natural service, and encourage general mass improvement is well recognized.

## SEX CONTROL

Sex ratios of dairy cattle at birth show that there is a slight deficiency in heifers; out of each 100 calves, on the average, 49 are heifer calves and 51 are bull calves.

Obviously, some method of controlling sex of offspring would have tremendous significance in the dairy field. Producers could then produce more needed heifer calves and fewer unwanted bull calves.

It appears that nature is about to yield to sex control. Predetermination of the sex of 6- to 12-day-old embryos is a reality, and progress is being made in the separation of sperm cells containing X chromosomes from those containing Y chromosomes.

## BIOTECHNOLOGY IN DAIRY CATTLE

Fig. 23–9. Mother cow and her "litter" of calves produced by embryo transfer. (Courtesy, Holstein-Friesian Assn. of America, Brattleboro, VT)

Biotechnology will revolutionize the dairy industry. Genetically engineered embryos will produce higher yielding cows that are more resistant to stress and disease. Through cloning, it will be possible for all animals in an entire herd to look alike, be genetically alike, have the same nutritive requirements, and produce the same quantity of milk of the same composition. In the next century, artificial insemination will be replaced by *in vitro* (test tube) artificially fertilized embryos.

## QUESTIONS FOR STUDY AND DISCUSSION

1. What is the objective of dairy cattle breeding?

2. Discuss the economic justification of improved dairy cattle breeding as illustrated in Fig. 23–1.

3. What is the difference between (a) breed improvement, and (b) herd improvement?

4. List, in descending order, the three main reasons for cows leaving dairy herds, and discuss each of these reasons for culling.

5. Name the four main reproductive diseases of dairy cattle and give the (a) symptoms and (b) prevention of each.

6. At what period in the heat cycle should a cow be bred in order to get the highest conception rate?

7. How are chromosome numbers maintained constant from one generation to the next?

8. Which is more important in determining the productivity of a dairy cow, heredity or environment?

9. What traits of dairy cattle have the highest heritability? What traits have the lowest heritability?

10. List the past methods of genetic evaluation of dairy cattle, and describe each of them briefly.

11. Define (a) Predicted Difference (PD), and (b) Repeatability. Of what value is each of them?

12. Describe the genetic evaluation of dairy cattle by the most commonly used present method known as Modified Contemporary Comparison (MCC).

13. Rank in descending order the three most accurate kinds of estimates of transmitting ability of bulls.

14. Rank in descending order the three most accurate kinds of phenotype information for estimating the transmitting ability of cows.

15. List reasonable goals for cow breeding, health, and lifetime records.

16. Discuss the design and the results of the New Zealand study on the effect of environment on dairy cattle.

17. Discuss each of the correction factors commonly used in dairy performance. Why are correction factors used?

18. List and discuss the three common *bases* of selection.

19. List and discuss the three most common *methods* of selection.

20. How would you go about selecting cow and dairy heifer replacements?

21. What three sources of herd sires does a dairy producer generally have? What are the advantages and the disadvantages of each source?

22. Under what circumstances might it be desirable to use a young sire rather than a proven sire? If a young sire is used, how should he be used?

23. Explain the difference between, and the advantages of (a) closebreeding, (b) linebreeding, (c) outcrossing, and (d) crossbreeding. Under what circumstances would you use each?

24. Outline a sound breeding program for a dairy herd.

25. Would you use artificial insemination or natural service in a dairy herd?

26. Of what value would sex control have for a dairy producer?

27. Of what value would genetically engineered embryos, cloning, and *in vitro* artificially fertilized embryos have in a dairy herd?

## SELECTED REFERENCES

| Title of Publication | Author(s) | Publisher |
| --- | --- | --- |
| *Artificial Insemination and Embryo Transfer of Dairy and Beef Cattle, The*, Seventh Edition | H. A. Herman<br>F. W. Madden | The Interstate Printers & Publishers, Inc., Danville, IL, 1987 |
| *Artificial Insemination of Farm Animals, The* | E. J. Perry | Rutgers University Press, New Brunswick, NJ, 1968 |
| *Cattle Fertility and Sterility* | S. A. Asdell | Little, Brown and Company, Boston, MA, 1968 |
| *Dairy Cattle Breeding* | L. O. Gilmore | J. B. Lippincott Co., Philadelphia, PA, 1951 |
| *Dairy Cattle; Principles, Practices, Problems, Profits*, Third Edition | D. L. Bath, *et al.* | Lea & Febiger, Philadelphia, PA, 1985 |
| *Dairy Cattle Science*, Second Edition | M. E. Ensminger | The Interstate Printers & Publishers, Inc., Danville, IL, 1980 |
| *Dairy Cattle Sterility* | H. D. Hafs<br>L. J. Boyd | W. D. Hoard and Sons Co., Fort Atkinson, WI |
| *Dairy Guide* | Staff | The Ohio State University Cooperative Extension Service, Columbus, OH, 1979 |
| *Illinois-Iowa Dairy Handbook* | Staff | University of Illinois at Urbana-Champaign, Iowa State University at Ames Cooperative Extension Service, 1983 |
| *Improving Cattle by the Millions* | H. A. Herman | University of Missouri Press, Columbia and London, 1981 |
| *Principles of Dairy Farming, The*, Sixth Edition | K. Russell<br>S. Williams | Farming Press Ltd., Ipswitch, England, 1972 |
| *Principles of Dairy Science* | G. H. Schmidt<br>L. D. Van Vleck | W. H. Freeman and Co. Publishers, San Francisco, CA 1974 |

Three top Guernsey females, products of heredity and environment. (Courtesy, American Guernsey Assn., Reynoldsburg, Ohio)

Healthy, hungry, and curious. (Courtesy, Holstein-Friesian Assn. of America, Brattleboro, VT)

# FEEDING AND MANAGING DAIRY CATTLE

## Chapter 24

Fig. 24-1. The attractive Maddox Dairy Headquarters, Riverdale, California, center of feeding and managing more than 3,600 lactating cows with a rolling herd average of 20,850 lb of milk and 3.72% fat on 3x day milking. (Photo by A. H. Ensminger)

Feed, more than any other one factor, determines the productivity and profitability of dairy cows. Within a herd, approximately 25% of the difference in milk production between cows is due to heredity; the remaining 75% is determined by environmental factors, with feed making up the largest portion. Feed accounts for about 55% (with a range from 45 to 65%) of the cost of milk production. Therefore, a good feeding program is necessary for profitable milk production.

It costs more to feed high producers than low producers. But high producers generally return more net income over feed cost than low producers. Fig. 24-2 shows how income over feed cost improves as production per cow increases.

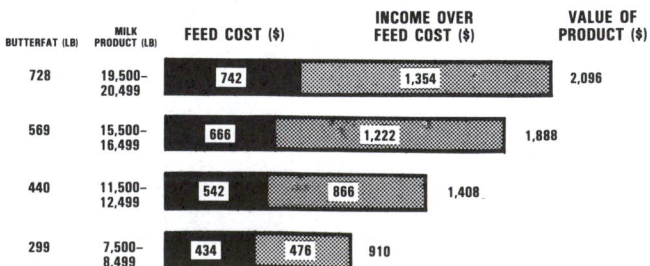

| BUTTERFAT (LB) | MILK PRODUCT (LB) | FEED COST ($) | INCOME OVER FEED COST ($) | VALUE OF PRODUCT ($) |
|---|---|---|---|---|
| 728 | 19,500-20,499 | 742 | 1,354 | 2,096 |
| 569 | 15,500-16,499 | 666 | 1,222 | 1,888 |
| 440 | 11,500-12,499 | 542 | 866 | 1,408 |
| 299 | 7,500-8,499 | 434 | 476 | 910 |

Fig. 24-2. It costs more to feed high-producing cows—but it pays. The reason: feed and overhead costs for maintenance are practically the same, regardless of level of production.

## DIGESTIVE SYSTEM OF THE COW

An understanding of the principal parts and functions of the digestive system is essential to intelligent feeding of dairy cattle. Fig. 24-3 shows the location of the parts of the ruminant's stomach and the route of digestion followed by most feed.

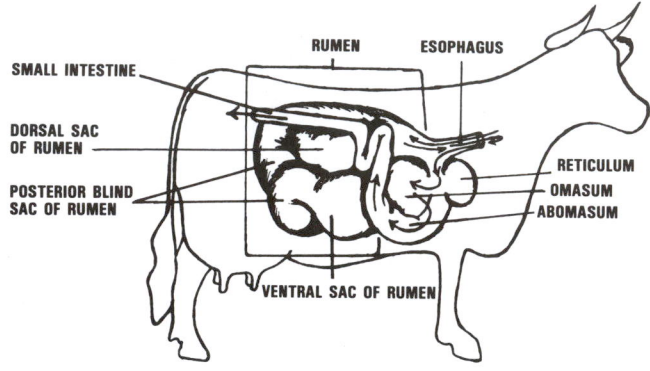

Fig. 24-3. Location and parts of the ruminant stomach (four compartments), with pathway of feeds indicated by arrows.

Further elucidation of the subject of digestion, including the parts and functions of the ruminant stomach is contained in Chapter 4 "Feeding Livestock"; hence, the reader is referred thereto.

## NUTRITIVE NEEDS OF DAIRY CATTLE

*A nutrient is any substance that aids in the support of life.* The first consideration in any dairy feeding program, therefore, is to determine the nutritive need. Dairy cattle require nutrients for growth, body maintenance, pregnancy or reproduction, and milk production

The nutritive needs for growth, body maintenance, and pregnancy generally are provided for before milk production can take place in quantity. For this reason, it does not pay to underfeed. A cow produces more economically when worked near full capacity.

The nutritive requirements for dairy cattle have been established by the National Academy of Sciences—NRC (*Nutrient Requirements of Dairy Cattle,* sixth revised edition, update 1989). Tables 24-1 and 24-2 were adapted by the author from this source. For additional categories—for weights and levels of production not given in Tables 24-1 and 24-2—and for nutrient requirements of rations, the reader is referred to the original publication. These requirements were adapted for, and presented in, *Feeds & Nutrition* and *Feeds & Nutrition Digest,* books by M. E. Ensminger, *et al.,* published by the Ensminger Publishing Company; hence, the reader is referred thereto. In using these tables, cognizance should be taken of the fact that the nutritive requirements given in them do not allow for any margin of safety; that is, they do not provide for animal differences, feed differences, and losses of certain nutrients in storage. Accordingly, in the formulation of rations, certain margins of safety should be provided.

**TABLE 24-1**
**DAILY NUTRIENT REQUIREMENTS OF GROWING DAIRY CATTLE[1]**

| Live Weight | | Gain | | Dry Matter Intake[2] | | Energy | | | | | Protein | | | Minerals | | Vitamins | |
|---|---|---|---|---|---|---|---|---|---|---|---|---|---|---|---|---|---|
| | | | | | | TDN | | DE | ME | $NE_m$ | $NE_g$ | CP | DIP[3] | UIP[4] | Ca | P | A | D |
| (lb) | (kg) | (lb) | (kg) | (lb) | (kg) | (lb) | (kg) | (Mcal) | (Mcal) | (Mcal) | (Mcal) | (g) | (g) | (g) | (g) | (g) | (1,000 IU) | (1,000 IU) |
| Growing Large-Breed Calves Fed Only Milk or Milk Replacer |
| 88 | 40 | 0.4 | 0.2 | 1.06 | 0.48 | 1.37 | 0.62 | 2.73 | 2.54 | 1.37 | 0.41 | 105 | — | — | 7 | 4 | 1.70 | 0.26 |
| Growing Small-Breed Calves Fed Only Milk or Milk Replacer |
| 55 | 25 | 0.4 | 0.2 | 0.84 | 0.38 | 1.08 | 0.49 | 2.16 | 2.01 | 0.96 | 0.37 | 84 | — | — | 6 | 4 | 1.10 | 0.16 |
| Growing Veal Calves Fed Only Milk or Milk Replacer |
| 331 | 150 | 2.4 | 1.1 | 6.00 | 2.72 | 4.74 | 2.15 | 9.60 | 8.46 | 3.69 | 2.29 | 598 | — | — | 24 | 15 | 6.40 | 0.99 |
| Large-Breed Growing Females |
| 441 | 200 | 1.8 | 0.8 | 10.96 | 4.97 | 7.36 | 3.34 | 14.71 | 12.62 | 4.57 | 2.25 | 796 | 295 | 294 | 22 | 15 | 8.48 | 1.32 |
| 992 | 450 | 1.3 | 0.6 | 21.15 | 9.59 | 12.59 | 5.71 | 25.18 | 21.12 | 8.40 | 2.53 | 1,151 | 686 | 176 | 28 | 19 | 19.08 | 2.97 |

[1]Adapted by the author from Nutrient Requirements of Dairy Cattle, 6th rev. ed., update 1989, NRC, National Academy Press, Washington, DC, pp. 81–84, Table 6–2.

**TABLE 24-2**
**DAILY NUTRIENT REQUIREMENTS OF LACTATING AND PREGNANT COWS[1]**

| Live Weight | | Energy | | | | Total Crude Protein | Minerals | | Vitamins | |
|---|---|---|---|---|---|---|---|---|---|---|
| | | TDN | | DE | ME | $NE_{lc}$ | | Ca | P | A | D |
| (lb) | (kg) | (lb) | (kg) | (Mcal) | (Mcal) | (Mcal) | (g) | (g) | (g) | (1,000 IU) | (1,000 IU) |
| Maintenance of Mature Lactating Cows[2] |
| 882 | 400 | 6.90 | 3.13 | 13.80 | 12.01 | 7.16 | 318 | 16 | 11 | 30 | 12 |
| 1,433 | 650 | 9.94 | 4.51 | 19.86 | 17.29 | 10.30 | 428 | 26 | 19 | 49 | 20 |
| 1,764 | 800 | 11.60 | 5.26 | 23.21 | 20.20 | 12.03 | 486 | 32 | 23 | 61 | 24 |
| Maintenance Plus Last 2 Months of Gestation of Mature Dry Cows[3] |
| 882 | 400 | 9.15 | 4.15 | 18.23 | 15.26 | 9.30 | 875 | 26 | 16 | 30 | 12 |
| 1,433 | 650 | 13.16 | 5.97 | 26.23 | 21.96 | 13.39 | 1,120 | 43 | 26 | 49 | 20 |
| 1,764 | 800 | 15.39 | 6.98 | 30.65 | 25.66 | 15.64 | 1,254 | 53 | 32 | 61 | 24 |

| Fat | Energy | | | | | | | Total Crude Protein | | Minerals | | | | Vitamins | |
|---|---|---|---|---|---|---|---|---|---|---|---|---|---|---|---|
| | TDN | | DE | | ME | | $NE_{lc}$ | | | Ca | | P | | A | D |
| (%) | (lb) | (kg) | (Mcal/lb) | (Mcal/kg) | (Mcal/lb) | (Mcal/kg) | (Mcal/lb) | (Mcal/kg) | (g/lb) | (g/kg) | (g/lb) | (g/kg) | (g/lb) | (g/kg) | | |
| Milk Production—Nutrients/2.2 lb or/kg of Milk of Different Fat Percentages |
| 3.0 | 0.616 | 0.280 | 0.56 | 1.23 | 0.49 | 1.07 | 0.29 | 0.64 | 35 | 78 | 1.24 | 2.73 | 0.76 | 1.68 | — | — |
| 3.5 | 0.662 | 0.301 | 0.60 | 1.33 | 0.52 | 1.15 | 0.31 | 0.69 | 38 | 84 | 1.35 | 2.97 | 0.83 | 1.83 | — | — |
| 4.0 | 0.708 | 0.322 | 0.64 | 1.42 | 0.56 | 1.24 | 0.34 | 0.74 | 41 | 90 | 1.46 | 3.21 | 0.90 | 1.98 | — | — |
| 4.5 | 0.755 | 0.343 | 0.69 | 1.51 | 0.60 | 1.32 | 0.35 | 0.78 | 44 | 96 | 1.57 | 3.45 | 0.98 | 2.13 | — | — |
| 5.0 | 0.801 | 0.364 | 0.73 | 1.61 | 0.64 | 1.40 | 0.38 | 0.83 | 46 | 101 | 1.68 | 3.69 | 1.04 | 2.28 | — | — |
| 5.5 | 0.847 | 0.385 | 0.77 | 1.70 | 0.69 | 1.48 | 0.40 | 0.88 | 49 | 107 | 1.78 | 3.93 | 1.10 | 2.43 | — | — |
| Live Weight Change During Lactation—Nutrients/kg of Weight Change[4] |
| Weight loss | -0.99 | -2.17 | -4.34 | -9.55 | -3.75 | -8.25 | -2.25 | -4.92 | -145 | -320 | — | — | — | — | — | — |
| gain | 1.03 | 2.26 | 4.52 | 9.96 | 3.88 | 8.55 | 2.32 | 5.12 | 145 | 320 | — | — | — | — | — | — |

[1]Adapted by the author from Nutrient Requirements of Dairy Cattle, 6th rev. ed., update 1989, NRC, National Academy Press, Washington, DC, p. 84, Table 6–3.

[2]To allow for growth of young lactating cows, increase the maintenance allowances for all nutrients except vitamins A and D by 20% during the first lactation and 10% during the second lactation.

[3]Values for calcium assume that the cow is in calcium balance at the beginning of the last 2 months of gestation. If the cow is not in balance, then the calcium requirement can be increased from 25 to 33%.

[4]No allowance is made for mobilized calcium and phosphorus associated with live weight loss or with live weight gain. The maximum daily nitrogen available from weight loss is assumed to be 30 g or 234 g of crude protein.

## ENERGY

Lack of energy is the most common deficiency of dairy rations. Cows cannot produce milk at peak levels if their rations are too low in energy.

Most of the energy required is supplied by carbohydrates and fats in forage and grain. All cows, except low-producing ones—those producing less than 15 to 20 lb of milk per day, need some grain if they are to yield at top levels.

The NRC energy requirements for dairy cattle, which are presented in Tables 24–1 and 24–2 are expressed as digestible energy (DE), metabolizable energy (ME), net energy for maintenance ($NE_m$), net energy for body gain ($NE_g$), net energy for lactation ($NE_{lc}$), and total digestible nutrients (TDN). Separate net energy values for each maintenance ($NE_m$) and gain ($NE_g$) are given because animals use energy for maintenance more efficiently than for growth. However, the efficiency of energy use by lactating cows for maintenance, pregnancy, and milk production is similar; so, only one energy value, net energy for lactation ($NE_{lc}$), is used for these functions.

The energy value of a feed may be separated into: (1) the losses that occur in digestion and metabolism, and (2) the net energy (NE) that is available to the animal for maintenance and production. The total energy in feed, which is determined by complete oxidation (burning) of the feedstuff and measurement of the heat produced, is known as *gross energy* and is expressed as calories. Common feedstuffs are similar in gross energy content, but differ in feeding value because of variations of digestibility. About 60% of the total energy in grain and 80% of the total energy in roughage is lost in feces, urine, gases, and heat.

In Tables 24–1 and 24–2, energy is also expressed as total digestible nutrients (TDN). TDN is comparable to digestible energy. It has been in use longer than the net energy system and more values are available for feedstuffs. TDN is computed as follows:

**TDN = digestible nitrogen-free extract (carbohydrate) + digestible crude fiber + digestible protein + (digestible ether extract × 2.25)**

NE of lactation can be calculated from TDN as follows:

**$NE_{lc}$(Mcal/lb DM) = (TDN, % of DM × .01114) − .054**

The energy requirement for maintaining a lactating cow is affected by a number of factors, especially the following: (1) *body size*—the larger the animal, the higher the maintenance energy requirement; (2) *activity*—to support grazing activity, the maintenance allowance may be increased by 10% on good pasture and up to 20% on poor pasture; and (3) *cold temperature*—under severe winter conditions without access to dry shelter, the maintenance feed allowance may be increased up to 8%. Also, during the first lactation, when a heifer is still growing, her energy needs are about 20% greater than a mature cow; and during the second lactation, her energy needs are 10% greater than a mature cow. The energy requirement for gestation is about 30% of that required for maintenance alone, with most of the increase during the last 8 weeks of pregnancy.

Table 24–2 includes allowances for liveweight changes during lactation. These values will aid the user in identifying the extent of dietary energy insufficiency during weight loss in early lactation and in estimating feed required to regain body weight in later lactation. The desired rate of liveweight gain will depend on the animal's body condition and stage of pregnancy.

## PROTEIN

Protein is essential for dairy cattle maintenance, growth, milk production, and the development of the fetus. Also, it is required for the formulation of enzymes and certain hormones that control or regulate chemical reactions in the body. The protein requirement is really a requirement for amino acids.

The protein composition of feeds, and the protein requirements of dairy cattle, may be expressed as crude protein, digestible protein, degraded intake protein, undegraded intake protein, and/or nonprotein nitrogen (NPN).

1. **Crude protein.** Chemically, most proteins contain 16% nitrogen; so, crude protein is determined by finding the nitrogen content, then multiplying the result by 6.25 (100 ÷ 16 = 6.25). It is called crude protein because not all nitrogen in feeds is in the form of protein; rather, it is a combination of true protein and nonprotein nitrogen.

2. **Digestible protein.** This is the amount of crude protein consumed less the crude protein excreted in the feces.

3. **Degraded intake protein (DIP).** This refers to the intake crude protein that is broken down (degraded) by microorganisms in the rumen.

4. **Undegraded intake protein (UIP).** This is the crude protein that is not broken down in the rumen; instead, it is swept out of the rumen into the abomasum and small intestine for breakdown there and absorption as peptides and amino acids. Undegraded protein is also known as bypass protein, protected protein, and escaped protein. (Also see Chapter 4 of this book, section on "Slow-Release and Rumen Bypass Treatments.")

5. **Nonprotein nitrogen (NPN).** Feedstuffs which contain nitrogen in a form other than proteins or peptides are termed nonprotein nitrogen (NPN).

The amount of protein needed in the total ration of lactating cows is determined primarily by the amount of milk produced. Milk is a rich source of high-quality protein; so, as milk production increases, a substantial amount of dietary protein is necessary. Thus, a high-producing 1,320-lb cow yielding 88 lb of 3.5% protein milk daily secretes 3.08 lb of milk protein. A deficiency of protein results in lowered milk production and may depress the protein content of milk. Excess protein usually results in high cost rations.

The amount of protein needed in the concentrate mix depends on the kind and quality of forage fed. As the amount of legume increases, the percentage of protein in the concentrate can be lowered. For most lactating cows, the total ration (forage plus grains and protein and energy supplements) should have 19% crude protein during the first ⅓ of lactation, lowered to 14% in midlactation and 12% during the dry period.

When more protein is fed than needed, the excess is used as a source of energy. Because protein feeds are generally more expensive than carbohydrate feeds, it usually is more economical to feed only the amount needed. Besides, a large excess of dietary protein may decrease the energy supply because excess protein must be deaminated to ammonia and, for the most part transformed back into urea for excretion. Most cows fed good quality alfalfa hay (fed free-choice), along with grain fed according to production by one of the recommended systems, will not need any supplemental protein until they produce more than 50 to 60 lb of milk daily. As production increases above this amount, protein intake must be increased gradually, usually by decreasing the hay intake and replacing it with grain and protein concentrates. So long as the hay consumption does not fall below 12 to 15 lb daily, the supplemental concentrate need not contain more than 15 to 16% protein.

## MINERALS

Minerals make up about 5% of the weight of the dairy cow. This is mostly calcium and phosphorus, found chiefly in the skeleton.

But the lactating cow has additional mineral requirements, over and above those needed for her own body or for the developing fetus.

Milk contains about 0.7% minerals. Thus, one cow producing 15,000 lb of milk gives 105 lb of mineral per year. By way of comparison, it is noteworthy that 3 steers produce only 120 lb of minerals by the time they reach 1,000 lb in weight (at 18 months of age). This means that in one lactation a cow produces in her milk nearly as much minerals as 3 steers store in their bodies in 54 months (3 × 18). Additionally, a milk cow needs minerals for body maintenance (which requirements are about the same as those of a steer), for development of the unborn calf, and for growth if she is a young cow.

Dairy cattle of all ages and stages of production are more apt to suffer from a lack of phosphorus in their feed than from a deficiency of any other mineral element. Major changes in dairy cattle feeding have accentuated the phosphorus deficiency in recent years. Among these changes are (1) increased crop yields as a result of improved varieties and heavy nitrogen fertilization, which have depleted the phosphorus of the soil; (2) alfalfa constituting more of the roughage component, and alfalfa is always a rich source of calcium (in alfalfa, calcium-phosphorus ratios of 6 to 8:1 are not uncommon); (3) more urea is being fed (and urea does not contain any minerals), whereas protein-rich oil meal supplements are a valuable source of phosphorus; and (4) high-level feeding and increased milk production, which carries with it a built-in stress factor that tends to emphasize any difficulty that may be encountered if any nutrient is deficient or supplied in excess. Generally speaking, the calcium-phosphorus ratio of the total ration should not be wider than 2:1.

It is also good business to guard against any trace mineral deficiencies by providing cobalt, copper, iodine, iron, manganese, molybdenum, selenium, and zinc. These trace minerals may be provided in the mineral mix, in trace-mineralized salt, or in the ration itself.

Salt and other minerals may be added to the concentrate mix, usually at the rate of about 1% salt and 1% other minerals. Even so, they should always be available free choice.

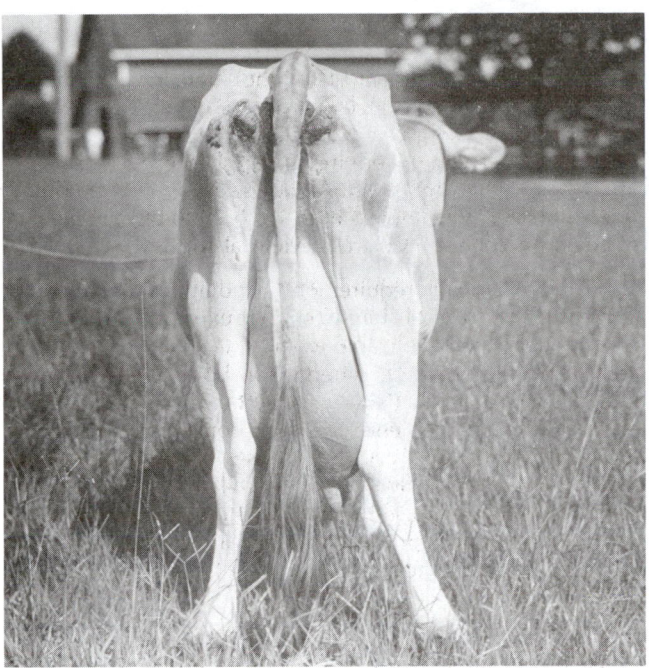

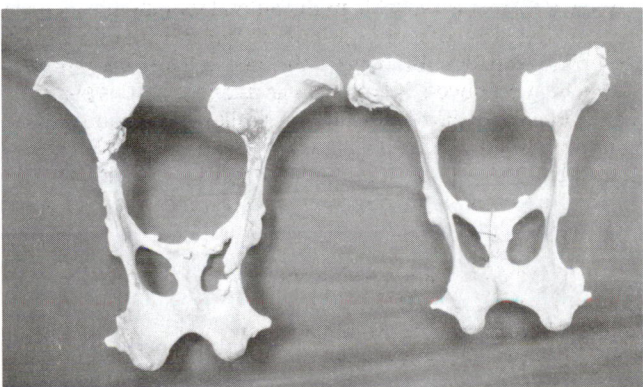

Fig. 24–4. Calcium deficiency. Lactating cows need calcium. Both hips of the cow shown above have been broken (knocked down) as a result of feeding a low-calcium ration. At lower left, the pelvis of a cow which suffered three breaks while the cow received a low-calcium ration. At lower right, the pelvis of the cow pictured above, showing the breaks involving both hip bones. (From Florida Ag. Exp. Sta. Tech. Bull. 262, through the courtesy of R. B. Becker.)

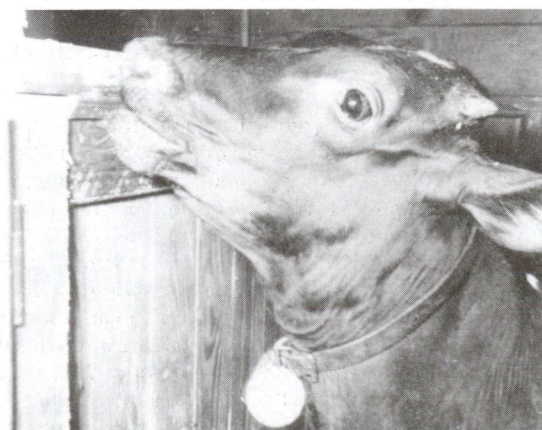

Fig. 24–5. Phosphorus-deficient calf chewing wood, a manifestation of depraved appetite. (Courtesy, Dr. S. E. Smith, Department of Animal Science, Cornell University)

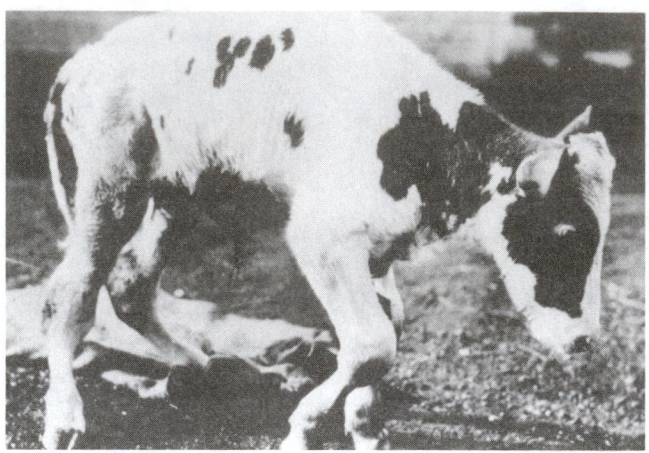

Fig. 24–6.  Calf with severe rickets. Note the emaciation, humping of back, swelling of joints, knuckling of pasterns, and bowing of legs. Rickets may be caused by a lack of either calcium, phosphorus, or vitamin D, or by an incorrect ratio of the two minerals. (From Michigan State University Bull. 150, through the courtesy of The Fertilizer Institute)

## VITAMINS

Dairy cattle, like other animals, require vitamins. Of the known vitamins, only A and D, and perhaps E under certain conditions, are likely to be lacking in the average dairy ration. Vitamin K and the B-complex vitamins are synthesized by the body tissues.

Lack of vitamin A causes a breakdown of the nervous system, skin, and body linings. Calves may be born weak, dead, or partially blind. Vitamin A deficiency is most likely to occur when hays which are bleached or badly weathered, and low in carotene, are fed. Thus, where average or poor quality hay is fed, a rich source of carotene or synthetic vitamin A should be added to the ration. The addition of about 5% dehydrated alfalfa or dehydrated grass to the grain ration will usually suffice.

The recommended vitamin A content of the total ration is as follows: lactating cows, 1,453 IU/lb; dry pregnant cows, 1,816 IU/lb; calf milk replacer, 1,725 IU/lb; and growing heifers, 999 IU/lb.

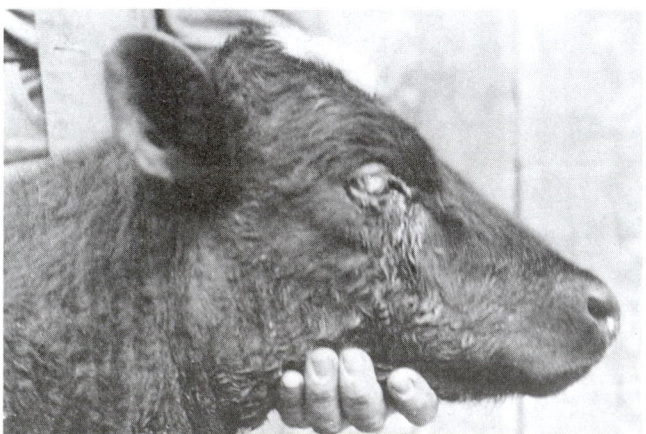

Fig. 24–7.  A three-year-old heifer showing typical advanced eye lesions accompanying a vitamin A deficiency. (From Calif. Ag. Exp. Sta. Bull. 560)

Dairy calves kept indoors may suffer from rickets, a vitamin D deficiency. Because it is difficult to predict the vitamin D content of forage, since it depends on the length of time forage is exposed to the sun's rays while growing and other factors, it is usually well to add vitamin D to most calf rations.

The recommended vitamin D content of the total ration is as follows: lactating cows, 454 IU/lb; dry pregnant cows, 545 IU/lb; calf milk replacer, 272 IU/lb; and growing heifers, 136 IU/lb.

Added vitamin E may be necessary under certain conditions because of its relationship to vitamin A utilization and the prevention of white muscle disease. When supplemental vitamin E is needed, dl-alpha-tocopherol acetate may be added to the dry ration at the following levels: lactating and dry pregnant cows, 7 IU/lb; calf milk replacer, 18 IU/lb; and growing heifers, 11 IU/lb.

## WATER

Large amounts of water are essential if a cow is to produce to her maximum capacity. Cows drink an average of 100 to 200 lb of water per day, with heavy producers drinking up to 300 lb per day. The amount of water a cow will drink depends on her size and milk yield, the temperature and relative humidity of the air, the temperature of the water, and the amount of moisture in her feed.

In extremely cold weather, it is a good idea to have a tank heater to keep the water from freezing. Also, frequency of watering is important. Cows stabled in a stanchion-type barn produce 3½ to 4% more milk if they have drinking cups available than if they are watered twice daily. Contrary to some opinions, cows do not produce more milk from softened than from normal hard water.

## OTHER FEED REQUIREMENTS

In addition to being nutritionally complete, the following factors should receive consideration in dairy cattle rations:

1. **Palatability.** If they don't eat it, they won't produce. Hence, palatability is important. The feeder should avoid mature, moldy, musty, coarse, or weed infested hay; finely ground hay or grain; and silages which are moldy, slimy, or too mature.

2. **Preparation.** Grain for dairy cows should either be steam rolled or medium ground. Calves under six months of age can be fed whole corn.

Pelleted concentrates are more compact and less dusty than ground grains. Also, cows will consume pelleted feeds faster than ground or flaked feeds. Cows fed pelleted grain produce slightly more milk, with a slightly lower fat content, than those fed unpelleted grain.

Cows produce as well on long hay as on chopped or ground hay. Finely ground, pelleted hay affects the amount and proportion of volatile fatty acids in the ruminant, with the result that the percentage fat content of the milk is lowered. Wafering, on the other hand, has no effect on fat percentage. Both pelleting and wafering lessen the transportation, storage, and handling charges on hay in comparison with long hay. (Also see Chapter 4, section entitled, "Feed Processing," including Table 4–10 therein; hence, the reader is referred thereto.)

3. **Variety.** Some variety in the ration is desirable, but palatability and nutritive content of individual feeds are more important than the number of ingredients. Cows crave some dry forage in their rations.

4. **Bulk.** Some bulk in the feed is desirable. However, the proportion of grain and roughage should be determined largely by comparative price.

5. **Laxativeness.** Cows that receive average amounts of legume hay and/or silage seldom become constipated. But grass hay or straw may cause some trouble. Constipation can be corrected by feeding such feeds as alfalfa, wheat bran, linseed meal, or molasses.

6. **Cost.** Cost is important. But even more important is net returns; hence, it may well be said that it's net returns rather than cost per ton, or per bag, that counts.

## FEEDS FOR DAIRY COWS

For convenience, the commonly used dairy cattle feeds are herewith classified as (1) roughages, (2) concentrates, and (3) special feeds.

## ROUGHAGES

Cows can produce up to 70% of their ability when fed good quality roughage alone. But with high-producing cows, a greater percentage of the total feed intake must be in the form of concentrates. Even so, large amounts of high-quality roughage should be the basis for feeding on most dairy farms.

In using any kind of roughage, three important points should be kept in mind: (1) To obtain the most nutrients from roughage, it must be of good quality; (2) the better the roughage, the smaller the requirement for grains; (3) the cow is, by nature, a good consumer of forage.

A high-quality roughage is one that possesses the physical and chemical characteristics commonly associated with palatability and an abundance of feed nutrients. The most important physical factors of quality that may be gauged in a practical way are (1) stage of maturity when cut, (2) percentage of leaves, (3) green color, (4) amount of heating, spoilage, etc., (5) pliability of stems, (6) aroma, and (7) freedom from foreign material.

1. **Hay.** Legumes make the best hay for dairy cattle. For best quality, alfalfa should be cut before one-tenth bloom. Table 24-3 data (from experiments conducted by the USDA) show the relationship of maturity to nutritive value. This table shows that protein decreases as hay matures, and that cows produce considerably less milk when fed very mature hay.

Fig. 24-8. High quality alfalfa hay is relished by lactating cows. (Courtesy, Holstein-Friesian Assn. of America, Brattleboro, VT)

Producers can afford to pay a higher price for good hay. Calculations show that if hay cut at full bloom is worth $80.00 per ton, hay cut at $1/10$ bloom is worth $140.00 per ton.

2. **Silage.** The feeding value of silage is no better than the material put in the silo. Good silage is easy to make if the crop is cut at the proper stage of growth, cut fine and ensiled as rapidly as possible at 55 to 70% moisture, evenly and well-packed, and stored in a silo free from air leaks.

In regions where rainy weather frequently makes haymaking difficult, it is wise to harvest forage as silage. This especially applies to the first cutting of grasses and/or legumes, which may normally be used as hay.

Usually, it is possible to produce more milk per acre of land when the crops are harvested as silage rather than as hay. Also, silage helps alleviate the week to week variation that often goes with pastures or green chop.

Many forages can be, and are, successfully used as silage for dairy cows, including corn silage, grass silage, oat silage, sorghum silage, and many others.

If *carefully mixed,* urea can be added to corn silage, at the rate of 10 lb of urea per ton of silage. Corn silage of 70% moisture content contains about 2.5% protein, or 7.5% protein on air-dry (90% dry matter) basis. The addition of 10 lb of urea per ton of corn silage will increase the protein content from 8 to 12% on a dry matter basis.

3. **Haylage (low-moisture silage).** Haylage is made from grass and/or legume that is wilted to 40 to 50% moisture content before ensiling. Properly made haylage is good smelling, palatable, high-quality feed. Cows usually receive more dry matter and net feed value in haylage than in silage made from the same cut.

Haylage is easy to prepare and preserve in a gas-tight silo where air is excluded. But it can be made in a conventional silo provided certain precautions are taken—precautions designed to keep out the air.

Haylage is growing in popularity. Its feed value depends on the stage of the growth of the crop when filled and the percentage of dry matter in the silage.

## TABLE 24-3
### ALFALFA MATURITY AND NUTRITIVE VALUE[1]

|  | Initial Bloom | Half Bloom | Full Bloom |
|---|---|---|---|
|  | (lb per acre) | | |
| Dry matter | 7,896 | 7,778 | 6,061 |
| Total digestible nutrients | 4,660 | 4,413 | 3,269 |
| Digestible protein | 1,106 | 1,016 | 722 |
| Production of 4% fat-corrected milk | 6,330 | 5,254 | 3,970 |

[1]USDA data.

4. **Green chop (soilage).** Many dairy producers harvest and feed green chop daily. It reduces wastage as compared to pasture. With tall-growing crops, 50% more feed value may be realized from a given area. However, green chop requires special equipment and harvesting every day. Also, there are harvesting problems in wet weather, and there is inevitable change in feed quality as the season progresses.

Fig. 24–9. Harvesting alfalfa as green chop for more than 4,000 lactating cows at Arizona Dairy Co., Higley, Arizona. (Courtesy, James Tappan)

5. **Pasture.** Good pastures provide a highly nutritious feed that milk cows relish, especially when fresh growth is maintained; and the cows do their own harvesting. However, pasturing results in considerable waste of the crop through trampling, with lower yield of nutrients per acre than from harvested crops. Also, it is sometimes difficult to maintain growth of even feed value because of variable weather conditions. This results in fluctuation in milk production. Additionally, in areas of high-priced land and where grain crops yield well, pastures must be exceedingly well managed to compete with other crops. For the latter reason, especially in the more fertile areas, dairy producers are shifting from pasture to continuous feeding of stored feed.

In recent years, the old system of continuous grazing on one field all season has been largely abandoned. The new systems of pasture management include rotation grazing, strip grazing (or daily rotation grazing), and zero grazing.

## CONCENTRATES

Concentrate feeds are those which are high in energy and low in fiber. Many different kinds of concentrate feeds can be, and are, used in dairy cattle feeding. They are usually classed according to total crude protein content as (1) low-protein, (2) medium-protein, and (3) high-protein feeds. The chemical analysis of feeds can be obtained from feed composition tables (such as those shown in *Feeds & Nutrition,* by M. E. Ensminger, et al, Ensminger Publishing Company, 1990; or from *Nutrient Requirements of Dairy Cattle,* sixth revised edition, update 1989, NRC National Academy Press).

Three factors besides chemical composition are important in evaluating concentrates for milk cows—palatability, quality of milk produced, and cost. The most infallible way in which

to appraise the first two factors is through actual feeding trials. Consideration of the third factor necessitates that producers be keen students of values. They must change the formulations of their rations in keeping with comparative feed prices.

## SPECIAL FEEDS AND FEED ADDITIVES FOR DAIRY CATTLE

Certain feeds and additives are especially adapted to dairy cattle, primarily to increase milk production and/or to affect milk composition; among such producers are (1) fats and oils, (2) fiber, (3) urea, and (4) the following additives: antibiotics, bovine somatotropin (BST), buffers, ionophores, isoacid (branched-chain fatty acids and valeric acid), and thyroprotein.

1. **Fats and oils.** Fat serves the following functions when added to dairy rations: it (a) increases the caloric density of the ration without lowering the forage (fiber) content; (b) controls dust; (c) lessens the wear and tear on feed mixing equipment; (d) facilitates pelleting of feeds; (e) increases palatability; (f) helps to homogenize and stabilize certain feed additives, especially those of a very fine particle size; and (g) increases the total amount of milk, butterfat, and SNF, but results in a slight decrease in the percentage of both butterfat and SNF.

Most forages and grains are low in lipids—they contain less than 2 to 3% fat. In general, dairy cows should be able to utilize 1 to 1½ lb of fat per day in addition to the fat present in natural feedstuffs. This means that about 3% more fat can be added to the total ration (forage plus concentrate), or that 5 to 6% fat can be added to the grain ration.

Added fat is especially effective in early lactation. Because of the increased caloric concentration provided by dietary fat and because high-producing cows are usually in negative energy balance during early lactation, fat is frequently added to the ration to increase the cow's energy intake and provide fatty acids to the udder. It may also be beneficial to provide supplemental fat when the capacity of the gastrointestinal tract limits energy intake.

An important consideration in the successful feeding of fats is that they may be used to provide increased energy without lowering the forage (fiber) intake. In early lactation, substituting fat for a portion of the starch obtained in cereal grains in the ration of cows is a way in which to maintain high energy concentrations and high fiber intakes. The substitution of fat for grain alleviates the low milk fat syndrome caused by inadequate fiber and excessive grain. A ration containing a high proportion of forage helps maintain normal rumen function and provides an environment in which fat is less inhibitory to rumen fermentation and nutrient digestion.

The type of fat (saturated or unsaturated) added to the ration greatly influences the animal's nutrient utilization, milk production, feeding behavior, ration acceptability, the amount of fat that can be fed, and milk composition. Unsaturated fats are less desirable for dairy cows because of their inhibitory effects on rumen fermentation and digestion. Animal fats (which are more saturated) and blended animal-vegetable fats have generally given the most positive responses in animal performance.

Because vegetable oils are high in unsaturated fat, they are less satisfactory than saturated fats as ration supplements. Whole seeds, such as cottonseed, soybeans, and sunflower seeds, have been used successfully, but the added fat derived from them should not exceed 1.0 lb/cow/day. Unsaturated fats that contain high levels of oleic acid apparently exceed the hydrogenation ability of the rumen microorganisms and result in the greatest milk fat depression. An increase in long-chain fatty acids in the ration (a) increases the secretion of milk and (b) inhibits the synthesis of short- and medium-chain fatty acids in mammary tissue.

Fig. 24–10. A complete (all-in-one) ration containing whole cottonseed, being distributed by a self-unloading truck, at Arizona Dairy Co., Higley, Arizona. Note the white cottonseeds in the feed. (Courtesy, Arizona Dairy Co., Higley, AZ)

Until the rumen becomes functional, young dairy calves require some fat in the diet. A level of 10% fat in milk replacers appears to be sufficient to supply essential fatty acids and carry fat-soluble vitamins, but insufficient to supply adequate energy for normal gains under optimum environmental temperatures. For veal production, a higher fat milk replacer (15 to 20% or more), will increase fat deposition in the carcass and is desirable. Also, 15 to 20% fat in milk replacers is needed for normal gains when calves are exposed in cold environmental temperatures.

(Also see Chapter 4 of this book, section on "Fats and Oils.")

2. **Fiber.** Fiber is important in dairy rations. Excessive fiber levels limit intake and energy concentrations, while a shortage of fiber reduces rumen digestibility and milk fat test.

The amount of fiber to include in the ration of dairy cattle is influenced by the body condition and level of production of the animal, the type of fiber, the particle size, the amount of total DM consumed and its bulk density, the buffering capacity of the forage, the frequency of feeding, and the economics. Lactating cows that are fed to produce large amounts of milk, or young animals that are fed to achieve rapid growth, should receive more energy and less fiber than lower producing animals. Forages that are finely ground (processed to small particle size) are rapidly consumed and fermented in the rumen, which reduces (a) the animal's chewing time, (b) ruminal fluid, (c) the acetate-to-propionate ratio in ruminal fluid. The result of these effects is a depression in milk fat percentage. Chopped alfalfa should average about ¼ in. in length to maintain a normal milk fat percentage. Feed factors, such as small particle size of forages, that reduce the pH of ruminal fluid, decrease the number and activity of fiber-degrading bacteria and cause a depression in fiber degradation. Feeding an insufficient amount of fiber or feeding forages that have a poor buffering capacity in the rumen may have undesirable effects on rumen

fermentation, fiber degradation, and milk fat percentage that are similar to those caused by reducing the particle size of the forage.

So, the general recommendation is that lactating dairy cows should receive at least one-third of the total ration dry matter as long hay or as its DM equivalent in medium-to-coarse chopped silage or other forage. A minimum of 5 lb of forage dry matter measuring 1 to 2 in. in length will meet the fiber need of most lactating cows.

The values for NDF and ADF are more accurate measures of the fiber component of feeds than are values for crude fiber. Yet, because both chemical and physical properties of feeds are involved in determining fiber quality and the energy value of feeds, there is currently no one fiber analysis that can accurately predict fiber quality and energy values for all feeds. NDF content is negatively correlated with dry matter intake and apparent digestibility of forages, but it is positively correlated with chewing time. ADF is more negatively correlated with apparent digestibility than is NDF. NDF and bulk density are positively related, which may explain the negative relationship between dry matter intake and the NDF content of the ration. According to University of Wisconsin researchers, NDF is a better predictor than ADF of dairy cow feed intake and milk production.

The optimum amount of NDF and ADF to include in the ration varies with the level of milk production and the type of forage that is fed to dairy cattle. A minimum of 21% of ADF and 28% of NDF is recommended for cows during the first 3 weeks of lactation. During times of high milk production, however, ADF and NDF contents of the ration are usually reduced to 19 and 25%, respectively, so that adequate dietary energy can be included to meet the cow's requirement. The ADF and NDF contents of the ration should be increased in later lactation to help prevent milk fat depression and because less energy is required for milk production. Seventy-five percent of the NDF in the ration should be supplied as forage.

3. **Urea.** Due to the increasing shortage and higher price of oilseed proteins, more and more urea is being fed to dairy cows. Urea is manufactured synthetically from the nitrogen of the air. When properly used, it is a safe, low-cost source of protein for dairy cattle; when improperly used, it becomes a hazard. It is recommended that urea in a concentrate mixture not exceed 2%, and that it not exceed 1% of the total hay and grain ration. It is further recommended that not more than one-third of the protein requirements be met through urea. Under experimental conditions, much higher quantities of urea have been fed to dairy cows. For example, beginning in 1958, Dr. A. I. Virtanen, the Finnish Nobel Prize winner, successfully fed dairy cows on a protein-free diet, utilizing small amounts of ammonium salts combined with urea as practically the sole source of nitrogen.[1]

Yet, it is emphasized that this was done under carefully controlled experimental conditions. Further, most state laws limit the amount of urea that can be put in commercial feeds.

One pound of 45% nitrogen (281% protein) urea provides as much protein value as 6.8 lb of 41% soybean meal or cottonseed meal (281 ÷ 41 = 6.8). However, urea does not provide energy, minerals, or vitamins, with the result that these must be provided through other sources when urea is substituted for protein meals.

---

[1]*Science,* Vol. 153, No. 3744, Sept. 30, 1966, pp. 1603–1614.

When added to the dairy cattle ration, urea must be mixed thoroughly to ensure even distribution, and used according to directions. It may be toxic when improperly used, or when fed in too large amounts.

Normally, dairy producers limit urea to 1.5 to 2.0% of the concentrate ration. Higher levels (up to 2.75% of the concentrate) are unpalatable and depress appetite. However, the unpalatability may be alleviated by pelleting the urea with alfalfa.

(Also, see Chapter 4 of this book, section on "Urea.")

4. **Feed additives.** Many additives are used by dairy producers to increase milk production, affect milk composition, and/or improve feed efficiency; and new products are constantly evolving. Among such additives are the following: antibiotics, bovine somatotropin (BST), buffers, ionophores, isoacids, (branched-chain fatty acids and valeric acid), and thyroprotein.

- **Antibiotics**—Antibiotics, which are widely used in the diet of young dairy calves, are especially beneficial for calves exposed to adverse conditions of housing, sanitation, and disease. However, they should not be used as a substitute for good management and a clean, sanitary environment. The greatest benefits from feeding antibiotics accrue when calves are started on the antibiotic as soon as possible after birth, then continued on it during the milk or milk replacer feeding period. Antibiotics are mainly effective in increasing feed intake and growth, along with preventing diarrhea. Generally, antibiotics are fed at the following concentrations: In the milk replacer (dry basis), or in an equivalent amount of whole milk, 20 to 40 ppm; and in the starter ration, 10 to 20 ppm. For the prevention and control of disease, higher concentrations may be necessary: 50 to 100 ppm in the milk replacer, and 25 to 50 ppm in the starter.

Some studies have shown a slight increase in milk production when low levels of antibiotics are fed to lactating cows. However, the practice is not recommended because of (1) the presence of residual antibiotic in milk, and (2) the possibility of drug resistance in the animal.

Those using antibiotics should always read and follow the label directions on any antibiotic container before slaughtering animals or selling milk from cows treated with antibiotics.

(Also, see Chapter 4 of this book, section on "Additives, Implants, and Injections.")

- **Bovine Somatotropin (BST)**—Experimentally, milk, lactose, milk fat, and protein yields have been increased significantly when exogenous bovine somatotropin (BST) has been injected into lactating cows. To meet this higher production, cows consume more total feed when BST is administered. However, the efficiency of milk production by cows is improved because a smaller proportion of the ingested nutrients in feed is needed to take care of maintenance requirements.

(Also, see Chapter 4 of this book, section on "Additives, Implants, and Injections.")

- **Buffers (Mineral Salts)**—Buffers are used primarily to improve the feed intake, rumen function, milk production, milk composition, and health of lactating cows. When used for young calves, buffers have given inconsistent results; they will likely be most beneficial when the calf diet being fed results in higher than normal acidity.

The common buffers are: sodium bicarbonate ($NaHCO_3$) magnesium oxide (MgO), sodium bentonite, sodium sesqui-

carbonate, and calcium carbonate or limestone ($CaCO_3$). Buffers function to maintain the hydrogen ion concentration in the rumen, intestines, tissues, and body fluids, or to increase the rate of passage of liquids from the rumen, or both.

Buffers are of greatest benefit to cows in the following situations: (a) during early lactation; (b) when large amounts of rapidly fermentable carbohydrates are fed, especially when they are fed at infrequent intervals; (c) when fermented forage, primarily corn silage, is the major or only forage in the ration; (d) when the concentrate and forage are fed separately; (e) when the particle size of the total ration dry matter has been reduced by chopping, grinding, or pelleting to the extent that it increases the rate of ruminal fermentation and depresses salivary secretion and buffering capacity; (f) when cows are abruptly switched from high-forage to high-concentrate rations, especially during early lactation; (g) when the animal's milk fat content is low; or (h) when off-feed problems resulting from feeding rapidly fermentable feeds are encountered.

(Also, see Chapter 4 of this book, section on "Additives, Implants, and Injections.")

- **Ionophores**—Ionophores are feed additives that change the metabolism within the rumen by altering the rumen microflora to favor propionic acid production. Currently, two ionophores—Bovatec (lasalocid) and Rumensin (monensin)—are FDA-approved for replacement heifers. Both are antibiotics. Feeding Bovatec or Rumensin to replacement heifers improves liveweight gains and the efficiency of feed utilization.

(Also, see Chapter 4 of this book, section on "Additives, Implants, and Injections.")

- **Isoacids (Branched-Chain Fatty Acids and Valeric Acid)**— The isoacids provide three branch-chain fatty acids (isobutyric acid, isovaleric acid, and 2-methyl butyric acid)—the same fatty acids that are made by ruminant bacteria and are present naturally in the rumen of cattle. Isoacids are essential for the growth of some rumen organisms that digest fiber. The use of isoacids may boost milk production by 8 to 10%, or 4 to 6 lb per day, with little or no increase in feed consumption. The mode of action of isoacids is not entirely clear, but it appears to be due to enhancing fiber digestion and acetate production without stimulating insulin secretion. Not all dairy cattle benefit from the use of isoacids. Moreover, the benefits of a profitable response are delayed 30 to 60 days following the initiation of isoacids.

(Also, see Chapter 4 of this book, section on "Additives, Implants, and Injections.")

- **Thyroprotein (Iodinated Casein/Hormone)**—Feeding materials with hormonal activity to lactating cows for the purpose of increasing milk production has fascinated scientists for many years, beginning with thyroprotein.

Cows that are given thyroprotein must receive additional feed to produce extra milk. Not all cows will respond to thyroprotein. Because of the problems involved in feeding the product, it is not widely used. Also, since thyroprotein is classified as a drug, milk and butterfat production records of cows fed thyroprotein are not accepted under DHIA (Dairy Herd Improvement Association) rules.

## COMMERCIAL DAIRY FEEDS

Commercial dairy feeds are just what the name implies—feeds mixed by commercial manufacturers who specialize in the business, rather than feeds that are farm mixed. In 1988, 103.1 million tons of primary feeds (complete feeds) were manufactured in the United States, of which 16.9% was fed to dairy cattle.

Several different types of commercial feeds are available for dairy cattle; among them, (1) complete dairy concentrates, (2) dry cow rations, (3) fitting rations, (4) growing or young stock rations, (5) calf starters, (6) milk replacer feeds, and (7) protein supplements.

Enlightened producers will know how to determine what constitutes the best in commercial feeds for their specific needs. They will not rely solely on how the feed looks and smells.

(Also, see Chapter 4 of this book, section on "How to Select a Commercial Feed.")

## RATIONS

Dairy producers must put together the available feeds so as to achieve the most profitable production. At its best, developing a dairy ration involves combining the art and the science of feeding. For small herds, individual animal response may be satisfactory. With large commercial herds, formulating rations must be more precise, because small costs per cow become large costs when multiplied by many cows. Yet, the most sophisticated computer must be augmented by the good judgment of the manager if the rations are to be successful in meeting the nutrient needs of individual cows and of the herd as a whole. Producers must always keep in mind that the best formula on paper is not always the best feed. A feed is of no value if it is not actually consumed.

Also, there should be a specific ration for every need—for lactating cows, dry cows, calves, replacement heifers, dairy beef, and show or sale animals.

Fig. 24–11.   Acres of dairy cattle at Arizona Dairy Co., Higley, Arizona, where more than 4,000 lactating cows plus young stock are fed and managed. Each year, U.S. dairy herd size, mechanization, and corral (group) feeding increase. (Courtesy, James Tappan, Arizona Dairy Co.)

## FEEDING LACTATING COWS

Few animal stresses are as great as those involved in the production of a large volume of milk. For each gallon of milk produced, 400 to 500 gal of blood must pass through the udder. Thus, if a cow is producing 10 gal (86 lb) of milk daily, 15 to 20 tons of blood course through the udder each 24 hours. This 10 gal of milk contains more than 3 lb of fat, more than 3 lb of protein, more than 4 lb of lactose (milk sugar), and more than ½ lb of minerals. All these must be supplied in the ration over and above the nutrients needed for the body processes, wastes, and energy to sustain the whole operation.

Fig. 24–12.   Production of a large volume of milk is very stressful. The stress of weather is alleviated by an environmentally controlled barn such as this one. (Courtesy, The American Jersey Cattle Club, Reynoldsburg, OH)

Also, producers realize greatest profits from feeding when cows convert the maximum proportion of their feed into milk. The nutrient requirements for production depend primarily on the amount and composition of the milk.

Additional considerations in feeding lactating cows include palatability of the ration; physical form, protein and mineral content of concentrates; proportion of concentrate to roughage; relative prices of ingredients; voluntary feed intake; and frequency and regularity of feeding. Thus, the proper feeding of lactating cows necessitates that producers have sufficient knowledge relative to basic nutrient requirements and principles to plan an efficient feeding program, and the experience and management ability to apply it.

Dry matter consumption is very important in feeding dairy cows. The best ration formulation on paper will not make for profitable production if the cows either fail to eat it or are given insufficient amounts of it. Also, high-producing cows must consume very large amounts of a balanced ration if they are to produce to their maximum.

### THUMB RULES FOR FEEDING LACTATING COWS

The feed requirements of lactating cows are significantly influenced by the volume and composition of the milk that they produce. Although knowledge of the nutrient requirements of the animals and of the composition of feeds is essential in order to feed properly, the ability of the cows to consume sufficient volume of the feed complicates adequate feeding. Table 24–4 may be used as a guide for dry matter intake; and the two sections that follow give some thumb rules relative to the amount and kind of forage and the amount and kind of concentrate to feed.

### TABLE 24–4
### DAILY DRY MATTER INTAKE GUIDELINES [1]

| Live Wt.: (lb) / *(kg)* | 900 / *409* | 1,100 / *499* | 1,200 / *545* | 1,300 / *590* | 1,500 / *681* |
|---|---|---|---|---|---|
| **Milk** [2] | **Percent of Body Weight** [3] | | | | |
| (lb/day) *(kg/day)* | (%) | (%) | (%) | (%) | (%) |
| 20 *9.1* | 2.6 | 2.3 | 2.2 | 2.1 | 2.0 |
| 30 *13.6* | 3.0 | 2.7 | 2.6 | 2.5 | 2.3 |
| 40 *18.2* | 3.4 | 3.1 | 2.9 | 2.8 | 2.5 |
| 50 *22.7* | 3.8 | 3.4 | 3.2 | 3.1 | 2.8 |
| 60 *27.2* | 4.1 | 3.7 | 3.5 | 3.4 | 3.1 |
| 70 *31.8* | 4.6 | 4.0 | 3.8 | 3.6 | 3.3 |
| 80 *36.3* | 5.1 | 4.3 | 4.1 | 3.8 | 3.5 |
| 90 *40.9* | | 4.7 | 4.4 | 4.1 | 3.7 |
| 100 *45.4* | | 5.0 | 4.7 | 4.4 | 3.9 |

[1]Adapted by the author from: Linn, J. G., M. F. Hutjens, W. T. Howard, L. H. Kilmer, and D. E. Otterby, *Feeding the Dairy Herd*, Cooperative Extension Services, Universities of Illinois, Iowa State, Minnesota, and Wisconsin, 1988, p.27, Table 19.

[2]Fat-corrected milk = (milk lb × .4) + (fat lb × 15).

[3]Intakes may be up to 18% less for cows in early lactation.

## AMOUNT AND KIND OF FORAGE TO FEED

The common thumb rules for forage feeding of lactating cows follow.

1. **Forage dry matter and intake.** The forage should constitute a minimum of 40% of the total dry matter of the ration and account for an intake of approximately 1.5% of the body weight daily.

2. **Acid detergent fiber (ADF).** The ADF should constitute 19% of the ration dry matter, increased to 21% during the first 3 weeks of lactation.

3. **Neutral detergent fiber (NDF).** The NDF should constitute 25% of the ration dry matter, increased to 28% during the first 3 weeks of lactation.

4. **Hay consumption.** If good quality hay only is fed, a cow will eat about 3 lb per 100 lb of body weight.

5. **Silage.** Depending on the moisture content, 2.5 to 4.5 lb of silage are equal to (and may replace) 1 lb of hay; the lower feeding value of silage is due to its high moisture content—hay runs 10 to 15% moisture, whereas silage runs 65 to 75% moisture.

6. **Hay/grain equivalent.** It takes about 3 lb of good hay to supply the same amount of usable energy as 2 lb of grain.

7. **Pasture (grass) consumption.** Cows will consume 100 to 200 lb of pasture per day; since pasture normally contains 70 to 85% moisture, that's 15 to 60 lb of dry matter per day.

8. **Yearly hay consumption.** Except for cows fed high grain rations, it takes 5 to 6 tons of hay (or an equivalent amount in dry matter from pasture or silage) to feed 1 cow for 1 year.

9. **Forage:concentrate ratio.** If forage is very high in quality, cows will eat more of it, with the result that the grain requirement will be lessened. However, over and above meeting the minimum requirement, the proportion of forage to concentrate should be determined primarily by the economics of the situation—that is, it should be decided on the basis of the relative price of available forage and concentrate, the milk production, and the net returns.

## AMOUNT AND KIND OF CONCENTRATE (GRAIN) TO FEED

The common thumb rules for concentrate feeding of dairy cows follow.

1. **Amount of concentrate (grain).** The concentrate (grain) should constitute a maximum of 60% of the total dry matter of the ration and account for an intake of not to exceed 2.3% of the body weight daily. Table 24–5 can be used as a guide for feeding concentrate (grain) according to milk production.

### TABLE 24–5
### AMOUNT OF CONCENTRATE (GRAIN) TO FEED BY PERIODS (1,400 LB *[636 KG]* COW, 4% MILK) [1]

| | Milk Production Ability of the Cow [2] | | | |
|---|---|---|---|---|
| Average Daily 1st Period .. (lb) | 50 | 60 | 80 | 90–100 |
| *Average Daily 1st Period . (kg)* | *23* | *27* | *36* | *41–45* |
| Lactation Total .......... (lb) | 10,000 | 12,000 | 15,000 | 18,000 |
| *Lactation Total ......... (kg)* | *4,540* | *5,448* | *6,810* | *8,172* |
| **Phase of Lactation** | **Grain to Milk Ratio** | | | |
| 1 (1st 10 weeks) ............... | 1:4 | 1:3 | 1:3 | 1:2.5 |
| 2 (2nd 10 weeks) .............. | 1:4 | 1:3 | 1:3 | 1:3 |
| 3 (last 24 weeks) ............... | 1:4 | 1:4 | 1:2.5 | 1:2.5 |
| | Daily | Daily | Daily | Daily |
| 4 (dry, 6–8 weeks) ........... (lb) | 0–4 | 0–4 | 0–4 | 0–6 |
| *(dry, 6–8 weeks) ........... (kg)* | *0–1.8* | *0–1.8* | *0–1.8* | *0–2.7* |
| Total grain (approximate) ....... (lb) | 3,000 | 4,000 | 5,000 | 6,000 |
| *Total grain (approximate) ....... (kg)* | *1,362* | *1,816* | *2,270* | *2,724* |

[1]Adapted by the author from: Linn, J. G., M. F. Hutjens, W. T. Howard, L. H. Kilmer, and D. E. Otterby, *Feeding the Dairy Herd*, Cooperative Extension Services, Universities of Illinois, Iowa State, Minnesota, and Wisconsin, 1988, p.26, Table 18.

[2]Ratios based on 100% dry matter basis, grain containing 80 Mcal, and forage 60 Mcal of $NE_{lc}$ per 100 lb *(45 kg)*.

2. **Amount and kind of protein.** Feed protein according to requirements (19% in early lactation, decreased thereafter according to milk production). A low rumen degradable protein source is recommended for high-producing cows in early lactation. Limit urea to 0.4 lb per day, and preferably to 0.2 lb per day, in phases 1 and 2.

3. **Added fat.** In addition to the fat present in natural feedstuffs, lactating cows may be fed 1 to 1½ lb of *added fat* per day; which translates into about 6% added fat to the concentrate (grain) ration, or 3% added fat to the total mixed ration (grain and forage combined). Fats in oilseeds (soybeans or whole cottonseed) should be considered as added fat. When feeding added fat, increase the calcium to 0.9 to 1.0%, the magnesium to 0.3%, and the acid detergent fiber to 20%.

4. **Salt.** Include 1% salt in the concentrate (grain) mix, or 0.5% salt in the total ration (concentrate and forage combined); which will provide for a salt intake of 2 to 3 oz per cow per day.

5. **Calcium/phosphorus and trace minerals.** A calcium/phosphorus mineral source should constitute 1 to 2% of the grain mix, or be fed at a rate of 1 oz per 10 lb of milk. Trace minerals should be incorporated in the ration or self-fed in trace mineralized salt to meet the requirements.

6. **Vitamins.** Vitamins A, D, and E should be added to the ration to meet the requirements.

Any thumb rules, and even calculated values (calculated by hand or computer), are estimates to be used with some judgment by the feeder. The rule of the successful feeder is to increase concentrates so long as cows respond with extra milk at a profit. The commonly used profit indicator is the "milk-feed price ratio," which is the pounds of 16% protein dairy concentrate equal in value to 1 lb of milk.

Generally the cost of 16% protein dairy concentrate per pound is about 70% of the price received for milk; of course there are yearly, seasonal, and area variations. For the United States as a whole, the milk-feed price ratio was $1.31 in 1975 and $1.79 in 1986. Thus, the cost per pound of 16% protein dairy concentrate was 76% the price received per pound of milk in 1975, and 56% in 1986.

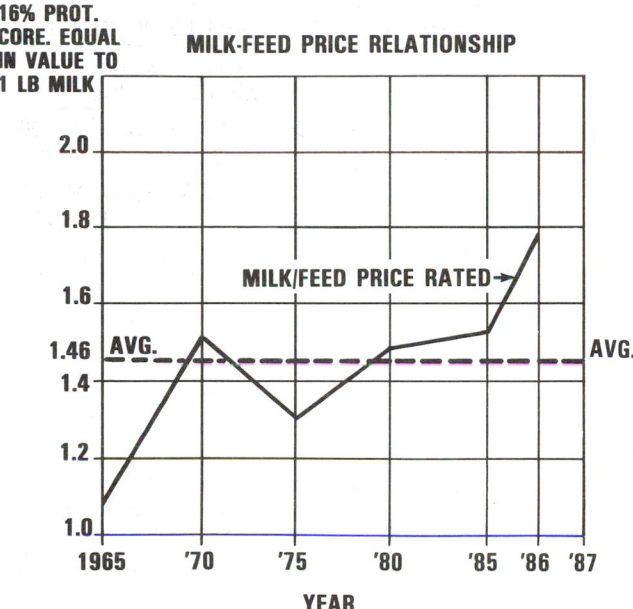

Fig. 24–13. Pounds of 16% protein dairy concentrate equal in value to 1 lb of milk. (Adapted by the author from USDA sources)

## RATIONS FOR LACTATING COWS

Forages constitute the primary basis of most dairy rations, with alfalfa hay as the preferred roughage. Frequently, corn silage is used to replace part of the hay, but, regardless of the kind of forage successful dairy producers balance the complete ration (forages and concentrates combined) so that all nutri-

ents—energy, protein, minerals, and vitamins—are fed in sufficient amounts to meet the needs of the cow. For maximum intake, the complete ration must contain over 55% dry matter, except when pasture or green chop forage is fed.

Grain provides energy in concentrated form; for example, 5 lb of barley contains as much energy as 8 lb of hay or 25 lb of silage.

Cows fed an all-forage ration produce about half as much milk as cows fed concentrates in average amounts (see Fig. 24–14). This is because a cow's stomach simply isn't big enough to hold all the forage necessary to get the amount of energy needed. To provide the needed energy, grain must be added. Grain provides energy in concentrated form.

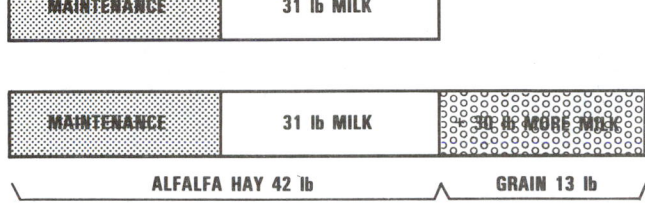

Fig. 24–14. A mature Holstein cow can consume sufficient good-quality alfalfa hay to meet the energy required for maintenance plus about 31 lb of milk daily. An additional 13 lb of grain daily will allow the cow to produce 30 lb more milk. Thus, the milk:grain ratio for the additional gain is 2:3.

Alfalfa hay of good quality will supply much of the protein required by lactating dairy cows. Table 24–6 shows that the percentage total protein needed in the grain ration depends upon the type of roughage fed. As shown, with high quality legume hay, a 10% protein grain mix will suffice. With a low quality grass forage, however, the protein level of the grain mix should be much higher—up to 18 to 22%.

### TABLE 24–6
### PERCENT PROTEIN NEEDED IN GRAIN MIX

| Forage Fed | Protein Needed in Grain Mix |
|---|---|
| | (%) |
| **All legume** | 10–16 |
| **Mixed** (part legume and part grass) | 14–20 |
| **All grass** | 18–22 |

When fed an all-hay ration, cows can sustain a production of 30 to 40 lb of milk daily, in addition to maintaining the other vital functions. The higher the quality of forage, the greater the amount that will be consumed, and the higher its percent digestibility. Cows can consume about 3% of their body weight in dry hay, which means that a 1,400-lb cow can consume more than 40 lb daily if hay is her only feed. As other feeds are introduced into the ration, hay consumption on a free-choice basis decreases.

Corn silage or hay-crop silage may replace hay at the rate of 2½ to 3 lb for each 1 lb of hay, based on dry matter and nutrient content. In tests with lactating cows, this was also the rate to which they voluntarily adapted when using various combinations of hay and silage. A daily silage allowance of 3 to 4½% of body weight can be fed to a lactating cow without interfering with the total nutrient intake. This means that a 1,400-lb cow can be fed 40 to 60 lb of silage daily. Experience has shown that above these levels, total nutrient intake is decreased regardless of the types of feeds used. This means that corn silage can replace about ⅓ of the hay when it is economical to make such a substitution.

Well-matured corn silage has about 30 to 45% of its weight in grain (dry basis); hence, on a moisture free basis it is higher in energy than the hay that it replaces, but lower in protein.

Table 24–7 shows what happens to the DM, energy, protein, and minerals contributed by the forage component of the ration when corn silage replaces about ⅓ of the alfalfa hay. Note that the DM and energy remain about the same, but that there is a decrease in the total protein, calcium, and phosphorus. Note, too, that there is little difference in energy values whether TDN or $NE_{lc}$ are used as the measure, and that the Ca:P ratio decreases from 7.5:1 with the hay alone to 5.3:1 for the combination of hay and corn silage.

### TABLE 24–7
### NUTRIENTS SUPPLIED BY ALFALFA HAY AND CORN SILAGE (AS-FED BASIS)[1]

| | DM | Energy | | | | Total Protein | | Calcium | Phos-phorus |
|---|---|---|---|---|---|---|---|---|---|
| | | TDN | | $NE_{lc}$ | | | | | |
| | | (lb) | (kg) | (Mcal/lb) | (Mcal/kg) | (lb) | (kg) | (g) | (g) |
| **Forage A:** | | | | | | | | | |
| Alfalfa hay (40 lb)[2] | 36.4 | 20.8 | 9.44 | 22.0 | 48.5 | 7.2 | 3.27 | 269 | 36 |
| **Forage B:** | | | | | | | | | |
| Alfalfa hay (25 lb)[2] | 22.7 | 13.0 | 5.90 | 13.7 | 30.2 | 4.5 | 2.04 | 168 | 23 |
| Corn silage (40 lb)[3] | 12.0 | 8.8 | 4.00 | 8.0 | 17.6 | 1.0 | 0.45 | 4.0 | 9.6 |
| Total forage | 34.7 | 21.8 | 9.90 | 21.7 | 47.8 | 5.5 | 2.49 | 172 | 32.6 |

[1]Values from Section V, Composition of Feeds of *Feeds & Nutrition.*

[2]Early bloom hay.

[3]Mature corn silage.

The concentrate ration needed to supplement the available roughage on the dairy farm may either be home mixed or commercially manufactured. Home-mixing involves the mixing of homegrown grains and such purchased feeds as necessary to balance the ration. Commercially manufactured feeds are generally nutritionally complete, concentrate feeds. On small dairies where grain is home-grown or abundantly available locally, home mixing of the concentrates is a widely used practice. However, there has been an increase in the use of commercial feeds on small dairies. Large specialized dairies purchase individual feed ingredients in bulk, then mix and feed total mixed rations.

Not all dairy producers, in the United States or in other countries, who home-mix feeds balance rations (1) on the basis of chemical analyses of their feed ingredients, (2) with the use of computers, or (3) by combining the concentrates and forage into a complete ration. For these producers, Table 24–8, Feed Mixing Guide for Lactating Cows, may serve as a useful guide. It shows how ingredients partitioned into four approximate protein levels (columns 2, 3, 4, and 5) may be combined to make concentrates suitable for feeding with three different qualities of roughages—excellent, medium, and poor.

Variations can and should be made in the rations listed in Table 24–8. Producers should give consideration to the supply of homegrown feeds, and to the availability and price of ingredients. Feeds of similar nutritive properties can and should be interchanged as price relationships warrant. Thus, the cereal grains may consist of corn, barley, wheat, oats, and/or sorghum; the protein supplements may consist of soybean, cottonseed, peanut and/or linseed meal; and a vast array of by-product feeds may be utilized.

Here is how to use Table 24–8: Let's assume that a producer has (1) medium quality forage, and (2) both low- and medium-high protein (columns 2 and 4) ingredients from which

Fig. 24–15. Many large and sophisticated dairies store concentrates as individual ingredients in bins such as these at Maddox Dairy. Then, they mix precise amounts of these ingredients to make balanced rations. (Courtesy, Maddox Dairy, Riverdale, CA)

## TABLE 24–8
## FEED MIXING GUIDE FOR LACTATING COWS (AS-FED BASIS)[1]

**Note:** *This shows how ingredients of 4 protein levels may be combined to make different concentrate mixes of approximate protein content to match 3 different qualities of roughages.*

| (1) Suggested Grain Mix, Based on Kind of Roughage Available | (2) Low Protein (under 12%) Ingredients | | (3) Low-Medium Protein (12–18%) Ingredients | | (4) Medium-High Protein (18–28%) Ingredients | | (5) High Protein (over 32%) Ingredients | |
|---|---|---|---|---|---|---|---|---|
| **Feeds** | **(% protein)** | | **(% protein)** | | **(% protein)** | | **(% protein)** | |
| | Barley, all analyses ...... 11.7 | | Dairy feed, 16% ......... 16.0 | | Brewers' dried grains* .... 27.3 | | Dairy feed, 32–34% .. 32–24 | |
| | Beet pulp w/molasses, dried . 9.3 | | Wheat bran ............ 15.5 | | Copra (coconut) meal ..... 21.3 | | Corn gluten meal ..... 60.8 | |
| | Corn-and-cob meal ....... 7.8 | | Wheat middlings ........ 16.4 | | Corn gluten feed ........ 23.0 | | Cottonseed meal* ..... 41.2 | |
| | Corn #2 ................ 8.9 | | | | Dairy feed, 18–24% .... 18–24 | | Linseed meal ......... 35.7 | |
| | Dairy feed, 12% ........ 12.0 | | | | Distillers' dried grains ..... 27.3 | | Peanut meal ........ 49.0 | |
| | Hominy feed ........... 10.3 | | | | Malt sprouts ............ 22.9 | | Soybean meal ....... 44.4 | |
| | Molasses, cane* ........ 4.3 | | | | Peas, field* ............ 23.2 | | | |
| | Oats, all analyses ....... 11.9 | | | | | | | |
| | Rye, all analyses* ....... 12.0 | | | | | | | |
| | Sorghum (milo) ........ 10.1 | | | | | | | |
| | Wheat, all analyses ...... 13.1 | | | | | | | |
| | **(lb)** | **(kg)** | **(lb)** | **(kg)** | **(lb)** | **(kg)** | **(lb)** | **(kg)** |
| **Excellent roughage**—High protein forage, 18%: (1) legume, or (2) legume and nonlegume mixed forages of *high quality;* consisting of dry forages and/or silage. | | | | | | | | |
| Mix No. 1 | 1,000 | 454 | | | | | | |
| Mix No. 2 | 900 | 409 | | | | | 100 | 45 |
| Mix No. 3 | 800 | 363 | | | 200 | 91 | | |
| Mix No. 4 | 850 | 386 | 100 | 45 | | | 50 | 23 |
| **Medium roughage**—Medium protein forage, 15–17%: (1) legume, or (2) legume and nonlegume mixed forages of *medium quality;* consisting of dry forages and/or silage. | | | | | | | | |
| Mix No. 5 | 800 | 363 | | | | | 200 | 91 |
| Mix No. 6 | 650 | 295 | | | 350 | 159 | | |
| Mix No. 7 | 700 | 318 | 100 | 45 | 100 | 45 | 100 | 45 |
| Mix No. 8 | Straight 16% dairy feed, or ½ Mix No. 9 and ½ 16% dairy feed | | | | | | | |
| **Poor roughage**—Low protein forage, under 14%: nonlegume forage; consisting of dry forages and/or silage. | | | | | | | | |
| Mix No. 9 | 700 | 318 | 300 | 136 | | | | |
| Mix No. 10 | 600 | 272 | | | 200 | 91 | 200 | 91 |
| Mix No. 11 | 600 | 272 | 100 | 45 | 100 | 45 | 200 | 91 |
| Mix No. 12 | 500 | 227 | | | | | 500 | 227 |

[1]The protein compositions in columns 2 to 5 were obtained from Section V, Composition of Feeds of *Feeds & Nutrition.*

**Comments:**

**Add**—To all rations (1) 1% iodized or trace-mineralized salt; (2) 1% steamed bone meal, dicalcium phosphate, or the equivalent (use monosodium phosphate or a high-phosphorus commercial mineral where alfalfa is fed liberally); (3) 1,000 IU of vitamin A/lb *(2,205 IU of vitamin A/kg)* of concentrate and, unless cows are in sunlight, add 150 IU of vitamin D/lb *(331 IU of vitamin D/kg)* of concentrate.

***Limitations**—Wheat, not more than 50% of the ration; dried molasses beet pulp, 20%; molasses, 15%; peas and brewers' dried grains, 30%; rye, 10%; and cottonseed meal, 20% of the mix for calves, but as needed for mature cows.

to choose. How many pounds each of the low- and medium-high protein ingredients will be required in a 1,000-lb concentrate mix? Step by step, here is the answer:

1. Look under "Medium roughage—medium protein forage" 15–17% (column to the left).
2. Mix No. 6, containing 650 lb of low protein ingredients

(under column 2: under 12% ingredients) and 350 lb of medium-high protein ingredients (column 4: 18 to 28% protein), will meet the needs. The concentrates may be chosen from among those listed at the top of the respective columns of Table 24–8—the low protein concentrates from column 2 (under 12%) and the medium-high protein concentrates from column 4 (18 to 28%).

## HOW TO BALANCE A DAIRY RATION

Good dairy producers have nutrient analyses made of all major ration ingredients and use computers to balance their rations for at least 20 nutrients.

It is recognized that rations should vary with conditions, and that many times they should be formulated to meet the conditions of a specific dairy farm. Also, a good producer should know how to balance a ration. Complete instructions on how to balance a ration (including an example of balancing a dairy ration) are given in Chapter 4 of this book, under the heading "How to Balance a Ration"; hence, the reader is referred thereto. By (1) following these instructions, and (2) using nutrient requirement tables similar to Tables 24–1 and 24–2 at the first of this chapter, it is possible to balance rations for specific weights and levels of production.

## FEEDING SYSTEMS

Traditional individual feeding of lactating cows in stanchioned barns or milking parlors is giving way to new feeding systems. Although the newer methods are not as effective as feeding cows individually, they are much more economical than feeding all cows in the herd the same amount of grain, regardless of production. Additionally, they make for considerable saving in labor and facilities.

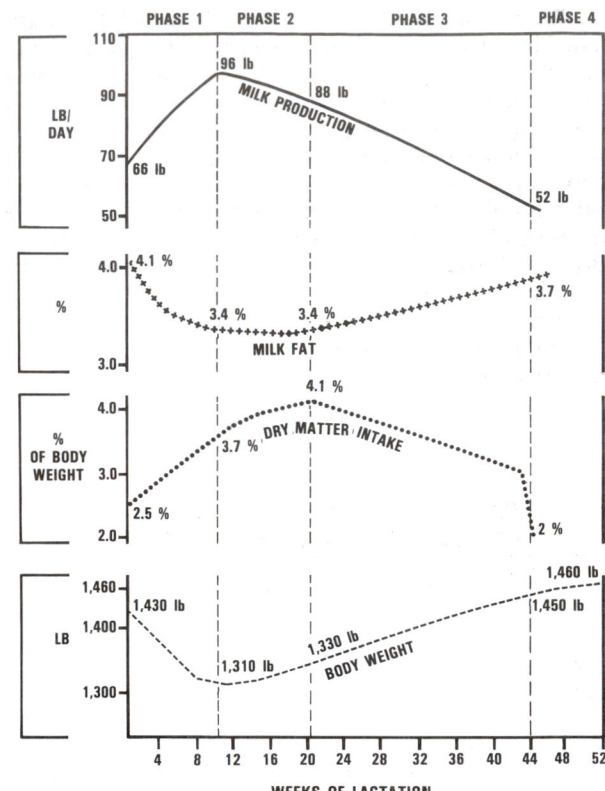

Fig. 24–17. Lactation cycle phases with corresponding changes in milk production, milk fat percentage, dry matter intake, and body weight. (Source: Linn, J. G., M. F. Hutjens, W. T. Howard, L. H. Kilmer, and D. E. Otterby, *Feeding the Dairy Herd*, Cooperative Extension Services, Universities of Illinois, Iowa State, Minnesota, and Wisconsin, 1988, p. 15, Fig. 6)

Fig. 24–16. Traditional individual feeding of lactating cows is giving way to corral (group) feeding, which is more economical and makes for savings in labor and facilities. (Courtesy, James Tappan, Arizona Dairy Co., Higley, AZ)

## PHASE FEEDING

*Phase feeding is a feeding program that is divided into periods based on milk production, milk fat percentage, feed intake, and body weight.* Fig. 24–17 illustrates the shape and relationship of curves for milk production, fat percentage, dry matter intake, and body weight. Based on these curves, four distinct feeding phases of lactating cows can be identified.

Producers should formulate rations to match each of these phases in order to optimize milk yield, minimize metabolic disorders, increase longevity, and increase profits. The four phases are:

1. **Phase 1, early lactation, 0 to 70 days postpartum.** During this period, milk production increases rapidly, peaking at 4 to 6 weeks after calving. But feed intake does not keep pace with nutrient needs (especially energy needs) for milk production, so body tissues are mobilized to meet these needs. During this phase, adjusting the cow to the lactation ration is an important management practice. After calving, the grain should be increased by 1 to 1.5 lb daily to meet increased nutrient demands and minimize off-feed problems and acidosis.

2. **Phase 2, peak dry matter intake, second 10 weeks postpartum.** During this phase, cows should be fed to maintain peak milk production as long as possible. Feed intake is near maximum and can supply nutrient needs. Cows should be maintaining weight or making slight gains.

Grain intake may reach, but should not exceed, 2.3% of the cow's body weight (dry matter basis). High quality forage should be provided, with a minimum intake of 1.5% of the cow's body weight (dry matter basis) in order to maintain rumen function and normal fat test.

To maximize nutrient intake (a) feed forage and grain 3 or more times daily, (b) feed high quality feeds, (c) limit urea to 0.2 lb/cow/day, (d) minimize stress, and (e) use a total mixed ration.

3. **Phase 3, mid- to late-lactation, 140 to 305 days postpartum.** This is the easiest phase to manage. During this period, milk production is declining, the cow is pregnant, and nutrient intake will easily meet or exceed requirements. The level of grain feeding should be adequate to meet production requirements, and to begin to replace body weight lost during early lactation. Lactating cows require less feed to replace a pound of body tissue than dry cows; hence, it is more efficient to have cows gain body weight near the end of lactation than during the dry period. Young cows should receive additional nutrients for growth; 2-year-old heifers should receive 20% more than for maintenance, and 3-year-olds should receive 10% more.

The example rations presented in Table 24–9 are suitable for the three phases.

### TABLE 24–9
### EXAMPLE RATIONS
### FOR VARIOUS MILK PRODUCTION PHASES,
### 1,350 LB *(613 KG)* COW, 3.8% FAT TEST[1]

| Item | | Phase 1 | Phase 2 | Phase 3 |
|---|---|---|---|---|
| Milk | (lb/day) | 90 | 80 | 50 |
| Milk | *(kg/day)* | *40.9* | *36.3* | *22.7* |
| DM intake[2] | (lb/day) | 49 | 51 | 38 |
| DM intake[2] | *(kg/day)* | *22.2* | *23.2* | *17.3* |

| | As-Fed | | | | | |
|---|---|---|---|---|---|---|
| | (lb/day) | *(kg/day)* | (lb/day) | *(kg/day)* | (lb/day) | *(kg/day)* |
| **Ration 1** | | | | | | |
| Alfalfa hay (88% DM), 140 RFV, 20% crude protein | 28 | *12.71* | 34 | *15.44* | 27 | *12.26* |
| Corn-oats[3] | 21 | *9.53* | 24 | *10.90* | 16 | *7.26* |
| Soybean meal, 44% | 5.0 | *2.27* | | | | |
| Dical, 18% phosphorus | 0.5 | *0.23* | 0.45 | *0.20* | 0.30 | *0.14* |
| Salt, vitamins, trace mineralized | 0.30 | *0.14* | 0.25 | *0.11* | 0.25 | *0.11* |
| Weight change | −1.5 | *−0.68* | — | — | + .5 | *+ .23* |
| **Ration 2** (corn silage limit fed) | | | | | | |
| Alfalfa hay, 140 RFV, 20% CP | 19 | *8.63* | 34 | *15.44* | 23 | *10.44* |
| Corn silage (35% DM) | 25 | *11.35* | 25 | *11.35* | 25 | *11.35* |
| Corn-oats | 18 | *8.17* | 12 | *5.45* | 10 | *4.54* |
| Soybean meal, 44% | 7.5 | *3.41* | 0.3 | *0.14* | — | — |
| Dical, 18% phosphrous | 0.45 | *0.20* | 0.50 | *0.23* | 0.3 | *0.14* |
| Salt, vitamins, trace mineralized | 0.30 | *0.14* | 0.25 | *0.11* | 0.25 | *0.11* |
| Weight change | −1.2 | *−0.54* | — | — | + .5 | *+ .23* |
| **Ration 3** (hay limit fed)[4] | | | | | | |
| Alf-grass hay, 113 RFV, 16% CP | 10 | *4.54* | 10 | *4.54* | 10 | *4.54* |
| Corn silage | 41 | *18.61* | 70 | *31.78* | 57 | *25.88* |
| Corn-oats | 16 | *7.26* | 11 | *4.99* | 6 | *2.72* |
| Soybean meal, 44% | 11.5 | *5.22* | 8.2 | *3.72* | 4.5 | *2.04* |
| Dical, 18% phosphrus | 0.40 | *0.18* | 0.30 | *0.14* | 0.25 | *0.11* |
| Limestone | 0.40 | *0.18* | 0.30 | *0.14* | 0.15 | *0.07* |
| Salt, vitamins, trace mineralized | 0.30 | *0.14* | 0.25 | *0.11* | 0.25 | *0.11* |
| Weight change | −1.4 | *−0.64* | + .7 | *+ .32* | + .5 | *+ .23* |
| **Ration 4** | | | | | | |
| Alf-grass hay, 113 RFV, 16% CP | 23 | *10.44* | 32 | *14.53* | 24 | *10.90* |
| Corn-oats | 22 | *9.99* | 22 | *9.99* | 19 | *8.63* |
| Soybean meal, 44% | 8.5 | *3.86* | 3.5 | *1.59* | 1.1 | *0.50* |
| Dical, 18% phosphorus | 0.45 | *0.20* | 0.40 | *0.18* | 0.25 | *0.11* |
| Limestone | 0.20 | *0.09* | | | | |
| Salt, vitamins, trace mineralized | 0.30 | *0.14* | 0.25 | *0.11* | 0.25 | *0.11* |
| Weight change | −1.9 | *−0.86* | — | — | + .5 | *+ .23* |

[1]*Source:* Linn, J. G., M. F. Hutjens, W. T. Howard, L. H. Kilmer, and D. E. Otterby, *Feeding the Dairy Herd,* Cooperative Extension Services, Universities of Illinois, Iowa State, Minnesota, and Wisconsin, 1988, p. 16, Table 6.

[2]Estimated average intake during the phase.

[3]85% corn–15% oats mix.

[4]Feed amounts may have to be limited during phase 2 and 3 to avoid over-conditioning.

4. **Phase 4, dry period, 45 to 60 days before parturition.** The dry phase is important. A good dry cow program can minimize metabolic problems at or immediately following calving and increase milk yield during the subsequent lactation. (See Table 24–10 for example dry cow rations.)

### TABLE 24–10
### EXAMPLE DRY COW RATIONS, 1,400 LB *(636 KG)* DRY COW[1]

| Forage | As-Fed | |
|---|---|---|
| | (lb/day) | *(kg/day)* |
| **Grass forage** | | |
| Orchard grass hay, 12% crude protein | 25.0 | *11.35* |
| Corn | 3.0 | *1.36* |
| Soybean meal | 0.5 | *0.23* |
| Limestone | 0.15 | *0.07* |
| Trace mineralized salt and vitamins | 0.1 | *0.05* |
| **Limited legume forage**[2] | | |
| Alfalfa hay, RFV 140, 20% crude protein | 12.0 | *5.45* |
| Corn silage | 43.0 | *19.52* |
| Monosodium phosphate | 0.1 | *0.05* |
| Trace mineralized salt and vitamins | 0.1 | *0.05* |
| **Limited corn silage** | | |
| Alfalfa-grass hay, RFV 113, 16% crude protein | 21 | *9.53* |
| Corn silage | 20 | *9.08* |
| Dicalcium phosphate | 0.1 | *0.05* |
| Trace mineralized salt and vitamins | 0.1 | *0.05* |

[1]*Source:* Linn, J. G., M. F. Hutjens, W. T. Howard, L. H. Kilmer, and D. E. Otterby, *Feeding the Dairy Herd,* Cooperative Extension Services, Universities of Illinois, Iowa State, Minnesota, and Wisconsin, 1988, p. 17, Table 7.

[2]Ration contains excess energy as formulated and may over-condition cows in some situations.

## CHALLENGE FEEDING (LEAD FEEDING)

*Challenge feeding, or lead feeding, refers to feeding the lactating cow so that she is challenged to reach her peak (summit) production level early in lactation.*

Because of the strong relationship between peak (summit) milk yield and the total milk production for the entire lactation period, emphasis should be placed on attaining maximum yield between weeks 3 and 8.

For every 5 lb increase in the summit peak, it is estimated that rolling herd average will increase 1,000 lb.

Preparation for challenge feeding begins during the dry cow period; (1) by having the dry cow in proper condition; and (2) by making the transition from the dry cow ration to the lactating ration, thereby preparing the rumen bacteria.

After calving, challenge feeding calls for increasing the grain allowance several pounds per day above the cow's exact requirements at the time. The objective is to allow each cow to reach peak production at or near her genetic potential. The advantage of achieving maximum peak production will be maintained throughout the remaining months of lactation.

Challenge feeding helps a cow reach peak production earlier than she otherwise might, thus taking advantage of the fact that her system, at the time, is physiologically adapted to heavy production.

After peak production is reached, the amount of concentrate fed should be determined by a concentrate feeding guide based on body weight, milk production, and fat test.

## CORRAL (GROUP) FEEDING

Individual feeding of lactating cows has largely given way to mechanized group feeding. The latter was developed for convenience and saving of labor, rather than for improved animal well-being or feed efficiency. Today, lactating herds with several hundred cows are common; and some herds number several thousand. In order to design a nutritional program for such large numbers that can be adapted to the specific needs of the cows, they are separated into groups according to production (and, therefore, nutritional needs).

Fig. 24–18.   Drive-through barn, with adjacent corrals, at Maddox Dairy, Riverdale, CA, in which lactating cows are group fed from self-unloading trucks. (Photo by A. H. Ensminger)

When producers decide to go to group feeding, they must decide on the number of groups into which to divide the herd. To answer this question, consideration should be given to the following: (1) herd size; (2) types and costs of available feeds; (3) current type of housing, feeding, and milking system; and (4) overall economic integration of the operation—for example, labor, machinery, etc.

In large herds (more than 250 milking cows), a commonly used system is one in which a minimum of 5 groups are established: (1) high-production cows (about 90 lb of milk/head/day), (2) medium-production cows (about 65 lb of milk/head/day), (3) low-production cows (about 45 lb of milk/head/day), (4) dry cows, and (5) first calf heifers. More groups are desirable in very large herds if corrals and facilities are available. Because of feeding and social considerations, a maximum of 100 cows per group is advisable. With this program, there can be a maximum of two moves during the lactation cycle. In many cases, only one move is necessary; and in a few cases, no moves are required. This system allows each group to be fed according to need. The high-producing groups should be fed the highest quality ingredients at maximum levels. The middle-producing cows should be fed in such a way as

to reduce feed costs, increase butterfat test, improve rumen function, and promote lactation persistency. The same holds true for the low-producing cows as for the medium producers except that considerable care must be exercised to avoid excessive fattening.

One of the problems inherent in group feeding concerns the behavioral adaptation of a newly introduced cow to a group. Group acceptance—*pecking order*—can pose occasional problems with a new cow, but the magnitude of the problem is usually not very great. One means of reducing this is to move several cows into a new group at the same time and just before feeding, rather than individually.

When group feeding programs are followed, grain is seldom fed in the milking parlor. This is commonly referred to as corral or bunk feeding since feeding generally takes place in bunks along the fenceline of the corrals or pens. Studies have demonstrated that cows fed their grain as a group in a common manger do as well as those fed individually in the milking parlor, but some cows may not always come into the parlor as easily when there is no grain to attract them. Some producers offer a minimum amount of feed in the parlor and the remaining amount in the corral with good success. The high producers seem to be more aggressive than the low producers; hence, they usually eat more when group fed.

Group feeding can be easily adapted to the use of complete feeds when the concentrates, roughages, and supplements are mixed into one feed rather than being fed separately. Some producers who use complete feeds prefer to feed dried roughages—especially long-stemmed hay—separately in order to enhance stimulation of the rumen and to facilitate mixing, because long hay does not lend itself to mixing in a mixer.

• **Automatic grain feeders**—The advent of electric and computerized grain feeders was heralded as a way in which producers could keep their cows in group housing but feed them individually. However, some producers report (1) that such feeders may result in overfeeding grain, accompanied by health problems and lower profits; and (2) that such feeders may not function properly when the grain ration contains adequate fiber content.

• **Example rations for group feeding**—Example rations for group feeding mature lactating cows at each of 3 milk production levels (90, 65, and 45 lb) are presented in Tables 24–11, 24–12, and 24–13; and example rations suitable for feeding dry cows are presented in Table 24–14.

When group feeding, first-calf heifers should be handled in a separate group and fed for both milk production and growth. Their nutrient requirements for milk production are similar to the requirements of their older counterparts producing milk at the same level, but, because of their growth, they should receive about 20% more nutrients than are required for maintenance.

Although variations can and should be made in Tables 24–11, 24–12, 24–13, and 24–14 rations, they are excellent guides. When milk yields and/or body weights differ from those used in these tables, a suitable computer program or hand calculations should be employed to obtain amounts to be fed, and to give consideration to cost of alternative feeds.

**TABLE 24-11**
**COMPLETE RATIONS FOR 1,300 TO 1,400 LB *(591 TO 636 KG)* COWS,**
**IN EARLY LACTATION—HIGH-PRODUCTION GROUP (90-LB MILK, 3.6% FAT AVERAGE)[1]**

| Feeds | All Alfalfa (lb) | (kg) | ¾ Alfalfa ¼ Corn Silage (lb) | (kg) | ⅔ Alfalfa ⅓ Corn Silage (lb) | (kg) | ½ Alfalfa ½ Corn Silage (lb) | (kg) | ⅓ Alfalfa ⅔ Corn Silage (lb) | (kg) | ¼ Alfalfa ¾ Corn Silage (lb) | (kg) |
|---|---|---|---|---|---|---|---|---|---|---|---|---|
| **All Amounts on Dry Matter Basis** | | | | | | | | | | | | |
| Alfalfa, 17% crude protein, 59% TDN | 24.2 | 10.99 | 19.2 | 8.72 | 17.28 | 7.85 | 13.59 | 6.17 | 9.24 | 4.19 | 7.03 | 3.19 |
| Corn silage, 8% crude protein, 68.7% TDN | — | — | 6.4 | 2.91 | 8.63 | 3.92 | 13.59 | 6.17 | 18.51 | 8.40 | 21.10 | 9.58 |
| Corn, 10% crude protein, 88% TDN | 18.9 | 8.58 | 16.1 | 7.31 | 15.32 | 6.96 | 13.11 | 5.95 | 11.22 | 5.09 | 10.16 | 4.61 |
| 44% crude protein supplement, 49% crude protein, 86% TDN | 5.5 | 2.50 | 6.6 | 3.00 | 7.00 | 3.18 | 7.83 | 3.55 | 8.82 | 4.00 | 9.34 | 4.24 |
| Fat, 182% TDN | 1.2 | 0.54 | 1.2 | 0.54 | 1.19 | 0.54 | 1.19 | 0.54 | 1.18 | 0.54 | 1.19 | 0.54 |
| Dicalcium phosphate, 22% calcium, 19% phosphorus | 0.42 | 0.19 | 0.52 | 0.24 | 0.51 | 0.23 | 0.49 | 0.22 | 0.46 | 0.21 | 0.44 | 0.20 |
| Limestone, 34% calcium | — | — | 0.07 | 0.03 | 0.14 | 0.06 | 0.25 | 0.11 | 0.41 | 0.19 | 0.48 | 0.22 |
| Trace mineralized salt | 0.25 | 0.11 | 0.25 | 0.11 | 0.25 | 0.11 | 0.25 | 0.11 | 0.25 | 0.11 | 0.25 | 0.11 |
| Mineral-vitamin mix | 0.08 | 0.04 | 0.08 | 0.04 | 0.08 | 0.04 | 0.09 | 0.04 | 0.10 | 0.05 | 0.11 | 0.05 |
| **Ration Nutrient Information** | | | | | | | | | | | | |
| Dry matter | 50.5 | 22.93 | 50.4 | 22.88 | 50.4 | 22.88 | 50.4 | 22.88 | 50.2 | 22.79 | 50.1 | 22.75 |
| Crude protein | 8.72 | 3.96 | 8.71 | 3.95 | 8.72 | 3.96 | 8.72 | 3.96 | 8.72 | 3.96 | 8.72 | 3.96 |
| Crude protein (%) | 17.26 | | 17.29 | | 17.29 | | 17.30 | | 17.36 | | 17.40 | |
| TDN | 37.52 | 17.03 | 37.39 | 16.98 | 37.43 | 16.99 | 37.39 | 16.98 | 37.33 | 16.95 | 37.30 | 16.93 |
| TDN (%) | 74.29 | | 74.19 | | 74.26 | | 74.18 | | 74.36 | | 74.446 | |
| NE$_{lc}$ (Mcal) | 38.72 | | 38.66 | | 38.72 | | 38.72 | | 38.72 | | 38.72 | |
| NE$_{lc}$ (Mcal/lb or kg) | 0.767 | 1.69 | 0.767 | 1.69 | 0.768 | 1.69 | 0.768 | 1.69 | 0.771 | 1.70 | 0.773 | 1.70 |
| Calcium (g) | 208 | | 206 | | 206 | | 206 | | 205 | | 205 | |
| Calcium (%) | 0.91 | | 0.90 | | 0.90 | | 0.90 | | 0.90 | | 0.90 | |
| Phosphorus (g) | 103 | | 114 | | 114 | | 114 | | 114 | | 114 | |
| Phosphorus (%) | 0.45 | | 0.50 | | 0.50 | | 0.50 | | 0.50 | | 0.50 | |
| Acid detergent fiber (%) | 18.04 | | 18.12 | | 18.07 | | 18.24 | | 18.09 | | 18.05 | |
| Ether extract (%) | 5.66 | | 5.54 | | 5.50 | | 5.41 | | 5.30 | | 5.28 | |
| Forage:grain ratio | 48:52 | | 51:49 | | 51:49 | | 54:46 | | 55:45 | | 56:44 | |

[1]Ration calculations based on 1988 NRC recommendations, and on the use of either alfalfa or alfalfa and corn silage (CS) forage. Table 24-11 was prepared by D. E. Otterby, Professor, and J. G. Linn, Extension Animal Scientist, Department of Animal Science, University of Minnesota, St. Paul; with metric added by the author.

**TABLE 24-12**
**COMPLETE RATIONS FOR 1,300 TO 1,400 LB *(591 TO 636 KG)* COWS,**
**IN MID-LACTATION—MEDIUM-PRODUCTION GROUP (65-LB MILK, 3.6% FAT AVERAGE)[1]**

| Feeds | All Alfalfa (lb) | (kg) | ¾ Alfalfa ¼ Corn Silage (lb) | (kg) | ⅔ Alfalfa ⅓ Corn Silage (lb) | (kg) | ½ Alfalfa ½ Corn Silage (lb) | (kg) | ⅓ Alfalfa ⅔ Corn Silage (lb) | (kg) | ¼ Alfalfa ¾ Corn Silage (lb) | (kg) | All Corn Silage (lb) | (kg) |
|---|---|---|---|---|---|---|---|---|---|---|---|---|---|---|
| **All Amounts on Dry Matter Basis** | | | | | | | | | | | | | | |
| Alfalfa, 17% crude protein, 59% TDN | 21.93 | 9.96 | 17.93 | 8.14 | 16.43 | 7.46 | 11.98 | 5.44 | 8.07 | 3.66 | 6.15 | 2.79 | — | — |
| Corn silage, 8% crude protein, 68.7% TDN | — | — | 5.98 | 2.71 | 8.20 | 3.72 | — | — | — | — | — | — | — | — |
| Urea-corn silage, 12% crude protein, 70% TDN | — | — | — | — | — | — | 11.98 | 5.44 | 16.16 | 7.34 | 18.46 | 8.38 | 26.39 | 11.98 |
| Corn, 10% crude protein, 91% TDN | 17.12 | 7.77 | 14.29 | 6.49 | 13.23 | 6.01 | 13.25 | 6.02 | 12.48 | 5.67 | 11.80 | 5.36 | 6.28 | 2.85 |
| 44% CP supplement, 49% CP, 86% TDN | 2.33 | 1.06 | 3.21 | 1.46 | 3.54 | 1.61 | 3.62 | 1.64 | 4.10 | 1.86 | 4.33 | 1.97 | 7.34 | 3.33 |
| Limestone | 0.34 | 0.15 | 0.32 | 0.15 | 0.31 | 0.14 | — | — | 0.01 | 0.01 | 0.07 | 0.03 | 0.28 | 0.13 |
| Trace mineralized salt | 0.21 | 0.10 | 0.21 | 0.10 | 0.21 | 0.10 | 0.21 | 0.10 | 0.21 | 0.10 | 0.21 | 0.10 | 0.20 | 0.09 |
| Mineral-vitamin mix | 0.08 | 0.04 | 0.07 | 0.03 | 0.08 | 0.04 | 0.08 | 0.04 | 0.07 | 0.03 | 0.10 | 0.05 | 0.10 | 0.05 |
| **Ration Nutrient Information** | | | | | | | | | | | | | | |
| Dry matter | 42.0 | 19.07 | 42.0 | 19.07 | 42.0 | 19.07 | 41.5 | 18.84 | 41.5 | 18.84 | 41.5 | 18.84 | 40.9 | 18.57 |
| Crude protein | 6.60 | 3.00 | 6.60 | 3.00 | 6.60 | 3.00 | 6.60 | 3.00 | 6.60 | 3.00 | 6.60 | 3.00 | 6.60 | 3.00 |
| Crude protein (%) | 15.72 | | 15.72 | | 15.72 | | 15.91 | | 15.91 | | 15.91 | | 16.14 | |
| TDN | 29.90 | 13.57 | 29.85 | 13.55 | 29.84 | 13.55 | 29.56 | 13.42 | 29.73 | 13.50 | 29.70 | 13.48 | 29.60 | 13.44 |
| TDN (%) | 71.18 | | 71.08 | | 71.05 | | 71.23 | | 71.63 | | 71.58 | | 72.37 | |
| NE$_{lc}$ (Mcal) | 30.73 | | 30.73 | | 30.73 | | 30.73 | | 30.73 | | 30.73 | | 30.73 | |
| NE$_{lc}$ (Mcal/lb or kg) | 0.73 | 1.61 | 0.73 | 1.61 | 0.73 | 1.61 | 0.74 | 1.63 | 0.74 | 1.63 | 0.74 | 1.63 | 0.75 | 1.65 |
| Calcium (g) | 146 | | 129 | | 122 | | 139 | | 122 | | 122 | | 122 | |
| Calcium (%) | 0.77 | | 0.68 | | 0.64 | | 0.74 | | 0.65 | | 0.65 | | 0.66 | |
| Phosphorus (g) | 86 | | 86 | | 86 | | 85 | | 85 | | 85 | | 80 | |
| Phosphorus (%) | 0.45 | | 0.45 | | 0.45 | | 0.45 | | 0.45 | | 0.45 | | 0.43 | |
| Acid detergent fiber (%) | 19.00 | | 19.57 | | 19.78 | | 19.73 | | 19.59 | | 19.69 | | 19.03 | |
| Ether extract (%) | 3.51 | | 3.36 | | 3.34 | | 3.00 | | 2.84 | | 2.74 | | 2.95 | |
| Forage:grain ratio | 52:48 | | 57:43 | | 59:41 | | 58:42 | | 58:42 | | 59:41 | | 65:35 | |

[1]Ration calculations based on 1988 NRC recommendations, and on the use of either alfalfa or alfalfa and corn silage (CS) forage. Table 24-12 was prepared by D. E. Otterby, Professor, and J. G. Linn, Extension Animal Scientist, Department of Animal Science, University of Minnesota, St. Paul; with metric added by the author.

## TABLE 24–13
### COMPLETE RATIONS FOR 1,300 TO 1,400 LB *(591 TO 636 KG)* COWS, IN LATE LACTATION—LOW-PRODUCTION GROUP (45–LB MILK, 3.8% FAT AVERAGE)[1]

| Feeds | All Alfalfa | | ¾ Alfalfa ¼ Corn Silage | | ⅔ Alfalfa ⅓ Corn Silage | | ½ Alfalfa ½ Corn Silage | | ⅓ Alfalfa ⅔ Corn Silage | | ¼ Alfalfa ¾ Corn Silage | | All Corn Silage | |
|---|---|---|---|---|---|---|---|---|---|---|---|---|---|---|
| | All Amounts on Dry Matter Basis | | | | | | | | | | | | | |
| | (lb) | *(kg)* | (lb) | *(kg)* | (lb) | *(kg)* | (lb) | *(kg)* | (lb) | *(kg)* | (lb) | *(kg)* | (lb) | *(kg)* |
| Alfalfa, 17% crude protein, 59% TDN ......... | 26.14 | *11.87* | 21.53 | *9.77* | 19.74 | *8.96* | 14.99 | *6.81* | 10.46 | *4.75* | 7.98 | *3.62* | — | — |
| Corn silage, 8% crude protein, 68.7% TDN ..... | — | — | 7.18 | *3.26* | 9.86 | *4.48* | — | — | — | — | — | — | — | — |
| Urea-corn silage, 12% crude protein, 66% TDN .. | — | — | — | — | — | — | 14.99 | *6.81* | 20.95 | *9.51* | 23.94 | *10.87* | 31.10 | *14.12* |
| Corn, 10% crude protein, 91% TDN ........... | 10.76 | *4.89* | 8.21 | *3.73* | 6.97 | *3.16* | 6.74 | *3.06* | 4.79 | *2.17* | 3.97 | *1.80* | 2.60 | *1.18* |
| 44% CP supplement, 49% CP, 81% TDN ..... | — | — | — | — | 0.34 | *0.15* | 0.13 | *0.06* | 0.63 | *0.29* | 0.94 | *0.43* | 2.20 | *1.00* |
| Dicalcium phosphate, 22% Ca, 19% P ....... | 0.27 | *0.12* | 0.26 | *0.12* | 0.25 | *0.11* | 0.30 | *0.14* | 0.31 | *0.14* | 0.31 | *0.14* | 0.30 | *0.14* |
| Limestone, 34% calcium ................... | — | — | — | — | — | — | — | — | — | — | — | — | 0.12 | *0.05* |
| Trace mineralized salt ................... | 0.19 | *0.09* | 0.19 | *0.09* | 0.19 | *0.09* | 0.19 | *0.09* | 0.19 | *0.09* | 0.19 | *0.09* | 0.18 | *0.08* |
| Mineral-vitamin mix .................... | 0.06 | *0.03* | 0.06 | *0.03* | 0.06 | *0.03* | 0.07 | *0.03* | 0.08 | *0.04* | 0.09 | *0.04* | 0.20 | *0.09* |
| | Ration Nutrient Information | | | | | | | | | | | | | |
| Dry matter ........................ | 37.42 | *16.99* | 37.42 | *16.99* | 37.42 | *16.99* | 37.42 | *16.99* | 37.42 | *16.99* | 37.42 | *16.99* | 36.7 | *16.66* |
| Crude protein ..................... | 5.52 | *2.51* | 5.11 | *2.32* | 5.09 | *2.31* | 5.09 | *2.31* | 5.09 | *2.31* | 5.09 | *2.31* | 5.09 | *2.31* |
| Crude protein ................ (%) | 14.75 | | 13.65 | | 13.60 | | 13.60 | | 13.60 | | 13.60 | | 13.87 | |
| TDN ............................. | 24.91 | *11.31* | 24.85 | *11.28* | 24.83 | *11.27* | 24.78 | *11.25* | 24.73 | *11.23* | 24.70 | *11.21* | 24.58 | *11.16* |
| TDN ......................... (%) | 66.52 | | 66.40 | | 66.35 | | 66.22 | | 66.10 | | 66.02 | | 66.98 | |
| NE$_{lc}$ ..................... (Mcal) | 25.37 | | 25.37 | | 25.37 | | 25.37 | | 25.37 | | 25.37 | | 25.37 | |
| NE$_{lc}$ ............ (Mcal/lb or kg) | 0.677 | *1.49* | 0.678 | *1.50* | 0.678 | *1.50* | 0.678 | *1.50* | 0.678 | *1.50* | 0.678 | *1.50* | 0.691 | *1.52* |
| Calcium ......................... (g) | 195 | | 173 | | 165 | | 148 | | 129 | | 117 | | 96 | |
| Calcium ......................... (%) | 1.15 | | 1.02 | | 0.97 | | 0.87 | | 0.76 | | 0.69 | | 0.58 | |
| Phosphorus ...................... (g) | 64 | | 64 | | 64 | | 64 | | 64 | | 64 | | 64 | |
| Phosphorus ...................... (%) | 0.38 | | 0.38 | | 0.38 | | 0.38 | | 0.38 | | 0.38 | | 0.38 | |
| Acid detergent fiber ............ (%) | 23.92 | | 24.63 | | 24.91 | | 25.42 | | 26.02 | | 26.16 | | 25.4 | |
| Ether extract ................... (%) | 3.47 | | 3.38 | | 3.32 | | 2.86 | | 2.59 | | 2.45 | | 2.09 | |
| Forage:grain ratio ............... | 70:30 | | 77:23 | | 79:21 | | 80:20 | | 84:16 | | 85:15 | | 85:15 | |

[1]Ration calculations based on 1988 NRC recommendations, and on the use of either alfalfa or alfalfa and corn silage (CS) forage. Table 24–13 was prepared by D. E. Otterby, Professor, and J. G. Linn, Extension Animal Scientist, Department of Animal Science, University of Minnesota, St. Paul; with metric added by the author.

## TABLE 24–14
### COMPLETE RATIONS FOR 1,300 TO 1,400 LB *(591 TO 636 KG)* DRY COWS[1]

| Feeds | Alfalfa Corn Silage | | Alfalfa-Grass Hay, Corn Stover | | Alfalfa-Grass Hay | | Oatlage | | Urea-Corn Silage, Grass Hay | | Grass Hay | |
|---|---|---|---|---|---|---|---|---|---|---|---|---|
| | All Amounts on Dry Matter Basis | | | | | | | | | | | |
| | (lb) | *(kg)* | (lb) | *(kg)* | (lb) | *(kg)* | (lb) | *(kg)* | (lb) | *(kg)* | (lb) | *(kg)* |
| Alfalfa, 17% CP, 59% TDN ........... | 11.09 | *5.03* | — | — | — | — | — | — | — | — | — | — |
| Alfalfa-grass hay, 16.5% CP, 58% TDN . | — | — | 14.59 | *6.62* | 22.52 | *10.25* | — | — | — | — | — | — |
| Grass hay, 12% CP, 60% TDN ....... | — | — | — | — | — | — | — | — | 13.50 | *6.13* | 17.93 | *8.14* |
| Oatlage, 12.8% CP, 59% TDN ....... | — | — | — | — | — | — | 25.74 | *11.69* | — | — | — | — |
| Corn silage, 8.7% CP, 68.7% TDN .... | 12.66 | *5.75* | — | — | — | — | — | — | — | — | — | — |
| Urea-corn silage, 12% CP, 66% TDN ... | — | — | — | — | — | — | — | — | 12.28 | *5.58* | — | — |
| Corn stover, 6.7% CP, 66% TDN ..... | — | — | 11.59 | *5.26* | — | — | — | — | — | — | — | — |
| Corn, 10% CP, 88% TDN ............. | — | — | — | — | — | — | 3.47 | *1.58* | 1.09 | *0.49* | — | — | 5.46 | *2.48* |
| 44% CP supp., 49.9% CP, 81% TDN ... | — | — | — | — | — | — | — | — | — | — | 0.53 | *0.24* |
| Dicalcium phosphate, 22% Ca, 19% P . | — | — | — | — | — | — | 0.01 | *0.01* | — | — | — | — |
| Limestone, 34% Ca ................. | — | — | — | — | — | — | 0.05 | *0.02* | 0.05 | *0.02* | 0.09 | *0.04* |
| Monosodium phosphate, 22.5% P .... | 0.05 | *0.02* | 0.10 | *0.05* | 0.03 | *0.01* | — | — | — | — | — | — |
| Trace mineralized salt ............. | 0.07 | *0.03* | 0.07 | *0.03* | 0.07 | *0.03* | 0.07 | *0.03* | 0.07 | *0.03* | 0.07 | *0.03* |
| Mineral-vitamin mix ................ | 0.03 | *0.01* | 0.02 | *0.01* | 0.02 | *0.01* | 0.06 | *0.03* | 0.06 | *0.03* | 0.06 | *0.03* |
| | Ration Nutrient Information | | | | | | | | | | | |
| Dry matter ......................... | 25.07 | *11.38* | 26.37 | *11.97* | 26.11 | *11.85* | 27.00 | *12.26* | 25.96 | *11.79* | 24.14 | *10.96* |
| Crude protein ..................... | 3.05 | *1.38* | 3.24 | *1.47* | 4.07 | *1.85* | 3.40 | *1.54* | 3.09 | *1.40* | 2.97 | *1.35* |
| Crude protein ................ (%) | 12.18 | | 12.29 | | 15.58 | | 12.60 | | 11.92 | | 12.31 | |
| TDN ............................. | 16.17 | *7.34* | 16.10 | *7.31* | 16.09 | *7.30* | 16.14 | *7.33* | 16.08 | *7.30* | 15.99 | *7.26* |
| TDN ......................... (%) | 64.49 | | 61.08 | | 61.64 | | 59.78 | | 61.98 | | 66.27 | |
| Calcium ......................... (g) | 88 | | 91 | | 87 | | 48 | | 48 | | 48 | |
| Calcium ......................... (%) | 0.78 | | 0.77 | | 0.73 | | 0.39 | | 0.41 | | 0.44 | |
| Phosphorus ...................... (g) | 30 | | 30 | | 30 | | 30 | | 30 | | 37 | |
| Phosphorus ...................... (%) | 0.27 | | 0.25 | | 0.25 | | 0.25 | | 0.27 | | 0.34 | |
| Acid detergent fiber ............ (%) | 30.16 | | 37.31 | | 28.96 | | 33.48 | | 35.05 | | 31.32 | |
| Forage:grain ratio ............... | 99:1 | | 99:1 | | 87:13 | | 95:5 | | 99:1 | | 74:26 | |

[1]Ration calculations based on 1988 NRC recommendations. Table 24–14 was prepared by D. E. Otterby, Professor, and J. G. Linn, Extension Animal Scientist, Department of Animal Science, University of Minnesota, St. Paul; with metric added by the author.

# FEEDING ON PASTURE

Problems of milk production are at a minimum during the early pasture season, when plant growth is lush. However, when the weather gets hot, it is a different story; there is the period known as the *summer slump*. High temperatures actually affect pasture growth more than the well-being of the cows. Many dairy producers have discontinued pasture grazing for two reason: (1) It is difficult to keep milk production uniform when cows are on pasture, because of changing temperatures and pasture growth; and (2) with larger herds, it is not possible to have sufficient pastures in close proximity to headquarters.

Pasture of high quality has a value that is intermediate between concentrate and hay. Thus, on good pasture, it is possible to sustain a high level of production with less grain than is needed for conventional winter forage supplementation.

The following summer feeding program is recommended for most dairy cows:

1. Have good pastures and follow good pasture management.

2. Consider ensiling the early pasture growth, thereby avoiding wasted grass early in the season.

3. Bunk feed hay with pasture, regardless of pasture quality. If the quality of the pasture is comparable to the roughage that was used in the winter ration, continue with the concentrate mix that was used in the winter. On the other hand, if summer pastures are considerably better or considerably poorer than the quality of the winter roughages, the grain mix should be changed accordingly. In other words, the grain mix used when cows are on pasture should be formulated so as to balance out the deficiencies of the grass, just the same as the winter concentrate balances out the deficiencies of the winter roughages.

Fig. 24–19. Pasture scenes of contented cows on pasture still characterize family dairy farms. (Courtesy, The Holstein-Friesian Assn. of America, Brattleboro, VT)

4. When pastures are poor and/or the weather is very hot, feed more grain and limit the hay, thereby providing needed added energy and avoiding the excess heat of high-fiber rations. But do not restrict roughage to the point that butterfat test drops.

5. Provide adequate shade for cows on summer pasture.

6. Consider supplementing summer pastures with silage.

# SPECIAL FEEDING PROGRAMS

The nutritive requirements of dairy cattle vary according to age, and according to whether they are lactating or dry. For this reason, more than one ration is necessary.

## DRY COWS

Dry cows have three important jobs: (1) recovering from a heavy milk producing period and resting the mammary glands, (2) developing the unborn calf (more than half the fetal growth occurs during the last two months of lactation), and (3) storing up body reserves for the next milking period. This necessitates that they be properly fed.

The following routine is recommended for dry cows:

1. Turn high-producing cows and cows milked 3 times daily dry 60 to 65 days before expected calving. Turn cows more than 4 years of age dry 50 to 60 days before freshening.

2. Stop grain feeding and milking abruptly to hasten drying off. Examine the udder at intervals, but do not milk it out. (See Chapter 25, section on "Drying Off Cows").

3. Feed only good quality roughage during the first 2 to 3 weeks of the dry period. Beginning 2 to 6 weeks before expected calving, start feeding concentrates twice daily. The amount of concentrate to feed should be determined primarily by the cow's condition and the quality of the roughage. It may vary from 4 to 20 lb per head daily. A guide to desirable gains follows:

| Weight of Cows | Pounds of Gain |
|---|---|
| 800–1,000 | 75–150 |
| 1,000–1,200 | 100–200 |
| 1,200–1,400 | 125–250 |

4. A special dry cow concentrate mix can be fed during this period. The dry cow mix may contain 2 to 4% less protein than the grain mixture normally fed to the milking herd. Also, if milk fever is a problem, as it is in most herds, it is desirable that the dry cow ration (a) be balanced out for calcium and phosphorus (a 1:1 ratio preferably, and not to exceed a 2:1 calcium to phosphorus ratio), and (b) contain added vitamin D.

Toward the end of the dry period, many successful producers follow a program of challenge or lead feeding. They reach a feeding level of 1 to 1½ lb of grain to each 100 lb of liveweight about one week before freshening, and continuing at this rate right up to freshening. This precalf feeding gets the rumen, and the cow's appetite and eating habits, adjusted to liberal feeding before freshening. Also, a cow freshening in good condition starts off better and maintains a higher level of production; her milk is usually higher in total solids; and the incidence of milk fever and ketosis is usually reduced. (See earlier section in this chapter on "Challenge Feeding [Lead Feeding]."）

## FEEDING AT CALVING TIME

Until recent years, cows were not fed much grain prior to freshening; and, immediately after calving. They were given all the roughage they wanted but the grain allowance was sharply reduced. Today, dairy producers feed appreciably more grain to cows prior to calving—from 12 to 20 lb per day before freshening; and feed is not withheld at calving. Instead, on the first day after calving, cows are fed the same amount of grain that they were used to before calving, followed by an increase of 2 to 3 lb per day according to the cow's appetite. The experienced caretaker is in the best position to determine how much, and what, to feed each individual cow at calving time.

## SHOW AND SALE ANIMALS

Dairy animals intended for show or sale should be fed so as to achieve a certain amount of finish or bloom, but they should not be too fat. Linseed meal, beet pulp, oats, barley, and wheat bran are popular feeds in a fitting and showing ration. Likewise, good roughages are always very important.

## DAIRY CALVES

One of the most important phases of dairy production is that of feeding and managing the dairy calves raised for replacement purposes. Statistics reveal that more than 20% of the dairy calves die of sickness or disease before reaching maturity. With good management, many of these losses may be reduced to 3 to 5%. Many of these deaths are caused by faulty nutrition and/or poor housing and management.

Fig. 24–20. Brown Swiss calves in individual pens. (Courtesy, *Brown Swiss Bulletin*, Beloit, WI)

A carefully planned and executed feeding program is necessary to produce growthy, vigorous, and healthy calves. The following feeding program is recommended:

| | |
|---|---|
| **Day 1** ............ | Dam's colostrum |
| **Day 2** ............ | Dam's colostrum |
| **Day 3** ............ | Dam's colostrum |
| **Day 4** ............ | Liquid feed of choice, introduce starter and water |
| **Day 5 to weaning** .... | Continue feeding program |
| **Weaning to 12 weeks** . | Starter (up to 5 lb daily), introduce forage |

Fig. 24–21. Calf barns at Arizona Dairy Co., Higley, Arizona. All calves (more than 4,000 of them each year) are taken from their mothers "wet" and fed *first* colostrum for 2 days, followed by *pooled* colostrum for the next 8 days. Calf losses after birth are only 1½%, which is extremely low. (Courtesy, James Tappan, Arizona Dairy Co.)

Since milk is the primary product of dairy production, it is necessary to switch the young calf to cheaper feeds as expeditiously as possible. At the same time, it is important that the diet promote good health, growth, and development. Four feeds that are routinely fed to calves are (1) colostrum, (2) milk, (3) milk replacers, and (4) calf starters.

### COLOSTRUM

*Colostrum is the milk which is high in antibodies, and which is secreted by cows, and other mammalian females, for the first few days following parturition.*

Colostrum (either dam's colostrum or mixed colostrum from first milking of older cows) should be fed to calves as soon after birth as possible (ideally within 15 minutes and certainly within 4 hours) to protect against disease. Some successful producers remove calves from their mothers *wet* and without nursing, so as to minimize infection from nursing. Then, as soon as possible, they are offered colostrum from a nippled bottle. If they fail to nurse naturally, the colostrum is hand-fed, using a specially designed tube to force it into the abomasum.

Early feeding of colostrum is necessary because:

1. Newborn calves have no antibodies to provide natural protection against disease until colostrum is received.

2. The calves' ability to absorb immunoglobulins (the disease protecting component) is substantially reduced after 24 to 36 hours.

3. Calves may become infected with highly pathogenic bacteria immediately after birth.

Calves should receive a total of 10 to 12% of their birth weight as first milking colostrum, with half of this amount received 4 to 6 hours after birth.

Surplus colostrum can be frozen and stored for a period of 1 year or longer without losing its antibody value. It may be thawed, warmed to about 100°F, and fed as needed.

Excess colostrum is a very nutritious feed, but it has little or no immune benefit (antibody protection) to a calf beyond the first 24 hours of life. Colostrum contains about a third more solids than milk or reconstituted milk replacer, and is highly digestible. So, storage and subsequent feeding of excess colostrum is highly desirable. It may be fed fresh; frozen, then thawed prior to feeding; or stored as sour (fermented) colostrum. Since it is higher in solids than normal milk, it should be diluted 25 to 50% when fed to other than newborn calves, in order to avoid overfeeding and scours.

The composition of colostrum changes rapidly after calving. The first six milkings are higher in nutrients than normal milk or most reconstituted milk replacers.

## MILK REPLACERS

Milk replacers vary in quality, so the buyer/user should study the feed tag. The best milk replacers contain 22% protein, all derived from milk products—skim milk powder, buttermilk powder, dried whole whey, de-lactosed whey, casein, and/or milk albumen. Chemically modified soy protein, soy isolates, and soy concentrates are good, but as plant proteins they are less digestible than milk protein. Meat solubles, fish protein concentrate, distillers' dried solubles, brewers' dried yeast, oat flour, and wheat flour are inferior as protein sources in milk replacers.

A good milk replacer powder should contain a minimum of 15% fat, and it may contain more than 20%. The higher fat level tends to reduce the severity of diarrhea and produce additional energy for growth. Good quality animal fats are preferable to most vegetable fats. However, soy lecithin, especially when homogenized, is an acceptable fat source and improves mixing qualities of the replacer.

The calf can use two carbohydrate sources in milk replacers: lactose (milk sugar) and dextrose. However, two other carbohydrate sources, starch and sucrose (table sugar), are not satisfactory and should be excluded from milk replacers.

Mastitic milk and discard milk (unmarketable milk from cows that were treated for mastitis, metritis, or other health problems) may be fed fresh in the same manner as whole milk, or they may be fermented or preserved with an organic acid. Milk collected for three to six milkings after antibiotic treatment will ferment normally. Extremely abnormal milk (bloody or watery) should not be fed.

Properly managed feeding of mastitic milk to calves will not increase mortality or cause mastitis in these animals when they freshen. Neither will it cause diarrhea. Mastitic milk should not be fed to calves less than 2 days old, as the intestine is permeable to large protein molecules. Calves fed milk containing antibiotics should not be marketed for meat unless the required withdrawal period is observed prior to slaughter.

## CALF STARTERS

A high quality palatable calf starter should be offered when the calf is 4 days old, and not later than 10 to 12 days of age. The best starters are high in energy, contain 16 to 18% protein (20% if calves are weaned before 4 weeks of age), and are free of excessive fines. To encourage consumption, starters should consist of whole, coarsely ground, cracked, or rolled grains. Up to 5% molasses improves palatability and minimizes fines and dust. Whole grains, especially oats, can be fed with starter rations to calves up to 3 months of age. Calf starters should be fed until calves are about 12 weeks of age, with intake limited to 5 to 7 lb per calf daily.

Fig. 24–22. Calf starter and water (note two pails) provided to calf in clean and well-bedded individual pen. (Courtesy, Maddox Dairy, Riverdale, CA)

Many good commercial starters are on the market. Also, calf starters may be home-mixed. Table 24–15 presents examples of some good grain calf starters.

## TABLE 24–15
### GRAIN STARTER RATIONS FOR CALVES[1]

| | | Grain Starters[2] | |
|---|---|---|---|
| | 1 | 2 | 3 |
| **Ingredients (air dry basis)** | | | |
| Corn (cracked or coarse ground) ...... (%) | 50 | 30 | |
| Ear corn (coarse ground) ............ (%) | | | 50 |
| Oats (rolled or crushed) ............. (%) | 22 | 18 | |
| Barley (rolled or coarse ground) ...... (%) | | 20 | 21 |
| Wheat bran ..................... (%) | | 8 | |
| Soybean meal .................... (%) | 20 | 16 | 21 |
| Molasses ....................... (%) | 5 | 5 | 5 |
| Dicalcium phosphate ............... (%) | 0.5 | 0.5 | 0.5 |
| Limestone ...................... (%) | 1.5 | 1.5 | 1.5 |
| Trace mineralized salt and vitamins ... (%) | 1 | 1 | 1 |
| **Composition (dry matter basis)** | | | |
| ADF ......................... (%) | 7.0 | 6.9 | 9.1 |
| Crude protein .................. (%) | 18.1 | 18.0 | 18.4 |
| TDN ......................... (%) | 80.0 | 78.8 | 78.0 |
| Calcium (Ca) ................... (%) | 0.80 | 0.80 | 0.82 |
| Phosphorus (P) ................. (%) | 0.48 | 0.56 | 0.47 |
| Vitamin A ..................... (IU/lb) | 1,000 | 1,000 | 1,000 |
| Vitamin A ..................... (IU/kg) | 2,205 | 2,205 | 2,205 |
| Vitamin D ..................... (IU/lb) | 150 | 150 | 150 |
| Vitamin D ..................... (IU/kg) | 331 | 331 | 331 |
| Vitamin E ..................... (IU/lb) | 11 | 11 | 11 |
| Vitamin E ..................... (IU/kg) | 24 | 24 | 24 |

[1]Source: Linn, J. G., M. F. Hutjens, W. T. Howard, L. H. Kilmer, and D. E. Otterby, *Feeding the Dairy Herd*, Cooperative Extension Services, Universities of Illinois, Iowa State, Minnesota, and Wisconsin, 1988, p. 20, Table 11.

[2]Hay may be offered free choice with grain starters.

## HAY OR SILAGE FOR CALVES

While calves may begin nibbling on good quality hay as early as 5 to 10 days of age, it is not necessary to feed forage before 8 to 10 weeks of age. If the housing and management system makes it inconvenient to provide forage, it may be desirable to incorporate a forage factor (more fiber) in the starter ration. Table 24–16 presents examples of suitable rations for calves not receiving hay or silage. Corn silage or pasture should not be fed before 3 months of age because of their high moisture content which can limit intake and growth. Low moisture haylage is acceptable if it is kept fresh.

### TABLE 24–16
### COMPLETE STARTER RATIONS FOR CALVES[1]

| | Complete Starters | | |
|---|---|---|---|
| | 1 | 2 | 3 |
| **Ingredients (air dry basis)** | | | |
| Corn (cracked or coarse ground) ...... (%) | 40 | 25 | 30 |
| Oats (rolled or crushed) ............. (%) | 14.5 | 8 | 18 |
| Beet pulp ........................ (%) | | 25 | 25 |
| Alfalfa hay (ground) ............... (%) | | 10 | |
| Corn cobs (ground) ................ (%) | 15 | | |
| Soybean meal .................... (%) | 23 | 18 | 20 |
| Molasses ........................ (%) | 5 | 5 | 5 |
| Dried whey ...................... (%) | | 7 | |
| Dicalcium phosphate ............... (%) | 0.5 | 0.5 | 0.5 |
| Limestone ....................... (%) | 1 | 0.5 | 0.5 |
| Trace mineralized salt and vitamins ... (%) | 1 | 1 | 1 |
| **Composition (dry matter basis)** | | | |
| ADF ............................. (%) | 13.3 | 15.8 | 14.2 |
| Crude protein .................... (%) | 18.3 | 18.0 | 18.2 |
| TDN ............................. (%) | 75.5 | 78.0 | 79.4 |
| Calcium (Ca) ..................... (%) | 0.63 | 0.72 | 0.58 |
| Phosphorus (P) ................... (%) | 0.45 | 0.44 | 0.43 |
| Vitamin A ....................... (IU/lb) | 1,000 | 1,000 | 1,000 |
| *Vitamin A* ..................... *(IU/kg)* | *2,205* | *2,205* | *2,205* |
| Vitamin D ....................... (IU/lb) | 150 | 150 | 150 |
| *Vitamin D* ..................... *(IU/kg)* | *331* | *331* | *331* |
| Vitamin E ....................... (IU/lb) | 11 | 11 | 11 |
| *Vitamin E* ..................... *(IU/kg)* | *24* | *24* | *24* |

[1]*Source:* Linn, J. G., M. F. Hutjens, W. T. Howard, L. H. Kilmer, and D. E. Otterby, *Feeding the Dairy Herd,* Cooperative Extension Services, Universities of Illinois, Iowa State, Minnesota, and Wisconsin, 1988, p. 20, Table 11.

## WATER

Clean fresh water in clean pails may be offered free choice starting on day four. Calves fed limited liquid (such as when fed once-a-day) should receive supplemental water, especially during warm weather. Calves offered water during the liquid feeding period (birth to 4 weeks) tend to consume more starter and perform better than calves fed liquid only.

## AMOUNT AND METHOD OF FEEDING, FREQUENCY OF FEEDING, AND AGE OF WEANING

Calves may be separated from their dams at birth, or within 12 to 24 hours after birth. In any case, they should receive their dam's colostrum for the first 3 days of life, following which they may be shifted to a liquid feed of the feeder's choice.

In order to obtain proper growth, calves must be provided adequate dry matter. For an 80- to 100-lb calf, this calls for 1 lb of dry matter (solids) daily from milk, surplus colostrum, or milk replacer, from birth to weaning at 4 weeks.

Milk or milk replacer may be fed by open pail, by nipple feeding from a pail or bottle, or by automated feeding equipment. Each method of feeding is satisfactory, provided it is accompanied by cleanliness and sanitation.

Most calf raisers feed twice daily. Weak or unthrifty calves may benefit from more frequent feedings. Calves need the same amount of dry matter daily, but liquid amounts may have to be reduced to avoid digestive upsets. Dry milk replacer can be added to whole or mastitic milk to increase solid content without increasing the volume of liquid fed. Once-a-day feeding of milk-fed calves has proven successful except when calves are housed in a very cold environment. If once-a-day feeding is practiced, calves should be checked for health and well being at least once in addition to feeding time.

When calves are housed in hutches and the weather is extremely cold, they should be fed a 20% fat replacer 3 times daily and the daily feed allowance should be increased by 1¼ to 1½ times in order to meet their increased energy requirements. Young calves that are doing poorly should be moved to warmer quarters.

Most producers wean calves between 4 to 8 weeks of age. When weaned at 3 weeks of age, calves may have slightly depressed growth rates, temporarily; however, by 12 weeks of age, their weights will be about the same as their later weaned mates. Weaning later than 8 weeks of age may lead to fat calves. A good practice is to wean according to starter intake—wean when the starter intake is 1 to 1½ lb/day. Starter intake can be encouraged by placing a little dry feed in the pail immediately after the liquid has been consumed. In general, early weaning (at 21 to 35 days of age) will reduce feed and labor costs. Calves eating less than 1 lb of starter per day or calves doing poorly should be fed liquid until performance improves and dry feed is consumed in satisfactory amounts.

## VEAL CALF PRODUCTION

*A veal calf is defined as a young bovine animal, usually not over 4 months of age, that has subsisted largely on milk or milk replacers, thus making the color of the lean meat light, grayish pink.* The majority of veal calves are of dairy breeding, consisting of bull calves and heifer calves not retained as replacements. Producers receive a premium if the lean meat of veal is light grayish pink in color (due to reduced muscle myoglobin), characteristic of feeding milk, which is naturally low in iron. Some producers attempt to enhance the desired grayish pink color by restricting iron intake and exercise. Research has shown that a dietary iron concentration of 11.4 to 13.6 mg per pound of dry matter in milk replacers is sufficient for the well being of veal calves and produces desirable grayish pink carcasses, with or without exercise.

A conversion rate of 10 lb of whole milk for 1 lb of body weight gain is normal. If milk replacer is used, the conversion rate is generally about 1.3 to 1.5 lb of dry replacer per pound of gain.

Profitable veal production depends on: (1) a low mortality rate, (2) economical housing, (3) plenty of inexpensive labor, and (4) an established market.

## PREVENTING CALF SCOURS

To prevent calf scours, the producer must prevent primary infection of the newborn. This rests on strict sanitary measures and isolation, along with other preventive measures. Observance of the following practices will lessen the incidence of calf scours:

1. **Make certain that the calf gets first-milk colostrum.** Never assume that the calf has nursed. Many newborn calves never receive sufficient colostrum to protect them from calfhood diseases. So, colostrum should be fed, preferably by hand, soon after birth; large breed calves should receive a minimum of 2 to 3 qt, small breed calves should receive about 3 pt.

2. **Augment natural resistance with vitamins.** Supplementation with vitamins A, D, and E (oral or injectable) immediately after birth is helpful in increasing the calf's natural resistance to scours, especially if colostrum is low in vitamin A content.

3. **Avoid overfeeding and irregularity of feeding.** Overfeeding and irregularity are common causes of calf scours.

4. **Keep feeding utensils clean and sanitary.** Clean the feeding utensils thoroughly after each feeding, then store them upside down to drain and dry.

5. **Don't overcrowd.** In bedded areas, provide 24 to 28 sq ft per calf. In confined, elevated stalls, provide about 20 sq ft per calf.

6. **Provide adequate ventilation.** Provide a minimum of 4 air exchanges per hour in winter and 15 air exchanges per hour in summer.

7. **Avoid damp, wet calves.** Provide plenty of dry bedding in maternity stalls; and provide adequate bedding and ventilation in the calf quarters.

• **Use of electrolytes**—Feeding an oral electrolyte solution usually is beneficial when a calf has a mild case of scours (not off feed, not depressed, and no fever). Recommended treatment follows:

1. Delete or drastically reduce the amount of milk or milk replacer fed.

2. Feed only water containing an electrolyte for 3 to 6 feedings, depending on how soon the feces become firm. Frequent feeding of small amounts is advantageous. A 100-lb calf should consume about 5 qt (10% of body weight) daily.

Oral electrolytes solution can be purchased commercially. If not readily available, a suitable electrolyte mixture can be made by combining the following ingredients:

> 4 tsp of table salt
> 3 tsp of baking soda
> ½ cup of light corn syrup
> 1 gal of water

## DAIRY HEIFER REPLACEMENTS

Fig. 24-23. Replacement Jersey heifers in individual pens. (Courtesy, The American Jersey Cattle Club, Reynoldsburg, OH)

Between weaning and calving (12 weeks to 2-year-olds), the nutrition of heifer replacements is often neglected. At its best, the feeding and management program during this period involves 3 distinct phases: (1) weaning (about 12 weeks of age) to 1 year; (2) 1 year to 2 months before calving at 2 years; and (3) 2 months before calving to calving.

• **Replacement heifers, weaning (about 12 weeks of age) to 1 year**—During this period, replacement heifers may be fed forage free-choice and limited grain. The amount, and the protein content, of the grain mix needed will be determined by the quality of the forage being fed. Pasture can be used successfully in the feeding program of replacement heifers, provided it is supplemented with a grain mix and some dried forage, along with suitable minerals (incorporated in the grain mix or offered free-choice). Also, there should be access to clean, fresh water.

During the yearling stage, replacement heifers should not be overfed and become too fat. Overconditioning has an inhibitory effect on mammary secretory tissue development during the critical period of its maximum development between 3 and 9 months of age and results in lower milk production later in life. Overconditioning of heifers after 15 months of age does not affect mammary secretory tissue.

Table 24-17 (see next page) lists some grower rations for 400-lb calves. If the protein content of the forage is good, little protein supplement will be required in the grain mix. Monensin or Lasalocid can be fed as directed on the label to heifers to improve growth and rate of gain.

**TABLE 24–17**
**RATIONS FOR LARGE BREED DAIRY HEIFERS OF DIFFERENT WEIGHTS[1]**

| Weight (lb) | (kg) | Rate of Gain (lb/day) | (kg/day) | Ration (As-fed) | (lb) | (kg) | Per cent (%) |
|---|---|---|---|---|---|---|---|
| 400 | 182 | 1.7 | 0.8 | Alfalfa hay, 140 RFV, 20% CP | 8.5 | 3.9 | |
| | | | | Grain mix, 12.0% CP .. | 3.0 | 1.4 | |
| | | | | coarse ground barley . | | | 71.0 |
| | | | | rolled or ground oats . | | | 23.0 |
| | | | | molasses | | | 5.0 |
| | | | | trace mineral salt | | | 0.5 |
| | | | | dicalcium phosphate | | | 0.4 |
| | | | | vitamin premix | | | 0.1 |
| | | 1.7 | 0.8 | Alfalfa-grass hay, 113 RFV, 16% CP | 7.0 | 3.2 | |
| | | | | Grain mix, 14.5% CP .. | 4.5 | 2.0 | |
| | | | | coarse ground corn | | | 83.1 |
| | | | | soybean meal | | | 15.5 |
| | | | | trace mineral salt | | | 0.5 |
| | | | | dicalcium phosphate | | | 0.8 |
| | | | | vitamin premix | | | 0.1 |
| | | 1.7 | 0.8 | Orchardgrass hay | 7.0 | 3.2 | |
| | | | | Grain mix, 16.5% CP .. | 5.0 | 2.3 | |
| | | | | rolled or coarse ground barley | | | 65.8 |
| | | | | molasses | | | 4.0 |
| | | | | soybean meal | | | 27.5 |
| | | | | trace mineral salt | | | 0.5 |
| | | | | limestone | | | 2.1 |
| | | | | vitamin premix | | | 0.1 |
| | | 1.7 | 0.8 | Orchardgrass hay | 4.5 | 2.0 | |
| | | | | Corn silage | 12.0 | 5.4 | |
| | | | | Grain mix, 19.7% CP .. | 2.5 | 1.1 | |
| | | | | coarse ground corn | | | 13.5 |
| | | | | rolled or coarse ground oats | | | 6.5 |
| | | | | soybean meal | | | 74.8 |
| | | | | trace mineral salt | | | 1.0 |
| | | | | limestone | | | 4.0 |
| | | | | vitamin premix | | | 0.2 |
| 700 | 318 | 1.7 | 0.8 | Alfalfa hay, 140 RFV, 20% CP | 18.0 | 8.1 | |
| | | | | Grain mix | 2.0 | 0.9 | |
| | | | | coarse ground corn | | | 95.3 |
| | | | | trace mineral salt | | | 1.5 |
| | | | | dicalcium phosphate | | | 3.0 |
| | | | | vitamin premix | | | 0.2 |
| | | 1.7 | 0.8 | Alfalfa hay, 140 RFV, 20% CP | 10.0 | 4.5 | |
| | | | | Corn silage | 15.0 | 6.8 | |
| | | | | Grain mix | 1.0 | 0.5 | |
| | | | | ground ear corn | | | 88.5 |
| | | | | trace mineral salt | | | 4.0 |
| | | | | dicalcium phosphate | | | 7.0 |
| | | | | vitamin premix | | | 0.5 |
| 700 | 318 | 1.7 | 0.8 | Alfalfa-grass hay, 113 RFV, 16% CP | 12.5 | 5.7 | |
| | | | | Grain mix | 6.0 | 2.7 | |
| | | | | ground ear corn | | | 95.9 |
| | | | | soybean meal | | | 2.5 |
| | | | | trace mineral salt | | | 0.6 |
| | | | | dicalcium phosphate | | | 0.9 |
| | | | | vitamin premix | | | 0.1 |
| | | 1.7 | 0.8 | Orchardgrass hay | 4.0 | 1.9 | |
| | | | | Corn silage | 30.0 | 13.6 | |
| | | | | Grain mix | 2.5 | 1.1 | |
| | | | | soybean meal | | | 91.8 |
| | | | | trace mineral salt | | | 2.0 |
| | | | | dicalcium phosphate | | | 1.0 |
| | | | | limestone | | | 5.0 |
| | | | | vitamin premix | | | 0.2 |
| 1,000 | 454 | 1.7 | 0.8 | Alfalfa hay, 113 RFV, 16% CP | 13.0 | 5.9 | |
| | | | | Corn silage | 35.0 | 15.9 | |
| | | | | Mineral-vitamin supplement | 0.5 | 0.2 | |
| | | | | soybean meal | | | 80.0 |
| | | | | trace mineral salt | | | 12.0 |
| | | | | dicalcium phosphate | | | 6.0 |
| | | | | vitamin premix | | | 2.0 |
| | | 1.7 | 0.8 | Orchardgrass hay | 10.0 | 4.5 | |
| | | | | Corn silage | 38.0 | 17.3 | |
| | | | | Supplement | 2.0 | 0.9 | |
| | | | | soybean meal | | | 89.0 |
| | | | | trace mineral salt | | | 3.0 |
| | | | | limestone | | | 7.5 |
| | | | | vitamin premix | | | 0.5 |
| | | 1.7 | 0.8 | Oatlage, 13% CP | 50.0 | 22.7 | |
| | | | | Grain mix | 4.0 | 1.8 | |
| | | | | coarse ground corn | | | 93.9 |
| | | | | trace mineral salt | | | 1.5 |
| | | | | limestone | | | 3.0 |
| | | | | dicalcium phosphate | | | 1.0 |
| | | | | vitamin premix | | | 0.6 |
| | | 1.7 | 0.8 | Urea-corn silage, 0.5% urea | 41.0 | 18.6 | |
| | | | | Grass hay | 10.0 | 4.5 | |
| | | | | Supplement | 0.5 | 0.2 | |
| | | | | soybean meal | | | 64.0 |
| | | | | trace mineral salt | | | 10.0 |
| | | | | limestone | | | 24.0 |
| | | | | vitamin premix | | | 2.0 |

[1] *Source:* Linn, J. G., M. F. Hutjens, W. T. Howard, L. H. Kilmer, and D. E. Otterby, *Feeding the Dairy Herd,* Cooperative Extension Services, Universities of Illinois, Iowa State, Minnesota, and Wisconsin, 1988, pp. 20–21, Table 12.

• **Replacement heifers, 1 year to 2 months before calving at 2 years**—If good quality forage is available, it may be the only feed required for heifers over 1 year of age. A suitable mineral mix should be provided on a free-choice basis. Heifers should gain 1.6 to 1.8 lb per day. If growth is not satisfactory, some grain should be provided. The rations shown in Table 24–17 indicate the amounts to feed when various forage and grain combinations are offered to 700- to 1,000-lb heifers.

Heifers on good pasture require no grain or other forage. As pastures mature, dry out, or are heavily grazed, supplemental feed should be provided. Heifers that are seriously deficient in energy, phosphorus, or vitamin A may not exhibit estrus.

First estrus in heifers is dependent on size and weight, primarily weight. A general guide is that heifers will show their first estrus at 40% of their mature weight, which should be before 12 months of age. Heifers fed high planes of nutrition will show estrus at an earlier age than heifers grown at recommended rates, but underfeeding of heifers will delay estrus. Underfed or very slow growing heifers may ovulate, but estrus signs are often suppressed. Heifers in good condition and gaining weight at breeding time generally show more definite signs of estrus and have improved conception rates over heifers in poor condition and/or losing weight. Overconditioned heifers require more services per conception than heifers of normal size and weight. Table 24–18 shows desirable weights for first breeding at 15 months of age along with weights for other age categories.

### TABLE 24–18
### NORMAL HEART GIRTH MEASUREMENT AND WEIGHT OF CALVES AND HEIFERS DURING THE GROWING PERIOD[1]

| Age in Months | Holstein | | Ayrshire | | Guernsey | | Jersey | |
|---|---|---|---|---|---|---|---|---|
| | (in.) | (lb) | (in.) | (lb) | (in.) | (lb) | (in.) | (lb) |
| Birth | 31 | 96 | 29½ | 72 | 29 | 66 | 24½ | 56 |
| 1 | 33½ | 118 | 32 | 98 | 31½ | 90 | 29½ | 72 |
| 2 | 37 | 161 | 35½ | 132 | 34½ | 122 | 32½ | 102 |
| 4 | 43½ | 272 | 42¾ | 236 | 41¼ | 217 | 38¼ | 181 |
| 6 | 50 | 396 | 48¼ | 340 | 47 | 304 | 44½ | 277 |
| 12 | 62½ | 714 | 59 | 583 | 58¼ | 549 | 56½ | 520 |
| 15 | 65¼ | 805 | 63 | 703 | 61¾ | 640 | 59 | 585 |
| 18 | 68½ | 912 | 66 | 781 | 65 | 727 | 61½ | 660 |
| 21 | 71½ | 1,025 | 68½ | 885 | 67½ | 816 | 64 | 740 |

[1]Body weight for Holsteins and Jerseys from *USDA Tech. Bull. 1098 and 1099*. Heart girth measurements for these weights taken from *Res. Bull. 194* (1960). Nebraska Ag. Exp. Sta. Weights and heart girth measurements for Ayrshires and Guernseys calculated from data furnished by Professor H. P. Davis, University of Nebraska, Lincoln.

• **Two months before calving time**—The feeding of heifers during this period can affect milk production during the first lactation. During the last 2 months of gestation, heifers should make daily gains of about 2.0 lb per day, in comparison with 1.7 lb during early pregnancy. Heifers that are growing rapidly at calving time, and continuing to grow during the first lactation, are more persistent milkers than full-sized heifers at calving.

The amount of grain to feed before calving will depend on forage quality, size, and condition of the heifer. A good thumb rule is to feed grain at 1% of body weight starting about 6 weeks before calving. The ration should have adequate protein, minerals, and vitamins. Excess salt intake can contribute to udder edema and should be avoided the last 2 weeks before calving.

Well-grown heifers will have a minimum of problems at calving time. But plane of nutrition can affect ease of calving in two ways: (1) calf size, and (2) fatness of the dam. Fat heifers have higher incidents of dystocia because of small pelvic openings and usually a larger than normal sized calf at birth. Underfed or poorly grown heifers will require more assistance at calving and have a higher death rate at calving than normal sized heifers.

Fig. 24–24. A Jersey bull (left) in vigorous breeding condition, along with a Jersey cow. (Courtesy, The American Jersey Cattle Club, Reynoldsburg, OH)

## DAIRY BULLS

Bull calves raised for breeding purposes should be fed and handled much the same as heifers. But, since they grow slightly faster than heifers, they should receive somewhat more feed than heifers of the same age.

Older bulls should be kept in thrifty, vigorous condition, but they should not be permitted to become too fat. Mature bulls can be fed the same grain ration as the lactating cows. Depending on the quality of the roughage, usually about ½ lb of grain per 100 lb of body weight will suffice for the mature bull. Also, individual differences must be considered, for some bulls are easier keepers than others.

## DAIRY BEEF

Dairy beef is beef derived from cattle of dairy breeding. Dairy heifers and steers finished at market weights comparable to the finished market weights of heifers and steers of beef breeding are commonly referred to as dairy beef. Generally speaking, there are two different finishing programs and market weights for dairy beef:

1. **High-energy rations; light market weights.** These animals are full fed a high-concentrate ration from about 300 lb to market weights of 750 to 950 lb.

2. **High-roughage; heavy market weights.** These animals are grown on a maximum of roughage to 600- to 750-lb weight, following which the proportion of concentrate is increased; and they are marketed at weights of 1,150 to 1,400 lb.

Economic conditions favor growing and finishing dairy steers and heifers, rather than marketing veal calves. The subject of "Dairy Beef" is fully covered in Chapter 18, "Finishing Cattle: Feedlot; Pasture." Hence, the reader is referred thereto.

## MILK AND BUTTERFAT RECORDS; TESTING PROGRAMS

Individual cow records are a must in any progressive dairy production program. Producers use records as a guide for feeding, for locating and culling out the least profitable cows, and for maintaining a permanent, detailed record of each cow. Records necessitate that each cow be individually identified, and that there be milk and butterfat production records.

The three testing programs sponsored by federal and state research extension services follow.

### DAIRY HERD IMPROVEMENT ASSOCIATION (DHIA)

In this program, supervisors or testers, employed by local or state testing associations, visit the herds one day each month. They identify all cows in the herd, and they weigh and take representative samples of the milk from all animals in the herd for 2 consecutive milkings (3 milkings on herds on 3-times-daily milkings). They then combine the milk samples and test them for butterfat. Sometimes, a test is also made for Solids-Not-Fat (SNF) or protein. In some testing associations, somatic cells or the California Mastitis Test (CMT) is made as an aid in monitoring udder health. Records are also obtained on the amount of grain fed to each cow and to the entire herd. Roughage consumption is calculated on a herd basis. Additional information on breeding dates, calving dates, dry dates, and other factors affecting cow productivity, is recorded. Costs, or value data, are obtained on feed used and milk or other products sold.

The above information is fed into a computer, programmed to provide monthly summaries on (1) individual cows , and (2) the herd, which are sent to each producer.

### OWNER-SAMPLER RECORDS (OS)

Under the owner-sampler plan, the owner, rather than the supervisor, weighs and samples the milk. The samples are then tested at a central laboratory.

### WEIGH-A-DAY-A-MONTH (WADAM)

In this program, the owner weighs the milk from each cow one day each month and enters the weight and feeding information on the forms provided. Then the information and forms are mailed to the supervisor, or a central office, where calculations are completed, following which summaries are returned to the owner.

As can be surmised, DHIA records are considered much more official than the other two, for the reasons that the tester represents an independent agency. Certainly, the other two programs are less costly than the DHIA program; and, if well kept, they can be nearly as valuable in herd management. However, because there is no one solely responsible for the testing and record, the hazard in both the owner-sampler and the weigh-a-day-a-month programs is that they will be neglected.

## OTHER FEED AND MANAGEMENT ASPECTS

There are innumerable other feed and management matters of great importance to dairy production. When disregarded, many of them will materially lessen production and make the enterprise unprofitable, no matter how good the breeding of the animals or the feed being used. Still others are important tools from the standpoint of enhancing good management. Some of these pointers will be discussed in the sections that follow.

### FEEDS AFFECTING MILK FLAVOR

Consumers want milk to taste like milk—not like silage, grass, or weeds.

Although feeds are not the only cause of milk flavors, they are major contributors. Feed flavors enter the milk through the digestive system, respiratory system, and by direct absorption. Research indicates that most feed flavors are detectable in the milk 20 minutes after the feed is consumed, and that they are usually most pronounced at the end of two hours.

Feed flavors that enter the milk through the respiratory system can usually be detected much sooner than those entering through the digestive system. For example, if a cow breathes air reeking with silage odors, these flavors can be detected in the milk almost immediately. Flavors that are directly absorbed by milk are less common, but they appear if the milk is left exposed for a long enough period.

The following control measures are recommended to alleviate feed flavors:

1. **Avoid sudden change to fresh, lush pasture.** Cows should be shifted from winter feeding, or old pasture, to new and lush pastures on a gradual basis. Also, cows should be taken out of such pastures two to three hours before milking. For the same reasons, freshly cut grass should not be fed immediately before milking.

2. **Control and avoid undesirable weeds.** Many weeds when eaten by cows will impart a strong flavor to milk; among them are wild onions, skunk cabbage, some members of the mustard family, bitterweed, carrot weed, ragweed, and others. It is easier to get rid of these weeds today than formerly, so they should be eliminated from pasture and hayfields utilized by milk cows.

3. **Silage flavor.** Silage flavor is both common and objectionable. It can be avoided by feeding all silages after milking, never before or during milking. Usually one will be safe if silage is not fed within two to four hours of milking time, but it's safer to feed it shortly after milking. This permits the flavor-causing material to pass through the cow's digestive system before the next milking.

If cows breathe the odor of silage, it will appear as flavor in the milk. Thus, silage should never be left in the mangers or feed alleys. In fact, it is preferable that it be fed in the corral, and not in the area where the cows are milked.

## TIME AND FREQUENCY OF FEEDING AND WATERING

Where complete rations are fed, cows are generally fed twice or three times daily. Where the forage and grain are fed separately, most dairy farms provide hay free-choice at all times, feed silage once or twice daily, and feed concentrates twice daily.

Hay may be put out in large amounts once daily, or in small amounts several times a day, depending on the type and preparation of the hay and the labor situation. Cows will eat more if moderate amounts are fed at frequent intervals, but the labor requirement may be prohibitive. At least twice-a-day feeding of hay is desirable to reduce wastage. If wastage exceeds 10%, there is need for improvement in the manner of feeding or the quality of the hay, or both.

Silage is usually fed once daily, because of the added cost of more frequent feeding. However, some of the automated systems do not have the added labor cost factor, with the result that they may be used to feed twice daily, or more frequently. Silage should be fed soon after milking, so that any residual feed flavors will disappear before the next milking. Cows should not have access to silage, or other feeds that cause off-flavored milk, for at least 2 to 3 hours prior to milking.

Grain should be fed twice, or more frequently, daily. When high-producing cows are fed in the milking parlor, they are not in the milking parlor long enough to eat all the grain that they need; hence, part of the concentrate of high producers should be fed in the manger along with the hay and/or silage. Group or corral feeding of grain has evolved, as a means of lowering labor costs. Under this system, lactating cows are grouped according to the level of production. Grain is fed twice daily in the manger, right after milking, either (1) as a top dressing on the silage and/or hay, or (2) mixed with the silage and/or hay, with the grain allocation for each group or corral determined by the average level of production of each cow in the group. The advent of electronic/computerized grain feeders gives the producer the opportunity to feed concentrates frequently and to control the amount of feed that cows consume.

Calves should be fed milk or milk replacer once or twice daily, at regular intervals. It is more harmful to overfeed than to underfeed a young calf.

Water should be clean, fresh, and available at all times. If there is too little water or if the cows must stand in line to get it, milk production will decrease. In cold weather, it may be necessary to protect water from freezing, as cows cannot get sufficient water by licking ice.

## FREQUENCY OF MILKING

Most cows are milked 2X per day. Increasing milking frequency to 3X per day increases milk production by 10 to 25% and milking 4X per day will stimulate milk yield another 5 to 15%. Whether these increases in milk production are worth the extra expense in labor, feed, utilities, and milking supplies depends upon economic conditions on each particular dairy farm. Also, with increasing genetic capability of cows, more frequent milking (3X and 4X) appears to improve udder health and lower the incidence of mastitis.

## FEED TERMS/CONVERSIONS

The following terms are generally used in analyzing dairy feed costs and practices:

• **Animal Unit Month (AUM)**—is the amount of feed required for one mature cow for one month. It is equivalent in nutrients to 0.4 ton of average hay, or 320 Mcal of energy.

• **Fat Corrected Milk (FCM)**—is a term used to compare milks of different composition on a standard energy basis. Four percent FCM is calculated by multiplying the actual milk yield by 0.4, then adding to this product the actual fat yield multiplied by 15. For example, if a cow produces 60 lb of 3.5% milk (60 × 3.5%) or 2.1 lb fat, her Fat Corrected Milk production is (60 × .4) plus (2.1 × 15), or 55.5 lb of 4% milk.

• **Estimating TDN from chemical analysis**—frequently, dairy producers have a chemical analysis made of a mixed feed. The results thereof are usually reported in terms of crude protein, crude fat (ether extract), crude fiber, ash, moisture, and N.F.E. Such a reading needs to be augmented by TDN, or energy, values. Where a mixed feed is involved, this poses a very difficult question because digestibility is affected by many things—level of feed intake, particle size, condition of animal, and individuality.

If the formulation of a mixed feed is known, the TDN values can be calculated as follows:

1. Obtain the digestible nutrients of each ingredient by multiplying the percentage of each nutrient by the digestion coefficient. For example, dent corn contains 8.9% protein of which 77% is digestible. Therefore, the percent of digestible protein is 6.9% This same procedure is applied to each ingredient and each nutrient.

2. Then, the TDN is the sum of all the digestible organic nutrients—protein, fiber, nitrogen-free extract, and fat (× 2.25).

This procedure is rather tedious and laborious.
A simple and quick rule-of-thumb method for arriving at the TDN of a mixed ration follows:

(a) 70% of the crude protein, plus
(b) fat times 2¼, plus
(c) N.F.E.

This rule-of-thumb method for determining TDN is close enough for most purposes. However, it is not recommended where considerable amounts of such by-product feeds as almond hulls, raisin stems, or grape seeds are included in the ration.

• **Converting crude protein to digestible protein**—in most good mixed dairy feeds, approximately 80% of the total protein is digestible. Hence, where the crude protein value is given, the digestible protein can be estimated by simply taking 80% of the crude protein. Where the digestible protein is given, and it is desired to convert it to crude protein, this may be done by dividing the digestible protein by 80 and then multiplying by 100.

(Also, see Chapter 4, "Feeding Livestock," section headed "Energy Definitions and Conversions.")

## HEALTH DISORDERS OF DAIRY COWS

Milk production is a very intensive and demanding form of production, accompanied by much stress. Therefore, certain health problems are associated with it; among them, acidosis, bloat, displaced abomasum, fat cow syndrome, grass tetany, hardware disease, ketosis, milk fever, retained placenta, and udder edema.

### ACIDOSIS (LACTIC ACIDOSIS)

*Acidosis is a metabolic disease of cattle and sheep.*
When ruminants are switched from a high-roughage ration to a high-grain ration too rapidly, ruminal acidosis and atony may occur. The microflora of the rumen do not have time to adapt to the new ration, resulting in serious digestive disturbances. The pH of the rumen contents falls, and lactic acid production increases dramatically. Many of the microorganisms cannot live in this environment resulting in radical changes in the bacterial and protozoal populations. Rumen motility, then, becomes static. Afflicted animals show signs of weakness, diarrhea, and abdominal discomfort, and in many cases die.

Prevention consists in starting animals on a high-roughage ration and gradually reducing the roughage and increasing the grain; avoiding erratic feeding; and avoiding abrupt ration changes. Also, the addition of buffers to a high-grain ration aid in the prevention of acidosis.

Various treatments have been used with different degrees of success. Perhaps the most successful treatment consists in decreasing both the amount and kinds of feeds fed (lessening the total amount of the ration), then returning to a higher forage mix.

(Also, see Chapter 4 of this book, Table 4–13.)

### BLOAT

Bloat affects all ruminants, dairy cattle included. When dairy animals are first introduced to concentrates or lush, legume pastures, bloat problems arise. Normally, the eructation process allows for the expulsion of gases that are produced in the rumen. In cases of bloat, frothing (the trapping of gas in the ingesta) prevents this process and intraruminal pressure builds, thereupon distending the left side of the abdomen until the animal is thrown off feed and goes down due to pain and the buildup of toxic metabolites.

Puncturing of the rumen of bloated cattle should be a last resort. **NOTE WELL:** Treatment, control, and prevention of bloat are detailed in the table referred to in the parentheses that follow.

(Also, see Chapter 4 of this book, Table 4–13.)

### DISPLACED ABOMASUM

In recent years, dairy producers have encountered increasing numbers of cows with displaced abomasums. The practice of feeding dry cows liberal amounts of grain has been pointed to as one contributing factor. It has been suggested that the lack of bulk and rumen fill promotes flabby muscle tone of the rumen, thereby permitting the abomasum to become displaced. The feeding of some effective roughage is recommended.

### FAT COW SYNDROME (FATTY LIVER SYNDROME)

Cows afflicted with fat cow syndrome have lowered level of liver function due to enlarged liver infiltrated with fat. General herd signs include very fat cows in the dry cow group, decreased resistance to infection, increased incidence of metabolic diseases such as ketosis, reduced feed intake, and reduced milk production and body weight.

Normally, the amount of fat in the liver is quite low (i.e. 1 to 2%), although it may increase to 4 to 10% precalving. However, cows with the fat cow syndrome may have more than 20% fat accumulation in the liver.

The incidence of the fat cow syndrome may be reduced by avoiding overconditioning of cows during late lactation and the dry period and by formulating rations that maximize feed intake after calving.

### GRASS TETANY (HYPOMAGNESEMIA)

Grass tetany is a metabolic condition that affects cows (especially lactating cows) grazing lush pasture high in nitrogen, resulting in low absorption of magnesium, and is most common in the spring. Afflicted animals develop tetany, walk with a stiff gait, go into convulsions, and may die. During the danger period, cows grazing lush pasture should be supplemented with magnesium oxide.

(Also, see Chapter 4 of this book, Table 4–13.)

### HARDWARE DISEASE

The condition in which the collection of foreign objects irritates or punctures the reticulum is called reticulitis, or *hardware disease*. Ruminants are grazers by nature and are sometimes indiscriminate in their selection of feed. Often they will consume nails, pieces of wire, and other foreign objects. Due to the motility patterns of the gastric region, these objects tend to accumulate in the reticulum; and the presence of these sharp objects can pose serious problems, especially if the reticulum should be punctured.

### KETOSIS

*Ketosis is a metabolic disease characterized by a drop in milk production, hypoglycemia, ketonuria, and a rapid loss of weight.* In general, the disorder develops within the first 30 days of lactation. While no preventative measures have proven to be 100% effective, the feeding of propylene glycol or sodium propionate has been successful in some cases; and the inclusion of niacin in the ration at a level of 6 g per head per day in the last 2 weeks of the dry period and in the fresh cow ration may aid in reducing the incidence of ketosis. Additionally, starting limited grain feeding during the latter part of the dry period is helpful in preventing ketosis.

If a cow comes down with ketosis, a glucose solution can be administered intravenously to promote rapid recovery.

(Also, see Chapter 4 of this book, Table 4–13.)

## MASTITIS

*Mastitis is an infection of the mammary gland caused by any one of several bacterial organisms, most frequently staphylococcus or streptococcus.* Symptoms vary with degree of inflammation. Acute cases show a swollen and painful udder, and frequently cause the cow to go off feed. Chronic cases result in slightly swollen udders and small flakes in the milk.

No feed is known to cause or cure mastitis. However, the sudden addition of nutrients may result in a marked increase in milk production and cause more stress; in turn, this may cause subclinical cases. Also, feeding recommended levels of selenium and vitamin E may be helpful in preventing mastitis.

## MILK FEVER

At or soon after calving (generally within 48 to 72 hours), a sharp decrease in blood calcium (hypocalcemia) occurs in some cows, resulting in loss of appetite, subnormal temperature and an unsteady gait. This is followed by nervousness, and, finally, collapse or complete loss of consciousness. The head is usually turned back. The name *milk fever* is a misnomer, because the body temperature is below normal.

The triggering mechanism for this drop in blood calcium is the onset of lactation—an intensive mobilization of calcium.

Feeding practices involving dry cows can markedly reduce the incidence of milk fever. When certain cows are known to have a history of milk fever, excessive calcium intake during the dry period should be avoided. If the problem persists, some nutritionists recommend the limited feeding of a high-energy, low-calcium (less than 15 g of calcium per day) ration. After calving, the calcium levels should be raised rapidly to meet the high requirements of lactation.

Recent studies have indicated that the addition of certain anions (negatively charged ions) to the ration may reduce the incidence of milk fever by aiding calcium absorption and mobilization. But more experimental work is needed relative to this method.

(Also, see Chapter 4 of this book, Table 4–13)

## RETAINED PLACENTA

Normally, the placenta is expelled within 3 to 6 hours after parturition. If it is retained as long as 12 hours after calving, competent assistance should be obtained.

Retained placenta occurs in about 10% of dairy cattle. It is more common following abnormally short or abnormally long pregnancies, among older cows, and following twinning. Experimentally, it has been found that a high incidence of retained afterbirth occurs when premature calving is induced by the administration of glucocorticoid drugs.

While infections such as brucellosis, vibriosis, and others have been associated with abortion and retained afterbirth, these are by no means the only causes. Its incidence increases with parturient hypocalcemia and appears to be related to the fat cow syndrome. Nutritionally, deficiencies of vitamin A, selenium, copper, and iodine have been incriminated. The prepartum injection of selenium at low doses has been shown to reduce the incidence of retained placenta. Also, it appears that fewer cases of retained placenta occur (1) when calves stay with their dams and nurse for 12 to 24 hours, and (2) when cows are kept on pasture year-round. Among cows which have previously retained the placenta, 20% are likely to do so again.

Calves born when the placenta is retained are likely to be weak. A retained placenta may cause pathological conditions resulting in uterine tissue destruction. This condition may or may not affect milk production, but it very likely will result in 5 to 10% lower fertility than for normal cows.

When a retained placenta is encountered, appropriate treatment should be administered by the veterinarian. The usual treatment consists of either antibiotics or sulfonamides, by direct infusion into the uterus or by other routes (or both).

It is seldom advisable to attempt removal of retained placenta. If the membranes are dragging in the ground, they should be cut off at the hocks. But never, never tie bricks or other objects to it. In most instances, the membranes will fall out by themselves in 1 to 2 weeks.

It is desirable to have all cows which have had retained afterbirth examined at about 30 days after calving. If pus is present, they may be treated with estrogenic hormones to induce heat and then the uterus can be infused with an antibiotic solution or perhaps with a dilute Lugol's (iodine) solution. Such examination and treatment may save considerable time with regard to the onset of normal cycles and may result in a higher conception rate.

## UDDER EDEMA

Udder edema, characterized by excessive accumulation of fluid in the intercellular spaces of the udder and forward of it, is sometimes of serious magnitude before calving. The cause is not well understood, but a reduction of blood proteins at calving time and increased blood flow without compensatory lymph removal have been suggested. It appears that high intakes of sodium chloride or potassium chloride increase the severity of udder edema, and that restriction of the salt intake will reduce the severity.

Severe edema may reduce milk production and may be one of the causes of pendulous udder.

## DRUGS AND PESTICIDES

Many drugs used in the treatment of cattle diseases, along with many pesticides, are excreted in milk. Such milk should be discarded to prevent the drugs from entering the human food supply. The presence of antibiotics, sulfas, and pesticides in milk is illegal. Dairy producers should follow a residue avoidance program.

# QUESTIONS FOR STUDY AND DISCUSSION

1. Discuss the relationship between, and the importance of, each of the following factors in the productivity and profitability of dairy cows: (a) heredity, (b) environment, and (c) feed.

2. Since it costs slightly more to feed high producers than low producers, how can high producers return the most profit?

3. Define the term *nutrient*. What nutrient needs of a lactating cow must generally be met before milk will be produced in quantity?

4. What are the most common sources of energy for cows? In Tables 24–1 and 24–2, why are separate net energy values given for each maintenance ($NE_m$) and gain ($NE_g$), whereas only one net energy value is given for the maintenance, pregnancy, and milk production involved in lactation ($NE_{lc}$)?

5. What factors affect the energy requirements of a lactating cow?

6. For what purposes is protein essential for dairy cattle? Define the following protein terms as they apply to cattle feeds and requirements: (a) degraded intake protein (DIP), (b) undegraded intake protein (UIP), and (c) nonprotein nitrogen (NPN).

7. Why are dairy cattle of all ages and stages of production more apt to suffer from lack of phosphorus in their feed than from a deficiency of any other mineral element?

8. In order to guard against any trace mineral deficiencies in the dairy ration, what trace minerals should be provided? How should they be provided?

9. What vitamins do dairy cattle require? What are the usual sources for each of the required vitamins?

10. Discuss the importance of each of the following factors in dairy cattle rations: palatability, preparation, variety, bulk, laxativeness, and cost.

11. List and discuss each of the common sources of roughages for dairy cattle. What factors should determine the proportion of roughage to grain fed to lactating cows?

12. Discuss the importance and sources of each of the following special feeds for dairy cattle: (a) fats and oils, (b) fiber, and (c) urea.

13. Many additives are used by dairy producers to increase milk production, affect milk composition, and/or improve feed efficiency. Why and how are each of the following additives used by some dairy producers: (a) antibiotics, (b) bovine somatotropin (BST), (c) buffers (mineral salts), (d) ionophores, (e) isoacids, and (f) thyroprotein? If a product contributes to overproduction of milk, should its use be banned by the federal government?

14. What is a *commercial dairy feed*? What different types of commercial dairy feeds are commonly available? Why are large dairies trending to buying separate ingredients and doing their own mixing?

15. The feed requirements of lactating cows are significantly influenced by the stress of milk production, including both volume and composition. How may Table 24–4 be used as a guide for dry matter intake?

16. Give the pertinent and common thumb rules for forage feeding of lactating cows.

17. Give the pertinent and common thumb rules for concentrate feeding of lactating cows.

18. Of what importance is the milk-feed price relationship portrayed in Fig. 24–13 to a practical dairy producer?

19. Evaluate the usefulness of Table 24–8, Feed Mixing Guide for Lactating Cows (As Fed Basis), for the small family farm dairy producer in the United States or abroad.

20. Discuss the concept and the method of applying each of the following feeding systems: (a) phase feeding, (b) challenge feeding (lead feeding), and (c) group feeding.

21. Give the two primary reasons for many dairy producers discontinuing pasture grazing. Outline a summer feeding program for most dairy cows.

22. Outline a special feeding program for (a) dry cows, and (b) cows at calving time.

23. Discuss the formulation and role/use of each of the following in raising dairy calves: (a) colostrum, (b) milk replacers, (c) calf starters, and (d) hay or silage.

24. Discuss the amount and method of feeding, frequency of feeding, and age of weaning dairy calves.

25. What's a veal calf? Will the percentage of dairy calves vealed increase or decrease in the future? Justify your opinion.

26. Outline a special feeding program for each of the following: (a) dairy heifer replacements, (b) dairy bulls, and (c) dairy beef.

27. Discuss how each of the following three dairy testing programs are conducted: (a) Dairy Herd Improvement Association (DHIA), (b) owner-sampler records, and (c) weigh-a-day-a-month.

28. List the most common feeds and feeding methods that affect milk flavor, and outline measures to alleviate each of them.

29. Discuss the following feed and management aspects as they apply to lactating cows: (a) time and frequency of feeding and watering, and (b) frequency of milking.

30. What is the meaning of and/or the method of computing each of the following feed terms/conversions: (a) animal unit month (AUM), (b) fat corrected milk (FCM), (c) estimating NDN from chemical analysis, and (c) converting crude protein to digestible protein?

31. Discuss the cause, symptoms, and treatment/prevention of disorders of dairy cows: (a) acidosis, (b) bloat, (c) displaced abomasum, (d) fat cow syndrome, (e) grass tetany, (f) hardware disease, (g) ketosis, (h) mastitis, (i) milk fever, (j) retained placenta, (k) udder edema, and (l) drugs and pesticides.

## SELECTED REFERENCES

| Title of Publication | Author(s) | Publisher |
|---|---|---|
| *Animal Nutrition* | L. A. Maynard, *et al.* | McGraw-Hill Book Company, New York, NY, 1979 |
| *Applied Animal Feeding and Nutrition*, Third Edition | M. H. Jurgens | Kendall/Hunt Publishing, Dubuque, IA, 1974 |
| *Basic Animal Nutrition and Feeding* | D. C. Church<br>W. G. Pond | O & B Books, Corvallis, OR, 1974 |
| *Dairy Cattle: Principles, Practices, Problems, Profits*, Third Edition | D. L. Bath, *et al.* | Lea & Febiger, Philadelphia, PA, 1985 |
| *Dairy Farm Management* | T. Quinn | Delmar Publishers, Inc., Albany, NY, 1980 |
| *Digestive Physiology and Nutrition of Ruminants* | D. C. Church | O & B Books, Corvallis, OR, 1974 |
| *Energy Metabolism of Ruminants, The* | K. L. Baxter | Hutchinson & Co., Ltd., London, England, 1962 |
| *Feed Formulations*, Third Edition | T. W. Perry | The Interstate Printers & Publishers, Inc., Danville, IL, 1982 |
| *Feeds and Feeding*, Third Edition | A. Cullison | Reston Publishing Company, Inc., Reston, VA, 1982 |
| *Feeds and Feeding*, 22nd Edition | F. B. Morrison | Morrison Publishing Co., Ithaca, NY, 1956 |
| *Illinois-Iowa Dairy Handbook* | Cooperative Extension Service | University of Illinois, Urbana, and Iowa State University, Ames, 1983 |
| *Large Dairy Herd Management* | C. J. Wilcox, *et al.* | University Press of Florida, 1978 |
| *Livestock Feeds & Feeding*, Second Edition | D. C. Church | O & B Books, Inc., Corvallis, OR, 1984 |
| *Nutrient Requirements of Dairy Cattle*, 6th Revised Edition | NRC, R. W. Henken, Chairman, Subcommittee, Dairy Cattle Nutrition | National Academy Press, Washington, DC, 1988 |
| *Principles and Practices of Feeding Dairy Cows* | W. H. Broster, *et al.* | College of Estate Management, Reading University, 1986 |
| *Science of Providing Milk for Man, The* | J. R. Campbell<br>R. T. Marshall | McGraw-Hill Book Company, New York, NY, 1975 |
| *Stockman's Handbook, The*, Sixth Edition | M. E. Ensminger | The Interstate Printers & Publishers, Inc., Danville, IL, 1983 |
| *United States-Canadian Tables of Feed Composition* | J. H. Conrad, Chairman of Subcommittee | National Academy Press, Washington, DC, 1982 |

Typical New England dairy farm. (Courtesy, United Cooperative Farmers, Inc.)

Contented cows. (Photo by J.C. Allen and Son, West Lafayette, IN)

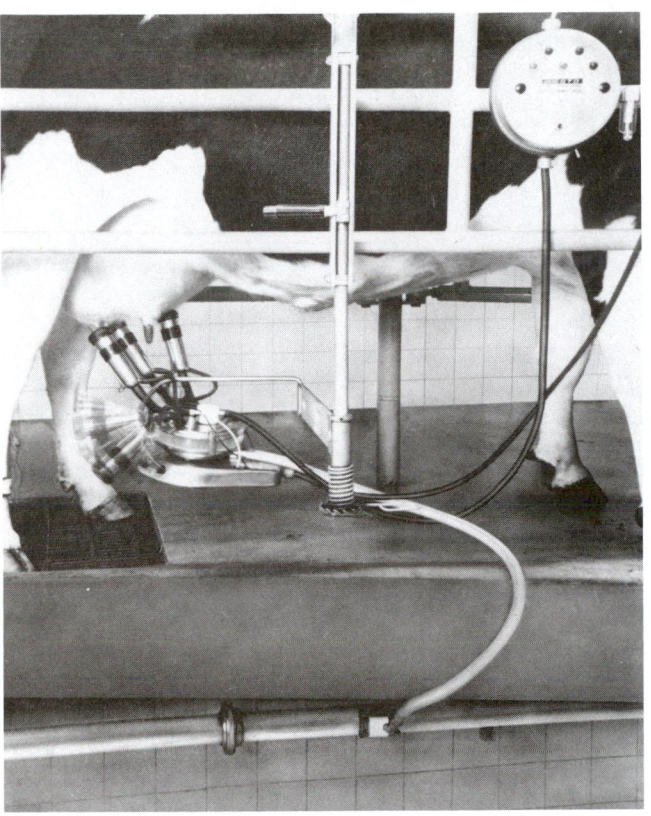

Modern milking machine, with (1) quarter take-off (QTO) milker, showing one QTO automatically dropping from right, front quarter; and (2) low pipeline. (Courtesy, Babson Bros. Co., Oak Brook, IL)

# MILK SECRETION
# AND HANDLING

# Chapter 25

Zoologically, cattle belong to the class *Mammalia*; and all mammalia are warmblooded, hairy animals that produce their young alive and suckle them for a variable period on a secretion from the mammary glands called *milk*.

The number of mammary glands and their position on the body is peculiar to each species. For example, the cow has four glands (quarters), each with a passageway (teat) to the outside; whereas in the sow and the bitch there are generally 10 or more mammary glands.

The ability of dairy cattle to produce large amounts of milk is the principal reason they are accorded a prominent place in American agriculture. It is important, therefore, that the physiology of milk production and the methods of "milk harvesting" be fully understood.

Fig. 25–1. Cows in holding area getting automatic udder wash before entering parlor for milking. (Courtesy, Babson Bros. Co., Oak Brook, IL)

## CHEMICAL COMPOSITION OF MILK

Contrary to popular belief, all milk is not alike. Chemically, it varies in composition by species (see Table 25–1). Also, the composition of milk differs according to breeds. Table 25–2 shows the average composition of milk of each of five major breeds of cattle. It is not claimed that these figures are true breed averages; rather, they give some indication of the levels of each component for each breed and the differences between breeds.

TABLE 25–1
AVERAGE PERCENTAGE COMPOSITION OF MILK
OF THE COW AND OTHER MAMMALS[1]

|  | Cow | Goat | Human | Sheep | Mare | Pig | Ass |
|---|---|---|---|---|---|---|---|
| Water | 87.70 | 86.0 | 88.2 | 81.3 | 89.8 | 81.9 | 90.1 |
| Fat | 3.61 | 4.6 | 3.3 | 6.9 | 1.2 | 6.8 | 1.3 |
| Lactose | 4.65 | 4.2 | 6.8 | 5.2 | 6.9 | 5.5 | 6.5 |
| Protein (N × 6.38) | 3.29 | 4.4 | 1.5 | 5.6 | 1.8 | 5.1 | 1.6 |
| Ash | 0.75 | 0.8 | 0.2 | 1.0 | 0.3 | 0.7 | 0.5 |

[1]Source: Pearson, D., *The Chemical Analysis of Foods*, The Chemical Rubber Co., Cleveland, OH, 1971, p. 411, Table 12.1, with pig added by the author.

TABLE 25–2
COMPOSITION OF MILK FROM DIFFERENT BREEDS[1]

| Breed | Fat | Protein | Lactose | Ash | Total Solids | SNF |
|---|---|---|---|---|---|---|
|  | (%) | (%) | (%) | (%) | (%) | (%) |
| Ayrshire | 4.1 | 3.6 | 4.7 | 0.7 | 13.1 | 8.52 |
| Brown Swiss | 4.0 | 3.6 | 5.0 | 0.7 | 13.3 | 8.99 |
| Guernsey | 5.0 | 3.8 | 4.9 | 0.7 | 14.4 | 9.01 |
| Holstein | 3.7 | 3.1 | 4.9 | 0.7 | 12.4 | 8.45 |
| Jersey | 5.1 | 3.9 | 4.9 | 0.7 | 14.6 | 9.21 |

Source: *Dairy Guide*, The Ohio State University, Columbus, OH.

Also, the composition of milk is greatly affected by both physiological and environmental factors, which will be discussed later in this chapter.

The first secretion of the mammary gland following parturition is known as colostrum. Colostrum is nature's product, designed to give young a good start in life. It is higher than milk in dry matter, protein, vitamins, and minerals. Additionally, it contains antibodies that give newborn animals protection against certain diseases.

# UDDER DEVELOPMENT

The development of the udder starts early in the growth of the fetus. At birth the udder consists of the teats, teat cisterns, gland cisterns, and structures which will later develop to form the duct system. From birth to puberty, little mammary change takes place, although there may be some deposition of fat in well-fed animals.

After puberty, there is some growth of the duct system with each recurrence of the estrus cycle. This growth is stimulated by estrogen, a hormone secreted by the ovarian follicle. The secretory tissue develops under the influence of another hormone, progesterone, from the corpus luteum of the ovary. However, there is little apparent enlargement of the udder until the secretion of colostrum begins, a short time before the birth of the calf.

# STRUCTURE OF THE UDDER

Mammary glands vary considerably among species in arrangement of glands and nipples. Udders also vary among cows—in sizes, shapes, quality, and udder attachments.

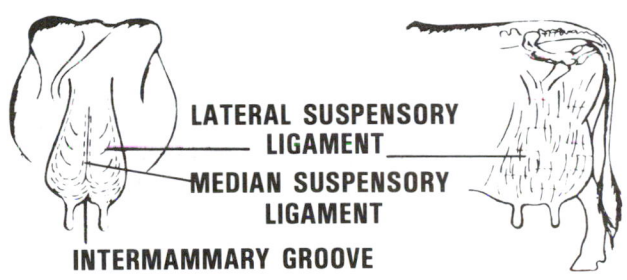

Fig. 25-2. How the udder is suspended.

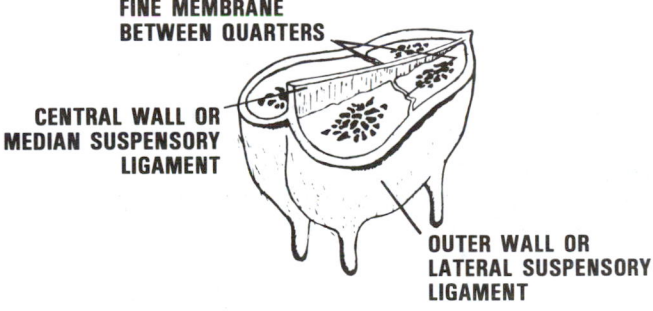

Fig. 25-3. Cross section of the udder.

The udder consists of the following parts:

1. **Suspension.** On a mature cow, the udder weighs from 25 to 60 lb, and it may hold 50 or more lb of milk; thus, good suspension is imperative. The main supporting structures of the udder are the median and lateral suspensory ligaments, and the skin.

A well-attached udder fits snugly against the abdominal wall in front and on the sides, and extends high between the thighs in the rear. "Breaking away" of the udder from the body occurs when the supporting ligaments weaken or stretch.

2. **Four glands.** The cow's udder consists of four separate quarters; and there is no way for the milk to move from one half to the other, or from one quarter to the other. The left and right halves are separated by a distinct membrane and the main supporting ligaments. The front and rear quarters are also completely separated, but the division is less marked.

Because of their greater depth, the rear quarters usually produce about 60% of the milk. This partially explains why the forequarters often are milked out before the rear quarters.

3. **Teats.** The teat is a tube of skin which hangs down from the udder. It is hollow and more or less closed at the top and at the bottom. The bottom of the teat is closed by a circular (sphincter) muscle. If this muscle is tight, the cow may be a *hard* milker. If it is loose, the cow is an *easy* milker.

The opening in the lower end of each teat is known as the teat canal.

4. **Milk-collecting system.** The duct system of the udder consists of the teat cistern, the gland cistern, and many collecting ducts. The udder also contains the nerve and lymph system and tissues that support and connect the basic parts.

5. **Alveolus.** The basic milk-producing unit in the udder is a very small bulb-shaped structure with a hollow center called the alveolus. It is estimated that each cubic inch of udder tissue contains one million alveoli; hence, in total there are billions of alveoli in the udder. The alveoli are lined with epithelial cells which manufacture milk from the blood circulating through the udder. The milk is then stored in the alveoli until it is forced out by hormone action on small muscle tissues around the alveoli. When this forcing action takes place, the milk enters small ducts, passes through the larger ducts, and collects in the milk cistern just above the teat.

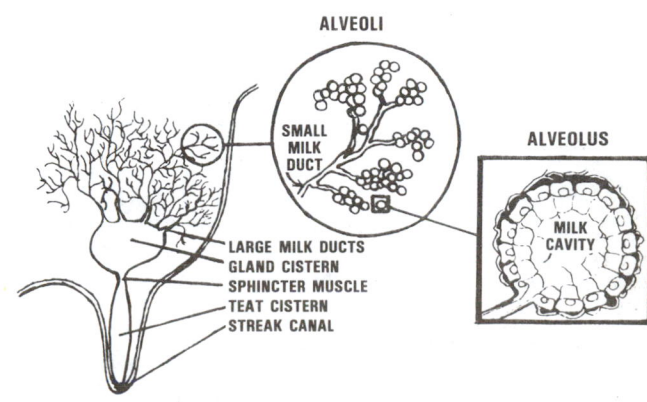

Fig. 25-4. Milk is produced and stored in the alveoli. By action on the musclelike cells surrounding the alveoli, the hormone, oxytocin, forces milk into the large ducts and cisterns of the udder. At milking time, it passes through the streak canal.

## MILK SECRETION

The phenomenon of milk secretion is very complex. The tiny epithelial cells in the alveoli are the "factories" in which the constituents carried by the blood are taken out of the blood capillaries and synthesized (when necessary) into the component parts of milk—fat, lactose, protein, minerals, vitamins, and the other constituents of milk. These cells have the unique ability of changing or manufacturing materials entirely different from those from which they came. Further, it is believed that each cell in the alveoli can manufacture all the milk constituents, and that there are no specialized cells for each type of compound.

The efficiency of milk secretion becomes apparent when it is realized that a cow that produces 14,500 lb of milk during one year manufactures 523 lb of milk fat, 674 lb of milk sugar, 477 lb of milk protein, and 109 lb of minerals and vitamins, or a total of over 1,783 lb of food. (Since milk contains approximately 87.7% water and 12.3% solids, 14,500 lb of milk contains 1,783 lb of solids—12.3% of 14,500.) That's equivalent to the carcass weight produced by 2½ steers in 18 months' time. Moreover, the cow is still alive and can repeat the productivity again and again, whereas the 2½ steers must be slaughtered or "spent."

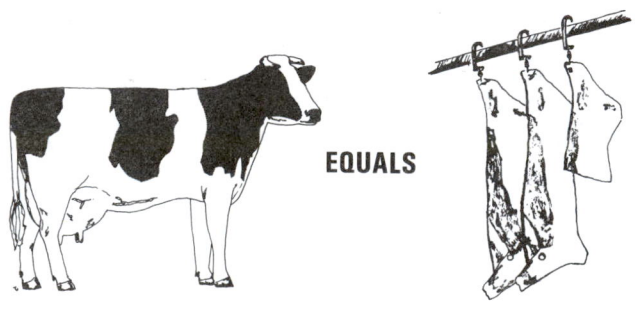

**EQUALS**

Fig. 25-5. It takes 2½ steers 18 months' time (for each steer) to produce as much carcass weight as one cow produces in milk in one year. And the cow remains alive to do it over again!

Approximately 500 lb of blood must pass through the udder to supply the raw materials for 1 lb of milk. This means that a cow producing 60 lb of milk in a day pumps 15 tons of blood through her udder daily. Hence, the blood supply to the udder is extremely important. In fact, milk production may be limited by the nutrients in the blood and the amount of blood available to milk-secreting cells of the udder.

Onset of milk production, shortly before parturition, is brought about by the action of the hormone, prolactin, secreted by the anterior pituitary. Other hormones are necessary for the maintenance of milk production; specifically, the thyroxin, secreted by the thyroid gland, and the adrenal gland secretions.

Milk secretion takes place more or less continuously until stopped by the mounting pressure within the alveoli.

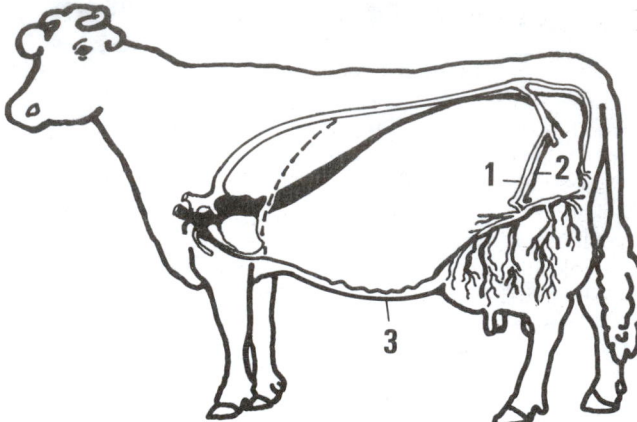

Fig. 25-6. Blood circulation to and from the udder. The external pudenal artery (1) supplies blood carrying the components from which milk is produced to each half of the udder. The blood and its components that are not used for milk production are drained away from the udder by the external pudenal vein (2) and the subcutaneous abdominal vein (3). (Source: *Dairy Guide*, The Ohio State University, Columbus)

## MILK EJECTION OR LET-DOWN

As has already been noted, the milk is stored in the alveoli. Before it is available to the calf or milker, it has to be forced from the alveoli into the larger ducts and cisterns. This process is known as the *let-down* of the milk. Here is how it works (see Fig. 25-7): When the udder (especially the teats) is stimulated by the calf or a milker, (1) impulses are conducted along the nerves to the posterior pituitary at the base of the cow's brain; (2) the posterior pituitary stores and releases the hormone oxytocin into the bloodstream; (3) the blood transports oxytocin back to the udder; and (4) oxytocin causes the smooth, muscle-like cells surrounding each alveolus to contract, thereby forcing the milk out of them into the large ducts and cisterns of the udder.

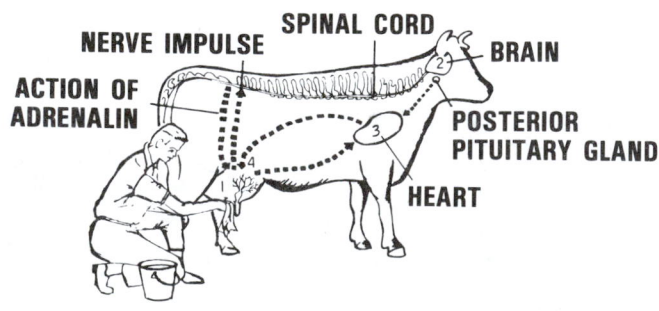

Fig. 25-7. Steps in milk let-down. See narrative for details.

The stimulation of the udder lasts for a limited time only—less than one minute in a fresh cow, since oxytocin is destroyed in the bloodstream. Hence, once the let-down has occurred, it is important that the milk be removed within approximately five minutes to obtain the greatest amount. This is so because a second stimulation cannot be obtained soon after the first.

## DISTURBANCES OR HOLD UP

When a cow is frightened or angry (from being hit, chased, shouted at, barked at, or for other reasons), she may not let down her milk. This is because of an overriding hormone action. Upon such occasions, another hormone, adrenalin, is released into the bloodstream. This hormone interferes with the action of oxytocin by reducing blood circulation to the alveoli.

Hold up of milk may also result from a poorly operated milking machine or from a poor hand milker.

## MILKING THE COW; MANAGED MILKING

Milking is the act of removing milk from the udder. The calf sucks milk, and, simultaneously, massages the teat. In hand milking, the teat is grasped between thumb and forefinger; then, by applying pressure with the other fingers, milk is forced from the teat cistern through the streak canal. In machine milking, the milk is drawn out by vacuum in much the same way that a nursing calf does.

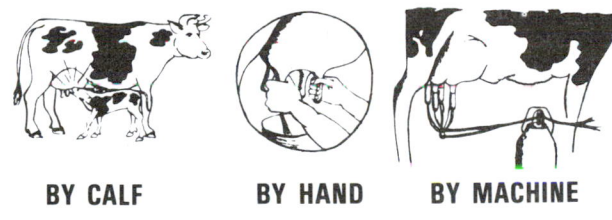

**BY CALF**  **BY HAND**  **BY MACHINE**

Fig. 25–8. Three ways to milk a cow.

Milking is the most important single job to be done on the dairy farm. Some individuals are excellent milkers, others are very poor—and this statement applies to both machine milking and hand milking. Also, it is important that the cow be milked at regular times, preferably by the same milker; and that each milking be a pleasant experience. Cows like to be milked—if it is done properly.

The physiology of the discharge of milk is a delicate process, and it requires the close cooperation of the milker and the cow if it is to be successful. A managed milking program is made up of the following coordinated steps:

1. **Preparing the equipment.** Prior to milking, the equipment to be used in the milking process should be assembled and sanitized. Also, it should be checked and adjusted if necessary.

2. **Preparing the cow.** Under natural conditions, the cow is primed or stimulated by the suckling of the calf. This process can be stimulated by washing the cow's teats and udder with warm water (120 to 130°F), then massaging and drying them with a paper towel. Following this process, remove 2 or 3 streams of milk from each quarter into a strip cup (never strip milk onto the floor) and examine for visible evidence of mastitis.

Also, this (a) washes out any debris adhering to the end of the teat, and (b) enhances the let-down effect.

About 45 seconds after the priming stimulus, the udder becomes full and firm (especially in early lactation), and milk occasionally will leak from the teat. This is evidence that the cow has let down her milk and is ready for the next step.

3. **Attaching the teat cup and beginning.** About 1 minute after washing the udder, and not more than 1½ minutes, the teat cups should be attached and milking should begin. Most cows will milk out in 3 to 6 minutes, depending upon the amount of milk and the characteristics of the cow. Also, and most important, each quarter should be milked individually, because some quarters milk out faster than others.

4. **Stripping by machine.** When it is apparent that the cow is about to be milked out, she should be machine stripped. This consists of pulling down on the teat cups with one hand, and massaging the udder downward with the other. This process should not take over about 20 seconds.

5. **Removing the teat cup.** Both incomplete and overmilking should be avoided. The greatest cause of machine injury is leaving the teat cups on too long. Incomplete milking usually results because one or more quarters are more difficult to milk than the others.

As soon as the udder is empty, and before the teat cups crawl up, they should be removed, properly and gently. Then, dip the teat with a fresh disinfectant solution (100 ppm idophor or chlorine, or other sanitizing agent). This will remove the milk from the ends of the teats and prevent the invasion of bacteria into the udder. Also, it will avoid attracting flies.

As soon as the teat cups have been removed from the udder of the cow, they should be cleaned. First, dip them in clean, cold water to remove milk inside the liners; then put them in a clean, warm, approved sanitizing solution. Change the solution after each five to seven cows.

6. **Cleaning up equipment.** After milking the last cow, all milking equipment should be thoroughly cleaned and put away.

7. **Milking time.** The actual milking time per cow will range from 3 to 6 minutes, with an average time of 3½ minutes for cows in mid-lactation. But additional time must be allowed for let-down, adjustments, and interval between cows.

The number of machines one milker can manage successfully depends upon the type of barn, the type of milking equipment, the ability of the milker, and the jobs the milker has other than milking. In a stanchion barn, 2 units per milker works best. In a herringbone system, 4 units per milker is the upper limit. Depending on the individual parlor installation, the number of cows milked per hour per milker usually ranges between 30 and 44. However, additional time must be allowed to bring the cows in from the outside, setting up, cleaning up, as well as milking problem cows.

8. **Milking order.** Cows that have mastitis or a history of chronic mastitis are a source of infection to noninfected cows. Hence, it is well to milk *clean* cows first. A desirable milking order in stanchion barns is:

    a. First-calf heifers that have been free of mastitis.

    b. Older cows that have been free of mastitis.

    c. Cows that have a previous history of mastitis, but which no longer show symptoms.

    d. Cows with quarters producing abnormal milk.

## MILKING PARLORS

The milking parlor is for the purpose of improving labor efficiency, working conditions, and sanitation surrounding the milking operation. Parlors are of various designs; seven of the most common arrangements of which are shown in Fig. 25–9.

# VARIOUS MILKING PARLORS

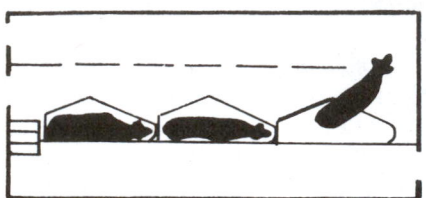

**SINGLE SIDE OPENING**

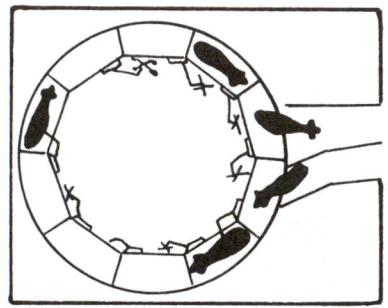

**ROTARY TANDEM**

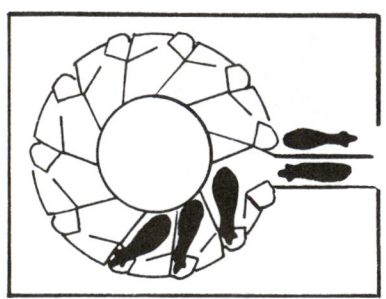

**ROTARY HERRINGBONE**

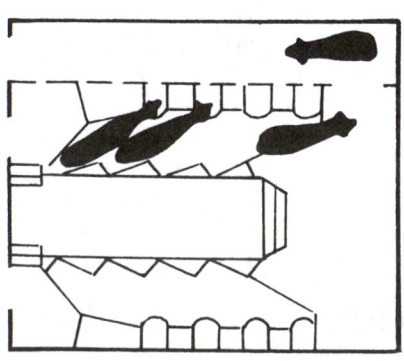

**DOUBLE HERRINGBONE**

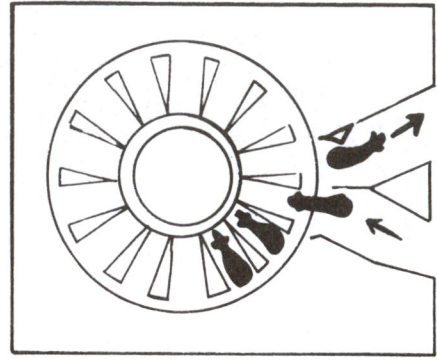

**TURNSTYLE**

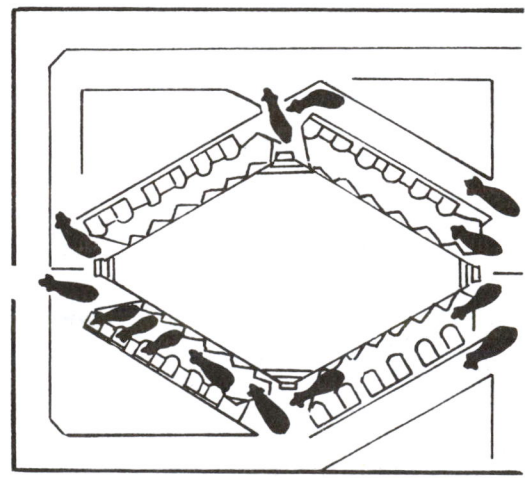

**POLYGON HERRINGBONE**

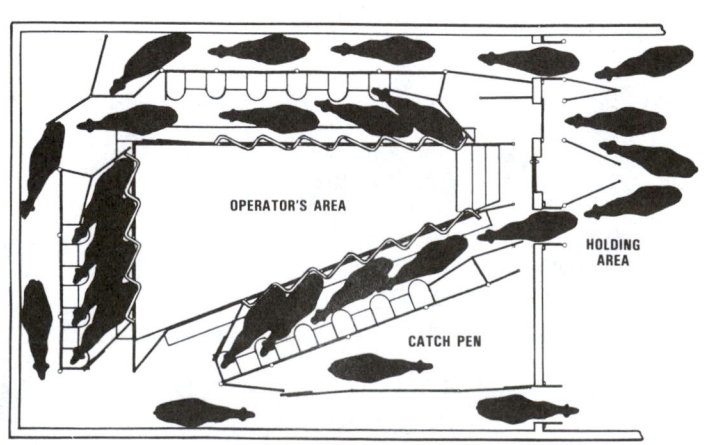

**TRIGON**

Fig. 25–9. Diagrams of different milking parlors.

## MILKING EQUIPMENT

Basically, there are two types of milking equipment: (1) the bucket system, and (2) the pipeline system.

In the bucket system, the milk is received directly into a nearby vacuumized portable bucket, which may be either of two types: (1) floor type, or (2) suspended type.

Fig. 25-10. Bucket system milking equipment in use on stanchioned cows in a stall barn. (Courtesy, Holstein-Friesian Assn. of America, Brattleboro, VT)

Conventional pipeline systems use a rigid heat-resistant glass or stainless sanitary pipe for carrying vacuum from the milk receiver to the individual milking units, and for carrying the milk from the units to the receiver. Pipeline milkers may be used in any of the following types of facilities: (1) stanchion barn, (2) herringbone milking parlor, (3) side-opening milking parlor, (4) walk-through milking parlor, or (5) rotary parlor.

Regardless of make, the mechanical milking systems can be separated broadly into four major parts: (1) vacuum supply, (2) milk flow, (3) pulsation, and (4) milking unit.

## BULK SYSTEMS OF HANDLING

There are two characteristics of milk which make it ideal for the development of bacteria: (1) It is a well-balanced food in which bacteria thrive, and (2) as it comes from the cow, the temperature is ideal for bacterial growth. For these reasons, milk must be cooled to at least 50°F (preferably to 40°F) as soon as possible in order to inhibit bacterial growth.

Milk may be handled by either of two systems: the can system, or the bulk system. Until 1939, when the bulk system was first introduced in California, all milk was handled in cans. Today, most modern dairies use bulk tanks. Although the initial cost is greater than where cans are used, the greater returns over a period of time justify the expense. Further, many producers are facing the situation of being forced into going to bulk tanks if they are to retain a market outlet.

Fig. 25-12. A stainless steel bulk tank in the milk room of an Illinois dairy. (Courtesy, J. C. Allen & Son, West Lafayette, IN)

Generally speaking, the following advantages accrue to the use of bulk tanks, in comparison with cans: (1) a saving in labor, (2) less loss in milk, (3) alleviating 10 gal cans, (4) higher butterfat tests (due to butterfat being left on lids of cans), (5) a saving in hauling costs, and (6) a premium paid by the plant.

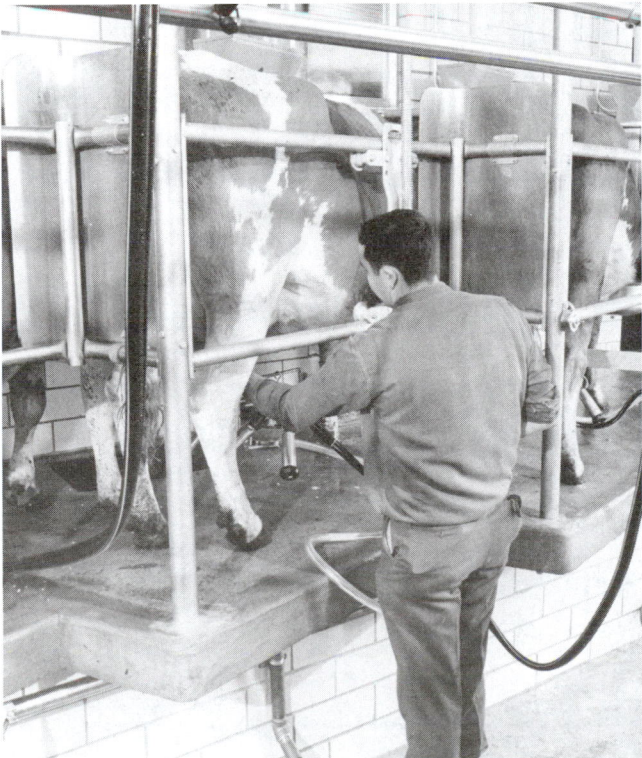

Fig. 25-11. One side of a double herringbone parlor with QTO milk units. Parlor has glazed tile and a low milk line. (Courtesy, Babson Bros. Co., Oak Brook, IL)

Fig. 25-13. The bulk truck handling milk. (Courtesy, Milk Marketing, Inc., Strongsville, OH)

## PRODUCE QUALITY MILK

Consumers and health departments all have a distinct interest in the quality of milk.

Quality milk can be produced only when the producer pays special attention to a number of factors:

1. **Health of the herd.** The herd should be free from disease that might be spread to human beings through the milk. Bacteria in milk coming from cows must be eliminated. Mastitis is the most important herd health problem at the present time.

2. **Clean animals.** The milker should clean the flanks and udders of cows just prior to milking to prevent dirt from getting into the milk. Clean floors and bedding and a well-drained yard make the cleaning job easier.

3. **Clean equipment.** All milking equipment should be kept as clean and free from bacteria as possible. Bacteria grow in cracks and rough spots on equipment if it is not washed properly.

4. **Cool and store milk properly.** Proper cooling and storage of milk on the dairy farm requires facilities which will cool the milk promptly from a temperature of about 90°F down to 40°F, and then hold it at that temperature until it is collected. Bacteria will reproduce (divide) once every 30 minutes in 70 to 90 degree temperature; thus, in 12 hours, one bacterium can reproduce 16 million. Cooling will control this growth.

5. **Keep barn and milk house clean.** The milking barn should be clean and should have a concrete floor. Barn odors may be eliminated by having a building well ventilated.

A milk room is important for the convenience of the operator, and is an aid to the production of high-quality milk.

6. **Control flies.** Fly control measures are important to milk producers. Flies add to the bacterial count of milk; cases are on record of flies carrying as many as 1,250,000,000 bacteria. They can carry typhoid, dysentery, and other contagious diseases.

Breeding places for flies, such as manure piles and mudholes, should be eliminated.

7. **Control bacteria.** In summary, here is how the bacterial count in milk can be kept down:

    a. Rinse the utensils and equipment with hot water after cleaning so they dry off quickly.

    b. Remove all milkstone[1] from the equipment, as bacteria must have food.

    c. Cool the milk as quickly as possible to 40°F, as bacteria like high temperatures.

    d. Wash and sanitize with proper cleaning and sterilizing material.

    e. Have a well-lighted barn and milk house, as bacteria like darkness rather than light.

## PHYSIOLOGICAL FACTORS AFFECTING AMOUNT AND COMPOSITION OF MILK

The variation in the butterfat composition of milk at the plant has puzzled dairy producers. And since the fat content of the milk has a bearing on the paycheck, it's an economical factor, too.

A number of physiological factors affect the amount and composition of milk:

1. **Breed and individual inheritance.** Variation in the ability of cows to produce total milk, fat, and solids-not-fat is an inherited characteristic. There is both a breed difference (see Table 25–2) and an individual difference. In general, total milk production decreases and butterfat content increases by breeds in the following order: Holstein, Brown Swiss, Ayrshire, Guernsey, and Jersey.

With the Holstein breed, a range in butterfat from 2.6 to 6.0% has been reported; and within the Jersey breed, from 3.3 to 8.4%. Similar variation between breeds and individuals exists in total milk production.

2. **Stage of lactation.** The greatest variation in the composition of milk takes place immediately following parturition, within the first five days after freshening. The secretory product known as colostrum, found in the udder at the time of calving and produced for a short time thereafter, is not milk as such. It contains more globulins, vitamins A and D, iron, calcium, magnesium, chlorine, and phosphorus than does milk; but it contains less lactose and potassium than milk. (See Table 25–3).

### TABLE 25–3
### COMPOSITION OF COLOSTRUM AND NORMAL MILK OBTAINED TWO WEEKS AFTER CALVING[1]

|  | Colostrum | Normal Milk |
|---|---|---|
|  | (%) | (%) |
| Total solids | 23.9 | 12.9 |
| Minerals | 1.1 | 0.7 |
| Protein | 14.0 | 3.1 |
| Fat | 6.7 | 4.0 |
| Lactose | 2.7 | 5.0 |

[1]Source: *Dairy Guide*, The Ohio State University, Columbus, OH.

Total milk production generally increases for the first month following freshening, then decreases gradually thereafter. Conversely, the butterfat test is usually higher toward the end of the lactation period than soon after freshening.

3. **Persistency.** This refers to the level at which milk production is maintained as lactation progresses. Generally speaking, following the peak lactation period, about a month after freshening, the total milk production each month is approximately 90% of that of the previous month (see Fig. 25–14).

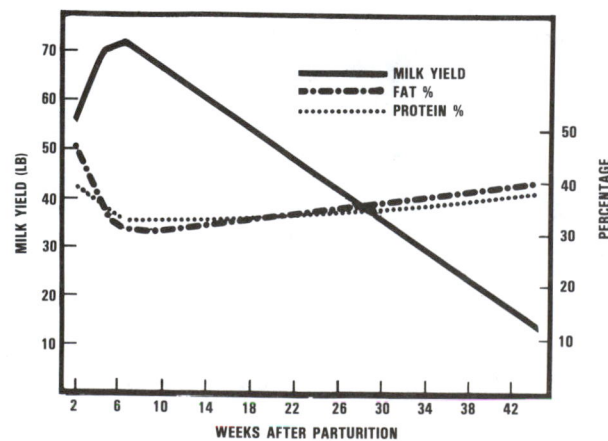

Fig. 25–14. Lactation curves of milk yield and milk fat and protein percentages of Holstein cows. (Source: *Dairy Guide*, The Ohio State University, Columbus)

---

[1]Milkstone is a complex mixture of milk and water minerals with entrapped fat, protein, soil particles, and microorganisms, plus cleaner and sanitizer residues. This film adheres tightly to the surface of milk-handling equipment and requires special acid treatment for removal.

4. **Estrus; pregnancy.** Milk and butterfat production may fluctuate, usually downward, on the day of or the day following the heat period. Pregnancy seems to have little effect on milk composition. However, beginning about the fifth month of pregnancy, total production of gestating cows declines more rapidly than that of nonpregnant cows. It has been estimated that the energy requirement of the fetus is equivalent to about 400 to 600 lb of milk.

5. **Calving interval.** Research indicates that it is most profitable for cows to calve at 12-month, rather than longer, intervals. With an 8-week dry period, this means a lactation period of 10 months.

6. **First- and last-drawn milk.** The percentage of fat in last-drawn milk is higher than that in first-drawn milk. The reasons for this are not known.

7. **Age.** The age of a cow has a definite effect on production. Most cows reach maturity and maximum milk production at about 6 years of age, following which there is a decline in production. Records indicate that cows produce approximately 25% more milk at maturity than they do as 2-year-olds. Also, after passing their prime—after 6 years of age—butterfat gradually decreases with advancing age.

Age adjustment factors have been developed to standardize 305-day lactation records to a mature equivalent basis and to minimize environmental variation due to month of the year in which the record began. These factors remove, with considerable accuracy, recent environmental effects from age and month of calving in individual breeds and regions. Table 23–5 in Chapter 23 of this book shows the milk and fat age adjustment factors for cows in the United States calving in the month of May, by breed, for selected ages. A complete list of adjustment factors for milk and fat by breeds, by regions (and for the United States), by month of calving, and by age, is given in the following report: *USDA-DHIA Factors for Standardizing 305-Day Lactation Records for Age and Month of Calving,* ARS-NE-40, U.S. Department of Agriculture.

8. **Size.** Within a breed, large cows usually produce more milk than small cows. However, according to Brody of the Missouri station, for each 100 lb increase in body weight, production increases only 70% of the proportional increase in body size.

## ENVIRONMENTAL FACTORS AFFECTING AMOUNT AND COMPOSITION OF MILK

All animals, including dairy animals, are the result of two forces—heredity and environment. Because of this, the maximum development of dairy cattle characteristics of economic importance—particularly total milk production—cannot be achieved unless there are optimum conditions of environment. Among the environmental factors affecting amount and composition of milk are the following:

1. **Feed.** If milk cows are not fed, or if they do not eat, they will not produce. There are a number of ways in which feed may affect the quantity and/or composition of milk. Among them—

    a. **Underfeeding.** By underfeeding we usually refer to not providing sufficient energy. The degree of milk reduction therefrom is related to the extent of underfeeding and the length of time it exists.

    b. **Challenge or lead feeding in early lactation.** One of the most critical periods for proper feeding is immediately following freshening. It is very difficult for high-producing cows to consume enough feed to supply the energy needs for production at this time. As a result, most cows lose weight during this period. The current system of increasing the concentrates, beginning 2 to 3 weeks before freshening until the cow is consuming 1.0 to 1.5 lb of concentrate per hundred pounds of body weight at calving, is known as challenge or lead feeding. In this system, after freshening, cows are fed to their inherited capacity for milk production as determined by profitability; in other words, at the point where the added milk produced does not pay for the added feed, it is time to discontinue further feed increases.

    c. **Deficiency of nutrients.** A deficiency of any essential nutrient required by the cow will lower milk production and feed efficiency, rather than make for significant changes in the composition of milk.

    d. **Some feed ingredients and rations influence feed composition.** Some feeds reduce the fat percentage of milk. Among such feeds are: cod-liver oil and other fish oils, certain pasturages (especially lush spring pastures), and pearl millet. Also, fine grinding of forage, too small an amount of roughage, or heated starch will lower the butterfat content of milk. On the other hand, such feeds as whole cottonseed, soybeans, and coconut oil result in an increase in the fat content of milk.

The amounts of fat-soluble vitamins A, D, and E in milk are influenced by the amounts of these particular vitamins in the ration, and in the case of vitamin D, exposure to sunlight is a factor, also.

2. **Length of dry period.** A dry period of approximately 60 days is recommended following each lactation period. This is important because it permits the cow's body to store up reserves so as to meet the rigorous demand of the next lactation, and it permits proper involution and conditioning of the udder. A short dry period usually results in lower milk production.

3. **Conditioning at calving time.** Cows that are in a thin, run-down condition at calving time produce less milk than cows in good condition. Excessive condition will also lower milk production after freshening, but it should be added that this seldom happens in good producing dairy cows.

Cows in good flesh at calving time have been observed to start their lactation with 25% more milk production than those calving in poor condition. Generous feeding of thin cows following freshening may eliminate some of this difference, but it is questionable that thin, high-producing cows can ever consume enough to catch up.

4. **Frequency of milking.** Frequency of milking does result in more total milk produced; cows milked 3 times a day consistently produce more milk than those milked twice a day, and cows milked 4 times a day produce more milk than those milked 3 times daily. It has also been observed that cows milked more frequently are more persistent in their production throughout the lactation; that is, milk production declines less rapidly as lactation progresses. Of course, a decision as to whether or not it pays to milk more than twice daily will depend on whether the additional milk more than covers the added labor and other costs of obtaining it. In a limited number of herds, managed for intensive production, 3 daily milkings have been profitable.

Frequency of milking has no effect on butterfat percentage.

5. **Irregular feeding and milking.** Unequal intervals between milkings affects both quantity and composition of milk; more milk of slightly lower fat content is obtained following the longer intervals.

6. **Change of milkers.** High-producing dairy cows may be under stress, with the result that they are usually very sensitive to any changes, including that of the caretaker. Creating a pleasant, quiet, and comfortable environment causes a cow to perform more efficiently.

7. **Environmental temperatures; season.** Butterfat percentage of milk varies with the season, being higher in the fall and winter and lower in the spring and summer. It may vary up and down seasonally by an average of 0.3 to 0.5%. Solids-not-fat also show a seasonable variation, with the low point in the spring and summer. The reasons for these changes are not known; it may be due to temperature and humidity, changes in body weight, or kinds and amounts of feeds may be reflected.

Severe weather conditions usually decrease the amount of total milk produced and may influence the fat test either up or down. Temperatures above 85°F greatly affect cows and the situation is accentuated when high temperatures are accompanied by high humidity.

It is also noteworthy that cows calving in the fall months consistently produce more than those calving at other times of the year. Cows calving in the spring produce the least. This difference may be as much as 10 to 15%. This phenomenon may be due in part to the temperature; but more than likely available feeds, including spring pastures to which fall-calving cows respond so well, may be a factor.

8. **Day-to-day variation.** Research has shown that day-to-day butterfat tests vary from 0.1 to 2.0%.

9. **Disease.** Disease does affect milk secretion, in both total production and composition, with the degree of the effect determined by the kind and severity of disease. Mastitis will, for example, lower both the total production of milk and the composition thereof.

10. **Drugs.** Many types of drugs have been used in an effort to increase milk production and affect its composition. Most of them have no effect, so it is questionable that they can be used on a practical basis.

When added to the feed at certain levels, thyroprotein (thyroxin) stimulates the cow to produce more milk of a higher percentage fat. However, to be effective, it must be added at a specific time during the lactation period and cows must be fed more when they are receiving the drug.

Oxytocin will, on a temporary basis, increase yields of both milk and fat. This is because it permits greater release of milk from the udder. but it must be administered just after each milking in order to get the residual milk, which makes its administration both expensive and time-consuming. Hence, it is not considered a practical procedure.

11. **Prepartum milking.** Prepartum milking is the practice of milking cows 10 days to 2 weeks before they are due to freshen. Those who follow this practice usually do so because they believe it will lessen congestion and swelling of the udder and belly of the cow. Among some, the feeling also persists that it will lessen the incidence of both mastitis and udder edema. It is known that prepartum milking will result in cows producing normal milk at the time of freshening, rather than colostrum. Thus, where prepartum milking is done, it is necessary to save (freeze) the early milk in order to have colostrum available for the newborn calf.

## MASTITIS

*Mastitis is an inflammatory reaction of udder tissue to bacterial, chemical, thermal, or mechanical injury.* The term *mastitis* is from the Greek word *mastos,* for breast, and *itis* refers to inflammation of.

According to the National Mastitis Council, nearly 40% of all dairy cows have some form of mastitis, which causes a yearly loss of $225 per afflicted cow.

It has been said that producers themselves are responsible directly, or indirectly, for 90% of their mastitis troubles; however, most producers blame their milking machine. The three main routes through which mastitis comes are: (1) dirty, or poorly adjusted, milking equipment; (2) poor milking practices; and (3) injuries to cows because of their surroundings.

Several species or groups of microorganisms may cause mastitis, but over 95% of all cases are caused by the following species of streptococci and staphylococci: *Streptococcus agalactiae, Streptococcus dysgalactiae, Streptococcus uberis,* and *Staphylococcus aureus.* Infection with any of these organisms is usually chronic, with flare-ups occurring at regular intervals. No amount of drugs given to cows today can prevent another attack next month under the same conditions.

Although mastitis is usually apparent, it may be a hidden disease. Therefore, several different tests have been developed for detecting the presence of the causative microorganisms; among them (1) *screening tests,* or *presumptive tests,* made either at the side of the cow or at the bulk tank, of which the California Mastitis Tests (CMT) is the most widely used one, and (2) *specific laboratory tests* designed to detect the causative organism. A reasonable goal, based on using the CMT test, is to have at least 75% of the bucket milk samples score negative or trace; less than 75% negative (−) and trace (T) bucket readings indicates a milking management problem. On an individual quarter basis, 90% of the samples scoring negative or trace indicates a well-managed herd.

Through the years, many different kinds of drug therapy have been used—including dyes, chemicals, sulfas, antibiotics, and nitrofurans. Many times such drugs have been effective, at least temporarily; in any event, acute cases of mastitis should be treated by a veterinarian.

In summary, it may be said that the producers themselves are unwittingly setting the stage for mastitis flare-ups in their herds, by providing the ideal conditions—poor milking practices, poor milking equipment, and improper surroundings—through managed milking and sanitation—producers can reduce or eliminate mastitis.

## MILK FLAVOR

Most consumers base the quality of any product on its flavor; and milk is no exception. They want milk that "tastes good." The flavors most often found in milk, and their cause and prevention, are:

1. **Feed and weed flavors.** These have been covered under the section on "Feeds Affecting Milk Flavor" in the chapter of "Feeding and Managing Dairy Cattle," so repetition is unnecessary.

2. **Oxidized flavor.** This has been described as a cardboard flavor. Some causes of oxidized flavor are (a) metallic contamination from copper and iron, which may be alleviated by using stainless steel; (b) exposure to sunlight or just daylight; (c) foaming; and (d) dry lot feeding. Feeding vitamin E to the milking herd will reduce or eliminate oxidized flavors.

3. **Rancid flavor.** This flavor is caused by a breakdown of the butterfat which releases strong-flavored acids. This action is caused by the enzyme lipase, which is present in all milk. The primary causes of rancid milk are (a) stripper cows (those well advanced in lactation), (b) excessive agitation of milk, due to high lifts and sharp turns in pipeline milking; and (c) slow cooling with foaming.

4. **Barney.** This flavor(s) is caused by dirty stables, poor ventilation, unclean milking, and unclean cows—all of which can be alleviated.

5. **Salty.** This flavor, which masks the slightly sweet flavor of milk, is caused by mastitis, stripper cows, or certain individual cows. Milk from cows that have mastitis, or from strippers, should not be marketed.

6. **Malty.** Malty flavor is primarily due to high bacteria count. The remedy is to keep bacteria out of milk as much as possible, and to prevent growth of those that do get into it. Clean and cold milk will practically eliminate malty flavor. Also, milk handlers should pick up all the milk and not leave any of it in the farm bulk tank.

7. **High-acid, sour milk.** This is due to a very high bacterial count. In these days of mechanical refrigeration, there is no excuse for sour milk; simply cool it as rapidly as possible from the 90°F temperature of the milk pail to 40°F.

8. **Unnatural or foreign.** This refers to flavors that come from medicinal agents and disinfectants. The control of such off-flavors consists in (a) handling medicines and disinfectants so that the flavor or odor from them will not get into the milk, and (b) using chemical sanitizers only in the concentrations indicated by the directions. Do not market milk from drug treated cows for at least 72 hours after last treatment, or longer if so prescribed on the drug label or by the veterinarian.

For good tasting milk, the producer should keep it clean, keep it cold, feed silage after milking (not before), use good quality feed, and not ship milk from problem cows.

## DRYING OFF COWS

There are several methods of drying off cows, ranging from complete cessation of milking (see Chapter 24, under section on "Dry Cows") to intermittent milking. Most experienced producers dry off cows as follows: When a cow is producing less than 35 lb of milk daily, they remove the grain from her ration and stop milking abruptly. After 2 to 3 days, the cow can be milked out for the final time and treated for mastitis.

High-producing cows may not dry-off easily. So, production should be reduced before milking is stopped. To change the environment, move such cows to a stall or pen, stop feeding grain, use poor quality forage, and limit the intake of water. If necessary, milk such cows once daily for 2 to 3 days. When production has dropped to 35 lb, quit milking the cow abruptly. Two days later, milk her out and treat her for mastitis.

## QUESTIONS FOR STUDY AND DISCUSSION

1. What characteristics are common to all mammalia?

2. What are the primary species differences in milk composition?

3. What are the primary breed differences in milk composition? How do you account for breed differences in milk composition?

4. Scientists have been able to make milk replacers. In fact, it is claimed that many such products are superior to milk, because the synthetic products are fortified with minerals, vitamins, and/or antibiotics. Despite this achievement, why have scientists not been able to make colostrum in the laboratory?

5. What are the main parts of the udder, and what are the functions of each part?

6. Where and how is milk formed in the cow's udder?

7. Cite the efficiency of a cow that produces 14,500 lb of milk in a year in terms of the quantity and type of food that she produces. How does this efficiency compare with a feedlot steer?

8. Describe the magnitude and the route/role of blood circulation and milk production in the cow.

9. What hormones are involved in the onset and the maintenance of milk production? How are they involved?

10. Outline in order the steps in milk let down.

11. Why is a rapid milker likely to get more milk from a cow than a slow milker?

12. Discuss the hormone action that may cause milk hold up.

13. List, in order, the coordinated steps of a well-managed milking program. Under favorable conditions, how many cows can be milked per hour per milker?

14. Diagram and describe four of the seven different milking parlor arrangements that are presented in Fig. 25–9.

15. Name and describe the two basic types (systems) of milking equipment. For what type of dairy is each system best adapted?

16. What advantages does the bulk tank system of handling milk have in comparison with the can system?

17. List and discuss the factors to which the producer must pay special attention in order to produce quality milk.

18. On the whole, what physiological factors have the most effect on the amount and composition of milk?

19. What do the lactation curves in Fig. 25–14 show for milk yield, milk fat, and protein show for Holstein cows?

20. List and discuss the environmental factors that may affect the quantity and/or composition of milk.

21. What is mastitis? What is the economic importance of mastitis? What may producers do to lessen the incidence of mastitis?

22. List the flavors most often found in milk, and give the cause and prevention of each.

23. Describe how to dry off cows.

## SELECTED REFERENCES

| Title of Publication | Author(s) | Publisher |
|---|---|---|
| *Biology of Lactation* | G. H. Schmidt | W. H. Freeman and Co. Publishers, San Francisco, CA, 1971 |
| *Dairy Cattle: Principles, Practices, Problems, Profits*, Third Edition | D. L. Bath, et al. | Lea & Febiger, Philadelphia, PA, 1985 |
| *Dairy Science* | W. E. Petersen | J. B. Lippincott Co., Philadelphia, PA, 1950 |
| *Harvesting Your Milk Crop* | C. W. Turner | Babson Bros. Co., Oak Brook, IL, 1973 |
| *Lactation—A Comprehensive Treatise*, Vols. I, II, and III | B. L. Larson<br>V. R. Smith | Academic Press, Inc., New York, NY, 1974 |
| *Mammary Gland, The* | C. W. Turner | Lucas Brothers Publishers, Columbia, MO, 1952 |
| *Principles of Dairy Science* | G. H. Schmidt<br>L. D. Van Vleck | W. H. Freeman and Co. Publishers, San Francisco, CA, 1974 |
| *Principles of Milk Production* | W. B. Nevens | McGraw-Hill Book Company, New York, NY, 1951 |
| *Science of Providing Milk for Man, The* | J. R. Campbell<br>R. T. Marshall | McGraw-Hill Book Company, New York, NY, 1975 |
| *Secretion of Milk* | D. Espe<br>V. R. Smith | Iowa State University Press, Ames, IA, 1952 |

Stanchioned Jersey cows in a stall barn. This California herd posted an annual lactation average of 15,289 lb of milk, 734 lb fat, and 582 lb protein. (Courtesy, The American Jersey Cattle Club, Reynoldsburg, OH)

How it used to be done! Milk delivered via horse-drawn milk wagon, in quart glass bottles, to the doorsteps of consumers. (Courtesy, The Bettmann Archive, Inc., New York, NY)

# MARKETING MILK AND DAIRY PRODUCTS

## Chapter 26

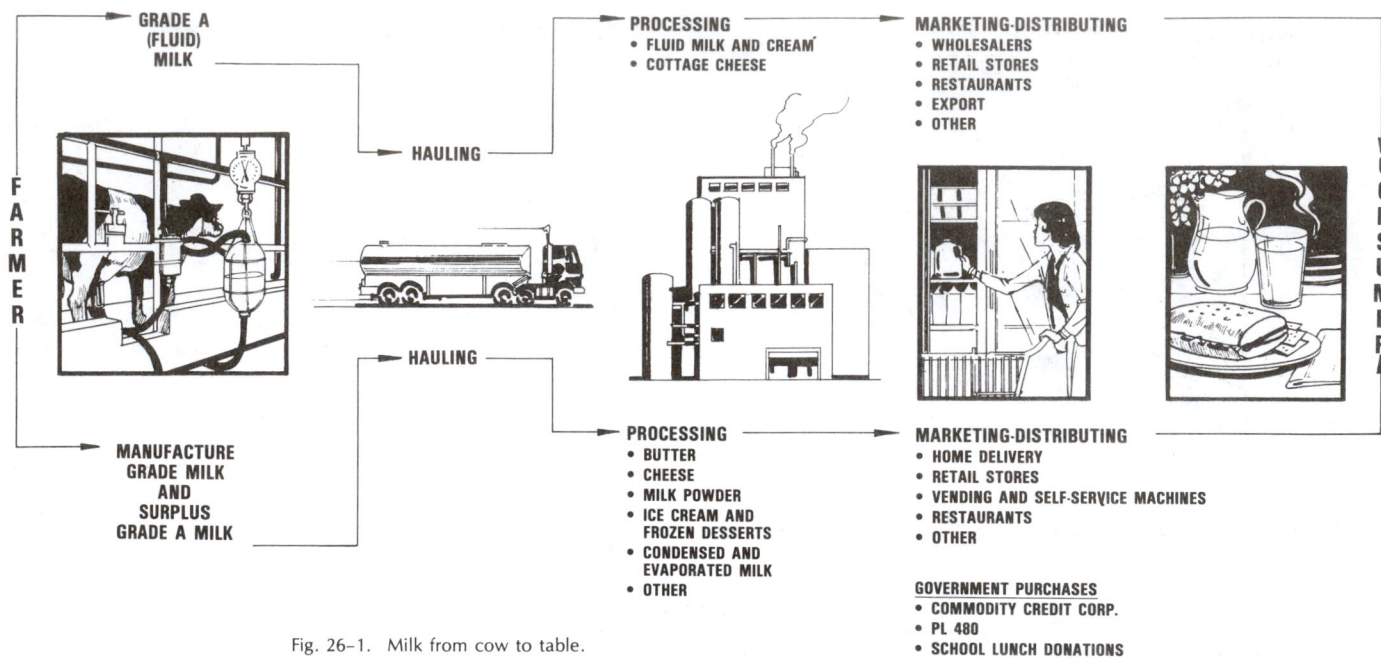

Fig. 26–1. Milk from cow to table.

Marketing is that all important end of the line; it's that which gives point and purpose to all that has gone before.

In our present system, the marketing of milk and dairy products is handled largely by specialists, usually under a multitude of regulations and controls. However, successful milk producers must understand milk markets and the factors affecting them if they are to take full advantage of their opportunities.

## MARKET IMPORTANCE OF MILK AND DAIRY PRODUCTS

The farm value of dairy products in 1986 was $17.8 billion. The total marketing bill—the cost of transporting, processing, and distributing dairy products, or the difference between consumer expenditures and farm value—that same year came to $34.2 billion. Upon being retailed, consumers spent $52 billion for these products (see Table 26–1).

### TABLE 26–1
### FARM VALUE, MARKETING BILL, AND CONSUMER EXPENDITURES FOR DAIRY PRODUCTS, 1986[1]

| Item | Value |
|------|-------|
|  | (million $) |
| Farm value | 17,800 |
| Marketing bill (cost for transporting, processing, and distributing) | 34,200 |
| Consumer expenditures (retail cost) | 52,000 |

[1]Source: *1988 Dairy Producer Highlights*, National Milk Producers Federation, Arlington, VA, pp. 6 and 30.

Other noteworthy statistics are: In 1986, (1) dairy products accounted for 13.1% of the cash income of the nation's farmers, and (2) consumers spent about 12.0% of their food dollar for dairy products.

## FARM PRODUCTION AND HANDLING OF MILK

Satisfactory milk marketing necessitates one basic ingredient—quality milk; and, ultimately, this means more income for the producer.

The difference between Grade A milk and the lower grades is considerable. But it goes beyond this; quality can mean increased consumer demand.

Buyers, consumers, and health departments all have a distinct interest in the quality of milk marketed.

Quality milk can be produced only when producers pay special attention to a number of factors; among them, herd health, the layout and structure of the barn and milk house, clean cows, care of the utensils, cooling and storage of milk, and transportation of milk to market.

## HOW MILK IS SOLD BY DAIRY PRODUCERS

Milk producers sell most of their product in the form of whole milk. At one time, they marketed a considerable amount of their product as farm separated cream, but the proportion of this product has been declining. In 1940, farm separated cream accounted for 38% of their total marketing; today, marketing of cream is negligible.

In 1987, only 1.6% of the total milk production was used on farms, compared with 15% in 1950. Obviously, the point has been reached where little additional marketing of milk by farmers can be expected through decreased use on the farm. Future increases in milk marketings will have to come from increased production.

Fig. 26–2 shows the milk supply, use, and stocks from 1975 to 1987. It is noteworthy that both total supply and government stocks have spiraled since 1975.

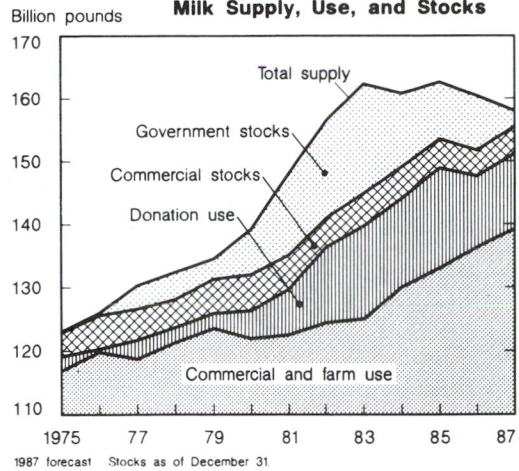

**Milk Supply, Use, and Stocks**

Billion pounds

1987 forecast Stocks as of December 31.

Fig. 26–2. Milk supply, use, and stocks, 1975 to 1987. (Source: *1988 Agricultural Chartbook*, Ag. Hdbk. No. 673, USDA, p. 91, Chart 198)

## HOW MILK IS USED

About 37.4% of the milk marketed by dairy farmers today is consumed in fluid form (Table 26–2, 36.6% sold by dealers + 0.8% sold by producers directly). Fluid milk is retailed as pasteurized milk, homogenized milk, fortified milk (vitamin D), skimmed milk, flavored milk (whole milk with flavor added), or flavored milk drink (skimmed milk with flavor added).

In 1960, 23.9% of the milk supply was used for butter, whereas only 14.8% was so used in 1987. However, it is noteworthy that the decline in the use of milk for butter has slowed in recent years.

In 1987, farmers sold 98.4 lb out of every 100 lb of milk they produced. Of the milk which they sold, 89% of it was eligible for the fluid market, as Grade A; 11% of it was manufacturing milk, or Grade B.

### TABLE 26–2
### UTILIZATION OF U.S. MILK SUPPLY, SELECTED YEARS 1965–1987[1]

| Year | Fluid Milk and Cream | | Butter, Creamery | Cheese | Evaporated and Condensed Milk | Frozen Products | Other[2] | Used on Farms Where Produced[3] |
|------|-----------------------|------------------------------------------|------------------|--------|-------------------------------|-----------------|----------|----------------------------------|
|      | Sold by Dealers | Sold by Producers Directly to Consumers | | | | | | |
| | (Percent of Total Milkfat Supply) | | | | | | | |
| 1960 | 41.4 | 1.7 | 23.9 | 10.9 | 4.4 | 7.7 | 2.6 | 7.4 |
| 1965 | 43.1 | 1.5 | 23.0 | 12.6 | 3.7 | 8.5 | 2.8 | 4.8 |
| 1970 | 42.8 | 1.5 | 20.4 | 16.6 | 2.8 | 9.4 | 3.1 | 3.4 |
| 1975 | 42.9 | 1.4 | 17.2 | 20.7 | 2.3 | 10.3 | 2.5 | 2.7 |
| 1980 | 38.4 | 1.1 | 17.7 | 26.4 | 1.6 | 9.3 | 3.7 | 1.8 |
| 1985 | 35.4 | 0.9 | 17.0 | 29.1 | 1.6 | 9.0 | 5.3 | 1.7 |
| 1987 | 36.6 | 0.8 | 14.8 | 30.0 | 1.5 | 9.4 | 5.3 | 1.6 |

[1] *1988 Milk Facts*, Milk Industry Foundation, p. 30. Supply of milkfat includes U.S. production, ingredient imports of milkfat and solids from sources outside the U.S., and net change in storage cream. Computations made by the Milk Industry Foundation based on data from the U.S.D.A.

[2] Dry whole milk, creamed cottage cheese, and other miscellaneous uses.

[3] Milk fed to calves, consumed on farms as milk and cream, and used for farm-churned butter.

## PER CAPITA CONSUMPTION

Table 26–3 shows the per capita consumption of each of the leading dairy products and the changes that have occurred since 1950.

### TABLE 26–3
### DAIRY PRODUCTS: PER CAPITA CIVILIAN CONSUMPTION
### UNITED STATES, SELECTED YEARS 1950–1987[1]

| Year | Fluid Milk Products | | | | | | | | | | | | | | | | | | Manufactured Milk Products | | | | | | | | | | | | | | | | | | | | |
|------|------|------|------|------|------|------|------|------|------|------|------|------|------|------|------|------|------|------|------|------|
| | Fluid Whole Milk | | Cream | | Low-Fat Milk | | Total Product | | Whole Milk Equivalent of Butterfat | | Butter | | Total Cheese | | Cottage Cheese | | Ice Cream | | Evaporated & Condensed Milk | | Nonfat Dried Milk | | Yogurt | | All Milk Equivalent | |
| | (lb) | (kg) | (lb) | (kg) | (lb) | (kg) | (lb) | (kg) | (lb) | (kg) | (lb) | (kg) | (lb) | (kg) | (lb) | (kg) | (lb) | (kg) | (lb) | (kg) | (lb) | (kg) | (lb) | (kg) | (lb) | (kg) |
| 1950 | 278 | 126 | 11.1 | 5.0 | 15.6 | 7.1 | 304 | 138 | 321 | 146 | 10.7 | 4.9 | 7.7 | 3.5 | 3.1 | 1.4 | 17.2 | 7.8 | 20.5 | 9.3 | 3.7 | 1.7 | — | — | 740 | 336 |
| 1955 | 290 | 132 | 9.6 | 4.4 | 20.0 | 9.1 | 320 | 145 | 326 | 148 | 9.0 | 4.1 | 7.9 | 3.6 | 3.9 | 1.8 | 18.0 | 8.2 | 16.2 | 7.3 | 5.5 | 2.5 | .1 | .0 | 706 | 320 |
| 1960[2] | 276 | 125 | 9.1 | 4.1 | 23.8 | 10.8 | 309 | 140 | 309 | 140 | 7.5 | 3.4 | 8.3 | 3.8 | 4.8 | 2.2 | 18.3 | 8.3 | 13.7 | 6.2 | 6.2 | 2.8 | .3 | .1 | 653 | 296 |
| 1965 | 264 | 120 | 7.6 | 3.4 | 34.0 | 15.4 | 306 | 139 | 294 | 133 | 6.4 | 2.9 | 9.6 | 4.4 | 4.7 | 2.1 | 18.5 | 8.4 | 10.7 | 4.9 | 5.6 | 2.5 | .3 | .1 | 620 | 281 |
| 1970 | 229 | 104 | 5.6 | 2.5 | 57.5 | 26.1 | 292 | 132 | 260 | 118 | 5.3 | 2.4 | 11.5 | 5.2 | 5.1 | 2.3 | 17.7 | 8.0 | 7.1 | 3.2 | 5.4 | 2.4 | .9 | .4 | 561 | 254 |
| 1975 | 195 | 88 | 5.9 | 2.7 | 84.7 | 38.4 | 286 | 130 | 244 | 111 | 4.8 | 2.2 | 14.5 | 6.6 | 4.7 | 2.1 | 18.6 | 8.4 | 5.3 | 2.4 | 3.3 | 1.5 | 2.1 | 1.0 | 546 | 248 |
| 1980 | 148 | 67 | 5.7 | 2.6 | 97.8 | 44.5 | 251 | 114 | 231 | 105 | 4.1 | 1.9 | 17.1 | 7.8 | 4.6 | 2.1 | 17.6 | 8.0 | 7.2 | 3.3 | 3.0 | 1.3 | 2.5 | 1.1 | 554 | 252 |
| 1985 | 120 | 55 | 6.0 | 2.7 | 107.0 | 48.6 | 239 | 109 | 186 | 85 | 4.9 | 2.2 | 22.6 | 10.0 | 4.1 | 1.9 | 18.0 | 8.2 | 7.9 | 3.6 | 2.2 | 1.1 | — | — | 593 | 270 |
| 1987 | 103 | 47 | 6.0 | 2.7 | 123.0 | 55.9 | 236 | 107 | 182 | 83 | 4.6 | 2.1 | 24.0 | 11.0 | 3.9 | 1.8 | 18.3 | 8.3 | 7.9 | 3.6 | 2.5 | 1.1 | — | — | 598 | 272 |

[1] *1978, 1980, and 1988 Dairy Producer Highlights*, National Milk Producers Federation.

[2] Includes Alaska and Hawaii beginning with 1960.

Fig. 26–3 shows change in U.S. per capita dairy product sales from 1977 to 1987. It is noteworthy that declines in per capita consumption were registered in the following products: ice milk, butter, buttermilk, flavored milk and drinks, evaporated and condensed whole milk, creamed cottage cheese, whole milk, and nonfat dry milk. It is noteworthy, too, that yogurt and low fat cottage cheese led the increases. Increase in yogurt reflects the rising interest in health foods, whereas increased consumption of low fat cottage cheese reflects weight watching.

## MARKET STAGES FOR MILK AND DAIRY PRODUCTS

Fig. 26–4. The bulk system of handling milk. This truck can haul 6,000 gal of milk. The insulated stainless steel tank holds milk within a few degrees of its original temperature. (Courtesy, Dairymen, Inc., Lexington, KY)

### PERCENT CHANGE IN PER CAPITA SALES 1977–87

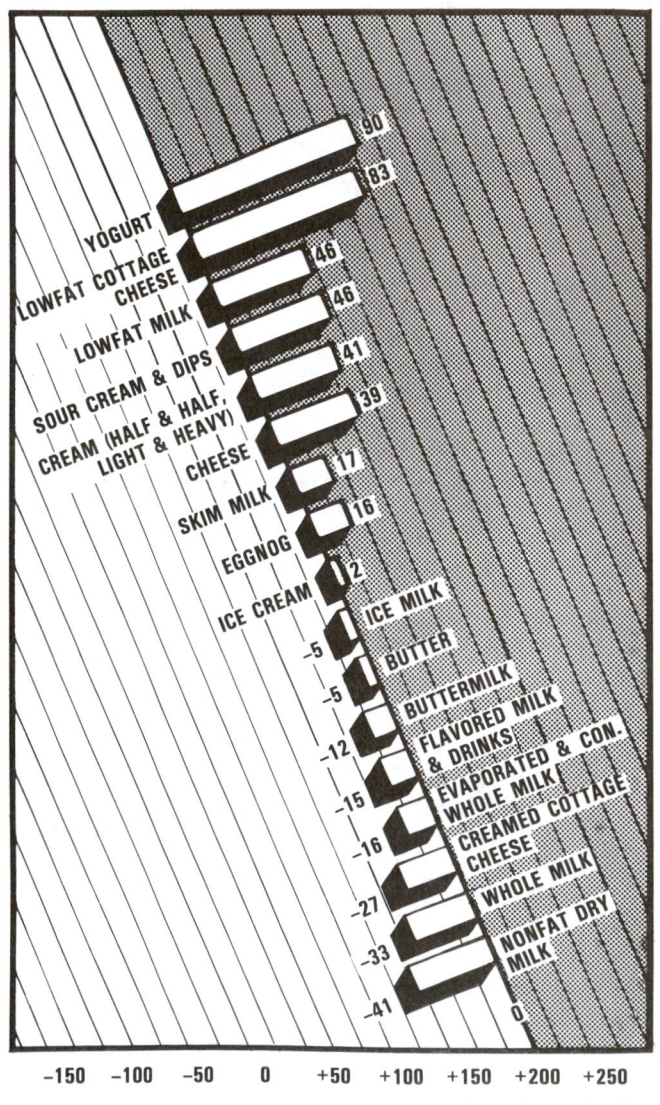

Fig. 26–3. Change in U.S. per capita consumption of milk and dairy products 1977 to 1987, based on sales. Bars to the left of the vertical line represent decreases; bars to the right represent increases. (Source: *1988 Milk Facts*, Milk Industry Foundation, Washington, DC, p. 16)

Milk moves from the farm to the consumer in the following three stages:

1. Assembly and transportation from farms to processing plants.

2. Processing and packaging or manufacturing into various dairy products.

3. Distribution of packaged milk and manufactured milk products to consumers.

Also, producers market their milk as (1) Grade A milk, or (2) manufacturing milk (Grade B).

## HOW MILK IS PRICED AND REGULATED

Chaotic conditions in milk marketing, resulting from the breakdown of private controls and the serious economic plight of farmers during the depression years of the early 1930s, brought requests from organized producers and distributors for government control. Out of this evolved two forms of government controls—those established by the federal government, and those established by the state governments; both were designed to bring more stability into the marketing of milk. Today, federal and state agencies, directly or indirectly, affect the pricing of milk marketed by dairy farmers in the United States. It has been estimated that nearly two-thirds of all milk eligible for fluid markets is affected by milk orders.

### FEDERAL MILK MARKETING ORDERS

Federal milk marketing orders are established and administered by the Secretary of Agriculture under acts of Congress passed in 1933 and 1937. They are legal instruments, and they are very complex. However, stated in simple terms, they are designed to stabilize the marketing of fluid milk and to assist farmers in negotiating with distributors for the sale of their milk. Prices paid to farmers are controlled, but there is no direct control of retail prices.

Federal orders are not concerned with sanitary regulations. These are administered by state and local health authorities. In 1987, there were 42 different federal order markets, each with a market administrator and provision for setting minimum farm prices and regulating transactions between farmers and milk dealers in their area.

Prices in other Grade A markets are influenced by prices established under federal orders or state control programs. Additionally, dairy support programs directly affect the prices of both manufacturing grade milk marketed by farmers and the milk farmers sell as farm separated cream.

## STATE MILK CONTROL

Through State Orders, 17 states have authority to set minimum farm prices and/or retail prices at the wholesale and retail levels. In some states, milk control commissions determine not only what farmers are to be paid but what price the stores can charge customers.

In setting minimum farm prices, state control agencies often operate in a manner similar to federal milk orders. Classified pricing principles are used, and prices are set for a particular market and not necessarily the whole state. Retail prices are based on the cost of processing and distribution. It is noteworthy that fewer and fewer state milk commissions set retail milk prices. The foes of retail pricing point out that, on the average, retail price setting results in a lower price to the farmer than when retail prices are let alone.

Because of their inability to cope with out-of-state milk, state milk controls will likely decline in importance in the future; they will be replaced by federal milk orders.

## COOPERATIVES

The practice of dealing separately with a large number of producers led to dissatisfaction in a number of cases. To rectify this situation, cooperatives were organized. These cooperative associations are of two general types:

1. Bargaining associations which do not handle any milk, but make all business arrangements.

2. Associations which process and distribute milk or assemble it for fluid use.

About 75% of the total deliveries of milk to plants and dealers in the United States is handled by cooperatives. In 1986, the net volume of 354 U.S. dairy cooperatives was $14.8 billion.

## OTHER REGULATORY PROGRAMS

Because of the essential nature of milk, plus the fact that it is easily contaminated and a favorable medium for bacterial growth, it is inevitable that numerous programs have evolved around it—federal, state, and local, some having been designed to control prices and assure a reasonably uniform flow of milk, and others for sanitary reasons.

### SANITARY REGULATIONS

The sanitation of milk and dairy products is assured by the enforcement of sanitary regulations by federal, state, and local authorities.

There are more than 15,000 state, county, local, and municipal health and sanitation jurisdictions in the U.S. Inspectors from these agencies regularly visit farms, plants, and stores, making sure dairy products keep their high quality. Unfortunately, from area to area, there are a bewildering number of different regulations, with the result that milk going to more than one city market is often subjected to duplication and confusion in inspection. Also, sanitary and health regulations have sometimes been used as barriers to keep milk out of a certain area for competitive reasons.

In 1923, the U.S. Public Health Service (USPHS) established an Office of Milk Investigations, and in 1924, the USPHS published its first Grade A pasteurized milk ordinance. Subsequently, this regulation has been revised several times.

Producers are issued permits allowing them to ship Grade A milk. The permit is revoked if either the bacteria count of raw milk exceeds 100,000 per milliliter or the cooling temperature exceeds 40°F in three of the last five samples.

The standard plate count of Grade A pasteurized milk may not exceed 20,000 per milliliter nor the coliform count 10 per milliliter in three of the last five samples or the processor's permit be revoked.

In addition to cleanliness and freedom from mastitis, temperature is important in processing quality milk. Bacteria cannot reproduce effectively below 40°F; so, dairy farmers should cool milk below 40°F as quickly as possible. By law, all fluid milk sold for human consumption must be pasteurized; so, at dairy processing plants, milk is pasteurized to kill disease-causing organisms. It may be pasteurized at either (1) 145°F for 30 minutes, or (2) 161°F for 15 seconds. Additionally, milk should be refrigerated while in the store or in the home.

The Food and Drug Administration (FDA) is charged with inspecting dairy products and processing plants for contamination and adulteration.

Presently, many cooperatives and some milk dealers pay a premium to dairy farmers for producing high quality milk. A variety of premiums and penalty bases are in use. The following standards for milk to qualify for the highest premiums are proposed as reasonable goals:

1. Standard plate count (SPC) or plate loop count (PLC), less than 10,000 per ml.
2. Preliminary incubation count (PIC), less than 20,000 per ml.
3. Somatic cell count (SCC), less than 200,000 per ml.
4. Antibiotic and chemicals, no detectable levels.
5. Temperature, 40°F or lower.
6. Odors and flavors, none objectionable.
7. Acid degree value, 1.0 or below.
8. Milkfat, 3.25% or above.
9. Protein, 3.2% or above.
10. Farm inspection score, 90 or above.

### STANDARDS AND GRADES

The U.S. Department of Agriculture has responsibility for the development of standards and grades for milk and dairy products. Milk is graded as Certified, Class I (Grade A), or Class II (Grade B).

The major dairy products for which the USDA has established grades, and the proportion graded are shown in Table 26–4.

It is expected that more and more dairy products will be federally graded.

### TABLE 26–4
### SELECTED DAIRY PRODUCTS GRADED BY USDA[1]

| Product | Volume | Share of U.S. Production |
|---------|--------|--------------------------|
|         | (million lb) | (%) |
| **B**utter | 107.7 | 9.75 |
| **C**heese | 5.3 | 0.1 |
| **N**onfat dried milk | 16.2 | 1.53 |

[1]Source: USDA.

## STATE TRADE PRACTICE LAWS

For almost 30 years, there has been considerable concern about competitive practices in the sale of fluid milk products and ice cream. Among the unfair trade practices sometimes observed or suspected in the marketing of milk are: discriminatory price cutting, secret rebates, loans, advertising rebates, furnishing and servicing equipment, and the giving of gifts and free signs.

Without doubt, more states will enact dairy fair trade practice laws, and this approach will be used by the state as a substitute for complete milk control.

## METHODS OF PRICING OR PAYING FOR FLUID MILK

Economists refer to the different systems of paying for milk as *price plans*. These plans, which in actual practice generally involve two or more plans—for example, pricing based on (1) class, (2) grade, and (3) base-surplus—are:

1. **Flat price plan.** This was the common method up to World War I. The milk producer was paid a uniform price for all milk sold, regardless of quality or the use made of it.

2. **Use classification plan.** Most marketing orders established two use classes—Class I and Class II.

Class I milk generally includes milk used in fluid form such as whole fluid milk, or milk for creamed drinks which must be made from milk approved by local health authorities. Generally speaking, Class I prices are 10 to 15% higher than Class II prices.

Class II milk usually includes milk in excess of fluid needs, which is used to make manufactured dairy products—primarily butter, nonfat dry milk, and cheese.

On some markets, a further division is made, primarily for milk going into cottage cheese, with the result that there are three classes of milk—Class I, Class II, and Class III.

3. **Blend price.** When dealers buy according to classification prices, they may pay producers a blend price. The blend is an average of class prices weighted by the volume of milk in each class, usually quoted at a specific point and for a specific test of milk.

4. **Quality grade plan.** Frequently, the terms Grade A and Grade B (Grade B is usually called *Manufacturing Grade Milk*) are encountered in milk marketing. Although there may be some local variations in their use, Grade A usually refers to milk produced under conditions which make it acceptable for fluid use in a given market. Grade B often refers to milk produced under conditions which do not make it acceptable for fluid milk use—it's manufacturing milk.

The production of Grade A milk relative to that of Grade B milk has been increasing in recent years (see Fig. 26–5). In 1987, U.S. farmers marketed 139,058 million lb of milk, of which 123,768 million lb, or 89%, was Grade A, and 15,290 million lb, or 11%, was Grade B.

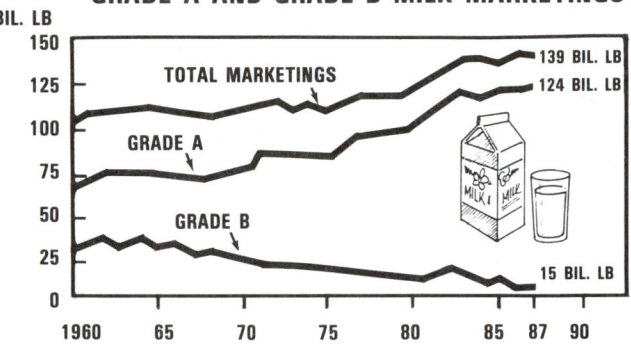

Fig. 26–5. Grade A and Grade B milk marketings, 1960 to 1987. (Source: USDA)

5. **Base surplus plan.** The base surplus plan (or base rating plan) is designed to encourage that a uniform supply of milk be available. It compensates the producer who maintains a high fall production, when more milk is needed. The base period is established during the lowest production months, usually over a period of three to six months. Then, a producer's base is established by the average amount of milk delivered during the base period. The base may be modified from time to time.

6. **Butterfat test price plan.** The butterfat test of milk affects the price. The common practice is to establish a price for 100 lb of milk of a specified butterfat test. Usually 3.5% butterfat is the basis for pricing, although several markets have established their base as high as 4.0% butterfat. Then, a price differential (per point or 0.1%) is set up for milk testing above or below this amount.

7. **Solids-not-fat price plan.** Today, the emphasis on the food value of milk is shifting from fat content to the other solids, especially protein. This is feasible because tests for solids-not-fat have been devised, and these are proving practical for field use. It is anticipated that this system of pricing milk will expand in the future.

On the average, whole milk contains about 2¼ lb of solids-not-fat for each pound of milk fat. Thus, milk testing 4% butterfat contains approximately 9 lb of solids-not-fat, to a total of 13 lb of solids per hundredweight.

8. **Component pricing plans.** Although various component pricing plans being formulated or adopted are not uniform, they continue to give price credit for butterfat in farm milk;

in addition, they give credit for solids-not-fat, including protein. Some also involve end product pricing in which farm milk prices are based on the yield and market value of cheese and other dairy products that can be manufactured from the milk. Most also either (1) establish a maximum somatic cell count at which component premiums will be paid, or (2) pay a premium of 6 to 12 cents per hundredweight of milk for minimum somatic cell counts.

On the average, farm milk contains about 3.7% butterfat and 8.55% solids-not-fat, including about 3.2% protein. A one point (0.1%) change in milk fat test is normally associated with a 0.4 point (0.04%) change in solids-not-fat. However, considerable variation in this average relationship does occur. Component pricing takes into consideration the value of variations in solids-not-fat as well as fat; thus, its advocates feel that it would correct an inequity to dairy farmers under current butterfat differential pricing.

In 1988, the Federal Milk Marketing Administrator reported that 48% of all Federal Order Milk is eligible for multiple component pricing (MCP), meaning that if the milk meets the minimum component, quality, and/or other requirements established by producers, cooperatives, or plants, payment may be based upon two or more components. If we include the non-Federal Order milk eligible for MCP, then approximately 60% of U.S. milk production is eligible for MCP. More and more component pricing will be implemented.

9. **Gallon or quart plan.** Occasionally, a producer supplies milk to a distributor on a per gallon or per quart basis. Since average milk weighs 2.15 lb to the quart and 8.6 lb to the gallon, 100 lb of milk would be equivalent to 46.5 qt or 11.6 gal. Thus, one can easily compute the possible returns from selling milk by different methods.

10. **Special milks.** Certain milks are sold under special labels. Among them are:

a. **Certified milk.** This is milk that is produced under special sanitary conditions prescribed by the American Association of Medical Commissioners. It is sold at a higher price than ordinary milk.

b. **Golden Guernsey milk.** Golden Guernsey milk is produced by owners of purebred Guernsey breeds who comply with the regulations of The American Guernsey Cattle Club. Such milk is sold under the trade name "Golden Guernsey," at a premium price.

c. **All-Jersey milk.** This is produced by registered Jersey herds whose owners comply with the regulations of The American Jersey Cattle Club. It is sold at a premium price under the trademark of "All-Jersey."

## THE PRICE SUPPORT PROGRAM

Some of the price support programs pertaining to surpluses since World War II had their origin in wartime programs designed to increase production. Following the war, the demand for dairy products for military and foreign use declined sharply. Thereupon, the Agricultural Act of 1948 extended the price support authorization at 90% of parity for milk and butterfat; and, beginning the following year, the Agricultural Act of 1949 authorized and directed the Secretary of Agriculture to support manufacturing milk prices to producers at between 75 and 90% of parity. The Act has been amended several times,

with the support level of manufacturing milk based on a percentage of parity equivalent basis. With each amendment, the support price has been lowered. Thus, in 1979–80, the percentage of parity equivalent on manufacturing milk was 79; in 1984–85, it was 58; and in 1987–88, it was 53.

In 1987, U.S. farmers received $12.30 per cwt for all milk sold to plants; with a price differentiation of $12.40 per cwt for fluid (Grade A) milk, and $11.10 per cwt for manufacturing (Grade B) milk.

In several of the postwar years, the government made substantial purchases of dairy products—through CCC and other purchase programs—to support prices at announced levels.

## MANUFACTURED MILK PRODUCTS

The production of manufacturing grade milk is primarily centered in the Midwest and the Great Lakes area. Thus, with the exception of ice cream making, the processing of manufactured dairy products—butter, nonfat dry milk powder, cheese, evaporated and condensed milk, and other products of minor importance—is concentrated near those areas of production.

## NUMBER AND SIZE OF PLANTS

Since the 1940s, the number of milk-manufacturing plants in the United States has been decreasing while the output per plant has been increasing (see Table 26–5). As shown, between 1948 and 1987, milk manufacturing plants became fewer and bigger. The statistics: During the 39-year period (1948 to 1987), 27.2 million more lb of milk was manufactured, the number of plants decreased from 9,737 to 1,933—7,804, or 80%, and the plants became an average of 7.5 times larger.

**TABLE 26–5**
**TOTAL MILK MANUFACTURED AND NUMBER OF PLANTS MANUFACTURING MILK, 1948–1987**

| Item | 1948 | 1987 | Percentage Change |
|---|---|---|---|
| | | | (%) |
| Total milk manufactured[1] ......million lb | 54.7 | 81.9 | + 50 |
| Number of plants[2] ........... | 9,737 | 1,933 | – 80 |
| Average annual volume per plant (whole milk equivalent) .......thousand lb | 5,622 | 42,392 | +654 |

[1]USDA sources.
[2]*1988 Dairy Producer Highlights,* National Milk Producers Federation, p. 14.

## USES OF MILK

The uses of milk have already been covered (see Table 26–2). Manufactured dairy products utilized 61% of U.S. milk production in 1987. A few pertinent points relative to each of the manufactured products will be presented in the sections which follow.

## BUTTER

Fig. 26–6. Fresh bread and butter with broccoli cream soup. (Courtesy, American Dairy Assn., Rosemont, IL)

Butter is made from cream. As marketed, it consists of about 80% milk fat. The remainder is water, salt, and traces of other substances.

Fig. 26–7 shows the per capita consumption of butter and margarine, from 1910 to 1987. As noted, the per capita consumption of margarine surpassed butter in 1957. In 1987, the per capita consumption of butter was 4.7 lb, whereas the per capita consumption of margarine was 10.5 lb.

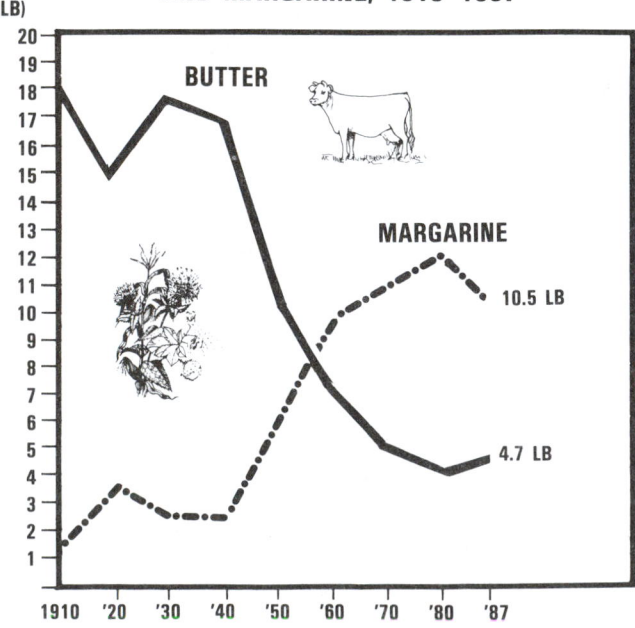

Fig. 26–7. The per capita consumption of butter and margarine from 1910 to 1987. (Source: USDA)

State and national laws and regulations bar additives to or changes in butter. Nevertheless, experiments are being conducted on low butterfat spreads. Besides their appeal to homemakers, these products would be better able to compete with oleomargarine.

Wisconsin is the leading butter-producing state; California ranks second.

## CASEIN

Casein, which is the major protein of milk, is found only in milk. It is obtained by acid or rennet coagulation of defatted milk. Casein contains a minimum of 80% crude protein. It gives milk its white color.

Casein is used as an ingredient of coffee whitener and whipped toppings, in baked goods, and as the main source of protein in the manufacture of meat analogs and in the protein supplementation of some meat products.

In 1987, the U.S. imported 108,136 metric tons of casein, which far exceeded imports of dried milk (1,301 metric tons), and of butter (905 metric tons).

## CHEESE

Fig. 26–8. A trio of cheeses—Parmesan, Provolone, and Cheddar—extends Italian sausage for a popular pizza. When served with peanut butter cookies, fruit, and chilled milk, it makes a delicious pizza supper. (Courtesy, United Dairy Industry Assn., Rosemont, IL)

Cheese is made by (1) exposing milk to specific bacterial fermentation, or (2) treating with enzymes, or both methods, to coagulate some of the proteins.

Milk can be, and is, processed into many different varieties of cheese. Some are made from whole milk, others from milk that has had part of the fat removed, and still others from skimmed milk. American types of cheese (Cheddar, Colby, washed curd, stirred, curd, Monterey, and Jack) make up 60% of the nation's cheese output. The most important variety produced from skimmed milk is cottage cheese. Other important types of cheese are Italian (mostly soft varieties), Swiss, Muenster, brick, blue, and processed cheese.

In 1987, 30% of all the milk used in manufactured dairy products was processed into cheese (exclusive of cottage cheese). The rising popularity of pizza in the U.S. accounts for much of the increase in cheese production and consumption in recent years.

The leading states in the production of cheese, excluding cottage cheese, by rank are: Wisconsin (with 35% of the total production), Minnesota, New York, California, Iowa, Pennsylvania, Missouri, Idaho, and South Dakota.

## CONDENSED AND EVAPORATED MILK

The primary products within this category are evaporated milk and condensed milk packed in cans for consumer use, and condensed whole and skimmed milk shipped in bulk. Condensed and evaporated milk are manufactured by removing a major portion of the water from the whole milk in a machine called a vacuum pan. Condensed milk is further treated by the addition of large amounts of sugar.

Candy manufacturers, especially bakers and ice cream processors, are large users of condensed milk.

The production of condensed and evaporated milk is declining. In 1945, 10.8% of the milk marketed in the United States was used in condensed and evaporated milk, compared with only 1.5% in 1987.

## CREAM

Cream is made by concentrating the fat portion of milk. Prior to the advent of the cream separator, this was accomplished by gravity separation. Today, it is done by passing milk through a cream separator. In commerce, whipping cream contains about 40% fat; coffee or table cream, 18 to 20%; and half-and-half, 12%.

## DRIED MILK (WHOLE MILK, SKIMMED MILK, AND WHEY)

Among the dried products produced from milk are nonfat dried milk (skimmed milk), for both human food and animal feed; dried whey, for both human food and animal feed; and dried whole milk.

In 1987, the following quantities of these dried products were produced in the United States: 1,067 million lb of dry whole milk, 1,059 million lb of nonfat dried milk, and 1,034 million lb of dried whey.

Dried milk products have many uses, principally as ingredients in other dairy and food products, although their use in the home has grown considerably in recent years. Despite its wide variety of uses, nonfat dried milk has been surplus much of the time.

## ICE CREAM AND SIMILAR FROZEN DESSERTS

Fig. 26-9. Meringue nut ice cream torte—a frozen dessert. (Courtesy, American Dairy Assn., Chicago, IL)

Currently, 99% of all frozen desserts in the United States consist of ice cream, ice milk, sherbet, and mellorine (made with a vegetable fat base). Other frozen desserts include frozen custard, frozen malted milk, artificially sweetened ice cream and ice milk, and water ices.

## IMPORTS AND EXPORTS

Imports of a number of dairy products are restricted by specific import quotas; among such products are several types of cheese, butter, butteroil, butterfat mixtures, ice cream, frozen cream, nonfat dry milk, dried buttermilk and whey, evaporated milk, condensed milk, chocolate crumb, and animal feed with milk solids. Although not formally restricted, certain other dairy products may be limited by agreement between the United States and the exporting country. The three main U.S. dairy exports in 1987, ranked in descending order of tonnage, were: cheese, 120,126 metric tons; casein, 108,136 metric tons; and butter, 905 metric tons.

As long as domestic prices are above world prices, and world supplies are ample, exporting countries will look to the United States as a possible market. As a result, import pressure will persist; yet, it is expected that imports of many commodities will continue to be limited by quotas.

Exports of dairy products are rather small. In 1987, U.S. exports of dairy products on a milk equivalent basis amounted to only 0.96% of the total U.S. milk supply. The three main

U.S. dairy product exports in 1987, ranked in descending order of tonnage, were: nonfat dry milk, 311,852 metric tons; cheese, 19,560 metric tons; and butter, 7,472 metric tons. The exports of nonfat dry milk were greater than the exports of all other dairy products combined.

In recent years, imports of dairy products have exceeded exports in monetary value, resulting in a trade deficit of $221 million in 1985 to a high of $277 million in 1989. The trade deficit is pointed up in Fig. 26–10 and Table 26–6.

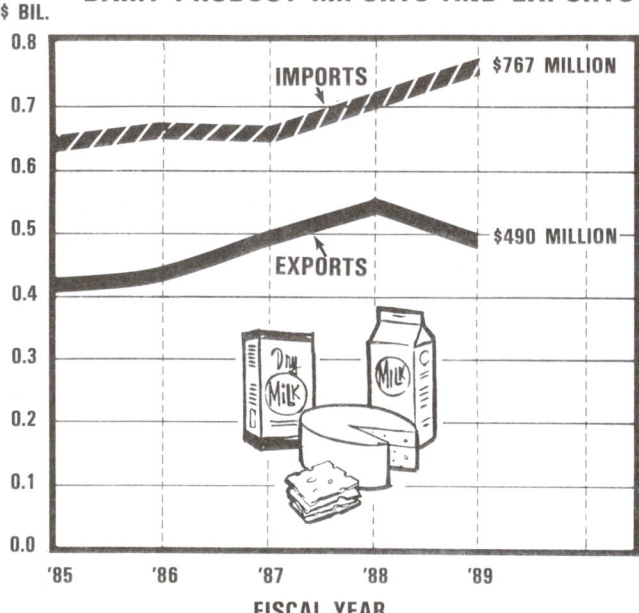

Fig. 26–10. Dairy product imports and exports. (Source: *Dairy, Livestock, and Poultry: U.S. Trade and Prospects*, USDA, Foreign Agricultural Service, FDLP 12–89, December 1989, pp. cover and #2.

## TABLE 26–6
### DAIRY PRODUCT IMPORTS AND EXPORTS[1]

| Year | Imports | Exports | Balance |
|------|---------|---------|---------|
|      | ($) | ($) | ($) |
| 1985 | 643 | 422 | −221 |
| 1986 | 666 | 434 | −232 |
| 1987 | 659 | 496 | −163 |
| 1988 | 712 | 541 | −171 |
| 1989 | 767 | 490 | −277 |

[1]Source: *Dairy, Livestock, and Poultry: U.S. Trade and Prospects*, USDA, Foreign Agricultural Service, FDLP, 12–89, December 1989, p. 2.

Exports of dairy products will continue to be influenced by the availability of surplus products and foreign policy. A more active role in meeting food deficiencies in the less developed countries of the world could increase total demand for dairy products and demand for export products.

## OUTLOOK

Per capita consumption of milk and most dairy products appears to have reached a rather stable stage. The most important factors affecting the future demand for dairy products will continue to be changes in population, income, consumer preferences, and new products.

Fig. 26–11. Consumers will determine the future of dairy products. This shows milk served with kabobs in outdoor cooking. (Courtesy, United Dairy Assn., Rosemont, IL)

It is expected that per capita consumption of various dairy products will follow the trend of recent years; products high in fat will decline in per capita consumption, while those low in fat will increase. The proportion of milk consumed in fluid form will increase; per capita consumption of butter and evaporated milk will likely decline on a gradual basis; and low fat fluid milk, nonfat dried milk, low fat frozen desserts, cheese, and sour cream dressings are likely to increase in per capita consumption.

Without doubt, new low fat dairy products and nondairy substitutes will replace some of the consumption of similar products higher in fat. The use of synthetic milk will increase.

Increasingly, the U.S. pricing system will (1) differentiate and price milk on a component basis, and (2) pay a premium for higher quality milk.

The operations involved in producing and marketing milk and dairy products will continue the current trend to bigness, fewer numbers, more mechanization, environmental control, and higher quality products.

# QUESTIONS FOR STUDY AND DISCUSSION

1. In 1986, the dairy farmers of the U.S. received $17.8 billion for their milk. But the retail cost to consumers was $52 billion. What is the explanation of this great difference between farm value and retail cost amounting to $34.2 billion?

2. In 1986, (a) dairy products accounted for 13.1% of the cash income of the nation's farmers, and (b) consumers spent about 12% of their food dollar for dairy products. Bear in mind that if these figures are increased, the figures for other items must be decreased. Can you justify increasing the percentages given above?

3. In 1987, only 1.6% of the total milk production was used on farms, compared with 15% in 1950. Why has "home use and fed" milk declined in recent years?

4. Fig. 26–2 shows that both total supply of milk and government stocks have spiraled since 1975. Interpreted from a practical standpoint, this means that U.S. dairy farmers have overproduced, and that the federal government has bought more and more surplus dairy products. Can such overproduction and government purchases and storage of surpluses be justified?

5. How do you account for each of the following facts relative to the utilization of the milk supply in 1987 (see Table 26–2):
   a. Only 37.4% of the milk and cream market is consumed in fluid form (36.6% sold to dealers, and 0.8% sold by producers directly to consumers).
   b. Of the milk sold by farmers, 89% of it was eligible for Grade A, and only 11% was Grade B.
   c. Only 14.8% of the milk supply was used for making butter.
   d. A total of 30% of the milk supply was used for making cheese.

6. How do you account for the following changes in per capita consumption of dairy products in recent years (see Table 26–3 and Fig. 26–3):
   a. Declines in butter, evaporated and condensed whole milk, and whole milk.
   b. Increases in yogurt and low fat cottage cheese.

7. In what three stages does milk move from the farm to the consumer?

8. How is milk priced and regulated today?

9. What is the primary purpose of federal milk marketing orders? Who administers them? How many different order markets are there?

10. What are State Orders? How do they differ from federal milk marketing orders?

11. What are milk cooperatives? What is their magnitude in volume of milk handled, and in volume (dollars) of business?

12. Discuss the importance of each of the following from the standpoints of sanitation and quality of milk:
   a. Dairy farm cooling of milk.
   b. Pasteurization.
   c. Grade A.
   d. Food and Drug Administration.
   e. Standard plate count, preliminary incubation count, and somatic cell count.

13. What are the standard grades of milk? Of what value are the federal grades of butter, cheese, and nonfat dried milk?

14. List and discuss each of the several price plans for paying for milk.

15. Why is so much more Grade A than Grade B milk marketed?

16. Do you favor pricing milk on (a) the traditional butterfat basis, or (b) the component basis? Justify your choice.

17. Do you favor a milk price support program? Justify your answer. Why has the support level of manufacturing milk gone down and down since 1984?

18. Why have milk manufacturing plants become fewer and bigger in recent years (see Table 26–5)?

19. For what is casein used?

20. Fig. 26–7 shows the per capita consumption of butter and margarine from 1910 to 1987. Why has the consumption of butter gone down and down while the consumption of margarine has gone up and up?

21. Why has the processing of cheese increased so greatly in recent years? List the five leading cheese producing states by rank.

22. Fig. 26–10 and Table 26–6 show that U.S. imports of dairy products exceeded exports from 1985 to 1989, creating a considerable trade deficit. During this same period (1985–1989), milk was federally subsidized and there was overproduction. Would you change this situation? If so, how would you change it?

23. What do you see ahead relative to the demand for, and marketing of, milk and dairy products? Are you optimistic or pessimistic about the future of the U.S. dairy industry?

## SELECTED REFERENCES

| Title of Publication | Author(s) | Publisher |
|---|---|---|
| *Federal Milk Marketing Order Program, The*, Mkt. Bull. No. 27 | | U.S. Department of Agriculture, Washington, DC, 1963 |
| *Fluid Milk Marketing* | G. M. Beal<br>H. H. Bakken | Mimir Publishers, Madison, WI, 1956 |
| *Grade "A" Pasteurized Milk Ordinance*, Pub. 229 | | U.S. Public Health Service, Washington, DC, 1965 |
| *Milk Production and Processing* | H. J. Judkins<br>H. A. Keener | John Wiley & Sons, Inc., New York, NY, 1963 |
| *Organization and Competition in the Dairy Industry*, Tech. Study No. 3 | | National Commission on Food Marketing, Washington, DC, 1966 |
| *Organization and Competition in the Midwest Dairy Industries* | S. W. Williams, *et al.* | Iowa State University Press, Ames, IA, 1970 |

How it used to be done! This shows homemade ice cream being made in a hand-turned freezer. (Courtesy, The Bettmann Archive, New York, NY)

A pastoral scene. (Courtesy, Soil Conservation Service)

# THE SHEEP AND GOAT INDUSTRY

## Chapter 27

Sheep and goats were first domesticated in the New Stone Age. The first Egyptian portrayal of sheep appears on one of the earliest sculptures known, dating back to 4000 B.C. Some subsequent sculptures showed one of the early uses of sheep—that of being driven across the freshly sown fields in the Nile Valley to tread in the grain. Other historical records show that sheep provided primitive pastoral peoples with meat, wool, tallow, skins, and milk.

Sheep belong to the genus *Ovis*, and goats and their wild relatives make up the genus *Capra*. These two genera of the family *Bovidae* are closely related; so closely related, in fact, that a naturalist never speaks lightly of "separating the sheep from the goats." Goats may be distinguished from sheep, however, by the presence of a beard, by the absence of foot glands (which sheep have), by the strong smell of the bucks, and by differences in horns and skeleton. Goats also are more intelligent, independent, and possess greater ability to fight and fend for themselves.

## ORIGIN AND DOMESTICATION OF SHEEP

Fig. 27-1. People have herded sheep and woven wool fibers since the Stone Age. Sheep have played an important part in world history. (A copper engraving from a 17th century edition of Vergil's "Bucolica." Photo obtained through the courtesy of The Wool Bureau, Inc.)

There is more confusion and disagreement about the ancestry and classification of sheep than with any other species. This difficulty arises from the bewildering number of breeds and the marked changes produced by domestication. There are more than 200 distinct breeds of sheep scattered throughout the world. Although differing widely in body form and wool character, domestic sheep of all breeds are universally timid and defenseless, and the least intelligent and the least teachable of all domestic four-footed animals. These traits are plainly the result of selection under domestication and are connected with the herding of sheep in large bands where independence of behavior is a disadvantage. As a result, domestic sheep have

become completely dependent on caretakers. Unlike other farm animals, they are unable to return to a wild life, which we refer to as becoming feral. Though this dependence is a logical final result of domestication, it does appear that the evolution of sheep in this direction may have gone too far, for they are pitifully helpless in emergencies.

It is certain that domestic sheep came from the wild sheep of Europe and Asia. The confusion and disagreement is over the number of species and the identity of the wild stocks mixed up in their ancestry. One of the chief difficulties in the way of tracing the ancestry of our domestic sheep lies in the fact that most of them are long-tailed, whereas all of the wild species from whence they came are short-tailed. It appears, however, that lengthening of the tail is a characteristic which appeared with domestication.

Domestic sheep are thought to descend mainly from two wild stocks: (1) The Moufflons (*Ovis musimon* and *Ovis orientalis*), and (2) the Asiatic Urial (*Ovis vignei*).[1]

There is, however, considerable evidence to indicate that the wild big-horned sheep of Asia may be at least partial progenitors of the fat-rumped sheep of central Asia. Perhaps, too, some modern breeds trace back to other wild stocks than those herein indicated.

## THE MOUFFLON

There are two wild stocks of the Moufflon, the Asiatic Moufflon (*O. orientalis*), a wild sheep still found in Asia Minor and the Caucasus, and the European Moufflon (*O. musimon*), which is native to Europe and still found in Sardinia and Corsica. These two relatives are closely allied, but the Asiatic Moufflon is redder and has a somewhat different twist to the horns. Both of the Moufflon stocks are considered as ancestors of domestic sheep.

Fig. 27-2. The European Moufflon (*Ovis musimon*), one of the ancestors of domestic sheep. Like all the wild species from which sheep descended, the Moufflon is short tailed. It appears, therefore, that lengthening of the tail is a characteristic which came with domestication. (Courtesy, New York Zoological Society)

---

[1]Based on difference in chromosome number, acceptance of the Asiatic Urial as the direct ancestor of domestic sheep may be questioned. But, based strictly on chromosome number, the wild ancestry of modern breeds of sheep remains unknown. This leads to the thinking that the Asiatic Urial is the primary ancestor of most of our modern breeds of sheep, but that mutations may have occurred along the way. Nevertheless, the chromosome theory of ancestry is presented in *Sheep & Goat Science*, pp. 4–6, a book by the same author and publisher as *Animal Science*.

Even today relatively unimproved Moufflon-like short-tailed domestic sheep exist in different sections of northern Europe. The least modified of these primitive types is the semiferal race of sheep on the uninhabited island of Soay, northwest of Scotland. The only essential difference between the Moufflon and the feral Soay sheep is the shorter wool of the latter. The island of Soay is visited once or twice each year by the residents of St. Kilda who hunt down the Soay sheep with dogs and shear them.

## THE ASIATIC URIAL

The Asiatic Urial (*O. vignei*), which is a smaller race of sheep than the Moufflon, is native to the grassy open plains of central Asia. It lives in large flocks and is much less a mountain animal than the Moufflon. Most of our familiar breeds of sheep are thought to be descendants of this wild stock. For example, the Merino seems to have originated in Asia Minor about the 8th century B.C., and to have been spread by the Phoenicians into North Africa and Spain. Likewise, the fat-tailed sheep in western Asia and Africa, the long-tailed African and Arabian breeds, and perhaps the fat-rumped sheep of central Asia, are descendants of the Asiatic Urial.

Fig. 27–3. A wild sheep (*Ovis vignei*), native to the province of Punjab in northern India. This is a member of the Asiatic Urial stock, the smaller of the two wild ancestors of domestic sheep. Most of our familiar U.S. breeds of sheep are thought to have descended chiefly, if not entirely, from the Asiatic Urial. (Courtesy, New York Zoological Society)

## ORIGIN AND DOMESTICATION OF GOATS

The ancestry of the goat is far less confused than is that of the sheep, but a report on the wild relatives of this genus will not be given here. Suffice it to say that goats have not produced nearly so many breeds nor, except for some of the milk-producing types, such extremely modified breeds. Unlike sheep, goats easily return to a wild state if given the opportunity. In fact, only the domestic cat can equal the goat in returning promptly and successfully to the independent life of a wild creature.

Like sheep, goats were probably among the first animals to be domesticated. Goatlike remains are found in the Swiss Lake dwellings of the New Stone Age, and the goat was well-known in Biblical days.

## POSITION OF SHEEP IN THE ZOOLOGICAL SCHEME

The following outline shows the basic position of the domesticated sheep in the zoological scheme:

Kingdom *Animalia*: Animals collectively; the Animal Kingdom.
  Phylum *Chordata*: One of approximately 21 phyla of the animal kingdom, in which there is either a backbone (in the vertebrates) or the rudiment of a backbone, the chorda.
    Class *Mammalia*: Mammals are warm-blooded, hairy animals that produce their young alive and suckle them for a variable period on a secretion from the mammary glands.
      Order *Artiodactyla*: Even-toed, hoofed mammals.
        Family *Bovidae*: Ruminants having polycotyledonary placenta; hollow, nondeciduous, unbranched horns; and nearly universal presence of a gallbladder.
          Genus *Ovis*: The genus consisting of the domestic sheep and the majority of wild sheep. The horns form a lateral spiral.
            Species *Ovis aries*: Domesticated sheep.

## WOOL; PRECIOUS FIBER THROUGH THE AGES

Wool was first spun and made into cloth many years before the beginning of recorded history. According to historians, fabrics of wool have been discovered in the ruins of the Swiss Lake villages, which were inhabited during the Neanderthal Age, between 10,000 and 20,000 years age. Moreover, wool was used as a clothing material by the Babylonians about 4000 B.C., and sheep are pictured on the earliest Egyptian monuments, which date some time between 5000 and 4000 B.C.

Fig. 27–4. Ancient Babylonian loom in operation. Wool was first used as a clothing material by the Babylonians beginning about 4000 B.C. (Courtesy, The Bettmann Archive)

The ancient Egyptians, Babylonians, Greeks, and Hebrews did hand spinning and weaving in the home. The wool industry, like most others, first developed as a household craft, rather than as a primitive factory system.

Sheep raising was known as the earliest pastoral industry, and reference is frequently made to it in Old Testament literature. For example, we are told that Abraham, the patriarch of the Old Testament, thrived and prospered through his great flocks and herds. Subjects of the King of Israel were taxed according to the number of their rams. The Bible also refers to Adam and Eve's son, Abel, as a "keeper of sheep"; and it was shepherds watching over their flocks by night who first saw the star over Bethlehem. Sheep raising was recognized as an early agricultural pursuit, and the early herds and flocks served as a medium of exchange.

Fig. 27–5. Ancient Egyptian spinning and weaving. As a household craft, the wool industry had its simple beginning among the Egyptians some time between 5000 and 4000 B.C. (Courtesy, The Metropolitan Museum of Art)

Sheep were treated with marked respect in Greece. Individual names were given to them and shepherds would proudly call out their favorites. To protect their fleeces from inclement weather, skins were spread over the animals, while the keepers of the flocks had to content themselves with loincloths.

When Rome was in its glory, the wealthy and refined citizens of the empire boasted of their achievement in producing the finest quality wool in the world. Sheep were given extraordinary care, and they were even blanketed so that a luster and gloss might be imparted to the wool. At frequent intervals, the fleece was parted, combed, and moistened with the rarest oils, oftentimes with wine. Surplus stock was usually killed at two years of age, for the Romans believed that the fleece was in its best condition at that period. The distinctive toga, a loose outer garment which was worn by officials of ancient Rome when appearing in public in time of peace, was made from woolen fabrics.

Before the year 1000, both Spain and England attached great importance to their flocks; and, by the year 1500, they were recognized as the two greatest sheep-producing countries of the world. Although the Spanish wools were much finer, for several hundred years Spain and England were regarded as competitors in the great wool markets of Flanders.

In Spain, the powerful nobility and clergy engaged in the lucrative sheep industry. In an attempt to produce the finest staple possible, the early Spanish keepers of flocks drove their sheep from southern to northern pastures in the spring and returned them in the fall. In this manner, it was possible to secure the most favorable grazing and climatic conditions for the flocks. The early laws of the kingdom stipulated that the owners of large flocks should be allowed a path 90 paces wide through all enclosed lands. In the migration process, any animal that failed to keep with the band was left by the wayside. Presumably, this accounts for the flocking or gregarious instinct of the Merino sheep as well as their hardiness. With the repeal of the migration laws and consequent prohibition of seasonal migration, Spanish shepherds blanketed their sheep through the colder months—their object being that of keeping an equitable temperature, thus producing a more uniform and higher quality product.

During the Middle Ages (500 to 1500 A.D.) in England, sheep were the sheet anchor of farming. Their chief product was not meat, milk, nor hides, but wool. Unlike the flocks of Spain, those in England were small; the sheep were not in the hands of a very few powerful owners as they were in Spain, and they were not compelled to travel across the country. The great problem of the English sheep farmers, therefore, was to procure sheep that were adapted to their particular locality. This largely accounts for the development of the many types and breeds in that country. The cold winters, scarcity of winter feed, and the presence of scab and foot rot made the early sheep raising of England a risky venture. The shepherd's position was highly respected, and lame shepherds were in great demand because lame shepherds were not so likely to overdrive their sheep. None of the wools from the English breeds was as fine as that of the Merino. Nevertheless, there was a ready market for the English wools because they were more suitable for a variety of uses than those from Spain.

## BAKEWELL'S IMPROVEMENT OF ENGLISH SHEEP

By the time Robert Bakewell of Dishley (1726 to 1795) entered the livestock-breeding business of England, wool had declined in price until—with the rapidly advancing values of English lands—it alone would no longer justify the keeping of sheep. Bakewell, however, had the foresight to picture the future needs of a growing population in terms of meat, and he set about improving the Leicestershire sheep of the day. He was successful in creating a low-set, blocky, quick-maturing type of animal. He paid little or no attention to fancy points; no animal met with his favor unless it had utility value, as measured by meat over the block. Bakewell gradually transformed the large, heavy-boned and heavy-framed sheep, that had little or no propensity to finish quickly, to a shorter-legged, blocky form with finer bone and quick-finishing propensities. Other breeders followed suit.

Fig. 27-6. Leicester sheep, the breed that Robert Bakewell improved. (Courtesy, *Farmer and Stock-Breeder*, London, England)

## EARLY IMPORTATIONS AND IMPROVEMENTS OF SHEEP IN THE UNITED STATES

The Bighorn or Rocky Mountain sheep, prevalent in the Rocky Mountain region from Alaska to California, were native to this continent, but the ancestors of our present-day domestic sheep were imported. Columbus brought sheep and goats to the West Indies on his second voyage in 1493. Cortez brought Merino sheep into Mexico in 1519. The Spaniards who founded old Santa Fe, New Mexico, were thought to have brought in the multicolored sheep from which the flocks of the Navajo Indians have descended. If, as is generally supposed, these early importations were of Spanish Merino extraction, special permission to take them out of Spain must have been granted by the king.

Fig. 27-7. Coarse-wooled, multicolored, unimproved Navajo sheep on the Navajo Reservation, New Mexico. It is thought that the ancestors of these sheep were brought to America from Spain by the Spaniards who founded old Santa Fe. (Courtesy, Southwestern Range and Sheep Breeding Laboratory, Fort Wingate, NM)

The first sheep of the British breeds to be introduced in this country are said to have been brought into Virginia by the London Company in 1609.[2]

Two decades later, there were as many as 400 sheep in Charleston, now a part of Boston. We are told that these early importations represented very poor specimens of British sheep, the imperfections of which were generally acknowledged. Furthermore, because of the lack of care, inadequate shelter, and promiscuous breeding, the sheep yielded a wool that had lost all pretense to fineness. Predatory animals played havoc with the flocks, and for this reason many of the early sheep were herded on small islands off the coast. Even today, sheep are found on the islands off the coast of Maine and Massachusetts, especially on Martha's Vineyard and Nantucket.

Sheep husbandry was promoted by the colonists primarily in order to furnish wool rather than for purposes of increasing the food supply. In addition to the demand for wool in most new countries, the product is well adapted to marketing in an undeveloped area—being (1) light in weight, value considered, and (2) imperishable with respect to time involved in getting it to market.

As early as 1662, there was a woolen mill at Watertown, Massachusetts. At about this time, the exportation of ewes or lambs, except to other colonies, was forbidden. In 1670, in order to encourage sheep growing, Connecticut required every person to labor for one day each year at clearing the underwood to make pasturage. Drastic legislation was passed by all the New England colonies to protect the sheep from dogs. A dog that bit or killed a sheep was often hanged as though it were a human malefactor. The execution was usually carried out in some nearby swamp, whence comes the name *Hang-Dog Swamp*, given to several localities in colonial Massachusetts and Connecticut.

The town common was open to all kinds of livestock, but because sheep were the most defenseless of domestic animals, and the hardest to raise, the town regulations concerning their care were numerous. Moreover, each town had one or more shepherds who were considered very important persons.

Owners identified their stock by marks or brands (ear notches or earholes were the most common means of marking)—a medieval European device revived in the New World. These marks were registered with the town clerk, and sometimes the description was embellished with a crude picture.

In due time, early importation to America and the improvements made by the colonists were to become the sturdy basis for some of our great American flocks. Breeding stock was imported from some of the foreign countries; and the early flock owners practiced rigid selection in order to improve the quality of wool produced, employing sight and touch, ordinary scales for weighing the fleeces, and in some instances a ruler for measuring fiber length. These empirical methods served well, however, and remarkable improvements in wool were brought about. Fleeces became heavier, and the fibers were longer and more uniform. It is also interesting to note that in the year 1836 fleeces produced in this country averaged only about 2 lb in weight and were satisfactory only for the coarser woolen fabrics. Half a century later the average weight had been increased to 5½ lb. Almost simultaneously,

---

[2]In 1607, a shipment of sheep came over on the *Susan Constant*, but these were consumed during the famine of the ensuing winter, thus having no permanent effect.

the growing importance of the lamb and mutton trade changed the character of the sheep from one with a small carcass and with a short, fine fleece to a heavier and larger animal growing a longer-stapled and coarser grade wool. Thus, a dual-purpose type of animal was developed.

Fig. 27-8. Three Merinos, two ewes and one ram, were imported to the United States from Spain in 1793, but these were butchered. *Don Pedro* (above) was then imported in 1801, by du Pont de Nemours at a cost of $1,000. It is claimed that this ram left a larger impression on the sheep of America than any other sheep ever imported. *Don Pedro* was acclaimed the father of the fine-wool industry in the United States. He weighed 138 lb. After careful washing in cold water, his fleece weighed 8½ lb, and it had a staple length of 1¾ in. (Courtesy, *National Wool Grower*, Salt Lake City, UT)

## THE WESTWARD MOVEMENT OF SHEEP

In 1810, the census figures clearly indicated that the northeastern part of the United States—New England and New York—was the sheep-producing center of the nation. At this date, there were an estimated 7 million head of sheep in the United States. By 1840, sheep numbers in the United States had increased to 19 million head. At that time there were no appreciable numbers of sheep in the Far West except those owned by the Navajo Indians in northern New Mexico. In fact, the only state west of the Mississippi having sheep in considerable numbers was Missouri. Ten years later, the densest sheep population was centered in the Ohio Valley and Great Lakes region. When this area became somewhat thickly settled and land values rose, many sheep producers, desiring to operate on a large scale, moved farther west where range was cheap and extensive. It must be remembered that during this period sheep were maintained primarily for wool production and that the market-lamb business, as we know it today, was practically unknown. It was not at all surprising, therefore, to find that with the opening up of cheaper rangelands in the West, there was also an immediate and marked shift of the sheep population from east to west.

Like other wars, the Civil War caused sharply increased sheep numbers as a result of the demand for and the high price of wool. Following the war, wool prices fell because of lower demand and increased competition from cotton and imported wool. Yet the westward expansion and the opening up of cheap lands continued.

## GROWTH, DECLINE, AND TRANSITION OF THE U.S. SHEEP INDUSTRY

Fig. 27-9 shows that sheep numbers have fluctuated rather widely from time to time. Although it is not shown in Fig. 27-9, they peaked in 1884; and they nearly peaked again in 1942 during World War II, when they reached 56 million head. On January 1, 1988, there were 10.8 million sheep and lambs, and during the year 1988, there were 1.61 million head of sheep and lambs on feed, in the United States. Also, it is noteworthy that there were more than 115,000 sheep operations in the United States in 1988.

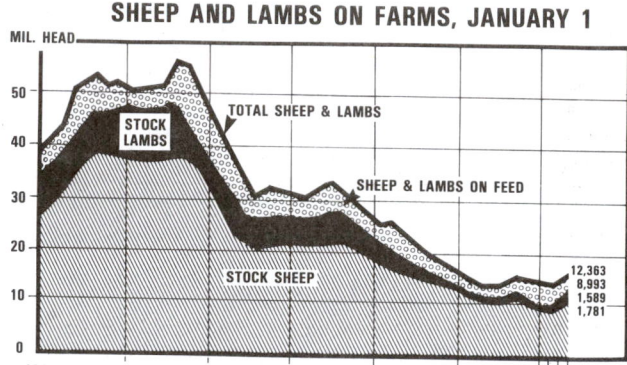

Fig. 27-9. Growth and decline of the U.S. sheep industry. (From USDA sources)

Several factors have contributed to the decline in sheep numbers since 1942, including (1) lower returns and higher risks from sheep than from cattle and some crops in many areas; (2) increased losses due to predators; (3) scarcity and high wages of competent sheep herders; (4) uncertainties in tariff levels and imports, wool incentive payments, and grazing allotments on public domain; and (5) application of more science and technology in competing meat and fiber industries.

Also, the geography of sheep production has shifted. The Rocky Mountain region had become the dominant sheep section of the country by 1900, but sheep numbers declined more sharply in the range states than in the native states during the forties. Since 1950, sheep numbers have declined proportionally in both farm flock and range band areas; at that time (1950), the 11 western states and Texas had 68.17% of the nation's stock sheep and lambs, and in 1988, these same 12 states still had 69.2% of the breeding sheep.

## WORLD DISTRIBUTION OF SHEEP

Despite the obscurity surrounding their domestication and the disagreement about their ancestry and classification, it is known that sheep raising followed the conquest and colonization of the Western Hemisphere to Australia, New Zealand, South Africa, and other countries. Today, the sheep industry is worldwide, with numerous breeds providing needed adaptation. Like other industries, sheep raising is affected by wars; national and international policies and politics; supply and demand; wool substitutes; competition for land, labor, and capital; and many other factors.

Table 27–1 ranks the 10 leading sheep producing countries of the world and presents related facts about the size and density of each country. The United States now ranks twenty-seventh among the nations of the world in sheep numbers.

**TABLE 27–1**
**LEADING SHEEP-PRODUCING COUNTRIES OF THE WORLD**

| Country | Sheep Population[1] | Human Population[2] | Size of Country[2] | | Sheep Per Capita | Density of Sheep | |
|---------|---------------------|---------------------|--------------------|--------|------------------|------------------|--------|
| | | | (sq mi) | (sq km) | | (sq mi) | (sq km) |
| Australia | 164,000,000 | 16,090,000 | 2,966,200 | 7,682,458 | 10.19 | 55.3 | 21.3 |
| U.S.S.R. | 140,783,000 | 287,015,000 | 8,649,496 | 22,402,195 | 0.49 | 16.3 | 6.3 |
| China | 102,655,000 | 1,069,628,000 | 3,705,390 | 9,596,960 | 0.10 | 27.7 | 10.7 |
| New Zealand | 64,970,000 | 3,397,000 | 103,736 | 268,672 | 19.13 | 626.3 | 241.8 |
| India | 51,684,000 | 833,422,000 | 1,266,595 | 3,280,481 | 0.06 | 40.8 | 15.8 |
| Turkey | 40,000,000 | 55,377,000 | 301,381 | 780,577 | 0.72 | 132.7 | 51.2 |
| Iran | 34,500,000 | 51,005,000 | 636,293 | 1,647,999 | 0.68 | 54.2 | 20.9 |
| United Kingdom | 27,820,000 | 56,648,000 | 94,226 | 244,045 | 0.49 | 289.5 | 114.0 |
| South Africa | 29,800,000 | 35,625,000 | 472,359 | 1,223,410 | 0.84 | 63.1 | 24.4 |
| Argentina | 29,202,000 | 32,617,000 | 1,065,189 | 2,758,840 | 0.90 | 27.4 | 10.6 |
| **World Total** | 1,172,828,000 | 5,055,000,000 | 57,800,000 | 149,702,000 | 0.23 | 20.3 | 7.8 |

[1]*FAO Production Yearbook 1988*, United Nations, Rome, Italy, Vol. 42, p. 247, Table 90. Data for 1988.

[2]*The World Almanac and Book of Facts, 1989.*

The majority of the world's sheep are concentrated in a relatively small number of countries; the 10 leading countries have over 58% of the world's sheep. Australia, the U.S.S.R., and China, account for 35% of the world total. Although the industry is worldwide, it is of greatest importance in those countries which have (1) vast frontier land areas that are sparsely settled, and (2) temperate climates. These conditions prevail in the Southern Hemisphere, and it is there that most of the world's sheep are located. On the other hand, there are areas within countries of the Northern Hemisphere with many great flocks and numerous small ones.

Since World War II, sheep numbers have increased in most countries. World sheep numbers increased from 987,400,000 in 1963 to 1,172,828,000 in 1988, an increase of 18.8%. By contrast, during this same period of time, U.S. sheep numbers declined from 30,170,000 to 10,744,000, a decline of 64%. This difference can be attributed to (1) the world's rising aggregate demand for food and fiber in relation to U.S. demands, and (2) the fact that sheep are an excellent subsistence occupation

for people in nations with a high proportion of nonarable land and relatively low living standards.

Favorable prices for wool and lambs, and satisfactory grazing and weather conditions usually make for stable or increased sheep numbers in those countries to which they are adapted. Historically, wars have always played a prominent part in stimulating the sheep business.

## WORLD WOOL PRODUCTION

Inspection of Table 27–2 reveals that, generally speaking, there is quite a close correlation between sheep numbers and wool production. It is noteworthy, however, that the U.S.S.R. has a low wool production per sheep; only 4.5 lb (scoured) are produced in comparison with 9 lb (scoured) in New Zealand. Australia produces 29% of the world's wool at the present time.

**TABLE 27–2**
**LEADING WOOL-PRODUCING COUNTRIES OF THE WORLD**

| Country | Wool Production[1] | | Number of Sheep[2] | Wool Produced/Sheep | |
|---------|--------------------|--------|--------------------|--------------------|--------|
| | (lb) | (kg) | | (lb) | (kg) |
| Australia | 1,190,000,000 | 541,000,000 | 164,000,000 | 7.3 | 3.3 |
| U.S.S.R. | 629,000,000 | 286,000,000 | 140,783,000 | 4.5 | 2.0 |
| New Zealand | 582,000,000 | 265,000,000 | 64,970,000 | 9.0 | 4.1 |
| China | 280,000,000 | 127,000,000 | 102,655,000 | 2.7 | 1.2 |
| Argentina | 203,000,000 | 92,000,000 | 29,202,000 | 7.0 | 3.2 |
| Uruguay | 119,000,000 | 54,000,000 | 26,049,000 | 4.6 | 2.1 |
| South Africa | 102,000,000 | 46,000,000 | 29,800,000 | 3.4 | 1.5 |
| United Kingdom | 100,000,000 | 45,000,000 | 27,820,000 | 3.6 | 1.6 |
| Turkey | 75,000,000 | 34,000,000 | 40,000,000 | 1.9 | 0.9 |
| Pakistan | 72,000,000 | 33,000,000 | 27,479,000 | 2.6 | 1.2 |
| **World Total** | 4,140,000,000 | 1,882,000,000 | 1,172,828,000 | 3.5 | 1.6 |

[1]*FAO Production Yearbook 1988*, United Nations, Rome, Italy, Vol. 42, p. 283, Table 104.

[2]*Ibid.*, p. 247, Table 90.

The Southern Hemisphere countries of Australia, New Zealand, South Africa, Argentina, and Uruguay are the main wool surplus producing areas. Two countries alone—Australia and New Zealand—account for 59.4% of all wool exports.

Wool production is definitely a frontier industry, thriving in those areas where there is an abundance of cheap range area and where the human population is sparse. On the other hand, wool consumption is greatest in centers of dense population, especially the temperate zone of the Northern Hemisphere. The leading importing countries, by rank, are: Japan, China, U.K., U.S.S.R., and France.

World production and consumption of wool has been steadily increasing since about 1976 (see Fig. 27–10), possibly due to (1) the manufacture of fabrics with a mixture of fibers, (2) the great desirability for wool fabrics in winter, and (3) the ever-increasing population growth.

## World Production and Consumption of Raw Wool

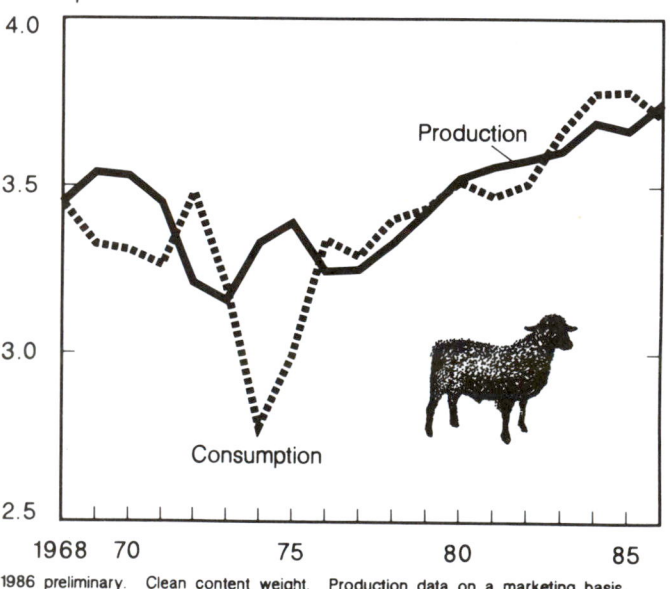

Billion pounds

1986 preliminary. Clean content weight. Production data on a marketing basis.

Fig. 27–10. Worldwide production and consumption of wool declined from 1968 to about 1976, but has increased since 1976. (*1988 Agricultural Chartbook*, USDA, Ag. Handbook, No. 673, p. 100, Chart 234)

## SHEEP RAISING IN AUSTRALIA

Australia holds undisputed claim to being the leading sheep country of the world, having about 14% of the world's sheep, and producing 29% of the world's wool. It is a large country which is best suited to a pastoral type of agriculture and in which sheep seem to have been the animals best adapted to grazing lands.

Although sheep existed in Australia at an early date, the Merino was first introduced in 1789. Other Merino importations followed, most of which came from Saxony in Germany. The early sheep industry of Australia was financed with English capital, the original aim being to render that country indepen-

Fig. 27–11. Mustering (called herding in the United States) sheep for shearing in Australia. Note that horses and dogs are used. Australia has wide, flat grasslands. (Courtesy, Australian News and Information Bureau)

dent of Spanish, German, and other foreign sources of supply. The wool from the Australian Merino flocks met with a ready demand on the part of English manufacturers, thus giving great encouragement to the further expansion of the sheep industry of Australia.

Merino blood still predominates in Australian flocks—accounting for 75% of the total, although the mutton breeds are gradually increasing in favor and greater attention is being given to the mutton qualities of the Merino. Most of the wool is marketed in England. Likewise, Australia disposes of most of its surplus mutton by shipping frozen carcasses to England.

Most of the flocks of Australia are kept in fenced holdings, rather than being herded as is the most common practice on the western ranges of the United States. The Australian owners prefer ranging on enclosed lands, contending that (1) the sheep make better use of the range under this system, the animals scattering out in contrast to each sheep regularly maintaining a fairly definite position in the band as happens in herding; (2) less driving is required as, in herding, the animals must be rounded into camp at night, driven to water, and kept from other bands of sheep; (3) the fences cost less than the added labor in herding; and (4) the fences give considerable protection against predatory animals, especially against the dingo—the wild dog of Australia, and help protect the forage from the ravages of rabbits (the fences being rabbit proof). Experienced operators contend that handling sheep in fenced holdings is satisfactory provided that the band can be kept under these conditions the year round, but that it will not work if the animals must be removed from fenced range at intervals and herded, for the band will then be untrained and unmanageable.

## SHEEP RAISING IN NEW ZEALAND

New Zealand is a small country, less than twice the area of the state of Illinois. However, there are over 65 million sheep in New Zealand, or 626 sheep per square mile, the densest sheep population of any country in the world. The sheep of this country run more to the mutton type than do the animals of Australia; about three-fourths of the sheep of New Zealand are Romneys. Practically all the flocks are kept in fenced holdings without herders, in a manner similar to the method followed in Australia. Year-round grazing is available. Cattle and sheep share many areas, to the advantage of each other, with cattle utilizing the coarser vegetation and sheep the finer grasses and legumes. The best lambs in New Zealand are produced by using Southdown rams on Romney ewes.

Fig. 27–13.  Sheep scene in South Africa. These are Dorpers, a mutton breed, developed from crossing Black Head Persian and Dorset Horn, adapted to the drier areas—with 7 to 11 in. of rainfall. (Courtesy, Embassy of South Africa, Washington, DC)

Fig. 27–12.  Romney sheep grazing fenced pasture in New Zealand. (Courtesy, Department of Scientific and Industrial Research, Wellington, New Zealand)

## SHEEP RAISING IN SOUTH AMERICA

There is a considerable sheep industry in Argentina, Chile, Uruguay, Brazil, and Peru.

Without doubt, the finest sheep country in South America, and one of the finest in the world, is the La Plata River area of Argentina and Uruguay, where sheep compete with cattle for the lush pasture of the Pampas region. Predatory animals are few, winter feeding is seldom necessary, and diseases are rare. In brief, it has been said that there is probably no other comparable area in the world where the shepherd's life is easier than in this particular territory. Sheep raising in other areas and countries of South America does not compare with the La Plata River section, either in terms of favorable conditions or quality of flocks.

In all the South American countries, Merino breeding was used as a foundation and in effecting improvement, but many subsequent importations of the mutton breeds have been made, especially the long-wooled breeds. Much of the coarse wool produced in South America is used in the manufacture of rugs.

## SHEEP RAISING IN SOUTH AFRICA

South Africa's chief claim to fame as a sheep country is in terms of wool production. Merino blood predominates, although a considerable number of representatives of the mutton breeds have been introduced in more recent years. The major handicaps to sheep production in South Africa are (1) prevalence of diseases and parasites, especially sheep scab; (2) unreliable labor; (3) predatory animals, especially jackals; and (4) frequent droughts.

## SHEEP AND WOOL PRODUCTION IN THE UNITED STATES

Table 27–3 gives the numbers and value of sheep and goats in the United States. As shown, sheep numbers totaled 10.8 million in 1988. This was only 19% of the 56 million head of the peak wartime year of 1942.

**TABLE 27–3**
**NUMBERS AND VALUE (TOTAL VALUE AND VALUE/HEAD) OF SHEEP ON UNITED STATES FARMS, AND GOATS IN TEXAS, 1988[1]**

| Class | Number | Farm Value | |
|---|---|---|---|
| | | Value | Total Value |
| | | ($/head) | (thousand $) |
| Sheep and lambs ........ | 10,774,000 | 89.90 | 968,918 |
| Goats and kids (1987) .... | 2,950,000[2] | 65.80[3] | 194,110[3] |

[1]Agricultural Statistics 1988, USDA, p. 282, Table 419; and p. 298, Table 440.

[2]The 2,950,000 goats in the United States consists of about 1,600,000 Angoras, 850,000 dairy goats, and 500,000 Spanish goats.

[3]Estimates made by the author of this book.

The United States is not a great sheep- and wool-producing nation. We rank twenty-seventh in world sheep numbers and fourteenth in world wool production. In terms of total farm value, both cattle and swine exceed sheep by a considerable sum.

The vast majority of U.S. sheep are still in the western range area (69.2% of the breeding sheep are in the 11 western states and Texas), where most of them graze on arid and semiarid pastures. For the most part, farm flocks utilize untillable areas and waste feeds.

Although the range area still dominates U.S. sheep production, many range operators have either retired or switched to cattle, and few young operators have entered the business.

The United States does not produce enough lamb or wool to meet its requirements; it is an importer. By 1987, lamb and mutton imports were equivalent to 11.1% of our production, and about 70.3% of our wool requirements came from abroad.

But sheep numbers alone do not tell the whole story. It's the production of lamb and mutton that counts. Fig. 27–14 shows that U.S. sheep have become more efficient through the years.

### Sheep Numbers, Lamb and Mutton Production

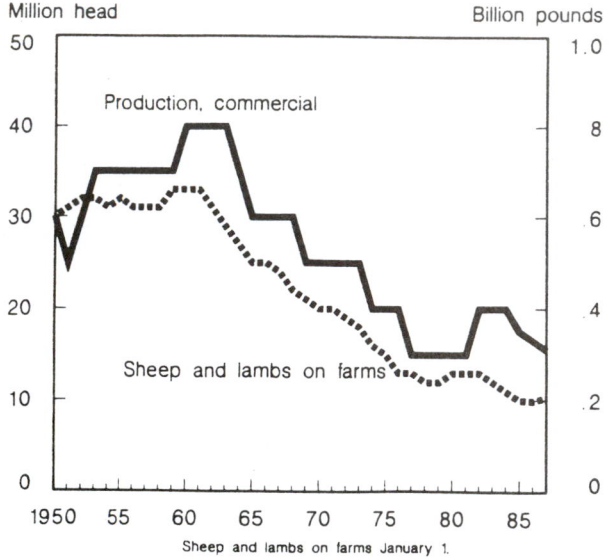

Fig. 27–14. Lamb and mutton production has gone up faster than sheep numbers, which means that sheep have become more productive and efficient through the years. (*1988 Agricultural Chartbook,* USDA, Ag. Hdbk. No. 673, p. 89, Chart 191)

## AREAS OF SHEEP PRODUCTION

Sheep raising in the United States may be divided into two areas: (1) the range sheep states, and (2) the farm flock sheep states. Each of these areas will be discussed separately.

### THE RANGE SHEEP STATES

About 69.2% of the breeding sheep (or stock sheep) of the nation is in the 11 western range states and Texas. For the most part, this type of sheep husbandry is characterized by large bands of from 1,000 to 1,500 ewes under the care of a herder. In general, these bands are run on unenclosed land. As more than half of the ranges of the West are publicly owned and are likely to continue so, fences will not be built simply because it is not wise to go to so much expense on other than private holdings. On the other hand, scarcity and high cost of labor have resulted in fencing a considerably amount of private holdings in the southwestern United States. As most of these ranges are used the year round and the bands are not herded on mountain ranges during the summer months, fenced range has been satisfactory.

Fig. 27–15. A range band of sheep.

Within the western range, there are great variations in topography, rainfall, and vegetation. Thus, in most cases, the sheep production of this territory is a migratory type of enterprise, in which deserts, plains, foothills, and mountains may be used during different seasons of the year in such manner as to obtain as nearly year-round foraging as possible. In general, the ranges of the southwestern states do not afford sufficiently good grazing to produce milk-fat market lambs.

Wool production with Rambouillet-type sheep, therefore, is of comparatively greater importance in this area than it is in the Northwest. On the better ranges of the Northwest, however, the objective is always that of producing a fat market lamb at weaning time, and feeder lambs result only because the ranges proved inadequate (perhaps through lack of moisture) or because of the usual rejects in culling. Because the vegetation is more abundant in the Northwest, a larger, crossbred type of sheep predominates in this area. This type produces a coarser fleece than the Rambouillet-type ewes of the Southwest and yields lambs of more desirable type and quality.

Most of the better range operators provide supplemental feeding during a part of the year, especially in the lambing season and during times when the range is covered with snow. Alfalfa from the irrigated areas is the chief hay crop; and corn, oats, barley, and various protein supplements may be used on occasion. In the trade, wools produced in the range states are known as territory wools.

## THE FARM FLOCK SHEEP STATES

This includes sheep production throughout the United States, except in the western range states. It embraces the farm flocks, ranging from a few head to a thousand or more, which are kept on the farms of the East, South, and Midwest. In the trade, the wool coming from farm flocks is known as native wool.

Fig. 27-16. Beside the still waters! A farm flock of sheep. (Courtesy, Dr. Clair E. Terrill, Staff Scientist, Agricultural Research Service, USDA)

Many of the smaller farm flocks of this area are primarily kept as scavengers, and, unfortunately, they are often accorded the neglect of a minor enterprise. In general, the lambs from farm flocks are sold directly for slaughter, with only a few finding their way to feedlots for further finishing. Farm flocks carry a heavier infestation of parasites than do range bands, and the fleeces lack the care and uniformity accorded to range wool production. Also, some lambs coming from farm flocks are not docked or castrated, a neglect which is a rarity on the western range where sheep production is more of a specialty.

Purebred flocks as well as commercial enterprises characterize farm sheep flocks. Purebred producers market their surplus stock (1) to other purebred breeders, and (2) as breeding stock for commercial enterprises.

## LEADING STATES IN SHEEP PRODUCTION

Some idea of the relative importance of sheep production in the leading states may be obtained through studying Table 27-4.

**TABLE 27-4**
**TEN LEADING STATES IN SHEEP AND LAMB NUMBERS, BY RANK, 1988[1]**

| State | Number |
|---|---|
| Texas | 1,960,000 |
| California | 1,015,000 |
| Wyoming | 865,000 |
| Colorado | 860,000 |
| South Dakota | 610,000 |
| Montana | 538,000 |
| Oregon | 490,000 |
| Utah | 478,000 |
| New Mexico | 451,000 |
| Iowa | 405,000 |
| **U.S. Total** | **10,774,000** |

[1]*Statistical Abstract of the United States 1989*, U.S. Dept. of Commerce, p. 653, Table 1136.

As will be noted, Texas has a substantial lead in sheep numbers, a position which it also holds in cattle, horse, and goat numbers. In fact, the leading sheep states are predominantly in the range sheep area of the United States.

# FACTORS FAVORABLE TO SHEEP PRODUCTION

Fig. 27–17. Sheep grazing a sage brush range. No other species is so well adapted to utilize arid areas. (Courtesy, U.S. Forest Service, Washington, DC)

As compared to other classes of livestock, sheep possess the following natural advantages:

1. They are unexcelled in the utilization of the more arid type of grazing.

2. In the farm flock states, they frequently utilize what would otherwise be wasteland. They are also excellent scavengers, gleaning fields and destroying weeds.

3. Compared to cattle, they produce more liberally in proportion to what they consume.

4. They produce two products, lamb and wool, which are available for market at two different periods of the year; and it seldom happens that both products sell at bottom prices the same year.

5. Their returns come quickly; lambs may be marketed eight months after the ewes are bred.

6. Their habit of bedding down on the highest areas of the field or range leaves the larger part of the droppings at the places where they are most needed. Moreover, the form in which sheep manure is dropped and the way it is tramped into the soil ensures a smaller waste than is possible under any other system of stock farming.

7. Their wool clip is easily stored and shipped, thus making wool production ideally suited to a frontier type of agriculture.

# FACTORS UNFAVORABLE TO SHEEP PRODUCTION

There are, however, many factors which are quite unfavorable to sheep production, as many sheep producers will testify to their regret. Some of these are:

1. Wool has always been in politics, and it is apt so to remain. It is rather difficult, therefore, to predict prices over a long period of time.

2. Consumption of lamb has declined to the point where it is a minor product, with the result that both on-foot and retail marketing problems have increased.

3. Sheep are very much subject to attack from numerous predatory animals, including dogs.

4. When disease or injury strikes, they have less resistance than other classes of livestock.

5. Herding is not a particularly attractive profession, thus resulting in a scarcity of satisfactory herders.

6. Sheep are quite susceptible to a number of very devastating parasites.

7. The competition from synthetic fibers affects the demand for wool.

# PRESENT POSITION; FUTURE OUTLOOK

The *present position* of the sheep industry is herewith summarized:

1. **U.S. sheep population has not kept pace with world trends.** World sheep numbers totaled 987,400,000 in 1963 vs 1,172,828,000 in 1988, an increase of 18.8%. By contrast, during this same period of time (1963 to 1988), U.S. sheep numbers declined from 30,170,000 to 10,774,000, a decline of 64.3%.

2. **Per capita consumption of lamb and mutton has declined.** Prior to 1945, U.S. per capita consumption of lamb and mutton—and most of it was lamb—was about 7 lb per year. In 1987, it was 1.5 lb.

3. **Synthetic fibers have increased.** In the period 1953 to 1986, U.S. per capita consumption of scoured wool declined from 3.1 to 1.7 lb. During this same 33-year period, per capita consumption of synthetic fibers increased from 9.5 to 36.3 lb.

4. **Imports of lamb, mutton, and wool increased.** Prior to 1957, imports represented less than 1% of this nation's lamb and yearling production; but by 1987 this figure had increased to 11.7%. In 1960, 55.5% of U.S. wool consumption was imported. By 1987, 70.3% of our wool consumption was imported.

5. **Fewer sheep on national forests.** In 1944, approximately 4.9 million sheep and goats grazed on national forests. By 1987, this number had been cut to 1.1 million head.

The *future outlook* for the U.S. sheep and wool industry, based on all available facts and figures and the considered judgment of the author, follows:

1. **Per capita consumption of lamb may decline slightly.** The future of lamb will be determined primarily on how well producers (a) meet consumer preference—and in a greater number of stores, (b) provide larger and year-round supplies of uniform quality, and (c) apply more science and technology.

2. **Per capita consumption of wool faces stiff competition.** Synthetic fiber producers will continue to invest large sums in research and promotion; and there is every reason to believe that these efforts will be as effective and productive as in the past.

3. **Imports of lamb and wool may increase.** A number of factors, especially improved transportation for carcass meat and the impact of the European Common Market, point to greater importation of lamb in the years ahead. If properly controlled, increased importation of lamb could be a help, thereby assuring year-round availability and stimulating consumption.

Domestic synthetic fiber production and imported fabrics (rather than fiber) will continue to constitute more of a threat to domestic wool production than foreign imports of wool.

4. **Some shift in the geography of production.** It is expected that sheep production will gradually shift from range bands to farm flocks—from west to east, primarily because of economics of production. Large range bands will, for the most part, be limited to those grazing areas that are not well suited to cattle and which are marginal or submarginal for most other agricultural products.

5. **Producer profits will be dependent on management.** As in all agriculture, increased production costs appear inevitable, but no comparable longtime rise in the price of lamb and wool is in the offing. Thus, superior management holds the key to profits in the years ahead.

6. **Science and technology will increase.** It is predicted that the sheep and wool industry will accelerate experimental work and the application of science and technology. Although those competing meat and fiber producers with a head start will not pull up the reins of progress, it is inevitable that, percentage-wise the sheep and wool industry can make greater scientific and technological strides in the immediate future. For example, considerable encouragement can be derived from recent advances in parasite control, thus lessening one of the major handicaps to sheep production. Also, from the standpoint of producing and marketing quality products, it appears that the U.S. sheep industry will emulate, in part, Australian wool production and New Zealand lamb production.

In brief, the future of the sheep business in America is largely dependent upon the industry itself. Significant breakthroughs are needed (1) in increased efficiency—in lambs raised and wool production per ewe; (2) in quality, merchandising, and promotion of lamb; and (3) in marketing and processing of wool.

## QUESTIONS FOR STUDY AND DISCUSSION

1. How may goats be distinguished from sheep?

2. There are more than 200 distinct breeds of sheep scattered throughout the world. How do you explain the fact that there are more breeds of sheep than there are of cattle and swine?

3. Sheep are unable to return to wildlife, which we refer to as becoming feral. Does this indicate that their domestication has gone too far? Justify your answer.

4. Apparently the wild stocks of sheep were short-tailed and the lengthening of the tail followed domestication. How can you explain the selection of long-tailed sheep?

5. From what two wild stocks are domestic sheep thought mainly to descend? Describe each of them.

6. Unlike sheep, goats easily return to the wild state if given the opportunity. Why the specie difference?

7. Christmas cards and other Biblical scenes depicting shepherds watching over their flocks by night, usually show more than one shepherd because the shepherds brought their flocks together at night. Why did they get together at night?

8. Briefly describe sheep raising (a) when Rome was in its glory, (b) in early-day Spain, and (c) in England during the Middle Ages.

9. Why and how did Robert Bakewell change English sheep? With what breed did he work?

10. Describe the early importations and improvements of sheep in the United States.

11. What forces caused the westward movement of sheep during the 1800s?

12. What forces caused the growth, decline, and transition of the U.S. sheep industry in the 1900s?

13. Name and rank the four leading sheep countries of the world. Why do these countries have so many sheep? What factors cause New Zealand to have more sheep per square mile than any other country in the world?

14. Why have sheep numbers increased in many parts of the world since World War II but, simultaneously, declined in the United States?

15. Name and rank the four leading wool producing countries of the world. Why do these countries produce so much wool?

16. What factors account for Australia holding undisputed claim to being the leading sheep country of the world?

17. Describe sheep raising in each (a) New Zealand, (b) South America, and (c) South Africa.

18. Describe sheep and wool production in the United States.

19. Discuss the primary differences in sheep production which characterize the two U.S. areas known as (a) the range sheep states, and (b) the farm flock sheep states.

20. Name and rank the 10 leading states of the U.S. in sheep and lamb numbers. Why do these states have so many sheep?

21. List the factors that are favorable to sheep production.

22. List the factors that are unfavorable to sheep production.

23. Describe the present position of the U.S. sheep industry.

24. Discuss the future outlook of the U.S. sheep and wool industry.

25. Where, and under what circumstances, would you recommend (a) that one enter sheep production, or (b) that one discontinue sheep production?

# SELECTED REFERENCES

| Title of Publication | Author(s) | Publisher |
|---|---|---|
| *America's Sheep Trails* | E. N. Wentworth | Iowa State University Press, Ames, IA, 1945 |
| *Approved Practices in Sheep Production*, Fourth Edition | E. M. Juergenson | The Interstate Printers & Publishers, Inc., Danville, IL, 1981 |
| *History of Livestock Raising in the United States, 1607–1860* | J. W. Thompson | U.S. Department of Agriculture, Washington, DC, 1942 |
| *Our Friendly Animals and Whence They Came* | K. P. Schmidt | M. A. Donohue & Co., Chicago, IL, 1938 |
| *Productive Sheep Husbandry* | W. C. Coffey | J. B. Lippincott Co., New York, NY, 1937 |
| *Profitable Sheep* | S. B. Collins R. F. Johnson | The Macmillan Company, New York, NY, 1956 |
| *Profitable Sheep Farming* | M. M. Cooper R. J. Thomas | Farming Press Ltd., Ipswich, England, 1971 |
| *Sheep and Wool* | M. P. Botkin R. A. Field C. L. Johnson | Prentice Hall, Inc., Englewood Cliffs, NJ, 1989 |
| *Sheep Book, The* | J. McKinney | John Wiley & Sons, Inc., New York, NY, 1959 |
| *Sheep Breeding* | Edited by G. L. Lomes D. E. Robertson R. J. Lightfoot | Butterworth & Co., Ltd., London, England, 1970 |
| *Sheepman's Production Handbook, The* | Edited by G. E. Scott | Sheep Industry Development Program, Inc., Denver, CO, 1986 and updated at intervals. |
| *Sheep Production* | R. V. Diggins C. E. Bundy | Prentice-Hall, Inc., Englewood Cliffs, NJ, 1958 |
| *Sheep Production and Management* | C. V. Ross | Prentice Hall, Inc., Englewood Cliffs, NJ, 1988 |
| *Sheep Science*, Revised Edition | W. G. Kammlade, Sr. W. G. Kammlade, Jr. | J. B. Lippincott Co., New York, NY, 1955 |
| *U.S. Sheep and Goat Industry; Products, Opportunities and Limitations, The* | C. S. Menzies, Chairman, Task Force | CAST, 250 Memorial Union, Ames, IA, 1982 |
| *Wild Sheep of the World, The* | R. Valdez | Wild Sheep and Goat International, Mesilla, NM, 1982 |

They also serve those who watch and wait. (Courtesy, Charles J. Belden, Pitchfork, WY)

Navajo sheep. This first American breed of sheep was molded by the first Americans—the Indians. The old-type Navajo sheep, developed by the Navajo Indians, produced the desired wool which the Navajo tribe wove into beautiful blankets and rugs. This shows old-type Navajo sheep, with a Navajo pueblo in the background.

# TYPES AND BREEDS OF SHEEP[1]

## Chapter 28

[1]Sometimes folks construe the write-up of a breed of livestock in a book or in a U.S. Department of Agriculture bulletin as an official recognition of the breed. Nothing could be further from the truth, for no person or office has authority to approve a breed. The only legal basis for recognizing a breed is contained in the Tariff Act of 1930, which provides for the duty-free admission of purebred breeding stock provided they are registered in the country of origin. But the latter stipulation applies to imported animals only.

In this book, no *official* recognition of any breed is intended or implied. Rather, the author has tried, earnestly and without favoritism, to present the factual story of the breeds in narrative and picture. In particular, such information relative to the new and/or less widely distributed breeds is needed, and often difficult to obtain.

In no other class of farm animals have so many types and breeds evolved as in sheep. As domestic sheep were improved in various parts of the world, the producers within different geographical areas soon became convinced that the animals under their care possessed special attributes not found in more distant flocks. Out of this thinking arose the approximately 200 different breeds of sheep that exist today. Many of these breeds are of little importance in commercial production, with more than three-fourths of the industry of the world based on the use of not more than six breeds. Even so, breed enthusiasts are usually vociferous about the relative merits of their particular breed, no matter how small the numbers.

## CLASSES OF SHEEP

Breeds of sheep may be and are classified on several different bases, including (1) their degree of suitability for mutton or wool production (mutton or wool type), (2) color of face (white or black face), (3) presence or absence of horns (horned or polled), (4) topography of the area in which they originated (mountain, upland, or lowland), (5) type of wool produced, (6) use in breeding, and (7) exotic uses. Each system of classification has its special merits, but perhaps a classification based on type of wool produced is as good as any.

## CLASSES OF SHEEP BY TYPE OF WOOL PRODUCED

Table 28–1 shows the most common U.S. breeds of sheep classed according to type of wool produced.

### TABLE 28–1
### BREEDS OF SHEEP CLASSIFIED ACCORDING TO THE TYPE OF WOOL PRODUCED

| Fine-Wool Type | Medium-Wool Type | Long-Wool Type | Crossbred-Wool Type | Carpet-Wool Type | Fur Type |
|---|---|---|---|---|---|
| American Merino | Cheviot | Cotswold | Columbia | Navajo | Karakul |
| Booroola Merino | Clun Forest | Leicester | Cormo | Scotch Highland | Romanov |
| Debouillet | Dorset | Lincoln | Corriedale | | |
| Delaine-Merino | Finnsheep | Romney | Panama | | |
| Rambouillet | Hampshire | | Polypay | | |
| | Montadale | | Targhee | | |
| | North Country Cheviot | | | | |
| | Oxford | | | | |
| | Shropshire | | | | |
| | Southdown | | | | |
| | Suffolk | | | | |
| | Tunis | | | | |

In general, all of the breeds listed within each of the six wool-type categories produce wool of a similar character, especially from the standpoint of diameter and length of fibers.

### FINE-WOOL BREEDS

The common fine-wool breeds of the United States are the American Merino, Debouillet, Delaine Merino, and Rambouillet. The Booroola Merino is a newcomer to the U.S. All of these breeds are of Spanish Merino extraction and really represent different types or ideals of Merinos brought about through selection. Because of their common ancestry, therefore, the five breeds still possess many characteristics in common. All of them are noted for fineness of wool and a great amount of yolk; the fleece sometimes loses over 70% in weight in scouring. In general, modern purebred animals of these breeds are of more acceptable mutton conformation than formerly, although they are not equal to the mutton breeds in this respect. The fine-wool breeds are hardy, gregarious, long-lived, and well suited to production under range management methods throughout the world. Also, like the Dorset and Tunis breeds, ewes of the fine-wool breeds will conceive out of season.

### MEDIUM-WOOL BREEDS

The Southdown, Shropshire, Oxford, Hampshire, and Suffolk breeds are collectively referred to as the *down* breeds, because of the nature of the country in which they were developed. This area in southern England is a country of hills or downs. The down breeds came into prominence during Bakewell's time. From the beginning, they were bred primarily for mutton, with special emphasis on those characteristics considered important to the nature of the particular grazing lands and climatic conditions as well as on the market demands of the regions in which they were developed. The face and leg color of all the down breeds is some shade of brown or black, and the fleece occupies a middle position between the length and coarseness of the long wools and the extreme fineness and density of the fine wools.

The medium-wool breeds have been popular in the farm-flock regions of the United States, and rams of the larger breeds have been extensively used in market-lamb production on the western ranges.

### LONG-WOOL BREEDS

The long-wool breeds, bred chiefly for mutton, are the largest of all sheep. They originated in Great Britain in an era when producers and consumers regarded with favor a large, coarse, slow-maturing sheep that produced long, coarse wool, and which when liberally fed would become very fat. These conditions gave rise to such important breeds as the Cotswold, Leicester, Lincoln, and Romney. All of the long-wool breeds are large framed, have square bodies, and are somewhat rangy in build with conspicuously broad backs. As compared with the fleeces of the fine- or medium-wool breeds, those of the long-wool sheep are open, coarse, and very long.

As their size would indicate, these breeds were developed in level-lying country where feeds were abundant and could be obtained without too much travel.

At the present time, most purebred sheep of the long-wool breeds are considered too slow in maturing and too big to satisfy the demands of the lamb market. Moreover, their carcasses are quite coarse and overlaid with fat. They are, however, of great value in crossbreeding to improve the weight of wool in other breeds, to increase the size of little sheep, and to increase the finishing qualities of some breeds. It is claimed that the long-wool breeds, especially the Romney, will thrive in regions of excessive rainfall, as the long wool carries the water off the body and does not soak it up as a more dense fleece will do. It is noteworthy, however, that most practical sheep producers caution against keeping sheep on marshy ground.

## CROSSBRED WOOL BREEDS[2]

The crossbred breeds, which are descended from a long-wool × fine-wool foundation, produce medium-fine wool and therefore often are classed with the medium-wool breeds. In general, however, the crossbred wool breeds are better adapted to the western range than are their respective parent stocks or any of the medium-wool breeds. Under range conditions, they produce better market lambs and heavier fleeces than the Rambouillet, and they are more active and have superior herding tendencies in comparison with either the long-wool or medium-wool breeds.

For many years, commercial ranchers crossed long-wool rams on grade fine-wool ewes in an effort to secure larger ewes that would yield more wool and produce heavier and superior market lambs. The results were often variable. Sometimes the mutton qualities would be improved, but the wool would be coarse; whereas at other times the wool yield would be greater, but the mutton qualities would be disappointing.

Despite the lack of uniformity, most commercial producers of the West preferred this method to the system of alternating in the use of black-faced and fine-wool rams. Topping the band with black-faced rams produced an excellent market lamb; but because of lack of the herding instinct, poor fleeces, and other deficiencies, the resulting crossbred ewe lambs were not suitable for flock replacements. In order to get desirable replacements, therefore, fine-wool rams had to be used at intervals, with the result that the wether lambs of this breeding did not meet market demands. Thus, the need was for a type of sheep which would eliminate the ram problem that invariably plagued producers in the alternative use of black-faced and fine-wool rams and which would produce lambs suitable for either market or replacement purposes. Such was the need, and out of this need arose the Columbia, Cormo, Corriedale, Panama, Polypay, and Targhee breeds of sheep—all descended from long-wool × fine-wool foundations.

## CARPET-WOOL BREEDS

Wools used in the manufacture of carpets and rugs in the United States are imported from New Zealand and Argentina, where the native sheep possess a coarse, wiry, tough fleece. Most American wools are too fine to be used in carpets. If so

used, they mat down and wear very rapidly. Carpet wools are quite variable, ranging from 1 to 13 in. in length and from 15 to 70 microns in diameter. In addition, carpet wools show a tremendous range of luster, strength, crimp, and resilience. Although several foreign breeds of sheep produce carpet wools, the only carpet-wool breeds known to most people in this country are the Navajo and the Scotch Highland.

## FUR SHEEP BREEDS

New processes have resulted in the use for fur of pelts obtained from many breeds of sheep. Except for the Karakul and the Romanov, however, all the other breeds of sheep are still kept primarily for lamb and wool production; and any fur production is secondary.

Karakul sheep are bred primarily because of the suitability of the lamb pelts for fur production. The majority of Karakul lambskins are produced in the regions of Bokhara and Bessarabia in the U.S.S.R.; in the cities of Herat, Afghanistan; Shiraz, Iran; and Baghdad and Salzfelle, Iraq; and in the countries of Namibia (formerly South West Africa) and India. The best-grade lambskins come from Bokhara, where the Karakul breed originated.

In order of their value, the pelts are classified as follows:

1. **Broadtail.** This is the most valuable, although its production is comparatively small. It is produced from prematurely born or stillborn lambs, and in some instances from those killed within a few hours after birth. The hair is undeveloped, grows in different directions, and reflects light, giving the pelt its moire appearance (moire means *watery design*).

Such premature deaths are not forced but are the result of accidents, such as abortions.

2. **Persian lamb.** This type of fur is next in value and comes from Karakul lambs 3 to 10 days old. It has a tight, lustrous curl that must be watched carefully from the time the lamb is born, for the curl is likely to open rapidly after the fifth day; and while the value of the pelt increases with size, it is essential that the curl remain tight.

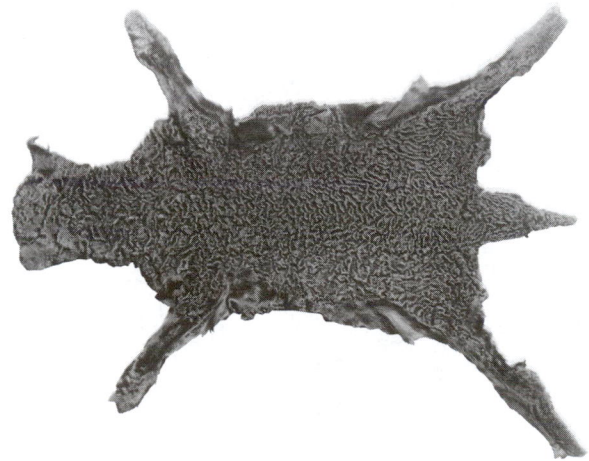

Fig. 28–1. A Persian lamb pelt properly dried (not shaped) and cleaned for shipment and sale. (Courtesy, Lowry Hagerman, Santa Fe, NM)

3. **Caracul.** Caracul, spelled with a "C," is a trade name given to the lustrous open-type of fur that shows a wavy moire pattern free from close curls.

---

[2]The listing of the crosses which produced each of the crossbred wool breeds is given in Table 28–4 for purposes of breed history. In no sense does it imply any lack of purity of the respective breeds, or that all of them are new breeds.

Caracul skins are light in weight and are best if removed when lambs are not more than two weeks old. But this type of pelt does not deteriorate so rapidly with growth as do those of either the broadtail or Persian lamb types. Caracul is the least valuable of the three types of pelts.

Within each of these three groups, there are different grades of pelts, depending upon quality, tightness of curl, luster, pattern, color, and general appearance. As would be expected, the price varies with the type and quality of the fur and with the supply and demand.

As is true in making any fur garment when the smaller furs are involved, large numbers of pelts are required for proper matching. For this reason, it is often difficult to sell small numbers of pelts to advantage.

The Romanov, the other fur or pelt sheep breed listed in Table 28–1, also originated in the U.S.S.R. Although the Romanov is well known in the U.S.S.R., it is a new breed to the United States, imported primarily because of its high lambing rate.

## CLASSES OF SHEEP BY BREEDING USE

Another means of classifying sheep is based on their breeding use—whether (1) their primary use is to produce replacement ewes or breeding rams; or (2) they are dual-purpose animals, used to produce both replacement ewes and breeding rams. Within this system, the breeds are classified into (1) ewe breeds, (2) ram breeds, (3) dual-purpose breeds, and (4) other breeds. Table 28–2 summarizes the classification of sheep based on their breeding use. Details follow.

### TABLE 28–2
### BREEDS OF SHEEP CLASSIFIED ACCORDING TO THEIR BREEDING USE[1]

| Ewe Breeds | Ram Breeds | Dual-Purpose Breeds | Other Breeds |
|---|---|---|---|
| Booroola Merino | Hampshire | Cheviot[2] | Cotswold |
| Corriedale | Oxford | Clun Forest[2] | Karakul |
| Debouillet | Shropshire | Columbia | Leicester[2] |
| Delaine-Merino | Southdown | Cormo[2] | Lincoln[2] |
| Finnsheep | Suffolk | Dorset | Navajo |
| Polypay[2] | | Montadale | Romanov |
| Rambouillet | | North Country Cheviot[2] | Scotch Highland |
| Targhee | | Panama[2] | Tunis |
| | | Romney | |

[1]Based on information from the American Wool Council except where noted.
[2]Author's classification.

## EWE BREEDS

These are the white-faced breeds; they produce fine, medium, or long wool, or crosses among these types. Ewe breeds are selected for adaptability to the environment, reproductive efficiency, wool production, size, milking ability, and longevity. Replacement ewe lambs should be raised from these breed types, or from crosses among these breeds.

## RAM BREEDS

These are the meat-type breeds. Ram breeds, or crosses of two ram breeds, are (1) raised for the production of rams to cross on ewe breeds, and (2) selected for growth rate and carcass quality.

## DUAL-PURPOSE BREEDS

These are the breeds that may be used to produce replacement ewes and/or breeding rams, depending on the production situation. For example, the Columbia is popular as a ewe breed in the mountain regions of the West, but it is used as a ram breed in the Midwest and South.

## OTHER BREEDS

For one reason or another, these breeds fail to fit into one of the other classifications. After each breed has been studied, the reasons for its exclusion will be apparent.

## EXOTIC CLASSES AND BREEDS OF SHEEP

Several exotic classes and breeds of sheep are being promoted today—some old, others new; some registered breeds, others without breed registries. There is some overlapping, with some breeds appearing in more than one class. Among the exotic classes and breeds being promoted today, are the following:

• **Colored sheep**—In the past, range sheep herders kept a few colored sheep to use as *markers*—usually, about one marker for each 100 sheep. Thus, with a band of 2,000 sheep, there would be 20 markers. So, upon checking and finding 20 markers present, the herder assumed that all 2,000 sheep were present and accounted for. But the woolen mills complained that the colored wool contaminated the white wool of the other sheep in the flock, especially at shearing time. Markers became taboo and colored sheep were bred out of the flock. Now, there is demand for colored wool for spinning and weaving for growing handcraft cottage industries; and this has created renewed interest in colored sheep.

Modern colored sheep are of many types and breeds, but all of them have one trait in common—they are colored; they are black, gray, red, tan, brown, silver, yellow, or any combination of these. Among the colored breeds are Barbados, Moufflon, Romney, Karakul, Scottish Highland, Lincoln, Cotswold, Jacob Sheep, Hungarian Racka, Teeswater, Salish, Navajo, and Tunis. These colored sheep raisers meet every five years as members of the World Congress of Colored Sheep.

• **Natural colored sheep (black sheep)**—The only requirement of this class is that they be black; so, it follows that there is great variation in type. *Natural colored sheep* are registered by the Natural Colored Wool Growers Association, 18150 Wild Flower Drive, Penn Valley, California 95946.

• **Jacob sheep**—These multi-colored sheep were brought to the United States from England, about 1900. Jacob sheep are always white, with large well-defined spots or patches of black or brown. Their faces are black and white.

• **Hair sheep**—There are over 100 million domesticated hair sheep in the world. Approximately 90 million of them are in Africa, 6 million in Brazil, 2 million in the Caribbean area, and most of the balance in Mexico, Colombia, Venezuela, India, Southeast Asia, and the United States. There are over 100 hair sheep breeds in the world; among them, the Barbados Blackbelly, Texas Barbado, St. Croix, Katahdin, Wiltshire Horn, Santa Ines, and Pelibuey.

• **Milking sheep**—In Europe, sheep have long been used to produce milk and cheese. For example, in the French province of Roquefort, 4,000 farmers own and milk 600,000 dairy ewes which produce 163,000,000 lb of milk annually, much of which is processed into Roquefort cheese by 14 cheese plants.

The increase in sheep's milk in the United States appears to be linked to the back-to-nature trend. As a result of this interest, the American Milking Sheep Society was founded. Dairy sheep in the United States consist chiefly of two breeds—Dorset and Tunis.

## POPULARITY OF BREEDS

Table 28–3 shows the 1988/89 and total registrations to date of the common breeds of sheep. Although the annual figures for 1988/89 are probably more meaningful at the present time than all-time registrations, it is recognized that one year's data fail to show trends.

**TABLE 28–3**
**REGISTRATIONS OF SHEEP IN U.S. BREED REGISTRY ASSOCIATIONS**

| Breed | 1988 or '89 Registrations | Total Registrations To Date |
|---|---|---|
| Suffolk[1] | 72,743 | 1,833,634 |
| Dorset | 18,600 | 440,000 |
| Hampshire | 18,189 | 1,742,812 |
| Rambouillet | 14,886 | 882,473 |
| Polypay (1989) | 12,000 | 49,000 |
| Columbia | 9,004 | 400,303 |
| Southdown | 5,517 | 466,870 |
| Corriedale | 4,883 | 517,463 |
| Shropshire | 3,472 | 1,385,445 |
| Montadale | 3,214 | 110,243 |
| Cheviot (1989) | 2,807 | 218,016 |
| Romney (1989) | 2,697 | 80,000 |
| Targhee | 2,537 | 33,615 |
| Oxford | 1,820 | 262,930 |
| Finnsheep | 1,172 | 14,631 |
| North Country Cheviot | 827 | 16,135 |
| Tunis | 561 | 15,786 |
| American and Delaine Merino | 478 | 81,350 |
| Cotswold (1989) | 321 | 13,733 |
| Clun Forest | 194 | 2,490 |

[1]Includes: National Suffolk Sheep Association, 1988 equals 52,843 and total equals 1,208,134; American Suffolk Sheep Society, 1988 equals 19,900 and total equals 625,500 registrations.

## CHARACTERISTICS OF BREEDS

The characteristics of the different breeds of sheep are summarized in Table 28–4.

**TABLE 28–4**
**BREEDS OF SHEEP AND THEIR CHARACTERISTICS, CLASSIFIED BY TYPE OF WOOL PRODUCED[1]**

| Breed | Place of Origin | Color; Face, Ears, and Legs | Head Characteristics | Other Distinguishing Characteristics | Disqualifications; Comments |
|---|---|---|---|---|---|
| **Fine-Wool Breeds** | | | | | |
| **American Merino** | Spain. | White. Reddish-brown spots may occasionally appear on lips, ears, and pasterns. | Most rams have horns, but there are some polled strains. | Distinguished from the Delaine Merinos by more skin wrinkles; the more wrinkled American Merinos being the "A" and "B" types. Strong flocking instinct. Ewes will breed out of season. | |
| **Booroola Merino** | Australia, (CSIRO) at a government research station. | White. | Similar to Merino, from which it is descended. | Very prolific. At the Holden Station in New Zealand, ewes wean an average of 2.9 lambs per ewe lambing. A single gene, known as the F gene, controls the Booroola's prolificacy; and the gene is transferable. | In addition to this prolificacy, Booroola Merino sheep produce high quality wool. |
| **Debouillet** | U.S.A.; on the Amos Dee Jones Ranches of Roswell, and Tatum, New Mexico, beginning in 1927–30. Association organized in 1954. | White. | Rams may have horns, but there are also polled strains; open face. | Comparatively smooth body; long staple. | Overshot or undershot jaw; broken-down pasterns; undersized; brown hair on face, ears, or legs; black spots in the fleece; too light fleece. |

*(Continued)*

**TABLE 28–4** (Continued)

| Breed | Place of Origin | Color; Face, Ears, and Legs | Head Characteristics | Other Distinguishing Characteristics | Disqualifications; Comments |
|---|---|---|---|---|---|
| **Delaine Merino** (the "C" Type, or Texas Delaine) | Spain. The Delaine Merino and Texas Delaine are of similar origin and appearance. | White. Reddish-brown spots may occasionally appear on lips, ears, and pasterns. | Most rams have horns, but there are some polled strains. The females should be free of horns or scurs. | Comparatively smooth body; of the "C" type. Strong flocking instinct. Ewes will breed out of season. | Texas Delaine Sheep Assn. disqualifies: abnormal testicles; swayback; close horns; black spots in the fleece or on the body; overshot or undershot jaw; weak pasterns. |
| **Rambouillet** | France; from Spanish Merino parent stock imported from Spain. | Cream to white. | Most rams have horns, but there are some polled strains. Ewes are hornless. | Largest fine-wool breed. Strong flocking instinct. Ewes will breed out of season. | Abnormal development of testicles or only one testicle descended in scrotum; unsound udder or inverted teats; overshot or undershot jaws; black spots in the fleece; rolled under eyelids; weak pasterns. |
| **Medium-Wool Breeds** | | | | | |
| **Cheviot** | Scotland; in the Cheviot Hills between Scotland and England. | White face with a black nose. Often black spots are on the ears. | Both sexes are polled. | Stylish, alert, and active. Head and legs are free from wool. | Black spots other than ears. Overshot or undershot jaw. |
| **Clun Forest** | In England, in the Clun Forest—a mountainous region. | Dark brown. | Bare faces, at least below the eyes. They have wool over the forehead that extends down to eye level. | Clun ewes are prolific, easy lambers, and good milkers. They are a medium-sized breed. | The Clun Forest registry association prohibits competitive showing of the breed. |
| **Dorset** | England; especially in the southern counties of Dorset and Somerset. Polled Dorsets originated in early 50s at North Carolina State University. | White and practically free from wool. | There are horned and polled strains, both of which are registered by the Continental Dorset Club. Except for the presence or absence of horns, the two strains are identical. | Ewes will breed out of season. Also, the ewes milk well. | All solid black hooves; black or dark colored septum (dividing tissue between nostrils); black or dark colored lining of the nose or mouth. |
| **Finnsheep** (Finnish Landrace) | Finland. Finnsheep were brought to the U.S. in 1968. An earlier importation from Finland was made to the University of Manitoba, in Canada. | White. | Head is free of wool. Usually, both sexes are hornless, but a few rams have light horns. | Prolificacy. In Finland, ewes (all ages included) average 2.41 lambs/birth. | The wool is usually 50–54 spinning count with a fleece weight of about 9.5 lb. Finnsheep are being used in breeding programs to increase multiple births. |
| **Hampshire** | England; in the south-central county of Hampshire. | Rich deep brown, approaching black. | Both sexes are hornless, although rams sometimes have scurs. | Large size; early maturity. | Undesirable traits are: crooked legs and poor feet; inverted eyelids; abnormal sex organs; black fibers; wool blindness; broken woolcap; horns; abnormal teeth or jaw development. |
| **Montadale** (Columbia × Cheviot) | U.S.A.; by E. H. Mattingly, St. Louis, Missouri. | White. | Both sexes are polled. Head free from wool. | Black hoofs and nose; black spots in ears. | Horns. Black spots in wool. Pink nose. |
| **North Country Cheviot** | Scotland; from the old Long Hill sheep, but with infusion of Merino, Ryeland, and Southdown blood in formative period. | White. | Nose straight to slightly Roman. Rams are sometimes horned. | Wool grades 50s and 56s; mature rams weigh up to 300 lb and mature ewes up to 200 lb. | |
| **Oxford** | England; in the south-central county of Oxford. The Oxford stems from a Hampshire × Cotswold cross, made in the 1830s. | Variable, from gray to brown. | Both sexes are polled. Topknot of wool. | Largest of the down breeds. | Black fiber; stub horns. |
| **Shropshire** | England; in the central-western counties of Shropshire and Stafford. | Dark face, but a gray nose is not objectionable. | Both sexes are polled, although rams frequently have scurs. | Covering of dense wool well over the poll. | Such lack of breed type as to render the identity of the breed doubtful; horns; overshot or under-shot jaw; severely splayed hoofs; very long, weak pasterns; severely post-legged or sickle-hocked; evidence of existing or surgically altered faults such as inverted eyelids, prolapse, hernia, or abnormal sexual organs. |

(Continued)

**TABLE 28-4** (Continued)

| Breed | Place of Origin | Color; Face, Ears, and Legs | Head Characteristics | Other Distinguishing Characteristics | Disqualifications; Comments |
|---|---|---|---|---|---|
| Southdown | England; in the south-eastern county of Sussex. | Light or mouse brown color preferred. | Both sexes are polled, although rams sometimes have scurs. | Superior conformation and quality of carcass. | Horns; speckled markings; black spots; color white or approaching black; one or both testicles not descended; incisor teeth not meeting dental pad. Ewes that need frequent tagging due to too much fat or wool on the rear should be discriminated against by breeders. |
| Suffolk | England; in the south-eastern counties of Suffolk, Essex, and Norfolk. | Very black head, ears, and legs. | Both sexes are polled, although rams frequently have scurs. | Head and ears are entirely free from wool. | |
| Tunis (or American Tunis) | Northern Africa; from the province of Tunis. First imported into U.S. in 1799. | Reddish brown to bright tan. | Both sexes are polled; long drooping ears; head free from wool. | Originally, it was fat-tailed sheep, which means that the tail was distinctly broad and fat. However, breeders have selected away from this trait. Pendulous ears. Will mate almost any season of the year; and the ewes milk well. | Horns; red or black wool; one testicle; undershot or overshot jaw. Newborn lambs are red or tan, but they gradually turn white. |
| **Long-Wool Breeds** | | | | | |
| Cotswold | England; in the Cotswold hills of Gloucestershire. | White, although grayish specks and bluish tinge are common. | Both sexes are polled, although rams frequently have scurs. | Natural wavy ringlets or curls in which the fleece hangs all over the body. Tuft of wool on the forehead. Second only to the Lincoln in size. | Unsound animals. |
| Leicester | England; in the central county of Leicester. | White, but may have bluish tinge or black spots. | Both sexes are polled. | The Border Leicester is a strain of Leicester sheep. It is distinguished from the English Leicester by being open faced and bare legged and having a shorter fleece. | |
| Lincoln | England; along the eastern coast of England and bordering the North Sea, in Lincolnshire. | White. Black spots may be present but are discriminated against. | Both sexes are polled. | Largest of all breeds of sheep. Rams weigh 250 to 375 lb; ewes 200 to 275 lb. Produces the heaviest fleece of any mutton breed. | |
| Romney | England; in the Romney Marsh region of the County of Kent. | White. | Both sexes are polled. Open face. | In comparison with other long-wool breeds; the Romney is shorter legged, more rugged, and its fleece is shorter, finer, and less open. | Black fleece, or black spots. |
| **Crossbred Wool Breeds**[1] | | | | | |
| Columbia (Lincoln rams, Rambouillet ewes) | U.S.A.; in Wyoming and Idaho. | White. | Both sexes are polled. | Open faced, with no tendency to wool blindness. | Horns or scurs; wool blindness; uneven or light fleece; overshot or undershot jaw; colored wool; excessive folds. |
| Cormo (Corriedale × Merino) | Australia; in Tasmania. | White. | Polled, open faced, with silky, translucent hair on their faces. | Cormo ewes are noted for fertility, twinning, good mothering, and strong herding instincts. | Yellow grease wool, and short staple wool. |
| Corriedale (Lincoln and Leicester rams, Merino ewes) | New Zealand. | White, with dark points. Black spots are sometimes present. | Both sexes are polled. | | Black or brown spots. Wool blindness. Malformed mouth. Horns or scurs. |
| Panama (Rambouillet rams, Lincoln ewes) | U.S.A.; by Laidlaw and Brockie of Muldoon, Idaho. | White. | Both sexes are polled. | | Horns, scurs, or knobs; overshot or undershot mouth; excessive folds or wrinkles; colored wool; colored spots larger than ¾ in. in diameter on clear areas; any unsound hereditary factor. |

(Continued)

TABLE 28–4 (Continued)

| Breed | Place of Origin | Color; Face, Ears, and Legs | Head Characteristics | Other Distinguishing Characteristics | Disqualifications; Comments |
|---|---|---|---|---|---|
| **Polypay** (Dorset × Targhee) × (Finnsheep × Rambouillet) | U.S. Sheep Expt. Station, Dubois, Idaho. | White. | Open faced. | Ewes mature early, lamb with ease, and are prolific. Mature ewes weigh 135 to 150 lb. | The Polypay breed is well adapted to twice-a-year lambing. |
| **Targhee** (Rambouillet rams, Lincoln-Rambouillet-Corriedale ewes) | U.S.A.; by the USDA at Dubois, Idaho. | White. | Both sexes are polled. Open faced. | Moderately low set. Long productive life. | Black or brown color in the fleece. Horns or scurs. |
| ***Carpet-Wool Breeds*** | | | | | |
| **Navajo** | In the U.S., from the Churro—a Spanish breed. | White or colored, generally a single color. | Navajo rams are horned (some are multi-horned), but ewes are usually polled. | Ewes frequently breed out of season. Navajo sheep produce a very coarse wool which is easy to spin into blankets and rugs. | They are of all colors, with many showing spots and other markings on the face and legs. From 1935 to 1966, the U.S. Department of Interior, at the South- |
| | | | western Range and Sheep Laboratory, Fort Wingate, New Mexico, attempted to improve the Navajo sheep by infusing improved breeding. But this may have been a disfavor to the Navajo Indians, because the crossbreds lost much of the hardiness, flocking instinct, and spinning and weaving qualities required for Navajo blankets and rugs. | | | |
| **Scotch Highland** (Black-faced Highland) | Scotland; in the highland country. | Black or mottled. | Both sexes have horns. | Striking stylish appearance. Fleece consists of long coarse outer-coat and a finer inner-coat. | |
| ***Fur Sheep Breeds*** | | | | | |
| **Karakul** | Asia; in the region of Bokhara (U.S.S.R.) | Black or brown. | Rams have horns, but ewes are hornless. | Drooping ears. Fat-tailed. Lamb pelts suitable for fur production. | |
| **Romanov** | In the Soviet Union, in the Volga Valley. | | High lambing rate, ranging from 184 to 320% in the U.S.S.R., early maturity, and the ability to breed out of season. | In the U.S.S.R., the Romanov is raised for its fur or pelt. There are more than a half million Romanov sheep in the U.S.S.R. But, in 1987, there were only 400 animals of this breed in Canada and 20 in the U.S., where the breed is being evaluated. | |
| ***Hair Sheep Breed*** | | | | | |
| **Katahdin** | Piel Farm, Abbot, Maine; from Virgin Island sheep, with infusion of some British breeds. | Most are white faced. But can be any color or pattern. | Polled animals preferred. | A woolless meat-type sheep. Does not require shearing. | Horned and scoured individuals are so designated, but not disqualified. |

[1]The listing of the crosses which produced each of the Crossbred Wool Breeds is given for breed history purposes only, and does not imply any lack of purity of the respective breeds. Nor does it indicate that all of them are new breeds; for example, the Corriedale, which is an old breed, was originated in New Zealand about 1880.

# FINE-WOOL TYPE

Fig. 28–2. American Merino, B-type. (Courtesy, USDA)

Fig. 28–3. Booroola Merino ewe and her triplet lambs. (Courtesy, USDA).

Fig. 28–4. Debouillet ram. (Courtesy, Amos Dee Jones Ranches, Tatum, NM)

Fig. 28–5. Delaine Merino ewe. (Courtesy, Texas Delaine Sheep Assn., Burnet, TX)

Fig. 28–6 Rambouillet ewe. (Courtesy, The American Rambouillet Sheep Breeders Assn., San Angelo, TX)

# MEDIUM-WOOL TYPE

Fig. 28–7. Cheviot ram. (Courtesy, *The Sheep Breeder and Sheepman,* Columbia, MO)

Fig. 28–8.   Clun Forest ram. (Courtesy, North American Clun Forest Assn., Holmen, WI)

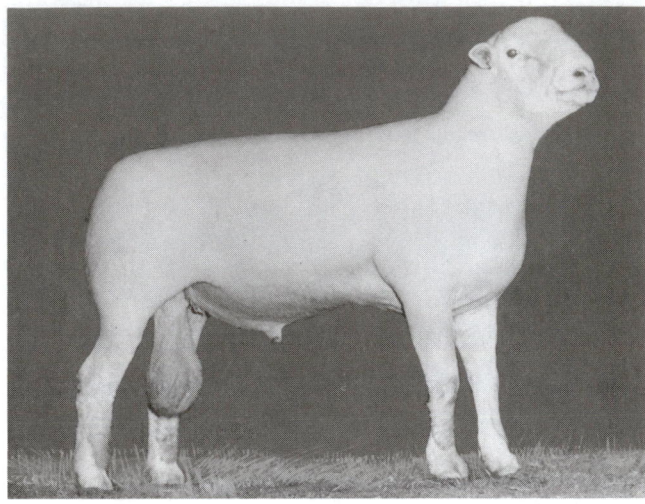

Fig. 28–9.   Dorset (Polled) ram. (Courtesy, *The Sheep Breeder and Sheepman*, Columbia, MO)

Fig. 28–10.   Finnsheep ram lamb. (Courtesy, Finnsheep Breeders Assn., Inc., Zionsville, IN)

Fig. 28–11.   Hampshire ram. (Courtesy, American Hampshire Sheep Assn., Ashland, MO)

Fig. 28–12.   Montadale yearling ewe. (Courtesy, Montadale Sheep Breeders Assn., Inc., Indianapolis, IN)

Fig. 28–13.   North Country Cheviot ewe. (Courtesy, American North Country Cheviot Sheep Assn., Longview, WA)

Fig. 28-14. Oxford ewe lamb. (Courtesy, *The Sheep Breeder and Sheepman*, Columbia, MO)

Fig. 28-15. Shropshire ram. (Courtesy, The American Shropshire Registry Assn., Inc., Schoolcraft, MI)

Fig. 28-16. Southdown yearling ram. (Courtesy, American Southdown Breeders' Assn., Fredonia, TX)

Fig. 28-17. Suffolk ewe. (Courtesy, National Suffolk Sheep Assn., Columbia, MO)

## LONG-WOOL TYPE

Fig. 28-18. Tunis yearling ram. (Courtesy, James McGuire, Oakfield, NJ)

Fig. 28-19. Cotswold ram. (Courtesy, American Cotswold Record Assn., Rochester, NH)

Fig. 28–20. Leicester (Border Leicester) yearling ram, bred in Scotland. (Courtesy, Canadian Sheep Breeders' Assn.)

Fig. 28–21. Lincoln ram. (Courtesy, National Lincoln Sheep Breeders' Assn., Decatur, IL)

Fig. 28–22. Romney yearling ram. (Courtesy, American Romney Breeders Assn., Corvallis, OR)

## CROSSBRED-WOOL TYPE

Fig. 28–23. Columbia ram. (Courtesy, *The Sheep Breeder and Sheepman*, Columbia, MO)

Fig. 28–24. Cormo ram. (Courtesy, American Cormo Sheep Assn., Montrose, CO)

Fig. 28–25. Corriedale ewe. (Courtesy, American Corriedale Assn., Inc., Seneca, IL)

Fig. 28-26.  Panama ram lamb. (Courtesy, University of Idaho)

Fig. 28-27.  Polypay ewe and twin lambs. (Courtesy, Russ Beattle, President, American Polypay Sheep Assn., Rexburg, ID)

## CARPET-WOOL TYPE

Fig. 28-28.  Targhee ram. (Courtesy, U.S. Targhee Sheep Assn., Jordan, MT)

Fig. 28-29.  Navajo ram. (Courtesy, USDA)

## FUR TYPE

Fig. 28-30.  Scotch Highland ram, bred by Mr. David Provan, Gateside, Douglas, Lanark, Scotland; and sold by auction in the Lanark Ram sale to Messrs. Ben Wilson, John Lammie, and James Anderson for $9,000. (Courtesy, David Provan of Scotland)

Fig. 28-31.  Karakul ewe lamb at one day of age, showing small to medium Persian lamb curl. (Courtesy, Lowry Hagerman, Santa Fe, NM)

Fig. 28-32. Romanov ewe and her triplet lambs. (Courtesy, USDA)

## HAIR TYPE

Fig. 28-33. Katahdin hair rams. A woolless meat-type sheep. (Courtesy, Laura Callan, Perryville, AR)

## QUESTIONS FOR STUDY AND DISCUSSION

1. Why have more types and breeds of sheep evolved than of other species of farm animals?

2. List the six different bases on which breeds of sheep may be classified.

3. What characteristics in common do the fine-wool breeds of sheep possess?

4. What breeds of sheep are known as the *down breeds*? What is the significance of the term *down breeds*?

5. With the current emphasis on size and tallness, will the long-wool breeds increase? Justify your opinion.

6. What need(s) prompted the development of the Columbia, Corriedale, Panama, Polypay, and Targhee breeds of sheep?

7. Why do we not use wool from the fine-wool breeds of sheep for making carpets? What kind of wool is best for making carpets?

8. List in order of value and describe the three classes of Karakul pelts.

9. Give the four classes of sheep by breeding use, and describe each class.

10. Describe and tell what forces prompted interest in each of the following exotic classes of sheep: (a) colored sheep, (b) natural colored sheep (black sheep), (c) Jacob sheep, (d) hair sheep, and (e) milking sheep.

11. What transitions within people were back of the increased interest in (a) colored sheep, and (b) milking sheep in the late 1900s?

12. Why are hair sheep so popular in Africa? Why are milking sheep so popular in the province of Roquefort France?

13. List by rank the five most populous breeds of sheep at the present time based on annual registrations. Why has each of these leading breeds achieved this status?

14. Give the name, place of origin, and distinguishing characteristics of three fine-wool breeds of sheep.

15. Give the name, place of origin, and distinguishing characteristics of six medium-wool breeds of sheep.

16. Give the name, place of origin, and distinguishing characteristics of two long-wool breeds of sheep.

17. Give the name, place of origin, and distinguishing characteristics of four crossbred breeds of sheep.

18. Give the name, place of origin, and distinguishing characteristics of two carpet-wool breeds of sheep.

19. What traits do the Booroola Merino, Romanov, and Cormo breeds of sheep possess that caused them to be imported to the United States in the late 1900s?

20. Was the infusion of improved breeding in the Navajo breed of sheep a disfavor to the Navajo Indians? If so, why?

21. Justify any preference that you may have for one particular breed of sheep.

22. Obtain breed registry association literature and a sample copy of a magazine of your favorite breed of sheep. (See Appendix Tables A–8 and A–9 for addresses.) Evaluate the soundness and value of the material that you receive.

## SELECTED REFERENCES

| Title of Publication | Author(s) | Publisher |
|---|---|---|
| *Breeds of Livestock, The* | C. W. Gay | The Macmillan Company, New York, NY, 1916 |
| *Breeds of Livestock in America* | H. W. Vaughan | R. G. Adams and Company, Columbus, OH, 1937 |
| *Karakul Sheep, The* | E. B. Bertone *et al.* | Food and Agriculture Organization of the United Nations, Rome, Italy, 1968 |
| *Modern Breeds of Livestock*, Fourth Edition | H. M. Briggs<br>D. M. Briggs | The Macmillan Company, New York, NY, 1980 |
| *Productive Sheep Husbandry* | W. C. Coffey | J. B. Lippincott Co., New York, NY, 1937 |
| *Profitable Sheep* | S. B. Collins<br>R. F. Johnson | The Macmillan Company, New York, NY, 1956 |
| *Profitable Sheep Farming* | M. M. Cooper<br>R. J. Thomas | Farming Press Ltd., Ipswich, England, 1971 |
| *Sheep & Goat Science*, Fourth Edition | M. E. Ensminger<br>R. O. Parker | The Interstate Publishers, Inc., Danville, IL, 1986 |
| *Sheep and Wool* | M. P. Botkin<br>R. A. Field<br>C. L. Johnson | Prentice Hall, Englewood Cliffs, NJ, 1988 |
| *Sheep Book, The* | J. McKinney | John Wiley & Sons, Inc., New York, NY, 1959 |
| *Sheep Book, The* | R. Parker | Charles Scribner's Sons, New York, NY, 1983 |
| *Sheep Breeds of the Mediterranean, The* | I. L. Mason | Commonwealth Agricultural Bureaux, Farnham Royal, Bucks, England, 1967 |
| *Sheepman's Production Handbook, The* | Edited by<br>G. E. Scott | Sheep Industry Development Program, Inc., Denver, CO, 1986 and updated at intervals |
| *Sheep of the World in Color* | K. Ponting | Bradford Press, Ltd., Poole, Dorset, England, 1980 |
| *Sheep Production* | L. J. Horlacher | McGraw-Hill Book Company, New York, NY, 1927 |
| *Sheep Production and Management* | C. V. Ross | Prentice Hall, Englewood Cliffs, NJ, 1989 |
| *Sheep Science*, Revised Edition | W. G. Kammlade, Sr.<br>W. G. Kammlade, Jr. | J. B. Lippincott Co., New York, NY, 1955 |
| *Stockman's Handbook, The*, Sixth Edition | M. E. Ensminger | The Interstate Printers & Publishers, Inc., Danville, IL, 1983 |
| *Study of Breeds in America, The* | T. Shaw | Orange Judd Co., New York, NY, 1900 |
| *Types and Breeds of Farm Animals* | C. S. Plumb | Ginn and Co., Boston, MA, 1920 |
| *Wool Handbook* | W. Von Bergen | Interscience Publishers (John Wiley & Sons, Inc.), New York, NY, 1963 |
| *World Dictionary of Breeds, Types, and Varieties of Livestock, A* | I. L. Mason | Commonwealth Agricultural Bureaux, Slough, Bucks, England, 1951 |

Also, breed literature pertaining to each breed may be secured by writing to the respective breed registry associations (see Table A–8, Appendix, for the name and address of each association).

New Zealand Merino sheep, with majestic Mt. Dark in the background. This shows a New Zealand musterer (counterpart of the U.S. western sheep herder), along with his horse and dogs, gathering up sheep from the windswept ridges. (Courtesy, National Publicity Studios, Wellington, New Zealand)

There is no greater thrill than selecting, fitting, and showing a champion, an animal that has been produced through intelligent breeding, then fitted to the height of perfection. This shows the champion Southdown wether lamb at the Junior Livestock Show, Oklahoma City, shown by Darrell Howard, Mulhall, OK. (Courtesy, Howard's Southdowns, Mulhall, OK)

# ESTABLISHING THE FLOCK; SELECTING AND JUDGING SHEEP

# Chapter 29

Whether a large range operation or a small farm flock is being established or maintained, consideration must be given to certain factors if the venture is to be successful. In the final analysis, sheep are maintained for the production of market lambs and wool. This means that each individual in the flock should possess those characteristics making for maximum and efficient production of these two products. Furthermore, if progress is to be made in the breeding program, each succeeding generation must represent an improvement over the parent stock.

## FACTORS TO CONSIDER IN ESTABLISHING THE FLOCK

The factors to consider in establishing a flock of sheep are not unlike those that the farmer must face in establishing a herd of cattle, hogs, or horses. They are, however, somewhat more confused because (1) two major products are involved—lamb and wool—instead of one, and (2) there are more breeds from which to select, and the practice of crossbreeding is more prevalent.

Currently, lamb represents about 80% of the gross income from sheep, while wool represents only about 20%. It follows that, when establishing a flock, the greater emphasis should be on lamb production.

## PUREBREDS, CROSSBREDS, OR GRADES

Generally speaking, only the experienced breeder should undertake the production of purebreds with the intent eventually of furnishing foundation or replacement stock to other purebred breeders or purebred rams to the commercial producer. Unless prices are unusually favorable, the beginner should start with crossbred or grade ewes and a purebred ram. The vast majority of range operators, most of whom are capable sheep specialists, elect to keep bands of high-grade ewes that are mated to purebred rams.

## NATIVE OR WESTERN EWES

Native ewes are those that are produced outside of the western range area, and that show a predominance of mutton-type breeding. Western ewes are those that are produced in the range area, and that either show a predominance of fine-wool breeding or represent fine-wool × long-wool crossbred types.

In starting a farm flock, the question often arises as to whether native or western ewes should be purchased. Both types of ewes are found in flocks throughout the country and may well be considered by the beginner. In general, western ewes are more uniform, smaller in size, less costly, and less

likely to be infested with parasites. On the other hand, native ewes are larger, and produce a larger lamb. If they are bought locally, a saving in price may be effected. In purchasing native ewes, however, buyers should make very certain that they are not obtaining the cull ewes (shy breeders and ewes with unsound udders or other defects) from one or several flocks. From this standpoint, the purchase of yearlings affords the best protection.

For range operations, where a large number of animals are to be handled in one or more bands, grade ewes of fine-wool breeding or ewes of the fine-wool × long-wool crossbred type are essential.

## SELECTION OF THE BREED

Within certain limitations, preference is usually the deciding factor in the selection of a breed. As already noted, however, there is good reason that most commercial bands of the West should carry considerable fine-wool breeding (chiefly because of the herding instinct in the animals). Except for this situation, sheep producers of this country have given little consideration to possible special-area adaptations of the different breeds. English sheep breeders, on the other hand, have long contended that some breeds are peculiarly adapted to certain conditions. This conviction springs from the fact that the several English breeds were developed in different geographical areas, each of which had different climatic, soil, and crop conditions.

According to English breeders, the small and more active breeds of sheep that were developed in the hill or down country of England are not adapted to the lowland region, whereas the large and more sluggish breeds developed in the fertile lowland areas will not thrive on the uplands. Moreover, the Romney breed is said to thrive better than any other breed in marsh areas similar to those of its native home in southern England.

Unfortunately, there is little experimental work either to substantiate or refute these contentions relative to environmental adaptations of breeds. It is noteworthy, however, that on the USDA Experiment Station at Dubois, Idaho, three crossbreds of sheep have been developed—the larger Columbias for the lush ranges of the West and the smaller Polypays and Targhees for the average ranges. It is also a well-known fact that for the production of hothouse lambs, the ewes must breed out of season and milk well. This limits the production of hothouse lambs to ewes of Dorset, Tunis, Rambouillet, or Merino breeding.

In so large and variable a country as the United States, it would appear quite probable that some breeds would be better adapted to certain conditions than others. It is also quite likely that rather wide differences in environmental adaptations may exist within breeds. Perhaps there is need for experimental work to determine these differences, if any, but it must be recognized that (1) breed preference is, and should remain, a powerful factor in the selection of a breed, and (2) it is difficult, if not impossible, to obtain a representative cross section of a widely disseminated breed of sheep for experimental study.

# THE *DOWN* BREEDS OF SHEEP IN THE LAND OF THEIR ORIGIN

The Hampshire, Oxford, Shropshire, Southdown, and Suffolk breeds of sheep are collectively referred to as the *down* breeds, because the area in southern England where all of them originated is a country of hills and downs. English breeders contend that these five breeds were bred and selected for their peculiar adaptation to a hilly country.

Fig. 29–1. *Above:* Hampshire sheep on the *downs.* They originated in the county of Hampshire. (Courtesy, *The Field,* London, England)

Fig. 29–2. *Right inset:* Oxford sheep on the *downs.* They originated in the county of Oxford. (Courtesy, *Farmer & Stock-Breeder,* London, England)

Fig. 29–3. *Right:* Shropshire sheep on the *downs.* They originated in the counties of Shropshire and Stafford. (Photo by Otto Karminski, in Shropshire, England)

Fig. 29–4. *Left:* Southdown sheep on the *downs.* They originated in the southeastern county of Sussex. (Courtesy, Arnold Nicholson, Philadelphia, PA)

Fig. 29–5. *Right:* Suffolk sheep on the *downs.* They originated in the southeastern counties of Suffolk, Essex, and Norfolk. (Courtesy, *Farmer & Stock-Breeder,* London, England)

# OTHER HISTORIC BREEDS OF SHEEP

Fig. 29-6. *Left:* Lincoln sheep, the largest of all breeds of sheep, originated in Lincolnshire, England, a low-lying area characterized by abundant feed and necessitating little travel; hence, they are well adapted to such areas. (Courtesy, *Farmer & Stock-Breeder,* London, England)

Fig. 29-7. *Right:* Romney sheep. They originated in the county of Kent, in England, a high rainfall, marshy area; hence, it is claimed that they are well adapted to wet areas. (Courtesy, Department of Scientific and Industrial Research, Wellington, New Zealand)

Fig. 29-8. *Left:* Scotch Highland sheep. They originated in the highland country of Scotland, a rugged, hilly area; hence, they are well adapted to mountainous areas. (Courtesy, Canadian Department of Agriculture, Ottawa, Canada)

Fig. 29-9. *Right:* Columbia ewes, originated in the United States, from a Lincoln × Rambouillet cross, a large breed adapted to the more lush ranges of the West. (Courtesy, Washington State University)

## SIZE OF THE FLOCK OR BAND

The unit of sheep management varies all the way from a small farm flock consisting of a few head to a large range band with as many as 3,500 head. Usually, the farm flock is merely a part of a diversified type of farm enterprise, whereas several range bands may constitute a highly specialized type of operation with little or no diversification.

Small farm flocks are variable in size, ranging from a few head to as many as 300 to 500 or more. In determining the size of the farm flock, it is to be noted that the labor and equipment cost, except at lambing time, differs very little whether the flock numbers 12 or 50. The smaller flock will require the services of one ram and practically the same amount of fencing to provide rotation of pastures or suitable corrals away from dogs. Furthermore, fewer sheep are much more likely to receive the neglect often accorded a minor enterprise.

On the other hand, the beginner can acquire valuable, practical experience with a very small flock without subjecting a large flock to the possible hazards that frequently accompany inexperience. In range operations, the size of the band varies somewhat according to the method of management, the general character of the country, and the season of the year. The number of bands run by a given operator is usually determined by the amount of range and capital available.

## TIME TO START

Late summer, after the lambs have been weaned and before the ewes are bred, is usually the best time in which to start in the sheep business. At this season of the year, it is generally possible to buy some of the surplus ewes from neighboring farmers or ranchers or to obtain a wide selection of ewes from the terminal markets. Ewes can usually be purchased at reasonable prices at this period, and for the beginner there is a period of valuable training prior to lambing time. Also, there is usually an abundance of meadows, grain stubble land, and other forage available on the farm or ranch that will make it possible to get the ewes in good thrifty condition prior to breeding time. Then, too, the purchase of a uniform flock of ewes makes possible the intelligent selection of the ram and the production of lambs with uniformity of type and quality.

## UNIFORMITY

In order to produce uniform market lambs and wool, it is first desirable that the ewe flock selected be uniform—preferably from the standpoint of breeding, size, body conformation, and fleece grade and quality. Such uniformity is of decided advantage in the marketing of products at premium prices. Also, the constructive breeder is able to select more wisely a ram or rams for most successful mating with a uniform flock or band of ewes.

## HEALTH

All ewes selected should be in a thrifty, vigorous condition. They should have every appearance of a life of usefulness ahead of them and give every evidence of being capable of producing a good fleece and raising strong, healthy lambs.

Animals showing dark blue skins, paleness or lack of coloring in the lining of the nose and eyelids, listlessness or a lack of vigor, and a general run-down condition should be regarded with suspicion.

## AGE

The vast majority of sheep producers prefer to establish a flock by acquiring yearling ewes that are bred to lamb when approximately 24 months of age. When it can be made certain that culls are not being secured, however, older ewes are frequently the best buy. Sometimes it is to the advantage of a farmer to secure range ewes that have one to two years of usefulness left, provided that they are placed under conditions where (1) less traveling is required, (2) the pastures are more abundant, and (3) more attention is given to winter feed.

## SOUNDNESS OF UDDER

In selecting ewes, particular attention should be given to the udder and teats. The udder should be soft and pliable and the teats normal. Ewes having hard or pendulous udders, teats that have been removed through careless shearing, or meaty or abnormally large teats should be rejected.

## SIZE

There is great variation in the size of the different breeds of sheep, extending all the way from the small, refined Southdowns to the large, ponderous Lincolns. The rank and file of commercial sheep producers in this country prefer an animal with considerable size, because, based on experience, it has been found that such animals are more profitable from the standpoint of the net receipts derived from the usual combination of lamb and wool production. This is due to the fact that, in general, the cost of handling sheep is on a per-head rather than on a per-pound basis, and our markets are not sufficiently discriminating in the purchase of lamb and wool to warrant any great sacrifice in quantity in favor of quality production. On the other hand, an exception exists when a premium is paid for early-maturing, high-quality lambs, such as in hothouse lamb production. Under the latter conditions, it is preferable to use early-maturing, mutton-type rams on medium-sized ewes. Also, where feed is sparse and sheep are run with a minimum of care, smaller sheep may be preferable, as they often are more efficient than larger animals.

In an effort to secure more size on the lush ranges of the northwestern United States, many commercial operators are keeping crossbred type ewes, based on long-wool × fine-wool crosses. Moreover, larger grade ewes of Rambouillet extraction have always been more popular on the western ranges than the smaller ewes of Merino breeding.

## PRICE

The price of breeding sheep must be based upon the projected price of the products of production—lamb and wool. Although it is usually sound business to pay a premium for quality foundation stock, it must be remembered that the ultimate objective of all sheep production is profitable market lamb and wool production; thus, the price should be governed accordingly.

## SELECTION BASES

The criteria used in selecting an individual or group of sheep or lambs will vary according to the use for which the animal or animals are intended. In general, commercial farmers or ranchers select sheep or lambs for the following purposes: (1) breeding sheep, with market lambs as the primary objective; (2) breeding sheep, with wool production as the chief objective; (3) feeder lambs for further finishing; or (4) finishing lambs to send to the market. Thus, where market lamb production is the main consideration, primary attention is given to those factors that will result in the production of a heavy, finished lamb of acceptable type and grade. When sheep are maintained chiefly for wool production, the clean weight and quality of fleece are of paramount importance. Feeder lambs are selected on the basis of desirable type and quality, health and thrift, and projected gains and finish with feeding. Slaughter lambs are selected out for the market, using finish, muscling, and quality as the criteria in determining value. The experienced operator can determine the degree of finish of sheep by touch, by rapidly applying the palm of the hand to the animals being examined as they file through a cutting chute.

Purebred breeders may add certain breed fancy points and highly prized pedigrees to the above considerations, and market specialists find it necessary to add other considerations in arriving at the rather imposing list of market classes and grades of sheep. It is not intended, therefore, that the amateur may become proficient as a sheep judge through merely reading about the subject. There is no shortcut or substitute for long years of patient practice.

In establishing a new flock or improving an old one, however, there are four bases of selection: (1) selection based on type or individuality, (2) selection based on pedigree, (3) selection based on show-ring winnings, and (4) selection based on production testing.

## SELECTION BASED ON TYPE OR INDIVIDUALITY

A vast majority of sheep, both purebred and commercial animals, are selected on the basis of type or individuality. In limited instances, three additional criteria may be available and invoked effectively; namely, pedigree, show-ring, and production selection. Although production selection offers a new and modern approach to animal improvement and pedigree selection may be a valuable guide in certain instances, perhaps for many years to come selection will continue to be based largely on individuality. Without doubt, however, the progressive sheep producer of the future will make increased use of production tested rams. Despite the greater certainty and more rapid progress afforded through selection based on production, it is well to note in retrospect that we have travelled a long way in sheep improvement—from the rather poor mutton type and average 2-lb clips of the sheep characteristic of General Washington's time to the modern sheep of today.

The judging of sheep—their selection based on individuality—differs materially from the judging of cattle, hogs, or horses because of the presence of the fleece. Not only must an additional characteristic of economic importance be considered, but the presence of the fleece makes it more difficult to determine the body type. This is especially true when the animal has been subjected to the blocking art of a clever shepherd, but the problem exists whenever there is any ap-

preciable wool growth. Even though complete production records are not available, progressive producers can arrive at sufficient evaluation to enable them to (1) cull the dry ewes, (2) remove wool-blind ewes and rams, and (3) sort out the light-fleeced animals by the *touch method* or according to actual fleece weights. This type of selection directly affects the income, even though it may not be very important genetically.

## SELECTION BASED ON PEDIGREE

Without doubt, less attention is paid to pedigree selection and family names in sheep than in any other class of farm animals. It is a rare occurrence, indeed, to find commercial producers who have any concern about the ancestry of the rams that they are contemplating purchasing. Although most purebred breeders are interested in the pedigree of stud rams, there is comparatively little interest in the ancestry of the flock ewes. In some respects, this relative lack of interest in pedigrees on the part of sheep breeders is fortunate, for breeding programs have not been subjected to the hazards of worshiping family names or of selecting a breeding animal chiefly because its ancestry traces to some noted ancestor many generations removed. On the other hand, it must be recognized that fancy pedigrees may have sales value from the standpoint of the purebred breeder.

## SELECTION BASED ON SHOW-RING WINNINGS

There can be no question that the great livestock shows of the land have been a profound influence in establishing type in the different breeds of sheep. It might well be added, however, that when utilitarian considerations have been ignored, their influence has not always been for the good. A prime example of the latter point was the face covering that was once bred into certain of the mutton breeds of sheep. This caused the affected breeds to diminish in importance and numbers, because, through sad experience, the practical operator had found that ewes subject to wool blindness produced less market lamb and wool. Utility value, therefore, should always come first and breed fancy points, second.

Fig. 29–10. A wool-blind sheep, an undesirable trait, once characteristic of show champions of certain breeds of sheep.

Winning rams and ewes and their progeny are usually in great demand, with the result that they generally bring premium prices. Provided that the type has been established wisely and is based on utilitarian considerations, the purebred breeder may find it desirable and profitable to select some animals on the basis of show-ring winnings.

## SELECTION BASED ON PRODUCTION TESTING

The relative merits of (1) performance testing, and (2) progeny testing have been presented in Chapter 3; and traits of importance in testing sheep and suggested records are given in Chapter 31. At this point, it may be well to emphasize that the importance attached to each trait will vary in different areas. For example, in the southwestern United States, where wool production is more important than elsewhere, greater stress should be placed upon selecting animals whose progeny possess the maximum in fleece weight and quality. On the other hand, in those areas where feed is more abundant and two-thirds, or more, of the annual income is derived from the sale of lamb, evaluations should be based largely on the market weight, type, and finish of the offspring.

Production selection should be given greater emphasis in animal improvement, for, in most instances, it is production—of which individuality or type is merely a part—that produces the income. Also, again and again, it should be emphasized that selection based on either performance testing or progeny testing is far more accurate than any other method of selection.

## FLOCK IMPROVEMENT THROUGH SELECTION

Once the flock or band has been established, improvement can be obtained only through constant, rigid culling and careful selection of replacements. Such procedure makes the flock more profitable from the standpoint of quantity and quality lamb and wool production and affords a means of accomplishing genetic gain in the next generation.

## JUDGING SHEEP

The judging of sheep differs from the judging of cattle or swine in that two products of economic importance are involved instead of one. That is, in addition to meat production, wool is a valuable product. The situation is made more difficult because body conformation is often difficult to determine because of the wool under which it is hidden. The latter situation is accentuated when sheep are subjected to the art of blocking by a clever shepherd.

## PARTS OF A SHEEP

The first step in preparation for judging sheep consists of mastering the language that describes and locates the different parts of the animal. Fig. 29–11 sets forth this information.

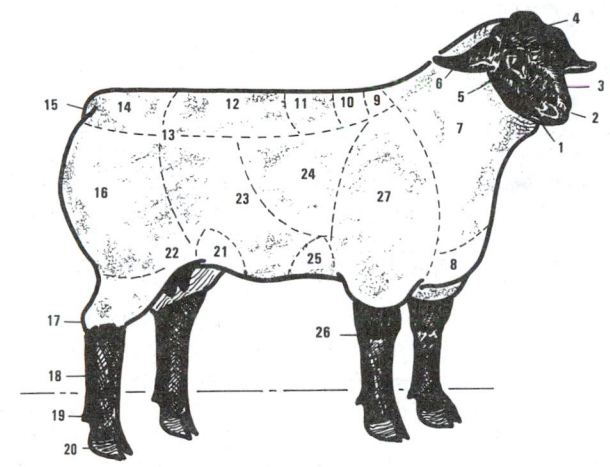

Fig. 29–11. Parts of a sheep. The first step in preparation for judging sheep consists in mastering the language that describes and locates the different parts of the animal. (Drawing by R. F. Johnson)

| | | |
|---|---|---|
| 1. Mouth | 10. Crops | 19. Dew claw |
| 2. Nostril | 11. Back | 20. Foot |
| 3. Face | 12. Loin | 21. Hind flank |
| 4. Forehead | 13. Hip | 22. Stifle |
| 5. Eye | 14. Rump | 23. Belly or paunch |
| 6. Ear | 15. Dock | 24. Ribs |
| 7. Neck | 16. Thigh or leg | 25. Fore flank |
| 8. Breast | 17. Hock | 26. Foreleg |
| 9. Top of shoulder | 18. Hind leg | 27. Shoulder |

## IDEAL MEAT TYPE AND CONFORMATION

The second requisite of successful judges of any class of livestock is that they have clearly in mind an ideal or standard of perfection. Thus, sheep producers, regardless of whether they are producing purebred or commercial animals or whether they are farm-flock or range-band operators, should have a type or ideal in mind and make their selections accordingly. For the mutton-type breeds, and when the production of market lambs is the primary objective, this ideal means plenty of size and growthiness; heavy muscling—especially in the leg and loin; a long body; trimness and freedom from waste; straight, widely set legs; a fleece of acceptable weight and quality; and a pink skin.

If purebred, the animals should show the characteristics of the breed represented. Rams should show boldness and masculinity, and ewes should be feminine.

Fig. 29–12. The successful sheep judge must have clearly in mind an ideal or standard of perfection. The ideal meat type (*left*) shows plenty of size, growthiness, muscling, trimness, and freedom from waste. The ideal fleece type (*right*) is more angular and produces a dense, clean, bright fleece.

## IDEAL FLEECE TYPE AND CONFORMATION

The greater the proportion of income derived from wool, the greater the emphasis on weight and quality of fleece. With wool-type sheep, the fibers should be long, fine, of good crimp; and the fleece should be dense, clean, and bright. Animals with fleeces that show black fibers or any tendency of the fleece to be hairy, loose, or open should be rejected. From the standpoint of body form, the typical wool-type sheep is quite different from those animals representing the ultimate in mutton type. The former are more angular, with considerably less muscling throughout. It is to be noted, however, that the present-day smooth-bodied Rambouillet sheep are of much more acceptable mutton type than the ideal extreme wool-type of former years.

## RECOGNIZING AND EVALUATING COMMON FAULTS

No animal is perfect. In judging, therefore, one must be able to recognize and appraise the common faults. Likewise, credit must be given for the good points. Finally, judges must be able to weigh and evaluate the relative importance of the characteristics that they have observed, and the degree to which they are good or bad. Skill and accuracy in this art can be achieved only through patient study and long experience. Fig. 29–13 shows the ideal meat type vs some common faults.

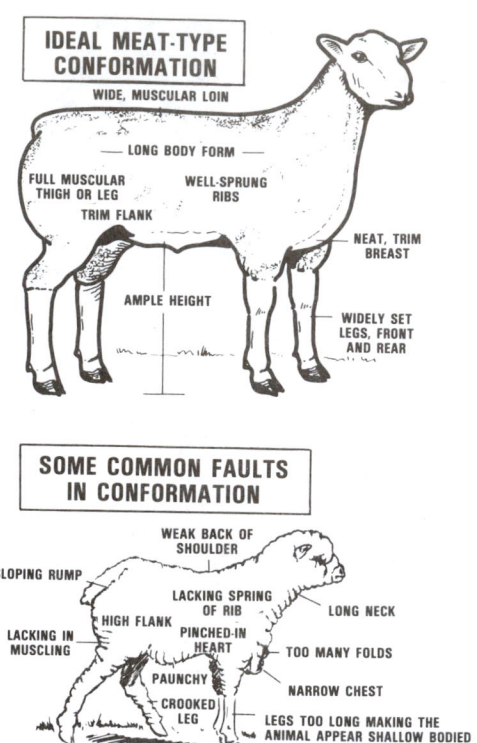

Fig. 29–13. Ideal meat type vs common faults. Successful sheep judges must know what they are looking for, and be able to recognize and appraise both the good points and the common faults.

## METHOD OF EXAMINING

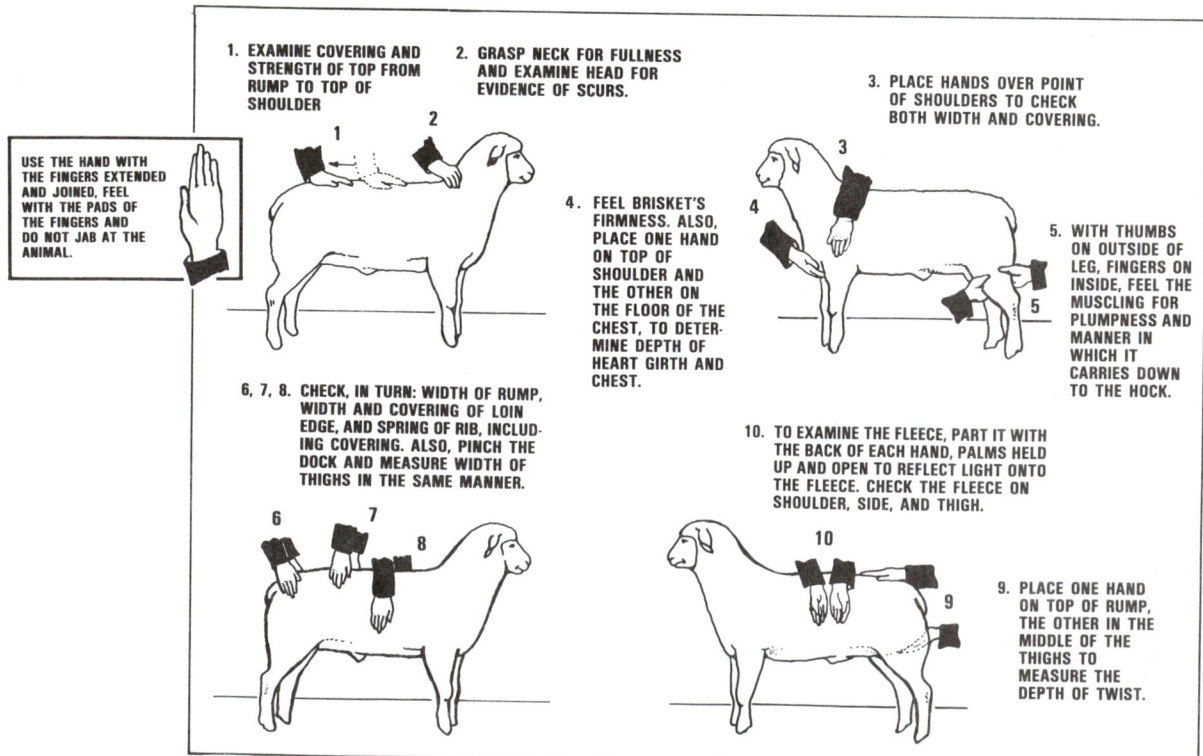

# A PROCEDURE FOR EXAMINING SHEEP

USE THE HAND WITH THE FINGERS EXTENDED AND JOINED, FEEL WITH THE PADS OF THE FINGERS AND DO NOT JAB AT THE ANIMAL.

1. EXAMINE COVERING AND STRENGTH OF TOP FROM RUMP TO TOP OF SHOULDER

2. GRASP NECK FOR FULLNESS AND EXAMINE HEAD FOR EVIDENCE OF SCURS.

3. PLACE HANDS OVER POINT OF SHOULDERS TO CHECK BOTH WIDTH AND COVERING.

4. FEEL BRISKET'S FIRMNESS. ALSO, PLACE ONE HAND ON TOP OF SHOULDER AND THE OTHER ON THE FLOOR OF THE CHEST, TO DETERMINE DEPTH OF HEART GIRTH AND CHEST.

5. WITH THUMBS ON OUTSIDE OF LEG, FINGERS ON INSIDE, FEEL THE MUSCLING FOR PLUMPNESS AND MANNER IN WHICH IT CARRIES DOWN TO THE HOCK.

6, 7, 8. CHECK, IN TURN: WIDTH OF RUMP, WIDTH AND COVERING OF LOIN EDGE, AND SPRING OF RIB, INCLUDING COVERING. ALSO, PINCH THE DOCK AND MEASURE WIDTH OF THIGHS IN THE SAME MANNER.

10. TO EXAMINE THE FLEECE, PART IT WITH THE BACK OF EACH HAND, PALMS HELD UP AND OPEN TO REFLECT LIGHT ONTO THE FLEECE. CHECK THE FLEECE ON SHOULDER, SIDE, AND THIGH.

9. PLACE ONE HAND ON TOP OF RUMP, THE OTHER IN THE MIDDLE OF THE THIGHS TO MEASURE THE DEPTH OF TWIST.

Fig. 29–14. A good procedure for examining sheep and some of the things for which to look.

The examination of sheep should be systematic and thorough, as it should be when examining any class of livestock. This is especially true in selecting breeding animals or in close competitive judging such as is encountered in the show-ring, but it is neither practical nor essential in handling large numbers of market sheep.

The sheep being examined should first be looked over from a distance, so that views from the front, side, and rear may be secured. It does not make any difference which view of the animal is noted first, but it is important that the same procedure be followed each time. This general inspection should furnish a good idea as to the size, balance, length of body, and muscling; the length of neck and the makeup of the head; the straightness and set of legs; and the breed type and character of the animal. Next, the impression gained through distant inspection should be verified by handling with the hands, and the fleece should be examined. Good judges differ as to whether they start handling the animal from the front or rear. Perhaps as good a method as any is illustrated in Fig. 29–14.

With market sheep, most of the examining that is necessary in arriving at the market classes and grades is done by observation. Even so, these specialists usually like to get their hands on slaughter lambs; and when part of a drove of lambs is fat enough to go the slaughter route—whereas others must go as feeders—they will make the cut after handling each of them as they file through a cutting chute.

## CATCHING AND HANDLING SHEEP

If sheep are to be caught and handled for any reason, they should first be confined to a small corral or shed. Sheep may best be caught around the neck, by the hind leg, or by the rear flank. Never should they be caught by the wool. Such rough treatment results in badly injured skin and tissue, which may require weeks or even months to heal. As a result of such mishandling, market sheep will exhibit a damaged carcass, and the fleece of breeding animals will lack uniformity because of the disturbance in the injured area.

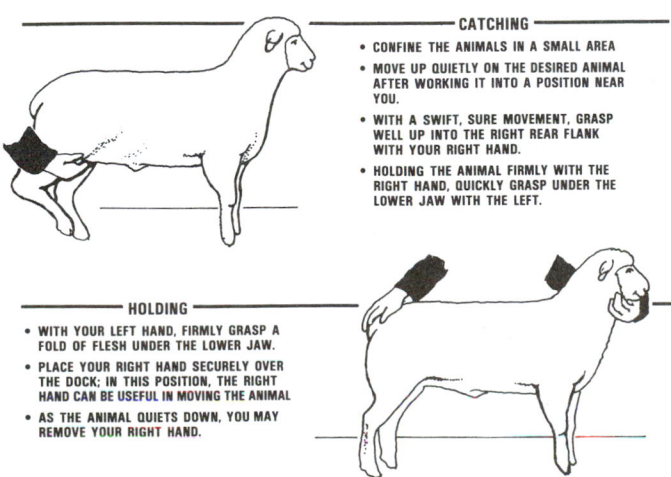

Fig. 29–15. Diagram showing the correct method of catching and holding a sheep.

In handling sheep, the fingers should be kept together. In this way, the correct touch is obtained in the palm of the hand, and the wool is not disarranged. In observing the fleece and skin, the wool should be parted well down on the shoulder, side, and leg. Opening the fleece on the back should be avoided, as it will allow water to run in.

## DETERMINING THE AGE OF SHEEP

Until sheep are 4 years of age, the front teeth of the lower jaw furnish a fairly reliable guide as to their development. The lamb has small, narrow teeth that are known as milk teeth. At 12 to 14 months of age, the 2 center incisors are replaced by 2 large, broad, permanent teeth. Each succeeding year, an additional pair of permanent teeth appears, one on either side of the first pair, until 4 years of age, when there is a full mouth. It must be remembered, however, that teeth represent the degree of development rather than the exact age according to birth, and, therefore, they are not an infallible indication of age.

Fig. 29–16. Diagram showing how to determine the age of sheep by the teeth.

After the sheep has a solid mouth (at 4 years), it is impossible to tell the exact age. With more advanced age, the teeth merely wear down and spread apart, and the degree of wearing or spreading is an indication of age. The normal number of teeth may be retained until 8 or 9 years, but often some are lost after about the fifth or sixth year, resulting in a "broken mouth." When most of the teeth have disappeared, animals are known as *gummers*.

Ewes of mutton breeding usually start on the decline at about five years of age, whereas ewes of the fine-wool breeds do not begin to decline until about a year later. Even so, outstanding producers should not be disposed of just because they have reached this age, especially in a purebred flock. Also, animals that have reached the age where it may no longer be wise to expose them to the rigors of range handling may often be retained satisfactorily and profitably in a farm flock for 1 to 2 years longer.

With market sheep, the break-joint or lamb-joint is also an important criterion of age. For further information on this point, the reader is referred to Chapter 33.

## SCORECARD JUDGING OF SHEEP

A scorecard is a listing of the different parts of an animal, with a numerical value assigned to each part according to its relative importance. It is a standard of excellence. The use of the scorecard involves studying each part, then assigning a score to each.

Different methods of scoring individual animals have evolved. All of them are based on visual appraisal. This point bears emphasis because producers and students often get the erroneous impression that just because some visual scoring system (scoring system based on visual appearance, in contrast to actual weights, measurements, etc.) is recommended for or used in conjunction with a production testing program, it must be more accurate than all other scoring systems. This is not true. All are visual methods, and the score resulting from the use of any of them is no better than the person making it. Some method of selecting all animals by score, preferably on a systematic and written basis, is the important thing.

A breeding sheep scorecard is shown in Fig. 29–17, while a market lamb scorecard is shown in Fig. 29–18.

As noted, the scorecard gives each of several traits a value, which total 100 for a perfect score.

A scorecard is a valuable teaching aid for beginners. It systematizes judging and avoids any part of the animal being overlooked. However, a scorecard has the following limitations: (1) It is not adapted to evaluating a great number of animals, or to comparative or show-ring judging, because of the time involved in using it; (2) a nearly worthless animal may score quite high—for example, an animal that is so structurally unsound that it can hardly walk may have a rather high total score; (3) it evaluates each part of an animal, rather than the system—the skeletal system, the muscle system, etc.; (4) it is based almost entirely on consumer needs (for example, on the end product—meat, in meat animals); and (5) it accords precious little consideration as to whether, or how, an animal can be better changed to conform to human needs and desires.

| | Perfect Score | No. 1 | No. 2 | No. 3 | No. 4 | Etc. |
|---|---|---|---|---|---|---|
| **CONFORMATION:** | 73 | | | | | |
| **General appearance**—(25 points) | | | | | | |
| Size and scale—big for age, roomy, heavy bone. (15) | | | | | | |
| Type—straight lined, balanced, deep ribbed, long, stylish. (10) | | | | | | |
| **Hindquarters**—(26 points) | | | | | | |
| Leg—muscular, plump, thick, deep. (9) | | | | | | |
| Rump—long, level, full, square dock. (7) | | | | | | |
| Loin—wide, strong, meaty. (9) | | | | | | |
| Twist—deep, full. (1) | | | | | | |
| **Forequarters**—(22 points) | | | | | | |
| Back—wide, straight, strong. (8) | | | | | | |
| Ribs—bold spring, deep ribbed. (6) | | | | | | |
| Shoulders—muscular, smooth. (4) | | | | | | |
| Chest—deep, wide chest floor. (3) | | | | | | |
| Neck—short, thick. (1) | | | | | | |
| **BREEDING QUALITIES:** | 27 | | | | | |
| **Head**—clean-cut, bright eyes, feminine or masculine, proper color of face, free from wool blindness. (5) | | | | | | |
| **Underpinning**—strong pasterns, legs correctly and squarely placed, rugged bone. (15) | | | | | | |
| **Fleece**—dense, uniform crimp, long staple, pink skin, fineness according to standard of the breed, free from black fiber. (7) | | | | | | |
| TOTAL | 100 | | | | | |

Fig. 29–17. Breeding sheep scorecard.

| | Perfect Score | ANIMAL No. 1 | No. 2 | No. 3 | No. 4 | Etc. |
|---|---|---|---|---|---|---|
| **CONFORMATION:** ................................. | 55 | | | | | |
| **General appearance**—(10 points) | | | | | | |
| Straight top and underline, muscular, thick, legs set wide apart, stylish, well balanced, adequate size for age. | | | | | | |
| **Hindquarters**—(26 points) | | | | | | |
| Legs—straight, set wide apart. ........................... (2) | | | | | | |
| Twist—clean, muscular. .................................. (1) | | | | | | |
| Leg—meaty, plump, long, deep, thick. .................... (8) | | | | | | |
| Rump—long, level, thickly muscled. ...................... (6) | | | | | | |
| Loin—meaty, thick, deep loin edge, straight. ............. (9) | | | | | | |
| **Forequarters**—(13 points) | | | | | | |
| Back—thick, straight. .................................. (6) | | | | | | |
| Ribs—bold spring, deep forerib. ......................... (3) | | | | | | |
| Shoulders—muscular, smooth. ............................ (2) | | | | | | |
| Neck—short, thick. ..................................... (1) | | | | | | |
| Breast—wide, deep chest floor, trim. .................. (0.5) | | | | | | |
| Legs—straight, set wide apart. ........................ (0.5) | | | | | | |
| **Middle**—(6 points) | | | | | | |
| Middle—trim, free from wastiness. | | | | | | |
| **FINISH:** ............................................. | 40 | | | | | |
| Uniformly covered with the correct amount of finish over back, ribs, loin, rump. ................... (30) | | | | | | |
| Covering over shoulder, dock. ......................... (5) | | | | | | |
| Trim in flanks, cod, etc. .............................. (5) | | | | | | |
| **QUALITY:** ........................................... | 5 | | | | | |
| Smooth pelt, head trim and refined, ample bone. | | | | | | |
| TOTAL ........................................... | 100 | | | | | |

Fig. 29–18. Market lamb scorecard.

# QUESTIONS FOR STUDY AND DISCUSSION

1. In establishing a flock of sheep, wherein do the factors that should be considered differ from the factors to consider in establishing a herd of cattle, hogs, or horses?

2. If you were establishing a new sheep operation, would you choose purebreds, crossbreds, or grades?

3. If you were establishing a new sheep operation, would you choose native or western ewes?

4. If you were establishing a new sheep operation, what breed would you choose? Justify your choice.

5. What factors should be considered in determining (a) the size of a farm flock, and (b) the size of a range band?

6. When establishing a flock of sheep, what importance would be accorded to each of the following: (a) time to start, (b) uniformity, (c) health, (d) age, and (e) soundness of udder?

7. In recent years, the trend has been to large sheep. Why has this been so?

8. When buying sheep to establish a flock, how can you determine if the price is right?

9. For what purposes do commercial sheep farmers or ranchers select sheep or lambs? When selecting sheep, what special considerations may a purebred breeder give thereto?

10. Discuss sheep selection based on type or individuality.

11. Discuss sheep selection based on pedigree. Why have sheep breeders accorded less importance to pedigrees than breeders of other animal species?

12. Discuss sheep selection based on show-ring winnings. Cite an example of selection based on show-ring winnings proving harmful.

13. Discuss sheep selection based on production testing.

14. Discuss the importance of flock improvement through selection.

15. When judging sheep, why is it important to know the parts?

16. Wherein do ideal meat-type and fleece-type sheep differ? Have these two types moved closer together—become more alike—in recent years; if so, why?

17. Name and describe what you consider to be the five worst faults of meat-type sheep.

18. When examining sheep, why is it important that it be systematic? Outline a good system of examining and tell of the things for which to look.

19. Describe the proper way of catching and holding sheep.

20. Why are sheep producers more conscious about, and familiar with, the determination of the age of sheep by the teeth than they are in applying this method to cattle, hogs, or horses?

21. What is a scorecard? Under what circumstances is a scorecard valuable? What are the primary differences between (a) a breeding sheep scorecard, and (b) a market lamb scorecard?

## SELECTED REFERENCES

| Title of Publication | Author(s) | Publisher |
|---|---|---|
| *Approved Practices in Sheep Production*, Fourth Edition | E. M. Juergenson | The Interstate Printers & Publishers, Inc., Danville, IL, 1981 |
| *Breeding Better Livestock* | V. A. Rice<br>F. N. Andrews<br>E. J. Warwick | McGraw-Hill Book Company, New York, NY, 1953 |
| *Livestock Judging, Selection and Evaluation*, Second Edition | R. E. Hunsley<br>W. M. Beeson<br>J. E. Nordby | The Interstate Printers & Publishers, Inc., Danville, IL, 1978 |
| *Profitable Sheep* | S. B. Collins<br>R. F. Johnson | The Macmillan Company, New York, NY, 1956 |
| *Selecting, Fitting and Showing Sheep*, Sixth Edition | J. E. Nordby<br>H. E. Lattig | The Interstate Printers & Publishers, Inc., Danville, IL, 1962 |
| *Sheep & Goat Science*, Fifth Edition | M. E. Ensminger | The Interstate Printers & Publishers, Inc., Danville, IL, 1986 |
| *Sheep and Wool* | M. P. Botkin<br>R. A. Field<br>C. L. Johnson | Prentice Hall, Inc., Englewood Cliffs, NJ, 1988 |
| *Sheep Book, The* | J. McKinney | John Wiley & Sons, Inc., New York, NY, 1959 |
| *Sheep Book, The* | R. Parker | Charles Scribner's Sons, New York, NY, 1983 |
| *Sheepman's Production Handbook, The* | Edited by<br>G. E. Scott | Sheep Industry Development Program, Inc., Denver, CO, 1986 and updated at intervals |
| *Sheep Production and Management* | C. V. Ross | Prentice Hall, Inc., Englewood Cliffs, NJ, 1989 |
| *Sheep Science*, Revised Edition | W. G. Kammlade, Sr.<br>W. G. Kammlade, Jr. | J. B. Lippincott Co., New York, NY, 1955 |
| *Stockman's Handbook, The*, Sixth Edition | M. E. Ensminger | The Interstate Printers & Publishers, Inc., Danville, IL, 1983 |

A farmer's pride, a nation's wealth! This is a pastoral scene of sheep in Australia, the leading sheep-producing country of the world.

Shepherd and dog watching over their flock. This shows Leonard Vanderhoop, former shepherd at Washington State University, and his Border Collie sheep dog, *Jan,* with a small flock of Hampshire ewe lambs. (Courtesy, Washington State University, Pullman)

# *Chapter 30*

It is logical that different systems of sheep production should exist in a country as large and diverse as the United States. This is fortunate, both from the standpoints of (1) the most efficient utilization of production factors, especially feeds, and (2) the meeting of consumer needs. For purposes of convenience and discussion, the different systems of sheep production are grouped herein into the following categories: (1) the farm flock method, (2) the purebred flock method, (3) the range-band method, (4) the confinement method, and (5) lamb feeding. Although the system of sheep production and the size of the enterprise may introduce new problems, the fundamental nature of sheep remains the same. For this reason, there is neither as much difference nor as many secrets to success in different systems of sheep production as some would have us believe. Essentially, success in any area or system depends upon maintaining a healthy and highly productive flock that is economically managed, followed by the advantageous marketing of the lamb and wool crop.

# DOWN ON THE FARM

Fig. 30–1. *Left:* Farm flock on pasture in Kentucky. (Courtesy, University of Kentucky, Lexington)

Fig. 30–2. Farm flock on the move. (Courtesy, *Sheep Breeder and Sheepman*, Columbia, MO)

Fig. 30–3. Farm flock at winter feed. (Courtesy, J. C. Allen and Son, West Lafayette, IN)

Fig. 30–4. Farm flock orphaned lambs being fed milk replacer. (Courtesy, Ralston Purina Co., St. Louis, MO)

## THE FARM FLOCK METHOD

The farm flock method is the common system of sheep production in the humid farming areas of central, southern, and eastern United States. In general, it accompanies a diversified and intensive system of farming. Since it stresses market lamb production, with wool production of secondary importance, most of the sheep in the farm flock states are of mutton type. Farm flocks range in size from a few head to several hundred head. The commercial farm flock is not under the care of a special shepherd or herder, but its handling is entrusted to a farm worker who has other responsibilities. For this reason, it is often neglected as a minor enterprise, especially when the flock is very small and the caretaker has other assignments which are considered more important and remunerative.

During the grazing season, farm flocks usually compete with beef and dairy cattle for the use of permanent or seeded pastures. On many farms, sheep are considered scavengers; as such they are given the assignment of keeping down the weeds and grass of fence rows, lanes, and draws. For best results, especially from the standpoint of greater efficiency of production and parasite control, however, the farm flock should be given better pastures, and a system of pasture rotation should be followed. Also, confinement or partial-confinement production should be considered.

## THE PUREBRED FLOCK METHOD

Fig. 30–5. Purebred flock of Debouillets. (Courtesy, Amos Dee Jones Ranches, Tatum, NM)

Fig. 30–6. Purebred flock of Hampshires. (Courtesy, *Sheep Breeder and Sheepman,* Columbia, MO)

For the most part, purebred flocks are comparatively small, and the vast majority of purebred breeders are located in the farm flock states. There are, however, a sizable number of purebred flocks in the range states, some of which are quite large and are handled by range-band methods.

In general, purebred sheep breeders have as their objectives the sale of rams to commercial producers and the sale of both rams and ewes to established or new purebred breeders. Over a period of years, the most successful purebred breeders are keen students of commercial sheep production, keeping ever in mind that the ultimate objective in all sheep production is the sale of market lambs and wool, thus gauging their type or ideal accordingly.

Most purebred flocks are given much closer attention than the average commercial flock. Frequently, a full-time caretaker is assigned. Also, in keeping with the general requirements for successful purebred production—regardless of the class or breed of farm animals—more attention is given to (1) the location of the farm, (2) individual records, (3) the careful study and selection of individual sheep rather than of flocks or bands, (4) matings that will produce animals with great inherent possibilities, (5) the maximum development of animals through feeding, and (6) different methods of advertising, including showing. In brief, the production of purebred sheep is a highly specialized business, and only a few producers should attempt this system of production.

# ON THE RANGE

Fig. 30-7. *Right:* Home on the range for the herder—a camp wagon; the herder; and the herder's horse. (Courtesy, Utah State University, Logan)

Fig. 30-8. *Left:* Winter grazing on the range, and in the snow. (Courtesy, U.S. Forest Service, Washington, DC)

Fig. 30-9. *Above:* A band of sheep on winter range, just before moving to spring range. (Courtesy, Utah State University, Logan)

Fig. 30-10. *Left:* Herder and dogs watching over a band of western ewes. (Courtesy, Ralston Purina Co., St. Louis, MO)

## THE RANGE-BAND METHOD

Nearly 70% of the sheep of the United States are located on the western range, with the vast majority of these animals handled according to the range-band method. Each band is under the care of an experienced herder who moves the animals over a comparatively large area of unenclosed land. In the southwestern United States, however—in Texas, Oklahoma, and much of northern New Mexico—nearly all of the range is fenced. Under the latter conditions, the management of sheep is very similar to that existing in Australia and New Zealand.

The relative emphasis on lamb and wool production in the range area varies according to the rainfall and vegetation. In the arid and semiarid regions of the Southwest, where feed is not sufficient for satisfactory grass finished lamb production, the production of wool is of greater relative importance than in those areas where the vegetation is more abundant. Even so, at the present time, lambs constitute about 80% of the gross income from sheep in these areas. In general, however, the sheep of this area are smaller, produce finer quality wool, and practically all the lambs go the feeder route for finishing in more distant areas where feeds are more abundant.

In the semiarid and subhumid areas of the West and Northwest, where the ranges are more lush, many of the lambs go the slaughter route at weaning time, being finished entirely on milk and grass. Most operators in this area produce feeder lambs only because the range is poorer than normal. The main source of income, therefore, is derived from the sale of lambs, with the income from wool being of secondary importance. The sheep of this area are larger—large Rambouillets or long-wool × fine-wool crossbreds predominating—produce coarser wool, and yield lambs of more desirable shape and meat quality.

## THE CONFINEMENT METHOD

Fig. 30–11.   Ewes in confinement barn, on slotted floors with a manure pit underneath. (Courtesy, Ohio Agricultural Research and Development Center, Wooster, OH)

Fig. 30–12.   Ewes on metal slotted floor. (Courtesy, *Sheep Breeder and Sheepman*, Columbia, MO)

Sheep and grass have long been associated together. But altogether too often this relationship has lessened the efficiency of the sheep industry. It has resulted in (1) the relegation of sheep to the least productive areas of the farm, (2) small and inefficient flocks, (3) poor management, (4) too many parasites, and (5) unprofitable operations. Some producers feel that these ills could be alleviated by raising sheep under intensive conditions and in confinement.

Without a doubt, parasite control ranks as the number one reason why people are interested in raising sheep in confinement. Also, interest in confinement sheep production has been accentuated by (1) the extent and success of confinement production in poultry, swine, and cattle; (2) scarcity of good pastures; (3) rise in land prices; (4) the growth of larger and more specialized farming operations; (5) the use of slotted floors and environmental control; and (6) the increase in multiple births and number of lambings per year.

This does not mean that the use of pasture for sheep production is antiquated. Rather, there now exist two alternatives for the producer, instead of just one; and the able manager will choose wisely between the two, or combine them.

Although more cost studies are needed, it appears that, on a per sheep basis, the total cost of confinement and pasture sheep production does not differ greatly. Despite this fact, a number of motivating forces have caused, and will continue to cause, more and more confinement sheep production. Among the factors **favorable** to confinement sheep production are:

1. It practically eliminates the problem of internal parasites.

2. It permits substituting automation for labor, especially in feeding and watering. Also, the added shelter and equipment costs of confinement production may be less than the cost of maintaining pasture fences and extensive water systems, plus the labor entailed in combating parasites, dogs, and bloat.

3. It frees land for more remunerative uses; for example, most Corn Belt farmers can make more money from growing corn and soybeans than from sheep pastures.

4. It makes for higher per acre yields of such crops as silage and dry forage than farmers can obtain by pasturing them with sheep.

5. It allows sheep producers to intensify their operations—to carry more sheep—without enlarging their acreage.

6. It results in more rapid gains and in lambs reaching market weight at an earlier age, thereby decreasing labor, risk, and production expense. Also, it results in improved carcass grade.

7. It requires less time in docking and castrating, vaccinating, controlling parasites, and loading for the market.

8. It makes for less distance to transport feed.

9. It facilitates sorting and lotting groups for size, uniformity, and single *vs* multiple births.

10. It facilitates environmental control.

11. It makes it possible to employ superior caretakers, because such operations are usually sufficiently large to warrant same.

12. It enhances the application of the latest in research, and the best in breeding, feeding, management, marketing, and business.

13. It lessens the chance of sheep being accorded the neglect of a minor enterprise, because it is a sizable business, and treated as such.

14. It generally makes for more favorable feed prices, due to purchase of quantity lots.

15. It makes it possible to feed ewes according to their requirements, rather than according to their appetites. This is important for, in general, pastures that will finish lambs provide more feed than ewes need.

On the other hand, the following factors are **favorable** *to conventional sheep production on pasture*, all or in part, over confinement production:

1. Lower building and equipment costs.

2. The use of pasture often makes for the most desirable land use, crop rotation, and soil conservation.

3. Pasture sheep operations are more flexible than confinement programs—an important consideration where renters are involved.

4. Pastures are especially valuable for the breeding flock—providing a combination of needed exercise, forage, and nutrients.

5. Fewer cases of wool eating, primarily caused by the stress and boredom of confinement, occur among sheep on pasture.

6. Sheep production on pasture does not require as high levels of skill and management as are necessary to make confinement production work.

Also, the monetary value of pastures varies rather widely, being affected by (1) the quality and quantity of the forage; (2) the length of the pasture season; (3) the relative price of protein supplements, grains, and dry forages; and (4) the class and age of the sheep.

Without doubt, many sheep producers can advantageously combine pasture and confinement production—using pastures for the breeding flock and confinement production for young lambs. Therefore, sheep confinement systems will not put good pasture systems out of business; rather, confinement production, all or in part, will increase and replace more and more of the conventional practice of grazing all sheep throughout the pasture season.

• **Partial-confinement method**—For those sheep producers who wish neither (1) to continue the conventional practice of grazing ewes and lambs together until the lambs are marketed, nor (2) to switch strictly to confinement raising, the following

alternative pasture systems are being evaluated, both experimentally and by practical operators:

1. Keep ewes and lambs together in drylot at night, but place them on separate pastures during the day.

2. Keep ewes and lambs together in drylot during the day, but place them on separate pastures at night.

3. Keep ewes and lambs together in drylot at night, place ewes on pasture during the day, and keep lambs in the barn on creep feeders.

4. Keep ewes and lambs together in drylot during the day, place ewes on pasture at night, and keep lambs in the barn on creep feeders.

5. Keep lambs in confinement until early weaned (8 to 12 weeks of age), then either (a) place lambs on separate (apart from the ewes), clean pastures until ready for market, or (b) finish lambs in drylot.

Where pastures are to be utilized for sheep, the particular system decided upon will depend upon the lambing season, temperature, quantity and quality of available pasture, and parasite situation.

## LAMB FEEDING

Fig. 30–13. Lambs on self-feeder. (Courtesy, J. C. Allen and Son, West Lafayette, IN)

Lamb feeding is a highly specialized system of sheep production. In general, where pasture or range conditions are sufficiently good, the farmer or rancher strives to produce grass finished lambs at weaning time. When the vegetation is not good or if for other reasons the lambs lack finish at weaning time, they are usually sold for further finishing. Seldom do producers attempt to finish out their own feeders. Most lamb feeders are large, specialized operators whose feedlots are located in close proximity to adequate and economical feed supplies, such as irrigated valleys, mill centers, or areas where winter wheat fields may be grazed. A full discussion of lamb feeding methods and problems is contained in Chapter 32.

Fig. 30–14. Lambs being finished in an Iowa feedlot. (Courtesy, Peterson Sheep Co., Spencer, IA)

## QUESTIONS FOR STUDY AND DISCUSSION

1. List the five different systems of sheep production.

2. Describe the typical farm flock method.

3. Describe the typical purebred flock method. In comparison with commercial production, to what factors or things is more attention given in a purebred operation?

4. Describe the typical range-band method.

5. How do you account for the fact that about 70% of ewe-lamb operations are by the range-band method, rather than the farm flock method?

6. What factors or forces caused greater interest in raising sheep in confinement?

7. What factors are favorable, and what factors are unfavorable, to the confinement method?

8. List five alternative partial confinement methods.

9. Select a certain farm or ranch (your home farm or ranch, or one with which you are familiar). Assume that there are no sheep on this establishment at the present time. Which of the systems of sheep production covered in this chapter would you elect to follow? Justify your decision.

10. Describe a typical lamb feeding operation.

11. Do lamb feeders fare best when range conditions are poor? Justify your answer.

## SELECTED REFERENCES

| Title of Publication | Author(s) | Publisher |
|---|---|---|
| *Approved Practices in Sheep Production*, Fourth Edition | E. M. Juergenson | The Interstate Printers & Publishers, Inc., Danville, IL, 1981 |
| *Profitable Sheep* | S. B. Collins<br>R. F. Johnson | The Macmillan Company, New York, NY, 1956 |
| *Sheep & Goat Science*, Fifth Edition | M. E. Ensminger | The Interstate Printers & Publishers, Inc., Danville, IL, 1986 |
| *Sheep and Wool* | M. P. Botkin<br>R. A. Field<br>C. L. Johnson | Prentice Hall, Inc., Englewood Cliffs, NJ, 1988 |
| *Sheep Book, The* | J. McKinney | John Wiley & Sons, Inc., New York, NY, 1959 |
| *Sheep Book, The* | R. Parker | Charles Scribner's Sons, New York, NY, 1983 |
| *Sheepman's Production Handbook, The* | Edited by<br>G. E. Scott | Sheep Industry Development Program, Inc., Denver, CO, 1986 and updated at intervals |
| *Sheep Production and Management* | C. V. Ross | Prentice Hall, Inc., Englewood Cliffs, NJ, 1989 |
| *Sheep Science*, Revised Edition | W. G. Kammlade, Sr.<br>W. G. Kammlade, Jr. | J. B. Lippincott Co., New York, NY, 1955 |
| *Wool Science* | W. D. McFadden | Pruett Press, Boulder, CO, 1967 |

The lost sheep! (Courtesy, New Zealand Embassy, Washington, DC)

A *litter* of six lambs, the lamb crop from one Finnsheep ewe. (Courtesy, USDA)

# BREEDING SHEEP

## Chapter 31

As with other classes of farm animals, sheep breeders are of two types: (1) commercial producers, and (2) purebred breeders. The commercial producers are primarily concerned with securing a high-percentage lamb crop, and with the utility features of the animals and their ability to make efficient use of feeds. They direct most of their effort toward improvement through the selection of purebred rams.

Purebred breeders are interested in producing rams for sale to the commercial sheep producers and, at the same time, in further improving their own flocks and selling both ewes and rams to other purebred breeders. Although breed fancy points are often given consideration, for success over a long period of years the purebred breeders must be ever aware of the fact that the ultimate goal in sheep production is the sale of lambs and wool.

Fig. 31–1.   Being born and born alive is the most important requisite of sheep breeding.

## NORMAL BREEDING HABITS OF EWES

Perhaps there is as much scientific information about the normal breeding habits of the ewe as there is about any class of farm animals. Even so, not all of the phenomena are clear, and much work in the field of sheep breeding remains to be done.

### AGE OF PUBERTY

Sheep reach puberty at five to seven months of age. As evidence of the fact that the sex organs of ram lambs become functional at this age, it might be pointed out that it is not unusual to find that they have bred a few of the ewes prior to the normal weaning time.

Ewe lambs are probably somewhat slower than ram lambs in reaching sexual maturity, but it must be realized that in sheep full development of the female may be reached before the onset of estrus or heat, for in sheep there are long periods (anestrus) when the female organs are not active. Generally, the first estrus

in ewe lambs of the mutton breeds occurs during the fall of their first year, when they are 8 to 10 months of age; whereas in the more slow-maturing Merinos, it may be delayed until they are 16 to 20 months of age.

### AGE TO BREED EWES

Most ewes are bred during the first breeding season after they are one year of age, producing their first lambs when approximately 24 months old.

As a rule, ewe lambs are not used for breeding purposes. Range operators almost never follow the practice, and only comparatively few farm flock owners breed ewe lambs so that they will drop their first lambs when approximately 12 months of age. Limited experimental work plus practical observations would indicate that the following results may be expected from the practice of breeding ewe lambs:

1. Growth of bred lambs is retarded temporarily, but they are not stunted permanently.
2. Wool yield is not affected by early breeding.
3. Some ewe lambs do not conceive; birth weights of lambs born to ewe lambs are lighter; and more troubles are encountered at lambing time. (More ewes require assistance at lambing and more lambs are disowned.)
4. If computed on the basis of total lifetime production up to five or six years of age, ewes bred as ewe lambs will show more lambs and total pounds of lambs produced per ewe than ewes bred at the normal yearling age.

Many good sheep producers are of the opinion that Suffolk rams should be used if ewe lambs are bred to black-faced rams, for the reason that the smaller head and shoulders of the progeny will cause less difficulty at lambing time.

### HEAT PERIODS

The duration of estrus in ewes ranges from 20 to 42 hours, with an average of 30 hours. Unlike other farm animals, the ewe shows no visible external indications of heat, the acceptance of the ram (or teaser with an apron on) being the only method of external detection. Ovulation seems to occur late in the heat period, usually from about 24 to 30 hours after the onset of estrus. If the ewe is not bred, or if she fails to get in lamb, estrus recurs after an interval of 14 to 20 days, with an average of 16 to 17 days.

### GESTATION PERIOD

The period of gestation of sheep varies between breeds and individuals, with the range from 144 to 152 days and the average 148 days. Medium-wool breeds, including the down breeds, have short gestation periods of from 144 to 148 days, whereas the fine-wool breeds, such as the Merino and Rambouillet, have long gestation periods ranging from 148 to 152 days. The long-wool breeds, such as the Lincoln and Romney, have gestation periods intermediate between the medium- and fine-wools; they average 146 to 149 days. Individual gestation periods within a breed may vary up to a range of 15 days.

## FERTILITY AND PROLIFICACY IN SHEEP

In a summary of all available data, two Swedish investigators, Johannson and Hansson, found that among all breeds of sheep and types of environment about 176 pairs of twins and 10 sets of triplets occur in every 1,000 births; and one set of quadruplet lambs is born in every 5,000 births.[1] This is a 119.6% lamb crop.

In general, sheep breeders have long considered that twinning is important. This is well illustrated in the following old English adage: "Ewes yearly by twinning rich masters do make; the lambs of such twinners for breeders go take." There is now substantial evidence to indicate that this adage was well founded.

Though twinning is inherited, it is not highly so in comparison with certain other traits. Experimental work has shown that the heritability of twinning is about 15%, whereas the heritability of face covering is about 56%.

Prolificacy in sheep is largely determined by the number of eggs liberated by the ovary at the heat period, and by the amount of fetal atrophy. If only one egg is released and fertilized, a single lamb will result unless this egg should divide so that twins are produced. Such division does not seem to occur with much frequency, it undoubtedly being true that most twins and triplets are due to the shedding of a like number of eggs, which are fertilized and complete their development.

Under natural conditions, the number of eggs shed depends on heredity, environment, and age. It is generally recognized that, because of heredity, some breeds, such as the Booroola Merino, Finnsheep, Polypay, and Romanov breeds, produce a higher percentage of twins and triplets than others. Also, under certain conditions, the number of eggs shed may be increased by flushing; that is, through providing improved nutritive conditions prior to the breeding season. Practical observation would indicate that age is a factor in twinning, with middle-aged ewes producing a higher percentage lamb crop than either ewe lambs or very old ewes.

Twins are desirable, for they greatly increase the weight of lambs sold per ewe. The annual maintenance requirements of ewes are not far different, regardless of whether they are producing twins or singles.

In some localities, rams of at least some of the breeds go through a period of lowered semen volume and quality during the summer months. The reasons for this phenomenon are not fully understood, and limited experimental studies to date have given conflicting results. Nevertheless, there appears to be a definite seasonal difference in the fertility of rams. Where early lambs are desired, this factor may be of considerable economic importance.

## FLUSHING

Flushing is that practice of feeding thin ewes more generously during the period of 2 to 3 weeks immediately prior to breeding. This may be accomplished either by providing more lush pasture or range or by grain feeding.

[1]Reeve, E. C., and F. W. Robertson, "Factors Affecting Multiple Births in Sheep," *Animal Breeding Abstracts*, Vol. 21, No. 3, 1953, pp. 211–224.

Fig. 31–2. A large percentage of twins is desirable. Apparently this factor is affected by heredity, environment, and age.

Although it is not likely that all of the benefits ascribed to flushing will be fully realized under all conditions, experiments and experiences show that flushing will result in a 15 to 30% increase in the lamb crop and that the ewes will both breed earlier and more nearly at the same time. Fat ewes are best conditioned for breeding by increasing the exercise.

## PREPARATION OF EWES AND RAMS FOR MATING

Several practices pertaining to both ewes and rams are important during the breeding season. Details follow.

### TRIMMING AND TAGGING THE EWES

Tagging is the removal of tags or locks of wool and dirt about the dock. It is important that this job be done prior to the breeding season in order to prevent the ewes from befouling themselves and to remove obstacles for the service of the ram.

### PREPARATION OF THE RAM

As the weather is usually rather warm at the time of the breeding season, shearing the ram just prior to this will make him more active. This is especially true of old show or sale rams. Where rams are not sheared completely, they should at least have the wool clipped from the neck and from the belly in the region of the penis, for this will result in copulation with greater ease. It is also important to see that the hoofs of the ram are properly trimmed prior to the breeding season.

### MARKING THE RAM

When several rams are turned in with a large band of ewes, it is impossible to detect individual rams that may be failing to settle ewes. Moreover, it is quite likely that a different ram will serve the ewe should there be a recurrence of heat, or perhaps more than one ram may serve the ewe at the time of estrus. When only one ram is being used on a small flock, however, it is important to know whether the ewes are getting with lamb. Then, too, with a purebred flock individual breeding records are rather important.

A breeding record can best be kept by using a marking harness (breeding harness), containing a crayon (different colored crayons are available), on the ram, or by smearing the breast of the ram and the area between his forelegs every day or two with a thick paste. Then, as the ram serves the ewe a mark will be left on her rump. Paint or tar should never be used for this purpose.

The color of the crayon or paste should be changed every 16 days (the approximate estrus cycle of the ewe) so that one can determine whether ewes that have been bred are returning in heat. For example, during the first 16-day interval, the thick paste used on the ram might well be a mixture of ordinary lubricating oil and yellow ochre; for the second 16-day interval, it might be lubricating oil and venetian red; and for the third 16-day interval (if there is still some question about some of the ewes having settled), it might be a paste made by using lubricating oil and lamp black (thus producing from light to dark colors).

Naturally, if a good percentage of the ewes are found coming in heat for a second time, the ram should be regarded with suspicion, and perhaps another ram should be obtained. The sterility in some instances may be temporary because of high condition and lack of exercise.

## CARE AND MANAGEMENT OF THE RAM

Fig. 31–3. Hampshire stud rams on pasture. (Courtesy, *Sheep Breeder and Sheepman*, Columbia, MO)

If possible, the rams should be secured considerably in advance of the breeding season. At this time, better rams are available and the prices are more favorable. Then, too, the ram will have an opportunity to become acclimated before being placed in service. In case of show or sale rams, it may also be advisable to remove some of their surplus flesh gradually.

Stud rams are usually kept separate from the ewes except during the breeding season. Their quarters need not be elaborate or expensive. Usually, a dry shelter that will provide protection during times of inclement weather is all that is necessary. Plenty of exercise should be provided at all times.

Rams may subsist largely on pasture and dry roughage. If the pasture has been scanty prior to the breeding season, the rams may be conditioned by feeding a little grain, usually not more than one lb daily. Rams are usually fed some grain when being fitted for show or sale, but it must be remembered that excess fat may actually be harmful from a breeding standpoint.

## AGE AND SERVICE

The number of ewes a ram will serve in a season depends on his age, vigor, and method of handling. Table 31–1 gives pertinent information relative to the use of the ram, including consideration that should be given to age and method of mating.

Generally speaking, it is best not to use a ram lamb, but a vigorous, well-grown, early-maturing lamb may be used on 20 to 25 ewes with no apparent harm. Many good breeders believe that there is a definite breed difference in early reproductive capacity. A vigorous ram 1 to 4 years old that is run with the flock during the breeding season is sufficient for 35 to 60 ewes. When the ram is turned with the flock for only a limited period daily—perhaps an hour in the morning and an hour in the evening—or when a teaser is used for the purpose of locating ewes that are in heat, one mature ram may be sufficient for 50 to 75 ewes. Unless the ram becomes extremely nervous and restless when removed, his energy will be conserved through keeping him away from the flock. By so handling, he will be available for heavier breeding service. Often he will remain contented if 1 or 2 wether lambs or bred ewes are kept with him. With a heavy ram and warm weather, a good plan is to allow him to run with the flock at night and to remove him to separate quarters during the day. Regardless of the system followed, the ewes should be checked daily and accurate records kept of the breeding dates. Though there is considerable variation, most range sheep operators usually plan to have 3 to 4 active mature rams with each hundred ewes.

**TABLE 31–1**
**HANDY RAM MATING GUIDE**

| Age | No. of Matings/Yr. | | Comments |
|---|---|---|---|
| | Hand Mating | Pasture Mating | |
| Lamb ........... | 20–25 | — | Most range operators use 1 ram to 25 to 35 ewes. |
| Yearling or older | 50–75 | 35–60 | A ram should remain a vigorous and reliable breeder up to 6 to 8 years of age. |

## ARTIFICIAL INSEMINATION (A.I.) AND EMBRYO TRANSFER

In the United States, artificial insemination is more extensively practiced with dairy cattle than with any other class of farm animals. Most experience with A.I. in sheep in the United States has been on an experimental basis. In the U.S.S.R., Eastern and Central Europe, and some areas of South America where labor costs are low, however, many sheep are also bred through artificial means. So, people are prone to ask why more sheep are not bred artificially in this country. The answer is that, to date, most sheep producers simply feel that the disadvantages outweigh the advantages. For some producers, however, A.I. provides a tool for meeting the demand for better efficiency, and they must consider the use of A.I. as it relates to their particular physical, economic, and genetic needs.

The procedures in artificial insemination of sheep are similar to those used in cattle and horses. Further, the equipment and supplies used for sheep are similar to those used for cattle, with the exception of a vaginal speculum and light source which aid the placement of the semen in the uterus.

Fig. 31–4. Semen tanks. Shipping tank (left). Regular semen storage tank (right). (Courtesy, American Breeders Service, DeForest, WI)

Embryo transfer is another tool that is becoming more readily available to livestock producers. It allows breeders to obtain more progeny from superior males and females within a shorter time span. Numerous embryos that are the result of matings of superior males and superior, superovulated females are removed from the uteri of the superior females and then placed in the uteri of commercial females. For some time to come, however, embryo transfer will probably be used only for top purebred breeders because of the cost.

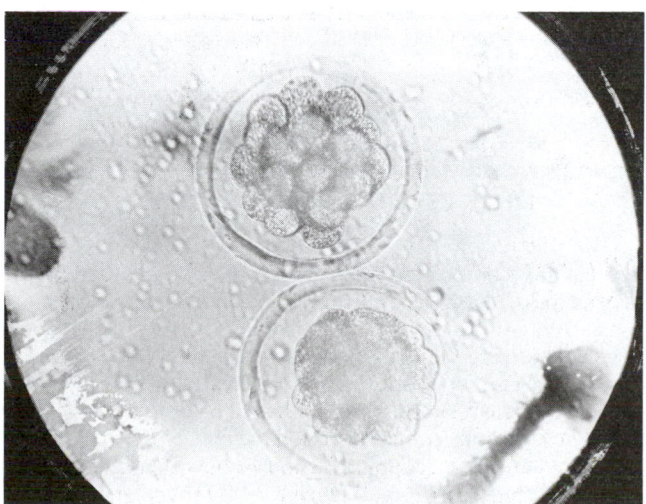

Fig. 31–5. Embryos recovered from a ewe. Numerous embryos flushed from the uterus of a superior ewe, bred to a superior ram, may be placed in the uteri of commercial ewes where they develop fully. The commercial ewes become surrogate mothers. (Courtesy, C. E. Lindley, Mississippi State University)

It is noteworthy that the first embryo transfers in farm animals were performed on sheep, but that the commercial interest now is in cattle and horses.

(Also, see Chapter 3, sections on "Artificial Insemination," and "Embryo Transfer.")

## CONTROLLED REPRODUCTION

Considerable research is underway to (1) synchronize heat, (2) increase ovulation rate and lambing crop, (3) shorten the lambing interval/breeding twice a year, and (4) control light, to boost the efficiency of production from sheep, by producing more lambs. As a result of the success of this research, sheep breeders will need to choose methods which are best suited for their operations.

• **Estrous synchronization**—A common procedure for synchronizing estrus in cycling ewes is the administration of progestogen for 12 to 14 days. Upon withdrawal, a majority of the ewes demonstrate estrus within 2 to 3 days.

Prostaglandin (prostaglandin $F_2$) can also be used for estrous synchronization during the breeding season. It may be used alone or in combination with a progestogen, but it is only effective when ewes are normally cycling.

Successful out-of-season breeding and synchronization can be accomplished by the administration of a progestogen and then the injection of PMSG (pregnant mare serum gonadotropin) which stimulates follicle growth and ovulation.

• **Increased ovulation rate and lamb crop**—Increased ovulation means increased lambing rate, and a larger lamb crop marketed means more profit. For example, records show that, on the average, range ewes that drop and rear twin lambs wean about 40 lb more lamb than ewes that have single lambs. Because most of the costs of sheep production are about the same regardless of the size of the lamb crop, multiple births are important.

The lamb crop as a percentage of ewes 1 year old and over runs from 106 to 108%.[2] However, lamb crop potential of 125 to 200% exists, depending on the production system.

Ovulation rate is affected by season (highest in the fall), by the level of feeding (flushing), and by breeding. Some breeds are more prolific than others. Also, fertility, prolificacy, and lamb livability are increased by crossbreeding.

The Booroola Merino, Finnsheep, Polypay, and Romanov breeds seldom have fewer than twins or triplets, and mature ewes of these breeds may have more. Crossbreeding with such prolific breeds can very quickly lead to high twinning rates. Also, as much as a 2% per generation (approximately 1% per year) increase in lambing rate can be realized when twin replacements are selected.

• **Shortened lambing interval/breeding twice a year**—Most ewes freeload half the year. They are bred in the fall, have a 5-month pregnancy period, and suckle lambs for another 4 to 5 months. They spend a whole year to produce 1 lamb crop. This is inefficient, and costly, too.

Using breeds of sheep with long breeding seasons (Dorsets, Rambouillets, and Merinos), along with proper management, can achieve lambing intervals of approximately 8 months—

[2]*Agricultural Statistics 1988,* USDA, p. 286, Table 424.

without the use of hormones. This means that 3 lamb crops can be produced in 2 years (1 crop each 8 months). It is possible, however, from both time and physiological standpoints, for a ewe to produce 2 lamb crops in 12 months. Research has shown that through the use of hormones, estrus and ovulation can be induced early enough so that ewes can be rebred once or twice after lambing, and yet produce another lamb crop in the 6-month period following lambing. Currently, the most hopeful efforts to achieve twice-a-year lambing (and 3 times in 2 years) seem to be selection for ewes with a longer natural breeding season, early postpartum fertility, and a shorter gestation.

Shortening the lambing interval will make for lower costs of production. Also, the lessening of seasonal restrictions will (1) contribute to a more even supply of lamb throughout the year, and (2) allow for greater flexibility in production.

• **Controlling reproduction with light**—Modification of the length of exposure to light can be used to induce out-of-season breeding in sheep. Artificially reducing the amount of light per day to which the animals are exposed may induce changes in the nervous and endocrine systems which initiate out-of-season cycles. However, under most production systems, it is difficult to provide a practical way of placing animals under decreasing light.

(Also, see Chapter 3, sections headed ''Estrous Cycle Manipulation'' and ''Superovulation.'')

## CARE OF THE PREGNANT EWE

Fig. 31-6.   Pregnant Polypay ewes in excellent breeding condition. Note the snow. (Courtesy, American Polypay Sheep Assn., Sidney, MT)

The requirements of the pregnant ewe are neither exacting nor difficult to meet. These needs are feed, water, exercise, and shelter.

## PREGNANCY TESTING

Barren ewes can no longer return feed costs for their wool production. Moreover, returns from lambs are of increasing importance to the sheep industry; today, about ¾ of the income is from lamb, and only ¼ from wool. These factors make pregnancy diagnosis of great economic importance. Detection of

multiple fetuses would also be of significance, because feed costs could then be lowered on ewes carrying only one lamb.

Several methods for pregnancy testing have been developed and used experimentally or practically in recent years; among them, the following: (1) rectal-abdominal palpation, (2) ultrasound scanning, and (3) intrarectal Doppler technique.

• **Rectal-abdominal palpation**—This is a relatively simple technique developed at the U.S. Sheep Experiment Station, Dubois, Idaho. The only equipment required is a 16-inch hollow, plastic rod (called a palpation rod), some lubricating material, and a device (such as a cradle) for holding a ewe on her back. The rectal-abdominal method of pregnancy detection is illustrated in Fig. 31-7. In the hands of an experienced operator, detection of ewes at mid-pregnancy (65 to 70 days postbreeding) with the pregnancy rod is virtually 100% accurate. With proper handling equipment, a technician and 3 assistants can pregnancy test 200 ewes per hour.

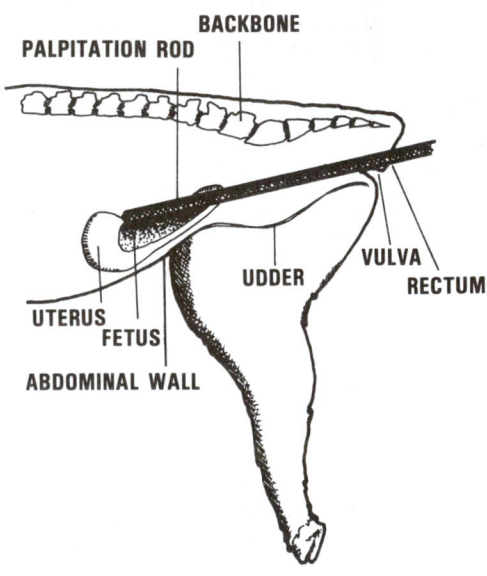

Fig. 31-7.   The palpation rod is inserted in the rectum, close to the spine. The objective is to (1) position the rod dorsal to the fetus, and (2) elevate and hold the fetus against the abdominal wall. Then, the free hand is used to feel the fetus through the abdominal wall. **Note:** To facilitate comprehension, the above diagram shows the ewe in an upright position. However, in a pregnancy examination, the ewe is placed on her back in a comfortable horizontal position.

In addition to determining pregnancy, this method has been used with a high degree of accuracy for determining the number of fetuses (twins, triplets, etc.) in a pregnant ewe.

• **Ultrasonic scanning**—The pregnancy checking equipment used in this method is similar to that used for determining back fat and loin eye area in cattle and swine. A transducer transmits ultrasonic sound waves through the body of the ewe and receives an echo based on the Doppler effect. If the ewe is pregnant, the presence of the fetus registers on the machine either by light or sound. This method appears to have a high degree of accuracy after 60 days of pregnancy; some claim 98% accuracy at six weeks. It cannot differentiate between single and multiple pregnancy.

• **Intrarectal Doppler technique**—This equipment detects the fetal heartbeat, which is 130 to 160 beats per minute compared to the ewe's 90 to 110 beats per minute. In preparation for this test, the ewe is placed on her back in a cradle with the hindlegs secured (see Fig. 31–8). A rectal probe is lubricated and inserted in the rectum on a horizontal plane below the uterus. As the probe is inserted, sounds are amplified on the machine—first the maternal heartbeats (90 to 110), and finally the very rapid (130 to 160) heartbeats of the fetus. When used by an experienced operator, this method is 98% accurate. The machine cannot differentiate between a single or multiple pregnancy.

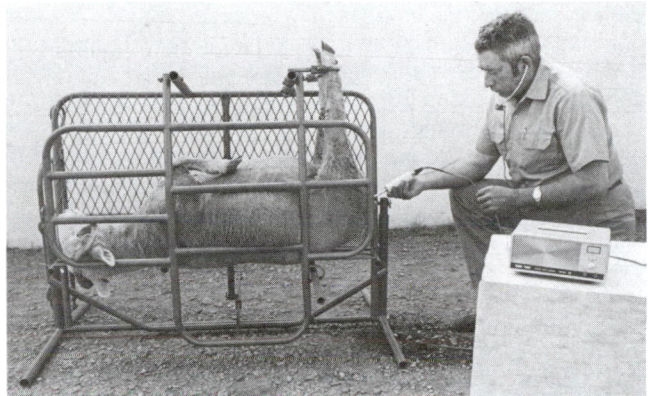

Fig. 31–8. The intrarectal Doppler technique for pregnancy determination in ewes. It detects the fetal heartbeat, which is 130 to 160 beats per minute compared to the ewe's 90 to 110 beats per minute. (Courtesy, Sheepman Supply Co., Barboursville, VA)

## FEED AND WATER

For successful sheep production, the ewes must be economically and properly fed and watered during the pregnancy period. As these requirements are fully covered in Chapter 32, no further discussion is necessary at this point.

## EXERCISE

During periods of inclement weather and when feed is brought into the barn, ewes quite often exercise entirely too little. As a result they become sluggish, and their blood circulation is poor. Forced exercise may be brought about by scattering a palatable roughage some distance from the shed or by driving the ewes at a moderate walk. Above all, overexertion, such as wading through a deep snow or being chased by dogs, should be avoided. With a good winter pasture and open weather conditions, no other arrangements for exercise will be necessary.

## SHELTER

The shelter should be of such nature as to protect the flock from becoming soaked with rain or wet snow. Dry snow or bitter cold has no harmful effect, and up until lambing time, a shelter open to the south on well-drained ground may be entirely satisfactory.

## CARE OF THE EWE AT LAMBING SEASON

As previously stated (under the section headed "Increased ovulation rate and lamb crop"), there is an average lamb crop of 106 to 108%. However, there is a mortality rate of 20 to 25% from birth to weaning. Most of these losses occur in the first few days of life.

The careful and observant shepherd or herder realizes the importance of having everything in readiness for the lambing season. If pregnant ewes have been so fed and managed as to drop a crop of strong, vigorous lambs, the next problem is that of saving the newborn animals.

As lambing time approaches, unsheared ewes should be tagged. This consists of shearing the wool from around the udder, flank, and dock. The ewe should also be placed where she has plenty of room, away from any jamming or crowding. The grain allowance should be materially reduced, but the roughage allowance may be continued, if it is certain that it is of good quality and palatable. Careless feeding at this time is likely to result in milk fever following parturition. At this time, the wool around the udder should be clipped short in order to allow the lamb to find the teats readily. If breeding records have not been kept, the signs of approaching parturition must be relied upon. A nervous, uneasy disposition; a sinking in front of the hips; and fullness of udder are such indications.

## THE LAMBING PEN

Fig. 31–9. Ewe and day old lambs in a lambing pen with slotted floor. Note the heat lamp.

Just before lambing, or immediately thereafter, the ewe should be placed in a lambing pen. These pens are usually 4 ft square and are made by placing together two hinged hurdles, which are then set against the walls of the sheep barn. Use of the lambing pen prevents other sheep from trampling on the newborn lamb; eliminates the possibility of the lamb wandering away and becoming chilled; and, through keeping the dam and offspring together, lessens the danger of disowned lambs.

Lambing pens should be clean, dry, well bedded, and well ventilated, and should be located so as to be free from drafts. During extremely cold weather, additional warmth may be provided for the first few hours after birth by throwing a blanket over the top of the pen.

## NORMAL PRESENTATION

A good rule for the shepherd to follow is to be near during parturition but not to disturb the ewe unless she needs help. Normal presentation of the lamb consists of having the forelegs extended with the head lying between them, although some lambs are delivered hind legs first. Even though the lambs are born in clean quarters, tincture of iodine should be applied to the navel soon after birth. The latter precaution may not be necessary when lambs are dropped on an uncontaminated pasture or range, although many range operators report that they have found it necessary to apply iodine to the navel of lambs born on the range as well as in the shed.

## TAKING THE LAMB

If the ewe has labored for some time with little progress or is laboring rather infrequently, it is usually time to give assistance. If the lamb is not in the proper position, such assistance consists of inserting the hand and arm in the vulva and turning the lamb so that the forefeet and head are in position to be delivered first. Delivery may then be helped by pulling the young outward and downward as the ewe strains. Before doing this, however, the fingernails should be trimmed closely and the hands and arms should be thoroughly washed with soap and warm water, disinfected, and then lubricated with Vaseline or linseed oil.

## CHILLED AND WEAK LAMBS

Lambs arriving during cold weather may become chilled before they have dried. One of the most effective methods of reviving a chilled lamb is to immerse the body, except for the head, in water that is as warm as one's elbow can bear. The lamb should be kept in this for a few minutes and then removed and rubbed vigorously with cloths. It then should be wrapped in an old blanket, a sheepskin, or other heavy material and should be given some warm milk as soon as possible. Another convenient and effective method of drying and warming a chilled lamb consists of putting it into a box containing a light bulb or an electric heater.

When strong, healthy ewes have been properly fed and cared for during pregnancy, there will be a minimum of weak lambs. The shepherd should first make certain that the membrane has been removed from the nostrils and that breathing has started. Blowing into the mouth, lifting the body and dropping it a short distance, working the legs, and pressing the sides are artificial methods to start respiration which may revive lambs that at first appear lifeless.

After breathing has started and the navel cord has been painted with iodine, an attempt should be made to get the lamb to nurse. Quite often even a very weak lamb will nurse the ewe if it is held to the teat. If it refuses to nurse in this manner, some of the colostrum of the ewe should be milked into a sterilized bottle, and the lamb should be fed a few teaspoonsful each hour by means of the bottle and nipple, until it gains strength.

If the ewe has no milk, an attempt should be made to obtain milk from another ewe that has just lambed, and perhaps in a few hours the normal flow of milk will start.

## DISOWNED LAMBS

When lambing pens are used, the number of disowned lambs is kept to a minimum. For the most part, disowning of lambs is due to improper feeding during pregnancy or because of a poor milk supply, an inflamed udder, or a maternal instinct that is not sufficiently developed, as is often true in ewes with their first lambs.

For the first few days, a ewe seems to recognize her young by scent or sense of smell. When difficulty is encountered in getting a ewe to own her own lamb or when it is desired to transfer or "graft" a lamb (as may be necessary with the loss of a lamb or when there are twins on an old ewe), deception in the sense of smell is an effective approach. One of the most common practices is to milk some of the ewe's milk on the rump of the lamb and then to smear some of it on the nose of the ewe. Many good shepherds take some of the mucus from the mouth and nose of the newborn lamb and smear it over the nose of the ewe. If these methods fail and the ewe persists in fighting the lamb away, blindfold her so that she cannot see the lamb. As a last resort, and when all other methods have failed, tie a dog in an adjoining pen. Sometimes the latter method will cause latent maternal instincts to rise to a surprising degree.

Occasionally a ewe will fail to own one of a pair of twin lambs. When this condition exists, about all that can be done is that the shepherd be patient in training the disowned lamb to nurse at the same time as its mate. Both lambs are usually kept from the ewe and turned with her at intervals.

## THE ORPHAN LAMB

A lamb may be orphaned through the death of its mother or because of the inability of the mother to suckle it. The most satisfactory arrangement for the orphan is to provide a foster mother. The good shepherd will try to have every ewe raise a lamb. There may be a ewe that has just lost her lamb or a strong, healthy ewe with just one lamb. When a lamb dies at birth and it is desired to transfer or "graft" another lamb on the ewe, two procedures are common. Sometimes a ewe will accept another lamb provided that the lamb to be adopted is first rubbed with the body of the dead lamb that it is to replace. Though a bit more bothersome, a more effective approach consists of removing the skin from the dead lamb and tying it over the lamb to be adopted. After 2 or 3 days, the skin may be removed gradually, a piece at a time. The latter method is commonly used in the range bands of the West.

When it is impossible to transfer an orphan lamb to another ewe, it may be raised either on cow's milk or on milk replacer. Of course, the problem will be simplified if the lamb has received some colostrum (the first milk) from its mother or from another ewe.

If cow's milk is used, it should not be diluted, because cow's milk is lower in butterfat and total solids than ewe's milk. Milk replacer should be mixed according to the manufacturer's directions.

Both cow's milk and milk replacer should be warmed to 100°F and fed in sterilized bottles. During the first few days, the orphan should be fed about one oz of milk (or milk replacer) every two hours. Gradually, the quantity may be increased and the intervals spaced further apart.

## FEED AND WATER AFTER LAMBING

Following parturition, the ewe is in a feverish condition and should be handled carefully. She may be watered immediately after lambing, and at frequent intervals thereafter, but she should never be allowed to gorge. It is also a good plan to take the chill off the water before giving it to her. In general, feeds of a bulky and laxative nature should be provided during the first few days. A mixture of equal parts of oats and wheat bran may be fed in very limited quantities, with all the hay that can be consumed. Heavy grain feeding at this time may cause udder trouble in the ewe and digestive disturbances in the lamb. The feed may be gradually increased until the ewe is on full feed in about a week.

## EXAMINATION OF THE UDDER

During the first two days following lambing, the udder should be examined night and morning. Sometimes a lamb will nurse one side only. If all the milk is not being taken by the lamb, the udder should be milked out and the ration lessened accordingly. If the udder becomes swollen and feverish, it should be milked out, bathed with warm water, and then dried. Following this, it should be painted with tincture of iodine. This treatment should be repeated once or twice daily, as necessary. Lambs should not be allowed to suckle their mothers when the udder is in such a condition. It is also a good plan to isolate the affected ewes from the rest of the flock.

## FACTORS AFFECTING REPRODUCTION

Normally, ewes come in heat during late summer or early fall, though there is both an area and a breed difference. The breeding season is usually restricted to about four months. Among the factors that affect reproduction in sheep are:

1. **Daylight.** The initiation of the breeding season is affected by the number of hours of light in the day. Sheep will generally begin cycling when the number of daylight hours drops below 14. This is the reason most breeds of sheep come into heat during the fall months—in September, October, and November. To initiate estrus, however, it does appear that the shorter days must be preceded by longer days. Ewes that come into heat before or after the normal breeding season demonstrate very erratic cycle lengths.

2. **Temperature.** Most breeds of sheep begin cycling with the coming of cooler fall weather, when night temperatures drop to around 74°F. Some breeds, such as the dark-faced mutton breeds, are particularly sensitive to heat levels. Other breeds, such as the Dorset, Merino, and Tunis, will cycle the year around and appear to be affected little by higher temperatures.

There is increasing research evidence that more ewes cycle and conceive in hot weather than has been previously thought. However, there is a high rate of embryo mortality during this time, and lambs born from ewes pregnant during hot weather are generally weak and much smaller than those born in cool weather. Death loss from such lambs is high.

Temperature also affects the male reproductive system. High temperatures will cause lower semen quality in rams. During prolonged periods of excessive heat (temperatures of 100°F or more), rams may become sterile. This damage is not permanent, however, and the ram usually becomes sexually sound after 4 to 6 weeks of cooler weather.

3. **Nutrition.** Proper nutrition is necessary for good reproductive performance.

Energy, in the form of carbohydrates, is important. Thus, the practice of flushing induces the shedding of a larger number of ova and consequently higher fertility.

Vitamin A is of particular importance to the ewe, for the maintenance of the germinal epithelium of the ovary. This portion of the ovary gives rise to the egg or ovum of the female. Under prolonged drought conditions (low carotene intake), the ewe will cease to produce ova, so reproduction becomes impossible.

## SYSTEMS OF LAMB RAISING

In general, four systems of lamb raising are practiced in the United States: (1) hothouse lambs, (2) Easter lambs, (3) spring lambs, and (4) lambs raised on grass.

## HOTHOUSE LAMBS

These are lambs produced out of season—principally for the Boston and New York markets—and sold from December to April. The lambs are dropped in the fall or early winter months and are ready for market in 6 to 12 weeks, weighing 30 to 60 lb liveweight. As the ewes must be bred out of season for this type of market lamb production, those individuals carrying a predominance of Dorset, Tunis, Rambouillet, or Merino breeding are usually used in the production of hothouse lambs. Generally, these ewes are mated to rams of the mutton breeds.

This system represents a highly specialized business, and the producers entering it must, first of all, have assurance of their market. Then, in addition to having heavy-milking ewes of the proper breeding, they must properly house and skillfully feed the lambs and ewes in order to produce the fancy carcass in demand. Hothouse lambs should be castrated; but, peculiarly enough, on certain markets, notably the Boston market, the consumer demand is for lambs that have not been docked. The usual explanation of the demand for undocked hothouse lambs is the fact that the city consumer associates the tailless condition with older sheep only, thus having considerable question about the age of a hothouse lamb should it be docked. Because of the high moisture content of the young carcass, the pelts are usually left on hothouse lambs until the carcass is ready for delivery to the consumer.

## EASTER LAMBS

The Easter lamb trade has been a variable one, the demand being within a rather wide range both as to weight and quality of lambs. During the past several years, there has been a considerable Easter demand for light lambs, 20 to 30 lb in weight and not carrying much finish. At the same time, there is always a rather constant demand for heavier lambs of higher quality for which good prices are paid.

## SPRING LAMBS

Young lambs marketed in the spring of the year, and prior to July first, are referred to as *spring lambs*. This class is not to be confused with *lambs* born the previous year but which are marketed the following spring, perhaps following a period in the feedlot. The first spring lambs usually come from the southwestern states and California.

Fig. 31-10.  Spring lambs on spring range. (Courtesy, USDA)

After July first, the market classification of spring lambs no longer exists, animals of similar birth simply being known as lambs. Likewise, after July first, those previously designated as lambs are yearlings (wethers) or yearling ewes.

## LAMBS RAISED ON GRASS

The fourth general system of market lamb production is that of having the lambs arrive in the spring and producing them to marketable weight entirely on milk and grass. This is the most common system of lamb production throughout the United States and the nearly universal method on the range. Lambs handled in this manner must have the benefit of lush pastures; otherwise, it will be necessary to market them via the feeder route. Also, they are generally more subject to parasites than earlier lambs, and they sell on a somewhat lower market; but the cost of production is relatively low.

## CROSSBREEDING IN SHEEP

Fig. 31-11.  The best lambs in New Zealand are crossbreds like these, produced by using Southdown lambs on Romney ewes. (Courtesy, Department of Scientific and Industrial Research, Wellington, New Zealand)

Although the common systems used in breeding sheep are not unlike those applied to other classes of farm animals, there appears to be more crossbreeding among sheep because of (1) the fact that sheep are called upon to produce two products, lamb and wool; (2) the many diverse conditions under which they are produced; and (3) the conviction on the part of many producers that the hybrid vigor of crossbreeding accounts for increased vigor and livability in the lamb crop. Crossbreeding, therefore, is extensively followed in the commercial sheep production on the western range. The ewe bands are predominantly of Rambouillet extraction; whereas, for market lamb production, Suffolk or Hampshire rams are generally used. The Rambouillet ewe bands are desired because of their (1) gregarious or flocking instinct, (2) great hardiness, and (3) superior shearing qualities. On the other hand, lambs of this breeding are not so desirable for market lambs. Thus, mutton-type rams are used in order to get large, fast-growing lambs that will attain a good market finish on milk and range vegetation or that can be readily sold to go into feedlots for further finishing. As black-faced crossbred lambs of this type are not suitable as flock replacements, both ewe and wether lambs are marketed. Replacement females are obtained by (1) outright purchase from a producer who has used white-faced rams (Rambouillets, Columbias, Targhees, or Panamas) for purposes of raising animals for sale as replacements, (2) using white-faced rams on the band every third year and retaining the ewe lambs (some producers with several bands simply use certain bands for producing lambs for replacement purposes), or (3) using both white-faced and black-faced rams simultaneously on the same ewe band. In the latter type of program, the better white-faced ewe lambs—which are easily recognized as the offspring of the white-faced rams—are selected out for breeding purposes.

As can be readily surmised, crossbreeding in sheep does make for a considerable problem from the standpoint of producing or purchasing replacement animals. Also, it often makes the ram problem a difficult one. This practice, however, was born of necessity, there being few or no existing breeds or types possessing all the desirable features needed. In recent years, considerable effort has been made toward developing breeds of sheep better adapted to the needs, with the hope of alleviating the necessity of crossbreeding. The Columbia, Targhee, and Panama breeds evolved out of this need.

In addition to the crossbreeding common to the western range, most hothouse lambs are produced through using this system of breeding. Usually grade Merino or Dorset ewes are topped with a Southdown ram. Ewes of this extraction will breed out of season, and they are excellent milkers; and Southdown rams impart to their progeny the ultimate in early maturity and mutton type. Crossbreeding has also gained in popularity in Kentucky where crossbred Hampshire-Rambouillet ewes are frequently bred to Southdown rams for the production of grass-fat lambs.

Crossbreeding in sheep is still extensively followed in England and New Zealand. Here again, this practice is attributed to the diverse areas in which sheep are produced and the great differences with respect to the demand for the products of sheep.

• **Complementarity**—In a crossbreeding program, breeds that complement each other should be selected, thereby maximizing the desirable traits and minimizing the undesirable traits. Complementarity in a crossbreeding system is incorporating additive genetic differences between breeds, recognizing that rams and ewes do not contribute equally to the performance of their offspring. For this reason the breeds of sheep are classified as ewe breeds and ram breeds.

1. **Ewe breeds.** These are the white-faced breeds which produce fine, medium, or long wool, or crosses among these types. Ewe breeds are selected for adaptability to environmental conditions, reproductive efficiency, wool production, size, milking ability, and longevity. Replacement ewe lambs should be raised from these breed types, or from crossing among these breeds. Ewe breeds include: Booroola Merino, Corriedale, Debouillet, Delaine-Merino, Finnsheep, Polypay, Rambouillet, and Targhee. (See Table 28–2.)

2. **Ram breeds.** These are the meat-type breeds, or crosses of two of these breeds. Ram breeds are raised for production of rams to be crossed on ewe breeds and selected for growth rate and carcass qualities. Ram breeds include: Hampshire, Oxford, Shropshire, Southdown, and Suffolk. (See Table 28–2.)

• **Factors affecting magnitude of advantages from crossbreeding**—Many examples of each of the advantages of crossbreeding could be cited. However, the total magnitude of the advantage of these factors—achieving the 15 to 25% potential immediate increase in yield per female unit through continuous crossbreeding compared to continuous straight breeding—depends upon the following:

1. **Making wide crosses.** The wider the cross, the greater the heterosis.

2. **Selecting breeds that are complementary.** A crossbreeding program should involve breeds that possess the favorable expression of traits desired in the crossbred offspring that will be produced.

3. **Using high-performing stock.** Once a crossbreeding program is initiated, further genetic improvement is primarily dependent upon the use of superior production-tested males.

4. **Following a sound crossbreeding system.** For a continuous high expression of heterosis and maximum output per female, a sound system of crossbreeding must be followed. This should include the use of crossbred females, for research clearly indicates that over one-half the higher profits from a crossbreeding program results therefrom.

5. **Tapping purebreds constantly.** Purebreds must be constantly tapped to renew the vigor of crossbreds. Otherwise, the vigor dissipates.

• **Systems of crossbreeding**—Without a planned breeding program, minimum benefits of crossbreeding will be realized. There are numerous different systems of crossbreeding. Among them are the following:

1. **Two-breed cross.** This consists of mating purebred rams to purebred or high-grade ewes of another breed. The two-breed cross is simple, but it does not take advantage of heterosis for maternal traits, since the resulting crossbred ewes are not saved for breeding purposes.

2. **Backcross or crisscross.** This system involves the use of rams of Breed A on ewes of Breed B. Then, the resulting crossbred females are kept for breeding and are mated to nonrelated males of Breed A or B. The advantage of this system is that the ewes and the lambs are crossbreds and will possess hybrid vigor.

3. **Three-breed cross.** This system uses males from each of three breeds in succession on crossbred females. For example, ewes of Breed B are mated to rams of Breed A; selected crossbred ewes (AB) are mated to breed C rams and all off-spring are sold for slaughter. In a rotational three-breed cross, selected crossbred ewes of each generation are bred in rotation to each of the three breeds. Three-breed crosses also take advantage of heterosis in the ewes and the lambs.

To produce market lambs from a three-breed cross, ram breeds should be mated to the crossbred ewes as a terminal cross. Such ram breeds should possess growthiness, carcass quality, sexual aggressiveness, and fertility. An example of a crossbreeding program of this type would be:

a. **Foundation ewe breeds:** Rambouillet, Merino, Columbia, Corriedale, Targhee. Two of these breeds to be selected and crossed to produce the $F_1$ females.

b. **$F_1$ females:** The $F_1$ females resulting from the above cross to be bred to Suffolk or Hampshire rams as a terminal cross. All lambs to be marketed.

(Also, see Chapter 3, section on "Crossbreeding.")

## LETHALS AND OTHER HEREDITARY ABNORMALITIES IN SHEEP

Lethal characters in sheep are caused by the presence of hereditary factors in the germ plasm that produce an effect so serious as to cause the death of the individual either at birth or later in life. In recent years, a disturbing condition known as *spider lamb syndrome* has appeared in several breeds (see Fig. 31–12). Though strictly nonlethal, other hereditary abnormalities of practical importance in sheep include overshot and undershot jaws. *An overshot jaw is one in which the lower jaw*

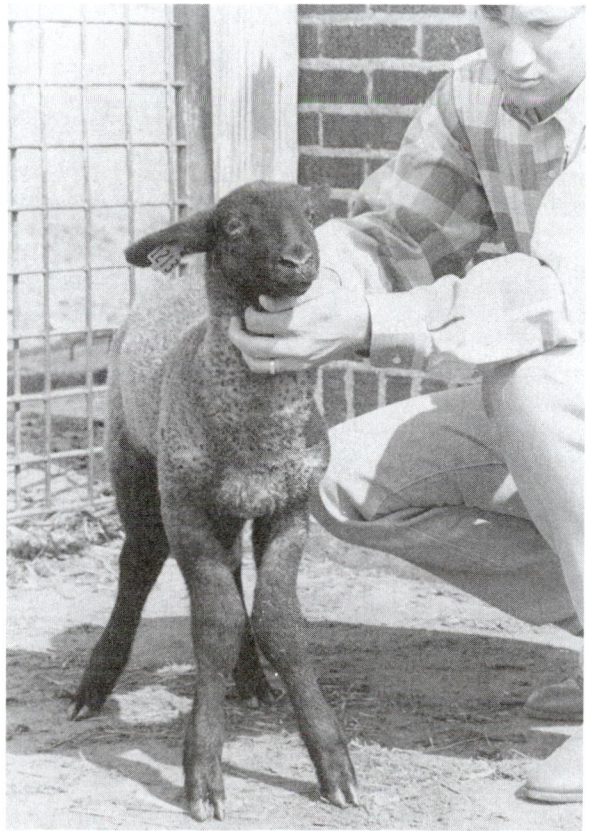

Fig. 31–12. Spider lamb syndrome. Note crooked front legs, along with extreme height and long legs. Apparently, the condition is caused by a single autosomal recessive gene. (Courtesy, Department of Animal Science, Oklahoma State University, Stillwater)

*is shorter than the upper jaw (see Fig. 31–13). An undershot jaw is one in which the lower jaw is longer than the upper jaw.* Either condition is of practical significance because affected animals cannot graze well. Genetically, these conditions are thought to be due to the interaction of several gene pairs, possibly recessive in nature. Affected animals should be culled from the breeding flock.

Other inherited defects, for which selection should be against, are rectal prolapse, entropion (inverted eyelids), and cryptorchidism.

Fig. 31–13. Sheep with overshot jaw, or parrot mouth, meaning that the lower jaw is too short. (Courtesy, USDA)

## PRODUCTION TESTING SHEEP

In Chapter 3, it was emphasized that (1) to make for breed improvement, either performance testing or progeny testing is imperative, and (2) to select intelligently on the basis of a testing program, the breeder must have adequate records and use them.

In comparison with that of chickens or even swine, production testing of sheep is slow, and like most investigational work with large animals, it is likely to be expensive. Even so, realizing that such testing is absolutely necessary if maximum improvement is to be made, the progressive purebred sheep breeder will wish to make a start.

Under practical conditions, progeny testing of sheep is usually confined to the ram, as a male produces during his lifetime many more offspring than a female. Because of the relatively small numbers of offspring, progeny testing with females (except with litter-bearing sows) is somewhat more difficult; and as a rule, a female cannot be considered adequately progeny tested until she has spent half or more of her lifetime in a flock. Certainly, intelligent breeders should study their females and cull those which consistently fail to produce good offspring. If possible, the females should also be progeny tested—especially those that may become the dams of stud rams.

A good plan in progeny testing rams consists of retaining and mating two or more ram lambs—the numbers depending upon the size of the flock—on a limited number of females during their first season of breeding. The progeny are then tested and evaluated, and only those rams that prove to be best on the basis of their progeny are retained for further breed purposes. If fertile ram lambs are each mated to 10 to 15 females,

lambs will be born 5 months later, and the progeny can be tested. Thus, with good fortune, it is possible to have progeny data on a ram when he is approximately 17 months of age. By comparison, with dairy bulls, whose daughters must be in lactation in order to make the test, it is not possible to make an evaluation until the animals are 6 to 7 years of age.

- **Using flock records in selection**—On-the-farm records are the first step toward successful production testing and selection. All too often a breeder remembers the good individuals produced by a given animal and forgets those which are mediocre or culls.

A prerequisite for any production data is that each animal be positively identified—by means of ear-notches, ear tags, or tattoos. For purebred breeders, who must use a system of animal identification anyway, this does not constitute an additional detail. But the recording of weights and grades does require additional time and labor—an expenditure which is highly worthwhile, however.

In order not to be burdensome, the record forms should be relatively simple. Furthermore, they should be in a form that will permit easy summarization; for example, the record of one ewe should be on one sheet if possible. Figs. 31–14a and 31–14b (see pages 631 and 632) show an individual ewe or ram record form.

Information on the productivity of *close relatives* (the sire and the dam and the brothers and the sisters) can supplement that on the animal itself and thus be a distinct aid in selection. This is especially important when traits, such as milk production, lambing rate, and semen production, cannot be measured in both sexes. The production records of more distant relatives are of little significance because, individually, due to the sampling nature of inheritance, they contribute only a few genes to an animal many generations removed.

Flock records have little value unless they are intelligently used in culling operations and in replacement decisions. Also, most breeders can and should use production records for estimating the rate of progress and for determining the relative emphasis to place on each character.

- **Determining relative emphasis to place on each character**—A replacement animal seldom excels in all of the economically important characters. The breeder must decide, therefore, how much importance shall be given to each factor. Thus, the producer will have to decide how much emphasis shall be placed on birth weight, weaning weight, body type, rate of gain, weight and quality of fleece and carcass characteristics.

Perhaps the relative emphasis to place on each character should vary according to the circumstances. Under certain conditions, some characters may even be ignored. Among the factors determining the emphasis to place on each character are the following:

1. **The economic importance of the character to the producer.** By economic importance is meant their dollar-and-cent value. Thus, those characters which have the greatest effect on profits should receive the most attention. In the west, wool currently represents about 20% of the gross income from the sheep enterprise.

2. **The heritability of the character.** It stands to reason that the more highly heritable characters should receive higher priority than those which are less heritable, for more progress can be made thereby.

## INDIVIDUAL EWE OR RAM RECORD

Breed _____ Reg. No. _____ Ear Nick _____ Tattoo _____ Birth Date _____

_____ Type of Birth (Single, Twin) _____ Date _____

Sire _____ Bred by _____ Temperament _____
(Gentle, nervous)

_____ Bought from _____

_____ Address _____ Face Covering[6] _____
(As a lamb)

Dam _____ Date Purchased _____ Face Covering[6] _____
(As a yearling)

_____ Type, Weaned[1] _____ Date _____

Type, Yearling[1] _____ Date _____ Disposed to _____ Date _____

Back[2] __ Rump[3] __ Leg[4] __ Why Disposed[7] _____

Defects & Abnormalities[5] _____

LAMBS (Use one line for each lamb for ewe's offspring; use one line for the average of a ram's progeny for each year.)

| Date of Birth | Ear Nick and No. | Vigor at Birth | Type of Birth[8] | Type of Rearing[9] | Sex | Birth Wt. | Defects and Abnormalities[5] | Sire | Milking Ability— Ewe[10] | Weaning Age, Days | 120-Day Weight | Weaning Condition[11] | Weaning Type[1] | Disposition[7] or Remarks |
|---|---|---|---|---|---|---|---|---|---|---|---|---|---|---|
| | | | | | | | | | | | | | | |
| | | | | | | | | | | | | | | |
| | | | | | | | | | | | | | | |
| | | | | | | | | | | | | | | |
| | | | | | | | | | | | | | | |
| | | | | | | | | | | | | | | |
| | | | | | | | | | | | | | | |
| | | | | | | | | | | | | | | |
| | | | | | | | | | | | | | | |
| | | | | | | | | | | | | | | |
| | | | | | | | | | | | | | | |
| | | | | | | | | | | | | | | |

[1]Trueness to breed appearance and desired meat conformation:
   "1," Excellent; "2," Good; "3," Medium; "4," Fair; "5," Poor.
[2]Straightness, strength, and spring of rib.
   "1," Excellent; "2," Good; "3," Medium; "4," Fair; "5," Poor.
[3]Width and levelness: 1-2-3-4-5 as above.
[4]Plumpness of thigh: 1-2-3-4-5 as above.
[5]Including overshot or undershot jaw, scurs, black fiber, etc.
[6]"1," Not covered beyond poll; "2," Covered to eyes; "3," Covered slightly below eyes, but open-faced; "4," Covered partially below eyes, but not subject to wool blindness; "5," Face covered and subject to wool blindness.

[7]Cause of death, reason for disposal, kept for breeding purposes, to whom sold.
[8]S—Single; T—Twin; Tr—Triplet.
[9]S—Single; T—Twin; Tr—Triplet; Gr—Grafted on foster mother and given her number.
[10]Good, medium, poor.
[11]Condition or degree of fatness:
   "1," Excellent; "2," Good; "3," Medium; "4," Fair; "5," Poor.

Fig. 31–14a.  Individual ewe or ram record form (see Fig. 31–14b for reverse side of record form).

3. **The genetic correlation between traits.** One trait may be so strongly correlated with another that selection for one automatically selects for the other. For example, rate of gain and economy of gain in meat animals are correlated to the extent that selection for rate of gain tends to select for the most economical gains as well; thus, economy of gain may be largely disregarded if rate of gain is given strong consideration. Conversely, one trait may be negatively correlated with another so that selection for one automatically selects against the other.

4. **The amount of variation in each character.** Obviously, if all animals were exactly alike in a given character, there could be no selection for that character. Likewise, if the amount of variation in a given character is small, the selected animals cannot be very much above the average of the entire flock, and progress will be slow.

5. **The level of performance already attained.** If a flock has reached a satisfactory level of performance for a certain character, there is not much need for further selection for that character. Thus, if sheep are almost free of wrinkles, there is little need to select against them; in fact, if sheep are entirely free of wrinkles, it would be impossible to select against them since there would be no wrinkled individuals to cull.

## WEIGHT RECORD OF EWE OR RAM

| Date | Age | Weight | Condition[1] | Remarks[2] |
|------|-----|--------|-----------|----------|
|  |  |  |  |  |
|  |  |  |  |  |
|  |  |  |  |  |
|  |  |  |  |  |
|  |  |  |  |  |
|  |  |  |  |  |
|  |  |  |  |  |
|  |  |  |  |  |
|  |  |  |  |  |
|  |  |  |  |  |
|  |  |  |  |  |
|  |  |  |  |  |
|  |  |  |  |  |
|  |  |  |  |  |
|  |  |  |  |  |
|  |  |  |  |  |

## REMARKS

(For example: bad udder, poor mother, aborted, veterinary treatment, and nature of ailment.)

Date:          Remarks:

### FLEECE
(Use one line for each year.)

| Length Side[3] | Fineness[4] | | | Date of Shearing | Days' Growth | Grease Weight | Percent of Yield | Clean Weight | Color of Skin | Purity[5] | Remarks About Fleece |
|----------------|-----------|------|-------|------------------|--------------|---------------|------------------|--------------|---------------|--------|----------------------|
|  | Shoulders | Side | Thigh |  |  |  |  |  |  |  |  |
|  |  |  |  |  |  |  |  |  |  |  |  |
|  |  |  |  |  |  |  |  |  |  |  |  |
|  |  |  |  |  |  |  |  |  |  |  |  |
|  |  |  |  |  |  |  |  |  |  |  |  |
|  |  |  |  |  |  |  |  |  |  |  |  |
|  |  |  |  |  |  |  |  |  |  |  |  |
|  |  |  |  |  |  |  |  |  |  |  |  |
|  |  |  |  |  |  |  |  |  |  |  |  |
|  |  |  |  |  |  |  |  |  |  |  |  |
|  |  |  |  |  |  |  |  |  |  |  |  |
|  |  |  |  |  |  |  |  |  |  |  |  |
|  |  |  |  |  |  |  |  |  |  |  |  |

[1]Condition or degree of fatness: "1," Excellent; "2," Good; "3," Medium; "4," Fair; "5," Poor.
[2]Factors affecting weight: *e.g.*, just shorn, soon lamb, etc.
[3]Length of staple, middle at side, to nearest 0.2 cm, just before shearing.
[4]Numerical grade as determined by USDA samples, just before shearing.
[5]Kemp, black fibers, etc.

Fig. 31–14b.  Reverse side of individual ewe or ram record form (for front side of record form, see Fig. 31–14a).

Sufficient selection pressure should be exerted to maintain the flock at the desired level of excellence, as low-producing animals retained in the flock lower the average production. Furthermore, under certain conditions relaxed selection may result in a regression toward the average of the breed.

- **Economically important traits in sheep and their heritability**—The following characteristics appear to be of greatest importance in evaluating the profitability of sheep:

1. **Multiple birth** (which has a heritability of approximately 15%).[3] Where adequate feeds are available, twin lambs are desirable because (a) they greatly increase the weight of lambs sold per ewe, and (b) the annual maintenance requirements of ewes is not far different, whether they are producing twins or singles.

2. **Birth weight of lambs** (which has an approximate heritability of 30%). The larger lambs at birth are generally more vigorous and make faster gains.

3. **Weaning weight** (which has an approximate heritability of 30% when over 100 days old at weaning). Heavy weaning weights are especially important in those areas where cost of production is largely on a per head rather than on a per pound basis, such as the western range.

4. **Rate of gain** (which has an approximate heritability of 30%). Daily rate of gain is important because (a) it is highly correlated with efficiency of gain, and (b) it makes for a shorter time in reaching market weight and condition, thus effecting a saving in labor, making for less exposure to risk and disease, and allowing for more rapid turnover in capital.

5. **Type** (which has an approximate heritability of 10% at weaning, and 40% at yearling stage). Type is important because it is a major factor in reflecting market values.

6. **Finish or condition at weaning** (which has an approximate heritability of 17%). Finish at weaning is largely determined by available feed and is not highly heritable. Yet it is most important because milk-fat lambs suitable for slaughter at weaning time almost always bring more per pound than thinner lambs that are sold as feeders. For the range area as a whole, about 25% of the lambs lack sufficient finish for slaughter at weaning time.

7. **Wrinkles or skin folds** (which are heritable by approximately the following percentages: neck folds at weaning age, 39%; and body folds at yearling age, 40%). Sheep with smooth bodies are preferred. Wrinkled sheep are difficult to shear and lack fiber uniformity.

8. **Face covering** (which has an approximate heritability of 56%). Wool-blind ewes do not graze well, require more labor if they are clipped around the eyes, and wean fewer pounds of lamb. At the Western Sheep Breeding Laboratory, ewes with open faces produced 11 lb more lamb per ewe bred than those with covered faces.

9. **Grease fleece weight** (which has an approximate heritability of 38%). Grease wool is valued on the basis of its clean weight. Of course, the heavier fleeces bring more money, because wool is sold by the pound.

10. **Clean fleece weight** (which has an approximate heritability of 40%). Clean fleece weight is most important, for the fiber is far more important than the materials scoured from grease wool.

11. **Staple length at weaning age** (which has an approximate heritability of 39%). Fiber length is important because it is a major factor in determining fleece weight and grade.

12. **Fleece grade** (which has an approximate heritability of 35%). The grade of fleece—which is based primarily on fiber diameter, but with consideration given to length, also—is important because it determines the use and price of wool.

13. **Fat thickness over loin eye** (which has an approximate heritability of 23%). Fat thickness is a measure of meatiness; excess fat results in an increase in fat trim and a decrease in percent lean cuts.

14. **Loin eye area** (which has an approximate heritability of 53%). Loin eye area is a good indicator of muscling.

The relative economic importance of lamb and wool production varies between breeds, areas, and years. But, for the nation as a whole, approximately ¾ of the income comes from lamb and ¼ from wool; hence, the emphasis is on market lamb production rather than on wool production.

- **Preweaning lamb performance test**—The following preweaning lamb performance test program is recommended for use in all purebred or commercial farm flocks:

**Step 1: Record individual data.** *Minimum data* should include (a) identification—lamb ear tag number; (b) sire number; (c) dam number; (d) age of dam, in years, at lambing time; (e) birth date of lamb; (f) sex of lamb; (g) type of birth—single, twin, triplet; and (h) how reared—single, twin, triplet, artificially.

*Optional data* may include (a) whether or not lamb was creep-fed; (b) slaughter grade; (c) 200-day yearling body weight; and (d) grease fleece weight and staple length to nearest ¹⁄₁₀ in.

**Step 2: Wean and weigh.** In advance, decide on weaning age—usually 90, 120, or 140 days. Wean and weigh as near to the intended age as possible.

**Step 3: Type score.** Even though the heritability of type, or conformation, is low at weaning (10%), it is a factor in determining today's market values; hence, all performance records should be augmented by type scores at weaning time. Also, and even more important because of the higher heritability (40%), all animals retained for breeding purposes should be type scored as yearlings.

Type can include any or all of the following: (a) characteristics that influence an animal's ability to live and perform in its environment—such as feet and legs, teeth, udder, lethals and sublethals; (b) traits that indicate meatiness; and/or (c) breed type.

Also, type score may include face cover score, wrinkle score, and record of the presence or absence of scurs or horns.

The determination of optimum type, the evaluation of it, and the use made of it should remain the responsibility of the individual breeder.

**Step 4: Adjust for certain environmental factors.** Adjust records for certain environmental factors such as age, sex, type of birth and rearing, and age of dam.

Weaning weights of lambs within a flock may be adjusted to 90, 120, or 140 days of age by finding the weight per day of age, then multiplying by the standardized age desired. Thus, the following formula may be used to provide the estimated 120-day weight of lambs:

$$\text{Adjusted 120-day weight (lb)} = \frac{\text{Actual weaning weight}}{\text{Actual days of age}} \times 120$$

---

[3]The rest is due to environment. The heritability figures given herein are averages based on large numbers; thus, some variation from these may be expected in individual flocks.

Additional adjustment factors are given in Table 31–2.

**TABLE 31–2**
**ADJUSTMENT FACTORS[1]**

| | Age of Dam | | |
|---|---|---|---|
| | **1-Year-Old** | **2-Years-Old or Over 6-Years-Old** | **3- to 6-Years-Old** |
| **Ewe Lamb** | | | |
| Single .................... | 1.22 | 1.09 | 1.00 |
| Twin—raised as twin ....... | 1.33 | 1.20 | 1.11 |
| Twin—raised as single ....... | 1.28 | 1.14 | 1.05 |
| Triplet—raised as triplet ..... | 1.46 | 1.33 | 1.22 |
| Triplet—raised as twin ...... | 1.42 | 1.28 | 1.17 |
| Triplet—raised as single ...... | 1.36 | 1.21 | 1.11 |
| **Wether** | | | |
| Single .................... | 1.19 | 1.06 | 0.97 |
| Twin—raised as twin ....... | 1.30 | 1.17 | 1.08 |
| Twin—raised as single ....... | 1.25 | 1.11 | 1.02 |
| Triplet—raised as triplet ..... | 1.43 | 1.30 | 1.19 |
| Triplet—raised as twin ...... | 1.39 | 1.25 | 1.14 |
| Triplet—raised as single ...... | 1.33 | 1.18 | 1.08 |
| **Ram Lamb** | | | |
| Single .................... | 1.11 | 0.98 | 0.89 |
| Twin—raised as twin ....... | 1.22 | 1.09 | 1.00 |
| Twin—raised as single ....... | 1.17 | 1.03 | 0.94 |
| Triplet—raised as triplet ..... | 1.35 | 1.22 | 1.11 |
| Triplet—raised as twin ...... | 1.31 | 1.17 | 1.06 |
| Triplet—raised as single ...... | 1.25 | 1.10 | 1.00 |

[1]Scott, G. E., edition, *The Sheepman's Production Handbook*, 2nd ed., Sheep Industry Development Program, Denver, CO, p. 27.

To use Table 31–2, multiply the 90-, 120-, or 140-day weight by the appropriate factor. For example, to find the adjusted 120-weight of a twin-born-and-reared lamb from a 2-year-old ewe that weighed 90 lb at 110 days of age, make the following calculations:

$$\frac{90 \text{ lb}}{110 \text{ days (of age)}} = 0.82 \text{ lb} \times 120 = 98 \text{ lb} \times \underset{\text{(adjustment factor)}}{1.09} = 107 \text{ lb}$$

The adjusted 120-day weight of the lamb would be 107 lb.

**Step 5: Cull.** Cull the lambs that fail to measure up in the preweaning performance test.

• **Postweaning lamb performance test**—The postweaning lamb performance test may be conducted either on the farm or at a central test station, as follows:

**Step 1: Conduct postweaning feed test.** Determine the growth rate by placing weaned lambs on a uniform feeding test for approximately 90 days. Make selection on the basis of growth rate during the feeding test.

**Step 2: Cull.** Cull the lambs that were poor gainers in the postweaning lamb performance test.

• **Ewe production test**—Reproductive traits in ewes should be measured as part of the on-the-farm performance record scheme. For this purpose, the record form shown in Figs. 31–14a and 31–14b will suffice for most flocks.

The following ewe performance test program is recommended for use in all purebred flocks:

**Step 1: Record individual data.** *Minimum data* should include (a) ewe number; (b) sire number; (c) dam number; (d) age of dam in years; (e) birth date of ewe; (f) type of birth of ewe—single, twin, triplet; (g) how reared—single, twin, triplet, artificially; (h) number of lambs born; (i) number of lambs weaned; and (j) adjusted weight of lambs weaned.

**Step 2: Type score yearlings.** At the yearling stage, type score and cull rigidly (for type score as a yearling is 40% heritable).

**Step 3: Evaluate fleece.** Record the shearing date, grease fleece weight, staple length to nearest $\frac{1}{10}$ in., and fleece grade.

**Step 4: Compute productivity for each ewe in the flock as—**

a. *Lamb productivity per ewe*, in total adjusted weight of lambs produced. It is intended that this should put strong emphasis on twinning.

b. *Ewe combined productivity score*, by combining both lamb and wool as follows:

Ewe combined productivity score = total pounds of lamb produced (using adjusted weights) + 3 times the 12-month wool total.

c. *Ewe productivity weight ratio or index*, to show the performance of an individual in relation to the average of all animals of the same group. Thus, the average productivity (total lamb and/or wool production) of the flock would be considered 100%. Individual ewes can then be rated based on their production above or below the flock average. Thus, a ewe producing 20% more lamb and/or wool than the average would have a productivity index of 120.

• **Progeny testing rams**—Progeny testing of rams can be very reliable if carefully planned and executed. For a valid progeny test, ewes must be assigned at random to the rams being tested, and sufficient ewes must be allotted to each ram to allow an accurate test. Progeny testing for rate of gain and carcass merit, for example, requires a minimum of 10 ewes per ram.

Progeny tests can be slow and expensive. Thus, if rams are tested as yearlings, they will be 2½ years old when the test is completed. Of course, testing ram lambs would speed up the process. Fortunately, research shows that selecting sires on the basis of their own growth rate (selecting them on the basis of the own performance tests) will result in more than half as much economic gain as selecting them on the basis of progeny test. Apparently, the added gain from progeny testing, in comparison with performance testing, is not sufficient to justify progeny testing of most rams that are to be used in natural service. If artificial insemination with frozen semen becomes practicable in sheep, progeny testing will become more important.

The following procedure is recommended for progeny testing rams:

**Step 1: Record individual data.** Record the following for each ram being progeny tested: (a) ram number and breed; (b) sire number; (c) dam number; (d) age of dam in years; (e) birth date of ram; (f) type of birth—single, twin, triplet; (g) how reared—single, twin, triplet, artificially; (h) lambs born from dam per ewe-year; and (i) adjusted 90-, 120-, or 140-day weight of lambs from dam per ewe-year (weight per day of age to 200 days of age, or to 12 to 16 months of age, is desirable; but, of course, this would not be available where ram lambs are being tested).

**Step 2: Conduct postweaning feed test.** Wean prospective stud rams at 60 to 90 days of age and place them on uniform feed test for 90 days. The heritability of growth rate is increased as maternal influence is decreased; hence, selection for growth should be based on the growth rate during the postweaning feeding test.

**Step 3: Select top ram lambs.** Select ram lambs that have excelled in both preweaning and postweaning performance.

If further testing—progeny testing—is not to be made, select future sires that have a high postweaning growth rate, that are well-muscled and have a minimum of fat, and that are out of ewes producing multiple births with maximum growth rate.

Where elite stud rams are desired, they should be progeny tested—proceed to Steps 4, 5, and 6.

**Step 4: Mate to randomly chosen ewes.** Mate each ram lamb to a minimum group of 10 randomly chosen ewes.

**Step 5: Test lambs.** Wean the lambs early (60 to 90 days), feed on uniform test for approximately 90 days, and slaughter.

**Step 6: Use best rams as yearlings.** The progeny test for gain and carcass merit can be computed in time to select and use the top progeny-tested rams in their yearling breeding season.

**Step 7: Evaluate fleece.** Record the shearing date, grease fleece weight, staple length to the nearest $\frac{1}{10}$ in., and fleece grade.

**Step 8: Type score.** At the yearling stage, type score and cull rigidly (for type score as a yearling is 40% heritable).

**Step 9: Cull.** Cull those rams that fail to measure up in the progeny test.

• **Production testing by state associations and colleges—** More states need to initiate production testing programs, and more breeders need to utilize them for flock improvement.

Also, production testing programs need to be standardized, or made uniform, from state to state; and where central stations are available and used, breeders need to be admonished that on the farm or ranch testing is the eventual goal. In view of this, some of the land-grant universities offer computerized on-the-farm performance testing. Two of the most used programs are maintained by the University of Wisconsin and The Ohio State University. For a small fee, these programs, using data provided by the sheep breeder, calculate ewe and lamb indexes based on multiple births, lamb growth, and ewe fleece weights and provide an annual printout.

Ohio has the only other on-the-farm testing program. It is patterned after the Wisconsin program—the original.

Most of the state programs are conducted in cooperation with state colleges of agriculture. For information relative to these programs, producers should contact their county agents or colleges of agriculture.

• **Production testing by sheep registry associations—**Some of the sheep registry associations have excellent production testing programs for their respective breeds and members. Literature pertaining to these may be obtained by writing directly to the breed registry association of the breed of special interest. Among such production testing programs are the following:

1. The Ram Certification program of the American Cheviot Sheep Society.
2. The Production Testing and Ram Certification programs of the American Hampshire Sheep Association.
3. The Certified Rams and Registry of Merit Rams of the American Rambouillet Sheep Breeders Association.
4. The Performance Registry and Ram Certification programs of the American Shropshire Registry Association, Inc.
5. The Performance Record of the Montadale Sheep Breeders Association, Inc.
6. The Production Registry and Ram Certification programs of the National Suffolk Sheep Association.

(Also, see Chapter 3, section headed "Selection Based on Production Testing.")

## QUESTIONS FOR STUDY AND DISCUSSION

1. What are the primary differences in the concerns and interests of commercial producers *vs* purebred breeders?

2. What is the normal age of puberty of (a) ram lambs, and (b) ewe lambs?

3. Under what conditions would you recommend that ewe lambs be bred? Justify your answer.

4. What are the normal (a) heat periods, and (b) gestation periods of sheep? What breed differences exist in the gestation period?

5. What can a sheep breeder do to increase twinning in a flock? What is *flushing*; and how is it accomplished?

6. What management techniques in preparing ewes and rams for mating are unique to sheep, and are not applied to cattle, hog, and horse breeding?

7. Discuss the care and management of the ram.

8. How do you account for the fact that artificial insemination in sheep has been little used in the United States, whereas it has been extensively used in the U.S.S.R., other countries in Eastern and Central Europe, and countries in South America?

9. Although the first embryo transfer of farm animals was performed with sheep, the commercial interest now is centered with cattle and horses. Why not sheep?

10. Discuss the techniques and economic importance of controlled reproduction in sheep by (a) estrous synchronization, (b) increased ovulation rate and lambing crop, (c) shortened lambing interval and breeding twice a year, and (d) controlling reproduction with light.

11. Discuss the age and service of the ram.

12. Why is pregnancy testing of sheep of great economic importance? Describe the rectal abdominal palpation technique of pregnancy testing.

13. Discuss the following needs of pregnant ewes: (a) feed and water, (b) exercise, and (c) shelter.

14. What is the average U.S. lamb crop percentage? How can the appalling mortality rate of 20 to 25% from birth to weaning be lessened?

15. What is the importance of each of the following as related to the proper care of the ewe at lambing time: (a) the lambing pen, (b) normal presentation, (c) taking the lamb, and (d) chilled and weak lambs?

16. What techniques are commonly tried to overcome disowned lambs?

17. How would you recommend that a beginner care for orphaned lambs?

18. What is the importance of each of the following as related to the care of the ewe after lambing: (a) feed and water, and (b) examination of the udder?

19. Discuss the effect of each of the following factors on reproduction of sheep: (a) daylight, (b) temperature, and (c) nutrition.

20. List the essential factors for success in the production of hothouse lambs.

21. Describe each of the following, and tell how they differ: (a) Easter lambs, (b) spring lambs, and (c) lambs raised on grass.

22. In a crossbreeding program, how may the matching of breeds that complement each other be applied through the use of (a) ewe breeds, and (b) ram breeds?

23. What factors affect the magnitude and advantages from crossbreeding sheep?

24. Describe each of the following systems of crossbreeding: (a) two-breed cross, (b) backcross or crisscross, and (c) three-breed cross.

25. Give an example of a three-breed cross designed to produce market lambs.

26. Define each of the following: (a) spider lamb syndrome, (b) overshot jaw, and (c) undershot jaw.

27. Why is it important that the breeder have adequate records and use them in order (a) to make for breed improvement, and (b) to select intelligently?

28. List and discuss the five factors which should determine the relative emphasis to place on each character in a sheep production testing program.

29. Discuss the economic importance and heritability of each of the following characteristics: (a) multiple birth, (b) rate of gain, (c) type, (d) face covering, (e) clean fleece weight, and (f) loin eye area.

30. Detail the recommended preweaning lamb performance test program.

31. Detail the recommended ewe production test.

32. Detail the recommended progeny testing of rams.

33. Why are not more sheep production tested?

## SELECTED REFERENCES

| Title of Publication | Author(s) | Publisher |
|---|---|---|
| *Control of the Ovarian Cycle in Sheep, The* | Edited by T. J. Robinson | Sydney University Press, Sydney, Australia, 1967 |
| *Profitable Sheep* | S. B. Collins R. F. Johnson | The Macmillan Company, New York, NY, 1956 |
| *Profitable Sheep Farming* | M. M. Cooper R. J. Thomas | Farming Press Ltd., Fenton House, Ipswich, England, 1971 |
| *Sheep and Wool* | M. P. Botkin B. A. Field C. L. Johnson | Prentice Hall, Inc., Englewood Cliffs, NJ, 1988 |
| *Sheep Book, The* | J. McKinney | John Wiley & Sons, Inc., New York, NY, 1959 |
| *Sheep Book, The* | R. Parker | Charles Scribner's Sons, New York, NY, 1983 |
| *Sheepman's Production Handbook, The* | Edited by G. E. Scott | Sheep Industry Development Program, Inc., Denver, CO, 1986 and updated at intervals |
| *Sheep Production and Management* | C. V. Ross | Prentice Hall, Inc., Englewood Cliffs, NJ, 1989 |
| *Sheep Science*, Revised Edition | W. G. Kammlade, Sr. W. G. Kammlade, Jr. | J. B. Lippincott Co., New York, NY, 1955 |

Feed my sheep!

# FEEDING AND MANAGING SHEEP

## Chapter 32

Sheep consume a higher proportion of forages than any other class of livestock, it being estimated that 94% of the total feed supply of the U.S. sheep production is derived from roughages. They are naturally adapted to grazing on pastures and ranges which supply a variety of forage plants, and they thrive best on forage that is short and fine rather than high and coarse. Although sheep will eat considerable quantities of weeds and brush, they prefer choice grasses and legumes.

Fig. 32–1. Sheep consume a higher proportion of forages than any other class of livestock. (Courtesy, U.S. Forest Service)

Except at lambing season, sheep seldom receive much grain. In the northern latitudes, farm-flock ewes are frequently given from ½ to 1 lb daily of a grain ration in addition to the roughage allowance from about 6 weeks before lambing to the time that they are turned to spring pasture. Higher levels of grain are fed during the suckling period than during gesta-

tion. Many of the farm flocks of the South and range bands of the Southwest, however, are kept in good thrifty condition, and the lambs are raised to the marketing stages, without the feeding of any grain. In still other areas, the ewes are fed only during periods of deep snows or extended droughts. The range bands in the colder regions of the West are normally fed alfalfa hay and grain during the period of about 3 to 4 weeks that they are confined to the lambing camp.

In general, for practical reasons, the ration of ewes should consist of as nearly year-round pastures as possible, with well-cured hay and other forages available the balance of the year, plus a limited grain allowance under certain conditions. Good quality sun-cured hay and lush pastures will not only provide most of the necessary proteins, but they are excellent sources of most of the minerals and vitamins, also.

## DIGESTIVE SYSTEM

The digestive systems of sheep and cows are very similar—both are ruminants. Of course, the digestive tract of sheep is much smaller than that of the cow and, too, the relative size of each of the four compartments of the sheep and the cow differ. (Also, see Chapter 4, Fig. 4–2; and Chapter 15, Fig. 15–2.)

## NUTRITIVE NEEDS OF SHEEP

The nutritive needs for sheep have been established by the National Academy of Sciences (and are presented in *Nutrient Requirements of Sheep,* Sixth Revised Edition, 1985). Tables 32–1, 32–2, 32–3, 32–4, and 32–5 present selected categories of the NRC nutritive requirements. Tables 32–6 and 32–7 present the NRC mineral requirements, and 32–8 presents the NRC vitamin E requirements of growing lambs. For additional categories—for animal weights not given in Tables 32–1, 32–2, 32–3, and 32–5, the reader is referred to the original NRC publication.

### TABLE 32–1
### DAILY NUTRIENT REQUIREMENTS OF SHEEP (PER ANIMAL)[1] (See footnotes at end of table.)

| Body Weight (lb) | (kg) | Weight Gain/Loss Per Day (lb) | (g) | Daily Consumption As-Fed[2] (lb) | (kg) | Moisture-Free Dry Matter[3] (lb) | (kg) | % Body Weight (%) | Energy[4] TDN (lb) | (kg) | DE (Mcal) | ME (Mcal) | Crude Protein (lb) | (g) | Ca (g) | P (g) | Vitamin A Activity (IU) | Vitamin E Activity (IU) |
|---|---|---|---|---|---|---|---|---|---|---|---|---|---|---|---|---|---|---|
| EWES[5] — Maintenance |||||||||||||||||||
| 154 | 70 | 0.02 | 10 | 2.9 | 1.3 | 2.6 | 1.2 | 1.7 | 1.5 | 0.66 | 2.9 | 2.4 | 0.25 | 113 | 2.5 | 2.4 | 3,290 | 18 |
| 198 | 90 | 0.02 | 10 | 3.4 | 1.6 | 3.1 | 1.4 | 1.5 | 1.7 | 0.78 | 3.4 | 2.8 | 0.29 | 131 | 2.9 | 3.1 | 4,230 | 21 |
| Flushing—2 Weeks prebreeding and first 3 weeks of breeding |||||||||||||||||||
| 154 | 70 | 0.22 | 100 | 4.4 | 2.0 | 4.0 | 1.8 | 2.6 | 2.3 | 1.06 | 4.7 | 3.8 | 0.36 | 164 | 5.7 | 3.2 | 3,290 | 27 |
| 198 | 90 | 0.22 | 100 | 4.9 | 2.2 | 4.4 | 2.0 | 2.2 | 2.6 | 1.18 | 5.1 | 4.2 | 0.39 | 177 | 6.1 | 3.9 | 4,230 | 30 |
| Nonlactating—First 15 weeks gestation |||||||||||||||||||
| 154 | 70 | 0.07 | 30 | 3.4 | 1.6 | 3.1 | 1.4 | 2.0 | 1.7 | 0.77 | 3.4 | 2.8 | 0.29 | 130 | 3.5 | 2.9 | 3,290 | 21 |
| 198 | 90 | 0.07 | 30 | 3.9 | 1.8 | 3.5 | 1.6 | 1.8 | 1.9 | 0.87 | 3.8 | 3.2 | 0.33 | 148 | 4.1 | 3.6 | 4,230 | 24 |
| Last 4 weeks gestation (130–150% lambing rate expected) or last 4–6 weeks lactation suckling singles[6] |||||||||||||||||||
| 154 | 70 | 0.40 (0.10) | 180 (45) | 4.4 | 2.0 | 4.0 | 1.8 | 2.6 | 2.3 | 1.06 | 4.7 | 3.8 | 0.42 | 193 | 6.2 | 5.6 | 5,950 | 27 |
| 198 | 90 | 0.40 (0.10) | 180 (45) | 4.9 | 2.2 | 4.4 | 2.0 | 2.2 | 2.5 | 1.18 | 5.1 | 4.2 | 0.47 | 212 | 6.4 | 6.5 | 7,650 | 30 |

*(Continued)*

### TABLE 32–1 (Continued)

| Body Weight (lb) | (kg) | Weight Gain/Loss Per Day (lb) | (g) | Daily Consumption As-Fed[2] (lb) | (kg) | Moisture-Free Dry Matter[3] (lb) | (kg) | % Body Weight (%) | Energy[4] TDN (lb) | (kg) | DE (Mcal) | ME (Mcal) | Crude Protein (lb) | (g) | Ca (g) | P (g) | Vitamin A Activity (IU) | Vitamin E Activity (IU) |
|---|---|---|---|---|---|---|---|---|---|---|---|---|---|---|---|---|---|---|
| EWES (Continued)[5] |||||||||||||||||||
| Last 4 weeks gestation (180–225% lambing rate expected) |||||||||||||||||||
| 154 | 70 | 0.50 | 225 | 4.7 | 2.1 | 4.2 | 1.9 | 2.7 | 2.8 | 1.24 | 5.4 | 4.4 | 0.47 | 214 | 7.6 | 4.5 | 5,950 | 28 |
| 198 | 90 | 0.50 | 225 | 5.1 | 2.3 | 4.6 | 2.1 | 2.3 | 3.0 | 1.37 | 6.0 | 5.0 | 0.51 | 232 | 8.9 | 5.7 | 7,650 | 32 |
| First 6–8 weeks lactation suckling singles or last 4–6 weeks lactation suckling twins[6] |||||||||||||||||||
| 154 | 70 | −0.06 (0.20) | −25 (90) | 6.1 | 2.8 | 5.5 | 2.5 | 3.6 | 3.6 | 1.63 | 7.2 | 5.9 | 0.73 | 334 | 9.3 | 7.0 | 5,950 | 38 |
| 198 | 90 | −0.06 (0.20) | −25 (90) | 6.6 | 3.0 | 5.9 | 2.7 | 3.0 | 3.8 | 1.75 | 7.6 | 6.3 | 0.78 | 353 | 9.6 | 7.8 | 7,650 | 40 |
| First 6–8 weeks lactation suckling twins |||||||||||||||||||
| 154 | 70 | −0.13 | −60 | 6.9 | 3.1 | 6.2 | 2.8 | 4.0 | 4.0 | 1.82 | 8.0 | 6.6 | 0.92 | 420 | 11.0 | 8.1 | 7,000 | 42 |
| 198 | 90 | −0.13 | −60 | 7.8 | 3.6 | 7.0 | 3.2 | 3.6 | 4.6 | 2.08 | 9.2 | 7.5 | 0.99 | 450 | 11.4 | 9.0 | 9,000 | 48 |
| EWE LAMBS |||||||||||||||||||
| Nonlactating—First 15 weeks gestation |||||||||||||||||||
| 110 | 50 | 0.30 | 135 | 3.7 | 1.7 | 3.3 | 1.5 | 3.0 | 1.9 | 0.88 | 3.9 | 3.2 | 0.35 | 159 | 5.2 | 3.1 | 2,350 | 22 |
| 154 | 70 | 0.28 | 125 | 4.1 | 1.9 | 3.7 | 1.7 | 2.4 | 2.2 | 1.00 | 4.4 | 3.6 | 0.36 | 164 | 5.5 | 3.7 | 3,290 | 26 |
| Last 4 weeks gestation (100–120% lambing rate expected) |||||||||||||||||||
| 110 | 50 | 0.35 | 160 | 3.9 | 1.8 | 3.5 | 1.6 | 3.2 | 2.2 | 1.00 | 4.4 | 3.6 | 0.42 | 189 | 6.3 | 3.4 | 4,250 | 24 |
| 154 | 70 | 0.33 | 150 | 4.4 | 2.0 | 4.0 | 1.8 | 2.6 | 2.5 | 1.14 | 5.0 | 4.1 | 0.43 | 194 | 6.8 | 4.2 | 5,950 | 27 |
| Last 4 weeks gestation (130–175% lambing rate expected) |||||||||||||||||||
| 110 | 50 | 0.50 | 225 | 3.9 | 1.8 | 3.5 | 1.6 | 3.2 | 2.3 | 1.06 | 4.7 | 3.8 | 0.45 | 204 | 7.8 | 3.9 | 4,250 | 24 |
| 154 | 70 | 0.47 | 215 | 4.4 | 2.0 | 4.0 | 1.8 | 2.6 | 2.5 | 1.14 | 5.0 | 4.1 | 0.46 | 210 | 8.2 | 4.7 | 5,950 | 27 |
| First 6–8 weeks suckling singles (wean by 8 weeks) |||||||||||||||||||
| 110 | 50 | −0.22 | −100 | 5.7 | 2.6 | 5.1 | 2.3 | 4.6 | 3.5 | 1.59 | 7.0 | 5.7 | 0.71 | 321 | 8.7 | 6.0 | 5,000 | 34 |
| 154 | 70 | −0.22 | −100 | 6.7 | 3.0 | 6.0 | 2.7 | 3.9 | 4.1 | 1.85 | 8.1 | 6.6 | 0.77 | 351 | 9.3 | 6.9 | 7,000 | 40 |
| REPLACEMENT EWE LAMBS[7] |||||||||||||||||||
| 88 | 40 | 0.40 | 182 | 3.4 | 1.6 | 3.1 | 1.4 | 3.5 | 2.0 | 0.91 | 4.0 | 3.3 | 0.39 | 176 | 5.9 | 2.6 | 1,880 | 21 |
| 154 | 70 | 0.22 | 100 | 3.7 | 1.7 | 3.3 | 1.5 | 2.1 | 1.9 | 0.88 | 3.9 | 3.2 | 0.29 | 132 | 4.6 | 2.8 | 3,290 | 22 |
| REPLACEMENT RAM LAMBS[7] |||||||||||||||||||
| 132 | 60 | 0.70 | 320 | 5.9 | 2.7 | 5.3 | 2.4 | 4.0 | 3.4 | 1.5 | 6.7 | 5.5 | 0.58 | 263 | 8.4 | 4.2 | 2,820 | 26 |
| 176 | 80 | 0.64 | 290 | 6.9 | 3.1 | 6.2 | 2.8 | 3.5 | 3.9 | 1.8 | 7.8 | 6.4 | 0.59 | 268 | 8.5 | 4.6 | 3,760 | 28 |
| LAMBS FINISHING—4 TO 7 MONTHS OLD[8] |||||||||||||||||||
| 66 | 30 | 0.65 | 295 | 3.2 | 1.4 | 2.9 | 1.3 | 4.3 | 2.1 | 0.94 | 4.1 | 3.4 | 0.42 | 191 | 6.6 | 3.2 | 1,410 | 20 |
| 110 | 50 | 0.45 | 205 | 3.9 | 1.7 | 3.5 | 1.6 | 3.2 | 2.7 | 1.23 | 5.4 | 4.4 | 0.35 | 160 | 5.6 | 3.0 | 2,350 | 24 |
| EARLY WEANED LAMBS—MODERATE GROWTH POTENTIAL[8] |||||||||||||||||||
| 22 | 10 | 0.44 | 200 | 1.2 | 0.6 | 1.1 | 0.5 | 5.0 | 0.9 | 0.40 | 1.8 | 1.4 | 0.38 | 127 | 4.0 | 1.9 | 470 | 10 |
| 66 | 30 | 0.66 | 300 | 3.2 | 1.4 | 2.9 | 1.3 | 4.3 | 2.2 | 1.00 | 4.4 | 3.6 | 0.42 | 191 | 6.7 | 3.2 | 1,410 | 20 |
| 110 | 50 | 0.66 | 300 | 3.7 | 1.7 | 3.3 | 1.5 | 3.0 | 2.6 | 1.16 | 5.1 | 4.2 | 0.40 | 181 | 7.0 | 3.8 | 2,350 | 22 |
| EARLY WEANED LAMBS—RAPID GROWTH POTENTIAL[8] |||||||||||||||||||
| 22 | 10 | 0.55 | 250 | 1.4 | 0.7 | 1.3 | 0.6 | 6.0 | 1.1 | 0.48 | 2.1 | 1.7 | 0.35 | 157 | 4.9 | 2.2 | 470 | 12 |
| 66 | 30 | 0.72 | 325 | 3.4 | 1.6 | 3.1 | 1.4 | 4.7 | 2.4 | 1.10 | 4.8 | 4.0 | 0.48 | 216 | 7.2 | 3.4 | 1,410 | 21 |
| 110 | 50 | 0.94 | 425 | 4.1 | 1.9 | 3.7 | 1.7 | 3.4 | 2.8 | 1.29 | 5.7 | 4.7 | 0.53 | 240 | 9.4 | 4.8 | 2,350 | 25 |

[1] Adapted by the author from *Nutrient Requirements of Sheep*, Sixth Revised Edition, NRC-National Academy of Sciences, 1985, pp. 45–47.

[2] As-fed was calculated using an average figure of 90% dry matter. When using silages, roots, and other wet feeds, these feeds should be converted to a moisture-free basis and the ration calculated using the moisture-free data.

[3] To convert dry matter to an as-fed basis, divide dry matter values by the percentage of dry matter in the particular feed.

[4] One kilogram TDN (total digestible nutrients) = 4.4 Mcal DE (digestible energy); ME (metabolizable energy) = 82% of DE. Because of rounding numbers, values in Table 1 and Table 2 may differ.

[5] Values are applicable for ewes in moderate condition. Fat ewes should be fed according to the next lower category and thin ewes at the next higher weight category.

[6] Values in parentheses are for ewes suckling lambs the last 4–6 weeks of lactation.

[7] Lambs intended for breeding; thus, maximum weight gains and finish are of secondary importance.

[8] Maximum weight gains expected.

## TABLE 32-2
## NUTRIENT CONCENTRATION IN RATIONS FOR SHEEP [1] [2](See footnotes at end of table.)

| Body Weight (lb) | (kg) | Weight Gain/Loss Per Day (lb) | (g) | Moisture Basis[3] A-F (as-fed) M-F (moisture-free) | Energy TDN[5] (%) | DE (Mcal/) (lb) | (kg) | ME (Mcal/) (lb) | (kg) | Example Diet Proportions Concentrate (%) | Forage (%) | Crude Protein (%) | Calcium (%) | Phosphorus (%) | Vitamin A Activity (IU/) (lb) | (kg) | Vitamin E Activity (IU/) (lb) | (kg) |
|---|---|---|---|---|---|---|---|---|---|---|---|---|---|---|---|---|---|---|
| EWES[6] |||||||||||||||||||
| Maintenance |||||||||||||||||||
| 154 | 70 | 0.02 | 10 | A-F | 50 | 4.8 | 2.2 | 4.0 | 1.8 | 0 | 100 | 8.5 | 0.18 | 0.18 | 5,442 | 2,468 | 31 | 14 |
|  |  |  |  | M-F | 55 | 5.3 | 2.4 | 4.4 | 2.0 | 0 | 100 | 9.4 | 0.20 | 0.20 | 6,046 | 2,742 | 33 | 15 |
| Flushing—2 weeks prebreeding and first 3 weeks of breeding |||||||||||||||||||
| 154 | 70 | 0.22 | 100 | A-F | 53 | 5.1 | 2.3 | 4.1 | 1.9 | 15 | 85 | 8.2 | 0.29 | 0.16 | 3,627 | 1,645 | 31 | 14 |
|  |  |  |  | M-F | 59 | 5.7 | 2.6 | 4.6 | 2.1 | 15 | 85 | 9.1 | 0.32 | 0.18 | 4,031 | 1,828 | 33 | 15 |
| Nonlactating—First 15 weeks gestation |||||||||||||||||||
| 154 | 70 | 0.07 | 30 | A-F | 50 | 4.8 | 2.2 | 4.0 | 1.8 | 0 | 100 | 8.4 | 0.23 | 0.18 | 4,664 | 2,115 | 31 | 14 |
|  |  |  |  | M-F | 55 | 5.3 | 2.4 | 4.4 | 2.0 | 0 | 100 | 9.3 | 0.25 | 0.20 | 5,182 | 2,350 | 33 | 15 |
| Last 4 weeks gestation (130–150% lambing rate expected) or last 4–6 weeks lactation suckling singles[7] |||||||||||||||||||
| 154 | 70 | 0.40 | 180 | A-F | 53 | 5.1 | 2.3 | 4.1 | 1.9 | 15 | 85 | 9.6 | 0.32 | 0.21 | 6,561 | 2,975 | 31 | 14 |
|  |  | (0.10) | (0.45) | M-F | 59 | 5.7 | 2.6 | 4.6 | 2.1 | 15 | 85 | 10.7 | 0.35 | 0.23 | 7,290 | 3,306 | 33 | 15 |
| Last 4 weeks gestation (180–225% lambing rate expected) |||||||||||||||||||
| 154 | 70 | 0.50 | 225 | A-F | 59 | 5.8 | 2.6 | 4.6 | 2.1 | 35 | 65 | 10.2 | 0.36 | 0.22 | 6,215 | 2,819 | 31 | 14 |
|  |  |  |  | M-F | 65 | 6.4 | 2.9 | 5.1 | 2.3 | 35 | 65 | 11.3 | 0.40 | 0.24 | 6,906 | 3,132 | 33 | 15 |
| First 6–8 weeks lactation suckling singles or last 4–6 weeks lactation suckling twins[7] |||||||||||||||||||
| 154 | 70 | -0.06 | -25 | A-F | 59 | 5.8 | 2.6 | 4.8 | 2.2 | 35 | 65 | 12.1 | 0.29 | 0.23 | 4,723 | 2,142 | 31 | 14 |
|  |  | (0.20) | (90) | M-F | 65 | 6.4 | 2.9 | 5.3 | 2.4 | 35 | 65 | 13.4 | 0.32 | 0.26 | 5,248 | 2,380 | 33 | 15 |
| First 6–8 weeks lactation suckling twins |||||||||||||||||||
| 154 | 70 | -0.13 | -60 | A-F | 59 | 5.8 | 2.6 | 4.8 | 2.2 | 35 | 65 | 13.5 | 0.35 | 0.26 | 4,962 | 2,250 | 31 | 14 |
|  |  |  |  | M-F | 65 | 6.4 | 2.9 | 5.3 | 2.4 | 35 | 65 | 15.0 | 0.39 | 0.29 | 5,513 | 2,500 | 33 | 15 |
| EWE LAMBS |||||||||||||||||||
| Nonlactating—First 15 weeks gestation |||||||||||||||||||
| 121 | 55 | 0.30 | 135 | A-F | 53 | 5.1 | 2.3 | 4.1 | 1.9 | 15 | 85 | 9.5 | 0.32 | 0.20 | 3,310 | 1,501 | 31 | 14 |
|  |  |  |  | M-F | 59 | 5.7 | 2.6 | 4.6 | 2.1 | 15 | 85 | 10.6 | 0.35 | 0.22 | 3,678 | 1,668 | 33 | 15 |
| Last 4 weeks gestation (100–120% lambing rate expected) |||||||||||||||||||
| 121 | 55 | 0.35 | 160 | A-F | 57 | 5.6 | 2.5 | 4.6 | 2.1 | 30 | 70 | 10.6 | 0.35 | 0.20 | 2,550 | 5,622 | 31 | 14 |
|  |  |  |  | M-F | 63 | 6.2 | 2.8 | 5.1 | 2.3 | 30 | 70 | 11.8 | 0.39 | 0.22 | 6,247 | 2,833 | 33 | 15 |
| Last 4 weeks gestation (130–175% lambing rate expected) |||||||||||||||||||
| 121 | 55 | 0.50 | 225 | A-F | 59 | 5.8 | 2.6 | 4.8 | 2.2 | 40 | 60 | 11.5 | 0.43 | 0.23 | 5,622 | 2,550 | 31 | 14 |
|  |  |  |  | M-F | 66 | 6.4 | 2.9 | 5.3 | 2.4 | 40 | 60 | 12.8 | 0.48 | 0.25 | 6,247 | 2,833 | 33 | 15 |
| First 6–8 weeks lactation suckling singles (wean by 8 weeks) |||||||||||||||||||
| 121 | 55 | 0.22 | -50 | A-F | 59 | 5.8 | 2.6 | 4.8 | 2.2 | 40 | 60 | 11.8 | 0.27 | 0.20 | 4,217 | 1,913 | 31 | 14 |
|  |  |  |  | M-F | 66 | 6.4 | 2.9 | 5.3 | 2.4 | 40 | 60 | 13.1 | 0.30 | 0.22 | 4,686 | 2,125 | 33 | 15 |
| First 6–8 weeks lactation suckling twins (wean by 8 weeks) |||||||||||||||||||
| 121 | 55 | -0.22 | -100 | A-F | 62 | 5.9 | 2.7 | 5.0 | 2.3 | 50 | 50 | 12.3 | 0.33 | 0.23 | 4,549 | 2,063 | 31 | 14 |
|  |  |  |  | M-F | 69 | 6.6 | 3.0 | 5.5 | 2.5 | 50 | 50 | 13.7 | 0.37 | 0.26 | 5,054 | 2,292 | 33 | 15 |
| REPLACEMENT EWE LAMBS[8] |||||||||||||||||||
| 110–154 lb 50–70 kg |  | 0.25 | 115 | A-F | 53 | 5.1 | 2.3 | 4.1 | 1.9 | 15 | 85 | 8.2 | 0.28 | 0.15 | 3,110 | 1,410 | 31 | 14 |
|  |  |  |  | M-F | 59 | 5.7 | 2.6 | 4.6 | 2.1 | 15 | 85 | 9.1 | 0.31 | 0.17 | 3,455 | 1,567 | 33 | 15 |
| REPLACEMENT RAM LAMBS[8] |||||||||||||||||||
| 132 | 60 | 0.70 | 320 | A-F | 57 | 5.6 | 2.5 | 4.6 | 2.1 | 30 | 70 | 9.9 | 0.32 | 0.16 | 3,292 | 1,493 | 31 | 14 |
|  |  |  |  | M-F | 63 | 6.2 | 2.8 | 5.1 | 2.3 | 30 | 70 | 11.0 | 0.35 | 0.18 | 3,658 | 1,659 | 33 | 15 |
| LAMBS FINISHING—4 TO 7 MONTHS OLD[9] |||||||||||||||||||
| 88 | 40 | 0.60 | 275 | A-F | 68 | 6.6 | 3.0 | 5.4 | 2.4 | 75 | 25 | 10.4 | 0.38 | 0.19 | 2,332 | 1,058 | 31 | 14 |
|  |  |  |  | M-F | 76 | 7.3 | 3.3 | 6.0 | 2.7 | 75 | 25 | 11.6 | 0.42 | 0.21 | 2,591 | 1,175 | 33 | 15 |

*(Continued)*

**TABLE 32-2** *(Continued)*

| Body Weight | | Weight Gain/Loss Per Day | | Moisture Basis[3] A-F (as-fed) M-F (moisture-free) | Energy[4] | | | | | | Example Diet Proportions | | Crude Protein | Cal- cium | Phos- phorus | Vitamin A Activity (IU/) | | Vitamin E Activity (IU/) | |
|---|---|---|---|---|---|---|---|---|---|---|---|---|---|---|---|---|---|---|---|
| | | | | | TDN[5] | DE (Mcal/) | | ME (Mcal/) | | | Con- centrate | Forage | | | | | | | |
| (lb) | (kg) | (lb) | (g) | | (%) | (lb) | (kg) | (lb) | (kg) | | (%) | (%) | (%) | (%) | (%) | (lb) | (kg) | (lb) | (kg) |
| **EARLY WEANED LAMBS—MODERATE AND RAPID GROWTH POTENTIAL[8]** | | | | | | | | | | | | | | | | | | | |
| 22 | 10 | 0.55 | 250 | **A-F** | 72 | 6.9 | 3.2 | 5.8 | 2.6 | | 90 | 10 | 23.4 | 0.74 | 0.34 | 1,866 | 846 | 40 | 18 |
| | | | | M-F | 80 | 7.7 | 3.5 | 6.4 | 2.9 | | 90 | 10 | 26.2 | 0.82 | 0.38 | 2,073 | 940 | 44 | 20 |
| 66 | 30 | 0.72 | 325 | **A-F** | 70 | 6.6 | 3.0 | 5.4 | 2.4 | | 85 | 15 | 13.6 | 0.46 | 0.22 | 2,153 | 977 | 31 | 14 |
| | | | | M-F | 78 | 7.3 | 3.3 | 6.0 | 2.7 | | 85 | 15 | 15.1 | 0.51 | 0.24 | 2,392 | 1,085 | 33 | 15 |
| 88–132 lb | | 0.88 | 400 | **A-F** | 70 | 6.6 | 3.0 | 5.4 | 2.4 | | 85 | 15 | 13.1 | 0.50 | 0.25 | 2,487 | 1,128 | 31 | 14 |
| 40–60 kg | | | | M-F | 78 | 7.3 | 3.3 | 6.0 | 2.7 | | 85 | 15 | 14.5 | 0.55 | 0.28 | 2,763 | 1,253 | 33 | 15 |

[1]Adapted by the author from *Nutrient Requirements of Sheep*, Sixth Revised Edition, NRC-National Academy of Sciences, 1985, p. 48.

[2]Values in Table 2 are calculated from daily requirements in Table 1 divided by DM intake. The exception, vitamin E daily requirements/head, are calculated from vitamin E/kg diet × DM intake.

[3]As-fed was calculated using an average figure of 90% dry matter. When using silages, roots, and other wet feeds, these feeds should be converted to a moisture-free basis and the ration calculated using the moisture-free data.

[4]One kilogram TDN = 4.4 Mcal DE (digestible energy); ME (metabolizable energy) = 82% of DE. Because of rounding numbers, values in Table 1 and Table 2 may differ.

[5]TDN calculated on following basis: hay DM, 55% TDN and on as-fed basis 50% TDN; grain DM, 83% TDN and on as-fed basis 75% TDN.

[6]Values are for ewes in moderate condition. Fat ewes should be fed according to the next lower weight category and thin ewes at the next higher weight category. Once desired or moderate weight condition is attained, use that weight category through all production stages.

[7]Values in parentheses are for ewes suckling lambs the last 4–6 weeks of lactation.

[8]Lambs intended for breeding; thus, maximum weight gains and finish are of secondary importance.

[9]Maximum weight gains expected.

## ENERGY

Lack of energy—hunger—is probably the most common nutritional deficiency of sheep. It may result from lack of feed or from the consumption of poor quality feed.

**Fig. 32-2.** Energy is needed for the production of meat, milk, and wool. This shows a band of western ewes "nooning down" (resting at noon) on the trail at Sun Valley, Idaho. (Courtesy, Union Pacific R.R., Omaha, NE)

Inadequate amounts of feed may result from overgrazing, droughts, snow covering the feed, or from a low dry matter content of lush, washy feeds. Also, poorly digested low-quality forage leads to reduced feed intake.

The energy needs of sheep are largely met through the consumption and digestion of roughages—pasture, hay, and silage. Grains, such as corn, barley, milo, wheat, and oats, are used to raise the energy level of the ration during periods when supplementation is necessary. In general, sheep subsist on an even higher proportion of roughages to concentrates than do beef cattle, and this applies to finishing lambs. The bacterial action in the paunch of the sheep efficiently converts roughages into suitable sources of energy.

It is generally recognized that the energy requirements of sheep are affected by size, age, pregnancy, lactation, growth, and protein content of the ration. It is also affected by environment, shearing, and sex. The net energy requirement for lambs of small, medium, and large mature weight genotypes are presented in Table 32-3 (see page 642); and the net energy requirements of ewes carrying different numbers of fetuses at various stages of gestation are given in Table 32-4 (see page 642).

*Symptoms of energy deficiency:* An energy deficiency is characterized by slowing and cessation of growth, loss of weight, reduced fertility or reproductive failure, lowered milk production and shortened lactation period, reduced quantity and quality of wool (including breaks in the fiber), lowered resistance to infection with internal parasites, and increased mortality.

**TABLE 32–3**
**NET ENERGY REQUIREMENTS FOR LAMBS OF SMALL, MEDIUM, AND LARGE MATURE WEIGHT GENOTYPES[1][2]**

| Body Weight[3] | Lb | 22 | 44 | 55 | 66 | 77 | 88 | 99 | 110 |
|---|---|---|---|---|---|---|---|---|---|
| | *Kg* | *10* | *20* | *25* | *30* | *35* | *40* | *45* | *50* |
| Daily Gain[3] | | kcal/d | kcal/d | kcal/d | kcal/d | kcal/d | kcal/d | kcal/d | kcal/d |
| (lb) | (*g*) | | | | | | | | |
| | | | | | ***NE_m* REQUIREMENTS**[4] | | | | |
| | | 315 | 530 | 626 | 718 | 806 | 891 | 973 | 1,053 |
| | | | | | ***NE_g* REQUIREMENTS** | | | | |
| | | | | | **Small mature weight lambs**[5] | | | | |
| .22 | *100* | 178 | 300 | 354 | 406 | 456 | 504 | 551 | 596 |
| .44 | *200* | 357 | 600 | 708 | 812 | 912 | 1,008 | 1,102 | 1,192 |
| .66 | *300* | 535 | 900 | 1,064 | 1,219 | 1,368 | 1,513 | 1,652 | 1,788 |
| | | | | | **Medium mature weight lambs**[6] | | | | |
| .22 | *100* | 155 | 261 | 309 | 354 | 397 | 439 | 480 | 519 |
| .55 | *250* | 388 | 653 | 771 | 884 | 993 | 1,097 | 1,199 | 1,297 |
| .88 | *400* | 621 | 1,044 | 1,234 | 1,415 | 1,589 | 1,756 | 1,918 | 2,076 |
| | | | | | **Large mature weight lambs**[7] | | | | |
| .22 | *100* | 132 | 221 | 262 | 300 | 337 | 372 | 407 | 439 |
| .55 | *250* | 329 | 553 | 654 | 750 | 842 | 930 | 1,016 | 1,099 |
| .88 | *400* | 526 | 885 | 1,046 | 1,200 | 1,347 | 1,489 | 1,626 | 1,760 |

[1]Adapted by the author from *Nutrient Requirements of Sheep*, Sixth Revised Edition, NRC-National Academy of Sciences, 1985, p. 49.

[2]Approximate mature ram weights of 209 lb (*95 kg*), 254 lb (*115 kg*), and 298 lb (*135 kg*), respectively.

[3]Weights and gains include fill.

[4]$NE_m = 56$ kcal • $W^{0.75}$ • $d^{-1}$.

[5]$NE_g = 317$ kcal • $W^{0.75}$ • LWG, kg • $d^{-1}$.

[6]$NE_g = 276$ kcal • $W^{0.75}$ • LWG, kg • $d^{-1}$.

[7]$NE_g = 234$ kcal • $W^{0.75}$ • LWG, kg • $d^{-1}$.

**TABLE 32–4**
**$NE_{PREG}$ ($NE_Y$) REQUIREMENTS OF EWES CARRYING DIFFERENT NUMBERS OF FETUSES AT VARIOUS STAGES OF GESTATION[1]**

| Number of Fetuses Being Carried | Stage of Gestation (days)[2] | | | | | |
|---|---|---|---|---|---|---|
| | 100 | %[3] | 120 | %[3] | 140 | %[3] |
| | $NE_{preg}$ Required (kcal/day) | | | | | |
| 1 .......... | 70 | 100 | 145 | 100 | 260 | 100 |
| 2 .......... | 125 | 178 | 265 | 183 | 440 | 169 |
| 3 .......... | 170 | 243 | 345 | 238 | 570 | 219 |

[1]Adapted by the author from *Nutrient Requirements of Sheep*, Sixth Revised Edition, NRC-National Academy of Sciences, 1985, p. 49. The ($NE_Y$) refers to reproductive process.

[2]For *gravid uterus* (plus contents) and mammary gland development only.

[3]As a percentage of a single fetus's requirement.

## PROTEIN

Sheep need protein, as do other classes of animals, for maintenance, growth, reproduction, and finishing. Additionally, sheep need protein for the production of wool—a protein product. Wool is especially rich in the sulfur-containing amino acid, cystine, but this requirement is usually amply met by the cystine of feeds or by methionine, another amino acid which is also rather widely distributed in natural sources and which is derived from rumen synthesis.

Green pastures and legume hays (alfalfa, clover, soybeans, lespedeza, etc.) are excellent and practical sources of proteins for sheep in most areas. Where the ranges are bleached and dry for an extended period, or legume hays cannot be produced for winter feeding, however, it may be desirable to provide sheep with such protein-rich supplements as soybean meal, cottonseed meal, linseed meal, canola meal, peanut meal, sunflower meal, or a commercial protein supplement, at the rate of about ¼ to ⅓ lb per ewe per day.

The protein requirements of sheep are affected by growth, pregnancy, lactation, mature size, weight for age, body condition, rate of gain, and protein-energy ratio. Though correspondingly less because of their smaller body size and lower milk production, the protein requirements of ewes nursing lambs are much like those of lactating cows. The crude protein requirements for lambs of small, medium, and large mature weight genotypes are presented in Table 32–5.

*Symptoms of protein deficiency:* A protein deficiency is characterized by reduced appetite, lowered feed intake, and poor feed efficiency. In turn, this makes for poor growth, poor muscular development, loss of weight, reduced reproductive efficiency, and reduced wool production. Under extreme conditions, there are severe digestive disturbances, nutritional anemia, and edema.

## TABLE 32–5
### CRUDE PROTEIN REQUIREMENTS FOR LAMBS OF SMALL, MEDIUM, AND LARGE MATURE WEIGHT GENOTYPES[1][2]

| Body Weight[3] | Lb | 22 | 44 | 55 | 66 | 77 | 88 | 99 | 110 |
|---|---|---|---|---|---|---|---|---|---|
| | Kg | 10 | 20 | 25 | 30 | 35 | 40 | 45 | 50 |
| Daily Gain[3] | | g/d | g/d | g/d | g/d | g/d | g/d | g/d | g/d |
| (lb) | (g) | | | | | | | | |
| Small mature weight lambs | | | | | | | | | |
| .22 | 100 | 84 | 112 | 122 | 127 | 131 | 136 | 135 | 134 |
| .44 | 200 | 123 | 145 | 152 | 154 | 156 | 158 | 154 | 151 |
| .66 | 300 | 162 | 178 | 182 | 181 | 180 | 180 | 174 | 168 |
| Medium mature weight lambs | | | | | | | | | |
| .22 | 100 | 85 | 114 | 125 | 130 | 135 | 140 | 139 | 139 |
| .55 | 250 | 147 | 167 | 174 | 175 | 177 | 179 | 175 | 171 |
| .88 | 400 | 209 | 221 | 224 | 221 | 219 | 217 | 210 | 202 |
| Large mature weight lambs | | | | | | | | | |
| .22 | 100 | 94 | 128 | 134 | 139 | 145 | 144 | 150 | 156 |
| .55 | 250 | 157 | 186 | 188 | 190 | 192 | 189 | 192 | 195 |
| .88 | 400 | 221 | 243 | 242 | 241 | 240 | 234 | 234 | 234 |

[1]Adapted by the author from *Nutrient Requirements of Sheep*, Sixth Revised Edition, NRC-National Academy of Sciences, 1985, p. 50.

[2]Approximate mature ram weights of 209 lb (*95 kg*), 254 lb (*115 kg*), and 298 lb (*135 kg*), respectively.

[3]Weights and gains include fill.

## MINERALS

Although the body contains many mineral elements, only 16 have been demonstrated to be essential for sheep—7 major mineral constituents, and 9 trace elements. These minerals, along with the requirements of each, are presented in Tables 32–6 and 32–7. Where known, the toxic levels of the microminerals are given also.

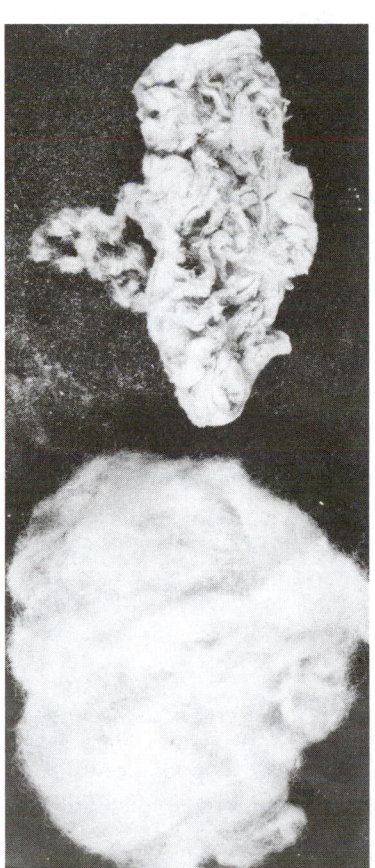

Fig. 32-2a. Limestone made the difference! Two samples of scoured wool from lambs fed different lespedeza hays. *Top:* Wool from lamb fed hay grown on soil given phosphate alone. *Bottom:* Wool from lamb fed hay grown on soil given both phosphate and limestone (calcium). (Courtesy, University of Missouri, Columbia)

## TABLE 32–6
### MACROMINERAL REQUIREMENTS OF SHEEP (PERCENTAGE OF RATION)[1]

| Nutrient | Requirement | |
|---|---|---|
| | As-fed[2] | Moisture-free |
| | (%) | (%) |
| Sodium | 0.08–0.16 | 0.09–0.18 |
| Chlorine | — | |
| Calcium | 0.18–0.74 | 0.20–0.82 |
| Phosphorus | 0.14–0.34 | 0.16–0.38 |
| Magnesium | 0.11–0.16 | 0.12–0.18 |
| Potassium | 0.45–0.72 | 0.50–0.80 |
| Sulfur | 0.13–0.23 | 0.14–0.26 |

[1]Adapted by the author from *Nutrient Requirements of Sheep*, Sixth Revised Edition, NRC-National Academy of Sciences, 1985, p. 48.

[2]As-fed was calculated using 90% dry matter (moisture-free).

## TABLE 32–7
### MICROMINERAL REQUIREMENTS OF SHEEP AND MAXIMUM TOLERABLE LEVELS (PPM OR MG/KG OF RATION)[1]

| Nutrient | Requirement | | Maximum Tolerable Level | |
|---|---|---|---|---|
| | As-fed[2] | Moisture-free | As-fed | Moisture-free |
| | (ppm or mg/kg) | (ppm or mg/kg) | (ppm or mg/kg) | (ppm or mg/kg) |
| Cobalt | 0.09–0.18 | 0.1–0.2 | 9 | 10 |
| Copper | 6–10 | 7–11[3] | 23 | 25[4] |
| Fluorine | — | — | 54–135 | 60–150 |
| Iodine | 0.09–0.72 | 0.10–0.80[5] | 45 | 50 |
| Iron | 27–45 | 30–50 | 450 | 500 |
| Manganese | 18–36 | 20–40 | 900 | 1,000 |
| Molybdenum | 0.45 | 0.5 | 9 | 10[4] |
| Selenium | 0.09–0.18 | 0.1–0.2 | 1.8 | 2 |
| Zinc | 18–30 | 20–33 | 675 | 750 |

[1]Adapted by the author from *Nutrient Requirements of Sheep*, Sixth Revised Edition, NRC-National Academy of Sciences, 1985, p. 50.

[2]As-fed was calculated using 90% dry matter (moisture-free).

[3]Requirement when dietary Mo concentrations are <1 mg/kg DM.

[4]Lower levels may be toxic under some circumstances.

[5]High level for pregnancy and lactation in rations not containing goitrogens; should be increased if rations contain goitrogens.

## MAJOR OR MACROMINERALS

The major or macrominerals involved in sheep nutrition are salt (sodium chloride), calcium, phosphorus, magnesium, potassium, and sulfur.

- **Salt (Sodium and Chlorine)**—Sheep are particularly fond of salt and consume considerably more of it per hundred pounds body weight than do cattle.

    The total salt requirement of growing lambs approximates 0.40% of the dry matter of the ration.

    Range operators commonly provide 0.5 lb to 0.75 lb salt per ewe per month. Mature sheep in drylot may consume more. Finishing lambs consume about 0.6 lb per head per month. Loose salt, rather than block salt, should be provided, for the reason that sheep bite at salt blocks, rather than lick, with the result that their teeth may be broken. In iodine-deficient areas, stabilized iodized salt should always be provided.

    When salt is added to mixed feeds, it is customary to add 0.5% to the complete ration or 1.0% to the concentrate portion. In the alkaline districts of the West, the water may contain enough salt to meet the requirements, and supplemental salt may not be needed.

    *Symptoms of salt deficiency:* A deficiency of salt may result in a depraved appetite, with the sheep trying to satisfy their craving by chewing wood, licking dirt, or eating toxic amounts of poisonous plants; decreased feed consumption; and decreased efficiency in the utilization of nutrients.

- **Calcium and phosphorus**—Calcium and phosphorus utilization depend on the presence of vitamin D and magnesium.

    Most pasture and range forage contains adequate amounts of calcium, although calcium supplementation of pastures is required in Florida, Louisiana, Nebraska, Virginia, and West Virginia.

    Legumes are an excellent source of calcium; hence, the requirements for this mineral are usually met when the winter forage for breeding ewes or the roughage for finishing lambs consists of one-third or more of a good quality legume hay. Corn silage is a poor source of calcium. Finishing lamb rations based on low-quality roughage, or high in concentrates, may require calcium supplementation.

    Mature pasture and range forage in North America is almost always deficient in phosphorus. Because of the high phosphorus content of the cereal grains, finishing lambs usually secure an adequate allowance of this mineral, unless a high proportion of beet by-products or other low-phosphorus feeds are fed.

    *Symptoms of calcium and phosphorus deficiency:* Rations that are decidedly lacking in calcium or phosphorus result in abnormal bone development, known as rickets in young animals and osteomalacia in adults.

    Signs of calcium deficiency due to low intake of calcium develop slowly because the body draws on the store of calcium in the bones until it is greatly reduced. In extreme cases, which may occur in lambs on high grain diets, low levels of calcium may result in tetany. Blood levels of calcium below 9 mg per 100 ml of serum indicate a calcium deficiency (hypocalcemia).

Fig. 32–3.  Lamb fed a ration deficient in phosphorus. Note the knock-kneed conformation. (Courtesy, University of Idaho)

Phosphorus deficiency symptoms are depraved appetite, slow growth, unthrifty appearance, listlessness, knock-knees, and low level of phosphorus in the blood (less than 4 mg/ 100 ml of plasma).

- **Magnesium**—Magnesium is a constituent of bone. Also, it is necessary for many enzyme systems and for proper functioning of the nervous system. A deficiency of magnesium may result in grass tetany in sheep, particularly on wheat pasture.

    The requirement for magnesium on a moisture-free (M-F) basis is 0.12, 0.15, and 0.18% for growing lambs, ewes in late pregnancy, and ewes in early lactation, respectively. Where ewes in early lactation are grazing forage with high nitrogen and potassium content, the minimum level of magnesium in the ration is 0.2%.

    The use of intraruminal magnesium alloy pellets (bullets) weighing 30 g has effectively prevented hypomagnesemic tetany in lactating ewes.

    *Symptoms of magnesium deficiency:* Acute tetany, characterized by stiff legs and head retraction, may occur as a result of insufficient dietary magnesium.

- **Potassium**—The potassium requirement for growth in lambs appears to be about 0.5% (M-F basis) of the ration. During periods of stress and during lactation, 0.7 to 0.8% may be required.

- **Sulfur**—Sulfur functions in the synthesis of the sulfur-containing amino acids, cystine and methionine, in the rumen and various compounds of the body. Also, wool is high in sulfur; hence, this element is closely related to wool production.

It is recommended that a dietary nitrogen-sulfur ratio of 10:1 be maintained. In percentage of dry matter, the requirements are as follows: mature ewes, 0.14–0.18%; young lambs, 0.18–0.26%.

Most feedstuffs contain more than 0.1% sulfur. However, mature grass and grass hays are sometimes low in sulfur and may not furnish enough of this element for optimum performance.

Where forages are low in sulfur, or where high-urea rations are fed, weight gains and growth of wool can be increased by feeding a sulfur supplement, such as sulfate sulfur, elemental sulfur, or sulfur-containing proteins or amino acids.

*Symptoms of sulfur deficiency:* Loss of appetite, reduced weight gains and feed conversion efficiency, and reduced wool growth. Also, excessive salivation and tears, and shedding of wool.

## TRACE OR MICROMINERALS

The trace or microminerals involved in sheep nutrition are cobalt, copper, fluorine (because of toxicity to sheep), iodine, iron, manganese, molybdenum, selenium, and zinc.

• **Cobalt**—Cobalt is essential for the synthesis of vitamin B–12 in the rumen. Indicators of the cobalt status of sheep are the levels of vitamin B–12 in the rumen contents, in the blood and liver, and in the feces.

Cobalt should be ingested frequently, preferably daily. This may be accomplished by adding cobalt to the salt, by adding cobalt to the soil, by placing cobalt pellets into the rumen, or by daily doses of cobalt.

The recommended amount of cobalt in the ration DM is 0.1 to 0.2 ppm. However, young rapidly growing lambs may have a slightly higher requirement.

Cobalt-deficient areas have been widely reported in the United States and Canada. In known deficient areas, it is recommended that cobalt be added to the salt at the rate of 1.4 grams per 100 lb of salt as cobalt chloride or cobalt sulfate.

*Symptoms of cobalt deficiency:* Affected sheep show loss of appetite, lack of thrift, severe emaciation, weakness, anemia, decreased fertility, and decreased milk production.

• **Copper**—A copper deficiency may exist alone or in combination with deficiencies of other trace minerals. In practice, copper deficiency is frequently induced by excess molybdenum in forages.

Copper is found in adequate amounts in most feeds throughout the United States. The NRC copper requirements vary depending on the molybdenum content of the feed as follows:

| | Recommended Cu Allowance | |
|---|---|---|
| | As-Fed | Moisture-Free |
| | (ppm) | (ppm) |
| **Mo content of diet ppm < 1.0:** | | |
| Growth | 8.9–9.0 | 8–10 |
| Pregnancy | 8.1–9.9 | 9–11 |
| Lactation | 6.3–7.2 | 7–8 |
| **Mo content of diet ppm > 3.0:** | | |
| Growth | 15.3–18.9 | 17–21 |
| Pregnancy | 17.1–20.7 | 19–23 |
| Lactation | 12.6–15.3 | 14–17 |

Merino sheep are less efficient in absorbing copper from feedstuffs than British breeds; so, they need an additional 1 to 2 ppm in their ration.

Copper-deficient areas have been reported in Florida, in the coastal plains region of the Southeast, and in Nevada, Oregon, and other western states. In such areas, it is recommended that copper sulfate be added to the salt at the rate of 0.5%. Copper is stored in the body; reserves may last as long as 4 to 6 months when animals are grazing copper-deficient forage.

*Symptoms of copper deficiency:* Lambs may be born weak and may die because of their inability to nurse. Suckling lambs show muscular incoordination and partial paralysis of the hindquarters.

Copper-deficient sheep produce "steely" wool, lacking in crimp, tensile strength, affinity for dyes, and elasticity. With a severe deficiency, the wool of black sheep is depigmented.

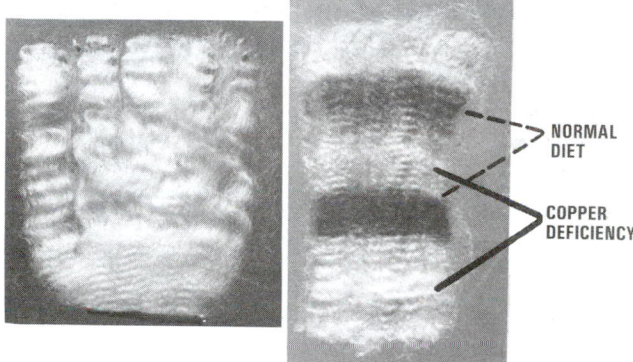

Fig. 32–4. Two samples of Australian wool, both of which show what may happen when sheep are on a copper-deficient diet. *Left:* The outer (bottom) two-thirds of this sample was produced by a sheep on a copper-deficient ration, resulting in hairlike or "steely" wool. Then copper was added to the sheep's diet, and normal, well-crimped wool was produced. *Right:* Wool sample from a normally black sheep. The white bands appeared at intervals when copper was deficient in the ration, because copper is essential for melanin or pigment production. Where such deficiencies occur under natural conditions, it is recognized that copper deficiencies result in the production of wool of lowered elasticity, tensile strength, and affinity for dyes.

• **Fluorine**—Fluorine is a cumulative poison. It occurs in some parts of the world as a result of consuming water high in fluoride or using rock phosphate that contains fluorine in amounts sufficient to be toxic—3 to 4%. Finishing lambs can tolerate up to 150 ppm of fluorine in the diet on a dry matter basis. Acute toxicity occurs at 200 ppm.

*Symptoms of fluorine toxicity:* Affected animals exhibit loss of appetite and weight, change in bone color from ivory to chalky white, thickened bones, and pitted and eroded teeth—especially the incisors.

• **Iodine**—Iodine is necessary for the formation of thyroxin, the iodine-containing hormone of the thyroid gland. Northwestern United States and the Great Lakes region are well-known iodine deficient areas, but other deficient areas are widely scattered throughout the United States.

The iodine requirement is 0.09 to 0.72 ppm in the ration on an as-fed basis, or 0.1 to 0.8 ppm moisture-free basis, with the higher levels indicated for pregnancy and lactation. When goitrogens such as kale or rape are fed, the dietary iodine should be increased.

Lamb losses can be prevented by feeding gestating ewes iodized salt containing 0.0078% iodine. The salt should be stabilized to prevent losses from exposure to sunlight and moisture.

Iodized salt should not be used in a feed mixture to govern feed intake, as the animals may consume too much iodine.

*Symptoms of iodine deficiency:* Lambs born with goiter (an enlarged thyroid gland) is the most common deficiency symptom. If the condition is not too advanced, afflicted lambs may survive. Other iodine deficiency signs are lambs born without wool, weak, or dead. An iodine deficiency in mature sheep may result in reduced wool yield and conception.

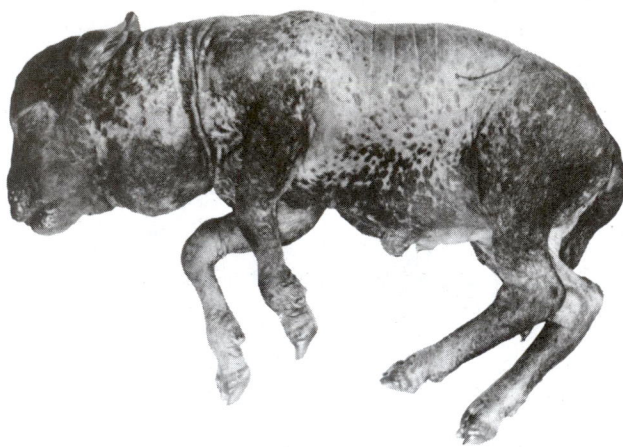

Fig. 32–5. Goiter. Woolless, goitered (big-necked) lamb stillborn due to iodine deficiency. (Courtesy, Montana State University, Bozeman)

• **Iron**—Iron deficiency anemia sometimes occurs in lambs raised in confinement, on slotted floors. It can be prevented (1) by giving the lambs 2 intramuscular injections of iron-dextran, each of 150 mg of iron, 2 to 3 weeks apart, or (2) by allowing free access to a commercial iron compound in the creep area.

*Symptoms of iron deficiency:* Iron deficiency is characterized by anemia, poor growth, lethargy, increased respiration rate, decreased resistance to infection, and, in severe cases, high mortality.

• **Manganese**—Although the exact requirements of sheep for manganese are unknown, it appears that 18 ppm in as-fed rations, or 20 ppm in moisture-free rations, will be adequate. It is needed for skeletal development and reproduction, as it is for various other species. A deficiency of manganese was produced in early-weaned lambs fed for a 5-month period on a ration containing less than 1 ppm of manganese. When a ration containing 8 ppm of manganese was fed to 2-year-old ewes for a 5-month period prior to breeding and throughout gestation, more services per conception (2.5 vs 1.5) were required than for ewes fed a ration containing 60 ppm manganese.

The manganese content of wool or mohair appears to be a good indicator of the manganese status of sheep.

*Symptoms of manganese deficiency:* A manganese deficiency results in impaired growth, skeletal abnormalities and lack of coordination of the newborn, and depressed or disturbed reproductive function.

• **Molybdenum**—Feeds containing small amounts of molybdenum are more effectively digested and produce more rapid gains in weight than rations without molybdenum.

A relationship exists between molybdenum and copper. A high molybdenum intake can induce copper deficiency in sheep even when the copper content of the pasture is quite high, the effect of which can be prevented by an increased copper intake. On the other hand, when pastures provide a low intake of molybdenum, excess copper accumulates in the body and results in a fatal jaundice, which can be prevented by increasing the molybdenum of animals.

*Molybdenum toxicity and copper toxicity:* Molybdenum toxicity (excess molybdenum) has been reported in California, Nevada, and England, where it causes a scouring disease. Sheep start to scour within a few days after being turned to high-molybdenum pasture (5–20 ppm on a dry matter basis; or on forage containing as little as 1 or 2 ppm molybdenum provided the dietary copper level is low and the sulfate level is high); the feces become soft, the fleece becomes stained, and the animals lose weight rapidly. Molybdenum toxicity can be controlled by increasing the copper level in the diet by 5 ppm.

Copper toxicity—characterized by hemolysis (dissolution of red corpuscles with liberation of their hemoglobin), jaundice (easily detected in the eyes), hemoglobinuria, and very dark-colored liver and kidneys—may result when sheep are fed diets high in copper and low in molybdenum. Copper toxicity may be prevented by lowering the copper level in the diet (normal is 8–11 ppm), or by a high-zinc diet (100 ppm on a dry matter basis). Recommended treatments for copper toxicity are (1) administering molybdenum and sulfate, or (2) drenching each lamb daily with 100 mg of ammonium molybdate and 1 g of sodium sulfate in 20 ml of water, for about 3 weeks. The Food and Drug Administration does not recognize molybdenum as safe; hence, it is not legal to add it to the feed of sheep.

• **Selenium**—A minimum level of 0.1 ppm of selenium in feeds is considered adequate for preventing a deficiency in sheep. There are extensive areas in northwestern, northeastern, and southeastern United States where the selenium content of crops is below this level. On the other hand, parts of South Dakota, Wyoming, and Utah produce forage so high in selenium that it causes selenium toxicity.

White muscle disease (stiff-lamb disease) in lambs, the main manifestation of selenium deficiency, can be prevented by (1) adding 0.3 ppm of selenium to the complete feed of ewes during gestation through weaning, or (2) adding 90 ppm of selenium to a salt-mineral mixture fed free-choice.

*Symptoms of selenium deficiency:* Selenium deficiency has serious effects on lamb production. The main signs of such deficiency are white muscle disease, sometimes called *stiff-lamb disease*, and reduced growth. If the muscle lesions are in the heart, lambs may die suddenly if subjected to exercise. Additional signs of inadequate selenium are unthriftiness, infertility, early embryonic death, and periodontal disease.

*Selenium toxicity:* Chronic selenium toxicity occurs when sheep consume plants containing more than 3 ppm selenium over a prolonged period. The extent to which plants take up selenium varies greatly by species. The most practical way in which to prevent livestock losses in high-selenium areas is to rotate the grazing between high-selenium and low-selenium areas. Although selenium is a cumulative poison, mild chronic signs can be readily overcome by feeding selenium-low forage. Also, small amounts of arsanilic acid are effective in reducing the toxicity of selenium.

- **Zinc**—Zinc is essential for sheep. A dietary level of 18 ppm of zinc appears to be adequate for growth, but a level of 30 ppm is necessary for normal reproduction in the ram. There is also evidence that the utilization of protein is impaired in zinc-deficient sheep.

*Symptoms of zinc deficiency:* Deficiency signs include impaired growth, excessive salivation, parakeratosis, wool loss, and delayed wound healing. Ram lambs show reduced testicular development and defective spermatogenesis. In females, all phases of the reproductive process from estrus to parturition and lactation may be adversely affected.

## VITAMINS

Many phenomena of vitamin nutrition are related to solubility—vitamins are soluble in either fat or water. Consequently, it is important that both nutritionists and sheep producers be well informed about solubility differences in vitamins and make use of such differences in programs and practices. Thus, in the discussion that follows, vitamins are grouped as either (1) fat-soluble vitamins, or (2) water-soluble vitamins.

### FAT-SOLUBLE VITAMINS

The fat-soluble vitamins are vitamin A (carotene), vitamin D, vitamin E, and vitamin K.

- **Vitamin A**—Dietary vitamin A or its precursor, carotene, is necessary for maintaining normal epithelial tissues.

Sheep do not convert carotene to vitamin A in the same ratio as rats. Thus, for sheep it is suggested that 1 mg of feed carotenes be considered as the equivalent of 400–500 IU of vitamin A. (For the rat, 1 mg of B-carotene = 1,667 IU of vitamin A.)

Vitamin A is fat-soluble and is stored in the body. It takes about 200 days to deplete entirely the liver storage of vitamin A of lambs previously pastured on green feed. This explains why animals which graze green forage during the normal growing season are able to do reasonably well on a low-carotene ration for periods of 4 to 6 months.

Both vitamin A and carotene are subjected to loss by oxidation. Sun-cured hay is usually lower in carotene than dehydrated hay. Feedstuffs that contain little green or yellow material or that have been badly weathered, heated, or stored for long periods are low in carotene. Stabilized vitamin A, which is resistant to oxidation, may be added to rations low in carotene.

The vitamin A requirements of the National Academy of Science are given in Tables 32–1 and 32–2.

Sheep that are deficient in vitamin A and weigh 70 lb or more should receive 100,000 IU of vitamin A by injection, followed by adjusting their rations to provide recommended levels of vitamin A or carotene. Ewes deficient in vitamin A should be given vitamin A either orally or by injection prior to breeding.

*Symptoms of vitamin A deficiency:* Vitamin A-deficient sheep develop keratinization of the respiratory, alimentary, reproductive, urinary systems, and ocular epithelia, accompanied by lowered resistance to infections. Also, a deficiency of vitamin A interferes with normal development of bone; may result in the birth of weak, malformed, or dead lambs; and causes night blindness, the appearance or nonappearance of which is the most common means of determining the vitamin A status of animals.

- **Vitamin D**—Vitamin D is required, in addition to calcium and phosphorus, for the prevention of rickets in sheep. The addition of vitamin D to lamb rations low in this nutrient has resulted in increased growth even when there are no signs of rickets.

Since vitamin D is fat-soluble and stored, it is less important in mature animals, except during pregnancy, when the demands are greater. If pregnant ewes have adequate storage of vitamin D, they provide their newborn lambs with sufficient of it to prevent rickets for 4–6 weeks.

Sheep exposed to sunlight obtain some vitamin D through irradiation, which may be sufficient to meet their requirements. Animals with white skin and/or short wool receive more vitamin D activity through irradiation than do animals with black skin and/or long wool.

Sheep on pasture seldom need additional vitamin D. But the question of adequacy arises during extended cloudy periods or when sheep are kept indoors. Under these circumstances, it is especially important that adequate vitamin D be included in the ration of fast-gaining lambs.

Early-weaned lambs require 300 IU of vitamin D per 100 lb body weight, daily. All other sheep require 250 IU/100 lb body weight, daily. Sheep use either $D_2$ or $D_3$ equally well.

Sun-cured hays are fairly good sources of vitamin D. Dehydrated hays, green feeds, seeds, and by-products of seeds are poor sources. Vitamin D is oxidized, but with greater difficulty than vitamin A. However, when mixed with minerals, especially calcium carbonate, its stability is poor.

*Symptoms of vitamin D deficiency:* A deficiency of vitamin D results in rickets in young lambs. Extreme deficiencies may cause congenital malformations in newborn lambs.

Fig. 32–6.  Bilateral bent leg in yearling Rambouillet ram due to a deficiency of vitamin D. (Courtesy, Utah Agricultural Experiment Station)

• **Vitamin E**—Vitamin E is essential for all sheep, especially for young lambs. Unlike vitamin A, it is not stored in the body in appreciable quantities.

Vitamin E functions as an important biological antioxidant and helps to prevent white muscle disease (stiff-lamb disease) in lambs, through its association with dietary selenium in metabolism. Some signs of deficiency, such as white muscle disease or nutritional muscular dystrophy, may respond to either selenium or vitamin E, or may require both. Although

Fig. 32–7. Lamb with stiff-lamb disease, caused by a deficiency of vitamin E. (Courtesy, Cornell University, Ithaca, NY)

vitamin E is a dietary requirement for young nursing lambs, experiments have failed to relate a deficiency of it to reproductive failure in sheep.

The need for vitamin E in the diet of young nursing lambs is related to the selenium level in the diet. Selenium has a sparing effect on the vitamin E requirement; the higher the selenium level in the diet, the lower the vitamin E requirement, and vice versa. White muscle disease in lambs is prevented by adding alpha-tocopherol and selenium to the diet. The suggested dietary levels of vitamin E are as follows: Lambs under 44 lb in weight should receive 8 IU/lb of as-fed ration; lambs over 44 lb in weight and pregnant ewes should receive 6 IU/lb of as-fed ration. (The IU is defined as 1 mg of dl-alpha-tocopherol acetate; 1 mg dl-alpha-tocopherol has the biological potency of 1.5 IU of vitamin E activity.) The above recommendations assume that dietary selenium levels are <0.05 ppm.

Values for the vitamin E requirements of sheep are presented in Tables 32–1, 32–2, and 32–8. The values presented in Table 32–1 were calculated from values per kilogram of dry feed consumed given in Table 32–2. Table 32–8 presents daily vitamin E requirements for lambs and the suggested amounts of alpha-tocopherol acetate to add to the rations to provide 100% of the requirements.

### TABLE 32–8
### VITAMIN E REQUIREMENTS OF GROWING-FINISHING LAMBS
### AND SUGGESTED LEVELS OF FEED FORTIFICATION TO PROVIDE 100% OF REQUIREMENTS[1]

| Body Weight | | Alpha-Tocopherol Acetate | | | Feed Intake Per Lamb | | Amount of Vitamin E Added to Concentrate | | | Amount of Vitamin E Added to Protein Supplement[2] | | |
|---|---|---|---|---|---|---|---|---|---|---|---|---|
| (lb) | (*kg*) | (mg/lamb/day)[3] | (mg/lb ration) | (*mg/kg ration*) | (lb) | (*kg*) | (mg/lb) | (*mg/kg*) | (mg/ton) | (mg/lb) | (*mg/kg*) | (mg/ton) |
| 22 | *10* | 5.0 | 44 | *20* | 0.50 | *0.23* | 9.1 | *20* | 18,200 | 133 | *60* | 120,000 |
| 44 | *20* | 10.0 | 44 | *20* | 1.00 | *0.45* | 9.1 | *20* | 18,200 | 60 | *133* | 120,000 |
| 66 | *30* | 15.0 | 33 | *15* | 2.10 | *0.96* | 6.8 | *15* | 13,600 | 45 | *100* | 90,000 |
| 88 | *40* | 20.0 | 33 | *15* | 2.86 | *1.30* | 6.8 | *15* | 13,600 | 45 | *100* | 90,000 |
| 110 | *50* | 25.0 | 33 | *15* | 3.50 | *1.60* | 6.8 | *15* | 13,600 | 45 | *100* | 90,000 |

[1]Adapted by the author from *Nutrient Requirements of Sheep*, Sixth Revised Edition, NRC-National Academy of Sciences, 1985, p. 51.

[2]Assumes the concentrate diet contains 15% protein supplement.

[3]Rounded values based on approximate diet intake containing recommended vitamin E levels.

Based on the average alpha-tocopherol content of feedstuffs generally used in growing-finishing rations (corn, soybean meal, and alfalfa hay), the typical ration may contain less than 6.8 mg of alpha-tocopherol/lb, which could result in inadequate intake of vitamin E. In addition, preintestinal destruction of vitamin E of an orally administered dose increases from 8 to 42% as the corn content of the ration increases from 20 to 80%. So, many sheep rations heretofore believed to be adequate in vitamin E may be inadequate, thus explaining the sporadic outbreaks of white muscle disease in areas considered adequate in selenium.

*Symptoms of vitamin E deficiency:* White muscle disease in nursing lambs is characterized by a stiffness (especially in the rear quarters), tucked-up rear flanks, and arched back. On autopsy, the disease is shown as white striations in the muscles, characterized by bilateral lesions. Affected lambs often die of pneumonia and starvation.

• **Vitamin K₁ and K₂**—These are fat-soluble vitamins, either of which is necessary in the blood-clotting mechanism. All green, leafy materials, fresh or dry, are good sources of vitamin $K_1$. Vitamin $K_2$ is usually synthesized in large amounts in the rumen, so no need for dietary supplementation of it has been established.

### WATER-SOLUBLE VITAMINS

All of the water-soluble vitamins except vitamin C are known as B vitamins. These vitamins are not stored.

• **B Vitamins**—Digestion in the young lamb is more like that of a non-ruminant animal than that of a ruminant. Thus, up to about 2 months of age, early-weaned lambs have dietary need for biotin, choline, folacin, niacin, pantothenic acid, riboflavin, thiamin, vitamin B–6 (pyridoxine), and vitamin B–12, which are usually supplied in the ewe's milk.

Normally, the B vitamins are not required in the ration of sheep with functioning rumens (older than about 2 months of age), because the microorganisms synthesize them in adequate amounts. It is noteworthy, however, that polioencephalomalacia (PEM), a noninfectious disease of sheep, the highest incidence of which occurs in lambs, responds to a parenteral injection of 0.5 g of thiamin hydrochloride, repeated at 2-day intervals if necessary. Apparently, PEM is caused by an antithiamin enzyme in the rumen.

It is necessary that cobalt be present for the synthesis of vitamin B–12 in the rumen.

• **Vitamin C**—Vitamin C is synthesized rapidly enough by the tissues to meet the animal's needs; hence, it is not a required dietary constituent for sheep.

## WATER

Sheep get water by drinking, and from snow, dew, and feed. The amount of water that sheep voluntarily consume is affected by temperature, rainfall, snow and dew covering, age, breed, stage of production, number of lambs carried, wool covering, respiratory rate, frequency of watering, kind and amount of feed, and exercise. On the average, mature animals consume approximately a gallon of water per day, whereas feeder lambs require about half this amount. However, sheep may go for weeks without drinking water when foraging on grasses and other feeds of high moisture content. This condition often prevails on desert ranges in the early spring and on many of the mountain ranges during the summer months.

## FEEDS FOR SHEEP

Sheep are adapted to the consumption of a great variety of feeds. Most of the common feeds used by them are of plant origin and bulky in nature. The feeding of concentrates is usually limited to the finishing of lambs and for use by the breeding flock at such special periods as the lambing season or just before and during the breeding season. Roughages constitute 100% of the ration of the vast majority of sheep during most seasons of the year. Most lambs are marketed as milk-fat lambs directly off pastures or ranges without having had any grain.

## PASTURES

No other class of farm animals is so well adapted to the utilization of maximum quantities of pasture as sheep. Although cattle compete with sheep for many of the same grazing areas and are also ruminants, sheep are unique in that the vast majority of the young are marketed as milk-fat animals directly off pastures. Also, in their grazing habits, sheep differ from cattle in that (1) they show a decided preference for short, fine forages, and (2) they have the gregarious or flocking instinct.

Although there are great differences in plants, sheep are able to utilize the various grasses, legumes, weeds, forbs (broadleafed herbaceous plants commonly called weeds by ranchers), and browse (broadleafed, woody plants, or shrubs, bushes, or small trees) that grow on millions of acres of cultivated and uncultivated land in this and other countries. This characteristic, plus the imperishable nature of wool from the standpoint of storage and transportation, has made sheep raising a frontier industry throughout the world.

Sheep are adapted to the grazing of both fenced and unfenced holdings. In this country, most farm flocks and a limited number of range bands of the Southwest are confined to fenced areas with no herder being necessary. On the other hand, most of the range pastures of the West are utilized by migratory bands under the supervision of a herder.

Regardless of the location of the area, year-round grazing is desired. In order to obtain succulent and palatable pastures, the range bands of the West are frequently travelled to different altitudes at different seasons of the year. In the mountainous sections, the summer ranges are usually at high altitudes, the spring-fall grazing at intermediate altitudes, and much of the winter range is on the desert or lowland areas. On some ranges, such traveling is not possible, the ranges being used on a year-long basis.

During the winter months and following periods of extended drought, pastures usually become leached and bleached. Although sheep can and do utilize these forages, it must be realized that, in comparison with green growing grasses, they are lacking in nutrients, being especially low in protein and carotene (provitamin A) content. Consideration of these facts should be given when providing supplemental feeds.

Fig. 32–8. Sheep on winter range. Note the snow. (Courtesy, Ralston Purina Co., St. Louis, MO)

Abundant and succulent pastures are ideal for stimulating milk production in ewes. Moreover, pastures of this type are desirable from the standpoint of the limited digestive capacity of the young lamb. Accordingly, the degree of finish carried by lambs at market time is an accurate reflection of the amount and quality of the forage available on the pasture or range.

Sheep may successfully utilize either permanent or cultivated pastures. In the range area, the vast majority of pastures are of a permanent type; whereas both types of pastures are used by native sheep. Where a choice is possible, improved grass-legume pastures are preferred. Where adapted, birdsfoot trefoil is the legume of choice because it is bloat-free and will remain productive for a long period, provided it is adequately fertilized and managed.

Range sheep graze nearly all the year and receive most of their sustenance from native range plants. Pronounced deficiencies (singly or in combination) of range forages in energy, protein, phosphorus, and carotene (provitamin A) sometimes occur on ranges. These deficiencies are most apt to happen when the forage is mature or dormant, during overgrazing, or in periods of drought. Also, they are most marked in ewes during gestation or lactation.

## HAYS AND OTHER DRY ROUGHAGES

Inclement weather, extreme droughts, over-stocked pastures and ranges, and leached and bleached pastures make it necessary that dry roughages be provided for sheep. They are fond of good roughage and make good use of it. In general, however, they cannot effectively use as much coarse roughage as cattle.

Hays are the standard winter feed for sheep when they cannot be out on the pasture or range or when the condition of the pastures is such as to require supplemental feeding. The choicest hay for sheep is a legume which has been produced on fertile soil, cut at the proper stage, and well cured. Such hay is palatable and rich in protein (and the quality of protein is good), calcium, and vitamins A and D. If legume hay cannot be secured, a high-quality grass-legume mixed hay will be entirely satisfactory and much superior to a straight grass hay. Sheep may do very well for a considerable period of time when they are fed no feed other than a good-quality legume hay, plus salt and water.

Fig. 32–10. Ewes in winter quarters feeding on hay. (Courtesy, *Sheep Breeder and Sheepman*, Colombia, MO)

## SILAGES AND ROOT CROPS

Silages for sheep may be made from a great variety of plants, including corn, sorghums, cereal grains, legumes, grasses, cannery refuse, pea vines, potatoes, beets, beet tops, sunflowers, and other materials. When properly preserved and fed, silages made from any of these materials are quite satisfactory.

Most practical sheep producers prefer to limit the silage allowance to about 4 to 6 lb per head per day, with the balance of the roughage ration consisting of hay. If a nonlegume silage is used, it is important that the hay be a legume. If a legume hay cannot be provided when a nonlegume silage is fed, it is very necessary that a suitable protein concentrate and minerals be provided.

Roots include all plants whose roots, tubers, bulbs, or other underground vegetative parts are used for feed. The important root crops for sheep are: mangels (stock beets), rutabagas (swedes), turnips, and carrots. These feeds are very succulent in nature, containing from 85 to 90% or more of water. They are highly relished by sheep and have a peculiarly beneficial effect upon the digestion and general thrift of these animals. The only objection to their general use is the cost and difficulty of growing, harvesting, and storing them. For the latter reason, roots are usually (1) a by-product from food produced for human consumption, (2) grown as a second crop and harvested by grazing sheep on them, or (3) grown and harvested mainly by producers of purebred animals and by shepherds in Europe. Presently, some sheep producers are using turnips as a fall and early winter grazing crop.

Roots are generally fed sliced, although some are fed whole. Where the teeth are good, practical sheep producers feel that little is gained by slicing roots. In this country, roots are usually limited to 5 to 6 lb per head daily, but in other countries up to 12 to 14 lb are often allowed daily per head.

Fig. 32–9. Ewes on winter pasture eating hay off the ground. (Courtesy, Ralston Purina Co., St. Louis, MO)

Although legume hays are preferable for sheep, nonlegume hays are fed extensively in the sheep-raising and lamb-feeding areas. If straight grass hays must be fed, they should be cut at an early stage of maturity. Even then, they will be lower in protein, calcium, and vitamins than the legumes; hence, for best results, protein and mineral supplements should be provided.

Bright, early-cut corn or sorghum fodder, or stover, and early-cut, green cereal hays, and straws of many kinds are fed to sheep in different areas. The feeding value of these coarser roughages varies considerably according to the stage of maturity at which they are cut, the amount of leaves, and the green coloration. Although these forages are not satisfactory as the sole roughage, especially during the latter part of gestation and in the suckling period, they may be successfully used when mixed with liberal quantities of good-quality legume hays. Where nonlegume roughages are fed, special attention should be given to providing a suitable protein concentrate and minerals, especially calcium.

## QUALITY OF FORAGE

The quality of forage—pasture, hay, or other dry roughage, or silage—greatly affects its consumption. High-quality forage is more digestible and passes through the digestive tract more rapidly than low-quality forage; hence, sheep will consume more of it.

The most favorable nutritional response is usually obtained by feeding forage harvested before the protein content has decreased and before lignification of the fiber content has increased. Loss of leaves, weather damage, fermentation, and leaching losses reduce the value of harvested forage.

When it is necessary to feed sheep low-protein forage, intake and utilization can be increased by adding a suitable protein supplement fortified with needed minerals and vitamins.

## CONCENTRATES

The concentrates include those feeds which are low in fiber and high in nutritive value. For purposes of convenience, concentrates are often further classified as (1) carbonaceous feeds, and (2) nitrogenous feeds.

Ordinarily, sheep are fed few concentrates, except immediately before and after lambing, in conditioning ewes and rams for breeding, or when finishing lambs. During these periods, the most frequently used concentrates consist of the common farm grains—oats, corn, barley, wheat, rye, and the grain sorghums. Numerous by-product feeds are also utilized for sheep, including those from the flour- and corn-milling industries, beet by-products, and oil meal cakes made from soybeans, cottonseed, flaxseed, or canola seed.

## ADDITIVES

Table 32–9 summarizes the growth stimulants that are presently available and can be used. All of these products have been shown to improve gain and feed efficiency of sheep. The information presented in Table 32–9 is the most recent available. But feed additives and implants do change from time to time; new products are developed, and sometimes old products are banned by the Food and Drug Administration. So, those using additives should always confer with local authorities and read and follow manufacturer's label directions for more complete details on the use of a specific drug or combination of drugs.

### TABLE 32–9
### SHEEP FEED ADDITIVES AND IMPLANTS

| Type of Additive | Method of Administering | Dosage | Effect On | | | Comments |
| | | | Daily Rate of Gain | Feed Efficiency | Carcass Quality | |
|---|---|---|---|---|---|---|
| | | | (% increase) | (% increase) | | |
| **Antibiotics** (chlortetracycline and oxytetracycline) | Feeding (oral) | **Aureomycin** (chlortetracycline) 10 to 25 mg/lb of feed. **Terramycin** (oxytetracycline) 5 to 10 mg/lb of feed. | Range: 0–31 Average: 11 | Range: 4–27 Average: 10 | No effect to slight improvement. | Antibiotics (especially chlortetracycline and oxytetracycline) may improve performance when added to creep and lamb finishing rations. Response to antibiotics varies markedly according to differences in management and degree of stress to which lambs are subjected. There is some evidence that antibiotics reduce the incidence of enterotoxemia. |
| **Bovatec** (lasalocid) | Feeding (oral) | 10–15 mg/lb complete feed, fed at rate of 15–70 mg lasalocid /day. | Range: 0–20 Average: 6–8 | Range: 5–15 Average: 8–10 | No effect. | Bovatec is an ionophore. In addition to increasing rate of gain and feed efficiency, Bovatec reduces rumen protein degradation and increases the amount of bypass protein. Greatest response is obtained where coccidiosis is a problem, for which purpose Bovatec was initially approved by FDA. |
| **Ralgro** (zeranol) | Implant | 12 mg/head | Range: 0–25 Average: 10 | Average: 6 | | Do not implant animals within 40 days of slaughter. Do not implant breeding animals. |

Antibiotics may improve performance when added to creep and lamb-finishing rations. Chlortetracycline and oxytetracycline are especially effective. The response to antibiotics seems to be affected by differences in management and the amount of stress to which the lambs are subjected. There is some evidence that antibiotics reduce the incidence of enterotoxemia.

In addition to the additives listed in Table 32–9, a number of feed additives are approved for treatment of specific diseases.

## FEED PREPARATION

The preparation of feeds for sheep is fully covered in Chapter 4, section entitled, "Feed Processing," including Table 4–10 therein; hence, the reader is referred thereto.

## HOW TO BALANCE A SHEEP RATION

Generally speaking the rations given in this chapter will suffice. But rations need to be changed from time to time in keeping with the supply and price of feed ingredients. So, sheep producers should know how to balance a ration. Complete instructions on how to balance a ration are given in Chapter 4, "Feeding Livestock," under the heading "How to Balance a Ration"; hence, the reader is referred thereto.

## FEEDING AND MANAGING BREEDING EWES

Success in the sheep business is largely measured by the percentage lamb crop raised and the pounds of lamb marketed per ewe. The most important factor affecting these criteria is the feed of the ewe. Also, the yearly feed of the ewe represents about 50% of all production costs. For purposes of convenience, the feeding of ewes will be discussed under the following headings: (1) drylot (confinement) feeding, (2) flushing ewes, (3) feeding pregnant ewes, (4) feeding at lambing time, (5) feeding lactating ewes, and (6) feeding ewes in accelerated lambing.

## DRYLOT (CONFINEMENT) FEEDING

The vast majority of the nation's sheep utilize pasture in season. However, some ewe-lamb producers are drylotting all or part of the year. So, now there are two alternatives, and the producer may choose between the two.

The **advantages** of drylot (confinement) production are:

1. The virtual elimination of losses from predators.

2. Freedom from the most harmful internal parasites.

3. Lowering of the energy requirement due to limited activity.

4. The opportunity to feed ewes according to their productivity and nutrient requirements rather than their appetites.

5. It results in more rapid gains, in lambs reaching market weight at an earlier age, and in improved carcass grade.

The **disadvantages** are:

1. A higher initial capital investment, especially in buildings and equipment.

2. It requires superior management.

3. All nutritive requirements must be met.

4. External parasites and contagious diseases may be increased.

5. Animal manure disposal and bedding costs will be greater.

The Table 32–14 rations are satisfactory for ewes in confinement, and the Table 32–12 and 32–13 rations are excellent for creep feeding lamb raised in confinement.

The following steps are necessary for successful drylot feeding:

1. Use a big purebred ram so that the lambs inherit rapid growth potential.

2. Have the lambs born so that they will be ready for the intended market in 4 to 4½ months.

3. After 2 months lactation, cut the ration of the ewes to the same amount that they were receiving 1 month before lambing.

4. Self-feed lambs; and wean them at 90 days of age. Do not turn lambs to pasture.

5. Market the lambs at 100 to 110 lb.

6. Vaccinate all lambs against enterotoxemia (overeating disease).

(Also, see Chapter 30, section headed ''Confinement Production Method.'')

## FLUSHING EWES

*Flushing is the practice of conditioning or having thin ewes gain in weight just prior to breeding.* Its purpose is to increase the ovulation rate and, consequently, the lambing rate.

This special feeding usually begins 2 to 3 weeks prior to breeding and continues into the breeding season. It may be accomplished by turning the ewes to a fresh luxuriant pasture 2 to 3 weeks before breeding time; or if such a pasture is not available, satisfactory results may be brought about by feeding a grain allowance of ½ to ¾ of a pound daily over a like period of time. Oats alone are excellent, or a mixture of equal parts of oats and corn is very satisfactory. Pumpkins, broken and scattered over the pasture, are also relished and are excellent for flushing purposes. Some shepherds like to feed cabbage at this season.

Although it is not likely that all of the benefits ascribed to flushing will be fully realized under all conditions, the general feeling persists that the practice will result in a 15 to 20% increase in the lamb crop, and that the ewes will breed both earlier and more nearly at the same time. Hence, it follows that the lamb crop will be earlier and more uniform in age and size.

Mature ewes appear to respond better than yearling ewes. Also, flushing may be more beneficial early and late in the breeding season than during the peak, when the ovulation rate is highest.

Fat ewes will not benefit from flushing. Instead, they should be conditioned for breeding by stepping up the exercise.

## FEEDING PREGNANT EWES

Fig. 32–11.   Feed conveyed by auger to pregnant ewes. (Courtesy, *The Sheepman's Production Handbook,* Denver, CO)

If a strong, healthy crop of lambs is to be expected, the ewes must be properly fed and cared for throughout the period of pregnancy. In general, this means the feeding of a suitable and well-balanced ration, together with the necessary minerals and vitamins as required for maintenance (and growth, if the ewe is not fully mature), growth of the fleece, and development of the fetus. In addition, plenty of exercise must be provided. Suitable shelter should be made available during inclement weather, and the animals should be given access to an abundance of fresh air and sunshine at all times. Ewes should gain in weight during the entire period of pregnancy, making a total gain of 20 to 30 lb for the period. They should enter the nursing period with some reserve flesh, because the lactation requirements are much more rigorous than those of the gestation period.

After the ewes are bred, they should have access to pastures as long as they are available and open. When the ground is firm, winter pasture or range, stalk, or stubble fields may be pastured to advantage. Green rye or wheat pastures furnish a very succulent feed and valuable exercise for the flock. Where winter pastures are either unavailable or inadequate, supplemental feeds must be provided. The most satisfactory forage is a good-quality legume hay—alfalfa, clover, lespedeza, or soybeans. The sheep producer, however, often seems to find it difficult to grow such roughage at satisfactory prices. Where grass hay, such as native hays or timothy, is used, every effort should be made to cut it at an early stage of maturity and to have it properly cured. Even then, a protein supplement should be provided, together with suitable minerals. Because of the known value of legumes from the standpoint of quality of proteins, minerals, and vitamins and the fact that grass hays are not recognized as too desirable for sheep, every effort should be made to supply at least a third of a good-quality legume roughage to pregnant ewes. A 150-lb ewe will eat about 4 lb of hay daily. In order to prevent waste and protect the wool from chaff and hay seeds, suitable racks should be provided.

Such succulent feeds as roots and silage are desirable in keeping the ewes healthy and doing well. Of the root crops, turnips seem to be preferable for pregnant ewes. Silage made from corn, milo, legumes, or grasses, and which is not frozen, spoiled, nor moldy, may be fed quite safely and is excellent feed. Ordinarily, the daily ration of roots or silage should not exceed 5 lb, which means that hay is usually fed in addition to the succulent feed.

During the last 4 to 5 weeks of pregnancy, the fetus develops very rapidly and the demands on the ewe are rather heavy. Also, ewes carrying twins or triplets, especially if they are a bit fat, are very prone to ketosis (lambing paralysis), which can be prevented by feeding a high-energy ration. So, during the last 4 to 5 weeks of pregnancy, ewes should be fed 0.5 to 1.0 lb of grain per head daily and gain 8 to 15 lb. Besides, ewes so fed milk better. The concentrate given to the farm flock usually consists of homegrown grains, whereas range bands are often given pelleted or cubed protein supplements.

## FEEDING AT LAMBING TIME

As lambing time approaches, or immediately after lambing, each ewe should be placed in an individual holding or lambing pen. At this time, the grain allowance should be materially reduced, but dry roughage may be fed free choice, when it is certain that it is of good quality and palatable. Usually, some five to seven days should elapse before ewes are placed on full feed following parturition. In general, feeds of a bulky and laxative nature should be provided during the first few days. A mixture of equal parts of oats and wheat bran is excellent. Soon after lambing, the ewe should be given water with the chill removed but should not be allowed to gorge.

## FEEDING LACTATING EWES

Following lambing, the feed allowance of the ewe should be increased according to her capacity and needs. Although there is great individual and breed variation, ewes will yield from 1 to 4 qt of milk per day. In comparison with cow's milk, ewe's milk is richer in protein and fat and higher in ash. It must also be borne in mind that, in addition to producing milk and maintaining her body, the ewe is growing wool, which is protein in character. Immature ewes are also growing. Under these circumstances, it is but natural and normal to expect ewes to lose in condition during the suckling period. The loss in weight is primarily determined by the inherent milking qualities of the individual and by the kind and amount of feed.

In general, it is considered good practice to feed lactating ewes rather liberally, for lambs make the most economical gains when suckling. It is a good plan to separate the ewes with twins from those with singles, giving the former more liberal rations or the benefit of the better pastures or ranges. In fact, some large sheep operators find this practice so advisable that they regularly separate out the twin bands.

Milk production can be greatly stimulated through the proper selection of feeds. If there is not sufficient high-quality roughage for the entire winter, the most palatable and succulent portion should be reserved for use during the suckling period. Pastures should be provided as soon as possible, but in the meantime a high-quality legume hay or, better yet, a combination of hay and silage, will take care of the roughage needs. Though varying somewhat with the size and condition of the ewe and whether there are twins or a single, an adequate ration for a lactating ewe may consist of approximately 4 lb of high-quality alfalfa hay plus 1 to 2 lb of grain daily. If neither a legume hay nor legume silage is available, a protein supplement should be included in the grain ration.

As soon as the spring pasture season has arrived, the use of harvested feeds should be discontinued, being both uneconomical and unnecessary.

## FEEDING EWES IN ACCELERATED LAMBING

*Accelerated lambing involves ewe lambs dropping their first lambs at 1 year of age, and lambing at intervals of 6 to 8 months thereafter.*

Experimental studies and practical observation indicate (1) that it is feasible and profitable to have ewe lambs drop their first lambs at 1 year of age, *provided* they are well fed, well grown, and early dropped; and (2) that it is possible to achieve a lambing interval of 6 to 8 months, *provided* breeds with long breeding seasons are used (Dorsets, Rambouillets, or Merinos), there is superior nutrition, and hormones are used when a 6-month interval is planned.

Ewe lambs that are to be bred so that they lamb at 12 months of age should be liberally fed (1) from birth, using one of the creep rations given in Tables 32–12 or 32–13; and (2) during pregnancy and lactation, using one of the suggested rations in Table 32–14.

Accelerated lambing can (1) make for lower cost of production, (2) contribute to a more even supply of lamb throughout the year, and (3) allow for greater flexibility in production.

## FEEDING RAMS

Fig. 32–12. Polypay rams in strong breeding condition. (Courtesy, American Polypay Sheep Assn., Sidney, MT)

The rams should be fed so as to remain in vigorous, active breeding condition. In general, rams should be fed the same kind of feed as ewes but in slightly larger quantities. They need a generous allowance of relatively high-quality feed just before and during the breeding season, when pasture is not available. During the balance of the year, pasture is usually adequate when available; otherwise, the ration may be comparable to that of ewes.

## FEEDING AND MANAGING RANGE SHEEP

Various geographical divisions are assumed in referring to the western range area—the native pasture area. Sometimes reference is made to the 17 range states, embracing a land area of 1.16 billion acres. At other times this area is broken down, chiefly on the basis of topography, into (1) the Great Plains area, and (2) the 11 western states. Much of the latter is federally owned. (See Chapter 2, section on "Land Area and Number of Farms and Ranches Devoted to Animal Production.")

In the early days of the range sheep industry, the animals were usually moved to lower winter ranges and expected to get their feed as best they could. There was precious little supplemental feeding. If the winter happened to be mild, and if a reasonable amount of grass was cured on the stalk, the band came through in pretty good shape. During an exceedingly cold winter, particularly when there was much snow, losses were severe and often disastrous. Today, the practical and successful range sheep producer winter feeds. The progressive rancher is also equipped to meet emergency feeding periods, of which droughts are the most common.

Ewes are normally maintained on winter grazing areas, with or without supplemental feeds, as long as possible. Usually these ranges are located at the lower altitudes and the vegetation consists of rather mature and bleached grasses or brush and browse. When the vegetation is sparse or covered by deep snow, supplemental feeds of hays, preferably alfalfa, some other

legume, or concentrates are provided. Often protein supplements in the form of pellets or cubes are used, for these may be scattered about the feeding grounds, neither being blown away nor difficult for the sheep to find. Usually such expensive protein supplements are fed only when native grass hays are being utilized, high-quality alfalfa not requiring a protein supplement.

Because of the magnitude of the range sheep industry and the fact that it is a highly specialized type of operation, in the sections that follow special discussion is devoted to the feeding and management of sheep on the range.

## NUTRIENT DEFICIENCIES OF RANGE FORAGE

Hunger, due to the lack of feed, is the most common deficiency on the western range. In particular, there may be a shortage of energy during droughts, late in the season, or early in the spring when grass is washy. Under such energy-deficient conditions, sheep lose weight and condition and lambs fail to grow. Also, reproduction is adversely affected.

Mature, weathered native range grass is almost always deficient in protein—being as low as 3% or less. Protein-leaching losses due to fall and winter rains may range from 37 to 73%.

Phosphorus deficiencies are rather common among range sheep, but calcium deficiency is seldom encountered.

Of the vitamins, vitamin A is most likely to be deficient in range forage, because dry, bleached range grass is very low in carotene (the precursor of vitamin A).

## RANGE SUPPLEMENTS

Four suggested range supplements, ranging from high to low protein, are given in Table 32–10.

Sheep on poor or weathered range grass should be supplemented by feeding the high protein formulation in Table 32–10, to correct the protein and phosphorus deficiencies. Of course, the supplements in Table 32–10 may be modified in keeping with the availability and cost of feeds, and yet meet known deficiencies. For example, if phosphorous is the only deficiency, it may be corrected by feeding a phosphorus supplement free choice.

There is no one best and most practical range supplement for any and all conditions. Many different feeds may be, and are, used; among them, (1) ranch or locally produced hay, (2) alfalfa pellets or cubes, with or without fortification, and (3) supplements of various kinds.

Also, producers can lessen the labor attendant to the daily feeding of a pasture or range supplement by (1) using protein blocks, or (2) self-feeding salt-feed mixtures.

Where salt is used for the purpose of governing consumption, the proportion of salt to feed may vary anywhere from 5 to 40% (with 30 to 33⅓% salt content being most common).

**TABLE 32–10**
**FORMULAS FOR RANGE SHEEP SUPPLEMENTS[1]**

| Feed[2] | Recommended Level of Protein | | | |
|---|---|---|---|---|
| | High | Medium-High | Medium-Low | Low |
| | (%) | | | |
| **B**arley, grain or corn, dent yellow, grain, grade 2 US, minimum 54 lb (24.5 kg)/bu | 5 | 40 | 75 | 65 |
| **B**eet, sugar, molasses, or sugar cane molasses, 48% invert sugar, minimum 79.5° Brix | 5 | 5 | 5 | 5 |
| **C**ottonseed with some hulls, solvent extracted, ground, minimum 41% protein, maximum 14% fiber, minimum 0.5% fat (cottonseed meal) | 66 | 36 | — | 16 |
| **S**oybean, seeds, solvent extracted, ground, maximum 7% fiber, 44% protein (soybean meal) | 10 | 10 | 10 | 10 |
| **U**rea, technical, 282% protein equivalent | — | — | 5 | — |
| **A**lfalfa, aerial parts, dehydrated, ground, minimum 17% protein or alfalfa, hay, sun-cured, early bloom | 10 | 5 | — | — |
| **V**itamin A ............................... (IU/lb) | — | 1,818 | 3,636 | 3,636 |
| *Vitamin A ............................... (IU/kg)* | — | *4,000* | *8,000* | *8,000* |
| **C**alcium phosphate, monobasic, commercial | 1 | 1 | 2 | 1 |
| **S**odium phosphate, monobasic, technical | 2 | 2 | 2 | 2 |
| **S**alt or trace mineralized salt | 1 | 1 | 1 | 1 |
| **Total** | 100 | 100 | 100 | 100 |

| Composition[3] | As-Fed[4] | M-F | As-Fed[4] | M-F | As-Fed[4] | M-F | As-Fed[4] | M-F |
|---|---|---|---|---|---|---|---|---|
| Digestible energy ......................... (Mcal/lb) | **1.4** | 1.5 | **1.4** | 1.5 | **1.4** | 1.5 | **1.3** | 1.4 |
| *Digestible energy ......................... (Mcal/kg)* | **3.0** | *3.3* | **3.0** | *3.3* | **3.0** | *3.3* | **2.8** | *3.1* |
| Protein (N × 6.25) ........................ (%) | **30.4** | 33.8 | **21.9** | 24.3 | **23.6** | 26.2 | **15.9** | 17.7 |
| Phosphorus ............................... (%) | **1.8** | 2.0 | **1.4** | 1.5 | **0.8** | 0.9 | **1.1** | 1.2 |
| Carotene ................................. (mg/lb) | **9.0** | 10.0 | **4.1** | 4.5 | — | — | — | — |
| *Carotene ................................. (mg/kg)* | **19.8** | *22.0* | **9.0** | *10.0* | — | — | — | — |
| Vitamin A ................................ (IU/lb) | — | — | **1,636.0** | 1,818.0 | **3,273.0** | 3,636.0 | **3,273.0** | 3,636.0 |
| *Vitamin A ................................ (IU/kg)* | — | — | **3,600.0** | *4,000.0* | **7,200.0** | *8,000.0* | **7,200.0** | *8,000.0* |
| Rate of feeding ........................... (lb/day) | **0.20–0.40** | 0.22–0.44 | **0.20–0.40** | 0.22–0.44 | **0.20–0.40[5]** | 0.22–0.44[5] | **0.20–0.40[5]** | 0.22–0.44[5] |
| *Rate of feeding ........................... (kg/day)* | **0.09–0.18** | *0.1–0.2* | **0.09–0.18** | *0.1–0.2* | **0.09–0.18[5]** | *0.1–0.2[5]* | **0.09–0.18[5]** | *0.1–0.2[5]* |

[1]Adapted by the author from *Nutrient Requirements of Sheep*, Sixth Revised Edition, NRC-National Academy of Sciences, 1985, p. 52, Table 11.

[2]Feeds mixed and fed in meal or pellet form.

[3]Molasses and alfalfa hay, sun-cured, early bloom not included.

[4]Estimated 90% dry matter.

[5]In emergency situations, up to 1.1 lb (0.5 kg) may be fed.

## RATE OF SUPPLEMENTAL FEEDING

The time and rate of supplemental feeding is determined by the reason for feeding supplements. Supplements are fed for two primary purposes: (1) to balance diets by adding small quantities of a nutrient (such as protein, a mineral, or a vitamin) or a combination of nutrients; and (2) to provide nutrients during short-term emergencies. As an example of the latter, a supplement may be needed to prevent sheep from eating poisonous plants during periods when they are on the trail or when forage is covered with snow.

Supplemental feeding should be timed to start when it is needed. If phosphorus supplementation is required, it should be provided continuously, perhaps by free-choice feeding. Where energy, protein, and /or vitamin A supplementation is involved, it takes a unique skill to recognize the nutritional state of the sheep, the range condition, and the need for supplement—both in kind and amount. The successful manager develops a grazing plan that minimizes the need for supplements, yet provides the proper supplement at the proper time and in the proper amounts.

The normal range of supplementation for sheep is ¼ to ½ lb per head per day. Rates above ½ lb approach a level that will result in reduced intake of range forage. Where range vegetation is so short as to require supplementation in excess of ½ lb per head per day, consideration should be given either to moving the sheep into drylot or to moving them to a better grazing area.

Some managers divide their sheep according to age, condition, and twins vs single lambs. Of course, this is facilitated where there are several bands. By so doing, it is possible (1) to give the animals that require the highest level of nutrition the best pasture or range, and/or (2) to supplement according to need.

## OTHER FEEDING AND MANAGEMENT ASPECTS

There are innumerable other feed and management aspects of great importance in sheep production; among them, those discussed in the sections that follow.

### POISONOUS PLANTS

The heaviest sheep losses from poisonous plants occur on the western range because (1) there has been less cultivation and destruction of poisonous plants in range areas; and (2) the frequent overgrazing on some of the western ranges has resulted in the elimination of some of the more nutritious and desirable plants, and these have been replaced by increased numbers of the less desirable and poisonous species.

The list of poisonous plants is so extensive that no attempt is made herein to list and discuss all of them. The most common poisonous plants of the intermountain ranges to which sheep are susceptible at certain times of the grazing season are listed in Table 32–11.

TABLE 32–11
COMMON POISONOUS PLANTS OF THE INTERMOUNTAIN REGION
TO WHICH SHEEP ARE SUSCEPTIBLE

| Plant | Time of Year |
|---|---|
| Broomweed | Spring and Summer |
| Chokecherry | Spring |
| Cooperweed | Summer |
| Death camas | Spring |
| Desert parsley | Spring |
| Greasewood | Fall |
| Halogeton | All Year |
| Horsebrush | Spring |
| Loco | Spring |
| Lupine | Summer and Fall |
| Milkweed | Summer |
| Rubberweed | Summer |
| Sneezeweed | Summer |
| Veratrum | Summer |

Halogeton and greasewood, poisonous plants that grow in the arid and semiarid saline regions of the West, cause especially heavy losses of range sheep. Both are toxic because of their high content of soluble oxalate (up to 30%). Toxicity occurs when large amounts of the oxalate are absorbed into the bloodstream where it produces its effect by (1) interference with certain steps in energy metabolism, (2) affinity for calcium, which produces hypocalcemia, and (3) mechanical damage to rumen and kidney tissues as a result of crystal formation. The toxic signs of halogeton and greasewood are: dullness, general weakness, incoordination, prostration, coma, and death.

## URINARY CALCULI (WATER BELLY, KIDNEY STONES, UROLITHIASIS)

Urinary calculi are mineral deposits which occur in the urinary tract. These deposits may block the flow of urine, followed by rupture of the urinary bladder and death. Affected animals stand with the back arched and strain to pass urine. They may kick at the belly, prefer to lie down, and become rather dull and disinterested in feed or water. In severe cases of some duration, water swellings (edema) of the lower abdomen may develop.

In feeder lambs, the disease appears to have a nutritional or metabolic origin; the affected animals excrete an alkaline urine that has a high phosphorus content. In range sheep, the disease is associated with the consumption of forages having a high silica content.

Salt (sodium chloride), fed at a level equivalent to 3 to 4% of the total ration, helps to prevent urinary calculi, especially in range sheep. The salt may be incorporated in the protein supplement provided adequate water is available. It increases water consumption and urine excretion.

The incidence of urinary calculi in finishing lambs can be greatly reduced by (1) preventing an excessive intake of phosphorus, and (2) maintaining a calcium:phosphorus ratio within the range of 2:1 to 2.5:1. Also, reducing the alkalinity of urine by feeding acid-forming salts is effective. For the latter purpose, ammonium chloride may be fed at a level of 1 oz per head per day.

## PREGNANCY DISEASE (KETOSIS, ACETONEMIA, PREGNANCY TOXEMIA)

Pregnancy disease is associated with undernourishment in late pregnancy. It is more common in ewes carrying twins and triplets than in those carrying singles.

Symptoms include grinding of the teeth, sluggishness, loss of appetite and weight, staggering gait, nervousness, and frequent urination. In the final stages, vision is impaired, and the animal collapses (unable to stand because of weakness, stiffness, or partial paralysis). Ewes that give birth during the early stages of the disease usually recover.

Pregnancy disease is a metabolic disorder—a disturbance of carbohydrate metabolism—which results in reduced blood glucose and elevated blood ketones. It can be prevented (1) by ensuring adequate feed intake in late pregnancy, and (2) by avoiding any change in feeding or management that might reduce the plane of nutrition or cause stress. At the onset of symptoms, a drench of propylene glycol (3 to 4 oz, 3 times daily) can be used as an energy source for ewes refusing to eat sufficient feed.

## FEEDING GROWING-FINISHING LAMBS

The growing-finishing stage of lambs refers to that period extending from birth to weaning at 4 to 6 months of age. At no other period in the life of the sheep is the promotion of growth and prevention of disease so important.

Where succulent pastures are available, most practical sheep producers, including producers with both farm flocks and range bands, consider that a combination of such green forage plus the ewe's milk is ample. In fact, lambs are unique among farm animals, inasmuch as they may be marketed at top prices off grass. Although young cattle may be sold off grass without having any other feed, they will usually fail to get sufficiently fat to bring top prices.

Frequently, farm-flock lambs are creep fed grain in addition to receiving their mother's milk and pastures. Usually creep feeding on the western range is too difficult. Should the range forages not be sufficiently abundant or lush to produce fat lambs, range sheep producers usually elect to sell their animals as feeders at weaning time.

Good pastures for lambs are those that are rather succulent and that are composed of plants that are palatable and nutritious. This means green, actively growing pastures in contrast to dormant or dried forages.

Hothouse lambs are born out of season, in the fall or early winter, when pastures are usually unavailable. It is necessary that these animals be crowded for slaughter at 2 to 4 months of age, when they should weigh from 40 to 60 lb. In addition to the right breeding for this specialty, therefore, hothouse lambs must be carefully fed. In the first place, the ewes should be given liberal quantities of a good succulent ration in order to stimulate the milk flow. Secondly, the lambs should be creep fed with a palatable and suitable ration from the time they are 2 weeks of age until marketing.

## EARLY WEANING

*Early weaning refers to the practice of weaning lambs earlier than usual—to weaning at 6 to 8 weeks of age or earlier.* There is much interest in early weaning because—

1. Of lambing out of season, multiple births, and more than one lamb crop per year.

2. Lactating ewes usually reach a peak in milk production 3 to 4 weeks after lambing, then decline thereafter. By 3 to 4 months after lambing, many ewes will be producing very little milk.

3. Fewer parasite problems accompany an early weaning program.

4. Increased knowledge of nutrition now makes it possible for scientists to improve upon milk (except for colostrum), chiefly by reinforcing it with certain vitamins and minerals.

5. Young gains are cheap gains, due to (a) the higher water and lower fat content of young animals in comparison with older animals, and (b) the higher feed consumption per unit weight of young animals.

6. Following weaning, ewes can be maintained on a limited feed allowance, thereby effecting a saving in cost.

For successful early weaning, superior nutrition and management are essential; and the earlier the weaning age the more exacting these requirements.

Early weaning of lambs is, to a considerable extent, a matter of preparation, rather than the abrupt separation of lambs from their mothers. Lambs that are to be early weaned should be creep fed from the time they are old enough to eat. At weaning time, the separation should be made by removing the ewes from the lambs, rather than *vice versa.* By keeping the lambs in familiar surroundings, stress is minimized.

An early-weaned lamb ration should meet the following specifications: contain a minimum of 16% crude protein; be fortified with supplemental iron if the lambs are raised on slotted floors; and have a calcium:phosphorus ratio of at least 1:1 (2:1 if urinary calculi has been experienced).

Milk replacers containing approximately 30% fat and 24% protein have been used successfully in feeding lambs receiving colostrum and weaned at one day of age. Replacers with reduced lactose content (from 42 to 27% on a dry matter basis) give improved performance. The milk is fed (1) cold at 36 to 40°F, rather than warm, to reduce overeating and bacterial contamination, and (2) free choice. From the beginning, lambs are offered a very palatable solid feed in addition to the milk. The milk replacer is discontinued when the lambs are eating sufficient quantities of the dry feed, usually at 21 to 35 days of age.

## CREEP FEEDING

*The practice of supplemental feeding of nursing lambs in a separate enclosure away from their dams is known as creep feeding.* Lambs will usually consume some creep feed at 10 to 14 days of age.

Creep rations can either be hand-fed or self-fed. Many sheep producers hand-feed until the lambs begin to eat regularly, then self-feed from this point on.

The amount of creep feed consumed is inversely proportional to the ewe's milk production. For this reason, (1) twin lambs usually consume more than single lambs, and (2) significant amounts of creep feed are consumed at 6 to 8 weeks of age, at which time the ewe's milk production usually drops.

Until lambs are 6 weeks old, the grain should be crimped, cracked, or rolled, unless a pelleted ration is used. After this age, whole grain may be fed unless it is extremely hard (like millet).

It is important that the creep ration be very palatable. For this reason, rolled oats, wheat bran, soybean meal, and molasses are important ingredients in a creep ration. Even then, if lambs have access to lush pasture, they may prefer it to the creep feed.

Suggested creep rations are given in Tables 32–12 and 32–13.

### TABLE 32–12
### SOME EXCELLENT CREEP RATIONS (AS-FED BASIS)[1]

| | Unpelleted | | Pelleted | |
|---|---|---|---|---|
| | First 2 Months | 2 Months to Market | First 2 Months | 2 Months to Market |
| | (%) | (%) | (%) | (%) |
| Ground corn ......... | 80 | 60 | 40 | 50 |
| Ground oats ......... | — | 20 | 15 | — |
| Soybean meal ....... | 20 | 10 | 20 | 10 |
| Alfalfa hay .......... | — | — | 10 | 35 |
| Bran .............. | — | 10 | 10 | 10 |
| Molasses ........... | — | — | 5 | 5 |
| Trace mineral salt .... | .5 | .5 | .5 | .5 |
| Limestone ........... | 1.0 | 1.0 | 1.0 | 1.0 |
| Antibiotic .... (mg/lb)[2] | 50 | 20 | 50 | 15 |
| Vitamin A ..... (IU/lb) | 1,000 | 1,000 | 1,000 | 1,000 |
| Vitamin D ..... (IU/lb) | 200 | 200 | 200 | 200 |
| Vitamin E ..... (mg/lb) | 20 | 20 | 20 | 20 |

[1]The addition of 0.25 to 0.50% ammonium chloride will minimize urinary calculi.

[2]Chlortetracycline (Aureomycin) or oxytetracycline (Terramycin).

**Feeding Directions:**

1. Lambs should be started on creep feed about 10 days after birth. Although they will not consume significant amounts of feed until 3–4 weeks of age, the small amounts consumed at earlier ages are critical for establishing both rumen function and the habit of eating.

2. Feed high quality legume hay in a separate rack. Feed hay and creep ration twice daily to keep them fresh.

3. The amount of creep feed consumed by lambs 2 to 6 weeks of age is affected by the palatability of the ration (ration composition and ration form) and the location and environment of the creep area. A well-bedded, well-lighted area located close to where the ewes congregate is preferred.

### TABLE 32–13
### SOME SIMPLE CREEP RATIONS (AS-FED BASIS)[1]

| | Unpelleted | | Pelleted | |
|---|---|---|---|---|
| | First 2 Months | 2 Months to Market | First 2 Months | 2 Months to Market |
| | (%) | (%) | (%) | (%) |
| Ground corn ......... | 49 | 89 | 64 | 59 |
| Crushed oats ........ | 30 | — | — | — |
| Soybean meal ....... | 20 | 10 | 20 | 10 |
| Limestone ........... | 1.0 | 1.0 | 1.0 | 1.0 |
| Trace mineral salt .... | .5 | .5 | .5 | .5 |
| Alfalfa ............. | — | — | 10 | 25 |
| Molasses ........... | — | — | 5 | 5 |

[1]The addition of 0.25 to 0.50% ammonium chloride will minimize urinary calculi.

**Feeding Directions:**

Same as presented with Table 32–12; so, see the latter.

## FEEDING ORPHAN ("BUMMER") LAMBS (ARTIFICIAL REARING)

Sheep producers estimate that about 10% of their lamb crop dies from starvation during the first week after birth. Some starvation results from newborn lambs sucking the scrotum and/or navel of another lamb. But most starved lambs are orphans (bummers) resulting from (1) the mother dying at lambing, (2) rejection by the mother, (3) the mother not being able to suckle the lamb because of mastitis or some similar problem, or (4) multiple births beyond the ewe's nursing capacity. Whatever the cause, the most satisfactory arrangement for the orphan is to provide a foster mother—to transfer (graft) the lamb to another ewe. The alternate to a foster mother arrangement is artificial rearing.

Observance of the following principles and practices will increase the chances of raising orphan lambs artificially:

• **Give colostrum**—Colostrum makes for a good start in life. A newborn lamb needs 3.2 oz of colostrum per pound body weight during the first 18 hours after birth, according to the Modern Research Institute in Scotland. Colostrum contains antibodies which impart immunity to infections for the first few weeks of life. This is important because the lamb's own immune system does not develop until it is 3 to 4 weeks old. Colostrum may be stored for this purpose. If a ewe either drops a stillborn lamb or loses her lamb within a day of birth, she may be milked and the colostrum frozen. Then it can be warmed to 100°F and fed as needed. If ewe colostrum is not available, colostrum from a cow or a goat may be used, although it will not impart immunity to certain infections that are specific to sheep.

• **Inject orphans**—When orphan lambs are placed in the nursery, inject them with (1) vitamins A, D, and E, (2) iron-dextran, and (3) selenium in selenium-deficient areas.

Also, enterotoxemia should be prevented. If the ewes were vaccinated with Type D toxoid prior to lambing, orphan lambs will receive colostral protection for 2 to 3 weeks; then, they should be vaccinated at 4 to 6 weeks of age. If the ewes were not vaccinated, the lambs should be vaccinated with the toxoid at 3 weeks of age and again at 5 weeks of age.

• **Use milk replacer**—A number of commercially prepared milk replacers are on the market. Best results will be obtained by using a replacer containing 25 to 30% fat, 20 to 25% protein provided by spray-dried milk products, and not to exceed 30 to 35% lactose; with the milk replacer diluted, mixed, and fed according to the manufacturer's directions.

Where several orphans are being fed, up to 12 lambs of similar size and age may be grouped together in a small pen and self-fed from a multiple nipple container, allowing one nipple for each 2 to 4 lambs. When self-feeding, cool milk (50° to 60°F) should be fed because (1) it does not sour as quickly, and (2) the lambs are not apt to engorge on it. However, hand-fed lambs can safely be given warm milk fed twice daily.

Do not use cow's milk or calf milk replacer for lambs. They contain too much lactose (milk sugar) for lambs and will cause scours.

Because milk replacer is expensive, the liquid-feeding period should be as short as possible. Lambs can be successfully weaned from milk replacer at 18 to 28 days of age.

• **Provide a good starter feed and water from day one**—From day one, orphan lambs should be provided access to a palatable dry ration to accustom them to eating dry feed and

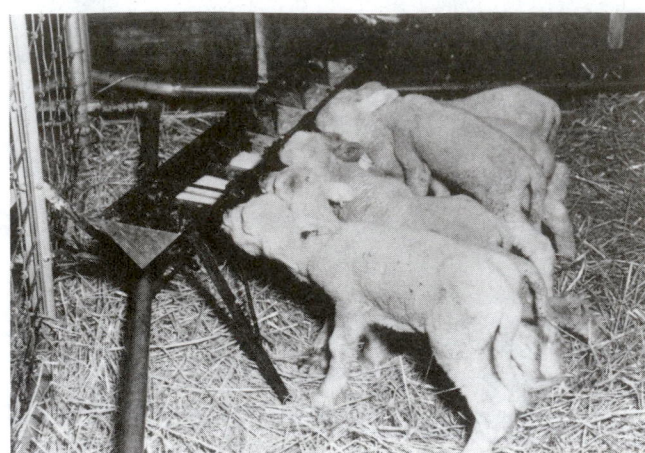

Fig. 32–13. Orphan lambs being self-fed milk replacer from a multiple-nipple container. (Courtesy, Ralston Purina Co., St. Louis, MO)

stimulate rumen development. This ration should be (1) palatable, (2) high in energy, (3) high in protein (22 to 24% crude protein on as-fed basis), (4) reinforced with minerals and vitamins, and (5) fed in finely ground (mash) form. A good starter ration follows:

**Lamb Starter Rations:**

| Ingredients | % |
| --- | --- |
| Soybean meal (49% CP) | 40.0 |
| Ground corn | 27.0 |
| Alfalfa meal | 15.0 |
| Dextrose (corn sugar) | 10.0 |
| Fat (e.g. vegetable oil) | 5.0 |
| Limestone | 2.0 |
| Trace mineral salt | 0.7 |
| Vitamin premix | 0.3 |
| Total | 100.0 |

Once the lambs have fully adjusted to the starter ration, they can be slowly switched onto the regular creep or grower ration. (See Tables 32–12 and 32–13.)

• **Maximize sanitation, observation, and TLC**—The successful artificial rearing of orphan lambs necessitates that the caretaker maximize sanitation, observation, and TLC—tender loving care.

## FINISHING LAMBS[1]

The primary objective of the sheep producer is that of producing milk-fat lambs suitable for slaughter at weaning time. Only when pasture is inadequate are lambs sold via the feeder route. Almost all feeder lambs come from the range area. Some range areas produce only a small percentage of lambs which are classed as feeders, whereas in other areas almost all the lambs must be sold as feeders because the vegetation is not sufficient to promote rapid growth and finishing. It is estimated that, for the range area as a whole, an average of at least 50% of all lambs produced in one year receive additional feed after they are removed from the range prior to slaughter.

[1]Many helpful suggestions for this section were received from the following authority: Dr. Clair Acord, Ph.D., Utah Wool Growers Assn., Salt Lake City, UT.

## AREAS AND TYPES OF LAMB FEEDING

Feeder lambs are generally sold to go into districts where grains and other concentrates are abundant or where fall and winter pastures are available. Such areas include (1) the irrigated districts of the Far West where the feeding of sugar beet by-products—beet pulp, beet tops, and molasses—predominates, (2) the wheat-raising sections of Oklahoma, Kansas, Nebraska, and Texas, where fall grazing of the wheat fields is practiced, and (3) the Corn Belt where stubble fields and meadows are gleaned and the corn crop may be harvested by lambs.

Colorado is the leading lamb-feeding state of the nation, finishing about one-sixth of the sheep and lambs fed in the United States. Here, locally grown alfalfa, sugar beet by-products, and barley are used extensively, along with considerable corn and protein supplements which are shipped in from outside areas. California ranks second in lamb feeding, followed by Texas, Wyoming, and Oregon.

Numerous feeding practices and a great variety of feeds are used in lamb-finishing operations. In general, however, all methods may be classified as either (1) field finishing, or (2) drylot finishing.

## FIELD FINISHING

Fig. 32–14. "Sheeping down" corn on an Iowa farm. First the crabgrass, weeds, and lower corn leaves are eaten. (Photo by A. M. Wettach, Mt. Pleasant, IA)

This method of finishing lambs is somewhat comparable to the pasture finishing of cattle, except that a greater variety of feeds is used by lamb feeders. Field feeding requires relatively little labor and equipment, and the manure is dropped back on the land where it will do the most good. Death losses run higher than in drylot feeding because the feed consumption cannot be controlled.

The kind of field feeding varies from area to area and even between farms within the same locality. Most of the feeder lambs are shipped to these feeding areas in August and September at the time the lambs are normally weaned from range ewes. Usually these field-fed lambs are ready for market in November and early December, though it is not uncommon for a small percentage of thin lambs to be held back for additional drylot finishing of 30 to 60 days.

Fig. 32–15. Lambs finishing on winter wheat in western Kansas. (Courtesy, Rufus F. Cox, Kansas State University)

Throughout the Corn Belt, feeder lambs are usually used as scavengers during the early part of the field feeding process. Frequently, the lambs are pastured in the stubble fields or on the meadow until all these feeds are consumed, after which they are turned into corn fields.

In Kansas, Oklahoma, Nebraska, and Texas, thousands of lambs are finished primarily by fall pasturing of the wheat fields. In the Pacific Northwest, a limited number of lambs are finished by gleaning pea stubble.

## DRYLOT FEEDING

Drylot feeding is, as the name indicates, feeding under restricted conditions. This may be either (1) shelter or barn feeding, or (2) open-yard feeding.

### SHELTER OR BARN FEEDING

Because of inclement weather in the fall and early winter, many of the lamb-feeding operations in the central and eastern states are in drylots which afford shelter. In some instances, the lambs are kept under cover without an exercising lot. These barns may consist of anything from open sheds to more costly and elaborate structures, including slotted floors. In the vast majority of instances, the feeds are locally grown, corn being the chief concentrate used in these feeding operations. Lambs are finished as a means of marketing the grain, conserving the fertility of the soil, and furnishing gainful work during the winter months. Most of these lambs are finished by farmers who feed one or more carloads, rather than by large operators who feed thousands of lambs. Practically all of the feeder lambs used in these operations come from the western ranges, either directly or via markets handling feeder lambs. After getting on full feed, these lambs are either hand-fed twice daily or self-fed.

### OPEN-YARD FEEDING

Open-yard feeding is the common method of finishing lambs in the irrigated areas of the West, though a few eastern lamb-feeding operations are in open yards. In this system, equipment costs are kept to a minimum—the facilities merely consisting of an enclosure and well-drained yard which may or may not have a natural or constructed windbreak, and the necessary feed bunks. Open-yard feeding is often used by large operators who feed thousands of lambs.

Fig. 32–16. Open-yard, self-feeding of finishing lambs. (Courtesy, *The Sheepman's Production Handbook,* Denver, CO)

Large quantities of alfalfa and sugar beet by-products are utilized in these yards. As it is often most economical to use a maximum of roughages, the feeding period may extend for as long as 4 to 5 months. Also, because of the desire to use a large proportion of roughage to grain, the lambs in these western feedlots are usually hand-fed twice daily rather than self-fed. In order to save labor, however, the practice of self-feeding is increasing. Lambs may be self-fed successfully, but it is recommended that the following precautions be taken in order to lessen the incidence of overeating disease: (1) good management, (2) vaccination against enterotoxemia, and (3) more roughage and less concentrate be used.

## BASIC CONSIDERATIONS IN FINISHING LAMBS

Although no rules of success are applicable to any and all conditions, the following basic considerations in finishing lambs are worth noting:

1. In lamb-feeding operations, the purchase price of the lambs represents 60–70% of all costs. This indicates the importance of keeping death losses to a minimum.

2. Experienced feeders normally expect to lose about 1 to 2% of lambs on feed. This is about twice the loss that occurs in commercial cattle-feeding operations.

3. Lamb feeding is seasonal in nature, usually extending from August to about the following May. This seasonal condition is due to the fact that (a) suitable feeder lambs are not available until the late summer and fall months, and (b) following the growing and harvesting seasons, the feeders have available quantities of marketable and unmarketable feeds which may be utilized by lambs.

4. As in cattle feeding, feedlot gains are expensive, usually costing more per pound than the selling price on the market. Thus, a reasonable margin or difference between the cost and selling price per hundredweight is necessary.

5. Feed accounts for approximately 66% of the cost of finishing feedlot lambs, exclusive of the initial purchase price of the feeder lambs.

6. Though the situation varies according to the kind of feed and the age of animals, it requires about the following amounts of feed per pound gain:

| Age | Lb Feed/Lb Gain |
|---|---|
| Preweaned lambs .......... | 2.0 to 2.5 |
| Early weaned lambs ........ | 2.5 to 4.0 |
| Late weaned lambs ......... | 6.0 to 8.0 |

7. In a 250-mile shipment, lambs will shrink about 5%. If properly fed, watered, and cared for *en route,* lambs may be shipped 1,500 to 2,000 miles without much greater shrinkage than this.

8. Most feeder lambs weigh between 60–80 lb when placed on feed and from 105–125 lb following a 90- to 120-day feeding period.

9. Wool is of importance in selecting feeder lambs because it has a bearing on their market value, the pelt being the most valuable slaughter by-product.

10. Range feeder lambs are more plentiful than native feeders, thus allowing for greater selection; and usually they are more uniform and are less heavily infected with parasites.

11. Lambs are frequently fed on a contract basis, with many and varied agreements being used.

12. Wether lambs appear to make slightly more rapid gains than ewe lambs, but they do not finish quite so early as ewe lambs.

13. Where western lambs have undergone a long shipment immediately after being taken from their mothers, special care is necessary in starting them on feed. After rest following shipment, lambs are usually started on grain by feeding about ¼ lb per head daily. Gradually this allowance is increased so that the lambs are getting a full feed of about 2 lb of grain per head daily and about the same amount of hay when on full feed 4 weeks later.

14. A great variety of feeds can be used in lamb feeding. In general, the successful feeder balances out the ration by selecting those feeds which are most readily available at the lowest possible price.

15. Unless such extremely hard seeds as millet are included in the ration, it does not pay to grind feeds for finishing lambs.

## FEED ALLOWANCE AND SOME SUGGESTED RATIONS FOR SHEEP

Sheep rations vary with the section of the country, depending chiefly on available local feeds. Fortunately, many feeds of similar nutritive properties can be interchanged in the ration as price relationships warrant. This makes it possible at all times to obtain a balanced ration at the lowest cost.

Except at lambing time or when emergencies occur as a result of drought or inclement weather, western bands receive little supplemental feed. Even with farm flocks, a minimum of grain is fed to breeding animals. Grain feeding usually is limited to the latter part of gestation and to the lactation period prior to turning to pasture.

Tables 32–14 and 32–15 contain some rations that have been used by successful sheep operators in various sections of the country.

### TABLE 32–14
### DAILY RATIONS FOR BREEDING EWES AT VARIOUS STAGES OF PRODUCTION

| Ration No. | Moisture Basis¹ A-F (as-fed) M-F (moisture-free) | Hay² (lb) | Hay² (kg) | Corn Silage (lb) | Corn Silage (kg) | Haylage (lb) | Haylage (kg) | Corn Straw (lb) | Corn Straw (kg) | Stover (Stalks) (lb) | Stover (Stalks) (kg) | Grain³ (lb) | Grain³ (kg) | Protein Supplement⁴ (lb) | Protein Supplement⁴ (kg) |
|---|---|---|---|---|---|---|---|---|---|---|---|---|---|---|---|
| **Maintenance** | | | | | | | | | | | | | | | |
| 1 | A-F | 3.0 | 1.4 | | | | | | | | | | | | |
|   | M-F | 3.3 | 1.5 | | | | | | | | | | | | |
| 2 | A-F | | | 6.0 | 2.7 | | | | | | | | | 0.20 | 0.09 |
|   | M-F | | | 6.7 | 3.0 | | | | | | | | | 0.22 | 0.10 |
| 3 | A-F | | | | | 6.0 | 2.7 | | | | | | | | |
|   | M-F | | | | | 6.7 | 3.0 | | | | | | | | |
| 4 | A-F | | | | | | | 3.0 | 1.4 | | | | | 0.40 | 0.18 |
|   | M-F | | | | | | | 3.3 | 1.5 | | | | | 0.44 | 0.20 |
| **Gestation, early (first 15 weeks)** | | | | | | | | | | | | | | | |
| 1 | A-F | 3.5 | 1.6 | | | | | | | | | | | | |
|   | M-F | 3.9 | 1.8 | | | | | | | | | | | | |
| 2 | A-F | 2.0 | 0.9 | | | | | | | | | 1.0 | 0.45 | | |
|   | M-F | 2.2 | 1.1 | | | | | | | | | 1.1 | 0.49 | | |
| 3 | A-F | 1.8 | 0.8 | | | | | | | | | 0.6 | 0.27 | 0.20 | 0.09 |
|   | M-F | 2.0 | 0.9 | | | | | | | | | 0.7 | 0.31 | 0.22 | 0.10 |
| 4 | A-F | | | 8.0 | 3.6 | | | | | | | | | 0.20 | 0.09 |
|   | M-F | | | 8.9 | 4.0 | | | | | | | | | 0.22 | 0.10 |
| 5 | A-F | | | | | 7.0 | 3.2 | | | | | 0.20 | 0.09 | | |
|   | M-F | | | | | 7.8 | 3.5 | | | | | 0.22 | 0.10 | | |
| 6 | A-F | 2.0 | 0.9 | | | | | | | 2.0 | 0.9 | 0.5 | 0.23 | | |
|   | M-F | 2.2 | 1.0 | | | | | | | 2.2 | 1.0 | 0.6 | 0.27 | | |
| 7 | A-F | 1.0 | 0.45 | | | | | | | 2.0 | 0.9 | 0.5 | 0.23 | 0.30 | 0.14 |
|   | M-F | 1.1 | 0.49 | | | | | | | 2.2 | 1.0 | 0.6 | 0.27 | 0.33 | 0.15 |

**Gestation, late (last 4 weeks): Add 0.5–1.0 lb (0.23–0.45 kg) grain per ewe daily to any of the above rations.**

| Ration No. | Moisture Basis | Hay² (lb) | Hay² (kg) | Corn Silage (lb) | Corn Silage (kg) | Haylage (lb) | Haylage (kg) | Corn Straw (lb) | Corn Straw (kg) | Stover (Stalks) (lb) | Stover (Stalks) (kg) | Grain³ (lb) | Grain³ (kg) | Protein Supplement⁴ (lb) | Protein Supplement⁴ (kg) |
|---|---|---|---|---|---|---|---|---|---|---|---|---|---|---|---|
| **Lactation** | | | | | | | | | | | | | | | |
| 1 | A-F | 4.0 | 1.8 | | | | | | | | | 2.0–3.0 | 0.9–1.4 | — | — |
|   | M-F | 4.4 | 2.0 | | | | | | | | | 2.2–3.3 | 1.1–1.5 | — | — |
| 2 | A-F | | | 10.0 | 4.5 | | | | | | | 1.5 | 0.7 | 0.25 | 0.11 |
|   | M-F | | | 11.1 | 5.0 | | | | | | | 1.7 | 0.8 | 0.28 | 0.13 |
| 3 | A-F | 1.0 | 0.45 | 8.0 | 3.6 | | | | | | | 1.5 | 0.7 | 0.20 | 0.09 |
|   | M-F | 1.1 | 0.49 | 8.9 | 4.0 | | | | | | | 1.7 | 0.8 | 0.22 | 0.10 |
| 4 | A-F | | | | | 8.9 | 3.6 | | | | | 2.0–3.0 | 0.9–1.4 | | |
|   | M-F | | | | | 8.9 | 4.0 | | | | | 2.2–3.3 | 1.1–1.4 | | |

¹As-fed was calculated using an average figure of 90% dry matter. When using silages, roots, and other wet feeds, these feeds should be converted to a moisture-free basis and the ration calculated using the moisture-free data.

²Alfalfa hay, midbloom, preferred.

³Grain may consist of corn, barley, wheat, oats, and/or grain sorghum.

⁴Protein supplement may consist of soybean, cottonseed, linseed, sunflower, safflower, or rapeseed meal.

**Feeding Directions:**

1. These rations are formulated to meet the requirements of a 154 lb (70 kg) ewe in average condition, and are designed for hand-feeding. The daily feed allowance can be increased or decreased, depending on the actual size of the ewe and the body condition.

2. Some of these rations are deficient in calcium and/or phosphorus; therefore, a supplement containing 50% trace mineral salt (for sheep) and 50% dicalcium phosphate should be fed free choice. The consumption of 0.05 lb (0.02 kg) per sheep per day of this mixture will provide the amounts of calcium and phosphorus needed for maintenance and the first 15 weeks of gestation; and 0.10 lb (0.05 kg)/day will provide the needed Ca and P for late gestation and lactation. Vitamins A and E should be added to the salt-mineral mix when sheep are fed the wheat straw and corn stover rations.

3. Ewes should gain 15 to 25 lb (6.8 to 11.4 kg) during gestation. During early gestation (first 15 weeks), they should gain 0.05 lb (0.02 kg)/day. During late gestation (last 4 weeks), they should gain 0.5 lb (0.23 kg)/day. If, during the last 4 weeks of pregnancy, ewes are fed 0.5 to 1.0 lb (0.23 to 0.45 kg) grain per head daily and gain 8 to 15 lb (3.6 to 6.8 kg), ketosis (lambing paralysis) can be prevented almost entirely.

4. During maintenance and early gestation, each ewe should have 14 in. (36 cm) of bunk feed space. In late gestation and during lactation, bunk space should be increased to 15 to 18 in. (38 to 46 cm).

## TABLE 32–15
## GROWING–FINISHING RATIONS FOR LAMBS[1]

| Ingredient | Moisture Basis[2] A-F (as-fed) M-F (moisture-free) | Rations Using Corn/Alfalfa Hay/Soybean Meal | | | | | Rations Using Milo/Cottonseed Hulls/Cottonseed Meal | | | | |
|---|---|---|---|---|---|---|---|---|---|---|---|
| | | 1 | 2 | 3 | 4 | 5 | 1 | 2 | 3 | 4 | 5 |
| | | (%) | (%) | (%) | (%) | (%) | (%) | (%) | (%) | (%) | (%) |
| Corn grain (dent yellow) | | 31.0 | 41.5 | 51.7 | 63.0 | 73.3 | — | — | — | — | — |
| Sorghum grain (milo) | | — | — | — | — | — | 19.5 | 32.7 | 46.2 | 60.7 | 73.7 |
| Alfalfa hay (mature) | | 55.0 | 45.0 | 35.0 | 25.0 | 15.0 | 15.0 | 15.0 | 15.0 | 15.0 | 15.0 |
| Cottonseed hulls | | — | — | — | — | — | 40.0 | 30.0 | 20.0 | 10.0 | — |
| Soybean meal (solvent 44% CP) | | 7.0 | 6.5 | 6.0 | 5.5 | 5.0 | — | — | — | — | — |
| Cottonseed meal (solvent 41% CP) | | — | — | — | — | — | 17.5 | 14.0 | 10.5 | 7.0 | 4.0 |
| Molasses (cane) | | 6.0 | 6.0 | 6.0 | 5.0 | 5.0 | 6.0 | 6.0 | 6.0 | 5.0 | 5.0 |
| Calcium carbonate | | — | — | .3 | .5 | .7 | 1.0 | 1.3 | 1.3 | 1.3 | 1.3 |
| Trace mineral salt (sheep) | | .5 | .5 | .5 | .5 | .5 | .5 | .5 | .5 | .5 | .5 |
| Ammonium chloride | | .5 | .5 | .5 | .5 | .5 | .5 | .5 | .5 | .5 | .5 |
| **Nutritional content** | | | | | | | | | | | |
| Dry matter (%) | A-F | 87.5 | 87.5 | 87.5 | 87.6 | 87.6 | 89.0 | 88.9 | 88.7 | 88.7 | 88.5 |
| | M-F | 100 | 100 | 100 | 100 | 100 | 100 | 100 | 100 | 100 | 100 |
| TDN (%) | A-F | 75.3 | 80.1 | 84.6 | 89.2 | 93.1 | 67.9 | 72.3 | 77.2 | 82.0 | 86.8 |
| | M-F | 65.9 | 70.1 | 74.0 | 78.1 | 82.0 | 60.4 | 64.3 | 68.5 | 72.7 | 76.8 |
| Net energy for maintenance (Mcal/lb) | A-F | .80 | .86 | .91 | .98 | 1.04 | .71 | .76 | .82 | .88 | .94 |
| | M-F | .70 | .75 | .80 | .86 | .91 | .63 | .68 | .73 | .78 | .83 |
| *Net energy for maintenance (Mcal/kg)* | *A-F* | *.36* | *.39* | *.41* | *.44* | *.47* | *.32* | *.35* | *.37* | *.40* | *.43* |
| | *M-F* | .32 | .34 | .36 | .39 | .41 | .29 | .32 | .33 | .35 | .38 |
| Net energy for gain (Mcal/lb) | A-F | .40 | .47 | .53 | .59 | .66 | .33 | .38 | .45 | .52 | .59 |
| | M-F | .35 | .41 | .46 | .52 | .58 | .29 | .34 | .40 | .46 | .52 |
| *Net energy for gain (Mcal/kg)* | *A-F* | *.18* | *.21* | *.24* | *.27* | *.30* | *.15* | *.17* | *.20* | *.24* | *.27* |
| | *M-F* | .16 | .19 | .21 | .24 | .26 | .13 | .15 | .18 | .21 | .24 |
| Crude protein (%) | A-F | 17.1 | 16.6 | 16.0 | 15.4 | 14.8 | 17.0 | 16.3 | 15.8 | 15.2 | 14.8 |
| | M-F | 15.0 | 14.5 | 14.0 | 13.5 | 13.0 | 15.1 | 14.5 | 14.0 | 13.5 | 13.1 |
| Protein bypass (%) | A-F | 42.3 | 44.1 | 45.8 | 47.9 | 49.7 | 44.6 | 47.6 | 51.0 | 54.7 | 57.9 |
| | M-F | 37.0 | 38.6 | 40.1 | 42.0 | 43.5 | 39.7 | 42.3 | 45.2 | 48.5 | 51.2 |
| Calcium (%) | A-F | .86 | .71 | .70 | .67 | .66 | .85 | .96 | .94 | .91 | .89 |
| | M-F | .75 | .62 | .61 | .59 | .58 | .76 | .85 | .83 | .81 | .79 |
| Phosphorus (%) | A-F | .29 | .30 | .31 | .31 | .32 | .37 | .37 | .36 | .36 | .36 |
| | M-F | .25 | .26 | .27 | .27 | .28 | .33 | .33 | .32 | .32 | .32 |

[1]Adapted by the author from *The Sheepman's Production Handbook*, published by the Sheep Industry Development Program, Inc., Denver, CO, 1986, p. N-44, Table 13.

[2]As-fed was calculated using an average figure of 90% dry matter. When using silages, roots, and other wet feeds, these feeds should be converted to a moisture-free basis and the ration calculated using the moisture-free data.

**Feeding Directions:**

1. These rations can be fed once daily in troughs or bunks if there is capacity for a day's feed. They can also be self-fed if the feeders are designed to handle such feed without bridging.

2. Offering lambs a good quality hay for 1–3 days along with rations 1 or 2 (provided free choice) can be used to start lambs on feed.

3. About 3 in. *(7.6 cm)* of self-feeder or trough space must be provided per lamb for self-feeding and about 12 in. *(30.5 cm)* if hand-fed.

4. Gradually adapt the lambs to the higher energy rations by allowing 4–7 days on a ration before switching to the ration with the next higher energy level.

5. Complete mixing to prepare a uniform ration is important.

6. Lambs must not be allowed to be without feed even for a short period of time.

7. The mineral and vitamin mixture given in Table 32–16 may replace the trace mineral salt in all Table 32–15 rations.

The rations in Table 32–15 are nutritionally adequate and balanced with respect to Ca:P and N:S ratios. The mineral and vitamin mixture given in Table 32–16 may replace the trace mineral salt in all the Table 32–15 rations.

The rations in Table 32–15 are nutritionally adequate and balanced with respect to Ca:P and N:S ratios. The mineral and vitamin mixture given in Table 32–16 may replace the trace mineral salt in all the Table 32–15 rations.

Some suggested grain fitting rations are given in Tables 32–17 and 32–18. To each of these grain rations, good quality roughage should be added.

### TABLE 32–16
### MINERAL AND VITAMIN MIXTURE FOR LAMB RATIONS[1]

| Ingredient | Lb/Ton | Kg/Ton | Contribution to Complete Ration[2] |
|---|---|---|---|
| Salt, plain fine mixing ...... | 1,729.613 | 785.244 | .43% salt |
| Sulfur, elemental[3] .......... | 200.00 | 90.80 | .05% S |
| Cobalt carbonate (CaCO₃) .... | 0.087 | 0.039 | .1 ppm Co |
| Ethylenediamine dihydro-iodide (EDDI) ........... | 0.100 | 0.045 | .2 ppm I |
| Manganese oxide (MnO) ..... | 10.300 | 4.676 | 20 ppm Mn |
| Zinc oxide (ZnO) ........... | 10.300 | 4.676 | 20 ppm Zn |
| Vitamin A[4] ................. | 17.6 | 8.0 | 600 IU/lb |
| Vitamin E[5] ................. | 32.0 | 14.5 | 10 IU/lb |

[1]Adapted by the author from *The Sheepman's Production Handbook,* published by the Sheep Industry Development Program, Inc., Denver, CO, 1986, p. N-45, Table 14.

[2]Contribution to the complete ration when 10 lb (*4.5 kg*) of the mineral and vitamin mixture is added to 1 ton of complete lamb ration.

[3]In complete rations containing ammonium sulfate instead of ammonium chloride for prevention of urinary calculi, sulfur should not be added to the mineral and vitamin premix (.5% NH₄SO₄ contributes .12% S to the ration).

[4]Contains 13,607,700 IU of vitamin A per pound.

[5]Contains 125,000 IU of vitamin E per pound.

### TABLE 32–17
### FITTING CONCENTRATE MIX FOR SHOW LAMBS

| Ingredient | % |
|---|---|
| Cracked corn ................................... | 50 |
| Whole or rolled oats ............................ | 35 |
| Soybean meal .................................. | 10 |
| Molasses ...................................... | 4 |
| Mineral (limestone/sheep salt-mineral mix) ................. | 1 |
| **Total** ...................................... | **100** |

### TABLE 32–18
### RATIONS FOR FITTING YEARLING AND MATURE SHEEP

| Ingredient | Ration Number | | | | | | | |
|---|---|---|---|---|---|---|---|---|
| | 1 | | 2 | | 3 | | 4 | |
| | (lb) | (kg) | (lb) | (kg) | (lb) | (kg) | (lb) | (kg) |
| Barley, rolled ........ | — | — | 40 | 18.2 | — | — | 10 | 4.5 |
| Corn, cracked ........ | — | — | — | — | 40 | 18.2 | — | — |
| Oats, rolled ......... | 50 | 22.7 | 40 | 18.2 | 40 | 18.2 | 60 | 27.2 |
| Peas (split) ......... | 40 | 18.2 | — | — | — | — | 10 | 4.5 |
| Protein supplement[1] ... | — | — | 10 | 4.5 | 10 | 4.5 | 10 | 4.5 |
| Wheat bran ......... | 10 | 4.5 | 10 | 4.5 | 10 | 4.5 | 10 | 4.5 |
| **Total** .......... | **100** | **45.4** | **100** | **45.4** | **100** | **45.4** | **100** | **45.4** |

[1]Cottonseed, linseed, rapeseed (canola), soybean, and/or sunflower meal.

## FITTING RATIONS

Fig. 32–17. Young Columbia rams fitted for sale. (Courtesy, *Sheep Breeder and Sheepman,* Columbia, MO)

In addition to being reasonably economical (mostly home-grown) and well-balanced, the ration for show and sale sheep must be palatable. Many feed combinations meet these specifications. The ration selected is usually determined by (1) the availability and price of feed in the area, and (2) the preference and judgment of the feeder.

Lambs (either creep fed or weaned lambs that are being fitted for show) will eat about 2½ lb of Table 32–17 fitting concentrate, per head daily when on full feed.

Yearlings will eat about 3 lb per head daily of one of the rations shown in Table 32–18 when on full feed, whereas mature sheep will eat about 3½ lb of one of these rations.

Exhibitors prefer to feed steam rolled oats and barley and nutted (pea-sized) old process, linseed meal.[2] Corn is usually cracked or coarsely ground, and peas are split or cracked. When pastures are not available, alfalfa is the most popular hay. But any good legume is quite satisfactory.

In fitting animals for show or sale, most successful shepherds feed a limited quantity of sliced carrots, cabbage, mangels (stock beets), rutabagas (swedes), or turnips. These succulent feeds are highly relished by sheep and appear to help their digestion and general thrift.

The following points also are pertinent in feeding sheep:

1. All classes and ages of sheep should be allowed free access to a double compartment mineral box, with loose salt in one compartment and a mixture of ⅓ salt and ⅔ steamed bone meal, or other suitable mineral, in the other.

2. Unless grains are unusually hard, they need not be ground for sheep. The animals prefer to do their own grinding, and the feeds are no more effectively utilized when ground.

[2]Among experienced shepherds, old-process linseed meal is especially popular for fine-wool sheep because of its conditioning effect on the fleece.

# QUESTIONS FOR STUDY AND DISCUSSION

1. In what ways may a farmer or rancher take practical advantage of the fact that sheep consume a higher proportion of forages than any other class of livestock?

2. Of what value are the nutritive needs of sheep as established by the National Academy of Sciences?

3. How are the energy needs of sheep usually met? What factors affect the energy requirements of sheep?

4. How are the protein needs of sheep usually met? What factors affect the protein requirements of sheep?

5. What are the deficiency symptoms in sheep of (a) energy, and (b) protein? How could you distinguish between them?

6. Why do the crude protein requirements of lambs of small, medium, and large mature weight genotypes differ?

7. How are the salt requirements of sheep usually provided? What are the symptoms of salt deficiency in sheep?

8. What are the common sources of calcium and phosphorus for sheep? What are the symptoms of (a) calcium deficiency in sheep, and (b) phosphorus deficiency?

9. Describe the deficiency symptoms in sheep of each of the following trace minerals, then tell how you could distinguish between them: cobalt, copper, fluorine, iodine, iron, manganese, molybdenum, selenium, and zinc.

10. What are the manifestations of toxicity in sheep of each of the following: molybdenum, copper, selenium, and fluorine?

11. Describe the deficiency symptoms in sheep of each of the following vitamins, then tell how you could distinguish between them: vitamin A, vitamin D, vitamin E.

12. What vitamin treatment is commonly used for polio-encephalomalacia?

13. Most lambs are marketed as milk-fat lambs directly off pastures or ranges without having any grain. What are the advantages of this from an economic standpoint? Can this practice be applied to cattle or swine?

14. How do sheep differ from cattle in their grazing habits?

15. Discuss the feeding of hays and other dry roughages to sheep.

16. Discuss the feeding of silages and root crops to sheep.

17. What is the physiological explanation of why sheep can consume more pounds of a high-quality forage than of a low-quality forage?

18. Discuss the feeding of additives to sheep.

19. List the steps that are necessary for successful drylot feeding of sheep.

20. What is the purpose of flushing? How would you flush a range band of ewes?

21. Discuss the feeding of pregnant ewes.

22. How does feeding at lambing time differ from feeding pregnant ewes?

23. How does feeding lactating ewes differ from feeding pregnant ewes?

24. What is *accelerated lambing*? How is accelerated lambing accomplished?

25. Discuss the proper feeding of rams.

26. Describe typical range grazing and management operations.

27. What are the most common nutrient deficiencies of range forage for sheep?

28. Discuss range supplements, both (a) as to kind, and (b) rate of feeding.

29. The heaviest sheep losses from poisonous plants occur in range areas rather than in farming areas. Why is this?

30. How would you lessen the incidence in sheep of (a) urinary calculi, and (b) pregnancy disease?

31. What is early weaning? Why is there much current interest in early weaning?

32. What is creep feeding? What are the primary requisites of a good creep ration?

33. What are the leading lamb-feeding states of the nation? Why are they adapted to lamb feeding?

34. Describe each of the following types of lamb feeding: (a) field finishing, (b) shelter or barn feeding, and (c) open yard feeding.

35. Lambs may be self-fed successfully, provided (a) they are vaccinated against enterotoxemia (overeating disease), and (b) considerable roughage and/or other bulky feeds are incorporated in the ration. What advantages does self-feeding have over hand-feeding?

36. Why is lamb feeding so seasonal in nature; more so than cattle feeding?

37. It requires 6 to 8 lb of feed to produce 1 lb of on-foot late weaned lamb. How does this efficiency of feed utilization compare with cattle and hogs?

38. Under what conditions might it be preferable that a feedlot operator feed lambs instead of cattle?

39. Evaluate the rations that are presented in Tables 32–14 and 32–15.

40. Evaluate the fitting rations that are presented in Tables 32–17 and 32–18.

## SELECTED REFERENCES

| Title of Publication | Author(s) | Publisher |
| --- | --- | --- |
| *Beginning Shepherd's Manual* | B. Smith | Iowa State University Press, Ames, IA, 1983 |
| *Handbook For Woolgrowers* | Edited by G. R. Moule | Australian Wool Board, Melbourne, Australia, 1972 |
| *Management and Diseases of Sheep, The* | Edited by the British Council | The British Council, London, and The Commonwealth Agricultural Bureaux, Slough, England, 1979 |
| *Nutrient Requirements of Sheep*, Sixth Revised Edition | National Research Council | National Academy Press, Washington, DC, 1985 |
| *Of Sheep & Shows* | J. C. P. Kroge | Paddock Publishing Co., Boulder, CO, 1972 |
| *Profitable Sheep Farming* | M. McG Cooper R. J. Thomas | Farming Press Ltd., Fenton House, Ipswich, England, 1971 |
| *Sheep and Goat Handbook* | Guest Professors | International Stockmen's School, Clovis, CA, annually since 1980 |
| *Sheep & Goat Science* | M. E. Ensminger R. O. Parker | The Interstate Printers & Publishers, Inc., Danville, IL, 1986 |
| *Sheep and Wool* | M. P. Botkin R. A. Field C. L. Johnson | Prentice Hall, Englewood Cliffs, NJ, 1988 |
| *Sheep: Applied and Basic Research Information Pub. During 1979* | Edited by D. R. Lincicome | International Goat and Sheep Research, Scottsdale, AZ, 1983 |
| *Sheep Book, The* | R. Parker | Charles Scribner's Sons, New York, NY, 1983 |
| *Sheepman's Production Handbook, The* | Edited by G. E. Scott | Sheep Industry Development Program, Inc., Denver, CO, 1985 and updated at intervals |
| *Sheep Production and Management* | C. V. Ross | Prentice Hall, Englewood Cliffs, NJ, 1989 |
| *Sheep Raisers Manual* | W. K. Kruesi | Williamson Publishing Co., Charlotte, VT, 1985 |
| *U.S. Sheep and Goat Industry, The* | C. S. Menzies, Chairman, Task Force | Council for Agricultural Science and Technology (CAST), Report No. 94, Ames, IA, 1982 |

New Zealand pastoral scene. New Zealand is noted for fine pastures, Romney sheep, and the densest sheep population in the world. (Courtesy, Department of Scientific and Industrial Research, Wellington, New Zealand)

Australian pastoral scene. Australia is the leading sheep country of the world; it has 14% of the world's sheep and produces 29% of the world's wool. (Courtesy, Australian News and Information Bureau)

Serve an elegant Spring Lamb Crown Roast with Vegetable Stuffing to celebrate the arrival of spring. (Courtesy, National Live Stock & Meat Board, Chicago, IL)

# MARKETING AND SLAUGHTERING SHEEP AND LAMBS

# Chapter 33

Marketing sheep and lambs is the job of pricing, assembling, sorting, transporting, and processing them, and of distributing their subsequent products. Marketing sheep and lambs is a complex and costly job, complicated by two situations: (1) The animals are widely dispersed over about 115,000 farms and ranches located in every state; and (2) the decline in slaughter sheep numbers and quantity of retail product put lamb at a disadvantage relative to other meats.

The trend toward fewer markets handling live sheep and lambs, and fewer, larger, and more specialized plants killing and processing them, makes the decision of where and how to market them a major consideration.

Also, the marketing of sheep and lambs differs from the marketing of cattle and hogs as follows:

1. In general, there is a greater transportation problem, because a preponderance of lambs are raised west of the Mississippi while most of the lamb is consumed east of the Mississippi.

2. Age is a greater factor in determining relative market values of sheep than it is in determining that of cattle and hogs; thus, 92 to 94% of the sheep slaughter is from lambs and yearlings.

3. Because of the wool or pelt, the by-products of sheep are more valuable than are the by-products obtained in cattle or hog slaughter.

4. Sheep producers take more breeders back to the country than do cattle or hog producers.

Fig. 33–1. Lamb chop mixed grill—appetite appeal and essential food nutrients. (Courtesy, American Meat Institute, Washington, DC)

## MARKETING CONSIDERATIONS PECULIAR TO SHEEP AND LAMBS

Although sheep and lambs are marketed through much the same channels as cattle and hogs (see Chapter 8, Marketing Livestock), the following differences exist:

• **Marketing on carcass weight and grade**—It is expected that the trend toward increased marketing on a carcass grade and weight basis will continue unabated. When marketing lambs on a carcass weight basis, the USDA guidelines call for weights to be taken and payments to be made on the basis of hot carcass weights without pencil shrink.

• **Pencil shrink**—In direct sales, it is common to deal on the basis of a pencil shrink: 3 to 4% for feeder lambs, and 4% for slaughter lambs. Producers should know how to convert or compare "asking or offer" prices for various shrinkage levels. For example, an offer of $60.00 with 4% pencil shrink is equivalent to a price of $57.60 without this reduction.

• **Spot (cash) sales**—The term *spot sales* appears to be more of the lexicon of sheep marketing than of cattle or hog marketing. Simply stated, it refers to on the spot negotiated price between the buyer and seller. In the marketing of sheep and lambs, the term *spot sales* is used primarily for the purpose of differentiating cash sales from contractual sales. Spot sales occur in the traditional market channels—direct, terminals, auctions, dealers, etc.—both on a live and a carcass basis. Among the **advantages** of spot sales are the following:

1. The price is determined quickly.
2. The title is transferred promptly.
3. The marketing costs, if any, are known and paid for soon after the transaction.
4. The seller is paid at the time of the sale or soon thereafter.

• **Ownership and control of lambs**—Most breeding sheep are owned by producers. However, an increasing number of lambs on feed are either (1) owned by packers, or (2) controlled through packers by some form of contractual or custom feeding arrangement. The vast majority of this ownership-control is centered in about a dozen firms. Their stated reasons for involvement in lamb feeding are (1) it increases plant efficiency, and (2) it improves merchandising programs.

Also, contracting for future delivery of either feeder lambs or slaughter lambs is common in the sheep industry. For example, a lamb feeder may contract for feeder lambs in July for delivery in September, or a packer may contract for slaughter lambs to be delivered the following week or later.

As a result of this ownership and control of lambs being centered in few packers, and of fewer lambs moving through traditional price-determining and price-reporting channels (like terminals and auctions), many knowledgeable sheep producers feel that a suitable market channel price barometer no longer exists.

• **Higher marketing costs**—Although the commission and yardage charges on a per head basis for lambs at central markets are about one-fourth that of cattle, on a per hundredweight basis for sheep and lamb costs are about three times higher than for cattle. This is attributed to low volume and problems in handling, as compared to other classes of livestock.

• **Discount plans for heavyweight live lambs**—More than 98% of all slaughter lambs grade USDA Prime or Choice. Consequently, there is less concern about lamb prices being discounted for low quality than for being discounted due to heavy live lambs or carcasses.

Western packers and dealers use the following four plans to discount live lambs at such times as wholesale markets discount heavyweight lamb carcasses:

1. **Sliding scale.** Under the sliding scale plan, the price decreases a specified amount per hundredweight for each pound that a shipment averages over a stipulated weight. For example, a 50-cent per hundredweight discount might be made for each pound that a shipment exceeds 110 lb. Thus, if the contract price is $60.00, and if lambs average 115 lb, the settlement or paying price would be $57.50.

2. **Stop weight.** Under this plan, the packer or dealer specifies an upper weight limit for the average liveweight of a lot, with no payment made for any weight in excess of this limit. For example, if the stop weight is 110 lb, and if the lot averages 115 lb, no payment is made for the 5 lb overweight.

3. **Guaranteed yield.** Under this plan, the buyer specifies that the shipment shall yield a certain dressing percentage; for example, 52%. If the lambs yield better than the guarantee, the payment is based on the actual at-plant liveweight. However, if the lambs fail to yield the specified percentage, the packer determines the pay weight by dividing the shrunk carcass weight by the guaranteed yield. Generally, packers weigh the carcass while hot and use an arbitrary calculate cooler shrink (usually 2.5 to 3%).

The following example will show how the guaranteed yield plan works: Lambs averaging 110 lb were bought on 52% guaranteed yield basis, equivalent to an average weight of 57.2 lb carcass weight shrunk. If the carcasses only average 56 lb after the pencil cooler shrink is applied to the hot weight, the packer would pay the seller for lambs averaging 107.69 liveweight ($56 \div 0.52 = 107.69$).

4. **Double dressed.** The double dressed weight plan is a variation of the guaranteed yield. The shrunk dressed weight of a lot is doubled. This is the live pay weight, regardless of the actual liveweight. The double dressed method assumes a 50% carcass yield. Thus, for lambs whose carcasses average 54 lb after the cooler shrink adjustment, payment is made on the basis of an average weight of 108 lb for the live lambs.

Although heavyweight lambs may be overfat and wasty, and, therefore, worth less per hundredweight than lightweight lambs, carefully selected and properly finished heavyweight lambs may actually have higher cutability carcasses and be more valuable per hundredweight than many lightweight lambs. This is so because:

    a. The carcass may yield a higher percentage of retail cuts.

    b. More pounds of retail cuts may be processed from heavyweight lambs for each work minute. It takes no longer to prepare a larger, higher cutability carcass for the meat case than for a light one; yet, the merchandiser has more meat to sell.

    c. The cuts may be a more desirable size for processing and merchandising and more preferred by the consumer.

## LEADING PUBLIC STOCKYARDS

Public markets have declined in importance as a market channel for sheep and lambs, as well as for cattle and hogs. Nevertheless, they do indicate the areas of market action and provide an established outlet for sheep producers, especially smaller operators. Although public stockyards vary from year to year in total receipts, the five largest sheep and lamb public stockyards in 1987, by rank, and their receipts, are given in Table 33–1.

**TABLE 33–1**
**SHEEP AND LAMB RECEIPTS OF FIVE LEADING PUBLIC STOCKYARDS, BY RANK[1]**

| Market | Receipts |
|---|---|
| **S**an Angelo, TX | 445,532 |
| **S**o. St. Paul, MN | 73,787 |
| **B**illings, MT | 68,720 |
| **S**ioux Falls, SD | 58,413 |
| **N**orfolk, NE | 53,559 |
| **United States total**, public stockyards | 898,156 |

[1]*Livestock and Meat Statistics, 1984–88*, USDA, p. 61, Table 57. Data for 1987.

## MARKET CLASSES AND GRADES OF SHEEP

Fig. 33–2. Blackfaced slaughter lambs, heavy weight—125 lb, Prime grade. (Courtesy, *Sheep Breeder and Sheepman*, Columbia, MO)

While no official grading of live sheep and lambs is done by the U.S. Department of Agriculture, grades do form a basis for uniform reporting of livestock marketing. Also, grades of live animals are intended to be directly related to the grades of the carcasses they produce.

The market classes and grades of sheep (see Table 33–2) follow closely the pattern for the classes and grades of cattle and swine. One notable difference is that a sizable number of sheep are sold as breeders. For the most part, this class is made up of mature western ewes that are sold to country buyers for the purpose of producing one or two more crops of lambs before again being returned to the market. Usually such ewes can be acquired at a lower cost than ewe lambs. Another difference between sheep and other species is found in the fact that one feeder class, namely the shearers, is based on wool value as well as adaptability for further feeding. (See next page.)

**TABLE 33-2**

**MARKET CLASSES AND QUALITY GRADES OF SHEEP**

| Sheep or Lambs | Use Selection | Sex Classes | Age | Weight Division | (pounds) | (kilograms) | Commonly Used Grades |
|---|---|---|---|---|---|---|---|
| Sheep | Slaughter sheep | Ewes | Yearling | Light | 120 down | 54.5 down | Prime, Choice, Good, Utility[1] |
| | | | | Medium | 120–140 | 54.5–63.6 | |
| | | | | Heavy | 140 up | 63.6 up | |
| | | | Mature (2-year-old or older) | Light | 140 down | 63.6 down | Choice, Good, Utility, Cull[1] |
| | | | | Medium | 140–160 | 63.6–72.7 | |
| | | | | Heavy | 160 up | 72.7 up | |
| | | Wethers | Yearling | Light | 130 down | 59.0 down | Prime, Choice, Good, Utility[1] |
| | | | | Medium | 130–150 | 59.0–68.2 | |
| | | | | Heavy | 150 up | 68.2 up | |
| | | | Mature (2-year-old or older) | Light | 150 down | 68.2 down | Choice, Good, Utility, Cull[1] |
| | | | | Medium | 150–170 | 68.2–77.3 | |
| | | | | Heavy | 170 up | 77.3 up | |
| | | Rams | Yearling | All weights | | | Prime, Choice, Good, Utility[1] |
| | | | Mature (2-year-old or older) | All weights | | | Choice, Good, Utility, Cull[1] |
| | Feeder sheep | Ewes and wethers | Yearlings | All weights | | | Fancy, Choice, Good, Medium, Cull |
| | | Ewes | Mature (2-year-old or older) | All weights | | | Choice, Good, Medium, Cull |
| | Breeding sheep | Ewes (rams occasionally purchased as breeders, but not listed in market reports) | Yearlings, 2-, 3-, or 4-year-old or older | All weights | | | Fancy, Choice, Good, Medium, Cull |
| Lambs | Slaughter lambs | Ewes, wethers, and rams | Hothouse lambs | 60 down | | | Prime, Choice, Good, Utility[1] |
| | | Ewes, wethers, and rams | Spring lambs | Light | 100 down | 45.4 down | Prime, Choice, Good, Utility[1] |
| | | | | Medium | 100–110 | 45.4–50.0 | |
| | | | | Heavy | 110 up | 50.0 up | |
| | | Ewes, wethers, and rams | Lambs | Light | 105 down | 47.7 down | Prime, Choice, Good, Utility[1] |
| | | | | Medium | 105–120 | 47.7–54.5 | |
| | | | | Heavy | 120 up | 54.5 up | |
| | Feeder lambs | Ewes and wethers | All ages | All weights | | | Fancy, Choice, Good, Medium, Cull |
| | Shearer lambs | Ewes and wethers | All ages | All weights | | | Choice, Good, Medium |

[1]In addition to the above quality grades, there are five yield grades applicable to all lamb and mutton carcasses, denoted by numbers 1 through 5, with the Yield Grade 1 representing the highest degree of cutability. Thus, slaughter sheep and lambs may be graded for (1) quality alone, (2) yield grade alone, or (3) both quality and yield grades.

# FACTORS DETERMINING MARKET CLASSES

The disposition or use to be made of sheep is determined by (1) whether they are sheep or lambs, (2) the use selection, (3) sex, (4) age, and (5) weight.

## SHEEP AND LAMBS

The first major market subdivisions of the ovine species separates the animals into sheep and lambs. Lambs include those animals that are approximately one year old and under. When there is any question as to whether animals should be classified as sheep or lambs, a final decision is based upon an examination of the teeth. If the first pair of larger, broader permanent teeth is about fully developed, the animal is classified as a yearling, for this change in the teeth takes place at about 12 months of age.

At the present time, lambs and yearlings make up to 92 to 94% of the total sheep slaughter.

## USE SELECTION OF SHEEP AND LAMBS

Sheep and lambs are divided into six market groups based on the uses made of them or the purposes for which they are best suited: slaughter sheep, feeder sheep, breeding sheep, slaughter lambs, feeder lambs, and shearer lambs. A brief description of each of these classes follows:

**Slaughter sheep**—Yearlings or older animals intended for immediate slaughter.

**Feeder sheep**—Yearlings or older animals best suited for further finishing. More than 60% of the lambs in the West sell as feeders.

**Breeding sheep**—Largely mature western ewes that are returned to the country for further reproduction. In addition to market grade in considering the suitability of ewes for breeding purposes, it is important that attention be given to the condition of the teeth and the breed, health, and general potentialities as a breeder.

**Slaughter lambs**—Young animals under one year of age that are sufficiently fat for immediate slaughter.

**Feeder lambs**—Young animals under one year of age that carry insufficient finish for slaughter purposes but which show indications of making good gains if placed on feed.

**Shearer lambs**—Those intended for shearing and further finishing prior to slaughtering. This classification is of importance on certain markets during the late winter and early spring months. Typical shearer lambs carry nearly a full year's growth of wool, but are not fat enough to be market-topping slaughter lambs. Such lambs are usually shorn out and finished before returning to market. The term *shorn lamb* is used to designate those lambs that have had their fleece removed within 60 days prior to marketing. Those fed lambs that have not been shorn are usually differentiated from shorn lambs by adding the term *wooled* to the class name.

## THE SEX CLASSES

The sex classes for sheep and lambs are: ewes, wethers, and rams. At the lamb and yearling stage, ewes and wethers are equally suitable for slaughter purposes. Ram lambs are usually somewhat discounted in price, and they are almost never used for feeder purposes. A definition of each sex class follows.

**Ewe**—A female sheep or lamb.

**Wether**—A male ovine animal that was castrated at an early age, before reaching sexual maturity and before developing the physical characteristics peculiar to rams.

**Ram**—An uncastrated male ovine animal of any age. The term *buck* is sometimes applied to animals of this sex class.

## AGE GROUPS

Each of the major age divisions is further separated into more exacting age groups. Thus, mature sheep may be designated as yearlings, 2-year-olds, 3-year-olds, 4-year-olds, or mature sheep. Yearlings are much more acceptable for slaughter purposes than mature sheep.

In a general way, age groups in lambs are indicated by the terms *hothouse lambs, spring lambs,* and *lambs.* Each of these age groups may be described as follows:

**Yearlings**—Young sheep between approximately one and two years of age. They may be identified by the fact that they have cut their first pair of permanent incisor teeth but not the second pair.

**Two-year-olds**—Sheep that are between 24 and 36 months old and which have cut their second pair of permanent incisor teeth.

**Three-year-olds**—Sheep that are between 36 and 48 months old and which have cut their third pair of permanent incisor teeth.

**Four-year-olds**—Sheep that are between 48 and 60 months old and which have a full set of permanent incisors.

**Mature sheep**—Usually animals that are two years old or over. With further age and the loss of teeth, they are referred to as broken mouthed. If all the incisors are missing or worn down to the gums, they are known as *gummers.* If the teeth are long and spread apart at the surface, the ewes are called *spreaders.* All gummers and broken-mouthed ewes should be rejected in buying sheep, for such animals are not likely to hold up in flesh on winter feeds.

**Hothouse lambs**—Considered by epicureans as being the most delectable of the lamb age groups. They are very young lambs—usually less than 3 months of age at slaughter—which are born and marketed out of season. Such milk-fat lambs are usually marketed during the period from Christmas to the Easter holidays at weights ranging from 30 to 60 lb. Hothouse lambs may consist of ewe, wether, or ram lambs. Although sex class is unimportant, these lambs should be undocked if sold on the Boston market.

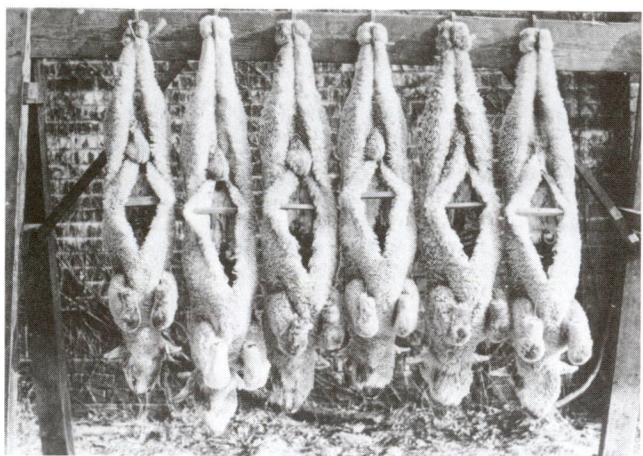

Fig. 33–3. Hothouse lambs "hog dressed" (meaning head and pelt on, but front feet and viscera removed). The method holds shrinkage to a minimum and maintains a pink carcass color in young lambs. The pluck is left in. (In this case, the pluck consists of the liver, heart, lungs, gullet, and windpipe.) (Courtesy, Pennsylvania State University)

**Spring lambs**—New-crop lambs arriving at the market in the spring of the year. Usually these are lambs born in the late fall or the early winter and marketed prior to July 1. After July 1, animals of like birth are simply designated as lambs on the market. The most desirable spring lambs range from 3 to 7 months in age and weigh from 70 to 90 lb. Because of the young age, sex class is unimportant in spring lambs—they may be ewe, wether, or ram lambs.

**Lambs**—All young ovine animals that do not classify as either hothouse or spring lambs. This is by far the most numerous class of market sheep under one year of age. These animals are usually born in the later winter or spring of the year and are marketed at 7 to 12 months of age. In general, they subsist on milk and grass. Lambs fed grain prior to marketing are designated as *fed lambs* in order to differentiate them from lambs.

## WEIGHT DIVISIONS

Weight is an especially important price factor in the case of slaughter lambs. Lambs weighing 105 to 120 lb are considered ideal. Heavy lambs (120 lb and up) are in less demand on the market than lighter-weight animals of similar grade. This is largely due to the consumer preference for small- or medium-sized cuts of lamb. Thus, weight is an important factor in both feeder and slaughter lambs. Heavy, fat ewes usually sell at a discount; they are not considered desirable for either slaughter or breeding purposes.

## MARKET GRADES OF SHEEP AND LAMBS

Before significant progress can be made in improving lamb production and marketing, it is necessary for producers and those engaged in marketing to have a clear understanding of the meaning and relative importance of the various live and carcass grades.

Fig. 33–4. Good-quality, crossbred, black-faced lambs on feed in Montana, finished and ready for market. (Photo: Osborn Studios, Sidney, MT)

Sheep and lambs may be graded either, or both, of two ways: (1) quality grade, or (2) yield grade. Quality grades relate to eating characteristics of the meat, whereas yield grades describe the volume or total proportion of trimmed, retail cuts a carcass contains.

The quality grades are designed to identify differences in (1) palatability and cooking characteristics, such as tenderness, juiciness, and flavor; and (2) carcass shape or conformation. Evaluation of lamb carcass quality is based on flank streaking in relation to maturity (see Fig. 33–5). Older lambs require more fat streaking in the flank than do younger animals to qualify for a specific grade. A minimum level of lean and fat firmness is specified for each grade. Also, conformation is as important as flank streaking and maturity in determining quality grades. Lamb is produced from animals less than a year old. Meat from older sheep is called yearling or mutton; where graded, these words must be stamped on the meat along with the grade. Grades for yearling and mutton are the same as for lamb, except that there is no Prime grade for mutton. About 70% of all lambs slaughtered commercially are quality graded. Of this number, about 80% grade Choice or higher.

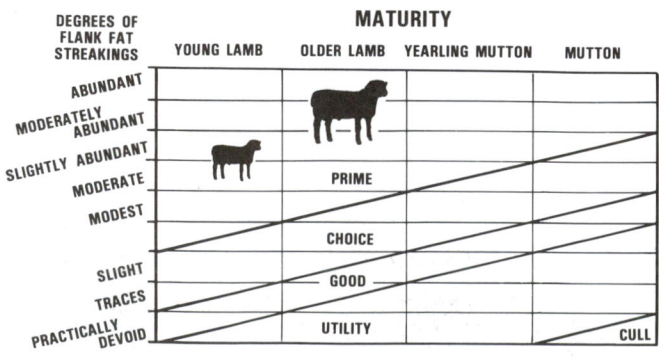

Fig. 33–5. Relationship between flank fat streakings, maturity, and quality.

The quality grades of slaughter sheep and lambs are: Prime, Choice, Good, Utility, and Cull (see Fig. 33–6). The corresponding grades of feeder sheep and lambs are: Fancy, Choice, Good, Medium, and Cull.

## PRIME

## CHOICE

## GOOD

## UTILITY

Fig. 33–6. Market grades of slaughter lambs. The lowest grade, Cull, is not pictured in Fig. 33–6. Names changed by the author to conform to present U.S. Department of Agriculture grades. (Courtesy, USDA)

In addition to the quality grades, yield grades were added in 1969. Both lamb and mutton carcasses may be yield graded. Yield grades provide a nationally uniform method of identifying carcasses for differences in cutability, or the percent of trimmed retail cuts. Five yield grades, numbered 1 through 5, cover the range in cutability, the Yield Grade 1 indicating the highest yield. A carcass which is typical of its yield grade would be expected to yield 3.5% more in total retail cuts than the next lower yield grade, when U.S. Department of Agriculture cutting and trimming methods are followed. The yield grade is determined by considering 3 factors: (1) the amount of external fat, (2) the amount of kidney and pelvic fat, and (3) the conformation grade of the leg. The leg is evaluated because U.S. Department of Agriculture studies show that, among lambs of the same degree of finish, those that have the higher leg conformation grades also have higher yields of retail cuts. Yield grades are applicable to wholesale and primal cuts as well as to carcasses.

A study conducted by Texas A&M University showed a $15.86 per hundred weight difference in carcass value between a Yield Grade 1 lamb and a Yield Grade 5 lamb. More astounding yet, nearly 30% of the weight of Yield Grade 5 lamb was waste fat.

To date, yield grades for lamb have not been used commercially. Emphasis on low calorie meat, coupled with the need for producers to be paid for producing consumer-preferred lamb, may eventually result in the commercial use of yield grades.

## OTHER SHEEP MARKET TERMS AND FACTORS

In addition to the rather general terms used in designating the different classes and grades of sheep, the following terms and factors are frequently of importance.

### CYCLICAL AND SEASONAL PRICE VARIATIONS

Sheep and lamb prices tend to vary both cyclically over the years and seasonally with the year. A knowledge of cyclical and seasonal price variations may be useful in planning production and marketing so that animals are available at the most advantageous time. Chapter 8, Fig. 8–26 shows cyclical trends.

A noteworthy aspect of cyclical price variations is that when sheep and lamb production is sufficiently profitable to induce increased production, breeding stock to increase production must come from either (1) a reduction of the rate of ewe slaughter, or (2) a holding back of ewe lambs to build breeding flocks. In either case, slaughter sheep and lamb supply is further reduced and even higher prices occur in the short run. Conversely, when prices are sufficiently low to cause producers to sell off breeding stock and/or keep fewer replacement lambs, this further adds to the sheep and lamb supply and results in lower prices.

Lamb supply and slaughter is seasonal, due to the breeding pattern of sheep. Spring lambing is common in the mountain, upper midwestern, and plains states with most marketing taking place in September and October. By contrast, about half of Texas' lamb crop is marketed between March and July; and along the Pacific Coast, May and June are the principle marketing months. The seasonality of supply is reflected in slaughter lamb prices (Fig. 33–7). Prices are highest in May and June, and lowest in the fall (September, October, and November).

It must be remembered that both cyclical and seasonal price information is based on past history and should be used only as rough guides.

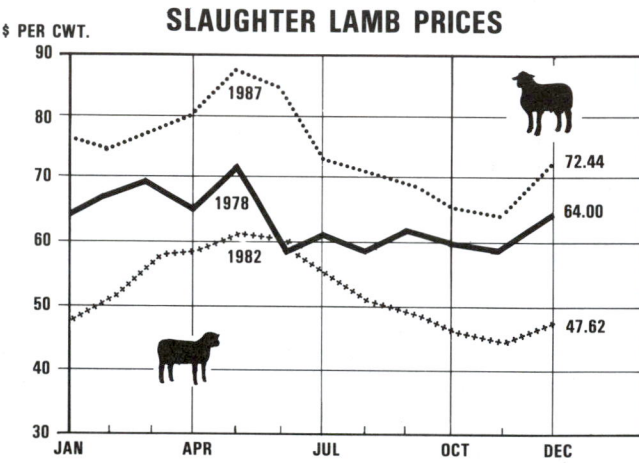

Fig. 33–7.  Slaughter lamb prices vary seasonally, largely because of seasonal supply. This shows slaughter lamb prices at Omaha for the years 1978, 1982, and 1987. (Source: *Livestock and Meat Statistics, 1984–88,* Stat. Bull. No. 784, USDA, p. 215, Table 125)

### NATIVE AND WESTERN SHEEP

Native sheep are predominantly of mutton breeding and are produced in the central, eastern and southern states, in the mixed farming areas. Westerns are predominantly of grade Rambouillet breeding or Rambouillet × mutton breed crossbreds that come from the western ranges. In the case of western sheep, their state of origin is frequently used as a designation rather than the broad classification as westerns.

### WHITEFACE AND BLACKFACE SHEEP

*Whiteface* and *blackface* are sometimes used to designate the general type and breeding of sheep and lambs. In general, whiteface sheep and lambs are predominantly of fine-wool breeding and do not possess quite the muscling and general excellent mutton conformation of the blackface sheep. Blackface sheep are either predominantly of mutton breeding or are sired by a black-faced ram of one of the mutton breeds.

## DISCOUNTING LONG-TAILED AND RAM LAMBS

All market lambs from commercial flocks should be docked and castrated at an early age. Long-tailed and buck lambs are frequently docked up to $2 to $5 per hundredweight on a discriminating market. Thin lambs, which should go the feeder route, are discounted more than slaughter animals. Despite the dock in price of ram lambs that are subsequently culled out and marketed for slaughter, purebred breeders can often afford to retain all male lambs as rams until weaning age, at which time more intelligent selections may be made. Of course, this can only be justified when some of the ram lambs are valuable for breeding purposes.

## VALUE OF THE WOOL OR PELT IN MARKET SHEEP

Pelts are the most valuable by-product of lamb slaughter, often accounting for 5 to 6% of the live lamb value. Most pelts are used as leather in the garment industry.

The value of the wool, or fleece, will depend upon (1) whether the animal has recently been shorn or is carrying a full fleece, and (2) the price of wool. Pelt prices are difficult to obtain, but they are generally reported in major industry publications.

## PACKER SLAUGHTERING AND DRESSING OF SHEEP AND LAMBS

Table 33–3 shows the proportion of sheep and lambs dressed commercially (meaning that they were slaughtered in federally inspected and other wholesale and retail establishments) and the proportion slaughtered on farms. The total figure refers to the number dressed in all establishments and on farms. It is noteworthy, too, that 95% of the lamb and yearling slaughter in 1987 was under federal inspection.

### TABLE 33–3
### PROPORTION OF SHEEP AND LAMBS SLAUGHTERED COMMERCIALLY, 1983 TO 1987[1]

| Year | Commercial | | | Farm | Total | % Federally Inspected | % Commercial |
| | Federally Inspected | Other | Total | | | | |
|---|---|---|---|---|---|---|---|
| | | | | | | (%) | (%) |
| 1983 | 6,412,000 | 207,000 | 6,619,000 | 171,000 | 6,790,000 | 94 | 97 |
| 1984 | 6,549,000 | 210,000 | 6,759,000 | 141,000 | 6,900,000 | 95 | 98 |
| 1985 | 5,976,000 | 189,000 | 6,165,000 | 135,000 | 6,300,000 | 95 | 98 |
| 1986 | 5,464,000 | 171,000 | 5,635,000 | 127,000 | 5,762,000 | 95 | 98 |
| 1987 | 5,042,000 | 158,000 | 5,200,000 | 112,000 | 5,312,000 | 95 | 98 |
| 5-year average | 5,889,000 | 187,000 | 6,076,000 | 137,000 | 6,213,000 | 95 | 98 |

[1]Agricultural Statistics 1988, USDA, p. 290, Table 430.

Today, 15 large U.S. packing plants slaughter 95% of all sheep and lambs slaughtered. This indicates a high degree of concentration of sheep and lamb slaughtering in relatively few, but large, plants. The trend toward fewer, but larger and more efficient, plants will continue.

Whether sheep are slaughtered in a large federally inspected packing plant, in a small local slaughterhouse, or on the farm, the slaughtering procedure is much the same. Because most sheep are slaughtered commercially rather than on farms, only packer slaughtering procedure will be discussed.

In order to avoid undue excitement and make for ease in handling, gregarious sheep and lambs are usually led to the packer's slaughtering pens by an old goat, commonly referred to as "Judas." A reasonable fast and quiet prior to slaughtering are especially important for sheep.

Fig. 33–8. "Judas" goat leading lambs to slaughter. (Courtesy, Monfort of Colorado, Greeley, CO)

## STEPS IN SLAUGHTERING AND DRESSING SHEEP AND LAMBS

Modern sheep and lamb slaughtering establishments use highly mechanized equipment, including a moving rail. The steps in packinghouse slaughter procedure are as follows:

1. **Rendering insensible.** The sheep are rendered insensible by use of a captive bolt stunner, gunshot, electric current, or carbon dioxide; a shackle is place around the hind leg just above the foot; and the animals are delivered to an overhead rail by means of a wheel hoist.

2. **Bleeding.** A double-edged knife is inserted into the neck just below the ear so that it severs the large blood vessel in the neck.

3. **Removing feet.** After bleeding, the front feet are removed. Lambs and most yearlings will break at the break-joint or lamb-joint, a temporary joint characteristic of young sheep which is located immediately above the ankle. In dressing mature sheep, the front feet are removed above the ankle, leaving a round joint on the end of the shank bone.

4. **Removing the pelt.** The pelt is next removed. Caution is taken to prevent damage to the fell (the thin tough membrane covering the carcass immediately under the pelt).

5. **Removing hind feet and head.** Next the hind feet and head are removed.

6. **Opening of carcass and eviscerating.** The carcass is opened down the medial line; the internal organs, windpipe, and gullet are removed; and the breastbone is split. The kidneys are left intact.

7. **Shaping.** The forelegs are folded at the knees and are held in place by a skewer. A spread-stick is inserted in the belly to allow for proper chilling and to give shape to the carcass.

8. **Washing.** Finally the carcass is washed, wiped, and promptly sent to the cooler. Some of the better-grade carcasses are wrapped in special coverings for marketing.

## THE BREAK-JOINT OR LAMB-JOINT

In general, the packer classes as lamb all carcasses in which the forefeet are removed at the break-joint or lamb-joint. This point—which can be severed on all lambs, most yearling wethers, and some yearling ewes (ewes mature earlier than wethers or rams)—is a temporary cartilage located just above the ankle. In lambs, the break-joint has four well-defined ridges that are smooth, moist, and red. In yearlings, the break-joint is more porous and dry. In mature sheep, the cartilage is knit or ossified and will no longer break, thus making it necessary

to take the foot off at the ankle instead. This makes a round-joint (commonly called spool-joint). All carcasses possessing the round-joint are sold as mutton rather than lamb.

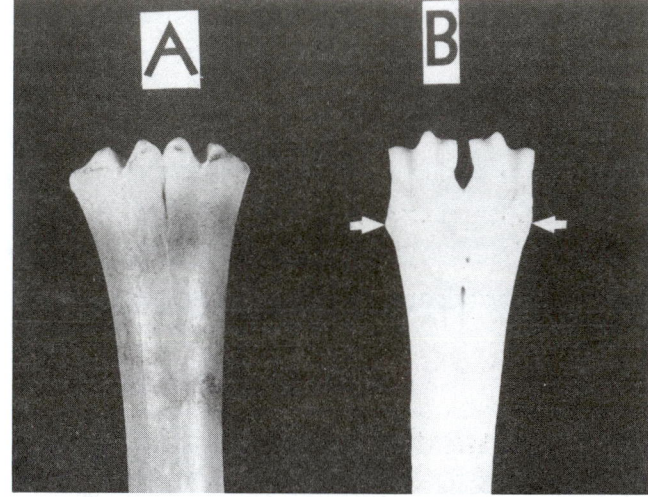

Fig. 33–9. The two joints of the foreleg of a sheep: A, the break-joint or lamb-joint, and B, the round-joint or spool-joint. Arrow indicates the location of the break-joint or ossification. All carcasses possessing the round-joint are sold as mutton rather than lamb. (Courtesy, Washington State University)

## THE DRESSING PERCENTAGE

In order to yield a high percentage of carcass, sheep must be (1) light in pelt, (2) well finished and heavily muscled, and (3) free from paunchiness. Because of the higher value of the offal of sheep, especially the high value of the wool or pelt, a high dressing percentage in sheep is not so important as in cattle or hogs. In fact, the wool is usually worth more per pound than the carcass of mutton or lamb. Wool yield, therefore, is usually an important item in slaughter return, and dressing return, and dressing percentage lowered by heavier wool yield may actually mean a greater total return. For this reason, the fleece should be as heavy as is consistent with the production of mutton or lamb of high quality.

Table 33–4 gives the dressing percentages that may be expected from the different grades of sheep and lambs. As would be expected, the highest dressing percentage is obtained when animals are slaughtered following shearing. Lambs of mutton breeds yield a somewhat higher percentage of carcass than those of the so-called wool breeds. The offal and by-products from the slaughter of sheep and lambs consist of the blood, pelt, feet, head, and viscera.

### TABLE 33–4
### DRESSING PERCENTAGE OF LAMBS AND SHEEP (MUTTON) BY GRADE[1]

| Lambs (wooled) | | | Sheep (excludes yearlings) | | |
|---|---|---|---|---|---|
| Grade | Range | Average | Grade | Range | Average |
| Prime | 49–55 | 52 | Choice | 49–54 | 52 |
| Choice | 47–52 | 50 | Good | 47–52 | 49 |
| Good | 45–49 | 47 | Utility | 44–48 | 46 |
| Utility | 43–47 | 45 | Cull | 40–46 | 43 |
| Cull | 40–45 | 42 | | | |

[1]From USDA sources.

The average liveweight of sheep and lambs dressed by federally inspected meat-packing plants and their percentage of yield in meat for the year 1987 was as shown in Table 33–5.

| | Average Liveweight | | Average Dressed Weight | | Average Dressing Percentage |
|---|---|---|---|---|---|
| | (lb) | *(kg)* | (lb) | *(kg)* | (%) |
| Sheep and lambs ...... | 120 | *55* | 60 | *27* | 50 |

[1]Source: *Livestock and Meat Statistics, 1984–88,* Stat. Bull. No. 784, USDA, pp. 119, 135, 136.

## AGING LAMB AND MUTTON

Prime and Choice lamb and mutton carcasses are best if aged two or three weeks. On the other hand, the medium and plain carcasses carrying less fat should be merchandised a few days after slaughter.

## DISPOSITION OF THE LAMB AND MUTTON CARCASS AND CUTS

Fig. 33–10. Lamb in the meat counter.

The lamb industry is turning to boxed lamb—the fabricating or cutting of lamb carcasses in the plant where they are slaughtered, with lamb primal cuts vacuum-sealed and shipped in cardboard boxes to retail outlets. The efficiency of the move becomes apparent when it is realized that 20% or more of an average lamb carcass winds up as waste at retail. Through central cutting, slaughterers and retailers save millions of dollars in transportation costs alone by not shipping fat and bones. In addition, in a program of central cutting where cuts are sealed in a plastic skintight container, the shrinkage is reduced from the normal 6 to 7% to less than 1%. Storage life is also increased, reducing perishability. To gain optimum economies, packaging, as well as cutting, must be moved outside the retail store. Also, it is noteworthy that block-ready and counter-ready cuts can help solve butcher resistance to lamb at the store level. Butchers are more apt to display lamb, can put a better product in the meat case, and can buy only the cuts needed.

About two-thirds of all U.S. meats are sold as fresh meats, a highly perishable commodity. The proportion of lamb sold fresh is even higher. For efficiency reasons, the longtime, but gradual, trend is toward prepackaged frozen meats.

The retail cuts of lamb are similar to those of other species. The legs and shoulders are used largely as roasts; the loin and rack are sources of chops; and the shanks, breasts, and flanks are boned for stewing or ground lamb.

Carcasses or wholesale cuts that are not considered desirable for the trade, or for which there is not sufficient demand, are processed at the packing plant. Eventually they are sold as prepared meats and meat food products.

In recent years, the following marketing methods and innovations have given a new look to the sheep industry:

• **Cuts and names**—The popular lamb cuts are: leg, chops, steaks, shanks, lamb-kabob meat, stew meat, ground lamb, sirloin roast, boned cutlets, rack, trimmed loin, spareribs, rack roast, crown roast, precarved shoulders, boneless shoulders, riblets, rolled breast, kidneys, and variety meats.

In the case of lamb, names may be mitigating against merchandising. Meat derived from lamb is called lamb, whereas meat from cattle becomes beef, meat from calves becomes veal, meat from hogs becomes pork, and meat from chickens becomes broiler. Not only that, but many lamb cuts are the actual names of live animals parts—leg, shoulder, breast, flank, etc. By contrast, the corresponding parts of other species generally bear alluring names which are not associated with the live animal part, such as chuck for neck of beef.

• **Boneless lamb and convenient forms**—Boneless cuts, such as boneless leg and boneless shoulder, may be cooked in smaller containers, and they are easier to carve. In certain areas, these boneless cuts have been merchandised in the form of netted roasts. The netting reduces the labor required for tying roasts and holds them in very uniform shape.

• **Frozen cuts**—Frozen lamb cuts are being promoted in some quarters, as a means of expanding the market for lamb and making it available to consumers throughout the year.

• **Specialty forms**—As a specialty product, with a high margin, some processors are interested in such items as canned leg of lamb, frozen shish kebabs, ground lamb combinations with other meats, and other similar items.

Also, processors are moving steadily in the direction of completely prepared, frozen products that require a minimum of preparation; many of these products are combined with sauces, gravies, or other foods. It is likely that greater quantities of precooked and frozen meats, including lamb, will be available in the future.

• **Retail display**—Retail stores normally allocate meat case display space proportionate to the sales volume of the product. Because lamb is a low-volume item, it usually has the smallest display allocation of any kind of meat.

• **Promotion**—Generally speaking, lamb is handicapped by lack of promotion, due to being a minor product. However, stores in higher lamb consumption areas tend to promote lamb as a gourmet item.

• **Foreign competition**—There are two schools of thought relative to importation of lamb and mutton products: (1) Imports may be beneficial in expanding the market in areas of low lamb consumption and in supplementing the domestic supply in seasons of low production; and (2) imports are an additional supply, which results in a lower price to domestic producers. More than likely, there is some truth in both arguments and the actual truth may lie somewhere between these two extremes.

It appears likely that increasing pressures to increase U.S. imports of lamb and mutton will be exerted by (1) Australian and New Zealand producers, anxious to find a market outlet for their very considerable surplus production of lamb and mutton, and (2) U.S. consumers, interested in low cost foods.

• **Future of lamb**—The future success of lamb appears to depend on the following:

1. Lamb being more competitively priced, due to greater efficiency of production, processing, and distribution.

2. Having lamb more readily available everywhere.

3. Improving lamb merchandising.

4. Maintaining a consistently high-quality product.

## LAMB CUTS AND HOW TO COOK THEM

The two major wholesale cuts of lamb are the (1) hindsaddle, and (2) foresaddle. The division into hindsaddle and foresaddle is made between the twelfth and thirteenth rib, with one pair of ribs remaining in the hindsaddle. Each of these two larger cuts comprises about 50% of the carcass weight.

The hindsaddle is further divided into the leg and loin. The foresaddle is subdivided into the shoulder, rib, foreshank, and breast.

Fig. 33–12 shows the common retail cuts of lamb, where they come from, and how to cook them.

## LAMB PROMOTION

The American Sheep Industry Association promotes and advertises lamb and wool for the American sheep industry. The needs: (1) to make consumers aware of lamb and its desirable qualities; (2) to teach food shoppers to think *lamb* when planning daily menus and preparing weekly food purchases; (3) to have high-quality lamb available throughout the year; and (4) to have lamb priced competitively with alternative protein sources. Promotion and advertising can give a big assist in meeting the first two needs, but consistent supplies, quality, and comparable pricing must be achieved primarily by producers.

## GOAT MEAT (CHEVON)

Goat meat is sold under the trade name of Chevon. However, most goat meat goes into processed meats. It is a perfectly wholesome food, though lacking in the conformation and degree of finish possessed by the better grades of lamb and mutton. Chevon from older goats is likely to possess a strong flavor. Only a limited number of goats are slaughtered for meat.

Fig. 33–11. Broiled shoulder lamb chops. Parsleyed potatoes fill the center of this lamb chop platter. Round bone lamb shoulder chops, shown here, are equally as flavorful and less expensive than loin chops. Add a salad of sliced tomatoes and cucumbers, warm yeast rolls, and a rhubarb pineapple pie to the meal featuring broiled chops. (Courtesy, American Meat Institute)

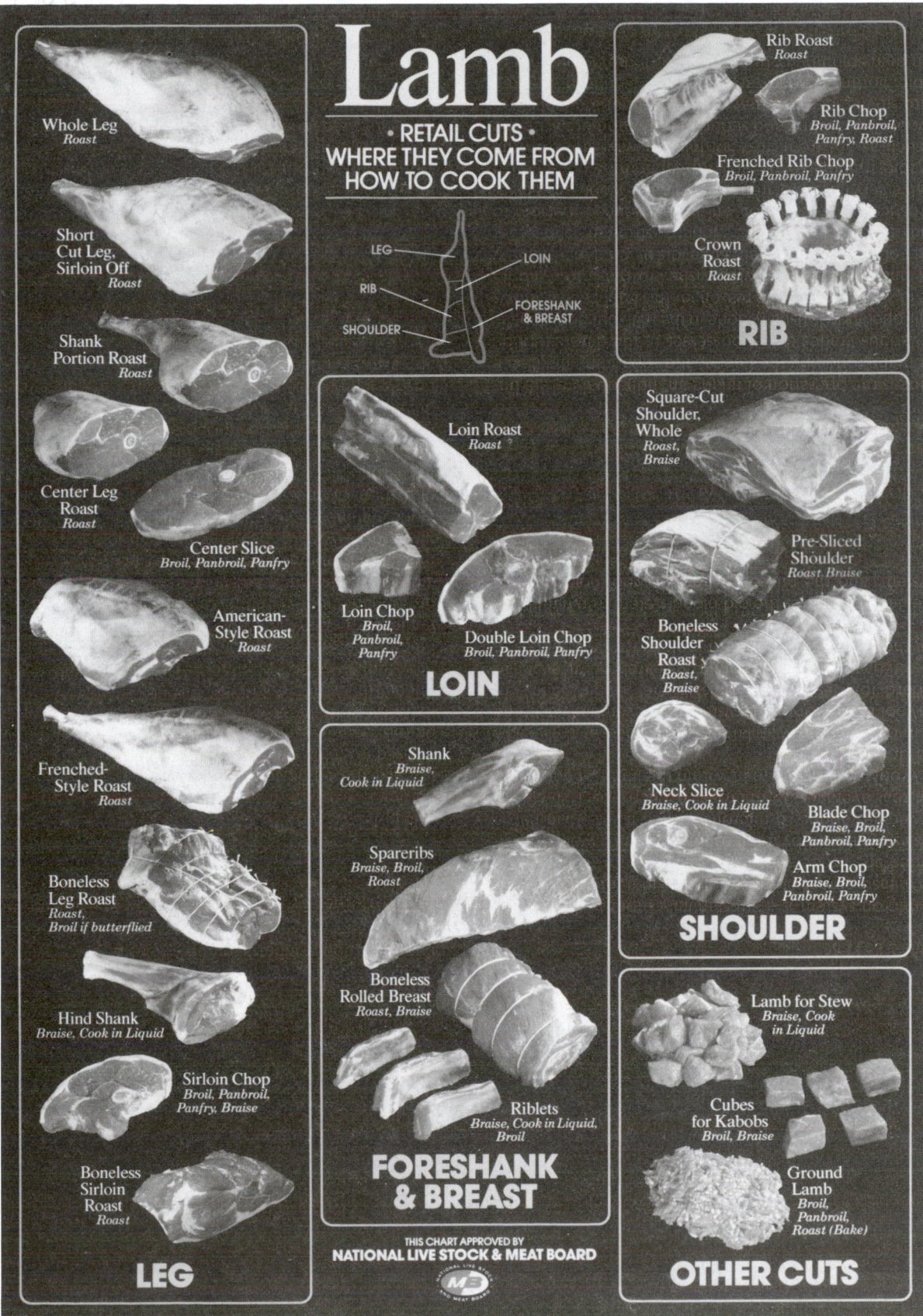

Fig. 33–12.   Retail cuts of lamb—where they come from, and how to cook them. (Courtesy, National Live Stock & Meat Board, Chicago, IL)

# QUESTIONS FOR STUDY AND DISCUSSION

1. What two situations complicate the marketing of sheep and lambs, and make their marketing complex and costly?

2. What primary differences exist between the marketing of sheep and cattle or hogs? Why does 92 to 94% of the sheep slaughter consist of lambs and yearlings, and only 6 to 8% of the slaughter consist of older sheep?

3. Define and discuss each of the following marketing considerations that are peculiar to sheep and lambs: (a) spot (cash) sales, and (b) discount plans for heavyweight live lambs.

4. List, by rank, the five leading sheep and lamb public stockyards.

5. Identify and describe two important differences between the market classes and grades of sheep and the market classes and grades of cattle and swine.

6. Define each of the following terms: (a) sheep, and (b) lambs. How can one be sure as to whether an animal should be classed as a sheep or lamb?

7. Name the market groups of sheep and lambs based on the use made of them, and describe each of the groups or classes.

8. Name the three sex classes of sheep and lambs, and define each of them.

9. List the market age groups of sheep, and describe each group.

10. List the market age groups of lambs, and describe each group.

11. Give the current weight divisions (including pounds) of slaughter lambs. What is the ideal weight range for slaughter lambs? Why are heavy lambs (weighing more than 125 lb on foot) in less demand on the market than light lambs?

12. What are the quality grades of sheep and lambs designed to identify? What carcass traits determine the quality grade? Name the quality grades of sheep and lambs in descending order.

13. What are yield grades of sheep and lambs designed to identify? Give the yield grades in descending order. What three factors determine yield grade? Why have yield grades not been used commercially to date?

14. What is the difference between (a) longtime cyclical trends (as shown in Chapter 8 of this book, Fig. 8–26) and (b) seasonal price variations (as shown in Fig. 33–7)?

15. What is meant by (a) native and western sheep, and (b) whiteface and blackface sheep?

16. Why are long-tailed and ram lambs discounted on the market?

17. How important is the pelt in determining the price of market sheep and lambs?

18. What is a *Judas goat*? List and briefly describe the steps in slaughtering and dressing sheep and lambs.

19. What is the break-joint or lamb-joint? Name, describe, and tell what each of the two types of joints indicate.

20. In order to have a high dressing percentage (a high carcass yield), what traits must sheep possess? Why is high dressing percentage of less importance in sheep than in cattle and hogs?

21. Discuss each of the following marketing methods and innovations from the standpoint of the disposition of the lamb and mutton carcass and cuts:
    a. Cuts and names.
    b. Boneless lamb and convenient forms.
    c. Frozen cuts.
    d. Specialty forms.
    e. Retail display.
    f. Promotion.
    g. Foreign competition.

22. Sketch a lamb carcass, and identify the location of the wholesale cuts (see Fig. 33–12).

23. What is Chevon?

# SELECTED REFERENCES

| Title of Publication | Author(s) | Publisher |
| --- | --- | --- |
| *Livestock and Meat Marketing*, Second Edition | J. H. McCoy | Avi Publishing Co., Westport, CT, 1979 |
| *Meat Handbook*, Fourth Edition | A. Levie | Avi Publishing Co., Westport, CT, 1979 |
| *Meat We Eat, The*, Twelfth Edition | J. R. Romans<br>P. T. Ziegler | The Interstate Printers & Publishers, Inc., Danville, IL, 1985 |
| *Sheep and Wool* | M. P. Botkin<br>R. A. Field<br>C. L. Johnson | Prentice Hall, Englewood Cliffs, NJ, 1988 |
| *Sheepman's Production Handbook, The*, Second Edition | Edited by<br>G. E. Scott | Sheep Industry Development Program, Inc., Denver, CO, 1986 and updated at intervals |
| *Sheep Production and Management* | C. V. Ross | Prentice Hall, Englewood Cliffs, NJ, 1989 |

Made of wool—fabric of beauty and history. (Courtesy, The Wool Bureau, New York, NY)

# WOOL AND MOHAIR

# Chapter 34

Sheep yield two products—lamb or mutton and wool, with the relative emphasis on each varying according to the remuneration derived therefrom. This same thinking applies wherever mutton and lamb are produced, which is throughout the world.

It is noteworthy that the income to U.S. sheep raisers from wool amounts to about 20%, with a range of 10 to 40%, of the gross income from sheep, lambs, and wool. Thus, about 80% of the total income to sheep producers is derived from meat (chiefly lamb).

The discussion that follows is designed to bridge the gap between producers and manufacturers. Although each group has particular problems, both are working toward a common goal. Producers breed, feed, and manage their flocks to supply the raw material, whereas manufacturers scour, comb, spin, and weave the fiber into cloth.

## WOOL IS THE NATURAL CLOTHING OF SHEEP

With all the perfection and modification in the fleece that has been wrought through centuries of domestication, breeding, selection, and improved environmental conditions, it must not be forgotten that wool is the natural hair as well as clothing of sheep. A covering of hair or feathers performs a thermoregulatory function for warm-blooded animals, the original intent being to protect the body from heat or cold. As wool fibers are poor conductors of heat, they serve to prevent any abnormal loss of heat from the body.

Fig. 34–1. Wool is the natural clothing of sheep. The Merino ram pictured is a member of the breed that produces the world's finest wool. (Courtesy, The Wool Bureau, Inc., New York, NY)

## VIRTUES OF WOOL

Long before domestication of sheep, some 12,000 years ago, Stone Age hunters observed that the fleeces of wild sheep were softer, warmer, and thicker than the hairy skins of other animals—and by this discovery they made a major contribution to human progress.

The unique characteristics and virtues of wool are:

1. It is porous and will absorb water more readily than any other textile fiber. It can absorb as much as 18% of its own weight in moisture without even feeling damp, and up to 50% of its weight without becoming saturated. This is an important health factor in clothing because body perspiration and outer dampness are prevented from clinging to the body in heat or cold, thus removing the chill line from the body.

2. It generates heat in itself.

3. It is a superior insulator, keeping the heat of the body from escaping and the cold air from entering. Because of this quality, wool is as effective as a protection from tropical heat and sun as it is against the gale-driven storms of winter.

4. It is light.

5. It is very elastic; the average fiber will stretch 30% of its normal length and still spring back in shape. Because of this resilience, wool garments resist wrinkling, stretching, or sagging during wear. Wool can be bent 20,000 times without breaking (silk breaks after 1,800 bends, rayon after 75).

6. It transmits the health-giving ultraviolet rays.

7. It dyes well.

8. It is durable.

9. It is strong. Diameter for diameter, a wool fiber is stronger than steel.

10. It is almost nonflammable. It will stop burning almost as soon as it is taken away from a flame.

11. It can be felted or matted easily.

## USES OF WOOL

About 91% of the wool consumed in the United States is apparel wool and 9% is carpet wool.

About 2% of apparel wool is used in batting and in the manufacture of pressed felt, mostly for hat bodies. The other 98% is consumed in the spinning of woolen and worsted yarn. About 15% of the woolen and worsted yarn is used in the production of knit goods, including sweaters, hosiery, underwear, gloves, and mittens. The remainder is used in the weaving of fabrics, including such apparel fabrics as suitings, trouserings, dress fabrics, and coatings and such nonapparel fabrics as blanketing, upholstery, draperies, and woven industrial felts.

The greater part of carpet wool is used in the manufacture of floor coverings, although small quantities are used in the manufacture of press cloth, knit and felt boots, and heavy fulled socks.

## WORLD WOOL PRODUCTION

World wool production totals more than 4 billion lb annually (4,140,000,000 lb in 1987). Australia alone produces 29% of the world's wool at the present time.

The Southern Hemisphere countries of Australia, New Zealand, Argentina, Uruguay, and the United Kingdom are the main wool surplus producing nations. Two countries, Australia and New Zealand, account for 59.4% of all wool exports. The Merino breed supplies at least a third of the world's wool.

(Also, see Chapter 27, section headed "World Wool Production," including Table 27–2.)

# MAGNITUDE OF THE U.S. WOOL AND TEXTILE INDUSTRY

As shown in Table 34–1, the farmers and ranchers of the United States receive millions of dollars annually for their wool clip. This does not include additional pulled wool, nor does it tell the story relative to the huge imports. Table 34–2 gives the wool production, imports, and consumption, and percent produced domestically. It is not expected that annual production plus annual imports will exactly equal annual consumption, due to stockpiling and certain other factors. However, among other things, the following noteworthy facts can be deducted from Table 34–2: (1) Virtually all of our carpet wool is imported; (2) today, we produce only about 35% of our apparel wool; and (3) in 1987, 73.6% of our total wool requirement (apparel and carpet) was imported.

### TABLE 34–1
### QUANTITY, PRICE PER POUND, AND TOTAL CASH VALUE OF WOOL PRODUCED IN THE U.S.[1]

| Year | Shorn Wool Produced | | Price per Pound | Cash Value |
|------|---------|---------|---------|---------|
| | (lb) | (kg) | (cents) | ($) |
| 1970 | 161,587,000 | 73,295,864 | 35.4 | 57,162,000 |
| 1975 | 119,535,000 | 54,221,077 | 44.8 | 53,505,000 |
| 1980 | 105,452,000 | 47,833,027 | 88.1 | 92,862,000 |
| 1985 | 87,941,000 | 39,973,182 | 63.3 | 55,657,000 |
| 1987 | 85,757,000 | 38,980,455 | 91.7 | 78,156,000 |

[1]*Agricultural Statistics 1981*, USDA, p. 334, Table 493. *Ibid.*, 1988, p. 297. Data 1970–1987.

### TABLE 34–2
### U.S. PRODUCTION, IMPORTS, AND CONSUMPTION OF WOOL, AND PERCENT PRODUCED DOMESTICALLY, CLEAN BASIS[1]

| Year | Production[2] | | | Imports[3] | | | Consumption[4] | | | Domestic Production as Percent of Consumption | | | Imports as Percent of Consumption[3] | | |
|------|---------|--------|-------|---------|--------|-------|---------|--------|-------|---------|--------|-------|---------|--------|-------|
| | Apparel | Carpet | Total | Apparel | Carpet | Total | Apparel | Carpet | Total | Apparel | Carpet | Total | Apparel | Carpet | Total |
| | (mil. lb) | (mil. lb) | (mil. lb) | (mil. lb) | (mil. lb) | (mil. lb) | (mil. lb) | (mil. lb) | (mil. lb) | (%) | (%) | (%) | (%) | (%) | (%) |
| 1960 | 144.6 | — | 144.6 | 74.3 | 153.9 | 228.2 | 246.4 | 164.6 | 411.0 | 58.7 | — | 35.8 | 30.2 | 93.6 | 55.5 |
| 1965 | 113.1 | — | 113.1 | 162.6 | 108.9 | 271.5 | 274.7 | 112.3 | 387.0 | 41.2 | — | 29.2 | 59.2 | 97.0 | 70.2 |
| 1970 | 88.2 | — | 88.2 | 79.8 | 73.3 | 153.1 | 163.7 | 76.6 | 240.3 | 53.9 | — | 36.7 | 48.7 | 95.7 | 63.7 |
| 1975 | 67.5 | — | 67.5 | 16.6 | 17.0 | 33.6 | 94.1 | 15.9 | 110.0 | 71.7 | — | 61.4 | 17.6 | 104.4 | 30.5 |
| 1980 | 56.4 | — | 56.4 | 30.5 | 26.0 | 56.5 | 113.4 | 9.1 | 122.5 | 49.7 | — | 46.0 | 26.9 | 285.7 | 46.1 |
| 1985 | 46.4 | — | 46.4 | 50.2 | 29.3 | 79.5 | 106.0 | 10.6 | 116.6 | 43.8 | — | 39.8 | 47.4 | 276.4 | 68.2 |
| 1987 | 45.3 | — | 45.3 | 74.0 | 31.1 | 105.1 | 129.7 | 13.1 | 142.8 | 34.9 | — | 43.1 | 57.0 | 237.4 | 73.6 |

[1]*Agricultural Statistics 1970*, p. 340, Table 497, and p. 341, Table 498; *Agricultural Statistics 1981*, p. 334, Table 493, and p. 338, Table 498; *Ibid.*, 1988, pp. 295. Data 1960–1987.
[2]Total wool production = shorn wool + pulled wool. Reported on basis of clean fiber, using conversion factor of 47.7% from 1966–1971, 52.8% from 1972 to 1980, and 72.9% for pulled wool production.
[3]Apparel wool includes dutiable wool. Carpet wool includes all duty-free wool.
[4]Consumption on the woolen and worsted systems.

Fig. 34–2 shows that wool production and imports, have declined through the years but that wool consumption has been on the increase in recent years. This has been due to (1) increased use of manufactured fibers (see Fig. 34–18); and (2) increased imports of manufactured wool textiles, which reflects that manufacturing labor costs are cheaper abroad than in the United States.

It is noteworthy that expenditures for clothing and accessories consumed 5.0% of the expenditures of the average American family in 1987.[1] In addition to the indispensability of its product, the wool textile industry is nationally important because it is one of the largest industries in the United States.

[1]*Statistical Abstracts of the United States 1989*, Bureau of the Census, U.S. Department of Commerce, p. 426, Table 693.

## U.S. Production, Imports, and Consumption of Raw Wool

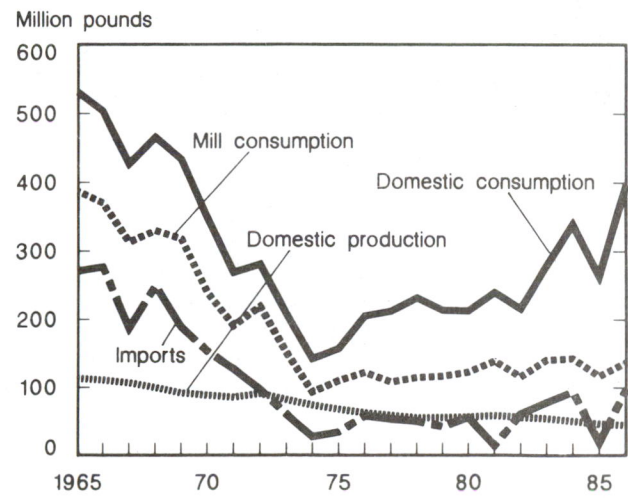

Fig. 34–2. Historic account of U.S. production, imports, and consumption of raw wool (clean basis). Production includes shorn and pulled wool. Imports include duty-free and dutiable wool. Mill consumption includes apparel and carpet wool. Domestic consumption includes mill consumption plus raw wool equivalent of net textile trade balance. (Source: *1988 Agricultural Chartbook*, USDA, Ag. Hdbk. No. 673, p. 100, Chart 233)

## THE WOOL FIBER

Wool, the natural protective covering of sheep, is the most complex of textile fibers. It is a protein material of highly organized structure. Wool differs from other animal fibers by having a serrated surface; a crimpy, wavy appearance; and excellent degree of elasticity; and an internal structure composed of numerous minute cells. In contrast, hair has a comparatively smooth surface, lacks in crimp or waviness, and will not stretch. As a product of the skin or cuticle of vertebrate animals, wool is similar in origin and general composition to the various other skin tissues found in animals—horns, nails, and hoofs.

## THE THREE CELL LAYERS

From the standpoint of structure, a microscope reveals that all wool fibers consist of two distinct cell layers, and some fibers have a third layer. According to their positions, these layers are known as (1) the epidermis or outer layer, (2) the cortex, and (3) the medulla. Although differing in characteristics, these same three cell layers are found in most hair. The chief characteristics of these three layers in wool fiber will be discussed briefly.

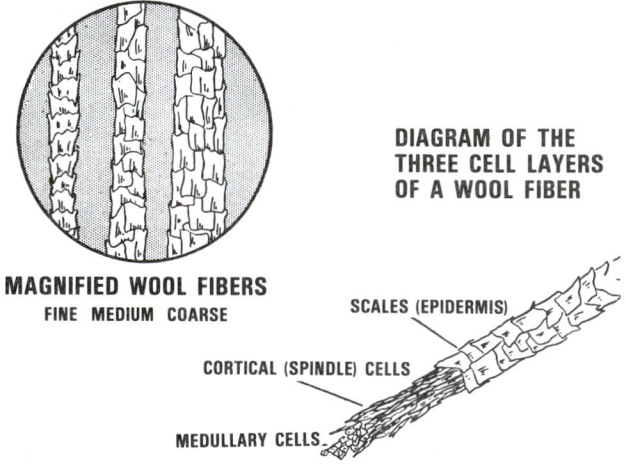

**DIAGRAM OF THE THREE CELL LAYERS OF A WOOL FIBER**

**MAGNIFIED WOOL FIBERS**
FINE   MEDIUM   COARSE

SCALES (EPIDERMIS)
CORTICAL (SPINDLE) CELLS
MEDULLARY CELLS

Fig. 34–3. Diagram showing the structure of wool fibers. Upper picture shows the epidermis—the outside cell layer—of fine, medium, and coarse fibers. Note that fine wool has a greater number of serrations. Bottom picture shows the three cell layers (1) the scales or epidermis, (2) the cortex, and (3) the medulla.

## CHEMICAL COMPOSITION OF WOOL

Chemically, wool is chiefly keratin, which is also the primary constituent of hair, nails, hoofs, horns, and feathers. Keratin is a mixture of nitrogen and sulfur compounds and amino acids. A typical chemical breakdown of wool is as follows: carbon, 50%; oxygen, 22 to 25%; nitrogen, 16 to 17%; hydrogen, 7%; and sulfur, 3 to 4%.

## SOME FLEECE CHARACTERISTICS

The chief fleece characteristics of interest and importance to both the producer and manufacturer are (1) grease, (2) length, (3) density, (4) diameter, and (5) variations of different body areas. It is well recognized that there are wide differences between breeds and individuals in such wool characters as fineness, length, density, and yield.

## GREASE

In a broad sense, grease refers to all the impurities found in unscoured wool, including the yolk, suint, and soluble foreign matter but not the vegetable matter. Shrinkage of fleeces varies widely—from 30 to 75%, with many factors affecting it, but on the average and with all grades included, U.S. shorn grease wool shrinks about 52.3%.[2] The following example shows how shrinkage is computed.

> 10.00 lb grease wool before scouring
> 4.77 lb clean wool after scouring
> 5.23 lb loss in wool scouring
> 5.23 lb = 52.3% = shrinkage
> 4.77 lb = 47.7% = yield

The commercial value of a clip of wool is largely determined by the amount of clean wool fiber that it yields. Although a part of the impurities found in grease wool are essential for the growth and well-being of the fleece and the animal, the manufacturer is primarily concerned with securing the highest possible yield of clean wool of the finest quality.

Many wool growers now sell their wool on a clean basis following a core test for clean yield.

## LENGTH

The length of the wool in fleece is a matter of much importance to both the producer and the manufacturer. Together with quality, it constitutes the principal basis of classification and grading in buying and selling and largely determines the use to which wool will be put. The wool producer regards good length as a desirable attribute, for it gives a greater weight of wool. In judging sheep, fiber length is based on an appraisal of the annual growth as determined by parting the fleece at three body areas—the shoulder, side, and britch. Fiber length varies anywhere from 1 to 20 in.

## DENSITY

Density refers to the closeness or compactness of the fibers in the fleece and is often defined as the number of fibers per unit area of skin. Experimental studies have revealed very clearly that density differences exist between breeds, individuals, and body areas. It has been estimated that the number of wool fibers per animal varies from a low of about 16,000,000 for some individuals of the medium or coarse wool breeds to a high of 120,000,000 fibers for individuals of the Australian Merino breed. Fleece density is an attribute in determining fleece weight. In judging, fleece density is determined by grasping the wool on the side to feel its fullness and compactness and by parting the fleece to examine the apparent closeness of the fibers. Experimental evidence shows that often this method is misleading, apparently being affected by the grease and dirt content.

---

[2]Based on entire 1946 clip purchased by the Commodity Credit Corporation which included virtually all the domestic wool clip that year. These wools showed the following shrinkage by grades: Fine, 56–65%; ½ Blood; 51–60%; ⅜ Blood, 41–50%; and ¼ Blood, 41–45%. (From *The Domestic Wool Clip*, Production and Marketing Administration, USDA, 1951, p. 14, Table I) Although definite information on proportion of various grades in the U.S. wool clip has not been available since 1946, it appears that a lower proportion of fine wool is now produced than in 1946. However, there has probably been little change in average shrinkage.

## DIAMETER

The fineness of wool is very important because the character of the yarns and fabrics produced is determined to a very great extent by the variations in the diameter of the fiber. Wool sorting is based on fineness of fiber, and this is considered to be the soundest basis on which wool and top qualities can be classified. In the trade, the experienced wool grader is able to estimate the fineness by visual inspection and handling. It is a well-known fact, however, that an ordinary sample of wool that is estimated by a wool grader as representing a certain fineness will, on close examination, usually show two or three other finenesses. That is, it is really a composite mixture. If placed under a microscope, it is also found that the shape or contour of fibers varies greatly. As a rule, fibers are irregular and possess varying degrees of ovality or ellipticity.

In judging sheep, the number of crimps is usually accepted as an index of fineness; that this is a good criterion is borne out by experiments in which more refined techniques have been used. The diameter of wool fibers varies anywhere from 0.0008 to 0.002 in. and the number of crimps, from 5 to 36 per in.

## PRODUCTION AND HANDLING OF THE WOOL CLIP

If U.S. sheep producers are to survive the inroads of imports and synthetic fibers, it is imperative that they market a higher quality product—one that does not require unnecessary processing expenditures in the textile mills.

Despite remarkable improvement in wool wrought through improved breeding, handlers, buyers, and processors of the domestic wool clip are in general agreement that the overall quality of the nation's wool clip has declined in recent years, and that the primary reason for this decline is the generally poor preparation of the clip. The most common explanations or excuses back of this are (1) carelessness and indifference on the part of sheep producers, and (2) lack of skilled and dedicated shearers. Whatever the cause, all are agreed that a change must be made if the domestic wool clip is to meet its increasing competition from imported wool and fabrics and from man-made fibers.

Observance of the following wool production and handling practices will result in marketing a higher quality product:

1. Producing superior fleeces through (a) feeding properly, (b) protecting the on-the-back fleece from foreign material, and (c) tagging sheep at intervals.
2. Using a scourable branding material, where branding or identifying is necessary.
3. Shearing in a clean place.
4. Using skilled shearers.
5. Packing properly.
6. Shipping in clean trucks or cars, and keeping the wool bags dry.

## SHEARING

Up to and during the early part of the 1900s, most shearing was done by hand shears. Then clippers, similar to barber's clippers, were developed; hand-powered at first, then electrically powered. A skilled shearer can clip 200 or more sheep per day using electric clippers. In the range states, the shearers travel in crews from ranch to ranch, staying on each ranch just long enough to shear the band.

High shearing cost and a scarcity of sheep shearers have spurred interest in finding an easier and less costly way of removing the fleece from sheep. The following three innovative approaches are in various stages of experimentation and application: chemical shearing, laser beam shearing, and computerized shearing.

• **Chemical shearing**—The chemical approach, which is still subject to U.S. Food and Drug Administration approval as safe, involves the use of the chemical, Cyclophosphamide (CPA). This synthetic drug, discovered by German scientists doing cancer research, was observed to cause patients' hair to fall out.

When the CPA pill is given to sheep orally (with a balling gun), it temporarily stops cell growth, constricting the fiber at the skinline and causing it to break. From 7 to 12 days after administering the drug, sheep may be sheared with the bare hands. They can be stripped as naked as the human body—they can be sheared without nicks, second cuts, or shearing skill.

To date, no harmful side effects have been reported from the use of CPA, including its use on pregnant ewes. However, even if FDA approval is forthcoming, the following problems will deter any widescale use of chemical shearing:

1. The necessity of handling sheep twice—when administering the pill and when defleecing 7 to 12 days later.
2. The susceptibility of "bald" sheep to sunburn or cold.
3. The loss of wool by rubbing on post and brush.
4. The variability of different body areas in response to the chemical; the wool on the back, shoulders, and sides is usually removed rather easily, whereas the wool from around the face and legs comes off with difficulty.
5. Variations in dosage levels, apparently due to individual and body weight differences.
6. Variations between animals in the time interval required from dosing to defleecing.
7. The regrowth of wool in Suffolks may come in black.

• **Laser beam shearing**—A group of Australians, headed by a former sheep shearer, has patented a laser beam for shearing. These ingenious inventors reasoned that lasers, which had already been used to cut woolen cloth and steel, could be adapted to shearing sheep. The laser actually severs the wool by burning. However, the Australian developers feel that it can be governed so that it will selectively cut only wool—that it can be designed so that it will automatically switch off when the beam strikes tissue or any other material differing in density from wool.

• **Computerized shearing**—This consists of a mechanical hand, guided by a computer (Fig. 34–8, page 687). The computer first creates a contour drawing—a computer map—of the sheep's body. This map is used by the computer to guide the mechanical hand, which holds the shears, over the animal's body. To avoid nicking, the shearer head retracts in $5/1000$ of a second when the slightest movement is detected. An experimental version at the University of Western Australia shears a sheep in 3 minutes, about the same as a human. Development of the computerized sheep shear was financed by the Australian Wool Corporation at a cost of $1 million.

# HARVESTING, SORTING, AND MARKETING OF WOO

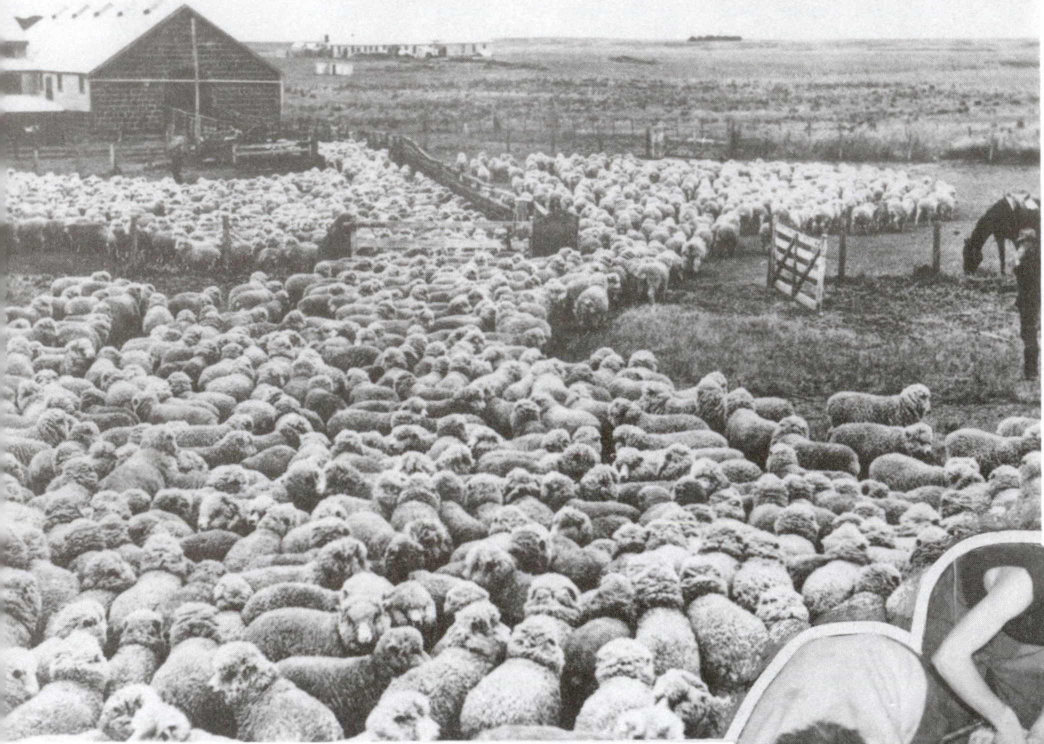

Fig. 34–4. *Above:* Yarding sheep for shearing in Australia. (Courtesy, Wool Bureau, Inc., New York, NY)

Fig. 34–5. *Below:* Shearing sheep with electric clippers just like the clippers of a barber.

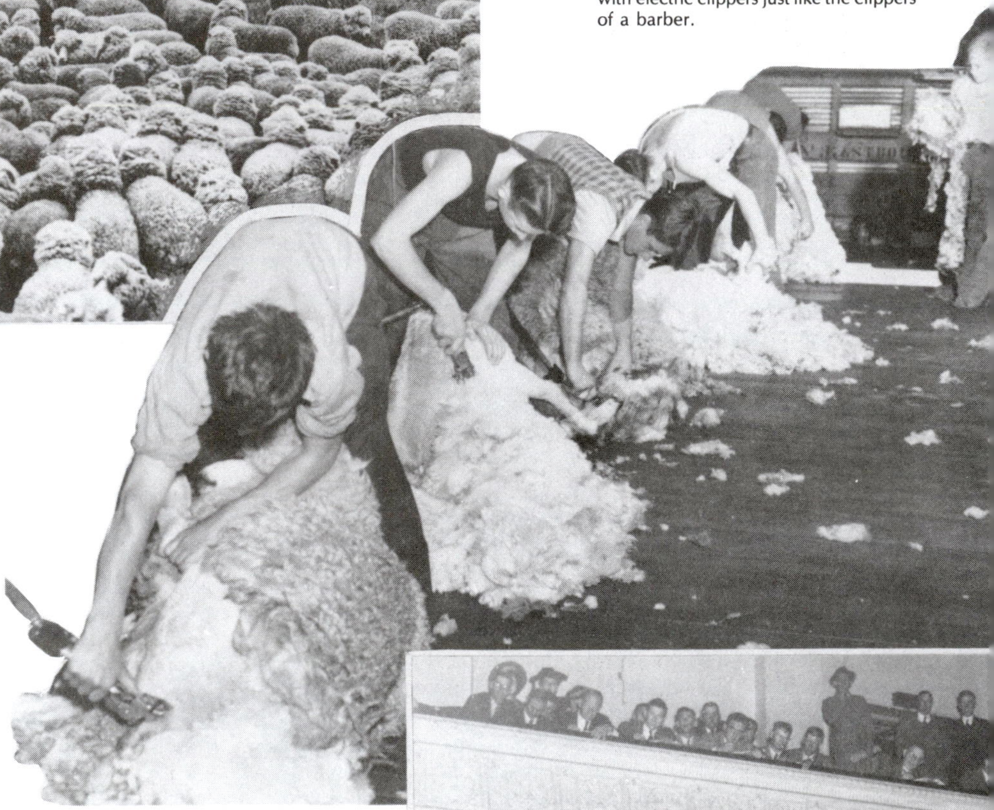

Fig. 34–6. *Below:* Wool sorting—the separation of fleeces into the various grades. (Courtesy, Pendleton Woolen Mills, Portland, OR)

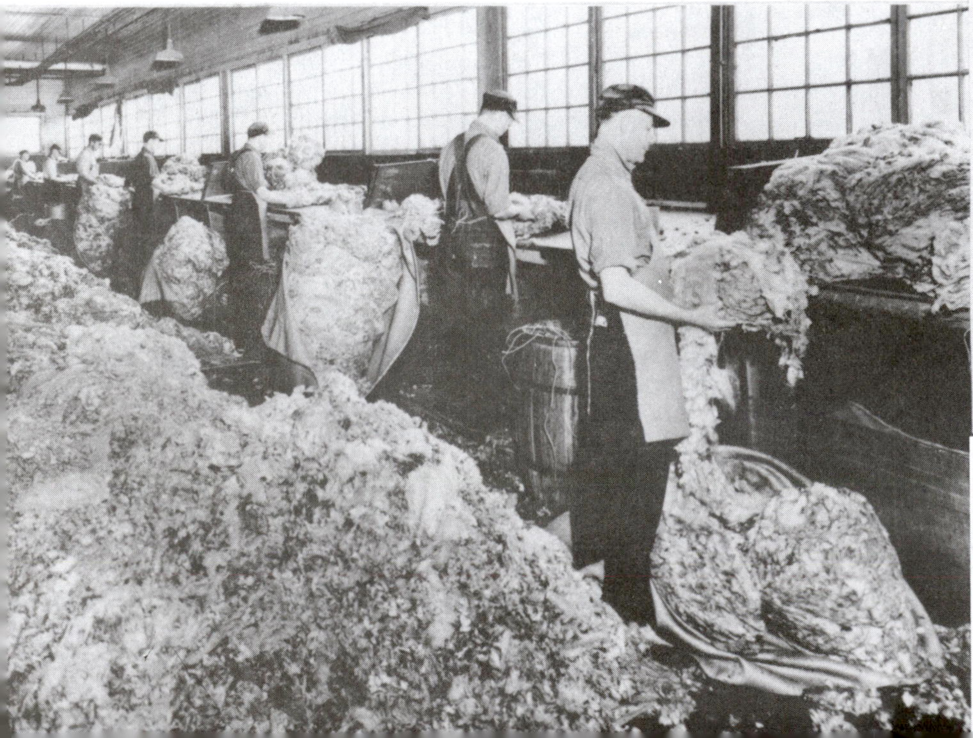

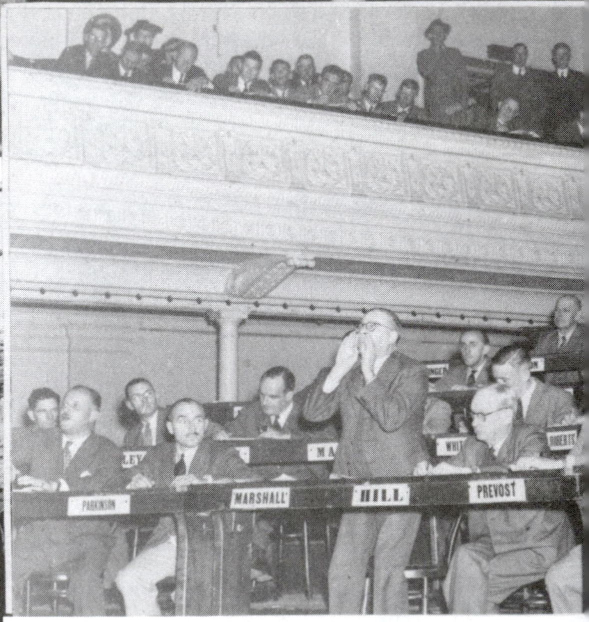


Fig. 34–7. Wool auction in Wellington, New Zealand, showing buyers bidding. (Courtesy, Department of Scientific and Industrial Research, Wellington, New Zealand)

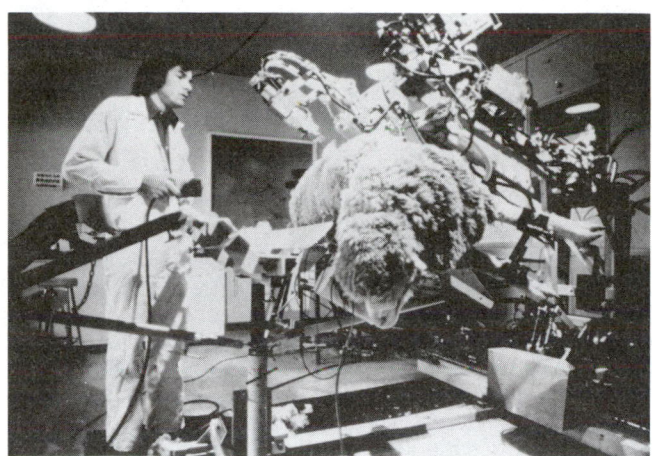

Fig. 34–8. Shearing sheep with a mechanical hand guided by a computer. The computer creates a contour drawing of the surface of the sheep and uses this to guide the shears over the animal's body. As technology improves, it may shear faster than a person. (Photo by Cary Wolinsky, through Stock, Boston, Inc., Boston, MA)

Despite the problems, chemical shearing, laser beam shearing and computerized shearing merit further study. The stakes are high, especially for the small flock owner who has difficulty in obtaining shearers.

## REQUISITES OF WOOL

Good wool possesses character, purity, strength of fiber, good condition, cleanliness, low shrinkage, adequate but not excessive grease, uniformity, and a bright white color. These requisites apply to all wools, regardless of the class or grade.

## CLASSES OF WOOL

The wool trade recognizes two major classes of wool—apparel wool and carpet wool. As the names imply, most apparel wools are those suitable for manufacture into yarns and fabrics for human clothing, whereas most wools of the carpet class are used in making floor covering. In 1987, 91% of the wool consumed in the United States was apparel wool, and 9% was carpet wool.

Apparel wools are further classified according to use as (1) combing wool or staple wool—the long-fibered wools within the class, (2) French combing wool—the wools of intermediate length, and (3) clothing wool—the short-fibered wools. Although these three classes are based largely on length of fiber, other factors—such as supply and demand, fiber diameter, purity, condition, etc—are important in determining the use made of wools. Thus, many wools used by the woolen industry are longer than some used in worsted manufacture; and a considerable amount of wool classed as clothing is used in the worsted industry. In general, however, the manufacturer can realize the greatest profit by utilizing apparel wools according to their best adaptation as indicated by the three classes. Further, carpet wool is not suited for use as apparel wool.

## COMBING OR STAPLE WOOL

Combing or staple wools are usually referred to as the highest priced and best wool obtained from sheep. Both fineness and length are requisite. For example, a 64s combing wool should be 3 in. or more in length, with the length varying according to grade as shown in Table 34–3 (see page 688). By and large, combing wools are used for making worsted fabrics. They take their name from the fact that one of the main processes in worsted manufacturing is the combing operation, which separates the long fibers from the short ones. The long fibers are used to make worsted cloths, and the short fibers (called noil) are used in the making of woolen cloths. In the former, the fibers are laid parallel to each other; whereas in the latter, the shorter wool fibers that are used in making woolen cloths and felts are laid in every direction—in fact, the more mixing the better in woolens. These differences are of importance to the consumer. Among other things, they explain why worsted suits hold their press better than woolen suits.

In the United States, wools with sufficiently long fibers and otherwise adapted to the making of worsted cloth are commonly combed on the Noble or Bradford comb.

Prior to World War II, approximately 70% of the apparel wool used in the United States was processed on the worsted system of manufacture. However, since that time there has been a gradual decline in the use of worsteds and an increase of woolens. In 1987, 53% of the mill consumption of apparel wool was processed on the worsted system and 47% on the woolen system.

## FRENCH COMBING WOOL

French combing wools are in between the combing wools and the clothing wools in length. These wools are manufactured on the French or Heilman comb, which is designed to use shorter wools and still produce worsted fabrics. Thus, the French system utilizes much wool that is not long enough for manufacture on the regular worsted system known as Noble combing. This system of combing is becoming more popular.

## CLOTHING WOOL

Clothing wool is the name usually given to the shortest wool. This wool is too short to be manufactured on the worsted system, but it can be used successfully on the woolen system. Although longer fibers can be used in making woolens, they are usually more expensive than short fibers and hence are reserved for making worsteds which usually sell at a slightly higher price than woolens. The term *clothing wool*, however, does not mean that the wool is suitable only for fabrics to be made into clothing. This type of wool is also used to make felts.

## CARPET WOOL

Carpet wools, which are usually the coarsest wools, are of low quality because they (1) contain mixtures of very coarse, hairy fibers and finer fibers, and (2) vary markedly in fiber length. The chief requisite of carpet wool is resilience, the quality that makes it resistant to matting down and to wear under the constant scuffing of passing feet. Most of this wool comes from long-wooled sheep and from sheep that show lack of breeding. Most carpet wools are imported because our flocks, except for sheep kept by the Navajo Indians, have been improved to the point where the vast majority of wool grades as apparel wool. In order to encourage importation to fill the need, carpet wool is duty-free. More than 90% of U.S. imports of carpet wool come from three countries: New Zealand, United Kingdom, and Argentina.

## WOOL GRADING

Wool grading is based primarily on fiber diameter or fineness, but consideration is also given to length. Many manufacturers desire wool of a certain fineness only. This means that the wool must be separated at the warehouse and like fleeces must be piled by themselves. This process is called *wool grading,* and it is done by highly trained wool graders.

A graded pile or bale of wool does not infer that all the wool therein is of one diameter. This is so because any single fleece of wool as it comes from the sheep may possess several different grades. Thus, a 60/62s combing wool simply means that the greater part of the wool on the fleece is of that fineness and length. The manufacturer knows that some wool in these fleeces, especially on the shoulder part of the fleece, will be finer; and also that some wool, as on the britch, will be considerably coarser. Because of this, a further separation, known as *sorting,* follows. The ability to grade wool, which is acquired only with considerable experience, requires a keen sense of sight combined with the sense of touch and rare good judgment.

## GRADES OF WOOL

The average diameter of fiber and the limits for the variation in diameter for the various grades are shown in Table 34–3. Maximum limits to the variation allowed for each grade are expressed by the statistical term—*standard deviation.* In application, if there is too much variation in fiber diameter, the wool is assigned to the next coarser grade. Wool can be separated roughly, after a little experience, into three broad market grades according to its diameter: (1) fine wool, (2) medium wool, and (3) coarse or braid wool. More accurately speaking, however, there are three distinct methods of grading wool according to diameter with several grades in each. The older method is called the blood system; the newer methods are the *numerical count system* and the *micron system.* A comparison of these three systems is contained in Table 34–3.

## TABLE 34–3
## COMPARATIVE WOOL GRADES AND CLASSES

| Type of Wool | Old Blood Grade | Numerical Count Grade | Micron System[3] Limit for Average Fiber Diameter | Micron System[3] Variability Limit[4] | Combing Wool over (in.) | Combing Wool over (cm) | French Combing Wool (in.) | French Combing Wool (cm) | Clothing Wool under (in.) | Clothing Wool under (cm) |
|---|---|---|---|---|---|---|---|---|---|---|
| | | | (microns)[5] | (microns)[5] | (in.) | (cm) | (in.) | (cm) | (in.) | (cm) |
| Fine | Fine | Finer than 80s | Under 17.70 | 3.59 | — | — | — | — | — | — |
| Fine | Fine | 80s | 17.70–19.14 | 4.09 | 2.75 | 6.99 | 1.25–2.75 | 3.18–6.99 | 1.25 | 3.18 |
| Fine | Fine | 70s | 19.15–20.59 | 4.59 | 2.75 | 6.99 | 1.25–2.75 | 3.18–6.99 | 1.25 | 3.18 |
| Fine | Fine | 64s | 20.60–22.04 | 5.19 | 2.75 | 6.99 | 1.25–2.75 | 3.18–6.99 | 1.25 | 3.18 |
| Medium | ½ blood | 62s | 22.05–23.49 | 5.89 | 3.0 | 7.62 | 1.5–3.0 | 3.81–7.62 | 1.5 | 3.81 |
| Medium | ½ blood | 60s | 23.50–24.94 | 6.49 | 3.0 | 7.62 | 1.5–3.0 | 3.81–7.62 | 1.5 | 3.81 |
| Medium | ⅜ blood | 58s | 24.95–26.39 | 7.09 | 3.25 | 8.26 | 2.0–3.25 | 5.08–8.26 | 2.0 | 5.08 |
| Medium | ⅜ blood | 56s | 26.40–27.84 | 7.59 | 3.25 | 8.26 | 2.0–3.25 | 5.08–8.26 | 2.0 | 5.08 |
| Medium | ¼ blood | 54s | 27.85–29.29 | 8.19 | 3.5 | 8.89 | 2.5–3.5 | 6.35–8.89 | 2.5 | 6.35 |
| Medium | ¼ blood | 50s | 29.30–30.99 | 8.69 | 3.5 | 8.89 | 2.5–3.5 | 6.35–8.89 | 2.5 | 6.35 |
| Coarse | Low ¼ | 48s | 31.00–32.69 | 9.09 | 4.0 | 10.16 | — | — | 4.0 | 10.16 |
| Coarse | Low ¼ | 46s | 32.70–34.39 | 9.59 | 4.0 | 10.16 | — | — | 4.0 | 10.16 |
| Coarse | Common[6] | 44s | 34.40–36.19 | 10.09 | 5.0 | 12.70 | — | — | 5.0 | 12.70 |
| Very coarse | Braid | 40s | 36.20–38.09 | 10.69 | 5.00 | 12.70 | — | — | 5.0 | 12.70 |
| Very coarse | Braid | 36s | 38.10–40.20 | 11.19 | 5.00 | 12.70 | — | — | — | — |
| Very coarse | Braid[6] | Coarser than 36s | Over 40.20 | — | — | — | — | — | — | — |

[1]Standards for grades of wool, as published by the USDA, August 20, 1965. *Federal Register* (7 CFR Part 31). These standards became effective January 1, 1966.

[2]There are no USDA official lengths for the different classes. The lengths given herein are in keeping with trade practices and were provided for use in this book by the Livestock Division Wool Laboratory, Standardization Branch, USDA, Denver, CO 80225.

[3]Beginning January, 1976, the unit designation terminology for wool prices changed to microns.

[4]Standard deviation maximum.

[5]A micron is 1/25,400 of an inch.

[6]Common and braid are not classified according to length because these wools are practically always of combing length. Carpet wool includes all those not suited to the three classes listed.

An experienced grader determines the grade of wool by the senses of sight and touch. However, for use in more objective grade determination and for arbitration purposes where there may be a dispute as to grade before final settlement, there is a scientific method of test prescribed. A copy of this method of test, which explains micro-projector equipment recommended and also sampling and testing procedures, may be obtained from the U.S. Department of Agriculture, Federal Center, Standardization Branch, Wool and Mohair Laboratory, Denver, Colorado. Testers and research workers may also use calipers or photographic or air-flow equipment for grade determination.

The grades of wool produced vary widely between areas. Thus, about 75% of the wool produced in Texas, New Mexico, Arizona, and Nevada, grades Fine and ½ Blood; whereas, ⅜ Blood and ¼ Blood wool predominates in the North Atlantic, East North Central, West North Central, and South Atlantic states.

The U.S. standard grades for wool are presented in Fig. 34–9.

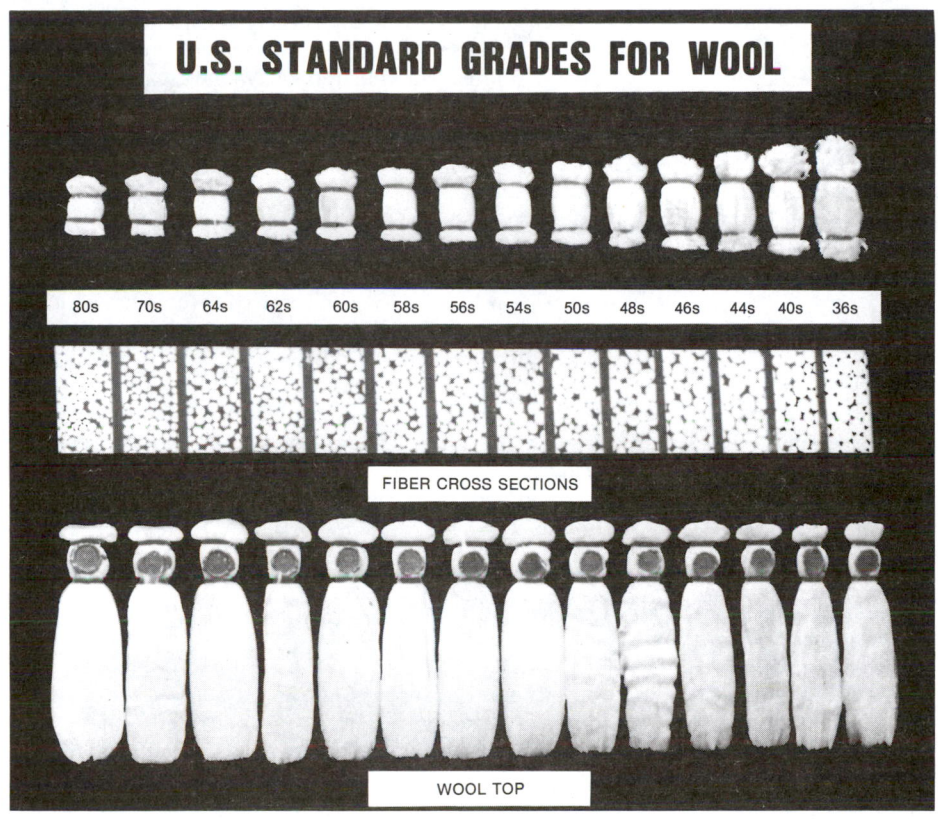

## U.S. STANDARD GRADES FOR WOOL

| 80s | 70s | 64s | 62s | 60s | 58s | 56s | 54s | 50s | 48s | 46s | 44s | 40s | 36s |

FIBER CROSS SECTIONS

WOOL TOP

Fig. 34–9. *Left:* U.S. Standard Grades of wool. (Courtesy, American Sheep Industry Assn., Englewood, CO)

## THE BLOOD SYSTEM

The blood system divides all wool, from finest to coarsest, into six market grades. These are (1) Fine, (2) ½ Blood, (3) ⅜ Blood, (4) ¼ Blood, (5) Low ¼ Blood, and (6) Common and Braid. Originally, these fractional Blood names denoted the amount of Merino blood in the sheep producing the wool. At the present time, these names indicate wool of a certain diameter only and have no connection whatsoever with the amount of Merino blood in the sheep. As a matter of fact, it is possible to have ⅜ Blood wool from a sheep with no Merino blood at all. The blood grades, therefore, are merely trade names identifying the different grades of wool, without relationship to the breeding of the sheep and are rapidly being replaced with the numerical count.

## NUMERICAL COUNT SYSTEM

The numerical count system divides all wool into 14 grades, and each grade is designated by a number. The numbers range from 80s for the finest wool down to 36 for the coarsest. This method gives more grades, and thus finer divisions can be made; and this is more satisfactory to the wool dealers and manufacturers. Table 34–3 shows the correlation between the two grade systems.

## WORSTED SPINNING COUNT

Theoretically, the numerical count system is based on the number of hanks of yarn (each hank representing 560 yards) that can be spun from one pound of such wool in the form of top. Wool of 50s quality, therefore, should spin 50 × 560 yards per pound of top, if spun to the maximum on the worsted system of manufacture. Unfortunately, this is not always true; the lower grades will not spin up to their number. Moreover, it is noteworthy that, in actual practice, wools are rarely spun to their maximum limit. Furthermore, spinning count is not determined by diameter alone; such factors as fiber length, moisture conditions, and the skill of the workers influence the count that may be spun. It may be concluded, therefore, that neither the blood system nor the numerical count system denotes accurately what it is supposed to indicate according to derivation of the respective term.

## MICRON SYSTEM

The micron system is a substantially more technical and accurate measurement of the wool fiber in a lot of wool. Sixteen grades are used, and are based on the average fiber thickness as measured by a micrometer. An 80s wool, for example, averages about 18 microns, which is less than half a 36s wool that averages 39 microns. A micron is 1/25,400 of an inch. Wool too variable to fit within the limits of one grade is placed down a grade.

The micron system was largely developed at the USDA Denver Wool Laboratory. This system may eventually become the standard for describing wools in the United States. In January, 1976, the U.S. Economic Research Service began reporting the unit designation terminology for wool prices in microns.

Table 34–3 compares the old blood system, the numerical count system, and the micron system of wool grades.

## WOOL SORTING

Sorting is the operation of taking an individual fleece, untying the twine, opening the fleece, and separating the fleece into the various grades that it possesses in the different body areas. This operation is usually done in the mill, but occasionally it is done in a warehouse. The reason for this is that a mill knows exactly what qualities of wool it wishes to put into a fabric. The object of sorting is to obtain large lots of wool that are very even and uniform in diameter, length, strength, and other characteristics. It is easy for an inexperienced person to distinguish a very fine wool from a very coarse wool, but it takes considerable training to be able to separate 2 consecutive grades, such as 56s from 58s. Sorting is always done on the grease wool. The dusting and scouring operation break up the fleece into small pieces, so that sorting of scoured wool is impracticable. Sorting is necessary on wool if a uniform worsted yarn with high spinning count is desired. If the wool is not to be spun to the maximum count, then only a superficial sorting is necessary. The thoroughness of the sorting varies according to the type of fabric into which the wool is to be made.

## THE MARKETING OF WOOL

Like most industries, the wool and textile business has progressed from the status of a family enterprise. In the early days of this country—and the same pattern held true in other nations—virtually every family owned a few sheep and produced sufficient wool to meet its own needs. Under the family system, carding, spinning, and weaving were carried on by members of the household for the purpose of supplying the family with clothing. Under these conditions, there was little or no marketing. With the concentration of population in the cities, the coming of artisans, and the bringing of wool from more distant points, however, markets were a necessity.

Today, wool is one of the most important commodities of world commerce, and it might well be added that the marketing operations connected therewith are among the most intricate. In the first place, it is one of the most difficult items of commerce to classify and grade for the benefit of the trade. Secondly, few items have to be transported greater distances from producers to consumers. Wool production is a frontier type of industry, with the surplus-producing areas in those regions that are relatively undeveloped. On the other hand, wool consumption is greatest in the more populated regions.

## INTERNATIONAL TRADE IN WOOL

Fig. 34–10 and Table 34–4 show the international trade in wool, in exports and in imports. About two-thirds of the world's wool production is in the Southern Hemisphere, with the five leading export countries, by rank, being Australia, New Zealand, Argentina, Uruguay, and the United Kingdom. However, the great wool-importing nations are in the Northern Hemisphere, with the five leading import countries, by rank, being Japan, China, United Kingdom, U.S.S.R., and Italy.

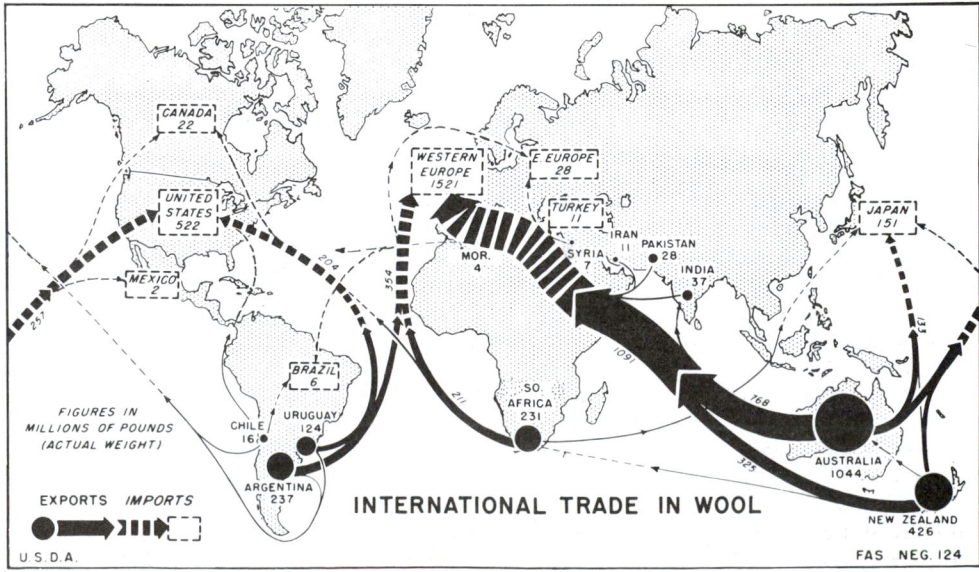

Fig. 34–10. The international trade in wool, exports and imports. (Courtesy, USDA)

**TABLE 34–4**
**MAJOR WOOL-EXPORTING AND WOOL-IMPORTING COUNTRIES OF THE WORLD[1]**

| | Exports | | | | Imports | | |
|---|---|---|---|---|---|---|---|
| Country | Greasy | Degreased | Total[2] | Country | Greasy | Degreased | Total[2] |
| | (MT) | (MT) | (MT) | | (MT) | (MT) | (MT) |
| Australia ............ | 677,197 | 105,098 | 782,295 | Japan ............. | 123,930 | 80,423 | 204,353 |
| New Zealand ......... | 121,717 | 181,358 | 303,075 | China ............. | 122,510 | 81,230 | 203,740 |
| Argentina ............ | 35,000 | 26,400 | 61,400 | United Kingdom ..... | 89,197 | 49,713 | 138,910 |
| Uruguay ............. | 29,994 | 15,090 | 45,084 | U.S.S.R. ........... | 133,999 | — | 133,999 |
| United Kingdom ...... | 13,429 | 29,089 | 42,518 | Italy ............. | 84,419 | 37,758 | 122,177 |
| France .............. | 24,638 | 16,822 | 41,460 | West Germany ...... | 55,301 | 24,375 | 79,676 |
| South Africa ........ | 29,000 | 6,000 | 35,000 | Belgium-Luxembourg . | 48,206 | 15,836 | 64,042 |
| Mongolia ............ | 14,000 | — | 14,000 | U.S.A. ............ | 31,698 | 15,960 | 47,658 |
| China ............... | 3,050 | 10,250 | 13,300 | South Korea ....... | 22,398 | 22,344 | 44,742 |
| Hong Kong ......... | 1,866 | 9,100 | 10,966 | Hong Kong ........ | 30,000 | 10,999 | 40,999 |
| World Total ....... | 1,018,300 | 482,003 | 1,500,303 | World Total ...... | 987,321 | 444,961 | 1,432,282 |

[1]FAO Trade Yearbook, 1987, United Nations, Rome, Italy, Vol. 41, pp. 271–274. Data for 1987.

[2]Because of the country to country variations in the proportion of greasy and degreased wools, and in the shrinkage of greasy wools, no claim is made that the country to country total exports, or the country to country total imports, are scientifically comparable. But they are the best, and only, totals available (1) for comparing wool-exporting countries, and (2) for comparing wool-importing countries.

London has been displaced as the greatest wool-marketing center of the world by Sydney, Australia. Today, Australia's wool auctions are the standard by which all others are judged.

In order to protect American sheep producers, there is an import duty on apparel wool of the type generally grown in this country. Wools of the type not grown in this country—carpet wools—are imported duty-free when used for floor coverings, press cloths, knit or felt boots, or heavy fulled socks.

The Hawley-Smoot Tariff Act of 1930, with periodic adjustments, continues in effect. A detailed current list of duties (Schedule 3, Textile Fibers and Textile Products) for each class of wool or wool product may be obtained by writing the Department of the Treasury, U.S. Customs Service, Washington, DC 20229.

## METHODS OF MARKETING WOOL IN THE UNITED STATES

There are several differences between the marketing of animals on foot and the marketing of wool. In the first place, the average livestock producer is usually familiar with more than one of the several avenues through which live animals may be disposed; whereas, except for the larger wool growers, there is generally little knowledge concerning possible market outlets for wool.

Unlike most farm products, there are no open- or auction-markets for wool. Most wool is bought and sold by private treaty. Buyers may be representatives from woolen mills or brokers who represent the mills.

There are several channels through which growers may sell their clips. Producers in the major wool growing areas of the United States (Texas and the western states) usually have more options than those in the farm flock states. The predominant market channels are:

1. Private treaty.
2. Wool pools.
3. Sealed bid.
4. Consignment.
5. Direct selling.
6. Wool warehouses.

## THE MANUFACTURING PROCESSES

After wool is purchased by the manufacturer, it must pass through a number of intricate processes before the final product evolves. Wool fabrics are of two types—worsteds and woolens, each requiring different manufacturing and finishing processes. Woolens are made from the shorter wool fibers—carded but not combed—with the result that the little fibers in the yarn are in a crisscross position, giving a soft, fuzzy yarn, as in broadcloth. The weave of a woolen cloth is more or less concealed by the fuzzy finish. A woolen also is usually softer and less firm than a worsted. Common worsted fabrics include serge, gabardine, rep, and coverts. Common woolen fabrics are cheviot, tweed, flannel, broadcloth, melton, kersey, cassimere, and mackinaw.

The longer fibers are used for worsted, and they are combed in addition to being carded. The combing process places the fibers parallel, giving a yarn that is spun more tightly than the woolen yarns, as in serge. A worsted also has a clear-cut weave and a smoother surface than a woolen.

The steps in manufacturing worsteds and woolens vary according to the available equipment, kind of wool, and the finished product desired, but in general they consist of the following: (1) sorting; (2) dusting and opening; (3) scouring; (4) drying; (5) carbonizing or bur picking; (6) blending, oiling, and mixing; (7) carding; (8) combing; (9) spinning; (10) weaving or knitting; and (11) dyeing and finishing.

# HOW IT USED TO BE DONE!

Fig. 34–11. *Above:* Unimproved sheep in rural France, the kind of sheep that produce carpet wool. (Courtesy, Ewing Galloway, New York, NY)

Fig. 34–12. *Above:* Hand-shearing sheep, the common method of shearing sheep in the U.S. until about 1925.

Fig. 34–13. *Left:* Spinning in colonial New England. (Courtesy, American Sheep Industry Assn., Englewood, CO)

# HOW IT IS DONE TODAY!

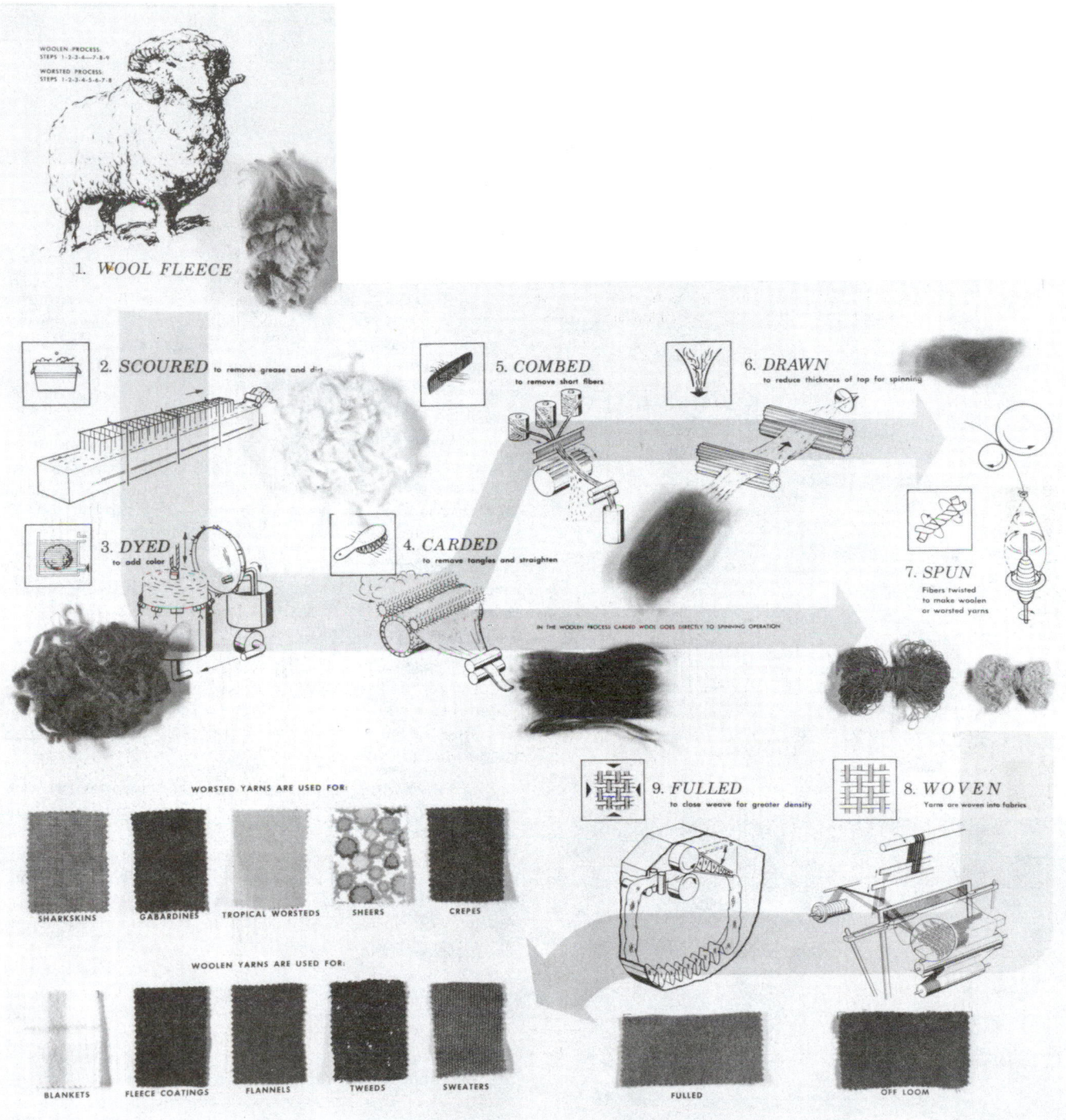

Fig. 34-14. How wool is made into fabric today. Note the two processes—worsted and woolen. (Courtesy, The Wool Bureau, Inc., New York, NY)

## *WOOL, FROM FLEECE TO FABRIC*

Fig. 34–15 shows the approximate amounts of grease wool required to make each of two common garments.

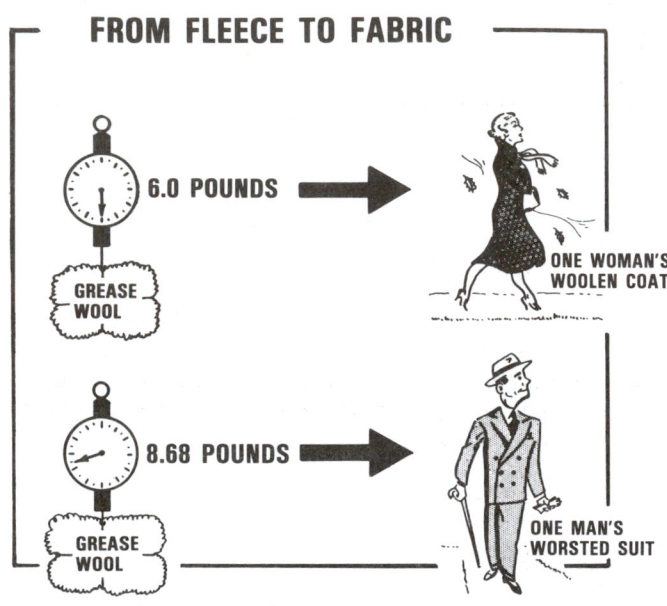

**FROM FLEECE TO FABRIC**

6.0 POUNDS → ONE WOMAN'S WOOLEN COAT

GREASE WOOL

8.68 POUNDS → ONE MAN'S WORSTED SUIT

GREASE WOOL

Fig. 34–15. Wool, from fleece to fabric. It requires 6.0 lb of grease wool (nearly ¾ of an average fleece) to make one woman's woolen coat. It requires 8.68 lb of grease wool (slightly more than one fleece) to make one man's worsted suit. **Note:** In 1987, the average weight per fleece in the U.S. was 7.78 lb.

Fig. 34–16. Wool garments, made from precious fiber for precious children. (Courtesy, The Wool Bureau, New York, NY)

The numerous operations from fleece to fabric are responsible for the often considered elusive reasons why there is so much difference in the price of grease wool and a suit. To be more specific, it is estimated that, out of each dollar which the consumer spends for apparel and household goods made of wool, the wool producer gets only 5 to 10 cents.

## *WOOL PRODUCTS LABELLING ACT OF 1939*

Few people, including the experts, can, on the basis of appearance, determine the composition of a fabric. In order to protect the consumer from fraudulent operators, federal legislation, known as the Wool Products Labelling Act of 1939, became law on October 14, 1940. This law merely stipulates that fabrics must be labeled under one of the following categories:

1. **Wool.** The act defines wool as including the fiber from the fleece of sheep or lamb, the hair of the Angora or the Cashmere goat, and the fibers from the camel, alpaca, llama, or vicuna—provided they are being used for the first time in the complete manufacture of a wool product.

2. **Reprocessed wool.** This includes wool fiber that has been woven or felted into a product and that, without ever having been utilized in any way by the ultimate consumer, is subsequently made into a fibrous state.

3. **Reused wool.** This consists of the resulting fiber when wool or reprocessed wool has been spun, woven, knitted, or felted into a wool product, which, after having been used in any way by the ultimate consumer, subsequently has been made into a fibrous state.

4. **Virgin wool.** In addition to the three categories established by the Wool Products Labelling Act, the Federal Trade Commission, under whose direction enforcement of the law is vested, has created a fourth category by defining virgin wool or new wool as material composed wholly of new or virgin wool which has never been used or reclaimed, reworked, reprocessed, or reused from any spun, woven, knitted, felted, manufactured, or used product.

Contrary to the opinion held by some, the act does not prohibit the use of any fibers whatsoever that the manufacturer may care to use; but it does not require that the contents be made known on the label, which must give the percentage of the total fiber weight represented by each kind of wool and also the percentage of other fibers present, if any.

In general, the consumer is quite correct in assuming that virgin wool is superior to reprocessed or reused wool. The very highest grades of reprocessed or reused wool, however, may be superior to low-grade virgin wool. In other words, virgin wool may be of very poor quality because of faulty breeding or feeding. Also, work quality is a major factor in determining the value of a fabric. While the Wool Products Labelling Act cannot give assurance relative to the many factors that enter into the determination of fabric quality other than the past history of the fabric, it does lend confidence to the consumer. It also benefits the wool grower, and in the final analysis, it benefits the manufacturer and the distributor.

## MOHAIR

Mohair, known as the most versatile of fibers, is produced by the Angora goat, one of the oldest animals known. Yet, few citizens of the United States are more than casually aware of the Angora goat or its existence, despite the fact that these animals graze millions of acres of land and the hardwearing fabrics made from their lustrous coats are used and admired from coast to coast. Mohair possesses qualities all its own, found in no other animal fiber. It has less crimp and smoother surface scales than sheep's wool. These qualities add luster, softness, and dust resistance to the other fine qualities mohair shares with wool. Mohair has remarkable resistance to wrinkles, great strength, and unequalled affinity to brilliant, deep colors that resist time, the elements, and hard wear.

## WORLD MOHAIR PRODUCTION

At the present time, the goat is bred on a commercial basis for fiber production in five countries: South Africa, Turkey, the United States, Argentina, and Lesotho. The mohair production of each of these nations is shown in Table 34-5.

**TABLE 34-5**
**LEADING MOHAIR-PRODUCING COUNTRIES[1]**

| Country | Production | |
|---|---|---|
| | (mil. lb) | (mil. kg) |
| South Africa[2] | 25.0 | 11.4 |
| United States | 16.0 | 7.3 |
| Turkey | 4.0 | 1.8 |
| Argentina | 2.0 | 0.9 |
| Australia | 1.5 | 0.7 |
| New Zealand | 0.8 | 0.4 |
| Total | 49.3 | 22.5 |

[1]Source: Mohair Council, San Angelo, TX. Data for 1989.
[2]Includes Lesotho.

## U.S. MOHAIR PRODUCTION AND CONSUMPTION

Although goats are rather widely distributed throughout the United States, the production of goats and mohair is of economic importance in a comparatively few states only. Table 34-6 summarizes important data relative to the mohair industry of the United States.

**TABLE 34-6**
**LEADING MOHAIR-PRODUCING STATES**
**(AND CASH RECEIPTS)[1]**

| State | Number Goats Clipped | Average Clip per Goat | | Total Production | | Average Price per Pound | Cash Receipts |
|---|---|---|---|---|---|---|---|
| | | (lb) | (kg) | (lb) | (kg) | ($) | ($) |
| Texas | 2,140,000 | 7.2 | 3.3 | 15,400,000 | 7,000,000 | 1.61 | 24,794,000 |
| Oklahoma | 92,000 | 7.6 | 3.5 | 700,000 | 318,000 | 1.46 | 1,022,000 |
| New Mexico | 125,000 | 5.0 | 2.3 | 625,000 | 284,000 | 1.27 | 794,000 |
| Arizona | 90,000 | 4.1 | 1.9 | 370,000 | 168,000 | 0.99 | 366,000 |
| Michigan | 20,000 | 8.0 | 3.6 | 160,000 | 73,000 | 1.39 | 222,000 |
| Total (5 states) | 2,467,000 | 7.0 | 3.2 | 17,255,000 | 7,843,000 | 1.58[2] | 27,198,000 |

[1]Wool and Mohair 1990, National Agricultural Statistics, USDA, March 23. Data for 1989.

[2]Although the average market price of mohair in 1989 was only $1.58 per lb, it is noteworthy that the incentive price for mohair in 1989 was $4.59 per lb. The incentive basis is explained in the later section in this chapter headed "The National Wool Act."

As noted, Texas is by far the leading state in Angora numbers and mohair production. It accounts for about 87% of U.S. mohair, with most of the goats located in the Edwards Plateau—a rough and broken area, with much brush and some grass.

U.S. goat numbers declined in the 1960s, primarily in response to low prices for mohair, along with rising labor costs and a resurgence of the animal predator problem. Other reasons for the drastic reduction in numbers in Texas were higher revenues from cattle raising, oil rights, and hunting permits, along with the high death losses caused by severe snowfalls and freezes in 1973. It is expected that U.S. goat numbers and mohair production will increase in response to a rise in the price of mohair.

Today, the United States is the second largest producer of mohair. In 1989, the goat raisers of this country produced 16 million lb of this fiber.

The United States is a major exporter of mohair, the bulk of which goes to Great Britain and the European continent.

## MOHAIR CHARACTERISTICS

As may be noted in Table 34-6, the Angora goats of Texas sheared an average annual clip of 7.2 lb of unscoured fleece per animal in 1989. Purebred herds often clip double this amount. Much of the domestic mohair, especially that produced in the Southwest, is taken off in two clips per year, in the spring and fall, whereas Turkish mohair is usually allowed a full year's growth prior to shearing.

The three types of fleeces, based on the type of lock and ranked according to desirability are: the tight or spiral lock, the flat lock, and the fluffy fleece. The tight lock hangs from the body in ringlets and is associated with the finest fibers. The flat lock is usually wavier and coarser, but it is associated with heavy shearing weight. The fluffy fleece is objectionable because it is easily broken and is torn out by brush to a greater extent than the other types.

The length of fiber averages about 12 in. for a full year's growth and 6 in. when the animals are shorn semiannually. Sometimes, with animals that do not have a tendency to shed and when special attention is given to tying the fleece up, mohair up to 3 ft long is produced in a period of 3 years. Such exceptionally long fibers are used in making ladies' switches, dolls' hair, and theatrical wigs.

In fineness, or diameter of fiber, mohair is somewhat coarser than wool. Length and luster are sought more than fineness. The fibers are usually very strong, high in luster, whitish in shade, fairly soft to the touch, and straight in staple appearance. Unfortunately, most mohair contains considerable kemp, which is highly undesirable from the standpoint of the manufacturer. Without doubt, the amount of kemp can be lessened through breeding and selection.

Mohair shrinkage in scouring averages 15 to 17% and does not depend on fineness, as does wool, since adult mohair shrinks as much as does kid mohair.

## PRODUCTION, HANDLING, AND MARKETING OF MOHAIR

Although mohair is usually accorded more neglect than wool, the principles involved in the economical production and advantageous marketing of a high-quality product are the same with both fibers. For practical reasons, chiefly as a means of lessening fleece losses caused by shedding or brush, more goats are shorn twice per year than is the case with sheep. Also, in the Southwest, goats are shorn twice each year because of the warm weather. Except for this difference, and the fact that it is not recommended that the mohair fleece be tied at shearing time, the discussion already presented relative to the production, handling, and marketing of wool is equally applicable to mohair. The market channels and leading market centers for wool and mohair are identical.

It is unfortunate that a large amount of the mohair produced in this country continues to be marketed by placing all grades in a single bag, with little attention given to sorting. On most shearing floors, altogether too little attention is given to keeping the fleece intact and rolling it together in order that an intelligent job of grading and sorting may be done later. So long as these careless production methods are followed, mohair will neither meet the highest requirements of the manufacturer nor command a top price for the grower.

Most of the mohair is exported through local buyers, representing English and European firms. Much of it is exported in original bags, then graded and processed abroad. However, some graded mohair is exported. Some mohair is graded and processed into *top* for domestic use.

## CLASSES AND GRADES OF MOHAIR

Mohair has certain physical and chemical properties which are basic to its commercial value as a textile fiber. The average fiber diameter is the major consideration as this characteristic determines, to a large degree, the type of fabric or product for which the mohair may be used. Other characteristics affecting grease mohair value are its length, yield of clean mohair, strength, luster, color, and character. Grease mohair standards (grades) became effective in July 1971, and mohair *top* grades became effective in January 1973.

The official grades of grease mohair and the specifications of each are given in Table 34-7.

### TABLE 34-7
### SPECIFICATIONS FOR THE OFFICIAL GRADES OF GREASE MOHAIR

| | Fiber Diameter | | Approximate Number of Fiber Measurements[1] |
|---|---|---|---|
| | Limits for Average | Maximum Standard Deviation | |
| | (microns) | (microns) | |
| Finer than 40s ... | Under 23.01 | 7.2 | 1,000 |
| 40s | 23.01–25.00 | 7.6 | 1,000 |
| 36s | 25.01–27.00 | 8.0 | 1,200 |
| 32s | 27.01–29.00 | 8.4 | 1,200 |
| 30s | 29.01–31.00 | 8.8 | 1,400 |
| 28s | 31.01–33.00 | 9.2 | 1,400 |
| 26s | 33.01–35.00 | 9.6 | 1,600 |
| 24s | 35.01–37.00 | 10.0 | 1,600 |
| 22s | 37.01–39.00 | 10.5 | 1,800 |
| 20s | 39.01–41.00 | 11.0 | 2,200 |
| 18s | 41.01–43.00 | 11.5 | 2,200 |
| Coarser than 18s .. | 43.01 and over | | 2,600 |

[1]The number of fibers to measure for each test shall be the number needed to attain confidence limits of the mean within ± 0.40 micron at a probability of 95%. Measurement of the approximate number of fibers for the grades listed above may serve as a guide to meet the required confidence limits. The numbers indicated are based on mohair matchings.

Kid hair is finest and is especially sought by mills. The fleeces from adults—especially bucks and old wethers—are the coarsest; and that from yearlings is intermediate between the other classes. These classes can be recognized by the grower and should be packed separately at shearing time. In addition, those fleeces that are extremely coarse, weak, and shorter than 6 in. or those having an excess of kemp, burs, or other foreign matter should be kept separate from clean, strong fleeces of desirable length and fineness.

The current U.S. Department of Agriculture grade standards (1) for grease mohair, and (2) for mohair top are pictured in Fig. 34-17, along with cross sections of each. The grades are based on average fiber diameter (fineness) and fiber diameter dispersion. Grease mohair refers to the fleece as it comes from Angora goats, and before processing; mohair top is the processed fiber obtained after raw mohair has been scoured, carded, and combed.

As with wool, the grades of mohair are based primarily on the presumed spinning count obtainable on the Bradford system (or the number of 560-yard hanks to the pound). In practice, fineness is associated with softness and is recognized by the experienced touch when handled between the thumb and fingers.

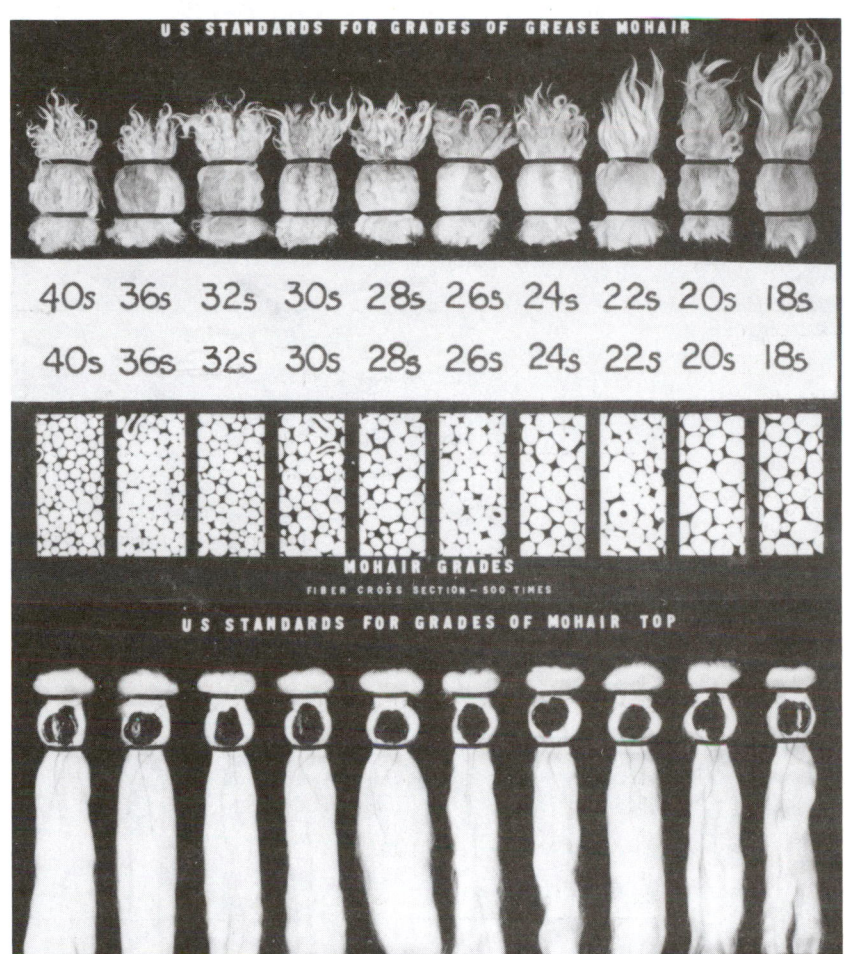

Fig. 34–17. Grades of mohair. *Upper:* Grades of grease mohair. *Lower:* Grades of mohair top. (Courtesy, USDA, Livestock Division Wool and Mohair Laboratory, Standardization Branch, Federal Center, Denver, CO)

## USES OF MOHAIR

Mohair is used for car upholstery, portieres, robes, rugs, braids, and artificial furs; and there is considerable use of superior mohair in suit linings and for men's summer suitings. The long-fibered mohair is also in demand for use in manufacturing wigs and switches for theatrical purposes.

## THE NATIONAL WOOL ACT

Through passage, and subsequent extension, of the National Wool Act (first passed in 1954), Congress recognizes wool as an essential and strategic commodity which is not produced in the United States in sufficient quantity to meet domestic needs.

The incentive payments are financed from the duties collected on the imports of wool. Also, the Act authorized an industry self-help program for the purpose of developing and conducting advertising and sales promotion programs for lamb and wool.

In the 1988 marketing year, the incentive price per pound of wool was 178¢.

In order to secure the most benefit from this Act, the wool grower should (1) sell for the highest price possible, and (2) obtain complete sale records. For example, let us assume that the national average wool price is 138¢ per pound. To bring the national average price of 138¢ to the incentive level of 178¢, each producer's price would need to be increased by 29% (178 − 138 = 40; then 40 ÷ 138 = 29%). Therefore, if you sell 1,000 lb of wool for 138¢ per pound, you will get (1) $1,380 from the buyer, and (2) $400.20 (29% more) from the U.S. Department of Agriculture, making a total of $1,780.20.

But if you sell your wool for 185¢ per pound, instead of 138¢, the story is as follows: You will get (1) $1,850 from the buyer, and (2) $536.50 (29% of $1,850) from the U.S. Department of Agriculture, making a total return of $2,386.50. This shows how the returns of 1,000 lb of wool could be increased by $606.30 through careful marketing.

In 1989, the incentive price for mohair was $4.59 per lb, while the 1989 average market price was $1.58 per lb.

## MANUFACTURED FIBERS

Manufactured fibers are of two kinds: (1) regenerated, and (2) synthetic. Both are used in blends with wool.

Regenerated fibers are made from the great quantities of cellulose and protein materials which are built up into polymers through natural processes, but which exist in a form that cannot be used as textile fibers. Today, we have regenerated fibers made from these long molecules of nature; among them, rayon fibers made from tree trunks and protein fibers made from peanuts.

Three important groups of synthetic (noncellulose) fibers have been developed—nylons (polyamides), polyesters, and acrylics. Basically, synthetic fibers are made from simple molecules, which, under certain conditions, have a coupling at each end. These molecules can be joined into long chains (called polymers) similar to boxcars being hitched together to form a train. Synthetic fibers are a product of fundamental research, conducted by duPont.

The impact of manufactured fibers on the U.S. textile market is evident from the following statistics: In the period 1945 to 1988, U.S. per capita consumption of manufactured fibers increased from 5.5 to 41.7 lb. During the same period, U.S. per capita consumption of scoured wool declined from 4.6 to 1.4 lb (see Fig. 34–18).

### U.S. Per Capita Consumption of Fibers

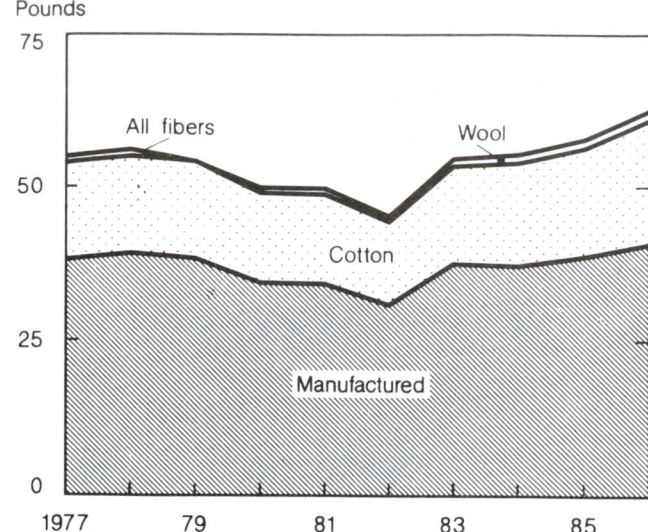

1986 preliminary. Mill consumption adjusted for fiber equivalent of trade balance in textile manufactures. All fibers do not include flax and silk.

Fig. 34–18. Per capita consumption of all fibers, with a breakdown into manufactured, wool, and cotton. Note that on a per capita basis the consumption of manufactured fiber has increased, the consumption of cotton has remained about the same, and the consumption of wool has increased very slightly. (Source: *1988 Agricultural Chartbook*, Ag. Hdbk. No. 673, USDA, p. 99, Chart 228)

## QUESTIONS FOR STUDY AND DISCUSSION

1. The income of U.S. sheep raisers from wool amounts to about 20% of the gross income from sheep, lambs, and wool. Should producers place greater emphasis on wool?

2. List five unique characteristics and virtues of wool.

3. What is apparel wool, and what is carpet wool? What percent of the wool consumed in the United States is apparel wool and what percent is carpet wool?

4. What factors in Australia cause it to produce 29% of the world's wool?

5. Table 34–2 reveals that (a) virtually all of our carpet wool is imported, (b) we produce only 34.9% of our apparel wool, and (c) more than 70% of our total wool requirement (carpet and apparel) is imported. Should the U.S. become self-sufficient in wool production?

6. Fig. 34–2 shows that production and imports have declined through the years, but that wool consumption has been on the increase in recent years. What forces caused these changes?

7. The average American family spends about 5% of its income for clothing and accessories. Should Americans spend more on clothing and be better dressed?

8. Discuss the structure and chemical composition of wool.

9. Describe, and discuss the importance to both the producer and manufacturer, of each of the following fleece characteristics: (a) grease, (b) length, (c) density, (d) diameter, and (e) variations of different body areas.

10. The observance of certain wool production and handling practices will result in marketing of a higher quality product. List these practices.

11. Describe each of the following methods of shearing sheep: (a) clipper shearing, (b) chemical shearing, (c) laser beam shearing, and (d) computerized shearing.

12. List the requisites of good wool.

13. List and describe the classes of wool, including the three classes of apparel wool.

14. How is wool graded? Describe each of the following systems of grading: (a) the blood system, (b) the numerical count system (and worsted spinning count), and (c) the micron system.

15. How and why is wool sorted?

16. List the five leading wool-exporting countries by rank. List the five leading wool-importing countries by rank. What conditions or forces have caused these countries to achieve their respective ranks?

17. Why are there no open- or auction-markets for wool in the United States? List the six predominant market channels for wool.

18. Describe the difference in the manufacturing process of worsted and woolen yarns, and tell how this imparts certain qualities to the finished fabrics.

19. List, in order, the common steps in manufacturing worsteds and woolens.

20. It is estimated that out of each dollar which the consumer spends for apparel and household goods made of wool, the wool producer gets only 5 to 10¢. Who gets the rest; and is the rest too big?

21. What is the *Wool Products Labelling Act of 1939*? Under what categories must fabrics be labeled? Describe each of these labels.

22. What is mohair? List the five leading mohair producing countries by rank.

23. Texas is by far the leading state of the U.S. in Angora numbers and mohair production. What conditions are so favorable to Angora goats and mohair production in Texas?

24. Mohair sells at more than double the price of wool on a per pound basis. Yet, the production, handling, and marketing of mohair is much neglected in comparison with wool. Why?

25. List the classes and grades of mohair.

26. What are the uses of mohair?

27. What is *The National Wool Act*? Why was such legislation enacted?

28. What are *manufactured fibers*?

29. Name the important groups of synthetic fibers. From 1945 to 1988, U.S. consumption of manufactured fibers increased from 5.5 to 41.7 lb. What forces/factors caused this phenomenal increase?

## SELECTED REFERENCES

| Title of Publication | Author(s) | Publisher |
| --- | --- | --- |
| *Handbook for Woolgrowers* | Edited by G. R. Moule | Australian Wool Board, Melbourne, Australia, 1972 |
| *Sheep & Goat Science*, Fifth Edition | M. E. Ensminger R. O. Parker | Interstate Publishers, Inc., Danville, IL, 1986 |
| *Sheep and Wool* | M. P. Botkin R. A. Field C. L. Johnson | Prentice Hall, Englewood Cliffs, NJ, 1988 |
| *Sheepman's Production Handbook, The* | Edited by G. E. Scott | Sheep Industry Development Program, Denver, CO, 1986 and updated at intervals |
| *Sheep Production and Management* | C. V. Ross | Prentice Hall, Englewood Cliffs, NJ, 1989 |
| *Textiles* | M. S. Woolman E. B. McGowan | The MacMillan Company, New York, NY, 1926 |
| *Wool Handbook* | W. Von Bergen | Interscience Publishers, New York, NY, 1963 |
| *Wool Handling* | | Eavenson & Levering Co., Camden, NJ, 1936 |
| *Wool Quality* | S. G. Barker | His Majesty's Stationery Office, London, England, 1931 |
| *Wool Science* | W. D. McFadden | Pruett Publishing Co., Boulder, CO, 1967 |

Also, valuable reference material on wool and mohair may be obtained from the following sources:

American Sheep Industry Association
6911 S. Yosemite Street
Englewood, CO 80112–1414

American Textile Manufacturers Institute
1101 Connecticut Avenue, NW
Suite 300
Washington, DC 10036

Mohair Council of America
516 Central National Bank Building
San Angelo, TX 76901

The Wool Bureau, Inc.
360 Lexington Avenue
New York, NY 10017

Made of wool! *Left:* A sporty threesome—pants, sweater, and scarf—made of 100% pure wool. *Right:* A pure wool suit, stylish and in impeccable good taste. (Courtesy, The Wool Bureau, New York, NY)

Mama, don't move! (Courtesy, Jodi Frediani, Santa Cruz, CA)

**GOATS**

═══════════════════════════════

# Chapter 35

═══════════════════════════════

Goats were probably among the first animals domesticated. Goat-like remains were found in the ruins of the Swiss lake dwellings of the New Stone Age; and, according to historical records, domestic goats and sheep were driven over the damp fields of the Nile Valley to trample the seed into the soil for the farmers of ancient Egypt. Also, according to Old Testament literature, goats were very versatile helpmates to humans in Biblical days. They furnished milk for food, fiber for clothing, skin for bottles, and served as the object of many a sacrificial offering for Jehovah.

Although there are many breeds of goats in the Old World, only a few have been introduced into America. These are divided into four groups: (1) Angora (mohair-bearing) goats, (2) dairy (milk) goats, (3) meat (Spanish) goats, and (4) pygmy goats.

## ANGORA (MOHAIR-BEARING) GOATS

The vast majority of the goats in America belong to the Angora (mohair-bearing) breed. Although there are more than 1 million head of these strange-looking, heavy-coated creatures in this country, few people outside the Angora goat districts know what they look like.

The breed derives its name from Angora, a province in Turkey, in which land it originated. Angora is a high plateau, lying from 1,000 to 4,000 ft above sea level. In 1881, the Sultan of Turkey passed an edict prohibiting the exportation of Angoras, expecting thereby to confine the mohair industry to Asia Minor and forever after hold a monopoly upon the mohair trade. Thirty years later, in 1910, South Africa followed suit, passing a law for the same purpose. Subsequent events proved that both Turkey and South Africa were too late in their efforts to hold a monopoly on the mohair trade, for some of the choicest Angora blood had already been brought to the United States.

Fig. 35–1. Angora goats in Turkey, the land of their origin.

The Angora goat was first introduced into this country by Dr. James B. Davis, of Columbia, South Carolina. Dr. Davis had been sent to Turkey by President Polk in answer to a request made by the Sultan for someone to experiment in the production of cotton in that country. Upon returning to the United States in 1849, 32 years before the Sultan's edict, Dr. Davis brought with him 9 choice Angora goats, including 7 does and 2 bucks. These and subsequent importations founded the Angora industry in the United States, which continued to thrive in this new land despite later restrictions imposed by Turkey.

## DISTRIBUTION AND ADAPTATION

Table 35–1 shows U.S. Angora goat numbers and mohair sales, for the leading states and for the nation.

**TABLE 35–1**
**LEADING ANGORA GOAT-PRODUCING STATES, PLUS MOHAIR SALES[1]**

| State | Angora Goats | | | Mohair Sales | | | |
|---|---|---|---|---|---|---|---|
| | Farms | No. of Goats | Average Number per Farm | Farms | Thousands of Pounds | Thousands of Dollars | Average Pounds per Farm |
| Texas | 2,852 | 1,419,310 | 498 | 2,877 | 11,394 | 39,520 | 3,960 |
| Arizona | 80 | 109,529 | 1,369 | 75 | 566 | 2,225 | 7,547 |
| New Mexico | 142 | 81,064 | 571 | 127 | 466 | 1,380 | 3,669 |
| Oklahoma | 166 | 37,501 | 226 | 160 | 298 | 999 | 1,863 |
| Michigan | 275 | 14,795 | 538 | 268 | 144 | 544 | 537 |
| Missouri | 129 | 5,982 | 46 | 98 | 42 | 135 | 429 |
| Ohio | 203 | 5,471 | 27 | 164 | 45 | 237 | 274 |
| Kansas | 55 | 3,978 | 72 | 43 | 43 | 160 | 1,000 |
| California | 152 | 2,694 | 18 | 89 | 18 | 57 | 202 |
| Wyoming | 18 | 2,443 | 136 | 11 | — | — | — |
| **U.S. Total** | 5,352 | 1,702,166 | 318 | 4,691 | 13,181 | 45,882 | 2,810 |

[1]Source: Bureau of the Census, U.S. Department of Commerce, Volume 1, Geographic Area Series, 1989. Data for 1987.

It is noteworthy that Texas alone accounts for the majority of the Angora goats found in the United States. The huge goat population of Texas is due principally to the fact that the Lone Star State has a large area of rugged grazing land which is well adapted to utilization by this species. The center of goat raising in Texas is in the south central part of the state, a region generally known as the Edwards Plateau. The area is charac-terized by rolling hills, somewhat rough and broken, covered by grasses, cedar and oak trees, and a considerable amount of brush. The elevation is between 1,500 and 3,000 ft, and the rainfall varies from 15 to 25 in. per year.

Generally speaking, the densest Angora goat populations in this country are found in those areas in which the grazing is too scanty, rough, or brushy for cattle or sheep. On the other hand, a very large percentage of Texas Angora goats are handled on the same ranges that support cattle and sheep. The goats utilize browse which would be of little or no value to sheep and cattle, and keep the brush from crowding out the natural grasses. On cut-over lands of the Pacific Coast states and in the Ozarks of the central states, Angora goats are frequently used for clearing land of brush; but if this end is to be accomplished, the area must be heavily stocked and closely grazed.

## CHARACTERISTICS

Fig. 35–2. Angora goats, showing long locks of close-curled, high-luster mohair. (Courtesy, Texas A&M, College Station, TX)

At the present time, the Angora goat of this country is considerably larger and more rugged than its Turkish ancestor. This transformation in type has been accomplished through long-continued selective breeding and some infusion of common or so-called Mexican goats. In range condition, mature bucks weigh from 125 to 175 lb and does from 80 to 90 lb. The breeders of this country prefer considerable size as long as it is not necessary to sacrifice fleece quality, because the added size gives a larger surface area upon which to produce a heavy fleece.

Angoras are almost always pure white, although a black one appears occasionally. Red kids will shed their hair and produce white mohair later, but it is recommended that these animals be culled out and sent to slaughter.

The outer coat of the animal is made up of long locks or strands of hair, known commercially as mohair, which covers the animal's body. The fleece should cover all parts of the body except the face and should be characterized by fine quality, a close curl, and high luster. It should be as free from kemp as possible.

Under range conditions, does and kids shear 5 to 7 lb and wethers 6 to 7 lb of mohair annually, which is usually removed in 2 clips. The best fleeces possess ample quantities of natural oil, and the best strains of Angoras show no tendency to shed.

The body conformation should be symmetrical and denote a good constitution. Both sexes are usually horned, but polled individuals occur. The rather thin, long, and pendulous ears droop out of the hair.

Fig. 35–3. Typical Angora flock on west Texas ranch. Texas is the leading Angora state of the nation. (Courtesy, *Sheep and Goat Raiser,* San Angelo, TX)

## REGISTRATION

The first U.S. breed registry association for Angora goats, known as the American Angora Goat Breeders' Association, was established in 1900. From that time until 1924, all animals registered by this association were either the original inspected stock and their progeny or goats subsequently imported. A second registry, known as the National Angora Record Association, was organized in 1918; but 6 years later, in 1924, it was merged with the American Angora Breeders' Association. Today, the lineage of registered Angora goats in this country is recorded in the American Association only.

As in sheep production, the breeding of registered Angoras is considered a specialty business, and few large range goat operators engage therein. Yet, most practical commercial goat producers prefer to use registered bucks and patronize the purebred breeders in order to secure bucks of the type that they hope will improve the breed.

## MANAGEMENT OF ANGORA GOATS

The proper care of goats differs little from that which should be accorded sheep under similar conditions. Like sheep, goats pay dividends for good management. It is unfortunate, therefore, that there is a widespread and common belief that goats will thrive despite neglect. This popular conception is not true. The successful goat raisers apply the same careful care and management to goat raising as is given to any other profitable livestock enterprise. Success with goats can be achieved in no other way. In the discussion which follows, particular attention is given to principles or systems of management wherein goats differ markedly from sheep. Needless repetition on such matters as fencing, castration, watering, etc., is omitted.

### HERDING

A large number of Angora goats in the United States are maintained under range conditions, where they are grazed in herds much like range bands of sheep. These herds vary in numbers from a few hundred to over 2,000 head, with an average herd numbering about 1,200. On rough range or in thick brush, a larger number than 1,200 can seldom be properly controlled by a single herder.

The principles and practices of good herding with goats are almost identical to those with sheep. There is one distinct difference: Rarely do sheep herders work ahead of the band; whereas it is common practice for a goat herder to work in front, turning the lead goats back to avoid unnecessary travel.

At the present time, most of the goats of Texas and other southwestern states are "loose grazed" (unherded) in wolf-proof fenced ranges, in the same manner that sheep are handled in this area.

## BREEDING

Fig. 35–4. Angora bucks. (Courtesy, American Angora Breeders' Assn., Rocksprings, TX)

The normal breeding habits of goats and sheep are similar, with the former having a slightly longer estrus period and a gestation period of 1 to 2 days longer (the gestation period of goats varies from 140 to 160 days, with an average of 151 days). The does are usually bred so that their first kids are dropped when they are approximately 2 years of age.

Like sheep, Angora does reproduce once each year, coming into heat in the late summer or early fall. The time of breeding is determined largely by the projected climatic conditions, feed supply, and shelter at kidding time. Most range goats of this country are bred to kid in March and April, for milk, settled weather and spring feed may normally be expected at this time.

## FEEDING

Unfortunately, there is a widely prevailing popular belief that goats will eat and do well on anything from newspapers to rusting tin cans. This is erroneous. Like other animals that are hungry or suffering from mineral or vitamin deficiencies, they will develop depraved appetites and chew on many things; but they prefer good quality, wholesome feeds and will pay dividends when so fed.

Goats are naturally browse animals; but a good goat range, in addition to furnishing abundant palatable evergreen brush (not cedar or other coniferous vegetation), should provide a mixture of grasses and broadleaved herbs. On the most desirable goat ranges, this feed combination is available the year round, although browse is usually the principal winter feed.

When supplemental feed is necessary, those feeds that are also suitable for sheep are used. The mineral (including salt) and water requirements for goats are similar to those of sheep.

## SHELTER

It is essential that newly sheared goats be protected from cold rains and storms, no matter how simple the shelter may be; otherwise, losses are sometimes disastrous. Goat sheds are generally low structures covered with a metal roof and boarded up on one or two sides. When well fed and not newly sheared, goats seldom succumb to cold and, therefore, do not need shelter.

## CARE AT KIDDING TIME

Goats require much more care than sheep at the time the newborn are arriving. Young kids are much more delicate than young lambs—neither being able to endure as much cold or damp weather nor to follow their mothers to the range as early in life. It is not surprising, therefore, to discover that an 80% kid crop is considered excellent for range bands.

If heavy losses of young kids are to be averted, a safe system of kidding must be followed. The two most common systems of kidding followed by progressive and successful goat raisers of today are (1) the toggle (or staking) system, and (2) the pen (or corral) system.

### TOGGLE SYSTEM

Under the toggle system, the young kids are staked with about 15 in. of rope attached to a swivel on the stake end, with a loop on the other end being attached to the fetlock joint of the kid. This loop should be changed to another leg once daily. Each kid is provided with a box, usually a small A-shaped structure made of 12-in. boards, with one end open, which furnishes protection from the elements. The camp should have sufficient stakes and boxes to allow for as many kids as may be dropped within any 10-day period. Ten feet between stakes is considered about the right distance, providing sufficient space to avoid confusion, affording room for handling, and minimizing quarreling of the does. Whenever a kid is staked, it and its mother are branded with corresponding marks, thus avoiding any possibility of confusion in identity. Some ranchers clip the doe's tail and place the number there, whereas others place brand marks on similar body areas of the does and kids.

The ideal toggle camp should have a good slope for drainage. With such a slope, the older does will spend the night on the higher side; so their kids should be staked there, with the kids of the younger does being placed in the lower part of the camp. Disturbances will be avoided if the kids or does that fight or habitually overturn the kid boxes are placed at one end or corner of the camp. On cold rainy nights, the does should be confined or otherwise kept out of the toggle camp, for each time a doe passes through the yard to her kid, other kids will come out of their boxes and suffer possible exposure.

A doe that disowns her kid should be staked with it. Kids may usually be grafted if the foster mother and orphan kid are staked or penned together. The presence of "dogies," as orphan kids are known, is evidence that such details have not been given adequate attention around a kidding camp.

The careful herder passes through the toggle camp each morning and evening, ascertaining whether each kid has nursed. If a kid is gaunt or restless, its mother should be brought to the stake; if a doe has a distended udder, she should be taken to the kid or kids that carry the corresponding number or mark. Does with extra big teats may have to be hand milked until the kids have learned to nurse the abnormally large teats.

Generally, kids should not be staked in the toggle camp for more than 10 days. At that time, they should be herded with their mothers. Some of the large operators, especially those in southwestern United States, release the kids directly from the toggle camp to the range with the does. These ranchers insist that kids handled in this manner develop into better foragers and possess more muscle and bone. Others prefer to make the change from the toggle camp to the range more gradually. Usually the latter operators first transfer the does and their kids in small numbers from the toggle camp to a small field. The size of the field and herd is then gradually increased until the kid band may be turned on the range when the young average about 6 weeks of age. A modification of the latter system consists of confining the kids to a corral for a few weeks while the does depart via a *jump board* (a structure about 18 in. high) to forage on the range in the daytime.

Fig. 35-5. A toggle camp (or stake kidding), showing the young kids tied with a swivel near an A-shaped box (kidding box) into which they go for protection from rain and the sun. (Courtesy, American Angora Goat Breeders' Assn., Rocksprings, TX)

## PEN (OR CORRAL) SYSTEM

The pen or corral system is most common today, because less work is involved in handling the animals and the results are as good as those obtained by using the toggle system. For a herd of 1,200 does, the pen system usually involves having about 8 corrals of ample size to enclose 50 does and their kids, a dozen small pens each large enough to hold an individual doe and her kid, and 1 or 2 larger corrals that may be adequate for 400 to 500 does and their kids.

Even the toggle system should be supplemented with a few individual kidding pens, which may be used as "bum pens" to force does to accept disowned kids of their own or orphan kids that are to be grafted. In general, in the pen or corral system

about 50 does and their kids of about the same age group are kept in one pen or corral for the first 2 or 3 weeks; the kids are confined to the corral, whereas the does are permitted to go out to the range in the daytime. As the kids get older, groups are combined and turned to larger areas, eventually to travel to the range in a kid herd consisting of 1,000 to 1,200 does and their kids.

## MOHAIR AND GOAT MEAT

The production, handling, and marketing of mohair have been fully covered in Chapter 34; hence, the reader is referred thereto for information on this subject.

Goats contribute 5.7% of the world meat supply.

Goat meat tastes much like mutton or lamb of similar finish and quality, and it is equally appetizing and nutritious. In general, however, goat carcasses are not so well finished and do not yield so high a dressing percentage as average sheep carcasses.

Reputable meat dealers sell goat meat under the trade name of chevon.[1] Because of the unwarranted prejudice against goat meat by consumers, however, it may be true that some of this product is passed over the counter as mutton or lamb. There is no federal restriction against marketing it in the latter way, but in some cities and in the state of Oregon this practice is prohibited. Most chevon is marketed in the West, as efforts to dispose of this product in the East have met with little enthusiasm. More than any other kind of meat, goat meat needs effective and deserved promotion.

## DAIRY (MILK) GOATS

Goats produce 1.8% of the total world milk supply. In some countries, goat milk accounts for up to 50% of the total milk production.

The goat has long been a popular milk animal in the Old World, where it is often referred to as the poor man's cow. When they are travelling or vacationing, Asiatics frequently take their goats with them in order to be assured of a supply of milk. Thus, it is interesting to recall that the late Mahatma Gandhi of India took two milk goats with him on his last visit to England.

Dairy goats were first introduced to America in early times—the first settlers in the Virginia colonies bringing their milk goats from the mother country. However, improved strains were not imported until many years later, for the first purebreds said to have been brought into the United States were four Toggenburgs imported to Ohio in 1893. Today, the dairy goat industry is growing, and these small animals are supplying nature's finest food—milk—to many children who would otherwise be undernourished. They are especially well adapted for furnishing a milk supply for low-income families in small towns and the suburbs of large cities where there is not enough feed available for a cow. A doe can often secure much of her feed from lawn clippings or garden and kitchen waste or by grazing in vacant lots.

---

[1]In Texas, meat from 5- to 6-month-old kids, weighing 30 to 40 lb on foot, is known as *cabrito.* It is regarded as a delicacy and is highly prized for barbecuing in the Southwest.

A good milking doe will average 5 lb of milk, or more, per day over a lactation period of 10 months, with superior animals producing 10 lb or more. The highest official milk production on record in the United States was made by a Saanen doe that produced 6,850 lb of milk in 305 days, in 1984 (see Fig. 35–6).

In comparison with cow's milk, goat's milk has smaller fat globules, a higher mineral content, and a sweeter flavor. Goat's milk forms a fine, soft curd during digestion, thus making it more easily digestible than cow's milk for some children and for older people who cannot use cow's milk. If the does are milked in clean quarters and away from the bucks, goat's milk will not have any unpleasant flavor or odor. The strong odor of the buck is quickly absorbed by warm milk.

Fig. 35–6. All-time, all-breed world record milk production holder. This Saanen doe produced 6,850 lb milk and 296 lb fat, in 305 days, in 1984. Bred by Gary and Sharon Swanson, Renton, WA; owned by Gary Lee Cox, Eagle Point, OR. (Courtesy, T. H. Teh, Ph.D., Prairie View A&M University Research Center, Prairie View, TX)

## IMPORTANCE OF DAIRY GOATS

Table 35–2 shows U.S. dairy (milk) goat numbers, and goat and milk sales, for the leading U.S. states and for the nation. It is noteworthy (1) that California has a commanding lead in all categories; and (2) that for the United States there were a total of 129,225 dairy goats on a total of 15,433 farms, with an average of only 8 goats per farm.

**TABLE 35–2**
**LEADING DAIRY GOAT-PRODUCING STATES, AND SALES OF GOATS AND MILK[1]**

| | Inventory | | | Sales | | | | |
|---|---|---|---|---|---|---|---|---|
| State | Farms | No. Dairy Goats | Goats per Farm | Farms | No. Milk Goats Sold | Farms | Gallons of Milk Sold | Total Sales |
| | (no.) | | | (no.) | (no.) | (no.) | (gal) | ($) |
| California | 1,049 | 16,055 | 15 | 401 | 8,242 | 243 | 973,291 | 2,693,000 |
| Texas | 1,011 | 10,559 | 10 | 315 | 3,322 | 145 | 253,156 | 776,000 |
| Ohio | 890 | 6,109 | 7 | 257 | 1,985 | 139 | 317,428 | 818,000 |
| Wisconsin | 450 | 5,562 | 12 | 139 | 2,910 | 93 | 409,938 | 841,000 |
| New York | 622 | 5,234 | 8 | 212 | 2,082 | 94 | 188,204 | 927,000 |
| Tennessee | 463 | 4,680 | 10 | 121 | 1,294 | 47 | 126,075 | 460,000 |
| Missouri | 673 | 4,512 | 7 | 172 | 1,470 | 66 | 69,822 | 227,000 |
| Oregon | 511 | 4,321 | 8 | 216 | 2,712 | 122 | 203,704 | 556,000 |
| Washington | 411 | 4,049 | 10 | 155 | 1,490 | 79 | 129,860 | 441,000 |
| Michigan | 533 | 3,844 | 7 | 173 | 1,488 | 51 | 141,975 | 364,000 |
| Colorado | 418 | 3,705 | 9 | 156 | 1,437 | 72 | 129,564 | 324,000 |
| Pennsylvania | 635 | 3,512 | 6 | 183 | 1,490 | 125 | 92,490 | 264,000 |
| Illinois | 495 | 3,432 | 7 | 150 | 1,312 | 78 | 107,074 | 335,000 |
| Indiana | 563 | 3,383 | 6 | 151 | 1,068 | 64 | 54,000 | 154,000 |
| Arkansas | 293 | 3,341 | 11 | 84 | 1,129 | 44 | 183,582 | 338,000 |
| Iowa | 425 | 3,195 | 8 | 105 | 909 | 52 | 123,669 | 268,000 |
| Florida | 283 | 3,183 | 11 | 92 | 1,169 | 35 | 95,648 | 325,000 |
| Arizona | 178 | 2,974 | 17 | 74 | 1,533 | 35 | 67,298 | 179,000 |
| Oklahoma | 427 | 2,878 | 7 | 110 | 773 | 54 | 22,692 | 69,000 |
| Minnesota | 388 | 2,838 | 7 | 110 | 821 | 64 | 57,249 | 155,000 |
| Virginia | 288 | 2,601 | 9 | 110 | 838 | 28 | 21,320 | 106,000 |
| Kentucky | 443 | 2,388 | 5 | 98 | 506 | 28 | 28,934 | 77,000 |
| Kansas | 438 | 2,297 | 5 | 106 | 552 | 79 | 36,773 | 104,000 |
| Georgia | 230 | 1,926 | 8 | 75 | 568 | 29 | 14,644 | 135,000 |
| Idaho | 244 | 1,624 | 7 | 89 | 655 | 40 | 29,566 | 88,000 |
| U.S. Total | 15,433 | 129,225 | 8 | 4,770 | 49,795 | 2,378 | 4,369,866 | 12,845,000 |

[1]Source: Bureau of the Census, U.S. Department of Commerce, Volume 1, Geographic Area Series, 1989. Data for 1987.

Table 35–3 shows the registrations of dairy goats by the American Dairy Goat Association. Of course, many dairy goats are not registered.

**TABLE 35–3**
**GROWTH IN REGISTRATIONS OF DAIRY GOATS[1]**

| Year | Number Registered by American Dairy Goat Association |
|------|------|
| 1960 | 4,041 |
| 1965 | 3,487 |
| 1970 | 6,792 |
| 1975 | 26,644 |
| 1980 | 46,683 |
| 1985 | 44,090 |
| 1989 | 36,481 |

[1]Registration numbers provided by American Dairy Goat Assn.

## U.S. BREEDS OF DAIRY GOATS

There are many distinct breeds of dairy goats in the Old World, but only a few breeds are important in the dairy business in America. Table 35–4 gives the characteristics of selected breeds.

**TABLE 35–4**
**BREEDS OF DAIRY GOATS AND THEIR CHARACTERISTICS[1]**

| Breed | Place of Origin | Color; Face, Ears, and Legs | Head Characteristics | Other Distinguishing Characteristics | Disqualification |
|-------|-----------------|------------------------------|----------------------|---------------------------------------|------------------|
| **Alpine** (including several varieties: British, French, Rock, and Swiss. French Alpine is most common) | Switzerland. All Alpines stem from Swiss foundation stock. | All colors and combinations of colors. | Upright ears. Straight face, Roman nose. | Medium to large size. The only breed with upright ears that offers all colors and combinations of colors, giving them distinction and individuality. | Toggenburg color and markings, or all-white, are discriminated against. |
| **La Mancha** | U.S.A., from a short-eared Spanish breed crossed on leading purebred breeds. | Any color or combination of colors. | Short ears or no ears; straight face; hornless or neatly disbudded. | The different-type ears are known as "gopher" or "elf-ear." The hair is short, fine, and glossy. Excellent dairy temperament. | Anything other than "gopher ears" in males. Ears other than true La Mancha type in females. |
| **Nubian** | The Nubian in the U.S. is of mixed origin. It evolved out of crossing Indian Jumna Pari and Egyptian Zariby types on British dairy goats. | They may be any color or colors, solid or patterned. Common colors are: black, gray, cream, white, shades of tan, brown, and rich reddish-brown. Common markings include lighter ears, facial stripes, muzzle, crown, and/or undertrim; overall light- or dark-colored spots or patches of any size are often found. | Some born with horns and disbudded, others are hornless. Long, drooping ears. Roman nose and prominent forehead. Does are beardless. | Relatively large breed. Noted for high milk and butterfat production. | Upright ears. Dished face. |
| **Oberhasli** | Switzerland. | Chamois, which is a bay—ranging from light to deep red, with the latter preferred. Black markings. A few white hairs. | Straight face. | Medium size. Alert. | Roman nose is discriminated against. |
| **Saanen** | Switzerland, in the Saanen Valley. | Pure white or creamy white. The cream color may vary from light to dark fawn. | Hornless animals preferred; straight or dished face; erect ears. | Medium to large size. | Large (1½ in. diameter or more) dark spot in hair; pendulous ears. Tendency to Roman nose discriminated against. |
| **Sable** (currently, the Sable Breeders Assn. is maintaining a separate registry) | Switzerland, from Saanen parents. | Dark- or sable-colored. Color results from Saanen being heterozygous for white. Thus, 25% of offspring from heterozygous Saanens will be Sable. Sable × Sable will always produce Sable. | Same as Saanen, except for color. | Same as Saanen, except for color. | White or light cream colors are disqualified; animals must be sable-colored. |
| **Toggenburg** | Switzerland, in the Toggenburg Valley; but they originally came from the Swiss Alps. | Light fawn to dark chocolate, with two white strips on the face and white on the legs below the knees. | Hornless or disbudded. Erect ears, carried forward. Face straight or dished, never Roman. | Medium size, sturdy, and vigorous. Alert appearance. | Tricolor or piebald; large (1½ in. or more) white spot in males; pendulous ears. |

[1]In addition to the specific breed disqualifications given in the right-hand column, the American Dairy Goat Assn. lists the following as disqualifications in any breed: total blindness; permanent lameness or difficulty in walking; blind or non-functioning half of udder; blind teat; double teats; extra teats that interfere with milking; hermaphrodism; navel hernia; crooked face in bucks; and extra teats; teats cut off, or double orifice in bucks.

Fig. 35-7.  Alpine doe. (Courtesy, *Dairy Goat Journal,* Scottsdale, AZ)

Fig. 35-8.  La Mancha doe. (Courtesy, American Dairy Goat Assn., Spindale, NC)

Fig. 35-9.  Nubian doe. (Courtesy, *Dairy Goat Journal,* Scottsdale, AZ)

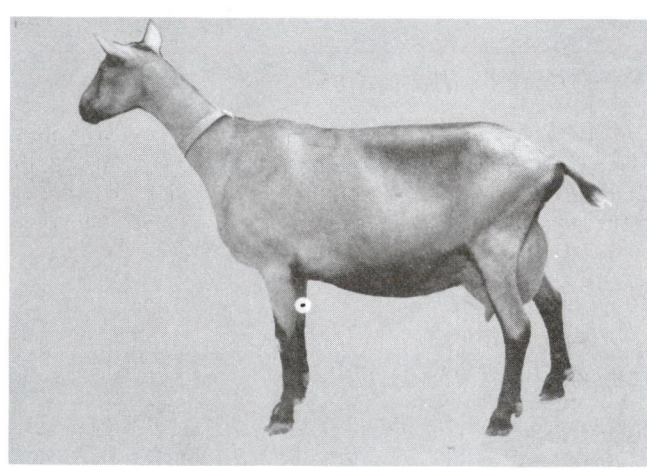

Fig. 35-10.  Oberhasli doe. (Courtesy, Dorthea Custer, Oberhasli Breeders of America, Manheim, PA)

Fig. 35-11.  Saanen doe. (Courtesy, Christine S. D. Williams, Michigan State University, East Lansing, MI)

Fig. 35-12.  Sable doe. (Courtesy, Donald West, Sable Breeders Assn., Cave Junction, OR)

Fig. 35–13. Toggenburg doe. (Courtesy, American Milk Goat Record Assn.)

## FEEDING DAIRY GOATS

The feed requirements of dairy goats are similar to those for milk cows. These are set forth in Chapter 24. Although the allowance should vary according to production, the concentrates for lactating does may consist of 1½ to 2 lb daily, plus either (1) pasture in season, or (2) hay and/or silage.

## MEAT (SPANISH) GOATS

Fig. 35–14. Typical multicolored meat (Spanish) goats. (Courtesy, USDA)

The terms *meat* or *Spanish goat* are used in the United States to refer to goats of mixed-breed origin. Since they are kept largely for meat production, the terms *meat goat* and *Spanish goat* are used. These terms are employed to distinguish these goats from the Angora and dairy breeds. Most meat goats are of the same origin as the Mexican Criollo—a breed derived from the Granada, Murcia, and Malaga breeds of Spain.

There are about 500,000 meat (Spanish) goats in the United States, most of which are located in Texas.

Meat goats are highly variable in appearance and performance. Some show traces of Nubian and Toggenburg breeding. Others lack external ears or have very small ears. As would be expected, the colors and markings of meat goats vary widely—there being no ideal or accepted color or marking. Colors range from solid black, brown, and white to fawn and brown with black points and a black stripe down the back. There are also many combinations of spotting—black and white, brown and white, black and brown, and some blue-gray. Most of these animals, both males and females, are horned. The horns of the males grow much larger and heavier than those of the females. A few meat goats are polled. Indeed, meat goats are mongrels.

Spanish goats are adaptable and unsurpassed in their ability to exist largely upon brush and still yield acceptable quantities of edible meat.

## CASHMERE GOATS

The Cashmere goat is a double-fibered goat which produces a fine undercoat known as *down*. It embraces several varieties rather than a distinct breed; one authority claims that 68 breeds of goats in 12 different countries have been identified as producing cashmere fiber.

The term *cashmere* refers to the fine down undercoat fiber produced by the Cashmere goat. Cashmere is the finest animal fiber used in commercial trade. Clothing made from cashmere is exceptionally light, soft, warm, and in the luxury class. On a comparable weight basis, cashmere is reported to have three times the insulating value of wool. The more common uses of cashmere are for sweaters, ladies' dress goods, shawls, and coatings.

Because Europeans first became aware of the fine undercoat (or down) produced by the goats native to the Central Asian mountains in Kashmir, they named it *cashmere* (after the old spelling of the country) and the goats from which it came were called *Cashmere goats*. But there are few Cashmere goats in Kashmir. Today, the three major producers of Cashmere goats and cashmere fiber are China, Mongolia, and Iran, in that order. But Cashmere goats have been transplanted to many parts of the world. Also, it is noteworthy that a few of the Spanish or meat goats of the United States have the ability to produce cashmere.

There appears to be a bright future in the United States for Cashmere goats and cashmere fiber in the decade ahead (and perhaps beyond), without flooding the market. However, U.S. producers face two major problems in developing a cashmere industry: (1) The small amount of cashmere fiber produced by a goat—about ½ lb annually in the better cashmere-producing areas of the world; and (2) combing (shearing) is a tedious, time-consuming job, and labor costs may be too high for it to become practical in this country.

## PYGMY GOATS

These miniature goats have been known by a variety of names—West African Dwarfs, African Pygmies, and Cameroons. They have proven to be very adaptable animals and valuable research animals. Pygmy goats are hardy, alert, animated, good-natured, gregarious, docile, responsive pets.

These small goats originated in the western part of Africa, particularly in the Cameroons. The original Pygmy goats reached zoos in the United States via imports from zoos in Sweden and West Germany sometime during the 1950s. In 1961, Dr. James Metcalfe established a colony at the University of Oregon Medical School at Portland by purchasing goats from two zoos. In 10 years, the colony produced over 400 offspring.

The Pygmy goat is small, cobby, and compact. It is full-barreled and well-muscled, and the body circumference in relation to height and weight is proportionately greater than that of other breeds. Also, the head and legs are short relative to body length. Hornlessness is considered a disqualifying fault.

Preferred colors range from white to gray and black in a predominantly grizzled (agouti) pattern. Muzzle, forehead, eyes, and ears are accented in lighter tones. Front and rear hoofs and cannons (socks) are black, as are the crown and dorsal stripe, or martingale. Coat length and density vary with climates.

Pygmy goats are precocious breeders, bearing 1 to 4 young every 9 to 12 months after a 5-month gestation period. Normally 1 or 2 kids are produced, weighing an average of 4 lb each. Does are usually bred for the first time when they are about 9 months of age, though they may conceive as early as 3 months if care is not taken to separate them early from the bucks. Mature females may produce 4 or more lb of milk of 6 to 9% butterfat at the peak of a 4- to 6-month lactation period.

Fig. 35–15. Pygmy goat. True to their name, mature goats of this breed measure only 16 to 23 in. at the withers. (Courtesy, Cottage Farm, Sherborn, MA)

## QUESTIONS FOR STUDY AND DISCUSSION

1. Cite the evidence that goats were probably among the first animals domesticated.

2. Name the four groups into which goats may be divided.

3. Detail the origin of Angora goats and their introduction to America.

4. To what may the huge goat population of Texas be attributed? How do the grazing areas commonly utilized by goats and sheep compare from the standpoints of (a) topography, and (b) type of vegetation?

5. Describe the characteristics of Angora goats.

6. Compare the (a) herding, (b) breeding, and (c) feeding of goats and sheep.

7. Compare the (a) shelter and (b) care at parturition time of goats and sheep.

8. Describe and compare the handling of young kids (a) by the toggle system, and (b) by the pen (or corral) system.

9. What percentage of the world meat supply is contributed by goat meat? Under what name is goat meat marketed? Compare goat carcasses and meat with lamb carcasses and meat.

10. What percentage of the world milk supply is contributed by goat milk? Why has the goat been more popular as a milk animal in the Old World than in the New World?

11. What breed of dairy goat holds the highest official milk production record in the U.S.? How much milk did she produce in 305 days?

12. Compare goat's milk with cow's milk. Why is goat's milk more easily digested than cow's milk by some children and older people?

13. List, by rank, the five leading U.S. dairy (goat) states in numbers. For the nation as a whole, what is the average number of dairy goats per farm?

14. Name the seven most common breeds of dairy goats; and for each breed give (a) the place of origin; (b) the color of the face, ears, and legs; (c) the head characteristics; and (d) other distinguishing characteristics.

15. What are *meat (Spanish) goats*? How do meat goats differ from Angora goats in characteristics and use?

16. What are Cashmere goats, and what is cashmere? What unique properties does cashmere fiber possess, and for what is it commonly used?

17. What are *pygmy goats*? How do they differ from Angora, dairy, and meat goats in characteristics and use?

## SELECTED REFERENCES

| Title of Publication | Author(s) | Publisher |
|---|---|---|
| *Angora Goat, The,* Farmers' Bull. No. 1203 | | U.S. Department of Agriculture, Washington, DC |
| *Dairy Goats* | B. E. Colby *et al.* | American Dairy Goat Association, Spindale, NC, 1972 |
| *Goat Production* | Edited by C. Gall | Academic Press, Inc., New York, NY, 1981 |
| *Milk Goats,* Farmers' Bull. No. 920 | | U.S. Department of Agriculture, Washington, DC |
| *Modern Milk Goats* | I. Richards | J. B. Lippincott Co., Philadelphia, PA, 1921 |
| *Selecting Angora Goats* | J. Gray<br>J. L. Groff | Texas A&M University, College Station, TX, 1969 |
| *Texas Angora Goat Production* | J. Gray<br>J. L. Groff | Texas A&M University, College Station, TX, 1970 |
| *U.S. Sheep and Goat Industry: Products, Opportunities and Limitations, The* | C. S. Menzies, Chairman, Task Force | CAST, 250 Memorial Union, Ames, IA, 1982 |

Also, breed literature pertaining to each breed of goats may be secured by writing to the respective breed registry associations (see Table A–8, Appendix, for the name and address of each association).

Angora goats on the Edwards Plateau of Texas. (Courtesy, *Sheep & Goat Raiser*)

Dairy goats on pasture. (Courtesy, *Dairy Goat Guide*, Countryside Publications, Ltd., Waterloo, WI)

Looking and listening. (Photo by J. C. Allen and Son, West Lafayette, IN)

# THE SWINE INDUSTRY

## Chapter 36

| Contents | Page |
|---|---|

| Contents | Page |
|---|---|

Nomadic people could not move swine about with them as easily as they could cattle, sheep, or horses. Moreover, close confinement was invariably accompanied by the foul odors of the pigsty. For this reason, the early keepers of swine were often regarded with contempt. This may have been the origin of the Hebrew and Moslem dislike of swine, later fortified by religious precept. As swine do not migrate great distances under natural conditions and the early nomadic peoples could not move them about easily, there developed in these animals, more than in most stock, a differentiation into local races that varied from place to place. It also appears that swine were domesticated in several different regions and that each region or country developed a characteristic type of hog.

## ORIGIN AND DOMESTICATION OF SWINE

Most authorities agree that two wild stocks contribute to American breeds of swine; namely, the European wild boar (*Sus scrofa*) and the East Indian pig (primarily *Sus vittatus*).[1] Both were gregarious, often forming large herds. Their feed consisted mostly of roots, mast (especially acorns and beechnuts), and such forage as they could glean from the fields and forests. Because of their roving nature, disease and parasites were almost unknown.

In addition to these two stocks, there exist certain wild types of piglike animals which, to this day, have never been domesticated. Included in the latter group are the brightly colored tropical river pigs and the giant forest hog of Africa; the hideous warthog of the African plains; the native American pigs (known as peccaries or javelinas); and the babirussa of Celebes, whose tusks resemble horns more than they do teeth.

Pigs were first domesticated in China in Neolithic times, about 4900 B.C. Biblical writings mention them as early as 1500 B.C., and legendary and historical accounts refer to the keeping of swine in Great Britain as early as 800 B.C. Swine seem to have been especially variable under domestication and especially amenable to human selection. As evidence of this assertion, one need but observe the difference in length of snout and size of ears of modern breeds of hogs.

When given the opportunity, pigs promptly revert within only a few generations to a wild or feral state in which they acquire the body form and characteristics of their wild progenitors many generations removed. The self-sustaining razorback of the United States is an example of this reversion.

## EUROPEAN WILD BOAR (SUS SCROFA)

The European wild boar (*Sus scrofa*) still lives in some of the forests of Europe. Although much reduced in numbers in the last few hundred years, it appears unlikely that the famous wild boar will become extinct like the Aurochs (the chief progenitor of domestic cattle). In comparison with the domestic pig, this race of hogs is characterized by its coarser hair (with an almost manelike crest along the back), larger and longer

---

head, larger feet, longer and stronger tusks, narrower body, and greater ability to run and fight. The color of mature animals is nearly black, with a mixture of gray and rusty brown on the body. Very young pigs are striped. The ears are short and erect.

Fig. 36–1. European wild boar (*Sus scrofa*). Note the coarse hair, long head and snout, large feet, and long tusks. (Courtesy, New York Zoological Society)

These sturdy ancestors of domestic swine are extremely courageous and stubborn fighters and are able to drive off most of their enemies, except hunters. If attacked, they will use their tusks with deadly effect, although normally they are as shy as most wild animals and prefer to avoid people. The wild boar hunt has been regarded as a noble sport throughout history. Custom decrees that the hunt shall be on horseback and with dogs, and that the quarry shall be killed with a spear.

The European wild boar will cross freely with domestic swine, and the offspring are fertile. It was domesticated somewhere around the Baltic Sea in Neolithic times.

## EAST INDIAN PIG

This broad classification includes a number of wild stocks of swine that were native to the East Indies and southeastern Asia. Though a bewildering number of domestic races are derived from the East Indian pig, including the domestic pig of China, all of them are smaller and more refined than the European wild boar. The East Indian pig is further distinguished by the absence of the crest of hair on the back and the white streak along the sides of the face. It is thought that *Sus vittatus* was the chief, if not the only, race or species of the East Indian pig that was crossed with the descendants of the European wild boar in founding the American breeds of swine.

---

[1]Some authorities now hold that the East Indian *Sus vittatus* is actually connected with *Sus scrofa* by a chain of intermediates.

Fig. 36–2. Malayan or Philippine pig (*Sus philippinensis*). This animal is rather typical of all East Indian pigs, the wild stock which, along with European wild boar, contributed to American breeds of swine. In comparison with the European wild boar, East Indian pigs are smaller and more refined. (Courtesy, Chicago Natural History Museum)

## POSITION OF THE HOG IN THE ZOOLOGICAL SCHEME

The following outline shows the basic position of the domesticated hog in the zoological scheme:

Kingdom *Animalia*: Animals collectively; the animal kingdom.
  Phylum *Chordata*: One of approximately 21 phyla of the animal kingdom, in which there is either a backbone (in the vertebrates) or the rudiment of a backbone, the chorda.
    Class *Mammalia*: Mammals, or warm-blooded hairy animals that produce their young alive and suckle them for a variable period on a secretion from the mammary glands.
      Order *Artiodactyla*: Even-toed, hoofed mammals.
        Family *Suidae*: The family of nonruminant, artiodactyl ungulates, consisting of wild and domestic swine but, in modern classifications, excluding the peccaries.
          Genus *Sus*: The typical genus of swine, formerly comprehensive but now restricted to the European wild boar and its allies, with the domestic breeds derived from them.
            Species *Sus scrofa* and *Sus vittatus*: *Sus scrofa* is a wild hog of continental Europe from which most domestic swine have been derived. *Sus vittatus* was the chief, if not the only, race or species of the East Indian pig that contributed to present-day domestic swine.

## INTRODUCTION OF SWINE TO AMERICA

Although many wild animals were widely distributed over the North American continent prior to the coming of the Spanish explorers, the wild boar was unknown to the native American Indian.

Columbus first brought hogs to the West Indies on his second voyage in 1493. According to historians, only eight head were landed as foundation stock. However, these hardy animals must have multiplied at a prodigious rate, for, 13 years later, the settlers of this same territory found it necessary to hunt the ferocious wild swine with dogs; they had grown numerous, and they were killing cattle.

Although swine were taken to other Spanish settlements following the early explorations of Columbus, pigs first saw America when crossing the continent with Hernando de Soto. The energetic Spanish explorer arrived in Tampa Bay (now Florida) in 1539. Upon his several vessels (between 7 and 10), he had 600 or more soldiers, some 200 to 300 horses, and 13 head of hogs.

This hardy herd of squealing, scampering pigs traveled with the army of the brave Spanish explorer. The hazardous journey stretched from the Everglades of Florida to the Ozarks of Missouri. In spite of battles with hostile Indians, difficult travel, and other hardships, the herd thrived so well that at the time of de Soto's death on the upper Mississippi, three years after the landing at Tampa, the hog herd had grown to 700. De Soto's successor, Moscoso, then ordered that the swine be auctioned off among the men.

Fig. 36–3. De Soto discovers the Mississippi. When in 1539, the Spanish explorer landed at Tampa Bay (now Florida), he brought with him 13 head of hogs. At the time of de Soto's death on the upper Mississippi, 3 years after the landing at Tampa, the hog herd had grown to 700. (Courtesy, The Bettmann Archive)

It is reasonable to assume, therefore, that the cross-country journey of de Soto's herd of pigs was the first swine enterprise in America. No doubt some of de Soto's herd escaped to the forest, and perhaps still others were traded to the Indians. At any rate, this sturdy stock served as foundation blood for some of our early American razorbacks.

## THE CREATION OF AMERICAN BREEDS OF SWINE

The most thoroughly American domestic animal is the pig. In no other class of animals have so many truly American breeds been created. These facts probably result from (1) the suitability of native Indian corn (maize) as a swine feed, (2) the ease with which pork could be cured and stored prior to the days of refrigeration, and (3) the need for fats and high-energy foods for laborers engaged in the heavy development work of a frontier country.

Unlike the beef and dairy producers, who sent their native cattle to slaughter and imported whole herds of blooded cattle from England, American hog raisers were content to use mongrel sows descended from colonial ancestry as a base, upon which they crossed imported Chinese, Neapolitan, Berkshire, Tamworth, Russian, Suffolk Black, Byfield, and Irish Grazier boars. These importations began as early as the second quarter of the 19th century. Out of the different crosses, which varied from area to area, were created the several genuinely American breeds of swine.

Structurally, the creation of the modern hog has been that of developing an animal that would put flesh on the sides and quarters, instead of running to bone and a big head. Physiologically, breed improvement has resulted in an elongation of the intestine of the hog, thus enabling it to consume more feed for conversion into meat. According to naturalists, the average length of the intestine of the wild boar compared with its body is in the proportion of 9 to 1; whereas, in the improved American breeds, it is in the proportion of 13.5 to 1.

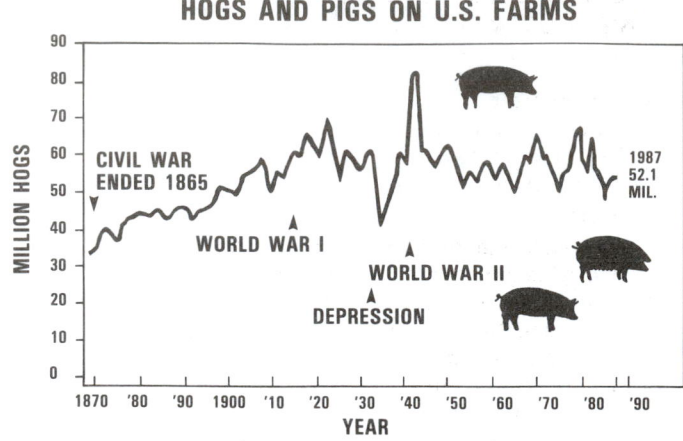

Fig. 36–5. Hogs and pigs on U.S. farms, 1867 to 1987. (From USDA sources)

Fig. 36–4. A typical Arkansas Razorback. This 2-year-old sow weighed 180 lb. (Courtesy, United Duroc Swine Registry)

## MAGNITUDE OF THE U.S. SWINE INDUSTRY

Hog numbers parallel very closely the production of corn in the north central or Corn Belt states; and these states produce nearly three-fourths of the corn grown in the country. Thus, when corn yields are down, the price of feed is up and swine production decreases. The opposite occurs when corn yields are high.

As shown in Fig. 36–5, hog numbers change sharply from year to year. Except for the sharp increase during the war years (with an all-time peak of 83,741,000 head on January 1, 1944), there has been a tendency for hog numbers to level out, rather than increase. This is probably the result, in part at least, of decreased export demands for pork and pork by-products and of the increased competition that lard has encountered from vegetable oils. Also, beef has moved ahead of pork in per capita consumption, and per capita poultry production has increased.

Another measure of the magnitude of the U.S. swine industry—and perhaps a more meaningful measure than the total count of hogs taken once a year, from the standpoint of reflecting the number of hogs marketed—is the size of the pig crop. This is shown in Fig. 36–6. Normally, the spring crop averages about five million head larger than the fall crop. This is a reflection of the more favorable weather in the spring than in the fall, from the standpoint of farrowing and raising pigs. But spring and fall farrowings do not exactly parallel each other. The latter situation is primarily due to the fact that the number of sows bred is a reflection of hog and feed prices at breeding time.

### Market Hogs and Pig Crops

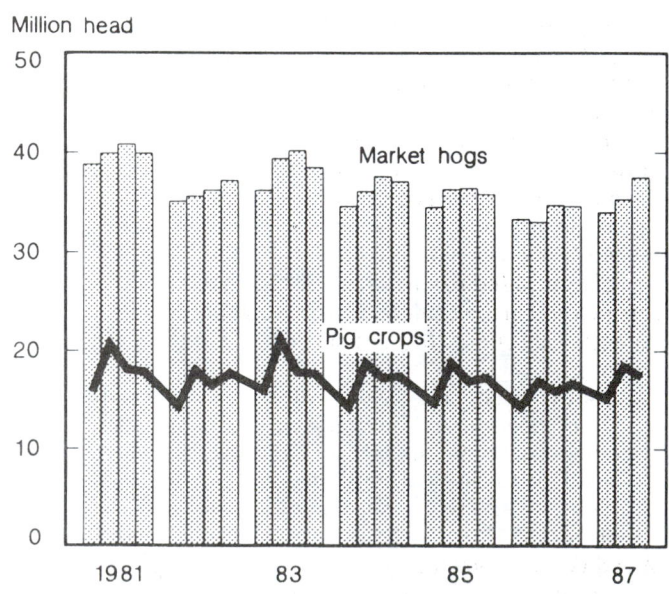

Fig. 36–6. Pig crops: Dec.-Feb., Mar.-May, June-Aug., and Sept.-Nov. Market hogs on farms: Dec. 1 previous year, Mar. 1, June 1, Sept. 1, and Dec. 1. (From: *1988 Agricultural Chartbook*, Ag. Hdbk. No. 673, USDA, p. 89, Chart 193)

# WORLD SWINE PRODUCTION AND DISTRIBUTION

Swine are produced most numerously in the temperate zones and in those areas where the population is relatively dense. There is reason to believe that these conditions will continue to prevail.

Table 36–1 gives data pertaining to the 10 leading swine-producing countries of the world. China has long had the largest hog population, but, because of the large human population, production in that country is largely on a domestic basis, with very negligible quantities of pork entering the world trade. It must also be remembered that in China pigs are primarily scavengers and that the value of the manure produced is one of the main incentives for keeping them.

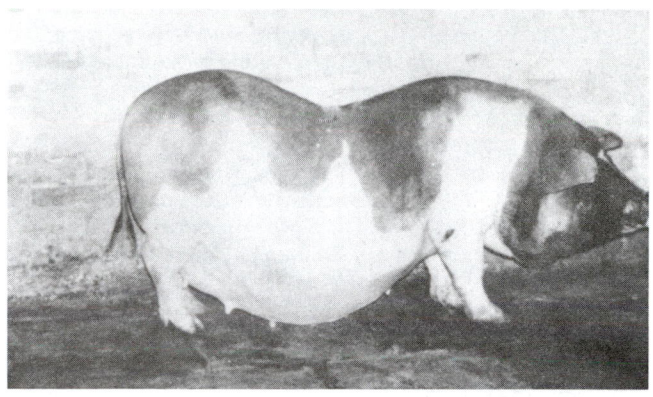

Fig. 36–7 A Chinese sow. She's black and white and swaybacked. But Chinese swine are very prolific. Most sows farrow 10 to 14 pigs, and litters of 23 to 25 are not uncommon in northern China. (Photo by A. H. Ensminger)

## TABLE 36–1
### LEADING SWINE-PRODUCING COUNTRIES OF THE WORLD

| Country | Swine Population[1] | Human Population[2] | Size of Country[2] | | Swine per Capita | Swine per | |
| --- | --- | --- | --- | --- | --- | --- | --- |
| | | | (sq mi) | (sq km) | | (sq mi) | (sq km) |
| China, People's Republic of ........ | 334,862,000 | 1,069,628,000 | 3,705,390 | 9,596,960 | 0.3 | 90.4 | 34.9 |
| U.S.S.R. ................. | 77,403,000 | 287,015,000 | 8,649,496 | 22,402,195 | 0.3 | 8.9 | 3.5 |
| United States ................. | 42,845,000 | 247,498,000 | 3,618,770 | 9,372,614 | 0.3 | 11.8 | 4.6 |
| Brazil ................. | 32,700,000 | 153,992,000 | 3,286,470 | 8,511,957 | 0.2 | 9.9 | 3.8 |
| West Germany ................. | 23,670,000 | 60,162,000 | 95,975 | 248,575 | 0.4 | 246.6 | 95.2 |
| Poland ................. | 19,605,000 | 38,389,000 | 120,727 | 312,683 | 0.5 | 162.4 | 62.7 |
| Spain ................. | 16,941,000 | 39,784,000 | 194,896 | 504,781 | 0.4 | 86.9 | 33.6 |
| Mexico ................. | 16,500,000 | 88,087,000 | 761,604 | 1,972,554 | 0.2 | 21.7 | 8.4 |
| Romania ................. | 15,224,000 | 23,155,000 | 91,699 | 237,500 | 0.7 | 166.0 | 64.1 |
| Netherlands ................. | 14,266,000 | 14,689,000 | 15,770 | 40,844 | 1.0 | 904.6 | 349.3 |
| World Total ................. | 823,403,000 | 5,055,000,000 | 57,800,000 | 149,702,000 | 0.2 | 14.2 | 5.5 |

[1]FAO Production Yearbook 1988, United Nations, Rome, Italy, Vol. 42, p. 247.
[2]The World Almanac and Book of Facts 1989, The Fresno Bee, 1989.

Hog production in the U.S.S.R. has risen sharply in recent years. As shown in Table 36–1, the U.S.S.R. ranks a solid second among the leading swine-producing countries of the world.

In general, in the European countries, hog numbers are closely related to the development of the dairy industry and the production of barley and potatoes—in much the same manner as the distribution of swine in the United States is closely related to the acreage of corn.

Corn is raised extensively in the La Plata region of South America and in the Danube Basin of southern Europe. In these corn-growing areas, hog production is a dominant type of farming.

Dairy by-products—skimmed milk, buttermilk, and whey—have long been important swine supplements in the northeastern United States and also in Denmark, the Netherlands, Canada, Ireland, Sweden, and Latvia. In Germany and Poland, potatoes have always been extensively used in swine feeding.

Since 1923, except for the increases occurring during World War II, there has been a downward trend in the exports of pork and lard from the United States. This has been due to a marked increase in production in Canada and in the European countries, particularly in Denmark, Germany, and Ireland and to various trade restrictions imposed by the importing countries. In no sense has this decrease been due to any lack of capacity to produce on the part of the American farmer.

Hog numbers fluctuate rather sharply on the basis of available feed supplies. Also, the annual per capita consumption of pork in different countries of the world varies directly with production and availability, cost, the taste preference of the people, and in some cases with the religious beliefs that bar the use of pork as food.

# SWINE PRODUCTION IN THE UNITED STATES

The contribution of humble pigs to American agriculture is expressed by their undisputed title as the "mortgage lifters." No other animal has been of such importance to the farmer. Hogs are produced on about a third of a million U.S. farms (in 1987, 328,640 farms, each with an average inventory of about 160 head, according to the USDA). These farms have about 5.2% of the hog population of the world, yet they produce 11.4% of the world's supply of pork.

In 1990, more than 60% of U.S. hogs were produced in confinement facilities; and 12 to 13% of all hogs were produced under contract.

On September 1, 1989, there were 58,400,000 head of hogs in the United States.[2] Hogs ranked second only to cattle in animal population. Also, they rank as one of the most important sources of farm income; in 1987, hogs accounted for $10.3 billion in farm income, which was 13.5% of the farm income derived from all livestock.

The geographical distribution of swine in the United States closely coincides with the acreage of corn, the principal swine feed. Normally, about ¾ of U.S. corn production is fed to livestock. Of the portion fed to animals, approximately ⅓ is fed to hogs. It is not surprising, therefore, to find that nearly ⅔ of the hog production is centered in the seven Corn Belt states: Iowa, Illinois, Indiana, Ohio, Missouri, Nebraska, and Kansas. From this it should not be concluded that sections other than the Corn Belt are not well adapted to pork production. As a matter of fact, an area that produces dairy by-products, small grains, and forage is admirably adapted to the production of bacon of the highest quality.

## LEADING STATES IN SWINE PRODUCTION

The state of Iowa has held undisputed lead in hog numbers since 1880, but the rank of the other states has shifted about considerably. A ranking of the 10 leading states, together with total numbers for the United States, is given in Table 36–2.

**TABLE 36–2**
**TEN LEADING STATES IN HOG NUMBERS, BY RANK, 1987[1]**

| State | Number of Hogs |
|---|---|
| Iowa | 13,800,000 |
| Illinois | 5,300,000 |
| Indiana | 4,600,000 |
| Minnesota | 4,350,000 |
| Nebraska | 4,000,000 |
| Missouri | 2,950,000 |
| North Carolina | 2,500,000 |
| Ohio | 2,150,000 |
| South Dakota | 1,520,000 |
| Kansas | 1,420,000 |
| **U.S. Total** | 53,795,000 |

[1]*Agricultural Statistics 1988*, USDA, p. 272.

Growing corn and producing pork have contributed largely toward making the farmers of the upper Mississippi Valley the wealthiest agricultural people on the globe.

## FACTORS FAVORABLE TO SWINE PRODUCTION

The important position that the hog occupies in American agriculture is due to certain factors and economic conditions favorable to swine production. These may be enumerated as follows:

1. Swine excel all other farm animals in the economy with which they convert concentrated farm feed into meat and meat products.

Experiments and experience indicate that the feed required to produce gain in the drylot during the market-finishing period of cattle, lambs, and pigs is about as shown in Table 36–3.

[2]*Hogs and Pigs*, Statistical Reporting Service, USDA, Sept. 29, 1989.

**TABLE 36–3**
**FEED EFFICIENCY OF CATTLE, SHEEP, AND SWINE**

| Age and Class of Animals | Days on Feed | Feed Required per 100 Lb (45.4 kg) Gain | | | |
|---|---|---|---|---|---|
| | | Grain | | Hay | |
| | | (lb) | (kg) | (lb) | (kg) |
| Yearling steers | 140 | 600 | 273 | 300 | 136 |
| Feeder lambs | 90 | 400 | 182 | 400 | 182 |
| Growing-finishing pigs | 90 | 400 | 181.4 | — | — |

2. Swine are prolific, commonly farrowing from 7 to 12 pigs, and producing 2 litters per year.

3. Swine excel in dressing percentage, yielding 65 to 80% of their liveweight when dressed packer style—with head, leaf fat, kidneys, and ham facings removed. On the other hand, cattle dress only 50 to 60%, and sheep and lambs 45 to 55%. Moreover, because of the small proportion of bone, the percentage of edible meat in the carcass of hogs is greater.

4. Pork is most nutritious. Because of the higher content of fat and the slightly lower content of water, the energy value of pork is usually higher than that of beef or lamb.

5. Hogs are efficient converters of wastes and by-products into pork. This includes grain wasted by finishing cattle, garbage, garden waste, and such dairy by-products as skim milk.

6. Since hogs are well adapted to the practice of self-feeding, labor is kept to a minimum.

7. Swine require a small investment for buildings and equipment.

8. The pig is adapted to both diversified and intensified agriculture.

9. The initial investment in getting into the business is small, and the returns come quickly. A gilt may be bred at 8 months of age, and the pig can be marketed 5 to 6 months after farrowing.

10. The spread in price in market hogs is relatively small—much smaller, for example, than the spread which usually exists between the price of Prime steers and Canner cows. Hogs may be sold at weights ranging from 150 to 275 lb without any great penalty in price. Also, old sows that have outlived their usefulness in the breeding herd may be disposed of without difficulty.

11. Hogs are unexcelled as a source of farm meats. This is due to their ease of dressing and the superior curing and keeping qualities of pork.

12. The hog excels all other farm animals in fat storing ability, and pork fat is more valuable than fats produced by other domestic animals with the exception of the dairy cow.

## FACTORS UNFAVORABLE TO SWINE PRODUCTION

It is not recommended that hogs be raised under any and all conditions. There are certain limitations that should receive consideration if the venture is to be successful. Some of the reservations are as follows:

1. Because of the nature of the digestive tract, the growing-finishing pig must be fed a maximum of concentrates and a minimum of roughages. Where or when grains are scarce and high in price, this may result in high production costs.

2. Because of the nature of their diet and their rapid growth

rate, hogs are extremely sensitive to unfavorable rations and to careless management.

3. Swine are very susceptible to numerous diseases and parasites.

4. Fences of a more expensive kind are necessary in hog raising.

5. Sows should have skilled attention at farrowing time.

6. Because of their rooting and close-grazing habits, hogs are hard on pasture.

7. Hogs are not adapted to a frontier type of agriculture where grazing areas are extensive and vegetation is sparse. Neither are they well suited to the utilization of permanent pasture areas.

## FUTURE OF THE AMERICAN SWINE INDUSTRY

Some of the factors that will determine the future of the American swine industry are:

1. **Foreign competition.** Several European and Scandinavian countries are pork-exporting nations. Some of the South American countries are potential pork-producing and pork-exporting nations, an encouraging market being the only needed incentive. Canada is making great progress in swine production, in both quality and quantity. Only tariffs, quotas, and embargoes enacted by our federal government can prevent future and serious competition from foreign imports. However, with our huge corn production and improved swine production methods, it is not anticipated that pork will ever have the potential foreign competition that exists with beef.

The potential pork export situation is not encouraging.

2. **The lard situation.** Lard was a ''drug on the market'' prior to 1941, and, soon after World War II, this status returned. Satisfactory vegetable oils can now be produced at lower cost. When processed lard sells for less than the price of hogs on foot, it should be perfectly evident that the product is lacking in demand. In order to alleviate the surplus lard situation, the soundest approach consists of (a) breeding a type of hog that is less lardy in conformation, (b) feeding so as to produce less excess fat, (c) marketing at lighter weights, and (d) purchasing hogs on a quality basis (preferably rail graded).

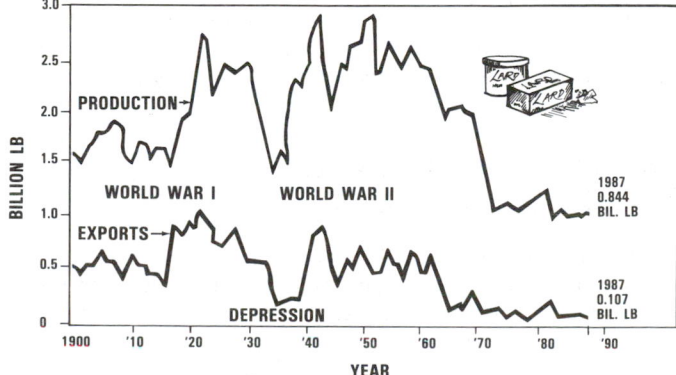

Fig. 36–8.   Production and exports of lard from the United States, 1900 to 1987. Note that lard exports increased sharply during both World War I and World War II, but they were very small between 1935 and 1940. Most authorities agree that future lard exports will be negligible. (From: *Agricultural Statistics 1988*, p. 281, Table 417)

3. **Increased human consumption.** Without doubt, some increased human consumption of pork could be brought about through the production of a higher quality product. Many folks, for example, would relish more high-quality breakfast bacon. It is not anticipated, however, that any considerable opportunity for increased demand for hogs will come from any marked increase in per capita consumption of pork. Rather, modest increases will accrue from our expanding population.

4. **Competition from other farm animals and between areas.** Despite the efficiency with which pigs convert grain to pork, they must continue to compete with all classes of animals for the available concentrates. For the nation as a whole, it appears, however, that there exists a fair balance of production with consumption of the several products of domestic animals. Thus, no immediate or sizable shift that would be either favorable or unfavorable to increased pork production is anticipated. On the other hand, certain minor shifts in production from area to area may be expected.

Because of the year-round grazing, interest in diversification, the development of vertical integration and the diverse and comparatively cheap feeds in the southern states, increased swine production in this area is likely.

5. **Confinement production.** Swine are better adapted to confinement rearing than other classes of four-footed animals. Moreover, confinement production facilitates manure handling and automatic feeding/watering, thereby lessening labor. Thus, more and more confinement production is in the future of the American swine industry.

Fig. 36–9.   The future of the American swine industry: More confinement production on slotted floors, self-feeding and automatic watering. (Photo by J. C. Allen and Son, West Lafayette, IN)

6. **Bigger operations.** In 1986, swine operations with 1,000 or more hogs accounted for 75% of the hogs marketed. That was a 120% increase over 1980. In 1990, more than 90% of the market hogs were controlled by producers raising 1,000 or more hogs per year.

7. **Contract production.** In 1990, 12 to 13% of U.S. hogs were produced under contract. It is predicted that contract production will double by the late '90s.

# QUESTIONS FOR STUDY AND DISCUSSION

1. Why couldn't nomadic people take swine with them as easily as other animal species? How did the trait of not migrating great distances contribute to many local races or breeds of swine?

2. Where and when were pigs first domesticated?

3. How did the razorback of the U.S. evolve?

4. Name and describe the two wild stocks that contributed to American breeds of swine.

5. Outline the basic position of the domestic hog in the zoological scheme.

6. When (in what year) did each Columbus and de Soto bring swine to America? What prompted them to bring swine with them?

7. During the 3-year journey from the Everglades of Florida to the Ozarks of Missouri, de Soto's hog herd grew from 13 to 700 head. How do you suppose this increase in numbers was accomplished with a traveling herd?

8. More truly American breeds were created in swine than was the case with cattle, sheep, and horses. How do you explain this situation?

9. Structurally and physiologically, what changes were made in creating modern hogs from their wild ancestors?

10. Fig. 36–5 shows (a) that hog numbers change sharply from year to year, and (b) that there is a trend for hog numbers to level out, rather than increase. Why is this?

11. How does the size pig crop indicate the number of hogs marketed?

12. Name and rank the six leading swine-producing countries of the world. Discuss the factors which account for each of them holding their respective rank.

13. What factors affect the annual per capita consumption of pork in different countries of the world?

14. United States farms have about 5.2% of the hog population of the world, yet they produce 11.4% of the world's supply of pork. What do these statistics tell about the American hog producer's efficiency?

15. How do you account for the fact that nearly two-thirds of the hog production is centered in the seven Corn Belt states?

16. List by rank the five leading U.S. states in hog numbers. What factors cause Iowa to have more than twice as many hogs as second ranking Illinois?

17. List, and rank according to importance, the five most important factors favorable to swine production.

18. List, and rank according to importance the five most limiting factors unfavorable to swine production.

19. Discuss the future of the American swine industry as determined by each of the following factors: (a) foreign competition, (b) the lard situation, and (c) increased human consumption of pork.

20. Discuss the future of the American swine industry as determined by (a) competition from other farm animals and between areas, (b) confinement production, and (c) bigger operations.

21. Select a certain farm or ranch (your home farm or ranch, or one with which you are familiar). Then, (a) discuss the relation of swine production on this farm or ranch to the type of agriculture, and (b) list the factors favorable and unfavorable to swine production on this particular farm or ranch.

## SELECTED REFERENCES

| Title of Publication | Author(s) | Publisher |
|---|---|---|
| Ancestor for the Pigs: Taxonomy and Phylogeny of the Genus "Sus" | C. Groves | Australian National University Press, Canberra, Australia, 1982 |
| Domesticated Animals from Early Times | J. Clutton-Brock | British Museum, London, and University of Texas Press, Austin, TX, 1981 |
| Encyclopaedia Britannica | | Encyclopaedia Britannica, Chicago, IL |
| Grzimek's Animal Life Encyclopedia, Vol. 13, Mammals IV | Edited by B. Grzimek | Van Nostrand Reinhold Company, New York, NY, 1972 |
| History of Livestock Raising in the United States, 1607–1860, Ag. History Series No. 5 | J. W. Thompson | U.S. Department of Agriculture, Washington, DC, 1942 |
| Natural History of the Pig, The | I. M. Mellen | Exposition Press, New York, NY, 1952 |
| Our Friendly Animals and Whence They Came | K. P. Schmidt | M. A. Donohue & Co., Chicago, IL, 1938 |
| Pigs from Cave to Corn Belt | C. W. Towne E. N. Wentworth | University of Oklahoma Press, Norman, OK, 1950 |
| Pork Production | W. W. Smith | The Macmillan Company, New York, NY, 1952 |
| Principles of Classification and a Classification of Mammals, The, Vol. 85 | G. G. Simpson | American Museum of Natural History, New York, NY, 1945 |
| Swine Production, Fifth Edition | J. L. Krider J. H. Conrad W. E. Carroll | McGraw-Hill Book Company, New York, NY, 1982 |
| Swine Science, Fifth Edition | M. E. Ensminger R. O. Parker | The Interstate Publishers, Danville, IL, 1984 |

From whence they came. The European Wild Boar, after a painting by Ernest Griset. (Courtesy, Smithsonian Institute)

# TYPES AND BREEDS OF SWINE[1]

## Chapter 37

[1]Sometimes folks construe the write-up of a breed of livestock in a book or in a USDA bulletin as an official recognition of the breed. Nothing could be further from the truth, for no person or office has authority to approve a breed. The only legal basis for recognizing a breed is contained in the Tariff Act of 1930, which provides for the duty-free admission of purebred breeding stock provided they are registered in the country of origin. But the latter stipulation applies to imported animals only.

In this book, no *official* recognition of any breed is intended or implied. Rather, the author has tried earnestly, and without favoritism, to present the factual story of the breeds in narrative and picture. In particular, such information relative to the new and/or less widely distributed breeds is needed, and often difficult to obtain.

In the hands of skilled livestock breeders, swine are the most plastic of any species of farm animals. This is due to their early maturity, multiple rate of reproduction, and short time between generations. A farmer who produces a total of 100 spring farrowed pigs yearly needs only 14 gilts to raise a crop of the same size the following spring. There will be approximately 50 gilts from which to select the 14 brood sows needed. Thus, there is a wide choice of keeping the meaty type of gilts or of picking others that are shorter and thicker. Continued selection each year with emphasis upon the same characteristics and the purchase of boars of the same type can completely alter the conformation of the hogs in a herd in the short period of 4 to 5 years.

Despite this fact, progress in producing meat-type hogs was often slow and painful throughout the thirties and forties. As a result, (1) pork gradually lost its place as the preferred meat, with beef taking the lead in the early fifties; and (2) a number of new American breeds of swine evolved, most of them carrying some Landrace breeding.

## TYPES OF HOGS

Swine types are the result of three contributing factors: (1) the demands of the consumer, (2) the character of the available feeds, and (3) the breeding and pursuit of type by breeders.

Historically, three distinct types of hogs have been recognized: (1) lard type, (2) bacon type, and (3) meat type.

### THE LARD TYPE

Originally, breeders of hogs stressed immense size and scale and great finishing ability. This general type persisted until the latter part of the 19th century. Beginning about 1890, breeders turned their attention to the development of early maturity, great refinement, and a very thick finish. In order to

obtain these desired qualities, animals were developed that were smaller in size, thick, compactly built, and very short of leg. In the Poland China breed, this fashionable fad was carried to the extreme. It finally culminated in the development of the "hot bloods." Hogs of this chuffy type were notoriously lacking in prolificacy. They often farrowed twins and triplets; and when they were carried to weights in excess of 200 lb, their gains were very expensive.

In order to secure increased utility qualities, breeders finally, about 1915, began the shift to the big-type strains. Before long, the craze swept across the nation, and again the pendulum swung too far. Breeders demanded great size, growthiness, length of body, and plenty of bone. The big-type animal was rangy in conformation and slow in maturity. Many champions of the show-ring included as their attributes long legs, weak loins, and "cat hams." One popular champion of the day was advertised as being "so tall that it makes him dizzy to look down." Inasmuch as this type failed most miserably to meet the requirements of either the packer or the producer—being too slow to reach maturity and requiring a heavy weight in order to reach market finish—another shift in ideals became necessary.

Fig. 37-2  A Poland China boar pig of the rangy type. Animals of this type dominated the American show-ring from 1915 to 1925.

### THE BACON TYPE

Bacon-type hogs are more common in those areas where the available feeds consist of dairy by-products, peas, barley, wheat, oats, rye, and root crops. As compared with corn, such feeds are not so fattening. Thus, instead of producing a great amount of lard, they build sufficient muscle for desirable bacon. The countries of Denmark, Canada, and Ireland have long been noted for the production of high-quality bacon. In the past, the surplus pork produced in these countries has found a ready market in England, largely selling as Wiltshire sides.

In emphasizing the importance of character of feeds as a factor influencing the production of bacon-type hogs, it is not to be inferred that there is no hereditary difference. That is to say, when bacon-type hogs are taken into the Corn Belt and fed largely on corn, they never entirely lose their bacon qualities.

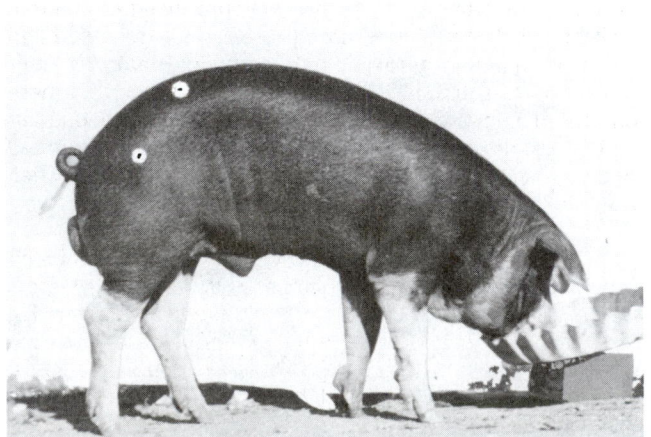

Fig. 37-1.  A Poland China gilt of the chuffy type. Small, refined animals of this type dominated the American show-ring from 1890 to 1910. (Courtesy, University of Illinois)

## THE MEAT TYPE

Since about 1925, American swine breeders have been striving to produce meat-type hogs—animals that are intermediate between the lard and bacon types. The best specimens of the meat type combine muscling, length of body, balance, and the ability to reach market weight and finish without excess fat. In achieving the meat type, the selection and breeding programs of producers have been augmented by meat certification programs, livestock shows, and swine-type conferences.

Fig. 37-3   Ideal meat-type barrow of 1990, a Chester White Champion barrow. (Courtesy, Chester White Assn., Peoria, IL)

Today's commercial producers of meat-type hogs have little allegiance to a breed or a color. Their primary concerns are carcass quality, efficiency of production, and herd health. Nevertheless, the eight major purebred breeds of swine are still very important in producing meat-type hogs in the United States. According to a recent survey, 57.8% of the nation's swine is sired by boars secured from traditional purebred breeders, while 27.7% is sired by boars obtained from *specialized breeding stock suppliers*.[2] The remaining 14.5% is sired by home-raised boars.

## POPULARITY OF BREEDS

Table 37-1 shows the 1988 and total registrations to date of the common breeds of hogs. Although the annual figures for 1988 are probably more meaningful at the present time than the all-time registrations, it is recognized that one year's data fail to show trends.

[2]Survey conducted by Michigan State University, financed by *National Hog Farmer*, reported in *National Hog Farmer*, September 15, 1989, pp. 52–58.

**TABLE 37-1**
**REGISTRATIONS OF SWINE IN U.S. BREED REGISTRY ASSOCIATIONS**

| Breed | 1988 Registrations (individual) | Total Registrations |
|---|---|---|
| Yorkshire | 213,427 | 3,239,107 |
| Duroc | 189,146 | 7,081,093 |
| Hampshire | 147,963 | 6,987,900 |
| Spotted | 62,179 | 2,338,448 |
| Chester White | 60,582 | 2,922,477 |
| Landrace | 30,422 | 829,533 |
| Berkshire | 17,611 | 1,192,016 |
| Poland China (1989) | 15,975 | 854,520 |
| Hereford | 1,044 | 121,220 |

## CHARACTERISTICS OF BREEDS

With the exception of the Berkshire, Landrace, Tamworth, and Yorkshire, the breeds of swine common to the United States are strictly American creations. This is noteworthy in view of the fact that few of our better-known breeds of beef cattle, dairy cattle, sheep, and draft horses were American creations. With the exception of the Hereford breed and some of the newer commercial and other specialized breeds of swine, the American breeds came into being in the period from 1800 to 1880—an era which was characterized by the production of an abundance of corn for utilization by hogs and by consumer demand for fat, heavy cuts of pork. The European breeds did not seem to meet these requirements.

It must be remembered, however, that the American breeds of swine were not developed without recourse to foreign stock. Prior to de Soto's importation, no hogs were found on the continent. The offspring of de Soto's sturdy razorbacks, together with subsequent importations of European and Oriental hogs, served as the foundation stock for the American breeds which followed. Out of these early-day multiple-colored and conglomerate types of swine, the swine producers of different areas of the Untied States, through the tools of selection and controlled matings, gradually molded uniform animals, later to be known as breeds. It is to be noted, however, that these foundation animals carried a variable genetic composition. This made them flexible in the hands of breeders, and accounted for the radical subsequent shifts in swine types that have been observed within the pure breeds.

Although there are many breed differences and most breed associations are constantly extolling the virtues of their respective breeds, it is perhaps fair to say that there is more difference within than between breeds from the standpoint of efficiency of production and carcass quality. Without doubt, the future and enduring popularity of each breed will depend upon how well it fulfills these two primary requisites.

The common U.S. breeds of swine and their characteristics are presented in Table 37-2 (see next page).

**TABLE 37-2**
**BREEDS OF SWINE AND THEIR CHARACTERISTICS**

| Breed | Place of Origin | Color | Distinctive Head Characteristics | Other Distinguishing Characteristics | Disqualifications |
|---|---|---|---|---|---|
| Berkshire | England. Chiefly in the south central counties of Berkshire and Wiltshire. | Black with 6 white points, 4 white feet, some white on the face, and a white switch on the tail. Any or all white points may be missing. | Medium short nose and erect ears. | Striking style and carriage. | A swirl on upper half of body; fewer than 12 teats; cryptorchid; or infertile vulva. Extreme pug nose is objectionable. |
| Chester White | U.S.A.; chiefly in Chester and Delaware counties of Pennsylvania. | White. Small bluish spots are sometimes found on the skin, but are discriminated against. | | | Off-colored hair, spots on hide larger than a silver dollar, cryptorchidism in males, hernia in males or females, or swirls on body above flanks. |
| Duroc | U.S.A.; chiefly in New York and New Jersey. | Red, varying from light to dark. | Medium-sized ear, tipping forward. | | White feet or white spots on any part of body, any white on end of nose, black spots larger than 2 in. in diameter, swirls on upper half of the body or neck, ridge-ling (one testicle) boars, or fewer than 6 udder sections on either side. |
| Hampshire | U.S.A.; in Boone County, Kentucky. | Black, with a white belt around the shoulders and body, including the front legs. | Longer and straighter in the face than most breeds; ears carried erect. | | Too much white on ham, hind legs, or head; or incomplete belt. |
| Hereford | U.S.A.; by R. U. Webber of La Plata, Missouri. | Red body color, with white face, legs, and switch similar to Hereford cattle. | A white face, and not less than two-thirds red, exclusive of face and ears. | | A white belt extending over shoulders, back, or rump; more than 1/3 white markings; no white markings on face; fewer than 2 white feet; swirl on any part of body; no marks of identification; ridgeling boar; permanent deformities of any kind; unsound underlines or fewer than 6 teats on each side. |
| Lacombe (55% Landrace, 23% Berkshire, and 22% Chester White) | Canada; at the Experimental Farm, Lacombe, Alberta, beginning in 1947. | White. | Medium-sized flop ears and a medium length, slightly dished face. | Of the 3 parent breeds, it resembles the Landrace most closely. | |
| Landrace | Denmark. | White, although small black skin spots are common. | Medium lop ears, straight snout, and trim jowl. | Very long side. | Black in the hair coat; fewer than 6 teats on either side; erect ears, with no forward break. |
| Poland China | U.S.A.; in Ohio, in the Miami Valley of Warren and Butler counties | Black or black with white spots, with 6 white points—the feet, face, and tip of the tail. | Drooping ears. | | Fewer than 6 teats on a side, a swirl on upper half of body, hernia, or cryptorchidism. |
| Spotted | U.S.A.; chiefly in Indiana. | Spotted black and white, about 50% each. | | Must have at least 6 prominent teats on each side to be eligible for show or sale. | Brown or sandy spots; swirl on the upper one-third of the body; ridgeling (one testicle) boar; solid black hair from ears forward; a combination black leg and black hoof. |
| Tamworth | England; in the central counties of Stafford, Leicester, Warwick, and Northamptom | Red, varying from light to dark. Black spots may occur, but are objectionable. | Wide between the ears, snout moderately long and straight, neat jowl, and medium-sized erect ears. | | Swirls. More than 5% black. |
| Yorkshire (known as the Large White in England) | England. | White, although black freckles may appear. | Slightly dished face, and erect ears. | Long bodied; prolific. | Swirls on upper third of body, hernia, hair color other than white, cryptorchidism, one testicle or any pronounced abnormal condition of the testicles, hermaphrodite, total blindness, fewer than 6 teats on each side, excessive amount of black or dark pigment of the skin, or extra dew claws. |

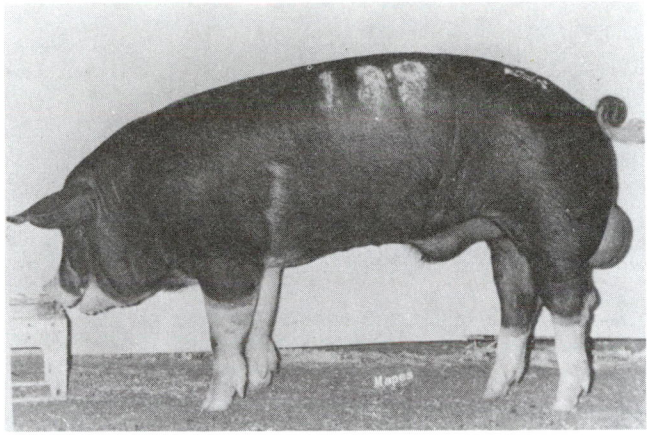

Fig. 37-4. Berkshire boar. (Courtesy, American Berkshire Assn., West Lafayette, IN)

Fig. 37-5. Berkshire gilt. (Courtesy, American Berkshire Assn., West Lafayette, IN)

Fig. 37-6. Chester White gilt. (Courtesy, Chester White Swine Record Assn., Peoria, IL)

Fig. 37-7. Chester White gilt. (Courtesy, Chester White Swine Record Assn., Peoria, IL)

Fig. 37-8. Duroc boar, bred by Sam Sparger, DeLeon, Texas. Sold to International Boar Semen, Eldora, Iowa, at $20,000. (Courtesy, United Duroc Swine Registry, Peoria, IL)

Fig. 37-9. Duroc sow, *HP Miss America*, bred by Howard Parrish, Eldon, Ohio. (Courtesy, United Duroc Swine Registry, Peoria, IL)

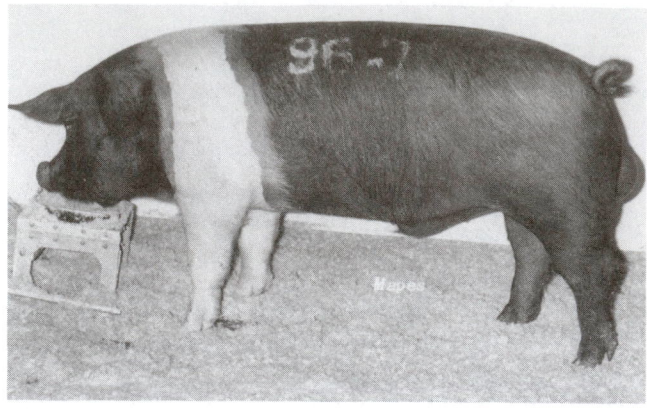

Fig. 37–10. Hampshire boar. (Courtesy, Compart's Boar Store, Nicollet, MN)

Fig. 37–11. Hampshire sow. (Courtesy, Lone Willow Genetics, Roanoke, IL)

Fig. 37–12. Champion Hereford boar. (Courtesy, National Hereford Hog Record Assn., Flandreau, SD)

Fig. 37–13. Grand Champion Hereford gilt, bred by Bernard C. Schulte, Norway, Iowa. (Courtesy, National Hereford Hog Record Assn., Flandreau, SD)

Fig. 37–14. Landrace boar, Grand Champion Boar at National Landrace Conference. (Courtesy, American Landrace Assn., West Lafayette, IN)

Fig. 37–15. Landrace female, *JMF Georgia Peach*, top selling Landrace female in the 1989 Winter Type Conference. (Courtesy, American Landrace Assn., West Lafayette, IN)

Fig. 37-16. Poland China boar. Champion at the 1990 National Barrow Show, Austin, Minnesota. (Courtesy, The Poland China Record Assn., West Lafayette, IN)

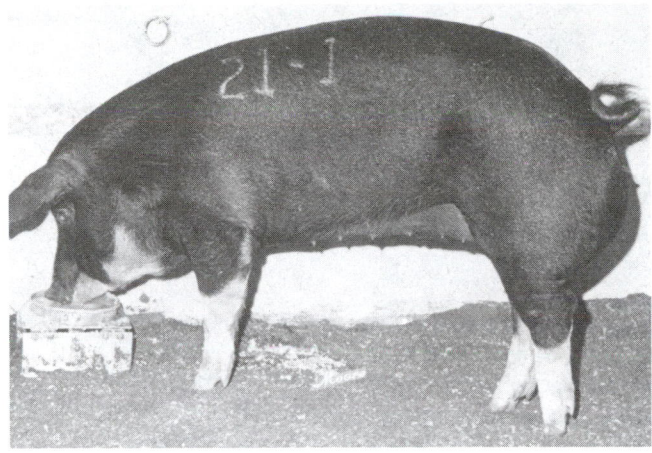

Fig. 37-17. Poland China gilt. (Courtesy, The Poland China Record Assn., West Lafayette, IN)

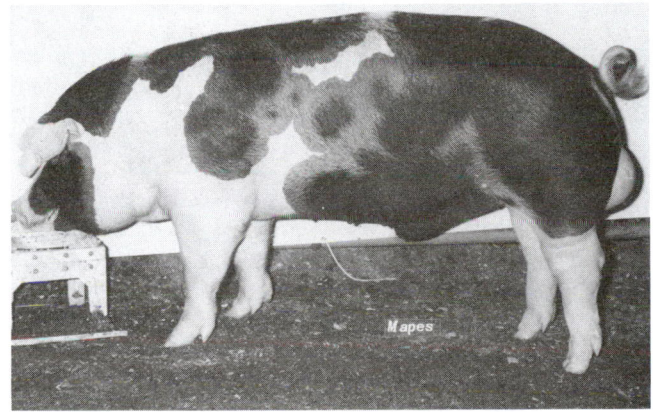

Fig. 37-18. Spotted boar. (Courtesy, National Spotted Swine Record, Inc., West Lafayette, IN)

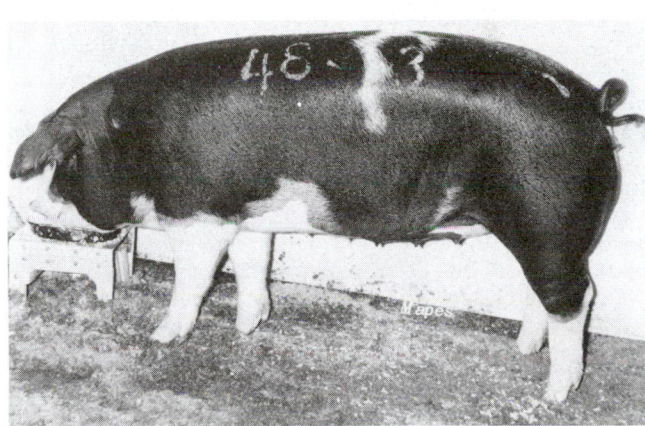

Fig. 37-19. Spotted gilt. (Courtesy, National Spotted Swine Record, Inc., West Lafayette, IN)

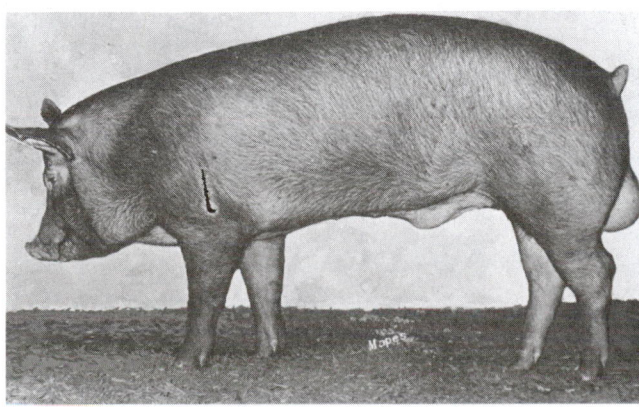

Fig. 37-20. Tamworth boar. (Courtesy, Tamworth Swine Assn., Winchester, OH)

Fig. 37-21. Tamworth gilt. (Courtesy, Tamworth Swine Assn., Winchester, OH)

Fig. 37–22. Yorkshire boar. (Courtesy, American Yorkshire Club, West Lafayette, IN)

Fig. 37–23. Yorkshire gilt. (Courtesy, American Yorkshire Club, West Lafayette, IN)

## INBRED BREEDS OF SWINE[3]

Beginning in the 1930s and continuing through the 1950s, there was interest in developing new breeds of swine that would be prolific, gain rapidly and efficiently, and produce a high quality carcass. New breeds were developed at the USDA Agricultural Research Center at Beltsville, Maryland, at several of the state agricultural experiment stations, by private breeders and in Canada. The goal was to bring together the best characteristics of several breeds to form new breeds. To do this, two or more of the established breeds were crossed with some backcrossing to one or more parent breeds. As the numbers of these inbred (new breeds) swine increased, a demand for registration arose. So, in 1946, the Inbred Livestock Registry was organized. Subsequently, it registered the following breeds: Beltsville No. 1, Beltsville No. 2, Connor Prairie, Lacombe, Maryland No. 1, Managra, Minnesota No. 1, Minnesota No. 2, Minnesota No. 3, Montana No. 1, Palouse, and San Pierre. Currently, very few, if any, of these breeds are being bred in the Untied States, Also, the Inbred Livestock Registry is dormant. But several of these inbred breeds were used in forming the specialized breeds (see next section); and the Lacombe breed seems to be well established in Canada.

While most of these newer breeds of swine did not survive, it is to their credit that they shook their older counterparts out of their lethargy. Thus, they more than justified the effort and cost back of them. Likewise, it is to the credit of the established breeds that they met the challenge and expedited the transition in type. As a result, pork that satisfies the consumer demands is once again being produced throughout America.

Fig. 37–24. Lacombe gilt. (Courtesy, Agriculture Canada, Research Branch, Lacombe, Alberta)

## SPECIALIZED BREEDING STOCK SUPPLIERS

Until about 1950, typical U.S. commercial swine producers of market hogs selected most replacement female breeding stock from their own herds, while almost all boars were purchased from purebred breeders. But the entry of *specialized breeding stock suppliers* has changed many breeding programs. A recent survey showed that *specialized breeding stock suppliers* were the source of 14.1% of the nation's gilts, and of 27.7% of the nation's boars.[4] Also, the survey revealed a trend of large commercial market hog producers to buy breeding stock from these new *specialized breeding stock suppliers*.

---

[3]The sections on "Inbred Breeds of Swine" and "Specialized Breeding Stock Suppliers" were authoritatively reviewed by, and helpful suggestions were received from, Dr. William G. Luce, Extension Swine Specialist, Oklahoma State University, Stillwater, Oklahoma; and Dr. Robert W. Seerley, Professor, Department of Animal and Dairy Science, University of Georgia, Athens.

[4]Survey conducted by Michigan State University, financed by *National Hog Farmer*, reported by *National Hog Farmer*, September 15, 1989, pp. 52–58.

# SPECIALIZED BREEDING STOCK

Fig. 37–25. *Blue Roan* gilt, developed by mating Large White (English Yorkshire) and American Yorkshire; then, mating the result of this cross to a Maternal Hampshire Line. (Courtesy, Lone Willow Genetics, Roanoke, IL)

Fig. 37–26. Duroc boar, class winner at the Nebraska State Fair. The breeder, Waldo Farms, has the largest Duroc herd in the United States. (Courtesy, Waldo Farms, DeWitt, NE)

Fig. 37–27. Hampshire × Duroc (F), Terminal Sire. (Courtesy, Genetic Improvement Services, Burlington, IN)

Fig. 37–28. Landrace boar. (Courtesy, Compart's Boar Store, Nicollet, MN)

Fig. 37–29. Large White (English Yorkshire) boar. (Courtesy, Lieske Genetics, Henderson, MN)

Fig. 37–30. Spot × 1 Boar; 50% Minnesota No. 1, and 50% Spotted Poland. (Courtesy, McLean County Hog Service, Inc., LeRoy, IL)

Most of the *specialized breeding stock suppliers* breed and market one or more of the traditional purebred breeds. In addition, they usually produce and market, often under their own genetic names, one or more hybrids, crossbreds, inbreds, maternal female lines, and/or terminal sire lines. They have well developed sales and testing programs, which emphasize performance and carcass quality. Some of them also provide financial assistance to their customers.

Among the *specialized breeding stock suppliers* are the following:[5]

Babcock Swine, Inc.
P.O. Box 759
Rochester, MN 55903

DeKalb Swine Breeders, Inc.
3100 Sycamore Road
DeKalb, IL 60115

Farmers Hybrid
P.O. Box 4528
Des Moines, IA 50306

Genetic Improvement Services
P.O. Box 447
Burlington, IN 46915

Lieske Genetics
Route 1, Box 211–A
Henderson, MN 56044

Lone Willow Farm
R.R. 2, Box 49
Roanoke, IL 61516

McLean County Hog Service, Inc.
Route 2, Box 96
LeRoy, IL 61752

Pig Improvement Co. (PIC)
P.O. Box 348
Franklin, KY 42134

Waldo Farms, Inc.
DeWitt, NE 68341

## MINIATURE SWINE

During the 1950s, the need developed for a smaller breed of swine for use in biomedical studies. For example, an awareness of the need arose from some of the work of the U.S. Atomic Energy Commission. Swine were selected as the animals of choice for certain studies. But the large size and the amount of waste (feces) produced by standard swine created a disposal problem and discouraged their use for long-term studies. In 1949, the Hormel Institute of the University of Minnesota initiated the development of genetically small pigs. These were used in some of the first Atomic Energy Commission studies. Later, other small strains were developed such as the Pitman-Moore, the Hanford Miniature Swine, and the Gottingen Miniature.[6] These miniatures are much smaller than standard swine. They are good experimental animals for biomedical studies, offering economic and convenience advantages over standard swine; they require less housing space, eat less, are easier to handle, and produce less waste (feces).

Vietnamese pigmy pigs, which are even smaller than miniature swine, are used as pets by some people.

---

[5]No claim is made that this list is complete. Neither should a listing be construed as an endorsement by the author.

[6]Dr. M. E. Ensminger, the author of this book, served as Consultant in the development and use of the Hanford Miniature at the Hanford Laboratories operated by the General Electric Company, Nucleonics Department (Atomic Energy Commission).

## QUESTIONS FOR STUDY AND DISCUSSION

1. Trace shifting of swine types throughout the years, including the factors that prompted such shifts.

2. What forces have caused Denmark, Canada, and Ireland to produce high quality bacon?

3. Describe a modern meat-type hog.

4. Rank the five leading breeds of swine based on annual registrations. Is the popularity of the leading breeds due to their meeting needs better than their rivals, or due to better promotion?

5. List the breeds of swine that are presented in Table 37–2, and give the place of origin, color, and distinguishing characteristics of each.

6. What circumstances caused the development of the *inbred breeds of swine* during the period 1930 to 1960?

7. Except for the Lacombe breed in Canada, the *inbred breeds of swine* did not survive; and the Inbred Livestock Registry became dormant. Why did this happen? Under such circumstances, can the effort and cost of their development be justified?

8. What are the *specialized breeding stock suppliers*? Most of them produce and market one or more *hybrids, crossbreds, inbreds, maternal female lines,* and/or *terminal sire lines.* Define each of these terms.

9. The trend of large market swine producers is to buy breeding stock from specialized breeding stock companies. Why this trend?

10. Is there a place (need) for miniature swine? If so, for what purpose?

11. Why have U.S. swine types shifted more rapidly and more drastically than cattle and sheep types?

12. Justify any preference that you may have for one particular breed of swine.

13. Obtain breed registry association literature and a sample copy of a magazine of your favorite breed of swine. (See Appendix Tables A–8 and A–9 for addresses.) Evaluate the soundness and value of the material that you receive.

## SELECTED REFERENCES

| Title of Publication | Author(s) | Publisher |
|---|---|---|
| *Breeds of Livestock, The* | C. W. Gay | The Macmillan Company, New York, NY, 1916 |
| *Breeds of Livestock in America* | H. W. Vaughan | R. G. Adams and Company, Columbus, OH, 1937 |
| *Modern Breeds of Livestock*, Fourth Edition | H. M. Briggs<br>D. M. Briggs | The Macmillan Company, New York, NY, 1980 |
| *Pigs from Cave to Corn Belt* | C. W. Towne<br>E. N. Wentworth | University of Oklahoma Press, Norman, OK, 1950 |
| *Pork Production* | W. W. Smith | The Macmillan Company, New York, NY, 1952 |
| *Raising Swine* | G. P. Deyoe<br>J. L. Krider | McGraw-Hill Book Company, New York, NY, 1952 |
| *Southern Hog Growing* | C. C. Scarborough | The Interstate Printers & Publishers, Inc., Danville, IL, 1958 |
| *Stockman's Handbook, The*, Sixth Edition | M. E. Ensminger | The Interstate Printers & Publishers, Inc., Danville, IL, 1983 |
| *Story of Duroc's, The* | B. R. Evans<br>G. G. Evans | United Duroc Record Association, Peoria, IL, 1946 |
| *Study of Breeds, The* | T. Shaw | Orange Judd Company, New York, NY, 1912 |
| *Swine Production* | C. E. Bundy<br>R. V. Diggins<br>V. W. Christensen | Prentice-Hall, Inc., Englewood Cliffs, NJ, 1976 |
| *Swine Production*, Fifth Edition | J. L. Krider<br>J. H. Conrad<br>W. E. Carroll | McGraw-Hill Book Company, New York, NY, 1982 |
| *Swine Production and Nutrition* | W. G. Pond<br>J. H. Maner | AVI Publishing Company, Inc., Westport, CT, 1984 |
| *Swine Science*, Fifth Edition | M. E. Ensminger<br>R. O. Parker | The Interstate Printers & Publishers, Inc., Danville, IL, 1984 |
| *Types and Breeds of Farm Animals* | C. S. Plumb | Ginn and Company, Boston, MA, 1920 |
| *World Dictionary of Breeds, Types, and Varieties of Livestock, A* | I. L. Mason | Commonwealth Agricultural Bureaux, Farnham House, Farnham Royal, Slough, Bucks, England, 1951 |

Also, breed literature pertaining to each breed may be secured by writing to the respective breed registry associations (see Table A–8, Appendix, for the name and address of each association).

# HOW BACON GOT ITS NAME

Trial of the Dunmow flitch (or side) of bacon. Custom decreed that the married couple kneel on pointed stones while swearing that they hold told the truth about their happy marital life. The traditional English ceremony of Dunmow, Essex, England, is—so the story goes—responsible for the origin of the word *bacon*. It was the custom to award a flitch of bacon to each couple who, after a year of marital life, could prove to the satisfaction of a judge and jury (composed of spinsters and bachelors) that they had been happy and had not wished themselves unwed. This homey 700-year-old tradition was started in the 12th century by a young lord who married a commoner, and having proved after a year that he was happy with her, was given the flitchof bacon by the Prior of Dunmow. (Courtesy, Picture Post Library, London, England)

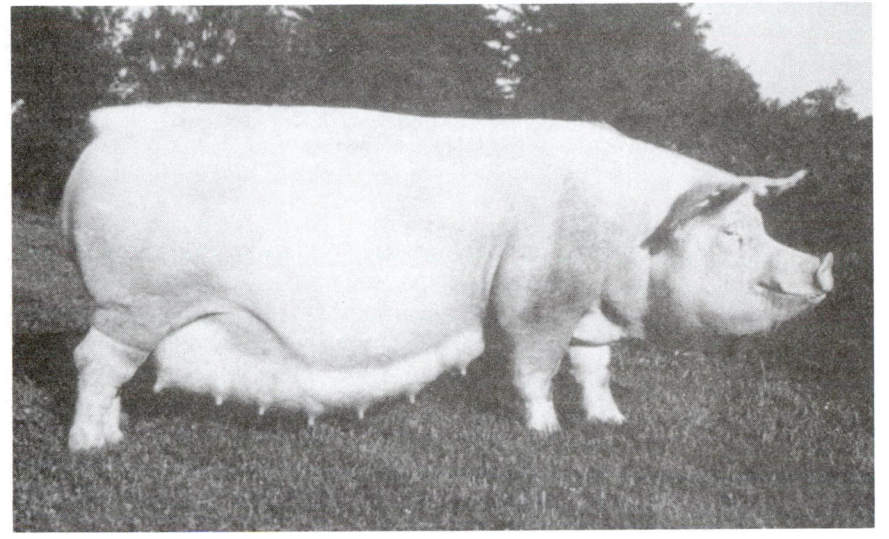

*White Diamond* gilt. The future of seedstock production will be as good as breeders make it. (Courtesy, Lieske Genetics, Henderson, MN)

# ESTABLISHING THE HERD; SELECTING AND JUDGING SWINE

## Chapter 38

The problems encountered and the principles employed in establishing the swine herd and in selecting and judging hogs are very similar to those for beef cattle and sheep. In general, however, one can establish a swine herd at a lower cost and more quickly than is possible with other classes of farm animals. Until recently hog markets were less discriminating than cattle or sheep markets, with the result that the average commercial producers gave far less attention to becoming proficient in selecting and judging swine.

## FACTORS TO CONSIDER IN ESTABLISHING THE HERD

At the outset, it should be recognized that the vast majority of the swine producers of this nation keep hogs simply because they expect them to be profitable. That hogs have usually lived up to this expectation is attested by their undisputed claim to the title of "the mortgage lifter." For maximum profit and satisfaction, in establishing the herd the individual swine producer must give consideration to the breeds and breeding, health, adaptation to the environment, and size of herd.

## BREEDS AND BREEDING

The U.S. Department of Agriculture estimated the total U.S. swine breeding stock inventory in 1987 to be 7,079,500 head. Although accurate data on breeding stock sales is not available, industry sources estimate (1) the annual boar sales from all sources to range between 250,000 and 300,000 and (2) the annual replacement gilt sales at 350,000 to 450,000. These estimates exclude boars and gilts retained by hog producers for their own use. In most commercial herds, 1 boar is selected for each 15 to 20 females.

Based on a nationwide survey, in 1989 Michigan State University reported that 57.8% of the nation's swine is sired by boars secured from traditional purebred breeders, while 27.7% is sired by boars obtained from *specialized seedstock suppliers* (suppliers of hybrids, crossbreds, inbreds, maternal female lines, and/or terminal sire lines). The latter *suppliers* generally advocate crossing selected lines to achieve heterosis, or hybrid vigor, leading to improved performance of breeding stock and improved carcasses of market animals. The survey also revealed that the trend is for more and more large commercial market hog producers to buy breeding animals from *specialized seedstock suppliers.*[1] Generally speaking, only experienced breeders should undertake the production of seedstock, whether it be purebreds or hybrids/crossbreds.

• **Selection of a purebred breed**—No one breed of hogs can be said to excel all others in all points of swine production and for all conditions. It is true, however, that particular breed characteristics may result in a certain breed being better adapted to given conditions; for example, hogs of light color are subject to sun scald in the deep South. Usually, however, there is a greater difference among individuals within the same breed than among the different breeds; this applies both to type

and efficiency of production. In the end, therefore, the selection of a particular breed is most often a matter of personal preference, and usually the breed that the individual producer likes is the one with which the greatest degree of success will be achieved. Where no definite preference exists, however, it is well to choose the breed that is most popular in the community. This consideration allows for greater latitude in the choice of foundation stock and makes the problem of securing herd boars less difficult. The producer should also give some thought to the local market demands and initial costs.

• **Selection of specialized seedstock**—The development of unique genetic lines of hybrids, crossbreds, inbreds, maternal female lines, and terminal sire lines requires a substantial investment in parent stock, facilities, and performance testing. In addition, it takes 6 to 10 years time before improved lines begin to make any return on investment. So, the development of specialized lines should be left to individuals or corporations in a very strong financial position.

The only sure way to determine the combining or "nicking" ability of lines is to cross them and performance test the offspring. Here again, the testing had best be left to *specialized seedstock suppliers.*

It follows that commercial hog producers wishing to use specialized seedstock (boars or gilts) must rely on the integrity of, and secure them from, seedstock suppliers that are in a strong financial position, that can withstand delayed returns, and that have the expertise to develop and test unique lines.

## HEALTH

Breeding animals that are in a thrifty, vigorous condition and that have been raised under a system of swine sanitation (by breeders who have exercised care in the control of disease and parasites) should have a decided preference.

A "Survey of Swine Breeding Systems of the U.S.," conducted by Michigan State University and funded by *National Hog Farmer* magazine, showed that *health* headed the list of factors considered by producers when choosing their source of purchased seedstock. One-third of the survey respondents indicated that they buy from Specific Pathogen-Free (SPF) herds—further emphasizing their concern about herd health.[2]

*Specific Pathogen-Free (SPF) pigs are pigs that are free of disease at birth.* Primary SPF herds are those originating from surgically derived stock and maintained in strict isolation. Any new blood added to the herd must also be obtained by surgical means. Primary herds are used to supply secondary multiplying herds, which, in turn supply breeding stock to commercial swine producers. The specific diseases and conditions which the SPF program is designed to eliminate and prevent are:

1. Brucellosis
2. Leptospirosis
3. Lice
4. Mange
5. Pneumonia lesions, including those caused by Mycoplasmal Pneumonia and Hemophilus Pneumonia
6. Pseudorabies
7. Swine Dysentery
8. Turbinate Citrophy and Snout Distortion.

---

[1]*National Hog Farmer*, September 15, 1989, pp. 52–58.

[2]*National Hog Farmer*, September 15, 1989, pp. 53–58.

## ADAPTATION TO THE ENVIRONMENT

The selection and breeding program should involve the propagation of pigs with the genetic capacity to respond favorably to a particular set of environmental conditions that are economically favorable in a specific setting. For example, if confinement raising on slotted floors is of importance as a means of saving labor in areas where labor represents a large proportion of the production costs, it is important that the swine breeder raise and select breeding animals that are able to thrive and produce efficiently in confinement and on slotted floors. Unfortunately, it is difficult to select swine on traits that do not lend themselves to objective measurements such as daily weight gain or feed efficiency. Nevertheless, if progress is to be made, it is necessary to simulate the commercial environment in which the animals will be produced and to discard the genes not useful therein.

Fig. 38–1. Seedstock being raised in confinement and on slotted floors, similar to most big commercial operations. Those that do not remain sound are culled. (Courtesy, Birchwood Genetics, Inc., West Manchester, OH)

## SIZE OF HERD

In 1975, a farm producing 5,000 hogs a year was considered a big operation. In 1990, it took 50,000 hogs a year to qualify as *big.*

In 1986, hog operations with 1,000 head or more accounted for 75% of the hogs marketed. By 1990, 90% of the hogs were controlled by producers raising 1,000 or more hogs a year. Hand in hand with the trend to bigness, the under-100-head hog operation was disappearing. In 1982, there were 84,620 such producers in the United States; in 1987, their numbers had dwindled to 60,303.

The trend to bigger hog operations is accompanied by a trend toward fewer hog operations. According to the Agricultural Census, there were 1.3 million hog farms in the nation in 1959. In 1987, U.S. hog operations were down to

239,000—only 19% of the 1959 figure. During this time, the total hog population was about the same.

It is predicted that operations will continue to increase in size as long as there are economies to be achieved by greater scale. But it appears that most of the economies of scale in production are in place when marketing about 10,000 hogs per year.

Hogs multiply more rapidly than any other class of farm animals. They also breed at an early age, produce twice each year, and bear litters. It does not take long, therefore, to get into the hog business.

The eventual size of the herd will be determined primarily by the following factors: (1) available capital, (2) kind and amount of labor, (3) the disease and parasite situation, (4) the probable market, and (5) comparative profits from hogs and other types of enterprises.

## SELECTION BASES

Generally, the selection of foundation hogs is made on the basis of one or more of the following considerations: (1) type or individuality, (2) pedigree, (3) show-ring winnings, or (4) production testing.

## SELECTION BASED ON TYPE OR INDIVIDUALITY

Selection based on type or individuality implies the selection of those animals that approach the ideal or standard of perfection most closely, and the culling out of those that fall short of these standards or that possess genetic abnormalities.

### TESTING FOR MEATINESS; MEASURING BACKFAT

The importance of backfat as a measure of meatiness becomes apparent when it is realized that each additional 0.1 in. of fat in a 140-lb carcass results in a 1.5% (2 lb) increase in fat trim and a decrease of about 5% (7 lb) in percent lean cuts. This is reflected in the marketplace. Also, it requires less feed to produce a pound of lean gain than a pound of fat gain.

Thickness of backfat, which has a heritability of 50%, has long been recognized as an important measure of meatiness in hogs. For many years, visual appraisal was the only method of estimating backfat thickness on live animals. However, even the most skilled are oftentimes wrong in their visual measurements.

Today, three mechanical methods are available and may be used by producers in determining backfat on live hogs; namely, the probe, the lean meter, and ultrasonics. Each of these methods requires hog restraint.

The probe and the lean meter were developed for the purpose of obtaining objective measures of backfat. Ultrasonics is used to determine loin eye area as well as backfat.

• **Probing**—Fig. 38–2 shows the probing sites on the live hog. These three locations correspond to the three locations where backfat determinations are made on the carcass.

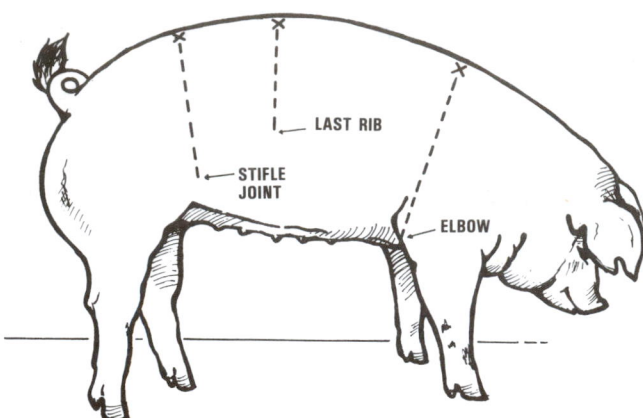

Fig. 38–2.   Probe each hog at three locations: (1) midpoint of shoulder above elbow; (2) middle of back where last rib joins the vertebrae; and (3) rump, straight above the stifle joint.

The only equipment needed for probing is a snare to restrain the hog, a sharp knife or scalpel blade, and a narrow 6-in. metal ruler with $\frac{1}{10}$-in. graduations. The steps and technique in probing are:

1. Wrap the knife or scalpel with tape about $\frac{3}{8}$ in. from the tip (in order to keep the blade from going too deep).
2. Weigh the hog.
3. Restrain the hog with a nose snare.
4. Jab the knife through the skin at a right angle to the hog's body at 1.5 to 2 in. to one side of the midline.
5. Insert the probe in the cut and slant it so that it points toward the center of the hog's body.
6. Force the probe through the fat down to the loin muscle. When the probe reaches the loin muscle, a firm resistance will be noted.
7. Push the clip on the probe down to the skin line. Remove the ruler and read the measurement.

In order to make valid comparisons and selections, backfat thickness should be adjusted to a common liveweight basis such as 200, 220, or 230 lb. This may be accomplished by calculating the average of the three readings, dividing by the hog's weight in pounds, and then multiplying the result by the selected common liveweight (200, 220 or 230 lb). Also, there are tables of average values designed for this purpose.

Where pinpoint accuracy is not considered essential, measuring backfat at a single probe site is suggested. The recommended site where one probe only is taken is the seventh rib, located at a distance approximately four fingers wide behind the shoulder-probing location and 1 in. off the midline of the back.

• **Lean meter**—This tool is more sophisticated than the probe, but also more expensive. The method is based on the difference in electrical conductivity of fat and muscle; fat is a relatively poor conductor, whereas muscle and blood are good conductors. The locations for fat determinations with the lean meter are the same as for the probe (see Fig. 38–2); and, like the probe, fat depth is read in tenths of inches.

• **Ultrasonic**—This electronic equipment employs the *pulse echo* technique. Basically, this is the generation of very short bursts of high-frequency (nondestructive, inaudible) sound into the animal, detecting the reflection of the pulses, and measuring the elapsed time between introduction of the sound pulse into the animal and the return of the reflected pulse. When the machine is properly calibrated, the fat depth can be read directly from the scale.

The ultrasonic backfat reading is made at the last rib, usually at the midline of the back. However, the reading may also be made 2 in. from the midline.

The ultrasonic may also be used to estimate loin eye area at the last rib.

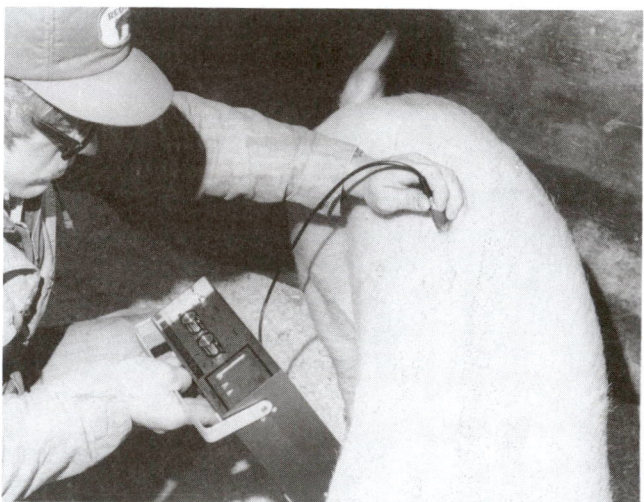

Fig. 38–3.   Backfat testing, using ultrasonic equipment. (Courtesy, International Livestock Services, Corp., Ames, IA)

Any one of these three mechanical methods for determining backfat thickness of the live animal is a valuable adjunct to visual appraisal and scales in the selection of meat-type breeding animals and the production of higher quality pork carcasses. Hence, backfat measurements should be used in the selection program of both purebred and commercial producers.

## TESTING FOR PORCINE STRESS SYNDROME (PSS)

The Porcine Stress Syndrome (PSS) is a nonpathological disorder that is of major concern to swine producers. When present, the disorder is usually associated with heavily muscled animals and results in sudden unexplained death losses. Fortunately, two different reliable tests may be made for the presence of PSS for the purpose of evaluating animals for the breeding herd. These are:

1. **The creatine phosphokinase (CPK) test.** This involves catching a small drop of blood on a special card and sending the card to a chemical laboratory to be analyzed for the activity of creative phosphokinase (CPK), a serum enzyme that is abnormally high in PSS swine. The producer may obtain blood for this test by making a small cut on the pig's ear.

2. **The halothane test.** This involves anesthetizing the animals with halothane, to which PSS swine respond by showing extreme muscle rigidity within five minutes. This test provides immediate results, but the equipment involved is expensive and must be used under the direction of a trained technician. The halothane test is generally regarded safe for only young pigs, since older PSS animals are likely to die after halothane exposure.

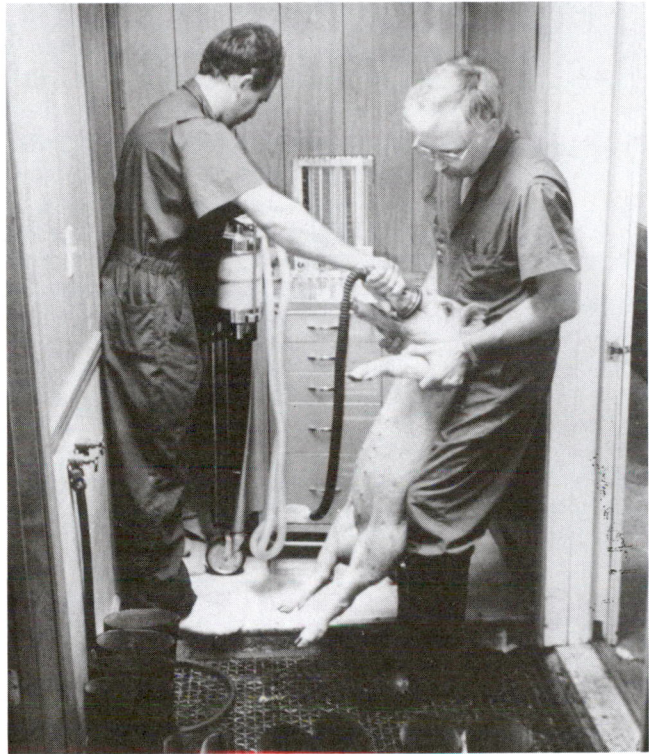

Fig. 38–4. The halothane test, showing a pig being anesthetized with halothane. (Courtesy, Genetic Improvement Services, Burlington, IN)

Both of these tests give reliable results for animals that are severely prone to the PSS condition. However, animals that are carriers of the trait and are not highly stress susceptible themselves may not react to either test.

## AVOIDING GENETIC DEFECTS

Anatomical abnormalities, or defects, occur in at least 1% of newborn pigs. These defects may be caused by either genetic or environmental factors. Among the environmental factors that may cause abnormalities are viral infections, dietary deficiencies, and ingestion of certain drugs, chemicals, and pesticides during pregnancy.

It is important to determine as nearly as possible if a defect is genetic or environmental. If it is due to an autosomal recessive, neither the parents nor littermates of affected individuals should be kept for breeding. If it is a sex-linked recessive, only the dam need be culled. Common inherited (genetic) defects are: atresia ani, cryptorchidism, hermaphrodites, porcine stress syndrome (PSS), scrotal hernia, and umbilical hernia.

## SELECTION BASED ON PEDIGREE

In the selection of breeding animals, the pedigree is a record of the individual's heredity or inheritance. If the ancestry is good, it lends confidence in projecting how well young animals may breed. It is to be emphasized, however, that mere names and registration numbers are meaningless. A pedigree may be considered desirable only when the ancestors close up in the lineage—the parents and grandparents—were superior individuals and outstanding producers. Too often, purebred hog breeders are prone to play up one or two outstanding animals back in the third or fourth generation. If pedigree selection is to be of any help, one must be familiar with the individual animals listed therein.

The boar should always be of known ancestry. This alone is not enough, for he should also be a good representative of the breed selected; and his pedigree should contain an impressive list of good animals. Likewise, it is important that the sows be of good ancestry, regardless of whether they are purebreds, grades, or crossbreds. Such ancestry and breeding gives more assurance of the production of high-quality pigs that are uniform and true to type.

• **Complementary**—In a crossbreeding program, breeds that complement each other should be selected, thereby maximizing the desirable traits and minimizing the undesirable traits. Thus, it is a matching of breeds so that they compensate each other. Providing seedstock that will permit the commercial producers to maximize heterosis and utilize the superior characteristics of each breed or strain through a systematic crossing program is a primary goal of seedstock producers.

## SELECTION BASED ON SHOW-RING WINNINGS

Swine producers have long looked favorably upon using show-ring winnings as a basis of selection. Purebred breeders have been quick to recognize the appeal and to extol their champions through advertising. In most instances, the selection of foundation or replacement hogs on the basis of show-ring winnings and standards has been for the good. On many occasions, however, purebred and commercial breeders alike have come to regret selections based on show-ring winnings. This was especially true during the eras when the chuffy or rangy types were sweeping shows from one end of the country to the other. This would indicate that some scrutiny should be given to the type of animals winning in the show, especially to ascertain whether such animals are of a type that are efficient from the standpoint of the producer, and whether, over a period of years, they will command a premium on a discriminating market.

Perhaps the principal value of selections based on show-ring winnings lies in the fact that shows direct the attention of the amateur to those types and strains of hogs that at the moment are meeting with the approval of the better breeders and judges.

# SELECTION BASED ON PRODUCTION TESTING

Fig. 38–5. Production records being used in selecting replacement gilts. (Courtesy, American Yorkshire Club, West Lafayette, IN)

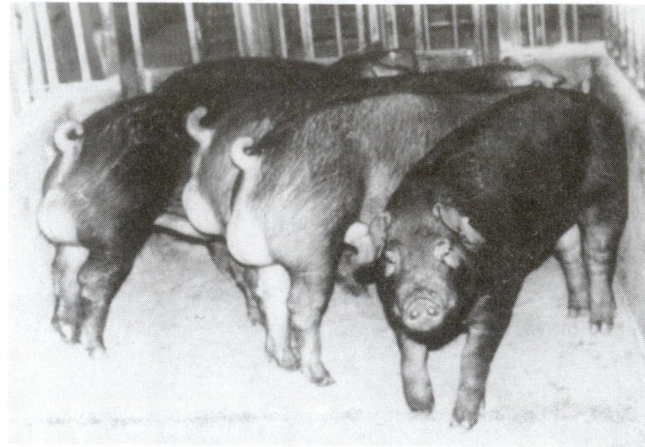

Fig. 38–6. Aggressive Duroc boars of breeding age. (Courtesy, United Duroc Swine Registry, Peoria, IL)

No criterion that can be used in selecting an animal is as accurate or important as past performance. It is recommended, therefore, that one purchase tried sows and a proved herd boar when such animals of the right kind can be secured at reasonable prices. Unfortunately, breeding animals of known merit are not usually available at a figure that a beginner can afford to pay. Sometimes, however, established breeders make the error of sacrificing brood sows of proved performance in favor of younger sows or find it necessary to sell proved boars because they can no longer be used in the breeding program. Animals of this kind constitute one of the soundest purchases that a young breeder can make.

Several of the purebred swine registry associations now have (1) production registry, and/or (2) meat certification programs. Also, many traditional purebred swine breeders, and most of the specialized breeding stock suppliers, production test. Although these programs are not perfect, they do represent a progressive step in the right direction. Without doubt, both purebred and commercial swine producers of the future will make increasing use of such records as a basis for selection.

Several states and local associations operate central swine testing stations, to assist producers in evaluating their breeding stock and to improve the performance and quality of market hogs. Most of these stations follow procedures outlined by the National Swine Improvement Federation (NSIF). (Production records in swine are discussed in Chapter 39.)

## BOAR SELECTION

Boar selection is very important, because the boar becomes the parent of more offspring than the female. In most commercial herds, 1 boar is selected for each 15 to 20 females. The following traits should be considered when selecting boars:

1. **Behavioral traits.** This includes docility, temperament, and the complex of traits associated with reproductive potential (*sexual*: development, maturity, aggressiveness).

2. **Sow productivity traits.** This includes reproductive ability, litter size, milking ability, and mothering ability. Some of these traits are not expressed in boars e.g., milk production and mothering ability, but they are expressed in the female offspring of the boar. Number of pigs farrowed and weaned, and individual pig and litter birth weights are the most common measures. Litter weight at 21 days/sow exposed to boars is probably the best single measure of sow productivity. When selecting boars for these traits, use the records of sire and dam, litter records, records of other relatives, and any records available on the boar being selected.

3. **Feedlot performance traits.** This includes growth rate—measured as gain/day from weaning to market, gain per day from 60 to 230 lb, or age at 230 lb; and feed conversion (lb feed/lb gain). When selecting for these traits, place more emphasis on the boar's own record and less emphasis on records of relatives.

4. **Carcass merit.** The boar's carcass merit can be evaluated by taking measurements of backfat thickness, loineye area, and percent muscle. Of these measurements, backfat is the single most important and best measure of leanness.

5. **Soundness and conformation type.** Traits associated with soundness include underline (teat spacing, number, and presentation), physical soundness of feet and legs, bone size and strength, genetic abnormalities, and mating ability (tagged penis, short penis, limp penis).

The conformation traits include those used in visually evaluating boars such as body length, depth, height, and skeleton size (ruggedness, frame, and body capacity), muscle size and shape, and boar masculinity characteristics, and testicular development.

6. **Age of boar.** Select and purchase boars at 6 to 7 months of age for use beginning at a minimum of 8 months of age.

• **Production performance records**—Performance records of boars or their littermates are important in boar selection. When selecting boars on the basis of performance records, consider (if possible) only those in the top 50% of the herd or test group, only those from litters of 10 or more pigs farrowed and 8 or more weaned, and only those that were raised under production conditions similar to those that exist on your farm—e.g., in confinement on concrete or slotted floor, or on pasture.

• **Health of boar**—The health history of the seller's herd and of the potential herd boar is one of the most important factors to consider. So, boars should be bought from reputable breeders who will give written health management records for the boars.

• **Central test stations**—Central test stations provide an excellent source of superior boars, with the ultimate in test records. The leading breeders put their best boars in these tests. These tests help in locating genetically superior boars since all boars are tested under a uniform environment. Since only limited numbers of test station boars are available, producers may look for boars in herds doing well in test stations.

Fig. 38–7. Swine testing station at the University of Missouri. (Courtesy, University of Missouri, Columbia)

• **Production records are important**—Daily gain, feed efficiency, and backfat are three major economic traits that can be easily measured on boars, regardless of whether they are tested in a central test station or on a breeder's farm.

• **Suggested selection standard for boars**—If possible, always select boars from the top 50% of the best group of both test station and on-farm sources. However, boars meeting the following standards should receive serious consideration as potential herd sires:

| Trait | Standard |
| --- | --- |
| Litter Size | 10 or more farrowed, and 8 or more weaned. |
| Underline | 12 or more fully developed, well-spaced teats. |
| Feet and Legs | Medium to large bone, wide stance both front and rear, free in movement, good cushion to both front and rear feet, equal sized toes. |
| Age at 230 lb | 155 days or less. |
| Feed/cwt gain, boar basis (60–230 lb) | 275 lb/cwt gain or less. |
| Daily gain (60–230 lb) | 2.00 lb/day or higher. |
| Backfat probe (adjusted 230 lb) | 1.0 in. or less. |

## FEMALE SELECTION

Fig. 38–8. The goal in female selection is the birth and raising of large litters of fast growing meat-type pigs. (Courtesy, Chester White Assn., Peoria, IL)

The productivity of the sow herd is the foundation of commercial pork production. The sow herd also contributes half of the genetic makeup of growing-finishing hogs. Together, these factors indicate the importance of the careful selection of replacement gilts and wise decisions on the culling of the sow herd. The following guidelines should be considered when making gilt selections and culling sows:

1. **Soundness.** Soundness in replacement females means being free from flaws or defects which would interfere with normal reproductive and maternal functions. Three areas of particular concern are reproductive soundness, mammary soundness, and skeletal soundness.

   a. **Reproductive soundness.** Replacement gilts should exhibit normal reproductive development, both anatomically and behaviorally. The external genitalia should be normally developed.

   b. **Mammary soundness.** Replacement gilts should have a sufficient number of functional teats to nurse a large litter of pigs—at least six functional teats on each side. Gilts with inverted or scarred nipples should not be saved. New

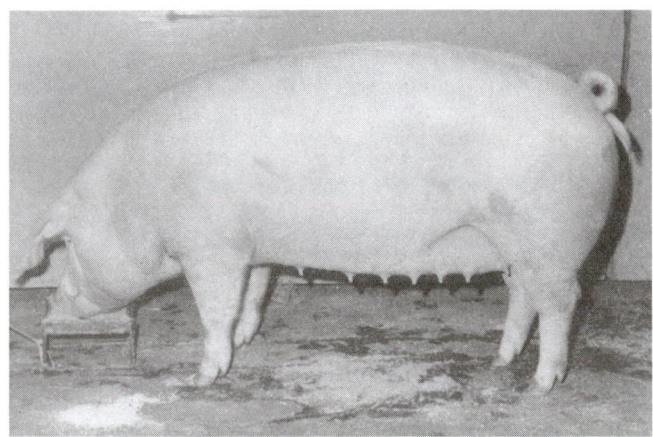

Fig. 38–9. Landrace gilt showing excellent mammary development, with seven functional teats on each side. (Courtesy, American Landrace Assn., West Lafayette, IN)

concrete, rough floors, and corrosive chemical compounds on the floors of farrowing houses can cause abrasion to the underside of gilts which result in nonfunctional teats. As the gilt approaches puberty, her underline should become more prominent, indicating normal development.

c. **Skeletal soundness.** Gilts with feet and leg problems which interfere with normal breeding, farrowing, and nursing functions should not be saved. Sows that are unable to get up and down in farrowing crates should be culled.

2. **Select the fastest growing, leanest gilts, which are sound and from large litters.** These are the kind to save for replacement purposes. Sows which fail to rebreed, or which had small litters, failed to milk, or had problems farrowing should be culled.

This selection and culling program requires identification of potential replacement gilts at birth. Also, it necessitates that gilts be weighed and backfat probed as they approach market weight. At that time, gilts should also be appraised for indications of normal reproductive development, functional appearance of the underline, and skeletal soundness. When gilts reach 180–200 lb, they should be evaluated, selected, and placed on a restricted diet.

Sufficient sow identification and farrowing house records should be kept in order to cull the right sow.

## SEEDSTOCK SELECTION

*Seedstock producers are those that produce specialized breeding stock (hybrids, crossbreds, inbreds, maternal female lines, and terminal sire lines) for commercial hog producers.* The primary goals of *seedstock producers* should include the following:

1. Supplying the genetic material for the production of health, fast growing, efficient, high quality, lean pigs.

2. Providing seedstock with the structural soundness necessary to breed and perform under a wide range of environmental conditions.

3. Providing animals capable of conceiving and raising large litters of uniform, thrifty pigs.

4. Maintaining sufficient volume to insure year-round availability of stock.

5. Providing seedstock that will permit the commercial producer to maximize heterosis and utilize the superior characteristics of each breed or strain, through systematic use of them, in a crossing program.

• **Traits to measure**—Seedstock producers should measure the following traits:

1. **Sow productivity traits.** Litter size, number reared per litter, total litter weight at 21 days, and litters per sow per year are lowly heritable traits. But these traits are of great economic importance, and seedstock producers can improve them in seedstock populations by (a) culling families that are extremely low in performance, (b) selecting sows with superior records, and (c) keeping the rate of inbreeding low.

2. **Production traits.** Growth rate and feed efficiency are economically important in most swine enterprises, and their heritabilities are of sufficient magnitude to respond to selec-

tion. Every seedstock producer should have a scale with which to measure weight for age. Growth rate may be expressed as (a) days required to reach a given weight, (b) average daily gain, or (c) weight at a fixed age.

Feed costs generally account for 65 to 75% of the total production costs of the commercial producer. Either of the following methods may be used by the seedstock producer to improve feed efficiency:

**Method 1:** Testing individual pigs for gain and backfat without feed records.

**Method 2:** Testing individual pigs or littermate pigs with feed efficiency records.

Research data show that Method 1 is about 70% as good as Method 2 on three littermate boars fed together. Of course, Method 1 is far less costly than Method 2.

3. **Carcass traits.** Ham-loin and lean cut percentages of pork carcasses are highly related to carcass value. Fortunately, they can be reliably predicted by backfat measured on live animals by either a metal ruler or ultrasonics.

4. **Pale, soft, and exudate (PSE) pork carcasses and porcine stress syndrome (PSS).** These problems can be severe in herds selected for extremes in muscling and/or those not utilizing objective tests to select against the malady. Most PSS pigs produce carcasses that are PSE, although other causes of poor quality muscle are known. Preventing PSS pigs from becoming herd replacements will lower the incidence of affected offspring. Either of the following tests for PSS may be used: (a) the creatine phosphokinase (CPK) test, or (b) the halothane test.

5. **Structural soundness.** There is great need for more sound, durable animals capable of withstanding the rigors of confinement rearing and breeding. Some seedstock producers raise their breeding animals in confinement, similar to the manner used by most commercial producers, then cull the unsound ones. Soundness is an easy trait for seedstock producers to improve through visual selection and culling.

6. **Visual traits.** Some traits, such as structural soundness, length of body, underline, general conformation, and presence of physical defects, can only be evaluated visually on the live animal. Thus, seedstock producers should use both performance records and "eyeball appraisal" in selecting breeding animals.

• **Basis for Seedstock Selection**—The primary objective of seedstock producers is to produce what commercial producers need; selling to other seedstock producers is secondary.

• **Performance Test Under Comparable Environment**—It is important that seedstock producers obtain performance records under a comparable environment. This calls for all test animals on a given farm being treated alike, without any preferential treatment of some animals.

• **Exerting Selection Pressure**—Selection pressure is largely a function of percent animals kept for breeding purposes. Only boars in the upper 10% of a breed in performance can be expected to be real improvers.

• **Rapid Generation Interval**—Turning over the herd as rapidly as possible while keeping the very best animals for replacements constitutes an optimum program for the seedstock producer. This rule applies to both boars and sows.

# PROGRESSIVE SEEDSTOCK PRODUCERS
## HAVE EVOLVED WITH UNIQUE GENETIC MATERIAL
## FOR COMMERCIAL SWINE PRODUCERS

Fig. 38-10. *Blue Roan* gilt, produced by mating Large White (English Yorkshire) × American Yorkshire, then mating to a Maternal Hampshire. (Courtesy, Lone Willow Genetics, Roanoke, IL)

Fig. 38-11. The X Y L line resulting from a Yorkshire × Lacombe cross. (Courtesy, McLean County Hog Service, Inc., LeRoy, IL)

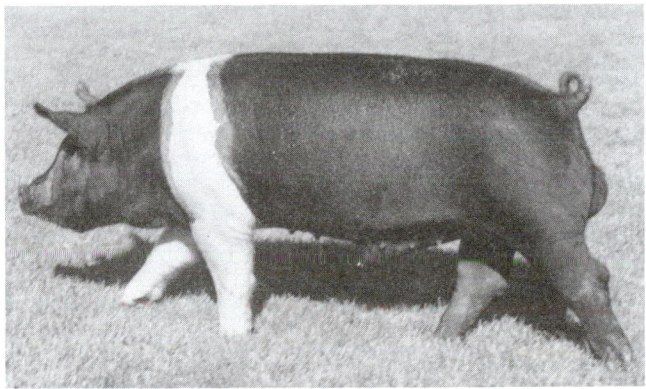

Fig. 38-12. *Superstuff*, a Hampshire × Duroc boar, for terminal planned mating to produce market hogs. (Courtesy, Lone Willow Farm, Roanoke, IL)

Fig. 38-13. Yorkshire boar, bred and used extensively by Compart's Boar Store. (Courtesy, Compart's Boar Store, Nicollet, MN)

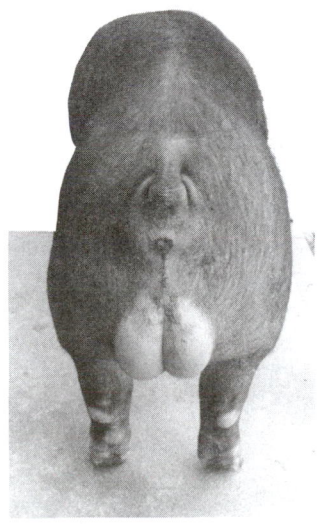

Fig. 38-14. Well-muscled (high percent lean) Duroc boar, bred by Waldo Farms, Inc., DeWitt, NE

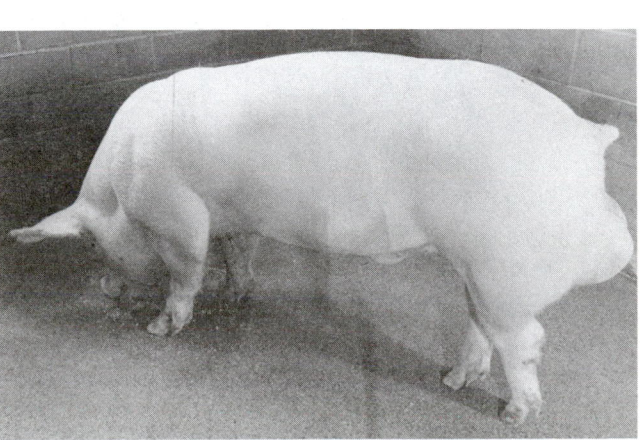

Fig. 38-15. Large White (English Yorkshire) boar. (Courtesy, Lieske Genetics, Henderson, MN)

## HERD IMPROVEMENT THROUGH SELECTION

Constructive breeders are those who effect improvement through breeding and selection, and this applies to both purebred and commercial producers. Such selection must be based on production factors of economic importance and market price determined by carcass quality on a discriminating market.

## JUDGING SWINE

Since the general requisites in judging, which are similar for all classes of livestock, are fully covered in Chapter 13, repetition at this point is unnecessary.

As previously indicated, until recently the small price spread in market classes and grades of swine offered little incentive to commercial swine producers to become proficient in judging. It is to the credit of purebred swine breeders, however, that they have been very progressive in this respect. The swine-type conferences sponsored by the various breed associations have made a unique contribution. Through bolstering live-animal work with a liberal amount of carcass data, these contests have soundly set fashions for both the producer and the packer.

## PARTS OF A HOG

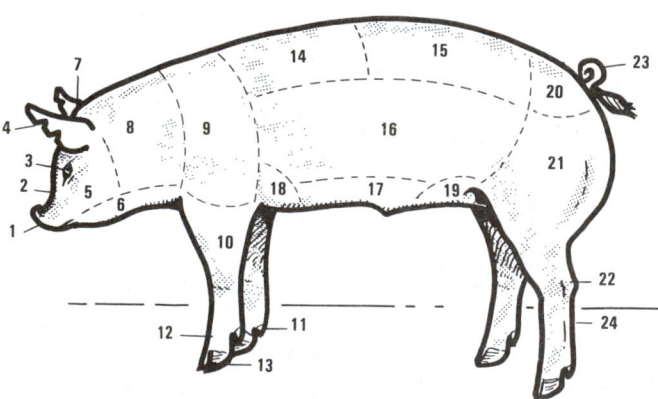

Fig. 38–16. Parts of a hog. The first step in preparation for judging hogs consists in mastering the language that describes and locates the different parts of the animal. (Drawing by Prof. R. F. Johnson)

| | | |
|---|---|---|
| 1. Snout | 9. Shoulder | 17. Belly |
| 2. Face | 10. Foreleg | 18. Foreflank |
| 3. Eye | 11. Dew claw | 19. Rear flank |
| 4. Ear | 12. Pastern | 20. Rump |
| 5. Cheek | 13. Toes | 21. Ham |
| 6. Jowl | 14. Back | 22. Hock |
| 7. Poll | 15. Loin | 23. Tail |
| 8. Neck | 16. Side | 24. Rear leg |

Most successful hog producers know the parts of a hog. In addition, they are aware of the possible merits of maximum development of these different parts from the standpoint of economical production and market value. Fig. 38–16 shows the parts of a hog.

## IDEAL TYPE AND CONFORMATION

Fig. 38–17 shows the ideal meat-type hog vs some of the common faults. Since no animal is perfect, the proficient swine judge must be able to recognize, weigh, and evaluate both the good points and the common faults. In addition, the judge must be able to arrive at a decision as to the degree to which the given points are good or bad.

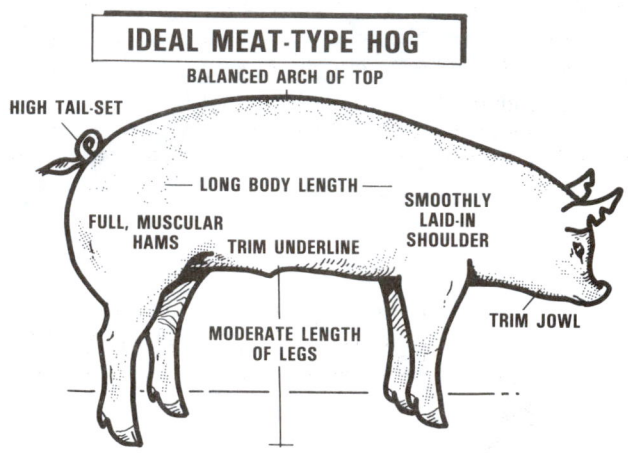

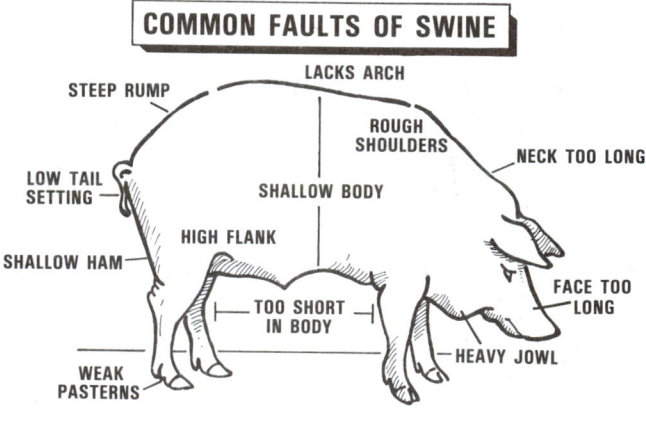

Fig. 38–17. Ideal meat type vs common faults. The successful hog judge must know what to look for and be able to recognize and appraise both the good points and the common faults. (Drawing by Prof. R. F. Johnson)

The next requisite in judging or selection is to have clearly in mind a standard or ideal. Presumably, this ideal should be based on a combination of (1) the efficient performance of the animal from the standpoint of the producer, and (2) the desirable carcass characteristics of market animals as determined by the consumer.

The most approved meat-type breeding animals combine size, smoothness, and quality, and the offspring possess the ability to finish during the growing period without producing an excessive amount of lard. The head and neck should be trim and neat; the back well arched and of ample width; the sides long, deep, and smooth; and the hams well developed and deep. The legs should be of medium length, straight, true, and squarely set; the pasterns should be short and strong; and the bone should be ample and show plenty of quality. With this splendid meat type, there should be style, balance, and symmetry and an abundance of quality and smoothness.

The most approved bacon-type breeding hogs differ from meat-type animals chiefly in that greater emphasis is placed on length of side and the maximum development of the primal cuts with the minimum of lard. Also, bacon-type hogs generally have less width over the back and have a squarer type ham, and show more trimness throughout.

With both meat- and bacon-type animals, the brood sows should show great femininity and breediness; and the udder should be well developed, carrying from 10 to 12 teats. The herd boar should show great masculinity as indicated by strength and character in the head, a somewhat crested neck, well-developed but smooth shoulders, a general ruggedness throughout, and an energetic disposition. The reproductive organs of the boar should be clearly visible and well developed. A boar with one testicle should never be used.

## METHOD OF EXAMINING

Since a pig will neither stand still nor remain in the same vicinity long, it is not possible to arrive at a set procedure for examining swine. In this respect, the judging of hogs is made more difficult than the judging of other classes of livestock. Where feasible, however, the steps for examining as illustrated in Fig. 38–18 are very satisfactory, and perhaps as good as any.

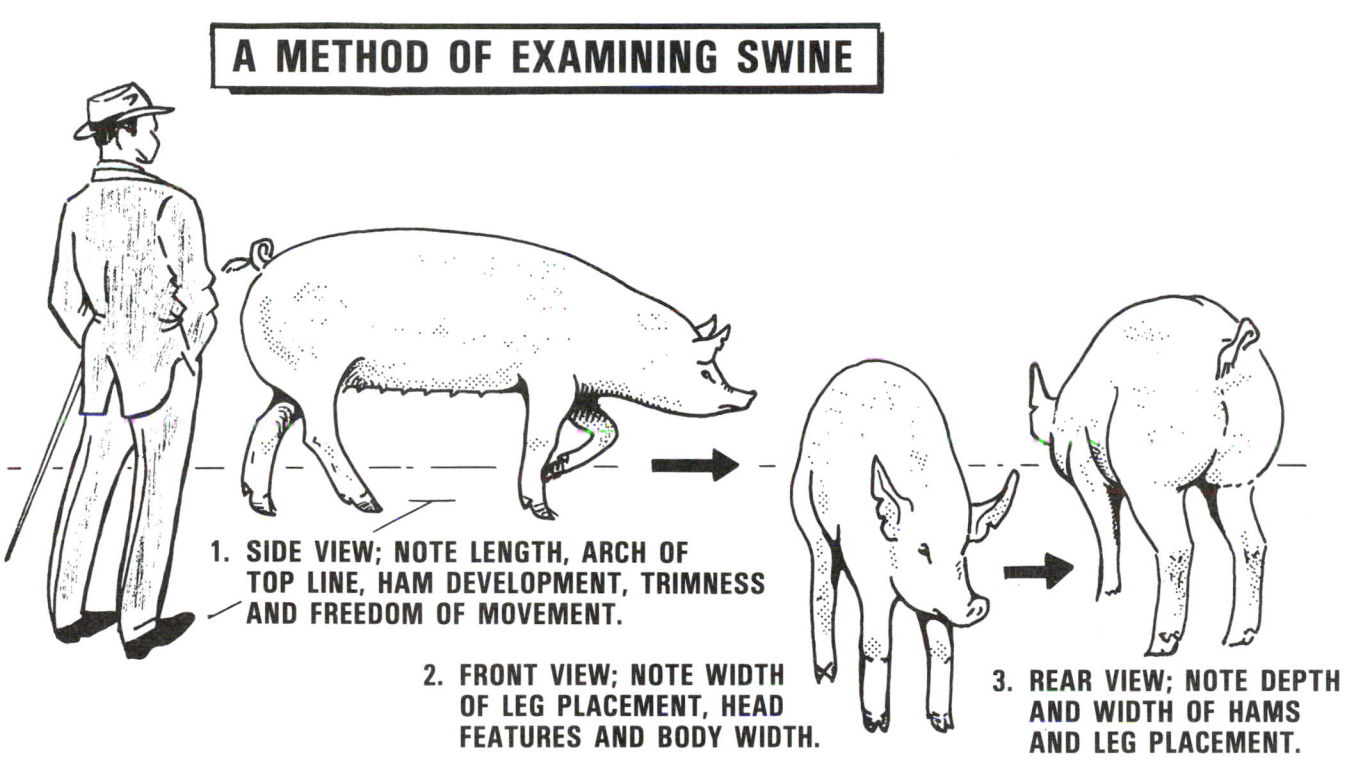

**A METHOD OF EXAMINING SWINE**

1. SIDE VIEW; NOTE LENGTH, ARCH OF TOP LINE, HAM DEVELOPMENT, TRIMNESS AND FREEDOM OF MOVEMENT.

2. FRONT VIEW; NOTE WIDTH OF LEG PLACEMENT, HEAD FEATURES AND BODY WIDTH.

3. REAR VIEW; NOTE DEPTH AND WIDTH OF HAMS AND LEG PLACEMENT.

Fig. 38–18. A good procedure for examining a hog and some of the things for which to look. (Drawing by Prof. R. F. Johnson)

## SCORECARD JUDGING

A scorecard is a listing of the different parts of a hog, with a numerical value assigned to each part according to its relative importance. It is a standard of excellence. The use of the scorecard involves studying each part, then assigning a score to it.

Figs. 38–19 and 38–20 are new and modern scorecards for swine; Fig. 38–19 is for scoring breeding swine, while Fig. 38–20 is for market hogs.

As noted, the scorecard gives each of several traits a value. These points total 100 for a perfect score. A scorecard is a valuable judging aid. It systemizes judging and avoids any part of the animal being overlooked.

# BREEDING SWINE SCORECARD

| | Perfect Score | No. 1 | No. 2 | No. 3 | No. 4 | Etc. |
|---|---|---|---|---|---|---|
| | | ANIMAL | | | | |

**GENERAL APPEARANCE:** .................................................................................. 25

  **Type**—heavy muscled, lean, trim, firm, smooth, long bodied, ham and rump should be wider than rest of body, moderately deep forerib, uniformly arched top, well balanced and stylish with a high degree of development in the valuable region of the hindquarters, same standards as for the market barrow. .........(12)

  **Size**—ample size, scale, and ruggedness for age. ...........................................(13)

**CONFORMATION:** ..................................................................................... 45

  **Hindquarters**—(26 points)

  Hind legs—set wide apart, out on the corners giving an indication of abundant muscling. .................(3)
  Ham—wide, deep, long, full, firm, meaty, deep in the seam. ...........................................(9)
  Rump—long, wide, uniformly turned, high tail setting, meaty. .........................................(6)
  Loin—muscular turn, long, lean. .....................................................................(8)

  **Forequarters**—(14 points)

  Back—muscular turn, long, lean, uniformly arched, full spring of rib. .................................(7)
  Shoulders—smooth, muscular, free from fatty creases and wrinkles, no evidence of fat deposits at the elbow. ...........................................................................................(4.5)
  Head—clean-cut, trim, firm jowl. ..................................................................(2)
  Neck—short. ......................................................................................(0.5)

  **Middle**—(5 points)

  Deep, roomy middle but not loose or wasty, belly trim and firm, deep ribbed.

**BREEDING QUALITIES:** ............................................................................... 30

  **Underpinning**—(13 points)

  Legs—straight as viewed from side, front, and rear, squarely set under corners of body, strong, straight toes. ...........................................................................................(8)
  Pasterns—strong, short, straight, but not buckled over. ..............................................(1)
  Action—free, easy, unhindered walk, not stiff or "peggy." ............................................(4)

  **Mammary system**—(12 points)

  Six (6) sound nipples on a side, nipples prominent and evenly spaced; no evidence of inverted or blind teats.

  **Breed character**—(5 points)

  Head—varies with the breed, wide between the eyes, sows feminine and boars masculine. ...............(3)
  Ears—relatively small and refined, not large and coarse thereby hindering vision; breeds having erect ears should show no tendency for drooping; breeds having drooping ears should show no tendency for being erect. ...........................................................................................(2)

  TOTAL ..........................................................................................100

Fig. 38-19. Breeding swine scorecard. (Adapted by the author from *Animal Science and Industry Laboratory Manual,* by Able, Bill V., *et al.*, Kansas State University, Kendall/Hunt Publishing Company, Dubuque, IA)

# MARKET BARROW SCORECARD

| | Perfect Score | ANIMAL | | | | |
|---|---|---|---|---|---|---|
| | | No. 1 | No. 2 | No. 3 | No. 4 | Etc. |
| **CONFORMATION:** . . . . . . . . . . . . . . . . . . . . . . . . . . . . . . . . . . . . . . . . . . . . . . . . . . . . . . . . . . . . . | 60 | | | | | |
| **General appearance**—(12 points) | | | | | | |
| Heavy muscled, lean, trim, firm, long bodied, ham and rump should be wider than the rest of the body, uniformly arched top, well balanced and stylish with a high degree of development in the valuable region of the ham and loin, adequate size for age. | | | | | | |
| **Hindquarters**—(28 points) | | | | | | |
| Hind legs—set wide apart, out on the corners, giving an indication of abundant inner muscling, straight. . .(2) | | | | | | |
| Ham—wide, deep, long, full, firm, meaty, deep in the seam. . . . . . . . . . . . . . . . . . . . . . . . . . . . . . .(10) | | | | | | |
| Rump—long, wide, uniformly turned, high tail setting, meaty. . . . . . . . . . . . . . . . . . . . . . . . . . . . .(7) | | | | | | |
| Loin—muscular turn, lean, long, uniformly arched. . . . . . . . . . . . . . . . . . . . . . . . . . . . . . . . . . . .(9) | | | | | | |
| **Forequarters**—(14 points) | | | | | | |
| Back—muscular turn, long, lean, uniformly arched, uniform width, full spring of rib. . . . . . . . . . . . . .(7) | | | | | | |
| Shoulders—smooth, muscular. . . . . . . . . . . . . . . . . . . . . . . . . . . . . . . . . . . . . . . . . . . . . . . . .(5) | | | | | | |
| Neck—short. . . . . . . . . . . . . . . . . . . . . . . . . . . . . . . . . . . . . . . . . . . . . . . . . . . . . . . . . . . .(0.5) | | | | | | |
| Head—clean-cut, refined, trim firm jowl. . . . . . . . . . . . . . . . . . . . . . . . . . . . . . . . . . . . . . . . . .(1.5) | | | | | | |
| **Middle**—(6 points) | | | | | | |
| Side—long, moderately deep forerib, smooth, free from wrinkles. . . . . . . . . . . . . . . . . . . . . . . . . .(3) | | | | | | |
| Underline (belly)—trim, firm, no evidence of looseness or wastiness. . . . . . . . . . . . . . . . . . . . . . .(3) | | | | | | |
| **FINISH:** . . . . . . . . . . . . . . . . . . . . . . . . . . . . . . . . . . . . . . . . . . . . . . . . . . . . . . . . . . . . . . . . . . | 36 | | | | | |
| Ham—firm and free from wrinkles at the base, firm in the crotch. . . . . . . . . . . . . . . . . . . . . . . . .(5) | | | | | | |
| Rump—no evidence of a counter sunk tail setting. . . . . . . . . . . . . . . . . . . . . . . . . . . . . . . . . . . .(5) | | | | | | |
| Back and loin—lean, meaty turn, evidence of abundant muscling accompanied by minimum amount of backfat. . . . . . . . . . . . . . . . . . . . . . . . . . . . . . . . . . . . . . . . . . . . . . . . . . . . . . . . . . . . . . . .(8) | | | | | | |
| Shoulders—firm and smooth, free from fatty creases and wrinkles, no evidence of fat deposit at the elbow. . .(8) | | | | | | |
| Jowl—trim and firm. . . . . . . . . . . . . . . . . . . . . . . . . . . . . . . . . . . . . . . . . . . . . . . . . . . . . . .(5) | | | | | | |
| Belly—trim and firm. . . . . . . . . . . . . . . . . . . . . . . . . . . . . . . . . . . . . . . . . . . . . . . . . . . . . . . .(5) | | | | | | |
| **QUALITY:** . . . . . . . . . . . . . . . . . . . . . . . . . . . . . . . . . . . . . . . . . . . . . . . . . . . . . . . . . . . . . . . . . | 4 | | | | | |
| Smooth throughout, not creased or wrinkled. . . . . . . . . . . . . . . . . . . . . . . . . . . . . . . . . . . . . . .(2) | | | | | | |
| Bone—ample substance of bone, definitely not fine but not overly coarse either. . . . . . . . . . . . . . . .(2) | | | | | | |
| TOTAL . . . . . . . . . . . . . . . . . . . . . . . . . . . . . . . . . . . . . . . . . | 100 | | | | | |

Fig. 38–20.  Market barrow scorecard. (Adapted by the author from *Animal Science and Industry Laboratory Manual*, by Able, Bill V., *et al.*, Kansas State University, Kendall/Hunt Publishing Company, Dubuque, IA)

# QUESTIONS FOR STUDY AND DISCUSSION

1. In establishing the herd, and in selecting and judging, what primary differences exist in swine as compared to beef cattle and sheep?

2. In 1989, Michigan State University reported that 57.8% of the nation's swine is sired by boars secured from traditional *purebred breeders*, while 27.7% is sired by boars obtained from *specialized seedstock suppliers*. What are the primary differences in the breeding of the swine from the two different sources?

3. Further, the Michigan State University survey revealed that the trend is for large commercial market hog producers to buy breeding animals from *specialized seedstock suppliers*. What is the reason for this trend?

4. What factors should be considered when selecting a purebred breed of swine? Justify any breed preference that you may have.

5. Discuss the capital requirements and delayed returns of *specialized seedstock production* in comparison with *commercial production*.

6. *Health* heads the list of factors considered by producers when choosing their source of purchased seedstock. Why is swine health so important?

7. What are Specific Pathogen-Free (SPF) pigs? Name the specific diseases and conditions which the SPF program is designed to eliminate and prevent.

8. How may a swine breeder raise and select animals that are adapted to confinement and slotted floors?

9. What forces have caused big operations to get bigger?

10. Discuss the importance of backfat as a measure of meatiness.

11. Name and describe each of the three methods available for determining backfat of live hogs.

12. What is the porcine stress syndrome (PSS)? Name and describe the two tests for detecting the presence of PSS.

13. List five genetic defects that should be avoided when selecting animals for breeding purposes.

14. Discuss swine selection based on pedigree. What is complementary?

15. What is the principal value of swine selection based on show-ring winnings?

16. Discuss swine selection based on production testing.

17. List five traits that should be considered when selecting boars; and describe and discuss the importance of each trait.

18. When selecting herd sires, what minimum standards should be established for each of the following traits: (a) age at 230 lb, (b) feed/cwt gain, (c) daily gain, and (d) backfat probe adjusted to 230 lb?

19. List and discuss the three areas of particular concern relative to soundness when selecting replacement gilts or culling sows.

20. When selecting gilts, for what traits should selection be made?

21. List the primary goals of seedstock producers.

22. List the traits which seedstock producers should measure.

23. Why should seedstock producers select only boars in the upper 10% of a herd in performance; and why should they turn over the herd as rapidly as possible?

24. Why is it important to know the parts of a hog? Describe the location of the following parts: jowl, dew claw, loin, rump, and ham.

25. Describe a modern meat-type breeding animal.

26. Why is it difficult to arrive at a set procedure for examining a pig?

27. What is a judging scorecard?

28. What are the primary differences between the two scorecards presented in this chapter; one for breeding swine, and the other for market barrows?

# SELECTED REFERENCES

| Title of Publication | Author(s) | Publisher |
|---|---|---|
| *Breeding Better Livestock* | V. A. Rice<br>F. N. Andrews<br>E. J. Warwick | McGraw-Hill Book Company, New York, NY, 1953 |
| *Determining the Age of Farm Animals by Their Teeth* | | U.S. Department of Agriculture, Washington, DC |
| *Elements of Livestock Judging, The* | W. W. Smith | J. B. Lippincott Co., Philadelphia, PA, 1930 |
| *Judging Livestock* | E. T. Robbins | University of Illinois, Urbana-Champaign, IL |
| *Livestock Judging, Selection and Evaluation,* Third Edition | R. E. Hunsley<br>W. M. Beeson | Interstate Publishers, Inc., Danville, IL, 1988 |
| *Livestock and Meat Manual, A* | | Washington State University, Pullman, WA |
| *Pork Production* | W. W. Smith | The Macmillan Company, New York, NY, 1952 |
| *Selecting, Fitting and Showing Swine* | J. E. Nordby<br>H. E. Lattig | The Interstate Printers & Publishers, Inc., Danville, IL, 1961 |
| *Stockman's Handbook, The,* Sixth Edition | M. E. Ensminger | The Interstate Printers & Publishers, Inc., Danville, IL, 1983 |

Born and born alive—the first and most important requisite in breeding swine. (Courtesy, Lone Willow Genetics, Roanoke, IL)

# BREEDING SWINE

# Chapter 39

The laws of heredity apply to swine breeding exactly as they do to all classes of farm animals. But the breeding of swine is more flexible in the hands of breeders because (1) hogs normally breed at an earlier age, thus making for a shorter interval between generations, and (2) they are litter-bearing animals. Because of these factors and because of the available feeds and the type of pork products demanded by the consumer, the American farmer has created more new breeds and made more rapid shifts in type in hogs than in any other class of farm animals.

## NORMAL BREEDING HABITS OF SWINE

The pig lends itself very well to experimental study in confined conditions. It is reasonable to expect, therefore, that we should have a considerable store of knowledge relative to the normal breeding habits of swine, perhaps more then we have of any other class of farm animals.

### AGE OF PUBERTY

The age of puberty in swine varies from 4 to 8 months. This rather wide range is due to differences in breeds and lines, sex, and environment—especially nutrition. Most gilts weigh 180 lb or more before the onset of puberty. In general, boars do not reach puberty quite so early as gilts.

### AGE TO BREED GILTS

Reasonably early breeding has the advantage of establishing regular and reliable breeding habits and reducing the cost of the pigs at birth.

Gilts that are well developed may, as a general rule, be bred to farrow at 11 to 12 months of age. It is to be emphasized, however, that this depends primarily upon development rather than age; thus it is recommended that gilts weigh at least 225 lb before breeding. Proper development is essential in order that animals may be able to withstand the strain of lactation, the demands of which are much more rigorous than gestation.

The breeding of show gilts is often delayed until after the show season. This practice frequently results in difficult conception and temporary, if not permanent, sterility, which appears to be due to the fitting.

### HEAT (ESTRUS) PERIODS

The heat period—the time during which the sow will accept the boar—lasts from 1 to 5 days, with an average of 2 to 3 days. Older sows generally remain in heat longer than gilts.

The external signs of heat in the sow are restless activity, swelling or enlargement and discharge from the vulva (although these signs are not always present), frequent mounting of other sows, frequent urination, and occasional loud grunting. However, all signs are not always present.

Ovulation occurs 38 to 42 hours after the start of heat. Live sperm must be in the female reproductive tract a few hours before ovulation occurs; otherwise, litter size will be reduced. Optimal breeding is based on the number of times per day that a producer checks the females for standing heat. With once-a-day detection, females should be bred each day they will accept the boar. With twice-a-day detection, females should be

bred at 12 and 24 hours after they are first detected in heat. Gilts will sometimes have heat periods less than 2 days long and may have to be bred as soon as they are detected in heat and then each succeeding 12 hours they will stand for the boar. Heat detection should always be accomplished in the presence of a boar, since his presence maximizes the chances of detecting all possible females in heat.

When the heat period lasts longer than 3 days, continued breeding is likely a waste of boar power since conception is doubtful. If not bred, the heat period normally recurs at intervals of 16 to 25 days, with an average of 21 days.

### GESTATION PERIOD

The average gestation period of sows is 114 days, although extremes of 98 to 124 days have been reported.

### BREEDING AFTER FARROWING

Sows will often come in heat during the first few days after farrowing, but they rarely conceive if bred at this time—for the reason that they fail to ovulate.

Following the early postfarrowing heat period, sows normally will not come in heat again until near the end of the suckling period. However, they may be brought into heat during the lactation period by removing the pigs for several consecutive nights. Some producers follow this practice and rebreed about the fifth week after farrowing. Nevertheless, it is common practice first to wean the litter and let the animals have a few days' rest before rebreeding. Normally, some will come in heat 3 to 10 days, with an average of about 7 days, after weaning.

When pigs are weaned under two weeks of age, it is recommended that sows be bred on the second heat period after weaning. However, when pigs are weaned at three or more weeks of age, mating on the first postweaning estrus will result in reproductive performance comparable to normal practices in terms of farrowing rate, size of litter farrowed, and subsequent lactation performance.

The most natural breeding season for sows seems to be in the early summer and late autumn, although they will breed any time of the year.

## FERTILITY AND PROLIFICACY IN SWINE

Under domestication and conditions of good care, a high degree of fertility is desired. The cost of carrying a litter of 10 pigs to weaning time is little greater than the cost of producing a litter of only 5 or 6. In other words, the maintenance costs on both the sow and the boar remain fairly constant. It must be remembered, however, that in the wild state high fertility may not have been characteristic of swine. Survival and natural selection were probably in the direction of smaller litters, but nature's plan has been reversed through planned matings and selection.

Low fertility in swine is most commonly attributed to hereditary and environmental factors. Maximum prolificacy depends upon having a large number of eggs shed at the time of estrus, upon adequate viable sperm present for fertilization at the proper time, and upon a minimum of embryonic and fetal mortality.

It is a well-known fact that some breeds and strains of swine are much more prolific than others. For example, litters of 12 are considered normal, rather than exceptional, among Chinese swine. Also, through selection, more prolific strains of swine can be developed. Furthermore, the number of pigs produced increases with the age of the sow. It also appears that flushing, or conditioning, of sows exerts an influence on the number of eggs shed. All in all, it appears that more can be accomplished through proper management to increase litter size than can be done through selection.

Practical swine producers generally associate type with prolificacy. To substantiate their theory, they point out that the fat, chuffy hogs in vogue during the early part of the present century were not prolific, but were prone to farrow twins and triplets. Experimental work substantiates this opinion.

Even though many eggs may be shed and fertilized, the size of the litter may be affected materially by embryonic and fetal mortality, which ranges from 5 to 40%. This condition is usually attributed to (1) hereditary factors, perhaps recessive lethals; (2) overcrowding resulting from a large number of pigs and a consequent limited uterine surface area available for the nourishment of individual embryos; (3) nutritionally incomplete rations prior to and during gestation; (4) old sperm or eggs at fertilization; (5) diseases or parasites; (6) accidents or injuries; or (7) hormone imbalances. Additional studies need to be made relative to the cause and prevention of embryonic and fetal mortality.

Certainly, the boar cannot affect the number of eggs shed, and under usual circumstances fertilization is very much an all or none phenomenon. Still, some evidence suggests that the boar can have a marked effect on litter size, primarily because embryos and fetuses sired by certain boars are less likely to survive to term. This further supports the contention that embryonic and fetal mortality are due to genetic and fertilization errors (aged sperm or eggs). In this regard, conception rate and litter size can be increased by using more than one boar on each female.

It is recommended that, in advance of regular use, new boars should be test mated to a few gilts. During these test matings, the producer should observe the boar for aggressiveness and desire to mate, and if necessary, give the boar assistance the first service or two. Also, the producer should check the boar's ability to enter the gilt, which may be hindered by a limp, infantile, or tied penis. Most importantly, serviced gilts should become pregnant. A semen evaluation conducted by a veterinarian or qualified technician will complement the test matings. While there is no absolute test for fertility, test matings and semen evaluation can often detect a sterile boar or one of questionable fertility.

## FLUSHING SOWS

The practice of feeding sows more liberally so that they gain in weight from 1 to 1½ lb daily from about 1 to 2 weeks before the opening of the breeding season until they are safely in pig is known as flushing. Some of the beneficial effects attributed to this practice are (1) more eggs are shed, and this results in larger litters; (2) the sows come in heat more promptly; and (3) conception is more certain.

If sows are already overfat, the best preparation for breeding consists of conditioning by providing plenty of exercise and access to a lush pasture, while decreasing the heavy gain ration.

## THE BREEDING PROCEDURE; HAND MATING; THE BREEDING CRATE

Hand mating is more generally practiced in swine than with either cattle or sheep. In fact, it is almost the universal procedure in purebred swine herds, and most all large commercial producers follow the same practice.

When a mature, heavy boar is to be bred to gilts or when a boar pig is to be bred to big sows, the use of a breeding crate is recommended. Animals that have formed the habit of breeding without the crate may be rather obstinate in accepting the new method, or may refuse service altogether. Commercial swine producers usually use active, young boars; in many instances they allow the boar to run with the sows to be bred.

## ARTIFICIAL INSEMINATION (A.I.)

Artificial insemination of swine in the United States is limited. Yet, there is great interest in it, and it is a valuable tool for some situations.

Fig. 39–1. A cutaway view of a cryogenic tank for storing frozen boar semen. (Courtesy, International Boar Semen, Eldora, IA)

Two primary reasons for using A.I. in swine are (1) disease control, and (2) genetic improvement. A.I. allows for the development of a closed herd, but it is possible to bring in new genetic material with a minimum of risk. Moreover, A.I. will allow better use of superior boars in large operations. For example, a mature boar should not breed more than 2 or 4 females per day. However, in a large unit where sows are weaned in groups, A.I. makes it possible to breed 10 or more sows in 1 day from one ejaculate. To accomplish this, boars have to be collected, their semen diluted, and then used to inseminate those females in heat that day. The disadvantage to A.I. in swine is the high level management and added labor that is required.

Advances have been made in freezing boar semen, and some is commercially available, but, on the average, conception rates will be lower and litter sizes smaller when frozen semen is used. For the most part, the use of fresh semen offers the greatest benefits for swine producers, though frozen semen is a means of introducing new genetic material into a herd.

Although A.I. requires a greater managerial input, a minimum amount of specialized equipment and specialized training is needed to carry out a successful on-farm program. Furthermore, by following a few precautions, litter size and conception rates will be equal to natural service while expanding the use of superior boars.

## PREGNANCY TESTING

With ultrasonic detectors, pregnancy diagnosis is a reality. Producers can determine with a 90 to 95% accuracy the number of females that have settled. These detectors are most accurate and yield the best return per dollar invested when they are used between 30 and 45 days after mating.

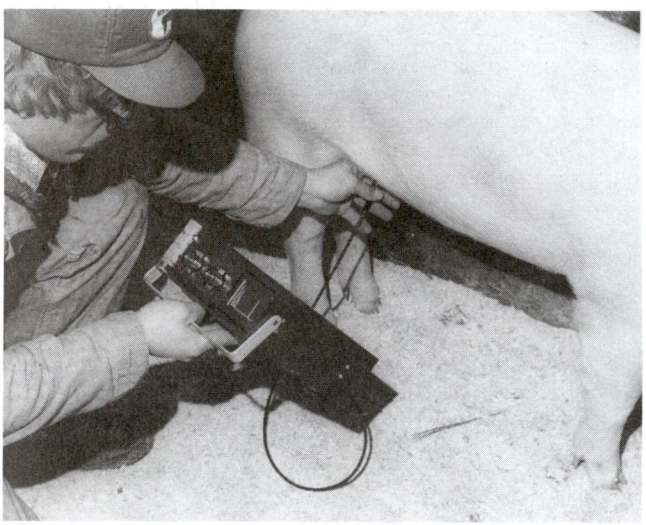

Fig. 39–2. Ultrasonic testing for pregnancy. (Courtesy, International Livestock Improvement Services Corp., Ames, IA)

The principle of ultrasonic pregnancy detectors is an ultrasonic echo from fluid in the uterus. Uterine fluid increases rapidly following conception and reaches detectable levels 25 to 30 days after breeding. It remains detectable for 80 to 90 days after breeding, following which the mass of pigs in the uterus exceeds the fluid content.

Among the several advantages of early pregnancy detection in sows and gilts are (1) it makes it possible to cull or rebreed nonpregnant, feed-wasting females; (2) it allows closer grouping of a number of sows for a farrowing period; (3) it gives early warning of breeding troubles, such as infertile boars and cystic ovaries of sows; (4) it enables producers to make more effective use of their breeding facilities and to plan more adequately for farrowing, nursing, and finishing; and (5) it makes it possible to guarantee pregnancy on females that are for sale.

## CARE AND MANAGEMENT OF THE BOAR

Proper care and management of the herd boar is most essential for successful swine production. Too frequently the boar is looked upon as a necessary evil and is neglected. Under such conditions, he is usually confined to a small, filthy pen—a typical pigsty; exercise is discouraged; and the feeding practices are anything but intelligent.

## FEED, SHELTER, AND EXERCISE

Outdoor exercise throughout the year is one of the first essentials in keeping the boar in a thrifty condition and virile. This may usually be accomplished by providing a well-fenced pasture. Even then, the caretaker may find it necessary to walk old boars or boars that are being fitted for the shows. In addition to the valuable exercise that is obtained in the pasture lot, green succulent pasture furnishes valuable nutrients for the herd boar. The amount of feed provided should be such as to keep the boar in a thrifty vigorous condition at all times. He should be neither overfat nor in a thin, run-down condition. The concentrate allowance should be varied with the age, development, and temperament of the individual; breeding demands; roughage consumed; etc. Feeding the boar is more fully covered in Chapter 40.

Boars of the same age or size can be run together during the off-breeding season, but boars of different ages should not be kept together.

In hot weather boars need to be kept cool with shade, fogging, a concrete hog wallow, or evaporative cooling. Boars subjected to high temperatures may demonstrate reduced fertility 4 to 6 weeks later.

Many producers successfully maintain boars in confinement, on slotted floors or concrete. Boars in confinement should be individually quartered in pens about 8 ft by 8 ft.

## RANTING

Some boars pace back and forth along the fence, often chopping their jaws and slobbering. Such action is called ranting. Young boars that take to excessive ranting may go off feed, become "shieldy" (hard in the shoulders), and fail to develop properly. Although this condition will not affect their breeding ability, it is undesirable from the standpoint of appearance. Isolation from other boars and from the sow herd is usually an effective means of quieting such boars. Should the boar remain off feed, placing a barrow or a bred sow in the pen with him will help to get him back on feed.

## CLIPPING THE TUSKS

It is never safe to allow the boar to have long tusks, for they may inflict injury upon other boars or even prove hazardous to the caretaker. Above all, such tusks should be removed well in advance of the breeding season, at which time it is necessary to handle the boar considerably. The common procedure in preparation for removing the tusks consists of drawing a strong rope over the upper jaw and tying the other end securely to a post or other object. As the animal pulls back and the mouth opens, the tusks may be cut with a bolt clipper.

## AGE AND SERVICE OF THE BOAR

The number of services allowed will vary with the age, development, temperament, health, breeding condition, distribution of services, and system of mating (hand coupling or pen breeding). No standard number of services can be recommended for any and all conditions. Yet the practices followed by good swine producers are not far different. Such practices are summarized in Table 39–1.

### TABLE 39–1
### HANDY BOAR MATING GUIDE

| Age | No. of Matings/Yr. | | Comments |
|-----|------|------|----------|
| | Hand Mating | Pen Mating | |
| **8** to 12 mo. of age .. | 24 | 12 | **B**oar pigs should be limited to one service/day; older boars to 2 services/day. |
| **Y**earling or older ..... | 50 | 35–40 | **A** boar should remain a vigorous and reliable breeder up to 6 to 8 years of age. |

For best results, the boar should be at least eight months old and well grown before being put into service. Even then, he should be limited to one service per day and a maximum total of two dozen services during his first breeding season, unless the mating period covers more than one month.

When fed and cared for by an experienced person, a strong, vigorous boar from 1 to 4 years of age (the period of most active service) may serve two sows per day during the breeding season provided hand coupling is practiced. With pen mating, fewer sows can be bred.

Excessive service will result in the release of a decreased concentration of sperm as well as immature sperm.

A boar should remain a vigorous and reliable breeder up to 6 or 8 years of age or older, provided that he has been managed properly throughout his lifetime.

## CARE OF THE PREGNANT SOW

Without attempting to review the discussion on feeding the gestating sow as found in Chapter 40, it may be well to reemphasize that there are two cardinal principles which the feeder should keep in mind when feeding sows during the pregnancy period. These are: to provide a ration which will ensure the complete nourishment of the sow and her developing fetal litter, and to choose the feeds and adopt a method of feeding which will proved economical and adaptable to local farm conditions.

Today, an increasing number of producers are confining sows, individually or in groups, as a means of lessening labor and automating, freeing land for crops, and controlling environment.

Some **advantages** of confinement sow housing include: (1) better control of mud, dust, and manure; (2) reduced labor for feeding, breeding, and moving to farrowing house; (3) improved control of internal and external parasites; (4) smaller land requirements; (5) better supervision of herd at breeding time; (6) use of existing buildings; (7) improved operator comfort and convenience; and (8) opportunities for better all-around management.

Some **disadvantages** of confinement sow housing include: (1) higher initial investment; (2) possible delayed sexual maturity and breeding age, lower conception rate in gilts, and lower rebreeding efficiency in sows; (3) requirement of better management and daily attention to details; and (4) increase in feet and leg problems.

Because of the initial capital investment, many producers prefer to remodel existing out-of-date farm structures for sow confinement rather than building new housing.

## CARE OF THE SOW AT FARROWING TIME

The careful and observant caretaker realizes the importance of having everything in readiness for farrowing time. If the pregnant sows have been so fed and managed as to give birth to a crop of strong, vigorous pigs, the next problem is that of saving the pigs at farrowing time.

It has been estimated that approximately 30% of the pigs farrowed never reach weaning age, and that an additional loss of 5% occurs after weaning. This means that only 65% of the pig crop reaches market age.

## SIGNS OF APPROACHING PARTURITION

The immediate indications that the sow is about to farrow are extreme nervousness and uneasiness, an enlarged vulva, and a possible mucous discharge. She usually makes a nest for her young, and milk is present in the teats.

## PREPARATION FOR FARROWING

About two weeks prior to farrowing, the sow should be dewormed. She should also be treated for external parasites at least twice within a few days of moving to the farrowing facility.

When farrowing in crates or pens, sows should be moved to these no later than the 110th day of gestation.

### SANITARY MEASURES

Before being moved into the farrowing quarters, the sow should be thoroughly scrubbed with soap and warm water, especially in the region of the sides, udder, and undersurface of the body. This removes adhering parasite eggs (especially the eggs of the common roundworm) and bacteria that are potential diarrhea-causing agents.

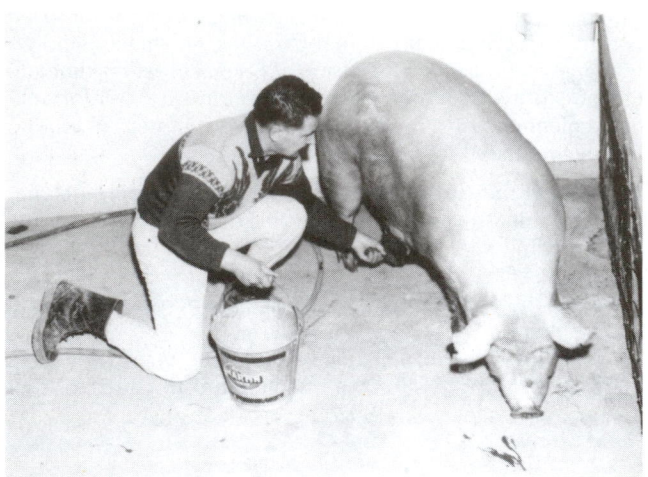

Fig. 39–3. W.S.U.'s swine herdsman washing a sow prior to moving her into the farrowing quarters. This removes adhering parasite eggs. (Courtesy, Washington State University, Pullman)

The farrowing unit should be thoroughly cleaned to reduce possible infection. If possible, the entire unit should be cleaned completely, disinfected, and left unused for 5 to 7 days. When this is not practical, the individual pen, stall, or crate should be cleaned and disinfected before a new sow is placed in the unit. Cleaning can be accomplished be scraping, high pressure cleaners, steam cleaners, and/or a stiff scrub brush. A complete job is necessary; otherwise, the use of a disinfectant is useless. Many good commercial disinfectants are available, including the quaternary ammonium compounds, iodoform compounds, and lye.

## THE QUARTERS

Hogs are sensitive to extremes of heat and cold and require more protection than any other class of farm animals. This is especially true at the time of parturition. It is recommended that the farrowing house temperature be maintained at 60° to 70°F, and that it not go below 40°F or above 85°F. In the cold areas and during winter months, use heat lamps or pig brooders when the farrowing house temperature falls below 60°F. Along with this temperature there should be adequate ventilation at all times.

The main requirements for satisfactory housing are that the quarters be dry, sanitary, and well ventilated and that they provide good protection from heat, cold, and winds. The buildings should be economical and durable.

## FARROWING CRATES OR PENS

Many producers use farrowing crates because they reduce the number of piglets crushed by the sow. Essentially, crates completely restrain sows. An additional advantage is that the operator is protected. Most farrowing crates are 5 ft wide by 7 ft long. The width includes an 18-in. piglet area on both sides of the 24-in. sow stall. Commercial crates adjust to accommodate very large or very small females. Crates may have slotted or solid floors.

When open pen farrowing is practiced, a guard rail around the farrowing pen is an effective means of preventing sows from crushing their pigs. The importance of this simple protective measure may be emphasized best by pointing out that approximately one-half of the young pig losses are accounted for by those pigs that are over-laid by their mothers. The rail should be raised 8 to 10 in. from the floor and should be 8 to 12 in. from the walls. It may be constructed of two-by-fours, two-by-sixes, or strong poles or steel pipe.

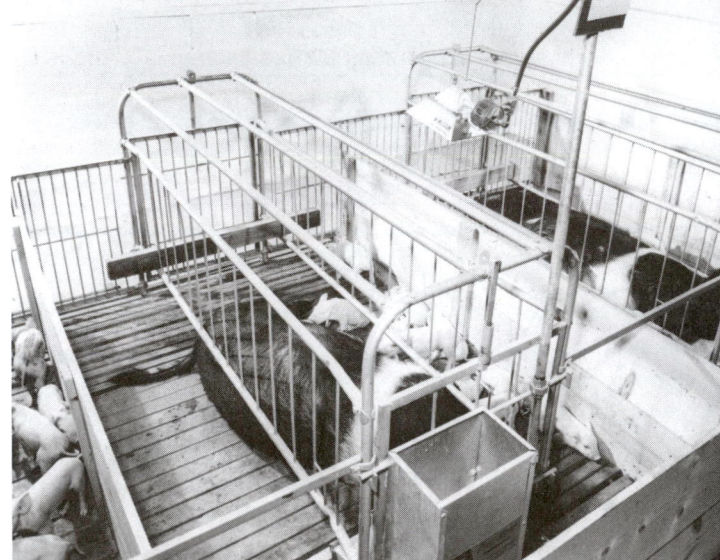

Fig. 39–4.   Farrowing crates. (Photo by J. C. Allen and Son, West Lafayette, IN)

## BEDDING

In open pen farrowing, the quarters should be bedded lightly with clean, fresh material. Any good absorbent that is not too long and coarse is satisfactory. Wheat, barley, rye, or oat straw; short or chopped hay; ground corncobs; peanut hulls; cottonseed hulls; shredded corn fodder; and shavings are most commonly used.

## THE ATTENDANT

Attending sows at farrowing decreases the incidence of stillborn pigs that die during the birth process, and the incidence of pigs dying within the first few hours after birth. Moreover, care given during this time improves survival the first few days after farrowing. The caretaker, therefore, should be on the job.

On the average, the interval between the birth of pigs is about 15 minutes and the whole birth process may last 30 minutes to 5 hours. Normal presentation is either head first or tail first. An attendant can (1) free piglets from the membranes, (2) help piglets reach a teat, (3) possibly revive some piglets which are not breathing, and (4) treat the navel cord with tincture of iodine. If farrowing is proceeding normally but slow, oxytocin may be used to speed the rate of delivery. Oxytocin should not be used if there is any suggestion that a pig is blocking the birth canal.

Continued strong labor for an extended period without the birth of piglets indicates the need for manual assistance by the attendant. A well lubricated gloved hand and arm should be inserted in the vulva and up the vagina as far as is needed to find the piglet blocking the birth canal. Then the piglet should be grasped and gently but firmly pulled. Since manual assistance increases the chances of complication, it is advisable to use an antibacterial solution as a lubricant and as an infusion following farrowing.

As soon as the afterbirth is expelled, it should be removed from the pen and burned or buried in lime. This prevents the sow from eating the afterbirth and avoids the development of bacteria and foul odors. Many good swine producers are convinced that eating the afterbirth encourages the development of the pig-eating vice. Dead pigs should be removed for the same reason.

If bedding is used, it is also well to work over the bedding; remove wet, stained, or soiled bedding and provide clean, fresh material.

## CHILLED AND WEAK PIGS

Pigs arriving during cold weather are easily chilled. Under such conditions, it may be advisable to take the pigs from the mother as they are born and place them in a half barrel or basket lined with straw or rags. In extremely cold weather, a few hot bricks or a jug of warm water (properly wrapped to prevent burns) may be placed in the barrel or basket; or the pigs may be taken to a warm room until they are dry and active.

One of the most effective methods of reviving a chilled pig is to immerse the body, except for the head, in water as warm as one's elbow can bear. The pig should be kept in this for a few minutes, then removed and rubbed vigorously with cloths.

## ORPHAN PIGS

Pigs may be orphaned through either sickness or death of their mother. In either event, the most satisfactory arrangement for the orphan is to provide a foster mother. When it is impossible to transfer the pigs to another sow, they may be raised on cow's milk or milk replacer. The problem will be simplified if the pigs have received a small amount of colostrum (the first milk) from their mother. Colostrum can be hand-milked if necessary.

If cow's milk is used, it is preferable that it be from a low-testing cow. Do not add cream or sugar; however, skim milk powder, at the rate of a tablespoonful to a pint of fluid milk may be added, if available. Milk replacer should be mixed according to the directions found on the container. The first 2 or 3 days the orphan should be fed regularly every 2 hours, and the milk should be at 100°F. Thereafter, the intervals may be spaced farther apart. All utensils (pan feeding or a bottle and nipple may be used) should be clean and sterilized.

Orphan pigs should be started on prestarter or starter ration when they are one week old. Also, a source of iron should be provided (in keeping with instructions given in Chapter 40).

## RUNTS

Small pigs or *runts* present a problem for producers. Larger litters have more runts, and these runts are often some of the last pigs born; thus, forcing them to compete for a teat with the larger earlier born pigs. Since runts often perish, in the past many producers have sacrificed them rather than try to save them. With today's economy, the effort to save these pigs may be worthwhile. Some research indicates that supplemental feeding of underweight (under 2 lb) newborn pigs (runts) can reduce their mortality. Supplemental feedings may consist of a commercial milk replacer or a mixture of 1 qt milk, ½ pt "half and half," and 1 raw egg, which is administered once or twice daily in 15 to 20 ml portions with a soft plastic tube attached to a syringe.[1]

## ARTIFICIAL HEAT

Raising the air temperature of the farrowing unit is important to prevent the chilling of newborns.

Artificial heat usually must be provided, especially for pigs farrowed in the northern United States. Most farrowing houses are equipped with a heating unit for use in winter farrowing, designed to maintain the temperature at 70° to 75°F.

Furthermore, providing an area of supplemental heat (heat lamps, gas brooders and/or heat in the floor) is necessary to maintain baby pigs in their thermoneutral zone. This heat zone should be 85° to 95°F. A major factor adversely affecting piglet survival is its difficulty in maintaining body temperature due to the high ratio between body surface area and size, sparse hair covering, and limited body fat. Even at thermoneutral temperatures, substantial amounts of metabolic energy are required to maintain body functions. A few days after birth, age-related changes occur which markedly improve the thermoneutral stability of the piglet.

Individual houses may be insulated by banking with straw and other insulating materials. Also, a heater may be suspended from the top of the house. It must be remembered, however, that there is considerable fire hazard with this practice.

## THE SOW AND LITTER

Fig. 39–5. A Landrace sow and her litter of 13 pigs. (Courtesy, American Landrace, Assn., West Lafayette, IN)

[1]Moody, N. W., et al., "Effects of Supplemental Milk on Growth and Survival of Suckling Pigs," *Journal of Animal Science*, Vol. 25, 1966, p. 1250.

The care and management given the sow and litter should be such as to get the pigs off to a good start. As is true of other young livestock, young pigs make more rapid and efficient gains than older hogs. Strict sanitation and intelligent feeding are especially important for the well-being of the young pig.

## SANITATION

No class of animals, with the possible exception of poultry, is as subject as swine to heavy losses from diseases and parasites. Each year, thousands of swine growers are put out of business by parasites and filth-borne infections.

Consciously or unconsciously, the successful swine producer follows a sanitation program similar to that which was worked out many years ago for the control of the common roundworm. The plan was given its first field test in McLean County, Illinois, in 1919; hence, it became known as the "McLean County System of Swine Sanitation."

But the importance of the McLean County System of Swine Sanitation extends far beyond roundworm control and historical significance. It consists of four simple steps: (1) cleaning and disinfecting the farrowing quarters, (2) washing the sow before placing her in the farrowing quarters, (3) hauling the sow and pigs to clean pasture, and (4) keeping the pigs on clean pasture

until they are at least four months old. The application of the same principles of sanitation is effective in reducing young pig troubles from other parasites and in disease control—thereby making for more profitable pork production; and the application of the same principles of sanitation is effective for both pasture and confinement production.

## THE NEEDLE TEETH

Newborn pigs have eight small, tusklike teeth (so-called needle or black teeth), two one each side of both the upper and lower jaw. As these are of no benefit to the pig, most swine producers prefer to cut them off soon after birth. This operation may be done with a small pair of pliers or with forceps made especially for the purpose. In removing the teeth, care should be taken to avoid injury to the jaw or gums, for injuries may provide an opening for bacteria. The needle teeth are very sharp and are often the cause of pain or injury to the sow, particularly if the udder is tender. Moreover, the pigs may bite or scratch each other, and infection may start and cause serious trouble.

## EAR NOTCHING THE LITTERS

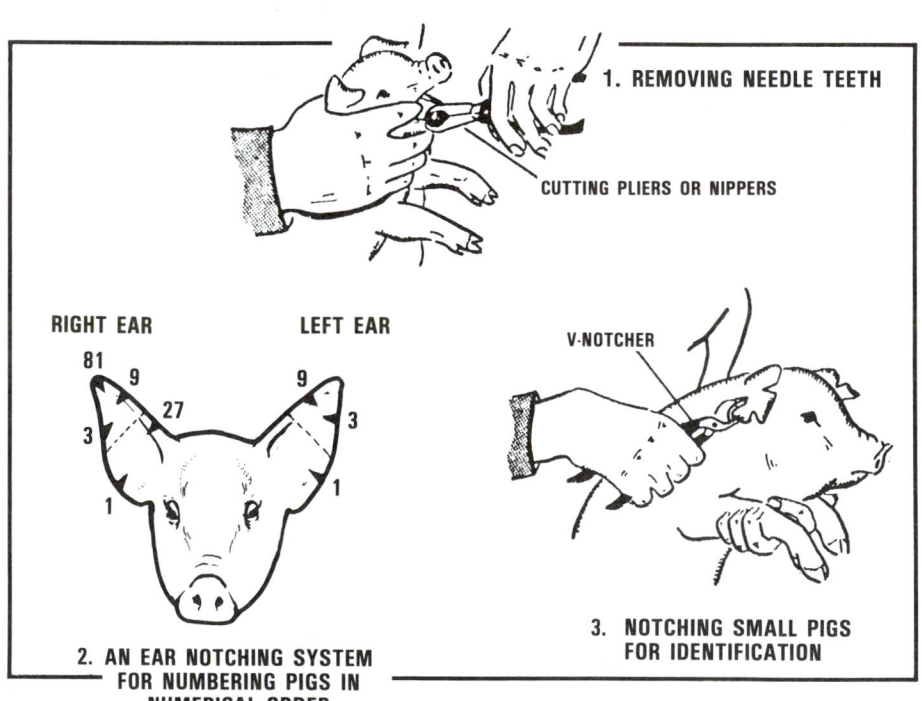

Fig. 39–6.   Diagram showing some important swine management practices: (1) removing needle teeth, (2) an ear-notching system for numbering pigs in numerical order (the right ear is used for the litter marks, and the left ear is used for the individual number), and (3) notching small pigs for identification. (Drawing by Prof. R. F. Johnson)

The common method of marking or identifying swine consists of ear notching the litters. Pigs are generally marked at the same time that the needle teeth are removed. Purebred breeders find it necessary to employ a system of marking so that they may determine the parentage of the individuals for purposes of registration and herd records. Even in the commercial herd, a system of identification is necessary if the gilts are to be selected from the larger and more efficient litters. The

ear notches are usually made with a special V-notcher. Most of the breed associations are in a position to recommend a satisfactory marking system; many require the Universal Ear Notching System shown in Fig. 39–6, (2). This is the most common system, though there are others.

Plastic ear tags, branding, and tattooing are also used for identifying swine.

## CASTRATION

All male pigs that are not to be used for breeding purposes should be castrated while they are still suckling their dam. They should be castrated early enough to allow plenty of time for the wounds to heal thoroughly before weaning. It should not be done at the same time that the pigs are vaccinated. In preparation for the operation, the pigs should be kept off feed 10 to 12 hours.

Before starting the castrating, the hands should be thoroughly washed with soap and water and rinsed in a good disinfectant. The knife should be disinfected both before and between operations; and the scrotum of the pig should be washed before and after making the incision. The knife should be sharp, and the slit should extend well down so as to allow for proper drainage. In young pigs, the cord should be pulled out or broken off well forward.

Pigs with an undescended testicle(s) or rupture (scrotal hernia) should be operated by a veterinarian.

## CONFINEMENT VS PASTURE

In recent years, confinement rearing, in which pigs are confined in buildings from birth to market, has increased. The main reasons given for going to confinement are savings in labor and land. The main problems encountered are high investment in buildings, more disease troubles, rations become more critical, manure disposal and odor control problems are greater, and sow fertility is lowered.

Contrary to the opinion held by some, pastures for sows and litters are not obsolete; rather, there now exists two alternatives for the breeding herd—confinement vs pasture—and the wise manager will choose between them.

On a per hog basis, the total cost of confinement and pasture production are about the same. Despite this fact, a number of motivating forces have caused, and will continue to cause, more and more confinement production. In particular, large operations do not lend themselves to the use of pastures.

## FEED REQUIREMENTS OF THE NURSING SOW

Until pigs are about 3 days old, the feed intake of the sow should be limited, and she should be fed plenty of bran. Thereafter, the feed allowance may be increased gradually so as to arrive at full feeding 10 days to 2 weeks later. Throughout the lactation period, the sow should be fed liberally with feeds that will stimulate milk production. The most essential ingredients in the brood sow's ration during this period are an ample amount of protein, vitamin, and mineral substances.

## NORMAL BREEDING SEASON AND TIME OF FARROWING

The season in which the sows are bred and the question of raising one or two litters a year or multiple farrowing depend primarily on the facilities at hand. The location of the producer (particularly the weather conditions in the area), availability and price of feeds, condition and growth of the sows, equipment for handling pigs during the winter months, available labor, and the type of production (purebred or commercial) should be taken into consideration. No positive advice can be given, therefore, for any and all conditions. Sows will breed any time of the year; but, as in other farm animals, the conception rate is much higher during those seasons when the temperature is moderate and the nutritive conditions are good. For the country as a whole, spring pigs are preferred, as is shown by the size of the spring pig crop in comparison with the fall pig crop.

## SHOW-YARD CLASSIFICATIONS MAY DETERMINE FARROWING DATES

The purebred breeder who exhibits breeding hogs should plan the breeding program so that the maximum advantage will be taken of the various age groups.

## TWO LITTERS VS ONE LITTER A YEAR

Most small swine producers follow a "two litters or one litter a year" system rather than multiple farrowing.

Whether to have each sow farrow two litters or one litter per year is a problem that each individual swine producer must decide. Where climatic conditions, facilities, and feeds are favorable to the two-litter system, it has the following distinct **advantages**:

1. Maximum use in made of the capital invested in facilities and equipment.

2. More certain and rapid improvement can be effected in the breeding herd through maintaining outstanding tried sows throughout their useful lives. The two-litter system is conducive to retaining such sows.

3. There is a better distribution of labor.

4. Pigs are marketed at two different time of the year, thus distributing risks.

5. There is a better distribution of farm income.

On the other hand, in the northern latitudes where suitable facilities and feeds may not always be available, the following disadvantages often apply to the two-litter system:

1. It is necessary that the spring farrowed litter arrive reasonably early, often during inclement weather. Young pig losses, under such conditions, are higher than is usually encountered in a one-litter system.

2. Except in the South where nearly year-round pastures are available, fall farrowed pigs require more concentrates and high-priced protein supplements than pigs born earlier in the year and run on pastures. Because of inadequate nutrition, most fall pigs also are less thrifty and make slower gains than pigs run on pasture.

Because of the high cost in maintaining tried sows for a whole year to raise one litter of pigs, the one-litter system is usually based chiefly or entirely on the use of gilts that are finished and marketed soon after weaning their litters.

The **advantages** of the one-litter system are:

1. There are fewer management problems; it is easier to keep on schedule.

2. Less total capital is tied up in hog buildings and equipment.

3. Less grain storage is required.

4. The weather is usually more favorable at farrowing time.

5. Less labor and hard work are required.

The **disadvantages** of the one-litter system are:

1. Buildings and equipment are not used to the maximum.
2. It limits the rapidity of improvement that can be made in the breeding herd; sows are usually sold after one litter only, or if retained there is only one litter per year.
3. The labor requirements are not distributed throughout the year.
4. Farm income is not distributed throughout the year.

## MULTIPLE FARROWING

*Multiple farrowing refers to that type of program in which there is a scheduling of breeding so that the litters arrive in a greater number of farrowing periods throughout the year than is the case in the conventional one- and two-litter systems.*

There is nothing mysterious or complicated about multiple farrowing. It does, however, entail some planning and close attention to management details. In practice, it generally means that the sow herd is split into either 2 or 3 groups, with each group farrowing twice each year. If 2 groups of sows are used, pigs are farrowed every 3 months. If 3 groups of sows are used, pigs can be farrowed every other month. Should a sow fail to conceive, she can be set back to another group.

As shown in Chapter 8, Fig. 8–27, Seasonality of Livestock Receipts, the lowest receipts of hogs occurs in February, and the highest in October. Increased multiple farrowing will help alleviate these rather sharp fluctuations in hog receipts and prices, and prove beneficial both to producers and packers.

Among the factors **favorable** to multiple farrowing are:

1. It makes for a more stable hog market, with fewer high and low market receipts and price fluctuations.
2. It distributes the work load for the swine producer.
3. It makes for better use of existing buildings and equipment; for example, the farrowing house can accommodate more litters because of the longer farrowing season.
4. It provides a more sustained flow of hogs to market, which, from the standpoint of the packer, is desirable because—(a) it makes for more complete use of labor and plant capacity, and (b) it enables the processor more nearly to meet the demands of the retailer and consumer. Also, the producer's income is distributed throughout the year.
5. It provides retailers with a steady supply of pork for their trade.
6. It avoids sharp price rises, which the consumer dislikes.

Among the factors **unfavorable** to multiple farrowing are:

1. The swine enterprise is more confining for a longer period of time, for the reason that competent help must be available over a more prolonged farrowing season.
2. The possibility of a disease break may be increased because of the likely build-up of pathogenic organisms. With multiple farrowing, it is not possible to clean out and air out for a long period of time.

No one expects the seasonal pattern of hog production to be completely eliminated, but, because of the several recognized advantages of multiple farrowing to both the processor and the producer, along with more confinement production, it will be used increasingly and make for a lessening of some of the market gluts of the past.

Also, because of the several advantages of multiple farrowing in big operations—especially more even distribution of labor, more complete use of facilities, and improved cash flow—multiple farrowing increases with the size of the operation.

## GILTS VS OLDER SOWS

Controlled experiments at the North Dakota and Wisconsin Experiment Stations, along with practical observations, in which gilts have been compared with older (tried) sows, bear out the following facts: (1) Gilts have fewer pigs in their litters than do older sows, and (2) the pigs from gilts average slightly smaller in weight at birth and also tend to make somewhat slower gains.

Despite these disadvantages, gilts have certain advantages, especially for the commercial pork producer. Their chief superiority lies in the fact that they continue to grow and increase in value while in reproduction. Although over-weight young sows bring less on the market than prime barrows, from this standpoint alone they generally return a handsome profit when the price of pork is sufficiently favorable. Many practical commercial producers, who probably are less close to the herd at farrowing time than purebred breeders, are of the opinion that the smaller gilts crush fewer pigs than do older and heavier sows.

In no case is it recommended that the purebred breeder rely only on gilts. Tried sows that are regular producers of large litters with a heavy weaning weight and that are good mothers and producers of progeny of the right type should be retained in the herd as long as they are fertile.

## CROSSBREEDING SWINE

*Crossbreeding is the mating of different breeds.*
Today, crossbreeding is being used by swine producers to (1) increase productivity over straightbreds because of the resulting hybrid vigor or heterosis; (2) produce commercial hogs with a desired combination of traits not available in any one breed; and (3) produce foundation stock for developing new breeds.

The motivating forces back of increased crossbreeding in farm animals are (1) more artificial insemination, thereby simplifying the rotation of sires of different breeds; and (2) the necessity for swine producers to become more efficient in order to meet their competition, both from within their industry and from without.

Crossbreeding will play an increasing role in the production of market animals in the future, because it offers the several advantages discussed in the sections which follow.

### HYBRID VIGOR OR HETEROSIS

*Heterosis, or hybrid vigor, is the name given to the biological phenomenon which causes crossbreds to outproduce the average of their parents.* For numerous traits, the performance of the cross is superior to the average of the parental breeds. This phenomenon has been well known for years and has been used in many breeding programs. The production of hybrid seed corn by developing inbred lines and then crossing them is probably the most important attempt to take advantage of hybrid vigor. Today, heterosis is also being used extensively in commercial swine, sheep, layer, and broiler production. An estimated 80% of market lambs and layers are crossbred; 95% of broilers are crosses; and about 90% of the hogs raised for slaughter are crossbred.

The genetic explanation for the hybrid's extra vigor is basically the same, whether it be cattle, hogs, sheep, layers, broilers, hybrid corn, hybrid sorghum, or whatnot. Heterosis is produced by the fact that the dominant gene of a parent is usually more favorable than its recessive partner. When the genetic groups differ in the frequency of genes they have and dominance exists, then heterosis will be produced.

Heterosis is measured by the amount the crossbred offspring exceeds the average of the two parent breeds or inbred lines for a particular trait, using the following formula for any one trait:

$$\frac{\text{Crossbred average} - \text{Purebred average}}{\text{Purebred average}} \times 100 = \text{Percent hybrid vigor}$$

Thus, if the average of the two parent populations for litter weaning weight is 284 lb and the average of their crossbred offspring is 336 lb, application of the above formula shows that the amount of heterosis is 52 lb.

Traits high in heritability—like carcass length, backfat thickness, and loin eye area—respond consistently to selection but show little response to hybrid vigor. Traits low in heritability—like litter size, litter weaning weight, and survival rate—usually demonstrate good response to hybrid vigor.

## COMPLEMENTARY

*Complementary refers to the advantage of a cross over another cross or over a purebred, resulting from the manner in which two or more characters combine or complement each other.* It is a matching of breeds so that they compensate each other, the objective being to get the desirable traits of each. Thus, in a crossbreeding program, breeds that complement each other should be selected, thereby maximizing the desirable traits and minimizing the undesirable traits. Since breeds which are selected because they tend to express a maximum of some trait will have some undesirable traits, different breeds must be selected for different purposes.

No one breed has a monopoly on all the desired characteristics. Therefore, producers must study their operation and the merits of different breeds before choosing a breed.

## INTRODUCE NEW GENES QUICKLY

Crossbreeding programs provide a way in which to introduce new and desired genes quickly—at a faster rate than can be achieved by selection within a breed. As undesirable qualities are often recessive, crossbreeding offers the best way in which to improve certain characteristics merely by hiding them with dominants.

## GET HYBRID VIGOR EXPRESSED IN THE FEMALE

Except for a two-breed cross, crossbreeding offers an opportunity to have hybrid vigor expressed in breeding females.

This is most important in the swine herd where it results in increased fertility, survivability of piglets, litter weaning size, and pig growth rate—all factors that mean more profit for the producer.

## SYSTEMS OF CROSSBREEDING

Crossbreeding has been widely used in swine because it increases production and profit. About 90% of all commercial hogs are crossbreds.

The superiority of crosses over purebreds is due to heterosis, or hybrid vigor, which results in the following increases in reproductive and performance traits:

| Trait | Heterosis, Advantage Over Purebreds |
|---|---|
| | (%) |
| Born alive . . . . . . . . . . . . . . | 8 |
| Litter size raised . . . . . . . . . | 23 |
| Litter weaning weight . . . . . . | 27 |
| Days to 230 lb . . . . . . . . . . | 7 |
| Five month litter weight . . . . | 18 |

However, heterosis gives little or no improvement in feed efficiency and carcass merit. These traits are obtained through the selection of the parents.

The availability of distinct purebred breeds provides an opportunity for maximizing heterosis in commercial herds. Three different rotational schemes follow:

1. **Rotational cross system** (Fig. 39–7). This system, which involves two or more breeds, can achieve nearly optimum heterosis levels and offers a simple, flexible alternative for the smaller operator or for those wishing to keep their own gilts and turn the generations quickly.

## ROTATIONAL CROSS SYSTEM

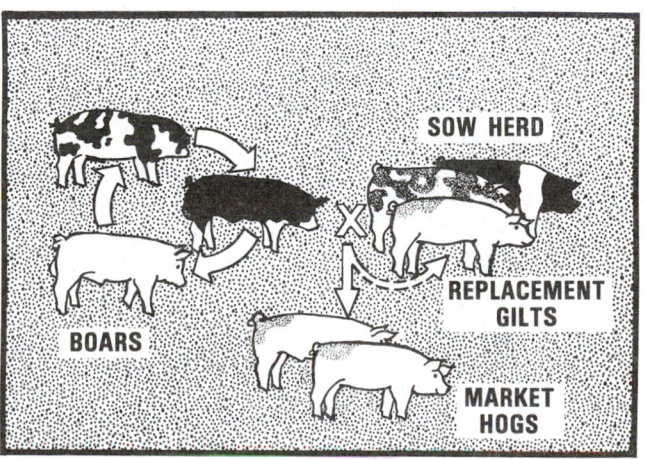

Fig. 39–7. Rotational cross system.

2. **Rototerminal cross system** (Fig. 39–8). When properly designed and implemented, this system produces 100% of maximum pig heterosis and a high proportion of maternal heterosis. Through optimum positioning of the breeds, such a system makes for superior matching (complementary) of the breeds involved, and is conducive to within-herd gilt replacement that can lower the cost of providing replacements and reduce the risk of disease.

## ROTOTERMINAL CROSS SYSTEM

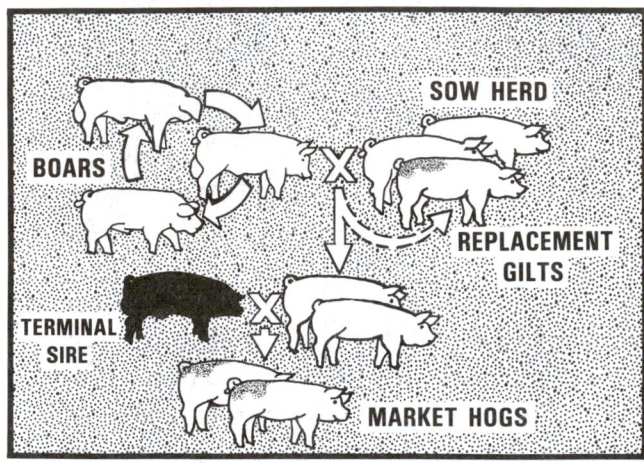

Fig. 39–8.   Rototerminal cross system.

3. **Terminal cross system** (Fig. 39–9). This is rapidly becoming the preferred system in larger units. Its success is due primarily to the great genetic diversity that exists between the pure breeds. With planned positioning of the pure breeds, this system will provide maximum hybrid vigor in both the pigs and the dam, make for superior matching (complementary) among the breeds utilized, and result in a final product of consistent performance and composition.

## TERMINAL CROSS SYSTEM

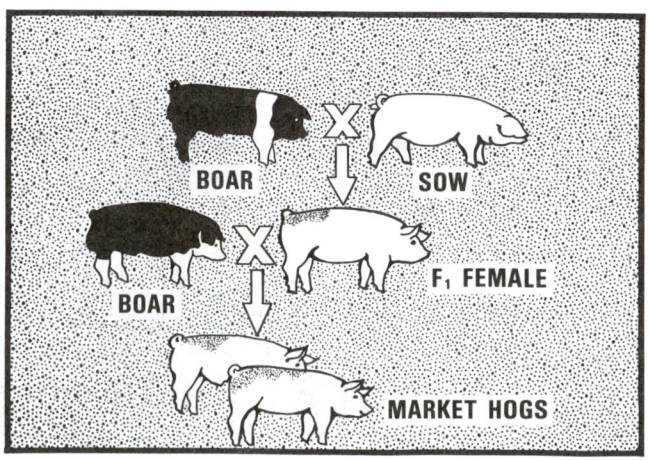

Fig. 39–9.   Terminal cross system.

Before starting a crossbreeding program, swine producers should know what is involved and what to expect. Also, they should realize that sound management and the selection of superior breeding stock on performance, potential carcass characteristics, and overall productivity are requisites for success. Moreover, the breeds used should complement each other. There is no one best system, breed, or source of breeding stock. Producers must evaluate their total program, and then make the most profitable choices.

## LETHALS AND OTHER HEREDITARY ABNORMALITIES IN SWINE

Lethal characters in swine, or in any class of animals, are caused by the presence of hereditary factors in the germ plasm that produce an effect so serious as to cause the death of the individual either at birth or later in life. Breeding animals possessing hereditary lethals should be culled from the herd.

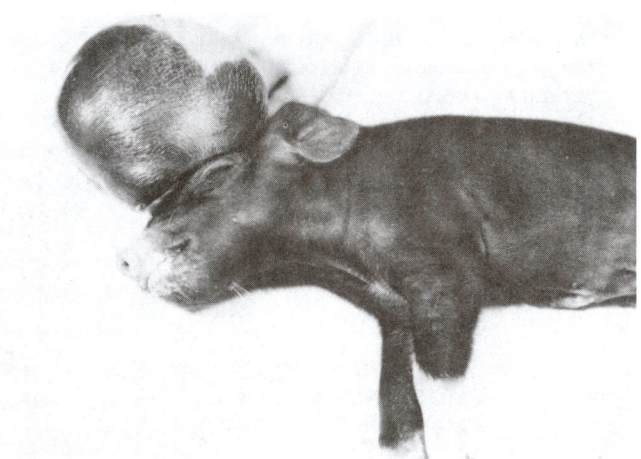

Fig. 39–10.   Hydrocephalic (literally meaning water in the head) pig. Pigs affected with this condition die soon after birth. It is inherited as a simple recessive. (Courtesy, Purdue University)

## PRODUCTION TESTING; HERITABILITY

Fig. 39–11.   The world's first swine production testing station, established in 1907 at Elsesminde, Denmark. (Courtesy, Danish Embassy)

# MODERN PRODUCTION TESTING OF SWINE

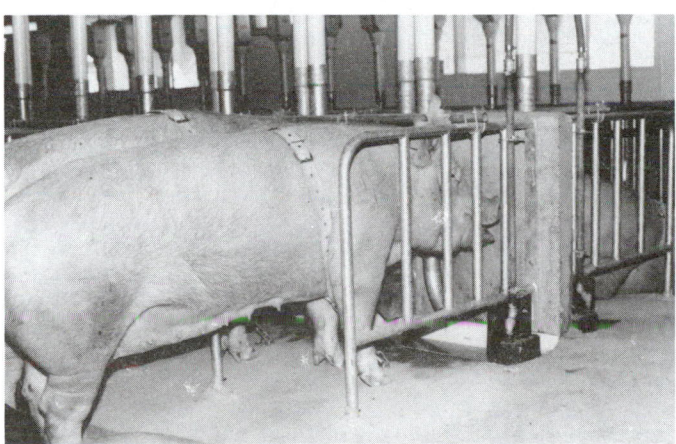

Fig. 39-12. New Ulm Swine Evaluation Station, New Ulm, Minnesota, a national center for swine testing. (Courtesy, University of Minnesota, St. Paul)

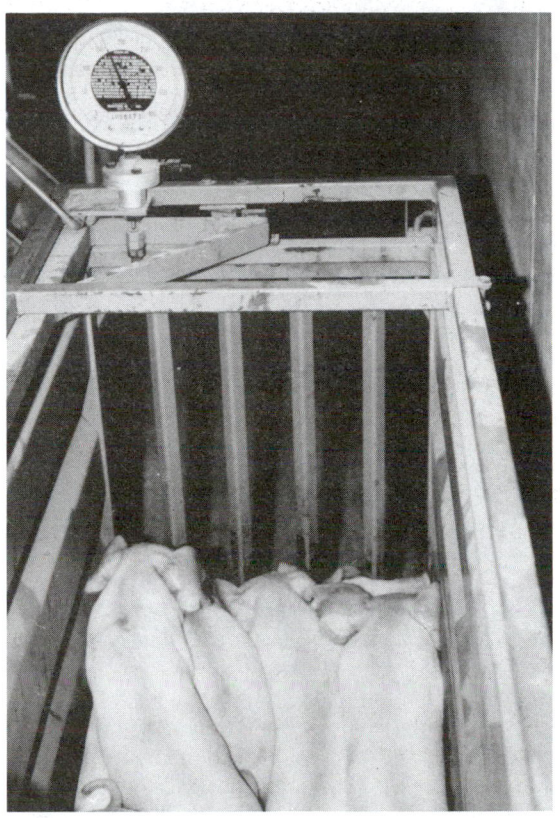

Fig. 39-14. Tethered sows (note belt). This makes it possible to determine individual feed efficiency. (Courtesy, Lone Willow Genetics, Roanoke, IL)

Fig. 39-13. Litter weight at 21 days of age—a production testing essential. (Courtesy, American Landrace Assn., West Lafayette, IN)

Fig. 39-15. Ultrasonic testing for backfat. (Courtesy, University of Minnesota, St. Paul)

Fig. 39-16. A very superior Yorkshire boar. His production test record: 121 days to reach 220 lb; 0.67 in. backfat; 6.25 sq in. loin eye; from a litter of 13. (Courtesy, Waldo Farms, DeWitt, NE)

As pointed out in Chapter 3, the effectiveness of selection can be increased provided that it is based upon carefully taken records rather than upon casual observation. Naturally, it would be illogical to expect upstanding, narrow-bodied, shallow sows and boars to beget meaty barrows that would be market toppers. Breeding animals of acceptable meat type can only transmit these qualities unfailingly to all their offspring when they themselves have been rendered relatively homozygous or pure for the necessary genes—a process that can be gradually accomplished through judgment by the eye method, but which can be made more rapid and certain through securing and intelligently using production records.

The following characteristics in swine are relatively easy to recognize and measure and are a fairly accurate guide in estimating the economy of production:

1. **Litter size at birth** (which has a heritability of 15%[2]). On the average, a sow will have consumed a total of 3/4 to 1 ton of feed during the period between breeding and the date her litter is weaned. Thus, if this quantity of feed must be charged against a litter of 4 or 5 pigs, the chance of eventual profit is small.

2. **Litter size at weaning** (which has an approximate heritability of 12%). Although greatly influenced by the caretaker, litter survival to weaning is a measure of the mothering ability of the sow.

3. **Birth weight of pigs** (which has a heritability of 5%). Very light pigs usually lack vigor.

4. **Litter weight at weaning** (which has a heritability of 17%). Weaning weight is important, for it has been shown that the pigs that are heaviest at weaning time reach market weight more quickly. The low heritability of this factor indicates that it is largely a function of the nursing ability of the sow rather than genetic.

5. **Daily rate of gain from weaning to marketing** (which has a heritability of 30%). Daily rate of gain from weaning to marketing is important because (a) it is highly correlated with efficiency of gain, and (b) it makes for a shorter time in reaching market weight and condition, thus effecting a saving in labor, making for less exposure to risk and disease, and allowing for a more rapid turnover in capital.

Rate of gain and lardiness may be correlated to some degree. Thus, one should not let this be the only factor upon which selection is based.

6. **Efficiency of feed utilization** (which has a heritability of 30%). Where convenient, accurate litter feed records should be kept, for the most profitable animals generally require less feed to make 100 lb of gain.

7. **Conformation score** (which has a heritability of 29%). This heritability figure is likely to be considerably higher in a herd of low quality.

8. **Carcass characteristics:**

   a. **Length** (which has a heritability of 60%). Carcass length is perhaps the most highly hereditary trait in hogs. This accounts for the rapid shifts that frequently have been observed; for example, in changing from chuffy to rangy hogs.

   b. **Backfat thickness** (which has a heritability of 50%). The probe, lean meter, or ultrasonic equipment can be used to measure backfat thickness on prospective breeding animals.

   c. **Loin lean area** (which has a heritability of 50%). Loin area is an indication of muscling or red meat.

   d. **Percent ham, based on carcass weight** (which has a heritability of 58%). Ham is a high-priced cut; hence, the aim is to get as large a ham as possible.

   e. **Percent lean cuts, based on carcass weight** (which has a heritability of 50%). A high yield of lean cuts means less trimable fat and more edible meat.

That swine show variations in these characteristics is generally recognized. The problem is to measure these differences from the standpoint of discovering the most desirable genes and then increasing their concentration and, at the same time, to purge the herd of the less desirable characters.

A prerequisite for any production data is that each animal be positively identified—by means of ear notches. For purebred breeders, who must use a system of animal identification anyway, this does not constitute an additional detail. But the taking of weights and grades does require additional time and labor—an expenditure which is highly worthwhile, however.

In order not to be burdensome, the record forms should be relatively simple. Fig. 39–17 is a litter record form for use in recording detailed information on one litter, whereas Fig. 39–18 is an individual sow record designed to use in recording the lifetime production record of one sow.

A good plan for progeny testing boars consists of retaining and mating one or more boar pigs—the number depending upon the size of the herd—to a limited number of females during their first season of breeding. The progeny are then tested and evaluated, and only those boars that prove to be the best on the basis of their progeny are retained for further breeding purposes. If boar pigs are each mated to 6 or 8 sows, pigs should be born 114 days later, and the progeny can be tested. Thus, with good fortune, it is possible to have progeny data on a boar when he is approximately 12 months of age.

Animals too young to progeny test may be evaluated by performance testing.

Herd records are, however, of little value unless they are intelligently used in culling operations and in deciding upon replacements. Also, most swine producers can and should use production records for purposes of estimating the rate of progress and for determining the relative emphasis to place on each character.

---

[2]The rest is due to environment. The heritability figures given herein are averages based on large numbers; thus, some variation from these may be expected in individual herds.

# SWINE

## LITTER RECORD

Breed _____  Litter No. _____
(notch, tattoo)

Data on Dam:
  Pedigree _____  { _____
        (name, reg. no., and ear notch)              (Sire)

  Birth date _____  _____
        (date and year)               (Dam)

  Litter mate carcass data, if any:
    No. carcasses _____  Av. backfat _____;  loin eye _____;  length _____
                            (in.)              (sq in.)          (in.)

  Sow's _____ litter.
    (1st, 2nd, etc.)

Data on Sire:
  Pedigree: _____  { _____
        (name, reg. no., and ear notch)               (Sire)

  Birth date _____  _____
                                 (Dam)

  Litter mate carcass data, if any:
    No. carcasses _____;  Av. backfat _____;  loin eye _____;  length _____
                            (in.)              (sq in.)          (in.)

Date of Birth _____  Health Services:

No. Pigs Born:

  Alive _____    Date cholera vaccinated _____

  Dead _____    Date erysipelas vaccinated _____

  Mummies _____    Date wormed _____

  Total _____    Other, including iron pills or shots (list) _____

No. Pigs Weaned _____    _____

## INDIVIDUAL PIG RECORD

| Pig's No. | Sex | No. Teats | Birth Wt. | Off-Color Markings | Defects & Abnormalities | Weaning Wt. ___ days (fill in) | Date Castrated | Date & Cause of Death | Disposal Date & To Whom | Remarks |
|---|---|---|---|---|---|---|---|---|---|---|
| | | | | | | | | | | |
| | | | | | | | | | | |
| | | | | | | | | | | |
| | | | | | | | | | | |
| | | | | | | | | | | |
| | | | | | | | | | | |
| | | | | | | | | | | |
| | | | | | | | | | | |
| | | | | | | | | | | |
| | | | | | | | | | | |
| | | | | | | | | | | |
| | | | | | | | | | | |
| | | | | | | | | | | |
| | | | | | | | | | | |
| | | | | | | | | | | |
| | | | | | | | | | | |
| | | | | | | | | | | |

Fig. 39–17.  Litter record form.

## SWINE

### Individual Sow Record

Breed _____ Name and registration no _____

Date farrowed _____ Identification _____
                                                                        (ear notch, tattoo)
Bred by _____
                                        (Name and address)

Sow's pedigree  _____  {  _____
                          (Sire)             {  _____

                _____  {  _____
                          (Dam)              {  _____

Record of litter of which the sow was a member

    No. in litter _____ No. of pigs weaned _____

    Weaning wt at _____ days of age
                              (fill in)
      Her own wt _____ Av. wt. of litter _____

Litter mate carcass record, if any

    No. carcasses _____; av. back fat _____; loin eye _____; length _____
                                              (in.)              (sq. in.)            (in.)

Number of teats _____

### Production Record of Sow

| | 1 | 2 | 3 | 4 | 5 | 6 | 7 | 8 |
|---|---|---|---|---|---|---|---|---|
| Litter no _____ | | | | | | | | |
| Sire _____ | | | | | | | | |
| No. services _____ | | | | | | | | |
| Farrowing data _____ | | | | | | | | |
|    Date _____ | | | | | | | | |
|    Temperament of sow (Gentle, nervous, cross) _____ | | | | | | | | |
|    No. pigs born: Alive _____ | | | | | | | | |
|       Dead _____ | | | | | | | | |
|       Mummies _____ | | | | | | | | |
|       Total _____ | | | | | | | | |
|    Av. birth weight _____ | | | | | | | | |
| No. functioning teats _____ | | | | | | | | |
| Weaning data   Age _____ | | | | | | | | |
|    No. weaned _____ | | | | | | | | |
|    Av. weaning wt. _____ | | | | | | | | |
| Offspring saved for breeding   No. gilts _____ | | | | | | | | |
|    No. boars _____ | | | | | | | | |

### DISPOSAL OF SOW

Date _____ Reasons _____

Sold to _____
                                        (Name and address)
Price $ _____

Fig. 39-18. Individual sow record form.

## PRODUCTION TESTING BY SWINE RECORD ASSOCIATIONS

The organized performance testing programs of the eight major purebred breeds of swine in the U.S. are under the "umbrella" of the National Association of Swine Records (NASR). This joint effort was first initiated in the central swine test station program in the 1950s. The new and current state-of-the-art genetic evaluation by the purebred breeds of swine is the STAGES program of testing, which embraces both on-farm test programs, and central test station programs.

- **On-farm test programs**—In the on-farm STAGES programs of purebred seedstock producers, the traits evaluated include reproduction (number born alive, number weaned, litter weight, and sow productivity index), growth (days to 230 lb), composition (backfat and loin muscle area), and the resulting indexes of these important traits.

The on-farm genetic evaluation from the STAGES program are used by purebred breeders (1) to cull the poorer producing sows, and (2) to select replacement gilts; and they may be considered by potential buyers of breeding animals.

- **Central test station programs**—There are more than 30 central test stations across the nation that genetically evaluate purebred boars. These test stations, which are owned and operated by impartial technicians, assemble boars from different purebred breeders and compare them under the same environment. The boars are evaluated for growth, composition, and feed efficiency. Then, within each station they are ranked on their performance, following which only the most superior animals are offered for sale. Similar records are accumulated by all test stations, and the genetically superior lines of pigs from across the nation are identified. This ranking of sires (Sire Summary) is made available to anyone interested in purchasing breeding stock.

These performance testing programs make for uniform tests, speed genetic progress, and make it possible for buyers to select genetically superior animals.

## QUESTIONS FOR STUDY AND DISCUSSION

1. Why is swine breeding more flexible than cattle or sheep breeding in the hands of breeders?

2. Discuss the normal breeding habits of swine relative to (a) age of puberty, (b) age to breed gilts, (c) heat periods, (d) gestation period, and (e) breeding after farrowing.

3. What can swine breeders do to increase the fertility and prolificacy of swine?

4. What is meant by flushing sows? How can sows be flushed?

5. What are the two primary reasons for using artificial insemination in swine? Why is A.I. not used more extensively in swine?

6. How are sows and gilts pregnancy tested? List the advantages of early pregnancy detection.

7. Give guidelines relative to the age and service of the boar.

8. List (a) the advantages, and (b) the disadvantages of confinement sow housing.

9. What sanitary measures should be taken in preparation for farrowing?

10. Discuss the use of farrowing crates versus open pens at farrowing time.

11. Detail the raising of orphan pigs.

12. Detail the handling of runts.

13. Why and how are needle teeth removed?

14. Why and how are litters ear notched?

15. What are the main reasons for increased confinement rearing?

16. For the small swine producer, what are the advantages and the disadvantages of (a) two litters a year *versus* (b) one litter a year?

17. What is meant by multiple farrowing? Why have big commercial swine producers gone to multiple farrowing?

18. What is hybrid vigor or heterosis? What traits give the greatest response to hybrid vigor? What traits respond the least to hybrid vigor?

19. What is *complementary*? How should complementary be used in a crossbreeding program?

20. Describe each of the following rotational schemes of crossbreeding: (a) rotational cross system, (b) rototerminal cross system, and (c) terminal cross system.

21. Discuss lethals and other hereditary abnormalities in swine.

22. List the economically important characteristics of swine, and divide them into (a) highly, (b) moderately, or (c) lowly heritable.

23. Discuss the importance of (a) individual animal identity and (b) record forms in production testing swine.

24. Discuss the production testing by swine record associations in the STAGES program (a) on-farm, and (b) in central test stations.

## SELECTED REFERENCES

| Title of Publication | Author(s) | Publisher |
|---|---|---|
| *All About Pigs*, Third Edition | | Farming Press, Ltd., Lloyds Chambers, Ipswich, England, 1970 |
| *Animal Reproduction—Principles and Practices* | A. M. Sorensen, Jr. | McGraw-Hill Book Co., New York, NY, 1979 |
| *Biology of the Pig, The* | W. G. Pond<br>K. A. Houpt | Cornell University Press, Ithaca, NY, 1978 |
| *Managing Swine Reproduction* | L. H. Thompson | University of Illinos, Urbana-Champaign, IL, 1981 |
| *Pig Husbandry*, Fourth Edition | J. R. Luscombe | Farming Press, Ltd., Fenton House, Ipswich, England, 1972 |
| *Pork Industry Handbook* | | Cooperative Extension Service, Purdue University, West Lafayette, IN |
| *Profitable Pig Farming*, Third Edition | E. G. Johnson | Farming Press, Ltd., Lloyds Chambers, Ipswich, England, 1970 |
| *Stockman's Handbook, The* | M. E. Ensminger | The Interstate Printers & Publishers, Inc., Danville, IL, 1983 |
| *Swine Production* | C. E. Bundy<br>R. V. Diggins<br>V. W. Christensen | Prentice-Hall, Inc., Englewood Cliffs, NY, 1976 |
| *Swine Production*, Fifth Edition | J. L. Krider<br>J. H. Conrad<br>W. E. Carroll | McGraw-Hill Book Company, New York, NY, 1982 |
| *Swine Production and Nutrition* | W. G. Pond<br>J. H. Maner | AVI Publishing Company, Inc., Westport, CT, 1984 |
| *Swine Science*, Fifth Edition | M. E. Ensminger<br>R. O. Parker | The Interstate Printers & Publishers, Inc., Danville, IL, 1983 |

Phosphorus made the difference! *Left:* Phosphorus-deficient pig, showing weak and crooked leg bones. *Right:* This pig received the same ration as the pig on the left, but with added phosphorus. (Courtesy, Purdue University, West Lafayette, IN)

# FEEDING AND MANAGING SWINE

# Chapter 40

In the natural state, the wild boar and his kind and kin roved through forests, gleaning the feeds provided by nature. On a modern farm, the range is restricted and sometimes entirely devoid of vegetation. By 1990, the author estimated that more than 60% of the market hogs in the United States were raised from farrow to finish in some type of confinement system—ranging all the way from simple shelters to environmentally controlled pig palaces. As a result of this confinement, domestic swine have less choice in their selection of feed than any other class of four-footed animals. For the most part they are able to consume only what the caretaker provides. This consists largely of concentrated feeds with only a small proportion of roughage. These conditions are made more critical because hogs grow much faster in proportion to their body weight than the larger farm animals, and they produce young at an earlier age. Thus, a knowledge of the nutritional needs of swine is especially important.

Fig. 40–1. More than 60% of market hogs in the U.S. are raised from farrow to finish in confinement. (Photo by J. C. Allen and Son, West Lafayette, IN)

Extensive surveys indicate that about 25 to 30% of all pigs farrowed fail to live to weaning age. Although these heavy losses are due to many and variable factors, certainly nutritional deficiencies play a major role.

Knowledge of feeding swine is also important from an economic standpoint, because feed accounts for approximately 65 to 75% of the total cost of producing pork.

## DIGESTIVE SYSTEM OF THE PIG

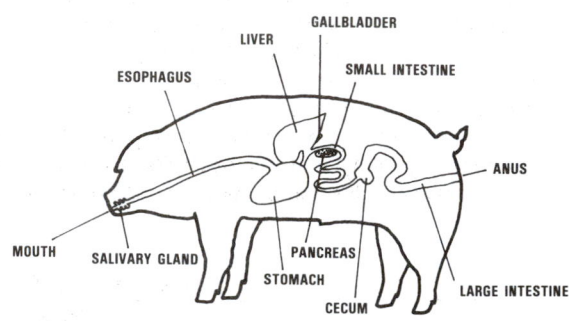

Fig. 40–2. The digestive tract of the pig—a nonruminant.

To grow rapidly and efficiently, swine must receive a high-energy concentrated grain ration, low in fiber. Cattle and sheep, on the other hand, can digest large quantities of fibrous feeds such as hay and pasture. This is largely due to differences in their digestive tracts.

Meat animals can be divided into two broad classifications—ruminants and nonruminants. Pigs are nonruminants. They have a single stomach, in contrast to ruminants which have a stomach divided into four compartments. Cattle and sheep are ruminants.

The digestive tract (or gastrointestinal tract) can be considered a continuous hollow tube—open at both ends—with the body built around it. It's a factory assembly line in reverse; instead of building something, it takes things apart. The digestive tract of the pig includes five main parts: the mouth, esophagus, stomach, small intestine, and large intestine.

Although ruminants and nonruminants differ in their physical makeup, the job of the digestive tract is the same in all animals. It breaks down feedstuffs into simple chemical components so that the animal can absorb and rearrange them into its own characteristic body composition.

## NUTRITIVE NEEDS OF SWINE

The science of animal nutrition has made great progress in unraveling some of the mysteries of the conversion of feed to food. However, this is a complex subject and much remains to be learned.

Plants have the ability to transform raw materials of the soil and air into forage and grain. For many years, livestock producers relied solely on these materials for livestock feeds. As the science of animal husbandry progressed, many supplemental sources of nutrients were discovered.

Swine differ in the kinds and amounts of nutrients needed. The need is influenced by age, function, disease level, nutrient interaction, environment, etc. It has been established that the pig has a requirement for over 40 individual nutrients. Fortunately, not all of them are of practical concern. It should be emphasized that the pig has a requirement for nutrients and not for particular ingredients.

A swine producer has a wide variety of ingredients from which to choose in formulating a ration. Each ingredient may contain several nutrients in varying amounts. Because ingredients vary in price, and in amount and quality of nutrients contained, judgment must be exercised in the choice made.

The nutrient present in the largest amount determines how an ingredient is classified. Thus, soybean meal is classified as a protein supplement because it is high in protein content. There are five main categories of nutrients: (1) energy (carbohydrates and fats), (2) protein (amino acids), (3) minerals, (4) vitamins, and (5) water. Their functions may be described as energy producing, structural, or regulatory.

No one nutrient is more important than another—and all are essential. Each nutrient has one or more particular and specific functions to perform in the body. If the nutrient is not supplied by the ration in proper amounts, the functions (growth, reproduction, lactation, etc.) will be impaired. Since the modern swine producer is interested in maximum performance, it follows that the input of nutrients must be ample to bring this about. For this reason, the recommended allowances of the various important nutrients that should be included in ration formulation for optimum performance are given in Table 40–1, rather than minimum requirements.

Sometimes nutritional requirements, or standards, like those in Table 40–1, impart the erroneous impression that such figures are absolute, final and unchangeable. Nothing could be further from the truth. Rather, Table 40–1 figures are guides, prepared by swine specialists of Iowa State University, the Land Grant University in the leading swine producing state in the nation.

In using Table 40–1, the following pertinent points should be recognized:

1. Feedstuffs produced in various parts of the country vary in nutritive value.

2. The environment in which pigs are produced can modify the requirements.

3. Animals bred for high performance have nutritional needs that are quite different from average performers.

## TABLE 40–1
## RECOMMENDED NUTRIENT ALLOWANCES FOR SWINE[1][2][3]

| | | Sows, Gilts and Boars | | | | Young Pigs | | Grower-Finisher Pigs | |
| --- | --- | --- | --- | --- | --- | --- | --- | --- | --- |
| | | Pregestation, Breeding, and Gestation | | | Lactation | Prestarter, Nursing to 12 lb | Starter, Creep to 40 lb | Grower, Re-placements 40–120 lb | Finisher 120–240 lb |
| | | 3 lb/day | 4 lb/day | 5 lb/day | | | | | |
| **Protein, amino acids:** | | | | | | | | | |
| Protein[4] | (%) | 13.00 | 12.00 | 11.00 | 13.00 | 20–24 | 18–20 | 15–17 | 13–15 |
| Lysine[4] | (%) | 0.60 | 0.45 | 0.35 | 0.60 | 1.40 | 1.15 | 0.75–0.85 | 0.60–0.70 |
| Threonine | (%) | 0.40 | 0.30 | 0.24 | 0.45 | 0.80 | 0.70 | 0.50 | 0.40 |
| Tryptophan | (%) | 0.13 | 0.10 | 0.08 | 0.12 | 0.20 | 0.18 | 0.13 | 0.11 |
| **Major or Macrominerals:** | | | | | | | | | |
| Salt (NaCl) | (%) | 0.50 | 0.40 | 0.30 | 0.50 | 0.25 | 0.25 | 0.25 | 0.25 |
| Calcium (Ca) | (%) | 1.00 | 0.75 | 0.60 | 0.75 | 0.90 | 0.80 | 0.60 | 0.55 |
| Phosphorus (P) | (%) | 0.75 | 0.60 | 0.50 | 0.60 | 0.70 | 0.65 | 0.50 | 0.45 |
| **Trace or Microminerals, added:[5]** | | | | | | | | | |
| Copper (Cu) | (ppm) | 7 | 5 | 4 | 5 | 8 | 8 | 4 | 2 |
| Iodine (I) | (ppm) | 0.20 | 0.14 | 0.11 | 0.14 | 0.14 | 0.14 | 0.07 | 0.04 |
| Iron (Fe) | (ppm) | 100 | 80 | 60 | 80 | 100 | 100 | 50 | 25 |
| Manganese (Mn) | (ppm) | 13 | 10 | 8 | 10 | 4 | 4 | 2 | 1 |
| Selenium (Se) | (ppm) | 0.20 | 0.15 | 0.12 | 0.15 | 0.30 | 0.30 | 0.15 | 0.08 |
| Zinc (Zn) | (ppm) | 65 | 50 | 40 | 50 | 100 | 100 | 50 | 25 |
| **Vitamins, added:[5]** | | | | | | | | | |
| Vitamin A | (IU/lb) | 2,500 | 2,000 | 1,500 | 2,000 | 2,000 | 2,000 | 1,000 | 500 |
| Vitamin D | (IU/lb) | 250 | 200 | 150 | 200 | 200 | 200 | 100 | 50 |
| Vitamin E | (IU/lb) | 13 | 10 | 8 | 10 | 10 | 10 | 5 | 2.5 |
| Biotin[6] | (mg/lb) | 0.13 | 0.10 | 0.08 | 0.10 | 0 | 0 | 0 | 0 |
| Niacin (Nicotinic Acid, Nicotinamide) | (mg/lb) | 7 | 5 | 4 | 5 | 10 | 10 | 5 | 2.5 |
| Pantothenic Acid (Vitamin B-3) | (mg/lb) | 8 | 6 | 5 | 6 | 8 | 8 | 4 | 2 |
| Riboflavin (Vitamin B-2) | (mg/lb) | 3 | 2 | 1.5 | 2 | 2 | 2 | 1.5 | 0.8 |
| Vitamin B-12 (Cobalamins) | (mcg/lb) | 10 | 7.5 | 6 | 7.5 | 10 | 10 | 5 | 2.5 |
| **Feed additives[7]** | (g/ton) | 0–300 | 0–300 | 0–300 | 100–300 | 100–300 | 100–300 | 0–100 | 0–50 |

[1]Adapted by the author from *Life Cycle Swine Nutrition*, Iowa State University, Ames, PM-489, June, 1988.

[2]The nutrient allowances are suggested for maximum performance, not as minimum requirements. They are based on research work with natural feedstuffs and have been found to give satisfactory results. Trace mineral and vitamin levels listed should be added to the ration in addition to those occurring in natural feedstuffs.

[3]To convert lb to kg, divide by 2.2. To convert IU/lb to IU/kg, multiply by 2.2. To convert mg/lb to mg/kg, multiply by 2.2. To convert mcg/lb to mcg/kg, multiply by 2.2. To convert g/ton (short) to g/ton (metric), divide by 0.907.

[4]Sow protein recommendations are based on corn-soybean meal rations. Other feedstuffs may require more protein to meet the amino acid requirement. Protein and lysine ranges for growing and finishing hogs allow for least cost formulation per unit of gain.

[5]Trace mineral and vitamin recommendations for finishing pigs are 50% of grower values. To convert trace or microminerals from ppm to mg/lb, divide by 2.2.

[6]Biotin additions are not needed in corn-soybean meal based rations.

[7]The feed additives may be antibiotics, arsenicals or other chemotherapeutics or combinations. Levels and combinations used and stage of production for which they are used must comply with Food and Drug Administration regulations. High levels for sows may be beneficial just before and after breeding and at farrowing. They are not recommended during the entire gestation-lactation period unless specific diseases are present. The feed additive and the level used during growing and finishing phases should be primarily for growth promotion and improvement of feed efficiency.

## ENERGY

Energy is the body's fuel supply. Every movement and activity of the pig's life involves the expenditure of fuel—energy for breathing, heart action, digestion, muscular movement, as well as heat to keep the body warm. If more energy is consumed than necessary to carry on vital functions, the excess is stored as body fat. In fact, this is what is done in finishing hogs. More energy is eaten than is needed for growth and body maintenance, with the result that the animal lays down fatty tissue with the excess.

The main nutrients supplying energy are carbohydrates. There are several forms of carbohydrates in plants. In feed analysis these forms are identified as either nitrogen-free-extract (NFE) or crude fiber. The NFE fraction includes the more soluble carbohydrates—sugars, starch, and some hemicellulose. All but hemicellulose are very digestible. Crude fiber, however, contains cellulose, hemicellulose, and lignin, all of which are highly digestible by the pig.

The kind of carbohydrate a feed contains determines its value as a source of energy for the pig. Cereal grains are widely used in swine feeding because of their very high NFE (60–70%) and low crude fiber content.

Another group of energy nutrients is the fats and oils. Fat, which is abundant in such common hog feeds as peanuts and soybeans, is a very concentrated source of fuel. It supplies approximately 2.25 times as much metabolizable energy as an equal weight of carbohydrates. Therefore, a feed high in fat, or a ration containing added fat, is much higher in energy value than a feed or ration low in fat. It is emphasized, however, that liberal qualities of either soybeans or peanuts will produce soft pork.

Although roughages are a good source of energy for ruminants, because of their bulky nature and the restricted size of the digestive tract of hogs in comparison with ruminants, only limited quantities of them are contained in normal swine rations. Roughages (pastures and ground legumes and hay) are added to swine rations because of their protein, minerals, and vitamins, rather than for energy purposes.

Energy values are generally expressed in feed tables as total digestible nutrients (TDN), digestible energy (DE), or metabolizable energy (ME). Since energy expenditure can be measured as heat, modern nutritionists measure the energy needs of the animal in calories (units of heat). TDN values may be converted to DE by assuming that 1 lb of TDN is equivalent to 2,000 kilocalories of DE. ME values are calculated from the formula—

$$ME = DE \frac{(96 - 0.2 \times \% \text{ crude protein})}{100}$$

This formula indicates that ME values are somewhat less than 96% of DE values.

*Symptoms of energy deficiency are:* slow or interrupted growth, lowered reproduction, and offspring dead or weak at birth.

- **Essential fatty acids**—In addition to supplying energy, certain dietary fats supply essential fatty acids, most commonly linoleic acid.

Practical swine rations contain adequate amounts of essential fatty acids. After the essential fatty acid requirement has been met, additions of fat increase the energy of the ration.

A deficiency of essential fatty acids in the young, growing pig results in a dull, dry hair coat and a scaly, dandrufflike dermatitis. In later stages, a brownish, gummy exudate and necrotic areas appear about the ears and under the flanks. Also, retarded sexual maturity, an underdeveloped digestive system, and an abnormally small gallbladder have been reported.

## PROTEIN

While carbohydrates and fats are the principal sources of energy, proteins supply the building materials from which body tissue and many body regulators, such as enzymes and hormones, are made. Each protein is made up of several nitrogen compounds called amino acids.

The pig has a specific requirement for each of the essential amino acids. (See Chapter 4, section entitled, "Quality of Proteins," for a list of amino acids.) Since they are needed for the formation of every new cell, the need is most critical when growth is rapid. This makes ration formulation for the young pig very important, because the protein provided at this time must supply the amino acids for muscle growth (lean meat), internal organs, blood, bone, and all other parts associated with growth and development.

Fig. 40–3. Lysine deficiency. *Top:* Pig gained 25 lb in 28 days after lysine was added to the basal diet (2.0% DL-lysine). *Bottom:* A lysine-deficient pig that received the basal diet only. This pig lost 2.0 lb in 28 days. (Courtesy, Purdue University, West Lafayette, IN)

*Quality of protein* is a term used to describe the amino acid balance of proteins. A protein is said to be of good quality when it contains all the essential amino acids in proper proportions and amounts, and to be of poor quality when it is deficient in either content or balance of essential amino acids. From this it is evident that the usefulness of a protein source depends upon its amino acid composition, because the real need of the pig is for amino acids and not for protein as such.

Although it is common practice to refer to *percent protein* in a ration, this term has little significance in swine nutrition unless there is information about the amino acids present. For swine, quality is just as important as quantity. It is possible for pigs to perform better on a 12% protein ration, well-balanced for amino acids, than on a 16% protein ration having a poor amino acid balance.

From a practical standpoint, the problem of building a balanced ration for swine is centered around correcting the deficiencies of the cereal grains. Although corn, wheat, and barley may contain from 8 to 12% protein, this protein is seriously deficient in the essential amino acid, lysine; corn is also deficient in tryptophan. Moreover, the digestive tract of the pig is not adapted to extensive synthesis of proteins by microorganisms like the paunch of ruminants. Also, since protein supplements are more expensive than grain, the tendency is to feed too little of them.

Previously, it was stated that when an excess of energy is consumed by the pig it is stored in the form of fat. Protein is not stored in the body in appreciable amounts. If an excess of protein is fed, the unused nitrogen portion is discarded as urea in the urine and the carbon fraction is used as a source of energy. From an economic standpoint, it is unprofitable to feed more protein than needed to meet the nutritional requirements of the pig.

*Symptoms of protein (amino acid) deficiency are:* reduced feed intake, stunted growth, poor hair and skin condition, and lowered reproduction.

# MINERALS

Of all common farm animals, the pig is most likely to suffer from mineral deficiencies. This is due to the following peculiarities of swine husbandry:

1. Hogs are fed principally upon cereal grains and their by-products, all of which are relatively low in mineral matter, particularly calcium.

2. The skeleton of the pig supports greater weight in proportion to its size than that of any other farm animal.

3. Hogs do not normally consume great amounts of roughage (pasturage or dry forage), which would tend to balance the mineral deficiencies of grains.

4. Hogs are fed to grow at a maximum rate for an early market, before they are mature.

5. Hogs reproduce when less mature than other classes of livestock.

6. Hogs raised in confinement are without access to soil or forage, which would tend to balance the mineral deficiencies of the grains.

Salt (sodium chloride), calcium and phosphorus are the supplemental minerals needed in largest quantities by the pig. Other minerals are required in small amounts and are known as *trace minerals.* The latter include cobalt, copper, iodine, iron,

manganese, selenium, and zinc. Although minerals constitute a small percentage of the swine ration, their importance to the health and well-being of the pig cannot be minimized.

## MAJOR OR MACROMINERALS

Salt (sodium and chlorine), calcium, phosphorus, magnesium, potassium, and sulfur are the major or macrominerals required by swine.

### SALT (SODIUM CHLORIDE)

Salt contains both sodium and chlorine, vital elements found in the fluids and soft tissues of the body. It improves the appetite, promotes growth, helps regulate body pH, and is essential for hydrochloric acid formation in the stomach.

Although swine require less salt than other classes of farm animals, it is generally advantageous to supply them with it, particularly if the protein supplement is not derived from tankage or fish meal (two feeds which supply salt). A lack of salt is marked by poor and depraved appetite, unthrifty condition, and failure to grow.

### CALCIUM AND PHOSPHORUS

A deficiency of either calcium or phosphorus in the diet of the pig can result in poor and inefficient gains, rickets or osteomalacia, broken bones, and posterior paralysis.

A large excess of either calcium or phosphorus interferes with the absorption of the other. Thus, it is important to have a suitable ratio between the two minerals. The most favorable calcium to phosphorus ratio is 1.2:1 to 1.5:1. Also, vitamin D is necessary for the proper utilization of these two minerals.

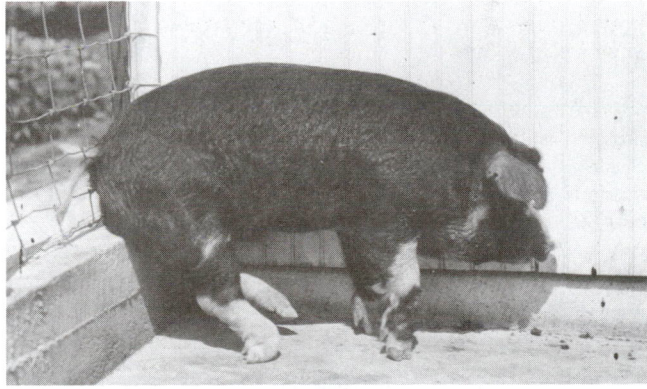

Fig. 40–4. Calcium deficiency. Note abnormal bone development and rachitic condition in advanced stage of deficiency. Lack of calcium retards normal skeletal development, but it does not usually depress total gain. (Courtesy, USDA)

It is important to supplement swine rations with both calcium and phosphorus. Cereal grains, which make up the bulk of swine rations, are quite low in calcium and are only fair sources of phosphorus. Moreover, about ½ to ⅔ of the phosphorus in cereal grains is present in the form of phytin phosphorus, a form of phosphorus which may be poorly utilized by swine.

(For a list of the commonly used calcium and phosphorus supplements, see Chapter 4, Table 4–4.)

## MAGNESIUM

Magnesium is a cofactor in many enzyme systems and a constituent of bone. Apparently, the magnesium requirement of swine is met by grain-soybean meal rations, or by rations containing grain and protein supplements.

## POTASSIUM

Potassium is the most abundant mineral in muscle tissue. Grain-soybean meal rations normally contain enough potassium to meet the requirements for all classes of swine.

The dietary potassium requirement for the pig is increased by high levels of dietary chloride, sulfate, and other anions, as it is for the chick.

## SULFUR

Sulfur is an essential element. However, the sulfur-containing amino acids (cystine and methionine) appear adequate to meet the pig's need for synthesis of sulfur-containing compounds. The addition of inorganic sulfate to low-protein swine rations has not been beneficial.

## TRACE OR MICROMINERALS

Minerals that are required in small amounts are known as trace or microminerals. These include cobalt, copper, iodine, iron, manganese, selenium, and zinc.

## COBALT

Cobalt is a component of vitamin B–12. The intestinal microflora of the pig are capable of synthesizing vitamin B–12 provided sufficient cobalt is present. But only a minimum level of dietary cobalt is necessary for this process. Intestinal synthesis is of greater importance if preformed vitamin B–12 is limiting.

## COPPER

The pig requires copper for the synthesis of hemoglobin and for the synthesis and activation of several oxidative enzymes necessary for normal metabolism.

The deficiency of copper leads to poor iron mobilization. A level of 6 ppm in the ration is adequate for baby pigs.

## IODINE

Hogs require a small amount of iodine. A deficiency of this mineral results in poor hair and skin condition, impaired reproduction, dead or weak offspring at birth, birth of hairless pigs, and goiter.

Fig. 40–5. Hairlessness in pigs caused by a deficiency of iodine. In iodine-deficient areas, farm animals should receive iodized salt throughout the year. (Courtesy, Dept. of Veterinary Pathology and Hygiene, College of Veterinary Medicine, University of Illinois, Urbana)

## IRON

Iron is necessary for the formation of hemoglobin in the red blood cells and the prevention of nutritional anemia. Hemoglobin serves as a carrier of oxygen throughout the body.

As the unborn pig develops, a supply of iron is stored in its body. The amount stored varies greatly between pigs of the same litter. In no case is the amount of iron adequate to keep the pig growing at its maximum for more than 10 days or two weeks after birth unless soil or some supplemental source of iron is available during the suckling period.

Sow's milk is very low in iron; and, to date, research has not uncovered any way of increasing its iron content. Thus, if suckling pigs are confined with no access to soil or feed, serious losses from anemia are likely. Once a pig begins to consume natural feedstuffs, the danger of anemia is practically nil because most feeds contain sufficient amount of iron to meet the pig's requirement.

Fig. 40–6. Suckling pig with nutritional anemia, caused by a lack of iron, characterized by listlessness and wrinkled skin. (Courtesy, University of Florida, Gainesville)

Anemic pigs lose their appetites and become weak and inactive. In more advanced stages of the deficiency, the pig's breathing becomes labored, a condition that is sometimes called *thumps*. In this condition they are more susceptible to other diseases and parasites. Death may occur in severe cases.

For the prevention or treatment of anemia in young pigs, either (1) place a little uncontaminated sod (topsoil, from an area where hogs have not run for years) in the corner of the pen daily, (2) inject a suitable iron preparation at a level of 100 to 200 mg into baby pigs at 1 to 3 days of age, (3) swab the sow's udder with iron solution, (4) give an iron-copper pill, or (5) allow access to oral iron preparations. In addition, the pigs should be encouraged to eat a pig starter ration as soon as they are old enough.

## MANGANESE

Manganese functions with many enzymes in soft tissue metabolism and also in bone development. Deficiency symptoms are: lameness, weakened bone structure, irregular estrus, offspring born dead or weak, and increased backfat. While manganese is usually present in adequate amounts without supplementation in most swine rations, it may not be adequate for the optimum reproductive performance of sows.

## SELENIUM

Selenium functions with gluthathione peroxidase, an enzyme which enables the tripeptide gluthathione to perform its role as a biological antioxidant in the body. This explains why deficiencies of selenium and vitamin E result in similar signs—loss of appetite and slow growth. However, high levels of vitamin E do not completely eliminate the need for selenium. Selenium is now approved by the FDA for addition as sodium selenite or sodium selenate at a level of 0.3 ppm of selenium in all swine rations.

## ZINC

The requirement for zinc in swine rations is very low, but when high levels of calcium are fed, zinc utilization is impaired and the requirement is increased. A zinc deficiency results in a mangelike skin condition called *parakeratosis*. Other symptoms are poor growth, inefficient feed conversion, gilts producing fewer and smaller pigs, boars with retarded testicular development, and young pigs with retarded thymic development.

Fig. 40–7. Zinc deficiency. *Left:* Pig received 17 ppm of zinc and gained only 3 lb in 74 days. Note severe dermatosis ("mangy" look), or parakeratosis. *Right:* Pig received the same diet as the pig on the left, except that the diet contained 67 ppm of zinc. this pig gained 111 lb in 74 days. (Courtesy, Purdue University, West Lafayette, IN)

## FEEDS AS A SOURCE OF MINERALS

The most satisfactory source of minerals for hogs is in the feed consumed. Thus, it is important to know whether the minerals in the ration are of the right kind and sufficient in amount. Certain general characteristics of feeds in regard to calcium and phosphorus (the two predominating mineral elements of the body) are worth noting:

1. The cereal grains and their by-products and protein supplements of plant origin are low in calcium but fairly good in phosphorus. However, as mentioned earlier, the phosphorus in plants is not fully utilized by swine.

2. The protein supplements of animal origin (skim milk, buttermilk, tankage, meat scraps, fish meal), legume forage (pasturage and hay), and rape, are all rich in calcium.

3. Most protein-rich supplements are high in phosphorus.

## METHOD OF FEEDING THE MINERAL SUPPLEMENT

Generally, minerals are incorporated in the ration (see Table 40–1). Additionally, it is recommended that hogs be given free access to a suitable mineral mix. This is cheap insurance against possible needs beyond what's provided in the ration. It's a way of hedging against (1) individual pig differences in mineral requirements, and (2) feed varying in both mineral content and availability.

## VITAMINS

Vitamins are complex organic compounds needed in minute amounts, which are essential for health and normal body functions. Like amino acids, each vitamin has a specific function to perform. Vitamins are classified into two groups—fat-soluble and water-soluble. The body can store reserves of the fat-soluble vitamins for a considerable period of time. But stores of the water-soluble vitamins are depleted quite rapidly.

Because of the greater prevalence of confinement feeding, swine are more likely to suffer from vitamin deficiencies than any other class of four-footed animals.

## FAT SOLUBLE VITAMINS

The primary fat-soluble vitamins of practical importance for swine are vitamins A, D, and E. Vitamin K may be of concern under some circumstances.

## VITAMIN A

The vitamin A needs of swine can be met by either vitamin A or carotene. Vitamin A does not occur in plants. However, green plants and yellow corn contain a yellow pigment called carotene which can be converted into vitamin A by the animal body. The combination of vitamin A and carotene present in the ration is referred to as its vitamin A activity.

Carotene is easily destroyed by the ultraviolet rays of the sun and by heat. The carotene content of corn and legume hay usually deteriorates quite rapidly in storage. Therefore, a synthetic concentrate is a more practical and reliable source of vitamin A than are natural sources. Most commercial feed companies fortify their feeds with a stabilized form of vitamin A, which is active over a considerable period of time.

Vitamin A is essential for vision, reproduction, growth, and the maintenance of differentiated epithelia and mucous secretions of swine.

Swine are able to store vitamin A in the liver, and to draw from this storage during periods of low intake.

Vitamin A deficiency signs in growing pigs are incoordination of movement, loss of control of the hind legs, weakness of the back, and night blindness. Sows may fail to come into heat, they may resorb their fetuses, or they may have young born dead with various deformities and defects. Vitamin A is also needed for normal vision and growth of new cells which line the respiratory, digestive, and reproductive tracts.

## VITAMIN D

Vitamin D is sometimes referred to as the "sunshine" vitamin, since the action of sunlight on a compound in the skin will produce it. As long as hogs are exposed to the sun, there is no danger of a deficiency. Living plants do not contain vitamin D. Plants that are mature or are cut and cured in the sun contain some vitamin D as a result of irradiation by sunlight. Pigs can utilize equally well either vitamin $D_2$ (from plant products) or vitamin $D_3$ (from animal products). Irradiated yeast is a good source of vitamin $D_2$.

Vitamin D is needed for the efficient assimilation of calcium and phosphorus; hence, it is required for the growth of strong bones. A lack of vitamin D will result in stiffness and lameness, rickets, broken or deformed bones, enlargement of joints, and general unthriftiness.

It is noteworthy that the vitamin D requirement is less when a proper balance of calcium and phosphorus exists in the ration.

Fig. 40–8. Rickets (advanced case) caused by a deficiency of vitamin D. The pig was fed indoors, without exposure to sunlight. Because of leg abnormalities, it was unable to walk. Later the pig responded to vitamin D. (Courtesy, University of Saskatchewan, Saskatoon, Canada)

## VITAMIN E

*Vitamin E is a biological oxidant which protects unsaturated fat against oxidation.* A number of compounds called tocopherols have vitamin E activity, the most active of which is alpha-tocopherol. Since cell membranes in the animal body contain unsaturated fat, a vitamin E deficiency may result in oxidative damage to the cell. This is manifested in the pig by liver necrosis, pale muscle, mulberry heart, edema, and sudden death.

Vitamin E is widely distributed in plants, and leafy forages (especially alfalfa) are a good source. Grains (particularly their germs) are fair sources of vitamin E. Most of the vitamin E is removed from soybeans with the oil in the manufacture of soybean meal; thus, soybean meal is a poor source of this vitamin.

The trace element selenium also functions with vitamin E in protecting the body against oxidative damage. The need for vitamin E is more acute when swine feeds are low in selenium. Thus, in areas where feed ingredients are low in selenium and where a majority of swine are raised in confinement without access to forages, supplemental vitamin E or selenium, or both are important.

## VITAMIN K

Vitamin K exists in three forms: phylloquinone ($K_1$), menaquinone ($K_2$), and menadione ($K_3$). Menadione is the synthetic form of vitamin K which has the same cyclic structure as vitamin $K_1$ and $K_2$. Vitamin $K_1$ occurs naturally in green plants. Vitamin $K_2$ is present in microorganisms and is formed by intestinal bacteria.

Vitamin K is one of the essential factors necessary for proper blood clotting. Generally, sufficient amounts of vitamin K are synthesized by bacteria in the digestive tract to meet the needs of swine. However, this synthesis may be inadequate in situations where high antibiotic levels are used, where clotting inhibitors (dicoumarol) may be present from molds in the feed, or where there is excess calcium.

The deficiency symptoms of vitamin K are: slow growth, hemorrhage, prolonged blood-clotting time, and hyperirritability.

## WATER-SOLUBLE VITAMINS

Biotin, niacin, pantothenic acid, riboflavin, and vitamin B–12 are the water-soluble vitamins most likely to be deficient in swine rations. However, occasionally the other water-soluble vitamins are deficient. So, choline, folacin, thiamin, vitamin B–6, and vitamin C are also discussed in the sections that follow.

## BIOTIN

Biotin is important metabolically as a cofactor for several enzymes. It is present in adequate amounts in most common feedstuffs, but its bioavailability varies greatly among ingredients.

In general, biotin supplementation has not improved the performance of baby pigs or of growing-finishing hogs fed a variety of feedstuffs. However, biotin supplementation of sow rations has significantly improved reproductive performance, including the number of pigs farrowed and weaned, litter weaning weight, and number of days from weaning to estrus. Also, biotin supplementation improves hoof-hardness of sows.

## CHOLINE

Choline is included with the vitamins, but does not qualify as a true vitamin because it is required at far greater levels than true vitamins and is not known to participate in any enzyme systems.

Choline functions as a *methyl donor* in metabolism and can lower the requirement of methionine to the extent that methionine is used for this function. It is also a constituent of some important phospholipids in the body. While choline is probably present in adequate amounts in most practical swine rations, studies have shown that more live pigs are born and weaned when sows receive supplemental choline throughout gestation. Also, since proteins supply most of the choline in swine rations, a reduction in ration protein when synthetic lysine is incorporated in the ration may at the same time create a choline-deficient situation.

Choline deficiency symptoms are: lowered reproduction, pigs weak at birth, lack of coordination, and spraddled legs.

## FOLACIN (FOLIC ACID)

Folacin includes a group of compounds with folic acid activity. A deficiency of folacin causes a disturbance in the metabolism of single-carbon compounds, including the synthesis of methyls groups, serine, purines, and thymine. Folacin is involved in the metabolic conversion of amino acids: of serine to glycine, and of homocysteine to methionine.

Folacin deficiency in pigs leads to slow weight gain, fading hair color, and anemia.

The folacin in feedstuffs commonly fed to swine, along with bacterial synthesis within the intestinal tract, usually adequately meets the requirement for all classes of swine.

## NIACIN (NICOTINIC ACID, NICOTINAMIDE)

This vitamin plays an important role in body metabolism as a constituent of two coenzymes, nicotinamide-adenine dinucleotide (NAD) and nicotinamide-adenine dinucleotide phosphate (NADP). These coenzymes in the pig are essential for the metabolism of carbohydrates, proteins and lipids.

A niacin deficiency results in *pig pellagra*. This is characterized by diarrhea, rough skin, and retarded growth.

Recent research has shown that the niacin in cereal grains is in a bound form, and that the niacin of corn may be almost completely unavailable to swine. It should be assumed that all niacin in cereal grains and their by-products is completely unavailable. The protein source and content of the ration can also affect the niacin requirement since tryptophan, an amino acid, can be converted to niacin.

## PANTOTHENIC ACID

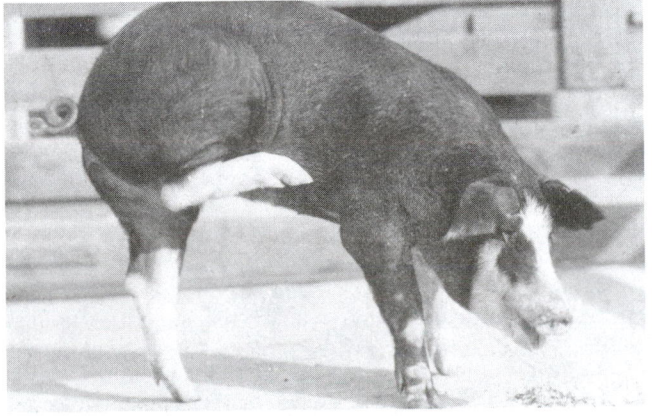

Fig. 40–9. Pig showing pantothenic acid deficiency symptoms. Note high-goose stepping gait. (Courtesy, University of California)

Pantothenic acid is a constituent of coenzyme A, which plays a key role in energy metabolism.

The biological availability of pantothenic acid is high from barley, wheat, and soybean meal, but low from corn and grain sorghum.

A lack of this vitamin may result in poor growth, diarrhea, loss of hair, and high-stepping gait of the hind legs—often called *goose stepping.*

Dried milk products, condensed fish solubles, and alfalfa meal are good natural sources of pantothenic acid. It is also available in synthetic form as calcium pantothenate.

## RIBOFLAVIN (B–2)

Riboflavin is sometimes referred to as vitamin B–2. It functions in the body as a constituent of two coenzymes, flavin mononucleotide and flavin adenine dinucleotide. Riboflavin is important in the metabolism of proteins, fats, and carbohydrates.

In growing swine, a deficiency may cause loss of appetite, stiffness, dermatitis, and eye problems. Poor conception and reproduction have been noted in gilts fed riboflavin-deficient rations. Pigs may be born prematurely, dead, or too weak to survive.

Milk products and other animal proteins, alfalfa meal, and distillers' solubles are good natural sources of riboflavin. (Also, synthetic riboflavin is available.)

## THIAMIN (VITAMIN B–1)

Fig. 40–10. Thiamin deficiency in littermate pigs. *Right:* Pig received no thiamin. *Left:* Pig received the equivalent of 2 mg thiamin/100 lb liveweight. Otherwise, their diets were the same. (Courtesy, USDA)

Thiamin is essential for carbohydrate and protein metabolism. The coenzyme thiamin pyrophosphate is essential for the oxidative decarboxylation of alpha-keto acids.

Pigs must have a dietary source of thiamin, because unlike ruminants, they cannot synthesize sufficient of it in the digestive tract. Normally, the thiamin content of feeds is sufficient to meet the needs of swine since cereal grains, which are major swine feeds, are good sources. However, deficiencies may result from: (1) heating feed ingredients excessively in processing, because thiamin is heat labile; (2) treating feedstuffs with sulfur dioxide, which inactivates the thiamin; or (3) feeding unprocessed fish or fish scraps of certain types of fish that contain the antithiamin factor known as thiaminase.

## VITAMIN B–6 (PYRIDOXINE, PYRIDOXAL, PYRIDOXAMINE)

Vitamin B–6 occurs in feedstuffs as pyridoxine, pyridoxal, pyridoxamine, and pyridoxal phosphate. Pyridoxal phosphate is an important cofactor for many amino acid enzyme systems. Vitamin B–6 plays a key role in central nervous system function.

Supplementation of grain—soybean meal rations with vitamin B–6 is generally unnecessary, because the concentration and availability of vitamin B–6 in the feed ingredients will meet the pig's requirement.

A deficiency of vitamin B–6 will reduce appetite and growth rate. Advanced deficiency will result in exudate around the eyes, convulsions, lack of coordination, coma, and death.

## VITAMIN B–12 (COBALAMINS)

Vitamin B–12 was discovered in 1948. Originally, it was known as the *animal protein factor* because of being associated with sources of animal protein—being particularly rich in fish solubles.

Vitamin B–12 stimulates the appetite, increases rate of growth, improves feed efficiency, and is necessary for normal reproduction.

Pigs require B–12 but responses to supplementation have been variable, primarily because of the synthesis of vitamin B–12 by microorganisms in the environment and within the intestinal tract, along with the pig's inclination toward coprophagy (feces eating).

## VITAMIN C (ASCORBIC ACID, DEHYDROASCORBIC ACID)

Vitamin C (ascorbic acid) is an antioxidant that is involved in a variety of metabolic processes. It is also essential for hydroxylation of proline and lysine, which are integral constituents of collagen. Collagen is essential for growth of cartilage and bone. Vitamin C enhances the formation of intercellular material, the formation of bone matrix, and the formation of tooth dentin.

A dietary source of vitamin C is essential for primates and guinea pigs, but domestic swine can synthesize this vitamin. However, some claims have been made that, under certain conditions—as during periods of excessive stress—pigs may not be able to synthesize vitamin C fast enough to meet their requirement. Yet, the conditions under which supplemental vitamin C may be beneficial are not well defined, so no recommendation for vitamin C is given for the pig.

### UNIDENTIFIED FACTORS

Some unidentified factor or factors may, under certain circumstances, be involved in securing optimum results during the critical periods (early growth and gestation-lactation). Sources of the unknown factor or factors are: distillers' dried solubles, fish solubles, dried whey, grass juice concentrate, soil, alfalfa meal, brewers' dried yeast, liver, and pasture.

## WATER

Water is so common that it is seldom thought of as a nutrient. However, it is the largest single part of nearly all living things. The body of a baby pig is about three-fourths water.

Water performs many tasks in the body. It makes up most of the blood which carries nutrients to the cells and carries waste products away; it is necessary in most of the body's chemical reactions; it is the body's built-in cooling system—it regulates the temperature; and it serves as a lubricant.

Life on earth would not be possible without water. An animal can live longer without feed than without water.

In general, swine will consume ¼ to ⅓ gal. of water for every pound of dry feed. The higher the temperature, the greater the water consumption. It is preferable that swine have access to automatic waterers, with cool, clean water available at all times. Otherwise, they should be hand watered at least twice daily. During winter, the drinking water should not be permitted to fall below 40°F. The estimated water needs of various classes of swine are:

| Class of Swine | Water Consumption (gal/head/day) |
|---|---|
| Gestating sows | 2–3 |
| Lactating sows | 4–5 |
| Weaned pigs (15–40 lb) | 0.5–1 |
| Growing pigs (40–110 lb) | 1 |
| Finishing pigs (110–240 lb) | 1.5–2 |

## FEEDS FOR SWINE

Throughout the world, swine are raised on a great variety of feeds, including numerous by-products. Except when on pasture or when ground dry forages are incorporated in the ration, they eat relatively little roughage; only 4.3% of the total feed consumed by swine in the United States is derived from roughages (see Chapter 2, Table 2–5).

Although corn is the chief concentrate fed to swine, the agriculture of the 50 states is very diverse, and the diet of the pig is readily adapted to the feeds produced locally. A similar adaptation in feeding practices is found in other countries. Thus, in most sections of the world, swine are fed predominantly on homegrown feeds. Ireland depends largely upon potatoes and dairy by-products; the swine industry of Denmark has been built up to augment the dairy industry, with milk and whey supplementing homegrown and imported cereals (mostly barley); in Germany, the pig is fed on such crops as potatoes, sugar beets, and green forage; and in China, the pig is primarily a scavenger, competing very little for grains suitable for human consumption.

## CONCENTRATES

Because of their simple monogastric stomach, swine consume more concentrates and less roughages than any other class of large farm animals. This characteristic gives pigs limited opportunity to consume large quantities of calcium and of vitamin-rich and better quality protein roughages, with the result that they suffer from more nutritional deficiencies than any other species except poultry.

Although most concentrate feeds are not suitable as the sole ration for hogs, it must be realized that swine can utilize a larger variety of feeds to greater advantage than other farm animals. In general, the grain crops—corn, barley, wheat, oats, rye, and the sorghums—constitute the major component of the swine ration. However, sweet potatoes and peanuts are successfully and extensively used in the South, soybeans in the central states, and peas in the Northwest. In those districts where they are grown, potatoes (cull) also are utilized in considerable quantities in feeding hogs. In addition, in almost every section of the country one or more by-product feeds are fed to hogs—including the by-products of the fishing, meat-packing, milling, and dairy industries. Human food wastes, such as refuse or garbage, are also fed extensively.

The protein and vitamin requirements of the monogastric pig differ very greatly from those of the ruminant, for the latter improves the quality of proteins and creates certain vitamins through bacterial synthesis.

Despite all this, it is possible to meet the nutritive needs of the pig on concentrated feeds by keeping in mind the following facts when balancing the ration:

1. The cereal grains and their by-products are relatively good in phosphorus, but low in calcium and other minerals.

2. Except for the carotene content of yellow corn and green peas, the grains are very poor sources of vitamins.

3. Most cereal grains supply proteins of poor quality.

4. Protein supplements of animal origin and soybean meal generally supply proteins of high quality, whereas proteins of plant origin, other than soybean meal, generally supply proteins of low quality.

5. Because of the inadequacies of most concentrates, it is usually necessary to rely on fortifications with minerals and vitamins.

## ENERGY FEEDS

Carbohydrates and fats may be classed together as energy feeds for swine. Three essential fatty acids—linoleic, linolenic, and arachidonic acid—are required by swine, but cereal grains contain adequate quantities of fat to meet the fatty acid requirements. The ingredients commonly used as sources of energy are corn, barley, sorghum, wheat, and fats and oils. Full replacement of corn with barley, oats, sorghum, or wheat will decrease dressing percentage by approximately 1% and decrease backfat by about 0.1 in.

### CORN

In this country, corn and swine production have always gone hand in hand. Normally, about one-fourth of the U.S. corn crop is fed to hogs. Because of the dominant position of corn as a swine feed, it is herein singled out for further elucidation.

Corn is an excellent energy feed for all classes of swine. It is an ideal finishing feed because it is high in digestible carbohydrate (starch), low in fiber, and very palatable. Also, it can be fed in a variety of ways. It may be fed shelled, ground, mixed, or free choice, or even as ear corn. It may be dry or high moisture.

In spite of its virtues, corn alone will not keep pigs alive. It contains 7 to 9% protein, but the protein is deficient in some of the essential amino acids required by the weaning pig, especially lysine and tryptophan (see Fig. 40–11). It is also so deficient in calcium and other minerals, and so inadequate in vitamin content, that pigs will die if they are limited to a ration containing only corn. So, corn must be supplemented with a protein that makes up its amino acid deficiencies. Equally important are the needed minerals and vitamins. When properly supplemented, corn is an excellent energy feed for all classes of swine.

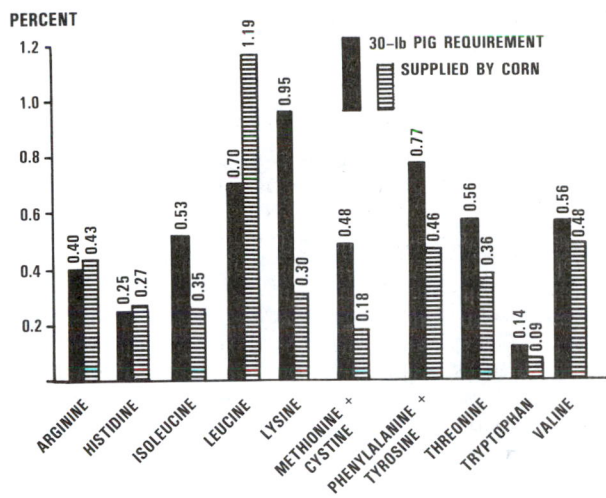

Fig. 40–11. Hogs cannot live by corn alone! This shows the amino acid content of corn (in %) in comparison with the ration requirements (in %) of a 30-lb pig. Note that corn is especially deficient in lysine and tryptophan. (Sources of data: Amino acid ration requirements [in % as-fed] of a 30-lb pig from *Nutrient Requirements of Swine*, 9th rev. ed., National Research Council, National Academy Press, 1988, p. 50, Table 5–1.)

It should be noted that corn is higher in fat than barley or wheat (4% vs less than 2%). Fat not only contributes to the high-energy content of corn, but it also improves its palatability and feeding properties in general.

- **High-moisture corn**—High-moisture corn may be substituted for dry corn on a dry matter basis with little effect on overall performance, in either feed conversion or rate of gain. It is usually fed free choice. Protein, vitamins, and minerals can be supplemented by either hand feeding daily or feeding free choice in separate feeders. High-moisture corn is usually very palatable, with the result that when fed free choice, pigs will often overeat on corn and undereat on the protein supplement. When the two are metered together, separation is likely to occur. If high-moisture corn is fed in a complete ration, the feed should be prepared frequently (every 1 or 2 days) to prevent spoilage. The feed should also be prevented from bridging in the feeders.

Thus, it may be said that the choice between high-moisture grain storage vs dry storage should be based largely on economics. The cost per bushel of storage capacity and the operating cost should also be compared in reaching a decision. Also, remember that when high-moisture corn is put in high-moisture storage facilities, it is suitable for feeding only and cannot be sold readily on the open market as dry corn.

- **High-lysine corn (Opaque-2)**—Corn is now being bred that is much higher than normal corn in lysine and tryptophan; hence, it has a better balance of the amino acids for swine. Also, the high-lysine corn (called *Opaque-2*), which was discovered by Purdue University in 1963, is higher in total protein than normal corn. However, Opaque-2 is lower in leucine than regular corn.

Currently, little high-lysine corn is available commercially. However, corn breeders are striving to develop an improved high-lysine corn for it would make for improved nutrition of both animals and people throughout the world, wherever corn (maize) is grown.

Subsequent to the discovery of Opaque-2, Purdue researchers found another high-lysine strain, which was named *Floury-2*. Then, in 1989, University of Minnesota researchers announced the discovery of a gene in corn that controls the level of protein produced, which can be used to produce corn with 3% more protein and 20% more methionine and lysine than normal corn, without lowering the yields or producing soft kernels.

## FATS AND OILS

The following types of fats and oils are available and may be added to swine rations as high-energy feeds: animal fat, poultry fat, tallow, lard, corn oil, soybean oil, and other plant seed oils. When of comparable quality, there is little difference in swine performance resulting from the type of fat or oil used.

When fat is included in growing-finishing rations, it is usually added at a level of 4 to 5%. As a result of added fat, daily feed intake usually decreases, daily gain increases slightly, and feed efficiency improves. Each 1% added fat produces about 2% improvement in feed efficiency..

When fat is added to rations, the caloric density of the ration increases. So, dietary nutrient density should increase in proportion to the increase in caloric density to maximize swine performance.

When fat is added to brood sow rations at a level of 7½% or more 10 to 14 days before farrowing, it increases the fat content of colostrum and milk and improves baby pig survival by 2 to 3% in herds where survival rates are less than 80%. However, added fat has little effect on litter size, birth weight, or weaning weight.

When oils are fed to pigs, they tend to make the body fat softer. To alleviate this problem, the level of oil must be limited and the oil must be deleted from the ration for a "hardening off" period prior to slaughter. Feeding fish oils gives a disagreeable fishy taste to the pork.

## OTHER ENERGY FEEDS FOR SWINE

Yellow corn is usually the cheapest source of energy over much of the United States. But price fluctuations frequently justify consideration of other feeds.

Although barley, oats, wheat, milo, and rye are higher in protein than corn, it is noteworthy that their protein is generally of the same poor quality as corn. It is noteworthy, too, that all of the cereal grains have about the same vitamin and mineral deficiencies.

The relative values, maximum quantities, preparation, and other pertinent information relative to various energy feeds that may be used as replacements for corn in swine rations are summarized in Table 40–2.

**TABLE 40–2**
**ENERGY SUBSTITUTION TABLE FOR SWINE**

| Grain | Relative Feeding (energy) Value Compared with Corn, Pound for Pound Basis | Maximum Percentage of the Corn It May Replace in a Typical Corn-Supplement Formula | Preparation | Remarks |
|---|---|---|---|---|
| **Corn** (yellow) | 100 | 100 | Ground or shelled. | Must be ground if under 15% moisture. |
| **Bakery waste** | 95 | 50 | | They average about 10% protein. Protein, mineral, and vitamin supplementation should be same as for the cereal grains. |
| **Barley** | 90–95 | 100 | Ground and mixed. | Feeding value varies considerably with test weight. Should be ground medium fine or rolled. |
| **Beans** (cuff) | 90 | 33–66 | | Cook thoroughly. Supplement with good quality protein. Limit to ⅓ of the grain ration (dry weight before cooking) for pigs up to 100 lb. For pigs above 100 lb, may be increased to ⅔ the grain ration when the price is right. |
| **Fat** (tallow, greases) | 240 | 5 | | There are no proteins (amino acids), minerals, or vitamins in fats. |
| **Lard** | 230 | 5 | | |
| **Molasses** (cane or beet) | 60–70 | 5 | | Too much molasses will cause scours. |
| **Oats** | 70–80 | 33 | Ground and mixed. | Highly variable test weights on oats. Due to their bulk, oats are a better feed for breeding animals than for young pigs or finishing hogs. Oat groats (dehulled and rolled oats) are an excellent feed for baby pigs. |
| **Potatoes** (cull, Irish) | 25–28 | 25–50 | | Contain only 23% as much dry matter as shelled corn. Not palatable in raw state; must be cooked. When cooked and fed in a ration of 3 lb of potatoes to 1 lb of grain, they are worth 25–28% as much as corn. |

*(Continued)*

**TABLE 40–2** (Continued)

| Grain | Relative Feeding (energy) Value Compared with Corn, Pound for Pound Basis | Maximum Percentage of the Corn It May Replace in a Typical Corn-Supplement Formula | Preparation | Remarks |
|---|---|---|---|---|
| Rye | 90 | 30 | Ground and mixed. | Not higly palatable.<br>Rye is often infested with the fungus *ergot*, which may cause abortion in sows. |
| Sorghum (milo) | 95 | 100 | Whole or ground. | Grain sorghums are the counterpart of corn in the Southwest.<br>If grain is very dry, it should be ground. Bird resistant varieties usually should be ground.<br>If gains are below normal, increase protein percentage in ration 2 percentage points. |
| Spelt | 65–80 | 25 | Ground and mixed. | Value influenced by amount of hulls in the feed. |
| Triticale | 80–90 | 50 | Ground and mixed. | More experimental work needed. |
| Wheat | 100–105 | 100 | Ground and mixed. | Wheat-corn mixtures are more efficient than wheat alone.<br>Wheat should be ground coarsely or rolled for hogs. Fine grinding makes it pasty and less palatable. |

## PROTEIN FEEDS

Protein is made up of nitrogenous compounds called amino acids. Protein feeds vary in the kind and amount of amino acids they contain. During the digestion process, the protein in feed is broken down into the various amino acids and the pig recombines them into the kind of protein needed for muscle development, repair of worn-out tissue, etc. Thus, the real need of the pig is for amino acids, not protein as such.

The pig can synthesize some of the amino acids, with the result that they are not required in the diet. However, 10 of the amino acids are termed *essential,* because the body cannot manufacture them in sufficient quantity to permit maximum growth and performance. It is important that ingredients rich in the essential amino acids be used in formulating the ration.

Although it is a common practice to refer to *percent protein* in a ration, this term has little meaning in swine feeds unless there is knowledge concerning the amino acids present. A protein feed is considered to be of good quality when it contains all the essential amino acids in the proportions and amounts needed by the pig.

## SOYBEAN MEAL

Soybean meal is by far the leading high-protein supplement of the United States. About 75% total high-protein animal feed (including both oilseed meals and animal proteins) consists of soybean meal.

Although soybean meal is marginal in methionine, it is otherwise very well balanced in amino acids. It must be supplemented with minerals and vitamins. Usually, it is not fed free choice because of its high palatability, which results in pigs eating more than is needed to meet their protein needs.

Producers generally have a choice of buying soybean meal of different protein content, usually 44 or 49%. The higher protein content meal is the most desirable to use in prestarter and starter rations because much of the hull has been removed. Thus, it is lower in fiber (not more than 3%), higher in energy,

and more palatable. Growing-finishing pigs can utilize both meals about equally well; hence, the choice should be on the basis of which is the best buy as determined by the price per unit of protein.

## FULL-FAT SOYBEANS

The protein of raw soybeans is poorly utilized by young, growing pigs due to the presence of antitrypsin, a powerful growth inhibitor that affects young swine.

Cooking or roasting at the proper temperature (250°F for 2½ to 3½ minutes in a roaster) destroys this factor and makes soybeans a satisfactory feed for young pigs. However, cooking whole soybeans for brood sows is not necessary; comparable performance results when raw soybeans replace soybean meal on an equal protein basis in rations fed to gestating and/or lactating sows.

Research has shown that whole cooked soybeans can be used to replace soybean meal or other forms of protein supplement in growing-finishing rations. They will increase daily gain up to 5%, and improve feed efficiency by 5 to 10% due to the higher fat content of the whole soybeans (17 to 18%), which makes for a higher energy ration. However, this improvement in feed efficiency may be offset by the lower protein content of whole soybeans; whole cooked beans average about 37% crude protein, whereas soybean meal usually runs 44%. Also, due to the higher energy of the beans, the protein content of the ration must be 1 to 2% higher than in a soybean meal ration in order to maintain the same protein-to-energy ratio.

The feeding of full-fat, cooked soybeans to growing-finishing pigs has little effect on grade and yield if the proper protein-to-energy ratio is maintained, but softer pork may be produced.

It is noteworthy that hogs fed cooked whole soybeans have a softer carcass than those fed a soybean meal ration. Whether this condition will influence the price packers are willing to pay for live hogs should be determined.

## OTHER PROTEIN FEEDS FOR SWINE

Table 40-3 gives the comparative value of the various ingredients commonly used as protein supplements for swine, along with the recommended percent of each to use (1) in a complete ration, and (2) in a protein supplement. Of course, a mixed protein supplement, like those given in Table 40-17 may be used.

### TABLE 40-3
### PROTEIN SUBSTITUTION TABLE FOR SWINE

| Protein Sources | Relative Feeding Value (lb for lb) in Comparison with Soybean Meal | Percent to Use in— | | Comments |
|---|---|---|---|---|
| | | Ration | Supplement | |
| **Soybean meal** (44%) ....... | 100 | 2–25 | 50–100 | **W**ell-balanced amino acids, very palatable. |
| **Blood meal** ................. | 123 | 0–9 | 0–20 | **H**igh in protein (above 80%). But the protein is lower in digestibility and quality than most other animal protein feeds. Also, it is not very palatable. |
| **Cottonseed meal** (44%) ..... | 85 | 0–9 | 0–25 | **G**ossypol which is toxic is present. But gossypol-free varieties of cottonseed are being developed. Cottonseed meal is low in lysine. |
| **Fish meal, Menhaden** (63%) .. | 115 | 0–10 | 0–25 | **E**xcellent balance of amino acids, expensive, good source of calcium and phosphorus. |
| **Linseed meal** (35%) ........ | 80 | 0–9 | 0–25 | **L**ow lysine, slightly laxative. |
| **Meat and bone meal** (50%) .. | 100 | 0–10 | 0–30 | **L**ow tryptophan, good source of calcium and phosphorus. |
| **Peanut meal** (44%) ......... | 95 | 0–15 | 0–50 | **L**ow lysine, very palatable. |
| **Skim milk, dried** .......... | 100 | 0–10 | 0–20 | **E**xcellent quality, very palatable, expensive. Especially good for use in prestarter and starter rations. |
| **Tankage** (60%) ............. | 110 | 0–10 | 0–30 | **L**ow tryptophan, good source of calcium and phosphorus. |

## PASTURES

Pasture was formerly thought to be an absolute essential for a successful swine operation. However, in recent years producing hogs in confinement has become a reality because of vastly improved rations, along with greater disease and parasite control. But it is still possible to utilize large amounts of forage effectively for the breeding herd.

Research reports on feed savings by swine on pasture vary considerably, depending on type of pasture, class and age of hogs, and management system. On the average, however, good pasture for growing-finishing hogs will effect a saving of 3 to 10% of the grain and as much as 33% of the supplement, in comparison with confinement production. Bred sows and gilts on legume pasture require about half as much grain and much less supplemental protein than those in drylots. Thus, the decision on whether to raise swine on pasture or in drylot should be based primarily on (1) net returns, and (2) whether the land can be put to a more profitable alternative use.

In general, temporary pastures are preferable to permanent pastures for swine, especially from the standpoint of disease and parasite control.

Over much of the country, alfalfa, the clovers (principally ladino, red, alsike, and sweet clover), rape, and oat mixtures make excellent swine pastures. Other plants which find use as hog pastures in certain areas and under certain conditions include bluegrass, bromegrass, orchardgrass, lespedeza, carpetgrass, rye, wheat, soybeans, cowpeas, field beans, sorghum, and Sudangrass.

Fig. 40-12. Pigs on pasture. Although more than 60% of the market hogs in the U.S. are raised in confinement, the use of pasture is still a viable alternative. (Courtesy, National Spotted Swine Record, Inc., Bainbridge, IN)

Little pollution potential exists from pasture systems with low animal densities or numbers, or where pastures are rotated. However, in high-density pasture systems involving a large number of hogs, a pollution potential does exist. Frequently there is little vegetation in these pasture lots, and rainwater falling on the lot drains away, carrying some solids with it. Where the lots are steeply sloping or where a water-course runs adjacent to or through the lots, the problem can be serious.

Where a pollution problem exists from pasture lots, the site should be abandoned or all runoff should be caught in a retention reservoir. After each rain, liquid must be removed from the reservoir by pumping and/or evaporation to ensure sufficient volume to capture all runoff from the next rain.

## DRY FORAGES

The favorable results obtained from feeding a corn-soybean meal ration, properly fortified with minerals and vitamins, in an era of relatively cheap grain, caused many researchers and swine producers to question the wisdom of using alfalfa meal in the ration. There was never any doubt about alfalfa meal being an excellent source for quality protein, carotene (vitamin A), B vitamins (riboflavin, pantothenic acid, and niacin), vitamin D if sun-cured, calcium, and unidentified factors. But these nutrients could be purchased more cheaply from other sources. Besides, it has not been proven that alfalfa meal contains some unknown nutrient essential for improved performance; and its high fiber (25 to 30%) and low palatability limit the amounts of it that can be used in baby pig and in growing-finishing rations. As a result, the swing was away from the use of alfalfa meal in swine rations.

But scarce and high-priced grains caused the return of alfalfa as a favorite feedstuff, particularly in rations for gestating sows. If the price of alfalfa is right—if it is a cheaper source of protein and energy than corn and other grains—it may, to advantage, be incorporated in swine rations at the following levels: up to 5% for grower-finishing hogs, up to 50% for gestating sows, and up to 10% for lactating sows. These levels serve as a safety factor to ensure the presence of certain minerals, vitamins, and possibly unidentified factors.

Alfalfa is especially well suited for winter feeding of swine. Because of its low energy and high fiber, it generates considerable heat when digested. The extra heat can be utilized to help maintain body temperature during the winter. Thus, alfalfa is more cost effective as a feed for swine during winter than during summer.

## SILAGES

Good quality silage is an excellent feed for brood sows; hence, it may be used to advantage where it is available on dairy and beef farms. Unless the sow herd is very large, however, it will not likely pay to construct a silo especially for hogs.

Silage should be fed fresh, daily, in amounts the sows will clean up in 2 to 3 hours. Sows will usually eat 10 to 15 lb, and gilts, 8 to 12 lb. Some wastage can be expected. It is important that the silage be supplemented with 1.0 to 1.5 lb of good protein supplement per head daily. Additional grain should be fed if necessary to keep sows in proper condition.

When feeding silage to hogs, the following precautions should be observed.

1. Never feed silage alone, as a poor pig crop will result.
2. Silage causes digestive upsets (diarrhea) in baby pigs. Do not feed it to lactating sows.
3. Avoid feeding moldy or frozen silage. It can cause sows to abort.

Even though considerable silage may be fed to brood sows to advantage, it is important that it be of good quality and that it be properly supplemented from a nutritional standpoint.

## HOGGING DOWN CROPS

Sometimes pigs are permitted to do their own harvesting. Corn is the principal crop so used, the animals being turned into the field when the grain is in the dent stage. Also, small grain crops that have been badly lodged or otherwise damaged, or that cannot be harvested because of weather, may be harvested by hogs. Soybeans and field peas may be hogged off. In the south, such crops as peanuts, sweet potatoes, chufas, and other root and tuber crops, are often harvested by hogs.

Fig. 40–13. Hogging down corn. The animals are usually turned into the field when the grain is in the dent stage. (Courtesy, USDA)

Space will not permit a full discussion of the method of utilizing the various crops. As corn is the main feed hogged down, comments will be limited to this crop; but the same general principles apply to other crops, when and if they are so utilized.

Some of the **advantages** of hogging down corn are:

1. It saves labor at a busy season of the year.
2. The maximum fertility value of the manure is conserved.
3. There is less danger of infesting swine with diseases and parasites than in drylot finishing.
4. Corn that is down or badly lodged is difficult to harvest, but it may be utilized through hogging down.

Some of the **disadvantages** of hogging down corn are:

1. During wet weather, a considerable amount of corn is lost by being tramped into the ground.

2. During wet weather, the tramping of animals puddles the soil and lowers its tilth. This is especially noticeable in heavy clay soils.

3. It usually requires additional fencing, for it is desirable to fence off a small area, then make the pigs clean it up before moving to a new area.

4. Early pigs cannot be used in hogging down corn, for they will be too far advanced.

5. When corn is hogged down, wheat cannot follow corn in the rotation.

## GARBAGE

Municipal garbage has long been fed to finishing hogs, but, following World War II, the practice declined because of (1) a gradual lowering in the feeding value of garbage, and (2) other competition for garbage—notably its manufacture into lawn, greenhouse, and garden fertilizer. The recent development of garbage recycling processes, along with high-priced grains, has created renewed interest in garbage as a hog feed.

As a rule of thumb, about 4 lb of heavy garbage may be considered as equivalent to 1 lb of concentrate.

Garbage may be utilized either as a feed for a sow and pig enterprise or for finishing feeder pigs that are obtained from other sources. Usually, the venture seems most successful when a combination of grain and garbage feeding is practiced.

It is also observed that the most successful garbage feeders use concrete feeding floors, practice rigid sanitation, and take every precaution to prevent diseases and parasites. Unless considerable grain is fed to market hogs, especially after weights are over 100 lb, soft pork and paunchiness will result in garbage-fed hogs.

All states now have laws requiring that commercial garbage be cooked.

The claim is frequently made that the greatest number of cases of trichinosis in humans occurs in communities where garbage is fed to hogs. It should be noted, however, that there is little or no danger in transmitting the disease in this way provided pork and pork products are thoroughly cooked.

## ADDITIVES

Certain additives have become standard ingredients of swine rations, especially for pigs from birth to market weight. They are not nutrients as such; hence, they should not be considered as dietary essentials. Although many different additives are used in swine rations, antibiotics and sulfas are most common.

• **Antibiotics**—Antibiotics are widely used as feed additives to stimulate growth, improve feed efficiency, secure uniformity of performance, and control infections. The response secured from their use depends on (1) the age of the pig, (2) the sanitary conditions, (3) the level fed, (4) the health and environment of the animal, (5) the type of ration, and (6) the season of the year. When fed to young, unthrifty pigs, antibiotics have increased growth rate by over 200%. For growing-finishing hogs under good sanitary conditions, antibiotics generally result in about 10% faster gains on 5% less feed. Pigs up to 100 lb weight give the greatest response to antibiotic feeding. Experimental results have been inconsistent relative to the value of antibiotics in brood sow rations, but it appears that breeding herds with a high disease level may show a favorable response.

In addition to antibiotics, certain other antimicrobial compounds can be used as feed additives, either (1) at low levels for promotion of growth and improvement in feed efficiency, or (2) for treatment and prevention of disease. Among these are nitrofurans, sulfonamides, copper, and arsenicals. Such compounds, alone or in combination with other antimicrobial compounds, should be used only at approved levels and for the specific purpose for which they are authorized.

*CAUTION:* Toxic residues in meat, milk, and/or eggs may result from the improper use of certain antimicrobial compounds, pesticides, tranquilizers, and other chemicals.

Regardless of the material used, the following safety precautions should be observed:

1. Use only those materials specifically authorized for use in swine. Regulations change frequently, so check them often.

2. Carefully follow label directions.

3. Do not mix unauthorized combinations.

4. Use minimal effective amounts.

5. Add or apply precise quantities.

6. Observe the proper interval between use and slaughter (or use for food).

• **Sulfonamides (sulfas)**—*Sulfonamides are organic compounds with bacterial and growth promotant properties similar to those of the antibiotics.* But, unlike the antibiotics, they are produced chemically rather than microbiologically. Also, the therapeutic use of sulfonamides preceded that of the antibiotics; they have been widely used in human and veterinary medicine since the mid-1930s. By 1955, two sulfas—sulfamethazine and sulfathiazole—were found to be effective in the treatment of atrophic rhinitis in swine. Soon, they were also used to treat or prevent swine bacterial enteritis, bacterial pneumonia, salmonellosis, and dysentery. Also, it was found that, in subtherapeutic doses, they would make for more rapid and efficient swine gains. They proved to be effective disease fighters and growth promotants. All went well until the 1960s and early 1970s, when the U.S. Department of Agriculture (USDA) discovered residues of sulfa drugs in many pork carcasses. The immediate concern was that a small percentage of the human population is hypersensitive to sulfas and may develop an allergic reaction after consuming low levels of sulfas in pork (or in other foods). Then, in 1980, a preliminary Food and Drug Administration (FDA) study showed that massive doses of sulfamethazine caused thyroid tumors in laboratory mice.

As a result of finding sulfa residues in pork, followed by an experiment showing that sulfamethazine produced thyroid tumors in mice, in 1987 the USDA and FDA launched a campaign to alleviate sulfa residues in pork.

Animal products are deemed to be in violation of sulfa residue regulations if levels of 0.1 ppm, or more, sulfa are found in the muscle, liver, kidney, eggs, or milk. Research at the University of Kentucky showed that as little as 1 g of sulfa per ton of complete feed can cause 100% violative liver residues and 63% kidney contaminations. This means that as little as

¼ tsp of sulfa per ton of complete feed can result in pork carcasses that are in violation. The sulfonamides are also capable of premise contamination.

Of the more than 5,000 sulfa compounds that have been synthesized, only two—sulfamethazine and sulfathiazole—are approved for use in swine feeds. Two additional sulfa drugs, sulfamerazine and sulfapyridine, are approved, along with sulfamethazine and sulfathiazole, for use as water medicants for swine.

Sulfamethazine and sulfathiazole are approved for use in swine feeds on the following basis: They must be used at one level only—100 g per ton; they must be used in legal combinations approved by the FDA (it is illegal to use sulfas alone for swine); they are approved for the prevention and treatment of enteritis, atrophic rhinitis, cervical abscesses, and leptospirosis (Although sulfas will improve feed efficiency and rate of gain of swine, they are not approved by the FDA as growth promotants); and there is a minimal 15-day withdrawal period for feed and water containing sulfamethazine, and a minimal 7-day withdrawal period for feed and water containing sulfathiazole.

*When used properly and withdrawn at the specified time before marketing, sulfa drugs have been shown to be safe.*

- **Porcine somatotropin (PST) for swine**—Porcine somatotropin (PST) is the scientific name for the growth hormone in swine. It is normally produced by the anterior pituitary at the base of the brain. It stimulates protein synthesis and growth in most tissues of the body, and it causes breakdown of fat deposits in adipose tissue.

Scientists have known for some time that providing extra somatotropin to a pig will cause it to grow more rapidly and produce leaner pork. But it was not feasible to apply this knowledge practically, because the only way to get porcine somatotropin was to isolate it from the pituitary glands of slaughtered hogs, which was very expensive because of the very low yield of hormone from each hog. Recently, it has become economically feasible to mass produce this compound by recombinant DNA procedures using *E. coli* bacteria. So, large supplies of PST are now available. The major remaining barrier to the use of PST is a delivery system—a method of administering. Presently, daily injections are required to capture the benefits of growth and carcass changes. PST is a naturally occurring protein that is broken down immediately by digestion; so, an orally active feed ingredient won't work.

Experiments have consistently shown that PST can make the following great leaps forward in the swine industry: 15–20% higher average daily gain, 20–30% improvement in feed conversion efficiency, 10–15% improvement in muscle mass, and 25–30% less backfat.

Before PST can be used, it must have FDA approval. Then its acceptance and use will be determined by a practical method of administering the product, and producer and consumer acceptance.

(Also, see Chapter 4, section on "Additives, Implants, and Injections.")

## FEED PROCESSING

The processing of feeds for swine is fully covered in Chapter 4, section entitled, "Feed Processing," including Table 4–10 therein; hence, the reader is referred thereto.

## CREEP FEEDING PIGS

Fig. 40–14. Creep for young pigs, located in a light, warm, dry, and draft-free area. (Photo by J. C. Allen and Son, West Lafayette, IN)

Baby pigs should have access to a creep feed beginning at 7 to 10 days of age. Commercial milk replacers (prestarters) and starters are readily available, or farm-mixed creep rations can be used (see Table 40–12 [p. 790] for suggested prestarter rations, and see Table 40–13 [p. 791] for suggested starter rations). A prestarter ration is normally fed to pigs weaned before 3 weeks of age and/or until they weigh approximately 12 lb, after which they can be switched to a starter ration. Also, a prestarter ration may be used for orphan pigs, when extreme disease outbreaks occur, or when the sow fails to produce sufficient milk. They should be fed a starter ration until they weigh about 40 lb. Pigs should receive a total of 3 to 5 lb of milk replacer (prestarter), after which they should be switched to starter until they weigh about 40 lb. Early availability of a quality prestarter and starter rations to young pigs will result in:

1. More uniform pigs with fewer runts.
2. Heavier weaning weights.
3. Less mortality of baby pigs.
4. Decrease in incidence and severity of baby pig scours.
5. Less setback to young pigs when weaned from the sow. The earlier pigs are to be weaned, the more important it is that they be eating dry feed at an early age.
6. Lower weight loss by the sow.

For successful creep feeding, the following pointers are pertinent:

1. Begin by giving the baby pigs a mere handful of creep feed, replenished daily. The creep feed should not be allowed to become stale or contaminated. Place feed in flat pans.

2. Once the pigs have started to eat readily, place the ration in a creep feeder so that they have access to the ration at all times. One linear foot of feeder space should be provided for each five pigs. The edge of the feeder trough should not be more than 4 in. above the floor. For maximum consumption of the creep ration, the feeder should be located close to the waterer for the baby pigs.

3. Make clean, fresh water available to the young pigs in a waterer. It is not sufficient to rely on the waterer for the sow to furnish water to the baby pigs.

4. The creep area should be light, warm, dry and draft-free. It should be located in an area where the pigs are the least

disturbed. Excitement, noise, and a change in feeding routine affect eating habits and subsequent feed consumption. Having the creep area near the sleeping area encourages more frequent eating. Arrange the creep area in such a way that it can be easily cleaned, and so that feed and water can be supplied conveniently without the producer getting into the area.

5. Individual litter creep areas are preferable. Where several litters have access to one creep area, it is advisable to limit the number of pigs to about 40 per creep.

6. If postweaning scours are encountered, the substitution of 200 to 400 lb of ground oats for a like amount of corn in the ration may be helpful.

7. By the time the pigs reach a weight of 40 lb, they will have consumed abut 54 lb of feed (4 lb of prestarter and 50 lb of starter).

## WEANING PIGS

The optimum age to wean pigs varies considerably, depending on nutritional programs, facilities, environment, health, and management. The average age at weaning for pigs in the United States is about 4 weeks, with a range of 2 to 7 weeks. The earlier pigs are weaned, the better the required feeding and management practices. Advantages of early weaning (3 to 4 weeks of age) are:

1. Heavier pigs at 8 weeks of age, with fewer runts.
2. Lower sow feed costs.
3. More litters per year.
4. Less weight loss of the sow.
5. Greater flexibility in rebreeding or selling sows.
6. Greater turnover of sows through the farrowing unit; hence, less total farrowing area space required and lower facility cost per sow.

Successful early weaning encompasses the following:

1. A sound breeding and feeding program of gestating sows to ensure large, healthy pigs at birth; there is a high positive relationship between birth weight and weight at weaning.

2. Good milking sows that supply plenty of nutrients to get the pigs off to a good start.

3. Good baby pig and sow management during lactation to ensure strong, uniform, and healthy pigs at weaning.

4. A good creep feeding program, beginning with a good quality prestarter ration (see Table 40–12 [p. 790]) when suckling pigs are 7 to 10 days of age, and switching to good quality starter ration (see Table 40–13 [p. 791]) after weaning.

For best results, the guidelines given in Table 40–4 should be observed when planning for early weaning.

**TABLE 40–4**
**GUIDELINES TO SUCCESSFUL EARLY WEANING[1]**

| Guideline | Age in Weeks | | | | |
|---|---|---|---|---|---|
| | 1 | 2 | 3 | 4 | 5 |
| **M**inimum pig weight .............................. (lb) | 5 | 9 | 12 | 15 | 21 |
| **N**ursery temperature at pig level ...................... (°F) | 85 | 85 | 83 | 81 | 79 |
| **M**inimum floor space per pig ...................... (sq ft)[2] | 3 | 3 | 3 | 3 | 3 |
| **M**aximum number of pigs per linear foot of feeding space .... | 5 | 5 | 5 | 5 | 5 |
| **M**aximum number of pigs per nipple waterer[3] .............. | 8 | 8 | 8 | 8 | 8 |
| **M**aximum number of pigs per group ...................... | 10 | 10 | 10 | 15 | 25 |

[1]See Appendix for conversion of U.S. customary to metric.

[2]The figures given herein are for solid floors. On slotted floors, this may be lowered 2 sq ft per pig from 1 to 5 weeks of age.

[3]Where bowls are used instead of nipples, there should be one bowl for each 12 pigs.

The age of pigs at weaning may vary from herd to herd, according to the facilities available, intensity of operation, and managerial skills of the producer. Generally, pigs can be weaned over a wide age range; however, the younger the pigs, the more demanding the management required to do it successfully. Observance of the following guides will reduce the stress at weaning:

1. Wean only pigs weighing more than 12 lb.

2. Wean over a 2- to 3-day period, weaning the larger pigs in the litter first.

3. For 3-week-old pigs provide an environmental temperature of 80–85°F.

4. Group pigs according to size.

5. Limit numbers in a pen to 30.

6. Limit feed intake for 48 hours if post-weaning scours are a problem.

7. Provide 1 feeder hole for 4 to 5 pigs and 1 waterer for each 20 to 25 pigs.

8. Medicate drinking water if scours develop.

## FEEDING GROWING-FINISHING PIGS

Fig. 40–15. Growing-finishing pigs self-fed on partially slotted floor. (Courtesy, National Hog Farmer, St. Paul, MN)

In the practical swine enterprise, growing-finishing generally refers to that period from weaning to market weight of about 240 lb. Because hogs are finished at an early age, the process really consists of both growing and finishing. In a general way, there are two methods of finishing hogs for market: (1) full feeding all the time until the animals attain market weight, and (2) limited feeding early in the period, with full feeding the last 60 to 70 days of the period before marketing.

For the production of lean (bacon) carcasses, the rate of gain should be restricted to about 1½ lb daily after a liveweight of 100 to 125 lb. This is easily accomplished by using a lighter, bulkier finishing ration (made by including 10 to 20% bran, oats, alfalfa, or other suitable bulky feed in the ration). Level of protein has no direct effect on carcass excellence, though it does affect the growth of the pig.

Neither system, full feeding nor limited feeding can be recommended as being best for any and all conditions. The plan to follow should be determined by (1) market conditions, (2) type and breeding of the pigs, (3) price of feeds, (4) feeds available on the farm, (5) kind and extent of pastures available, (6) available labor, and (7) capital invested in facilities. Self-feeders are well adapted to a system of full feeding, but hand-feeding or interval feeding are necessary in any plan for limiting the ration.

Research indicates that there is little difference in the feed efficiency of group-fed pigs between (1) limited feeding of 5 lb per day from 125 lb to market, and (2) self-feeding. However, limited feeding results in slower gains, increased labor or mechanization, variable performance, and increased supervision. For these reasons, the practice can be justified only when sufficient premium is paid for the modest increase achieved in the lean-to-fat ratio. It should be added that the selection of hogs with bred-in meat-type carcasses has largely alleviated the need for restricting rate of gain and using bulky rations as a means of getting leaner carcasses.

When on full feed, finishing pigs will consume 5 to 6 lb of feed daily per 100 lb liveweight up to 120 lb in weight. From 120 lb to a finished weight of 240 lb, pigs on full feed consume about 4 lb of feed daily for each 100 lb of liveweight. With good feed and management, about 700 lb of feed are required to produce 200 lb of gain during the growing-finishing period from 40 lb to 240 lb weight. This means that 3.5 lb of feed are required to produce 1 lb of gain during the growing-finishing stage, from 40 lb weight to 240 lb weight (700 ÷ 200 = 3.5 lb).

As previously indicated, pigs can utilize a great variety of concentrates. The chief ingredients of a growing-finishing ration, therefore, are usually, for practical reasons, those most readily available at the lowest possible price.

Suggested growing-finishing rations are given in Table 40–16 (p. 793). Note that provision is made for meeting the nutritional needs at two different stages of growth—40 to 120 lb, and 121 to 240 lb—with three different rations suggested for each stage.

## FEEDING PROSPECTIVE BREEDING GILTS

Fig. 40–16. Selecting prospective breeding gilts, following which they will be limit-fed. (Courtesy, American Yorkshire Club, West Lafayette, IN)

Prospective breeding gilts should be kept from getting too fat. Meat-type animals can usually be left on a high-energy ration until they reach 175 to 200 lb without becoming too fat. It is neither necessary nor desirable that females intended for breeding purposes carry the same degree of finish as market animals. After selecting replacement gilts, they should be fed as follows:

1. Give about 5 lb per day through their second heat period.
2. Flush—full feed—after the second heat period until breeding on the third heat period.
3. After breeding, limit the feed intake to 3 to 5 lb per day. Overfeeding during gestation can cause embryonic death and thus decrease litter size.

## FEEDING BOARS

The feed allowance of young boars should vary according to condition of the animals, the climatic condition, and the individuality. If the animals are inclined to get too fat, which is likely to happen in self-feeding, the ration may well contain a considerable amount of bulky feeds; otherwise, limited feeding may be necessary.

The feed requirements of the herd boar are abut the same as those of a female of equal weight. He should always be kept in thrifty, vigorous condition and virile. In no case should boars be overfat, nor should they be in a thin run-down condition. Normally, the following feed allowances will suffice: for boars weighing 120 to 150 lb, 6 to 9 lb of feed daily; for mature boars, 5 to 7 lb of feed daily. A more liberal ration must be provided in the wintertime and when the sire is in heavy service. The feed allowance should be varied with the age, development, temperament, breeding demands, and roughage consumed.

Boars and pregestating/gestating sows may be fed the same rations. Rations, formulated for feeding 3, 4, or 5 lb per head daily, are presented in Tables 40–8, 40–9, and 40–10 (pp. 788 to 789).

## FEEDING BROOD SOWS

The nutrition of brood sows is critical, for it may materially affect conception, reproduction, and lactation. Proper feeding of sows should begin with replacement gilts and continue through each stage of the breeding cycle—finishing, gestation, farrowing, and lactation.

### FLUSHING SOWS

*The practice of conditioning or having the sows gain in weight just prior to breeding is known as flushing.* The purpose of flushing is to increase the number of ova shed during estrus. About 10 to 14 days prior to expected breeding, the female should be fed a ration that will make for gains of 1 to 1¼ lb per day. Generally 6 to 8 lb per day of a high-energy, 14 to 16% protein feed that is well balanced in minerals and vitamins, is adequate. Immediately after breeding, the females should be put back on limited feeding. Continuation of high level of feeding after breeding will result in a higher embryo mortality.

## GESTATION PERIOD

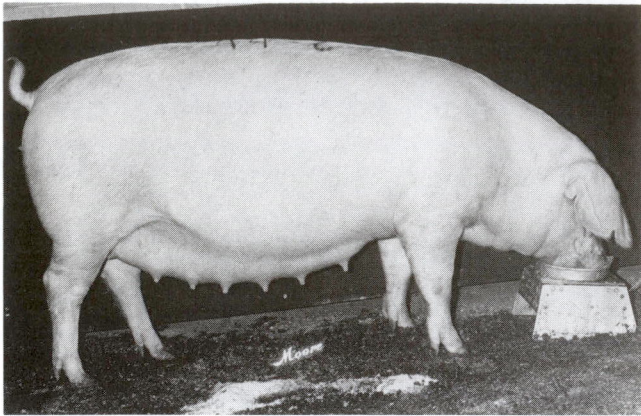

Fig. 40–17. Gestating gilt. (Courtesy, American Landrace Assn., West Lafayette, IN)

The nutrients fed the pregnant gilt or sow must first take care of the usual maintenance needs. If the gilt is not fully mature, nutrients are required for both maternal growth and growth of the fetus. Quality and quantity of proteins, minerals, and vitamins become particularly important in the ration of young pregnant gilts, for their requirements are much greater and more exacting than those of the mature sow.

Approximately two-thirds of the growth of the fetus is made during the last month of the gestation period. It may be said, therefore, that the demands resulting from pregnancy are particularly accelerated during the latter third of the gestation period. Again, the increased needs are primarily for proteins, vitamins, and minerals.

During gestation, it is also necessary that the body reserves be stored for subsequent use during lactation. With a large litter and a sow that is a heavy milker, the demands for milk production are generally greater than can be supplied by the ration fed at the time of lactation. Although desired gains will vary somewhat according to the initial condition, mature sows are generally fed to gain about 70 lb during the pregnancy period, and first litter pregnant gilts are fed to gain about 90 lb. This means that from the day of mating until farrowing time gilts should be fed to gain about 0.9 lb per day and mature sows about 0.7 lb per day. This calls for a daily feed allowance of approximately 4 to 4.5 lb per head, with variation according to environmental conditions.

It is important that the condition of dry sows should be regulated so that they are neither too fat nor too thin at farrowing time. Overly fat sows may have difficulty in farrowing and give birth to weak or dead pigs. Sows that are too thin at farrowing tend to become suckled down during lactation. Thus, one way or another, limited feeding is a must for gestating gilts and sows. This may be accomplished by any one of the following feeding systems (these feeding systems are further detailed later in this chapter in the section headed ''Feeding Systems'').

1. By adding sufficient bulk.
2. By interval feeding.
3. By group hand feeding.
4. By individual feeding.

In addition to the above limited feeding systems, the use of pasture should be considered. Where available, a leguminous pasture is the ideal way in which to limit-feed gestating gilts and sows. Dry sows on good legume pasture are usually fed ½ lb less supplement and 2 lb less corn per day. In addition to limiting the feed intake, the pasture system provides valuable quality protein, minerals, vitamins, and exercise.

Suggested gestation rations, formulated for feeding 3, 4, or 5 lb per head daily, are given in Tables 40–8, 40–9, and 40–10 (pp. 788 to 789).

## FARROWING TIME

It is considered good practice to feed lightly and with bulky laxative feeds from 4 to 5 days before and after farrowing. Wheat bran or oats may constitute half of the limited ration, and a small amount of linseed meal may be added.

The sow may be watered at frequent intervals before or after farrowing, but in no event should she be allowed to gorge. It is also a good plan to take the chill off the water in the wintertime.

## LACTATION PERIOD

The nutritive requirements of a lactating sow are more rigorous than those during gestation. They are very similar to those of a milk cow, except they are more exacting relative to the quality of proteins and the B vitamins because of the absence of rumen synthesis in the pig. A good lactating sow will produce an average of 10 to 15 lb of milk daily during the suckling period. A sow's milk is also richer than cow's milk in all nutrients, especially in fat. Thus, sows suckling litters need a liberal allowance of concentrates rich in protein, calcium, phosphorus, and vitamins.

Fig. 40–18. Lactation period, where life begins—usually in the safety of a farrowing crate. (Courtesy, Lone Willow Genetics, Roanoke, IL)

It is essential that suckling pigs receive a generous supply of milk, for at no other stage in life will they make such economical gains. The gains made by pigs from birth to weaning are largely determined by the milk production of the sows; and this in turn is dependent upon the ration fed and the sow's inherent ability to produce milk. The lactating sow should be provided with a liberal feed allowance—ranging from 2½ to 4½ lb daily for each 100 lb weight. Generous feeding during lactation, with a small shrinkage in weight, is more economical than a stingy allowance of feed, for the nutrients in milk must come either from the feed or from the sow's back. Lactating sows are commonly self-fed, because even when hand fed they are practically on full feed.

Suggested lactation rations are given in Table 40–11 (p. 790).

## FEEDING ORPHAN PIGS

There is no replacement for the sow's colostrum. If the newborn pig does not receive colostrum, it has a lesser chance for survival. An orphan pig can obtain colostrum by being placed with another sow (a foster sow) that has just farrowed. If no such sow is available, the orphan can be fed a commercial milk replacer, several good ones of which are on the market. A homemade milk replacer can be prepared by mixing the following ingredients:

1 quart milk
1 pint half-and-half
1 raw egg
Oral, water-soluble antibiotic

Portions of this mixture should be warmed to 98 to 100°F and fed about every 3 hours, with each pig receiving about ¼ cup per feeding. The use of a shallow pan for feeding is recommended. Immersing the baby pig's nose in the milk a few times will result in its drinking readily. It is extremely important that all feeding utensils be kept clean and sanitary; otherwise, scouring will occur.

Orphan pigs can be fed a dry 22 to 23% crude protein prestarter from 5 to 7 days of age until about 2 to 3 weeks of age (see Table 40–12, p. 790). At this time, they can be switched to a 20 to 21% crude protein pig starter (see Table 40–13, p. 791).

## CORN-HOG RATIO

The corn-hog ratio refers to the number of bushels of corn required to be equivalent in value to 100 lb of live hogs at local markets, based on average prices received by farmers for corn and hogs. During the 14-year period 1973 through 1987, the corn-hog ratio averaged 19.3. This means that the price relationship was such that 19.3 bushels of corn equalled in value 100 lb of hogs.

### HOG-CORN RATIO 1973–87

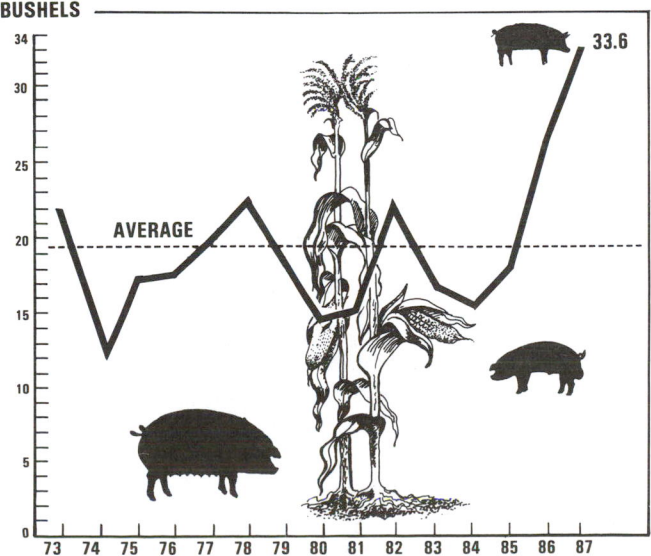

Fig. 40–19. The corn-hog ratio, 1973 to 1987. As shown, it averaged 19.3. (Based on data from *Agricultural Statistics, 1988*, USDA, p. 279, Table 414.)

A high corn-hog ratio—one above 19.3 in recent years—indicates cheap corn and high-priced hogs and likely profit to the producer—conditions that stimulate more breeding and more feeding to heavier weights. On the other hand, a low ratio, one which is below 19.3, means high-priced corn and low-priced hogs—conditions that result in less breeding and feeding of swine.

It is noteworthy that the corn-hog ratio averaged 12.6 for the period 1930 to 1981, whereas it averaged 19.3 for the period 1973 to 1987.

## SOFT PORK

Feed fats are laid down in the body without undergoing much change. Thus, when finishing hogs are liberally fed on high fat content feeds in which the fat is liquid at ordinary temperatures, soft pork results. This condition prevails when hogs are liberally fed such feeds as soybeans, peanuts, mast, or garbage. The fat of the cereal grains is also liquid at ordinary temperatures, but fortunately the fat content in these feeds is relatively low. When such feeds are liberally fed to swine, most of the pork fat is actually formed from the more abundant carbohydrates in these feeds.

Soft pork is undesirable from the standpoint of both the processor and the consumer. It remains flabby and oily even under refrigeration. In soft pork, there is a higher shrinkage in processing; the cuts do not stand up and are unattractive in the showcase; it is difficult to slice the bacon; and the cooking losses are higher through loss of fat. For these reasons, hogs that are liberally fed on those feeds known to produce soft pork are heavily discounted on the market.

The firmness of pork carcasses may be judged by (1) grasping the flank below the ham, (2) lifting one end of the cut while permitting the other end to rest on the table (a firm pork cut will not bend readily), or (3) applying a slight pressure of the thumb (not gouging) on a cut surface. Experimentally, either the iodine number or the refractive index is used in determining the degree of softness; this is a measure of the degree of unsaturation (see Chapter 4, section headed "Fats" for a discussion of unsaturation).

Unless the producer is willing to take the normal reduction in price (about $1.00 per hundredweight), it is recommended that feeds which normally produce soft pork be fed liberally only to pigs under 85 lb in weight and to the breeding herd. For growing-finishing pigs over 85 lb in weight, soybeans and peanuts should not constitute more than 10% of the ration if a serious soft pork problem is to be averted.

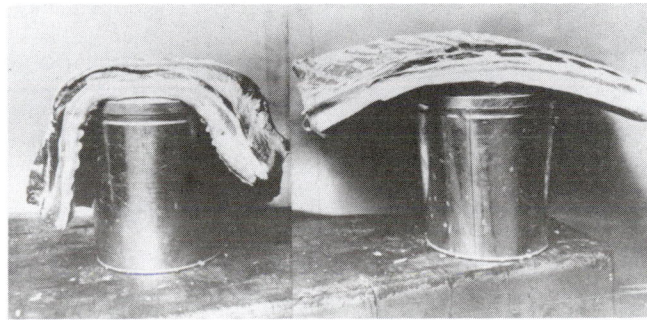

Fig. 40–20. Soft pork. Feed fats do affect body fats. The bacon belly on the left came from a hog liberally fed on soybeans. (Courtesy, University of Illinois, Urbana)

Experimental evidence and practical observation have shown, however, that when a ration producing hard fat is given following a period of feeds rich in unsaturated fats, the body fat gradually becomes harder. It has also been found that this process takes place more rapidly if the animals are first fasted for a period before the change in ration is made. This practice is called *hardening off*. Thus, many hogs that are, for practical reasons, finished primarily on such feeds as soybeans, peanuts, or garbage, are hardened off with a ration of corn or some other suitable grain.

## FEED ALLOWANCES AND SOME SUGGESTED RATIONS FOR SWINE

In most instances, the nutrient allowances should be higher than the minimum requirements established by the national Research Council. This is desirable to obtain maximum performance and reduce the risk of nutrient deficiencies that might occur because of the differences in ingredient quality, environment, health, genetics, and performance of individual animals.

From a nutritional standpoint, there is no *best* swine ration in terms of the ingredients used. So, ingredients should be selected on the basis of availability, price, and the quality of the nutrients that they contain. Corn, grain sorghum, barley, and wheat are the primary energy supplying ingredients in rations for swine. Yet, each of these feed grains is deficient in certain indispensable amino acids, minerals, and vitamins. Soybean meal, other oilseed meals or animal protein meals are commonly added as sources of supplemental amino acids to the grain. Thus, corn and soybean meal, which are the primary ingredients in the Corn Belt, may be replaced by other ingredients in other areas as determined by availability and price; then, balanced nutritionally, especially to meet the amino acid, mineral, and vitamin requirements.

Following the selection of feed ingredients, they may be fed to swine in either combined (complete) form or cafeteria style, and they may be purchased commercially or home mixed.

• **Recommended nutrient allowances of the swine specialists at Iowa State University**—In this chapter, the author presents the nutrient allowances, along with premixes and rations, of the swine specialists of Iowa State University, the Land Grant University in the state of Iowa, the leading swine state in the nation.

The recommended allowances and rations formulated by the swine specialists at Iowa State University consist of three premixes (Tables 40–5, 40–6, 40–7 [pp. 786 to 787]), and nine complete rations (Tables 40–8 to 40–16 [pp. 788 to 793]). This makes for a popular and convenient method of building swine rations by blending the proper premix with the main ingredients of the ration.

Normally, premixes should be purchased from a commercial company, because they have much better quality control and mixing facilities to handle the small quantities of minerals and/or vitamins required.

Vitamin potency may decrease with extended storage and can be completely destroyed when vitamins are in contact with minerals over a prolonged period of time. If the minerals and vitamins are purchased in one premix (such as shown in Table 40–7, p. 787), they should be used within 30 days of purchase for optimum vitamin potency. Also, mineral-vitamin premixes should be stored in a cool, dry, and dark place. Stabilizing agents will increase the shelf-life of mineral-vitamin premix combinations.

Fig. 40–21. Two different rations stored in outside steel bins and augured into environmentally controlled building. (Courtesy, Iowa State University, Ames)

Special care should be taken when blending the premix with the primary ingredients to obtain a thorough dispersion throughout the feed. A common method of blending is to mix enough premix for 1 ton of feed with 20 to 50 lb of finely ground corn or soybean meal, then add this to the mixer.

### TABLE 40–5
### COMPOSITION AND ANALYSIS OF TRACE MINERAL PREMIX[1]

| Element | Source[2] | Amount (lb) | Amount (kg) | Percent in Premix (%) | Parts Per Million When Added to a Complete Ration At The Following Pounds Per Ton: 2 (ppm) | 3 (ppm) | 4 (ppm) | 5 (ppm) |
|---|---|---|---|---|---|---|---|---|
| Copper (Cu) | Copper sulfate | 1.500 | 0.681 | 0.38 | 4 | 6 | 8 | 10 |
| Iodine (I) | Potassium iodide[3] | 0.010 | 0.005 | 0.008 | 0.08 | 0.11 | 0.15 | 0.19 |
| Iron (Fe) | Ferrous sulfate | 25.000 | 11.350 | 5.03 | 50 | 75 | 101 | 126 |
| Manganese (Mn) | Manganese sulfate | 2.500 | 1.135 | 0.57 | 6 | 9 | 11 | 14 |
| Selenium (Se) | Sodium selenite[3] | 0.025 | 0.011 | 0.011 | 0.11 | 0.17 | 0.23 | 0.29 |
| Zinc (Zn) | Zinc sulfate | 25.000 | 11.350 | 5.68 | 57 | 85 | 114 | 142 |
| | Carrier | 45.965 | 20.868 | | | | | |
| | Total | 100.000 | 45.400 | | | | | |

[1]Adapted by the author from *Life Cycle Swine Nutrition*, Iowa State University, Ames, PM–489, June, 1988.

[2]Other sources of trace minerals may be substituted. Iodine may be omitted if iodized salt is used.

[3]Iodine and selenium probably will be added in a separate premix form.

## TABLE 40-6
## COMPOSITION OF VITAMIN PREMIX[1][2]

| Vitamins | Amount | Unit | Units Per Pound of Complete Ration When Added At The Following Pounds Per Ton: | | | |
|---|---|---|---|---|---|---|
| | | | 3 | 5 | 8 | 10 |
| **Essential[3]** | | | | | | |
| Vitamin A ....................................... (million IU) | 5.0 | (IU) | 750.00 | 1,250.00 | 2,000.00 | 2,500.00 |
| Vitamin D ....................................... (million IU) | 0.6 | (IU) | 90.00 | 150.00 | 240.00 | 300.00 |
| Vitamin E ....................................... (thousand IU) | 26.0 | (IU) | 3.90 | 6.50 | 10.40 | 13.00 |
| Niacin (nicotinic acid, nicotinamide) ........................... (g) | 25.0 | (mg) | 3.75 | 6.25 | 10.00 | 12.50 |
| d-Pantothenic acid (vitamin B-3) ........................... (g) | 20.0 | (mg) | 3.00 | 5.00 | 8.00 | 10.00 |
| Riboflavin (vitamin B-2) ........................... (g) | 6.0 | (mg) | 0.90 | 1.50 | 2.40 | 3.00 |
| Vitamin B-12 (cobalamins) ........................... (mg) | 25.0 | (mcg) | 3.75 | 6.25 | 10.00 | 12.50 |
| **Optional[4]** | | | | | | |
| Biotin ........................... (g) | 0.3 | (mg) | 0.05 | 0.08 | 0.12 | 0.15 |
| Menadione (source of vitamin K) ........................... (g) | 4.0 | (mg) | 0.60 | 1.00 | 1.60 | 2.00 |
| **Carrier** ........................... | ? | | | | | |
| **Total[5]** ........................... (lb) | 10.0 | | | | | |

[1]Adapted by the author from *Life Cycle Swine Nutrition,* Iowa State University, Ames, PM-489, June, 1988.

[2]A feed additive may be included in the vitamin premix.

[3]Most natural feedstuffs contain very little vitamin D or B-12. The amount of provitamin A (beta-carotene) in feedstuffs will depend on processing and storage, while niacin in most grains is relatively unavailable for swine. Riboflavin and pantothenic acid in natural feedstuffs can meet part of the requirement.

[4]Supplemental biotin is not necessary with corn-soybean meal based rations. It should be included in sow rations based on other grains. The vitamin K requirement is normally met by the level present in natural feedstuffs and by intestinal synthesis. A hemorrhagic or bleeding syndrome has been diagnosed which is probably due to a vitamin K antimetabolite. The antimetabolite is thought to be produced by mold occuring in one or more of the ration ingredients. When this has occurred, adding menadione has been helpful in preventing or overcoming the problem.

[5]If this premix is used in Table 40-7, dilute to 2 lb (0.9 kg) only.

## TABLE 40-7
## COMPLETE MINERAL-VITAMIN PREMIXES FOR CORN-SOYBEAN MEAL RATIONS[1][2][3]

| Ingredients | 1 | | 2 | |
|---|---|---|---|---|
| | (lb) | (kg) | (lb) | (kg) |
| Calcium carbonate ....................... | 540 | 245 | 300 | 136 |
| Dicalcium phosphate ....................... | 1,030 | 468 | | |
| Defluorinated phosphate ....................... | | | 1,060 | 481 |
| Salt ....................... | 250 | 114 | 250 | 114 |
| Trace mineral premix (Table 40-5) .......... | 80 | 36 | 80 | 36 |
| Vitamin premix (Table 40-6)[4] ........... | 40 | 18 | 40 | 18 |
| Carrier (corn, middlings, or grain by-products) ... | 60 | 27 | 270 | 123 |
| **Total** ....................... | 2,000 | 908 | 2,000 | 908 |

**Calculated Analyses:**

| | (%) | (lb) | (kg) |
|---|---|---|---|
| Salt (NaCl) ....................... | 12.5 | | |
| Calcium (Ca) ....................... | 22.8 | | |
| Phosphorus (P) ....................... | 9.5 | | |
| Copper (Cu) ....................... | 0.015 | | |
| Iodine (I) ....................... | 0.00031 | | |
| Iron (Fe) ....................... | 0.201 | | |
| Manganese (Mn) ....................... | 0.023 | | |
| Selenium (Se) ....................... | 0.00046 | | |
| Zinc (Zn) ....................... | 0.227 | | |
| Vitamin A ....................... (IU) | | 50,000 | 110,000 |
| Vitamin D ....................... (IU) | | 6,000 | 13,200 |
| Vitamin E ....................... (IU) | | 260 | 572 |
| Vitamin K (optional) ....................... (mg) | | 40 | 88 |
| Biotin (optional) ....................... (mg) | | 3 | 7 |
| Niacin (Nicotinic Acid, Nicotinamide) .. (mg) | | 250 | 550 |
| Pantothenic Acid (Vitamin B-3) ..... (mg) | | 200 | 440 |
| Riboflavin (Vitamin B-2) ........... (mg) | | 60 | 132 |
| Vitamin B-12 (Cobalamins) ........ (mcg) | | 250 | 550 |

**Mixing Directions**

| Stage | Premix | | Soybean Meal, 44% | | Corn | |
|---|---|---|---|---|---|---|
| | (lb) | (kg) | (lb) | (kg) | (lb) | (kg) |
| **Gestation:** | | | | | | |
| 3 lb (1.4 kg)/day ...... | 100 | 45 | 270 | 123 | 1,630 | 740 |
| 4 lb (1.8 kg)/day ...... | 80 | 36 | 160 | 73 | 1,760 | 799 |
| 5 lb (2.3 kg)/day ...... | 65 | 30 | 100 | 45 | 1,835 | 833 |
| **Lactation** ............... | 80 | 36 | 270 | 123 | 1,650 | 749 |
| **Starter rations** ........ | 65 | 30 | 690 | 313 | 1,245 | 565 |
| **Grower-finisher** | 50 | 23 | 300– | 136– | 1,700– | 772– |
| rations ............... | | | 450 | 204 | 1,500 | 681 |

[1]Adapted by the author from *Life Cycle Swine Nutrition,* Iowa State University, Ames, PM-489, June, 1988.

[2]Due to the instability of vitamins in the presence of trace minerals this premix should be used within 30 days of preparation.

[3]These premixes can be fed free-choice to sows on pasture or in other instances where free-choice minerals and vitamins are needed. Do not add or feed free-choice additional minerals or vitamins with any of the ration formulas in Tables 40-8 to 40-16, because they contain sufficient minerals and vitamins.

[4]Table 40-6 premix diluted to 2 lb (0.9 kg) instead of 10 lb (4.5 kg). See footnote 5 under Table 40-6.

Suggested rations are shown for each of the following classes of swine:

Table 40-8, Pregestation, Breeding, and Gestation Rations—for boars, sows, or gilts fed 3 lb per day.

Table 40-9, Pregestation, Breeding, and Gestation Rations—for boars, sows, or gilts fed 4 lb per day.

Table 40-10, Pregestation, Breeding, and Gestation Rations—for boars, sows, or gilts fed 5 lb per day

Table 40-11, Lactation Rations.

Table 40-12, Prestarter Rations—for baby pigs before 3 weeks of age.

Table 40-13, Starter Rations.

Table 40-14, Recommended Rations for Performance Testing of Boars.

Table 40-15, Swine Conditioner Rations—for newly received feeder pigs, for stress periods and convalescence.

Table 40-16, Grower-Finisher Rations—for 40 to 240 lb.

As previously indicated, the rations shown in Tables 40-8 to 40-16 call for the use of the mineral and vitamin premixes shown in Tables 40-5 to 40-7.

## TABLE 40-8
### PREGESTATION, BREEDING, AND GESTATION RATIONS (FOR BOARS, SOWS OR GILTS FED 3 LB *[1.4 KG]* PER DAY)[1][2][3]

| Ingredient | Ration # | | | | | | | |
|---|---|---|---|---|---|---|---|---|
| | 1 | | 2 | | 3 | | 4 | |
| | (lb) | *(kg)* | (lb) | *(kg)* | (lb) | *(kg)* | (lb) | *(kg)* |
| **Corn,** yellow (8.4% protein)[4] | 1,636 | *743* | 1,559 | *708* | 1,640 | *744* | 1,562 | *709* |
| **Soybean meal,** solvent extracted (44.0% protein)[5] | 270 | *123* | 250 | *113* | 194 | *88* | 175 | *80* |
| **Alfalfa meal,** dehydrated (17.0% protein) | | | 100 | *45* | | | 100 | *45* |
| **Meat and bone meal** (50.0% protein) | | | | | 100 | *45* | 100 | *45* |
| **Dicalcium phosphate** | 50 | *22* | 51 | *23* | 31 | *14* | 31 | *14* |
| **Limestone** | 19 | *8* | 15 | *7* | 10 | *5* | 7 | *3* |
| **Iodized salt** | 10 | *5* | 10 | *5* | 10 | *5* | 10 | *5* |
| **Trace mineral premix** (Table 40-5) | 5 | *2* | 5 | *2* | 5 | *2* | 5 | *2* |
| **Vitamin premix** (Table 40-6) | 10 | *5* | 10 | *5* | 10 | *5* | 10 | *5* |
| **Feed additives**[6] | | | | | | | | |
| **Total** | 2,000 | *908* | 2,000 | *908* | 2,000 | *908* | 2,000 | *908* |
| **Calculated analyses:** | | | | | | | | |
| Metabolizable energy (kcal/lb) | 1,470 | | 1,434 | | 1,469 | | 1,433 | |
| Metabolizable energy *(kcal/kg)* | *3,234* | | *3,155* | | *3,232* | | *3,153* | |
| Protein (%) | 12.81 | | 12.90 | | 13.66 | | 13.76 | |
| Lysine (%) | 0.60 | | 0.60 | | 0.60 | | 0.60 | |
| Threonine (%) | 0.49 | | 0.49 | | 0.50 | | 0.51 | |
| Tryptophan (%) | 0.14 | | 0.15 | | 0.13 | | 0.14 | |
| Calcium (Ca) (%) | 1.01 | | 1.01 | | 1.00 | | 1.01 | |
| Phosphorus (P) (%) | 0.75 | | 0.75 | | 0.75 | | 0.75 | |

[1]Adapted by the author from *Life Cycle Swine Nutrition,* Iowa State University, Ames, PM-489, June, 1988.

[2]These rations can be used for sows on pasture since they require only supplemental minerals or at the most 2 to 3 lb *(0.9 to 1.4 kg)* of complete feed. If less than 3 lb *(1.4 kg)* is fed per day, free-choice minerals should be available.

[3]These rations can also be used as a silage balancer. Gestating sows will eat 5 to 7 lb *(2.3 to 3.2 kg)* of corn silage daily which should be supplemented with 2 to 3 lb *(0.9 to 1.4 kg)* of one of these rations.

[4]Ground oats can replace corn up to 20% of the total ration. Ground milo, wheat, or barley can replace the corn.

[5]To replace 44% soybean meal with 47% soybean meal or whole soybeans, use the following ratios:
Each 100 lb *(45 kg)* SBM (44%) = 93 lb *(42 kg)* SBM (47%) + 7 lb *(3 kg)* corn.
Each 100 lb *(45 kg)* SBM (44%) + 35 lb *(16 kg)* corn = 135 lb *(61 kg)* whole soybeans.

[6]Feed additives are not generally recommended during gestation or for gilts during the developer period after selection unless specific disease problems exist. High levels (100-300 g/ton) may be beneficial when fed 2 weeks before and after breeding and 2 weeks before farrowing.

Fig. 40-22. Open front, low-profile shelter with concrete apron for gestating gilts. (Courtesy, Iowa State University, Ames)

## TABLE 40-9
## PREGESTATION, BREEDING, AND GESTATION RATIONS (FOR BOARS, SOWS, OR GILTS FED 4 LB *[1.8 KG]* PER DAY)[1][2]

| Ingredient | Ration # | | | | | | | |
|---|---|---|---|---|---|---|---|---|
| | 1 | | 2 | | 3 | | 4 | |
| | (lb) | *(kg)* | (lb) | *(kg)* | (lb) | *(kg)* | (lb) | *(kg)* |
| Corn, yellow (8.4% protein)[3] | 1,769 | *803* | 1,692 | *768* | 1,767 | *802* | 1,692 | *768* |
| Soybean meal, solvent extracted (44.0% protein)[4] | 160 | *72* | 140 | *64* | 90 | *41* | 68 | *31* |
| Alfalfa meal, dehydrated (17.0% protein) | | | 100 | *45* | | | 100 | *45* |
| Meat and bone meal (50.0% protein) | | | | | 100 | *45* | 100 | *45* |
| Dicalcium phosphate | 36 | *16* | 36 | *16* | 16 | *7* | 16 | *7* |
| Limestone | 15 | *7* | 12 | *5* | 7 | *3* | 4 | *2* |
| Iodized salt | 8 | *4* | 8 | *4* | 8 | *4* | 8 | *4* |
| Trace mineral premix (Table 40-5) | 4 | *2* | 4 | *2* | 4 | *2* | 4 | *2* |
| Vitamin premix (Table 40-6) | 8 | *4* | 8 | *4* | 8 | *4* | 8 | *4* |
| Feed additives[5] | | | | | | | | |
| **Total** | 2,000 | *908* | 2,000 | *908* | 2,000 | *908* | 2,000 | *908* |
| **Calculated analyses:** | | | | | | | | |
| Metabolizable energy (kcal/lb) | 1,493 | | 1,457 | | 1,492 | | 1,456 | |
| Metabolizable energy *(kcal/kg)* | *3,285* | | *3,205* | | *3,282* | | *3,203* | |
| Protein (%) | 10.95 | | 11.04 | | 11.90 | | 11.95 | |
| Lysine (%) | 0.45 | | 0.45 | | 0.46 | | 0.46 | |
| Threonine (%) | 0.41 | | 0.41 | | 0.43 | | 0.43 | |
| Tryptophan (%) | 0.11 | | 0.12 | | 0.10 | | 0.11 | |
| Calcium (Ca) (%) | 0.75 | | 0.76 | | 0.75 | | 0.76 | |
| Phosphorus (P) (%) | 0.60 | | 0.60 | | 0.60 | | 0.60 | |

[1]Adapted by the author from *Life Cycle Swine Nutrition,* Iowa State University, Ames, PM–489, June, 1988.

[2]These rations can be used for gilts on pasture during gestation since they require 3 to 4 lb *(1.4 to 1.8 kg)* of feed daily. These rations can also be used for interval-fed sows or gilts if the average daily intake is approximately 4 lb *(1.8 kg)*.

[3]Ground oats can replace corn up to 20% of the total ration. Ground milo, wheat, or barley can replace the corn.

[4]To replace 44% soybean meal with 47% soybean meal or whole soybeans, use the following ratios:
Each 100 lb *(45 kg)* SBM (44%) = 93 lb *(42 kg)* SBM (47%) + 7 lb *(3 kg)* corn.
Each 100 lb *(45 kg)* SBM (44%) + 35 lb *(16 kg)* corn = 135 lb *(61 kg)* whole soybeans.

[5]Feed additives are not generally recommended during gestation or for gilts during the developer period after selection unless specific disease problems exist. High levels (100–300 g/ton) may be beneficial when fed 2 weeks before and after breeding and 2 weeks before farrowing.

## TABLE 40-10
## PREGESTATION, BREEDING, AND GESTATION RATIONS (FOR BOARS, SOWS, OR GILTS FED 5 LB *[2.3 KG]* PER DAY)[1]

| Ingredient | Ration # | | | | | | | |
|---|---|---|---|---|---|---|---|---|
| | 1 | | 2 | | 3 | | 4 | |
| | (lb) | *(kg)* | (lb) | *(kg)* | (lb) | *(kg)* | (lb) | *(kg)* |
| Corn, yellow (8.4% protein)[2] | 1,845 | *838* | 1,764 | *801* | 1,853 | *841* | 1,824 | *828* |
| Soybean meal, solvent extracted (44.0% protein)[3] | 100 | *45* | 85 | *38* | 20 | *9* | | |
| Alfalfa meal, dehydrated (17.0% protein) | | | 100 | *45* | | | 50 | *23* |
| Meat and bone meal (50.0% protein) | | | | | 100 | *45* | 100 | *45* |
| Dicalcium phosphate | 26 | *12* | 26 | *12* | 6 | *3* | 7 | *3* |
| Limestone | 14 | *6* | 10 | *5* | 6 | *3* | 4 | *2* |
| Iodized salt | 6 | *3* | 6 | *3* | 6 | *3* | 6 | *3* |
| Trace mineral premix (Table 40-5) | 3 | *1* | 3 | *1* | 3 | *1* | 3 | *1* |
| Vitamin premix (Table 40-6) | 6 | *3* | 6 | *3* | 6 | *3* | 6 | *3* |
| Feed additives[4] | | | | | | | | |
| **Total** | 2,000 | *908* | 2,000 | *908* | 2,000 | *908* | 2,000 | *908* |
| **Calculated analyses:** | | | | | | | | |
| Metabolizable energy (kcal/lb) | 1,508 | | 1,473 | | 1,507 | | 1,489 | |
| Metabolizable energy *(kcal/kg)* | *3,318* | | *3,241* | | *3,315* | | *3,276* | |
| Protein (%) | 9.95 | | 10.13 | | 10.72 | | 10.59 | |
| Lysine (%) | 0.38 | | 0.38 | | 0.37 | | 0.36 | |
| Threonine (%) | 0.37 | | 0.38 | | 0.38 | | 0.37 | |
| Tryptophan (%) | 0.10 | | 0.10 | | 0.08 | | 0.08 | |
| Calcium (Ca) (%) | 0.61 | | 0.60 | | 0.61 | | 0.61 | |
| Phosphorus (P) (%) | 0.50 | | 0.50 | | 0.50 | | 0.50 | |

[1]Adapted by the author from *Life Cycle Swine Nutrition,* Iowa State University, Ames, PM–489, June, 1988.

[2]Ground oats can replace corn up to 20% of the total ration. Ground milo, wheat, or barley can replace the corn.

[3]To replace 44% soybean meal with 47% soybean meal or whole soybeans, use the following ratios:
Each 100 lb *(45 kg)* SBM (44%) = 93 lb *(42 kg)* SBM (47%) + 7 lb *(3 kg)* corn.
Each 100 lb *(45 kg)* SBM (44%) + 35 lb *(16 kg)* corn = 135 lb *(61 kg)* whole soybeans.

[4]Feed additives are not generally recommended during gestation or for gilts during the developer period after selection unless specific disease problems exist. High levels (100–300 g/ton) may be beneficial when fed 2 weeks before and after breeding and 2 weeks before farrowing.

## TABLE 40–11
## LACTATION RATIONS[1][2]

| Ingredient | Ration # 1 (lb) | (kg) | 2 (lb) | (kg) | 3 (lb) | (kg) | 4 (lb) | (kg) |
|---|---|---|---|---|---|---|---|---|
| Corn, yellow (8.4% protein) | 1,658 | 752 | 1,545 | 701 | 1,580 | 717 | 1,469 | 667 |
| Soybean meal, solvent extracted (44.0% protein)[3] | 270 | 122 | 283 | 128 | 177 | 80 | 260 | 118 |
| Fat or oil source | | | 100 | 45 | | | | |
| Meat and bone meal (50.0% protein) | | | | | 100 | 45 | | |
| Oats (11.5% protein) | | | | | | | 200 | 90 |
| Beet pulp (8.0% protein) | | | | | 100 | 45 | | |
| Dicalcium phosphate | 35 | 16 | 35 | 16 | 15 | 7 | 33 | 15 |
| Limestone | 15 | 7 | 15 | 7 | 6 | 3 | 16 | 7 |
| Iodized salt | 10 | 5 | 10 | 5 | 10 | 5 | 10 | 5 |
| Trace mineral premix (Table 40–5) | 4 | 2 | 4 | 2 | 4 | 2 | 4 | 2 |
| Vitamin premix (Table 40–6) | 8 | 4 | 8 | 4 | 8 | 4 | 8 | 4 |
| Feed additives[4] | | | | | | | | |
| Total | 2,000 | 908 | 2,000 | 908 | 2,000 | 908 | 2,000 | 908 |
| **Calculated analyses:** | | | | | | | | |
| Metabolizable energy (kcal/lb) | 1,487 | | 1,409 | | 1,468 | | 1,457 | |
| Metabolizable energy (kcal/kg) | 3,271 | | 3,100 | | 3,230 | | 3,205 | |
| Protein (%) | 12.90 | | 12.72 | | 13.43 | | 13.04 | |
| Lysine (%) | 0.60 | | 0.60 | | 0.60 | | 0.60 | |
| Threonine (%) | 0.49 | | 0.49 | | 0.48 | | 0.49 | |
| Tryptophan (%) | 0.14 | | 0.14 | | 0.13 | | 0.15 | |
| Calcium (Ca) (%) | 0.75 | | 0.76 | | 0.76 | | 0.76 | |
| Phosphorus (P) (%) | 0.61 | | 0.60 | | 0.60 | | 0.60 | |

[1]Adapted by the author from *Life Cycle Swine Nutrition,* Iowa State University, Ames, PM–489, June, 1988.

[2]These rations may be limit-fed from a few days before farrowing and full-fed during lactation.

[3]To replace 44% soybean meal with 47% soybean meal or whole soybeans, use the following ratios:
   Each 100 lb *(45 kg)* SBM (44%) = 93 lb *(42 kg)* SBM (47%) + 7 lb *(3 kg)* corn.
   Each 100 lb *(45 kg)* SBM (44%) + 35 lb *(16 kg)* corn = 135 lb *(61 kg)* whole soybeans.

[4]High levels of feed additives (100–300 g/ton) may be beneficial when fed 2 weeks before and after breeding and 2 weeks before farrowing.

## TABLE 40–12
## PRESTARTER RATIONS (FOR BABY PIGS BEFORE 3 WEEKS OF AGE)[1][2]

| Ingredient | Ration # 1 (lb) | (kg) | 2 (lb) | (kg) | 3 (lb) | (kg) | 4 (lb) | (kg) | 5 (lb) | (kg) |
|---|---|---|---|---|---|---|---|---|---|---|
| Corn, yellow (8.4% protein) | 917 | 416 | 878 | 399 | 866 | 393 | 621 | 282 | 774 | 351 |
| Soybean meal, solvent extracted (44.0% protein) | | | | | 635 | 288 | | | | |
| Soybean meal, solvent extracted, dehulled (47.0% protein) | 680 | 309 | 580 | 263 | | | 580 | 263 | 725 | 329 |
| Oat groats (dehulled oats) (16.0% protein) | | | | | | | 200 | 91 | | |
| Skim milk, dried (33.0% protein) | 100 | 45 | 100 | 45 | 200 | 91 | 200 | 91 | | |
| Whey, dried (12.0% protein) | 200 | 91 | 300 | 136 | 200 | 91 | 200 | 91 | 400 | 182 |
| Fish meal, menhaden (62.0% protein) | | | 50 | 23 | | | | | | |
| Sugar | | | | | | | 100 | 45 | | |
| Fat or oil source | 40 | 18 | 40 | 18 | 40 | 18 | 40 | 18 | 40 | 18 |
| Dicalcium phosphate | 26 | 12 | 19 | 9 | 25 | 12 | 25 | 12 | 25 | 12 |
| Limestone | 19 | 9 | 15 | 7 | 16 | 7 | 16 | 7 | 18 | 8 |
| Iodized salt | 5 | 2 | 5 | 2 | 5 | 2 | 5 | 2 | 5 | 2 |
| Trace mineral premix (Table 40–5) | 5 | 2 | 5 | 2 | 5 | 2 | 5 | 2 | 5 | 2 |
| Vitamin premix (Table 40–6) | 8 | 4 | 8 | 4 | 8 | 4 | 8 | 4 | 8 | 4 |
| Feed additives[3] | ◄————————————— 100 to 300 g per ton —————————————► | | | | | | | | | |
| Total | 2,000 | 908 | 2,000 | 908 | 2,000 | 908 | 2,000 | 908 | 2,000 | 908 |
| **Calculated analyses:** | | | | | | | | | | |
| Metabolizable energy (kcal/lb) | 1,458 | | 1,459 | | 1,441 | | 1,387 | | 1,441 | |
| Metabolizable energy (kcal/kg) | 3,208 | | 3,210 | | 3,170 | | 3,051 | | 3,170 | |
| Protein (%) | 22.68 | | 22.32 | | 22.11 | | 22.34 | | 22.64 | |
| Lysine (%) | 1.40 | | 1.40 | | 1.40 | | 1.40 | | 1.40 | |
| Threonine (%) | 0.94 | | 0.95 | | 0.94 | | 0.93 | | 0.96 | |
| Tryptophan (%) | 0.29 | | 0.28 | | 0.29 | | 0.29 | | 0.30 | |
| Calcium (Ca) (%) | 0.91 | | 0.91 | | 0.91 | | 0.90 | | 0.91 | |
| Phosphorus (P) (%) | 0.70 | | 0.70 | | 0.70 | | 0.71 | | 0.70 | |

[1]Adapted by the author from *Life Cycle Swine Nutrition,* Iowa State University, Ames, PM–489, June, 1988.

[2]The prestarter ration is normally fed in only limited amounts. It should be fed to pigs weaned before 3 weeks of age until they reach approximately 12 lb *(5.4 kg).* They can then be switched to a starter ration. These are good rations for orphan pigs, when extreme disease outbreaks (TGE) occur, or when the sow fails to produce sufficient milk.

[3]The feed additive may be part of the vitamin premix, or if a separate premix, it should replace an equal amount of corn.

Fig. 40-23. Early-weaned pigs in double-deck nursery pens. Early weaning requires special facilities and special rations. (Courtesy, University of Illinois, Urbana)

## TABLE 40-13
## PIG STARTER RATIONS [1] [2]

| Ingredient | Ration # 1 (lb) | (kg) | 2 (lb) | (kg) | 3 (lb) | (kg) | 4 (lb) | (kg) | 5 (lb) | (kg) |
|---|---|---|---|---|---|---|---|---|---|---|
| Corn, yellow (8.4% protein) | 1,072 | 487 | 939 | 426 | 1,032 | 468 | 1,273 | 578 | 865 | 393 |
| Soybean meal, solvent extracted (44.0% protein) | | | | | 610 | 277 | 689 | 313 | | |
| Soybean meal, solvent extracted, dehulled (47.0% protein) | 570 | 259 | 580 | 263 | | | | | | |
| Soybeans, full-fat, cooked (37.0% protein) | | | | | | | | | 800 | 363 |
| Oat groats, dehulled (16.0% protein) | | | 200 | 91 | | | | | | |
| Whey, dried (12.0% protein) | 300 | 136 | 200 | 91 | 300 | 136 | | | 300 | 136 |
| Fat or oil source | | | 20 | 9 | | | | | | |
| Dicalcium phosphate | 24 | 11 | 25 | 12 | 25 | 12 | 3 | 1 | 2 | 1 |
| Limestone | 16 | 7 | 18 | 8 | 15 | 7 | 17 | 8 | 15 | 7 |
| Iodized salt | 5 | 2 | 5 | 2 | 5 | 2 | 5 | 2 | 5 | 2 |
| Trace mineral premix (Table 40-5) | 5 | 2 | 5 | 2 | 5 | 2 | 5 | 2 | 5 | 2 |
| Vitamin premix (Table 40-6) | 8 | 4 | 8 | 4 | 8 | 4 | 8 | 4 | 8 | 4 |
| Feed additives [3] | ◄————————— 100 to 300 g per ton —————————► | | | | | | | | | |
| **Total** | 2,000 | 908 | 2,000 | 908 | 2,000 | 908 | 2,000 | 908 | 2,000 | 908 |
| **Calculated analyses:** | | | | | | | | | | |
| Metabolizable energy (kcal/lb) | 1,483 | | 1,472 | | 1,460 | | 1,473 | | 1,525 | |
| Metabolizable energy (kcal/kg) | 3,263 | | 3,238 | | 3,212 | | 3,241 | | 3,355 | |
| Protein (%) | 19.70 | | 20.37 | | 19.55 | | 20.39 | | 20.13 | |
| Lysine (%) | 1.15 | | 1.16 | | 1.15 | | 1.15 | | 1.20 | |
| Threonine (%) | 0.82 | | 0.82 | | 0.82 | | 0.81 | | 0.85 | |
| Tryptophan (%) | 0.25 | | 0.25 | | 0.26 | | 0.26 | | 0.28 | |
| Calcium (Ca) (%) | 0.80 | | 0.81 | | 0.81 | | 0.80 | | 0.84 | |
| Phosphorus (P) (%) | 0.65 | | 0.65 | | 0.65 | | 0.65 | | 0.68 | |

[1]Adapted by the author from *Life Cycle Swine Nutrition*, Iowa State University, Ames, PM-489, June, 1988.

[2]The pig starter ration can be used as a creep ration before weaning and fed after weaning until the pigs reach approximately 40 lb *(18 kg)*. They can then be switched to a grower-finisher ration.

[3]The feed additive may be part of the vitamin premix, or if a separate premix, it should replace an equal amount of corn.

**TABLE 40-14**
**RECOMMENDED RATIONS**
**FOR PERFORMANCE TESTING OF BOARS** [1] [2]

| Ingredient | Conditioner Ration (lb) | Conditioner Ration (kg) | Test Ration (lb) | Test Ration (kg) |
|---|---|---|---|---|
| **Corn,** yellow (8.4% protein) [3] | 1,115 | 506 | 1,359 | 617 |
| **Soybean meal,** solvent extracted (44.0% protein) [4] | 500 | 227 | 550 | 250 |
| **Wheat middlings** (15.5% protein) | 200 | 91 | | |
| **Whey,** dried (12.0% protein) | 100 | 45 | | |
| **Dicalcium phosphate** | 50 | 23 | 60 | 27 |
| **Limestone** | 15 | 7 | 11 | 5 |
| **Iodized salt** | 5 | 2 | 5 | 2 |
| **Trace mineral premix** (Table 40-5) | 5 | 2 | 5 | 2 |
| **Vitamin premix** (Table 40-6) | 10 | 5 | 10 | 5 |
| **Feed additives** (g/ton) | (100-300) | | (0-100) | |
| **Total** | 2,000 | 908 | 2,000 | 908 |

| Calculated analyses: | | | | |
|---|---|---|---|---|
| Metabolizable energy (kcal/lb) | 1,438 | | 1,459 | |
| Metabolizable energy (kcal/kg) | 3,164 | | 3,210 | |
| Protein (%) | 17.83 | | 17.81 | |
| Lysine (%) | 0.98 | | 0.97 | |
| Threonine (%) | 0.71 | | 0.70 | |
| Tryptophan (%) | 0.23 | | 0.22 | |
| Calcium (Ca) (%) | 1.01 | | 1.01 | |
| Phosphorus (P) (%) | 0.87 | | 0.89 | |

[1] Adapted by the author from *Life Cycle Swine Nutrition,* Iowa State University, Ames, PM-489, June, 1988.

[2] These rations normally would be used for growing boars from 40 to 250 lb *(18 to 114 kg)* body weight.

[3] If the ration is to be pelleted, 25 to 50 lb *(11 to 23 kg)* of molasses or binder can replace an equal amount of corn.

[4] To replace 44% soybean meal with 47% soybean meal or whole soybeans, use the following ratios:
Each 100 lb *(45 kg)* SBM (44%) = 93 lb *(42 kg)* SBM (47%) + 7 lb *(3 kg)* corn.
Each 100 lb *(45 kg)* SBM (44%) + 35 lb *(16 kg)* corn = 135 lb *(61 kg)* whole soybeans.

**TABLE 40-15**
**SWINE CONDITIONER RATIONS** [1] [2]

| Ingredient | Ration # 1 (lb) | 1 (kg) | 2 (lb) | 2 (kg) | 3 (lb) | 3 (kg) |
|---|---|---|---|---|---|---|
| **Corn,** yellow (8.4% protein) | 882 | 401 | 840 | 382 | 979 | 444 |
| **Oats** (11.5% protein) | 300 | 136 | 600 | 272 | 600 | 272 |
| **Wheat middlings** (15.5% protein) | 300 | 136 | | | | |
| **Soybean meal,** solvent extracted (44.0% protein) [3] | 170 | 77 | 300 | 136 | 350 | 159 |
| **Whey,** dried (12.0% protein) | 200 | 91 | 200 | 91 | | |
| **Alfalfa meal,** dehydrated (17.0% protein) | 50 | 23 | | | | |
| **Fish meal,** menhaden (62.0% protein) | 50 | 23 | | | | |
| **Dicalcium phosphate** | 14 | 6 | 26 | 12 | 30 | 14 |
| **Limestone** | 14 | 6 | 14 | 6 | 16 | 7 |
| **Iodized salt** | 5 | 2 | 5 | 2 | 10 | 5 |
| **Trace mineral premix** (Table 40-5) | 5 | 2 | 5 | 2 | 5 | 2 |
| **Vitamin premix** (Table 40-6) | 10 | 5 | 10 | 5 | 10 | 5 |
| **Feed additives** [4] | ◄─────── 100 to 300 g/ton ───────► | | | | | |
| **Total** | 2,000 | 908 | 2,000 | 908 | 2,000 | 908 |
| **Calculated analyses:** | | | | | | |
| Metabolizable energy (kcal/lb) | 1,398 | | 1,387 | | 1,391 | |
| Metabolizable energy (kcal/kg) | 3,076 | | 3,051 | | 3,060 | |
| Protein (%) | 14.67 | | 14.78 | | 15.26 | |
| Lysine (%) | 0.75 | | 0.75 | | 0.75 | |
| Threonine (%) | 0.59 | | 0.60 | | 0.58 | |
| Tryptophan (%) | 0.18 | | 0.18 | | 0.19 | |
| Calcium (Ca) (%) | 0.73 | | 0.71 | | 0.71 | |
| Phosphorus (P) (%) | 0.61 | | 0.60 | | 0.60 | |

[1] Adapted by the author from *Life Cycle Swine Nutrition,* Iowa State University, Ames, PM-489, June, 1988.

[2] These rations are recommended as the first ration fed to newly received feeder pigs, for stress periods and convalescence.

[3] To replace 44% soybean meal with 47% soybean meal use the following ratio:
Each 100 lb *(45 kg)* SBM (44%) = 93 lb *(42 kg)* SBM (47%) + 7 lb *(3 kg)* corn.

[4] Be certain that only approved feed additives and levels are used for therapy. The feed additive may be a part of the vitamin premix, or if a separate premix, it should replace an equal amount of corn.

## TABLE 40-16
### GROWER-FINISHER RATIONS (FOR PIGS FROM 40 TO 240 LB [18 TO 109 KG])[1][2][3][4]

| Ingredient | For Pigs 40–120 Lb (18–54 Kg) | | | | | | For Pigs 121–240 Lb (55–109 Kg)[5] | | | | | |
|---|---|---|---|---|---|---|---|---|---|---|---|---|
| | 1 | | 2 | | 3 | | 4 | | 5 | | 6 | |
| | (lb) | (lb) | (lb) | (lb) | (lb) | (lb) | (lb) | (lb) | (lb) | (lb) | (lb) | (lb) |
| **Corn,** yellow (8.4% protein)[6][7] | 1,571 | 1,497 | 1,422 | 1,343 | 1,561 | 1,492 | 1,692 | 1,613 | 1,575 | 1,481 | 1,694 | 1,620 |
| **Soybean meal,** solvent extracted (44.0% protein)[8] | 380 | 455 | | | 310 | 380 | 265 | 345 | | | 190 | 264 |
| **Soybeans,** full-fat, cooked (37.0% protein)[9] | | | 530 | 610 | | | | | 380 | 475 | | |
| **Meat and bone meal** (50.0% protein) | | | | | 110 | 110 | | | | | 100 | 100 |
| Dicalcium phosphate | 21 | 19 | 19 | 18 | | | 18 | 17 | 20 | 20 | | |
| Limestone | 15 | 16 | 16 | 16 | 6 | 5 | 15 | 15 | 15 | 14 | 6 | 6 |
| Iodized salt | 5 | 5 | 5 | 5 | 5 | 5 | 5 | 5 | 5 | 5 | 5 | 5 |
| Trace mineral premix (Table 40–5) | 3 | 3 | 3 | 3 | 3 | 3 | 2 | 2 | 2 | 2 | 2 | 2 |
| Vitamin premix (Table 40–6) | 5 | 5 | 5 | 5 | 5 | 5 | 3 | 3 | 3 | 3 | 3 | 3 |
| Feed additives[10] | ◄————— 0 to 100 g per ton —————► | | | | | | ◄————— 0 to 50 g per ton —————► | | | | | |
| **Total** | 2,000 | 2,000 | 2,000 | 2,000 | 2,000 | 2,000 | 2,000 | 2,000 | 2,000 | 2,000 | 2,000 | 2,000 |
| **Calculated analyses:** | | | | | | | | | | | | |
| Metabolizable energy ......... (kcal/lb) | 1,500 | 1,497 | 1,543 | 1,547 | 1,498 | 1,495 | 1,510 | 1,507 | 1,538 | 1,543 | 1,508 | 1,505 |
| Metabolizable energy ......... (kcal/kg) | 3,300 | 3,293 | 3,395 | 3,403 | 3,296 | 3,289 | 3,322 | 3,315 | 3,384 | 3,395 | 3,318 | 3,311 |
| Protein ......... (%) | 14.96 | 16.30 | 15.78 | 16.93 | 16.13 | 17.38 | 12.94 | 14.36 | 13.65 | 15.01 | 13.79 | 15.11 |
| Lysine ......... (%) | 0.75 | 0.85 | 0.81 | 0.90 | 0.77 | 0.86 | 0.60 | 0.70 | 0.65 | 0.76 | 0.60 | 0.70 |
| Threonine ......... (%) | 0.58 | 0.63 | 0.61 | 0.66 | 0.60 | 0.65 | 0.49 | 0.55 | 0.52 | 0.58 | 0.51 | 0.56 |
| Tryptophan ......... (%) | 0.17 | 0.20 | 0.20 | 0.21 | 0.17 | 0.19 | 0.14 | 0.17 | 0.16 | 0.18 | 0.13 | 0.15 |
| Calcium (Ca) ......... (%) | 0.60 | 0.60 | 0.61 | 0.61 | 0.61 | 0.60 | 0.55 | 0.55 | 0.59 | 0.58 | 0.55 | 0.56 |
| Phosphorus (P) ......... (%) | 0.50 | 0.50 | 0.51 | 0.51 | 0.51 | 0.53 | 0.46 | 0.46 | 0.49 | 0.51 | 0.47 | 0.49 |

[1]Adapted by the author from *Life Cycle Swine Nutrition,* Iowa State University, Ames, PM–489, June, 1988.

[2]Feed the ration with the higher level of soybean meal (lower level of corn) to lighter pigs in each group and decrease the soybean meal (increase the corn) until you reach the lower level as pig weights increase. If preferred, one level of protein can be fed from 40 to 240 lb *(18 to 109 kg)* with similar results as with varying the levels. To accomplish this, use the lower protein formulations from rations 1, 2, or 3 (for example, in ration No. 1 use 1571 lb *[713 kg]* of corn and 380 lb *[173 kg]* of soybean meal).

[3]If barrows and gilts are separated, use the higher range for soybean meal for the gilts and the lower range for the barrows.

[4]To convert lb to kg, divide by 2.2. To convert g/ton (short) to g/ton (metric), divide by 0.907.

[5]For potential replacement gilts, the level of dicalcium phosphate should be increased by 10 lb *(4.5 kg)* per ton. This will provide a minimum dietary level of 0.67% calcium and 0.55% phosphorus.

[6]Ground milo, wheat, or barley can replace the ground corn. Ground oats can replace corn up to 20% of the total ration.

[7]If the ration is to be pelleted, 25 to 50 lb *(11 to 23 kg)* of molasses or binder can replace an equal amount of corn.

[8]To replace 44% soybean meal with 47% soybean meal or synthetic lysine, use the following ratios:
  Each 100 lb *(45 kg)* SBM (44%) = 93 lb *(42 kg)* SBM (47%) + 7 lb *(3 kg)* corn.
  Each 100 lb *(45 kg)* SBM (44%) = 3 lb *(1.4 kg)* 98% lysine hydrochloride + 1 lb *(0.45 kg)* dicalcium phosphate + 96 lb *(43.6 kg)* corn.

[9]The fat content of whole soybeans increases the energy content of the ration. For maximum utilization of the ration, the protein content has been increased to maintain a similar energy-to-protein ratio.

[10]The feed additive may be part of the vitamin premix, or if it is a separate premix, it should replace an equal amount of corn.

Fig. 40–24. (Courtesy, *Hog Farm Management,* Minnetonka, MN)

• **Formulation and use of complete protein supplements**—In addition to the formulations given in Tables 40–5 to 40–16, Table 40–17 (see page 794) gives the formulas for complete protein supplements, ranging from 34.49 to 43.64% protein content, which may be used to make growing-finishing, gestation, and lactation rations; and Table 40–18 (see page 794) gives the mixing directions for incorporating the Table 40–17 protein supplements in complete rations.

**TABLE 40–17**
**COMPLETE PROTEIN SUPPLEMENTS[1 2 3 4]**

| Ingredient | 1 | 2 | 3 | 4 | 5 | 6 | 7 | 8 |
|---|---|---|---|---|---|---|---|---|
| | (lb) | (lb) | (lb) | (lb) | (lb) | (lb) | (lb) | (lb) |
| Wheat middlings (15.5% protein)[5] | 192 | 102 | 59 | 202 | | | | |
| Soybean meal, solvent extracted (44.0% protein) | 1,500 | 1,495 | 1,150 | 1,200 | | | | |
| Soybean meal, solvent extracted, dehulled (47.0% protein) | | | | | 1,636 | 1,415 | 1,420 | 1,193 |
| Alfalfa meal, dehydrated (17.0% protein) | | 100 | 200 | | | 100 | | |
| Meat and bone meal (50.0% protein)[6] | | | 400 | 400 | | | 300 | 500 |
| Fish meal, menhaden (62.0% protein) | | | | | | 150 | | 100 |
| Dicalcium phosphate | 145 | 145 | 70 | 70 | 169 | 153 | 111 | 59 |
| Limestone | 75 | 70 | 33 | 40 | 90 | 77 | 64 | 43 |
| Iodized salt | 40 | 40 | 40 | 40 | 45 | 45 | 45 | 45 |
| Trace mineral premix (Table 40–5) | 16 | 16 | 16 | 16 | 20 | 20 | 20 | 20 |
| Vitamin premix (Table 40–6) | 32 | 32 | 32 | 32 | 40 | 40 | 40 | 40 |
| Feed additives[7] (g/ton) | | | | | | | | |
| **Total** | 2,000 | 2,000 | 2,000 | 2,000 | 2,000 | 2,000 | 2,000 | 2,000 |
| **Calculated analyses:** | | | | | | | | |
| Metabolizable energy (kcal/lb) | 1,228 | 1,203 | 1,167 | 1,222 | 1,260 | 1,243 | 1,249 | 1,254 |
| Metabolizable energy (kcal/kg) | 2,702 | 2,647 | 2,567 | 2,688 | 2,772 | 2,735 | 2,748 | 2,759 |
| Protein (%) | 34.49 | 34.53 | 37.46 | 37.97 | 38.45 | 38.75 | 40.87 | 43.64 |
| Lysine (%) | 2.24 | 2.24 | 2.22 | 2.26 | 2.54 | 2.59 | 2.54 | 2.65 |
| Threonine (%) | 1.40 | 1.41 | 1.44 | 1.46 | 1.55 | 1.57 | 1.59 | 1.66 |
| Tryptophan (%) | 0.49 | 0.50 | 0.45 | 0.45 | 0.52 | 0.52 | 0.49 | 0.48 |
| Salt, added (%) | 2.00 | 2.00 | 2.00 | 2.00 | 2.25 | 2.25 | 2.25 | 2.25 |
| Calcium (%) | 3.39 | 3.36 | 3.39 | 3.39 | 3.94 | 3.94 | 3.94 | 3.95 |
| Phosphorus (%) | 1.87 | 1.84 | 1.86 | 1.91 | 2.09 | 2.10 | 2.10 | 2.10 |

[1]Adapted by the author from *Life Cycle Swine Nutrition,* Iowa State University, Ames, PM–489, June, 1988.

[2]These supplements can be used to make growing-finishing, gestation, or lactation rations. See Table 40–15, including the table footnote, for mixing directions.

[3]Supplements with meat and bone meal may be self-fed free-choice with shelled corn for growing-finishing pigs.

[4]To convert lb to kg, divide by 2.2. To convert g/ton (short) to g/ton (metric), divide by 0.907.

[5]The wheat middlings may be replaced with corn, corn distillers' grains with solubles, or other grain by-products.

[6]The meat and bone meal was assumed to contain 8.10% calcium and 4.10% phosphorus. If meat and bone meal with a higher concentration of calcium and phosphorus is used, the amount of dicalcium phosphate should be reduced accordingly.

[7]The concentration of feed additives will depend on the type of ration in which the supplement will be used. The concentration should be 3 to 5 times higher in supplements 1 to 4 and 4 to 6 times higher in supplements 5 to 8 than desired in the complete ration.

**TABLE 40–18**
**COMPLETE RATIONS (USING SUPPLEMENTS IN TABLE 40–17)[1 2 3]**

| Ingredient | Protein Level in Complete Rations | | | | | | | | | |
|---|---|---|---|---|---|---|---|---|---|---|
| | 13% | | 14% | | 15% | | 16% | | 17% | |
| | (lb) | (lb) | (lb) | (lb) | (lb) | (lb) | (lb) | (lb) | (lb) | (lb) |
| Corn, yellow (8.4% protein) | 1,635 | 1,685 | 1,555 | 1,625 | 1,480 | 1,550 | 1,400 | 1,490 | 1,320 | 1,425 |
| Supplements 1–4 (See Table 40–17) | 365 | | 445 | | 520 | | 600 | | 680 | |
| Supplements 5–8 (See Table 40–17) | | 315 | | 375 | | 450 | | 510 | | 575 |
| **Total** | 2,000 | 2,000 | 2,000 | 2,000 | 2,000 | 2,000 | 2,000 | 2,000 | 2,000 | 2,000 |
| **Calculated analyses:[4]** | | | | | | | | | | |
| Metabolizable energy (kcal/lb) | 1,485 | 1,506 | 1,470 | 1,497 | 1,456 | 1,485 | 1,440 | 1,475 | 1,425 | 1,465 |
| Metabolizable energy (kcal/kg) | 3,267 | 3,313 | 3,234 | 3,293 | 3,203 | 3,267 | 3,168 | 3,245 | 3,135 | 3,223 |
| Protein (%) | 13.04 | 13.13 | 14.05 | 14.04 | 15.01 | 15.16 | 16.02 | 16.06 | 17.04 | 17.04 |
| Lysine (%) | 0.61 | 0.61 | 0.68 | 0.68 | 0.76 | 0.76 | 0.83 | 0.83 | 0.91 | 0.91 |
| Threonine (%) | 0.50 | 0.50 | 0.54 | 0.54 | 0.58 | 0.58 | 0.62 | 0.62 | 0.67 | 0.66 |
| Tryptophan (%) | 0.14 | 0.13 | 0.15 | 0.15 | 0.17 | 0.16 | 0.18 | 0.17 | 0.19 | 0.19 |
| Salt, added (%) | 0.37 | 0.35 | 0.45 | 0.42 | 0.52 | 0.51 | 0.60 | 0.57 | 0.68 | 0.65 |
| Calcium (%) | 0.62 | 0.63 | 0.76 | 0.75 | 0.88 | 0.89 | 1.02 | 1.01 | 1.15 | 1.14 |
| Phosphorus (%) | 0.54 | 0.54 | 0.60 | 0.60 | 0.66 | 0.66 | 0.72 | 0.72 | 0.78 | 0.78 |

[1]Adapted by the author from *Life Cycle Swine Nutrition,* Iowa State University, Ames, PM–489, June, 1988.

[2]Suggested stages of production for using the above rations:

| | | | | | | | | | | |
|---|---|---|---|---|---|---|---|---|---|---|
| Grower | ( ——— Protein and lysine low ——— ) | | | | + | | + | | ( ———— Calcium high ———— ) | |
| Finisher | + | + | + | + | + | | + | | ( ———— Calcium high ———— ) | |
| Gestation: | | | | | | | | | | |
| 3 lb (1.4 kg) per day | – | – | – | – | – | | – | (Phosphorus marginal) | + | + |
| 4 lb (1.8 kg) per day | – | – | (Phosphorus marginal) | | + | | + | + | + | + |
| 5 lb (2.3 kg) per day | + | + | + | + | + | | + | + | + | + |
| Lactation | – | – | (Phosphorus marginal) | | + | | + | + | + | + |

[3]To convert lb to kg, divide by 2.2.

[4]Expected analysis using minimum analyses for each nutrient in the supplements from Table 40–17.

Also, commercial supplements (usually a combined protein-mineral-vitamin-additive supplement) may be bought and mixed with the locally available grain. Table 40–19 shows the proportion of grain for 8.4% protein to mix with supplements ranging from 30 to 45% protein to obtain finished rations ranging from 10 to 18% protein content.

Where commercial supplements are bought to use with farm-grown grains, they may be utilized in the following ways:

1. Mixed with ground, farm-grown grain in the approximate amounts shown in Table 40–19 to make a complete ration (see the Table 40–19 footnote example for mixing directions).

2. Self-fed in separate feeders, with the ground or whole grain also being self-fed in separate self-feeders.

3. Hand-fed; with the supplement and the grain each being hand-fed in the proportions recommended in Table 40–19.

### TABLE 40–19
### GRAIN AND SUPPLEMENT COMBINATIONS (POUNDS)
### NEEDED TO FORMULATE RATIONS OF DIFFERENT PROTEIN LEVELS (GRAIN VALUED AT 8.4% PROTEIN)[1][2][3]

| Protein in Supplement | | Percent Protein in Total Ration | | | | | | | | |
|---|---|---|---|---|---|---|---|---|---|---|
| | | 10 | 11 | 12 | 13 | 14 | 15 | 16 | 17 | 18 |
| (%) | | (lb) | (lb) | (lb) | (lb) | (lb) | (lb) | (lb) | (lb) | (lb) |
| 30 | Grain | 1,852 | 1,759 | 1,667 | 1,574 | 1,481 | 1,389 | 1,296 | 1,204 | 1,111 |
| | Supplement | 148 | 241 | 333 | 426 | 519 | 611 | 704 | 796 | 889 |
| 31 | Grain | 1,858 | 1,770 | 1,681 | 1,593 | 1,504 | 1,416 | 1,327 | 1,239 | 1,150 |
| | Supplement | 142 | 230 | 319 | 407 | 496 | 584 | 673 | 761 | 850 |
| 32 | Grain | 1,864 | 1,780 | 1,695 | 1,610 | 1,525 | 1,441 | 1,356 | 1,271 | 1,186 |
| | Supplement | 136 | 220 | 305 | 390 | 475 | 559 | 644 | 729 | 814 |
| 33 | Grain | 1,870 | 1,789 | 1,707 | 1,626 | 1,545 | 1,463 | 1,382 | 1,301 | 1,220 |
| | Supplement | 130 | 211 | 293 | 374 | 455 | 537 | 618 | 699 | 780 |
| 34 | Grain | 1,875 | 1,797 | 1,719 | 1,641 | 1,563 | 1,484 | 1,406 | 1,328 | 1,250 |
| | Supplement | 125 | 203 | 281 | 359 | 438 | 516 | 594 | 672 | 750 |
| 35 | Grain | 1,880 | 1,805 | 1,729 | 1,654 | 1,579 | 1,504 | 1,429 | 1,353 | 1,278 |
| | Supplement | 120 | 195 | 271 | 346 | 421 | 496 | 571 | 647 | 722 |
| 36 | Grain | 1,884 | 1,812 | 1,739 | 1,667 | 1,594 | 1,522 | 1,449 | 1,377 | 1,304 |
| | Supplement | 116 | 188 | 261 | 333 | 406 | 478 | 551 | 623 | 696 |
| 37 | Grain | 1,888 | 1,818 | 1,748 | 1,678 | 1,608 | 1,538 | 1,469 | 1,399 | 1,329 |
| | Supplement | 112 | 182 | 252 | 322 | 392 | 462 | 531 | 601 | 671 |
| 38 | Grain | 1,892 | 1,824 | 1,757 | 1,689 | 1,622 | 1,554 | 1,486 | 1,419 | 1,351 |
| | Supplement | 108 | 176 | 243 | 311 | 378 | 446 | 514 | 581 | 649 |
| 39 | Grain | 1,895 | 1,830 | 1,765 | 1,699 | 1,634 | 1,569 | 1,503 | 1,438 | 1,373 |
| | Supplement | 105 | 170 | 235 | 301 | 366 | 431 | 497 | 562 | 627 |
| 40 | Grain | 1,899 | 1,835 | 1,772 | 1,709 | 1,646 | 1,582 | 1,519 | 1,456 | 1,392 |
| | Supplement | 101 | 165 | 228 | 291 | 354 | 418 | 481 | 544 | 608 |
| 41 | Grain | 1,902 | 1,840 | 1,779 | 1,718 | 1,656 | 1,595 | 1,534 | 1,472 | 1,411 |
| | Supplement | 98 | 160 | 221 | 282 | 344 | 405 | 466 | 528 | 589 |
| 42 | Grain | 1,905 | 1,845 | 1,786 | 1,726 | 1,667 | 1,607 | 1,548 | 1,488 | 1,429 |
| | Supplement | 95 | 155 | 214 | 274 | 333 | 393 | 452 | 512 | 571 |
| 43 | Grain | 1,908 | 1,850 | 1,792 | 1,734 | 1,676 | 1,618 | 1,561 | 1,503 | 1,445 |
| | Supplement | 92 | 150 | 208 | 266 | 324 | 382 | 439 | 497 | 555 |
| 44 | Grain | 1,910 | 1,854 | 1,798 | 1,742 | 1,685 | 1,629 | 1,573 | 1,517 | 1,461 |
| | Supplement | 90 | 146 | 202 | 258 | 315 | 371 | 427 | 483 | 539 |
| 45 | Grain | 1,913 | 1,858 | 1,803 | 1,749 | 1,694 | 1,639 | 1,585 | 1,530 | 1,475 |
| | Supplement | 87 | 142 | 197 | 251 | 306 | 361 | 415 | 470 | 525 |

[1]Adapted by the author from *Life Cycle Swine Nutrition*, Iowa State University, Ames, PM–489, June, 1988. To convert lb to kg, divide by 2.2.

[2]The grain common to the area may be substituted in Table 40–19, with the 8.4% protein content changed in keeping with the protein content of the grain used, and the proportions of grain and supplement adjusted to obtain the desired percent protein in the total ration.

[3]**Example**: In order to obtain a total ration with 15% protein, each 2,000 lb *(908 kg)* of feed should contain 1,389 lb *(631 kg)* of the 8.4% protein grain and 611 lb *(277 kg)* of the 30% supplement.

## POINTERS IN FORMULATING RATIONS AND FEEDING SWINE

In formulating rations and in feeding swine, the following points are noteworthy:

1. Feeds of similar nutritive properties can be interchanged in the ration as price relationships warrant.

2. If wheat, barley, oats, or grain sorghum is used instead of corn as the grain in a ration, the protein supplement may be slightly reduced.

3. When proteins of animal origin predominate, adequate mineral protection can be obtained by allowing hogs free access to a 2-compartment box or self-feeder with (a) salt (trace mineralized) in one side, and (b) a mixture of ⅓ salt (salt added for purposes of palatability) and ⅔ monosodium phosphate or other phosphorus supplement, in the other side. When supplements of plant origin constitute most of the source of proteins, add a third compartment to the mineral box and place in it a mixture of ⅓ salt (trace mineralized) and ⅔ ground limestone or oystershell flour.

4. When hogs are not exposed to sunlight or when dehydrated alfalfa meal is fed, vitamin D should be added in keeping with the recommended allowances (see Table 40–1).

5. Where the ration consists chiefly of white corn, barley, wheat, oats, rye, kafir, or by-products of these grains, there may be a deficiency of vitamin A (see Table 40–1 for recommended allowances).

6. Except for gestating sows and boars of breeding age, hogs are generally self-fed. All of the ingredients may be mixed together and placed in the same self-feeder or the grain may be placed in one self-feeder (or compartment) and the protein supplements (including any ground alfalfa) in another. If the (a) cereal grains and (b) protein supplements (including ground alfalfa) are hand-fed, the grain and supplement should be fed separately, in the proportions indicated in the suggested rations.

7. An exception should be made to the cafeteria-style feeding when the grain ration consists of barley, oats, rye, or kafir. These feeds are higher in protein content than corn, and for this reason are generally fed as a mixed ration. Otherwise, the pigs will often eat more protein supplement than is necessary to balance the ration. Likewise, when corn is fed as the grain, sometimes such protein supplements as (a) roasted soybeans, (b) soybean meal, and (c) peanut meal are too palatable to be fed separately from the corn, especially if the corn is not of good quality.

8. Full-fed finishing hogs will consume 4 to 5 lb of feed daily per 100 lb liveweight until they weigh 100 lb. They will eat 3 to 4 lb daily per 100 lb weight from this stage until marketing.

## FITTING RATIONS FOR SHOW AND SALE SWINE

Any of the rations listed in Tables 40–8, 40–9, 40–10, 40–14, or 40–16 for the respective classes and ages of swine are suitable for use in fitting show animals of similar classification. Because of the high cost of labor, the recent trend has been toward self-feeding both young breeding animals and market barrows and gilts that are being fitted for show. Many of them are left on self-feeders right up to show time, others are hand-fed only during the last month or two of the fitting period. However, most experienced exhibitors feel that they can get superior bloom and condition by either (1) hand-feeding, or (2) using a combination of hand-feeding and self-feeding (hand-feeding twice daily and allowing free access to

a self-feeder). When hand-feeding, they also prefer mixing the ration with skim milk, buttermilk, or condensed buttermilk, and feeding the entire ration in the form of a slop.

Adding milk to a ration that is already properly balanced makes for a higher protein content than necessary. On the other hand, most experienced caretakers prefer using rations of higher protein content for fitting purposes. They feel they get more bloom that way. In general, however, when skim milk or buttermilk is used in slop feeding, the protein feeds of the ration may be reduced by one-half without harm to the animal.

In fitting show barrows, it may be necessary to decrease or discontinue slop feeding 2 to 4 weeks before the show to avoid paunchiness and lowering of the dressing percentage.

When oatmeal (oat groats, rolled hulled oats) is not too high priced, many successful hog caretakers replace up to 50% of the grain (corn, wheat, barley, oats, and/or sorghum) in the ration with oatmeal. They do this especially when fitting hogs—both breeding animals and barrows—in the younger age groups. Oatmeal is highly palatable, lighter, and less fattening than corn.

Suitable minerals and vitamins should always be provided.

## FEEDING SYSTEMS

The choice of the feeding system(s) and the choice of the ration(s) must go hand in hand. For example, if the grain and the protein supplements are to be self-fed in separate feeders or compartments, it is important that they be of equal palatability; otherwise, pigs will consume too much of one and too little of the other. A listing and discussion of each of the common feeding systems follows.

• **Complete self-fed rations**—The trend is toward the use of complete, self-fed rations for baby pigs and growing-finishing hogs, because, in comparison with free-choice feeding, they (1) lend themselves better to automation, (2) provide better control of nutrient intake, and (3) result in faster gains.

Complete rations may be formulated either by ''building from the ground up'' (by adding each ingredient, one by one), or by mixing a complete supplement, a base mix, or a premix with ground grain.

• **Floor or drop feeding**—Floor or drop feeding is particularly suited to the controlled feeding of growing-finishing swine or the breeding herd. Feeding in the sleeping area encourages cleanliness, since pigs are less inclined to defecate where they

Fig. 40–25. National Barrow Show Champion Truckload of barrows; bred, fitted, and shown by Waldo Farms. (Courtesy, Waldo Farms, DeWitt, NE)

eat. Feed wastage is reduced to a minimum when the animals do not have more feed available than they will consume at one eating. Even though automated, restricted feeding requires close attention, because the daily feed intake of pigs is affected by the weather.

- **Free-choice**—Grain and protein supplements may be fed separately and free-choice. Generally, pigs fed free-choice rations in separate feeders or compartments will not make as uniform or as fast gains as pigs fed a complete mixed ration. The free-choice system requires more supervision, as the palatability of the grain or the protein supplement may vary and the pigs will then overeat or undereat the supplement or the grain. There is very little, if any, difference in economy of gain between feeding a free-choice or a complete ground mixed ration. Free-choice feeding may be the best feeding system for the small producer who does not have mixing equipment.

- **Liquid feeding**—Liquid feeding usually involves mixing predetermined amounts of feed and water prior to, or at the time of, feeding. When properly used, this method can practically eliminate feed dust in the feeding area and minimize wastage. Ratios of feed and water can be varied to produce a free-flowing liquid or a thick paste. In some cases, feed is automatically dropped into the water in the feed trough. Research has shown no difference in the rate of gain of pigs full-fed on liquid or dry feeds. Neither does liquid feeding have any effect on dressing percentage, carcass measurements, or carcass quality. However, pigs full-fed liquid rations generally require more feed per pound of gain than pigs full-fed dry rations.

- **Limit feeding**—With gestating sows, limit feeding to 4 to 6 lb per head daily is a must in order to keep them from getting too fat. Overly fat sows have difficulty in farrowing and give birth to weak or dead pigs. With growing-finishing pigs, it is a way in which to increase slightly the proportion of lean to fat in the carcass. A discussion of limit feeding of (1) gilts and sows, and (2) growing-finishing pigs follows.

1. **Gilts and sows.** Replacement gilts should be started on a limited feeding program at 180 to 200 lb; and all gestating sows and gilts should be limit fed. Limit feeding may be accomplished by any one of the following methods:

a. **By feeding bulky, fibrous feeds**, such as silage, haylage, or alfalfa, with such feed constituting at least one-third of the ration. Actually, this is a way in which to lower the energy content of the ration. Although bulky feeds will hold the weight down, they usually do not lower feed cost.

b. **By interval feeding**, in which gilts or sows are turned to self-feeders for 2 to 8 hours every second or third day. Under this system, gilts will usually eat around 12 lb of feed at a time (or an average of 4 lb per day) and older sows will consume around 15 lb (or an average of 5 lb per day). The amount of feed consumed in interval feeding may be controlled either (1) by varying the interval, from every other day to twice a week, (2) by varying the length of time that the gilts and sows are left on the self-feeders (from 2 to 8 hours), or (3) by hand-feeding.

c. **By group hand-feeding** a limited ration to several sows. This is apt to result in the "bossy" sows getting too much and the timid sows getting too little. This problem can be partially alleviated by feeding over a large area.

d. **By individual feeding** in either individual stalls or in tie stalls, tethered by a neck collar or belt.

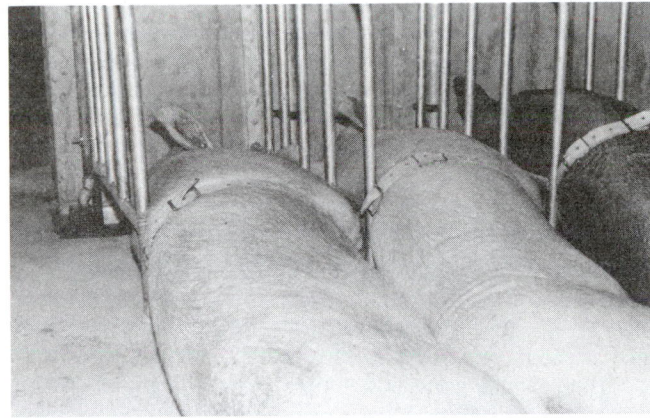

Fig. 40–26. Individually fed sows, comfortably tethered. (Courtesy, Lone Willow Genetics, Roanoke, IL)

2. **Growing-finishing pigs.** Sometimes growing-finishing pigs are limit-fed in order to produce leaner carcasses. Usually, it is started when pigs weigh around 100 lb and feed is limited to about 85 to 95% of what pigs of comparable age consume when self-fed. Limit feeding of market hogs results in slower gains, increased labor, and more mechanization. Thus, unless sufficient premium is paid for the modestly leaner carcasses, it cannot be justified.

- **Pelleted complete rations**—The use of pelleted complete rations for growing-finishing hogs will increase the average daily gain by 2 to 5% and improve the feed efficiency by approximately 5 to 10%. Thus, when a complete ration is purchased, buying a pelleted feed may be more economical than buying a meal. But the advantage of pelleting will usually not be sufficient to offset the cost of hauling grain to the mill and having a pelleted ration made. Also, pellet machines are costly; hence, the purchase of such equipment cannot be justified with the volume of hogs handled by most swine producers.

## FEEDER PIG PRODUCTION

*Feeder pig production refers to the production and sale of immature pigs weighing 30 to 60 lb, usually throughout the year, for growing and finishing on other farms.* It makes for a two-phase system in swine production, similar to the two-phase system so well known in the cattle industry where some operators specialize in the cow-and-calf system (the production of feeder cattle) and others in finishing cattle. Until recent years, it was generally assumed that a two-phase system lent itself more logically to beef cattle than to swine because of the western range and of fewer disease problems. But several important scientific and technological developments which occurred in the swine industry in the 1950s and early 1960s ushered in considerable two-phase production of hogs. Among such developments were: (1) specific pathogen-free (SPF) herds and other improved disease control measures; (2) confined and continuous production—which increased specialization in breeding, in farrowing, and in finishing; and (3) increased mechanization.

## CONTRACT HOG PRODUCTION (CUSTOM FEEDING/LEASING)

*A contract is an agreement between two or more persons to do or refrain from doing certain things.* In recent years, swine producers have shown increasing interest in contract hog production due to (1) the high cost of capital, (2) the difficulty of many producers in obtaining adequate financing, and (3) the desire to forego the possibility of large profits for the assurance of more reliable returns. In the late 1980s, an estimated 8 to 10% of U.S. hogs were under some kind of production contract, with a much smaller number under a marketing contract.

## OTHER FEED/MANAGEMENT RELATED ASPECTS

In addition to the subject matter covered earlier in this chapter, there are other feed and management aspects of great importance in swine production. Some of these will be discussed in the sections that follow.

## FEED REQUIRED TO PRODUCE A POUND OF MARKET PIG

Nationally, it has been estimated that it requires 4.0 lb of feed to produce 1 lb of on-foot hog (live) from birth to market weight (see Chapter 1, Table 1–7, exclusive of the feed required by sows and boars to produce pigs. This is high. But remember that 25 to 30% of all pigs farrowed die before weaning. Remember too, that many swine producers are inefficient.

Table 40–20 shows realistic goals for well-managed swine operations.

**TABLE 40–20**
**ESTIMATED FEED REQUIRED TO PRODUCE 240–LB MARKET PIG[1]**

| Stage of Production | Feed Required per 240–Lb Market Pig |
|---|---|
| | (lb) |
| Sow gestation ration (includes pregestation and breeding) | 110 |
| Boar ration | 8 |
| Lactation ration | 45 |
| Starter ration (creep to 40 lb) | 54 |
| Grower-finisher ration (40 to 240 lb) | 700 |
| Total, lb | 917 |

Per 100 lb of pork produced $\frac{917}{240} \times 100 = 382$ lb

[1]See Appendix for conversion of U.S. customary to metric.

The values given in Table 40–20 are estimates based on realistic standards for apportioning the quantities of sow and boar feed to each pig and the feed conversion normally attained during the starter and grower-finisher periods. Although these data do not provide for pig deaths after weaning, normal milling losses, and feed wastage, it is assumed that these losses are not considered in the pounds of hog produced. Data obtained

in commercial herds where accurate records have been kept indicate that 382 lb of feed per 200 lb live hog produced is a realistic goal for a practical swine operation. However, to achieve this level of efficiency, a sound feeding and management program must be followed, including limit feeding of pregnant sows, high conception rates, large litters weaned, early weaning and rebreeding, low death losses, minimal disease problems, balanced rations, and minimal feed wastage.

## EFFECT OF SEX ON PERFORMANCE OF GROWING-FINISHING PIGS

When full fed, boars consume 10 to 15% less feed daily than barrows and gilts and are 10 to 15% more efficient in feed conversion. Also, boars gain faster than barrows and gilts. Barrows gain approximately $\frac{1}{10}$ lb faster per day than gilts, which reduces their age at slaughter by 10 days. Feed per pound of gain is similar for barrows and gilts. Gilts yield carcasses having .11 in. less backfat, .52 sq in. larger loin area, and 1.8% more lean cuts than barrows. Dressing percentage usually favors barrows, which is consistent with their greater depth of backfat.

## SCOURING (DIARRHEA, ENTERITIS)

Scouring is one of the major problems facing swine producers. It is estimated that about 40% of U.S. swine herds are affected with scouring, and that about 20% of pig losses between farrowing and weaning are caused by scouring. Many different etiological agents cause scouring, including nutrition, management, environment, stress, bacteria, viruses, and parasites. Although the nutritional aspects of scouring are frequently discussed, they are poorly understood. However, it is generally recognized that good nutrition, along with good management and minimal stress, will lessen the incidence of scouring and be effective treatment when an outbreak of the disease occurs.

## NUTRITIONAL ANEMIA

*Anemia is a blood condition in which there is a deficiency of hemoglobin which transports oxygen to various parts of the body.* It is caused by a deficiency of iron and/or copper in the diet, and it is most likely to occur in nursing pigs that do not have access to soil.

Anemic pigs show listlessness, rough hair coat, wrinkled skins, drooping ears and tails, pale membranes around the mouth and eyes, and labored breathing.

Any of the following anemia prevention measures may be used:

1. Inject intramuscularly in the neck or ham muscle 100 to 200 mg of iron, in the form of iron dextran, iron dextrin, or gleptoferron into baby pigs at 2 to 3 days of age. If pigs remain in confinement and do not have access to creep feed at an early age, a second injection at 2 to 3 weeks of age is desirable. Injection is the method of choice, for it assures that every pig receives its requirement.

2. Orally administer iron from iron chelates within the first few hours of life; early administration before gut closure to large molecules is crucial. To ensure daily intake by all pigs, it is important to have a preparation that is palatable and readily consumed. Also, placement of the oral preparation at the right location in the creep area is important.

3. Give the pigs iron tablets or paste at 2 to 3 days of age. Repeat the treatment every 7 to 10 days until the pigs are eating their creep ration adequately. If pills are given, it is important to see that the pigs swallow them and not spit them out.

4. Place clean soil in the farrowing pen daily. Soil should not be contaminated with parasite eggs and other disease organisms. Iron sulfate can be sprinkled over the soil.

5. Swab sow's udder daily with a solution of 1 lb ferrous sulfate dissolved in 1 gal of warm water.

## POISONS AND TOXINS

Swine are susceptible to a number of poisons, any one of which may be disastrous in a herd. Among them are the following: moldy feed, including three species of mycotoxins—aflatoxins, ergot poisoning, and estrogenic syndrome; pitch poisoning; lead poisoning; mercury poisoning; pesticides, plant poisoning, involving several plants that are toxic to swine; salt poisoning; and blue-green algae.

## DOCKING TAILS TO STOP TAIL BITING

Tail biting accompanies close confinement. It results when pigs are prevented from rooting, nibbling, and chewing—from disturbing the pig's normal behavior pattern. Tail docking has become a common management practice to prevent subsequent tail biting of pigs in confinement. It should be done on all market hogs. Tails should be cut ¼ to ½ in. from the body with side-cutting pliers or another blunt instrument; the crushing action stops bleeding. The tail stump should be disinfected with a good antiseptic, and the instrument should be disinfected after docking each pig.

## FEEDING AND WATERING SPACE

It is especially important that hogs raised in confinement have adequate feeding and watering space. To this end, the following recommendations are made:

1. **Self-feeder space—**

| Class of Swine | No. of Animals/linear ft |
| --- | --- |
| Sows . . . . . . . . . . . . . . . . . . . . . . | 2 |
| Gilts . . . . . . . . . . . . . . . . . . . . . . | 3 |
| Boars . . . . . . . . . . . . . . . . . . . . . | 1 |
| Sows with litters . . . . . . . . . . . . . . | 1 |
| Weaning to 75 lb . . . . . . . . . . . . . | 4 |
| 75 lb to market . . . . . . . . . . . . . . | 3 |

2. **Automatic watering cups—**

| Class of Swine | No. of Animals/cup |
| --- | --- |
| Sows . . . . . . . . . . . . . . . . . . . . . . . . | 10 |
| Gilts . . . . . . . . . . . . . . . . . . . . . . . . | 2 |
| Boars . . . . . . . . . . . . . . . . . . . . . . . | 2 |
| Sows with litters . . . . . . . . . . . . . . . | 4 |
| Weaning to 75 lb . . . . . . . . . . . . . . . | 20 |
| 75 lb to market . . . . . . . . . . . . . . . | 20 |

## ENVIRONMENTAL EFFECTS OF SWINE

Pollution potential, affecting the environment of both people and swine, increased as the U.S. swine industry moved toward specialization, mechanization, high animal density, and confinement.

When manure and urine are stored and undergo anaerobic digestion, dangerous and disagreeable gases are produced. The ones of primary concern are: hydrogen sulfide, ammonia, carbon dioxide, and methane.

High levels of dust particles resulting from automated dry feed handling systems, dander and hair from hogs, and dried manure particles can occur inside swine buildings. Manure gases can cling to those dust particles in such a way that inhaling these gas-laden particles is uncomfortable and objectionable. Particulate matter also includes viral, bacterial, and fungal agents from the building environment and carries them into the respiratory system of people and hogs.

To operate compatibly within the community, to provide maximum self-protection, and to avoid neighbor complaints and legal actions seeking either monetary damages or court injunctions, swine producers must be aware of some basic information and strategy concerning pollution and apply pollution control measures appropriate to the location.

# QUESTIONS FOR STUDY AND DISCUSSION

1. How has the shift to confinement rearing affected the nutritional well-being of pigs? From an economic standpoint, how important is a knowledge of swine feeding?

2. What are the chief differences between pigs and ruminants, anatomically and in kind of feed consumed?

3. Why aren't nutritional requirements, or standards, like those in Table 40–1, absolute, final, and unchangeable?

4. What are the primary sources of energy for swine? What are the symptoms of energy-deficient swine?

5. Why is protein most frequently the limiting factor in the ration of swine, both from the standpoint of quantity and quality? What are the symptoms of protein-deficient swine?

6. What peculiarities of swine husbandry are conducive to swine suffering from mineral deficiencies?

7. What are the symptoms of deficiencies in swine of (a) salt, and (b) calcium/phosphorus?

8. What are the symptoms of deficiencies in swine of (a) cobalt, (b) copper, (c) iodine, and (d) iron?

9. What are the symptoms of deficiencies in swine of (a) manganese, (b) selenium, and (c) zinc?

10. What are vitamins?

11. How do fat-soluble and water-soluble vitamins differ? Classify the vitamins of practical importance for swine as (a) fat-soluble or (b) water-soluble.

12. What are the primary sources and the deficiency symptoms in swine of each of the following vitamins: (a) vitamin A, (b) vitamin D, (c) vitamin E, and (d) vitamin K?

13. What are the primary sources and the deficiency symptoms in swine of each of the following vitamins: (a) biotin, (b) pantothenic acid, (c) thiamin, and (d) vitamin B-12?

14. What are *unidentified factors*? List rich sources of unidentified factors.

15. Only 4.3% of the total feed consumed by swine in the U.S. is derived from roughage. Is this good or bad?

16. List twelve important concentrates that are used as swine feeds. When balancing rations for swine, what nutritional characteristics of concentrates should be kept in mind?

17. Why is so much corn fed to hogs in the U.S., despite the fact that it is deficient in both quantity and quality of protein?

18. Discuss the role of each of the following high energy feeds for swine: (a) high-moisture corn, (b) high-lysine corn, and (c) fats and oils.

19. How do you account for the fact that soybean meal is the leading high-protein supplement in the U.S.?

20. Discuss the preparation and feeding value of full-fat soybeans for swine.

21. Is the use of pastures for swine outmoded in the U.S.?

22. Discuss the role of (a) dry forages and (b) silages for swine.

23. Under what circumstances would you recommend (a) hogging down crops, and (b) feeding garbage?

24. Discuss the role of (a) antibiotics, and (b) sulfonamides as feed additives.

25. What is porcine somatotropin (PST)? What do experiments show relative to its value as a swine additive?

26. Why and how is creep feeding practiced?

27. List and discuss the advantages and disadvantages of early weaning. Give guidelines for weaning at 3 weeks of age.

28. Discuss the feeding of (a) growing-finishing pigs, (b) prospective breeding gilts, and (c) boars.

29. Why and how may sows be flushed? What is unique and important relative to feeding sows/gilts (a) during the gestation period, and (b) at farrowing time?

30. Present a program for feeding orphan pigs.

31. What is the corn-hog ratio? What does it average? What is the value of the corn-hog ratio?

32. What is soft pork; what causes it; and why is it objectionable?

33. How may a swine producer use, and benefit from, the feed allowances and suggested rations given in this chapter?

34. List and discuss the "pointers in formulating rations" presented in this chapter.

35. Discuss fitting rations for show and sale swine.

36. List and describe swine feeding systems.

37. What is (a) feeder pig production and (b) contract hog production?

38. Discuss Table 40-20, showing the estimated feed required to produce a 240-lb market hog.

39. What is nutritional anemia? How may it be prevented?

40. Why dock the tails of pigs? Is it necessary to dock the tails of pigs raised on pasture?

41. Discuss the environmental effects of swine.

## SELECTED REFERENCES

| Title of Publication | Author(s) | Publisher |
|---|---|---|
| *All About Pigs*, Third Edition | | Farming Press, Ltd., Lloyds Chambers, Ipswich, England, 1970 |
| *Approved Practices in Swine Production*, Sixth Edition | J. K. Baker E. M. Juergenson | The Interstate Printers & Publishers, Inc., Danville, IL, 1979 |
| *Digestion in the Pig* | D. E. Kidder M. J. Manners | Scientechnica Bristol Kingston Press, Oldfield Park, Bath, England, 1978 |
| *Life Cycle Nutrition* | P. Holden, et al. | Iowa State University, Ames, IA, 1988 |
| *Nutrient Requirements of Swine*, Ninth Revised Edition | National Research Council | National Research Council, National Academy Press, Washington, DC, 1988 |
| *Pig Husbandry*, Fourth Edition | J. R. Luscombe | Farming Press, Ltd., Fenton House, Warfedale Road, Ipswich, England, 1972 |
| *Pig Nutrition—recent developments in* | D. J. A. Cole W. Haresign | Butterworths, London, England, 1985 |
| *Pork Industry Handbook* | Staff | University of Illinois, Cooperative Extension Service, 1988 |
| *Practical Pig Nutrition* | C. T. Whittemore F. W. H. Elsley | Farming Press Limited, Fenton House, Ipswich, Suffolk, England, 1976 |

*(Continued)*

## SELECTED REFERENCES (Continued)

| Title of Publication | Author(s) | Publisher |
|---|---|---|
| *Profitable Pig Farming*, Third Edition | E. G. Johnson | Farming Press, Ltd., Lloyds Chambers, Ipswich, England, 1970 |
| *Stockman's Handbook, The*, Sixth Edition | M. E. Ensminger | The Interstate Printers & Publishers, Inc., Danville, IL, 1983 |
| *Swine Feeding and Nutrition* | T. J. Cunha | Academic Press, Inc., New York, NY, 1977 |
| *Swine Production*, Fifth Edition | J. L. Krider<br>J. H. Conrad<br>W. E. Carroll | McGraw-Hill Book Company, New York, NY, 1982 |
| *Swine Production and Nutrition* | W. G. Pond<br>J. H. Maner | AVI Publishing Co., Westport, CT, 1984 |
| *Swine Production in Temperate and Tropical Environment* | W. G. Pond<br>J. H. Maner | W. H. Freeman, San Francisco, CA, 1974 |
| *Swine Science*, Fifth Edition | M. E. Ensminger<br>R. O. Parker | The Interstate Printers & Publishers, Inc., Danville, IL, 1984 |

Distribution facility of Lieske Genetics, Inc., Henderson, Minnesota. This firm markets approximately 3,000 boars annually. (Courtesy, Lieske Genetics, Henderson, MN)

Why worry?

How bacon got its name. One theory is that the word *bacon* is derived from the crest of Lord Bacon (1561–1626), noted English viscount, lawyer, statesman, and politician, which depicts a pig. (Courtesy, Picture Post Library, London, England)

# *Chapter 41*

## MARKETING AND SLAUGHTERING HOGS

Hogs are marketed through the several channels open to all classes of livestock. In comparison with the marketing of cattle and sheep, however, the marketing of hogs differs in the following respects: (1) A higher proportion of hogs is marketed through nonpublic markets than of cattle and sheep; (2) relatively few hogs are taken out of the market for further feeding because of parasite and disease problems; and (3) the proportion of hogs slaughtered on farms exceeds that of cattle or sheep because of the greater ease of slaughtering and the adaptation of pork to home curing.

Hog slaughtering is unique in that much more pork is cured than is the case with beef or lamb; and pork fat (lard) is not classed as a by-product, although the surplus fats of beef, veal, mutton, and lamb are in the by-product category.

## MARKET CHANNELS FOR HOGS

In 1987, packers purchased their hogs through the following market channels:

|  | Percentage |
|---|---|
| Nonpublic markets | 88.8 |
| Terminal markets | 6.3 |
| Auction Markets | 4.9 |
| Total | 100.0 |

*Nonpublic markets is defined as all purchases from sources except public markets—terminals and auctions. A terminal market is defined as a public market at which two or more independent commission firms (selling agencies) operate. An auction market is defined as a public market at which only one selling agency operates.*

More hogs than cattle or sheep are purchased at nonpublic markets; and fewer hogs than cattle or sheep are marketed on a grade and weight basis. In 1987, nonpublic markets accounted for the following percentages of total marketings: hogs, 88.8%; sheep, 81.4%; and cattle, 80.2%. In 1987, marketings on a grade and weight basis accounted for the following percentages of the total marketings: sheep and lambs, 35.9%; cattle, 30.4%; and hogs, 13.5%.

## LEADING PUBLIC STOCKYARDS

Public stockyards have declined in importance as market outlets for hogs, just as they have for cattle and sheep. Nevertheless, they still indicate the areas of market action and provide an established outlet.

As would be expected, the leading hog markets of the United States are located in or near the Corn Belt—the area of densest hog population. With the advent of the motor truck and improved highways, the pork packing plants were moved nearer the areas of hog raising. Thus, since World War II, local or interior packers have increased in numbers. In order to meet this added competition, the large packers at more distant points resorted to direct buying and the purchase and construction of interior plants.

The rank of the major public hog markets shifts considerably according to feed supplies and general economic conditions. However, the six largest hog markets and their receipts are listed in Table 41-1.

**TABLE 41-1**
**RECEIPTS AT LEADING PUBLIC HOG MARKETS[1]**

| Market | Year Average 1983–1987 | 1987 |
|---|---|---|
| Sioux City, Iowa | 1,096,000 | 858,000 |
| National Stockyards, E. St. Louis, Illinois | 1,011,000 | 843,000 |
| South St. Paul, Minnesota | 861,000 | 745,000 |
| Omaha, Nebraska | 731,000 | 557,000 |
| South St. Joseph, Missouri | 621,000 | 498,000 |
| Kansas City, Missouri | 184,000 | 139,000 |
| U.S. Total[2] | 8,322,000 | 6,857,000 |

[1]*Agricultural Statistics 1988*, USDA, p. 276, Table 409.

[2]The number of stockyards reporting varies from 41 to 68.

## MARKET CLASSES AND GRADES OF HOGS

The market classes of grades of hogs are summarized in Table 41-2.

**TABLE 41-2**
**THE MARKET CLASSES AND GRADES OF HOGS**

| Hogs or Pigs | Use Selection | Sex Class | Weight Divisions (lb) | | Weight Divisions (kg) | | Commonly Used Grades |
|---|---|---|---|---|---|---|---|
| Hogs | Slaughter hogs | Barrows and Gilts (often called butcher hogs) | 120–140<br>140–160<br>160–180<br>180–200<br>200–220<br>220–240 | 240–270<br>270–300<br>300–330<br>330–360<br>360–400<br>400 lb up | 55–64<br>64–73<br>73–82<br>82–91<br>91–100<br>100–109 | 109–123<br>123–136<br>136–150<br>150–163<br>163–182<br>182 kg up | U.S. No. 1<br>U.S. No. 2<br>U.S. No. 3<br>U.S. No. 4<br>U.S. Utility |
| | | Sows (or packing sows) | 270–300<br>300–330<br>330–360<br>360–400 | 400–450<br>450–500<br>500–600<br>600 lb up | 123–136<br>136–150<br>150–163<br>163–182 | 182–204<br>204–227<br>227–272<br>272 kg up | U.S. No. 1<br>U.S. No. 2<br>U.S. No. 3<br>Medium, Cull |
| | | Stags | All weights | | | | Ungraded |
| | | Boars | All weights | | | | Ungraded |
| | Feeder hogs | Barrows and Gilts | 120–140<br>140–160<br>160–180 | | 55–64<br>64–73<br>73–82 | | U.S. No. 1, U.S. No. 2,<br>U.S. No. 3, U.S. No. 4,<br>U.S. Utility, Cull |
| Pigs | Slaughter pigs | Barrows, Gilts and Boars | Under 30<br>30–60 | | Under 13.6<br>13.6–27.2 | | Ungraded |
| | | Barrows and Gilts | 60–80<br>80–100<br>100–120 | | 27.2–36.3<br>36.3–45.4<br>45.4–54.5 | | Ungraded |
| | Feeder pigs | Barrows and Gilts | 80–100<br>100–120 | | 36.3–45.4<br>45.4–54.5 | | U.S. No. 1, U.S. No. 2<br>U.S. No. 3, U.S. No. 4,<br>U.S. Utility, Cull |

The market classes and grades of swine were developed in much the same manner as the classifications of cattle were developed and brought into use. They also serve much the same purpose. Swine classes and grades do differ from those used in cattle and sheep in that (1) there are no age divisions by years (e.g., cattle are classified as yearling and two-year-old and over), (2) only a limited number of hogs are returned to the country as feeders for further growth or finishing, and (3) rarely are hogs of any kind purchased on the market for use as breeding animals. As in the classification of market cattle, the class of market hogs indicates the use to which the animals are best adapted, whereas the grade indicates the degree of perfection within the class.

## FACTORS DETERMINING MARKET CLASSES

The market class of hogs is determined by the following factors: (1) hogs vs pigs, (2) use selection, (3) sex, and (4) weight.

### HOGS AND PIGS

All swine are first divided into two major groups according to age—hogs and pigs. Although actual ages are not observed, the division is made largely by weight in relation to the apparent age of the animal. Young animals weighing under 120 lb (under 4 mos. of age) are generally known as pigs, whereas those weighing over 120 lb are called hogs.

### USE SELECTION OF HOGS AND PIGS

Hogs and pigs are each further divided into two subdivisions as slaughter animals and feeders. Slaughter swine are hogs and pigs that are suitable for immediate slaughter. The demand for lightweight slaughter pigs is greatest during the holiday season when they are in demand as roasting pigs for hotels, clubs, restaurants, steamships, and other consumers. Such pigs weigh from 30 to 60 lb, are dressed shipper style (with the head on), and must produce a plump and well-proportioned carcass. Slaughter hogs (the older animals) are in demand throughout the year.

Feeder swine include those animals that show ability to take on additional weight and finish. The feeder group is relatively small. Moreover, because of the greater disease hazard with hogs, this class is under very close federal supervision. Before being released for return to the country, feeder swine must be inspected from a health standpoint, and then either sprayed or dipped as a precautionary measure to prevent the spread of diseases and parasites.

### THE SEX CLASSES

The sex class is used only when it affects the usefulness and selling price of animals. In hogs, this subdivision is of less importance than in cattle. Thus, barrows and gilts are always classed together in the case of both slaughter and feeder hogs.

This is done because the sex condition affects their usefulness so little that a price differentiation is not warranted. In addition, because the carcass is not affected, no sex differentiations are made for slaughter pigs under 60 lb in weight. The terms *barrow, gilt, sow, boar,* and *stag* are used to designate the sex classes of hogs. The definition of each of these terms follows:

**Barrow**—A castrated male swine animal that was castrated at an early age—before reaching sexual maturity and before developing the physical characteristics peculiar to boars.

**Gilt**—A female swine that has not produced pigs and which has not reached an evident stage of pregnancy.

**Sow**—A female swine that shows evidence of having produced pigs or which is in an evident stage of pregnancy. Piggy sows are usually docked 40 lb, but they may be docked from 0 to 50 lb, depending on the market.

**Boar**—An uncastrated male swine animal of any age. Mature boars should always be stagged and fed a month longer before being sent to market. The market value of boars is necessarily low, for a considerable number are condemned as unfit for human consumption, primarily because of odor.

**Stag**—A stag is a male swine animal that was castrated after developing the physical characteristics of a mature boar. Because of relatively thick skins, coarse hair, and heavy bones, stags are subject to a dockage. They are usually docked 70 lb, but they may be docked from 40 to 80 lb, depending on the market. When marketed direct, stags are usually not docked in weight but are purchased at a price that reflects their true value from a meat standpoint.

### WEIGHT DIVISIONS

Occasionally, the terms *light, medium,* and *heavy* are used to indicate the approximate weights, but most generally the actual range in weight in pounds is specified both in trading and in market reporting. Moreover, hogs are usually grouped according to relatively narrow weight ranges, because variations in weight affect (1) the dressing percentages, (2) the weight and desirability of the cuts of meat, and (3) the amount of lard produced (heavier weights produce more lard). Boars and stags are not usually subdivided according to weights.

## THE FEDERAL GRADES OF HOGS AND THEIR CARCASSES

The market grade of swine, as for other kinds of livestock, is a specific indication of the degree of excellence within a given class. While no official grading of live animals is done by the U.S. Department of Agriculture, market grades do form a basis for uniform market hog reporting. Also, it is intended that the grade of slaughter hogs on foot be correlated with the carcass grade.

Tentative grades for pork carcasses were first issued in 1931. Subsequently, the standards have been revised from time to time, better to reflect producer and consumer needs at the time. In January 1985, the current standards for grades of barrow and gilt carcasses and slaughter barrows and gilts became effective. These grades are based on two factors: (1) last rib backfat thickness, and (2) muscling.

If a carcass qualifies as acceptable in quality of lean and in belly thickness, and is not soft and oily, it is graded U.S. No. 1, 2, 3, or 4, based entirely on projected carcass yields of the 4 lean cuts (ham, loin, picnic shoulder, and Boston butt). The expected yields of each of the grades in the 4 lean cuts, based on using the U.S. Department of Agriculture standard cutting and trimming methods, are as given in Table 41–3.

### TABLE 41–3
### EXPECTED YIELDS OF THE FOUR LEAN CUTS BASED ON CHILLED CARCASS WEIGHT, BY GRADE[1]

| Grade | Yield |
|---|---|
| U.S. No. 1 | 60.4 and over |
| U.S. No. 2 | 57.4–60.3 |
| U.S. No. 3 | 54.4–57.3 |
| U.S. No. 4 | Less than 54.4 |

[1]USDA Source. These yields will be approximately 1% lower if based on hot carcass weight.

Carcasses vary in their yields of the four lean cuts because variations in their degree of fatness and in their degree of muscling (thickness of muscling in relation to skeletal size).

The degree of muscling specified for each of the four grades decreased progressively from U.S. No. 1 grade through the U.S. No. 4 grade. This reflects the fact that, among carcasses of the same weight, fatter carcasses normally have lesser degree of muscling. Three degrees of muscling are recognized: thick (superior), average, and thin (inferior).

The grade is determined by calculating a preliminary grade based on the backfat thickness over the last rib according to the schedule shown in Table 41–4, then adjusting it up or down one grade for thick or thin muscling.

### TABLE 41–4
### PRELIMINARY CARCASS GRADE BASED ON BACKFAT THICKNESS OVER THE LAST RIB

| Preliminary Grade | Backfat Thickness Range |
|---|---|
| U.S. No. 1 | Less than 1.00 in. |
| U.S. No. 2 | 1.00 to 1.24 in. |
| U.S. No. 3 | 1.25 to 1.49 in. |
| U.S. No. 4 | 1.50 in. and over[1] |

[1]Carcasses with last rib backfat thickness of 1.75 in. or over cannot be graded U.S. No. 3, even with thick muscling.

Thus, the on-foot and carcass federal grades of *slaughter swine* are: U.S. No. 1, U.S. No. 2, U.S. No. 3, U.S. No. 4, and U.S. Utility (see Fig. 41–1). The market grades of swine differ from the grades of cattle in that (1) hogs possess 5 grades instead of the 8 common to cattle, (2) the top grade of hogs is U.S. No. 1 instead of Prime, and (3) no Cutter or Canner designations are used in hogs—instead the lowest grade is known as U.S. Utility.

As a rule, slaughter pigs that weigh under 60 lb are not graded, because they have not reached sufficient maturity for variations in their carcass traits to affect the market value materially.

The five market grades of *feeder pigs* are shown in Fig. 41–2.

# SLAUGHTER SWINE
## U.S. GRADES

**U.S. NO. 1**

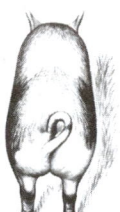

**U.S. NO. 2**

**U.S. NO. 3**

**U.S. NO. 4**

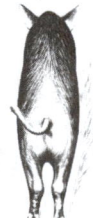

**U.S. UTILITY**

Fig. 41–1. The five market grades of slaughter swine, as adapted by the author. (Courtesy, USDA)

# FEEDER PIGS
## U.S. GRADES

 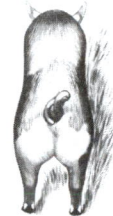

### U.S. NO. 1

### U.S. NO. 2

### U.S. NO. 3

### U.S. NO. 4

### U.S. UTILITY

Fig. 41-2. The five market grades of feeder pigs, as adapted by the author. (Courtesy, USDA)

## PRODUCTION AND PRICE MOVEMENTS

The following three types of production and price movements are of concern to livestock market analysts and hog producers:

1. **Secular trends.** These are longtime trends that persist over a period of several cycles. The long-run trend in hog numbers from 1920 to 1980 has been unchanged. However, the longtime trend in pork production (weight of hogs slaughtered) has been slightly upward. The difference between hog numbers and pork production is attributed to improved breeding, feeding, and management, which have increased productivity per head and resulted in marketing hogs at younger ages. Despite the increased total production, however, supplies of pork per person have tended slightly downward.

2. **Cyclical movement.** These are movements that follow a pattern that repeats itself. Hog cycles average about 4 years—2 years of expansion, and 2 years of liquidation. Basically, cycles are the response of producers to prices; and prices reflect quantity produced. In the past, the hog-corn ratio served as a barometer; when it was favorable, an expansion in hog numbers followed; when it was unfavorable, a cutback followed. Today, the corn-hog ratio is not as much a factor as formerly. Also, large swine units are inclined to maintain production near optimum levels for the size unit without regard to cyclical movements. So, cycles have become less important.

3. **Seasonal variation.** These are movements that tend to follow a more or less uniform pattern within the year, and which conform to this pattern over a period of years. Hogs follow a distinctive seasonal pattern which is directly related to farrowing and marketing. Spring farrowing is concentrated in March and April, and fall farrowing is concentrated in September. As a result, slaughter barrows and gilts are usually highest in price in July, August, and September, and lowest in November to May.

Feeder pig prices generally have a decided peak in early spring, then decline in a July-August low. This is followed by a partial recovery in September and October and another dip in December. This general seasonal pattern reflects the number of feeder pigs on the market.

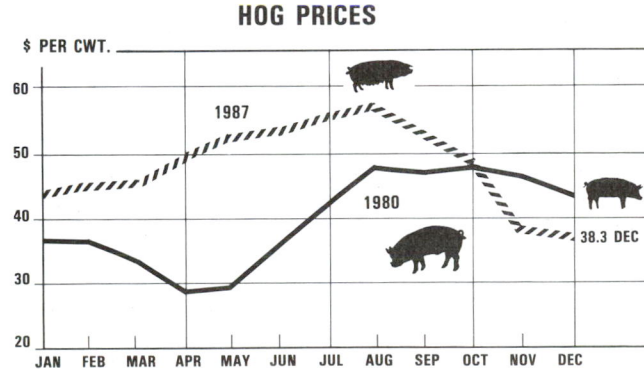

Fig. 41-3. Hog prices vary seasonally; this is directly related to farrowing and marketing. This shows slaughter barrow and gilt prices at seven Midwest markets combined, by months and for each of the years 1980 and 1987. (Source: *Livestock and Meat Statistics,* 1984–88, USDA, Bull. No. 784, p. 205, Table 120.)

## OTHER HOG MARKET TERMS AND FACTORS

Additional market terms and factors of importance in marketing hogs are: roasters (roasting pigs), suspects (governments), cripples, and dead hogs.

### ROASTERS (ROASTING PIGS)

Roasters are fat, plump, suckling pigs, weighing 30 to 60 lb on foot. These are dressed shipper style (with the head on); and they are not split at the breast or between the hams. When properly roasted and attractively served with the traditional apple in the mouth, roast pig is considered a great delicacy for the holiday season.

### SUSPECTS (GOVERNMENTS)

Suspects, or governments, are suspicious animals that federal inspectors tag at the time of the antemortem inspection to indicate that more careful scrutiny is to be given in the postmortem inspection. If the carcass is deemed unfit for human consumption, it is condemned and sent to the inedible tank.

### CRIPPLES

Cripples are hogs that are not able to walk and that must be hauled to the packing plants in cripple carts. They are docked heavily on a live basis because some of them must be condemned, and others may be partially condemned.

### DEAD HOGS

Dead hogs are those that arrive dead at the market. They have practically no salvage value. Their carcasses are sent to the tanks for conversion into inedible grease, fertilizer, etc.

## PACKER SLAUGHTERING AND DRESSING OF HOGS

The page headed, "How It Used to Be Done" depicts, more vividly than words, the history of market swine from feeding corn with a scoop shovel to packing pork in a barrel.

Table 41–5 shows the proportion of hogs slaughtered commercially (meaning slaughtered in federally inspected and other wholesale and retail establishments) and the proportion on farms. The total figure refers to the number dressed in all establishments and on farms.

Although modern equipment may be lacking, the slaughtering procedure on the farm or in a small local plant is much the same as that followed in a large federally inspected plant. Only the procedure used in a modern commercial packing plant will be discussed, however.

After purchase, hogs are driven from the holding pens to the packing plant where they are given a shower and are held temporarily in a small pen while awaiting slaughter.

**TABLE 41–5**
**U.S. HOGS SLAUGHTERED, BY CLASS, 1970–1987[1]**

| Year | Farm Slaughter | Commercial | | | Total | Percent Commercial |
| | | Federally Inspected | Other | Total | | |
| | | (million lb) | | | | (%) |
| 1970 | 178 | 12,114 | 1,134 | 13,248 | 13,426 | 98.7 |
| 1975 | 189 | 10,733 | 581 | 11,314 | 11,503 | 98.4 |
| 1980 | 195 | 15,742 | 689 | 16,431 | 16,626 | 98.8 |
| 1985 | 79 | 14,312 | 414 | 14,726 | 14,805 | 99.5 |
| 1987 | 62 | 13,954 | 358 | 14,312 | 14,374 | 99.6 |

[1]*Livestock and Meat Statistics 1984–88*, USDA, Stat. Bull. No. 784, p. 138.

## STEPS IN SLAUGHTERING AND DRESSING HOGS

The slaughtering and dressing of hogs consists in carrying out the following steps in rapid, and almost rhythmic, succession:

1. **Rendering insensible.** The hogs are rendered insensible[1] by use of a captive bolt stunner, gunshot, electric current, or carbon dioxide.

2. **Shackling and hoisting.** The hogs are shackled just above the hoof on the hind leg and are then hoisted to an overhead rail.

3. **Sticking.** The sticker sticks the hog just under the point of the breastbone, severing the arteries and veins leading to the heart. The animal is allowed to bleed for a few minutes.

4. **Scalding.** The animals are next placed in a scalding vat for about four minutes. By means of automatically controlled steam jets, the temperature of the water in the vats is maintained at about 145°F. The scalding process loosens the hair and scurf.

---

[1]See footnote No. 2 under "Steps in Slaughtering and Dressing Cattle," Chapter 19, relative to humane slaughter methods.

# HOW IT USED TO BE DONE

Fig. 41-4. *Above:* Prior to 1900, hogs were fed chiefly on corn and marketed at 10 to 16 months of age; and sows farrowed once a year. But what was time to a hog! (Courtesy, American Feed Manufacturers Assn.)

Fig. 41-5. *Above:* Hogs being driven to market in the early 1900s, before truck transportation which was first started in 1911. (Photo by J. C. Allen and Son, West Lafayette, IN)

Fig. 41-6. *Above:* Slaughtering and dressing hogs in a meat-packing plant in 1860. (Courtesy, Bettmann Archive, New York, NY)

Fig. 41-7. *Above:* Dry-curing bacon, with a mixture of 8 lb of table salt and 3 lb of sugar.

Fig. 41-8. *Right:* Sweet-pickle curing hams in a hardwood barrel, using a mixture of 7 lb table salt, 3 lb sugar, and 5 gal water per 100 lb meat.

5. **Dehairing.** After scalding, the animals are elevated into a dehairing machine which scrapes them mechanically. In modern packing plants, a single dehairing machine will handle up to 500 hogs per hour. Most large plants are equipped with twin machines which provide a scraping capacity of a thousand hogs per hour.

6. **Returning to overhead tracks.** As the animals are discharged from the dehairing machine, the gam cords of the hind legs are exposed; and gambrel sticks are inserted in the cords. Then the carcass is again hung from the rail.

Fig. 41-9. On the rail. Pork carcasses enroute through a meat packing plant.

7. **Dressing.** A conveyor then moves the carcass slowly along a prescribed course where attendants perform the following tasks:

    a. Washing and singeing.

    b. Removing the head.

    c. Opening the carcass and eviscerating.

    d. Splitting or halving the carcass with a cleaver or electric saw.

    e. Removing the leaf fat.

    f. Exposing the kidneys for inspection and facing the hams (removing the skin and fat from the inside face or cushion of the ham).

    g. Washing the carcass and sending it to the coolers where the temperature is held around 34°F. (Rapid chilling is desirable.)

During the dressing process, federal inspectors carefully examine the head, viscera, and carcass. If the carcass gives no evidence of disease, it is stamped *U.S. Inspected and Passed* prior to being sent to the coolers. If it shows evidence of a diseased condition and is not considered wholesome, it is stamped *U.S. Condemned* and is sent to the inedible tank room along with the viscera. There the carcass is cooked under steam pressure in sealed tanks until all the disease germs are destroyed.

## PACKER VS SHIPPER STYLE OF DRESSING

The two common styles of dressing hogs in packing plants are (1) packer style, and (2) shipper style. In general, the packer style is followed, and this system is almost exclusively used when carcasses are to be converted into the primal cuts. In packer-style dressing, the backbone is split full length through the center; the head, without the jowl, is removed; and the kidneys and the leaf fat are removed and the hams faced.

The shipper style is ordinarily limited to light-weight slaughter pigs that are sold as entire carcasses to the wholesale trade. In this style of dressing, the carcass is merely opened from the crotch to the tip of the breastbone; the backbone is left intact; the leaf fat is left in; and the entire head is left attached. Roasting pigs are dressed shipper style, and prior to cooling are placed in a trough with front legs doubled back from the knee joints and the hind legs extending straight back from the hams.

When dressed packer style, the carcass yields are four to eight percent less than can be obtained shipper style. The lower yield is due to the removal of the head, leaf fat, kidneys, and ham facings.

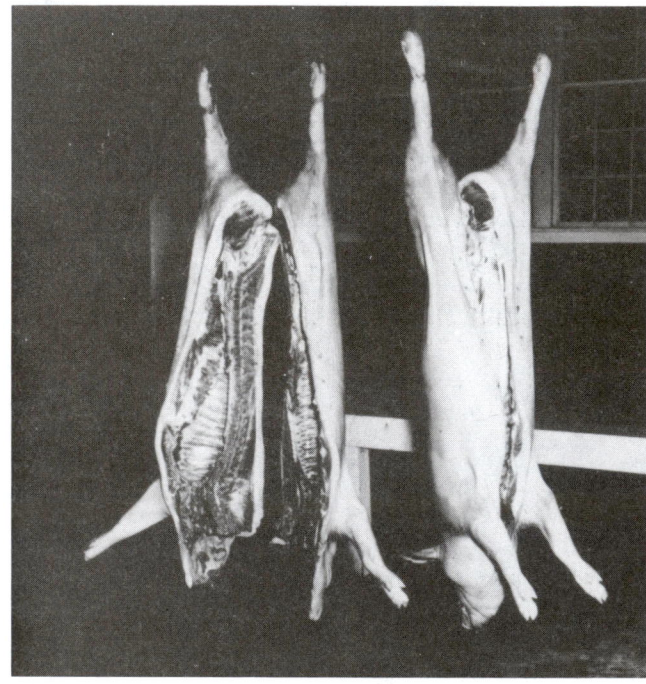

Fig. 41-10. Packer (left) vs shipper (right) style of dressing. Both hogs dressed 75% shipper style, but subsequent removal of the head, leaf fat, and kidneys from the carcass to the left—in preparing packer style—lowered the dressing percentage of this carcass 4% further. (Courtesy, Washington State University, Pullman)

## THE DRESSING PERCENTAGE

*Dressing percentage may be defined as the percentage yield of chilled carcass in relation to the weight of the animal on foot.*

Because hogs have a smaller digestive capacity, fill is less important in determining their dressing percentage than is the case with cattle. The degree of finish and style of dressing are the important factors affecting dressing percentage in hogs. U.S. No. 1 hogs dressed packer style (with head, leaf fat, and kidneys removed) dress about 70%, whereas hogs dressed shipper style (head left on, and leaf fat and kidneys in) dress 4 to 8% higher.

It is generally recognized that fat, lardy-type hogs give a higher dressing percentage than can be obtained with meat- or bacon-type animals. Because lard frequently sells at a lower price than is paid for hogs on foot, an excess yield of lard very obviously represents an economic waste of feed in producing the animals and is undesirable from the standpoint of the processor. Accordingly, attaching great importance to the projected dressing percentage of hogs is outmoded. The more progressive buyers are now focusing their attention on the cutout value of the carcass, especially on the maximum yield of the more sought primal cuts of high quality.

Hogs have a relatively smaller barrel and chest cavity than cattle and sheep. In addition they are dressed with their skin and shanks on. Consequently, they dress higher than other classes of slaughter animals.

Generally, when barrows and gilts are finished for market, they weigh approximately 240 lb. After slaughter, the resulting carcass represents approximately 70% of the live weight. So, a 240-lb live hog will produce a carcass weighing approximately 168 lb.

The average liveweight of hogs, dressed packer style by federally inspected meat-packing plants, and their percentage yield in meat for the year 1980 was as shown in Table 41–6.

### TABLE 41–6
### AVERAGE LIVEWEIGHT, CARCASS YIELD, AND DRESSING PERCENTAGES OF ALL HOGS COMMERCIALLY SLAUGHTERED IN THE U.S.[1]

| Hogs | Average Liveweight | | Average Dressing Weight[2] | | Dressing |
|------|------|------|------|------|------|
| | (lb) | (kg) | (lb) | (kg) | (%) |
| 1977 | 237 | 107 | 170 | 77 | 71.7 |
| 1987 | 248 | 113 | 177 | 80 | 71.4 |

[1]*Livestock and Meat Statistics, 1984–88*, USDA, Stat. Bull. No. 784, pp. 107, 135, 136.
[2]Packer style.

## DISPOSITION OF THE PORK CARCASS

Fig. 41–11.  Hams in a modern packing house. (Courtesy, *The National Provisioner*, Chicago, IL)

Almost all hog carcasses are cut up at the slaughtering plant and are sold in the form of wholesale and retail cuts. The whole-carcass trade is largely confined to roasting and slaughter pigs.

The handling of pork differs further from that of beef and lamb in that only a relatively small percentage, about 30%, of the pork is sold fresh. The remaining 70% is either cured by various methods, rendered into lard, or manufactured into meat products. In general, loins, shoulders, and spareribs are most likely to be sold as fresh cuts. But it must be remembered that practically every pork cut may be cured, and, under certain conditions, is cured. Because pork is well adapted to curing, it has a decided advantage over beef and mutton, which are sold almost entirely in the fresh state. The hog market is stabilized to some extent by this factor.

## LARD

*Lard is fat rendered (melted out) from fresh, fatty pork tissue.* It is considered a primary product of hog slaughter and not a by-product.

The proportion of lard varies with the type, weight, and finish of the hogs, and the relative price of lard and the cuts of meat.

Lard production per slaughtered hog has decreased sharply in recent years, with the shift away from fat, lardy-type hogs to lean, muscular animals. In 1950, an average of 33.2 lb of lard per hog slaughtered was produced in the United States; in 1983, only 11.1 lb of lard per hog was produced in the United States.

## KINDS OF LARD

Lard is classified as follows according to the part of the animal from which the fat comes and the method of rendering:

- **Open kettle rendered leaf lard**—Kettle rendered leaf lard is made from the leaf fat only. It is rendered in a steam-jacketed open kettle at a temperature of 230° to 250°F. It is very white in color, fine textured, and possesses excellent keeping qualities and a pleasing flavor. It is the highest grade of lard.

- **Kettle rendered lard**—Kettle rendered lard is made from backfat and leaf fat, usually in equal amounts. It is also rendered in a steam-jacketed open kettle, but at a temperature of 240° to 260°F. Kettle rendered lard has good keeping qualities, but it is likely to be somewhat darker than kettle rendered leaf lard, to which it ranks second in grade.

- **Steam rendered lard (prime steam lard)**—Probably 85% of packinghouse lard is of the steam rendered type. It is made from killing and cutting fats that are rendered in a closed tank under steam pressure of from 30 to 50 lb and at a temperature of 285°F. It is somewhat milder in flavor and odor and lighter in color than lard produced by dry rendering.

- **Neutral lard**—Neutral lard is made entirely from leaf fat that is rendered at a very low temperature of about 126°F. It is used almost entirely in the manufacture of oleomargarine.

- **Dry-process rendered lard**—Dry-process rendered lard is essentially a kettle rendered lard, except that the fat is melted in a closed container, usually under reduced pressure. This method of rendering gives a product that has a fine flavor and excellent keeping quality.

- **Lard substitutes**—Lard substitutes, which are sometimes used in place of lard, are made from (1) lard and other animal fats (lard compound); (2) vegetable oils with animal fats; and (3) hydrogenated vegetable oils, the most common of which are cottonseed, soybean, peanut, and coconut oil.

- **Lard oil and stearin**—When fat is stored at high temperatures of 90° to 100°F, the liquid (lard oil) separates from the solid (stearin). Lard oil, which is made from prime steam lard, consists mainly of olein. It is used in the manufacture of margarine, as a burning oil, and as a lubricant for thread-cutting machines.

Stearin is the white solid material composed of glycerin and stearic acid left after the pressing operation forces out the lard oil.

- **Modern lard (or deodorized or hydrogenated lard)**—Modern lard is deodorized lard to which extra atoms of hydrogen and a stabilizing agent (inhibitors of vegetable origin) have been added. Such lard is bland, has a high melting point, and is less subject to rancidity than ordinary lard. This type of lard was first placed on the market in 1941.

## PORK CUTS AND HOW TO COOK THEM

A minimum of 24 hours chilling at temperatures ranging from 33° to 38°F is necessary to remove the animal heat properly and give the carcasses sufficient firmness to make possible a neat job of cutting. After chilling, the carcasses are brought to the cutting floor where they are reduced to the wholesale cuts.

The method of cutting varies somewhat according to the value of lard and the relative demand for different cuts. Despite some variation, the most common wholesale cuts of pork are ham, bacon, loin, picnic shoulder, Boston butt, jowl, spareribs, and feet.

Market hogs weighing from 240 lb will yield from 54% of their liveweight in the four primal cuts: the ham, loin, picnic shoulder, and Boston butt. However, because of the relatively higher value per pound of these cuts, they make up three-quarters of the value of the entire carcass.

Fig. 41-12, *PORK* shows retail cuts, where they come from, and how to cook them. Also, the wholesale cuts of pork are shown in the drawing immediately below the heading.

# QUESTIONS FOR STUDY AND DISCUSSION

1. What primary differences exist between the marketing of hogs and the marketing of cattle and sheep?

2. Through what three market channels do packers purchase their hogs? Packers purchase the vast majority of their hogs (88.8% of them in 1987) through nonpublic markets. Is this good or bad?

3. Why are the leading public hog markets located in or near the Corn Belt? List by rank the three leading public hog markets of the United States.

4. Compare the market classes and grades of hogs with the market classes and grades of cattle and sheep.

5. Explain how each of the following factors affects the market class of hogs: (a) hogs and pigs, (b) use selection of hogs and pigs, (c) sex classes, and (d) weight divisions.

6. Define each of the following terms: (a) barrow, (b) gilt, (c) sow, (d) boar, and (e) stag.

7. In 1985, new grades of barrow and gilt carcasses and slaughter barrows and gilts became effective. On what two factors are these grades based?

8. Name the four lean pork cuts on which carcass grades are based. For a U.S. No. 1 chilled carcass, what minimum yield of these four cuts is specified?

9. The degree of muscling specified for each of the four carcass grades decreases progressively from U.S. No. 1 grade through U.S. No. 4 grade. Why is this?

10. Name in descending order (from highest to lowest) the on-foot and carcass grades of slaughter barrows and gilts. Name in descending order (from highest to lowest) the five market grades of feeder pigs.

11. Discuss each of the following three types of price movements as they affect hog producers and pork consumers: (a) secular trends, (b) cyclical movements, and (c) seasonal variations. Why have cycles become less important in recent years?

12. Define each of the following market terms and factors: (a) roasters, (b) suspects, (c) cripples, and (d) dead hogs.

13. Table 41-5 shows the farm slaughter of hogs is becoming a lost art. Why is this so?

14. List in chronological order the steps in slaughtering and dressing hogs.

15. How do packer and shipper styles of dressing differ?

16. Why is attaching great importance to the projected dressing percentage of hogs outmoded?

17. If a market hog weighed 240 lb on foot and yielded a carcass weighing 168 lb, what was the dressing percentage?

18. Why is so much more pork than beef or lamb cured before selling, rather than sold fresh?

19. Define the term *lard*. Why is lard considered a primary product of hog slaughter and not a by-product, whereas the trimmed fats of beef and lamb are classed as by-products?

20. What is *modern lard*?

21. How may a meat shopper or consumer benefit from a chart such as Fig. 41-12?

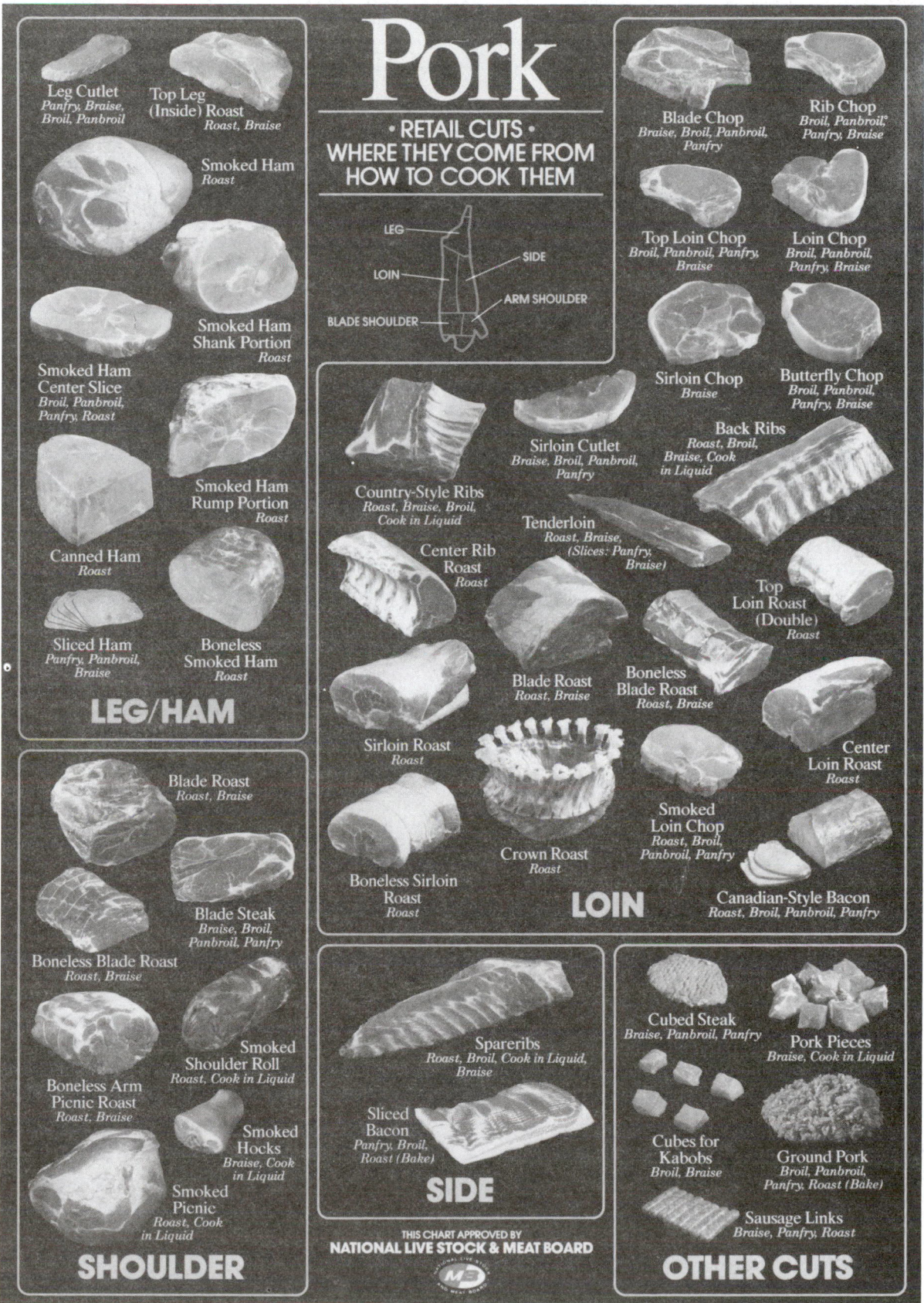

Fig. 41–12. PORK, retail cuts, where they come from, and how to cook them. Note, too, that the wholesale cuts are shown in the drawing immediately below the heading. (Courtesy, National Live Stock and Meat Board, Chicago, IL)

## SELECTED REFERENCES

| Title of Publication | Author(s) | Publisher |
|---|---|---|
| *Livestock and Meat Marketing*, Second Edition | J. H. McCoy | Avi Publishing Co., Westport, CT, 1979 |
| *Meat Handbook*, Fourth Edition | A. Levie | Avi Publishing Co., Westport, CT, 1979 |
| *Meat We Eat, The*, Twelfth Edition | J. R. Romans<br>P. T. Ziegler | The Interstate Printers & Publishers, Inc., Danville, IL, 1985 |
| *Pork Industry Handbook* | Staff | Cooperative Extension Service, Purdue University, West Lafayette, IN, updated at intervals |
| *Stockman's Handbook, The* | M. E. Ensminger | The Interstate Publishers, Inc., Danville, IL, 1983 |
| *Swine Production* | J. L. Krider<br>J. H. Conrad<br>W. L. Carroll | McGraw-Hill Book Company, New York, NY, 1982 |

Pork roast. (Courtesy, National Live Stock & Meat Board, Chicago, IL)

# THE POULTRY
# INDUSTRY

Life begins! Good breeding and good feeding make for a good start in life. (Courtesy, Maple Leaf Mills, Ltd., Ontario, Canada)

# Chapter 42

815

The term *poultry* covers a rather wide variety of birds of several species, and it refers to them whether they are alive or dressed. It includes chickens, turkeys, ducks, geese, swans, guineas, pigeons, peafowl, ostriches, pheasants, and other game birds.

The study of birds which are not classed as poultry is known as *ornithology*.

There are about 600,000 species of animals in the world, of which 10,000 species are birds. The most highly developed animals are mammals, which include humans and the 4-footed farm animals, and which are distinguished by the presence of hair and mammary glands. Birds are distinguished by the covering of feathers.

## DOMESTICATION AND EARLY USE OF POULTRY

Ancient people persuaded chickens to live and produce near their abodes. It is not known exactly when this happened, but it's obvious that chickens were domesticated at a remote period. The keeping of poultry was probably contemporary with the keeping of sheep by Abel and the tilling of the soil by Cain. Chickens were known in ancient Egypt, and they had already achieved considerable status in the time of the Pharaohs, because artificial incubation was then practiced in crude ovens resembling some still in use in that country.

The use of poultry and eggs as food goes back to very early times in the history of the human race. Methods of slaughter and preparation for consumption have varied with succeeding civilizations and cultures. Not until fairly recent times did these operations become a matter of great commercial importance, or of serious concern to consumers, public health officials, and government alike.

## IMPORTANCE OF THE POULTRY INDUSTRY

Several criteria may be used to measure or gauge the importance of the poultry industry; among them, the following:

1. **As source of farm income.** U.S. poultry producers sold eggs, broilers, and turkeys valued at $11.5 billion in 1987, representing 8.3% of the total cash farm income that year (see Fig. 2–11). Actually, the percentage of total farm income derived from poultry has changed very little during the past quarter century, through the years accounting for 8 to 10% of the cash farm income. But, this cash income has been received by fewer and fewer poultry producers, because the big have gotten bigger. Further, the production of different kinds of poultry has gradually been concentrated in certain geographic areas where farmers have the greatest comparative advantage in the production of one or more kinds of poultry or poultry products.

Fig. 42–1. Roast turkey. Turkeys produce a higher percentage of edible meat to liveweight (on-foot) than any other species, and compare favorably with other meats as a source of amino acids. (Courtesy, National Turkey Federation, Mount Morris, IL)

2. **As a food.** Poultry meat and eggs are used chiefly as human food. In 1987, the U.S. per capita consumption of chicken and turkey on a ready-to-cook basis totaled 62.8 and 15.1 lb, respectively. Additionally, in that same year, 249 eggs were consumed per person.

Fig. 42–2 shows that the per capita consumption of poultry meat and eggs from 1976 to 1986. As noted, the consumption of turkeys and broilers, on a per capita basis, has increased very sharply. On the other hand, the per person consumption of eggs has declined, primarily due to a change in eating habits of the average urban dweller.

## PER CAPITA CONSUMPTION OF POULTRY AND EGGS

### POUNDS OF POULTRY                    NUMBER OF EGGS

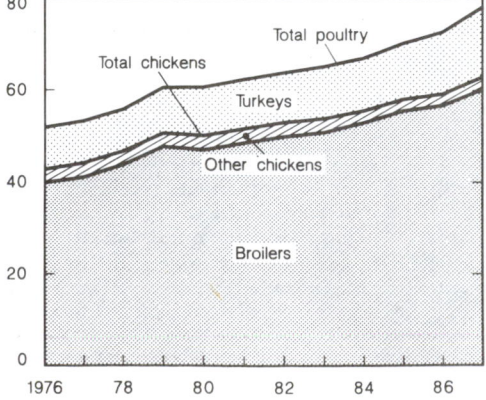

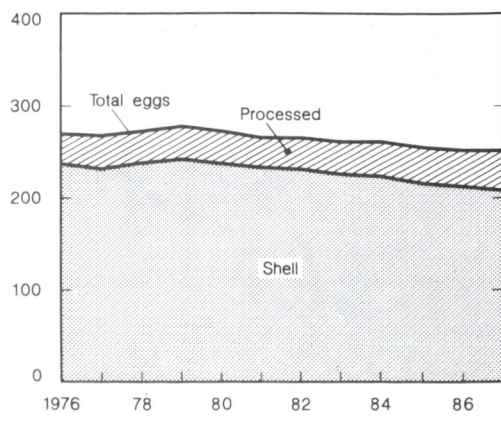

Fig. 42–2. Per capita consumption of poultry and eggs. (Source: *1988 Agricultural Chartbook,* Ag. Hdbk. No. 673, USDA, p. 92, Chart 203)

Eggs contain an abundance of proteins, vitamins, and minerals. Also, they have large amounts of high-quality and easily digestible proteins. Moreover, the proteins are complete proteins; that is, they contain all the essential amino acids to maintain life and promote growth. Additionally, eggs are a rich source of iron; phosphorus; trace minerals; vitamins A, E, K; and all the B vitamins, including vitamin B–12. Eggs are second only to fish-liver oils as a natural source of vitamin D. Eggs are second only to fish-liver oils as a natural source of vitamin D. Eggs are moderate from the standpoint of calorie content, a medium-size egg containing about 77 calories. Table 42–1 points up the nutritive value of eggs.

Notwithstanding the high nutritive value of eggs, millions of Americans routinely eat eggs for breakfast each morning simply because they like them.

Poultry meat is supplied chiefly by chickens and turkeys, although ducks, geese, guinea and other fowls contribute thereto.

Poultry meat is economical, quick, and easy to prepare and serve. Also, it has a number of desirable nutritional properties (see Table 42–2).

Nutritionally, people eat meat primarily for its protein content. Turkey and chicken meat is higher in protein than beef and other red meats (see Table 42–2). Additionally, poultry meat contains high-quality protein—it's a rich source of all the essential amino acids (see Table 42–3). The close resemblance of the amino acid content of poultry meat to the amino acids of milk and eggs (top-quality proteins) serves to emphasize the latter point.

### TABLE 42–1
### SOME OF THE ESSENTIAL NUTRIENTS IN TWO MEDIUM EGGS (108 gm) WITHOUT SHELL[1]

| Nutrient | Function | Percentage of Daily Dietary Recommendations[2] |
|---|---|---|
| Protein ......... | To build muscles and body tissues. | 16.3% |
| Iron ............ | To build red blood cells and help promote good health. | 17.3% |
| Vitamin A ....... | To help give normal vision and clear, healthy skin. | 22.0% |
| Thiamin ......... (Vitamin B–1) | To promote growth, good appetite, and a healthy nervous system. | 8.3% |
| Riboflavin ....... (Vitamin B–2) | To promote growth and good health. | 14.8% |
| Vitamin B–12 .... | To help prevent and cure pernicious anemia. | 1.0% microgram[3] |
| Vitamin D ....... | To help calcium in building bones and teeth. | 25.0% |

[1]From the Poultry and Egg National Board, Chicago, IL.
[2]Recommendations for girls, ages 16 to 19.
[3]No daily recommendations established.

### TABLE 42–2
### COMPARISON OF NUTRIENT COMPOSITION OF COOKED TURKEY, CHICKEN, AND BEEF[1]

| Kind of Meat | Protein | Fat | Moisture | Food Energy Calories/Lb |
|---|---|---|---|---|
| | (%) | (%) | (%) | |
| **Turkey** (mature, roasted and boned) | | | | |
| Breast (white meat) ................................... | 34.2 | 7.5 | 58 | 925 |
| Leg (dark meat) ...................................... | 30.5 | 11.6 | 56.5 | 1,029 |
| **Chicken** (16 weeks old, roasted and boned) | | | | |
| Breast (white meat) ................................... | 31.5 | 1.3 | 66 | 625 |
| Leg (dark meat) ...................................... | 25.4 | 7.3 | 67 | 761 |
| **Beef** (cooked and boned) | | | | |
| Round steak ......................................... | 27.0 | 13.0 | 59 | 1,056 |
| Rump roast .......................................... | 21.0 | 32.0 | 46 | 1,714 |
| Hamburger ........................................... | 22.0 | 30.0 | 47 | 1,650 |

[1]From Poultry Meat, Pub. 9, Ontario Department of Agriculture, p. 20, Table 16.

### TABLE 42–3
### COMPARISON OF AMINO ACID COMPOSITION OF VARIOUS ANIMAL FOODS[1]

| Amino Acid | Percentage of the Carcass Proteins | | | | | |
|---|---|---|---|---|---|---|
| | Turkey[2] | Chicken | Beef | Pork | Milk | Eggs |
| Arginine ............ | 6.5 | 6.7 | 6.4 | 6.7 | 4.3 | 6.4 |
| Cystine[3] ............ | 1.0 | 1.8 | 1.3 | 0.9 | 1.0 | 2.4 |
| Histidine[3] .......... | 3.0 | 2.0 | 3.3 | 2.6 | 2.6 | 2.1 |
| Isoleucine[3] .......... | 5.0 | 4.1 | 5.2 | 3.8 | 8.5 | 8.0 |
| Leucine[3] ........... | 7.6 | 6.6 | 7.8 | 6.8 | 11.3 | 9.2 |
| Lysine[3] ............ | 9.0 | 7.5 | 8.6 | 8.0 | 7.5 | 7.2 |
| Methionine[3] ......... | 2.6 | 1.8 | 2.7 | 1.7 | 3.4 | 4.1 |
| Phenylalanine[3] ...... | 3.7 | 4.0 | 3.9 | 3.6 | 5.7 | 6.3 |
| Threonine[3] .......... | 4.0 | 4.0 | 4.5 | 3.6 | 4.5 | 4.9 |
| Tryptophan[3] ......... | 0.9 | 0.8 | 1.0 | 0.7 | 1.6 | 1.5 |
| Tyrosine ............ | 1.5 | 2.5 | 3.0 | 2.5 | 5.3 | 4.5 |
| Valine[3] ............. | 5.1 | 6.7 | 5.1 | 5.5 | 8.4 | 7.3 |

[1]From Poultry Meat, Pub. 9, Ontario Department of Agriculture, p. 19, Table 15.  [2]Average of values for whole turkey (breast and leg).  [3]Essential amino acids.

3. **In industrial uses.** Science and technology have teamed up to make for many uses of poultry and eggs, and their by-products. Among such industrial uses are: fertile eggs used in the preparation of vaccines; inedible eggs used in the preparation of animal feed and fertilizers; egg whites used in the making of pharmaceuticals, paints, varnishes, adhesives, printer's ink, photography, book binding, wine clarification, leather tanning, and textile dyeing; egg yolks used in the making of cake mixes, soap, paints, shampoos, leather finishing, and book binding; eggshells used in making mineral mixes and in fertilizer; feathers used in animal feed, fertilizer, millinery goods, pillows, cushions, mattresses, dusters, and insulation material; poultry offal used in animal and mink feeds; and endocrine glands used in making biological products.

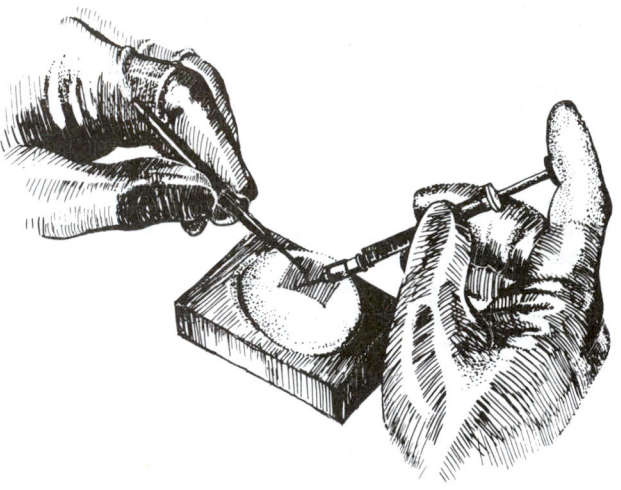

Fig. 42-3. Fertilized eggs are required for the production of many vaccines.

4. **In research.** The chick is much more sensitive to the lack of several substances in the diet than is the laboratory rat. At first, this great sensitivity was a handicap and made it very hard to keep chicks alive on the kinds of diets which would support a rat fairly well. Eventually, the higher sensitivity of the chick to many dietary factors proved a great advantage in bringing to light new information on vitamins, minerals, and amino acids, and in their more accurate estimation. Also, chicks have the advantages of being cheap and readily available, and large numbers of them can be hatched at the same time so as to provide the accuracy which goes with numbers of animals. Additionally, many scientists feel that the nutritional needs of the human are more like that of the chicken than the rat.

## TRANSFORMATION OF THE AMERICAN POULTRY INDUSTRY

The American poultry industry had its humble beginning when chickens were first brought to this continent by the early settlers. Small home flocks were started at the time of the establishment of the first permanent homes in Jamestown in 1607. For many years thereafter, chickens were tenderly cared for by the farmer's wife, who fed them on table scraps and the unaccounted-for grain from the crib.

As villages and towns were established, and increased in size, the nearby farm flocks were sold or bartered for groceries and other supplies in the nearby towns. Eventually, grain production to the West, the development of transportation facili-

ties, the use of refrigeration, and artificial incubation further stimulated poultry production in the latter part of the 1800s.

Since World War II, changes in poultry and egg production and processing have paced the whole field of agriculture. Practices in all phases of poultry production—breeding, feeding, management, housing, marketing and processing—have become very highly specialized. The net result is that more products have been made available to consumers at favorable prices, comparatively speaking, and per capita consumption has increased. This is shown in Fig. 42-4.

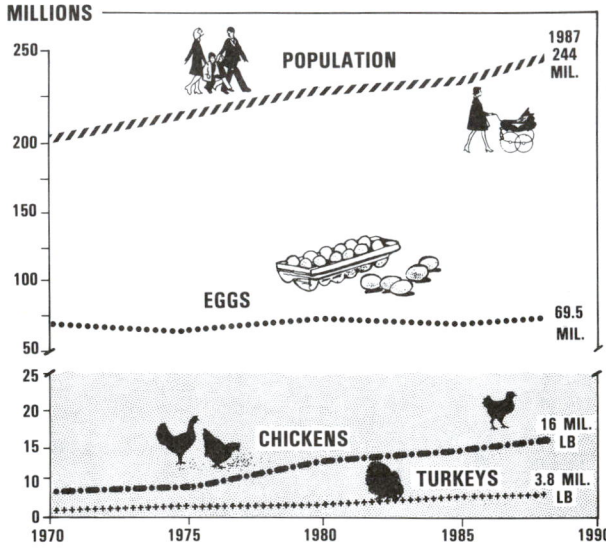

Fig. 42-4. Poultry and egg production and population. (Source: *Agricultural Statistics 1983/1988,* USDA)

A discussion of each of the most important changes in the poultry industry follows:

1. **Changes in breeding methods.** Standard-bred chickens decreased as modern breeding methods were applied.

The poultry geneticist discovered that family, as well as individual bird records, are needed to develop high egg production. From this base, breeders created certain strains for high egg production and feed efficiency.

When breeding for broiler production, hybrid vigor is obtained by systematic matings that may involve crossing different breeds, different strains of the same breed, or the crossing of inbred lines. Many of the strains used as sires trace their ancestry to the broad-breasted Cornish breed. The main objective is the improvement of broiler growth rate to eight weeks of age, although improvements in other economic factors is sought.

Breeding for egg production differs from breeding for broiler production in that the individual methods most useful for improving growth rate have little value in selection to improve egg production, because egg production is of low heritability. In breeding for egg production, high-producing families are selected. Then, either of two types of crosses is made: (a) crossing of inbred lines, or (b) using strains which are not inbred.

2. **Changes in hatcheries.** In the beginning, hatching was done according to nature's way—by a setting hen hovering over eggs. Then came the first American incubator, patented in 1844, followed by the U.S. Post Office acceptance of chicks for shipment by mail in 1918. Hatcheries became larger in size and fewer in number. In 1934, a half billion chicks were hatched

in 11,000 hatcheries. In 1989, 6.3 billion chicks, including both broiler and egg type, were hatched in 376 hatcheries each with an average capacity of 1.4 million eggs; and 289 million turkey poults were hatched in 82 hatcheries each with an average capacity of 527,000 eggs.

Fig. 42–5. Many modern hatcheries have a capacity of more than 1,000,000 eggs. (Courtesy, Hubbard Farms, Walpole, NH)

3. **Changes in egg production.** A hundred years ago, a hen produced about 100 eggs per year. In 1987, the U.S. average was 248 eggs per hen (see Fig. 42–6). Formerly, eggs were sold largely on an ungraded basis. Today most of them are candled for inferior quality, weighed, cartoned, and sold according to size and quality. In many modern egg-grading plants, efficient power operated weighing machines speed grading and move the eggs through the marketing system with dispatch.

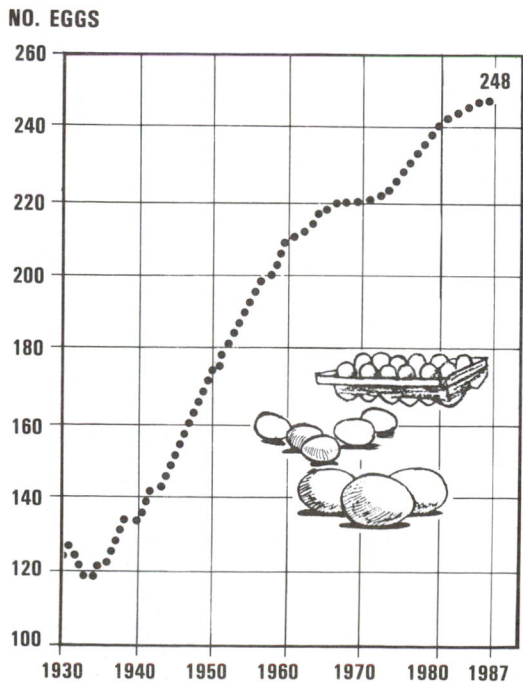

Fig. 42–6. Changes in U.S. annual egg production per hen. (From USDA sources)

Beginning in the early 1930s, three important changes took place in relation to egg production: (a) with the greater emphasis on commercial size flocks, the light breeds and strains of chickens gradually replaced general or dual-purpose breeds for egg production; (b) as the technique for "sexing" chicks became perfected, only the female chicks of egg-type breeds were sold by the hatcheries to layer operations; and (c) feeding, breeding, management, and disease control practices were improved so that more eggs were produced per layer, thereby requiring fewer layers to provide the eggs necessary to supply the market demands.

Changes in the number of layers, total egg production, and rate of lay are shown in Fig. 42–7.

## Eggs: Rate of Lay, Production, and Number of Layers

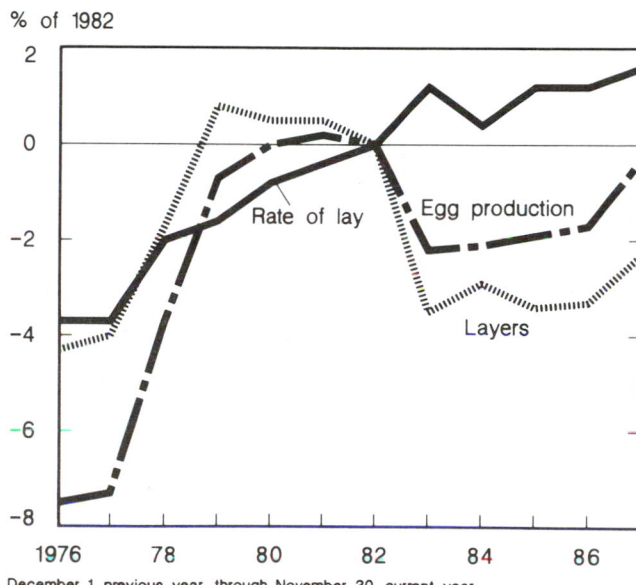

December 1, previous year, through November 30, current year.

Fig. 42–7. The number of layers, total egg production, and rate of lay, 1976–1987. (Source: *1988 Agricultural Chartbook*, Ag. Hdbk. No. 673, USDA, p. 93, Chart 205)

4. **Changes in chicken and meat production.** Prior to 1930, chicken meat was mainly the by-product of egg production. Birds which were no longer producing eggs at a satisfactory rate were sold for meat purposes, mainly in the fall of the year. Cockerels raised with the pullets were disposed of as fryers, or roasters, at weights of 3 to 8 lb.

In 1934, 34 million broilers were produced in the United States. By 1987, 4.65 billion broilers were being produced, or 136.6 times as many as 53 years earlier. Modern broiler production is so concentrated, and so highly commercialized, that the industry might properly be classed as a poultry meat factory, rather than a farming operation. Very little land is required beyond the space necessary for a broiler house and a driveway. Chicks, feed, and other items used in production are purchased, or are obtained from another division of an integrated operation of which the broiler production is a part. The operation is highly specialized, mechanized, and carried on within the limits of the broiler house. Also, the operation is characterized by large numbers.

Similar progress has been made in the processing of poultry. In most areas of the country, poultry processing has become a highly industrialized, large-scale operation using modern mechanical equipment and sanitary methods.

5. **Changes in turkey production.** Like the situation in chicken production, the production of turkeys was mostly a small sideline enterprise until 1910. At that time, 870,000 farmers raised 3.66 million turkeys or an average of 4 turkeys per farm. In 1987, farmers raised 240 million turkeys, and flocks of more than 50,000 birds were common.

6. **Changes in number and size of poultry farms.** In 1910, more than 5.5 million farms in the United States (88% of the 6.4 million farms in the nation) kept chickens. The average size flock in the United States numbered 50 laying hens.

In 1987, there were 144,438 layer operations in the United States with an average of 2,586 layers per unit. In 1987, there were 27,645 broiler operations, marketing an average of 157,785 broilers per unit during the year. In 1987, there were 7,347 turkey units, marketing an average of 33,210 turkeys per unit during the year.

7. **Changes in ownership and organization.** As poultry operations grew in size and efficiency, they vertically integrated (chiefly with feed companies and processors) to secure more credit; and contractual production became commonplace. Today, 92% of the broilers are produced under contract, and the remaining 8% are raised on integrator-owned farms. Also, it is estimated that 90% of the turkeys, and 89% of the eggs are produced under some kind of integrated or contract arrangement.

8. **Changes in the labor requirements.** Poultry producers have achieved remarkable efficiency, primarily through increased confinement production and mechanization. In 1935 to 1939, it required 8.5 labor-hours to produce 100 lb of broiler; in 1982 to 1986, it required only 0.1 labor-hour to produce 100 lb of broiler; and in this same period of time, the labor requirements to produce 100 lb of turkey were lowered from 23.7 labor-hours to 0.2 labor-hour. (See Chapter 2, Table 2–7.) But no such progress has been made in lowering the labor requirements for the production of red meats, with the result that poultry producers have achieved a very real advantage.

9. **Changes in feed efficiency.** Fig. 42–8 and Table 42–4 show the marked lowering of feed required to produce a unit of eggs, turkeys, and broilers since 1940. In 1940, it required 4.7 lb of feed to produce 1 lb of weight gain of broilers; in 1990, it took only 1.9 lb. But no such progress has been made in red meat or milk animals. It requires about 9 lb of feed to produce 1 lb of beef; and the feed efficiency of beef cattle has changed very little since 1900. Also, it requires 4.0 lb of feed to produce a pound of gain in hogs compared to 6.5 lb needed in the early 1900s. Likewise, feed efficiency of dairy cows has changed very little—in 1910, it was 1.2 lb of feed per pound of milk; today, it still takes 1.11 lb of feed per pound of milk. (See Chapter 1, Table 1–7)

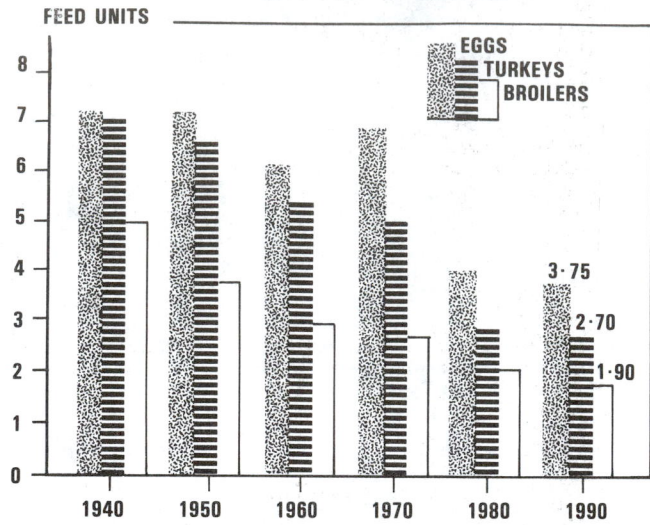

Fig. 42–8. Feed per unit of eggs and poultry produced. (1940 to 1970 from USDA. 1980 and 1990 provided by the author)

**TABLE 42–4**
**FEED UNITS REQUIRED PER UNIT**
**OF EGGS AND POULTRY PRODUCED[1]**

| Year | Per Dozen Eggs | Per Pound Liveweight | |
| | | Turkey | Broiler |
| --- | --- | --- | --- |
| | (feed units) | (feed units) | (feed units) |
| 1940 | 7.4 | 7.2 | 4.7 |
| 1950 | 7.2 | 5.6 | 3.7 |
| 1960 | 6.4 | 5.4 | 3.0 |
| 1970 | 4.6 | 5.2 | 2.6 |
| 1980 | 4.10 | 2.9 | 2.1 |
| 1990 | 3.75 | 2.7 | 1.9 |

[1]Feed units used per dozen eggs or per lb of liveweight turkey or boiler produced (1940 to 1970 from USDA, 1980 and 1990 provided by the author).

10. **Changes in geography of production and processing.** A rising proportion of broilers, turkeys, and eggs are being produced in the Southeast. Processors also shifted to the proximity of production centers.

11. **Changes in marketing.** In the movement of poultry and eggs from the producer to the consumer, fewer agencies are being used and there is more direct marketing.

12. **Changes in proportion of federally inspected slaughter.** Prior to passage of the Poultry Inspection Act of 1957, relatively little poultry was slaughtered under federal inspection. Today, all commercial poultry slaughter is under federal inspection.

In 1990, the U.S. Department of Agriculture initiated a two-year study to determine the optimum process for implementing the *Hazard Analysis and Critical Control Point (HACCP)* system of meat and poultry inspection—a system for monitoring sanitation and process control. The HACCP system is designed to benefit consumers, the livestock and poultry industry, and the Food Safety and Inspection Service of the U.S. Department of Agriculture as follows:

　　a. Focusing inspection activities on the critical areas of product safety, wholesomeness, and adulteration prevention.

　　b. Focusing industry responsibilities and action to produce safe and wholesome food.

　　c. Increasing the scientific basis for inspection operations.

13. **Changes in processing.** In 1947, New York dressed chicken (only the blood and feathers removed) accounted for 80% of the chickens and 75% of the turkeys marketed. By 1963, only about 11% of the chickens and 5% of the turkeys were marketed that way. By the mid-1960s, 88% of the broilers were ice-packed and ready-to-cook, whereas 86% of the turkeys were marketed frozen. Today, practically all chickens and turkeys are eviscerated. Not only that, much of the meat is cut up or further processed—changes designed to lessen labor in

the kitchen and to add variety to menus. Also, there is a big trend toward ready-to-eat items and TV dinners.

## PRESENT STATUS OF THE POULTRY INDUSTRY

As is true in the production of all other commodities, it is important that poultry producers be well informed concerning worldwide poultry production in order to know which countries are potential competitors. Likewise, producers within the country need to know where both their competition and markets are located.

## WORLD POULTRY DISTRIBUTION

Tables 42–5, 42–6, and 42–7 show the leading poultry-producing countries of the world.

### TABLE 42–5
### LEADING CHICKEN-PRODUCING COUNTRIES OF THE WORLD

| Country | Chickens[1] | Hen Egg Production[1] | Human Population[2] | Size of Country[2] | | Chickens per Capita | Chickens per | | Hen Eggs per Capita[3] |
|---|---|---|---|---|---|---|---|---|---|
| | | (metric tons) | | (sq mi) | (sq km) | | (sq mi) | (sq km) | |
| China | 1,849,000,000 | 6,685,000 | 1,069,628,000 | 3,705,390 | 9,596,960 | 1.7 | 499 | 193 | 110 |
| U.S.A. | 1,540,000,000 | 4,045,600 | 247,498,000 | 3,539,289 | 9,166,759 | 6.2 | 435 | 168 | 288 |
| U.S.S.R. | 1,129,000,000 | 4,656,000 | 287,015,000 | 8,649,496 | 22,402,195 | 3.9 | 131 | 50 | 286 |
| Brazil | 550,000,000 | 1,280,000 | 153,992,000 | 3,286,473 | 8,511,953 | 3.6 | 167 | 65 | 147 |
| Indonesia | 410,000,000 | 434,000 | 187,726,000 | 735,268 | 1,904,344 | 2.2 | 558 | 215 | 41 |
| Japan | 334,000,000 | 2,409,000 | 123,231,000 | 145,856 | 377,767 | 2.7 | 2,290 | 884 | 345 |
| India | 260,000,000 | 990,000 | 833,422,000 | 1,266,595 | 3,280,481 | 0.3 | 205 | 79 | 21 |
| Mexico | 224,000,000 | 928,000 | 88,087,000 | 761,604 | 1,972,554 | 2.5 | 294 | 114 | 186 |
| France | 189,000,000 | 912,000 | 55,813,000 | 220,668 | 571,530 | 3.4 | 856 | 331 | 288 |
| Pakistan | 150,000,000 | 194,600 | 110,358,000 | 310,403 | 803,944 | 1.4 | 483 | 187 | 31 |
| **World Total** | 10,215,000,000 | 34,879,920 | 5,052,000,000 | 57,800,000 | 149,702,000 | 2.0 | 177 | 68 | 122 |

[1]*FAO Production Yearbook 1988*, United Nations, Rome, Italy, Vol. 42, pp. 250 and 280. Data for 1988.

[2]*The World Almanac and Book of Facts 1989*, The Fresno Bee, Fresno, CA.

[3]There are 9 medium-sized eggs per pound.

### TABLE 42–6
### LEADING TURKEY-PRODUCING COUNTRIES OF THE WORLD

| Country | Turkeys[1] | Human Population[2] | Size of Country[2] | | Turkeys per Capita | Turkeys per | |
|---|---|---|---|---|---|---|---|
| | | | (sq mi) | (sq km) | | (sq mi) | (sq km) |
| U.S.A. | 78,000,000 | 247,498,000 | 3,539,289 | 9,166,759 | 0.32 | 22.0 | 8.5 |
| U.S.S.R. | 48,000,000 | 287,015,000 | 8,649,496 | 22,402,195 | 0.17 | 5.5 | 2.1 |
| Italy | 23,000,000 | 57,439,000 | 116,303 | 301,225 | 0.40 | 197.8 | 76.4 |
| France | 20,000,000 | 55,813,000 | 220,668 | 571,530 | 0.36 | 90.6 | 35.0 |
| Mexico | 12,000,000 | 88,087,000 | 761,604 | 1,972,554 | 0.14 | 15.8 | 6.1 |
| United Kingdom | 9,000,000 | 56,648,000 | 94,226 | 244,045 | 0.16 | 95.5 | 36.9 |
| Israel | 7,000,000 | 4,477,000 | 7,847 | 20,324 | 1.56 | 892.1 | 344.4 |
| Canada | 6,000,000 | 25,334,000 | 3,851,790 | 9,976,136 | 0.24 | 1.6 | 0.6 |
| Brazil | 5,000,000 | 153,992,000 | 3,286,473 | 8,511,953 | 0.03 | 1.5 | 0.6 |
| Madagascar | 4,000,000 | 11,148,000 | 226,657 | 587,042 | 0.36 | 17.6 | 6.8 |
| **World Total** | 233,000,000 | 5,052,000,000 | 57,800,000 | 149,702,000 | 0.05 | 4.0 | 1.6 |

[1]*FAO Production Yearbook 1988*, United Nations, Rome, Italy, Vol. 42, p. 250. Data for 1988.

[2]*The World Almanac and Book of Facts 1989*, The Fresno Bee, Fresno, CA.

**TABLE 42–7**
**LEADING DUCK-PRODUCING COUNTRIES OF THE WORLD**

| Country | Ducks[1] | Human Population[2] | Size of Country[2] | | Ducks per Capita | Ducks per | |
|---|---|---|---|---|---|---|---|
| | | | (sq mi) | (sq km) | | (sq mi) | (sq km) |
| China | 325,000,000 | 1,069,628,000 | 3,705,390 | 9,596,960 | 0.30 | 88.0 | 34.0 |
| Bangladesh | 32,000,000 | 112,757,000 | 55,598 | 143,999 | 0.30 | 576.0 | 222.0 |
| Indonesia | 29,000,000 | 187,726,000 | 735,268 | 1,904,344 | 0.20 | 39.0 | 15.0 |
| Vietnam | 27,000,000 | 66,708,000 | 128,401 | 332,559 | 0.40 | 210.0 | 81.0 |
| U.S.A. | 21,600,000 | 247,498,000 | 3,539,289 | 9,166,759 | 0.09 | 6.1 | 2.4 |
| Thailand | 16,000,000 | 55,017,000 | 198,456 | 514,001 | 0.30 | 81.0 | 31.0 |
| France | 11,000,000 | 55,813,000 | 220,668 | 571,530 | 0.20 | 50.0 | 19.0 |
| Mexico | 7,000,000 | 88,087,000 | 761,604 | 1,972,554 | 0.08 | 9.0 | 4.0 |
| Brazil | 6,000,000 | 153,993,000 | 3,286,473 | 8,511,953 | 0.04 | 2.0 | 0.7 |
| Philippines | 6,000,000 | 61,971,000 | 115,831 | 300,002 | 0.10 | 52.0 | 20.0 |
| **World Total** | 519,000,000 | 5,052,000,000 | 57,800,000 | 149,702,000 | 0.10 | 9.0 | 3.5 |

[1]FAO Production Yearbook 1988, United Nations, Rome, Italy, Vol. 42, p. 250. Data for 1988. Except U.S.A. data, which was provided by the author, and is for 1985.
[2]The World Almanac and Book of Facts 1989, The Fresno Bee, Fresno, CA.

Pertinent facts pointed up by the preceding tables follow:

1. **Table 42–5, Chickens.** China leads in chicken numbers. The United States ranks seconds, and the U.S.S.R. ranks third. In chickens per capita, the United States leads, with the U.S.S.R. ranking second, and Brazil ranking third. Japan produces the highest number of hen eggs per capita—345—of any nation in the world.

2. **Table 42–6, Turkeys.** The United States ranks first in turkey numbers, and the U.S.S.R. ranks second. However, Israel is far in the lead in turkeys per capita, with 1.56 turkeys per person.

3. **Table 42–7, Ducks.** China is far in the lead in duck numbers. It is noteworthy, too, that the Southeast Asian countries produce the largest number of ducks.

## U.S. POULTRY PRODUCTION

Tables 42–8, 42–9, and 42–10 show the leading states of the United States in numbers of eggs, broilers, and turkeys produced.

**TABLE 42–8**
**LEADING U.S. EGG-PRODUCING STATES[1]**

| State | Produced |
|---|---|
| | (no.) |
| California | 8,023,000,000 |
| Indiana | 5,750,000,000 |
| Pennsylvania | 4,853,000,000 |
| Georgia | 4,476,000,000 |
| Ohio | 4,351,000,000 |
| Arkansas | 3,874,000,000 |
| Texas | 3,424,000,000 |
| North Carolina | 3,251,000,000 |
| Alabama | 2,605,000,000 |
| Florida | 2,564,000,000 |
| **U.S. Total** | 69,492,000,000 |

[1]Agricultural Statistics 1988, USDA, p. 365. Data for 1987.

From these tables and other facts, the following deductions can be made:

1. From an overall standpoint, the Southeast ranks high as a poultry area.

2. California is the leading egg-producing state, by a considerable margin. Indiana ranks second.

3. The Southeast—Arkansas, Georgia, Alabama, North Carolina, and Mississippi—completely dominates broiler production.

4. North Carolina, Minnesota, and California hold a sizable lead in number of turkeys produced. It's difficult to understand why these three states should lead the nation in turkey production, for few states could be more dissimilar in climate, crops grown, and population density.

**TABLE 42–9**
**LEADING U.S. BROILER-PRODUCING STATES[1]**

| State | Raised |
|---|---|
| | (no.) |
| Arkansas | 878,574,000 |
| Georgia | 733,417,000 |
| Alabama | 666,538,000 |
| North Carolina | 477,700,000 |
| Mississippi | 343,395,000 |
| Maryland | 264,196,000 |
| Texas | 259,000,000 |
| Delaware | 209,818,000 |
| California | 196,120,000 |
| Virginia | 154,036,000 |
| **U.S. Total** | 5,002,934,000 |

[1]Agricultural Statistics 1988, USDA, p. 357, Table 510. Data for 1987.

**TABLE 42–10**
**LEADING U.S. TURKEY-PRODUCING STATES[1]**

| State | Produced |
|---|---|
| | (no.) |
| North Carolina | 48,350,000 |
| Minnesota | 40,500,000 |
| California | 25,500,000 |
| Arkansas | 18,000,000 |
| Virginia | 16,200,000 |
| Missouri | 15,500,000 |
| Indiana | 13,000,000 |
| Iowa | 8,500,000 |
| Pennsylvania | 8,000,000 |
| Wisconsin | 5,450,000 |
| **U.S. Total** | 240,349,000 |

[1]Agricultural Statistics 1988, USDA, p. 363. Data for 1987.

# THE MODERN POULTRY INDUSTRY

Fig. 42–9. *Left:* Layers in a modern facility, with automated feeding and egg collecting. (Courtesy, Ralston Purina Company, St. Louis, MO)

Fig. 42–10. *Above:* Meat-type hens being flock-mated. (Photo by J. C. Allen and Son, West Lafayette, IN)

Fig. 42–11. *Above:* Broilers with automatic feed and water facilities. (Photo by J. C. Allen and Son, West Lafayette, IN)

Fig. 42–12. *Above:* Market turkeys in semiconfinement in Wisconsin. (Courtesy, *Turkey World*, Mount Morris, IL)

Fig. 42–13. *Above:* Pekin ducks. (Photo by J. C. Allen and Son, West Lafayette, IN)

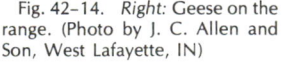

Fig. 42–14. *Right:* Geese on the range. (Photo by J. C. Allen and Son, West Lafayette, IN)

## RELATIVE IMPORTANCE OF DIFFERENT SPECIES OF POULTRY

Table 42–11 shows the relative importance of different species of poultry. This shows that both broiler and turkey production have dominated poultry production in recent years. Although small in numbers when compared to broilers and turkeys, it is noteworthy that U.S. duck numbers nearly doubled from 1980 to 1985.

**TABLE 42–11**
**U.S. POULTRY PRODUCTION¹**

| Year | Species of Poultry | | | |
| --- | --- | --- | --- | --- |
| | Chickens (broilers) | Turkeys | Ducks² | Geese² |
| | (millions) | | | |
| 1930 | 34³ | 17 | 11.3 | 3.9 |
| 1940 | 142 | 33 | 11.9 | 1.3 |
| 1950 | 631 | 44 | 11.5 | 1.1 |
| 1960 | 1,795 | 85 | 11.1 | .9 |
| 1970 | 4,151 | 116 | 10.7 | .7 |
| 1980 | 3,963 | 165 | 15.1 | .6 |
| 1987 | 4,646 | 240 | 21.6 | .5 |

¹*Agricultural Statistics* 1949, 1962, 1984, 1988, USDA.
²Calculated from available data.
³Data for 1934.

## ALLIED INDUSTRIES

No other livestock industry has spawned so many allied industries as has the poultry business. In addition to actual breeding, hatching, growing, marketing, processing, and distributing poultry meat and eggs, there are numerous closely related industries; among them, the commercial feed industry that produced 45.9 million tons of primary poultry feeds (complete feeds) in 1988, which represented 44.5% of the total primary feed produced for all U.S. livestock that year (see Chapter 4, section headed "Commercial Feeds"). Also, there's pullorum testing and vaccinating against certain diseases; the production of drugs; communications; poultry futures market; fertilizer; and numerous items of equipment—incubators, brooders, feeders and waterers, egg cases and cartons, chick boxes, leg bands, time clocks, laying cages, and refrigeration.

## FUTURE OF THE POULTRY INDUSTRY

Science and technology will continue to be the great multipliers in the poultry industry. The following transitions will usher the U.S. poultry industry into the 21st century—and beyond.

1. **Increased mass production, economical products, and year-round availability of products.** Because poultry products can be batched on schedule, handled as a flock, and adapted to confinement, mass production methods can be applied economically to produce poultry meat and eggs on a continuous basis, very efficiently and on a highly competitive price

basis. Today, the industry is so efficient that more food in the form of poultry products can be produced with less raw materials and energy than is required by any animal source other than some phases of aquaculture.

2. **More contract and integrated production, and fewer and bigger integrators.** Poultry production has become so highly specialized that farmers contract with large integrated companies to produce one kind or phase of production only; for example, the production of started pullets, eggs, broilers, or turkeys. The integrators supply chicks, poults, feed, litter, fuel, and medication, and the contract growers provide housing and the care of the birds. Most contract growers are on an incentive basis; they receive a graduated fee that compensates them for superior management and bird performance.

Fig. 42–15. White Leghorn replacement pullets, a highly specialized phase of production, produced under contract. (Courtesy, University of Georgia, Athens)

Because capital requirements for a modern poultry operation are so great and the margin of profit per bird so small, individual growers cannot afford the financial risk of selling poultry or eggs on the open market. As a result, in 1990 essentially 100% of the broilers, and about 90% of the turkeys and eggs were produced under some kind of contract or integrated arrangement. In the future, further integration of turkeys and eggs will likely occur, and there will be fewer and bigger integrators.

3. **Increased laborsaving devices and mechanization.** Higher priced labor, along with more sophisticated equipment, is making, and will continue to make, for increased mechanization all the way along the line, from production through processing and marketing.

Fig. 42–16. Modern egg handling is highly automated. This shows an automatic vacuum lift placing eggs in a filler flat. (Courtesy, DeKalb Poultry Research, Inc., DeKalb, IL)

a way as to bring about improved houses, brooders, incubators, and laborsaving equipment.

5. **Increased biotechnology.** The application of biotechnology, along with improved breeding, feeding, and management, will make for increased production and efficiency of layers, broilers, and turkeys.

6. **Improved quality of products.** No other country in the world produces as high-quality poultry meat and eggs as the United States. Yet, further improvements can and will be made.

7. **Increased consumption of poultry products.** Per capita consumption of poultry meat will increase because it (a) is available fresh or frozen, (b) is easy to prepare, (c) is easily combined with other foods, (d) is lean, nutritious, tender, and easy to chew, (e) is mild in flavor, (f) is easy to digest, and (g) is a good buy when compared with other animal products.

Eggs are also easy to prepare, nutritious, digestible, and economical. Following a decline in consumption, it appears that consumption has leveled off at about 250 eggs per capita annually.

In the future, it is predicted that the consumption of poultry meat will increase due to both increased human population and increased per capita consumption; and that there will be a modest increase in total egg consumption due to population increase, rather than increased per capita consumption. Increased per capita consumption of poultry meats will be achieved with the development of more further processed products such as poultry hot dogs and weiners; more deboned, filleted, smoked, or formed-into-patties meats; more convenience foods; and increased shelf life.

4. **Improved housing and environmental control.** Physics, engineering, and physiology are being combined in such

## QUESTIONS FOR STUDY AND DISCUSSION

1. Define the terms *poultry* and *ornithology*. How are birds and mammals distinguished?

2. Discuss the domestication and early use of poultry.

3. Discuss the importance of the poultry industry (a) as a source of farm income, (b) as a food, (c) in industrial uses, and (d) in research.

4. Why has poultry production changed more rapidly, and more completely, than production in any other class of animals?

5. Discuss each of the following changes in the poultry industry: (a) changes in breeding methods, (b) changes in hatcheries, (c) changes in egg production, and (d) changes in chicken meat production.

6. Discuss each of the following changes in the poultry industry: (a) changes in turkey production, (b) changes in number and size of poultry farms, (c) changes in ownership and organization, and (d) changes in labor requirements.

7. Discuss each of the following changes in the poultry industry: (a) changes in feed efficiency, (b) changes in marketing, (c) changes in federally inspected slaughter, and (d) changes in processing.

8. Name the three leading chicken-producing countries of the world. What factors account for the dominant position of these countries?

9. Name the three leading turkey-producing countries of the world. What factors account for the dominant position of these countries?

10. Name the three leading duck-producing countries of the world. What factors account for the dominant position of these countries?

11. Name the three leading states of the United States in numbers of (a) eggs produced, (b) broilers, and (c) turkeys. What deductions can be made from Tables 42–8, 42–9, and 42–10?

12. Discuss the relative importance of different species of poultry, and the changes in numbers of each species, as presented in Table 42–11.

13. Discuss the future of the poultry industry from the standpoints of (a) increased mass production, economical products, and year-round availability of products; (b) more contract and integrated production, and fewer and bigger integrators; and (c) increased labor saving devices and mechanization.

14. Discuss the future of the poultry industry from the standpoints of (a) improved housing and environmental control, (b) increased biotechnology, (c) improved quality of products, and (d) increased consumption of poultry products.

15. Is there a hazard of monopolies evolving as poultry production gets into fewer and fewer hands?

16. If the per capita consumption of poultry meats increases in the future, is it likely to be due to increasing total per capita meat consumption or to decreasing the consumption of some of the red meats?

17. How may young men and women train for and get employment in the large poultry units of today?

## SELECTED REFERENCES

| Title of Publication | Author(s) | Publisher |
|---|---|---|
| *American Poultry History 1823–1973* | O. A. Hanke<br>J. L. Skinner<br>J. H. Florea | American Printing & Publishing Co., Madison, WI, 1974 |
| *Commercial Chicken Production Manual,* Third Edition | M. O. North | Avi Publishing Co., Westport, CT, 1984 |
| *Poultry Husbandry* | M. A. Jull | McGraw-Hill Book Company, New York, NY, 1951 |
| *Poultry Husbandry I* | C. J. Price | Food and Agriculture Organization of the United Nations, Rome, Italy, 1969 |
| *Poultry Husbandry II* | C. J. Price<br>J. E. Reed | Food and Agriculture Organization of the United Nations, Rome, Italy, 1971 |
| *Poultry Keeping in Tropical Areas* | W. Thomann | Food and Agriculture Organization of the United Nations, Rome, Italy, 1968 |
| *Poultry Meat and Egg Production* | C. R. Parkhurst<br>G. L. Mountney | Van Nostrand Reinhold Company, New York, NY, 1988 |
| *Poultry Production,* Twelfth Edition | L. E. Card<br>M. C. Nesheim | Lea & Febiger, Philadelphia, PA, 1979 |
| *Poultry Products Technology* | G. J. Mountney | Avi Publishing Co., Westport, CT, 1966 |
| *Poultry Science,* Second Edition | M. E. Ensminger | The Interstate Printers & Publishers, Inc., Danville, IL, 1980 |
| *Poultry Science and Practice* | A. R. Winter<br>E. M. Funk | J. B. Lippincott Co., New York, NY, 1960 |
| *Practical Poultry Management* | J. E. Rice<br>H. E. Botsford | John Wiley & Sons, Inc., New York, NY, 1956 |
| *Processing of Poultry* | G. C. Mead | Elsevier Applied Science, London, England, 1989 |
| *U.S. Broiler Industry, The* | F. A. Lasley, *et al.* | USDA, Economic Research Service, Ag. Econ. Report No. 591, 1988 |

# POULTRY BREEDS
# AND BREEDING;
# SELECTING
# AND CULLING

From whence they came! *Gallus gallus,* one of the ancestors of domestic chickens. The male weighed about 2½ lb.

# Chapter 43

827

Breeding of poultry differs from the breeding of four-footed farm animals primarily in that (1) it is more flexible due to greater numbers and more rapid reproduction; (2) it has passed through the total presently known cycle of breeding methods more than any other species of animals; and (3) it is concentrated in fewer hands.

## BREEDS, AND BREEDING CHICKENS

It is improbable that all of the present-day breeds and varieties of chickens sprang from a common origin. The habits of the varieties in the Asiatic class indicate an ancestry which roosted on the ground and nested on a mound of earth. Such breeds as the Leghorn probably had tree roosting ancestors.

Authorities generally agree that the red jungle fowl, *Gallus bankiva*, was one of the ancestors of domestic chickens. But more recent investigations suggest that at least four species of jungle fowl may have contributed to the development of domestic fowl.

## TYPES AND CLASSES

Type refers to the general shape and form, without regard to breed. Commercially speaking, chickens are of two types:

the egg type, which is bred for egg production, and the meat type which is bred for meat production.

The term *class* is used to designate groups of breeds which have been developed in certain regions; thus, the class names—American, English, Asian, etc.

## BREEDS AND VARIETIES

The term *breed* refers to an established group of fowls, related by breeding, possessing a distinctive shape and the same general weight.

A *variety* is a subdivision of a breed, distinguished by color, color and pattern, or comb. Hence, a breed may embrace a number of varieties, distinguished by different color—white, black, buff, etc.; or by color and markings—as Light Brahma, Dark Brahma, etc.; or by different combs—as single comb, rose comb, etc.

The *Standard of Perfection* lists nearly 200 varieties of chickens. However, only two breeds (White Leghorn and White Rock) are really important today, with three other breeds (Rhode Island Red, Barred Rock, and New Hampshire) of negligible importance.

Table 43-1 lists some representative breeds and varieties of chickens and gives their more important characteristics.

### TABLE 43-1
### SOME BREEDS AND VARIETIES OF POULTRY AND THEIR CHARACTERISTICS

| Breed and Variety | Plumage Color | Standard Weight | | Type of Comb | Color of Ear Lobe | Color of Skin | Color of Shank | Shanks Feathered? | Color of Egg |
|---|---|---|---|---|---|---|---|---|---|
| | | Cock | Hen | | | | | | |
| | | (lb) | (lb) | | | | | | |
| **American** | | | | | | | | | |
| Jersey Black Giant | Black | 13 | 10 | Single | Red | Yellow | Black | No | Brown |
| New Hampshire | Red | 8½ | 6½ | Single | Red | Yellow | Yellow | No | Brown |
| Rhode Island Red | Red | 8½ | 6½ | Single and rose | Red | Yellow | Yellow | No | Brown |
| White Plymouth Rock | White | 9½ | 7½ | Single | Red | Yellow | Yellow | No | Brown |
| White Wyandotte | White | 8½ | 6½ | Rose | Red | Yellow | Yellow | No | Brown |
| **Asiatic** | | | | | | | | | |
| Black Langshan | Greenish-black | 9½ | 7½ | Single | Red | White | Bluish-black | Yes | Brown |
| Brahma (light) | Columbian pattern | 12 | 9½ | Pea | Red | Yellow | Yellow | Yes | Brown |
| Cochin (buff) | Buff | 11 | 8½ | Single | Red | Yellow | Yellow | Yes | Brown |
| **English** | | | | | | | | | |
| Australorp | Black | 8½ | 6½ | Single | Red | White | Dark slate | No | Brown |
| Buff Corpington | Buff | 10 | 8 | Single | Red | White | White | No | Brown |
| Silver-gray Dorking | Black body with silver white markings | 9 | 7 | Single | Red | White | White | No | White |
| Sussex | Speckeled/Red/Light | 9 | 7 | Single | Red | White | White | No | Brown |
| White Cornish | White | 10½ | 8 | Pea | Red | Yellow | Yellow | No | Brown |
| **Mediterranean** | | | | | | | | | |
| Andalusian (blue) | Bluish slate (laced) | 7 | 5½ | Single | White | White | Slaty-blue | No | White |
| Anocona | Black, may be white tipped | 6 | 4½ | Single and rose | White | Yellow | Yellow | No | White |
| Minorca (S.C. black) | Black | 9 | 7½ | Single | White | White | Dark slate | No | White |
| White Leghorn | White | 6 | 4½ | Single and rose | White | Yellow | Yellow | No | White |

A total of 342 breeds and varieties of domesticated land fowl and waterfowl are listed in the *Standard of Perfection*, published by the American Poultry Association. The latter named association was first organized in 1873 by representatives of different sections of the United States and Canada. Its primary objective was to standardize the breeds and varieties of domestic fowl shown in exhibition.

In the poultry industry, there are no breed registry associations like those for four-footed farm animals. Hence, in a sense, the *Standard of Perfection* takes the place of such a registry association by recognizing as *purebred* only those individuals that show characteristics conforming to those given in the *Standard of Perfection*.

Not all commercially important breeds of chickens have been officially recognized by the *Standard of Perfection*. One such breed is the California Grey, an egg-laying strain which was once widely used by hatchery producers for crossing. Without doubt, the California Grey would have been eligible for recognition by the American Poultry Association. However, the developers of the strain did not request such recognition.

Fig. 43–1.  Single Comb White Leghorns. (Courtesy, *Poultry Tribune,* Mount Morris, IL)

Fig. 43–2.  Barred Plymouth Rocks. (Courtesy, *Poultry Tribune,* Mount Morris, IL)

Fig. 43–3.  White Plymouth Rocks. (Courtesy, *Poultry Tribune,* Mount Morris, IL)

Fig. 43–4.  White Cornish. (Courtesy, *Poultry Tribune,* Mount Morris, IL)

## SYSTEMS OF BREEDING

The breeding of chickens has passed through the total presently known systems of breeding; in fact, each method has been, and still is being, used successfully. But the vast majority of chickens in America today are hybrids of one form or another—either strain crosses, breed crosses, or crosses between inbred lines. A discussion of each method of breeding follows:

1. **Standard breed and variety.** This system of poultry breeding is similar to purebred breeding of four-footed farm animals. It consists in either (a) mass mating of phenotypically selected individuals of the same standard breed and variety,

or (b) mating birds based on some measurement of performance, either individual, parental, sib, or progeny. Despite the fact that hybrids produce more eggs than average commercial layers, it is noteworthy that egg-laying tests show that Single Comb White Leghorns compete on even terms with hybrids under test conditions. Certainly the hybrids can equal, but the point is that they do not excel, the best Leghorns.

2. **Pure strains.** So-called pure strains, which are produced commercially today, are no more pure than the standard breeds and varieties (point 1 above). A purebred strain of chickens generally takes the name of the breeder who developed it. It is called pure strain for the reason that the developer has closed the flock to outside blood for several years. Pure strains are now used extensively as parents of commercial crosses. Generally speaking, such pure strains are not for sale; they are the exclusive property of the developer, who promotes them.

3. **Strain crosses.** These are crosses of two or more different strains within the same breed; hence, they are still purebreds. The crossing of two inbred strains may result in some favorable hybrid vigor in egg production, or in other economically valuable characteristics. Thus, this method of breeding has some advantages over pure-line breeding for the production of commercial stock. Most of the chickens produced today for commercial egg production are Leghorn strain crosses.

4. **Breed crosses or crossbreds.** This consists in crossing different breeds or varieties that combine well, based on the performance of the progeny. Such crossbreds usually show hybrid vigor or heterosis for egg or meat production. Among the more important commercial types of crossbreds are the following:

a. **Sex-linked cross.** When we speak of sex-linked traits, we refer to those traits that are determined by genes carried (linked) on the X chromosome, one of the two sex chromosomes (the other is known as the Y chromosome). A common sex-linked cross in chickens consists in the use of either a Rhode Island Red or a New Hampshire male on a Barred Plymouth Rock female. The male progeny from such a cross are barred like their mother, but the females are nonbarred like their father. The chicks from such a cross can be easily distinguished at hatching, with the result that the cockerels are raised for meat production, and the pullets are kept as layers.

It is noteworthy that the reciprocal cross—that is, a Barred Rock male mated to a Rhode Island Red or a New Hampshire female—produces all barred crossbred progeny.

b. **Leghorn-Red cross.** The offspring of a Leghorn male and a Rhode Island Red female is known as a Leghorn-Red cross. This cross, which was used rather extensively in the Midwest in layer production, results in medium-weight layers which produce eggs of an intermediate color between the chalk white of the Leghorn and the brown of the Rhode Island Red. In addition to being good layers, when the hens are marketed as meat they usually command a better price than Leghorns.

The reciprocal cross—that is, a Rhode Island Red male mated to a Leghorn female—is quite different; the offspring

mature more slowly, and are more inclined to broodiness, but they do have a lower adult mortality rate.

c. **Austra-White.** This cross, which results from mating an Australorp male and a White Leghorn female, has been used somewhat as layers in the Midwest and in southern California. Although they are good layers, they do tend toward excessive broodiness; and sometimes the dark shanks of the pullets and the tinted eggs are discriminated against on the market.

The reciprocal cross—Leghorn male on an Australorp female—results in a cross known as the White-Austra. These crossbreds are less inclined to broodiness than their counterparts, the Austra-White, yet they are apparently quite resistant to respiratory diseases.

5. **Inbred hybrid.** The use of inbreeding and hybridization to produce commercial hybrids is more or less patterned after the methods of the corn breeder. The actual details of commercially inbred hybrids are trade secrets of the companies producing them. Nevertheless, the general principles used by all are the same—the development and testing of many small inbred lines. Three or more generations of brother-sister matings, or sufficient generations of less intense inbreeding to produce an equivalent degree of inbreeding, may be required to produce a genetically stable stock. Some lines fall by the wayside in the process, due to reduction in hatchability, egg production, and viability under intense inbreeding, with the result that they must be discarded. The more viable and productive lines are mated with one another, and the cross-line progeny tested. Then, if the cross-line progeny is superior, further tests are conducted to find four-line (or more) combinations which will produce outstanding, uniform progeny in commercial volume. Because the inbred population are, within themselves, not generally outstanding, only sufficient numbers of the inbred lines are maintained to produce the cross-line progeny which are mated to produce four-line combinations for commercial distribution.

In recent years, as a result of performance entries in random sample tests, inbred hybridization has lost some of its popularity, because strain-cross entries frequently prove superior to some of the inbred-hybrid entries. Thus, at this time, there is considerable debate as to the virtues of inbred hybrids. Eventually, the issue will be settled by commercial producers for, over a period of time, they will follow that system of breeding which is most profitable to them.

It is generally recognized that hybrid stocks show quicker recovery from disease outbreaks and greater resistance to unfavorable environment than standardbred stocks. On the other hand, some hybrids have suffered higher mortality than standardbred stocks from the avian leukosis complex. But hybrids with average or better resistance to leukosis have been developed.

6. **Recurrent reciprocal selection.** The most interesting development in systems of mating since inbred hybrids has been the breeding for cross-line performance by using the system of recurrent reciprocal selection. This system is much like strain crossing, but differs from it in the way the pure strains are multiplied by using those individuals which cross best with the other line, rather than on the basis of their own performance.

Fig. 43-5. H & N "nick chick" Leghorn, produced by recurrent reciprocal selection. (Courtesy, Heisdorf & Nelson, Inc., Kirkland, WA)

use, copyrighted trade names. The larger ones have well-trained geneticists, physiologists, pathologists, and veterinarians on their staffs. Also, they make use of computers to handle the thousands of records which they must process. The foundation breed furnishes eggs to hatcheries.

Fig. 43-6. Progeny testing. Different combinations of lines and families within lines being tested by collecting data on progeny—egg numbers, egg size, body size, livability, etc. (Courtesy, Colonial Poultry Farms, Inc., Pleasant Hill, MO)

## BUSINESS ASPECTS OF POULTRY BREEDING

Broadly speaking, commercial chicken producers are of two kinds; they're either egg producers or broiler producers. In either event, it's net return that counts. This means that the commercial producer is interested in increasing product output per bird, increasing the efficiency of producing the product, and improving the quality of the product produced. Improvements in fertility, hatchability, growth rate, body conformation, egg yield, meat yield, feed conversion, egg quality, meat quality, and viability (for example, PPLO-free) are all important in that they contribute to the main goal—increased net returns.

No industry has done a better job than the poultry industry in combining science and technology. Literally speaking, these two have upped the ounce to the pound in broiler production, and upped the dozen to the gross in egg production. Hand in hand with the transformation of the poultry industry, there has come a very high degree of specialization. From a breeding standpoint, today's poultry breeding is centered in two types of business enterprise: (1) the foundation breeder, and (2) the hatchery (or multiplier).

### FOUNDATION BREEDERS

Most modern foundation breeders are large, well financed, and well managed. In fact, many of them have the characteristics of any other large business. Generally they are incorporated, departmentalized—breeding and development, sales, advertising, office management, purchases, etc. Many of them have sales representatives abroad, as well as throughout the United States. Most of them recognize the importance of, and

## HATCHERIES

Fig. 43-7. Hatchery trays full of baby chicks. (Courtesy, Indian River International, Division of Hy-Line Indian River Company, Nacogdoches, TX)

Hatcheries multiply the stock supplied by foundation breeders, through hatching eggs. In turn, they sell chicks or poults to the farmer or producer, who then grows them out as commercial egg layers, broilers, or market turkeys.

Until about 1950, foundation breeders and hatcheries operated independently of each other. Today, most of them are associated together through a franchise. A contract, signed by the two parties, specifies that the breeder will provide the hatchery with the breeding stock for the hatchery supply flock. Such supply flocks are commonly referred to as parent flocks, because the hatchery uses the chickens as parents of the commercial chicks that they sell.

Such franchise arrangements give breeders virtual control over the stock sold by associated hatcheries. On the other hand, hatcheries find the franchise agreement desirable because they do not have to spread themselves so thinly; they can leave the breeding problems to the foundation breeders and concentrate their efforts on hatching and selling chicks.

## RANDOM SAMPLE PERFORMANCE TEST; MULTIPLE UNIT POULTRY TEST

The idea for random sample testing was first proposed by Hagedoorn, a geneticist of the Netherlands, at the 1927 World's Poultry Congress held in Ottawa, Canada. Twenty years later, the first U.S. random sample laying test was conducted at Pomona, California.

Random Sample Performance Tests are designed to provide information on the performance of commercial chicks and poultry under uniform testing conditions. In these centralized tests, stocks from several growers are hatched at the test, the chicks raised, and the resultant pullets maintained to a fixed age (500 days or older) to determine egg production. Broiler tests are conducted for the purpose of evaluating growth, quality and pounds of feed required to produce a pound of liveweight to the completion of the test. Entries are kept on the performance test for a period determined by the test management—usually 9 to 12 weeks.

Theoretically, centralized random sample performance tests of this kind should give a true comparative evaluation of the stocks tested. However, the following serious criticisms have been leveled at them: (1) that stocks sampled for testing are not truly random—which biases comparisons, (2) that such tests lead to monopolistic tendencies and to a serious reduction of the world's supply of potentially valuable genetic stock by the elimination of small breeders and (3) that such tests are mainly a tool of breeders used to promote sales.

In order to alleviate some of the disadvantages and criticisms of the Random Sample Performance Test, the Multiple Unit Poultry Test evolved. In this program, as implied by the name, more than one type of test is carried out at more than one farm or location.

Whatever the merits or demerits of Random Sample Performance Tests or of Multiple Unit Poultry Tests, they declined. But even the most vociferous critics of these tests agree that performance tests of this kind have exerted a powerful influence in developing high genetic merit of commercially sold poultry.

## INHERITANCE OF SOME CHARACTERS IN CHICKENS

The hereditary material of a chicken is located in 39 pairs of chromosomes. The turkey has 41 pairs of chromosomes.

Some of the Mendelian principles were first applied to poultry, following the rediscovery of Mendel's laws about 1900. Fig. 43–8 shows the inheritance of a pair of characters; the result of mating a White Wyandotte hen and a White Plymouth Rock male. Wyandottes have rose combs and Plymouth Rocks have single combs. The gene for rose comb is designated by R, and that for single comb by r. The gametes from the pure Wyandotte will carry only Factor R for rose comb and those from the Plymouth Rock will carry only factor r for single comb (Fig. 43–8). When birds are mated, the offspring, or $F_1$ generation, will have rose combs. The character (rose comb) is called dominant while the single comb that does not appear is termed recessive.

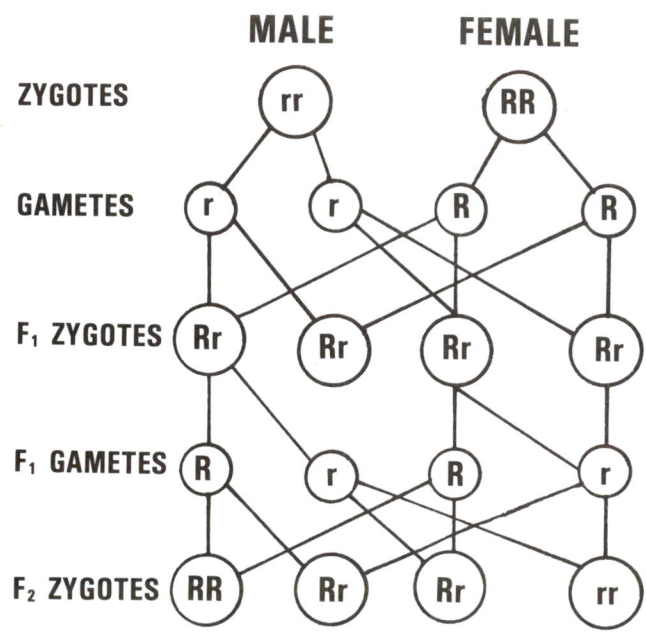

Fig. 43–8. Inheritance of a pair of characters. When a rose comb (R) female is crossed with a single comb (r) male, the first generation birds have rose comb (dominant character). When the $F_1$ birds are mated among themselves, about 75% of the $F_2$ generation will have rose combs and the remainder will have single combs.

An example of the inheritance of two pairs of characters may be obtained by crossing a Black Wyandotte with a White Plymouth Rock. In this case, rose comb (R) is dominant to single (r), and black (B) is dominant to white (b). All of the first generation crosses from such a mating will be rose combed and black. However, the $F_2$ generation will consist of 9 rose comb-black, 3 rose comb-white, 3 single comb-black, and 1 single comb-white, or a 9:3:3:1 ratio.

1. **Plumage color.** White, or light-colored, feathers have become an important factor in the breeding of broilers because they are easier to pick clean than chickens with dark-colored feathers. Colored chickens often have pigmented pinfeathers at broiler age which have not broken through the skin and, therefore, cannot be readily removed with the result that they detract from the appearance of the carcass and cut-up parts. Hence, White Rock and specially developed *dominant white* meat-type male lines are preferred for broiler production. Also,

white turkeys and white ducks are preferred for the same reason. Bronze turkeys have been replaced with white turkeys; and most ducks produced commercially are of the White Pekin breed.

Fig. 43–9. Shaver Redbro parents, for producing red-feathered broilers. Not all consumers want white chickens! (Courtesy, Shaver Poultry Breeding Farm, Ltd., Cambridge, Ontario, Canada)

Geneticists are well aware of the inheritance of plumage color and make use of it in the production of broilers. Also, in some cases, they make use of it where sexing of chicks is desired.

2. **Skin and shank color.** The several different shank colors found in fowl result from different combinations of pigments in the upper and lower layers of skin. Yellow shanks are due to the presence of carotenoid pigments in the epidermis, and the absence of melanic pigment. Black shanks are due to the presence of melanic pigments in the epidermis. White shanks are the result of complete absence of both types of pigments.

Blue, or slaty blue, shanks occur when melanic pigment is present in the dermis, but neither type of pigment is present in the epidermis. Green shanks occur when there is black in the dermis and yellow in the overlying epidermis.

3. **Rate of feather development.** Early feathering is essential for minimizing pinfeathers on the dressed carcass. To meet this requisite, all modern broiler strains now carry the sex-linked early feathering gene. This is a recessive gene. Early feathering chicks can be identified at hatching by the length of the covert feathers of the wing in proportion to the length of the primary feathers. Also, at about 10 days of age, rapid-feathering chicks show well-developed tail feathers, whereas slow-feathering chicks show no tails.

Since the early feathering gene is sex-linked, it may be used in determining sex at hatching time, thereby alleviating the necessity of vent sexing. An early-feathering male mated to a late-feathering female produces slow-feathering male progeny and early feathering pullets. This makes it possible to select pullets at hatching time.

4. **Egg production.** Perhaps it is fair to say that geneticists have been more successful in increasing broiler (meat) production than they have in egg production. Most of the income from layers is from the sale of eggs for food or for hatching purposes. Thus, the higher the egg production secured, the lower the feed and the total cost per dozen eggs. Consequently, producers are interested in high egg production. Egg production is affected by feeding. Likewise, management and environment exert considerable influence on egg production; for example, a bird will lay more eggs than it would otherwise if the eggs are removed from the nest daily.

Egg yield is the product of two forces: rate of laying and length of the laying period before molting.

a. **Sexual maturity** (approximately 30% heritability). A pullet is said to be sexually mature when she lays her first egg. The earlier a pullet commences laying, the longer will be her laying year, with the result that there is more possibility of her producing more eggs. In turn, this means lower cost per egg.

Age at maturity is hereditary. On the average, Leghorns become sexually mature at between 170 and 185 days of age; the dual-purpose breeds, such as the White Wyandotte, reach sexual maturity about 2 weeks later. However, there are differences between strains, and selection can be made accordingly.

Sexual maturity can also be greatly advanced or delayed by environment, especially by the lighting program followed during growing. Feeding programs and disease can also delay age at first egg.

If birds of the light breeds, such as Leghorns, are to lay approximately 250 eggs during the pullet year, they should come into production when about 5 months old. The general-purpose breeds, such as Rhode Island Reds, should start laying when they are about 5½ to 6 months old.

In a flock of birds of about the same age, the first 75% to come into production will be the best layers.

b. **Egg size** (approximately 50% heritability). Size of eggs is correlated with a number of factors, among them (1) body size—the larger breeds generally produce larger eggs, (2) age of pullets—egg size increases from the time pullets start to lay until somewhere about six months later, (3) weather—the size of eggs declines during the hot summer months, (4) second year—the eggs produced during the second year are larger than those produced the first year, (5) period of time within the clutch—those laid at the beginning of the clutch are larger than those laid at the end, and (6) total eggs laid—there is a tendency toward a decline in egg size with the total number of eggs laid in a year.

Shell color is also heritable. As is well known, some breeds produce white eggs, others produce brown eggs. Varying shades of color may be expected among brown eggs, but tinted shells should be avoided among white eggs, as these are not desired by consumers. For this reason, white eggs that have tinted shells should not be set.

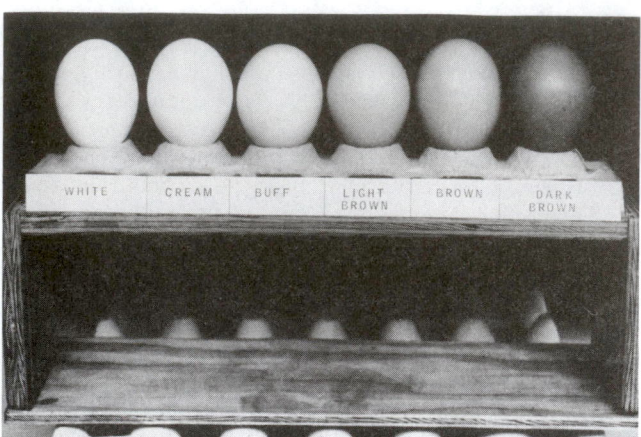

Fig. 43–10. Eggshell color can be varied by breeding. (Courtesy, DeKalb Ag Research, Inc., DeKalb, IL)

c. **Intensity** (approximately 10% heritability). Intensity, or rate of production, which refers to the number of eggs laid by a hen during a given period of time, is most important from a profit standpoint. Thus, if a flock peaks at 85% production, the average hen in the flock must be laying an egg, on the average, more than 8 out of every 10 days. Such a flock is said to have a high intensity of lay.

d. **Hatchability** (approximately 12% heritability). Hatchability refers to the percentage of fertile eggs which hatch under artificial incubation. It is largely influenced by feeding and management. Even so, the poultry breeder should select against lethal genes which either reduce or prevent hatching, and against such indirect effects as large and small egg size and poor shell texture.

e. **Broodiness.** When the birds are brooding, they aren't laying. Hence, a minimum of broodiness is desired.

Broodiness is partially determined by a dominant sex-linked character. Breed differences exist; the light breeds, such as the Leghorn, are less broody than general-purpose breeds, such as the White Wyandotte. However, within a given breed or variety, there are strain differences in broodiness.

There is also evidence that broodiness is determined by complementary effects of genes. For this reason, when certain breeds are crossed, the progeny generally show more broodiness than that shown by either of the parent breeds.

Of course, broodiness was nature's way of aiding propagation. After laying a certain number of eggs, wild birds set on them until they are hatched out. But, since this type of behavior is no longer needed in modern poultry production, it has been practically eliminated in today's egg-laying strains.

5. **Body weight and growth rate** (approximately 60% and 35% heritability, respectively). It is generally recognized that there are wide differences in body size among breeds—the extremes being illustrated by the difference in body size between bantams and Brahmas. Large body size is of importance to broiler and turkey breeders, chiefly because mature body size is correlated with rate of growth and efficiency of feed utilization.

Because weight of broilers is highly heritable, progress has been excellent through the selection of the larger individuals

at broiler age. Further, there is some question whether heterosis, as obtained through hybrid breeding, contributes substantially to broiler weight; for example, the best New Hampshire strains generally equal the most rapid-gaining crosses. Despite the latter fact, crossbreeding is usually practiced in broiler production for, among other reasons, more uniformity of growth and fewer runts and culls are found in crossbreds than in standardbred stocks. Also, the crossbreds are generally more viable.

Body conformation is especially important in turkeys, because, in comparison with broilers, turkeys are marketed at higher weights and the carcass is usually marketed whole rather than cut up. With broiler producers, on the other hand, conformation is of secondary importance because of the increasing practice of marketing broiler meat in cut-up and packaged form. However, the poultry breeder has produced more desirable broiler carcasses through the infusion of broad-breasted, heavily muscled, Cornish breeding.

Body conformation is of little consequence in layers, for the reason that after they have finished their usefulness to produce eggs, their carcasses are usually manufactured into chicken soup and other prepared foods.

Growth rate and feed efficiency are highly correlated; hence, growth rate is of great importance in the breeding of both meat chickens and turkeys. Rapid growth makes for a saving in time, labor, feed consumption, and overhead in the production of meat.

6. **Viability.** Viability, or livability, is influenced greatly by feeding and management practices. Experimental evidence also shows that there are family differences in susceptibility to and resistance against pullorum, fowl typhoid, range paralysis, roundworm infestation, crooked keels, and reproductive troubles. Hence, poultry geneticists have concentrated, and will continue to concentrate, on developing strains of higher livability.

The possibility of developing genetically resistant strains to certain diseases was clearly demonstrated by Cornell University in 1955. In their studies, resistant strains of White Leghorns showed only 2 to 3% mortality from leukosis, whereas susceptible strains showed a mortality of 25%.

As has already been pointed out, crossbreeding generally results in increased vigor and higher livability. On the other hand, inbreeding usually results in reduced vigor and increased mortality.

## METHODS OF MATING

The two most common methods of mating are flock and pen mating, although stud mating and artificial insemination are sometimes used.

1. **Flock mating.** Flock or mass mating means that a number of males are allowed to run with the entire flock of hens. Other things being equal, better fertility is obtained from flock mating than from pen mating.

The number of hens per male will vary with the size and age of the birds. With the light breeds, such as Leghorns, it is customary to use one male for 15 to 20 hens. In the case of the general-purpose breeds, such as the White Wyandotte, it is customary to use one male with 10 to 15 hens. With the heavy breeds, one male is placed with each 8 to 12 hens. Also, age is a factor: young males are more active than older ones. Males past 3 years old are not too satisfactory as breeders.

Fig. 43-11. Flock mating or mass mating, in which a number of males are allowed to run with an entire flock of hens. (Courtesy, The Cobb Breeding Corporation, Concord, MA)

Fertility can be improved by using cockerels with older hens, and by using older males with pullets.

2. **Pen mating.** In pen mating, a pen of hens is mated with one male. If the birds are trap nested and the hen's leg-band number recorded on the egg, this system makes it possible to know the parents of every chick hatched from a pen mating.

About the same numbers of hens are mated with one male in pen mating as in flock mating. However, fertility is generally not so good in pen mating as in flock mating because (a) there is no opportunity for the birds to mate with the ones they choose, and (b) there is no competition between males.

3. **Stud mating.** Stud mating is comparable to what is called hand mating among four-footed farm animals. In this method, the females are mated individually with a male that is kept by himself in a coop or pen. This system makes it possible to mate more females to a male than can be accomplished in either pen or flock mating. However, stud mating involves more labor than the other two systems because birds should be mated at least once each week in order to maintain good fertility.

Sometimes stud mating is used when a very valuable male is being used as a breeder, and it is desired to use him to the maximum.

4. **Artificial insemination.** Artificial insemination is frequently used in turkey breeding where poor fertility is encountered. Also, it may be used in experimental work where hens are kept in cages.

Fig. 43-12. Pen mating, in which a pen of hens is mated to one male. (Photo by J. C. Allen and Son, West Lafayette, IN)

Fig. 43-13. Artificial insemination of caged turkey breeders. (Courtesy, *Turkey World*, Mount Morris, IL)

## SELECTING AND CULLING CHICKENS

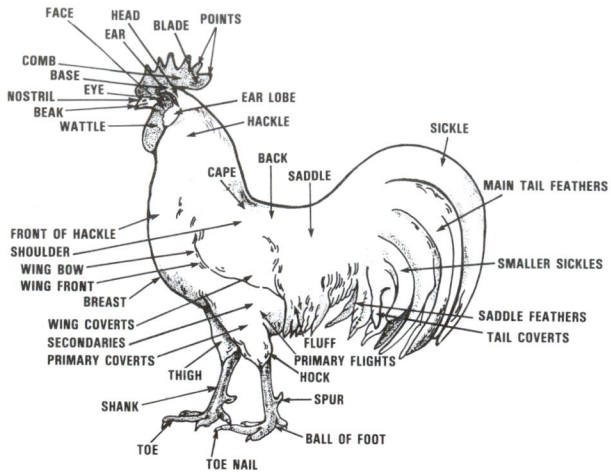

Fig. 43–14. Parts of a chicken. Before attempting to select and cull chickens, it is necessary to master the nomenclature that describes and locates each part, and to know the relative importance of each part.

The terms *selection* and *culling* carry opposite connotations. Selection aims at progress; it deals with retaining the best in the flock, seldom more than the top 20 to 25%, and generally not more than 10 to 15%, for carrying forward. On the other hand, culling refers to the removal of the least productive part of the flock. It is aimed at prevention of retrogression.

## METHODS OF SELECTION

Except for the poultry fancier, who is interested in breeds and varieties from the standpoint of the *Standard of Perfection*, poultry are seldom exhibited today.

The following methods of selection are used in poultry breeding:

1. **Individual or mass selection.** In this, selection of individuals is based on physical appearance. Many geneticists and poultry producers maintain that this system has little or no value. Yet, two facts must be recognized: (a) A great deal of the poultry improvement to date has been the result of widely applied mass selection practices, particularly from the standpoint of lessening undesirable individuals that are slow gainers, poor layers, and which detract from the uniformity of the flock; and (b) mass selection may be effective in those cases where the flock is considerably below average in some trait—for example, in a flock where the average egg production is only 150 to 160 during the pullet year.

On the other hand, where a flock is average or above average in productivity, and for those characters of economic importance which are influenced by many different genes, individual or mass selection is ineffective.

2. **Pedigree selection.** Pedigree selection is of special importance when production data is not available, or when selection is being made between two males that are comparable in all other respects. For example, pedigree selection is of value where one is selecting breeding cockerels whose dams have very different production records, as between 250 eggs and 175 eggs. In making use of pedigree selections, however, it must

be remembered that the nearest ancestors in the pedigree are much more important than those many generations removed.

3. **Family selection.** Family selection refers to the performance or appearance of the rest of the members of the family, particularly the bird's sibs (sisters and brothers). Sons of a 250-egg hen which was 1 of 10 sisters laying between 240 and 270 eggs each are more likely to transmit desirable genes for egg production to their daughters than are the sons of a 290-egg hen whose sisters finished the year with records of 180 to 240 eggs. Family testing has been the key to maximizing genetic progress in producing high-laying strains of chickens.

Indeed, individuality, pedigree, and family are important, and all should be used as tools in selection, but the only really sure basis concerning the ability of an individual to transmit genes for the desired characters to most of its progeny is based on a breeding test or progeny test.

4. **Blood group system.** Research work in genetics and animal breeding has demonstrated that in a population there may be more than two alternative genes that can occupy the loci on chromosomes. Such genes are called multiple alleles. The best known multiple alleles are those involved in blood types in humans. Three different genes, called genes A, B, and O, are known to be involved in humans. So far, seven blood group systems have been identified in chickens, and each group is controlled by a group of genes forming an allelic series.

To date, blood typing has not proven to be a simple and quick way for poultry improvement; someday it may. However, blood typing is valuable as a tool for obtaining greater insight concerning genetic mechanisms and for determining parentage.

## MEASURING EGG PRODUCTION

With layers, annual egg production is, without doubt, the most important single criterion to invoke when it comes to selection. Yet, it is not the only trait of importance to poultry breeders; they must select for body size and general appearance, livability, rapid growth and proper feathering, and egg size, shape, color, shell texture, and interior quality. It must be remembered, however, that selection for two or more characters automatically cuts down the effectiveness of selection for any one of them. Also, annual egg production is of low heritability, probably about 20%. All of this means that improvement in egg production through selection is relatively slow, a fact which is confirmed by egg records of the past.

1. **Trapnests.** Trapnests differ from regular nests in that they are provided with trapdoors by which birds shut themselves in when they enter. This was the invention of Professor James E. Rice, of Cornell University, about 1895.

There are two primary shortcomings to trapnesting chickens: (a) The female is in her second year of production before she is used as a breeder—hence, it lengthens the time between generations; and (b) the males are themselves untested—usually they are the sons of phenotypically superior females. Additionally, trapnesting requires more labor, particularly if it is done 7 days a week and each week of the year. However, studies have shown that a satisfactory measure of the egg-producing ability of a hen is possible by trapnesting 3 days a week. This reduces labor costs.

2. **Banding.** Some poultry producers give each pullet a numbered band, or make use of a series of colored bands of different color, in lieu of a complete quantitative trapnest record. By establishing a *key of colors*, and using both the right and the left leg, it is possible to mark the birds so as to indicate such important characteristics as (a) time of starting to lay, (b)

winter pause and molt, (c) broodiness, and (d) time of stopping lay in the fall.

3. **Examining the birds.** If one cares to do so, the hens may be examined early in the morning, and those which are going to lay on that day can be detected by feeling the egg in the shell gland. By going through this procedure on three successive days each month, a highly accurate record of the relative laying ability of the hens that make up the flock can be obtained, as well as an estimate of the actual number of eggs laid by each hen.

## MEASURING BROILER PRODUCTION

Growth rate of broilers to market weight has an estimated heritability of 35 to 40%. This trait is of major importance in the breeding of meat chickens as well as turkeys, because rapid growth means a saving in time, labor, feed consumption, and overhead in production costs. Also, it is known that feed efficiency is highly correlated with growth rate.

Because of the high heritability of growth rate, progress in breed improvement for weight at broiler age is an excellent way in which to make progress.

Feed efficiency is also highly correlated with growth rate, and it is known that feathering is highly heritable. Because these traits are highly heritable, rapid progress in broiler improvement has been made through mass selection. Also, carcass quality and uniformity have been improved through crossbreeding.

## NATIONAL POULTRY IMPROVEMENT PLAN

The National Poultry Improvement Plan was established in 1935, with the approval of the Secretary of Agriculture and under the authority of an appropriation made by Congress for the United States Department of Agriculture, for the improvement of poultry, poultry products, and hatcheries. The National Turkey Improvement Plan became operative in 1943. Then, in 1970, the two plans were consolidated into one National Poultry Improvement Plan with general provisions applicable to problems and conditions peculiar to particular classes of such fowl.

The objective of the National Poultry Improvement Plan is to provide a cooperative state-federal program through which new technology can be effectively applied to the improvement of poultry and poultry products throughout the country. The provisions of the plan, developed jointly by industry members and state and federal officials, establish standards for the evaluation of poultry breeding stock and hatchery products with respect to production qualities and freedom from hatchery disseminated diseases. Products conforming to specific standards are identified by authorized terms that are uniformly applicable to all parts of the country.

The provisions of the plan are changed from time to time to conform with the development of the industry and to utilize new information as it becomes available.

Initially, this program aided in identifying the bacterial disease transmitted from the hen to the chick through the egg caused by the bacterium *Salmonella pullorum*. Today, breeders are not only carefully screened for this disease, but they are also screened or vaccinated against a host of other diseases, including pullorum, Arizona infection, fowl typhoid, infectious bronchitis, Newcastle, *Mycoplasma gallisepticum*, and *Mycoplasma synoviae*.

Acceptance of the plan is optional with the states and individual members of the industry within the states. Table 43–2 shows the participation of chickens and turkeys in the National Poultry and Turkey Improvement Plans, by leading states.

### TABLE 43–2
### PARTICIPATION OF CHICKENS AND TURKEYS IN THE NATIONAL
### POULTRY AND TURKEY IMPROVEMENT PLANS, BY LEADING STATES[1]

| State | National Poultry Improvement Plan | | | | State | National Turkey Improvement Plan | | | |
|---|---|---|---|---|---|---|---|---|---|
| | Hatcheries | | Supply Flocks | | | Hatcheries | | Supply Flocks | |
| | (number) | (egg capacity) | (flocks) | (birds) | | (number) | (egg capacity) | (flocks) | (birds) |
| Alabama | 57 | 72,316,000 | 417 | 4,602,000 | Minnesota | 7 | 8,124,000 | 118 | 808,000 |
| Georgia | 37 | 61,838,000 | 677 | 7,685,000 | North Carolina | 8 | 4,875,000 | 174 | 1,078,000 |
| North Carolina | 34 | 48,212,000 | 890 | 8,610,000 | Missouri | 4 | 2,520,000 | 15 | 188,000 |
| Arkansas | 65 | 42,311,000 | 477 | 3,951,000 | Ohio | 3 | 2,148,000 | 48 | 249,000 |
| Mississippi | 88 | 31,074,000 | 317 | 3,047,000 | Virginia | 3 | 1,788,000 | 43 | 213,000 |
| Pennsylvania | 33 | 15,684,000 | 130 | 804,000 | Arkansas | 2 | 1,204,000 | 20 | 122,000 |
| Florida | 35 | 15,561,000 | 146 | 1,330,000 | Wisconsin | 3 | 976,000 | 18 | 98,000 |
| U.S. Total | 925 | 423,633,000 | 6,181 | 37,324,000 | U.S. Total | 73 | 29,789,000 | 660 | 3,918,000 |

[1]*Agricultural Statistics 1988*, USDA, p. 355. Data for 1986.

Information relative to the plan is available in *The National Poultry Improvement Plan*, a copy of which may be secured from the U.S. Department of Agriculture, Agriculture Research Service, BARC—East Building 265, Beltsville, MD 20705.

Proponents of the plan point out that it has been highly effective in increasing egg production and lowering pullorum disease. By 1980, pullorum disease had practically ceased to exist in the United States. Of the 35.4 million chickens tested that year, only .000002% were reactors; and of the 3.2 million turkeys tested in 1980, there were no reactors. This is no small achievement when it is realized that, in 1920, when pullorum testing was first started in the United States, there were 11% pullorum reactors in chicken flocks.

## CULLING

The attitude toward culling has changed as American laying flocks have increased in size. Today, extensive culling is no longer practiced in many high egg-producing commercial flocks. Because only healthy and well-developed birds are placed in commercial laying housing, these flocks normally contain a very small number of poor layers. Hence, it is not considered practical to disturb high producers and reduce egg production merely to identify a few culls. On the other hand, proper culling can increase efficiency of egg production in most farm flocks; and the poorer the flock, the greater the need to cull.

Culling is particularly valuable if the producer keeps hens for a second year of egg production. However, it is questionable whether or not it pays to keep a flock longer than the pullet year, since a hen lays 20 to 25% fewer eggs in her second year than in her first year.

Identifying and removing nonlaying and low-producing birds from the flock accomplishes the following: (1) keeps the egg production rate of the flock high; (2) saves the cost of feeding unproductive birds (and hens eat about 7 lb of feed per month whether laying or not); (3) reduces the spread of disease from hens to young birds; and (4) provides more space for the remaining birds.

Culling should take place throughout the year. When chicks are started, all obviously weak birds should be culled. As the flock gets older, runty and slow-growing pullets should be eliminated. When the flock is put into the laying house, slow-maturing birds should be removed. During the laying year, flock owners should remove the sick, lame, or injured birds. Generally speaking, heavy culling should not be necessary during the first eight to nine months of lay. However, if egg production slumps badly, it may be desirable to locate the cause and remedy it.

It is possible to secure an indication of a hen's present, past, and future egg production by physical examination of the bird. These characteristics, and their significance, are summarized in Table 43–3, Culling Chart.

### TABLES 43–3
### CULLING CHART[1]

| Separating Layers and Nonlayers | | |
|---|---|---|
| **Character** | **Layer** | **Nonlayer** |
| Comb ......... | Large, smooth, bright red, glossy. | Dull, dry, shriveled, scaly. |
| Face ......... | Bright red. | Yellowish tint. |
| Vent .......... | Large, smooth, moist. | Shrunken, puckered, dry. |
| Pubic bones ..... | Thin, pliable, spread apart. | Blunt, rigid, close together. |
| Abdomen ....... | Full, soft, pliable. | Contracted, hard, fleshy. |
| Skin ........... | Soft, loose. | Thick, underlaid with fat. |

| Separating High and Low Producers | | |
|---|---|---|
| **Character** | **High Producer (continuous laying)** | **Low Producer (brief laying)** |
| Vent .......... | Bluish-white. | Yellow or flesh color. |
| Eye-ring ........ | White. | Yellow. |
| Earlobe ........ | White. | Yellow. |
| Beak .......... | White. | Yellow. |
| Shanks ........ | White, flattened. | Yellow, round. |
| Plumage ....... | Worn, soiled. | Not much worn. |
| Molting ........ | Late, rapid. | Early, slow. |

| Characteristics of Desirable Producers | |
|---|---|
| Time of maturity ........... | Leghorns begin to lay at 5 to 5½ months; Rhode Island Reds, Plymouth Rocks, and similar breeds, at 5 to 6½ months. |
| Rate of production .......... | Hens lay at least 220 eggs a year. |
| Broodiness ................. | Birds are seldom broody. |
| Persistence of production ...... | Good producers lay consistently for 12 to 15 months. |

[1]From *Culling Hens*, USDA Farmers' Bull. No. 2216, p. 10.

## BREEDS, AND BREEDING TURKEYS

Turkeys are native to America. They were found in great numbers by the pioneer settlers, and a limited number of wild turkeys still exist in certain remote areas. Turkeys were also plentiful in Mexico.

Fig. 43–15. Turkey hens in a breeder house showing trap nests on the left. Note ramp for easy access of the turkeys to the upper level nests. Breeders and facilities of Cuddy Farms, Ltd., North Carolina. (Photo courtesy James Strawser, University of Georgia, Athens)

It is reported that turkeys were taken from this continent to Spain in 1498, and to England soon thereafter. Later, some of these European stocks were brought back to this country where, along with the native wild turkey, they were used in developing our present varieties.

Only one breed of turkeys is recognized by the *Standard of Perfection*; hence, correctly speaking, we should refer to varieties rather than breeds.

Fig. 43–16. Bronze gobbler. (Courtesy, National Turkey Federation, Mount Morris, IL)

The varieties of turkeys of historical importance are listed in Table 43–4. Today, only the Large White and Bronze are commercially important in the United States.

## TABLE 43–4
### VARIETIES OF TURKEYS AND THEIR CHARACTERISTICS

| Variety | Adult Tom | Adult Hen | Plumage Color | Beak | Color of Throat Wattle | Beard | Shank & Toes | Comments |
|---|---|---|---|---|---|---|---|---|
| | (lb) | (lb) | | | | | | |
| Black | 33 | 18 | Metallic black. | Slaty black. | Red, changeable to bluish white. | Black. | Pink in adults. | Black color evolved from selecting darker birds in the population. Black turkeys were popular in Spain, France, and Italy. |
| Bourbon Red | 33 | 18 | Brownish red, with white wing and tail markings. | Light horn at tip, dark at base. | Red, changeable to bluish white | Black. | Reddish pink in adults. | Developed in Bourbon County, Kentucky. Very attractive. |
| Bronze | 36 | 20 | Black; with an iridescent sheen of red, green, and bronze. | Light horn at tip, dark at base. | Red, changeable to bluish white. | Black. | Dull black in young; smoky pink in mature birds. | The Broad-Breasted Bronze is a sub-variety. Of all meat animals, the Broad-Breasted Bronze most uniformly produces a well-fleshed carcass. |
| Narragansett | 33 | 18 | Dull black, with white markings. | Horn. | Red, changeable to bluish white. | Black. | In adults, deep salmon. | Developed in Narragansett Bay area of Rhode Island, by crossing domestic stock on wild turkeys. |
| White Broad Breasted | 33 | 18 | Pure white. | Light pinkish horn. | Red, changeable to pinkish white. | Deep black. | Pinkish white. | Developed from crossing Broad-Breasted Bronze with white feathered variety. Very similar to Bronze; only white, and slightly higher fertility. |
| White Holland | 33 | 18 | Pure white. | Light pinkish horn. | Red, changeable to pinkish white. | Black. | Pinkish white. | White Holland is thought to be a sport from bronze turkeys. |

The White Broad Breasted strain listed in Table 43–4 refers to strains that have been developed since 1950 by first crossing Bronze and White Holland varieties and then backcrossing the second generation white progeny to Bronze males. This procedure is repeated for several generations so that the resulting Large White is essentially a Bronze turkey with white feathering and broad-breasted in conformation. These birds have been developed in recent years in response to processors' objections to the dark pins of the Broad-Breasted Bronze. This accounts for the fact that Broad-Breasted Bronze turkeys have decreased in number since 1950, while the White Broad Breasted variety has increased. In 1934, the USDA started the development of a small white turkey variety, which was released under the name of Beltsville Small White in 1941. However, turkey growers did not like the Small White, because the tonnage of meat produced was low. One person can take care of about the same number of big turkeys as of small turkeys, so labor cost per pound is higher on the smaller varieties. As a result, the Beltsville Small White has practically ceased to exist.

Fig. 43–17. White Broad Breasted turkeys; male (left) and female. (Courtesy, *Turkey World*, Mount Morris, IL)

Most turkeys are bred as standardbreds; that is, bred pure rather than hybridized. Also, individual selection and mass matings are the usual practices. Color, size, and conformation are highly heritable, with the result that they have responded well to these simple methods.

A few breeders have trapnested their birds and selected for egg production and hatchability.

In the case of turkeys, it appears that the breeding accomplishments in developing fast-growing, broad-breasted birds have left the breeder with a good market bird but with breeding populations seriously lacking in reproductive qualities. Part of this is attributed to the fact that heavily fleshed males are clumsy in mating. For this reason, many breeders have successfully used artificial insemination.

## BREEDS, AND BREEDING DUCKS

Fig. 43–18.  Duck breeding house in which there are more than 3,000 female breeders. (Courtesy, Cherry Valley Farms, Ltd., Rothwell, England)

The wild mallard duck, *Anas boschas,* is the ancestor of all domestic breeds of ducks. Ducks must have been domesticated a long time, because the Romans referred to them as early as 2,000 years ago. Also, it is believed that commercial duck raising has been practiced longer in China than in any other country.

In 1985, 21.6 million ducks were marketed in the United States. Duck production was once centered on Long Island, New York. Although Long Island still produces many ducks, large numbers are now raised in Indiana, Michigan, North Carolina, and Minnesota.

The choice of a breed of ducks should depend upon the market that is to be supplied. White Pekin, Aylesbury, and Muscovy ducks are excellent meat producers, Rouen Cayuga, Swedish, and Call ducks reach market weights that make them valuable as meat producers; but poor egg production and to some extent colored plumage makes them unsatisfactory for mass commercial production.

Khaki Campbells and Indian Runners are excellent egg-laying breeds. Accordingly, where special duck egg markets exist, the choice of either of these breeds would be wise.

Fig. 43–19.  White Pekin duck. (Courtesy, USDA)

Since the vast majority of ducks that are raised commercially in the United States are of the White Pekin variety, only it will be described; White Pekins reach market weight (7 lb) in 8 weeks. The breed originated in China and was introduced in the United States in the late 1870s. White Pekins are large, white-feathered birds, with orange-yellow bills, reddish-yellow shanks and feet, and yellow skin. Their eggs are tinted white. Adult drakes weigh 9 lb and adult ducks (females) weigh 8 lb.

White Pekins average approximately 160 eggs per year, but they are not good setters and they seldom bother to raise a brood. They are nervous, with the result that they should be treated gently.

## BREEDS, AND BREEDING GEESE

Fig. 43–20.  White Chinese geese. (Courtesy, USDA)

The goose was first domesticated 4,000 years ago in Egypt, where it was regarded as a sacred bird. History also records that the Romans learned to value goose liver as a delicacy, with the result that they placed large numbers of geese in pens and fattened them to increase the size of the liver. We are also told that they learned to use the feathers for filling mattresses, cushions, etc. Geese became well distributed over Europe during the Christian era, and even today goose raising is an important enterprise throughout eastern and western Europe.

In the United States, geese are raised primarily for meat production. However, several varieties of geese are bred by poultry fanciers. Also, a considerable number of geese are used for weeding crops.

The principal meat-producing varieties of geese in the United States are the Toulouse, Emden, and Pilgrim. As is true in chickens, turkeys, and ducks, a white or near-white goose can most easily and attractively be dressed.

Geese differ somewhat from ducks in their mating habits. The large breeds of geese mate best in pairs or trios, although ganders of some lighter breeds will mate satisfactorily with as many as five females. Canada wild geese are largely monogamous and will usually mate that way for life.

## BREEDING OTHER POULTRY

There are many other species, breeds, and varieties of poultry, some bred for fancy and show, some bred for game, others bred for racing, and still others bred to fight to the death. A few of these will be discussed:

1. **Game birds.** Game birds are raised by and for those who like to hunt, and those who like to eat them. Among such game birds are pheasants, quail, grouse, chukars (or Chukar partridge), wild ducks, and game birds bred for cockfighting.

2. **Guinea fowl.** Guinea fowl are native to Africa, but they were brought to Europe during the Middle Ages. They are sometimes used as a substitute for game birds. It is thought that they might be more popular were it not for their harsh and seemingly never-ending cry, and their nervous disposition.

There are three domesticated varieties of Guineas; the Pearl, the White, and the Lavender. The Pearl is by far the most popular. It has a purplish-gray plumage, dotted or "pearled" with white.

Like quail and most other wild birds, Guinea fowls have a tendency to mate in pairs. However, one male may be mated with three or four hens.

3. **Ornamental birds.** Peafowls and swans are kept chiefly for ornamental purposes.

Peafowls are native to India. They like the habitat of shrubbery or trees. Four or five hens may be mated with one cock bird.

Swans are more common in Europe than in the United States. They live in pairs and remain faithful to each other until death. Swans live to be very old; the females will breed for 30 years, and males have been known to live for more than 60 years.

4. **Ostriches.** The ostrich is the largest bird in the world. At maturity, it may stand 10 ft tall and weigh more than 330 lb. Young ostriches grow very rapidly; they reach full size in about 6 months, but they do not attain sexual maturity until 3 to 4 years of age. They may live to 70 years of age. The ostrich is the only bird that eliminates its urine and feces separately.

Fig. 43–21. Ostrich.

Ostriches are valued primarily for their skins, which are made into fine quality leather. Spasmodically, the plumes are popular for decorations and accessories. The eggs may be used for human food, but the meat is seldom consumed because it is tough and has an unpleasant taste.

Ostrich farms in the United States are relatively new; so, it remains to be seen whether ostriches in this country will be a passing fad or an infant industry.

5. **Pigeons.** Pigeons are kept in all parts of the United States for squab production, for racing and messengers, and for exhibit.

There is a demand for squabs, especially in large cities, to take the place of game. Pigeons are the most rapid growing of all kinds of poultry. Squabs exceed the normal adult weight at the time they leave the nest, when 30 to 35 days of age. Flight and activity soon slim them down, however.

There are many varieties of pigeon, but the Homer, White King, and Swiss Mondaines are the most popular.

Pigeons mate in pairs and usually remain with their mates throughout life, although the mating may be changed if desired by placing the male and female in a coop together and leaving them there for 6 to 14 days, or until such time as they become settled.

Fig. 43-22. *Above:* Old English Game Chicken. (Courtesy, Watt Publishing Co., Mt. Morris, IL)

# A PROUD
# HERITAGE
## paintings by famous artists

Fig. 43-23. *Above:* Dark Cornish. (Courtesy, Watt Publishing, Co., Mt. Morris, IL)

Fig. 43-24. *Above:* A sex-linked cross (New Hampshire male and Barred Rock female with progeny). *Note:* Male progeny are barred. (Courtesy, Watt Publishing Co., Mt. Morris, IL)

Fig. 43-25. *Right:* Broad Breasted Bronze turkeys. (Courtesy, Watt Publishing Co., Mt. Morris, IL)

# QUESTIONS FOR STUDY AND DISCUSSION

1. What are the primary differences of poultry breeding and the breeding of four-footed animals?

2. Why is poultry breeding more flexible in the hands of breeders than the breeding of four-footed animals?

3. Discuss and describe the wild ancestry of chickens.

4. Define each of the following terms: *Type, class, breed,* and *variety.*

5. What is the *Standard of Perfection?* Historically, what role has the *Standard of Perfection* played in the poultry industry?

6. For each of the White Plymouth Rock, New Hampshire, Brahma (light), White Cornish, and White Leghorn, list their characteristics under the following categories: (a) plumage color, (b) weight, (c) type of comb, and (d) color of egg.

7. Discuss each of the following methods of breeding poultry: (a) standard breed and variety, (b) pure strains, and (c) strain crosses.

8. Discuss each of the following methods of breeding poultry: (a) breed crosses or crossbreds, (b) inbred hybrids, and (c) recurrent reciprocal selection.

9. Why has crossbreeding been used more extensively in poultry production than in the production of four-footed animals?

10. Records indicate that (a) Single Comb White Leghorns compete on even terms with hybrids in egg production, and (b) the best New Hampshire strains generally equal the most rapid-growing crosses in broiler weights. Why crossbreed?

11. Will the standardbred breeds of chickens be completely eliminated in the United States?

12. Why has crossbreeding been used more extensively with chickens than with turkeys, ducks, and geese?

13. Describe the role, or function (a) of foundation breeders, and (b) of hatcheries.

14. Describe each of the following: (a) Random Sample Performance Test, and (b) Multiple Unit Poultry Test. Why didn't they continue and expand?

15. Discuss the inheritance and economic importance of each of the following characters in chickens: (a) plumage color, (b) skin and shank color, and (c) rate of feather development.

16. Discuss the inheritance and economic importance of each of the following characters in chickens: (a) egg production, (b) body weight and growth rate, and (c) viability.

17. Discuss each of the following methods of mating: (a) flock mating, (b) pen mating, (c) stud mating, and (d) artificial insemination.

18. Define the terms *selection* and *culling.* Is it necessary that modern chicken producers, of either layers or broilers, know anything about judging and culling chickens?

19. Describe, and give the advantages and disadvantages of, each of the following methods of selection used in poultry breeding: (a) individual or mass selection, (b) pedigree selection, (c) family selection, and (d) blood group selection.

20. Discuss each of the following methods of measuring broiler production: (a) growth rate, and (b) feed efficiency.

21. What is the National Poultry Improvement Plan? How effective has it been?

22. What attitude toward culling would you recommend to a layer operator?

23. Name six varieties of turkeys, and tell about the origin of each of them.

24. For Bronze and White Broad Breasted turkeys, list their characteristics under the following categories: (a) weight, and (b) plumage color.

25. The modern turkey breeder still makes most selections on the basis of physical appearance, whereas the modern chicken producer has largely discarded this basis of selection. Why the difference?

26. Describe White Pekin ducks. In what states are most of the ducks in the United States raised?

27. Some breeds of ducks are excellent layers. Why haven't they been selected and used for this purpose, in competition with laying flocks of chickens?

28. List the three principal meat-producing varieties of geese in the United States. For what purposes are geese raised in the United States?

29. For what purpose(s) is each of the following birds bred: (a) guinea fowl, (b) ostriches, (c) peafowls, and (d) pigeons?

# SELECTED REFERENCES

| Title of Publication | Author(s) | Publisher |
| --- | --- | --- |
| *Commercial Chicken Production Manual,* Third Edition | M. O. North | AVI Publishing Co., Westport, CT, 1984 |
| *Poultry Breeding and Genetics* | Edited by R. D. Crawford | Elsevier Science Publishing Co., Inc., New York, NY, 1990 |
| *Poultry Meat and Egg Production* | C. R. Parkhurst G. J. Mountney | Van Nostrand Reinhold Co., New York, NY, 1988 |
| *Poultry Production,* Twelfth Edition | M. C. Nesheim R. E. Austic L. E. Card | Lea & Febiger, Philadelphia, PA, 1979 |
| *Standard of Perfection* | M. C. Wallace | American Poultry Association, Crete, NE, 1966 |

Turkeys on the range and about ready to go to market. Produced by Cuddy Farms, Ltd., North Carolina. (Photo courtesy James Strawser, University of Georgia, Athens)

It takes a lot of feed for a big turkey operation. This shows the feed mill, elevators for feed ingredients, and rail access for shipping corn, soybean meal, and other feed ingredients. Feed facilities of Cuddy Farms, Ltd., North Carolina. (Photo courtesy James Strawser, University of Georgia, Athens)

# FEEDING AND MANAGING POULTRY[1]

# Chapter 44

Contents | Page
---|---
**Digestive System of Poultry** | 846
**Nutritive Needs of Poultry** | 846
   Energy | 846
   Protein | 847
   Minerals | 847
   Vitamins | 848
   Water | 851
**Evaluating Poultry Feeds** | 851
**Feeds Used in Poultry Rations** | 852
**Nonnutritive Additives** | 853
**Poultry Ration Pointers** | 853
   Factors Involved in Formulating Poultry Rations | 853
      Nutrient Requirements | 853
      Availability, Nutrient Content, and Cost of Feedstuffs | 853
      Acceptability and Physical Condition of Feedstuffs | 854

[1]The author is very grateful to the following poultry scientists for their authoritative review and suggestions for improvement of this chapter: Dr. Leo S. Jensen, Professor, Department of Poultry Science, College of Agriculture, The University of Georgia, Athens; and Dr. Peter R. Ferket, Extension Poultry Nutritionist, School of Agriculture and Life Sciences, North Carolina State University, Raleigh.

845

Poultry feeding has changed more than the feeding of any other species. Originally, it was strictly a backyard enterprise; the mother hen did her own incubating and raised her young, and the farmer's wife fed the chickens on table scraps and the unaccounted-for grain from the crib. Reproduction was confined to the spring months when green feeds, insects, and sunshine were all available to contribute to the nutrition of the baby chicks. Feeding was largely an art rather than a science, and such commercial feeds as were sold were largely "secret formulas and patented potions." But all this has changed. Today, the vast majority of commercial poultry is produced in large units wherein the maximum of science and technology exist. Confinement production is commonplace, and well-balanced rations containing adequate sources of all known nutrient materials are fed for maximum production. The current trend in poultry production is toward controlled environment, which usually results in lowered feed consumption. Under such conditions, the daily feed consumption must be taken into consideration and the nutrient content of the feed (energy, amino acids, vitamins, and minerals) increased so as to compensate for the reduced feed intake and meet the requirements.

Poultry nutrition is more critical than that of other farm animals with regard to a number of factors. This is so because birds are quite different from four-footed animals; their digestion is more rapid, their respiration and circulation are faster, their body temperature is 8 to 10 degrees higher (about 107°F), they are more active, they are more sensitive to environmental influences, growth takes place at a more rapid rate, and birds mature at an earlier age. Also, egg production is an all-or-none phenomenon—that is, birds must have enough nutrients to produce an egg, otherwise no egg is produced.

The economic importance of poultry feeding becomes apparent when it is realized that 65 to 75% of the total production cost of poultry is from feed, with the production of eggs toward the lower side of this range and the production of broilers and turkeys toward the upper side. For this reason, the efficient use of feed is extremely important to the poultry producer.

The major objective of poultry feeding is the conversion of feedstuffs into human food. In this respect, the domestic fowl is quite efficient (see Chapter 1, Table 1–7).

## DIGESTIVE SYSTEM OF POULTRY

An understanding of the principal parts and functions of the digestive system of poultry is requisite to intelligent feeding. Fig 44–1 shows the organs and structure of the digestive tract as they would appear after being removed from the chicken and arranged in a functional sequence.

The rate of digestion in the fowl is rapid. If the alimentary tract is empty, feed will pass through it in about 3½ hours. When feeding is continuous (as in self-feeding), feed will pass from the mouth to the cloaca in about 12 hours. The transfer is quicker during daylight hours than at night.

The end products of digestion and metabolism are excreted in the feces, the urine, and as carbon dioxide and water eliminated by way of respiration. The mixture of feces and urine voided by birds is known as manure.

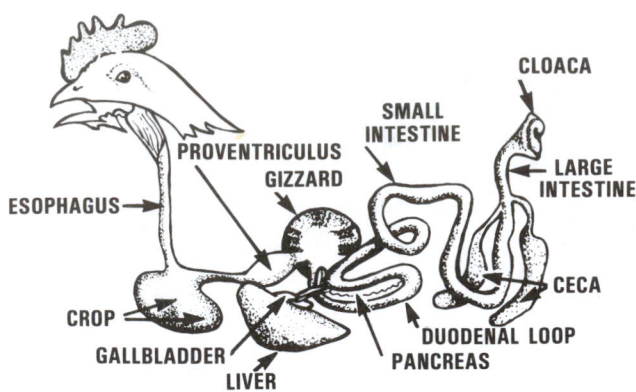

Fig. 44–1. Schematic diagram of the digestive tract of a chicken.

On the average, a laying hen excretes about .37 of a pound of water each day through the intestinal waste and kidneys, and about .09 of a pound is lost through the lungs. A hen will produce about 90 lb (wet weight) of manure each year.

## NUTRITIVE NEEDS OF POULTRY

Nutrients, the chemical substances found in feed materials, are needed by the bird in definite amounts, with the quantities varying according to the kind of bird and the purpose for which it is being fed. A deficiency in a nutrient can be, and often is, a limiting factor in egg production or growth.

The nutrient composition of chickens and eggs shown in Table 44–1 gives some idea of the relative importance of these nutrients as body constituents.

**TABLE 44–1**
**NUTRIENT COMPOSITION OF CHICKENS AND EGGS[1]**

| Nutrient | Chicken | Egg |
|---|---|---|
| | (%) | (%) |
| **W**ater | 55–78[2] | 66 |
| **P**rotein | 21 | 13[3] |
| **F**at | 17 | 10[4] |
| **M**inerals | 3.5 | 11[5] |

[1]From *Poultry Nutrition and Feeding*, Pub. MM–214, The Ohio State University, p. 5, Table 1.
[2]Depends on age of bird; younger birds have a higher water content.
[3]Chiefly in egg white.
[4]Chiefly in the yolk.
[5]Nearly all is calcium in shell.

## ENERGY

Carbohydrates, which constitute about 75% of the dry weight of plants and grain, make up a large part of poultry rations. They serve as a source of heat and energy in the bird's body. A surplus taken into the body may be transformed into fat and stored as a reserve supply of heat and energy.

In poultry feeds, the term *nitrogen-free extract* (NFE) is often used to refer to the soluble and digestible portion of carbohydrates, whereas the term *fiber* is used to refer to the insoluble and undigestible carbohydrates that are the structural components of plants.

Fats constitute about 17% of the dry weight of the market broiler and about 40% of the dry weight of a whole egg.

Food fat affects body fat. Thus, poultry consuming soft fat, such as most vegetable oils, may accumulate fat that is somewhat soft or oily.

Because the primary function of both carbohydrates and fats is to serve as a source of energy for the body, an insufficient supply of these nutrients results in reduced growth rate or egg production in poultry.

Because animals eat primarily to satisfy their energy needs, the energy concentration of the diet determines the amount of feed consumed. Therefore, the proper ratio of amino acids and other nutrients to energy must be maintained to insure an adequate intake of these nutrients for growth and egg production.

## PROTEIN

Typical broiler starter rations contain from 21 to 24% protein, and typical laying rations from about 16 to 17% protein. Grain and millfeeds supply approximately one-half of the protein needs for most poultry rations. Additional protein is supplied from high protein concentrates of either animal or vegetable origin.

From the standpoint of poultry nutrition, the amino acids that make up proteins are really the essential nutrients, rather than the protein molecule itself. Hence, protein content as a measure of the nutritional value of a feed is becoming less important, and each amino acid is being considered individually. The essential amino acid requirements of chickens and turkeys are given in Table 44–4, 44–10, and 44–13.

In practice, the amino acid requirements of growing chickens and turkeys, and of laying hens, are met by proteins from plant and animal sources. Protein supplements that most nearly supply the essential amino acids of the bird are known as *high-quality* supplements. Usually it is necessary to choose more than one source of dietary protein, then combine them in such a way that the amino acid composition of the mixture meets the requirements of the bird.

In poultry nutrition, special attention needs to be given to supplying the amino acids lysine, methionine and cystine, and tryptophan. These are sometimes referred to as the *critical amino acids* in poultry nutrition.

When formulating poultry rations, they must be so designed as to supply all the essential amino acids in ample amounts. Additionally, there must be sufficient total nitrogen for the chickens to synthesize the other amino acids needed. An amino acid deficiency always results in slow growth or poor egg production. Also, feathering is often poor, and usually the fat content of the carcass of protein-deficient birds is higher than that of adequately nourished chicks.

It has been determined that the chick requires dietary sources of protein to furnish 13 different amino acids. These amino acids are referred to as *essential,* since the chicken cannot produce them in sufficient amounts for maximum growth or egg production, and because a dietary deficiency of any one of them interferes with body protein formation and affects growth or egg production. The primary object of protein feeding, therefore, is to furnish the bird with protein which, upon digestion, will yield sufficient quantities of the 13 essential amino acids needed for top performance.

Any excess protein consumed by the bird can be burned in the body to yield energy in somewhat the same manner as carbohydrates and fats. In practical feeding of poultry, it is seldom wise to use excessive protein because carbohydrates and fats are generally more economical sources of energy.

## MINERALS

Fig. 44–2.   Perosis, or slipped tendon, due to manganese deficiency. (Courtesy, Virginia Poultry Federation, Harrisburg, VA)

The minerals which have been shown to be essential for chickens and turkeys are: sodium, chlorine, potassium, calcium, phosphorus, magnesium, copper, iodine, iron, manganese, molybdenum, selenium, sulfur, and zinc. Of these, sodium, chlorine, calcium, phosphorus, manganese, selenium, and zinc are considered to be of most importance since outside sources of them must be added to practical feed formulation for chickens and turkeys. Pertinent facts relative to poultry minerals are summarized in Table 44–2 (see next page).

The present consensus among poultry nutritionists is that the diet of laying hens should have a minimum allowance of 3 to 4% calcium, with the higher level used in high energy formulations (1,300 to 1,400 kilocalories metabolizable energy per pound).

For growing chickens and turkeys, a calcium to available phosphorus ratio of 2:1 is considered acceptable.

**TABLES 44-2**
**POULTRY MINERAL CHART**

| Mineral | Functions of Mineral | Some Deficiency Symptoms | Practical Sources of Mineral | Types of Poultry Rations Usually Requiring Supplementation | | | | Comments |
|---|---|---|---|---|---|---|---|---|
| | | | | Start-ing | Grow-ing | Lay-ing | Breed-ing | |
| **Major or Macrominerals** | | | | | | | | |
| **Sodium, chlor-ine, and potas-sium** | Maintain balance in tis-sue fluids; serve as buffers, preventing body from be-coming acid or alkaline. | Deficiency of sodium and chlorine causes poor growth, a nervous condition which usual-ly results in cannibalism and poor feed use. Deficiency of potassium reduces growth and increases nitrogen excretion. | **Sodium and chlorine:** Products of animal origin, and common salt. **Potassium:** Products of plant origin. | Yes | Yes | Yes | Yes | Potassium is not deficient in nor-mal rations, due to large amounts of plant products in poultry feeds. |
| **Calcium** | Formation of bone and eggshell. Clotting of blood. Muscular action. | Bones of growing birds be-come soft and rubbery, called rickets. Thin-shelled eggs, drop in egg production, and lowered hatch-ability. | Oystershell. Limestone. | Yes | Yes | Yes | Yes | Most poultry feeds are deficient in Ca and P, thus they must be added. |
| **Phosphorus** | Structural part of bone, proteins, and certain fats. | Rickets in growing birds. Reduced egg production. | Dicalcium phosphate. Defluorinated phosphate. Steamed bone meal. Monosodium phosphate. | Yes | Yes | Yes | Yes | Organic phosphorus (present in plants) is poorly utilized by grow-ing birds, but is satisfactory for adult birds. |
| **Magnesium** | Regulation of cellular fluids. | Muscular collapse. | | Seldom necessary to add. | | | | Most poultry rations contain ade-quate magnesium. Do not use dolemitic (high magnesium) limestones in poultry feeds, as high magnesium interferes with Ca utilization. |
| **Trace or Microminerals** | | | | | | | | |
| **Iodine** | | Goiter. | Calcium iodate. Iodized salt. | | | | | Iodine, iron, and zinc are some-times partially deficient in poultry rations unless added. An extreme iodine deficiency may affect hatchability. |
| **Iron** | | Anemia. | Ferrous sulfate. | | | | | |
| **Zinc** | | Retarded growth and poor feather development. | Zinc oxide, carbonate, or sulfate. | Yes | Yes | Yes | Yes | |
| **Manganese** | Necessary for normal bone and tendon develop-ment. | Slipped tendon or perosis. Poor egg production, shell quality, and hatchability. | Manganese sulfate or oxide. | Yes | Yes | Yes | Yes | |
| **Selenium** | Prevents uncontrolled oxidation. | Edema, anemia, and muscular dystrophy. | Sodium selenite. | Yes | Yes | Yes | Yes | Corn and soybeans grown in the U.S. are generally deficient in selenium. |
| **Copper, sulfur, cobalt, molyb-denum, fluor-ine** | | Anemia. | | Generally adequate in common poultry feeds. | | | | These trace minerals are essen-tial, but do not have to be added to most poultry rations. |

## VITAMINS

The vitamins required by poultry, along with their deficien-cy symptoms and dietary sources, are shown in Table 44-3.

The next to last column of Table 44-3 indicates the types of poultry rations in which special attention must be paid to the inclusion of special dietary sources of the vitamins. As shown, vitamins A, D, B-12 and riboflavin are commonly low in most poultry rations. It is also to be emphasized that vitamin

$D_3$, the animal form (made by the irradiation of 7-dehydro-cholesterol) is more active for poultry, and should, therefore, be used instead of vitamin $D_2$, the plant form of the vitamin.

The fat-soluble vitamins (A, D, E, and K) can be stored and accumulated in the liver and other parts of the body, while only very limited amounts of the water-soluble vitamins (biotin, choline, folacin, niacin, pantothenic acid, riboflavin, thiamin, vitamin B-6, and vitamin B-12) are stored. For this reason, it is important that the water-soluble vitamins be fed regularly in the ration in adequate amounts.

**TABLE 44–3**
**POULTRY VITAMIN CHART**

| Vitamin | Some Deficiency Symptoms | Practical Sources of Vitamin | Type of Poultry Ration Usually Requiring Supplementation | | | | Comments |
|---|---|---|---|---|---|---|---|
| | | | Start-ing | Grow-ing | Lay-ing | Breed-ing | |
| **Fat-Soluble Vitamins** | | | | | | | |
| **Vitamin A** | **Chicks:** Wobbly gait, uric acid deposits in ureters and kidneys, and general unthriftiness. **Hens:** Reduced egg production and poor hatchability. | Green forage, alfalfa meal, corn gluten meal, yellow corn, fish oils, synthetic vitamin A. | Yes | Yes | Yes | Yes | |
| **Vitamin D** | **Chicks:** Leg deformities, soft bones (rickets), reduced growth. **Hens:** Poor eggshell formation, reduced egg production, and hatchability. | Irradiated animal sterols, fish liver oils, vitamin A and D feeding oils, synthetic Vitamin D₃. | Yes | Yes | Yes | Yes | Vitamin $D_3$ is more than 30 times as efficient for preventing rickets in chickens as vitamin $D_2$. Requirements for poultry are expressed in International Chick Units. |
| **Vitamin E** | **Chicks:** Encephalomalacia or "crazy chick disease," edema, or muscular dystrophy. **Hens:** Poor hatchability. | Alfalfa meal, vegetable oils, wheat germ, and pure vitamin concentrates. | Yes | No | No | Yes | |
| **Vitamin K** | **Chicks:** Hemorrhages due to failure of blood to clot. **Hens:** Same as for chicks except that condition is rarely seen. | Green pasture, alfalfa meal, synthetic vitamin K (menadione sodium bisulfite). | Yes | No | No | No | |
| **Water-Soluble Vitamins** | | | | | | | |
| **Biotin** | **Chicks:** Cracking and degeneration of skin on feet, around beak, slipped tendon, and fatty liver and kidney syndrome. **Hens:** Reduced hatchability. | Soybean meal, alfalfa meal, dried yeast, milk products, green pasture, synthetic biotin. | Yes | No | No | Yes | The biotin in wheat and barley is unavailable. |
| **Choline** | **Chicks:** Retarded growth and slipped tendon. **Hens:** No deficiency known. | Fish products and choline chloride. | Yes | Yes | No | No | |
| **Folacin** (Folic acid) | **Chicks:** Poor growth, poor feathering, perosis, and anemia. **Hens:** Reduced hatchability. | Alfalfa meal, wheat, soybean meal, liver preparations, and synthetic folic acid. | Yes | No | No | Yes | |
| **Niacin** (Nicotinic, acid, nicotinamide) | **Chicks:** Enlargement of hock joints and perosis, retarded growth, and inflammation of mouth and tongue. **Hens:** No symptoms observed in hen except on protein-deficient diet. | Chemically synthesized nicotinic acid, liver, yeast, wheat bran and middlings, fermentation products, and most grasses. | Yes | No | No | No | |
| **Panthothenic acid** | **Chicks:** Poor growth, ragged feather development, and degeneration of skin around beak, eyes, and vent. **Hens:** Reduced hatchability. | Pure calcium pantothenate, alfalfa meal, dried milk products, and fermentation residues. | Yes | No | No | Yes | |
| **Riboflavin** (B–2) | **Chicks:** Curled toe paralysis and reduced growth. **Hens:** Poor hatchability with many embryos dying during second week of incubation. | Alfalfa meal, milk products, distillers' solubles, fermentation products, and pure vitamin. | Yes | Yes | Yes | Yes | |
| **Thiamin** (B–1) | **Chicks:** Loss of appetite, head retractions, loss in body weight. **Hens:** Same as for chicks, and egg-laying stops. | Grains and grain products, oilseed meals, milk products, and pure vitamin. | No | No | No | No | |
| **Vitamin B–6** (Pyridoxine, Pyridoxal, Pyridoxamine) | **Chicks:** Poor growth, lack of coordination, and convulsions. **Hens:** Reduced body weight, egg production, and hatchability. | Milk products, meat and fish by-products, soybean meal. | No | No | No | No | A deficiency in ordinary rations is not likely. |
| **Vitamin B–12** (Cobalamins) | **Chicks:** Reduced growth. **Hens:** Poor hatchability. | Fish meal, fish solubles, meat scrap, liver preparations, fermentation products, and synthetic vitamin B–12. | Yes | Yes | No | Yes | |

In addition to the vitamins listed in Table 44–3, certain unidentified or unknown factors are important in poultry nutrition. They are referred to as *unidentified* or *unknown* because they have not yet been isolated or synthesized in the laboratory. Nevertheless, rich sources of these factors and their effects have been well established. A diet that supplies the specific levels of all the known nutrients, but which does not supply the unidentified factors is inadequate for best performance. There is evidence that the growth factors exist in dried whey, marine and packinghouse by-products, distillers' solubles, antibiotic fermentation residues, alfalfa meal, and certain green forages. There is also evidence that at least one unknown hatchability factor is in fish solubles and green forage. Most of the unidentified factor sources are added to the diet at a level of 1 to 3%, although, antibiotic fermentation residues may be used at levels ranging from 7 to 10 lb per ton.

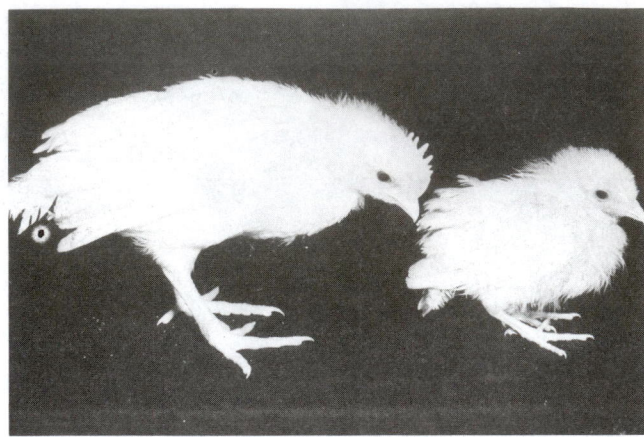

Fig. 44–5. Effect of niacin deficiency on chick growth. (Courtesy, Department of Poultry Science, University of Wisconsin, Madison)

Fig. 44–3. A chick deficient in vitamin D, showing ungainly manner of balancing body. The beak is also soft and rubbery. (Courtesy, Department of Poultry Science, Cornell University, Ithaca, NY)

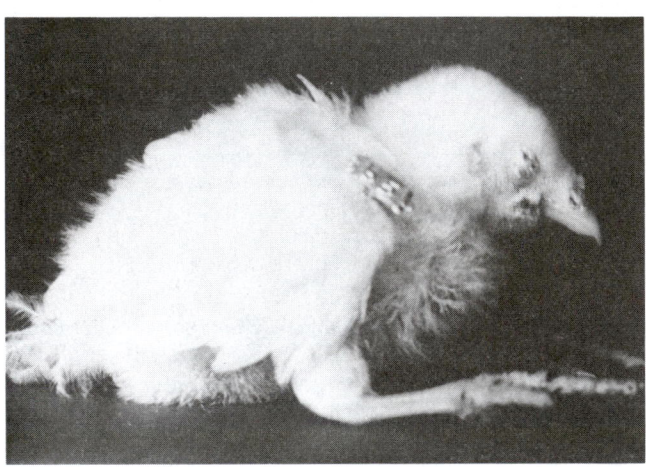

Fig. 44–6. An advanced stage of pantothenic acid deficiency. Note the lesions at the corners of the mouth and on the eyelids and feet. (Courtesy, Department of Poultry Science, Cornell University, Ithaca, NY)

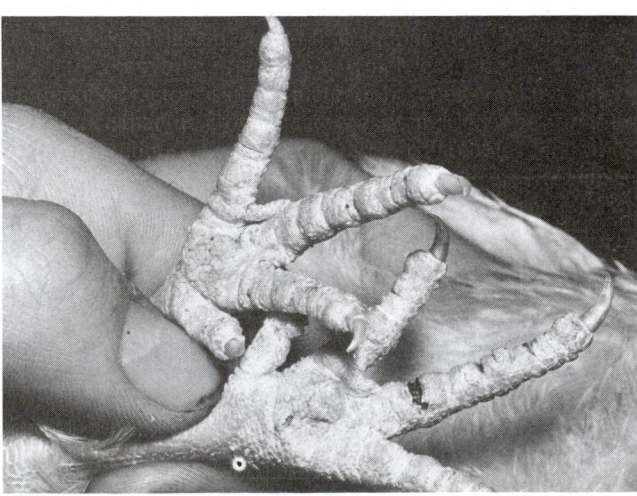

Fig. 44–4. Biotin deficiency. Note the severe lesions on the bottom of the feet. (Courtesy, H. R. Bird, Department of Poultry Science, University of Wisconsin, Madison)

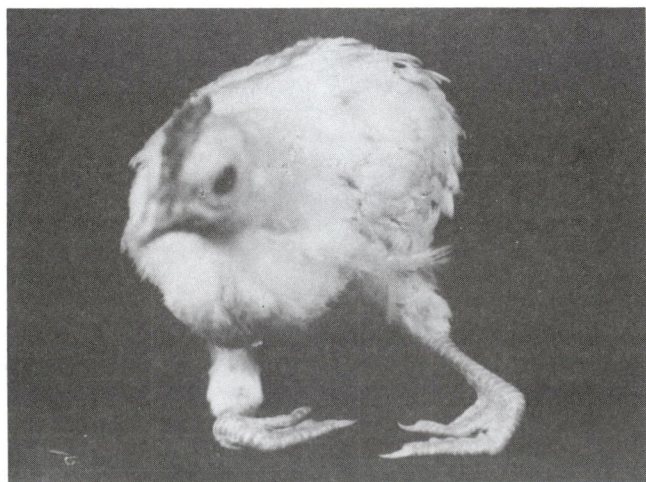

Fig. 44–7. Riboflavin deficiency in a young chick. Note the curled toes and tendency to squat on hocks. (Courtesy, Department of Poultry Science, Cornell University, Ithaca, NY)

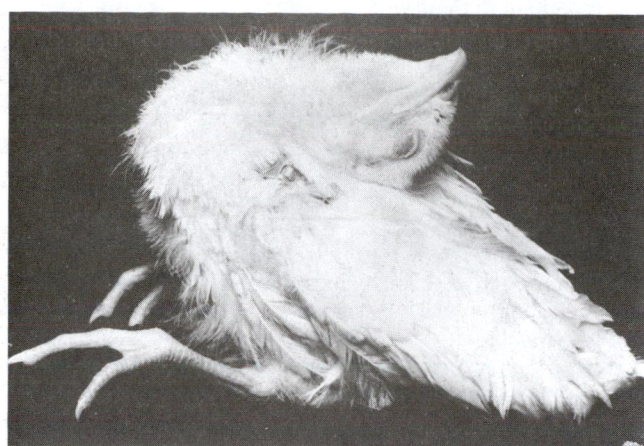

Fig. 44–8. Chick in the acute state of polyneuritis, due to a thiamin deficiency. Note characteristic head retraction. (Courtesy, Department of Poultry Science, University of Wisconsin)

## WATER

Poultry should have free access to clean, fresh water at all times. A mature chicken will consume about ½ lb of water daily under average conditions. The ratio of water to feed consumption of layers averages about 2.2, with a range from just under 2.0 to more than 3.0 to 1. Of course, water consumption varies according to the kind of feed, temperature, humidity, and activity of the chicken. During hot weather, chickens will consume 2 to 3 times as much water as they do under conditions of average temperature.

## EVALUATING POULTRY FEEDS

It is not expected that poultry producers conduct experiments to determine the nutrient requirements of poultry or that they evaluate the different feeds that they use; that is, unless they are very large operators. It is important, however, that they have a working knowledge of the value of different poultry feeds from the standpoint of purchasing and utilizing them.

The nutritive requirements for a specific substance are determined by finding the minimum amount of that particular nutrient or substance that will permit maximum development of the physiological function or economic characteristics of concern. In general, the economic characteristics of importance in poultry are growth, efficiency of feed utilization, egg production, and hatchability. For example, if the need of a certain nutrient for growth is being determined, groups of birds must be fed on an experimental ration containing different levels of the nutrient in question until it is known that increasing the quantity of the test nutrient beyond a particular level will not result in further increases in growth. If the test ration is complete in all other respects, then the nutrient requirement will be equal to the minimum supplemental level found to give maximum growth.

Some feeds are more valuable than others; hence, measures of their relative usefulness are important. Among such methods of evaluating the usefulness of poultry feeds are the following:

1. **Chemical analyses of feed.** While the biological response of animals (feeding trials) is the ultimate indicator of nutritive adequacy in a ration, tests of this type are difficult to perform, require extended periods of time, and are usually expensive. Thus, certain chemical analyses have been developed which are rough indicators of the value of a feedstuff or ration with regard to specific nutrient substances. The usual chemical analysis of feeds includes crude protein, ether extract or crude fat, crude fiber, ash or mineral, and moisture. It is recognized, however, that such proximate analysis of poultry feeds leaves much to be desired because in many cases the protein and nitrogen-free extract indicated may not be available to poultry.

In addition to the so-called proximate analysis, specific chemical and microbiological determinations can be made from many of the vitamins and individual mineral elements.

2. **Biological tests.** Most chemical and microbiological tests for nutrient substances give information about the total amount of nutrient present in a particular feedstuff or ration. However, these tests do not tell anything about the digestibility and utilization of the feedstuff or ration in the digestive tract of the animal. Hence, biological tests directly involving the bird are required to establish the true usefulness of feed supplying the nutrient needs of the bird. These biological tests are particularly important in evaluating protein and energy-yielding nutrients like carbohydrates and fats.

a. **Biological measure of protein utilization.** The amount of protein or nitrogen digested by the bird can be determined by a balance experiment in which a measured intake of protein is compared to the measured undigested protein in the feces of the bird. The biological value of a protein source is defined as the amount of protein retained in the body expressed as a percentage of the digestible protein available. Thus, this expression is a reflection of the kinds and amounts of amino acids available to the bird after digestion. If the amino acids available to the bird closely match those needed for body protein formation, the biological value of the protein is high. If, on the other hand, there are excesses of certain amino acids and deficiencies of other amino acids as a result of digestion, the biological value of the protein is low because of the increased number of amino acids which must be excreted via the kidney.

b. **Biological measure of energy utilization.** The total energy content of a feed can be measured by completely burning the feed in an apparatus known as a bomb calorimeter (see Chapter 4, Fig. 4–30). Birds, like other animals, are not able to extract all of the energy present in feeds. Hence, the term *digestible energy* is used to describe the total energy of the feed minus that which remains undigested. Metabolizable energy is the total energy in the feed minus both fecal and urinary energy; it represents all the available energy for any use in the animal. The net energy value of a feed is the metabolizable energy content minus the energy employed in utilizing it; thus, net energy may be used for body storage or the production of heat and muscular activity. Metabolizable energy values are used to describe the energy content of poultry feedstuffs and rations. Metabolizable energy values are relatively easy to measure in poultry where the feces and urine are voided together and are little affected by various physiological conditions. The energy terms used in this section, and their relationship to one another, are given in Chapter 4.

3. **Cost factor.** From the standpoint of a poultry producer the most important measurement of a feed's usefulness is in terms of net returns. Cost per pound or per ton of feed, and pounds of feed required to produce a pound of broiler or a dozen eggs, are important only as they reflect or affect the cost per unit of poultry products produced. For example, if the cost of a broiler ration is 9¢ a pound and 1.9 lb of the ration are required to produce 1 lb body weight, then the feed cost per pound of body weight can be arrived at by multiplying the above figures (9 × 1.9), which gives a feed cost of 17.1¢ per pound. Obviously when rations are compared, the ration that produces a unit of poultry product at the lowest total feed cost is the most desirable from an economic point of view.

## FEEDS USED IN POULTRY RATIONS

A wide variety of feedstuffs can be, and are, used in poultry rations. Broadly speaking, these may be classed as energy feedstuffs, protein supplements, mineral supplements, and vitamin supplements.

1. **Energy feedstuffs.** The major energy sources of poultry feeds are the cereal grains and their by-products and fats. Corn is the most important grain used by poultry, supplying about one-third of the total feed which they consume. Wheat is used when the price is right; and the sorghum grains (milo and kafir) are important in the southern states.

Fig. 44–9. Corn, the most important grain used in poultry. Ear corn like this is shelled, ground, and incorporated in a mash, crumbles, or pellets. (Courtesy, USDA)

Animal and vegetable fats are now used extensively in poultry feed. In addition to their high energy value, fats reduce the dustiness of feed mixtures, increase their palatability, and improve the texture and appearance of the feed. However, the use of fats in poultry feeds requires good mixing equipment. Also, it is necessary that the fat be properly stabilized in order to prevent rancidity.

Many other energy feedstuffs, including a great array of milling by-products, are used in poultry feeds.

2. **Protein and amino acid supplements.** The usefulness of a protein feedstuff depends upon its ability to furnish the essential amino acids required by the bird, the digestibility of the protein, and the presence or absence of toxic substances. As a general rule, several different sources of protein produce better results than single protein sources. Both animal and vegetable protein supplements are used for poultry. Most of the protein supplements of animal origin contribute minerals and vitamins which significantly affect their value in poultry rations, but they are generally more variable in composition than the vegetable protein supplements.

Among the animal protein supplements commonly used in poultry rations are meat by-products, milk by-products, marine products, and such miscellaneous animal by-products as blood meal, hydrolyzed poultry feathers, and poultry by-product meal.

Fig. 44–10. Menhaden herring. After oil extraction, the dried residue—*fish meal*—provides a high quality protein for poultry. (Courtesy, National Fisheries Institute, Inc., Washington, DC)

The common vegetable protein supplements used in poultry feeding include the oilseed meals (soybean meal, canola meal, cottonseed meal, peanut meal, and occasionally a limited amount of linseed meal), corn gluten meal, and alfalfa meal and other legume meals.

Use of synthetic amino acid supplements is now common in poultry diets and greatly facilitates the balancing of diets for meeting the amino acid requirements. Chemically made DL-methionine and microbially fermented L-lysine are used extensively. Other synthetic amino acids such as thiamin and tryptophan are available as feed supplements. Methionine hydroxy analog is also a common supplement and birds are able to convert it to methionine.

3. **Mineral supplements.** Mineral supplements are required by poultry for skeletal development in growing birds, for eggshell formation in laying hens, and for certain other regulatory processes in the body.

The common calcium supplements used in poultry feeding are ground limestone, crushed oyster shells or oystershell flour, bone meal, calcite, chalk, and marble.

Most of the phosphorus in plant products is in organic form and not well utilized by young chicks or turkey poults. Hence, for poultry, emphasis is placed upon inorganic phosphorus sources in feed formulation. Bone meal, dicalcium phosphate, defluorinated phosphate, and raw rock phosphate are used where both calcium and phosphorus are needed in the ration.

Salt is added to most poultry rations at 0.2 to 0.5% level, depending primarily upon the amount of sodium required. Too much salt will result in increased water consumption and wet droppings.

4. **Vitamin supplements.** Over the past three decades, a great many vitamins have been discovered that are important in present-day poultry feed formulation. Formerly, a wide variety of crude feedstuffs were added to poultry formulas primarily for their vitamin content. Today, many of these have been replaced by special vitamin supplements, which in many cases are chemically pure sources that need to be used only in very minute amounts. In modern poultry feed formulation and production, premixes often represent the commonsense approach to providing both vitamin and mineral needs for poultry.

## NONNUTRITIVE ADDITIVES

Modern poultry feeds commonly contain one or more non-nutritive additives. These additives are used for a variety of reasons. They are not nutrients, but some of them improve production under certain circumstances, others prevent rancidity in the feed, etc. There is no evidence of a nutritional deficiency when they are omitted from a ration. Among such additives are the following:

1. **Antibiotics.** The primary reason for using antibiotics in poultry feeds is for their growth-stimulating effect, for which purpose they are generally used in both broiler and market turkey rations. The reasons for this still remain obscure, but the best explanation for the growth-stimulating activity of antibiotics is the so-called disease level theory. This theory is based on the fact that antibiotics have failed to show any measurable effect on animals maintained under germ-free conditions.

Antibiotics are generally fed to poultry at levels of 5 to 50 g per ton of feed, depending upon the particular antibiotic used. Higher levels of antibiotics (100 to 400 g per ton of feed) are used for disease control purposes.

The antibiotics most commonly used in poultry rations are bacitracin, virginiamycin, bambermycins, and lincomycin.

2. **Arsenicals and nitrofurans.** These products exert many of the same effects as the antibiotics; hence, they are often added to poultry feeds to improve performance. It would appear that the action of arsenicals and that of antibiotics are very similar, since the effects of the two are not considered to be additive. For broilers, arsanilic acid (or sodium arsanilate) is used at 45 to 90 g per ton, and "3-nitro" is used at 22.5 to 45.0 g per ton of ration; but, in keeping with the recommendation made relative to the use of any drug, the manufacturer's directions should be followed.

3. **Drugs.** Poultry rations frequently contain drugs designed to prevent specific diseases. For example, a wide variety of chemical substances, sold under many trade names, are available for use in the prevention of coccidiosis. These drugs are known as coccidiostats.

4. **Antioxidants.** Antioxidants are used to prevent rancidity in poultry feeds. The antioxidants which are presently accepted for addition to fat in poultry feeds are butylated hydroxyanisole (BHA), butylated hydroxytoluene (BHT), and ethoxyquin. These antioxidants may be used to prolong the induction period in fats and to prevent oxidation in mixed feeds. They are used at a level of 0.25 lb per ton. Antioxidants are chemical compounds that are capable of temporarily inhibiting the destructive effects of oxygen on sensitive feeding ingredients—the unsaturated fats, fat-soluble vitamins, and other constituents. These chemicals are normally incorporated in the vitamin-trace mineral premix to prevent vitamins A and E from oxidative destruction. Some are added to feed fats to stabilize them

against rancidity. BHT and BHA are commonly used to stabilize fat. Ethoxyquin is also a common antioxidant in poultry feed.

5. **Grit.** Grit is a controversial subject. Some research indicates that hens on an all-mash ration do not need grit, but there is growing evidence that as a component of or supplement to all mash it will improve feed utilization and increase production under some conditions. The purpose of grit is primarily to help the gizzard grind food materials that pass through it. It is definitely needed when birds consume whole grains or coarse, fibrous feedstuffs. Crushed granite or other hard, insoluble material can be used for grit.

(Also, see Chapter 4 of this book, section on "Additives, Implants, and Injections," for additional additives.)

## POULTRY RATION POINTERS

In 1988, 103.1 million tons of commercially prepared primary (complete) livestock and poultry feeds were produced in the United States, of which 44.5%, or 45.9 million tons, was fed to poultry.

The poultry producer has the following alternatives for purchasing and preparing feeds:

1. Purchase of a commercially prepared complete feed.
2. Purchase of a commercially prepared protein supplement, reinforced with vitamins and minerals, which may be blended with local or homegrown grain.
3. Purchase of a commercially prepared vitamin-mineral premix which may be mixed with an oil meal, and then blended with local or homegrown grain.
4. Purchase of individual ingredients (including vitamins and minerals) and mixing the feed from the ground up.

Today, very few large poultry producers purchase a commercially prepared mixed feed. Instead, most of them choose either option 3 or 4 from the preceding list, with option Number 3 being most common.

## FACTORS INVOLVED IN FORMULATING POULTRY RATIONS

Before anyone can intelligently formulate a poultry ration, it is necessary to know (1) the nutrient requirements of the particular birds to be fed; (2) the availability, nutrient content, and cost of feedstuffs; (3) the acceptability and physical condition of feedstuffs; and (4) the presence of substances harmful to product quality.

### NUTRIENT REQUIREMENTS

Through experimentation, the requirements of most of the nutrients needed by poultry have been established. These requirements differ according to the type and age of poultry being fed, as well as the purpose for which they are kept.

### AVAILABILITY, NUTRIENT CONTENT, AND COST OF FEEDSTUFFS

Generally speaking, it is most practical to use feedstuffs which are readily available at the lowest cost—provided they are satisfactory in other respects. Substitutes can be found and used when certain ingredients become scarce or unavailable.

The nutrient content of different feed ingredients can be obtained from feed composition tables.

The price of feedstuffs must also be taken into consideration. When the price of any one ingredient gets out of line, based on the nutrient that it contributes, substitutions can usually be made. For example, wheat can be used to replace corn, and vice versa. Likewise, soybean meal can be substituted for cottonseed meal or peanut meal when the prices of these protein supplements change.

## ACCEPTABILITY AND PHYSICAL CONDITION OF FEEDSTUFFS

Some feeds cannot be used in poultry feed formulations because they lack acceptability to the birds, as determined by palatability, color, and particle size. Other feed materials may be processed in such a way that their physical condition makes them undesirable for feeding.

Feeds containing milo and rye can be markedly improved by pelleting and spraying hot fat over the exterior of the pellet.

Most feeds for broilers and growing turkeys are fed in pelleted form because it results in improved growth and feed efficiency.

## PRESENCE OF SUBSTANCES HARMFUL TO PRODUCT QUALITY

The composition of the feed can affect the product. The color of the skin or shanks of a broiler or of the yolk of an egg is primarily due to the carotenoid pigments consumed in the feed. Corn, alfalfa meal, and corn gluten meal are the main feeds used to contribute these pigments.

Alfalfa meal is seldom used in broiler diets because of its high fiber content. Extracts of marigold petals are now used as a feed supplement to provide carotenoid pigments.

Screw process cottonseed meal, which is high in gossypol, when fed to laying hens may cause egg yolk discoloration in stored eggs. Some fish products may impart fishy flavors to poultry meat or eggs. Thus, certain feedstuffs may be undesirable simply because of the effect they produce on the end product.

## FORMULATING POULTRY RATIONS

The increasing complexity of poultry rations, along with larger and larger enterprises, makes it imperative that producers who choose to mix feeds be absolutely sure that they will have nutritionally balanced and adequate rations.

The larger commercial feed companies, and the larger poultry producers who do their own mixing or formulating, generally rely on the services of a nutritionist and the use of a computer in formulating their rations. Even though they are more time-consuming, and fewer factors can be considered simultaneously, a good job can be done in formulating rations by the hand method. Instructions on balancing rations are given in Chapter 4 of this book, under the heading, "How to Balance a Ration"; hence, the reader is referred thereto.

## FEED PREPARATION

The preparation of feeds for poultry is fully covered in Chapter 4, section entitled, "Feed Processing," including Table 4–10 therein; hence, the reader is referred thereto.

## FEEDING SYSTEMS

The vast majority of large poultry enterprises feed complete rations, in which all of the nutrients needed by the bird are provided in the quantities necessary. Little or no grain is fed along with this ration because such feeding would destroy the balance of nutrients provided by the complete ration. Laying or breeding hens are sometimes provided with a supplementary source of calcium, to which they are given free access.

Sometimes poultry producers, especially farm flock owners, feed whole grain and a fortified protein supplement in separate hoppers, cafeteria style. The supplement provides the extra amounts of protein, vitamins, and minerals lacking in the grain. By estimating the approximate consumption of each the grain and the supplement, and fortifying the supplement accordingly, an overall nutrient intake that is reasonably well balanced is achieved. This system enables the producer to make maximum use of farm-grown or local grains. However, it generally results in the birds eating too much grain and not enough supplement.

## SPECIAL FEEDING PROGRAMS; SUGGESTED RATIONS

The nutritive requirements of poultry vary according to species (between chickens, turkeys, ducks, and geese), according to age, and according to the type of production—whether the birds are kept for layers, breeders, or for meat production. For this reason, many different rations are required.

To be successful, rations must meet the nutritive requirements of the birds to which they are fed. Also, in using nutrient requirement tables to formulate practical rations, it must be remembered that they are minimum requirements, which means that they do not provide for any margins of safety. Further, the protein-energy relationships shown therein should be retained.

Also, rations must be reformulated from time to time, in keeping with new developments and changing prices. The larger commercial feed companies and poultry producers rely on their nutritionist to make such changes as necessary. Perhaps the suggested rations that follow will serve as useful guides, especially for the smaller operator who does not have a nutritionist on the staff.

Special feeding programs and suggested rations for each poultry species and type of production are presented in the following sections:

1. Feeding Layers and Breeders.
2. Feeding Broilers.
3. Feeding Turkeys.
4. Feeding Ducks.
5. Feeding Geese.
6. Feeding Bobwhite Quail.

## FEEDING LAYERS AND BREEDERS

Fig. 44-11.   Replacement Leghorn pullets. (Courtesy, Don Bell, Riverside County Poultry Farm Advisor, Riverside, CA)

Fig. 44-12.   Layers, cage system, four decks, with automated feeding, watering, and egg gathering. (Courtesy, DeKalb Poultry Research, Inc., DeKalb, IL)

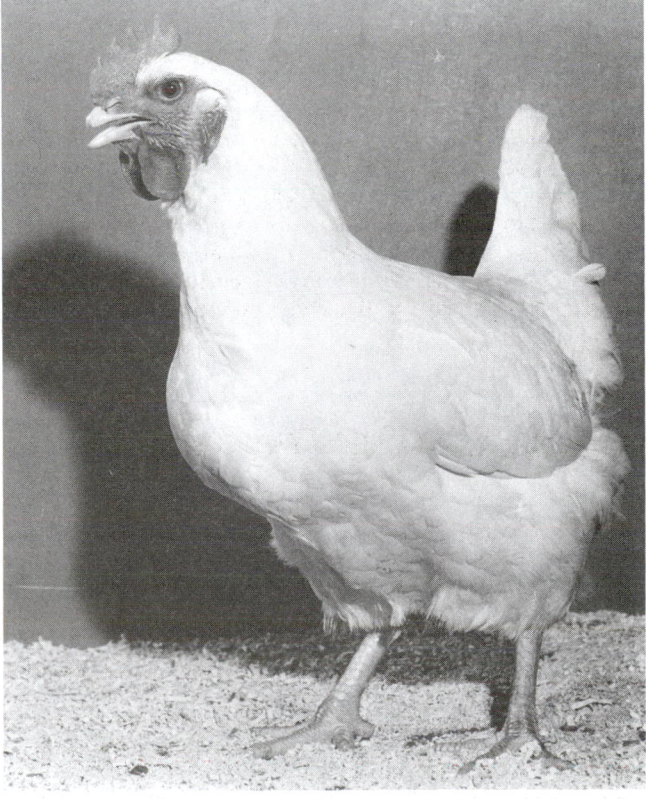

Fig. 44-13.   Meat-type breeder. (Courtesy, Hubbard Farms, Walpole, NH)

The nutrient requirements of, and suggested rations for, chicken layers and breeders are presented in the following tables:

Table 44–4, Nutrient Requirements of Leghorn-type Chickens For Growing, Laying, and Breeding.

Table 44–5, Replacement and Breeder Rations.

Footnotes To Tables 44–5, –6, and –11.

Table 44–6, Layer Rations.

Table 44–7, Metabolizable Energy Required Daily By Layers In Relation To Body Weight and Egg Production.

Table 44–8, Body Weights and Feed Requirements of Leghorn-type Pullets and Hens.

Table 44–9, Body Weights and Feed Allowances of Meat-type Replacement Chickens.

### TABLE 44–4
### NUTRIENT REQUIREMENTS OF LEGHORN-TYPE CHICKENS FOR GROWING, LAYING, BREEDING[1]

| Energy Base: | | Growing | | | Laying | | Breeding |
|---|---|---|---|---|---|---|---|
| | | 0–6 Weeks | 6–14 Weeks | 14–20 Weeks | | Daily Intake | |
| kcal ME/lb Diet[2] | | 1,315 | 1,315 | 1,315 | 1,315 | Per Hen (mg)[3] | 1,315 |
| kcal ME/kg Diet[2] | | 2,900 | 2,900 | 2,900 | 2,900 | | 2,900 |
| Protein | (%) | 18 | 15 | 12 | 14.5 | 16,000 | 14.5 |
| **Amino acids:** | | | | | | | |
| Arginine | (%) | 1.00 | 0.83 | 0.67 | 0.68 | 750 | 0.68 |
| Glycine and serine | (%) | 0.70 | 0.58 | 0.47 | 0.50 | 550 | 0.50 |
| Histidine | (%) | 0.26 | 0.22 | 0.17 | 0.16 | 180 | 0.16 |
| Isoleucine | (%) | 0.60 | 0.50 | 0.40 | 0.50 | 550 | 0.50 |
| Leucine | (%) | 1.00 | 0.83 | 0.67 | 0.73 | 800 | 0.73 |
| Lysine | (%) | 0.85 | 0.60 | 0.45 | 0.80 | 750 | 0.68 |
| Methionine + cystine | (%) | 0.60 | 0.50 | 0.40 | 0.60 | 660 | 0.60 |
| Methionine | (%) | 0.30 | 0.25 | 0.20 | 0.32 | 350 | 0.32 |
| Phenylalanine + tyrosine | (%) | 1.00 | 0.83 | 0.67 | 0.80 | 880 | 0.80 |
| Phenylalanine | (%) | 0.54 | 0.45 | 0.36 | 0.40 | 440 | 0.40 |
| Threonine | (%) | 0.68 | 0.57 | 0.37 | 0.45 | 500 | 0.45 |
| Tryptophan | (%) | 0.17 | 0.14 | 0.11 | 0.17 | 187 | 0.17 |
| Valine | (%) | 0.62 | 0.52 | 0.41 | 0.55 | 600 | 0.55 |
| **Linoleic acid** | (%) | 1.00 | 1.00 | 1.00 | 1.00 | 1,100 | 1.00 |
| **Major or macrominerals:** | | | | | | | |
| Calcium (Ca) | (%) | 0.80 | 0.70 | 0.60 | 3.40 | 3,750 | 3.40 |
| Chlorine (Cl) | (%) | 0.15 | 0.12 | 0.12 | 0.15 | 165 | 0.15 |
| Magnesium (Mg) | (mg) | 600 | 500 | 400 | 500 | 55 | 500 |
| Phosphorus (P, available | (%) | 0.40 | 0.35 | 0.30 | 0.32 | 350 | 0.32 |
| Potassium (K) | (%) | 0.40 | 0.30 | 0.25 | 0.15 | 165 | 0.15 |
| Sodium (Na) | (%) | 0.15 | 0.15 | 0.15 | 0.15 | 165 | 0.15 |
| **Trace or microminerals:** | | | | | | | |
| Copper (Cu) | (mg) | 8 | 6 | 6 | 6 | 0.88 | 8 |
| Iodine (I) | (mg) | 0.35 | 0.35 | 0.35 | 0.30 | 0.03 | 0.30 |
| Iron (Fe) | (mg) | 80 | 60 | 60 | 50 | 5.50 | 60 |
| Manganese (Mn) | (mg) | 60 | 30 | 30 | 30 | 3.30 | 60 |
| Selenium (Se) | (mg) | 0.15 | 0.10 | 0.10 | 0.10 | 0.01 | 0.10 |
| Zinc (Zn) | (mg) | 40 | 35 | 35 | 50 | 5.50 | 65 |
| **Fat-soluble vitamins:** | | | | | | | |
| Vitamin A | (IU) | 1,500 | 1,500 | 1,500 | 4,000 | 440 | 4,000 |
| Vitamin D | (ICU) | 200 | 200 | 200 | 500 | 55 | 500 |
| Vitamin E | (IU) | 10 | 5 | 5 | 5 | 0.55 | 10 |
| Vitamin K | (mg) | 0.50 | 0.50 | 0.50 | 0.50 | 0.055 | 0.50 |
| **Water-soluble vitamins:** | | | | | | | |
| Biotin | (mg) | 0.15 | 0.10 | 0.10 | 0.10 | 0.011 | 0.15 |
| Choline | (mg) | 1,300 | 900 | 500 | ? | ? | ? |
| Folacin (Folic Acid) | (mg) | 0.55 | 0.25 | 0.25 | 0.25 | 0.0275 | 0.35 |
| Niacin (Nicotinic Acid, Nicotinamide) | (mg) | 27.0 | 11.0 | 11.0 | 10.0 | 1.10 | 10.0 |
| Pantothenic Acid (Vitamin B–3) | (mg) | 10.0 | 10.0 | 10.0 | 2.20 | 0.242 | 10.0 |
| Riboflavin (Vitamin B–2) | (mg) | 3.60 | 1.80 | 1.80 | 2.20 | 0.242 | 3.80 |
| Thiamin (Vitamin B–1) | (mg) | 1.8 | 1.3 | 1.3 | 0.80 | 0.088 | 0.80 |
| Vitamin B–6 (Pyridoxine, Pyridoxal, Pyridoxamine) | (mg) | 3.0 | 3.0 | 3.0 | 3.0 | 0.33 | 4.50 |
| Vitamin B–12 (Cobalamins) | (mg) | 0.009 | 0.003 | 0.003 | 0.004 | 0.00044 | 0.004 |

[1]Adapted and updated by the author from *Nutrient Requirements of Poultry,* 8th rev. ed., NRC, National Academy Press, Washington, DC, 1984, p. 12, Table 4.

[2]These are typical dietary energy concentrations.

[3]Assumes an average daily intake of 110 g of feed/hen daily.

**TABLE 44-5**
**REPLACEMENT AND BREEDER RATIONS[1]**

| | Replacement Rations | | | | Breeder Rations | |
| | Starter | | Grower[2] | | | |
| Ingredient | 20% Protein | 18% Protein | 14% Protein | 12% Protein | Body Weight (3½-5 lb) | Body Weight (5-8 lb) |
|---|---|---|---|---|---|---|
| | (lb) | (lb) | (lb) | (lb) | (lb) | (lb) |
| Ground yellow corn (2, 3) .......... | 1,267 | 1,310 | 1,438 | 1,481 | 1,313 | 1,366 |
| Wheat middlings ................. | 130 | 200 | 254 | 323 | — | 100 |
| Alfalfa meal (17%) ............... | 25 | 25 | 25 | 25 | 50 | 50 |
| Soybean meal (dehulled) .......... | 422 | 309 | 217 | 104.8 | 310 | 208 |
| Fish meal, herring (65%) (4, 5) ..... | 50 | 50 | — | — | 60 | 75 |
| Meat and bone meal (47%) (5) ..... | 50 | 50 | — | — | 60 | 50 |
| Lysine ........................ | — | — | 1 | 1.2 | — | — |
| Dicalcium phosphate (6) ........... | 10 | 9 | 30 | 29 | 2 | — |
| Ground limestone (7) ............. | 19 | 20 | 28 | 29 | 157 | 144 |
| DL-Methionine, or equivalent ........ | — | — | — | — | 0.4 | — |
| Stabilized yellow grease, or equivalent ................... | 20 | 20 | (16) | (16) | 51 | (16) |
| Iodized salt (4) ................... | 7 | 7 | 7 | 7 | 7 | 7 |
| Antibiotic supplement ..... (footnote) | (8) | (8) | — | — | — | — |
| Antioxidant ............ (footnote) | (9) | (9) | (9) | (9) | (9) | (9) |
| Zinc (17) .................... (g) | — | — | — | — | 16 | 16 |
| Coccidiostat ............ (footnote) | (10) | (10) | (10) | (10) | — | — |
| Manganese (11) ............... (g) | 52 | 52 | 52 | 52 | 52 | 52 |
| Selenium .............. (footnote) | (25) | (25) | (25) | (25) | (25) | (25) |
| Vitamin supplements (12) | | | | | | |
| Vitamin A .............. (USPU) | 3,000,000 | 3,000,000 | 3,000,000 | 3,000,000 | 4,000,000 | 4,000,000 |
| Vitamin D₃ .............. (ICU)[3] | 1,000,000 | 1,000,000 | 1,000,000 | 1,000,000 | 2,000,000 | 2,000,000 |
| Vitamin E ................. (IU) | — | — | — | — | 2,000 | 2,000 |
| Vitamin K (20) ............ (mg) | — | — | — | — | — | — |
| Vitamin B-12 .............. (mg) | 6 | 6 | 6 | 6 | 6 | 6 |
| Riboflavin ................. (mg) | 1,500 | 1,500 | 1,500 | 1,500 | 3,000 | 3,000 |
| Niacin ................... (mg) | 10,000 | 10,000 | 10,000 | 10,000 | 10,000 | 10,000 |
| Calcium pantothenate ........ (mg) | 4,000 | 4,000 | 3,000 | 3,000 | 6,000 | 6,000 |
| Choline ................. (mg) | 213,000 | 298,000 | 125,000 | 209,000 | 168,000 | 197,000 |
| **Total** (lb) (21) ................... | 2,000 | 2,000 | 2,000 | 2,000 | 2,000.4 | 2,000 |
| **Calculated Analysis** (27) | | | | | | |
| Metabolizable energy ....... (kcal/lb) | 1,361 | 1,362 | 1,341 | 1,342 | 1,342 | 1,293 |
| Protein ...................... (%) | 20.03 | 18.01 | 14.01 | 12.01 | 17.01 | 16.01 |
| Lysine ..................... (%) | 1.04 | 0.89 | 0.63 | 0.49 | 0.87 | 0.79 |
| Methionine ................. (%) | 0.34 | 0.32 | 0.24 | 0.21 | 0.33 | 0.30 |
| Methionine + cystine ........ (%) | 0.64 | 0.59 | 0.46 | 0.41 | 0.59 | 0.55 |
| Fat ......................... (%) | 4.48 | 4.70 | 3.54 | 3.74 | 5.89 | 3.72 |
| Fiber ...................... (%) | 2.67 | 2.83 | 3.00 | 3.15 | 2.42 | 2.72 |
| Calcium .................... (%) | 0.90 | 0.90 | 0.90 | 0.90 | 3.24 | 3.01 |
| Total phosphorus ............. (%) | 0.66 | 0.66 | 0.66 | 0.65 | 0.51 | 0.53 |
| Available phosphorus ........... (%) | 0.41 | 0.41 | 0.40 | 0.40 | 0.45 | 0.46 |
| Vitamins (units or mg/lb) | | | | | | |
| Vitamin A activity ......... (USPU) | 4,188 | 4,237 | 4,381 | 4,430 | 5,990 | 6,050 |
| Vitamin D₃ added ........... (ICU) | 500 | 500 | 500 | 500 | 1,000 | 1,000 |
| Riboflavin ................. (mg) | 1.78 | 1.76 | 1.64 | 1.62 | 2.51 | 2.55 |
| Niacin .................... (mg) | 19.08 | 20.49 | 20.53 | 21.84 | 15.34 | 18.01 |
| Pantothenic acid ........... (mg) | 5.34 | 5.27 | 4.72 | 4.64 | 5.70 | 5.81 |
| Choline ................... (mg) | 600.20 | 600.11 | 420.20 | 419.76 | 500.02 | 500.39 |

[1]From *NECC Chicken and Turkey Rations 1980*, pp. 3 and 6, prepared cooperatively for distribution by the New England Land-Grant Universities. In using Table 44–5, also see the footnotes that follow under the heading "Footnotes to Tables 44–5, –6, and –11." The footnote numbers are those indicated in parentheses in this table. To convert pounds to kilograms, divide by 2.2.

[2]For those persons wanting a restricted feeding program, feed the grower at 80% or less of the amount normally consumed.

[3]ICU = International Chick Unit.

## TABLE 44–6
## LAYER RATIONS[1]

| Ingredient | Protein Level of Rations[2] | | | | |
|---|---|---|---|---|---|
| | 15% | 16% | 17% | 18% | 19% |
| Ground yellow corn (2, 3) ............... (lb) | 1,457 | 1,403 | 1,339 | 1,242 | 1,177 |
| Alfalfa meal (17%) ..................... (lb) | 25 | 25 | 25 | 25 | 25 |
| Soybean meal (dehulled) ............... (lb) | 292.2 | 340.6 | 393.6 | 451.6 | 504.6 |
| Meat and bone meal (47%) (5) ......... (lb) | 50 | 50 | 50 | 50 | 50 |
| DL-Methionine, or equivalent ......... (lb) | 0.8 | 0.4 | 0.4 | 0.4 | 0.4 |
| Dicalcium phosphate (6) ............... (lb) | 9 | 8 | 8 | 7 | 7 |
| Ground limestone (7) ................... (lb) | 159 | 159 | 159 | 174 | 174 |
| Iodized salt (4) ........................ (lb) | 7 | 7 | 7 | 7 | 7 |
| Stabilized yellow grease, or | | | | | |
| equivalent ......................... (lb) | — | 7 | 18 | 43 | 55 |
| Antioxidant ..................... (footnote) | (9) | (9) | (9) | (9) | (9) |
| Zinc (17) ............................... (g) | 16 | 16 | 16 | 16 | 16 |
| Manganese (11) ....................... (g) | 52 | 52 | 52 | 52 | 52 |
| Vitamin supplements (12) | | | | | |
| Vitamin A ...................... (USPU) | 6,000,000 | 6,000,000 | 6,000,000 | 6,000,000 | 6,000,000 |
| Vitamin D$_3$ ................... (ICU)[3] | 2,000,000 | 2,000,000 | 2,000,000 | 2,000,000 | 2,000,000 |
| Vitamin K (20) .................. (mg) | — | — | — | — | — |
| Vitamin B-12 .................... (mg) | 6 | 6 | 6 | 6 | 6 |
| Riboflavin ...................... (mg) | 2,000 | 2,000 | 2,000 | 2,000 | 2,000 |
| Niacin .......................... (mg) | 12,000 | 12,000 | 12,000 | 12,000 | 12,000 |
| Calcium pantothenate ........... (mg) | 5,000 | 4,500 | 4,500 | 4,500 | 4,000 |
| Choline ......................... (mg) | 274,000 | 231,000 | 184,000 | 140,000 | 94,000 |
| **Total** (21) ........................... (lb) | 2,000 | 2,000 | 2,000 | 2,000 | 2,000 |
| **Calculated Analysis** (27) | | | | | |
| Metabolizable energy ............... (kcal/lb) | 1,306.2 | 1,303.9 | 1,303.4 | 1,304.1 | 1,304.5 |
| Protein .............................. (%) | 15.07 | 16.00 | 17.01 | 18.00 | 19.01 |
| Lysine ........................... (%) | 0.68 | 0.75 | 0.83 | 0.91 | 0.98 |
| Methionine ...................... (%) | 0.29 | 0.29 | 0.30 | 0.31 | 0.32 |
| Methionine + cystine ............ (%) | 0.53 | 0.54 | 0.56 | 0.59 | 0.61 |
| Fat .................................. (%) | 3.29 | 3.54 | 3.98 | 5.05 | 5.54 |
| Fiber ................................ (%) | 2.20 | 2.20 | 2.21 | 2.18 | 2.18 |
| Calcium ............................. (%) | 3.25 | 3.24 | 3.24 | 3.50 | 3.50 |
| Total phosphorus .................... (%) | 0.52 | 0.52 | 0.53 | 0.52 | 0.53 |
| Available phosphorus ................ (%) | 0.45 | 0.45 | 0.45 | 0.45 | 0.45 |
| Vitamins (units or mg/lb) | | | | | |
| Vitamin A activity .................. (USPU) | 5,904 | 5,842 | 5,770 | 5,660 | 5,586 |
| Vitamin D$_3$ added .................... (ICU) | 1,000 | 1,000 | 1,000 | 1,000 | 1,000 |
| Riboflavin .......................... (mg) | 1.84 | 1.85 | 1.86 | 1.86 | 1.87 |
| Niacin .............................. (mg) | 15.40 | 15.48 | 15.55 | 15.50 | 15.56 |
| Pantothenic acid .................... (mg) | 500.13 | 4.88 | 4.99 | 5.07 | 4.95 |
| Choline ............................. (mg) | | 500.34 | 500.05 | 500.39 | 500.48 |

[1]From *NECC Chicken and Turkey Rations 1980*, pp. 3 and 6, prepared cooperatively for distribution by the New England Land-Grant Universities. In using Table 44–6, also see the footnotes that follow under the heading "Footnotes to Tables 44–5, –6, and –11." The footnote numbers are those indicated in parentheses in this table. To convert pounds to kilograms, divide by 2.2.

[2]Five rations varying in protein levels are presented. Actual feed consumption of the hens must be known to use properly the 15%, 16%, 17%, 18%, and 19% layer rations. When feed consumption is not known, it is suggested that not less than a 17% protein ration be used.

[3]ICU = International Chick Unit.

# FOOTNOTES TO TABLES 44–5, –6, and –11

The following footnotes are applicable to each of the example rations given in Tables 44–5, 44–6, and 44–11. Thus, instead of footnoting each of three tables, the footnotes are given in this one section. **NOTE WELL:** In the tables, the footnote references are enclosed in parentheses. Thus, footnote No. 1 which follows refers to (1) in the tables.

(1) Wherever substitutions are made in the rations, the total nutrient content should be adjusted to meet established requirements.

(2) Two to 400 lb of coarsely ground wheat or yellow hominy may be used to replace an equal amount of corn. If wheat is used, add 200,000 IU of vitamin A for each 100 lb of corn removed.

(3) There is usually some loss of provitamin A activity in corn and alfalfa meal during storage. If stored ingredients are used, it may be advisable to increase the added vitamin A level of the ration by 1,000 or 2,000 IU/lb. This can be accomplished by increasing the recommended supplement by 2,000,000 or 4,000,000 IU/ton of feed.

(4) The added salt level should be reduced by the amount supplied by the fish meal and other by-product ingredients.

(5) Poultry by-product meal may be substituted for all of the meat and bone scrap and up to 50% of the fish meal. Correct for calcium and phosphorus loss due to substitution of poultry by-product meal.

(6) Based on an 18.5% phosphorus product: Steamed bone meal or defluorinated rock phosphate may replace the dicalcium phosphate on a phosphorus basis.

(7) Based on 35% calcium, low-magnesium limestone.

(8) An antibiotic may be used in these rations at the level recommended by the manufacturer.

(9) The chick starter, 1, 2-dihydro-6-ethoxy-2, 2, 4-trimethylquinoline (ethoxyquin) may be used in broiler and breeder rations at the 0.0125% level to help prevent the appearance of encephalomalacia (crazy chick disease). If desired, it, or an equivalent antioxidant, may be added to help prevent the oxidation of dietary components. Total ethoxyquin from all sources must not exceed 0.25 lb per ton.

(10) A coccidiostat or antihistomonal may be used in these rations, as required, at levels recommended by the manufacturer.

(11) This amount of manganese will be furnished by 0.5 lb of manganese sulfate or 0.21 lb manganous oxide (70% feeding grades). An equivalent amount of manganese may be added from other acceptable sources.

(12) Caution should be used when high-potency vitamin mixes are involved. It is recommended that 10 lb be the minimum amount of any item added to a ton of feed to ensure proper mixing. Thus, high-potency vitamin, mineral, or drug mixes should be premixed with a carrier (such as corn meal) to such a dilution that 10 lb of the final mix will be added for each ton of feed mixed. Minerals and vitamins should not be premixed together.

(13) Available phosphorus has been taken as 30% of total phosphorus from plant sources for chicks, and 75% of total phosphorus from plant sources for adult birds. Phosphorus from other than plant sources is considered to be 100% utilized.

(14) For those persons wanting a restricted feeding program, feed the grower at 80% or less of the amount normally consumed.

(15) The amount of manganese will be furnished by 0.7 lb manganese sulfate or 0.3 lb manganous oxide (70% feeding grades). An equivalent amount of manganese may be added from other acceptable sources.

(16) Stabilized fats may replace an equal amount of cereal grains to provide a higher energy level, control dust, and aid pelleting. Where maintaining body weight in layers is a problem, increase fat by 1% or 2% during the winter months by replacing an equal amount of cereal grains.

(17) Approximately this amount of zinc will be furnished by 29 grams of zinc carbonate or 20 grams of zinc oxide. An equivalent amount of zinc may be used from other acceptable sources.

(18) Feed starting ration until birds are 35 days old.

(19) Based on 3-nitro-4-hydroxyphenylarsonic acid at a level of 45 grams (0.1 lb) per ton. Other compounds that may be used at a level recommended by the manufacturer are sodium arsanilate or arsanilic acid.

(20) In the absence of alfalfa or if the birds are raised on wire, 2 grams of vitamin K activity should be added. Values in the broiler rations are based on menadione. Other compounds supplying equivalent levels of vitamin K may be used.

(21) If an even 2,000 lb is desired, adjust by removing or adding ground yellow corn.

(22) May be fed with grain after 20 weeks.

(23) This amount of manganese will be furnished by approximately 0.3 lb of manganese sulfate or 0.13 lb of manganous oxide (70% feeding grades). An equivalent amount of manganese may be added from other acceptable sources.

(24) This amount of zinc will be furnished by approximately 53 grams of zinc carbonate or 37 grams of zinc oxide. An equivalent amount of zinc may be added from other acceptable sources.

(25) Federal law, which strictly regulates the addition of selenium to poultry rations, should be consulted. Selenium, as sodium selenite or sodium selenate, may be added to complete feed for poultry at a level not to exceed 0.3 part per million.

(26) For heavy caged layer pullets, the NECC suggests feeding 18% protein 0–6 weeks, 14% protein 7–12 weeks, and 12% protein 13–20 weeks of age.

(27) Any discrepancies in calculated analysis that occur in the decimal part of the figures are due to rounding errors built in by the computers used in the calculations.

Data showing the approximate quantities of feed required for the production of eggs are presented in Tables 44–7 and 44–8.

### TABLE 44–7
### METABOLIZABLE ENERGY REQUIRED DAILY BY LAYERS IN RELATION TO BODY WEIGHT AND EGG PRODUCTION[1][2]

| Body Weight | | Rate of Egg Production (%) | | | | | |
|---|---|---|---|---|---|---|---|
| | | 0 | 50 | 60 | 70 | 80 | 90 |
| (lb) | (kg) | ← | | (Metabolizable Energy/Hen Daily (kcal)[3]) | | | → |
| 2.2 | 1.0 | 130 | 192 | 205 | 217 | 229 | 242 |
| 3.3 | 1.5 | 177 | 239 | 251 | 264 | 276 | 289 |
| 4.4 | 2.0 | 218 | 280 | 292 | 305 | 317 | 330 |
| 5.5 | 2.5 | 259 | 321 | 333 | 346 | 358 | 371 |
| 6.6 | 3.0 | 296 | 358 | 370 | 383 | 395 | 408 |
| 7.7 | 3.5 | 333 | 395 | 408 | 420 | 432 | 445 |

[1]Adapted by the author from *Nutrient Requirements of Poultry*, 8th rev. ed., NRC, National Academy Press, Washington, DC, 1984, p. 15, Table 10.

[2]A number of formulas have been suggested for prediction of the daily energy requirements of chickens. The formula used here was derived from that in *Effect of Environment on Nutrient Requirements of Domestic Animals* (NRC, 1981).

$$ME/hen\ daily = W^{0.75} (173 - 1.95T) + 5.5 \triangle W + 2.07 EE$$

where:    $W$ = body weight (kg),
          $T$ = ambient temperature (°C),
          $\triangle W$ = change in body weight in g/day, and
          $EE$ = daily egg mass (g).

[3]Temperature of 22°, egg weight of 60 g, and no change in body weight were used in calculations.

### TABLE 44–8
### BODY WEIGHTS AND FEED REQUIREMENTS OF LEGHORN-TYPE PULLETS AND HENS[1]

| Age | Body Weight[2] | | Feed Consumption[3] | | Typical Egg Production (Hen/Day) |
|---|---|---|---|---|---|
| (weeks) | (lb) | (kg) | (lb/week) | (kg/week) | (%) |
| 0 | 0.07 | 0.03 | 0.11 | 0.05 | — |
| 2 | 0.31 | 0.14 | 0.20 | 0.09 | — |
| 4 | 0.60 | 0.27 | 0.40 | 0.18 | — |
| 6 | 0.99 | 0.45 | 0.57 | 0.26 | — |
| 8 | 1.37 | 0.62 | 0.73 | 0.33 | — |
| 10 | 1.74 | 0.79 | 0.86 | 0.39 | — |
| 12 | 2.09 | 0.95 | 0.95 | 0.43 | — |
| 14 | 2.34 | 1.06 | 1.01 | 0.46 | — |
| 16 | 2.56 | 1.16 | 1.01 | 0.46 | — |
| 18 | 2.78 | 1.26 | 1.01 | 0.46 | — |
| 20 | 3.00 | 1.36 | 1.11 | 0.50 | 3 |
| 22 | 3.13 | 1.42 | 1.33 | 0.60 | 40 |
| 24 | 3.31 | 1.50 | 1.60 | 0.73 | 84 |
| 26 | 3.48 | 1.58 | 1.65 | 0.75 | 90 |
| 30 | 3.81 | 1.73 | 1.70 | 0.77 | 94 |
| 40 | 4.01 | 1.82 | 1.72 | 0.78 | 89 |
| 50 | 4.12 | 1.87 | 1.68 | 0.76 | 84 |
| 60 | 4.19 | 1.90 | 1.67 | 0.76 | 79 |
| 70 | 4.19 | 1.90 | 1.64 | 0.75 | 73 |

[1]Adapted and updated by the author from *Nutrient Requirements of Poultry*, 8th rev. ed., NRC, National Academy Press, Washington, DC, 1984, p. 13, Table 5.

[2]Pullets and hens of Leghorn-type strains are generally fed *ad libitum* but are occasionally control-fed to limit body weights. Values shown are typical but will vary with strain differences, season, and lighting. Specific breeder guidelines should be consulted for desired schedules of weights and feed consumption.

[3]Based on rations containing 1,315 ME kcal/lb *(2,900 ME kcal/kg)*. Consumption will vary depending upon the caloric density of the ration, environmental temperature, and rate of production (see Table 44–9).

Table 44–9 is a yardstick of feed and time required to obtain certain average liveweights of chickens. Such a guide is useful in setting up budgets and measuring results.

### TABLE 44–9
### BODY WEIGHTS AND FEED ALLOWANCES OF MEAT-TYPE REPLACEMENT CHICKENS[1][2]

| Age | Male Body Weight[3] | | Male Feed Consumption[4] | | Female Body Weight[3] | | Female Feed Consumption[4] | | Typical Egg Production |
|---|---|---|---|---|---|---|---|---|---|
| (weeks) | (lb) | (g) | (lb/week) | (g/week) | (lb) | (g) | (lb/week) | (g/week) | (hen/day %) |
| 0 | 0.09 | 40 | 0.22 | 100 | 0.09 | 40 | 0.17 | 75 | — |
| 2 | 0.55 | 250 | 0.55 | 250 | 0.50 | 225 | 0.56 | 225 | — |
| 4 | 1.20 | 545 | 0.77–0.85 | 350–385 | 1.00 | 455 | 0.69–0.73 | 315–330 | — |
| 6 | 1.75 | 795 | 0.86–0.94 | 390–425 | 1.46 | 660 | 0.73–0.77 | 330–350 | — |
| 8 | 2.25 | 1,020 | 0.89–1.03 | 405–475 | 1.85 | 840 | 0.77–0.88 | 350–400 | — |
| 10 | 2.98 | 1,250 | 1.03–1.21 | 475–550 | 2.20 | 1,000 | 0.85–0.98 | 385–445 | — |
| 12 | 3.26 | 1,480 | 1.19–1.38 | 540–625 | 2.60 | 1,180 | 0.94–1.06 | 425–480 | — |
| 14 | 3.75 | 1,700 | 1.27–1.54 | 575–700 | 3.00 | 1,360 | 1.01–1.21 | 460–550 | — |
| 16 | 4.25 | 1,930 | 1.38–1.69 | 625–765 | 3.42 | 1,550 | 1.09–1.32 | 495–600 | — |
| 18 | 4.74 | 2,150 | 1.47–1.82 | 665–825 | 3.81 | 1,730 | 1.16–1.48 | 525–670 | — |
| 20 | 5.29 | 2,400 | —[5] | —[5] | 4.25 | 1,930 | 1.26–1.61 | 570–730 | — |
| 22 | 5.82 | 2,640 | — | — | 4.65 | 2,110 | 1.40–1.75 | 635–795 | 10 |
| 24 | 7.05 | 3,200 | — | — | 5.40 | 2,450 | 1.76–2.04 | 800–925 | 15 |
| 26 | 7.80 | 3,540 | — | — | 6.02 | 2,730 | 2.09–2.31 | 950–1,050 | 30 |
| 28 | 8.27 | 3,750 | — | — | 6.35 | 2,880 | 2.38–2.52 | 1,078–1,141 | 56 |
| 30 | 8.60 | 3,900 | — | — | 6.61 | 3,000 | 2.38–2.52 | 1,078–1,141 | 75 |
| 32 | 9.02 | 4,090 | — | — | 6.81 | 3,090 | 2.38–2.52 | 1,078–1,141 | 80 |
| 34 | 9.30 | 4,220 | — | — | 6.90 | 3,130 | 2.38–2.52 | 1,078–1,141 | 78 |
| 36 | 9.57 | 4,340 | — | — | 6.97 | 3,160 | 2.38–2.52 | 1,078–1,141 | 76 |
| 38 | 9.81 | 4,450 | — | — | 7.01 | 3,180 | 2.36–2.50 | 1,071–1,134 | 73 |
| 40 | 10.01 | 4,540 | — | — | 7.01 | 3,180 | 2.35–2.48 | 1,064–1,127 | 72 |

[1]Adapted by the author from *Nutrient Requirements of Poultry*, 8th rev. ed., NRC, National Academy Press, Washington, DC, 1984, p. 15, Table 9.

[2]Broiler-breeder strains must be grown on a controlled feeding program to limit weight. Values shown are typical but will vary according to strain. Specific breeder guidelines should be consulted for desired schedule of weights and feed allotments.

[3]Values are typical for fall-hatched chicks. Spring-hatched chicks will have decreasing natural daylight during the time of sexual maturity and usually need to be heavier to attain sexual maturity at the desired age.

[4]Adjust as required to maintain desired body weight.

[5]Males and females intermingled.

• Pointers pertinent to **feeding replacement chicks (pullets):**

1. Replacement chicks are usually fed a ration lower in energy than broiler chicks. Also, feed and daily light periods may be restricted, so as to permit the pullets to reach larger body size before they start to lay than would be the case were they full-fed, fully lighted pullets.

2. About 18 to 23 lb of starting and growing ration are required to develop an egg production type pullet to the age of sexual maturity.

3. Always use complete starter feeds for chicks (see Table 44–5), and give chicks starter feeds without grain supplement until they are six weeks old.

4. When chicks are 6 weeks old, change to growing ration (see Table 44–5).

5. When pullets are 17 to 19 weeks old, replace the growing mash with laying mash (see Table 44–6).

• Pointers pertinent to **feeding layers:**

1. In the final analysis, the objective of feeding laying hens is to produce a dozen eggs of good quality at the lowest possible feed cost. Thus, the actual cost of the feed that a layer eats in producing a dozen eggs—not the price per pound of feed—determines the economy of the ration.

2. Feed consumption per bird varies primarily with egg production and body size (Table 44–8). It is also influenced by the health of the birds and the environment, especially the temperature.

3. Normally, a mature Leghorn, or other light-weight bird, eats about 80 lb of feed per year and produces about 22 dozen eggs in that same period of time. Hence, it requires 3.6 lb of feed to produce 1 dozen eggs. With light weight layers, the producer should aim for a feed efficiency of 3.5 to 4.0 lb, or less, of feed per dozen eggs. A bird of the heavier breeds eats 95 to 115 lb of feed per year; hence, they are not as efficient egg producers.

• Pointers pertinent to **feeding breeder chickens:**

1. The nutritive requirements of the breeding flock are more rigorous than those for commercial laying flocks (see Table 44–4 for the nutritive requirements of breeding hens). Breeders require certain nutrients that are not required for laying hens, and they require greater amounts of certain nutrients than do laying flocks. Rations with these added ingredients in the right proportion give high hatchability and good development of chicks. Such rations cost more than normal layer rations.

2. Birds intended as breeders should be started on breeder rations about one month before hatching eggs are to be saved.

• **Phase feeding**—*Phase feeding,* has become part of the poultry producer's language. As usually used, phase feeding refers to changes in the laying hen's diet (1) to adjust to age and stage of production of the hen, (2) to adjust for season of the year and for temperature and climatic changes, (3) to account for differences in body weight and nutrient requirements of different strains of birds, and (4) to adjust one or more nutrients as other nutrients are changed for economic or availability reasons. Research has shown, for example, that a hen laying at the rate of 60% has different nutritional requirements than one laying at the rate of 80%; hens have different requirements in summer and in winter; a 24-week-old layer has different needs than one 54 weeks old. The main objective, therefore, of phase feeding is to reduce the waste of nutrients caused by feeding more than a bird actually needs under different sets of conditions. In this way, feed efficiency can be improved and the cost of producing a dozen eggs reduced.

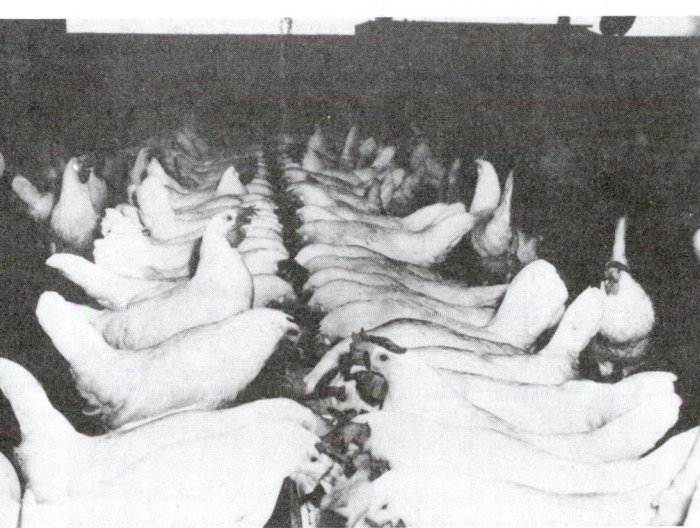

Fig. 44–14. Feeding layers went ultramodern with the development of the concept of *phase feeding,* which is the practice of changing ration formulations to meet changing nutritional requirements due to such factors as rate of lay, environmental temperature, etc. (Courtesy, DeKalb Poultry Research, Inc., DeKalb, IL)

A phase feeding program for laying hens generally calls for use of a rather high-protein feed (usually 17 to 18%) from the onset of egg production through the peak production period. Thereafter, a lower level of protein (about 16%) is fed for the next 5 or 6 months, followed by still lower levels (usually 15%) until the laying period is completed. This general plan takes age into consideration, but for greatest benefits other factors will also need to be considered.

Although phase feeding has its advantages, it does present some problems; it is a complicated procedure, it necessitates a knowledgeable producer, and it requires more bulk bins, closer check on feed deliveries, etc.

Phase feeding is being practiced, and it is increasing in use. Although it does not promise to bring about large increases in egg production, it can help production reach a higher peak and sustain it longer if other conditions are right. In the latter respect, it is very much like *lead* or *challenge* feeding in dairy

cows. Most of all, phase feeding offers a good potential for lowering costs and increasing income. Like many other developments, it favors the larger operator.

## FEEDING BROILERS

Fig. 44–15. Broilers, showing automated feeding and watering. (Courtesy, Gold Kist—Communications, Atlanta, GA)

The nutrient requirements of, and suggested rations for, broilers are presented in the following tables:

Table 44–10, Nutrient Requirements of Broilers.

Table 44–11, Broiler Rations.

Table 44–12, Body Weights and Feed Requirements of Broilers/Roasters.

## TABLE 44–10
## NUTRIENT REQUIREMENTS OF BROILERS[1]

| Energy Base: | | Weeks 0–3 | Weeks 3–6 | Weeks 6–8 |
|---|---|---|---|---|
| kcal ME/lb Diet[2] | | 1,452 | 1,452 | 1,452 |
| *kcal ME/kg Diet[2]* | | *3,200* | *3,200* | *3,200* |
| **Protein** | (%) | 23.0 | 20.0 | 18.0 |
| **Amino acids:** | | | | |
| Arginine | (%) | 1.30 | 1.05 | 0.95 |
| Glycine and serine | (%) | 1.50 | 1.00 | 0.70 |
| Histidine | (%) | 0.35 | 0.30 | 0.26 |
| Isoleucine | (%) | 0.80 | 0.70 | 0.60 |
| Leucine | (%) | 1.35 | 1.18 | 1.00 |
| Lysine | (%) | 1.20 | 1.00 | 0.85 |
| Methionine + cystine | (%) | 0.93 | 0.78 | 0.72 |
| Methionine | (%) | 0.50 | 0.40 | 0.36 |
| Phenylalanine + tyrosine | (%) | 1.34 | 1.17 | 1.00 |
| Phenylalanine | (%) | 0.72 | 0.63 | 0.54 |
| Threonine | (%) | 0.80 | 0.74 | 0.68 |
| Tryptophan | (%) | 0.23 | 0.18 | 0.17 |
| Valine | (%) | 0.82 | 0.72 | 0.62 |
| **Linoleic acid** | (%) | 1.00 | 1.00 | 1.00 |
| **Major or macrominerals:** | | | | |
| Calcium (Ca) | (%) | 1.00 | 0.90 | 0.80 |
| Chlorine (Cl) | (%) | 0.15 | 0.15 | 0.15 |
| Magnesium (Mg) | (mg) | 600 | 600 | 600 |
| Phosphorus (P), available | (%) | 0.45 | 0.40 | 0.35 |
| Potassium (K) | (%) | 0.40 | 0.35 | 0.30 |
| Sodium (Na) | (%) | 0.15 | 0.15 | 0.15 |
| **Trace or microminerals:** | | | | |
| Copper (Cu) | (mg) | 8.0 | 8.0 | 8.0 |
| Iodine (I) | (mg) | 0.35 | 0.35 | 0.35 |
| Iron (Fe) | (mg) | 80.0 | 80.0 | 80.0 |
| Manganese (Mn) | (mg) | 60.0 | 60.0 | 60.0 |
| Selenium (Se) | (mg) | 0.15 | 0.15 | 0.15 |
| Zinc (Zn) | (mg) | 40.0 | 40.0 | 40.0 |
| **Fat-soluble vitamins:** | | | | |
| Vitamin A | (IU) | 1,500 | 1,500 | 1,500 |
| Vitamin D | (ICU) | 200 | 200 | 200 |
| Vitamin E | (IU) | 10 | 10 | 10 |
| Vitamin K | (mg) | 0.50 | 0.50 | 0.50 |
| **Water-soluble vitamins:** | | | | |
| Biotin | (mg) | 0.15 | 0.15 | 0.10 |
| Choline | (mg) | 1,300 | 850 | 500 |
| Folacin (Folic Acid) | (mg) | 0.55 | 0.55 | 0.25 |
| Niacin (Nicotinic Acid, Nicotinamide) | (mg) | 27.0 | 27.0 | 11.0 |
| Pantothenic Acid (Vitamin B–3) | (mg) | 10.0 | 10.0 | 10.0 |
| Riboflavin (Vitamin B–2) | (mg) | 3.60 | 3.60 | 3.60 |
| Thiamin (Vitamin B–1) | (mg) | 1.80 | 1.80 | 1.80 |
| Vitamin B–6 (Pyridoxine, Pyridoxal, Pyridoxamine) | (mg) | 3.0 | 3.0 | 2.5 |
| Vitamin B–12 (Cobalamins) | (mg) | 0.009 | 0.009 | 0.003 |

[1]Adapted and updated by the author from *Nutrient Requirements of Poultry*, 8th rev. ed., NRC, National Academy Press, Washington, DC, 1984, p. 13, Table 6.

[2]These are typical dietary energy concentrations.

## TABLE 44–11
## BROILER RATIONS (1)[1]

| Ingredient | | Starter (18) | Finisher[2] |
|---|---|---|---|
| Ground yellow corn (3) | (lb) | 1,104 | 1,258 |
| Soybean meal (dehulled) | (lb) | 605 | 420 |
| Corn gluten meal (60%) | (lb) | 50 | 75 |
| Fish meal, herring (65%) (4, 5) | (lb) | 50 | 50 |
| Meat and bone meal (47%) (5) | (lb) | 50 | 50 |
| Dicalcium phosphate (6) | (lb) | 10 | 9 |
| Ground limestone (7) | (lb) | 16 | 14 |
| DL-Methionine, or equivalent | (lb) | 3.2 | — |
| Stabilized yellow grease, or equivalent | (lb) | 106 | 115 |
| Iodized salt (4) | (lb) | 7 | 7 |
| Antibiotic supplement | (footnote) | (8) | (8) |
| Antioxidant | (footnote) | (9) | (9) |
| Coccidiostat | (footnote) | (10) | (10) |
| Zinc (24) | (g) | 30 | 30 |
| Manganese (15) | (g) | 75 | 75 |
| Selenium | (footnote) | (25) | (25) |
| Organic arsenical supplement (19) | | 0.1 | 0.1 |
| Vitamin supplements (12) | | | |
| Vitamin A | (USPU) | 4,000,000 | 4,000,000 |
| Vitamin D₃ | (ICU)[3] | 1,000,000 | 1,000,000 |
| Vitamin E | (IU) | 2,000 | 2,000 |
| Vitamin K (20) | (mg) | 1,000 | 1,000 |
| Vitamin B–12 | (mg) | 12 | 12 |
| Riboflavin | (mg) | 3,000 | 3,000 |
| Niacin | (mg) | 20,000 | 20,000 |
| Calcium pantothenate | (mg) | 5,000 | 5,000 |
| Choline | (mg) | 503,000 | 672,000 |
| **Total** (lb) (21) | | 2,001.3 | 1,998.1 |
| **Calculated Analysis** (27) | | | |
| Metabolizable energy | (kcal/lb) | 1,462 | 1,500 |
| Protein | (%) | 24.08 | 21.09 |
| Lysine | (%) | 1.30 | 1.05 |
| Methionine | (%) | 0.57 | 0.38 |
| Methionine + cystine | (%) | 0.93 | 0.81 |
| Fat | (%) | 8.20 | 8.92 |
| Fiber | (%) | 1.97 | 2.11 |
| Calcium | (%) | 0.84 | 0.81 |
| Total phosphorus | (%) | 0.64 | 0.60 |
| Available phosphorus (13) | (%) | 0.40 | 0.38 |
| Vitamins (units or mg/lb) | | | |
| Vitamin A activity | (USPU) | 3,769 | 6,049 |
| Vitamin D₃ (added) | (ICU) | 500 | 500 |
| Riboflavin | (mg) | 2.44 | 2.49 |
| Niacin | (mg) | 21.36 | 21.33 |
| Pantothenic acid | (mg) | 5.69 | 5.51 |
| Choline | (mg) | 800.03 | 800.48 |
| Xanthophyll | (mg) | 9.50 | 14.05 |

[1]From *NECC Chicken and Turkey Rations 1980*, pp. 3 and 6, prepared cooperatively for distribution by the New England Land-Grant Universities. In using Table 44–11, also see the footnotes under the heading "Footnotes to Tables 44–5, –6, and –11." The footnote numbers are those indicated in parentheses in this table. To convert pounds to kilograms, divide by 2.2.

[2]Comply with Food and Drug Administration regulations for withdrawal times of feed additives.

[3]ICU = International Chick Unit.

The body weights and feed requirements of male and female broilers/roasters are given in Table 44–12. As shown, males are slightly heavier at 9 weeks of age, and require slightly more feed per pound of gain than females. At 9 weeks of age, males weigh about 7.8 lb, with a feed efficiency of 2.28 lb of feed per pound of gain. At 9 weeks of age, females weigh about 6.4 lb, with a feed efficiency of 2.2 lb of feed per pound of gain.

### TABLE 44-12
### BODY WEIGHTS AND FEED REQUIREMENTS
### OF BROILERS/ROASTERS[1] [2]

| Age | Body Weights | | | | Weekly Food Consumption | | | |
|---|---|---|---|---|---|---|---|---|
| | Male | | Female | | Male | | Female | |
| (weeks) | (lb) | (g) | (lb) | (g) | (lb) | (g) | (lb) | (g) |
| 1 | .33 | 150 | .32 | 147 | .29 | 132 | .29 | 130 |
| 2 | .85 | 385 | .79 | 360 | .64 | 290 | .58 | 265 |
| 3 | 1.56 | 700 | 1.40 | 640 | 1.04 | 470 | .94 | 430 |
| 4 | 2.40 | 1,100 | 2.16 | 980 | 1.50 | 670 | 1.30 | 600 |
| 5 | 3.40 | 1,560 | 3.00 | 1,360 | 2.00 | 900 | 1.70 | 770 |
| 6 | 4.50 | 2,060 | 3.90 | 1,760 | 2.45 | 1,115 | 2.10 | 940 |
| 7 | 5.60 | 2,560 | 4.80 | 2,160 | 2.90 | 1,310 | 2.40 | 1,100 |
| 8 | 6.70 | 3,060 | 5.60 | 2,550 | 3.30 | 1,500 | 1.74 | 1,240 |
| 9 | 7.80 | 3,540 | 6.40 | 2,910 | 3.70 | 1,670 | 3.00 | 1,370 |

[1]Most broilers are marketed at 6 to 8 weeks of age, at 3 to 5 lb live weight. Roasters weigh 6 to 9 lb.

[2]The data presented in this table was provided by Dr. Leo S. Jensen, Professor, Department of Poultry Science, College of Agriculture, The University of Georgia, Athens.

• Pointers pertinent to **feeding broilers:**

1. Feed is the largest cost item in broiler production, representing 65 to 75% of the total cost.

2. The starting ration of replacement chicks usually contains about 20% protein, whereas the starting ration of broilers usually contains about 24% protein for the first 6 weeks. The exact percentage of protein needed depends on the feed's energy content.

3. At about 6 weeks of age, broilers should be placed on a finishing mash that has an increased energy level and reduced protein level. Pellets are often used for broilers from 6 weeks until they reach market weight.

## FEEDING TURKEYS

Fig. 44-16. Feed is the largest single item in the cost of growing turkeys; hence, the selection of the ration greatly influences the opportunity for profit in a turkey enterprise. (Courtesy, *Turkey World*, Mount Morris, IL)

Fig. 44-17. Turkeys on the range. Note self-feeders on the right. (Courtesy, *Turkey World*, Mount Morris, IL)

The nutrient requirements of, and suggested rations for, turkeys are presented in the following tables:

Table 44–13, Nutrient Requirements of Turkeys For Growing, Holding, Breeding.

Table 44–14, Turkey Prestarter, Starter, Grower, and Finisher Rations.

Table 44–15, Turkey Finisher and Holding Rations.

Table 44–16, Turkey Breeder Rations.

Table 44–17, Vitamin-Mineral Premixes For Turkey Rations in Tables 44–14, –15, and –16.

Table 44–18, Body Weights and Feed Consumption of Turkeys During Holding and Breeding Periods.

Table 44–19, Body Weights, Feed Consumption, and Feed-to-Gain Ratios of Turkey Toms.

Table 44–20, Body Weights, Feed Consumption, and Feed-to-Gain Ratios of Turkey Hens.

## TABLE 44–13
### NUTRIENT REQUIREMENTS OF TURKEYS FOR GROWING, HOLDING, BREEDING[1]

| | | Age (Weeks) | | | | | | | |
|---|---|---|---|---|---|---|---|---|---|
| Male | | 0–4 | 4–8 | 8–12 | 12–16 | 16–20 | 20–24 | | Breeding |
| Female | | 0–4 | 4–8 | 8–11 | 11–14 | 14–17 | 17–20 | Holding | Hens |
| **Energy Base:** | | | | | | | | | |
| kcal ME/lb Diet[2] | | 1,270 | 1,315 | 1,361 | 1,406 | 1,452 | 1,497 | 1,315 | 1,315 |
| *kcal ME/kg Diet[2]* | | *2,800* | *2,900* | *3,000* | *3,100* | *3,200* | *3,300* | *2,900* | *2,900* |
| **Protein** | (%) | 28 | 26 | 22 | 19 | 16.5 | 14 | 12 | 14 |
| **Amino acids:** | | | | | | | | | |
| Arginine | (%) | 1.60 | 1.5 | 1.25 | 1.1 | 0.95 | 0.8 | 0.6 | 0.6 |
| Glycine and serine | (%) | 1.0 | 0.9 | 0.8 | 0.7 | 0.6 | 0.5 | 0.4 | 0.5 |
| Histidine | (%) | 0.58 | 0.54 | 0.46 | 0.39 | 0.35 | 0.29 | 0.25 | 0.3 |
| Isoleucine | (%) | 1.1 | 1.0 | 0.85 | 0.75 | 0.65 | 0.55 | 0.45 | 0.5 |
| Leucine | (%) | 1.9 | 1.75 | 1.5 | 1.3 | 1.1 | 0.95 | 0.5 | 0.5 |
| Lysine | (%) | 1.6 | 1.5 | 1.3 | 1.0 | 0.8 | 0.65 | 0.5 | 0.6 |
| Methionine + cystine | (%) | 1.05 | 0.9 | 0.75 | 0.65 | 0.55 | 0.45 | 0.4 | 0.4 |
| Methionine | (%) | 0.53 | 0.45 | 0.38 | 0.33 | 0.28 | 0.23 | 0.2 | 0.2 |
| Phenylalanine + tyrosine | (%) | 1.8 | 1.65 | 1.4 | 1.2 | 1.05 | 0.9 | 0.8 | 1.0 |
| Phenylalanine | (%) | 1.0 | 0.9 | 0.8 | 0.7 | 0.6 | 0.5 | 0.4 | 0.55 |
| Threonine | (%) | 1.0 | 0.93 | 0.79 | 0.68 | 0.59 | 0.5 | 0.4 | 0.45 |
| Tryptophan | (%) | 0.26 | 0.24 | 0.2 | 0.18 | 0.15 | 0.13 | 0.1 | 0.13 |
| Valine | (%) | 1.2 | 1.1 | 0.94 | 0.8 | 0.7 | 0.6 | 0.5 | 0.58 |
| **Linoleic acid** | (%) | 1.0 | 1.0 | 0.8 | 0.8 | 0.8 | 0.8 | 0.8 | 1.0 |
| **Major or macrominerals:** | | | | | | | | | |
| Calcium (Ca) | (%) | 1.2 | 1.0 | 0.85 | 0.75 | 0.65 | 0.55 | 0.5 | 2.25 |
| Chlorine (Cl) | (%) | 0.15 | 0.14 | 0.14 | 0.12 | 0.12 | 0.12 | 0.12 | 0.12 |
| Magnesium (Mg) | (mg) | 600 | 600 | 600 | 600 | 600 | 600 | 600 | 600 |
| Phosphorus (P), available | (%) | 0.6 | 0.5 | 0.42 | 0.38 | 0.32 | 0.28 | 0.25 | 0.35 |
| Potassium (K) | (%) | 0.7 | 0.6 | 0.5 | 0.5 | 0.4 | 0.4 | 0.4 | 0.6 |
| Sodium (Na) | (%) | 0.17 | 0.15 | 0.12 | 0.12 | 0.12 | 0.12 | 0.12 | 0.15 |
| **Trace or microminerals:** | | | | | | | | | |
| Copper (Cu) | (mg) | 8 | 8 | 6 | 6 | 6 | 6 | 6 | 8 |
| Iodine (I) | (mg) | 0.4 | 0.4 | 0.4 | 0.4 | 0.4 | 0.4 | 0.4 | 0.4 |
| Iron (Fe) | (mg) | 80 | 60 | 60 | 60 | 50 | 50 | 50 | 60 |
| Manganese (Mn) | (mg) | 60 | 60 | 60 | 60 | 60 | 60 | 60 | 60 |
| Selenium (Se) | (mg) | 0.2 | 0.2 | 0.2 | 0.2 | 0.2 | 0.2 | 0.2 | 0.2 |
| Zinc (Zn) | (mg) | 75 | 65 | 50 | 40 | 40 | 40 | 40 | 65 |
| **Fat-soluble vitamins:** | | | | | | | | | |
| Vitamin A | (IU) | 4,000 | 4,000 | 4,000 | 4,000 | 4,000 | 4,000 | 4,000 | 4,000 |
| Vitamin D[3] | (ICU) | 900 | 900 | 900 | 900 | 900 | 900 | 900 | 900 |
| Vitamin E | (IU) | 12 | 12 | 10 | 10 | 10 | 10 | 10 | 25 |
| Vitamin K | (mg) | 1.0 | 1.0 | 0.8 | 0.8 | 0.8 | 0.8 | 0.8 | 1.0 |
| **Water-soluble vitamins:** | | | | | | | | | |
| Biotin | (mg) | 0.2 | 0.2 | 0.15 | 0.125 | 0.100 | 0.100 | 0.100 | 0.15 |
| Choline | (mg) | 1,900 | 1,600 | 1,300 | 1,100 | 950 | 800 | 800 | 1,000 |
| Folacin (Folic Acid) | (mg) | 1.0 | 1.0 | 0.8 | 0.8 | 0.7 | 0.7 | 0.7 | 1.0 |
| Niacin (Nicotinic Acid, Nicotinamide) | (mg) | 70.0 | 70.0 | 50.0 | 50.0 | 40.0 | 40.0 | 40.0 | 30.0 |
| Pantothenic Acid (Vitamin B–3) | (mg) | 11.0 | 11.0 | 9.0 | 9.0 | 9.0 | 9.0 | 9.0 | 16.0 |
| Riboflavin (Vitamin B–2) | (mg) | 3.6 | 3.6 | 3.0 | 3.0 | 2.5 | 2.5 | 2.5 | 4.0 |
| Thiamin (Vitamin B–1) | (mg) | 2.0 | 2.0 | 2.0 | 2.0 | 2.0 | 2.0 | 2.0 | 2.0 |
| Vitamin B–6 (Pyridoxine, Pyridoxal, Pyridoxamine) | (mg) | 4.5 | 4.5 | 3.5 | 3.5 | 3.0 | 3.0 | 3.0 | 4.0 |
| Vitamin B–12 (Cobalamins) | (mg) | 0.003 | 0.003 | 0.003 | 0.003 | 0.003 | 0.003 | 0.003 | 0.003 |

[1]Adapted by the author from *Nutrient Requirements of Poultry*, 8th rev. ed., NRC, National Academy Press, Washington, DC, 1984, p. 17, Table 11.

[2]These are typical ME concentrations for corn-soya rations. Different ME values may be appropriate if other ingredients predominate.

[3]These concentrations of vitamin D are satisfactory when the dietary concentrations of calcium and available phosphorus conform with those in this table.

## TABLE 44-14
### TURKEY PRESTARTER, STARTER, GROWER, AND FINISHER RATIONS[1]

| Ingredient | Prestarter (0–3 weeks) | Starter (3–6 weeks) | Grower I (6–9 weeks) | Grower II (9–12 weeks) | Grower III (12–15 weeks) | Finisher (15–18 weeks) |
|---|---|---|---|---|---|---|
| | (%) | (%) | (%) | (%) | (%) | (%) |
| Yellow corn | 43.38 | 45.72 | 53.76 | 64.88 | 67.90 | 71.59 |
| Soybean meal (48%) | 43.94 | 40.71 | 30.77 | 18.33 | 16.77 | 11.00 |
| Fish meal (60%) | 3.00 | — | — | — | — | — |
| Poultry meal | 2.86 | 4.44 | 7.00 | 8.00 | 8.00 | 8.00 |
| Poultry fat | 1.68 | 4.38 | 4.73 | 5.25 | 4.80 | 7.19 |
| Dicalcium phosphate (21% Ca) | 2.90 | 2.63 | 2.03 | 1.74 | 0.92 | 0.59 |
| Limestone | 1.23 | 1.14 | 0.95 | 1.01 | 0.86 | 0.88 |
| DL-methionine | 0.23 | 0.24 | 0.08 | 0.09 | 0.01 | 0.03 |
| L-lysine-HCl | 0.02 | 0.14 | 0.03 | 0.14 | 0.10 | 0.07 |
| Choline chloride (50%) | 0.20 | 0.20 | 0.20 | 0.20 | 0.20 | 0.20 |
| Salt (NaCl) | 0.30 | 0.20 | 0.25 | 0.16 | 0.24 | 0.25 |
| Vitamins and Minerals[2] | 0.20 | 0.20 | 0.20 | 0.20 | 0.20 | 0.20 |
| Total | 100.00 | 100.00 | 100.00 | 100.00 | 100.00 | 100.00 |
| **Calculated Analysis** | | | | | | |
| Protein (%) | 28.50 | 26.00 | 24.00 | 20.00 | 19.00 | 16.50 |
| M.E. (Kcal/lb) | 1,300 | 1,370 | 1,420 | 1,475 | 1,520 | 1,600 |
| Lysine (%) | 1.70 | 1.60 | 1.30 | 1.10 | 1.00 | 0.80 |
| TSAA (%) | 1.14 | 1.10 | 0.90 | 0.90 | 0.70 | 0.65 |
| Calcium (%) | 1.50 | 1.30 | 1.30 | 1.16 | 1.12 | 0.80 |
| Available phosphorus (%) | 0.85 | 0.72 | 0.65 | 0.60 | 0.45 | 0.38 |
| Crude fat (%) | 5.58 | 8.07 | 8.59 | ? | 9.19 | 11.52 |

[1]Rations provided by Dr. Peter Ferket, Extension Poultry Nutritionist, North Carolina State University, School of Agriculture and Life Sciences, Raleigh, NC.

[2]Vitamins and minerals premix supplied per kg of diet: 13,200 IU vitamin A; 4,400 ICU vitamin D₃; 33 IU vitamin E; 22 mg vitamin B-12; 13 mg riboflavin; 66 mg niacin; 22 mg d-pantothenic acid; 2.2 mg menadione; 1.2 mg folic acid; 6.6 mg pyridoxine; 2.2 mg thiamin; 165 mg d-biotin; 150 mg manganese (MnSO₄ H₂O); 180 mg zinc (ZnO); 40 mg iron (FE₃ (SO₄)₂ 7H₂O); 6 mg copper (CuSO₄).

## TABLE 44-15
### TURKEY FINISHER AND HOLDING RATIONS[1] [2]

| Ingredient | Finisher | | Holding Rations | | |
|---|---|---|---|---|---|
| | No. 25 | No. 26 | No. 27 | No. 28 | No. 29 |
| | (kg) | (kg) | (kg) | (kg) | (kg) |
| Ground yellow corn | 811 | 817 | — | 103 | 233 |
| Ground wheat | — | — | 203 | 168 | — |
| Pulverized oats | — | — | 227 | 168 | 263 |
| Ground barley | — | — | 230 | 260 | 250 |
| Wheat middlings | — | — | 140 | 130 | — |
| Wheat shorts | — | — | 140 | 130 | — |
| Wheat bran | — | — | — | — | 150 |
| Soybean meal (48%) | 103 | 82 | 10 | 10 | 57 |
| Meat meal (50%) | — | 22 | — | 10 | — |
| Fish meal (60%) | — | — | — | 10 | — |
| Fat (stabilized) | 50 | 50 | 10 | 10 | 10 |
| Ground limestone | 12 | 8 | .5 | 12 | 12 |
| Calcium phosphate (20% P) | 15.5 | 10.5 | 12 | 8 | 13 |
| Salt (iodized) | 3.5 | 3.5 | 5 | 5 | 5 |
| Vitamin:mineral premix No. 7 and No. 8 | 10 | 10 | 10 | 10 | 10 |
| DL-methionine | .6 | .6 | .75 | .6 | .75 |
| Choline chloride | 1 | 1 | — | — | .25 |
| Total (kg) | 1,000 | 1,000 | 1,000 | 1,000 | 1,000 |

(Add methionine and choline at the expense of corn or wheat)

| Calculated analyses | | | | | |
|---|---|---|---|---|---|
| Crude protein (%) | 12.0 | 12.1 | 13.3 | 13.3 | 13.2 |
| Digestible protein (%) | 10.9 | 11.0 | 11.5 | 11.4 | 11.0 |
| Crude fat (%) | 8.1 | 8.3 | 4.0 | 4.1 | 4.1 |
| Crude fiber (%) | 2.3 | 2.3 | 6.7 | 6.6 | 7.5 |
| Metabolizable energy (kcal/kg) | 3,349 | 3,363 | 2,680 | 2,730 | 2,656 |
| Calcium (%) | 0.81 | 0.77 | 0.90 | 0.86 | 0.81 |
| Available phosphorus (%) | 0.40 | 0.40 | 0.38 | 0.37 | 0.40 |
| Sodium (%) | 0.19 | 0.20 | 0.28 | 0.27 | 0.28 |
| Potassium (%) | 0.57 | 0.54 | 0.60 | 0.55 | 0.66 |
| Choline equivalents (mg/kg) | 1,570 | 1,547 | 2,044 | 1,713 | 1,672 |
| Methionine (%) | 0.29 | 0.30 | 0.27 | 0.28 | 0.28 |
| Methionine + cystine (%) | 0.15 | 0.46 | 0.47 | 0.48 | 0.48 |
| Lysine (%) | 0.49 | 0.50 | 0.50 | 0.53 | 0.51 |
| Tryptophan (%) | 0.50 | 0.15 | 0.20 | 0.19 | 0.22 |
| Threonine (%) | 0.53 | 0.52 | 0.44 | 0.46 | 0.47 |

[1]Rations provided by Dr. Peter Ferket, Extension Poultry Nutritionist, North Carolina State University, School of Agriculture and Life Sciences, Raleigh, NC.

[2]To convert kilograms to pounds, multiply by 2.2.

**TABLE 44–16**
**TURKEY BREEDER RATIONS[1] [2]**

| Ingredient | No. 30 | No. 31 | No. 32 | No. 33 | No. 34 | No. 35 |
|---|---|---|---|---|---|---|
| | (kg) | (kg) | (kg) | (kg) | (kg) | (kg) |
| Ground yellow corn | 600 | — | 292 | 671 | — | 303 |
| Ground wheat | — | 720 | 245 | — | 753 | 250 |
| Ground barley | 155 | 112 | 250 | 100 | 100 | 250 |
| Soybean meal (48%) | 165 | 88 | 133 | 122 | 40 | 82 |
| Meat meal (50%) | — | — | — | 20 | 20 | 20 |
| Fish meal (60%) | — | — | — | 20 | 20 | 20 |
| Fat (stabilized) | 10 | 10 | 10 | 10 | 10 | 10 |
| Ground limestone | 45 | 45 | 45 | 40 | 40 | 40 |
| Calcium phosphate (20% P) | 15 | 15 | 15 | 7.5 | 7.5 | 7.5 |
| Salt (iodized) | 3.5 | 3.5 | 3.5 | 3.5 | 3.5 | 3.5 |
| Vitamin:mineral premix No. 8 | 10 | 10 | 10 | 10 | 10 | 10 |
| DL-methionine | 0.75 | 1.13 | 0.75 | 0.50 | 0.75 | 0.60 |
| Choline chloride | 0.75 | — | 0.50 | 0.75 | — | 0.50 |
| **Total**  (kg) | 1,000.00 | 1,004.63 | 1,004.75 | 1,005.25 | 1,004.75 | 997.1 |

(Add methionine and choline at the expense of corn or wheat)

| Calculated analyses | | | | | | |
|---|---|---|---|---|---|---|
| Crude protein (%) | 14.9 | 14.9 | 15.0 | 15.0 | 15.1 | 15.0 |
| Digestible protein (%) | 13.4 | 13.3 | 13.8 | 13.4 | 13.5 | 13.3 |
| Crude fat (%) | 2.7 | 2.4 | 2.9 | 4.0 | 2.5 | 3.2 |
| Crude fiber (%) | 3.2 | 3.3 | 3.7 | 3.0 | 3.2 | 3.9 |
| Metabolizable energy (kcal/kg) | 2,929 | 2,886 | 2,858 | 3,000 | 2,931 | 2,899 |
| Calcium (%) | 2.1 | 2.1 | 2.1 | 2.05 | 2.06 | 2.07 |
| Available phosphorus (%) | 0.40 | 0.42 | 0.41 | 0.41 | 0.43 | 0.42 |
| Sodium (%) | 0.18 | 0.20 | 0.19 | 0.20 | 0.23 | 0.21 |
| Potassium (%) | 0.73 | 0.66 | 0.72 | 0.65 | 0.58 | 0.63 |
| Choline chloride (mg/kg) | 1,610 | 1,971 | 1,785 | 1,558 | 1,980 | 1,757 |
| Methionine (%) | 0.34 | 0.34 | 0.33 | 0.34 | 0.32 | 0.33 |
| Methionine + cystine (%) | 0.56 | 0.58 | 0.56 | 0.56 | 0.56 | 0.55 |
| Lysine (%) | 0.71 | 0.70 | 0.71 | 0.74 | 0.72 | 0.73 |
| Tryptophan (%) | 0.21 | 0.23 | 0.22 | 0.90 | 0.22 | 0.21 |
| Threonine (%) | 0.63 | 0.51 | 0.57 | 0.64 | 0.51 | 0.58 |

[1]Rations provided by Dr. Peter Ferket, Extension Poultry Nutritionist, North Carolina State University, School of Agriculture and Life Sciences, Raleigh, NC.

[2]To convert kilograms to pounds, multiply by 2.2.

**TABLE 44–17**
**VITAMIN-MINERAL PREMIXES FOR TABLES 44–14, –15, AND –16[1]**

| Vitamin:Mineral Premix No.[2] | Turkey | | | |
|---|---|---|---|---|
| | Starter 5 | Grower 6 | Finisher 7 | Breeder 8 |
| Vitamin A (millions IU) | 10.0 | 8.0 | 7.5 | 10.0 |
| Vitamin D (millions ICU) | 1.7 | 1.5 | 1.2 | 1.7 |
| Vitamin E (thousand IU) | 20.0 | 15.0 | 12.0 | 30.0 |
| Riboflavin (g) | 6.0 | 5.0 | 5.0 | 6.0 |
| Calcium pantothenate (g) | 15.0 | 12.0 | 10.0 | 15.0 |
| Vitamin B-12 (mg) | 0.015 | 0.012 | 0.010 | 0.015 |
| Niacin (g) | 40.0 | 40.0 | 30.0 | 50.0 |
| Vitamin K (g) | 1.5 | 1.5 | 1.5 | 1.5 |
| Folic acid (g) | 1.0 | 0.5 | 0.5 | 1.0 |
| Biotin (g) | 0.3 | 0.2 | 0.1 | 0.5 |
| Manganese (g) | 70.0 | 70.0 | 70.0 | 70.0 |
| Zinc (g) | 60.0 | 60.0 | 60.0 | 60.0 |
| Copper (g) | 8.0 | 8.0 | 8.0 | 8.0 |
| Selenium (g) | 0.2 | 0.2 | 0.2 | 0.2 |
| Iron (g) | 90 | 90 | 90 | 90 |

**For supplements multiply the above values by the factors shown below:**

| | | | | |
|---|---|---|---|---|
| Diets using 35% protein supplement | — | — | — | — |
| Diets using 40% protein supplement | — | 3.3 | 6.7 | 5.0 |
| Diets using special protein supplement | — | 20.0 | 20.0 | 20.0 |

[1]Rations provided by Dr. Peter Ferket, Extension Poultry Nutritionist, North Carolina State University, School of Agriculture and Life Sciences, Raleigh, NC.

All premixes should be made up to 20 kg by the addition of a carrier with the exception of the special premixes which should be made up to 50 kg before being mixed in a diet.

The amounts shown are the *levels of nutrients* to be added per 1,000 kg of diet.

[2]Note the values shown are amounts of pure ingredients.

Fig. 44–18. Baby turkey poult. (Courtesy, National Turkey Federation, Reston, VA)

## TABLE 44-18
### BODY WEIGHTS AND FEED CONSUMPTION OF TURKEYS DURING HOLDING AND BREEDING PERIODS[1][2]

| Age | Hens | | | | | Toms | | | |
|---|---|---|---|---|---|---|---|---|---|
| | Weight | | Egg Production | Feed | | Weight | | Feed | |
| (weeks) | (lb) | (kg) | (%) | (lb/day) | (g/day) | (lb) | (kg) | (lb/day) | (g/day) |
| 20 | 15.4 | 7.0 | — | 0.44 | 200 | 26.5 | 12.0 | 0.88 | 400 |
| 25 | 17.6 | 8.0 | — | 0.47 | 215 | 29.8 | 13.5 | 0.93 | 420 |
| 30 | 19.8 | 9.0 | Start light Stimulation | 0.51 | 230 | 35.3 | 16.0 | 0.97 | 440 |
| 35 | 20.9 | 9.5 | 66 | 0.57 | 260 | 37.5 | 17.0 | 0.99 | 450 |
| 40 | 20.5 | 9.3 | 63 | 0.56 | 255 | 39.7 | 18.0 | 1.01 | 460 |
| 45 | 20.1 | 9.1 | 60 | 0.55 | 250 | 40.1 | 18.2 | 1.06 | 480 |
| 50 | 19.8 | 9.0 | 50 | 0.53 | 240 | 40.8 | 18.5 | 1.10 | 500 |
| 55 | 19.8 | 9.0 | 40 | 0.51 | 230 | 41.5 | 18.8 | 1.12 | 510 |
| 60 | 19.8 | 9.0 | 35 | 0.49 | 220 | 41.9 | 19.0 | 1.15 | 520 |

[1]Adapted by the author from *Nutrient Requirements of Poultry*, 8th rev. ed., NRC, National Academy Press, Washington, DC, 1984, p. 18, Table 13.

[2]These values are based on experimental data involving *in-season* egg production (*i.e.*, November through July) of commercial stock. It is estimated that summer breeders would produce 70–90% as many eggs and consume 60–80% as much feed, respectively, as in-season breeders.

Tables 44–19 and 44–20 give, for 1989, the body weight, feed consumption, and feed-to-gain ratio of turkey toms and hens, respectively. Nationwide, toms were marketed at 123 days of age and an average weight of 26.35 lb, and hens were marketed at 100 days of age and an average weight of 14.68 lb. At 18 weeks of age, toms averaged 27.06 lb live weight, with a feed/pound turkey of 2.80. At 15 weeks of age, hens averaged 15.51 lb live weight, with a feed/pound turkey of 2.55.

## TABLE 44-19
### BODY WEIGHTS, FEED CONSUMPTION, AND FEED-TO-GAIN RATIOS OF TURKEY TOMS[1]

| Age | Live Weight | | Cumulative Feed/Tom on the Basis of National Average[2] | Feed/Lb of Turkey on the Basis of National Average | Metabolizable Energy Intake, kcal/Lb of Turkey[3] |
|---|---|---|---|---|---|
| | Average | Gain for Period | | | |
| (weeks) | (lb) | (lb) | (lb) | (lb) | (kcal/lb) |
| 1 | .32 | .26 | .32 | 1.24 | 1,605 |
| 2 | .66 | .34 | .75 | 1.27 | 1,645 |
| 3 | 1.15 | .49 | 1.27 | 1.31 | 1,720 |
| 4 | 1.86 | .71 | 2.43 | 1.35 | 1,810 |
| 5 | 2.79 | .93 | 3.85 | 1.41 | 1,920 |
| 6 | 3.93 | 1.14 | 5.80 | 1.50 | 2,040 |
| 7 | 5.27 | 1.34 | 8.49 | 1.60 | 2,170 |
| 8 | 6.74 | 1.47 | 11.49 | 1.71 | 2,310 |
| 9 | 8.34 | 1.60 | 15.07 | 1.82 | 2,460 |
| 10 | 10.09 | 1.75 | 19.36 | 1.92 | 2,610 |
| 11 | 11.96 | 1.87 | 24.16 | 2.03 | 2,765 |
| 12 | 13.99 | 2.03 | 29.81 | 2.14 | 2,920 |
| 13 | 16.07 | 2.18 | 36.02 | 2.25 | 3,080 |
| 14 | 18.26 | 2.19 | 42.95 | 2.36 | 3,245 |
| 15 | 20.46 | 2.20 | 50.39 | 2.47 | 3,415 |
| 16 | 22.67 | 2.21 | 58.56 | 2.58 | 3,605 |
| 17 | 24.87 | 2.20 | 66.74 | 2.69 | 3,805 |
| 18 | 27.06 | 2.19 | 75.87 | 2.80 | 4,005 |
| 19 | 29.24 | 2.18 | 85.21 | 2.92 | 4,210 |
| 20 | 31.40 | 2.16 | 94.96 | 3.04 | 4,420 |
| 21 | 33.51 | 2.11 | 105.37 | 3.16 | 4,635 |
| 22 | 35.54 | 2.03 | 116.02 | 3.28 | 4,855 |
| 23 | 37.51 | 1.97 | 126.96 | 3.40 | 5,075 |
| 24 | 39.41 | 1.90 | 138.55 | 3.53 | 5,295 |

[1]Source: *Turkey World*, January-February, 1990, p. 12, article entitled, "Faster Growing, More Efficient Turkeys in 1989," by Jerry L. Sell, Department of Animal Science, Iowa State University; reproduced with the permission of Dr. Sell. To convert pounds to kilograms, divide by 2.2.

[2]Feed-to-gain ratios calculated on the basis of plant weights minus condemnation weights.

[3]Estimated on the basis of national average feed-to-gain ratios.

### TABLE 44–20
### BODY WEIGHTS, FEED CONSUMPTION, AND FEED-TO-GAIN RATIOS OF TURKEY HENS[1]

| Age | Live Weight | | Cumulative Feed/Hen on the Basis of National Average[2] | Feed/Lb of Turkey on the Basis of National Average | Metabolizable Energy Intake, kcal/Lb of Turkey[3] |
|---|---|---|---|---|---|
| | Average | Gain for Period | | | |
| (weeks) | (lb) | (lb) | (lb) | (lb) | (kcal/lb) |
| 1 | .30 | .24 | .30 | 1.24 | 1,605 |
| 2 | .63 | .33 | .71 | 1.27 | 1,640 |
| 3 | 1.11 | .48 | 1.34 | 1.31 | 1,710 |
| 4 | 1.77 | .66 | 2.33 | 1.36 | 1,805 |
| 5 | 2.58 | .81 | 3.65 | 1.45 | 1,915 |
| 6 | 3.55 | .97 | 5.37 | 1.55 | 2,035 |
| 7 | 4.65 | 1.10 | 7.57 | 1.65 | 2,165 |
| 8 | 5.88 | 1.23 | 10.24 | 1.76 | 2,315 |
| 9 | 7.18 | 1.30 | 13.31 | 1.87 | 2,480 |
| 10 | 8.56 | 1.38 | 16.83 | 1.98 | 2,655 |
| 11 | 9.96 | 1.40 | 20.79 | 2.10 | 2,840 |
| 12 | 11.40 | 1.44 | 25.06 | 2.21 | 3,030 |
| 13 | 12.81 | 1.41 | 29.71 | 2.33 | 3,230 |
| 14 | 14.18 | 1.37 | 34.45 | 2.44 | 3,435 |
| 15 | 15.51 | 1.33 | 39.42 | 2.55 | 3,645 |
| 16 | 16.79 | 1.28 | 44.84 | 2.67 | 3,860 |
| 17 | 18.00 | 1.21 | 50.59 | 2.82 | 4,095 |
| 18 | 19.13 | 1.13 | 56.83 | 2.98 | 4,335 |
| 19 | 20.19 | 1.06 | 63.61 | 3.16 | 4,585 |
| 20 | 21.17 | .98 | 70.51 | 3.34 | 4,850 |

[1]Source: *Turkey World*, January-February, 1990, p. 14, article entitled, "Faster Growing, More Efficient Turkeys in 1989," by Jerry L. Sell, Department of Animal Science, Iowa State University; reproduced with the permission of Dr. Sell. To convert pounds to kilograms, divide by 2.2.

[2]Feed-to-gain ratios calculated on the basis of plant weights minus condemnation weights.

[3]Estimated on the basis of national average feed-to-gain ratios.

- Pointers pertinent to **feeding turkeys:**

1. Prevent poult "starve-out." Upon arrival, poults should be encouraged to eat feed and drink water as soon as possible. Using colored feed, or placing brightly colored marbles in the feed and waterers, may help. It may be necessary to dip the beaks of some of them in feed and water to start them eating and drinking. Some may even have to be force fed.

2. Turkeys grow faster than chickens; hence, they have relatively higher feed and protein requirements.

3. As they approach maturity, turkeys fed for market purposes should be fed rations that are quite different from those that are fed to turkey breeders.

4. There is a tendency among turkey breeders (egg producers) to provide a high-fiber holding ration, for use beginning at 16 weeks of age. This type of ration retards sexual maturity and may result in some desirable effects upon later reproductive performance. The holding ration limits energy intake, but should not limit protein, vitamins, and minerals. Where a holding ration is used, the birds should be switched to the breeder ration one month prior to egg production. Table 44–18 shows the feed consumption of turkeys during the holding and breeding periods.

5. Good range provides green feed and tends to reduce feed costs. However, it may make for higher losses from blackhead and other diseases, and from predators; and range turkey operations may make the neighbors unhappy because of dust, odors, and noise. So, fewer and fewer turkeys are being range raised.

## FEEDING DUCKS

Fig. 44–19. Growing ducklings. (Photo by J. C. Allen and Son, West Lafayette, IN)

The commercial duck industry of the United States is focused around the production of young ducklings for meat purposes, the vast majority of which are of the White Pekin breed. Ducklings are fed a high-energy, high-protein starter feed for the first two weeks. The feed is fed in the form of pellets or crumbles. Feed in this form is easier to consume, results in less waste, and makes for improved feed conversion. A grower feed consisting of ½ in. pellets is fed until marketing.

Suggested duck rations are presented in the following tables:

Table 44-21, Duck Rations.

Table 44-22, Suggested Macronutrient Requirements Of Ducks.

Table 44-23, Micronutrients For Duck Rations In Table 44-21.

## TABLE 44-21
### DUCK RATIONS[1]

| Ingredient | Percentage of Complete Ration | | | | |
|---|---|---|---|---|---|
| | Starter | Grower | Finisher | Developer | Layer[2] |
| | (%) | (%) | (%) | (%) | (%) |
| Yellow corn, #2 dent | 70.00 | 73.58 | 77.25 | 39.50 | 59.00 |
| Barley | — | — | — | 15.00 | 15.04 |
| Oats | — | — | — | 11.20 | — |
| Soybean meal (48% protein) | 18.18 | 19.70 | 16.13 | 12.40 | 13.95 |
| Alfalfa meal (17% protein) | 2.00 | — | — | — | — |
| Fish meal (60% protein) | 7.50 | — | — | — | — |
| Meat & bone meal (50% protein) | — | 5.00 | 5.00 | — | 5.00 |
| Wheat bran | — | — | — | 10.00 | — |
| Wheat middlings | — | — | — | 8.00 | — |
| DL-methionine | 0.17 | 0.22 | 0.16 | 0.15 | 0.14 |
| Dicalcium phosphate (18.5% protein) | 0.55 | 0.28 | 0.15 | 1.30 | 0.18 |
| Ground limestone | 0.75 | 0.77 | 0.86 | 2.00 | 6.24 |
| Iodized salt | 0.25 | 0.25 | 0.25 | 0.25 | 0.25 |
| Vitamin-mineral package | 0.20[3] | 0.20[4] | 0.20[4] | 0.20[4] | 0.20[5] |
| Chlortetracycline - 50 | 0.40 | — | — | — | — |
| **Calculated Analysis** | | | | | |
| Protein ... (%) | 20.0 | 18.3 | 17.0 | 15.0 | 16.0 |
| Metabolizable energy ... (kcal/lb) | 1,400 | 1,410 | 1,426 | 1,200 | 1,312 |
| Calcium ... (%) | 0.90 | 0.85 | 0.80 | 0.75 | 2.90 |
| Available phosphorus ... (%) | 0.45 | 0.40 | 0.35 | 0.38 | 0.35 |
| Lysine ... (%) | 1.12 | 0.90 | 0.80 | 0.70 | 0.75 |
| Methionine + cystine ... (%) | 0.90 | 0.80 | 0.70 | 0.65 | 0.65 |

[1]Rations provided by Dr. Peter Ferket, Extension Poultry Nutritionist, North Carolina State University, School of Agriculture and Life Sciences, Raleigh, NC.

[2]Layer diet may be supplemented with free choice ground oyster shell.

[3]Supplies/pound of complete ration the vitamins and minerals in the amounts listed in package No. 1 (see Table 44-23).

[4]Supplies/pound of complete ration the vitamins and minerals in the amounts listed in package No. 2 (see Table 44-23).

[5]Supplies/pound of complete ration the vitamins and minerals in the amounts listed in package No. 3 (see Table 44-23).

## TABLE 44-22
### SUGGESTED MACRONUTRIENT REQUIREMENTS OF DUCKS[1] [2]

| Breeder Nutrient Layer | Starter 0-2 weeks | Grower 2-6 weeks | Finisher 6-8 weeks | Developer | Breeder |
|---|---|---|---|---|---|
| Metabolizable energy[4] ... (kcal/lb) | 1,400 | 1,400 | 1,400 | 1,175 | 1,300 |
| Protein ... (%) | 20.0 | 18.0 | 16.0 | 14.5 | 16.0 |
| Lysine ... (%) | 1.1 | 0.9 | 0.8 | 0.65 | 0.75 |
| Arginine ... (%) | 1.1 | 1.0 | 0.9 | 0.7 | 0.85 |
| Methionine + cystine ... (%) | 0.9 | 0.8 | 0.7 | 0.6 | 0.65 |
| Calcium ... (%) | 0.9 | 0.8 | 0.8 | 0.7 | 2.9 |
| Available phosphorus ... (%) | 0.45 | 0.4 | 0.4 | 0.35 | 0.35 |
| Linoleic acid ... (%) | 1.0 | 1.0 | 1.0 | 0.8 | 1.0 |

[1]Rations provided by Dr. Peter Ferket, Extension Poultry Nutritionist, North Carolina State University, School of Agriculture and Life Sciences, Raleigh, NC.

[2]Nutrients shown in this table apply only to the energy level specified.

[3]Begin feeding breeder layer feed one month before the first egg is laid.

[4]The energy concentration is only an example. The energy concentration may vary from 1,000 to 7,500 kcal/lb, provided the concentration of each nutrient per unit of energy remains the same.

## TABLE 44–23
## MICRONUTRIENTS FOR DUCK RATIONS IN TABLE 44–21[1]

| Nutrient | Vitamin-Mineral Package[2] | | |
|---|---|---|---|
| | 1<br>0–2 Wks. | 2<br>2 Wks.-Adult | 3<br>Breeder |
| **Minerals** | | | |
| Potassium[3] ........ (%) | 0.7 | 0.6 | 0.6 |
| Sodium .......... (%) | 0.17 | 0.14 | 0.14 |
| Chlorine .......... (%) | 0.12 | 0.12 | 0.12 |
| Magnesium ..... (mg/lb) | 230 | 230 | 230 |
| Manganese ..... (mg/lb) | 25 | 25 | 25 |
| Zinc .......... (mg/lb) | 32 | 25 | 30 |
| Iron .......... (mg/lb) | 35 | 20 | 30 |
| Copper ........ (mg/lb) | 4 | 3 | 3 |
| Iodine ........ (mg/lb) | 0.18 | 0.14 | 0.20 |
| Cobalt ........ (mcg/lb) | 90 | 90 | 90 |
| Selenium ...... (mcg/lb) | 70 | 70 | 70 |
| **Vitamins** | | | |
| Vitamin A ........ (IU/lb) | 4,000 | 2,500 | 4,000 |
| Vitamin D₃ ..... (ICU/lb) | 500 | 400 | 400 |
| Vitamin E ....... (IU/lb) | 10 | 5 | 10 |
| Vitamin K ...... (mg/lb) | 1.0 | 0.5 | 1.0 |
| Riboflavin ....... (mg/lb) | 3.0 | 1.5 | 3.0 |
| d-pantothenic acid (mg/lb) | 6 | 4 | 5 |
| Niacin ......... (mg/lb) | 25 | 20 | 25 |
| Vitamin B–12 .. (mcg/lb) | 4 | 2 | 4 |
| Choline ....... (mg/lb) | 900 | 450 | 450 |
| Biotin .......... (mg/lb) | 0.05 | 0.05 | 0.05 |
| Folic acid ....... (mg/lb) | 0.6 | 0.4 | 0.5 |
| Thiamin ........ (mg/lb) | 1.6 | 1.5 | 1.4 |
| Pyridoxine ...... (mg/lb) | 1.4 | 1.4 | 1.4 |
| Ethoxyquin ...... (mg/lb) | 60 | 60 | 60 |

[1]Rations provided by Dr. Peter Ferket, Extension Poultry Nutritionist, North Carolina State University, School of Agriculture and Life Sciences, Raleigh, NC.

[2]Vitamin-Mineral Package(s) should provide the following levels/pound of complete feed.

[3]Not needed in commercial Vitamin-Mineral premixes.

• Pointers pertinent to **feeding ducks:**

1. Ducks should be fed pellets rather than mash; use ³⁄₃₂- or ³⁄₁₆-in. pellets. Pellets will make for a saving of 15 to 20% in the feed required to produce a market duck.

2. Ducks are nearly as good foragers as geese.

3. Ducks should be ready for market between 7½ and 8 weeks of age.

4. The holding rations are designed to maintain breeding ducks from about 8 weeks of age until the breeding season commences, without them getting too fat. It is recommended that birds on holding rations be limited to about ½ lb per bird per day.

5. A breeder diet should be substituted for the holding diet about 4 weeks before eggs are desired for hatching purposes.

## FEEDING GEESE

Fewer than a million geese are raised commercially in the United States, annually. However, geese are more popular in Canada, and they are the holiday bird of choice in Poland, Hungary, Czechoslovakia, France, and Belgium.

Fig. 44–20. Farm geese. (Photo by J. C. Allen and Son, West Lafayette, IN)

The nutrient requirements of and suggested rations for geese are presented in the following tables:

Table 44–24, Nutrient Requirements Of Geese.
Table 44–25, Goose Rations.

**TABLE 44–24**
**NUTRIENT REQUIREMENTS OF GEESE[1] [2]**

| | Starting (0–6 Weeks) | Growing (After 6 Weeks) | Breeding | | Starting (0–6 Weeks) | Growing (After 6 Weeks) | Breeding |
|---|---|---|---|---|---|---|---|
| **Energy Base:** | | | | **Energy Base:** | | | |
| kcal ME/lb Diet[3] ............. | 1,315 | 1,315 | 1,315 | kcal ME/lb Diet[3] ................. | 1,315 | 1,315 | 1,315 |
| *kcal ME/kg Diet[3]* ............ | *2,900* | *2,900* | *2,900* | *kcal ME/kg Diet[3]* ................. | *2,900* | *2,900* | *2,900* |
| **Protein** ..................... (%) | 22.0 | 15.0 | 15.0 | **Fat-soluble vitamins:** | | | |
| **Amino acids:** | | | | Vitamin A ........................... (IU) | 1,500 | 1,500 | 4,000 |
| Lysine ...................... (%) | 0.9 | 0.6 | 0.6 | Vitamin D ........................... (ICU) | 200 | 200 | 200 |
| Methionine + cystine .............. (%) | 0.75 | — | — | **Water-soluble vitamins:** | | | |
| **Major or macrominerals:** | | | | Niacin (Nicotinic Acid, Nicotinamide) .. (mg) | 55.0 | 35.0 | 20.0 |
| Calcium (Ca) ...................... (%) | 0.8 | 0.6 | 2.25 | Pantothenic Acid (Vitamin B–3) ............. (mg) | 15.0 | — | — |
| Phosphorus (P), available ........... (%) | 0.4 | 0.3 | 0.3 | Riboflavin (Vitamin B–2) ................. (mg) | 4.0 | 2.5 | 4.0 |

[1]Adapted by the author from *Nutrient Requirements of Poultry*, 8th rev. ed., NRC, National Academy Press, Washington, DC, 1984, p. 19, Table 14.

[2]For nutrients not listed, see requirements for chickens as a guide.

[3]These are typical dietary energy concentrations.

**TABLE 44–25**
**GOOSE RATIONS[1] [2]**

| | All-Mash Formulas | | | Mash and Grain Formulas | |
|---|---|---|---|---|---|
| | Age in Weeks | | | Age in Weeks | |
| | 0–3 (confinement) | 3– Market (range) | Breeding (confinement) | 3– Market (range) | Breeding (confinement) |
| | (lb/ton) | | | | |
| **Ingredient:** | | | | | |
| Ground yellow corn ......................... | 975 | 920 | 835 | 680 | 610 |
| Wheat shorts .............................. | 100 | 200 | 100 | 100 | 100 |
| Wheat middlings ........................... | 100 | 200 | 200 | 200 | 200 |
| Ground barley ............................. | 200 | 400 | 400 | 400 | 200 |
| Dehydrated green feed ..................... | 60 | 20 | 100 | 40 | 140 |
| Meat meal (50%) .......................... | 40 | 40 | 40 | 50 | 40 |
| Fish meal (60%) ........................... | 40 | — | 40 | — | 80 |
| Dried whey ............................... | 40 | — | 30 | 40 | 50 |
| Soybean meal (50%) ....................... | 400 | 175 | 150 | 420 | 365 |
| Ground limestone .......................... | 10 | 10 | 65 | 20 | 150 |
| Dicalcium phosphate ....................... | 10 | 10 | 15 | 15 | 30 |
| Salt (iodized) ............................. | 10 | 10 | 10 | 20 | 20 |
| Trace mineral premix ...................... | 5 | 5 | 5 | 5 | 5 |
| Vitamin premix ............................ | 10 | 10 | 10 | 10 | 10 |
| **Total** ........................... (lb) | 2,000 | 2,000 | 2,000 | 2,000 | 2,000 |
| **Trace mineral premix:** | | | | | |
| Manganous oxide (56% Mn) ............... (oz) | 3.0 | 3.0 | 3.0 | 6.0 | 6.0 |
| Zinc oxide (80% Zn) ..................... (oz) | 2.0 | 2.0 | 2.0 | 4.0 | 4.0 |
| Ground limestone .......................... (oz) | 75.0 | 75.0 | 75.0 | 70.0 | 70.0 |
| **Total** ........................... (lb) | 5.0 | 5.0 | 5.0 | 5.0 | 5.0 |
| **Vitamin premix:** | | | | | |
| Vitamin A ................. (millions of USPU) | 3.0 | 2.0 | 3.0 | 4.0 | 10.0 |
| Vitamin D₃ ................. (millions of ICU) | 0.6 | 0.4 | 1.0 | 0.8 | 2.0 |
| Riboflavin ................................ (g) | 4.0 | 2.0 | 3.0 | 4.0 | 8.0 |
| d-calcium pantothenate .................... (g) | 4.0 | 2.0 | 5.5 | 4.0 | 12.0 |
| Vitamin B–12 ............................ (mg) | 6.0 | 4.0 | 4.0 | 8.0 | 8.0 |
| Niacin .................................. (g) | 30.0 | 15.0 | 15.0 | 30.0 | 25.0 |
| Menadione sodium bisulfate .............. (g) | 1.0 | 1.0 | 1.0 | 2.0 | 2.0 |
| Vitamin E ................. (thousands of IU) | 2.5 | 2.0 | 5.0 | 4.0 | 10.0 |
| Ethoxyquin ............................. (oz) | — | — | — | — | — |
| Ground yellow corn to 10 lb ................. | + | + | + | + | + |
| **Total** ........................... (lb) | 10.0 | 10.0 | 10.0 | 10.0 | 10.0 |

[1]*Duck and Goose Raising*, Ontario (Canada) Department of Agriculture, Pub. 532, p. 79. Practical formulas of this type can be adapted to any area by judicious feed substitutions based on available feeds and prices.

[2]To convert pounds to kilograms, divide by 2.2.

- Pointers pertinent to **feeding geese:**

1. Geese are normally allowed access to pasture. Depending on the pasture, up to 20 to 25 adult geese, and up to 50 young geese can be grazed per acre.

2. Geese rations should be pelleted, in 3/32 in. or 3/16 in. pellets. Crumbles result in too much feed wastage, and should not be used.

3. Geese should be ready for market at about 15 weeks of age.

## FEEDING BOBWHITE QUAIL

Raising bobwhite quail for release in hunting or conservation areas or for sale to the gourmet food market is becoming increasingly popular. Although bobwhite quail have been raised domestically for many years, they are still *wild* birds compared to other domestic fowl. Consequently, a combination of good management and sound nutrition is essential to raising these birds successfully.

Good commercially prepared game bird feed, available at most local feed stores, will usually meet the nutritional needs of the bobwhite quail. However, a ration may be home-mixed.

The nutrient requirements of and suggested rations for bobwhite quail are presented in the following tables:

Table 44-26, Nutrient Requirements Of Bobwhite Quail.

Table 44-27, Bobwhite Quail Rations.

Table 44-28, Micronutrients For Bobwhite Quail Rations in Table 44-27.

**TABLE 44-26**
**NUTRIENT REQUIREMENTS OF BOBWHITE QUAIL[1]**

| Nutrient | Age (weeks) | | | | Breeder[2] |
|---|---|---|---|---|---|
| | 0–4 | 4–6 | 6–12 | 12–Adult | |
| Metabolizable energy ........ (kcal/lb) | 1,300–1,400 | 1,300–1,430 | 1,300–1,430 | 1,200–1,430 | 1,150–1,300 |
| Protein .................... (%) | 28.00 | 24.00 | 18.00 | 18.00 | 20.00 |
| Lysine .................... (%) | 1.60 | 1.40 | 1.20 | 1.00 | 1.00 |
| Methionine + cystine ........ (%) | 1.10 | 0.90 | 0.70 | 0.70 | 0.70 |
| Glysine ................... (%) | 1.60 | 1.45 | 1.10 | 1.10 | 1.15 |
| Calcium ................... (%) | 1.10 | 1.00 | 0.90 | 0.70 | 2.85 |
| Total phosphorus ........... (%) | 1.00 | 0.75 | 0.80 | 0.65 | 0.70 |
| Available phosphorus ......... | 0.60 | 0.50 | 0.45 | 0.40 | 0.45 |
| Linoleic acid .............. (%) | 1.00 | 1.00 | 0.80 | 0.80 | 1.00 |

[1]Rations provided by Dr. Peter Ferket, Extension Poultry Nutritionist, North Carolina State University, School of Agriculture and Life Sciences, Raleigh, NC.

[2]Begin feeding breeder feed two weeks before the first egg is laid.

**TABLE 44-27**
**BOBWHITE QUAIL RATIONS[1]**

| Ingredient | Percentage of Complete Ration | | | | |
|---|---|---|---|---|---|
| | 0–4 Weeks | 4–6 Weeks | 6–12 Weeks | 12 Weeks–Adult | Breeder |
| | (%) | (%) | (%) | (%) | (%) |
| Yellow corn, #2 dent ............ | 44.15 | 55.85 | 71.85 | 65.40 | 59.00 |
| Wheat middlings (std.) .......... | — | — | — | 5.00 | 4.35 |
| Soybean meal (48% protein) ...... | 43.50 | 34.80 | 18.70 | 24.00 | 24.50 |
| Alfalfa meal (17% protein) ....... | 2.50 | 2.50 | 2.00 | 2.65 | — |
| Fish meal (60% protein) ......... | 5.00 | 4.00 | — | — | — |
| Meat and bone meal ............ | — | — | 5.00 | — | 5.00 |
| Animal fat ................... | 1.70 | — | — | — | — |
| DL-methionine ................ | 0.18 | 0.10 | 0.15 | 0.13 | 0.10 |
| L-lysine-HCl ................. | — | — | 0.41 | 0.07 | — |
| Dicalcium phosphate (18.5% P) ... | 1.60 | 1.30 | 0.70 | 1.50 | 0.60 |
| Limestone ................... | 0.90 | 1.00 | 0.74 | 0.80 | 6.00 |
| Iodized salt ................. | 0.25 | 0.25 | 0.25 | 0.25 | 0.25 |
| Vitamin-mineral pack ........... | 0.20[2] | 0.20[2] | 0.20[3] | 0.20[3] | 0.20[4] |
| Amprolium ................... | 0.025 | 0.025 | — | — | — |
| Bacitracin[5] ................ | 0.020 | 0.020 | — | — | — |
| **Calculated Analysis** | | | | | |
| Protein .................... (%) | 28 | 24 | 18 | 18 | 20 |
| Metabolizable energy ........ (kcal/lb) | 1,300 | 1,320 | 1,380 | 1,350 | 1,275 |
| Calcium ................... (%) | 1.10 | 1.00 | 0.95 | 0.75 | 2.95 |
| Available phosphate .......... (%) | 0.60 | 0.50 | 0.45 | 0.40 | 0.45 |
| Lysine .................... (%) | 1.70 | 1.40 | 1.20 | 1.00 | 1.00 |
| Methionine + cystine ......... (%) | 1.10 | 0.90 | 0.70 | 0.75 | 0.75 |

[1]Rations provided by Dr. Peter Ferket, Extension Poultry Nutritionist, North Carolina State University, School of Agriculture and Life Sciences, Raleigh, NC.

[2]Supplies per pound of complete ration the vitamins and minerals in the amounts listed in package No. 1 (see Table 44-28).

[3]Supplies per pound of complete ration the vitamins and minerals in the amounts listed in package No. 2 (see Table 44-28).

[4]Supplies per pound of complete ration the vitamins and minerals in the amounts listed in package No. 3 (see Table 44-28).

[5]Zinc bacitracin or bacitracin methylene disalicylate.

- Pointers pertinent to **feeding bobwhite quail:**

1. Bobwhite prefer bite-sized particles. They avoid eating particles that are too big or too small.

2. Fine crumbles are best for young bobwhite under 8 weeks of age, whereas coarse crumbles or pellets are best for older birds.

**TABLE 44–28**
**MICRONUTRIENTS FOR BOBWHITE QUAIL RATIONS IN TABLE 44–27[1]**

| Nutrient | Vitamin-Mineral Package | | |
| --- | --- | --- | --- |
| | 1 0–6 Wks. | 2 6 Wks.-Adult | 3 Breeder |
| **Minerals** | | | |
| Potassium[3] ........ (%) | 0.2 | 0.16 | 0.16 |
| Sodium .......... (%) | 0.15 | 0.15 | 0.15 |
| Sodium chloride .... (%) | 0.20 | 0.20 | 0.20 |
| Chlorine .......... (%) | 0.11 | 0.12 | 0.12 |
| Magnesium ..... (mg/lb) | 300 | 300 | 300 |
| Manganese ..... (mg/lb) | 40 | 40 | 40 |
| Zinc .......... (mg/lb) | 35 | 25 | 30 |
| Iron .......... (mg/lb) | 45 | 40 | 30 |
| Copper ........ (mg/lb) | 3.6 | 3.0 | 3.6 |
| Iodine ........ (mg/lb) | 0.15 | 0.15 | 0.20 |
| Cobalt ....... (mcg/lb) | 90 | 90 | 90 |
| Selenium ...... (mcg/lb) | 45 | 45 | 45 |
| **Vitamins** | | | |
| Vitamin A ....... (IU/lb) | 3,000 | 2,000 | 3,000 |
| Vitamin D ...... (ICU/lb) | 600 | 500 | 600 |
| Vitamin E ....... (IU/lb) | 8 | 6 | 12 |
| Vitamin K₁ ...... (mg/lb) | 1.1 | 1.1 | 1.1 |
| Riboflavin ....... (mg/lb) | 4.0 | 2.0 | 4.0 |
| d-pantothenic acid (mg/lb) | 7.0 | 5.5 | 8.0 |
| Niacin ......... (mg/lb) | 30 | 20 | 20 |
| Vitamin B-12 .. (mcg/lb) | 8 | 5 | 11 |
| Choline ....... (mg/lb) | 1,000 | 800 | 600 |
| Biotin .......... (mg/lb) | 0.14 | 0.10 | 0.10 |
| Folic acid ....... (mg/lb) | 1.5 | 1.0 | 1.0 |
| Thiamin ........ (mg/lb) | 1.5 | 1.5 | 2.0 |
| Pyridoxine ..... (mg/lb) | 2.0 | 1.5 | 1.8 |
| Ethoxyquin ...... (mg/lb) | 60 | 60 | 60 |

[1]Rations provided by Dr. Peter Ferket, Extension Poultry Nutritionist, North Carolina State University, School of Agriculture and Life Sciences, Raleigh, NC.

Fig. 44–21. Bobwhite quail.

Fig. 44–22. Japanese quail.

# QUESTIONS FOR STUDY AND DISCUSSION

1. Trace the transition of the poultry industry from a backyard enterprise to modern confinement production.

2. Why is poultry nutrition more critical than that of other farm animals?

3. What proportion of the total production cost of poultry is due to feed?

4. How does the digestive system of poultry differ from the digestive system of ruminants?

5. What are the main differences in the nutrient composition of chicken meat and eggs?

6. Discuss the energy needs of poultry.

7. Discuss the protein needs of poultry.

8. For each calcium and phosphorus, briefly summarize pertinent poultry nutrition information under the following headings: (a) functions of mineral, (b) some deficiency symptoms, and (c) practical sources.

9. For each vitamin A, vitamin D, and vitamin E, summarize pertinent poultry nutrition information under the following headings: (a) some deficiency symptoms, and (b) practical sources.

10. For each biotin, niacin, riboflavin, and vitamin B–12, summarize pertinent poultry nutrition information under the following headings: (a) some deficiency symptoms, and (b) practical sources of the vitamin.

11. How may a poultry producer evaluate poultry feeds?

12. What are the major energy feed sources for poultry?

13. What are the major protein feed sources for poultry?

14. Discuss the reasons for using each of the following non-nutritive additives: (a) antibiotics, (b) arsenicals and nitrofurans, (c) drugs, (d) antioxidants, and (e) grit.

15. How do you account for the fact that more than 40% of commercially prepared primary (complete) feed in the United States is fed to poultry?

16. Before anyone can intelligently formulate a poultry ration, what four factors should be known?

17. Why are distinct and separate feeding programs and suggested rations presented for each of the following: chicken layers and breeders, broilers/roasters, turkeys, ducks, geese, and bobwhite quail?

18. List pointers which are pertinent to feeding replacement pullets.

19. List pointers which are pertinent to feeding breeder chickens.

20. List pointers which are pertinent to feeding layers.

21. What is *phase feeding* of layers; and what is it designed to accomplish?

22. List pointers which are pertinent to feeding broilers.

23. What do Tables 44–19 and 44–20 show relative to the market age and weight of turkey (a) toms, and (b) hens?

24. List pointers which are pertinent to feeding turkeys.

25. Why do the protein requirements of turkeys from one day of age to marketing differ so widely?

26. List pointers which are pertinent to feeding ducks.

27. Few geese are produced commercially in the United States; yet, goose is the holiday bird of choice in Poland, Czechoslovakia, France, and Belgium. What is the reason for this difference in popularity?

28. List pointers which are pertinent to feeding geese.

29. How does the feeding of bobwhite quail differ from the feeding of chickens?

## SELECTED REFERENCES

| Title of Publication | Author(s) | Publisher |
| --- | --- | --- |
| Commercial Chicken Production Manual, Third Edition | M. O. North | AVI Publishing Company, Inc., Westport, CT, 1984 |
| Feeding Poultry | G. F. Heuser | John Wiley & Sons, Inc., New York, NY, 1955 |
| Feeds and Feeding, 22nd Edition | F. B. Morrison | The Morrison Publishing Company, Ithaca, NY, 1956 |
| Nutrient Requirements of Poultry, No. 1 | | National Academy of Sciences, National Research Council, Washington, DC, 1971 |
| Nutrition of the Chicken | M. L. Scott<br>M. C. Nesheim<br>R. J. Young | M. L. Scott & Associates, Ithaca, NY, 1982 |
| Nutrition of the Turkey | M. L. Scott | M. L. Scott of Ithaca, Ithaca, NY, 1987 |
| Poultry: Feeds and Nutrition, Second Edition | H. Patrick<br>P. J. Schaible | AVI Publishing Company, Inc., Westport, CT, 1980 |
| Poultry Meat and Egg Production | C. R. Parkhurst<br>G. J. Mountney | Van Nostrand Reinhold Company, New York, NY, 1988 |
| Poultry Nutrition, Fifth Edition | W. R. Ewing | The Ray Ewing Company, Pasadena, CA, 1963 |
| Poultry Nutrition Handbook | J. D. Summers<br>S. Leeson | Department of Animal and Poultry Science, Ontario Agricultural College, University of Guelph, Guelph, Ontario, Canada, 1985 |
| Poultry Production, Twelfth Edition | M. C. Nesheim<br>R. E. Austic<br>L. E. Card | Lea & Febiger, Philadelphia, PA, 1979 |
| Scientific Feeding of Chickens, The, Fifth Edition | H. W. Titus<br>J. C. Fritz | The Interstate Printers & Publishers, Inc., Danville, IL, 1971 |
| Stockman's Handbook, The, Sixth Edition | M. E. Ensminger | The Interstate Printers & Publishers, Inc., Danville, IL, 1983 |
| United States-Canadian Tables of Feed Composition | J. H. Conrad, Chairman of Subcommittee | National Academy Press, Washington, DC, 1982 |

Futuristic housing for futuristic layers. (Courtesy, Alabama Poultry and Egg Assn., Cullman, AL)

# POULTRY HOUSES AND EQUIPMENT

## Chapter 45

Properly designed poultry buildings and equipment should provide for housing, feeding, and handling of poultry in accordance with recommended production practices.

No standard set of poultry buildings and equipment can be expected to be well adapted to such diverse conditions and systems of poultry production as exist in the United States. In presenting the following discussion, therefore, it is intended that it be considered as a guide only. Detailed plans and specifications for buildings and equipment can be obtained from a local architect or through contacting the college of agriculture of the state.

## SPACE REQUIREMENTS OF BUILDINGS AND EQUIPMENT FOR POULTRY

Fig. 45–1, Breeder males influence fertility, hatch, and progeny performance; hence, they should have adequate space and proper nutrition. (Courtesy, Jamesway Division, Butler Manufacturing Company, Kansas City, MO)

One of the first, and frequently one of the most difficult, problems confronting the poultry producer who wishes to construct a building or item of equipment is that of arriving at the proper size or dimensions. Less space than needed may jeopardize the health and well-being of the birds, whereas too much space will make the buildings and equipment more expensive than necessary. Suggested space requirements are given in Table 45–1 (see facing page).

## REQUISITES OF POULTRY HOUSING

There are certain general requisites of all animal buildings, regardless of the class of livestock, that should always be considered; among them, reasonable construction and maintenance costs, reduced labor, and utility value. These and other general requisites are fully covered in Chapter 5 of this book, so repetition at this point is unnecessary. In the case of poultry, however, increased emphasis needs to be placed on the following features of buildings:

1. **Temperature.** Feathers give some protection against cold. However, the bird's efficiency in egg production, meat production, and feed utilization is greatly lowered when it must endure temperatures appreciably below the zone of comfort.

Birds have a very poor defense against heat, and their cooling system is not very efficient because they do not have sweat glands. They attempt to adjust to heat (a) by panting or breathing rapidly with the mouth open, (b) by eating less and drinking more, (c) by holding their wings away from the body, and (d) by resting against a cool surface such as the damp earth or a concrete floor.

The optimum temperature for layers is 55° to 70°F and for broilers, 75°F.

Tests conducted by the U.S. Department of Agriculture showed that hens kept at 55°F laid at a rate of 75% and consumed 3.5 lb of feed for each pound of eggs, whereas those maintained at 85°F laid at a rate of only 50% and consumed 4 lb for each pound of eggs, while those kept at 23°F laid at a rate of only 26% and ate 12.3 lb of feed for each pound of eggs. Experiments conducted at the University of Connecticut showed a 12.5% increase in feed efficiency in broilers grown in a 75°F house compared to those grown at 45°F.

Humidity influences are tied closely to temperature effects. For laying houses, the relative humidity should be within the range of 50 to 75%.

2. **Insulation.** The term *insulation* refers to materials which have a high resistance to the flow of heat. Such materials are commonly used in the walls and ceiling of poultry houses. Proper insulation makes for a more uniform temperature—cooler houses in the summer and warmer houses in the winter—and makes for a substantial fuel saving in brooder houses.

Heat produced by layers varies with their weight and the environmental temperature. White Leghorn layers will produce 9 BTUs per hen per hour at a housing temperature of 55°F. This is approximately 40 BTUs per hour for a typical 4½-lb White Leghorn hen.

3. **Vapor barrier.** There is much moisture in poultry houses; it comes from open water fountains, wet litter, the respiration of the birds, and from the droppings. When the amount of water vapor in the house is greater than in the outside air, the vapor will tend to move from inside to outside. Since warm air holds more water vapor than cold air, the movement of vapor is more pronounced during the winter months. The effective way to combat this problem in a poultry house is to use a vapor barrier with the insulation. It should be placed on the warm side or inside of the house.

**TABLE 45–1**
**SPACE REQUIREMENTS FOR POULTRY[1]**

| Type of Bird | Type of Facility | Age of Birds (wk) | Space/Bird (sq ft) | Space/Bird (sq cm) | Feeder Space/Bird Linear (in.) | Feeder Space/Bird Linear (cm) | Water Space/Bird Linear (in.) | Water Space/Bird Linear (cm) |
|---|---|---|---|---|---|---|---|---|
| **Commercial layers** (Leghorn-type hybrids) | Floor | 0–4 | 0.30 | 279 | 1.0 | 2.5 | 0.2 | 0.5 |
| | Floor | 4–8 | 0.60 | 557 | 1.0 | 2.5 | 0.4 | 1.0 |
| | Floor | 9–16 | 1.25 | 1,161 | 1.5 | 3.8 | 0.6 | 1.5 |
| | Floor | 16 over | 1.50 | 1,394 | 1.5 | 3.8 | 1.0 | 2.5 |
| | Cage, individual | Adult | 0.50 | 465 | 3.0 | 7.6 | 1.5 | 3.8 |
| | Cage, colony | Adult | 0.50 | 465 | 3.0 | 7.6 | 1.5 | 3.8 |
| **Broiler-breeder pullets** | Floor | 0–8 | 0.80 | 743 | 1.0 | 2.5 | 0.5 | 1.3 |
| | Floor | 9–16 | 1.30 | 1,208 | 3.0 | 7.6 | 0.6 | 1.5 |
| | Floor | 16 over | 2.00 | 1,858 | 4.0 | 10.2 | 1.0 | 2.5 |
| **Broiler-breeder hens** | Floor | Adult | 2.50 | 2,322 | 4.0 | 10.2 | 2.0 | 5.0 |
| | Slat floor | Adult | 2.00 | 1,858 | 4.0 | 10.2 | 2.0 | 5.0 |
| **Broilers** | Floor | 0–4 | 0.30 | 279 | 1.0 | 2.5 | 0.2 | 0.5 |
| | Floor | 4–8 | 0.75 | 697 | 1.0 | 2.5 | 0.2 | 0.5 |
| **Roasters** | Floor | 0–8 | 0.80 | 743 | 1.0 | 2.5 | 0.5 | 1.3 |
| | Floor | 9–16 | 1.50 | 1,394 | 2.0 | 5.0 | 1.0 | 2.5 |
| **Turkey-breeders** | Confinement on floor | Adult | 4.50 | 4,181 | 3.0 | 7.6 | 1.0 | 2.5 |
| | Confinement and range | Adult | 2.50 in house plus range | 2,322 | 3.0 | 7.6 | 1.0 | 2.5 |
| **Turkeys-market** | Confinement on floor | 0–4 | 1.25 | 1,161 | 1.0 | 2.5 | 0.5 | 1.3 |
| | Confinement on floor | 4–16 | 2.50 | 2,322 | 2.0 | 5.0 | 1.0 | 2.5 |
| | Confinement on floor | 16–29 | 4.00 | 3,716 | 2.5 | 6.4 | 1.0 | 2.5 |
| **Duck-breeders** | Confinement-yard | Adult | 2.50 in house plus yard | 2,322 | 2.0 | 5.0 | 1.5 | 3.8 |
| **Ducklings-market** | Wire | 0–3 | 0.50 | 465 | 1.0 | 2.5 | 0.5 | 1.3 |
| | Floor | 0–3 | 1.00 | 929 | 1.0 | 2.5 | 0.5 | 1.3 |
| | Floor | 3–5 | 1.50 | 1,394 | 1.5 | 3.8 | 1.0 | 2.5 |
| | Floor | After 5 | 2.00 | 1,858 | 2.0 | 5.0 | 1.5 | 3.8 |
| **Geese** | Floor | 0–2 | 1.25 | 1,161 | 1.5 | 3.8 | 1.0 | 2.5 |
| | Floor | After 2 | 2.50 | 2,322 | 2.5 | 6.4 | 2.0 | 5.0 |

[1]"Report of the Committee on Avian Facilities," *Poultry Science*, 1974, 53(6):2257, with turkeys, ducks, and geese added by the author.

4. **Ventilation.** Ventilation refers to the changing of air—the replacement of foul air with fresh air. Poultry houses should be well ventilated, but care must be taken to avoid direct drafts and coldness. Good poultry house ventilation saves feed and helps make for maximum production.

Three factors are essential for good ventilation: (a) fresh air moving into the poultry house, (b) insulation to keep the house temperatures warm, and (c) removal of moist air.

In most poultry houses, easily controlled electric fans do the best job of putting air where it is needed. Gravity flow—or relying on air movement without fans—is suitable only for narrow houses and few birds.

A complete ventilation system has three parts: (a) a fan, or fans, to move fresh air through the house; (b) enough inlets to let plenty of fresh air in; and (c) enough outlets to let stale, moisture-laden air out. All these parts are necessary for success.

Exhaust-type ventilation systems are usually used. Wall fans are generally cheaper and easier to install than ceiling fans. However, ceiling fans provide gravity ventilation if electricity should fail.

The water-holding capacity of air increases with rising temperature (Fig. 45-2).

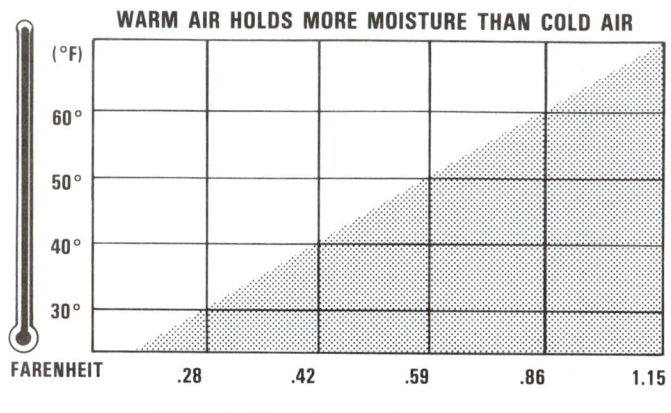

Fig. 45-2. Influence of air temperature on its water holding capacity. (From Washington State University Ext. Bull. No. 529)

Fans are rated in cubic feet per minute (cfm) of air they move. To arrive at fan capacity, allow (a) 3 cfm per bird for layers of the light breeds, (b) 4 cfm per bird for layers of the medium-weight breeds, and (c) ½ cfm per lb of body weight for broilers and turkeys.

5. **Lighting.** It has long been known that light stimulates egg production in the domestic fowl and other birds. Records show that the ancient Chinese made their canaries sing more by placing a lighted candle by their cage at night. Much later, early in the 1900s, poultry farmers in the state of Washington found that they could increase winter egg production by placing a lighted lantern in the chicken house for a few hours each evening. At the time, however, it was thought that the role of light was primarily a matter of increasing the "workday" of the bird. Today, the action of light is known to be physiological; light enters the eye of the bird and stimulates the pituitary gland. In turn, the pituitary gland releases certain hormones which cause ovulation. Because of this phenomenon, artificial lighting in the poultry house is extremely important. For pullets, an increase in the day length during the growing period will stimulate early maturity, whereas a decrease in day length will delay the age of maturity. For mature layers, an increase in day length will stimulate egg production, whereas a decrease in day length will suppress egg production.

Automatic time switches are available at moderate cost and should be installed in poultry houses for pullets or layers.

6. **Manure management.** The handling of manure is probably the single most important problem confronting commercial poultry producers today. It must be removed for sanitary reasons, and there is a limit to how much of it can be left to accumulate in pits or other storage areas. Also, to the urbanite, manure odor is taboo. Hence, it has two primary outlets: (a)

as a fertilizer, and (b) as a feed for ruminants (see Chapter 47 section on "Manure"). Its value for both purposes has increased in recent years—as a replacement for high-priced, petroleum-based chemical fertilizer, and as a substitute for scarce and high-priced livestock feed.

## REQUISITES OF POULTRY EQUIPMENT

The successful poultry producer must have adequate equipment with which to provide feed, water, and care of birds.

Certain features are desirable in all types of animal equipment, regardless of the species for which it is used. These points are discussed in Chapter 5 of this book; hence, the reader is referred thereto. But, because poultry are subject to numerous diseases and parasites, equipment for them should be constructed for easy cleaning and disinfection.

There are many types and designs of the various pieces of poultry equipment, and some producers will introduce their own ideas.

## SPECIALIZED BUILDINGS AND EQUIPMENT FOR SPECIALIZED PURPOSES

With the decline in farm flocks, fewer and fewer general-purpose type poultry houses and equipment are seen. The large and highly specialized poultry enterprises have developed buildings and equipment adapted to their highly specialized purposes. Although there is hardly any limit to the number of different styles, sizes, and colors of poultry buildings and equipment, within specialty types of operations certain principles are similar; that is, certain principles are observed in the facilities and equipment for layers, for brooding of chicks, for replacement pullets, for broilers, for turkey poults, for market turkeys, and for breeding turkeys. Building and equipment requisites for each of these specialty areas follow.

## HOUSING AND EQUIPMENT FOR LAYERS

The design and construction of houses and equipment for layers should be such as to provide for top performance of the layers, optimum environmental control, functional arrangement of equipment, maximum labor efficiency, satisfactory waste disposal, and minimum housing and care costs per dozen eggs produced. Also, futuristic housing and equipment must be big. It is estimated that the average egg firm owned 150,000 layers in 1990. By the year 2000, it is projected that this figure will exceed 250,000 per firm.

### HOUSES

Layer houses may be colony houses, multiple-unit houses, or multiple-story houses; some are permanent and others are movable.

The starting point in designing a layer house is the selection of the type of laying system. Presently, layer houses are being arranged in the following ways, or according to the following systems:

1. **Cage system.** Wild hens were caught and kept in cages by primitive people in order to facilitate egg gathering. Then, beginning in the 1930s, a limited number of commercial layers were kept in cages. Today, more than 90% of the layers in the United States are kept in cages.

Concrete floors are widely used in caged layer houses. In the cage system, small wire cages placed side by side, in long rows, hold the birds. These cages stand in rows on each side of an aisle (usually about 30 to 36 in. wide), placed at a convenient working height for the operator. The cage arrangement may range from single-deck to five-decks. Where multiple decks are used, the second row (and subsequent rows) of cages may be located directly above the lower row, or placed so that the bottom row projects forward about one-half the cage's depth, giving a stair-step effect.

Originally, 1 or 2 hens were kept in individual wire cages. To reduce the cost per bird, however, *colony cages* which hold up to 20 to 25 hens have been developed. Three or four hens per cage appears to be the most popular arrangement at the present time.

Cage sizes are not standardized, but Table 45–2 will serve as a useful guide.

Fig. 45–3. Cage system of housing layers, with three decks of cages, (Courtesy, DeKalb Agricultural Association, DeKalb, IL)

**TABLE 45–2**
**CAGE SIZES, BIRDS PER CAGE, AND FEED AND WATER SPACE**

| Cage Size | No. Birds | Total Sq. Inches | Sq. Inches per Bird | Inches Feed and Water Space per Bird |
|---|---|---|---|---|
| 8″ × 16″ | 1 | 128 | 128 | 8 |
| 8″ × 16″ | 2 | 128 | 64 | 4 |
| 12″ × 16″ | 3 | 192 | 64 | 4 |
| 12″ × 18″ | 3 | 216 | 72 | 4 |
| 12″ × 18″ | 4 | 216 | 53 | 3 |
| 24″ × 18″ | 7 | 432 | 62 | 3.5 |
| 24″ × 18″ | 8 | 432 | 54 | 3 |
| 12″ × 20″ | 4 | 240 | 60 | 3 |
| 2′ × 2½′ | 10 | 720 | 72 | 3 |
| 3′ × 4′ | 20 | 1,728 | 86 | 2.4 |
| 3′ × 4′ | 25 | 1,728 | 69 | 1.9 |

While the cage system (a) accommodates more birds in a given floor area than the litter floor system and (b) eliminates many internal parasite troubles, it does give rise to problems. High initial investment and high labor requirements have been experienced in the hen-per-cage system. With the colony-type cage arrangement, cannibalism and similar social problems have often been encountered.

Controlling flies and removing manure have been particularly difficult with the cage system.

2. **Slat floor.** This means just what the name indicates—the use of slats or wire over the entire floor. Slats, usually of metal or wood, are placed on edge. In slat or wire floors, the droppings collect in the space beneath the wire floors, and can be removed with a mechanical cleaner, or a floor section can be taken out once each year and the droppings removed with tractor-mounted equipment.

Fig. 45–4. New slatted floor layer house, with White Leghorn pullets. (Photo by J. C. Allen and Son, West Lafayette, IN)

The **advantages** of slat floors are: they require less floor space per bird than when the birds are kept on a litter floor, no litter is needed, there is better control of bacterial diseases, and it is not necessary to clean during the laying year—except to eliminate moisture.

The **disadvantages** of slat floors are: high humidity, no place where the birds can relax, birds appear more nervous, feather conditions become rougher, more egg breakage, more feather picking and cannibalism, and lower egg production.

3. **Combination slat and litter floor.** A slat-and-litter combination floor is popular for birds producing hatching eggs, particularly meat-type breeders. Such a system provides good fertility and keeps hatching eggs cleaner than other systems. In this system, slats are generally placed over two-thirds of the floor area, with the feeders and waterers located on the slats. The other third is covered with litter. Generally, the slats/litter are in either of two locations: (1) half the slats on each side of the house, with the litter area down the center; or (2) the slats across the center of the house, with half the litter area in front of the house and half in back of the house.

Fig. 45–6. Layers in solid floor, litter-type house. (Courtesy, Babcock Industries, Ithaca, NY)

Fig. 45–5. Combination slat and litter floor, used for breeders. Note slatted floor on left and litter floor and nest boxes on right. (Courtesy, Dr. M. J. Wineland, North Carolina State University, Raleigh)

As would be expected, each of these systems has its strong advocates, and the ardent supporters of each system can cite experiences and experiments to substantiate their claims. After studying the overall results obtained in the form of net income per hen or per laborer by these four main types of systems, the author came to the conclusion that there is no clear-cut advantage for any one type—that among the many factors influencing profit in the egg business, the type or system of housing is not paramount. The most important things are to design the building to fit the local climate; to provide adequate protection, ventilation, and cooling for the birds; and to make for good working conditions for the operator. It is further recommended that anyone planning to construct a layer house should inspect all types or systems, and confer with those who have used them; then, reach a decision.

In addition to the layer house, a building for handling, cooling, and holding eggs under refrigeration on the egg farm is essential.

Most poultry feed is purchased in bulk and delivered in bulk tanks from which it is withdrawn as used, by gravity or mechanical conveyors. Where whole grain is used, it may be practical to have sufficient storage facilities to permit purchase of grain at harvest time, when prices are usually more favorable.

A service building for storing supplies and small equipment, and for making repairs, is also needed. It may be a separate building, or it may be a part of the garage or egg room.

4. **Floor or litter-type house.** This is the oldest of the systems or arrangements, and at one time it was the exclusive type of layer house. It consists of litter covering the entire floor. Feeders and waterers are located on the litter, and nests usually line one or both sides of the house. This arrangement calls for a minimum amount of equipment. Also, fly control is simple. But it requires a well-insulated house with a good ventilation system.

Litter floor houses may be used in climates where slotted or wire floor houses or cages are unsatisfactory. They are expensive to construct, but maintenance costs are low. Also, they are more flexible; for example, additional brooding facilities are not required since the chicks may be brooded in floor-laying type houses which have been cleared of hens.

## EQUIPMENT

Good laying-house equipment is essential for satisfactory production. It should be simple in construction, movable, and easily cleaned. The nests, roosts, feeders, and waterers are of particular importance.

1. **Nests.** Nests should be roomy, movable, easily cleaned, cool and well ventilated, dark, and conveniently located. They are usually about 14 in. square, 6 in. deep, and with 15 in. head room. All-metal nests are preferred to wooden nests because of ease of cleaning and less chance of becoming infected by mites.

Nesting material should consist of small particles that are highly moisture absorbent, such as shavings, oat hulls, sawdust, or excelsior pads.

Roll-away nests, without nesting material, are becoming more common. They consist of plastic-covered wire bottoms, sloped to a covered egg tray.

Fig. 45-7. Nest box in a broiler breeder house. (Courtesy, Dr. M. J. Wineland, North Carolina State University, Raleigh)

Trapnests differ from regular nests in that they are provided with trap doors by which the birds shut themselves in when they enter. They are the accepted means of securing accurate individual egg records, and they are essential where pedigree breeding is practiced—that is, where more than one female is continuously mated with one male.

2. **Roosts.** Where roosts are used, they are commonly made of 2 in. × 3 in. or 2 in. × 4 in. lumber, placed sideways, with the edges rounded off.

3. **Feeders.** Since feed cost is the major item in egg production, it is necessary that there be adequate feeder space and that the feeders be good. They should be easy to fill, easy to clean, built to avoid waste, arranged so that the birds cannot roost on them, high enough so that the birds cannot scratch litter into them, and so constructed that as long as they contain any feed the birds can reach it.

Automatic (mechanical) feeders are standard equipment in large commercial egg operations.

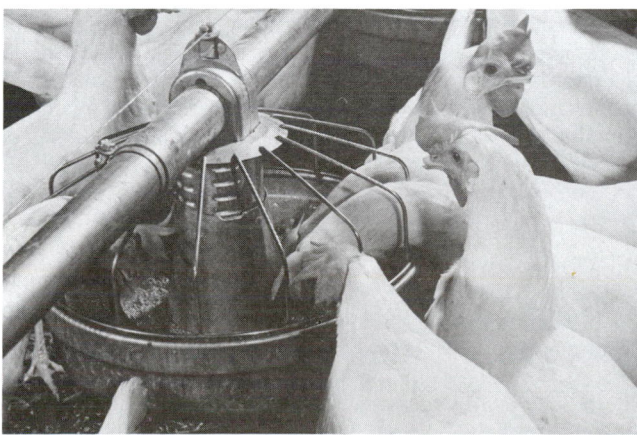

Fig. 45-8. Layers at feeder pan of an automatic feeder line. (Courtesy, Hy-Line International, Division of Hy-Line Indian River Company, Johnston, IA)

4. **Waterers.** Laying hens drink 2 to 3 lb of water for each pound of feed they eat. Watering devices should keep the water clean, be easily cleaned, and prevent spillage of water around the vessels or containers. Also, it is important that the waterers be distributed throughout the laying house so that a hen never has to travel more than 15 ft for a drink.

Many cage installations for layers are equipped with either drip-type or small-cup waterers that minimize cleaning problems as well as spillage.

## CARE OF THE HOUSE

Good husbandry and housekeeping are essential for optimum production and high egg quality. Also, it is necessary to minimize the spread of diseases. Accordingly, the following practices should be a part of the regular chores of the caretaker: inspect the birds daily, clean waterers daily, keep nests clean, keep light bulbs clean, clean windows regularly, control flies and rodents, and inspect all equipment routinely. Additionally, the following management and environmental factors should receive special attention.

1. **Litter.** Where litter is used, 6 to 8 in. of it should be provided. Litter absorbs moisture from droppings and then gives this moisture to the air brought in by ventilation. A good litter is highly absorbent and fairly coarse, so as to prevent packing. The litter should be free of mold and contain a minimum amount of dust. Availability and cost will determine the type of litter used. Common litter materials include shredded cane pulp, soft wood shaving, peanut hulls, ground corncobs, rice hulls, and peat moss.

2. **Cooling.** In warm areas, summer heat may cause retarded egg production and even result in death losses. Well-ventilated houses help, but when temperatures become extreme, artificial cooling is necessary. During extremely hot weather, the house can be kept more comfortable by increasing the movement of air, cleaning the fan blades and screens, painting the roof white, and sprinkling the roof. Foggers, controlled by thermostats, may also be used to produce a fine spray. In an emergency, sprinkling water over the hens with a garden hose, using a fine spray, will help cut down death due to heat prostration if enough breeze is available to evaporate the water and cool the birds.

3. **Manure removal.** A manure removal or holding system should be planned before the operation is started. The importance of this becomes apparent when it is realized that 100,000 layers will produce over 12 tons of manure a day, or well over 4,000 tons a year.

## BROODER HOUSES AND EQUIPMENT FOR CHICKS

Wherever chicks are raised—whether for broilers or replacement layers, and whether for continued rearing in confinement or on the range—artificial brooding of some kind is necessary. No phase of the poultry business is so important as brooding—it's the part that makes for a proper start in life.

## BROODER HOUSES

Until recent years, brooder houses for chicks were, for the most part, either portable or stationary buildings that were used for other purposes the balance of the year—part of the laying house, or the garage, or perhaps one end of the machine shed was used for brooding purposes. Housing arrangements of this type are still common among farm flock owners. However, in large commercial installations, where it is not uncommon to find 30,000 to 50,000 or more chicks being brooded together as a unit, special brooder houses or arrangements are common.

Fig. 45–9. Brooder house, with baby chicks. (Courtesy, Gold Kist-Communications, Atlanta, GA)

## BROODING EQUIPMENT

Heating, feeding, and watering equipment are the three main items of equipment needed for the brooding of chicks.

1. **Heating equipment.** The heat requirements for artificially brooding chicks may be supplied by a wide variety of devices and methods, among them the following:

a. **Portable brooders.** These units are, as indicated by the name, portable or movable. Although they come in a wide variety of styles and sizes, they generally consist of a central heating unit surrounded by a hover. They may be heated by gas, oil, or electricity. Portable brooders cost less to install than central heating systems, but they cost more per chick to operate. Also, they require more attention and labor, and there is more fire hazard from using them than central systems.

b. **Infrared lamps.** In this method, infrared lamps are suspended 18 to 27 in. above the floor litter. These lamps do not heat the surrounding air, but they warm the chicks in the same manner as direct rays from the sun. One 250-watt infrared bulb will suffice for 60 to 100 chicks.

c. **Battery brooders.** Commercial battery brooders may be either (1) unheated brooders made for use in warm rooms, or (2) those equipped with heating units and warm compartments for use in rooms held at 60° to 70°F. Most batteries of today have heating units warmed by electricity and equipped with thermostat regulators.

d. **Central heating.** Large commercial operations, which handle 5,000 to 25,000 or more chicks per house, need a central heating system. Several different heating systems have been developed for such use. Most of them are highly automated, thermostatically controlled, and fueled by oil, gas, or electricity. Central heating may provide warmth through either hot water (pipes or radiant heating) or hot air (direct or indirect).

2. **Feeders and waterers.** Feeding and watering equipment which is suited to six-week-old birds is not satisfactory for day-old chicks. Hence, special feeding and watering equipment

must be provided, and the chicks must learn to eat and drink when they are first placed in the brooder.

## HOUSING AND EQUIPMENT FOR REPLACEMENT (STARTED) PULLETS

Regular replacement of the laying flock is one of the most important, and most expensive, essentials of egg production. It ranks next to feed in the cost of egg production, and it exceeds labor, housing, equipment, interest, taxes, and other costs.

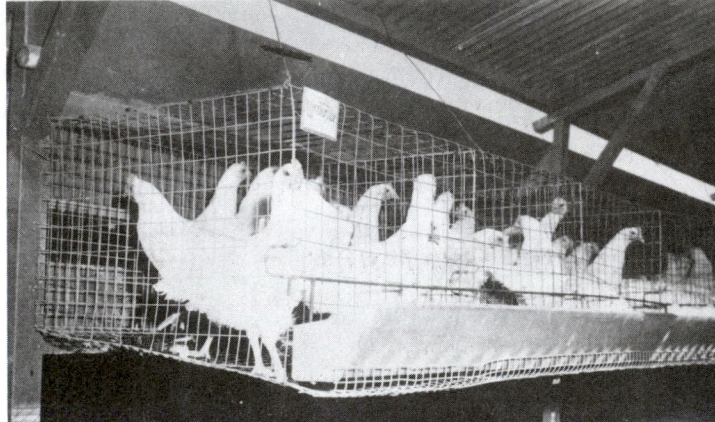

Fig. 45–10. Cage-reared replacement pullets. (Courtesy, Don Bell, Farm Advisor, University of California, Riverside)

Producers may choose between growing their own pullets or purchasing started pullets. In any event, the building and equipment requirements are the same, whether pullets are raised for sale or raised by the one who will retain them as layers.

Usually it is wise to provide pullets with levels of environment—housing, feed, and general management—considerably above minimum standards, but consistent with realistic cost consideration. This calls for housing that gives reasonable protection from heat and cold and provides good ventilation; and it means adequate feed and water space and facilities.

## HOUSING

Effective isolation of growing stock from older birds is important for disease control, particularly during the early stages. Hence, separate housing should be provided for replacement pullets, completely separated from layers. Housing that is suitable for brooding chicks, or for housing a laying flock, will be satisfactory for rearing pullets. Usually, they are started out in the brooder house, then switched to the layer house.

## FEEDERS AND WATERERS

Water space is less important than a good distribution of waterers through the house with a constant supply of fresh water. The same principle applies to some degree to feeders. Both feeders and waterers should be located to permit birds in any part of the house to feed or drink conveniently, without having to find their way around barriers or to travel more than 15 ft.

## OTHER EQUIPMENT

Hoppers for grit should be provided for replacement pullets throughout the growing period, and, as they near maturity, additional hoppers should be provided for oystershell or other calcium supplement.

Roosts are not always used where pullets are reared, but they are a desirable addition to the equipment. When used over enclosed pits, roosts aid in sanitation. Also, they add to the comfort of the birds in hot weather and help to establish desirable roosting habits for the laying period that follows. Some producers also feel that flocks with good roosting habits are less likely to present severe problems of floor eggs.

Nesting equipment should be available by the time the pullets begin to lay. Of course, only a few nests are necessary if pullets are moved from rearing quarters to their laying quarters before production becomes heavy.

Except for automatic feeders and waterers, little mechanization exists in the vast majority of pullet rearing facilities, especially when compared to layer facilities.

## HOUSES AND EQUIPMENT FOR BROILERS

No other segment of agriculture is as well suited to assembly-line production techniques as broiler production.

In modern commercial broiler production, the bird spends its entire life in one house; that is, it is not brooded in a special brooder house, then moved to a house for growing, for broiler raising is basically a brooding operation. Instead, brooder houses are thoroughly cleaned and disinfected between flocks, preferably with the quarters left idle one to two weeks before starting a new group.

Today, the vast majority of the nation's broilers are produced in large production units; and virtually 100% of them are grown under some type of vertical integration or contractual arrangement. Hence, in the discussion that follows, only the buildings and equipment common to these larger establishments will be described.

## HOUSING

A broiler house should provide clean, dry, comfortable surroundings for birds throughout the year. The house should be kept warm enough, but not too warm; the litter should be kept reasonably dry; provision should be made to modify the air circulation as broilers grow; and fresh air should be circulated, but the house should be free from drafts. In short, the broiler house should not be too cold, too hot, too wet, or too dry.

Most of the new broiler houses being constructed today are 24 to 40 ft wide, with gable-type roofs. Truss-type (of either wood or steel) construction is replacing pole-type due to lower labor costs in construction and greater ease of cleaning with a tractor. The length varies from 200 to 600 ft, with most of them averaging 300 to 400 ft. Capacity varies, but in the newer houses they generally range from 7,200 to 20,000 broilers. All of the birds may be in one large pen, but the newer trend is to pen units of 1,200 to 2,500 birds. There is increasing interest in controlled environment housing for broilers. When the broiler house is insulated and environmentally controlled, a centrally heated brooding system may be difficult to justify. However, where broiler houses are not environmentally controlled, a central heating system, of either hot or cold air or

Fig. 45-11. Typical curtain-sided broiler houses. (Courtesy, Dr. S. E. Scheideler, North Carolina State University, Raleigh)

hot water, is preferred. Where a chick hover is used, large—1,000 capacity—units are most popular.

Although maintaining adequate temperature is of great importance, constant attention is needed to ventilate the broiler house properly. This calls for a building that is properly insulated and in which there is a forced ventilation system.

## FEEDERS

Baby chicks should be started eating from new, cut-down chick boxes or box lids placed at floor level, allowing one feeder lid per 100 chicks. These feeders allow the chicks to find the feed easily.

Following the *box lid stage*, broilers may be fed from trough feeders, hanging tube-type feeders, or from mechanical feeders. Bulk feed bins and mechanical feeders are the most costly of the various types of feeding equipment, but they also make for more saving in labor. As a rule of thumb, installation of a mechanical feeder or feeders is worthwhile if the investment is no more than five times the labor saved per year.

## WATERERS

Clean water at 55°F should be available at all times. Gallon fountains should be provided for baby chicks, but these should be replaced by automatic, hanging waterers as soon as the chicks have learned to drink from the latter.

All waterers should be cleaned and washed daily.

## HOUSES AND EQUIPMENT FOR TURKEYS

Although there are some species differences between turkeys and chickens, primarily because of a difference in size, the general principles relative to buildings and equipment for turkeys and chicks are very similar.

## FACILITIES FOR BREEDER TURKEYS

Breeding turkeys may be kept (1) on restricted range, (2) in confinement housing, or (3) in semiconfinement.

Open range without shelters should not be used unless the flock can be protected from cold winds by trees and/or sloping hillsides. Turkey range should be enclosed by a well-constructed, permanent fence that is at least 5 ft high, preferably with an electric wire around the outside further to discourage predators. A 4-acre area will accommodate about 600 breeders, with provision for some rotation of pens and 2 night pens. Night pens should be equipped with roosts and lights arranged to ensure the entire area is lighted.

Day pens should be equipped with waterers, feeders, and nests.

Buildings for strict confinement are usually 40 to 50 ft in width and covered with wire on all sides. Most operators enclose 3 sides of the shelter with plastic to give protection during the winter months. Strict confinement lends itself to automatic feeding and watering.

Semiconfinement gives the protection advantage of complete confinement during the bad weather, with the added yard space for improved sanitation and mating. Buildings for semiconfinement are usually of lower cost construction than those used in strict confinement; generally they consist of an open-front, pole-type building, with a fenced yard.

Choice between the system—restricted range, strict confinement, and semiconfinement—depends to a very large extent upon the physical characteristics of the building site on the farm, the cost involved, and the weather.

## FACILITIES FOR BROODING TURKEY POULTS

Turkey poults should be brooded and reared separately from all ages of chickens and adult turkeys. In large commercial turkey operations, they're usually brooded in permanent-type houses. Some growers prefer to brood in batteries for the first 2 or 3 weeks before placing the poults on the floor. Regardless of the method of brooding employed, at least 1 sq ft of floor space is required to 8 weeks of age. More floor space must be provided as the poults grow older.

Fig. 45–12. Turkey poults in a brooder house with access to feed, water, and heat from a gas brooder. Integrated operation of Cuddy Farms, Ltd., North Carolina. (Photo courtesy, James Strawser, University of Georgia, Athens)

Any of the types of eating equipment commonly used for baby chicks can be used for brooding poults. To begin with, poults should be maintained at a temperature of 95° to 100°F, with the temperature lessening following the first week. A guard should be placed around the brooder for the first two weeks to prevent crowding and smothering in the corners of the house.

The watering and feeding facilities used by turkey poults are very similar to that used by baby chicks, but generally they are somewhat larger and, of course, more space needs to be provided per bird.

## FACILITIES FOR MARKET TURKEYS

For the first 8 to 10 weeks of life, the brooding of turkey poults is the same regardless of whether they are subsequently to be raised in confinement or moved to the range. Although a limited number of market turkeys are reared on the range, there is an increasing tendency to grow them in confinement in pole-type shelters, or even in environmentally controlled buildings. However, breeder turkeys are usually only partially confined.

Fig. 45–13. Typical turkey grow-out facility in southeastern U.S. (Photo courtesy James Strawser, University of Georgia, Athens)

Where turkeys are to be range raised under a rotation system, portable range shelters, roosts, feeders, and waterers are necessary. These may be placed on runners or wheels and moved where and when needed.

When confinement reared, turkeys are generally provided with a pole-type shelter, although more costly environmentally controlled units are being used, particularly for turkey-fryer production. When raised in strict confinement, 3 to 5 sq ft of floor space should be provided to carry each bird to market age and weight, the amount depending on the size of the strain of turkeys. Automatic waterers and feeders, or bulk feeders, help reduce labor costs. Care should be taken to provide adequate ventilation and occasional additions to the litter to reduce dampness and dirty litter conditions.

Fig. 45-14. Inside view of grow-out facility with 7-week-old turkeys. Integrated operation of Cuddy Farms, Ltd., North Carolina. (Photo courtesy James Strawser, University of Georgia, Athens)

## HOUSES AND EQUIPMENT FOR DUCKS AND GEESE

Elaborate and expensive housing facilities are not necessary for ducks or geese. Except during storms, they prefer to be outdoors. A fenced-in area, in which there is a colony house or open shed to provide protection during inclement weather, is all that is needed.

Most commercial producers of ducks and geese allow breeders access to water for swimming. However, ducklings and goslings can be reared successfully without swimming facilities, provided they have a constant supply of readily available fresh drinking water.

Commercially produced ducks and geese are brooded much like chicks and poults. Brooding temperatures should be 85° to 90°F under the hover at the start, but reduced to 80°F by the end of the first week and 70°F by the end of the third week.

Fig. 45-15. Breeder ducks in breeding house on the world's largest duck farm. Note the nest boxes in the foreground. (Courtesy, Cherry Valley Farms Ltd., Rothwell, Lincoln, England)

Many different types of feeders and waterers are used for ducks and geese. The main requisites are that feeders handle pellets without wastage, and that the waterers be of a type which birds cannot get into and splash.

## OTHER POULTRY FACILITIES AND EQUIPMENT

In modern large-scale poultry operations, two other types of facilities and equipment, which for the most part are used in all types of poultry production operations, merit mentioning. Namely, (1) facilities and equipment for handling manure, and (2) an emergency warning system.

### FACILITIES AND EQUIPMENT FOR HANDLING MANURE

The facilities and equipment for handling poultry manure will vary according to the disposition made of it. Among the common manure disposal systems and equipment are the following:

1. **Dry spreading.** In most places, the spreading of dry manure on crop land is still the most common and economical way to dispose of it. The major problems or drawbacks to this method are the land acreage required and the odors produced. A manure spreader, a manure pit where storage is planned, and a field constitute the necessary facilities and equipment for this system.

2. **Wet spreading.** This refers to manure to which water has been added for the purpose of facilitating handling. Liquid manure is stored in large water-tight storage pits or tanks (a 30,000 gal storage facility is required to store manure from 10,000 laying hens for 8 weeks), with 90 to 93% water where a sludge pump is used and 80 to 85% water for a vacuum pump. For conveying the manure away from the storage tank or pit, either a watertight manure spreader or an irrigation pipe must be available.

3. **Lagoon.** Outdoor lagoons should be between 3 and 5 ft deep, and 1 acre in size for each 1,000 to 2,000 laying hens. Indoor lagoons should have a minimum of 3.5 cu ft of water per bird, proper inlet and outlet mechanisms to control water depth, and provision for supplying oxygen.

4. **Dehydration.** Drying equipment and marketing require a sizable investment for this type of operation. Dehydration equipment to handle manure is expensive. Also, there is an odor problem, which precludes the possibility of locating a manure drying operation near any population center.

5. **Incinerators.** Incinerators are rather expensive, require the use of considerable fuel to consume manure, and create an odor problem; hence, the use of incinerators for manure disposal does not appear to be the answer.

6. **Processing for feed.** Poultry house litter may be used as either an energy or a protein source for beef cattle. As an energy source, it has 10 to 40% the feeding value of No. 2 corn, and it may replace 15 to 25% of the grain ration. As a protein supplement, it has 50 to 55% the value of soybean meal (41%), and it may replace 25% of the oil meal supplement of a ration.

### EMERGENCY WARNING SYSTEMS

The confinement of large flocks of birds in a mechanically controlled environment entails considerable risk because of the possibility of (1) power or equipment failure or (2) fire or abnormal temperatures. To guard against such troubles, an emergency warning system should be installed. Such a warning system may save its cost during just one power interruption, fire, or undesirable temperature in an incubator, brooder house, layer house, broiler house, egg storage room, or furnace room. The continuation of the poultry enterprise as the result of a timely warning can be far more valuable than any monetary insurance settlement after the operation has failed.

## QUESTIONS FOR STUDY AND DISCUSSION

1. How would you prepare detailed building and equipment plans and specifications for a large and modern layer or broiler operation?

2. What are the likely consequences of (a) allowing too much space for housing and equipment, or (b) allowing too little space for housing and equipment?

3. Discuss each of the following special requisites of poultry houses: (a) temperature, (b) insulation, and (c) vapor barrier.

4. Why are (a) proper lighting, and (b) manure disposal so important in poultry production? Discuss the phenomenon of lighting from the standpoint of egg production.

5. Describe each of the following systems or arrangements of layer houses, and give the advantages and disadvantages of each: (a) cage system, (b) slat floor, (c) combination slat and litter floor, and (d) floor or litter-type house.

6. It is estimated that more than 90% of layers in the United States are kept in cages today. Why have cages become so popular?

7. Discuss the types, along with the importance, of each of the following kinds of laying-house equipment: (a) nests, (b) roosts, (c) feeders, and (d) waterers.

8. Discuss the special attention that should be accorded to each of the following management and environmental factors for optimum production and high egg quality: (a) litter, (b) cooling, and (c) manure removal.

9. For chicks, discuss the kind, along with the importance, of modern (a) brooder houses, (b) heating equipment, and (c) feeders and waterers.

10. For replacement pullets, discuss the kind, along with the importance, of modern (a) housing, (b) feeders and waterers, and (c) other equipment.

11. For broilers, discuss the kind, along with the importance, of modern (a) housing, (b) feeders, and (c) waterers.

12. For turkeys, discuss the kind, along with the importance, of modern (a) facilities for breeder turkeys, (b) facilities for breeding turkey poults, and (c) facilities for market turkeys.

13. Discuss the kind, along with the importance, of modern houses and equipment for ducks and geese.

14. Name six methods of disposing of manure, and give the primary advantages and disadvantages of each.

15. Why should there be an emergency warning system to guard against power failure in an environmentally controlled poultry house?

## SELECTED REFERENCES

| Title of Publication | Author(s) | Publisher |
| --- | --- | --- |
| Commercial Chicken Production Manual, Third Edition | M. O. North | AVI Publishing Company, Inc., Westport, CT, 1984 |
| Farm Building Design | L. W. Neubauer H. B. Walker | Prentice-Hall, Inc., Englewood Cliffs, NJ, 1961 |
| Farm Buildings, Second Edition | J. C. Wooley | McGraw-Hill Book Company, New York, NY, 1946 |
| Farm Buildings, Third Edition | D. G. Carter W. A. Foster | John Wiley & Sons, Inc., New York, NY, 1947 |
| Farm Service Buildings | H. E. Gray | McGraw-Hill Book Company, New York, NY, 1955 |
| Farm Structures | H. J. Barre L. L. Sammet | John Wiley & Sons, Inc., New York, NY, 1950 |
| Poultry Handbook | L. D. Schwartz F. W. Hicks | The Pennsylvania State University Press, University Park, PA, 1972 |
| Poultry Meat and Egg Production | C. R. Parkhurst G. J. Mountney | Van Nostrand Reinhold Company, New York, NY, 1988 |
| Poultry Production, Twelfth Edition | M. C. Nesheim R. E. Austic L. E. Card | Lea & Febiger, Philadelphia, PA, 1979 |
| Principles of Animal Environment | M. L. Esmay | Avi Publishing Co., Westport, CT, 1969 |

Picture of health! (Courtesy, Hubbard Farms, Walpole, NH)

# POULTRY HEALTH, DISEASE PREVENTION, AND PARASITE CONTROL

## Chapter 46

Fig. 46–1. When birds are placed in close confinement in great numbers under forced production and eat and sleep in close contact with their own body discharges, the control of diseases and parasites becomes of paramount importance. This shows market turkeys in confinement housing in Wisconsin. (Courtesy, *Turkey World*, Mount Morris, IL)

The annual report of the U.S. Department of Agriculture, Poultry Inspection Branch, also shows problems associated with certain poultry diseases. Each year, the USDA inspectors reject for human consumption about 2% of the total weight of poultry inspected. The most important causes of rejection are: leukosis, mycoplasma air sacculitis, and septicemia. (The latter includes birds that are rejected because of anemia, edema, dehydration, inflammatory lesions, cyanosis, hyperemia, or other indications of disease.)

Losses from some diseases have been cut to low levels because effective control procedures have been developed. Thus, infectious bronchitis and Newcastle disease do not show up as major causes of loss, primarily because of control measures that are now available. It is recognized, however, that relaxation of these control measures would quickly restore these diseases to the forefront of causes of economic loss.

Laying flock death losses of about 1% per month, or 10 to 12% per year, are considered normal. In broilers, the average mortality may be about 4%, but as low as 1 to 2% deaths up to market age can be obtained. Death losses in turkeys are high; normally, 8 to 9% of poults do not survive, and 5 to 6% of breeder hens die each year.

Not all poultry losses can be prevented, but they can be reduced; and the more knowledgeable producers and those who counsel with them are, the more successful they will be in instituting and carrying out programs of poultry health, disease prevention and parasite control. To the latter end, this chapter, along with Chapter 6, is presented.

U.S. poultry producers recognize the importance of healthy birds, and realize that unhealthy birds cause financial losses. Yet, death and condemnation losses take a tremendous toll. Even greater economic losses among the living result from decreased growth and egg production and lowered feed efficiency. The U.S. Department of Agriculture estimates that poultry disease losses approach $2 billion annually, with respiratory diseases (primarily Newcastle disease, infectious bronchitis, mycoplasmosis, laryngotracheitis, pasteurellosis, and turkey corza) accounting for one fourth of the total, or $500 million.

Since great numbers of broilers, layers, turkeys, ducks, and geese are concentrated in confined areas in modern production systems, potential losses from a severe disease outbreak are great. Of course, the prevention of an outbreak of disease is the key to minimizing losses. With few exceptions, treatment after a disease outbreak has occurred is costly, often unsuccessful, and the recovered flock does not return to peak performance. Thus, management of poultry must be done with disease control and prevention clearly in mind.

The *Journal of Avian Diseases* publishes an annual report from diagnostic laboratories in several regions of the United States that summarizes the diseases diagnosed at these laboratories the previous year. Approximately 100 separate conditions involving infectious agents, nutritional diseases, parasites, or mismanagement are listed in these reports.

## A PROGRAM OF POULTRY HEALTH, DISEASE PREVENTION, AND PARASITE CONTROL

Although the exact program will and should vary according to the specific conditions existing on each individual poultry farm, the basic principles will remain the same. With this in mind, the following program of poultry health, disease prevention, and parasite control is presented. The producers may use it (1) as a yardstick with which to compare their existing programs, and (2) as a guidepost so that they and their veterinarians and other advisers, may develop a similar and more specific program for each enterprise.

1. **Get clean stock.** Clean stock simplifies the problem of disease control by reducing the number and severity of problems that are present when the flock is established. Disease-free stock is best assured by purchasing stock from breeders who have conscientiously participated in organized disease control programs.

2. **Avoid bringing infection in.** Visitors should be kept out of poultry houses; such equipment as trucks, shipping crates, and chick boxes should be restricted from the standpoint of area; and rats, mice, and birds should be controlled.

Fig. 46-2. Guarding against infectious diseases is important on a big turkey operation. This shows a vehicle washing station on a turkey breeder operation in which all vehicles are washed before being allowed to enter the farm. (Photo courtesy James Strawser, University of Georgia, Athens)

3. **Follow vaccination program.** Vaccination is cheap insurance against heavy losses from certain diseases. Vaccines are available for the prevention of many diseases (see Table 46-1).

4. **Control internal and external parasites; reduce stresses.** Feed is always too costly to feed parasites. Also, The control of external and internal parasites, along with keeping stresses to a minimum, helps maintain the birds in good condition so that they can resist disease organisms.

5. **Recognize diseases early.** The best way in which to recognize the early stages when trouble is about to strike is through record keeping. A slump in feed and water consumption is usually one of the best early indicators. Thus, it pays to keep daily records on feed and water consumption, egg production, and mortality. Any major change from day to day, or over a period of time, may mean that a disease is present in the flock.

Also, caretakers should set aside a certain time each day for the purpose of observing the flock. At that time, they should note the birds' actions, how they're eating and drinking, and whether there are any unusual sounds—any sneezing or rattling.

6. **Use diagnostic laboratory.** Modern poultry farms represent a large investment. Thus, heavy losses may accrue if the wrong medication is given. Producers should not attempt to identify all diseases on the farm. When in doubt, they should always use the laboratory facilities of their area to get an accurate diagnosis.

7. **Dispose of carcasses properly.** Sanitary disposal of dead birds is becoming an increasing problem, particularly in large commercial operations. Satisfactory dead bird disposal helps control disease.

8. **Periodically vacate and clean.** Periodically, all poultry buildings should be vacated, thoroughly cleaned, and disinfected. This is the most effective way in which to prevent the development of disease cycles.

## NORMAL SICKNESS AND LOSSES

Certain rules of thumb are helpful in evaluating the importance of a developing disease problem; among them are the following:

1. **More than 1% of the birds sick at one time.** If the producer finds that more than 1% of the flock is sick at any one time, it should be obvious that a disease is present and that the problem should receive immediate attention.

2. **First 3 weeks' losses; 2% chickens, 3% poults.** During the first 3 weeks of a chick's life, mortality losses generally average about 2%. For turkey poults, they run about 1% higher. If losses are greater than these figures, there may be cause for alarm.

3. **After 3 weeks, 1% per month.** The normal mortality rate after 3 weeks of age should not exceed 1% per month. However, a slight rise in mortality can be expected as adult flocks come into egg production.

## DIAGNOSTIC LABORATORY

Most states have one or more poultry pathology laboratories, operated by the state, available to their producers and the state's poultry industry. Additionally, many veterinarians in private practice provide limited laboratory services for areas adjacent to their practice.

Accurate identification of poultry diseases is often difficult for even the best pathologists if they must work without the aid of a properly equipped laboratory. A wrong diagnosis often results in faulty recommendations with improper medication and unsatisfactory results. This can prove to be quite costly due to ineffective medication and a continuance of high mortality and increased numbers of diseased birds.

When a disease outbreak is suspected, live birds showing typical symptoms of the sick birds, along with a complete history of the flock, should be submitted immediately to a poultry diagnostic laboratory for examination. Such laboratories are equipped to identify the disease problem and make recommendations for control. Practicing veterinarians, industry service representatives, and trained extension personnel working with producers and a diagnostic laboratory can bring about a reduction in losses due to disease.

## VACCINATION PROGRAM

In a concentrated poultry production area, it is necessary to vaccinate poultry against certain diseases, to prevent (1) depressed growth, (2) lowered egg production, (3) death, and/or (4) decreased carcass or egg quality. The choice of vaccines and the decisions in respect to application procedures often are difficult. A single vaccination program cannot be designed that is perfect for all enterprises.

Technical problems attendant to immunization make it necessary that even the best programs be a compromise. The most satisfactory approach is to develop a program that recognizes the disease hazards of the farm and make allowances for the housing and management practices employed.

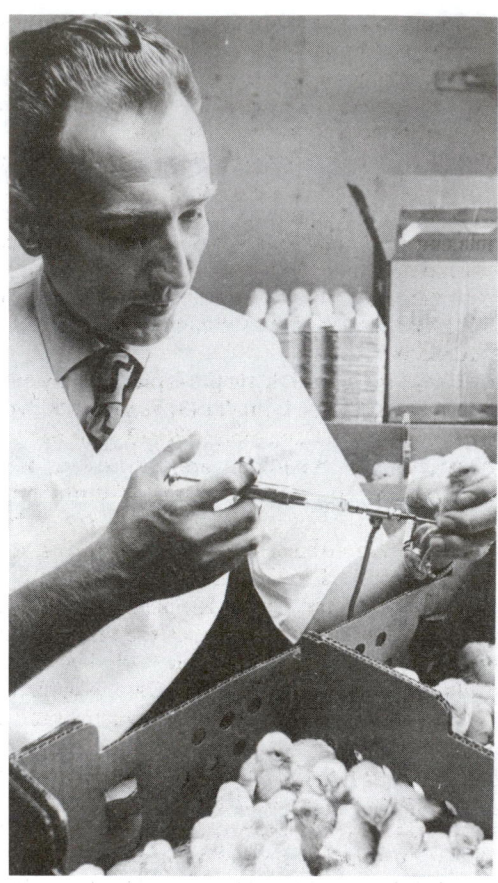

Fig. 46–3. One-day-chicks being vaccinated against Marek's disease.

Immunization is not the only way to control and prevent disease. With many diseases, sanitation procedures and isolation measures are relatively effective. When disease breaks, drugs may be available to provide control. With some diseases, such as coccidiosis, while immunization is possible, most operators find drug medication can prevent problems with less demand on management skill.

Before vaccinating, two points are noteworthy: (1) Certain diseases seldom occur in some areas; and (2) most vaccines act as a stress on poultry and their use introduces a hazard. In general, vaccination should be attempted when the following criteria are met: (1) the cost of the vaccine and the stress on the birds are appreciably less than the cost of the other means of control; and (2) the chances of occurrence of the disease are so great as to make the risk of not vaccinating too hazardous. Appraisal of the degree of risk should include the size and type of flock, the degree of isolation from other poultry, the nature of the disease, and the prevalence of the disease on the farm and in the area in recent years.

The particular vaccination program and schedule followed should vary according to the area and from flock to flock; and vaccines should always be used according to the instructions of the manufacturer, regarding time and method. Also, they should be kept cool before being opened and used immediately after they are opened.

In no case should the vaccination program be used as a substitute for good management. Likewise, no vaccination program is entirely successful without strict management practices to limit possible spread of infection.

## CARCASS DISPOSAL

Producers find it increasingly important to dispose of dead poultry by methods which are acceptable in an expanding urbanization, as well as economically feasible. The following three methods are commonly used:

1. **Rendering plant.** Where there is sufficient concentration of poultry and the service is available, a rendering plant is a safe, rapid, convenient, and economical method of carcass disposal. Usually, renderers are required to use trucks, equipment, and practices that will avoid hazards to animals and people.

2. **Disposal pit.** A disposal pit is simply an air-tight underground pit covered by boards and earth or concrete, with an opening through which dead birds can be dropped. Disposal pits should not be used in sandy soils and where there is a hazard of polluting ground water. With the exception of this precaution, they are entirely satisfactory.

3. **Incinerator.** Cremation, or burning, is the most satisfactory form of flesh disposal. If done properly, there is no odor, disease, rodent, fly, or water pollution problem. Incinerators may be home constructed, or a steel-jacketed commercial unit may be purchased. The incinerator should be equipped with an afterburner. Propane, gas, or oil may be used as the fuel.

## DISINFECTANTS AND THEIR USE

Fig. 46–4. Many disease problems can be avoided by thorough cleaning and disinfecting between broods. (Courtesy, DeKalb Poultry Research, Inc., DeKalb, IL)

The use of disinfectants, substances which have the power to kill microorganisms, is an important part of disease prevention and control in the poultry establishment. A section on "Disinfectants and Their Use" is given in Chapter 6 of this book.

Fig. 46–5. Worker scrubbing boots with disinfectant before entering a facility, to prevent transmitting disease. (Courtesy, Arbor Acres Farm, Inc., Glastonbury, CT)

## HATCHERY FUMIGATION

Fumigation of eggs and incubators is an essential part of a hatchery sanitation program. When improperly done, fumigation can be a hazard; hence, it should always be done by, or under the supervision of, an experienced person.

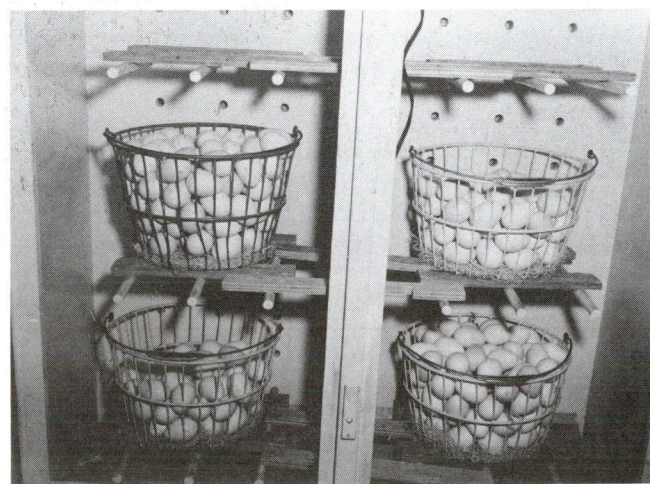

Fig. 46–6. Hatching eggs in fumigation cabinet. (Courtesy, *Turkey World*, Mount Morris, IL)

Routine preincubation fumigation of hatching eggs on the farm is highly recommended for eliminating Salmonella infection from poultry flocks. Each egg entering the hatchery should have been subjected to preincubation fumigation as soon as possible after its collection from the nest.

High levels of formaldehyde gas will destroy Salmonella organisms on shell surfaces if used as soon as possible after the eggs are laid. An inexpensive cabinet for enclosing the eggs is required. Fans circulate the gas during the fumigation process and then exhaust it from the cabinet. Eggs for fumigation should be placed on racks so that the gas can reach the entire surface. Plastic trays used for washing market eggs are ideal for this purpose.

A high level of formaldehyde gas is provided by mixing 1.2 cubic centimeters (cc) of formalin (37% formaldehyde) with 0.6 g of potassium permanganate ($KMnO_4$) for each cu ft of space in the cabinet. An earthenware, galvanized, or enamelware container having a capacity at least 10 times the volume of the total ingredients should be used for mixing the chemicals. The gas should be circulated within the enclosure for 20 minutes, then expelled to the outside.

Humidity for this method of preincubation fumigation is not critical; but temperature should be maintained at approximately 70°F. Extra humidity may be provided in dry weather. Eggs should be set as soon as possible after fumigation and extra care taken to ensure that they are not exposed to new sources of contamination.

Eggs should be routinely refumigated after transfer to the hatchery to destroy organisms that may have been introduced as a result of handling. Recommendations for loaded-incubator fumigation vary widely, depending upon the make of the machine. Therefore the method, concentration, and duration of fumigation should be in accordance with the manufacturer's instructions.

## DISEASES AND PARASITES OF POULTRY

For purposes of convenience and discussion, the diseases of poultry are herein classified as (1) nonnutritional diseases/parasites, (2) nutritional diseases and ailments, and (3) other ailments and health problems.

### NONNUTRITIONAL DISEASES/PARASITES

Ask poultry producers to name the most serious disease and parasite problems and more than likely they will name the ones that are afflicting their flocks at the moment. There are a host of poultry diseases and parasites. Fortunately, most of them can be controlled. But constant vigilance is necessary to keep the old ones under control, and to be on the alert for new ones.

An alphabetically listed tabular summary of the common nonnutritional diseases of poultry is presented in Table 46–1; and an alphabetically arranged summary of internal and external parasites of poultry is presented in Table 46–2.

TABLE 46-1

| Disease | Species Affected | Cause | Symptoms and Signs | Postmortem |
|---|---|---|---|---|
| **Arizona Infection** | Chiefly turkeys. | *Salmonella arizona.* | Poults unthrifty, and a good many develop eye opacity and blindness.<br>Mortality usually confined to first 3 to 4 weeks of age. | Yolk sacs slowly absorbed.<br>Liver enlarged and mottled.<br>Infected intestinal tract. |
| **Aspergillosis** (brooder pneumonia) | Almost all birds and animals, including humans. | Mold (*Aspergillus*). | Fever, difficult breathing, nervous symptoms. | Nodules in lungs and air sacs, pus in air sacs. |
| **Avian encephalomyelitis** (epidemic tremor) | All birds. | Virus. | Unsteadiness, sitting on hocks, inability to move, muscular tremors of head, neck, and limbs. | No gross lesions of the nervous system seen. |
| **Bluecomb disease** (coronaviral enteritis) | Turkeys. | Virus. | Poults appear cold and seek heat, stop eating and lose weight, and have frothy or watery droppings.<br>In growing turkeys, the appearance of the disease is sudden with a concurrent drop in feed and water consumption. Sick birds show darkening of the head and skin. | Birds show few lesions. The contents of the duodenum, jejunum, and ceca are watery and gaseous. |
| **Botulism** (limber-neck; food poisoning) | All birds except vultures. | Toxin produced by the anaerobic bacterium *Clostridium botulinum*. | Convulsions, paralysis, and sudden death. | Enteritis. |
| **Chronic respiratory disease** (C.R.D. or air sac disease) | Chickens.<br>Turkeys. | Various mycoplasms (*E. coli* causes secondary infection) | Coughing, gurgling, sneezing, nasal exudate, slow spread, loss of weight. | Mucus in trachea, air sacs thickened and containing yellow pus, thickened membrane over heart. |
| **Coryza** (roup) | Primarily chickens. | Bacterium (*Hemophilus gallinarum*) | Gasping, swollen eyes, nasal discharge, offensive odor. | White to yellow pus in eyes and sinuses. |
| **Duck plague** (duck virus enteritis) | Ducks.<br>Geese.<br>Swans. | Virus. | Spreads rapidly.<br>Can cause heavy mortality.<br>Affected birds reluctant to walk. | Hemorrhages of internal organs. |
| **Duck virus hepatitis** | Ducks.<br>Geese. | Virus. | Acute disease. Especially affects ducklings up to 3 weeks of age.<br>Mortality ranges from 5 to 90% | Characteristic liver lesions. |
| **Epidemic tremor** (avian encephalomyelitis) | All birds, but primarily chickens. | Virus. | Tremors of head and neck, muscular incoordination.<br>Temporary drop in egg production of layers. | None. |
| **Erysipelas** | Turkeys primarily, but other fowl affected.<br>Also humans and swine. | Bacterium (*Erysipelothrix*) | Sudden losses, swollen snood, discoloration of parts of face, droopy. | Hemorrhages in muscles, mucus in mouth; reddened intestines. |
| **Fowl cholera** | Chickens.<br>Turkeys.<br>Water fowl.<br>Other birds. | Bacterium (*Pasteurella multocida*). | Fever, purplish head, greenish-yellow droppings, sudden deaths. | Enlarged liver, hemorrhages in heart and in other organs. |
| **Fowl pox** (avian pox; avian diptheria; contagious epithelioma; sore head) | Chickens.<br>Turkeys.<br>Other birds. | Virus. | Small clear to yellow blisters on comb and wattles that soon scab over; decreased egg production. | May have lesions in throat and trachea. |
| **Fowl typhoid** | Chickens.<br>Turkeys.<br>Ducks.<br>Pigeons.<br>Pheasants. | Bacterium (*Salmonella gallinarium*) | Inactive, fever, greenish-yellow droppings. | Liver and spleen enlarged, bronze or greenish colored liver with some lesions. |
| **Gumboro** (infectious bursal disease) | Chickens. | Virus. | Sleepy.<br>White, watery diarrhea. | Enlarged bursa, hemorrhages. |
| **Hemorrhagic enteritis** (enteritis) | Turkeys. | Unknown, probably a virus. | Usually the only sign is one or more dead birds. | Severe hemorrhagic inflammation of intestinal lining from gizzard to ceca. |

## NONNUTRITIONAL DISEASES OF POULTRY

| Distribution and Losses Caused By | Treatment | Prevention | Remarks |
|---|---|---|---|
| Widespread. | Nitrofurans. | Elimination of infected breeder flocks. Hatchery fumigation and sanitation. | It is an egg-transmitted infection. |
| The incidence of the disease is not great. | No treatment. Remove source of infection. | Avoid musty and moldy feed and litter, provide good ventilation. | Young fowl most susceptible. |
| Worldwide. | Affected birds are usually destroyed. | Vaccination of breeder pullets at 10 to 15 weeks of age, repeated at molting if held second year. | Differentiated from Vitamin E deficiency by history, signs, and histological study. |
| Heaviest losses are in condition and production, but death losses may be high in young turkey poults. | Antibiotics or nitrofurans according to directions may be helpful in reducing the mortality. | Sanitation. | |
| Botulism occurs worldwide. | Flush exposed birds that do not show symptoms with Epsom salts. Move birds to clean facilities. Antitoxin may be used on valuable birds, but it is difficult to obtain and expensive. | Do not feed spoiled or decomposing feed. Prompt disposal of dead birds and rodents. Avoid wet spots in litter. | The toxin is very potent, being 17 times more deadly than cobra venom for the guinea pig. |
| Worldwide. | Antibiotics in feed or water according to directions. In severe outbreaks, inject birds with appropriate antibiotics. | Secure mycoplasma-free stock. | |
| Worldwide. | Sulfathiazole or antibiotics according to directions. | Keep age groups separate. Periodic complete depopulation. | Do not expose susceptible birds to recovered birds. Latter are lifetime carriers. |
| Worldwide. | No successful treatment. | A modified live virus vaccine. | This is a reportable disease. |
| Worldwide. | No known treatment. | Modified live virus vaccine. Strict isolation. | |
| Worldwide. Morbidity in affected flocks averages 5 to 10%. | None. | Vaccination of breeders. | Vaccination of commercial laying flocks is of questionable value. |
| The disease is of economic concern to turkey growers throughout the world. | Use antibiotics according to recommendations. | Vaccinate. | Transmitted via wounds or skin abrasions. |
| Worldwide. At times it causes high mortality; at other times the losses are nominal. | Sulfonamides and antibiotics according to directions. | Sanitation, disposal of sick birds, isolation of new stock. Vaccination. | When there is an outbreak, work fast in treatment. |
| Worldwide. Mortality is not high. Economic loss is in reduced feed efficiency and production. | Treatment is of little value. | Vaccination. | Control mosquitoes. |
| Worldwide. Mortality of affected birds ranges from 1 to 40% if treatment not instituted promptly. The cost of fowl typhoid is primarily in testing under the National Poultry improvement plan. | Nitrofurans or sulfa drugs according to directions. But every effort should be made to eradicate the disease. | Get stock from disease-free sources. | Egg and mechanical transmission. |
| Disease occurs in most of concentrated poultry-producing areas of the world. Heaviest losses in chicks up to 6 weeks. | None. | Live and dead vaccines. | The disease damages the birds' immune processes. Thus, vaccines to other disease are less effective. |
| The disease has been reported in the U.S., Canada, Japan, Australia, India, and Israel. | Injection of convalescent antiserum. | Vaccine. | |

(Continued)

| Disease | Species Affected | Cause | Symptoms and Signs | Postmortem |
|---|---|---|---|---|
| Infectious bronchitis | Chickens only. | Virus. | **Young birds:** Gasping, wheezing, nasal discharge.<br>**Older birds:** Sharp and prolonged drop in egg production, and soft-shelled eggs. | Yellowish mucus or plugs in lower trachea and air passages of lungs. |
| Infectious sinusitis | Turkeys. | Specific causative agent under investigation. | Nasal discharge, swollen sinuses, labored breathing, coughing. | Exudate in sinuses, cheesy material in air sacs. |
| Infectious synovitis | Chickens.<br>Turkeys. | *Mycoplasma synoviae* | The disease occurs primarily in growing birds from 4 to 12 weeks of age.<br>Enlarged hocks, foot pads, lame. Breast blisters. | Exudate at joints.<br>Enlarged liver, spleen. |
| Laryngotracheitis (gapes, chicken flu) | Chickens.<br>Pheasants. | Virus. | Gasping, coughing, loss of egg production, soft-shelled eggs, extending of neck outward on inhalation and slumping on exhaling, weeping of eyes. | Blood-stained mucus in trachea. |
| Leukosis (big liver disease; lymphoid leukosis) | Chickens.<br>Turkeys.<br>Other fowl. | Virus. | Loss of weight, diarrhea, thickened bones, gray eyes. | Enlarged liver and spleen, tumors in various parts of body. |
| Marek's disease (range paralysis, acute leukosis) | Chickens.<br>Other fowl. | Herpes virus. | Sudden death, loss of weight, diarrhea, paralysis of legs or wings.<br>Skin lesions in young birds. May occur as early as 5 to 8 weeks of age. | Enlarged liver, spleen, kidney, ovary, and testicles; or nodular tumors on these organs. Enlarged nerves in wings or legs.<br>Skin lesions in young birds. |
| Mycoplasma gallisepticum | All poultry. | Bacterium. | Infection of the air passages.<br>Coughing, sneezing, nasal discharge. | Thickened air sacs filled with exudates. |
| Mycoplasma synovial | All poultry. | Bacterium. | Affects the joints, producing an exidative tendonitis and bursitis. | |
| Mycosis (crop) | Chickens.<br>Turkeys.<br>Other fowl. | Pathogenic fungus (*Candida albicans*) | Crop distention. | Cheesy scum on crop lining. |
| Mycotoxicosis | All poultry. | Ingestion of toxic substances caused by molds growing on feeds and possibly litter.<br>*Aspergillus flavis*, which produces aflatoxins, is of most concern. The B-1 toxin is the most toxic and of greatest concern to the poultry industry. | Reduced growth and egg production and high mortality. | Hemorrhages.<br>Pale, fatty liver, kidneys. |
| Newcastle disease | Chickens.<br>Turkeys.<br>Other fowl. | Virus. | Gasping, wheezing, twisting of neck, paralysis, severe drop in egg production, soft-shelled eggs. | Often none. Sometimes mucus in trachea and thickened air sacs containing yellow exudate. |
| Paratyphoid | Chickens.<br>Turkeys.<br>Waterfowl and other birds. | Bacteria; Salmonella species other than *S. pullorum* and *S. gallinarum*. | Seen mainly in poults. | Enteritis, nodules in wall of intestines. |
| Pullorum | Chickens.<br>Turkeys.<br>Other domestic and wild fowl. | Bacteria (*Salmonella pullorum*). | Sleepy, pasted up, inactive, high mortality in young birds. | Lesions on lungs, liver, and intestines.<br>Unabsorbed egg yolks. |

*(Continued)*

| Distribution and Losses Caused By | Treatment | Prevention | Remarks |
|---|---|---|---|
| Worldwide. Economic loss is in lowered production and quality of eggs, and in mortality and lowered gains and feed efficiency of young chickens. | No specific treatment. | Inactivated and live vaccines. | |
| It may cause significant mortality in poults. | Drain sinuses and inject antibiotics into sinuses. | Secure poults from disease-free breeders and keep isolated from chickens. | |
| Probably worldwide. Mortality varies from 2 to 75% with 5 to 15% being most usual. | Antibiotics. | Test breeders and purchase clean chicks and poults. Heat treatment of eggs. | |
| Laryngotracheitis has been identified in most countries. | No drug treatment is effective. | Vaccination. | Farm eradication can be accomplished if security measures are superior. |
| With few exceptions, leukosis virus infection occurs in all chicken flocks; by sexual maturity most birds have been exposed. | None. | No vaccine. Buy birds from complement fixation avian leukosis (Cofal) flocks. Raise birds in isolation away from old or adult stock. | Now virtually eliminated from major breeding units. |
| Worldwide. Prior to development of a vaccine, losses ranged from 25 to 60%. | None. | Vaccination of day-old chicks at the hatchery. | |
| Worldwide. | Antibiotics. | Eradication from breeding flocks. An oil-emulsion vaccine is available. | |
| Worldwide. | Antibiotics. | Eradication from breeding flocks. | |
| | Fungicidal drugs. | Sanitation. Do not overcrowd. Nystatin: 50 gm/ton continuously. | |
| The disease first became prominent in 1960 when 100,000 turkeys died, later found to be caused by a toxin in moldy peanut meal. | Remove source of aflatoxin from the diet. | Avoid feed spoilage. Treat high moisture grains with propionic or acetic acids. | Is on increase. Young are more susceptible than mature birds. |
| Worldwide. Mortality of affected chickens varies from 0 to nearly 100%. | There is no effective treatment. | Vaccination. | It was first recognized in England in 1926; and it is named after the town of Newcastle. Newcastle disease was first reported in the U.S. in 1944. Symptoms in turkeys are mild; reduction of egg production of turkey breeders is main economic loss. This is a notifiable disease. |
| Worldwide. Mortality in turkey poults usually 1 to 20%. Outbreaks in ducks (keel disease) often run very high. | Sulfonamides, antibiotics, and nitrofurans may be employed to reduce mortality. | Hatchery and flock sanitation are the most important factors in paratyphoid prevention. | Egg and mechanical transmission; and through "blow-up" of infected eggs during incubation, and through some feeds. |
| Presently, the main economic loss from pullorum in the U.S. is due to the necessity of testing breeding flocks of chickens and turkeys to ensure freedom from infection. | Sulfonamides, nitrofurans, and antibiotics may be used to check mortality, but there is no substitute for a sound eradication program. | Eggs from disease-free breeders, hatched in disease-free incubators. Eradication from breeding stock. Buy chicks and poults only from pullorum typhoid-clean breeding stock. | Primarily egg-transmitted, but transmission may be by other means. |

*(Continued)*

**TABLE 46–1**

| Disease | Species Affected | Cause | Symptoms and Signs | Postmortem |
|---|---|---|---|---|
| Tuberculosis | All poultry. | Bacterial (*Mycobacterium avium*) | Unthriftiness, lowered egg production, and finally death. | Characteristic grayish-white or yellowish nodules of varying sizes in the liver, spleen, and intestines. |
| Turkey corza | Turkeys. | Bacterium. | Snicking, rales, and discharge of excessive nasal mucus. | Lesions and excessive mucus in the upper respiratory system. |
| Turkey veneral disease | Turkeys. | *Mycoplasma meleagridis*. | Cull poults, lowered fertility and hatchability. | Infected air sacs. |
| Ulcerative enteritis (quail disease) | Chickens. Turkeys. Game birds. | Bacterium (*Clostridium colinum*) | Sleepy, loss of appetite. | Ulcers on intestines, enteritis. |
| Vibrionic hepatitis | Chickens. | Bacterium. | Large number of culls. Decreased egg production. | Enlarged liver, spleen. |

Fig. 46–7. Bluecomb afflicted hen (left) vs normal hen (right). Note the dark and dried comb and wattles of the affected bird. This darkening starts at the outer edge, and progresses toward the base. (Courtesy, Salsbury Laboratories, Charles City, IA)

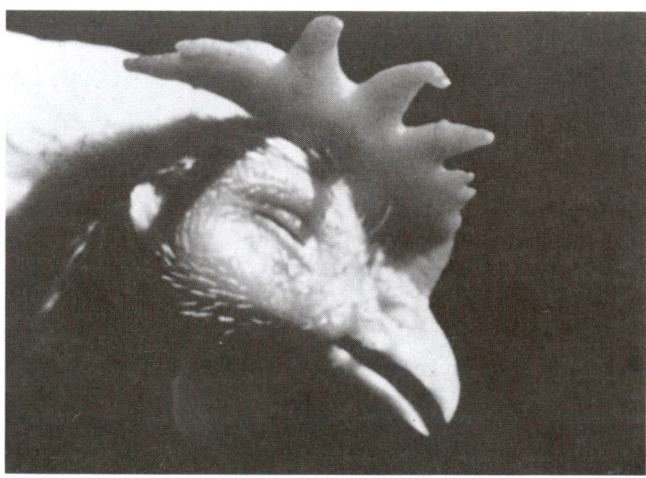

Fig. 46–8. Corza, showing swelling of tissue around the eye. (Courtesy, A. S. Rosenwalk, DVM, University of California, Davis)

Fig. 46–9. Ducklings with *duck viral hepatitis (DVH)*. Symptoms: Ducklings lie on side with heads drawn back and paddle feet spasmodically. (Courtesy, Salsbury Laboratories, Charles City, IA)

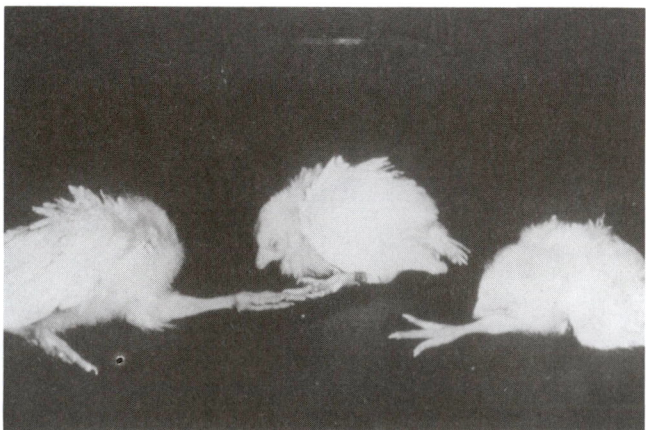

Fig. 46–10. Epidemic tremor (avian encephalomyelitis), a virus disease which can kill chicks and cause a lowering of egg output in laying flocks. Note that the chicks rest on their haunches or on their sides, and appear too weak to move about. The lack of muscle coordination prevents them from reaching feed and water. (Courtesy, Salsbury Laboratories, Charles City, IA)

*(Continued)*

| Distribution and Losses Caused By | Treatment | Prevention | Remarks |
|---|---|---|---|
| Worldwide, but occurs most frequently in the North Temperate Zone.<br>In 1972, tuberculosis was the cause for condemnation of 0.04% of the 186.9 million mature chickens slaughtered under federal inspection in the U.S. | None. | Sanitation; put disease-free birds in a clean house or on clean ground. | Avian tuberculosis is transmissible to swine, so keep swine and chickens separated. |
| Worldwide. | No treatment is entirely effective, although use of antibiotics in the early stages of the disease may be helpful. | A commercial vaccine is available. | Morbidity is generally 100%, while mortality ranges from 5 to 75%. |
| | | Dip eggs in antibiotics. | |
| Death losses may be high in replacement pullets and quail. | Antibiotics. | Sanitation.<br>Raising birds on wire is an effective preventive measure. | Increasing problem in chickens. |
| Vibrionic hepatitis has been identified in the U.S., Canada, Netherlands, Germany, Italy, and Switzerland.<br>Egg production may drop as much as 35%. Mortality is usually low, but may be as high as 10 to 15%. | Nitrofurans, antibiotics, or sulfas. | There is no successful immunization. | Affects young and immature chickens. |

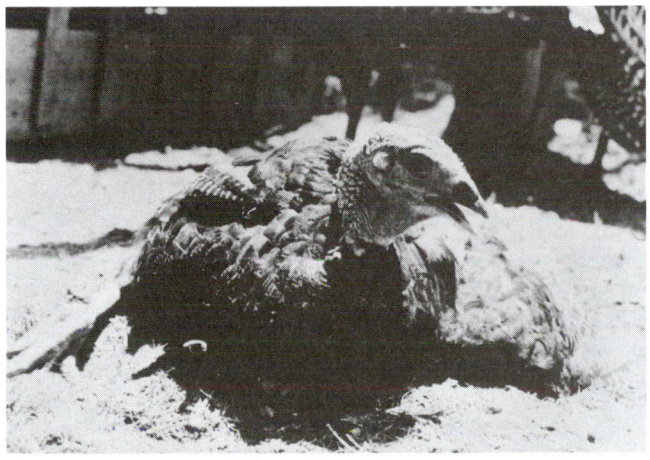

Fig. 46–11. Fowl cholera afflicted turkey. Symptoms: fever, purplish head, greenish-yellow droppings, ruffled feathers, and sudden death. (Courtesy, Virginia Poultry Federation, Harrisburg, VA)

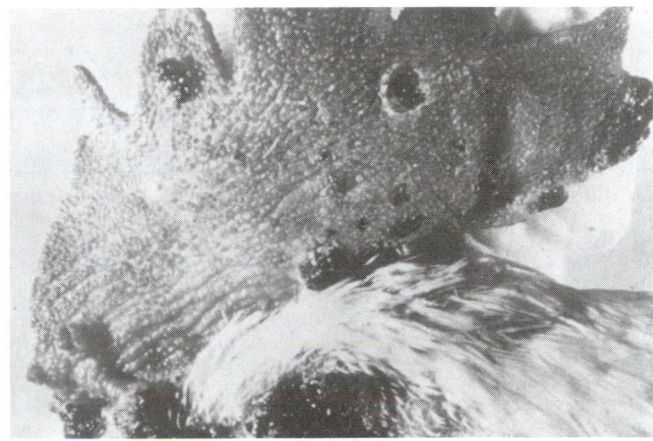

Fig. 46–12. Fowl pox—the dry or skin form. This common virus-caused disease may produce brownish scabs on the comb, face, and wattles. It usually occurs in the fall or early winter and may appear in the same locality from year to year. Pox scabs are wartlike, deep seated, and won't peel off. (Courtesy, Salsbury Laboratories, Charles City, IA)

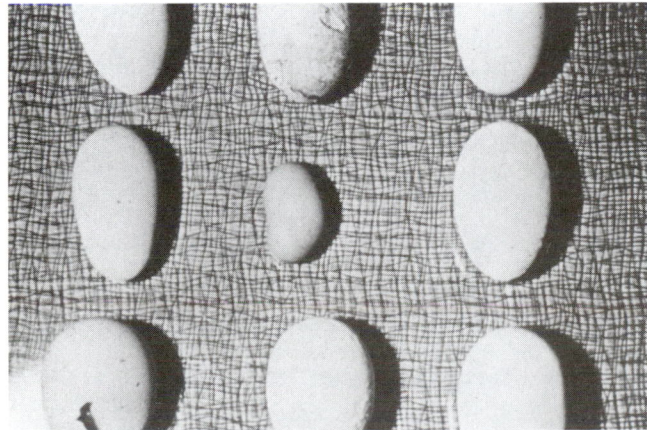

Fig. 46–13. Infectious bronchitis in the laying or breeding flock causes egg production to drop quickly; and the few eggs laid are often misshapen, rough or soft-shelled, and may vary widely in size. Egg quality may never return to normal. (Courtesy, Salsbury Laboratories, Charles City, IA)

Fig. 46–14. Laryngotracheitis (gapes), a highly contagious virus disease. The gasping or cawing position is typical, and blood may fleck from the mouth. Birds die suddenly. (Courtesy, Salsbury Laboratories, Charles City, IA)

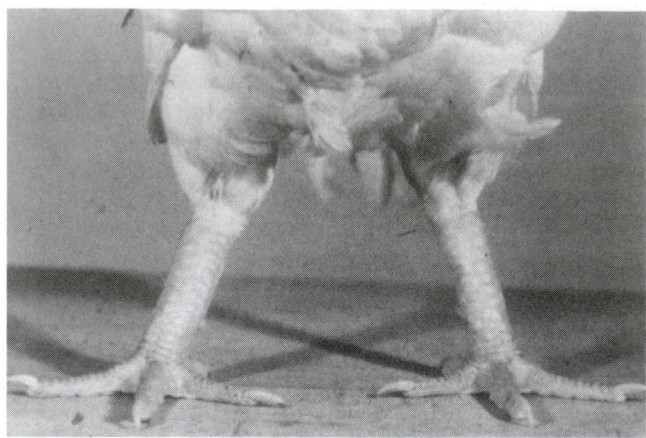

Fig. 46–15. Leukosis, showing a close-up view of leg bone enlargement caused by the bone form of leukosis. (Courtesy, Salsbury Laboratories, Charles City, IA)

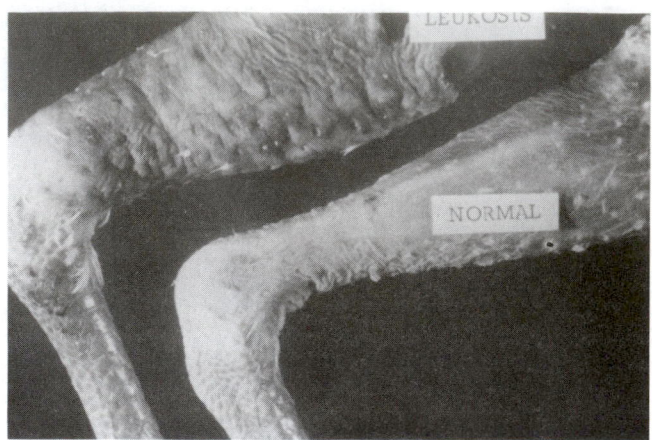

Fig. 46–16. Marek's disease, showing the skin form, sometimes referred to as skin leukosis. (Courtesy, Salsbury Laboratories, Charles City, IA)

**TABLE 46–2**

| Parasite | Species Affected | Cause | Symptoms and Signs | Postmortem |
|---|---|---|---|---|
| **Blackhead** (Histomoniasis) | Chickens, but more resistant than turkeys.<br>Turkeys. | Protozoa (*Histomonas meleagridis*). | Droopiness, loss of weight, sulfur-colored diarrhea, and darkened heads. | Lesions in liver and ceca, ceca enlarged, liver enlarged and spotted with dark red or yellow circular areas. |
| **Coccidiosis** | Chickens.<br>Turkeys.<br>Ducks.<br>Geese.<br>Game birds. | Protozoa (*Eimeria* species).<br>Six species of *Eimeria* cause the disease in chickens and three in turkeys.<br>Other species cause the disease in ducks, geese, and game birds. | Bloody droppings (usually cecal-type only), sleepy, pale, ruffled feathers, unthrifty. | Bloody or cheesy plugs in ceca (cecal-type).<br>Intestinal wall thickened with small white or reddish areas (intestinal-type). |
| **Hexamitiasis** (infectious catarrhal enteritis) | Turkeys.<br>Ducks.<br>Quail.<br>Partridges.<br>Pigeons.<br>But not chickens. | *Turkeys:* The one-celled parasite *Hexamita meleagridis*. | Listlessness, foamy and watery diarrhea, and convulsions. | Dehydration, emaciation, thin and watery intestinal contents, and bulbous areas in intestines. |
| **Large roundworms** | Chickens.<br>Turkeys. | Ascarid infestation (*Ascaridia galli*). | Droopiness, emaciation, and diarrhea. | Roundworms, 1½ to 3 in. long, in intestines. |
| **Leucocytozoorosis** | Turkeys.<br>Ducks. | Protozoa (*Leucocytozoon*). | Symptoms are not apparent in older birds.<br>Loss of appetite, droopiness, weakness, increased thirst and rapid, labored breathing in young birds. | |
| **Lice** | All poultry. | More than 40 species of lice. | Frequent picking, pale head and legs, loss of weight. | |
| **Mite infestation** (mites on body only during night) | Chickens are the commonest hosts, but these mites may occur on all poultry. | Common red, or roost, mites (*Dermanyssus gallinae*). | Reduced egg production, retarded growth, lowered vitality, damaged plumage, and even death. | |

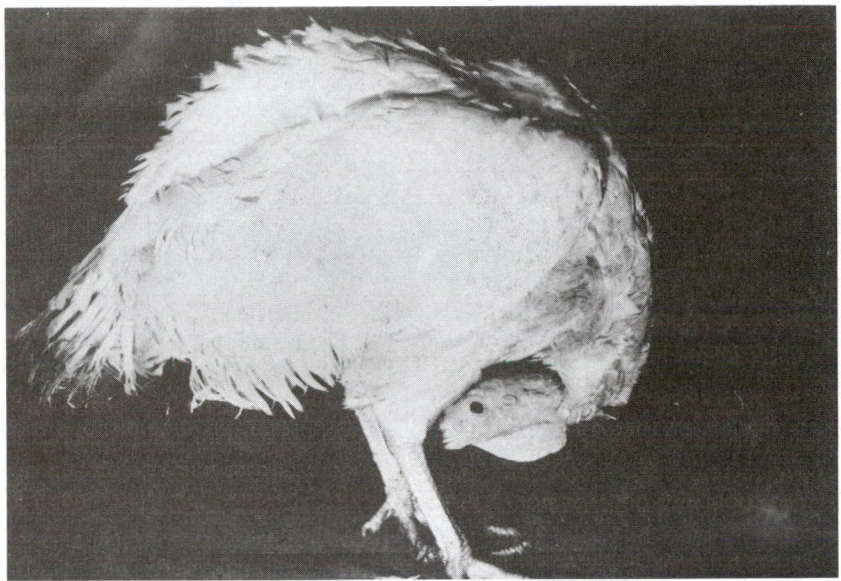

Fig. 46–17. Newcastle disease in 12-week-old turkey. (Courtesy, Virginia Poultry Federation, Harrisonburg, VA)

## PARASITES OF POULTRY

| Distribution and Losses Caused By | Treatment | Prevention | Remarks |
|---|---|---|---|
| Average annual losses in turkeys estimated to exceed $2 million. | A number of drugs are on the market, from which a selection may be made and used according to manufacturer's directions. | Sanitation, frequent range rotation. Do not crowd. Preventive medication in feed and water according to directions. Do not keep chickens and turkeys on the same premises. | Transmitted by droppings from infected birds. Control cecal worms. |
| Coccidia are found wherever poultry are raised. Estimates of annual looses in the U.S. range up to $200 million. More than $80 million spent on preventive medication, annually. | Anticoccidials in feed or soluble form of drugs in water according to directions. | Preventive medication in feed or water according to directions. Vaccination is useful in certain types of operations, but seek expert advice before using it. | Transmitted by droppings of infected birds. |
| Reported in the U.S., Canada, Scotland, England, and Germany. Hexamitiasis is not of major importance today. | The disease does not respond well to treatment. Furazolidone, aureomycin, and terramycin have been used with some success. | Segregation of age groups, and sanitation. | |
| | Several deworming drugs are on the market. Select and use according to manufacturer's directions. | Sanitation and rotation of range and yards. Careful use of old litter. Cage rearing and housing. | See Fig. 46–21, p. 900, for life cycle. |
| The disease occurs most frequently in southern and southeastern U.S. | Drug treatment of leucocytozoorosis has had limited success. Clopidol may be used as a treatment or preventive. | Exterminate black fly population, and do not raise turkeys near running streams. Segregate breeding and brooding operations. Brooding in houses with cheesecloth over openings during black fly season. Sulfadimethoxine and sulfaquinoxaline prevent infection of certain types of the parasite. | |
| Worldwide. Heavy infestations affect bird performance. | Select an approved insecticide and use according to manufacturer's directions. | Buy louse-free birds and never add lousy birds to clean stock. | |
| Worldwide. Can cause reduced performance. | Select an approved insecticide and use according to manufacturer's directions. | Sanitation. Examine birds frequently for signs of mites. Preventive insecticide treatment of quarters. Control sparrows. | |

*(Continued)*

TABLE 46-2

| Parasite | Species Affected | Cause | Symptoms and Signs | Postmortem |
|---|---|---|---|---|
| **Mite infestation** (mites always on body) | **A**ll poultry. | **N**orthern fowl mites or tropical feather mites. | **D**roopy, pale condition and listlessness. | |
| **Tapeworms** | **C**hickens.<br>**T**urkeys.<br>**O**ther birds. | **S**everal species of tapeworms. | **P**ale head and legs, poor flesh. | **T**apeworms in intestines. |
| **Trichomoniasis** | **C**hickens.<br>**T**urkeys.<br>**P**igeons.<br>**Q**uail. | **P**rotozoa (*Trichomonas gallinae*). | **L**oss of appetite, droopiness, loss of weight, and darkened head. | **L**esions—necrotic ulcerations—in the upper digestive tract, affecting the crop in particular. |

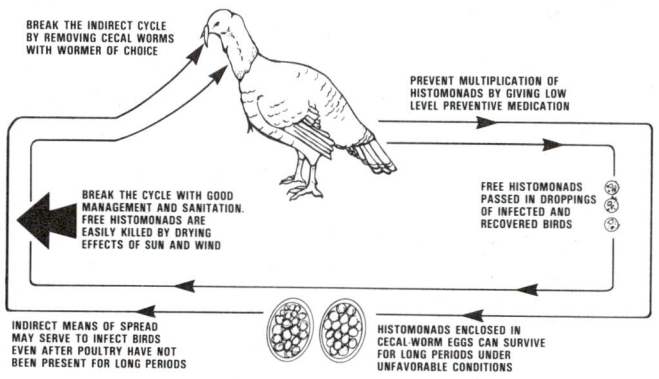

Fig. 46–18.  Blackhead (Histomoniasis), to which turkeys are especially susceptible. Preventive measures are: (1) good sanitation; (2) control of cecal worms; (3) turkeys kept on wire, away from chickens; and (4) a histomonastat administered continuously in feeds to turkeys over 6 weeks old.

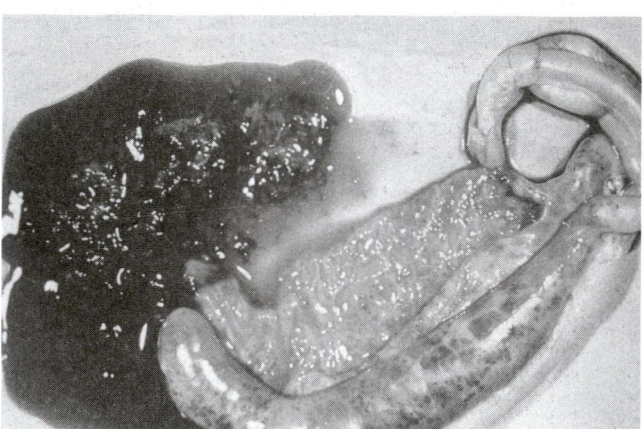

Fig. 46–19.  Cecal, or bloody, coccidiosis in a chicken, showing accumulation of blood in ceca, caused by the protozoan organism, *Eimeria tenella*. (Courtesy, C. F. Hall, DVM, College of Veterinary Medicine, Texas A&M University, College Station)

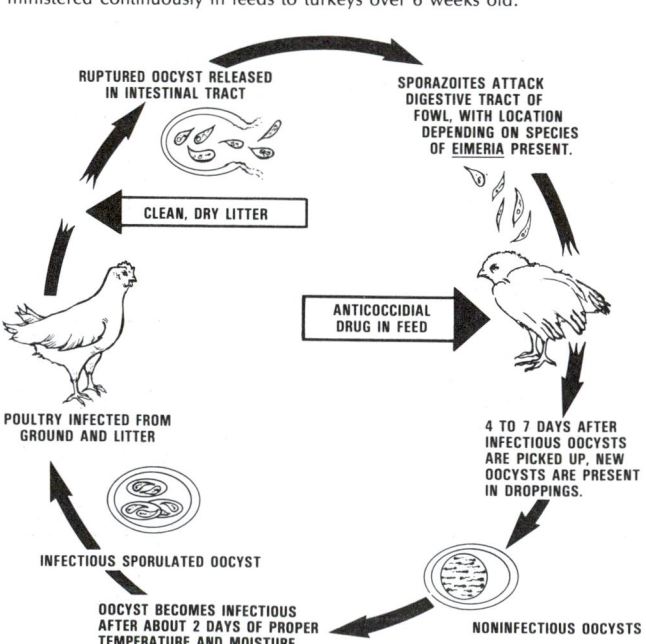

Fig. 46–20.  Life cycle of *E. tenella*, one of the genus containing many species causing coccidiosis in poultry, showing (1) the typical 7- to 9-day cycle, and (2) the stage of development at which anticoccidial drugs kill the organism.

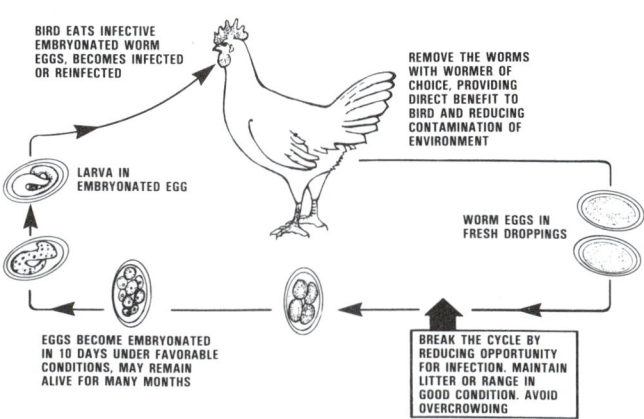

Fig. 46–21.  Life cycle of the common, large roundworm. *Ascaridia galli*, the most prevalent of the worm parasites of chickens and the cause of heavy economic losses.

*(Continued)*

| Distribution and Losses Caused By | Treatment | Prevention | Remarks |
|---|---|---|---|
| The Northern fowl mite, *Ornithonyssus sylviarum*, is rated as the most important permanent parasite in all major producing areas of the U.S. | Select an approved insecticide and use according to manufacturer's directions. | Examine birds frequently for evidence of mite infestation. Preventive insecticide treatment of quarters. Control sparrows. | |
| | Butynorate is the most widely used product for treatment. Use according to manufacturer's directions. | Eliminate the intermediate insect hosts. Control snails, earthworms, beetles, and flies. | See Fig. 46–24 for life cycle. |
| Worldwide. | Copper sulfate (bluestone) in drinking water; 1:2,000 dilution for 4 to 7 days. Enheptin, used according to manufacturer's directions, has been of value in treatment of this disease. | Sanitation, clean feed and water. Eliminate recovered or carrier birds. | |

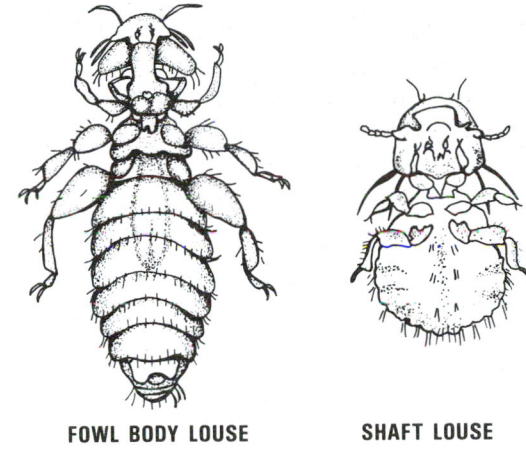

**FOWL BODY LOUSE**        **SHAFT LOUSE**

Fig. 46–22. Lice: Fowl body louse, and shaft louse.

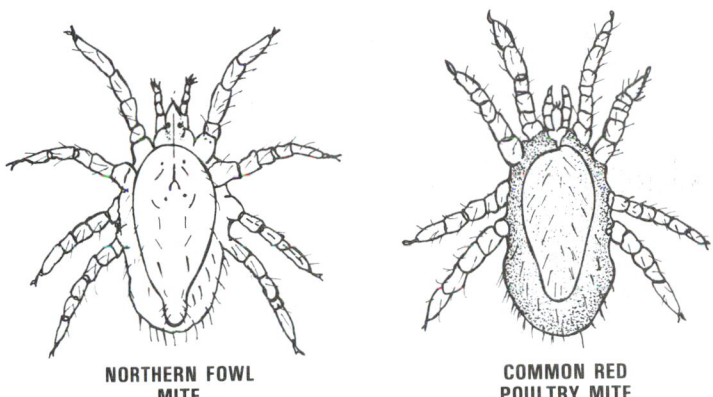

**NORTHERN FOWL MITE**        **COMMON RED POULTRY MITE**

Fig. 46–23. *Mites: Northern fowl mite. Arnithonyssus sylviarum* (left), and common red mite. *Dermanyssus gallinae* (right).

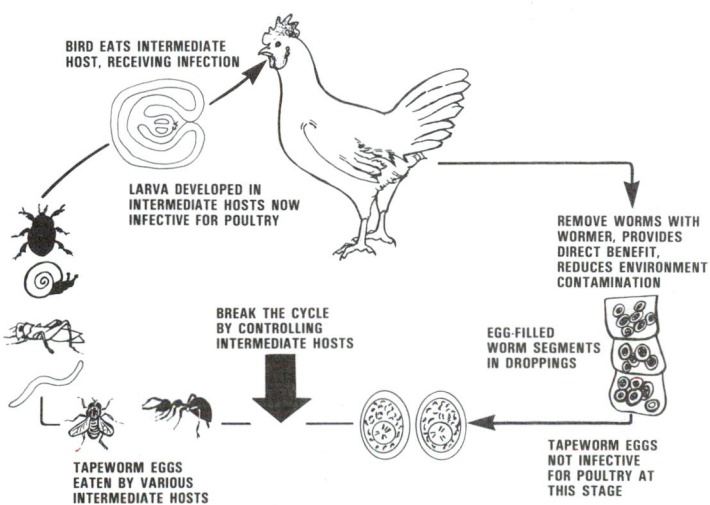

BIRD EATS INTERMEDIATE HOST, RECEIVING INFECTION

LARVA DEVELOPED IN INTERMEDIATE HOSTS NOW INFECTIVE FOR POULTRY

REMOVE WORMS WITH WORMER, PROVIDES DIRECT BENEFIT, REDUCES ENVIRONMENT CONTAMINATION

EGG-FILLED WORM SEGMENTS IN DROPPINGS

BREAK THE CYCLE BY CONTROLLING INTERMEDIATE HOSTS

TAPEWORM EGGS NOT INFECTIVE FOR POULTRY AT THIS STAGE

TAPEWORM EGGS EATEN BY VARIOUS INTERMEDIATE HOSTS

Fig. 46–24. *Right:* Life cycle of tapeworms, several species of which infect chickens and turkeys. Poultry become infected by eating the intermediate host.

## NUTRITIONAL DISEASES AND AILMENTS

The vast majority of nutritional deficiency diseases and ailments of poultry are brought about by the deficiency of one or more of the minerals or vitamins. A summary of these deficiency diseases is given in Tables 44–2 and 44–3, in Chapter 44.

## OTHER AILMENTS AND HEALTH PROBLEMS

Although not properly considered diseases, there are several other vices, ailments, or problems which can cause serious problems, and even mortality in a flock. Some of these follow.

1. **Cannibalism.** Cannibalism may be encountered among birds of all ages. Among baby chicks, the trouble is usually confined to toe and tail picking. With mature birds, the vent, tail, and comb are the regions most frequently picked.

The cause of cannibalism is not fully understood. It is known that it is more frequent under confined conditions, and that it increases as space is reduced. Without doubt, it may be accentuated by deficiencies in management and nutrition. Also, it may be brought on by just plain boredom.

The best way in which to control cannibalism is by debeaking. Many broiler producers, and some egg producers, have their chicks debeaked at the hatchery. Additionally, layers can be debeaked at the time of moving into the laying house. Reducing light will help, also.

2. **Cage fatigue.** Cage fatigue is a paralytic condition observed in birds held in cages. The disease is most common among high-producing young pullets during the summer. The exact cause is not understood; however, the disorder is considered to be a disturbance in mineral metabolism. Treatment

with Vitamin $D_3$ may help. Affected birds usually make a spontaneous recovery if placed on the floor or if the cage bottom is covered with newspaper or other such material.

3. **Fatty liver syndrome.** The *fatty liver disease* continues to be a serious anomaly in commercial egg operations. It is characterized by deranged fat metabolism resulting in the deposition of excess fat in the liver (yellow to hemorrhagic liver) and body cavities. It is seen most commonly in caged, fat hens, but on occasion it may strike floor birds, particularly in the heavy breeds. Fatty liver syndrome usually occurs in overweight birds, in high-producing flocks, and in hot weather. The cause is unknown. However, factors which predispose the condition include reduced activity in a cage operation and use of high-energy feeds. Although the cause of fatty liver disease is not known, it does appear that the incidence of the malady is lessened by (a) lower energy diets, or (b) higher levels of choline, vitamin E, and vitamin B–12 than normally fed.

4. **Aplastic anemia.** This disease, sometimes referred to as hemorrhagic anemia syndrome, is characterized by hemorrhage and anemia. It is considered to be a disease of chickens only, although there have been reports of its occurrence in turkeys. The condition may affect birds of all ages, but usually those between 4 and 12 weeks of age. The economic losses are in mortality and retarded growth. Little information is available regarding the cause, treatment, and prevention of *aplastic anemia*.

5. **Pendulous crop.** This condition, sometimes known as *baggy crop* or *drop crop*, may be found among chickens and among growing poults two to three months of age. The condition may be caused by irregular feeding and by overconsumption of feed or water at any given time. It also seems that there are fewer baggy crops where turkeys are in houses which are not overheated, and when there is adequate shade for poults on the range.

## QUESTIONS FOR STUDY AND DISCUSSION

1. It is estimated that poultry disease losses approach $2 billion annually. Who absorbs this loss?

2. How do the normal death losses of layers and broilers, as given in the introductory section of this chapter, compare with the normal mortality of four-footed animals? (See the introductory section in Chapter 6 of this book.)

3. Select a specific poultry farm (either your own or one with which you are familiar) and outline (in 1, 2, 3 order) a program of poultry health, disease prevention, and parasite control.

4. What might be considered the normal (a) percentage of sick birds at one time, and (b) percentage of mortality losses during the first 3 weeks of life of chicks and of poults?

5. Why are diagnostic laboratories more frequently used by poultry producers than by producers of four-footed animals?

6. What consideration should be given to each of the following as related to vaccinations: (a) alternatives to immunization; (b) the fact that some diseases seldom occur, and the stress/hazard of vaccination; and (c) the program and schedule adaptations resulting from area and flock differences?

7. List and discuss three methods of disposing of dead birds.

8. Define the term *disinfectant*; and list five things upon which effective disinfection depends (see Chapter 6, section on "Disinfectants and Their Use").

9. What is the primary purpose of hatchery fumigation? How is it accomplished?

10. In your area, what are the six most serious disease problems of poultry?

11. For each of the six diseases listed in answer to question No. 10, give the following: (a) species affected, (b) cause, (c) symptoms and signs, (d) treatment, and (e) prevention.

12. In your area, what are the six most troublesome parasites of poultry?

13. For each of the six parasites listed in answer to question No. 12, give the following: (a) species affected, (b) cause, (c) symptoms and signs, (d) treatment, and (e) prevention.

14. How may a producer make practical use of life cycles of parasites such as those shown in Figs. 46–18, –20, –21, and –24?

15. Present pertinent information about each of the following conditions: (a) cannibalism, (b) cage fatigue, (c) fatty liver syndrome, (d) aplastic anemia, and (e) pendulous crop.

16. Assume that a specific contagious disease (you name it) has broken out in a flock. What steps would you take to meet the situation? (List in 1, 2, 3 order; be specific.)

17. Assume that a specific parasite (you name it) has become troublesome in a particular flock. What steps would you take to meet the situation? (List in 1, 2, 3 order; be specific.)

## SELECTED REFERENCES

| Title of Publication | Author(s) | Publisher |
| --- | --- | --- |
| *Diseases of Poultry*, Eighth Edition | Edited by M. S. Hofstad *et al.* | Iowa State University Press, Ames, IA, 1984 |
| *Merck Veterinary Manual, The*, Sixth Edition | Edited by C. M. Fraser *et al.* | Merck & Co., Inc., Rahway, NJ, 1986 |
| *Poultry Handbook* | L. D. Schwartz F. W. Hicks | The Pennsylvania State University Press, University Park, PA, 1972 |
| *Poultry Health Handbook* | L. D. Schwartz | The Pennsylvania State University Press, University Park, PA, 1972 |
| *Serviceman's Poultry Health Handbook* | E. H. Peterson | Better Poultry Health Company, Fayetteville, AR, 1975 |

In addition to these selected references, valuable publications on different subjects pertaining to poultry diseases, parasites, disinfectants, and related matters can be obtained from the following sources: Division of Publications, Office of Information, U.S. Department of Agriculture, Washington, D.C.; your state agricultural college; and several biological, pharmaceutical, and chemical companies.

A sign of good health—a turkey gobbler *strutting*. (Courtesy, USDA)

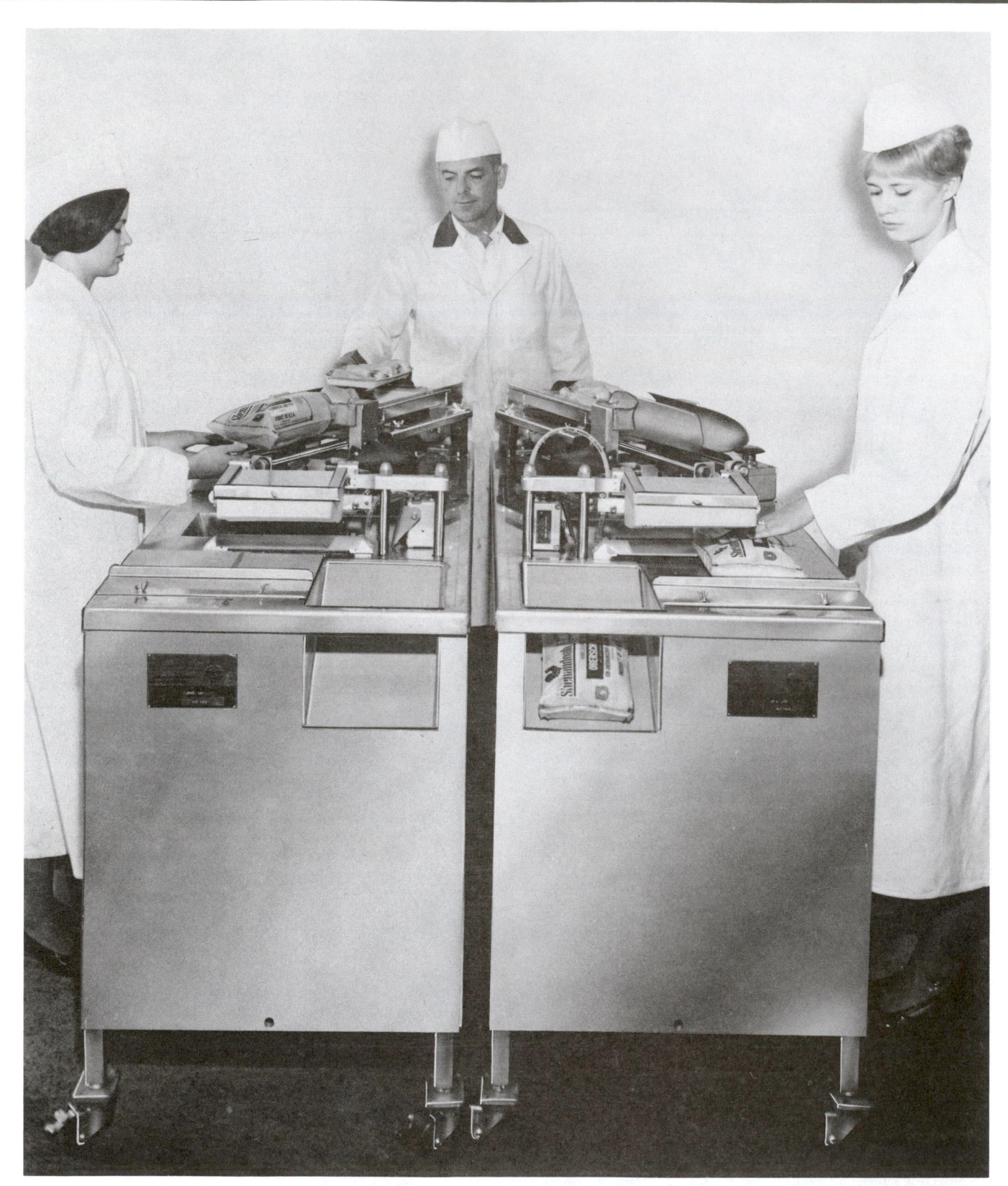

Kitchen-ready poultry—safest in the world. (Courtesy, Gordon Johnson Company)

Superlayer perched on the 448 eggs that she laid nonstop. (Courtesy, University of Missouri, Columbia)

# SPECIALIZATION; BUSINESS MANAGEMENT IN POULTRY PRODUCTION

## Chapter 47

No phase of American Agriculture has undergone such pronounced changes since World War II as has the poultry industry. Broiler and turkey production have skyrocketed in numbers; and poultry has shifted from small farm flocks, often accorded the neglect of a minor enterprise, to large commercial flocks. Hand in hand with this transition has come a high degree of specialization. For example, commercial egg producers no longer hatch their own replacement stock. Instead, they rely upon highly specialized chicken breeders to produce the foundation stock, and on hatcheries to multiply and incubate. More than likely they even rely on another specialized enterprise to raise their replacement pullets.

With the growth of highly specialized enterprises, business methods have also changed. Big enterprises require more financing. In turn, this has spawned vertical integration and contractual arrangements.

The general principles of poultry breeding, feeding, health, buildings and equipment, and marketing are covered elsewhere in this book; hence repetition at this point is unnecessary. Instead, this chapter is devoted to the practical application of these principles to each of the highly specialized types of poultry production.

## CHICKEN PRODUCTION

Modern chicken production embraces the following specialized types of operations: chicken breeding, hatcheries, growing replacement pullets, commercial egg production, and broiler production.

Fig. 47-1. Meat-type breeder hen and 1-day-old broiler chicks. (Courtesy, Euribrid B.V., Boxmeer, Holland)

## CHICKEN BREEDING

The breeding of chickens has become a highly specialized industry. Today, from a commercial standpoint, the small farm flock poultry breeder is almost a thing of the past. The number of chicken breeding farms has been greatly reduced, and the chicken breeders have become very large. Today's chicken breeding industry is a complex, specialized, and highly competitive aspect of poultry production.

Chicken breeding is dominated by the foundation breeder and the hatchery, usually joined together under a franchise ar-

rangement. The foundation breeder is concerned with genetic improvement, whereas the hatchery multiplies the stock supplied by the foundation breeder, hatches the eggs, and sells the chickens.

Foundation breeders generally employ well-trained geneticists, veterinarians, and physiologists to direct their breeding programs. They recognize the importance of *name lines* or *trade names,* which they copyright and promote. These trade names are as identifying as a make of a car or a brand of shoes.

Modern foundation chicken breeding farms are organized along the lines of any other big and successful business. They are generally departmentalized—breeding, sales, foreign markets, advertising, business management, purchases, etc.— with specially trained personnel in each division. In their breeding programs, they employ advanced techniques and make use of data-processing equipment.

## BREEDING METHODS

Commercial egg and broiler producers are not interested in breed fancy points. No bird meets with their favor unless it efficiently produces meat or eggs, as the case may be. As a result of this thinking, interest has shifted from the *Standard of Perfection,* or so-called Standard breeding, to the breeding of crosses or strains for efficient production of meat and eggs. Individual birds in the flock have value only insofar as they increase the total egg or meat production of the flock. That is, they have no value as a Standard breed as such. This change of emphasis from the individual bird (Standard breeding) to the flock performance (flock breeding) has greatly changed both breeds and methods of breeding. Each modern breeder establishment produces its own strain and selects birds only for their value to produce eggs or meat.

### BREEDING FOR EGG PRODUCTION

Fig. 47-2. Egg-type breeder hen—a White Leghorn. (Courtesy, Hubbard Farms, Walpole, NH)

Breeders for egg production must supply stock that will lay well. In addition, the eggs must be of desirable size and shape, shell texture, and interior quality. Also, the birds must have good livability.

Breeding for egg production differs from breeding for broiler production primarily because egg production as a trait is of low heritability. By contrast, the rate and efficiency of gain of broilers are of high heritability. Hence, in breeding for egg production, family selection rather than individual merit of a particular bird in the flock is considered.

## BREEDING FOR BROILER PRODUCTION

Most of the effort in broiler breeding in recent years has been directed toward increasing the rate of growth. To this end, chickens used for breeding to improve broiler growth rate are selected on a weight for age basis. Those which weigh the most in each brood are the ones chosen for breeders.

Hybrid vigor is obtained by systematic matings that may involve crossing of different breeds, different strains of the same breed, or the crossing of inbred lines. In addition to hybrid vigor, improvement in economic factors often results from these crosses, provided the mating includes stock having superior qualities of genetic origin.

The end product—modern white, yellow-shanked broiler—is often obtained by crossing the male line from specialized breeders with female lines produced likewise by specialists. Each line may be the result of crossing two or more strains. The male lines usually have dominant white feathers and are selected for rapid growth; meat characteristics such as breast width, body depth, live market grade, and dressing yield; and rapid feathering. The female line also must have outstanding growth rate, high hatchability, and good, but not outstanding, production of eggs of desirable size and texture. In recent years, considerable attention has been given to selection for livability, feed efficiency, skin texture, skin and shank pigmentation, and, in the case of large broilers and roasters, feathering on the breast or absence of breast blisters.

It is noteworthy that in breeding for broiler production, inbred lines have not been used.

## BREEDING TESTS

Public tests designed to measure the egg or meat production of the various breeds and commercial crosses have declined in numbers and importance in recent years. This subject is discussed in Chapter 43, under the heading, "Random Sample Performance Test; Multiple Unit Poultry Test"; hence, the reader is referred thereto. Today, large poultry breeding companies are staffed with scientific personnel and conduct their own progeny testing, the results of which are used to generate greater sales.

## CHICKEN BREEDING MANAGEMENT PRACTICES

Every condition for the efficient production of table eggs is required for the production of hatching eggs. In addition, there are certain management practices, and costs, which are peculiar to chicken breeding operations and which do not apply to commercial egg production. Among these, are the following:

1. **Cost of production.** Breeders of egg-type chicks are retained for 12 months or more of production, whereas breeders of broiler-type chicks are seldom profitable after 9 to 10 months of laying. The hatching-egg yield of broiler supply flocks averages between 40 and 60% during this 9- to 10-month period, while egg-type supply flocks average 60 to 70% or even

higher for the longer period. Broiler hatching eggs are, therefore, more costly to produce.

The extra cost involved in producing eggs for broiler hatches arise from the greater feed cost per dozen eggs due to larger bodies and lower egg production per hen. Also, more floor space (about 3 sq ft per bird) is required than the normal 2 to 2½ sq ft of floor space used for the small egg-type hens.

2. **Methods of mating.** Pedigree matings are used where information is desired on family performance. This consists of one male and about 15 females housed in a separate pen. The females are trapnested so that each egg may be identified for pedigree hatching. When the eggs are transferred to the hatching trays on the eighteenth day of incubation, each hen's eggs are placed in a separate basket. Following hatching, the chicks in each basket are wing banded to identify their parentage.

Flock matings are more common. The females are housed in large flocks varying from 100 to 1,000 or more per pen. In broiler-type stock, about 8 males per 100 females are placed together; in egg-type chickens, fewer males are used—usually about 6 per 100 females.

3. **Selecting and handling hatching eggs.** Certain physical characteristics of eggs are related to hatchability. For best results, hatching eggs should be neither too large nor too small. They should be of normal shape with strong clean shells.

Usually a reasonably high percentage of fertile eggs can be expected a week after the males are put in the flock, but maximum fertility will not be reached until approximately two weeks after adding males.

Eggs are suitable for hatching any time after the flock has been laying longer than 6 weeks, or about a month after the flock has reached 50 eggs per 100 hens. Pullets are about 7 months of age at this time.

Improper care of hatching eggs can greatly reduce their hatchability (see Chapter 48, section on "Handling Hatching Eggs").

## HATCHERIES

Fig. 47–3. Hatchery. (Courtesy, Chick Master, Morton & Associates, Gainesville, GA)

Few phases of the poultry industry have undergone more changes than has the hatchery business. In the early 1900s, the "old cluck" was forced to yield her pleasant job of spending three weeks on the nest "expecting." The incubator in the small hatchery gradually took over the job. As hatcheries developed, they grew in size and improved in efficiency, and the old setting hen was practically eliminated. Today, virtually all chicks raised, either farm chickens or commercial broilers, are bought from hatcheries as baby chicks.

Perhaps hatcheries have exerted more influence on improving the general level of the poultry industry than any other segment of the business. This has been achieved through their own breeding programs or through the breeding programs of the poultry breeders with whom they have a franchise agreement.

## TRENDS IN THE HATCHING INDUSTRY

The major changes or trends in the hatching industry in recent years are:

1. **Fewer and larger hatcheries.** Back in 1934, there were 11,405 chicken hatcheries in the United States; in 1989, there were 376—only 3.3% remained. But, in this same period of time, the average size of hatchery increased by nearly ninefold, so that there was a substantial excess of capacity in the hatchery industry. The following figures from the U.S. Department of Agriculture point up the trend toward fewer and larger chicken hatcheries:

| Year | Number of Hatcheries | Total Egg Capacity | Average Egg Capacity per Hatchery |
|---|---|---|---|
| 1934 | 11,405 | 276,287,000 | 24,000 |
| 1953 | 8,233 | 616,976,000 | 80,000 |
| 1965 | 2,365 | 471,318,000 | 199,000 |
| 1975 | 797 | 416,000,000 | 521,000 |
| 1981 | 538 | 466,096,000 | 466,000 |
| 1989 | 376 | 524,615,000 | 1,395,000 |

The increased size of hatcheries becomes obvious when it is realized that on January 1, 1989, U.S. chick hatcheries with incubators capable of holding 500,000 or more eggs accounted for 60.4% of the nation's capacity. A similar trend has occurred in turkey poult hatchery capacity. On January 1, 1989, 38 of

Fig. 47-4. Processing baby chicks in a modern hatchery. (Courtesy, Indian River International, Division of Hy-Line Indian River Company, Nacogdoches, TX)

the 82 turkey hatcheries had a capacity of 500,000 or more eggs, and these 38 hatcheries accounted for 46.3% of the poult hatching capacity in the United States.

2. **Year-round business.** Formerly, most hatcheries operated in the spring of the year only. In 1937–38, slightly under 2.0 hatches of chicks were taken off per year. In 1989, more than 11 chicken hatches and more than 6 turkey hatches were obtained during the year. (This is known as turnover; it is obtained by dividing hatch by capacity.) The primary reasons for extending the hatching season to a more nearly year-round business are (a) less seasonality in the production of eggs and turkeys, and (b) the development of a year-round commercial broiler industry.

3. **Franchise agreements have come in.** In 1937–38, franchising was practically unknown in the hatchery business. Today, many hatcheries have franchise arrangements with breeders who use advanced methods to produce superior strains of birds. Under these agreements, the breeders furnish the eggs and the hatcheries agree to sell breeders' strains of birds in specified areas.

4. **Less custom-hatching.** Today, very little custom-hatching is done. Most of that which is done is for the hatching of turkey poults.

5. **More integration.** Many hatcheries have become a part of broader businesses which involve poultry growing and processing, or sale of supplies. This may involve the selling of poultry feeds, medicines, equipment, farm supplies, and farm products.

6. **Fewer started chicks sold.** With the growth of larger and more specialized poultry farms and their improved brooding technology, fewer and fewer started chicks (chicks two to four weeks of age) are being sold.

7. **Geographical shift.** With the shift of the broiler industry to the South and South Atlantic states, the total output of hatcheries has also increased in this area, which means that hatcheries have followed broiler production, geographically speaking.

## GROWING REPLACEMENT PULLETS

A recent trend in poultry production involves specialization within a specialization. For example, some commercial egg producers are only concerned with producing market eggs—leaving the rearing of replacement pullets to other specialists. This means that commercial chicken producers have two alternatives of specializing—raising replacement pullets or producing eggs. Likewise, commercial egg producers have two alternatives in procuring replacement stock: (1) growing their own baby chicks, or (2) buying started pullets. The majority of large commercial egg producers today prefer to buy started pullets rather than raise their own. In the final analysis, the choice of the system should be based on the cost of pullets of the desired quality, with added consideration given to available facilities and labor.

Many factors influence the cost of pullets, whether they are grown by the layer operator or purchased; and these vary from area to area, year to year, and farm to farm. Approximate percentages of the different items are: feed, 60%; chick, 15%; labor, 15%; and all other costs, 10%.

Growers of starter pullets should be specialists—operators who raise pullets to sell. They should not have any adult birds on their farms unless they are large enough to keep the two operations well separated.

## STARTER PULLET MANAGEMENT

Well-bred and well-grown, healthy, vigorous started pullets are requisite to success in the laying house. Proper feeding, care, and management during the pullet starting period can develop the potential productive egg capacity of the birds, whereas poor practices and management may reduce or obliterate the genetic potential that has been bred into them. The following management pointers in raising started pullets are recommended:

1. **Sanitation.** A strict sanitation program is a must in raising started pullets. Recommended rules are (a) do not mix ages or strains of birds in the same house; (b) do not allow visitors in the house; (c) screen out birds and control rats, mice, and other rodents; (d) do not permit contaminated equipment to come on the farm; (e) use renderer, incinerator, or disposal pit for dead birds; (f) keep birds free from external and internal parasites; and (g) post all dead birds.

2. **Vaccination.** The vaccination program should be in keeping with the needs of the area. Also, it should be completed far enough ahead of moving the pullets so that they will not be going through a vaccination reaction during or soon after moving.

3. **Debeaking and dubbing.** Pullets should be debeaked at 9 to 16 weeks of age, perhaps at the same time that they are vaccinated for fowl pox. This will make for a permanent job.

Dubbing (removal of the comb) should be done when the birds are a day old. Dubbing alleviates a source of injury, especially to caged birds, and results in 2 to 4% more eggs.

4. **Lights.** The lighting system used will depend on the type of building and the season of the hatch. It should be designed to avoid an increasing photo-period during the last half to three-fourths of the growing period.

5. **Moving the pullets.** The pullets should be moved in such a manner that there will be a minimum of stress. This calls for clean (preferably steamed) coops and truck, not moving in inclement weather, providing fresh air but avoiding drafts during the move, and gentle handling.

## COMMERCIAL EGG PRODUCTION

Fig. 47–5. Caged laying birds, showing egg collection conveyor belt and automatic feeder in the foreground. (Courtesy, Dr. T. A. Carter, North Carolina State University, Raleigh)

Specialization and business are the key words in modern egg production. The industry is changing rapidly and providing new opportunities for those who can meet its requirements.

The time has passed when egg production can be considered a small sideline operation, worthy only of the leftovers of the producer's time and money. Current commercial egg production units represent major enterprises in which success or failure depends largely on good production and business management practices. Of course, an important tool in business management is the maintenance and use of accurate records.

### CURRENT TRENDS IN EGG PRODUCTION

In recent years, the number of laying flocks has become fewer and the size of the flocks larger. In 1989, *Egg Industry* reported that 59 U.S. firms each owned a million or more commercial layers; that the 10 largest U.S. egg companies owned 61.2 million layers, which represented 27% of the nation's laying flock; and that one firm—Cal-Maine Foods, Inc., Jackson, Mississippi—owned 15 million laying hens.[1]

These big flocks have been described as "egg factories," where inputs of housing, equipment, pullets, feed, supplies, and labor are combined to produce eggs—hopefully for a profit.

### FINANCING

By 1990, 89% of the nation's market eggs were produced under contract.

Most egg contracts require that the operator furnish housing, equipment, litter, electricity, and water, and that the contractor supply the pullets, feed, medication, and some supervision. For facilities and services, the owner usually receives a fixed rate per dozen commercial eggs sold plus a bonus for superior egg production and feed conversion.

### SIZE OF FLOCK

The size of flocks has been increasing steadily over the years. The determining factors in size of flock are (1) amount of income which the owner wishes to derive from the business, (2) how the egg-producing business fits in with the other farm enterprises or other employment, (3) method of marketing eggs, and (4) available finances and labor.

Generally speaking, a flock of 10,000 or more hens is more efficient than one of smaller size.

### HOUSING AND EQUIPMENT

By 1980, more than 90% of the U.S. layers were in cage houses. However, houses with slat floors, combination slat and litter floors, and litter floors still have their advocates—and a place.

Modern cage houses vary from 36 to 40 ft in width and 360 to 500 ft in length; and they're equipped with automatic feeders and egg belts. Some new environmental houses have no windows, which makes them completely dependent on electricity to run fans and lights.

[1]*Egg Industry*, a Watt Poultry Publication, November/December, 1989, p. 20.

The space requirements for different types of houses are given in Chapter 45; hence the reader is referred thereto. In addition to space needs for the birds, size and construction of cages, ventilation, type of feeding equipment, and manure disposal should be considered in determining the width and length of poultry houses.

(Also, see Chapter 45, Poultry Houses and Equipment.)

Fig. 47–6. A modern commercial laying house. Note the 3 tiers of cages, the automatic feeder, and the automatic egg collecting belt. (Courtesy, Poultry Department, California Polytechnic University, San Luis Obispo)

## BREEDS

Unless there is a premium market for brown eggs, it is recommended that either White Leghorns or Leghorn crosses be used for commercial production. Hens require about 8 lb of feed per year per pound of body weight just for maintenance. Heavy breeds average from 1 to 1½ lb more body weight than Leghorns and require 8 to 12 lb more feed per year. If feed costs 8¢/lb this means that it will cost 50¢ to $1 more to keep a heavy hen for a year than a Leghorn. Of course, the heavy hen brings more as a cull, but the difference does not compensate for the extra feed consumed. Egg production is about the same, or perhaps even a little lower for the heavy hen.

It is more important to get a high-producing strain than to get a particular breed or cross. It pays to make a careful check on the stock before purchasing.

## MANAGEMENT PRACTICES AND GUIDES

Bigness alone does not assure success and profit in commercial egg production. Rather, it makes it imperative that there be superior management. Among the management practices and guides followed by successful commercial egg producers are the following:

1. An average of 20 eggs per bird per month is a good standard for high-producing strains.
2. Mortality of about 1% per month for layers is normal.
3. Lights should be provided so as to make about a 14-hour working day for the hens. Lights may also be used to help bring slow-maturing pullets into lay.
4. Culling should be done to remove diseased or low-vitality birds. With high-producing strains, little, if any, culling may be required. Accurate records should be kept on percentage of lay, feed intake, mortality, and other pertinent facts.

5. Eggs should be gathered 3 to 5 times daily, cleaned immediately, and stored at 55°F and 75 to 80% relative humidity.
6. Where the cage system is followed, it is important that there be a regular spray schedule, along with proper dropping management to control flies, that birds be debeaked if multiple caging is used, and that cages and waterers be cleaned once each week.
7. Ventilation should provide fresh air and remove moisture, without producing drafts. The condition of the litter and the amount of fumes (ammonia fumes) are good indicators of the adequacy of ventilation.

## PRODUCTION GOALS

The producer of commercial eggs strives for the maximum of egg production at a minimum cost—yet, bearing in mind that net returns are more important than costs as such.

The largest item of cost in the production of eggs is feed. It normally constitutes 65 to 75% of the total cost. Labor costs and cost of replacement pullets are the other two major cost items in production.

Generally speaking, higher egg production means lower costs per dozen eggs. This is so because the feed required for maintenance is constant for hens of any given weight and bears no relation to the number of eggs laid. Hence, it is important that the commercial producer strive for high egg production.

The following production goals are suggested for the commercial egg producer:

1. Production of 260 eggs per hen per year.
2. Feed conversion of 3.5 to 4.0 lb of feed per dozen eggs for light weight layers, with the heavy breeds slightly less efficient.
3. A laying house mortality of less than 10%.
4. Seventy-five percent or more extra large and large Grade A eggs.
5. Ninety-five percent or more marketable eggs.
6. On-farm egg breakage under 2%.
7. No layers over 19 months old.
8. Mortality of less than 5% from 1 day old to 5 months.

## BROILER PRODUCTION

The terms *broilers, fryers,* and *young chickens* are used interchangeably. Most broilers weigh 3 to 5 lb live at 6 to 8 weeks of age.

Fig. 47–7. Broilers in a typical commercial house, showing feeder line on the left and nipple drinker line on the right. (Courtesy, Dr. S. E. Scheideler, North Carolina State University, Raleigh)

Commercial broiler production is a highly specialized and complex business enterprise. During the past three decades, it has been the fastest-growing segment of the poultry industry. The total output increased from 34 million broilers in 1934 to 5 billion in 1987. In 1987, broilers accounted for over $6.2 billion in the $11.5 billion total income from poultry and poultry products in the United States.

No other segment of agriculture is as well suited to assembly-line production techniques as broiler production.

## CURRENT TRENDS IN BROILER PRODUCTION

The broiler industry has been characterized by rapid changes. Among the pertinent trends affecting broiler production are the following:

1. **Shift in production centers.** In recent years, the center of broiler production has moved from the Eastern Seaboard—the Delmarva Peninsula, Maine, and Connecticut—to the southern and southeastern states of Arkansas, Georgia, Alabama, North Carolina, and Mississippi. In 1987, the 10 leading states in broiler production produced 4.2 billion birds, or 84% of all the broilers in this nation. Area competition is primarily dependent on production and transportation costs of processed broilers.

2. **Integration.** By 1990, 92% of the broilers of the United States were produced under contract, and the remaining 8% were raised on integrator-owned farms.

In the beginning, most commercial broiler production was by independent growers. They paid cash for everything and took all the profit. However, as margins became smaller, and flocks became larger, there was need for more credit. At first, the local feed dealer was the source of credit. As the industry grew, feed dealers began to depend on feed manufacturers as a source of funds. Then, to spread their risks, both feed dealers and feed manufacturers integrated vertically with hatcheries and processors.

Integration may include one or more phases, such as financing, feed, hatching, breeding stock, production, processing, and retail marketing. In most cases, the production of breeding stock and retailing have not been included.

3. **Financing.** Today, banks and the production credit associations largely finance the broiler industry through credit extended to feed manufacturers, dressing plants processors, hatcheries, and integrated operations. Much of this money is lent at the local level to growers with mortgages as collateral.

One of the principal reasons for the rapid expansion of the poultry industry in the South has been that financing is easier in that area, poultry house construction is cheaper, labor is cheaper, and more capital is available to invest in poultry enterprises.

4. **Contracts.** A contract is a legal agreement between the grower and the integrated operator. The numerous types of broiler contracts now being used indicate the need for those involved to have a clear understanding of the various specifications and their possible impact on any present or future method of operation.

## SIZE OF ENTERPRISE

In 1987, there were 27,645 broiler operations in the United States, marketing an average of 157,785 broilers per unit during the year. By 1990, the eight largest broiler companies held over half of the market.

A competent person on a full-time basis, with a really efficient operation, including mechanical feeders and waterers, can care for 75,000 broilers at a time. Also, some caretakers handle more than 100,000 birds, with some extra help the first week when the labor requirements are high. But most units fall short of this maximum. Even with houses fully automated, family units on a full-time basis generally run in the 30,000 to 50,000 range. However, not all broiler operations are mechanized. Further, not all operators devote full time to the enterprise. As a result, there are many small operations. But even most part-time operators will not undertake a broiler enterprise unless it will provide at least a quarter of the income. This means that most of them will have around 10,000 broilers.

## HOUSING AND EQUIPMENT

There are, of course, many different styles and designs of houses, and even more variations in equipment. The important thing is that broiler houses and equipment provide comfortable conditions, including adequate feed and water, so that the birds can perform at the highest level of which they are genetically capable. A satisfactory broiler house must protect against heat and cold, high winds, and inclement weather. A typical broiler growout house is 40 ft wide by 300 ft long and houses between 65,000 and 80,000 birds.

## CONTROLLED ENVIRONMENT

There is a tendency for the larger and better financed broiler operators to construct environmentally controlled housing. These are windowless, fan ventilated, and insulated. With this type of arrangement, it is possible to provide a more uniform temperature, along with a proper supply of clean fresh air without drafts. The final result is an improved market-quality bird and higher income.

## SELECTING CHICKS

The selection of commercial broiler or roaster stock can make the difference between profit and loss. Accordingly, it is most important that the broiler producer select stock which, from a genetic standpoint, is capable of producing a high-quality product as efficiently as possible.

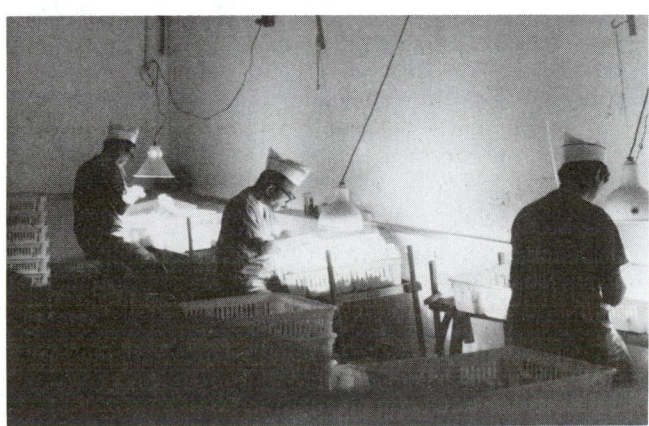

Fig. 47–8. Chicks being sexed at the hatchery soon after they are removed from the hatchers. (Courtesy, Dr. M. J. Wineland, North Carolina State University, Raleigh)

Males are slightly heavier at 9 weeks of age and require slightly more feed per pound of gain than females. At 9 weeks of age, males weigh about 7.8 lb, with a feed efficiency of 2.28 lb of feed per pound of gain. At 9 weeks of age, females weigh about 6.4 lb, with a feed efficiency of 2.2 lb of feed per pound of gain. (See Chapter 44, Table 44–12.)

There is a sex difference—and this difference becomes more marked at heavier weights. The producer should determine whether the advantages to accrue from separating the sexes will offset the cost of sexing day-old chicks.

## MANAGEMENT PRACTICES AND GUIDES

The management practices and guides followed by most successful broiler producers include the following:

1. Only one age on the farm at any one time.
2. Debeak chicks when necessary.
3. Use all-night lights.
4. Adapt vaccination schedule to local needs.
5. Keep visitors out of houses; lock doors.
6. With automatic feeders, not more than 18 minutes' labor should be required per 1,000 broilers per day; without automatic equipment, labor should not exceed 30 minutes.
7. Use 2 to 4 in. of litter—less in hot weather.

Fig. 47–9.  Turkey breeders. (Courtesy, Jamesway Division of Butler Manufacturing Co., Kansas City, MO)

## PRODUCTION GOALS

Broiler producers should aim for broilers with an average live weight of 3 to 5 lb at 6 to 8 weeks of age. Feed conversion should be less than 2.2, and mortality should be under 1%.

# TURKEY PRODUCTION

Turkey has been considered traditional Thanksgiving and Christmas fare since the Pilgrims hunted wild turkeys to grace their tables that first Thanksgiving Day.

Today, turkey production and marketing is a highly efficient process, and turkey meat consumption is substantial in every month of the year. Yet, more than a third of our annual turkey consumption occurs during the two holiday months.

In 1946, 39.7 million turkeys were produced in this country, while the total production sold in 1987 was in excess of 240 million. Gross from turkeys in 1987 was $1.7 billion.

In 1987, there were 7,347 turkey units in the United States, marketing an average of 33,210 turkeys per unit during the year.

## TURKEY BREEDING

The breeding of turkeys is the most specialized phase of turkey production. The breeder must have a good knowledge of genetics in order to develop a sound breeding program. Further, all traits of economic importance must be taken into consideration at selection time.

## SELECTION

Most turkeys for breeding are selected from flocks being raised for market. However, there is much merit in raising turkeys especially for breeding purposes, and selecting the birds intended for breeding purposes therefrom.

The initial selection for future breeding stock should be made at 12 to 16 weeks of age. An additional and final selection should be made at 22 to 26 weeks of age, or before the remainder of the flock is sent to market. Among the many traits of economic importance for which the breeder should make selection are body type and conformation, livability, health and vigor, early market finish, and rapid feathering. Additionally, selection for reproductive traits—egg production, fertility, hatchability—is most important. In the case of the reproductive traits, it is necessary to select on a family basis in order to achieve the maximum rate of improvement, while for the physical traits selection of the outstanding individuals is sufficient.

The selection of a variety is largely determined by intended market, along with any preference that the producer may have. In general, the family size turkey most preferred weighs about 15 lb or less. Until recently, large size turkeys were used primarily for the hotel and restaurant trade. However, the further processing of large turkeys is increasing rapidly and turkey is now available in many different forms—canned, cut up, rolled, pies, and weiners.

## BREEDING SYSTEMS

Our present turkey breeding stock was developed much like broiler breeder stock. Most breeders used phenotype selection to improve physical traits. Some used pedigree selection with trapnesting to develop pure lines with improved reproductive traits. Live and dressed turkey shows focused attention on the need for improved meat characteristics; consumers demanded more meaty breasts on the turkeys that they purchased; and the hatchery industry pressured for improvement in number of eggs and hatchability.

Crossbreeding of bronze-feathered and white-feathered strains was used to develop the Broad White variety. However, even within this variety, no further crossbreeding was followed after establishing the foundation. To date, therefore, crossbreeding has been little used in turkey breeding.

As has been previously pointed out, chicken breeders have made extensive use of the inbreeding and hybridization techniques of the corn breeder. To date, hybridization breeding in turkeys has been confined largely to the experimental stage.

## MALE:FEMALE RATIO

One tom to 10 hens of the large varieties, and 1 tom to 15 hens of the small-type strains of turkeys, is recommended.

## ARTIFICIAL INSEMINATION

Fig. 47–10. Turkey poults produced by artificial insemination. (Courtesy, USDA)

Artificial insemination in turkeys is considered practical in flocks where natural fertility is below 80% during the early season or below 70% late in the season. In some flocks, a combination of artificial insemination and natural mating is used, with each supplementing the other.

Generally speaking, inseminating the hens every two weeks is considered sufficient. On the other hand, fertility in broadbreasted hens may be improved slightly by inseminating at weekly intervals.

## ARTIFICIAL LIGHTING

Turkeys are generally light stimulated when they are between 29 and 32 weeks of age. A 13- to 15-hour day should be provided about one month before hatching eggs are wanted.

A 60-watt lamp with reflectors 6 ft above the floor, every 10 ft in the house, will provide adequate light to stimulate turkeys into egg production. Males should be lighted 2 to 3 weeks before the hens to ensure maximum fertility.

## FACILITIES

Turkey hens are generally housed in total confinement or semiconfinement, with 3 to 7 sq ft of floor space per hen. The trend is toward total confinement in order to permit optimum control of lights for off-season egg production.

Breeder toms are housed in separate buildings from the hens, in small groups of 10 to 15 birds per pen to reduce fighting, and with 5 to 10 sq ft of floor space per tom.

Both hens and toms are kept on litter on concrete, asphalt, or dirt floors.

## MANAGEMENT PRACTICES AND GUIDES

Among the pertinent management practices and guides peculiar to, or of particular importance in, the breeding of turkeys are the following:

1. **Saddles.** These are just what the name implies—canvas saddles fitted over the hens that are to be mated by natural service, prior to the onset of the mating. This prevents much of the loss from hens being torn or cut by the toenails or spurs of the toms during mating.

2. **Care of broodies.** Broodiness is responsible for much of the poor egg production in turkeys. Broody hens will produce 20 to 30 fewer eggs than nonbroody hens. Consequently, it is very important to remove the broody hens from the rest of the flock as soon as they exhibit such behavior. A good practice is to remove all the hens that are on the nest at dark; then, place them in other quarters, but give them the same care and feed as the other hens in the flock in order to get them back into production as soon as possible. It also helps to move the nests to various locations to discourage broodiness in turkeys.

3. **Debeaking and wing clipping.** Debeaking of turkeys is commonly done to prevent cannibalism, just as it is with chickens. With turkeys, it is usually done at two to three weeks of age, with at least one-half of the upper beak removed with an electric debeaker.

Wing clipping is done to prevent flight. When practiced, it is done during the first three weeks of life. Poults are wing clipped by cutting off the tip of one wing just outside the last joint. It can be done with an electric debeaker, which cauterizes the cut and prevents excessive bleeding.

## MARKET TURKEY PRODUCTION

The marketing of turkeys is big business. In 1987, a total of 4.9 billion lb liveweight of turkeys was marketed in the United States. Another noteworthy statistic is that in 1987, the per capita consumption of turkey in the United States was 15.1 lb, exceeding both lamb (1.5 lb) and veal (1.8 lb). These figures clearly indicate that the production and marketing of turkeys are no longer confined to providing a bird for Thanksgiving. Instead, turkey meat is consumed throughout the year.

Fig. 47–11. Market turkeys raised in confinement. (Courtesy, *Turkey World*, Mount Morris, IL)

## MARKET CONTRACTS

In 1990, 90% of the nation's turkeys were produced under contract. That same year, the comparable figure for broilers was 100% produced under contract. In general, the types of contracts used in turkey production are very similar to those used in broiler production. Turkey contracts and returns are discussed in Chapter 49; hence, the reader is referred thereto.

## BUILDINGS AND EQUIPMENT

Poults are usually brooded in permanent-type houses. Some growers prefer to brood in batteries for the first 2 to 3 weeks before placing the poults on the floor. After 8 to 10 weeks of age, heat is no longer needed. At that time, the poults may be either moved to the range or placed in their confinement quarters for further growing. If properly managed, either system of rearing is satisfactory, although there is an increasing tendency to grow market turkeys in confinement in pole-type shelters. In Georgia, pole-type houses generally have sheet metal styrofoam insulated roofs, side curtains, and dirt litter floors, the houses are generally 50 ft wide and 400–500 ft long. About 2.0 to 2.5 sq ft of floor space is accorded per hen, and about 3.0 to 3.5 sq ft is allowed per tom. Birds are fed by mechanical feeders.

Fig. 47–12. Curtain-sided turkey house with sun-porch. (Courtesy, C. E. Brewer, North Carolina State University, Raleigh)

Some of the advantages claimed for confinement rearing are: (1) better protection against thieves, predators, disease and adverse weather conditions, (2) lower land costs, (3) lower labor costs because of automatic feeding and watering, and (4) better control of overall operations. Among the disadvantages: (1) higher costs for housing and equipment, (2) greater risk of respiratory diseases and cannibalism, and (3) danger from overcrowding.

## MANAGEMENT PRACTICES AND GUIDES

The following management practices and guides are pertinent to raising market turkeys:

1. **Sexed and unsexed poults.** Some producers buy day-old sexed poults. Others buy straight-run poults, then separate the sexes at 12 to 14 weeks of age. In any event, most producers raise the sexes apart because slightly better weight gain can be obtained by separating the sexes, either on the range or in confinement.

2. **Beaks blunted/desnooded/toes removed.** In addition to vent sexing the poults, hatcheries will usually render the following services for a small fee: Blunt the beaks to prevent cannibalism; remove the snood to prevent swelling and injuries from fighting; remove the toes just behind the nails to prevent the birds from scratching each other when they become frightened during loading and when they trample each other.

Fig. 47–13. Turkeys being debeaked at the hatchery. (Courtesy, C. E. Brewer, North Carolina State University, Raleigh)

## DUCK PRODUCTION

Ducks have for ages been popular with connoisseurs of good food.

In 1985, 21.6 million ducks were marketed in the United States. Duck production was once centered on Long Island, New York. Although Long Island still produces many ducks, large numbers of ducks are now raised in Indiana, Michigan, North Carolina, and Minnesota.

Most meat ducks are marketed as ducklings, at 7 to 8 weeks of age and at a liveweight of about 6.25 lb. They are generally frozen and ready to cook after thawing.

There is little demand for duck eggs in the United States.

Fig. 47-14. Ducklings at 16 days of age. (Courtesy, Cherry Valley Farms, Ltd., Rothwell, England)

## GEESE PRODUCTION

Geese are very hardy, are the closest grazers known, and can live almost entirely on good pasture. Yet the production of geese for meat purposes has never enjoyed the popularity in the United States that it has in some European countries. In recent years, considerable attention has been focused on raising geese for weeding purposes.

Weeder geese are used with great success to control and eradicate troublesome grass and certain weeds in a great variety of crops and plantings, including cotton, hops, onions, garlic, strawberries, nurseries, corn, orchards, groves, and vineyards. The geese eat grass and young weeds as quickly as they appear, but they do not touch certain cultivated plants. They will work continuously from daylight to dark, seven days a week (even on bright moonlight nights) nipping off the grass and weeds as promptly as new growth appears.

Fig. 47-15. Geese working their way down a cotton field, eating grass and weeds along the way. (Courtesy, *The Fresno Bee*, Fresno, CA)

At the end of the weeding season, geese are generally brought from the field and placed in pens for fattening for 3 to 4 weeks, until they weigh 10 to 12 lb or more. Markets are highest during the 4 to 6 weeks prior to Thanksgiving and Christmas.

The carrying of geese over from one season to the next for weeding purposes is not recommended, because older geese are less active in hot weather than young birds.

## GAME AND ORNAMENTAL BIRD PRODUCTION

Game and ornamental birds are sometimes raised for pleasure. A limited number of producers also raise them for profit, on a full-time or part-time basis. Game birds are raised for sale to game preserves or for shooting preserves. For these outlets, there is a limited, but satisfactory market. Also, there is a limited market for the sale of ornamental birds.

Fig. 47-16. Some of the 6,000 pheasants that this farmer hatches annually to keep the farm well stocked for the 21 hunters who lease the exclusive hunting rights on the farm. (USDA, Soil Conservation Service photo)

## BUSINESS AND MANAGEMENT ASPECTS OF POULTRY PRODUCTION

With the increase in specialization and size of enterprises, the business and management aspects of poultry production have become more important. More capital is required, competent management is in demand, records are essential, computers have come in, futures trading has increased, and such things as tax management, estate planning, and liability have taken on a new look. The general principles of many of these business aspects are similar for all types of animal production. These are covered in Chapter 10 of this book; hence, they will not be repeated. Instead, those business and management aspects that are peculiar to and/or particularly important to the poultry industry will be covered in the sections that follow.

## BROILER CONTRACTS

Broilers were one of the first animal industry commodities to shift from the conventional merchant-producer credit arrangement to a contractual-type program.

For maximum protection and minimum misunderstanding, all contracts should be in writing. Also, those involved should know and understand what constitutes a good contract.

## BASIC CONSIDERATIONS FOR CONTRACTS

All contracts should make provision for the following:

1. **Tenure.** The contract should be specific as to starting and ending dates per brood or time basis. A time contract usually specifies three to four broods per year.

2. **Renewal.** Each party should retain the same right for continuing or closing the program. This could minimize hardships for growers who have used credit in providing housing and equipment.

3. **Cancellation.** This should be specific and clearly understood, with equal rights and privileges.

4. **Management.** The contract should make known who is responsible for decisions; and details of the management program should be spelled out.

5. **Production and credit resources.** Detail as to who is to furnish what, when, amount (under what provisions), and security for credit should be given.

6. **Payment or settlement.** The contract should be clear as to method of computing rate, time, incentives, penalties, and losses including condemnation.

7. **Assignment of interest.** There should be mutual agreement as to assignment privileges as dictated by the situation.

8. **Arbitration.** There should be procedures providing for binding settlement to avoid court proceedings.

9. **Legal relationship of contracting parties.** It should be clearly stated whether or not the contract is a partnership, employer-employee situation, or arranged on an independent basis. This is important for social security and income tax purposes.

## MAJOR SPECIFICATIONS TO BE SPELLED OUT

In most broiler contracts, the following matters are spelled out: number and size of broods; disease; parasite and cannibalism control; ownership, facilities, and labor; technical assistance and supervision; taxes and insurance; marketing method, price, age, weight, loading, weighing, and transportation; feeding program and its control; records and accounting; changing or adjusting contract; responsibility and division of losses; credit arrangements, notes and chattels; other poultry on premises and visitors; method of figuring costs; and method of settlement and payment for broilers at time of marketing, leftover supplies, and division of interest.

## TYPES OF CONTRACTS

There are numerous types of contracts. But most of them fall into one of the following seven categories.

1. **Conventional-type credit plan or *open account* agreement.** The producer-owner retains title, makes all decisions, assumes all risks, takes any profit and/or loss (except where dealer provides a "no-loss" clause), and pays credit account at time of sale of birds. Credit may be secured by note or chattel.

2. **Share.** Under this type of contract, the producer furnishes the house, equipment, labor, and may or may not fur-

nish the litter and fuel; shares in the proceeds above costs as designated by terms of the contract; and shares in condemnation losses.

The contractor retains title, provides all production supplies, medication, supervision, makes necessary decisions, and usually assumes losses.

3. **Flat fee.** The grower receives a guaranteed fixed amount on a per bird or on a per pound basis. Some contracts include other incentives and provisions on feed conversion, gains, market price, etc.

4. **Feed conversion.** The grower is paid on specific rates based on feed conversion. Variations of incentives in addition to feed conversion are: share, market price and share, and production cost and share.

5. **Market price.** Payment to the grower is calculated on a specific schedule of rates based on the various market prices.

6. **Guaranteed price.** The dealer guarantees the grower a specific (prearranged) market price per pound for the broilers sold. The dealer may provide a *no-loss* clause as further protection for the grower.

7. **Salary.** Payment is made to the grower on a regular wage scale, paid on a weekly or monthly basis. If the grower owns buildings and equipment, he may increase income through rental or lease to contractors.

## TURKEY AND EGG CONTRACTS

Turkey contacts are very similar to broiler contracts.

The usual contract provisions for egg production cover the following: management of laying house (feeding, space, etc.); strain and number of birds; months of lay (production and quality); egg gathering and delivery; egg cleaning (who does it, methods used); cooling (temperature, humidity, and time); oiling specifications, if used; disease, parasite, and cannibalism control; quantity and quality of eggs to be delivered; restriction of other poultry and visitors; record keeping and maintenance; pullet replacement program; sale and removal of fowl at end of lay period; provision for disposal of manure; provision for use of eggs and fowl; obligation of assets to insure the contract; credit (amount and repayment); labor (present and future); incurring additional expense; share of fixed cost during the down period.

The main kinds of egg contracts are:

1. Shell and/or breaker egg.

2. Hatching egg.

3. Hatching egg foundation stock, lease basis.

4. Pullet growing contracts for laying flock.

Many adaptations of methods and techniques are used in the various kinds of egg contracts.

## MANURE

Most poultry manure is utilized as a fertilizer. However, a limited amount of it is being fed to beef cattle.

1. **Fertilizer.** In pure form, without added litter, poultry manure is produced in about the quantities shown in Table 47–1.

**TABLE 47–1**
**QUANTITY AND VALUE OF PURE MANURE FROM VARIOUS FLOCKS[1]**

| Birds | Type of Flock | Average Bird Weight | Quantity Manure Produced (dry basis) | Time Period | Estimated Value of Manure[2] |
|---|---|---|---|---|---|
| | | (lb) | (lb) | | ($) |
| 100 | Laying hens | 4.5 | 2,400 | 12 mos. | 14.78 |
| 1,000 | Broilers (chickens) | 4.0 | 2,700 | 9 wks. | 16.63 |
| 1,000 | Broilers (turkeys) | 8.0 | 4,320 | 16 wks. | 26.61 |
| 1,000 | Heavy turkeys | 20.00 | 35,000 | 24 wks. | 215.60 |

[1]From University of Maryland Extension Service, Fact Sheet 39.

[2]Assuming a value of $12.32 per ton (see Table 5–9).

Using the values shown in Table 5–9, a ton of poultry manure (free of bedding) contains an average of 31.2 lb of nitrogen (N), 8.0 lb of phosphorus (18.3 lb $P_2O_5$), and 7.0 lb of potassium (8.4 lb $K_2O$). Assuming that mixed commercial fertilizer sells at 29¢ per lb for N, 30¢ per lb for $P_2O_5$, and 12¢ per lb for $K_2O$, pure poultry manure has a value of $15.55 per ton. Of course, these values are based on pure dry manure only (droppings as they are cleaned from cage houses) with no litter added and no allowance for loss of nutrients through leaching, heating, addition of dry agents, or evaporating changes.

Where litter is used, the value of the manure for fertilizer varies according to the chemical composition of the litter, and the amount of litter used.

2. **Feeding manure (poultry house litter).** There are a number of reports of U.S. beef cattle operations using poultry manure as a feed. Also, several experiment stations have conducted, or are conducting, experimental work on the feeding value of this product. Of course, the nutritive content of manure varies appreciably, according to the type of ration fed, and more particularly according to the type of litter used. But, in general, the economic efficiencies of feeding poultry manure, in properly balanced rations, in the wintering rations of cows, or in the finishing ration of cattle appears to be worthwhile.

As an energy source, in comparison with No. 2 corn which is arbitrarily assigned a feeding value of 100, poultry manure has a relative feeding value, pound for pound, ranging from 10 to 40; and it may replace up to 15 to 25% of the grain ration. As a protein source, in comparison with 41% soybean meal which is arbitrarily assigned a feeding value of 100, poultry manure has a relative feeding value, pound for pound, ranging from 50 to 55; and it may replace up to 25% of the protein supplement.

It should be pointed out, however, that there are certain possible hazards to the feeding of poultry manure. For example, much remains to be known about the transmission of diseases from poultry to other species. Also, it is not generally known whether certain drugs administered to poultry are toxic to cattle when consumed at the levels found in litter. Neither is the effect of certain molds common in litter known. Finally, there is the esthetic angle, but this does not appear insurmountable within itself.

Cattle fed poultry manure are, in common with all other slaughtered animals, subject to Food and Drug Administration regulations governing the amount of residue in the meat. It is conceivable, therefore, that high carcass residues might result from litter containing a high percentage of medicated poultry feeds. Currently, the Food and Drug Administration does not permit the sale of chicken manure as livestock feed. However, poultry producers may feed the manure produced by their birds to their cattle.

## RECORDS

The key to good business and management is records. The historian, Santayana, put it this way, "Those who are ignorant of the past are condemned to repeat it." Also, good records help the poultry producer to overcome the banker's traditional fear of feathers.

The record forms will differ somewhat according to the type of enterprise. For example, with layers, cost per dozen eggs is the important thing, whereas in broiler and market turkey production, it's cost per pound of bird. Net returns are important, but it is also necessary that records show all the items of cost and income—egg production, feed consumption, and mortality of layers; rate of growth, pounds of chicken per 100 lb of feed, mortality, and quality of broilers produced. Good records, properly analyzed and used, will increase net earnings and serve as a basis for sound management and husbandry.

## QUESTIONS FOR STUDY AND DISCUSSION

1. What factors have caused the development of such a high degree of specialization in the poultry industry; for example, in the chicken business, there are breeders, hatchery operators, broiler producers, growers of replacement pullets, and commercial egg producers?

2. Are poultry breeders any more highly specialized than breeders of four-footed animals? Bear in mind, for example, that in the beef cattle industry there are specialized purebred breeders, commercial cow-calf operators, feeders of grower-type rations (those who background cattle) and cattle finishers or feeders.

3. Describe modern chicken breeding.

4. How do you account for the fact that hybrid breeding has been more extensively used in broiler production than in layer production; and that hybrid production in chickens has been more prevalent than in turkeys?

5. List and discuss management practices and costs which are peculiar to chicken breeding operations and which do not apply to commercial egg production.

6. List and discuss the major changes or trends in the hatching industry in recent years.

7. What are *replacement pullets*? List the factors that influence the cost of pullets, and give the approximate percentages that each of these factors contribute to the total cost of pullets.

8. List and discuss five management pointers pertinent to raising starter pullets.

9. Discuss the current trends in egg production in (a) size of laying flock, (b) proportion produced under contract, (c) housing, and (d) breeds.

10. List the management practices and guides followed by successful commercial egg producers.

11. List reasonable and achievable production goals for commercial egg producers.

12. Give statistical evidence of the growth of the broiler industry in recent years.

13. What economic factors have caused the recent shifts in broiler production to the southeastern part of the United States? List and discuss other important trends affecting the broiler industry, including (a) integration, (b) financing, and (c) contracts.

14. Discuss the current size of broiler enterprises. Describe the size and capacity of a typical, modern broiler house.

15. List the management practices and guides that are followed by most broiler producers.

16. List reasonable and achievable production goals for broilers in (a) weight for age, (b) feed conversion, and (c) mortality.

17. Present statistics to show the growth and current magnitude of the turkey industry.

18. Discuss selection of turkeys for breeding.

19. Describe the breeding systems used in turkeys.

20. Why is artificial insemination used in turkeys?

21. Why and how is artificial lighting used with turkey breeders?

22. Describe and discuss the following management practices and guides which are peculiar to, and important in turkeys: (a) saddles, (b) care of broodies, and (c) debeaking and wing clipping.

23. Discuss each of the following aspects as they apply to market turkey production: (a) magnitude, (b) market contracts, and (c) buildings and equipment.

24. List and discuss the management practices and guides pertinent to raising market turkeys.

25. Describe modern duck production.

26. Describe modern geese production.

27. List and discuss the basic considerations of broiler contracts. What are the common types of broiler contracts.

28. List and discuss the usual provisions of egg contracts. List the main kinds of egg contracts.

29. Discuss (a) the quantity of manure produced from different types of flocks; and (b) the value and use of manure as a fertilizer and as a feed.

30. Discuss the value and kind of records to keep.

## SELECTED REFERENCES

| Title of Publication | Author(s) | Publisher |
|---|---|---|
| *Chicken Broiler Industry, The: Structure, Practices, and Costs*, Mktg. Res. Rpt. No. 930 | F. L. Faber<br>R. J. Irvin | Economic Research Service, USDA, Washington, DC, 1971 |
| *Hatchery Production 1989 Summary* | Staff | U.S. Department of Agriculture, Washington, DC, March 1990 |
| *Layers and Egg Production* | Staff | U.S. Department of Agriculture, Washington, DC, January 1990 |
| *Market Structure of the Food Industries*, Mktg. Res. Rpt. No. 971 | D. F. Dunham *et al.* | Economic Research Service, USDA, Washington, DC, 1972 |
| *Raising Ducks*, Farmers' Bull. No. 2215 | W. J. Ash | U.S. Department of Agriculture, Washington, DC, 1969 |
| *Raising Geese* | R. A. Ernst<br>W. S. Coates | Cooperative Extension, University of California, USDA, Berkeley, CA, 1975 |
| *Turkey Industry, The: Structure, Practices, and Costs*, Mktg. Res. Rpt. No. 1000 | W. W. Gallimore<br>R. J. Irvin | Economic Research Service, USDA, Washington, DC, 1973 |
| *Turkey Raising*, Farmers' Bull. No. 1409 | S. J. Marsden | U.S. Department of Agriculture, Washington, DC, 1952 |
| *U.S. Broiler Industry* | F. A. Lasley *et al.* | U.S Department of Agriculture, Washington, DC, Nov. 1988 |

Turkey eggs and newly hatched poult. (Courtesy, Nicholas Turkey Breeding Farm, Sonoma, CA)

# THE EGG

---

## Chapter 48

---

| Contents | Page |
|---|---|

| Contents | Page |
|---|---|

The bird egg is a marvel of nature. It's one of the most complete foods known to man, as evidenced by the perfect balance of proteins, fats, carbohydrates, minerals, and vitamins which it provides during that 20-day in-the-shell period when it serves as the developing chick's only source of food. Also, the egg is one of the few foods that is produced in prepackaged form. Not only that, it is the reproductive cell (ovum) of the hen. Upon fertilization by the male's reproductive cell (sperm), the egg will develop into a chick when incubated properly.

## PARTS OF AN EGG

A schematic side view of an egg is shown in Fig. 48–1, with the various parts labeled in their normal position.

The protective covering, known as the shell, is composed primarily of calcium carbonate, with 6,000 to 8,000 microscopic pores permitting transfer of volatile components. The air cell, located in the large end of the egg, is formed when the cooling egg contracts and pulls the inner and outer shell membranes apart. The cordlike chalaza holds the yolk in position in the center of the egg. As shown, the yolk is surrounded by membrane, known as the vitelline membrane. The germinal disc, a normal part of every egg, is located on the surface of the yolk. Embryo formation begins here only in fertilized eggs.

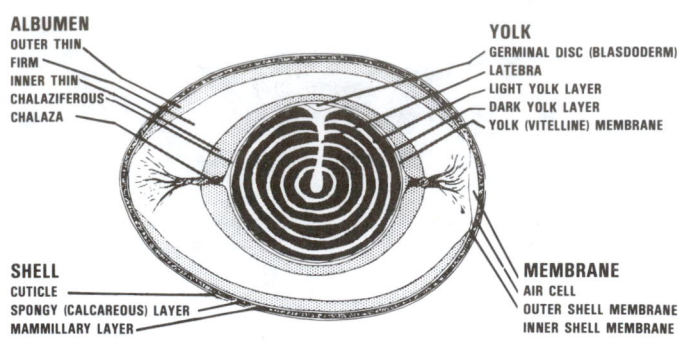

Fig. 48–1. Parts of an egg.

## COMPOSITION OF AN EGG

The chemical composition of the egg is given in Table 48–1.

### TABLE 48–1
### CHEMICAL COMPOSITION OF THE EGG

|  | Percent | Water | Protein | Fat | Ash |
|---|---|---|---|---|---|
|  | (%) | (%) | (%) | (%) | (%) |
| Whole egg ............ | 100 | 65.5 | 11.8 | 11.0 | 11.7 |
| White ............... | 58 | 88.0 | 11.0 | 0.2 | 0.8 |
| Yolk ................ | 31 | 48.0 | 17.5 | 32.5 | 2.0 |
| Shell ................ | 11 | — | — | — | 96.0 |

## FORMATION OF THE EGG

The "egg-making" machinery of the hen consists of two main parts: the ovary and the oviduct. These are shown in Fig. 48–2.

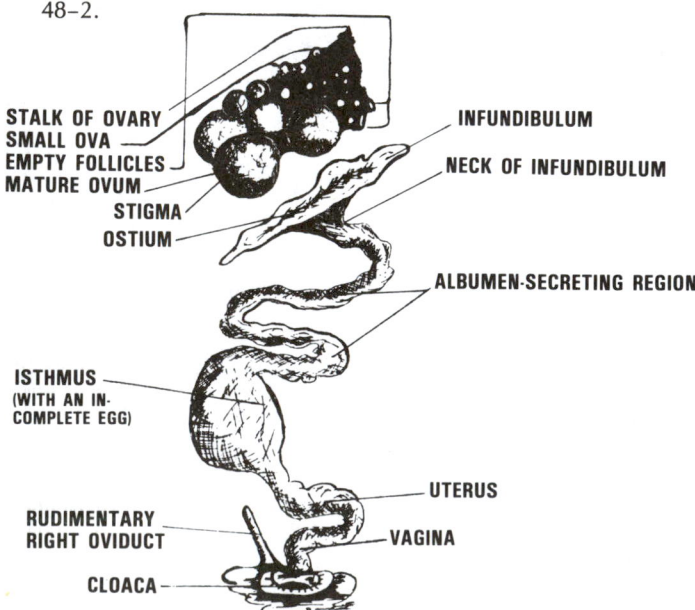

Fig. 48–2. Reproductive organs of the hen.

## OVARY

Females of most animals have two functional ovaries—a right one and a left one. But the hen has only one functional ovary, the left one, which is situated in the body cavity near the backbone. At the time of hatching, the female chick's left ovary contains up to approximately 3,600 to 4,000 tiny ova from which full-sized yolks may develop when the hen matures.

Each yolk (ovum) is enclosed in a thin-walled sac (follicle) which is attached to the ovary by a stalk. This sac contains the vast network of blood vessels which supply the yolk materials.

When a pullet reaches sexual maturity, or comes into egg production, some of the ova develop to mature yolks. When mature, the yolk is released from the follicle by rupture of the follicle wall along a line called the stigma. Soon after its release, the yolk is picked up, or engulfed, by the funnel of the oviduct.

## OVIDUCT

The oviduct is a coiled, folded tube about 20 to 30 in. long occupying a part of the left side of the abdominal cavity. The oviduct is divided into 5 rather clearly defined regions, each of which plays a specific role in the completion of the whole egg. A normal hen requires slightly over 24 hours to complete an egg. Within 30 minutes after the egg is laid, another yolk will be released from the ovary for laying the following day.

The functions of each of the 5 parts of the oviduct are set forth in Table 48–2.

**TABLE 48–2**
**FUNCTIONS OF THE OVIDUCT**

| Part | Approximate Time Egg Spends in Section | Functions |
|------|------|------|
| **Infundibulum** (funnel) | 15 min. | Picks up yolk from the body cavity after it is released from the follicle.<br>If live sperm are present, fertilization occurs in this section. |
| **Magnum** (albumen secreting region) | 3 hrs. | Thick white (albumen) is desposited around the yolk. This layer later forms the chalaziferous layer, the chalaza, and inner thin and thick white. |
| **Isthmus** | 1¼ hrs. | Inner and outer shell membranes are added, with some water and mineral salts. These membranes give some protection to the egg contents from outside contamination. |
| **Shell Gland** (uterus) | 21 hrs. | During the first part of the egg's stay in the shell gland, water and minerals pass through the shell membranes into the white, inflating the egg and giving rise to the outer layer of thin white. Soon after the egg is inflated, the shell gland starts to add calcium over the shell membranes, continuing this process until just prior to laying. If the shell is going to be colored, pigment is added in this section. |
| **Vagina** | Entire time from ovulation to laying is slightly more than 24 hrs. | The egg passes into this section just prior to laying. Its function is not known. Some believe it adds a protein sealer to the shell to seal the pores which then functions as a protective layer. |

## EGG QUALITY, AND FACTORS AFFECTING IT

The quality of the eggs produced by today's commercial egg-laying flocks is becoming increasingly superior through improved breeding, feeding, management, and marketing. However, eggs are a perishable product, and they should be handled as such. It is of utmost importance, therefore, that producers, jobbers, and retailers maintain superior quality.

Consumers want eggs with fresh-laid appearance, good flavor, and high nutritive value. The shells should be strong, regular, and clean; the white (or albumen) should be thick, clear, and firm; the yolk should be light colored, well centered, and free from blood and meat spots. The factors which determine final grade are summarized in Chapter 49, Table 49–2.

## PRODUCING QUALITY EGGS

Contrary to popular belief, not all eggs gathered from the nest are necessarily of first-rate quality. However, a large percentage of top quality eggs can be produced by adopting the following practices:

1. **Select a strain of birds noted for its ability to lay eggs of high quality.** Inherent capacity for producing high initial egg quality is important. Egg shape, shell color, shell strength, albumen quality, and incidence of blood and meat spots are quality factors which can be improved through selective breeding. Most breeders are giving considerable attention to this possibility. Data from random sample tests are highly useful for this purpose.

2. **Feed well-balanced rations.** Deficiencies of calcium, phosphorus, manganese, and vitamin D₃ lead to poor shell quality. Yolk color is almost entirely dependent on the bird's diet. Low vitamin A levels may increase the incidence of blood spots.

3. **Keep the flock disease free.** Certain diseases, especially Newcastle and infectious bronchitis, often cause birds returning to production to lay eggs of poor shape, poor shell quality, and low interior quality.

4. **Replace birds in the laying flock when they are 18 to 20 months old.** The finest quality eggs are laid by pullets. Older hens lay eggs lacking in acceptable shell quality and albumen firmness.

5. **Productive infertile market eggs.** Keep males out of the laying flock.

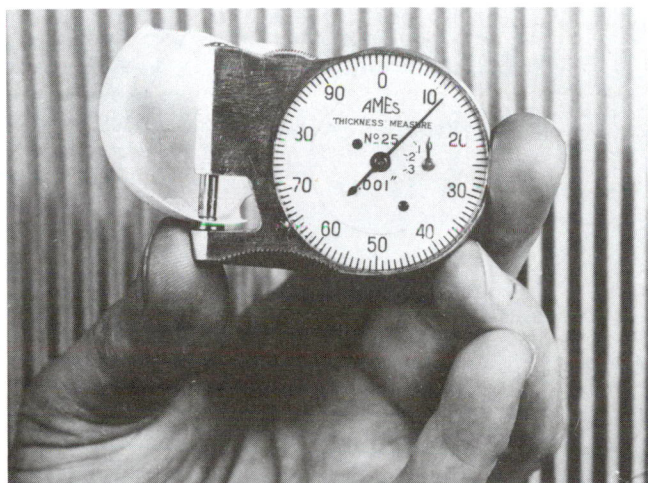

Fig. 48–3. Shell thickness is an important factor considered by breeders. Thickness is correlated with shell strength. (Courtesy, DeKalb Poultry Research, Inc., DeKalb, IL)

## HANDLING TABLE EGGS

Observance of the following rules will aid in maintaining top quality all the way to the consumer:

1. **Gather eggs frequently.** Eggs should be gathered in well-ventilated baskets or filler-flats three to four times daily. Frequent gatherings reduce the amount of body heat to which eggs are exposed. Also, the number of broken and cracked eggs will be reduced.

2. **Produce clean eggs.** Eggs are usually cleanest when they are laid. If plenty of nonstaining, dry, nesting material is provided and changed when needed, and the eggs are gathered three times a day, very few eggs will become dirty or stained.

3. **Clean soiled eggs.** To avoid excessive handling, clean all eggs rather than take time to separate the dirty eggs. The

essential items for properly cleaning shells include water, detergent-sanitizer, and an egg-washing machine. Use clean water for washing eggs and maintain the water temperature between 110° and 120°F. Add detergent-sanitizer to water according to manufacturer's directions.

4. **Coat shell.** To preserve the high initial quality, coat the eggshell with a thin, odorless, colorless, and tasteless mineral oil to prevent the egg's carbon dioxide from escaping into the surrounding atmosphere. Apply oil, either automatically with in-line egg-processing facilities or with a hand-type aerosol spray.

5. **Cool eggs properly.** Cooling eggs immediately after gathering removes the animal heat and retards any reaction which might be conducive to deterioration of quality. The egg cooler should be large enough to accommodate the daily production plus eggs held until they are marketed. An egg cooler operating at 55°F or lower, with a relative humidity of 70 to 80%, is considered adequate.

Eggs cooled prior to packing will sweat if removed from the cooler for candling or sizing. Therefore, eggs should be processed in coolers or in an adjoining air-conditioned room.

6. **Candle eggs.** Candling is the most practical way to determine interior quality of shell eggs. The object of candling is to discover and cull out eggs with blood spots and checks.

The shell should be sound and free from checks or cracks, and it should be of good texture.

7. **Separate eggs into weight classes.** After candling, eggs should be separated into weight classes. In large commercial operations, this is done by conveying the eggs over a series of scales to obtain uniform weight for a given pack. When so classified, eggs have eye appeal. Weight classes of eggs based on minimum net weight per dozen are given in Chapter 49, Table 49-4.

8. **Pack eggs properly.** Eggs should be packed with small ends down. It is possible for the air cell to break loose and move to the small end when the large end is packed down.

It is also important that cartons and cases be kept clean. This prevents the possibility of mold formation which may pass off-flavors to the eggs.

9. **Make frequent deliveries.** Eggs should be moved to market as quickly as possible—at least twice a week. Also, it is essential to keep all eggs cool enroute to market.

After eggs are delivered to the jobber or retailer, maintenance of quality is the responsibility of marketers. Generally speaking, they have adequate facilities to protect high-quality products that producers have delivered to them.

Refrigeration is a must for both short-time holding and displaying of eggs.

## HANDLING HATCHING EGGS

The fertile egg is usually in a fairly advanced stage of development from an early embryological standpoint at the time of laying. Accordingly, ideal handling would consist of setting the egg at once, so that development could proceed without being checked. Obviously, this isn't practical under most conditions, for hatching eggs must be held for varying lengths of time. The practical problem, therefore, is to hold these eggs in a suspended state of development without destroying the developing embryo. To this end, the following handling practices are recommended:

1. **Gather frequently.** Generally, hatching eggs are gathered more frequently than eggs intended for table use.

When temperatures are normal, three to four gatherings a day will suffice. However, when temperatures are extremely hot or cold, hatching eggs should be gathered every hour.

2. **Length of holding.** Hatchery eggs should be held for as short a period of time as possible, for hatchability decreases as the time of holding is increased. At the most, they should not be held longer than 10 days. Commercial hatcheries usually set twice a week.

3. **Holding temperature.** As soon as possible, hatching eggs should be cooled to a temperature between 50° and 60°F with a relative humidity of about 70 to 80%.

When it is necessary to hold hatching eggs more than 7 days, it is recommended that they be warmed to 100°F for 1 to 5 hours early in the holding period.

4. **Position and turning.** Hatching eggs are usually packed with the large end up, just as is the situation with market eggs. However, they may be placed in trays in a horizontal position.

If it is necessary to hold hatching eggs for more than five days, they should be turned. This prevents the yolk from sticking to the shell. Turning can be done by tipping the egg cases sharply. It is recommended that the cases be turned in this manner twice daily.

## FERTILIZATION OF THE EGG

The hen can produce an egg without mating with a male bird. But such an egg will not hatch.

Each male chicken has two testes, located within the body, about midway of the back. These testes discharge sperm.

If the rooster mates with and fertilizes the hen, the male sperm unites with the ova, which is found on the yolk. Such an egg will hatch.

## INCUBATION

Fig. 48–4. Incubator with thousands of turkey eggs. Note that the eggs are turned at a 45° angle. Integrated operation of Cuddy Farms, Ltd., North Carolina. (Photo courtesy James Strawser, University of Georgia, Athens)

Eggs have been incubated by artificial means for thousands of years. Both the Chinese and the Egyptians are credited with having originated artificial incubation procedures. The Chinese developed a method whereby they burned charcoal to supply the heat. They also used the hotbed method in which decomposing manure furnished the heat. The Egyptians constructed large brick incubators which they heated with fires right in the rooms where the eggs were incubated. These ancient methods were crude when compared to our present-day mammoth incubators that hold from a few thousand up to 100,000 or more eggs, and in which the temperature, humidity, ventilation, and turning are automatically controlled.

There are four factors of major importance in incubating eggs artificially: temperature, humidity, oxygen, and turning. Of these, temperature is the most critical. In natural conditions, heat is furnished by the body of the setting hen. This temperature is usually slightly lower than that of the nonbroody hen's average temperature of 106°F.

1. **Temperature.** Maintenance of the proper temperature is of prime importance for good hatchability of fertile eggs. Depending on the type of incubation, optimum temperatures range from 99° to 103°F. In the usual forced-air machine, the temperature should be maintained at about 99.5°F.

Overheating is much more critical than underheating; it will speed up rate of development, cause abnormal embryos in the early stages, and lower the percentage of hatchability. Fig. 48–5 shows the effect of temperature on the percentage of fertile eggs.

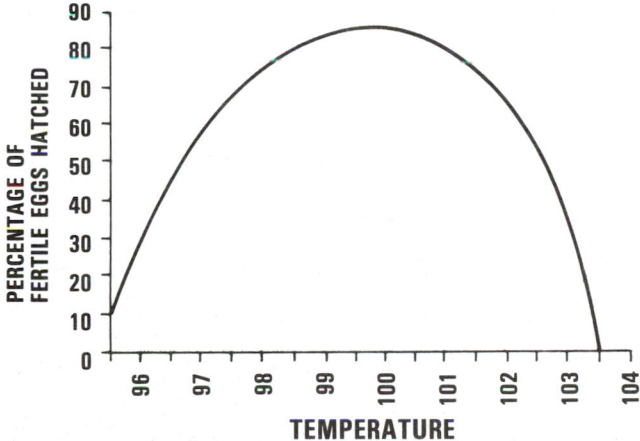

Fig. 48–5. The effect of incubation temperature on percentages of fertile eggs hatched. Relative humidity 60%, oxygen 21%, carbon dioxide below 0.5%. (From the University of Connecticut College of Agriculture Extension Service publication entitled, *Incubation and Embryology of the Chick*, p.3)

2. **Humidity.** Humidity is of great importance for normal development of the chicken embryo. Although a variation of 5 to 10% is acceptable, the relative humidity of the air within an incubator for the first 18 days should be about 60%. During the last 3 days, or the hatching period, it should be nearer 70%. Lower humidity causes excess evaporation of water, while high humidity prevents the evaporation of sufficient amounts of water from the egg. In both cases, hatchability is reduced.

3. **Egg turning.** The embryo head must occupy a position in the large end of the egg for proper hatching. Thus, the egg must be incubated large end up as gravity orients the embryo with its head uppermost. Somewhere between the fifteenth and sixteenth day, the head of the embryo is near the air cell.

Eggs should be turned from three to five times a day between the second and the eighteenth day. The purpose of this turning is to prevent the germ spot from migrating through the albumen and becoming fastened to the shell membrane. That is, turning the egg prevents an adhesion between the chorion and the shell membrane.

Proper turning consists of rotating the egg back and forth, not in one direction (a 30- to 45-degree angle is best).

4. **Oxygen utilization.** As the embryo develops, it uses oxygen and gives off carbon dioxide. Thus, sufficient ventilation within the incubator is required to assure an adequate supply of oxygen and the proper removal of carbon dioxide.

The best hatching results are obtained with 21% oxygen in the air—the normal oxygen level in the atmosphere. The embryo will tolerate a carbon dioxide level of .5%, but it will die if this level reaches 5%.

5. **Incubation time.** The normal incubation periods for several species of birds follow:

| Common Name | Incubation Period |
|---|---|
| | (in days) |
| Chicken . . . . . . . . . . . . . . . . . . | 21 |
| Turkey . . . . . . . . . . . . . . . . | 28 |
| Duck . . . . . . . . . . . . . . . . . . | 28 |
| Muscovy duck . . . . . . . . . . . . . | 33–35 |
| Goose . . . . . . . . . . . . . . . . | 28–32 |
| Guinea . . . . . . . . . . . . . . . . | 26–28 |
| Pigeon . . . . . . . . . . . . . . . . . | 18 |
| Pheasant . . . . . . . . . . . . . . . | 24 |
| Quail . . . . . . . . . . . . . . . . | 24 |
| Peafowl . . . . . . . . . . . . . . . . | 28 |
| Ostrich . . . . . . . . . . . . . . . . | 42 |
| Swan . . . . . . . . . . . . . . . . . . | 35–40 |

## EMBRYONIC DEVELOPMENT

The development which takes place in the egg from the time of fertilization to the time of hatching is one of the wonders of nature. Our knowledge of this phenomenon has been gained largely through the use of the microscope. The complexity of the development cannot be understood without some thorough training in embryology. In this book, we shall confine further discussion to outlining the highlights of development. (The reader who is interested in more detailed discussion of the complexities of embryology should consult some of the references listed at the end of this chapter.)

Since the fertilized germinal disc, or blastoderm, spends about 24 hours in the warmth of the hen's body (at about 107°F) while the egg is being completed, certain stages of embryonic development occur during that time. About 3 hours after fertilization, the newly formed single cell divides and makes 2 cells. Then there are 4, 8, 16, and more. Cell division continues until there are many cells grouped in a small, whitish spot visible on the upper surface of the egg yolk.

When the egg is laid and its temperature drops below about 80°F, shell development ceases. Cooling at ordinary temperatures will not kill the embryo, and it will begin to develop again when the egg is placed in the incubator. Keeping eggs at temperatures above 80°F prior to incubation will cause a slow growth which leads to a weakening and eventual death of the embryo.

During incubation, various processes occur. They are mainly respiration, excretion, nutrition, and protection.

The stages of embryonic development are listed in summary form in Table 48-3, and the important events are pictured in Fig. 48-6.

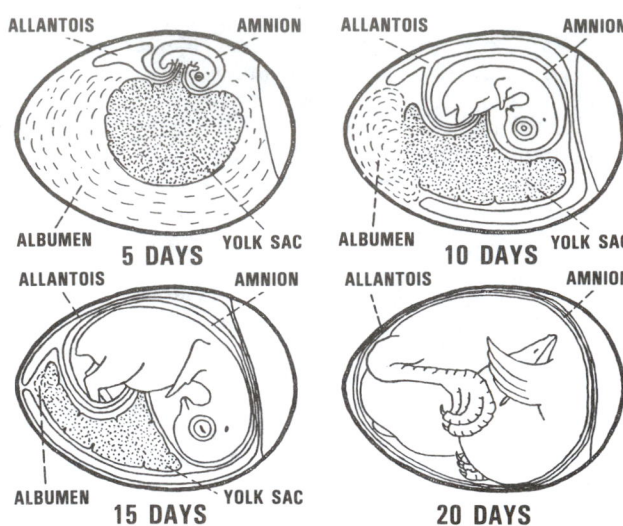

Fig. 48-6. Successive changes in the position of the chick embryo and its embryonic membranes. (From A. L. Romanoff, Cornell Rural School Leaflet, September 1939)

### TABLE 48-3
### IMPORTANT EVENTS IN EMBRYONIC DEVELOPMENT

| Stage or Period | What Takes Place |
|---|---|
| **Before Egg Laying** | Fertilization, division, and growth of living cells, segregation of cells into groups of special function. |
| **Between Laying and Incubation** | No growth; stage of inactive embryonic life. |
| **During Incubation** | |
| First day: | |
| 16 hours ......... | First sign of resemblance to chick embryo. |
| 18 hours ......... | Appearance of alimentary tract. |
| 20 hours ......... | Appearance of vertebral column. |
| 21 hours ......... | Beginning of formation of nervous system. |
| 22 hours ......... | Beginning of formation of head. |
| 23 hours ......... | Appearance of blood islands—vitelline circulation. |
| 24 hours ......... | Beginning of formation of eye. |
| Second day: | |
| 25 hours ......... | Beginning of formation of heart. |
| 35 hours ......... | Beginning of formation of ear. |
| 42 hours ......... | Heart begins to beat. |
| Third day: | |
| 50 hours ......... | Beginning of formation of amnion. |
| 60 hours ......... | Beginning of formation of nasal structure. |
| 62 hours ......... | Beginning of formation of legs. |
| 64 hours ......... | Beginning of formation of wings. |
| 70 hours ......... | Beginning of formation of allantois. |
| Fourth day ........ | Beginning of formation of tongue. |
| Fifth day .......... | Beginning of formation of reproductive organs and differentiation of sex. |
| Sixth day ......... | Beginning of formation of beak and egg-tooth. |
| Eighth day ........ | Beginning of formation of feathers. |
| Tenth day ......... | Beginning of hardening of beak. |
| Thirteenth day ..... | Appearance of scales and claws. |
| Fourteenth day ..... | Embryo turns its head toward the blunt end of egg. |
| Sixteenth day ...... | Scales, claws, and beak becoming firm and horny. |
| Seventeenth day .... | Beak turns toward air cell. |
| Nineteenth day ..... | Yolk sac begins to enter body cavity. |
| Twentieth day ..... | Yolk sac completely drawn into body cavity; embryo occupies practically all the space within the egg except the air cell. |
| Twenty-first day .... | Hatching of chick. |

Newly hatched chicks can be shipped long distances (up to 17 hours of travel time) without food or water. The yolk is largely unused by the embryo and is drawn into the body of the chick on the nineteenth day, just before it hatches. The yolk is highly nourishing and provides proteins, fats, vitamins, minerals, and water for the first several hours of the chick's life. The yolk is gradually used up during the first 10 days of life of the chick.

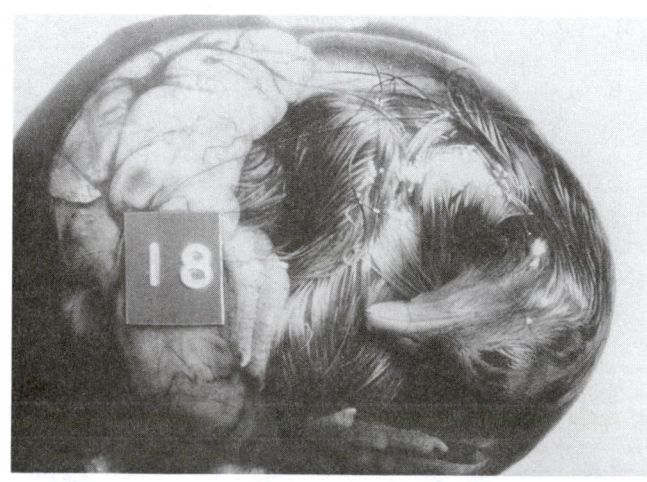

Fig. 48-7. Eighteen-day embryo. (Courtesy, Euribrid B.V., Boxmeer, Holland)

# QUESTIONS FOR STUDY AND DISCUSSION

1. What characteristics of a bird egg cause it to be referred to as a marvel of nature?

2. Sketch a cross-section of an egg and label the parts.

3. Nutritionally, or chemically, how do 2 eggs compare with 1 glass of milk or 8 oz of beef?

4. Sketch the reproductive system of the hen, name each part, and trace the formation of an egg.

5. Describe a quality egg as desired by the consumer.

6. What practices should a producer adopt in order to secure quality eggs?

7. In what ways do the handling of table eggs and hatching eggs differ?

8. Why shouldn't eggs intended for table use be fertilized?

9. How do you account for the ancient Chinese and Egyptians originating artificial incubation procedures, rather than relying on nature's method—the setting hen?

10. For proper incubation, what is considered desirable temperature, humidity, egg turning, and oxygen utilization; and what is the function of each of these environmental factors?

11. Outline the occurrence of the most prominent changes in the development of the chick.

# SELECTED REFERENCES

| Title of Publication | Author(s) | Publisher |
| --- | --- | --- |
| Avian Egg, The | A. L. Romanoff<br>A. J. Romanoff | John Wiley & Sons, Inc., New York, NY, 1963 |
| Avian Embryo, The | A. L. Romanoff | The Macmillan Company, New York, NY, 1960 |
| Early Embryology of the Chick | D. M. Patten | The Blakiston Co., New York, NY, 1952 |
| Fertility and Hatchability of Chicken and Turkey Eggs | L. W. Taylor | John Wiley & Sons, Inc., New York, NY, 1949 |
| Hatchery Operation and Management | E. M. Funk<br>M. R. Irwin | John Wiley & Sons, Inc., New York, NY, 1955 |
| Lillie's Development of the Chick | H. L. Hamilton | Henry Holt and Co., New York, NY, 1952 |

Newly hatched baby chicks. (Courtesy, Foster Farms, Livingston, CA)

Cuddy Farm turkey hatchery/North Carolina. This illustrates the transfer of turkey eggs from the setter to the hatcher components. Turkey eggs are incubated for 25 days in the setter and placed in the hatcher for the last 3 days before hatching. (Photo courtesy, James Strawser, University of Georgia, Athens)

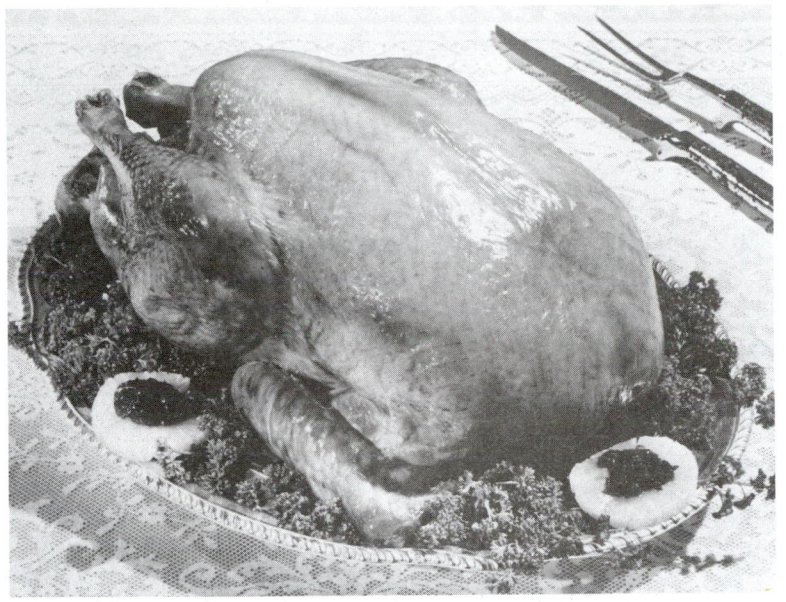

The end of the marketing line. This shows roast turkey. (Courtesy, National Turkey Federation, Mt. Morris, IL)

# MARKETING EGGS AND POULTRY

---

## Chapter 49

---

| Contents | Page |
|---|---|

| Contents | Page |
|---|---|

Marketing is selling. Practically everything that is done to eggs and poultry is done with the hope of improving salability.

The changes in the marketing of eggs and poultry have been particularly striking in recent years. These changes encompass processing and distribution as well as farm production. Technological advances in poultry breeding, nutrition, housing, disease control, and other phases of production have brought with them organizational changes which have lowered production costs and transformed traditional poultry farming into a factory-type operation. The transformation of the industry structure has advanced furthest for broilers, but turkeys and eggs are now closing the gap.

Fig. 49–1. Loading market turkeys. (Courtesy, *Turkey World,* Mount Morris, IL)

## MARKET VALUE OF EGGS AND POULTRY

U.S. poultry producers sold eggs and poultry valued at a total of $11.5 billion in 1987. This represented 8.3% of the total cash farm income. Eggs account for about 28% of the value of farm poultry sales, broilers about 54%, turkeys around 15%, and other poultry (ducks, geese, guineas, pigeons, pheasants, and turkey hatching eggs)—the remainder.

The U.S. cash receipts from poultry products are given in Fig. 49–2.

### U.S. CASH RECEIPTS FROM POULTRY PRODUCTS

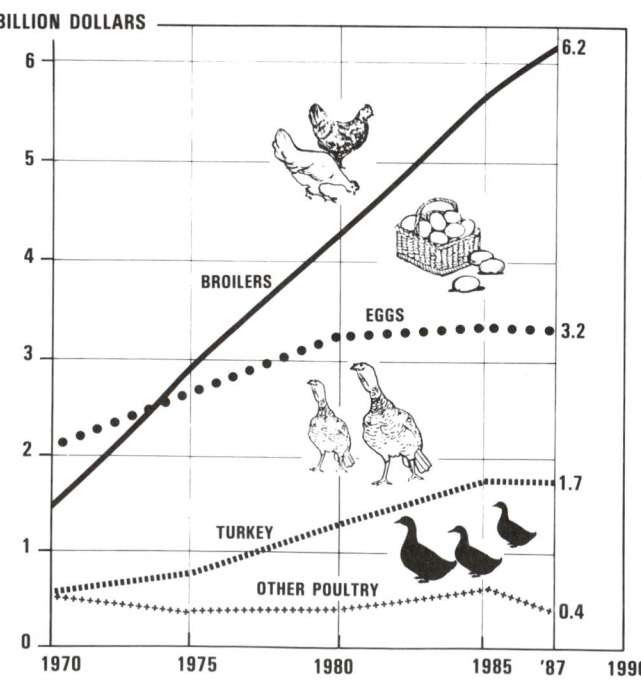

Fig. 49–2. Cash receipts from poultry products 1970–1987. (Source: *Agricultural Statistics,* USDA, 1981, p. 459 and 1988, p. 409)

## EGG AND POULTRY CONSUMPTION

Since 1960, chicken and turkey consumption, on a per capita basis, has gone up, while egg consumption per person has dropped. Consumption figures are given in Table 49–1.

### TABLE 49–1
### CONSUMPTION PER PERSON—EGGS, CHICKEN, AND TURKEY, 1960–1987[1]

| | Annual Consumption per Person | | | | | | |
|---|---|---|---|---|---|---|---|
| | 1960 | 1965 | 1970 | 1975 | 1980 | 1985 | 1987 |
| Eggs (farm basis), number | 335 | 314 | 311 | 279 | 272 | 290 | 284 |
| Chicken, ready-to-cook pounds | 27.8 | 33.3 | 40.4 | 40.1 | 50.0 | 57.6 | 62.7 |
| Turkey, ready to cook pounds | 6.2 | 7.4 | 8.0 | 8.5 | 10.5 | 12.1 | 15.1 |

[1]From USDA sources.

Consumers have been willing to increase their consumption of chicken and turkey meat as advances in production and marketing technology make for lower prices in relation to red meats (see Fig. 49–3).

### RETAIL PRICES OF SELECTED LIVESTOCK PRICES

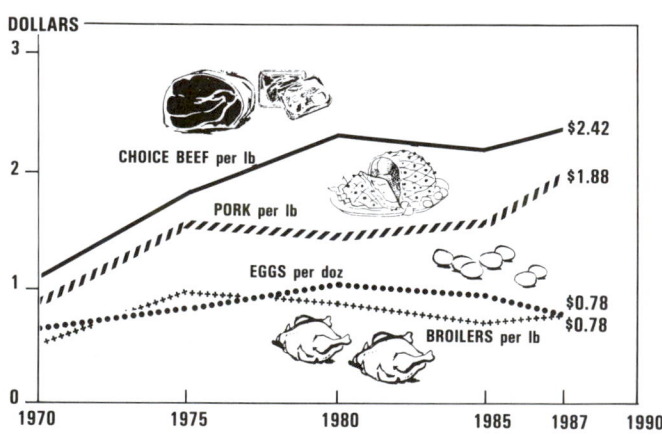

Fig. 49–3. Consumer food prices. (Source: *Statistical Abstract of the United States,* U.S. Department of Commerce, 1977, p. 486, and 1989, p. 476)

But technological advances and lower prices have brought no such consumption increases in eggs. Even at lower prices, the average person uses 15% fewer eggs than in 1960. The reasons for eggs not faring so well as poultry meat are not fully understood.

## MARKET CHANGES, AND THE FORCES BACK OF THEM

Changes in marketing poultry and eggs have extended all the way from the producer to the consumer, and they have involved technology, organization, and location. It is impossible to separate the changes from the forces back of them.

Among the pertinent changes affecting the marketing of eggs and poultry are the following:

1. **Fewer and larger producers.** All poultry-producing units—layers, broilers, and turkeys—have become fewer in number and larger in size in recent years. Many layer and broiler units now have more than 1,000,000 birds, and there are market turkey operations with 500,000, or more, birds. These operations are large enough (a) to command competent farm management know-how, (b) to justify labor and cost-saving facilities such as automatic watering and feeding equipment, and (c) to warrant the cooling facilities and other equipment necessary to maintain quality.

2. **Increased production efficiency.** This is best measured on the basis of pounds of feed required to produce a unit of product. In 1940, it required 7.4 lb of feed to produce a dozen eggs; in 1990, 3.75 lb of feed produced a dozen eggs in the most efficient operations. In 1940, it required 4.7 lb of feed to produce a pound of broiler; in 1990, 1.9 lb of feed produced a pound of broiler in the most efficient operations. The ratio for turkeys was 7.2 in 1940, and 2.7 in 1990 in the most efficient operations. (See Chapter 42, Fig. 42–8 and Table 42–4.)

3. **Shifts in production areas.** Hand in hand with changes in efficiency and industry organization, there have been shifts in the principal production areas. From 1950 to 1987, the 10 leading layer-producing states increased their proportion of the U.S. egg production from 46 to 62%. From 1950 to 1987, the 5 top broiler states—Arkansas, Georgia, Alabama, North Carolina, and Mississippi—increased their dominance from 27 to 62% of the commercial production. In 1987, the 3 major turkey-producing states of North Carolina, Minnesota, and California accounted for 48% of U.S. turkey production.

4. **More vertical integration, contracts, and direct marketing.** Rapid technical advances in poultry production following World War II made it possible for a dozen eggs or a pound of poultry meat to be produced with decreasing amounts of feed and other production costs. The incentive was strong to achieve these new efficiencies as quickly as possible. But capital was necessary. Feed manufacturers stepped in, providing capital in order to have a market for their feeds without generally involving added feed selling costs. This set off a whole chain of events involving vertical integration all the way from the producer to the consumer. Also, contractual arrangement and direct marketing followed; and all three developments—integration, contracts, and direct marketing—became commonplace among egg, broiler, and turkey producers.

a. **Egg integration.** Integration and contracts among egg producers were prompted because of the formation of large-scale specialized egg producing units, resembling factory production systems. These require much capital.

In 1990, 89% of the nation's market eggs were under some kind of integrated or contract arrangement. This means that 89% of the eggs were under contract, either in production or in marketing, and that only 11% of the eggs were sold after being produced—with no contract or prior arrangement.

b. **Broiler integration.** Initially, feed manufacturers served as the integrators in broiler production. To protect their interests, processors followed—integrating both with feed companies and producers. Integration spread quickly to all stages of broiler production and distribution except two: (1) development of the basic breeding stock, and (2) distribution to consumers.

In order to protect their financial investment, and to exercise certain controls over management phases, con-

tracts evolved. These contracts proved particularly attractive to those who had land and/or facilities and a surplus of labor, but who were unable to finance the purchase of birds and feed. From the standpoint of the integrator—the feed manufacturer and the processor—they were also attractive because they involved no social security, worker's compensation, or other similar employee fringe benefits. Likewise, the integrators could use their capital to earn higher returns in other ways.

By 1990, 92% of all broilers were produced under contract, and the remaining 8% were raised on integrator-owned farms; so, virtually 100% of the broilers were integrated.

c. **Turkey integration.** Vertical coordination and control have not advanced quite as far in turkey as in broiler production. This is attributed to the fact that turkeys need a longer growing period, the market is seasonal, and capital and management requirements have been greater and disease risk higher than in broiler production. Also, cooperatives have played a more important role in turkey than in broiler production.

In 1990, 90% of the nation's turkeys were produced or marketed under some kind of integrated or contract agreement. This means that only 10% of the turkeys were sold after being produced—with no contract or prior arrangement.

Hand in hand with increased vertical integration and contracts, eggs, broilers, and turkeys have moved through shorter and more direct marketing channels. Increasingly, direct movement from packing plants to retailers is bypassing wholesale distributors, with the result that the latter have declined in volume and relative importance.

5. **Fewer and larger processors.** In recent years, there has been a marked trend to larger processors and they have shifted their location near the production rather than the consumption area.

Because commercial egg production is more widely dispersed over the United States, the concentration of egg handling and processing isn't as great in the market egg business as in poultry meat processing. Processing is higher in liquid and frozen egg processing than in the handling of fresh eggs.

6. **More further processing.** *Further processing of poultry products is the conversion of poultry carcasses and eggs into products with added value, appeal, and convenience to the consumer, and with added utilization of products and services for the producer.*

• **Further processing of eggs**—Further processing of shell eggs is done to maintain quality. Eggs are generally cooled and held at 60°F (or lower) and 70% relative humidity. Quality can be maintained a little longer by spraying the eggs with a mist of mineral oil soon after they are gathered, to prevent the loss of carbon dioxide.

Eggs that are stored in 3 dozen cartons may be processed as follows: Removed from the cooler, picked up by vacuum lifter, transferred to a rubber conveyor line, and moved through an egg washer (containing water at about 90°F, a detergent for cleaning, and a sanitizer to control microorganisms); then, rinsed with warm water, sprayed with a thin coating of oil, flash candled, and sized; and packed—most commonly in dozen-sized cartons. Some eggs are broken. Still others are prepared as scrambled egg mixes and various egg and fruit juice drink combinations and sold as liquid eggs. Other eggs are dried.

• **Further processing of chickens and turkeys**—Historically, broilers have been marketed in five forms: (a) live, (b) New York dressed (picked, but not eviscerated), (c) ready-to-cook whole, (d) ready-to-cook parts, and (e) further processed products. Traditionally, broilers are chilled, but not frozen (only 7% of broilers were frozen in 1987). For many years, most turkeys were marketed whole and frozen.

In the 1970s and 1980s, many broilers were marketed as parts. As the demand for preferred parts (breasts, thighs) increased, a surplus of the less popular parts—backs, necks, wings—was created. But, gradually, deboning was perfected, backs, necks, wings, and whole carcasses were ground and the bone fragments and cartilage were removed by sieving, and made into hot dogs and other products. Today, chicken and turkey carcasses are converted into cut portions and further processed; they're marketed battered, breaded (enrobed), precooked, cooked, as cooked and frozen dinners, as cold cuts (such as luncheon meats and bologna) or nuggets, and made into burger patties, hot dogs, turkey rolls, roasts, and other forms.

In 1988, the USDA reported that 22.1% of the nation's broilers certified as slaughter were further processed.[1]

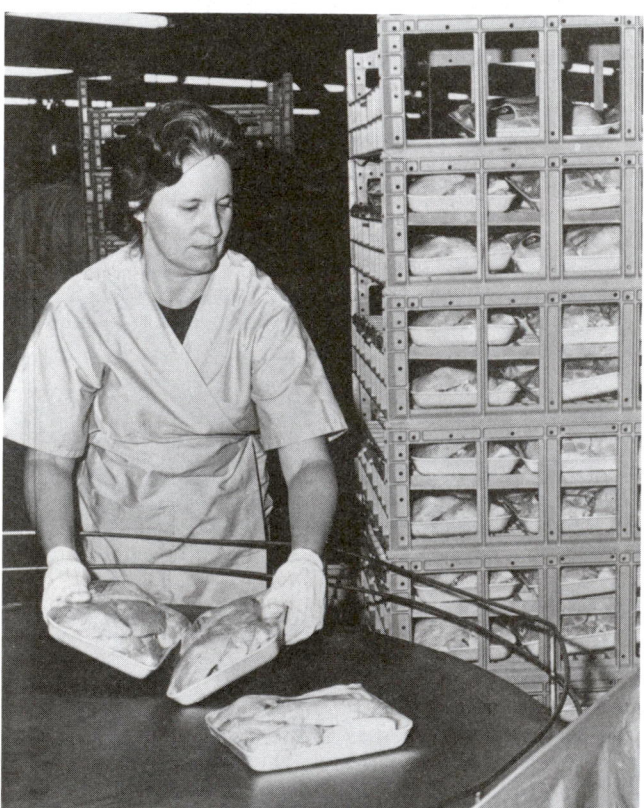

Fig. 49–4. Parts of fryer being stacked on racks. (Courtesy, Foster Farms, Livingston, CA)

• **Factors favorable to further processing of poultry**—Further processing of poultry carcasses has grown, and will continue to grow, for the following reasons:

1. **It is generally profitable to processors.** Further processing is usually profitable because competition is limited, due

[1]Lasley, F. A., et al., *The U.S. Broiler Industry,* USDA, ERS, Agricultural Economic Report No. 591, Nov. 1988, p. 61.)

to the high cost of establishing such a business, with the result that the profits are usually good.

2. **Brand name products impart consumer confidence.** Consumers are able and willing to pay more for brand name products because they impart assurance of repeatability and quality.

3. **Consumers demand convenience foods.** With two-thirds of American women between the ages of 20 and 64 years of age working outside the home, and having less time but more money, ready-to-eat foods have become popular. Likewise, such institutions as schools and hospitals favor further processed poultry as a means of circumventing high labor costs and inefficiency; they can buy ready-to-cook and ready-to-eat products, prepared by the processor with maximum automation, at a lower cost than they can prepare it in their own kitchens.

4. **Retailers like it, too.** Retailers like having further processed poultry products arrive prepackaged, weighed, dated, and even priced if they so desire. This limits their labor in handling the product primarily to stocking the counter.

The trend to further processing of poultry products will continue unabated, with more convenience and time saving built into the products and with the processor able to add value to the initial products. Thus, in the future more deboned, ready-to-cook, and ready-to-serve poultry products will become available under alluring brand names.

## MARKET CLASSES AND GRADES

The U.S. Department of Agriculture has established specifications for different kinds, classes, and grades of poultry. They define kind as referring to the different species of poultry, such as chickens, turkeys, ducks, geese, guineas, and pigeons. *Class* refers to kinds of poultry by groups which are essentially of the same physical characteristics, such as broilers or hens. These physical characteristics are associated with age and sex. The kinds and classes of live, dressed, and ready-to-cook poultry listed in the U.S. classes, standards, and grades are in general use in all segments of the poultry industry.

## EGGS

The grading of eggs involves their sorting according to quality, size, and weight, and other factors that determine their relative value. U.S. standards for quality of individual shell eggs have been developed on the basis of such interior quality factors as condition of the white and yolk, the size of the air cell, and the exterior quality factors of cleanliness and soundness of the shell. These standards cover the entire range of edible eggs.

Eggs are also classified according to weight (or size), expressed in ounces per dozen.

Egg grading, then, is the grouping of eggs into lots according to similar characteristics as to quality and weight. Although color is not a factor in the U.S. standards of grades, eggs are sometimes sorted for color and sold as either *whites* or *browns.*

Four sets of grades, based on the quality standards for individual shell eggs, are used in this country: (1) consumer grades—used in the sale of eggs to individual consumers, (2) wholesale grades—used in the wholesale channels of trade, (3)

U.S. Procurement Grades—used for institutional buying and Armed Forces purchases, and (4) U.S. Nest Run Grade—which is also used in wholesale channels of trade.

The U.S. standards for quality of individual shell eggs are applicable only to eggs of the domesticated chicken that are in the shell. These are given in Table 49-2.

### TABLE 49-2
### SUMMARY OF U.S. STANDARDS FOR QUALITY OF INDIVIDUAL SHELL EGGS[1]

| Quality Factor | Specifications for Each Quality Factor | | | |
| --- | --- | --- | --- | --- |
| | **AA Quality** | **A Quality** | **B Quality** | **C Quality** |
| **Shell** | Clean. Unbroken. Practically normal. | Clean. Unbroken. Practically normal. | Clean; to slightly stained. Unbroken. May be slightly abnormal. | Clean; to moderately stained. Unbroken. May be abnormal. |
| **Air Cell** | ⅛ in. or less in depth. May show unlimited movement and may be free or bubbly. | 3/16 in. or less in depth. May show unlimited movement and may be free or bubbly. | ⅜ in. or less in depth. May show unlimited movement and may be free or bubbly. | May be over ⅜ in. in depth. May show unlimited movement and may be free or bubbly. |
| **White** | Clear. Firm (72 Haugh units or higher). | Clear. May be reasonably firm (60 to 72 Haugh units). | Clear. May be slightly weak (31 to 60 Haugh units). | May be weak and watery. Small blood clots or spots may be present (less than 31 Haugh units).* |
| **Yolk** | Outline slightly defined. Practically free from defects. | Outline may be fairly well defined. Practically free from defects. | Outline may be well defined. May be slightly enlarged and flattened. May show definite but not serious defects. | Outline may be plainly visible. May be enlarged and flattened. May show clearly visible germ development but no blood. May show other serious defects. |

*If they are small (aggregating not more than ⅛ in.) in diameter.

For eggs with dirty or broken shells, the standards of quality provide three additional qualities. These are:

| Dirty | Check | Leaker |
| --- | --- | --- |
| Unbroken. May be dirty. | Checked or cracked but not leaking. | Broken so contents are leaking. |

[1]*United States Standards, Grades, and Weight Classes for Shell Eggs*, Agricultural Marketing Service, Poultry Division, USDA, effective July 1, 1974.

The basis for the egg grades given in Table 49-2 is resemblance to normal new-laid eggs.

Consumer grades are those used for lots of eggs that have been carefully candled and graded for retail trade. These are given in Table 49-3.

### TABLE 49-3
### SUMMARY OF U.S. CONSUMER GRADES FOR SHELL EGGS[1]

| U.S. Consumer Grade (origin) | Quality Required[1] | Tolerance Permitted[2] | |
| --- | --- | --- | --- |
| | | **Percent** | **Quality** |
| Grade AA or Fresh Fancy Quality | 85% AA | Up to 15 | A or B |
| | | Not over 5 | C or Check |
| Grade A | 85% A or better | Up to 15 | B |
| | | Not over 5 | C or Check |
| Grade B | 85% B or better | Up to 15 | C |
| | | Not over 10 | Checks |

| U.S. Consumer Grade (destination) | Quality Required[1] | Tolerance Permitted[3] | |
| --- | --- | --- | --- |
| | | **Percent** | **Quality** |
| Grade AA or Fresh Fancy Quality | 80% AA | Up to 20 | A or B |
| | | Not over 5 | C or Check |
| Grade A | 80% A or better | Up to 20 | B |
| | | Not over 5 | C or Check |
| Grade B | 80% B or better | Up to 20 | C |
| | | Not over 10 | Checks |

[1]*United States Standards, Grades, and Weight Classes for Shell Eggs*, Agricultural Marketing Service, Poultry Division, USDA, July 1, 1974, p. 3.

[2]For the U.S. Consumer grades (at origin), a tolerance of 0.30% Leakers or loss (due to meat or blood spots) in any combination is permitted. No Dirties or other type loss is permitted.

[3]For the U.S. Consumer grades (destination), a tolerance of 0.50% Leakers, Dirties, or loss (due to meat or blood spots) in any combination is permitted, except that such loss may not exceed 0.30%. Other types of losses are not permitted.

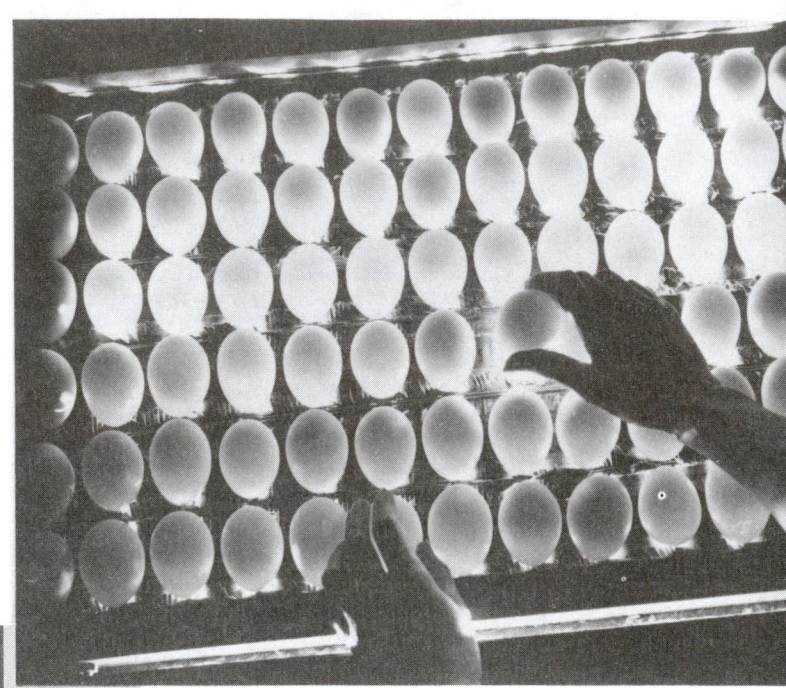

# PROCESSING EGGS

Fig. 49-5. *Above:* Washing eggs. (Courtesy, DeKalb Poultry Research, DeKalb, IL)

Fig. 49-6. *Above:* Candling eggs. (Courtesy, DeKalb Poultry Research, DeKalb, IL)

Fig. 49-7. *Above:* Grading eggs by automatic machine. (Courtesy, DeKalb Poultry Research, DeKalb, IL)

Fig. 49-8. *Right:* Racking eggs for storage. (Courtesy, Don Bell, Riverside County Poultry Farm Advisor, Riverside, CA)

Wholesale grades differ from consumer grades in the tolerance of lower quality eggs permitted and in possible inclusion of some loss or inedible eggs. The grade designations of wholesale grades are U.S. Specials, U.S. Extras, U.S. Standards, U.S. Trades, U.S. Dirties, and U.S. Checks.

The Nest Run Grade is for the purpose of expediting trading by anticipating the grade yield of eggs before processing—before washing, grading, and sizing.

In the marketing of eggs, weight classes are also provided. These are given in Table 49–4.

It is to be emphasized that weight is separate and distinct from egg quality.

## TABLE 49–4
## U.S. WEIGHT CLASSES FOR CONSUMER GRADES FOR SHELL EGGS[1]

| Size or Weight Class | Minimum Net Weight per Dozen | Minimum Net Weight per 30 Dozen | Minimum Weight for Individual Eggs at Rate per Dozen |
|---|---|---|---|
| | (oz) | (lb) | (oz) |
| Jumbo | 30 | 56 | 29 |
| Extra large | 27 | 50½ | 26 |
| Large | 24 | 45 | 23 |
| Medium | 21 | 39½ | 20 |
| Small | 18 | 34 | 17 |
| Peewee | 15 | 28 | — |

[1]*United States Standards, Grades and Weight Classes for Shell Eggs*, effective July 1, 1974.

## POULTRY

The commonly used market classes of chickens are:

1. **Rock Cornish game hen or Cornish game hen.** A Rock Cornish game hen or Cornish game hen is a young immature chicken (usually about 4 weeks of age) weighing not more than 2 lb ready-to-cook weight, which was prepared from a Cornish chicken or the progeny of a Cornish chicken crossed with another breed of chicken.

2. **Broiler or fryer.** A broiler or fryer is a young chicken (usually under 8 weeks of age), of either sex, that is tender meated with soft, pliable, smooth-textured skin and flexible breastbone cartilage.

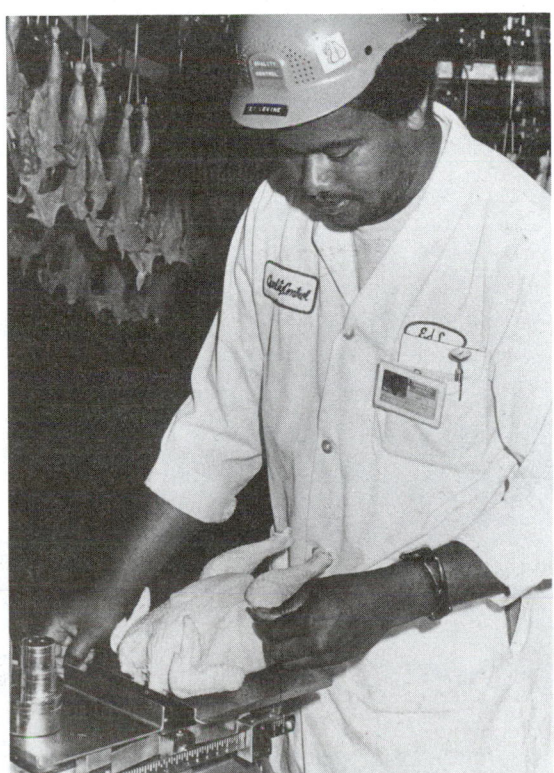

Fig. 49–9. Weighing a broiler. (Courtesy, Foster Farms, Livingston, CA)

3. **Roaster.** A roaster is a young chicken (usually under 10 weeks of age), of either sex, that is tender meated with soft, pliable, smooth-textured skin and breastbone cartilage that may be somewhat less flexible than that of a broiler or fryer.

4. **Capon.** A capon is a surgically unsexed male chicken (usually under 5 months of age) that is tender meated with soft, pliable, smooth-textured skin.

5. **Stag.** A stag is a male chicken (usually under 10 months of age) with coarse skin, somewhat toughened and darkened flesh, and considerable hardening of the breastbone cartilage. Stags show a condition of fleshing and a degree of maturity intermediate between that of a roaster and a cock or rooster.

6. **Hen or stewing chicken or fowl.** A hen or stewing chicken or fowl is a mature female chicken (usually more than 10 months of age) with meat less tender than that of a roaster, and nonflexible breastbone tip.

7. **Cock or rooster.** A cock or rooster is a mature male chicken (usually more than 10 months of age) with coarse skin, toughened and darkened meat, and hardened breastbone tip.

Classes are also provided for turkeys, ducks, geese, guineas, and pigeons.

• **Grades**—The grades of individual live birds are: A or No. 1 Quality, B or No. 2 Quality, and C or No. 3 Quality. The criteria used in determining grade are health and vigor, feathering, conformation, fleshing, fat covering, and defects.

Dressed and ready-to-cook poultry are graded for class, condition, and quality. These are most important since they are the grades used at the retail level. These grades are: U.S. Grade A, U.S. Grade B, and U.S. Grade C. These grades apply to dressed and ready-to-cook chickens, turkeys, ducks, geese, guineas, and pigeons.

Additionally, there are the U.S. Procurement Grades, which are designed primarily for institutional use. These grades are: U.S. Procurement Grade 1, and U.S. Procurement Grade 2. In procurement grades, more emphasis is placed on meat yield than on appearance.

The factors determining the grade of carcasses, or ready-to-cook poultry parts therefrom, are: conformation, fleshing, fat covering, pinfeathers, exposed flesh, discoloration, disjoined bones, broken bones, missing parts, and freezing defects.

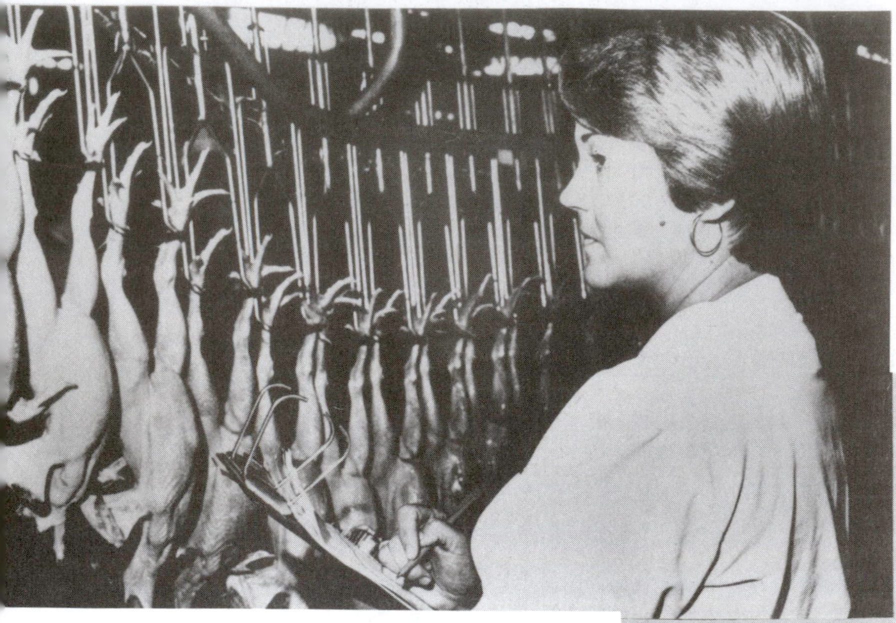

# PROCESSING BROILERS

Fig. 49-10. *Above:* Inspecting broilers—quality assurance. (Courtesy, Foster Farms, Livingston, CA)

Fig. 49-11. *Right:* Eviscerating broilers. Note the stainless steel trough and sanitation. (Courtesy, Gordon Johnson Co.)

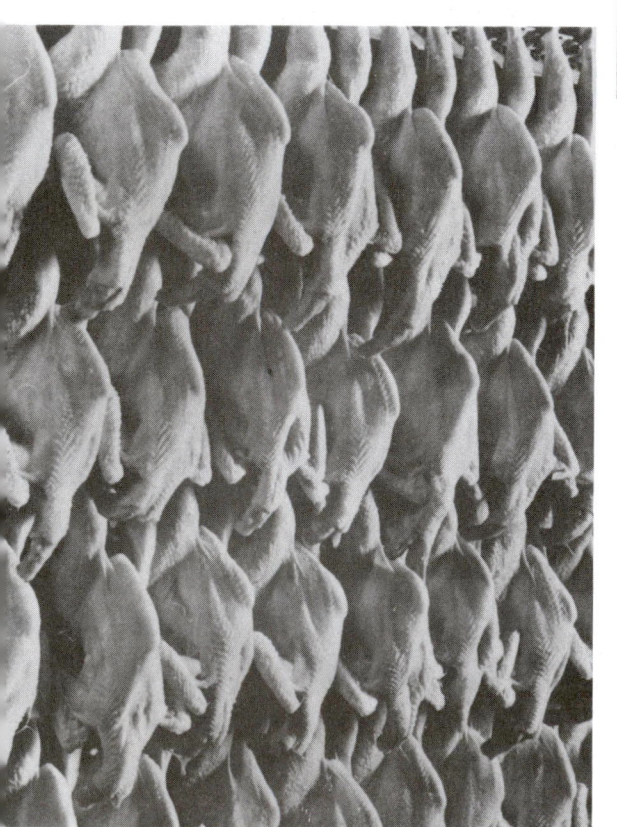

Fig. 49-12. *Left:* Broilers ready for market. (Photo by J. C. Allen and Son, West Lafayette, IN)

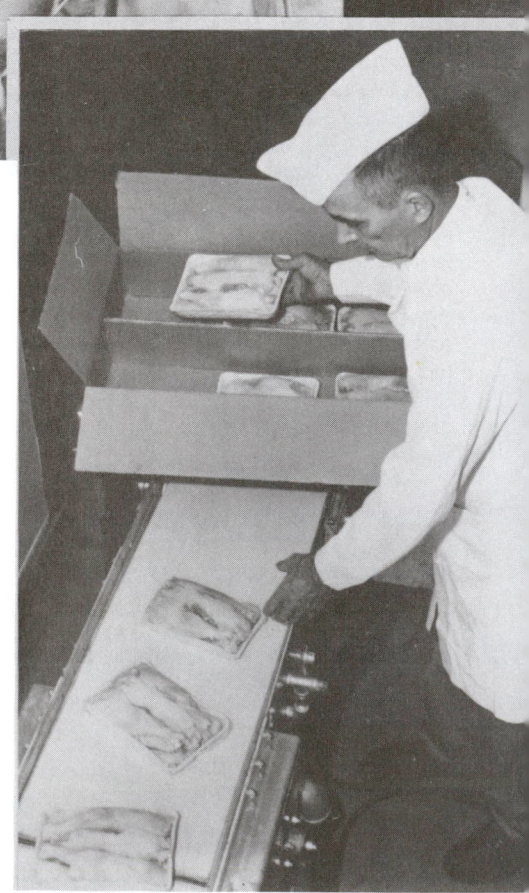

Fig. 49-13. *Right:* Broiler parts, packaged and chilled. (Courtesy, Gordon Johnson Co.)

## FEDERAL INSPECTION

The Poultry Products Inspection Act, Public Law 85–175, was enacted on August 28, 1957, and became fully effective January 1, 1959. The Animal and Plant Health Inspection Service in the U.S. Department of Agriculture is charged with the responsibility of administering this Act. The law requires inspection for wholesomeness of all poultry-processing plants shipping any of their products in interstate or foreign commerce. Personnel and supervisory cost of the service required by the Act except for necessary overtime are borne by appropriated federal funds.

Under the provisions of this Act, the USDA has four major responsibilities: (1) to determine that the poultry being processed is fit for human food—as determined by antemortem and postmortem inspection, (2) to make sure that the processing is done in a sanitary manner, (3) to protect poultry and poultry products from adulteration, and (4) to require that poultry and poultry products are properly labeled in compliance with the requirements of the law.

The grading services for both poultry meat and eggs are on a voluntary basis, with charges made for those requesting the service. However, inspection of poultry and egg products is mandatory in those poultry plants processing and shipping any of their products in interstate or foreign commerce.

But the current method of examining birds individually by federal inspectors is outmoded and inefficient. Imagine having sufficient trained inspectors to examine individually more than 5 billion broilers annually! So, the National Academy of Science, along with producer and consumer groups, has recommended that the U.S. Department of Agriculture develop a state-of-the-art meat and poultry inspection service. Currently (1990), the Food Safety Inspection Service (FSIS) of the U.S. Department of Agriculture is studying a system which it calls *Hazard Analysis and Critical Control Point (HACCP)* for meat and poultry operations. This system identifies what it calls Critical Control Points (CCP), or places, such as where thousands of birds are processed together; for example, (1) scalding freshly killed chickens in a common bath to loosen feathers; (2) defeathering many birds with the same rubber fingers; (3) eviscerating many birds with an improperly adjusted machine; and (4) cooling defeathered chickens in a common chill bath. But, until, and unless, the U.S. Department of Agriculture develops an improved inspection system, the only solution is more quality control inspectors—both company and federal.

## POULTRY IRRADIATION

Factory-style broiler production has spread salmonella, which was once either rare or easily isolated in individual birds and flocks. Thus, from 1973 to 1987, poultry accounted for 9.8% of U.S. food-borne illnesses, with these attributed to salmonella and other pathogens, according to the Centers for Disease Control, Atlanta. Moreover, the inspectors of USDA's Food Safety and Inspection Service are not supposed to be looking for bacteria such as salmonella, which cannot be detected with the naked eye. This set of circumstances prompted the U.S. Department of Agriculture to petition the Food and Drug Administration (FDA) for approval of irradiation to control bacteria which may be present in chickens, turkeys, and other fresh or frozen uncooked poultry.

In 1990, the FDA announced that on May 1, it approved the use of irradiation to control bacteria in uncooked poultry; and described the process as a system to "pasteurize" solid foods. FDA added (1) that, as in the heat pasteurization of milk, the irradiation process greatly reduces, but does not eliminate, all bacteria; and (2) that irradiation does not make the food radioactive and, therefore, does not expose consumers to radiation. Treated food must be labeled, stating that is was irradiated; and the packaging must carry an international symbol representing the irradiation process.

Although the safety of irradiated foods has been carefully researched and thoroughly tested, at this point and period of time consumers haven't accepted it. Fortunately, the same results can be achieved by cooking poultry.

## MARKET CHANNELS AND SELLING ARRANGEMENTS

Most eggs and poultry are marketed through retail food stores. The institutional markets, government purchases, and exports account for the remainder. In all cases, market channels have become more direct.

Until the mid-1940s, most eggs moved from the producers to county buying stations, or to hucksters and peddlers, thence to central assembling plants and shippers, thence to city wholesalers and jobbers, and finally to store warehouses or retail stores. In recent years, the desire to sell to premium outlets has had a major impact on the channels used in areas close to markets, with the result that a substantial proportion of eggs are moving from the producers either directly to consumers or to retail stores.

The marketing of poultry meat has also changed, largely as a result of a substantial increase in commercial broiler production and a shift from New York dressed (blood and feathers removed only) to ready-to-cook birds or parts and further processed products. In the heavy producing broiler areas, birds are moved directly from producers to the processing plants, thence to chain store warehouses or direct to retail stores. In the less populous poultry areas, the buying station type of agency assembles small quantities of poultry from individual producers, then sells them in larger quantities to city or country processing plants.

In summary, it may be said that, in both egg and broiler production, larger operations have resulted in processors locating near production and the elimination of some of the people who formerly performed a needed service where many small producers were involved.

As integration and various forms of vertical coordination continue to spread within the poultry industries, fewer actual purchases and sales of products occur and fewer genuine negotiated prices are generated. Also, formula pricing reduces further the fraction of total supply entering the market price formation. Consequently, it is increasingly difficult for a poultry producer to determine what is the "going market," and markets are more vulnerable from the standpoint of manipulation.

Fortunately, two major market news services remain: (1) The U.S. Department of Agriculture market news service, and (2) the Urner Barry report.

• **The U.S. Department of Agriculture market news service—** The USDA operates a nationwide news service in cooperation with state agencies. Egg reports are made on frozen and dried eggs. These reports include data on supply and demand, movement, cold storage stocks, trading activity, price activities, and quality ranges. Prices and statistics are collected and disseminated nationwide on broilers/fryers, fowl, roasters, turkeys, ducks, and other miscellaneous poultry and rabbits.

Information on eggs and poultry is gathered by professionally trained staff from producers, major consuming centers, shipping points, and other sources. The information is disseminated nationwide by leased wire service and to the news media.

• **The Urner Barry publications**—Urner Barry reports egg prices on a regional basis.

## PRICING EGGS

The egg-pricing system in use today is still operating largely on the basis of wholesale trading in terminal markets, despite the fact that wholesalers, who formerly handled most of the nation's eggs, have largely been displaced by assembler-packers. As a result, the egg-pricing system has become increasingly controversial and problem laden.

Today, there is no satisfactory price barometer for eggs. The volume of eggs traded on the New York Mercantile Exchange and the Chicago Mercantile Exchange is too small. Also, mercantile exchange trading in both New York and Chicago has been criticized for too frequent and too wide fluctuations in price, although egg production, nationwide, is increasingly uniform.

There is no easy solution to the complex problem of egg pricing. Among the alternative pricing methods or systems proposed are the following:

1. **Computerized buying and selling.** This could be accomplished through an existing or new organization of traders who would agree to conduct transactions according to prescribed trading rules. The results of trading could be used directly, or they could become a major indicator for base price quotations.

2. **Using prices paid by retailers.** This could replace the wholesale level. The Los Angeles and San Francisco markets have moved in this direction.

3. **Committee pricing.** This could best be carried out under specific legislation. A group of designated individuals, supported by a staff to gather and analyze market information, could suggest prices which they consider to reflect supply and demand for specific locations, grades and sizes, and time periods. Such a committee could encourage more and better information and quickly adapt to changing industry structure and practices.

4. **Decentralized pricing.** This is the method used in the pricing of live meat animals. An objection to the use of this method in eggs is that the latter are more homogeneous in quality than live animals; hence, lots from various areas may be readily substituted as necessary, making the market national in scope.

5. **Administered pricing.** This would be operated almost entirely by private industry. This method would be a distinct possibility if the egg industry becomes more integrated or coordinated than at present. Presumably, this would require stricter industry determination and scheduling of quantities produced than exist at present.

6. **Futures oriented pricing.** This would involve derivation of cash market prices by adjusting from values for the nearest futures option. For this system to work, active futures markets would be required throughout the year, and on several grades of eggs.

In summary, it may be said that the egg-pricing system which evolved in the egg industry served it long and well—as long as the industry remained relatively static. But as the basic structure of the industry began to undergo rapid and extensive changes, the egg-pricing system did not change with it. As a result, progress in pricing bypassed egg marketing. Although there is no easy solution to the problem, the egg industry recognizes that pricing is vital to a market economy. Gradually, improved egg pricing will evolve and close the gap between needs and performance.

## PRICING BROILERS

For broilers and turkeys, there is a great deal of formula pricing. Originally, the typical broiler formula was based on live price divided by 73% (the approximate yield of ready-to-cook from liveweight) plus 3 to 7 cents to cover processing costs. However, as more and more broilers were produced under contract, fewer and fewer live broilers changed hands. As a result, the trend is away from farm base calculations to ready-to-cook prices.

## PRICING TURKEYS

Since most turkeys are sold frozen, there is less urgency in either selling or buying them than in the case of broilers or fresh meat. The Urner Barry daily report of wholesale turkey prices in the New York area is widely used as a guide in price negotiations. In the Chicago area, for example, the typical rule of thumb for turkey transactions is to apply the Urner Barry New York quotation, five-day average for the week in which delivery was made, less 1.25 cents per pound.

## FINANCIAL ASPECTS

Most people are in business to make money, and poultry producers are people; hence, they are no exception in this regard, nor should they be. The following sections reveal, to the extent that the facts and figures are available, the financial aspects of each of the main segments of the poultry industry, and tell how well or how poorly they are doing from the standpoint of profits.

## EGG CONTRACTS AND RETURNS

Contract egg producers usually own the buildings and equipment and furnish labor, litter, electricity, and water in return for a fixed rate per dozen eggs. Contractors usually furnish hens, feed, medication, packing materials, and service people to supervise the management and care of the flocks. In some areas, the contractors haul the eggs, in others the producers do the hauling. Generally, the producer only washes the dirty and badly stained eggs.

The terms of contracts vary among areas and are affected by the amount of competition and the level of commercial egg prices. The following types of contracts are rather typical:

• **Contracts most frequently used:**

1. $0.08 per dozen for Grade A and $0.05 per dozen for undergrades.[2]

2. $0.07 per dozen for Grade A and $0.04 per dozen for undergrades.

3. $0.07 per dozen for all marketable eggs.

4. $0.08 per dozen for all marketable eggs.

5. Fifteen percent of gross egg sales: Eggs are valued according to the average farm price for eggs marketed nest run on a grade yield basis to packing plants.

6. Molted flocks: $0.0175 per hen per week until molted hens obtain 50% production, then $0.08 per dozen for all marketable eggs.

7. $0.10 per dozen for all marketable eggs (sometimes offered for molted flocks).

• **Contracts sometimes used:**

1. $0.0175 per week per hen until the flock reaches 50% production, then $0.07 per dozen for Grade A and $0.04 per dozen for undergrades. (This contract is usually given when the flock is placed prior to 20 weeks of age.)

2. Sliding price scale plan: (a) When the average price is $0.60 per dozen or more, the contractor pays $0.08 per dozen for Grade A and $0.05 per dozen for undergrades; and (b) when the average price is below $0.60 per dozen, the contractor pays $0.07 per dozen for Grade A and $0.04 per dozen for undergrades.

3. $0.07 per dozen for Grade A and $0.04 per dozen for undergrades until pullet, feed, and medication expenses are covered; then contractor and producer divide profit equally.

4. $0.075 per dozen for all marketable eggs.

5. A contract rate of 2.4 cents per layer per week for heavy-type layers (brown-egg layers) versus 1.8 cents (75% of 2.4) for Leghorns.

Returns are usually quoted on a per hen per year basis. They vary according to the stipulations of the contract, the type of housing and equipment, and whether or not hired labor is used. In Southern California in 1989, the net return was 10.5 cents a dozen eggs, or $2.10 per hen for the year.

## BROILER CONTRACTS AND RETURNS

A typical broiler contract is one in which the grower (usually a farmer) provides the housing and grow-out equipment—feeders, waterers, brooders—and such other items as water, electricity, fuel, litter, and labor. The contractor (the company) provides the chicks, feed, necessary medication, supervision, and labor and equipment for catching and hauling the birds to market.

The company retains title to the birds and the grower is paid a return for labor and facilities, usually a bonus or incentive arrangement. Some contractors pay growers on a per pound basis—usually 4 cents per pound of broiler marketed, whereas others pay on a per bird basis—usually 12 to 16 cents per bird marketed. Bonuses are earned by having lower mortality than the average grower, by producing more pounds of broiler on a given amount of feed, by having fewer birds condemned during processing, or by a combination of part or all these accomplishments. Grower income less operating ex-

penses, depreciation, taxes, and return on capital invested has been estimated at between $1.00 and $5.00 per hour, the amount depending on efficiencies achieved in producing the birds and how fully a grower's labor and capital can be kept employed.

## TURKEY CONTRACTS AND RETURNS

Fig. 49–14. Turkeys on the range about ready to go to market, produced by a large and successful integrator—Cuddy Farms, Ltd., North Carolina. By 1990, 90% of the market turkeys in the U.S. and Canada were produced by integrators or under contract, and most of them were produced in confinement. (Photo courtesy James Strawser, University of Georgia, Athens)

When turkeys are grown under contract, contractors generally provide poults, feed, and services, while producers provide land, facilities, fuel, litter, and labor. Two common types of turkey contracts follow:

• **Profit-sharing contract**—The contractor provides the following inputs at cost: (1) feed, (2) services, (3) medication, (4) insurance, and (5) interest charges. The costs of these inputs plus hauling charges are first deducted from the gross receipts from the sale of turkeys. Seventy percent of the remaining receipts are distributed to the contractor and 30% to the grower until the cost of the poults is paid, then the balance of the receipts (if any remain) is distributed 70% to the grower and 30% to the contractor.

• **Base pay and bonuses contract**—The contractor provides poults, feed, medication, and services, and pays the grower according to the number of weeks in production plus bonuses for livability, feed conversion, and grade yield. The basic payments are provided for all marketable turkeys as follows:

1. $0.05 per turkey per week for the first 8 weeks (brooding).

2. $0.025 per turkey per week for 9 through 25 weeks (range).

3. $0.05 per turkey per week from 26 weeks until marketed (though turkeys are usually marketed before 26 weeks).

---

[2]Undergrades refer to Peewee, Grade B, Cracks, Checks, Stains, and other marketable eggs other than Grade A.

A Georgia study showed that average base payments to growers were 3.6 cents per lb for marketable liveweight for toms and 4.1 cents per lb for hens. Additionally, an efficiency bonus was paid for feed cost savings.[3]

## PRODUCER'S SHARE OF THE CONSUMER'S DOLLAR

The producer's share of the consumer's dollar is shown in Table 49-5. Beef has been added for comparative purposes.

### TABLE 49-5
### FARMER'S SHARE OF
### CONSUMER'S POULTRY PRODUCT DOLLAR[1]
(Beef for Comparison)

| Item | Eggs (Grade A) | Broiler | Turkey | Beef (Choice) |
|---|---|---|---|---|
| | (/doz) | (/lb) | (/lb) | (/lb) |
| Farm value ............ $ | 0.45 | 0.34 | 0.37 | 1.47 |
| Marketing costs ......... $ | 0.17 | 0.51 | 0.59 | 1.08 |
| Retail price ............ $ | 0.62 | 0.85 | 0.96 | 2.55 |
| Farmer's share of retail price ............... % | 73 | 40 | 39 | 58 |

[1]*Livestock and Poultry Situation and Outlook Report*, USDA, LPS-33, February 1989, pp. 16, 27, 29, 33. Data for 1988.

Without doubt, the smaller retail markup shown in Table 49-5 for broilers and turkeys than for beef can be attributed largely to the practice of many retail outlets using poultry as a *loss-leader*. This refers to the practice of selling poultry during special sale periods at a relatively low price, and on a small markup basis. Table 49-5 also reveals that marketing costs for poultry products are much lower than for beef. This is due to the fact that poultry units have tended to become more con-

centrated in certain areas, and that operations have become larger, with the result that the cost of assembling the products has lessened.

## FUTURES TRADING

Most egg, broiler, and turkey producers operate on a contract basis, which means that they are guaranteed so much per dozen eggs or per pound of broiler or turkey produced; hence, their risk is minimal.

But big integrators take big risks; risks from disease, tornados/fires, fluctuating feed prices, and volatile egg, broiler, and turkey prices. Although most integrators are in a strong position financially, many of them prefer to protect themselves by *hedging* in the futures market.

Futures trading is conducted by the Chicago Mercantile Exchange in both eggs and broilers. Egg futures are in units of 750 cases, and broiler futures are in 30,000 lb.

(Also, see Chapter 8, section of "Futures and Options Trading.")

## FOREIGN TRADE IN POULTRY PRODUCTS

Prior to 1950, exports of poultry products amounted to no more than 1% of domestic production. However, beginning in the late 1950s, exports of broilers increased rapidly, and by 1962 they peaked at 262 million pounds and accounted for 3.4% of domestic production. Beginning in 1962, however, the European Economic Community (EC) imposed high levies on chickens and turkeys. As a result, chicken exports to these countries declined. But exports to other countries have expanded, with the result that, in 1987, chicken exports represented 5% of U.S. production, and turkey exports represented 0.9% of our production. Fig. 49-15 shows that there is a rather sizable, and a stable, export market for poultry products.

**U.S. Exports of Poultry Products**

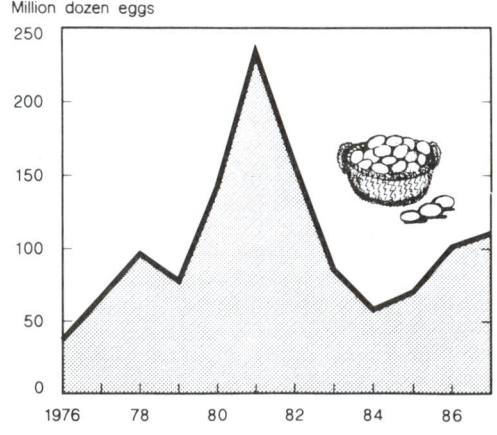

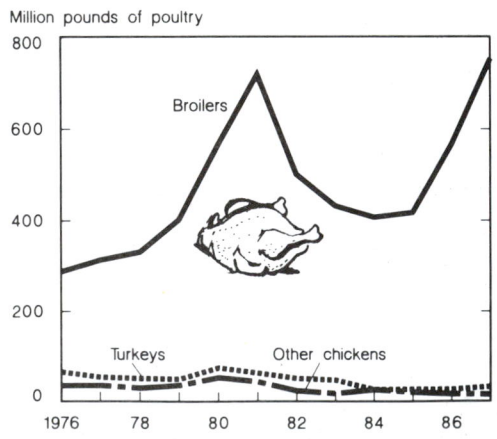

Fig. 49-15. Exports of eggs and broilers both rose in 1987, reflecting a weaker dollar, lower U.S. prices, and increased sales through the Export Enhancement Program. (Source: *1988 Agricultural Chartbook*, p. 92, Chart 202)

[3]Lance, C. C., *Production and Marketing Practices and Contracts for Georgia Turkey Producers*, The University of Georgia, Athens, Georgia, 1982, p. 8.

In 1987, the major importers of U.S. poultry and poultry products, by rank, were: West Germany, Japan, Saudi Arabia, U.S.S.R., and Hong Kong.

U.S. imports of poultry products, mainly eggs (in the shell) from Canada, have been small compared with exports of poultry products. Total product imports amounted to $4,187,000 in 1987, 1% the value of U.S. exports of poultry products.

In 1990, the United States had an import duty of 3½ cents per dozen on shell chicken eggs, 3 cents per pound on plucked chicken, 5 cents per pound on eviscerated chicken, 8½ cents per pound on plucked turkey, and 5 cents per pound on eviscerated turkey.

## TRENDS IN MARKETING AHEAD

There is every indication that egg, broiler, and turkey production will each become more highly specialized, larger, and concentrated in fewer farms, and more vertically integrated.

Contract production and owner integration are expected to expand further in market eggs and turkeys. The motivating forces back of increasing bigness are the advantages in management, marketing, and distribution. Also, with increasing mechanization and smaller labor requirements, it is quite likely that more and more of the production units will be owned by integrators, rather than controlled on a contract basis.

Further gains in marketing efficiency are needed and likely.

Market reporting and bases for establishing the going market price of eggs, broilers, and turkeys leave much to be desired. Perhaps the ultimate solution lies in the "giants" in the poultry industry taking a page out of the book of the automobile manufacturers—learning to live together in competition, but cutting back or closing down production at such intervals as necessary to avoid overproduction and ruinous prices. Of course, there remain two great differences between poultry producers and automobile manufacturers: (1) Consumers don't eat cars, and (2) biologically controlled animals cannot be turned on and off as can the manufacture of a car.

## QUESTIONS FOR STUDY AND DISCUSSION

1. What forces have caused the marketing of eggs and poultry to be particularly striking in recent years?

2. The U.S. cash receipts from eggs and broilers were about the same in 1975. However, in 1987, the cash receipts from broilers were nearly double the receipts for eggs. (See Fig. 49–2.) What caused the phenomenal increase in cash receipts from broilers?

3. Table 49–1 shows that per capita egg consumption has gone down since 1960. During this period of time, the per capita consumption of chickens and turkeys has more than doubled. How do you explain this situation?

4. How does the retail price of broilers compare with the retail price of beef and pork? (See Fig. 49–3.) If this gap is narrowed, will it be accomplished by broilers increasing in price, or by beef and pork prices falling?

5. Discuss each of the following pertinent changes as factors affecting the marketing of eggs and poultry: (a) number and size of units, (b) increased production efficiency, and (c) shifts in production areas.

6. Explain how more vertical integration, contracts, and direct marketing in egg, broiler, and turkey production affected marketing.

7. Discuss each of the following pertinent changes as factors affecting the marketing of eggs and poultry: (a) fewer and larger processors, and (b) more further processing.

8. Discuss the market classes and grades of eggs (a) for quality of individual shell eggs (Table 49–2), (b) consumer grades for shell eggs (Table 49–3), and (c) weight classes for consumer grades for shell eggs (Table 49–4).

9. List and define the commonly used market classes and grades of chickens.

10. List the major responsibilities of the USDA under the provisions of the Poultry Products Inspection Act.

11. How would you recommend that the USDA examine individually more than 5 billion broilers annually? Describe the feed safety system known as the *Hazard Analysis and Critical Control Point (HACCP)*.

12. Poultry irradiation was approved for use effective May 1, 1990. What is it designed to accomplish? Is the irradiation of poultry good or bad?

13. Discuss how the market channels and selling arrangements have changed in recent years relative to (a) eggs, and (b) poultry.

14. Identify and describe the two remaining market news services.

15. Describe the current pricing system for (a) eggs, (b) broilers, and (c) turkeys.

16. Discuss the pertinent financial aspects of (a) egg contracts and returns, (b) broiler contracts and returns, and (c) turkey contracts and returns.

17. Discuss the producer's share of the consumer's dollar. Define *loss-leader*, and explain how it is supposed to work.

18. What is *Futures Trading*? Explain how it is supposed to work.

19. Discuss the following aspects of foreign trade in poultry products: (a) exports as a percent of U.S. production, (b) major importers of U.S. poultry and poultry products, and (c) U.S. import duties.

20. What trends in marketing may be expected in the decades to come?

21. What is the ultimate solution relative to (a) establishing market price for eggs, broilers, and turkeys, and (b) avoiding over-production and ruinous prices?

## SELECTED REFERENCES

| Title of Publication | Author(s) | Publisher |
|---|---|---|
| *Chicken Broiler Industry, The: Structure, Practices, and Costs*, Mktg. Res. Rpt. No. 930 | F. L. Faber<br>R. J. Irvin | Economic Research Service, USDA, Washington, DC, 1971 |
| *Market Structure of the Food Industries*, Mktg. Res. Rpt. No. 971 | D. F. Dunham *et al.* | Economic Research Service, USDA, Washington, DC, 1972 |
| *Marketing Poultry Products*, Fifth Edition | E. W. Benjamin *et al.* | John Wiley & Sons, Inc., New York, NY, 1960 |
| *Poultry Meat and Egg Production* | C. R. Parkhurst<br>G. J. Mountney | Van Nostrand Reinhold Company, New York, NY, 1988 |
| *Poultry Production*, Twelfth Edition | L. E. Card<br>M. C. Nesheim | Lea & Febiger, Philadelphia, PA, 1979 |
| *Poultry Products Technology* | G. J. Mountney | AVI Publishing Co., Westport, CT, 1966 |
| *Processing of Poultry* | Edited by<br>G. C. Mead | Elsevier Applied Science, London, England, 1989 |
| *Readings on Egg Pricing* | Edited by<br>G. B. Rogers<br>L. A. Voss | College of Agriculture, University of Missouri, Columbia, MO, 1971 |
| *U.S. Broiler Industry, The* | F. A. Lasley *et al.* | USDA, ERC, Agricultural Economic Report No. 591, Nov. 1988 |

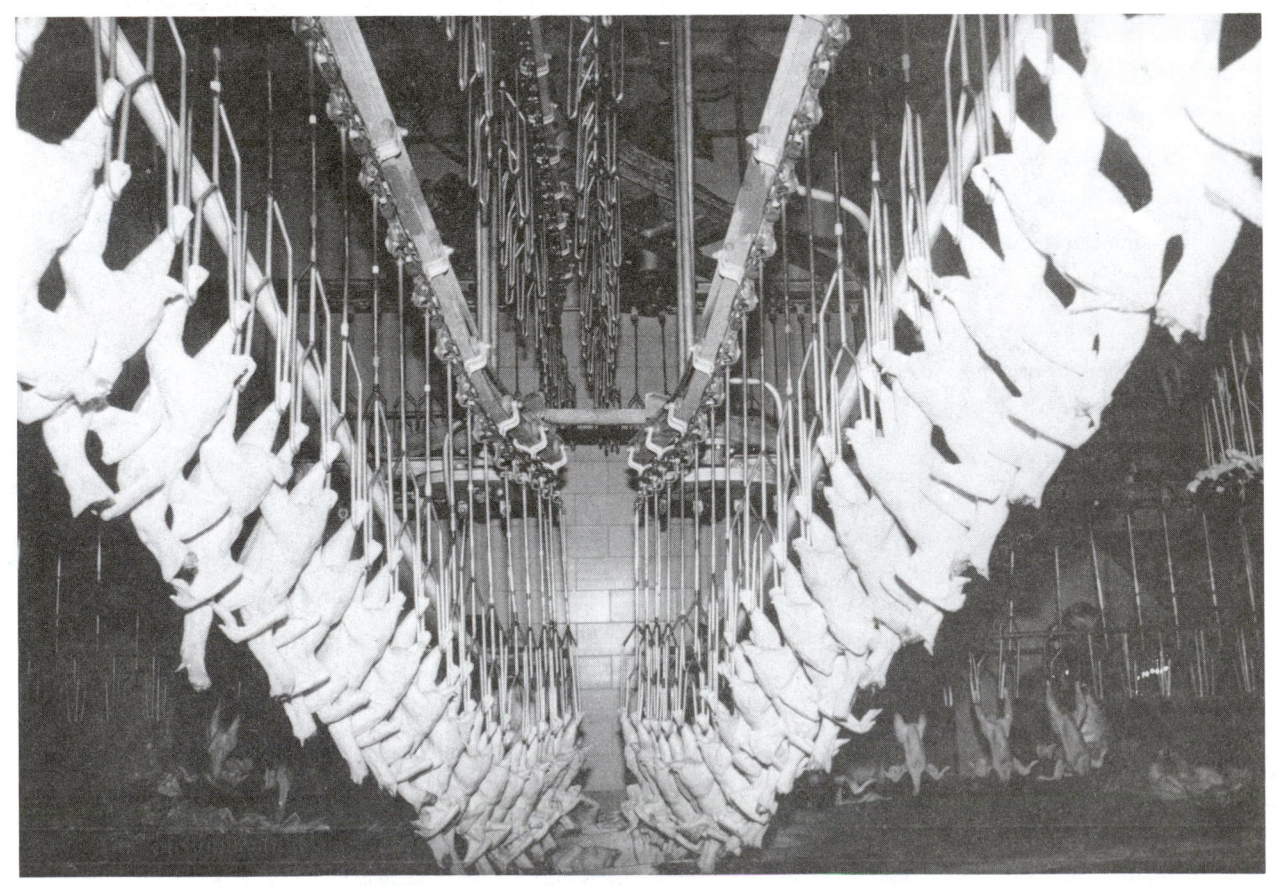

Processing broilers. (Courtesy, Gold Kist-Communications, Atlanta, GA)

Curious yearlings on a Kentucky horse farm. (Courtesy, Kentucky Department of Travel Development, Frankfort)

# THE HORSE INDUSTRY

---

# *Chapter 50*

---

The evolution and transformation of the horse from the early-day wild forms, its subsequent domestication, and the overlapping uses made of it in both war and peace is a fascinating story.

Likewise, the rise and fall of the horse as a beast of burden—the replacement of muscle power by the relentless wheels of progress—constitutes one of the most thrilling chapters of history. A century ago, muscles provided 94% of the world's energy needs; coal, oil, and waterpower provided the other 6%. Today, the situation is reversed in the developed nations. They now obtain 94% of their energy needs from coal, oil, natural gas, and waterpower, and only 6% from the muscle power of people and animals.

The unique thing about the horse business, not found in any other industry, is the human values back of it. It's a people business, and a way of life for many.

## EVOLUTION OF THE HORSE

Fossil remains prove that members of the horse family roamed the plains of America (especially what is now the Great Plains area of the United States) during most of Tertiary time, beginning about 58 million years ago. Yet no horses were present on this continent when Columbus discovered America in 1492. Why they perished, only a few thousand years before, is still one of the unexplained mysteries of evolution. As the disappearance was so complete and so sudden, many scientists believe that it must have been caused by some contagious disease or some fatal parasite. Others feel that perhaps it was due to multiple causes; including (1) climatic changes, (2) competition, and/or (3) failure to adapt. Regardless of why horses disappeared, it is known that conditions in America were favorable for them at the time of their reestablishment by the Spanish Conquistadores less than 500 years ago.

Through fossil remains, it is possible to reconstruct the evolution of the horse, beginning with the ancient 4-toed ancestor, the *Eohippus* (meaning dawn horse). This was a small animal, scarcely more than a foot high, with 4 toes on the front feet and 3 toes on the hind feet, and with slender legs, a short neck, and even teeth. It was well adapted to travelling in, and feeding on, the herbage of the swamps. Gradually, the descendants of *Eohippus* grew in size and changed in form, evolving into a 3-toed animal known as *Mesohippus*, which was about 24 in. in height, or about the size of a Collie dog. Further changes continued, transforming the animal from a denizen of the swamp to a creature capable of surviving in the forest and finally to one adapted to the prairie. In terms of conformation, the animal grew taller. The teeth grew longer, stronger, and more roughened to suit the gradual changes to grazing on the prairie. The cannon bones (metacarpals and metatarsals) lengthened; the middle toe (or third toe) grew longer and stronger, forming a hoof; and the other toes (second and fourth toes) gradually disappeared except for vestiges, the slender bones known as splints under the skin. The transformation in length and structure of foot made for greater speed over the prairie-type terrain, thereby enabling the animal to feed farther and farther from water, and providing for greater safety in its struggle to survive. The horse is an excellent example, therefore, of the slow adaptation of animal life to changing conditions in environment, climate, food, and soil. The animal was transformed from one adapted to a swamp type of environment to one adapted to the prairie.

Though all horses eventually perished in the New World and none were present on the continent when America was discovered, fortunately some of these animals had long before emigrated to Asia and Europe at a time when there was a land bridge connecting Alaska and Siberia (now the Bering Strait). These emigrants formed the sturdy wild European stock from which the horse family of today has descended, and this stock also populated Africa with its asses and zebras.

In summary, it is noteworthy that the evolution of horses covered a period of approximately 58,000,000 years, but that people hunted them as recently as 25,000 years ago, and that the Spanish Conquistadores returned them to the New World less than 500 years ago.

## PRZEWALSKI HORSE

Fig. 50–1. Przewalski horse. This is the only surviving species of original wild horses—not feral or escaped from domestication—known to exist at the present time. Note that the animal is small and stockily built, with an erect mane and no forelock. (Courtesy, New York Zoological Society, New York)

The Przewalski horse, named after the Russian explorer, Przewalski, who rediscovered it in the northwestern corner of Mongolia, in 1879, maintained itself during some 60 million years of evolution, in Europe and Asia. It was last seen in the wild in 1968 in southwestern Mongolia, along the border with China. Today, it is almost certainly extinct in the wild. However, about 700 Przewalski horses are maintained in captivity. It is a small stockily built, and distinctly yellowish horse, with an erect mane and no forelock. There is usually a dark stripe on the shoulders and down the middle of the back. In the wild, Przewalski horses separated into bands, seldom more than 40 in summer, with a stallion leader in each group. When crossed on domestic horses, the hybrids are fertile, which proves that Przewalski horses are very closely related to the domestic horse.

It is not intended to imply that Przewalski horse was the foundation stock of any or all of the present-day improved breeds throughout the world. Rather, this wild horse of Asia is extremely interesting because it is the only one known to have survived some 60 million years of evolution.

## ORIGIN AND DOMESTICATION OF THE HORSE

The horse was probably the last of present-day farm animals to be domesticated. According to early records, after subduing the ox, the sheep and the goat, primitive people domesticated the ass and then the camel; and finally, the horse.

Horses were first domesticated by the nomads that roamed the steppes (plains) of Central Asia, whom the Greeks and Chinese called the Scythians. They were bearded men, with dark, deep-set eyes, weather-beaten faces and long wind-snarled hair. They are credited with the invention, or at least the improvement, of the chief items of riding gear—the reins and the saddle. The Scythians used horses for mounts and draft animals, for milk and meat, and for waging war. They were fierce fighters; they charged at the gallop, shooting deadly arrows from their bows. Also, they adhered to the custom of scalping their foes as trophies.

From Central Asia, horses spread westward through southern Europe in the time of the Lake Dwellers. They were reported in Babylonia as early as 2000 B.C., perhaps coming into the country via neighboring Persia.

Although the Egyptians—the most advanced civilization of the day—had domesticated and used the ass from the earliest times, horses were wholly unknown to them until the dynasty of the Shepherd Kings, who entered Egypt from Asia in 1680 B.C. It is reported that, thereafter, the horse was much favored in Egypt.

Presence of the horse seems to have prompted the invention of the chariot, a type of vehicle drawn by horses that the Egyptians used in war and other pursuits. The Bible also relates that when Joseph took his father's remains from Egypt back to Canaan "there went up with him both chariots and horsemen."[1] It is probable that the Egyptians were largely responsible for the spread of domesticated horses to other countries.

Fig. 50-2. Joseph using horses in his move to Egypt (about 1500 B.C.), from a miniature painting in the Bible of the Counts of Toggenburg, 16th century. (Courtesy, Bettmann Archive)

Certainly, Greece was not even peopled, and there were no horses in Arabia during the early period when they were flourishing in Egypt. But horses and chariots were in use in Greece at least a thousand years before Christ, to judge from

[1]Genesis 50:9.

the account of their use in the siege of Troy. It is also interesting to note that the first and most expert equestrians of Greece, the Thessalians, were colonists from Egypt. As evidence that the Greeks were accomplished equestrians it might be pointed out that they developed the snaffle bit at an early period. One of their number is also said to have originated the axiom: "No foot, no horse." Yet, the use of the saddle and the stirrups appears to have been unknown in Greece at this time.

Fig. 50-3. Distinguished young Greek in fashionable riding habit. Bowl painting, 500 B.C. Though the Greeks were accomplished equestrians at this time the use of the saddle and stirrups appears to have been unknown to them. (Courtesy, The Bettmann Archive)

From Greece, the horse was later taken to Rome and from there to other parts of Europe. The Romans proved to be master equestrians. They invented the curb bit. According to historians, when Caesar invaded Britain, about 55 B.C., he took horses with him. Although there were other horses in Britain at the time of the Roman occupation, eastern breeding was probably greatly infused at this time—thus laying the foundation for the Blood Horse of today.

The Arabs, strangely enough, did not use horses to any extent until after the time of Mohammed (570 to 632 A.D.), depending on camels before that time. As evidence of this fact, it is noted that in the seventh century after Christ, when Mohammed attacked the Koreish near Mecca, he had but 2 horses in his whole army; and, at the close of his murderous campaign, although he drove off 24,000 camels and 40,000 sheep and carried away 24,000 ounces of silver, not one horse appeared in his list of plunder. This would seem to indicate rather conclusively that Arabia, the country whose horses have done so much to improve the horses of the world, was not the native home of the horse and that the Arabs did not use horses until after the time of Christ.

Of course, it seems incredible that all the various breeds, colors, and types of draft, light, and pony horses should have descended from a common wild ancestor. Rather, there were probably many different wild stocks giving descent to domestic horses.

Fig. 50–4. Horses vary in size and use. The Shetland Pony foal (left) is thought to have descended from the small, shaggy wild stock of northern Europe, whereas the draft horse (right) is thought to have descended primarily from the ponderous wild black horse of Flanders. (Courtesy, Iowa State University)

## ORIGIN AND DOMESTICATION OF THE DONKEY

The two species of the horse family that have been tamed by humans are *Equus caballus*, the horse and *Equus asinus*, the ass or donkey. The history of the domestic donkey is as clear as that of the horse is obscure. Donkeys were first domesticated in Egypt, where they served people from earliest times. Good figures of them appear on slates of the First Dynasty, about 3400 B.C. Domestic donkeys are descended from the wild donkey (the Nubian wild ass) of North Africa, a species which is now

Fig. 50–5. A Mongolian wild ass in the Gobi desert in Asia. (Courtesy, American Museum of Natural History, New York)

almost extinct. Because of the frequent tendency to stripes on the legs, however, some zoologists also think that the domestic donkey is related to the Somali wild ass of Africa.

From Egypt, the use of the domestic donkey spread into southwestern Asia sometime prior to the year 1000 B.C. The Bible first refers to the ass in relating how Abraham, the patriarch of the Old Testament, rode one of these animals from Beersheba to Mount Mordah. Every child is familiar with the fact that Jesus rode into Jerusalem on an ass. This mode of transportation was not unusual at the time of Christ, for donkeys were then the common saddle animals throughout the Near East.

As is generally known, the donkey is commonly used in this country in the production of mules.[2] Mules have been known from very ancient times, as we learn from accounts of the Trojan War.

## POSITION OF THE HORSE IN THE ZOOLOGICAL SCHEME

The following outline shows the basic position of the domesticated horse in the zoological scheme:

Kingdom *Animalia*: Animals collectively; the Animal Kingdom.
  Phylum *Chordata*: One of the approximately 21 phyla of the animal kingdom, in which there is either a backbone (in the vertebrates) or the rudiment of a backbone, the chorda.
    Class *Mammalia*: Mammals, or warm-blooded, hairy animals that produce their young alive and suckle them for a variable period on secretions from the mammary glands.
      Order *Perissodactyla*: Nonruminant hoofed mammals, usually with an odd number of toes, the third digit the largest and in line with the axis of the limb. This suborder includes the horse, tapir, and rhinoceros.
        Family *Equidae*: The members of the horse family may be distinguished from the other existing *perissodactyla* (rhinoceros and tapir) by their comparatively more slender and agile build.
          Genus *Equus*: Includes horses, asses, and zebras.
            Species *Equus caballus*: The horse is distinguished from asses and zebras by the longer hair of the mane and tail, the presence of the *chestnut* on the inside of the hind leg, and by other less constant characteristics such as larger size, larger hoofs, more arched neck, smaller head, and shorter ears.

## USES OF HORSES

The name *horse* is derived from the Anglo-Saxon *hors*, meaning swiftness; and the word *horseman* comes from the Hebrew root *to prick or spur*.[3] These early characterizations of the horse, within themselves, tell somewhat of a story. Perhaps the very survival of the wild species was somewhat

---

[2]In recent years, some miniature donkeys are being used as children's pets in the U.S.

[3]The Jews were forbidden to use horses by divine authority. In fact, they were required to hamstring horses captured in war.

Fig. 50–6. Chariot driven through Pompeii. The horse-drawn chariot was used by the sports-loving Greeks in chariot races, as well as in war and other pursuits. (Courtesy, The Bettmann Archive)

dependent upon its swiftness, which provided escape from both beast and people. The Hebrew description of an equestrian was obviously assigned after the horse had been domesticated and ridden.

The various uses that people have made of the horse down through the ages, in order of period of time, follow: (1) as a source of food, (2) for military purposes, (3) in the pastimes and sports of the nations, (4) in agricultural and commercial pursuits, and (5) for recreation and sport.

## INTRODUCTION OF HORSES AND MULES TO, AND EARLY HISTORY IN, THE UNITED STATES

It has been established that most of the evolution of the horse took place in the Americas, but this animal was extinct in the Western World at the time of Columbus' discovery, and apparently extinct even before the arrival of the Indians some thousands of years earlier.

Columbus first brought horses to the West Indies on his second voyage in 1493. Cortez brought Spanish horses with him to the New World in 1519 when he landed in Mexico (16 animals were in the initial contingent, but approximately 1,000 head more were subsequently imported during the 2-year conquest of Mexico). Horses were first brought directly to what is now the United States by de Soto in the year 1539. Upon his vessels, he had 237 horses. These animals traveled with the army of the explorer in the hazardous journey from the Everglades of Florida to the Ozarks of Missouri. Following de Soto's death and burial in the upper Mississippi 3 years later, his followers returned by boats down the Mississippi, abandoning many of their horses.

One year following de Soto's landing in what is now Florida, in 1540, another Spanish explorer, Coronado, started an expedition with an armed band of horsemen from Mexico, penetrating to a point near the boundary of Kansas and Nebraska.

Beginning about 1600, the Spaniard established a chain of Christian missions among the Indians in the New World. This chain of missions extended from the eastern coast of Mexico up the Rio Grande, thence across the mountains to the Pacific Coast. Each mission brought animals from the mother country, including horses.

There are two schools of thought relative to the source of the foundation stock of the first horses of the American Indians, and the hardy bands of mustangs—feral horses of the Great Plains. Most historians agree that both groups were descended from animals of Spanish (Arabian) extraction. However, some contend that their foundation stock came from the abandoned and stray horses of the expeditions of de Soto and Coronado, whereas others claim that they were obtained chiefly from Santa Fe, an ancient Spanish mission founded in 1606. It is noteworthy that Santa Fe and other early Spanish missions were the source of Spanish Longhorn cattle, thus lending credence to the theory that the missions were the source of the foundation horses for the Indian and the wild bands of mustangs.

Much romance and adventure is connected with the mustang, and each band of wild horses was credited with leadership by the most wonderful stallion ever beheld by humans. Many were captured, but the real leaders were always alleged to have escaped by reason of speed, such as not possessed by a domesticated horse. The mustangs multiplied at a prodigious rate. In one high luxuriant bunchgrass region in the state of Washington, wild horses thrived so well that the region became known as *horse heaven*, a name it bears today.

The coming of the horse among the Indians increased the strife and wars between tribes. Following the buffalo on horseback led to greater infringement upon each other's hunting grounds, which had ever been a cause for war. From the time the Indians came into possession of horses until the country was taken over by the white people there was no peace among the tribes.

Later, animals of both light- and draft-horse breeding were introduced from Europe by the colonists. For many years, however, sturdy oxen continued to draw the plows for turning the sod on many a rugged New England hillside. Horses were largely used as pack animals, for riding, and later for pulling wagons and stagecoaches. It was not until about 1840 that the buggy first made its appearance.

Six mares and two stallions were brought to Jamestown in 1609, these being the first European importations. Some of these animals may have been eaten during the period of near starvation at Jamestown, but importations continued; and it was reported in 1611 that a total of 17 horses had been brought to this colony.

The horse seems to have been much neglected in early New England, as compared with cattle and sheep. This is not surprising, inasmuch as oxen were universally used for draft purposes. Roads were few in number; speed was not essential; and the horse had no meat value like that of cattle. Because of the great difficulty in herding horses on the commons, they were usually hobbled. Despite the limited early-day use of the horse, the colonists must have loved them, because, very early, the indiscriminate running of stallions among the mares upon the commons was recognized as undesirable. Massachusetts, before 1700, excluded from the town commons all stallions "under 14 hands high and not of comely proportion."[4]

Even before horses found much use in New England, they became valuable for export purposes to the West Indies for work in the sugar mills. In fact, this business became so lucrative that horse stealing became a common offense in New England in the 18th century. Confiscation of property, public whippings, and banishment from the colony constituted the common punishment for a horse thief.

[4]Thompson, J. W., *History of Livestock Raising in the United States, 1607–1860*, Agricultural History Series No. 5, USDA, Nov. 1942.

As plantations materialized in Virginia, the need for easy-riding saddle horses developed, so that the owners might survey their broad estates. Racing also became a popular sport among the Cavaliers in Virginia, Maryland, and the Carolinas—with the heat races up to four miles being common events. The plantation owners took considerable pride in having animals worthy of wearing their colors. So great was the desire to win that by 1730 the importation of English racehorses began.

George Washington maintained an extensive horse- and mule-breeding establishment at Mount Vernon. The President was also an ardent race fan, and riding to hounds was a favorite sport with him. As soon as Washington's views on the subject of mules became known, he received some valuable breeding stock through gifts. In 1787, the Marquis de Lafayette presented him with a jack and some jennets of the Maltese breed. The jack, named *Knight of Malta*, was described as a superb animal, of a black color, with the form of a stag and the ferocity of a tiger. In 1795, the King of Spain gave Washington a jack and two jennets that were selected from the royal stud at Madrid. The Spanish jack, known as *Royal Gift*, was 16 hands high, of a gray color, heavily made, and of a sluggish disposition. It was said that Washington was able to combine the best qualities of the two gift jacks, especially through one of the descendants named *Compound*. General Washington was the first to produce mules of quality in this country, and soon the fame of these hardy hybrids spread throughout the South.

The Dutch, Puritan, and Quaker colonists to the north adhered strictly to agricultural pursuits, frowning upon horse races. They imported heavier types of horses. In Pennsylvania, under the guidance of William Penn, the farmers prospered. Soon their horses began to improve, even as the appearance and fertility of their farms had done. Eventually, these large horses were hitched to enormous wagons and used to transport freight overland to and from river flatboats and barges along the Ohio, Cumberland, Tennessee, and Mississippi Rivers. Both horses and wagons were given the name Conestoga, after the Conestoga Valley, a German settlement in Pennsylvania. The Conestoga[5] wagon was the forerunner of the prairie schooner,

Fig. 50-7. Conestoga freight wagon drawn by six Conestoga horses, in front of a country inn. These improved horses and large wagons were both given the name Conestoga, after the Conestoga Valley, a German settlement in Pennsylvania. The advent of the railroads drove the Conestoga horses into oblivion, and the Conestoga wagon was succeeded by the prairie schooner. (Courtesy, The Bettmann Archive)

[5]It is noteworthy that the American custom of driving to the right on the road, instead of to the left as is the practice in most of the world, is said to have originated among the Conestoga wagon drivers of the 1750s. The drivers of these 4- and 6-horse teams either sat on the left wheel horse or on the left side of the seat, the better to wield their whip hand (the right hand) over the other horses

and before the advent of the railroad it was the freight vehicle of the time. It was usually drawn by a team of six magnificent Conestoga horses, which were well groomed and expensively harnessed. At one time, Conestoga horses appeared likely to become a new breed—a truly American creation. However, the railroads replaced them, eventually driving them into permanent oblivion. Other breeds were developed later, but that is another story.

## GROWTH AND DECLINE OF U.S. HORSE AND MULE PRODUCTION

The golden age of the horse extended from the Gay Nineties to the mechanization of agriculture—to the advent of the automobile, truck, and tractor. During this era, everybody loved the horse. The town livery stable, watering trough, and hitching post were trademarks of each town and village. People wept when a horse fell down on the icy street, and jailed people who beat or mistreated horses. The oat bag, carriage, wagon, buggy whip, axle grease, horseshoe, and horseshoe nail industries were thriving and essential parts of the national economy. Every school child knew and respected the village blacksmith.

Bobtailed Hackneys attached to high-seated rigs made a dashing picture as they pranced down the avenue; they were a mark of social prestige. A few memorable dinner parties of the area were even staged on horseback, with the guests lining up in exclusive restaurants astride their favorite mounts, and drinking and eating to the merriment of music, while their steeds munched oats and costumed lackeys cleaned up behind them.

In 1900, the automobile was still the plutocrat's plaything, and the truck and tractor were unknown. Most of the expensive 8,000 cars in the country at the time were either imported or custom-built. Tires cost about $40.00 each, and lasted only 2,000 miles. Few really loved the auto. People complained of the noise that they made, enacted laws against them going through the city parks, and split their sides with laughter when they had to be pushed uphill or got stuck in the mud.

Then, in 1908, Henry Ford produced a car to sell at $825. The truck, the tractor, and improved highways followed closely in period of time. Old Dobbin did not know it at the time, but his days were numbered. As shown in Fig. 50-8 and Table 50-1, the passing of the horse age and the coming of the machine age went hand in hand; as tractor and truck numbers increased, horse and mule numbers declined.

The number of horses in the United States increased up to 1915, at which time there was a record number of 21,431,000 head (horses only; not including mules). Horse production expanded with the growth and development of farms.

Mules on farms slowly but steadily increased in numbers for 10 years after horses began their decline, reaching a peak in 1925 at 5,680,897 head. Mule numbers decreased proportionally less rapidly than horses because of their great use in the deep South where labor was cheaper and more abundant and the farms smaller in size.

In 1915—the peak year, there were 26,493,000 horses and mules, combined, on farms and ranches in the United States and an additional 2,000,000 head in cities. By January 1, 1960,

in the team. Also, when two Conestoga drivers met, they pulled over to the right so that, sitting on the left wheel horse or on the left side of the seat they could see that the left wheels of their wagons cleared each other. Lighter vehicles naturally followed the tracks of the big Conestoga wagons.

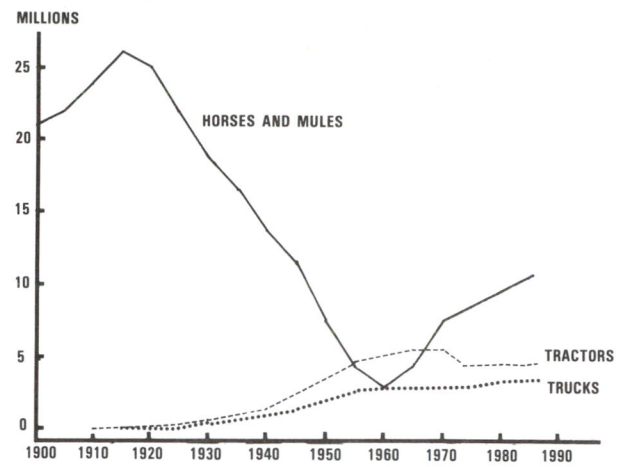

Fig. 50–8. Growth and decline of U.S. horse and mule population. The period of decline in horse and mule numbers coincided closely with the advent of mechanized power, especially the tractor and truck. (Source: USDA)

## TABLE 50–1
### U.S. FARM AND RANCH (1) HORSE AND MULE POPULATION (2) TRACTOR NUMBERS, AND (3) TRUCK NUMBERS, FROM 1900 TO 1986

| Year | Horses and Mules[1] | Tractor Nos. (including garden)[2] | Truck Nos.[3] |
|------|------|------|------|
| 1900 | 21,531,635 | — | — |
| 1905 | 22,077,000 | — | — |
| 1910 | 24,042,882 | 1,000 | 0 |
| 1915 | 26,493,000 | 25,000 | 25,000 |
| 1920 | 25,199,552 | 246,083 | 139,169 |
| 1925 | 22,081,520 | 503,933 | 459,000 |
| 1930 | 18,885,856 | 920,021 | 900,385 |
| 1935 | 16,676,000 | 1,048,000 | 890,000 |
| 1940 | 13,931,531 | 1,567,430 | 1,047,084 |
| 1945 | 11,629,000 | 2,422,000 | 1,490,000 |
| 1950 | 7,604,000 | 3,610,000 | 2,207,000 |
| 1955 | 4,309,000 | 4,692,000 | 2,701,000 |
| 1960 | 3,089,000 | 5,138,000 | 3,110,000 |
| 1965 | 4,580,000 | 5,486,000 | 3,030,000 |
| 1970 | 7,668,000 | 5,424,000 | 2,984,000 |
| 1975 | 8,568,000 | 4,469,000 | 3,032,000 |
| 1980 | 9,663,000 | 4,752,000 | 3,344,000 |
| 1985 | 10,581,000 | 4,676,000 | 3,380,000 |
| 1986 | 10,601,000 | 4,670,000 | 3,380,000 |

[1]From 1900 to 1960, USDA sources. From 1960 to 1986, *FAO Production Yearbooks*, United Nations.
[2]USDA sources. Beginning in 1975, garden tractors excluded.
[3]USDA sources.

the census showed that there were only 3,089,000 head of horses and mules on the nation's farms and ranches (not counting suburban owned horses and those kept on parcels under 10 acres in size)—the lowest number ever recorded.

Ironical as it may seem, the development of manufacturing and commerce was responsible for both the rise and the fall of the horse and mule industry of the United States. The early growth of American industry created a large need for horses to transport the raw and manufactured goods and to produce needed agricultural products for those people who lived in the cities and villages. With further scientific develop-

ments—especially the invention of the tractor and truck—the horse was replaced, first ever so slowly, but then rapidly and drastically.

Today, very few horses are found on city streets. The old-time livery stable has long since passed out of existence; draft horses are seldom hitched to large dray wagons; and horses hitched to a delivery wagon or to a plow are almost a novelty.

But, the horse came up fast in the fields of recreation and sport. In 1986, there were 10,600,000 horses in the United States, most of which were light horses; and there were only 1,000 mules. Draft horses and mules were the victims of mechanization—farming changed. From 1915 to 1986, farm tractors increased from 25,000 to 4,670,000—a 187-fold increase; and, during this same period of time, farm trucks increased from 25,000 to 3,380,000 (Table 50–1).

A survey conducted by the American Horse Council in 1985, showed that the leading states in horse numbers at that time, by rank, were: Texas, California, Oklahoma, Colorado, New York, Ohio, Michigan, Pennsylvania, Washington, and Kentucky.[6]

## WORLD DISTRIBUTION OF HORSES AND MULES

At a very early date and throughout the world, the versatility and adaption of horses were recognized. They were unexcelled in carrying riders comfortably and swiftly on long journeys; possessed a long life of usefulness; and, above all, they were intelligent. Despite all these virtues, in some areas horses have been unable to replace patient *roughage* burning oxen and water buffalos. To this day, oxen are still the main source of power on farms in such densely populated countries as India, Pakistan, and China, in many Near Eastern and African countries, and in some countries of Latin America; and water buffalos are the main source of power in rice-producing areas, because of their ability to work in muddy paddy fields. In the more isolated portions of the New England states, oxen are occasionally used, and stoneboat-pulling contests are a great attraction at the New England fairs.

Members of the ass family (mules and donkeys) are distributed in the warmer regions of the world, where they still occupy a rather important place among the animals used for both pack and draft purposes.

Table 50–2 (see next page) shows the size and density of the horse population of the important horse countries of the world. As noted, world horse numbers totaled 65,064,000 in 1986. This was far below the 1934–38 prewar average of 96.4 million head. The decline in horse numbers since 1938 can be attributed chiefly to the mechanization of agriculture. For example, the number of tractors in use in agriculture in the world in 1986 totaled 25,284,480 compared with fewer than 2 million in 1939.

Fig. 50–9 (see next page) shows the 10 leading horse countries of the world in 1986. As noted, by rank they are: China, United States, Mexico, U.S.S.R., Brazil, Argentina, Mongolia, Columbia, Ethiopia, and Poland.

[6]*The Economic Impact of the U.S. Horse Industry*, prepared by The American Horse Council, Washington, DC, 1987, p. 4, Table 2.

## TABLE 50-2
## LEADING HORSE-PRODUCING COUNTRIES OF THE WORLD

| Country | Horse Population[1] | Human Population[2] | Size of Country | | Horses Per Capita | Horses Per | |
|---|---|---|---|---|---|---|---|
| | | | (sq. mi.) | *(sq. km.)* | | (sq. mi.) | *(sq. km.)* |
| China, P.R. of | 11,000,000 | 1,069,628,000 | 3,705,390 | *9,596,960* | 0.01 | 3.0 | *1.1* |
| United States | 10,840,000 | 247,498,000 | 3,539,289 | *9,166,759* | 0.04 | 3.1 | *1.2* |
| Mexico | 6,135,000 | 88,087,000 | 761,604 | *1,972,554* | 0.07 | 8.1 | *3.1* |
| U.S.S.R. | 5,800,000 | 287,015,000 | 8,649,496 | *22,402,195* | 0.02 | 0.7 | *.3* |
| Brazil | 5,500,000 | 153,992,000 | 3,286,470 | *8,511,957* | 0.04 | 1.7 | *0.6* |
| Argentina | 3,000,000 | 32,617,000 | 1,065,189 | *2,758,840* | 0.09 | 2.8 | *1.1* |
| Mongolia | 1,971,000 | 2,093,000 | 604,247 | *1,565,000* | 0.94 | 3.3 | *1.2* |
| Colombia | 1,950,000 | 31,821,000 | 439,735 | *1,138,914* | 0.06 | 4.4 | *1.7* |
| Ethiopia | 1,590,000 | 47,709,000 | 471,776 | *1,221,900* | 0.03 | 3.4 | *1.3* |
| Poland | 1,272,000 | 38,389,000 | 120,727 | *312,683* | 0.03 | 10.5 | *4.1* |
| **World Total** | 65,064,000 | 5,052,000,000 | 57,800,000 | *149,702,000* | 0.01 | 1.1 | *0.4* |

[1]*Production Yearbook 1986*, Vol. 40, FAO of the United Nations, Rome, pp. 195–197. The horse figures are for 1986.

[2]*The World Almanac and Book of Facts 1989*, pub. by Newspaper Enterprise Assn., Inc.

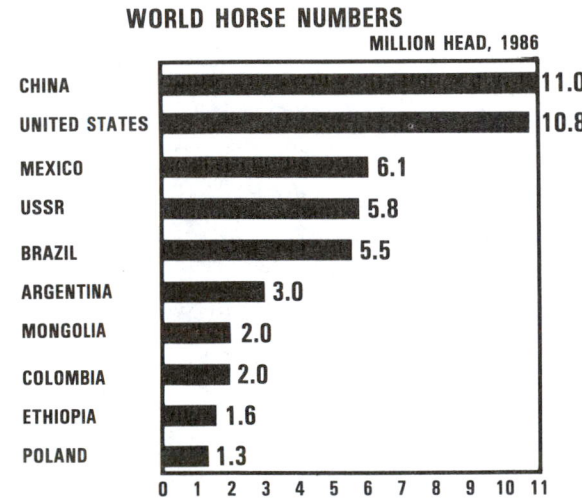

Fig. 50-9. Ten leading horse countries of the world, 1986. (Based on data from *FAO Production Yearbook, 1986* Vol. 40, pp. 195–197.)

## PRESENT STATUS OF THE U.S. HORSE INDUSTRY

The unique thing about the horse business, not found in any other industry, is the human values back of it. It's a people's business, and a way of life for many.

Also, the U.S. horse industry is big and important business, and it will get bigger. The following facts and figures attest to the magnitude and importance of the industry:

• There are more than 10½ million horses in the United States.

• The United States horse industry is a $15 billion industry.

• Annual expenditures for horse feed, drugs, tack, and equipment average about $1,000 per horse.

• Annual wages paid in the horse industry exceed $1 billion.

• There are more than twice as many 4-H Club horse projects as beef cattle projects. These figures are presented in Table 50-3.

## TABLE 50-3
## 4-H CLUB BEEF CATTLE AND HORSE PROJECTS[1]

| | 1965 | 1966 | 1970 | 1974 | 1975 | 1980 | 1985 | 1987 |
|---|---|---|---|---|---|---|---|---|
| Beef cattle | 160,914 | 157,949 | 150,056 | 160,846 | 164,208 | 129,905 | 123,617 | 107,376 |
| Horses | 146,541 | 165,510 | 231,206 | 320,767 | 320,050 | 248,527 | 225,843 | 224,903 |

USDA.

• Horse shows are increasing in size and numbers. Currently, horse shows generate total revenue from attendees, participants, and sponsorships of about ⅓ billion dollars annually. (See Table 50-4). The five leading states in horse show revenues, by rank, are: Texas, California, Kentucky, Pennsylvania, and New Jersey.

• The horse show data reported in Table 50-4 are the latest and most complete available. Although the U.S. economy, inflation, and human population increase will make for year-to-year changes, it is anticipated that future horse shows will continue to be of like magnitude with similar relationship between items, as reported in Table 50-4.

## TABLE 50-4
## HORSE SHOWS[1]

| Item | Nationally Sanctioned | Locally Sanctioned | Total |
|---|---|---|---|
| Total shows ........ No. | 7,348 (20%) | 29,392 (80%) | 36,740 (100%) |
| Spectators ......... No. | 4,577,094 | 19,898,384 | 24,475,478 |
| Prizes ............... $ | 29,886,067 | 61,517,456 | 91,403,523 |
| Total Expenses ........ $ | 84,832,882 | 187,303,694 | 272,136,576 |
| Admissions ........... $ | 5,984,587 | 39,414,672 | 45,399,259 |
| Total Revenues ....... $ | 96,059,559 | 221,615,680 | 317,675,239 |
| Charity (donations) .... $ | 8,618,425 | 19,104,800 | 27,723,225 |

[1]From *Economic Impact of Horse Shows 1979*, a report based on a survey conducted by the American Horse Council, Washington, DC. Note that 20% of the horse shows were *nationally sanctioned* and 80% were *locally sanctioned.*

• Horse racing ranks third as a spectator sport. In 1988, 74,158,269 people went to horse races (Thoroughbred and harness racing).

The rankings and figures of the seven leading U.S. spectator sports are given in Table 50–5.

**TABLE 50–5**
**U.S. SPORTS ATTENDANCE 1988[1]**

| Leading Sports | Fan Attendance |
|---|---|
| 1. Baseball | 89,525,211 |
| 2. Auto racing | 85,968,814 |
| 3. Horse racing (Thoroughbred and harness) | 74,158,269 |
| 4. Football | 51,024,820 |
| 5. Basketball | 46,555,845 |
| 6. Hockey | 26,875,143 |
| 7. Greyhound racing | 26,618,552 |

[1]*1988 Survey of Sports Attendance*, Triangle Publications, Inc., Hightstown, NJ.

Fig. 50–10. Lipizzan stallion at the Spanish Riding School, Wels, Austria, literally flying through the air with perfectly seated stirrupless rider. The horse is decorated with gold trappings, and the rider is smartly dressed in an old-time Napoleonic military uniform. The stallion is pictured doing the Capriole, one of the several intricate movements resembling the leaping, twisting, fighting, and frolicking of high-spirited horses in pasture. (Courtesy, Spanish Riding School, Wels, Austria)

• Saddle clubs are springing up everywhere, and more people are riding horses for pleasure than ever before. According to a survey of the U.S. Department of the Interior, 27 million people ride at least once a year, more than half of them on a regular basis.

• On the western range, cow ponies are still used in the traditional manner; mechanized replacement for them has not yet been devised.

• Horses are benefactors of people in numerous other ways. Limited numbers of them are used by the Forest Service; others are used as pack animals into remote areas not otherwise accessible by surface travel. They are still the show in many motion pictures and in parades. Those responsible for law enforce-

ment have found that mounted patrols are one of the most effective ways in which to handle crowds and riots.

In the laboratory, horses serve as a factory for the manufacture of antitoxins that are used for rendering animals and people immune to certain diseases, such as tetanus. Also, medical doctors use equine produced estrogens (female sex hormones), obtained from the urine of pregnant mares, to relieve the menopause (change of life) of women.

Despite the magnitude of the industry, U.S. horse owners suffer appalling losses. They are—

1. Spending millions of dollars on needless concoctions and using unbalanced and deficient rations.

2. Producing only a 50% foal crop, which means that they are keeping two mares a whole year to produce one foal.

3. Keeping too many stallions for producing too few foals.

4. Maintaining horse-breeding establishments that return little or nothing on investment.

5. Retiring an appalling number of horses from tracks, shows, and other uses due to unsoundnesses.

6. Losing through inefficiency and deaths millions of dollars due to disease and parasites.

Such wanton losses prompt the question: If the horse industry is so good, why not better?

Chapters 50 to 54 of this book are designed to impart sound, scientific information relative to breeding, feeding, care, and management of horses; to give instruction in the arts of selecting and judging; to give impetus to the rapidly expanding horse industry; and to make for greater knowledge and enjoyment in the grandest sport of all.

Fig. 50–11. An attractive horse farm. (Courtesy, Windfields Farm, Chesapeake City, MD)

## FUTURE OF THE HORSE INDUSTRY

This generation has more money to spend and more leisure time in which to spend it than any population in history. A shorter workweek, increased automation, more suburban and rural living, and the continued recreation and sports surge, with emphasis on physical fitness and the out-of-doors, will require more horses and support more racetracks, shows, and other horse events.

It is expected that the estimated 500,000 horses in the 17 western range states will continue to hold their own. Even the

Fig. 50-12. Horses will continue to be used for recreation and sport. This shows a hunter in competition. (Courtesy, American Quarter Horse Assn., Amarillo, TX)

Jeep is not sufficiently versatile for use in roping a steer on the range. It is reasonable to assume, therefore, that the cow pony will continue to furnish needed assistance to ranchers in the West.

Fig. 50-13. The cow pony of the West will continue to furnish needed assistance in handling the range herd. (Courtesy, *The Quarter Horse Journal*, Amarillo, TX)

Horse racing will continue to be a popular spectator sport, although there will be increased competition for the recreation and sports dollar in the years ahead.

In the final analysis, the dominant factors that will determine the future of the horse industry are (1) the need for the cow pony, and (2) the use of horses for recreation and sport.

Horse production will, in common with most businesses, encounter increasing competition in the years ahead. Competition will be keen for land, labor, and capital, as well as from other sports.

Skilled management and production programs geared to produce horses that meet more exacting market demands will be the two essential ingredients for success. Also, it will require greater skill and understanding of fundamental relationships to take care of highly bred, sensitive animals in forced production.

Never has there been so much reason to have confidence in, and to be optimistic about, the future. The years ahead will be the most rewarding in the history of the horse industry.

## HORSE RESEARCH

The Age of Research was ushered in with World War II. Now we are in the biotechnology age, and all industry, big and little—including the horse business—must be geared to it. Other animal industries have long been cognizant of new frontiers possible through research. But horse research has lagged, with the result that we have just begun to apply science, automation, and technology to light horses. In 1986, the scientific man years (a man year is defined as one person devoting full time to research for one year) devoted to research on each class of livestock by USDA and college personnel was as follows:

| Class of Animal | Scientific Years Devoted to Research in 1986 |
| --- | --- |
| Beef Cattle | 446.3 |
| Dairy Cattle | 338.5 |
| Poultry | 264.4 |
| Swine | 210.0 |
| Sheep and Wool | 101.8 |
| Horses, Ponies, and Mules | 73.5 |
| Laboratory Animals | 57.1 |
| Pets | 28.1 |
| Goats and Mohair | 27.5 |
| Other Animals | 5.6 |

There is every reason to believe that today's research will be reflected in a host of tomorrow's advances—that many of today's horse problems will be solved through research. Indeed, horse research should be expanded. More specifically, and among other things, we need to know the following in the horse business.

1. We need to know how to modernize rations and effect savings in costs; we need to eliminate needless concoctions and unbalanced and deficient rations.

2. We need to know how to rectify appalling and costly sterility and reproductive failures; we need to produce more than a 50% foal crop.

3. We need to know how to bring mares in heat at will.

4. We need to improve artificial insemination of horses.

5. We need to be able to transplant fertilized horse eggs.

6. We need to know more about the relationship between soil fertility, plant nutrients, and horses.

7. We need to know how to provide laborsaving buildings and equipment—how to automate the horse business. Seventy-five percent of horse work is still hand labor, one-third of which could be eliminated by mechanization and modernization.

8. We need to know how to improve upon the control of diseases and parasites.

9. We need to know how to increase the durability and useful life of a horse—in racing, in showing, and in breeding; we need to lessen the appalling number of horses that we are retiring from tracks, shows, and other uses due to unsoundnesses.

10. We need to know how to make a fair return on capital invested in horse breeding establishments.

We must remember, however, (1) that horse research is both slow and costly, and (2) that other industries have long liberally supported research costs with no assistance from the taxpayer, simply including them as a normal part of their operating costs. In addition to individual owners contributing to the support of research programs, the time has arrived when equestrians should review where racing dollars go. Perhaps a liberal proportion of racing revenue which now goes into the treasuries of the 30 states having pari-mutuel betting should be earmarked for horse research, teaching, and extension.

Otherwise, there is grave danger of starving "the goose that laid the golden egg."

Finally, it should be emphasized that research can make the information available, but it is still up to each individual to secure and apply the results. "You can lead a horse to water, but you can't make it drink." Nevertheless, in the years ahead equestrians will not be able to cling to horse-and-buggy methods while the rest of the industry forges ahead. For sheer survival, they must use science and technology.

## QUESTIONS FOR STUDY AND DISCUSSION

1. What caused the shift in the developed nations from muscle power to fossil fuel power over the past century?

2. Trace the evolution of horses from the standpoints of (a) the body changes which transformed them from denizens of the swamp to creatures capable of surviving in the forest, and finally, to animals adapted to the prairie, and (b) the period of years covering their evolution, domestication, and return to America.

3. Conjecture why Przewalski Horses are the only wild horses that survived 60 million years of evolution.

4. Trace the domestication and use of the horse prior to the time of Christ.

5. Trace the origin and domestication of the donkey.

6. Name and discuss the family, genus, and species of the horse in the zoological scheme.

7. In order of period of time, what uses have been made of horses down through the ages?

8. To which school of thought do you subscribe relative to the source of the foundation stock of the first horses of the American Indians: (a) that they came from the abandoned and stray horses of de Soto and Coronado, or (b) that they were obtained chiefly from Santa Fe, an ancient Spanish mission? Justify your answer.

9. What is said to be the origin of the American custom of driving to the right of the road, instead of to the left as is the practice in most of the world?

10. Discuss the rise and fall of the horse population of the United States from the standpoints of (a) the forces which caused such shifts, and (b) the economic effects thereof.

11. How do you account for the rise in the number of horses used for recreation and sport in the U.S.? List the three leading states in numbers of horses.

12. What forces make for (a) China being the leading horse-producing nation of the world, and (b) Texas being the leading state in number of horses in the United States?

13. Discuss the magnitude and importance of the U.S. horse industry from the standpoint of (a) money invested; (b) expenditures for feed, tack, medication, and services; (c) number of 4-H Club projects; (d) horse shows; (e) racing attendance; (f) saddle clubs; and (g) cow ponies.

14. Discuss horses as benefactors of mankind in the following ways: (a) law enforcement, and (b) the manufacture of antitoxins.

15. What is your prediction as to the future of the horse industry (a) in the United States and (b) throughout the world?

## SELECTED REFERENCES

| Title of Publication | Author(s) | Publisher |
| --- | --- | --- |
| Agricultural Statistics | | U.S. Department of Agriculture, Washington, DC, annual |
| Animals and Men | H. Dembeck | The American Museum of Natural History, The Natural History Press, Garden City, NY |
| Asiatic Wild Horse, The | E. Mohr, translated by D. M. Goodall | J. A. Allen & Co. Ltd., London, England, 1971 |
| Breeding and Raising Horses, Ag. Hdbk. No. 394 | M. E. Ensminger | Agricultural Research Service, USDA, Washington, DC, 1972 |
| Economic Impact of the U.S. Horse Industry, The | | The American Horse Council, Washington, DC, 1987 |
| Evolution of the Horse | W. D. Matthew S. H. Chubb | American Museum of Natural History, New York, NY, 1921 |
| FAO Yearbook | | Food and Agriculture Organization of the United Nations, Rome, Italy, annual |

*(Continued)*

## SELECTED REFERENCES *(Continued)*

| Title of Publication | Author(s) | Publisher |
|---|---|---|
| First Horsemen, The | F. Trippett | Time-Life Books, New York, NY, 1974 |
| Harper's Encyclopedia for Horsemen: The Complete Book of the Horse | L. Taylor | Harper & Row Publishers, New York, NY, 1973 |
| History of Domesticated Animals, A | F. E. Zuener | Harper & Row Publishers, Inc., Great Britain |
| Horse, The | J. W. Evans<br>A. Borton<br>H. F. Hintz<br>L. D. Van Vleck | W. H. Freeman and Company, San Francisco, CA, 1977 |
| Horse, The | P. D. Rossdale | The California Thoroughbred Breeders Association, Arcadia, CA, 1972 |
| Horse, The, through fifty centuries of civilization. | A. Dent | Phaidon Press Limited, London, England, 1974 |
| Horse Breeding Farm, The | L. C. Willis | A. S. Barnes & Co., Inc., Cranbury, NJ, 1973 |
| Horse Science Handbook, Vols. 1, 2, and 3 | Edited by M. E. Ensminger | Agriservices Foundation, Clovis, CA, 1963, 1964, and 1966 |
| Horses | G. G. Simpson | Oxford University Press, New York, NY, 1951 |
| Horses and Horsemanship, 6th Edition | M. E. Ensminger | The Interstate Publishers, Inc., Danville, IL, 1990 |
| Horses and Horsemanship Through the Ages | L. Gianoli | Crown Publishers, Inc., New York, NY, 1969 |
| Horseman's Encyclopedia | M. C. Self | A. S. Barnes & Co., Inc., Cranbury, NJ, 1963 |
| Horsemanship and Horse Care, Ag. Info. Bull. No. 353 | M. E. Ensminger | Agricultural Research Service, USDA, Washington, DC, 1972 |
| Horses, Their Selection, Care & Handling | M. C. Self | A. S. Barnes & Co., Inc., Cranbury, NJ, 1943 |
| Kinships of Animals and Man | E. Adams | McGraw-Hill Book Company, Inc., New York, NY, 1955 |
| Light Horses, Farmers' Bull. No. 2127 | M. E. Ensminger | U.S. Department of Agriculture, Washington, DC, 1965 |
| Natural History of the Horse, The | J. Clabby | Weidenfeld and Nicolson, London, England, 1976 |
| Our Friendly Animals and Whence They Came | K. P. Schmidt | M. A. Dohohue & Co., Chicago, IL, 1938 |
| Principles of Classification and a Classification of Mammals, The, Vol. 85 | G. G. Simpson | American Museum of Natural History, New York, NY, 1945 |
| Statistical Abstracts of the United States | | U.S. Department of Commerce, Bureau of the Census, Washington, DC, annual |
| Stud Managers Course, Lectures | | Stud Managers Course, University of Kentucky, Lexington, KY, intermittent years since 1951 |
| Stud Managers' Handbook, The | Edited by M. E. Ensminger | Agriservices Foundation, Clovis, CA, pub. annually since 1965 |
| Summerhays' Encyclopedia for Horsemen | R. S. Summerhays | Frederick Warne and Co., Ltd., New York, NY, 1966 |
| World Almanac and Book of Facts | | Scripps Howard Company, New York, NY, annual |

The future use of a foal is determined primarily by heredity and training. (Photo by Ernst Peterson, Courtesy, Bitterroot Stock Farm, Hamilton, MT)

# CLASSES, TYPES, USES, AND BREEDS OF HORSES

## Chapter 51

Contents          Page

Contents          Page

In no class of animals have so many diverse and distinct classes, types, and uses been developed as in the horse. The descendants of the Oriental light-legged horse have, for generations, been bred and used for riding and driving purposes—first as the chariot and riding horses of Egypt, Persia, Greece, and Arabia; later, as the running horses of England; and finally, for purposes of recreation and sport in the United States and throughout the world. In due time, further refinements in breeding light horses were made, and these animals were adapted for more specific purposes.

As with light horses, similar distinctive types—though smaller in number—evolved in the draft horses. From the ponderous beast of Flanders, used as foundation stock, the Great War Horse of the Middle Ages was developed; the Great War Horse, in turn, served as the forerunner of the draft horse of commerce and agriculture. Further and eventual refinements through breeding and selection adapted draft animals to many and diverse uses, some of which have subsequently passed into oblivion with mechanization.

## CLASSES OF HORSES

Horses may be classified as light horses, ponies, or draft horses, according to size, build, and use.

**Light horses** stand 14–2 to 17 hands high, weigh 900 to 1,400 lb, and are used primarily for riding, driving, racing, or for utility purposes. Light horses generally are more rangy and are capable of more action and greater speed than draft horses.

**Ponies** stand under 14–2 hands high and weigh 500 to 900 lb.

**Draft horses** stand 14–2 to 17–2 hands high, weigh 1,400 lb or more, and are suited for drawing loads and for other heavy work. Today, draft horses are of negligible importance and primarily of historical interest only.

## TYPES AND USES OF LIGHT HORSES

In attempting to produce horses for specific uses, different types were developed. In some cases, however, the particular use or performance is so exacting that only one breed appears to be sufficiently specialized. For example, harness racing is now synonymous with the Standardbred breed. The several types, uses, and breeds of light horses are summarized in Table 51–1.

### RIDING HORSES AND PONIES

Riding horses and ponies have many and varied uses, but as the name indicates, they are all ridden. They may have a very definite utility value, as is true of stock horses, or they may be used chiefly for purposes of recreation and sport. For the latter use, training, manners, and style are of paramount importance, although durability and efficiency are not to be overlooked in any horse.

### THREE- AND FIVE-GAITED SADDLE HORSES

Long after the development of the New England town, the opening up of roads along the Eastern Seaboard, the development of the buggy and the popularity of the roadster type of horse, the states of Virginia, West Virginia, Kentucky, Tennessee, and Missouri still consisted of large plantations under the ownership of southern gentlemen. Roads were few and far

between, and travel was largely on horseback over the most natural paths that could be found. Thus, there was need for a horse that would carry the plantation owners with dignity befitting their station in life and with the least distress possible to both rider and horse. As the plantation owners rode over their broad estates, easy gaits were a necessity. Such was the need, and out of this need arose the beautiful American Saddlebred Horse.

Animals qualifying as either 3- or 5-gaited saddle horses in the leading American horse shows are generally of American Saddlebred breeding, a truly American creation.[1] Occasionally, however, animals of the other light horse breeds are trained to execute the 5 gaits. It must also be remembered that the vast majority of American horses of all breeds are of the 3-gaited variety and that only a relatively small proportion of these animals are ever exhibited. Instead, most of the 3-gaited horses are used for pleasure riding.

The gaits of a 3-gaited horse are: the walk, the trot, and the canter. In addition to performing these same gaits, the 5-gaited horse must possess a slow gait and the rack. The slow gait may be either the running walk, fox trot, or stepping pace (slow pace); but, for show purposes, only the stepping pace is accepted. In the show-ring, generally the judge requests the 5-gaited horses execute the gaits in the following order: the walk, the trot, the slow gait, the rack, and the canter.

Whether an animal is 3-gaited or 5-gaited is primarily a matter of training. Custom decrees that 3-gaited horses be shown with their manes roached or clipped short and their tails clipped or sheared for a short distance from the base; whereas 5-gaited horses are shown with flowing manes and full-length tails. Also, because of the speed at which 5-gaited horses are expected to perform at the trot and the rack, they are permitted to wear quarter boots to protect the heels of the front feet, a practice which is forbidden in 3-gaited classes.

Fig. 51–1. Three-gaited saddle horse, *Steppin On*, American Saddlebred mare, a many times champion. Note that the mane is roached (clipped short), and that the tail is clipped for a short distance from the base—as custom decrees. (Courtesy, Sunnyslope Farms, Scott City, KS)

---

[1]Herein reference is made to the Saddlebred Horse Division as described by The American Horse Shows Association, and not to the several performance classes in which three-gaited horses of various breeds compete.

**TABLE 51–1**
**LIGHT HORSE SUMMARY**

| Type | Primary Use | Breeds | |
|---|---|---|---|
| **R**iding Horses and Ponies | **T**hree-gaited saddle horses .......... | American Saddlebred<br>Andalusian<br>Appaloosa<br>Arabian<br>Hanoverian<br>Hungarian Horse<br>Lipizzan<br>Morab<br>Morgan | National Show Horse<br>Paint Horse<br>Palomino<br>Pinto<br>Quarter Horse<br>Spanish-Barb<br>Spotted Saddle Horse<br>Thoroughbred<br>Trakehner |
| | **G**aited horses ..................... | American Saddlebred<br>Missouri Fox Trotting Horse<br>National Show Horse<br>Paso Fino<br>Peruvian Paso<br>Tennessee Walking Horse | |
| | **S**tock horses ..................... | Grades, crossbreds, or following purebreds:<br>American Mustang<br>Appaloosa<br>Buckskin<br>Chickasaw<br>Galiceno<br>Hungarian Horse | Morgan<br>Paint Horse<br>Pinto Horse<br>Quarter Horse<br>Spanish-Barb<br>Spanish Mustang<br>Thoroughbred |
| | **E**quine sports horses .............. | Purebreds, crossbreds, and graders<br>  of the following breeds:<br>Akhal-Teke<br>American Creme/American White<br>Lipizzan<br>Thorcheron<br>Thoroughbred<br>Trakehner | |
| | **P**onies for riding ................. | American Walking Pony<br>Connemara Pony<br>Gotland Pony<br>National Appaloosa Pony<br>Pony of the Americas | Quarter Pony<br>Shetland Pony<br>Welara Pony<br>Welsh Pony and Cob |
| **R**acehorses[1] | **R**unning racehorses ............... | Thoroughbred | |
| | **Q**uarter racehorses ............... | Quarter Horse | |
| | **H**arness racehorses<br>  (trotters and pacers) .............. | Standardbred<br>Trottingbred | |
| **D**riving Horses and Ponies | **D**riving horses:<br>Heavy harness horses ............<br>Fine harness horses .............<br>Roadsters ..................... | Hackney<br>American Saddlebred (predominantly, although other breeds are also used)<br>Standardbred | |
| | **D**riving ponies:<br>Harness show ponies ............<br>Heavy harness ponies ............ | Hackney<br>Shetland Pony<br>Welsh Pony and Cob | |
| **A**ll-Purpose Horses and Ponies | **F**amily horses/ponies .............. | American Bashkir Curly<br>American Gotland Horse<br>Haflinger<br>Norwegian Fjord Horse | |
| **M**iniature Horses and Donkeys | **D**riving ......................<br>**P**ets ......................... | Miniature Horse<br>Miniature Donkey | |

[1]In a few states, Appaloosa and Arabian horses are also being raced under saddle.

Fig. 51-2. Five-gaited saddle horse, *Wing Commander*, American Saddlebred stallion, six times World's Grand Champion. Note the flowing mane and full-length tail—as custom decrees. (Photo by Horst. Courtesy, American Saddlebred Horses Assn., Lexington, KY)

Both 3- and 5-gaited horses are shown under saddle; and each may be shown in combination classes, in which they must perform both in harness and under saddle. Also, 5-gaited horses (but not 3-gaited horses) may be shown in a third division; namely, in fine harness classes.

In combination classes, the entries enter the ring hitched to an appropriate four-wheeled vehicle, with the saddle and bridle hidden in the back of the rig. The judge works the class both ways of the ring, then lines them up in the center for inspection and backs each horse in order to test its manners. Next the judge orders that the entries be unhitched, unharnessed, saddled, bridled and worked under saddle both ways of the ring. Finally, the horses are again lined up in the center of the ring, and each animal is backed under saddle.

A fine harness horse is exactly what the name implies—a fine horse presented in fine harness. The entire ensemble is elegant, and represents the ultimate in grace and charm.

Fine harness horses are penalized if driven at excessive speed. Combination horses, especially five-gaited ones, should be driven at a more speedy trot than fine harness horses.

In addition to executing the gaits with perfection, both three-gaited and five-gaited animals should possess the following characteristics:

1. **Superior conformation,** in which the principal requirements are:

    a. Graceful lines obtained through a fairly long, arched neck; short, strong back and loin with a good seat; a nicely turned croup; a smartly carried, flowing tail; and a relatively long underline.

    b. A shapely and smart head.

    c. Nicely sloping shoulders and pasterns.

    d. Symmetry and blending of all parts.

    e. Quality, as evidenced by a clean-cut, chiseled appearance throughout, and soundness.

    f. Style, alertness, and animation, sometimes said to be comparable to that of a "peacock."

2. **Perfect manners,** which include form, training, and obedience—those qualities that make for a most finished performance.

3. **Superior action,** including an elastic step, high action, and evidence of spirit and dash.

## WALKING HORSES

This particular class of horses is largely comprised of one breed—the Tennessee Walking Horse.

Fig. 51-3. Walking horse, *Son's Shadow*, Tennessee Walking Horse stallion, owned by Mr. and Mrs. E. Carl Hengen, Lawn Vale Farm, Gainsville, VA. (Courtesy, *The National Horseman*, Middleton, KY)

Horses of this type were first introduced into Tennessee by the early settlers from Virginia and the Carolinas. For many years, the plantation owners of middle Tennessee—persons who spent long hours daily in supervising labor from the saddle—selected and bred animals for their easy, springy gaits, good dispositions, and intelligence. Particular stress was placed upon the natural gait known as the running walk and upon the elimination of the trot. Thus, the three gaits that evolved in the walking horse (also called Plantation Walking Horse) were: the walk, the running walk, and the canter.

At the running walk there is a characteristic nodding of the head. Sometimes there is also a flopping of the ears and a snapping of the teeth while the animal is in this rhythmic movement. Walking horses are also noted for their wonderful dispositions. Their easy gaits and a superb disposition make them an ideal type of horse for the amateur rider or the professional society person who rides infrequently.

## STOCK HORSES

Stock horses constitute the largest single class of light horses of this country; there are approximately 500,000 of them in use in the 17 range states. They are the cow ponies of the West.

Usually, stock horses are of mixed breeding. Most generally they are descendants from the Mustang—the feral horse of the United States. Subsequently, Mustang mares were mated to sires of practically every known light horse breed—especially Thoroughbreds and Quarter Horses. Stallions of the Palomino, Morgan, Arabian, and other breeds have also been used. Such grading-up has improved the size, speed, and perhaps the appearance of the cow pony, but most equestrians will concede that no amount of improved breeding will ever produce a gamier, hardier, and more durable animal than the Mustang. In addition to being game and hardy, the stock horse must be agile, surefooted, fast, short coupled, deep, powerfully muscled, durable, and possess good feet and legs. Above all, cowhands insist that their ponies be good companions and possess *cow sense.*

Fig. 51–4.  The superbly trained Quarter Horse, *Lucky Penny,* working without a bridle. (Courtesy, *The Western Horseman,* Colorado Springs, CO)

## EQUINE SPORTS HORSES

In addition to racing, horses are used for many equine sports, including dressage, endurance trials, hunters/jumpers, polo, and parades. Purebreds, crossbreds, and grades of the following breeds are especially well adapted to the equine sports described in the sections that follow: Akhal-Teke, American Creme and American White, Lipizzan, Thorcheron, Thoroughbred, and Trakehner.

• **Dressage**—*Dressage is the guiding of a horse through natural maneuvers without emphasis on the use of reins, hands, and feet.*

The term *dressage* comes from the French verb meaning *to train.* After the horse has learned to respond to the simple directions of moving forward or backward, turning, changing gait, and halting, equestrians who wish to develop the horse's strength, willingness, and agility as much as possible continue training by giving special exercises to develop these traits. This training is called dressage.

All of the movements described as dressage or training movements are based on natural movements of the horse while at liberty. Thus, by watching horses (especially young horses) in a corral, they will be seen to execute with ease changes of the leading leg at the canter when changing direction, and such intricacies of the *Haute Ecole* as the pirouette, the piaffe, the passage, and the pesade. Why, then, must a horse be trained to do these things if it already knows how? There are two reasons: (1) the horse must learn to balance itself under the weight of the rider, and (2) the horse must learn to do these movements when so requested by the rider.

• **Endurance trials (rides)**—*Competitive tests designed to test the stamina of horses are known as endurance trials.* The riders must take their horses over a prescribed course, which is usually

of rugged terrain, and which may require anywhere from one to three days to cover. The time for the different courses varies according to the topography, elevation, and footing. Regular tests (temperature, pulse, respiration, etc.) are made at intervals.

One of the best known endurance rides is the Tevis Cup Ride, held at Auburn, California each August. It is 100 miles over extremely rough terrain. The time limit is 24 hours, and there are three mandatory rest stops of one hour each. During the rest stops, the veterinarians check the pulse, respirations, and temperature of each horse, both at the beginning and the end of the hour, so as to determine whether its rate of recovery is satisfactory. There are two kinds of competition in the Tevis Cup Ride: (1) a race over the 100-mile course, and (2) the best-conditioned horse that completes the ride within the 24-hour period.

Generally over 200 competitors start on the Tevis Cup Ride and about 160 (or 80%) finish. The winning horse makes the ride in approximately 11 hours.

• **Hunters and jumpers**—*The hunter is that type of horse used in following the hounds in fox hunting.* The sport is traditional in England, and each year it is sharing its glamour with more people in the United States.

The hunter is not necessarily of any particular breeding, but Thoroughbred blood predominates. The infusion of some cold blood (draft breeding) is often relied upon in order to secure greater size and a more tractable disposition.

In addition to being of ample size and height, hunters must possess the necessary stamina and conformation to keep up with the pack. They must be able to hurdle with safety such common field obstacles as fences and ditches. The good hunter, therefore, is rugged, short coupled, and heavily muscled throughout.

Fig. 51–5.  A jumper in action. (Courtesy, Al-Marah Arabians, Tucson, AZ)

All hunters are jumpers to some degree, but high jumpers are not necessarily good hunters. To qualify as hunters, horses must do more. They must execute many and varied jumps over a long period of time.

*Jumpers are a nondescript group, consisting of all breeds and types; the only requisite is that they can jump.* In the show-ring, an unsoundness does not penalize a jumper unless it is sufficiently severe to be considered an act of cruelty.

- **Polo**—As the name would indicate, *polo mounts are horses that are particularly adapted for use in playing the game of polo.* This game, which was first introduced into this country in 1876, is played by four mounted players on each team. The object is to drive a wooden ball between two goalposts at either end of a playing field 300 yards long and 120 to 150 yards wide. Long-handled regulation mallets are used to drive the ball.

At the time the game was first introduced into the United States, there was a decided preference for ponies under 13–2 hands in height. Later, horses up to 14–2 hands were accepted, and more recently horses up to 15–2 and over have been used.

Although very similar to the hunter in type, polo mounts are generally smaller in size. They must be quick and clever in turning, and they must be able to dodge, swerve, or wheel while on a dead run. They must like the game of polo and be able to follow the ball.

The polo mount is trained to respond to the pressure of the reins on the neck, so that the rider may be free to guide it with only one hand. Up to 5 or 6 years is required to complete the schooling of a polo horse, and as many as 4 to 6 mounts may be used by each player in a single game—all of which contributes to the expensiveness of the sport.

Polo ponies are usually of mixed breeding, but most of them are predominantly Thoroughbred. Type and training, together with native ability and intelligence, are the primary requisites.

- **Parade horses**—*Parade horses are horses of any breed, cross, or color used under elaborate Western, Mexican, or Spanish equipment in parades.* Attractive colors and good manners are important. Parade horses are shown at an animated walk and a parade gait. The latter is a prancing cadenced trot at about five miles per hour.

Fig. 51-6. A parade horse, showing elaborate tack. (Courtesy, Willard Beanland, Canoga Park, CA)

## PONIES FOR RIDING

Several pony breeds are used for riding; among them, the Connemara, Gotland Horse, Pony of America, Shetland, and Welsh. Also, the smaller animals failing to meet the minimum height requirements of the American Saddlebred, Appaloosa, Quarter Horse, and Tennessee Walking Horse breeds are registered as Saddlebred Pony, Appaloosa Pony, Quarter Pony, and Walking Pony, respectively.

In the 1950s, the Welsh Pony and the Tennessee Walking Horse were crossed to create the American Walking Pony. In the 1960s, the Shetland and Hackney breeds were crossed to produce the Saddlebred type of pony. In 1980, the Welara Pony Society was formed to register the breed created by crossing the Arabian and Welsh breeds.

Ponies are children's mounts. So, in addition to their small size, they should possess the following characteristics: (1) gentleness, (2) sound feet and legs, (3) symmetry, (4) good eyes, (5) endurance, (6) intelligence, (7) patience, (8) faithfulness, and (9) hardiness. Above all, they must be kind and gentle in disposition.

Fig. 51-7. Pony for riding, Shetland breed. (Courtesy, Sir Richard Musgrave)

## RACEHORSES

The term *racehorse* refers to a horse that is bred and trained for racing.

According to historians, the Greeks introduced horse racing in the Olympic Games in 1450 B.C. Also, it is reported that a planned horse race of consequence was run in England in 1377 A.D., between animals owned by Richard II and the Earl of Arundel. The sporting instinct of people being what it is, it is reasonable to surmise, however, that a bit of a contest was staged the first time that two proud mounted equestrians chanced to meet.

The development of horse racing in Britain dates from the 17th century, although it is known to have taken place much earlier. Records exist of racing during the Roman occupation; and during the reign of Henry II races took place at Smithfield, which was the great London horse market at the time. But it was in the reign of James I that racing first began to be an organized sport. He took a great liking to Newmarket, where he had a royal palace and a racecourse built. Also, he established public races in various parts of the country.

The famous Rowley Mile Course in Newmarket, the home of English flat racing, is named after Charles II. "Old Riley" was his nickname, after his favorite riding horse by that name. Charles II loved racing; he rode in matches, founded races called the Royal Plates, and sometimes adjudicated in the disputes.

The Jockey Club came into existence at Newmarket in 1752, with many rich and influential people among its members. It gradually became the governing body of English racing.

Today, three types of horse races are run: (1) running races (including steeplechase races), (2) quarter races, and (3) harness races. For the most part, each type of race is dominated by one breed. Thus, in running races, it's Thoroughbreds; in quarter races, it's Quarter Horses; and harness races, it's Standardbreds. However, on a limited basis, and in a few states, Appaloosa and Arabian horses are now being raced under saddle.

## *RACING ATTENDANCE*

Horse racing ranks third as a U.S. spectator sport. In 1988, 74,158,269 people went to horse races (Thoroughbred and harness racing). Baseball (major and minor leagues) ranked first, with an attendance of 89.6 million; auto racing ranked second with 86 million fans; football (pro and college) ranked fourth, with 51 million fans; and basketball (pro and college) stood fifth with an attendance of 46.6 million.[2]

## *RACING RECORDS*

In running races (Thoroughbreds), the records at some of the popular American distances are: **For 1 mile**—*Dr. Fager* set the world's record at 1:32$\frac{1}{5}$ minutes, as a 4-year-old and carrying 134 lb, at Arlington Park, Chicago, in 1968; **for 1$\frac{3}{8}$ miles on the dirt**—*Man O'War* set the American record at 2:14$\frac{1}{5}$ minutes, as a 3-year-old and carrying 126 lb, at Belmont Park, New York, in 1920; and **for 1$\frac{3}{8}$ miles on the turf**—*Cougar 2nd* set the world's record at 2:11 minutes, as a 6-year-old and carrying 126 lb, at Hollywood Park, California, in 1972. It is noteworthy that the fastest mile (run by *Dr. Fager*) at 1:32$\frac{1}{5}$ was equal to a speed of 45.4 miles per hour.

In quarter races (Quarter Horses), *Truckle Feature* set the U.S. record for a quarter mile (440 yards) at :21.02 in 1969.

In harness racing (Standardbreds), *Mack Lobell* holds the world's trotting record for a mile at 1:52$\frac{1}{5}$ minutes, which was established in 1987; and *Matts Scooter* set the world's pacing record for a mile at 1:48$\frac{2}{5}$ minutes in 1988.

## *RUNNING RACEHORSES*

Racehorses used for running (an extended gallop) under the saddle are now confined almost exclusively to one breed, the Thoroughbred. On the other hand, the Thoroughbred breed (including both purebreds and crossbreds) has been used widely for other purposes, especially as polo mounts and hunters.

Although trials of speed had taken place between horses from earliest recorded history, the true and unmistakable foundation of the Thoroughbred breed, as such, traces back only to the reign of Charles II, known as the "father of the English turf."

Although the length of the race, weight carried, and type of track have undergone considerable variation in recent years, the running horse always has been selected for speed and more speed at the run. The distinguishing characteristics of the running horse, as represented by the Thoroughbred, are the extreme refinement, oblique shoulders, well-made withers, heavily muscled rear quarters, straight hind legs, and close travel to the ground.

Fig. 51-8. A running race, of Thoroughbreds. (Photo by Bob Coglianese. Courtesy, *The Maryland Horse,* Timonium, MD)

## *QUARTER RACEHORSES*

Quarter racing has become an increasingly popular sport. For the most part, races of this type are confined to animals of the Quarter Horse breed, which animals derived their name and initial fame for their extraordinary speed at distances up to a quarter mile. Although the great majority of Quarter Horses are used to work cattle and never appear on the racetrack, the proponents of quarter racing advocate the racetrack as a means of proving animals. Performance, so they argue, is the proof of whether or not a horse can do the job for which it is bred. Thus, quarter racing is used as a breed proving ground for the Quarter Horse, for in racing the fundamental quality of speed can be accurately measured and recorded in such a way that the performance of horses in all parts of the country can be compared.

Fig. 51-9. Quarter racehorses rounding the final turn at Los Alamitos. (Photo by Milt Martinez. Courtesy, Mrs. Frank Vessels, Jr., Los Alamitos, CA)

[2]Statistics courtesy of Triangle Publications, Inc., Hightstown, NJ.

## HARNESS RACEHORSES (TROTTERS AND PACERS)

Prior to the advent of improved roads and the automobile, but following the invention of the buggy, there was need for a fast, light-harness type of horse. This horse was used to draw vehicles varying in type from the light roadster of the young gallant to the dignified family carriage. In the process of meeting this need, two truly American breeds of horse evolved—the Morgan and the Standardbred. The first breed traces to the foundation sire, *Justin Morgan*; and the latter to *Hambletonian 10*, an animal which was line bred to imported *Messenger*.

As horse-and-buggy travel passed into permanent oblivion, except for recreation and sport, the Standardbred breeders wisely placed greater emphasis on the sport of racing; whereas the Morgan enthusiasts directed their breeding programs toward transforming their animals into a saddle breed.

The early descendants of *Messenger* were sent over the track, trotting (not galloping) under the saddle; but eventually the jockey races in this country came to be restricted to a running type of race in which the Thoroughbred was used. With this shift, qualifying standards—a mile in 2:30 at the trot and 2:25 at the pace when hitched to a sulky—were set up for light harness races; and those animals so qualifying were registered.[3] The pneumatic-tire racing vehicles, known as sulkies, were first introduced in 1892. With their use that year, the time was reduced nearly 4 seconds below the record of the previous year. Thus, were developed harness racing and the Standardbred breed of horses, which today is the exclusive breed used for this purpose.

Fig. 51–10. A harness race horse, Standardbred breed. This shows the World's Champion trotting filly as a three-year-old. (Courtesy, The United States Trotting Assn., Columbus, OH)

Trotters and pacers are of similar breeding and type, the particular gaits being largely a matter of training. In fact, many individuals show speed at both the trot and the pace. It is generally recognized, however, that pacers are handicapped in the mud, in the sand, or over a rough surface.

The Standardbred breed—like the Thoroughbred—also finds other uses, as driving horses in roadster classes, delivery horses, and general utility horses. By way of comparison with

---

[3]On January 1, 1933, registration on performance alone was no longer granted, and registration of both sire and dam was required. The qualifying standards were initiated in 1879.

the Thoroughbred, the Standardbred possesses a more tractable disposition, is smaller, longer bodied, closer to the ground, heavier-limbed, and sturdier in build. The latter characteristic is very necessary because harness races are usually *heat races*—for example, the best two out of three races.

In the beginning, horses of this type found their principal use in harness races at county and state fairs. However, in recent years pari-mutuel harness racing has been established at a number of tracks. Today, harness racehorses are almost exclusively of the Standardbred breed.

## DRIVING HORSES AND PONIES

At the present time, driving horses and ponies are used chiefly for purposes of recreation. According to the specific use made of them, driving horses are classified as heavy harness horses, fine harness horses, roadsters, or ponies. Driving ponies are of two kinds; harness show ponies and heavy harness ponies.

### HEAVY HARNESS HORSES

These are also known as carriage horses. At the present time, this type of horse has very little place in the utility field, its use being largely confined to the show-ring. As the name implies, the heavy harness horse of the show-ring wears heavier leather than the fine harness horse or the roadster, though it in no way approaches the draft harness. The heavy leather used on these animals was first decreed by fashion in England, the idea being that to drive handsomely one must drive heavily. The vehicles drawn were of heavy construction and elegant design and logically and artistically the harness has to be in proportion thereto.

Heavy harness horses were especially popular during the Victorian era, and the ownership of a handsome pair was an indication of social prestige. In this country during the Gay Nineties, bobtailed Hackneys attached to high-seated rigs made a dashing picture as they pranced down the avenue.

Fig. 51–11. Handsome and heavy. A Hackney heavy harness horse in action. (Courtesy, Mrs. Dean J. Briggs, Garden Plain, KS)

At one time, there were several heavy harness breeds, but at present all except the Hackney have practically ceased to exist in America. In this country, therefore, the Hackney is now the heavy harness breed; and the American Horse Shows

Association officially refers to show classifications as Hackneys rather than Heavy Harness Horses.

Hackneys should possess the following distinguishing characteristics:

1. **Beauty.** Beauty is obtained through graceful, curved lines; full-made form; and high carriage. Showing style decrees that heavy harness horses be docked and have their manes pulled.

2. **High action.** Animals of this type are bred for high hock and knee action, but skilled training, bitting, and shoeing are necessary for their development. In the show-ring, heavy harness horses must be able to fold their knees, flex their hocks, and set their chins. "Woodenlegged" horses cannot take competition.

3. **Manners and temperament.** Perfection in manners and disposition of pleasure horses of this type is a requisite of first rank.

4. **Color.** Seal brown, brown, bay, and black colors are preferred in heavy harness horses. White stockings are desired for purposes of accentuating high action.

5. **Height.** For horse show purposes, the maximum height of Hackney ponies shall be 14-2 hands.

## FINE HARNESS HORSES

A fine harness horse is exactly what the name implies—a fine horse presented in fine harness. The entire ensemble is elegant and represents the ultimate in grace and charm.

In the show-ring, fine harness horses are, according to the rules of the American Horse Shows Association, limited to the American Saddlebred breed. In some shows, however, other breeds are exhibited in fine harness classes. Fashion decrees that fine harness horses shall be shown wearing long mane and tail and drawing a four-wheeled vehicle, preferably a small buggy with wire wheels, but without top, or with top drawn. Light harness with a snaffle bit is required. Fine harness horses are shown at an animated park trot and at an animated walk.

## ROADSTERS

The sport of showing a roadster originated in the horse and buggy era. It was founded upon the desire to own an attractive horse that possessed the necessary speed to pass any of its rivals encountered upon the city or county thoroughfares.

In the show-ring, roadsters are generally shown in either or both (1) roadster to bike, or (2) roadster to road wagon or buggy classes. The latter are hitched singly or in pairs. Some shows also provide a class or classes[4] for roadsters under saddle. In all divisions—whether shown to bike or buggy, or under saddle—entries must trot; pacing is barred.

Originally, roadster classes included animals of both Standardbred and Morgan extraction. In recent years, however, the Morgan has developed in the direction of the saddler. Today the *Rule Book* of the American Horse Shows Association lists two divisions for roadsters: (1) Roadster Division, for Standardbred and Standardbred-type horses, and (2) Roadster Ponies, for ponies under 12-2 hands (50 in. and under).

Particular stress is placed in roadster show classes upon the manners, style, and beauty of conformation, combined with

[4]In many of the larger shows, a roadster appointment class is provided. Appointments are listed in the A.H.S.A. Rule Book.

Fig. 51-12.  Roadster to buggy.

speed. In striking contrast to heavy harness classes, the roadster is shown hitched to very light vehicles permitting fast travel.

Custom decrees that roadsters shall enter the ring at a jog, and work the wrong way (clockwise) of the track first. After jogging for a brief time, usually the judge asks that they perform the road gait, then jog again (all clockwise of the ring). Then, in succession, the judge asks them to reverse, jog, road gait, and turn on or trot at speed. Lastly, they are called to the center of the ring for inspection in a standing position, at which time the judge usually tests their manners by asking the drivers, in order, to back their horses.

## PONIES FOR DRIVING

Fig. 51-13.  Pony for driving—harness pony. (Courtesy, *The Hackney Journal,* Crawfordsville, IA)

In horse shows, ponies for driving are either harness show ponies or heavy harness ponies.

• **Harness show ponies**—The best harness show ponies are vest-pocket editions of fine harness horses; that is, they possess the same desirable characteristics, except that they are in miniature. According to the rules of the American Horse Shows Association, harness show ponies may be of any breed or combination of breeds; the only requisite is that they must be under 12-2 hands in height. Three breeds produce more animals that qualify under this category; namely, the Shetland, Welsh, and Hackney breeds.

• **Heavy harness ponies**— These must be under 14–2 hands. Generally they are either purebred Hackneys, or predominantly of Hackney breeding.

Three breeds produce animals that qualify as ponies; namely, the Hackney, Welsh, and Shetland breeds. The Hackney is generally exclusively of the harness type, but the Welsh and Shetland breeds are used either under saddle or in harness. In the major horse shows of the land, the latter two breeds may be exhibited in harness, but in practical use they are children's mounts.

## ALL-PURPOSE HORSES AND PONIES

All-purpose horses are horses that are used for riding, driving, and/or draft purposes. They may be used for pleasure riding, carriage horses, jumpers, hunters, trail riders, stock horses, or for light draft purposes. they are family-type horses, suitable for both children and adults. Because of their versatility, they are especially well adapted for use by suburban and part-time farm families. The following breeds are extolled as all-purpose horses and ponies; American Bashkir Curly, American Gotland Horse, Haflinger, and Norwegian Fjord Horse.

Fig. 51–14. All purpose Norwegian Fjord mares competing in the Rocky Mountain Championship Pulling Contest, Rifle, CO. (Courtesy, Norwegian Fjord Association of North America, Grayslake, IL)

## MINIATURE HORSES AND DONKEYS

The Miniature Horse is a scaled-down model of a full-sized horse, and not a dwarf. The American Miniature Horse Association, which was organized in 1978, stipulates that a Miniature Horse cannot exceed 34 in. at the withers. Some people keep Miniature Horses as pets. Others exhibit them as driving horses, in single pleasure and roadster driving classes. Still others exhibit them in multiple hitches, pulling miniature wagons, stage coaches, carriages and other vehicles. Although they can pull a pretty good load, because of their small size only a small child can ride them.

Donkeys, which are small members of the ass family, *Equus asinus*, were domesticated in Egypt, where they were first ridden and used as pack animals about 3400 B.C. Like the horse, donkeys vary in size. Miniature Mediterranean donkeys, native of Sardinia and Sicily, are under 38 in. high, and are used as pets. Sometimes, they are hitched to specially made small vehicles.

Fig. 51–15 Miniature Horse, *Blue Araby Princess*, 31 in. at the withers. (Courtesy, Flying W Farms, Piketon, OH)

Fig. 51–16. Miniature donkey. (Courtesy, Danby Farm, Millard, NE)

## BREEDS OF HORSES AND DONKEYS[5]

*A breed of horses may be defined as a group of horses having a common origin and possessing certain well-fixed, distinctive, uniformly transmitted characteristics that are not common to other horses.* There are about 300 breeds of horses in the world, of which 35 breeds of light horses, 10 breeds of ponies, and 6 breeds of draft horses are found in the United States.

There is scarcely a breed of horses that does not possess one or more distinctive breed characteristics in which it excels all others. Moreover, any one of several breeds is often well adapted to the same use. Certainly, if any strong preference exists, it should be an important factor in determining the choice of a breed, although it is recognized that some breeds are better adapted to specific purposes than others.

---

[5]No person or department has authority to approve a breed. The only legal basis for recognizing a breed is contained in the Tariff Act of 1930, which provides for the duty-free admission of purebred breeding horses provided they are registered in the country of origin. But this applies to imported animals only. In this book, therefore, no official recognition of any breed is intended or implied. Rather, every effort has been made to present the factual story of the breeds. In particular, information about the new or less widely distributed breeds is needed and often difficult to obtain.

# BREED NUMBERS

Table 51–2 shows (1) the 1987 or 1988, and (2) the total registration to date of the breeds of light horses and ponies, draft horses, and asses and mules. The recent annual figures reflect the current popularity of the different breeds, although it is recognized that only one year's data fails to show trends.

## TABLE 51–2
## ANNUAL (1987 OR 1988) AND TOTAL REGISTRATIONS
## OF LIGHT HORSES IN U.S. BREED REGISTRIES

| Breed | Annual (1987 or 1988) Registrations | Total Registrations (since breed registry started) | Breed | Annual (1987 or 1988) Registrations | Total Registrations (since breed registry started) |
|---|---|---|---|---|---|
| Quarter Horse | 128,352 | 2,777,860 | Welsh Pony and Cob | 577 | 32,298 |
| Thoroughbred | 50,375 | 1,158,803 | Shetland Pony | 568 | 137,360 |
| Arabian | | | National Spotted Saddle Horse | 436 | 3,996 |
| Purebred | 24,570 | 422,829 | Andalusian[5] | 355 | 1,903 |
| Half-Arabian | 6,500 | 274,200 | Haflinger | 298 | 2,000 |
| Anglo-Arabian | 188 | 7,353 | American Bashkir Curly | 163 | 875 |
| Standardbred | 17,393 | 632,000 | National Appaloosa Pony | 150 | 5,729 |
| Ridden Standardbred[1] | 300 | 5,000 | American Part-Blooded | 147 | 14,450 |
| Paint Horse | 15,518 | 144,076 | Lipizzan | 124 | 928 |
| Appaloosa | 12,317 | 492,511 | Welara Pony | 93 | 423 |
| Tennesse Walking Horse | 8,000 | 245,000 | Spanish Mustang | 70 | 1,510 |
| American Saddlebred | 3,811 | 207,381 | American White and American Creme[6] | 34 | 3,698 |
| Half Saddlebred[2] | 182 | 1,440 | Norwegian Fjord Horse | 21 | 424 |
| Morgan | 3,526 | 88,000 | Spanish-Barb | 21 | 381 |
| Palomino[3] | 3,493 | 78,080 | American Walking Pony | 15 | 300 |
| Trottingbred | 3,441 | 10,578 | Miniature Horse | — | 30,000 |
| Paso Fino | 2,060 | 13,500 | Hanoverian | — | 1,811 |
| Peruvian Paso[4] | 1,813 | 12,072 | American Mustang | — | 1,000 |
| Missouri Fox Trotting Horse | 1,800 | 32,800 | Hungarian | — | 545 |
| Buckskin | 852 | 12,540 | Morab | — | 320 |
| Pony of the Americas | 709 | 36,000 | American Warmblood | — | 71 |
| Trakehner | 625 | 5,987 | Thorcheron | — | 57 |

[1]Registred in the Ridden Standardbred Association, 1578 Fleet Rd., Troy, OH 49373

[2]The Half Saddlebred Registry was founded in May, 1971 for the registration of horses of which one parent is an American Saddlebred.

[3]Palomino Horse Breeders of America: 1987 registrations, 1,746; total registrations, 53,000. The Palomino Horse Association, Inc: 1988 registrations, 1,747; total registrations, 25,080.

[4]American Association of Owners & Breeders of Peruvian Paso Horses: 1987 registrations, 675; total registrations, 6,500. Peruvian Paso Horse Registry of North America: 1987 registrations, 1,138; total registrations, 5,572.

[5]There are two Andalusian registries. From inception to 1988, the American Andalusian Association reported a total of 703 registrations. From inception to 1988, the International Andalusian Association reported a total of 1,200 registrations.

[6]The American White and the American Creme are registered in separate divisions by the International American Albino Association, Naper, NE. In 1987, a total of 34 American Creme and American White horses were registered. The number of animals registered by the association from its formation through 1987 totaled 3,698.

# CHARACTERISTICS OF BREEDS

A summary of the U.S. breeds of light horses, ponies, and draft horses, and their characteristics, is given in Table 51–3. It is noteworthy that most U.S. breeds of light horses are American creations. There are two primary reasons for this: (1) the diverse needs and uses for which light horses have been produced, and (2) the fact that many people of wealth have bred light horses. (See pages 964 to 971.)

## TABLE 51–3
## BREEDS OF LIGHT HORSES, PONIES, AND DRAFT HORSES; AND THEIR CHARACTERISTICS

| Breed | Place of Origin | Color | Other Distinguishing Characteristics | Primary Uses | Disqualifications; Comments |
|---|---|---|---|---|---|
| *Light Horses and Ponies* | | | | | |
| Akhal-Teke | The U.S.S.R., in Southern Turkmenia, by the *Teke* tribe. | The prevailing color is gold; either a golden dun, golden bay, or golden chestnut, although other colors do occur. | The Akhal-Teke is a desert-bred horse, with a light, elegant build. It has a long, tapering face, wide nostrils, large eyes, mobile ears; a long, straight neck; a short, sparse mane and forelock; sloping shoulders; prominent withers; a long, lean, narrow, and sinewy body; a pronounced croup; long legs, muscular forearms, strong hocks, and dense hooves; and gliding elastic action. Akhal-Tekes are 15 to 15–2 hands high; weigh 900 to 1,000 lb; and are of quiet temperament. | All competitive equine sports, including endurance trials, dressage, and jumping. | |
| American Bashkir Curly | In Bashkiria, on the eastern slopes of the Ural Mountians, in the U.S.S.R.; hence, the name Bashkir. Modern history of curly horses in America began in 1898, when Peter Damele of Ely, Nevada, cut three curly animals from a herd of wild horses in the Peter Hanson Mountain Range. Most of today's curly horses trace to Damele ranch breeding. | All colors are accepted. | Curly coat, with corkscrew mane and wavy tail. In build, Curlies are medium size and chunky, somewhat resembling the early-day Morgan in conformation. The breed is noted for small nostrils, a gentle disposition, and heavy milking. Many of the animals have a natural fox-trot gait. | As a pleasure horse; for utility purposes—including light draft work; family trail horses; and children's mounts. | Weight in excess of 1,350 lb; faulty conformation. |
| American Creme[1] | U.S.A. Pale cream horses have existed for a very long time, primarily in Oregon and Washington. They were not given breed status until 1970, when the American Albino Assn., Inc., established a separate American Creme Horse Division for its registration. | The following color classifications of American Creme horses are registered: A–Body ivory white, mane white (lighter than body), eyes blue, skin pink. B–Body cream, mane darker than body, cinnamon buff to ridgeway, eyes dark. C–Body and mane of the same color, pale cream; eyes blue; skin pink. D–Body and mane of same color, sooty cream; eyes blue; skin pink. Combinations of the above classifications are also accepted. | Both horses and ponies are accepted for registry; animals above 14–2 hands are classed as horses, those below 14–2 are classed as ponies. | Pleasure and stock horses; for exhibition purposes; as parade and flag-bearer horses. | Pink eyes, or any body color other than ivory, white or cream, are disqualifications. |
| American Gotland Horse | Baltic Island of Gotland, a part of Sweden. | Bay, brown, black, dun, chestnut, palomino, roan, and some leopard and blanket markings. | Average to 51 in. high, with a range of 11 to 14 hands. | An all-purpose horse, suitable for children and medium-sized adults. | Pintos and animals with large markings are disqualified. |
| American Mustang | Along the Barbary Coast of North Africa. From here, they were taken to Spain by the conquering Moors, propagated in Anadalusia, and brought to America by the conquistadors. The American Mustang Assn., Inc., was formed in 1962. | Any color. | Historic origin, along with hardiness and versatility. | Pleasure riding, show, trail riding, endurance trials, stock horses, and jumping. | They must be between 13–2 and 15 hands high. |

*(Continued)*

**TABLE 51–3** (Continued)

| Breed | Place of Origin | Color | Other Distinguishing Characteristics | Primary Uses | Disqualifications; Comments |
|---|---|---|---|---|---|
| *Light Horses and Ponies* (Continued) | | | | | |
| American Saddle-bred | U.S.A.; in Fayette County, Kentucky. | Bay, brown, chestnut, gray, roan, black, or golden. Gaudy white markings are frowned upon. | Ability to furnish an easy ride, with great style and animation. Long, graceful neck and proud action. | Three- and five-gaited saddle horses, fine harness horses, pleasure horses, and stock horses. | |
| American Walking Pony | U.S.A.; near Macon, Georgia, in 1968. | No color coat stipulation. Since it is a cross between the Welsh Pony and Tennessee Walking Horse, the colors of both parent breeds occur. | They range in height from 13 to 14–2 hands. Ponies that perform the running walk gait. | Pleasure riding; children's or small adults' mounts. | Standing over 14–2 hands; not being able to perform at the running walk. |
| American White[1] | U.S.A.; White Horse Ranch, Naper, Nebraska. | Snow-white hair, as white as clean snow; pink skin; light blue, dark blue (near black), brown, or hazel eyes. | Both horses and ponies are accepted for registry; animals above 14–2 hands are classed as horses, those below 14–2 are classed as ponies. | Pleasure horses; trained horses for exhibition purposes; parade and flag-bearer horses. | Pink eyes. |
| Andalusian | Spain. From desert-bred Barbs (introduced by the invading Moors) crossed on the light, agile horses of southern Spain. | White, grays, and bays most common. Also, there are a few blacks, roans, and chestnuts. | Andalusians stand 14–2 to 16 hands and weigh from 1,000 to 1,200 lb. | Bullfighting, parade horses, dressage, jumping, and pleasure riding. | Animals not tracing to the Spanish Registry, which is supervised by the Army in Spain, are not eligible for registry. |
| Appaloosa | U.S.A.: in Oregon, Washington, and Idaho; from animals originating in Fergana, Central Asia. | Variable, but usually white over the loin and hips, with dark round or egg-shaped spots thereon. | The eye is encircled by white, the skin is mottled, and the hoofs are striped vertically black and white. | Stock horses, pleasure horses, parade horses and racehorses. | Not having Appaloosa characteristics; draft horse, pony, Albino, Paint, or Pinto breeding; gray or nonappaloosa roan, or progeny of these colors; cryptorchids; or under 14 hands at 5 years of age. |
| Arabian | Arabia | Bay, gray, and chestnut with an occasional white or black. White marks on the head and legs are common. The skin is always dark. | A beautiful head, short coupling, docility, great endurance, and airy way of going. | Saddle horses, show purposes, pleasure horses, stock horses, and racing. | |
| Buckskin | U.S.A. | Buckskin, red dun, grulla (mouse dun). International Buckskin Horse Assn.: Dorsal strip, leg barring, shoulder stripe or shadowing; black ear tips; hazel eyes; cobwebbing on face, frosted mane and tail. | American Buckskin Registry Assn., Inc.: All types accepted. Horses must be over 54 in., ponies under 54 in. | Stock horses, pleasure horses, and show purposes. | American Buckskin Registry Assn., Inc., disqualifications: Palominos, chestnuts, sorrels, or bays with dorsal stripe; draft type; blue or glass eyes; white spots on body (indicating Pinto or Appaloosa blood) or white markings above the knees or hocks. International Buckskin Horse Assn. disqualifications: Excessive white; showing paint, pinto, Appaloosa, draft type, or pony characteristics. |
| Chickasaw | U.S.A. Developed by the Chickasaw Indians (hence, the name of the breed) of Tennessee, North Carolina, and Oklahoma, from horses of Spanish extraction. | Bay, black, chestnut, gray, roan, sorrel, or palomino. | Short head and ears; a short back; a short neck; square, stocky hips; a low-set tail; a wide chest; and great width between the eyes. | Cow ponies. | Preferred height: 53 to 59 in. |
| Connemara Pony | Ireland; along the west coast. | Gray, black, bay, brown, dun, cream, with occasional roans and chestnuts. | Range in height. So, the society registers in 2 sections: (1) *pony*, 13 to 14–2 hands; and (2) *small horse*, over 14–2 hands. | Jumpers, showing under saddle and in harness; for both adults and children. | Piebalds, skewbalds, and cream with blue eyes not accepted. |
| Galiceno | Galicia, a province in northwestern Spain. | Solid colors prevail. Bay, black, chestnut, dun, gray, brown, and palomino are most common. | Intermediate in size; at maturity they stand 12 to 13–2 hands and weigh 625 to 700 lb. | Riding horses. | Albinos, pintos, and paints are ineligible for registry. Cryptorchids or monorchids. |

*(Continued)*

## TABLE 51-3 (Continued)

| Breed | Place of Origin | Color | Other Distinguishing Characteristics | Primary Uses | Disqualifications; Comments |
|---|---|---|---|---|---|
| *Light Horses and Ponies* (Continued) | | | | | |
| **Hackney** | England; on the eastern coast, in Norfolk and adjoining counties. | Chestnut, bay, and brown most common; roans and blacks are seen. White marks are common and are desired. | In the show-ring, custom decrees that Hackney horses and ponies be docked and have their manes pulled. High natural action. | Heavy harness or carriage horses and ponies. For cross-breeding purposes to produce hunters and jumpers. | Animals of piebald or skewbald color not eligible for registry. |
| **Haflinger** | Near the village of Hafling, which was a part of Austria to the end of World War I, now a part of Italy. | Chestnut, ranging from honey blond to dark chocolate, with white or flaxen mane and tail. | The Haflinger is a relatively small, stocky horse, standing from 13 to 14-3 hands high. It is noted for beauty, strength, vitality, intelligence, and a good disposition. | Haflingers are adapted to, and used for, either work or pleasure. In the U.S., they are used for riding and other light horse purposes. | Foals born to registered Haflingers are ineligible for registration if either sire or dam had not achieved the age of three years at the time of breeding. |
| **Hanoverian** | Germany, in the Hanover section, beginning in 1732. They were developed for the purpose of providing a superior horse for military use, with emphasis on size, intelligence, and temperament. In 1978, the American Hanoverian Society was incorporated. | Variable. | Hanoverian horses are big and powerful. Many of them stand 16½ hands or better and weigh 1,200 lb or more. They combine nobility, size, and strength in a unique way. | In Europe, they are used for riding, driving (carriage horses), hunting, jumping, dressage, and utility purposes. In the U.S., the breed is used for all light horse purposes, especially for hunting, jumping, and dressage. | Without doubt, the breeding and selection program followed with Hanoverian horses in Germany is the finest equine production-testing program in the world. |
| **Hungarian Horse** | Hungary | All colors, either solid or broken. | Unique combination of style and beauty with ruggedness. | Stock horses, cutting horses, pleasure horses, trail riding, hunters, and jumpers. | Cryptorchids; glass eyed. The Hungarian Horse Association of the U.S. was formed in Sept., 1966. |
| **Lipizzan** | In Lippiza, Yugoslavia, the town from which the breed takes its name. | Most mature animals are white, but foals are born dark brown or gray, then turn white at 4 to 6 years of age. About 1 in 600 remains black or brown throughout life. When the latter happens, it is considered good luck. | An elastic walk, with considerable knee action. | Dressage (for which purpose they are without a peer), harness horses, pleasure horses, hunters, jumpers, and parade horses. | Glass eyes, extreme Roman nose, or deformed or crooked limbs. |
| **Miniature Horse** | The Miniature Horse is a new breed, with an old history. Small horses were used in England and Northern Europe to pull ore carts in the coal mines as early as 1765. They were also used as pets by some of the royal families of Europe. They were brought to the U.S. in the late 19th century and used in the mines of West Virginia and Ohio. | All colors are accepted. | Miniature Horses cannot exceed 34 in. at the withers. They are of two types: (1) the more refined Arabian type, and (2) the heavier Quarter Horse type. | Miniature Horses are used as pets, and as driving horses—pulling various small vehicles. | Animals standing more than 34 in. in height when measured from the base of the last hairs of the mane to the ground while standing squarely are not eligible for registration. A dwarf is not eligible for registration. |
| **Missouri Fox Trotting Horse** | U.S.A.; in the Ozark Hills of Missouri and Arkansas. | Sorrels predominate, but any color is accepted. | The fox-trot gait. | Pleasure horses, stock horses, and trail riding. | If animal cannot fox trot. In 1948, the Missouri Fox Trotting Horse Breed Assn. was formed. |
| **Morab** | Although Morgan × Arabian crosses were made in the early 1800s, the name *Morab* was coined in the 1920s when William Randolph Hearst crossed Arabian stallions on Morgan mares to produce horses for his San Simeon ranch in California. | Bay, black, brown, buckskin, chestnut, dun, gray grulla, palomino, or roan. | The breed possesses the muscular strength and ruggedness of the Morgan and the refinement and beauty of the Arabian. | Show purposes, pleasure riding, endurance rides, and ranch work. | In order to be eligible for registry, Morabs must (1) not show more than 75%–25% Morgan or Arabian bloodlines; and (2) not show any patterns, spots, or similar white above the knee or hock. |
| **Morgan** | U.S.A.; in the New England states. | Bay, brown, black, and chestnut; extensive white markings are uncommon. | Easy keeping qualities, endurance, and docility. | Saddle horses, and stock horses. | Wall-eye (lack of pigmentation of the iris), or natural white markings above the knee or hock except on the face. |

*(Continued)*

**TABLE 51–3** *(Continued)*

| Breed | Place of Origin | Color | Other Distinguishing Characteristics | Primary Uses | Disqualifications; Comments |
|---|---|---|---|---|---|
| *Light Horses and Ponies* (Continued) | | | | | |
| **National Appaloosa Pony** | U.S.A.; near Rochester, Indiana. | Vari-colored, but leopard, blanket-type, snow-flake, and roan are most popular. The skin, nose, and area around the eyes are mottled. White sclera encircles the eyes. | Appaloosa color, and standing under 14–2 hands. | Working ponies; show ponies; and for trail riding, jumping, and racing. | Albino-, pinto-, or paint-colored animals not eligible for registration. |
| **National Show Horse** | In the U.S., the National Show Horse Registry was conceived and incorporated by Gene La Croix, a prominent Arabian breeder, located in Arizona, in 1981. | All colors common to Arabians and Saddlebreds, the parent breeds of the National Show Horse. | The National Show Horse combines the beauty, refinement, and stamina of the Arabian and the size and high-stepping action of the Saddlebred. The ideal National Show Horse is long-necked, pretty-headed, and relatively short-backed, with long legs and a high, animated trot. | The goal of the breed founders was to produce the ideal show horse. National Show Horses are shown in the following classes: English pleasure, three-gaited, five-gaited, pleasure driving, fine harness, equitation, and country pleasure. | |
| **Norwegian Fjord Horse** | In Norway, where they have been selectively bred for more than 2,000 years. | Dun colored (ranging from light to dark), with a dorsal stripe, dark bars on the legs, and dark hooves, eyes, and ear tips. | The body is compact and muscular. They range in height from 13 to 14–2 hands. | Throughout Europe, the Norwegian Fjord Horse is used for riding, driving, and draft purposes. In the U.S., Fjord Horses are used for draft, carriage, pleasure riding, jumping, dressage, and trail riding. | The following infraction of rules will disqualify Norwegian Fjord Horses for registry: 1. Horses resulting from inbreeding (mother to son, father to daughter, brother to sister, or ½ sister to ½ brother) after Jan. 2, 1986. 2. A Fjord Horse stallion or mare bred to a non-Fjord will lose its registered status. |
| **Paint Horse** | U.S.A. | White plus any other color. Must be a recognizable paint. Paint Horses are distinguished by two color patterns—overo or tobiano. | No discrimination is made against glass, blue, or light-colored eyes. | Stock horses, pleasure horses, show purposes, and racing. | Lack of natural white markings above the knees or hocks except on the face; horses with Appaloosa color or blood; adult horses under 14 hands; 5-gaited horses. |
| **Palomino** | U.S.A.; from animals of Spanish extraction. | Golden (the color of a newly minted gold coin, or 3 shades lighter or darker), with a light-colored mane and tail (white, silver, or ivory, with not more than 15% dark or chestnut hair in either). White markings on the face or below the knees are acceptable. Dark skin. | | Stock horses, parade horses, pleasure horses, saddle horses, and fine harness horses. | The Palomino Horse Breeders of America lists the following disqualifications: a dorsal stripe of brown or black along the spine, zebra stripes around the legs or across the wither, or patches of white hair with underlying pink or light skin, unless caused by injury. The Palomino Horse Assn., Inc., lists the following disqualifications: buckskin, chocolate, or sorrel color; blue or chalk eyes; a dorsal stripe along the spine; horses with white or dark spots to indicate Pinto, Appaloosa, or Paint background; horses with known Albino breeding; horses with characteristics of draft horses. |
| **Paso Fino** | In the Caribbean area; mostly Puerto Rico, Colombia, the Dominican Republic and Cuba. | Any color, although solid colors are preferred. Bay, chestnut, and black with white markings are most common. Occasionally, palominos and pintos appear. | The paso fino gait may be described as a broken pace. The legs on the same side move together, but the hind foot strikes the ground a fraction of a second before the front foot, producing a 4-beat gait. They range in height from 13–2 to 15–2 hands. | Pleasure, cutting, and parade horses; and for endurance riding and drill team work. | Animals that do not possess the paso fino gait or do not trace directly to the purebred paso fino ancestry are not eligible for registration. |

*(Continued)*

## TABLE 51-3 *(Continued)*

| Breed | Place of Origin | Color | Other Distinguishing Characteristics | Primary Uses | Disqualifications; Comments |
|---|---|---|---|---|---|
| *Light Horses and Ponies* (Continued) | | | | | |
| **Peruvian Paso** | Peru. | Any color, although solid colors are preferred. | Breed is characterized by the piso, or gait, a natural four-beat, lateral gait—a broken pace. Breed does only two acceptable gaits—the paso llano and sobreandando. | Pleasure horses, parade horses, and endurance horses. | Peruvian Paso Horse Registry of North America: cryptorchids, monorchids, and albino animals. American Assn. of Owners & Breeders of Peruvian Paso Horses disqualifies horses with more than 25% body white, including on legs and head. |
| **Pinto Horse** | U.S.A.; from horses brought in by the Spanish conquistadors. | Preferably half color or colors and half white, with many spots well placed. The two distinct pattern markings are: overo and tobiano. | Glass eyes are not discounted. Assn. has separate registry for ponies and/or horses under 14 hands. | Any light horse purpose, but especially for show, parade, novice, pleasure purposes, stock horses. | Animals with Appaloosa ancestry or color, are of known draft horse breeding, are ineligible for registry. |
| **Pony of the Americas** | U.S.A.; Mason City, Iowa. | Similar to appaloosa; white over the loin and hips, with dark, round or egg-shaped spots. | Happy medium of Arabian and Quarter Horse in miniature, ranging in height from 46 to 54 in., with appaloosa color. | Children's western type using pony. | Ponies possessing any of the following characteristics are disqualified for registry: pinto or paint color, characteristics, or parents; white markings with underlying light skin (1) above the knees or hocks, (2) behind a line running from the center of each ear to the corner of each side of the mouth, or (3) on the lower lip above a line running from one corner of the mouth to the other corner. Animals that mature (at 6 years of age) over 56 in. high or under 46 in. are eligible for transfer to breeding stock only category. |
| **Quarter Horse** | U.S.A. | Chestnut, sorrel, bay, and dun are most common, although they may be palomino, black, brown, roan, or copper colored. | Well muscled and powerfully built. Small, alert ears; sometimes heavily muscled cheek and jaw. | Stock horses, racing, and pleasure horses. | Paint, Pinto, Appaloosa, or albino colors are ineligible for registry; also, white markings beyond prescribed lines, excessive white markings or spots shall be subject to cancellation. |
| **Quarter Pony** | In the United States, in 1975. | The same colors as Quarter Horses, from which they descend. | Quarter Ponies possess the same characteristics as Quarter Horses, only they are smaller. | Mounts for children and juniors; and used for all the purposes for which their larger counterparts—Quarter Horses—are used. | Quarter Ponies cannot measure more than 14 hands at two years of age. Registration is not permitted of pinto, paint, or Appaloosa coloring, or for other excessive spotting from the ears back. |
| **Shetland Pony** | Shetland Isles | All colors, either solid or broken. | Small size. Two types are promoted by the Shetland Pony Club: (1) the classic type, which is short and chunky, and (2) the modern type, which is fine boned, long necked and high going. | Children's mounts, harness horses, roadsters, and racing. | Over 46 in. in height. |
| **Spanish Barb** | Barb horses were taken from Africa to Spain with the conquest of Spain by the Moors in 711 A.D. From Spain, they were taken to Cuba in 1511, to Mexico in 1519, to southwestern U.S. in 1540, and to Florida in 1565. The Spanish Barb Breeders Assn. was organized in 1972. | All colors are represented in the breed; but dun, grulla, sorrel, and roan are most common. Most animals are solid colored. A dorsal stripe and zebra markings occur in all duns and grullas and in some sorrels. | Spanish Barbs are small horses (the standard height is 13-3 to 14-3 hands), with short coupling (usually 5 lumbar vertebrae; at times 17 thoracic vertebrae), deep bodies, good action, and without extreme muscling. | Cow ponies, western riding, English riding, and packhorses. | |

*(Continued)*

**TABLE 51-3** *(Continued)*

| Breed | Place of Origin | Color | Other Distinguishing Characteristics | Primary Uses | Disqualifications; Comments |
|---|---|---|---|---|---|
| *Light Horses and Ponies* (Continued) | | | | | |
| **Spanish Mustang** | U.S.A. Trace to feral and semiferal (Indian-owned) horses of Barb and Andalusian ancestry brought to America by the Spanish in the early 1500s and 1600s. Beginning about 1925, Robert E. Brislawn, Sr., and Ferdinand L. Brislawn began gathering pure Spanish Mustangs. Breed registry was founded in 1957, at Sundance, WY. | The whole gamut of colors, including all the solid colors and all the broken colors. | Some have 5 to 5½ lumbar vertebrae. Stand 13-2 to 15-2 hands. | Cow ponies, western riding, English riding, and packhorses. | |
| **Spotted Saddle Horse** | In the United States, primarily in Middle Tennessee. | Spotted. | Animals of this breed must be spotted, and they must be saddle horses. In addition to color, the breed is characterized for its ruggedness, intelligent eyes, fox-type ears, surefootedness, and good dispositions. | Pleasure horses, jumpers, and as mounts for bird hunters and for coon hunters. | Animals are disqualified if they are not spotted, and if they are not saddle horses. |
| **Standardbred** | U.S.A. | Bay, brown, chestnut, and black are most common; but grays, roans, and duns are found. | Smaller, less leggy and with more substance and ruggedness than the Thoroughbred. | Harness racing, either trotting or pacing; and harness horses in horse shows. | The Ridden Standardbred Association was formed in 1983 and is located at Troy, Ohio. It promotes Standardbreds for uses other than sulky racing; and it registers purebred, half-bred, and partial-bred Standardbreds. |
| **Tennessee Walking Horse** | U.S.A.; in the Middle Basin of Tennessee. | Sorrel, chestnut, black, roan, white, bay, brown, gray, or golden. White markings on the face and legs are common. | The running walk gait. | Plantation walking horses, pleasure horses, and show horses. | |
| **Thorcheron** | In the United States, in Michigan, beginning in 1970, with Thoroughbred mares mated to a Percheron stallion. | | The Thorcheron is a hunter/jumper in type, ranging in size from 16 to 17 hands and averaging about 1,250 lb. Thorcherons are short-coupled and good movers, with the refinement of Thoroughbreds and the muscle and good nature of Percherons. They resemble the big Irish hunters. | Hunters and jumpers. | |
| **Thoroughbred** | England. The first edition of the *General* (English) *Stud Book* was published in 1793. | Bay, brown, chestnut, and black; less frequently, roan and gray. White markings on the face and legs are common. | Fineness of conformation. Long, straight, and well-muscled legs. | Racing, stock horses, saddle horses, polo mounts, and hunters. | All Thoroughbreds in the world trace to 3 stallions; The Darley Arabian, the Byerly Turk, and the Godolphin Barb. |
| **Trakehner** | In Trakehnen, East Prussia, in 1732. It evolved from the blending of the indigenous Prussian horses, Thoroughbreds, and Arabians. | | Horses with the size of the Thoroughbred, but more rugged, and possessing the elegance of the Arabian. | | |
| **Trottingbred** | In the United States, in the 1960s. | | Trottingbreds are miniature Standardbreds, with a dash of pony breeding. They must not measure more than 13-1 hands. | Harness racing. Trottingbred harness racing is do-it-yourself and family oriented. | Animals measuring more than 51½ in. are disqualified. Full Standardbreds are not allowed to compete in Trottingbred races. |

*(Continued)*

TABLE 51–3 (Continued)

| Breed | Place of Origin | Color | Other Distinguishing Characteristics | Primary Uses | Disqualifications; Comments |
|---|---|---|---|---|---|
| **Light Horses and Ponies** (Continued) | | | | | |
| **Welara Pony** | In the United States, from Welsh × Arabian crosses, with the breed registry opened in 1981. | All the colors common to the two parent breeds, Welsh and Arabian. Any color except Appaloosa is acceptable. | It resembles a miniature coach horse. They range from 46 to 58 in. in height. The Welara is noted for its beauty. | Fine harness, English pleasure, halter, hunter, and native costume classes. Also, in competitive trail rides. | Appaloosa color. |
| **Welsh Pony and Cob** | Wales. | Any color except piebald and skewbald. Gaudy white markings are not popular. | Small size; intermediate between Shetland Ponies and other light horse breeds. The Welsh Pony and Cob Society of America maintains the following divisions, according to height stipulations: 1. The Welsh Mountain Pony, Section A of the Stud Book, for Welsh not exceeding 12–2 hands. 2. The Welsh Pony, Section B of the Stud Book, for Welsh over 12–2 but under 14–2. 3. The Welsh Pony (Cob Type) and Welsh Cob, respectively. Section C is For Welsh not exceeding 13–2 hands, without a lower limit. Section D is for Welsh exceeding 13–2 hands, with no upper limit on height. | Mounts for children and small adults; racing, roadsters, trail riding, parade horses, stock cutting, and hunting. | Piebald or skewbald. |
| **Draft Horses:** | | | | | |
| **American Cream** | U.S.A. | Cream, with white mane and tail and pink skin. Some white markings. | Medium size. Good disposition. | Farm work horses and exhibition purposes. | |
| **Belgian** | Belgium. | Bay, chestnut, and roan are most common, but browns, grays, and blacks are occasionally seen. Many Belgians have a flaxen mane and tail and a white-blazed face. | Lowest set and most massive of all draft breeds. | Farm work horses, and exhibition purposes. | Side bones or curbs disqualify an animal for registry. |
| **Clydesdale** | Scotland; along the River Clyde. | Bay and brown with white markings are most common; but blacks, grays, chestnuts, and roans are occasionally seen. | Superior style and action. Feather or hair on the legs. | Farm work horses, and exhibition purposes. | |
| **Percheron** | France; in the northwestern district of La Perche. | Mostly black or gray; but bays, browns, chestnuts, and roans are seen. | In comparison with other draft breeds, noted for its handsome clean-cut head. | Farm work horses, and exhibition purposes. | |
| **Shire** | England; primarily in the east central counties of Lincolnshire and Cambridgeshire. | Common colors are bay, brown, and black with white markings; although grays, chestnuts, and roans are occasionally seen. | Taller than any other draft breed. Feather or hair on the legs. | Farm work horses, and exhibition purposes. | |
| **Suffolk** | England; in the eastern county of Suffolk. | Chestnut only. | They are the smallest of the draft breeds. Close-to-the-ground and chunky build. | Farm work horses, and exhibition purposes. | Any color other than chestnut. |
| **Jacks, Donkeys, and Mules** | | | | | |
| **Mammoth Jack** (or American Standard Jack) | Southern Europe, mostly from the area bordering on the Mediterranean. | Variable, but mostly black with white nose, gray, or red/sorrel. | Mammoth is the largest of the asses. Jacks should be 56 in. or over in height, and jennets should be 54 in. or over in height. | Production of mules. | Mammoth is a blend of several breeds of jack stock imported from Southern Europe in the 1800s. |

*(Continued)*

**TABLE 51–3** *(Continued)*

| Breed | Place of Origin | Color | Other Distinguishing Characteristics | Primary Uses | Disqualifications; Comments |
|---|---|---|---|---|---|
| *Jacks, Donkeys, and Mules* (Continued) | | | | | |
| **Large Standard Donkey** (Spanish Donkey) | Southern Europe. | Variable, but mostly black with white nose, gray, or red/sorrel. | Males stand from 48.01 to 56 in. high and females from 48.01 to 54 in. high. | Production of mules. | |
| **Standard Donkey** (Burro) | From donkeys throughout the world. | Variable, but many are mouse colored. | They range in height from 36.01 in. to 48.0 in. at the withers. | Driving, riding, working, and as pets. | The term *burro* refers to an un-improved member of the ass family. |
| **American Spotted Ass** | Throughout the world. | Spotted. | A blending of all breeds and sizes. | Pets and show. | They must be spotted, and they must be registered with the American Council of Spotted Asses. |
| **Miniature Mediterranean Donkey** | Sicily and Sardinia. | Dorsal stripe, forming a cross with stripe over withers and down shoulders. | They must be under 36 in. at the withers to be registered by the American Donkey and Mule Society as miniature. | Pets. | Over 36 in. high. Without the cross. |
| **Mule** (hinny) | The mule is a hybrid, a cross between the horse and ass. | Variable color, but sorrel is the preferred color. | The mule resembles his sire, the jack, more than the mare. But the desired conformation is identical to that of a horse for similar use, with more stress placed upon size, and set and quality of ear. | Driving, riding, pack-horses, working, and show purposes. | |

[1]Both the American Creme and the American White are registered by the International American Albino Assn., Naper, NE, with separate divisions provided for each.

Fig. 51–17.  Akhal-Teke stallion, *Senetir.* (Courtesy, The Akhal-Teke Assn. of America, Staunton, VA)

Fig. 51–18.  American Bashkir Curly mare. (Photo by Georgia Cheer, Madera, CA)

Fig. 51-19. American Creme stallion, *Polar Bar,* owned by Edward Bales, Prineville, OR. (Courtesy, International American Albino Assn., Naper, NE)

Fig. 51-20. American Gotland mare, *Gerta,* owned by Krona Horse Farms, Columbia, MO. (Courtesy, Krona Horse Farms)

Fig. 51-21. American Mustang mare, *Lochinvar's Jubilation,* owned by Nancy Bunge, Norco, CA. (Courtesy, American Mustang Assn., Yucaipa, CA)

Fig. 51-22. American Saddlebred stallion, *Kalarama Rex,* one of the great stallions of the breed. (Courtesy, American Saddle Horse Breeders Assn., Louisville, KY)

Fig. 51-23. American Walking Pony stallion, *BT Golden Splendor,* owned by Browntree Stables, Macon, GA. (Courtesy, American Walking Pony Assn. Macon, GA)

Fig. 51-24. American White Stallion, *Westwind Zepherous.* (Courtesy, Karen S. Wales, Taylor Ridge, IL)

Fig. 51-25. Andalusian stallion, *Ofendido VII*, owned by Karen Jenkins, Nashville, TN. (Courtesy, American Andalusian Horse Assn., Springfield, OH)

Fig. 51-26. Appaloosa stallion, *Romanno*, owned by James and Linda McKay, Purcell, OK. (Courtesy, Appaloosa Horse Club, Moscow, ID)

Fig. 51-27. *AM Canadian Beau\**, prominent stallion at Al-Marah Arabians, winner of many championships in both halter and performance. (Photo by Louise L. Serpa, Tucson, AZ. Courtesy, Mrs. Garvin Tankersley, Al-Marah Arabians, Tucson, AZ)

Fig. 51-28. Buckskin stallion, *Leo Reno*, a golden buckskin with dorsal stripe and dark ear tips and ear outline, owned by Holiday Farm, Dyer, IN. (Courtesy, International Buckskin Horse Assn., Shelby, IN)

Fig. 51-29. Chickasaw gelding, *Johnny Reb*, national high point Chickasaw horse, owned by Mr. and Mrs. Roy Hughes, Clarinda, IA. (Courtesy, Mr. and Mrs. Hughes)

Fig. 51-30. Connemara Pony mare, *Oak Hills Miss Independence.* (Courtesy, Mavis Connemara Farm, Rochester, IL)

Fig. 51-31. Galiceno mare, *Vestido Azal,* owned by Glenn H. Bracken, Tyler, TX. (Courtesy, Galiceno Horse Breeders Assn., Godley, TX)

Fig. 51-32. Hackney stallion, *Creation's King.* (Courtesy, USDA)

Fig. 51-33. Haflinger yearling. (Courtesy, Haflinger Assn. of America, Hemlock, MI)

Fig. 51-34. Hungarian Horse stallion, *Hungarian Kallo,* owned by Bitterroot Stock Farm, Hamilton, MT. (Photo by Ernst Peterson, Hamilton, MT. Courtesy, Bitterroot Stock Farm)

Fig. 51-35. Lipizzan stallion, *Maestoso Bregova II.* (Courtesy, Evelyn L. Dreitzler, Snohomish, WA)

Fig. 51-36. Miniature Horse stallion, *FWF Little Blue Boy.* He measures 30 in. at the withers. (Courtesy, Fredericka Wagner, Flying W Farms, Piketon, OH)

Fig. 51-37. Missouri Fox Trotting Horse stallion, *Yankee Digger Joe*, owned by Eddie Atchison, Mountain Grove, MO. (Courtesy, Missouri Fox Trotting Horse Breed Assn., Ava, MO)

Fig. 51-38. Morab mare, *B.C.M. Qizan Star*, owned by Andrea Rowland, Walnut Creek, CA. (Courtesy, North American Morab Horse Assn., Hilbert, WI)

Fig. 51-39. Morgan stallion, *Rex's Major Monte*. (Courtesy, American Morgan Horse Assn., Shelburne, UT)

Fig. 51-40. Norwegian Fjord stallion, *Montano*, with band of Norwegian Fjord mares in the adjacent corral. (Courtesy, Norwegian Fjord Assn. of North America, Grayslake, IL)

Fig. 51-41. Palomino mare, *Palolena*, owned by Tom Abbot, Jr., Fort Worth, TX. (Courtesy, Palomino Horse Breeders of America, Tulsa, OK)

Fig. 51-42. Paint Horse mare, *Baby Doll*, a chestnut and white *Tobiano*, owned by Bill and Bonnie Bailey, Kent, OH. (Courtesy, American Paint Horse Assn., Fort Worth, TX)

Fig. 51-43.   Paso Fino stallion, *Latines de A-USA,* owned by Angel and Esther Usateaui, Miami, FL. (Courtesy, Paso Fino Horse Assn., Bowling Green, FL)

Fig. 51-44.   Peruvian Paso stallion.

Fig. 51-45.   Pinto Horse. (Courtesy, Pinto Horse Assn. of America, Ft. Worth, TX)

Fig. 51-46.   Pony of the Americas stallion, *Siri Chief,* owned by Paula Cooper, Willcox, AZ. (Courtesy, Pony of the Americas Club, Indianapolis, IN)

Fig. 51-47   Quarter Horse halter champion, *Devious Skip.* (Courtesy, *The Quarter Horse Journal,* Amarillo, TX)

Fig. 51-48.   Shetland Pony mare, *T.J.'s Take A Chance,* owned by Jean Brown, Singletree Farm, Rocky Ridge, MD. (Courtesy, Jean Brown)

Fig. 51-49. Spanish Barb stallion, *Little Bit*. (Courtesy, Wild Horse Research Farm, Porterville, CA)

Fig. 51-50. Spanish Mustang stallion, *Syndicate*, owned by Cayuse Ranch, Oshoto, WY. (Courtesy, Cayuse Ranch)

Fig. 51-51. *Adios*, all-time great Standardbred sire. (Photo by George Smallsreed, Jr. Courtesy, United States Trotting Assn., Columbus, OH)

Fig. 51-52. Tennessee Walking Horse stallion, *Go Boy's Shadow*, twice champion at the Walking Horse National Celebration. (Courtesy, USDA)

Fig. 51-53. Thorcheron mare, *Kinsorbia*, owned by Thorcheron Breeding Farm, Kalamazoo, MI. (Courtesy, Thorcheron Breeding Farm)

Fig. 51-54. Thoroughbred yearling by *Sea Bird II*, sold for $405,000. (Photo by J. Moye, Versailles, KY. Courtesy, *Thoroughbred Record*, Lexington, KY)

Fig. 51-55. Welsh Pony stallion, *Sleight of Hand,* imported from United Kingdom, owned by Mr. and Mrs. J. D. Morris, Hughes, AZ. (Courtesy, Welsh Pony & and Cob Society of America, Winchester, VA)

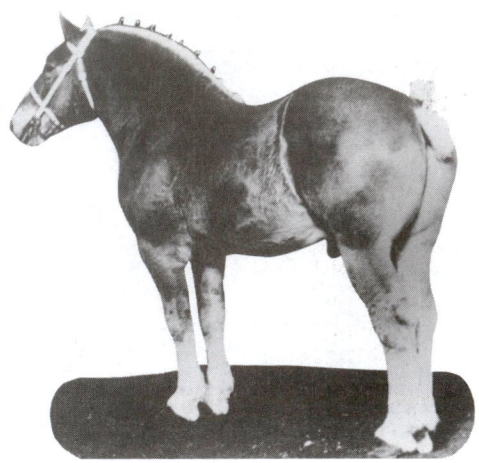

Fig. 51-56. Belgian stallion. *Congressman,* shown as a 6-year-old, standing 18-2 hands and weighing 2,200 lb. *Congressman* was many times Grand Champion Belgian Stallion through the United States and Canada, including the Royal Agricultural Winter Fair, Toronto, Canada, in 1974. Bred and owned by Mr. Harold Clark, Millersburg, IN. (Courtesy, Harold Clark)

Fig. 51-57. Clydesdale stallion, *Benefactor,* imported from Scotland. (Courtesy, Clydesdale Breeder's Assn. of United States, Pecatonica, IL)

Fig. 51-58. Percheron stallion, *Maverick,* winner of 10 Grand Championships, including the National Percheron Show and the Canadian Royal Winter Fair. He stands 18 hands high and weighs 2,200 lb. *Maverick* is owned by Art Bast, Hartford, WI. (Courtesy, Art Bast and family)

Fig. 51-59. Shire stallion breed champion.

Fig. 51-60. Suffolk stallion, imported from England. (Courtesy, American Suffolk Horse Assn., Inc., Ledbetter, TX)

Fig. 51-61. Mammoth Jack (or American Standard Jack) of fine type, with a smoothly turned body.

Fig. 51-62. Large Standard Donkey (Spanish Donkey), *Fairview's Mister Jack.* (Courtesy, American Donkey and Mule Society, Denton, TX)

Fig. 51-63. Miniature donkeys. (Courtesy, Danby Farm, Millard, NE)

# QUESTIONS FOR STUDY AND DISCUSSION

1. Compare horses from the standpoint of the diverse and distinct types that have been developed with each: beef cattle, sheep, and swine.

2. Describe each of the following classes of horses according to size, build, and use: light horses, ponies, and draft horses.

3. Define a breed of horse.

4. What one breed of light horses do you consider to be best adapted to each of the specific uses listed in Table 51-1, page 955?

5. What accounts for the fact that so many light horse breeds were American creations, but that only one draft breed originated in this country?

6. In outline form, list the (a) place of origin, (b) distinguishing characteristics, and (c) disqualifications of each

breed of light and draft horses; then discuss the importance of each of these listings.

7. Is there need to develop one or more new breeds of light horse for certain specific uses? If so, to what uses does this apply?

8. What specific uses of light horses have no utility value; that is, they are for show only?

9. How do you account for the fact that some folks prefer donkeys or mules to horses—for riding, driving, showing, and as pets?

10. Justify any preference that you may have for one particular breed of horse.

11. Obtain breed registry association literature and a sample copy of a magazine of your favorite breed of horses. (See Appendix Tables A-8 and A-9 for addresses.) Evaluate the soundness and value of the material that you receive.

# SELECTED REFERENCES

| Title of Publication | Author(s) | Publisher |
|---|---|---|
| *America's Quarter Horses* | P. Laune | Doubleday & Company, Inc., Garden City, NY, 1973 |
| *Appaloosa* | F. Haines | Amon Carter Museum of Western Art, Fort Worth, TX, 1963 |
| *Appaloosa Horse, The* | F. Haines G. B. Hatley R. Peckinpah | R. G. Bailey Printing Company, Lewiston, ID, 1957 |
| *Arabian Horse Breeding* | H. H. Reese | Borden Publishing Co., Los Angeles, CA, 1953 |
| *Breeding and Raising Horses, Ag. Hdbk. No. 394* | M. E. Ensminger | Agricultural Research Service, USDA, Washington, DC, 1972 |
| *Breeding and Rearing of Jacks, Jennets and Mules, The* | L. W. Knight | The Cumberland Press, Nashville, TN, 1902 |
| *Encyclopedia of the Horse, The* | Edited by C. E. Hope G. N. Jackson | The Viking Press, New York, NY, 1973 |
| *Foundation Sires of the American Quarter Horse* | R. M. Denhart | University of Oklahoma Press, Norman, OK, 1976 |
| *Gentle Giants* | R. Whitlock | Lutterworth Press, Guildford and London, England, 1976 |
| *Harper's Encyclopedia for Horsemen: The Complete Book of the Horse* | L. Taylor | Harper & Row, Publishers, New York, NY 1973 |
| *History of American Jacks and Mules* | F. C. Mills Edited by H. L. Hall | Hutch-Line, Inc., Hutchinson, KS, 1971 |
| *History of Horse Racing, The* | R. Longrigg | Stein and Day Publishers, New York, NY, 1972 |
| *History of the Percheron Horse, A* | A. H. Sanders W. Dinsmore | Breeder's Gazette Print, Chicago, IL 1917 |
| *History of Thoroughbred Racing in America* | W. H. P. Robertson | Prentice-Hall, Inc., Englewood Cliffs, NJ, 1965 |
| *Horse America Made, The* | L. Taylor | American Saddle Horse Breeders Association, Louisville, KY, 1961 |
| *Horseman's Encyclopedia, The* | M. C. Self | A. S. Barnes & Co., Inc., Cranbury, NJ, 1963 |
| *Horsemanship and Horse Care, Ag. Info. Bull. No. 353* | M. E. Ensminger | Agricultural Research Service, USDA, Washington, DC, 1972 |
| *Horsemanship and Horsemastership* | Edited by G. Wright | Doubleday & Company, INc., Garden City, NY, 1962 |
| *Horses* | M. C. Self | A. S. Barnes & Co., Inc., Cranbury, NJ, 1953 |
| *Horses and Horsemanship, Sixth Edition* | M. E. Ensminger | The Interstate Publishers, Inc., Danville, IL, 1990 |
| *Horses and Horsemanship* | L. E. Walraven | A. S. Barnes & Co., Inc., New York, NY, 1970 |
| *Horses of Today* | H. H. Reese | Wood & Jones, Pasadena, CA, 1956 |
| *Horse Shows* | A. N. Phillips | The Interstate Printers & Publishers, Inc., Danville, IL, 1956 |
| *Horses, Their Selection, Care, and Handling* | M. C. Self | A. S. Barnes & Co., Inc., New York, NY, 1943 |
| *Horse, The* | D. J. Kays Rev. by J. M. Kays | A. S. Barnes & Co., Inc., Cranbury, NJ, 1969 |
| *Kellogg Arabians, The* | H. H. Reese G. B. Edwards | Borden Publishing Co., Los Angeles, CA, 1958 |
| *King Ranch Quarter Horses, The* | R. M. Denhardt | University of Oklahoma Press, Norman, OK, 1978 |
| *Light Horse Breeds, The* | J. W. Patten | A. S. Barnes & Co., Inc., New York, NY, 1960 |

*(Continued)*

## SELECTED REFERENCES (Continued)

| Title of Publication | Author(s) | Publisher |
|---|---|---|
| *Light Horses*, Farmers' Bull. No. 2127 | M. E. Ensminger | U.S. Department of Agriculture, Washington, DC, 1965 |
| *Modern Breeds of Livestock*, Fourth Edition | H. M. Briggs | The Macmillan Company, New York, NY, 1980 |
| *Morgan Horse Handbook, The* | J. Mellin | Stephen Green Press, Brattleboro, VT, 1973 |
| *People with Long Ears* | R. Borwick | Cassell & Company Ltd., London, England, 1970 |
| *Percheron Horse, The* | M. C. Weld | O. Judd Co., New York, NY, 1886 |
| *Pinto, The* | | Yearbook and Studbook of the Pinto Horse Association of America, 1958-59 |
| *Rule Book* | | The American Horse Shows Association, Inc., New York, NY, annual |
| *Shetland Pony, The* | L. F. Bedell | The Iowa State University Press, Ames, Iowa, 1959 |
| *Steeplechasing* | J. Hislop | J. A. Allen & Co., Ltd., London, England, 1970 |
| *Stockman's Handbook, The*, Sixth Edition | M. E. Ensminger | The Interstate Printers & Publishers, Inc., Danville, IL, 1983 |
| *Summerhays' Encyclopaedia for Horsemen* | R. S. Summerhays | Frederick Warne & Co., Inc., New York, NY, 1966 |
| *Trotting Horse of America, The* | H. Woodruff | University Press; Welch, Bigelow, & Co., Cambridge, MA, 1871 |
| *Using the American Quarter Horse* | L. N. Sikes | The Saddlerock Corporation, Houston, TX, 1958 |
| *Western Horse, The* | J. A. Gorman | The Interstate Printers & Publishers, Inc., Danville, IL, 1967 |
| *World Dictionary of Breeds, Types, and Varieties of Livestock, A* | I. L. Mason | Commonwealth Agricultural Bureaux, Farnham House, Farnham Royal, Slough, Bucks, England, 1951 |
| *World of Pinto Horses, A* | Edited by R. D. Greene | The Pinto Horse Association of America, San Diego, CA, 1970 |

Fig. 51-64. A hunter in action, and well ridden. (Courtesy, A Mackay-Smith, Middleburg, VA)

Pause for the drink that refreshes. (Photo by Charles J. Belden, Pitchfork, WY. Courtesy, *The Western Horseman*, Colorado Springs, CO)

Select early in life! This foal displays natural aptitude for racing. (Photo by Ernst Peterson. Courtesy Bitterroot Stock Farm, MT)

# SELECTING AND
# JUDGING HORSES

$$\text{Chapter 52}$$

The great horse shows throughout the land have exerted a powerful influence in molding the types of certain breeds of light horses. Other breeds have been affected primarily through selections based on performance, such as the racetrack. It is realized, however, that only comparatively few animals are subjected annually to the scrutiny of experienced judges or to trial on the racetrack. Rather, the vast majority of them are evaluated by horse users—by those who take pride in owning a good horse and who conduct their own buying and selling operations.

Experienced livestock producers generally agree that horses are the most difficult to judge of all classes of farm animals. In addition to considering conformation—which is the main criterion in judging other farm animals—action and numerous unsoundnesses are of paramount importance. For this reason, the amateur should enlist the help of a competent equestrian when buying a horse.

## HOW TO SELECT A HORSE

When selecting a horse, the buyer must first decide what kind of horse is needed. This means that consideration should be given to the following points:

1. The mount should be purchased within a price range that the buyer can afford.

2. The amateur or child should have a quiet, gentle, well-broken horse that is neither headstrong nor unmanageable. The horse should never be too spirited for the rider's skill.

3. The size of the horse should be in keeping with the size and weight of the rider. A small child should have a small horse or pony, but a heavy person should have a horse of the weight-carrying type. Also, an extremely tall person looks out of place if not mounted on a horse of considerable height.

4. Usually the novice will do best to start with a three-gaited horse and first master the three natural gaits before attempting to ride a horse executing the more complicated gaits.

5. Other conditions being equal, the breed and color of horse may be decided on the basis of performance.

6. The mount should be suited to the type of work to be performed.

After deciding on the kind of horse needed and getting an ideal in mind, the buyer is ready to select the individual horse. Selection on the basis of body conformation and performance is the best single method of obtaining a good horse. Of course, when animals are selected for breeding purposes, two additional criteria should be considered. These are (1) the record of the horse's progeny if the animal is old enough and has reproduced, and (2) the animal's pedigree. Also, show-ring winnings may be helpful.

Proficiency in judging horses necessitates a knowledge of (1) the parts of a horse; (2) correct conformation and stance;

Fig. 52-1. Horses may be bought at private treaty or at an auction. This shows a young American Saddlebred being sold at auction at the famous Tattersalls, Lexington, Kentucky. (Courtesy, American Saddlebred Horse Assn., Kentucky Horse Park, Lexington, KY)

(3) the gaits; (4) defects in action; (5) blemishes and unsoundnesses; (6) the proper value assigned to each part (a scorecard may be used for this purpose); (7) vices; (8) how to determine age; (9) how to measure a horse; and (10) colors and markings.

## PARTS OF A HORSE

In selecting and judging horses, parts are usually referred to rather than the individual as a whole. Nothing so quickly sets a real equine expert apart from a novice as a thorough knowledge of the parts and the language commonly used in describing them. Fig. 52-2 shows the parts of a horse.

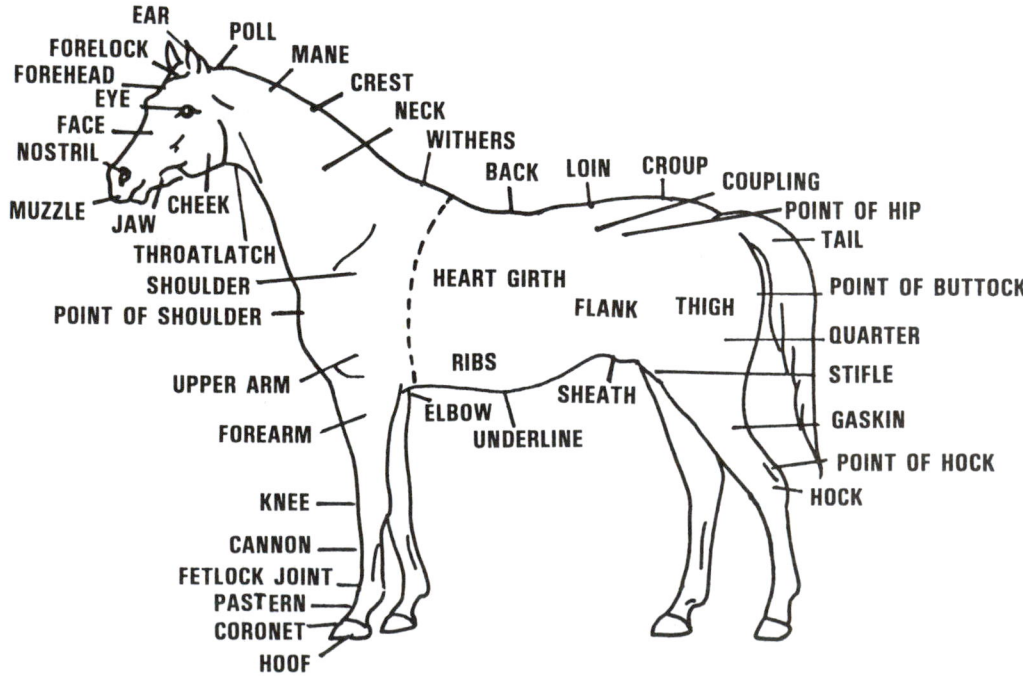

Fig. 52-2. Parts of a horse.

It is noteworthy that the knee joint in the horse is the counterpart of the wrist joint in the human; the stifle joint in the horse is the counterpart of the knee joint in the human; and the hock joint in the horse is the counterpart of the ankle joint in the human.

## CONFORMATION, STANCE, AND ACTION

A good horse must conform to the specific type which fits it for the function that it is expected to perform. Secondly, it should be true to the characteristics of the breed that it represents.

Anyone selecting a horse should have clearly in mind the ideal, recognizing full well that few animals meet this high standard and that it may be necessary to settle for less. Regardless of type or breed, however, the following desirable characteristics are sought in all horses: correct form; correct legs, feet, and pasterns; good action; and sound.

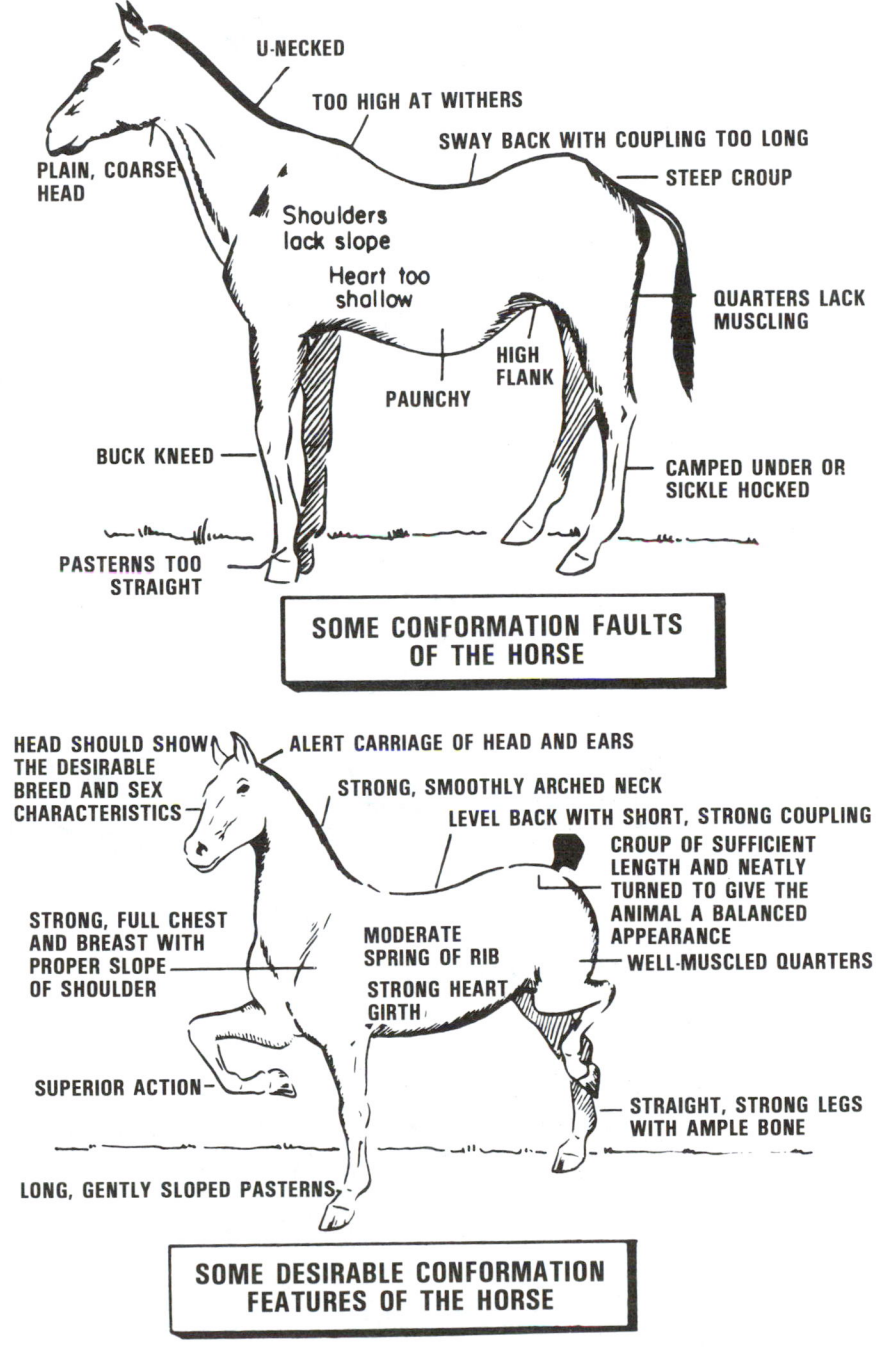

Fig. 52-3. Ideal type vs common faults. Regardless of type or breed, certain desirable characteristics should be present in all horses. The successful horse judge must be able to recognize both the desirable characteristics and the common faults, and the relative importance of each. (Drawing by Prof. R. F. Johnson)

## CORRECT FORM

Correct form calls for the following:

1. Good head, neck, and shoulders.
2. Strong, heavily muscled topline; short back and loin; level croup.
3. Ample chest and middle.
4. Well-muscled arm, forearm, and gaskins. (Also see the All-Breed Horse Scorecard at the end of this chapter.)

## CORRECT LEGS, FEET, AND PASTERNS

There has long been a saying "no foot, no horse." After all, the value of a horse lies chiefly in its ability to move; hence, the necessity of good underpinning. The legs should be straight, true, and squarely set; the bone should be well placed and clearly defined. The pasterns should be sloping; the feet large and wide at the heels and tough in conformation.

The hock should be large, clean, wide from front to back, deep, clean-cut, and correctly set. The knee should be deep from front to rear, wide when viewed from the front, straight, and taper gradually into the leg. Since the hock and knee joints of the horse are subject to great wear and are the seat of many unsoundnesses, they should receive every attention.

## GOOD ACTION

Although the degree of action of the horse will vary somewhat with the type (draft, speed, show, and saddle), the usefulness of all horses is dependent upon their action and their ability to move in various types of racing, driving, hunting, riding, polo, etc. In all types and breeds, the motion should be straight and true with a long, swift, and elastic stride.

## THE GAITS

*The gait is a particular natural or acquired way of going, characterized by a distinctive rhythmic movement of the feet and legs.*

In proper show-ring procedure, horses are brought back to a walk each time before they are called upon to execute a different gait. An exception sometimes is made in five-gaited classes—the rack may be executed from the slow gait. The gaits are:

**Walk**—A natural, slow gait of four beats in which each foot leaves and strikes the ground at separate intervals. The walk should be springy, regular, and true.

**Trot**—A natural, rapid, two-beat diagonal gait in which the front foot and the opposite hind foot take off together and strike the ground simultaneously.

All four feet are off the ground at the same time for a brief moment; the trotting horse thus seems to float through the air.

This gait varies considerably with different breeds. The trot of the Standardbred is characterized by length and rapidity of individual strides; the trot of the Hackney shows extreme flexion of the knees and hocks that produces a high-stepping show gait.

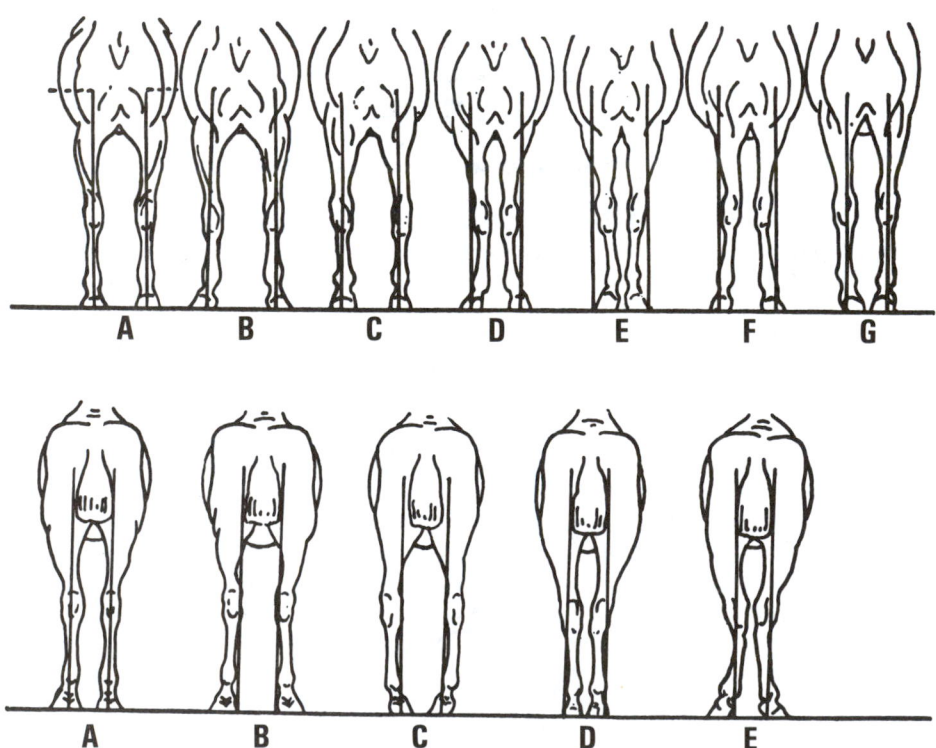

Fig. 52–4.   The proper and faulty conformation of the forelegs (top) when viewed from the front and the hind legs (bottom) when viewed from the rear. *The forelegs*: A represents correct conformation; B, splay-footed or base-narrow forefeet, toe cut out, heels in; C, bowed legs; D, knock-kneed, set close together with toes pointing outward; E, conformation predisposing to interfering; F, knees set close together; G, pigeon-toed or toe-narrow—a conformation which will cause the animal to wing or throw out the feet as they are elevated. *The hind legs*: A represents correct conformation; B, hind legs set too far apart; C, bandy-legged—wide at the hocks and hind feet toe in; D, hind legs set too close together; E, cow-hocked. The direction of the leg and the form of the foot are very important in the horse. (Courtesy, USDA)

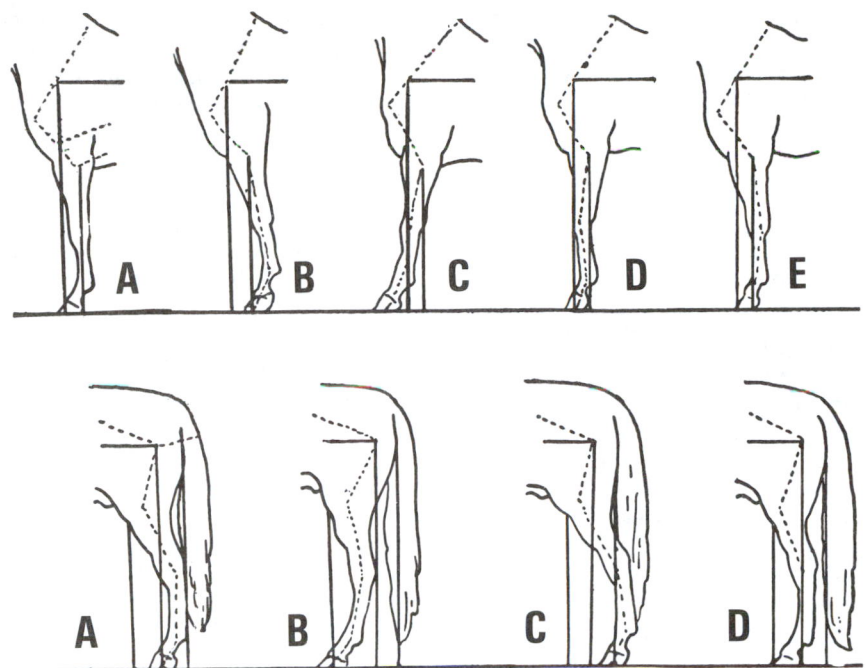

Fig. 52–5. The proper and faulty conformation of the forelegs (top) when viewed from the side and the hind legs (bottom) when viewed from the rear. *The forelegs:* A, correct conformation; B, forelegs too far under the body; C, forelegs too far advanced; D, knee-sprung or buck kneed—over in the knees; E, calf-kneed—standing with knees too far back. *The hind legs:* A, correct conformation; B, sickle hocked—hind legs too far under the body; C, legs set too far back; D, hock joint is too straight. The direction of the legs and the form of the foot are very important in the horse. (Courtesy, USDA)

**Run or gallop**—A fast, 3-beat gait during which 2 diagonal legs are paired and strike the ground together between the successive beats of the other 2 unpaired legs. All 4 feet are off the ground for a brief interval. The 2 unpaired legs that act independently—the forefoot with which the horse leads and the diagonal hindfoot—naturally bear more weight and are subject to more fatigue than the paired legs that act jointly.

In the gallop, propulsion is chiefly in the hindquarters, although the forequarters sustain a tremendous jar as the horse lands. The gallop is the fast natural gait of horses.

**Canter**—A slow, restrained gallop or run. Like the gallop, it is a three-beat gait, and it puts unusual wear on the leading forefoot and its diagonal hindfoot. It is important frequently to change the lead. A well-trained horse will do this easily at the will of the rider.

In the show-ring, the lead should be toward the inside of the ring, and the lead is changed by reversing direction of travel (when the ringmaster calls for "reverse and canter"). This gait should be executed in such a slow collected manner that the animal may perform in a relatively small circle.

**Lope**—The lope is the western adaptation of a very slow canter. It is a smooth, slow gait at which the head is carried low.

**Pace**—A fast, two-beat gait at which the front and hind feet on the same side start and stop simultaneously. The feet rise just above the ground level. All four feet are off the ground for a split second and the horse appears to float forward.

The pace is faster than the trot but slower than the run or gallop. It allows for a quick getaway, but it produces an objectional side or rolling type of motion. This gait is not suited to travel in mud or snow; a smooth, hard footing and easy draft are necessary for its best execution.

The pace was once popular in England, but it lost in favor soon after the development of the Thoroughbred early in the 18th century.

**Stepping pace (or slow pace)**—A modified pace in which the objectionable side or rolling motion is eliminated because the 2 feet on each side do not move exactly together. Instead, it is a 4-beat gait with each of the 4 feet striking the ground separately. The hind and front feet start almost together in the take off, but the hind foot touches the ground slightly ahead of the front foot on the same side. This is the preferred slow gait for 5-gaited show horses.

**Fox trot**—A slow, short, broken type of trot in which the head usually nods. In executing the fox trot, the horse brings each hind foot to the ground an instant before the diagonal forefoot. This gait is accepted as a slow gait, but it is not as popular as the stepping pace.

**Running walk**—A slow, 4-beat gait, intermediate between the walk and rack. The hind foot oversteps the front foot from 2 or 3 to as many as 18 in., giving the motion a smooth, gliding effect. This gait is characterized by a bobbing or nodding of the head, a flopping of the ears, and a snapping of the teeth in rhythm with the movements of the legs.

The running walk is easy on both horse and rider; it is the all-day working gait of the South, executed at a speed of six to eight miles per hour. It is a necessary gait for Tennessee Walking Horses.

**Rack**—A fast, brilliant, flashy, unnatural, four-beat gait in which each foot strikes the ground separately at equal intervals; known originally as the *single foot*. The rack is easy on the rider, hard on the horse. It is, undoubtedly, the most popular gait in the American show-ring. On the tanbark, greater speed at the rack is requested with the command "rack on."

**Traverse** or *side step*—The traverse or side step is simply a lateral movement of the animal without forward or backward movement. This step often helps the rider in opening and closing gates, lining up horses in the show-ring, and taking position in a mounted drill or posse.

## COMMON DEFECTS IN WAY OF GOING

The feet of a horse should move straight ahead parallel to an imaginary center line drawn in the direction of travel. Any deviation from this way of going constitutes a defect—some defects are:

**Cross-firing**—A scuffing on the inside of the diagonal forefeet and hindfeet; generally confined to pacers.

**Dwelling**—A noticeable pause in the flight of the foot, as though the stride were completed before the foot reaches the ground; most noticeable in trick-trained horses.

**Forging**—Striking forefoot with toe of hindfoot.

**Interfering**—Striking fetlock or cannon with the opposite foot; most often done by base-narrow, toe-wide, or splay-footed horses.

**Lameness**—A defect detected when the animal favors the affected foot when standing. The load on the ailing foot in action is eased and a characteristic bobbing of the head occurs as the affected foot strikes the ground.

**Paddling**—Throwing the front feet outward as they are picked up; most common in toe-narrow or pigeon-toed horses.

**Pointing**—Perceptible extension of the stride with little flexion; likely to occur in the long-strided Thoroughbred and Standardbred breeds—animals bred and trained for great speed.

**Pounding**—Heavy contact with ground, instead of desired light, springy movement.

**Rolling**—Excessive lateral shoulder motion; characteristic of horses with protruding shoulders.

**Scalping**—The hairline at the top of the hindfoot hits the toe of the forefoot as it breaks over.

**Speedy cutting**—The inside of diagonal fore and hind pasterns makes contact; sometimes seen in fast-trotting horses.

**Stringhalt**—Excessive flexing of hind legs; most easily detected when a horse is backed.

**Trappy**—A short, quick, choppy stride; a tendency of horses with short, straight pasterns and straight shoulders.

**Winding or rope-walking**—A twisting of the striding leg around in front of supporting leg, which results in contact like that of a rope-walking artist; often occurs in horses with very wide fronts.

**Winging**—An exaggerated paddling particularly noticeable in high-going horses.

## SOUND

An integral part of selecting a horse lies in the ability to recognize common blemishes and unsoundnesses and to rate the importance of each.

A thorough knowledge of normal, sound structure makes it easy to recognize imperfection.

*Any abnormal deviation in the structure or function of a horse constitutes an unsoundness.* From a practical standpoint, however, a differentiation is made between abnormalities that do and those that do not affect serviceability.

*Blemishes include abnormalities that do not affect serviceability*—such as wire cuts, rope burns, nail punctures, shoe boils, or capped hocks.

*Unsoundnesses include more serious abnormalities that affect serviceability.*

Fig. 52–6 shows the location of common blemishes and unsoundnesses.

One should consider the use to which it is intended to put the animal before buying a blemished or unsound horse.

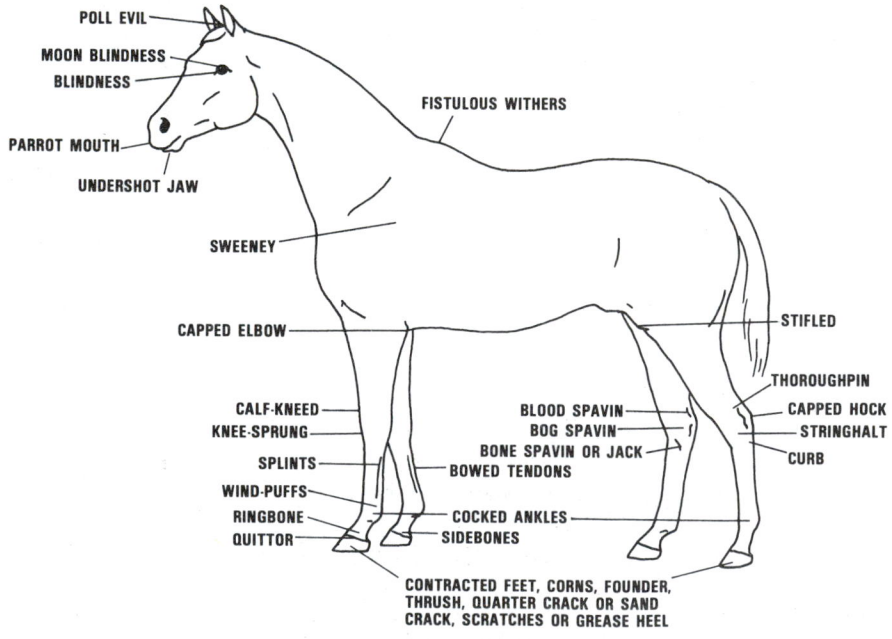

Fig. 52–6.  Location of common blemishes and unsoundnesses in horses.

## OTHER CONSIDERATIONS IN BUYING A HORSE

In addition to the desirable qualities in conformation already enumerated, there should be style and beauty, balance and symmetry, an abundance of quality, an energetic yet manageable disposition, freedom from vices, good wind, suitable age, freedom from disease, and proper condition. The buyer should also be on the alert for possible misrepresentation.

Also, price is always of importance. Although a high price may be justified for superior breeding and performing horses, sound judgment should always prevail.

## HOW TO DETERMINE AGE BY THE TEETH

The life-span of horses averages about 25 years—about one-third that of a human; hence, a 25-year-old horse is equivalent to a person 75 years old. Horses generally are at their best between 3 and 12 years of age. This may vary because of individual differences in animals or because of the differences in the kind of work they do. The age of a horse is, therefore, important to the breeder, seller, and buyer.

The approximate age of a horse can be determined by noting the time of appearance, shape, and degree of wear of temporary and permanent teeth. Temporary, or milk teeth are easily distinguishable from permanent ones because they are smaller and whiter.

The best way to learn to determine age in horses is by examining the teeth of individual horses of known ages.

A mature male horse has 40 teeth; a mature female has 36.[1] A foal of either sex has 24. The mare does not have tushes as a rule.

Table 52–1 may be used a guide to determining age of horses by their teeth. The gradual wearing and disappearance of the cups (the inside or center of the tooth) according to a rather definite pattern in period of time enables the experienced equestrian to judge the age of an animal with a fair degree of accuracy up to 12 years.

Even experienced equestrians cannot determine the age of an animal accurately after it is 12 years old. After this age, the teeth change from oval to triangular, and they project or slant forward more and more as the horse becomes older.

Side views of the mouth of 5-, 7-, and 20-year-old horses are shown in Fig. 52–7 (see next page).

---

[1]Quite commonly, a small, pointed tooth, known as a *wolf tooth*, may appear in front of each first molar tooth in the upper jaw, thus increasing the total number of teeth to 42 in the male and 38 in the female. Less frequently, two more *wolf teeth* in the lower jaw increase the total number of teeth in the male and female to 44 and 40, respectively.

## TABLE 52–1
## A SUMMARY OF THE CHANGES OF THE TEETH OF HORSES

| Age | Condition of Teeth | |
|---|---|---|
| At birth or before 10 days of age | First or central upper and lower incisors appear. | } Appearance of temporary teeth |
| 4 to 6 weeks of age | Second or intermediate upper and lower incisors appear. | |
| 6 to 10 months of age | Third or corner upper and lower incisors appear. | } Wear of temporary teeth |
| 1 year of age | Crowns of central incisors show wear. | |
| 1½ years of age | Intermediate incisors show wear. | |
| 2 years of age | All temporary incisors show wear. | } Appearance of permanent teeth |
| 2½ years of age | First or central incisors appear. | |
| 3½ years of age | Second or intermediate incisors appear. | |
| 4½ years of age | Third or corner incisors appear. | |
| 4 to 5 years of age (in male) | Canines appear. | } Wear of permanent teeth |
| 5 years of age | Cups in all incisors. | |
| 6 years of age | Cups worn out of lower central incisors. | |
| 7 years of age | Cups worn out of lower intermediate incisors. | |
| 8 years of age | Cups worn out of all lower incisors, and dental *star* appears on upper central and intermediate pairs. | |
| 9 years of age | Cups also worn out of upper central incisors, and dental *star* appears on upper central and intermediate pairs. | |
| 10 years of age | Cups also worn out of upper intermediate incisors, and dental *star* is present on all incisors, both upper and lower. | |
| 11 years of age | Cups worn out of all upper and lower incisors, and dental *star* approaches center of cups. | |
| 12 years of age | No cups. *Smooth mouthed.* | |

An animal's environment can affect wear on teeth material- ly. Teeth of horses raised in dry, sandy areas, for example, will show more than normal wear; a 5-year-old western horse may have teeth that would be normal in a 6- to 8-year-old horse

raised elsewhere. The teeth of cribbers also show more than normal wear. It is hard to determine the age of such animals. The age of a horse with a parrot mouth, or undershot jaw, also is difficult to estimate.

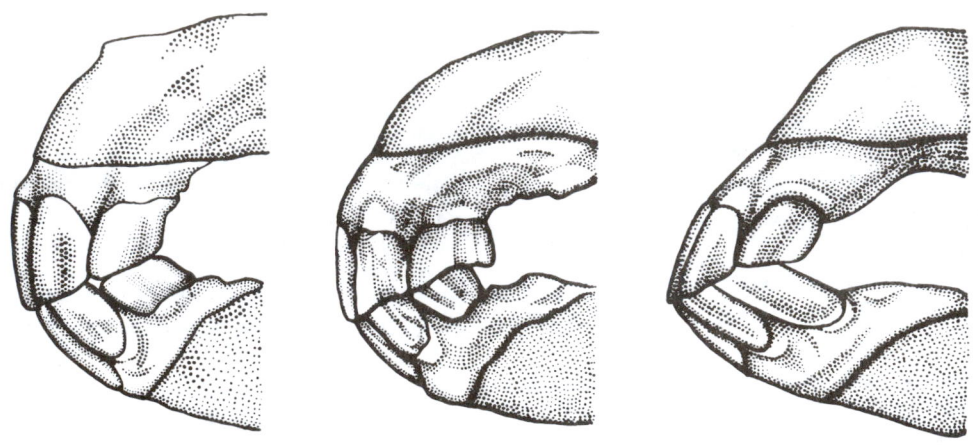

Fig. 52–7. Side view of 5-, 7-, and 20-year-old mouth. Note that as the horse advances in age, the teeth change from nearly perpendicular to slanting sharply toward the front.

## HOW TO MEASURE A HORSE

Normal pertinent measurements are height, weight, girth, and bone.

**Height**—The height of a horse is the vertical distance from the highest point of its withers to the ground when the animal is standing squarely on a level area. The unit of measurement used in expressing height is the *hand*, which is 4 in. A horse measuring 62 in. is said to be 15–2 hands high (15 hands and 2 in.).

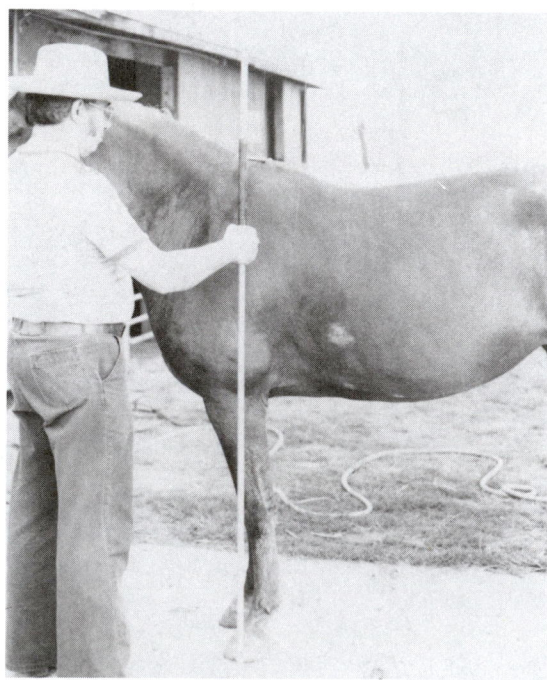

Fig. 52–8. Measuring the height of a horse. (Courtesy, American Mustang Assn., Yucaipa, CA)

People can estimate a horse's height if they know the ex- act number of inches from the level of their eyes to the ground. Knowing this, all they need do is stand beside the animal's front limbs and look at the highest point of the withers; they can then estimate the horse's height rather closely.

**Weight**—Although weight may be estimated by means of a tape measure, it is best to use scales.

**Girth**—Girth is a measurement of the circumference of the chest behind the withers and in front of the back.

**Bone**—Size of bone usually is determined by placing a tape measure around the cannon bone halfway between the knee and fetlock joints. This measurement is in inches.

## COLORS AND MARKINGS

Within certain breeds, some colors are preferred or re- quired, and others are undesirable or constitute disqualifica- tion for registration. A good equestrian needs a working knowledge of horse colors and patterns because they are the most conspicuous features by which a horse can be described or identified.

The five basic body colors of horses are:

1. **Bay**—Bay is a mixture of red and yellow. It includes many shades, from a light yellowish tan (light bay) to a dark, rich shade that is almost brown (dark bay). A bay horse usual- ly has a black mane and tail and black points.

2. **Black**—A black horse is completely black, including the muzzle and flanks. If in doubt whether the horse is dark brown or black, note the color of the fine hairs on the muzzle and the hair on the flanks; tan or brown hairs at these points in- dicate the horse is not a true black, but a seal brown.

3. **Brown**—A brown horse is almost black but can be distinguished by the fine tan or brown hairs on the muzzle and flanks.

4. **Chestnut (sorrel)**—A chestnut horse is basically red. The shades vary from light washy yellow (light chestnut) to a dark

liver color (dark chestnut). Between these come the brilliant red-gold and copper shades. Normally, the mane and tail of a chestnut horse are the same shade as the body, although they may be lighter. When they are lighter, they are referred to as *flaxen mane and tail*. Chestnut color is never accompanied by a black mane and tail.

5. **White**—A true white horse is born white and remains white throughout life. White horses have snow-white hair, pink skin, and brown eyes (rarely blue).

Besides the five basic colors, horses have five major variations to these coat colors. The variations are:

1. **Dun (buckskin)**—Dun is a yellowish color of variable shading from pale yellow to a dirty canvas color. A dun horse has a stripe down the back.

2. **Gray**—This is a mixture of white and black hairs. Sometimes a gray horse is difficult to distinguish from a black horse at birth, but gray horses get lighter with age.

3. **Palomino**—This is a golden color. Palomino horses have a light-colored mane and tail of white, silver, or ivory.

4. **Pinto (calico or paint)**—Pinto is a Spanish word that means painted. The pinto color is characterized by irregular colored and white areas in either piebald or skewbald patterns. Piebald horses are black and white, and skewbald horses are white and any other color except black.

5. **Roan**—Roan is a mixture of white hairs with one or more base colors. White with bay is red roan; white with chestnut is strawberry roan; and white with black is blue roan.

When identifying an individual horse, it is generally necessary to include more identification than just body color. Head marks and leg marks are often used to describe a horse. For example, it may be necessary to identify the dark sorrel as the one with the blaze face and the white stocking on the right front foot.

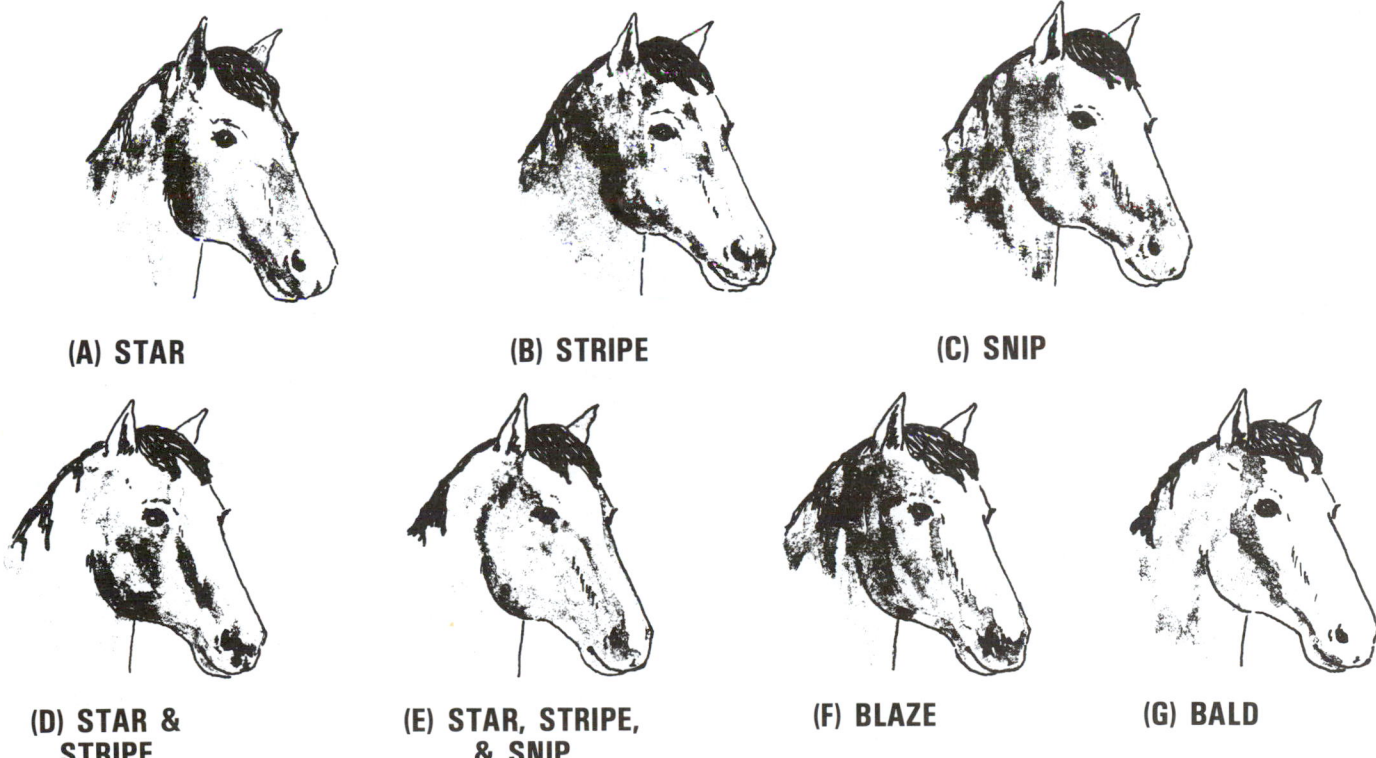

**(A) STAR**  **(B) STRIPE**  **(C) SNIP**

**(D) STAR & STRIPE**  **(E) STAR, STRIPE, & SNIP**  **(F) BLAZE**  **(G) BALD**

Fig. 52-9.   The head marks of horses. (A) Star is any white mark on the forehead located above a line running from eye to eye; (B) stripe is a narrow white marking that extends from about the line of the eyes to the nostrils; (C) snip is a white mark between the nostrils or on the lips; (D) star and stripe includes both a star and a stripe; (E) star, stripe, and snip includes all three of these marks—star, stripe, and snip; (F) blaze is a broad, white marking covering almost all the forehead but not including the eyes or nostrils; (G) bald is a bald, or white, face including the eyes and nostrils, or a partially white face.

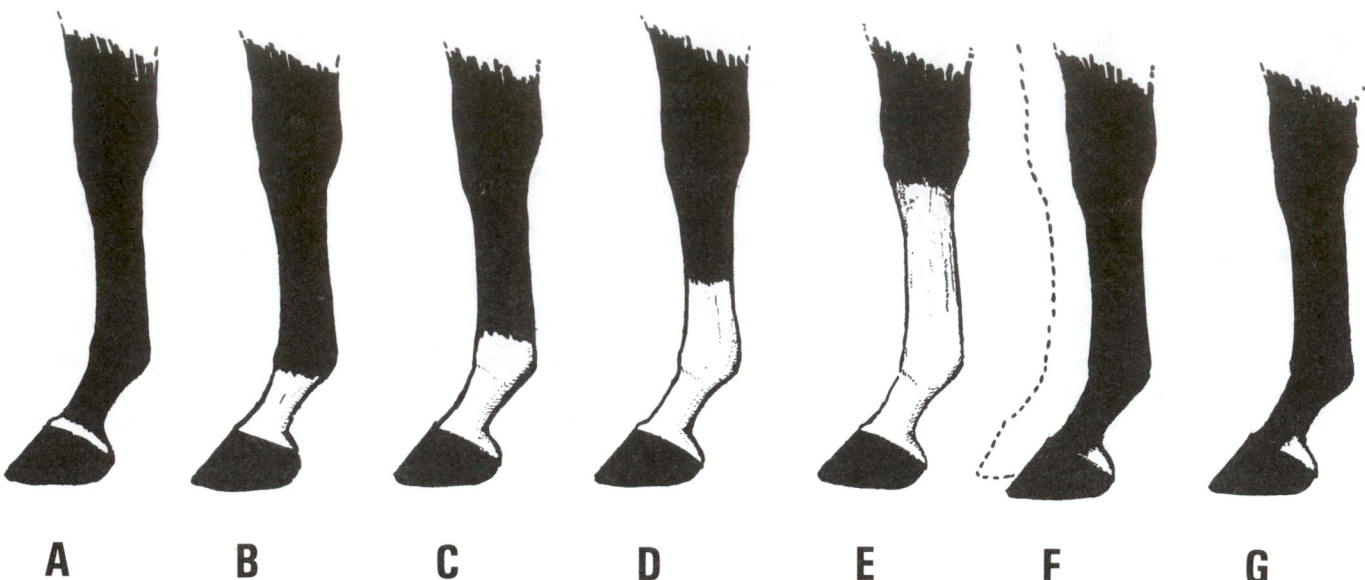

Fig. 52–10. Most common leg markings. A, *coronet*—a white strip covering the coronet band; B, *pastern*—white extends from the coronet to and including the pastern; C, *ankle*—white extends from the coronet to and including the fetlock; D, *half stocking*—white extends from the coronet to the middle of the cannon; E, *stocking*—white extends from the coronet to the knee. When the white includes the knee, it is known as a full stocking; F, *white outside heels*—both heels are white; G, *white outside heel*—outside heel only is white.

## METHOD OF EXAMINING

Fig. 52–11. A halter class. (Photo by W. W. Dexter)

Fig. 52–12. A performance class. (Courtesy, Round Robbin Farm, East Pepperell, MA)

Custom decrees somewhat different procedures in examining or judging horses. In the show-ring, halter classes usually are first examined while lined up side by side, and later inspected while moved (first, at the walk, then at the trot) one at a time; whereas performance classes are first examined with the entire class in action, and later lined up for close inspection. After a judge has inspected a light horse performance class, both in action and when lined up, it is considered entirely pro-

per to request that certain animals be pulled out and again put through their paces.

When selecting or buying a horse, the procedure followed is essentially the same as in show-ring judging. Young stock and breeding animals are examined at the halter—first when standing, then when led at the walk and trot. Performance animals are observed in action (in their intended use—jumping, as pleasure horses, or whatnot), followed by a close inspection.

# HORSE SCORECARD

A horse must conform to the specific type that is needed for the function it is to perform. Second, the horse should conform to the characteristics of the breed that it represents. The use of a scorecard is a good way to make sure that no part is overlooked and a proper value is assigned to each part.

A scorecard is a listing of the different parts of an animal, with a numerical value assigned to each part according to its relative importance. Also, breed characteristics may be considered in a scorecard. An all-breed horse scorecard follows.

| | POINTS or % | NAME and/or No. of Horse | NAME and/or No. of Horse | NAME and/or No. of Horse | NAME and/or No. of Horse |
|---|---|---|---|---|---|
| **BREED TYPE** .................................................... | 15 | | | | |
| The breed is distinguished by its unique combination of style and beauty. | | | | | |
| COLOR: In keeping with the breed. | | | | | |
| HEIGHT AT MATURITY: Proper height; extremes undesirable. | | | | | |
| WEIGHT AT MATURITY: Proper weight; extremes undesirable. | | | | | |
| **FORM** .......................................................... | 35 | | | | |
| STYLE AND BEAUTY: Attractive, good carriage, alert, refined, symmetrical, and all parts nicely blended together. | | | | | |
| BODY: Nicely turned, long, well-sprung ribs; heavily muscled. | | | | | |
| BACK AND LOIN: Short and strong, wide, well muscled, and short coupled. | | | | | |
| CROUP: Long, level, wide, muscular, with a high-set tail. | | | | | |
| REAR QUARTERS: Deep and muscular. | | | | | |
| GASKIN: Heavily muscled. | | | | | |
| WITHERS: Prominent, and of the same height as the high point of the croup. | | | | | |
| SHOULDERS: Deep, well laid in, and sloping (about a 45° angle). | | | | | |
| CHEST: Fairly wide, deep, and full. | | | | | |
| ARM AND FOREARM: Well muscled. | | | | | |
| **FEET AND LEGS** ................................................ | 15 | | | | |
| LEGS: Correct position and set (when viewed from front, side, and rear). | | | | | |
| PASTERNS: Long, sloping (about a 45° angle). | | | | | |
| FEET: In proportion to size of horse, good shape, wide and deep at heels, dense texture of hoof. | | | | | |
| HOCKS: Deep, clean-cut, and well supported. | | | | | |
| KNEES: Broad, tapering gradually into cannon. | | | | | |
| CANNONS: Clean, flat, with tendons well defined. | | | | | |
| **HEAD AND NECK** ................................................ | 10 | | | | |
| Alertly carried, showing style and character. | | | | | |
| HEAD: Well proportioned to rest of body, refined, clean-cut, with chiseled appearance; broad, full forehead with great width between eyes; ears medium sized, well carried and attractive; eyes large and prominent. | | | | | |
| NECK: Long and nicely arched, clean-cut about the throatlatch; with head well set on, gracefully carried. | | | | | |
| **QUALITY** ....................................................... | 10 | | | | |
| Clean, flat bone; well defined and clean joints and tendons, and fine skin and hair. | | | | | |
| **ACTION** ........................................................ | 15 | | | | |
| WALK: Easy, springy, prompt, balanced, a long step, with each foot carried forward in a straight line, feet lifted clear of the ground. | | | | | |
| TROT: Prompt, straight, elastic, balanced, with hocks carried closely, and high flexion of knees and hocks. | | | | | |
| **DISCRIMINATION:** Any abnormality that affects the serviceability of the horse. | | | | | |
| **DISQUALIFICATION:** Blindness (except by injury), bone spavin, stifled, stringhalt, cryptorchid. | | | | | |
| **TOTAL SCORE** ......................................... | 100 | | | | |

# QUESTIONS FOR STUDY AND DISCUSSION

1. In judging horses, what primary differences exist in comparison with beef cattle, sheep, and hogs?

2. What points should a buyer consider in deciding what kind of horse is needed?

3. After deciding on the kind of horse needed and getting an ideal in mind, what selection basis is the best method of obtaining a good horse?

4. Describe the location of the following parts of a horse: poll, crest, withers, coupling, stifle, gaskin, coronet, fetlock, and arm.

5. Describe the conformation, stance, and action of an ideal horse.

6. Classify horse gaits as (a) 2-beat gaits, (b) 3-beat gaits, or (c) 4-beat gaits.

7. Describe the following gaits; canter, lope, pace, and rack.

8. Describe eight common defects in way of going.

9. Prepare a list of common horse blemishes and describe each of them.

10. Rank in order of their severity six common horse unsoundnesses and describe each of them.

11. How does the life-span of horses and humans compare?

12. What is involved in determining the age of horses by the teeth (a) up to 5 years of age, (b) from 6 to 12 years of age, (c) beyond 12 years of age?

13. Of what importance are the following measurements: (a) height, (b) weight, (c) girth, and (d) bone? How is each of them determined?

14. List (a) the five basic body colors of horses, and (b) the five variations of the basic body colors.

15. Name and describe three common head marks and three common legs marks of horses.

16. In the show-ring, what different procedure is followed in examining (a) a halter class, and (b) a performance class?

17. What is a horse scorecard? What are the chief virtues of a scorecard?

# SELECTED REFERENCES

| Title of Publication | Author(s) | Publisher |
| --- | --- | --- |
| *Anatomy and Conformation of the Horse* | G. B. Edwards | Dreenan Press Ltd., Croton-on-Hudson, NY, 1973 |
| *Breeding and Raising Horses*, Ag. Hdbk. No. 394 | M. E. Ensminger | Agricultural Research Service, USDA, Washington, DC, 1972 |
| *Color of Horses, The* | B. K. Green | Northland Press, Flagstaff, AZ, 1974 |
| *Determining the Age of Farm Animals by Their Teeth*, Farmers' Bull. No. 1721 | | U.S. Department of Agriculture, Washington, DC |
| *Fair Exchange* | H. S. Finney | Charles Scribner's Sons, New York, NY, 1974 |
| *Horse Buyer's Guide, The* | J. K. Posey | A. S. Barnes & Co., Inc., Cranbury, NJ, 1973 |
| *Horsemanship and Horse Care*, Ag. Info. Bull. No. 353 | M. E. Ensminger | Agricultural Research Service, USDA, Washington, DC, 1972 |
| *Horses: Their Selection, Care & Handling* | M. C. Self | A. S. Barnes & Co., Inc., New York, NY, 1943 |
| *Judging Manual for American Saddlebred Horses* | J. Foss | American Saddle Horse Breeders Association, Louisville, KY, 1973 |
| *Lameness in Horses*, Second Edition | O. R. Adams | Lea & Febiger, Philadelphia, PA, 1966 |
| *Livestock Judging and Evaluation: A Handbook for the Student* | W. M. Beeson R. E. Hunsley J. E. Nordby | The Interstate Printers & Publishers, Inc., Danville, IL, 1970 |
| *Points of the Horse*, Seventh Revised Edition | M. H. Hayes | Arco Publishing Co., Inc., New York, NY, 1969 |
| *Rule Book* | | The American Horse Shows Association, Inc., New York, NY, annual |
| *Selecting, Fitting and Showing Horses* | J. E. Nordby H. E. Lattig | The Interstate Printers & Publishers, Inc., Danville, IL, 1963 |
| *Stockman's Handbook, The*, Sixth Edition | M. E. Ensminger | The Interstate Printers & Publishers, Inc., Danville, IL, 1983 |

Because a stallion can have so many more offspring than a mare, he is more important than any one mare from a hereditary standpoint. This shows the Morgan stallion, *Ultimate Command*. (Photo by H. R. Hoover. Courtesy, *The Morgan Horse*, Shelburne, VT)

# BREEDING HORSES

# Chapter 53

Horses have the poorest reproductive performance of all domestic animals, due to the intervention of people with the natural breeding season.

However, they will continue to be bred to (1) provide recreation and sport, and (2) serve the livestock industry in the West. It is important, therefore, that both the student and the progressive horse breeder be familiar with the breeding of them.

## NORMAL BREEDING HABITS OF MARES

Perhaps at the outset it should be emphasized that strictly normal breeding habits of the horse do not exist under domestication. In the wild state, each band of 30 to 40 mares was headed by a stallion leader who sired all of the colts in that band. With plenty of outdoor exercise on natural footing, superior nutrition derived from plants grown on unleached soils, regular production beginning at an early age, little possibility of disease or infection, and frequent services during the heat period, 90% or higher foaling rates were commonplace. Under domestication, the average conception rate is less than 50%, and only the better establishments exceed 70%. Thus, the low fertility usually encountered under domestication must be caused to a large extent by the relatively artificial conditions under which horses are mated.

## AGE OF PUBERTY

Fillies generally start coming in heat when 12 to 15 months of age.

## AGE TO BREED MARES

Only exceptionally well-grown fillies should be bred as late 2-year-olds, so as to foal at 3 years of age. Under a system of early breeding, the fillies must be fed exceptionally well in order to provide growth for their own immature bodies as well as for the developing fetus. Furthermore, they usually should not be bred the following year. Generally speaking, it is best to breed the mare as a 3-year-old so that she will foal when 4. Not only will the 3-year-old be better grown, but there will not be the handicap of training her while she is heavy in foal.

If they are properly cared for, it is not uncommon for broodmares to produce regularly up to 14 to 16 years of age; and, of course, in the more exceptional cases they may produce up to 25 years of age.

In selecting a broodmare, it is usually advisable either to obtain a young 3- or 4-year-old or to make certain of the sure and regular breeding habits of any old mares.

## HEAT PERIODS

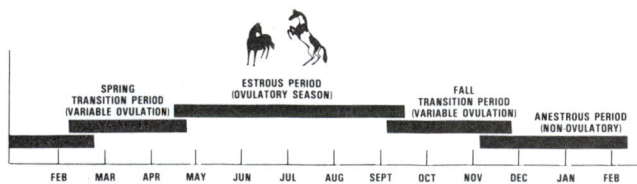

Fig. 53–1. Mare Seasonality. From about mid-April to mid-September, the mare cycles and exhibits sexual receptivity to the stallion every 15 to 19 days.

The heat periods of mares recur at approximately 21- to 23-day intervals, and last 3 to 7 days, with the mare exhibiting sexual receptivity to the stallion every 15 to 19 days. The highest rate of conception may be obtained by serving the mare daily or every other day during the heat period, beginning with the third day.

## SIGNS OF ESTRUS

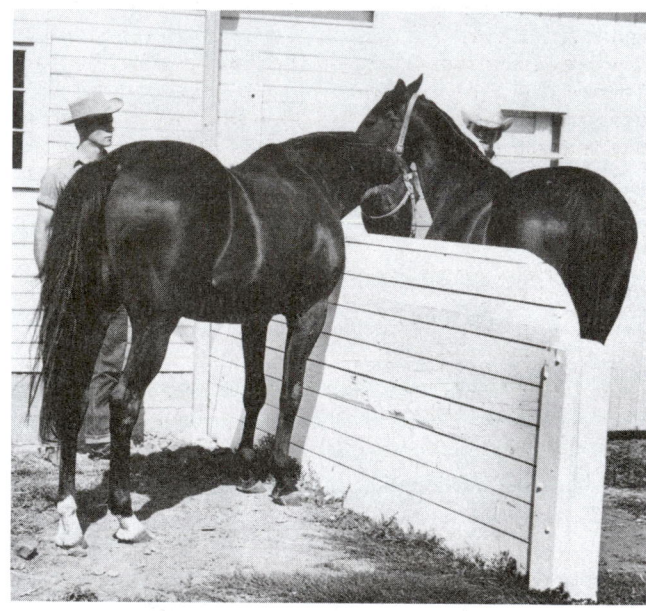

Fig. 53–2. Teasing a mare, using a solid fence for separation. Breeding a mare that is not in season is a wasteful practice, and more important, it may result in damaging future breeding efficiency. (Courtesy, Washington State University)

The experienced breeder who is familiar with a band of mares can usually detect those that are in season by observing (1) the relaxation of the external genitals, (2) more frequent urination, (3) the teasing of other mares, (4) the apparent desire for company, and (5) a slight mucous discharge from the vagina. In shy breeders or when there is any question about the mare being in season, she should be tried.

Palpation and teasing are the two most common management tools used in detecting heat. Most managers use a combination of palpation and teasing for estrus detection and breeding determination. Mares are teased and those showing signs of estrus are palpated better to define reproductive status.

Above all, precaution should be taken against false heats of mares in foal, the breeding of which may result in abortion.

## FERTILIZATION

Generally, the egg is liberated during the period of one day before to one day after the end of heat. Unfortunately, there is no reliable way of predicting the length of heat nor the time of ovulation, although an expert technician can predict the time of ovulation by palpation—by feeling the ovary (follicle) with the hand through the rectal wall.

The sperm (or male germ cells) are deposited in the uterus at the time of service and from there ascend the reproductive tract. Under favorable conditions, they meet the egg, and one of them fertilizes it in the upper part of the oviduct near the ovary.

A series of delicate time relationships must be met, however, or the egg will never be fertilized. The sperm cells live only 24 to 30 hours in the reproductive tract of the female, and it probably requires 4 to 6 hours for them to ascend the female reproductive tract. Moreover, the egg is viable for an even shorter period of time than the sperm, probably for not more than 4 to 6 hours after ovulation. For conception, therefore, breeding must take place within 20 to 24 hours before ovulation.

As mares usually stay in heat for 3 to 7 days, perhaps the highest rate of conception may be obtained by serving the mare daily or every other day during the heat period, beginning with the third day. When many mares are being bred and heavy demands are being made upon a given stallion, this condition may be obtained by reinforcing a natural service with subsequent daily artificial inseminations as long as heat lasts. In no case should the mare be bred twice in the same day.

## GESTATION PERIOD

The average gestation period of mares is 336 days, or a little over 11 months. This will vary, however, with individual mares and may range from 310 to 370 days.

A handy rule-of-thumb method which may be used to figure the approximate date of foaling is to subtract one month and add one day to the date the mare was bred. Hence, a mare bred May 20 should foal April 21 the following year.

## BREEDING AFTER FOALING

Mares usually come in heat 7 to 10 days after foaling, with individual mares varying from 3 to 13 days after foaling. Provided that foaling has been entirely normal and there is no discharge or evidence of infection at this time, many good breeders plan to rebreed the mare during this first recurrence of heat after foaling on or about the ninth day. They believe that mares so handled are more likely to conceive than if bred at a later period. Mares suffering from an infection of the genital tract are seldom settled in service; and, even if they do conceive, there is danger of the foal being undersized and poorly developed. Also, the infection may be needlessly spread to the stallion and other mares by allowing such a practice. Mares not bred at this time or not conceiving will come in heat between the twenty-fifth and thirtieth day from foaling.

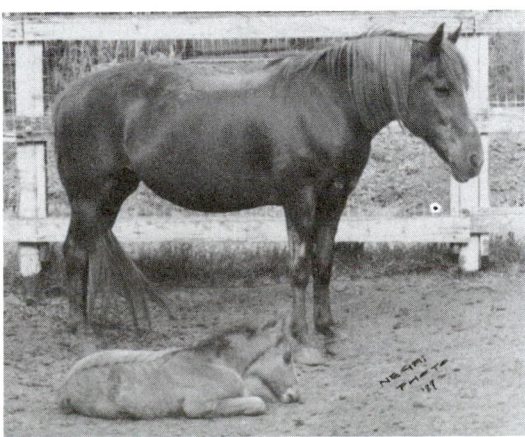

Fig. 53-3.  If foaling is normal and there is no discharge, mares may be rebred, about the ninth day after foaling. (Courtesy, American Mustang Assn., Yucaipa, CA)

The usual reasons advanced in favor of rebreeding at the first heat period following foaling are:

1. It gives an added chance to rebreed the mare in an effort to get her in foal.
2. Occasionally a mare will not again show signs of heat during the breeding season.
3. If the mare conceives, she will foal about 20 days earlier the following year. This may be important where an early foal is desired.

Some arguments against rebreeding on the ninth day following foaling are:

1. If only one service is given on the ninth day following foaling, it is estimated that not more than 25% of the mares conceive.
2. During the period extending up to two weeks following foaling, the broodmare is more susceptible to genital infection than during any other period of her life.
3. Older mares that have been raising foals regularly each year may require a longer period of rest between pregnancies.
4. If the chances of conception are not too great, it may be unwise to overwork the stallion deliberately.

## PERCENT OF MARES BRED PRODUCING FOALS

Fig. 53-4.  Breeding horses necessitates that foals be born—and born alive. (Courtesy, United States Trotting Assn., Columbus, OH)

Without question, more difficulty is experienced in breeding mares than any other kind of livestock. The percentage of mares bred that actually conceive each year will vary from 40% to a high of 85%, with an average probably running less than 50%, and some of this number will fail to produce living foals.

This means that, on the average, two mares are kept a whole year in order to produce one foal. By contrast, nationally, 88% of all beef cows that are bred calve; 95% of all ewes that are bred lamb; and 85% of all sows that are bred farrow. The lower percentage conception in mares than in other classes of livestock is due primarily to the following: (1) Research in the field has lagged, (2) we try to get mares bred in about 4 months instead of 12, and (3) we have arbitrarily limited our breeding season (late winter and early spring) to a period that at its best is only about 50% in agreement with nature.

In the Bluegrass Region of Kentucky, where there are both good and experienced horse breeders and as desirable conditions for breeding as can be secured under domestication, 60% foaling is considered as average for the area.

Recognition of the following facts may help to increase the percentage of foals produced:

1. Mares bred in the late spring of the year are more likely to conceive. If mares are bred out of season, spring conditions should be duplicated as nearly as possible.

2. Mares bred as three- and four-year-olds and kept in regular production thereafter are more likely to conceive and produce living foals.

3. Infections or other unhealthy conditions of either the mare or the stallion are not favorable for production.

4. More conceptions will occur if the mare is bred at the proper time within the heat periods. Usually mares bred just before going out of heat are more likely to conceive.

5. Returning the mares to the stallion for retrial or rebreeding is important.

6. Mares in foal should be fed and cared for properly, so as to develop the young. Balance of proteins, minerals, and vitamins is important.

7. It must also be remembered that old mares, overfat mares, or mares in a thin, run-down condition are less likely to be good breeders. Unfortunately, these conditions frequently apply to mares that are bred following retirement from the racetrack or the show-ring.

A shift of the date of birth (the January 1 birthday, for purposes of racing and showing) to somewhere between March 1 and May 1 would improve conception rate and foaling percentage, simply because mares would be bred under more natural and ideal spring conditions. Thus, it would have considerable virtue from the standpoint of the horse producer. On the other side of the ledger, however, it would create problems in racing and in registration, both here and abroad. Also, such deep-rooted tradition would be difficult to change; in fact, much consideration has been given to this matter from time to time. In the final analysis, therefore, stepping up breeding research is the primary avenue through which the deplorably low percentage foal crop may be improved.

## STERILITY OR BARRENNESS IN MARES

Sterility is a condition of infertility. Whatever the cause, there are no cure-alls for the condition. Rather, each individual case requires careful diagnosis and specific treatment for what is wrong. Also, it should be recognized that there are two types of sterility—temporary and permanent—although no sharp line can be drawn between the two.

Regardless of the cause of sterility, it is well to give a word of caution against the so-called *opening up* of mares, which is the practice of inserting the hand and arm into the genital organs for the purpose of rearranging the organs in order to ensure conception. Few caretakers, no matter how expert they may classify themselves, have either sufficient knowledge of the anatomy of the mare or appreciation of the absolutely sterile methods necessary in such procedure to be probing about. Moreover, it is only rarely that the reproductive organs are out of place. Unless the *opening up* is recommended and conducted by a competent veterinarian, it should not be permitted. When performed by an amateur, or even most would-be experts, it is a dangerous practice that is to be condemned.

## TEMPORARY STERILITY

Some common causes of temporary sterility are:

1. Lack of exercise, irregular work, and overfeeding accompanied by extremely high condition.

2. Overwork, underfeeding, and an extremely thin and run-down condition.

3. Nutritional deficiencies.

4. Infections of various kinds.

5. Some types of physiological imbalance characterized by such things as cystic ovaries or failure to ovulate at the proper time.

Temporary sterility can be reduced by removing the cause and correcting the difficulty.

## PERMANENT STERILITY

Naturally, permanent sterility is much more serious to the horse breeder. Perhaps the most common causes of permanent sterility:

1. Old age, which is usually accompanied by irregular breeding and eventual total sterility.

2. Infections in the reproductive tract, usually in the cervix, uterus, or fallopian tubes.

3. Some types of physiological imbalances characterized by such things as cystic ovaries or failure to ovulate at the proper time.

4. Closure of the female genital organs.

Sometimes a veterinarian is able to correct the latter two conditions; and on an extremely valuable breeding mare, it may be worthwhile to obtain such professional service in an effort to bring about conception.

Retained afterbirth or other difficulties encountered in foaling may cause inflammation and infection that will prevent conception as long as the condition exists. There is real danger of spreading the infection if the mare is bred while in such a condition.

## FERTILITY OF THE STALLION

Any stallion of breeding age that is purchased should be a guaranteed breeder; this is usually understood among reputable horse breeders.

The most reliable and obvious indication of potency is a large number of healthy, vigorous foals from a season's service. As an added protection, or in order to follow the horse during the midst of a heavy breeding program, a microscopic examination of the semen may be made by an experienced person. As the stallion dismounts from service, some of the semen is collected in a sterilized funnel by holding the penis over the plugged funnel. A sample of the semen is then placed on a slide for examination. A great number of active sperm cells is an indication, although not definite assurance, that the stallion is fertile. Some establishments make a regular practice of making such a microscopic examination twice each week during the breeding season. If it is desired to examine a stallion's semen after the breeding season or when a mare is not in season, an artificial vagina may be used. When an entire ejaculate is available for study, the four main criteria of quality are (1) semen volume, (2) spermatazoan count, (3) progressive movement, and (4) morphology.

If the stallion is a shy breeder or lacks fertility even though the feed and exercise have been up to standard, masturbation should be suspected.

## CONDITIONING THE MARE FOR BREEDING

Proper conditioning of the mare prior to breeding is just as important as in the stallion. Such conditioning depends primarily upon adequate and proper feed and the right amount of exercise.

For the highest rate of conception, mares should neither be too thin nor too fat; a happy medium in condition makes for best results. It is especially important that one avoid the natural tendency of barren or maiden mares to get too fat.

Time permitting, mares of the light-horse breeds may best be exercised and conditioned by riding under the saddle or driving in harness. When these methods are not practical or feasible, permitting a band of mares to run in a large pasture will usually provide a satisfactory amount of exercise.

## THE BREEDING OPERATIONS

No phase of horse production has become more unnatural or more complicated with domestication than the actual breeding operations. It is important, therefore, that those responsible for breeding conduct each operation thereof in such a manner as to maximize conception.

### HAND BREEDING, CORRAL BREEDING, AND PASTURE BREEDING

Hand mating is undoubtedly the best way in which to breed mares; it is the accepted practice in the better breeding establishments throughout the world. It guards against injury to both the stallion and the mare

Although leaving much to be desired, corral breeding is next best to hand breeding. In this system, after first ascertaining that the mare is in heat, she and the stallion are turned loose together in a small, well-fenced corral. The attendants should remain out of the corral, where they can see but not be seen by the animals, until service is completed, following which the stallion and the mare are returned to their respective quarters.

Pasture breeding simply consists of turning the stallion into a pasture with the band of mares which it is intended that he serve. Except on the ranges of the Far West, this method of breeding is seldom practiced with domestic horses. With valuable animals, both corral and pasture breeding are too likely to cause injury, and the practices should be condemned. In pasture breeding, a stallion will handle fewer mares because of the repeated services of a mare, and he may even become sterile toward the end of the breeding season. Moreover, in pasture breeding, accurate breeding records are impossible.

### EXAMINATION OF THE MARE

Before accepting a mare for service, the stallion owner should check every possible condition with care. The stallioner should (1) examine the mare closely and question the owner concerning her health, last foaling date, breeding record, and similar matters; and (2) be well acquainted with the symptoms

of dourine and other venereal diseases. Even though these diseases are not common in this country, there is always danger of finding them in imported stallions and mares.

When mares have been barren over an extended period or when there is the slightest suspicion of infection, it is good protection to require a veterinarian's certificate to the effect that the mare is in a healthy breeding condition.

## SIGNS AND TESTS OF PREGNANCY

In order to produce as high a percentage of foals as possible and to have them arrive at the time desired, the horse breeder should be familiar with the signs of and tests for pregnancy. This is doubly important when it is recognized that a great many mares may either be shy breeders or show signs of heat even when well advanced in gestation. The signs and tests of pregnancy follow:

1. **The cessation of the heat period.** One of the simplest determinations of pregnancy is the cessation of the heat period—the mare does not exhibit any signs of heat 18 to 20 days after her last ovulation. This may be difficult to determine as well as misleading. Some mares will continue to exhibit the characteristic heat symptoms when in foal; sometimes, they show such pronounced signs of heat that they are given the service of a stallion, which may result in abortion. Other mares may not cycle due to follicular or corpora luteal abnormalities or have silent heat periods in which external signs of estrus are not evident. Because of these situations, other methods of pregnancy determination are commonly used.

2. **Rectal palpation.** The most widely used method of pregnancy determination is by rectal palpation. An experienced technician can determine pregnancy (or barrenness) of mares at 98 to 100% accuracy by feeling with the hand through the rectal wall. Normally, the test is made 43 to 45 days after breeding, but pregnancy in maiden mares often can be detected within 35 to 40 days following conception. When performed by an experienced person, the manual test is quite reliable.

The following procedure is employed by most experienced technicians in making the manual test:

a. The examination is made in surroundings familiar to the mare and in an unhurried manner, as this makes for a minimum of restraint and avoids roughness.

b. Two helpers are needed; one to twitch the mare, and the other to hold the tail to one side. If the mare objects to the twitch, it is left off.

c. The latex obstetrical sleeve with glove attached is slipped on and lubricated. The rectum is entered and evacuated, the arm is inserted nearly to the shoulder, reaching forward and downward until the ovaries are located (the left ovary is most accessible for right-handed operators; the right ovary for left-handed operators); and the uterus is gently palpated and massaged. If the mare is 43 to 45 days pregnant, an enlargement approximately the size of a large orange can be located along the bottom of one uterine horn.

3. **Blood tests.** Blood tests, involving blood samples taken 20 to 120 days following breeding, give an indication of pregnancy.

4. **Ultrasonography.** In recent years, ultrasonography has increased in use in pregnancy determination. This technique can be used to obtain a visual image of the mare's reproductive tract, and thus detection of pregnancy, before palpation

is normally performed. Ultrasonography can detect pregnancies as early as 10 days postovulation.

The ultrasound can also be used to detect follicular development, to detect estrus, and to detect reproductive abnormalities.

## CARE AND MANAGEMENT OF THE STALLION

Although certain general recommendations can be made, it should be remembered that each stallion should be studied as an individual, and his care, feeding, exercise, and handling should be varied accordingly.

Fig. 53–5. *Secretariat*, great Thoroughbred stallion, shown in breeding condition, at Claiborne Farm, Paris, Kentucky, with Mr. Lawrence Robinson, Head Stud Groom. (Photo by J. Noye, Versailles, KY. Courtesy, Dr. W. C. Kaufman, DVM, Claiborne Farm, Paris, KY)

## QUARTERS

The most convenient arrangement for the stallion is a roomy box stall which opens directly onto a 2- or 3-acre pasture paddock, preferably separated from the other horses by a double fence. A paddock fence made of heavy lumber is safest. The stall door opening into such a paddock may be left open except during extremely cold weather; this will give the stallion plenty of fresh air, sunshine, and additional exercise.

## FEEDING

The feed and water requirements of the stallion are adequately discussed in Chapter 54. In addition to this, it may be well to reemphasize that, in season, clean, lush pastures produced on fertile soils are excellent for the stallion. Grass is the horse's most natural feed, and it is a rich source of the vitamins

that are so necessary for vigor and reproduction. Perhaps the ideal arrangement in providing pasture for the stallion is to give him access to a well-sodded paddock.

## EXERCISE

Regular exercise daily for the stallion is important. It is one of the best means of keeping the horse in a thrifty, natural condition and virile.

Stallions of the light-horse breeds are most generally exercised under the saddle or hitched to a cart. Thus, Standardbred stallions are usually jogged 3 to 5 miles daily while drawing a cart. Thoroughbred stallions and saddle stock stallions of all other breeds are best exercised under the saddle for from 30 minutes to 1 hour daily, especially during the breeding season.

Exercise should not be hurried or hard; the walk and the trot are the best gaits to use for this purpose. After the stallion is exercised, he should be rubbed down and cooled off before he is put up, especially if he is hot. Better yet, the ride should be so regulated at the end that the horse will be brought in cool, in which case he can be brushed off and turned into his corral.

Frequently, in light horses, bad feet exclude exercise on roads, and faulty tendons exclude exercise under the saddle. Under such a condition, one may have to depend upon (1) exercise taken voluntarily by the stallion in a large paddock, (2) longeing or exercising on a 30- to 40-foot rope, or (3) leading.

Longeing should be limited to a walk and a trot; and, if possible, the stallion should be worked on both hands; that is, made to circle both to the right and to the left. It is also best that this type of exercise be administered within an enclosure. Two precautions in longeing are (1) do not longe a horse when the footing is slippery, and (2) do not pull the animal in such a manner as to make him pivot too sharply with the hazard of breaking a leg.

Leading is a satisfactory form of exercise for some stallions if it is not practical to ride them. In leading, a bridle should always be used—never a halter—and one should keep away from other horses and be careful that the horse being ridden is not a kicker.

The objection to relying upon paddock exercise alone is that the exercise cannot be regulated, especially during inclement weather. Some animals may take too much exercise and others too little. Moreover, merely running in the paddock will seldom, if ever, properly condition any stallion.

A 2- or 3-acre grassy paddock should always be provided, even for horses that are regularly exercised. Stallions that are worked should be turned out at night and on idle days. While the pasture paddock is much superior to close confinement and no exercise in a stall, the stallion not being used regularly at useful work will be benefitted by any additional forced exercise, even though it does become somewhat monotonous.

## GROOMING

Proper grooming of the stallion is necessary, not only to make the horse more attractive in appearance, but to assist in maintaining the best of health and condition. Grooming serves to keep the functions of the skin active. It should be thorough, with special care taken to keep all parts of the body clean and free from any foulness, but not so rough nor severe as to cause irritation either of the skin or temper.

## AGE AND SERVICE OF THE STALLION

It should be remembered that the number and kind of foals that a stallion sires in a given season is more important than the total number of services. The number of services allowed during a season will vary with the age, development, temperament, health, and breeding condition of the stallion and the distribution of services. Therefore, no definite best number of services can be recommended for any and all conditions. Yet, the practices followed by good stud managers are not far different. All are agreed that excessive service of the stallion may reduce his fertility.

Table 53–1 contains recommendations relative to the number of services for stallions, with consideration given to age and type of mating.

Because of their more naturally nervous temperaments, stallions of the light-horse breeds are usually more restricted in services than stallions of the draft-horse breeds. Also, there is a difference between breeds.

The most satisfactory arrangement for the well-being of the stallion is to allow not more than one service each day. With proper handling, however, the mature, vigorous stallion may with certainty, and apparently without harm, serve two mares in a single day. During the heavy spring breeding season, this may often be necessary. It is a good plan to allow a stallion to rest at least one day a week.

### TABLE 53–1
### HANDY STALLION MATING GUIDE[1]

| Age | No. of Matings/Yr. | | Comments |
|---|---|---|---|
| | Hand Mating | Pasture Mating | |
| 2-year-old .......... | 10–15 | Preferably no pasture mating | Limit the 2-year-old to 2 to 3 services/week; the 3-year-old to 1 service/day; and the 4-year-old or over to 2 to 3 services/day. |
| 3-year-old .......... | 20–40 | | |
| 4-year-old .......... | 40–60 | | |
| Mature horse ...... | 50–70 | | A stallion should remain a vigorous and reliable breeder up to 20 to 25 years of age. |
| Over 18 years old .. | 20–40 | | |

[1]There are breed differences. Thus, when first entering stud duty, the averge 3-year-old Thoroughbred shold be limited to 20 to 25 mares per season, whereas a Standardbred of the same age may breed 25 to 30 mares; and the 4- or 5-year-old Thoroughbred should be limited to 30 to 40 mares, whereas a Standardbred of the same age may breed 40 to 50 mares. Mature stallions of the draft breeds may and do breed up to 70, or more, mares in a season.

In order to secure a higher conception of the mares and yet avoid overwork of the stallion with an excessive number of natural services, many breeders now reinforce each natural service with one artificial insemination.

Stallions often remain virile and valuable breeders until 20 to 25 years of age, especially if they have been properly handled. However, it is usually best to limit the number of services on a valuable old sire in order to preserve his usefulness and extend his longevity as long as possible.

Occasionally, Thoroughbred and Standardbred stallions are used to a limited extent before retirement to the stud, although many good breeders seem to feel that it is not best to use them until it is time for them to be retired. Saddle horses may be bred to a few mares and still be used in the show-ring. However, sometimes it makes them more difficult to handle.

It frequently happens that a wonderful horse is injured in the midst of his racing career, and, while awaiting the next racing season, he is bred to a few mares.

If two services a day are planned with the mature stallion, one should be rather early in the morning and the other late in the afternoon. It is also best not to permit teasing or service immediately before or after feeding the stallion, for this may result in digestive disturbances, particularly in the nervous, fretful individual.

## CARE OF THE PREGNANT MARE

Barren and foaling mares are usually kept separately because pregnant mares are sedate, whereas barren mares are more likely to run, tease, and kick. Precautions in handling the pregnant mare will be covered in the discussion that follows.

### QUARTERS

If mares are worked under the saddle or in harness, they may be given quarters like those accorded to the rest of the horses used similarly, at least until near parturition time. Idle mares may best be turned to pasture. Even in the wintertime, a simple shelter is adequate. In some sections of the country, an open shed is satisfactory.

### FEEDING

The feed and water requirements of the pregnant mare are adequately discussed in Chapter 54, so repetition is unnecessary.

### EXERCISE

The pregnant mare should have plenty of exercise. This may be obtained by allowing a band of broodmares to roam over large pastures in which shade, water, and minerals are available. Mares of the light-horse breeds may be exercised for an hour daily under the saddle or hitched to a cart. When handled carefully, the broodmare may be so exercised to within a day or two of foaling. Above all, when not receiving forced exercise or on idle days, she should not be confined to a stable or a small dry lot.

## CARE AT FOALING TIME

A breeding record should be kept on each mare so that it will be known when she is due to foal. As has been previously indicated, the period of gestation of a mare is about 336 days, but it may vary as much as a month in either direction. Therefore, the careful and observant caretaker will be ever alert and make certain definite preparations in ample time.

The period of parturition is one of the most critical stages in the life of the mare. Through carelessness or ignorance, all of the advantages gained in selecting genetically desirable and healthy parent stock and in providing the very best of environmental and nutritional conditions through gestation can be quickly dissipated at this time. Generally speaking, less difficulty at parturition was encountered in the wild state, when the females of all species brought forth their young in the fields and glens.

## WORK AND EXERCISE

Saddle or light-harness mares should be exercised moderately in the accustomed manner. If they are not used one or two days prior to foaling, other gentle exercise, such as leading, should be provided. This is especially important if she has not been accustomed to being on pasture and if it is desired to avoid any abrupt changes in feeding at this time.

## SIGNS OF APPROACHING PARTURITION

Perhaps the first sign of approaching parturition is a distended udder, which may be observed 2 to 6 weeks before foaling time. About 7 to 10 days before the arrival, there will generally be a marked shrinking or falling away of the muscular parts of the top of the buttocks near the tailhead and a falling of the abdomen. Although the udder may have filled out previously, the teats seldom fill out to the ends more than 4 to 6 days before parturition. About this time, the vulva becomes full and loose. As foaling time draws nearer, milk will drop from the teat; and the mare will show restlessness, break into a sweat, urinate frequently, lie down and get up, etc. It should be remembered however, that there are times when all signs fail and a foal may be dropped when least expected. Therefore, it is well to be prepared as much as 30 days in advance of the expected foaling date.

## PREPARATION FOR FOALING

Approximately one month ahead of foaling, consideration should be given to vaccinating gestating mares. At this time, one prominent California horsebreeder routinely gives each mare a four-way booster shot against tetanus, influenza, and Western and Eastern encephalomyelitis (sleeping sickness), along with a vaccine for the prevention of strangles. The tetanus booster shot administered at this time forms sufficient antibodies to protect both the mare and foal (the latter via colostrum). Also, at this time, sutures are removed from sutured (Caslicked) mares; and the feet and teeth receive such attention as necessary.

When signs of approaching parturition seem to indicate that the foal may be expected within 7 to 10 days, arrangements for the place of foaling should be completed. Thus, the mare will become accustomed to the new surroundings before the time arrives.

During the spring, summer, and fall months when the weather is warm, the most natural and ideal place for foaling is a clean, open pasture away from other livestock. Under these conditions, there is decidedly less danger of either infection or mechanical injury to the mare and foal. Of course, in following this practice, it is important that the ground be dry and warm. Small paddocks or lots that are unclean and foul with droppings are unsatisfactory and may cause such infectious troubles as navel-ill.

During inclement weather, the mare should be placed in a roomy, well-lighted, well-ventilated, comfortable, quiet box stall, free of projections, which should first be carefully cleaned, disinfected, and bedded for the occasion. It is best that the mare be stabled therein at nights a week to 10 days before foaling so that she may become accustomed to the new surroundings. The foaling stall should be at least 12 ft square and free from any low mangers, hayracks, or other obstructions that might cause injury either to the mare or to the foal. After the foaling

Fig. 53–6.   An Arabian mare and her newborn foal. (Courtesy, Arabian Horse Registry of America, Inc., Westminster, CO)

stall has been thoroughly cleaned, it should be disinfected to reduce possible infection. This may be done by scrubbing with boiling-hot lye water, made by using 8 oz of lye to 20 gal. of water (one-half this strength of solution should be used in scrubbing mangers and grain boxes). The floors should then be sprinkled with air-slaked lime. Plenty of clean, fresh bedding should be provided at all times.

A foaling stall somewhat away from the other horses and with a smooth, well-packed clay floor is preferred. The clay floor may be slightly more difficult to keep smooth and sanitary than concrete or other such surface materials, but there is less danger to the mare and the newborn foal from slipping and falling; and it is decidedly better for the hoofs.

## FEEDING AT FOALING TIME

Shortly before foaling, it is usually best to decrease the grain allowance slightly and to make more liberal use of light and laxative feeds, especially wheat bran. If there are any signs of constipation, a wet bran mash should be provided.

## THE ATTENDANT

A good rule for the attendant is to *be near but not in sight*. Some mares seem to resent the presence of an attendant at this time, and they will delay foaling as long as possible under such circumstances. Mares that have foaled previously and which have been properly fed and exercised will usually not experience any difficulty. However, young mares foaling for the first time, old mares, or mares that are either in an overfat or in a thin, run-down condition may experience considerable difficulty. The presence of the attendant may prevent possible injury to the mare and foal; and, when necessary, the attendant may aid the mare or call a competent veterinarian.

## PARTURITION

The immediate indications that the mare is about to foal are extreme nervousness and uneasiness, lying down and getting up, biting of the sides and flanks, switching of the tail, sweating in the flanks, and frequent urination.

The first actual indication of foaling is the rupture of the outer fetal membrane, followed by the escape of a large amount of fluid. This is commonly referred to as the rupture of the *water bag*. The inner membrane surrounding the foal appears next, and labor than becomes more marked.

With normal presentation, a mare foals rapidly, usually not taking more than 15 to 30 minutes. Usually, when the labor pains are at their height, the mare will be down; and it is in this position that the foal is generally born, while the mare is lying on her side with all legs stretched out.

In normal presentation, the front feet, with heels down, come first, followed by the nose which is resting on them, then the shoulders, the middle (with the back up), the hips, and then the hind legs and feet. If the presentation is other than normal, a veterinarian should be summoned at once, for there is great danger that the foal will smother if its birth is delayed. If the feet are presented with the bottoms up, it is a good indication that they are the hind ones, and there is likely to be difficulty.

If after reasonable time and effort have been expended, a mare appears to be making no progress in parturition, it is advisable that an examination be made and assistance be rendered before the animal has completely exhausted her strength in futile efforts at expulsion. In rendering any such assistance, the following cardinal features should exist:

1. Cleanliness.
2. Quietness.
3. Gentleness.
4. Perseverance.
5. Knowledge, skill, and experience.

When parturition is unduly delayed or retarded, the fetus often dies from twists or knots in the umbilical cord, or from remaining too long in the passage. In either case, there may be stoppage of fetal circulation or lack of oxygen for the fetus, or both.

If foaling has been normal, the attendant should enter the stable to make certain that the foal is breathing and that the membrane has been removed from its mouth and nostrils. If the foal fails to breathe immediately, artificial respiration should be applied. This may be done by blowing into the mouth of the foal, working the ribs, rubbing the body vigorously and permitting the foal to fall around. Then after the navel has been treated, the mare and foal should be left to lie and rest quietly as long as possible so that they may gain strength.

## THE AFTERBIRTH

To aid in afterbirth (placenta) expulsion and involution of the uterus, immediately after foaling some horse breeding farm managers give each foaling mare, except old mares or mares that have had difficult foaling, 100 USP units of the hormone oxytocin.

If the afterbirth is not expelled as soon as the mare gets up, it should either be tied up in a knot or be tied to the tail of the mare. This should be done so that the foal or mare will not step on it, thereby increasing the danger of inflammation of the uterus and foal founder in the mare. Usually the afterbirth will be expelled within one to six hours after foaling.

If it is retained for a longer period or if lameness is evident, the mare should be blanketed, and an experienced veterinarian should be called. Retained afterbirth often causes laminitis, which is recognized by lameness in the mare. This is usually treated by feeding easily digested feed for a period of 36 hours and by applying cold applications to the mare's feet until the condition is relieved.

The afterbirth should always be examined for evidence of infection and to ensure that none of it remains in the mare.

To prevent development of bacteria and foul odors, the afterbirth should be removed from the stall and burned or buried in lime.

A soon as possible after the afterbirth has passed, mares that were previously sutured are resutured (Caslicked).

## CLEANING THE STALL

Once the foal and mare are up, the stall should be cleaned. Wet, stained, or soiled bedding should be removed. The floor should be sprinkled with lime; and clean, fresh bedding should be provided. Such sanitary measures will be of great help in preventing the most common type of joint-ill.

If the weather is extremely cold and the mare is hot and sweaty, she should be rubbed down, dried, and blanketed soon after getting on her feet.

## FEED AND WATER AFTER FOALING

Fig. 53–7. When the mare begins lactating, her nutritional requirements increase significantly. (Courtesy, Bobbie Lieberman, Gaithersburg, MD)

Following foaling, the mare usually is somewhat hot and feverish. She should be given small quantities of lukewarm water at intervals but should never be allowed to gorge. It is also well to feed lightly and with laxative feeds for the first few days. The very first feed might well be a wet bran mash with a few oats or a little oatmeal soaked in warm water. About one-half the usual amount should be fed. Usually, for the first week, no better grain ration can be provided than bran and oats. The quantity of feed given should be governed by the milk flow, the demands of the foal, and the appetite and condition of the mare. Usually the mare can be back on full feed within a week or 10 days after foaling.

## BE OBSERVANT

The good caretaker will be ever alert to discover difficulties before it is too late. If the mare has much temperature (normal for the horse is about 100.5°F) something is wrong, and the veterinarian should be called. As a precautionary measure, the mare's temperature should be taken a day or two after foaling. Any discharge from the vulva should also be regarded with suspicion.

## HANDLING THE NEWBORN FOAL

Immediately after the foal has arrived and breathing has started, it should be thoroughly rubbed and dried with warm towels. Then it should be placed in one corner of the stall on clean, fresh straw. Usually the mare will be less restless if this corner is in the direction of her head. The eyes of a newborn foal should be protected from bright light.

### THE NAVEL CORD

At the time the umbilical cord is ruptured, there is a direct communication from outside the body to some of the vital organs and the blood of the foal. Usually this opening is soon closed by the ensuing swelling and final drying and sloughing-off process. Under natural conditions, in the wild state, there was little danger of navel infection, but domestication and foaling under confined conditions have changed all this.

To reduce the danger of navel infection (which causes a disease known as joint-ill or navel-ill) the navel cord of the newborn foal should be treated at once with tincture of iodine (Metaphen or Merthiolate may be used). This may be done by placing the end of the cord in a widemouthed bottle nearly full of tincture of iodine while pressing the bottle firmly against the abdomen. This is best done with a foal lying down. The cord should then be dusted with a good antiseptic powder. Dusting with powder should be continued daily until the stump dries up and drops off and the scar heals, usually in three to four days. If an antiseptic powder is not available, air-slaked lime may be used. Any foreign matter that accumulates on the navel should be pressed out, and a disinfectant should be applied.

If left alone, the navel cord of the newborn foal usually breaks within 2 to 4 in. from the belly. Under such conditions, no cutting is necessary. However, if it does not break, it should be severed about 2 in. from the belly with clean, dull shears,

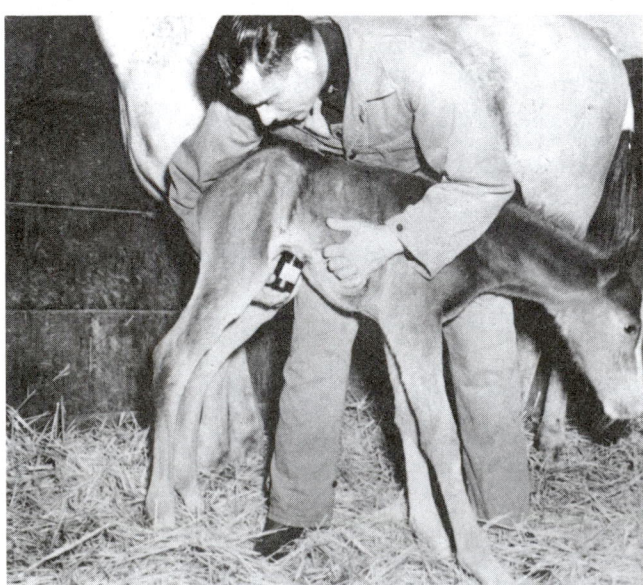

Fig. 53–8.   Treating the navel cord of the newborn foal with iodine.

or it may be scraped in two with a knife. Never cut diagonally across. A torn or broken blood vessel will bleed very little; whereas, one that is cut directly across may bleed excessively. If severing of the cord is resorted to, it should be immediately treated with iodine.

### NAVEL INFECTION (JOINT-ILL OR NAVEL-ILL)

Although most newborn foal infections are referred to as navel infection—implying that the infection is postnatal, with entrance to the body gained through the umbilical cord after birth—many such troubles are of prenatal origin. In the latter type, infection of the foal takes place in the uterus (womb) of the dam before the foal is born. The infection may either be present in the dam before she is bred, or it may be introduced by the stallion, if he is infected or if he has previously bred other infected mares. If prenatal infection does not result in abortion and the mare carries the fetus to normal term, the foal is often born weak or develops navel-ill within a few days and dies; or if it does not die, it becomes a hopeless cripple that must be destroyed.

Fig. 53–9.   Foal with navel infection (joint-ill or navel-ill). The disease is fatal in about 50% of the cases. Also, a large proportion of the animals that survive are left with deformed joints, like the foal shown above. (Courtesy, College of Veterinary Medicine, University of Illinois)

Under unsanitary conditions, there is also great danger from infection that may enter the bloodstream through the opening of the navel cord prior to the time it has dried up and the scar has healed over. When weather conditions permit foaling on a clean pasture in the fresh air and sunshine, danger of such infection is held to a minimum. On the other hand, foaling in a filthy paddock or stall and with no precautions taken is very likely to result in infection and navel-ill. For this reason, when it is necessary to have mares foal in the stall, every precaution should be taken. The stall should be thoroughly cleaned, disinfected, and bedded; and the navel should be treated with iodine immediately after the foal arrives, followed by dusting with a good antiseptic powder several times daily.

Navel infection (joint-ill or navel-ill) may be recognized by a loss of appetite, soreness and stiffness in the joints, and a general listlessness of the foal. If this is recognized in the early stages and a veterinarian is called at once, the infected foal may be treated and may recover. If, however, the disease has reached the pus-forming stage, very likely it will be fatal. Blood transfusions from the dam to foal have been given in all types of foal infections, usually with good results. With certain specific types of infections, antibiotics, sulfanilamide, serums, or bacterins may be used successfully; but these should always be administered by a veterinarian. Prevention is decidedly the best protection.

In summary, it may be stated that the practice of sanitation and hygiene, starting with the stallion and broodmare at the time of mating and continuing with the broodmare and the young foal at foaling time, usually prevents the most common type of joint-ill. In certain areas, particularly those known to be goiterous or semigoiterous, such as the Pacific Northwest, the feeding of stabilized iodized salt to in-foal mares appears to reduce losses from joint-ill.

## THE COLOSTRUM

The colostrum is the milk that is secreted by the dam for the first few days following parturition. It differs from ordinary milk in the following aspects:

1. It is more concentrated.
2. It is higher in protein content, especially globulins.
3. It is richer in vitamin A.
4. It contains more antibodies.
5. It has a more stimulating effect on the alimentary tract.

Because of these beneficial qualities of colostrum, the caretaker should make very certain that the newborn foal secures this first milk.

The strong, healthy foal will usually be up on its feet and ready to nurse within 30 minutes to 2 hours after birth. Occasionally, however, a big, awkward foal will need a little assistance and guidance during its first time to nurse. The stubborn foal should be coaxed to the mare's teats (forcing is useless). This may be done by backing the mare up on additional bedding in one corner of the stall and coaxing the foal with a bottle and nipple. The attendant may hold the bottle while standing on the opposite side of the mare from the foal. The very weak foal should be given the mare's first milk even if it must be drawn in a bottle and fed by nipple for a time or two. Sometimes these weak individuals will nurse the mare if steadied by the attendant.

Aside from the difference in chemical composition, the colostrum (the milk yielded by the mother for a short period following the birth of the young) has the following functions:

1. It contains antibodies that temporarily protect the foal against certain infections, especially those of the digestive tract.
2. It serves as a natural purgative, removing fecal matter that has accumulated in the digestive tract.

This, therefore, explains why mares should not be milked out prior to foaling and why colostrum is important to the newborn foal.

Before allowing the foal to nurse for the first time, it is usually good practice to wash the mare's udder with a mild disinfectant and to rinse it with clean, warm water.

## BOWEL MOVEMENT OF THE FOAL

The regulation of the bowel movement in the foal is very important. Two common abnormalities are constipation and diarrhea or scours.

Impaction in the bowels of the excrement accumulated during the development prior to birth—material called meconium—may prove fatal if not handled promptly. Usually a good feed of colostrum will cause elimination, but not always—especially when foals are from stall-fed mares.

Bowel movement of the foal should be observed within 4 to 12 hours after birth. If by this time there has been no discharge and the foal seems rather sluggish and fails to nurse, it should be given an enema. This may be made by using 1 to 2 quarts of water at blood heat, to which a little glycerin has been added; or warm soapy water is quite satisfactory. The solution may be injected with a baby syringe (one having about a 3-inch nipple) or a tube and can. This treatment may be repeated as often as necessary until the normal yellow feces appear.

If the foal is scouring, the ration of the mare should be reduced, and a part of her milk should be taken away by milking her out at intervals.

Diarrhea or scours in foals may be associated with infectious diseases or may be caused by unclean surroundings. Any of the following conditions may bring on diarrhea: contaminated udder or teats; non-removal of fecal matter from the digestive tract; fretfulness or temperature above normal in the mare; and excess of feed affecting the quality of the mare's milk; cold, damp bed; or continued exposure to cold rains. As treatment is not always successful, the best practice is to avoid the undesirable conditions.

Diarrhea is caused by an irritant in the digestive tract that should be removed if recovery is to be expected. Only in exceptional cases should an astringent be given with the idea of checking the diarrhea; and such treatment should be prescribed by a veterinarian.

## RAISING THE ORPHAN FOAL

Occasionally a mare dies during or immediately after parturition, leaving an orphan foal to be raised. Also, there are times when mares fail to give a sufficient quantity of milk for the newborn foal. Sometimes there are twins. In such cases, it is necessary to resort to other milk supplies. The problem will be simplified if the foal has at least received the colostrum from the dam, for it does play a very important part in the well-being of the newborn young.

If at all possible, the foal should be shifted to another mare. Some breeding establishments regularly follow the plan of breeding a mare that is a good milk producer but whose foal is expected to be of little value. Her own foal is either destroyed or raised on a bottle, and the mare is used as a foster mother or nurse mare.

Most nurseries usually keep a supply of colostrum on hand. They remove colostrum from mares that (1) have had dead foals, or (2) produce excess milk, which is then stored in a freezer for future use for foals that do not receive colostrum from their dams. When needed, it can be removed from the freezer, heated, and fed. This is an excellent practice.

If no colostrum is available, the foal should be placed on either (1) cow's milk made as nearly as possible of the same composition as mare's milk, or (2) a synthetic milk replacer.

A comparison of cow's milk and mare's milk is given in Table 53–2.

**TABLE 53–2**
**COMPOSITION OF MILK FROM COWS AND MARES[1]**

| Source | Water | Protein | Fat | Sugar | Ash |
|--------|-------|---------|-----|-------|-----|
|        | (%)   | (%)     | (%) | (%)   | (%) |
| Cow .............. | 87.18 | 3.55 | 3.69 | 4.88 | 0.75 |
| Mare .............. | 90.78 | 1.99 | 1.21 | 5.67 | 0.35 |

[1]USDA, Farmers' Bull. No. 803.

As can be observed, mare's milk is higher in percentage of water and sugar than cow's milk and is lower in other components.

For best results in raising the orphan foal, milk from a fresh cow, low in butterfat, should be used. To about a pint of milk, add a tablespoon of sugar and from 3 to 5 tablespoonsful of lime water. Warm to body temperature, and for the first few days feed about ¼ pint every hour. After 3 to 4 weeks the sugar can be stopped, and at 5 to 6 weeks skimmed milk can be used entirely.

Orphan foals may also be raised on milk replacer, fed according to the directions of the manufacturer. Here again the situation is simplified if the foal has first received colostrum.

For the first few days, the milk (either cow's milk or milk replacer) may be fed by using a bottle and a rubber nipple. Later, the foal should be taught to drink from a pail. It is important that all receptacles be kept absolutely clean and sanitary (clean and scald each time), and that feeding be at regular intervals. Grain feeding should be started at the earliest possible time with the orphan foal.

## NORMAL BREEDING SEASON AND THE TIME OF FOALING

The most natural breeding season for the mare is in the spring of the year. Usually mares are gaining flesh at this time; the heat period is more evident; and they are more likely to conceive. Furthermore, the spring-born foal may be dropped on pasture—with less danger of infection and with an abundance of exercise, fresh air, and sunshine to aid in its development. Also there will be good, green, succulent pasture for the mare. Such conditions are ideal.

However, when the demands for using the mares are such that spring foaling interferes and fall or perhaps late winter foaling are desired, plans may be changed accordingly. Under such circumstances, spring conditions should be duplicated at the breeding season. That is, the mare should be fed to gain in flesh and, if necessary, should be blanketed for comfort.

Also, it must be remembered that the exhibitors will want to give consideration to having the foals dropped at such time that they may be exhibited to the best advantage. The same applies to the person who desires to sell well-developed yearlings of the light-horse breeds or to race two-year-olds. It is noteworthy, however, that the percentage of barren mares that conceive at an early breeding is markedly lower than is obtained later in the season. Nevertheless, some mares do conceive early in the year, and even a small percentage is advantageous to some breeders.

## ARTIFICIAL INSEMINATION

Stallion semen is now being collected, processed, and frozen somewhat like bull semen. As a result, there is renewed interest in artificial insemination (AI) in horses. The main advantage of AI is that it allows more mares to be bred from a single ejaculate, about 17 may be bred by AI compared to only one when natural service is used. When properly used, AI will result in over 90% conception after three estrous cycles, compared to 80 to 85% using natural service. Consequently, it is not uncommon for some of the more popular stallions to breed as many as 400 to 500 mares in a single season by being collected only once daily or every other day. By comparison, when using natural service the stallion should not be booked for more than 75 to 80 mares, and he would have to serve as much as three times per day almost every day of the breeding season.

## HYBRIDS WITH THE HORSE AS ONE PARENT

The mule, representing a cross between the jack (male of the ass family) and the mare (female of the horse family) is the best-known hybrid in the United States. The resulting offspring of the reciprocal cross of the stallion mated to a jennet is known as the hinny.

Rarely have mules proven fertile; only five authentic cases of mare mules producing foals have been reported in the United States. This infertility of the mule is probably due to the fact that the chromosomes will not pair and divide equally in the reduction division.

The offspring of fertile mules are generally horselike in appearance,[1] showing none of the characteristics of the mule's sire (or ass). For the most part, therefore, the eggs (ova) which produce them do not carry chromosomes from the ass; they are pure horse eggs without any inheritance from their maternal grandfathers. This indicates that in the production of eggs in mare mules the reduction division is such that all of the horse chromosomes go to the egg and none to the polar bodies.

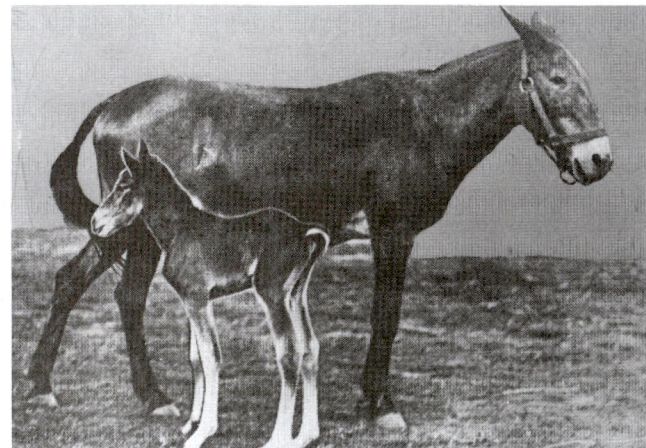

Fig. 53–10. It has been said that the mule has no pride of ancestry and no hope of posterity, but *Old Beck*—a bay *cotton-type* mule owned by Texas A&M University—was an exception. Shown above is *Old Beck* and her second living foal, a *horselike* offspring sired by a stallion. The colt was fertile and sired a number of living foals. (Courtesy, Texas A&M University)

[1]Not all are horselike, however. Thus, one of *Old Beck's* (fertile mule owned by Texas A&M University) three offspring was mulelike in appearance.

The zebroid, a zebra × horse hybrid, is rather popular in certain areas of the tropics because of its docility and resistance to disease and heat.

## LETHALS IN HORSES

*The term lethal refers to a genetic factor that causes death of the young, either during prenatal life or at birth.* Lethals which have been reported in horses include the following: abnormal sex ratio, atresia coli, lethal white, and stiff forelegs. Breeding animals carrying these or other lethal factors should be culled.

## PERFORMANCE TESTING HORSES

The breeders of racehorses have always followed a program of mating animals of proven performance on the track. For example, it is interesting to note that the first breed register which appeared in 1791—known as ''An Introduction to the General Stud Book,'' recorded the pedigrees of all the Thoroughbred horses winning important races. In a similar way, the Standardbred horse—which is an American creation—takes its name from the fact that, in its early history, animals were required to trot a mile in 2 minutes and 30 seconds, or to pace a mile in 2 minutes and 25 seconds, before they could be considered eligible for registry. The chief aim, therefore, of the early-day breeders of racehorses was to record the pedigree of outstanding performers rather than all members of the breed.

The simplest type of progeny testing in horses consists of the average record of merit of an individual stallion's or mare's offspring. Thus, the offspring of Thoroughbred or Standardbred animals bred for racing may be tested by timing on the track. Less satisfactory tests for saddle horses and harness horses have been devised. However, it is conceivable that actual exhibiting on the tanbark in the great horse shows of the country may be an acceptable criterion for saddle- and harness-bred animals. Also, the dynamometer might conceivably be used for testing animals of draft-horse breeding, although it has not been so used in the past.

## HOW TO LOWER THE COST OF RAISING HORSES

Some management principles that should receive consideration in lowering the cost of raising horses are:

1. Attain higher fertility in both mares and stallions; secure a higher percent foal crop. With a 50% foal crop, two mares are kept a whole year to raise one foal.

2. Eliminate unnecessary concoctions, and drugs, if they are not needed.

3. Begin using horses moderately at two years of age, at which time their use should more than compensate for the feed cost.

4. Keep all horses of usable age earning their way. Animals that are not necessary or that do not increase in value at a profitable rate are a needless expense.

5. Utilize pastures to the maximum. Such a practice will supply nutritious feeds at a low cost, save time in feeding, reduce labor in caring for the horses, and do away with bedding the stalls and cleaning the barn.

6. Utilize the less salable roughage as much as possible, particularly during the second and third years.

Fig. 53–11. The cost of raising horses can generally be lowered by utilizing pastures to the maximum. This shows Quarter Horse mares and foals on pasture in Oklahoma. (Courtesy, *The Quarter Horse Journal*, Amarillo, TX)

7. Do not construct or maintain costly quarters for the young, growing horse.

8. Keep animals free from parasites, both internal and external. Feeding parasites is always too costly.

9. Provide a balanced ration, including a balance of proteins, necessary minerals, and vitamins. Plenty of good, clean water should be available at all times.

## BUYING HORSES OR RAISING FOALS

Where horses are needed, either they must be purchased or foals must be raised. The primary factors to consider in determining whether horses will be bought or foals raised are (1) the experience of the individual, (2) comparative cost, and (3) risks surrounding the introduction of horses.

## EXPERIENCE OF THE BUYER/BREEDER

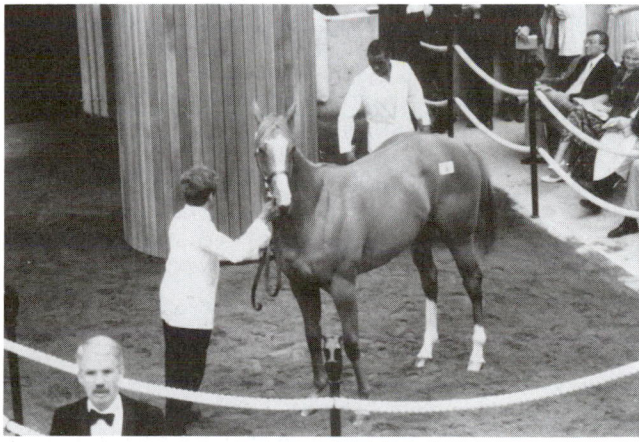

Fig. 53–12. A filly, sired by *Mr. Prospector* and out of *Larida*, which sold in the 1988 Saratoga Selected Yearling Sale, Saratoga Springs, New York, for $1.5 million. (Courtesy, Fasig-Tipton Company, Lexington, KY)

Certainly it must be recognized that the person who would attempt to raise replacements must have more knowledge of horse production than the person buying mature horses. In addition to knowing the regular care and management aspects of horse production, the person who raises replacements must be somewhat familiar with the breeding of horses and the rearing of foals.

## COMPARATIVE COST

In determining whether horses will be bought or foals raised, the comparative cost of the two methods should be computed. In arriving at such comparative cost figures, the following factors should be remembered:

1. Such figures should be on the basis of animals of equal merit and usefulness for the purpose desired. Consideration should also be given to age and future depreciation.

2. Computing the purchase price on horses should be on the basis of price delivered to the farm. Commission, freight or trucking, and insurance charges should not be overlooked.

3. In computing the cost of raising a foal to usable age, feed price should be figured on the basis of farm values rather than on actual grain market values. Also, consideration should be given to the fact that cheap and somewhat unusable roughages may often be used. Further, such items as service fees, manure produced, and handling charges should be considered.

## RISKS SURROUNDING THE INTRODUCTION OF HORSES

After giving full consideration to the experience of the buyer/breeder and the comparative cost of the two methods, there are still some rather perplexing problems encountered in introducing horses. These difficulties may be summarized as follows:

1. **Misrepresentations**—The inexperienced buyer especially, is likely to encounter misrepresentations as to age, soundness, vices, and the training and usefulness of the horse.

2. **Diseases**—In moving a horse, there is always a possible exposure to the many ills. Sometimes these are of sufficiently serious nature as to make the use of the animal impossible at a time when it is most needed; occasionally, they even prove fatal. Also, it must be remembered that such diseases as are contracted very likely may spread to the other horses on the farm and even to those in the community, thus exposing them to the same risk.

3. **Acclimating**—Horses coming from a distance usually need time to become acclimated before being most useful.

4. **Condition**—In all too many instances, horses brought in for sale and speculative purposes have been made fat for the occasion. Usually such liberal feeding has been made even more harmful through accompanying lack of work and confinement to a stall. Such horses are soft and require a period of gradual fitting for work. Also, it must be remembered that fat will cover up a multitude of defects.

## BLOOD TYPING

Horse blood typing was developed at the University of California at Davis during the period 1958–64. It involves a study of the components of the blood which are inherited according to strict genetic rules that have been established in the research laboratory. By determining the genetic *markers* in each blood sample and then applying the rules of inheritance, parentage can be affirmed or denied. To qualify as the offspring of a given mare and stallion, a foal must not possess any genetic markers not present in the alleged parents. If it does, it constitutes grounds for illegitimacy.

Horse blood typing is used for the following purposes:

• **To verify parentage**—The test is used in instances where the offspring may bear some unusual color or markings or carry some undesirable recessive characteristic. It may also be used to verify a registration certificate. Through blood typing, parentage can be verified with 90% accuracy.[2] Although this means that 10% of the cases cannot be settled, it is not possible to do any better than this in human blood typing.

• **To determine which of two sires**—When a mare has been served by two or more stallions during one breeding season, blood typing can exclude the incorrect stallion and include the correct stallion in over 90% of the cases.

• **To provide a permanent blood type record for identification purposes**—Two samples of blood are required from each animal to be studied; and the samples must be taken in tubes and in keeping with detailed instructions provided by the laboratory. In parentage cases, this calls for blood samples from the offspring and both parents; in paternity cases, samples must be taken from the offspring, the dam, and all the sires.

• **To substitute for fingerprinting**—Much attention is now being given to the idea of utilizing blood typing as a positive means of identification of stolen animals, through proving their parentage.

• **Blood typing laboratories**—The following laboratories are capable of determining equine parentage:

### In the United States

Department of Veterinary Sciences, University of Kentucky
102 Animal Pathology Building
Lexington, KY 40546

Serology Laboratory, School of Veterinary Medicine
University of California
Davis, CA 95616

Stormont Laboratories, Inc.
1237 E. Beamer St., Suite D
Woodland, CA 95695

### In Canada

Mann Equitest, Inc.
550 McAdam Road
Mississauga, Ontario L4Z 1P1

---

[2]In a personal communication to the author, Dr. Clyde Stormont, Professor Emeritus, University of California-Davis, presently owner/operator of Stormont Laboratories, Woodland, California, reported that approximately 91% of all horse parentage cases can be solved by blood typing.

# QUESTIONS FOR STUDY AND DISCUSSION

1. What is the normal age of puberty in fillies, and what is the normal age to breed mares?

2. Describe the normal heat period of mares. How would you breed (serve) the mare in order to secure the highest conception rates?

3. What are the pros and cons of rebreeding the mare at the first heat period following foaling?

4. Why is the percentage of mares bred producing foals so much lower than the percentage of young crop of cows, ewes, and sows?

5. List the things that may be done to increase the percentage of foals produced.

6. Discuss the pros and cons of a shift of the date of birth of foals from January 1 (a) to April 1, or (b) to the actual date when born—as is done with other classes of animals.

7. Describe each of the following types of breeding operations: hand breeding, corral breeding, and pasture breeding.

8. List the signs and tests of pregnancy of mares. Which pregnancy test would you recommend? Justify your choice.

9. Give recommendations relative to the number of services for stallions, with consideration given to age and type of mating.

10. Describe the signs of approaching parturition of the mare.

11. How should the owner of a pregnant mare prepare for foaling?

12. How should the mare be fed and watered after foaling?

13. How should the navel cord of the newborn foal be treated?

14. What is colostrum? How does it differ from ordinary milk? Why is it important for the newborn foal, and why is it important that it be consumed soon after birth?

15. Outline a program for raising an orphan foal.

16. Why do not the horse breed registrations liberalize their rules and regulations governing the registration of foals produced by A.I.?

17. How can you explain the following: (a) rarely have mules proven fertile, and (b) the offspring of mules are generally horselike in appearance?

18. How would you go about performance testing the breed of horses of your choice? List and discuss each step.

19. List and discuss some management principles that should receive consideration in lowering the cost of raising horses.

20. List and discuss the primary factors to consider in determining whether horses will be bought or foals raised.

21. What is blood typing? For what purposes may it be used?

# SELECTED REFERENCES

| Title of Publication | Author(s) | Publisher |
|---|---|---|
| Animal Breeding Plans | J. L. Lush | Collegiate Press, Inc., Ames, IA, 1963 |
| Arabian Horse Breeding | H. H. Reese | Bordon Publishing Company, Los Angeles, CA, 1953 |
| Breeding and Improvement of Farm Animals | V. A. Rice, et al. | McGraw-Hill Book Company, New York, NY, 1967 |
| Breeding and Raising Horses, Ag. Hdbk. No. 394 | M. E. Ensminger | U.S. Department of Agriculture, Washington, DC, 1972 |
| Equine Genetics & Selection Procedures | Staff of Equine Research Publications | Equine Research Publications, Dallas, TX, 1978 |
| Equine Reproducton | Edited by I. W. Rowlands W. R. Allen P. D. Rossdale | Blackwell Scientific Publications, Oxford, England, 1975 |
| Equine Reproduction II | Edited by I. W. Rowlands W. R. Allen | Journals of Reproduction and Fertility, Ltd., Cambridge, England, 1979 |
| Farm Animals | J. Hammond | Edward Arnold & Company, London, England, 1952 |

*(Continued)*

## SELECTED REFERENCES *(Continued)*

| Title of Publication | Author(s) | Publisher |
|---|---|---|
| *Genetic Principles in Horse Breeding* | J. F. Lasley | John F. Lasley, Columbia, MO, 1970 |
| *Genetics Is Easy* | P. Goldstein | Lantern Press, New York, NY, 1967 |
| *Genetics of the Horse* | W. E. Jones<br>R. Bogart | Edwards Brothers, Inc., Ann Arbor, MI, 1971 |
| *Genetics of Livestock Improvement*, Third Edition | J. F. Lasley | Prentice-Hall, Inc., Englewood Cliffs, NJ, 1978 |
| *Hammond's Farm Animals* | J. Hammond, Jr.<br>I. L. Mason<br>T. J. Robinson | Butler & Tanner, Ltd., Frome and London, England, 1971 |
| *Horse Science Handbook*, Vols. 1-3 | Edited by<br>M. E. Ensminger | Agriservices Foundation, Clovis, CA, 1963, 1964, 1966 |
| *Horsemanship and Horse Care*, Ag. Info. Bull. No. 353 | M. E. Ensminger | U.S. Department of Agriculture, Washington, DC, 1972 |
| *How Life Begins* | J. Power | Simon and Schuster, Inc., New York, NY, 1965 |
| *Lectures, Stud Managers Course* | | Stud Managers Course, Lexington, KY, intermittent since 1951 |
| *Light Horses*, Farmers' Bull. No. 2127 | M. E. Ensminger | U.S. Department of Agriculture, Washington, DC, 1965 |
| *Reproduction in Farm Animals*, Fourth Edition | E. S. Hafez | Lea & Febiger, Philadelphia, PA, 1980 |
| *Reproductive Physiology* | A. V. Nalbandov | W. H. Freeman and Co., Publishers, San Francisco, CA, 1958 |
| *Stud Farm Diary, A* | H. S. Finney | J. A. Allen & Co., Ltd., London, England, 1973 |
| *Stud Managers' Handbooks* | Edited by<br>M. E. Ensminger | Agriservices Foundation, Clovis, CA, annually since 1965 |
| *Studies on Reproduction in Horses* | Y. Nishakawa | Japan Racing Association, Tokyo, Japan, 1959 |

A horse breeding farm. (Courtesy, Velma Morrison, Boise, ID)

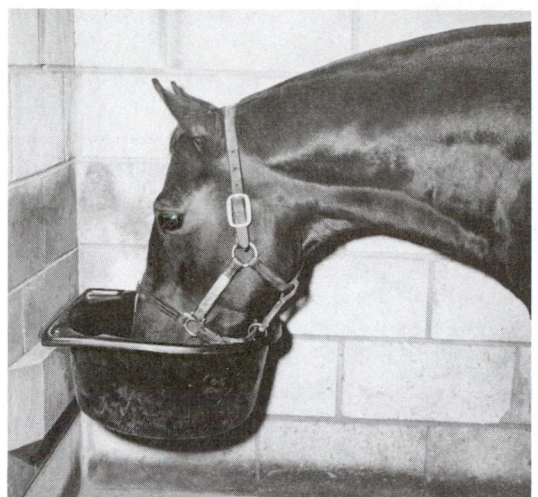

Feeding time!

# FEEDING AND MANAGING HORSES

## Chapter 54

Feed is the most important influence in the environment of the horse. Unless horses are fed properly, their maximum potential in reproduction, growth, body form, speed, endurance, style, and attractiveness cannot be achieved.

The following conditions make it imperative that the nutrition of horses be the best that science and technology can devise:

1. **Confinement.** Many horses are kept in stables or corrals most of the time.

2. **Fitting yearlings.** When forcing young equines, it is important to their development and soundness that the ration be nutritionally balanced.

3. **Racing two-year-olds.** In the United States, we race more two-year-olds than any other nation in the world; our richest races are for them. If the nutrient content of the ration is not adequate, there is bound to be more breakdown on the track than with older horses—this is costly.

4. **Stress.** Stress is affected by excitement, temperament, fatigue, number of horses together, previous nutrition, breed, age, and management. Race and show horses are always under stress; and the more tired they are and the greater the speed, the greater the stress. Thus, the ration for race and show horses should be scientifically formulated, rather than based on fads, foibles, and trade secrets. The greater the stress the more exacting the nutritive requirements.

5. **Horses are unique.** They differ from other farm animals from the standpoint of use and should not be fed the same feeds. They have greater value; are kept for recreation, sport, and work; are fed for a longer life of usefulness; have a smaller digestive tract; should not carry surplus weight; and are fed for nerve, mettle, animation, and character of muscle.

Also, feed constitutes the greatest single item in the horse business.

## DIGESTIVE SYSTEM OF THE HORSE

The digestive system of the horse differs anatomically and physiologically from that of the cow (ruminant) as follows:

1. It is smaller (see Fig. 54–1), with the result that the horse cannot eat as much roughage as cattle. Not only that, it functions best at two-thirds capacity. Because of its small size, if a horse is fed too much roughage, labored breathing and quick tiring may result. Actually, the horse's stomach is designed for almost constant intake of small quantities of feed (such as happens when a horse is grazing on pasture), rather than large amounts at one time.

2. Without feed, the horse's stomach will empty completely in 24 hours, whereas it takes about 72 hours (3 times as long) for the cow's stomach to empty. At the time of eating, feed passes through the horse's stomach very rapidly—so much so that the feed eaten at the beginning of the meal passes into the intestine before the last part of the meal is completed.

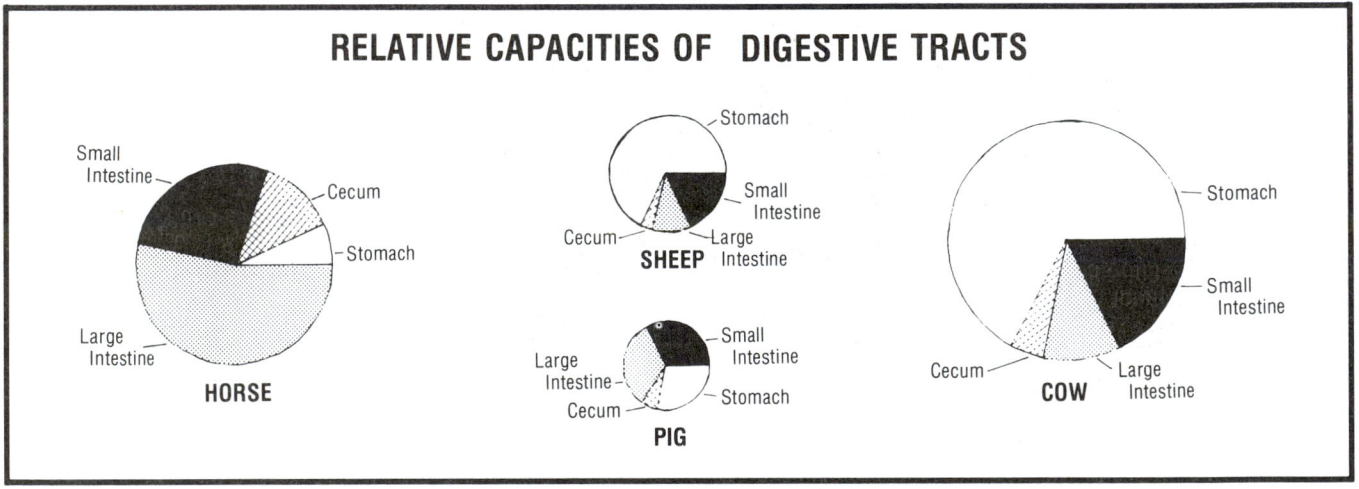

Fig. 54–1. Diagram showing comparative capacities of digestive tracts of different classes of farm animals. Note that the horse has a relatively small stomach.

3. The cow has four compartments (rumen, reticulum, omasum, and the abomasum or true stomach), whereas the horse has one.

4. There is comparatively little microbial action in the stomach of the horse, but much such action in the stomach (rumen) of the cow. As a result, the horse does not break down more than about 30% of the cellulose of feed, whereas the ruminant breaks down 60 to 70%. Hence, horses cannot handle as much roughage as can ruminants. Also, higher quality (lower cellulose content) forages must be fed to horses.

5. The primary seats of microbial activity in ruminants and horses occupy different locations in the digestive system in relation to the small intestine. In cows (and sheep), the rumen precedes the small intestine; in horses, the cecum follows it.

As a result, the efficiency of absorption of nutrients synthesized by the microorganisms is likely to be lower in a horse than in a ruminant.

The limited protein synthesis in the horse (limited when compared with ruminants), and the lack of efficiency of absorption due to the cecum being on the lower end of the gut (thereby not giving the small intestine a chance at the ingesta after it leaves the cecum), clearly indicate that the horse rations should contain high-quality proteins, adequate in amino acids.

In comparison to a cow, therefore, a horse should be fed less roughage, more and higher quality protein (no urea), and added B vitamins. Actually, the nutrient requirements of a horse more nearly parallel those of a pig than a cow.

Further elucidation of the digestive system is contained in Chapter 4, Feeding Livestock; hence, the reader is referred thereto.

## NUTRIENT REQUIREMENTS VS ALLOWANCES

In ration formulation, two words are commonly used—*requirements* and *allowances*. Requirements do not provide for margins of safety. Thus, to feed a horse on the basis of meeting the bare requirements would not be unlike building a bridge without providing margins of safety for heavier than average loads or for floods. No competent engineer would be so foolish as to design such a bridge. Likewise, knowledgeable horse nutritionists provide for margins of safety—they provide for the necessary nutritive allowances. They allow for variations in feed composition; possible losses during storage and processing; day-to-day, and period-to-period, differences in needs of animals; age and size of animal; stage of gestation and lactation; the kind and degree of activity; the amount of stress; the system of management; the health, condition, and temperament of the animal; and the kind, quality, and amount of feed—all of which exert a powerful influence in determining nutritive needs.

## RECOMMENDED NUTRIENT ALLOWANCES

Presently available information indicates that the recommended nutrient allowances given in Table 54–1 will meet the minimum requirements for horses and provide reasonable margins of safety.

## TABLE 54–1
### RECOMMENDED NUTRIENT ALLOWANCES FOR HORSES (TOTAL RATION/AS-FED BASIS)[1] (See footnotes at end of table.)

| | Mature Horses (Consuming 25 lb feed/horse/day. Idle horses require less feed and/or consume more roughage than heavily worked horses or lactating mares.) | | | | | Young Horses, Based on Mature Weight 1,000 lb | | | | |
|---|---|---|---|---|---|---|---|---|---|---|
| | Idle Horses/ Light Work/ Moderate Work (1,000 lb Wt.) | Heavy Training/ Heavy Work (1,000 lb Wt.) | Stallions in Breeding Season (1,000 lb Wt.) | Mares, Last 90 Days Gestation (1,000 lb Wt.) | Mares, Peak of Lactation (1,000 lb Wt.) | Creep Feed (250 lb Body Wt/11 lb Feed Daily) | Weanlings (450 lb Body Wt/12 lb Feed Daily) | Yearlings (650 lb Body Wt/13 lb Feed Daily) | 2-Yr-Olds & 3-Yr-Olds (800 lb Body Wt/14 lb Feed Daily) | 2-Yr-Olds in Light Training (800 lb Body Wt/15 lb Feed Daily) |
| **Digestible Energy:** | | | | | | | | | | |
| TDN[2] (%) | 55 | 62.50 | 75 | 62.50 | 75 | 75 | 75 | 70 | 60 | 65 |
| Mcal per (lb) | 0.8 | 1.2 | 1.0 | 0.90 | 1.10 | 1.25 | 1.25 | 1.15 | 1.00 | 1.10 |
| Mcal per (kg)[3] | 1.80 | 2.55 | 2.15 | 2.0 | 2.35 | 2.60 | 2.60 | 2.50 | 2.20 | 2.40 |
| **Crude Protein** (%) | 9.0 | 11.0 | 14.0 | 13.0 | 14.0 | 18.0 | 16.0 | 14.0 | 13.0 | 13.0 |
| Lysine (%) | 0.25 | 0.36 | 0.30 | 0.32 | 0.41 | 0.54 | 0.55 | 0.48 | 0.38 | 0.41 |
| **Major or Macrominerals:** | | | | | | | | | | |
| Salt (%) | 0.75 | 0.75 | 0.75 | 0.75 | 0.75 | 0.75 | 0.75 | 0.75 | 0.75 | 0.75 |
| Calcium (%) | 0.21 | 0.31 | 0.26 | 0.29 | 0.47 | 0.62 | 0.55 | 0.40 | 0.28 | 0.31 |
| Phosphorus (%) | 0.15 | 0.23 | 0.19 | 0.30 | 0.30 | 0.34 | 0.30 | 0.22 | 0.15 | 0.17 |
| Magnesium (%) | 0.08 | 0.12 | 0.10 | 0.10 | 0.09 | 0.07 | 0.07 | 0.07 | 0.08 | 0.09 |
| Potassium (%) | 0.27 | 0.39 | 0.33 | 0.33 | 0.38 | 0.27 | 0.27 | 0.27 | 0.27 | 0.29 |
| Sulfur (%) | 0.15 | 0.15 | 0.15 | 0.15 | 0.15 | 0.15 | 0.15 | 0.15 | 0.15 | 0.15 |
| **Trace or Microminerals:** | | | | | | | | | | |
| Cobalt (ppm)[4] | 0.11 | 0.11 | 0.11 | 0.11 | 0.11 | 0.11 | 0.11 | 0.11 | 0.11 | 0.11 |
| Copper (ppm) | 25 | 25 | 25 | 25 | 30 | 40 | 40 | 30 | 25 | 25 |
| Iodine (ppm) | 0.11 | 0.11 | 0.11 | 0.11 | 0.11 | 0.11 | 0.11 | 0.11 | 0.11 | 0.11 |
| Iron (ppm) | 40 | 60 | 90 | 90 | 90 | 90 | 80 | 60 | 60 | 60 |
| Manganese (ppm) | 46 | 46 | 46 | 46 | 46 | 46 | 46 | 46 | 46 | 46 |
| Selenium (ppm) | 0.11 | 0.11 | 0.11 | 0.11 | 0.11 | 0.11 | 0.11 | 0.11 | 0.11 | 0.11 |
| Zinc (ppm) | 80 | 90 | 90 | 100 | 100 | 100 | 100 | 100 | 90 | 90 |
| | (/lb) | (/lb) | (/lb) | (/lb) | (/lb) | (/lb) | (/lb) | (/lb) | (/lb) | (/lb) |
| **Fat-soluble Vitamins in Feed:** | | | | | | | | | | |
| Vitamin A (IU) | 1,045 | 1,045 | 1,045 | 1,569 | 1,569 | 1,045 | 1,045 | 1,045 | 1,045 | 1,045 |
| Vitamin D (IU) | 156 | 156 | 156 | 314 | 314 | 419 | 419 | 419 | 419 | 419 |
| Vitamin E (IU) | 26 | 41 | 41 | 41 | 41 | 41 | 41 | 41 | 41 | 41 |
| Vitamin K (mg) | 0.32 | 0.32 | 0.32 | 0.32 | 0.32 | 0.30 | 0.30 | 0.30 | 0.30 | 0.30 |
| **Water-soluble Vitamins in Feed:** | | | | | | | | | | |
| Biotin (mg) | 0.1 | 0.1 | 0.1 | 0.1 | 0.1 | 0.1 | 0.1 | 0.1 | 0.1 | 0.1 |
| Choline (mg) | 20 | 30 | 30 | 30 | 30 | 62.5 | 62.5 | 62.5 | 62.5 | 62.5 |
| Folacin (mg) | 0.8 | 1.2 | 1.2 | 1.2 | 1.2 | 3.0 | 3.0 | 3.0 | 3.0 | 3.0 |
| Niacin (mg) | 10 | 20.8 | 10 | 10 | 10 | 10 | 10 | 10 | 10 | 10 |
| Pantothenic acid (mg) | 10 | 20.8 | 10 | 10 | 10 | 10 | 10 | 10 | 10 | 10 |
| Riboflavin (mg) | 1.6 | 1.6 | 1.6 | 1.6 | 1.6 | 1.6 | 1.6 | 1.6 | 1.6 | 1.6 |
| Thiamin (B-1) (mg) | 1.57 | 2.61 | 1.57 | 1.57 | 1.57 | 1.57 | 1.57 | 1.57 | 1.57 | 1.57 |
| Vitamin B-6 (mg) | 1.0 | 1.0 | 1.0 | 1.0 | 1.0 | 0.5 | 0.5 | 0.5 | 0.5 | 0.5 |
| Vitamin B-12 (mg) | 0.005 | 0.006 | 0.006 | 0.006 | 0.006 | 0.007 | 0.007 | 0.007 | 0.007 | 0.007 |
| Vitamin C (Ascorbic acid) (mg) | 2.4 | 4.0 | 4.0 | 4.0 | 4.0 | 3.75 | 3.75 | 3.75 | 3.75 | 3.75 |

[1]Where hay is fed separately, double the amounts shown in this table should be added to the concentrate.

[2]1 lb TDN = 2 Mcal or 2,000 Kcal.

[3]1 kg = 2.2 lb or 1,000 g.

[4]1 ppm (parts per million) = 1 mg/kg.

• **Stress affects nutritive needs**—Stress may be caused by excitement, temperament, fatigue, number of horses together, previous nutrition, breed, age, and management. Race and show horses are always under stress; and the more tired they become and the greater the speed, the greater the stress, Thus, the ration for race and show horses should be scientifically formulated. The greater the stress, the more exacting the nutritive requirements.

• **Other factors affect nutritive needs**—The feed requirements of horses do not necessarily remain the same from day-to-day or from period-to-period. The age and size of the animal; the stage of gestation or lactation of a mare; the kind and degree of activity; climatic conditions; the kind, quality, and amount of feed; the system of management; and the health, condition, and temperament of the animal are all continually exerting a powerful influence in determining its nutritive needs. How well the caretaker understands, anticipates, interprets, and meets these requirements usually determines the success or failure of the ration.

No set of instructions, calculator, or book of knowledge can substitute for experience and born horse intuition. Skill and good judgment are essential.

## NUTRITIVE NEEDS OF HORSES

The proper nutrition of horses is a major factor in determining their efficiency and years of service.

The various nutritive needs of the horse will be discussed under the following headings: (1) energy (carbohydrates and fats), (2) protein, (3) minerals, (4) vitamins, and (5) water.

### ENERGY

The energy requirements of horses for various activities are hard to develop because it is difficult to express quantitatively the type of exercise, the intensity and duration of work, the condition and training of the animals, the ability and weight of the rider and driver, the degree of fatigue, and the environmental temperature—all of which influence energy requirements. Based on Cornell studies by Pagan, Hintz reported the energy requirements given in Table 54–2. The Cornell researchers found that the amount of energy expended was proportional to the body weight of the riderless horse or the combined weight of the horse plus the rider, and that the amount of energy expended was exponentially related to speed. Additional studies are needed to determine the energy expenditures at speeds faster than the 13 miles per hour reported in Table 54–2.

**TABLE 54–2**
**DIGESTIBLE ENERGY REQUIREMENTS FOR VARIOUS ACTIVITIES OF LIGHT HORSES[1]**

| Gait | Speed (Miles/Hour)[2] | DE/Hour (Kcal/kg of Wt.)[3] |
|---|---|---|
| Slow walk | 2.2 | 1.7 |
| Fast walk | 3.6 | 2.5 |
| Slow trot | 7.5 | 6.5 |
| Medium trot | 9.3 | 9.5 |
| Fast trot/slow canter | 11.2 | 13.7 |
| Medium canter | 13.0 | 19.5 |

[1]Hintz, H. F., Energy Requirements of Horses, *Feed Management*, Vol. 37, No. 2, Feb. 1986, p. 15.
[2]To convert to metric, see Appendix, Weights and Measures.
[3]Body weight of horse plus weight of rider and tack.

Fig. 54–2. When racing, horses may require up to 100 times more energy when running than at rest. (Courtesy, *The Quarter Horse Journal*, Amarillo, TX)

A lack of energy intake may cause slow and stunted growth in foals and loss of weight, poor condition, and excessive fatigue in mature horses. Excess energy may result in obese horses, which are more susceptible to stress and founder and have lowered reproductive efficiency and decreased longevity.

It is common knowledge that a ration must contain proteins, fats, and carbohydrates. Although each of these has specific functions in maintaining a normal body, they can all be used to provide energy for maintenance, for work, or for fattening. From the standpoint of supplying the normal energy needs of horses, however, the carbohydrates are by far the most important, more of them being consumed than any other compound, whereas the fats are next in importance for energy purposes. Carbohydrates are usually more abundant and cheaper, and they are very easily digested, absorbed, and transformed into body fat. Also, carbohydrate feeds may be more easily stored in warm weather and for longer periods of time. Feeds high in fat content are likely to become rancid, and rancid feed is unpalatable, if not actually injurious in some instances. Thus, when fat is added to the ration, it is important that it be stabilized. Recent, experimental work indicates that horses will readily consume 10 to 20% added fat to the ration, without difficulty—and even with benefit.

Generally, increased energy for horses is met by increasing the grain and decreasing the roughage.

The fiber of growing pasture grass, fresh or dried, is more digestible than the fiber of most hay. Likewise, the fiber of early cut hay is more digestible than that of hay cut in the late bloom or seed stages. The difference is due to both chemical and physical structure, especially the presence of certain encrusting substances (notably lignin) which are deposited in the cell wall with age. This is understandable when it is recognized that lignin is the principal constituent of wood, for no one would think of feeding wood to horses.

Young equines and working (or running) horses must have rations in which a large part of the carbohydrate content of the ration is low in fiber, and in the form of nitrogen-free extract.

### PROTEIN

Horses of all ages and kinds require adequate amounts of protein of suitable quality for maintenance, growth, finishing, reproduction, and work. Of course, the protein requirements for growth and reproduction are the greatest and most critical.

A deficiency of proteins in the horse may result in the following deficiency symptoms; depressed appetite, poor growth, loss of weight, reduced milk production, irregular estrus, lowered foal crops, loss of condition, and lack of stamina.

Since the vast majority of protein requirements given in feeding standards meet minimum needs only, the allowance for race, show, breeding, and young animals should be higher.

In the case of ruminants (cattle and sheep), there is tremendous bacterial action in the paunch. These bacteria build body proteins of high quality from sources of inorganic nitrogen that nonruminants (humans, rats, chickens, swine, poultry, and dogs) cannot. Farther on in the digestive tract, the ruminant digests the bacteria and obtains good proteins therefrom. Although the horse is not a ruminant, apparently the same bacterial process occurs to a more limited extent in the cecum, that greatly enlarged blind pouch of the large intestine of the horse. However, it is much more limited than in ruminants, and the cecum is located beyond the small intestine, the main area for digestion and absorption of nutrients. This points up the fallacy of relying on cecum synthesis in the horse; above all, it must be remembered that little cecum synthesis exists in young equines.

The limited protein synthesis in the horse (limited when compared with ruminants) and the lack of efficiency of absorption due to the cecum being on the lower end of the gut (thereby not giving the small intestine a chance at the ingesta after it leaves the cecum), clearly indicate that horse rations should contain high-quality proteins, adequate in amino acids. This is especially important for young equines, because cecum synthesis is very limited early in life. In practical horse feeding, foals should be provided with some protein feeds of animal origin in order to supplement the protein found in grains and forages. In feeding mature horses, a safe plan to follow is to provide plant protein from several sources.

There is some evidence that nonprotein nitrogen (urea) can be substituted for protein in the diet of the horse, but the conversion to protein is inefficient. Up to 5% of urea in the total ration does not appear to be harmful. Nevertheless, in recognition of the more limited bacterial action in the horse and the hazard of toxicity, most state laws forbid the use of such nonprotein nitrogen sources as urea in horse rations.

The extent to which the horse's ration is supplemented with protein depends on the age of the horse and on the quality of the forage fed. Growing or lactating animals require somewhat more protein than horses that are idle, gestating, or working. Also, grass hays are generally low in quality and quantity of proteins and require more supplementation than legumes.

## PROTEIN POISONING

Some opinions to the contrary, protein poisoning as such has never been documented. There is no proof that heavy feeding of high-protein feeds to horses is harmful, provided (1) the ration is balanced out in all other respects, (2) the animal's kidneys are normal and healthy (a large excess of protein in terms of body needs increases the work of the kidneys for the excretion of the urea), (3) any ration change to high-protein feed is made gradually, as is recommended in any change in feed, and (4) there is adequate exercise and normal metabolism.

Some horses do appear to be allergic to certain proteins or to excesses of specific amino acids, as a result of which they may develop *protein bumps*.

It is recognized that protein in excess of what the body can use tends to be wasted insofar as its specific functions are concerned, since it cannot be stored in any but very limited amounts and must be catabolized. Nevertheless, some wastage of protein in terms of its known functions may be both physiologically and economically desirable in order to (1) maintain the protein reserves, (2) provide an adequate protein-calorie ratio for efficient energy utilization, and (3) assure that protein quality needs are met, despite the marked difference of quality needs among commonly fed rations. Generally speaking, high protein feeds are more expensive than high-energy feeds (feeds high in carbohydrates and fats), with the result that there is the temptation to feed too little of them.

## MINERALS[1]

When we think of mineral for the horse, we instinctively think of bones and unsoundnesses. This is so because (1) a horse's skeleton is very large, weighing 100 lb or more in a full-grown horse, of which more than half consists of organic matter and minerals, and (2) experienced trainers estimate that one-third of the horses in training require treatments for unsoundnesses, in one form or another. But in addition to furnishing structural material for the growth of bones, teeth, and tissues, minerals regulate many of the life processes.

In an amazingly short time after birth, a healthy foal can run almost as fast as its mother—and on legs almost as long. In fact, the cannon bones (the lower leg bones extending from the knees and hocks to the fetlocks) are as long at the time of birth as they will ever be. This indicates that important development of the skeleton takes place in the fetus, before the foal is born. it is evident, therefore, that adequate minerals must be provided the broodmare if the bones of her offspring are to be sound.

The mineral requirements of mares in lactation are even more rigorous than those during gestation. Mares weighing about 1,000 lb will produce an average of 2 gal., or more, of milk per day throughout the 7-month suckling period. That's a total of 3,612 lb of milk. Since fresh mare's milk contains 0.7% ash, this amount of mare's milk contains 25.8 lb of mineral (3,612 × 0.7% = 25.28.) Here's how this phenomenon works: The mare's skeleton is like a bank—people deposit money in a bank, then draw out or write checks on their reserves as needed. So, when properly fed before breeding, in early pregnancy, and when barren, mineral deposits are made in the mare's skeleton. Then at those times when the mineral demands are greater than can be obtained from the feed—the last of pregnancy, and during lactation—the mare draws from the stored reserves in her skeleton. Of course, if there hasn't been proper storage in the mare's skeleton, something must "give"— and that something is the mother. Nature has ordained that growth of the fetus, and the lactation that follows, shall take priority over the maternal requirements. Hence, when there is a mineral deficiency, the mare's body will be deprived, or even stunted if she is young, before the developing fetus or milk production will be materially affected.

Eighteen mineral elements are known to be required by at least some animal species. They can be divided into the following two groups based on the relative amounts needed in the ration:

---

[1]In this section, when reference is made to a National Academy of Sciences recommendation, this implies the following source: *Nutrient Requirements of Horses*, No. 6, 5th rev. ed., National Academy of Sciences, 1989.

| Major or Macrominerals | Trace or Microminerals |
|---|---|
| Calcium (Ca) | Iodine (I) |
| Phosphorus (P) | Manganese (Mn) |
| Sodium (Na) | Iron (Fe) |
| Chlorine (Cl) | Zinc (Zn) |
| Potassium (K) | Copper (Cu) |
| Magnesium (Mg) | Molybdenum (Mo) |
| Sulfur (S) | Fluorine (F) |
| | Chromium (Cr) |
| | Selenium (Se) |
| | Silicon (Si) |
| | Cobalt (Co) |

Approximately 70% of the mineral content of the horse's body consists of calcium and phosphorus. About 99% of the calcium and over 80% of the phosphorus are found in the bones and teeth.

Although acute mineral deficiency diseases and actual death losses are relatively rare, inadequate supplies of any one of the essential mineral elements may result in lack of thrift, poor gains, inefficient feed utilization, lowered reproduction, and decreased performance in racing, showing, riding, or whatnot.

The typical horse ration of grass hay and farm grains is usually deficient in calcium, but adequate in phosphorus. Also, salt is almost always deficient; and many horse rations do not contain sufficient iodine and certain other trace elements. Thus, horses usually need special mineral supplements. But they should not be fed either more or less minerals than needed. Also, it is recognized that mineral allowances given with the ration or in a mineral mix should vary according to the mineral content of the soil on which feeds are grown.

Fig. 54-3. Rabbit with bowed legs and enlarged joints resulting from eating alfalfa produced on low-phosphorus soils. There is reason to believe that the same thing happens to horses. (Courtesy, Washington State University, Pullman)

The proper development of the bone is particularly important in the horse, as evidenced by the stress and strain on the skeletal structure of the racehorse, especially when racing the 2-year-old. Since the greatest development of the skeleton takes place in the young, growing animal, it is evident that adequate minerals must be provided at an early age if the bone is to remain sound.

• **Metabolic bone disease (MBD)**— In recent years, there has been a greater increase in metabolic bone disease in growing horses, especially epiphysitis, contracted tendons, and osteochondritis dissecans (OCD). A brief description of each of these conditions follows:

1. **Epiphysitis.** This is an inflammation of the growth plate of the long bones, primarily found at the lower end of the radius above the knee, but it may be noticeable at the distal tibial and the distal metacarpal and metatarsal bones. Epiphysitis results in a firm and painful swelling.

2. **Contracted tendons.** This involves the shortening of the flexor tendons, causing the heels to be raised and the pasterns to be straight or, in severe cases, to knuckle forward with the horse walking on its toe. Contracted tendons may be present at birth, or they may be acquired during growth.

3. **Osteochondritis dissecans (OCD).** This is a condition in which the cartilage in a growing foal does not properly convert into bone. It may appear in either of two forms: (a) The form in which it is localized in one or a few joints (most commonly the stifle and hock joints, although any joint may be involved), usually without any clinical signs; and (b) the second and less common form, which most commonly affects the more distal joints such as pastern and fetlock, although it may affect any joint, including those of the back.

At this time, the cause of the increase in the incidence of bone diseases is not entirely clear. However, it appears that the major factors are: (1) rapid growth and excess weight, (2) injury to the epiphysis, (3) nutritional imbalances, (4) genetic predisposition, (5) limited forced exercise, (6) exercise on hard ground, and (7) faulty conformation.

Based on field observations and a study conducted by Ohio State University, involving 384 yearlings raised on 19 breeding farms in Ohio and Kentucky, and including the Thoroughbred, Standardbred, Arabian, and Quarter Horse breeds, there is strong evidence that calcium, phosphorus, copper, and zinc deficiencies/imbalances and/or masking are involved. They reported that the average calcium content of the rations on farms with fewest skeletal problems was 1.16% ± 0.09 and the phosphorus content was 0.72% ± 0.08.[2]

On the other hand, Krook and Maylin of Cornell University theorize that overfeeding of calcium to the growing horse or the pregnant mare is the primary cause of osteochondrosis in the foal. Their explanation: A dietary calcium overload causes an excessive secretion of the hormone calcitonin which acts directly on the bone of growing horses; calcitonin inhibits (1) the conversion of cartilage to bone, and (2) the resorption of calcium from bone. In the pregnant mare, calcitonin can be transferred through the placenta to the fetus; which might explain how a foal could be born with osteochondrosis. The Cornell scientists also theorize that osteochondrosis would predispose racehorses to fractures. They recommend that alfalfa

[2]Knight, Debra A., et al, Correlation of Dietary Minerals to Incidence and Severity of Metabolic Bone Disease in Ohio and Kentucky, College of Veterinary Medicine, The Ohio State University; paper presented at the 1985 American Association of Equine Practitioners meeting in Toronto, Canada.

hay not be fed to pregnant mares and pregnant and growing horses because of its high calcium content, and that calcium be limited to 34 g per horse per day.[3]

Both the Ohio State and Cornell scientists recognize the seriousness of metabolic bone disease. But they differ markedly as to the cause. Ohio State researchers submit strong evidence that deficiencies of calcium, phosphorus, copper, and zinc are the primary causes, whereas Cornell scientists theorize that high calcium is the main cause.

Further experimental studies are needed. In the meantime, horse owners and caretakers are admonished to practice the old adage: "Use moderation in all things."

Based on experiments (including unpublished work at both Ohio State and Cornell) and experiences, the author recommends (1) that breeders continue to feed alfalfa hay to pregnant mares and growing horses, and (2) that the levels of calcium, phosphorus, copper, iron, manganese, and zinc be in keeping with the recommendations given in Table 54–1.

## MAJOR OR MACROMINERALS

The major or macrominerals required by the horse are salt (sodium chloride), calcium, phosphorus, magnesium, potassium, and sulfur.

## SALT (SODIUM CHLORIDE)

Salt, which serves as both a condiment and a nutrient, is needed by all classes of animals, but more especially by herbivora (grass-eating animals, like the horse). It may be provided in the form of granulated, rock, or block salt. In general, the form selected is determined by price and availability. It is to be pointed out, however, that it is difficult for horses to eat very hard block and rock salt. This often results in inadequate consumption. Also, if there is much competition for the salt block, the more timid animals may not get their requirements.

The horse requires both sodium and chlorine. Generally, the chlorine requirement will be met if the sodium needs are met.

A deficiency of sodium over a long period of time results in depraved appetite, rough coat, reduced growth, and lowered milk production.

The salt requirement is greatly increased under conditions which cause heavy sweating, thereby resulting in large losses of this mineral from the body. Unless it is replaced, fatigue will result. For this reason, when engaged in hard work and perspiring profusely, horses should receive liberal allowances of salt.

On the average, a horse needs about 3 oz of salt daily or 1½ lb per week, although salt requirements vary with work and temperature. Iodized salt should be used in iodine-deficient areas.

Salt can be fed free choice to horses, provided they have not been salt starved. That is, if the animals have not previously been fed salt for a considerable length of time, they may overeat, resulting in digestive disturbances and even death from salt cramps. Salt starved animals should first be hand fed salt, and the daily allowance should be increased gradually until they start leaving a little in the mineral box. When this point is reached, self-feeding may be followed.

When added to the concentrate ration, salt should be added at a level of 0.5 to 1.0%.

---

[3]Sellnow, L., "Linking Breakdowns and Diet," *The Blood-Horse*, May 3, 1986, p. 3162.

## CALCIUM AND PHOSPHORUS

Horses are more apt to suffer from a lack of calcium and phosphorus than from any of the other minerals except salt. These two minerals comprise about ¾ the ash of the skeleton and from ⅓ to ½ of the minerals of milk.

A deficiency of either calcium or phosphorus will cause rickets in foals.

The availability to the horse of calcium and phosphorus in common feedstuffs is unknown. But the availability of calcium is assumed to be 55 to 75%, and the availability of phosphorus is assumed to be 35 to 55%. Several factors account for the poor absorption, including the Ca:P ratio, level of intake, source of calcium and phosphorus, and the presence of organic inhibitors such as oxalate and phytate. Also, due to poor utilization, the calcium and phosphorus requirement of aged animals (animals over 20 years of age) are higher than for younger animals.

In considering the calcium and phosphorus requirements of horses, it is important to realize that proper utilization of these minerals by the body is dependent upon three factors: (1) an adequate supply of calcium and phosphorus in an available form; (2) a suitable ration between them; and (3) sufficient vitamin D to make possible the assimilation and utilization of the calcium and phosphorus. If plenty of vitamin D is present (as provided either by sunlight or through the ration), the ration of calcium to phosphorus becomes less important. Also, less vitamin D is needed when there is a desirable calcium phosphorus ratio.

Normally, the calcium to phosphorus ratio should be about 1.1:1. However, the ratio varies according to age. For example, older horses can have a calcium-phosphorus ratio of 2:1. Provided adequate phosphorus is fed, weanling foals will *tolerate* a 3:1 ratio and mature horses a 5:1 ratio. It is important, however, to have more calcium than phosphorus—but not too much calcium. Feeding excessive calcium interferes with the utilization of magnesium, manganese, and iron—and perhaps with the utilization of zinc.

During pregnancy, the mare deposits an amount equivalent to 10 to 12% of her body weight in products of conception; and these products contain approximately 1.2% calcium and 0.6% phosphorus. Since most of these minerals are deposited in the bones, and approximately 90% of the bone development occurs during the last 90 days of gestation, about 6 g of calcium and 3 g of phosphorus per day will be deposited during this period in a 1,100-lb pregnant mare.

The calcium and phosphorus requirements for lactation depend on level of production. Milk production varies among mares and with stage of lactation. During peak production, mares will give up to 4½ gal. (38.7 lb/17.6 kg) of milk per day, with each pound containing 0.45 g of calcium and 0.2 g of phosphorus. Thus, a mare producing 4½ gal. of milk per day will deposit 17.4 g of calcium and 7.7 g of phosphorus per day.

Lack of either or both calcium and phosphorus can result in bone disorders, with the type and severity of the disorder dependent upon the age of the animal and the degree and duration of the deficiency. Deficiency in young horses is generally characterized by poorly formed, soft bones, which may bend or bow; and deficiency in older animals, by porous, fragile bones. Because these conditions are not completely reversible, prevention is imperative.

A deficiency of either calcium or phosphorus will cause rickets in foals. Also, there is substantial evidence that lack of calcium and phosphorus, along with deficiencies of copper and zinc, causes epiphysitis, contracted tendons, and osteochondritis dissecans (OCD) in young horses.

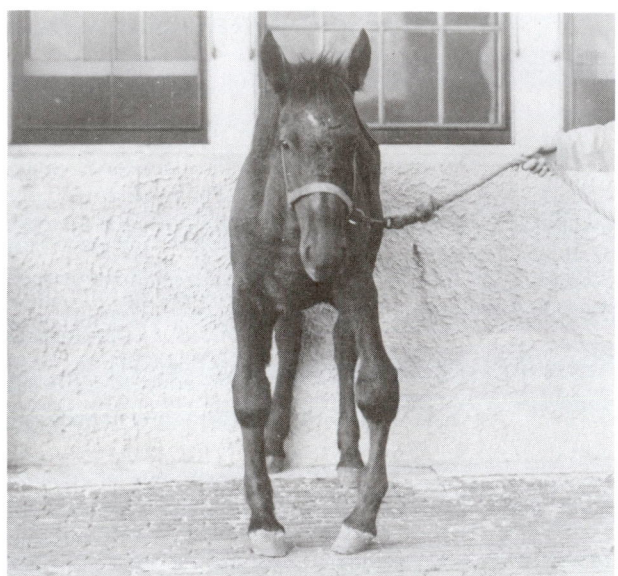

Fig. 54-4. Foal with severe rickets. Note the enlarged joints and crooked legs. (Courtesy, Department of Veterinary Medicine, University of Illinois, Urbana)

Bone disturbances (called osteodystrophia tebrosa, nutritional secondary hyperparathyroidism, osteomalacia, osteoporosis, and Miller's disease) develop in adult horses fed rations containing limited calcium and high phosphorus. The disease develops when rations with a calcium-phosphorus ratio of 0.8:1 are fed for 6 to 12 months, and it progresses rapidly when the ratio is 0.6:1.

After giving consideration to all the above mentioned factors, and after considering the experiments, the author recommends that calcium and phosphorus be provided to lactating mares and young horses at the levels in the total ration given in Table 54-1.

Generally speaking, legume forages, such as alfalfa hay or pasture, are rich in calcium; cereal grains and their byproducts—oats, corn, barley, and wheat bran—are fair to good sources of phosphorus; and the protein supplements—linseed meal, soybean meal, and dried skim milk—are good sources of both calcium and phosphorus. So, by selecting and combining the common horse feeds properly, the maintenance needs of most horses can be met. (See Fig. 54-5.)

Where both calcium and phosphorus are needed, the author favors the use of high-quality steamed bone meal for horses, because bone meal contains many ingredients in addition to calcium and phosphorus. It is a good source of iron, manganese, and zinc, and it contains such trace minerals as copper and cobalt. However, it is increasingly difficult to get good bone meal. Some of the imported products are high in fat, rancid, and/or odorous and unpalatable. Where good bone meal is not available, dicalcium phosphate is generally recommended.

Where phosphorus alone is needed, defluorinated rock phosphate, sodium monophosphate, or sodium polyphosphate are the minerals of choice. Sodium monophosphate and sodium polyphosphate are not palatable, hence, it is important that they be combined with more palatable products.

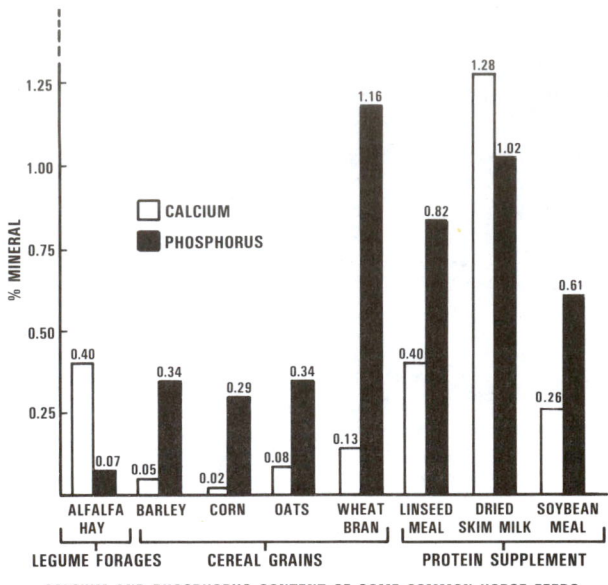

Fig. 54-5. Calcium and phosphorus content of some common horse feeds (as-fed basis).

Fig. 54-6. Calcium-phosphorus imbalance. Horse with "big head disease" (nutritional secondary hyperparathyroidism) resulting from feeding a ration low in calcium and high in phosphorus. Note that the upper jaw is enlarged because calcium is replaced by fibrous connective tissue. (Courtesy, National Academy of Sciences, and College of Veterinary Medicine, Texas A&M University, College Station)

## MAGNESIUM

Rations containing 50% forage will likely contain sufficient magnesium for unstressed horses, unless the forage is known to be deficient in magnesium. But horses at hard work (as in racing and showing) consume more grain (which is low in magnesium) and less forage. Also, horses being raced or shown, or otherwise stressed, are frequently keyed up, high-strung, and jumpy, similar to the nervousness that characterizes animals and humans known to be suffering from a magnesium deficiency.

In view of the above, it would appear prudent that ½ to ⅔ of the recommended daily magnesium allowance of the horse be added to the ration.

## POTASSIUM

Forage-consuming animals generally require about 0.3% of potassium in their rations. A ration that contains at least 50% forage can be expected to meet potassium requirements. However, a horse ration that does not contain roughage, molasses, or oil meals may be deficient in potassium.

Significant amounts of potassium are lost during heavy sweating.

A reduced appetite is an early sign of potassium deficiency. A severe deficiency may cause muscle tremors and erratic heart beat.

## SULFUR

Inorganic sulfur is not known to be an essential dietary constituent of the horse. If the protein requirement of the ration is met, the sulfur intake will usually be at least 0.15%, which appears to be adequate.

## TRACE OR MICROMINERALS

The need for trace minerals may be inferred from the many reports on the value of blackstrap molasses (a good source of trace minerals) for horses fed low-quality hays.

Trace minerals may be supplied (1) as part of either the concentrate or complete ration, and/or (2) in a trace mineralized (TM) salt. When incorporated in the concentrate mix, TM salt should be added at the rate of 1%. When TM salt is fed free-choice, it should be placed in a conveniently located covered mineral box in amounts that will be consumed in not more than 1 to 2 weeks. When remaining in a mineral box longer than this period of time, it may become unpalatable and there may be losses of some elements.

A discussion of each of the trace minerals follows. They are listed alphabetically, and not necessarily in order of importance.

• **Cobalt**—Cobalt is required for the synthesis of vitamin B-12 in the intestinal tract of the horse. A lack of cobalt and/or B-12 will result in anemia. However, the cobalt requirement of the horse is very low, for horses have remained in good health while grazing pastures so low in cobalt that ruminants confined to them have died. This means that the cobalt requirement, if any, of the horse is lower than that of ruminants. However, it is noteworthy that anemia in horses has responded to vitamin B-12 treatment; and, of course, B-12 contains cobalt in the molecular structure. Thus, inclusion of cobalt in the ration of horses is in the nature of good insurance.

• **Copper**—A copper deficiency has been reported in Australia in horses grazing on pastures low in copper. Also, mare's milk (along with milk from all species) is low in copper; and its copper concentration decreases greatly during the first weeks of lactation. The presence of 5 to 25 ppm of molybdenum in forages causes disturbances in copper utilization in horses.

Copper is of special interest to nutritionists because, in addition to its effect on iron metabolism, it is closely associated with normal bone development in young growing animals of all species. Abnormal bone development has been reported in foals on low copper rations.

There are wide species differences in tolerance of copper. Horses are very tolerant to copper, whereas sheep are very sensitive to it. The maximum tolerable levels of copper for growing animals in ppm, according to the National Academy of Sciences, are: horse 800 ppm; chicken, 300 ppm; swine, 250 ppm; cattle, 100 ppm; and sheep, 25 ppm.

The author's recommended copper allowances are: 30 ppm of the total ration for lactating mares, and 40 ppm for young horses. (See Table 54–1.)

In high molybdenum areas, it is recommended that the copper level for horses be about five times higher than the normal level.

• **Iodine**—Pregnant mares are very susceptible to iodine deficiency. Where such a deficiency exists, the foals are usually stillborn or so weak that they cannot stand and suck. There is also some evidence to indicate that the incidence of navel ill in foals may be lessened by feeding iodine to broodmares.

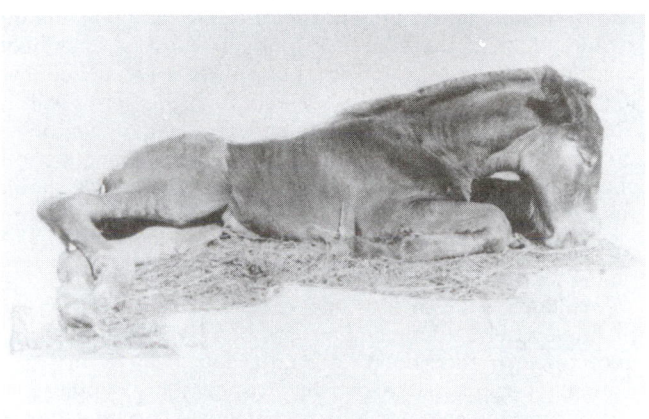

Fig. 54–7. Newborn weak colt affected with simple goiter due to deficiency of iodine during prenatal period. (Courtesy, Western Washington University Agricultural Experiment Station, Puyallup)

Other facts pertinent to iodine for horses follow:

1. **Amount in body and function.** It is estimated that the mature animal body contains less than 0.00004 percent iodine, but if this minute amount is not maintained in the diet, disaster results. More than half of the total iodine content of the body is located in the thyroid gland of the neck. Iodine, which is secreted by the thyroid gland in the form of thyroxin (an iodine-containing hormone), controls the rate of metabolism of the body.

2. **Iodine deficiencies.** If the soil—and the water and feed crops coming therefrom—is low in iodine, the body is likely to show deficiency symptoms in the form of simple goiter, unless an adequate source of iodine is provided artificially. A goiter is simply an enlargement of the thyroid gland, which is nature's way of trying to make enough thyroxin when there is insufficient iodine in the feed.

In general, it may be said that goiter is an advanced symptom of iodine deficiency, but that the chief loss is from interference with reproductive processes and the birth of weak, deformed offspring that fail to survive.

Iodine deficiencies are worldwide. In the United States, the northwestern states, the Pacific Coast, and the Great Lakes region are classed as goiter areas.

3. **Recommended iodine supplements.** The simplest method of supplying iodine in deficient areas is through use of salt containing (a) 0.01 percent potassium iodine (0.0076% iodine), or (b) calcium iodate. Most of the salt companies now manufacture stabilized iodized salt. If iodized salt is fed, addi-

tional iodine supplementation is unnecessary. Organic iodine supplements, such as kelp, are not any better than iodized salt.

4. **Precautions in feeding iodized salt.** Although iodized salt is an effective preventive measure, no satisfactory treatment has been developed for animals which have developed pronounced deficiency symptoms.

Excessive and continued iodine supplementation, such as feeding kelp meal in addition to iodized salt or trace mineralized salt, may cause goiter.

Iodized salt should always be kept in a dry place and it should be kept fresh. It should also be provided in such form and quantities as to ensure an adequate intake of iodine.

• **Iron**—If horses are fed diets that are too low in iron, or in iron and copper, nutritional anemia results.

The National Academy of Sciences estimates the maintenance requirements of the horse for iron at 40 ppm, and the requirements of the foal at 50 ppm. However, it has been reported that horses which are subjected to pressure from racing, showing, or other heavy use, require higher levels of iron. To be on the safe side, approximately one-half of the iron requirement of the horse should be added to the ration; and it should be in a biologically available form (iron oxide should not be used as a source of iron for horses because it is poorly absorbed).

Other facts pertinent to iron for horses follow:

1. **Body store of iron and copper at birth.** Nature has planned wisely. Young equines are born with a store of iron and copper in their bodies, which usually suffices until they normally begin to eat feeds which supply these constituents. This is most fortunate, as milk is very low in iron and copper. When young animals are continued on a milk diet for a long period of time, particularly under confined conditions and with little or no supplemental feeds, nutritional anemia will likely develop.

2. **Natural sources of iron.** In obtaining sources of iron, it is well to remember that simple inorganic iron salts, such as ferric chloride, are readily utilized, whereas the iron in the complex organic compounds in the hemoglobin of the blood is much less readily available, if at all. Also, though certain small amounts of iron are very essential, too much of this element in the diet may actually be deleterious—interfering with phosphorus absorption by forming an insoluble phosphate—and rickets may thus result from a diet otherwise adequate.

• **Manganese**—Feeds containing 60 to 70 ppm of manganese are recommended, with the higher levels fed to foals, stressed horses, and breeding animals.

Since most natural feedstuffs are rich in manganese, it can be assumed that part of the requirement for this element will be met by the normal ration.

• **Selenium**—Selenium is an essential mineral for horses. Deficient animals have muscle disorders and lowered serum selenium.

It is recommended that horse rations contain 0.1 ppm of selenium in the complete feed. Excess selenium about 5 ppm results in selenium poisoning, or alkali disease.

• **Zinc**—A level of 80 to 100 ppm of zinc is recommended, with young and highly stressed animals receiving the upper level.

Zinc is necessary for the maintenance and development of skin and hair. Since beautiful hair coats are important in horses, fortifying the daily ration with zinc will prevent any possibility

of a zinc deficiency; and if the zinc in the feed is on the low side, the supplemental zinc should improve the hair coat.

## CHELATED MINERALS

The word chelate is derived from the Greek *chelae*, meaning a claw or pincerlike organ. Those selling chelated minerals generally recommend a smaller quantity of them (but at a higher price per pound) and extoll their "fenced-in" properties.

When it comes to synthetic chelating agents, much needs to be learned about their selectivity toward minerals, the kind and quantity most effective, their mode of action, and their behavior with different species of animals and with varying rations.

It is possible that their use may actually create a mineral imbalance. These answers, and more, should be forthcoming through carefully controlled experiments.

## MINERAL IMBALANCES

Having the right balance and forms of minerals can be very important. The more calcium you feed, the more phosphorus you need. The more copper you feed, the more manganese you need.

Also, minerals can be fed in several different forms. For example, iron can be fed as oxide, sulfite, sulfate, or as a proteinate. Oxides may be absorbed at about 2 to 5%, while sulfites may be absorbed at up to 10%, and sulfates at 25%.

Thus, the requirements of any mineral may be modified (1) by another mineral which enhances or interferes with its utilization, or (2) by the form of the mineral.

From the above, it is apparent that excess fortification of the horse's ration with one or more mineral elements may prove more detrimental than helpful. Thus, caretakers who know and care will avoid harmful imbalances; they will provide minerals on the basis of *recommended allowances* (see Table 54-1). Also, when fortifying rations with minerals, consideration should be given to the minerals provided by the ingredients of the normal ration, for it is the total composition of the feed that counts.

## FEEDING MINERALS

With the exception of sodium, the self-feeding of the major minerals cannot be relied upon to meet the needs of horses. This is so because horses consume such supplements on the basis of palatability, rather than because of dietary need. As a result, the free-choice intake of minerals among individual horses will vary from too little to too much. Sometimes minerals are incorporated in a salt mix, but salt consumption is erratic and variable according to the sodium content of the feedstuffs being fed. So, the only way to ensure that each horse receives the needed major minerals is to incorporate the proper amounts in the animal's feed and/or water.

Trace minerals may be added to the ration and/or incorporated in the salt. In either case, the amounts and proportions of trace minerals should be selected with care because the improper use of trace minerals can lead to induced deficiencies. Theoretically, the total ration (grain plus forage) should be balanced in trace mineral content, with the trace mineral mix providing only the minerals needed and with each one in the right amount. Of course, this is not practical. Therefore, a trace mineral mix must contain an array of minerals in adequate levels to meet a wide variety of conditions. Fortunately, the horse is tolerant to most trace mineral excesses.

When horses are on pasture and no grain or protein supplement is being fed, minerals may be self-fed, usually as either a commercially manufactured mineral block or as a mineral mixture. A suitable home-mixed mineral for self-feeding on pasture may be prepared as follows:

1. **Where the pasture is primarily grass.** Prepare a mixture containing two parts of calcium to one part of phosphorus.
2. **Where the pasture is primarily a legume.** Prepare a mixture containing one part of calcium to one part of phosphorus.

To each of the above mixes, add one-third trace-mineralized salt to provide the microminerals and improve the palatability.

## VITAMINS[4]

Until early in the 20th century, if a ration contained proteins, fats, carbohydrates, and minerals, together with a certain amount of fiber, it was considered to be a complete ration.

The lack of vitamins in a horse ration may, under certain conditions, be more serious than a short supply of feed. Deficiencies may lead to failure in growth or reproduction, poor health, and even characteristic disorders known as deficiency diseases.

Unfortunately, there are no warning signals to tell a caretaker when a horse is not getting enough of certain vitamins. But a continuing inadequate supply of any one of several vitamins can produce illness which is very hard to diagnose until it becomes severe, at which time it is difficult and expensive—if not too late—to treat. The important thing, therefore, is to insure against such deficiencies occurring. But caretakers should not shower a horse with mistaken kindness through using shotgun-type vitamin preparations. Instead, the quantity of each vitamin should be based on available scientific knowledge.

It has long been known that the vitamin content of feeds varies considerably according to soil, climatic conditions, and curing and storing.

Deficiencies may occur during periods (1) of extended drought or of restriction in diet, (2) when production is being forced, or during stress, (3) when large quantities of highly refined feeds are being fed, or (4) when low-quality forages are utilized.

Although the occasional deficiency symptoms are the most striking results of vitamin deficiencies, it must be emphasized that in practice mild deficiencies probably cause higher total economic losses than do severe deficiencies. It is relatively uncommon for a ration, or diet, to contain so little of a vitamin that obvious symptoms of a deficiency occur. When one such case does appear, it is reasonable to suppose that there must be several cases that are too mild to produce characteristic symptoms, but that are sufficiently severe to lower the state of health and the efficiency of production.

Certain vitamins are necessary for the growth, development, health, and reproduction of horses. Deficiencies of vitamins A and D are sometimes encountered. Also, indications are that vitamin E and some of the B vitamins are required by horses. Further, it is recognized that single, uncomplicated

vitamin deficiencies are the exception rather than the rule.

High-quality, leafy, green forages plus plenty of sunshine generally give horses most of the vitamins they need. Horses get carotene (which they can convert to vitamin A) and riboflavin from green pasture and green hay not over a year old, and they get vitamin D from sunlight and sun-cured hay. If plenty of green forage and sunlight are not available, the caretaker should get the advice of a nutritionist or veterinarian on the use of vitamin additives to the feed.

Table 54–1 lists the vitamins most commonly involved in horse nutrition. Although there is no evidence of deficiencies of certain vitamins, it is possible that more of them may be destroyed or used by horses during stress or strain than can be obtained through normal feeds or synthesized by the intestinal microflora of the horse; hence, adding them to the ration may assure maximum performance.

### FAT-SOLUBLE VITAMINS

The fat-soluble vitamins, which are stored in the body in appreciable quantities, are vitamin A (carotene), vitamin D, vitamin E, and vitamin K.

### VITAMIN A

Vitamin A is strictly a product of animal metabolism, no vitamin A being found in plants. The counterpart in plants is known as carotene, which is the precursor of vitamin A. Because the animal body can transform carotene into vitamin A, this compound is often spoken of as *provitamin A*.

Carotene is the yellow-colored, fat-soluble substance that gives the characteristic color to carrots and to butterfat (vitamin A is nearly a colorless substance). Carotene derives its name from the carrot, from which it was first isolated over 100 years ago. Although its empirical formula was established in 1906, it was not until 1919 that Steenbock, of the University of Wisconsin, discovered its vitamin A activity. Though the yellow color is masked by the green chlorophyll, the green parts of plants are rich in carotene and have high vitamin A value. Also, the degree of greenness in a roughage is a good index of its carotene content, provided it has not been stored too long. Early cut, leafy green hays are very high in carotene.

Vitamin A is not synthesized in the cecum. Thus, it must be provided in the feed, either (1) as vitamin A, or (2) as carotene, the precursor of vitamin A. For horses, 1 mg of beta-carotene is equivalent to 400 IU of vitamin A.

Aside from yellow corn, practically all of the cereal grains used in horse feeding have little carotene or vitamin A value. Even yellow corn has only about one-tenth as much carotene as well-cured hay. Dried peas of the green and yellow varieties and carrots are also valuable sources of carotene.

Severe deficiency of vitamin A may cause night blindness (impaired adaptation to darkness), lacrimation (tears), keratinization of the cornea and skin, reproductive difficulties, poor or uneven hoof development, difficulty in breathing, incoordination, convulsive seizures, progressive weakness, and poor appetite. There is also some evidence that deficiency of this vitamin may cause or contribute to certain leg bone weaknesses. When vitamin A deficiency symptoms appear, the caretaker should add a stabilized vitamin A product to the ration.

A considerable margin of safety in vitamin A and carotene is provided in the recommended allowances due to the ox-

---

[4]In this section, when reference is made to a National Academy of Sciences recommendation, this implies the following source: *Nutrient Requirements of Horses*, No. 6, 5th rev. ed., National Academy of Sciences, 1989.

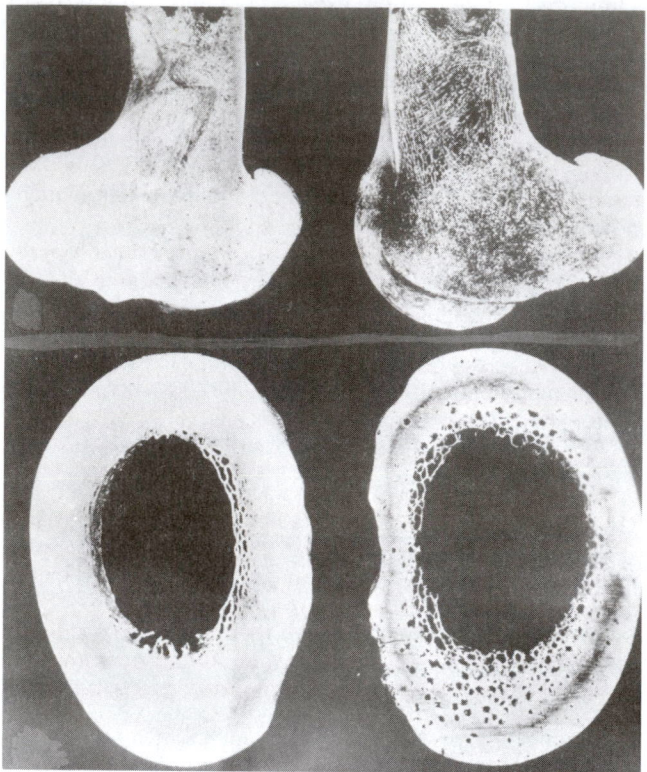

Fig. 54–8. Vitamin A made the difference! *Upper:* On the right is shown the sagittal section of the distal end of the femur of a vitamin A-deficient horse compared to normal bone (left). *Lower:* On the right is shown the cross-section of the cannon bone from a vitamin A-deficient horse compared to normal bone (left). (Courtesy, G. E. Howell, California Agricultural Experiment Station, Davis)

idative destruction of these materials in feeds during storage. But, it is wasteful to feed more vitamin A than is needed. Also, feeding excessively high levels of vitamin A over an extended period of time may cause bone fragility, hyperostosis, and exfoliated epithelium. When fed as directed, the vast majority of horse feeds won't provide excesses of vitamin A.

Other facts pertinent to vitamin A for horses follow:

1. **Circumstances conducive to vitamin A deficiencies.** The circumstances most conducive to vitamin A deficiencies are (a) extended periods of drought, resulting in the pastures becoming dry and bleached; (b) a long winter feeding period on bleached hays or straws, especially overripe cereal hays and straws; and (c) using feeds which have lost their vitamin A potency as a result of either heat or extended storage (for example, it has been found that alfalfa may lose nine-tenths of its vitamin A value in a year's storage). There is reason to believe that mild deficiencies of vitamin A, especially in the winter and early spring, are fairly common.

Fortunately, horses are able to store vitamin A, primarily in the liver, during periods of abundance to tide them through periods of scarcity. Thus, horses that have been consuming green forage for 4 to 6 weeks usually store sufficient vitamin A in the liver to maintain adequate levels of plasma vitamin A for 3 to 6 months.

It is noteworthy (1) that the absorption of vitamin A is adversely affected by the presence of parasites in the intestinal tract, and (2) that the presence of enough protein of good quality enhances the conversion of carotene to vitamin A.

It is generally believed that stressed horses have a higher vitamin A requirement than those not under stress. Among such stress factors are: racing, showing, fatigue, hot weather, confinement, excitement, and number of animals run together.

The vitamin A requirements for gestating mares may be five times the minimum maintenance requirements. therefore, unless properly fed, broodmares may become almost depleted of their vitamin A reserves by the end of the winter—at a time when vitamin A deficiency could be critical to the rapid development of the fetus.

2. **Measurement of vitamin A potency.** The vitamin A potency (whether due to the vitamin itself, to carotene, or to both) of feeds is usually reported in terms of IU or USP units. These two units of measurement are the same. They are based on the growth response of rats. The carotene or vitamin A content of feeds is commonly determined by colorimetric or spectroscopic methods.

## VITAMIN D

For horses, both $D_2$ (the plant form) and $D_3$ (the animal form) are equally effective, so there is no need to use some of each.

Foals sometimes develop rickets because of insufficient vitamin D, calcium, or phosphorus. Rickets is characterized by reduced bone calcification, stiff and swollen joints, stiffness of gait, irritability, and reduction in serum calcium and phosphorus. It can be prevented by exposing the animal to direct sunlight as much as possible, by allowing free access to a suitable mineral mixture, or by providing good quality suncured hay or luxuriant pasture grown on well-fertilized soil. In northern areas that do not have adequate sunshine, many caretakers provide the foal with a vitamin D supplement.

With vitamin D, as with vitamin A, there is need for adequacy without harmful excesses. Too much vitamin D may harm a horse. Vitamin D toxicity is characterized by calcification of the blood vessels, heart, and other soft tissues, and by bone abnormalities. Also, there is general weakness and loss of body weight. Although the toxic level of vitamin D in the horse has not been established, a level of 50 times the requirement may be harmful.

The vitamin D requirement is less when a proper balance of calcium and phosphorus exists in the ration.

Other facts pertinent to vitamin D for horses follow:

1. **Vitamin D, and cholesterol and ergosterol.** Most of the commonly used feeds contain little or no vitamin D, yet there is no widespread need for special supplements containing this factor. Fortunately, the skin of horses and many feeds contain provitamins in certain forms of cholesterol and ergosterol, respectively, which, through the action of ultraviolet light (light of such short wave length that it is invisible) from the sun, are converted into vitamin D. These certain forms of cholesterol and ergosterol themselves have no antirachitic effect.

2. **Vitamin D limited in feeds.** Of all the known vitamins, vitamin D has the most limited distribution in common feeds. Very little of this factor is contained in the cereal grains and their by-products, in roots and tubers, in feeds of animal origin, or in growing pasture grasses. The only important natural sources of vitamin D are sun-cured hay and other roughages. The chief vitamin D rich concentrates include vitamin $D_2$, vitamin $D_3$, sun-cured hay, cod and certain other fish liver oils, irradiated cholesterol and ergosterol, and irradiated yeast.

3. **Effectiveness of sunlight in producing vitamin D.** The effectiveness of sunlight is determined by the lengths and intensity of the ultraviolet rays which reach the body. It is more

potent in the tropics than elsewhere, more potent at noon than earlier or later in the day, more potent in the summer than in the winter, and more potent at high altitudes. The ultraviolet rays are largely screened out by clothing, window glass, clouds, smoke, or dust. Also, some biochemists theorize that the color of the skin of humans is nature's way of regulating the manufacture of vitamin D—that the dark skin of races near the equator filters out excess ultraviolet light. Perhaps color of hair and skin in horses exercise a similar control, although this is not known.

## VITAMIN E

Vitamin E, or tocopherol, is associated with reproduction. Also, it prevents and corrects anhidrosis, a condition characterized by a dry, dull hair coat, elevated temperature, high blood pressure, and labored breathing. Anhidrosis has been successfully treated by the oral administration of 1,000 to 3,000 IU of vitamin E daily for 1 month.

Most practical rations contain liberal quantities of vitamin E, perhaps enough except under conditions of work, stress, or reproduction, or where there is interference with its utilization. Green forages, especially alfalfa, are good sources.

It is now preferable to use milligrams of alpha-tocopherol equivalents as a summation term for all vitamin E activity. However, feed composition tables generally give values in IU, and IU is still used for labeling most feed products.

The requirements for vitamin E are influenced by interrelationships with other essential nutrients—increased by the presence of interfering substances, and spared by the presence of other substances that may be protective or that may assume part of its functions. The recommended allowances of vitamin E are given in Table 54-1.

## VITAMIN K

When vitamin K is deficient, the coagulation time of the blood is increased and the prothrombin level is decreased. This is the main justification for adding this vitamin to the ration of the horse. Also, vitamin K has value in veterinary medicine as an aid in controlling hemorrhages.

## WATER-SOLUBLE VITAMINS

The large amounts of water which pass through the horse's body daily tend to carry out the water-soluble vitamins, thereby depleting the supply. Thus, they must be supplied in the horse's ration on a day-to-day basis. All of the water-soluble vitamins except C are known as B vitamins.

Vitamins of the B-complex, particularly biotin, choline, folacin (folic acid), niacin, pantothenic acid, riboflavin, thiamin (B-1), vitamin B-6 (pyridoxine) and vitamin B-12 may be essential, especially for (1) young horses before the synthesis of the B-complex vitamins by the microflora begins, and (2) horses that are under stress, as in racing and showing.

However, it is not clear which ones are needed, in what quantities they are needed, and as to their status from the standpoint of synthesis and absorption in the horse. Healthy horses usually get enough of them either in natural rations or by synthesis in the intestinal tract. However, when neither green pasture nor high-quality dry forage is available, it may be in the nature of good insurance to provide them, especially for horses that are under stress.

Although some of the B vitamins and unidentified factors are synthesized in the cecum of the horse, it is doubtful that microbial activity is sufficient to meet the need during the critical periods—growth, reproduction, and when animals are subjected to great stress as in showing or racing. Also, there is reason to question the efficacy of absorption this far down the digestive tract; for, in comparison with that of humans and other animals, the cecum is on the wrong end of the digestive tract. Moreover, it is known that horses fed thiamin-deficient rations lose weight, become nervous, and show incoordination in the hindquarters; then, when thiamin is added to the ration, this condition is cured. For these reasons, in valuable horses it is not wise to rely solely on bacterial synthesis. The B vitamins, along with unidentified factors, may be provided by adding to the ration such ingredients as distiller's dried solubles, dried brewer's yeast, dried fish solubles, or animal liver meal; usually through a reputable commercial feed.

- **Biotin**—Ordinary equine rations probably contain ample biotin, or horses synthesize all they need. But, in recognition that biotin is required by all species, and that it plays an important role in metabolism of carbohydrates, fats, and proteins, it is possible that adding biotin to the ration of the horse may assure maximum performance. Also, there is some indication that biotin is essential for sound hooves; but it should not be concluded that all hoof problems are due to biotin deficiency. They are not. Actually, several nutrients are known to influence hoof growth—biotin among them. Studies indicate that a complete balanced ration is essential for proper hoof growth. Without doubt, heritability is also a factor.

- **Choline**—Choline is a metabolic essential for building and maintaining cell structure and in the transmission of nerve impulses. Choline deficiency has been produced in rats, dogs, chickens, pigs, and other species. Slow growth is a nonspecific symptom.

The dietary requirement for choline depends on the level of methionine (an amino acid) in the ration. Also, it is noteworthy that all naturally occurring fats contain some choline; however, normal horse feeds are low in fat. Hence, the addition to the ration of 500 mg of choline per day is the recommended allowance for a 1,000-lb horse.

- **Folacin (Folic Acid)**—There is no single compound vitamin with the name folacin; rather, the term *folacin* is used to designate folic acid and a group of closely related substances which are essential for vertebrates.

Folacin is widely distributed in horse feeds. Also, it is synthesized in the lower digestive tract of the horse. Hence, it is unlikely that a dietary source is required, although a small amount may be in the nature of cheap insurance.

- **Niacin (Nicotinic Acid, Nicotinamide)**—Niacin is a collective term which includes nicotinic acid and nicotinamide, both natural forms of the vitamin with equal niacin activity.

Some evidence indicates that niacin is synthesized by the horse. Also, the horse can convert the essential amino acid tryptophan into niacin. Hence, it is important to make certain that the ration is adequate in niacin; otherwise, the horse will use tryptophan to supply niacin needs. Niacin is widely distributed in feeds; fermentation solubles, and certain oil meals are especially good sources. Only a modest addition of niacin to the ration is indicated.

- **Pantothenic Acid (Vitamin B-3)**—Intestinal synthesis of pantothenic acid has been found to occur in all species studied. In the case of the horse, such synthesis appears to be sufficiently extensive to meet body needs, at least in part. However, of all the B vitamins, pantothenic acid is most likely to be deficient under stable (confinement) conditions. A daily allowance of 250 mg of pantothenic acid is recommended for a 1,000-lb gestating or lactating mare.

• **Riboflavin (Vitamin B-2)**—A deficiency of riboflavin may cause periodic ophthalmia (moon blindness), characterized by catarrhal conjunctivitis in one or both eyes, accompanied by photophobia, and lacrimation. Repeated attacks affect the retina, lens, and ocular fluids and cause impaired vision or blindness. But it is known that lack of this vitamin is not the only factor causing this condition. Sometimes moon blindness follows leptospirosis in horses, and it may be caused by a localized hypersensitivity or allergic reaction. Periodic ophthalmia caused by lack of riboflavin may be prevented by feeding green hay and green pasture, supplying feeds high in riboflavin, or by adding crystalline riboflavin to the ration.

Two properties of riboflavin lend support to riboflavin supplementation for the horse: (1) it is destroyed by light, and it is destroyed by heat in an alkaline solution; and (2) body storage is very limited, so day-to-day needs must be provided in the ration.

• **Thiamin (Vitamin B-1)**—Vitamin B-1 is synthesized in the lower gut of the horse by bacterial action, but there is some doubt as to its sufficiency and as to the amount absorbed always meeting the full requirements.

A thiamin deficiency has been produced experimentally. It is characterized by loss of appetite, loss of weight, anemia, incoordination (especially of the hind legs), lower blood thiamin, elevated blood pyruvic acid, and dilated and hypertrophied heart.

Vitamin B-1 is required for normal carbohydrate metabolism. Since carbohydrate metabolism is increased during physical exertion, it is important that B-1 be available in quantity at such times.

• **Vitamin B-6 (Pyridoxine, Pyridoxal, Pyridoxamine)**—There is no evidence that deficiencies of vitamin B-6 occur in horses on commonly fed rations; and it is not expected that deficiencies should occur in view of the widespread distribution of vitamin B-6 in feedstuffs and the probable synthesis of B-6 in the cecum. Yet, these sources may not be adequate to assure maximum performance of the horse. So, the daily supplementation of 25 mg of vitamin B-6 per horse may be in the nature of cheap insurance, especially because of the important role of this vitamin.

Vitamin B-6, in its coenzyme forms, is involved in a large number of physiologic functions, particularly the metabolism of protein, carbohydrate, and fat. Also, it is involved in clinical problems, including (1) anemia that is not iron responsive, (2) kidney stones, and (3) the physiologic demands of pregnancy.

• **Vitamin B-12 (Cobalamins)**—It has been reported that horses in poor nutritional condition showing anemia respond to the administration of vitamin B-12. An allowance of 0.084 mg of B-12 per day is recommended for weanlings.

Vitamin B-12 injections are frequently given to horses to improve performance and to prevent or cure anemia. There is no experimental evidence that such shots are either helpful or harmful.

• **Vitamin C (Ascorbic Acid, Dehydroascorbic Acid)**—A dietary need for ascorbic acid is limited to humans, monkeys, guinea pigs, fruit-eating bats, and bulbul birds.

The vitamin is probably required by all other species, in-

cluding the horse, but is likely synthesized adequately in the body. It is noteworthy, however, that catfish and trout require dietary sources of vitamin C when they are stressed by being raised in crowded conditions. This theory of stress has been carried over to human nutrition. So, it is conjectured that heavily stressed horses may not be able to synthesize sufficient vitamin C for maximum performance, hence, adding the vitamin to the ration may make for added assurance.

## VITAMIN IMBALANCES

Experiments have shown that the amounts needed of certain vitamins may be affected by the supply of another vitamin or of some other nutritive essential. Also, it is known that excess fortification of the horse's ration with certain vitamins may prove more detrimental than helpful. Thus, caretakers should avoid harmful imbalances; they should provide vitamins on the basis of recommended allowances. Also, when fortifying with vitamins, consideration should be given to the vitamins provided by the ingredients of the normal ration, for it is the total composition of the feed that counts.

## UNIDENTIFIED FACTORS

Since the U.S. foal crop is only around 50%, and since horses under stress (racing, showing, etc.) frequently become temperamental in their eating habits, it is obvious that there is room for improvement in the ration somewhere along the line. Perhaps unidentified factors are involved.

Unidentified factors include those vitamins which the chemist has not yet isolated and identified. For this reason, they are sometimes referred to as the vitamins of the future. There is mounting evidence of the importance of unidentified factors for animals, including humans. Among other things, they lower the incidence of ulcers in humans and swine. For horses, they appear to increase growth and improve feed efficiency and breeding performance when added to rations thought to be complete with regard to known nutrients. The anatomical and physiological mechanism of the digestive system of the horse, plus the stresses and strains to which modern horses are subjected, would indicate the wisdom of adding unidentified factor sources to the ration of the horse. Unidentified factors appear to be of special importance during breeding, gestation, lactation, and growth.

Three highly regarded unidentified factor sources are dried whey product, corn fermentation solubles, and dehydrated alfalfa meal.

## WATER

Water is one of the most vital of all nutrients. In fact, horses can survive for a longer period without feed than they can without water. The loss of 10% body water will result in disorders; the loss of 20% body water will cause death. But, fortunately, under ordinary conditions water can be readily provided and at little cost.

Water is one of the largest single constituents of the animal body, varying in amount with condition and age. The younger the animal, the more water it contains. Also, the fatter the animal, the lower the water content. Thus, as an animal matures, it requires proportionately less water on a weight basis, because it consumes less feed per unit of weight and the water content of the body is being replaced by fat.

Surplus water is excreted from the body, principally in the urine, and to a slight extent in the perspiration, feces, and water vapor from the lungs.

The average horse will drink 10 to 12 gal. of water daily, the amount varying according to the weather, amount of work done, (sweating), rations fed, and size of horse.

Free access to water is desirable. When this is not possible, horses should be watered at approximately the same times daily. Opinions vary among caretakers as to the proper times and method of watering horses. All agree, however, that regularity and frequency are desirable. Most caretakers agree that water may be given before, during, and after feeding.

Frequent, small waterings between feedings are desirable during warm weather or when the animal is being put to hard use. Do not allow a horse to drink heavily when it is hot, because it may founder; and do not allow a horse to drink heavily just before being put to work.

Fig. 54-9. Automatic waterer. (Courtesy, *Sunset Magazine*)

Automatic waterers are the modern way to provide clean, fresh water at all times—as nature intended. Also, frequent but small waterings prevent gorging. All waterers should have drains for easy cleaning, and should be heated to 40 to 45°F during the winter months in cold regions. Waterers should be available in both stalls and corrals.

## FEEDS FOR HORSES

More than one kind of hay makes for appetite appeal. In season, any good pasture can replace part or all of the hay unless work or training conditions make substitution impractical.

Good quality oats and timothy hay always have been considered standard feeds for light horses. However, feeds of similar nutritive properties can be interchanged in the ration as price relationships warrant; among them, the grains—oats, corn, barley, wheat, and sorghum; the protein supplements—linseed meal, soybean meal, and cottonseed meal; and hays of many varieties. Feed substitution makes it possible to obtain a balanced ration at lowest cost.

During the winter months, it is well to add a few sliced carrots to the ration, an occasional bran mash, or a small amount of linseed meal. Also a bran mash or linseed meal may be used to regulate the bowels.

The proportion of concentrates must be increased and the roughages decreased as energy needs rise with the greater amount, severity, or speed of work. A horse that works at a trot needs considerably more feed than one that works at a walk. For this reason, riding horses in medium to light use require somewhat less grain and more hay in proportion to body weight than light horses that are racing. Also, from an esthetic standpoint, large, paunchy stomachs are objectionable on horses that are used for recreation and sport.

In addition to making for a nutritionally complete ration, the following factors should be considered when choosing horse feeds: cost, palatability, preparation, variety, bulk, and laxativeness.

For purposes of convenience in the discussion that follows, the author has classed feeds as (1) pasture, (2) hay, (3) concentrates, (4) protein supplements, (5) special feeds and additives, and (6) treats.

## PASTURES

The great horse-breeding centers of the world are characterized by good pastures. Thus, the bluegrass area of Kentucky is known for its lush pastures produced on residual limestone soils. In short, good equine caretakers, good pastures, and good horses go hand in hand—good pasture is the cornerstone of successful horse production. Yet, it is becoming increasingly difficult to provide good pastures for many horses, especially those in suburban areas. Also, it is recognized that many folks are prone to overrate the quality of their grass.

Fig. 54-10. Good pastures make for success in the horse business. (Photo by Ernst Peterson. Courtesy, Bitterroot Stock Farm, Hamilton, MT)

In season, there is no finer forage for horses than superior pastures—pastures that are much more than gymnasiums. This is especially true for idle horses, broodmares, and young stock. In fact, pastures have a very definite place for all horses, with the possible exception of animals at heavy work or in training. Even with the latter groups, pastures may be used with discretion. Horses in use may be turned to pasture at nights or over the weekend. Certainly, the total benefits derived from pasture are to the good, although pasturing may have some laxative effects and produce a greater tendency to sweat.

In addition to the nutritive value of the grass, pasture provides invaluable exercise on natural footing—with plenty of sunshine, fresh air, and lowered feed costs as added benefits. Feeding on pasture is the ideal existence for young stock and breeding animals.

The use of temporary seeded pasture grown in regular crop rotation is recommended instead of a permanent pasture that may become infested with parasites. However, the parasites in horse pastures can be reduced dramatically by picking up the manure twice a week. An Ohio State University study showed that a pasture routinely cleaned in this manner has 18 times fewer parasites than an uncleaned pasture. Manure can be removed manually or mechanically. In England, a power sweeper is available, consisting of a small tractor fitted with a vacuum pump powered by a tractor or small engine.

Horse pastures should be well drained and not too rough or stony. All dangerous places—such as pits, stumps, poles, and tanks—should be guarded. Shade, water, and suitable minerals should be available in all pastures.

Most horse pastures can be improved by seeding new and better varieties of grasses and legumes, and by fertilizing and management. Also, caretakers need to give attention to balancing pastures. Early-in-the-season pastures are of high water content and lack energy. Mature, weathered grass is almost always deficient in protein (being as low as 3% or less) and low in carotene (the precursor of vitamin A). But these deficiencies can be corrected by proper supplemental feeding.

## HAYS

Through mistaken kindness or carelessness, horses are often fed too much hay or other roughage, with the result that they breathe laboriously and tire quickly. With cattle and sheep, on the other hand, it is usually well to feed considerable roughage. This difference between horses and ruminants is due primarily to the relatively small size of the simple stomach of the horse in comparison with the fourfold stomach of the ruminant.

When limiting the allowance of roughage, it is sometimes necessary to muzzle greedy horses (gluttons) to prevent them from eating the bedding.

Usually, young horses and idle horses can be provided with an unlimited allowance of hay. In fact, much good will result from feeding young and idle horses more roughage and less grain. But one should gradually increase the grain and decrease the hay as work or training begins.

The hay should be early cut, leafy, green, well cured, and free from dust and mold. Hay native to the locality is usually fed. However, horse owners and caretakers everywhere prefer good quality timothy. With young stock and breeding animals especially, it is desirable that a sweet grass-legume mixture of alfalfa hay be fed. The legume provides a source of high-quality proteins and certain minerals and vitamins.

Horses like variety. Therefore, if at all possible, it is wise to have more than one kind of hay in the stable. For example, timothy may be provided at one feeding and a grass-legume mixed hay at the other feeding. Good caretakers often vary the amount of alfalfa fed, for increased amounts of alfalfa in the ration will increase urination and give a softer consistency to the bowel movements. This means that elimination from kidneys and bowels can be carefully regulated by the amount and frequency of alfalfa feedings. Naturally, such regulation becomes more important with irregular use and idleness. On the other hand, in some areas alfalfa is fed as the sole roughage with good results.

- **Timothy**—Timothy is the preferred hay by most caretakers. Although it may be grown alone, it is commonly seeded in mixtures with medium red or alsike clover.

Timothy is easy to harvest and cure. However, in comparison with hay made from the legumes, it is low in crude protein and minerals, particularly calcium.

As with all forages, the feeding value of timothy is affected by the stage of growth of the plants at the time of cutting. With increasing maturity, (1) the percentage of crude protein decreases, (2) the percentage of crude fiber increases, (3) the hay becomes less palatable, and (4) the digestibility decreases. However, delaying cutting until timothy has reached the full bloom stage, or later, usually results in the highest yields. When both yield and quality are considered, the best results are obtained when timothy is cut for hay at the early bloom stage.

- **Alfalfa (lucerne)**—Alfalfa is an important, perennial, leguminous forage plant with trifoliate leaves and bluish-purple flowers. It is grown widely, principally for hay. Alfalfa is capable of surviving dry periods because of its extraordinarily long root system, and it is adapted to widely varying conditions of climate and soil. It yields the highest tonnage per acre and has the highest protein content of the legume hays.

Good quality alfalfa hay is excellent for horses. It averages 15.3% protein, which is of high quality; and it is a good source of certain minerals and vitamins. In addition to being used as hay, alfalfa is an ingredient of most all-pelleted feeds.

- **Oat hay**—Oat hay is an excellent feed for horses. It is easy to cure, and horses like it. Early cutting (in the soft dough stage) greatly increases its feeding value, due to the higher protein content. Even though considerable energy is stored in the kernels at maturity, shattering of the grains during harvesting of mature oats results in energy losses and decreased feeding value compared with early-cut hay.

Oat hay is low in protein; hence, its feeding value is greatly enhanced when it is fed with alfalfa or some other legume.

## FORAGE SUBSTITUTIONS

Forages of similar nutritive properties may be interchanged in the ration as availability and price relationships warrant, thereby making it possible at all times to obtain a balanced ration at the lowest cost. Table 54–3 presents, in summary form, some roughage substitution guidelines.

**TABLE 54–3**
**FORAGE SUBSTITUTION TABLE**

| Forage | Value Compared to Timothy Hay, Which Is Designated As Equal To 100 | Maximum Percent of Timothy Hay Which It Can Replace |
|---|---|---|
| Timothy hay ........ | 100 | 100 |
| Alfalfa hay ......... | 133⅓ | 100 |
| Corn silage ......... | 44-55 (wet basis) | 33⅓-50 |
| Oat hay ........... | 100 | 100 |
| Prairie hay ........ | 100 | 100 |
| Sorghum fodder ...... | 100 | 50 |

# CONCENTRATES

Of all the concentrates, heavy oats most nearly meet the needs of horses; and, because of the uniformly good results obtained from their use, they have always been recognized as the leading grain for horses. Corn is also widely used as a horse feed, particularly in the central states. Despite occasional prejudice to the contrary, barley is a good horse feed. As proof of the latter assertion, it is noteworthy that the Arabs—who were good equestrians—fed barley almost exclusively. Also, wheat, wheat bran, molasses, and commercial mixed feeds are extensively used. Milk by-products and milk replacers are commonly fed to young stock. It is to be emphasized, therefore, that careful attention should be given to the prevailing price of feeds available locally, for many feeds are well suited to horses. Often substitutions can be made that will result in a marked saving without affecting the nutritive value of the ration. When corn or other heavy grains are fed, it is important that a little linseed meal or wheat bran be used, in order to regulate the bowels.

## BRAN MASH

Feeding a bran mash is the traditional way of regulating the bowels of horses on idle days and at such other times as required.

The mash is prepared by filling a 2- to 2½ gal. bucket with wheat bran, pouring enough boiling hot water over it to make it the consistency of breakfast oatmeal, covering the bucket with a blanket and allowing it to steam until cool, then feeding it to the horse.

Occasionally, when horses are offered a bran mash for the first time, they may refuse to eat it. When this occurs, the animal may be enticed to eat the mash by either (1) introducing a little of it by hand, or (2) sprinkling some sugar, or some other well-liked feed over it.

## CONCENTRATE SUBSTITUTIONS

Table 54–4, Grain Substitution Table, shows the value for horses of certain cereal grains compared to oats.

### TABLE 54–4
### GRAIN SUBSTITUTION TABLE

| Energy Feeds | Relative Value Compared to Oats, Which Is Designated As Equal to 100 | Maximum Percent of Oats Which It Can Replace |
|---|---|---|
| Oats ............ | 100 | 100 |
| Barley ............. | 110 | 100 |
| Corn .............. | 115 | 100 |
| Milo (sorghum) ....... | 110–115 | 85 |
| Molasses ........... | 80–95 | 10 |
| Wheat ............. | 115 | 50 |

# PROTEIN SUPPLEMENTS

Grass hays and farm grains are low in quality and quantity of proteins. Hence, they must be supplemented with other sources of protein.

In practical horse feeding, foals should be provided with some protein feeds of animal origin in order to supplement the proteins found in grains and forages. In feeding mature horses, a safe plan to follow is to provide plant protein from several sources.

In general, feeds of high protein content are more expensive than those high in carbohydrates or fats. Accordingly, there is a temptation to feed too little protein. On the other hand, when protein feeds are the cheapest—as is often true of cull peas in certain sections of the West—excess quantities of them may be fed as energy feeds without harm, provided the ration is balanced in all other respects. Any amino acids that are left over, after the protein requirements have been met, are deaminated or broken down in the body. In this process, a part of each amino acid is turned into energy, and the remainder is excreted via the kidneys.

The following oil meals are most commonly used as protein supplements for horses: linseed meal, soybean meal, cottonseed meal, rapeseed meal (canola meal), and sunflower meal.

## UREA FOR THE HORSE

It is recognized that horses frequently consume urea-containing cubes and blocks intended for cattle and sheep, particularly on the western range. Moreover, it appears that mature horses are able to do so without untoward effects. The latter observation was confirmed in one limited experiment[5] in which 4 horses consumed an average 4.57 lb per day of a urea-containing supplement, or 0.55 lb/head/day of feed urea (262%) for 5 months. Also, the Louisiana Station[6] did not find urea detrimental or toxic to horses when it constituted up to 5% of the grain ration, with up to 0.5 lb per day of urea consumed. There are reports, however, of urea toxicity in foals, in which bacterial action is more limited than in older horses.

Thus, there is some evidence that nonprotein nitrogen (urea) can be substituted for protein in the diet of the horse, but the conversion to protein is inefficient. Up to 5% of urea in the total ration does not appear to be harmful to mature horses. Nevertheless, in recognition of the more limited bacterial action in the horse and hazard of toxicity—especially to young equines, most state laws forbid the use of such nonprotein nitrogen sources as urea in horse rations.

# SPECIAL FEEDS AND ADDITIVES

Special horse feeds may be needed from time to time for promoting growth of young stock, preventing disease, or imparting bloom and attractiveness.

• **Antibiotics**—Antibiotics are not nutrients; they are drugs. They are chemical substances, produced by molds or bacteria, which have the ability to inhibit the growth of or to destroy other microorganisms.

The author was a member of the research team that conducted the first U.S. study on feeding antibiotics to foals, which study was subsequently used in obtaining Food and Drug Administration (FDA) approval for feeding Aureomycin to foals. This experiment revealed that an 85 milligram level of Aureomycin, fed to foals from 5 days of age to 5 months, produced 22 lb more weight.[7]

---

[5]*Veterinary Medicine*, Vol. 58, No. 12, Dec, 1963, pp. 945–946.

[6]"Non-Toxicity of Urea Feeding to Horses," *Veterinary Medicine/Small Animal Clinician*, Nov. 1965.

[7]Wash. Ag. Exp. Sta. Circ. 263, April 1955.

Certain antibiotics, at stipulated levels, are approved by the FDA for growth promotion and for the improvement of feed efficiency of young equines up to one year of age. Unless there is a disease level, however, there is no evidence to warrant the continuous feeding of antibiotics to mature horses. Such practice may even be harmful, Hence, where antibiotics are needed for therapeutic purposes, it is best to seek the advice of a veterinarian.

It appears that antibiotics may be especially helpful for young foals which suffer setbacks from infections, digestive disturbances, inclement weather, and other stress factors. Also, horses may benefit from antibiotics (1) when being transported from one location to another—for example, when being moved to a new show or track; (2) when there is a low disease level in the herd; or (3) when mares are foaling.

The poorer the feed, the greater the response from antibiotics; and the poorer the management, the greater the response from antibiotics. It follows, therefore, that there is a temptation to use antibiotics as a "crutch," rather than improve the regimen.

When added to feed, the level of antibiotics should be in keeping with the directions of the manufacturer and with the Food and Drug Administration regulations.

- **Bloom-imparting feeds**—Bloom or gloss is important in horses. But sometimes they lack this desired quality—their hair is dull and dry. Feeding a well-balanced ration will usually rectify this situation. Also feeding the following products will make for an attractive, shiny coat:

1. **Corn oil or safflower oil.** Feed at the rate of 2 oz (2 Tbsp) per horse twice a day.
2. **Whole flaxseed soaked.** Put a handful of whole flaxseed in a teacup, cover it with water, let it stand overnight, then pour it over the morning feed. Repeat twice each week.

Unless the horse is afflicted with lice, mange, or some other ailment, either of the above treatments will impart bloom or gloss to the coat.

- **Lysine**—Protein quality is important for horses. Because of more limited amino acid synthesis in the horse than in ruminants, plus the fact that the cecum is located beyond the small intestine—the main area for digestion and absorption of nutrients, it is generally recommended that high quality protein rations, adequate in amino acids, be fed to equines. This is especially important for young equines, because cecal synthesis is very limited in early life.

Fortunately, the amino acid content of proteins from various sources varies. Thus, the deficiencies of one protein may be improved by combining it with another, and the mixture of the two proteins often will have a higher feeding value than either one alone. It is for this reason, along with added palatability, that a considerable variety of feeds in the horse ration is desirable.

Cornell University reported that the addition of lysine to the diet of growing horses increased weight gains, feed consumption, and feed efficiency.

In recognition that lysine is the first limiting amino acid of horses and is thus an indicator of the quality of protein which horses require, the recommended lysine allowance for horses is given in Table 54–1.

- **Milk by-products**—The superior nutritive values of milk by-products are due to their high-quality proteins, vitamins, a good mineral balance, and the beneficial effect of the milk sugar,

lactose. In addition, these products are palatable and highly digestible. They are an ideal feed for young equines and for balancing out the deficiencies of the cereal grains. Most foal rations contain one or more milk by-products, primarily dried skim milk, with some dried whey and dried buttermilk included at times. The chief limitation to their wider use is price.

- **Milk replacer**—As indicated by the name, a milk replacer is a replacement for milk. Such replacers generally contain the following composition: animal or vegetable fat, 17–20%; crude soybean lecithin, 1–2%; skimmed milk solids, 78–82% (10–15% dried whey powder can be included in place of an equivalent amount of skimmed milk solids); plus fortification with minerals and vitamins.

Foals suckling their dams generally develop satisfactorily up to weaning time. But the most critical period in the entire life of a horse is that space from weaning time (about six months of age) until one year of age. This is especially so in the case of the young horse being fitted for shows or sales, where condition is so important. Thus, where valuable weanlings or yearlings are to shown or sold, the use of a milk replacer may be practical.

## TREATS

Horses are fed a great variety of treats. On a government horse-breeding establishment in Brazil, the author saw a large and well-manicured vegetable garden growing everything from carrots to melons, just for horses. Also, trainers recognize that most racehorses, which are the *prima donnas* of the equine world, don't "eat like a horse;" they eat like people—and sometimes they're just as finicky. Their menus may include a choice of carrots or other roots, apples and other fruits, pumpkins, squashes, melons, molasses, sugar, honey, or innumerable other goodies.

Ask the average equestrians why they feed treats to their horses and you'll get a variety of answers. However, high on the list of reasons will be (1) as appetizers, (2) as a source of nutrients and as conditioners, (3) as rewards, (4) as a means of alleviating obesity (dieting the horse), or (5) folklore.

But treats can be overdone. Hence, horses should not be permitted to eat too much of any treat, simply because they like it.

## PALATABILITY

Palatability is important, for horses must eat their feed if it is to do them any good. But many horses are finicky simply because they are spoiled. For the latter, stepping up the exercise and halving the ration will usually effect a miraculous cure.

Also, it seems possible that well-liked feeds are digested somewhat better than those which are equally nutritious, but less palatable.

Palatability is particularly important when feeding horses that are being used hard, as in racing or showing. Unless the ration is consumed, such horses will not obtain sufficient nutrients to permit maximum performance. For this reason, lower quality feeds, such as straw or stemmy hay, should be fed to idle horses.

Familiarity and habit are important factors concerned with the palatability of horse feeds. For example, horses have to learn to eat pellets, and very frequently they will back away from feeds with new and unfamiliar odors. For this reason, any change in feeds should be made gradually.

Occasionally, the failure of horses to eat a normal amount of feed is due to a serious nutritive deficiency. For example, if horses are fed a ration made up of palatable feeds, but deficient in one or more required vitamins or minerals, they may eat normal amounts for a time. Then when the body reserves of the lacking nutrient(s) are exhausted, they will usually consume less feed, due to an impairment of their health and consequent lack of appetite. If the deficiency is not continued so long that the horses are injured permanently, they will usually recover their appetites if some feed is added which supplies the nutritive lack and makes the ration complete.

## FEED PREPARATION

The physical preparation of cereal grains for horses has been practiced by equestrians for a very long time. Basically, grain is soaked, cooked, ground, or rolled (wet or dry), and hay is fed long, or pelleted, or cubed.

For horses, flaking is the preferred method of grain preparation; it makes for a light ration and fewer digestive disturbances. For animals with good teeth, the value of oats is increased only 5% by processing.

Hay for horses is usually fed long or incorporated in an all-pelleted ration (with the grain and hay combined).

Further elucidation of feed preparation for horses is contained in Chapter 4, section entitled, "Feed Processing"; hence, the reader is referred thereto.

## RATIONS

Correctly speaking, a ration is the amount of feed given to a horse in a day, or a 24-hour period. To most caretakers however, the word implies the feeds fed to an animal without limitation of the time in which they are consumed.

To supply all the needs—maintenance, growth, reproduction and lactation, and work (running)—horses must receive feeds in quality and quantity to furnish the necessary energy (carbohydrates and fats), proteins, minerals, vitamins, and perhaps unknown factors and additives. Such rations are said to be balanced. Moreover, the feed must be palatable—horses must like it. The rations listed in Table 54–5 meet these standards. Also, liberal margins of safety have been provided to compensate for variations in feed composition, environment, possible losses of nutrients during storage, and differences in individual animals.

## HOME MIXED FEEDS

A horse feeding guide is given in Table 54–5. In selecting rations, compare them with commercial feeds. If only small quantities are required or little storage space is available, it may be more satisfactory to buy ready-mixed feeds.

When home mixed feeds are used, feeds of similar nutritive properties can be interchanged in the ration as price relationships warrant. This makes it possible to obtain a balanced ration at lowest cost.

The quantities of feeds recommended in Table 54–5 are intended as guides only. For example, the caretaker should increase the feed, especially the concentrates, when the horse is too thin and decrease the feed if it gets too fat.

Sudden changes in the diet should be avoided, especially when changing from a less concentrated ration to a more concentrated one. When this rule of feeding is ignored, digestive disturbances result and the horse goes *off feed*. In either adding or omitting one or more ingredients, the change should be made gradually. Likewise, caution should be exercised in turning horses to pasture or in transferring them to more lush grazing.

In general, horses may be given as much non-legume roughage as they will eat. But they must be accustomed gradually to legumes because legumes may be laxative.

In feeding horses, as with other classes of livestock, it is recognized that nutritional deficiencies (especially deficiencies of certain vitamins and minerals) may not be of sufficient proportions to cause clear-cut deficiency symptoms. Yet, such deficiencies without outward signs may cause great economic losses because they go unnoticed and unrectified. Accordingly, sufficient additives (especially minerals and vitamins) should always be present, but care should be taken to avoid imbalances.

## SUGGESTED RATIONS

Table 54–5 contains some suggested rations for different classes of horses. This is merely intended as a general guide. The feeder should give consideration to (1) the quality, availability, and cost of feeds; (2) the character and severity of the work; and (3) the age and individuality of the animal.

**TABLE 54–5**
**LIGHT HORSE FEEDING GUIDE[1]**

| Age, Sex, and Use | Daily Allowance | Kind of Hay | Suggested Grain Rations | | |
| --- | --- | --- | --- | --- | --- |
| | | | Rations No. 1 | Rations No. 2 | Rations No. 3 |
| | | | (lb) | (lb) | (lb) |
| **Stallions in breeding season** (weighing 900 to 1,400 lb) | ¾ to 1½ lb grain per 100 lb body weight, together with a quantity of hay within same range. | Grass-legume mixed; or ⅓ to ½ legume hay, with remainder grass hay. | Oats ........ 55 <br> Wheat ........ 20 <br> Wheat bran ........ 20 <br> Linseed meal ........ 5 | Corn ........ 35 <br> Oats ........ 35 <br> Wheat ........ 15 <br> Wheat bran ........ 15 | Oats ........ 100 |
| **Pregant mares** (weighing 900 to 1,400 lb) | ¾ to 1½ lb grain per 100 lb body weight, together with a quantity of hay within the same range. | Grass-legume mixed; or ⅓ to ½ legume hay, with remainder grass hay (straight grass hay may be used first half of pregnancy). | Oats ........ 80 <br> Wheat bran ........ 20 | Barley ........ 45 <br> Oats ........ 45 <br> Wheat bran ........ 10 | Oats ........ 95 <br> Linseed meal ........ 5 |

*(Continued)*

**TABLE 54-5** *(Continued)*

| Age, Sex, and Use | Daily Allowance | Kind of Hay | Suggested Grain Rations | | |
|---|---|---|---|---|---|
| | | | Rations No. 1 | Rations No. 2 | Rations No. 3 |
| | | | (lb) | (lb) | (lb) |
| **Foals before weaning** (weighing 100 to 350 lb with projected mature weights of 900 to 1,400 lb) | ½ to ¾ grain per 100 lb body weight, together with a quantity of hay within same range. | Legume hay. | Oats ............. 50<br>Wheat bran ......... 40<br>Linseed meal ........ 10 | Oats ............. 30<br>Barley ............. 30<br>Wheat bran ......... 30<br>Linseed meal ........ 10 | Oats ............. 80<br>Wheat bran .......... 20 |
| | | | Rations balanced on basis of following assumption: Mares of mature weights of 600, 800, 1,000, and 1,200 lb may produce 36, 42, 44 and 49 lb of milk daily. | | |
| **Weanlings** (weighing 350 to 450 lb) | 1 to 1½ lb grain and 1½ to 2 lb hay per 100 lb body weight. | Grass-legume mixed; or ½ legume hay, with remainder grass hay. | Oats ............. 30<br>Barley ............. 30<br>Wheat bran ......... 30<br>Linseed meal ........ 10 | Oats ............. 70<br>Wheat bran ......... 15<br>Linseed meal ........ 15 | Oats ............. 80<br>Linseed meal .......... 20 |
| **Yearlings, second summer** (weighing 450 to 700 lb) | Good, luxuriant pastures. (If in training for other reasons without access to pastures, the ration should be intermediate between the adjacent upper and lower groups.) | | | | |
| **Yearlings, or rising 2-year-olds, second winter** (weighing 700 to 1,000 lb) | ½ to 1 lb grain and 1 to 1½ lb hay per 100 lb body weight. | Grass-legume mixed; or ⅓ to ½ legume hay, with remainder grass hay. | Oats ............. 80<br>Wheat bran ......... 20 | Barley ............. 35<br>Oats ............. 35<br>Bran ............. 15<br>Linseed meal ........ 15 | Oats ............. 100 |
| **Light horses at work; riding, driving, and racing** (weighing 900 to 1,400 lb) | Hard use—1¼ to 1⅓ lb grain and 1 to 1¼ lb hay per 100 lb body weight. Medium use—¾ to 1 lb grain and 1 to 1¼ lb hay per 100 lb body weight. Light use—⅜ to ½ lb grain and 1¼ to 1½ lb hay per 100 lb body weight. | Grass hay. | Oats ............. 100 | Oats ............. 70<br>Corn ............. 30 | Oats ............. 70<br>Barley ............. 30 |
| **Mature idle horses; stallions, mares, and geldings** (weighing 900 to 1,400 lb) | 1½ to 1¾ lb hay per 100 lb body weight. | Pasture in season; or grass-legume mixed hay. | (With grass hay, and ¾ lb of a high-protein supplement daily.) | | |

[1] With all rations and for all classes and ages of horses, provide free access to a mineral box as follows: (1) *Where the pasture or hay is primarily grass*, use a mixture containing 2 parts of calcium to 1 part of phosphorus; and (2) *where the pasture or hay is primarily a legume*, use a mixture containing 1 part of calcium to 1 part of phosphorus. To each of these mixes, add ⅓ salt (trace mineralized) to improve acceptability. If preferred, a good commercial mineral may be used. Self-feed salt separately.

## ROUGHAGES

Actually, a horse does not have to have any hay. Also, more horses receive too much roughage than not enough, as evidenced by hay bellies (distended digestive tracts), quick tiring, and labored breathing.

Under most conditions, the roughage requirement of horses ranges from 0.5% to 1.0% of body weight, or from 5 to 10 lb of roughage daily for a 1,000-lb horse.

Racehorses should receive a minimum of roughage, since they need a maximum of energy. Sometimes it is necessary to muzzle greedy horses to keep them from eating bedding when their roughage allowance is limited.

## COMMERCIAL HORSE FEED

Commercial horse feeds are feeds mixed by manufacturers who specialize in the feed business.

Commercial feed manufacturers are able to purchase feed in quantity lots, making possible price advantages and the scientific control of quality. Many horse owners have found that because of the small quantities of feed usually involved, and the complexities of horse rations, they have more reason to rely on good commercial feeds than do owners of other classes of farm animals.

The nutritive requirements of horses vary according to age, weight, use or demands, growth, stage of gestation or lactation, and environment. Also, part of the horse ration may be homegrown.

## SWEET FEED

*Sweet feed* refers to a feed to which has been added one or more ingredients that are sweet. Most commonly, it is considerable molasses (approximately 10%); although brown sugar (about 5%) is sometimes used, and occasionally honey.

The horse has a *sweet tooth*; hence, it's not easy to switch him from a sweet feed to what may be a more nutritious ration. Of course, the manufacturer of the sweet feed would have it that way. Also, it must be remembered that sweet feeds are a way in which a feed manufacturer may make poor quality feed ingredients more appetizing. Remember, too, that most boys and girls would rather eat candy than foods that are more nutritious. But doctors and parents know best!

## RATION FORMULATION BY IMITATION

Sometimes horse trainers pattern their ration after what some great horse is getting—they get some of the same "stuff."

Of course, it is difficult to argue with success. Also, it is well known that horse owners and caretakers as a whole are great imitators. The author has known them to pay $50 for a gallon of a mysterious concoction, in a green jug, made in a little hamlet in Kentucky. Of course, the fallacy of such imitation—of feeding what the "great horse" got—is that the "name" horse might have been even greater had it been fed properly, and there must be a reason why there are so few truly great horses. Also, the following searching question might well be asked: Why do many horses start training in great physical shape, only to slow down and lose appetite, and be taken out of training for some rest?

## RESULTS MORE IMPORTANT THAN COST PER BAG

As is true when buying anything—whether it be a suit of clothes, a dinner, or whatnot—horse feed should be bought on a quality basis, rather than what is cheapest—results are more important than cost per bag. If this were not so, one might well buy and feed many cheap products, including sawdust.

Consideration should be given to meeting the specific needs of the horse, with special attention given to providing adequate quantity and quality proteins, minerals, vitamins, unidentified factors, and palatability.

## ART OF FEEDING

Feeding horses is both an art and a science. The art is knowing how to feed and how to take care of each horse's individual requirements. The science is meeting the nutritive requirements with the right combination of ingredients.

## AMOUNT TO FEED

The main qualities desired in horses are trimness, action, spirit, and endurance. These qualities cannot be obtained with large, paunchy stomachs or lack of energy, which may result from excessive use of roughage. Moreover, a healthy condition is desired, but excess fat is to be avoided. The latter is especially true with horses used for racing, where the carrying of any surplus body weight must be avoided.

The quantity of grain and hay required by horses depends primarily upon the following:

1. The individuality—horses vary in keeping qualities, just as people do. Some horses simply utilize their feed more efficiently than others. A hard keeper will require considerably more feed than an easy keeper when doing the same amount of work.

2. The age, size, and condition of the animal.

3. The kind, severity, regularity, amount, and speed of work performed. With greater speed, the horse requires proportionately greater energy; hence, considerably more concentrate is required when performing work at a trot than at a walk.

4. The weather; for example, under ideal October weather conditions in Missouri, a horse may require 14 lb of 60% TDN feed daily, whereas in the same area, the same horse requires 16 lb daily of the same feed in July and August, and 20 lb in the winter.

5. Kind, quality, and amount of feed.

6. System of management.

7. Health, condition, and temperament of the animal.

When given all the feed that they will consume, mature horses will generally eat an amount equivalent to about 2.5% of their body weight. Growing foals and lactating mares eat more heartily—they will consume up to 3% of their body weight.

Because the horse has a rather limited digestive capacity, the amount of concentrates must be increased and the roughage decreased when the energy needs rise with greater amount, severity, or speed of work. The following are general guides for the daily ration of horses under usual conditions

1. **For horses at light work** (1 to 3 hours per day of riding or driving) allow ⅔ to ½ lb of grain and 1¼ to 1½ lb of hay per day per 100 lb liveweight.

2. **For horses at medium work** (3 to 5 hours per day of riding or driving), allow ¾ to 1 lb of grain and 1 to 1¼ of hay per 100 lb of liveweight.

3. **For horses at hard work** (5 to 8 hours per day of riding or driving), allow about 1¼ to 1⅓ lb of grain and 1 to 1¼ lb of hay per 100 lb of liveweight.

The recommended feed allowance on the basis of animal weight are equally applicable to equines of all sizes, including ponies and donkeys; simply vary as necessary according to the work performed and the individuality of the animal.

As will be noted from these recommendations, the total allowance of concentrates and hay should be within the range of 2.0 to 2.5 lb daily per 100 lb of liveweight. No grain should be left from one feeding to the next, and all edible forage should be cleaned up at the end of each day.

About 6 to 12 lb of grain daily is an average grain ration for a light horse at medium or light work. Racehorses in training usually consume 10 to 16 lb of grain per day—the exact amount varying with the individual requirements and the amount of work. The hay allowance averages about 1 to 1¼ lb daily per 100-lb liveweight, but it is restricted as the grain allowance is increased. Light feeders should not be overworked.

It is to be emphasized that the quantities of feeds recommended above are intended as guides only. The feeder will increase the allowance, especially the concentrates, when the horse is too thin, and decrease the feed when the horse is too fat.

The regular practice of turning horses to pasture at night, on idle days, and in off-work seasons is good for the health and well-being of the animals and decreases the quantity of grain and hay required. If the horse must be confined to the stall on idle days, the grain ration should be reduced by 50% in order to avoid azoturia or other digestive disturbances. When idle, it is also advisable to add some wheat bran to the ration. A mixture of ⅔ grain and ⅓ bran is quite satisfactory. Many good caretakers regularly give a feeding of bran, either dry or as a wet mash, on Saturday night.

## OVERFEEDING

Overfeeding may result in two consequences: if done suddenly, it may cause founder (laminitis, colic, or enterotoxemia); if prolonged, it will likely result in obesity (too fat). Both are bad.

## STARTING HORSES ON FEED

Horses must be accustomed to changes in feed gradually. In general, they may be given as much nonlegume roughage as they will consume. But they must be accustomed gradually to high-quality legumes, which may be very laxative. This can be done by slowly replacing the nonlegume roughage with greater quantities of legumes. Also, as the grain ration is increased, the roughage is decreased.

Starting horses on grain requires care and good judgment. Usually, it is advisable first to accustom them to a bulky type of ration; a starting ration with considerable rolled oats is excellent for this purpose.

The keenness of the appetite and the consistency of the droppings are an excellent index of a horse's capacity to take more feed. In all instances, scouring should be avoided.

## FREQUENCY, ORDER, AND REGULARITY OF FEEDING

The grain ration usually is divided into 3 equal feedings given morning, noon, and night. Because a digestive tract distended with hay is a hindrance to hard work, most of the hay should be fed at night. The common practice is to feed ¼ of the daily hay allowance at each of the morning and noon feedings and the remaining ½ at night when the animals have plenty of time to eat leisurely.

Usually the grain ration is fed first and then the roughage. This way, the animals can eat the bulky roughages more leisurely.

Horses learn to anticipate their feed. Accordingly, they should be fed at the same time each day. During warm weather, they will eat better if the feeding hours are early and late, in the cool of the day.

## AVOID SUDDEN CHANGES

Sudden changes in diet should be avoided, especially when changing from a less concentrated ration to a more concentrated one. If this rule of feeding is ignored, horses have digestive disturbances and go *off-feed*. When ingredients are added or omitted, the change should be made gradually. Likewise, caution should be exercised in turning horses to pasture or in transferring them to a more lush grazing.

Sometimes caretakers experience difficulty in switching horses from an overly sweet or highly flavored feed to a more nutritious ration. But the end result usually justifies the effort.

## ATTENTION TO DETAILS PAYS

The successful equestrian pays great attention to details, in addition to maintaining the health and comfort of animals, he/she should also give consideration to their individual likes and temperaments.

It is important to avoid excessive exercise to the point of fatigue and undue stress. Also, rough treatment, excitement, and noise usually result in nervous and inefficient use of feed.

## BOLTING FEED

Horse that eat too rapidly are said to be bolting their feed. It can be lessened by spreading the concentrate thinly over the bottom of a large grain box, so that the horse cannot get a large mouthful; or by placing in the grain box a few smooth stones about the size of baseballs, so that the horse has to work to get feed.

## EATING BEDDING

Sometimes gluttonous animals eat their bedding. This is undesirable because (1) most bedding materials are low in nutritional value, and (2) feces-soiled bedding adds to the parasite problem. The problem can be alleviated by muzzling the horse.

## FEEDING SYSTEMS

Most horses are hand-fed. The grain ration usually is divided into three equal feeds given morning, noon, and night. Because a digestive tract distended with hay is a hindrance in hard work, most of the hay should be fed at night. The common practice is to feed one-fourth of the daily allowance at each of the morning and noon feedings and the remaining one-half at night when the animals have plenty of time to eat leisurely.

A few caretakers self-feed high-energy rations, but, sooner or later, they usually founder a valuable horse. Except for the use of reasonable hard salt-protein blocks, salt-free mixes in meal form (never in pellet form), or high-roughage rations, the self-feeding of horses is not recommended.

## GENERAL FEEDING RULES

In addition to the guides already mentioned, observance of the following general rules will help avoid some of the common difficulties:

1. Know the approximate weight and age of each animal.
2. Feed by weight of feed, not by volume (volume as determined by a coffee can or marked bucket). Horses do not require a certain volume of feed; rather, they require a certain weight of nutrients based on their body weights.
3. Avoid sudden changes in the ration.
4. Never feed moldy, musty, dusty, or frozen feed.
5. Feed regularly. Horses anticipate their feed.
6. Look for problems at feeding time; don't just dump the feed and run. Look for injuries and abnormalities.
7. Check the feces. Any change in quantity, odor, color, or composition may presage trouble.
8. Inspect the feedbox frequently to see if the horse goes off feed. Feed refusal means (1)the horse was over-fed, (2) something is wrong with the feed, or (3) the horse is sick.
9. Keep the feed and water containers clean. Scrub them periodically to insure proper sanitation.
10. Do not overfeed. Some horses suffer from obesity, while others suffer from deficiency. Fat horses not receiving adequate exercise are predisposed to colic and founder. An old Arab proverb cautions: "The greatest enemies of horses are fat and rest."
11. Force aggressive eaters to slow down. Some horses may bolt their feed when fed in deep narrow feed boxes. Their eating may be slowed by scattering the feed in a larger box, or by placing large round stones, bricks, or salt blocks in the feed container.
12. Accord timid eaters solitude to eat. Feed them where it is quiet and they will not be disturbed.
13. Do not feed from the hand; this can lead to *nibbling*.
14. Exercise stalled horses daily. It improves their appetite, digestion, and overall well being. This may be accomplished by riding, longeing, walking, ponying, swimming, or treadmilling.
15. Avoid excessive exercise (to the point of fatigue and stress), rough treatment, noise, and excitement.

16. Do not feed concentrates 1 hour before or within 1 hour after hard work.

17. Feed horses as individuals; consider their likes and temperaments. Learn the peculiarities and desires of each animal because each one is different.

18. Gradually decrease the condition of horses that have been fitted for show or sale. Many caretakers accomplish this difficult task, and yet retain strong vigorous animals, by cutting down gradually on the feed and increasing the exercise.

19. Prevent wood chewing. This habit usually results from boredom, lack of exercise, lack of adequate roughage, or lack of phosphorus; so, alleviate the causes.

20. Make certain that the horse's teeth are sound.

21. Know the signs of a well-fed, healthy horse, any departure from which constitutes a warning signal.

## FEEDING PLEASURE HORSES

Keeping pleasure horses—horses used for recreation and sport—in peak condition makes for greater satisfaction when they are used.

It is difficult to feed pleasure horses properly because their exercise is often irregular. Sometimes they are used moderately; at other times they are idle; at still other times they are worked hard over the weekend or on a trail ride.

Most horses used for pleasure are worked lightly, perhaps 1 to 3 hours of riding per day. Others are worked medium hard, as when ridden 3 to 5 hours per day. Still others are worked very hard, as when raced or when ridden 5 to 8 hours per day. The recommended daily feed allowance per 100 lb body weight of pleasure horses in light, medium, and hard use follows:

| Lb Daily/100 Lb Weight of Horse | Light Use | Medium Use | Hard Use |
|---|---|---|---|
| Hay | 1¼–1½ | 1–1¼ | 1–1¼ |
| Grain | ⅖–½ | ¾–1 | 1¼–1⅓ |

As shown above, the roughage content of the ration decreases and the concentrate content increases as the amount of work increases. This is because the digestibility and the efficiency of conversion are greater for high-energy concentrates than for roughages.

Of course, horses differ in temperament and in ease of keeping. Also, no two horses will perform the same amount of work with an equal expenditure of energy, and no two equestrians will get the same amount of work out of the same horse. So, the feed allowance should be increased if the horse fails to maintain condition, and it should be decreased if the animal becomes too fat.

In season, pasture may replace hay, all or in part, according to the quality of the pasture. But the concentrate allowance of the working horse should remain about the same on pasture as in the stable or dry corral. There is a tendency of the pastured working horse to sweat and tire more easily (be *soft*), probably due to the high water content of green forage.

In addition to forage and grain, pleasure horses should have access to salt and suitable mineral mix, free choice. The mineral requirements of the working horse differ from the idle horse mainly in the salt requirements, due to the loss of salt in perspiration.

The vitamin requirements of working horses are approximately the same as those of idle horses, except for the increase in the B-complex requirements due to the greater carbohydrate metabolism of the working horse.

## FEEDING HORSES IN TRAINING

Horses in heavy training for specific purposes—such as training for racing, cutting, roping, jumping, or hunting—have a higher nutritional requirement than most pleasure horses. And the younger the animal in training, the higher the level of nutrition needed in order to develop and maintain sound legs and build a strong frame and body. Therefore, the level of work, the temperament of the individual, and the age of the horse determine the nutritional needs. For this reason, horses in training should be fed as individuals.

Horses in training will eat about 1½ lb of grain and 1 lb of hay per 100-lb liveweight.

## FEEDING RACEHORSES

Racehorses are equine athletes whose nutritive requirements are the most exacting, but the most poorly met, of all animals. This statement may be shocking to some, but it's true because racehorses are commonly:

• Started in training very shortly past 12 months of age, which is comparable to an adolescent boy or girl doing sweatshop labor.

• Moved from track to track under all sorts of conditions.

• Trained the year around, raced innumerable times each year, and forced to run when fatigued.

• Outdoors only a short time each day—usually before sunup, with the result that the sun's rays have little chance to produce vitamin D from the cholesterol in the skin.

• Without opportunity for even a few mouthfuls of grass—a rich, natural source of the B vitamins and unidentified factors.

• Fed oats, grass, hay, and possibly bran—produced in unknown areas, and on soils of unknown composition. Such an oats-grass-hay-bran ration is almost always deficient in vitamins A and D and the B vitamins, and lopsided and low in calcium and phosphorus.

• Given a potion of some concoction of questionable value—if not downright harmful.

By contrast, human athletes—college football teams and participants in the Olympics, for example—are usually required to eat at a special training table, supervised by nutrition experts. They are fed the best diet that science can formulate and technology can prepare. It is high in protein, rich in readily available energy, and fortified and balanced in vitamins and minerals.

It's small wonder, therefore, that so many equine athletes go unsound, whereas most human athletes compete year after year until overtaken by age.

Indeed, high strung and highly stressed racehorses need special rations just as human athletes do—and for the same reasons; and, the younger the age, the more acute the need. This calls for rations rich in protein, rich in readily available energy, fortified with vitamins, minerals, and unidentified factors—and with all nutrients in proper balance.

A racehorse is asked to develop a large amount of horsepower in a period of 1 to 3 minutes. The oxidations that occur in a racehorse's body are at a higher pitch than in an idle horse, and therefore, more vitamins are required.

Also, racehorses are the *prima donnas* of the equine world; most of them are temperamental, and no two of them can be fed alike. They vary in rapidity of eating, in the quantity of feed that they will consume, in the proportion of concentrate to roughage that they will take, and in the response to different caretakers. Thus, for best results, they must be fed as individuals.

Most racehorse rations are deplorably deficient in protein, simply because they are based on the minimum requirements of little stressed, slow, plodding horses.

During the racing season, the hay of a racehorse should be limited to 7 or 8 lb, whereas the concentrate allowance may range up to 16 lb. Heavy roughage eaters may have to be muzzled, to keep them from eating their bedding. A bran mash is commonly fed once a week.

## FEEDING BROODMARES

Regular and normal reproduction is the basis for profit on any horse breeding establishment. However, only 40 to 60% of mares bred produce foals. There are many causes of reproductive failure, but inadequate nutrition is a major one. The following pointers are pertinent to feeding a broodmare properly:

1. Condition the mare for breeding by providing adequate and proper feed and the right amount of exercise prior to the breeding season.

2. See that adequate proteins, minerals, and vitamins are available during the last third of pregnancy when the fetus grows most rapidly.

3. Feed and water with care immediately before and after foaling. For the first 24 hours after parturition, the mare may have a little hay and a limited amount of water from which the chill has been taken. A light feed of bran or a wet bran mash is suitable for the first feed and the following meal may consist of oats or a mixture of oats and bran. A reasonably generous allowance of good quality hay is permissible after the first day. If confined to the stable, as may be necessary in inclement weather, the mare should be kept on a limited and light grain and hay ration for about 10 days after foaling. Feeding too much grain at this time is likely to produce digestive disturbances in the mare and, even more hazardous, it may produce too much milk, which may cause indigestion in the foal. If weather conditions are favorable and it is possible to allow the mare to foal on a clean, lush pasture, she will regulate her own feed needs most admirably.

4. Provide adequate nutrition during lactation, because the requirements during this period are more rigorous than the requirements during pregnancy.

5. Make sure that young growing mares receive adequate nutrients; otherwise, the fetus will not develop properly or the dam will not produce milk except at the expense of her body tissue.

## FEEDING STALLIONS

The ration exerts a powerful effect on sperm production and semen quality. Successful breeders adhere to the following stallion feeding rules:

1. Feed a balanced ration, giving particular attention to proteins, minerals, and vitamins.

2. Regulate the feed allowance because the stallion can become infertile if he gets too fat. Also, increase the exercise when the stallion is not a sure breeder.

3. Provide pasture in season as a source of both nutrients and exercise.

## FEEDING FOALS

Growth is the very foundation of horse production. This is so because horses cannot perform properly, or possess the necessary speed and endurance if their growth has been stunted or their skeletons have been injured by inadequate rations during early age. Naturally, these requirements become increasingly acute when horses are forced for early use, such as the training and racing of the two-year-old. Also, unless foals are rather liberally fed when young, they never attain the much desired body form, even if they are well fed later in life; this point is especially important where young stock is sold or shown.

Fig. 54–11. Off to a good start in life! (Photo by Ernst Peterson. Courtesy, Bitterroot Stock Farm, Hamilton, MT)

The foal should be provided with a low-built grain box, or with a creep (a separate enclosure away from the dam) if on pasture.

The following pointers are pertinent to proper feeding of foals:

1. Start on feed early, which means at 10 days to 3 weeks of age. Rolled oats and wheat bran, to which a little brown sugar has been added, is especially palatable as a starting ration. Crushed or ground oats, cracked or ground corn, wheat bran, and a little linseed meal may be provided later with good result; or, a good commercial ration may be fed if desired and available.

2. Use a scientifically formulated ration even though it seems expensive, for usually it will represent a wise investment.

Supplemental feeding also affords a convenient way in which to improve upon milk, by reinforcing it with certain minerals (for example, milk is low in iron and copper), vitamins, and additives.

Provide a good hay, preferably a legume, or pasture, in addition to its grain ration.

3. Allow about ½ lb of grain daily per 100 lb of body weight at 4 to 5 weeks of age. This ration should be increased by weaning time to about ¾ lb or more per 100 lb of body weight. The exact amount of the ration will vary with the individual, the type of feed, and the development desired.

4. Obtain growth with durability and soundness, which calls for expert care and particular emphasis on the kind of ration, feed allowance, and exercise.

5. Simplify weaning and setback by feeding foals so that they rely less upon their mothers.

## FEEDING WEANLINGS

The most critical period in the entire life of a horse is that interval from weaning time (about six months of age) until one year of age. Foals suckling their dams and receiving no grain may develop vary satisfactorily up to weaning time. However, lack of preparation prior to weaning and neglect following the separation from the dam may prevent the animal from gaining proper size and shape. The primary objective in the breeding of horses is the economical production of a well-developed, sound individual at maturity. To achieve this result requires good care and management of weanlings.

No great setback or disturbances will be encountered at weaning time provided that the foals have developed a certain independence from proper grain feedings during the suckling period. Generally, weanlings should receive 1 to 1½ lb of grain and 1½ to 2 lb of hay daily per each 100 lb liveweight. The amount of feed will vary somewhat with the individuality of the animal, the quality of roughage, available pastures, the price of feeds, and whether the weanling is being developed for show, race, or sale. Naturally, animals being developed for early use or sale should be fed more liberally, although it is equally important to retain clean, sound joints, legs, and feet—a condition which cannot be obtained so easily in heavily fitted animals.

Because of the rapid development of bone and muscle in weanlings, it is important that, in addition to ample quantity of feed, the ration also provides quality of proteins, and adequate minerals and vitamins.

## FEEDING YEARLINGS

If young animals have been fed and cared for so that they are well grown and thrifty as yearlings, usually little difficulty will be experienced at any later date.

When on pasture, yearlings that are being grown for show or sale should receive grain in addition to grass. They should be confined to their stalls in the daytime during the hot days and turned out at night (because of not being exposed to sunshine, adequate vitamin D must be provided). This point needs to be emphasized when forced development is desired; for, good as pastures may be, they are roughages rather than concentrates.

The winter feeding program for the rising 2-year-old should be such as to produce plenty of bone and muscle rather than fat. From ½ to 1 lb of grain and 1 to 1½ lb of hay should be

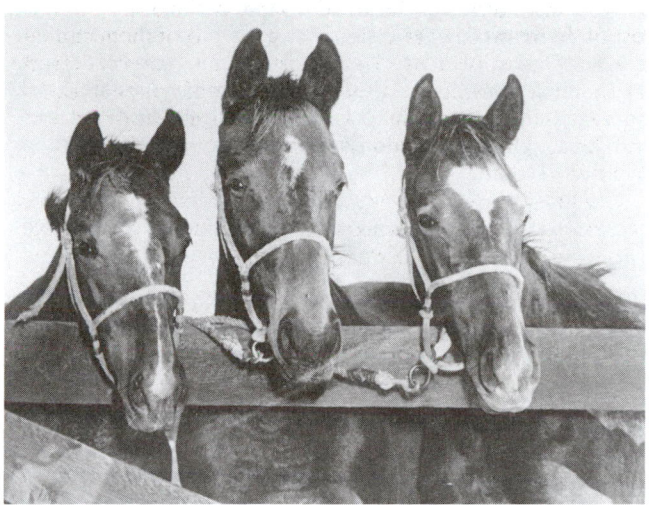

Fig. 54–12. Well-fed yearlings—curious, contented, and alert. (Courtesy, Theracon, Inc., Topeka, KS)

fed for each 100 lb of liveweight. The quantity will vary with the quality of the roughage, the individuality of the animal, and the use for which the animal is produced. In producing for sale, more liberal feeding may be economical. Access to salt and to a mineral mixture should be incorporated in the ration. An abundance of fresh, pure water must be available.

## FEEDING TWO- AND THREE-YEAR-OLDS

Except for the fact that the 2- and 3-year-olds will be larger, and, therefore, will require more feed, a description of their proper care and management would be merely a repetition of the principles that have already been discussed for the yearling.

## FITTING FOR SHOW OR SALE

Each year, many horses are fitted for shows or sales. In both cases, a fattening process is involved, but exercise is doubly essential.

For horses that are being fitted for shows, the conditioning process is also a matter of hardening, and the horses are used daily in harness or under saddle. Regardless of whether a sale or a show is the major objective, fleshing should be obtained without sacrificing action or soundness or without causing filling of the legs and hocks.

In fattening horses, the animals should be brought to a full feed rather gradually, until the ration reaches a maximum of about 2 lb of grain daily for each 100 lb of liveweight. When on full feed, horses make surprising gains. Daily weight gains of 4 to 5 lb are not uncommon. Such animals soon become fat, sleek, and attractive. This is probably the basis for the statement that "fat will cover up a multitude of sins in a horse."

Although exercise is desirable from the standpoint of keeping the animals sound, it is estimated that such activity decreases the daily rate of gains by as much as 20%. Because of the greater cost of gains and the expense involved in bringing about forced exercise, most feeders of sale horses limit the exercise to that obtained naturally from running in a paddock.

In comparison with finishing cattle or sheep, there is more risk in fattening horses. Heavily fed horses kept in idleness are likely to become blemished and injured through playfulness, and there are more sicknesses among liberally fed horses than in other classes of stock handled in a similar manner.

In fitting show horses, the finish must remain firm and hard, the action superb, and the soundness unquestioned. Thus, they must be carefully fed, groomed, and exercised to bring them to proper bloom.

Persons fitting and selling yearlings or younger animals may feed a palatable milk replacer or commercial feed to advantage.

## SOIL ANALYSIS

For horse farms that produce their own hay, a soil analysis can be very helpful; for example (1) the phosphorus content of soils affects the plant composition, (2) soils high in molybdenum and selenium affect the composition of feed produced, (3) iodine deficiency areas are important to horse nutrition, and (4) other similar soil-plant relationships are important.

No analysis is any better than the sample taken. So, make sure that you get a representative sample of soil. Ask your county agent (farm advisor) how to take the samples and where to send them. Some colleges of agriculture make soil analyses at nominal cost.

## NUTRITIONAL DISEASES AND AILMENTS

Nutritional deficiencies may be brought about either by (1) too little feed, or (2) rations that are too low in one or more nutrients. Also, forced production (such as racing 2-year-olds) and the feeding of forages and grains which are often produced on leached or depleted soils, have created many problems in nutrition. This condition has been further aggravated through the increased confinement of horses, many animals being confined to stalls or lots all or a large part of the year. Under these unnatural conditions, nutritional diseases and ailments have become increasingly common.

Chapter 4, "Feeding Livestock," Table 4–13, contains a summary of the important nutritional diseases and ailments affecting horses, as well as other classes of farm animals; hence, the reader is referred thereto.

## PICA; WOOD CHEWING

Horses, particularly those confined to stalls or lots, sometimes consume such materials as dirt, hair, bones, or feces.

Such depraved appetites are known as *pica*. This condition is usually caused by one or more of the following conditions:

1. **Boredom**, because they have nothing to do. The more limited the exercise, and the more quickly they consume their feed, the greater the unoccupied time available and the consequent boredom.

By contrast, little *Eohippus* (the dawn horse of 58 million years ago) was a denizen of the swamp. Later, through evolution, it became a creature of the prairies. Although its natural habitat shifted during this long predomesticated period, until people confined the horse, it gleaned the feeds provided by nature. Inevitably, this occupied its time and provided exercise.

2. **Nutritional inadequacies**, which may be due to (a) a deficiency of one or more nutrients, (b) an imbalance between certain nutrients, or (c) objection to the physical form of the ration—for example, it may be ground too finely.

3. **Psychological stress and habit**, which contribute to the behavior of horses, and which have been accentuated by the unnatural environment to which man has subjected them.

Whatever the reason(s) for pica, the suspected causative factor(s) should first be rectified. When and where needed, the exercise should be stepped up; the eating time should be prolonged, and the interval between feeding shortened; nutritional deficiencies, imbalances, and physical form of ration should be corrected; and stress should be minimized. Even after these conditions have been rectified, it may be disconcerting to find that wood chewing, and perhaps various other forms of pica, persist among certain horses—perhaps due to habit. Thus, in the final analysis, there is only one foolproof way in which to prevent wood chewing; namely, to have no wood on which they can chew—to use metal, or other similar materials for fences and barns. Of course, this isn't always practical. So, you can lessen, although you cannot entirely prevent, wood chewing through one or more of the following practices:

1. Stepping up the exercise.

2. Feeding three times a day, rather than twice a day, even though the total daily feed allowance remains the same.

3. Spreading out feed in a larger feed container, and/or placing a few large stones about the size of a baseball in the feed container, thereby making the horse work harder and longer to obtain its feed.

4. Providing 2 to 4 lb of straw or coarse grass hay per animal per day, thereby giving the horse something to nibble on during its spare time.

## QUESTIONS FOR STUDY AND DISCUSSION

1. What conditions pertaining to the care and use of horses make it imperative that their nutrition be the best that science and technology can devise?

2. How does the digestive track of the horse differ from that of a cow?

3 What's the difference between *nutritive requirements* and *nutritive allowances*?

4. How do you explain that the energy requirements of a horse increase with speed?

5. What's meant by *protein poisoning*? Is there such a thing?

6. How can a lactating mare produce milk that contains more minerals than the feed she is consuming at the time?

7. Discuss metabolic bone diseases, including cause and prevention.

8. List the minerals that are most apt to be deficient in horse rations; then, for each of them (a) list the deficiency symptoms, and (b) give practical sources.

9. What three factors determine the proper utilization of calcium and phosphorus?

10. List the vitamins that are apt to be deficient in horse rations; then, for each of them (a) list the deficiency symptoms,

and (b) give practical sources.

11. Explain how the action of ultraviolet light on the skin of horses and on some feeds may produce vitamin D.

12. At what point in the loss of body water may death result?

13. Why do horse feeders prefer oats and timothy hay to all other feeds?

14. Why do caretakers use each of the following: old process linseed meal, bran mash, treats, and sweet feed?

15. Under what circumstances would you recommend that a horse owner use (a) a home-mixed feed, or (b) a commercial feed?

16. Discuss the art and the science of feeding horses.

17. What would you do about a horse that (a) bolts his feed, or (b) eats his bedding?

18. Discuss the proper feeding of each of the following classes of horses: (a) pleasure horses, (b) horses in training, (c) racehorses, (d) broodmares, (e) stallions, (f) foals, (g) weanlings, (h) yearlings, (i) 2- and 3-year-olds, and (j) horses being fitted for show or sale.

19. Under what circumstances would you have the soil on a horse farm chemically analyzed?

20. What is pica? What causes it and what would you do to stop it?

21. Write down the ration that is fed to a certain horse (your horse or a friend's horse). Then, evaluate it from the standpoints of (a) rate of feeding; (b) protein content; (c) content of salt, iodine, calcium, and phosphorus; and (d) content of vitamin A, vitamin D, and riboflavin.

22. List all of the ways in which the nutritive requirements and the feeding of horses are different from cattle, sheep, swine, and poultry.

## SELECTED REFERENCES

| Title of Publication | Author(s) | Publisher |
|---|---|---|
| *Animal Science,* 8th Edition | M. E. Ensminger | The Interstate Printers & Publishers, Inc., Danville, IL, 1983 |
| *Breeding and Raising Horses,* Ag. Hdbk. No. 394 | M. E. Ensminger | Agricultural Research Service, USDA, Washington, DC, 1972 |
| *Complete Encyclopedia of Horses, The* | M. E. Ensminger | A. S. Barnes & Co., Inc., Cranbury, NJ, 1977. (Now, Oak Tree Publications, San Diego, CA) |
| *Feeding and Care of the Horse* | L. D. Lewis | Lea & Febiger, Philadelphia, PA, 1982 |
| *Feeding Ponies* | W. C. Miller | J. A. Allen & Co., London, England, 1968 |
| *Feeds & Nutrition,* 2nd Edition | M. E. Ensminger J. E. Oldfield W. W. Heinemann | The Ensminger Publishing Company, Clovis, CA, 1990 |
| *Feeds & Nutrition Digest,* 2nd Edition | M. E. Ensminger J. E. Oldfield W. W. Heinemann | The Ensminger Publishing Company, Clovis, CA, 1990 |
| *Horse Feeding and Nutrition* | T. J. Cunha | Academic Press, New York, NY, 1980 |
| *Horse, The* | P. D. Rossdale | The California Thoroughbred Breeders Assn., Arcadia, CA, 1972 |
| *Horse, The* | J. W. Evans, *et al* | W. H. Freeman and Co., San Francisco, CA, 1977 |
| *Horse Science Handbook,* Vols. 1, 2, and 3 | Edited by M. E. Ensminger | Agriservices Foundation, Clovis, CA, 1963, 1964, and 1966 |
| *Horsemanship and Horse Care,* Ag. Info. Bull. No. 353 | M. E. Ensminger | Agricultural Research Service, USDA, Washington, DC, 1972 |
| *Horses and Tack* | M. E. Ensminger | Houghton Mifflin Company, Boston, MA, 1990 |
| *Light Horse Management* | R. C. Barbalace | Caballus Publishers, Fort Collins, CO, 1974 |
| *Light Horses,* Farmers' Bull. No. 2127 | M. E. Ensminger | U.S. Department of Agriculture, Washington, DC, 1965 |
| *Nutrient Requirements of Horses,* No. 6, 5th Revised Edition | National Research Council | National Academy of Sciences, Washington, DC, 1989 |
| *Shetland Pony* | L. F. Bedell | Iowa State University Press, Ames, IA, 1959 |
| *Stud Managers' Handbook* | Edited by M. E. Ensminger | Agriservices Foundation, Clovis, CA, annually since 1965 |

Feed is the most important influence in the environment. These well-conditioned broodmares and their foals are on excellent pasture. (Photo by Ernst Peterson, Courtesy, Bitterroot Stock Farm, Hamilton, MT)

Californian doe and litter. Rabbit feed is very important nutritionally and economically. (Courtesy, Pel-Freez Rabbit Meat, Inc., Rogers, AR)

# Chapter 55

# RABBITS

Rabbits are members of the order *logomorpha,* which also includes the hares. Rabbits can be distinguished from hares by their young, which are born hairless, blind, and helpless; whereas young hares are born furred, with eyes open, and able to hop minutes after birth. The modern *logomorphs* consist of two families (*leporids* and *ochotonids*) with 12 genera. The two main genera of rabbits are the true rabbits (*oryctolagus*) and the cottontail rabbit (*sylvilagus*). Domestic rabbits and cottontails cannot be crossed. Neither can domestic rabbits and hares be crossed.

The origin and evolution of the rabbit is clouded in obscurity. Fossil remains trace the order *logomorpha* back about 45 million years to the late Eocene period. Originally inhabiting the western half of the Mediterranean basin, the rabbit spread throughout temperate Europe. Introduced into Australia for hunting, where natural enemies were few, it increased beyond bounds and became a nuisance, destroying much of the grazing land.

The rabbit has long been domesticated, and many variations have been produced. But all breeds of domestic rabbits are descendants of the European wild rabbit, *Orcyctolagus cuniculus.*

Rabbits are found in virtually every country of the world, and are raised for the following purposes:

1. **Meat.** The production of rabbits for meat has long been important in Europe, where estimated annual per capita consumption is as follows: Hungary, 8.8 lb; France, 7.9 lb, Spain, 7.9 lb, Italy, 6.2 lb; and Portugal, 4.4 lb. Traditionally, rabbits have been raised by small farmers in these countries to provide meat and supplementary income for the family.

Fig. 55–1. A 10,000-doe rabbitry in Hungary, which produces about 300,000 fryers annually. Hungary has the world's largest rabbitries. (Photo by David J. Harris; print courtesy of Pel-Freez Rabbit Meat, Inc., Rogers, AR)

2. **Wool.** The Angora rabbit produces wool which is used in the manufacture of luxury garments and in handicraft work.

3. **Fur.** The Rex breed has a unique fur, with short guard hairs and erect underfur, the pelts of which can be manufactured into high-quality coats, gloves, hats, and other garments.

4. **Research.** Rabbits are extensively used in biomedical teaching and research and as laboratory animals.

5. **Show or exhibition.** In some countries, including Germany, Great Britain, and the United States, fancy rabbits are raised for show or exhibition purposes.

6. **Pets and companionship.** Many rabbits are raised as pets and for companionship, especially for the young and the elderly.

Fig. 55–2. Pet rabbit being fed garden greens in the backyard. (Courtesy, David J. Harris, Siloam Springs, AR)

Regardless of the purpose for which rabbits are kept—meat, wool, fur, research, show, or pets—the principles of breeding, feeding, management, and health are much the same. Likewise, the objectives—healthy, well nourished animals—are the same.

## RABBIT PRODUCTION IN THE UNITED STATES

Commercial rabbit production in the United States began about 1900, when Belgian Hares were imported from England. During that period, rabbitries sprang up at a tremendous rate, centering primarily in southern California. The demand for quality breeding stock during this time was so great that rabbits were sold at highly inflated prices—in the thousands of dollars. As with most economic booms, the bubble burst, and rabbits dwindled to their present status.

Through the years, rabbits have been accorded little attention in American agriculture; yet, large numbers of them are being used in medicinal and biological research, and for meat, fur, pets, and show.

While rabbit production in the United States is rather limited in comparison with other animal industries, it warrants serious consideration because about 200,000 people are either directly or indirectly engaged in it, and because of its potential. Rabbits are raised in small rabbitries with three or four hutches and in large commercial operations. Rabbit raising lends itself to both types of production.

Commercial operations produce rabbits primarily for meat, with much of the demand centered around the ethnic eating habits of immigrants in urban areas. Prices for fur are currently so low that it should be considered only as a supplemental income. The vast majority of rabbits are raised by part-time backyard operators as a sideline or hobby.

Most full-time commercial operations involving a family with a minimum of hired help require 500 to 1,000 does in order to be an economic unit. In the past, rabbit producers routinely produced 4 litters per doe annually. Today, with the high costs of labor, feed and facilities, most commercial operations must produce 7 to 8 litters per doe, annually, to make a profit. To attain the 7 to 8 litters per doe level of production, young rabbits are weaned at 4 to 5 weeks of age and raised separately from their mother until they reach market weight, generally at 8 to 9 weeks of age. These intensive breeding programs require that does be rebred anywhere from 7 to 14 days after kindle (after giving birth). Since 1 doe can produce 8 litters per year, she has the ability to produce from 130 to 250 lb of liveweight per year through her reproductive ability. In large operations, 1 buck is kept for as many as 20 to 30 does and produces more than 1,000 offspring annually. From this, it may be concluded that 4 does can produce 700 lb liveweight, or 400 lb of dressed carcass weight, annually—as much as an average beef cow can produce in 1½ years—and produce it on less feed.

Commercial rabbit raising is an extremely intensive form of animal production. Californian and New Zealand Whites are the most widely used breeds for meat production because of their large litters, good mothering ability, large size, and the preference of many processors for white rabbits. New Zealand Whites and Florida Whites are the most widely used rabbits for research. The American Rabbit Breeders Association recognized 41 different breeds (in 1984).

Fig. 55–3.  New Zealand White fryers at 3 weeks of age. (Courtesy, Pel-Freez Rabbit Meat, Inc., Rogers, AR)

The meat of the domestic rabbit is white, high in protein (25%), low in fat (4%), and low in cholesterol and sodium—facts that should please the discriminating American consumer. When the public is made aware of the high quality of rabbit meat, consumption of rabbit should increase; but this will happen only after vigorous educational promotion by the industry, extolling the virtues of rabbit meat and telling how to prepare and serve it. Currently, Americans eat 25 to 30 million lb of domestic rabbit meat each year.

Rabbits are also used extensively in biomedical teaching and research. Over 600,000 rabbits are used annually for these purposes, helping researchers investigate a host of maladies, including cardiac diseases, hypertension, antibody production, endocrinology, venereal disease, and virology. Instructors find rabbits invaluable for laboratory demonstrations. Since large numbers of adult rabbits are needed for these purposes, a market exists. But producing for such an outlet is risky as the demand is unpredictable. The producer must have on hand enough rabbits to meet the sporadic demands, yet, at the same time, keep the number of adult breeder rabbits at as low a level as possible in order to minimize feed costs.

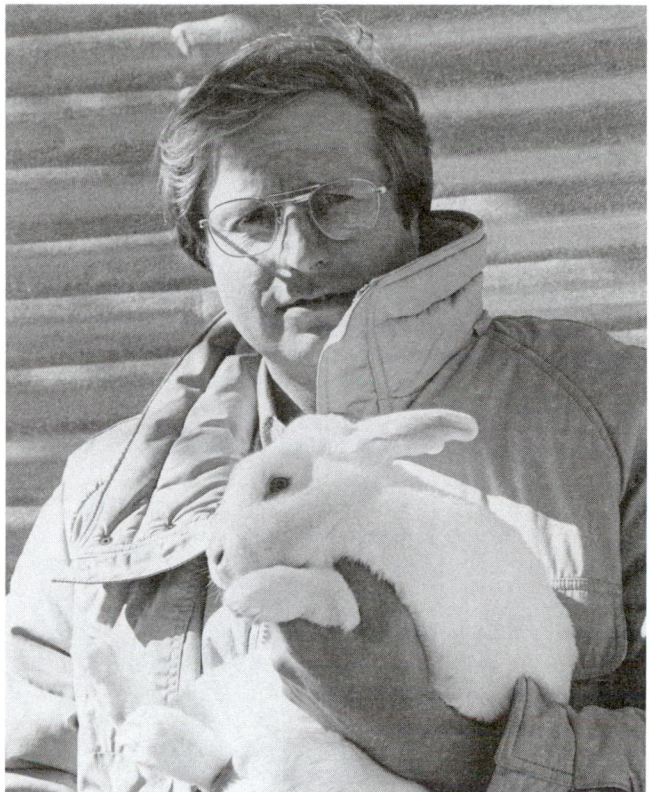

Fig. 55–4.  This rabbit produces antibodies for the pesticides assay tests called ELISA, or Enzyme Linked Immunosolvent Assay. It can analyze all the pesticides in the soil. (Courtesy, USDA)

Despite their prolificacy, rapid growth rate, good feed conversion, use of humanly inedible foods, and nutritious meat, there are many failures in the rabbit business; primarily, because of the intensive labor requirements, disease problems, and unstable markets.

## RABBITRY

The building and equipment needed for the rabbitry will depend on local building regulations, climatic conditions, the size of the operation, and the amount of money to be invested. Whatever the size of the enterprise, the rabbitry should be designed to save labor, to make for neatness and attractiveness, and to prevent pollution.

## HOUSING

In areas where the climate is mild, hutches can be placed outdoors in the shade of trees, or they can be placed under a lath superstructure. Sunlight in the rabbitry will help maintain sanitary conditions, but the rabbits should be given a choice between being in the shade or in the sun at all times.

During hot weather, some cooling measures must be provided in addition to shade. This can be accomplished by using overhead sprinklers, or foggers, placed within the building. Make sure that the building is adequately ventilated and that the rabbits receive the benefit of prevailing breezes.

Fig. 55-5. Curtain sided rabbitry. (Courtesy, David J. Harris, Siloam Springs, AR)

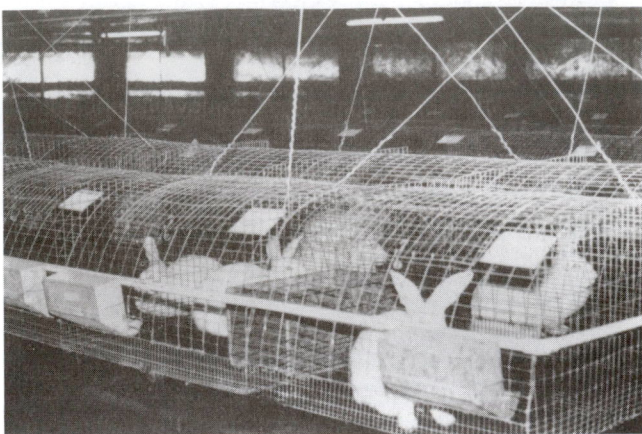

Fig. 55-7. Quonset style cages. (Courtesy, David J. Harris, Siloam Springs, AR)

Fig. 55-6. An outdoor rabbitry suitable for backyard production. (Courtesy, Peter R. Cheeke, Oregon State University, Corvallis)

In areas where strong winds and cold weather prevail, hutches can be protected by placing them in buildings that open to the south or east. During stormy weather these buildings can be closed by the use of curtains or panels. Where weather is extremely cold, extra protection may be needed.

## HUTCHES (CAGES)

Individual hutches (cages) should be provided for mature rabbits. They should be about 2 ft high and no more than 2½ ft deep. The length of hutches should be in keeping with breed size: For small breeds, 3 ft long; for medium breeds, 4 ft long; and for large breeds, 6 ft long. (All figures are inside measurements.)

Hutch construction varies from all-wire Quonset shaped hutches for use inside buildings to semienclosed hutches for use outdoors.

An inexpensive hutch suitable for small rabbitries may be home-made, using wood-frame construction and hardware cloth or wire mesh. It should be light, movable, and easy to build. During warm weather it can be placed under trees or on the protected side of a building. During cold weather it can be moved into a well-ventilated building.

Several types of metal and all-wire hutches (cages) are available on the market.

Several types of flooring can be used in hutches. Wire mesh flooring is used extensively in commercial rabbitries where self-cleaning hutches are desirable. Solid and slat flooring, or a combination of solid flooring at the front and a strip of wire mesh or slats at the back, can also be used in hutch construction.

If mesh flooring is used, examine the surface for sharp points. Always put the smooth surface on top. For slat flooring, use 1-in. hardwood slats and space them ⅝ or ¾ in. apart. To provide drainage, solid floors should slope slightly from the front to the rear of the hutch.

## NEST BOXES

Fig. 55-8. Common styles of wooden nest boxes. (Courtesy, David J. Harris, Siloam Springs, AR)

A good nest box should provide seclusion for the doe when she gives birth to her litter, and protection for the litter. It should be easy to clean and maintain, provide good drainage and ventilation, and be accessible to the young when they are large enough to leave and return to the nest.

Inexpensive nest boxes can be made from sturdy packing boxes—apple boxes are ideal. In order to provide easy access for the doe and young, an opening may be cut in one end of the box, or a portion of one end may be removed.

As an alternative, one end of the box may be removed and replaced with removable slats. The slats may be left in place until the young rabbits are large enough to leave and return to the nest.

Also nail kegs, with a board nailed across the open end to cover ⅓ to ½ the opening make suitable nest boxes. To keep the keg from rolling, the board covering the opening may be extended a few inches beyond the edge of the opening. Several 1-in. holes should be drilled in the closed end of the keg to provide ventilation.

During cold weather, young rabbits may need more protection than provided by the standard nest box. A simple winter nest box can be made by lining the inside of a standard nest box with 2 or 3 layers of corrugated cardboard. Drill 2 or 3 holes in the lid for ventilation, then fill the box with clean straw so that the doe can burrow a cavity for a nest.

## FEEDING EQUIPMENT

Crocks, troughs, or hoppers, along with hay mangers, are commonly used as feed containers. They should be (1) large enough to hold several feeds, and (2) designed to prevent waste and contamination of feed.

Crocks that are especially designed for rabbit feeding are available. They are sufficiently heavy that they cannot be easily tipped, and they have lips that prevent the rabbits from scratching out feed. Crocks can be used for feeding pellets or whole grain.

Fig. 55–9. Crocks for rabbit feeding. (Courtesy, Dr. Terry Reed, Markle, IN)

When hay or green feed is included in the ration, hay mangers should be incorporated into the hutches. Space can be saved by having one manger serve two hutches. Hay wastage can be reduced by placing troughs under the mangers. These troughs can also be used to feed supplemental grains.

Hoppers save considerable time and labor when they are designed for self- feeding. They may be used for feeding pregnant does, does with suckling litters, and rabbits being conditioned for market. Hoppers can be made from cans, hardboard, wood, or other suitable material. An inexpensive self-feeder can be made from a 5-gal can.

## WATERING EQUIPMENT

Rabbits need clean, fresh water at all times. During warm weather, a doe and her litter will drink about 1 gal of water per day.

Crocks and coffee cans are suitable watering devices for small rabbitries. Coffee cans are especially useful during cold weather because the ice can easily be broken and removed from the cans.

Automatic watering systems provide a steady supply of clean, fresh water. They are used extensively in commercial rabbitries to reduce labor, but they are not practical for use in most small rabbitries.

## BREEDS OF RABBITS

Fig. 55–10. The main breeds for meat production are the New Zealand White (left) and the Californian (right). The middle animal is a crossbred NZW × Californian. (Courtesy, Peter R. Cheeke, Oregon State University, Corvallis)

The American Rabbit Breeders Association (ARBA) recognizes 41 different breeds (in 1984) in its current *Official Guide*. Many of these breeds contain a number of color varieties. Details of each recognized breed are described in the *Standard of Perfection,* published by the ARBA. In addition to the breeds of rabbits in America, there are various other breeds in other parts of the world which are seldom, if ever, seen or recognized here. A summary of the most common breeds is presented in Table 55–1 (see next page).

The rabbit breeds are grouped into four fur types—normal, satin, rex, and angora. Most breeds of rabbits have normal-type fur.

**Normal** fur has a dense undercoat protected by longer guard hairs.

**Satin** fur is finer than normal fur, is composed of transparent hair, and has a sheen or luster.

**Rex** fur is very dense and the guard hairs are approximately the same length as the undercoat, giving a plush feeling.

**Angora** fur contains a long, wool-type fiber that is harvested and used in the manufacture of clothing.

Table 55-1 lists some common breeds of rabbits.

## TABLE 55-1
## STANDARD RABBIT BREEDS AND VARIETIES

| Breed | Variety | Ideal Mature Weight | | Utility Value | |
|---|---|---|---|---|---|
| | | Buck | Doe | Meat | Fur |
| | | (lb) | (lb) | | |
| American | Blue, white. | 9 | 10 | Good | Good |
| Angora, English | Black, blue, fawn, white. | 6 | 7 | Good | Good |
| Angora, French | Black, blue, fawn, white. | 8 | 8 | Fair | Fair |
| Belgian Hare | | 8 | 8 | Fair | Fair |
| Beveren | Black, blue, white. | 9 | 10 | Good | Good |
| Britannia Petite | Red-eyed white | 2¼ | 2¼ | | |
| Californian | | 9 | 9 | Very Good | Very Good |
| Champagne D'Argent | | 10 | 10½ | Good | Good |
| Checkered Giant, American | Black, blue. | 11 + | 12 + | Good | Fair |
| Chinchilla, American | | 10 | 11 | Good | Good |
| Chinchilla, Giant | | 13 to 14 | 14 to 15 | Good | Good |
| Chinchilla, Standard | | 6½ | 7 | Good | Good |
| Cinnamon | | 9 | 10 | Good | Fair |
| Creme D'Argent | | 9 | 10 | Good | Good |
| Dutch | Black, blue, chocolate, gray, steel gray, tortoise. | 4½ | 4½ | Small | Fair |
| English Spot | Black, blue, chocolate, gold, gray, lilac, tortoise. | 6 | 7 | Small | Fair |
| Flemish Giant | Black, blue, fawn, light gray, sandy, steel gray, white. | 14 + | 15 + | Good | Good |
| Florida White | | 5 | 5 | Small | Good |
| Harlequin | Japanese, magpie. | 7 | 7 | Fair | Good |
| Havana | Chocolate, blue, black. | 6 | 6 | Small | Good |
| Himalayan | Black, blue. | 3½ | 3½ | Small | Fair |
| Hotot, Dwarf | White, with thin black eye band around each eye. | 2¼ | 2¼ | | |
| Hotot | Frosty white, with thin black eye band around each eye. | 9 | 10 | | |
| Jersey Wooly | | 3 | 3 | | |
| Lilac | Light lilac. | 6 to 7 | 6½ to 7½ | Good | Fair |
| Lop, English | Broken, solid. | 10 | 11 | Good | Fair |
| Lop, French | Broken, solid. | 10 | 11 | Good | Fair |
| Lop, Holland | Solid and broken. | 3 | 3 | | |
| Lop, Mini | Solid and broken. | 5½ | 5½ | | |
| Netherland Dwarf | Agouti, patterned, selfs, shaded, tan, any other variety. | 2 | 2 | Small | Fair |
| New Zealand | Black, red, white. | 10 | 11 | Very Good | Very Good |
| Palomino | Golden, lynx. | 9 | 10 | Good | Good |
| Polish | Black, chocolate, blue-eyed white, ruby-eyed white. | 2½ | 2½ | Small | Fair |
| Rex | Black, blue, California, castor, chinchilla, chocolate, lilac, lynx, opal, red, sable, seal, white. | 8 | 9 | Good | Good |
| Rhinelander | Tri-colored. | 8½ | 9 | Fair | Fair |
| Sable | Sable color. | 8 | 9 | Good | Fair |
| Satin | Black, blue, Californian, chinchilla, chocolate, copper, red, Siamese, white. | 9½ | 10 | Good | Good |
| Satin Angora | | 8½ | 8½ | | |
| Silver | Brown, fawn, grey. | 6 | 6 | Small | Fair |
| Silver Fox | Black, blue. | 10½ | 10½ | Good | Good |
| Silver Marten | Black, blue, chocolate, sable. | 7½ | 8½ | Good | Good |
| Tan | Black, blue, chocolate, lilac. | 4½ | 5 | Small | Good |
| Vienna, Blue | | 9 | 9½ | Good | Good |

## SELECTING THE BREED

The most important decision that rabbit breeders need to make at the outset is whether they wish to raise animals for pleasure or to sell them for meat, pets, research, or to other breeders. If they wish to raise rabbits to sell they should find out about possible markets and marketing methods before they begin.

Breeders who wish to sell rabbits should determine the type and breed of rabbit in demand. They should visit several rabbit shows and confer with a number of breeders before buying their first rabbit.

Breeders who wish to raise meat animals should select one of the fast- growing, medium-size rabbit breeds, such as the New Zealand White or Californian. These breeds develop a quality carcass, and they usually produce large litters. White-furred rabbits are usually preferable since the rabbit fur market often brings a higher return for white pelts that can be dyed.

Breeders who wish to raise specialty-type rabbits for selling to breeders or as pets, should select a breed that they can successfully raise in their rabbitry, buy at a reasonable price, and that they think will sell well.

For the pet market, the preferred breeds are those that are small, come in various colors and patterns, are easy to raise, can be weaned early, and do not tend to become mean or flighty when handled. Pets must be capable of adjusting to unusual conditions. Many animals sold as pets are marketed soon after weaning or sometimes even before they are fully weaned.

Breeders who wish to sell rabbits to fanciers, should make certain that the rabbits fit the description for that breed as closely as possible. Fanciers usually decide on a breed because they admire specific qualities of that breed. Breeders often require animals that are registered or that can be registered.

Laboratories often desire a large-eared, white animal. However, commercially raised breeds that meet specific age, sex, pregnancy, and freedom from disease requirements are currently filling most laboratory needs through private contracts.

The *Standard of Perfection* describes all of the approved breeds of rabbits and cavies in the United States. It can be purchased from the American Rabbit Breeders Association, 2401 E. Oakland Avenue, Bloomington, IL 61701.

## SELECTING FOUNDATION STOCK

After deciding on the breed, the prospective rabbit breeder should consider the following five major points when selecting breeding stock:

1. **Large Litters.** Large litters are essential for maximum profits. Some individuals are more productive than others. Profitable does should consistently kindle 7 to 10 young, 5 to 6 times each year, and should average 6 to 8 young marketed per litter.

2. **Milk Production.** Heavy milk production of the doe is one of the most important factors in insuring that the young will live and grow rapidly to marketing age. Feeding can give a big assist in assuring an ample supply of milk, although there are good and poor producing does, just as there are good and poor producing cows and feed alone cannot make producers.

The milk production of the doe can be determined by checking the weight of the litter at the end of the third week.

3. **Consistent Reproduction.** Five to six litters per year is one of the outstanding characteristics of a good breeding doe. The tendency of some does to molt heavily in the fall or to miss their summer or winter breeding greatly reduces income from a rabbitry.

4. **Desirable Type.** The rabbit fryer at 8 weeks should be heavily meated and should weigh 4 lb or more, liveweight. Some breeds are definitely meat types; others are not. Foundation breeding stock should be free from such undesirable characteristics as buck teeth (malocclusion), yellow fat, woolliness or undesirable pelt color. These characteristics are inherited.

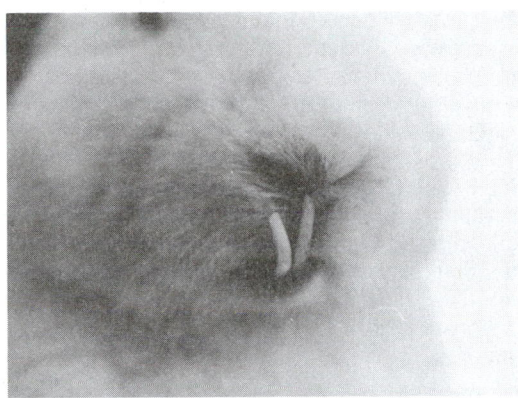

Fig. 55–11. Buck teeth (malocclusion). (Courtesy, David J. Harris, Siloam Springs, AR)

5. **Disease Resistance and Longevity.** These are inherited characteristics of vital importance to the rabbit raiser. Bucks and does capable of transmitting long life and disease resistance are worth many times more than stock lacking in these traits. Selection for these traits may be accomplished by retaining breeders from older does with large litters which survive to weaning age.

Thus, it is important that the foundation animals excel in those characteristics which the rabbit producer wishes to stress and perpetuate in future generations, such as (1) large litters, (2) heavy milk production, (3) consistent reproduction, (4) desirable type, and (5) disease resistance and longevity.

## BREEDING RABBITS

Rabbit breeders find a challenge in producing different types and varieties of rabbits. The development, production, and purification of different characteristics and colors make rabbit breeding particularly interesting. So, breeders exhibit their rabbits at fairs and shows where experienced judges evaluate and compare the animals to decide which ones most nearly meet the standard for that breed.

### SYSTEMS OF BREEDING

After deciding on the breed and selecting the foundation animals, the system of breeding should be charted. The three choices are:

• **Outcrossing**—The most common system of breeding is outcrossing, or the combining of two unrelated strains. It is a relatively safe system, for it is unlikely that two such unrelated animals will carry the same undesirable genes and pass them on to their offspring.

- **Inbreeding**—Inbreeding involves the crossing of closely related animals. For example, breeding a buck to his daughters, breeding full brothers to full sisters, breeding a son to its mother. Before instituting inbreeding, it is important to establish that there are no outstanding weaknesses in the participating families, for close breeding intensifies every family characteristic—good or bad. Therefore, if an organic weakness exists, if there is a tendency toward low milk production, or if there is a general family tendency toward some disease, the undesirable trait will be intensified in the offspring. On the other hand, if there is a tendency for heavy milk production, for long life, for excellent pelts, for greater size, for disease resistance, or for any other desirable quality, these characteristics will be intensified by inbreeding.

- **Line Breeding**—This is the safest system for most breeders. Line breeding is the combining of distantly related animals; for example, a buck might be crossed on his half-sister or on his grandmother, or a grandson might be bred to his grandmother. By such means, desirable characteristics may be intensified, but the chance of intensifying undesirable characteristics is minimized.

## MATING METHODS

The time to breed a young doe depends on its age and weight. Lightweight rabbits tend to reach sexual maturity at a younger age than do heavyweight rabbits. Lightweight rabbits can be bred for the first time at 5 to 6 months of age; heavyweight rabbits should not be bred until they are 6 to 8 months of age. Before mating, the breeder should make sure that the doe has reached the recommended weight for its age. A fast-maturing doe can be bred at a younger age. Once the breeding cycle is established, the doe should be kept in active production.

Natural mating is the most common method. For highest conception, the doe should be examined daily. She is at the peak for conception when the vulva is moist and a bright pink to a reddish color all the way to the tip. As the cycle is waning, the vaginal opening becomes bright purple.

The doe should always be taken to the buck's cage. If the buck is taken to the doe's cage, she may fight or resist him. The in-heat female usually accepts the buck within a few minutes. The caretaker should watch to see that the buck services the doe. If the doe urinates immediately, she may lose the sperm. If that happens, the doe should be taken back right away to be rebred. Also the caretaker should make sure that the buck does not lose his sperm before completing the act of breeding. Even if the first service went well, the doe should be returned to the buck's cage for rebreeding 6 to 12 hours after the first mating.

A first-time buck may be slow about breeding. So, the caretaker should make sure the doe does not scare or harm him. Also it is important that bucks not be used in service until they are mature enough; bucks of lightweight breeds may be used at 4 to 6 months of age; bucks of heavyweight breeds should not be used until 5 to 7 months of age.

Some does must be *force-mated*; they may have to be restrained so the buck can mount them. If a shy or hard-to-breed doe is caged next to the buck for 12 to 24 hours before breeding, she may prove easier to breed. If a doe continues to be difficult to mate, consider culling her.

*Colony breeding* is sometimes resorted to with a hard-to-breed doe. The doe is left with the buck in his cage for 28 days, then returned to her cage. This method ties up a buck for 28 days; and there is always the possibility that the doe may injure the buck, usually within the first 2 to 3 days.

## ARTIFICIAL INSEMINATION

Although artificial insemination of rabbits has been developed and used extensively in the laboratory, very few commercial breeders use artificial insemination in their rabbitries.

The semen from bucks averages about 0.5 cc in volume, with a range of 0.1 to 6 cc. It contains about 700 million to 2 billion sperm per cubic centimeter. The total number of sperm per ejaculate average about 250 million.

Semen is collected from the bucks by means of an artificial vagina. After the artificial vagina has been prepared, the collection is made by using a doe for a mounting animal. The doe is taken to the buck's cage; and when the buck mounts, the artificial vagina is placed between the buck and the doe. When the buck locates the artificial vagina, he will ejaculate into the open end with the same behavior as when breeding naturally. The operator must be alert to prevent the buck from breeding the doe. After the buck has been trained, a dummy made of a stuffed rabbit skin may be substituted for the doe.

A simple insemination tube is used for insemination of does. It consists of a glass tube and rubber bulb similar to a medicine dropper, with the last ½ in. bent at a 30° angle.

The number of does bred to a buck depends on many factors, so an extreme range is possible. For an average buck, collections can be made at least twice weekly and the number of does bred will depend on the motility, density, and volume of the semen produced.

In summary, artificial insemination is applicable to rabbits, but little practical use of it has been made to date.

## SEASONAL EFFECT ON BREEDING

There are seasons of the year when breeding is very difficult. During summer heat, most older bucks, especially the heavyweights, become temporarily sterile. So, during hot weather, young bucks about 6 to 8 months of age should be used.

In the fall and early winter, the does may not conceive. At this season, the caretaker will need patience and persistence to get does into production. During the summer and early fall months, rabbits should not be overfed. Overly fat animals are hard to breed and may have difficulty both during the gestation period and at kindling time.

## PREGNANCY CHECK

Does should be checked for pregnancy 12 to 14 days after breeding. The most commonly practiced method, although not the best, is to introduce the doe to the buck. If pregnant, she will whine and avoid the attention of the buck. However, she may also accept the buck even though she conceived on the first breeding date.

Palpating is the best means of checking for pregnancy. It is accomplished as follows: Use one hand to hold the ears and the loose skin over the shoulders, to restrain the doe. Place the other hand slightly in front of the pelvis so the doe is relaxed, and lightly and gently move this hand backwards and forwards. If the doe is pregnant, the embryos may be felt. At two weeks, they are about the size of marbles. This method should be used cautiously until pregnancy examiners are sure of what they are feeling for; otherwise, they may injure the growing embryos.

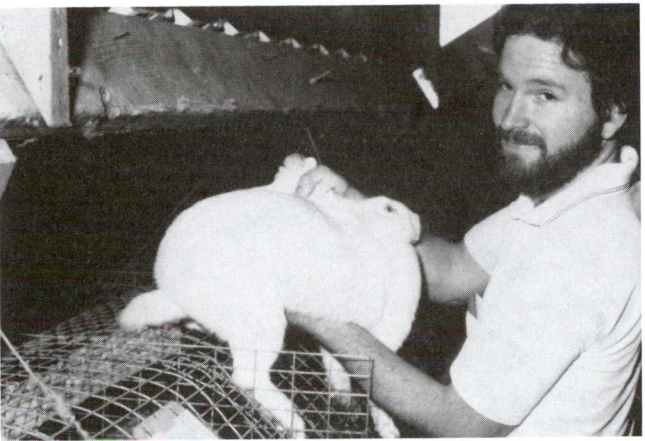

Fig. 55–12.  Confirming pregnancy by palpation. (Photo by D. J. Harris. Print courtesy of Wm. K. Sanchez, Manna Pro Corp., Portland, OR)

## KINDLING

The gestation period for rabbits (the time between breeding and kindling) usually ranges from 28 to 32 days, depending on the breed and its size. So, the nest box should be put in the hutch 25 to 28 days after mating the doe.

Fig. 55–13.  Doe making a nest in a subterranean style nest box. (Courtesy, David J. Harris, Siloam Springs, AR)

Most does take good care of their young if they have a good nest box, plenty of nesting material, and the protection of a good cage. Clean pine shavings or clean straw may be used as nesting materials. During cold weather, the nesting material should be 4 to 6 in. deep; in warm weather, it should be 1 to 2 in. deep. The doe will mix her fur with the nesting material and provide a comfortable nest for herself and her young. If hay is placed on top of the shavings, the doe will use it for food and for mixing in the nest. Alfalfa and oat hays are the best

for rabbits because they provide the roughage so vital to their good health. Also rice hulls, chopped straw, or cut-up paper may be used in nest boxes.

A day or two before kindling, the doe usually consumes less food than normally. At that time, she should not be disturbed, and she should be made as comfortable as possible. She may be tempted with small quantities of green feed. This will have a beneficial effect on her digestive system.

Most litters are kindled at night. After kindling, the doe may be restless. She should not be disturbed until she has quieted down.

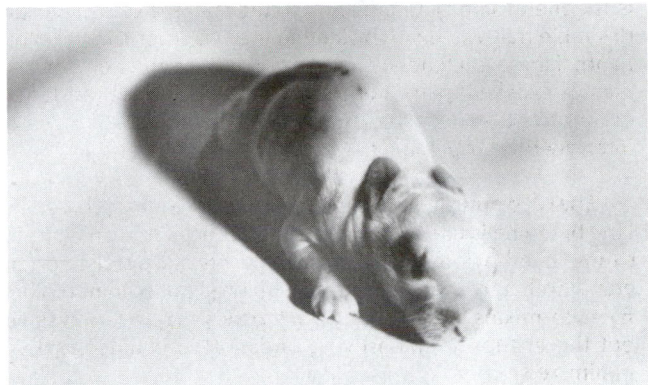

Fig. 55–14.  Three-day-old baby rabbit. The young are born hairless, blind, and helpless. (Courtesy, David J. Harris, Siloam Springs, AR)

## COMPLICATIONS AT KINDLING TIME

Anterior, or breech presentation, of young at birth is normal. If the doe is in proper condition for kindling, complications are rare. Pregnancy, however, makes a heavy demand on the doe and lowers her vitality, making her more susceptible to such complications as pneumonia, caked breast, and mastitis.

• **Pneumonia**—A few days before or several days following kindling, pneumonia may develop. For the successful treatment of pneumonia, early detection is essential. With pneumonia, the doe's head is held high and tilted backwards; and breathing is difficult. Treatment consists in making the doe comfortable and in adding a little green feed to the ration. Injections of penicillin or a penicillin-streptomycin mixture are effective in treating colds and pneumonia and reducing mortality.

• **Caked breast**—Caked breast may be caused by the milk not being removed from the breast, or by injuries. Early symptoms of caked breast are firm, pink breasts that feel feverish to the touch. As caked breast develops, the tissues around the involved teats become enlarged and hard. The skin turns dark, the ends of the teats become discolored and tender, and the doe refuses to allow the young to nurse. Treatment consists in rubbing lanolin on the teats and massaging the involved portion of the breast. The caretaker may restrain the doe to allow her own young or those from other litters to remove the milk. Milk may also be stripped from the teats, taking care not to use too much pressure. But the *tissues* should not be lanced.

• **Mastitis**—Mastitis, or "blue breast" is caused by bacterial infection and may be very contagious. The symptoms: The doe fails to consume her feed and is inactive; the breast is congested and feverish, turns red or purple, and the teats are discolored. The treatment: Reduce the feed, give some green feed, and inject penicillin intramuscularly in the thigh.

## CARE OF THE LITTER

On the day of kindling, or soon after, inspect the litter, and remove any deformed, undersized, or dead young. If the attendant is careful, and quiet when making the inspection, generally the doe will not object. *There is no danger of causing her to disown the young.* If she is nervous and irritable, distract her attention and calm her by placing some tempting feed in the hutch immediately after inspection.

Litters vary in size. The utility breeds usually average 8 young. Some may number 12 to 18. For commercial purposes 7, 8, or 9 may be left with the doe. Does from strains that have been developed for heavy production may care for 9 or 10.

Some of the baby rabbits can be transferred from a large litter to a foster mother that has a small litter. Adjusting the number of young to the capacity of the doe insures more uniform development and finish at weaning time. For best results, the young that are transferred should be within 3 or 4 days of the age of the foster mother's young.

When the babies are about 4 weeks old, remove the nest box; remove it sooner in summer and later in winter. Clean and sanitize the nest box so that it will be ready for the next litter. Leave the fryers with the doe until weaning age, at approximately 8 weeks of age, providing the doe was not bred less than 42 days after kindling.

Once the young are out of the nest, the doe and litter should be fed only what they will clean up in 24 hours; and they should have a continuous supply of fresh, clean water.

Occasionally a doe may have problems while she is raising a litter. If she is flighty after kindling, it may be best to leave her alone for a time. Also, the caretaker should talk to her, reassure her, and try distracting her attention by giving her some tempting feed, such as a handful of hay, a piece of bread, or a carrot. An extremely nervous doe should be left completely alone for the first week.

The doe should not be overfed after kindling because she may produce a surplus of milk, causing caked udder, which will make her uncomfortable. In very cold weather, it is wise to give a nursing doe warm water and even warm milk to keep her drinking liquids and producing milk. Warm tea is also excellent.

- **Sexing**—The sex of a rabbit can be determined at 1 day of age, but it is easier to do so at 3 days of age. However, sexing with accuracy may be done at 8 weeks of age as follows: Place the rabbit on its back with your right hand holding its head and ears. Place the index and middle fingers of the left hand over the tail; and gently pull the tail away from the body. Use your thumb to pull open the genital area and move it toward the upper body. The buck's opening will come up round and the doe's opening will form a slit and have a slight depression at the end next to the rectum.

- **Culling**—When choosing young rabbits to save for future breeders or show prospects, the undesirable animals should be culled at 8 weeks, again at 10 to 12 weeks, and finally at 6 months. At 10 weeks, each prospective breeding or show animal should be housed in an individual cage. Only those rabbits that exhibit qualities as good as or better than those of the parents should be saved. Individual animals should be thoroughly checked for any disqualification and breeding faults,

such as buck teeth, crooked bone, or any other weak points of body type or markings.

- **Tattooing**—Rabbits should be permanently identified by tattooing. The breeder may choose any numbering system, but the tattoo should always be in the left ear. The recommended procedure for tattooing follows: Clean the ear with alcohol; carefully select the tattoo and clamp it onto paper to be sure it is correct; then firmly and quickly insert it into the ear; rub the ink briskly into the tattoo with a toothbrush and allow it to dry for several days. Then, use Vaseline or alcohol to remove the excess ink. There is also a tattoo paste available that can be applied with a swab.

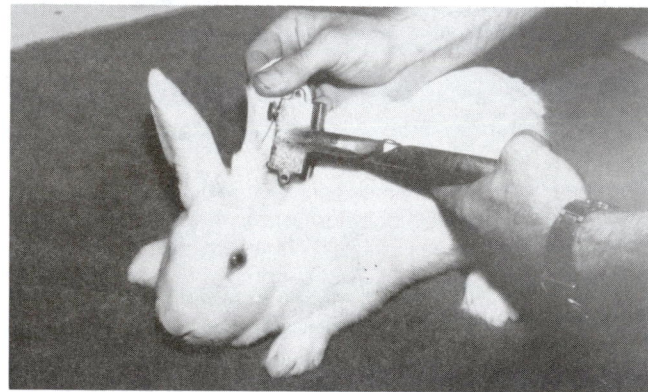

Fig. 55–15.   Tattooing in a rabbit's left ear. (Courtesy, David J. Harris, Siloam Springs, AR)

A tattoo box is helpful, but wrapping a rabbit in a towel or sheet also works well. The simplest way is to back the animal's rump into something solid, and have someone firmly hold the animal's back while the tattooing is being done. The job should be done right the first time. Care should be taken not to injure the ear and to avoid the blood vessel that is in the ear.

## REBREEDING DOES

Commercial breeders may rebreed does 14 days after kindling. They may also use other intervals that are multiples of 7 days. The most common intervals for rebreeding are 35, 42, and 56 days. In many situations, the most desirable time length may be 42 days, or 6 weeks after kindling. This schedule gives the doe 2 weeks more with her litter and 2 weeks alone before the next litter. With this schedule, each doe may be expected to produce 5 litters a year.

## FEEDING RABBITS

Confined production precludes the choice of feeds by rabbits. It follows that big differences in their performance can be expected as a result of different diet compositions and feeding methods.

Annually, an estimated 250,000 tons of commercial feeds are fed to the more than 10 million rabbits produced in the United States. Most rabbitries use commercially pelleted rabbit feeds.

As a meat-type animal, the rabbit compares very favorably in feed conversion with the more traditional animals. With a balanced ration, feed conversions of 3:1 can be obtained in fryer rabbits. While this conversion rate is not as favorable as the 2.4:1 achieved in broilers, it does compare very favorably with the steer (9:1). Protein efficiency in rabbits is approximately 6:1, compared to 1.9:1 in broilers and 10.6:1 in steers. (See Chapter 1, Food and Animals—a global perspective, Table 1–7, for the feed-to-food efficiency rating of rabbits in comparison with other animal species.)

Since the rabbit possesses an enlarged cecum, similar to that of the horse, approximately 40% of commercial rabbit rations consist of alfalfa or other roughage material. So, it follows that rabbits are not in direct competition with humans for foodstuffs. In the future, if competition for human food becomes more acute, the use of roughages will assume added importance in the livestock industry; and the rabbit will out-compete many other animals in the ability to utilize roughages.

## NUTRITIVE NEEDS OF RABBITS

Rabbits are nonruminant herbivores. The digestive anatomy and physiology of the rabbit closely resemble those of the horse, and in many ways the nutritive requirements are similar. Several differences, such as the habit of coprophagy (the ingestion of fecal material) and decreased fiber utilization in rabbits, alter the requirements somewhat. Nevertheless, the types of feeds used for rabbits and horses are very similar.

Rabbits excrete two types of feces—hard and soft. The hard feces (or day feces, containing about 40% water), which are produced in the large intestine, are the fecal pellets which are most commonly seen. The soft feces (or night feces, containing bout 70% water), are produced in the cecum, excreted in grape-like clusters, and consumed directly from the anus in a peculiar type of coprophagic behavior displayed by rabbits as

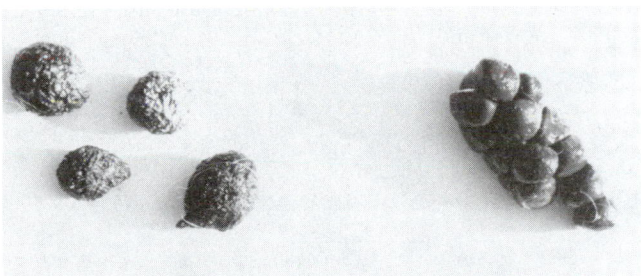

Fig. 55–16. Two types of feces. *Left:* Regular hard pellets. *Right:* Soft feces in a grapelike cluster. (Courtesy, Pel-Freez Rabbit Meat, Inc., Rogers, AR)

early as 3 weeks of age. **NOTE WELL:** Night feces is really a misnomer, because soft feces are often produced during the day.

Coprophagy by rabbits plays an important role in the modification of their nutritive requirements. It has been speculated that the act of coprophagy may aid in the absorption of some of the essential amino acids and certain vitamins—including vitamin K and the B complex vitamins. By recycling the digesta, certain feeds are digested and absorbed that were not utilized the first time. The small intestine is the main site of absorption for these nutrients, but a good deal of this synthesis takes place in the cecum. Since the cecum lies behind the small intestine, much of that which is digested in the cecum is not absorbed but is passed in the feces. Coprophagy enables the rabbits to recycle these nutrients which may not have been absorbed in the cecum or the large intestine. In addition to the microbial synthesis of many of these nutrients, some fiber digestion takes place in the cecum.

## ENERGY

Rabbit production is intensive in nature, and the energy demands on the doe are high if she is to produce up to eight litters a year. Likewise, growth of young juniors creates a high-energy requirement.

Carbohydrates (starch and cellulose) and fats are the primary sources of energy for rabbits. Starch is found in cereal grains and tubers; cellulose is the structural component (fiber) of plants.

Good quality legume hays and supplemental concentrates or commercial pellets are routinely used to supply a high level of energy during peak production periods. The energy requirements for maintaining dry does or young bucks not in service are low, with the result that good-quality hay should be sufficient.

While no requirement for dietary fat has been established, rations routinely contain 2 to 5% fat. It has been suggested that higher fat levels might be feasible, but caution should be exercised in order to prevent digestive disturbances, such as scours. Also, high levels of fat may reduce feed pellet quality, resulting in pellets that break apart easily.

## PROTEIN

All production parameters—fur, growth, reproduction, and lactation—require high levels of good-quality protein. Workers at the Arkansas Station recently completed research showing that gestating/lactating does have a very high dietary protein requirement; they found that does fed a diet that contained 26% crude protein the last 2 weeks of gestation and during early lactation produced kits that were 30% heavier and had higher quality fur than does fed an 18% crude protein ration during late gestation and early lactation.[1]

Fig. 55–17. Protein quality, affected by processing, made the difference! *Left:* rabbit fed raw soybeans. *Right:* rabbit fed solvent extracted soybean meal. (Photo by D. J. Harris, Pel-Freez Rabbit Meat, Inc., Rogers, AR. Print courtesy Wm. K. Sanchez, Manna Pro Corp., Portland, OR)

[1]Unpublished research. Reported in a personal communication from Professor L. B. Daniels, The University of Arkansas, to the author, 1986.

Rabbits require certain amino acids in the diet. The essential amino acid profile is very similar to that of the chick and the pig. Of the essential amino acids, lysine is most likely to be deficient in rabbit feeds.

Studies have shown that nonprotein nitrogen (NPN), such as urea, is of little value in rabbit rations. This is attributed to the fact that NPN sources are degraded and absorbed in the small intestine and subsequently eliminated as waste products before the NPN ever reaches the cecum where it might be transformed into bacterial protein.

Coprophagy has been shown to increase the biological value of certain low-quality proteins. There are indications that some increase in value is observed with the coprophagy of high-quality protein, but the magnitude of the increase is not large. Confirmation of the benefits of coprophagy in protein utilization in rabbits is presented in Table 55–2. Casein was compared to gelatin in two feeding systems; in one system the rabbits were allowed to consume their feces, in the other they were not. Rabbits on the casein diet ate more than those on the gelatin diet. Rabbits that were allowed to consume their feces demonstrated better protein utilization in both diets, but the resulting increase in apparent biological value of the gelatin diet as a result of coprophagy was quite dramatic.

**TABLE 55–2**
**EFFECT OF COPROPHAGY ON VOLUNTARY INTAKE AND PROTEIN PARAMETERS OF MATURE RABBITS[1]**

| Source of Protein[2] | Dry Matter Consumption | Digestion Coefficient of Nitrogen | Nitrogen Balance | Apparent Biological Value |
|---|---|---|---|---|
| | (g/day/$W_{kg}^{.75}$) | (%) | (mg/day/$W_{kg}^{.75}$) | |
| Casein—coprophagy allowed | 34.2 | 96.1 | 458.3 | 37.8 |
| Casein—coprophagy prevented | 27.2 | 88.3 | 236.1 | 26.8 |
| Gelatin—coprophagy allowed | 19.8 | 87.9 | 89.4 | 12.8 |
| Gelatin—coprophagy prevented | 17.1 | 70.7 | −186.2 | −48.6 |

[1]Adapted by the author from Kennedy, L. G., and T. V. Hershberger, "Protein Quality for the Nonruminant Herbivore," *Journal of Animal Science*, 1974, 39:510.

[2]*Casein diet:* casein, 22.25%; corn starch, 18.62%; glucose monohydrate, 25.62%; alphacel, 15.51%; corn oil, 10.00%; vitamin and mineral mix, 8%.

*Gelatin diet:* gelatin, 19.22%; corn starch, 18.62%; glucose monohydrate, 25.62%; alphacel, 18.54%; corn oil, 10.00%; vitamin and mineral mix, 8%.

**TABLE 55–3**

| Minerals Which May Be Deficient Under Normal Conditions | Conditions Usually Prevailing Where Deficiencies Are Reported | Functions of Mineral | Some Deficiency Symptoms |
|---|---|---|---|
| ***Major or macrominerals:*** | | | |
| Salt (NaCl) | Negligence, for salt is cheap. | Sodium and chlorine help maintain osmotic pressure in body cells, upon which depends the transfer of nutrients to the cells and the removal of waste materials. Also, sodium is important in making bile, which aids in the digestion of fats and carbohydrates; and chlorine is required for the formation of hydrochloric acid in the gastric juice so vital to protein digestion. | Depressed growth. |
| Calcium (Ca) | Rations of grass hay and farm grains. | Essential for development and maintenance of normal bones and teeth. Important in blood coagulation and lactation. Enables heart, nerves, and muscles to function. Regulates permeability of tissue cells. Affects availability of phosphorus and zinc. | Rickets; tetany; brittle bones. |
| Phosphorus (P) | High legume forage ration without supplemental phosphorous. | Essential for sound bones and teeth, and for the assimilation of carbohydrates and fats. A vital ingredient of the proteins in all body cells. Necessary for enzyme activation. Acts as a buffer in blood and tissue. Occupies a key position in biologic oxidation, and reactions requiring energy. | Rickets; tetany; brittle bones. |
| Magnesium (Mg) | | Necessary for many enzyme systems and for proper functioning of the nervous system. Closely associated with the metabolism of calcium and phosphorus. | Poor fur growth; fur chewing; hyperirritability. |
| Potassium (K) | | Essential for proper enzyme, muscle and nerve function, cecal mircroorganism activity, and appetite. | Muscular dystrophy. |
| ***Trace or microminerals:*** | | | |
| Cobalt (Co) | Feeds from cobalt-deficient areas. | Constituent of vitamin B–12. | Anemia. |
| Copper (Cu) | | Copper, along with iron, is necessary for hemoglobin formation, although it forms no part of the hemoglobin molecule of red blood cells. | Anemia; graying of hair. |
| Iodine (I) | Feeds grown in iodine-deficient areas. | Needed for the production of thyroxin, an iodine-containing hormone that regulates metabolic rate. | |
| Iron (Fe) | | Necessary for formation of hemoglobin, an iron-containing compound which enables the blood to carry oxygen. Also, important to certain enzyme systems. | Microcytic and hypochromic anemia. |
| Manganese (Mn) | Excess calcium and phosphorus decreases absorption of manganese. | Considered essential in utilization of calcium and phosphorus, for proper functioning of mammary glands and normal reproduction. | Abnormalities of the skeletal system. |
| Zinc (Zn) | | Component of several enzyme systems and also required for normal protein synthesis. | Weight loss, alopecia, graying of hair, dermatitis, low hematocrit, reproductive problems. |

[1]As used herein, the distinction between "mineral requirements" and "recommended allowances" is as follows: In mineral requirements, no margins of safety are included intentionally, whereas in recommended allowances, margins of safety are provided in order to compensate for variations in feed composition, environment, and possible losses during storage or processing. Where preceded by an asterisk, the mineral

Legumes are excellent sources of protein, and alfalfa is used extensively in rabbit rations. Oilseed meals are widely used as protein supplements when high-protein levels are required. Animal and fish products are seldom included in rabbit diets because of their high costs.

## MINERALS

Little research has been reported concerning the mineral requirements of rabbits. It is generally thought that the requirements involve the same elements as are required by other animals; but for many elements, quantitative requirements have not yet been established.

A summary of the various minerals, along with their respective requirements and nutritional considerations, is presented in Table 55–3.

Fig. 55–18. Rabbit with bowed legs and enlarged joints resulting from eating alfalfa produced on low-phosphorus soils. (Courtesy, Dr. Wilton W. Heinemann, Washington State University, Pullman)

## RABBIT MINERAL CHART

| Mineral Requirements[1] Mineral Content of Ration | Recommended Allowances[1] Daily Nutrients/ Animal | Percent of Total Ration | Practical Sources of the Mineral | Comments |
|---|---|---|---|---|
| *0.5%. | | 0.5–1.0 | Salt spools. Can be added to feed. | Sodium and chlorine are low in feeds of plant origin. There is little danger of overfeeding salt unless a salt-starved animal is suddenly given access to too much salt or if liberal amounts of water are not available. |
| Variable according to age and production. | | 0.4–1.0 | Ground limestone or oystershell flour. When both Ca and P are needed, use bone meal, dicalcium phosphate, or defluorinated phosphate. | The Ca:P ratio should be maintained close to 1:1, although 2:1 is acceptable when the higher calcium content is due to the presence of legume. Where there is a shortage of calcium in the ration, it is withdrawn from the bones. |
| Variable according to age and production. | | 0.22–0.50 | Monosodium phosphate. Monoammonium phosphate. When both Ca and P are needed, use bone meal, dicalcium phosphate, or defluorinated phosphate. | (Same as stated for Ca under "Comments" above.) If plenty of vitamin D is present, the ratio of Ca:P becomes less important. Apparently phosphorus cannot be withdrawn from the bone. |
| *300–400 ppm. | | 0.03–0.04 | Magnesium sulfate. Magnesium oxide. | In high Ca diets, Mg may become deficient due to interference from the Ca in absorption. |
| *0.6%. | | 0.6 | Potassium chloride. Roughages. | |
| | | | Cobalt chloride, cobalt sulfate, cobalt oxide, or cobalt carbonate. | Cobalt-deficient areas are: Fla., Mass., Mich., N.H., N.Y., N.C., Pa., S.C., and Wisc. Also, Australia and western Canada. |
| *3 ppm. | | 0.00003 | Copper carbonate. Copper sulfate. | Trace mineralized salt containing copper is satisfactory. |
| *0.2 ppm. | | 0.2 ppm | Iodized salt. | |
| | | 0.0050 | Trace mineralized salt. | |
| Variable according to age and production. | 1 mg for growing animals; 0.3 mg for mature animals. | 0.000025–0.000085 | Trace mineralized salt. | Manganese is needed for growth and reproduction of most animals. |
| | | | Zinc carbonate. Zinc sulfate. | |

requirements, allowances, and other facts presented herein were taken from *Nutrient Requirements of Rabbits*, 2nd rev. ed., NRC-National Academy of Sciences, 1977.

## VITAMINS

As with mineral nutrition in rabbits, the amount of information concerning the vitamin requirements is very limited.

Table 55–4, Rabbit Vitamin Chart, lists the vitamins that have been studied in rabbits, and gives pertinent information pertaining to each.

**TABLE 55–4**

| Vitamins Which May Be Deficient Under Normal Conditions | Conditions Usually Prevailing Where Deficiencies Are Reported | Functions of Vitamins | Some Deficiency Symptoms |
|---|---|---|---|
| **Fat-soluble vitamins:** | | | |
| A | When on high-forage rations that have been stored a long time and have lost their carotene value. | Promotes growth and stimulates appetite. Assists in reproduction and lactation. Keeps the mucous membranes of respiratory and other tracts in healthy condition. Makes for normal vision. Prevents night blindness. | Retarded growth, incoordination and paralysis, blindness, drooping ears, and hydrocephalus (enlarged head) of fetuses from vitamin A-deficient does. |
| D | In confinement rearing where the rabbits do not have access to sunlight or sun-cured hay. | Assimilation and utilization of calcium and phosphorus, necessary in normal bone development—including the bones of the fetus. | Rickets. |
| E | | Serves as insurance against destruction of vitamin A. Makes for improved reproduction. Protection of cellular lipids from oxidation. | Nutritional muscular dystrophy of skeletal and cardiac muscle, paralysis, and fatty livers. |
| K | Intestinal disorders or when antibiotics are used. | Concerned with blood coagulation. | Prolonged bleeding following a minor injury and abortion and placental hemorrhage in does. |
| **Water-soluble vitamins:** | | | |
| Biotin | The presence in feeds of avidin, a biotin antagonist. | Important in the metabolism of carbohydrates, fats, proteins. | Loss of hair and dermatitis. |
| Choline | Rations low in methionine, an amino acid. | Prevention of fatty livers, transmitting nerve impulses, and the metabolism of fat. | Depressed growth, fatty and cirrhotic liver, and necrotic kidneys. |
| Folacin (Folic Acid) | | Essential for normal growth and reproduction, for the prevention of blood disorders, and for important biochemical mechanisms within each cell. | |
| Niacin (Nicotinic Acid, Nicotinamide) | | Constituent of coenzymes which produce energy within the cells. | Loss of appetite, diarrhea, and emaciation. |
| Pantothenic Acid (Vitamin B–3) | No deficiency ever produced in rabbits. | Constituent of coenzyme A (CoA). It plays a key role in energy metabolism. | |
| Riboflavin (Vitamin B–2) | Destruction of riboflavin by light or by heat in an alkaline solution. | In the oxidative mechanisms in the cells. | Retarded growth and lowered feed efficiency. |
| Thiamin (Vitamin B–1) | Prolonged feeding of a thiamin-deficient diet. | A cofactor of certain enzymes involved in carbohydrate and fat metabolism. | Loss of appetite, muscle paralysis, and accumulation of pyruvic acid in the blood. |
| Vitamin B–6 (Pyridoxine, Pyridoxal, Pyridoxamine) | | Key constituent of cofactors involving amino acid and energy metabolism. | Depressed growth, acrodynia (dermatitis), convulsions, and paralysis. |
| Vitamin B–12 (Cobalamins) | Lack of cobalt. | Two coenzyme forms—coenzyme B–12, and methyl B–12. | Retarded growth. |

[1]As used herein, the distinction between "vitamin requirements" and "recommended allowances" is as follows: In vitamin requirements, no margins of safety are included intentionally, whereas in recommended allowances, margins of safety are provided in order to compensate for variations in feed composition, environment, and possible losses during storage or processing. Where preceded by an asterisk, the vitamin requirements, allowances, and other facts presented herein were taken from *Nutrient Requirements of Rabbits*, 2nd rev. ed., NRC-National Academy of Sciences, 1977.

## WATER

The water requirement of rabbits is influenced by a number of factors; among them, the following:

1. **Temperature and humidity.** When the temperature and humidity increase to levels above the zone of thermoneutrality, water within the body becomes an important means of heat dissipation.

2. **Stage of production.** A doe with a litter of seven can drink up to 1 gal of water per day.

3. **Composition of feed.** Feeds that are high in protein and fiber increase the need for water because of the increased need to excrete end products produced in the digestion and metabolism of these feed components. The inclusion of suc-

## RABBIT VITAMIN CHART

| Vitamin Requirements[1] Vitamin Content of Ration | Recommended Allowances[1] | | Practical Sources of the Vitamin | Comments |
|---|---|---|---|---|
| | Daily Nutrients Per Rabbit | Per Lb of Ration | | |
| Variable according to age and production. | 23 mcg of carotene per pound of ration has been shown to prevent symptoms of vitamin A deficiency. | For does in production, 4,545 IU/lb feed. | Stabilized vitamin A. Green grass or legume. | Vitamin A is not synthesized in the cecum. A vitamin A level of 86,355 IU/lb of diet, which is only 19 times the vitamin A requirement, is toxic to rabbits. |
| | No allowances have been recommended for vitamin D supplementation. | | Sun-cured hays. Exposure to sunlight. Commercially available supplements of either $D_2$ or $D_3$. | The vitamin D requirement is directly related to the Ca:P ratio and their respective levels. Vitamin D toxicity is of greater concern than deficiency in rabbits. Toxicity symptoms are loss of appetite, impaired movement, and calcification of soft tissues such as the kidneys and the arteries. |
| *18 mg/lb for rabbits in production. | 0.5 mg/lb of body weight. | 25 IU | Stabilized vitamin E. Germ or gum oils of plants. Green plants and hays. Cereals. | Selenium does not exert a vitamin E-sparing effect in rabbits; instead, rabbits depend entirely on vitamin E for protection against peroxides. |
| *0.1 mg/lb for gestating does. | No allowances have been recommended for vitamin K supplementation. | | Synthetic vitamin K. Green grass. Well-cured hays. | Studies indicate that dietary vitamin K is required for reproduction but not growth. |
| | | | Widely distributed in nature. | Deficiencies are not associated with normal rations. |
| *0.5 g/lb for growing rabbit. | | 0.12% | Alfalfa hay and meal. Choline chloride. Rice polishings. | |
| | | | Alfalfa hay/meal and oil seed meals. | |
| *82 mg/lb for growing rabbit. | 5 mg/lb of body weight. | | Nicotinamide (synthetic). Nicotinic acid (synthetic). Rice polishings. Yeast (brewers', torula). | Limited amounts of niacin can be synthesized from the amino acid tryptophan. Supplemental niacin in some cases can increase growth rate. |
| | | | Calcium pantothenate (synthetic), rice polishings, yeast (brewers', torula), alfalfa hay and meal. | |
| Not known. | | | Synthetic riboflavin, alfalfa hay, whey, yeast. | |
| Not known. | | | Thiamin hydrochloride, rice polishings. | |
| *18 mg/lb for growing rabbit. | | 0.5 mg | Pyridoxine hydrochloride. Rice polishings. Alfalfa hay/meal. | Normally, rabbit rations contain adequate vitamin B-6. |
| | | | Meat animal by-products. Marine by-products. | |

culent feeds into the ration provides an additional source of water. But care should be taken to restrict the use of this type of feed as the high water content may restrict the level of nutrient intake.

Plenty of clean, fresh water should be available to rabbits at all times. Because of the possible presence of parasites, water should not come from ponds to which dogs and wild animals have access. Also, well water should be tested for impurities.

Because water provides a medium for cross-contamination of bacteria and parasites between rabbits, water bowls should be disinfected routinely. If a drip system of watering is being used, the producer should clean the system periodically, especially when new stock is being introduced to the herd.

## NONNUTRITIVE FACTORS

Two nonnutritive factors—fiber and feed additives—are discussed with regards to their respective influences on rabbit nutrition.

### FIBER

Roughages make up a large percentage of rabbit rations, and the digestibility of fiber plays an important role in the utilization of feed by rabbits. Experiments have shown that rabbits do not digest fiber efficiently. In a study designed to compare the digestibility of an all-alfalfa ration and a ration containing both alfalfa and grain, rabbits were observed to digest the same amount of crude protein on a percentage basis as the horse and the pony; but fiber digestibility in the rabbit was less than one-half that of either the horse or the pony in both rations (see Table 55–5).

#### TABLE 55–5
#### COMPARISON OF DIGESTIBILITY OF ALFALFA AND ALFALFA-GRAIN RATION IN HORSES, PONIES, AND RABBITS[1]

|  | Organic Matter | Crude Protein | Crude Fiber |
|---|---|---|---|
|  | (%) | (%) | (%) |
| Alfalfa composition[2] | 90.1 | 19.7 | 25.2 |
| Digestion coefficients: |  |  |  |
| Horse | 60.4 | 74.0 | 34.7 |
| Pony | 62.5 | 76.2 | 38.1 |
| Rabbit | 54.3 | 73.7 | 16.2 |
| Alfalfa-grain composition[2][3] | 91.6 | 16.7 | 17.8 |
| Digestion coefficients: |  |  |  |
| Horse | 71.1 | 77.3 | 38.6 |
| Pony | 72.4 | 79.6 | 40.9 |
| Rabbit | 65.2 | 73.2 | 18.1 |

[1]Adapted by the author from Slade, L. M., and H. F. Hintz, "Comparison of Digestion in Horses, Ponies, Rabbits, and Guinea Pigs," *Journal of Animal Science*, 1969, 28:842–843.

[2]Composition based on a dry matter basis.

[3]Ration consisted of 50% alfalfa, 30% rolled barley, 9% wheat mill run, 10% molasses, and 1% salt.

Although fiber is not a useful energy source for rabbits, it is a very important component of rabbit feeds. Numerous studies have shown that low fiber diets cause increased enteritis. Also, when fiber is lacking in the ration, rabbits will sometimes resort to eating their own fur. Fur then accumulates in the digestive tract, resulting in obstruction. The inclusion of some roughage in the ration will remedy the problem in many cases.

Nonproducing animals can be fed rations that contain up to 16 to 22% fiber while animals in production should be limited to 12 to 14% fiber because of the greater need for digestible nutrients.

The form of fiber can affect its value to the rabbit. There is some evidence indicating that finely ground fiber can cause diarrhea. Thus, for proper utilization, fiber should be fed in a coarse form.

### FEED ADDITIVES

Feed additives in rabbit rations can be divided into three groups: coccidiostats, antibiotics, and antioxidants.

### Anticoccidial Drugs

Coccidiosis is the most prevalent parasitic disease in rabbits. Four species of this protozoan live in the intestine of the rabbit, while an additional species infiltrates the liver. The intestinal species cause diarrhea, loss of appetite, weight loss, and sometimes death. The liver species—the most pathogenic form of coccidiosis in rabbits—enters via the intestine and travels up the bile ducts to the liver. While this form may not be lethal, infected livers must be condemned. Treatment for either the intestinal or liver forms of coccidiosis in rabbits is as follows: sulfaquinoxaline (1) administered in the drinking water at 0.04% for 14 to 30 days; or (2) given in the feed at 0.025% for 20 days, or for 2 days out of every 8. Rabbits should not be treated for coccidiosis within 10 days of slaughter if they are to be used for food. Since most rabbit producers feed only one feed, or at the most two, the addition of anticoccidial drugs to the feed becomes impractical. The best method to administer these drugs is through the drinking water. Withdrawing these drugs can then be facilitated by merely changing the water.

### Antibiotics

The addition of low levels of antibiotics to the rations of various young animals has generally shown growth-promoting effects. Oxytetracycline is approved by the Food and Drug Administration as a feed additive for rabbits at a use level of 10 g/ton, as an aid in stimulating growth and improving feed efficiency. (Also see Chapter 4, section on "Additives, Implants, and Injections— •*Antibiotics*.")

### Antioxidants

Several antioxidants can be added to rabbit rations in order to prevent spoilage due to the autoxidation of fats in the feed. Also, vitamin E is an effective, natural antioxidant. Compared to the composition of rations for other livestock, rabbit rations are relatively low in fat; but the addition of antioxidants to the ration can ensure against feed spoilage and increase the storage life of the feed. (Also see Chapter 4, section on "Additives, Implants, and Injections— •*Antioxidants*.")

## FEEDS FOR RABBITS

Generally speaking, a good-quality legume hay is sufficient to meet most of the maintenance needs of the rabbit; but as production pressures intensify, there is need for additional feed sources that are higher in energy. A variety of feedstuffs has been used successfully in feeding rabbits.

### ENERGY FEEDS

Production demands necessitate high-energy levels in rations. Traditionally, cereal grains and their by-products have filled this need. Whole grains from barley, buckwheat, corn, oats, rye, wheat, and the grain sorghums have been widely used in rabbit rations.

The digestibility of certain grains can be increased with minimal processing. For example, corn is more easily digested when it is cracked.

Cereal by-products, such as red dog flour, wheat bran, mill run, and middlings, offer excellent alternatives to the feeding of whole grains; but since these by-products tend to have laxative properties, the level of usage in the ration should be carefully monitored, and sufficient fiber should be present. Economic considerations should be the determining factor as to which feed should be used.

## PROTEIN SUPPLEMENTS

Fish and animal by-products are seldom used as protein supplements in rabbit rations. But various oilseed meals are routinely added. Peanut meal, the protein supplement most readily eaten by rabbits, provides an excellent source of both protein and energy. Soybean meal, sunflower meal, rapeseed meal (canola meal), safflower meal, sesame meal, linseed meal, and cottonseed meal can all be used interchangeably to different extents.

Amino acid balance in dietary protein is important in the growing rabbit. The Rowett Research Institute, Bucksburn, Scotland, found that the addition of both lysine and methionine to a cereal-peanut meal diet improved the growth rate of weaned rabbits. The minimum requirements for normal growth were found to be 6.2 g methionine plus cystine and 9.4 g lysine/kg diet.[2]

## DRY FORAGES

Rabbits have the ability to utilize dry forages as a large portion of their diet. A good-quality hay will alleviate much of the expense of feeding more costly grains. When using hays, the producer must make certain that they are clean and free from dirt and mold, since it does not take much nutritionally related stress to cause digestive disturbances in rabbits. Once a young rabbit develops scours, precious weight is lost, lowering the efficiency of feed utilization, and the animal becomes highly susceptible to disease and other stresses.

Fig. 55–19. Supplementary hay being fed from the top of the cage. (Courtesy, Pel-Freez Rabbit Meat, Inc., Rogers, AR)

## LEGUME HAYS

Legumes are probably the best roughage feed for rabbits, with alfalfa preferred. These hays are high in protein and are extremely palatable to rabbits. A good-quality legume hay can

be the sole feed source for bucks not in service, dry does, and young, growing replacement stock. Up to 40% of the ration of pregnant and lactating does can consist of legume hay. Besides being excellent sources of nutrients, legume hays also supply much of the bulk and fiber that are needed to maintain healthy rabbits. The most commonly used legume hays in rabbit rations are: alfalfa, clover (except sweet clover), vetch, lespedeza, kudzu, cowpea, and peanut.

## GRASS HAYS

Typical commercial rabbit feeds contain about 40% alfalfa meal, or occasionally some other legume. However, when the price is right, and when incorporated in nutritionally balanced rations, high quality grass hays can be used successfully in rabbit rations, often at a saving in feed costs. Among such grass hays are Bermudagrass, bromegrass, fescue, Johnsongrass, orchardgrass, timothy, ryegrass, and Sudangrass. In comparison with legume hays, grass hays are generally lower in protein and calcium, lower in vitamin A activity, higher in fiber, and less palatable.

Research workers at the Arkansas Experiment Station reported (1) that Coastal Bermudagrass is equal to alfalfa for growing rabbits, and (2) that the Bermudagrass cost 40% less than alfalfa at the time and place of the study, thereby greatly reducing the cost of the ration[3].

## GRASSES (GREEN)

A wide variety of green grasses (and green legumes) can be fed to rabbits—from high-quality alfalfa to lawn clippings. Since fresh grasses are high in water and bulk, it is wise to use a supplemental feed to ensure against any nutritional deficiencies. A number of plants should not be used in rabbit rations because they are either of no nutritional value or they contain compounds that are toxic to rabbits.

## MISCELLANEOUS FEEDS

Many unusual and waste feeds can be used in rabbit rations. Often feedstuffs can come from the garden or table scraps—especially in backyard production. Fats, meats, and spoiled foods should not be used, but garden trimmings and vegetables are low cost and may be efficient.

### MISCELLANEOUS ROUGHAGES

Scraps and by-products from various fruits can be fed to rabbits. Potato peelings are readily eaten. The pulp and peelings of citrus fruits as well as apples are often good sources of cheap, supplemental feed. Caution should be exercised when using fruit pulp, lest the rabbits consume too much fiber.

### ROOTS AND TUBERS

The addition of roots and tubers to the rations of rabbits can be beneficial, especially in the winter when fresh greens are not available. They are highly palatable and are good sources of vitamins and minerals. However, the water content of roots and tubers tends to be extremely high (about 90%),

[2]Spreadbury, D., "A study of the protein and amino acid requirements of the growing New Zealand White rabbit with emphasis on lysine and the sulphur-containing amino acids," British Journal of Nutrition, 1978, 39, 601–6.

[3]Daniels, L. B., L. A. Shriver, and T. S. Nelson, Evaluation of Bermudagrass in Diets of Domestic Rabbits, Arkansas Farm Research, May-June, 1985, p. 2.

and the protein level is quite low (1 to 4%). Thus, the producer should not incorporate too high a level of them in the ration. A deficiency in some of the nutrients may result when feeding roots and tubers, because rabbits preferentially eat this type of feed first, subsequently neglecting the higher quality feeds. For this reason, the daily allowance of roots and tubers in a maintenance ration should be limited to 1.5% of body weight. Rabbits in production should not be fed any roots or tubers.

### SHRUBS AND TREES

Occasionally, twigs from woody plants can be given to rabbits. While the nutritional value of such feeds may be doubtful, they provide the rabbit with something on which to chew, and some additional fiber.

## FEED PREPARATION

The size, type, and intensity of operation will, in most cases, determine the type of feed preparation. Since most backyard operations are small and do not have access to feed mixing facilities, they generally use commercially available complete pelleted rations, along with some additional hay.

If rabbit production is a sideline to a farm operation, homegrown grains and hays are fed with very little processing involved. Commercial operations must obtain feed in large quantities; hence, the processing and preparation of feed for them warrants considerable attention.

Nutritionally complete pelleted rations are available commercially and are being used extensively and successfully. They generally come in 2 types: (1) production rations, which are high in protein and energy; and (2) maintenance rations, which are somewhat lower in protein and energy. The complete pelleted rations generally contain 50 to 60% concentrate and 40 to 50% roughage, with micronutrients supplied at nutritionally adequate levels. The size of the pellet is important. It should be ⅛ to ⁵⁄₃₂ in. in diameter and ⅛ to ¼ in. long. If the pellets are larger than these specifications, the rabbits will bite off pieces of them, with the result that there will be fines and considerable waste. Also, smaller pellets result in kits getting on dry feed more quickly.

Fig. 55-20. *Left:* pelleted feed. *Right:* unpelleted feed. (Courtesy, Pel-Freez Rabbit Meat, Inc., Rogers, AK)

Grains can be fed whole to rabbits, but oats and barley should be rolled and corn cracked in order to maximize digestibility. Rabbits are extremely sensitive to moldy and dusty feeds, so care should be taken to provide only clean, high-quality grains.

Protein supplements should be in cake, pelleted, or crumble form. If the supplement is fed in mash form, there is the possibility that it will settle out from the rest of the feed and be wasted. Since protein supplements are probably the most costly components of the ration, it is imperative that they be used as efficiently as possible. Cottonseed meal must be degossypolized before feeding it to rabbits, and soybean seeds must be heat-treated in order to make them palatable.

Forages should be clean, leafy, and free from mold. In order to maximize the utilization of forages, they should be cut in lengths of about 3 in. and fed in a hay rack. If forage is fed in a form longer than 3 in., rabbits will pull the feed from the hay rack, drag it out on the floor of the hutch, chew a piece from the end, then leave the remainder.

## FEED ALLOWANCES AND SUGGESTED RATIONS

Generally, the term *nutrient requirements* implies a rigid, inflexible level of a certain nutrient for a particular nutritional demand. In reality, the actual nutrient requirement for a particular animal in a fixed set of environmental influences is seldom as precise as has been stated by a group of scientists agreeing on a hypothetical figure. Rather, nutrient allowances are given in order that the producer might have a general idea as to the range in which the real nutrient requirement should fall.

When formulating rations for rabbits, one should consider the following:

• **Fat**—Maintenance rations should contain 2 to 4% fat. Production rations should have 3 to 6% fat.

• **Nitrogen-free extract**—This indicator of the carbohydrate portion of the ration should constitute 42 to 50% of maintenance rations and 44 to 52% of production rations.

• **Crude protein**—For adult rabbits not in production, a ration containing 12 to 15% crude protein will suffice. Rabbits 6 months or older that are growing and/or being fattened, should receive a 16 to 18% crude protein ration. During the last 2 weeks of gestation and early lactation, does should be fed a ration containing 24 to 26% crude protein.

• **Ash**—Ash may run from 5 to 6.5% in all rabbit rations.

• **Fiber**—Rabbits on maintenance rations, should be fed more fiber than rabbits on production regimens since fiber feeds tend to be cheaper than either energy or protein feeds. Most maintenance rations contain 16 to 22% crude fiber, whereas production rations contain 12 to 16%.

A listing of the nutrient allowances for rabbits is given in Table 55-6.

### TABLE 55-6
### NUTRIENT ALLOWANCES FOR RABBITS[1]

|  | Crude Protein[2] | Fat | Fiber | Nitrogen-Free Extract | Ash |
|---|---|---|---|---|---|
|  | (%) | (%) | (%) | (%) | (%) |
| **P**regnant does | 15–17 (16) | 3–6 | 12–16 | 44–52 | 5–6.5 |
| **D**oes with litters | 24–26 (25) | 3–6 | 12–16 | 44–52 | 5–6.5 |
| **D**ry does | 12–15 (13) | 2–4 | 16–22 | 42–50 | 5–6.5 |
| **H**erd bucks | 12–15 (13) | 2–4 | 16–22 | 42–50 | 5–6.5 |
| **D**eveloping young (after weaning) | 16–18 (17) | 3–6 | 12–16 | 44–52 | 5–6.5 |

[1]Allowances are on the basis of air-dried feed (as-fed basis).
[2]Values in parentheses represent what the author feels is optimal for most rations.

## FEEDING GUIDE AND RATIONS

A rabbit feeding guide, along with suggested rations, is presented in Table 55–7.

**TABLE 55–7**
**RABBIT FEEDING GUIDE AND SUGGESTED RATIONS[1]**

| Age, Sex, Production | Total Daily Feed Per Day As-fed (oz) | (g) | Total Daily Feed As % Live Weight As-fed (%) | Suggested Rations[2] Ingredient | Total Diet A (%) | B (%) |
|---|---|---|---|---|---|---|
| **Normal growth, does or bucks,** average 6.5 lb (2.95 kg) | 5.1 | 145 | 5.8 | Alfalfa hay | 60 | |
| | | | | Corn, grain | 21.5 | |
| | | | | Barley, grain | 15 | |
| | | | | Soybean meal | 3 | |
| | | | | Salt | 0.5 | |
| **Normal growth and fattening, does or bucks,[3]** at body weight of: | | | | | | |
| 4 lb (1.8 kg) | 4.0 | 113 | 6.2 | Alfalfa hay | 40 | 50 |
| 5 lb (2.3 kg) | 4.8 | 136 | 6.0 | Corn | — | 23.5 |
| 6 lb (2.7 kg) | 5.4 | 154 | 5.7 | Wheat bran | 5 | 5 |
| 7 lb (3.2 kg) | 6.1 | 172 | 5.4 | Barley, grain | 31.5 | 11 |
| | | | | Oats, grain | 18 | — |
| | | | | Soybean meal | 5 | 10 |
| | | | | Salt | 0.5 | 0.5 |
| **Maintenance, does or bucks,** at body weight of: | | | | | | |
| 5 lb (2.3 kg) | 3.2 | 91 | 4.0 | Alfalfa hay | 70 | — |
| 10 lb (4.5 kg) | 5.3 | 150 | 3.3 | Clover | — | 70 |
| 15 lb (6.8 kg) | 7.2 | 204 | 3.0 | Oats, grain | 19.5 | 29.5 |
| | | | | Wheat, grain | 10 | — |
| | | | | Salt | 0.5 | 0.5 |
| **Pregnant does,** at body weight of: | | | | | | |
| 5 lb (2.3 kg) | 4.0 | 113 | 5.0 | Alfalfa | — | 50 |
| 10 lb (4.5 kg) | 6.6 | 186 | 4.1 | Clover hay | 50 | — |
| 15 lb (6.8 kg) | 9.0 | 254 | 3.7 | Oats, grain | 43.5 | 45.5 |
| | | | | Soybean meal | 6 | 4 |
| | | | | Salt | 0.5 | 0.5 |
| **Lactating doe,** at body weight of 10 lb (4.5 kg) with a litter of 7 | 18.3 | 520 | 3.4 | Alfalfa hay | 40 | 40 |
| | | | | Wheat, grain | 25 | 25 |
| | | | | Sorghum, grain | 24.5 | 22.5 |
| | | | | Soybean meal | 10 | 12 |
| | | | | Salt | 0.5 | 0.5 |

[1]Adapted by the author from *Nutrient Requirements of Rabbits*, No. 9, 1st and 2nd rev. ed., NRC-National Academy of Sciences, 1966 and 1977.

[2]In iodine-deficient areas, use iodized salt.

[3]Ration can be used for pregnant and lactating does, also.

## FEEDING RABBITS

When good nutrition and proper management procedures are followed, growing rabbits should attain feed conversion rates of 2.8 to 4.0; that is, they should make 1 lb of gain from 2.8 to 4.0 lb of feed. Most producers estimate that it takes about 100 lb of feed to grow one litter to market weight of fryers. This figure includes the feed consumed by the doe from the time of mating to the weaning of the litter.

When including working does, their young, bucks, and replacement does, the average feed efficiency for rabbits ranges from 4.5:1 to 5:1.

Rabbits should be fed according to their level of production. Thus, lactating does have a much greater feed requirement than dry or pregnant does.

## FEEDING DRY DOES AND HERD BUCKS

Since these animals are not involved in an intensive type of production, they can be fed maintenance rations—rations that will supply just enough nutrients to keep them healthy and sound. A good-quality leafy, fine-stemmed legume hay plus trace mineral salt can maintain dry does and bucks not in service if the feed is available free-choice. If the legume is not of good quality or if a carbonaceous hay is to be offered free-choice, supplementation with a grain-protein mixture or an all-grain pellet is recommended. A general rule of thumb is to feed 2 oz of supplement for every 8 lb of liveweight. A buck that is in active service requires more feed. Four to six ounces of a complete pelleted ration, or a good-quality legume hay plus 2 oz of supplement per 8 lb of body weight, should suffice. If the sexually active buck begins to put on weight, the amount of feed should be reduced.

## FEEDING PREGNANT DOES

Immediately after mating, but before pregnancy is confirmed, the does should be kept on a maintenance level ration. Pregnancy should be confirmed before the plane of nutrition is increased, because if the doe does not conceive, she may become overweight.

The greatest nutritional demands do not occur until the latter stages of pregnancy. Many rabbit producers feed only commercial pellets to pregnant does to minimize disease and nutritional problems inherent in the feeding of hay.

A recommended practice for feeding pregnant does is as follows:

1. If a doe is not nursing, she should be fed 4 to 6 oz per day of a commercial feed from the time she is bred until she is ready to kindle, with the level of feeding varying according to the level of energy in the ration; generally, more of a low-energy ration must be fed in order to obtain performance comparable to a high-energy ration.

2. A day or two before kindling, the doe may eat less; her feed allowance should be reduced accordingly.

## FEEDING LACTATING DOES

Lactation creates extremely high nutrient demands on the doe. Either a complete pelleted ration or a combination of good-quality hay and supplement will provide the necessary energy and protein.

After kindling, gradually increase the allowance over the next few days to full feed. During lactation, the doe should continue to receive feed *ad libitum.*

When the young are weaned, the doe can be placed on a maintenance ration until she is rebred and diagnosed as pregnant.

## FEEDING JUNIORS

In the medium-size breeds, 2 to 4 oz of a pelleted supplement daily and free-choice of good-quality hay will be adequate. If a complete pelleted ration is to be used, the juniors should be fed 4 to 6 oz per day. When juniors are being grown for breeding stock, a ration of 99% alfalfa pellet (15 to 16% protein) and 1% salt may be used. Young does and bucks should not be fed to the point that they become obese, because obese rabbits often have reproductive problems. Besides, the practice is economically wasteful.

## FEEDING ORPHAN LITTERS

When a doe dies at or after kindling, the producer must make provisions for feeding the orphan litter. If a recently kin-

dled doe with a small litter is available, some of the young may be transferred (fostered) to her. This method is far more practical than feeding each kit by hand. However, since this practice is not always possible, a procedure like the following may be used:

1. For the first 2 weeks of life, feed cow's or goat's milk, or a commercial milk replacer. Heat the milk to body temperature, then feed it by using an eyedropper or a doll's nursing bottle. The eyes of the young rabbits will open at about day 10.

2. After the initial nursing period, solid food, such as fresh grass and rolled oats, can be offered in addition to milk. This will stimulate development of the gut.

3. When the young are about 17 days of age, they can be taught to drink from a pan and offered small quantities of a good growing ration.

4. Gradually, the quantity of solid feed can be increased.

## FEEDING METHODS

A standard rule is that rabbits should never be exposed to radical changes in their diets. If feed changes are to be made, they should be made gradually, taking as long as 5 to 10 days. If, for example, the ration is to be lowered from 100% alfalfa to 40% alfalfa, a gradual daily reduction in the alfalfa content of the feed is recommended until the desired feeding level has been attained. Occasionally, when making a slow transition in the ration, some rabbits will waste a large amount of feed by digging and searching for the old, familiar feed. When this happens, an abrupt switch to the new ration may be made.

### HAND-FEEDING

While this method of feeding involves considerable time and labor, it enables producers to keep a close watch relative to the general condition and feeding habits of their animals.

Rabbits can be fed once, twice, or three times daily. However, in most operations, a once-a-day feeding practice is the best. Less labor is involved; and the rabbits are more likely to maintain an active appetite if they are allowed to clean up their feed before the next feeding. Since rabbits eat about 2½ times as much at night as during the day, a once-a-day feeding should be offered in the evening. If more than one daily feeding is used, the last feeding of the day should provide the largest quantity of feed. The number of feedings is not critical, but it is important that the time of feeding be regular from day to day. Rabbits are creatures of habit, and any break of routine can cause digestive problems.

If rabbits are hand-fed only the amount of feed that they will clean up in a day, it is easy to detect an animal that is off feed; hence, the daily allowance may serve as an important health check.

### SELF-FEEDING

Growth tends to be more rapid and efficient in self-fed rabbits than those that are fed by hand. This improved growth rate is due to the fact that self-fed rabbits have access to feed at all times; consequently, they eat more frequently and chew their food more thoroughly.

Animals that are on maintenance rations may consume more feed than necessary if feed is provided *ad libitum.* Thus, only lactating does and market rabbits 4 to 8 weeks of age should be fed free-choice.

Filling self-feeders with enough pelleted feed to last several days may cause problems in areas with high humidity, resulting in pellets absorbing moisture, expanding, lodging in the feeders, and molding.

## MANAGING THE HERD

Success in raising rabbits depends on efficient management. Caretakers should become thoroughly acquainted with their animals—their characteristics and behavior, their likes and dislikes. Consideration for the welfare of animals is always necessary for success in raising them. Proper arrangement of equipment, hutches, and buildings is also essential to efficient management. When entering the rabbitry, caretakers and visitors should do so quietly and make their presence known by speaking in a low tone. Otherwise, the rabbits may become frightened, race around in the hutch and injure themselves, or jump into the nest boxes and injure the litters.

## METHODS OF HANDLING RABBITS

Never lift rabbits by the ears or legs. Handling in this manner may injure them.

Small rabbits can be lifted and comfortably carried by grasping the loin region gently and firmly with the heel of the hand toward the tail of the animal. This method prevents bruising the carcass or damaging the pelt.

To lift and carry a medium-weight rabbit, let the right hand grasp the fold of skin over the rabbit's shoulder; then, support the rabbit by placing the left hand under its rump.

## CAUSES OF LOSSES IN NEWBORN LITTERS

If the doe is disturbed, she may kindle on the hutch floor and the litter may die from exposure. Even if predators—cats, snakes, rats, weasels, minks, bobcats, coyotes, strange dogs—cannot gain access to the rabbitry, they may be close enough for the doe to detect their presence, and she may be frightened and kindle prematurely. If she is disturbed after the litter is born and jumps into the nest box, she may stamp with her back feet and injure or kill the newborn rabbits.

Occasionally a doe fails to produce milk. In such cases the young will starve within 2 or 3 days unless the condition is noted and the young transferred to foster mothers. The attendant should keep a close check on newborn litters for several days after birth to make sure they are being fed and cared for properly.

Does sometimes eat their young. This may result from a ration that is inadequate in either quantity or quality, or from the nervousness of a doe disturbed after kindling. It is also possible that the doe is of a strain that exhibits poor maternal instincts. Does usually do not kill and eat healthy young, but limit their cannibalism to young born dead, or those that are injured and have died. Proper feeding and handling during pregnancy will do more than anything else to prevent this tendency.

A good rule is to give another chance to a valuable doe that destroys her first litter; if she repeats the practice, dispose of her.

## WEANING

Under most management programs the young are weaned at 8 weeks of age. At that age, young meat rabbits should average 4 lb in weight and be ready for market. Some commercial producers leave the young with the doe for 9 or 10 weeks to get a 4½ to 5½-lb fryer. Small litters (fewer than 5 young) can be weaned at an earlier age, and the doe may be rebred at that time. Also, under accelerated breeding programs where does are bred less than 35 days following kindling, it is advisable to wean the young at 5, 6, or 7 weeks of age to allow the doe to prepare for her next litter. It is best to allow a few days between removal of one litter and birth of the next. For example: if a doe is bred 28 days after kindling, it is possible to leave the litter with her until they are 56 days of age, allowing for kindling about 3 days later. It depends upon the condition of the doe and her ability to stand up under this type of program. The breeder may wish to remove the young at 7 weeks of age and give the doe 7 to 10 days to prepare for the next kindling.

## CASTRATION

Castration of bucks may be desirable; for example, where Angoras are to be kept in colonies for wool production. In producing domestic rabbit meat for market, there are no advantages to be derived from castrating bucks for improving the rate of growth and condition, reducing the quantity of feed required to produce a pound of gain in liveweight, or improving the carcass and pelt. Probably the only advantage to be derived would be that it reduces fighting and makes possible the maintaining of a number of castrated bucks in one enclosure, thereby saving equipment, time, and labor. Castration is a simple operation, most easily performed when bucks are 3 to 4 months old.

To restrain an animal for the operation, have an assistant hold the buck's left forefoot and left hind foot with the left hand, and the right forefoot and right hind foot with the right hand, with the animal's back held firmly, but gently, against the caretaker's lap. Clip all the wool from the scrotum. Disinfect a sharp knife or razor blade. If a disinfecting agent is not used on the rabbit, it will lick the wound frequently, keeping it clean and the tissues soft, thus promoting healing.

Press one of the testicles out into the scrotum. Hold it firmly between the thumb and forefinger of the left hand. Make an incision parallel to the median line and well toward the back end of the scrotum to allow the wound to drain readily. To keep the testicle from being drawn up into the abdominal cavity, as soon as it comes from the incision pull it out far enough from the body for the cord to be severed just above it. To prevent excessive hemorrhage, sever the cord by scraping with a knife rather than by cutting. If too much tension is put on the cord and it is drawn too far from the body, injury may be brought about by internal hemorrhage or other complication.

After the second testicle has been removed in the same manner, lift the scrotum to make sure that the ends of the cord go back into the cavity.

Handle the animal gently. After the operation, place it in a clean hutch where it can be quiet and comfortable.

## CARE OF HERD DURING EXTREME TEMPERATURES

• **Heat**—In almost all sections of the United States, high summer temperatures necessitate some changes in the general care and management of rabbits. Adequate shade should be provided for the animals during the hottest part of the day. Good circulation of air throughout the rabbitry is necessary, but strong drafts should be avoided. An abundant supply of water should be provided at all times.

Newborn litters and does advanced in pregnancy are most susceptible to high temperatures. Heat suffering in the young is characterized by extreme restlessness; in older animals, by rapid respiration, excessive moisture around the mouth, and occasionally slight hemorrhages around the nostrils. Rabbits that show symptoms of suffering from the heat should be moved to a quiet, well-ventilated place and given a feed sack moistened with cold water to lie on. Water crocks and large bottles filled with cracked ice and placed in the hutch so that the rabbits can lie next to them make for comfort.

In well-ventilated rabbitries, wetting the tops of the hutches and the floors of the houses on a hot, dry day will reduce the temperatures 6 to 10 °F. The tops of hutches should be waterproof, as rabbits should be kept dry. Overhead sprinkling equipment can be used in houses with concrete or soil floors that drain readily, or sprinklers may be used above the roof of rabbit sheds. A thermostatically controlled sprinkler that will work automatically should be installed.

In hot, dry climates such as southwest United States, evaporative coolers on the roof or sides of the buildings can be used, drawing air over wet pads and distributing it through the building. This type of cooler is widely used in homes and can be adapted to use in rabbitries which are partially, or totally, enclosed.

In areas of high humidity, the use of sprinklers or extra water will aggravate the situation and add to the rabbit's discomfort. Under such conditions, it is advisable to install fans, or place the buildings to take advantage of all breezes, in order to get maximum movement of air. The use of refrigerant air conditioning is usually uneconomical and impractical, due to the high initial investment and operating costs.

• **Cold**—If kept out of drafts, mature rabbits suffer little from low temperatures. However, precautions should be taken to protect rabbits from direct exposure to rain, sleet, snow, and winds. If they are enclosed in a building, care must be taken to provide adequate ventilation and to prevent the accumulation of moisture. Cold weather, drafts, and high humidity are conducive to the spread of respiratory infections. Young litters should be provided with nest boxes and sufficient bedding to keep them warm.

• **Controlled Environments**—The use of controlled environment in rabbitries, where rabbits are maintained under more or less constant environmental conditions, is receiving increased attention. Several large commercial rabbitries in the western United States are changing to or are constructing this type of housing. The advantages of controlled environment are the elimination of extremes in weather and, perhaps, seasonal fluctuations in production.

## PREVENTING INJURIES

Paralyzed hindquarters in rabbits usually result from improper handling or from injuries caused by slipping in the hutch while exercising or attempting to escape predators, especially around kindling time. Such slipping usually occurs at night. Common injuries are dislocated vertebras, damaged nerve tissue, or strained muscles or tendons. If the injury is mild, the animal may recover in a few days. The injured animal should be made comfortable and fed a balanced diet. If it does not improve within a week, it should be destroyed to prevent unnecessary suffering. It is important, therefore, that rabbits be provided with quiet, comfortable surroundings and be protected from predators and unnecessary disturbances.

The toenails of rabbits confined in hutches do not wear normally. They may even become long enough to cause foot deformity. The nails may also catch in the wire mesh floor and cause injury and suffering. So, periodically the nails should be cut with side cutting pliers, with the cut made below the tip of the cone in the toenail. The cone can be observed by holding the foot up to daylight. This will not cause hemorrhaging or injury to the sensitive portion.

## PREVENTING SORE DEWLAPS

During warm weather the dewlap, or fold of skin under the rabbit's chin, may become sore. This is caused by drinking frequently from crocks and keeping the fur on the dewlap wet so long that it becomes foul and turns green. The skin on the dewlap and on the inside of the front legs becomes rough and the fur may be shed. The animal scratches the irritated area, causing abrasions and infection.

This condition may be alleviated by removing the cause—by placing a board or brick under the water crock to raise it so that the dewlap will not get wet when the rabbit drinks. If the skin becomes infected, clip off the fur and treat the area with a medicated ointment until the irritation clears up. The best solution to the problem is to use an automatic dewdrop watering system which eliminates the possibility of wet dewlaps.

## FUR-EATING HABIT

Rabbits that eat their own fur or bedding material, or gnaw the fur on other rabbits, usually do so because the diet is inadequate in quality or quantity. A common cause is a diet low in fiber or bulk. Sometimes the protein content of the diet is too low; adding more soybean, peanut, sesame, or linseed meal will correct the latter deficiency.

Fig. 55–21. Fur-chewed rabbits. (Courtesy, Pel-Freeze Rabbit Meat, Inc., Rogers, AR)

The experienced breeder notes the condition of each animal in the herd and regulates the quantity of feed to meet its individual requirement. Providing good-quality hay or feeding fresh, sound green feed or root crops as a supplement to the grain or pelleted diet also helps to correct an abnormal appetite.

## PREVENTING FUR BLOCK

In cleaning themselves by licking their coats, or when eating fur from other animals, rabbits swallow some wool or fur which is not digested. The only noticeable result may be droppings fastened together by fur fibers. However, if the rabbit swallows any appreciable amount, it may collect in the stomach and form a *fur block* that interferes with digestion. If it becomes large enough, it blocks the alimentary tract and the animal starves. The most satisfactory method of preventing this is to shear Angoras regularly, and lessen fur eating among rabbits by providing adequate roughage and protein in their diets. A block of wood or other material upon which the rabbit can chew may be used to reduce fur chewing.

## GNAWING WOOD HUTCH

Gnawing wood is natural for rabbits. So, the wooden parts of the hutch should be protected by placing wire mesh on the inside of the frame when constructing the hutch and by using strips of tin to protect exposed wooden edges. Placing twigs or pieces of soft wood in the hutch may protect it to some extent; rabbits may chew these instead of the hutch.

Rabbits that have access to good-quality hay and are receiving some fresh green feed or root crops are less likely to gnaw on their hutches.

## DISPOSAL OF RABBIT MANURE

Rabbit manure has a high nitrogen content when the rabbits are fed a well-balanced diet. It will not burn lawns or plants and is easy to incorporate in the soil. It will benefit gardens, lawns, flowering plants, shrubs, and trees.

The value of rabbit manure depends on how it is cared for and used. There will be less loss of fertilizing elements if the material is immediately incorporated into the soil. When manure is stored in piles and exposed to the weather, chemicals are lost through leaching and heat. Much of this loss can be prevented by keeping the manure in a compost heap or in a bin or pit.

## EARTHWORMS CONVERT DROPPINGS TO CASTS

Where earthworms are active throughout the year as in warm climates, they may be used to advantage under rabbit hutches to save labor in removing fertilizer. Make bins for confining the worms the same length and width as the hutch and 1 ft deep. Place bins on the ground, not on solid floors, and keep the fertilizer moist to insure the worms working throughout the bin.

Earthworms convert the rabbit droppings into *casts*—a convenient form of fertilizer for use on flowers, lawns, shrubs, trees, and other foliage. If a large population of worms is kept, there will be no objectionable odor. Very few flies will breed in the bins. By using earthworms in this manner, it will be necessary to remove the manure only at 5- to 6-month intervals.

## RECORDS

A convenient and simple system of records is essential for keeping track of breeding, kindling, and weaning operations. This information may be used in culling unproductive animals and in selecting replacement breeding stock. The essential features of a simple record system are illustrated in the hutch card and the buck breeding record card shown in Figs. 55-22 and 55-23.

Similar record forms may be obtained from firms dealing in supplies for the rabbitry, or may be home prepared. Some feed mills also furnish their customers with hutch cards and record forms.

### HUTCH CARD

ANIMAL NO. _____ BORN _____ BREED _____
SIRE _____ DAM _____ LITTER NO. _____

| DATE BRED | BUCK NO. | DATE KINDLED | NUMBER YOUNG BORN | | NUMBER YOUNG RETAINED | LITTER NO. | DATE WEANED | NUMBER WEANED |
| | | | ALIVE | DEAD | | | | |
|---|---|---|---|---|---|---|---|---|
| | | | | | | | | |
| | | | | | | | | |
| | | | | | | | | |
| | | | | | | | | |

### PRODUCTION RECORD

| LITTER NO. | WEANING | | | NOTES: |
| | NUMBER | AGE | WEIGHT | |
|---|---|---|---|---|
| | | | | |
| | | | | |
| | | | | |
| | | | | |

Fig. 55-22. Hutch card, a useful record form. *Top:* front. *Bottom:* back.

### BUCK BREEDING RECORD

BUCK NO. _____

BREED _____ SIRE _____
DATE BORN _____ DAM _____

| DOE | LOCATION | DATE BRED | RESULT OF BREEDING | | | | |
| | | | KINDLED | | PASSED | | |
| | | | ALIVE | DEAD | DATE | NUMBER | WEIGHT |
|---|---|---|---|---|---|---|---|
| | | | | | | | |
| | | | | | | | |
| | | | | | | | |
| | | | | | | | |

Fig. 55-23. Buck breeding record form.

## RABBIT HEALTH, DISEASE PREVENTION, AND PARASITE CONTROL

Disease is always an ominous threat when animals are kept in close confinement such as in a rabbitry. Nationwide, the United States' losses due to rabbit diseases and parasites averages 20 to 25%; this includes kits born dead, nest box deaths, fryer deaths, and the loss of bucks and does.

## HERD HEALTH PROGRAM

The control of diseases and parasites in the rabbitry should be a matter of prevention rather than cure. To prevent, or minimize, disease outbreaks the following rabbit herd health program is proposed:

1. **Buy healthy foundation stock.** Start with disease-free foundation stock obtained from a reputable breeder.

2. **Keep feed and water clean and fresh.** Disease can be spread through contaminated feed and water.

3. **Clean and disinfect the nest box after each litter.** Newborn rabbits are highly susceptible to disease and an infected nest box can be a real source of trouble.

4. **Learn to recognize rabbit diseases.** But remember that many diseases are quite similar. So, have a correct diagnosis before beginning treatment.

5. **Isolate all rabbits brought into rabbitry (show stock, new breeding stock, etc.)** Until positive that all danger of infecting the herd is over, all incoming rabbits should be isolated.

6. **Use wire floored hutches.** This will keep the animals out of contact with their droppings and make it easier to keep the rabbitry sanitary and free of disease.

7. **Allow no visitors in the rabbitry.** This applies especially to *pick-up* people and other rabbit raisers who might unknowingly transmit disease to the herd.

8. **Sear hutch wire and equipment.** After marketing each litter, use a blow torch flame. This will destroy most disease organisms and protect the next occupant of the hutch.

9. **Select disease-resistant animals for breeding stock.** They will pass on to their offspring greater ability to survive.

## DISEASES, PARASITES, AND AILMENTS OF RABBITS

The most serious disease of domestic rabbits is pasteurellosis. This disease manifests itself in a wide variety of conditions such as pneumonia, snuffles (sinusitis), and other respiratory infections; and septicemia, a generalized blood infection.

Another serious problem in rabbit health is enteritis, or bloat. Three types of enteritis are distinguished: diarrhea, mucoid, and hemorrhagic. The specific cause of enteritis is not known and there are no reliable measures for prevention or treatment.

Coccidiosis, both of the liver and intestines, is a serious problem in some areas but can be successfully treated.

The tapeworms that infest the rabbit are those which at a later stage infest dogs and cats, but the rabbits seem to suffer little harm from them.

Tularemia, the disease that has killed so many wild rabbits in recent years, is spread by ticks and fleas. If domestic rabbits are kept in clean conditions, free from ticks and fleas, they will not contract tularemia.

Domestic rabbits suffer from other ailments such as fungal infections, mange, sore hocks, and spirochetosis or vent disease, but these usually can be successfully treated and do not present a major problem.

Many of the most common diseases, parasites, and ailments of rabbits are presented in summary form in Table 55–8.

## TABLE 55–8
## DISEASES, PARASITES, AND AILMENTS OF RABBITS

| Disease | Symptoms | Cause | Treatment and Control |
|---|---|---|---|
| Abscesses | A "running" sore. | Most abscesses are caused by *P. multocida*, which invade a break in the skin caused by a scratch, cut, or sore. | Penicillin-streptomycin combination is an effective treatment. However, in most rabbitries it is probably best to cull afflicted rabbits immediately. Good ventilation and sanitation lessen the incidence of abscesses. |
| Buckteeth, malocclusion | The lower teeth protrude. The upper teeth curve into the mouth. Afflicted rabbits have difficulty eating. | This is a genetically inherited defect, although the mode of inheritance is not known. | The long teeth may be clipped with a sharp wire cutter. Rabbits with buck teeth should not be used for breeding purposes. |
| Cannibalism | Does that eat their young. | Generally caused by mismanagement, such as strangers in the rabbitry, predators, noise, or moving the doe just before kindling. | Rectify any causative mismanagement. If a doe cannibalizes two successive litters and there is no mismanagement, she should be culled. |
| Coccidosis | Mild intestinal cases, no symptoms; moderate cases, diarrhea and no weight gain. Severe cases have pot belly, diarrhea with mucus, and pneumonia is often secondary. Severe liver coccidosis can be fatal. | Parasitic infection of the intestinal tract or liver, caused by coccidia. Ten different coccidia species of the genus *Eimeria* infect the intestine. (Liver coccidiosis is caused by *Eimeria steidae*.) | Keep floor clean, dry; remove droppings frequently. Prevent fecal contamination of feed and water. Add feed grade sulfaquinoxaline so that level will be 0.025 percent, feed 3 to 4 weeks. Water soluble sulfaquinoxaline can be added to the drinking water at level of 0.04% and administered for 2 to 3 weeks. These treatments, combined with sanitation, will greatly reduce numbers of parasites and animals infected. |
| Conjunctivitis or weepy eye | Inflammation of the eyelids; discharge may be thin and watery or thick and purulent. Fur around the eye may become wet and matted. | Infection of the eyelids caused by several different bacteria; also may be due to irritation from smoke, dust, sprays, or fumes. | Early cases may be cleared up with eye ointments, argyrol, yellow oxide of mercury, or antibiotic. A combination of 400,000 units of penicillin combined with ½ gr. streptomycin to each 2 ml. For eye infections drop directly into eye. Protect animals from airborne irritants. |

*(Continued)*

**TABLE 55-8** *(Continued)*

| Disease | Symptoms | Cause | Treatment and Control |
|---|---|---|---|
| **Ear mange or canker** | Shaking of head, scratching of ears. Brown scaly crusts at base of inner ear. | Ear mites, *Psoroptes cuniculi.* | Mites can be treated by using any number of mite medications in keeping with manufacturer's directions. Most such products contain a mineral oil base, with a parasiticide such as malathion added. |
| **Enterotoxemia** | The clinical signs of enterotoxemia are profuse diarrhea, dehydration, reduced feed intake, and rough hair coat. Rabbits 4 to 8 weeks of age are susceptible. | Several bacteria including *Clostridium perfungens, C. spiroforme,* and *Escherichia coli.* | The most successful treatment is to change feed, to lower energy and higher fiber feed. Broad-spectrum antibiotics such as oxytetracycline in the water give short-term relief. There is no effective vaccine for rabbit enterotoxemia. |
| **Fur blocks, hairballs, trichobezoars** | Animals reduce feed intake or stop eating completely, fur becomes rough, and weight is lost. Stomach is filled with undigested fur, blocking passage to intestinal tract. Pneumonia may become secondary. | Caused by rabbits ingesting excess hair. Lack of sufficient fiber, bulk, or roughage in the diet. Junior does or developing does most susceptible. | Control: Increase fiber or roughage in the ration. Feed dry alfalfa or timothy hay. Treatment: Give rabbits with blockage 10 cc of fresh pineapple juice for three consecutive days. |
| **Heat prostration** | Rapid respiration, prostration, blood-tinged fluid from nose and mouth. Does that are due to kindle are most susceptible. | Extreme outside temperature. Degree varies with location and humidity. Rabbits have no sweat glands. | Reduce temperature with water sprays, foggers. Place wet burlap in hutch or wet the animal to help reduce body temperature. |
| **Hutch burn** | Inflammation of external sex organs and anus. Area may form crusts and bleed and, if severely infected, pus will be produced. | Bacterial infection of the membranes. | Keep hutch floors clean and dry. Pay particular attention to corners where animals urinate. Antibiotic salves and ointments are helpful. Generally it is best to cull animals with urine-hutch burn. |
| **Ketosis, pregnancy toxemia** | Sudden death of does shortly before or shortly after kindling. On post-mortem examination, a yellow or tan liver is observed, caused by an accumulation of fat in the liver cells. | The cause of the disease is unknown, but the feeding of high energy diets may be involved. | Treatment of the disease in rabbits is seldom attempted. |
| **Mange** | Reddened, scaly skin, intense itching and scratching, some loss of fur. | Mites, *Cheyletiella parasitivorax* and *Listraphorus gibbas* are the most common ones. | If only a few rabbits are involved, a cat flea powder will be effective. If many animals are affected, a 0.5% malathion dip, used according to manufacturer's directions, works well. |
| **Mastitis, blue bag, caked breasts, caked udder** | The mammary glands become feverish and pink, nipples red and dark. Temperature above normal, appetite poor, breasts turn black and purplish. | Bacterial infection of the breasts. It is often caused by the bacterium *Staphylococcus aureus,* but may be caused by several other bacteria. | Inject 100,000 units of penicillin intramuscularly twice each day for 3 to 5 days. Disinfect hutch and reduce feed concentrates. If severe case, destroy. NEVER transfer young from infected doe to another doe. |
| **Metritis and orchitis** | Metritis is an infection of the uterus, commonly characterized by a yellowish white discharge from the doe's vent. However, metritis may be present without a discharge. Orchitis is an infection of the male's testicles, causing one or both testicles to be enlarged and hot. | Metritis is caused by *P. multocida.* Orchitis is also caused by *P. multocida.* | The only practical treatment for metritis and orchitis is culling the infected rabbits from the herd. |
| **Mucoid enteritis** | This is a diarrheal disease of rabbits of any age. The clinical signs are: gelatinous or mucous-covered feces, loss of appetite, loss of weight, drinking much water, and often a bloated abdomen due to excess water. | Constipation—an impaction of the digestive tract. | There is no effective medical treatment. Changing to a new batch of feed will usually eliminate the disease, for unknown reasons. |
| **Paralyzed hindquarters** | Found in mature does, hind legs drag, cannot support weight of pelvis or stand. Urinary bladder fills but does not empty. | Injury, resulting in broken back, displaced disc, damage to spinal cord or nerves. | Protect animals from disturbing factors, predators, night prowlers, and visitors or noises that startle animals, especially pregnant does. |
| **Pasteurellosis** | The term *pasteurellosis* covers a host of clinical conditions all caused by *P. multocida,* including sniffles, pneumonia, abscesses, weepy eyes, metritis, orchitis, and wry neck. | Bacterial infection, *Pasteurella multocida.* | For treatment and control, see each of the several diseases caused by *P. multocida.* These are listed in the second column (Symptoms). |
| **Pinworms** | No specific symptoms in live animals. White threadlike worms found in cecum and large intestine cause slight local irritation. | Pinworms, *Passalurus ambiquus.* | Infection not considered of economic importance. If treatment is necessary, add piperazine citrate to the water, at dosage of 100 mg per 100 ml of water. |
| **Pneumonia** | Labored breathing with nose held high, bluish color to eyes and ears. Lungs show congestion, red, mottled, moist, may be filled with pus. Often secondary to enteritis. | Bacterial infection of the lungs. Organisms involved may be *Pasteurella multocida, Boredetella bronchiseptica,* and *Staphylococcus* and *Streptococcus* sp. | Broad-spectrum antibiotcs and sulfa drugs have been used with little success. |
| **Pseudotuberculosis** | A wasting type of disease characterized by poor appetite, depression, emaciation, and death. | The bacterium, *Yersinia pseudotuberculosis.* | Treatment is not effective. Prevention consists in good sanitation. |

*(Continued)*

**TABLE 55–8** *(Continued)*

| Disease | Symptoms | Cause | Treatment and Control |
|---|---|---|---|
| **Rabbit syphillus or spirochetosis, or vent disease** | Similar lesions as produced by urine or hutch burn. Raw lesions or scabs appear on sex organs; transmitted by mating. | Spirochete, *Treponema cuniculi.* | Inject intramuscularly 100,000 units of penicillin. Do not breed until lesions heal. If only a few animals infected, it is easier to cull than treat. Do not lend bucks. |
| **Ringworm or favus** | Circular patches of scaly skin with red, elevated crusts. Usually starts on head. Fur may break off or fall out. | Fungi, *Trichophyton* and *Microsporum.* | Griseofulvin given orally in the feed at the rate of 10 mg per pound body weight for 15 to 25 days is the best herd treatment.<br>Iodine or an ointment containing hexetidine is effective in treating isolated cases of ringworm. |
| **Salmonellosis** | Characterized by profuse diarrhea, with septicemia and death very common. Abortion is common. Mortality is greatest in young rabbits and pregnant does. | The bacterium, *Salmonella typhimurium* or *S. enteriditidis.* | Treatment with streptomycin or nitrofurazone is effective. Prevention is important. The disease can be transmitted to rabbits by other rabbits, humans, wild birds, and rodents. So, avoid other infected species as well as infected rabbits. |
| **Snuffles** | Sneezing, rubbing nose; nasal discharge may be thick or thin. Mats fur on inside of front feet. May develop into pneumonia, usually chronic type of infection. | Bacterial infection of the nasal sinuses, *Pasteurella multocida,* or *Bordetella bronchiseptica.* | Individual animals may be treated with a combination of 200,000 units of penicillin combined with ¼ gr. streptomycin per day injected in the muscle of the rabbit for three days. |
| **Sore hocks** | Bruised, infected, or abscessed areas on hocks. May be found on front feet in severe cases. Animal shifts weight to front feet to help hocks. | Bruised or chafed areas become infected. Caused by wet floors, irritation from wire or nervous "stompers."<br>Sore hocks are a major problem in Rex rabbits and in giant breeds such as the Flemish Giant. | Small lesions may be helped by placing animal on lath platform or on ground. Advanced cases are best culled. Medication is temporarily effective. |
| **Staphylococcosis** | This disease syndrome causes a number of different conditions, including mastitis, staphylococcal septicemia, conjunctivitis (weepy eyes), and abscesses. | Caused by the bacterium, *Staphylococcus aureus.* | Many antibiotics have been used in treatment, but with limited success.<br>The best defense is prevention by sanitation and disinfectants. |
| **Tapeworms** | White streaks in liver or small white cysts attached to membrane on stomach or intestines. Usually cannot detect in live animals. | Larval stage of the dog tapeworm, *Taenia pisiformis,* or of the cat tapeworm, *T. taeniaeformis.*<br>Adult tapeworms are rarely found in domestic rabbits. | Treatment is not practical.<br>Keep dogs and cats away from feed, water, and nest box material. Eggs of tapeworm occur in droppings of dogs and cats. |
| **Tularemia** | This is primarily a disease of wild rabbits. Few cases have been diagnosed in domestic rabbits. Afflicted rabbits become sluggish and die. | The bacterium, *Francisella tularensis.* | No treatment is known. |
| **Tyzzer's disease** | The clinical signs are profuse diarrhea and rapid death, especially in fryer rabbits. | The bacterium, *Bacillus piliformis,* and the disease is associated with poor sanitation and stress. | No treatment is effective.<br>In severe outbreaks, complete elimination of the herd, followed by thorough cleaning and disinfection of the rabbitry, has allowed repopulation.<br>Do not allow dogs access to area where feed and nesting material are stored. |
| **Wry neck** | Head twisted to one side, animals roll over, cannot maintain equilibrium. | Caused by *P. multocida* infection of middle ear. | No effective treatment, afflicted rabbits should be culled. |

Fig. 55–24. Wry neck. (Courtesy, David J. Harris, Siloam Springs, AR)

## TYPES OF PRODUCTION

From a commercial utility standpoint, rabbits are produced for meat (fryers and roasters), wool, and rabbit skins.

## FRYER PRODUCTION

According to the regulations governing the grading and inspection of domestic rabbits, issued by the Department of Agriculture, "A fryer or young rabbit is a young domestic rabbit carcass weighing not less than 1½ pounds and rarely more than 3½ pounds processed from a rabbit usually less than 12 weeks of age."

Rabbits raised for meat and fur usually are marketed when they reach fryer weight even though the pelts are not prime. In order to yield a carcass weighing from 1½ to 3½ lb, young rabbits should have a live weight of approximately 3 to 6 lb.

Best carcass yields are usually from young rabbits weighing from 4 to 4¾ lb, when weaned at 2 months of age. These should yield a carcass (including liver and heart) of 50 to 59% of the liveweight, 78 to 80% of which is edible.

For fryer production, medium-weight to heavyweight breeds are preferred. Their young are most apt to develop to the desired weight and finish by the time they are 2 months old.

A pound of marketable fryer rabbit will require 2½ to 3½ lb of feed, or a total of approximately 100 lb for a doe and litter of 8, from mating of the doe to marketing of the young when 2 months old. Good does nurse their litters 6 to 8 weeks. The young develop more rapidly if they are in the hutch with their mothers until they are 8 weeks of age. By that time, the milk supply will have decreased, the young will be accustomed to consuming other feed, and weaning will be less of a shock than if undertaken at an earlier age. Young that are weaned and held for several days before market may either fail to gain or actually lose weight. Therefore, it is usually best to leave the young with their mothers until they go to market.

## ROASTER PRODUCTION

According to the regulations governing the grading and inspection of domestic rabbits, issued by the Department of Agriculture, "A roaster or mature rabbit is a mature or old domestic rabbit carcass of any weight, but usually over 4 pounds processed from a rabbit usually 8 months of age or older."

Roasters consist of (1) rabbits that are developed to heavier weights than fryers, and (2) culls from the breeding herd that are in good condition. Such rabbits should yield a carcass that is 55 to 65% of the liveweight, with 87 to 90% of it edible. However, the quantity of feed required to produce a pound of gain, liveweight, increases with each pound of gain, and may amount to 12 to 14 lb to increase the liveweight from 9 to 10 lb. Therefore, the cost of feed required to produce these gains must be assessed against the value of the heavier rabbits. Unless a premium is paid for mature rabbits for their meat or better fur quality, it is doubtful if such production would be more profitable than that of rabbits of fryer weight.

## ANGORA RABBIT WOOL PRODUCTION

Fig. 55–25. Angora rabbits in China, used for wool production. The better types of Angoras will produce from 2 to 2½ in. of wool in approximately 11 weeks, and an annual growth of 8–10 in. weighing 12–16 oz. When fed properly balanced rations, 100 lb of feed will, on the average, produce 1 lb of wool. (Photo by A. H. Ensminger)

Angora rabbits are raised primarily for wool production. Wool on Angoras grows to a length of 2½ to 3½ in. each 3 months, or approximately 1 in. per month. A mature Angora that is not nursing young will shear 14 to 15 oz of wool a year. This wool is valued for its softness, warmth, and strength. It is used in blends with other fibers in the manufacture of children's clothing, sport clothes, garment trimmings, and clothes for general wear. When used alone, it is usually too light and fluffy; blends create better tensile strength and durability.

There are two main types of Angora rabbits—the English and the French. Present standards of the American Rabbit Breeders Association, however, make English and French types of wool synonymous. Usually, the typical French Angora is larger than the English. The wool fiber of the French is shorter and coarser than that of the English, but the wool yield is greater. Owing to competition with other fibers, both natural and synthetic, and competition with imported Angora rabbit wool, the market price is generally low; hence, it is advisable to use the Angora as a dual purpose animal for both meat and wool production. The commercial Angora weighs at least 8 lb and is being bred more and more to improve its quality for meat.

Wool should be harvested prior to breeding to prevent rough treatment and soiling of the wool.

Angoras are generally sheared or plucked every 10 to 11 weeks, though some producers pluck their animals monthly, and still others at intervals beyond 3 months.

The breeding, feeding, care and management of Angoras is similar to that of the other rabbit breeds.

## RABBIT SKINS

Domestic rabbit skins vary greatly in density and quality, depending on the degree of care that breeders take in breeding and in curing. Good fur can be produced on efficient meat-producing animals by selective mating. Better skins command higher prices.

White skins bring higher prices than colored skins because of the adaptability for use in the lighter shades of garments and hats.

All rabbit skins have some value in the fur trade. About 85% of domestic rabbit skins are from rabbits 8 to 10 weeks old. These skins are known in the trade as *fryer skins*. They are usually sold by the pound as butcher run, that is, ungraded. Five or six fryer skins usually weigh a pound. In full-fed rabbits weighing 4 to 12 lb, the poorest skins come from animals up to 134 days old. Older animals produce a higher percentage of better grade skins. The better grade skins from older domestic rabbits are usually sold by the piece, primarily because they are larger than fryer skins.

Raw-fur buyers usually grade rabbit skins as first, seconds, thirds, and hatters.

First are prime pelts that are large, properly shaped, and properly dried. They are used for making garments.

Fryer skins contain only a small percentage of fur usable for garments primarily because of shedding or molting marks and secondarily because of thin fur and leather.

Seconds are pelts that have shorter fur and less underfur than firsts. The unprime colored skin shows dark pigment spots or streaks and, sometimes, large black splotches on the leather side. These markings do not show on white skins since pigment is lacking. Seconds also include pelts that are improperly shaped and dried, have been damaged in shipment, or show poor spots where the skin has been pierced or the fur is short or missing.

Thirds are pelts with short fur and thin underfur and those from animals too young or those that are shedding. Thirds are of no value to the furriers. They are used in the manufacture of toys, specialty articles, and felt for hats.

All skins that do not meet requirements of the other grades are *hatters*. Pelts that are badly cut or otherwise mutilated, or poorly stretched and dried, also are classed as hatters. The underfur of such pelts is used in making felt. Since the denser skins yield more cut fur, the hat trade pays more for them.

The distribution of domestic rabbit skins into these several grades depends on the demand for each kind. The market may be such that practically all the rabbit skins at a given time will be sold as hatters. Under some conditions, there may be but little demand even in the hat trade.

## FITTING AND SHOWING RABBITS

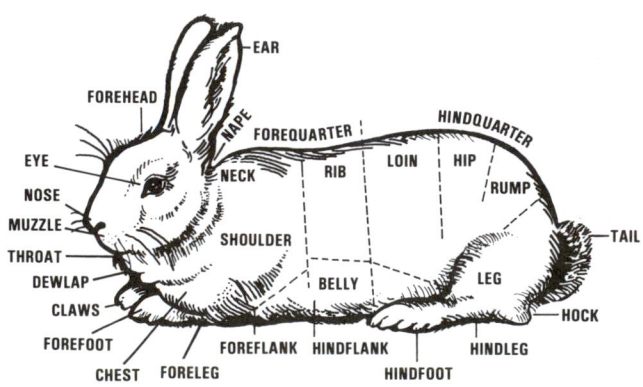

Fig. 55–26.   Parts of a rabbit.

For many rabbit producers, exhibiting rabbits in fairs and shows is an appealing adjunct to the overall project. Because of the contacts with individuals with a similar interest, the spirit of competition, and the opportunity to see high quality stock, exhibiting in shows often becomes a way of life to the rabbit producer. The 4-H member, the hobbyist, the breeder, and the commercial rabbit raiser can, through entering their rab-

bits in fairs and shows, see how their rabbits compare with those from other producers. These shows will range from local or county fairs, to state fairs, to breed association shows, to the national shows sponsored by the American Rabbit Breeders Association.

Breed and variety characteristics, condition of the fur, health of individual, and the ability of the animal to show off for the judge are qualities which are necessary for high placings in any show. The individual rabbit must exhibit breed type and variety markings according to the *Standard of Perfection*. For this reason, exhibitors must be quite selective in the animals that they prepare for exhibition. Those rabbits that fit the breed specifications should also be clean, tame, healthy, and possess fur and flesh that is in good condition; otherwise, it is a waste of time and money to exhibit them. Most fairs and shows have breed, age, and sex groupings.

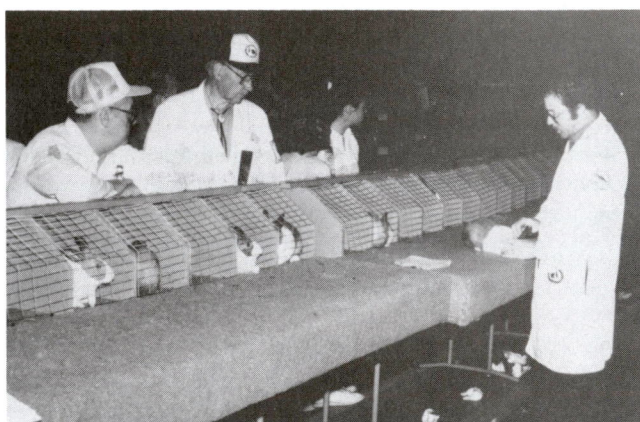

Fig. 55–27.   A class of rabbits being judged at an American Rabbit Breeders Assn. (ARBA) National convention. (Courtesy, David J. Harris, Siloam Springs, AR)

Rabbit producers who plan to exhibit should identify their animals by tattooing. Not only does this serve as identification for age verification, but it also can alleviate ownership problems among animals of similar appearance. Ear notching, tags, or similar identification methods are not desirable as they detract from the rabbit's appearance on the show bench, and may cause the rabbit to be disqualified.

Rabbits that are to be shown should be handled frequently so that they will show off to advantage on the show bench. The judge must be able to pose the various rabbits for best comparison. A rabbit that is nervous will not receive maximum consideration in spite of adherence to breed characteristics.

## QUESTIONS FOR STUDY AND DISCUSSION

1. Brief the origin and history of domestic rabbits. How may domestic rabbits be distinguished from hares?

2. List and describe the 6 purposes for which rabbits are raised.

3. Describe rabbit production in the United States.

4. Describe each of the following types of buildings and equipment in a modern rabbitry: (a) housing, (b) hutches, (c) nest boxes, (d) feeding equipment, and (e) watering equipment.

5. Name and give the color and ideal mature weight of (a) 4 meat-type breeds of rabbits, (b) 2 pelt-type breeds, and (c) 1 wool-type breed.

6. List and brief 5 major points which should be considered when selecting rabbit breeding stock.

7. List and describe 3 systems of breeding rabbits.

8. Describe the common mating methods of rabbits. Would more artificial insemination of rabbits be advantageous?

9. Discuss each of the following: (a) seasonal effect on breeding rabbits, and (b) pregnancy check of does.

10. Describe (a) normal kindling, and (b) some complications at kindling time.

11. Describe the care of the litter, including (a) sexing, (b) culling, and (c) tattooing.

12. What is coprophagy? Explain the important role of coprophagy in modifying the nutritive requirements of rabbits.

13. List the common sources for rabbits of (a) energy, and (b) protein.

14. For rabbits, list 2 major minerals which may be deficient and for each of them give (a) the functions, (b) some deficiency symptoms, and (c) practical sources.

15. For rabbits, list 2 trace minerals which may be deficient and for each of them give (a) the functions, (b) some deficiency symptoms, and (c) practical sources.

16. For rabbits, list 3 fat-soluble vitamins which may be deficient and for each of them give (a) the functions, (b) some deficiency symptoms, and (c) practical sources.

17. For rabbits, list 3 water-soluble vitamins which may be deficient and for each of them give (a) the functions, (b) some deficiency symptoms, and (c) practical sources.

18. Discuss the use of the following types of feeds for rabbits: (a) legume hays, (b) green grass, and (c) roots and tubers.

19. How should feeds be prepared for rabbits?

20. When formulating feeds for rabbits, what consideration should be given to each of the following: (a) fat, (b) nitrogen-free extract, (c) crude protein, (d) ash, and (e) fiber?

21. What are the primary differences in feeding rabbits in each of the following categories: (a) dry does, (b) pregnant does, (c) lactating does, and (d) juniors?

22. How would you feed and care for an orphan litter of rabbits?

23. What are the advantages, and the disadvantages of each (a) hand-feeding, and (b) self-feeding?

24. Discuss each of the following as they apply to managing a herd of rabbits: (a) method of handling rabbits, (b) weaning, and (c) care of the herd during extreme temperatures.

25. Discuss each of the following as they apply to managing a herd of rabbits: (a) fur-eating habit, (b) disposal of rabbit manure, and (c) records.

26. In outline form, present a rabbit herd health program.

27. Name 4 common diseases of rabbits and for each of them give the (a) symptoms, (b) cause, and (c) treatment and control.

28. Name 2 common parasites of rabbits and for each of them give the (a) symptoms, (b) cause, and (c) treatment and control.

29. Describe each of the following types of rabbit production: (a) fryer production, (b) roaster production, (c) wool production and (d) rabbit-skin production.

30. Of what value are rabbit shows? How should rabbits be fitted and shown?

## SELECTED REFERENCES

| Title of Publication | Author(s) | Publisher |
|---|---|---|
| *Commercial Rabbit Raising*, Ag. Hdbk. No. 309 | R. B. Casady *et al.* | Agricultural Research Service, USDA, Washington, DC, 1971 |
| *Domestic Rabbit, The*, Third Edition | J. C. Sandford | Granada Publishing, Ltd., London, England, 1979 |
| *Nutrient Requirements of Rabbits*, 2nd Revised Edition | National Research Council | National Academy of Science, Washington, DC, 1977 |
| *Offical Guide to a Progressive Program for Raising Better Rabbits and Cavies* | Staff | The American Rabbit Breeders Association, Inc., Bloomington, IL, 1984 |
| *Rabbits* | H. Dyson | Cassell & Collier MacMillan Pub., London, England, 1975 |
| *Rabbit Feeding and Nutrition* | P. R. Cheeke | Academic Press, Orlando, FL, 1987 |
| *Rabbit Production* | P. R. Cheeke *et al.* | The Interstate Printers and Publishers, Inc., Danville, IL, 1987 |
| *Raising Rabbits Successfully* | B. Bennett | Williamson Publishing Co., Carlotte, VT, 1984 |
| *Raising Small Meat Animals* | V. M. Giammattei | The Interstate Printers & Publishers, Inc., Danville, IL, 1976 |

Also, information on rabbits may be obtained from the American Rabbit Breeders Association, 1925 S. Main, P.O. Box 426, Dept. H7, Bloomington, IL 61701.

# RABBITS— A PEOPLE'S BUSINESS

A typical commercial rabbitry, with a side curtain used to control ventilation. (Courtesy, Dr. P. R. Cheeke, Oregon State University Rabbit Research Center, Corvallis, OR)

Interior of a large commercial rabbitry. (Courtesy, Pel-Freez Rabbit Meat, Inc., Rogers, AR)

Jennifer Harris feeding rabbits. (Courtesy, Dr. D. J. Harris, Siloam Springs, AR)

A young lady and her pet rabbit. People need pets, and pets need people. (Courtesy, USDA)

Weighing a rabbit at a rabbit show. (Courtesy, Dr. D. J. Harris, Siloam Springs, AR)

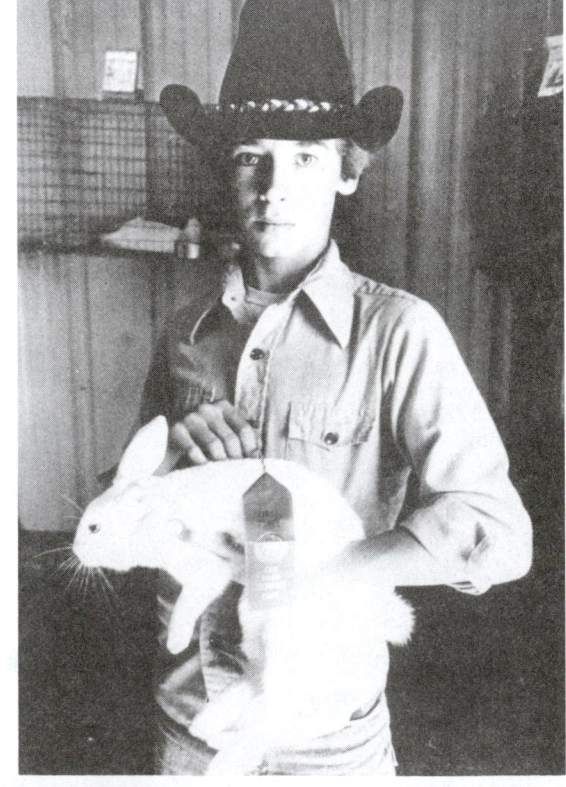

*Right:* 4-H Club member, with his blue-ribbon winner at the show. (Courtesy, USDA)

A large trout hatchery in Idaho. (Courtesy, C. E. Smith, Director, Fish Technology Center, U.S. Fish & Wildlife Service, Bozeman, MT)

# AQUACULTURE

## Chapter 56

Presently, seafood, fish, and shellfish contribute less than 1% of the world food supplies in terms of dietary energy, less than 5% of the total protein, and less than 14% of the animal protein. As health concerns and dieting increase throughout the world, more fish will be needed in human diets.

Seafood, fish, and shellfish originate from two primary sources: (1) The first and most important source is from natural or capture fisheries, and (2) the second source is aquaculture or farm raised.

Aquaculture is an old term. It is derived from the Latin word *aqua,* which means water, and from the word *culture,* which means to till, to cultivate, and to grow. So, *aquaculture is the controlled cultivation and harvest of aquatic animals and plants in water.* **Note well:** The definition includes the words *controlled cultivation.* It encompasses the farming of both ocean and freshwater plants and animals.

Aquaculture dates back to more than 4,000 years in China, Japan, and Egypt; to more than 3,000 years in India and Java; and to more than 2,500 years in Europe. By 2000 B.C., the Chinese were artificially hatching fish; and by the time of the Han Dynasty (201 B.C. to 190 A.D.), they were raising pond fish in commercial numbers.

Fig. 56–1. Carp have been harvested in China for 4,000 years. (Photo by A. H. Ensminger)

These early aquaculturalists recognized that fish could be farmed, just as hunters had previously come to realize that certain land animals could be domesticated. So, they constructed ponds and stocked them with fish. Legal documents of the time attest to the importance of these food reservoirs.

# WORLD AQUACULTURE

The world demand for fish and shellfish has been increasing steadily over the past few years, but the ocean catches have not been keeping up with demand. The Food and Agriculture Organization of the United Nations (FAO) estimates that world markets will fall 20% short of demand for fish and shell products by the year 2000. To meet this deficit, many countries are moving ahead in developing their aquaculture. But the concern is that their moves may be too little and too late.

Fig. 56–2. National Fisheries Research Center in Hungary—an FAO project. (Courtesy, Dr. J. E. Halver, School of Fisheries, University of Washington, Seattle)

Although aquaculture has been practiced throughout the world for many years, significant growth has occurred only since about 1975; with the increase attributed to (1) the capture fisheries not being able to expand, because commercial fishing has already reached its maximum sustainable yield; (2) increase in world population; (3) increased seafood consumption; and (4) the application of new technologies in aquaculture facilities.

The worldwide production of farmed, cage-reared, primarily Atlantic salmon, is the fastest growing segment of world aquaculture. The growth rate has been about 40% since 1980. By 1990, farm salmon accounted for more than 20% of the world's salmon supply. The leading countries in this industry, by rank, have been Norway, Japan, and Canada; with the three major markets being Japan, the United States, and European Common Market (EC).

The world's shrimp farmers also produced spectacularly in the 1980s. By 1990, farmed shrimp accounted for more than 25% of the total world shrimp market. The primary market for the world shrimp crop is the United States, Japan, and the EC.

Aquaculture will continue to be a growth industry, worldwide; (1) producing wholesome and healthful food for a growing population; (2) creating economic development; (3) reducing trade deficits, (4) enhancing a growing recreation industry; and (5) contributing to a sustainable agriculture.

## UNITED STATES AQUACULTURE

The aquaculture industry is the fastest growing segment of U.S. agriculture, increasing at a rate of 20% per year throughout the 1980s. The primary species contributing to this phenomenal growth include catfish, trout, crayfish, salmon, tilapia, and striped bass.

Despite this remarkable growth, aquaculture production accounts for only about 11% of the total supply of U.S. edible fishery products, which suggests that the growth potential of the U.S. aquaculture industry is substantial.

The catfish industry is the most important success story of the U.S. aquaculture industry. It has grown from a total of 3 million pounds processed in 1969 to more than 295 million pounds processed in 1988, valued at $225 million (consumption from live hauling fee fishing and direct purchases are not included in these catfish processing figures).

Fig. 56-3. Feeding channel catfish with a mechanical blower. (Courtesy, Dr. D. Gatlin, Dept. of Wildlife and Fisheries Sciences, Texas A&M University, College Station)

It is expected that U.S. fish farming will continue to grow, but at a slower pace in the years ahead. Catfish are efficient feed converters; besides, the feeds that they consume are not generally used for human consumption.

## TYPES OF AQUACULTURE

Aquaculture is of two types: (1) freshwater, or (2) saltwater (or marine water). Freshwater fish live in the lakes, streams, and ponds—on land, while saltwater fish live in the oceans that cover 70% of the earth's surface. Within these two types of

production the following four systems of management are practiced:

1. **Hatchery operation.** Young fish are hatched and raised to a size where they can be released into natural populations to grow and reproduce.

Fig. 56-4. Trout hatch at Hardy Hatchery, ID. (Courtesy, C. E. Smith, Director, Fish Technology Center, U.S. Fish & Wildlife Service, Bozeman, MT)

2. **Capture of young fish.** In this system, young fish are captured and grown to market weight on supplemental feeding or on natural feeds in the environment aided by active fertilization of the water.

3. **Growing of young fish to market weight.** In this system, the farmer is only concerned with growing fish and does not attempt to select breeding stock. Shrimp farming is an example of this system.

4. **Management of entire life cycle.** Catfish and trout production in the United States are examples of this system. Young fish are hatched and grown either to be marketed or kept as replacement breeder stock.

Freshwater fish production can be divided into two classifications—coldwater culture (40 to 60°F); and warmwater culture (70 to 100°F). In the United States, trout and Pacific salmon represent the bulk of coldwater fish production, while channel catfish is the predominant warmwater species. In other parts of the world, the carps are the major warmwater fish.

## SPECIES AND SPECIALIZED AQUACULTURE

U.S. aquaculture made a difference in the 1980s! In 1986, approximately 620 million pounds of aquaculture products were produced by private operators. The main contributors that year were catfish (327 million pounds), crawfish (98 million pounds), salmon (75 million pounds), and trout (5 million pounds). In addition to food fish, other aquaculture industries blossomed, including bait fish (mainly gold shiners, fathead minnows, and goldfish), hobby (ornamental) fishes, tropical fishes, and others.

Hand in hand with the growth of aquaculture, the industry expanded in area, method, and technology. In addition to traditional ponds, it included flowing water systems (called raceways), and enclosed systems ranging from rafts and cages, both floating and submerged, to closing off and farming fiords.

Research and development have resulted in improved genetic strains of fish, feed and nutrition, water quality management, disease control, and harvesting, processing, and marketing. Also, innovative management techniques such as sea ranching, double cropping, and polyculture evolved.

• **Sea ranching**—This method, which involves releasing hatchery-reared animals of various sizes into marine waters for rearing and the subsequent recapture of the adult fish upon their return to the point of release, is creating considerable interest.

Fig. 56-5. Coldwater species, trout and salmon, being produced in Scandinavia in sea pens equipped with automated, pneumatic feeding systems. (Courtesy, International Aquaculture Research Center, Hagerman, ID)

• **Double cropping**—This development involves the raising of channel catfish and rainbow trout in the same facility at different times of the year. Trout fingerlings are stocked in the fall (when water temperature drops to 70°F), then marketed in the spring, at 120 days or less. After the trout are harvested, channel catfish are stocked in the spring, then fed out and marketed in the fall.

• **Polyculture**—*Polyculture is the stocking of different species together.* Channel catfish and crayfish continue to be the major warmwater species farmed together, but polyculture involving carp (Chinese and common), tilapias, and buffaloes is increasing.

Today, there are many choices of species in aquaculture. Additionally, there are opportunities to specialize within a species; for example, among three trout farmers, one may produce eggs, the second may produce fingerlings, and the third may produce market fish.

## MAJOR AQUACULTURE SPECIES

A pertinent *brief* relative to each of the major aquaculture species follows:

### CHANNEL CATFISH

Catfish farming makes a significant contribution to the economies of many southern states, especially Mississippi (accounting for almost 80% of the aquaculture production of the state), Arkansas, Alabama, and Louisiana. The channel catfish is preferred over other species of catfish because of its tolerance for handling, good feed conversion, high dress-out percentage, and ease of spawning. Flat bottomed ponds filled by well water are the ideal situation for catfish production, but channel catfish adapt readily to raceways, cages, and tanks.

### TROUT AND SALMON

Culture of trout is a major form of aquaculture in the United States. The leading states in numbers of commercial growers are Idaho, Wisconsin, Colorado, Michigan, Pennsylvania, and North Carolina. Salmon are cultured for food in three states: Washington, Oregon, and Maine.

The outlook for trout farming is good. Interest in and public demand for trout fishing is projected to increase. Also, some water resources are still available for expansion in the leading trout states listed above.

The future expansion of the cultured salmon industry will be greatly influenced by the legality of sea ranching, the number of permits granted for pond and pen culture, and market prices of culture salmon.

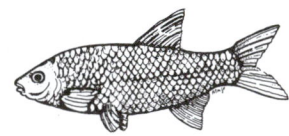

### BAIT FISH

At least 20 species of fish have been raised commercially for bait. Gold shiners, fathead minnows, and goldfish are among the most popular bait species, and all of them can be produced reliably in ponds. These species reproduce naturally and grow well on inexpensive feeds. A series of flat bottomed ponds that can be drained, seined, and refilled are needed so buyers can be supplied on a regular basis.

The commercial production of bait fish for use as bait in sport fishing is a multimillion dollar industry. The acreage devoted to minnow farming almost equals that used for channel cat fish farming.

# MAJOR AQUACULTURE SPECIES

Fig. 56–6.  Channel catfish (*Ictalurus punctatus*). (Courtesy, Dr. D. Gatlin, Dept. of Wildlife and Fisheries Sciences, Texas A&M University, College Station)

Fig. 56–7.  Brook trout. (Courtesy, C. E. Smith, Director, Fish Technology Center, U.S. Fish & Wildlife Service, Bozeman, MT)

Fig. 56–8.  Golden shiner (*Notemigonus crysoleucas*), a common baitfish. (Courtesy, Dr. D. Gatlin, Dept. of Wildlife and Fisheries Sciences, Texas A&M University, College Station)

## MINOR AQUACULTURE SPECIES

In addition to the major aquaculture species listed in the preceding sections, a *brief* relative to each of the less populous aquaculture species in America is presented in alphabetical order in the sections that follow.

### ALLIGATORS

The alligator (*Alligator mississippiensis*) is a large, aquatic reptile highly valued for its hide and meat. It was once abundant in the deep South, before overhunting and habitat destruction reduced its numbers throughout the region. However, extensive conservation efforts have restored alligators to pre-exploitation levels wherever suitable habitats remain, principally in southern Louisiana and south-central Florida.

The conservation effort led to the development of alligator culture techniques. Several commercial alligator farms are now being operated in Louisiana and Florida. The demand for alligator products is great and prices are high, with the result that alligator farming can be very lucrative.

A prospective alligator farmer can raise alligators extensively in fenced outdoor earthen ponds or in indoor sophisticated incubation and rearing containers. An alligator grows to a harvestable length of 5 to 6 ft in a period of 26 months during which temperatures range between 80 and 90°F. Inasmuch as growth stops when temperatures fall below 60°F, the effective growing season is 5 to 6 months per year in most of the South unless supplemental heating is provided. In unheated farms, 48 to 52 calendar months are required to raise alligators of marketable size.

Domesticated alligators should be used for brood stock. Wild alligators, even when available, are very aggressive, and no more than one adult male and three adult females can be stocked in a 4-acre brood enclosure. In contrast, as many as 75 domesticated female alligators 6 to 7 ft long, and 25 males 8 to 9 ft long, can be stocked in such an enclosure.

Brood enclosures can be about 4 acres in area and must be built of sturdy materials to prevent escape. The most practical and economical materials are welded wire and treated posts; the fences should be 6 ft high, and the mesh size either 2 × 4 in. or 1 × 2 in. To prevent escape beneath the fence, 1- by 6-in. treated boards should be installed underground along the base of the entire fence line. A poultry wire retainer must be installed at the top of the fence and angled inward to prevent escape of wild alligators, which are agile climbers. A 4-acre enclosure should contain a 1½-acre pond 4 to 6 ft deep.

Alligators are fed ground fish when they are newly hatched, chopped fish as they grow larger, and finally whole fish.

Although alligators fed fish have been shown to grow well, there is evidence that reproductive success is reduced unless the diet is changed to include meat products from warm-blooded mammals.

Breeding activity begins when air temperatures approach 70°F in spring. Bellowing and courtship displays are conspicuous activities. The females build nests from sedges, rushes, grasses, and mud, although some use only mud. Vegetation must be provided for nest-building if little is present in the breeding pen. Usually a 10- to 11-year-old domesticated alligator 7½ to 8½ ft long lays 38 to 40 eggs in a single laying.

It is advisable to remove alligator eggs from nests soon after all are laid, and to incubate them artificially.

Most prospective alligator farmers think of alligators in terms of hides; however, the demand for alligator meat is great, and it can be as valuable as the hides. Body parts such as teeth and skull can also be sold as curios.

### AQUARIUM FISH (HOBBY FISH/TROPICAL FISH)

This group includes representatives of several families, and over 100 species of small, colorful, and unique fishes.

Although aquarium fish are not cultured for food, they do represent a major segment of the pet industry. Also, they are used extensively in research; for example, in water quality testing. There are many commercial feeds for aquarium fish, in a variety of shapes and formulations. Most of these feeds are nutritionally adequate; but they are closed formulas and are subject to periodic change in composition, thereby limiting their use for experiments. Expensive carotenoids and other pigments are commonly included in the feed to promote bright coloration of the fish.

### BASS, STRIPED/HYBRID

Demand for striped bass and hybrid striped bass is high, especially on the east coast where the wild catch is declining and pollution problems worry consumers. The cross between white and striped bass has proven to be an excellent fish for pond and tank culture.

They do not reproduce well in reservoirs because the eggs and larvae require long stretches of flowing water; so, continuous restocking is necessary.

Bass add an element of size to fisheries because of their large size; and are highly prized by sportsmen. Striped bass may weigh up to 40 to 60 lb, or more. Hybrid bass weigh up to 25 lb.

Hybrid bass are hardier and faster growing than striped bass, but their life span is shorter—only 8 to 10 years.

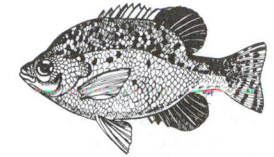

## BLUEGILL

The bluegill is the best known member of the sunfish family, originally ranging from the Mississippi, Great Lakes, and Eastern Seaboard drainage basin. However, because of its adaptability, the bluegill has been widely introduced across the United States.

The bluegill has a deep, compressed head and a very small mouth. Its ear flap is broad, short, and colored blue or black. A distinct, irregular dark spot is present on the posterior of the dorsal fin. Colors may vary with sex and water quality, but normally several vertical bars are visible along the sides. The bluegill is widely stocked as both a sport and forage fish.

Bluegill mature at lengths of 3 in. with females capable of laying between 10,000 and 60,000 eggs per spawn. Eggs hatch in 3 to 5 days depending on water temperature. No parental care is provided after hatching.

Young bluegill feed primarily on plankton, switching to a diet of insects and other small aquatic life as their size increases. Bluegill also feed on the eggs and fry of other fish species.

The bluegill is a sunfish preferring static, clear waters of ponds, reservoirs, and sluggish streams. It seldom strays far from shore and prefers structures such as weed beds, fallen timber, and pilings.

## BULLFROGS

Frog legs are gourmet items on the menus of many fine restaurants. Most come from the wild. However, the demand is greater than the supply, and availability is seasonal. Managed culture or production of bullfrogs could make the supply of frog legs more predictable and possibly lower the cost to the consumer. Then, too, there is a substantial market for live frogs for use in biological research.

Because of the high sale price for frog legs and live frogs, unscrupulous persons have tried to convince prospective producers that bullfrog culture is a lucrative business. Generally, it has not been possible to culture bullfrogs commercially because they have a complicated life cycle that includes completely aquatic, gill-bearing tadpoles and semi-aquatic, lung breathing juveniles and adults. Also, adult frogs are reported to consume food only if it is moving. This has led to some extremely innovative and labor intensive methods, including the dangling of food pellets from horsehair. Although bullfrogs breed readily in shallow, vegetation-choked ponds, controlled laboratory breeding and rearing have been achieved only within the past few years. Bullfrog culture is now technically feasible, but profitability is questionable.

The Japanese and Taiwanese practice open-pond culture of bullfrogs from eggs to adults, but their techniques have never been applied successfully in the United States.

Two markets are available for domestically produced frogs. Small, ¼-lb frogs, about 4 to 5 months post-metamorphosis, command favorable prices in the research market (e.g., biological supply houses). Larger ½-lb frogs, 6 to 10 months post metamorphosis, may be sold for food, but prices are not now high enough to make this market as attractive as the market for live frogs.

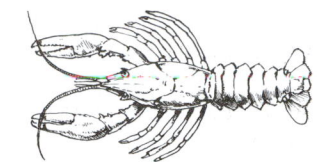

## CRAYFISH (CRAWFISH)

Large crayfish are prized as gourmet food, while small ones are in demand for bait. About 500 species are found throughout the world, and more than 250 species are found in the United States. Red swamp crawfish are the traditional cultured species in the United States.

Louisiana is the leading U.S. producer of crayfish, with approximately 125,000 acres devoted thereto, which is 90% of the nation's crawfish acreage. About 8% of the U.S. crayfish acreage is in Texas, and the remaining 2% is in the states of Mississippi, South Carolina, and Arkansas. Annual harvest range from 500 to 2,000 lb per acre. Some crayfish are farmed on rice land as a secondary crop.

Feeds for crayfish include natural food and agricultural by-products. The feeding of specific formulations manufactured for crayfish is neither necessary nor economically sound.

The simplest type of culture involves the trapping of crayfish from marginal wetlands, usually adjacent to crop or timber lands. However, higher levels of production are achieved by the construction and management of ponds specifically designed for crayfish.

## EEL (ELVERS)

Eel is considered a gourmet food in Japan, Taiwan, and in most European countries; and it commands high prices.

In various countries around the world, extensive production of eel in brackish-water ponds and inlets is being conducted.

Almost all eels produced in Japan are from pond units, and the techniques of production are becoming more and more refined. In the past, the natural feeds were supplemented only by fresh fish diets. As a result, feed conversion rates were extremely poor (5 to 15:1). Today, modern Japanese diets make for conversion rates of 1.2 to 1.6:1 using a paste feed, and 1.0 to 1.6:1 using a dry pellet. The diets generally range from 20 to 30% protein.

American eel culturists have been left to the mercy of unpredictable overseas markets because there are virtually no domestic sales in North America. Each year, 2,000 to 3,000 metric tons of wild eel are caught, primarily in Virginia, Maryland, and South Carolina, and shipped to Japan.

## LOBSTER

Lobsters are large hard-shelled marine crustaceans that are closely related to crabs, crayfish, and shrimp. The American Lobster, *Homarus americanus,* which is perhaps the largest lobster, lives along the Atlantic coast from Labrador to Virginia; it may reach a length of 3 ft and a weight of 44 lb. Most European lobsters are smaller.

Lobsters live on the bottom of the ocean near the shore and hide in holes or under rocks. A lobster is a predator that sits in its burrow all day, waving its feelers outside the entrance and holding its claws ready to capture any prey that comes near. Lobsters eat crabs, snails, small fish, and even other lobsters. At night, the lobster walks the ocean bottom seeking food.

Compared to farm animal and poultry nutrition, and even fish nutrition, lobster nutrition is in its infancy. Until recently, lobsters required natural food diets to grow and survive—such foods as crab, shrimp, or fish were used.

Attempts to substitute cereal-based or purified diets for these natural foods have usually resulted in poor growth and low survival. However, with the development of a successful purified diet in 1980 by researchers at the Bodega Marine Laboratory of the University of California, nutritional requirements can now be determined and thereby provide knowledge of diets from readily available feedstuffs. Although the commercial production of lobsters is not yet feasible, research continues.

## PADDLEFISH

Interest in paddlefish or ''spoonbill'' culture is mainly due to prices of $30 or more per pound for its eggs, which are substituted for Iranian and Russian caviar. The flesh has a small value at present, but is very fine for smoking.

Paddlefish culture is experimental so the level of investment should be kept low. Approximately 10 years are required before a female will produce eggs. Paddlefish feed by filtering very small animal life from the water. A production system that appears promising is to culture paddlefish in catfish ponds for 2 to 3 years and then stock them in large ponds or lakes for later capture by commercial fishing techniques. There are a few paddlefish fingerling producers, but production is not reliable. Paddlefish fingerlings are reported to be very susceptible to birds due to their habit of swimming near the surface.

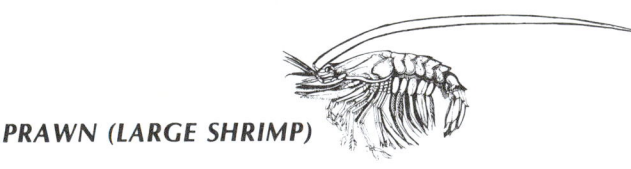

## PRAWN (LARGE SHRIMP)

Prawns and other crustaceans are in heavy demand in the United States and throughout much of the world.

Because larvae must spend some time in brackish water, seed stocks must be specially handled before being made available to the freshwater culturist. The requirement of water temperatures of 75° to 95°F for prawn culture greatly limits the length of the growing season in the United States.

Feeding need not begin until the small prawns reach a length of 1 to 1½ in.—usually in 30 to 40 days. Special shrimp rations or the cheaper sinking catfish feeds, offered at about 3% of the body weight of the prawns, are suitable.

Prawn farmers in the United States usually purchase their seed stock from growers in tropical areas. The larvae are unique in that they must live through several moltings in brackish water before they can be stocked, as post larvae, into fresh water for culture.

Prawns may be cultured with fish. The major problem with polyculture of prawns and fish is that prawns must be harvested by early fall and hand-sorted.

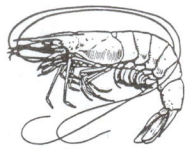

## SHRIMP

*Shrimp is the common name applied to about 2,000 species of crustaceans, whereas the term prawn may refer to any large shrimp.* Shrimp, which have a very high market value, are shellfish and are closely related to crabs, crayfish, and lobsters.

Shrimps are predominantly marine, but several families live in fresh or brackish waters, and one species that lives in swamps in South America is semiterrestrial. Most are bottom dwellers, some even burrow into the sea bottom; others swim in the open sea. Shrimps are filter feeders and scavengers. They commonly feed on small animal or plant organisms. Several genera form the basis of a major fishery, and some shrimps are commercially raised, with Japan leading in number of shrimp farms and volume of commercial shrimp production.

Because shrimp are slow feeders, feeds that remain stable in water for several hours must be used. The pellets should be small (0.08 to 0.16 in.). Shrimp farmers who culture shrimp in large tanks or small earthen ponds generally feed a mixture containing 40% dry pellets (formulations similar to broiler rations), 40% shrimp waste, and 20% clam waste.

Shrimp are a luxury food for which Americans seem to have an ever-growing hunger. To satisfy this hunger, major corporations and private investors have built shrimp farms in Central America, South America, and Asia, where the tropical climate is favorable.

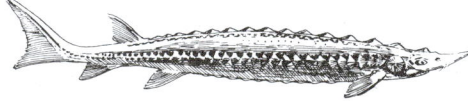

## STURGEON

The Russians have done considerable research on the production of sturgeon. Both wild and cultured fish are used in caviar production. Dry diets using fish and animal meals, distillers' solubles, yeast, and vitamins have been developed and are being used extensively.

## TILAPIA

Originally found in Africa and the Near East, many different species of tilapia have been introduced around the world because of their suitability for aquaculture.

Tilapia are fast growing, rapid-reproducing, and easy-to-manage fish. Since these omnivorous fish are extremely sensitive to cold temperature, tilapia culture is limited to tropical climates. They were originally introduced to the southern region of the United States as a means of controlling aquatic vegetation in lakes and ponds. They are still useful in this aspect of pond management. Today, they are also being produced in limited numbers as commercial food fish and sport fish.

Research has indicated that tilapia utilize many plant ingredients to about the same degree as trout utilize animal products. Thus, a program combining pond fertilization with the feeding of a dry diet and chopped leaves and grasses should provide a suitable ration for culturing tilapia.

# MINOR AQUACULTURE
# SPECIES

Fig. 56-9.   One-year-old alligators in incubation. (Courtesy, Dr. H. K. Dupree, Director, Fish Farming Experimental Laboratory, U.S. Dept. of Interior, Stuttgart, AR)

Fig. 56-10.   Aquarium fish. (Courtesy, *1984 Yearbook of Agriculture Animal Health Livestock and Pets*, USDA, p. VI)

Fig. 56-11.   Hybrid striped bass (*Morone saxatilis* × *M. chrysops.*) (Courtesy, Dr. D. Gatlin, Dept. of Wildlife and Fisheries Sciences, Texas A&M University, College Station)

Fig. 56-12.   Bullfrog. (Courtesy, Dr. H. K. Dupree, Director, Fish Farming Experimental Laboratory, U.S. Dept. of Interior, Stuttgart, AR)

Fig. 56-13.   Marketable cultured eel. (Courtesy, Dr. H. K. Dupree, Director, Fish Farming Experimental Laboratory, U.S. Dept. of Interior, Stuttgart, AR)

Fig. 56-14.   Blue tilapia (*Oreochromis aureua*). (Courtesy, Dr. D. Gatlin, Dept. of Wildlife and Fisheries Sciences, Texas A&M University, College Station)

## BREEDING FISH[1]

Substantial gains have been made in livestock production through selective breeding and hybridization. Improved cattle, sheep, swine, horses, and poultry are available to farmers worldwide. In fish farming, most fish stocks have undergone selection only for cultural traits that increase survival, such as tolerance to adverse water conditions and resistance to disease. Several species, primarily the rainbow trout, koi carp, goldfish, tropical aquarium fishes, and the common carp, have undergone some selective breeding for production or for ornamental traits.

All channel catfish now being cultured have been domesticated for intensive production on commercial fish farms and in government fish hatcheries. This does not mean that further improvement cannot be made in feed conversion efficiency, growth rate, and dress-out percentage. Techniques used to accomplish these goals include strain evaluation, selective breeding, crossbreeding, and hybridization. Strain evaluation involves determining the characteristics of various fish strains so that the appropriate strain can be matched with each production system. Long-term selective breeding enhances desired characteristics, thus developing a high-performance fish for each production system and geographical area. Short-term crossbreeding involves mating unrelated strains of the same species to enhance production or survival characteristics, or to avoid inbreeding. Hybridization involves the crossing of different species such as blue catfish and channel catfish to enhance desirable production or survival characteristics found in these species.

Unfortunately, the commercial fish industry has not had available selectively improved seed stock similar to that being used in other animal fields. Breeding programs, which require substantial commitments in personnel, funds, and facilities, have not been available to fish geneticists. Additionally, a complicating factor lies in the biology of the fish: it is difficult to observe because it lives in water and it usually requires roughly three years to produce a new generation.

Despite these impediments, some fish selection studies are in progress. These studies have led to several recommendations to the fish industry which follow.

1. **Avoidance of inbreeding.** Inbreeding, which involves the mating of closely related individuals, results in depressed growth and reduced survival. The selection of the largest individuals in a pond for brood fish may result in brother-sister matings, since these fish may have been the product of a single spawn. Keeping the same brood fish for 4 to 10 years, with replacement brood fish coming from the progeny produced on the same farm, may result in father-daughter and mother-son matings. Brood fish purchased from another producer may also contain the same narrow genetic base. Inbreeding can be avoided or eliminated by applying the following techniques:

a. **Enrichment of bloodlines.** The addition of unrelated stock may be effective in correcting deterioration in quality of brood stock common to inbreeding and disorderly selection. The need to enrich bloodlines might be suspected if a high percentage of deformed progeny, low hatchability of eggs, low survival of fry, or poor growth become evident.

b. **Crossbreeding of unrelated stocks.** Stocks originating from different river systems and different commercial lines are usually unrelated, and combining them may yield a desired combination of traits, especially in resistance to disease. If a hatchery stock shows susceptibility to disease (e.g., to channel catfish virus disease), crossbreeding should be tried. Branding enables the fish farmer to keep two or more stocks at the same facility at the same time. These stocks may then be outbred to correct the negative effects of inbreeding and to produce fingerlings of uniform performance year after year.

2. **Selection of brood fish.** It should be possible to improve the performance of fish by selecting the largest fish and saving these for breeding. However, several cautions must be noted and several pitfalls avoided. The fish must be of identical age. When fish are of unequal age, the oldest are usually the largest, and if they came from a single spawn, these largest fish would be brothers and sisters. They should be sexed to ensure a roughly equal number of males and females. Inasmuch as males usually are larger than females of equal age, the producer could be saving all or nearly all males for brood fish. Consequently, the males and females should be selected separately. That is, the largest males and the largest females should be selected—not just the largest fish. These selections should be made when the fish are 18 to 24 months old, and have reached marketable size.

## HYBRIDIZATION

The offspring of two different species may possess hybrid vigor, such as fast growth, high dress-out percentage, good tolerance to adverse water conditions, and increased resistance to disease. Hybrid crosses have been produced among channel catfish, blue catfish, white catfish, flathead catfish, brown bullheads, black bullheads, and yellow bullheads. Other crosses have resulted from pairing a hybrid to a species not used to produce the hybrid (out-crossing) and the crossing of the hybrid with one of the parental species (back-crossing).

Some hybrid crosses can be obtained by pairing the fish in pens or aquaria, and encouraging of spawning through special handling, use of hormones, and alterations of the environment. Other hybrid crosses have been obtained by fertilizing the hormone-induced ovulated eggs with surgically excised testicular tissue. Either method complicates fingerling production when compared with the techniques used to produce true species.

Some crosses—specifically the channel catfish × blue catfish hybrid—grow faster than either parent species. The major limitation to the use of this hybrid in commercial aquaculture appears to be the lack of sufficient seed stock at an affordable price.

## GENETICALLY ENGINEERED FISH

Genetic engineering wizardry is now being applied experimentally to fish. A transgenic carp (so-called because it contained genes of another species) became a reality on June 26, 1989, when it was released to swim in an experimental pond in Alabama. The stated objective: To improve rate and efficiency of gains. Success of the experiment could ultimately help overcome world shortage of fish and lead to improved nutrition around the world. Earlier, the Alabama Experiment Station workers succeeded in inserting rainbow trout, coho salmon, and human growth hormone into channel catfish in efforts to improve growth and efficiency of that commercial fish.

---

[1]The section on "Breeding Fish" was adapted by the author from *Third Report to the Fish Farmers*, edited by Harry K. Dupree and Jay V. Huner, published by U.S. Department of the Interior, Fish and Wildlife Service, 1984, pp. 69–72.

# BREEDING FISH

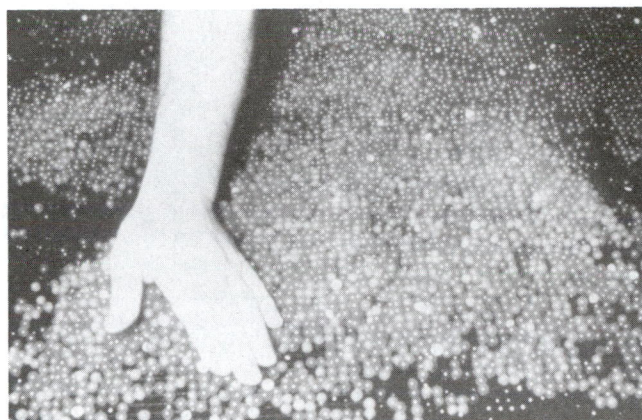

Fig. 56-15. Eyed eggs on hatching trays. (Courtesy, Dr. H. K. Dupree, Director, Fish Farming Experimental Laboratory, U.S. Department of Interior, Stuttgart, AR)

Fig. 56-16. Putting eggs in upwelling incubators. (Courtesy, C. E. Smith, Director, Fish Technology Center, U.S. Fish & Wildlife Service, Bozeman, MT)

Fig. 56-17. Hatching channel catfish eggs. (Courtesy, Dr. D. Gatlin, Dept of Wildlife and Fisheries Sciences, Texas A&M University, College Station)

Fig. 56-18. 1,000 Springs Hatchery—Idaho. (Courtesy, Dr. H. K. Dupree, Director, Fish Farming Experimental Laboratory, U.S. Department of Interior, Stuttgart, AR)

Fig. 56-19. Channel catfish fry. (Courtesy, Dr. D. Gatlin, Dept of Wildlife and Fisheries Sciences, Texas A&M University, College Station)

Fig. 56-20. Red drum (Sciaenops ocellatus fingerling. (Courtesy, Dr. D. Gatlin, Dept of Wildlife and Fisheries Sciences, Texas A&M University, College Station)

# AQUACULTURE SPECIES OF THE UNITED STATES

Table 56–1 lists the most common aquaculture species of the United States, including the description, habitat/fishing, processing, and uses of each of them.

## TABLE 56–1
## AQUACULTURE SPECIES OF THE UNITED STATES

| Common and Scientific Name(s) | Description | Habitat/Fishing | Processing/Uses | Remarks |
|---|---|---|---|---|
| **Bass, Sea** (striped bass) *Roccus saxatillis* | **U**sually dark olive-green vary to bluish or black above, paling to silver on the sides and white on the belly; 7 or 8 horizontal stripes; elongated body rather stout; 2 spiny dorsal fins; a slightly projecting lower jaw; large scales; a 5-year-old may weigh 6 to 7 lb *(2.7 to 3.2 kg)*. | **Habitat:** Native to the Atlantic coast of North America; range from St. Lawrence River to northern Florida and in streams along Gulf of Mexico from western Florida to Louisiana; successfully introduced on the Pacific coast and range from San Diego, California to the Columbia River, Oregon. **Fishing:** Shoal areas of open bays and sounds or in large tributaries along Atlantic coast; 50% caught with gill net, remainder with pound nets, and haul seines; some incidental catches on Pacific coast. | **Processing:** Filleted, steaked, chunked or whole dressed. **Uses:** Baking, broiling, pan frying, planking or poaching. | **S**ea bass were a vital fishery resource for the early colonists. **S**port fishermen catch striped bass by surf casting. **L**argest striped bass recorded weighed 125 lb *(56.7 kg)* and was 6 ft *(1.8 m)* long. Most fishing, however, is for pan-sized fish weighing about 1 lb *(0.45 kg)*. |
| **Carp** *Cyprinus carpio* | **S**cale carp, mirror carp, and leather carp; few scales on leather carp; scales golden-colored; average 3 to 5 lb *(1.4 to 2.3 kg)*, but may reach 80 lb *(36 kg)*. | **Habitat:** Inland lakes and streams in most parts of the Northern Hemisphere; introduced to U.S. from Asia in late 1800s, spread rapidly; prolific breeders. **Fishing:** Not a sport fish in U.S.; high-ranking sport fish in England; some farmed in Europe and China. | **Processing:** Fresh, smoked, or canned. **Uses:** Not a favorite of Americans; eaten by Europeans. | **T**he goldfish is a kind of carp. **C**arp flourish in muddy, stagnant or polluted water. They prefer large rivers, slow-moving streams, and backwaters. |
| **Catfish, Channel** *Ictalurus punctatus* | **L**ong barbels (whiskers) about the mouth for locating food; scaleless with heavy, sharp pectoral and dorsal spines; deeply forked tail; small irregular spots on sides. | **Habitat:** Warm, quiet, slow-moving waters; prefer large rivers and lowland lakes; originally found mainly in Mississippi basin waters, now in many waters throughout the U.S. **Fishing:** From catfish farming in states west from Florida to Texas, and extending north to Kansas and Missouri; some caught as sport fish. | **Processing:** Purchased as whole dressed, skinned dressed, steaks or fillets; fresh or frozen. **Uses:** Baked, broiled, grilled, barbecued, smoked, sauteed or stuffed. | **F**or many years, the catfish market was adequately supplied by fishermen who harvested wild catfish. **C**atfish farmers can get fish to reach 1½ lb *(0.7 kg)* in a 210 day season. |
| **Clam** (class: *Bivalvia*. There are more than 12,000 species.) | **T**he clam has two shells that open and close by muscular action; a muscular foot which is used to burrow in mud or sand; and gills that filter oxygen and food from water. | **Habitat:** Clams live on the bottoms of oceans, lakes, and streams in many parts of the world. They feed on tiny water plants and animals called *plankton*. **Fishing:** Dredges, tongs, rakes, forks, and shovels. | **Processing:** Marketed live in shell; fresh or frozen shucked; frozen breaded, raw, and cooked; canned. **Uses:** Deep fried; pan fried patties; deviled clams, clam chowder; clam cakes and rolls. | **C**ommercial harvesting. Also hobby harvesting. |
| **Crab** There are about 4,500 species. | **F**lattened, broad body covered by a shell. Crabs have up to 19 pairs of appendages, 5 pairs of which are developed into walking legs. | **Habitat:** Crabs are found throughout the world, chiefly in marine waters, but they also inhabit fresh water and land. Most crabs feed on small fish or worms, or scavenge along the shore or sea bottom. **Fishing:** Most crabs are taken in nets and wicker traps. | **Processing:** Marketed live in shell; fresh or frozen cooked; frozen breaded; raw or cooked; canned. **Uses:** Steamed (boiled), cakes, patties, deviled, stuffed, casseroles, salads, appetizers. | **S**everal crab are prized as food, including the *Alaska King Crab*, the *Blue Crab*, and the *Dungeness Crab*. Soft-shelled crabs are considered a delicacy. |
| **Crawfish** (crayfish) *Cambarus virilis* and *Cambarus bartoni* | **F**reshwater crustaceans which look like small lobsters; usually brownish green but may be white, pink, or blue; seldom over 6 in. *(15 cm)* long. | **Habitat:** Streams and ponds throughout the world. **Fishing:** Baited poles and traps. | **Processing:** None. **Uses:** Like lobster; thick soup, crayfish bisque in Louisiana. | **C**rayfish are eaten as food and used as live fish bait in many parts of the world. |
| **Lobster** (spiny lobster) *Homarus* sp or *Panulirus* sp | **M**ost shells, dark green or dark blue with spots; bright red when cooked; 12 to 24 in. *(30 to 61 cm)* long weighing 1 to 20 lb *(0.5 to 9 kg)*; 4 pair thin walking legs; 1 pair of large claws. | **Habitat:** Atlantic and Pacific oceans at depths of 6 to 120 ft *(1.8 to 37 m)*, under rocks and in holes. **Fishing:** Lobster pots. | **Processing:** Live in shell; fresh or frozen; cooked meat; tails raw; canned; usually sold in shell. **Uses:** Baked, broiled or simmered; variety of recipes for use. | **T**here are three main types of lobster: (1) Northern, (2) Rock, and (3) Spiny. **B**est lobster comes from the North Atlantic. |

*(Continued)*

**TABLE 56-1** *(Continued)*

| Common and Scientific Name(s) | Description | Habitat/Fishing | Processing/Uses | Remarks |
|---|---|---|---|---|
| **Oysters** (Eastern or Atlantic oyster; Pacific oyster; Western oyster) *Crassostrea virginica* (Eastern); *Crassostrea gigas* (Pacific); and *Ostrea lurida* (Western) | Dissimilar lower and upper shells; upper shell flat and lower concave; adductor muscle near center of oyster's body controls shell. | **Habitat:** In beds along the temperate and tropical coastlines of all continents; between tidal levels or in shallow waters of bays and estuaries. **Fishing:** Depends on area; handpicking at low tide or use of tongs; hand operated dredges; suction dredges; machine-hoisted dredge; some oyster farming. | **Processing:** Live in shell market; shucked fresh and frozen; frozen breaded, raw, or fried; canned; smoked. **Uses:** Steamed, baked, sauteed, or used in variety of dishes. | Oysters should be cooked just enough to heat through and remain plump and tender. Many oyster beds have been depleted by over-fishing. |
| **Perch, Yellow** *Perca flavescens* | Golden yellow color accentuated by greenish-black; pelvic fins yellow to red tinged; distinctly deeper than wide; seldom exceeds 14 in. *(35 cm)* or 1 lb *(0.45 kg)*. | **Habitat:** Native to freshwaters east of the Continental Divide in large lakes and rivers. **Fishing:** Commercially from the Great Lakes with trap, gill, fyke, hoop, and pound nets in deep waters. | **Processing:** Whole, dressed, or filleted; frozen, fresh or smoked. **Uses:** Baked, broiled, pan or deep fried, and planked. | In many areas, the prolific yellow perch are in constant danger of over-populating their environment. Sport fishing regulations are liberal in most bodies of water. |
| **Pompano** (cobblerfish; butterfish; palmeta) *Trachinotus carolinus* | Silvery body shading to metallic blue above and golden yellow below; deep-bodied with deeply forked caudal and dorsal fins; ranging from 1½ to 3 lb *(0.7 to 1.4 kg)*. | **Habitat:** Mediterranean; Gulf of Mexico; Atlantic coast from Massachusetts to Brazil. **Fishing:** Netted with trammel or gill nets; haul seines and otter trawls may be used. | **Processing:** Whole or dressed. **Uses:** Baked, broiled, planked, pan fried, or deep fried. | Florida pompano offer potential for farming because of their limited supply and great demand. |
| **Salmon** *Oncorhynchus* sp | Five important species: (1) Chinook, (2) Sockeye, (3) Pink, (4) Coho, and (5) Chum; coloration varies from red and green to bluish green with silvery sides; weighing 3 to 20 lb *(1.4 to 9.1 kg)* and reaching 3 ft *(1 m)* in length; deep salmon to almost white flesh. | **Habitat:** Range from Monterey Bay, California to Alaska; migrate from freshwater streams to ocean after hatch; spend most of life in ocean; some travel 3,000 miles to spawn in freshwater streams. **Fishing:** Purse seining, trolling, and gill netting; some with poles. | **Processing:** Dressed, steaked, filleted, fresh or frozen; smoked; canned. **Uses:** Baked, broiled, barbecued, fried, steamed, or poached; variety of recipes and dishes. | Salmon has nourished the human race since ancient times. The Roman, Pliny, wrote in 77 A.D. that "the river salmon is preferred to all fish that swim the sea." Salmon do not eat once they begin their spawning run. Salmon return to the streams in which they hatched, spawn, and die. |
| **Scallop** *Placepecten magellanicus* | Young shells pink and white with some darker color variations; fan-shaped shell with fluted (scalloped) edges; active swimmers by snapping shell together with oversized muscle called the *eye*; shell grows to 8 in. *(20 cm)* in diameter with *eye* of 2 in. *(5 cm)* in diameter. | **Habitat:** Middle and South Atlantic, New England, Gulf of Mexico; deep waters for sea scallops; bays and estuaries for bay scallops. **Fishing:** Dredging; dip nets, rakes or by hand in shallow waters. | **Processing:** Shucked soon after harvesting; fresh or frozen shucked, frozen breaded, raw or cooked; specialty products. **Uses:** Boiled or sauteed; cocktails. | Buildings in ancient Pompeii were ornamented with scallop shell designs. During the Crusades, scallop shells were the symbol of the holy pilgrimages. |
| **Shrimp** (Northern shrimp; North Pacific shrimp; Southern shrimp) *Penaeus* sp and *Pandalus* sp | Body shape resembles a small lobster or crayfish without pincers; color from gray, brown, white, pink, yellow, red, to blue; move rapidly backward by flipping tail. | **Habitat:** Offshore waters of Maine and Massachusetts for Northern shrimp; along the coastlines of California, Oregon, Washington, and Alaska for North Pacific shrimp; waters of Gulf and South Atlantic states for Southern shrimp. **Fishing:** Otter trawl along bottom. | **Processing:** Raw in shell; peeled, fresh or frozen; deveined, fresh or frozen, frozen raw or cooked breaded; cooked, peeled and deveined, fresh or frozen; canned. **Uses:** Simmered, baked, broiled, fried, or oven finish; cocktail; hundreds of uses such as casseroles, salads and sauces. | Three species of Southern shrimp are commercially important: (1) white shrimp;, (2) brown shrimp, and (3) pink or brown spotted shrimp. Southern shrimp are usually the largest and North Pacific shrimp are the smallest. |
| **Sturgeon** *Acipenser* sp | Slender bodies covered with rows of bony plates; long snout; barbels around small tubelike mouth; white sturgeon reach 286 lb *(130 kg)*. | **Habitat:** Marine species anadromous; some entirely freshwater; both Atlantic and Pacific types. **Fishing:** Nets and hook and line. | **Processing:** Flesh usually smoked; eggs (roe) for caviar; swim bladder for isinglass. **Uses:** Specialty item. | There are four genera of sturgeon but most North American sturgeon belong to *Acipenser*. Sturgeon are the only surviving member of their family. |
| **Trout, Lake** *Salvelinus namaycush* | Vary widely in color with shades of gray and olive predominating; body mottled; flesh varies from ivory to deep pink; for commercial purposes range from 4 to 5 lb *(1.8 to 2.3 kg)*; size varies from one body of water to the next. | **Habitat:** Principally the large, cool, freshwater lakes of North America. **Fishing:** Commercial fish in the Great Lakes with gill nets. | **Processing:** Whole, dressed, filleted; steaks; fresh or frozen. **Uses:** Baked, broiled, poached, fried, steamed, or sauteed. | One of the largest trout on record was 49.5 in. *(1.3 m)* long and weighed 102 lb *(46.3 kg)*. It was caught in Lake Athabaska, Saskatchewan in 1961. |
| **Trout, Rainbow** *Salmon gairdneri* | Broad reddish band or *rainbow* running along their sides; olive green on their backs and pure white or silvery on their bellies. | **Habitat:** Native to the Pacific slope of the Sierras from California to Alaska; now transplanted to almost every state; prefer cool, clear, freshwater. **Fishing:** Popular for sport fishing; commercial production from trout farms. | **Processing:** From trout farms, cleaned and packaged for fresh or frozen market; boned, and boned and breaded. **Uses:** Baked, broiled, pan fried, poached, or steamed. | Some rainbow trout migrate to the ocean and then return to their stream to spawn. These are called steelheads. |

## FEEDING FISH

Fish are very efficient feed converters. It makes only 1.5 to 1.7 lb of feed to produce 1 lb of fish; this compares with 2.1 lb of feed to produce 1 lb of broiler. Also, protein efficiency (pounds of protein in feed required to produce 1 lb of fish protein) is very excellent, 2.1:1, although it is exceeded by broilers (1.9:1). Besides, fish consume proteins which are not generally used for human consumption; for example, trash fish and animal by-products—they recycle protein. For these reasons, plus the fact that fish do not compete with crop production, it is expected that both freshwater and brackish-water fisheries, including fish production in small ponds and paddy fields, will receive increasing attention in the future.

Fish are poikilotherms (cold-blooded animals); that is, their body temperature changes to that of the environment. From a production standpoint, this offers both advantages and disadvantages when comparing fish farming with the production of homothermal (warm-blooded animal) species. Since the body temperature of fish is the same as the temperature of their environment, little or no energy is required to maintain their body temperature. By contrast, considerable energy is required to maintain the body temperature of warm-blooded animals. However, fish are very susceptible to stresses which result from environmental changes, especially rapid fluctuations in water temperature. It follows that the production of freshwater fish is more extensive than that of marine fish because it is easier to monitor and control the environmental stresses which affect their growth rate.

Freshwater fish production can be divided into two classifications—coldwater culture (40 to 60°F); and warmwater culture (70 to 100°F). In the United States, channel catfish is the predominant warmwater species, whereas trout and Pacific salmon represent the bulk of coldwater fish production. In other parts of the world, the carps are the major warmwater fish.

## NUTRITIVE NEEDS OF FISH

Fish in the wild seldom show signs of nutrient deficiency: natural foods are relatively well-balanced nutritionally and growth rate is proportional to quantity, not quality, of food available. In confinement, where natural food is limited or absent, nutritional requirements become critical. Although fish require essentially the same nutrients for the same metabolic functions as land animals, there are the following differences:

1. The energy requirements of fish are lower, resulting in superior feed efficiency.
2. Vitamin C (ascorbic acid) is synthesized poorly, or not at all, by most fish.
3. Most fish require dietary omega-3 fatty acids, whereas warm-blooded animals do not.
4. Fish can absorb some soluble minerals from the water, thereby reducing the need for those minerals in the diet.

The nutritional requirements of fish do not vary greatly among species. There are exceptions, however, usually associated with coldwater or warmwater fish, finfish or crustaceans, and freshwater or marine species. These differences generally relate to the requirement of essential fatty acids, and to carbohydrate utilization.

Fish can be divided into three types of eaters—carnivores, herbivores, and omnivores.

1. **Carnivores.** Consume primarily animal material. Foods consumed by this type of fish may be as small as a microscopic crustacean or insect or as large as an amphibian or a small mammal.
2. **Herbivores.** Consume primarily available vegetation and decayed organic materials in the environment.
3. **Omnivores.** Consume both animal organisms and plant materials.

Some anatomic adaptations in the mouth and digestive system, relating to their foods and feeding habits, are observed in some fishes.

Based on well-defined differences in the organization and structure of the mouth, it is possible to classify fish according to their feeding habits into the following categories:

1. **Predators.** These fish, of which trout and salmon are examples, feed on organisms that are generally large enough to be seen with the naked eye. Teeth are well developed and act as a means of grasping and holding the prey. Some predators rely primarily on sight to hunt while others rely on the sense of taste and touch or on lateral-line sense organs.
2. **Grazers.** These fish, of which the mullet is an example, graze in the same sense as grazing animals. Generally, they feed continuously at the bottom of the water habitat on either plants or small animal organisms. Ingested food is taken in well-defined bites.
3. **Strainers.** These fish, of which menhaden are an example, select food primarily by size, rather than by type. An adult menhaden can strain in excess of 6 gal of water per minute through its gill rakers. Through this process of rapid straining, the menhaden is able to concentrate a relatively large mass of plankton and other organisms.
4. **Suckers.** This group of fish, of which buffalo fish are an example, feeds primarily on the bottom of their habitat—sucking in mud, filtering, and extracting digestible material.
5. **Parasites.** Some fish, notably the sea lamprey, attach themselves to other animals and exist on the body fluids of the host.

In addition to the anatomical adaptations to eating, fish have developed behavioral feeding patterns which are sensitive to environmental stimuli. The fish farmer should recognize the influences of environment on feeding behavior if the efficiency of production is to be maximized. By knowing the behavioral patterns of the particular species of fish, the producer can adapt a system of feeding which will best utilize labor and feed. Some environmental influences on feeding behavior follow:

1. **Time of day.** Some fish depend largely on their sense of sight in locating food, while others rely primarily on their sense of taste, touch, and smell. One would expect the fish using sight for feed to be active feeders during the daylight hours. Night feeders rely on the other senses to locate food.
2. **Season of the year.** Some fish, such as largemouth bass, cease feeding activity during their spawning season. Most fish increase feed intake in the spring in the temperate climatic regions when the water temperature starts to rise. As a result, the peak growth period for most fish occurs in the spring and early summer.
3. **Rapid changes of light intensity.** Some fish, such as the yellow perch, show peak feed activity at dawn and dusk.
4. **Physical contact with the food.** Quite often the texture of a potential food source is felt before the fish will consume it.

5. **Water temperature.** Apart from feed and environmental qualities (dissolved oxygen, etc.), water temperature is probably the most important factor affecting appetite and amount of feed consumed. Sharply reduced feed consumption in trout occurs when the temperature decreases to 38°F. The lower limit for feed consumption by catfish is about 50°F.

Size and age of fish are also important factors in determining the nutrient requirements of fish.

Feeding fish is complex. Fish nutrition is truly "an art and a science" because of the complications imposed by the responsiveness of fish to their environment. As a result, fish farmers cannot merely turn to a reference source to find out how much, when, or what to feed. They must be able to recognize the numerous factors affecting feed utilization and adapt their management programs accordingly.

Even though the digestive anatomy, physiology, and feeding habits in fish are different from the warm-blooded domestic animal, the ultimate nutritional requirements can be expressed in the same terms—maintenance, growth, and reproduction. Since the water environment can supply a number of nutrients, and since such environments vary widely, it is not easy to identify the limiting nutrients to be included in a formulated feed. Most of the research relative to the nutrient requirements of fish has centered around salmon, trout, catfish, and carp. However, experiments and experiences indicate that the nutritional requirements of all coldwater fishes as a group, and of all warmwater fishes as a group, are similar. So, the information presented in this chapter for fish of each group can be used for formulating diets for other species within the group.

One very important factor in establishing the nutrient requirements of fish is the effect of water temperature. To deal with this problem, the National Research Council (NRC) reports Standard Environmental Temperatures (SET) for various species of fish in order that fish producers will have an idea of the applicability of the requirements to their particular systems. The Standard Environmental Temperatures (SET's) used by the National Research Council are: 59°F for chinook salmon, 50°F for rainbow trout, and 86°F for channel catfish. As water temperature deviates from these standards, nutrient requirements increase. If the temperature of the water is lower or higher than the SET for a particular species of fish, feed intake is usually reduced.

A discussion of the nutrient requirements of fish follows, with separate sections devoted to energy, protein, minerals, vitamins, water, and nonnutritional factors.

## ENERGY

The energy requirements of fish are lower than the energy requirements of warm-blooded animals due to the following differences in energy metabolism:

1. Fish have a low basal energy need, because, being cold-blooded, they do not expend energy to maintain body temperature. By not having to regulate body temperature, more energy is available for growth, fattening, and reproduction. It has been estimated that only 70% of the dietary calories are used for maintenance, thus leaving a sizable number of calories for growth, fattening, and reproduction.

2. Fish have a low energy need for locomotion and voluntary activity; they have no need for large antigravitation muscles because (1) of the buoyancy of their water environment, and

(2) a streamlined fish moving through water represents one of nature's most efficient modes of locomotion.

3. Fish use little energy requirement for protein catabolism and waste nitrogen excretion, since ammonia is the principal end product of their protein catabolism as contrasted with the more complex urea and uric acid formed by terrestrial homeotherms. Fishes excrete ammonia principally through the gills and expend little energy in the excretion process.

4. Fish use proteins and fats for energy, preferentially, and carbohydrates sparingly; however, carbohydrates are used more efficiently by warmwater fish than by coldwater fish. By contrast, warm-blooded farm animals and poultry use carbohydrates for energy, preferentially.

The amount of energy required by fish is affected by species, diet, size, age, sex, reproductive stage, water temperature, water quality, light exposure (darkness and the rest period that accompanies it, decreases the energy requirement in some species), and activity.

• **Energy value of feedstuffs for fish**—The standard methods of measuring and expressing energy value of feedstuffs for farm animals and poultry is fully covered in Chapter 4, Feeding Livestock. So, only the differences between fish and warm-blooded animals relative to the energy value of feedstuffs will be discussed in the rest of this section.

Digestible energy (DE) and metabolizable energy (ME) are used to express feed values and energy requirements of fishes. Theoretically, ME should be superior to DE, since, in fish, it accounts for energy losses from protein via urine and gills. Practically, however, ME offers little advantage over DE in evaluating useful energy in feeds for fish because there are no urinary or gill losses of the fat and carbohydrates; hence, the ME and DE of the fat and carbohydrate fractions of feed ingredients or mixed feeds are identical. Gill losses do affect the protein portion of the diet; fish excrete ammonia through the gills directly into the water with little energy expenditure, with the result that protein has a higher energy value for fish than for mammals. Because of their efficiency in eliminating nitrogenous wastes through the gill tissue directly into the water, fish are able to utilize more protein in their diet than is required for maximum growth. Nevertheless—protein is more expensive than fats and carbohydrates; hence, only the optimal amount of protein required for maintenance, growth, and reproduction should be included in the diet, and the less expensive fats and carbohydrates should supply the energy and spare the protein for growth.

From the above, it may be deduced that the energy value of feedstuffs for fish and warm-blooded animals is not far different; it is slightly higher in fish due to their greater efficiency in digesting protein.

• **DE versus ME values**—Insufficient data are available to ascertain whether DE values alone are sufficient or if the extra work and expense associated with determining ME is justified. Moreover, there is lack of agreement relative to the method to determine energy values of feedstuffs for fish; i.e., direct method or indirect method. The direct method involves confining fish in metabolism chambers, along with the difficulties encountered in simultaneously collecting fecal, gill, and urinary excretions. The procedures followed are very similar to the digestion trials with other organisms. Indirect methods commonly involve the use of chromic oxide as a marker and the partial collection of wastes.

Digestibility values are easier to determine; besides, the fish are not stressed when allowed to feed voluntarily. In order to determine whether DE or ME values should be used in evaluating fish feed, we need to know the effect of feeds on stress, species of fish, temperature, feeding level, activity, water oxygen content, and metabolite accumulation. These and many other questions need to be answered through research.

• **Energy values of fish**—Since only limited energy values for fish foods are available, it is suggested that the metabolizable energy values for poultry be used. As a result of a comparison of efficiencies of energy and protein retention between rainbow trout and broiler chickens, Cowley et al. concluded that the retention of dietary energy and protein by the rainbow trout is only moderately superior to that of broiler chickens.

As with traditional livestock, fish derive their energy from three sources—carbohydrates, fats, and proteins.

## CARBOHYDRATES

Carbohydrates are a major source of energy for humans and domestic animals, but not for fish. The primary sources of carbohydrates in fish feeds are plant feedstuffs, including cereal grains, wheat by-products, soybean meal, and cottonseed meal.

No carbohydrate requirement has been established for fish because carbohydrates do not supply any essential nutrients that cannot be obtained from other nutrients in the feed; i.e., the energy requirement of fish may be satisfied by fat or protein, as well as by carbohydrate. If sufficient energy nutrients are not available in the feed, the organism will catabolize protein for energy at the expense of growth and tissue repair. The use of carbohydrate for energy to save protein for other purposes is known as the *protein-sparing effect* of carbohydrate.

Carbohydrate in excess of the immediate energy need is converted into fat and deposited in various tissues as reserve energy for use during periods of less abundant feed.

In addition to being a source of energy, carbohydrates may also serve as precursors of nonessential amino acids. Starches also improve the pelleting quality of fish feeds.

Dietary fiber is not utilized by fish. Levels over 10% in salmonoid feeds and over 20% in catfish feeds reduce nutrient intake and impair the digestibility of feeds. Most of the fiber in the feed ultimately becomes a pollutant in the water.

• **Carbohydrates in salmonoid feeds**—Nutritionists have placed the maximum carbohydrate level of salmonoid diets at 12 to 20%. The gross energy from carbohydrates available to mammals is 4.15 kcal/g, whereas the value for trout is only 1.6 kcal/g, a 40% relative efficiency. It is noteworthy, however, that recent research has shown that the processing method affects the availability of starch; extrusion processing (as used in making floating pellets) increases the availability of carbohydrate dramatically in salmonoid diets.

• **Carbohydrates in feeds for warmwater fish**—Studies have shown that channel catfish and carp can utilize higher levels of carbohydrates than trout—they can use up to 25% carbohydrates as effectively as fats as an energy source. Starches are more readily utilized than sugars.

## FATS (LIPIDS)

In nature, fat is the major source of energy for fish. In addition to providing energy, fats serve several other functions

for fish such as reserve energy storage, insulation of the body, cushion for vital organs, lubrication, transport of fat-soluble vitamins, and maintenance of neutral buoyancy.

The average gross energy value of fat is 9.45 kcal/g, and the fish has the ability to use 8 kcal for every gram of fat, representing about 84% efficiency of energy utilization.

In nature, fish tend to deposit fat peculiar to the species. However, the diet of fish will alter the type of deposited fat; so, the fat of cultured fish tends to be similar to the fat ingested.

Polyunsaturated (soft) fats in fish feeds are digested easily in both warm and cold water, but unsaturated (hard) fats are digested efficiently only in warm water.

• **Fat requirements for coldwater fishes**—When there is little or no fat in the feed, trout form their own fat from carbohydrates and proteins. Feed manufacturers use fish oil and vegetable oil in fish feeds as the primary energy source.

Linolenic fatty acids (omega-3 type) are essential for trout and salmon and should be incorporated at a level of at least 1% of the ration for maximum growth response.

The level of dietary fat required for trout and salmon depends on the size of the fish, the protein level in the feed, and the kind of supplemental fat. The recommended percent of fat and protein in the ration dry matter for different ages of trout and salmon follows:

| Feed/Size of Fish | Protein | Fat |
|---|---|---|
| | (%) | (%) |
| Starter feeds (fry) . . . . . . . . . . . . | 50 | 19 |
| Grower feeds (fingerlings) . . . . . . | 40 | 15 |
| Production feeds (older fish) . . . . | 35 | 12 |

Some fatty acid deficiency signs reported for trout include poor growth, necrosis of the caudal fin, pale and fatty livers, dermal pigmentation, increased muscle water content, and reduced hemoglobin.

• **Fat requirements for warmwater fishes**—The available data suggest that warmwater fish feeds should contain 10 to 15% fat, and that more than 15% fat in the feed will not improve growth or increase protein deposition.

The level of omega-3 type fatty acids required by warmwater fishes is unknown, but it is believed to be between 0.3 and 0.5% of the dry diet.

Feed fats affect the taste and storage quality of catfish products. Rancid or off-flavor fish oil has an adverse effect on the flavor of fresh and frozen fish. Fish raised on soybean oil, safflower oil, or corn oil have a better flavor than those fed beef tallow or fish oil.

The principal EFA deficiency signs of warmwater fishes are reduced growth rate, reduced feed efficiency, and in some cases increased mortality.

## PROTEIN AS AN ENERGY SOURCE

In fish feeds, fats and carbohydrates are the primary sources of energy, but excess protein, or protein that is deficient in one or more essential amino acids, is also utilized for energy. In order to metabolize protein for energy, it must be deaminated and the carbon skeleton altered in such a manner as to permit it to enter the energy metabolic pathway. However, fish are relatively efficient in using protein, deriving 3.9 of the 4.65 gross kilocalories per gram from protein, for an 84% efficiency.

## PROTEIN

The primary objective of fish husbandry is to produce fish flesh, which contains more than 50% protein on a dry weight basis.

Weight gain of the fish is essentially proportional to the protein content of the diet at a range of 20 to 40%, but above 50% does not improve growth. Excess protein can be detrimental to fish if not properly balanced with adequate lipid levels; besides, excess protein makes for added feed cost. Fish digest the protein in feeds into amino acids, which are then absorbed into the blood and carried to the cells. Amino acids serve a threefold role in the nutrition of fish: (1) to meet the requirements for formation of the functional body proteins (hormones, enzymes, and products of respiration); (2) to provide tissue repair and growth; and (3) to provide energy.

Fish can synthesize some amino acids, but usually not in sufficient quantity to satisfy their total requirement. So, certain amino acids must be supplied in the feed due to the inability of fish to synthesize them. Fish require the same ten essential amino acids as higher mammals, and they require them in about the same amounts as do broilers. When fish are fed feeds lacking in one or more of the essential amino acids, they become inactive and lose both appetite and weight.

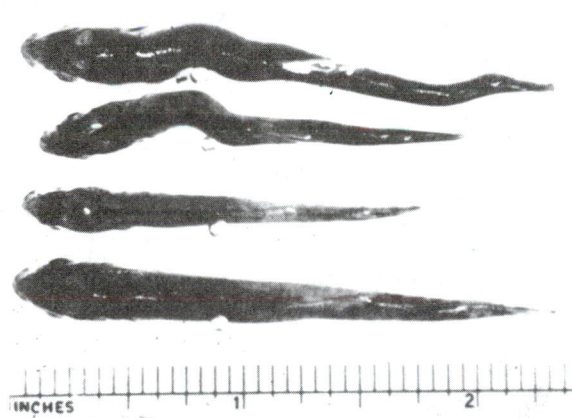

Fig. 56–21. Tryptophan deficiency (scoliosis) condition reversed after 7 to 14 days on a complete diet. (Courtesy, C. E. Smith, Director, Fish Technology Center, U.S. Fish & Wildlife Service, Bozeman, MT)

Several factors determine the requirement for protein in fish feeds; among them—

1. **Temperature.** The protein requirement increases with a rise in temperature.

2. **Age.** Older fish have a lower protein requirement than young fish.

3. **Species.** There are species differences; for example, young catfish require less crude protein than salmonoids (see Table 56–2).

4. **Energy content of the feed.** For optimal growth and feed efficiency, there should be a balance between the protein and energy content of the feed.

5. **Feeding Rate.** The feeding rate determines the daily amount of protein received by the fish. When the feed allowance is increased, the protein level in the feed can be reduced; when the feed allowance is decreased, the protein level in the feed should be increased to assure that fish receive sufficient dietary protein.

A number of protein deficiency symptoms can be documented. Eye problems are evident—resulting in a cloudy condition in the lens, as well as cataract formation. Osteoporosis and a dystrophy of the epiphyseal cartilages can often be seen. On occasion, tryptophan—an essential amino acid—is deficient in a particular ration, with the result that scoliosis and lordosis occur (reported in salmon and trout).

6. **Size.** The protein requirements for fish are size-related (see Table 56–2).

Table 56–2 gives the recommended protein levels for different species and sizes of fish.

**TABLE 56–2**
**PROTEIN REQUIREMENTS (PERCENT OF TOTAL RATION, BY WEIGHT) FOR RAPID GROWTH IN DIFFERENT KINDS OF FISH[1]**

| Species | Fry To Fingerlings | Fingerlings To Subadults | Adults and Brood Fish |
|---|---|---|---|
| | (%) | (%) | (%) |
| Trout and salmon | 50 | 35–40 | 30–32 |
| Channel catfish | 35–40 | 25–36 | 28–32 |
| Common carp | 43–47 | 37–42 | 28–32 |
| Largemouth bass | 40 | 40 | 35 |
| Striped bass | 40 | 36 | 35 |
| Eel | 50–60 | 45–50 | |

[1]Adapted by the author from: Dupree, H. K. and J. V. Huner, Editors, *Third Report to the Fish Farmers*, U.S. Department of the Interior, Fish and Wildlife Service, 1984, p. 142, Table 11.1.

The quality of protein, reflecting the amino acid content, is most important in optimizing the utilization of dietary proteins. If a ration is deficient in any of the ten essential amino acids, poor growth and decreased efficiency of feed conversion will result, despite a high total protein level in the feed. Animal protein sources are generally of higher quality than plant sources. Synthetic amino acids can be added to fish feed, but more research needs to be done to establish benefits and costs therefrom. However, amino acid balance at reasonable cost is best achieved by using a combination of animal proteins, particularly fish meal, and vegetable proteins.

Fish meal appears to be a highly desirable feed ingredient in fish formulas. Substitutions can be made for most of the ingredients in standard fish formulations, but whenever fish meal is left out poorer growth and feed conversion usually result. Starter diets usually contain at least 15% fish meal (starter trout and salmon diets generally contain at least 50% fish meal), and production and brood stock diets normally contain more then 5% fish meal. Other commonly used protein ingredients for fish diets are: soybean meal, cottonseed meal, corn gluten meal, poultry by-product meal, hydrolyzed feather meal, dried blood meal, and dried skimmilk.

Fish cannot utilize nonprotein nitrogen sources such as urea and diammonium citrate, which even nonruminant animals can utilize to a limited extent; thus, these have no value as a feed source for fish. They can even be toxic at high levels.

The chemical composition of fish tissue can be altered by the levels and ingredients in feeds. Within limits, there is a general increase in the percentage of protein in the carcass in relation to the amount in the feed. Also, there is a direct relation between the percentage of protein in the feed and the water content of the carcass. Fish fed low-protein feeds tend to have more fat, less water, and less protein.

• **Protein in salmonoid feeds**—The protein requirements for salmon and trout are similar (see Table 56-2).

The level of protein required in feed varies with the quality and proportion of natural proteins that make up the feed. Between 0.5 and 0.7 lb of dietary protein, in a balanced hatchery feed, is required to produce a pound of trout. The requirement for protein is also temperature-dependent. The optimal protein level in the feed for chinook salmon is 40% at 47°F and 55% at 58°F.

• **Protein in catfish feeds**—Protein utilization of catfish is affected by the protein source and water temperature.

Channel catfish convert the best protein source, fish meal, better than they do the best plant source, soybean meal. In catfish feeds, generally 20% of the dietary protein requirement consists of animal protein.

Better efficiency is obtained for all proteins when they are fed to catfish at temperatures between 75° and 88°F than at 65°F or below.

## MINERALS

Fish, like domesticated animals, require minerals. However, across the gills and skin, fish have the ability to absorb from, and excrete into, the water a number of minerals, thereby reducing the mineral requirement in the diet. For this reason, research concerning the dietary mineral requirements of fish has been difficult to conduct, and the results have been inconclusive. Most researchers agree that fish require all of the macro- and micro-elements required by other animals for enzymes and cofactors. Calcium and cobalt are readily absorbed and excreted through the gills. While phosphate, chloride, and sulfate ions may be absorbed from the water, they are more readily absorbed in the digestive tract.

Because of the difficulties in determining the mineral requirements of fish, along with the lack of sufficient fish mineral research, some nutritionists suggest that a trace mineral mixture suitable for poultry be included in fish feeds.

## FISH MINERAL CHART

More research is necessary on the requirements, functions, and interactions of minerals in the diet and water of fish. In the meantime, based on presently available information, Table 56-3 summarizes the mineral deficiency symptoms and gives (1) the *requirements* for coldwater fish, and (2) the recommended *allowances* for warmwater fish. **Note well:** The figures for coldwater fish are *requirements*, whereas the figures for warmwater fish are *allowances*.

### TABLE 56-3
### FISH MINERAL CHART

| | Deficiency Symptoms | Coldwater Fish Requirement in Feed[1] | Warmwater Fish Recommended Allowance in Feed | |
|---|---|---|---|---|
| | | (%) | (%) | (mg/kg)[2] |
| **Major or Macrominerals:** Calcium | *Trout:* Poor growth, appetite, and feed efficiency. *Catfish:* Reduced growth and lower carcass ash, calcium, and phosphorus. | 0.2–1.0 | 0.35–0.45[3] | 3,500–4,500[3] |
| Phosphorus | *Coldwater and warmwater fish:* Reduced growth, appetite, feed conversion, and bone ash. Additionally, salmonoids show skeletal abnormalities and bone deformities. | 0.7–0.8 (inorganic P) | 0.45[3] (available P) | 4,500[3] (available P) |
| Magnesium | *Coldwater and warmwater fish:* Poor growth, loss of appetite, sluggishness, muscle flacidity, high mortality, and depressed magnesium levels in the body. Additionally, trout show renal calculi, vertebral curvature, degeneration of the muscles, and high mortality. | > .006 | 0.04[3] | 400[3] |
| **Trace or Microminerals:** Cobalt | *Coldwater and warmwater fish:* Required for the synthesis of vitamin B-12 by gut bacteria. | | 0.000005[4] | 0.05[4] |
| Copper | *Warmwater fish:* Depressed growth. | | 0.0005[4] | 5.0[4] |
| Iodine | *Salmonoids:* Thyroid hyperplasia (goiter). | 0.00006–0.00011 | 0.0005[4] | 5.0[4] |
| Iron | *Coldwater and warmwater fish:* Hypochromic, microcytic anemia. | | 0.003[4] | 30.0[4] |
| Manganese | *Coldwater and warmwater fish:* Depressed growth and loss of appetite. Additionally, trout show abnormal tail growth and shortening of the body. | | 0.001[4] | 10.0[4] |
| Selenium | *Salmonoids:* Muscular dystrophy and exudative diathesis. | 0.1–0.35 | 0.1[4] | 1,000[4] |
| Zinc | *Salmonoids:* Cataracts, caudal fin erosion, and depressed growth. *Channel catfish:* Poor growth and appetite. | 0.0015–0.003 | 0.015[4] | 150[4] |

[1]From: Hilton, J. W. and S. J. Slinger, *Nutrition and Feeding of Rainbow Trout*, Department of Nutrition, College of Biological Science, University of Guelph, Guelph, Ontario; published by Department of Fisheries and Oceans, Ottawa, 1981, p. 6, Table 3.

[2]1 mg is the same as 1 ppm.

[3]From: Lovell, R. T., *Nutrition and Feeding of Channel Catfish* (Revised), Southern Cooperative Series Bul. No. 296, 1984, p. 28. These are dietary mineral requirements for catfish.

[4]From: Gatlin, D. M. III, University of Arkansas at Pine Bluff, Pine Bluff, AR, in a personal communication to the author, 1986. These are recommendations for channel catfish.

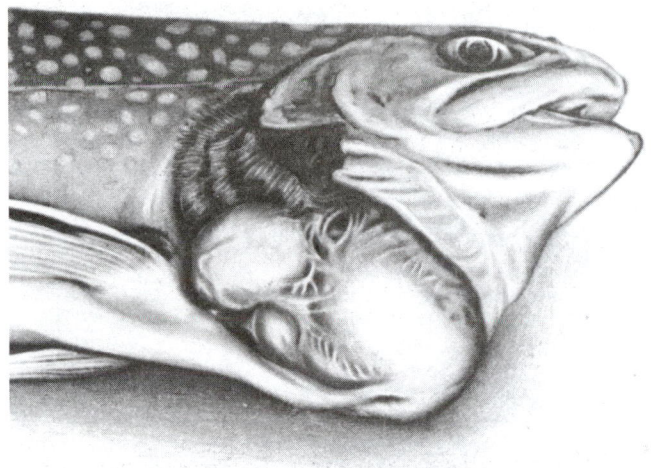

Fig. 56–22. Iodine deficiency caused goiter. When thyroid tumors in fish were first described in 1891, it was believed that the condition represented a form of throat cancer. After considerable debate and research, it was found that the tumors resulted from a dietary iodine deficiency—goiter. Today, iodine is routinely added to fish diets, and thyroid tumors are rarely seen in commercial fish production. (From Gaylord & Marsh; print provided by J. E. Halver, School of Fisheries, University of Washington, Seattle)

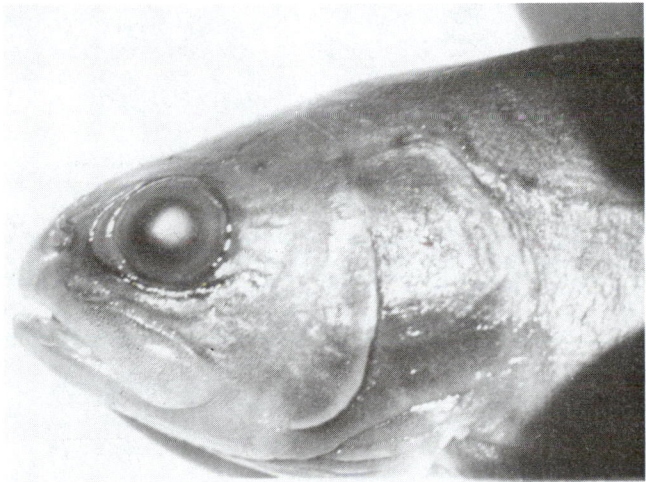

Fig. 56–23. Zinc-deficient diet produced the cataract. (Courtesy, C. E. Smith, Director, Fish Technology Center, U.S. Fish & Wildlife Service, Bozeman, MT)

to a large extent on liver storage of a particular vitamin. Additional studies are needed further to validate earlier requirements, using improved purified diets, and particularly, practical diets. In assessing the requirement for a particular vitamin, as many as possible of the following parameters should be employed: weight gain; feed efficiency; mortality; absence of deficiency symptoms; vitamin tissue storage levels; blood analysis—hemoglobin level, hematocrit value, and red blood cell count; carcass moisture, protein, and lipid content; and vitamin specific enzyme activities.

Fig. 56–24. Folic acid deficiency in the top fish showing pale gills as a result of anemia. Bottom fish is normal. (Courtesy, C. E. Smith, Director, Fish Technology Center, U.S. Fish & Wildlife Service, Bozeman, MT)

Fig. 56–25. Vitamin C deficiency symptoms in coho salmon (top and bottom). *Top:* SCOLIOSIS—a lateral curvature of the spine. *Bottom:* LORDOSIS—a forward curvature of the spine. *Center:* Normal (control) fish that received adequate vitamin C. (Courtesy, Dr. J. E. Halver, School of Fisheries, University of Washington, Seattle)

## VITAMINS

As the digestive system of the fish is monogastric in structure and function, there is definite need for vitamins in fish diets. In general, the vitamin requirements of fish resemble those of monogastric animals, with a few exceptions. Fish represent one of the few types of higher animals that have been shown to have a requirement for vitamin C. There is not enough bacterial activity in the gut of the fish to satisfy either the B complex or the vitamin K requirements.

Many of the presently available requirement levels for vitamins involve the use of semipurified diets and were based

## FISH VITAMIN CHART

The presently recommended vitamin requirements and allowances for fish are given in Table 56–4. Vitamin losses in fish feeds occur due to oxidation and reaction with other feed components. For this reason, amounts in excess of the requirements should be added—the latter are known as recommended allowances.

Prior to 1970, only *complete* fish feeds contained added vitamins. Today, almost all feeds for fish contain vitamin supplements. It is noteworthy that the vitamin requirements of fish differ between species and increase as the fish grows.

**TABLE 56–4**
**FISH VITAMIN CHART**

| | Coldwater Fishes[1] | | | | Warmwater Fishes[2] | | | |
|---|---|---|---|---|---|---|---|---|
| | Requirements[3] | | Recommended Allowances[4] | | Supplemental Allowances[5] | | Complete Diet Allowance[5] | |
| | (per lb body wt./day) | (per kg body wt./day) | (per lb dry feed) | (per kg dry feed) | (per lb dry feed) | (per kg dry feed) | (per lb dry feed) | (per kg dry feed) |
| **Fat-Soluble Vitamins:** | | | | | | | | |
| A ................... (IU) | 34 | 75 | 1,136 | 2,500 | 1,000 | 2,200 | 2,500 | 5,500 |
| D ................... (IU) | 33 | 72 | 1,091 | 2,400 | 100 | 220 | 454 | 1,000 |
| E ................... (IU) | 0.45 | 1 | 13.6 | 30 | 5 | 11 | 23 | 50 |
| K ................... (mg) | 0.05 | 0.1 | 4.5 | 10 | 2.3 | 5 | 4.5 | 10 |
| **Water-Soluble Vitamins:** | | | | | | | | |
| Biotin .............. (mg) | 0.02 | 0.05 | 0.45 | 1 | 0 | 0 | 0.05 | 0.1 |
| Choline ............. (mg) | 14–23 | 30–50 | 1,364 | 3,000 | 200 | 440 | 250 | 550 |
| Folacin (Folic Acid) ................. (mg) | 0.09 | 0.20 | 2.3 | 5 | 0 | 0 | 2.3 | 5 |
| Inositol ............. (mg) | 8–9 | 18–20 | 182 | 400 | 0 | 0 | 45 | 100 |
| Niacin (Nicotinic Acid, Nicotinamide) .......... (mg) | 1.4–2.7 | 3–6 | 68 | 150 | 7.7–12.7[6] | 17–28[6] | 45 | 100 |
| Pantothenic Acid (Vitamin B–3) ........... (mg) | 0.45 | 1 | 18 | 40 | 3.2–5[6] | 7–11[6] | 23 | 50 |
| Riboflavin (Vitamin B–2) .................. (mg) | 0.2–0.5 | 0.5–1.0 | 9 | 20 | 0.9–3.2[6] | 2–7[6] | 9.1 | 20 |
| Thiamin (Vitamin B–1) .................. (mg) | 0.07–0.09 | 0.15–0.20 | 4.5 | 10 | 0 | 0 | 9.1 | 20 |
| Vitamin B–6 (Pyridoxine, Pyridoxal, Pyridoxamine) ... (mg) | 0.09–0.18 | 0.2–0.4 | 4.5 | 10 | 5 | 11 | 9.1 | 20 |
| Vitamin B–12 (Cobalamins) ............. (mg) | 0.0003 | 0.0006 | 0.009 | 0.02 | 0.9–4.5 | 2–10 | 9.1 | 20 |
| Vitamin C (Ascorbic Acid, Dehydroascorbic Acid) ..... (mg) | 1.4–2.7 | 3–6 | 45 | 100 | 0–45[6] | 0–100[6] | 14–45[6] | 30–100[6] |

[1]From: *Nutrient Requirements of Coldwater Fishes*, NRC-National Academy of Sciences, 1981, p. 41, Table 7.

[2]From: *Nutrient Requirements of Warmwater Fishes*, NRC-National Academy of Sciences, 1977, p. 18; and *Third Report to the Fish Farmers*, edited by Dupree, H. K. and J. V. Huner, published by the U.S. Department of the Interior, Fish and Wildlife Service, 1984, p. 147, Table 11.5.

[3]Based on young fish.

[4]Total vitamin contribution from all sources. Other amounts may be more appropriate to offset losses resulting from the effects of formulation and storage, or when feeding other than small fish at SET.

[5]These amounts do not allow for processing or storage losses. Other amounts may be more appropriate for various species and under various environmental conditions.

[6]Highest amounts probably appropriate when "standing crop" of fish exceeds 500 kg/hectare of water surface.

## VITAMIN DEFICIENCY SYMPTOMS

The symptoms of vitamin deficiencies are given in Table 56–5.

### TABLE 56–5
### SYMPTOMS OF VITAMIN DEFICIENCIES IN FISHES[1]

| | Vitamin Deficiency (Hypovitaminosis) Symptoms |
|---|---|
| **Fat-Soluble Vitamins:** | |
| A | *Salmonoids:* Cataracts, intolerance to light, anemia.<br>*Catfish:* Exophthalmia, edema, hemorrhaged kidneys.<br>*Carp:* Faded color, exophthalmia, warped operculum, fin and skin hemorrhages. |
| D | *Salmonoids:* Lethargy, tetany-like contractions, fatty liver, muscles, and carcass, droopy tails.<br>*Catfish:* Low bone ash.<br>*Carp:* None tested. |
| E | *Salmonoids:* Anemia, depigmentation, heart abnormality.<br>*Catfish:* Depigmentation, fatty liver, anemia, mortality.<br>*Carp:* Muscular dystrophy, mortality. |
| K | *Salmonoids:* Hemorrhages, increased prothrombin time, pale gills.<br>*Catfish:* Skin hemorrhages.<br>*Carp:* Not tested. |
| **Water-Soluble Vitamins:** | |
| Biotin | *Salmonoids:* Loss of appetite, pale gills, high glycogen in liver.<br>*Catfish:* Depigmentation, anemia.<br>*Carp:* Poor growth. |
| Choline | *Salmonoids:* Hemorrhages, fatty livers, poor growth.<br>*Catfish:* Enlarged liver, hemorrhages in kidneys and intestines, reduced weight gain.<br>*Carp:* Fatty liver. |
| Folacin (Folic Acid) | *Salmonoids:* Anemia, fragility of caudal fin, lethargy, pale gills.<br>*Catfish:* Loss of appetite and lethargy.<br>*Carp:* None detected. |
| Inositol | *Salmonoids:* Loss of appetite, poor growth, poor feed efficiency.<br>*Catfish:* None detected.<br>*Carp:* Skin lesions. |
| Niacin (Nicotinic Acid, Nicotin-amide) | *Salmonoids:* Swollen gills, intestinal lesions, poor coordination, anemia.<br>*Catfish:* Skin and fin lesions, hemorrhages, prominate eyeball, tetany, lethargy, mortality.<br>*Carp:* Skin hemorrhages, mortality. |
| Pantothenic Acid (Vitamin B–3) | *Salmonoids:* Clubbed gills, anemia, sluggish, prostration.<br>*Catfish:* Clubbed gills, anemia, erosion of skin, barbels mortality.<br>*Carp:* Poor growth, anemia, skin hemorrhages, exophthalmia. |
| Riboflavin (Vitamin B–2) | *Salmonoids:* Cataracts, anemia, dark coloration.<br>*Catfish:* Short body, dwarfism, opaque eye lens.<br>*Carp:* Skin and fin hemorrhages, mortality. |
| Thiamin (Vitamin B–1) | *Salmonoids:* Convulsions, neuritis.<br>*Catfish:* Dark coloration, reduced gains, lethargy, mortality.<br>*Carp:* Fin congestion, nervousness, fading body color. |
| Vitamin B–6 (Pyridoxine, Pyridoxal, Pyridoxamine) | *Salmonoids:* Anemia, nervousness, fits, erratic swimming.<br>*Catfish:* Nervousness, tetany, erratic swimming, greenish blue color, mortality.<br>*Carp:* Nervous disorders. |
| Vitamin B–12 (Cobalamins) | *Salmonoids:* Anemia, fragmented and immature erythrocytes.<br>*Catfish:* Reduced hematocrit, reduced weight gains.<br>*Carp:* None detected. |
| Vitamin C (Ascorbic Acid, Dehydroascorbic Acid) | *Salmonoids:* Spinal deformities, anemia, lethargy, prostration.<br>*Catfish:* Scoliosis, lordosis, reduced growth, dislocated vertebrae, popeye, bloated belly, reduced bone collagen.<br>*Carp:* None tested. |

[1]In the preparation of this table, the authors adapted material from two primary sources: Salmonoids from *Nutrition and Feeding of Rainbow Trout*, Hilton, J. W. and S. J. Slinger, Department of Nutrition, College of Biological Science, University of Guelph, Guelph, Ontario, published by Department of Fisheries and Oceans, Ottawa, 1981, p. 5, Table 2; and catfish and carp from *Nutrient Requirements of Warmwater Fishes and Shellfishes*, NRC-National Academy of Sciences, 1983, p. 16, Table 8. **Note well:** In addition to catfish and carp, other warmwater species are listed in the NRC publication listed above.

## WATER

In addition to its role as a nutrient carrier, water serves additional functions in fish. In freshwater fish, the concentration of ions in the blood is stronger than in the water of the environment. Since water always diffuses from the area of weakest ionic concentration to the strongest, fresh water readily diffuses through the gills and digestive tract into the fish. In saltwater fish, the blood ion concentration is weaker than that of marine water, consequently forcing the fish to absorb nutrients from the environment.

Water also acts as a medium for carrying oxygen. The amount of dissolved oxygen in water depends upon the movement of air over the water's surface, the movement of the water itself, the population of aquatic plants, the amount of sunshine, the population and activity of aquatic animals, and water temperature. Water temperature is critical because less oxygen is dissolved as the temperature increases and, equally important, as the temperature increases, the metabolic rate of aquatic animals accelerates, increasing the oxygen demand. On a physiological basis, if the water temperature is increased and there is less oxygen in the water, the fish has to increase its respiratory rate; and the demand for energy in the fish to sustain this activity is increased.

Fig. 56–26. Aerators on inflowing water to the raceways. (Courtesy, C. E. Smith, Director, Fish Technology Center, U.S. Fish and Wildlife Service, Bozeman, MT)

## NONNUTRITIVE FACTORS

While nonnutritive factors do not directly contribute to the maintenance, growth, or reproduction of fish, they should be considered in the formulation of rations as they can affect feed efficiency or the quality of the final marketable product. Three nonnutritive factors—antioxidants, fiber, and pigment-producing factors—warrant discussion concerning fish nutrition.

## ANTIOXIDANTS

Due to the high fat content of fish diets, along with the highly unsaturated nature of the fats, the dangers of oxidation and subsequent spoilage of feeds can present major problems.

One means of controlling these problems is the addition of antioxidants to the feed, such as ethoxyquin, butylated hydroxyanisole (BHA), butylated hydroxytoluene (BHT), or vitamin E. The main disadvantage in the use of these chemicals is that the levels allowable in feed are adequate only for the prevention of oxidation of animal fats. Vegetable oils and fish oils are highly unsaturated, with the result that high levels of antioxidants are required—far higher than the regulatory guidelines will permit. Vitamin E functions in the fish's physiological system as well as in feed preservation. But the level of vitamin E in fish feeds must be adequate to prevent oxidation of oils and still meet the nutritional requirements of the fish.

Fig. 56–27. Feed storage bin and demand feeder. (Courtesy, Ann Gannam, UAPB Exp. Sta., Pine Bluff, AR)

## FIBER

Due to the simplicity in the structure of the gastrointestinal tract of fish, the digestibility of fiber is extremely low—about 10%. Very little microbial breakdown of fiber has been noted. Herbivore fish have the ability to tolerate higher amounts of fiber than carnivores; nevertheless, it is still recommended that crude fiber not exceed 10% of fish diets—preferably not more than 5 to 6%. Fiber does serve as a source of bulk, which facilitates the passage of ingesta through the digestive tract.

## PIGMENT-PRODUCING FACTORS

Quite often, producers wish to enhance the skin and tissue color in order to make the product more attractive to the consumer. This can be achieved through the addition of certain ingredients to the diet. Paprika fed at a level of 2% of the diet will improve the coloration of brook trout. Xanthophylls from alfalfa meal, corn gluten meal, and dried egg products will increase yellow pigmentation in brown trout skin. The use of shrimp or prawn wastes, which contain carotinoids, will produce a healthy "rosy" color when fed to trout. Also, commercial sources of coloring agents are now available for addition to fish feeds. Species differences with regard to the amount of color change have been observed, so it is possible that what works with one type of fish will not be effective with another.

## FEEDS FOR FISH

Feed for commercial fish production accounts for 30 to 50% of the total expense. So, if fish farmers wish to maximize production at the lowest cost, they must know the nutritional requirements of the fish species which they produce. Also, they recognize and, to the extent practical, maintain the environment to which fish are exposed. Producers have the good fortune of being able to select a type of fish which can adapt to their environment.

Good quality feeds, along with proper processing and feeding techniques, are essential in the intensive production of fish. Feeds that are not eaten do not produce fish; worse yet, they often reduce production and fish quality by contributing to oxygen demand and water quality deterioration. Several types of fish feeds are available, with the choice determined by the formulation and nutrient content, the cost of the feed, the behavioral characteristics of the species, and the size of the fish. Also, the feed preferences of different species often change during their life span, and are often influenced by the type of culture system being used and the level of production.

The processing methods used in preparing fish feeds are similar to those used in preparing feeds for domestic animals.

## NATURAL FOODS AND FEEDS

As the term implies, natural foods are obtained from the immediate environment. Small fish feed upon algae and zooplankton. As fish grow, they devour progressively larger natural foods—insects, worms, mollusks, crustaceans, small fish, tadpoles, frogs, and plants.

Pondfish culturalists take advantage of the natural foods present. The insects, worms, and forage fish which pond fish consume are high in water—containing 75 to 80%. The remaining components are: protein, 12–15%; fat, 3–7%; ash, 1–4%; and carbohydrate, less than 1%. During warm weather when insects hatch and bottom organisms are abundant, a pond can provide a considerable amount of food for fish. The food production can be increased by pond fertilization with chemical fertilizers, organic materials, and animal manures. Because the environment tends to be highly variable in its production of biomass, this method of providing food is inefficient unless the producer is utilizing large bodies of water. However, natural food organisms are relied upon to provide nutrients lacking in supplemental feeds used in pond culture.

• **Fertilizing farm ponds**—A fertilized pond will usually produce three to four times as many pounds of fish each year as a nonfertilized pond. The well-fertilized, well-managed pond will usually produce 200 to 400 lb of fish per acre per year.

Fertilizer may be applied to ponds to produce phytoplankton (tiny green plants or "bloom"). The increased phytoplankton provide more food for small fish and insects, which, in turn, are eaten by large fish.

Fertilizer recommendations vary from area to area, depending primarily on soil type, water quality, and temperature. Also, the use of organic matter such as chicken manure, hay, or oilseed protein meals in conjunction with the chemical fertilizer may be desirable. The fish farmer should obtain the fertilizer recommendation (kind of fertilizer, and rate and schedule of application) from the local County Extension Agent, or from the fish specialist(s) of the U.S. Department of the Interior Fish and Wildlife Service, the U.S. Department of Agriculture, or the university of the state in which the fish farm is located.

## ARTIFICIAL WET FEEDS

Artificial wet feeds contain various organs, meats, and by-products from animals, poultry, and fish. The most common ingredients in wet feeds are liver, spleen, ovaries, intestines, blood, testicles, condemned meat, trash fish, kidneys, fish scraps, mollusks, brain, meat trimmings, heart, poultry offal and by-products, and milk by-products. When feeding artificial wet diets, it is important to ensure that sufficient amounts of the omega-3 fatty acid (linolenic acid family) is present. Also, fish products contain thiaminase necessitating heat treatment of these fish ingredients or supplemental thiamin.

## FORMULATED FEEDS

Most commercial fish producers are now using dry formulated feeds, that contain a combination of both plant and animals ingredients. The most commonly used grains are wheat and corn. The most commonly used grain by-products are brewers' grains, corn gluten meal, cottonseed meal, peanut meal, soybean meal, rice bran, and various wheat by-products. The most commonly used animal by-products are blood meal, feather meal (hydrolyzed), fish meal, poultry by-product meal, shrimp meal, and whey.

Formulated feed may be either supplemental or complete rations.

• **Supplemental fish feeds**—These feeds are formulated to provide adequate energy and protein, but they may be deficient in minerals and vitamins which the fish are expected to obtain from natural foods. Such feeds are fed to fish reared in low densities in ponds.

• **Complete fish feeds**—These feeds are formulated to provide all the essential nutrients required by fish for optimal growth. If high densities of fish are being reared, a complete feed must be provided, as natural feeds will be limited or absent. Complete feeds must be (1) of a physical consistency that will allow them to be fed in the water with minimum leaching, yet ingested and digested by the fish; (2) properly sized for different sizes of fish; (3) palatable to fish so that it will be readily consumed and not left to dissipate into the water; and (4) relatively free from dust and fine particles, which are not well consumed, and which, in excess, may cause water pollution.

Formulated feeds are manufactured in compressed (sinking) pellets, expanding (floating) pellets, moist or semi-moist pellets, crumbles (granules), meals, or flakes.

Dry pelleted feeds have several advantages over other feeds; among them (1) availability at all times of the year in any quantity, (2) the size of the pellets can be altered so as to be suitable for the size of the fish consuming it, (3) they give improved feed conversions and lower feed cost per unit of weight gain, and result in less waste and contamination of the water than meals or wet feeds, and (4) they lend themselves to lower cost bulk handling, storage, and automatic feeding.

• **Compressed or sinking pellets**—These pellets are made by adding steam to the feed as it is pelleted. The steam increases the moisture content by 5 to 6% and raises the temperature to 150–180°F during processing. The feed mixture is forced through a die (dies are available in different sizes) to extrude a compressed, dense pellet. The pellets are cooled and air dried to no more than 10% moisture immediately after pelleting.

Fig. 56–28.  Pellets for fish: *Left:* compressed or sinking pellets. *Right:* expanded or floating pellets. (Courtesy, Department of Fisheries and Allied Aquaculture, Auburn University, Auburn, AL)

• **Expanded or floating pellets**—These pellets require higher temperatures and pressures than compressed or sinking pellets. Under these conditions, raw starch is gelatinized; and bonds are formed within the gelatinized starch to give a durable, water-stable pellet. The sudden release of pressure following extrusion allows water vapor to expand and the ensuing entrapment of gas creates a buoyant, floating food particle. The major disadvantages of floating pellets are the higher cost (8 to 15% higher than sinking pellets), greater bulk (which reduces feed intake), and possibly more vitamin destruction due to the high temperature required in processing. Many fish producers prefer floating feeds because they can observe the fish feeding, which aids in management and reduces feed wastage due to overfeeding. Also, extruded pellets are very durable and breakage of pellets with the production of fines and consequent wastage is considerably reduced. Recent studies with catfish have shown that feeding 15% of the ration as floating feed and 85% as sinking feed gives better feed utilization and is more economical than feeding either alone.

• **Moist and semi-moist pellets**—Moist pellets, which contain 30–50% moisture are made from variable amounts of either fresh or frozen pasteurized fish, together with some dry ingredients. No heat is required for pelleting moist feeds. Refrigeration must be used to protect moist feeds against spoilage. After extrusion, moist pellets should be quick-frozen and stored at –14°F. Moist pelleted feed spoils rapidly when thawed. Also, moist feeds cost more to manufacture, ship, and store than dry pelleted feeds because they must be kept frozen. Salmon producers are the major users of moist feeds.

Semi-moist pellets contain 20–25% moisture, which is intermediate in moisture between moist and dry pellets. They do not require refrigeration, but they must be protected by adequate mold inhibitors and preservatives.

• **Crumbles (granules)**—These are made by crushing pellets, followed by screening out the granules to the desired sizes. The finished feed should be sized and contain not more than 15% oversize or undersize granules. When adding oil or fat, not more than 3% fat should be included at the time of mixing; added fat may be sprayed on crumbles after manufacture.

• **Meals and flakes**—Meals are often coated with vegetable oils or animal fats to increase energy level and improve flotation. They are usually scattered over the surface of the water. Meals are fed to bait minnows, goldfish, and fry of striped bass, grass carp, and sunfish.

Flaked feed is prepared for aquarium fishes. It is usually sprinkled on the water surface.

## FEED HANDLING AND STORAGE

Two rules should be followed in the handling and storage of pelleted fry feed for fish. They are:

1. **Handle with care.** Pellets are fragile and easily broken. When breakage occurs, fines from the pellet represent lost feed and can cause water pollution.

In order to reduce the fines, avoid rough treatment of feed. If the feed is bagged, do not walk or stand on the bags. Use machinery that will not break the pellets. Screw-type augers are known to be extremely hard on pelleted feeds.

2. **Store feed in a cool, dry place and use it within 90 days.** Due to the high protein and fat levels in fish diets, the potential of ingredient spoilage is extremely great. The storage area should be kept clean and adequately ventilated. Protection from insects and rodents and from chemical contamination should receive high priority in feed storage. When storing bagged feed, use multiwalled paper and/or plastic bags instead of burlap. Besides affording extra protection from breakage, plastic-lined paper bags protect the feed from moisture and oxidation. Storage bins are probably best because the feed that has been stored for the longest time is the first to be used. Moist feeds should be refrigerated at temperatures of −10°F or lower.

## RATIONS FOR FISH

Because the cost of feed is the largest single item of expense in fish production, the fish farmer must utilize feed as efficiently as possible. An excess of a certain ingredient could well be as economically detrimental to overall production expenses as a deficiency. If producers are to obtain efficient feed conversions, they must adapt the feed to the particular needs of the fish. Water temperature, size and species of fish, and stocking density are probably the most critical factors to consider when formulating a ration.

### TROUT AND SALMON DIETS

In trout and salmon diets, protein levels should not be less than 45% in starter diets, not less than 40% in production diets, and not less than 35% in brood stock diets. Fat levels should be 15 to 20% in starter diets, 10 to 15% in production diets, and 10 to 15% in brood stock diets. Crude fiber should not exceed 4% in starter diets, and not exceed 5% in production and brood stock diets. The starter diet should not be fed longer than necessary to get the fish off to a good start; prolonged feeding of the starter can cause gill irritation, so it is important to switch to larger size particles as soon as possible.

Formulation specifications for trout are presented in Table 56-6; and diets for Pacific salmon are given in Table 56-7.

**TABLE 56-6**
**FORMULATION SPECIFICATIONS FOR TROUT DIETS[1]**

| | Percent of Diet | | |
|---|---|---|---|
| | Starter | Produc-tion (Grower) | Brood Stock |
| | (%) | (%) | (%) |
| **Fish meal (herring, anchovy, mackerel, capelin):** | | | |
| —minimum crude protein 65% | | | |
| —stabilized with ethoxyquin | | | |
| —maximum fat level 12% | 45–50 | 25–35 | 30–35 |
| —maximum moisture 10% | | | |
| —maximum salt level not to exceed 3% | | | |
| —maximum ash 15% | | | |
| **Wheat middlings:** | | | |
| —minimum crude protein 16% | 0–15 | 10–30 | 15–35 |
| —maximum crude fiber 9.5% | | | |
| **Wheat gluten meal:** | | | |
| —minimum crude protein 80% | 0–3 | 0–2 | 0–1 |
| **Soybean meal[2]:** | | | |
| —minimum crude protein 48% | 5–10 | 5–15 | 5–20 |
| **Corn gluten meal:** | | | |
| —minimum crude protein 60% | 0–10 | 0–10 | 0–10 |
| **Dehydrated alfalfa meal:** | | | |
| —minimum crude protein 17% | – | 0–3 | 0–5 |
| —maximum crude fiber 27% | | | |
| **Dried whey:** | | | |
| —minimum crude protein 13% | | | |
| —partially delactosed | 0–5 | 0–3 | 0–5 |
| **Yeast, dried brewers':** | | | |
| —45% crude protein | 0–5 | 0–5 | 0–5 |
| **Corn distillers' dried solubles:** | | | |
| —minimum crude protein 27% | 0–10 | 0–10 | 0–10 |
| **Animal by-product meals:** | | | |
| —hydrolyzed feather meal, minimum crude protein 85% | 0–5 | 0–7 | 0–7 |
| —Poultry by-product meal, minimum crude protein 60% | 0–5 | 0–7 | 0–7 |
| —blood meal, ring dried or spray dried, minimum crude protein 80% | 0–5 | 0–7 | 0–7 |
| —meat meal, minimum crude protein 50% | 0–5 | 0–7 | 0–7 |
| **Fat supplement:** | | | |
| —marine oil (salmon, capelin, herring, mackerel, etc.) | 5–15 | 5–15 | 5–15 |
| —vegetable oil (soybean oil, canola oil) | 0–5 | 0–8 | 0–5 |
| —animal fat and grease, all oils and fats to be stabilized | 0–5 | 0–5 | 0–5 |
| **Vitamin premix** | 2–4 | 2–4 | 2–4 |
| **Mineral premix** | 2–4 | 2–4 | 2–4 |

[1]Adapted by the author from *Nutrition and Feeding of Rainbow Trout*, by Hilton, J. W., and S. J. Slinger, Department of Nutrition, College of Biological Science, University of Guelph, Guelph, Ontario; published by Department of Fisheries and Oceans, Ottawa, 1981, p. 8.

[2]Soybean meal is not commonly used in Pacific salmon diets due to poor growth, probably because of unpalatability.

## TABLE 56-7
### DIETS FOR PACIFIC SALMON—OREGON MOIST AND ABERNATHY DRY FORMULATIONS[1]

| Ingredients | Starter Moist (%) | Starter Dry (%) | Small Pellets or Crumbles[2] Moist (%) | Small Pellets or Crumbles[2] Dry (%) | Large Pellets Moist (%) | Large Pellets Dry (%) | Ingredients | Starter Moist (%) | Starter Dry (%) | Small Pellets or Crumbles[2] Moist (%) | Small Pellets or Crumbles[2] Dry (%) | Large Pellets Moist (%) | Large Pellets Dry (%) |
|---|---|---|---|---|---|---|---|---|---|---|---|---|---|
| Herring meal (70% crude protein)[3] | 49.9 | 58.0 | 47.5 | 55.0 | 28.0 | 50.0 | Choline chloride (70% product) | 0.5 | 0.5 | 0.5 | 0.5 | 0.5 | 0.5 |
| Wheat germ meal (25% crude protein) | 10.0 | — | Remainder | 5.0 | Remainder | 5.0 | Binder | — | 2.0 | 3.0 | 2.0 | — | 2.0 |
| Dried whey (12% crude protein) | 8.0 | 5.0 | 4.0 | 5.0 | 5.0 | 5.0 | Cottonseed meal (47% crude protein) | — | — | — | — | 15.0 | — |
| Trace mineral premix | 0.1 | 0.1 | 0.1 | 0.1 | 0.1 | 0.1 | Corn distillers' solubles (25% crude protein) | — | — | — | — | 4.0 | — |
| Vitamin premix | 1.5 | 1.5 | 1.5 | 1.5 | 1.5 | 1.5 | Dried blood meal (spray or flash, 80% crude protein) | — | 10.0 | — | 10.0 | — | 10.0 |
| Wet fish (pasteurized) | 20.0 | — | 30.0 | — | 30.0 | — | Wheat middlings (15% crude protein) | — | Remainder | — | Remainder | — | Remainder |
| Fish oil (stabilized) | 10.0 | 12.0 | 7.0 | 9.0 | 7.75 | 9.0 | | | | | | | |
| Vitamin C | 0.15 | 0.1 | 0.15 | 0.1 | 0.15 | 0.1 | | | | | | | |

[1]Formulations provided for this book by L. G. Fowler, U.S. Fish and Wildlife Service, Longview, WA 98632

[2]Small pellets are moist feeds, and crumbles are dry feeds.

[3]Anchovy meal (65% crude protein) allowed in dry crumbles and large pellets.

## CATFISH DIETS

Catfish feeds are generally lower in protein than those of coldwater fish. Fingerlings require 35 to 40% protein, production (grower) fish 25 to 36%, and brood stock 28 to 32% protein. Examples of various catfish feeds are found in Tables 56-8 and 56-9.

### TABLE 56-8
### FEED INGREDIENTS AND PERCENT COMPOSITION OF SOME PRACTICAL FEED FORMULATIONS FOR CHANNEL CATFISH[1]

| Ingredient | Percent Protein | Fish Farming Experimental Station[2] | Texas A&M University 1 | Texas A&M University 2 | Auburn University Fingerlings (36% protein) | Auburn University Production (32% protein) |
|---|---|---|---|---|---|---|
| Alfalfa meal | 17.5 | 3.5 | | | | |
| Blood meal | 75.3 | 5.0 | | | | |
| Bone meal | 11.2 | | | | | |
| Brewers' grains | 25.0 | | | | | |
| Corn, distillers' grains | 27.1 | | | | | |
| Corn, distillers' solubles | 27.3 | 8.0 | | | 5.0 | |
| Corn grain, yellow | 9.6 | | 30.0 | 30.5 | 23.5 | 29.1 |
| Corn grain, flint | 9.9 | | | | | |
| Cottonseed meal | 40.8 | 10.0 | | | | |
| Cottonseed meal (without hulls) | 50.0 | | | | | |
| Dicalcium phosphate | | | 0.25 | 1.5 | 1.5 | 1.0 |
| Fats, animal | | | | 2.0 | 2.5 | 1.5 |
| Fish meal, catfish | 55.3 | | | | | |
| Fish meal, menhaden | 61.1 | 12.0 | | 9.0 | 10.0 | 8.0 |
| Fish meal, tuna | 59.4 | | | | | |
| Fish meal, white | 61.9 | | | | | |
| Grains, distillers' | 27.4 | | | | | |
| Liver meal, animal | 66.5 | | | | | |
| Meat meal, whole animal | 54.3 | | | | | |
| Meat meal (with bone) | 50.5 | | 15.0 | | | |
| Oats, cereal by-product | 14.6 | | | | | |
| Peanut meal | 44.0 | | | | 18.0 | |
| Poultry by-product meal | 57.8 | | | | | |
| Poultry feathers, hydrolyzed | 85.4 | 5.0 | | | | |
| Rice bran | 12.7 | 25.0 | | | | 10.0[3] |
| Rice polishings (dust) | 12.1 | 10.0 | | | | |
| Soybean meal | 44.0 | | 47.5 | | | |
| Soybean meal (without hulls) | 48.8 | 20.0 | | 54.4 | 37.0 | 48.3 |
| Wheat bran | 15.1 | | | | | |
| Wheat middlings | 16.7 | | 0.85 | | | |
| Wheat grain | 14.9 | | | | | |
| Wheat shorts | 16.4 | | | | | |
| Whey (low lactose) | 16.5 | | 2.5 | | | |
| Yeast (brewers') | 45.1 | | | | | |
| Vitamin premix | | [4] | [4] | [4] | [5] | [5] |
| Mineral premix | | | 0.50 | 0.50 | [6] | [6] |
| Limestone | | | 0.90 | | | |

[1]Adapted by the author from *Third Report to the Fish Farmers*, by Dupree, H. K., and J. V. Huner, U.S. Department of the Interior, Fish and Wildlife Service, 1984, p. 151, Table 11.6.

[2]Herring fish meal (70% protein) may be substituted at 10% of the formulation. Wheat shorts, wheat middlings, or cereal grains may be substituted for the rice bran.

[3]Wheat shorts may be substituted for the rice bran.

[4]Vitamin premixes as published by the National Research Council (footnote 1, above).

[5]Quantities of vitamins per ton (grams, unless otherwise indicated): vitamin A, 4 million IU; vitamin D, 2 million IU; vitamin B-12, 8 mg; vitamin E, 50; menadione, 10; choline chloride (70%), 500; niacin, 80; riboflavin, 12; pyridoxine, 10; thiamin, 10; pantothenic acid, 32; folic acid, 2; and ethoxyquin, 125. Ascorbic acid (335 g) is added to the feed during pelleting, or top-dressed on extruded feed.

[6]Quantities of minerals per ton (milligrams): manganese, 110; zinc, 105; iron, 36; copper, 4.5; iodide, 2.3; and cobalt, 0.5.

## TABLE 56-9
### EXTRUDED (FLOATING) AND HARD (SINKING) PELLET FORMULA FOR CATFISH IN PONDS, RACEWAYS, AND CAGES[1]
### (ALSO MINNOWS, GOLDFISH, CARP, AND BUFFALO)

| Ingredient | Percent |
|---|---|
| **Fish meal,** menhaden, minimum 61% protein . . . . . . . . . . . . . . | 10.0 |
| **Soybean meal,** solv extd, w/o hulls, minimum 49% protein . . | 35.0 |
| **Cottonseed meal,** solv extd, minimum 41% protein . . . . . . . . | 12.0 |
| **Wheat,** whole grain (ground)[2] . . . . . . . . . . . . . . . . . . . . . . . . | 31.7 |
| **Rice bran,** with germs, solv extd . . . . . . . . . . . . . . . . . . . . . | 3.5 |
| **Fat,** animal or plant[3] . . . . . . . . . . . . . . . . . . . . . . . . . . . . . . | 5.0 |
| **Dicalcium phosphate** . . . . . . . . . . . . . . . . . . . . . . . . . . . . . | 1.0 |
| **Vitamin premix[4]** . . . . . . . . . . . . . . . . . . . . . . . . . . . . . . . . | 0.8 |
| **Mineral premix[5]** . . . . . . . . . . . . . . . . . . . . . . . . . . . . . . . . | 1.0 |
| **Analysis:** | |
| Crude protein, more than . . . . . . . . . . . . . . . . . . . . . . . . . | 32.0 |
| Crude fiber, less than . . . . . . . . . . . . . . . . . . . . . . . . . . . . | 3.5 |
| Crude fat, more than . . . . . . . . . . . . . . . . . . . . . . . . . . . . . | 7.0 |

[1]Feed formula prepared for this book by H. Dupree, Fish Farming Experimental Station, U.S. Fish and Wildlife Service, Stuttgart, Ark.

[2]Wheat may be replaced up to 25% with corn.

[3]Sprayed on after manufacture.

[4]A. If the feed is to be used as a supplemental pond feed (*i.e.,* low intensity pond culture), the premix should be as follows (per ton of feed basis): vitamin A, 900,000–1,800,000 IU; vitamin D₃, 450,000–900,000 IU; vitamin E, 27 g; thiamin, 0.9 g; riboflavin, 8.1 g; pyridoxine, 2.7 g; pantothenic acid, 9–18 g; nicotinic acid, 12 g; and ascorbic acid, 55 g.

B. If the feed is to be used as a complete feed (*i.e.,* raceways, cages, and high intensity pond culture), the premix should be as follows (per ton of feed basis): vitamin A, 4,000,000 IU; vitamin D₃, 2,000,000 IU; vitamin E, 50 g; vitamin K, 10 g; thiamin, 10 g; riboflavin, 12 g; pyridoxine, 10 g; pantothenic acid, 32 g; nicotinic acid, 80 g; folic acid, 2.0 g; choline chloride, 500 g; ascorbic acid, 350 g; and vitamin B-12, 8 mg.

[5]Mineral premix should provide the following (per ton of feed basis): manganese, 100 g; iodide, 2.5 g; copper, 3.9 g; zinc, 80 g; iron, 40 g; and cobalt, 45 mg.

## FEEDING COLDWATER FISH

In the United States, trout and salmon are the two most common species of coldwater fish grown commercially.

The art and science of feeding coldwater fish has progressed dramatically over the last 100 years. Originally, producers relied on natural feeds to grow trout, followed years later by the addition of wet diet supplements to natural feeds, and finally to the recent development of complete diets.

Variations in feeding habits occur among the different types of coldwater fish. For example, rainbow trout are surface feeders whereas brown trout are bottom feeders. Therefore, the type of feed pellet to be used must be given careful consideration. Since trout consume their feed in about 5 to 10 minutes, the producer does not have to be concerned with the pellets falling apart unless too much feed is being used and remains uneaten.

The producer should feed according to the stocking rate, the size of the fish, the type of pond or culture facility, the water temperature, and the energy content of the feed. Feed consumption is markedly affected by water temperature, decreasing in cold weather. Also, feed intake is affected by the energy content of the feed, since fish, like terrestrial animals, eat to satisfy their energy needs. Feed consumption is reduced in polluted water.

The feeding rate for salmonoids is commonly expressed as a percentage of body weight fed per day. Cognizance is taken that (1) smaller fish require feed at a greater percentage of their body weight per day than larger fish, and (2) water temperature has a marked effect on feed requirement, being very low at 43°F. Very low and very high temperatures constitute a con-

siderable stress to salmonoids. But feeding guides are just that—guides. Thus, feed as a percentage of body weight for growing fish can vary between 0.5 to 10% depending on numerous factors. For example, in the spring when the water begins to warm and when the photoperiod is increasing, it is possible to feed up to twice the amounts shown in most feeding guide tables.

In addition to feeding rates, the following feeding practices are important:

1. Frequency of feeding, with swim-up fry being fed small amounts of feed 20 to 24 times per day, gradually reduced to one to three times per day.

2. Feed particle size, hardness and texture, palatability, and placement of feed in relation to fish size are important. Very small fish will not travel far for feed.

3. Changes in feed intake and feed size should be made gradually, extending over a few days.

4. Most salmonoids should be selected for brood stock at 2 to 3 years of age. At this time, they should be switched from the production or grower diet to a brood stock diet and fed only once a day. Depending on the temperature of the water, breeders should be taken off feed 3 to 6 weeks before spawning, then gradually brought back to full feed over a period of 2 to 4 weeks after spawning. Overfeeding before spawning reduces reproductive performance.

## FEEDING WARMWATER FISH

Warmwater fish production involves numerous species of fish, but the major industries are catfish and carp farming. There have been attempts to produce predacious fish commercially, but due to the food requirements and behavioral characteristics of the fish, success has been very limited.

Carp production is the oldest form of aquaculture—dating back 4,000 years to China. Today, different types of carp culture can be found throughout the world, quite often in conjunction with another type of agriculture production—for example, polyculture in rice fields.

The catfish industry in the United States has experienced phenomenal growth in recent years. In 1960, only 400 acres were under catfish culture, and the production was a mere 300,000 lb. By 1985, catfish farming was more than a $200 million-a-year industry producing over 290 million pounds of catfish on about 100,000 acres. Indications are that the industry will continue to expand as consumers become more familiar with the product.

## FEEDING CATFISH

Catfish raised for food and for fee-fishing in the Southwest are fed to liveweights of 2 to 2½ lb. Most catfish marketed in the Southeast—the area of greatest production—are processed at liveweights of ¾ to 2 lb. If good management practices are followed, fingerlings will have feed conversion rates of 1.5 to 1.7.

Catfish, being omnivores, have well-defined stomachs and can effectively digest meat; but in commercial production the use of fresh meat diets is extremely limited. Most feeds are in dry pellet form, which make for economical storage and handling. Catfish diets generally contain the oil meals, distillers' solubles, and fish meal, in addition to vitamin and mineral premixes.

Since a sizable portion of the diet of catfish can come from the natural feed chain, ponds containing fingerlings should be fertilized in order to establish an optimum level of plankton growth. But fingerling producers must avoid over fertilizing.

It is generally recommended that 50 lb of 16–20–4 or 16–20–0 inorganic fertilizer per acre to be applied every 10 days until an adequate plankton bloom is present. The producer can tell if the pond needs more fertilizer by sticking an arm in the water up to the elbow. If the hand can be seen, more fertilizer is needed.

In general, catfish are bottom feeders, but they can be taught to feed at the surface. While floating pellets are more expensive, some producers justify the added cost because they can routinely observe the condition of the fish at feeding.

The water temperature, fish weight, and water quality are the primary factors affecting the feed consumption of warm-water fish.

Table 56–10 presents a feeding guide for channel catfish in ponds. Table 56–11 gives a feeding guide for catfish in

**TABLE 56–10**

**TYPICAL SPRING-SUMMER-FALL FEEDING SCHEDULE FOR CHANNEL CATFISH IN PONDS STOCKED AT 2,000–3,000 FINGERLINGS PER ACRE AS 5-IN. FISH AND HARVESTED AS 1.1-LB FOOD FISH, IN SOUTHEASTERN UNITED STATES[1][2]**

| Date | Water Temperature (°F) | Water Temperature (°C) | Fish Weight (lb) | Fish Weight (kg) | Feed Allowance Per Day (% of fish wt.) |
|---|---|---|---|---|---|
| April  15 | 68 | 20.0 | 0.04 | 0.02 | 2.0 |
| April  30 | 72 | 22.2 | 0.06 | 0.03 | 2.5 |
| May  15 | 78 | 25.5 | 0.11 | 0.05 | 2.8 |
| May  30 | 80 | 26.6 | 0.16 | 0.07 | 3.0 |
| June  15 | 83 | 28.3 | 0.21 | 0.10 | 3.0 |
| June  30 | 84 | 28.8 | 0.28 | 0.13 | 3.0 |
| July  15 | 85 | 29.4 | 0.35 | 0.16 | 2.8 |
| July  30 | 85 | 29.4 | 0.42 | 0.19 | 2.5 |
| August  15 | 86 | 30.0 | 0.60 | 0.27 | 2.2 |
| August  30 | 86 | 30.0 | 0.75 | 0.34 | 1.8 |
| September  15 | 83 | 28.3 | 0.89 | 0.40 | 1.6 |
| September  30 | 79 | 26.1 | 1.01 | 0.46 | 1.4 |
| October  15 | 73 | 22.7 | 1.10 | 0.50 | 1.1 |

[1]Adapted by the author from *Third Report to the Fish Farmers*, by Dupree, H. K., and J. V. Huner, U.S. Department of the Interior, Fish and Wildlife Service, 1984, p. 156, Table 11.8.

[2]Feed allowances are based on data obtained with rations containing 36% protein and about 2.88 kcal of digestible energy per gram of protein. If feeds of lower protein and energy concentrations are used, daily allowances should be increased proportionally. Data adapted from R. T. Lovell, 1977. Feeding practices. Pages 50–55 in Nutrition and feeding of channel catfish. *Southern Cooperative Series Bul. 218.* Alabama Ag. Exp. Sta., Auburn University, Auburn, AL.

**TABLE 56–11**

**FEEDING RATES (PERCENT BODY WEIGHT PER DAY) FOR CHANNEL CATFISH FED A COMPLETE FEED (25% FLOATING, 75% SINKING FEED) IN RACEWAYS[1]**

| Water Temperature | Size and Weight 1–2 in. 0.001–0.004 lb | Size and Weight 2–5 in. 0.004–0.04 lb | Size and Weight over 5 in. over 0.04 lb |
|---|---|---|---|
| (°F) | (% body weight) | (% body weight) | (% body weight) |
| Below 55° | 1 | 1 | 1 |
| At 55° | 3 | 2 | 1.5 |
| Above 55° | 5 | 3 | 2 |

[1]Adapted by the author from *Fish Hatchery Management*, by Piper, R. G., I. B. McElwain, L. E. Orme, J. P. McCraren, L. G. Fowler, and J. R. Leonard, U.S. Department of the Interior, Fish and Wildlife Service, Washington, DC, 1982, p. 252, Table 29.

raceways. The pond feed is a supplemental, 36% protein diet; the raceway feed is a complete formulation.

Multiple daily feeding can increase growth rate. In such a production system, feed is offered in the maximum amounts that can be metabolized in the pond. Feeding frequency varies with water temperature (see Table 56–12). A rule of thumb is that 90% of the feed should be eaten in 15 minutes or less.

**TABLE 56–12**

**SUGGESTED MAXIMUM FEEDING RATES AND FREQUENCIES FOR CHANNEL CATFISH[1]**

| Water Temperature (°F) | Water Temperature (°C) | Feeding Frequency (times per day) | Feeding Rates (% of total fish wt.) |
|---|---|---|---|
| 90 | 32.2 | 1 | 1 |
| 80–86 | 26.6–30.0 | 2 | 3 |
| 68–80 | 20.0–26.6 | 1 | 2½ |
| 58–68 | 14.4–20.0 | 1 | 1½ |
| 50–58 | 10.0–14.4 | 0.5[2] | ¾–1 |
| 50 | 10.0 | 0.3[3] | ½–1 |

[1]Adapted by the author from *Third Report to the Fish Farmers*, by Dupree, H. K., and J. V. Huner, U.S. Department of the Interior, Fish and Wildlife Service, 1984, p. 156, Table 11.9.

[2]Feed once on alternate days.

[3]Feed once every 3 to 4 days.

• **Rules of thumb for feeding catfish**—The following rules of thumb for feeding catfish are generally followed:

1. **Be aware of water temperature.** Catfish make their most efficient gains in water temperatures of around 84°F. When the water temperature falls below 55°F, daily supplemental feeding should be reduced to about 1%, or less, of body weight. In the winter months, catfish will require about ¼ the amount of feed that would be fed in the summer. If the water temperature drops below 45°F, it is only necessary to feed at the rate of 0.5% of body weight every 4 to 5 days.

2. **Do not full-feed on rainy days or when it has been overcast for extended periods (more than four days).** It has been found that catfish will show a dramatic drop in feed consumption on rainy days. If the sky is overcast for several days, feeding activity decreases but will return to normal once the sky clears. This is due to the fact that on overcast days, plants and algae take up more oxygen than they give off, thereby decreasing the level of oxygen in the water.

3. **Until catfish reach a weight of 1 lb, the maximum feeding rate should be 3% of body weight per day.**

4. **The feeding rate of catfish weighing more than 1 lb should not exceed 2% of body weight per day.**

• **Feeding catfish fry and fingerlings**—Catfish feeding begins with the fry and fingerling stages. Newly hatched catfish, which are called fry, live on nutrients from the yolk sac for 3 to 10 days (depending on the water temperature), following which they accept food from a variety of sources. Fry can either be stocked in a rearing trough or moved directly to a specially prepared rearing pond. Feed for trough-feeding of fry should be small in particle size, high in animal protein, and high in fat. In troughs, fry are held until they have absorbed their yolk sacs and have developed a graying color. At the latter stage, they are called swim-up fry because they rise to the surface to seek feed. As soon as possible, the fry should be transferred to ponds with high zooplankton densities to utilize the natural food source. At this time, supplemental feeding should begin, using 36% high quality protein catfish feed for fry and fingerlings.

Catfish fingerlings have been shown to utilize feeds extremely efficiently—demonstrating feed conversions of 0.9 to 1.0. Of course, as the fish increase in size, the total efficiency will decrease because the natural foods will constitute smaller and smaller amounts of the total food intake.

## FEEDING CARP

Carp are the most extensively cultured fish in the world. They grow well under a variety of cultural conditions, use natural food efficiently, and respond well to supplemental feeding.

China produces 75–80% of the world's farm-raised carp, most of which are raised in mixed culture ponds and fed pig and poultry manure and grass clippings.

The feeding principles of carp production can be generally considered to be applicable to the less commonly produced warmwater food and bait fish—for example, minnows, buffalo fish, and barbs.

Carp grow fastest in water which is 77° to 86°F. When the temperature drops to 60°F, growth is inhibited; and if the temperature falls below 55°F, feeding activity is greatly decreased. However, carp still require supplemental feed in cold water to avoid excessive weight loss and debilitation. For every 18°F above 55°F, there is a two- to threefold increase in feed consumption by carp.

Some researchers recommend that at least 50% of the feed for carp consist of natural feeds. Carp eat plankton primarily, along with small animals close to the shore and the bottom. But they will also eat some of the natural vegetation. With emphasis on the utilization of natural feeds, pond fertilization becomes an important production technique.

Artificial feeding—supplementary feeding is probably a more appropriate term—is commonly practiced. Soybean, corn, and wheat are the most widely used feedstuffs; but barley, oats, rye, beans, potatoes, millet, rice bran, vetch, and grass seeds may be used.

Carp are slow eaters. It generally takes them 30 minutes to an hour to finish eating a dry feed as compared to 5 minutes for trout. The pellets should be water stable in order to prevent leaching of nutrients, wastage of feed, and possible reduction in water quality.

Feeds for fry carp and related species are generally in a fine, flourlike consistency. It is best to feed on all sides of the pond in order to ensure that all the young fish have adequate access to feed.

In the warm regions of the world, where rice is commonly grown, carp are sometimes raised in rice fields. The warm water provides an excellent environment for rapid growth of both the rice and carp. The waste products from the carp fertilize the rice, and the carp consume some of the aquatic organisms that can interfere with the growth of the rice. Carp grow most efficiently during the summer, which is the time that rice will have its greatest growth.

In most countries, fertilization of ponds with manure is a common practice. Manure and organic wastes can be used to fertilize ponds, but the producer should be aware of possible nitrate-nitrite toxicity and oxygen depletion in the ponds. By

using manure as a pond fertilizer, two purposes are served. First, it provides a convenient method of dealing with the potential pollution problem resulting from the feeding of livestock. Second, nutrients from the manure can be utilized to support the natural food system of the pond.

The Chinese have fed manure to fish since the days of the Ming Dynasty; hence, the use of manure in fish culture is of long standing in that country. In Kwang Tung Province, in the heart of the land of fish and rice, the author was fascinated with an integrated hog-fish production operation on a commune where, in 1 year, they produced 49,700 pigs and had a fish catch of 863,000 lb. The fish were fed solely on natural feeds produced in the ponds from the manure flushed from the nearby hog barns, plus grass clippings from the pond banks. This unique operation provided a method of handling manure plus pollution control and, at the same time, recycled and used the feed twice—first through the hogs, and second through the fish. It is noteworthy, too, that the Chinese were feeding manure to fish without the hazard of nitrate-nitrite toxicity and/or of oxygen starvation.

The Chinese Academy of Agriculture and Forestry Sciences recommends the following practices when feeding manure to fish: (1) mixing the manure with plant materials, then composting it before feeding (although manure may be, and is, fed to fish without composting, pollution may be lessened by first composting); (2) fertilizing and fermenting the composted manure; and (3) feeding manure to the little fingerlings, rather than to the big fish. But, according to the fish experts of China, the feed of fish should vary according to species (carp are especially well adapted to the use of animal waste) and age. They also report that, in addition to manure, freshwater fish of China are fed a great variety of feeds, including silage, the leaves of such crops as sweet potatoes and turnips, soybeans, bean cake and curd, distillers' grains, and wheat. It's noteworthy, too, that in addition to supplying needed food, the freshwater fish of China lessen the mosquito menace.

Fig. 56–29. A hog-fish combination in China, where, in 1 year, they produced 49,700 pigs and had a fish catch of 863,000 lb. (Photo by A. H. Ensminger)

## FEEDING SYSTEMS

In all feeding systems, the producer must compare the costs of labor to the costs of automation. In general, larger operations can justify more automation. The producer must also take into consideration the frequency of feeding to be followed. Small fish must be fed frequently—8 to 20 times daily for trout—

in order to prevent cannibalism and irregular growth. As the fish become large, the frequency of feeding can eventually be reduced to once or twice daily.

There are three methods of feeding fish—hand-feeding, semiautomatic feeding, and automatic feeding.

## HAND-FEEDING

This is the oldest form of feeding. It involves the fish culturist walking along the water banks and spreading the feed from a ladle or by hand. It is best to feed along all sides of the pond. This ensures that all the fish have access to feed with a minimum of competition; hence, more pounds per acre are yielded. The farmer must be attentive to the amount of feed that will be consumed. Overfeeding will reduce profits as well as cause pollution. On small operations, the expense of this additional labor is far less than the expense of converting to more automatic methods.

## SEMIAUTOMATIC FEEDING

In operations that involve intense production, a certain amount of automation is necessary. Fish can be fed from boats or by blowing feed from mechanical equipment traveling along the edge of the water. In extremely large operations, feed can be released from airplanes flying close to the water.

Fig. 56-30. Automatic feeders. (Courtesy, C. E. Smith, Director, Fish Technology Center, U.S. Fish & Wildlife Service, Bozeman, MT)

## AUTOMATIC FEEDING (MECHANICAL FEEDERS)

Automatic feeding can be divided into two classes of feeder systems: (1) the demand type, which is activated by the fish; and (2) the automatic type, which is activated by a time clock. In the demand feeder, fish can obtain the food by mechanically tripping a feed release trigger. This type of feeder can create

Fig. 56-31. Demand feeders on trout raceway. (Courtesy, C. E. Smith, Director, Fish Technology Center, U.S. Fish & Wildlife Service, Bozeman, MT)

some wastage of feed, but the producer must consider the cost of labor vs the cost of wasted feed. An important advantage of the trigger-type self-feeder is that it can be adapted to almost any size of operation. The second type of automatic feeder is designed to dispense fixed amounts of feed at present time intervals; it may offer more feed than is eaten or less than is required. Since these feeders are rather sophisticated, the cost cannot be justified unless the operation is large and intensive.

Many large operators find that the time and labor involved in filling self-feeders can be considerable; hence, most large operations use a truck or tractor equipped with feed blowers.

## OTHER FEED AND MANAGEMENT ASPECTS

When discussing the various nutritional aspects of fish culture, two additional areas should be considered—feeds affecting the quality and flavor of fish, and types of specialized aquatic production.

### FEED INGREDIENTS AFFECTING FISH QUALITY AND FLAVOR

In many cases, fish farmers can alter the texture and flavor of their product merely by altering the composition of the ration. With this in mind, they should be aware of the demands of the consumer and attempt to produce a uniform product.

Trout that are fed wet or moist diets tend to be high in fat content and have a soft meat texture, whereas those fed dry diets have a more desirable flavor with a firmer texture. If the wet diet contains fresh sea fish, there is a possibility that an off-flavor (sardine taste) will result. Also, off-flavor may be caused by spoiled feed, by feed containing oxidized (rancid) oils, or by algae in the water.

Carp fed bread and potatoes develop a wet, soft meat consistency which is considered undesirable. Supplemental feeding of cereal grains will produce carp that have a desirable flavor as well as a firm meat texture. If no supplemental feeding is provided—that is, if carp get all their required nutrients from natural sources—an undesirable wild flavor may result.

Catfish flavor is affected by several factors. Excessive plankton bloom, feeds high in fish oil, presence of muskgrass, overfeeding, chemicals, and organic debris can all produce an off-flavor in catfish. When the water temperature is high—as in late summer—there is a greater chance of off-flavors occurring in the meat.

## NUTRITIONAL TOXICANTS OF FISH

In addition to being susceptible to excesses and deficiencies of certain nutrients, fish are highly susceptible to a number of chemicals that often are found in water—either naturally or as a result of soil pollution due to past agricultural or industrial practices. Metal toxicity can result from the earth surrounding the pond. In the past, pesticides and industrial chemicals have been routinely dumped in waterways. Organisms, such as red tide, can grow in the water and poison fish.

### HEAVY METAL TOXICITIES

Several metals have been shown to be toxic to fish. In some cases, the toxicity may result from natural deposits of the metal in the environment while often it can be traced to needless pollution by humans. The amount of literature dealing with heavy metal toxicity in fish is limited. Mercury poisoning in fish has received the most attention, but cases of toxicities from arsenic, chromium, copper, and selenium have been documented.

Of all the forms of mercury found in the environment, methylmercury appears to be the most toxic. This is the most prevalent form found in the mercury-contaminated fish. Fish absorb mercury at an extremely rapid rate while they metabolize and excrete it very slowly; thus, there is a cumulative effect of the element when fish are exposed to a contaminated environment. The element is widely distributed through the body and the effects can be devastating. Cataracts result when mercury accumulates in the eye. Reproduction is impaired when the gonads are affected. In general, there is a decreased ability of the fish afflicted by heavy metal toxicants to absorb sodium; hence, the osmotic regulatory mechanism of the fish is adversely affected.

### ORGANIC COMPOUND TOXICITIES

Numerous naturally occurring and synthetic organic compounds elicit toxic responses in fish. Aflatoxin, nitrosamines, cyclopropenoid fatty acids, and tannic acid have all been shown to induce liver cancer in fish. Gossypol—a toxin present in glanded cottonseed—causes loss of appetite and ceroid accumulation in the livers. Phytic acid, which ties up zinc in feed, and growth inhibitors found in soybean meal can be destroyed by proper heating during processing. Chlorinated hydrocarbons occur as contaminants of fish meal and can cause mortality when present in fry feeds; and brood fish transfer these compounds from the feed to their eggs, resulting in low hatchability and high mortality of swim-up fry. Toxaphene affects the utilization of vitamin C in catfish and can cause the *broken back syndrome*.

## DISEASES AND PARASITES OF FISH[2]

Intensive production of fish increases the likelihood and severity of disease and parasite outbreaks. Many fish farmers have learned that some diseases and parasites can kill an entire fish population in a short time. Diseases cause economic losses not only from mortality, but also from treatment expense, growth reduction during and after an outbreak, and postponement or loss of the opportunity to ship or sell the fish.

Serious infectious diseases are also caused by bacteria, fungi, and viruses. Erratic behavior and syndromes characteristic of infectious diseases may be produced by environmental stressors (low oxygen, supersaturation of dissolved gases such as nitrogen or oxygen), toxins (hydrogen sulfide, ammonia, nitrite, many pesticides, and even therapeutic agents used to treat infectious diseases), and nutritional deficiencies. The parasites described in Table 56–13 include protozoans, trematodes, (flukes), cestodes (tapeworms), nematodes (roundworms), leeches, and crustaceans (fish lice and anchor parasites).

It is essential that a fish farmer observe the fish daily and be able to recognize the signs of fish diseases, and to notice problems when they first appear. Good management is the key to successful production of healthy fish.

## FISH DISEASE SIGNS

Diseased fish usually exhibit either physical or behavioral signs, or both. These signs can be helpful, but also misleading; in almost all instances an accurate evaluation can be made only at a diagnostic laboratory.

The following *behavioral signs* indicate that certain diseases may be present:

- **Failure to feed**—many diseases.

- **Swimming weakly, lazily, erratically, or in spirals**—many diseases.

- **Scratching, flashing, or rubbing against objects in the pond**—external parasite.

- **Twitching, darting, convulsions**—toxins; nutritional disease; external parasites.

- **Failure to flee when exposed to fright stimuli**—low oxygen; metabolic factors; many diseases.

- **Crowding or gathering in vegetation, shallow water, or at water inflow, hiding under objects to avoid light**—many diseases; low oxygen; toxins.

- **"Topping" or "piping" at water surface, floating head-up, moribund (dying)**—low oxygen; toxins; external parasites; bacteria.

Some *physical signs* suggest disease. The following associations are often made:

- **Dead or dying fish**—many diseases.

---

[2]The narrative in this section on "Diseases and Parasites of Fish" was adapted by the author from *Third Report to the Fish Farmers*, edited by Harry K. Dupree and Jay V. Huner, published by U.S. Department of the Interior, Fish and Wildlife Service, 1984, pp. 177–183.

- **Open lesions or sores, bloody or reddened areas—**bacteria; bacteria secondary to parasite infections; external parasites; toxins.

- **Gaping mouths—**low oxygen; diseased gills.

- **Scale loss—**Myxobolus notemigoni (milk scale disease); external parasites; fighting; predation; rough handling.

- **Gills pale, eroded, puffy, bloody, or brown, or gill covers flared—**anemia; vitamin deficiency; gill disease; environmental stress; toxins; external parasites; Branchiomyces (fungus); Flexibacter columnaris (bacterium).

- **Bleached skin color—**vitamin E deficiency; low oxygen.

- **Exophthalmia (popeye), stargazing—**bacterial dropsy; brain flukes; gas bubble disease; malnutrition; environmental contaminants.

- **Bloated belly (dropsy)—**bacteremia; white grubs (flukes); Ligula (tapeworm); catfish virus (affects fingerlings).

- **Excess mucus (light gray film), sloughing of skin, scratches on body—**external parasites; fungus; fighting; predation.

- **Spinal curvature—**vitamin C deficiency; pesticides; genetic deformities.

- **Folded fins or tail, pectoral fins pointed forward—**toxins; many diseases.

- **Nodules, pustules, white spots—**myxosporidian cysts (protozoans); larval trematodes (flukes); Ichthyophthirius or "Ich" (protozoan); yellow grub (fluke); larval nematodes.

- **Fluid in body cavity (cloudy, bloody, or clear)—**bacterial dropsy; channel catfish virus; malnutrition.

- **Bloody internal organs—**bacteria; virus; vitamin A or B deficiency.

- **White "fungus" patches—**external fungus; Epistylis (protozoan).

- **Frayed fins or tail, eroded tail—**external parasites; Flexibacter columnaris or other bacteria; chemical contaminants.

- **Emaciation (thin fish, pinheads), reduced growth—**any disease that causes fish to reduce feed intake or cease feeding; underfeeding; malnutrition; intestinal worms (helminths); vitamin deficiency.

- **Air bubbles under skin—**gas bubble disease (excessive nitrogen or oxygen in the water).

- **Cloudy eyes—**eye flukes; nutritional deficiencies.

- **Red spots near bases of fins—**larval Lernaea (copepod); external parasites; bacteria.

- **Gray, chalky white, or dull opaque yellowish ovaries or eggs in golden shiners—**Pleistophora ovariae (protozoan).

- **Ruptured abdomen—**toxic algae (in fry); Ligula (cestode); white grub (trematode).

- **Dirty gray or yellow lesions—**bacteria; external parasites, external fungi.

- **Foul-smelling lesions—**Edwardsiella tarda (bacterium).

- **Hole-in-the-head—**Edwardsiella ictaluri (bacterium).

- **Brown blood—**nitrate toxicity.

Farmers with major investments in fish culture should seriously consider arranging for training in parasite identification and disease diagnosis for themselves and key personnel, and setting up basic facilities for examining fish and monitoring water quality. This training may provide diagnosticians with vital information that could save time in verifying or selecting the proper treatment.

- **Submitting fish and samples for diagnosis—**After observing fish with disease signs, the farmer must provide the diagnostician with specimens and information to ensure an accurate evaluation of the problem.

## ENVIRONMENT AND DISEASE

Because fish are totally dependent on water, it is essential for the farm manager to know as much as possible about this complex environment—not only the water source and volume for each pond, but also such characteristics as dissolved oxygen concentration, total hardness, pH, alkalinity, nitrite and ammonia content, and temperature. Variations of any of these factors could affect the health of fish.

Fig. 56-32. *Top:* normal striped bass. *Bottom:* environmentally damaged striped bass. (Courtesy, Dr. H. K. Dupree, Director, Fish Farming Experimental Laboratory, U.S. Dept. of Interior, Stuttgart, AR)

Stress caused by extremes in environmental conditions plays an important role in fish diseases. Low-oxygen stress is considered to be a major cause in outbreaks of *Aeromonas* (a bacterial pathogen), and channel catfish virus (CCV) disease during summer. *Flexibacter columnaris* (a bacterial pathogen) is associated with stress due to high water temperature and organically rich water.

Various management practices such as handling, hauling, excessive chemical treatments, overcrowding, and the feeding of nutritionally inadequate feeds may also impair fish health. Overcrowding of fish is common and is associated with many bacterial and protozoan diseases. Disease organisms are nearly always present in the water—ready to become problem-causing pathogens when fish are weakened by stress. Holding stress within the tolerance limits of the particular fish species being raised is a role of good management. The profit motive often overrules good management techniques, however; consequently, coping with a host of problems caused by poor water conditions is often considered a necessary headache.

## NUTRITION AND DISEASE

The absence or a deficiency of essential food ingredients can cause nutritional diseases. Mineral and vitamin deficiencies are summarized in earlier tables in this chapter. (See Table 56–3, Fish Mineral Chart; and Table 56–4, Fish Vitamin Chart.)

## DISEASE PREVENTION

Only a few federally registered therapeutics are available for treatment of fish diseases, and it is generally recognized that disease prevention is more economical than treating sick fish with costly drugs and chemicals. Experience has shown that the following three management practices often prevent disease outbreaks.

1. **Suitable pond management.** Where infectious disease organisms are known to occur, most can be eliminated from the pond after the fish have been harvested. If possible, the pond should be drained and allowed to dry. If it cannot be completely dried, all remaining fish can be killed by applying just enough fresh calcium hydroxide (hydrated lime) to cover the entire pond area (about 2 tons per acre), and additional liberal amounts to the puddles and wet spots. This chemical leaves no final toxic residue. The application of calcium hypochlorite (HTH) to water at the rate of 40 lb per acre-foot (10 ppm available chlorine) has been used for killing fish, and also kills many types of disease organisms. The chemical should remain in the pond 2 or more days before the pond is drained or flushed. The HTH must be used with caution, because the chemical is corrosive and very irritating to the skin, eyes, and nose.

If all the fish are kept out of the pond for several weeks, many fish pathogens (but not their spores) will die. Treatment of brood stock or fingerlings to remove the external parasites is a simple precautionary measure. Also, the exclusion of wild fish that enter fish ponds may help avoid disease outbreaks.

2. **Maintenance of a disease-free water supply.** Ordinarily, no pathogens are introduced with well water. Water from springs, however, may contain wild fish that pass infectious organisms along to the cultured fish. If water from streams or reservoirs containing fish is used, one can expect pathogens to be introduced into the culture ponds. At some trout hatcheries and warmwater fish farms, the contaminated incoming water is passed through sand and gravel filters to remove certain protozoans and larger parasites. Reservoir water can be treated with rotenone to remove unwanted fish, or with HTH to remove predators and parasites. The chemicals must be dissipated or neutralized before fish are stocked into the pond. There is no practical way to remove bacteria and viruses from large, flowing supplies of surface water.

3. **Constant surveillance.** Where there is danger of transmitting infectious diseases through fish transfer, the fish should first be inspected by a diagnostician. The practice of disinfecting nets, seines, tubs, and other equipment before and after harvesting and transporting fish, even though no disease problems are observed, reduces the possibility of introducing or spreading infectious diseases.

## SUMMARY OF DISEASES AND PARASITES OF FISH

For convenience, the diseases and parasites of fish are grouped as follows:

- **Table 56–13, Bacterial Diseases of Fish**—*Bacteria are the smallest and simplest known form of plant life. They are microscopic, possess just one cell, vary in shape, multiply by transverse fission, and possess no chlorophyll.*

- **Table 56–14, Fungal Diseases of Fish**—*Fungi are distinguished by the formation of mycelium (a network of filaments or threads), or by spore masses.*

- **Table 56–15, Viral Diseases of Fish**—*Viruses are disease-producing agents that (a) are so small that they cannot be seen by using an electron microscope, (b) are capable of passing through the pores of special filters which retain ordinary bacteria, and (c) propagate only in living tissue.*

- **Table 56–16, Parasitic Diseases of Fish**—*Parasites are organisms living in, on, or at the expense of another living organism.* The parasites listed in Table 56–16 include nematodes (roundworms), trematodes (flukes), protozoa (internal and external), and crustacean (fish lice and anchor parasites).

**TABLE 56-13**
**BACTERIAL DISEASES OF FISH**

| Common Name | Distribution | Cause | Clinical Signs | Prevention/Treatment |
|---|---|---|---|---|
| **Bacterial gill disease** | Mainly in hatchery salmonoids and aquarium fish. | *Flavobacteria.* Associated with crowding, poor water quality, and high ammonia. | Impaired gill function, and breathing problem. | **Prevention:** 1. Improving water quality. 2. Avoiding over stocking. **Treatment:** Antibacterial drugs to control secondary infections. |
| **Coldwater disease/ fin and tail rot** | Most common in coldwater fish, but may occur in warmwater fish at low temperatures. | *Cytophaga psychrophila.* | Skin ulcerations; kidney involvement. | **Treatment:** 1. Oxytetracycline orally. 2. Sulfamethazine or sulfisoxasole orally. |
| **Columnaris disease** (cottonmouth disease, saddleback) | Worldwide occurrence. Affects all species of freshwater and aquarium fish. | *Flexibacter columnaris.* | Cottonmouth, skin lesions, some scale loss, tail rot. | **Treatment:** 1. Oxytetracycline orally. 2. Sulfamethazine or sulfisoxasole orally. Both treatments to be administered according to manufacturer's directions. |
| **Edwardsiella ictaluri** (hole-in-the-head) | *Edwardsiella ictalur.* Afflicts harvestable-size channel catfish. | | Lesions on the sides, redness at the base of the fins and on the belly, and sometimes a slit beween the eyes exposing the brain. | |
| **Edwardsiella tarda** | | *Edwardsiella tarda.* Attacks channel catfish in ponds and raceways. | Large deep lesions on the sides of fish, usually behind the pectoral fin, which emit a highly objectionable odor when cut open. | **Prevention and control:** Remove infected fish. Control by feeding oxytetracycline according to manufacturer's directions. |
| **Fish tuberculosis.** | Common in aquarium fish. | *Mycobacterium piscium: M. platypoecilus; M. fortuitum.* | Emaciation, skin ulceration and hemorrhages, and skeletal deformities. | Avoid infection; destroy infected fish. |
| **Furunculosis/ulcer disease** | Worldwide, except Australia and New Zealand. Chiefly disease of salmonoids and goldfish. | *Aeromonas salmonicida.* | Boils (furuncles) on skin; ulcer disease of goldfish. | **Prevention:** Avoid infected fish. **Treatment:** 1. Sulfamerazine in food. 2. Oxytetracycline in food. |
| **Hemorrhagic septicemia** | Worldwide occurrence in most species of fresh water fish. | *Aeromonas hydrophilia.* Associated with stressful freshwater, pond culture, or aquarium conditions. | Fish not feeding, redness at base of fins, bloated, and eyes may bulge (popeyed). | **Prevention:** Avoid infected fish. **Treatment:** Oxytetracycline or chloramphenicol orally or intraperitoneally. |
| **Kidney disease** (dee disease) | Economically important in cultured salmonoids. But occurs in other species. | *Corynebacterium.* | Blindness and emaciation. | Obtain disease-free stock and prevent contamination by infected wild stock. |
| **Nocardiosis** | | *Nocardia asteroides.* | Emaciation, skin hemorrhages, and skeletal deformities. | Avoid infection; destroy infected fish. |
| **Pseudomonas diseases** | All species of freshwater fish seem susceptible. | *Pseudomonas fluorescens.* Capsulated form very virulent. | Hemorrhagic septicemia with external reddening and hemorrhages are found in the peritoneum. | Antibiotics in food or water, according to manufacturer's directions. |
| **Ulcer disease** | Chiefly brook trout in the U.S. | *Hemophilus piscium.* | External ulcers; septicemia. | Oxytetracycline or chloramphenicol orally. |
| **Vibriosis** | Worldwide; mostly in marine and estuarine environment; also common in aquarium fish. | *Vibrio anguillarum.* | Skin, fin, and tail hemorrhages and ulcerations; hemorrhagic and degenerative changes of internal organs. | **Prevention:** 1. Avoid overcrowding. 2. Minimize stressful conditions. 3. Vaccination with formalin-killed *vibris.* **Treatment:** Sulfamerazine used according to manufacturer's directions. |

## TABLE 56-14
## FUNGAL DISEASES OF FISH

| Disease | Distribution | Cause | Clinical Signs | Prevention/Treatment |
|---------|--------------|-------|----------------|----------------------|
| Branchiomycosis (gill rot) | Common in cultured European foodfish. | *Branchiomyces sanguinis.* | The first sign is red spotting on the gills. Later, the gills become grayish-white and stop working, and the fish suffocate and die. | **Prevention:** Remove decaying organic material. Thin out fish, increase water flow, and clean ponds between stockings. |
| Ichthyophonus | A common fungal infection in wild fish and aged cultured and aquarium fish. | *Ilchthyosporidium hoferi.* | Characteristic spherical cyst stages observed microscopically in the smears of lesions of the heart, liver, spleen, kidney, skin, and muscle. The disease usually is chronic and progressive. | **Prevention:** Remove infected fish and avoid feeding fish products that contain the organism. There is no known treatment. |
| Saprolegnia infection | A common fungal infection of fish and fish eggs. | *Saprolegnia,* a fungus. | Grayish-white, cottonlike growths on the skin, gills, eyes, or fins, which may invade deeper tissues of the body. | **Prevention:** Remove the predisposing causes—dead and infected fish and decaying organic material. |

## TABLE 56-15
## VIRAL DISEASES OF FISH

| Disease | Distribution | Cause | Clinical Signs | Prevention/Treatment |
|---------|--------------|-------|----------------|----------------------|
| Carp pox | | Invasion of tissue by the virus particles. | Poxlike gelatinous lesions on the skin surface. | **Prevention:** Select breeding stock that test negative. |
| Channel catfish virus disease | | Herpesvirus. | Popeye, bloated abdomen, and redness of the belly. The gills are pale. Mortality can reach 80%. | No known treatment. |
| Golden shiner disease | | Virus. | A red back or red head, caused by expanded blood vessels. | Avoidance is recommended. There is no known cure. |
| Hepatoma (epithelioma) | | Invasion of tissue by the virus particles. | Nodules in the liver. | No known treatment. |
| Herpesvirus disease of salmonoids | Afflicts young rainbow trout and Kokanee salmon. | Virus of the rhabdovirus group. | Anemia, exophthalmia, and ascites. Mortality is at least 50%. | |
| Herpesvirus disease of turbot | Afflicts both wild and cultured turbot. | Virus. | Hypertrophy and fusion of epithelial cells of the skin and gills of young fish. Heavy mortality associated with heavy gill infections and poor water quality. | **Treatment:** Maintenance of high levels of oxygenation. |
| Infectious hematopoietic necrosis | An infection of salmonoids. | Rhabdovirus. | Produces high mortality in fry and fingerlings. Victims darken, show pale gills and exophthalmia, and shed thick fecal pseudocasts. | **Prevention:** Raise fry separately, and avoid introduction of virus by disinfecting nets, pails, and other implements. |
| Infectious pancreatic necrosis | A disease of salmonoids. | Virus. | High mortality in young trout. Occurs subclinically and chronically in other salmonoids. Signs of the disease include a dark color, abdominal swelling, popeye, and victims whirling about their long axis. | Stock fish from certified disease-free hatcheries. Raise fry separately from older fish. |
| Lymphocystis | Affects both wild and captive marine and freshwater species. | Invasion of tissue by the virus. | Whitish rounded lumps on the fins or skin, sometimes nearly covering the body surface. | No known treatment. Control should be directed to getting rid of the affected fish. |
| Pike fry rhabdovirus disease | An infection of young northern pike. | Virus. | Affected fry have pale gills, exophthalmia, and hydrocephalus. | |
| Spring viremia of carp. | It has occurred only in Europe and the U.S.S.R. | A rhabdovirus disease of cultured carp. | Petechiation of the skin, gills, and visceral mass. | Oxytetracycline administered according to manufacturer's directions, will lessen secondary infections. |
| Viral erythrocytic necrosis | It affects salmon, herring, cod, and other marine species. | Causative agent is thought to be an iridoviridae. | | |
| Viral hemorrhagic septicemia | Affects rainbow trout of all ages; confined to Europe. | A rhabdovirus disease. | Victims show signs of anemia, hemorrhages of the skin and fins, and are popeyed. | |

## TABLE 56-16
## PARASITIC DISEASES OF FISH

| Disease | Cause | Clinical Signs | Prevention/Treatment |
|---|---|---|---|
| **Worms** | | | |
| Cestodes | A large Asian tapeworm. | Greatly swollen belly. | Treat held-over brood stock with di-n-butyl tinoxide and drain/lime pond. |
| Leeches | *Trypanosma cryptobia.* | Worms attached to the body, fins, gills, and mouth. | Trichlorfon used according to manufacturer's directions. |
| Nematodes, internal (roundworms) | Ingestion of parasite at infective stage. | Weight loss or emaciation; fish may be bloated; the worms frequently extrude from the anus. | Piperazine citrate or Levamisole according to manufacturer's directions. |
| Trematodes | *C. complanatum,* the yellow grub. *P. minimum,* the white grub. | Edges of gills are gray and thickened; open opercula. May also appear on body as bluish white turbidity. Larvae develop in the eye of channel catfish and other warmwater fishes, causing opaque, white lens. Heavy infestations may cause blindness. | Destruction of snails, the intermediary host, is the best control. |
| **Protozoa** | | | |
| Coccidia | Ingestion of infective larvae. | Yellow, bloody feces; nodules on mucosa. | Sulfamethazine in the feed according to manufacturer's directions. |
| Hexamita | A small protozoan parasite. | Emaciation. | Metronidazole in fish food according to manufacturer's directions. |
| Myxobolus (nodular disease) | Ingestion of spores. | White elevated areas, often covering the entire body. | Partial drainage of ponds followed by addition of chlorine and eventual drying. |
| Myxosoma (whirling disease) | Ingestion of spores. | Whirling movements; mandibular and vertebral deformities. | The disease can be prevented by purchasing fingerlings or brood stock from hatcheries that are certified-free of the disease. |
| **External Protozoa** | | | |
| Chilodonella | Two species, *C. cyprini* and *C. hexasticha,* afflict channel catfish. | **Acute form:** Affects gills, sudden death. **Chronic form:** Causes respiratory signs, paleness, weakness, and uncoordinated swimming. | Reduce fish population, increase water flow, and decrease organic matter in water. |
| Flukes | 1. Body fluke. 2. Gill fluke. | Irritation and hemorrhage at base of fins and on body. Gill damage. | Trichlorfon used according to manufacturer's directions. |
| Henneguya | A sporozoan parasite. | White cyst formations in skin, muscles, and gills. Breathing difficulty. | No treatment for the parasite. Drainage, treatment with chlorine, and drying will reduce number of spores. |
| Ichthyopthirius (ich or white spot) | Invasion of skin by infective stage. | White spots on skin. Fish may appear sluggish and lie at the bottom of the pond, raceway, or aquarium. | The best control is prevention. No drug can safely penetrate and kill the encystated parasite in the skin, but each of the following drugs will kill the organism in the aquarium water: 1. Quinine hydrochloride, 10 ppm. 2. Sodium chloride, 30,000 ppm. 3. Nifuripirinol, 0.05–0.2 ppm. |
| Ichtyobodo (costia) | Invasion by the protozoan parasite. | Steel-gray discoloration of the skin. Breathing difficulty. | Increasing rate of flow and oxygenation of water. |
| Oodinium (velvet disease) | Invasion by infective stage of parasite. | Whitish-gray to dirty yellow nodules on skin and fins. | Copper sulfate according to manufacturer's directions. |
| Plistophora | A sporozoan parasite. | Reduced weight, growth, and spawn. | No effective cure. Replace brood stock annually. |
| Trichodinids | Variety of finfish pathogens. | Breathing difficulty. | No treatment for food fish is known. Prevention consists of reducing density and organic content of water. |
| Trichophyra | A protozoan parasite belonging to the class *Suctoria.* | Infected gills, breathing difficulty. | Copper sulfate, potassium permanganate, and formalin have been used to treat fish. |
| **Crustaceans** | | | |
| Copepods | Many genera of freshwater and marine parasitic copepods exist. | Barblike attachments to the skin or gills. | Trichlorfon is the drug of choice for treating infested fish, used according to manufacturer's directions. |

# AQUACULTURE AND MANAGEMENT

Aquaculture embraces all aquatic animals—from frogs to fishes, including alligators, crayfish, eel, lobsters, prawns, and shrimp. Each of the major species of aquatic animals is briefed in the earlier section headed "Species and Specialized Aquaculture." Although species make for differences in culture and management, the same principles apply to all of them. So, because of space limitations, the sections that follow will be limited to the culture and management of fish—by far the leading type of aquaculture.

Fish farming can be divided into two types of production—freshwater and marine water.

## KINDS OF WATER FACILITIES

The kind of water facility constructed will depend on the size of the farm and the type of fish farming planned. The most commonly used structures are: ponds, raceways, tanks, and aquariums.

● **Ponds**—This refers to any size of earthen structure for holding a standing body of water—water that does not flow or that has limited flow. Except for trout and salmon, ponds are the most common water-holding facility for fish.

Several kinds of ponds are employed in fish farming, with the kind of pond determined by the function or use; the common kinds are:

1. **Holding pond.** A holding pond is used for the brood fish between spawning seasons. It may range from ¼ to 1 acre in size.

2. **Spawning pond.** A spawning pond is a small pond in which the brood fish are placed for spawning; they may be placed in pens or in the open pond.

3. **Rearing pond.** A rearing pond is a small pond in which fry are placed for growing into fingerlings.

4. **Growing pond.** This is a pond in which fingerlings are grown to market fish size.

5. **Catch-out pond.** This is a pond in which fish are grown, or held, until caught by sports fishers.

Three types of ponds may be constructed for fish culture:

1. **Ravine pond.** Constructed in a gorge, gully, or similar watershed area.

2. **Excavated pond.** Constructed in a fairly level area and not subject to overflow from creeks or rivers.

3. **Levee pond.** Construction on flat land with the levee on all sides.

Fig. 56-34. Raceways, 6 ft × 60 ft at the Bozeman Fish Technology Center. (Courtesy, C. E. Smith, Director, Fish Technology Center, U.S. Fish & Wildlife Service, Bozeman, MT)

Fig. 56-33. Ponds. (Photo by A. H. Ensminger)

Fig. 56-35. Round tank in an Atlantic salmon hatchery. (Courtesy, C. E. Smith, Director, Fish Technology Center, U.S. Fish & Wildlife Services, Bozeman, MT)

Fig. 56-36. Troughs and diameter circular tanks at the Bozeman Fish Technology Center. (Courtesy, C. E. Smith, Director, Fish Technology Center, U.S. Fish & Wildlife Service, Bozeman, MT)

- **Raceways**—Raceways are usually rectangular structures of concrete, block, tile, bricks, or other durable material. A few raceways are circular, and a few are earthen. Raceways are used almost exclusively for trout, although a few are used for catfish. Flowing water is used in raceways, with the waste removed and oxygen replaced fast enough to intensify production greatly.

- **Tanks**—Tanks are generally circular, 3 to 30 ft in diameter, and constructed of concrete or fiberglass. Tanks are well suited for holding a small number of fish of varying ages and sizes for selective breeding, nutrition studies, and many types of research projects.

Raceways and tank culture is practical only if large quantities of cheap, high-quality water are available for a *once-through* or *open* system, or if the water can be recirculated and wastes efficiently removed for a *closed* system.

- **Cages**—These are small wire or fiber mesh enclosures over a supporting frame, which are attached to floats and anchored in rivers, lakes, or large ponds. Cages are used for catfish or rainbow trout. They are generally 39 in. deep, wide, and long.

- **Pens**—These are similar to cages, but they are much larger. In northwestern United States pens approximately 65 ft square × 33 ft deep are used in growing salmon.

## SELECTION OF FISH

In the United States, about 11% of the fish harvest is the product of fish farming. Catfish is the largest crop, grown primarily in Mississippi, Arkansas, and Louisiana. Catfish cultivation is relatively simple because the fish are omnivorous feeders and reproduce easily. Carp farming is also a possibility in the United States; it was an important business in some areas in the past.

Trout culture requires large quantities of pure, cool water. Of the cultured-trout crop, 90% is produced in Idaho, where there is an abundant supply of fast-running river water. Fertilized eggs are obtained from brood stock and the fish are raised in long, shallow tanks. They can be held there until it is time to harvest them, or they may be used as game fish for stocking local streams.

The production of Pacific salmon in hatcheries, and their release into the ocean to augment wild stocks, has been an industry in the United States for more than a century. Hatcheries raise the spawn to young smolts that are released into the ocean and caught when they return to spawn. They may also be raised in pens and harvested after one year, when they weigh about 1 lb.

Culture systems for other desirable aquatic species are also being developed. Freshwater shrimp are raised successfully in Hawaii, and the Japanese have established a small marine shrimp industry. In addition, attempts have been made to raise lobsters, abalone, crabs, and oysters, but commercial success has been limited. Some species of marine fish, especially yellowtail and sea bream, are farmed in Japan, but the high cost of raising these fish limits their sale to the luxury market.

## PARTS OF A FISH

The major parts of all fish perform the same functions, and they are located in about the same places on the fish's body. But the size, shape, and color are often different, which helps

to tell fish apart. Also, knowing how a healthy fish looks is important.

All fish have a tail consisting of the *caudal peduncle* and the *caudal fin*. The fish's fins help it steer through the water and hold it upright in the water. Often a sick fish cannot steer or flops over on its side. Other fins on the body include:

1. **Pectoral.** Usually located on the sides of the fish behind the head.

2. **Pelvic.** Usually located towards the rear of the body.

3. **Dorsal.** Runs along the top of the fish. May be single or double. The second dorsal fin is sometimes called the *soft dorsal fin*.

4. **Anal.** Usually located right behind the *anal vent* (anus) on the rear bottom end of the fish.

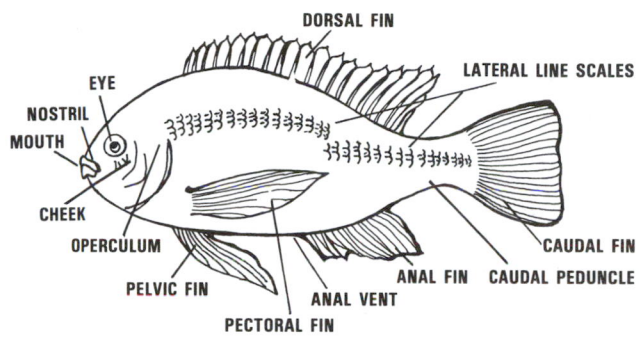

Fig. 56–37. Parts of the Tilapia, a pond fish.

Most fish have *eyes*, but even with eyes fish cannot see very well. All fish have *gills*. The gills are covered by a flap called the *operculum*. The gills are extremely important. Fish take in water through their mouths. The water is then passed through the gills which remove the oxygen and nutrients from the water. The water is then passed outside of the body of the fish through the gill slits.

It is possible to tell a lot about a fish's health and eating habits by looking at its gills. Fish with many, many feathery gill rakers and few if any teeth eat the smaller foods in the pond. Fish with few and larger gill filaments eat the larger particles from the pond. Healthy gills are a bright red color. When fish farmers see fish with gills that do not have this healthy red color, or have white spots all over, for example, they know that fish is not healthy and should not be bought or placed in their ponds. If the fish is already in their ponds, they know they must take steps to get rid of the disease before it afflicts more fish.

Other identifying parts that all fish have are the *mouth*, the *genital openings* (to reproductive organs), and the *lateral line*. The lateral line is a small line of nerve cells which runs along the length of the body about midway on the side of the body. Sometimes the lateral line is covered by a layer of scales; sometimes it is a different color from the rest of the body. In any case, the lateral line is an area of sensitivity that helps the fish feel pressure and temperature changes in the water around it.

Some fish, like catfish, have *barbels*, small projections that hang down from the sides of the mouth. Barbels help the catfish sense its surroundings, find food, and attract small fish to the catfish so that it can eat them.

## WATER QUALITY

Fish growth depends greatly on the quality of the water used in the pond; and the quality of the water depends upon where it comes from. Testing the water quality means making sure that all the factors which relate to the water are right for the fish. These factors are: temperature, oxygen content, pH, turbidity, hardness, alkalinity, and nutrient availability (source of food for the fish). Farmers do not need to know these particular words to raise fish well, but they do need a working knowledge of the factors that are part of the water world in which fish live.

Fig. 56-38. Aerator in operation. (Courtesy, Ann Gannam, UAPB Exp. Sta., Pine Bluff, AR)

## MANAGING THE POND

As defined at the beginning of this chapter, *Management* is the art of caring for, handling, or controlling. It follows that *managing the pond* involves (1) stocking properly, (2) servicing the pond, (3) managing fry and fingerlings, (4) managing brood stock, and (5) controlling predators.

### STOCKING PROPERLY

The *stocking rate* is the term used to refer to the number of one species which are put into a pond. Therefore, in a monoculture pond, the stocking rate is the same as the stocking density because there is only one kind of fish.

Stocking rate and density are important. There is only enough food and room in a pond for a certain number of fish. The good growth of fish depends upon putting the right number of fish into the pond.

The age of the fish must also be considered when stocking ponds. For example, more fingerlings can be placed in a pond than brood fish, because fingerlings require less food per fish than brood fish. If the food available in the pond is not supplemented, proper stocking rates and density are even more important.

Disease-free fish should be stocked regardless of whether fish are to be put into a pond, a raceway, or a home aquarium. Also, quarantine of newly purchased fish is a good practice, especially if the disease status of the fish is not known.

## SERVICING THE POND

After pond is stocked, ongoing management includes the following:

1. Feeding and fertilizing as necessary.
2. Keeping the pond in good condition.
3. Watching for trouble and disease.

Each pond, whether it is small or large, one pond or one of several, requires supervision in the above areas; and good management requires that checks of the condition of the fish and the pond be a regular part of the pond manager's day.

Ponds and the fish in them must be taken care of every day. It is a good idea to have the pond manager follow a checklist of things to do. Daily care will greatly lessen the chance that something will go wrong in the pond. A suggested checklist follows:

1. Check the pond for leaks.
2. Clean filters.
3. Watch fish behavior near the feeding area.
4. Feed the fish. (See earlier section on "Feeding Fish.")
5. Add fertilizer, if necessary. (See the earlier section on "Natural Foods and Feeds.")
6. Watch for predators.

**Note well:** Check the ponds at the same time each day. Early morning is the best time because oxygen levels in the water are lowest then, and the fish are more likely to have trouble at that time of day—if they are going to have trouble at all.

## MANAGING FRY AND FINGERLINGS

Large fish farmers usually find it more convenient and economical to produce their own fry, rather than purchase them from other farmers. However, smaller farmers usually purchase fingerlings from hatcheries that specialize in fingerling production.

Fig. 56-39. Fry in a nursing tank (about 10 days old). (Courtesy, Ann Gannam, UAPB Exp. Sta., Pine Bluff, AR)

Fry should be counted, then moved to either a rearing trough or pond.

Food fish farmers must secure good quality fingerlings at a reasonable price. Whether purchased or raised, fingerlings should be healthy and free from diseases and parasites.

Wherever the fry or fingerlings come from, it is important that a count be made before putting them into a pond, raceway, tank, or other container. This procedure can help farmers evaluate their management. Thus, if an excessive number of the fish dies between the fry or fingerling stage and the time of harvesting for market, the farmer should determine the reasons for the loss before restocking with fry or fingerlings.

## MANAGING BROOD FISH

*A brood fish is a fish that has reached its full growth and is able to reproduce.* The age at which this happens depends upon the kind of fish, the climate, and the quality and amount of food.

Fig. 56–40. Striped bass brood fish. (Courtesy, Dr. H. K. Dupree, Director, Fish Farming Experimental Laboratory, U.S. Dept. of Interior, Stuttgart, AR)

Good brood fish are essential for good fingerling production. Fingerlings can be no better than the brood fish used. This calls for the careful selection and proper management of breeding stock. Brood fish must receive year-round care if large, viable spawns are to be produced.

The specific characteristics of brood fish are basically the same for every fish species. In general, good brood fish are:

1. Well formed. The female should have a well-rounded belly, indicating spawning potential. The male should have a head wider than the body, and large and protruded genital papilla.
2. Free of parasites and disease.
3. Lively and active.
4. More than 36 months of age.
5. Sexually mature (so they can be separated by sex).

## CONTROLLING PREDATORS

*Predators are animals that live by preying on other animals.* The common fish predators are snakes, fish-eating birds, insects, and certain species of turtles. Sometimes, alligators become poachers. Otters are a problem in Florida. If present in large numbers, each of these may inflict considerable losses; hence, their populations should be monitored and controlled.

## HARVESTING AND MARKETING FISH

*Harvesting means capturing fish.* Two types of harvesting are used: (1) partial, or (2) complete.

Fig. 56–41. A basketful of farm-raised channel catfish ready for market, weighing 2 to 2½ lb. (Courtesy, Mississippi State University, Mississippi State)

Harvesting fish raised in tanks is relatively easy; they are lifted out with a dip net. Raceways are usually harvested by using a seine. Pond harvest is much more complex. In big ponds, fish are harvested by (1) draining, (2) seining, (3) trapping, and (4) using a fish holding bay (or live car). The latter is the newest technique. It consists of having a fish holding bag in the middle of the chute formed by the seine net. The fish are crowded into the live car, which holds about 5½ tons of fish. The filled car is detached and moved to the bank, and an empty car takes its place.

Marketing is the end of the line! It involves all those activities that occur between production and consumption. Marketing may be either (1) simple and direct to the consumer, as in a fish-out operation; or (2) complex, as where the product is changed in form and undergoes many buying and selling transactions before reaching the ultimate consumer.

For three freshwater products—catfish, trout, and bait fish—the amount of competition from wild fish sources varies. Cultured catfish make up over half of the quantity sold. Cultured trout represent all the trout marketed, including all the trout captured from the *wild*. For bait fish, the proportion of cultured and captured fish cannot be determined. In all cases the cultured and the wild fish are marketed through the same channels.

Knowing the market is the key to success in aquaculture. **The rule:** Do not put any fish in the water that cannot be sold when they are taken out. Commercial aquaculture is a business; hence, it should be treated as such.

To bolster consumer confidence, there is need for mandatory inspection of seafood products similar to the inspection accorded beef, pork, and lamb. The U.S. Congress is considering such legislation, but two major problems must first be resolved: (1) What federal agency will do the inspecting, and (2) who will pay for the inspection?

## GLOSSARY OF FISH

**Aeration:** Adding oxygen to water by spraying or bubbling air through it.

**Algae:** Small or large water plants from five classes of plants.

**Alkalinity:** The ability to combine with an acid to form a salt.

**Aquaculture:** The cultivation of animal and vegetable life in water.

**Backwashing:** Forcing water in the opposite direction from its normal flow.

**Barbels:** Sensitive organs that hang down on the sides of the mouth of certain fishes.

**Bloom:** A very good growth of algae in a pond that has a strong green color.

**Bottom feeders:** Fish that feed on bottom organisms (organisms that live in mud on the bottom of the pond).

**Breeding:** The cycle of reproduction in animals.

**Brine:** Water that is saturated with common salt, or the water from a salt water body (the ocean).

**Brood ponds:** Ponds where the fish used for breeding are kept.

**Brood stock:** The fish used for breeding in fish ponds.

**Cage:** An enclosure to hold fish in the water.

**Captivity:** The state of being held in a confined place (fish in ponds are captive).

**Carnivore:** An organism that eats animal products.

**Contaminant:** Something that makes something else impure; a pollutant.

**Dam:** The wall of a fish pond or other body of water.

**Debris:** Rubbish, garbage, anything that is not supposed to be in a certain area.

**Density:** The number of fish in a pond.

**Dike:** The wall of a fish pond or other body of water.

**Diversion channel:** A ditch that takes water from a stream or river to a fish pond.

**Elevation:** The height of land or of some object.

**Exotic species:** Species that are not native to the area.

**Fertility:** The productivity.

**Fertilizer:** Anything added to water or soil to make it more productive.

**Fingerling:** A fish that is about as long as a person's finger.

**Fish culture:** The breeding and cultivation of fish in ponds.

**Fry:** Fish from hatching to fingerling size.

**Genitals:** Reproductive organs.

**Genital opening:** The opening on the fishes' body where the eggs or sperm are released.

**Gills:** The part of a fish that allows it to breathe in the water.

**Gravity:** The gravitational attraction of the earth's mass for bodies at or near its surface as modified by the centrifugal force due to the earth's rotation.

**Herbivore:** An organism that eats only plants and plant products.

**Hypophysation:** Hormone injection to induce breeding of fish.

**Hypophysis:** The pituitary gland.

**Hormones:** Components that are secreted by glands of the body to cause certain changes in the body's functions.

**Impermeable:** A substance that nothing can leak through.

**Induced spawning:** Causing a fish to spawn by injecting it with hormones.

**Introduced species:** Fish not native to an area that are used in fish ponds of the area.

**Mortality rate:** The rate of death.

**Natural food:** Food that a fish eats in nature.

**Operculum:** The gill covering.

**Oxygen:** A gas that is necessary for all life.

**Pens:** Enclosures for fish culture in large bodies of water.

**Phytoplankton:** Tiny green or brown plants that are microscopic, free-floating in water, that are used as food by fish.

**Plankton:** The tiny plants and animals that grow in ponds that are eaten by fish.

**Ponds:** Any enclosure that holds water so that fish can be grown inside it.

**Predators:** Animals that prey on other animals.

**Productivity:** Ability to grow food in a pond, whether it is plankton or fish.

**Reproduction:** Producing offspring.

**Respiration:** Breathing.

**Serrations:** Rough edges, like on a fishes' fin.

**Spawning:** The release and fertilization of eggs and sperm.

**Stress:** Any change that is not normal in the environment that creates problems.

**Trash fish:** Fish not wanted in the pond, or fish that are too small to eat, or spoiled fish.

**Watertight:** Impermeable.

**Zooplankton:** Small animals in ponds that can be seen with the naked eye.

## QUESTIONS FOR STUDY AND DISCUSSION

1. Define aquaculture. Wherein do aquaculture and natural or capture fisheries differ?

2. Trace the history of aquaculture.

3. From a human nutrition standpoint, how important is world aquaculture in comparison with world fisheries catch?

4. List and discuss the two primary reasons the energetics of protein production favor expansion of the aquaculture industry in the decades to come.

5. To what can the growth of aquaculture since about 1975 be attributed?

6. Discuss the growth and current magnitude of U.S. aquaculture.

7. Identify the types of aquaculture in the U.S. What are the main species contributors?

8. Define (a) sea ranching, (b) double cropping, and (c) polyculture.

9. Describe each of the following major aquaculture species, including the magnitude of each: (a) channel catfish, (b) trout and salmon, and (c) bait fish.

10. Describe each of the following less populous aquacultural species: (a) alligators, (b) aquarium fish, and (c) bass, striped/hybrid.

11. Describe each of the following less populous aquacultural species: (a) crayfish, (b) lobster, (c) prawn, (d) shrimp, and (e) tilapia.

12. Compare the modern breeding methods of fish with the modern breeding methods of four-footed animals.

13. Describe the genetically engineered carp experiment that is reported in this chapter.

14. Present under the headings that follow pertinent information relative to each of these aquacultural species: (a) bass, (b) carp, (c) channel catfish, and (d) crayfish:
   Description
   Habitat/Fishing

15. Present under the headings that follow pertinent information relative to each of these aquacultural species: (a) lobster, (b) salmon, (c) shrimp, and (d) rainbow trout:
   Description
   Habitat/Fishing

16. How do fish and broilers compare in efficiency of feed conversion?

17. What is *poikilotherm*?

18. Give the approximate temperature range best suited for (a) coldwater fish, and (b) warmwater fish.

19. Fish may be divided into three types of eaters: (a) carnivores, (b) herbivores, and (c) omnivores. Describe each type of eater.

20. Describe the feeding habits of each of the following types of fish: (a) predators, (b) grazers, (c) strainers, (d) suckers, and (e) parasites.

21. Discuss the effect on the feeding behavior of fish of each of the following environmental influences: (a) time of day, (b) season of the year, (c) rapid changes of light intensity, (d) physical contact with food, and (e) water temperature.

22. Explain why the energy requirements of fish are lower than the energy requirements of warm-blooded animals.

23. Discuss the fat requirements of each (a) coldwater fishes, and (b) warmwater fishes.

24. List the factors that determine the requirements for protein in fish feeds.

25. Describe the deficiency symptoms of fish due to insufficiency of each of the following minerals: (a) calcium, (b) phosphorus, (c) iodine, and (d) selenium.

26. Describe the deficiency symptoms of fish due to insufficiency of each of the following vitamins: (a) vitamin A, (b) vitamin E, (c) biotin, and (d) vitamin C.

27. In addition to its role as a nutrient carrier, what additional functions does water serve for fish?

28. Discuss the economic importance, along with the sources, of pigment-producing factors for fish.

29. Discuss *natural foods and feeds* for fish. Why are not more ponds fertilized with chicken manure?

30. Describe, and discuss the role, of each of the following types of fish feeds: (a) compressed or sinking pellets, and (b) expanded or floating pellets.

31. How do the recommended dietary protein levels of (a) trout/salmon and (b) catfish compare?

32. Compare the feeding of (a) coldwater fish, and (b) warmwater fish.

33. Describe the feeding of carp. Describe the hog-fish feeding combination which is sometimes practiced in China.

34. Describe each of the following fish feeding systems: (a) hand-feeding, (b) semiautomatic feeding, and (c) automatic feeding.

35. List some feed ingredients which may affect fish quality and flavor.

36. List some common nutritional toxicants of fish.

37. Describe four fish behavioral signs which indicate that certain diseases (you name them) may be present.

38. Discuss the relationship of environment and disease.

39. Discuss the role of each of the following three management practices in disease prevention: (a) suitable pond management, (b) maintenance of a disease-free water supply, and (c) constant surveillance.

40. Name three important bacterial diseases of fish, and give for each of them the (a) cause, (b) clinical signs, and (c) prevention/treatment.

41. Name two important fungal diseases of fish, and give for each of them the (a) cause, (b) clinical signs, and (c) prevention/treatment.

42. Name three important viral diseases of fish, and give for each of them the (a) cause, (b) clinical signs, and (c) prevention/treatment.

43. Name four important parasitic diseases of fish, and give for each of them the (a) cause, (b) clinical signs, and (c) prevention/treatment.

44. List the four systems of management that are practiced within each freshwater and marine water production.

45. For what primary purpose(s) is each of the following types of ponds used: (a) holding pond, (b) spawning pond, and (c) catch-out pond?

46. Describe and tell how each of the following water-holding facilities is used: (a) raceways, (b) tanks, and (c) cages.

47. Diagram a fish, and label all its parts.

48. List the factors that constitute water quality.

49. Briefly describe what is involved in managing the pond from the standpoint of each of the following: (a) stocking properly, (b) servicing the pond, (c) managing fry and fingerlings, (d) managing brood stock, and (e) controlling predators.

50. Define the term *harvesting of fish*. How is it accomplished?

51. For commercial fish farmers, why is it so important that they know their markets?

52. Why, and how, will the mandatory inspection of seafood products bolster consumer confidence?

## SELECTED REFERENCES

| Title of Publication | Author(s) | Publisher |
|---|---|---|
| *Artificial Propagation of Marine Fish* | J. E. Shelbourne | T. F. H. Publications, Jersey City, NJ |
| *Commercial Catfish Farming* | J. S. Lee | The Interstate Printers & Publishers, Inc., Danville, IL, 1973 |
| *Finfish Nutrition and Fishfeed Technology*; Vols. 1 and 2 | J. E. Halver<br>K. Tiews | Heenemann, Berlin, Germany, 1979 |
| *Fishery Statistics, Catches and Landings*, Vol. 64 | Staff | Food and Agriculture Organization of the United Nations, Rome, Italy, 1987 |
| *Fishes of the World*, Second Edition | J. S. Nelson | John Wiley & Sons, New York, NY, 1984 |
| *Fish Farming Handbook* | E. E. Brown<br>J. B. Gratzek | AVI Publishing Co., Inc., Westport, CT, 1980 |
| *Fish Hatchery Management* | R. G. Piper *et al.* | U.S. Department of the Interior, Fish and Wildlife Service, Washington, DC, 1982 |
| *Fish Nutrition*, Second Edition | Edited by<br>J. E. Halver | Academic Press, New York, NY, 1989 |
| *Fish Physiology* | Edited by<br>W. S. Hoar<br>D. J. Randall | Academic Press, New York, NY, 1989 |
| *Foods & Nutrition Encyclopedia* | A. H. Ensminger *et al.* | Pegus Press, Clovis, CA, 1983 |
| *Ichthyology* | K. F. Lagler<br>J. E. Bardach<br>R. R. Miller | John Wiley & Sons, Inc., New York, NY, 1962 |
| *Manual of Fish Culture, Part 3, Section B* | A. M. Phillips | Bureau of Sport Fisheries and Wildlife, 1970 |
| *Merck Veterinary Manual, The* | Edited by<br>C. M. Fraser | Merck & Co., Inc., Rahway, NJ, 1986 |
| *Nutrient Requirements of Coldwater Fishes* | G. L. Rumsey *et al.* | National Research Council, National Academy Press, Washington, DC, 1981 |
| *Nutrient Requirements of Warmwater Fishes and Shellfishes* | R. R. Stickney *et al.* | National Research Council, National Academy Press, Washington, DC, 1983 |
| *Nutrition and Feeding in Fish* | Edited by<br>C. B. Crowley<br>A. M. Mackie<br>J. G. Bell | Academic Press, New York, NY, 1985 |
| *Nutrition and Feeding of Channel Catfish, SCSB–296* | Edited by<br>E. H. Robinson<br>R. T. Lovell | Texas Agricultural Experiment Station, Dept. of Agric. Communications, Texas A&M Univ., College Station, TX, 1984 |
| *Nutrition and Feeding of Rainbow Trout* | J. W. Hilton<br>S. J. Slinger | Department of Fisheries and Oceans, Ottawa, Canada, 1981 |
| *Physiology of Fishes, The* | M. E. Brown | Academic Press, New York, NY, 1957 |
| *Special Methods in Pond Fish Husbandry* | Edited by<br>J. E. Halver | Halver Corporation, Seattle, WA, 1984 |
| *Third Report To The Fish Farmers* | Edited by<br>H. K. Dupree<br>J. V. Huner | U.S. Fish and Wildlife Service, Washington, DC, 1984 |
| *World Fish Farming—cultivation & economics* | E. E. Brown | AVI Publishing Co., Inc., Westport, CT, 1977 |

1 HORSE (yearling)

In energy equivalents per 1,000 lb body weight ...

1 COW

5 SHEEP

7 GOATS

**Ruminants**

Ruminants—cattle, sheep and goats—have four-compartment stomachs; they are adapted to eating large quantities of bulky fibrous feeds.

56 RABBITS

**Nonruminant herbivore**

Nonruminant herbivore—horses and rabbits—have a large functional cecum and a big colon, a digestive system that is intermediate between ruminants and monogastrics; they are adapted to eating feeds that are intermediate in bulk and fiber.

3 SOWS

116 MINK

**Monogastrics**

Monogastrics—swine, mink, and fish—have a simple digestive system with limited capacity and limited microbial action and fiber digestion; they are adapted to eating concentrate feeds such as grains.

259 FISH

**Poultry**

Poultry—chickens and turkeys—have no teeth, store feed in a crop, and crush and grind feed in a gizzard; they are adapted to eating cereal grains and by-product feeds.

71 LAYERS

24 TURKEYS

with each species eating its most common feeds

Tom Phillips

# *Appendix*

## WEIGHTS AND MEASURES

Weights and measures are the standards employed in arriving at weights, quantities, and volumes. Even among primitive people, such standards were necessary; and with the growing complexity of life, they become of greater and greater importance.

Weights and measures form one of the most important parts of modern agriculture. This section contains pertinent information relative to the most common standards used in the U.S. livestock industry.

## METRIC SYSTEM[1,2]

The United States and a few other countries use standards that belong to the *customary*, or English, system of measurement. This system evolved in England from older measurement

___

[1]For further information on the federal government's metric activities contact: U.S. Dept. of Commerce, Office of Metric Programs, Room 4845, Washington, DC 20230, (202) 377-3036.

[2]For additional conversion factors, or for greater accuracy, see *Misc. Pub. 223,* the National Bureau of Standards.

standards, beginning about the year 1200. All other countries—including England—now use a system of measurements called the *metric system,* which was created in France in the 1790s. Increasingly, the metric system is being used in the United States. Hence, everyone should have a working knowledge of it.

The basic metric units are the *meter* (length/distance), the *gram* (weight), and the *liter* (capacity). The units are then expanded in multiples of 10 or made smaller by ⅒. The prefixes, which are used in the same way with all basic metric units, follow:

| | | | | |
|---|---|---|---|---|
| "milli-" | = | ⅟₁₀₀₀ | "deca-" | = | 10 |
| "centi-" | = | ⅟₁₀₀ | "hecto-" | = | 100 |
| "deci-" | = | ⅟₁₀ | "kilo-" | = | 1,000 |

The following tables will facilitate conversion from metric units to U.S. customary, and vice versa:

Table A–1   Weights and Measures—
Weight
Length
Surface/Area
Volume

Table A–2   Temperature

**TABLE A–1**
**WEIGHTS AND MEASURES**

| Weight | | |
|---|---|---|
| **Unit** | **Is Equal To** | |

| Unit | (metric) | (U.S. customary) |
|---|---|---|
| **Metric system:** | | |
| 1 microgram (mcg) | .001 mg | |
| 1 milligram (mg) | .001 g | .015432356 grain |
| 1 centigram (cg) | .01 g | .15432356 grain |
| 1 decigram (dg) | .1 g | 1.5432 grains |
| 1 gram (g) | 1,000 mg | .03527396 oz |
| 1 decagram (dkg) | 10 g | 5.643833 dr |
| 1 hectogram (hg) | 100 g | 3.527396 oz |
| 1 kilogram (kg) | 1,000 g | 35.274 oz; 2.2046223 lb |
| 1 ton | 1,000 kg | 2,204.6 lb; 1.102 tons (short or 0.984 ton (long) |
| **U.S. customary:** | **(U.S. customary)** | **(metric)** |
| 1 grain | .037 dr | 64.798918 mg; .064798918 g |
| 1 dram (dr) | .063 oz | 1.771845 g |
| 1 ounce (oz) | 16 dr | 28.349527 g |
| 1 pound (lb) | 16 oz | 453.5924 g or 0.4536 kg |
| 1 hundredweight (cwt) | 100 lb | |
| 1 ton (short) | 2,000 lb | 907.18486 kg or 0.907 (metric) ton |
| 1 ton (long) | 2,200 lb | 1,016.05 kg or 1.016 (metric) ton |
| 1 part per million (ppm) | 1 microgram/gram; 1 mg/l; 1 mg/kg | .4535924 mg/lb; .907 g/ton; .0001%; .00013 oz/gal |
| 1 percent (%) (1 part in 100 parts) | 10,000 ppm; 10 g/l | 1.28 oz/gal; 8.34 lb/100 gal |

| Weight Conversions | | | | |
|---|---|---|---|---|
| **U.S. Customary to Metric** | | | **Metric to U.S. Customary** | |
| **To Change** | | **Multiply By** | **To Change** | **Multiply By** |
| grains | to milligrams | 64.799 | | |
| ounces | to grams | 28.35 | grams                to ounces | 0.035 |
| pounds | to grams | 453.6 | | |
| pounds | to kg | 0.454 | kg                to pounds | 2.205 |
| tons | to metric tons | 0.9 | metric tons                to tons | 1.102 |

| Weight—Unit Conversion Factors | | | | |
|---|---|---|---|---|
| **To Change** | | **Multiply By** | **To Change** | **Multiply By** |
| mg/lb | to g/ton | 2 | mg/g                to mg/lb | 453.6 |
| g/lb | to g/ton | 2,000 | mg/kg                to mg/lb | 0.4536 |
| lb/ton | to g/ton | 453.6 | mcg/kg                to g/lb | 0.4536 |
| ppm | to mg/lb | 0.4536 | g/ton                to g/lb | 0.0005 |
| ppm | to % | move decimal 4 places to left | g/ton                to lb/ton | 0.0022 |
| mg/lb | to ppm | 2.2046 | g/ton                to % | 0.00011 |
| | | | %                to g/ton | 9,072.00 |
| ppm | to g/ton | 0.907 | g/ton                to ppm | 1.1 |

*(Continued)*

**TABLE A–1** *(Continued)*

## Length

| Unit | Is Equal To | |
|------|-------------|---|
| **Metric system:** | **(metric)** | **(U.S. customary)** |
| 1 millimicron (m ) | .000000001 m | .000000039 in. |
| 1 micron ( ) | .000001 m | .000039 in. |
| 1 millimeter (mm) | .001 m | .0394 in. |
| 1 centimeter (cm) | .01 m | .3937 in. |
| 1 decimeter (dm) | .1 m | 3.937 in. |
| 1 meter (m) | 1 m | 39.37 in.; 3.281 ft; 1.094 yd |
| 1 hectometer (hm) | 100 m | 328.08 ft; 19.8338 rd |
| 1 kilometer (km) | 1,000 m | 3,280.8 ft; 0.621 mi |
| **U.S. customary:** | **(U.S. customary)** | **(metric)** |
| 1 inch (in.) | 1 in. | 25 mm; 2.54 cm |
| 1 hand* | 4 in. | 10.16 cm |
| 1 foot (ft) | 12 in. | 30.48 cm; .305 m |
| 1 yard (yd) | 3 ft | .914 m |
| 1 fathom** (fath) | 6.08 ft | 1.829 m |
| 1 rod (rd), pole, or perch | 16½ ft; 5½ yd | 5.029 m |
| 1 chain | 792 in.; 66 ft; 22 yd | 20.116 m |
| 1 furlong (fur.) | 220 yd; 40 rd | 201.168 m |
| 1 mile (mi) | 5,280 ft; 1,760 yd; 320 rd; 8 fur. | 1,609.35 m; 1.609 km |
| 1 knot or nautical mile | 6,080 ft; 1.15 land miles | 1.85 km |
| 1 league (land) | 3 mi (land) | 4.827 km |
| 1 league (nautical) | 3 mi (nautical) | 4.827 km |

### Length Conversions

| U.S. Customary to Metric | | Metric to U.S. Customary | |
|--------------------------|---|--------------------------|---|
| To Change | Multiply By | To Change | Multiply By |
| inches ......... to millimeters | 25.4 | millimeters ......... to inches | 0.04 |
| inches ......... to centimeters | 2.54 | centimeters ......... to inches | 0.4 |
| feet ......... to centimeters | 30.5 | centimeters ......... to feet | 0.033 |
| feet ......... to meters | 0.305 | meters ......... to feet | 3.3 |
| yards ......... to meters | 0.914 | meters ......... to yards | 1.1 |
| miles ......... to kilometers | 1.609 | kilometers ......... to miles | 0.6 |

*Used in measuring height of horses.

**Used in measuring depth at sea.

*(Continued)*

**TABLE A-1** *(Continued)*

### Surface/Area

| Unit | Is Equal To | |
|---|---|---|
| **Metric system:** | **(metric)** | **(U.S. customary)** |
| 1 square millimeter (mm²) ..................... | .000001 m² ..................... | .00155 in.² |
| 1 square centimeter (cm²) ..................... | .0001 m² ..................... | .155 in.² |
| 1 square decimeter (dm²) ..................... | .01 m² | 15.50 in.² |
| 1 square meter (m²) ..................... | 1 centare (ca) ..................... | 1,550 in.²; 10.76 ft²; 1.196 yd² |
| 1 are (a) ..................... | 100 m² | 119.6 yd² |
| 1 hectare (ha) ..................... | 10,000 m² | 2.47 acres |
| 1 square kilometer (km²) ..................... | 1,000,000 m² | 247.1 acres; .386 mi² |
| **U.S. customary:** | **(U.S. customary)** | **(metric)** |
| 1 square inch (in.²) ..................... | 1 in. × 1 in. ..................... | 6.452 cm² |
| 1 square foot (ft²) ..................... | 144 in.²; 0.111 yd² ..................... | .093 m² |
| 1 square yard (yd²) ..................... | 1,296 in.²; 9 ft² ..................... | .836 m² |
| 1 square rod (rd²) ..................... | 272.25 ft²; 30.25 yd² ..................... | 25.29 m² |
| 1 rood ..................... | 40 rd² | 10.117 a |
| 1 acre ..................... | 43,560 ft²; 4,840 yd²; 160 rd²; 4 roods ............ | 4,046.87 m²; 0.405 ha |
| 1 square mile (mi²) ..................... | 640 acres; 1 section ..................... | 2.59 km² or 259 ha |
| 1 township ..................... | 36 sections; 6 miles square | |

### Surface/Area Conversions

| U.S. Customary to Metric | | Metric to U.S. Customary | |
|---|---|---|---|
| To Change | Multiply By | To Change | Multiply By |
| sq in. ............ to cm² ............ | 6.452 | cm² ............ to sq in. ............ | 0.155 |
| sq ft ............ to cm² ............ | 929.1 | cm² ............ to sq ft ............ | 0.001 |
| sq ft ............ to m² ............ | 0.09 | m² ............ to sq ft ............ | 10.764 |
| sq yd ............ to m² ............ | 0.836 | m² ............ to sq yd ............ | 1.196 |
| sq mi ............ to km² ............ | 2.6 | km² ............ to sq mi ............ | 0.4 |
| acres ............ to ha ............ | 0.4 | ha ............ to acres ............ | 2.5 |

### Weights/Measures/Unit Area

| Unit | Is Equal To |
|---|---|
| **Volume per unit area:** | |
| 1 liter/hectare ..................... | 0.107 gal/acre |
| 1 gal/acre ..................... | 9.354 liter/ha |
| **Weight per unit area:** | |
| 1 kilogram/cm² ..................... | 14.22 lb/in² |
| 1 kilogram/hectare ..................... | 0.892 lb/acre |
| 1 lb/sq in. ..................... | 0.0703 kg/cm² |
| 1 lb/acre ..................... | 1.121 kg/ha |
| **Area per unit weight:** | |
| 1 square centimeter/kilogram ..................... | 0.0703 in.²/lb |
| 1 sq in./lb ..................... | 14.22 cm²/kg |

*(Continued)*

**TABLE A–1** *(Continued)*

## Volume

| Unit | Is Equal To | | |
|------|------|------|------|
| **Metric system**<br>**liquid and dry:** | **(U.S. customary)**<br>**(liquid)** | | **(U.S. customary)**<br>**(dry)** |
| 1 milliliter (ml) .........................001 liter | .271 dram (fl) ........................ | | .061 in.³ |
| 1 centiliter (cl) .........................01  liter | .338 oz (fl) .......................... | | .610 in.³ |
| 1 deciliter (dl) ..........................1  liter | 3.38 oz (fl) | | |
| 1 liter ..............................1,000  cc | 1.057 qt or 0.2642 gal (fl) | | .908 qt |
| 1 hectoliter (hl) ......................100  liters | 26.418 gal | | 2.838 bu |
| 1 kiloliter (kl) .....................1,000  liters | 264.18 gal ........................... | | 1,308 yd³ |
| **U.S. customary**<br>**liquid:** | **(ounces)** | **(cubic inches)** | **(metric)** |
| 1 teaspoon (t) ...................... 60 drops | ⅛ ................................. | | 5 ml |
| 1 dessert spoon ..................... 2 t | | | |
| 1 tablespoon (T) .................... 3 t | ½ ................................. | | 15 ml |
| 1 fl oz ............................. 1 | 1 | 1.805 ................. | 29.57 ml |
| 1 gill (gi) ......................... ½ c | 4 | 7.22 | 118.29 ml |
| 1 cup (c) ........................... 16 T | 8 | 14.44 | 236.58 ml or 0.24 liter |
| 1 pint (pt) ......................... 2 c | 16 | 28.88 | .47 liter |
| 1 quart (qt) ........................ 2 pt | 32 | 57.75 | .95 liter |
| 1 gallon (gal) ...................... 4 qt | 8.34 lb | 231 | 3.79 liters |
| 1 barrel (bbl) ...................... 31½ gal | | | |
| 1 hogshead (hhd) .................... 2 bbl | | | |
| **Dry:** | | | |
| 1 pint (pt) ......................... ½ qt | | 33.6 | .55 liter |
| 1 quart (qt) ........................ 2 pt | | 67.20 | 1.10 liters |
| 1 peck (pk) ......................... 8 qt | | 537.61 | 8.81 liters |
| 1 bushel (bu) ....................... 4 pk | | 2,150.42 ........................ | 35.24 liters |

| Unit | Is Equal To | |
|------|------|------|
| **Solid**<br>**metric system:** | **(metric)** | **(U.S. customary)** |
| 1 cubic millimeter (mm³) ......................... | .001                            cc | |
| 1 cubic centimeter (cc) .......................... | 1,000 mm³ | .061 cu in. |
| 1 cubic decimeter (dm³) .......................... | 1,000 cc | 61.023 cu in. |
| 1 cubic meter (m³) ............................... | 1,000 dm³ | 35.315 ft³; 1.308 yd³ |
| **U.S. customary:** | **(U.S.customary)** | **(metric)** |
| 1 cubic inch (in.³) .............................. | ..................................... | 16.387 cc |
| 1 board foot (fbm) ............................... | 144 in.³ ............................. | 2,359.8 cc |
| 1 cubic foot (ft³) ............................... | 1,728 in.³ ........................... | .028 m³ |
| 1 cubic yard (yd³) ............................... | 27 ft³ ............................... | .765 m³ |
| 1 cord .......................................... | 128 ft³ .............................. | 3.625 m³ |

### Volume Conversions

| U.S. Customary to Metric | | Metric to U.S. Customary | |
|------|------|------|------|
| **To Change** | **Multiply By** | **To Change** | **Multiply By** |
| ounces (fluid) ............ to cc ............. | 29.57 | cc ............. to oz (fluid) ................. | 0.034 |
| ounces ............ to ml ................. | 29.57 | ml ............. to oz ........................ | 0.034 |
| qt ............ to liters ................. | 0.946 | liters ............. to qt ................... | 1.057 |
| cu in. ............ to cc ................. | 16.387 | cc ............. to cu in. ................... | 0.061 |
| cu yd ............ to cm ................. | 0.765 | cm ............. to cu yd ................... | 1.308 |

## TABLE A–2
## TEMPERATURE

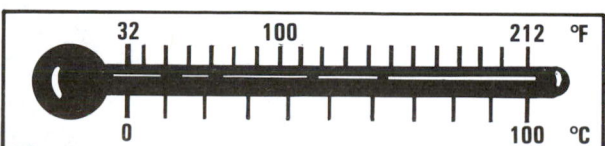

Fig. A–1. Fahrenheit-Centigrade scale for direct conversion and reading.

*One Fahrenheit (F) degree is 1/180 of the difference between the temperature of melting ice and that of water boiling at standard atmospheric pressure. One Fahrenheit degree equals 0.556°C.*

*One Centigrade (C) degree is 1/100 the difference between the temperature of melting ice and that of water boiling at standard atmospheric pressure. One Centigrade degree equals 1.8°F.*

| To Change | To | Do This |
|------|------|------|
| Degrees<br>Fahrenheit | Degrees<br>Centigrade | Subtract 32, then multiply by .556 (5⁄9) |
| Degrees<br>Centigrade | Degrees<br>Fahrenheit | Multiply by 1.8 (9⁄5) and add 32 |

## WEIGHTS AND MEASURES OF COMMON FEEDS

In calculating rations and mixing concentrates, it is usually necessary to use weights rather than measures. However, in practical feeding operations it is often more convenient for the farmer or rancher to measure the concentrates by volume. Table A–3 will serve as a guide in feeding by measure.

**TABLE A–3**
**WEIGHTS AND MEASURES OF COMMON FEEDS**

| Feed | Approximate Weight | |
|---|---|---|
| | Lb per Quart [1] | Lb per Bushel [1] |
| Alfalfa meal | 0.6 | 19 |
| Barley | 1.5 | 48 |
| Beet pulp (dried) | 0.6 | 19 |
| Brewers' grain (dried) | 0.6 | 19 |
| Buckwheat | 1.6 | 51 |
| Buckwheat bran | 1.0 | 32 |
| Corn, husked ear | — | 70 |
| Corn, cracked | 1.6 | 51 |
| Corn, shelled | 1.8 | 58 |
| Corn meal | 1.6 | 51 |
| Corn-and-cob meal | 1.4 | 45 |
| Cottonseed | 0.9–1.0 | 29–32 |
| Cottonseed meal | 1.5 | 48 |
| Cowpeas | 1.9 | 61 |
| Distillers' grain (dried) | 0.6 | 19 |
| Fish meal | 1.0 | 32 |
| Flax | 1.7 | 54 |
| Gluten feed | 1.3 | 42 |
| Linseed meal (old process) | 1.1 | 35 |
| Linseed meal (new process) | 0.9 | 29 |
| Meat scrap | 1.3 | 42 |
| Milo (grain sorghum) | 1.7 | 54 |
| Molasses feed | 0.8 | 26 |
| Oat middlings | 1.5 | 48 |
| Oats | 1.0 | 32 |
| Oats, ground | 0.7 | 22 |
| Peanut meal | 1.0 | 32 |
| Peas | 1.9 | 61 |
| Rice | 1.4 | 45 |
| Rice bran | 0.8 | 26 |
| Rye | 1.7 | 54 |
| Sorghum (grain) | 1.7 | 54 |
| Soybeans | 1.8 | 58 |
| Sunflower | 0.7 | 22 |
| Tankage | 1.6 | 51 |
| Velvet beans, shelled | 1.8 | 58 |
| Wheat | 1.9 | 61 |
| Wheat bran | 0.5 | 16 |
| Wheat middlings, standard | 0.8 | 26 |
| Wheat screenings | 1.0 | 32 |

[1] 32 qts per bushel.

## GRAIN WEIGHT IN A BIN

Sometimes farmers need to estimate the weight of grain in storage. Such estimates are difficult to make because of differences in moisture content, depth of material stored, and other factors. However, the following procedure will enable one to figure feed quantities fairly closely.

1. **Corn (shelled) or small grain in rectangular cribs or bins.** Multiply the width by the length by the average depth (all in feet) and multiply by 0.8 to get the number of bushels (multiplying by 0.8 is the same as dividing by 1¼, the number of cubic feet in a bushel).

2. **Ear corn in rectangular cribs or bins.** Multiply the width by the length by the average depth (all in feet) and multiply by 0.4 to get the number of bushels (multiplying by 0.4 is the same as dividing by 2½, the number of cubic feet in a bushel of ear corn).

3. **Round bins or cribs.** To find the cubic feet in a cylindrical bin, multiply the squared radius by 3.1416 by the depth.
Thus, the volume of a round bin 20 ft in diameter and 10 ft deep is determined as follows:

a. The radius is half the diameter, or 10 ft.

b. $10 \times 10 = 100$

c. $100 \times 3.1416 = 314.16$

d. $314.16 \times 10 = 3,141.6$ cu ft

e. Where shelled corn or small grain is involved, one would multiply $3,141.6 \times 0.8$, which equals 2,513.28 bu of grain that it would hold if full.

f. Where ear corn is involved, one would multiply $3,141.6 \times 0.4$ which equals 1,256.64 bu of ear corn that it would hold if full.

## HAY WEIGHT IN A BARN OR STACK

Livestock producers and hay dealers frequently buy and sell large quantities of hay in the stack or in the barn. This practice is especially prevalent in the western and Great Plains states where cattle and sheep are brought into the valleys to be wintered on hay bought from valley hay producers. Under such circumstances, the weight of hay is usually estimated, because (1) no scales are available, and/or (2) it is impractical to weigh the hay due to the time, labor, and wastage involved. In many such instances, the hay is fed directly from the stack or barn, in racks arranged about it. Under these and other circumstances, there is need for a simple and reasonably accurate method of estimating the weight of hay in a stack or in a barn.

In order to estimate the tonnage of hay in a stack or in a barn, it is necessary (1) to compute the volume of hay, and (2) to know the number of cubic feet per ton of hay. Table A–4 gives the latter information.

**TABLE A–4**
**CUBIC FEET PER TON OF HAY**

| Feed | Settled 1–2 Months | Settled Over 3 Months |
|---|---|---|
| | (cu ft) | (cu ft) |
| Alfalfa | 485 | 470 |
| Clover | 512 | 500 |
| Hay, baled (closely stacked) | 150–300 | 150–200 |
| Hay, chopped | 225 | 210 |
| Straw, baled | 200 | 200 |
| Straw, loose | 1,000 | 600–1,000 |
| Timothy | 640 | 625 |
| Wild hay | 600 | 450 |

In using Table A-4, it should be recognized that many factors—other than kind of hay, form (loose, chopped, or baled), and period of settling—affect the density of hay in a barn or in a stack, including (1) moisture content at haying time, and (2) texture and foreign material.

It is relatively simple to compute the volume of hay in a mow, but it is more difficult to determine the volume of a stack. Although different rules or formulas may be and are used, the following are recommended by the U.S. Department of Agriculture.[3]

1. **Volume of hay in barns.** Multiply the width by the length by the height, all in feet, and divide by the cubic feet per ton as given in Table A-4.

2. **Volume of hay in oblong stacks.** Three types of oblong stacks are common, as shown in Fig. A-2.

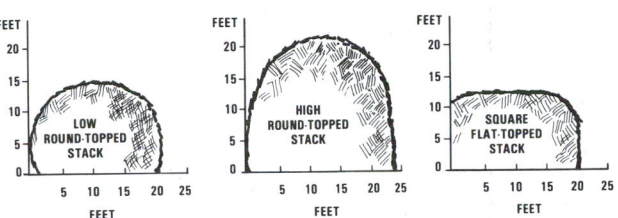

Fig. A-2.  Three common types of oblong stacks.

The volume of each type of oblong stack may be determined as follows:

   a. For low, round-topped stacks—

$$(0.52 \times O) - (0.44 \times W) \times W \times L$$

   b. For high, round-topped stacks—

$$(0.52 \times O) - (0.46 \times W) \times W \times L$$

   c. For square, flat-topped stacks—

$$(0.56 \times O) - (0.55 \times W) \times W \times L$$

In these formulas "O" is the "over" or "overthrow," which is the distance in feet from the ground on one side of the stack, up and over the stack and down to the ground on the other side; W is the width; and L is the length.

The application of this formula is illustrated as follows:

   **Example.** *It is desired to estimate the amount of alfalfa hay in a low, round-topped type of oblong stack that has settled for 4 months. The stack is 20 ft wide, 30 ft long, and has an over of 40 ft.*

The answer is secured as follows:

   a. Volume = $(0.52 \times 40) - (0.44 \times 20) \times 20 \times 30 = 7,200$ cu ft.

   b. Table A-6 shows that there are 470 cu ft per ton of settled alfalfa.

   c. $7,200 \div 470 = 15$ tons of hay.

[3]*Measuring Hay in Stacks,* USDA Leaflet No. 72.

3. **Volume of hay in round stacks.** The rules or formulas used for oblong stacks do not apply to round stacks. The volume of round stacks can be calculated by using the following formula:

$$\textbf{Volume} = \textbf{(0.04} \times \textbf{O)} - \textbf{(0.012} \times \textbf{C)} \times \textbf{C}^2$$

*In this formula, C equals the circumference or distance around the stack at the ground, and O equals the over or distance from the ground on one side over the peak to the ground on the other side (usually it is best to take 2 over measurements at right angles to each other, and to average them).*

Thus, the computation of the volume of a large round stack may be illustrated by the following example:

   **Example.** *It is desired to determine the amount of alfalfa hay in a round stack that is 100 ft in circumference and has an average over of 60 ft.*

The answer is secured as follows:

   a. Volume = $(0.04 \times 60) - (0.012 \times 100) \times (100)^2 = 12,000$ cu ft.

   b. Table A-6 shows that there are 470 cu ft per ton of settled alfalfa.

   c. $12,000 \div 470 = 25.5$ tons of hay.

## STORAGE SPACE REQUIREMENTS FOR FEED AND BEDDING

The space requirements for feed storage for the livestock enterprise—whether it be for cattle, sheep, hogs, or horses, or, as is more frequently the case, a combination of these—vary so widely that it is difficult to provide a standard method of calculating space requirements applicable to such diverse conditions. The amount of feed to be stored depends primarily upon (1) length of pasture season, (2) method of feeding and management, (3) kind of feed, (4) climate, and (5) the proportion of feeds produced on the farm or ranch in comparison with those purchased. Normally, the storage capacity should be sufficient to handle all feed grain and silage grown on the farm and to hold purchased supplies. Forage and bedding may or may not be stored under cover. In those areas where weather conditions permit, hay and straw are frequently stacked in the fields or near the barns in loose, baled, or chopped form. Sometimes poled, framed sheds or a cheap cover of waterproof paper, grass, or cereal straw grass are used for protection. Other forms of low-cost storage include temporary upright silos, trench silos, temporary grain bins, and open-walled buildings for hay.

Table A-5 gives the storage space requirements for feed and bedding. This information may be helpful to the individual operator who desires to compute the barn storage space required for a specific livestock enterprise. This table provides a convenient means of estimating the amount of feed or bedding in storage.

## TABLE A-5
## STORAGE SPACE REQUIREMENTS FOR FEED AND BEDDING

| Kind of Feed or Bedding | Pounds per Cubic Foot | Cubic Feet per Ton | Pounds per Bushel of Grain |
|---|---|---|---|
| **Hay-Straw:** [1] | | | |
| 1. Loose | | | |
|   Alfalfa | 4.4–4.0 | 450– 500 | |
|   Nonlegume | 4.4–3.3 | 450– 600 | |
|   Straw | 3.0–2.0 | 670–1,000 | |
| 2. Baled | | | |
|   Alfalfa | 10.0–6.0 | 200– 330 | |
|   Nonlegume | 8.0–6.0 | 250– 330 | |
|   Straw | 5.0–4.0 | 400– 500 | |
| 3. Chopped | | | |
|   Alfalfa | 7.0–5.5 | 285– 360 | |
|   Nonlegume | 6.7–5.0 | 300– 400 | |
|   Straw | 8.0–5.7 | 250– 350 | |
| **Corn:** | | | |
| 15½% moisture: | | | |
|   Shelled | 44.8 | | 56.0 |
|   Ear | 28.0 | | 70.0 |
|   Shelled, ground | 38.0 | | 48.0 |
|   Ear, ground | 36.0 | | 45.0 |
| 30% moisture: | | | |
|   Shelled | 54.0 | | 67.5 |
|   Ear, ground | 35.8 | | 89.6 |
| **Barley,** 15% moisture | 38.4 | | 48.0 |
|   Ground | 28.0 | | 37.0 |
| **Flax,** 11% moisture | 44.8 | | 56.0 |
| **Oats,** 16% moisture | 25.6 | | 32.0 |
|   Ground | 18.0 | | 23.0 |
| **Rye,** 16% moisture | 44.8 | | 56.0 |
|   Ground | 38.0 | | 48.0 |
| **Sorghum grain,** 15% moisture | 44.8 | | 56.0 |
| **Soybeans,** 14% moisture | 48.0 | | 60.0 |
| **Wheat,** 14% moisture | 48.0 | | 60.0 |
|   Ground | 43.0 | | 50.0 |

[1]Many factors—other than kind of hay-straw, form (loose, baled, chopped), and period of settling—affect the density of hay-straw in a stack or in a barn, including (a) moisture content at haying time, and (b) texture and foreign material.

## ANIMAL UNITS

An animal unit is a common animal denominator, based on feed consumption. It is assumed that 1 mature cow represents an animal unit. Then, the comparative (to a mature cow) feed consumption of other age groups or classes of animals determines the proportion of an animal unit which they represent. For example, it is generally estimated that the ration of one mature cow will feed 5 mature ewes, or that 5 mature ewes equal 1.0 animal unit.

The original concept of an animal unit included a weight stipulation—an animal unit referred to a 1,000–lb cow, with or without a calf at side. Unfortunately, in recent years, the 1,000–lb qualification has been dropped. Certainly, there is a wide difference in the daily feed requirements of a 900–lb cow and of a 1,500–lb cow. Both will consume dry matter on a daily basis at a level equivalent to about 2% of their body weight.

Hence, a 1,500–lb cow will consume 50% more feed than a 1,000–lb cow.

Also, the period of time to be grazed has an effect on the total carrying capacity. For example, if an animal is carried for 1 month only, it will take ¹⁄₁₂ of the total feed required to carry the same animal 1 year. For this reason, the term *animal unit months* is becoming increasingly important. So, in

addition to the weight factor, the time factor has a distinct bearing on the ultimate carrying capacity of a tract of land.

Table A–6 gives the animal units of different classes and ages of livestock.

## TABLE A-6
## ANIMAL UNITS

| Type of Livestock | Animal Units |
|---|---|
| **Cattle:** | |
| Cow, with or without unweaned calf at side, or heifer 2 years old or older | 1.0 |
| Bull, 2 years old or older | 1.3 |
| Young cattle, 1 to 2 years | 0.8 |
| Weaned calves to yearlings | 0.6 |
| **Horses:** | |
| Horse, mature | 1.3 |
| Horse, yearling | 1.0 |
| Weanling colt or filly | 0.75 |
| **Sheep:** | |
| 5 mature ewes, with or without unweaned lambs at side | 1.0 |
| 5 rams, 2 years old or over | 1.3 |
| 5 yearlings | 0.8 |
| 5 weaned lambs to yearlings | 0.6 |
| **Goats — 7** | 1.0 |
| **Swine:** | |
| Sow | 0.4 |
| Boar | 0.5 |
| Pigs to 200 lb | 0.2 |
| **Chickens:** | |
| 75 layers or breeders | 1.0 |
| 325 replacement pullets to 6 months of age | 1.0 |
| 650 8-week-old broilers | 1.0 |
| **Turkeys:** | |
| 35 breeders | 1.0 |
| 40 turkeys raised to maturity | 1.0 |
| 75 turkeys to 6 months of age | 1.0 |
| **Rabbits — 56** | 1.0 |
| **Fish — 259** | 1.0 |

Fig. A–3. White turkeys on the range. Forty turkeys raised to maturity equal one animal unit. (Courtesy, J. C. Allen & Son, West Lafayette, IN)

# CALCULATING ANIMAL WEIGHTS

Feeders who finish large numbers of animals have scales in their feedyards for use in determining in-weights, out-weights, and interim weight gains of animals while they are on feed. Likewise, both purebred and commercial breeders usually have scales. However, those with only one animal, or a few head—such as 4-H Club and FFA members, and part-time farmers—may not have scales. As a result, rations cannot be accurately evaluated, rate of gain cannot be calculated, and an animal's "weight readiness" for a livestock show or for market cannot be determined. Under such circumstances, a simple but reasonably accurate method of estimating body weight is very useful. Fortunately, animal weights may be determined with reasonable accuracy by taking two body measurements (body length and heart girth), then applying an appropriate formula.

## BEEF CATTLE WEIGHTS

Here is how to do it:

**Step 1.** Measure the circumference (heart girth), from a point slightly behind the shoulder blade, thence down over the foreribs and under the body, behind the elbow (distance C of Fig. A-4).

**Step 2.** Measure the length of body, from the point of the shoulder to the point of the rump (pinbone), in inches (distance A-B of Fig. A-4).

**Step 3.** Take the values obtained in Steps 1 and 2 and apply the following formula to calculate body weight:

**Heart girth × heart girth × body length ÷ 300 = weight in pounds**

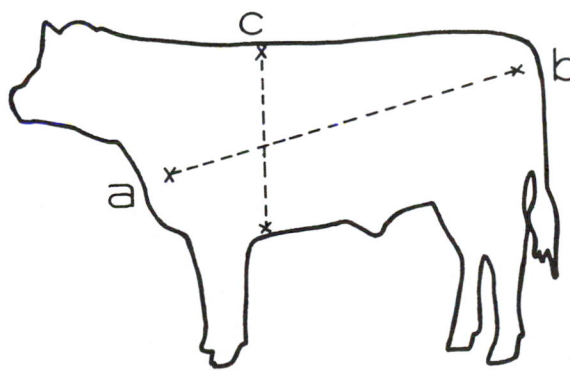

Fig. A-4. How and where to measure beef cattle.

**Example of a beef animal.** *Assume that the heart girth measures 76 in. and the body length, 66 in. How much does the animal weigh?*

$$76 \times 76 = 5,776$$
$$5,776 \times 66 = 381,216$$
$$381,216 \div 300 = 1,270 \text{ lb}$$

## DAIRY HEIFER WEIGHTS

Weight for age is important in dairy heifers from the standpoint of determining the growth progress made by herd replacements.

Table 24-18 of Chapter 24, Feeding and Managing Dairy Cattle, shows the weight and heart girth measurements of dairy calves or heifers at monthly intervals up to 22 months of age. If producers do not have scales, they can measure the heart girth with a tape (see Fig. A-5) and use Table 24-18 to estimate weight within 95% accuracy.

Fig. A-5. How to tape measure a dairy heifer.

## SHEEP AND GOAT WEIGHTS

The weight of sheep and goats is estimated in the same way as for beef cattle; hence, it involves making the measurements and applying the formula given for beef cattle. There is one important precaution, however; with unshorn sheep, be sure to part, or compress, the wool to ensure an accurate heart girth measurement.

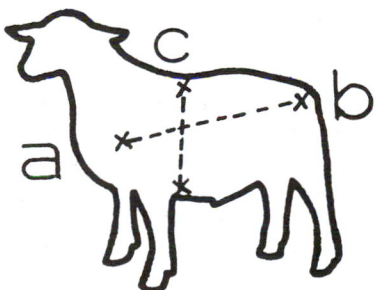

Fig. A-6. How and where to measure sheep.

## SWINE WEIGHTS

Hog weights can be calculated from body measurements, similar to beef cattle, but a different formula must be used. Here is how to estimate the weight of hogs:

**Step 1.** Measure the circumference (heart girth) of the animal (C in diagram).

**Step 2.** Measure the length of body (A-B in Fig. A-7). With the animal standing or restrained in the position shown in Fig. A-7, measure the distance from the poll (between the ears), over the backbone, to the base of the tail.

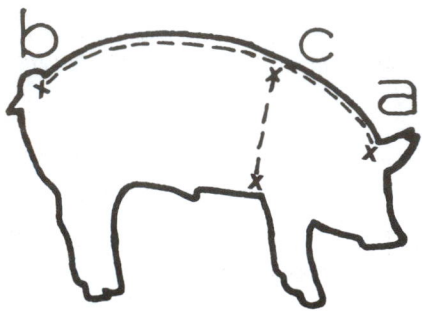

Fig. A–7. How and where to measure hogs.

**Step 3.** Apply the following formula:

**Heart girth × heart girth × length ÷ 400 = weight in pounds.**

**Note:** For hogs weighing less than 150 lb, add 7 lb to the weight figure obtained from the formula. For animals weighing 151 to 400 lb, no adjustment is necessary.

## HORSE WEIGHTS

It is easy to estimate the weight of a horse; and tests have shown that the results obtained this way are accurate within 3% of actual scale weight. This procedure is as follows:

**Step 1.** Measure the circumference (heart girth) of the body in inches (C in diagram).

**Step 2.** Measure the length of body from the point of the shoulder to the point of croup (A-B in the diagram).

**Step 3.** Apply the following formula to calculate the weight of the horse:

**Heart girth × heart girth × length ÷ 300 + 50 lb = weight of horse.**

**Example.** *Assume that the heart girth is 70 in. and the body length is 65 in. How much does the horse weigh?*

    70 × 70 × 65 ÷ 300 + 50 lb = weight
    4,900 × 65 = 318,500
    318,500 ÷ 300 = 1,061 lb
    1,061 + 50 = 1,111 lb body weight

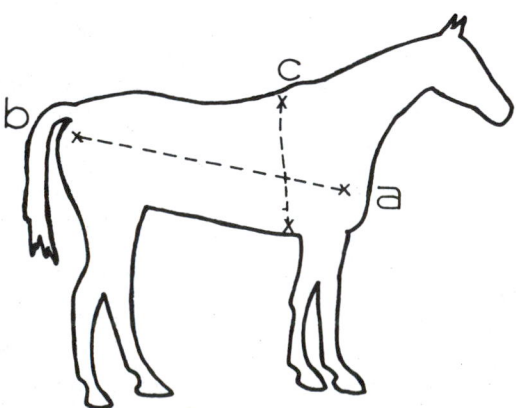

Fig. A–8. How and where to measure horses.

## GESTATION TABLE

The producer who has information relative to breeding dates can easily estimate parturition dates from Table A–7.

**TABLE A–7**
**GESTATION TABLE**

| Date Bred | Cow 283 Days | Ewe 148 Days | Sow 114 Days | Mare 336 Days |
|---|---|---|---|---|
| | (date due) | (date due) | (date due) | (date due) |
| Jan. 1 | Oct. 11 | May 29 | April 25 | Dec. 3 |
| Jan. 6 | Oct. 16 | June 3 | April 30 | Dec. 8 |
| Jan. 11 | Oct. 21 | June 8 | May 5 | Dec. 13 |
| Jan. 16 | Oct. 26 | June 13 | May 10 | Dec. 18 |
| Jan. 21 | Oct. 31 | June 18 | May 15 | Dec. 23 |
| Jan. 26 | Nov. 5 | June 23 | May 20 | Dec. 28 |
| Jan. 31 | Nov. 10 | June 28 | May 25 | Jan. 2 |
| Feb. 5 | Nov. 15 | July 3 | May 30 | Jan. 7 |
| Feb. 10 | Nov. 20 | July 8 | June 4 | Jan. 12 |
| Feb. 15 | Nov. 25 | July 13 | June 9 | Jan. 17 |
| Feb. 20 | Nov. 30 | July 18 | June 14 | Jan. 22 |
| Feb. 25 | Dec. 5 | July 23 | June 19 | Jan. 27 |
| Mar. 2 | Dec. 10 | July 28 | June 24 | Feb. 1 |
| Mar. 7 | Dec. 15 | Aug. 2 | June 29 | Feb. 6 |
| Mar. 12 | Dec. 20 | Aug. 7 | July 4 | Feb. 11 |
| Mar. 17 | Dec. 25 | Aug. 12 | July 9 | Feb. 16 |
| Mar. 22 | Dec. 30 | Aug. 17 | July 14 | Feb. 21 |
| Mar. 27 | Jan. 4 | Aug. 22 | July 19 | Feb. 26 |
| April 1 | Jan. 9 | Aug. 27 | July 24 | Mar. 3 |
| April 6 | Jan. 14 | Sept. 1 | July 29 | Mar. 8 |
| April 11 | Jan. 19 | Sept. 6 | Aug. 3 | Mar. 13 |
| April 16 | Jan. 24 | Sept. 11 | Aug. 8 | Mar. 18 |
| April 21 | Jan. 29 | Sept. 16 | Aug. 13 | Mar. 23 |
| April 26 | Feb. 3 | Sept. 21 | Aug. 18 | Mar. 28 |
| May 1 | Feb. 8 | Sept. 26 | Aug. 23 | April 2 |
| May 6 | Feb. 13 | Oct. 1 | Aug. 28 | April 7 |
| May 11 | Feb. 18 | Oct. 6 | Sept. 2 | April 12 |
| May 16 | Feb. 23 | Oct. 11 | Sept. 7 | April 17 |
| May 21 | Feb. 28 | Oct. 16 | Sept. 12 | April 22 |
| May 26 | Mar. 5 | Oct. 21 | Sept. 17 | April 27 |
| May 31 | Mar. 10 | Oct. 26 | Sept. 22 | May 2 |
| June 5 | Mar. 15 | Oct. 31 | Sept. 27 | May 7 |
| June 10 | Mar. 20 | Nov. 5 | Oct. 2 | May 12 |
| June 15 | Mar. 25 | Nov. 10 | Oct. 7 | May 17 |
| June 20 | Mar. 30 | Nov. 15 | Oct. 12 | May 22 |
| June 25 | April 4 | Nov. 20 | Oct. 17 | May 27 |
| June 30 | April 9 | Nov. 25 | Oct. 22 | June 1 |
| July 5 | April 14 | Nov. 30 | Oct. 27 | June 6 |
| July 10 | April 19 | Dec. 5 | Nov. 1 | June 11 |
| July 15 | April 24 | Dec. 10 | Nov. 6 | June 16 |
| July 20 | April 29 | Dec. 15 | Nov. 11 | June 21 |
| July 25 | May 4 | Dec. 20 | Nov. 16 | June 26 |
| July 30 | May 9 | Dec. 25 | Nov. 21 | July 1 |
| Aug. 4 | May 14 | Dec. 30 | Nov. 26 | July 6 |
| Aug. 9 | May 19 | Jan. 4 | Nov. 31 | July 11 |
| Aug. 14 | May 24 | Jan. 9 | Dec. 6 | July 16 |
| Aug. 19 | May 29 | Jan. 14 | Dec. 11 | July 21 |
| Aug. 24 | June 3 | Jan. 19 | Dec. 16 | July 26 |
| Aug. 29 | June 8 | Jan. 24 | Dec. 21 | July 31 |
| Sept. 3 | June 13 | Jan. 29 | Dec. 26 | Aug. 5 |
| Sept. 8 | June 18 | Feb. 3 | Dec. 31 | Aug. 10 |
| Sept. 13 | June 23 | Feb. 8 | Jan. 5 | Aug. 15 |
| Sept. 18 | June 28 | Feb. 13 | Jan. 10 | Aug. 20 |
| Sept. 23 | July 3 | Feb. 18 | Jan. 15 | Aug. 25 |
| Sept. 28 | July 8 | Feb. 23 | Jan. 20 | Aug. 30 |
| Oct. 3 | July 13 | Feb. 28 | Jan. 25 | Sept. 4 |
| Oct. 8 | July 18 | Mar. 5 | Jan. 30 | Sept. 9 |
| Oct. 13 | July 23 | Mar. 10 | Feb. 4 | Sept. 14 |
| Oct. 18 | July 28 | Mar. 15 | Feb. 9 | Sept. 19 |
| Oct. 23 | Aug. 2 | Mar. 20 | Feb. 14 | Sept. 24 |
| Oct. 28 | Aug. 7 | Mar. 25 | Feb. 19 | Sept. 29 |
| Nov. 2 | Aug. 12 | Mar. 30 | Feb. 24 | Oct. 4 |
| Nov. 7 | Aug. 17 | April 4 | Mar. 1 | Oct. 9 |
| Nov. 12 | Aug. 22 | April 9 | Mar. 6 | Oct. 14 |
| Nov. 17 | Aug. 27 | April 14 | Mar. 11 | Oct. 19 |
| Nov. 22 | Sept. 1 | April 19 | Mar. 16 | Oct. 24 |
| Nov. 27 | Sept. 6 | April 24 | Mar. 21 | Oct. 29 |
| Dec. 2 | Sept. 11 | April 29 | Mar. 26 | Nov. 3 |
| Dec. 7 | Sept. 16 | May 4 | Mar. 31 | Nov. 8 |
| Dec. 12 | Sept. 21 | May 9 | April 5 | Nov. 13 |
| Dec. 17 | Sept. 26 | May 14 | April 10 | Nov. 18 |
| Dec. 22 | Oct. 1 | May 19 | April 15 | Nov. 23 |
| Dec. 27 | Oct. 6 | May 24 | April 20 | Nov. 28 |

# BREED REGISTRY ASSOCIATIONS

A breed registry association consists of a group of breeders banded together for the purposes of: (1) recording the lineage of their animals, (2) protecting the purity of the breed, (3) encouraging further improvement of the breed, and (4) promoting the interest of the breed. A list of the breed registry associations is given in Table A–8.

**TABLE A–8**
**BREED REGISTRY ASSOCIATIONS[1]**

| Class of Animal | Breed | Association and Address | Class of Animal | Breed | Association and Address |
|---|---|---|---|---|---|
| *Beef and Dual-Purpose Cattle* | American | American Breed Assn. P.O. Box 10679 Midwest City, OK 73140 | *Beef and Dual-Purpose Cattle* (Continued) | Charolais/Charbray | American-International Charolais Assn. P.O. Box 20247 Kansas City, MO 64195 |
| | Amerifax | Amerifax Cattle Assn. P.O. Box 149 Hastings, NE 68900 | | Chianina | American Chianina Assn. P.O. Box 890 Platte City, MO 64079 |
| | Angus | American Angus Assn. 3201 Frederick Boulevard St. Joseph, MO 64501 | | Devon | Devon Cattle Assn., Inc. P.O. Box 61 The Plains, VA 22171 |
| | Ankina | Ankina Breeders 5803 Oakes Road Clayton, OH 45315 | | Dexter | American Dexter Cattle Assn. Route 1, Box 378 Concordia, MT 64020 |
| | Ankole-Watusi | Ankole-Watusi International Registry Box 319 Phippsburg, CO 80469 | | Galloway | American Galloway Breeders Assn. Route 1, Box 106A Athol, ID 83801 |
| | Barzona | Barzona Breeders Assn. P.O. Box 631 Prescott, AZ 86302 | | | Galloway Cattle Society of America Hennepin, IL 61327 |
| | Beefalo | American Beefalo World Registry 116 Executive Park Louisville, KY 40207 | | Gelbvieh | American Gelbvieh Assn. 5001 National Western Drive Denver, CO 80216 |
| | Beef Friesian | Beef Friesian Society 118 Livestock Exchange Building 4701 Marion Street Denver, CO 80216 | | | Canadian Gelbvieh Assn. Box 536, Marlborough P.O. Centre Calgary, Alta. T2A 7L4 Canada |
| | Beefmaster | Beefmaster Breeders Universal 6800 Park Ten Blvd. Suite 290 West San Antonio, TX 78213 | | Hereford | American Hereford Assn., The P.O. Box 014059 Kansas City, MO 64101 |
| | | Foundation Beefmaster Assn. Livestock Exchange Building, Suite 200 4701 Marion Street Denver, CO 80216 | | Indu-Brazil (Zebu) | International Zebu Breeders Assn. 783 N. Loop 337 New Braunfels, TX 78130 |
| | | National Beefmaster Assn. P.O. Box 368 Canton, TX 75103 | | Limousin | North American Limousin Foundation 100 Livestock Exchange Building P.O. Box 16767 Denver, CO 80216-0767 |
| | Belgian Blue | Canadian Belgian Beef Cattle Assn. R.R. #2 Orangeville, Ont. L9W 2Y9 Canada | | Lincoln Red | Canadian Lincoln Red Promotion Assn. Box 447 Richmond Hill, Ont. L4C 4Y8 Canada |
| | Blonde d'Aquitaine | Canadian Blonde d'Aquitaine Assn., 207 1606 Centre Street North Calgary, Alta. 72E 2R9 Canada | | Maine-Anjou | American Maine-Anjou Assn. 528 Livestock Exchange Building 1600 Genesee Street Kansas City, MO 64102 |
| | Braford | International Braford Assn., Inc. P.O. Box 2727 Fort Pierce, FL 34954 | | | Maine-Anjou International 334 9th Avenue NE Calgary, Alta. T2E 7A6 Canada |
| | Brahman[2] | American Brahman Breeders Assn. 1313 La Concha Lane Houston, TX 77054 | | Marchigiana | American Int'l. Marchigiana Society P.O. Box 198 Walton, KS 67151-0198 |
| | Bralers | American Bralers Assn. Star Route, Box 47 Ganado, TX 77962 | | Murray Grey | American Murray Grey Assn. 1222 N. 27th, Suite 208 Billings, MT 59101 |
| | Brangus | International Brangus Breeders Assn., Inc. 5750 Epsilon San Antonio, TX 78230 | | Normande | American Normande Assn. P.O. Box 350 Kearney, MO 64060 |

[1]No claim is made that all breed registries are listed.

[2]Includes three breeds of *Bos indicus* cattle that have contributed and are contributing to beef production in the U.S.; namely, American Brahman, Indu Brazil, and Africander.

*(Continued)*

**TABLE A-8** (Continued)

| Class of Animal | Breed | Association and Address | Class of Animal | Breed | Association and Address |
|---|---|---|---|---|---|
| *Beef and Dual-Purpose Cattle* (continued) | Piedmontese | Canadian Piedmontese Assn. Box 11 Admiral, Sask. SON OBO Canada | *Dairy Cattle* | Ayrshire | Ayrshire Breeders' Assn. 2 Union Street Brandon, VT 05733 |
| | | Piedmontese Assn. of the U.S. Route 1 Cost, TX 78614 | | Brown Swiss | Brown Swiss Cattle Breeders' Assn., The P.O. Box 1038 Beloit, WI 53511-1038 |
| | Pinzgauer | American Pinzgauer Assn. Route 1, Box 104E Kelly, IA 50134-9801 | | Dutch Belted | Dutch Belted Cattle Assn. of America, Inc. P.O. Box 358 Venus, FL 33960 |
| | | Canadian Pinzgauer Assn. 233 Stockman's Centre 2116 - 27th Avenue, N.E. Calgary, Alta. T2E 7A6 Canada | | Guernsey | American Guernsey Assn. P.O. Box 666 Reynoldsburg, OH 43068-0666 |
| | Polled Hereford | American Polled Hereford Assn. 4700 E. 63rd Street Kansas City, MO 64130 | | Holstein | Holstein-Friesian Assn. of America P.O. Box 808 Brattleboro, VT 05301 |
| | Red Angus | Red Angus Assn. of America 4201 I-35 North Denton, TX 76201 | | Illawarra | International Illawarra Assn. P.O. Box 449 Beloit, WI 53511 |
| | Red Brangus | American Red Brangus Assn. P.O. Box 1326 Austin, TX 78767 | | Jersey | American Jersey Cattle Club, The 6486 E. Main St. Reynoldsburg, OH 43068-2362 |
| | Red Poll | American Red Poll Assn. Box 35519 Louisville, KY 40232 | | Milking Shorthorn | American Milking Shorthorn Society P.O. Box 449 Beloit, WI 53511 |
| | Romagnola/Romark | Canadian Romark Assn. Box 177 Jarvie, Alta. TOG 1HO Canada | *Sheep* | Border Leicester | American Border Leicester Assn. 7594 S.R. 534 West Farmington, OH 44491 |
| | Salers | American Salers Assn. 5600 S. Quebec, Suite 220A Englewood, CO 80111 | | Cheviot | American Cheviot Sheep Society R.R. 1, Box 100 Clarks Hill, IN 47930 |
| | Santa Gertrudis | Santa Gertrudis Breeders International P.O. Box 1257 Kingsville, TX 78363 | | | American North Country Cheviot Sheep Assn. 833 Fall Creek Road Longview, WA 98632 |
| | Scotch Highland | American Scotch Highland Breeders' Assn. P.O. Box 81 Remer, MN 56672 | | Clun Forest | North American Clun Forest Assn. W. 5855 Mahlum Road Holmen, WI 54636 |
| | Senepol | North American Senepol Assn. P.O. Box 300168 Kansas City, MO 64130 | | Columbia | Columbia Sheep Breeders Assn. of America P.O. Box 272 Upper Sandusky, OH 43351 |
| | Shorthorn | American Shorthorn Assn. 8288 Hascall Street Omaha, NE 68124 | | Cormo | American Cormo Sheep Assn. |
| | | Canadian Shorthorn Assn. Gummer Bldg., 5 Douglas Street Guelph, Ont. H1H 2S8 Canada | | Corriedale | American Corriedale Assn., Inc. Box 29C Seneca, IL 61360 |
| | Simmental/Simbrah | American Simmental Assn. 1 Simmental Way Bozeman, MT 59715 | | Cotswold | American Cotswold Record Assn. 282 Meaderboro Road Rochester, NH 03867 |
| | South Devon | North American and International South Devon Assn. P.O. Box 68 Lynnville, IA 50153 | | Debouillet | Debouillet Sheep Breeders Assn. 300 S. Kentucky Avenue Roswell, NM 88201 |
| | Tarentaise | American Tarentaise Assn. P.O. Box 446 Reedpoint, MT 59069 | | Delaine Merino | American & Delaine Merino Record Assn. 1193 Township Road 346 Nova, OH 44859 |
| | Texas Longhorn | Texas Longhorn Breeders Assn. of America 2315 N. Main Street, Suite 402 Fort Worth, TX 76106 | | | Black Top & National Delaine Merino Sheep Assn. 290 Beech Street Muse, PA 15350 |
| | Welsh Black | United States Welsh Black Cattle Assn. Route 1, Box 76B Shelburn, IN 47879 | | | Texas Delaine Sheep Assn. Route 1 Burnet, TX 78611 |
| | White Park | White Park Cattle Assn. 419 N. Water Street Madrid, IA 50156 | | | |

**TABLE A-8** *(Continued)*

| Class of Animal | Breed | Association and Address | Class of Animal | Breed | Association and Address |
|---|---|---|---|---|---|
| *Sheep* (continued) | **Dorset** | Continental Dorset Club, Inc.<br>P.O. Box 506<br>Hudson, IA 50643 | *Sheep* (continued) | **Tunis** | National Tunis Sheep Registry<br>R.D. 1<br>Wayland, NY 14572 |
| | **Finnsheep** | Finnsheep Breeders Assn., Inc.<br>P.O. Box 512<br>Zionsville, IN 46077-0512 | *Goats* | **Angora** | American Angora Goat Breeders' Assn.<br>P.O. Box 195<br>Rocksprings, TX 78880 |
| | **Hampshire** | American Hampshire Sheep Assn.<br>Box 345<br>Ashland, MO 65010 | *Dairy Goats* | **All breeds** | Alpines International<br>Route 1, Box 2065<br>Ft. Pierce, FL 33451 |
| | **Katahdin** | Katahdin Registry<br>Piel Farm<br>P.O. Box 89<br>Abbot, ME 04406 | | | American Dairy Goat Assn.<br>P.O. Drawer 865<br>Spindale, NC 28160 |
| | **Lincoln** | National Lincoln Sheep Breeders' Assn.<br>R.R. 6, Box 24<br>Decatur, IL 62521 | | | International Nubian Breeders' Assn.<br>P.O. Box 130<br>Crewell, OR 97426 |
| | **Milking Sheep** | American Milking Sheep Society<br>Bethel, MO 63434 | | | National Pygmy Goat Assn.<br>5621 W. Michigan<br>Tucson, AZ 85746 |
| | **Montadale** | Montadale Sheep Breeders' Assn., Inc.<br>P.O. Box 44300<br>Indianapolis, IN 46244 | *Swine* | **Berkshire** | American Berkshire Assn.<br>1769 U.S. 52 North, Box 2436<br>West Lafayette, IN 47906 |
| | **Natural Colored** | Natural Colored Wool Growers Assn.<br>Route 2, Box 2382<br>Davis, CA 95616 | | **Chester White** | Chester White Swine Record Assn.<br>1803 W. Detweiller Drive<br>Peoria, IL 61615 |
| | **North American Clun Forest** | North American Clun Forest Assn.<br>High Meadow Farm<br>Ferryville, WI 54628 | | **Duroc** | United Duroc Swine Registry<br>1803 W. Detweiller Drive<br>Peoria, IL 61615 |
| | **North Country Cheviot** | American North Country Cheviot Sheep Assn.<br>833 Fall Creek Road<br>Longview, WA 98632 | | **Hampshire** | Hampshire Swine Registry<br>6748 N. Frostwood Pkwy.<br>Peoria, IL 61615 |
| | **Oxford** | American Oxford Down Record Assn.<br>Route 4<br>Ottawa, IL 61350 | | **Hereford Hog** | National Hereford Hog Record Assn.<br>Route 1, Box 37<br>Flandreau, SD 57028 |
| | **Panama** | American Panama Registry Assn.<br>HC 85, Box 297<br>Grandview, ID 83624 | | **Landrace** | American Landrace Assn., Inc.<br>P.O. Box 2340<br>West Lafayette, IN 47906 |
| | **Polypay** | American Polypay Sheep Assn.<br>1934 E. Rua Bronco<br>Sandy UT 84092 | | **Poland China** | Poland China Record Assn.<br>P.O. Box 2537<br>West Lafayette, IN 47906 |
| | **Rambouillet** | American Rambouillet Sheep Breeders' Assn., The<br>2709 Sherwood Way<br>San Angelo, TX 76901 | | **Spotted** | National Spotted Swine Record, Inc.<br>P.O. Box 2807<br>West Lafayette, IN 47906 |
| | **Romney** | American Romney Breeders Assn.<br>4375 N.E. Weslinn Drive<br>Corvallis, OR 97333 | | **Yorkshire** | American Yorkshire Club, Inc.<br>P.O. Box 2417<br>West Lafayette, IN 47906 |
| | **Shropshire** | American Shropshire Registry Assn., Inc.<br>6508 West "R" Ave.<br>Schoolcraft, MI 49087 | *Light Horses* | **Akhal-Teke** | Akhal-Teke Association of America, Inc., The<br>Shenandoah Farm, Route 5, Box 110<br>Staunton, VA 24401 |
| | **Southdown** | American Southdown Breeders' Assn.<br>Route 4, Box 14B<br>Bellefonte, PA 16283 | | **American Bashkir Curly** | American Bashkir Curly Registry<br>P.O. Box 453<br>Ely, NV 89301 |
| | **Suffolk** | American Suffolk Sheep Society<br>P.O. Box 256<br>Newton, UT 84327 | | **American Creme** | International American Albino Association<br>Box 194<br>Naper, NE 68755 |
| | | National Suffolk Sheep Assn.<br>P.O. Box 324<br>Columbia, MO 63205 | | **American Mustang** | American Mustang Association, Inc.<br>P.O. Box 338<br>Yucaipa, CA 92399 |
| | **Targhee** | U.S. Targhee Sheep Assn.<br>P.O. Box 34<br>Jordan, MT 59337 | | | |

*(Continued)*

## TABLE A–8 (Continued)

| Class of Animal | Breed | Association and Address | Class of Animal | Breed | Association and Address |
|---|---|---|---|---|---|
| **Light Horses** (Continued) | **American Saddlebred** | American Saddlebred Horse Association, Inc. 4093 Iron Works Pike Lexington, KY 40511 | | **Morab** | North American Morab Horse Association, Inc. W3174 Faro Springs Road Hilbert, WI 54129 |
| | **American White** | International American Albino Association Box 194 Naper, NE 68755 | | **Morgan** | American Morgan Horse Association, Inc., The P.O. Box 960 3 Bostwick Road Shelburne, VT 05482 |
| | **Andalusian** | American Andalusian Horse Association P.O. Box 68 Tamacacori, AZ 85640 | | | |
| | | International Andalusian Horse Association 256 S. Robertson, No. 9378 Beverly Hills, CA 90211 | | **National Show Horse** | National Show Horse Registry Plainview Triad North, Suite 237 10401 Linn Station Road Louisville, KY 40223 |
| | **Appaloosa** | Appaloosa Horse Club, Inc. P.O. Box 8403 Moscow, ID 83843 | | **National Spotted Saddle Horse** | National Spotted Saddle Horse Association, Inc. P.O. Box 898 Murfreesboro, TN 37130 |
| | **Arabian** | Arabian Horse Registry of America, Inc. 12000 Zuni Street Westminster, CO 80234 | | **Norwegian Fjord Horse** | Norwegian Fjord Association of North America 24570 W. Chardon Road Grayslake, IL 60030 |
| | | International Arabian Horse Registry of North America P.O. Box 325 Delphi Falls, NY 13051 | | **Paint Horse** | American Paint Horse Association P.O. Box 18519 Ft. Worth, TX 76118 |
| | **Buckskin** | International Buckskin Horse Association, Inc. P.O. Box 357 St. John, IN 46373 | | **Palomino** | Palomino Horse Association, Inc., The P.O. Box 324 Jefferson City, MO 65101 |
| | **Chickasaw** | Chickasaw Horse Association, Inc., The P.O. Box 8 Love Valley, NC 28677 | | | Palomino Horse Breeders of America, Inc. 15253 E. Skelly Dr. Tulsa, OK 74116-2620 |
| | | National Chickasaw Horse Association Route 2 Clarinda, IA 51232 | | **Paso Fino** | Paso Fino Horse Association, Inc. P.O. Box 600 Bowling Green, FL 33834 |
| | **Galiceno** | Galiceno Horse Breeders Association, Inc. 111 E. Elm Street Tyler, TX 75701 | | **Peruvian Paso** | American Association of Owners & Breeders of Pervuian Paso Horses 221 N. Alameda Ave. Burbank, CA 91502 |
| | **Hackney** | American Hackney Horse Society P.O. Box 174 Pittsfield, IL 62363 | | | Peruvian Paso Horse Registry of North America 1038 4th Street, Suite 4 Santa Rosa, CA 95404 |
| | **Haflinger** | Haflinger Association of America 14570 Gratiot Road Hemlock, MI 48626 | | | |
| | | Haflinger Registry of North America 14640 State Route 83 Coshocton, OH 32812 | | **Quarter Horse** | American Quarter Horse Association P.O. Box 200 Amarillo, TX 79168 |
| | **Half Saddlebred** | Half Saddlebred Registry of America, The 319 S. 6th Street Coshocton, Oh 43812 | | **Spanish-Barb** | Spanish-Barb Breeders Association 2888 Bluff St., Box 465 Boulder, CO 80301 |
| | **Hanoverian** | American Hanoverian Society, The Office 2–E, 831 Bay Ave. Capitola, CA 95010 | | **Spanish Mustang** | Spanish Mustang Registry Inc., The 8328 Stevenson Ave. Sacramento, CA 95828 |
| | **Hungarian Horse** | Hungarian Horse Association P.O. Box 98 Anselmo, NE 68813 | | **Standardbred** | United States Trotting Association 750 Michigan Avenue Columbus, OH 43215 |
| | **Lipizzan** | United States Lipizzan Registry 12479 Duncan Plains Road N.W. Johnstown, OH 43031 | | **Standardbred (Ridden)** | Ridden Standardbred Association[3] 1578 Fleet Road Troy, OH 49373 |
| | **Missouri Fox Trotting Horse** | Missouri Fox Trotting Horse Breed Association, Inc. P.O. Box 1027 Ava, MO 65608 | | **Tennessee Walking Horse** | Tennessee Walking Horse Breeders' and Exhibitors Association P.O. Box 286 Lewisburg, TN 37091 |

[3]The Ridden Standardbred Association accepts for registry purebred, half-bred, and partial-bred standardbreds.

(Continued)

TABLE A-8 *(Continued)*

| Class of Animal | Breed | Association and Address | Class of Animal | Breed | Association and Address |
|---|---|---|---|---|---|
| **Light Horses** (Continued) | **Thorcheron** | Thorcheron Hunter Association 3749 S. 4th St. Kalamazoo, MI 49009 | **Draft Horses** (Continued) | **Clydesdale** | Clydesdale Breeders Association of the United States Route 3 Waverly, IA 50677 |
| | **Thoroughbred** | Jockey Club, The 380 Madison Aveune New York, NY 10017 | | **Percheron** | Percheron Horse Association of America P.O. Box 141 Fredericktown, OH 43819 |
| | **Trakehner** | American Trakehner Association, Inc. 1520 West Church St. Newark, OH 43055 | | **Shire** | American Shire Horse Association Route 1, Box 10 Adel, IA 50003-9702 |
| | **Trottingbred** | International Trotting & Pacing Association, Inc. 575 Broadway Hanover, PA 17331 | | **Suffolk** | American Suffolk Horse Association, Inc. Route 1, Box 212 Ledbetter, TX 78946 |
| | **Welsh Cob** | Welsh Cob Society of America Grazing Field Farm Head of the Bay Road Buzzard Bay, MA 02532 | **Jacks, Donkeys, and Mules** | **Jack and Jennet** | Standard Jack and Jennet Registry of America P.O. Box 1155 Pulaski, TN 38478-1155 |
| **Ponies** | **American Gotland Horse** | American Gotland Horse Association R.R. 2, Box 181 Elkland, MO 65644 | | **Miniature Donkey** | Miniature Donkey Registry of the United States, Inc. 2901 N. Elm Denton, TX 76201 |
| | **American Walking Pony** | American Walking Pony Association Rt. 27, Box 605 Upper River Road Macon, GA 31211 | | **Donkey and Mule** | American Donkey and Mule Society, Inc. 2901 N. Elm Denton, TX 76201 |
| | **Connemara Pony** | American Connemara Pony Society R.D. 1 Hoshiekon Farm Goshen, CT 06756 | | **Mules** | American Mule Registry 2901 N. Elm Denton, TX 76201 |
| | **National Appaloosa Pony** | National Appaloosa Pony, Inc. Box 206 Gaston, IN 47342 | **All Horses and Half-Breeds** | **Any and all colors and types of horses** (including animals not eligible for registry, eligible but not registered, or registered in existing associations) including both light and draft horses. | International American Albino Association Box 194 Naper, NE 68755 |
| | **Pony of the Americas** | Pony of the Americas Club, Inc. 5240 Elmwood Ave. Indianapolis, IN 46203 | | | |
| | **Quarter Ponies** | National Quarter Pony Association 5131 Country Road, #25, Rt. 1 Marengo, OH 43334 | | **American Warmblood** | American Warmblood Society Route 5, Box 1219A Phoenix, AZ 85043 |
| | **Shetland Pony** | American Shetland Pony Club P.O. Box 3415 Peoria, IL 61614 | | | |
| | **Welara Pony** | American Welara Pony Society P.O. Box 401 Yucca Valley, CA 92284 | | | |
| | **Welsh Pony and Cob** | Welsh Pony and Cob Society of America P.O. Box 2977 Winchester, VA 22601 | | | |
| **Miniature Horse** | **Miniature Horse** | American Miniature Horse Association, Inc. P.O. Box 129 Burleson, TX 76028 | | | |
| | | A.S.P.C./A.M.H.R. P.O. Box 3415 Peoria, IL 61614 | | | |
| **Draft Horses** | **American Creme** | Route 1, Box 88 Hubbard, IA 50122 | | | |
| | **Belgian** | Belgian Draft Horse Corporation of America P.O. Box 335 Wabash, IN 46992 | | | |

The American Warmblood can be any breed or combination of breeds, except that they cannot be 100% hot-blooded (i.e. Arabian or Thoroughbred) or 100% cold-blooded (most draft breeds). Pedigree history is desirable, but not required for registration; the only requisite is that the horse must be a warmblood.

The American Warmblood Society, which was formed in the early 1980s, is strictly a performance registry for the following events: dressage, show jumping, combined training, and combined driving. The goal is to breed and train horses to compete internationally in the equestrian events sponsored by the United States Equestrian Team, Inc.

The program and policies of the American Warmblood Society are similar to their much older European counterparts; for example, (1) the *Royal Studbook of the Netherlands*, the government-financed agency that registers and promotes the Dutch Warmblood; and (2) the British Warmblood Society, which is privately (non-government) sponsored.

| Class of Animal | Breed | Association and Address | Class of Animal | Breed | Association and Address |
|---|---|---|---|---|---|
| **All Horses and Half-Breeds** (Continued) | **Half-bred Thoroughbreds** | American Remount Association, Inc. (Half-Thoroughbred Registry)[4] P.O. Box 1066 Perris, CA 92370 | **All Horses and Half-Breeds** (Continued) | **Half-bred Arabian** | International Arabian Horse Association P.O. Box 33696 Denver, CO 80233 |

**Half-bred Thoroughbreds** — American Remount Association, Inc. (Half-Thoroughbred Registry)[4]:

Section 1: *The American Remount Half-Thoroughbred*—Must have one Thoroughbred parent registered in the American (Jockey Club) Stud Book.

Section 2: *The American Remount Anglo*—Must have one Thoroughbred parent registered in the American (Jockey Club) Stud Book and the other parent registered in the Stud Book of a recognized breed.

Section 3: *The American Remount Thoroughbred Kind*—Must have one Thoroughbred parent of a recognized Foreign Registry or must have both parents registered in the American (Jockey Club) Stud Book but be ineligible for registry in the American (Jockey Club) Stud Book.

Section 4: *The American Remount Hunter-Jumper*—Must be a minimum of 36 months of age; be performance certified by an approved Equine Practitioner, a Master of Fox Hounds, an Official of the American Horse Show Association, or a Steward of the American Remount Association; and be ineligible for registry in the American (Jockey Club) Stud Book.

Section 5: *The American Remount Polo Pony*—Must be performance certified by an approved Equine Practitioner, a five-goal rated player, an Officer of the U.S. Polo Association, or a Steward of the American Remount Association; and be ineligible for registry in the American (Jockey Club) Stud Book.

Section 6: *The American Remount Endurance Horse*—Must be performance certified by an approved Equine Practitioner, an Official of the American Show Horse Association, or a Steward of the American Remount Association; and be ineligible for registry in the American (Jockey Club) Stud Book

Section 7: *The American Remount Record*—This is an identification Certificate issued to a horse that has apparent Thoroughbred ancestry but is not otherwise eligible for registry in the American (Jockey Club) Stud Book or any other recognized Stud Book.

**Half-bred Arabian** — International Arabian Horse Association:

1. *Anglo-Arabs* must carry not more than ¾ and not less than ¼ Arabian blood. May be either —
   (a) By Thoroughbred stallions and out of registered Arabian mares;
   (b) By registered Arabian stallions and out of registered Thoroughbred mares;
   (c) By registered Thoroughbred or Arabian stallions and out of registered Anglo-Arab mares; or
   (d) By Anglo-Arab stallions and out of either Anglo-Arab mares, registered Thoroughbred mares, or registered Arabian mares.
2. *Half-Arabians* are by registered Arabian stallions and out of mares that are not registered Thoroughbreds or Arabians.

| Class of Animal | Breed | Association and Address |
|---|---|---|
| | **Half-bred, grade, and crossbred horses**, involving American Saddlebred, Appaloosa, Hackney, Morgan, Quarter Horse, Standardbred, Tennessee Walking Horse, Welsh Pony, and certain other breeds. Grade horses must not be registered in another breed association. | American Part-Blooded Horse Registry 4120 S.E. River Drive Portland, OR 97222 — National Grade Horse Registry P.O. Box 338 10221 Slater Ave., #103 Fountain Valley, CA 92708 |

[4]Formerly the Half-Bred Stud Book operated by the American Remount Association, but now a privately owned registry. It records only foals sired by registerd Thoroughbred stallions and out of mares not registered in the American (Jockey Club) Stud Book, or in the Arabian Stud Book.

## BREED MAGAZINES

The livestock magazines publish news items and informative articles of special interest to producers. Also, many of them employ field representatives whose chief duty it is to assist in the buying and selling of animals.

In the compilation of the list herewith presented (see Table A–9) no attempt was made to list the general livestock magazines, of which there are numerous outstanding ones. Only the magazines which are devoted to a specific class or breed of animal are included.

| Class & Breed of Animal | Publication | Address | Class & Breed of Animal | Publication | Address |
|---|---|---|---|---|---|
| **Beef Cattle** **General** | Arkansas Cattle Business | 11701 I-30, Bldg. 4, Suite 412 Little Rock, AR, 72209 | **Beef Cattle (Continued)** **General** (Continued) | Calf News | 18345 Ventura Blvd., Suite 303 Tarzana, CA 91356 |
| | Beef | 1999 Shepard Road St. Paul, MN 55116 | | California Cattleman | 1221 H Street Sacramento, CA 95814 |
| | Beef Digest | Box 1009 Manhattan, KS 66502 | | Cattleman, The | 1301 W. Seventh Street Fort Worth, TX 76102 |
| | Beef Week | P.O. Box 4264 Macon, GA 31208 | | Cattlemen | 1750 Ellice Avenue Winnipeg, Man. R3H 0B6 Canada |
| | Better Beef Business | P.O. Box 7386 Louisville, KY 40207 | | El Ganadero Internacional | 11201 Morning Court San Antonio, TX 78213 |

[1]No claim is made that all magazines are listed.

(Continued)

**TABLE A-9** *(Continued)*

| Class & Breed of Animal | Publication | Address | Class & Breed of Animal | Publication | Address |
|---|---|---|---|---|---|
| **Beef Cattle** (Continued) | | | **Beef Cattle** (Continued) | | |
| **General** (Continued) | Florida Cattleman & Livestock Journal | P.O. Box 1403 Kissimmee, FL 32741 | **Simmental** | Simmental Country | No. 13–4101 195th Street, N.E. Calgary, Alta. T2E 7C4 Canada |
| | Gulf Coast Cattleman | 11201 Morning Court San Antonio, TX 78213 | | Simmental Shield Update, The | P.O. Box 522 Lindsborg, KS 67456 |
| | Ideal Beef Memo | Route 1, Box 79 Huxley, IA 50124 | **Tarentaise** | Tarentaise Times | Tongue River Route Miles City, MT 49301 |
| | The Ketch Pen | P.O. Box 96 Ellensburg, WA 98926 | **Texas Loghorn** | Texas Longhorn Journal | P.O. Box 1209 Monument, CO 80132 |
| | Oregon Cattleman | 1000 N.E. Multnomah Street Portland, OR 97232 | **Welsh Black** | Welsh Black Cattle World | Route 1 Wahkon, MN 56386 |
| | Southern Beef Producer | P.O. Box 843 Franklin, TN 37064 | **Dairy Cattle** | | |
| | World of Beef | 105 Stockman's Centre 2116 - 27th Avenue N.E. Calgary, Alta. T2E 7A6 Canada | **General** | California Agribusiness Dairyman | 1185 W. Hedges Fresno, CA 93728 |
| **Amerifax** | Amerifax, The | Box 149 Hastings, NE 68901 | | Dairy Contact | 11802 - 124th Street, Suite 214 Edmonton, Alta. T5L 0M3 Canada |
| **Angus** | Angus Journal | 3201 Frederick Blvd. St. Joseph, MO 64501 | | Dairy Illustratied | 1880 Country Farm Drive Naperville, IL 60540 |
| **Ankina** | Ankina Breeder | 4803 Oakes Road Clayton, OH 45315 | | Dairy Journal | Tulare, CA 93275 |
| **Ankole Watusi** | Watusi | 2810 East 3rd Amarillo, TX 79104 | | Dairyman, The | P.O. Box 299 Sandy Creek, NY 13145 |
| **Barzona** | The Barzonian | P.O. Box 631 Prescott, AZ 86302 | | Dairymen's Digest | P.O. Box 5040 Arlington, TX 76005 |
| **Beefmaster** | Beefmaster Cowman, The | 11201 Morning Court San Antonio, TX 78213 | | Farm & Dairy | Box 38 Salem, OH 44460 |
| **Brahman** | Brahman Journal, The | P.O. Box 220 Eddy, TX 76524 | | Hoard's Dairyman | 38 W. Milwaukee Avenue Fort Atkinson, WI 53538-0801 |
| **Brangus** | Brangus Journal | 5750 Epsilon San Antonio, TX 78249 | | Sunbelt Dairyman, The | P.O. Box 843 Franklin, IN 37064 |
| **Charolais** | Charolais Journal | 11700 N.W. Plaza Circle Kansas City, MO 64195 | **Ayrshire** | Ayrshire Digest | 2 Union Street Brandon, VT 05733 |
| **Chianina** | American Chianina Journal | P.O. Box 890 Platte City, MO 64079 | **Brown Swiss** | Brown Swiss Bulletin, The | P.O. Box 1038 Beloit, WI 53511 |
| **Gelbvieh** | Gelbvieh Country | 5001 National Western Drive Denver, CO 80216 | **Guernsey** | Guernsey Breeders Journal | P.O. Box 27410 Columbus, OH 43227 |
| **Hereford** | American Hereford Journal, The | P.O. Box 4059 Kansas City, MO 64101 | **Holstein** | Arizona Holstein Sun | Box 365 Marana, AZ 85238 |
| | Canadian Hereford Digest | 5160 Skyline Way, N.E. Calgary, Alta T2E 6V1 Canada | | California Holstein News | 1177 West Hedges Fresno, CA 93728 |
| | Texas Hereford | 4609 Airport Freeway Fort Worth, TX 76118 | | Holstein Journal | 335 Lesmill Road Don Mills, Ont. M3B 2V1 Canada |
| **Maine-Anjou** | Maine-Anjou International | 334 Ninth Avenue, N.E. Calgary, Alta. T2E 0V6 Canada | | Holstein World | P.O. Box 299 Sandy Creek, NY 13145 |
| **Murray Grey** | Murray Grey News | P.O. Box 30085 Billings, MT 59107 | | Texas Holstein News | Route 1 Buda, TX 78610 |
| **Polled Hereford** | Polled Hereford World | 4700 E. 63rd Street Kansas City, MO 64130 | **Jersey** | Canadian Jersey Breeder | 343 Waterloo Avenue Guelph, Ont. N1H 3K1 Canada |
| | Texas Polled Hereford News | Box 70 Rio Vista, TX 76093 | | Jersey Journal | 6486 E. Main Street Reynoldsburg, OH 43068-2362 |
| **Red Angus** | American Red Angus | 4201 I-35 North Denton, TX 76201 | **Milking Shorthorn** | Journal of the Milking Shorthorn and Illawarra Breeds | P.O. Box 449 Beloit, WI 53511 |
| **Red Poll** | Red Poll News | P.O. Box 35519 Louisville, KY 40232 | **Sheep** | | |
| **Santa Gertrudis** | Santa Gertrudis Journal, The | P.O. Box 7636 Fort Worth, TX 76111-0636 | **General** | California Sheepman's Quarterly | 3382 E. Camino Avenue #6 Sacramento, CA 95821 |
| **Shorthorn** | Shorthorn Country | 8288 Hascall Street Omaha, NE 68124 | | Montana Wool Grower | P.O. Box 1693 Helena, MT 59601 |
| | | | | Ranch Magazine | P.O. Box 2678 San Angelo, TX 76902 |

*(Continued)*

**TABLE A-9** (Continued)

| Class & Breed of Animal | Publication | Address | Class & Breed of Animal | Publication | Address |
|---|---|---|---|---|---|
| *Sheep* (Continued) | | | *Swine* (Continued) | | |
| **General** (Continued) | Sheep! | Box 329<br>Jefferson, WI 53549 | **Yorkshire** | Yorkshire Journal | P.O. Box 2417<br>West Lafayette, IN 47906 |
| | Sheep Breeder and Sheepman, The | P.O. Box 796<br>Columbia, MO 65205 | *Poultry* | | |
| | sheep! Magazine | W2997 Market Road<br>Helenville, WI 53137 | **General** | Broiler Industry | Watt Publishing Co.<br>Sandstone Building<br>Mount Morris, IL 61054-1497 |
| | Shepherd Magazine, The | 5696 Johnston Road<br>New Washington, OH 44854 | | Poultry Digest | Watt Publishing Co.<br>Sandstone Building<br>Mount Morris, IL 61054 |
| | Southern Sheep Producer | Route 2<br>Clinton, KY 42031 | | Poultry International | 18 Chapel St.<br>Petersfield, Hants GU32 3D2<br>England |
| **Columbia** | Speaking of "Columbias" | P.O. Box 272<br>Upper Sandusky, OH 43351 | | Poultry Press | P.O. Box 947<br>York, PA 17405 |
| **Corriedale** | Corriedale Extra, The | Box 29C<br>Seneca, IL 61360 | | Poultry Tribune | Watt Publishing Co.<br>Sandstone Building<br>Mount Morris, IL 61054-1497 |
| **Suffolk** | Suffolk Banner, The | Box AA<br>Cuba, IL 61427 | | Turkey World | Watt Publishing Co.<br>Sandstone Building<br>Mount Morris, IL 61054-1497 |
| *Sheep & Goats* | | | | World Poultry | Surrey House, 1 Throwley Way<br>Sutton, Surrey SM1 4QQ England |
| **General** | Better Goat Keeping | Harvard, MA 01451 | *Light Horses and Ponies* | | |
| | Dairy Goat Guide | Hwy 19 East<br>Waterloo, Wi 53594 | **General** | American Farriers Journal | 63 Great Road<br>Maynard, MA 01754 |
| | Marker, The | Star Route, Box 48<br>Brooks, CA 95606 | | Bluegrass Horseman, The | P.O. Box 389<br>Lexington, KY 40501 |
| | News Dispatch | Route 2, Box 112<br>DeLeon, TX 76444 | | Bridle & Bit | P.O. Box 54520<br>Phoenix, AZ 85078 |
| | Ranch Magazine, The | Box 2678<br>San Angelo, TX 76902 | | California Horseman's News | P.O. Box 474<br>San Marcos, CA 92069 |
| | United Caprine News | Drawer A<br>Rotan, TX 79546 | | California Horse Review | P.O. Box 2437<br>Fair Oaks, CA 95628 |
| *Swine* | | | | Canadian Rider | P.O. Box 7065<br>Ancaster, Ont. L9G 3L1 Canada |
| **General** | Hog Digest | Box 1009<br>Manhattan, KS 66502 | | Carriage Journal, The | R.D. #1, Box 115<br>Salem, NJ 08079 |
| | Hog Farm Management | 12400 Whitewater Drive<br>Minnetonka, MN 55343 | | Chronicle of the Horse, The | P.O. Box 46<br>Middleburg, VA 22117 |
| | Hogs Today | 230 W. Washington Square<br>Philadelphia, PA 19105 | | Corral, The | P.O. Box 151<br>Medina, OH 44256 |
| | National Hog Farmer | 7900 International Drive<br>Minneapolis, MN 55425 | | Cuttin' Hoss Chatter, The | 4704 Hwy. 377 South<br>Fort Worth, TX 76116-8805 |
| | Nebraska Pork Talk | P.O. Box 369<br>Madison, NE 68748 | | Dressage | P.O. Box 12130<br>Cleveland, OH 44112 |
| | Pig Farming | Fenton House, Wharfedale Road<br>Ipswich, Suffolk 1P1 4LG England | | Equestrian Trails | 13741 Foothil Blvd., 220<br>Sylmar, CA 91342 |
| | Pork | P.O. Box 2939<br>Shawnee Mission, KS 66201-1339 | | Equine Journal | P.O. Box 623<br>Keene, NH 03431 |
| | Southern Hog Producer | P.O. Box 843<br>Franklin, TN 37064 | | Equus | 656 Quince Orchard Road<br>Gaithersburg, MD 20760 |
| **Berkshire** | Purebred Picture | Route #2, Box 146<br>Prophetstown, IL 61277 | | Horse and Horseman | P.O. Box HH<br>Capistrano Beach, CA 92624 |
| **Chester White** | Chester White Journal | 1803 W. Detweiller Drive<br>Peoria, IL 61615 | | Horse & Pony | P.O. Box 2050<br>Seffner, IL 33584 |
| **Duroc** | Duroc News | 1803 W. Detweiller Drive<br>Peoria, IL 61615 | | Horse & Rider | 941 Calle Negocio<br>San Clemente, CA 92672 |
| **Hampshire** | American Hampshire Herdsman | 6748 Frostwood Pkwy.<br>Peoria, IL 61615 | | | |
| **Spotted** | Spotted News | P.O. Box 2807<br>West Lafayette, IN 47906 | | | |
| **Tamworth** | Tamworth News | 2656 Horner Road<br>Winchester, OH 45697 | | | |

(Continued)

**TABLE A-9** *(Continued)*

| Class & Breed of Animal | Publication | Address | Class & Breed of Animal | Publication | Address |
|---|---|---|---|---|---|
| *Light Horses and Ponies* (Continued) | | | *Light Horses and Ponies* (Continued) | | |
| **General** | *Horse Illustrated* | P.O. Box 6050<br>Mission Viejo, CA 92690 | **Appaloosa** | *Appaloosa Journal* | Box 8403<br>Moscow, ID 83843 |
| | *Horseman* | 25025 I-45 N., Suite 390<br>Spring, TX 77380 | | *Cal-Western Appaloosa* | 3097 Willow #15<br>Clovis, CA 93612 |
| | *Horsemen's Corral* | P.O. Box 110<br>New London, OH 44851 | **Arabian** | *Arabian Horse Country* | P.O. Box 4607-A<br>Portland, OR 97208 |
| | *Horsemen's Yankee Pedlar* | 785 Southbridge Street<br>Auburn, MA 01501 | | *Arabian Horse Express* | P.O. Box 845<br>Coffeyville, KS 67337 |
| | *Horseplay* | 11 Park Avenue, P.O. Box 130<br>Gaithersburg, MD 20884 | | *Arabian Horse Times, The* | R.R. 3<br>Waseca, MN 56093 |
| | *Horses—All* | P.O. Box 9<br>Babb, MT 59411-0009 | | *Arabian Horse World* | 409 Sherman Ave.<br>P.O. Box 60910<br>Palo Alto, CA 94306 |
| | *Horse Show* | 220 E. 42nd Street<br>New York, NY 10017-5806 | | *Arabian Visions* | Box 230<br>Platte City, MO 64079 |
| | *Horse World* | P.O. Box 1007<br>Shelbyville, TN 37160 | | *Crabbet Influence in Arabians Today, The* | 12723 Road 34¾<br>Madera, CA 93638 |
| | *Inside International* | P.O. Box 33696<br>Denver, CO 80233 | **Buckskin** | *Buckskin Country* | 4627 W. Frier Drive<br>Glendale, AZ 85301 |
| | *Lariat, The* | 12675 First Street, S.W.<br>Beaverton, OR 97005 | **Hackney** | *Hackney Journal, The* | P.O. Box 200<br>Crawfordsville, IA 52621 |
| | *Modern Horse Breeding* | 656 Quince Orchard Road<br>Gaithersburg, MD 20878 | **Haflinger** | *Haflinger Highlite* | 2061 Kenyon Ave., S.W.<br>Massillon, OH 44646 |
| | *National Horseman, The* | 11603 Shelbyville Road, Box 43397<br>Middletown, KY 40243 | **Lipizzan** | *Lipizzan Journal, The* | c/o United States Lipizzan Registry<br>12479 Duncan Plains Rd., N.W.<br>Johnstown, OH 93031 |
| | *Northeast Equine Journal* | P.O. Box 623<br>Keene, NH 03431 | **Miniature Horse** | *Miniature Horse World, The* | American Miniature Horse Association, The<br>P.O. Box 129<br>Burleson, TX 76028 |
| | *Pennmarva* | 1225 Industrial Hwy.<br>Southampton, PA 18966 | **Missouri Fox Trotting Horse** | *Journal, The* | P.O. Box 1027<br>Ava, MO 65608 |
| | *Practical Horseman* | Gum Tree Corner<br>Unionville, PA 19375 | | *Missouri Fox Trotter* | Box 191<br>West Plains, MO 65775 |
| | *Record Horseman* | P.O. Box 900<br>Winter Park, CO 80482 | **Morgan** | *Morgan Horse, The* | American Morgan Horse Association, Inc., The<br>P.O. Box 960<br>3 Bostwick Road<br>Shelburn, VT 05482 |
| | *Southern Horseman, The* | P.O. Box 71<br>Meridian, MS 39302 | | | |
| | *Spur* | P.O. Box 85<br>Middleburg, VA 22117 | | | |
| | *Trail Rider, The* | P.O. Box 387<br>Chatsworth, GA 30705 | **Morab** | *Morab World, The* | W3174 Faro Springs Road<br>Hilbert, WI 54129 |
| | *Turf and Sport Digest* | 26 West Pennsylvania Avenue<br>Towson, MD 21204 | **National Show Horse** | *National Show Horse* | 10401 Linn Station Road<br>Louisville, KY 40223 |
| | *USCTA News* | 292 Bridge Street<br>South Hamilton, MA 01982 | **National Spotted Saddle Horse** | *National Spotted Saddle Horse Journal* | P.O. Box 898<br>Murfreesboro, TN 37133-0898 |
| | *Western Horseman, The* | P.O. Box 7980<br>Colorado Springs, CO 80933 | **Norwegian Fjord Horse** | *Fjord Times, The* | Norwegian Fjord Association of North America<br>24570 W. Chardon Road<br>Grayslake, IL 60030 |
| **American Bashkir Curly** | *Curly Cues* | Box 453<br>Ely, NV 89301 | | | |
| **American Mustang** | *American Mustang World* | P.O. Box 338<br>Yucaipa, CA 92399 | | | |
| **American Saddlebred** | *American Saddlebred Magazine, The* | 4093 Iron Works Pike<br>Lexington, KY 40511 | **Paint Horse** | *Paint Horse Journal, The* | P.O. Box 18519<br>Ft. Worth, TX 76118 |
| | *Bluegrass Horseman, The* | P.O. Box 389<br>Lexington, KY 40501 | **Palomino** | *Palomino Horses* | P.O. Box 71<br>Meridian, MS 39302 |
| | *Saddle and Bridle* | 375 N. Jackson Ave.<br>St. Louis, MO 63130 | | *Palomino Parade* | P.O. Box 324<br>Jefferson City, MO 65101 |
| | *Saddle Horse Report* | P.O. Box 1007<br>Shelbyville, TN 37160 | **Paso Fino** | *Paso Fino Horse World* | P.O. Box 600<br>Bowling Green, FL 33834 |
| **Andalusian** | *The Spanish Bit* | 1941 Old Mill Road<br>Springfield, OH 45502 | | | |

**TABLE A-9** *(Continued)*

| Class & Breed of Animal | Publication | Address | Class & Breed of Animal | Publication | Address |
|---|---|---|---|---|---|
| ***Light Horses and Ponies*** (Continued) | | | ***Light Horses and Ponies*** (Continued) | | |
| **Peruvian Paso** | *Caballo Magazine* | P.O. Box 1959 Corona, CA 91718-1959 | **Thoroughbred** (Continued) | *Backstretch, The* | 19363 James Couzens Hwy. Detroit, MI 48235 |
| | *International Peruvian Paso* | 7228 Kentwood Ave. Westchester, CA 90045 | | *Blood-Horse, The* | Box 4038 Lexington, KY 40544-4038 |
| **Pinto Horse** | *Pinto Horse, The* | P.O. Box 71 Meridian, MS 39302 | | *Daily Racing Form* | 10 Lake Drive Hightstown, NJ 08520 |
| **Pony of the Americas** | *Pony of the Americas* | 5240 Elmwood Ave. Indianapolis, IN 46203 | | *Florida Horse, The* | P.O. Box 2106 Ocala, FL 32678 |
| **Quarter Horse** | *Canadian Quarter Horse Journal* | P.O. Box 7065 Ancaster, Ont. L9G 3L1 Canada | | *Maryland Horse, The* | P.O. Box 427 Timonium, MD 21093 |
| | *Intermountain Quarter Horse, The* | 2225 E 4800 So., Suite 110 Salt Lake City, UT 84117 | | *Texas Thoroughbred, The* | P.O. Box 14967 Austin, TX 78761 |
| | *Quarter Horse Journal, The* | P.O. Box 32470 Amarillo, TX 79120 | | *Thoroughbred of California, The* | 201 Colorado Place Arcadia, CA 91006 |
| | *Quarter Racing Journal, The* | P.O. Box 32470 Amarillo, TX 79120 | | *Thoroughbred Record, The* | 367 W. Short St. Lexington, KY 40533 |
| | *Quarter Week* | 10554 Progress Way, Suite G Cypress, CA 90630 | | *Thoroughbred Times* | P.O. Box 8237 Lexington, KY 40533 |
| **Shetland Pony** | *Pony Journal, The* | P.O. Box 3415 Peoria, IL 61614 | | *Washington Thoroughbred, The* | P.O. Box 88258 Seattle, WA 98178 |
| **Spanish-Barb** | *Spanish Barb World/ Journal, The* | 188 Springridge Rd. Terry, MS 39170 | **Trakehner** | *American Trakehner, The* | 5008 Pine Creek Dr., Suite B Westerville, OH 43081 |
| **Spanish Mustang** | *Spanish Mustang Registry* (annual) *Newsletter, The* (quarterly) | 8328 Stevenson Ave. Sacramento, CA 95828 | **Trottingbred** | *Trottingbred, The* | 575 Broadway Hanover, PA 17331 |
| **Spotted Saddle Horse** | *Spotted Saddle Horse News* | P.O. Box 1046 Shelbyville, TN 37160 | **Welara Pony** | *Welara Journal* | P.O. Box 401 Yucca Valley, CA 92284 |
| **Standardbred** | *Harness Horse, The* | P.O. Box 10779 Harrisburg, PA 17105 | **Welsh** | *Welsh Roundabout* | 5051 Townline Rd. East Troy, WI 53120 |
| | *Hoof Beats* | 750 Michigan Avenue Columbus, OH 43215 | ***Draft Horses*** (all breeds, and mules) | *Draft Horse Journal* | Box 670 Waverly, IA 50677 |
| | *Horseman and Fair World, The* | 904 N. Broadway Lexington, KY 40505 | **Belgian** | *Belgian Review* (annual) | P.O. Box 335 Wabash, IN 46992 |
| **Tennessee Walking Horse** | *Voice of the Tennessee Walking Horse* | P.O. Box 286 Lewisburg, TN 37091 | **Percheron** | *Percheron Notes* | P.O. Box 141 Fredericktown, OH 43019 |
| | *Walking Horse Report* | P.O. Box 1007 Shelbyville, TN 37160 | ***Donkeys and Mules*** | *Brayer, The* | 2901 N. Elm Denton, TX 76201 |
| **Thoroughbred** | *American Turf Monthly* | 438 W. 37th St. New York, NY 10018 | | *Mules* | American Mule Association 6725 Union Road Paso Robles, CA 93446 |
| | *Arizona Thoroughbred, The* | 1501 West Bell Rd. Phoenix, AZ 85069 | | *Mules & More* | P.O. Box 872 Carthage, MO 64836 |

# U.S. LAND-GRANT UNIVERSITIES AND CANADIAN PROVINCIAL UNIVERSITIES

U.S. producers can obtain a list of available bulletins and circulars, and other information by writing to (1) their state agricultural college (land-grant institution), and (2) the U.S. Superintendent of Documents, Washington, DC; or by going to the local county extension office (farm advisor) of the county in which they reside. Canadian producers may write to the Department of Agriculture of their province or to their provincial university. A list of U.S. land-grant institutions and Canadian provincial universities follows in Table A–10.

### TABLE A–10
### U.S. LAND-GRANT INSTITUTIONS AND CANADIAN PROVINCIAL UNIVERSITIES

| State | Address |
|---|---|
| Alabama | School of Agriculture, Auburn University, Auburn, AL 36830 |
| Alaska | Department of Agriculture, University of Alaska, Fairbanks, AK 99701 |
| Arizona | College of Agriculture, The University of Arizona, Tucson, AZ 85721 |
| Arkansas | Division of Agriculture, University of Arkansas, Fayetteville, AR 72701 |
| California | College of Agricultural and Environmental Sciences, University of California, Davis, CA 95616 |
| Colorado | College of Agricultural Sciences, Colorodao State University, Fort Collins, CO 80521 |
| Connecticut | College of Agriculture and Natural Resources, University of Connecticut, Storrs, CT 06268 |
| Delaware | College of Agricultural Sciences, University of Delaware, Newark, DE 19711 |
| Florida | College of Agriculture, University of Florida, Gainesville, FL 32611 |
| Georgia | College of Agriculture, University of Georgia, Athens, GA 30602 |
| Hawaii | College of Tropical Agriculture, University of Hawaii, Honolulu, HI 96822 |
| Idaho | College of Agriculture, University of Idaho, Moscow, ID 83843 |
| Illinois | College of Agriculture, University of Illinois, Urbana-Champaign, IL 61801 |
| Indiana | School of Agriculture, Purdue University, West Lafayette, IN 47907 |
| Iowa | College of Agriculture, Iowa State University, Ames, IA 50010 |
| Kansas | College of Agriculture, Kansas State University, Manhattan, KS 66506 |
| Kentucky | College of Agriculture, University of Kentucky, Lexington, KY 40506 |
| Louisiana | College of Agriculture, Louisiana State University and A&M College, University Station, Baton Rouge, LA 70803 |
| Maine | College of Life Sciences and Agriculture, University of Maiine, Orono, ME 04473 |
| Maryland | College of Agriculture, University of Maryland, College Park, MD 20742 |
| Massachusetts | College of Food and Natural Resources, University of Massachusetts, Amherst, MA 01002 |
| Michigan | College of Agriculture and Natural Resources, Michigan State University, East Lansing, MI 48823 |
| Minnesota | College of Agriculture, University of Minnesota, St. Paul, MN 55101 |
| Mississippi | College of Agriculture, Mississippi State University, Mississippi State, MS 39762 |
| Missouri | College of Agriculture, University of Missouri, Columbia, MO 65201 |
| Montana | College of Agriculture, Montana State University, Bozeman, MT 59715 |
| Nebraska | College of Agriculture, University of Nebraska, Lincoln, NE 68503 |
| Nevada | The Max C. Fleischmann College of Agriculture, University of Nevada, Reno, NV 89507 |
| New Hampshire | College of Life Sciences and Agriculture, University of New Hampshire, Durham, NH 03824 |
| New Jersey | College of Agriculture and Environmental Science, Rutgers University, New Brunswick, NJ 08903 |
| New Mexico | College of Agriculture and Home Economics, New Mexico State University, Las Cruces, NM 88003 |
| New York | New York State College of Agriculture, Cornell University, Ithaca, NY 14850 |
| North Carolina | School of Agriculture, North Caronlina State University, Raleigh, NC 27607 |
| North Dakota | College of Agriculture, North Dakota State University, State University Station, Fargo, ND 58102 |
| Ohio | College of Agriculture and Home Economics, The Ohio State University, Columbus, OH 43210 |
| Oklahoma | College of Agriculture and Applied Science, Oklahoma State University, Stillwater, OK 74074 |
| Oregon | School of Agriculture, Oregon State University, Corvallis, OR 97331 |

*(Continued)*

**TABLE A–10** *(Continued)*

| State | Address |
|---|---|
| Pennsylvania | College of Agriculture, The Pennsylvania State University, University Park, PA 16802 |
| Puerto Rico | College of Agricultural Sciences, University of Puerto Rico, Mayaguez, PR 00708 |
| Rhode Island | College of Resource Development, University of Rhode Island, Kingston, RI 02881 |
| South Carolina | College of Agricultural Sciences, Clemson University, Clemson, SC 29631 |
| South Dakota | College of Agriculture and Biological Sciences, South Dakota State University, Brookings, SD 57006 |
| Tennessee | College of Agriculture, University of Tennessee, P.O. Box 1071, Knoxville, TN 37901 |
| Texas | College of Agriculture, Texas A&M University, College Station, TX 77843 |
| Utah | College of Agriculture, Utah State University, Logan, UT 84321 |
| Vermont | College of Agriculture, University of Vermont, Burlington, VT 05401 |
| Virginia | College of Agriculture, Virginia Polytechnic Institute and State University, Blacksburg, VA 24061 |
| Washington | College of Agriculture, Washington State University, Pullman, WA 99163 |
| West Virginia | College of Agriculture and Forestry, West Virginia University, Morgantown, WV 26506 |
| Wisconsin | College of Agricultural and Life Sciences, University of Wisconsin, Madison, WI 53706 |
| Wyoming | College of Agriculture, University of Wyoming, University Station, P.O. Box 3354, Laramie, WY 82070 |

| Canada | Address |
|---|---|
| Alberta | University of Alberta, Edmonton, Alberta T6H 3K6 |
| British Columbia | University of British Columbia, Vancouver, British Columbia V6T 1W5 |
| Manitoba | University of Manitoba, Winnipeg, Manitoba R3T 2N2 |
| New Brunswick | University of New Brunswick, Fredericton, New Brunswick E3B 4Z7 |
| Ontario | University of Guelph, Guelph, Ontario N1G 2W1 |
| Quebec | Faculty d'Agriculture, L'Universite Laval, Quebec City, Quebec G1K 7D4; and Macdonald College of McGill University, Ste. Anne de Bellevue, Quebec H9X 1C0 |
| Saskatchewan | University of Saskatchewan, Saskatoon, Saskatchewan S7N 0W0 |

## POISON INFORMATION CENTERS

With the large number of chemical sprays, dusts, and gases now on the market for use in agriculture, accidents may arise because of operators being careless in their use. Also, there is always the hazard that a child may eat or drink something that may be harmful. Centers have been established in various parts of the country where doctors can obtain prompt and up-to-date information on treatment of such cases, if desired.

Local medical doctors have information relative to the Poison Information Centers of their area, along with some of the names of their directors, telephone numbers, and street numbers. When calling any of these centers, one should ask for the "Poison Information Center." If this information cannot be obtained locally, call the U.S. Public Health Service at Atlanta, Georgia; or Wenatchee, Washington.

Also, the *National Poison Control Center* is located at the University of Illinois, Urbana-Champaign. It is open 24 hours a day, every day of the week. The *hot line* number is: 217/333-3611. The toxicology group is staffed to answer questions about known or suspected cases of poisoning or chemical contaminations involving any species of animal. It is not intended to replace local veterinarians or state toxicology laboratories, but to complement them. Where consultation over the telephone is adequate, there is no charge to the veterinarian or producer. Where telephone consultation is inadequate or the problem is of major proportions, a team of veterinary specialists can arrive at the scene of a toxic or contamination problem within a short time. The cost of a personal visitation varies according to the distance traveled, personnel time, and laboratory services required.

# Index

## D

Agriculture is scientific from breeder to feedlot, and from processor to kitchen.

Angus and Brangus heifers near London, Texas. (Courtesy, *Livestock Weekly*, San Angelo, TX)

Beef cows winter grazing on cornstalks. (Courtesy, Kansas State University, Manhattan)

Cattle finishing operation, showing feed mill in the distance.

An attractive dairy. (Courtesy, USDA)

Holstein lactating cows feeding on alfalfa hay. (Photo by Holstein-Friesian Assn. of America, Brattleboro, VT)

Cows in comfortable free stalls. (Courtesy, Holstein-Friesian Assn. of America, Brattleboro, VT)

Merino rams, a fine wool breed and the most common breed in the world.

Angora goats in the Edwards Plateau area of Texas. (Courtesy, Texas A&M University Ag. Res. and Ext. Center, San Angelo, TX)

Toggenburg doe. She produced 5,750 lb of milk in one year. (Courtesy, *Dairy Goat Journal*, Scottsdale, AZ)

Shed opening onto concrete pens, used for replacement gilts and gestating sows. (Courtesy, American Yorkshire Club, West Lafayette, IN)

Sows in individual feeding stalls. (Courtesy, American Landrace Assn., West Lafayette, IN)

Sow and litter in a farrowing crate with a slotted floor.

White Leghorn. (Courtesy, Hy-Line Indian River Company, Johnstown, IA)

Stacking ready-to-cook fryer in racks. (Courtesy, Foster Farms, Livingston, CA)

Close up of turkey eggs in an incubator at Cuddy Farms. (Photo courtesy James Strawser, University of Georgia, Athens)

Forty-eight-day-old ducklings, weighing 6.6 lb, ready for market. (Courtesy, Cherry Valley Farms Ltd., Rothwell, Lincoln, England)

New Zealand White—at 8 weeks of age. New Zealand Whites are the most widely used breed for meat production. (Photo courtesy David J. Harris, Siloam Springs, AR)

Catfish being harvested for market, at liveweight of 2 to 2½ lb. (Courtesy, USDA)

(Left) Hungarian mares and foals at Bitter-root Stock Farm, Hamilton, Montana. (Photo by Ernst Peterson, Hamilton, MT)

(Below) Lipizzan stallion about to kick out in the capriole movement. (Courtesy, *Popular Horseman*, Harrisburg, PA)

(Below) Smelling the flower! (Photo by Ernst Peterson, Hamilton, MT)

(Right) The legendary *Man O' War*, one of the greatest racehorses of all time. (Courtesy, Kentucky Department of Public Information, Frankfort, KY)